Differentiation

$$(cu)' = cu' \qquad (c \text{ constant})$$

$$(u + v)' = u' + v'$$

$$(uv)' = u'v + v'u$$

$$\left(\frac{u}{v}\right)' = \frac{u'v - v'u}{v^2}$$

$$\frac{du}{dx} = \frac{du}{dy} \cdot \frac{dy}{dx} \qquad (\text{Chain rule})$$

$$(x^n)' = nx^{n-1}$$

$$(e^x)' = e^x$$

$$(a^x)' = a^x \ln a$$

$$(\sin x)' = \cos x$$

$$(\cos x)' = -\sin x$$

$$(\tan x)' = \sec^2 x$$

$$(\cot x)' = -\csc^2 x$$

$$(\sinh x)' = \cosh x$$

$$(\cosh x)' = \sinh x$$

$$(\ln x)' = \frac{1}{x}$$

$$(\log_a x)' = \frac{\log_a e}{x}$$

$$(\arcsin x)' = \frac{1}{\sqrt{1 - x^2}}$$

$$(\arccos x)' = -\frac{1}{\sqrt{1 - x^2}}$$

$$(\arctan x)' = \frac{1}{1 + x^2}$$

$$(\text{arc cot } x)' = -\frac{1}{1 + x^2}$$

Integration

$$\int uv' \, dx = uv - \int u'v \, dx$$

$$\int x^n \, dx = \frac{x^{n+1}}{n + 1} + c \qquad (n \neq -1)$$

$$\int \frac{1}{x} \, dx = \ln |x| + c$$

$$\int e^{ax} \, dx = \frac{1}{a} e^{ax} + c$$

$$\int \sin x \, dx = -\cos x + c$$

$$\int \cos x \, dx = \sin x + c$$

$$\int \tan x \, dx = -\ln |\cos x| + c$$

$$\int \cot x \, dx = \ln |\sin x| + c$$

$$\int \sec x \, dx = \ln |\sec x + \tan x| + c$$

$$\int \csc x \, dx = \ln |\csc x - \cot x| + c$$

$$\int \frac{dx}{x^2 + a^2} = \frac{1}{a} \arctan \frac{x}{a} + c$$

$$\int \frac{dx}{\sqrt{a^2 - x^2}} = \arcsin \frac{x}{a} + c$$

$$\int \frac{dx}{\sqrt{x^2 + a^2}} = \sinh^{-1} \frac{x}{a} + c$$

$$\int \frac{dx}{\sqrt{x^2 - a^2}} = \cosh^{-1} \frac{x}{a} + c$$

$$\int \sin^2 x \, dx = \tfrac{1}{2}x - \tfrac{1}{4} \sin 2x + c$$

$$\int \cos^2 x \, dx = \tfrac{1}{2}x + \tfrac{1}{4} \sin 2x + c$$

$$\int \tan^2 x \, dx = \tan x - x + c$$

$$\int \cot^2 x \, dx = -\cot x - x + c$$

$$\int \ln x \, dx = x \ln x - x + c$$

$$\int e^{ax} \sin bx \, dx$$
$$= \frac{e^{ax}}{a^2 + b^2} (a \sin bx - b \cos bx) + c$$

$$\int e^{ax} \cos bx \, dx$$
$$= \frac{e^{ax}}{a^2 + b^2} (a \cos bx + b \sin bx) + c$$

Advanced Engineering Mathematics

EIGHTH EDITION

Advanced Engineering Mathematics

ERWIN KREYSZIG
Professor of Mathematics
Ohio State University
Columbus, Ohio

JOHN WILEY & SONS, INC.
New York   Chichester   Brisbane   Toronto   Singapore

Publisher: Peter Janzow
Mathematics Editor: Barbara Holland
Marketing Manager: Audra Silveric
Freelance Production Manager: Jeanine Furino
        Lorraine Burke HRS Electronic Text Management
Designer: HRS Electronic Text Management
Illustration Editor: Sigmund Malinowski
Electronic Illustrations: Precision Graphics
Cover Photo: Chris Rogers/The Stock Market

This book was set in Times Roman by GGS Information Services
and printed and bound by Von Hoffmann Press.
The cover was printed by Phoenix Color Corp.

This book is printed on acid-free paper. ∞™

Kreyszig, Erwin.
        Advanced engineering mathematics / Erwin Kreyszig.—8th ed.
            p.   cm.
        Accompanied by instructor's manual.
        Includes bibliographical references and index.
        ISBN 0-471-15496-2 (cloth : acid-free paper)
        1. Mathematical physics.   2. Engineering mathematics.   1. Title.

Printed in the United States of America
10 9 8 7 6

PREFACE

See also http://www.wiley.com/college/mat/kreyszig154962/

Purpose of the Book

This book introduces students of engineering, physics, mathematics, and computer science to those areas of mathematics which, from a modern point of view, are most important in connection with practical problems.

The content and character of mathematics needed in applications are changing rapidly. Linear algebra—especially matrices—and numerical methods for computers are of increasing importance. Statistics and graph theory play more prominent roles. Real analysis (ordinary and partial differential equations) and complex analysis remain indispensable. The material in this book is arranged accordingly, in seven independent parts (see also the diagram on the next page):

A Ordinary Differential Equations   (Chaps. 1–5)
B Linear Algebra, Vector Calculus   (Chaps. 6–9)
C Fourier Analysis and Partial Differential Equations   (Chaps. 10, 11)
D Complex Analysis   (Chaps. 12–16)
E Numerical Methods   (Chaps. 17–19)
F Optimization, Graphs   (Chaps. 20, 21)
G Probability and Statistics   (Chaps. 22, 23)

This is followed by

References   (Appendix 1)
Answers to Odd-Numbered Problems   (Appendix 2)
Auxiliary Material   (Appendix 3 and inside of covers)
Additional Proofs   (Appendix 4)
Tables of Functions   (Appendix 5).

This book has helped to pave the way for the present development and will prepare students for the present situation and the future by a modern approach to the areas listed above and the ideas—some of them computer related—that are presently causing basic changes: Many methods have become obsolete. New ideas are emphasized, for instance, stability, error estimation, and structural problems of algorithms, to mention just a few. Trends are driven by supply and demand: supply of powerful new mathematical and computational methods and of enormous computer capacities, demand to solve problems of growing complexity and size, arising from more and more sophisticated systems or production processes, from extreme physical conditions (for example, those in space travel), from materials with unusual properties (plastics, alloys, superconductors, etc.), or from entirely new tasks in computer vision, robotics, and other new fields.

The general trend seems clear. Details are more difficult to predict. Accordingly, students need solid knowledge of basic principles, methods, and results, and a clear perception of what engineering mathematics is all about, in all three phases of solving problems:

- **Modeling:** Translating given physical or other information and data into mathematical form, into a mathematical *model* (a differential equation, a system of equations, or some other expression).

Parts of the Book and Corresponding Chapters

PART A
Chaps. 1–5
Ordinary differential equations
Chaps. 1–3 Basic material

Chap. 4 Series solutions, Special functions	Chap. 5 Laplace transforms

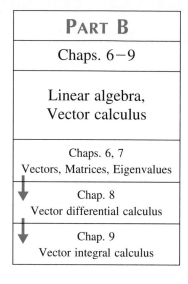

PART B
Chaps. 6–9
Linear algebra, Vector calculus
Chaps. 6, 7 Vectors, Matrices, Eigenvalues
Chap. 8 Vector differential calculus
Chap. 9 Vector integral calculus

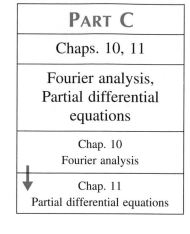

PART C
Chaps. 10, 11
Fourier analysis, Partial differential equations
Chap. 10 Fourier analysis
Chap. 11 Partial differential equations

PART D
Chaps. 12–16
Complex analysis
Chap. 12–15 Basic material
Chap. 16 Potential theory

PART E
Chaps. 17–19
Numerical methods

Chap. 17 General numerical methods	Chap. 18 Methods for linear algebra	Chap. 19 Methods for differential equations

PART F
Chaps. 20, 21
Optimization, Graphs

Chap. 20 Linear programming	Chap. 21 Graphs, Combinatorial optimization

PART G
Chaps. 22, 23
Probability, Statistics
Chap. 22 Probability theory
Chap. 23 Mathematical statistics

GUIDES AND MANUAL
MAPLE Computer Guide MATHEMATICA Computer Guide
STUDENT SOLUTIONS MANUAL (New!)

- **Solving:** Obtaining the solution by selecting and applying suitable mathematical methods, and in most cases doing numerical work on a computer. This is the main task of this book.

- **Interpreting:** Understanding the meaning and the implications of the mathematical solution for the original problem in terms of physics—or wherever the problem comes from.

It would make no sense to overload students with all kinds of little things that might be of occasional use. Instead, it is important that students become familiar with ways to think mathematically, recognize the need for applying mathematical methods to engineering problems, realize that mathematics is a systematic science built on relatively few basic concepts and involving powerful unifying principles, and get a firm grasp for the interrelation between theory, computing, and experiment.

The rapid ongoing developments just sketched have led to many changes and new features in the present edition of this book.

In particular, many sections have been rewritten in a more detailed and leisurely fashion to make it a simpler book.

This has also led to a still better balance between applications, algorithmic ideas, worked-out examples, and theory.

Big Changes in This Edition

PROBLEM SETS CHANGED

The new problems place more emphasis on qualitative methods and applications. There is a (slight) reduction of formal manipulations in favor of problems that require mathematical thinking and understanding, as opposed to a routine use of a CAS (Computer Algebraic System).

PROJECTS

Modern engineering work is team work, and TEAM PROJECTS will help the student to prepare for this. (These are relatively simple, so that they will fit into the time schedule of a busy student.) WRITING PROJECTS will help in learning how to plan, develop, and write coherent reports. CAS PROJECTS and CAS PROBLEMS will invite the student to an increased use of computers (and programmable calculators); these projects are not mandatory, simply because *this book can be used independently of computers or in connection with them* (see page x).

NUMERICAL ANALYSIS UPDATED

Details are given below.

Further Changes and New Features in Chapters

Ordinary Differential Equations (Chaps. 1–5)

▶ **First-Order Differential Equations** (Chap. 1). Qualitative aspects emphasized by discussing direction fields early (Sec. 1.2). Presentation streamlined by combining exact equations and integrating factors into one section (Sec. 1.5) and moving Picard's iteration to Sec. 1.9 on existence and uniqueness.

▶ **Linear Second and Higher Order Differential Equations** combined into one chapter (Chap. 2), to save some time by avoiding duplications.

▶ **Systems of Differential Equations** (Chap. 3). Lotka–Volterra predator–prey population model included.

▶ **Frobenius Method** (Chap. 4). Material on Bessel functions slightly reduced.

▶ **Laplace Transforms** (Chap. 5). New section on systems of differential equations included.

Linear Algebra, Vector Calculus (Chaps. 6–9)

▶ **Matrices, Linear Systems** (Chap. 6). Slightly more rapid start by combining the first two sections from the last edition. Applications appearing earlier. Cramer's rule absorbed into the (slightly condensed) section on determinants (Sec. 6.6). Inverse treated more compactly (Sec. 6.7).

▶ **Eigenvalue Problems** placed in a separate chapter (Chap. 7), to have the basic material on vectors and matrices in a chapter of its own.

Fourier Analysis and Partial Differential Equations (Chaps. 10, 11)

▶ **Fourier Series** (Chap. 10) streamlined by moving half-range expansions into the section on even and odd functions (Sec. 10.4).

▶ **Partial Differential Equations** (Chap. 11). Fourier transform method absorbed into the section on Fourier integrals for heat problems (Sec. 11.6).

Complex Analysis (Chaps. 12–16)

▶ **Complex Numbers and Functions** (Chap. 12). The old chapter on conformal mapping no longer exists. Its (slightly reduced) material has been distributed in Sec. 12.5 on conformality, Secs. 12.6–12.8 on special functions, and Sec. 12.9 on linear fractional transformations. Hence for a better understanding we now discuss geometric properties of functions simultaneously with their analytic formulas, as we do all the time in calculus.

▶ **Complex Integration** (Chap. 13). Integration methods right after the definition of the integral.

▶ **Laurent Series,** formerly in a chapter jointly with power series, combined with residue integration in Chap. 15.

Numerical Methods (Chaps. 17–19)

▶ **Numerical Methods in General** (Chap. 17). Updated in the light of computer requirements and developments. Idea of error estimation by halving. Changes in Sec. 17.4 on splines, in Sec. 17.5 on error estimates in integration. Adaptive integration and Romberg integration included (Sec. 17.5).

▶ **Methods for Differential Equations** (Chap. 19). Automatic variable step size selection in modern codes, Runge–Kutta–Fehlberg method (Sec. 19.1), extension of Euler and Runge–Kutta methods to systems and higher order equations (Sec. 19.3) included.

Optimization, Graphs (Chaps. 20, 21)

▶ **Linear Programming** (Chap. 20). Simplex method completely rewritten in terms of matrix language and techniques.

Probability and Statistics (Chaps. 22, 23)

▶ **Probability** (Chap. 22) beginning with a section on data analysis, explaining stem-and-leaf plots and boxplots and motivating probability by relative frequency (Sec. 22.1).

▶ **Statistics** (Chap. 23) beginning with a section on the use of random number generators (Sec. 23.1). Introduction to correlation added (Sec. 23.10).

Appendices

▶ **Appendix 1** (References) updated.

Suggestions for Courses: A Four-Semester Sequence

The material may be taken in sequence and is suitable for four consecutive semester courses, meeting $3-5$ hours a week:

First semester.	Ordinary differential equations (Chaps. 1–4 or 5)
Second semester.	Linear algebra and vector analysis (Chaps. 6–9)
Third semester.	Complex analysis (Chaps. 12–16)
Fourth semester.	Numerical methods (Chaps. 17–19)

For the remaining chapters, see below. Possible interchanges are obvious; for instance, numerical methods could precede complex analysis, etc.

Suggestions for Courses: Independent One-Semester Courses

The book is also suitable for various independent one-semester courses meeting 3 hours a week; for example:

Introduction to ordinary differential equations (Chaps. $1-2$)
Laplace transform (Chap. 5)
Vector algebra and calculus (Chaps. 8, 9)
Matrices and linear systems of equations (Chaps. 6, 7)
Fourier series and partial differential equations (Chaps. 10, 11, Secs. 19.4–19.7)
Introduction to complex analysis (Chaps. $12-15$)
Numerical analysis (Chaps. 17, 19)
Numerical linear algebra (Chap. 18)
Optimization (Chaps. 20, 21)
Graphs and combinatorial optimization (Chap. 21)
Probability and statistics (Chaps. 22, 23)

General Features of This Edition

The selection, arrangement, and presentation of the material has been made with greatest care, based on past and present teaching, research, and consulting experience. Some major features of the book are these:

The book is **self-contained,** except for a few clearly marked places where a proof would be beyond the level of a book of the present type and a reference is given instead. Hiding difficulties or oversimplifying would be of no real help to students.

The presentation is **detailed,** to avoid irritating readers by frequent references to details in other books.

The examples are **simple,** to make the book teachable—why choose complicated examples when simple ones are as instructive or even better?

The notations are **modern and standard,** to help students read articles in journals or other *modern* books and understand other mathematically oriented courses.

The chapters are largely **independent,** providing flexibility in teaching special courses (see above).

Computer Use and CASs (Computer Algebraic Systems)

The use of computers and (programmable) calculators is invited but not requested.

This technology may be helpful in solving many of the about 4000 problems in this book. ***Intelligent utilization*** of these amazingly powerful and versatile systems may give the student additional motivation, insight, and help in working in classes, tutorials, labs, and at home, as well as in the preparation for future jobs after graduation.

For these reasons we have included CAS projects as useful enrichments of the problem sets, which, however, remain complete entities without them.

Similarly, working through the text is possible without computer use.

A list of software is included before Chap. 17, the first chapter on numerical methods, on p. 829.

Acknowledgments

I am indebted to many of my former teachers, colleagues, and students who directly or indirectly helped me in preparing this book, in particular, the present edition of it. Various parts of the manuscript were copied and distributed to my classes and returned to me with suggestions for improvement. Discussions with engineers and mathematicians (as well as written comments) were of great help to me; I want to mention particularly Professors S. L. Campbell, J. T. Cargo, R. Carr, P. L. Chambré, V. F. Connolly, J. Delany, J. W. Dettman, D. Dicker, L. D. Drager, D. Ellis, W. Fox, R. B. Guenther, J. L. Handley, V. W. Howe, W. N. Huff, J. Keener, V. Komkow, H. Kuhn, G. Lamb, H. B. Mann, I. Marx, K. Millet, J. D. Moore, W. D. Munroe, A. Nadim, J. N. Ong, Jr., P. J. Pritchard, W. O. Ray, J. T. Scheick, L. F. Shampine, H. A. Smith, J. Todd, H. Unz, A. L. Villone, H. J. Weiss, A. Wilansky, C. H. Wilcox, H. Ya Fan, L. Zia, A. D. Ziebur, all from the U.S.A., Professors H. S. M. Coxeter, E. J. Norminton, R. Vaillancourt, and Mr. H. Kreyszig (whose computer expertise was of great help in Chaps. 17–19) from Canada, and Professors H. Florian, H. Unger, and H. Wielandt from Europe. I can offer here only an inadequate acknowledgment of my appreciation.

My very special cordial thanks goes to Privatdozent Dr. M. Kracht for the formidable task of checking all the details of the manuscript, resulting in many substantial improvements.

I also want to thank Ms. Barbara Holland, Editor, for her unusually great help and effort during the periods of preparing the manuscript and producing the book.

Furthermore, I wish to thank John Wiley and Sons (see the list on p. iv) as well as Mr. J. T. Nystrom and Ms. C. A. Elicker of GGS Information Services for their effective cooperation and great care in preparing this edition.

Suggestions of many readers were evaluated in preparing the present edition. Any further comments and suggestions for improvement of the book will be gratefully received.

ERWIN KREYSZIG

CONTENTS

Part A. Ordinary Differential Equations. 1

Chapter 1   First-Order Differential Equations. 2

1.1   Basic Concepts and Ideas, 2
1.2   Geometrical Meaning of $y' = f(x, y)$. Direction Fields, 10
1.3   Separable Differential Equations, 14
1.4   Modeling: Separable Equations, 19
1.5   Exact Differential Equations. Integrating Factors, 25
1.6   Linear Differential Equations. Bernoulli Equation, 33
1.7   Modeling: Electric Circuits, 41
1.8   Orthogonal Trajectories of Curves. *Optional,* 48
1.9   Existence and Uniqueness of Solutions. Picard Iteration, 52
Chapter Review, 59
Chapter Summary, 61

Chapter 2   Linear Differential Equations of Second and Higher Order 64

2.1   Homogeneous Linear Equations of Second Order, 64
2.2   Second-Order Homogeneous Equations with Constant Coefficients, 72
2.3   Case of Complex Roots. Complex Exponential Function, 76
2.4   Differential Operators. *Optional,* 81
2.5   Modeling: Free Oscillations (Mass–Spring System), 83
2.6   Euler–Cauchy Equation, 93
2.7   Existence and Uniqueness Theory. Wronskian, 97
2.8   Nonhomogeneous Equations, 101
2.9   Solution by Undetermined Coefficients, 104
2.10   Solution by Variation of Parameters, 108
2.11   Modeling: Forced Oscillations. Resonance, 111
2.12   Modeling of Electric Circuits, 118
2.13   Higher Order Linear Differential Equations, 124
2.14   Higher Order Homogeneous Equations with Constant Coefficients, 132
2.15   Higher Order Nonhomogeneous Equations, 138
Chapter Review, 142
Chapter Summary, 143

Chapter 3   Systems of Differential Equations, Phase Plane, Qualitative Methods. 146

3.0   Introduction: Vectors, Matrices, Eigenvalues, 146
3.1   Introductory Examples, 152
3.2   Basic Concepts and Theory, 159
3.3   Homogeneous Systems with Constant Coefficients. Phase Plane, Critical Points, 162
3.4   Criteria for Critical Points. Stability, 170
3.5   Qualitative Methods for Nonlinear Systems, 175
3.6   Nonhomogeneous Linear Systems, 184
Chapter Review, 190
Chapter Summary, 192

Chapter 4   Series Solutions of Differential Equations.
**   Special Functions** **194**
4.1   Power Series Method, 194
4.2   Theory of the Power Series Method, 198
4.3   Legendre's Equation. Legendre Polynomials $P_n(x)$, 205
4.4   Frobenius Method, 211
4.5   Bessel's Equation. Bessel Functions $J_\nu(x)$, 218
4.6   Bessel Functions of the Second Kind $Y_\nu(x)$, 228
4.7   Sturm–Liouville Problems. Orthogonal Functions, 233
4.8   Orthogonal Eigenfunction Expansions, 240
Chapter Review, 247
Chapter Summary, 248

Chapter 5   Laplace Transforms **250**
5.1   Laplace Transform. Inverse Transform. Linearity. Shifting, 251
5.2   Transforms of Derivatives and Integrals. Differential Equations, 258
5.3   Unit Step Function. Second Shifting Theorem. Dirac's Delta Function, 265
5.4   Differentiation and Integration of Transforms, 275
5.5   Convolution. Integral Equations, 279
5.6   Partial Fractions. Differential Equations, 284
5.7   Systems of Differential Equations, 291
5.8   Laplace Transform: General Formulas, 296
5.9   Table of Laplace Transforms, 297
Chapter Review, 299
Chapter Summary, 302

Part B.   Linear Algebra, Vector Calculus. **303**

Chapter 6   Linear Algebra: Matrices, Vectors, Determinants.
**   Linear Systems of Equations** **304**
6.1   Basic Concepts. Matrix Addition, Scalar Multiplication, 305
6.2   Matrix Multiplication, 311
6.3   Linear Systems of Equations. Gauss Elimination, 321
6.4   Rank of a Matrix. Linear Independence. Vector Space, 331
6.5   Solutions of Linear Systems: Existence, Uniqueness, General Form, 338
6.6   Determinants. Cramer's Rule, 341
6.7   Inverse of a Matrix. Gauss–Jordan Elimination, 350
6.8   Vector Spaces, Inner Product Spaces, Linear Transformations. *Optional,* 358
Chapter Review, 365
Chapter Summary, 367

Chapter 7   Linear Algebra: Matrix Eigenvalue Problems **370**
7.1   Eigenvalues, Eigenvectors, 371
7.2   Some Applications of Eigenvalue Problems, 376
7.3   Symmetric, Skew-Symmetric, and Orthogonal Matrices, 381
7.4   Complex Matrices: Hermitian, Skew-Hermitian, Unitary, 385
7.5   Similarity of Matrices. Basis of Eigenvectors. Diagonalization, 392
Chapter Review, 398
Chapter Summary, 399

Chapter 8   Vector Differential Calculus. Grad, Div, Curl. **400**
8.1   Vector Algebra in 2-Space and 3-Space, 401
8.2   Inner Product (Dot Product), 408

8.3   Vector Product (Cross Product), 414
8.4   Vector and Scalar Functions and Fields. Derivatives, 423
8.5   Curves. Tangents. Arc Length, 428
8.6   Curves in Mechanics. Velocity and Acceleration, 435
8.7   Curvature and Torsion of a Curve. *Optional,* 440
8.8   Review from Calculus in Several Variables. *Optional,* 443
8.9   Gradient of a Scalar Field. Directional Derivative, 446
8.10   Divergence of a Vector Field, 453
8.11   Curl of a Vector Field, 457
Chapter Review, 459
Chapter Summary, 461

Chapter 9   Vector Integral Calculus. Integral Theorems 464
9.1   Line Integrals, 464
9.2   Line Integrals Independent of Path, 471
9.3   From Calculus: Double Integrals. *Optional,* 478
9.4   Green's Theorem in the Plane, 485
9.5   Surfaces for Surface Integrals, 491
9.6   Surface Integrals, 496
9.7   Triple Integrals. Divergence Theorem of Gauss, 505
9.8   Further Applications of the Divergence Theorem, 510
9.9   Stokes's Theorem, 515
Chapter Review, 521
Chapter Summary, 522

Part C.   Fourier Analysis and Partial Differential Equations 525

Chapter 10   Fourier Series, Integrals, and Transforms 526
10.1   Periodic Functions. Trigonometric Series, 527
10.2   Fourier Series, 529
10.3   Functions of Any Period $p = 2L$, 537
10.4   Even and Odd Functions. Half-Range Expansions, 541
10.5   Complex Fourier Series. *Optional,* 547
10.6   Forced Oscillations, 550
10.7   Approximation by Trigonometric Polynomials, 553
10.8   Fourier Integrals, 557
10.9   Fourier Cosine and Sine Transforms, 564
10.10 Fourier Transform, 569
10.11 Tables of Transforms, 576
Chapter Review, 579
Chapter Summary, 580

Chapter 11   Partial Differential Equations 582
11.1   Basic Concepts, 583
11.2   Modeling: Vibrating String, Wave Equation, 585
11.3   Separation of Variables. Use of Fourier Series, 587
11.4   D'Alembert's Solution of the Wave Equation, 595
11.5   Heat Equation: Solution by Fourier Series, 600
11.6   Heat Equation: Solution by Fourier Integrals and Transforms, 610
11.7   Modeling: Membrane, Two-Dimensional Wave Equation, 616
11.8   Rectangular Membrane. Use of Double Fourier Series, 619
11.9   Laplacian in Polar Coordinates, 626
11.10 Circular Membrane. Use of Fourier–Bessel Series, 629

11.11 Laplace's Equation in Cylindrical and Spherical Coordinates. Potential, 636
11.12 Solution by Laplace Transforms, 643
Chapter Review, 647
Chapter Summary, 648

Part D.   Complex Analysis. 651

Chapter 12   Complex Numbers and Functions.
Conformal Mapping. 652
12.1   Complex Numbers. Complex Plane, 652
12.2   Polar Form of Complex Numbers. Powers and Roots, 657
12.3   Derivative. Analytic Function, 663
12.4   Cauchy–Riemann Equations. Laplace's Equation, 669
12.5   Geometry of Analytic Functions: Conformal Mapping, 674
12.6   Exponential Function, 679
12.7   Trigonometric Functions, Hyperbolic Functions, 682
12.8   Logarithm. General Power, 687
12.9   Linear Fractional Transformations. *Optional,* 692
12.10 Riemann Surfaces. *Optional,* 699
Chapter Review, 701
Chapter Summary, 702

Chapter 13   Complex Integration 704
13.1   Line Integral in the Complex Plane, 704
13.2   Cauchy's Integral Theorem, 713
13.3   Cauchy's Integral Formula, 721
13.4   Derivatives of Analytic Functions, 725
Chapter Review, 730
Chapter Summary, 731

Chapter 14   Power Series, Taylor Series 732
14.1   Sequences, Series, Convergence Tests, 732
14.2   Power Series, 741
14.3   Functions Given by Power Series, 746
14.4   Taylor Series and Maclaurin Series, 751
14.5   Uniform Convergence. *Optional,* 759
Chapter Review, 767
Chapter Summary, 768

Chapter 15   Laurent Series, Residue Integration 770
15.1   Laurent Series, 770
15.2   Singularities and Zeros. Infinity, 776
15.3   Residue Integration Method, 781
15.4   Evaluation of Real Integrals, 787
Chapter Review, 794
Chapter Summary, 796

Chapter 16   Complex Analysis Applied to Potential Theory. 798
16.1   Electrostatic Fields, 799
16.2   Use of Conformal Mapping, 804
16.3   Heat Problems, 808
16.4   Fluid Flow, 812
16.5   Poisson's Integral Formula, 819
16.6   General Properties of Harmonic Functions, 822

Chapter Review, 826
Chapter Summary, 827

Part E. Numerical Methods 828

Software 829

Chapter 17 Numerical Methods in General. 830
17.1 Introduction: Floating Point. Round-off, Error Propagation, etc., 831
17.2 Solution of Equations by Iteration, 838
17.3 Interpolation, 848
17.4 Splines, 861
17.5 Numerical Integration and Differentiation, 869
Chapter Review, 882
Chapter Summary, 884

Chapter 18 Numerical Methods in Linear Algebra. 886
18.1 Linear Systems: Gauss Elimination, 886
18.2 Linear Systems: LU-Factorization, Matrix Inversion, 894
18.3 Linear Systems: Solution by Iteration, 900
18.4 Linear Systems: Ill-Conditioning, Norms, 906
18.5 Method of Least Squares, 914
18.6 Matrix Eigenvalue Problems: Introduction, 917
18.7 Inclusion of Matrix Eigenvalues, 920
18.8 Eigenvalues by Iteration (Power Method), 925
18.9 Tridiagonalization and QR-Factorization, 929
Chapter Review, 938
Chapter Summary, 940

Chapter 19 Numerical Methods for Differential Equations. 942
19.1 Methods for First-Order Differential Equations, 942
19.2 Multistep Methods, 952
19.3 Methods for Systems and Higher Order Equations, 956
19.4 Methods for Elliptic Partial Differential Equations, 962
19.5 Neumann and Mixed Problems. Irregular Boundary, 971
19.6 Methods for Parabolic Equations, 976
19.7 Methods for Hyperbolic Equations, 982
Chapter Review, 984
Chapter Summary, 987

Part F. Optimization, Graphs. 989

Chapter 20 Unconstrained Optimization, Linear Programming 990
20.1 Basic Concepts. Unconstrained Optimization, 990
20.2 Linear Programming, 994
20.3 Simplex Method, 998
20.4 Simplex Method: Degeneracy, Difficulties in Starting, 1002
Chapter Review, 1007
Chapter Summary, 1008

Chapter 21 Graphs and Combinatorial Optimization 1010
21.1 Graphs and Digraphs, 1010
21.2 Shortest Path Problems. Complexity, 1015
21.3 Bellman's Optimality Principle. Dijkstra's Algorithm, 1020
21.4 Shortest Spanning Trees. Kruskal's Greedy Algorithm, 1024

21.5   Prim's Algorithm for Shortest Spanning Trees, 1028
21.6   Networks. Flow Augmenting Paths, 1031
21.7   Ford–Fulkerson Algorithm for Maximum Flow, 1038
21.8   Assignment Problems. Bipartite Matching, 1041
Chapter Review, 1046
Chapter Summary, 1048

Part G.   Probability and Statistics **1049**

Chapter 22   Data Analysis. Probability Theory **1050**
22.1   Data: Representation, Average, Spread, 1050
22.2   Experiments, Outcomes, Events, 1055
22.3   Probability, 1058
22.4   Permutations and Combinations, 1064
22.5   Random Variables, Probability Distributions, 1069
22.6   Mean and Variance of a Distribution, 1075
22.7   Binomial, Poisson, and Hypergeometric Distributions, 1079
22.8   Normal Distribution, 1085
22.9   Distributions of Several Random Variables, 1091
Chapter Review, 1100
Chapter Summary, 1102

Chapter 23   Mathematical Statistics **1104**
23.1   Introduction. Random Sampling, 1104
23.2   Estimation of Parameters, 1106
23.3   Confidence Intervals, 1109
23.4   Testing of Hypotheses, Decisions, 1118
23.5   Quality Control, 1128
23.6   Acceptance Sampling, 1133
23.7   Goodness of Fit. χ^2-Test, 1137
23.8   Nonparametric Tests, 1142
23.9   Regression Analysis. Fitting Straight Lines, 1145
23.10 Correlation Analysis, 1150
Chapter Review, 1153
Chapter Summary, 1155

Appendix 1   References **A1**
Appendix 2   Answers to Odd-Numbered Problems **A5**
Appendix 3   Auxiliary Material **A51**

A3.1   Formulas for Special Functions, A51
A3.2   Partial Derivatives, A57
A3.3   Sequences and Series, A60

Appendix 4   Additional Proofs **A65**
Appendix 5   Tables **A85**
Index **I1**

P A R T   A

Ordinary Differential Equations

Chapter 1   **First-Order Differential Equations**

Chapter 2   **Linear Differential Equations
of Second and Higher Order**

Chapter 3   **Systems of Differential Equations.
Phase Plane, Qualitative Methods**

Chapter 4   **Series Solutions of Differential Equations.
Special Functions**

Chapter 5   **Laplace Transforms**

Differential equations are of fundamental importance in engineering mathematics because many physical laws and relations appear mathematically in the form of such equations. In Part A, which consists of five chapters, we shall consider various physical and geometrical problems that lead to differential equations, and we shall explain the most important standard methods for solving such equations.

Modeling. We shall pay particular attention to the derivation of differential equations from given physical (or other) situations. This transition from the given physical problem to a corresponding "mathematical model" is called *modeling*. This is of great practical importance to the engineer, physicist, and computer scientist, and we shall illustrate it using typical examples.

Computers. Differential equations are very well suited for computers. Corresponding **NUMERICAL METHODS** for solving differential equations are explained in Secs. 19.1–19.3. These sections are independent of other sections on numerical methods, so that they can be studied directly after Chaps. 1 and 2, respectively.

Evaluating Results. We must make sure that we understand what a mathematical result means in physical or other terms in a given problem. If we obtained the result using a computer, we must check the result for reliability—the computer can sometimes give us nonsense. This applies to all the work with computers.

CHAPTER 1

First-Order Differential Equations

In this chapter we begin our program of studying ordinary differential equations and their applications. This includes the derivation of differential equations from physical or other problems (**modeling**), the solution of these equations by methods of practical importance, and the interpretation of the results and their graphs in terms of a given problem. We also discuss the questions of existence and uniqueness of solutions.

We start with the simplest equations. These are called **differential equations of the first order** because they involve only the *first* derivative of the unknown function. Our usual notation for the unknown function will be $y(x)$ or $y(t)$.

Numerical methods for these equations follow in Secs. 19.1 and 19.2, which are totally independent of other sections in Chaps. 17–19, and can be taken up immediately after this chapter.

Prerequisite for this chapter: integral calculus.
Sections that may be omitted in a shorter course: 1.7–1.9.
References: Appendix 1, Part A.
Answers to Problems: Appendix 2.

1.1 Basic Concepts and Ideas

An **ordinary differential equation** is an equation that contains one or several derivatives of an unknown function, which we call $y(x)$ and which we want to determine from the equation. The equation may also contain y itself as well as given functions and constants. For example,

$$(1) \qquad\qquad y' = \cos x,$$

$$(2) \qquad\qquad y'' + 4y = 0,$$

$$(3) \qquad\qquad x^2 y''' y' + 2e^x y'' = (x^2 + 2)y^2$$

are ordinary differential equations. The word *"ordinary"* distinguishes them from *partial* differential equations, involving an unknown function of two or more variables and its *partial* derivatives; these equations are more complicated and will be considered later (in Chap. 11).

Differential equations arise in many engineering and other applications as mathematical models of various physical and other systems. The simplest of them can be solved by remembering elementary calculus.

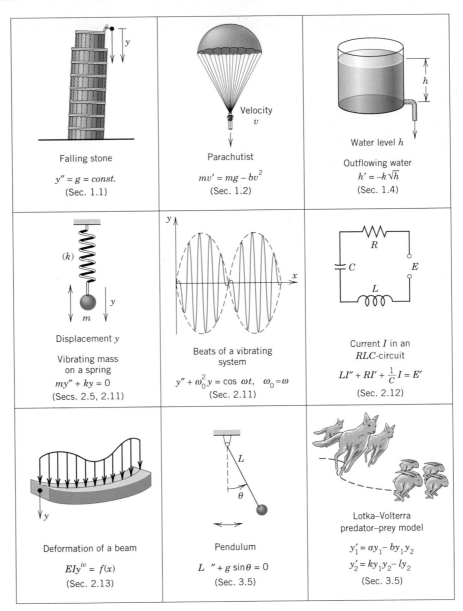

Some Applications of Differential Equations

For example, if a population (of humans, animals, bacteria, etc.) grows at a rate $y' = dy/dx$ (x = time) equal to the population $y(x)$ present, the population model is $y' = y$, a differential equation. If we remember from calculus that $y = e^x$ (or more generally $y = ce^x$) has the property that $y' = y$, we have obtained a solution of our problem.

As another example, if we drop a stone, then its acceleration $y'' = d^2y/dx^2$ (x = time, as before) is equal to the acceleration of gravity g (a constant). Hence the model of this problem of "free fall" is $y'' = g$, in good approximation, since the air resistance will not matter too much in this case. By integration we get the velocity $y' = dy/dx = gx + v_0$, where v_0 is the initial velocity with which the motion started (e.g., $v_0 = 0$). Integrating once more, we get the distance traveled $y = \frac{1}{2}gx^2 + v_0 x + y_0$, where y_0 is the distance from 0 at the beginning (e.g., $y_0 = 0$).

More complicated practical problems, as illustrated by various figures in this book, lead to differential equations whose solution needs more refined methods. We shall discuss such methods systematically. This begins with a classification of differential equations by "order."

The **order** of a differential equation is the order of the highest derivative that appears in the equation.

Thus **first-order differential equations,** to be considered in this chapter, contain only y' and may contain y and given functions of x. Hence we can write them

(4)
$$F(x, y, y') = 0$$

or sometimes

$$y' = f(x, y).$$

Examples are (1) and $y' = y$ just considered. Equations (2) and (3) are of second and third order, respectively; such higher order differential equations will be discussed in Chaps. 2–5.

Concept of Solution

A **solution** of a given first-order differential equation (4) on some open interval[1] $a < x < b$ is a function $y = h(x)$ that has a derivative $y' = h'(x)$ and satisfies (4) for all x in that interval; that is, (4) becomes an identity if we replace the unknown function y by h and y' by h'.

EXAMPLE 1   **Concept of solution. Verification of solution**

Verify that $y = x^2$ is a solution of $xy' = 2y$ for all x.

Indeed, by substituting $y = x^2$ and $y' = 2x$ into the equation we obtain $xy' = x(2x) = 2x^2 = 2y$, an identity in x.   ◀

Sometimes a solution of a differential equation will appear as an implicit function, that is, implicitly given in the form

$$H(x, y) = 0,$$

and is called an *implicit solution,* in contrast to an *explicit solution* $y = h(x)$.

EXAMPLE 2   **Implicit solution**

The function y of x implicitly given by $x^2 + y^2 - 1 = 0$ ($y > 0$) represents a semicircle of unit radius in the upper half-plane. This function is an implicit solution of the differential equation $yy' = -x$, on the interval $-1 < x < 1$, as the student may verify by differentiation.   ◀

We next observe that a differential equation may (and in general, will) have many solutions. This should not really surprise us because we know from calculus that integration introduces arbitrary constants.

---

[1]By definition, the concept of **interval** includes as special cases $a < x < \infty$ and $-\infty < x < b$ as well as the whole x-axis $-\infty < x < \infty$. All our intervals are **open,** that is, their endpoints are not regarded as points belonging to the interval.

EXAMPLE 3   Our equation (1) was $y' = \cos x$ and can be solved by calculus. Integration gives sine curves $y = \sin x + c$ with arbitrary c. Each c gives one of them, and these are all possible solutions, as we know from calculus. Figure 1 shows some of them, for $c = -3, -2, -1, 0, 1, 2, 3, 4$.   ◀

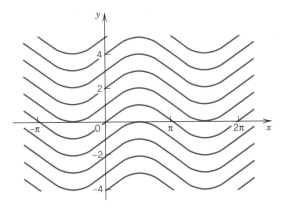

Fig. 1.   Solutions of $y' = \cos x$

This example, simple as it is, is typical of most equations of first order. It illustrates that all solutions are represented by a single formula involving an arbitrary constant c (additively as here, multiplicatively as in $y = ce^x$ above or in some other way). Such a function involving an arbitrary[2] constant is called a **general solution** of a first-order differential equation. Geometrically, these are infinitely many curves, one for each c. We call this a **family of curves.** And if we choose a specific c ($c = 2$ or 0 or $-5/3$, etc.), we obtain what is called a **particular solution** of that equation.

Thus, $y = \sin x + c$ is a general solution of $y' = \cos x$, and $y = \sin x$, $y = \sin x - 2$, $y = \sin x + 0.75$, etc. are particular solutions.

In the following sections we shall develop various methods for obtaining general solutions of first-order equations. For a given equation, a general solution obtained by such a method is unique, except for notation, and will then be called ***the*** general solution of that differential equation.

COMMENT   **Singular solutions**

A differential equation may sometimes have an additional solution that cannot be obtained from the general solution and is then called a **singular solution.** This is not of great engineering interest, and we mention it merely for completeness. For example,

(5) $y'^2 - xy' + y = 0$

has the general solution $y = cx - c^2$, as the student may verify by differentiation and substitution. This represents a family of straight lines, one line for each c. These are the particular solutions shown in Fig. 2. Substitution also shows that the parabola $y = x^2/4$ in Fig. 2 is also a solution. This is a singular solution of (5) because we cannot obtain it from $y = cx - c^2$ by choosing a suitable c.   ◀

We shall see that the conditions under which a given differential equation has solutions are fairly general. But we should note that there are simple equations that do not have

---

[2]The range of the constant may have to be restricted in some cases to avoid imaginary expressions or other degeneracies.

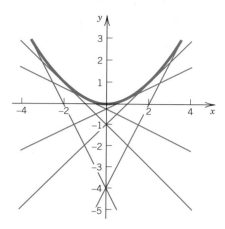

Fig. 2. Singular solution (parabola)
and particular solutions of (5)

solutions at all, and others that do not have a general solution. For example, the equation $y'^2 = -1$ does not have a solution for real y. (Why?) The equation $|y'| + |y| = 0$ does not have a general solution, because its only solution is $y \equiv 0$ (meaning that y is zero for all x).

Applications. Modeling. Initial Value Problems

Differential equations are of great importance in engineering and science because many physical laws and relations appear mathematically in the form of differential equations, for reasons that will soon become apparent.

To begin with, let us consider a basic physical application that will illustrate the typical steps of **modeling,** that is, the steps that lead from the physical situation (*physical system*) to a mathematical formulation (*mathematical model*) and solution, and to the physical interpretation of the result. This may be the easiest way of obtaining a first idea of the nature and purpose of differential equations and their applications.

EXAMPLE 4 **Radioactivity, exponential decay**

Experiments show that a radioactive substance decomposes at a rate proportional to the amount present. Starting with a given amount of substance, say, 2 grams, at a certain time, say, $t = 0$, what can be said about the amount available at a later time?

Solution. 1st Step. Setting up a mathematical model (a differential equation) of the physical process.
We denote by $y(t)$ the amount of substance still present at time t. The rate of change is dy/dt. According to the physical law governing the process of radiation, dy/dt is proportional to y:

(6)
$$\frac{dy}{dt} = ky.$$

Hence y is the unknown function, depending on t. The constant k is a definite physical constant whose numerical value is known for various radioactive substances. (For example, in the case of radium $_{88}Ra^{226}$ we have $k \approx -1.4 \cdot 10^{-11}$ sec^{-1}.) Clearly, since the amount of substance is positive and decreases with time, dy/dt is negative, and so is k. We see that the physical process under consideration is described mathematically by an ordinary differential equation of the first order. Hence this equation is the mathematical model of that physical process. *Whenever a physical law involves a rate of change of a function, such as velocity, acceleration, etc., it will lead to a differential equation. For this reason differential equations occur frequently in physics and engineering.*

2nd Step. Solving the differential equation.   We do not yet know any methods of solution, but calculus will help us here. Indeed, Eq. (6) tells us that if there is a solution $y(t)$, its derivative must be proportional to y. Now we remember from calculus that exponential functions have this property. Indeed, by differentiation and substitution we see that a solution for all t is $y(t) = e^{kt}$ because $y'(t) = (e^{kt})' = ke^{kt} = ky(t)$. More generally, a solution for all t is

$$(7) \qquad\qquad\qquad y(t) = ce^{kt}$$

with any constant c because $y'(t) = cke^{kt} = ky(t)$. Since c is arbitrary, (7) is the **general solution** of (6), by definition.

3rd Step. Determination of a particular solution from an initial condition.   Clearly, our physical process behaves uniquely. Hence we should be able to get from (7) a unique particular solution. Now the amount of substance at some time t will depend on the initial amount $y = 2$ grams at time $t = 0$, or, written as a formula,

$$(8) \qquad\qquad\qquad y(0) = 2.$$

This is called an **initial condition.** We use it to find c in (7):

$$y(0) = ce^0 = 2, \qquad \text{thus} \qquad c = 2.$$

With this c, Eq. (7) gives as the answer the particular solution

$$(9) \qquad\qquad\qquad y(t) = 2e^{kt} \qquad\qquad\qquad \text{(Fig. 3).}$$

Thus the amount of radioactive substance shows exponential decay (exponential decrease with time). This agrees with physical experiments.

4th Step. Checking.   From (9) we have

$$\frac{dy}{dt} = 2ke^{kt} = ky \qquad \text{and} \qquad y(0) = 2e^0 = 2.$$

We see that the function (9) satisfies the equation (6) as well as the initial condition (8).

  The student should never forget to carry out this important final step, which shows whether the function is (or is not) the solution of the problem. ◀

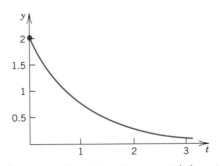

Fig. 3.   Radioactivity (Exponential decay)

A differential equation together with an initial condition, as in our example, is called an **initial value problem.** With x as the independent variable (instead of t) it is of the form

$$(\mathbf{10}) \qquad\qquad \boxed{y' = f(x, y), \qquad y(x_0) = y_0}$$

where x_0 and y_0 are given values. (In our example, $x_0 = t_0 = 0$ and $y_0 = y(0) = 2$.) The initial condition $y(x_0) = y_0$ is used to determine a value of c in the general solution.

Let us show next that geometrical problems also lead to differential equations and initial value problems.

EXAMPLE 5   **A geometrical application**

Find the curve through the point (1, 1) in the xy-plane having at each of its points the slope $-y/x$.

Solution. The function giving the desired curve must be a solution of the differential equation

(11)
$$y' = -\frac{y}{x}.$$

We shall soon learn how to solve such an equation. Meanwhile the student may verify that the general solution of (11) is (Fig. 4)

(12)
$$y = \frac{c}{x}$$
                                                            (c arbitrary).

Since we are looking for the curve that passes through (1, 1), we must have $y = 1$ when $x = 1$. This initial condition $y(1) = 1$ gives $c = 1$ in (12) and as the answer the particular solution $y = 1/x$.   ◀

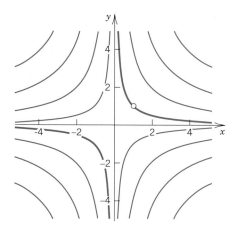

Fig. 4.   Solutions of $y' = -y/x$ (hyperbolas)

PROBLEM SET 1.1

Calculus. Solve the following differential equations.
   1. $y' = x^2$ **2.** $y' = \sin 3x$ **3.** $y'' = x^{-4}$ **4.** $y' = xe^{-x^2}$

Verification

State the order of the differential equation. Verify that the given function is a solution. (a, b, c are arbitrary constants.)
   5. $y' + y = x^2 - 2$, $y = ce^{-x} + x^2 - 2x$
   6. $y'' + y = 0$, $y = a \cos x + b \sin x$
   7. $y''' = e^x$, $y = e^x + ax^2 + bx + c$
   8. $y'' + 2y' + 2y = 0$, $y = e^{-x}(a \cos x + b \sin x)$
   9. $x + yy' = 0$, $x^2 + y^2 = 1$

What happens to the differential equation in Prob. 9:
10. If we change the solution to $x^2 - y^2 = 1$?
11. If we replace 1 by 2? By any number?

Initial Value Problems

Verify that y is a solution of the differential equation. Determine c so that the resulting particular solution satisfies the given initial condition. Graph this solution.

12. $x^3 + y^3 y' = 0$, $x^4 + y^4 = c \ (y > 0)$, $y = 1$ when $x = 0$

13. $y' + 2y = 2.8$, $y = ce^{-2x} + 1.4$, $y = 1.0$ when $x = 0$

14. $xy' = 3y$, $y = cx^3$, $y = 16$ when $x = -4$

15. $yy' = 2x$, $y^2 - 2x^2 = c \ (y > 0)$, $y(1) = \sqrt{3}$

16. $y' = y \tan x$, $y = c \sec x$, $y(0) = \pi/2$

17. $4yy' + x = 0$, $x^2 + 4y^2 = c \ (y > 0)$, $y(2) = 1$

18. What happens in Prob. 17 if we change the initial condition to $y(a) = 0$, where a is any constant?

Modeling, Applications

19. (Half-life) The half-life of a radioactive substance is the time in which half of the given amount will disappear; hence it measures the radioactive decay process. What is the half-life of $_{88}Ra^{226}$ (in years) in Example 4?

20. (Half-life) Radium $_{88}Ra^{224}$ has a half-life of about 3.6 days. Given 1 gram, how much will still be present after 1 day? After 1 year? First guess, then calculate.

21. (Half-life) What would be the effect on the answers in Prob. 20 if the half-life 3.6 were too small, say, by 1%? Would the results increase or decrease? By 1% or more? By less than 1%?

22. (Falling body) If we drop a stone or an iron ball, we can neglect air resistance. Experiments then show that the acceleration of the motion is constant (equal to $g = 9.80$ meters/sec^2 = 32 ft/sec^2; this is called the *acceleration of gravity*). Show that a corresponding model is $y''(t) = g$, where $y(t)$ is the distance fallen as a function of time t. If the motion starts at time $t = 0$ from rest (i.e., with velocity $v = y' = 0$), show that we obtain the familiar *law of free fall*

$$s = \tfrac{1}{2}gt^2.$$

23. (Falling body) How long does a fall of 100 meters take? Of 200 meters? (Guess first.) Why is the second answer less than twice the first? Graph t as a function of y.

24. (Exponential decay; airplane engines) The efficiency of the engines of (subsonic) airplanes depends on air pressure and usually is maximum near about 35 000 ft. Find the air pressure $y(x)$ at this height. *Physical information.* The rate of change $y'(x)$ is proportional to the pressure. At 18 000 ft it is half its value $y_0 = y(0)$ at sea level. *Hint.* Remember from calculus that if $y = e^{kx}$, then $y' = ke^{kx} = ky$. Can you see without calculation that the answer should be close to $y_0/4$?

25. (Exponential growth, a population model) Exponential decay (see Example 4 in the text) and exponential growth are quite important models in physics, biology, etc. If relatively small populations (of humans, animals, bacteria, etc.) are left undisturbed, they often grow according to **Malthus's law.**[3] This law states that the time rate of growth is proportional to the population $y(t)$ present. Model this by a differential equation. Show that the solution is $y(t) = y_0 e^{kt}$. Determine y_0 and k from the first two columns of the table, which contains data for the United States. Calculate values for 1860, 1890, $\cdots$, 1980 from your formula. Compare with the observed values. Comment.

t	0	30	60	90	120	150	180
Year	1800	1830	1860	1890	1920	1950	1980
Population (millions)	5.3	13	31	63	106	150	230

---

[3]THOMAS ROBERT MALTHUS (1766–1834), English social scientist, one of the leaders in classical national economics.

26. (Interest rates) Let $y(x)$ be the investment resulting from a deposit y_0 after x years at an interest rate r. Show that

$$y(x) = y_0[1 + r]^x \qquad \text{(interest compounded annually)}$$

$$y(x) = y_0[1 + (r/4)]^{4x} \qquad \text{(interest compounded quarterly)}$$

$$y(x) = y_0[1 + (r/365)]^{365x} \qquad \text{(interest compounded daily).}$$

Now recall from calculus that $[1 + (1/n)]^n \to e$ as $n \to \infty$, hence $[1 + (r/n)]^{nx} \to e^{rx}$, which gives

$$y(x) = y_0 e^{rx} \qquad \text{(interest compounded continuously).}$$

What differential equation does the last function satisfy? Let $y_0 = \$1000.00$ and $r = 8\%$. Compute $y(1)$ and $y(5)$ from each of the four formulas and confirm that there is not much difference between daily and continuous compounding.

27. WRITING PROJECT. Growth and Decay. Write a short report on exponential growth and decay and its importance. Try to find additional examples of your own. Discuss how the solution changes if k in the exponent is changed (increased or decreased).

1.2 Geometrical Meaning of $y' = f(x, y)$ Direction Fields

We shall now begin with a systematic study of first-order differential equations. Any such equation must contain the first derivative y' of the unknown function y, and it may contain y itself and given functions of x. Hence we can write any first-order differential equation in the form

(1) $$F(x, y, y') = 0.$$

(This is called the *implicit form.*) Not always but in most applications we can write a first-order differential equation in the **explicit form**

(2) $$\boxed{y' = f(x, y).}$$

Before we discuss solution methods, let us explain the simple geometrical meaning of (2).

From calculus we know that $y' = dy/dx$ is the slope of the curve of $y(x)$. Hence if (2) has a solution $y(x)$ passing through a point (x_0, y_0) of the xy-plane, it must have at (x_0, y_0) the slope $f(x_0, y_0)$, as we see from (2).

This suggests the idea of plotting approximate solution curves of a given differential equation (2) without actually solving the equation, so that we obtain a picture of the general behavior of these solution curves. This is of practical interest because many differential equations have complicated solution formulas or no explicit solution formulas at all. Then we can do the following.

Given a differential equation (2), at some points in the xy-plane we indicate the slope $f(x, y)$ by short segments (as illustrated in Fig. 5a on p. 11) called **lineal elements.** This is called a **direction field** or **slope field.** It is a field of tangent directions (slopes) of solution curves of the given equation (2). And we can use this for plotting (approximate) solution curves that have the given tangent directions. Each such curve represents a particular solution corresponding to some initial condition. Look at Fig. 5a and note how nicely the solution curve follows the tangent directions. Of course, we can plot as many solution curves as we want or need.

Plotting Direction Fields by Computer

A **computer algebra system (CAS)** will plot direction fields consisting of lineal elements at the points of a square grid. The mesh size of the grid can be suitably chosen. Subregions R of rapid changes of y' may often require a smaller mesh size. In such cases, a separate enlarged plot of R may be the simplest way of gaining more accuracy.

Plotting Direction Fields by Hand. Isoclines

This is the older method. It consists of three steps.

1st Step. Draw the curves along each of which the slope of the solution curves will be constant, $f(x, y) = k = const$. These are not yet solution curves—don't get confused! These curves $f(x, y) = const$ are also called **isoclines** (meaning curves of equal inclination).

2nd Step. Along each isocline $f(x, y) = k = const$ draw many lineal elements of slope k. Do this for one isocline after another, until your field is sufficiently covered with lineal elements. This is the direction field of (2).

3rd Step. In this direction field sketch approximate solution curves of (2) that have the directions of the lineal elements as their tangent directions.

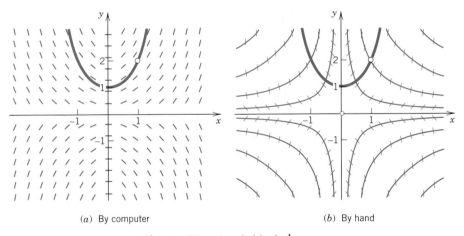

(a) By computer   (b) By hand

Fig. 5. Direction field of $y' = xy$

EXAMPLE 1   **Direction field. Isoclines**

Plot the direction field of the differential equation

(3) $$y' = xy$$

and an approximation to the solution curve through the point $(1, 2)$. Compare with the exact solution.

Solution by computer.   This is shown in Fig. 5a.

Solution by hand.   The isoclines are the equilateral hyperbolas $xy = k$ (because $f(x, y) = xy$), together with the two coordinate axes. We graph some of them (see Fig. 5b). Then we draw lineal elements by sliding a triangle along a fixed ruler. Figure 5b shows the result as well as the solution curve through the point $(1, 2)$. In the next section we shall see that differential equations such as (3) can easily be solved exactly. At present, this has the advantage that we can get an impression of the accuracy of the direction field method by comparing with exact solutions. So let us verify that the general solution of (3) is

$$y(x) = ce^{x^2/2}$$

with general constant c. Indeed, by differentiation (chain rule!)

$$y' = xce^{x^2/2} = xy.$$

The particular solution in Fig. 5 through $(x, y) = (1, 2)$ must satisfy $y(1) = 2$. Thus, $2 = ce^{1/2}$, $c = 2e^{-1/2}$. Substitution into $y(x)$ gives

$$y(x) = 2e^{-1/2}e^{x^2/2} = 2e^{(x^2-1)/2}. \qquad\blacktriangleleft$$

A famous equation for which we do need direction fields is

(4) $$y' = 0.1(1 - x^2) - \frac{x}{y}.$$

(It is related to the van der Pol equation of electronics, which we shall discuss in Sec. 3.5.) The direction field in Fig. 6 shows lineal elements generated by the computer. We have also added the isoclines for $k = -5, -3, \frac{1}{4}, 1$ as well as three typical solutions, one that is (almost) a circle and two spirals approaching it from inside and outside.

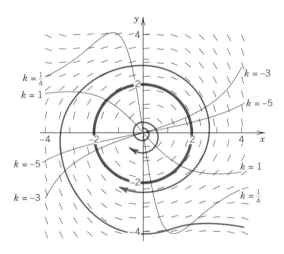

Fig. 6. Direction field of $y' = 0.1(1 - x^2) - \dfrac{x}{y}$

PROBLEM SET 1.2

1. Plot further solutions in Fig. 5, say, those satisfying $y(0) = \pm 1, \pm 2$, and compare your graph with the exact solutions.

Comparison of Accuracy. Plot a direction field (by a CAS or by hand). In it plot some approximate solution curves by hand. Then solve the differential equation and compare to get an impression of the accuracy of the direction field method.

2. $y' = 2x$ **3.** $y' = -y$ **4.** $y' = 2y$ **5.** $y' = \cos \pi x$

Direction Field and Approximate Solution Curves. Plot a direction field (by a CAS or by hand). In it plot several approximate solution curves by hand.

6. $y' = x^2$ **7.** $y' = -x/y$ **8.** $y' = x + y$ **9.** $y' = y^2$

Initial Value Problems. Plot a direction field (by a CAS or by hand). In it plot the particular solution satisfying the initial condition.

10. $y' = x, \quad y(-2) = 6$ **11.** $y' = -y, \quad y(0) = 3$

12. $y' = -xy, \quad y(0) = 1$ **13.** $y' = -x/y, \quad y(\sqrt{2}) = \sqrt{2}$

14. $y' + y^2 = 0, \quad y(5) = 0.25$ **15.** $9yy' + 4x = 0, \quad y(3) = -4$

16. **(Motion)** A body B moves on a straight line L. Let $s(t)$ be the body's distance from a fixed point O on L. Assume that at each instant the velocity of B equals $1/s(t)$, and $s = 1$ when $t = 0$. Find the model (the differential equation). Plot the direction field of it and the (approximate) solution curve $s(t)$ of the problem.

17. **(Skydiver)** Plot a direction field of the differential equation

$$mv' = mg - bv^2$$

with $g = 9.8$ (meters/sec^2; the acceleration of gravity) and $m = 1$, $b = 1$ (for simplicity). This equation models a parachutist of mass m (of the person plus the equipment). $v = v(t)$ is the velocity, t is time, and mv' is mass times acceleration. By Newton's second law, mv' equals the two forces, the attraction mg by the earth and $-bv^2$ ($b > 0$), the air resistance. Suppose that the parachute opens when $v = 10$ m/sec; call this instant $t = 0$; thus $v(0) = 10$. Plot this solution curve in the direction field. Can you see from the direction field that all solutions seem to have the same limit (about 3.13)? That they are all monotone increasing (for what initial conditions?) or monotone decreasing?

18. **(Verhulst's logistic population model)** Plot a direction field of the differential equation

$$y' = ay - by^2 \qquad\qquad (\textit{Verhulst's equation})$$

with $a = 4$ and $b = 1$. This equation was introduced by Verhulst as a model of human populations, whose rate of change y' is equal to ay (this alone would cause exponential growth of y) minus by^2 (a "braking term" that prevents unlimited growth). Draw the following conclusions directly from the direction field. All solution curves in the strip $0 < y < 4$ are monotone increasing. What about solution curves about the line $y = 4$? Can you see directly from the equation that $y = 4 = const$ must be a solution? (We discuss Verhulst's equation in more detail in Sec. 1.6.)

19. **WRITING PROJECT. Direction Fields.** Describe the practical use of direction fields. Compare the two methods of obtaining them, stating the advantages and disadvantages of each method and illustrating your arguments with examples of your own. Discuss the possibility of letting the computer draw the isoclines and then doing the rest of the work by hand. Would this be feasible?

 20. **CAS PROJECT. Direction Fields.** Graph the direction fields of a few important differential equations as follows.

(a) Graph portions of the direction field in Fig. 5, for instance, $-1 \leqq x \leqq 1$, $-1 \leqq y \leqq 1$, or the portion in the first quadrant. Explain what you have gained by this enlargement of portions of the field.

(b) Graph the direction field of $y' = -2y$ and some solutions of your own choice. How do they behave? Why do they decrease for $y > 0$?

(c) Make a conjecture about the solutions of $y' = -x/y$ from its direction field.

(d) Find a differential equation with the general solution $x^2 + 4y^2 = c$ ($y > 0$) by implicit differentiation and graph its direction field. Does it give the impression that the solutions may be semi-ellipses? Can you do similar work for circles? Hyperbolas? Parabolas? Other curves?

1.3 Separable Differential Equations

Many first-order differential equations can be reduced to the form

(1) $$g(y)y' = f(x)$$

by algebraic manipulations. Since $y' = dy/dx$, we find it convenient to write

(2) $$\boxed{g(y)\,dy = f(x)\,dx,}$$

but we keep in mind that this is merely another way of writing (1). Such an equation is called a **separable equation,** because in (2) the variables x and y are *separated* so that x appears only on the right and y appears only on the left.

To solve (1), we integrate on both sides with respect to x, obtaining

$$\int g(y)\,\frac{dy}{dx}\,dx = \int f(x)\,dx + c.$$

Now on the left we can switch to y as the variable of integration. By calculus, $(dy/dx)\,dx = dy$, so that we get

(3) $$\boxed{\int g(y)\,dy = \int f(x)\,dx + c.}$$

If we assume that f and g are continuous functions, the integrals in (3) will exist, and by evaluating these integrals we obtain the general solution of (1).

EXAMPLE 1 Solve the differential equation

$$9yy' + 4x = 0.$$

Solution. By separating variables we have

$$9y\,dy = -4x\,dx.$$

By integrating on both sides we obtain the general solution

$$\frac{9}{2}y^2 = -2x^2 + c^*, \qquad \text{thus} \qquad \frac{x^2}{9} + \frac{y^2}{4} = c \qquad \left(c = \frac{c^*}{18}\right).$$

The solution represents a family of ellipses. Some of them are shown in Fig. 7. ◀

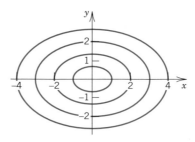

Fig. 7. General solution in Example 1

EXAMPLE 2   Solve the differential equation

$$y' = 1 + y^2.$$

Solution.   By separating variables and integrating we obtain

$$\frac{dy}{1 + y^2} = dx, \qquad \text{arc tan } y = x + c, \qquad y = \tan(x + c).$$

It is of great importance to introduce the constant of integration immediately when the integration is performed. $y = \tan x + c$ (with $c \neq 0$) would not be a solution! Verify this. ◀

EXAMPLE 3   **Exponential growth or decay**

Exponential growth or decay is governed by the differential equation

$$y' = ky$$

with positive or negative (fixed) k, respectively. This equation is separable. We obtain

$$\frac{dy}{y} = k\,dx, \qquad \ln|y| = kx + \tilde{c}$$

because by calculus, $(\ln|y|)' = y'/y$; indeed, when $y > 0$, then $(\ln|y|)' = (\ln y)' = y'/y$; and when $y < 0$, then $-y > 0$ and $(\ln|y|)' = (\ln(-y))' = -y'/(-y) = y'/y$. Taking exponentials and noting that $e^{a+b} = e^a e^b$, we get

$$|y| = e^{kx+\tilde{c}} = e^{kx}e^{\tilde{c}}, \qquad \text{thus} \qquad y = ce^{kx},$$

where $c = +e^{\tilde{c}}$ when $y > 0$ and $c = -e^{\tilde{c}}$ when $y < 0$, and we can also admit $c = 0$ (giving $y \equiv 0$). This is the general solution. ◀

EXAMPLE 4   **Initial value problem**

Solve the initial value problem

$$y' = -\frac{y}{x}, \qquad y(1) = 1.$$

Solution.   Separation of variables and integration gives

$$\frac{dy}{y} = -\frac{dx}{x}, \qquad \ln|y| = -\ln|x| + \tilde{c} = \ln\frac{1}{|x|} + \tilde{c}.$$

Taking exponentials, we get $y = c/x$. This is the general solution. From this and the initial condition, $1 = c/1$, so that $c = 1$. The answer is $y = 1/x$. This is shown in Fig. 4 in Sec. 1.1. ◀

EXAMPLE 5   **Initial value problem. Bell-shaped curves**

Solve the initial value problem

$$y' = -2xy, \qquad y(0) = 1.$$

Solution.   Separating variables, integrating, and taking exponentials, we obtain

$$\frac{dy}{y} = -2x\,dx, \qquad \ln|y| = -x^2 + \tilde{c}, \qquad |y| = e^{-x^2+\tilde{c}}.$$

Setting $e^{\tilde{c}} = +c$ when $y > 0$, and $e^{\tilde{c}} = -c$ when $y < 0$, and admitting also $c = 0$ (which gives the solution

$y \equiv 0$), we get the general solution

$$y = ce^{-x^2}.$$

This represents so-called **"bell-shaped curves,"** which play a role in heat conduction (Sec. 11.6) and in probability and statistics (Sec. 22.8 and Chap. 23). Figure 8 shows some of them for $c > 0$.

The student may show that the particular solution of our initial value problem is

$$y = e^{-x^2}.$$

This is shown as the second of the three curves in Fig. 8.   ◀

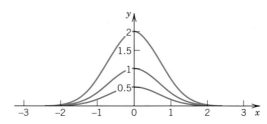

Fig. 8.   Solutions of $y' = -2xy$ ("bell-shaped curves")
in the upper half-plane

Reduction to Separable Form

Certain differential equations are not separable but can be made separable by the introduction of a new unknown function. We illustrate the idea of this method by some typical cases.

EXAMPLE 6   **Differential equations of the form[4]** $y' = g\left(\dfrac{y}{x}\right)$

Here g is any (differentiable) function of y/x, for instance, $(y/x)^3$, $\cos (y/x)$, and so on. The form of the equation suggests that we set $y/x = u$; thus,

(4) $\boxed{y = ux}$ and by product differentiation $\boxed{y' = u'x + u.}$

Substitution into $y' = g(y/x)$ then gives

$$u'x + u = g(u), \qquad \text{thus} \qquad u'x = g(u) - u.$$

We see that this can be separated,

(5) $$\frac{du}{g(u) - u} = \frac{dx}{x}.$$

For instance, solve

$$2xyy' = y^2 - x^2.$$

---

[4]These equations are sometimes called **homogeneous equations.** We shall not use this terminology but reserve the term "homogeneous" for a much more important purpose (see Sec. 1.6).

Solution.   We divide the given differential equation by $2xy$, obtaining

$$y' = \frac{y^2}{2xy} - \frac{x^2}{2xy} = \frac{1}{2}\left(\frac{y}{x} - \frac{x}{y}\right).$$

Now $y' = u'x + u$ by (4), so that we obtain

$$u'x + u = \frac{1}{2}\left(u - \frac{1}{u}\right),$$

hence by simplification,

$$u'x = -\frac{1}{2}\left(u + \frac{1}{u}\right) = -\frac{u^2 + 1}{2u}.$$

We see that separation of variables now gives

$$\frac{2u\,du}{1 + u^2} = -\frac{dx}{x}.$$

By integration,

$$\ln(1 + u^2) = -\ln|x| + c^* = \ln\frac{1}{|x|} + c^*, \qquad \text{thus} \qquad 1 + u^2 = \frac{c}{x}.$$

Replacing u by y/x yields $1 + (y/x)^2 = c/x$. We multiply by x^2 to get

$$x^2 + y^2 = cx, \qquad \text{thus} \qquad \left(x - \frac{c}{2}\right)^2 + y^2 = \frac{c^2}{4}.$$

This general solution represents the family of circles with centers on the x-axis and all passing through the origin, as illustrated in Fig. 9.   ◀

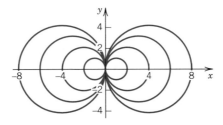

Fig. 9.   General solution (family of circles)
in Example 6

EXAMPLE 7   **Transformations $v = ay + bx + k$**

Sometimes a transformation $v = ay + bx + k$ may lead to a separable differential equation. We illustrate the idea of this method by a simple equation,

$$(2x - 4y + 5)y' + x - 2y + 3 = 0.$$

Solution.   The terms $2x - 4y$ and $x - 2y$ suggest that we try $v = x - 2y$. Then $2y = x - v$ and $y' = (1 - v')/2$. By substitution and simplification the given equation becomes

$$(2v + 5)v' = 4v + 11.$$

Multiplying by 2 and separating variables, we obtain

$$\frac{4v + 10}{4v + 11}\,dv = \frac{4v + 11 - 1}{4v + 11}\,dv = \left(1 - \frac{1}{4v + 11}\right)dv = 2\,dx.$$

By integration,

$$v - \frac{1}{4} \ln |4v + 11| = 2x + c^*.$$

Since $v = x - 2y$, this implicit general solution of the given equation may be written

$$4x + 8y + \ln |4x - 8y + 11| = c.$$

(Transformations $v = ay + bx + k$ are discussed in full in the older literature, e.g., in Ince's book [A5] listed in Part A of Appendix 1.)   ◀

PROBLEM SET 1.3

1. Why is it important to add the constant of integration immediately when the integration is performed?

General Solution

Solve the following differential equations. (In Probs. 7–10 use the indicated transformation.) Check your answer by substitution.

2. $yy' + 25x = 0$

3. $y' = 1 + 0.01y^2$

4. $y' + 3x^2y^2 = 0$

5. $y' = xy/2$

6. $y' = -ky^2$

7. $xy' = y^2 + y$ $(y/x = u)$

8. $xy' = x + y$ $(y/x = u)$

9. $y' = (x^2 + y^2)/xy$ $(y/x = u)$

10. $y' = (y + 4x)^2$ $(y + 4x = v)$

11. $y' + \csc y = 0$

Initial Value Problems

Solve the following initial value problems. (In Prob. 19, L, R, and I_0 are constants.) Show the details of your work.

12. $y' = -x/y,$ $y(1) = \sqrt{3}$

13. $xy' + y = 0,$ $y(2) = -2$

14. $y^3y' + x^3 = 0,$ $y(0) = 1$

15. $e^xy' = 2(x + 1)y^2,$ $y(0) = 1/6$

16. $y' = 1 + 4y^2,$ $y(0) = 0$

17. $y' \cosh^2 x - \sin^2 y = 0,$ $y(0) = \pi/2$

18. $dr/dt = -2tr,$ $r(0) = 2.5$

19. $L(dI/dt) + RI = 0,$ $I(0) = I_0$

Setting $y/x = u$, solve the following initial value problems.

20. $xy' = (y - x)^3 + y,$ $y(1) = 3/2$

21. $xy' = y + 3x^4 \cos^2 (y/x),$ $y(1) = 0$

22. $xy' = y + x^2 \sec (y/x),$ $y(1) = \pi$

23. $xyy' = 2y^2 + 4x^2,$ $y(2) = 4$

24. Differential equations $y' = f(ax + by + k)$ can be made separable by using as a new unknown function $v(x) = ax + by + k$. Using this method, solve $y' = (x + y - 2)^2$.

25. Solve $y' = \dfrac{1 - 2y - 4x}{1 + y + 2x}$. *Hint.* Use $y + 2x = v$.

26. TEAM PROJECT. Family of Curves. A family of curves can often be characterized as the general solution of a differential equation $y' = f(x, y)$.

   (a) Show that for the circles with center at the origin we get $y' = -x/y$.

   (b) Graph some of the hyperbolas $xy = c$. Find a differential equation for them.

   (c) Find a differential equation for the straight lines through the origin.

   (d) You will see that the product of the right sides of the differential equations in (a) and (c) equals -1. Do you recognize this as the condition for the two families to be **orthogonal** (i.e., to intersect at right angles)? Do your graphs confirm this?

   (e) Sketch families of curves of your own choice and find their differential equations. Can every family of curves be described by a differential equation?

27. CAS PROJECT. Graphing Solutions. A CAS can usually plot solutions, even if they are integrals that cannot be evaluated by the usual analytical methods of calculus.

(a) Show this for the five initial value problems $y' = e^{-x^2}$, $y(0) = 0, \pm 1, \pm 2$, graphing all five curves on the same axes.

(b) Graph approximate solution curves, using the first few terms of the Maclaurin series (obtained by termwise integration of that of y') and compare with the exact curves.

(c) Repeat the work in (a) for another differential equation and initial conditions of your own choice, leading to an integral that cannot be evaluated as indicated.

1.4 Modeling: Separable Equations

Modeling means setting up a mathematical model of a physical or other system. The model may be a function to be evaluated or plotted, or a differential or other equation to be solved, and so on. In this section we consider systems that can be modeled in terms of a separable differential equation (see Sec. 1.3).

EXAMPLE 1   **Radiocarbon dating**

Suppose that an archaeologist excavates a bone and measures its content of radioactive carbon $_6C^{14}$. If the result is 25% of the content present in bones of a *living* organism, what can be said about the age of the bone?

Idea of the method to be used.   In the atmosphere, the ratio of radioactive carbon $_6C^{14}$ (made radioactive by cosmic rays) and ordinary carbon $_6C^{12}$ is constant. The same holds for **living** organisms. When an organism dies, the absorption of $_6C^{14}$ by breathing and eating terminates. Hence one can estimate the age of a fossil by comparing the carbon ratio in the fossil with that in the atmosphere. This is W. Libby's idea of radiocarbon dating (Nobel Prize for chemistry, 1960). The half-life of $_6C^{14}$ has been found to be 5730 years (*CRC Handbook of Chemistry and Physics,* 54th edition, p. B251).

Solution.   As in Example 4 of Sec. 1.1, the mathematical model of the process of radioactive decay is

$$y' = ky, \qquad \text{solution} \qquad y(t) = y_0 e^{kt}.$$

Here, y_0 is the initial amount of $_6C^{14}$. By definition, the **half-life** (5730 years) is the time after which the amount of radioactive substance ($_6C^{14}$) has decreased to half its original value. Thus,

(1) $$y_0 e^{k \cdot 5730} = \tfrac{1}{2}y_0.$$

This is an equation for the unknown constant k. Division by y_0 and taking logarithms gives

(2) $$e^{5730k} = \tfrac{1}{2}, \qquad k = \frac{\ln 1/2}{5730} = -0.000\ 121.$$

The time after which 25% of the original amount of $_6C^{14}$ is still present can now be computed from

(3) $$y_0 e^{-0.000\ 121t} = \tfrac{1}{4}y_0, \qquad t = \frac{\ln 1/4}{-0.000\ 121} = 11\ 460\ [\text{years}].$$

Hence the mathematical answer is that the bone has an age of 11 460 years. Actually, the experimental determination of the half-life of $_6C^{14}$ involves an error of about 40 years. Also, a comparison with other methods shows that radiocarbon dating tends to give values that are too small, perhaps because the ratio of $_6C^{14}$ to $_6C^{12}$ may have changed over long periods of time. Hence 12 000 or 13 000 years is probably a more realistic answer to our present problem.

Have you noticed that the answer is twice the half-life? Is this just by chance?   ◀

EXAMPLE 2 **Mixing problem**

The tank in Fig. 10 contains 200 gal of water in which 40 lb of salt are dissolved. Five gal of brine, each containing 2 lb of dissolved salt, run into the tank per minute, and the mixture, kept uniform by stirring, runs out at the same rate. Find the amount of salt $y(t)$ in the tank at any time t.

Solution. 1st Step. Modeling. The time rate of change $y' = dy/dt$ of $y(t)$ equals the inflow of salt minus the outflow. The inflow is 10 lb/min (5 gal of brine, each containing 2 lb). We determine the outflow. $y(t)$ is the total amount of salt in the tank. The tank always contains 200 gal because 5 gal flow in and 5 gal flow out per minute. Thus 1 gal contains $y(t)/200$ lb of salt. Hence the 5 outflowing gallons contain $5y(t)/200 = y(t)/40 = 0.025y(t)$ lb of salt. This is the outflow. The time rate of change y' is the balance:

$$y' = \text{Salt inflow rate} \quad - \quad \text{Salt outflow rate}$$

Since the inflow of salt is 10 lb/min and the outflow is $0.025y(t)$ lb/min, this equation becomes

(4) $$y' = 10 - 0.025y.$$

Initially, $y(0) = 40$, by assumption. This initial value problem is our model to be solved.

2nd Step. Solution of the model and interpretation. By algebra and separation of variables we obtain from (4)

$$y' = -0.025(y - 400), \qquad \text{thus} \qquad \frac{dy}{y - 400} = -0.025 \, dt.$$

Integration gives

$$\ln |y - 400| = -0.025t + \tilde{c}.$$

Taking exponentials, we have

$$y - 400 = ce^{-0.025t}.$$

From this and the initial condition we obtain

$$y(0) - 400 = 40 - 400 = -360 = c.$$

Hence the amount of salt in the tank at time t is (Fig. 10)

(5) $$y(t) = 400 - 360e^{-0.025t} \; [\text{1b}].$$

We see that $y(t)$ increases with time. Can you explain this physically? We also note that the limit is 400 lb = 200×2 lb. Could you see this directly from the differential equation (4)? From the physics of the given problem? ◀

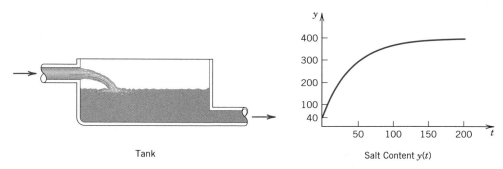

Tank   Salt Content $y(t)$

Fig. 10. Mixing problem in Example 2

EXAMPLE 3   **Heating problem (Newton's law of cooling[5])**

Suppose that you turn off the heat in your home at night 2 hours before you go to bed; call this time $t = 0$. If the temperature T at $t = 0$ is 66°F and at the time you go to bed ($t = 2$) has dropped to 63°F, what temperature can you expect in the morning, say, 8 hours later ($t = 10$)? Of course, this process of cooling off will depend on the outside temperature T_A, which we assume to be constant at 32°F.

Physical information.   Experiments show that the time rate of change dT/dt of the temperature T of a body is proportional to the difference between T and the temperature T_A of the surrounding medium. This is called **Newton's law of cooling.** Ideally, the body would be a copper ball that is heated and placed into cold water. (Copper is a good heat conductor.) Nevertheless, our application of this law will give us a good qualitative understanding of the process.

Solution. 1st Step. Modeling.   All we have to do is write Newton's law of cooling as an equation. Denoting the unknown constant of proportionality by k, we have

(6)
$$\frac{dT}{dt} = k(T - T_A) = k(T - 32).$$

2nd Step. General solution.   Separation of variables, integration, and taking exponentials leads to the general solution of (6),

$$\frac{dT}{T - 32} = k\, dt, \qquad \ln |T - 32| = kt + \tilde{c}, \qquad T(t) = 32 + ce^{kt} \quad (c = e^{\tilde{c}}).$$

3rd Step. Particular solution.   The initial condition is $T(0) = 66$. From this and the general solution we obtain (Fig. 11)

$$T(0) = 32 + c = 66, \qquad c = 34, \qquad T(t) = 32 + 34e^{kt}.$$

4th Step. Determination of k.   For this we use $T(2) = 63$, as given, and solve algebraically for k,

$$T(2) = 32 + 34e^{k \cdot 2} = 63, \qquad e^{2k} = \frac{63 - 32}{34} = 0.911\ 765,$$

thus $k = \frac{1}{2} \ln 0.911\ 765 = -0.046\ 187$.

5th Step. Answer and interpretation.   By inserting this value of k into the particular solution in Step 3 and taking $t = 10$ (10 hours after shutoff), we find that the temperature in the morning will be 53°F, approximately. Indeed,

$$T(10) = 32 + 34e^{-0.046\ 187 \cdot 10} = 53.4 \ [°F].$$

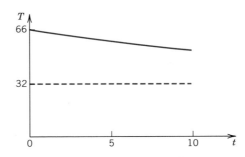

Fig. 11.   Temperature in Example 3, Step 3

---

[5]Sir ISAAC NEWTON (1642–1727), great English physicist and mathematician, became a professor at Cambridge in 1669 and Master of the Mint in 1699. He and the German mathematician and philosopher GOTTFRIED WILHELM LEIBNIZ (1646–1716) invented (independently) the differential and integral calculus. Newton discovered many basic physical laws and created the method of investigating physical problems by means of calculus. His *Philosophiae naturalis principia mathematica* (*Mathematical Principles of Natural Philosophy,* 1687) contains the development of classical mechanics. His work is of greatest importance to both mathematics and physics.

Is this answer reasonable? T dropped by 3°F within 2 hours. If T were a linear function, it should drop by 15°F from 66 to 51. Can you explain why the actual result is higher?

6th Step. Check the result.   ◀

EXAMPLE 4   **Velocity of escape from the earth**

In reaching other planets it is critically important that the velocity v of a space ship be so large that v does not become zero and the ship is not pulled back toward the earth.

   In this example we shall determine the minimum initial velocity of a projectile that is fired in a radial direction from the earth and is supposed to escape from the earth (without receiving any additional push while under way). We neglect the air resistance and the gravitational pull of other celestial bodies.

Solution. 1st Step. Modeling. By **Newton's law of gravitation** the gravitational force is proportional to $1/r^2$, where r is the distance from the center of the earth to the projectile. The corresponding acceleration is

(7) $$a(r) = -\frac{gR^2}{r^2},$$

where R is the radius of the earth. The minus sign occurs because the gravitational attraction acts in the negative r-direction, toward the center of the earth. For $r = R$ this becomes $a(R) = -g$, where g is the gravitational acceleration at the earth's surface.

   From (7) we get a differential equation for the velocity v by the **"chain-rule trick"** (worth remembering),

(8) $$a = \frac{dv}{dt} = \frac{dv}{dr}\frac{dr}{dt} = \frac{dv}{dr}v.$$

Substituting the last expression into (7), we obtain

(9) $$\frac{dv}{dr}v = -\frac{gR^2}{r^2}.$$

2nd Step. Solving the differential equation by separation and integration gives

(10)   (a) $v\,dv = -gR^2\dfrac{dr}{r^2}$ and   (b) $\dfrac{v^2}{2} = \dfrac{gR^2}{r} + c.$

3rd Step. c in terms of the initial velocity. On the earth's surface, $r = R$ and $v = v_0$, the initial velocity (to be suitably chosen). This gives in (10b)

$$\frac{v_0^2}{2} = \frac{gR^2}{R} + c, \qquad \text{so that} \qquad c = \frac{v_0^2}{2} - gR.$$

With this c we obtain from (10b), multiplied by 2,

(11) $$v^2 = \frac{2gR^2}{r} + v_0^2 - 2gR.$$

4th Step. Determination of the velocity of escape. If $v^2 = 0$, then $v = 0$; thus the projectile will stop and return to the earth. If we choose

(**12**) $$\boxed{v_0 = \sqrt{2gR},}$$

then $v_0^2 - 2gR = 0$ in (11) and v^2 remains positive because the first term on the right in (11) is positive, $2gR^2/r > 0$. The quantity v_0 in (12) is called the **velocity of escape from the earth.** A smaller v_0 would not suffice because then we would have $v^2 = 0$ in (11) for some r.

5th Step. Numerical values. The radius of the earth is $R = 6372$ km $= 3960$ mi. Furthermore, we have $g = 9.80$ meters/sec$^2 = 0.00980$ km/sec$^2 = 32.15$ ft/sec$^2 = 0.00609$ mi/sec^2 give

$$v_0 = \sqrt{2gR} = 11.2 \text{ km/sec} = 6.96 \text{ mi/sec.}$$

6th Step. Checking. Check the result.   ◀

PROBLEM SET 1.4

1. **(Exponential growth)** If in a culture of yeast the rate of growth $y'(t)$ is proportional to the amount $y(t)$ present at time t, and if $y(t)$ doubles in 1 day, how much can be expected after 3 days at the same rate of growth? After 1 week?

2. **(Airplane takeoff)** An airplane taking off from a landing field has a run of 2 kilometers. If the plane starts with a speed of 10 meters/sec, moves with constant acceleration, and makes the run in 50 sec, with what speed does it take off?

3. **(Airplane)** What happens in Prob. 2 if the acceleration is 1.5 meters/sec^2?

4. **(Rocket)** A rocket is shot straight up. During the initial stages of flight it has acceleration $7t$ meters/sec^2. The engine cuts out at $t = 10$ sec. How high will the rocket go? (Neglect air resistance.)

5. **(Radiocarbon dating)** What should be the $_6C^{14}$ content (in percent of y_0) of a fossilized tree that is claimed to be 3000 years old?

6. **(Dryer)** If a wet sheet in a dryer loses its moisture at a rate proportional to its moisture content, and if it loses half of its moisture during the first 10 minutes, when will it be practically dry, say, when will it have lost 99% of its moisture? First guess, then calculate.

7. **(Dryer)** Could you see practically without calculation that the answer in Prob. 6 must lie between 60 and 70 minutes? Explain.

8. **(Linear accelerator)** Linear accelerators are used in physics for accelerating charged particles. Suppose that an alpha particle enters an accelerator and undergoes a constant acceleration that increases the speed of the particle from 10^3 meters/sec to 10^4 meters/sec in 10^{-3} sec. Find the acceleration a and the distance traveled during this period of 10^{-3} sec.

9. **(Boyle–Mariotte's law for ideal gases[6])** Experiments show that for a gas at low pressure p (and constant temperature) the rate of change of the volume $V(p)$ equals $-V/p$. Solve the corresponding differential equation.

10. **(Velocity of escape)** At the earth's surface the velocity of escape is 11.2 km/sec; see Example 4. If the projectile is carried by a rocket and is separated from it at a distance of 1000 km from the earth's surface, what would be the minimum velocity at this point sufficient for escape from the earth? Why is it smaller than 11.2 km/sec?

11. **(Sugar inversion)** Experiments show that the rate of inversion of cane sugar in dilute solution is proportional to the concentration $y(t)$ of unaltered sugar. Let the concentration be 1/100 at $t = 0$ and 1/300 at $t = 4$ hours. Find $y(t)$.

12. **(Exponential decay) Lambert's law of absorption[7]** states that the absorption of light in a very thin transparent layer is proportional to the thickness of the layer and to the amount incident on that layer. Formulate this in terms of a differential equation and solve it.

13. **(Newton's law of cooling)** A thermometer, reading 5°C, is brought into a room whose temperature is 22°C. One minute later the thermometer reading is 12°C. How long does it take until the reading is practically 22°C, say, 21.9°C?

14. **(Mixing problem)** A tank contains 400 gal of brine in which 100 lb of salt are dissolved. Fresh water runs into the tank at the rate of 2 gal/min, and the mixture, kept practically uniform by stirring, runs out at the same rate. How much salt will there be in the tank at the end of 1 hour?

---

[6]ROBERT BOYLE (1627–1691), English physicist and chemist, one of the founders of the Royal Society; EDMÉ MARIOTTE (about 1620–1684). French physicist and prior of a monastery near Dijon.

[7]JOHANN HEINRICH LAMBERT (1728–1777), German physicist and mathematician, known for his contributions to cartography and astronomy.

15. (Curves) What curves in the xy-plane have the property that at each point (x, y) their tangent has the slope $-4x/y$?

16. (Mothball) Suppose that a mothball loses volume by evaporation at a rate proportional to its instantaneous area. If the diameter of the ball decreases from 2 cm to 1 cm in 2 months, how long will it take until the ball has practically gone, say, until its diameter is 1 mm?

17. (Curves) Find all curves in the xy-plane whose tangents all pass through the origin.

18. (Friction) If a body slides on a surface, it experiences friction F (a force against the direction of motion). Experiments show that $|F| = \mu|N|$ (*Coulomb's*[8] *law of kinetic friction without lubrication*), where N is the normal force (force that holds the two surfaces together; see Fig. 12) and the constant of proportionality μ is called the *coefficient of kinetic friction*. In Fig. 12 assume that the body weighs 45 nt (about 10 lb; see front cover for conversion), $\mu = 0.20$ (corresponding to steel on steel), $\alpha = 30°$, the slide is 10 meters long, the initial velocity is zero, and air resistance is negligible. Find the velocity of the body at the end of the slide.

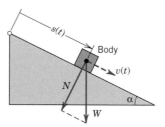

Fig. 12. Problem 18

19. PROJECT. Half-Life Formula. A standard method of determining the half-life t_H of a radioactive substance is based on the use of measurements of the amounts y_1 and y_2 of the substance present at times t_1 and t_2, respectively.

    (a) Find t_H from $y_1 = y(1) = 0.2$ gram and $y_2 = y(4) = 0.05$ gram, where t is measured in days.

    (b) Using your special formula in (a) as a guide, find a general half-life formula.

    (c) For what purpose could you use a third measurement if it is available.

20. TEAM PROJECT. Outflow from a Tank. Torricelli's Law.[9] Torricelli's law states that water issues from a hole in the bottom of a tank with velocity

(13)
$$v(t) = 0.600\sqrt{2gh(t)},$$

where $h(t)$ is the height of the water above the hole at time t, $g = 980$ cm/sec$^2 = 32.15$ ft/sec^2 is the acceleration of gravity at the surface of the earth, and 0.600 is the Borda contraction factor[10] (because the stream has a smaller cross section than the hole). See Fig. 13 on p. 26.

    (a) Show that $\sqrt{2gh}$ in (13) is equal to the velocity of free fall (without air resistance).

    (b) Using (13), show that a model of the outflow is

(14)
$$h' = -26.56 \frac{A}{B(h)} \sqrt{h},$$

---

[8]CHARLES AUGUSTIN DE COULOMB (1736–1806), French physicist and engineer.

[9]EVANGELISTA TORRICELLI (1608–1647), Italian physicist and mathematician, pupil and later successor of GALILEO GALILEI (1564–1642) at Florence.

[10]Suggested by J. C. BORDA in 1766.

where A is the area of the hole and $B(h)$ is the cross-sectional area of the tank. *Hint.* Equate the decrease of the water volume in the tank to the volume flowing out during a short time Δt.

(c) Solve (14) for a cylindrical tank of constant B.

(d) When will the tank in (c) be empty if $h(0) = 150$ cm, the tank has diameter 1 meter, and the hole has diameter 1 cm (see Fig. 13)?

(e) What happens in (c) if the hole is increased? If the cross-sectional area of the tank is increased? Can this be seen directly from (14) without solving it?

(f) Build a small tank from a can and check the theory by suitable experiments. Improve the mathematical model if you can.

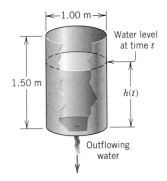

Fig. 13.   Tank in Team Project 20

1.5 Exact Differential Equations Integrating Factors

We remember from calculus that if a function $u(x, y)$ has continuous partial derivatives, its **differential**[11] is

$$du = \frac{\partial u}{\partial x}\, dx + \frac{\partial u}{\partial y}\, dy.$$

From this it follows that if $u(x, y) = c = const$, then $du = 0$.

   For example, if $u = x + x^2 y^3 = c$, then

$$du = (1 + 2xy^3)\, dx + 3x^2 y^2\, dy = 0$$

or

$$y' = \frac{dy}{dx} = -\frac{1 + 2xy^3}{3x^2 y^2},$$

a differential equation that we can solve by going backward. This idea gives a powerful solution method, as follows.

---

[11]Sometimes also called the *total differential* of $u(x, y)$.

A first-order differential equation of the form

(1)
$$M(x, y)\, dx + N(x, y)\, dy = 0$$

is called an **exact differential equation** if the **differential form** $M(x, y)\, dx + N(x, y)\, dy$ is **exact,** that is, this form is the differential

(2)
$$du = \frac{\partial u}{\partial x}\, dx + \frac{\partial u}{\partial y}\, dy$$

of some function $u(x, y)$. Then the differential equation (1) can be written

$$du = 0.$$

By integration we immediately obtain the general solution of (1) in the form

(3)
$$u(x, y) = c.$$

Comparing (1) and (2), we see that (1) is an exact differential equation if there is some function $u(x, y)$ such that

(4)   (a) $\dfrac{\partial u}{\partial x} = M,$ (b) $\dfrac{\partial u}{\partial y} = N.$

Now suppose that M and N are defined and have continuous first partial derivatives in a region in the xy-plane whose boundary is a closed curve having no self-intersections. Then from (4) (see Appendix 3.2 for notation)

$$\frac{\partial M}{\partial y} = \frac{\partial^2 u}{\partial y\, \partial x},$$

$$\frac{\partial N}{\partial x} = \frac{\partial^2 u}{\partial x\, \partial y}.$$

By the assumption of continuity the two second partial derivatives are equal. Thus

(5)
$$\frac{\partial M}{\partial y} = \frac{\partial N}{\partial x}$$

This condition is not only necessary but also sufficient[12] for (1) to be an exact differential equation.

If (1) is exact, the function $u(x, y)$ can be found by guessing or in the following systematic way. From (4a) we have by integration with respect to x

---

[12]We shall prove this fact at another occasion (Theorem 3 in Sec. 9.2); the proof can also be found in some books on elementary calculus; see Ref. [13] in Appendix 1.

$$(6) \qquad \boxed{u = \int M\, dx + k(y);}$$

in this integration, y is to be regarded as a constant, and $k(y)$ plays the role of a "constant" of integration. To determine $k(y)$, we derive $\partial u/\partial y$ from (6), use (4b) to get dk/dy, and integrate dk/dy to get k.

Formula (6) was obtained from (4a). Instead of (4a) we may equally well use (4b). Then instead of (6) we first have

$$(6^*) \qquad \boxed{u = \int N\, dy + l(x).}$$

To determine $l(x)$ we derive $\partial u/\partial x$ from (6*), use (4a) to get dl/dx, and integrate. We illustrate all this by the following typical examples.

EXAMPLE 1   **An exact equation**

Solve

$$(7) \qquad (x^3 + 3xy^2)\, dx + (3x^2 y + y^3)\, dy = 0.$$

Solution. 1st Step. Test for exactness.   Our equation is of the form (1) with

$$M = x^3 + 3xy^2, \qquad N = 3x^2 y + y^3. \qquad \text{Thus} \qquad \frac{\partial M}{\partial y} = 6xy, \qquad \frac{\partial N}{\partial x} = 6xy.$$

From this and (5) we see that (7) is exact.

2nd Step. Implicit solution.   From (6) we obtain

$$(8) \qquad u = \int M\, dx + k(y) = \int (x^3 + 3xy^2)\, dx + k(y) = \frac{1}{4} x^4 + \frac{3}{2} x^2 y^2 + k(y).$$

To find $k(y)$, we differentiate this formula with respect to y and use formula (4b), obtaining

$$\frac{\partial u}{\partial y} = 3x^2 y + \frac{dk}{dy} = N = 3x^2 y + y^3.$$

Hence $dk/dy = y^3$, so that $k = (y^4/4) + \tilde{c}$. Inserting this into (8) we get the answer

$$(9) \qquad u(x, y) = \tfrac{1}{4}(x^4 + 6x^2 y^2 + y^4) = c.$$

3rd Step. Checking. CAUTION!   Note well that the present method gives the solution in implicit form, $u(x, y) = c = const$, not in explicit form, $y = f(x)$. For **checking**, we can differentiate $u(x, y) = c$ implicitly and see whether this leads to $dy/dx = -M/N$ or $M\, dx + N\, dy = 0$, the given equation.

In the present case, differentiating (9) implicitly with respect to x, we obtain

$$\tfrac{1}{4}(4x^3 + 12xy^2 + 12x^2 yy' + 4y^3 y') = 0.$$

Collecting terms, we see that this equals $M + Ny' = 0$ with M and N as in (7); thus $M\, dx + N\, dy = 0$. This completes the check.   ◄

EXAMPLE 2   **An initial value problem**

Solve the initial value problem

$$(10) \qquad (\sin x \cosh y)\, dx - (\cos x \sinh y)\, dy = 0, \qquad y(0) = 3.$$

Solution. The student may verify that the equation is exact. From (6) we obtain

$$u = \int \sin x \cosh y \, dx + k(y) = -\cos x \cosh y + k(y).$$

From this, $\partial u / \partial y = -\cos x \sinh y + dk/dy$. Hence $dk/dy = 0$, and $k = const$. The general solution is $u = const$, that is, $\cos x \cosh y = c$. The initial condition gives $\cos 0 \cosh 3 = 10.07 = c$. Hence the answer is $\cos x \cosh y = 10.07$.

Figure 14 shows the six particular solutions satisfying $y(0) = 0, 1, 2, 3$ (thick), 4, 5. All of them approach infinity as $x \to \pm \pi/2$. Can you explain why?

Checking. $(\cos x \cosh y)' = -\sin x \cosh y + \cos x (\sinh y)y' = 0$, which gives (10). Also, $\cos 0 \cosh 3 = 10.07$ shows that the answer satisfies the initial condition.   ◀

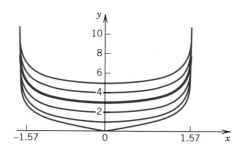

Fig. 14. Six particular solutions in Example 2

EXAMPLE 3   **WARNING! Breakdown in the case of nonexactness**

Consider the equation

$$-y \, dx + x \, dy = 0.$$

We see that $M = -y$, $N = x$, hence $\partial M / \partial y = -1$ but $\partial N / \partial x = 1$. Hence the equation is not exact. Let us show that in such a case, the present method does not work. From (6),

$$u = \int M \, dx + k(y) = -xy + k(y).$$

From this,

$$\frac{\partial u}{\partial y} = -x + k'(y).$$

This should equal $N = x$. But this is impossible, since $k(y)$ can depend only on y. Try (6*); it will also fail. Solve the equation by another method that we have discussed.   ◀

Reduction to Exact Form. Integrating Factors

The differential equation in Example 3,

$$-y \, dx + x \, dy = 0,$$

is not exact. But if we multiply it by $1/x^2$, we get an exact equation,

$$(11) \qquad \frac{-y \, dx + x \, dy}{x^2} = -\frac{y}{x^2} \, dx + \frac{1}{x} \, dy = d\left(\frac{y}{x}\right) = 0.$$

Integration of (11) then gives the general solution $y/x = c = const$.

Of course, exactness of (11) can be checked by (4) in the usual fashion,

$$M = -\frac{y}{x^2}, \qquad \frac{\partial M}{\partial y} = -\frac{1}{x^2}, \qquad N = \frac{1}{x}, \qquad \frac{\partial N}{\partial x} = -\frac{1}{x^2}.$$

Our differential equation illustrates the simple idea of reduction to exact form. All we have done was the multiplication of a given nonexact equation, say,

(12)
$$\boxed{P(x, y)\, dx + Q(x, y)\, dy = 0,}$$

by a function F, which in general will be a function of both x *and* y. The result was an equation

(13)
$$FP\, dx + FQ\, dy = 0$$

that is exact, so that we can solve it as just discussed. The function $F = F(x, y)$ is then called an **integrating factor** of (12).

EXAMPLE 4   **Integrating factors**

The integrating factor in (11) is $F = 1/x^2$. Hence in this case the exact equation (13) is

$$FP\, dx + FQ\, dy = \frac{-y\, dx + x\, dy}{x^2} = d\left(\frac{y}{x}\right) = 0. \qquad \text{Solution} \qquad \frac{y}{x} = c.$$

These are straight lines $y = cx$ through the origin.

It is remarkable that we can readily find other integrating factors for the equation $-y\, dx + x\, dy = 0$, namely $1/y^2$, $1/xy$, and $1/(x^2 + y^2)$, because

(14) $\dfrac{-y\, dx + x\, dy}{y^2} = -d\left(\dfrac{x}{y}\right)$, $\dfrac{-y\, dx + x\, dy}{xy} = -d\left(\ln\dfrac{x}{y}\right)$, $\dfrac{-y\, dx + x\, dy}{x^2 + y^2} = d\left(\arctan\dfrac{y}{x}\right)$. ◄

How to Find Integrating Factors

In simpler cases, integrating factors may be found by inspection or perhaps after some trials [keeping (14) in mind]. In the general case, the idea is this:

For $M\, dx + N\, dy = 0$ the exactness condition (4) is $\partial M/\partial y = \partial N/\partial x$. Hence for (13), $FP\, dx + FQ\, dy = 0$, the exactness condition is

(15)
$$\frac{\partial}{\partial y}(FP) = \frac{\partial}{\partial x}(FQ).$$

By the product rule, with subscripts denoting partial derivatives, this gives

$$F_y P + F P_y = F_x Q + F Q_x.$$

In the general case, this would be complicated and useless. So we follow the ***Golden Rule:*** If you cannot solve your problem, try to solve a simpler one—the result may be useful (and may also help you later on). Hence we look for an integrating factor depending only on **one** variable; fortunately, in many practical cases, there are such factors, as we shall see. Thus, let $F = F(x)$. Then $F_y = 0$ and $F_x = F' = dF/dx$, so that (15) becomes

$$FP_y = F'Q + FQ_x.$$

Dividing by FQ and reshuffling terms, we have

$$(16) \qquad \frac{1}{F}\frac{dF}{dx} = \frac{1}{Q}\left(\frac{\partial P}{\partial y} - \frac{\partial Q}{\partial x}\right).$$

This proves

THEOREM 1   **[Integrating factor $F(x)$]**

If (12) *is such that the right side of* (16), *call it R, depends only on x, then* (12) *has an integrating factor $F = F(x)$, which is obtained by integrating* (16) *and taking exponentials on both sides,*

$$(17) \qquad F(x) = \exp \int R(x)\,dx.$$

Similarly, if $F = F(y)$, then instead of (16) we get

$$(18) \qquad \frac{1}{F}\frac{dF}{dy} = \frac{1}{P}\left(\frac{\partial Q}{\partial x} - \frac{\partial P}{\partial y}\right)$$

and we have the companion

THEOREM 2   **[Integrating factor $F(y)$]**

If (12) *is such that the right side $\widetilde{R}$ of* (18) *depends only on y, then* (12) *has an integrating factor $F = F(y)$, which is obtained from* (18) *in the form*

$$(19) \qquad F(y) = \exp \int \widetilde{R}(y)\,dy.$$

EXAMPLE 5   **Application of Theorems 1 and 2. Initial value problem**

Find an integrating factor using either Theorem 1 or 2 and solve the initial value problem

$$2 \sin (y^2)\,dx + xy \cos (y^2)\,dy = 0, \quad y(2) = \sqrt{\pi/2}.$$

Solution. 1st Step. Check for exactness.   We have

$$P = 2 \sin (y^2) \quad \text{and} \quad Q = xy \cos (y^2).$$

The equation is not exact because

$$P_y = 4y \cos (y^2) \neq Q_x = y \cos (y^2).$$

2nd Step. Integrating factor.   We try Theorem 1. On the right side of (16) we obtain

$$R = \frac{1}{Q}(P_y - Q_x) = \frac{1}{xy \cos (y^2)}\left[4y \cos (y^2) - y \cos (y^2)\right] = \frac{3y}{xy} = \frac{3}{x}.$$

This shows that Theorem 1 applies. From (17) we thus obtain the integrating factor

$$F(x) = \exp \int R(x)\, dx = \exp \int \frac{3}{x}\, dx = x^3.$$

Multiplying the given equation by x^3, we get the new equation

(20) $2x^3 \sin (y^2)\, dx + x^4 y \cos (y^2)\, dy = 0.$

This equation is exact because

$$\frac{\partial}{\partial y} [2x^3 \sin (y^2)] = 4x^3 y \cos (y^2) = \frac{\partial}{\partial x} [x^4 y \cos (y^2)].$$

3rd Step. General solution. From (6) we obtain for the new equation (20), written as $u_x\, dx + u_y\, dy = 0,$

$$u = \int 2x^3 \sin (y^2)\, dx = \tfrac{1}{2} x^4 \sin (y^2) + k(y).$$

From this and the second term in the new equation (20),

$$u_y = x^4 y \cos (y^2) + k'(y) = x^4 y \cos (y^2).$$

Hence $k'(y) = 0$ and $k = const.$ This gives the general solution

$$u(x, y) = \tfrac{1}{2} x^4 \sin (y^2) = c = const.$$

4th Step. Particular solution of the initial value problem. Substituting the initial condition $y(2) = \sqrt{\pi/2}$ into $u(x, y)$, we have

$$\frac{1}{2} \cdot 2^4 \sin \frac{\pi}{2} = 8 = c.$$

Hence the desired particular solution is (Fig. 15)

$$\tfrac{1}{2} x^4 \sin (y^2) = 8 \qquad \text{or} \qquad x^4 \sin (y^2) = 16.$$

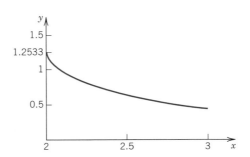

Fig. 15. Particular solution in Example 5

PROBLEM SET 1.5

Exact Equations for Given Solutions. Given $u(x, y)$, find the exact differential equation $du = 0$ and plot some of the solution curves $u(x, y) = const.$

1. $u = x^2 + 4y^2$
2. $u = x^2 - y^2$
3. $u = e^{x^2/y}$
4. $u = 1/(x^2 + y^2)$
5. $u = \tan (y^2 - x^3)$
6. $u = \sin x \cosh y$

Solution of Exact Equations. Show that the following equations are exact and solve them.

7. $2xy\,dx + x^2\,dy = 0$

8. $-yx^{-2}\,dx + x^{-1}\,dy = 0$

9. $\sinh x \cos y\,dx = \cosh x \sin y\,dy$

10. $e^{3\theta}(dr + 3r\,d\theta) = 0$

11. $e^{-2\theta}(r\,dr - r^2\,d\theta) = 0$

12. $(\cot y + x^2)\,dx = x \csc^2 y\,dy$

Test for Exactness. Initial Value Problems. Are the following equations exact? Solve the initial value problems by one of the methods discussed. (Show the details of your work.)

13. $3y^2\,dx + x\,dy = 0,\qquad y(1) = \frac{1}{2}$

14. $2y^{-1}\cos 2x\,dx = y^{-2}\sin 2x\,dy,\qquad y(\pi/4) = 3.8$

15. $2xy\,dy = (x^2 + y^2)\,dx,\qquad y(1) = 2$

16. $ye^x\,dx + (2y + e^x)\,dy = 0,\qquad y(0) = -1$

17. $[(x + 1)e^x - e^y]\,dx = xe^y\,dy,\qquad y(1) = 0$

18. $2\sin \omega y\,dx + \omega \cos \omega y\,dy = 0,\qquad y(0) = \pi/2\omega$

19. $2\sin 2x \sinh y\,dx - \cos 2x \cosh y\,dy = 0,\qquad y(0) = 1$

20. $(2xy\,dx + dy)e^{x^2} = 0,\qquad y(0) = 2$

On Exactness and Integrating Factors

21. Under what conditions is $(ax + by)\,dx + (kx + ly)\,dy = 0$ exact? (Here, a, b, k, l are constants.) Solve the exact equation.

22. Can you figure out what the solution curves in Example 1 look like? *Hint.* Introduce new variables by setting $x = s + t$, $y = s - t$.

23. Verify that y, xy^3, and x^2y^5 are integrating factors of $y\,dx + 2x\,dy = 0$ and solve.

24. **(Checking)** Checking solutions is always important. In connection with the method of integrating factors it is particularly essential since one may have to exclude the function $y(x)$ given by $F(x, y) = 0$. To see this, consider $(xy)^{-1}\,dy - x^{-2}\,dx = 0$; show that an integrating factor is $F = y$ and leads to $d(y/x) = 0$, hence $y = cx$, where c is arbitrary, but $F = y = 0$ is not a solution of the original equation.

Verifying Integrating Factors and Solving. Show that the given function is an integrating factor and solve:

25. $\sin y\,dx + \cos y\,dy = 0,\quad e^x$

26. $y\,dx + [y + \tan(x + y)]\,dy = 0,\quad \cos(x + y)$

27. $(a + 1)y\,dx + (b + 1)x\,dy = 0,\quad x^a y^b$

28. $3(y + 1)\,dx = 2x\,dy,\quad (y + 1)/x^4$

29. $(2y + xy)\,dx + 2x\,dy = 0,\quad 1/xy$

30. $2\cos y\,dx = \tan 2x \sin y\,dy,\quad \cos 2x$

Find an Integrating Factor by inspection or by either Theorem 1 or 2 and solve:

31. $2\cosh x \cos y\,dx = \sinh x \sin y\,dy$

32. $2xy\,dx + 3x^2\,dy = 0$

33. $(2\cos y + 4x^2)\,dx = x \sin y\,dy$

34. $2\cos y\,dx = \sin y\,dy$

35. $2x \tan y\,dx + \sec^2 y\,dy = 0$

36. $(y + 1)\,dx - (x + 1)\,dy = 0$

37. $x^{-1}\cosh y\,dx + \sinh y\,dy = 0$

38. **WRITING PROJECT. Differential Equations Solvable by Several Methods.** Give a short summary of the methods of solution that we have discussed so far. Using equations of your own choice or from the text or problem sets, show that certain differential equations can be solved by more than one method. Compare the amount of work involved in each case.

39. **PROJECT. Working Backward.** Working backward from the solution to the problem is useful in many areas. Euler and other great masters did it. To get additional insight into the idea of integrating factors, start from $u(x, y)$ of your own choice, find $du = 0$, destroy exactness by division by some $F(x, y)$, and see what equations you can get that are solvable by integrating factors. Can you proceed systematically, beginning with the simplest $F(x, y)$?

40. **CAS PROJECT. Plotting Particular Solutions.** This project concerns particular solutions of the differential equation

(21) $$dy - y^2 \sin x\,dx = 0.$$

(a) Show that (21) is not exact. Find an integrating factor using either Theorem 1 or 2. Solve (21).

(b) Solve (21) by separating variables. Is this simpler than (a)?

(c) Plot the seven particular solutions satisfying the following initial conditions $y(0) = 1$, $y(\pi/2) = \pm\frac{1}{2}, \pm\frac{2}{3}, \pm 1$ (see Fig. 16).

(d) Which solution of (21) do we not get in (a) or (b)?

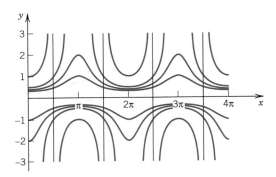

Fig. 16. Particular solutions in CAS Project 40

1.6 Linear Differential Equations Bernoulli Equation

A first-order differential equation is said to be **linear** if it can be written

$$(1) \qquad \boxed{y' + p(x)y = r(x).}$$

The characteristic feature of this equation is that it is linear in the unknown function y and its derivative y', whereas p as well as r on the right may be *any* given functions of x.

If the right side $r(x)$ is zero for all x in the interval in which we consider the equation (written $r(x) \equiv 0$), the equation is said to be **homogeneous;** otherwise it is said to be **nonhomogeneous.**

Let us find a formula for the general solution of (1) in some interval I, assuming that p and r are continuous in I. For the homogeneous equation

$$(2) \qquad y' + p(x)y = 0$$

this is very simple. Indeed, by separating variables we have

$$\frac{dy}{y} = -p(x)\,dx, \qquad \text{thus} \qquad \ln|y| = -\int p(x)\,dx + c^*$$

and by taking exponentials on both sides

$$(3) \qquad \boxed{y(x) = ce^{-\int p(x)\,dx}} \qquad (c = \pm e^{c^*} \quad \text{when} \quad y \gtrless 0);$$

here we may also take $c = 0$ and obtain the ***trivial solution*** $y(x) \equiv 0$.

The nonhomogeneous equation (1) will now be solved. It turns out that it has the pleasant property of possessing an integrating factor depending only on x. Indeed, we first write (1) as

$$(py - r)\, dx + dy = 0.$$

This is $P\, dx + Q\, dy = 0$, where $P = py - r$ and $Q = 1$. Hence (16) in Sec. 1.5 becomes simply

$$\frac{1}{F}\frac{dF}{dx} = p(x).$$

Since this depends only on x, Eq. (1) has an integrating factor $F(x)$, which we obtain directly by integration and exponentiation [as in (17), Sec. 1.5]:

$$F(x) = e^{\int p\, dx}.$$

Multiplying (1) by this F and observing the product rule of differentiation gives

$$e^{\int p\, dx}(y' + py) = (e^{\int p\, dx}y)' = e^{\int p\, dx}\, r.$$

We now integrate the second and the third of these three expressions with respect to x,

$$e^{\int p\, dx}\, y = \int e^{\int p\, dx}\, r\, dx + c.$$

We divide this equation by $e^{\int p\, dx}$ on both sides. Abbreviating $\int p\, dx$ by h, we obtain

(4)
$$y(x) = e^{-h}\left[\int e^{h} r\, dx + c\right], \qquad h = \int p(x)\, dx.$$

This represents the general solution of (1) in the form of an integral.[13] (The choice of the value of the constant of integration in $\int p\, dx$ does not matter; see Prob. 2.)

EXAMPLE 1 Solve the linear differential equation

$$y' - y = e^{2x}.$$

Solution. Here

$$p = -1, \qquad r = e^{2x}, \qquad h = \int p\, dx = -x$$

and from (4) we obtain the general solution

$$y(x) = e^{x}\left[\int e^{-x}e^{2x}\, dx + c\right] = e^{x}[e^{x} + c] = ce^{x} + e^{2x}.$$

In simpler cases, such as the present, we may not need the general formula (4), but may wish to proceed directly, multiplying the given equation by $e^{h} = e^{-x}$. This gives

$$(y' - y)e^{-x} = (ye^{-x})' = e^{2x}e^{-x} = e^{x}.$$

---

[13]If the integral cannot be integrated by the usual methods of calculus (as often happens in practice), we may have to use a numerical method for integrals (Sec. 17.5) or for the differential equation itself (Secs. 19.1, 19.2).

Integrating on both sides, we obtain the same result as before:

$$y e^{-x} = e^x + c, \qquad \text{hence} \qquad y = e^{2x} + ce^x. \qquad \blacktriangleleft$$

EXAMPLE 2   **Mixing problem**

The tank in Fig. 17 contains 1000 gal of water in which 200 lb of salt are dissolved. Fifty gallons of brine, each containing $(1 + \cos t)$ lb of dissolved salt, run into the tank per minute. The mixture, kept uniform by stirring, runs out at the same rate. Find the amount of salt $y(t)$ in the tank at any time t.

Solution.   ***1st Step. Modeling.***   This is similar to Example 2 in Sec. 1.4. The essential difference is the variability of the salt content in the inflow, which is $50(1 + \cos t)$. $y(t)$ is the amount of salt in the tank, which always contains 1000 gal; hence $y(t)/1000$ is the salt content per gallon, and $50y(t)/1000 = 0.05y(t)$ is the salt content in the outflow per minute. The rate of change $y' = dy/dt$ of y equals the balance,

$$(5) \qquad\qquad y' = \text{In} - \text{Out} = 50(1 + \cos t) - 0.05y.$$

2nd Step. Solution.   Eq. (5) is a nonhomogeneous linear differential equation, which we can write

$$(5) \qquad\qquad y' + 0.05y = 50(1 + \cos t).$$

Hence $p = 0.05$, $h = 0.05t$, and (4) gives the general solution

$$\begin{aligned}
y &= e^{-0.05t} \left(\int e^{0.05t} \, 50(1 + \cos t) \, dt + c \right) \\
&= e^{-0.05t} \left(e^{0.05t} \, (1000 + a \cos t + b \sin t) + c \right) \\
&= 1000 + a \cos t + b \sin t + ce^{-0.05t}
\end{aligned}$$

where $a = 2.5/(1 + 0.05^2) = 2.494$ and $b = 50/(1 + 0.05^2) = 49.88$, which we obtained by evaluating the integral. From this and the initial condition $y(0) = 200$ we have

$$y(0) = 1000 + a + c = 200, \qquad c = 200 - 1000 - a = -802.5.$$

Hence the solution of our problem is

$$y(t) = 1000 + 2.494 \cos t + 49.88 \sin t - 802.5 \, e^{-0.05t}.$$

3rd Step. Discussion of solution.   Figure 17 shows the solution $y(t)$. The last term in $y(t)$ is the only term that depends on the initial condition (because c does). It decreases monotone. As a consequence, $y(t)$ increases, but keeps oscillating about 1000 as the limit of the mean value.

This mean value is also shown in Fig. 17. It is obtained as the solution of the differential equation

$$y' + 0.05y = 50.$$

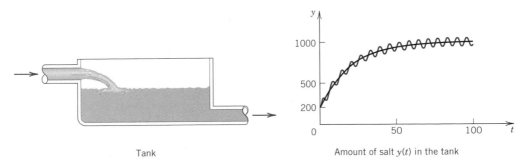

Tank   Amount of salt $y(t)$ in the tank

Fig. 17.   Mixing problem in Example 2

This equation is (5) with the right side replaced by the mean salt inflow of 50 lb/min. The solution satisfying the initial condition is

$$y = 1000 - 800\,e^{-0.05t}.$$

Compare this curve with that in Example 2 of Sec. 1.4 and comment.   ◄

EXAMPLE 3   Solve

$$y' + 2y = e^x(3 \sin 2x + 2 \cos 2x).$$

Solution.   Here $p = 2$, $h = 2x$, so that (4) gives

$$y = e^{-2x}\left[\int e^{2x}e^x(3 \sin 2x + 2 \cos 2x)\,dx + c\right]$$

$$= e^{-2x}[e^{3x} \sin 2x + c]$$

$$= ce^{-2x} + e^x \sin 2x.$$ ◄

EXAMPLE 4   **Initial value problem**

Solve the initial value problem

$$y' + y \tan x = \sin 2x, \qquad y(0) = 1.$$

Solution.   Here $p = \tan x$, $r = \sin 2x = 2 \sin x \cos x$, and

$$\int p\,dx = \int \tan x\,dx = \ln |\sec x|.$$

From this we see that in (4),

$$e^h = \sec x, \qquad e^{-h} = \cos x, \qquad e^h r = (\sec x)(2 \sin x \cos x) = 2 \sin x,$$

and the general solution of our equation is

$$y(x) = \cos x \left[2 \int \sin x\,dx + c\right] = c \cos x - 2 \cos^2 x.$$

From this and the initial condition, $1 = c \cdot 1 - 2 \cdot 1^2$; thus $c = 3$ and the solution of our initial value problem is $y = 3 \cos x - 2 \cos^2 x$.   ◄

Reduction to Linear Form. Bernoulli Equation

Certain nonlinear differential equations can be reduced to linear form, as we shall illustrate in the problem set. The practically most famous of these is the **Bernoulli equation**[14] (see also Prob. 31, etc.)

(6) $$\boxed{y' + p(x)y = g(x)y^a}$$ (a any real number).

---

[14]JAKOB BERNOULLI (1654–1705), Swiss mathematician, professor at Basel, also known for his contributions to elasticity theory and mathematical probability. The method for solving Bernoulli's equation was discovered by Leibniz in 1696. Jakob Bernoulli's students include his nephew NIKLAUS BERNOULLI (1687–1759), who contributed to probability theory and infinite series, and his youngest brother JOHANN BERNOULLI (1667–1748), who had profound influence on the development of calculus, became Jakob's successor at Basel, and had among his students GABRIEL CRAMER (see Sec. 6.6) and LEONHARD EULER (see Sec. 2.6). His son DANIEL BERNOULLI (1700–1782) is known for his basic work in fluid flow and the kinetic theory of gases.

If $a = 0$ or $a = 1$, Equation (6) is linear. Otherwise it is nonlinear. Then we set

$$u(x) = [y(x)]^{1-a}.$$

We differentiate this and substitute y' from (6), obtaining

$$u' = (1 - a)y^{-a}y' = (1 - a)y^{-a}(gy^a - py).$$

Simplification gives

$$u' = (1 - a)(g - py^{1-a}),$$

where $y^{1-a} = u$ on the right, so that we get the linear equation

(7) $$u' + (1 - a)pu = (1 - a)g.$$

EXAMPLE 5   **Bernoulli equation. Verhulst equation. Logistic population model**

Solve the special Bernoulli equation, called the **Verhulst equation**[15]:

(8) $$y' - Ay = -By^2 \qquad (A, B \text{ positive constants}).$$

Solution.   Here, $a = 2$, so that $u = y^{-1}$, and by differentiation and substitution of y' from (8),

$$u' = -y^{-2}y' = -y^{-2}(-By^2 + Ay) = B - Ay^{-1},$$

that is,

$$u' + Au = B.$$

From (4) with $p = A$, hence $h = Ax$, and $r = B$ we obtain

$$u = e^{-Ax}\left[\int Be^{Ax}\,dx + c\right] = e^{-Ax}\left[\frac{B}{A}e^{Ax} + c\right] = ce^{-Ax} + \frac{B}{A}.$$

This gives the general solution of (8),

(9) $$y = \frac{1}{u} = \frac{1}{(B/A) + ce^{-Ax}},$$

and directly from (8) we see that $y(x) \equiv 0$ is also a solution.

Equation (9) is called the **logistic law** of population growth, where x is time. For $B = 0$ it gives exponential growth $y = (1/c)e^{Ax}$ (**Malthus's law,** Prob. 25 in Sec. 1.1). $-By^2$ in (8) is a "braking term," preventing the population from growing without bound. Indeed, (9) shows that initially small populations $[0 < y(0) < A/B]$ increase monotone to A/B, whereas initially large populations $[y(0) > A/B]$ decrease monotone to the same limit A/B (Fig. 18).

The logistic law has useful applications to human populations (see CAS Project 50) and animal populations (see C. W. Clark, *Mathematical Bioeconomics,* New York, Wiley, 1976). ◀

---

[15]PIERRE-FRANÇOIS VERHULST, Belgian statistician, who introduced Eq. (8) as a model for human population growth in 1838.

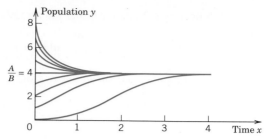

Fig. 18. Logistic population model.
Curves (9) in Example 5 with $A/B = 4$

Input and Output

Linear differential equations (1) have various applications, as we shall illustrate in the problem set as well as in the next section. Then the independent variable x will often be time; the function $r(x)$ on the right side of (1) may represent a force, and the solution $y(x)$ a displacement, a current, or some other variable physical quantity. In engineering mathematics $r(x)$ is frequently called the **input,** and $y(x)$ is called the **output** or *response to the input* (and the initial conditions). For instance, in electrical engineering the differential equation may govern the behavior of an electric circuit and the output $y(x)$ is obtained as the solution of that equation corresponding to the input $r(x)$. We shall discuss this idea in terms of typical examples in the next section (and for second-order equations in Secs. 2.5, 2.11, 2.12).

In the solution formula (4) of (1) the only quantity depending on the initial condition is the constant c. Writing (4) as a sum of two terms,

$$y(x) = e^{-h} \int e^h r \, dx + c e^{-h},$$

we see the following:

(10)   Total Output = Response to the Input + Response to the Initial Data.

Problem Set 1.6

1. Show that $e^{-\ln x} = 1/x$ (but not $-x$) and $e^{-\ln (\sec x)} = \cos x$.
2. Show that the choice of the value of the constant of integration in $\int p \, dx$ [see (4)] does not matter (so that we may choose it to be zero).

General Solution. In Probs. 3–14 find the general solutions of the following differential equations. (Show the details of your work.)

3. $y' - y = 4$ 4. $y' + 2y = 2.5$ 5. $y' + 3xy = 0$
6. $y' + xy = 4x$ 7. $y' + ky = e^{-kx}$ 8. $y' + 4y = \cos x$

9. $xy' = 2y + x^3 e^x$ **10.** $y' + y = e^{-x} \tan x$ **11.** $y' = (y - 2) \cot x$

12. $x^3 y' + 3x^2 y = 1/x$ **13.** $y' + y \sin x = e^{\cos x}$ **14.** $x^2 y' + 2xy = \sinh 5x$

Initial Value Problems. Solve the following initial value problems. (Show the details of your work.)

15. $y' + 4y = 20, \quad y(0) = 2$ **16.** $y' - (1 + 3x^{-1})y = x + 2, \quad y(1) = e - 1$

17. $y' = 2(y - 1) \tanh 2x, \quad y(0) = 4$ **18.** $y' = y \tan x, \quad y(\pi) = 2$

19. $y' + 3y = \sin x, \quad y(\pi/2) = 0.3$ **20.** $y' + 6x^2 y = e^{-2x^3}/x^2, \quad y(1) = 0$

21. $y' = 1 + y^2, \quad y(0) = 0$ **22.** $xy' + 4y = 8x^4, \quad y(1) = 2$

General Properties of Homogeneous Linear Differential Equations (2). Show that (2) has the following basic properties. Illustrate each with an example of your own.

23. $y \equiv 0$ is a solution of (2), called the **trivial solution.**

24. If y_1 is a solution of (2), so is cy_1 with any constant c.

25. The sum $y_1 + y_2$ of two solutions y_1 and y_2 of (2) is a solution of (2).

General Properties of Nonhomogeneous Linear Differential Equations (1) (with $r(x) \not\equiv 0$). Show that (1) has the following basic properties. Give an example for each.

26. If y_1 is a solution of (1) and y_2 is a solution of (2), then $y = y_1 + y_2$ is a solution of (1).

27. The difference $y = y_1 - y_2$ of two solutions y_1 and y_2 of (1) is a solution of (2).

28. If y_1 is a solution of (1), then $y = cy_1$ is a solution of $y' + py = cr$.

29. If y_1 is a solution of $y_1' + py_1 = r_1$ and y_2 is a solution of $y_2' + py_2 = r_2$ (with the same p), then $y = y_1 + y_2$ is a solution of $y' + py = r_1 + r_2$.

30. If $p(x)$ and $r(x)$ in (1) are constant, say, $p(x) = p_0$ and $r(x) = r_0$, then (1) can be solved by separating variables and the result will agree with that obtained from (4).

Reduction of Nonlinear Differential Equations to Linear Form. Reduce to linear form and solve the following equations. Show all the steps of your work, not merely the linear equation or the solution. (Some are Bernoulli equations. Others become linear if you take x as the unknown function and y as the independent variable.)

31. $y' + 2y = y^2$ **32.** $y' + y = -x/y$

33. $y' + \frac{1}{3}y = \frac{1}{3}(1 - 2x)y^4$ **34.** $y' = (\tan y)/(x - 1)$

35. $y' = 1/(6e^y - 2x)$ **36.** $y'(\sinh 3y - 2xy) = y^2$

37. $y' + xy = xy^{-1}$ **38.** $2xyy' + (x - 1)y^2 = x^2 e^x \quad (y^2 = z)$

Some Applications (More in the next section)

39. (Mixing problem) What will happen in Example 2 in the text if we replace $\cos t$ by $e^{-0.1t} \cos t$? First guess. Then calculate and plot.

40. Hormone secretion can be modeled by

$$y' = a - b \cos \frac{2\pi t}{24} - ky.$$

Here, t is time [in hours, with $t = 0$ suitably chosen, e.g., 8:00 A.M.], $y(t)$ is the amount of a certain hormone in the blood, a is the average secretion rate, $b \cos (\pi t/12)$ models the daily 24-hr secretion cycle, and ky models the removal rate of the hormone from the blood. Find the solution when $a = b = k = 1$ and $y(0) = 2$.

41. (Newton's law of cooling, Sec. 1.4) If the temperature of a cake is 300°F when it leaves the oven and is 200°F 10 minutes later, when will it be practically equal to the room temperature of 60°F, say, when will it be 61°F?

42. (Atomic waste disposal) If a sealed container with atomic waste is dumped into the ocean, Newton's second law,

$$\text{Mass} \times \text{Acceleration} = \text{Force},$$

gives as a model (v the speed)

(11)
$$m \frac{dv}{dt} = W - B - kv, \qquad v(0) = 0,$$

where W is the weight of the container (acting downward), B the buoyancy force of the water (acting upward), and $-kv$ the drag (acting against the motion). Solve the equation to obtain $v(t)$. Integrate to get $y(t)$ such that $y(0) = 0$. The container should not break when it hits the bottom of the ocean. Assume that it will not break if it hits the bottom with speed $v_{\text{crit}} = 12$ meters/sec. Determine the critical time t_{crit} when the container reaches v_{crit}, assuming that $W = 2254$ nt (about 507 lb), $B = 2090$ nt (about 470 lb), and $k = 0.637$ kg/sec. Show that breaking will occur at points where the ocean is deeper than 105 meters, approximately. (Linearity of the drag force as a function of v is confirmed experimentally for speeds that are not too large.)

43. What happens in Prob. 42 if both the weight and the buoyancy force are increased by 1000 nt (about 225 lb)?

Riccati and Clairaut Equations

44. A **Riccati equation**[16] is of the form $y' + p(x)y = g(x)y^2 + h(x)$. Verify that the Riccati equation $y' = x^3(y - x)^2 + x^{-1}y$ has the solution $y = x$ and reduce it to a Bernoulli equation by the substitution $w = y - x$ and solve it.

45. Show that the general Riccati equation in Prob. 44 (which is a Bernoulli equation when $h \equiv 0$) can be reduced to a Bernoulli equation if one knows a solution $y = v$, by setting $w = y - v$.

46. A **Clairaut equation**[17] is of the form $y = xy' + g(y')$. Solve the special Clairaut equation $y = xy' + 1/y'$. *Hint.* Differentiate the equation with respect to x.

47. Show that the general Clairaut equation in Prob. 46, with arbitrary $g(s)$ has as solutions a family of straight lines $y = cx + g(c)$ and a singular solution determined by $g'(s) = -x$, where $s = y'$. (Those lines are tangents to the latter.) *Hint.* Differentiate the equation with respect to x, as in Prob. 46.

48. Show that the straight lines whose segment between the positive x-axis and y-axis has constant length 1 are solutions of the Clairaut equation $y = xy' - y'/\sqrt{1 + y'^2}$, whose singular solution is the **astroid** $x^{2/3} + y^{2/3} = 1$. Make a sketch.

49. TEAM PROJECT. Transformation of Differential Equations. Such transformations have the purpose of extending solution methods to larger classes of equations. We have transformed equations to separable form, to exact form, and to linear form. Describe the key idea for each of these transformations. Illustrate each using two examples of your own choice, giving each step of your calculations (not just the transformed equations).

50. CAS PROJECT. Verhulst Population Model for the United States. In this project we study the role of the parameters in the solution (9) of the Verhulst population model (8) and the effect of changes by experimenting with plots obtained from a CAS. We shall improve the fit of (9) to the following (rounded) data for the United States:

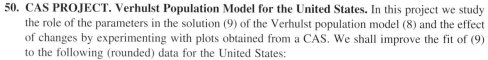

x	0	30	60	90	120	150	180	190
Year	1800	1830	1860	1890	1920	1950	1980	1990
Population (millions)	5.3	13	31	63	105	150	230	250

---

[16]JACOPO FRANCESCO RICCATI (1676–1754), Italian mathematician, who introduced his equation in 1723.

[17]ALEXIS CLAUDE CLAIRAUT (1713–1765), French mathematician, also known for his work in geodesy and astronomy.

(a) Verhulst predicted in 1845 the values $A = 0.03$ and $B = 1.6 \cdot 10^{-4}$ in (9), where x is measured in years (with $x = 0$ corresponding to 1800) and $y(x)$ in millions. Graph the data and the curve, using $y(0) = 5.3$. Describe the quality of fit in words. See Fig. 19a.

(b) Conclude from (a) that improvement of fit will be achieved by an increase of the limit. Figure 19b corresponds to a limit of 280, but you may wish to choose an even larger value. Determine and plot this new $y(x)$.

(c) If you want to proceed somewhat more systematically, determine A, B, c in (9) such that $y(0) = 5.3$, the limit is $L = 280$ (or whatever you choose) and the curve passes through a given point (x_m, y_m) somewhere in the middle. Derive the formula for A from these conditions,

$$A = \frac{1}{x_m} \ln \frac{1/y(0) - 1/L}{1/y_m - 1/L}.$$

Take $(x_m, y_m) = (120, 105)$, compute and graph $y(x)$ and the data, and compare the quality of fit with that in (b) and comment. See Fig. 19c.

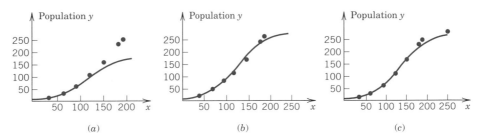

Fig. 19.   Verhulst population model. Data for the United States.
See CAS Project 50, parts (a), (b), (c).

1.7 Modeling: Electric Circuits

Recall from Secs. 1.1 and 1.4 that **modeling** means setting up mathematical models of physical or other systems. In this section we shall model electric circuits. Their models will be linear differential equations. Although of particular interest to students of electrical engineering, computer engineering, etc., our discussion will be profitable to *all* students because modeling skills can be acquired most successfully by considering practical problems from *various* fields. And we shall see that our present modeling will be relatively simple and straightforward.

Basic Elements of an Electric Circuit

To help *all* students, we first explain the basic concepts needed.

The simplest electric circuit is a series circuit in which we have a source of electric energy (**electromotive force**) such as a generator or a battery, and a resistor, which uses energy, for example an electric lightbulb (Fig. 20). If we close the switch, a current I will flow through the resistor, and this will cause a **voltage drop,** that is, the electric potential at the two ends of the resistor will be different; this potential difference or voltage drop can be measured by a voltmeter. Experiments show that the following law holds.

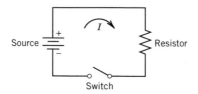

Fig. 20. Circuit

The **voltage drop E_R across a resistor** *is proportional to the instantaneous current I,* say,

(1) $$\boxed{E_R = RI}$$ **(Ohm's law)**

where the constant of proportionality R is called the **resistance** of the resistor. The current I is measured in *amperes,* the resistance R in *ohms,* and the voltage E_R in *volts.*[18]

The other two important elements in more complicated circuits are **inductors** and **capacitors.** An inductor opposes a change in current, having an inertia effect in electricity similar to that of mass in mechanics; we shall consider this analogy later (Sec. 2.12). Experiments yield the following law.

The **voltage drop E_L across an inductor** *is proportional to the instantaneous time rate of change of the current I,* say,

(2) $$\boxed{E_L = L\frac{dI}{dt}}$$

where the constant of proportionality L is called the **inductance** of the inductor and is measured in *henrys;* time t is measured in seconds.

A capacitor is an element that stores energy. Experiments yield the following law.

The **voltage drop E_C across a capacitor** *is proportional to the instantaneous electric charge Q on the capacitor,* say,

(3*) $$\boxed{E_C = \frac{1}{C}Q}$$

where C is called the **capacitance** and is measured in *farads;* the charge Q is measured in *coulombs.* Since

(3′) $$I(t) = \frac{dQ}{dt}$$

Eq. (3*) may be written

---

[18]These and the subsequent units are named after ANDRÉ MARIE AMPÈRE (1775–1836), French physicist; GEORG SIMON OHM (1789–1854), German physicist; ALESSANDRO VOLTA (1745–1827), Italian physicist; JOSEPH HENRY (1797–1878), American physicist; MICHAEL FARADAY (1791–1867), English physicist; and CHARLES AUGUSTIN DE COULOMB (1736–1806), French physicist and engineer.

(3)
$$E_C = \frac{1}{C} \int_{t_0}^{t} I(t^*) \, dt^*.$$

All this is summarized in Table 1.1.

Table 1.1
Elements of Electric Circuits

Name	Symbol		Notation	Unit	Voltage Drop
Ohm's resistor	⟿	R	Ohm's resistance	ohms (Ω)	RI
Inductor	⟿	L	Inductance	henrys (H)	$L \dfrac{dI}{dt}$
Capacitor	⟿	C	Capacitance	farads (F)	Q/C

Currents in Circuits Obtained by Kirchhoff's Voltage Law

The current $I(t)$ in a circuit may be determined by solving the equation (or equations) resulting from the application of the following physical law.

Kirchhoff's voltage law (KVL)[19]

The algebraic sum of all the instantaneous voltage drops around any closed loop is zero, or the voltage impressed on a closed loop is equal to the sum of the voltage drops in the rest of the loop.

EXAMPLE 1   **RL-circuit**

Model the "*RL*-circuit" in Fig. 21 and solve the resulting equation for (A) a constant electromotive force, (B) a periodic electromotive force.

Solution. 1st Step. Modeling. By (1) the voltage drop across the resistor is *RI*. By (2) the voltage drop across the inductor is *L dI/dt*. By KVL the sum of the two voltage drops must equal the electromotive force $E(t)$; thus

(4)
$$L \frac{dI}{dt} + RI = E(t).$$

2nd Step. Solution of the equation for a special circuit. Before we solve the general problem, let us solve the special case in which $L = 0.1$ henry, $R = 5$ ohms, and a 12-volt battery gives the electromotive force. Then (4) is

$$0.1 \frac{dI}{dt} + 5I = 12, \qquad \text{thus} \qquad \frac{dI}{dt} + 50I = 120.$$

We can read off an integrating factor e^{50t} and get

$$\frac{d}{dt}(e^{50t}I) = 120e^{50t}, \qquad e^{50t}I = \frac{120}{50}e^{50t} + c, \qquad I = 2.4 + ce^{-50t}.$$

---

[19]GUSTAV ROBERT KIRCHHOFF (1824–1887), German physicist. Later we shall also need **Kirchhoff's current law (KCL):**
   At any point of a circuit, the sum of the inflowing currents is equal to the sum of the outflowing currents.

Or we can use (4) in Sec. 1.6 with $x = t$, $y = I$, $p = 5/0.1$, $r = 12/0.1$, and get the same result:

$$I(t) = e^{-50t}\left[\int e^{50t}120\,dt + c\right] = \frac{120}{50} + ce^{-50t}.$$

Figure 22 shows the three kinds of solution, depending on the initial condition, namely: constant solution $I = E_0/R = 12/5 = 2.4$ (if $I(0) = 2.4$) or exponential approach to 2.4 from below (if $I(0) < 2.4$) or from above (if $I(0) > 2.4$).

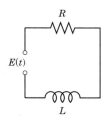

Fig. 21. *RL*-circuit

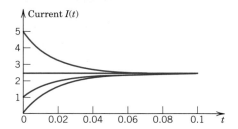

Fig. 22. Current in an *RL*-circuit due to a constant electromotive force

3rd Step. Solution of Eq. (4) in general. We get the general solution of (4) from (4) in Sec. 1.6, which was derived for an equation $y' + py = r$, with y' having coefficient 1; thus we must first divide (4) by L,

$$\frac{dI}{dt} + \frac{R}{L}I = \frac{E}{L}.$$

Then (4), Sec. 1.6, with $x = t$, $y = I$, $p = R/L$, and $r = E/L$ gives

$$(5) \qquad I(t) = e^{-\alpha t}\left[\int e^{\alpha t}\frac{E}{L}\,dt + c\right], \qquad\qquad \alpha = \frac{R}{L}.$$

4th Step. Case A. Constant electromotive force $E = E_0$. Since $\int e^{\alpha t}\,dt = e^{\alpha t}/\alpha$, for constant $E = E_0$, Eq. (5) gives the solution

$$(5^*) \qquad I(t) = e^{-\alpha t}\left[\frac{E_0}{L}\cdot\frac{L}{R}e^{\alpha t} + c\right] = \frac{E_0}{R} + ce^{-\alpha t}.$$

The last term goes to zero as $t \to \infty$; practically, after some time the current will be constant, equal to E_0/R, the value it would have immediately (by Ohm's law) had we no inductor in the circuit, and we see that this limit is independent of the initial value $I(0)$.

From (5*) we see that $I(0) = E_0/R + c$. Hence for the initial condition $I(0) = 0$ we get $c = -E_0/R$ and from (5*) the particular solution

$$(5^{**}) \qquad I(t) = \frac{E_0}{R}(1 - e^{-\alpha t}) = \frac{E_0}{R}(1 - e^{-t/\tau_L}).$$

Here $\tau_L = L/R\ (= 1/\alpha)$ is called the **inductive time constant** of the circuit (see Prob. 3). The lowest curve in Fig. 22 gives an impression of this solution.

5th Step. Case B. Periodic electromotive force $E(t) = E_0\sin\omega t$. For this $E(t)$, Eq. (5) is

$$I(t) = e^{-\alpha t}\left[\frac{E_0}{L}\int e^{\alpha t}\sin\omega t\,dt + c\right], \qquad\qquad \alpha = \frac{R}{L}.$$

Integration by parts yields

$$(6^*) \qquad I(t) = ce^{-(R/L)t} + \frac{E_0}{R^2 + \omega^2 L^2}(R\sin\omega t - \omega L\cos\omega t).$$

This may be written [see (14) in Appendix A3.1]

$$(6) \qquad I(t) = ce^{-(R/L)t} + \frac{E_0}{\sqrt{R^2 + \omega^2 L^2}} \sin(\omega t - \delta), \qquad\qquad \delta = \arctan \frac{\omega L}{R}.$$

The exponential term will approach zero as t approaches infinity. This means that after some time the current $I(t)$ will execute practically harmonic oscillations. (See Fig. 23.) Figure 24 shows the phase angle δ as a function of $\omega L/R$. If $L \approx 0$, then $\delta \approx 0$, and the oscillations of $I(t)$ are in phase with those of $E(t)$. ◀

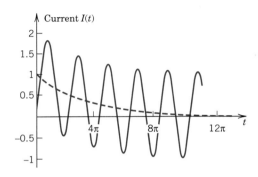

Fig. 23. Current (6) in an *RL*-circuit due to a sinusoidal electromotive force. (For simplicity, $I(t) = \exp(-0.1t) + \sin(t - \pi/4)$.) Dashed: the exponential term

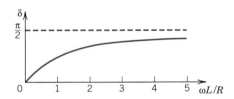

Fig. 24. Phase angle δ in (6) as a function of $\omega L/R$

An electrical (or dynamical) system is said to be in the **steady state** when the variables describing its behavior are periodic functions of time or constant, and it is said to be in the **transient state** (or *unsteady state*) when it is not in the steady state. The corresponding variables are called *steady-state functions* and *transient functions,* respectively.

In Example 1, Case A, the function E_0/R is a steady-state function or **steady-state solution** of (4), and in Case B the steady-state solution is represented by the last term in (6). Before the circuit (practically) reaches the steady state it is in the transient state. It is clear that such an interim or transient period occurs because inductors and capacitors store energy, and the corresponding inductor currents and capacitor voltages cannot be changed instantly. Practically, this transient state will last only a short time.

EXAMPLE 2   *RC*-circuit

Model the "*RC*-circuit" in Fig. 25 and find the current in the circuit for Cases A and B of the electromotive force $E(t)$ considered in Example 1.

Solution. 1st Step. Modeling.   From (1), (3), and KVL we get

(7)
$$RI + \frac{1}{C} \int I\, dt = E(t).$$

To get rid of the integral we differentiate the equation with respect to t, finding

(8)
$$R\frac{dI}{dt} + \frac{1}{C}I = \frac{dE}{dt}.$$

2nd Step. Solution of the equation.   Dividing (8) by R, we get from (4), Sec. 1.6, the general solution

(9)
$$I(t) = e^{-t/(RC)}\left(\frac{1}{R}\int e^{t/(RC)}\frac{dE}{dt}\, dt + c\right).$$

3rd Step. Case A. Constant electromotive force.   If $E = const$, then $dE/dt = 0$ and (9) is simply

(10)
$$I(t) = ce^{-t/(RC)} = ce^{-t/\tau_C} \qquad \text{(Fig. 26)}$$

where $\tau_C = RC$ is called the **capacitive time constant** of the circuit.

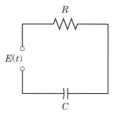

Fig. 25.   *RC*-circuit

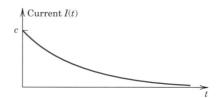

Fig. 26.   Current (10) in an *RC*-circuit due to a constant electromotive force

4th Step. Case B. Periodic electromotive force $E(t) = E_0 \sin \omega t$.   For this $E(t)$ we have

$$\frac{dE}{dt} = \omega E_0 \cos \omega t.$$

By inserting this in (9) and integrating by parts we find

(11)
$$I(t) = ce^{-t/(RC)} + \frac{\omega E_0 C}{1 + (\omega RC)^2}(\cos \omega t + \omega RC \sin \omega t)$$

$$= ce^{-t/(RC)} + \frac{\omega E_0 C}{\sqrt{1 + (\omega RC)^2}} \sin(\omega t - \delta),$$

where $\tan \delta = -1/(\omega RC)$. The first term decreases steadily as t increases, and the last term represents the steady-state current, which is sinusoidal. The graph of $I(t)$ is similar to that in Fig. 23.   ◀

   "*RLC*-circuits" containing all three kinds of elements lead to second-order differential equations and will be considered in Sec. 2.12; electrical networks leading to systems of differential equations will be discussed in Chap. 3.

PROBLEM SET 1.7

RL-Circuits

1. **(Limit)** Can the limit of (5*) as $t \to \infty$ be seen directly from the differential equation without actually solving it?

2. **(Phase angle)** How does the phase angle δ in (6) depend on L? Is this physically understandable?

3. **(Inductive time constant)** Show that $\tau_L = L/R$ is the time at which the current (5**) reaches about 63% of its final value.

4. **(Increase and decrease)** In an RL-circuit with constant $E(t) = E_0$, particular solutions increase if $I(0) < E_0/R$, and decrease if $I(0) > E_0/R$. Could you see this directly from (4) (with $E(t) = E_0$) without solving (4)? *Hint.* When will $I' > 0$? $I' < 0$?

5. **(Half of final value)** At what time will the current (5**) reach half of its theoretical maximum value?

6. **(Choice of R)** If $L = 10$ henrys, what R should we choose so that (5**) will reach 99% of its final value at $t = 1$ sec?

7. **(Choice of L)** What L should we choose in (4) with $E = E_0 = const$ and $R = 1000$ ohms if we want the current to grow from 0 to 25% of its final value within 10^{-4} sec?

8. **(Avoiding integration)** Derive the steady-state solution in (6) by substituting the function $I_p = A \cos \omega t + B \sin \omega t$ into (4) with $E(t) = E_0 \sin \omega t$ and determining A and B by equating the cosine and sine terms in the resulting equation. (This avoids integration by parts.)

9. Solve (4) with $E(t) = e^{-t}$ when (a) $R \neq L$ and (b) $R = L$.

10. **TEAM PROJECT. RL-Circuits with Discontinuous Electromotive Force.** In practice, discontinuous electromotive forces $E(t)$ may occur because of switching-on or off.

    (a) Show that to a jump J of $E(t)$ in (4) at some $t = a$ there corresponds a jump J/L of I'.

    (b) Let $R = 1$ ohm, $L = 1$ henry, $E(t) = 1$ volt when $0 \leq t \leq 4$ sec, and $E(t) = 0$ when $t > 4$ sec. Find $I(t)$, assuming $I(0) = 1/2$ ampere.

    (c) Using the experience that you have gained, generalize (b) to arbitrary R, L, $I(0) = I_0$, $E(t) = E_0 = const$ and time interval $0 \leq t \leq a$.

RC-Circuits

11. **(Verification)** Show that (11) is a solution of (8) with $E = E_0 \sin \omega t$.

12. **(Particular solution)** Obtain from (11) the solution satisfying the initial condition $I(0) = 0$.

13. **(Voltage)** A capacitor ($C = 0.2$ farad) in series with a resistor ($R = 200$ ohms) is charged from a source ($E_0 = 24$ volts); see Fig. 25 with $E(t) = E_0$. Find the voltage $V(t)$ on the capacitor, assuming that at $t = 0$ the capacitor is completely uncharged.

14. **(Current)** Find the current $I(t)$ in the RC-circuit shown in Fig. 25, assuming that $E = 100$ volts, $C = 0.25$ farad, R is variable according to $R = (200 - t)$ ohms when $0 \leq t \leq 200$ sec, $R = 0$ when $t > 200$ sec, and $I(0) = 1$ ampere.

15. **(Discharge of a capacitor)** Show that (7) can also be written

    $$(12) \qquad\qquad R\frac{dQ}{dt} + \frac{1}{C}Q = E(t).$$

    Solve this equation with $E(t) = 0$, assuming $Q(0) = Q_0$.

16. **(Discharge)** Find the time when the capacitor in Prob. 15 has lost 99% of his initial charge.

17. **(Discharge)** In (12) in Prob. 15, let $R = 10$ ohms and $C = 0.1$ farad, and let $E(t)$ be exponentially decaying, say, $E(t) = 30e^{-3t}$ volts. Assuming $Q(0) = 0$, find and graph $Q(t)$. At what time does $Q(t)$ reach a maximum? What is that maximum charge?

18. **(Change of data)** What happens in Prob. 17 if the resistance R is doubled? First guess, then calculate.

19. **(Periodic electromotive force)** Find the steady-state solution of (12) when $R = 50$ ohms, $C = 0.04$ farad, and $E(t) = 100 \cos 2t + 25 \sin 2t + 200 \cos 4t + 25 \sin 4t$.

20. **TEAM PROJECT. *RC*-Circuits with Discontinuous Electromotive Force.** In this project we shall see that *RC*- and *RL*-circuits (see Team Project 10) behave differently under discontinuous electromotive forces.

   (a) Show that if the initial charge in the capacitor in Fig. 25 is $Q(0)$, the initial current in the *RC*-circuit is $I(0) = E(0)/R - Q(0)/RC$.

   (b) Show that if $E(t)$ in Fig. 25 has a jump of magnitude J at $t = a$, then the current $I(t)$ in the *RC*-circuit has a jump of magnitude J/R at $t = a$.

   (c) Find the plot $I(t)$ when $R = 1$, $C = 1$ (for simplicity), and $E(t) = \frac{1}{2}t^2$ if $0 < t < 2$ and $E = 0$ thereafter, assuming zero initial charge on the capacitor.

   (d) Give a similar discussion of the differential equation of the *RC*-circuit for another discontinuous $E(t)$ of your own choice.

1.8 Orthogonal Trajectories of Curves. *Optional*

As another interesting application, we shall see in this section how to use differential equations for finding curves that intersect given curves at right angles,[20] a task that arises rather often in practice. The new curves are then called the **orthogonal trajectories** of the given curves (and conversely). Here, "orthogonal" is another word for "perpendicular."

For instance, the meridians and parallels on the globe are orthogonal trajectories of each other, and so are the curves of steepest descent and the contour lines on a map. In an electric field, the curves of electric force are the orthogonal trajectories of the equipotential lines (= curves of constant voltage) and conversely. A simple example of this is shown in Fig. 27, where the equipotential lines are concentric circles (cylinders in space appearing as circles in this cross section), and their orthogonal trajectories are straight lines (planes in space through the axis of the cylinders). Other important applications arise in fluid flow, heat conduction, and other fields of physics.

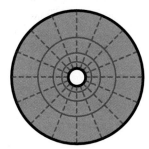

Fig. 27. Equipotential lines and lines of electric force (dashed) between two concentric cylinders

---

[20]Remember that the **angle of intersection** of two curves is defined to be the angle between the tangents of the curves at the point of intersection.

How to Find Orthogonal Trajectories

1st Step. We find a differential equation

$$(1) \qquad\qquad y' = f(x, y)$$

for which the given curves are solution curves.

2nd Step. We write down the differential equation of the orthogonal trajectories to be found, which is

$$(2) \qquad\qquad \boxed{y' = -\dfrac{1}{f(x, y)}}$$

with the same $f(x, y)$ as in (1). Why? Well, the slope of a given curve passing through a point (x_0, y_0) is $f(x_0, y_0)$, by (1). The trajectory through (x_0, y_0) has slope $-1/f(x_0, y_0)$, by (2). The product of these slopes is -1, as we see. This is the condition for perpendicularity (orthogonality, as we now say), known from calculus.

3rd Step. Solve the differential equation (2).

This needs some explanation. The differential equation (1) has infinitely many solution curves, one for each value of the arbitrary constant c in its general solution. Hence we can write this family of curves as

$$(3) \qquad\qquad F(x, y, c) = 0.$$

For example, a family of circles $x^2 + y^2 = c$ can be written

$$F(x, y, c) = x^2 + y^2 - c = 0.$$

This is called a **one-parameter family of curves,** and c is called the *parameter* of the family. From (3) we get the differential equation (1) by differentiation, as we shall explain. It is quite important that in this step the parameter c disappears—under no circumstances must it appear in (1)! The rest of the method is straightforward (except perhaps for difficulties in solving (2), which we may have with solving *any* differential equation). We illustrate all this with a typical example.

EXAMPLE 1   **Family of curves. Their orthogonal trajectories**

Given the curves

$$(4) \qquad\qquad y = cx^2,$$

where c is arbitrary. Find their orthogonal trajectories.

Solution. 1st Step. Differential equation of the family **(4).**   Equation (4) represents a one-parameter family of curves, where c is the parameter. These are parabolas. For each c we get one of them. For positive c they open upward and for negative c downward. For $c = 0$ we get a straight line (the x-axis); see Fig. 28. We can write (4) in the form (3),

$$F(x, y, c) = y - cx^2 = 0.$$

We derive a differential equation of the form (1) of the family (4). Differentiating (4), we have

$$y' = 2cx.$$

But this is no good because it still contains c, so we do something else. We first solve (4) algebraically for c,

$$\frac{y}{x^2} = c.$$

Then we differentiate this equation with respect to x. By calculus,

$$\frac{y'}{x^2} - 2\frac{y}{x^3} = 0.$$

Next we multiply by x^2 and take the last term to the right, obtaining

(5) $$y' = \frac{2y}{x}.$$

This is the differential equation (1) of the family (4).

2nd Step. *Differential equation of the orthogonal trajectories.* The differential equation (2) of the orthogonal trajectories of (4) is now obtained from (5) in the form

(6) $$y' = -\frac{x}{2y}.$$

Do not forget the minus sign!

3rd Step. *Orthogonal trajectories.* We find the orthogonal trajectories of (4) by solving (6). Separation of variables and integration give

$$2y\,dy = -x\,dx, \qquad y^2 = -\frac{x^2}{2} + c^*.$$

This is a one-parameter family of ellipses. Note that we must use another notation for the parameter, not c. We write this family in the more usual form

(7) $$\tfrac{1}{2}x^2 + y^2 = c^*.$$

Then we see that $c^* = 1$ gives the ellipse with semi-axes $\sqrt{2}$ (in the x-direction) and 1. For every positive c^* we obtain an ellipse with semi-axes $\sqrt{2c^*}$ and $\sqrt{c^*}$, as shown in Fig. 28. For $c^* = 0$ we obtain the origin and for $c^* < 0$ no (real) solution. This solves our problem. ◀

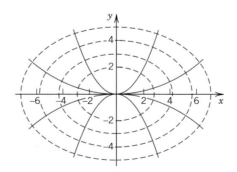

Fig. 28. Curves and orthogonal trajectories
in Example 1

PROBLEM SET 1.8

Families of Curves. Represent the following families in the form (1). Sketch some of the curves.

1. All ellipses with foci -2 and 2 on the x-axis
2. All circles of radius 2 with centers on the cubic parabola $y = x^3$
3. The catenaries obtained by translating the **catenary** $y = \cosh x$ in the direction of the straight line $y = -x$

Differential Equations of Families of Curves. Plot some of the given curves. Find the differential equation of the family.

4. $2y - x + c = 0$
5. $y = ce^{2x}$
6. $y = \tan(x + c)$
7. $y = cx + x$
8. $y = cx^4$
9. $c^2x^2 + y^2 = c^2$

Orthogonal Trajectories. Find the orthogonal trajectories. First guess from the plot of the given curves. Then apply the method used in Example 1. Plot or sketch some curves and trajectories. (Show the details of your work.)

10. $y = ce^{x^2}$
11. $y = ce^{-x}$
12. $y = \ln|x| + c$
13. $y = c\sqrt{x}$
14. $y = \sqrt{x + c}$
15. $y = cx^{3/2}$
16. $xy = c$
17. $y = c/x^2$
18. $(x - c)^2 + y^2 = c^2$

Applications

19. **(Electric field)** In the electric field between two concentric cylinders (Fig. 27) the **equipotential lines** (= curves of constant potential) are circles given by $U(x, y) = x^2 + y^2 = const$ [volts]. Use the method of Example 1 to get their trajectories (the curves of electric force).

20. **(Electric field)** Experiments show that the electric lines of force of two opposite charges of the same strength at $(-1, 0)$ and $(1, 0)$ are the circles through $(-1, 0)$ and $(1, 0)$. Show that these circles can be represented by the equation $x^2 + (y - c)^2 = 1 + c^2$. Show that the equipotential lines (orthogonal trajectories) are the circles $(x + c*)^2 + y^2 = c*^2 - 1$ (dashed in Fig. 29).

21. **(Fluid flow)** If the **streamlines** of the flow (= paths of the particles of the fluid) in the channel in Fig. 30 are $\Psi(x, y) = xy = const$, what are their orthogonal trajectories (called **equipotential lines,** for reasons explained in Sec. 16.4)?

22. **(Temperature field)** If the **isotherms** (= curves of constant temperature) in a body are $T(x, y) = 2x^2 + y^2 = const$, what are their orthogonal trajectories (the curves along which heat will flow in regions free of heat sources or sinks and filled with homogeneous material)?

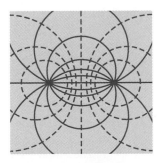

Fig. 29.   Electric field in Problem 20

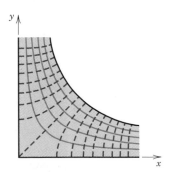

Fig. 30.   Flow around a corner in Problem 21

Other Forms of the Differential Equations

23. Show that the orthogonal trajectories of a family $g(x, y) = c$ can be obtained from the following differential equation and use it to solve Prob. 21:

$$\frac{dy}{dx} = \frac{\partial g/\partial y}{\partial g/\partial x}.$$

24. (Cauchy–Riemann equations). Show that for a family $u(x, y) = c = const$ the orthogonal trajectories $v(x, y) = c^* = const$ can be obtained from the following so-called *Cauchy–Riemann equations* (which are basic in complex analysis in Chap. 12). Use these equations to find the orthogonal trajectories of $e^x \cos y = const$.

25. WRITING PROJECT. Orthogonal Trajectories. Write a short report on the observation that if F is the family of orthogonal trajectories of a given family G, then G is the family of orthogonal trajectories of F. Illustrate this observation with two examples of your own, showing all the details of your calculations.

26. TEAM PROJECT. Conic Sections. (a) Give a brief description of the general method of obtaining orthogonal trajectories.

(b) Are the orthogonal trajectories of families of ellipses $x^2/a^2 + y^2/b^2 = c$ always conic sections? Find conditions under which this is the case. Make CAS plots or sketches by hand illustrating your result. What happens if $a \to 0$? If $b \to 0$?

(c) Investigate families of hyperbolas $x^2/a^2 - y^2/b^2 = c$ in a similar fashion.

(d) Can you think of more complicated curves for which you still get differential equations that you can solve? Give it a try.

1.9 Existence and Uniqueness of Solutions Picard Iteration

So far, for the differential equations considered there existed a general solution. Also, for an **initial value problem,** say,

(1)
$$\boxed{y' = f(x, y), \qquad y(x_0) = y_0}$$

consisting of a differential equation and an initial condition $y(x_0) = y_0$, we got a unique particular solution. However, this was just *one* of three possibilities illustrated by the following examples:

The initial value problem

$$|y'| + |y| = 0, \qquad y(0) = 1$$

has no solution because $y \equiv 0$ is the only solution of the differential equation. (Why?) The initial value problem

$$y' = x, \qquad y(0) = 1$$

has precisely one solution, namely, $y = \frac{1}{2}x^2 + 1$. The initial value problem

$$xy' = y - 1, \qquad y(0) = 1$$

has infinitely many solutions, namely, $y = 1 + cx$, where c is an arbitrary constant. From these three examples we see that an initial value problem may have no solutions, precisely one solution, or more than one solution. This leads to the following two fundamental questions.

Problem of existence. *Under what conditions does an initial value problem of the form (1) have at least one solution?*

Problem of uniqueness. *Under what conditions does that problem have at most one solution?*

Theorems that state such conditions are called **existence theorems** and **uniqueness theorems,** respectively.

Of course, our three examples are so simple that we can find the answer to these two questions by inspection, without using any theorems. However, it is clear that in more complicated cases—for example, when the equation cannot be solved by elementary methods—existence and uniqueness theorems may be of considerable practical importance. Even when you are sure that your physical or other system behaves uniquely, once in a while your model may be oversimplified and may not give a faithful picture of the reality. So make sure that your model has a unique solution before you try to compute that solution. The following two theorems take care of almost all conceivable practical cases.

The first theorem says that if $f(x, y)$ in (1) is continuous in some region of the xy-plane containing the point (x_0, y_0) (corresponding to the given initial condition), then the problem (1) has at least one solution.

The second theorem says that if, moreover, the partial derivative $\partial f/\partial y$ exists and is continuous in that region, then the problem (1) can have at most one solution; hence, by Theorem 1, it has precisely one solution.

Read again what you have just read—these are entirely new ideas in our discussion.

We now formulate our statements in a precise way.

THEOREM 1   Existence theorem

If $f(x, y)$ is continuous at all points (x, y) in some rectangle (Fig. 31)

$$R: \qquad |x - x_0| < a, \qquad |y - y_0| < b$$

and bounded[21] *in R, say,*

(2)
$$|f(x, y)| \leqq K \qquad \text{for all } (x, y) \text{ in } R,$$

then the initial value problem (1) has at least one solution $y(x)$. This solution is defined at least for all x in the interval $|x - x_0| < \alpha$ where α is the smaller of the two numbers a and b/K.

---

[21]A function $f(x, y)$ is said to be **bounded** when (x, y) varies in a region in the xy-plane if there is a number K such that $|f(x, y)| \leqq K$ when (x, y) is in that region. For example, $f = x^2 + y^2$ is bounded, with $K = 2$ if $|x| < 1$ and $|y| < 1$. The function $f = \tan (x + y)$ is not bounded for $|x + y| < \pi/2$.

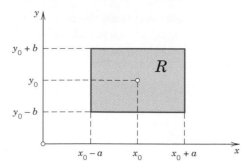

Fig. 31. Rectangle R in the existence and
uniqueness theorems

THEOREM 2 **Uniqueness theorem**

If $f(x, y)$ and $\partial f/\partial y$ are continuous for all (x, y) in that rectangle R and bounded, say,

(3)   (a) $|f| \leq K,$ (b) $\left|\dfrac{\partial f}{\partial y}\right| \leq M$ *for all (x, y) in R,*

then the initial value problem (1) has at most one solution $y(x)$. Hence, by Theorem 1, it has precisely one solution. This solution is defined at least for all x in that interval $|x - x_0| < \alpha$.

Understanding These Theorems

Proofs of these theorems are beyond the level of this book (see Ref. [A5] in Appendix 1). However, we want to include some remarks and examples that may help the student to better undertand these theorems.

Since $y' = f(x, y)$, the condition (2) implies that $|y'| \leq K$, that is, the slope of any solution curve $y(x)$ in R is at least $-K$ and at most K. Hence a solution curve that passes through the point (x_0, y_0) must lie in the colored region in Fig. 32 bounded by the lines l_1 and l_2 whose slopes are $-K$ and K, respectively. Depending on the form of R, two different cases may arise. In the first case, shown in Fig. 32a on p. 55, we have $b/K \geq a$ and therefore $\alpha = a$ in the existence theorem, which then asserts that the solution exists for all x between $x_0 - a$ and $x_0 + a$. In the second case, shown in Fig. 32b, we have $b/K < a$. Therefore $\alpha = b/K$, and all we can conclude from the theorems is that the solution exists for all x between $x_0 - b/K$ and $x_0 + b/K$. For larger or smaller x's the solution curve may leave the rectangle R, and since nothing is assumed about f outside R, nothing can be concluded about the solution for those larger or smaller x's; that is, for such x's the solution may or may not exist—we don't know.

Let us illustrate our discussion with a simple example. We shall see that our choice of a rectangle R with a large base (a long x-interval) will lead to the case in Fig. 32(b).

EXAMPLE 1   Consider the problem

$$y' = 1 + y^2, \qquad y(0) = 0$$

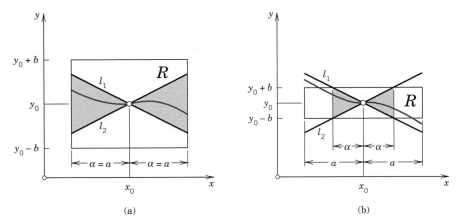

Fig. 32.   The condition (2) of the existence theorem.
(a) First case. (b) Second case

and take R: $|x| < 5$, $|y| < 3$. Then $a = 5$, $b = 3$ and

$$|f| = |1 + y^2| \leqq K = 10, \qquad |\partial f/\partial y| = 2|y| \leqq M = 6, \qquad \alpha = b/K = 0.3 < a.$$

Indeed, the solution of the problem is $y = \tan x$ (see Example 2 in Sec. 1.3). This solution is discontinuous at $x = \pm \pi/2$. And there is no *continuous* solution valid in the entire interval $|x| < 5$ from which we started. ◀

The conditions in the two theorems are sufficient conditions rather than necessary ones, and can be lessened. For example, by the mean value theorem of differential calculus we have

$$f(x, y_2) - f(x, y_1) = (y_2 - y_1) \left. \frac{\partial f}{\partial y} \right|_{y = \tilde{y}}$$

where (x, y_1) and (x, y_2) are assumed to be in R, and $\tilde{y}$ is a suitable value between y_1 and y_2. From this and (3b) it follows that

(4) $$\left| f(x, y_2) - f(x, y_1) \right| \leqq M|y_2 - y_1|.$$

It can be shown that (3b) may be replaced by the weaker condition (4), which is known as a **Lipschitz condition.**[22] However, continuity of $f(x, y)$ is not enough to guarantee the *uniqueness* of the solution. This may be illustrated with the following example.

EXAMPLE 2   **Nonuniqueness**

The initial value problem

$$y' = \sqrt{|y|}, \qquad y(0) = 0$$

has the two solutions

$$y \equiv 0 \qquad \text{and} \qquad y^* = \begin{cases} x^2/4 & \text{if } x \geqq 0 \\ -x^2/4 & \text{if } x < 0 \end{cases}$$

---

[22]RUDOLF LIPSCHITZ (1832–1903), German mathematician, professor at Bonn, who also contributed to algebra, number theory, potential theory, and mechanics.

although $f(x, y) = \sqrt{|y|}$ is continuous for all y. The Lipschitz condition (4) is violated in any region that includes the line $y = 0$, because for $y_1 = 0$ and positive y_2 we have

$$(5) \qquad \frac{|f(x, y_2) - f(x, y_1)|}{|y_2 - y_1|} = \frac{\sqrt{y_2}}{y_2} = \frac{1}{\sqrt{y_2}}, \qquad (\sqrt{y_2} > 0)$$

and this can be made as large as we please by choosing y_2 sufficiently small, whereas (4) requires that the quotient on the left side of (5) should not exceed a fixed constant M. ◀

Picard Iteration for Initial Value Problems. *Optional*

Picard's iteration method[23] gives approximate solutions of an initial value problem (1),

$$y' = f(x, y), \qquad y(x_0) = y_0.$$

The method is based on two ideas. The first idea is that we can integrate (1) on both sides, obtaining

$$(1^*) \qquad y(x) = y_0 + \int_{x_0}^{x} f[t, y(t)] \, dt.$$

y_0 is the constant of integration. We have chosen it and the limits of integration so that for $x = x_0$ the integral is zero, thus satisfying the initial condition. Since x occurs as the upper limit of integration, we had to choose another letter of integration, t.

The second idea of the method is to solve (1^*) by **iteration.** That is, in the first step we substitute $y(t) = y_0$ in the integrand in (1^*) and calculate a first approximation $y_1(x)$ of the unknown solution,

$$y_1(x) = y_0 + \int_{x_0}^{x} f(t, y_0) \, dt.$$

In the second step we substitute the function $y_1(x)$ in the integrand in the same way and calculate a presumably better second approximation

$$y_2(x) = y_0 + \int_{x_0}^{x} f[t, y_1(t)] \, dt,$$

etc. The nth step of this iteration gives an approximating function

$$(6) \qquad y_n(x) = y_0 + \int_{x_0}^{x} f[t, y_{n-1}(t)] \, dt.$$

In this way we obtain a sequence of approximations

$$(7) \qquad y_1(x), \qquad y_2(x), \cdots, \qquad y_n(x), \cdots.$$

---

[23]EMILE PICARD (1856–1941), French mathematician, professor in Paris since 1881, also known for his important contributions to complex analysis (see Sec. 15.2 for his famous theorem).

THEOREM 3   **(Convergence of Picard's iteration)**

Under the conditions of Theorems 1 and 2 the sequence (7) of functions defined by (6) (with $y_0(t) = y_0 = const$) converges to the solution $y(x)$ of the initial value problem (1).

EXAMPLE 3   **Picard iteration**

Find approximate solutions to the initial value problem

$$y' = 1 + y^2, \qquad y(0) = 0.$$

Solution.   In this case, $x_0 = 0$, $y_0 = 0$, $f(x, y) = 1 + y^2$, and (6) becomes

$$y_n(x) = \int_0^x \left[1 + y_{n-1}^2(t)\right] dt = x + \int_0^x y_{n-1}^2(t)\, dt.$$

Starting from $y_0 = 0$, we thus obtain (see Fig. 33)

$$y_1(x) = x + \int_0^x 0 \, dt = x$$

$$y_2(x) = x + \int_0^x t^2 \, dt = x + \frac{1}{3}x^3$$

$$y_3(x) = x + \int_0^x \left(t + \frac{t^3}{3}\right)^2 dt = x + \frac{1}{3}x^3 + \frac{2}{15}x^5 + \frac{1}{63}x^7$$

etc. Of course, we can obtain the exact solution of our present problem by separating variables (see Example 2 in Sec. 1.3), finding

(8) $$y(x) = \tan x = x + \frac{1}{3}x^3 + \frac{2}{15}x^5 + \frac{17}{315}x^7 + \cdots \qquad \left(-\frac{\pi}{2} < x < \frac{\pi}{2}\right).$$

The first three terms of $y_3(x)$ and of the series in (8) are the same. The series in (8) converges for $|x| < \pi/2$, and all we may expect is that our sequence $y_1, y_2, \cdots$ converges to a function that is the solution of our problem for $|x| < \pi/2$. This illustrates that the study of convergence is of practical importance.   ◀

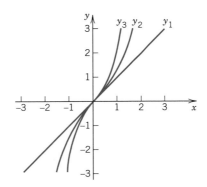

Fig. 33.   Approximate solutions in Example 3

Practical Significance of Picard's Iteration

Picard used his iteration method for proving Theorems 1–3. His method involves integrations, which generally become quite involved and lengthy after a few steps. Hence in precomputer times his method was of little *practical* value.

The situation has changed with the advent of computers, beginning around 1950. Iteration methods are very practical in general, because in each step they use the same sequence of operations, with new data that have just been generated in the preceding step (as in Example 3, or in several of the previous steps in other iterations). This makes programs relatively short. Your CAS may have a built-in program for Picard iteration that can be called by a single command. If not, you can write a program of your own. Don't forget to include a printing command for the approximate solution curve in each step—it is frequently quite exciting to watch how these curves make it that they converge to the exact solution curve. The MAPLE and MATHEMATICA MANUALS for this book contain nice examples of this.

Numerical methods on first-order differential equations can now be taken up if so desired. The corresponding Secs. 19.1 and 19.2 are completely independent of any other section in Chaps. 17–19.

PROBLEM SET 1.9

Existence and Uniqueness of Solutions

1. Show that the initial value problem $xy' = 4y$, $y(0) = 1$ has no solution. Does this contradict our present existence theorem?

2. What happens in Prob. 1 if we replace $y(0) = 1$ by $y(0) = 0$? Do we get a contradiction to our present theorems?

3. If the assumptions of Theorem 1 are satisfied not merely in a rectangle R but even in a vertical strip given by $|x - x_0| < a$, show that the solution of (1) exists for all x in the interval $|x - x_0| < a$.

4. Find all initial conditions such that the initial value problem $(x^2 - 2x)y' = 2(x - 1)y$, $y(x_0) = y_0$ has (a) no solution, (b) more than one solution, and (c) precisely one solution. Do your answers contradict Theorems 1 and 2?

5. Show that if $y' = f(x, y)$ satisfies the assumptions of Theorems 1 and 2 in a rectangle R, then any two solution curves of this equation in R cannot have a point in common in R.

6. Find all solutions of $y' = x|y|$.

7. **(Linear differential equation)** Write $y' + p(x)y = r(x)$ in the form (1). If p and r are continuous for all x such that $|x - x_0| \leq a$, show that $f(x, y)$ in this equation satisfies the conditions of our present existence and uniqueness theorems so that a corresponding initial value problem has a unique solution. [This also follows directly from (4) in Sec. 1.6, so that for the *linear* differential equation, we do not need these theorems.]

8. For many differential equations, solutions may exist in intervals larger than those given in Theorems 1 and 2. Illustrate this for the initial value problem $y' = y^2$, $y(1) = 1$ by finding *the best* possible α (choosing b optimally) and then by determining for what x the solution actually exists.

9. What best possible α can we achieve in Example 1 of the text by choosing a and b suitably?

10. **PROJECT. Lipschitz Condition. (a)** State the definition of a Lipschitz condition and explain its relation to the existence of a partial derivative. Explain its significance in our present context. Illustrate your statements using examples of your own.

(b) Show that $f(x, y) = |\sin y| + x$ satisfies a Lipschitz condition (4) with $M = 1$ on the whole xy-plane, but $\partial f / \partial y$ does not exist when $y = 0$. Explain the reason. Explain what this example illustrates.

(c) Show that for the *linear* differential equation $y' + p(x)y = r(x)$ with *continuous* $p(x)$ and $r(x)$ in $|x - x_0| \leq a$, a Lipschitz condition holds. This is quite remarkable! It means that for the *linear* differential equation the *continuity* of $f(x, y)$ guarantees the *uniqueness of the solution* of an initial value problem. (Of course, this also follows directly from (4) in Sec. 1.6.) This is not generally true for a *nonlinear* differential equation.

(d) For a few simple differential equations that you can solve, discuss uniqueness of solutions and find out whether a Lipschitz condition is satisfied.

Picard Iteration

Apply Picard's iteration to the following problems. Do three steps if you are working by hand. On the computer, do as many steps (at most 10) as can be done within a minute or two. Sketch or plot every approximate solution curve obtained. Find the exact solution. Compare.

11. $y' = y$, $y(0) = 1$. Show that the approximations approach $y = e^x$, the exact solution.

12. $y' = x + y$, $\quad y(0) = 0$

13. $y' = x + y$, $\quad y(0) = -1$

14. $y' = y^2$, $\quad y(0) = 1$

15. $y' = xy + 2x - x^3$, $\quad y(0) = 0$

16. $y' = 2\sqrt{y}$, $y(1) = 0$. Find all solutions. Which of them does Picard's iteration approximate?

17. $y' = y - y^2$, $\quad y(0) = \frac{1}{2}$

18. $y' = 3y/x$, $\quad y(1) = 1$

19. Show that if f in $y' = f(x, y)$ is independent of y, then the approximations obtained by Picard's method are identical to the exact solution. Why?

 20. CAS PROJECT. Picard Iteration. **(a)** Write a program for the Picard iteration that gives a printout of all approximations as well as a plot of them on common axes. Try your program on two initial value problems of your own choice.

(b) Find the Maclaurin series of the function

$$y = e^{x^2/2} \int_0^x e^{-t^2/2} \, dt$$

by deriving an initial value problem that y satisfies and applying to it the Picard method. Compare the amount of work with that of integrating the Maclaurin series of $e^{-x^2/2}$, multiplying it by that of $e^{x^2/2}$ and ordering the result in powers of x.

(c) Experiment with the conjecture that Picard's iteration converges to the solution of the problem for any initial choice of y in the integrand in (6) (leaving y_0 outside the integral as it is). Begin with a simple differential equation and see what happens. When you are reasonably sure, take a slightly more complicated equation and give it a try.

CHAPTER 1 REVIEW

1. What is the order of a differential equation?

2. What is the difference between an ordinary and a partial differential equation?

3. What is a general solution? A particular solution? What is the practical significance of these two concepts?

4. What do you know about the existence of solutions of a differential equation? About the uniqueness? Is this of practical interest?

5. Make a list of the main methods for solving differential equations that we have considered. Explain each in a few sentences and include a typical example of your own.

6. Are there differential equations that can be solved by more than one of our methods? By none of these standard methods?

7. Make a list of the applications in this chapter, in both the text and the problem sets.

8. What do you know about direction fields? About their practical importance?

9. What do you know about Picard's iteration method? About iteration methods in general?

10. Explain a typical application of orthogonal trajectories. How would you obtain orthogonal trajectories if a family of curves is given?

11. Why do electric circuits lead to differential equations?
12. What is the reason that differential equations appear as models in mechanics?
13. What is exponential growth? Exponential decay? What is the practical significance of these concepts?
14. What is the role of the computer in connection with differential equations?

General Solution. Find the general solution using one of the methods in this chapter. Indicate the method you are using. Show the details of your work.

15. $y' + 4y = 17 \sin x$
16. $y' = ay + by^2$ $(a \neq 0)$
17. $25yy' - 9x = 0$
18. $(x^2 + 1)y' + y^2 + 1 = 0$
19. $\sin x \sin 2y \, dx = 2 \cos x \cos 2y \, dy$
20. $(2xe^{x^2} \cosh y + 1) \, dx + e^{x^2} \sinh y \, dy = 0$
21. $4xyy' = y^2 - x^2$ $(y/x = u)$
22. $xy' = y + x^2 \sec (y/x)$
23. $(3xe^y + 2y) \, dx + (x^2 e^y + x) \, dy = 0$
24. $2x \tan y \, dx + \sec^2 y \, dy = 0$

Initial Value Problems. Solve the following initial value problems, indicating the method used and giving all steps in detail.

25. $y' = y \tanh x$, $y(0) = \pi$
26. $y' = \sqrt{1 - y^2}$, $y(0) = 1/\sqrt{2}$
27. $xy' + y = x^2 y^2$, $y(1) = \frac{1}{2}$
28. $y' + 4xy = e^{-2x^2}$, $y(0) = -4$
29. $9 \sec y \, dx + \sec x \, dy = 0$, $y(0) = 0$
30. $(2x + e^y) \, dx + xe^y \, dy = 0$, $y(2) = 0$
31. $3x^2 y \, dx + 2x^3 \, dy = 0$, $y(1) = 3$
32. $x \sinh y \, dy = \cosh y \, dx$, $y(3) = 0$
33. $y' + xy = xy^{-1}$, $y(0) = 2$
34. $y' + \frac{1}{2}y = y^3$, $y(0) = 1$

Direction Fields. Plot a direction field by hand or with a CAS and sketch some solution curves. Solve exactly and compare.

35. $y' = -3y$
36. $y' = -2xy$
37. $y' = y + 1.01 \sin 10x$
38. $y' = 2y^2$

Orthogonal Trajectories. Determine the orthogonal trajectories of the given family of curves. Plot or sketch the curves and their trajectories on common axes. Show each step of your calculation.

39. $y = x^2 + c$
40. $y = cx^{-3}$
41. $y = ce^{-x^2/2}$
42. $y = \sqrt{2 \ln |x| + c}$

Picard Iteration. Apply Picard's iteration by hand (3 iterates) or by CAS (5–10 iterates). Sketch or plot the iterates.

43. $y' = 2y$ $y(0) = 1$
44. $y' = 1 - y$, $y(0) = 2$

Applications

45. **(Exponential growth)** If the growth rate of a culture of bacteria is proportional to the number of bacteria present and after 1 day is 1.5 times the original number, within what interval of time will the number of bacteria (a) double, (b) triple?

46. **(Newton's law of cooling)** A metal bar whose temperature is 20°C is placed in boiling water. How long does it take to heat the bar to practically 100°C, say, to 99.9°C, if the temperature of the bar after 1 min of heating is 51.5°C? First guess, then calculate.

47. **(Half-life)** If in a reactor, uranium $_{92}U^{237}$ loses 10% of its weight within 1 day, what is its half-life? How long would it take for 99% of the original amount to disappear?

48. **(Electric circuit)** How should one choose R and L in an RL-circuit connected to a battery of 48 volts if one wants the steady-state current to be 10 amp and the time to practically reach this value (say, 9.99 amp) to be 10^{-2} sec after the instant when the circuit is connected to the battery?

49. **(Heart pacemaker)** Figure 34 on p. 61 shows a heart pacemaker consisting of a capacitor of capacitance C, a battery of voltage E_0, and a switch that is periodically moved from A (charging period $t_1 < t < t_2$ of the capacitor) to B (discharging period $t_2 < t < t_3$ during which the capacitor sends an electric stimulus to the heart, which acts as a resistor of resistance R). Find

the current during the discharging period, assuming that the charging period is such that the charge on the capacitor is 99% of its maximum possible value.

50. **(Mixing problem)** The tank in Fig. 35 contains 80 lb of salt dissolved in 500 gal of water. The inflow per minute is 20 lb of salt dissolved in 20 gal of water. The outflow is 20 gal/min of the uniform mixture. Find the time when the salt content $y(t)$ in the tank reaches 95% of its limiting value (as $t \to \infty$).

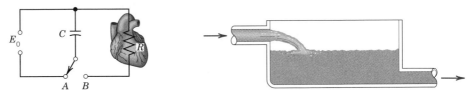

Fig. 34.   Heart pacemaker   **Fig. 35.**   Tank in Problem 50

51. **(Linear accelerator)** Linear accelerators are used in physics for accelerating charged particles. Suppose that an alpha particle enters an accelerator and undergoes a constant acceleration that increases the speed of the particle from 10^4 meters/sec to 10^6 meters/sec in 10^{-4} sec. Find the acceleration a and the distance traveled during this period of 10^{-4} sec.

52. **(Electric field)** What are the lines of electrical force between two elliptical copper plates $x^2 + 2y^2 = 1$ and $\frac{1}{4}x^2 + \frac{1}{2}y^2 = 1$ that are kept at different electric potentials?

53. **(Heat flow)** In a thin plate in the xy-plane, if heat flows along the curves $xy = const$, what are the isotherms (curves of constant temperature) in the plate?

54. **(Chemical reaction)** The **law of mass action** states that under constant temperature the velocity of a chemical reaction is proportional to the product of the concentrations of the reacting substances. A bimolecular reaction $A + B \to M$ combines a moles per liter of a substance A and b moles of a substance B. If $y(t)$ is the number of moles per liter that have reacted after time t, the rate of reaction is $dy/dt = k(a - y)(b - y)$. Solve this equation, assuming that $a \neq b$.

55. **(Lemniscate)** The solution curve of $dr/d\theta + (a^2/r) \sin 2\theta = 0$, $r^2(0) = a^2$, $a \neq 0$, is called a **lemniscate.** Find this solution curve and sketch it. (r and θ are polar coordinates.)

SUMMARY OF CHAPTER 1
FIRST-ORDER DIFFERENTIAL EQUATIONS

This chapter concerns **first-order differential equations** and their applications; these are equations

$$(1) \qquad F(x, y, y') = 0 \qquad \text{or in explicit form} \qquad y' = f(x, y)$$

involving the derivative y' of an unknown function y, given functions of x, and, perhaps, y itself. We started with the basic concepts (Sec. 1.1) and with the method of direction fields (Sec. 1.2), then studied solution methods and modeling (Secs. 1.3–1.8), and, finally, ideas on existence and uniqueness of solutions (Sec. 1.9).

Ordinarily such an equation has a **general solution,** that is, a solution involving an arbitrary constant, which we denoted by c. In most applications, however, one has to find a solution satisfying a given condition, resulting from the physical or other system of which the equation is a mathematical model. This leads to the concept of an **initial value problem**

$$(2) \qquad y' = f(x, y), \qquad y(x_0) = y_0 \qquad (x_0, y_0 \text{ given numbers})$$

in which the **initial condition** $y(x_0) = y_0$ is used to determine a **particular solution,** that is, a solution obtained from the general solution by specifying a value of c. Geometrically, a general solution represents a family of curves, and each particular solution corresponds to a curve of this family. See Sec. 1.2 on **direction fields,** which give information on the general behavior of such a family.

Perhaps the simplest equations are **separable equations,** those we can put into the form $g(y)\, dy = f(x)\, dx$ by algebraic manipulation (possibly combined with transformations, such as $y/x = u$). These equations can then be solved by integration on both sides (Secs. 1.3, 1.4).

An **exact equation**

$$M(x, y)\, dx + N(x, y)\, dy = 0$$

is one for which $M\, dx + N\, dy$ is the differential

$$du = \frac{\partial u}{\partial x}\, dx + \frac{\partial u}{\partial y}\, dy$$

of a function $u(x, y)$, so that we get the implicit solution $u(x, y) = c$ (Sec. 1.5). This method extends to nonexact equations that can be made exact by multiplying them by some function $F(x, y)$, called an **integrating factor** (Sec. 1.5).

Linear equations (Sec. 1.6)

$$(3) \qquad y' + p(x)y = r(x)$$

are very important. They have an integrating factor is $F(x) = \exp\left(\int p(x)\, dx\right)$, which leads to the solution formula (4), Sec. 1.6. Certain nonlinear equations can be reduced to linear form by substituting new variables. This holds for the **Bernoulli equation** $y' + p(x)y = g(x)y^a$ (Sec. 1.6).

Modeling is included at various places. Sections entirely devoted to **applications** are 1.4 on separable equations, 1.7 on linear equations for **electrical circuits,** and 1.8 on **orthogonal trajectories,** that is, curves that intersect given curves at right angles.

Picard's iteration method (Sec. 1.9) gives approximate solutions of initial value problems by iteration. It is important as the theoretical basis of **Picard's existence and uniqueness theorems** (Sec. 1.9).

Numerical methods on first-order differential equations can be taught from Secs. 19.1 and 19.2 (as indicated in Sec. 1.9).

CHAPTER 2

Linear Differential Equations of Second and Higher Order

The ordinary differential equations may be divided into two large classes, namely, **linear equations** and **nonlinear equations.** Whereas nonlinear equations are difficult in general, linear equations are much simpler because their solutions have general properties that facilitate working with them, and there are standard methods for solving many practically important linear differential equations.

We first consider linear differential equations of second order—homogeneous equations in Secs. 2.1−2.7 and nonhomogeneous equations in Secs. 2.8−2.12. Higher order equations follow in Secs. 2.13−2.15.

We concentrate on second-order equations for two main reasons. First, they have important applications in mechanics (Secs. 2.5, 2.11) and in electric circuit theory (Sec. 2.12). Second, their theory is typical of that of linear differential equations of any order n (but involves much simpler formulas), so that the transition to higher order n needs only very few new ideas.

Numerical methods for second-order differential equations are included in Sec. 19.3, which is independent of the other sections in Chaps. 17−19 and can be studied after Sec. 2.12 (or after the end of this chapter) if desired.

(Legendre's, Bessel's, and the hypergeometric equations will be considered in Chap. 4.)

Prerequisite for this chapter: Chap. 1, in particular Sec. 1.6.
Sections that may be omitted in a shorter course: 2.4, 2.7, 2.10, 2.12, 2.13−2.15.
References: Appendix 1, Part A.
Answers to problems: Appendix 2.

2.1 Homogeneous Linear Equations of Second Order

We have already discussed linear differential equations of first order, and we now define and discuss linear differential equations of second order.

Linear Differential Equation of Second Order

A second-order differential equation is called **linear** if it can be written

(1)
$$y'' + p(x)y' + q(x)y = r(x)$$

and **nonlinear** if it cannot be written in this form.

The characteristic feature of Eq. (1) is that it is linear in the unknown function y and its derivatives, whereas p and q as well as r on the right may be any given functions of x. If the first term is, say, $f(x)y''$, we have to divide by $f(x)$ to get the **"standard form"** (1), with y'' as the first term, which is practical.

If $r(x) \equiv 0$ (that is, $r(x) = 0$ for all x considered), then (1) becomes simply

(2)
$$y'' + p(x)y' + q(x)y = 0$$

and is called **homogeneous.** If $r(x) \not\equiv 0$, then (1) is called **nonhomogeneous.** This is similar to Sec. 1.6.

The functions p and q in (1) and (2) are called the **coefficients** of the equations.

An example of a nonhomogeneous linear differential equation is

$$y'' + 4y = e^{-x} \sin x.$$

An example of a homogeneous linear equation is

$$(1 - x^2)y'' - 2xy' + 6y = 0.$$

Examples of nonlinear differential equations are

$$x(y''y + y'^2) + 2y'y = 0$$

and

$$y'' = \sqrt{y'^2 + 1}.$$

We shall always suppose that x varies on some open interval I, and all our assumptions and statements will refer to such an I, which we need not specify in each case. (Recall from footnote 1 in Sec. 1.1 that I may be the entire x-axis.)

A **solution** of a second-order (linear or nonlinear) differential equation on some open interval $a < x < b$ is a function $y = h(x)$ that has derivatives $y' = h'(x)$ and $y'' = h''(x)$ and satisfies that differential equation for all x in that interval; that is, the equation becomes an identity if we replace the unknown function y and its derivatives by h and its corresponding derivatives.

Second-order linear differential equations have many basic applications, as we shall see. Some of them are very simple, their solutions being familiar functions from calculus. Others are more involved, their solutions being higher functions (for instance, Bessel functions) occurring in engineering problems.

Homogeneous Equations: Superposition or Linearity Principle

We now begin our discussion of second-order homogeneous linear differential equations (Secs. 2.1–2.7). Nonhomogeneous equations follow in Secs. 2.8–2.12.

EXAMPLE 1　**Solutions of a homogeneous linear differential equation**

$y = e^x$ and $y = e^{-x}$ are solutions of the homogeneous linear differential equation

$$y'' - y = 0$$

for all x because for $y = e^x$ we get $(e^x)'' - e^x = e^x - e^x = 0$ and similarly for $y = e^{-x}$, as the student should verify.

We can even go an important step further. We can multiply e^x and e^{-x} by different constants, say, -3 and

$\frac{2}{5}$ (or any other numbers) and then take the sum

$$y = -3e^x + \tfrac{2}{5}e^{-x}$$

and verify that this is another solution of our homogeneous equation for all x because

$$(-3e^x + \tfrac{2}{5}e^{-x})'' - (-3e^x + \tfrac{2}{5}e^{-x}) = -3e^x + \tfrac{2}{5}e^{-x} - (-3e^x + \tfrac{2}{5}e^{-x}) = 0. \qquad \blacktriangleleft$$

This example illustrates the very important fact that for a **homogeneous linear** equation (2), we can always obtain new solutions from known solutions by multiplication by constants and by addition. Of course, this is a great advantage because in this way we can get further solutions from given ones. Now from y_1 ($= e^x$) and y_2 ($= e^{-x}$) we have obtained a function of the form

$$(3) \qquad\qquad\qquad y = c_1 y_1 + c_2 y_2 \qquad\qquad (c_1, c_2 \text{ arbitrary constants}).$$

This is called a **linear combination** of y_1 and y_2. Using this concept, we can now formulate the result suggested by our example, often called the **superposition principle** or **linearity principle.**

THEOREM 1   **Fundamental Theorem for the homogeneous equation (2)**

For a homogeneous linear differential equation (2), any linear combination of two solutions on an open interval I is again a solution of (2) on I. In particular, for such an equation, sums and constant multiples of solutions are again solutions.

PROOF.   Let y_1 and y_2 be solutions of (2) on I. Then by substituting $y = c_1 y_1 + c_2 y_2$ and its derivatives into (2), and using the familiar rule $(c_1 y_1 + c_2 y_2)' = c_1 y_1' + c_2 y_2'$, etc., we get

$$\begin{aligned} y'' + py' + qy &= (c_1 y_1 + c_2 y_2)'' + p(c_1 y_1 + c_2 y_2)' + q(c_1 y_1 + c_2 y_2) \\ &= c_1 y_1'' + c_2 y_2'' + p(c_1 y_1' + c_2 y_2') + q(c_1 y_1 + c_2 y_2) \\ &= c_1(y_1'' + py_1' + qy_1) + c_2(y_2'' + py_2' + qy_2) = 0, \end{aligned}$$

since in the last line, $(\cdot \cdot \cdot) = 0$ because y_1 and y_2 are solutions, by assumption. This shows that y is a solution of (2) on I. $\blacktriangleleft$

Caution! Always remember this highly important theorem, but don't forget that it ***does not hold*** for *nonhomogeneous linear equations* or *nonlinear equations,* as the following two examples illustrate.

EXAMPLE 2   **A nonhomogeneous linear differential equation**

Substitution shows that the functions $y = 1 + \cos x$ and $y = 1 + \sin x$ are solutions of the nonhomogeneous linear differential equation

$$y'' + y = 1,$$

but the following functions are *not* solutions of this differential equation:

$$2(1 + \cos x) \qquad \text{and} \qquad (1 + \cos x) + (1 + \sin x). \qquad \blacktriangleleft$$

EXAMPLE 3   **A nonlinear differential equation**

Substitution shows that the functions $y = x^2$ and $y = 1$ are solutions of the nonlinear differential equation

$$y'' y - xy' = 0,$$

but the following functions are *not* solutions of this differential equation:

$$-x^2 \qquad \text{and} \qquad x^2 + 1.$$ ◀

Initial Value Problem. General Solution. Basis

For a *first-order* differential equation, a general solution involved one arbitrary constant c, and in an initial value problem we used one initial condition $y(x_0) = y_0$ to get a particular solution in which c had a definite value. The idea of a general solution was to get all possible solutions, and we know that for *linear* equations (Sec. 1.6) we succeeded (because those have no singular solutions). We now extend this idea to second-order equations:

For second-order **homogeneous** linear equations (2), a general solution will be of the form

(4)
$$\boxed{y = c_1 y_1 + c_2 y_2,}$$

a linear combination of two (suitable) solutions involving two arbitrary constants c_1, c_2. An **initial value problem** now consists of Eq. (2) and two **initial conditions**

(5)
$$\boxed{y(x_0) = K_0, \qquad y'(x_0) = K_1,}$$

prescribing values K_0 and K_1 of the solution and its derivative (slope of the curve) at the same given x_0 in the open interval considered. We shall use (5) to get from (4) a particular solution of (2), in which c_1 and c_2 have definite values.

Let us illustrate this by a simple example, which will also help us to see that we have to impose a condition on y_1 and y_2 in (4).

EXAMPLE 4   **Initial value problem**

Solve the initial value problem

$$y'' - y = 0, \qquad y(0) = 4, \qquad y'(0) = -2.$$

Solution. 1st Step. e^x and e^{-x} are solutions (by Example 1), and we take

$$y = c_1 e^x + c_2 e^{-x}.$$

(This will turn out to be a general solution as defined below.)

2nd Step.   From the initial conditions, since $y' = c_1 e^x - c_2 e^{-x}$, we get

$$y(0) = c_1 + c_2 = 4$$
$$y'(0) = c_1 - c_2 = -2.$$

Hence $c_1 = 1$, $c_2 = 3$. This gives the answer $y = e^x + 3e^{-x}$ shown in Fig. 36 on the next page. Note that the curve begins at $y = 4$, with a negative slope -2, so that the initial tangent intersects the x-axis at $x = 2$, as shown. (To save space, we have chosen different scales on the two axes.)

Observation.   Had we taken $y_1 = e^x$ and $y_2 = le^x$ (where l is any constant) and thus obtained

$$y = c_1 e^x + c_2 l e^x = (c_1 + c_2 l)e^x = y',$$

our solution would not have been general enough to satisfy the two initial conditions and solve the problem. Why? Well, our present y_1 and y_2 are proportional, $y_1/y_2 = 1/l$, whereas before they were not, $y_1/y_2 = e^x/e^{-x} = e^{2x}$. This is the point. It motivates the following definition as well as its importance in connection with initial value problems. ◀

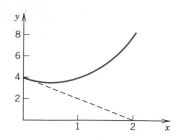

Fig. 36. Particular solution in Example 4

Definition (General solution, basis, particular solution)

A **general solution** of an equation (2) on an open interval[1] I is a solution (4) with y_1 and y_2 not proportional solutions of (2) on I and c_1, c_2 arbitrary[2] constants. These y_1, y_2 are then called a **basis** (or **fundamental system**) of (2) on I.

A **particular solution** of (2) on I is obtained if we assign specific values to c_1 and c_2 in (4).

As usual, y_1 and y_2 are called *proportional* on I if[3]

$$(6) \qquad \text{(a)} \quad y_1 = ky_2 \qquad \text{or} \qquad \text{(b)} \quad y_2 = ly_1$$

holds for all x on I, where k and l are numbers, zero or not. ◄

Actually, we can also formulate our definition of a basis in terms of "linear independence." We call two functions $y_1(x)$ and $y_2(x)$ **linearly independent** on an interval I where they are defined if

$$(7) \qquad k_1 y_1(x) + k_2 y_2(x) = 0 \qquad \text{on } I \text{ implies} \qquad k_1 = 0, \quad k_2 = 0,$$

and we call them **linearly dependent** on I if (7) also holds for some constants k_1, k_2 not both zero. Then, if $k_1 \neq 0$ or $k_2 \neq 0$, we can divide and solve, obtaining

$$y_1 = -\frac{k_2}{k_1} y_2 \qquad \text{or} \qquad y_2 = -\frac{k_1}{k_2} y_1.$$

Hence then y_1 and y_2 are proportional, whereas in the case of linear independence, they are not proportional. We thus have the following

Definition of a basis (Reformulated)

A **basis of solutions** of (2) on an interval I is a pair y_1, y_2 of linearly independent solutions of (2) on I. ◄

---

[1] See Sec. 1.1.

[2] The range of the constants may have to be restricted in some cases to avoid imaginary expressions or other degeneracies.

[3] If $k \neq 0$ in (a), it implies (b) with $l = 1/k$, but not when $k = 0$, that is, $y_1 = 0$.

EXAMPLE 5   **Basis, general solution, particular solution**

e^x and e^{-x} in Example 4 form a basis of the differential equation $y'' - y = 0$ for all x. Hence a general solution is $y = c_1 e^x + c_2 e^{-x}$. The answer in Example 4 is a particular solution of the equation. ◄

EXAMPLE 6   **Basis, general solution**

The student may verify that $y_1 = \cos x$ and $y_2 = \sin x$ are solutions of the differential equation

$$y'' + y = 0.$$

Now $\cos x$ and $\sin x$ are not proportional, $y_1/y_2 = \cot x \neq const.$ Hence they form a basis of our equation for all x, and a general solution is

$$y = c_1 \cos x + c_2 \sin x.$$ ◄

How to Obtain a Basis if One Solution Is Known. Reduction of Order

This is of practical interest because we can often guess a solution y_1 of (2) on some interval I or obtain it by some method. Then to obtain a basis we need a second linearly independent solution y_2 of (2) on I. We now show that we can get y_2 from a first-order differential equation by a **method of reduction of order**[4], which is applicable when one solution is known.

To get y_2, we set $y_2 = u y_1$ and determine u. For this we substitute $y_2 = u y_1$ and its derivatives (product differentiation!)

$$y_2' = u' y_1 + u y_1' \qquad \text{and} \qquad y_2'' = u'' y_1 + 2u' y_1' + u y_1''$$

into (2). This gives

(8) $$u'' y_1 + 2u' y_1' + u y_1'' + p(u' y_1 + u y_1') + q u y_1 = 0.$$

Collecting terms in u'', u', and u, we have

$$u'' y_1 + u'(2y_1' + p y_1) + u(y_1'' + p y_1' + q y_1) = 0.$$

Now comes the main point. Since y_1 is a solution of (2), the expression in the last parentheses is zero. Hence u is gone, and we are left with an equation in u' and u''. We divide this remaining equation by y_1 and set $u' = U$, $u'' = U'$,

$$u'' + u' \frac{2y_1' + p y_1}{y_1} = 0, \qquad \text{thus} \qquad U' + \left(\frac{2y_1'}{y_1} + p\right) U = 0.$$

---

[4]Credited to the great mathematician JOSEPH LOUIS LAGRANGE (1736—1813), who was born in Turin, of French extraction, got his first professorship when he was 19 (at the Military Academy of Turin), became director of the mathematical section of the Berlin Academy in 1766, and moved to Paris in 1787. His important major work was in the calculus of variations, celestial mechanics, general mechanics (*Mécanique analytique*, Paris, 1788), differential equations, approximation theory, algebra, and number theory.

This is the desired first-order equation, the reduced equation. Separation of variables and integration gives

$$\frac{dU}{U} = -\left(\frac{2y_1'}{y_1} + p\right) dx \qquad \text{and} \qquad \ln|U| = -2\ln|y_1| - \int p\,dx.$$

By taking exponentials we finally obtain

(9)
$$U = \frac{1}{y_1^2} e^{-\int p\,dx}.$$

Here $U = u'$, so that $u = \int U\,dx$. Hence the desired solution

$$y_2 = uy_1 \text{ is } y_2 = y_1 \int U\,dx.$$

The quotient $y_2/y_1 = u = \int U\,dx$ cannot be a constant (because $U \neq 0$), so that y_1 and y_2 form a basis.   ◄

EXAMPLE 7   **Reduction of order if a solution is known. Basis**

Find a basis of solutions for the following second-order homogeneous linear equation for positive x:

(10)
$$x^2 y'' - xy' + y = 0.$$

Solution.   A solution is $y_1 = x$ because $y_1' = 1$, $y_1'' = 0$, so that substitution gives $-x \cdot 1 + x = 0$. Now comes an important point. Equation (8) was derived for an equation (2) in **standard form.** Hence before we can apply (9), we have to find p for (10) in standard form. Division by x^2 gives the standard form

$$y'' - \frac{1}{x}y' + \frac{1}{x^2}y = 0.$$

We now see that $p = -1/x$, $-\int p\,dx = \ln x$, so that (9) gives

$$U = \frac{1}{x^2} e^{\ln x} = \frac{1}{x}.$$

Thus

$$y_2 = ux = x\int U\,dx = x\ln x.$$

Answer: A basis of (10) for $x > 0$ is $y_1 = x$, $y_2 = x\ln x$. Check this by substitution.   ◄

Practical Role of General Solutions

In practice, one mainly uses a general solution to get particular solutions from it, by imposing two initial conditions (5), because it is the *particular solution* that describes the unique behavior of a given physical or other system. Being interested in first gaining experience in that practical task, we give the underlying theory afterwards (in Sec. 2.7). For the time being, it suffices to know the following. If the coefficients p and q of (2) and the function r are continuous on some interval I, then (2) always has a general solution on I, from which one obtains the solution of any initial value problem (1), (5) on I, which is unique. Also, (2) does not have **singular solutions** (i.e., solutions not obtainable from a general solution).

PROBLEM SET 2.1

Reduction of Linear or Nonlinear Equations to First Order

A general second-order differential equation is of the form $F(x, y', y'') = 0$, involving y'', given functions of x, and, perhaps, y and y'. Of course, reduction to first order is of great practical interest. This is possible (without knowledge of a solution) if y does not occur explicitly (Prob. 1) or x does not occur explicitly (Prob. 2):

1. Show that $F(x, y', y'') = 0$ can be reduced to first order in $z = y'$ (from which y follows by integration). Give two examples of your own.

2. Show that $F(y, y', y'') = 0$ can be reduced to a first-order equation with y as the *independent* variable and $y'' = (dz/dy)z$, where $z = y'$; derive this by the chain rule. Give two examples.

Using the methods in Example 7 or in Probs. 1 and 2, reduce to first order and solve (showing each step of your calculation in detail):

3. $y'' = y'$

4. $2xy'' = 3y'$

5. $yy'' = 2y'^2$

6. $xy'' + 2y' + xy = 0$, $y_1 = (\sin x)/x$

7. $y'' + e^y y'^3 = 0$

8. $xy'' + y' = 0$

9. $x^2 y'' - 5xy' + 9y = 0$, $y_1 = x^3$

10. $y'' + (1 + y^{-1})y'^2 = 0$

11. $x^2 y'' + xy' + (x^2 - \frac{1}{4})y = 0$, $y_1 = x^{-1/2} \cos x$

12. $(1 - x^2)y'' - 2xy' + 2y = 0$, $y_1 = x$

Applications of Reducible Equations

13. **(Motion)** A small body moves on a straight line so that the product of its velocity and acceleration is constant, say, 1 meter^2/sec^3. If at $t = 0$ the body's distance from the origin is 2 meters and its velocity is 2 meters/sec, what are the distance and velocity at $t = 6$ sec?

14. **(Motion)** What happens in Prob. 13 if the acceleration equals the velocity (the other data being as before)? Will the distance at $t = 6$ be larger? First guess, then calculate.

15. **(Curve)** Find the curve through the origin in the xy-plane which satisfies $y'' = 2y'$ and whose tangent at the origin has slope 1.

16. **(Hanging cable)** It can be shown that the curve $y(x)$ of an inextensible flexible homogeneous cable hanging between two fixed points is obtained by solving $y'' = k\sqrt{1 + y'^2}$, where the constant k depends on the weight. This curve is called a *catenary* (from Latin *catena* = the chain). Find and graph $y(x)$, assuming that $k = 1$ and those fixed points are $(-1, 0)$ and $(1, 0)$ in a vertical xy-plane.

Initial Value Problems. (More in the next problem set.) Verify that the given functions form a basis of solutions of the given equation and solve the given initial value problem.

17. $y'' + 9y = 0$, $y(0) = 4$, $y'(0) = -6$; $\cos 3x$, $\sin 3x$

18. $y'' + 2y' + y = 0$, $y(0) = 1$, $y'(0) = 0$; e^{-x}, xe^{-x}

19. $4x^2 y'' - 3y = 0$, $y(1) = 3$, $y'(1) = 2.5$; $x^{-1/2}$, $x^{3/2}$

20. **WRITING PROJECT. General Properties of Solutions of Linear Differential Equations.** Write a short essay (with proofs and simple examples of your own) that includes the following.
    (a) The superposition principle.
    (b) $y \equiv 0$ is a solution of the homogeneous equation (2) (called the **trivial solution**).
    (c) The sum $y = y_1 + y_2$ of a solution y_1 of (1) and y_2 of (2) is a solution of (1).
    (d) Explore possibilities of making further general statements on solutions of (1) and (2) (sums, differences, multiples).

2.2 Second-Order Homogeneous Equations with Constant Coefficients

In this section and the next one, we show how to solve homogeneous linear equations

(1)
$$y'' + ay' + by = 0$$

whose coefficients a and b are constant. These equations have important applications, especially in connection with mechanical and electrical vibrations, as we shall see in Secs. 2.5, 2.11, and 2.12.

To solve (1), we remember from Sec. 1.6 that a *first-order* linear differential equation $y' + ky = 0$ with constant coefficient k has an exponential function as solution, $y = e^{-kx}$. This gives us the idea to try as a solution of (1) the function

(2)
$$y = e^{\lambda x}.$$

Substituting (2) and the derivatives

$$y' = \lambda e^{\lambda x} \qquad \text{and} \qquad y'' = \lambda^2 e^{\lambda x}$$

into our equation (1), we obtain

$$(\lambda^2 + a\lambda + b)e^{\lambda x} = 0.$$

Hence (2) is a solution of (1) if λ is a solution of the quadratic equation

(3)
$$\lambda^2 + a\lambda + b = 0.$$

This equation is called the **characteristic equation** (or *auxiliary equation*) of (1). Its roots are

(4) $\qquad \lambda_1 = \frac{1}{2}(-a + \sqrt{a^2 - 4b}), \qquad \lambda_2 = \frac{1}{2}(-a - \sqrt{a^2 - 4b}).$

Our derivation shows that the functions

(5) $\qquad y_1 = e^{\lambda_1 x} \qquad \text{and} \qquad y_2 = e^{\lambda_2 x}$

are solutions of (1). The student may check this by substituting (5) into (1).

Directly from (4) we see that, depending on the sign of the discriminant $a^2 - 4b$, we obtain

(Case I)	*two real roots if $a^2 - 4b > 0$,*
(Case II)	*a real double root if $a^2 - 4b = 0$,*
(Case III)	*complex conjugate roots if $a^2 - 4b < 0$.*

We discuss Cases I and II now and Case III in the next section.

Case I. Two Distinct Real Roots λ_1 and λ_2

In this case,

$$y_1 = e^{\lambda_1 x} \qquad \text{and} \qquad y_2 = e^{\lambda_2 x}$$

constitute a basis of solutions of (1) on any interval (because y_1/y_2 is not constant; see Sec. 2.1). The corresponding general solution is

(6)
$$\boxed{y = c_1 e^{\lambda_1 x} + c_2 e^{\lambda_2 x}.}$$

EXAMPLE 1   **General solution in the case of distinct real roots**

We can now solve $y'' - y = 0$ (in Example 1 of Sec. 2.1) in a systematic fashion. The characteristic equation is $\lambda^2 - 1 = 0$. Its roots are $\lambda_1 = 1$ and $\lambda_2 = -1$. Hence a basis is e^x and e^{-x} and, as before, gives the general solution

$$y = c_1 e^x + c_2 e^{-x}. \qquad \blacktriangleleft$$

EXAMPLE 2   **An initial value problem in the case of distinct real roots**

Solve the initial value problem

$$y'' + y' - 2y = 0, \qquad y(0) = 4, \qquad y'(0) = -5.$$

Solution. 1st Step. General solution.   The characteristic equation is

$$\lambda^2 + \lambda - 2 = 0.$$

Its roots are

$$\lambda_1 = \tfrac{1}{2}(-1 + \sqrt{9}) = 1 \qquad \text{and} \qquad \lambda_2 = \tfrac{1}{2}(-1 - \sqrt{9}) = -2,$$

so that we obtain the general solution

$$y = c_1 e^x + c_2 e^{-2x}.$$

2nd Step. Particular solution.   Since $y'(x) = c_1 e^x - 2c_2 e^{-2x}$, we obtain from the general solution and the initial conditions

$$y(0) = c_1 + c_2 = 4,$$
$$y'(0) = c_1 - 2c_2 = -5.$$

Hence $c_1 = 1$ and $c_2 = 3$. This gives the *answer* $y = e^x + 3e^{-2x}$. Figure 37 shows that the curve begins at $y = 4$ with a negative slope (-5, but note that the axes have different scales!), in agreement with the initial conditions. $\qquad \blacktriangleleft$

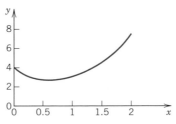

Fig. 37.   Solution in Example 2

Case II. Real Double Root $\lambda = -a/2$

If the discriminant $a^2 - 4b$ is zero, we see directly from (4) that we get only one root, $\lambda = \lambda_1 = \lambda_2 = -a/2$, hence only one solution,

$$y_1 = e^{-(a/2)x}.$$

To obtain a second independent solution y_2 (needed for a basis), we use the method discussed in the last section. That is, we set $y_2 = uy_1$. Substituting this and its derivatives $y_2' = u'y_1 + uy_1'$ and y_2'' into (1), we first have

$$(u''y_1 + 2u'y_1' + uy_1'') + a(u'y_1 + uy_1') + buy_1 = 0.$$

Collecting terms in u'', u', and u, as in the last section, we obtain

$$u''y_1 + u'(2y_1' + ay_1) + u(y_1'' + ay_1' + by_1) = 0.$$

The expression in the last parentheses is zero, since y_1 is a solution of (1). The expression in the first parentheses is zero, too, since

$$2y_1' = -ae^{-ax/2} = -ay_1.$$

We are thus left with $u''y_1 = 0$. Hence $u'' = 0$. By two integrations, $u = c_1x + c_2$. To get a second independent solution $y_2 = uy_1$, we can simply take $u = x$. Then $y_2 = xy_1$. Since these solutions are not proportional, they form a basis. Our result is that *in the case of a double root of* (3) *a basis of solutions of* (1) *on any interval is*

$$e^{-ax/2}, \qquad xe^{-ax/2}.$$

The corresponding general solution is

(7)
$$\boxed{y = (c_1 + c_2x)e^{-ax/2}.}$$

Warning. If λ is a *simple* root of (4), then $(c_1 + c_2x)e^{\lambda x}$ is **not** a solution of (1).

EXAMPLE 3   **General solution in the case of a double root**

Solve

$$y'' + 8y' + 16y = 0.$$

Solution. The characteristic equation is $\lambda^2 + 8\lambda + 16 = 0$. It has the double root $\lambda = -4$. Hence a basis is e^{-4x} and xe^{-4x} and the corresponding general solution is $y = (c_1 + c_2x)e^{-4x}$.   ◀

EXAMPLE 4   **An initial value problem in the case of a double root**

Solve the initial value problem

$$y'' - 4y' + 4y = 0, \qquad y(0) = 3, \qquad y'(0) = 1.$$

Solution. The characteristic equation $\lambda^2 - 4\lambda + 4 = (\lambda - 2)^2 = 0$ has the double root $\lambda = 2$, so that a general solution of the differential equation is

$$y(x) = (c_1 + c_2x)e^{2x}.$$

By differentiation we obtain

$$y'(x) = c_2e^{2x} + 2(c_1 + c_2x)e^{2x}.$$

From this and the initial conditions it follows that

$$y(0) = c_1 = 3, \qquad y'(0) = c_2 + 2c_1 = 1.$$

Hence $c_1 = 3$, $c_2 = -5$, and the answer is

$$y = (3 - 5x)e^{2x}.$$ ◄

The remaining Case III of **complex conjugate roots** of the characteristic equation (4) will be discussed in the next section.

PROBLEM SET 2.2

General Solution. Find a general solution. Check your answer by substitution.

1. $4y'' + 4y' - 3y = 0$
2. $y'' + 3.2y' + 2.56y = 0$
3. $2y'' - 9y' = 0$
4. $y'' - 8y = 0$
5. $y'' + 9y' + 20y = 0$
6. $16y'' - \pi^2y = 0$
7. $9y'' - 30y' + 25y = 0$
8. $10y'' + 6y' - 4y = 0$
9. $y'' + 2ky' + k^2y = 0$

Initial Value Problems. Solve the initial value problems. Check that your answer satisfies the equation as well as the initial conditions. (Show the details of your work.)

10. $y'' + y' - 6y = 0$, $\qquad y(0) = 10$, $\qquad y'(0) = 0$
11. $y'' + 4y' + 4y = 0$, $\qquad y(0) = 1$, $\qquad y'(0) = 1$
12. $y'' - y = 0$, $\qquad y(0) = 3$, $\qquad y'(0) = -3$
13. $8y'' - 2y' - y = 0$, $\qquad y(0) = -0.2$, $\qquad y'(0) = -0.325$
14. $4y'' - 25y = 0$, $\qquad y(0) = 0$, $\qquad y'(0) = -5$
15. $y'' + 2.2y' + 1.17y = 0$, $\qquad y(0) = 2$, $\qquad y'(0) = -2.6$
16. $y'' - k^2y = 0$, $(k \neq 0)$ $\qquad y(0) = 1$, $\qquad y'(0) = 1$
17. $4y'' - 4y' - 3y = 0$, $\qquad y(-2) = e$, $\qquad y'(-2) = -e/2$

Linear independence is of basic importance in this chapter in connection with general solutions, as explained in the text. Are the following functions linearly independent or dependent on the given interval? (Show the details of your work.)

18. e^{-x}, e^x, any interval
19. 0, $\tan x$ $(|x| < \pi/4)$
20. x^2, $x^2 \ln x$ $(x \geq 1)$
21. $\ln x$, $\ln (x^4)$ $(x > 1)$
22. $\ln x$, $(\ln x)^4$ $(1 \leq x \leq 2)$
23. $\sin^2 x$, $\sin (x^2)$ $(0 < x < \sqrt{\pi})$
24. $x|x|$, x^2 $(0 \leq x \leq 1)$
25. $x|x|$, x^2 $(|x| \leq 1)$
26. $\sin 2x$, $\cos x \sin x$ $(x < 0)$

27. (Nonsense) Why is it nonsense to talk about linear dependence or independence of two functions at a point?

28. (Subintervals) If f, g are linearly dependent on an interval I, show that this also holds for any subinterval J of I. Is the same true for linear independence? (Give reasons and examples.)

 29. CAS PROJECT. Linear Independence. Write a program for testing linear independence and dependence. Try it out on some of the problems in this problem set and on examples of your own.

30. TEAM PROJECT. General Properties of Solutions

    (a) Coefficient formulas. Show how a and b in (1) can be expressed in terms of λ_1 and λ_2. Explain how these formulas can be used in constructing equations for given bases.

    (b) Root zero. Solve $y'' + 4y' = 0$ (i) by the present method, and (ii) by reduction to first order. Can you explain why the result must be the same in both cases? Can you do the same for a general equation $y'' + ay' = 0$?

    (c) Double root. Verify directly that $xe^{\lambda x}$ with $\lambda = -a/2$ is a solution of (1) in the case of a double root. Verify and explain why $y = e^{-2x}$ is a solution of $y'' - y' - 6y = 0$ but xe^{-2x} is not.

    (d) Limits. Double roots should be limiting cases of distinct roots λ_1, λ_2 as, say, $\lambda_2 \to \lambda_1$. Experiment with this idea. (Remember L'Hôpital's rule from calculus.) Can you arrive at $xe^{\lambda_1 x}$? Give it a try.

2.3 Case of Complex Roots Complex Exponential Function

For homogeneous linear differential equations with constant coefficients

$$(1) \qquad\qquad y'' + ay' + by = 0$$

we now discuss the remaining case that the characteristic equation

$$(2) \qquad\qquad \lambda^2 + a\lambda + b = 0$$

has roots

$$(3) \qquad \lambda_1 = -\tfrac{1}{2}a + \tfrac{1}{2}\sqrt{a^2 - 4b}, \qquad \lambda_2 = -\tfrac{1}{2}a - \tfrac{1}{2}\sqrt{a^2 - 4b}$$

that are complex. Equation (3) shows that this happens if the discriminant $a^2 - 4b$ is negative. This is Case III of the last section.

For those students not too familiar with complex numbers, we begin with an important example.

EXAMPLE 1   **Complex roots**

Find a general solution of the equation

$$(4) \qquad\qquad y'' + y = 0.$$

Solution. We can write (4) as $y'' = -y$. Hence we want functions that come back under two differentiations, times a minus sign. Recall from calculus that $(\cos x)'' = -\cos x$ and $(\sin x)'' = -\sin x$. Thus we are done, and a general solution is

$$y = A \cos x + B \sin x$$

with arbitrary A and B.

This was a trick. Let us now turn to a method. The characteristic equation of (4) is $\lambda^2 + 1 = 0$. Thus, $\lambda^2 = -1$ and $\lambda = \pm\sqrt{-1} = \pm i$. Here, i is the standard notation for $\sqrt{-1}$ used in the general form of a complex number (e.g., $3 + 4i$). Hence we first get two complex solutions

$$e^{ix} \qquad \text{and} \qquad e^{-ix}.$$

What do they mean? Well, from the definition (and motivation) of the *complex* exponential function given below we shall see that these are related to the *real* cosine and sine by the **Euler formula**

(5)
$$\boxed{\begin{array}{ll} \text{(a)} & e^{ix} = \cos x + i \sin x \\ \text{(b)} & e^{-ix} = \cos x - i \sin x. \end{array}}$$

We add (5a) and (5b). Then $\sin x$ drops out. Division by 2 gives

(6a)
$$\boxed{\cos x = \tfrac{1}{2}(e^{ix} + e^{-ix}).}$$

Similarly, subtracting (5a) − (5b) we lose $\cos x$. Division by $2i$ then gives

(6b)
$$\boxed{\sin x = \frac{1}{2i}(e^{ix} - e^{-ix}).}$$

Since e^{ix} and e^{-ix} are solutions of (1), so are $\cos x$ and $\sin x$ by the superposition principle (Sec. 2.1) or by direct verification. Hence we get the same general solution as before, this time derived methodically.   ◀

Complex Exponential Function

To justify our example and to settle Case III in general, we define the **complex exponential function** e^z of a complex variable[5] $z = s + it$. The definition in terms of the real functions e^s, $\cos t$, and $\sin t$ is

(7)
$$\boxed{e^z = e^{s+it} = e^s e^{it} = e^s(\cos t + i \sin t).}$$

This is motivated as follows. For real $z = s$, the function e^z becomes the real exponential function e^s (because $\cos 0 = 1$ and $\sin 0 = 0$). It can be shown that $e^{z_1+z_2} = e^{z_1}e^{z_2}$, just as in real. (Proof in Sec. 12.6) Finally, the Maclaurin series of e^x with $x = it$ gives (since $i^2 = -1$, $i^3 = -i$, $i^4 = 1$, etc., and under the assumption that we can reorder the terms as shown—this can be proved) the following series and earlier result:

---

[5] We write $z = s + it$ instead of $z = x + iy$ since x and y are used as variable and unknown function in our equations.

$$e^{it} = 1 + it + \frac{(it)^2}{2!} + \frac{(it)^3}{3!} + \frac{(it)^4}{4!} + \frac{(it)^5}{5!} + \cdots$$

$$= 1 - \frac{t^2}{2!} + \frac{t^4}{4!} - + \cdots + i\left(t - \frac{t^3}{3!} + \frac{t^5}{5!} - + \cdots\right)$$

$$= \cos t + i \sin t.$$

(Look up these series in your calculus book if necessary.) We see that we have obtained the Euler formula (5a).

Case III. Complex Roots

We now apply (7) to (3). We begin by introducing a better notation. In Case III the radicand $a^2 - 4b$ in (3) is negative. To make it positive, we pull out -1 under the root and use $\sqrt{-1} = i$. Then we put $1/2 = \sqrt{1/4}$ under the root. This gives in (3)

$$\lambda_1 = -\tfrac{1}{2}a + \tfrac{1}{2}\sqrt{a^2 - 4b} = -\tfrac{1}{2}a + \tfrac{1}{2}i\sqrt{4b - a^2} = -\tfrac{1}{2}a + i\sqrt{b - \tfrac{1}{4}a^2}$$

and similarly for λ_2. Our result is

$$(8) \qquad\qquad \lambda_1 = -\tfrac{1}{2}a + i\omega, \qquad \lambda_2 = -\tfrac{1}{2}a - i\omega$$

where $\omega = \sqrt{b - \tfrac{1}{4}a^2}$.

     Using this, we apply (7):

$$e^{\lambda_1 x} = e^{-(a/2)x + i\omega x} = e^{-(a/2)x}(\cos \omega x + i \sin \omega x)$$

$$e^{\lambda_2 x} = e^{-(a/2)x - i\omega x} = e^{-(a/2)x}(\cos \omega x - i \sin \omega x).$$

We now add and divide by 2, as in Example 1. This gives y_1. Then we subtract and divide by $2i$. This gives y_2. Hence

$$(9) \qquad\qquad y_1 = e^{-ax/2} \cos \omega x, \qquad y_2 = e^{-ax/2} \sin \omega x.$$

These are solutions of (1), as follows by differentiation and substitution. They form a basis because they are linearly independent on any interval. Indeed, $y_2/y_1 = \tan \omega x$ is not constant because $\omega \neq 0$ (why?), so that y_1 and y_2 are not proportional. The corresponding general solution is

(10)
$$\boxed{y = e^{-ax/2}(A \cos \omega x + B \sin \omega x).}$$

EXAMPLE 2   **Complex roots. Initial value problem**

Solve the initial value problem

$$y'' + 0.2y' + 4.01y = 0, \qquad y(0) = 0, \qquad y'(0) = 2.$$

Solution. 1st Step. General solution.   The characteristic equation is $\lambda^2 + 0.2\lambda + 4.01 = 0$. It has the roots $-0.1 \pm 2i$. Hence $\omega = 2$, and a general solution (10) is

$$y = e^{-0.1x}(A \cos 2x + B \sin 2x).$$

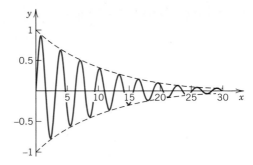

Fig. 38. Damped oscillations in Example 2

2nd Step. Particular solution.   The first initial condition gives $y(0) = A = 0$. There remains $y = Be^{-0.1x} \sin 2x$. We need the derivative (chain rule!)

$$y' = B(-0.1e^{-0.1x} \sin 2x + 2e^{-0.1x} \cos 2x).$$

From this and the second initial condition we get $y'(0) = 2B = 2$. Hence $B = 1$. Our solution is

$$y = e^{-0.1x} \sin 2x.$$

Figure 38 shows y and the curves of $e^{-0.1x}$ and $-e^{-0.1x}$ (dashed), between which the curve of y oscillates. Such ***"damped vibrations"*** (with $x = t =$ time) have important mechanical and electrical applications, as we shall soon see.   ◄

EXAMPLE 3   **Complex roots**

A general solution of the equation

$$y'' + \omega^2 y = 0 \qquad\qquad (\omega \text{ constant, not zero})$$

is

$$y = A \cos \omega x + B \sin \omega x.$$

With $\omega = 1$ this confirms Example 6 in Sec. 2.1.   ◄

This completes the discussion of all three cases, which we summarize as follows.

Summary of Cases I–III

Case	Roots of (2)	Basis of (1)	General Solution of (1)
I	Distinct real λ_1, λ_2	$e^{\lambda_1 x}, e^{\lambda_2 x}$	$y = c_1 e^{\lambda_1 x} + c_2 e^{\lambda_2 x}$
II	Real double root $\lambda = -\tfrac{1}{2}a$	$e^{-ax/2}, xe^{-ax/2}$	$y = (c_1 + c_2 x)e^{-ax/2}$
III	Complex conjugate $\lambda_1 = -\tfrac{1}{2}a + i\omega,$ $\lambda_2 = -\tfrac{1}{2}a - i\omega$	$e^{-ax/2} \cos \omega x$ $e^{-ax/2} \sin \omega x$	$y = e^{-ax/2}(A \cos \omega x + B \sin \omega x)$

It is very interesting that in applications to mechanical systems or electrical circuits, these three cases correspond to three different forms of motion or flows of current, respectively. We shall discuss this basic relation between theory and practice in detail in Sec. 2.5 (and again in Sec. 2.12).

Boundary Value Problems

Applications sometimes also lead to conditions of the type

(11)
$$y(P_1) = k_1, \qquad y(P_2) = k_2.$$

These are known as **boundary conditions,** since they refer to the endpoints P_1, P_2 (*boundary points P_1, P_2*) of an interval I on which the equation (1) is considered. Equation (1) and conditions (11) together constitute what is known as a **boundary value problem.** We confine ourselves to the discussion of a typical example.

EXAMPLE 4 **Boundary value problem**

Solve the boundary value problem

$$y'' + y = 0, \qquad y(0) = 3, \qquad y(\pi) = -3.$$

Solution. 1st Step. General solution. A basis is $y_1 = \cos x$, $y_2 = \sin x$. The corresponding general solution is

$$y(x) = c_1 \cos x + c_2 \sin x.$$

2nd Step. Solution of the problem. The left boundary condition gives $y(0) = c_1 = 3$. The right boundary condition gives $y(\pi) = c_1 \cos \pi + c_2 \cdot 0 = -3$. Now $\cos \pi = -1$ and $c_1 = 3$, so that this equation holds, and we see that it yields no condition for c_2. Hence as a solution of the problem on the interval $0 \leqq x \leqq \pi$ considered we obtain

$$y = 3 \cos x + c_2 \sin x.$$

Here c_2 is still arbitrary. This is a surprise. Of course, the reason is that $\sin x$ is zero at 0 and π. The reader may conclude and prove (Prob. 23) that the solution of a boundary value problem (1), (11) is unique if and only if no solution $y \neq 0$ of (1) satisfies $y(P_1) = y(P_2) = 0$. ◀

PROBLEM SET 2.3

Conversion to Real Form. Verify that the given function is a solution and derive the corresponding real general solution.

1. $y = c_1 e^{(1+i)x} + c_2 e^{(1-i)x}$, $\qquad y'' - 2y' + 2y = 0$

2. $y = c_1 e^{2\pi i x} + c_2 e^{-2\pi i x}$, $\qquad y'' + 4\pi^2 y = 0$

3. $y = c_1 e^{(-0.5+1.5i)x} + c_2 e^{(-0.5-1.5i)x}$, $\qquad 4y'' + 4y' + 10y = 0$

4. $y = c_1 e^{(-k+2i)x} + c_2 e^{(-k-2i)x}$, $\qquad y'' + 2ky' + (k^2 + 4)y = 0$

General Solution. State whether the given equation corresponds to Case I, Case II, or Case III. Find a real general solution. (Show each step of your derivation.)

5. $25y'' + 40y' + 16y = 0$ $\qquad$ **6.** $y'' + y' - 12y = 0$

7. $16y'' - 8y' + 5y = 0$ $\qquad$ **8.** $y'' + 4y' + (4 + \omega^2)y = 0$

9. $y'' - 9\pi^2 y = 0$ $\qquad$ **10.** $y'' - 2\sqrt{2}y' + 2.5y = 0$

11. $y'' - 2\sqrt{2}y' + 2y = 0$ $\qquad$ **12.** $y'' + 2ky' + (k^2 + k^{-2})y = 0$

Initial Value Problems. Solve the following problems. (Show each step.)

13. $9y'' + 6y' + y = 0$, $\qquad y(0) = 4$, $\qquad y'(0) = -13/3$

14. $4y'' + 16y' + 17y = 0$, $\qquad y(0) = -0.5$, $\qquad y'(0) = 1$

15. $y'' - 25y = 0$, $\qquad y(0) = 0$, $\qquad y'(0) = 20$

16. $y'' + 0.4y' + 0.29y = 0$, $\qquad y(0) = 1$, $\qquad y'(0) = -1.2$

17. $y'' - y' - 2y = 0$, $\qquad y(0) = -4$, $\qquad y'(0) = -17$

18. $y'' - 2y' + (4\pi^2 + 1)y = 0$, $\qquad y(0) = -2$, $\qquad y'(0) = 6\pi - 2$

Boundary Value Problems. Solve the following problems. (Show each step.)

19. $y'' + 4y = 0$, $y(0) = 3$, $y(\pi/2) = -3$

20. $y'' - 25y = 0$, $y(-2) = y(2) = \cosh 10$

21. $y'' + 2y' + 2y = 0$, $y(0) = 1$, $y(\pi/2) = 0$

22. $3y'' - 8y' - 3y = 0$, $y(-3) = 1$, $y(3) = 1/e^2$

23. Show that the solution of a boundary value problem (1), (11) is unique if and only if no solution $y \neq 0$ of (1) satisfies $y(P_1) = y(P_2) = 0$.

24. PROJECT. Damped Oscillations. (a) Investigate Fig. 38 by calculus, determining maxima and minima, inflection points, and points of contact with the dashed curves. Do the latter coincide with the extrema? (First guess.) Do the inflection points lie on the x-axis? (First guess.)

(b) What happens if you increase the damping by changing $e^{-0.1x}$ to, say, $e^{-0.2x}$? Could you make a qualitative statement about the effect of an arbitrary increase of the damping?

25. REVIEW PROJECT. Complex Numbers. Section 12.1 contains much more about complex numbers than you will presently need. Extract the information from Sec. 12.1 that you find to be useful and write a short report that includes examples of your own.

 26. CAS PROJECT. From Complex Roots to Double Root. Explore the transition from Case III to Case II graphically by choosing in (10) a sequence of positive values of ω tending to zero.
(a) Show that (1) with $a = 1$ can be written $y'' + y' + (\frac{1}{4} + \omega^2)y = 0$.
(b) Take the equation in (a) with, say, $\omega = 5, 0.5, 0.1, 0.01, \cdots$, and plot the solutions satisfying $y(0) = 1$, $y'(0) = -2$. Do the solution curves approach a limiting curve? Rapidly? Does it look like a curve in Case II?
(c) Does your CAS give you the limit? Give it a try. Find the limit analytically, showing all steps. (Remember l'Hôpital.)
(d) Find a problem in which the limiting curve does not intersect the x-axis.

2.4 Differential Operators. *Optional*

This section gives an introduction to differential operators. It will be used only once and for a minor purpose (in Sec. 2.14), so that it can be omitted without interrupting the flow of ideas.

By an **operator** we mean a transformation that transforms a function into another function. Operators and corresponding techniques are called **operational methods.**

Differentiation suggests an operator as follows. Let D denote differentiation with respect to x, that is, write

$$Dy = y'.$$

D is an operator. We say that D *operates on* y; that is, D transforms y (assumed differentiable) into its derivative y'. For example,

$$D(x^2) = 2x, \qquad D(\sin x) = \cos x.$$

Applying D twice, we obtain the second derivative $D(Dy) = Dy' = y''$. We simply write $D(Dy) = D^2 y$, so that

$$Dy = y', \qquad D^2 y = y'', \qquad D^3 y = y''', \cdots.$$

More generally,

(1)
$$L = P(D) = D^2 + aD + b$$

is called a **second-order differential operator.** Here a and b are constant. When L is applied to a function y (assumed twice differentiable), it produces

(2)
$$L[y] = (D^2 + aD + b)y = y'' + ay' + by.$$

L is a *linear* operator. By definition this means that we have

$$L[\alpha y + \beta w] = \alpha L[y] + \beta L[w]$$

for any constants α and β and any (twice differentiable) functions y and w.

The homogeneous linear differential equation $y'' + ay' + by = 0$ may now be simply written

(3)
$$L[y] = P(D)[y] = 0.$$

For example,

(4)
$$L[y] = (D^2 + D - 6)y = y'' + y' - 6y = 0.$$

Since

$$D[e^{\lambda x}] = \lambda e^{\lambda x}, \qquad D^2[e^{\lambda x}] = \lambda^2 e^{\lambda x},$$

we have from (2) and (3)

(5)
$$P(D)[e^{\lambda x}] = (\lambda^2 + a\lambda + b)e^{\lambda x} = P(\lambda)e^{\lambda x} = 0.$$

This confirms our result of Sec. 2.2 that $e^{\lambda x}$ is a solution of (3) if and only if λ is a solution of the characteristic equation $P(\lambda) = 0$. If $P(\lambda)$ has two different roots, we obtain a basis. If $P(\lambda)$ has a double root, we need a second independent solution. To obtain that solution, we differentiate

$$P(D)[e^{\lambda x}] = P(\lambda)e^{\lambda x}$$

[see (5)] on both sides with respect to λ and interchange differentiation with respect to λ and x. This gives

$$P(D)[xe^{\lambda x}] = P'(\lambda)e^{\lambda x} + P(\lambda)xe^{\lambda x}$$

where $P' = dP/d\lambda$. For a double root, $P(\lambda) = P'(\lambda) = 0$, so that we have $P(D)[xe^{\lambda x}] = 0$. Hence $xe^{\lambda x}$ is the desired second solution. This agrees with Sec. 2.2.

$P(\lambda)$ is a polynomial in λ, in the usual sense of algebra. If we replace λ by D, then we obtain the "operator polynomial" $P(D)$. The point of this **"operational calculus"** is that $P(D)$ can be treated just like an algebraic quantity. In particular, we can factor it.

EXAMPLE 1   **Factorization, solution of a differential equation**

Factor $P(D) = D^2 + D - 6$ and solve $P(D)[y] = 0$.

Solution. $D^2 + D - 6 = (D + 3)(D - 2)$. Now $(D - 2)y = y' - 2y$ by definition. Hence

$$(D + 3)(D - 2)y = (D + 3)[y' - 2y] = D(y' - 2y) + 3(y' - 2y)$$
$$= y'' - 2y' + 3y' - 6y = y'' + y' - 6y.$$

Hence our factorization is "permissible," that is, yields the correct result. Solutions of $(D + 3)y = 0$ and $(D - 2)y = 0$ are $y_1 = e^{-3x}$ and $y_2 = e^{2x}$. This is a basis of $P(D)[y] = 0$ on any interval. The student should verify that our method in Sec. 2.2 gives the same result. This is not unexpected, since we factored $P(D)$ in the same way as we factor the characteristic polynomial $P(\lambda) = \lambda^2 + \lambda - 6$.   ◄

It was essential that L in (2) had *constant* coefficients. Extension of operator methods to *variable-coefficient* equations is more difficult, and will not be considered here.

If operational calculus were limited to the simple situations illustrated in this section, it would perhaps not be worth mentioning. Actually, the power of the operator approach comes out in more complicated engineering problems, as we shall see in Chap. 5.

PROBLEM SET 2.4

Applications of Differential Operators. In each problem apply the given operator to the given functions. (Show the details of your work.)

1. $D^2 + 3D;$ $\cosh 3x,$ $e^{-x} + e^{2x},$ $10 - e^{-3x}$
2. $D - 4;$ $3x^2 + 4x,$ $4e^{4x},$ $\cos 2x - \sin 2x$
3. $(D - 2)(D + 1);$ $e^{2x},$ $xe^{2x},$ $e^{-x},$ xe^{-x}
4. $(D + 5)^2;$ $5x + \sin 5x,$ $xe^{5x},$ xe^{-5x}

General Solution. Find a general solution, using factorization (as in Example 1 in the text). (Show the details.)

5. $(D^2 - D - 2)y = 0$ 6. $(9D^2 + 6D + 1)y = 0$
7. $(D^2 - 4D)y = 0$ 8. $(25D^2 - 1)y = 0$
9. $(D^2 + 2kD + k^2)y = 0$ 10. $(D^2 + \pi(\pi - 1)D - \pi^3)y = 0$
11. $(64D^2 + 16D + 1)y = 0$ 12. $(2D^2 + D)y = 0$
13. $(10D^2 + 12D + 3.6)y = 0$

14. **(Linear operator)** Show that L in (2) is a linear operator.

2.5 Modeling: Free Oscillations (Mass–Spring System)

Homogeneous linear differential equations with constant coefficients have important engineering applications. In this section we discuss the motions of a basic mechanical system, a mass on an elastic spring ("mass–spring system," Fig. 39). We **model** the system, that is, we set up its mathematical equation (a differential equation), solve it, and discuss the types of motion. Very interestingly, the latter will correspond to Cases I, II, III in Secs. 2.2 and 2.3.

An equally important application from electrical engineering (a fundamental electric circuit) will be studied in Sec. 2.12.

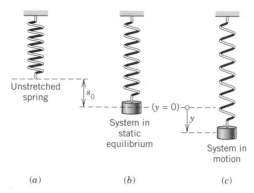

Fig. 39. Mechanical system under consideration

Setting up the Model of the Mass–Spring System

We take an ordinary spring that resists compression as well as extension and suspend it vertically from a fixed support (Fig. 39). At the lower end of the spring we attach a body of mass m. We assume m to be so large that we may disregard the mass of the spring. If we pull the body down a certain distance and then release it, it undergoes a motion. We assume that the body moves strictly vertically.

We want to determine the motion of our mechanical system. This motion is governed by **Newton's second law**

(1)
$$\boxed{\text{Mass} \times \text{Acceleration} = my'' = \text{Force}}$$

where "Force" is the resultant of all the forces acting on the body.[6] Here, $y'' = d^2y/dt^2$, where $y(t)$ is the displacement of the body and t is time.

We choose the ***downward direction as the positive direction,*** thus regarding downward forces as positive and upward forces as negative.

Consider Fig. 39. The spring is first unstretched. When we attach the body, the latter stretches the spring by an amount s_0 (see Fig. 39). This causes an upward force F_0 in the spring. Experiments show that this force F_0 is proportional to the stretch, say,

(2) $$F_0 = -ks_0$$ **(Hooke's law[7]).**

$k\,(>0)$ is called the **spring constant** (or spring modulus). The minus sign appears because F_0 points upward (recall that this is our *negative* direction!). A stiff spring has a large k; hence it gives a small s_0.

The extension s_0 is such that F_0 balances the weight $W = mg$ of the body (where $g = 980 \text{ cm/sec}^2 = 32.17 \text{ ft/sec}^2$ is the gravitational constant). Consequently, $F_0 + W = -ks_0 + mg = 0$. These forces will not affect the motion. Spring and body are again at rest. This is called the **static equilibrium** of the system (Fig. 39*b*). We let this

---

[6]For systems of units and conversion factors, see inside of front cover.

[7]ROBERT HOOKE (1635–1703), English physicist, a forerunner of Newton with respect to the law of gravitation.

position of the body be $y = 0$, that is, we measure the displacement $y(t)$ of the body from this position as the origin, positive downward and negative upward.

Now comes the main point. From the position $y = 0$ we pull the body downward. This further stretches the spring by some amount $y > 0$ (the distance we pull it down). By Hooke's law this causes an (additional) upward force F_1 in the spring,

$$F_1 = -ky.$$

F_1 is a **restoring force.** It has the tendency to *restore* the system, that is, to pull the body back to $y = 0$.

Undamped System: Equation and Solution

Every system has damping—otherwise it would keep moving forever. But practically, the effect of damping may often be negligible, for example, for the motion of an iron ball on a spring during a few minutes. Then F_1 is the only force in (1) causing motion. Hence $my'' = -ky$ from (1). We see that our model of the mechanical system without damping is the linear differential equation with constant coefficients

(3) $$\boxed{my'' + ky = 0.}$$

By the method in Sec. 2.3 (see Example 3) we get the general solution

(4) $$\boxed{y(t) = A \cos \omega_0 t + B \sin \omega_0 t} \qquad \omega_0 = \sqrt{k/m}.$$

The corresponding motion is called a **harmonic oscillation.** Figure 40 shows typical forms of (4). These correspond to some positive initial displacement $y(0)$ [which determines $A = y(0)$ in (4)] and different initial velocities $y'(0)$. Each of the latter determines a value of B in (4) because $y'(0) = \omega_0 B$.

By applying the addition formula for the cosine, the student may verify that (4) can be written [see also (13) in Appendix A3.1]

(4*) $$y(t) = C \cos (\omega_0 t - \delta) \qquad \left(C = \sqrt{A^2 + B^2}, \quad \tan \delta = \frac{B}{A} \right).$$

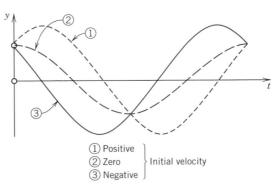

① Positive ⎫
② Zero ⎬ Initial velocity
③ Negative ⎭

Fig. 40.   Harmonic oscillations

Since the period of the trigonometric functions in (4) is $2\pi/\omega_0$, the body executes $\omega_0/2\pi$ cycles per second. The quantity $\omega_0/2\pi$ is called the **frequency** of the oscillation and is measured in cycles per second. Another name for cycles/sec is hertz (Hz).[8]

EXAMPLE 1   **Undamped system. Harmonic oscillations**

If an iron ball of weight $W = 89.00$ nt (about 20 lb) stretches a spring 10.00 cm (about 4 inches), how many cycles per minute will this mass–spring system execute? What will its motion be if we pull down the weight an additional 15.00 cm (about 6 inches)?

Solution.   Hooke's law (2) with W as the force and 10 cm $= 0.1$ meter as the stretch gives $W = 0.1k$, thus $k = W/0.1 = 89.00/0.1000 = 890.0 \left[\text{kg/sec}^2\right] = 890.0$ [nt/meter]. The mass is $m = W/g = 89.00/9.8000$ [kg], that is, 9.082 kg. This gives the frequency

$$\omega_0/2\pi = \sqrt{890.0/9.082}/2\pi = 9.899/2\pi = 1.576 \text{ [Hz]}$$

or 94.5 cycles per minute. From (4) and the initial conditions, $y(0) = A = 0.1500$ [meter] and $y'(0) = \omega_0 B = 0$. Hence the motion is

$$y(t) = 0.1500 \cos 9.899t \text{ [meters]} \qquad \text{or} \qquad 0.492 \cos 9.899t \text{ [ft]}.$$

If you have a chance of experimenting with a mass–spring system, don't miss it. You will be surprised about the good agreement between theory and experiment, usually within a fraction of one percent if you measure carefully.   ◀

Damped System: Equation and Solutions

If we connect the mass to a dashpot (Fig. 41), we have to take the corresponding viscous damping into account. The corresponding damping force has the direction opposite to the instantaneous motion. We assume that it is proportional to the velocity $y' = dy/dt$ of the body. This is generally a good approximation, at least for small velocities. Thus the damping force is of the form

$$F_2 = -cy'.$$

c is called the **damping constant.** Let us show that c is positive. If y' is positive, the body moves downward (in the positive y-direction) and $-cy'$ must be an upward force, that is,

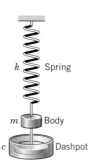

Fig. 41.   Damped system

---

[8]HEINRICH HERTZ (1857–1894), German physicist, who discovered electromagnetic waves and made important contributions to electrodynamics.

by agreement, $-cy' < 0$, which implies $c > 0$. For negative y' the body moves upward and $-cy'$ must be a downward force, that is, $-cy' > 0$, which implies $c > 0$.

    The resultant of the forces acting on the body is now

$$F_1 + F_2 = -ky - cy'.$$

Hence, by Newton's second law,

$$my'' = -ky - cy'.$$

This shows that the motion of the damped mechanical system is governed by the linear differential equation with constant coefficients

(5)
$$my'' + cy' + ky = 0.$$

The corresponding characteristic equation is

$$\lambda^2 + \frac{c}{m}\lambda + \frac{k}{m} = 0.$$

The roots are

$$\lambda_{1,2} = -\frac{c}{2m} \pm \frac{1}{2m}\sqrt{c^2 - 4mk}.$$

We use the short notations

(6)
$$\alpha = \frac{c}{2m} \quad \text{and} \quad \beta = \frac{1}{2m}\sqrt{c^2 - 4mk}.$$

Then we can write

$$\lambda_1 = -\alpha + \beta \quad \text{and} \quad \lambda_2 = -\alpha - \beta.$$

The form of the solution of (5) will depend on the damping. As in Secs. 2.2 and 2.3, we now have the following three cases:

Case I.	$c^2 > 4mk$.	*Distinct real roots λ_1, λ_2.*	*(**Overdamping**)*
Case II.	$c^2 = 4mk$.	*A real double root.*	*(**Critical damping**)*
Case III.	$c^2 < 4mk$.	*Complex conjugate roots.*	*(**Underdamping**)*

Discussion of the Three Cases

Case I. Overdamping

If the damping constant c is so large that $c^2 > 4mk$, then λ_1 and λ_2 are distinct real roots, and the general solution of (5) is

(7)
$$y(t) = c_1 e^{-(\alpha-\beta)t} + c_2 e^{-(\alpha+\beta)t}.$$

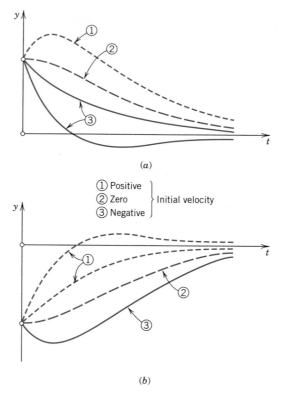

Fig. 42. Typical motions (7) in the overdamped case
(*a*) Positive initial displacement
(*b*) Negative initial displacement

We see that in this case the body does not oscillate. For $t > 0$ both exponents in (7) are negative because $\alpha > 0$, $\beta > 0$, and $\beta^2 = \alpha^2 - k/m < \alpha^2$. Hence both terms in (7) approach zero as t approaches infinity. Practically speaking, after a sufficiently long time the mass will be at rest at the static equilibrium position ($y = 0$). This is understandable since the damping takes energy from the system and there is no external force that keeps the motion going. Figure 42 shows (7) for some typical initial conditions.

Case II. Critical damping

If $c^2 = 4mk$, then $\beta = 0$, $\lambda_1 = \lambda_2 = -\alpha$, and the general solution is

(8)
$$y(t) = (c_1 + c_2 t)e^{-\alpha t}.$$

Since the exponential function is never zero and $c_1 + c_2 t$ can have at most one positive zero, it follows that the motion can have at most one passage through the equilibrium position ($y = 0$). If the initial conditions are such that both c_1 and c_2 are positive (or both negative), there is no such passage at all (because $t \geqq 0$). Figure 43 shows typical forms of (8).

Case II marks the border between nonoscillatory motions and oscillations; this explains its name "critical case."

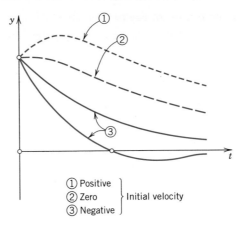

① Positive
② Zero   } Initial velocity
③ Negative

Fig. 43. Critical damping [see (8)]

Case III. Underdamping

This is the most interesting case. If the damping constant c is so small that $c^2 < 4mk$, then β in (6) is pure imaginary, say,

$$(9) \quad \beta = i\omega^* \qquad \text{where} \qquad \omega^* = \frac{1}{2m}\sqrt{4mk - c^2} = \sqrt{\frac{k}{m} - \frac{c^2}{4m^2}} \quad (> 0).$$

(We write ω^* to reserve ω for Sec. 2.11.) The roots of the characteristic equation are complex conjugate,

$$\lambda_1 = -\alpha + i\omega^*, \qquad \lambda_2 = -\alpha - i\omega^*$$

with α given in (6). Hence the corresponding general solution is

$$(10) \qquad y(t) = e^{-\alpha t}(A \cos \omega^* t + B \sin \omega^* t) = Ce^{-\alpha t} \cos(\omega^* t - \delta)$$

where $C^2 = A^2 + B^2$ and $\tan \delta = B/A$ [as in (4*)].

This solution represents damped oscillations. Since $\cos(\omega^* t - \delta)$ varies between -1 and 1, the curve of the solution lies between the curves $y = Ce^{-\alpha t}$ and $y = -Ce^{-\alpha t}$ in Fig. 44, touching these curves when $\omega^* t - \delta$ is an integer multiple of π.

The frequency is $\omega^*/2\pi$ cycles per second. From (9) we see that the smaller c (> 0) is, the larger is ω^* and the more rapid the oscillations become. As c approaches zero, ω^* approaches the value $\omega_0 = \sqrt{k/m}$ corresponding to the harmonic oscillation (4).

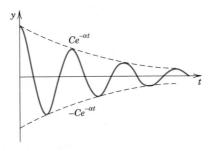

Fig. 44. Damped oscillation in Case III [see (10)]

EXAMPLE 2   The three cases of damped motion

How does the motion in Example 1 change if the system has damping given by

$$\text{(I)} \ c = 200.0 \text{ kg/sec}, \qquad \text{(II)} \ c = 179.8 \text{ kg/sec}, \qquad \text{(III)} \ c = 100.0 \text{ kg/sec?}$$

Solution.   It is instructive to study and compare these cases with the behavior of the system in Example 1.
   (I)   The problem is

$$9.082y'' + 200.0y' + 890.0y = 0, \qquad y(0) = 0.1500 \text{ [meter]}, \qquad y'(0) = 0.$$

The characteristic equation has the roots $\lambda_{1,2} = -\alpha \pm \beta = -11.01 \pm 4.822$, thus $\lambda_1 = -6.190$, $\lambda_2 = -15.83$. From (7) and the initial conditions, $c_1 + c_2 = 0.1500$, $\lambda_1 c_1 + \lambda_2 c_2 = 0$. The solution is

$$y(t) = 0.2463e^{-6.190t} - 0.0963e^{-15.83t}.$$

It approaches 0 as $t \to \infty$. The approach is very rapid. After a few seconds, it is practically 0; that is, the body is at rest.
   (II)   The problem is as before, with $c = 179.8$ instead of 200. Since $c^2 = 4mk$, we get the double root $\lambda = -9.899$. From (8) and the initial conditions, $c_1 = 0.1500$, $c_2 + \lambda c_1 = 0$, $c_2 = 1.485$. The solution is

$$y(t) = (0.150 + 1.485t)e^{-9.899t}.$$

It decreases rapidly to zero. (Actually, to get the critical case exactly, we would have to calculate with a large number of digits. Try it.)
   (III)   The problem is as before, with $c = 100$. This c is small enough for obtaining oscillations. Indeed, the roots are complex conjugate, $\lambda_{1,2} = -\alpha \pm i\omega^* = -5.506 \pm 8.227i$. From (10) and the initial conditions we obtain $A = 0.1500$, $-\alpha A + \omega^* B = 0$ or $B = 0.1004$. This gives the solution

$$y(t) = e^{-5.506t}(0.1500 \cos 8.227t + 0.1004 \sin 8.227t).$$

These damped oscillations have smaller frequency than the harmonic oscillations in Example 1 by about 17%. Their amplitude goes to zero very fast (why?).   ◀

In this section we have been concerned with **free motions** of mass–spring systems. These are governed by homogeneous differential equations, as we have seen. **Forced motions** under the influence of a "driving force" lead to nonhomogeneous equations and will be studied in Sec. 2.11, after we have learned how to solve such equations.

PROBLEM SET 2.5

Harmonic Oscillations (Motion without Damping)

1. Show that the harmonic oscillation (4) starting from initial displacement y_0 with initial velocity v_0 is $y(t) = y_0 \cos \omega_0 t + (v_0/\omega_0) \sin \omega_0 t$ and represent this in the form (4*).
2. If a spring is such that a weight of 20 nt (about 4.5 lb) would stretch it 2 cm, what would the frequency of the corresponding harmonic oscillation be? The period? (Proceed as in Example 1.)
3. How does the frequency of the harmonic oscillation change if we (i) double the mass, (ii) take a stiffer spring? Before you look at formulas first try to find qualitative answers by physical arguments.
4. Could you make a harmonic oscillation move faster by giving the body a greater initial push?
5. Show that the frequency of a harmonic oscillation of a body on a spring is $(\sqrt{g/s_0})/2\pi$, so that the period is $2\pi\sqrt{s_0/g}$, where s_0 is the elongation in Fig. 39.

6. If a body hangs on a spring of modulus $k_1 = 8$, which in turn hangs on a spring of modulus $k_2 = 12$, what is the modulus k of this combination of springs?

7. What are the frequencies of vibration of a mass $m = 5$ kg (i) on a spring with modulus $k_1 = 20$ nt/m, (ii) on a spring with $k_2 = 45$ nt/m, (iii) on the two springs in parallel? See Fig. 45.

8. Archimedes's principle states that the buoyancy force equals the weight of the water displaced by the body (partly or totally submerged). Figure 46 shows a cylindrical buoy 60 cm in diameter standing in water with its axis vertical. When depressed slightly and released, its period of vibration is 2 sec. Find the weight of the buoy.

9. If 1 liter of water is allowed to vibrate up and down in a U-shaped tube 2 cm in diameter (Fig. 47), what is the frequency? Neglect friction.

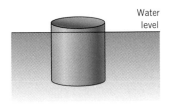

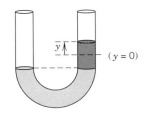

Fig. 45.   Problem 7   **Fig. 46.**   Buoy in Problem 8   **Fig. 47.**   Tube in Problem 9

10. TEAM PROJECT. Harmonic Motions of Similar Models. The unifying power of mathematical methods results to a large extent from the fact that different physical (or other) systems may have the same or very similar models. Illustrate this for the three systems shown in Figs. 48–50.

    (a) Pendulum (Fig. 48). Determine the frequency of oscillation of the pendulum of length L, neglecting air resistance and the weight of the rod. Assume θ to be so small that $\sin \theta$ practically equals θ.

    (b) Pendulum clock. A clock has a 1-meter pendulum. The clock ticks once for each time the pendulum completes a swing, returning to its original position. How many times a minute does the clock tick?

    (c) Flat spring (Fig. 49). The harmonic oscillations of a flat spring with a body attached at one end and horizontally clamped at the other are also governed by (3). Find its motions, assuming that the body weighs 8 nt (about 1.8 lb), the system has its static equilibrium 1 cm below the horizontal line, and we let it start from this position with initial velocity 10 cm/sec.

    (d) Torsional vibrations (Fig. 50). Undamped torsional vibrations (rotations back and forth) of a wheel attached to an elastic thin rod or wire are governed by the equation $I_0\theta'' + K\theta = 0$, where θ is the angle measured from the state of equilibrium. Solve this equation for $K/I_0 = 13.69 \text{ sec}^{-2}$, initial angle $30°$ ($= 0.5235$ rad) and initial angular velocity $20° \text{ sec}^{-1}$ ($= 0.349 \text{ rad} \cdot \text{sec}^{-1}$).

Fig. 48.   Pendulum   **Fig. 49.**   Flat spring   **Fig. 50.**   Torsional vibrations

Motions with Damping

11. (Overdamping) Show that for (7) to satisfy initial conditions $y(0) = y_0$ and $v(0) = v_0$ we must have $c_1 = [(1 + \alpha/\beta)y_0 + v_0/\beta]/2$ and $c_2 = [(1 - \alpha/\beta)y_0 - v_0/\beta]/2$.

12. (Overdamping) Show that in the overdamped case, the body can pass through $y = 0$ at most once (Fig. 42).

13. (Critical damping) Find the critical motion (8) that starts from y_0 with initial velocity v_0.

14. (Critical damping) Under what conditions does (8) have a maximum or minimum at some instant $t > 0$?

15. (Underdamping) Determine the values of t corresponding to the maxima and minima of the oscillation $y(t) = e^{-t} \sin t$. Check your result by graphing $y(t)$.

16. (Underdamping) Show that the maxima and minima of an underdamped motion occur at equidistant values of t, the distance between two consecutive maxima being $2\pi/\omega^*$.

17. (Damping constant) Consider an underdamped motion of a body of mass $m = 0.5$ kg. If the time between two consecutive maxima is 3 sec and the maximum amplitude decreases to $\frac{1}{2}$ its initial value after 10 cycles, what is the damping constant of the system?

18. (Logarithmic decrement) Prove that the ratio of two consecutive maximum amplitudes of a damped oscillation (10) is constant, the natural logarithm of this ratio being $\Delta = 2\pi\alpha/\omega^*$. ($\Delta$ is called the *logarithmic decrement* of the oscillation.) Find Δ in the case of $y = e^{-t} \cos t$ and determine the values of t corresponding to the maxima and minima.

19. (Frequency) Show that the frequency $\omega^*/2\pi$ of the underdamped motion decreases as the damping increases. Is this physically understandable?

20. CAS PROJECT. Transition between Cases I, II, III. Study this transition in terms of plots of typical solutions.

(a) *Delimitation* to a reasonable amount of work is quite important in any project. *Avoiding unnecessary generality* is part of good modeling. Decide that the initial value problems (A) and (B),

$$\text{(A)} \quad y'' + cy' + y = 0, \qquad y(0) = 1, \qquad y'(0) = 0$$

(B) the same with different c and $y'(0) = -2$ (instead of 0), will meet the objective, giving practically as much information as a problem with other m, k, $y(0)$, $y'(0)$.

(b) *Consider* (A). Choose suitable c, perhaps better ones than in Fig. 51, for the transition from Case III to II and I. Guess c for the curves in the figure.

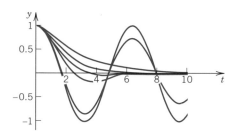

Fig. 51. CAS Project 20

(c) *Time to go to rest.* Theoretically, this time is infinite (why?). Practically, the system is at rest when its motions have become very small, say, less than 1% of the initial displacement (this choice being up to us), that is,

$$(11) \qquad |y(t)| < 0.01 \qquad \text{for all } t \text{ greater than some } t_1.$$

In engineering constructions, damping can often be varied without too much trouble. Experimenting with your plots, find empirically a relation between t_1 and c.

(d) ***Solve* (A)** ***analytically.*** Give a reason why the solution c of $y(t_2) = -0.01$, with t_2 the solution of $y'(t) = 0$, will give you the best possible c.

(e) ***Consider* (B)** empirically as in (a) and (b). What is the main difference between (B) and (A)?

2.6 Euler–Cauchy Equation

Exponential functions $e^{\lambda x}$ reproduce under differentiation, $(e^{\lambda x})' = \lambda e^{\lambda x}$, etc., and this property gave us the idea in solving linear constant-coefficient equations (Secs. 2.2, 2.3). Powers $y = x^m$ decrease by 1 when we differentiate, $y' = mx^{m-1}$, $y'' = m(m-1)x^{m-2}$. Hence they should solve linear differential equations in y, xy', x^2y'', where the x-power compensates for the loss. That is,

(1)
$$x^2y'' + axy' + by = 0 \qquad (a, b \text{ constant}).$$

This is called the **Euler–Cauchy equation.**[9] We try

(2)
$$y = x^m.$$

Substituting this and its derivatives into (1), we find

$$x^2m(m-1)x^{m-2} + axmx^{m-1} + bx^m = 0.$$

We have a common factor x^m, which is not zero when $x \neq 0$. We can drop it and arrange the terms in m^2, m, and m^0. This gives the auxiliary equation

(3)
$$m^2 + (a-1)m + b = 0$$

for determining m in (2).

---

[9]LEONHARD EULER (1707–1783) was an enormously creative Swiss mathematician. He studied in Basel under JOHANN BERNOULLI and in 1727 became a professor of physics (and later of mathematics) in St. Petersburg, Russia. In 1741 he went to Berlin as a member of the Berlin Academy. In 1766 he returned to St. Petersburg. He contributed to almost all branches of mathematics and its applications to physical problems, even after he became totally blind in 1771; we mention his fundamental work in differential and difference equations, Fourier and other infinite series, special functions, complex analysis, the calculus of variations, mechanics, and hydrodynamics. He is the exponent of a very rapid expansion of analysis. (So far, the first seventy (!) volumes of his Collected Works have appeared.)

This expansion of mathematics was followed by a period characterized by greater rigor, dominated by the great French mathematician AUGUSTIN-LOUIS CAUCHY (1789–1857), the father of modern analysis. Cauchy studied and taught mainly in Paris. He is the creator of complex analysis and exercised a great influence on the theory of infinite series and ordinary and partial differential equations (see the Index to this book for some of these contributions). He is also known for his work in elasticity and optics. Cauchy published nearly 800 mathematical research papers, many of them of basic importance.

Three Cases of Solutions

Case I. Distinct real roots. If (3) has distinct real roots m_1, m_2, we get a basis of solutions $y_1 = x^{m_1}$, $y_2 = x^{m_2}$ and a corresponding general solution of (1)

(4)
$$y = c_1 x^{m_1} + c_2 x^{m_2}$$
(c_1, c_2 arbitrary)

for all x for which y_1 and y_2 are defined.

EXAMPLE 1   **General solution in the case of different real roots**

Solve the Euler–Cauchy equation

$$x^2 y'' - 2.5xy' - 2.0y = 0.$$

Solution.   The auxiliary equation is

$$m^2 - 3.5m - 2.0 = 0.$$

(Note -3.5, not -2.5!) The roots are $m_1 = -0.5$ and $m_2 = 4$. This gives

$$y_1 = 1/\sqrt{x},$$

$$y_2 = x^4$$

and the general solution

$$y = \frac{c_1}{\sqrt{x}} + c_2 x^4$$

valid for all positive x.   ◀

Case II. Double root. If (3) has a double root, this root must be $\frac{1}{2}(1 - a)$ because

$$(m - \tfrac{1}{2}(1 - a))^2 = m^2 + (a - 1)m + b \qquad\qquad (b = \tfrac{1}{4}(1 - a)^2).$$

We then get a first solution

(5)
$$y_1 = x^{(1-a)/2}$$

and a second solution y_2 by the method of reduction of order (as in Sec. 2.2). Thus, substituting $y_2 = uy_1$ and its derivatives into (1), we obtain

$$x^2(u'' y_1 + 2u' y_1' + uy_1'') + ax(u' y_1 + uy_1') + buy_1 = 0.$$

Reshuffling terms gives

(6)
$$u'' x^2 y_1 + u' x(2xy_1' + ay_1) + u(x^2 y_1'' + axy_1' + by_1) = 0.$$

The last expression $(\cdots)$ is zero since y_1 is a solution of (1). From (5) we get in (6) for the first expression in parentheses

$$2xy_1' + ay_1 = (1 - a)x^{(1-a)/2} + ax^{(1-a)/2} = x^{(1-a)/2} = y_1.$$

This reduces (6) to $(u''x^2 + u'x)y_1 = 0$. We divide this by y_1 ($\neq 0$), separate variables, and integrate. Then for $x > 0$ we get

$$\frac{u''}{u'} = -\frac{1}{x}, \qquad \ln|u'| = -\ln x, \qquad u' = \frac{1}{x}, \qquad u = \ln x.$$

Thus $y_2 = y_1 \ln x$, which is not proportional to y_1. Hence in the case of a double root of (3), a basis of (1) for all positive x is y_1, $y_2 = y_1 \ln x$ with y_1 in (5). The corresponding general solution is

(7)
$$\boxed{y = (c_1 + c_2 \ln x)x^{(1-\alpha)/2}}$$
$(c_1, c_2$ arbitrary).

EXAMPLE 2   **General solution in the case of a double root**

Solve
$$x^2 y'' - 3xy' + 4y = 0.$$

Solution.   The auxiliary equation has the double root $m = 2$. Hence a basis of real solutions for all positive x is x^2, $x^2 \ln x$, and the corresponding general solution is

$$y = (c_1 + c_2 \ln x)x^2.$$
◀

Case III. Complex conjugate roots. This is of no great practical importance. We present it for completeness. If the roots of (3) are complex, they are conjugate, say, $m_1 = \mu + i\nu$, $m_2 = \mu - i\nu$. The trick is to write $x^{i\nu} = (e^{\ln x})^{i\nu} = e^{i\nu \ln x}$ and then to apply the Euler formula (5), Sec. 2.3. This gives

$$x^{m_1} = x^\mu x^{i\nu} = x^\mu e^{i\nu \ln x} = x^\mu [\cos(\nu \ln x) + i \sin(\nu \ln x)]$$

$$x^{m_2} = x^\mu x^{-i\nu} = x^\mu e^{-i\nu \ln x} = x^\mu [\cos(\nu \ln x) - i \sin(\nu \ln x)].$$

By addition the sine drops out and by subtraction the cosine drops out. Again this gives solutions because (1) is linear and homogeneous. We divide that sum by 2 and that difference by $2i$ (as in Sec. 2.3). We then obtain the real solutions

$$x^\mu \cos(\nu \ln x) \qquad \text{and} \qquad x^\mu \sin(\nu \ln x)$$

and the corresponding general solution for all positive x,

(8)
$$y = x^\mu [A \cos(\nu \ln x) + B \sin(\nu \ln x)].$$

EXAMPLE 3   **General solution in the case of complex conjugate roots**

Solve
$$x^2 y'' + 7xy' + 13y = 0.$$

Solution.   The auxiliary equation (3) is $m^2 + 6m + 13 = 0$. The roots of this equation are complex conjugate, $m_{1,2} = -3 \pm \sqrt{9 - 13} = -3 \pm 2i$. By (8) a general solution for all positive x is

$$y = x^{-3}[A \cos(2 \ln x) + B \sin(2 \ln x)].$$
◀

Euler–Cauchy equations occur in certain applications, and we illustrate this by a simple example from electrostatics.

EXAMPLE 4   **Boundary value problem. Electric potential field between two concentric spheres**

Find the electrostatic potential $v = v(r)$ between two concentric spheres of radii $r_1 = 4$ cm and $r_2 = 8$ cm kept at potentials $v_1 = 110$ volts and $v_2 = 0$, respectively.

Physical information. $v(r)$ is a solution of $rv'' + 2v' = 0$, where $v' = dv/dr$.

Solution.   The auxiliary equation $m^2 + m = 0$ has the roots 0 and -1. This gives the general solution $v(r) = c_1 + c_2/r$. From the "boundary conditions" (the potentials on the spheres),

$$v(4) = c_1 + c_2/4 = 110, \qquad v(8) = c_1 + c_2/8 = 0.$$

By subtraction, $c_2/8 = 110$, $c_2 = 880$, thus $c_1 = -110$. *Answer.* $v(r) = -110 + 880/r$ volts. Figure 52 shows that the potential is convex, not a straight line, as it would be for a potential between two parallel plates. For instance, on the sphere of radius 6 it is not $110/2 = 55$ volts, but considerably less. (What is it?) ◀

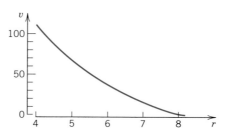

Fig. 52.   Potential in Example 4

PROBLEM SET 2.6

1. Verify directly by substitution that $x^{(1-a)/2} \ln x$ is a solution of (1) if (3) has a double root, but $x^{m_1} \ln x$ and $x^{m_2} \ln x$ are **not** solutions of (1) if the roots m_1 and m_2 of (3) are different.

General Solution. Find a real general solution. (Show the details of your work.)

2. $x^2 y'' - 4xy' + 6y = 0$
3. $x^2 y'' - 20y = 0$
4. $xy'' + 2y' = 0$
5. $10x^2 y'' + 46xy' + 32.4y = 0$
6. $x^2 y'' - xy' + 2y = 0$
7. $x^2 y'' + xy' + y = 0$
8. $(xD^2 + D)y = 0$
9. $(4x^2 D^2 + 12xD + 3)y = 0$
10. $(x^2 D^2 + 0.7xD - 0.1)y = 0$
11. $(x^2 D^2 + 1.25)y = 0$
12. $(x^2 D^2 - 0.2xD + 0.36)y = 0$
13. $(x^2 D^2 + 7xD + 9)y = 0$

Initial Value Problems. Solve and plot the solution. (Show your work.)

14. $x^2 y'' - 2xy' + 2y = 0$, $y(1) = 1.5$, $y'(1) = 1$
15. $4x^2 y'' + 24xy' + 25y = 0$, $y(1) = 2$, $y'(1) = -6$
16. $x^2 y'' + xy' + 9y = 0$, $y(1) = 2$, $y'(1) = 0$
17. $(x^2 D^2 - 3xD + 4)y = 0$, $y(1) = 0$, $y'(1) = 3$
18. $(x^2 D^2 + 3xD + 1)y = 0$, $y(1) = 3$, $y'(1) = -4$

19. **(Boundary value problem)** How does the potential in Example 4 change if the spheres have diameter 5 and 10 cm and potential 30 and 300 volts, respectively?

20. **(Relation of Euler–Cauchy Equations to Constant-Coefficient Equations)** These two large classes of differential equations have in common that we can solve them "algebraically" without actual integration. As a curiosity, show that they can even be transformed into each other. *Hint.* Set $x = e^t$.

2.7 Existence and Uniqueness Theory. Wronskian

In this section we give a general theory for homogeneous linear equations

(1)
$$y'' + p(x)y' + q(x)y = 0$$

with continuous, but otherwise arbitrary variable coefficients p and q. This will concern the existence of a general solution

(2)
$$y = c_1 y_1 + c_2 y_2$$

of (1) as well as initial value problems consisting of the differential equation (1) and two initial conditions

(3)
$$y(x_0) = K_0, \qquad y'(x_0) = K_1$$

with given x_0, K_0, and K_1.

Clearly, no such theory was needed for constant-coefficient or Euler–Cauchy equations because everything came out explicitly from our calculations.

Central to our present discussion is the following theorem.

THEOREM 1   **Existence and Uniqueness Theorem for Initial Value Problems**

If $p(x)$ and $q(x)$ are continuous functions on some open interval I (see Sec. 1.1) and x_0 is in I, then the initial value problem consisting of (1) and (3) has a unique solution $y(x)$ on the interval I.

The proof of existence uses the same prerequisites as those of the existence theorem in Sec. 1.9 and will not be presented here; it can be found in Ref. [A5] listed in Appendix 1. Uniqueness proofs are usually simpler than existence proofs. But in the present case, even the uniqueness proof is long, and we give it as an additional proof in Appendix 4.

Linear Independence of Solutions. Wronskian

Theorem 1 will imply very important properties of general solutions (2) of (1). As we know, these are made up of a **basis** y_1, y_2, that is, of a pair of linearly independent solutions. Remember from Sec. 2.1 that we call y_1 and y_2 **linearly independent** on an interval I if the equation

$$k_1 y_1(x) + k_2 y_2(x) = 0 \quad \text{on } I \qquad \text{implies} \qquad k_1 = 0, k_2 = 0.$$

And we call y_1 and y_2 **linearly dependent** on I if this equation also holds for k_1, k_2 not both zero. In this case, and only in this case, y_1 and y_2 are proportional on I, that is (see Sec. 2.1),

(4)   (a) $y_1 = k y_2$ or   (b) $y_2 = l y_1$.

For our discussion the following criterion of linear independence and dependence of solutions will be helpful. This criterion uses the so-called *Wronski determinant*[10] or, briefly, the **Wronskian,** of two solutions y_1 and y_2 of (1), defined by

(5)
$$W(y_1, y_2) = \begin{vmatrix} y_1 & y_2 \\ y_1' & y_2' \end{vmatrix} = y_1 y_2' - y_2 y_1'.$$

THEOREM 2   **(Linear dependence and independence of solutions)**

Suppose that (1) has continuous coefficients $p(x)$ and $q(x)$ on an open interval I. Then two solutions y_1 and y_2 of (1) on I are linearly dependent on I if and only if their Wronskian W is zero at some x_0 in I. Furthermore, if $W = 0$ for $x = x_0$, then $W \equiv 0$ on I; hence if there is an x_1 in I at which W is not zero, then y_1, y_2 are linearly independent on I.

PROOF.   **(a)** If y_1 and y_2 are linearly dependent on *I*, then (4a) or (4b) holds on *I*. If (4a) holds, then

$$W(y_1, y_2) = W(ky_2, y_2) = \begin{vmatrix} ky_2 & y_2 \\ ky_2' & y_2' \end{vmatrix} = ky_2 y_2' - y_2 ky_2' \equiv 0;$$

similarly if (4b) holds.

(b) Conversely, we assume that $W(y_1, y_2) = 0$ for some $x = x_0$ in *I* and show that then y_1, y_2 are linearly dependent. We consider the linear system of equations

(6)
$$k_1 y_1(x_0) + k_2 y_2(x_0) = 0$$
$$k_1 y_1'(x_0) + k_2 y_2'(x_0) = 0$$

in the unknowns k_1, k_2. This system is homogeneous. Its determinant is just the Wronskian $W[y_1(x_0), y_2(x_0)]$, which is zero by assumption. Hence the system has a solution k_1, k_2 where k_1 and k_2 are not both zero (see Theorem 4, Sec. 6.6). Using these numbers k_1, k_2, we introduce the function

$$y(x) = k_1 y_1(x) + k_2 y_2(x).$$

By Fundamental Theorem 1 in Sec. 2.1 this function $y(x)$ is a solution of (1) on *I*. From (6) we see that it satisfies the initial conditions $y(x_0) = 0$, $y'(x_0) = 0$. Now another solution of (1) satisfying the same initial conditions is $y^* \equiv 0$. Since p and q are continuous, Theorem 1 applies and guarantees uniqueness, that is, $y \equiv y^*$, written out,

$$k_1 y_1 + k_2 y_2 \equiv 0$$

on *I*. Now since k_1 and k_2 are not both zero, this means linear dependence of y_1, y_2 on *I*.

---

[10]Introduced by I. M. HÖNE (1778–1853), Polish mathematician, who changed his name to Wrónski. Second-order determinants should be familiar from elementary calculus; otherwise, consult the beginning of Sec. 6.6, which is independent of the other sections in Chap. 6.

(c) We prove the last statement of the theorem. If $W = 0$ at an x_0 in I, we have linear dependence of y_1, y_2 on I by part (b), hence $W \equiv 0$ by part (a) of this proof. Hence $W \neq 0$ at an x_1 in I cannot happen in the case of linear dependence, so that $W \neq 0$ at x_1 implies linear independence.   ◄

EXAMPLE 1   **Illustration of Theorem 2**

$y_1 = \cos \omega x$ and $y_2 = \sin \omega x$ are solutions of $y'' + \omega^2 y = 0$. Their Wronskian is

$$W(\cos \omega x, \sin \omega x) = \begin{vmatrix} \cos \omega x & \sin \omega x \\ -\omega \sin \omega x & \omega \cos \omega x \end{vmatrix} = \omega(\cos^2 \omega x + \sin^2 \omega x) = \omega.$$

Theorem 2 shows that they are linearly independent if and only if $\omega \neq 0$. Of course, we can see this directly because their quotient $y_2/y_1 = \tan \omega x$ is not constant if $\omega \neq 0$. For $\omega = 0$ we have $y_2 \equiv 0$, which implies linear dependence (why?).   ◄

EXAMPLE 2   **Illustration of Theorem 2 for a double root**

A general solution of $y'' - 2y' + y = 0$ on any interval is $y = (c_1 + c_2 x)e^x$. (Verify!) The corresponding Wronskian is not zero, which shows linear independence of e^x and xe^x on any interval.

$$W(e^x, xe^x) = \begin{vmatrix} e^x & xe^x \\ e^x & (x + 1)e^x \end{vmatrix} = (x + 1)e^{2x} - xe^{2x} = e^{2x} \neq 0.$$ ◄

A General Solution of (1) Includes All Solutions

THEOREM 3   **(Existence of a general solution)**

If $p(x)$ and $q(x)$ are continuous on an open interval I, then (1) has a general solution on I.

PROOF.   By Theorem 1, equation (1) has a solution $y_1(x)$ on I satisfying the initial conditions

$$y_1(x_0) = 1, \qquad y_1'(x_0) = 0$$

and a solution $y_2(x)$ on I satisfying the initial conditions

$$y_2(x_0) = 0, \qquad y_2'(x_0) = 1.$$

From this we see that the Wronskian $W(y_1, y_2)$ has at x_0 the value 1. Hence y_1, y_2 are linearly independent on I, by Theorem 2; they form a basis of solutions of (1) on I, and $y = c_1 y_1 + c_2 y_2$ with arbitrary c_1, c_2 is a general solution of (1) on I.   ◄

We now reach the final goal of this section by proving that a general solution of (1) is as general as it can be, namely, it includes *all* solutions of (1):

THEOREM 4   **(General solution)**

Suppose that (1) has continuous coefficients $p(x)$ and $q(x)$ on some open interval I. Then every solution $y = Y(x)$ of (1) on I is of the form

(7) $$Y(x) = C_1 y_1(x) + C_2 y_2(x)$$

where y_1, y_2 form a basis of solutions of (1) on I and C_1, C_2 are suitable constants.

*Hence (1) does not have **singular solutions** (i.e., solutions not obtainable from a general solution).*

PROOF.   By Theorem 3, our equation has a general solution

$$(8) \qquad\qquad y(x) = c_1 y_1(x) + c_2 y_2(x)$$

on I. We have to find suitable values of c_1, c_2 such that $y(x) = Y(x)$ on I. We choose any fixed x_0 in I and show first that we can find c_1, c_2 such that

$$y(x_0) = Y(x_0), \qquad y'(x_0) = Y'(x_0),$$

written out

$$(9) \qquad \begin{aligned} c_1 y_1(x_0) + c_2 y_2(x_0) &= Y(x_0), \\ c_1 y_1'(x_0) + c_2 y_2'(x_0) &= Y'(x_0). \end{aligned}$$

In fact, this is a linear system of equations in the unknowns c_1, c_2. Its determinant is the Wronskian of y_1 and y_2 at $x = x_0$. Since (8) is a general solution, y_1 and y_2 are linearly independent on I. From Theorem 2 it follows that their Wronskian is not zero. Hence the system has a unique solution $c_1 = C_1$, $c_2 = C_2$ (which can be obtained by elimination or by Cramer's rule in Sec. 6.6). By using these constants we obtain from (8) the particular solution

$$y^*(x) = C_1 y_1(x) + C_2 y_2(x).$$

Since C_1, C_2 are solutions of (9), from (9) we now see that

$$y^*(x_0) = Y(x_0), \qquad y^{*\prime}(x_0) = Y'(x_0).$$

From this and the uniqueness theorem (Theorem 1) we conclude that y^* and Y must be equal everywhere on I, and the proof is complete.   ◀

PROBLEM SET 2.7

Bases of Solutions. Wronskians. Find the Wronskian of the given bases and verify Theorem 2. (Show the details of your work.)

1. $e^{\lambda_1 x}$, $e^{\lambda_2 x}$
2. 1, e^x
3. $e^{-ax/2} \cos 3x$, $e^{-ax/2} \sin 3x$
4. x^{m_1}, x^{m_2}
5. x^4, $x^4 \ln x$
6. $e^{\lambda x}$, $xe^{\lambda x}$
7. $x^\mu \cos (2 \ln x)$, $x^\mu \sin (2 \ln x)$
8. $e^{-x} \cos \omega x$, $e^{-x} \sin \omega x$

Equations for Given Bases. Wronskians. Find a second-order homogeneous linear differential equation for which the given functions are solutions. Find the Wronskian and use it to verify linear independence by Theorem 2. (Show the details of your work.)

9. e^{3x}, xe^{3x}
10. x^5, x^{-5}
11. x^2, $x^2 \ln x$
12. $\cosh 2x$, $\sinh 2x$
13. x^2, $x^{1/2}$
14. 1, e^{-2x}
15. $\cos 2\pi x$, $\sin 2\pi x$
16. $\cos (\ln x)$, $\sin (\ln x)$
17. $x^{3/2}$, $x^{-3/2}$

18. **TEAM PROJECT. Some Consequences of the Present Theory.** This concerns some general properties of solutions that are worth noting. Assume that the differential equation (1) has continuous coefficients on some open interval I.

(a) Prove that the solutions of a basis cannot be zero at the same point.

(b) Prove that the solutions of a basis cannot have a maximum or minimum at the same point.

(c) Let y_1, y_2 be a basis. Prove that $z_1 = a_{11}y_1 + a_{12}y_2$, $z_2 = a_{21}y_1 + a_{22}y_2$ is a basis if and only if the determinant of the coefficients a_{jk} is not zero.

(d) Illustrate (c) with $y_1 = e^x$, $y_2 = e^{-x}$, $z_1 = \cosh x$, $z_2 = \sinh x$.

(e) How are the arbitrary constants in general solutions in (d) related?

2.8 Nonhomogeneous Equations

In this section we turn from homogeneous *to* **nonhomogeneous linear equations**

(1)
$$y'' + p(x)y' + q(x)y = r(x)$$

where $r(x) \not\equiv 0$. How can we solve such an equation? Before we consider methods, let us first explore what we need for proceeding from the corresponding homogeneous equation

(2)
$$y'' + p(x)y' + q(x)y = 0$$

to the nonhomogeneous equation (1). The key that relates (1) to (2) and gives us a plan for solving (1) is the following theorem, which will also motivate the next definition.

THEOREM 1 **[Relations between solutions of (1) and (2)]**

(a) *The difference of two solutions of* (1) *on some open interval I is a solution of* (2) *on I.*

(b) *The sum of a solution of* (1) *on I and a solution of* (2) *on I is a solution of* (1) *on I.*

PROOF. (a) Denote the left side of (1) by $L[y]$. Let y and $\tilde{y}$ be any solutions of (1) on I. Then $L[y] = r(x)$, $L[\tilde{y}] = r(x)$, and since we have $(y - \tilde{y})' = y' - \tilde{y}'$, etc., we obtain

$$L[y - \tilde{y}] = L[y] - L[\tilde{y}] = r(x) - r(x) \equiv 0.$$

(b) Similarly, for y as before and any solution y^* of (2) on I,

$$L[y + y^*] = L[y] + L[y^*] = r(x) + 0 = r(x). \qquad \blacktriangleleft$$

Definition (General solution, particular solution)

A **general solution** of the nonhomogeneous equation (1) on some open interval I is a solution of the form

(3)
$$y(x) = y_h(x) + y_p(x),$$

where $y_h(x) = c_1 y_1(x) + c_2 y_2(x)$ is a general solution of the homogeneous equation (2) on I and $y_p(x)$ is any solution of (1) on I containing no arbitrary constants.

A **particular solution** of (1) on I is a solution obtained from (3) by assigning specific values to the arbitrary constants c_1 and c_2 in $y_h(x)$. $\qquad \blacktriangleleft$

A General Solution of (1) Includes All Solutions

If the coefficients of (1) and $r(x)$ are continuous functions on I, then (1) has a general solution on I because $y_h(x)$ exists on I by Theorem 3, Sec. 2.7, and the existence of $y_p(x)$ will be shown in Sec. 2.10. Also, an initial value problem for (1) has a unique solution on I. This follows from Theorem 1, Sec. 2.7, once the existence of $y_p(x)$ has been established. Indeed, if initial conditions

$$y(x_0) = K_0, \qquad y'(x_0) = K_1$$

are given and a y_p has been determined, by that theorem there exists a unique solution $\tilde{y}$ of the homogeneous equation (2) on I satisfying

$$\tilde{y}(x_0) = K_0 - y_p(x_0), \qquad \tilde{y}'(x_0) = K_1 - y'_p(x_0),$$

and $y = \tilde{y} + y_p$ is the unique solution of (1) on I satisfying the given initial conditions.

Furthermore, justifying the terminology, we now prove that a general solution of (1) includes *all* solutions of (1); this is the same as for the homogeneous equation:

THEOREM 2   **(General solution)**

Suppose that the coefficients and $r(x)$ in (1) are continuous on some open interval I. Then every solution of (1) on I is obtained by assigning suitable values to the arbitrary constants in a general solution (3) of (1) on I.

PROOF.   Let $\tilde{y}(x)$ be any solution of (1) on I. Let (3) be any general solution of (1) on I; this solution exists because of our continuity assumption. Theorem 1(a) implies that the difference $Y(x) = \tilde{y}(x) - y_p(x)$ is a solution of the homogeneous equation (2). By Theorem 4 in Sec. 2.7, this solution $Y(x)$ is obtained from $y_h(x)$ by assigning suitable values to the arbitrary constants c_1, c_2. From this and $\tilde{y}(x) = Y(x) + y_p(x)$, the statement follows.   ◀

Practical Conclusion

To solve the nonhomogeneous equation (1) or an initial value problem for (1), we have to solve the homogeneous equation (2) and find any particular solution y_p of (1).

Methods for finding y_p and many applications follow in Secs. 2.9–2.12. At present we merely illustrate the basic technique and our notation with a simple example.

EXAMPLE 1   **Initial value problem for a nonhomogeneous equation**

Solve the initial value problem

$$y'' + 2y' + 101y = 10.4e^x, \qquad y(0) = 1.1, \qquad y'(0) = -0.9.$$

Solution. 1st Step. General solution of the homogeneous equation.   The characteristic equation of the homogeneous equation is $\lambda^2 + 2\lambda + 101 = 0$. It has the roots $-1 \pm \sqrt{1 - 101} = -1 \pm 10i$. Hence a real general solution of the homogeneous equation is (see Sec. 2.3)

$$y_h = e^{-x}(A \cos 10x + B \sin 10x).$$

2nd Step. General solution of the nonhomogeneous equation.   For this we need any particular solution y_p of the nonhomogeneous equation. Since e^x on the right has the derivatives e^x, e^x, we try

$$y_p = Ce^x.$$

Substitution gives $(1 + 2 + 101) Ce^x = 10.4e^x$. By comparison, $C = 0.1$. Hence a general solution of the nonhomogeneous equation is

$$y = y_h + y_p = e^{-x}(A \cos 10x + B \sin 10x) + 0.1e^x.$$

3rd Step. Particular solution satisfying the initial conditions. From the general solution y and the first initial condition we obtain $y(0) = A + 0.1 = 1.1$, hence $A = 1$. The derivative of y with $A = 1$ is

$$y' = e^{-x}(-\cos 10x - B \sin 10x - 10 \sin 10x + 10B \cos 10x) + 0.1e^x.$$

Thus $y'(0) = -1 + 10B + 0.1 = -0.9$ from the second initial condition. Hence $B = 0$. The solution of our initial value problem is (Fig. 53)

$$y = e^{-x} \cos 10x + 0.1e^x.$$

The first term, resulting from the homogeneous equation, is an oscillation with maximum amplitude decreasing to zero, so that y approaches the exponential curve $0.1e^x$ (dashed). The other dashed curves are $\pm e^{-x} + 0.1e^x$, between which the solution oscillates. ◀

Solution methods follow in the next sections.

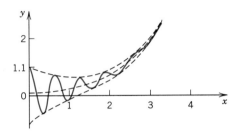

Fig. 53. Solution in Example 1

PROBLEM SET 2.8

Particular and General Solutions. Verify that y_p is a solution of the given differential equation and find a general solution. (Show the details of your work.)

1. $y'' - y = 8e^{-3x}$, $y_p = e^{-3x}$
2. $y'' - y = 8e^{-3x}$, $y_p = e^{-3x} - 3e^x$
3. $y'' + 3y' + 2y = 4x^2$, $y_p = 2x^2 - 6x + 7$
4. $y'' - 2y' + 5y = 5x^3 - 6x^2 + 6x$, $y_p = x^3$
5. $(D^2 + 3D - 4)y = 8 \cos 2x + 6 \sin 2x$, $y_p = -\cos 2x$
6. $(D^2 - 4D + 4)y = e^x \cos x$, $y_p = -\frac{1}{2}e^x \sin x$
7. $(D^2 + 1)y = -x^{-2} + \ln \pi x$, $y_p = \ln \pi x$
8. $(8D^2 - 6D + 1)y = 6 \cosh x$, $y_p = \frac{1}{5}e^{-x} + e^x$

Initial Value Problems. In Probs. 9–15 verify that y_p is a solution of the given equation. Solve the initial value problem. (Show the details of your work.)

9. $y'' + y = 2x$, $y(0) = -1$, $y'(0) = 8$; $y_p = 2x$
10. $y'' - y = 2 \cos x$, $y(0) = 0$, $y'(0) = -0.2$; $y_p = -\cos x$
11. $y'' - y = 2e^x$, $y(0) = -1$, $y'(0) = 0$; $y_p = xe^x$
12. $(D^2 + 4)y = -12 \sin 2x$, $y(0) = 1.8$, $y'(0) = 5.0$; $y_p = 3x \cos 2x$
13. $(x^2 D^2 - 3xD + 3)y = 3 \ln x - 4$, $y(1) = 0$, $y'(1) = 1$; $y_p = \ln x$

14. $(x^2D^2 - 2xD + 2)y = (3x^2 - 6x + 6)e^x$, $y(1) = 2 + 3e$, $y'(1) = 3e$; $y_p = 3e^x$

15. $(D^2 + 4D + 4)y = e^{-2x}/x^2$, $y(1) = 1/e^2$, $y'(1) = -2/e^2$; $y_p = -e^{-2x}\ln x$

16. TEAM PROJECT. Structure of General Solutions of the Nonhomogeneous Equation. Before we discuss solution methods for nonhomogeneous differential equations, it is important that we have a clear understanding of formula (3) for general solutions of the nonhomogeneous equation (1).

(a) What are the steps for solving an initial value problem for (1)?

(b) What can we say about the relation between y_p in (3) and the solution of an initial value problem for (1), call it y*?

(c) If we have two general solutions $y = y_h + y_p$ and $\tilde{y} = y_h + \tilde{y}_p$ of the same equation (1), what can we say about the relation between y_p and $\tilde{y}_p$?

(d) In (c), could we have $y = y_h + y_p$ and $\tilde{y} = \tilde{y}_h + y_p$, that is, different general solutions of the homogeneous equation even if y_p is the same? (Give a reason.)

(e) What happens in Prob. 14 if we replace the initial conditions by $y(0) = 3$, $y'(0) = 7$? Is the solution still unique? Explain.

2.9 Solution by Undetermined Coefficients

A general solution of a nonhomogeneous linear equation is a sum of the form

$$y = y_h + y_p$$

where y_h is a general solution of the corresponding homogeneous equation and y_p is any particular solution of the nonhomogeneous equation. This has just been shown. Hence our main task is to discuss methods for finding such y_p. There is a general method for this that always works and that we shall consider in the next section. There also is a much simpler special method of practical interest, which we discuss now. This method is called the **method of undetermined coefficients** and applies to equations

(1)
$$\boxed{y'' + ay' + by = r(x)}$$

with constant coefficients and special right sides $r(x)$, namely, exponential functions, polynomials, cosines, sines, or sums or products of such functions. These $r(x)$ have derivatives of a form similar to $r(x)$ itself. This gives the key idea: Choose for y_p a form similar to that of $r(x)$ and involving unknown coefficients to be determined by substituting that choice for y_p into (1). Example 1 in the last section illustrates this for an exponential function; the undetermined coefficient was C. The rules of the method are as follows.

Rules for the Method of Undetermined Coefficients

(A) Basic Rule. *If $r(x)$ in (1) is one of the functions in the first column in Table 2.1, choose the corresponding function y_p in the second column and determine its undetermined coefficients by substituting y_p and its derivatives into (1).*

(B) Modification Rule. *If a term in your choice for y_p happens to be a solution of the homogeneous equation corresponding to (1), then multiply your choice of y_p by x (or by x^2 if this solution corresponds to a double root of the characteristic equation of the homogeneous equation).*

(C) Sum Rule. *If $r(x)$ is a sum of functions in several lines of Table 2.1, first column, then choose for y_p the sum of the functions in the corresponding lines of the second column.* ◀

The Basic Rule tells us what to do in general. The Modification Rule takes care of difficulties that occur in the case indicated. Accordingly, we always have to solve the homogeneous equation first. The Sum Rule is obtained if we note that the sum of two solutions of (1) with $r = r_1$ and $r = r_2$, respectively, is a solution of (1) with $r = r_1 + r_2$. (Verify!)

The method corrects itself in the sense that a false choice of y_p or one with too few terms will lead to a contradiction, usually indicating the necessary correction, and a choice of too many terms will give a correct result, with superfluous coefficients coming out zero.

Table 2.1
Method of Undetermined Coefficients

Term in $r(x)$	Choice for y_p
$ke^{\gamma x}$	$Ce^{\gamma x}$
kx^n $(n = 0, 1, \cdots)$	$K_n x^n + K_{n-1} x^{n-1} + \cdots + K_1 x + K_0$
$k \cos \omega x$ $k \sin \omega x$	$K \cos \omega x + M \sin \omega x$
$ke^{\alpha x} \cos \omega x$ $ke^{\alpha x} \sin \omega x$	$e^{\alpha x}(K \cos \omega x + M \sin \omega x)$

Examples Illustrating Rules (A)−(C)

EXAMPLE 1 **Application of Rule (A)**

Solve the nonhomogeneous equation

$$(2) \qquad\qquad y'' + 4y = 8x^2.$$

Solution. Table 2.1 suggests the choice

$$y_p = K_2 x^2 + K_1 x + K_0. \qquad \text{Then} \qquad y_p'' = 2K_2.$$

Substitution gives

$$2K_2 + 4(K_2 x^2 + K_1 x + K_0) = 8x^2.$$

Equating the coefficients of x^2, x, and x^0 on both sides, we have $4K_2 = 8$, $4K_1 = 0$, $2K_2 + 4K_0 = 0$. Thus $K_2 = 2$, $K_1 = 0$, $K_0 = -1$. Hence $y_p = 2x^2 - 1$, and a general solution of (2) is

$$y = y_h + y_p = A \cos 2x + B \sin 2x + 2x^2 - 1.$$

Note well that although $r(x) = 8x^2$, a trial $y_p = K_2 x^2$ would fail. Try it. Can you see why it fails? ◀

EXAMPLE 2 **Modification Rule (B) in the case of a simple root**

Solve

$$(3) \qquad\qquad y'' - 3y' + 2y = e^x.$$

Solution. The characteristic equation $\lambda^2 - 3\lambda + 2 = 0$ has the roots 1 and 2. Hence $y_h = c_1 e^x + c_2 e^{2x}$. Ordinarily, our choice would be $y_p = Ce^x$. But we see that e^x is a solution of the homogeneous equation corresponding to a simple root (namely, 1). Hence Rule (B) suggests

$$y_p = Cxe^x. \qquad \text{We need} \qquad y_p' = C(e^x + xe^x), \qquad y_p'' = C(2e^x + xe^x).$$

Substitution gives

$$C(2 + x)e^x - 3C(1 + x)e^x + 2Cxe^x = e^x.$$

The xe^x-terms cancel out, and $-Ce^x = e^x$ remains. Hence $C = -1$. A general solution is

$$y = c_1 e^x + c_2 e^{2x} - xe^x.$$

Check it! Try $y_p = Ce^x$ to convince yourself that it does not work. ◀

EXAMPLE 3   **Modification Rule (B) in the case of a double root**

Solve the initial value problem

(4) $$y'' + 2y' + y = (D + 1)^2 y = e^{-x}, \qquad y(0) = -1, \qquad y'(0) = 1.$$

Solution.   The characteristic equation is $(\lambda + 1)^2 = 0$. It has the double root $\lambda = -1$. The corresponding general solution of the homogeneous equation is $y_h = (c_1 + c_2 x)e^{-x}$.

We need y_p. Ordinarily we would choose Ce^{-x}. But since e^{-x} is a solution of the homogeneous equation corresponding to a double root (-1), the Modification Rule (B) calls for the choice $y_p = Cx^2 e^{-x}$. Then $y_p' = C(2x - x^2)e^{-x}$, $y_p'' = (2 - 4x + x^2)e^{-x}$. Substituting all this into (4), we see that the x- and x^2-terms drop out and we obtain

$$C(2 - 4x + x^2)e^{-x} + 2C(2x - x^2)e^{-x} + Cx^2 e^{-x} = 2Ce^{-x} = e^{-x}.$$

Hence $C = 1/2$. This gives the general solution of (4),

$$y = (c_1 + c_2 x)e^{-x} + \tfrac{1}{2}x^2 e^{-x}.$$

Thus $y(0) = c_1 = -1$ from the first initial condition. By differentiation and from the second initial condition,

$$y' = (c_2 - c_1 - c_2 x)e^{-x} + (x - \tfrac{1}{2}x^2)e^{-x}, \qquad y'(0) = c_2 - c_1 = 1, \qquad c_2 = 0.$$

This gives the answer

$$y = (\tfrac{1}{2}x^2 - 1)e^{-x}.$$

Figure 54 shows that this curve begins at $y = -1$ with slope 1 (note that the units on the axes are different!), crosses the x-axis at $\sqrt{2}$, has a maximum at $1 + \sqrt{3} = 2.73$ (verify!), and then goes monotone to zero. ◀

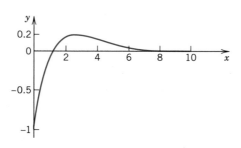

Fig. 54.   Solution in Example 3

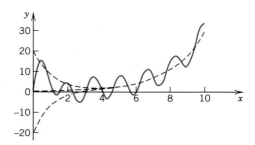

Fig. 55.   Solution in Example 4

EXAMPLE 4   **Sum Rule (C). Initial value problem**

Solve the initial value problem

(5) $$y'' + 2y' + 5y = 1.25e^{0.5x} + 40 \cos 4x - 55 \sin 4x, \qquad y(0) = 0.2, \qquad y'(0) = 60.1.$$

1st Step. General solution of the homogeneous equation. The characteristic equation is $\lambda^2 + 2\lambda + 5 = 0$. It has the roots $-1 + 2i$ and $-1 - 2i$. Hence a general solution of the homogeneous equation is (see Sec. 2.3)

$$y_h = e^{-x}(A \cos 2x + B \sin 2x).$$

2nd Step. Particular solution y_p. From Table 2.1 and the Sum Rule we see that we should choose

$$y_p = Ce^{0.5x} + K \cos 4x + M \sin 4x.$$

Then
$$y_p' = 0.5\,Ce^{0.5x} - 4K \sin 4x + 4M \cos 4x,$$

$$y_p'' = 0.25\,Ce^{0.5x} - 16K \cos 4x - 16M \sin 4x.$$

We substitute this into (5). Then the left side becomes

$$(0.25 + 1 + 5)Ce^{0.5x} + (-16K + 8M + 5K) \cos 4x + (-16M - 8K + 5M) \sin 4x.$$

This must equal the right side of (5). By equating the exponential terms on both sides, then the cosine terms, and finally the sine terms, we obtain

$$6.25C = 1.25, \qquad C = 0.2$$

$$-11K + 8M = 40, \qquad -8K - 11M = -55, \qquad K = 0, \qquad M = 5.$$

This gives the general solution of (5),

(6) $$y = e^{-x}(A \cos 2x + B \sin 2x) + 0.2e^{0.5x} + 5 \sin 4x.$$

3rd Step. Particular solution satisfying the initial conditions. From (6) and the first initial condition we have $y(0) = A + 0.2 = 0.2$, $A = 0$. Differentiating (6) with $A = 0$, we obtain

$$y' = e^{-x}(-B \sin 2x + 2B \cos 2x) + 0.1e^{0.5x} + 20 \cos 4x.$$

Hence by the second initial condition, $y'(0) = 2B + 0.1 + 20 = 60.1$, $B = 20$. We thus obtain the answer (Fig. 55)

$$y = 20e^{-x} \sin 2x + 0.2e^{0.5x} + 5 \sin 4x.$$

The first term approaches zero relatively fast. When $x = 4$, it is practically zero, as the dashed curves $\pm 20e^{-x} + 0.2e^{0.5x}$ show. From then on the last term, $5 \sin 4x$, causes an oscillation about $0.2e^{0.5x}$ (the monotone increasing dashed curve).   ◀

Basic applications follow in Secs. 2.11 and 2.12.

PROBLEM SET 2.9

General Solutions of Nonhomogeneous Equations

Find a (real) general solution. Which rule are you using? (Show each step of your calculation.)

1. $y'' + 4y = \sin 3x$

2. $y'' - y = 2e^x + 6e^{2x}$

3. $y'' + 3y' = 28 \cosh 4x$

4. $y'' - y' - 2y = 3e^{2x}$

5. $y'' + 2y' + 10y = 25x^2 + 3$

6. $3y'' + 10y' + 3y = 9x + 5 \cos x$

7. $y'' + y' - 6y = -6x^3 + 3x^2 + 6x$

8. $y'' + 6y' + 9y = 50e^{-x} \cos x$

9. $y'' + 2y' - 35y = 12e^{5x} + 37 \sin 5x$

10. $y'' - y' - \frac{3}{4}y = 21 \sinh 2x$

11. $y'' + 10y' + 25y = e^{-5x}$

12. $y'' + 3y' - 18y = 9 \sinh 3x$

13. $y'' + 8y' + 16y = 64 \cosh 4x$

14. $y'' - 4y' + 20y = 377 \sin x$

Initial Value Problems for Nonhomogeneous Equations

Solve the given initial value problem. Indicate the rules you are using. Show each step of your calculation.

15. $y'' + 1.5y' - y = 12x^2 + 6x^3 - x^4$, $y(0) = 4$, $y'(0) = -8$

16. $y'' - 6y' + 13y = 4e^{3x}$, $y(0) = 2$, $y'(0) = 4$

17. $y'' - 4y = e^{-2x} - 2x$, $y(0) = 0$, $y'(0) = 0$

18. $y'' + 9y = 6 \cos 3x$, $y(0) = 1$, $y'(0) = 0$

19. $y'' + 1.2y' + 0.36y = 4e^{-0.6x}$, $y(0) = 0$, $y'(0) = 1$

20. $y'' - 2.8y' + 1.96y = 2e^{1.4x}$, $y(0) = 0$, $y'(0) = 0$

21. $y'' + y' = 2 + 2x + x^2$, $y(0) = 8$, $y'(0) = -1$

22. $y'' + y' + 9.25y = 9.25(4 + e^{-x})$, $y(0) = 7$, $y'(0) = -2$

 23. CAS PROJECT. Structure of Solutions of Initial Value Problems. Using the present method, find, graph, and discuss the solutions y of initial value problems of your own choice. Explore how changes of the initial conditions affect the solution. Plot y as well as y_p and $y - y_p$ (resulting from y_h) separately, to see the separate effects. Find a problem in which (a) y_h decreases to zero, (b) y_h increases, and (c) y_h is not present in the answer. Study a problem with $y(0) = 0$, $y'(0) = 0$. Consider a problem in which the Modification Rule applies. Make sure that your problems cover all three Cases I, II, III.

24. TEAM PROJECT. Extension of the Method of Undetermined Coefficients. Table 2.1 includes exponential functions, powers of x, and cosines and sines. Extend the method to products of such functions. Comment on the practical significance of such extensions.

25. WRITING PROJECT. Initial Value Problem. Using your own words and formulations, write out all the calculations in Example 4 of the text in greater detail. Discuss Fig. 55 in more detail. Why does the first "half-wave" extend beyond the upper dashed curve? Why does the second not touch the lower dashed curve?

2.10 Solution by Variation of Parameters

The method in the last section is simple and has important engineering applications as we shall see in the next sections. But it applies only to constant-coefficient equations with special right sides $r(x)$. In this section we discuss the so-called **method of variation of parameters,**[11] which is completely general (but more complicated). That is, it applies to differential equations

(1) $$y'' + p(x)y' + q(x)y = r(x)$$

with arbitrary variable function p, q, and r that are continuous on some interval I. The method gives a particular solution y_p of (1) on I in the form

(2) $$y_p(x) = -y_1 \int \frac{y_2 r}{W} \, dx + y_2 \int \frac{y_1 r}{W} \, dx$$

where y_1, y_2 form a basis of solutions of the homogeneous equation

(3) $$y'' + p(x)y' + q(x)y = 0$$

---

[11]Credited to Lagrange (see footnote 4 in Sec. 2.1). The name of the method is explained below.

corresponding to (1) and

(4) $$W = y_1 y_2' - y_2 y_1'$$

is the Wronskian of y_1, y_2 (see Sec. 2.7).

CAUTION! Before applying (2), make sure that your equation is written in the standard form (1), with y'' as the first term; divide by $f(x)$ if it starts with $f(x)y''$. (The simple reason is that (2) is obtained under this assumption.)

The integrations in (2) may often cause difficulties. If you have a choice, apply the previous method. It is simpler. Let us first work an example to which it does not apply, so that we see how to use (2).

EXAMPLE 1   **Method of variation of parameters**

Solve the differential equation

$$y'' + y = \sec x.$$

Solution.   A basis of solutions of the homogeneous equation on any interval is

$$y_1 = \cos x, \qquad y_2 = \sin x.$$

This gives the Wronskian

$$W(y_1, y_2) = \cos x \cos x - \sin x\,(-\sin x) = 1.$$

Hence from (2), choosing the constants of integration to be zero, we get the particular solution

$$y_p = -\cos x \int \sin x \sec x\, dx + \sin x \int \cos x \sec x\, dx$$

$$= \cos x \ln |\cos x| + x \sin x$$

of the given equation. From this and the general solution $y_h = c_1 y_1 + c_2 y_2$ of the homogeneous equation we obtain the answer

$$y = y_h + y_p = [c_1 + \ln|\cos x|]\cos x + (c_2 + x)\sin x.$$

Had we included two arbitrary constants of integration $-c_1$, c_2, we would have obtained in (2) the additional $c_1 \cos x + c_2 \sin x = c_1 y_1 + c_2 y_2$, that is, a general solution of the given equation directly from (2). This will always be the case.   ◀

Idea of the Method. Derivation of (2)

What was Lagrange's idea? Where does the name of the method come from? How can we get (2)? Where do we use the continuity assumption?

The continuity of p and q implies that the homogeneous equation (3) has a general solution

$$y_h(x) = c_1 y_1(x) + c_2 y_2(x)$$

on I, by Theorem 3 in Sec. 2.7. The method of variation of parameters involves replacing the constants c_1 and c_2 (here regarded as "parameters" in y_h) by functions $u(x)$ and $v(x)$ to be determined so that we obtain a particular solution $y_p(x)$ of (1) on I given by

(5) $$y_p(x) = u(x)y_1(x) + v(x)y_2(x)$$

By differentiating (5) we obtain

$$y_p' = u'y_1 + uy_1' + v'y_2 + vy_2'.$$

Now (5) contains *two* functions u and v, but the requirement that y_p satisfy (1) imposes only *one* condition on u and v. Hence it seems plausible that we may impose a second arbitrary condition. Indeed, our further calculation will show that we can determine u and v such that y_p satisfies (1) and u and v satisfy as a second condition the relationship

$$(6) \qquad u'y_1 + v'y_2 = 0.$$

This reduces the expression for y_p' to the form

$$(7) \qquad y_p' = uy_1' + vy_2'.$$

By differentiating this function we have

$$(8) \qquad y_p'' = u'y_1' + uy_1'' + v'y_2' + vy_2''.$$

Substituting (5), (7), and (8) into (1) and collecting terms containing u and terms containing v, we readily obtain

$$u(y_1'' + py_1' + qy_1) + v(y_2'' + py_2' + qy_2) + u'y_1' + v'y_2' = r.$$

Since y_1 and y_2 are solutions of the homogeneous equation (3), this reduces to

$$u'y_1' + v'y_2' = r.$$

Equation (6) is

$$u'y_1 + v'y_2 = 0.$$

This is a linear system of two algebraic equations for the unknown functions u' and v'. The solution is obtained by Cramer's rule (Sec. 6.6) or as follows. Multiply the first equation by $-y_2$ and the second by y_2' and add to get

$$u'(y_1y_2' - y_2y_1') = -y_2r, \qquad \text{thus} \qquad u'W = -y_2r,$$

where W is the Wronskian (4) of y_1, y_2. Now multiply the first equation by y_1 and the second by $-y_1'$ and add to get

$$v'(y_1y_2' - y_2y_1') = y_1r, \qquad \text{thus} \qquad v'W = y_1r.$$

Since y_1, y_2 form a basis, we have $W \neq 0$ (by Theorem 2 in Sec. 2.7) and can divide by W, obtaining

$$(9) \qquad u' = -\frac{y_2r}{W}, \qquad v' = \frac{y_1r}{W}.$$

By integration,

$$u = -\int \frac{y_2 r}{W} \, dx, \qquad v = \int \frac{y_1 r}{W} \, dx.$$

These integrals exist because $r(x)$ is continuous. Substituting them into (5), we obtain (2). This completes the derivation. ◀

PROBLEM SET 2.10

General Solution of Nonhomogeneous Equations. Find a general solution, showing all details of your calculations.

1. $y'' - 4y' + 4y = e^{2x}/x$

2. $y'' + 9y = \sec 3x$

3. $y'' + 2y' + y = e^{-x} \cos x$

4. $y'' + 9y = \csc 3x$

5. $y'' - 2y' + y = e^x/x^3$

6. $y'' - 4y' + 5y = e^{2x} \csc x$

7. $(D^2 - 2D + 1)y = 3x^{3/2}e^x$

8. $(D^2 + 6D + 9)y = 16e^{-3x}/(x^2 + 1)$

9. $(D^2 + 4D + 4)y = 2e^{-2x}/x^2$

10. $(D^2 + 2D + 2)y = 4e^{-x} \sec^3 x$

Nonhomogeneous Euler–Cauchy Equations. Find a general solution, showing all details of your calculations. *Caution!* First divide the equation by the coefficient of y'' to get the standard form (1).

11. $x^2 y'' - 4xy' + 6y = 21x^{-4}$

12. $xy'' - y' = (3 + x)x^2 e^x$

13. $4x^2 y'' + 8xy' - 3y = 7x^2 - 15x^3$

14. $(x^2 D^2 - 4xD + 6)y = 7x^4 \sin x$

15. $(x^2 D^2 - 2xD + 2)y = x^3 \cos x$

16. $(x^2 D^2 + xD - 1)y = 1/x^2$

17. $(x^2 D^2 + xD - 9)y = 48x^5$

18. TEAM PROJECT. Comparison of Methods in Secs. 2.9 and 2.10. The method in Sec. 2.9 should be used whenever it applies because it is much simpler than the present method. Compare the two methods as follows.

(a) Solve $y'' + 4y' + 3y = 65 \cos 2x$ by both methods and compare the amount of work.

(b) Sometimes you may use one method for one part of the right side and the other for the other part. Do this in the best possible way for $y'' - 2y' + y = 35x^{3/2}e^x + x^2$.

(c) Can you invent an undetermined-coefficient method for certain nonhomogeneous Euler–Cauchy equations? (Look at the answers in this section and experiment.)

2.11 Modeling: Forced Oscillations. Resonance

Free motions of the mass–spring system in Fig. 56 on the next page are motions in the absence of external forces, and they are governed by the homogeneous differential equation

(1) $$my'' + cy' + ky = 0 \qquad \text{(Sec. 2.5)}.$$

Here, y (as a function of time t) is the displacement of the body from rest, m the mass of the body, my'' the force of inertia, cy' the damping force, and ky the spring force.

Forced motions are obtained if we let an external force $r(t)$ act on the body. To get the model, we simply have to add our new force $r(t)$ to those forces; this gives the

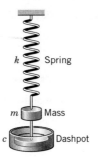

Fig. 56. Mass on a spring

nonhomogeneous differential equation

$$my'' + cy' + ky = r(t).$$

$r(t)$ is called the **input** or **driving force.** A corresponding solution is called an **output** or a **response** *of the system to the driving force.* (See also Sec. 1.6.)

Of particular interest are *periodic inputs,* and we shall consider a sinusoidal force, say,

$$r(t) = F_0 \cos \omega t \qquad\qquad (F_0 > 0,\ \omega > 0).$$

Then we have the equation

(2)
$$my'' + cy' + ky = F_0 \cos \omega t.$$

Its solution will familiarize us with further interesting facts fundamental in engineering mathematics, in particular with resonance.

Solving the Equation

A general solution of (2) is the sum of a general solution y_h of (1), which we know from Sec. 2.5, and a particular solution y_p of (2). We can best determine y_p by the method of undetermined coefficients (Sec. 2.9). Accordingly, we start from

(3)
$$y_p(t) = a \cos \omega t + b \sin \omega t.$$

By differentiating this function we have

$$y_p' = -\omega a \sin \omega t + \omega b \cos \omega t,$$
$$y_p'' = -\omega^2 a \cos \omega t - \omega^2 b \sin \omega t.$$

We substitute these expressions into (2) and collect the cosine and sine terms:

$$[(k - m\omega^2)a + \omega cb] \cos \omega t + [-\omega ca + (k - m\omega^2)b] \sin \omega t = F_0 \cos \omega t.$$

Equating the coefficients of the cosine and sine terms on both sides we have

$$(4) \qquad \begin{aligned} (k - m\omega^2)a &+ & \omega c b &= F_0 \\ -\omega c a &+ & (k - m\omega^2)b &= 0. \end{aligned}$$

This is a linear system of two algebraic equations in the two unknowns a and b. The solution is obtained in the usual way by elimination or by Cramer's rule (if necessary, see Sec. 6.6). We find

$$a = F_0 \frac{k - m\omega^2}{(k - m\omega^2)^2 + \omega^2 c^2}, \qquad b = F_0 \frac{\omega c}{(k - m\omega^2)^2 + \omega^2 c^2}$$

provided the denominator is not zero. If we set $\sqrt{k/m} = \omega_0 \ (> 0)$ as in Sec. 2.5, this becomes

$$(5) \qquad a = F_0 \frac{m(\omega_0^2 - \omega^2)}{m^2(\omega_0^2 - \omega^2)^2 + \omega^2 c^2}, \qquad b = F_0 \frac{\omega c}{m^2(\omega_0^2 - \omega^2)^2 + \omega^2 c^2}.$$

We thus obtain the general solution of (2) in the form

$$(6) \qquad y(t) = y_h(t) + y_p(t),$$

where y_h is a general solution of (1) and y_p is given by (3) with coefficients (5).

　　We shall now discuss the behavior of the mechanical system, distinguishing between the two cases $c = 0$ (no damping) and $c > 0$ (damping). These cases will correspond to two different types of output.

Case 1. Undamped Forced Oscillations. Resonance

If there is no damping, then $c = 0$. We first assume that $\omega^2 \neq \omega_0^2$ (where $\omega_0^2 = k/m$, as in Sec. 2.5). This is essential. We then obtain from (3) and (5) (where $b = 0$ because $c = 0$)

$$(7) \qquad y_p(t) = \frac{F_0}{m(\omega_0^2 - \omega^2)} \cos \omega t = \frac{F_0}{k[1 - (\omega/\omega_0)^2]} \cos \omega t.$$

From this and (4*) in Sec. 2.5 we have the general solution

$$(8) \qquad y(t) = C \cos (\omega_0 t - \delta) + \frac{F_0}{m(\omega_0^2 - \omega^2)} \cos \omega t.$$

*This output represents a superposition of two harmonic oscillations. Their frequencies are the "**natural frequency**" $\omega_0/2\pi$ [cycles/sec] of the system (that is, the frequency of the free undamped motion) and the frequency $\omega/2\pi$ of the input.*

　　From (7) we see that the maximum amplitude of y_p is

$$(9) \qquad a_0 = \frac{F_0}{k} \rho \qquad \text{where} \qquad \rho = \frac{1}{1 - (\omega/\omega_0)^2}.$$

a_0 depends on ω and ω_0. If $\omega \to \omega_0$, then ρ and a_0 tend to infinity. This phenomenon of excitation of large oscillations by matching input and natural frequencies ($\omega = \omega_0$) is known as **resonance,** and is of basic importance in the study of vibrating systems (see below). The quantity ρ is called the *resonance factor* (Fig. 57). From (9) we see that $\rho/k = a_0/F_0$ is the ratio of the amplitudes of the function y_p and of the input.

In the case of resonance, equation (2) becomes

(10)
$$y'' + \omega_0^2 y = \frac{F_0}{m} \cos \omega_0 t.$$

From the Modification Rule in Sec. 2.9 we conclude that a particular solution of (10) is of the form

$$y_p(t) = t(a \cos \omega_0 t + b \sin \omega_0 t).$$

By substituting this into (10) we find $a = 0$, $b = F_0/2m\omega_0$, and (Fig. 58)

(11)
$$y_p(t) = \frac{F_0}{2m\omega_0} t \sin \omega_0 t.$$

We see that y_p becomes larger and larger. In practice, this means that systems with very little damping may undergo large vibrations that can destroy the system. We shall return to this practical aspect of resonance later in this section.

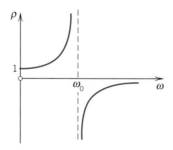

Fig. 57.   Resonance factor $\rho(\omega)$

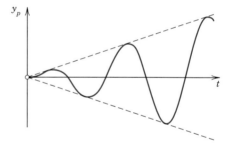

Fig. 58.   Particular solution in the case of resonance

Another interesting and highly important type of oscillation is obtained when ω is close to ω_0. Take, for example, the particular solution [see (8)]

(12)
$$y(t) = \frac{F_0}{m(\omega_0^2 - \omega^2)} (\cos \omega t - \cos \omega_0 t) \qquad (\omega \neq \omega_0)$$

corresponding to the initial conditions $y(0) = 0$, $y'(0) = 0$. This may be written [see (12) in Appendix A3.1]

$$y(t) = \frac{2F_0}{m(\omega_0^2 - \omega^2)} \sin\left(\frac{\omega_0 + \omega}{2} t\right) \sin\left(\frac{\omega_0 - \omega}{2} t\right).$$

Since ω is close to ω_0, the difference $\omega_0 - \omega$ is small, so that the period of the last sine function is large, and we obtain an oscillation of the type shown in Fig. 59. This is what musicians are listening to when they **tune** their instruments.

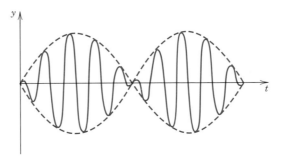

Fig. 59. Forced undamped oscillation when the difference of the input and natural frequencies is small (**"beats"**)

Case 2. Damped Forced Oscillations

If there is damping, then $c > 0$. From Sec. 2.5 we know that then the general solution y_h of the *homogeneous* equation (1) is (see (9), (10), Sec. 2.5)

$$y_h(t) = e^{-\alpha t}(A \cos \omega^* t + B \sin \omega^* t) \qquad \left(\alpha = \frac{c}{2m} > 0\right).$$

This solution approaches zero as t goes to infinity. Practically, it is zero after a sufficiently long time. Now the general solution (6) of the nonhomogeneous equation (2) is $y = y_h + y_p$. This is called the **transient solution.** It approaches the **steady-state solution** y_p. *Hence after a sufficiently long time, the output corresponding to a purely sinusoidal input will practically be a harmonic oscillation whose frequency is that of the input. This is what happens in practice, because no physical system is completely undamped.*

Whereas in the undamped case the amplitude of y_p approaches infinity as ω approaches ω_0, this will not happen in the damped case; *in this case the amplitude will always be finite,* but may have a maximum for some ω, depending on c. This may be called **practical resonance.** It is of great importance because it shows that some input may excite oscillations with such a large amplitude that the system can be destroyed. Such cases happened in practice, in particular in earlier times when less was known about resonance. Machines, cars, ships, airplanes, and bridges are vibrating mechanical systems, and it is sometimes rather difficult to find constructions that are completely free of undesired resonance effects.

Amplitude of y_p

To study the amplitude of y_p as a function of ω, we write (3) in the form

(13) $y_p(t) = C^* \cos (\omega t - \eta)$

where, according to (5), the amplitude C^* and the angle η are given by

(14)
$$C^*(\omega) = \sqrt{a^2 + b^2} = \frac{F_0}{\sqrt{m^2(\omega_0{}^2 - \omega^2)^2 + \omega^2 c^2}},$$

$$\tan \eta = \frac{b}{a} = \frac{\omega c}{m(\omega_0{}^2 - \omega^2)}.$$

Let us determine the maximum of $C^*(\omega)$. By setting $dC^*/d\omega = 0$ we find

$$\left[-2m^2(\omega_0{}^2 - \omega^2) + c^2\right]\omega = 0.$$

(Verify this!) The expression in brackets is zero when

(15)
$$c^2 = 2m^2(\omega_0{}^2 - \omega^2).$$

For sufficiently large damping ($c^2 > 2m^2\omega_0{}^2 = 2mk$) equation (15) has no real solution, and C^* decreases in a monotone way as ω increases (Fig. 60). If $c^2 \leqq 2mk$, equation (15) has a real solution $\omega = \omega_{\max}$, which increases as c decreases and approaches ω_0 as c approaches zero. The amplitude $C^*(\omega)$ has a maximum at $\omega = \omega_{\max}$, and by inserting $\omega = \omega_{\max}$ into (14) we find

(16)
$$C^*(\omega_{\max}) = \frac{2mF_0}{c\sqrt{4m^2\omega_0{}^2 - c^2}}.$$

We see that $C^*(\omega_{\max})$ is finite when $c > 0$. Since $dC^*(\omega_{\max})/dc < 0$ when $c^2 < 2mk$, the value of $C^*(\omega_{\max})$ increases as c ($\leqq \sqrt{2mk}$) decreases and approaches infinity as c approaches zero, in agreement with our result in Case 1. Figure 60 shows the **amplification** C^*/F_0 (ratio of the amplitudes of output and input) as a function of ω for $m = 1$, $k = 1$, and various values of the damping constant c.

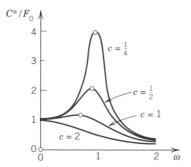 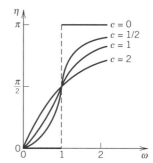

Fig. 60. Amplification C^*/F_0 as a function of ω for $m = 1$, $k = 1$, and various values of the damping constant c

Fig. 61. Phase lag η as a function of ω for $m = 1$, $k = 1$, thus $\omega_0 = 1$, and various values of the damping constant c

The angle η in (14) is called the **phase angle** or **phase lag** (Fig. 61) because it measures the lag of the output behind the input. If $\omega < \omega_0$, then $\eta < \pi/2$; if $\omega = \omega_0$, then $\eta = \pi/2$; and if $\omega > \omega_0$, then $\eta > \pi/2$.

PROBLEM SET 2.11

Steady-State Solutions. Find the steady-state oscillation of the mass–spring system governed by the given equation. Show the details of your calculation.

1. $y'' + 3y' + 2y = 20 \cos 2t$

2. $y'' + 2y' + 5y = -13 \sin 3t$

3. $y'' + 2y' + 4y = \sin 0.2t$

4. $(D^2 + 4D + 3)y = \cos t + \frac{1}{3} \cos 3t$

5. $(D^2 + D + 1)y = e^t$

6. $(D^2 + 5D + 10)y = 7 \cos t + 2 \sin t$

Transient Solutions. Find the transient motion of the mass–spring system governed by the given equation. (Show the details of your work.)

7. $y'' + 3y' + 2y = 170 \sin 4t$

8. $y'' + 3y = 11 \cos 0.5t$

9. $y'' + y = \cos \omega t, \qquad \omega^2 \neq 1$

10. $(D^2 + 6D + 9)y = 25 \sin t$

11. $(D^2 + 4D + 3)y = 26 \cos 2t$

12. $(D^2 + D)y = 1 + \cos t$

Initial Value Problems. Find the motion of the mass–spring system corresponding to the given equation and initial conditions. Sketch or plot the solution curve. State the time when the solution practically reaches the steady state. (Show the details of your work.)

13. $y'' + 25y = 24 \sin t, \qquad y(0) = 1, \qquad y'(0) = 1$

14. $y'' + 2y' + 2y = \cos t, \qquad y(0) = 1.2, \qquad y'(0) = 1.4$

15. $4y'' + 8y' + 3y = 425 \sin 2t, \qquad y(0) = -16, \qquad y'(0) = -26$

16. $(D^2 + 8D + 17)y = 474.5 \sin 0.5t, \qquad y(0) = -5.4, \qquad y'(0) = 9.4$

17. $(D^2 + 4)y = \sin t + \frac{1}{3} \sin 3t + \frac{1}{5} \sin 5t, \qquad y(0) = 1, \qquad y'(0) = \frac{3}{35}$

18. WRITING PROJECT. Free and Forced Vibrations. Write a condensed report of 2−3 pages on the most important facts about free and forced vibrations. *Hint.* First make a list of 5−6 points that you think are most relevant; this will help you to balance your essay without exceeding three pages.

19. TEAM PROJECT. Practical Resonance. (a) Give a detailed derivation of the crucial formula (16) for the maximum amplitude.

(b) Illustrate the situation with a differential equation of your own in which you vary the damping constant c, and sketch or plot corresponding curves as in Fig. 60.

(c) Consider your equation with a fixed c and a sum of two terms as the input, one term whose frequency is close to that of the practical resonance frequency and the other whose frequency is not. Discuss and sketch or plot the output.

(d) Describe other applications (not in the book) in which resonance plays an important role.

20. CAS PROJECT. Undamped Vibrations for Various Input Frequencies. (a) Solve the initial value problem $y'' + y = \cos \omega t, \omega^2 \neq 1, y(0) = 0, y'(0) = 0$. Show that the solution can be written

(17)
$$y(t) = \frac{2}{1 - \omega^2} \sin \left[\frac{1}{2}(1 + \omega)t \right] \sin \left[\frac{1}{2}(1 - \omega)t \right].$$

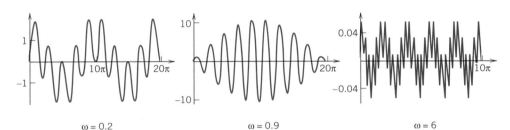

$\omega = 0.2$ $\qquad\qquad\qquad$ $\omega = 0.9$ $\qquad\qquad\qquad$ $\omega = 6$

Fig. 62. Typical solution curves in CAS Project 20

(b) Now experiment with plots of (17) for various ω, intelligently chosen so that you can see the change of the curves from those for small ω (> 0) to beats and on to resonance, and finally beyond resonance to high values of ω (Fig. 62). Note that (17) explains beats nicely, but for very high frequencies, formula (11) in Appendix A3.1 is more informative.

2.12 Modeling of Electric Circuits

The last section was devoted to the study of a mechanical system that is basic for the understanding of vibrations and resonance. We shall now consider a similarly important *electrical system,* which may be regarded as a basic building block in electrical networks. This consideration will also provide a striking example of the important fact that ***entirely different physical systems may correspond to the same mathematical model***—in the present case, to the same differential equation—so that they can be solved by the same methods. This is an impressive demonstration of the ***unifying power*** of mathematics.

Indeed, we shall obtain a *correspondence between mechanical and electrical systems that is not merely qualitative but strictly quantitative* in the sense that to a given mechanical system we can construct an electric circuit whose current will give the exact values of the displacement in the mechanical system when suitable scale factors are introduced.

The practical importance of such an ***analogy between mechanical and electrical systems*** is almost obvious. The analogy may be used for constructing an "electrical model" of a given mechanical system. In many cases this will be an essential simplification, because electric circuits are easy to assemble and currents and voltages are easy to measure, whereas the construction of a mechanical model may be complicated and expensive, and the measurement of displacements will be more time-consuming and less accurate than that of currents.

Setting up the Model

RL-circuits and *RC*-circuits were modeled in Sec. 1.7, which you may first wish to review. We now consider the ***RLC-circuit*** in Fig. 63. In it, an Ohm's resistor of resistance R [ohms], an inductor of inductance L [henrys], and a capacitor of capacitance C [farads] are connected in series to a source of electromotive force $E(t)$ [volts], where t is time. The equation for the current $I(t)$ [amperes] in the *RLC*-circuit is obtained by considering the three voltage drops

$$E_L = LI' \qquad\qquad \text{(Drop across the inductor)}$$

$$E_R = RI \qquad\qquad \text{(Drop across the resistor, by } Ohm's \ law)$$

$$E_C = \frac{1}{C} \int I(t)\, dt \qquad\qquad \text{(Drop across the capacitor)}.$$

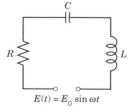

$$E(t) = E_0 \sin \omega t$$

Fig. 63. *RLC*-circuit

Their sum equals the electromotive force $E(t)$. This is **Kirchhoff's voltage law** (Sec. 1.7), the analog of Newton's second law (Sec. 2.5) for mechanical systems. For a sinusoidal $E(t) = E_0 \sin \omega t$ (E_0 constant), this law gives

$$(1') \qquad LI' + RI + \frac{1}{C} \int I \, dt = E(t) = E_0 \sin \omega t.$$

This process of modeling is the same as that in Sec. 1.7. Indeed, if we add $E_L = LI'$ to the equation (7) in Sec. 1.7 for the RC-circuit, we obtain our present equation $(1')$ for the RLC-circuit.

To get rid of the integral in $(1')$, we differentiate with respect to t, obtaining

$$(1) \qquad \boxed{LI'' + RI' + \frac{1}{C} I = E_0 \omega \cos \omega t.}$$

This is of the same form as (2), Sec. 2.11. Hence our RLC-circuit is the electrical analog of the mechanical system in Sec. 2.11. The corresponding analogy of electrical and mechanical quantities is shown in Table 2.2.

Table 2.2
Analogy of Electrical and Mechanical Quantities in
(1), This Section, and (2), Sec. 2.11

Electrical System	Mechanical System
Inductance L	Mass m
Resistance R	Damping constant c
Reciprocal $1/C$ of capacitance	Spring modulus k
Derivative $E_0 \omega \cos \omega t$ of electromotive force	Driving force $F_0 \cos \omega t$
Current $I(t)$	Displacement $y(t)$

Comment. We recall from Sec. 1.7 that $I = Q'$. Hence $I' = Q''$ and $\int I \, dt = Q$. We thus obtain from $(1')$ the following differential equation for the charge Q on the capacitor:

$$(1'') \qquad LQ'' + RQ' + \frac{1}{C} Q = E_0 \sin \omega t.$$

In most practical problems, the current $I(t)$ is more important than $Q(t)$, and for this reason we shall concentrate on (1) rather than on $(1'')$.

Solving Equation (1), Discussion of Solution

To obtain a particular solution of (1) we may proceed as in Sec. 2.11. We substitute

$$(2) \qquad I_p = a \cos \omega t + b \sin \omega t$$

$$I_p' = \omega(-a \sin \omega t + b \cos \omega t)$$

$$I_p'' = \omega^2(-a \cos \omega t - b \sin \omega t)$$

into (1). Then we collect the cosine terms and equate them to $E_0\omega \cos \omega t$ on the right, and we equate the sine terms to zero,

$$L\omega^2(-a) + R\omega b + a/C = E_0\omega \qquad \text{(Cosine terms)}$$

$$L\omega^2(-b) + R\omega(-a) + b/C = 0 \qquad \text{(Sine terms)}.$$

The solution is (verify!)

(3) $$a = \frac{-E_0 S}{R^2 + S^2}, \qquad b = \frac{E_0 R}{R^2 + S^2}$$

where S is the so-called **reactance,** given by the expression

(4) $$S = \omega L - \frac{1}{\omega C}.$$

In any practical case, $R \neq 0$, so that the denominator in (3) is not zero. The result is that (2), with a and b given by (3), is a particular solution of (1).

Using (3), we may write I_p in the form

(5) $$I_p(t) = I_0 \sin(\omega t - \theta)$$

where [see (14) in Appendix A3.1]

$$I_0 = \sqrt{a^2 + b^2} = \frac{E_0}{\sqrt{R^2 + S^2}}, \qquad \tan \theta = -\frac{a}{b} = \frac{S}{R}.$$

The quantity $\sqrt{R^2 + S^2}$ is called the **impedance.** Our formula shows that the impedance equals the ratio E_0/I_0. This is somewhat analogous to $E/I = R$ (Ohm's law).

A general solution of the homogeneous equation corresponding to (1) is

$$I_h = c_1 e^{\lambda_1 t} + c_2 e^{\lambda_2 t}$$

where λ_1 and λ_2 are the roots of the characteristic equation

$$\lambda^2 + \frac{R}{L}\lambda + \frac{1}{LC} = 0.$$

We can write these roots in the form $\lambda_1 = -\alpha + \beta$ and $\lambda_2 = -\alpha - \beta$, where

$$\alpha = \frac{R}{2L}, \qquad \beta = \frac{1}{2L}\sqrt{R^2 - \frac{4L}{C}}.$$

As in Sec. 2.11 we conclude that if $R > 0$ (which, of course, is true in any practical case), the general solution $I_h(t)$ of the homogeneous equation approaches zero as t approaches infinity (practically: after a sufficiently long time). Hence the transient current $I = I_h + I_p$ tends to the steady-state current I_p, and *after some time* **the output will practically be a harmonic oscillation,** *which is given by (5) and whose frequency is that of the input.*

EXAMPLE 1   *RLC-circuit*

Find the current $I(t)$ in an *RLC*-circuit with $R = 100$ ohms, $L = 0.1$ henry, $C = 10^{-3}$ farad, which is connected to a source of voltage $E(t) = 155 \sin 377t$ (hence 60 Hz = 60 cycles/sec), assuming zero charge and current when $t = 0$.

Solution. 1st Step. General solution.   Equation (1) is

$$0.1 I'' + 100 I' + 1000 I = 155 \cdot 377 \cos 377t.$$

We calculate the reactance $S = 37.7 - 1/0.377 = 35.0$ and the steady-state current

$$I_p(t) = a \cos 377t + b \sin 377t$$

where

$$a = \frac{-155 \cdot 35.0}{100^2 + 35^2} = -0.484, \qquad b = \frac{155 \cdot 100}{100^2 + 35^2} = 1.380.$$

Then we solve the characteristic equation

$$0.1\lambda^2 + 100\lambda + 1000 = 0.$$

The roots are $\lambda_1 = -10$ and $\lambda_2 = -990$ (more exactly, $-10.102\ 051$ and $-989.897\ 949$). This gives the general solution

(6) $$I(t) = c_1 e^{-10t} + c_2 e^{-990t} - 0.484 \cos 377t + 1.380 \sin 377t.$$

2nd Step. Particular solution.   We determine c_1 and c_2 from the initial conditions $Q(0) = 0$ and $I(0) = 0$. The second condition gives

(7) $$I(0) = c_1 + c_2 - 0.484 = 0.$$

How to use $Q(0) = 0$? Solving $(1')$ algebraically for I', we have

(8) $$I' = \frac{1}{L}\left[E(t) - RI(t) - \frac{1}{C}Q(t) \right]$$

since $\int I\, dt = Q$. Here $E(0) = 0$, $I(0) = 0$, and $Q(0) = 0$, so that $I'(0) = 0$. Differentiating (6), we thus obtain

$$I'(0) = -10c_1 - 990c_2 + 1.380 \cdot 377 = 0.$$

The solution of this and (7) is $c_1 = -0.042$, $c_2 = 0.526$. From (6) we thus have the answer

$$I(t) = -0.042 e^{-10t} + 0.526 e^{-990t} - 0.484 \cos 377t + 1.380 \sin 377t.$$

Figure 64 shows $I(t)$ as well as $I_p(t)$, which practically coincide, except for a very short time (hardly noticeable near $t = 0$) because the exponential terms go to zero very rapidly. Thus after a very short time the current will practically execute harmonic oscillations of frequency 60 Hz = 60 cycles/sec, which is the input frequency.

Note that by (5) we can write the steady-state current in the form

$$I_p(t) = 1.463 \sin (377t - 0.34). \qquad \blacktriangleleft$$

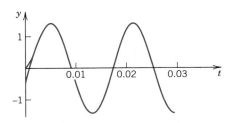

Fig. 64.   Transient and steady-state currents in Example 1

PROBLEM SET 2.12

RLC-circuits

1. **(Types of damping)** What are the conditions for an *RLC*-circuit to be overdamped (Case I), critically damped (Case II), and underdamped (Case III)? In particular, what is the critical resistance R_{crit} (the analog of the critical damping constant $2\sqrt{mk}$)?
2. **(Transient current)** Prove the claim in the text that if $R > 0$, then the transient current approaches I_p as $t \to \infty$.
3. **(Tuning)** In tuning a radio to a station we turn a knob on the radio that changes C (or perhaps L) in an *RLC*-circuit (Fig. 63) so that the amplitude of the steady-state current becomes maximum. For what C will this be the case?

Steady-State Current. Find the steady-state current in the *RLC*-circuit in Fig. 65 for the given data. (Show the details of your work.)

4. $R = 2$ ohms, $L = 1$ henry, $C = 0.5$ farad, $E = 50 \sin t$ volts
5. $R = 8$ ohms, $L = 2$ henrys, $C = 0.1$ farad, $E = 160 \cos 5t$ volts
6. $R = 4$ ohms, $L = 1$ henry, $C = 2 \cdot 10^{-4}$ farad, $E = 220$ volts

Transient Current. Find the current in the *RLC*-circuit in Fig. 65 for the given data. (Show the details of your work.)

7. $R = 40$ ohms, $L = 0.5$ henry, $C = 1/750$ farad, $E = 25 \cos 100t$ volts
8. $R = 20$ ohms, $L = 5$ henrys, $C = 10^{-2}$ farad, $E = 425 \sin 4t$ volts
9. $R = 10$ ohms, $L = 0.1$ henry, $C = 1/340$ farad, $E = e^{-t}(169.9 \sin t - 160.1 \cos t)$

Initial Value Problems. Solve the following problems, assuming zero initial current and charge. (Show the details of your work.)

10. $R = 80$ ohms, $L = 10$ henrys, $C = 0.004$ farad, $E = 240.5 \sin 10t$
11. $R = 8$ ohms, $L = 2$ henrys, $C = 0.1$ farad, $E = 10$ volts
12. $R = 3$ ohms, $L = 0.5$ henry, $C = 0.08$ farad, $E = 12 \cos 5t$ volts

LC-circuits

(In practice, these are *RLC*-circuits with negligibly small *R*.)

Find the current $I(t)$ in the *LC*-circuit in Fig. 66 with the following data, assuming zero initial current and charge. (For the use of $Q(0) = 0$ see Example 1.) (Show the details of your work.)

13. $L = 10$ henrys, $C = 0.1$ farad, $E = 10t$ volts
14. $L = 2$ henrys, $C = 5 \cdot 10^{-5}$ farad, $E = 110$ volts
15. $L = 2$ henrys, $C = 0.005$ farad, $E = 220 \sin 4t$ volts
16. **(Charge)** Find the charge Q in Prob. 14 (a) from the current, and (b) by solving (1'').

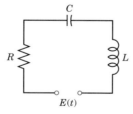

Fig. 65.   *RLC*-circuit

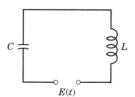

Fig. 66.   *LC*-circuit

17. **PROJECT. Analogy of Mass–Spring Systems and Electric Circuits.** (a) Write an essay of about 2–3 pages based on Table 2.2. Describe the analogy in more detail. Characterize its practical significance.

(b) What *RLC*-circuit with $L = 1$ henry is the analog of the mass–spring system with mass 2 kg, damping constant 20 kg/sec, spring constant 58 kg/sec², and driving force $110 \cos 5t$ nt?

(c) Illustrate the analogy with an example of your own choice.

18. TEAM PROJECT. Complex Method for Particular Solutions. Differentiating an exponential function, $(e^{ax})' = ae^{ax}$, is easier than differentiating a sine or cosine. This motivates the method. The idea of the method results from the Euler formula (Sec. 2.3)

$$(9) \qquad e^{i\omega t} = \cos \omega t + i \sin \omega t \qquad (i = \sqrt{-1}).$$

Indeed, engineers like to obtain steady-state currents for equations such as (1) by replacing $E_0\omega \cos \omega t$ with $E_0\omega e^{i\omega t}$, the real part of which is $E_0\omega \cos \omega t$. That is, they consider the complex equation

$$(10) \qquad \boxed{LI'' + RI' + \frac{1}{C}I = E_0\omega e^{i\omega t}} \qquad (i = \sqrt{-1}).$$

They determine a particular solution I_p of (10) and finally take the real part $\tilde{I}_p$ of I_p, which is the desired solution of the given equation (1). Work out these ideas as follows.

(a) Substitute

$$(11) \qquad I_p = Ke^{i\omega t}$$

and its derivatives into (10). Use $i^2 = -1$. Show that this gives

$$\left(-\omega^2 L + i\omega R + \frac{1}{C}\right) Ke^{i\omega t} = E_0\omega e^{i\omega t}.$$

Solve this algebraically for K to get [with S the reactance given by (4)]

$$(12) \qquad K = \frac{E_0}{-\left(\omega L - \dfrac{1}{\omega C}\right) + iR} = \frac{E_0}{-S + iR} = \frac{-E_0(S + iR)}{S^2 + R^2}.$$

The last equality should not make you unhappy. It is obtained by multiplying both the numerator and the denominator by $-S - iR$ (the conjugate of $-S + iR$), a standard trick of complex division (see also Sec. 12.1). Substitute (12) into (11). Using (9), decompose the result into the real and imaginary part (by multiplying out and again using $i^2 = -1$). Show that the real part is

$$(13) \qquad I_p = \frac{-E_0}{S^2 + R^2}(S \cos \omega t - R \sin \omega t),$$

in agreement with (2), (3).

(b) Show that by introducing the so-called **complex impedance**

$$(14) \qquad Z = R + iS = R + i\left(\omega L - \frac{1}{\omega C}\right)$$

we can write (12) simply as

$$(12^*) \qquad K = \frac{E_0}{iZ}.$$

Note that the real part of Z is R, the imaginary part is the reactance S, and the absolute value is the **impedance,** $|Z| = \sqrt{R^2 + S^2}$ (see Fig. 67).

(c) Solve $I'' + I' + 3I = 5 \cos t$ first by the real method and then by the complex method. Compare. (Show the details of your work.)

(d) Apply the complex method to an equation of your own choice.

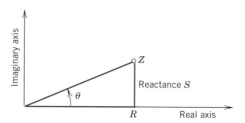

Fig. 67.   Complex impedance Z

2.13 Higher Order Linear Differential Equations

We turn from second-order equations to differential equations of arbitrary order n,

$$F(x, y, y', \cdots, y^{(n)}) = 0 \qquad \left(y^{(n)} = \frac{d^n y}{dx^n} \right).$$

Such an equation is called **linear** if it can be written

(1)
$$y^{(n)} + p_{n-1}(x)y^{(n-1)} + \cdots + p_1(x)y' + p_0(x)y = r(x).$$

(For $n = 2$ this is (1) in Sec. 2.1 with $p_1 = p$ and $p_0 = q$.) The **coefficients** $p_0, \cdots, p_{n-1}$ and r are any given functions of x, and y is unknown. $y^{(n)}$ has coefficient 1. We call this the **"standard form."** This is practical. (If you have $f(x)y^{(n)} + \cdots$, divide by $f(x)$ to get this form.) An equation that cannot be written in the form (1) is called **nonlinear.**

If $r(x)$ is identically zero, $r(x) \equiv 0$ (zero for all x considered), then (1) becomes

(2)
$$y^{(n)} + p_{n-1}(x)y^{(n-1)} + \cdots + p_1(x)y' + p_0(x)y = 0$$

and is called **homogeneous.** If $r(x)$ is not identically zero, the equation is called **nonhomogeneous.** This is as in Sec. 2.1.

Solution. General Solution. Linear Independence

A **solution** of an nth order (linear or nonlinear) differential equation on some open interval I is a function $y = h(x)$ that is defined and n times differentiable on I and is such that the equation becomes an identity if we replace the unknown function y and its derivatives in the equation by h and its corresponding derivatives.

We now turn to the homogeneous equation **(2)** *and begin with the following.*

THEOREM 1   **(Superposition principle or linearity principle)**

For the **homogeneous** *linear differential equation* (2), *sums and constant multiples of solutions on some open interval I are again solutions of* (2) *on I.*

The proof is a simple generalization of that in Sec. 2.1 and is left to the student. We repeat our *warning* that the theorem *does not hold* for the nonhomogeneous equation (1) or for a nonlinear equation.

Our further discussion parallels and extends that for second-order equations in Sec. 2.1. So we define next a general solution of (2), which will need an extension of linear independence from two to n functions, a concept of great general importance, far beyond our present purpose.

Definition (General solution, basis, particular solution)

A **general solution** of (2) on an open interval I is a solution of (2) on I of the form

$$(3) \qquad \boxed{y(x) = c_1 y_1(x) + \cdots + c_n y_n(x)} \qquad (c_1, \cdots, c_n \text{ arbitrary}^{12})$$

where $y_1, \cdots, y_n$ is a **basis** (or **fundamental system**) of solutions of (2) on I; that is, these solutions are linearly independent on I, as defined below.

A **particular solution** of (2) on I is obtained if we assign specific values to the n constants $c_1, \cdots, c_n$ in (3). ◀

Definition (Linear independence and dependence)

n functions $y_1(x), \cdots, y_n(x)$ are called **linearly independent** *on some interval I* where they are defined if the equation

$$(4) \qquad \boxed{k_1 y_1(x) + \cdots + k_n y_n(x) = 0 \quad \text{on } I}$$

implies that all $k_1, \cdots, k_n$ are zero. These functions are called **linearly dependent** on I if this equation also holds on I for some $k_1, \cdots, k_n$ not all zero. ◀

If and only if $y_1, \cdots, y_n$ are linearly dependent on I, we can express (at least) one of these functions on I as a **"linear combination"** of the other $n - 1$ functions, that is, as a sum of those functions, each multiplied by a constant (zero or not). This motivates the term "linearly dependent." For instance, if (4) holds with $k_1 \neq 0$, we can divide by k_1 and express y_1 as the linear combination

$$y_1 = -\frac{1}{k_1}(k_2 y_2 + \cdots + k_n y_n).$$

Note that when $n = 2$, these concepts reduce to those defined in Sec. 2.1.

EXAMPLE 1   **Linear dependence**

Show that the functions $y_1 = x$, $y_2 = 3x$, $y_3 = x^2$ are linearly dependent on any interval.

Solution. $y_2 = 3y_1 + 0y_3$. ◀

---

^{12}See footnote 2 in Sec. 2.1.

EXAMPLE 2   **Linear independence**

Show that $y_1 = x$, $y_2 = x^2$, $y_3 = x^3$ are linearly independent on any interval, for instance, on $-1 \leq x \leq 2$.

Solution.   Equation (4) is $k_1 x + k_2 x^2 + k_3 x^3 = 0$. Taking $x = -1$, 1, 2, we get

$$-k_1 + k_2 - k_3 = 0, \qquad k_1 + k_2 + k_3 = 0, \qquad 2k_1 + 4k_2 + 8k_3 = 0,$$

respectively, which implies $k_1 = 0$, $k_2 = 0$, $k_3 = 0$, that is, linear independence.

This calculation was not too pleasant and illustrates the need for a better method of testing linear independence, at least for solutions. We shall get to this soon.   ◄

EXAMPLE 3   **General solution, basis**

Solve the fourth-order differential equation

$$y^{\mathrm{iv}} - 5y'' + 4y = 0.$$

Solution.   As in Sec. 2.2, we try $y = e^{\lambda x}$. Then substitution and omission of the common (nonzero) factor $e^{\lambda x}$ gives the characteristic equation

$$\lambda^4 - 5\lambda^2 + 4 = 0,$$

which is a quadratic equation in $\mu = \lambda^2$,

$$\mu^2 - 5\mu + 4 = 0.$$

The roots are $\mu = 1$ and $\mu = 4$. Hence $\lambda = -2, -1, 1, 2$, which gives four solutions, so that a general solution on any interval is

$$y = c_1 e^{-2x} + c_2 e^{-x} + c_3 e^x + c_4 e^{2x}$$

provided those solutions are linearly independent. This is true, but will be shown later.   ◄

Initial Value Problem, Existence and Uniqueness

An **initial value problem** for equation (2) consists of (2) and n **initial conditions**

$$(5) \qquad y(x_0) = K_0, \qquad y'(x_0) = K_1, \qquad \cdots, \qquad y^{(n-1)}(x_0) = K_{n-1}$$

where x_0 is some fixed point in the interval I considered.

In extension of Theorem 1 in Sec. 2.7 we now have the following.

THEOREM 2   **Existence and Uniqueness Theorem for Initial Value Problems**

If $p_0(x), \cdots, p_{n-1}(x)$ are continuous functions on some open interval I and x_0 is in I, then the initial value problem (2), (5) has a unique solution $y(x)$ on the interval I.

Existence is proved in Ref. [A5] in Appendix 1, and uniqueness can be proved by a slight generalization of the uniqueness proof at the beginning of Appendix 4.

EXAMPLE 4   **Initial value problem for a third-order Euler–Cauchy equation**

Solve the initial value problem

$$x^3 y''' - 3x^2 y'' + 6xy' - 6y = 0, \qquad y(1) = 2, \qquad y'(1) = 1, \qquad y''(1) = -4$$

on any open interval I on the positive x-axis containing $x = 1$.

Solution. 1st Step. General solution.   As in Sec. 2.6 we try $y = x^m$. Differentiation and substitution gives

$$m(m - 1)(m - 2)x^m - 3m(m - 1)x^m + 6mx^m - 6x^m = 0.$$

Ordering terms and dropping the factor x^m, we obtain

$$m^3 - 6m^2 + 11m - 6 = 0.$$

If we can guess the root $m = 1$, we can divide by $m - 1$ and find as the other roots $m = 2$ and $m = 3$. (Without guessing, for orders higher than four, one has to use a numerical root-finding method, such as Newton's; see Sec. 17.2.) The corresponding solutions x, x^2, x^3 are linearly independent on I (see Example 2). Hence a general solution on I is

$$y = c_1 x + c_2 x^2 + c_3 x^3.$$

(Our interval I does not include 0, where the coefficients of our equation in standard form [the given form divided by x^3] are not continuous, but we see that, actually, y is a general solution on any interval.)

2nd Step. Particular solution.   We now also need the derivatives

$$y' = c_1 + 2c_2 x + 3c_3 x^2, \qquad y'' = 2c_2 + 6c_3 x.$$

From this and y and the initial conditions we get

$$y(1) = c_1 + c_2 + c_3 = 2$$
$$y'(1) = c_1 + 2c_2 + 3c_3 = 1$$
$$y''(1) = 2c_2 + 6c_3 = -4.$$

By elimination or Cramer's rule (Sec. 6.6) we obtain $c_1 = 2$, $c_2 = 1$, $c_3 = -1$. *Answer:* $y = 2x + x^2 - x^3$. ◀

Linear Independence of Solutions. Wronskian

Very few new ideas occur in the transition from order $n = 2$ to general n. Linear dependence of more than two functions is one of them. A criterion for it is desirable. Fortunately the criterion involving the Wronskian in Sec. 2.7 (not overly important when $n = 2$) extends to general n and becomes practically valuable. It uses the **Wronskian** W of n solutions defined as the nth order determinant

$$(6) \qquad W(y_1, \cdots, y_n) = \begin{vmatrix} y_1 & y_2 & \cdots & y_n \\ y_1' & y_2' & \cdots & y_n' \\ \cdot & \cdot & \cdots & \cdot \\ y_1^{(n-1)} & y_2^{(n-1)} & \cdots & y_n^{(n-1)} \end{vmatrix}$$

and can be stated as follows.

THEOREM 3   **(Linear dependence and independence of solutions)**

Suppose that the coefficients $p_0(x), \cdots, p_{n-1}(x)$ of (2) are continuous on some open interval I. Then n solutions $y_1, \cdots, y_n$ of (2) on I are linearly dependent on I if and only if their Wronskian is zero for some $x = x_0$ in I. Furthermore, if $W = 0$ for $x = x_0$, then $W \equiv 0$ on I; hence if there is an x_1 in I at which $W \neq 0$, then $y_1, \cdots, y_n$ are linearly independent on I.

PROOF.   **(a)** Let $y_1, \cdots, y_n$ be linearly dependent on I. Then, by definition, there are constants $k_1, \cdots, k_n$, not all zero, such that for all x in I,

$$(7) \qquad k_1 y_1 + \cdots + k_n y_n = 0.$$

By $n - 1$ differentiations of this identity (7) we obtain

$$
\begin{aligned}
k_1 y_1' + \cdots + k_n y_n' &= 0 \\
&\;\;\vdots \\
k_1 y_1^{(n-1)} + \cdots + k_n y_n^{(n-1)} &= 0.
\end{aligned}
$$

(8)

(7), (8) is a homogeneous linear system of algebraic equations with a nontrivial solution $k_1, \cdots, k_n$. Hence its coefficient determinant must be zero for every x on I, by Cramer's theorem (Sec. 6.6). But that determinant is the Wronskian W, as we see. Thus, $W = 0$ for every x on I.

(b) Conversely, let $W = 0$ for an x_0 in I. Then the system (7), (8) with $x = x_0$ has a solution $\tilde{k}_1, \cdots, \tilde{k}_n$, not all zero, by that same theorem. With these constants we define the solution $\tilde{y} = \tilde{k}_1 y_1 + \cdots + \tilde{k}_n y_n$ of (2). By (7), (8) it satisfies the initial conditions $\tilde{y}(x_0) = 0, \cdots, \tilde{y}^{(n-1)}(x_0) = 0$. But another solution also satisfying these conditions is $y \equiv 0$. Hence $\tilde{y} \equiv y$ on I by Theorem 2; that is, (7) holds identically on I. This means linear dependence of $y_1, \cdots, y_n$.

(c) If $W = 0$ at an x_0 in I, we have linear dependence by (b). Hence $W \equiv 0$ by (a). It follows that $W \neq 0$ at any x_1 in I implies linear independence of the solutions $y_1, \cdots, y_n$ on I. ◀

EXAMPLE 5 **Basis, Wronskian**

We can now prove that in Example 3 we do have a basis. In evaluating W, pull out the exponentials columnwise. In the result, subtract column 1 from columns 2, 3, 4. Then expand by row 1. In the resulting third-order determinant, subtract column 1 from column 2 and expand the result by row 2:

$$
W = \begin{vmatrix} e^{-2x} & e^{-x} & e^x & e^{2x} \\ -2e^{-2x} & -e^{-x} & e^x & 2e^{2x} \\ 4e^{-2x} & e^{-x} & e^x & 4e^{2x} \\ -8e^{-2x} & -e^{-x} & e^x & 8e^{2x} \end{vmatrix} = \begin{vmatrix} 1 & 1 & 1 & 1 \\ -2 & -1 & 1 & 2 \\ 4 & 1 & 1 & 4 \\ -8 & -1 & 1 & 8 \end{vmatrix} = \begin{vmatrix} 1 & 3 & 4 \\ -3 & -3 & 0 \\ 7 & 9 & 16 \end{vmatrix} = 72. \;\; ◀
$$

A General Solution of (2) Includes All Solutions

We first show that general solutions always exist. Indeed, Theorem 3 in Sec. 2.7 extends as follows.

THEOREM 4 **(Existence of a general solution)**

If the coefficients $p_0(x), \cdots, p_{n-1}(x)$ of (2) are continuous on some open interval I, then (2) has a general solution on I.

PROOF. We choose any fixed x_0 in I. By Theorem 2, equation (2) has n solutions $y_1, \cdots, y_n$, where y_j satisfies initial conditions (5) with $K_{j-1} = 1$ and all other K's equal to zero. Their Wronskian at x_0 equals 1; for instance, when $n = 3$, then $y_1(x_0) = 1$, $y_2'(x_0) = 1$, $y_3''(x_0) = 1$ and the other initial values are zero, so that

$$
W(y_1(x_0),\, y_2(x_0),\, y_3(x_0)) = \begin{vmatrix} y_1(x_0) & y_2(x_0) & y_3(x_0) \\ y_1'(x_0) & y_2'(x_0) & y_3'(x_0) \\ y_1''(x_0) & y_2''(x_0) & y_3''(x_0) \end{vmatrix} = \begin{vmatrix} 1 & 0 & 0 \\ 0 & 1 & 0 \\ 0 & 0 & 1 \end{vmatrix} = 1.
$$

Hence these solutions are linearly independent on I, by Theorem 3; they form a basis on I, and $y = c_1 y_1 + \cdots + c_n y_n$ with arbitrary constants $c_1, \cdots, c_n$ is a general solution of (2) on I. ◄

We can now prove the basic property that from a general solution of (2) every solution of (2) can be obtained by choosing suitable values of the arbitrary constants. Hence an nth order *linear* differential equation has no **singular solutions,** that is, solutions that cannot be obtained from a general solution.

THEOREM 5   **(General solution includes all solutions)**

Suppose that (2) *has continuous coefficients* $p_0(x), \cdots, p_{n-1}(x)$ *on some open interval I. Then every solution* $y = Y(x)$ *of* (2) *on I is of the form*

$$(9) \qquad\qquad Y(x) = C_1 y_1(x) + \cdots + C_n y_n(x),$$

where $y_1, \cdots, y_n$ *is a basis of solutions of* (2) *on I and* $C_1, \cdots, C_n$ *are suitable constants.*

PROOF.   Let $y = c_1 y_1 + \cdots + c_n y_n$ be a general solution of (2) on I and choose any fixed x_0 in I. We show that we can find values of $c_1, \cdots, c_n$ for which y and its first $n - 1$ derivatives agree with Y and its corresponding derivatives at x_0. Written out, this means that for $x = x_0$ we should have

$$
\begin{aligned}
c_1 y_1 + \cdots + c_n y_n &= Y \\
c_1 y_1' + \cdots + c_n y_n' &= Y' \\
&\;\;\vdots \\
c_1 y_1^{(n-1)} + \cdots + c_n y_n^{(n-1)} &= Y^{(n-1)}.
\end{aligned}
$$

$$(10)$$

But this is a linear system of equations in the unknowns $c_1, \cdots, c_n$. Its coefficient determinant is the Wronskian of $y_1, \cdots, y_n$ at $x = x_0$, which is not zero by Theorem 3 because $y_1, \cdots, y_n$ are linearly independent on I (they form a basis!). Hence (10) has a unique solution $c_1 = C_1, \cdots, c_n = C_n$ (by Cramer's theorem, Sec. 6.6). With these values we get from our general solution the particular solution

$$y^*(x) = C_1 y_1(x) + \cdots + C_n y_n(x)$$

on I. From (10) we see that y^* agrees with Y at x_0, and the same holds for the first $n - 1$ derivatives of y^* and Y. That is, y^* and Y satisfy at x_0 the same initial conditions. From the uniqueness theorem (Theorem 2) it now follows that $y^* \equiv Y$ on I, and the theorem is proved. ◄

This completes our theory of the homogeneous linear equation (2). Note that for $n = 2$ it is identical with that in Sec. 2.7, as had to be expected.

Application: Elastic Beams

Whereas second-order differential equations have various applications, higher order equations occur much more rarely in engineering work. An important fourth-order equation governs the bending of an elastic beam, such as a wooden or iron girder in a building or a bridge.

EXAMPLE 6   **Bending of an elastic beam under a load**

We consider a beam B of length L and constant (e.g., rectangular) cross section and homogeneous elastic material (e.g., steel); see Fig. 68. We assume that under its own weight the beam is bent so little that it is practically straight. If we apply a load to B in a vertical plane through the axis of symmetry (the x-axis in Fig. 68), B is bent. Its axis is curved into the so-called **elastic curve** C (or **deflection curve**). It is shown in elasticity theory that the bending moment $M(x)$ is proportional to the curvature $k(x)$ of C. We assume the bending to be small, so that the deflection $y(x)$ and its derivative $y'(x)$ (determining the tangent direction of C) are small. Then, by calculus, $k = y''/(1 + y'^2)^{3/2} \approx y''$. Hence

$$M(x) = EIy''(x).$$

EI is the constant of proportionality. E is *Young's modulus of elasticity* of the material of the beam. I is the moment of inertia of the cross section about the (horizontal) z-axis in Fig. 68.

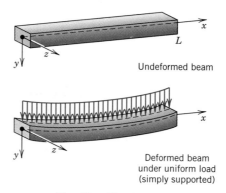

L

Undeformed beam

Deformed beam
under uniform load
(simply supported)

Fig. 68.   Elastic Beam

Elasticity theory shows further that $M''(x) = f(x)$, where $f(x)$ is the load per unit length. Together,

(11)
$$\boxed{EIy^{\text{iv}} = f(x).}$$

The practically most important supports and corresponding boundary conditions are as follows (see Fig. 69).

     (A)   Simply supported $y = y'' = 0$ at $x = 0$ and L

     (B)   Clamped at both ends $y = y' = 0$ at $x = 0$ and L

     (C)   Clamped at $x = 0$, free at $x = L$ $y(0) = y'(0) = 0$, $y''(L) = y'''(L) = 0$.

The boundary condition $y = 0$ means no displacement at that point, $y' = 0$ means a horizontal tangent, $y'' = 0$ means no bending moment, and y''' means no shear force.

Let us apply this to the uniformly loaded simply supported beam in Fig. 68. The load is $f(x) \equiv f_0 = const$. Then (11) is

(12)
$$y^{\text{iv}} = k, \qquad k = \frac{f_0}{EI}.$$

This can be solved simply by calculus. Two integrations give

$$y'' = \frac{k}{2} x^2 + c_1 x + c_2.$$

$y''(0) = 0$ gives $c_2 = 0$. Then $y''(L) = L(\frac{1}{2}kL + c_1) = 0$, $c_1 = -kL/2$ (since $L \neq 0$). Hence

$$y'' = \frac{k}{2}(x^2 - Lx).$$

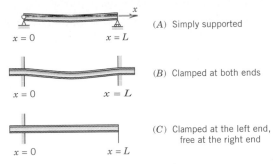

(A) Simply supported

$x = 0$ $x = L$

(B) Clamped at both ends

$x = 0$ $x = L$

(C) Clamped at the left end, free at the right end

$x = 0$ $x = L$

Fig. 69. Supports of a Beam

Integrating this twice, we obtain

$$y = \frac{k}{2}\left(\frac{1}{12}x^4 - \frac{L}{6}x^3 + c_3 x + c_4\right)$$

with $c_4 = 0$ from $y(0) = 0$. Then

$$y(L) = \frac{kL}{2}\left(\frac{L^3}{12} - \frac{L^3}{6} + c_3\right) = 0, \qquad c_3 = \frac{L^3}{12}.$$

Inserting the expression for k, we obtain as our solution

$$y = \frac{f_0}{24EI}(x^4 - 2Lx^3 + L^3 x).$$

Since the boundary conditions at both ends are the same, we expect the deflection $y(x)$ to be "symmetric" with respect to $L/2$, that is, $y(x) = y(L - x)$. Verify this directly or set $x = u + L/2$ and show that y becomes an even function of u,

$$y = \frac{f_0}{24EI}\left(u^2 - \frac{1}{4}L^2\right)\left(u^2 - \frac{5}{4}L^2\right).$$

From this we can see that the maximum deflection in the middle at $u = 0$ ($x = L/2$) is $5f_0 L^4/(16 \cdot 24EI)$. Recall that the positive direction points downward. ◀

Vibrations of beams will be considered in Sec. 11.4.

PROBLEM SET 2.13

Typical Examples of Bases. Initial Value Problems

To get an impression of what to expect for higher order linear equations, prove in Probs. 1–9 that the given functions form a basis of the corresponding given equation. Then solve the initial value problem. Show all details of your work. (Solution methods follow in the next sections.)

1. $1, x, x^2, x^3$; $y^{iv} = 0$, $y(0) = 1, y'(0) = 0, y''(0) = -1, y'''(0) = 30$

2. e^x, e^{-x}, e^{2x}; $y''' - 2y'' - y' + 2y = 0$, $y(0) = -2, y'(0) = -5, y''(0) = -11$

3. $e^{-3x}, xe^{-3x}, x^2 e^{-3x}$; $y''' + 9y'' + 27y' + 27y = 0$, $y(0) = 4, y'(0) = -13, y''(0) = 46$

4. $1, \cos x, \sin x$; $y''' + y' = 0$, $y(0) = 15, y'(0) = 0, y''(0) = -3$

5. $e^x \cos x, e^x \sin x, e^{-x} \cos x, e^{-x} \sin x$; $(D^4 + 4)y = 0$, $y(0) = 0, y'(0) = 2,$
$y''(0) = 0, y'''(0) = 4$

6. $\cosh x, \sinh x, \cos x, \sin x;$ $(D^4 - 1)y = 0,$ $y(0) = y'(0) = y''(0) = y'''(0) = 1$

7. $1, x, x^{-1};$ $xy''' + 3y'' = 0,$ $y(1) = 4, y'(1) = -8, y''(1) = 10$

8. $\cos x, \sin x, \cos 2x, \sin 2x;$ $(D^4 + 5D^2 + 4)y = 0,$ $y(0) = 1, y'(0) = 1, y''(0) = -1,$ $y'''(0) = -4$

9. $e^{2x}, e^{-2x}, \cos x, \sin x;$ $(D^4 - 3D^2 - 4)y = 0,$ $y(0) = 4, y'(0) = -4, y''(0) = 16,$ $y'''(0) = 4$

Linear Independence and Dependence

Since these concepts are of *general* importance, far beyond our present study, we add a few more problems on them.

Are the following functions linearly independent or dependent on the positive x-axis? (Give a reason.)

10. $\cos x, \sin x, \sin 2x$ **11.** $\cos^2 x, \sin^2 x, \pi$

12. $\ln x, -\ln (x^2), \ln (x^3)$ **13.** $\cosh 2x, e^x, e^{-x}, e^{2x}$

14. $\sinh 3x, e^{-3x}, e^{3x}$ **15.** $(x - 1)^2, (x + 1)^2, x$

16. $\ln x, (\ln x)^2, e^x, e^{-x}$ **17.** $e^x \cos x, e^x \sin x, e^x$

18. $\tan^2 x, \tan x, \cot x, 0$

19. TEAM PROJECT. General Properties of Solutions of Linear Differential Equations. These properties are of basic practical importance in obtaining new solutions from given ones, in constructing bases, and so on. For this reason extend Team Project 30 in Problem Set 2.2 to nth order equations. Explore statements on sums and multiples of solutions of (1) or (2) in a systematic way, with proofs. Be sure to recognize clearly that no new ideas are needed in this extension from $n = 2$ to general n.

20. TEAM PROJECT. Linear Independence and Dependence. (a) Prove the following basic facts about a set S on an interval I. Give an illustrative example for each.

(1) If S contains the zero function as an element, then S is linearly dependent.

(2) If S is linearly independent on a subinterval J of I, it is linearly independent on I. If S is linearly dependent on I, what can you say about S on J?

(3) If S is linearly dependent on I and T contains S, then T is linearly dependent on I. What about T if S is linearly independent on I?

(b) In what case can you use the Wronskian for testing for linear independence? What other means can you think of for such a test?

2.14 Higher Order Homogeneous Equations with Constant Coefficients

If a linear differential equation has *constant* coefficients, it can be solved by the methods for second-order constant-coefficient equations. We discuss this, beginning with the homogeneous equation, which we write in the form

$$(1) \qquad y^{(n)} + a_{n-1}y^{(n-1)} + \cdots + a_1 y' + a_0 y = 0.$$

Substituting $y = e^{\lambda x}$ (as in Sec. 2.2), we obtain the characteristic equation

$$(2) \qquad \lambda^n + a_{n-1}\lambda^{n-1} + \cdots + a_1\lambda + a_0 = 0$$

of (1). To get solutions of (1) we have to determine the roots of (2). This will generally require a numerical root-finding method (for example, Newton's method, Sec. 17.2), which your CAS may contain.

There are more cases than in Secs. 2.2 and 2.3. We discuss all of these cases and illustrate them with typical examples.

Distinct Real Roots

If all the n roots $\lambda_1, \cdots, \lambda_n$ of (2) are real and different, then the n solutions

$$(3) \qquad \boxed{y_1 = e^{\lambda_1 x}, \qquad \cdots, \qquad y_n = e^{\lambda_n x}}$$

constitute a basis for all x. The corresponding general solution of (1) is

$$(4) \qquad y = c_1 e^{\lambda_1 x} + \cdots + c_n e^{\lambda_n x}.$$

Indeed, the solutions in (3) are linearly independent, as we shall see after the example.

EXAMPLE 1   **Distinct real roots**

Solve the differential equation

$$y''' - 2y'' - y' + 2y = 0.$$

Solution.   The roots of the characteristic equation

$$\lambda^3 - 2\lambda^2 - \lambda + 2 = 0$$

are -1, 1, and 2, and the corresponding general solution (4) is

$$y = c_1 e^{-x} + c_2 e^x + c_3 e^{2x}.$$

Verify linear independence by using the Wronskian.   ◀

We prove linear independence of (3). Students familiar with nth order determinants may verify that by pulling out all exponentials from the columns, the Wronskian of $e^{\lambda_1 x}, \cdots, e^{\lambda_n x}$ becomes

$$(5) \quad W = \begin{vmatrix} e^{\lambda_1 x} & e^{\lambda_2 x} & \cdots & e^{\lambda_n x} \\ \lambda_1 e^{\lambda_1 x} & \lambda_2 e^{\lambda_2 x} & \cdots & \lambda_n e^{\lambda_n x} \\ \lambda_1^2 e^{\lambda_1 x} & \lambda_2^2 e^{\lambda_2 x} & \cdots & \lambda_n^2 e^{\lambda_n x} \\ \cdot & \cdot & \cdots & \cdot \\ \lambda_1^{n-1} e^{\lambda_1 x} & \lambda_2^{n-1} e^{\lambda_2 x} & \cdots & \lambda_n^{n-1} e^{\lambda_n x} \end{vmatrix}$$

$$= e^{(\lambda_1 + \cdots + \lambda_n)x} \begin{vmatrix} 1 & 1 & \cdots & 1 \\ \lambda_1 & \lambda_2 & \cdots & \lambda_n \\ \lambda_1^2 & \lambda_2^2 & \cdots & \lambda_n^2 \\ \cdot & \cdot & \cdots & \cdot \\ \lambda_1^{n-1} & \lambda_2^{n-1} & \cdots & \lambda_n^{n-1} \end{vmatrix}.$$

The exponential function is never zero. Hence $W = 0$ if and only if the determinant on the right is zero. This is a so-called **Vandermonde** or **Cauchy determinant**.[13] It can be shown that it equals

$$(6) \hspace{4cm} (-1)^{n(n-1)/2}V$$

where V is the product of all factors $\lambda_j - \lambda_k$ with $j < k \ (\leqq n)$; for instance, when $n = 3$ we get $-V = -(\lambda_1 - \lambda_2)(\lambda_1 - \lambda_3)(\lambda_2 - \lambda_3)$. This shows that the Wronskian is not zero if and only if all the n roots of (2) are different and thus gives the following.

THEOREM 1   **(Basis)**

Solutions $y_1 = e^{\lambda_1 x}, \cdots, y_n = e^{\lambda_n x}$ of (1) (with any real or complex λ_j's) form a basis of solutions of (1) if and only if all n roots of (2) are different.

Actually, Theorem 1 is an important special case of our more general result obtained from (5) and (6):

THEOREM 2   **(Linear independence)**

Any number of solutions $y_1 = e^{\lambda_1 x}, \cdots, y_m = e^{\lambda_m x}$ of (1) are linearly independent on an open interval I if and only if $\lambda_1, \cdots, \lambda_m$ are all different.

Simple Complex Roots

If complex roots occur, they must occur in conjugate pairs since the coefficients of (1) are real. Thus, if $\lambda = \gamma + i\omega$ is a simple root of (2), so is the conjugate $\bar{\lambda} = \gamma - i\omega$, and two corresponding linearly independent solutions are (as in Sec. 2.3, except for notation)

$$y_1 = e^{\gamma x} \cos \omega x, \hspace{2cm} y_2 = e^{\gamma x} \sin \omega x.$$

EXAMPLE 2   **Simple complex conjugate roots. Initial value problem**

Solve the initial value problem

$$y''' - y'' + 100y' - 100y = 0, \hspace{1cm} y(0) = 4, \hspace{1cm} y'(0) = 11, \hspace{1cm} y''(0) = -299.$$

Solution.   The characteristic equation is $\lambda^3 - \lambda^2 + 100\lambda - 100 = 0$. It has the root 1, as can perhaps be seen by inspection. Then division by $\lambda - 1$ shows that the other roots are $\pm 10i$. Hence a general solution and its derivatives (obtained by differentiation) are

$$(a) \hspace{2cm} y = c_1 e^x + A \cos 10x + B \sin 10x,$$

$$(b) \hspace{2cm} y' = c_1 e^x - 10A \sin 10x + 10B \cos 10x,$$

$$(c) \hspace{2cm} y'' = c_1 e^x - 100A \cos 10x - 100B \sin 10x.$$

---

[13]ALEXANDRE-THÉOPHILE VANDERMONDE (1735–1796), French mathematician, who worked on solution of equations by determinants. For CAUCHY, see footnote 9 in Sec. 2.6.

From this and the initial conditions we obtain by setting $x = 0$

(a) $c_1 + A = 4,$

(b) $c_1 + 10B = 11,$

(c) $c_1 - 100A = -299.$

Then (a) minus (c) gives $101A = 303$, $A = 3$, $c_1 = 1$. And $B = 1$ from (b). The solution is (Fig. 70)

$$y = e^x + 3 \cos 10x + \sin 10x.$$

This is a curve that oscillates about e^x (dashed). ◀

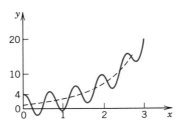

Fig. 70.   Solution in Example 2

Multiple Real Roots

If a real **double root** occurs, say, $\lambda_1 = \lambda_2$, then $y_1 = y_2$ in (3) and we take y_1 and $y_2 = xy_1$ as two linearly independent solutions corresponding to this root; this is as in Sec. 2.2.

If a **triple root** occurs, say, $\lambda_1 = \lambda_2 = \lambda_3$, then $y_1 = y_2 = y_3$ in (3) and three linearly independent solutions corresponding to this root are

(7) $y_1, \qquad xy_1, \qquad x^2y_1.$

More generally, *if λ is a* **root of order** ***m,*** *then m corresponding linearly independent solutions are*

(8) $e^{\lambda x}, \qquad xe^{\lambda x}, \qquad \cdots, \qquad x^{m-1}e^{\lambda x}.$

Linear independence of these functions on any open interval follows from that of $1, x, \cdots, x^{m-1}$, which in turn follows from Theorem 3 in Sec. 2.13 and the fact that these are solutions of $y^{(m)} = 0$ with a nonzero Wronskian W. How does one get (8)? We show this after the example.

EXAMPLE 3   **Real double and triple roots**

Solve the differential equation

$$y^{\text{v}} - 3y^{\text{iv}} + 3y''' - y'' = 0.$$

Solution. The characteristic equation

$$\lambda^5 - 3\lambda^4 + 3\lambda^3 - \lambda^2 = 0$$

has the roots $\lambda_1 = \lambda_2 = 0$ and $\lambda_3 = \lambda_4 = \lambda_5 = 1$, and the answer is

$$(9) \qquad\qquad y = c_1 + c_2 x + (c_3 + c_4 x + c_5 x^2)e^x. \qquad\qquad \blacktriangleleft$$

As promised, we now show how (8) is obtained [and that these functions are solutions of (1) in the present case]. To simplify formulas a little, we use operator notation (see Sec. 2.4), writing the left side of (1) as

$$L[y] = \left[D^n + a_{n-1}D^{n-1} + \cdots + a_0\right]y.$$

For $y = e^{\lambda x}$ we can perform the indicated differentiations and get

$$L\left[e^{\lambda x}\right] = (\lambda^n + a_{n-1}\lambda^{n-1} + \cdots + a_0)e^{\lambda x}.$$

Let λ_1 be an mth order root of the polynomial on the right, and let $\lambda_{m+1}, \cdots, \lambda_n$ be the other roots, all different from λ_1, when $m < n$. In product form we then have

$$L\left[e^{\lambda x}\right] = (\lambda - \lambda_1)^m h(\lambda)e^{\lambda x}$$

with $h(\lambda) = 1$ if $m = n$ and $h(\lambda) = (\lambda - \lambda_{m+1}) \cdots (\lambda - \lambda_n)$ if $m < n$. Now comes the key idea: We differentiate on both sides with respect to λ,

$$(10) \qquad \frac{\partial}{\partial\lambda} L[e^{\lambda x}] = m(\lambda - \lambda_1)^{m-1}h(\lambda)e^{\lambda x} + (\lambda - \lambda_1)^m \frac{\partial}{\partial\lambda}\left[h(\lambda)e^{\lambda x}\right].$$

Differentiations with respect to x and λ are independent and the occurring derivatives are continuous, so that we can interchange their order on the left:

$$(11) \qquad\qquad \frac{\partial}{\partial\lambda} L[e^{\lambda x}] = L\left[\frac{\partial}{\partial\lambda} e^{\lambda x}\right] = L[xe^{\lambda x}].$$

Now the right side of (10) is zero for $\lambda = \lambda_1$ because of the factors $\lambda - \lambda_1$ (and $m \geqq 2$). Hence (11) shows that $xe^{\lambda_1 x}$ is a solution of (1).

We can repeat this step and produce $x^2 e^{\lambda_1 x}, \cdots, x^{m-1}e^{\lambda_1 x}$ by another $m - 2$ such differentiations with respect to λ. Going one step further would no longer give zero on the right because the lowest power of $\lambda - \lambda_1$ would then be $(\lambda - \lambda_1)^0$, multiplied by $m!h(\lambda)$ and $h(\lambda_1) \neq 0$ because $h(\lambda)$ has no factors $\lambda - \lambda_1$; so we get *precisely* the solutions in (8). $\qquad \blacktriangleleft$

Multiple Complex Roots

In this case, real solutions are obtained as for complex simple roots above. Consequently, if $\lambda = \gamma + i\omega$ is a **complex double root**, so is the conjugate $\bar{\lambda} = \gamma - i\omega$. Corresponding linearly independent solutions are

$$(12) \qquad e^{\gamma x}\cos\omega x, \qquad e^{\gamma x}\sin\omega x, \qquad xe^{\gamma x}\cos\omega x, \qquad xe^{\gamma x}\sin\omega x.$$

The first two of these result from $e^{\lambda x}$ and $e^{\bar{\lambda} x}$ as before, and the second two from $xe^{\lambda x}$ and $xe^{\bar{\lambda} x}$ in the same fashion. Obviously, the corresponding general solution is

(13) $$y = e^{\gamma x}\left[(A_1 + A_2 x) \cos \omega x + (B_1 + B_2 x) \sin \omega x\right].$$

For *complex triple roots* (which hardly ever occur in applications), one would obtain two more solutions $x^2 e^{\gamma x} \cos \omega x$, $x^2 e^{\gamma x} \sin \omega x$, and so on.

PROBLEM SET 2.14

General Solution. Solve the following differential equations. (Show the details of your work.)

1. $y^{\mathrm{iv}} - 16y = 0$

2. $y''' + 9y'' + 27y' + 27y = 0$

3. $y^{\mathrm{iv}} - 2y'' + y = 0$

4. $y^{\mathrm{iv}} + 2y'' + y = 0$

5. $y''' - 2y'' - y' + 2y = 0$

6. $y^{\mathrm{iv}} + 5y'' + 4y = 0$

7. $(D^3 - D^2 - D + 1)y = 0$

8. $(D^3 - 3D + 2)y = 0$

9. $(16D^4 - 40D^2 + 9)y = 0$

10. $(D^3 + 6D^2 + 11D + 6)y = 0$

Initial Value Problems. Solve the following initial value problems and sketch or plot the solution. (Show your work.)

11. $y^{\mathrm{iv}} = 0$, $\quad y(0) = 1$, $y'(0) = 16$, $y''(0) = -4$, $y'''(0) = 24$

12. $y''' - 3y'' + 3y' - y = 0$, $\quad y(0) = 2$, $y'(0) = 2$, $y''(0) = 10$

13. $y''' - y'' - y' + y = 0$, $\quad y(0) = 2$, $y'(0) = 1$, $y''(0) = 0$

14. $(D^4 - 1)y = 0$, $\quad y(0) = -1$, $y'(0) = 7$, $y''(0) = -1$, $y'''(0) = 7$

15. $(D^3 + 5D^2 - D - 5)y = 0$, $\quad y(0) = 5$, $y'(0) = 0$, $y''(0) = 125$

16. $(D^4 + 10D^2 + 9)y = 0$, $\quad y(0) = 0$, $y'(0) = 0$, $y''(0) = 32$, $y'''(0) = 0$

17. $(D^4 + 4D^3 + 8D^2 + 8D + 4)y = 0$, $\quad y(0) = 1$, $y'(0) = 0$, $y''(0) = -2$, $y'''(0) = 2$

18. $(D^3 + 6D^2 + 12D + 8)y = 0$, $\quad y(0) = 1$, $y'(0) = -2$, $y''(0) = 6$

 19. CAS PROJECT. Wronskians. Euler–Cauchy Equations of Higher Order. (a) Write a program for calculating the value of Wronskians.

  (b) Apply the program to typical bases occurring in connection with third-order and fourth-order constant-coefficient equations.

  (c) Extend the solution method in Sec. 2.6 to Euler–Cauchy equations of any other n. Solve $x^3 y''' + x^2 y'' - 2xy' + 2y = 0$ and two other equations of your own choice, and in each case calculate the Wronskian.

20. PROJECT. Reduction of Order. This is of practical interest because a single solution can often be obtained by inspection or by some experimentation. **(a)** Explain how you would proceed for a constant-coefficient equation. (This is rather simple.)

  (b) For an equation with variable coefficients,

$$y''' + p_2(x)y'' + p_1(x)y' + p_0(x)y = 0,$$

this can be rather involved; however, if a solution $y_1(x)$ is known, another solution is $y_2(x) = u(x)y_1(x)$ with $u(x) = \int z(x)\,dx$ and z obtained by solving

$$y_1 z'' + (3y_1' + p_2 y_1)z' + (3y_1'' + 2p_2 y_1' + p_1 y_1)z = 0.$$

Derive this general formula. (*Hint.* Model your work after that in Sec. 2.1 for second-order equations.)

  (c) Apply the formula in (b) to $x^3 y''' - 3x^2 y'' + (6 - x^2)xy' - (6 - x^2)y = 0$ with $y_1 = x$ (perhaps obtainable by inspection).

2.15   Higher Order Nonhomogeneous Equations

We now turn to nonhomogeneous linear differential equations of nth order, which we write in standard form

(1)
$$y^{(n)} + p_{n-1}(x)y^{(n-1)} + \cdots + p_1(x)y' + p_0(x)y = r(x)$$

with $y^{(n)} = d^n y/dx^n$ as the first term, which is practical, and $r(x) \not\equiv 0$. As for second-order equations, a general solution of (1) on an interval I of the x-axis is of the form

(2)
$$y(x) = y_h(x) + y_p(x).$$

Here $y_h(x) = c_1 y_1(x) + \cdots + c_n y_n(x)$ is a general solution of the corresponding homogeneous equation

(3)
$$y^{(n)} + p_{n-1}(x)y^{(n-1)} + \cdots + p_1(x)y' + p_0(x)y = 0$$

on I. Also, y_p is any solution of (1) on I containing no arbitrary constants. If (1) has continuous coefficients and a continuous $r(x)$ on I, then a general solution of (1) exists and includes all solutions. Thus Eq. (1) has no singular solution.

An **initial value problem** for (1) consists of (1) and n initial conditions

(4)
$$y(x_0) = K_0, \qquad y'(x_0) = K_1, \qquad \cdots, \qquad y^{(n-1)}(x_0) = K_{n-1}$$

with x_0 in I. Under those continuity assumptions it has a unique solution. The ideas of proof are the same as those for $n = 2$ in Sec. 2.8.

Method of Undetermined Coefficients

Equation (2) shows that for solving (1) we have to determine a particular solution $y_p(x)$ of (1). For a constant-coefficient equation

(5)
$$y^{(n)} + a_{n-1}y^{(n-1)} + \cdots + a_1 y' + a_0 y = r(x)$$

($a_0, \cdots, a_{n-1}$ constant) and special $r(x)$ as in Sec. 2.9, such a $y_p(x)$ can be determined by the **method of undetermined coefficients,** as in Sec. 2.9, using these rules.

(A) Basic Rule as in Sec. 2.9.

(B) Modification Rule. *If a term in your choice for y_p is a solution of the homogeneous equation (3), then multiply $y_p(x)$ by x^k, where k is the smallest positive integer such that no term of $x^k y_p(x)$ is a solution of (3).*

(C) Sum Rule as in Sec. 2.9.

The practical application of the method is the same as that in Sec. 2.9. It suffices to illustrate the typical steps of solving an initial value problem and, in particular, the new Modification Rule, which includes the old Modification Rule as a particular case (with $k = 1$ or 2). We shall see that the technicalities are the same as for $n = 2$, perhaps except for the more involved determination of the constants.

EXAMPLE 1   **Initial value problem. Modification Rule**

Solve the initial value problem

(6) $\qquad\qquad y''' + 3y'' + 3y' + y = 30e^{-x}, \qquad y(0) = 3, \qquad y'(0) = -3, \qquad y''(0) = -47.$

Solution. 1st Step.   The characteristic equation is $\lambda^3 + 3\lambda^2 + 3\lambda + 1 = (\lambda + 1)^3 = 0$. It has the triple root $\lambda = -1$. Hence a general solution of the homogeneous equation is

$$y_h = c_1 e^{-x} + c_2 x e^{-x} + c_3 x^2 e^{-x}$$
$$= (c_1 + c_2 x + c_3 x^2)e^{-x}.$$

2nd Step.   If we try $y_p = Ce^{-x}$, we get $-C + 3C - 3C + C = 30$, which has no solution. Try Cxe^{-x} and $Cx^2 e^{-x}$. The Modification Rule calls for

$$y_p = Cx^3 e^{-x}.$$

Then
$$y_p' = C(3x^2 - x^3)e^{-x},$$
$$y_p'' = C(6x - 6x^2 + x^3)e^{-x},$$
$$y_p''' = C(6 - 18x + 9x^2 - x^3)e^{-x}.$$

Substitution of these expressions into (6) and omission of the common factor e^{-x} gives

$$C(6 - 18x + 9x^2 - x^3) + 3C(6x - 6x^2 + x^3) + 3C(3x^2 - x^3) + Cx^3 = 30.$$

The linear, quadratic, and cubic terms drop out, and $6C = 30$. Hence $C = 5$. This gives $y_p = 5x^3 e^{-x}$.

3rd Step.   We now write down $y = y_h + y_p$, the general solution of the given equation. From it we find c_1 by the first initial condition. We insert the value, differentiate, and determine c_2 from the second initial condition, insert the value, and finally determine c_3 from $y''(0)$ and the third initial condition:

$$y = y_h + y_p = (c_1 + c_2 x + c_3 x^2)e^{-x} + 5x^3 e^{-x}, \qquad y(0) = c_1 = 3$$
$$y' = [-3 + c_2 + (-c_2 + 2c_3)x + (15 - c_3)x^2 - 5x^3]e^{-x}, \qquad y'(0) = -3 + c_2 = -3, \qquad c_2 = 0$$
$$y'' = [3 + 2c_3 + (30 - 4c_3)x + (-30 + c_3)x^2 + 5x^3]e^{-x}, \qquad y''(0) = 3 + 2c_3 = -47, \qquad c_3 = -25.$$

Hence the answer of our problem is (Fig. 71)

$$y = (3 - 25x^2)e^{-x} + 5x^3 e^{-x}.$$

The dashed curve in Fig. 71 is y_p. Does the curve of y agree with what you can expect from the initial conditions? From the limit as $x \to \infty$? ◀

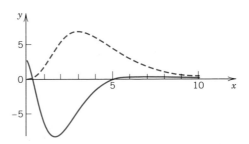

Fig. 71. y and y_p (dashed) in Example 1

Method of Variation of Parameters

The method of variation of parameters (see Sec. 2.10) also extends to arbitrary order n. It gives a particular solution y_p for the nonhomogeneous equation (1) (in standard form with $y^{(n)}$ as the first term!) by the formula

(7)
$$y_p(x) = y_1(x) \int \frac{W_1(x)}{W(x)} r(x)\, dx + y_2(x) \int \frac{W_2(x)}{W(x)} r(x)\, dx$$
$$+ \cdots + y_n(x) \int \frac{W_n(x)}{W(x)} r(x)\, dx$$

on an interval I on which the coefficients of (1) and $r(x)$ are continuous. In (7) the functions $y_1, \cdots, y_n$ form a basis of the homogeneous equation (3), with Wronskian W, and W_j ($j = 1, \cdots, n$) obtained from W by replacing the jth column of W by the column $[0 \ \ 0 \ \ \cdots \ \ 0 \ \ 1]^T$. Thus, when $n = 2$, this becomes identical with (2) in Sec. 2.10,

$$W = \begin{vmatrix} y_1 & y_2 \\ y_1' & y_2' \end{vmatrix}, \qquad W_1 = \begin{vmatrix} 0 & y_2 \\ 1 & y_2' \end{vmatrix} = -y_2, \qquad W_2 = \begin{vmatrix} y_1 & 0 \\ y_1' & 1 \end{vmatrix} = y_1.$$

The proof of (7) uses an extension of the idea of the proof of (2) in Sec. 2.10 and can be found in Ref. [A5] listed in Appendix 1.

EXAMPLE 2   **Variation of parameters. Nonhomogeneous Euler–Cauchy equation**

Solve the nonhomogeneous Euler–Cauchy equation

$$x^3 y''' - 3x^2 y'' + 6xy' - 6y = x^4 \ln x \qquad\qquad (x > 0).$$

Solution. 1st Step. General solution.   Substitution of $y = x^m$ and the derivatives into the homogeneous equation and deletion of the factor x^m gives

$$m(m - 1)(m - 2) - 3m(m - 1) + 6m - 6 = 0.$$

The roots are 1, 2, 3 and give as a basis of the homogeneous equation

$$y_1 = x, \qquad y_2 = x^2, \qquad y_3 = x^3.$$

2nd Step. Determinants needed in (7).   These are

$$W = \begin{vmatrix} x & x^2 & x^3 \\ 1 & 2x & 3x^2 \\ 0 & 2 & 6x \end{vmatrix} = 2x^3, \qquad W_1 = \begin{vmatrix} 0 & x^2 & x^3 \\ 0 & 2x & 3x^2 \\ 1 & 2 & 6x \end{vmatrix} = x^4,$$

$$W_2 = \begin{vmatrix} x & 0 & x^3 \\ 1 & 0 & 3x^2 \\ 0 & 1 & 6x \end{vmatrix} = -2x^3, \qquad W_3 = \begin{vmatrix} x & x^2 & 0 \\ 1 & 2x & 0 \\ 0 & 2 & 1 \end{vmatrix} = x^2.$$

3rd Step. Integration.   In (7) we also need the right side $r(x)$ of our equation in standard form, which we obtain by dividing the given equation by x^3 (the coefficient of y'''); this gives $r(x) = (x^4 \ln x)/x^3 = x \ln x$. From this and (7),

$$y_p = x \int \frac{x}{2} x \ln x \, dx - x^2 \int x \ln x \, dx + x^3 \int \frac{1}{2x} x \ln x \, dx$$

$$= \frac{x}{2}\left(\frac{x^3}{3} \ln x - \frac{x^3}{9} \right) - x^2\left(\frac{x^2}{2} \ln x - \frac{x^2}{4} \right) + \frac{x^3}{2}(x \ln x - x).$$

Simplification gives the answer (see Fig. 72)

$$y_p = \frac{x^4}{6}\left(\ln x - \frac{11}{6}\right).$$

Can you explain the shape of this curve? Its behavior near $x = 0$? The occurrence of a minimum? Its eventual rapid increase? Can you explain why the method of undetermined coefficients would not have produced the solution? ◀

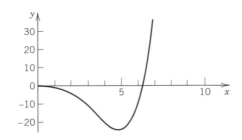

Fig. 72. Solution y of the nonhomogeneous Euler–Cauchy equation in Example 2

PROBLEM SET 2.15

General Solution. Find a general solution. (Show the details of your work.)

1. $y''' + 3y'' + 3y' + y = 8e^x + x + 3$

2. $x^3 y''' + x^2 y'' - 2xy' + 2y = x^3 \ln x$

3. $x^3 y''' + x^2 y'' - 2xy' + 2y = x^{-2}$

4. $y''' + 2y'' - y' - 2y = 1 - 4x^3$

5. $y''' - y'' - 4y' + 4y = 12e^{-x}$

6. $y''' - 6y'' + 12y' - 8y = \sqrt{x}\, e^{2x}$

7. $xy''' + 3y'' = e^x$

8. $4x^3 y''' + 3xy' - 3y = 4x^{11/2}$

Initial Value Problems. Solve the following initial value problems. (This is quite similar to the procedure for second-order equations in Sec. 2.9 and 2.10.) (Show your work.)

9. $y''' + 3y'' + 3y' + y = e^{-x}\sin x,$ $y(0) = 2,$ $y'(0) = 0,$ $y''(0) = -1$

10. $x^3 y''' + xy' - y = x^2,$ $y(1) = 1,$ $y'(1) = 3,$ $y''(1) = 3$

11. $x^3 y''' - 3x^2 y'' + 6xy' - 6y = 24x^5,$ $y(1) = 1,$ $y'(1) = 3,$ $y''(1) = 14$

12. $y^{iv} + 10y'' + 9y = 40\sinh x,$ $y(0) = 0,$ $y'(0) = 6,$ $y''(0) = 0,$ $y'''(0) = -26$

13. $y''' - 4y' = 10\cos x + 5\sin x,$ $y(0) = 3,$ $y'(0) = -2,$ $y''(0) = -1$

14. CAS PROJECT. Variation of Parameters Versus Undetermined Coefficients. Variation of parameters is complicated compared to the method of undetermined coefficients. Hence it seems worthwhile to gain experience in extending the latter by performing computer experiments. Find equations for which an extension seems feasible and others for which this is impossible. *Hint.* Work backward, solving equations with a CAS and then looking at the solution to decide whether and how it could be obtained by undetermined coefficients. For example, consider

$$y''' - 3y'' + 3y' - y = x^{1/2}e^x \quad\text{and}\quad x^3 y''' + x^2 y'' - 2xy' + 2y = x^3 \ln x.$$

15. WRITING PROJECT. Comparison of the Two Solution Methods in This Section. Write an essay on the method of undetermined coefficients and the method of variation of parameters. Discuss and compare the advantages and disadvantages of each method. Illustrate your findings with typical examples. Try to show that the method of undetermined coefficients, say, for an equation with constant coefficients and an exponential function on the right, can be derived from the method of variation of parameters.

CHAPTER 2 REVIEW

1. What is the superposition principle? Does it hold for nonlinear equations? For nonhomogeneous linear equations? For homogeneous linear equations? Why is it important?

2. How many arbitrary constants does a general solution of a nonhomogeneous linear equation involve? A general solution of a homogeneous linear equation? How many additional conditions does one need to determine them?

3. How is linear independence and dependence of n functions defined? Why are these concepts important in this chapter?

4. How would you practically test for linear independence of two functions? Of n solutions of a linear differential equation?

5. Why are differential equations of second order practically more important than differential equations of higher order?

6. Does it make sense to talk about linear dependence of functions at a single point? Explain.

7. What is a particular solution? Why are particular solutions generally more common as final answers to practical problems than general solutions?

8. What is the Wronskian? What role did it play in this chapter?

9. We considered two large classes of differential equations that can essentially be solved by "algebraic" manipulations. What are they?

10. What do you know about existence and uniqueness of solutions?

11. What is the Modification Rule and when did we need it?

12. Describe the idea and some details of the method of variation of parameters from memory, as best as you remember.

13. In modeling, one generally prefers linear over nonlinear differential equations whenever one can hope to get a faithful picture of the reality from a linear equation. What is the reason for this?

14. For second-order constant-coefficient equations, we distinguished three cases. What are they? What is their significance in connection with mass–spring systems? In RLC-circuits?

15. What do we mean by "resonance"? Where and under what conditions does it occur?

General Solution. Find a general solution. (Show the details of your calculations.)

16. $4y'' + 24y' + 37y = 0$

17. $2y'' - 3y' - 2y = 13 - 2x^2$

18. $x^2 y'' + xy' - 9y = 0$

19. $x^2 y'' - 3xy' + 4y = 12$

20. $y''' - 4y'' - y' + 4y = 30e^{2x}$

21. $x^3 y''' - 9x^2 y'' + 33xy' - 48y = 0$

22. $y'' - 2\pi y' + \pi^2 y = 2e^{\pi x}$

23. $y'' + 2y' + 2y = 3e^{-x} \cos 2x$

24. $(D^2 + 4D + 4)y = e^{-2x}/x^2$

25. $(D^4 - 5D^2 + 4)y = 40 \cos 2x$

26. $(D^3 - D)y = \sinh x$

27. $(D^3 + 3D^2 + 3D + 1)y = 8 \sin x$

28. $(D^2 + D + 1)y = \cos x + 13 \cos 2x$

29. $(x^2 D^2 - 0.4xD + 0.49)y = 0.07$

30. $(D^2 - 4D + 4)y = e^{2x}/x$

Initial Value Problems. Solve the following problems. (Show the details.)

31. $y'' + 16y = 17e^x$, $y(0) = 6$, $y'(0) = -2$

32. $y'' - 3y' + 2y = 10 \sin x$, $y(0) = 1$, $y'(0) = -6$

33. $x^2 y'' - 4xy' + 6y = \pi^2 x^4 \sin \pi x$, $y(1) = 5$, $y'(1) = 5 + \pi$

34. $y'' + 4y' + (4 + \omega^2)y = 0$, $y(0) = 1$, $y'(0) = \omega - 2$

35. $y'' + 4y = 8e^{-2x} + 4x^2 + 2$, $y(0) = 2$, $y'(0) = 2$

36. $(D^2 - 4D + 3)y = 10 \sin x$, $y(0) = 2$, $y'(0) = 1$

37. $(x^2 D^2 + xD - 1)y = 16x^3$, $y(1) = -1$, $y'(1) = 11$

38. $(D^3 - D^2 - D + 1)y = 0$, $y(0) = 2$, $y'(0) = 1$, $y''(0) = 0$

39. $(x^3 D^3 - 3x^2 D^2 + 6xD - 6)y = 12/x$, $y(1) = 5$, $y'(1) = 13$, $y''(1) = 10$

40. $(D^3 + 3D^2 + 3D + 1)y = 8 \sin x$, $y(0) = -1$, $y'(0) = -3$, $y''(0) = 5$

Applications

41. Find the steady-state current in the *RLC*-circuit in Fig. 73, assuming that $L = 1$ henry, $R = 2000$ ohms, $C = 4 \cdot 10^{-3}$ farad, and $E(t) = 110 \sin 415t$ (66 cycles/sec).

42. Find a general solution of the homogeneous equation corresponding to the equation in Prob. 41.

43. Find the steady-state current in the *RLC*-circuit in Fig. 73 when $R = 50$ ohms, $L = 30$ henrys, $C = 0.025$ farad, $E(t) = 200 \sin 4t$ volts.

44. Find the current in the *RLC*-circuit in Fig. 73 when $R = 20$ ohms, $L = 0.1$ henry, $C = 1.5625 \cdot 10^{-3}$ farad, $E(t) = 160t$ volts if $0 < t < 0.01$, $E(t) = 1.6$ volts if $t > 0.01$ sec, assuming that $I(0) = 0$, $I'(0) = 0$.

45. Find an electrical analog of the mass–spring system with mass 4 kg, spring constant 10 kg/sec^2, damping constant 20 kg/sec, and driving force $100 \sin 4t$ nt.

46. Using the complex method, find the steady-state current in the *RLC*-circuit in Fig. 73 with $L = 4$ henrys, $R = 20$ ohms, $C = 0.5$ farad, and $E = 10 \sin 10t$ volts.

47. Find the steady-state solution of the system in Fig. 74 when $m = 1$, $c = 2$, $k = 6$ and the driving force is $\sin 2t + 2 \cos 2t$.

48. Find the motion of the mass–spring system in Fig. 74 with mass 0.125 kg, damping 0, spring constant 1.125 kg/sec^2, and driving force $\cos t - 4 \sin t$ nt, assuming zero initial displacement and velocity. For what frequency of the driving force would you get resonance?

49. In Prob. 47, find the solution corresponding to initial displacement 1 and initial velocity 0.

50. In Fig. 74, let $m = 1$, $c = 4$, $k = 24$ and $r(t) = 10 \cos \omega t$. Determine ω such that you get the steady-state vibration of maximum possible amplitude. Determine this amplitude. Then find the general solution with this ω and check whether the results are in agreement.

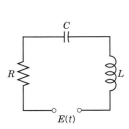

Fig. 73.   *RLC*-circuit

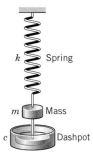

Fig. 74.   Mass–spring system

Summary of Chapter 2
Linear Differential Equations
of Second and Higher Order

Second-order linear differential equations are of great practical importance. Their theory and applications show all the typical features of linear differential equations of any order n. For this reason we studied them extensively in this chapter (Secs. 2.1–2.12). The transition to any n was then rather simple (Secs. 2.13–2.15).

A second-order differential equation is called **linear** if it can be written

$$(1) \qquad\qquad y'' + p(x)y' + q(x)y = r(x).$$

This **"standard form"** of a linear differential equation can always be achieved by division by $f(x)$ if such an equation begins with the term $f(x)y''$.

Equation (1) is called **homogeneous** if the function on the right side is zero for all x under consideration, written $r(x) \equiv 0$. Equation (1) is called **nonhomogeneous** if $r(x) \not\equiv 0$. Thus a homogeneous equation of second order is of the form

$$(2) \qquad\qquad y'' + p(x)y' + q(x)y = 0.$$

It has the very important property that a linear combination of solutions is again a solution (**superposition principle** or **linearity principle,** Sec. 2.1). Two linearly independent solutions y_1, y_2 of (2) on an open interval I form a **basis** (or **fundamental system**) of solutions on I, and $y = c_1 y_1 + c_2 y_2$ with arbitrary constants c_1, c_2 forms a **general solution.** From it we obtain a **particular solution** if we specify numerical values of c_1 and c_2, usually by prescribing two initial conditions.

For a nonhomogeneous equation (1) a **general solution** is of the form

$$(3) \qquad\qquad y = y_h + y_p \qquad\qquad \text{(Sec. 2.8)}$$

where y_h is a general solution of (2) and y_p any particular solution (i.e., a solution containing no arbitrary constants) of (1). The practical problem of determining such a y_p can be solved by the **method of variation of parameters** (Sec. 2.10).

There also exists the much simpler **method of undetermined coefficients,** valid for constant p, q and special r (powers of x, sine, cosine, etc.; see Sec. 2.9). In this case we write (1) using the notation

$$(4) \qquad\qquad y'' + ay' + by = r(x) \qquad\qquad (a,\ b \text{ constant}).$$

To solve the corresponding homogeneous equation $y'' + ay' + by = 0$ we substitute $y = e^{\lambda x}$. This gives a solution if λ is a root of the **characteristic equation**

$$(5) \qquad\qquad \lambda^2 + a\lambda + b = 0.$$

Hence there are three cases (Sec. 2.2):

Case	Type of Roots	General Solution
I	Distinct real λ_1, λ_2	$y = c_1 e^{\lambda_1 x} + c_2 e^{\lambda_2 x}$
II	Double $-\frac{1}{2}a$	$y = (c_1 + c_2 x)e^{-ax/2}$
III	Complex $-\frac{1}{2}a \pm i\omega$	$y = e^{-ax/2}(A \cos \omega x + B \sin \omega x)$

Equation (4) has important engineering applications in mechanics and electrical engineering (Secs. 2.5, 2.11, 2.12), which are fundamental in the study of **vibrations** and **resonance.**

Another large class of equations also solvable by an "algebraic" method consists of the **Euler–Cauchy equations** (Sec. 2.6)

$$(6) \qquad\qquad x^2 y'' + axy' + by = 0.$$

If we substitute $y = x^m$, we can determine m from the auxiliary equation

$$(7) \qquad\qquad m^2 + (a - 1)m + b = 0.$$

For second-order equations (1) or (2), an **initial value problem** consists of the equation and two initial conditions (Sec. 2.1)

$$(8) \qquad\qquad y(x_0) = K_0, \quad y'(x_0) = K_1 \qquad\qquad (x_0, K_0, K_1 \text{ given numbers}).$$

If p, q, and r are continuous on some open interval $I: \alpha < x < \beta$ and x_0 is in I, then (1) and (2) have a general solution on I, and the initial value problem (1), (8) or (2), (8) has a unique solution on I (which is a particular solution; thus (1) or (2) has no singular solutions). See Sec. 2.7.

All these facts and methods extend to equations of any order n. A **linear differential equation of nth order** is an equation that can be written in the form

$$(9) \qquad\qquad y^{(n)} + p_{n-1}(x)y^{(n-1)} + \cdots + p_1(x)y' + p_0(x)y = r(x)$$

with $y^{(n)} = d^n y/dx^n$ as the first term; we again call this the **standard form.** This extends (1). Equation (9) is called **homogeneous** if $r(x) \equiv 0$, **nonhomogeneous** if $r(x) \not\equiv 0$. For the homogeneous equation

$$(10) \qquad\qquad y^{(n)} + p_{n-1}(x)y^{(n-1)} + \cdots + p_1(x)y' + p_0(x)y = 0$$

the superposition principle holds, just as in the case $n = 2$. A basis of solutions of (10) on an open interval I consists of n linearly independent solutions $y_1, \cdots, y_n$ of (10) on I. A **general solution** of (10) on I is a linear combination of these,

$$(11) \qquad\qquad y = c_1 y_1 + \cdots + c_n y_n \qquad\qquad (c_1, \cdots, c_n \text{ arbitrary constants}).$$

A **general solution** of the nonhomogeneous equation (9) on I is of the form

$$(12) \qquad\qquad\qquad y = y_h + y_p \qquad\qquad\qquad \text{[see (3)]}$$

where y_h is a general solution of (10) on I and y_p is any particular solution of (9) on I. The latter can be obtained by the method of **variation of parameters** or [for constant-coefficient equations with special $r(x)$] by the method of **undetermined coefficients** (Sec. 2.15).

An **initial value problem** now consists of an equation (9) or (10) and n initial conditions

$$(13) \qquad y(x_0) = K_0, \quad y'(x_0) = K_1, \quad \cdots, \quad y^{(n-1)}(x_0) = K_{n-1}$$

with given $x_0, K_0, \cdots, K_{n-1}$. General solutions of (9) and (10) on an open interval I exist if the coefficients $p_0(x), \cdots, p_{n-1}(x)$ and the function $r(x)$ are continuous on I. Then the initial value problem (9), (11) or (10), (11) has a unique solution.

The solutions of linear differential equations with variable coefficients will generally be higher functions (as opposed to the "elementary functions" considered in calculus). We study the practically most important ones in Chap. 4.

CHAPTER 3

Systems of Differential Equations, Phase Plane, Qualitative Methods

Systems of differential equations occur in various applications (see Secs. 3.1 and 3.5). Their theory (outlined in Sec. 3.2) includes that of a single equation. Linear systems (Secs. 3.3, 3.4, 3.6) are best treated by the use of matrices and vectors, of which, however, only modest knowledge will be needed here (see Sec. 3.0).

In addition to actually *solving* linear systems of differential equations (Secs. 3.3, 3.6), there is a totally different approach, namely, the powerful method of discussing the general behavior of solutions in the **phase plane** (Sec. 3.4). This is called a **qualitative method** because it does not need actual solutions (which would make it *"quantitative"*), which for many practically important systems cannot be obtained analytically.

The qualitative method also gives information on **stability.** This concept is of general importance in engineering science (for instance, in control theory). It means that, roughly speaking, small changes of a physical system at some instant cause only small changes in the behavior of the system at all later times.

Phase plane methods can be extended to nonlinear systems (Sec. 3.5), for which they are particularly useful.

Notations for variables and functions. For the unknown functions we shall write $y_1 = y_1(t)$, $y_2 = y_2(t)$ (or sometimes $y_1 = y_1(x)$, $y_2 = y_2(x)$). This seems preferable to *suddenly* changing the *independent* variable x in $y = f(x)$ (Calculus!) and in $y = y(x)$ (Chaps. 1, 2) into a *dependent* variable ($x_1 = x_1(t)$, $x_2 = x_2(t)$), as it is sometimes done in systems of differential equations.

Prerequisites for this chapter: Secs. 1.2, 1.6, Chap. 2.
References: Appendix 1, Part A.
Answers to problems: Appendix 2.

3.0 Introduction: Vectors, Matrices, Eigenvalues

In discussing linear systems of differential equations we shall use matrices and vectors. This will simplify formulas and clarify ideas. It will require knowledge of some rather elementary facts probably at the disposal of most students, who therefore may immediately go to Sec. 3.1 and use this section *for reference* as needed. (The full treatment of matrices in Chaps. 6 and 7 will *not* be needed in this chapter.)

Most of our linear systems will consist of two differential equations in two unknown functions $y_1(t)$, $y_2(t)$ of the form

(1)
$$y_1' = a_{11}y_1 + a_{12}y_2,$$
$$y_2' = a_{21}y_1 + a_{22}y_2,$$

for example,

$$y_1' = -5y_1 + 2y_2$$
$$y_2' = 13y_1 + \tfrac{1}{2}y_2$$

(probably with an additional *known* function in each equation on the right).

Similarly, a linear system of n first-order differential equations in n unknown functions $y_1(t), \cdots, y_n(t)$ is of the form

(2)
$$y_1' = a_{11}y_1 + a_{12}y_2 + \cdots + a_{1n}y_n$$
$$y_2' = a_{21}y_1 + a_{22}y_2 + \cdots + a_{2n}y_n$$
$$\cdots\cdots\cdots\cdots\cdots\cdots\cdots\cdots\cdots$$
$$y_n' = a_{n1}y_1 + a_{n2}y_2 + \cdots + a_{nn}y_n$$

(probably with an additional *known* function in each equation on the right).

Some Definitions and Terms

In (1) the (constant or variable) coefficients form a **2 × 2 matrix A,** that is, an array

(3) $$\mathbf{A} = [a_{jk}] = \begin{bmatrix} a_{11} & a_{12} \\ a_{21} & a_{22} \end{bmatrix},$$ for example, $$\mathbf{A} = \begin{bmatrix} -5 & 2 \\ 13 & \tfrac{1}{2} \end{bmatrix}.$$

Similarly, the coefficients in (2) form an **n × n matrix**

(4) $$\mathbf{A} = [a_{jk}] = \begin{bmatrix} a_{11} & a_{12} & \cdots & a_{1n} \\ a_{21} & a_{22} & \cdots & a_{2n} \\ \cdot & \cdot & \cdots & \cdot \\ a_{n1} & a_{n2} & \cdots & a_{nn} \end{bmatrix}.$$

The $a_{11}, a_{12}, \cdots$ are called **entries,** the horizontal lines **rows** or *row vectors,* and the vertical lines **columns** or *column vectors.* Thus, in (3) the first row is $[a_{11} \quad a_{12}]$, the second row is $[a_{21} \quad a_{22}]$ and the first and second columns are

$$\begin{bmatrix} a_{11} \\ a_{21} \end{bmatrix}$$ and $$\begin{bmatrix} a_{12} \\ a_{22} \end{bmatrix}.$$

In the *"**double subscript notation**"* for entries, the first subscript denotes the *row* and the second the *column* in which the entry stands. Similarly in (4). The **main diagonal** is the diagonal $a_{11} \quad a_{22} \quad \cdots \quad a_{nn}$ in (4), hence $a_{11} \quad a_{22}$ in (3).

We shall also need vectors. A **column vector x** with n **components** $x_1, \cdots, x_n$ is

$$\mathbf{x} = \begin{bmatrix} x_1 \\ x_2 \\ \vdots \\ x_n \end{bmatrix}, \qquad \text{thus if } n = 2, \qquad \mathbf{x} = \begin{bmatrix} x_1 \\ x_2 \end{bmatrix}.$$

Similarly, a **row vector v** is of the form

$$\mathbf{v} = [v_1 \quad \cdots \quad v_n], \qquad \text{thus if } n = 2, \qquad \mathbf{v} = [v_1, \quad v_2].$$

Calculations with Matrices and Vectors

Equality. Two $n \times n$ matrices are *equal* if and only if corresponding entries are equal. Thus for $n = 2$,

$$\mathbf{A} = \begin{bmatrix} a_{11} & a_{12} \\ a_{21} & a_{22} \end{bmatrix} = \mathbf{B} = \begin{bmatrix} b_{11} & b_{12} \\ b_{21} & b_{22} \end{bmatrix}$$

if and only if

$$a_{11} = b_{11}, \qquad a_{12} = b_{12}$$
$$a_{21} = b_{21}, \qquad a_{22} = b_{22}.$$

Two column vectors (or two row vectors) are *equal* if and only if they both have n components and corresponding components are equal. Thus for column vectors with $n = 2$ components,

$$\mathbf{v} = \begin{bmatrix} v_1 \\ v_2 \end{bmatrix} = \mathbf{x} = \begin{bmatrix} x_1 \\ x_2 \end{bmatrix} \qquad \text{if and only if} \qquad \begin{aligned} v_1 &= x_1 \\ v_2 &= x_2. \end{aligned}$$

Addition is performed by adding corresponding entries (or components); here, matrices must both be $n \times n$, and vectors must both have the same number of components. Thus for $n = 2$,

$$(5) \qquad \mathbf{A} + \mathbf{B} = \begin{bmatrix} a_{11} + b_{11} & a_{12} + b_{12} \\ a_{21} + b_{21} & a_{22} + b_{22} \end{bmatrix}, \qquad \mathbf{v} + \mathbf{x} = \begin{bmatrix} v_1 + x_1 \\ v_2 + x_2 \end{bmatrix}.$$

Scalar multiplication (multiplication by a number c) is performed by multiplying each entry (or component) by c. For example, if

$$\mathbf{A} = \begin{bmatrix} 9 & 3 \\ -2 & 0 \end{bmatrix}, \qquad \text{then} \quad -7\mathbf{A} = \begin{bmatrix} -63 & -21 \\ 14 & 0 \end{bmatrix}.$$

If

$$\mathbf{v} = \begin{bmatrix} 0.4 \\ -13 \end{bmatrix}, \qquad \text{then} \quad 10\mathbf{v} = \begin{bmatrix} 4 \\ -130 \end{bmatrix}.$$

Matrix multiplication. The product $\mathbf{C} = \mathbf{AB}$ (in this order) of two $n \times n$ matrices $\mathbf{A} = [a_{jk}]$ and $\mathbf{B} = [b_{jk}]$ is the $n \times n$ matrix $\mathbf{C} = [c_{jk}]$ with entries

(6)
$$c_{jk} = \sum_{m=1}^{n} a_{jm}b_{mk}$$
$$j = 1, \cdots, n$$
$$k = 1, \cdots, n,$$

that is, multiply each entry in the *j*th *row* of $\mathbf{A}$ by the corresponding entry in the *k*th *column* of $\mathbf{B}$ and then add these n products. One says briefly that this is a "multiplication of rows into columns." For example,

$$\begin{bmatrix} 9 & 3 \\ -2 & 0 \end{bmatrix}\begin{bmatrix} 1 & -4 \\ 2 & 5 \end{bmatrix} = \begin{bmatrix} 9 \cdot 1 + 3 \cdot 2 & 9 \cdot (-4) + 3 \cdot 5 \\ -2 \cdot 1 + 0 \cdot 2 & (-2) \cdot (-4) + 0 \cdot 5 \end{bmatrix}$$

$$= \begin{bmatrix} 15 & -21 \\ -2 & 8 \end{bmatrix}.$$

Caution! Matrix multiplication is *not commutative,* $\mathbf{AB} \neq \mathbf{BA}$ in general. In our example,

$$\begin{bmatrix} 1 & -4 \\ 2 & 5 \end{bmatrix}\begin{bmatrix} 9 & 3 \\ -2 & 0 \end{bmatrix} = \begin{bmatrix} 1 \cdot 9 + (-4) \cdot (-2) & 1 \cdot 3 + (-4) \cdot 0 \\ 2 \cdot 9 + 5 \cdot (-2) & 2 \cdot 3 + 5 \cdot 0 \end{bmatrix}$$

$$= \begin{bmatrix} 17 & 3 \\ 8 & 6 \end{bmatrix}.$$

Multiplication of an $n \times n$ matrix $\mathbf{A}$ by a vector $\mathbf{x}$ with n components is defined by the same rule: $\mathbf{v} = \mathbf{Ax}$ is the vector with the n components

$$v_j = \sum_{m=1}^{n} a_{jm}x_m \qquad\qquad j = 1, \cdots, n.$$

For example,

$$\begin{bmatrix} 12 & 7 \\ -8 & 3 \end{bmatrix}\begin{bmatrix} x_1 \\ x_2 \end{bmatrix} = \begin{bmatrix} 12x_1 + 7x_2 \\ -8x_1 + 3x_2 \end{bmatrix}.$$

Differentiation. The *derivative* of a matrix (or vector) with variable entries (or components) is obtained by differentiating each entry (or component). Thus, if

$$\mathbf{y}(t) = \begin{bmatrix} y_1(t) \\ y_2(t) \end{bmatrix} = \begin{bmatrix} e^{-2t} \\ \sin t \end{bmatrix}, \qquad \text{then} \qquad \mathbf{y}'(t) = \begin{bmatrix} y_1'(t) \\ y_2'(t) \end{bmatrix} = \begin{bmatrix} -2e^{-2t} \\ \cos t \end{bmatrix}.$$

Transposition is the operation of writing columns as rows and conversely and is indicated by T. Thus the transpose $\mathbf{A}^{\mathsf{T}}$ of the 3×3 matrix $\mathbf{A}$ given by is

$$\mathbf{A} = \begin{bmatrix} a_{11} & a_{12} & a_{13} \\ a_{21} & a_{22} & a_{23} \\ a_{31} & a_{32} & a_{33} \end{bmatrix} = \begin{bmatrix} 4 & 0 & -8 \\ 1 & -6 & 7 \\ 9 & 9 & 5 \end{bmatrix}$$

$$\mathbf{A}^\mathsf{T} = \begin{bmatrix} a_{11} & a_{21} & a_{31} \\ a_{12} & a_{22} & a_{32} \\ a_{13} & a_{23} & a_{33} \end{bmatrix} = \begin{bmatrix} 4 & 1 & 9 \\ 0 & -6 & 9 \\ -8 & 7 & 5 \end{bmatrix}.$$

The transpose of a column vector, say,

$$\mathbf{v} = \begin{bmatrix} v_1 \\ v_2 \end{bmatrix}, \qquad \text{is a row vector,} \qquad \mathbf{v}^\mathsf{T} = [v_1 \quad v_2],$$

and conversely.

Inverse of a matrix. The $n \times n$ **unit matrix** $\mathbf{I}$ is the $n \times n$ matrix with main diagonal $1, 1, \cdots, 1$ and all other entries zero. If for a given $n \times n$ matrix $\mathbf{A}$ there is an $n \times n$ matrix $\mathbf{B}$ such that $\mathbf{AB} = \mathbf{BA} = \mathbf{I}$, then $\mathbf{A}$ is called **nonsingular** and $\mathbf{B}$ is called the **inverse** of $\mathbf{A}$ and is denoted by $\mathbf{A}^{-1}$; thus

$$(7) \qquad\qquad \mathbf{AA}^{-1} = \mathbf{A}^{-1}\mathbf{A} = \mathbf{I}.$$

If $\mathbf{A}$ has no inverse, it is called **singular.** For $n = 2$,

$$(8) \qquad\qquad \mathbf{A}^{-1} = \frac{1}{\det \mathbf{A}} \begin{bmatrix} a_{22} & -a_{12} \\ -a_{21} & a_{11} \end{bmatrix},$$

where the **determinant** of $\mathbf{A}$ is

$$(9) \qquad\qquad \det \mathbf{A} = \begin{vmatrix} a_{11} & a_{12} \\ a_{21} & a_{22} \end{vmatrix} = a_{11}a_{22} - a_{12}a_{21}.$$

(For general n, see Sec. 6.7, but this will not be needed in this chapter.)

Linear independence. r given vectors $\mathbf{v}^{(1)}, \cdots, \mathbf{v}^{(r)}$ with n components are called a *linearly independent set* or, more briefly, **linearly independent,** if

$$(10) \qquad\qquad c_1 \mathbf{v}^{(1)} + \cdots + c_r \mathbf{v}^{(r)} = \mathbf{0}$$

implies that all scalars $c_1, \cdots, c_r$ must be zero; here, $\mathbf{0}$ denotes the **zero vector,** whose n components are all zero. If (10) also holds for scalars not all zero (so that at least one of these scalars is not zero), then these vectors are called a *linearly dependent set* or, briefly, **linearly dependent,** because then at least one of them can be expressed as a **linear combination** of the others; that is, if, for instance, $c_1 \neq 0$ in (10), then we can obtain

$$\mathbf{v}^{(1)} = -\frac{1}{c_1} (c_2 \mathbf{v}^{(2)} + \cdots + c_r \mathbf{v}^{(r)}).$$

Systems of Differential Equations as Vector Equations

Using matrix multiplication and differentiation, we can now write (1) as

$$(11) \quad \mathbf{y}' = \begin{bmatrix} y_1' \\ y_2' \end{bmatrix} = \mathbf{Ay} = \begin{bmatrix} a_{11} & a_{12} \\ a_{21} & a_{22} \end{bmatrix} \begin{bmatrix} y_1 \\ y_2 \end{bmatrix}, \quad \text{e.g.,} \quad \mathbf{y}' = \begin{bmatrix} -5 & 2 \\ 13 & \frac{1}{2} \end{bmatrix} \begin{bmatrix} y_1 \\ y_2 \end{bmatrix}.$$

Similarly for (2) by means of an $n \times n$ matrix $\mathbf{A}$ and a column vector $\mathbf{y}$ with n components, namely, $\mathbf{y}' = \mathbf{Ay}.$ The vector equation (11) is equivalent to two equations for the components, and these are precisely the two equations in (1).

Eigenvalues, Eigenvectors

Eigenvalues and eigenvectors will be very important in this chapter (and, as a matter of fact, throughout mathematics).

Let $\mathbf{A} = [a_{jk}]$ be an $n \times n$ matrix. Consider the equation

$$(12) \qquad \boxed{\mathbf{Ax} = \lambda\mathbf{x}}$$

where λ is a scalar (a real or complex number) to be determined and $\mathbf{x}$ is a vector to be determined. Now for every λ a solution is $\mathbf{x} = \mathbf{0}$. A scalar λ such that (12) holds for some vector $\mathbf{x} \neq \mathbf{0}$ is called an **eigenvalue** of $\mathbf{A}$, and this vector is called an **eigenvector** of $\mathbf{A}$ corresponding to this eigenvalue λ.

We can write (12) as $\mathbf{Ax} - \lambda\mathbf{x} = \mathbf{0}$ or

$$(13) \qquad \boxed{(\mathbf{A} - \lambda\mathbf{I})\mathbf{x} = \mathbf{0}.}$$

These are n linear algebraic equations in the n unknowns $x_1, \cdots, x_n$ (the components of $\mathbf{x}$). For these equations to have a solution $\mathbf{x} \neq \mathbf{0}$, the determinant of the coefficient matrix $\mathbf{A} - \lambda\mathbf{I}$ must be zero. This is proved as a basic fact in linear algebra (Theorem 4 in Sec. 6.6). In this chapter we need this only for $n = 2$. Then (13) is

$$(14) \qquad \begin{bmatrix} a_{11} - \lambda & a_{12} \\ a_{21} & a_{22} - \lambda \end{bmatrix} \begin{bmatrix} x_1 \\ x_2 \end{bmatrix} = \begin{bmatrix} 0 \\ 0 \end{bmatrix};$$

in components,

$$(14^*) \qquad \begin{aligned} (a_{11} - \lambda)x_1 + \quad a_{12}x_2 &= 0 \\ a_{21}x_1 + (a_{22} - \lambda)x_2 &= 0. \end{aligned}$$

Now $\mathbf{A} - \lambda\mathbf{I}$ is singular if and only if its determinant $\det(\mathbf{A} - \lambda\mathbf{I})$, called the **characteristic determinant** of $\mathbf{A}$ (also for general n), is zero. This gives

$$\det (\mathbf{A} - \lambda \mathbf{I}) = \begin{vmatrix} a_{11} - \lambda & a_{12} \\ a_{21} & a_{22} - \lambda \end{vmatrix}$$

(15)
$$= (a_{11} - \lambda)(a_{22} - \lambda) - a_{12}a_{21}$$

$$= \lambda^2 - (a_{11} + a_{22})\lambda + a_{11}a_{22} - a_{12}a_{21} = 0.$$

This quadratic equation in λ is called the **characteristic equation** of $\mathbf{A}$. Its solutions are the eigenvalues λ_1 and λ_2 of $\mathbf{A}$. First determine these. Then use (14*) with $\lambda = \lambda_1$ to determine an eigenvector $\mathbf{x}^{(1)}$ of $\mathbf{A}$ corresponding to λ_1. Finally use (14*) with $\lambda = \lambda_2$ to find an eigenvector $\mathbf{x}^{(2)}$ of $\mathbf{A}$ corresponding to λ_2. Note that if $\mathbf{x}$ is an eigenvector of $\mathbf{A}$, so is $k\mathbf{x}$ for any $k \neq 0$.

EXAMPLE 1   **Eigenvalue problem**

Find the eigenvalues and eigenvectors of the matrix

(16)
$$\mathbf{A} = \begin{bmatrix} -4.0 & 4.0 \\ -1.6 & 1.2 \end{bmatrix}.$$

Solution.   The characteristic equation is the quadratic equation

$$\det [\mathbf{A} - \lambda \mathbf{I}] = \begin{vmatrix} -4 - \lambda & 4 \\ -1.6 & 1.2 - \lambda \end{vmatrix} = \lambda^2 + 2.8\lambda + 1.6 = 0.$$

It has the solutions $\lambda_1 = -2$ and $\lambda_2 = -0.8$. These are the eigenvalues of $\mathbf{A}$.
  Eigenvectors are obtained from (14*). For $\lambda = \lambda_1 = -2$ we have from (14*)

$$(-4.0 + 2.0)x_1 + \quad 4.0x_2 \quad = 0$$

$$-1.6x_1 \quad + (1.2 + 2.0)x_2 = 0.$$

A solution of the first equation is $x_1 = 2$, $x_2 = 1$. This also satisfies the second equation. (Why?). Hence an eigenvector of $\mathbf{A}$ corresponding to $\lambda_1 = -2.0$ is

(17)
$$\mathbf{x}^{(1)} = \begin{bmatrix} 2 \\ 1 \end{bmatrix}. \qquad \text{Similarly,} \qquad \mathbf{x}^{(2)} = \begin{bmatrix} 1 \\ 0.8 \end{bmatrix}$$

is an eigenvector of $\mathbf{A}$ corresponding to $\lambda_2 = -0.8$, as obtained from (14*) with $\lambda = \lambda_2$. Verify this. ◀

3.1 Introductory Examples

We first illustrate with a few typical examples that systems of differential equations have various applications, and that an explicit higher order equation [see (8) in this section] can always be reduced to a first-order system. This accounts for the practical importance of these systems.

EXAMPLE 1   **Mixing problem involving two tanks**

A mixing problem involving a single tank leads to a single differential equation, as we have shown in Example 2 of Sec. 1.4. You may want to review that example because the principle of modeling will be the same for a mixing problem involving two tanks. This will lead to a system of two first-order differential equations.

Tank T_1 in Fig. 75 on the next page contains initially 100 gal of pure water. Tank T_2 contains initially 100 gal of water in which 150 lb of fertilizer are dissolved. Liquid circulates through the tanks at a constant rate of 2 gal/min, and the mixture is kept uniform by stirring. Find the amounts of fertilizer $y_1(t)$ and $y_2(t)$ in T_1 and T_2, respectively, where t is time.

Solution. 1st Step. Modeling.   As for a single tank, the time rate of change $y_1'(t)$ of $y_1(t)$ (amount of fertilizer in T_1) equals inflow minus outflow. Similarly for tank T_2. From Fig. 75 we see that

$$y_1' = \text{Inflow/min} - \text{Outflow/min} = \frac{2}{100}y_2 - \frac{2}{100}y_1 \qquad (\text{Tank } T_1)$$

$$y_2' = \text{Inflow/min} - \text{Outflow/min} = \frac{2}{100}y_1 - \frac{2}{100}y_2 \qquad (\text{Tank } T_2).$$

Hence the mathematical model of our mixture problem is the system of first-order differential equations

$$y_1' = -0.02y_1 + 0.02y_2 \qquad (\text{Tank } T_1)$$

$$y_2' = 0.02y_1 - 0.02y_2 \qquad (\text{Tank } T_2).$$

As a vector equation with vector $\mathbf{y} = [y_1 \ \ y_2]^\mathsf{T}$ (a column vector because we are taking the transpose!) and matrix $\mathbf{A}$ this becomes

$$\mathbf{y}' = \mathbf{Ay}, \qquad \text{where} \qquad \mathbf{A} = \begin{bmatrix} -0.02 & 0.02 \\ 0.02 & -0.02 \end{bmatrix}.$$

2nd Step. General solution.   As for a single equation, we try an exponential function of t,

$$(1) \qquad\qquad \mathbf{y} = \mathbf{x}e^{\lambda t}. \qquad \text{Then} \qquad \mathbf{y}' = \lambda \mathbf{x}e^{\lambda t} = \mathbf{Ax}e^{\lambda t}.$$

Dividing by $e^{\lambda t}$ and interchanging the left and right sides, we obtain

$$\mathbf{Ax} = \lambda \mathbf{x}.$$

We need nontrivial solutions (solutions that are not identically zero). Hence we have to look for eigenvalues and eigenvectors of $\mathbf{A}$. The eigenvalues are the solutions of the characteristic equation

$$(2) \qquad \det(\mathbf{A} - \lambda\mathbf{I}) = \begin{vmatrix} -0.02 - \lambda & 0.02 \\ 0.02 & -0.02 - \lambda \end{vmatrix} = (-0.02 - \lambda)^2 - 0.02^2 = \lambda(\lambda + 0.04) = 0.$$

We see that $\lambda_1 = 0$ (which can very well happen—don't get mixed up—it is eigen*vectors* that must not be zero) and $\lambda_2 = -0.04$. Eigenvectors are obtained from (14*) in Sec. 3.0 with $\lambda = 0$ and $\lambda = -0.04$. For our present $\mathbf{A}$ this gives [we need only the first equation in (14*)]

$$-0.02x_1 + 0.02x_2 = 0 \qquad \text{and} \qquad (-0.02 + 0.04)x_1 + 0.02x_2 = 0,$$

respectively. Hence $x_1 = x_2$ and $x_1 = -x_2$, respectively, and we can take $x_1 = x_2 = 1$ and $x_1 = -x_2 = 1$. This gives the eigenvectors

$$\mathbf{x}^{(1)} = \begin{bmatrix} 1 \\ 1 \end{bmatrix} \qquad \text{and} \qquad \mathbf{x}^{(2)} = \begin{bmatrix} 1 \\ -1 \end{bmatrix}.$$

From (1) and the superposition principle (which continues to hold for systems of homogeneous linear equations) we thus obtain a solution

$$(3) \qquad\qquad \mathbf{y} = c_1\mathbf{x}^{(1)}e^{\lambda_1 t} + c_2\mathbf{x}^{(2)}e^{\lambda_2 t} = c_1 \begin{bmatrix} 1 \\ 1 \end{bmatrix} + c_2 \begin{bmatrix} 1 \\ -1 \end{bmatrix} e^{-0.04t}$$

where c_1 and c_2 are arbitrary constants. Later we shall call this a *general solution*.

3rd Step. Initial conditions. Answer. The initial conditions are $y_1(0) = 0$ (no fertilizer in tank T_1) and $y_2(0) = 150$. From this and (3) with $t = 0$ we obtain

$$\mathbf{y}(0) = c_1 \begin{bmatrix} 1 \\ 1 \end{bmatrix} + c_2 \begin{bmatrix} 1 \\ -1 \end{bmatrix} = \begin{bmatrix} c_1 + c_2 \\ c_1 - c_2 \end{bmatrix} = \begin{bmatrix} 0 \\ 150 \end{bmatrix}.$$

In components this is $c_1 + c_2 = 0$, $c_1 - c_2 = 150$. The solution is $c_1 = 75$, $c_2 = -75$. This gives the answer

$$\mathbf{y} = 75\mathbf{x}^{(1)} - 75\mathbf{x}^{(2)}e^{-0.04t} = 75 \begin{bmatrix} 1 \\ 1 \end{bmatrix} - 75 \begin{bmatrix} 1 \\ -1 \end{bmatrix} e^{-0.04t}.$$

In components,

$$y_1 = 75 - 75e^{-0.04t} \qquad \text{(Tank } T_1\text{, lower curve)}$$

$$y_2 = 75 + 75e^{-0.04t} \qquad \text{(Tank } T_2\text{, upper curve).}$$

Figure 75 shows the exponential increase of y_1 and the exponential decrease of y_2 to the common limit 75 lb. Did you expect this for physical reasons? Can you physically explain why the curves look "symmetric"? Would the limit change if T_1 initially contained 100 lb of fertilizer and T_2 contained 50 lb? ◀

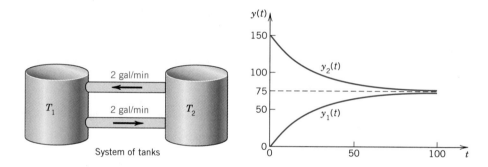

Fig. 75. Fertilizer content in Tanks T_1 (lower curve) and T_2

EXAMPLE 2 **Model of an electrical network**

Find the currents $I_1(t)$ and $I_2(t)$ in the network shown in Fig. 76a on p. 156, assuming that all charges and currents are zero at $t = 0$, the instant when the switch is closed.

Solution. 1st Step. Setting up the mathematical model. The mathematical model of this network is obtained from Kirchhoff's voltage law, as in Secs. 1.7 and 2.12 (where we considered single circuits). The left loop yields

$$I_1' + 4(I_1 - I_2) = 12$$

or

(4a) $$I_1' = -4I_1 + 4I_2 + 12,$$

where I_1 is the current in the left loop and I_2 is the current in the right loop, and $4(I_1 - I_2)$ is the voltage drop across the resistor because I_1 and I_2 flow through the resistor in opposite directions. Similarly, for the right loop we obtain

$$6I_2 + 4(I_2 - I_1) + 4 \int I_2 \, dt = 0$$

or, by differentiation and division by 10,

$$I_2' - 0.4I_1' + 0.4I_2 = 0.$$

Replacing $-0.4I_1'$ by using (4a) and ordering, we obtain

(4b) $$I_2' = -1.6I_1 + 1.2I_2 + 4.8.$$

In matrix form, (4) is (we write **J** since **I** is the unit matrix)

(5) $\quad$ **J' = AJ + g,** $\quad$ where $\quad$ $\mathbf{J} = \begin{bmatrix} I_1 \\ I_2 \end{bmatrix}$, $\quad$ $\mathbf{A} = \begin{bmatrix} -4.0 & 4.0 \\ -1.6 & 1.2 \end{bmatrix}$, $\quad$ $\mathbf{g} = \begin{bmatrix} 12.0 \\ 4.8 \end{bmatrix}$.

2nd Step. Solving (5). This is a nonhomogeneous linear system. No method of solution is yet available, but we could try to proceed as for a single equation, solving first the homogeneous system $\mathbf{J'} = \mathbf{AJ}$ (thus $\mathbf{J'} - \mathbf{AJ} = \mathbf{0}$) by substituting $\mathbf{J} = \mathbf{x}e^{\lambda t}$; this gives

$$\mathbf{J'} = \lambda\mathbf{x}e^{\lambda t} = \mathbf{Ax}e^{\lambda t}, \qquad \text{thus} \qquad \mathbf{Ax} = \lambda\mathbf{x}.$$

Hence, to obtain a nontrivial solution, we again need the eigenvalues and eigenvectors. For the present matrix **A** these are given in Example 1 in Sec. 3.0:

$$\lambda_1 = -2, \qquad \mathbf{x}^{(1)} = \begin{bmatrix} 2 \\ 1 \end{bmatrix}; \qquad \lambda_2 = -0.8, \qquad \mathbf{x}^{(2)} = \begin{bmatrix} 1 \\ 0.8 \end{bmatrix}.$$

Hence a "general solution" of the homogeneous system is

$$\mathbf{J}_h = c_1\mathbf{x}^{(1)}e^{-2t} + c_2\mathbf{x}^{(2)}e^{-0.8t}.$$

For a particular solution of (5), since **g** is constant, we try a constant vector $\mathbf{J}_p = \mathbf{a} = [a_1 \quad a_2]^\mathsf{T}$. Then $\mathbf{J}_p' = \mathbf{0}$. Substitution gives $\mathbf{Aa} + \mathbf{g} = \mathbf{0}$; in components,

$$-4.0a_1 + 4.0a_2 + 12.0 = 0$$

$$-1.6a_1 + 1.2a_2 + 4.8 = 0.$$

The solution is $a_1 = 3$, $a_2 = 0$; thus $\mathbf{a} = [3 \quad 0]^\mathsf{T}$. Hence

(6) $$\mathbf{J} = \mathbf{J}_h + \mathbf{J}_p = c_1\mathbf{x}^{(1)}e^{-2t} + c_2\mathbf{x}^{(2)}e^{-0.8t} + \mathbf{a};$$

in components,

$$I_1 = 2c_1e^{-2t} + c_2e^{-0.8t} + 3$$

$$I_2 = c_1e^{-2t} + 0.8c_2e^{-0.8t}.$$

The initial conditions give

$$I_1(0) = 2c_1 + c_2 = 0$$

$$I_2(0) = c_1 + 0.8c_2 + 3 = 0.$$

Hence $c_1 = -4$ and $c_2 = 5$. As the solution of our problem we thus obtain

(7) $$\mathbf{J} = -4\mathbf{x}^{(1)}e^{-2t} + 5\mathbf{x}^{(2)}e^{-0.8t} + \mathbf{a}.$$

In components (Fig. 76*b*),

$$I_1 = -8e^{-2t} + 5e^{-0.8t} + 3$$

$$I_2 = -4e^{-2t} + 4e^{-0.8t}.$$

We see that I_1 has the limit 3 amperes and I_2 zero. Did you expect this? Can you explain it physically?

Figure 76b shows $I_1(t)$ and $I_2(t)$ as two separate curves. Figure 76c shows these two currents as a single curve $[I_1(t), I_2(t)]$ in the I_1I_2-plane. This is a parametric representation with t as the parameter. It is often important

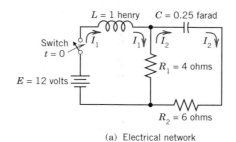

(a) Electrical network

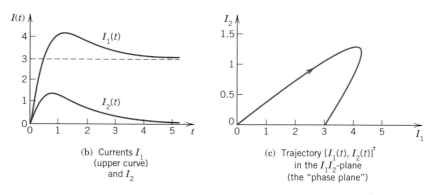

(b) Currents I_1
(upper curve)
and I_2

(c) Trajectory $[I_1(t), I_2(t)]^T$
in the $I_1 I_2$-plane
(the "phase plane")

Fig. 76. Electrical network and currents in Example 2

to know in what sense such a curve is traced. This can be indicated by an arrow in the direction of increasing t, as shown. The $I_1 I_2$-plane is called the **phase plane** of our system (5) and the curve is called a **trajectory.** We shall see that such "phase plane representations" are far more important than graphs as in Fig. 76b because they will give a much better qualitative overall impression of the general behavior *of whole families* of solutions, not merely of one solution as in the present case. ◀

Conversion of an *n*th Order Differential Equation to a System

We show that an nth order differential equation can be converted to a system of n first-order differential equations. This is practically and theoretically important. Practically, it permits the study and solution of single equations by methods for systems. Theoretically, it gives a possibility of including the theory of higher order equations into that of first-order systems. This conversion is another reason for the importance of systems, in addition to their use as models in various basic applications.

The idea underlying the conversion is simple. An nth order differential equation

$$(8) \qquad\qquad y^{(n)} = F(t, y, y', \cdots, y^{(n-1)})$$

can always be reduced to a system of n first-order differential equations simply by setting

$$(9) \qquad\qquad \boxed{y_1 = y, \quad y_2 = y', \quad y_3 = y'', \cdots, y_n = y^{(n-1)}.}$$

We then immediately obtain the first-order system

$$y_1' = y_2$$
$$y_2' = y_3$$
$$\vdots$$
$$y_{n-1}' = y_n$$
$$y_n' = F(t, y_1, y_2, \cdots, y_n).$$

(10)

The first $n - 1$ of these equations follow immediately from (9) by differentiation. Also, $y_n' = y^{(n)}$ by (9), so that the last equation in (10) follows from the given differential equation (8). This reduction is also a reason why one generally concentrates on *first-order* systems. To gain confidence, let us consider an old friend of ours.

EXAMPLE 3   **Mass on a spring**

For

$$y'' + \frac{c}{m} y' + \frac{k}{m} y = 0 \qquad\qquad \text{(Sec. 2.5)}$$

the system (10) is linear and homogeneous,

$$y_1' = y_2$$

$$y_2' = -\frac{k}{m} y_1 - \frac{c}{m} y_2$$

Setting $\mathbf{y}^\mathsf{T} = [y_1 \quad y_2]$, we get in matrix form

$$\mathbf{y}' = \begin{bmatrix} 0 & 1 \\ -\dfrac{k}{m} & -\dfrac{c}{m} \end{bmatrix} \mathbf{y}.$$

The characteristic equation is

$$\det(\mathbf{A} - \lambda\mathbf{I}) = \begin{vmatrix} -\lambda & 1 \\ -\dfrac{k}{m} & -\dfrac{c}{m} - \lambda \end{vmatrix} = \lambda^2 + \frac{c}{m}\lambda + \frac{k}{m} = 0.$$

It agrees with the characteristic equation in Sec. 2.5. For a simple illustrative computation, assume that $m = 1$, $c = 2$, and $k = 0.75$. Then $\lambda^2 + 2\lambda + 0.75 = 0$, and this gives the eigenvalues $\lambda_1 = -0.5$, $\lambda_2 = -1.5$. Eigenvectors are $\mathbf{x}^{(1)} = [2 \quad -1]^\mathsf{T}$ (from $0.5x_1 + x_2 = 0$) and $\mathbf{x}^{(2)} = [1 \quad -1.5]^\mathsf{T}$ (from $1.5x_1 + x_2 = 0$). Hence we get the solution

$$\mathbf{y} = c_1 \begin{bmatrix} 2 \\ -1 \end{bmatrix} e^{-0.5t} + c_2 \begin{bmatrix} 1 \\ -1.5 \end{bmatrix} e^{-1.5t}.$$

The first component of this vector equation is the expected solution

$$y = y_1 = 2c_1 e^{-0.5t} + c_2 e^{-1.5t}$$

and $y_2 = y_1'$ comes out as it should,

$$y_2 = -c_1 e^{-0.5t} - 1.5c_2 e^{-1.5t} = y_1'.$$

◄

PROBLEM SET 3.1

Mixing Problems. What would happen in Example 1:
1. If the exchange rate were doubled (4 gal/min)? First guess, then calculate.
2. If T_1 were replaced by a tank of 200 gal? First guess.
3. If the size of the tanks were reduced to 50 gal each?

Electrical Network. In Example 2, find the currents:
4. If the initial currents are $I_1(0) = 28$ amperes, $I_2(0) = 14$ amperes.
5. If the initial currents are $I_1(0) = 9$ amperes, $I_2(0) = 0$ ampere.
6. If the capacitance is changed to $C = 5/27$ farad. (General solution only.)

Conversion to Systems. Solve the given equation (a) by first converting it to a system, and (b) as given. (General solution. Show the details of your work.)

7. $y'' - y = 0$ 8. $y'' + 3y' + 2y = 0$
9. $y'' - 9y = 0$ 10. $4y'' - 15y' - 4y = 0$
11. $y'' - 4y' = 0$ 12. $y''' + 2y'' - y' - 2y = 0$

 13. **CAS PROJECT. Electrical Network.** (a) In Example 2 choose a sequence of values of C increasing beyond bound and compute the corresponding sequences of eigenvalues of **A.** What limits of these sequences do your numerical values (approximately) suggest?
     (b) Find these limits analytically.
     (c) Explain your result physically.
     (d) Below what value (approximately) must you decrease C to get vibrations?

14. **TEAM PROJECT. Conversion of Second-Order Systems. Masses on Springs.** (a) Model the (undamped) mechanical system in Fig. 77.
     (b) Solve the system of two second-order differential equations directly as obtained. (Try exponential functions $\mathbf{y} = \mathbf{x}e^{\omega t}$ and set $\omega^2 = \lambda$. Proceed as in Example 1 or 2.)
     (c) Convert your system into a first-order system of four equations and solve it. Convince yourself that in the present case, (b) seems more practical.

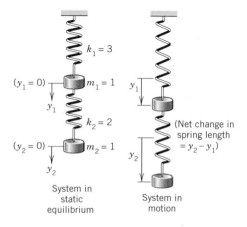

Fig. 77. Mechanical system in Team Project 14

3.2 Basic Concepts and Theory

In this section we discuss some basic concepts and facts about systems of differential equations that are similar to those for single equations.

The first-order systems in the last section were special cases of the more general system

(1)

$$
\begin{aligned}
y_1' &= f_1(t, y_1, \cdots, y_n) \\
y_2' &= f_2(t, y_1, \cdots, y_n) \\
&\quad \cdots \\
y_n' &= f_n(t, y_1, \cdots, y_n).
\end{aligned}
$$

Introducing the *column* vectors $\mathbf{y} = [y_1 \ \cdots \ y_n]^\mathsf{T}$ and $\mathbf{f} = [f_1 \ \cdots \ f_n]^\mathsf{T}$ ($^\mathsf{T}$ means transposition and saves us space!), we can write (1) as a vector equation

(1)

$$
\mathbf{y}' = \mathbf{f}(t, \mathbf{y}).
$$

This system (1) includes almost all cases of practical interest. For $n = 1$ it becomes $y_1' = f_1(t, y_1)$ or, simply, $y' = f(t, y)$, well known to us from Chap. 1.

A **solution** of (1) on some interval $a < t < b$ is a set of n differentiable functions

$$
y_1 = h_1(t), \cdots, y_n = h_n(t)
$$

on $a < t < b$ that satisfy (1) throughout this interval. In vector form, introducing the *"solution vector"* $\mathbf{h} = [h_1 \ \cdots \ h_n]^\mathsf{T}$ (a column vector!) we can write

$$
\mathbf{y} = \mathbf{h}(t).
$$

An **initial value problem** for (1) consists of (1) and n given initial conditions

(2) $y_1(t_0) = K_1,$ $y_2(t_0) = K_2,$ $\cdots,$ $y_n(t_0) = K_n,$

in vector form, $\mathbf{y}(t_0) = \mathbf{K}$, where t_0 is a specified value of t in the interval considered and the components of $\mathbf{K} = [K_1 \ \cdots \ K_n]^\mathsf{T}$ are given numbers. Sufficient conditions for the existence and uniqueness of a solution of an initial value problem (1), (2) are stated in the following theorem, which extends the theorems in Sec. 1.9 for a single equation. (For a proof, see Ref. [A3].)

THEOREM 1   **(Existence and uniqueness theorem)**

Let $f_1, \cdots, f_n$ in (1) be continuous functions having continuous partial derivatives $\partial f_1/\partial y_1, \cdots, \partial f_1/\partial y_n, \cdots, \partial f_n/\partial y_n$ in some domain R of $t y_1 y_2 \cdots y_n$-space containing the point $(t_0, K_1, \cdots, K_n)$. Then (1) has a solution on some interval $t_0 - \alpha < t < t_0 + \alpha$ satisfying (2), and this solution is unique.

Linear Systems

Extending the notion of a linear different equation, we call (1) a **linear system** if it is linear in $y_1 \cdots, y_n$; that is, if it can be written

(3)
$$
\boxed{
\begin{aligned}
y_1' &= a_{11}(t)y_1 + \cdots + a_{1n}(t)y_n + g_1(t) \\
&\quad\vdots \\
y_n' &= a_{n1}(t)y_1 + \cdots + a_{nn}(t)y_n + g_n(t).
\end{aligned}
}
$$

In vector form, this becomes

(3)
$$\boxed{\mathbf{y}' = \mathbf{A}\mathbf{y} + \mathbf{g}}$$

where
$$
\mathbf{A} = \begin{bmatrix} a_{11} & \cdots & a_{1n} \\ \cdot & \cdots & \cdot \\ a_{n1} & \cdots & a_{nn} \end{bmatrix}, \qquad
\mathbf{y} = \begin{bmatrix} y_1 \\ \vdots \\ y_n \end{bmatrix}, \qquad
\mathbf{g} = \begin{bmatrix} g_1 \\ \vdots \\ g_n \end{bmatrix}.
$$

This system is called **homogeneous** if $\mathbf{g} = \mathbf{0}$, so that it is

(4)
$$\mathbf{y}' = \mathbf{A}\mathbf{y}.$$

If $\mathbf{g} \neq \mathbf{0}$, then (3) is called **nonhomogeneous.** The system in Example 1 in the last section is homogeneous and in Example 2 nonhomogeneous.

For a linear system (3) we have $\partial f_1/\partial y_1 = a_{11}(t), \cdots, \partial f_n/\partial y_n = a_{nn}(t)$ in Theorem 1. Hence for a linear system we simply obtain the following.

THEOREM 2 **(Existence and uniqueness in the linear case)**

Let the a_{jk}'s and g_j's in (3) be continuous functions of t on an open interval $\alpha < t < \beta$ containing the point $t = t_0$. Then (3) has a solution $\mathbf{y}(t)$ on this interval satisfying (2), and this solution is unique.

As for a single homogeneous linear equation we have

THEOREM 3 **(Superposition principle or linearity principle)**

*If $\mathbf{y}^{(1)}$ and $\mathbf{y}^{(2)}$ are solutions of the **homogeneous linear** system (4) on some interval, so is any linear combination $\mathbf{y} = c_1\mathbf{y}^{(1)} + c_2\mathbf{y}^{(2)}$.*

PROOF. Differentiating and using (4), we obtain

$$
\begin{aligned}
\mathbf{y}' &= \left[c_1\mathbf{y}^{(1)} + c_2\mathbf{y}^{(2)}\right]' \\
&= c_1\mathbf{y}^{(1)\prime} + c_2\mathbf{y}^{(2)\prime} \\
&= c_1\mathbf{A}\mathbf{y}^{(1)} + c_2\mathbf{A}\mathbf{y}^{(2)} \\
&= \mathbf{A}(c_1\mathbf{y}^{(1)} + c_2\mathbf{y}^{(2)}) = \mathbf{A}\mathbf{y}.
\end{aligned}
$$

◀

The general theory of linear systems is quite similar to that of a single linear equation in Secs. 2.7 and 2.8. To see this, we shall explain the most basic concepts and facts. For proofs we refer to more advanced texts, such as [A3].

Basis. General Solution. Wronskian

By a **basis** or a **fundamental system** of solutions of the homogeneous system (4) on some interval[1] J we mean a linearly independent set of n solutions $\mathbf{y}^{(1)}, \cdots, \mathbf{y}^{(n)}$ of (4) on that interval. We call a corresponding linear combination

$$(5) \qquad \mathbf{y} = c_1\mathbf{y}^{(1)} \cdots + c_n\mathbf{y}^{(n)} \qquad (c_1, \cdots, c_n \text{ arbitrary})$$

a **general solution** of (4) on J. It can be shown that if the $a_{jk}(t)$ in (4) are continuous on J, then (4) has a basis of solutions on J, hence a general solution, which includes every solution of (4) on J.

We can write n solutions $\mathbf{y}^{(1)}, \cdots, \mathbf{y}^{(n)}$ of (4) on some interval J as columns of an $n \times n$ matrix

$$(6) \qquad \mathbf{Y} = \begin{bmatrix} \mathbf{y}^{(1)} & \cdots & \mathbf{y}^{(n)} \end{bmatrix}.$$

The determinant of $\mathbf{Y}$ is called the **Wronskian** of $\mathbf{y}^{(1)}, \cdots, \mathbf{y}^{(n)}$, written

$$(7) \qquad W(\mathbf{y}^{(1)}, \cdots, \mathbf{y}^{(n)}) = \begin{vmatrix} y_1^{(1)} & y_1^{(2)} & \cdots & y_1^{(n)} \\ y_2^{(1)} & y_2^{(2)} & \cdots & y_2^{(n)} \\ \cdot & \cdot & \cdots & \cdot \\ y_n^{(1)} & y_n^{(2)} & \cdots & y_n^{(n)} \end{vmatrix}.$$

The columns are these solutions, each in terms of components. These solutions form a basis on J if and only if W is not zero at any t_1 in this interval. W either is identically zero or is nowhere zero in J. (This is similar to Secs. 2.7 and 2.13.) If these solutions form a basis (a fundamental system), then (6) is often called a **fundamental matrix.**

We can relate (7) to Sec. 2.7. If y and z are solutions of a second-order homogeneous linear differential equation, their Wronskian is

$$W = \begin{vmatrix} y & z \\ y' & z' \end{vmatrix}.$$

If we write this equation as a system, we have to set $y = y_1$, $y' = y_1' = y_2$, and similarly for z (see Sec. 3.1). Thus, $W(y, z)$ becomes (7) with $n = 2$, except for notation.

---

[1] We write J because we need I to denote the unit matrix. Students familiar with vector spaces (Sec. 6.4) will notice that the solutions of a ***homogeneous*** linear system form an n-dimensional vector space. This will not be essential in what follows.

3.3 Homogeneous Systems with Constant Coefficients. Phase Plane, Critical Points

We continue our discussion of homogeneous linear systems

(1) $$\mathbf{y'} = \mathbf{Ay},$$

now assuming that the $n \times n$ matrix $\mathbf{A} = [a_{jk}]$ is constant, that is, its entries do not depend on t. We wish to solve (1). For this we remember that a single equation $y' = ky$ has the solution $y = Ce^{kt}$. Accordingly, we try

(2) $$\mathbf{y} = \mathbf{x}e^{\lambda t}.$$

By substitution into (1) we get

$$\mathbf{y'} = \lambda \mathbf{x}e^{\lambda t} = \mathbf{Ay} = \mathbf{Ax}e^{\lambda t}.$$

Dividing by $e^{\lambda t}$, we are left with the **eigenvalue problem**

(3) $$\mathbf{Ax} = \lambda\mathbf{x}.$$

Thus the nontrivial solutions of (1) are of the form (2), where λ is an eigenvalue of $\mathbf{A}$ and $\mathbf{x}$ is a corresponding eigenvector.

Let us further assume that $\mathbf{A}$ has a basis of n eigenvectors $\mathbf{x}^{(1)}, \cdots, \mathbf{x}^{(n)}$ corresponding to eigenvalues $\lambda_1, \cdots, \lambda_n$ (which may all be different or some of which—or all—may be equal). Then the corresponding solutions (2) are

(4) $$\mathbf{y}^{(1)} = \mathbf{x}^{(1)}e^{\lambda_1 t}, \qquad \cdots, \qquad \mathbf{y}^{(n)} = \mathbf{x}^{(n)}e^{\lambda_n t}.$$

Their Wronskian [(7) in Sec. 3.2] is

$$W(\mathbf{y}^{(1)}, \cdots, \mathbf{y}^{(n)}) = \begin{vmatrix} x_1^{(1)}e^{\lambda_1 t} & \cdots & x_1^{(n)}e^{\lambda_n t} \\ x_2^{(1)}e^{\lambda_1 t} & \cdots & x_2^{(n)}e^{\lambda_n t} \\ \cdot & \cdots & \cdot \\ x_n^{(1)}e^{\lambda_1 t} & \cdots & x_n^{(n)}e^{\lambda_n t} \end{vmatrix}$$

$$= e^{\lambda_1 t + \cdots + \lambda_n t} \begin{vmatrix} x_1^{(1)} & \cdots & x_1^{(n)} \\ x_2^{(1)} & \cdots & x_2^{(n)} \\ \cdot & \cdots & \cdot \\ x_n^{(1)} & \cdots & x_n^{(n)} \end{vmatrix}.$$

On the right, the exponential function is never zero, and the determinant is not zero either because its columns are the linearly independent eigenvectors that form a basis. This proves

THEOREM 1   **(General solution)**

If the constant matrix **A** *in the system* (1) *has a linearly independent set of n eigenvectors,*[2] *then the corresponding solutions* $\mathbf{y}^{(1)}, \cdots, \mathbf{y}^{(n)}$ *in* (4) *form a basis of solutions of* (1), *and the corresponding general solution is*

(5)
$$\mathbf{y} = c_1 \mathbf{x}^{(1)} e^{\lambda_1 t} + \cdots + c_n \mathbf{x}^{(n)} e^{\lambda_n t}.$$

How to Plot Solutions. Phase Plane

We shall now concentrate on homogeneous linear systems (1) with constant coefficients consisting of two equations:

(6)
$$\mathbf{y}' = \mathbf{Ay}; \quad \text{in components,} \quad \begin{aligned} y_1' &= a_{11}y_1 + a_{12}y_2 \\ y_2' &= a_{21}y_1 + a_{22}y_2. \end{aligned}$$

Of course, we can plot solutions

(7)
$$\mathbf{y}(t) = \begin{bmatrix} y_1(t) \\ y_2(t) \end{bmatrix}$$

of (6) as two curves over the t-axis, one for each component of $\mathbf{y}(t)$. (Figure 76b in Sec. 3.1 shows an example.) But we can also plot (7) as a single curve in the y_1y_2-plane. This is a parametric representation (parametric equation) with parameter t, known from calculus. (See Fig. 76c for an example. Many more to follow.) Such a curve is called a **trajectory** (or sometimes an *orbit* or *path*) of (6). The y_1y_2-plane is called the **phase plane**[3] of (1). If we fill the phase plane with trajectories of (6), we obtain the so-called **phase portrait** of (6).

EXAMPLE 1   **Trajectories in the phase plane (phase portrait)**

To see what is going on, as a first introductory example let us find and plot solutions of the system

(8)
$$\mathbf{y}' = \mathbf{Ay} = \begin{bmatrix} -3 & 1 \\ 1 & -3 \end{bmatrix} \mathbf{y}, \quad \text{thus} \quad \begin{aligned} y_1' &= -3y_1 + y_2 \\ y_2' &= y_1 - 3y_2. \end{aligned}$$

Solution.   By substituting $\mathbf{y} = \mathbf{x}e^{\lambda t}$ and $\mathbf{y}' = \lambda \mathbf{x}e^{\lambda t}$ and dropping the exponential function we get $\mathbf{Ax} = \lambda \mathbf{x}$. The characteristic equation is

$$\det(\mathbf{A} - \lambda\mathbf{I}) = \begin{vmatrix} -3 - \lambda & 1 \\ 1 & -3 - \lambda \end{vmatrix} = \lambda^2 + 6\lambda + 8 = 0.$$

---

[2]This holds if **A** is symmetric ($a_{jk} = a_{kj}$) or if it has *n different* eigenvalues. (See Theorems 3 and 4 in Sec. 7.5.)

[3]A name that comes from physics, where it is the y-(mv)-plane, used to plot a motion in terms of position y and velocity $y' = v$ (m = mass); but the name is now used quite generally for the y_1y_2-plane.

This gives the eigenvalues $\lambda_1 = -2$ and $\lambda_2 = -4$. Eigenvectors are then obtained from

$$(-3 - \lambda)x_1 + x_2 = 0.$$

For $\lambda_1 = -2$ this is $-x_1 + x_2 = 0$. Hence we can take $\mathbf{x}^{(1)} = [1 \quad 1]^\mathsf{T}$. For $\lambda_2 = -4$ this becomes $x_1 + x_2 = 0$, and an eigenvector is $\mathbf{x}^{(2)} = [1 \quad -1]^\mathsf{T}$. This gives the general solution

$$\mathbf{y} = \begin{bmatrix} y_1 \\ y_2 \end{bmatrix} = c_1 \mathbf{y}^{(1)} + c_2 \mathbf{y}^{(2)} = c_1 \begin{bmatrix} 1 \\ 1 \end{bmatrix} e^{-2t} + c_2 \begin{bmatrix} 1 \\ -1 \end{bmatrix} e^{-4t}.$$

Figure 78 shows a phase portrait of some of the trajectories (to which more trajectories could be added if so desired). The two straight trajectories correspond to $c_1 = 0$ and $c_2 = 0$ and the others to other choices of c_1, c_2. ◀

Studies of solutions in the phase plane have recently become quite important, along with advances in computer graphics, because a phase portrait gives a good general qualitative impression of the entire family of solutions.

Critical Points of the System (6)

The point $\mathbf{y} = \mathbf{0}$ in Fig. 78 seems to be a common point of all trajectories, and we want to explore the reason for this remarkable observation. The answer will follow by calculus. Indeed, from (6) we obtain

$$(9) \qquad \frac{dy_2}{dy_1} = \frac{dy_2/dt}{dy_1/dt} = \frac{y_2'}{y_1'} = \frac{a_{21}y_1 + a_{22}y_2}{a_{11}y_1 + a_{12}y_2}.$$

This associates with every point P: (y_1, y_2) a unique tangent direction dy_2/dy_1 of the trajectory passing through P, except for the point $P = P_0$: $(0, 0)$, where the right side of (9) becomes $0/0$. This point P_0, at which dy_2/dy_1 becomes undertermined, is called a **critical point** of (6).

Five Types of Critical Points

There are five types of critical points depending on the geometrical shape of the trajectories near them. They are called **improper nodes, proper nodes, saddle points, centers,** and **spiral points.** We define and illustrate them in Examples 1–5.

EXAMPLE 1 **(continued) Improper node (Fig. 78)**

An **improper node** is a critical point P_0 at which all the trajectories, except for two of them, have the same limiting direction of the tangent. The two exceptional trajectories also have a limiting direction of the tangent at P_0 which, however, is different.

The system (8) has an improper node at **0,** as its phase portrait Fig. 78 shows. The common limiting direction at **0** is that of the eigenvector $\mathbf{x}^{(1)} = [1 \quad 1]^\mathsf{T}$ because e^{-4t} goes to zero faster than e^{-2t} as t increases. The exceptional limiting tangent direction is that of $\mathbf{x}^{(2)} = [1 \quad -1]^\mathsf{T}$. ◀

EXAMPLE 2 **Proper node (Fig. 79)**

A **proper node** is a critical point P_0 at which every trajectory has a definite limiting direction and for any given direction d at P_0 there is a trajectory having d as its limiting direction.

The system

$$(10) \qquad \mathbf{y}' = \begin{bmatrix} 1 & 0 \\ 0 & 1 \end{bmatrix} \mathbf{y}, \qquad \text{thus} \qquad \begin{aligned} y_1' &= y_1 \\ y_2' &= y_2 \end{aligned}$$

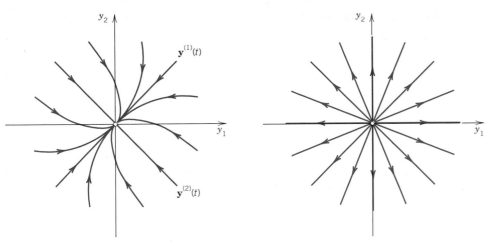

Fig. 78. Trajectories of the system (8)
(Improper node)

Fig. 79. Trajectories of the system (10)
(Proper node)

has a proper node at the origin (see Fig. 79) because a general solution is

$$\mathbf{y} = c_1 \begin{bmatrix} 1 \\ 0 \end{bmatrix} e^t + c_2 \begin{bmatrix} 0 \\ 1 \end{bmatrix} e^t \qquad \text{or} \qquad \begin{aligned} y_1 &= c_1 e^t \\ y_2 &= c_2 e^t \end{aligned} \qquad \text{or} \qquad c_1 y_2 = c_2 y_1. \qquad \blacktriangleleft$$

EXAMPLE 3   **Saddle point (Fig. 80)**

A **saddle point** is a critical point P_0 at which there are two incoming trajectories, two outgoing trajectories, and all the other trajectories in a neighborhood of P_0 bypass P_0.

   The system

$$(11) \qquad\qquad \mathbf{y}' = \begin{bmatrix} 1 & 0 \\ 0 & -1 \end{bmatrix} \mathbf{y}, \qquad \text{thus} \qquad \begin{aligned} y_1' &= y_1 \\ y_2' &= -y_2 \end{aligned}$$

has a saddle point at the origin because a general solution is

$$\mathbf{y} = c_1 \begin{bmatrix} 1 \\ 0 \end{bmatrix} e^t + c_2 \begin{bmatrix} 0 \\ 1 \end{bmatrix} e^{-t} \qquad \text{or} \qquad \begin{aligned} y_1 &= c_1 e^t \\ y_2 &= c_2 e^{-t} \end{aligned} \qquad \text{or} \qquad y_1 y_2 = \textit{const}.$$

This is a family of hyperbolas (and the coordinate axes); see Fig. 80. $\qquad \blacktriangleleft$

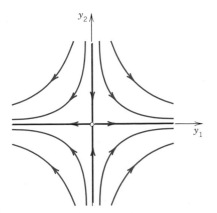

Fig. 80. Trajectories of the system (11)
(Saddle point)

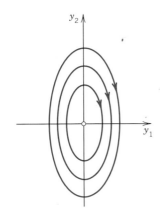

Fig. 81. Trajectories of the system (12)
(Center)

EXAMPLE 4   **Center (Fig. 81)**

A **center** is a critical point that is enclosed by infinitely many closed trajectories.

The system

(12) $$\mathbf{y}' = \begin{bmatrix} 0 & 1 \\ -4 & 0 \end{bmatrix} \mathbf{y}, \qquad \text{thus} \qquad \begin{aligned} y_1' &= y_2 \\ y_2' &= -4y_1 \end{aligned}$$

has a center at the origin, as we now show. The characteristic equation is $\lambda^2 + 4 = 0$. It has the eigenvalues $2i$ and $-2i$, $i = \sqrt{-1}$, and eigenvectors $[1 \quad 2i]^\mathsf{T}$ and $[1 \quad -2i]^\mathsf{T}$, respectively (verify!). Hence a complex general solution is

(12*) $$\mathbf{y} = c_1 \begin{bmatrix} 1 \\ 2i \end{bmatrix} e^{2it} + c_2 \begin{bmatrix} 1 \\ -2i \end{bmatrix} e^{-2it}, \qquad \text{thus} \qquad \begin{aligned} y_1 &= c_1 e^{2it} + c_2 e^{-2it} \\ y_2 &= 2ic_1 e^{2it} - 2ic_2 e^{-2it}. \end{aligned}$$

The next step would be the transformation of this solution to real form by the Euler formula (Sec. 2.3). But we were just curious to see what kind of eigenvalues we obtain in the case of a center. Accordingly, we do not continue, but start again from the beginning and use a shortcut. We rewrite the given equations in the form $y_1' = y_2$, $4y_1 = -y_2'$; then the product of the left sides must equal the product of the right sides,

$$4y_1 y_1' = -y_2 y_2'. \qquad \text{By integration,} \qquad 2y_1^2 + \tfrac{1}{2} y_2^2 = \textit{const.}$$

This is a family of ellipses (see Fig. 81 on the previous page) enclosing the center at the origin.   ◀

EXAMPLE 5   **Spiral point (Fig. 82)**

A **spiral point** is a critical point P_0 about which the trajectories spiral, approaching P_0 as $t \to \infty$ (or tracing these spirals in the opposite sense, away from P_0).

The system

(13) $$\mathbf{y}' = \begin{bmatrix} -1 & 1 \\ -1 & -1 \end{bmatrix} \mathbf{y}, \qquad \text{thus} \qquad \begin{aligned} y_1' &= -y_1 + y_2 \\ y_2' &= -y_1 - y_2 \end{aligned}$$

has a spiral point at the origin, as we show. The characteristic equation is $\lambda^2 + 2\lambda + 2 = 0$. It gives the eigenvalues $-1 + i$ and $-1 - i$. Corresponding eigenvectors are obtained from $(-1 - \lambda)x_1 + x_2 = 0$. For $\lambda = -1 + i$ this becomes $-ix_1 + x_2 = 0$ and we can take $[1 \quad i]^\mathsf{T}$ as an eigenvector. Similarly, an eigenvector corresponding to $-1 - i$ is $[1 \quad -i]^\mathsf{T}$. This gives the complex general solution

$$\mathbf{y} = c_1 \begin{bmatrix} 1 \\ i \end{bmatrix} e^{(-1+i)t} + c_2 \begin{bmatrix} 1 \\ -i \end{bmatrix} e^{(-1-i)t}.$$

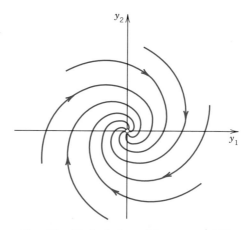

Fig. 82.   Trajectories of the system (13)
(Spiral point)

The next step would be the transformation of this complex solution to a real general solution by the Euler formula. But, as in the last example, we just wanted to see what eigenvalues to expect in the case of a spiral point. Accordingly, we start again from the beginning and instead of that rather lengthy systematic calculation we use a shortcut. We multiply the first equation in (13) by y_1, the second by y_2, and add, obtaining

$$y_1 y_1' + y_2 y_2' = -(y_1{}^2 + y_2{}^2).$$

Introducing polar coordinates r, t, where $r^2 = y_1{}^2 + y_2{}^2$, we see that this equation becomes $\frac{1}{2}(r^2)' = -r^2$. Now $(r^2)' = 2rr'$ by differentiation. Together,

$$\frac{1}{2}(r^2)' = rr' = -r^2$$

$$r' = -r$$

$$\ln r = -t + \tilde{c}.$$

Hence by taking exponentials we have

$$r = ce^{-t}.$$

For each real c this is a spiral, as claimed (see Fig. 82).   ◀

No Basis of Eigenvectors Available

When could this happen and what could we do? Well, it *cannot* happen if the matrix $\mathbf{A}$ is symmetric ($a_{jk} = a_{kj}$, as in Examples 1–3) or skew-symmetric ($a_{jk} = -a_{kj}$, thus $a_{jj} = 0$), and it also does not happen in many other cases (e.g., in Examples 4 and 5). This is the case for any n, not just for $n = 2$. If it happens, what can we do?

Suppose that an $n \times n$ matrix $\mathbf{A}$ has a double eigenvalue μ [that is, the product representation of $\det(\mathbf{A} - \lambda\mathbf{I})$ has a factor $(\lambda - \mu)^2$] with only one eigenvector (and its multiples) corresponding to it, instead of two linearly independent eigenvectors, so that we first get only one solution $\mathbf{y}^{(1)} = \mathbf{x}e^{\mu t}$. In this case we can obtain a second independent solution by substituting

(14)
$$\boxed{\mathbf{y}^{(2)} = \mathbf{x}te^{\mu t} + \mathbf{u}e^{\mu t}}$$

into (1). (The $\mathbf{x}t$-term alone, resembling what we did in Sec. 2.2 in the case of a double root, would not be enough. Try it.) This gives

$$\mathbf{y}^{(2)\prime} = \mathbf{x}e^{\mu t} + \mu\mathbf{x}te^{\mu t} + \mu\mathbf{u}e^{\mu t} = \mathbf{A}\mathbf{y}^{(2)} = \mathbf{A}\mathbf{x}te^{\mu t} + \mathbf{A}\mathbf{u}e^{\mu t}.$$

Since

$$\mu\mathbf{x} = \mathbf{A}\mathbf{x},$$

the terms $\mu\mathbf{x}te^{\mu t}$ and $\mathbf{A}\mathbf{x}te^{\mu t}$ cancel, and division by $e^{\mu t}$ gives

(15) $\mathbf{x} + \mu\mathbf{u} = \mathbf{A}\mathbf{u},$ thus $(\mathbf{A} - \mu\mathbf{I})\mathbf{u} = \mathbf{x}.$

Although $\det(\mathbf{A} - \mu\mathbf{I}) = 0$, this can always be solved for $\mathbf{u}$, as can be shown.

EXAMPLE 6   **No basis of eigenvectors available. Degenerate node (Fig. 83)**

Find a general solution of

$$\mathbf{y}' = \mathbf{A}\mathbf{y} = \begin{bmatrix} 4 & 1 \\ -1 & 2 \end{bmatrix}\mathbf{y}.$$

Solution.   The matrix $\mathbf{A}$ is *not* skew-symmetric! (Why?) Its characteristic equation is

$$\det(\mathbf{A} - \lambda\mathbf{I}) = \begin{vmatrix} 4 - \lambda & 1 \\ -1 & 2 - \lambda \end{vmatrix} = \lambda^2 - 6\lambda + 9 = (\lambda - 3)^2 = 0.$$

It has a double root $\lambda = 3$. Eigenvectors are obtained from $(4 - \lambda)x_1 + x_2 = 0$, thus from $x_1 + x_2 = 0$, say, $\mathbf{x}^{(1)} = [1 \quad -1]^T$ and multiples of this (which do not help). Now (15) is

$$(\mathbf{A} - 3\mathbf{I})\mathbf{u} = \begin{bmatrix} 1 & 1 \\ -1 & -1 \end{bmatrix}\mathbf{u} = \begin{bmatrix} 1 \\ -1 \end{bmatrix}, \qquad \text{thus} \qquad \begin{aligned} u_1 + u_2 &= 1 \\ -u_1 - u_2 &= -1 \end{aligned}$$

and we can take simply $\mathbf{u} = [0 \quad 1]^T$. This gives the answer (Fig. 83)

$$\mathbf{y} = c_1\mathbf{y}^{(1)} + c_2\mathbf{y}^{(2)} = c_1 \begin{bmatrix} 1 \\ -1 \end{bmatrix} e^{3t} + c_2 \left(\begin{bmatrix} 1 \\ -1 \end{bmatrix} t + \begin{bmatrix} 0 \\ 1 \end{bmatrix} \right) e^{3t}.$$

This critical point at the origin is often called a **degenerate node** (or sometimes an **improper node**, although it differs from that in Example 1). $c_1\mathbf{y}^{(1)}$ gives the heavy straight line, with $c_1 > 0$ corresponding to the lower part (in the fourth quadrant) and $c_1 < 0$ corresponding to the upper part. $\mathbf{y}^{(2)}$ gives the right part of the heavy curve from 0 through the second, first, and finally fourth quadrants. $-\mathbf{y}^{(2)}$ gives the other part of that curve.   ◄

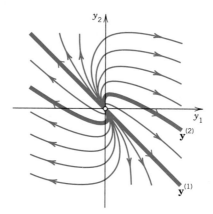

Fig. 83.   Degenerate node in Example 6

Now suppose that (1) consists of three or more equations and that $\mathbf{A}$ has a triple eigenvalue μ with only a single linearly independent eigenvector corresponding to it. Then we get a second solution (14) with a vector satisfying (15), as just discussed, and a third of the form

(16) $$\mathbf{y}^{(3)} = \tfrac{1}{2}\mathbf{x}t^2 e^{\mu t} + \mathbf{u}t e^{\mu t} + \mathbf{v}e^{\mu t}$$

with $\mathbf{u}$ satisfying (15) and $\mathbf{v}$ determined from

(17) $$(\mathbf{A} - \mu\mathbf{I})\mathbf{v} = \mathbf{u},$$

which can always be solved.

We finally mention that if $\mathbf{A}$ has a triple eigenvalue μ and *two* linearly independent eigenvectors $\mathbf{x}^{(1)}$, $\mathbf{x}^{(2)}$ corresponding to it, then three linearly independent solutions are

(18) $$\mathbf{y}^{(1)} = \mathbf{x}^{(1)}e^{\mu t}, \qquad \mathbf{y}^{(2)} = \mathbf{x}^{(2)}e^{\mu t}, \qquad \mathbf{y}^{(3)} = \mathbf{x}te^{\mu t} + \mathbf{u}e^{\mu t}$$

where $\mathbf{x}$ is a linear combination of $\mathbf{x}^{(1)}$ and $\mathbf{x}^{(2)}$ such that

(19) $$(\mathbf{A} - \mu\mathbf{I})\,\mathbf{u} = \mathbf{x}$$

is solvable for $\mathbf{u}$.

PROBLEM SET 3.3

General Solution. Find a real general solution of the following systems. (Show the details of your work.)

1. $y_1' = y_2$
$y_2' = y_1$

2. $y_1' = 2y_1 + 2y_2$
$y_2' = 5y_1 - y_2$

3. $y_1' = y_1 + y_2$
$y_2' = 3y_1 - y_2$

4. $y_1' = 6y_1 + 9y_2$
$y_2' = y_1 + 6y_2$

5. $y_1' = y_1 - y_2$
$y_2' = y_1 + y_2$

6. $y_1' = -8y_1 - 2y_2$
$y_2' = 2y_1 - 4y_2$

7. $y_1' = 10y_1 - 10y_2 - 4y_3$
$y_2' = -10y_1 + y_2 - 14y_3$
$y_3' = -4y_1 - 14y_2 - 2y_3$

8. $y_1' = -3y_1 - y_2 + 2y_3$
$y_2' = -4y_2 + 2y_3$
$y_3' = y_2 - 5y_3$

9. $y_1' = -y_1 - 4y_2 + 2y_3$
$y_2' = 2y_1 + 5y_2 - y_3$
$y_3' = 2y_1 + 2y_2 + 2y_3$

Initial Value Problems. Solve the following initial value problems for systems. (Show the details.)

10. $y_1' = 2y_1 + 2y_2$
$y_2' = 5y_1 - y_2$
$y_1(0) = 0, y_2(0) = -7$

11. $y_1' = y_2$
$y_2' = y_1$
$y_1(0) = 1, y_2(0) = 0$

12. $y_1' = y_1 + y_2$
$y_2' = 4y_1 + y_2$
$y_1(0) = 4, y_2(0) = 4$

13. $y_1' = 2y_1 + 5y_2$
$y_2' = -\frac{1}{2}y_1 - \frac{3}{2}y_2$
$y_1(0) = 10, y_2(0) = -5$

14. $y_1' = 2y_1 + 3y_2$
$y_2' = \frac{1}{3}y_1 + 2y_2$
$y_1(0) = 0, y_2(0) = 2$

15. $y_1' = -14y_1 + 10y_2$
$y_2' = -5y_1 + y_2$
$y_1(0) = -1, y_2(0) = 1$

16. **(Mixing problem, Fig. 84 on p. 170)** Each of the two tanks contains 200 gal of water in which initially 100 lb (Tank T_1) and 200 lb (Tank T_2) of fertilizer are dissolved. The inflow, circulation, and outflow are shown in Fig. 84 on the next page. The mixture is kept uniform by stirring. Find the fertilizer contents $y_1(t)$ (Tank T_1) and $y_2(t)$ (Tank T_2) in the tanks.

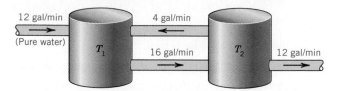

Fig. 84.   Tanks in Problem 16

17. (Network, Fig. 85) Show that a model for the currents $I_1(t)$ and $I_2(t)$ in Fig. 85 is

$$\frac{1}{C}\int I_1 \, dt + R(I_1 - I_2) = 0, \qquad LI_2' + R(I_2 - I_1) = 0.$$

Find a general solution, assuming that $R = 3$ ohms, $L = 4$ henrys, and $C = 1/12$ farad.

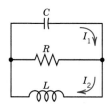

Fig. 85.   Network in Problem 17

18. Find the matrix **A** of the system in Fig. 85 for any R, L, C. When will **A** have real eigenvalues? Complex conjugate eigenvalues?

19. CAS PROJECT. Phase Portraits. Plot some of the figures in this section on the computer, in particular Fig. 83 on the degenerate node, in which the vector in $\mathbf{y}^{(2)}$ depends on t. In each figure highlight a trajectory that satisfies an initial condition of your own choice.

20. TEAM PROJECT. Conversion of Complex to Real Solutions. Replace the shortcuts in Examples 4 and 5 by systematic calculations as indicated, and compare the amount of work.

3.4 Criteria for Critical Points.   Stability

In the preceding section we discussed homogeneous linear systems in two equations with constant coefficients

$$(1) \qquad \mathbf{y}' = \mathbf{A}\mathbf{y}, \qquad \text{in components,} \qquad \begin{aligned} y_1' &= a_{11}y_1 + a_{12}y_2 \\ y_2' &= a_{21}y_1 + a_{22}y_2 \end{aligned}$$

and plots of their solutions as curves in the y_1y_2-plane, the **phase plane,** given by the parametric representation

$$(2) \qquad \qquad \mathbf{y}(t) = \begin{bmatrix} y_1(t) \\ y_2(t) \end{bmatrix}.$$

Such a solution curve is called a **trajectory** of (1). We have seen that in the case of *constant* coefficients a_{jk} in (1) this leads to **eigenvalue problems.** The simple reason is that we set

(3) $\qquad \mathbf{y}(t) = \mathbf{x}e^{\lambda t}.$ $\qquad$ Then $\qquad \mathbf{y}'(t) = \lambda \mathbf{x}e^{\lambda t} = \mathbf{A}\mathbf{y} = \mathbf{A}\mathbf{x}e^{\lambda t}$

and by dropping the common factor $e^{\lambda t}$ we get

(4) $\qquad\qquad\qquad\qquad\qquad\qquad \boxed{\mathbf{A}\mathbf{x} = \lambda\mathbf{x}.}$

This defines the *eigenvalues* of **A** as scalars λ such that (4) has a nonzero vector **x** as a solution, called an *eigenvector* of **A** corresponding to that eigenvalue λ.

Our examples in the last section indicate that the general character of the **phase portrait**—that is, the family of trajectories in the phase plane—is determined to a large extent by the type of **critical point** of the system (1). This is a point at which [see (9) in Sec. 3.3]

(5) $\qquad\qquad\qquad \dfrac{dy_2}{dy_1} = \dfrac{dy_2/dt}{dy_1/dt} = \dfrac{y_2'}{y_1'} = \dfrac{a_{21}y_1 + a_{22}y_2}{a_{11}y_1 + a_{12}y_2}$

becomes undetermined, 0/0. If you reconsider those examples, you will recognize that the type of a critical point is somehow related to the kind of eigenvalues of the system (1). This is the first new idea that we pursue in this section. The second is that of *stability*.

Criteria for Types of Critical Points

We know that the eigenvalues of **A** in (1) are the solutions λ_1, λ_2 of the characteristic equation

(6) $\qquad\qquad \det(\mathbf{A} - \lambda\mathbf{I}) = \begin{vmatrix} a_{11} - \lambda & a_{12} \\ a_{21} & a_{22} - \lambda \end{vmatrix}$

$$= \lambda^2 - (a_{11} + a_{22})\lambda + \det\mathbf{A} = 0.$$

We now introduce the standard notations

(7) $\qquad p = a_{11} + a_{22}, \qquad q = \det\mathbf{A} = a_{11}a_{22} - a_{12}a_{21}, \qquad \Delta = p^2 - 4q.$

Then from the right side of (6) and from the product representation we obtain the equation

(8) $\qquad \lambda^2 - p\lambda + q = (\lambda - \lambda_1)(\lambda - \lambda_2) = \lambda^2 - (\lambda_1 + \lambda_2)\lambda + \lambda_1\lambda_2.$

Hence p is the sum of the eigenvalues, q the product, and Δ the discriminant. This gives the following criteria (9), whose derivation we shall indicate on p. 173.

Criteria for Critical Points. The critical point P_0 of (1) is a:

(9)

> (a) *node* if $q > 0$ and $\Delta \geqq 0$,
>
> (b) *saddle point* if $q < 0$,
>
> (c) *center* if $p = 0$ and $q > 0$,
>
> (d) *spiral point* if $p \neq 0$ and $\Delta < 0$.

Stability

Criteria (9) classify critical points P_0 of (1) in terms of trajectories near P_0. Another classification is in terms of stability. Stability concepts are basic in many engineering and other applications. They are suggested by physics, where **stability** means, roughly speaking, that a small change (a small disturbance) of a physical system at some instant changes the behavior of the system only slightly at all future times t. For critical points, the following definitions are appropriate.

Definitions

P_0 is called a **stable**[4] **critical point** of (1) if, roughly speaking, all trajectories of (1) that at some instant are sufficiently close to P_0 remain close to P_0 at all future times; precisely: if for every disk D_ϵ of radius $\epsilon > 0$ with center P_0 there is a disk D_δ of radius $\delta > 0$ with center P_0 such that every trajectory of (1) that has a point P_1 (corresponding to $t = t_1$, say) in D_δ has all its points corresponding to $t \geqq t_1$ in D_ϵ. See Fig. 86.

P_0 is called **unstable** if P_0 is not stable.

P_0 is called a **stable and attractive critical point**[5] of (1) if P_0 is stable and every trajectory that has a point in D_δ approaches P_0 as $t \to \infty$. See Fig. 87.

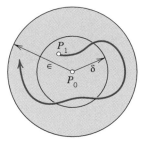

Fig. 86. Stable critical point P_0 of (1) (The trajectory initiating at P_1 stays in the disk of radius ϵ.)

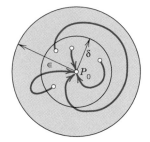

Fig. 87. Stable and attractive critical point P_0 of (1)

---

[4]More precisely: **stable in the sense of Liapunov.** There exist other definitions of stability, but Liapunov's definition—the only we shall consider—is probably the most useful one.

[5]Or an **asymptotically stable critical point.**

Stability Criteria for Critical Points. A critical point P_0 is:

(10)

(a) *stable and attractive* if $p < 0$ and $q > 0$,
(b) *stable* if $p \leqq 0$ and $q > 0$,
(c) *unstable* if $p > 0$ or $q < 0$.

The criteria in (9) and (10) are summarized in the **stability chart** in Fig. 88. In this chart, the region of instability is dark blue.

We indicate how these criteria are obtained. If $q = \lambda_1 \lambda_2 > 0$, both eigenvalues are positive or both negative or complex conjugates. If also $p = \lambda_1 + \lambda_2 < 0$, both are negative or have a negative real part. Hence P_0 is stable and attractive. The reasoning for the other two lines in (10) is similar.

If $\Delta < 0$, the eigenvalues are complex conjugates, say, $\lambda_1 = \alpha + i\beta$ and $\lambda_2 = \alpha - i\beta$. If also $p = \lambda_1 + \lambda_2 = 2\alpha < 0$, this gives a spiral point that is stable and attractive. If $p = 2\alpha > 0$, this gives an unstable spiral point.

If $p = 0$, then $\lambda_2 = -\lambda_1$ and $q = \lambda_1 \lambda_2 = -\lambda_1^2$. If also $q > 0$, then $\lambda_1^2 = -q < 0$, so that λ_1, and thus λ_2, must be pure imaginary. This gives periodic solutions, their trajectories being closed curves around P_0, which is a center.

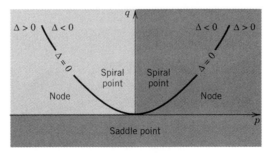

Fig. 88. Stability chart of the system (1) with p, q, Δ defined in (7).
Stable and attractive: The second quadrant without the q-axis.
Stability also on the positive q-axis (which corresponds to centers).
Unstable: Dark blue region

EXAMPLE 1   Application of the criteria (9) and (10)

The system (8) in the last section is

(11)
$$\mathbf{y}' = \mathbf{A}\mathbf{y} = \begin{bmatrix} -3 & 1 \\ 1 & -3 \end{bmatrix} \mathbf{y}.$$

We see that $p = a_{11} + a_{22} = -6$, $q = \det \mathbf{A} = 8$, and $\Delta = (-6)^2 - 4 \cdot 8 = 4$. Hence, by (9a), the critical point at the origin is a node, which, by (10a), is stable and attractive. This agrees with our previous result. The other examples in Sec. 3.3 can be discussed similarly. ◀

EXAMPLE 2   Free motions of a mass on a spring

What kind of critical point does the model of a mass on an elastic spring have? (Solution see next page.)

Solution. The equation in [see (5) in Sec. 2.5]

$$y'' + \frac{c}{m} y' + \frac{k}{m} y = 0.$$

To get a system, we set $y_1 = y$, $y_2 = y'$ (see Sec. 3.1), obtaining

$$\mathbf{y}' = \begin{bmatrix} 0 & 1 \\ -\dfrac{k}{m} & -\dfrac{c}{m} \end{bmatrix} \mathbf{y}, \qquad \text{thus} \qquad \begin{aligned} y_1' &= y_2 \\ y_2' &= -\frac{k}{m} y_1 - \frac{c}{m} y_2. \end{aligned}$$

For our criteria (9) and (10) we need $p = -c/m$, $q = k/m$, and $\Delta = (c/m)^2 - 4k/m$. This gives the following results.

No damping. $c = 0$, $p = 0$, $q > 0$, a center.
Underdamping. $c^2 < 4mk$, $p < 0$, $q > 0$, $\Delta < 0$, a stable and attractive spiral point.
Critical damping. $c^2 = 4mk$, $p < 0$, $q > 0$, $\Delta = 0$, a stable and attractive node.
Overdamping. $c^2 > 4mk$, $p < 0$, $q > 0$, $\Delta > 0$, a stable and attractive node. ◀

PROBLEM SET 3.4

Type and Stability of Critical Point

Determine the type and stability of the critical point. Then find a real general solution. Finally, sketch or plot some trajectories in the phase plane. (Show the details of your work.)

1. $y_1' = y_1$
$y_2' = 2y_2$

2. $y_1' = 2y_1 + y_2$
$y_2' = 5y_1 - 2y_2$

3. $y_1' = y_1 + 2y_2$
$y_2' = 2y_1 + y_2$

4. $y_1' = -6y_1 - y_2$
$y_2' = -9y_1 - 6y_2$

5. $y_1' = -2y_1 + 2y_2$
$y_2' = -2y_1 - 2y_2$

6. $y_1' = y_1 - 2y_2$
$y_2' = 5y_1 - y_2$

7. $y_1' = y_2$
$y_2' = -9y_1$

8. $y_1' = -y_1 + 4y_2$
$y_2' = 3y_1 - 2y_2$

9. $y_1' = -2y_1 - 6y_2$
$y_2' = -8y_1 - 4y_2$

10. $y_1' = -y_1$
$y_2' = -5y_1 - y_2$

11. $y_1' = 2y_1 + y_2$
$y_2' = 6y_1 + 2y_2$

Trajectories of Systems and Second-Order Differential Equations

12. (**Harmonic oscillations**) Solve $y'' + \frac{1}{9}y = 0$. Find the trajectories. Sketch and plot some of them.

13. (**Trajectories**) Solve $y'' + ay' = 0$ (a constant) and plot some of the trajectories.

14. (**Damped oscillations**) Solve $y'' + 2y' + 2y = 0$. What kind of curves do you get as trajectories?

15. (**Types of critical points**) Discuss the critical points in (10)–(13) in Sec. 3.3 by applying the criteria (9) and (10).

16. (**Transformation of parameter**) What happens to the critical point in Example 1 if you introduce $\tau = -t$ as a new independent variable?

17. (**Perturbation of center**) What happens in Example 4 in Sec. 3.3 if you change $\mathbf{A}$ to $\mathbf{A} + 0.1\mathbf{I}$, where $\mathbf{I}$ is the unit matrix?

18. **(Perturbation of center)** If a system has a center as its critical point, what happens if you replace its matrix $\mathbf{A}$ by $\tilde{\mathbf{A}} = \mathbf{A} + k\mathbf{I}$ with any real number k (where k may represent measurement errors in the diagonal entries)?

19. **(Perturbation)** The system in Example 4 in Sec. 3.3 has a center. Replace all four a_{jk} by $a_{jk} + b$ (where b may be an error of measurement). Find values of b for which you get a spiral point, saddle point, or node.

20. **WRITING PROJECT. Stability.** Stability concepts are of basic importance in physics and various engineering applications (in connection with fluid flow, cars, airplanes, machines, bridges, etc.). Write a two-part essay, about three pages for each part, devoting part (A) to general applications in which stability plays a role (be as precise as you can) and part (B) consisting of a summary of the material on stability in this section. Use your own formulations and examples of your own choice.

3.5   Qualitative Methods for Nonlinear Systems

Qualitative methods for differential equations are methods of investigating general properties of solutions in a qualitative fashion without actually solving the equation. For instance, all solutions $y(t)$ of $y' = 1 + y^2$ must be increasing with t because $1 + y^2 > 0$, which makes the derivative y' positive—this is a typical (albeit very simple) qualitative conclusion.

Qualitative methods become particularly valuable for systems whose solution by analytic methods is difficult or impossible, as is the case for many practically important first-order **nonlinear systems**

(1)
$$\mathbf{y}' = \mathbf{f}(\mathbf{y}), \qquad \text{thus} \qquad \begin{aligned} y_1' &= f_1(y_1, y_2) \\ y_2' &= f_2(y_1, y_2). \end{aligned}$$

Phase plane methods (as just discussed) are qualitative methods. In this section we extend phase plane methods to nonlinear systems (1).

We assume that (1) is an **autonomous system,** that is, the independent variable t does not occur explicitly. We shall see that the extended methods give a characterization of various general properties of solutions. As before, we shall exhibit entire families of solutions. This generally is an advantage over numerical methods, which give only one (approximate) solution at a time (albeit with much greater accuracy).

As before we call the $y_1 y_2$-plane the **phase plane,** a solution curve of (1) in the phase plane a **trajectory,** and a point P_0: (y_1, y_2) at which both $f_1(y_1, y_2) = 0$ and $f_2(y_1, y_2) = 0$ a **critical point.**

If (1) has several critical points, we discuss one after the other. Each time we move the point P_0: (a, b) to be discussed to the origin. This can be done by a translation

$$\tilde{y}_1 = y_1 - a, \qquad \tilde{y}_2 = y_2 - b.$$

Thus we can assume that P_0 is the origin $(0, 0)$, and we continue to write y_1, y_2 (instead of $\tilde{y}_1, \tilde{y}_2$), for simplicity. We also assume that P_0 is *isolated,* that is, it is the only critical point within a (sufficiently small) circular disk with center at the origin.

Linearization of Nonlinear Systems

How can we determine the type and stability property of a critical point P_0: $(0, 0)$ of (1)? In most practical cases this can be done by **linearization,** that is, by investigating a certain *linear system,* as follows.

We assume that f_1 and f_2 in (1) are continuous and have continuous partial derivatives in a neighborhood of P_0. Then (1) and a linear system by which we shall approximate (1) near P_0 will have the same kind of critical point at P_0 (with two exceptions; see below).

Since P_0 is a critical point, we have $f_1(0, 0) = 0$ and $f_2(0, 0) = 0$. Hence f_1 and f_2 have no constant terms. Their linear terms we write explicitly. Then (1) becomes

$$(2) \qquad \mathbf{y}' = \mathbf{A}\mathbf{y} + \mathbf{h}(\mathbf{y}), \qquad \text{thus} \qquad \begin{aligned} y_1' &= a_{11}y_1 + a_{12}y_2 + h_1(y_1, y_2) \\ y_2' &= a_{21}y_1 + a_{22}y_2 + h_2(y_1, y_2). \end{aligned}$$

Here $\mathbf{A}$ is constant (independent of t) because (1) is autonomous. One can prove that if $\det \mathbf{A} \neq 0$, then the type and stability of P_0 is the same as that of the critical point $(0, 0)$ of the *linear system* obtained by **linearization,** that is, by dropping $\mathbf{h}(\mathbf{y})$ from (2):

$$(3) \qquad \boxed{\mathbf{y}' = \mathbf{A}\mathbf{y}, \qquad \text{thus} \qquad \begin{aligned} y_1' &= a_{11}y_1 + a_{12}y_2 \\ y_2' &= a_{21}y_1 + a_{22}y_2. \end{aligned}}$$

Indeed, the above assumption on the derivatives implies that h_1 and h_2 are small near P_0. *Two exceptions* occur if the eigenvalues of $\mathbf{A}$ are equal or pure imaginary. Then, in addition to the type of critical points of the *linear* system, the nonlinear system may have a spiral point. For proofs, see Ref. [A3], pp. 375–388.

EXAMPLE 1 **Free undamped pendulum. Linearization**

Figure 89a shows a pendulum consisting of a body of mass m (the bob) and a rod of length L. Determine the locations and types of the critical points. Assume that the mass of the rod and air resistance are negligible.

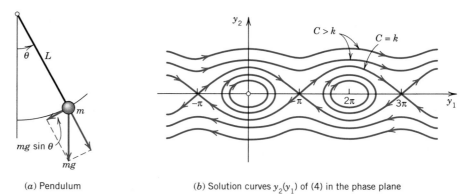

(a) Pendulum   (b) Solution curves $y_2(y_1)$ of (4) in the phase plane

Fig. 89.   Examples 1 and 2 (*C* will be explained in Example 4.)

Solution. 1st Step. Setting up the mathematical model.   Let θ denote the angular displacement, measured counterclockwise from the equilibrium position. The weight of the bob is mg (g the acceleration of gravity). It causes a restoring force $mg \sin \theta$ tangent to the curve of motion (circular arc) of the bob. By Newton's second law, at each instant this force is balanced by the force of acceleration $mL\theta''$, where $L\theta''$ is the acceleration;

hence the resultant of these two forces is zero, and we obtain as the mathematical model

$$mL\theta'' + mg \sin \theta = 0.$$

Dividing this by mL, we have

(4) $$\boxed{\theta'' + k \sin \theta = 0} \qquad \left(k = \frac{g}{L} \right).$$

When θ is very small, we could approximate $\sin \theta$ rather accurately by θ and obtain as an *approximate* solution $A \cos \sqrt{k}t + B \sin \sqrt{k}t$, but the *exact* solution for any θ is not an elementary function.

2nd Step. Discussion of critical points by linearization. To obtain a system of equations, we set $\theta = y_1$, $\theta' = y_2$. Then from (4) we obtain a nonlinear system (1) of the form

(4*) $$y_1' = f_1(y_1, y_2) = y_2$$
$$y_2' = f_2(y_1, y_2) = -k \sin y_1.$$

The right sides are both zero when $y_2 = 0$ and $\sin y_1 = 0$. This gives the infinitely many critical points $(n\pi, 0)$, where $n = 0, \pm 1, \pm 2, \cdots$. We consider $(0, 0)$. Since the Maclaurin series is

$$\sin y_1 = y_1 - \tfrac{1}{6}y_1^3 + - \cdots \approx y_1,$$

the linearized system at $(0, 0)$ is

$$\mathbf{y'} = \mathbf{Ay} = \begin{bmatrix} 0 & 1 \\ -k & 0 \end{bmatrix} \mathbf{y}, \qquad \text{thus} \qquad \begin{aligned} y_1' &= y_2 \\ y_2' &= -ky_1. \end{aligned}$$

In our criteria (9), (10) in Sec. 3.4 we need $p = a_{11} + a_{22} = 0$, $q = \det \mathbf{A} = k \ (> 0)$, and $\Delta = p^2 - 4q = -4k$. From this and (9c) in Sec. 3.4, we conclude that $(0, 0)$ is a center, which is always stable. Since $\sin \theta = \sin y_1$ is periodic, with period 2π, the critical points $(n\pi, 0)$, $n = \pm 2, \pm 4, \cdots$, are all centers.

We now consider the critical point $(\pi, 0)$, setting $\theta - \pi = y_1$ and $(\theta - \pi)' = \theta' = y_2$, so that in (4),

$$\sin \theta = \sin (y_1 + \pi) = -\sin y_1 = -y_1 + \tfrac{1}{6}y_1^3 - + \cdots \approx -y_1$$

and the linearized system at $(\pi, 0)$ is now

$$\mathbf{y'} = \mathbf{Ay} = \begin{bmatrix} 0 & 1 \\ k & 0 \end{bmatrix} \mathbf{y}, \qquad \text{thus} \qquad \begin{aligned} y_1' &= y_2 \\ y_2' &= ky_1. \end{aligned}$$

We see that $p = 0$, $q = -k \ (< 0)$, and $\Delta = -4q = 4k$, so that (9b) in Sec. 3.4 gives a saddle point, which is always unstable. By periodicity, the critical points $(n\pi, 0)$, $n = \pm 1, \pm 3, \cdots$, are all saddle points. These results agree with the impression we get from Fig. 89b. ◀

EXAMPLE 2   **Linearization of the damped pendulum equation**

To gain further experience in investigating critical points, as another practically important case, let us see how Example 1 changes when we add a damping term $c\theta'$ (damping proportional to the angular velocity) to equation (4), so that it becomes

(5) $$\boxed{\theta'' + c\theta' + k \sin \theta = 0}$$

where $k > 0$ and $c \geqq 0$ (which includes our previous case of no damping, $c = 0$). Setting $\theta = y_1$, $\theta' = y_2$, as before, we obtain the system

$$y_1' = y_2$$
$$y_2' = -k \sin y_1 - cy_2.$$

We see that the critical points have the same locations as before, namely, $(0, 0)$, $(\pm\pi, 0)$, $(\pm 2\pi, 0)$, $\cdots$. We consider $(0, 0)$. Linearizing $\sin y_1 \approx y_1$ as in Example 1, we get the linearized system at $(0, 0)$

(6) $$\mathbf{y'} = \mathbf{Ay} = \begin{bmatrix} 0 & 1 \\ -k & -c \end{bmatrix} \mathbf{y}, \qquad \text{thus} \qquad \begin{aligned} y_1' &= y_2 \\ y_2' &= -ky_1 - cy_2. \end{aligned}$$

This is identical with the system in Example 2 of Sec. 3.4, except for the (positive!) factor m (and except for the physical meaning of y_1). Hence for $c = 0$ (no damping) we have a center (see Fig. 89b), for small damping we have a spiral point (see Fig. 90), and so on.

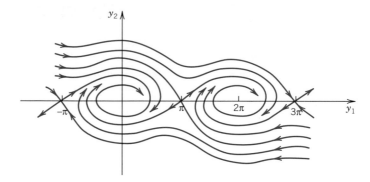

Fig. 90.   Trajectories in the phase plane for the damped pendulum

We now consider the critical point $(\pi, 0)$. We set $\theta - \pi = y_1$, $(\theta - \pi)' = \theta' = y_2$ and linearize

$$\sin \theta = \sin (y_1 + \pi) = -\sin y_1 \approx -y_1.$$

This gives the new linearized system at $(\pi, 0)$

$$(6^*) \qquad\qquad \mathbf{y}' = \mathbf{A}\mathbf{y} = \begin{bmatrix} 0 & 1 \\ k & -c \end{bmatrix} \mathbf{y}, \qquad \text{thus} \qquad \begin{aligned} y_1' &= y_2 \\ y_2' &= ky_1 - cy_2. \end{aligned}$$

For our criteria (9), (10) in Sec. 3.4 we calculate $p = a_{11} + a_{22} = -c$, $q = \det \mathbf{A} = -k$, and $\Delta = p^2 - 4q = c^2 + 4k$. This gives the following results.

   No damping. $c = 0$, $p = 0$, $q < 0$, $\Delta > 0$, a saddle point. See Fig. 89b.
   Damping. $c > 0$, $p < 0$, $q < 0$, $\Delta > 0$, a saddle point. See Fig. 90.

Since $\sin y_1$ is periodic with period 2π, the critical points $(\pm 2\pi, 0)$, $(\pm 4\pi, 0)$, $\cdots$ are of the same type as $(0, 0)$, and the critical points $(-\pi, 0)$, $(\pm 3\pi, 0)$, $\cdots$ are of the same type as $(\pi, 0)$, so that our task is finished.

Figure 90 shows the trajectories in the case of damping. What we see agrees with our physical intuition. Indeed, damping means loss of energy. Hence instead of the closed trajectories of periodic solutions in Fig. 89b we now have trajectories spiraling around one of the critical points $(0, 0)$, $(\pm 2\pi, 0)$, $\cdots$. Even the wavy trajectories corresponding to whirly motions eventually spiral around one of these points. Furthermore, there are no more trajectories that connect critical points (as there were in the undamped case).   ◀

Lotka–Volterra Population Model

EXAMPLE 3   **Predator–prey population model (Lotka, 1925; Volterra, 1931)[6]**

This model involves two species, say, rabbits and foxes. The foxes prey on the rabbits, and we assume the following.

   1. Rabbits have unlimited food supply. Hence if there were no foxes, their number $y_1(t)$ would grow exponentially, $y_1' = ay_1$.
   2. Actually, y_1 is decreased because of the kill by foxes, say, at a rate proportional to y_1y_2, where $y_2(t)$ is the number of foxes. Hence $y_1' = ay_1 - by_1y_2$, where $a > 0$ and $b > 0$.
   3. If there were no rabbits, then $y_2(t)$ would exponentially decrease to zero, $y_2' = -ly_2$. However, y_2 is increased by a rate proportional to the number of encounters between predator and prey; together we have $y_2' = -ly_2 + ky_1y_2$, where $k > 0$ and $l > 0$.

---

[6]ALFRED J. LOTKA (1880—1949), American biophysicist. VITO VOLTERRA (1860—1940), Italian mathematician, the initiator of functional analysis (1887).

This gives the **Lotka–Volterra system**

(7)

$$
\begin{aligned}
y_1' &= ay_1 - by_1y_2 \\
y_2' &= ky_1y_2 - ly_2.
\end{aligned}
$$

We see that the critical points are the solutions of

$$y_1(a - by_2) = 0, \qquad y_2(ky_1 - l) = 0.$$

The two solutions are $(y_1, y_2) = (0, 0)$ and $(l/k, a/b)$. At $(0, 0)$ the linearized system is

$$
\begin{aligned}
y_1' &= \quad ay_1 \\
y_2' &= -ly_2.
\end{aligned}
$$

The eigenvalues are $\lambda_1 = a > 0$ and $\lambda_2 = -l < 0$, and we obtain a saddle point.

At the critical point $(l/k, a/b)$ we set $y_1 = l/k + \tilde{y}_1$, $y_2 = a/b + \tilde{y}_2$. This gives

$$\tilde{y}_1' = \left(\tilde{y}_1 + \frac{l}{k}\right)\left[a - b\left(\tilde{y}_2 + \frac{a}{b}\right)\right] = \left(\tilde{y}_1 + \frac{l}{k}\right)(-b\tilde{y}_2)$$

$$\tilde{y}_2' = \left(\tilde{y}_2 + \frac{a}{b}\right)\left[k\left(\tilde{y}_1 + \frac{l}{k}\right) - l\right] = \left(\tilde{y}_2 + \frac{a}{b}\right)k\tilde{y}_1.$$

Dropping the two nonlinear terms we obtain the linearized system

(7*)

$$(a) \qquad \tilde{y}_1' = -\frac{lb}{k}\,\tilde{y}_2$$

$$(b) \qquad \tilde{y}_2' = \quad \frac{ak}{b}\,\tilde{y}_1.$$

We use the shortcut of Example 4 in Sec. 3.3, namely, the left side of (a) times the right side of (b) must equal the left side of (b) times the right side of (a), and by integration we obtain the ellipses

$$\frac{ak}{b}\,\tilde{y}_1{}^2 + \frac{lb}{k}\,\tilde{y}_2{}^2 = const.$$

Hence this critical point of the linearized system is a center (Fig. 91). It can be shown (by a rather complicated analysis) that the nonlinear system (7) also has a center at $(l/k, a/b)$, surrounded by closed trajectories (not ellipses), not a spiral point.

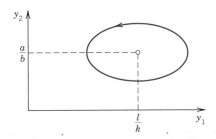

Fig. 91. Ecological equilibrium point and trajectory
of the linearized Lotka–Volterra system (7*)

We see that the predator and prey have a cyclic variation about the critical point. Let us move counterclockwise around the ellipse, beginning at the right vertex, where the rabbits have a maximum number. Foxes are sharply increasing in number until they reach a maximum at the upper vertex, and the number of rabbits is then sharply decreasing until it reaches a minimum at the left vertex, and so on. Cyclic variations of this kind have been observed in nature, for example, for lynx and snowshoe hare near the Hudson Bay, with a cycle of about 10 years.

Our model is based on assumptions that are too simple in many cases. For more refined models and a systematic discussion, see C. W. Clark, *Mathematical Bioeconomics* (Wiley, 1976).   ◀

Transformation to a First-Order Equation in the Phase Plane

Another phase plane method is based on the idea of transforming a second-order **autonomous equation** (a differential equation in which t does not occur explicitly)

$$F(y,\, y',\, y'') = 0$$

to first order by taking $y = y_1$ as the independent variable, setting $y' = y_2$ and transforming y'' by the chain rule,

$$y'' = y_2' = \frac{dy_2}{dt} = \frac{dy_2}{dy_1}\frac{dy_1}{dt} = \frac{dy_2}{dy_1}\,y_2.$$

Then the equation becomes of first order,

(8) $$F\!\left(y_1,\ y_2,\ \frac{dy_2}{dy_1}\,y_2\right) = 0$$

and can sometimes be solved or treated by direction fields. We illustrate this for the equation in Example 1 and shall gain much more insight into the behavior of solutions.

EXAMPLE 4   **An equation (8) for the free undamped pendulum**

If in (4) we set $\theta = y_1$, $\theta' = y_2$ (the angular velocity) and use

$$\theta'' = \frac{dy_2}{dt} = \frac{dy_2}{dy_1}\frac{dy_1}{dt} = \frac{dy_2}{dy_1}\,y_2,$$

we get from (4)

$$\frac{dy_2}{dy_1}\,y_2 = -k\sin y_1.$$

We can now separate variables, $y_2\,dy_2 = -k\sin y_1\,dy_1$, and integrate,

(9) $$\tfrac{1}{2}y_2{}^2 = k\cos y_1 + C$$ (*C* constant).

Multiplying this by mL^2, we get

$$\tfrac{1}{2}m(Ly_2)^2 - mL^2k\cos y_1 = mL^2C.$$

We see that these three terms are energies. Indeed, y_2 is the angular velocity, so that Ly_2 is the velocity and the first term is the kinetic energy. The second term (including the minus sign) is the potential energy of the pendulum, and mL^2C is its total energy, which is constant, as expected from the law of conservation of energy, because there is no damping (no loss of energy). The type of motion depends on the total energy, hence on C, as follows.

Figure 89b on p. 176 shows portions of trajectories for various values of C. These graphs continue periodically with period 2π to the left and to the right. We see that some of them are ellipse-like and closed, others are wavy, and there are two trajectories (passing through the saddle points $(n\pi, 0)$, $n = \pm 1, \pm 3, \cdots$) that separate those two types of trajectories. From (9) we see that the smallest possible C is $C = -k$; then $y_2 = 0$, and $\cos y_1 = 1$, so that the pendulum is at rest. The pendulum will change its direction of motion if there are points at which $y_2 = \theta' = 0$. Then $k \cos y_1 + C = 0$ by (9). If $y_1 = \pi$, then $\cos y_1 = -1$ and $C = k$. Hence if $-k < C < k$, then the pendulum reverses its direction for a $|y_1| = |\theta| < \pi$, and for these values of C with $|C| < k$ the pendulum oscillates. This corresponds to the closed trajectories in the figure. However, if $C > k$, then $y_2 = 0$ is impossible and the pendulum makes a whirly motion that appears as a wavy trajectory in the y_1y_2-plane. Finally, the value $C = k$ corresponds to the two "separating trajectories" in Fig. 89b connecting the saddle points. ◄

The phase plane method of deriving a single first-order equation (8) may be of practical interest not only when (8) can be solved (as in Example 4) but also when solution is not possible and we have to utilize direction fields (Sec. 1.2). We illustrate this with a very famous example:

EXAMPLE 5   **Self-sustained oscillations, van der Pol equation**

There are physical systems such that for small oscillations, energy is fed into the system, whereas for large oscillations, energy is taken from the system. In other words, large oscillations will be damped, whereas for small oscillations there is "negative damping" (feeding of energy into the system). For physical reasons we expect such a system to approach a periodic behavior, which will thus appear as a closed trajectory in the phase plane, called a **limit cycle.** A differential equation describing such vibrations is the famous **van der Pol equation**[7]

(10)
$$y'' - \mu(1 - y^2)y' + y = 0 \qquad (\mu > 0, \text{ constant}).$$

It first occurred in the study of electrical circuits containing vacuum tubes. For $\mu = 0$ this equation becomes $y'' + y = 0$ and we obtain harmonic oscillations. Let $\mu > 0$. The damping term has the coefficient $-\mu(1 - y^2)$. This is negative for small oscillations, namely, $y^2 < 1$, so that we have "negative damping," is zero for $y^2 = 1$ (no damping), and is positive if $y^2 > 1$ (positive damping, loss of energy). If μ is small, we expect a limit cycle that is almost a circle because then our equation differs but little from $y'' + y = 0$. If μ is large, the limit cycle will probably look different.

Setting $y = y_1$, $y' = y_2$ and using $y'' = (dy_2/dy_1)y_2$ as in (8), we have from (10)

(11)
$$\frac{dy_2}{dy_1} y_2 - \mu(1 - y_1{}^2)y_2 + y_1 = 0.$$

The isoclines in the y_1y_2-plane (the phase plane) are the curves $dy_2/dy_1 = K = const$, that is,

$$\frac{dy_2}{dy_1} = \mu(1 - y_1{}^2) - \frac{y_1}{y_2} = K.$$

Solving algebraically for y_2, we see that the isoclines are given by

$$y_2 = \frac{y_1}{\mu(1 - y_1{}^2) - K} \qquad \text{(Figs. 92, 93).}$$

We have included these curves in Figs. 92 and 93 on p. 182, although they look somewhat complicated. Figure 92 shows some of them for a small $\mu = 0.1$ as well as the limit cycle (almost a circle) and two trajectories approaching the limit cycle, one from the outside and one from the inside. The latter is a narrow spiral and only the initial part of it is shown in the figure. For larger μ the situation changes and the limit cycle no longer resembles a circle. Figure 93 illustrates this for $\mu = 1$. Note that the approach of the trajectories to the limit cycle is much more rapid than for $\mu = 0.1$. ◄

---

[7]BALTHASAR VAN DER POL (1889—1959). Dutch physicist.

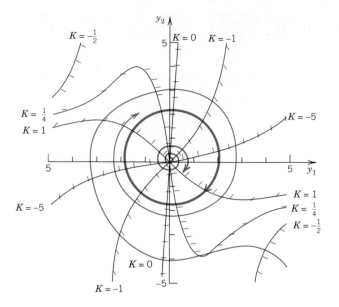

Fig. 92. Lineal element diagram for the van der Pol equation with $\mu = 0.1$ in the phase plane, showing also the limit cycle and two trajectories

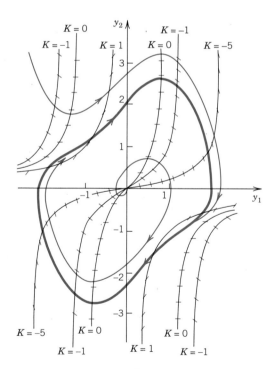

Fig. 93. Lineal element diagram for the van der Pol equation with $\mu = 1$ in the phase plane, showing also the limit cycle and two trajectories approaching it

PROBLEM SET 3.5

1. **(Limit cycle)** What is the essential difference between a limit cycle and one of the closed trajectories surrounding a center?

2. **(Harmonic oscillation)** What is the radius of a real general solution of $y'' + y = 0$ in the phase plane?

3. To what state (position, speed, direction of motion) of the oscillating pendulum do the four points of intersection of a closed trajectory with the axes in Fig. 89*b* correspond? The point of intersection of a wavy curve with the y_2-axis?

4. What type of critical points do we obtain in Example 1 if $k < 0$?

5. **(Linearization)** Show that $y'' - y + y^2 = 0$ has the critical points $(0, 0)$ and $(1, 0)$ and determine their type and stability.

Critical Points, Linearization. Determine the location and type of all critical points of the given equation. (Use linearization of the corresponding system. Show the details of your work.)

6. $y'' + y + y^2 = 0$ 7. $y'' + y - \frac{1}{2}y^2 = 0$

8. $y'' + y - y^3 = 0$ 9. $y'' - 9y + y^3 = 0$

10. $y'' + 4y - 5y^3 + y^5 = 0$ 11. $y'' + \cos y = 0$

12. **(Trajectories)** Convert $y'' - 4y + y^3 = 0$ to a system, solve it for y_2 as a function of y_1 and sketch or plot some of the trajectories in the phase plane.

13. **(Trajectories)** What kind of curves are the trajectories of $yy'' + y'^2 = 0$?

14. **(Trajectories)** In Prob. 12 add a linear damping term to get $y'' + y' - 4y + y^3 = 0$. Using mechanical arguments and a comparison with Prob. 12, guess the type of each critical point. Then determine these types by linearization, showing all details of the calculation.

 15. **CAS PROJECT. Van der Pol Equation.** Convert the van der Pol equation to a system and plot the limit cycle and some trajectories approaching it from inside and outside, for $\mu = 0.2$, 0.4, 0.6, 0.8, 1, 1.5, 2. Describe in words how the limit cycle is deformed with growing μ.

16. **TEAM PROJECT. Self-Sustained Oscillations. (a) Van der Pol equation.** Determine the type of critical point at $(0, 0)$ when $\mu > 0$, $\mu = 0$, $\mu < 0$. Show that if $\mu \to 0$, the isoclines approach straight lines through the origin. Why is this to be expected?

(b) Rayleigh[8] equation. Show that the so-called Rayleigh equation

$$Y'' - \mu(1 - \tfrac{1}{3}Y'^2)Y' + Y = 0 \qquad\qquad (\mu > 0)$$

also describes self-sustained oscillations and that by differentiating it and setting $y = Y'$ one obtains the van der Pol equation.

(c) Duffing equation. The Duffing equation is

$$y'' + \omega_0^2 y + \beta y^3 = 0$$

where usually $|\beta|$ is small, thus characterizing a small deviation of the restoring force from linearity. $\beta > 0$ and $\beta < 0$ are called the cases of a *hard spring* and a *soft spring,* respectively. Find the equation of the trajectories in the phase plane. (Note that for $\beta > 0$ all these curves are closed.)

---

[8]LORD RAYLEIGH (JOHN WILLIAM STRUTT) (1842—1919), English physicist and mathematician, professor at Cambridge and London, known by his important contributions to the theory of waves, elasticity theory, hydrodynamics, and various other branches of applied mathematics and theoretical physics.

3.6 Nonhomogeneous Linear Systems

In this last section of Chap. 3 we discuss methods for solving nonhomogeneous linear systems. From Sec. 3.2 we recall that such a system is of the form

(1)
$$\mathbf{y}' = \mathbf{A}\mathbf{y} + \mathbf{g}$$

with the vector $\mathbf{g}(t)$ not identically zero. We assume $\mathbf{g}(t)$ and the entries of the $n \times n$ matrix $\mathbf{A}(t)$ to be continuous on some interval J of the t-axis. From a general solution $\mathbf{y}^{(h)}(t)$ of the homogeneous system $\mathbf{y}' = \mathbf{A}\mathbf{y}$ on J and a **particular solution** $\mathbf{y}^{(p)}(t)$ of (1) on J [i.e., a solution of (1) containing no arbitrary constants], we get a solution of (1),

(2)
$$\mathbf{y} = \mathbf{y}^{(h)} + \mathbf{y}^{(p)},$$

called a **general solution** of (1) on J, because it includes every solution of (1) on J. This follows from Theorem 2 in Sec. 3.2 (see Prob. 1).

Having studied homogeneous linear systems in Secs. 3.1–3.4, our present task will be to explain methods for obtaining particular solutions of (1). We discuss the method of undetermined coefficients and the method of the variation of parameters; these have counterparts for a single equation, as we know from Secs. 2.9 and 2.10. As a third method of solution we also consider the reduction to diagonal form.

Method of Undetermined Coefficients

This method is suitable if the components of $\mathbf{g}$ are integer powers of t, exponential functions, or sines and cosines. We explain this in terms of an example.

EXAMPLE 1   **Method of undetermined coefficients**

Find a general solution of the nonhomogeneous linear system

(3)
$$\mathbf{y}' = \mathbf{A}\mathbf{y} + \mathbf{g} = \begin{bmatrix} 2 & -4 \\ 1 & -3 \end{bmatrix} \mathbf{y} + \begin{bmatrix} 2t^2 + 10t \\ t^2 + 9t + 3 \end{bmatrix}.$$

Solution.   The form of $\mathbf{g}$ suggests to assume $\mathbf{y}^{(p)}$ in the form

$$\mathbf{y}^{(p)} = \mathbf{u} + \mathbf{v}t + \mathbf{w}t^2$$

and to determine the vectors $\mathbf{u}$, $\mathbf{v}$, and $\mathbf{w}$. By substitution.

$$\mathbf{y}^{(p)\prime} = \mathbf{v} + 2\mathbf{w}t = \mathbf{A}\mathbf{u} + \mathbf{A}\mathbf{v}t + \mathbf{A}\mathbf{w}t^2 + \mathbf{g}.$$

In terms of components, this gives

$$\begin{bmatrix} v_1 \\ v_2 \end{bmatrix} + \begin{bmatrix} 2w_1 t \\ 2w_2 t \end{bmatrix} = \begin{bmatrix} 2u_1 - 4u_2 \\ u_1 - 3u_2 \end{bmatrix} + \begin{bmatrix} 2v_1 - 4v_2 \\ v_1 - 3v_2 \end{bmatrix} t + \begin{bmatrix} 2w_1 - 4w_2 \\ w_1 - 3w_2 \end{bmatrix} t^2 + \begin{bmatrix} 2t^2 + 10t \\ t^2 + 9t + 3 \end{bmatrix}.$$

Equating the t^2-terms on both sides, we get

$$0 = 2w_1 - 4w_2 + 2, \qquad 0 = w_1 - 3w_2 + 1, \qquad \text{thus} \qquad w_1 = -1, \quad w_2 = 0.$$

From the t-terms we get

$$2w_1 = 2v_1 - 4v_2 + 10, \qquad 2w_2 = v_1 - 3v_2 + 9, \qquad \text{thus} \qquad v_1 = 0, \quad v_2 = 3.$$

From the constant terms we finally get

$$v_1 = 2u_1 - 4u_2, \qquad v_2 = u_1 - 3u_2 + 3, \qquad \text{thus} \qquad u_1 = 0, \quad u_2 = 0.$$

From this and the general solution of the homogeneous system (verify!)

$$\mathbf{y}^{(h)} = c_1 \begin{bmatrix} 4 \\ 1 \end{bmatrix} e^t + c_2 \begin{bmatrix} 1 \\ 1 \end{bmatrix} e^{-2t}$$

we obtain the answer

$$\mathbf{y} = \mathbf{y}^{(h)} + \mathbf{y}^{(p)} = c_1 \begin{bmatrix} 4 \\ 1 \end{bmatrix} e^t + c_2 \begin{bmatrix} 1 \\ 1 \end{bmatrix} e^{-2t} + \begin{bmatrix} -t^2 \\ 3t \end{bmatrix}. \qquad \blacktriangleleft$$

A **modification** is necessary if a term $\mathbf{g}$ involves $e^{\lambda t}$ with λ being an eigenvalue of $\mathbf{A}$. Then instead of assuming in $\mathbf{y}^{(p)}$ a term $\mathbf{u}e^{\lambda t}$, we must start from $\mathbf{u}te^{\lambda t} + \mathbf{v}e^{\lambda t}$. (The first of these two terms is the analog of the modification in Sec. 2.9, but it would not be sufficient here. Try it out.)

EXAMPLE 2   Modification of the undetermined-coefficient method

Find a general solution of

$$(4) \qquad \mathbf{y}' = \mathbf{A}\mathbf{y} + \mathbf{g} = \begin{bmatrix} -3 & 1 \\ 1 & -3 \end{bmatrix} \mathbf{y} + \begin{bmatrix} -6 \\ 2 \end{bmatrix} e^{-2t}.$$

Solution.   A general equation of the homogeneous system is (see Example 1 in Sec. 3.3)

$$(5) \qquad \mathbf{y}^{(h)} = c_1 \begin{bmatrix} 1 \\ 1 \end{bmatrix} e^{-2t} + c_2 \begin{bmatrix} 1 \\ -1 \end{bmatrix} e^{-4t}.$$

Since $\lambda = -2$ is an eigenvalue of $\mathbf{A}$, we must assume $\mathbf{y}^{(p)} = \mathbf{u}te^{-2t} + \mathbf{v}e^{-2t}$ (rather than $\mathbf{u}e^{-2t}$). By substitution,

$$\mathbf{y}^{(p)'} = \mathbf{u}e^{-2t} - 2\mathbf{u}te^{-2t} - 2\mathbf{v}e^{-2t} = \mathbf{A}\mathbf{u}te^{-2t} + \mathbf{A}\mathbf{v}e^{-2t} + \mathbf{g}.$$

Equating the te^{-2t}-terms on both sides, we have $-2\mathbf{u} = \mathbf{A}\mathbf{u}$. Hence $\mathbf{u}$ is an eigenvector of $\mathbf{A}$ corresponding to $\lambda = -2$; thus [see (5)] $\mathbf{u} = a[1 \quad 1]^T$ with any $a \neq 0$. Equating the other terms gives

$$\mathbf{u} - 2\mathbf{v} = \mathbf{A}\mathbf{v} + \begin{bmatrix} -6 \\ 2 \end{bmatrix} \qquad \text{or} \qquad (\mathbf{A} + 2\mathbf{I})\mathbf{v} = \mathbf{u} - \begin{bmatrix} -6 \\ 2 \end{bmatrix} = a \begin{bmatrix} 1 \\ 1 \end{bmatrix} - \begin{bmatrix} -6 \\ 2 \end{bmatrix};$$

in components,

$$-v_1 + v_2 = a + 6$$
$$v_1 - v_2 = a - 2.$$

Hence $a = -2$ (to have a solution) and then $v_2 = v_1 + 4$, say, $v_1 = k$ and $v_2 = k + 4$; thus

$$\mathbf{v} = k \begin{bmatrix} 1 \\ 1 \end{bmatrix} + \begin{bmatrix} 0 \\ 4 \end{bmatrix}$$

and we can simply choose $k = 0$. This gives the answer

$$(6) \qquad \mathbf{y} = \mathbf{y}^{(h)} + \mathbf{y}^{(p)} = c_1 \begin{bmatrix} 1 \\ 1 \end{bmatrix} e^{-2t} + c_2 \begin{bmatrix} 1 \\ -1 \end{bmatrix} e^{-4t} - 2 \begin{bmatrix} 1 \\ 1 \end{bmatrix} te^{-2t} + \begin{bmatrix} 0 \\ 4 \end{bmatrix} e^{-2t}.$$

For other k we get other $\mathbf{v}$; for instance, $k = -2$ gives $\mathbf{v} = [-2 \quad 2]^T$, so that the answer becomes

$$(6^*) \qquad \mathbf{y} = c_1 \begin{bmatrix} 1 \\ 1 \end{bmatrix} e^{-2t} + c_2 \begin{bmatrix} 1 \\ -1 \end{bmatrix} e^{-4t} - 2 \begin{bmatrix} 1 \\ 1 \end{bmatrix} te^{-2t} + \begin{bmatrix} -2 \\ 2 \end{bmatrix} e^{-2t}, \qquad \text{etc.} \qquad \blacktriangleleft$$

Method of Variation of Parameters

This method can be applied to nonhomogeneous linear systems

$$(7) \qquad \boxed{\mathbf{y}' = \mathbf{A}(t)\mathbf{y} + \mathbf{g}(t)}$$

with variable $\mathbf{A} = \mathbf{A}(t)$ and general $\mathbf{g}(t)$. It yields a particular solution $\mathbf{y}^{(p)}$ of (7) on some interval J of the t-axis if a general solution

$$(8) \qquad \mathbf{y}^{(h)} = c_1\mathbf{y}^{(1)} + \cdots + c_n\mathbf{y}^{(n)}$$

of the homogeneous system on J is known. In components, (8) is

$$\mathbf{y}^{(h)} = \begin{bmatrix} c_1 y_1^{(1)} + \cdots + c_n y_1^{(n)} \\ \vdots \\ c_1 y_n^{(1)} + \cdots + c_n y_n^{(n)} \end{bmatrix} = \begin{bmatrix} y_1^{(1)} & \cdots & y_1^{(n)} \\ & \vdots & \\ y_n^{(1)} & \cdots & y_n^{(n)} \end{bmatrix} \begin{bmatrix} c_1 \\ \vdots \\ c_n \end{bmatrix} = \mathbf{Y}(t)\mathbf{c}.$$

We see that $\mathbf{Y}(t)$ is the fundamental matrix (see Sec. 3.2), whose columns are the basis vectors in (8), and $\mathbf{c} = [c_1 \;\; \cdots \;\; c_n]^{\mathsf{T}}$ is the vector of the arbitrary constants. As in Sec. 2.10, we replace the constant vector $\mathbf{c}$ by a variable vector $\mathbf{u}(t)$, so that we have

$$(9) \qquad \mathbf{y}^{(p)} = \mathbf{Y}(t)\,\mathbf{u}(t)$$

and we determine $\mathbf{u}(t)$ by substituting $\mathbf{y}^{(p)}$ into (7). This gives

$$(10) \qquad \mathbf{Y}'\mathbf{u} + \mathbf{Y}\mathbf{u}' = \mathbf{A}\mathbf{Y}\mathbf{u} + \mathbf{g}.$$

Now since $\mathbf{y}^{(1)}, \cdots, \mathbf{y}^{(n)}$ are solutions of the homogeneous system, we have

$$\mathbf{y}^{(1)\prime} = \mathbf{A}\mathbf{y}^{(1)}, \qquad \mathbf{y}^{(2)\prime} = \mathbf{A}\mathbf{y}^{(2)}, \qquad \cdots, \qquad \mathbf{y}^{(n)\prime} = \mathbf{A}\mathbf{y}^{(n)}.$$

We can write these n vector equations as a single matrix equation $\mathbf{Y}' = \mathbf{A}\mathbf{Y}$. (Make sure you understand this before you go on.) Hence $\mathbf{Y}'\mathbf{u} = \mathbf{A}\mathbf{Y}\mathbf{u}$ in (10), which thus reduces to $\mathbf{Y}\mathbf{u}' = \mathbf{g}$. Now the determinant of $\mathbf{Y}$ is the Wronskian (Sec. 3.2), which is not zero for a basis. Hence $\mathbf{Y}$ is nonsingular (see Sec. 3.0), so that its inverse $\mathbf{Y}^{-1}$ exists and we can solve

$$\mathbf{Y}\mathbf{u}' = \mathbf{g} \qquad \text{to get} \qquad \mathbf{u}' = \mathbf{Y}^{-1}\mathbf{g}.$$

Integrating from some t_0 in J to a variable t, we obtain

$$\mathbf{u}(t) = \int_{t_0}^{t} \mathbf{Y}^{-1}(\tilde{t})\,\mathbf{g}(\tilde{t})\,d\tilde{t} + \mathbf{c}.$$

The integrand is a vector with n components. Integration means that we integrate each component separately. This gives the vector $\mathbf{u}$ with n components. By leaving the constant vector $\mathbf{c}$ general we get a general solution

$$(11) \qquad \mathbf{y} = \mathbf{Y}\mathbf{u} = \mathbf{Y}\mathbf{c} + \mathbf{Y} \int_{t_0}^{t} \mathbf{Y}^{-1}(\tilde{t})\,\mathbf{g}(\tilde{t})\,d\tilde{t},$$

and for $\mathbf{c} = \mathbf{0}$ this is a particular solution (9) of (7).

EXAMPLE 3   **Solution by the method of variation of parameters**

For the system (4) in Example 2 we have from (5) and (4)

$$(12) \qquad \mathbf{Y} = [\mathbf{y}^{(1)} \quad \mathbf{y}^{(2)}] = \begin{bmatrix} e^{-2t} & e^{-4t} \\ e^{-2t} & -e^{-4t} \end{bmatrix}, \qquad \mathbf{g} = \begin{bmatrix} -6 \\ 2 \end{bmatrix} e^{-2t}.$$

From this and (8) in Sec. 3.0 we obtain the inverse

$$\mathbf{Y}^{-1} = \frac{1}{-2e^{-6t}} \begin{bmatrix} -e^{-4t} & -e^{-4t} \\ -e^{-2t} & e^{-2t} \end{bmatrix} = \frac{1}{2} \begin{bmatrix} e^{2t} & e^{2t} \\ e^{4t} & -e^{4t} \end{bmatrix}.$$

We multiply this by **g,** obtaining **u**$'$ (see above),

$$\mathbf{u}' = \mathbf{Y}^{-1}\mathbf{g} = \frac{1}{2} \begin{bmatrix} e^{2t} & e^{2t} \\ e^{4t} & -e^{4t} \end{bmatrix} \begin{bmatrix} -6e^{-2t} \\ 2e^{-2t} \end{bmatrix} = \frac{1}{2} \begin{bmatrix} -4 \\ -8e^{2t} \end{bmatrix} = \begin{bmatrix} -2 \\ -4e^{2t} \end{bmatrix}.$$

Integration now gives

$$\mathbf{u}(t) = \int_0^t \begin{bmatrix} -2 \\ -4e^{2\tilde{t}} \end{bmatrix} d\tilde{t} = \begin{bmatrix} -2t \\ -2e^{2t} + 2 \end{bmatrix}$$

(where $+2$ comes from the lower limit of integration). From this and (12) we obtain

$$\mathbf{Yu} = \begin{bmatrix} e^{-2t} & e^{-4t} \\ e^{-2t} & -e^{-4t} \end{bmatrix} \begin{bmatrix} -2t \\ -2e^{2t} + 2 \end{bmatrix} = \begin{bmatrix} -2te^{-2t} - 2e^{-2t} + 2e^{-4t} \\ -2te^{-2t} + 2e^{-2t} - 2e^{-4t} \end{bmatrix} = \begin{bmatrix} -2t - 2 \\ -2t + 2 \end{bmatrix} e^{-2t} + \begin{bmatrix} 2 \\ -2 \end{bmatrix} e^{-4t}.$$

The last term on the right is a solution of the homogeneous system. Hence we can absorb it into $\mathbf{y}^{(h)}$. We thus obtain as a general solution of the system (4), in agreement with (6*),

$$(13) \qquad \mathbf{y} = c_1 \begin{bmatrix} 1 \\ 1 \end{bmatrix} e^{-2t} + c_2 \begin{bmatrix} 1 \\ -1 \end{bmatrix} e^{-4t} - 2 \begin{bmatrix} 1 \\ 1 \end{bmatrix} te^{-2t} + \begin{bmatrix} -2 \\ 2 \end{bmatrix} e^{-2t}.$$

Method of Diagonalization

The idea of this method is to "decouple" the n equations of a linear system, so that each equation contains only one of the unknown functions $y_1, \cdots, y_n$ and thus can be solved independently of the other equations. This works for systems

$$(14) \qquad \boxed{\mathbf{y}' = \mathbf{Ay} + \mathbf{g}(t)}$$

for which **A** has a basis of eigenvectors $\mathbf{x}^{(1)}, \cdots, \mathbf{x}^{(n)}$. [This is true in the important cases of a symmetric matrix ($a_{jk} = a_{kj}$) or a skew-symmetric matrix ($a_{jk} = -a_{kj}$).] It can be shown that then

$$(15) \qquad \boxed{\mathbf{D} = \mathbf{X}^{-1}\mathbf{AX}}$$

is a diagonal matrix with the eigenvalues $\lambda_1, \cdots, \lambda_n$ of **A** on the main diagonal; here **X** is the $n \times n$ matrix with columns $\mathbf{x}^{(1)}, \cdots, \mathbf{x}^{(n)}$. (Proof in Sec. 7.5.) Note that $\mathbf{X}^{-1}$ exists because these columns are linearly independent. For instance, in Example 2 (verify!),

(16) $\mathbf{X} = \begin{bmatrix} 1 & 1 \\ 1 & -1 \end{bmatrix}$, $\mathbf{D} = \begin{bmatrix} \frac{1}{2} & \frac{1}{2} \\ \frac{1}{2} & -\frac{1}{2} \end{bmatrix} \begin{bmatrix} -3 & 1 \\ 1 & -3 \end{bmatrix} \begin{bmatrix} 1 & 1 \\ 1 & -1 \end{bmatrix} = \begin{bmatrix} -2 & 0 \\ 0 & -4 \end{bmatrix}$,

To apply diagonalization to (14), we define the new unknown function

(17)   (a) $\mathbf{z} = \mathbf{X}^{-1}\mathbf{y}$.   Then   (b) $\mathbf{y} = \mathbf{X}\mathbf{z}$.

Substituting this into (14), we have (note that $\mathbf{X}$ is constant!)

$$\mathbf{X}\mathbf{z}' = \mathbf{A}\mathbf{X}\mathbf{z} + \mathbf{g}.$$

We multiply this by $\mathbf{X}^{-1}$ from the left, obtaining

$$\mathbf{z}' = \mathbf{X}^{-1}\mathbf{A}\mathbf{X}\mathbf{z} + \mathbf{h}   \text{where}   \mathbf{h} = \mathbf{X}^{-1}\mathbf{g}.$$

Because of (15) we can write this

(18) $\mathbf{z}' = \mathbf{D}\mathbf{z} + \mathbf{h};$ in components, $z_j' = \lambda_j z_j + h_j,$

where $j = 1, \cdots, n$. We can now solve each of these n linear differential equations as in Sec. 1.6, to get

(19)
$$z_j(t) = c_j e^{\lambda_j t} + e^{\lambda_j t} \int e^{-\lambda_j t} h_j(t)\, dt.$$

These are the components of $\mathbf{z}(t)$, and from them we obtain the answer $\mathbf{y} = \mathbf{X}\mathbf{z}$ by (17b).

The present method has no calculational advantage over the other methods because we need the eigenvalues and eigenvectors, just as before, but it is interesting that solving a linear system of differential equations and diagonalizing a matrix are related problems.

EXAMPLE 4   **Method of diagonalization**

For the system (4) in Example 2 we get from (5) and (4) [see also (16)]

$$\mathbf{h} = \mathbf{X}^{-1}\mathbf{g} = \begin{bmatrix} \frac{1}{2} & \frac{1}{2} \\ \frac{1}{2} & -\frac{1}{2} \end{bmatrix} \begin{bmatrix} -6e^{-2t} \\ 2e^{-2t} \end{bmatrix} = \begin{bmatrix} -2e^{-2t} \\ -4e^{-2t} \end{bmatrix}.$$

Since the eigenvalues are $\lambda_1 = -2$ and $\lambda_2 = -4$, the diagonalized system is

$$\mathbf{z}' = \begin{bmatrix} -2 & 0 \\ 0 & -4 \end{bmatrix} \mathbf{z} + \mathbf{h}, \qquad \text{thus} \qquad \begin{aligned} z_1' &= -2z_1 - 2e^{-2t} \\ z_2' &= -4z_2 - 4e^{-2t}. \end{aligned}$$

From (19) we obtain the solutions $z_1 = c_1 e^{-2t} - 2te^{-2t}$ and $z_2 = c_2 e^{-4t} - 2e^{-2t}$. From this and (17b) we get the answer

$$\mathbf{y} = \mathbf{X}\mathbf{z} = \begin{bmatrix} 1 & 1 \\ 1 & -1 \end{bmatrix} \begin{bmatrix} c_1 e^{-2t} - 2te^{-2t} \\ c_2 e^{-4t} - 2e^{-2t} \end{bmatrix} = \begin{bmatrix} c_1 e^{-2t} - 2te^{-2t} + c_2 e^{-4t} - 2e^{-2t} \\ c_1 e^{-2t} - 2te^{-2t} - c_2 e^{-4t} + 2e^{-2t} \end{bmatrix}.$$

This is identical with (13) and (6*).   ◀

PROBLEM SET 3.6

1. (General solution) Prove that (2) includes every solution of (1).

General Solution. Find a real general solution of the following nonhomogeneous linear systems. (Show the details of your work.)

2. $y_1' = 2y_2 + t$

$y_2' = 2y_1 + 1$

3. $y_1' = y_2 + e^{3t}$

$y_2' = y_1 - 3e^{3t}$

4. $y_1' = 3y_1 - 4y_2 + 10 \cos t$

$y_2' = y_1 - 2y_2$

5. $y_1' = 3y_1 + y_2 - 3 \sin 3t$

$y_2' = 7y_1 - 3y_2 + 9 \cos 3t - 16 \sin 3t$

6. $y_1' = 4y_1 + y_2 + t$

$y_2' = 2y_1 + 3y_2 - t$

7. $y_1' = -2y_1 + y_2$

$y_2' = -y_1 + e^t$

Initial Value Problems. Showing all details of the calculation, solve:

8. $y_1' = 4y_2$

$y_2' = 4y_1 + 2 - 16t^2$ $y_1(0) = 3, \quad y_2(0) = 1$

9. $y_1' = y_2 + 6e^{2t}$

$y_2' = y_1 - 3e^{2t}$ $y_1(0) = 11, \quad y_2(0) = 0$

10. $y_1' = 4y_1 - 8y_2 + 2 \cosh t$

$y_2' = 2y_1 - 6y_2 + \cosh t + 2 \sinh t$ $y_1(0) = 0, \quad y_2(0) = 0$

11. $y_1' = 5y_2 + 23$

$y_2' = -5y_1 + 15t$ $y_1(0) = 1, \quad y_2(0) = -2$

12. $y_1' = -3y_1 - 4y_2 + 5e^t$

$y_2' = 5y_1 + 6y_2 - 6e^t$ $y_1(0) = 19, \quad y_2(0) = -23$

13. $y_1' = y_2 - 5 \sin t$

$y_2' = -4y_1 + 17 \cos t$ $y_1(0) = 5, \quad y_2(0) = 2$

14. $y_1' = y_1 + 4y_2 - t^2 + 6t$

$y_2' = y_1 + y_2 - t^2 + t - 1$ $y_1(0) = 2, \quad y_2(0) = -1$

15. $y_1' = 5y_1 + 4y_2 - 5t^2 + 6t + 25$

$y_2' = y_1 + 2y_2 - t^2 + 2t + 4$ $y_1(0) = 0, \quad y_2(0) = 0$

16. (**Undetermined coefficients: Modification**) Explain why in Example 2 of the text we have some freedom in choosing the vector **v**.

17. (**Network**) In Fig. 94 let $R_1 = 2$ ohms, $R_2 = 8$ ohms, $L = 1$ henry, $C = 0.5$ farad, $E = 200$ volts. Find the currents. Show the details of your work.

18. (**Network**) What is the solution in Prob. 17 if you change E to $440 \sin t$ volts, the other data remaining the same?

19. (**Network**) In Prob. 17 find the particular solution corresponding to zero currents and charge when $t = 0$.

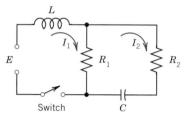

Fig. 94. Problems 17–19

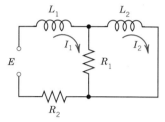

Fig. 95. Problem 20

20. (**Network**) Find the currents in Fig. 95, where $R_1 = 1$ ohm, $R_2 = 1.4$ ohms, $L_1 = 0.8$ henry, $L_2 = 1$ henry, $E = 100$ volts, and $I_1(0) = I_2(0) = 0$.

CHAPTER 3 REVIEW

1. What do we mean by a nonhomogeneous linear system of differential equations? A nonlinear system?

2. How can we transform a linear differential equation of second order into a system of equations? A fourth-order equation?

3. What do we mean by the phase plane? By a trajectory?

4. What is a critical point? Why is it important?

5. Without referring to the text, write down what you remember about the classification of critical points.

6. What do we mean by stability? Why is this basic in engineering?

7. What do we mean by linearizing a nonlinear system?

8. Review the undamped and damped pendulum equations and their linearizations.

9. What do we mean by the phase portrait? Compare the phase portraits in Prob. 8.

10. What are self-sustained oscillations? What is the van der Pol equation and its treatment in this chapter?

11. Why do eigenvalue problems come up in the present connection?

12. What is a basis of eigenvectors? How did we use it in a basis of solutions?

13. What can we do if a matrix has no basis of eigenvectors?

14. What do trajectories near a saddle point look like? Near a node?

15. What is the difference between a limit cycle and a closed trajectory surrounding a center?

General Solution. Critical Points. Find a general solution. Determine the type and stability of the critical point. (Show the details of your work.)

16. $y_1' = 3y_1 + 4y_2$

$y_2' = 3y_1 + 2y_2$

17. $y_1' = -3y_1 + 2y_2$

$y_2' = 4y_1 - y_2$

18. $y_1' = -2y_2$

$y_2' = 2y_1$

19. $y_1' = -2y_1 + 5y_2$

$y_2' = -y_1 - 6y_2$

20. $y_1' = y_1 + y_2$

$y_2' = -6y_1 - 4y_2$

21. $y_1' = 2y_1 + 4y_2$

$y_2' = 3y_1 + y_2$

22. $y_1' = y_1 + 4y_2$

$y_2' = y_1 + y_2$

23. $y_1' = 4y_1 - 2y_2$

$y_2' = 13y_1 - 6y_2$

Nonhomogeneous Systems. Find a general solution. (Show the details.)

24. $y_1' = 2y_1 + 2y_2 + e^t$

$y_2' = -2y_1 - 3y_2 + e^t$

25. $y_1' = 2y_1 + 3y_2 - 2e^{-t}$

$y_2' = -y_1 - 2y_2$

26. $y_1' = y_1 + 4y_2 - 2\cos t$

$y_2' = y_1 + y_2 - \cos t + \sin t$

27. $y_1' = 4y_2$

$y_2' = 4y_1 + 32t^2$

28. **(Critical point)** If $y' = Ay$ has a saddle point at $(0, 0)$, prove that $y' = A^2y$ has an unstable node at $(0, 0)$.

29. **(Critical point)** If A in $y' = Ay$ has the eigenvalues -4 and 3, what type of critical point is $(0, 0)$?

30. **(Network)** Find the currents in Fig. 96 when $R = 1$ ohm, $L = 1.25$ henrys, $C = 0.2$ farad, and $I_1(0) = I_2(0) = 1$ ampere.

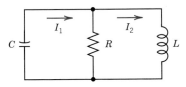

Fig. 96. Network in Problem 30

31. **(Network)** Find the currents in Fig. 97 when $R = 2.5$ ohms, $L = 1$ henry, $C = 0.04$ farad, $E(t) = 169 \sin t$ volts, and $I_1(0) = I_2(0) = 0$.

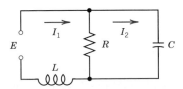

Fig. 97. Network in Problems 31, 32

32. (Network) Find the currents in Fig. 97 when $R = 1$ ohm, $L = 10$ henrys, $C = 1$ farad, $E = 100$ volts, and $I_1(0) = I_2(0) = 0$.

33. (Mixing problem) Tank T_1 in Fig. 98 contains initially 100 gal of pure water. Tank T_2 contains initially 100 gal of water in which 90 lb of salt are dissolved. Liquid is pumped through the system as indicated, and the mixtures are kept uniform by stirring. Find the amounts of salt $y_1(t)$ and $y_2(t)$ in T_1 and T_2, respectively.

34. (Comparison of methods) Find a general solution of $y_1' = y_2 + t$, $y_2' = -y_1$ by applying each of the three methods discussed in Sec. 3.6.

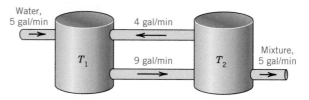

Fig. 98.   Tanks in Problem 33

Linearization of Systems. Determine the location and type of all critical points by using linearization.

35. $y_1' = y_2$

$y_2' = y_1 - y_1{}^3$

36. $y_1' = y_2$

$y_2' = -\tan y_1$

37. $y_1' = -4y_2$

$y_2' = \sin y_1$

38. $y_1' = 2y_2 + 2y_2{}^2$

$y_2' = -8y_1$

SUMMARY OF CHAPTER 3
SYSTEMS OF DIFFERENTIAL EQUATIONS
PHASE PLANE, QUALITATIVE METHODS

Whereas single electric circuits or single mass–spring systems are governed by single differential equations as their mathematical model (Chap. 2), networks consisting of several circuits, mechanical systems of several masses and springs, and other problems of engineering interest lead to **systems of differential equations** (Sec. 3.1). In such a system we simultaneously deal with several unknown functions, representing the currents in the various circuits, the displacements of those masses, etc. Of central interest are first-order systems (Sec. 3.2)

$$\mathbf{y}' = \mathbf{f}(t, \mathbf{y}), \qquad \text{in components,} \qquad \begin{aligned} y_1' &= f_1(t, y_1, \cdots, y_n) \\ &\vdots \\ y_n' &= f_n(t, y_1, \cdots, y_n), \end{aligned}$$

to which higher order equations and systems can be reduced. In this summary we let $n = 2$, for simplicity; thus

$$y_1' = f_1(t, y_1, y_2)$$

(1)

$$y_2' = f_2(t, y_1, y_2).$$

A **linear system** is of the form

$$y_1' = a_{11}y_1 + a_{12}y_2 + g_1$$

(2')

$$y_2' = a_{21}y_1 + a_{22}y_2 + g_2;$$

in vector form

(2) $\mathbf{y}' = \mathbf{A}\mathbf{y} + \mathbf{g}$,   where $\mathbf{A} = \begin{bmatrix} a_{11} & a_{12} \\ a_{21} & a_{22} \end{bmatrix}$, $\mathbf{y} = \begin{bmatrix} y_1 \\ y_2 \end{bmatrix}$, $\mathbf{g} = \begin{bmatrix} g_1 \\ g_2 \end{bmatrix}$.

A **homogeneous linear system** is of the form

(3) $\mathbf{y}' = \mathbf{A}\mathbf{y},$ written out, $\begin{aligned} y_1' &= a_{11}y_1 + a_{12}y_2 \\ y_2' &= a_{21}y_1 + a_{22}y_2. \end{aligned}$

A system (3) with *constant* $a_{11}, \cdots, a_{22}$ has solutions $\mathbf{y} = \mathbf{x}e^{\lambda t} \neq \mathbf{0}$, with the λ's being the solutions of the quadratic equation (Sec. 3.3)

$$\begin{vmatrix} a_{11} - \lambda & a_{12} \\ a_{21} & a_{22} - \lambda \end{vmatrix} = (a_{11} - \lambda)(a_{22} - \lambda) - a_{12}a_{21} = 0$$

and vector $\mathbf{x} \neq \mathbf{0}$ with components x_1, x_2 determined up to a multiplicative constant by

$$(a_{11} - \lambda)x_1 + a_{12}x_2 = 0.$$

(These λ's are called the **eigenvalues** and these vectors $\mathbf{x}$ **eigenvectors** of the matrix **A.** Further explanation of this is given in Sec. 3.0.)

The $y_1 y_2$-plane is called the **phase plane.** A **trajectory** is the curve of a solution $y_1(t)$, $y_2(t)$ in the phase plane. A **critical point** $P: (y_1, y_2)$ of a system (1) [or (2)] is one at which the right sides of the system are both zero. Depending on the behavior of the trajectories near P we call P a **node, saddle point, center,** or **spiral point,** and critical points are also classified in terms of **stability** (Secs. 3.3, 3.4). For nonlinear systems this is done by linearization (Sec. 3.5). Phase plane methods are primarily applied to **autonomous systems,** that is, systems in which t does not occur explicitly. They are **qualitative methods.** In applying them, we do not solve a system but discuss general properties of its solutions. This reveals a surprisingly large amount of information, particularly on nonlinear equations and systems that cannot be solved analytically. Three famous applications, namely, the pendulum and van der Pol equations and the Lotka–Volterra predator–prey population model, are included in Sec. 3.5.

Three methods for solving nonhomogeneous systems are discussed in Sec. 3.6.

CHAPTER 4

Series Solutions of Differential Equations. Special Functions

If a homogeneous linear differential equation has **constant coefficients,** it can be solved by algebraic methods, and its solutions are elementary functions known from calculus (e^x, cos x, etc.), as we know from Chap. 2. However, if such an equation has *variable coefficients* (functions of x), it must usually be solved by other methods. **Legendre's equation** (Sec. 4.3), the **hypergeometric equation** (Sec. 4.4), and **Bessel's equation** (Sec. 4.5) are very important equations of this type. Since these and other equations and their solutions play a basic role in applied mathematics, we devote an entire chapter to two standard methods of solution and their applications: the **power series method** (Secs. 4.1, 4.2), which yields solutions in the form of power series, and an extension of it, called the **Frobenius method** (Sec. 4.4).

The study of those solutions (and of other "higher" functions not discussed in calculus) is called the **theory of special functions.** Hence Chap. 4 will give the student a chance to become familiar with some of the methods in this area. This will include a discussion of **Sturm–Liouville theory** (Secs. 4.7, 4.8) based on **orthogonality of functions,** an idea whose significance to mathematical physics and its engineering applications can hardly be overestimated.

COMMENT. This chapter can also be studied directly after Chap. 2 because it uses no material from Chap. 3.

Prerequisite for this chapter: Chap. 2.
Sections that may be omitted in a shorter course: 4.2, 4.6−4.8.
References: Appendix 1, Part A.
Answers to problems: Appendix 2.

4.1 Power Series Method

The **power series method** is the standard basic method for solving linear differential equations with *variable* coefficients. It gives solutions in the form of power series; this explains the name. These series can be used for computing values of solutions, for exploring their properties, and for deriving other kinds of representations of those solutions, as we shall see. In this section we begin by explaining and illustrating the basic idea of the method.

Power Series

We first remember from calculus that a **power series**[1] (in powers of $x - x_0$) is an infinite series of the form

(1)
$$\sum_{m=0}^{\infty} a_m(x - x_0)^m = a_0 + a_1(x - x_0) + a_2(x - x_0)^2 + \cdots.$$

$a_0, a_1, a_2, \cdots$ are constants, called the **coefficients** of the series. x_0 is a constant, called the **center** of the series, and x is a variable.

If in particular $x_0 = 0$, we obtain a **power series in powers of x**

(2)
$$\sum_{m=0}^{\infty} a_m x^m = a_0 + a_1 x + a_2 x^2 + a_3 x^3 + \cdots.$$

We shall assume in this section that all variables and constants are real.

Familiar examples of power series are the Maclaurin series

$$\frac{1}{1 - x} = \sum_{m=0}^{\infty} x^m = 1 + x + x^2 + \cdots \qquad (|x| < 1, \textit{ geometric series}),$$

$$e^x = \sum_{m=0}^{\infty} \frac{x^m}{m!} = 1 + x + \frac{x^2}{2!} + \frac{x^3}{3!} + \cdots,$$

$$\cos x = \sum_{m=0}^{\infty} \frac{(-1)^m x^{2m}}{(2m)!} = 1 - \frac{x^2}{2!} + \frac{x^4}{4!} - + \cdots,$$

$$\sin x = \sum_{m=0}^{\infty} \frac{(-1)^m x^{2m+1}}{(2m + 1)!} = x - \frac{x^3}{3!} + \frac{x^5}{5!} - + \cdots.$$

Idea of the Power Series Method

The idea of the power series method for solving differential equations is simple and natural. We begin by describing the practical procedure and illustrate it for simple equations whose solutions we know, so that we can see what is going on. The mathematical justification of the method follows in the next section.

For a given differential equation

$$y'' + p(x)y' + q(x)y = 0$$

we first represent $p(x)$ and $q(x)$ by power series in powers of x (or of $x - x_0$ if solutions

---

[1]The term "power series" alone usually refers to a series of the form (1), but ***does not include*** series of negative powers of x such as $a_1 x^{-1} + a_2 x^{-2} + \cdots$ or series involving fractional powers of x. Note that in (1) we write, for convenience, $(x - x_0)^0 = 1$, even when $x = x_0$.

We use m as the summation letter, reserving n as a standard notation for the parameters in the Legendre and Bessel equations for integer values.

in powers of $x - x_0$ are wanted). Often, $p(x)$ and $q(x)$ are polynomials, and then nothing needs to be done in this first step. Next we assume a solution in the form of a power series with unknown coefficients,

$$(3) \qquad y = \sum_{m=0}^{\infty} a_m x^m = a_0 + a_1 x + a_2 x^2 + a_3 x^3 + \cdots$$

and insert this series and the series obtained by termwise differentiation,

$$(4) \quad \text{(a)} \quad y' = \sum_{m=1}^{\infty} m a_m x^{m-1} = a_1 + 2a_2 x + 3a_3 x^2 + \cdots$$

$$\text{(b)} \quad y'' = \sum_{m=2}^{\infty} m(m-1) a_m x^{m-2} = 2a_2 + 3 \cdot 2a_3 x + 4 \cdot 3a_4 x^2 + \cdots$$

into the equation. Then we collect like powers of x and equate the sum of the coefficients of each occurring power of x to zero, starting with the constant terms, the terms containing x, the terms containing x^2, etc. This gives relations from which we can determine the unknown coefficients in (3) successively.

We illustrate this for some simple equations that can also be solved by elementary methods.

EXAMPLE 1   Solve

$$y' - y = 0.$$

Solution.   In the first step, we insert (3) and (4a) into the equation:

$$(a_1 + 2a_2 x + 3a_3 x^2 + \cdots) - (a_0 + a_1 x + a_2 x^2 + \cdots) = 0.$$

Then we collect like powers of x, finding

$$(a_1 - a_0) + (2a_2 - a_1)x + (3a_3 - a_2)x^2 + \cdots = 0.$$

Equating the coefficient of each power of x to zero, we have

$$a_1 - a_0 = 0, \qquad 2a_2 - a_1 = 0, \qquad 3a_3 - a_2 = 0, \cdots.$$

Solving these equations, we may express $a_1, a_2, \cdots$ in terms of a_0, which remains arbitrary:

$$a_1 = a_0, \qquad a_2 = \frac{a_1}{2} = \frac{a_0}{2!}, \qquad a_3 = \frac{a_2}{3} = \frac{a_0}{3!}, \cdots.$$

With these coefficients the series (3) becomes

$$y = a_0 + a_0 x + \frac{a_0}{2!} x^2 + \frac{a_0}{3!} x^3 + \cdots,$$

and we see that we have obtained the familiar general solution

$$y = a_0 \left(1 + x + \frac{x^2}{2!} + \frac{x^3}{3!} + \cdots \right) = a_0 e^x. \qquad \blacktriangleleft$$

EXAMPLE 2   Solve

$$y' = 2xy.$$

Solution.   We insert (3) and (4a) into the equation:

$$a_1 + 2a_2x + 3a_3x^2 + \cdots = 2x(a_0 + a_1x + a_2x^2 + \cdots).$$

We must perform the multiplication by $2x$ on the right and can write the resulting equation conveniently as

$$a_1 + 2a_2x + 3a_3x^2 + 4a_4x^3 + 5a_5x^4 + 6a_6x^5 + \cdots$$
$$= \qquad 2a_0x + 2a_1x^2 + 2a_2x^3 + 2a_3x^4 + 2a_4x^5 + \cdots.$$

From this we see that

$$a_1 = 0, \qquad 2a_2 = 2a_0, \qquad 3a_3 = 2a_1, \qquad 4a_4 = 2a_2, \qquad 5a_5 = 2a_3, \qquad \cdots.$$

Hence $a_3 = 0$, $a_5 = 0$, $\cdots$ and for the coefficients with even subscripts,

$$a_2 = a_0, \qquad a_4 = \frac{a_2}{2} = \frac{a_0}{2!}, \qquad a_6 = \frac{a_4}{3} = \frac{a_0}{3!}, \cdots;$$

a_0 remains arbitrary. With these coefficients the series (3) gives the following solution, which you should confirm by the method of separating variables.

$$y = a_0 \left(1 + x^2 + \frac{x^4}{2!} + \frac{x^6}{3!} + \frac{x^8}{4!} + \cdots \right) = a_0 e^{x^2}. \qquad \blacktriangleleft$$

EXAMPLE 3   Solve

$$y'' + y = 0.$$

Solution.   By inserting (3) and (4b) into the equation we obtain

$$(2a_2 + 3 \cdot 2a_3x + 4 \cdot 3a_4x^2 + \cdots) + (a_0 + a_1x + a_2x^2 + \cdots) = 0.$$

Collecting like powers of x, we find

$$(2a_2 + a_0) + (3 \cdot 2a_3 + a_1)x + (4 \cdot 3a_4 + a_2)x^2 + \cdots = 0.$$

Equating the coefficient of each power of x to zero, we have

$$2a_2 + a_0 = 0 \qquad\qquad\qquad \text{coefficient of } x^0$$
$$3 \cdot 2a_3 + a_1 = 0 \qquad\qquad\qquad \text{coefficient of } x^1$$
$$4 \cdot 3a_4 + a_2 = 0 \qquad\qquad\qquad \text{coefficient of } x^2$$

etc. Solving these equations, we see that a_2, a_4, $\cdots$ may be expressed in terms of a_0; and a_3, a_5, $\cdots$ may be expressed in terms of a_1:

$$a_2 = -\frac{a_0}{2!}, \qquad a_3 = -\frac{a_1}{3!}, \qquad a_4 = -\frac{a_2}{4 \cdot 3} = \frac{a_0}{4!}, \cdots;$$

a_0 and a_1 are arbitrary. With these coefficients the series (3) becomes

$$y = a_0 + a_1x - \frac{a_0}{2!}x^2 - \frac{a_1}{3!}x^3 + \frac{a_0}{4!}x^4 + \frac{a_1}{5!}x^5 + \cdots.$$

Reordering terms (which is permissible for a power series), we can write this in the form

$$y = a_0 \left(1 - \frac{x^2}{2!} + \frac{x^4}{4!} - + \cdots \right) + a_1 \left(x - \frac{x^3}{3!} + \frac{x^5}{5!} - + \cdots \right)$$

and we recognize the familiar general solution

$$y = a_0 \cos x + a_1 \sin x.$$

◀

Do we need the power series method for these or similar equations? Of course not; this was just to explain the method. So what happens if we apply the method to an equation not of the kind considered in Chap. 2, even an innocent-looking one such as $y'' + xy = 0$ ("Airy's equation")? We most likely end up with new functions given by power series. And if such an equation and its solutions are of practical (or theoretical) interest, they are given names and are thoroughly investigated. This is what happened to Legendre's, Bessel's, and Gauss's hypergeometric equations, to mention just the most prominent ones. However, before we can discuss these highly important equations, we must first explain the power series method (and an extension of it) in more detail.

PROBLEM SET 4.1

Technique of the Power Series Method. Apply the power series method to the following differential equations. Show the details of your work. (More problems of this kind follow in Problem Set 4.2.)

1. $y' = 3y$
2. $y' = ky$
3. $y' + 2y = 0$
4. $(1 - x)y' = y$
5. $y' = 2xy$
6. $(1 + x)y' = y$
7. $y' = xy$
8. $y'' = 4y$
9. $y'' + 9y = 0$
10. $y'' = y'$
11. $y' = 3x^2 y$
12. $y'' = y$

13. **WRITING PROJECT. Power Series.** Write a concise review (2–3 pages) on power series as they are discussed in the usual calculus courses. Do not just copy from calculus texts—use your own formulations and include a small number of *simple* illustrative examples.

4.2 Theory of the Power Series Method

In the last section we saw that the power series method gives solutions of differential equations in the form of power series. In this section we first review a few relevant facts on power series from calculus, then list the operations on power series needed in the method (differentiation, addition, multiplication, etc.), and finally say a word about the existence of power series solutions.

Basic Concepts

Recall from calculus that a **power series** is an infinite series of the form

(1)
$$\sum_{m=0}^{\infty} a_m (x - x_0)^m = a_0 + a_1(x - x_0) + a_2(x - x_0)^2 + \cdots.$$

As before, we assume the variable x, the **center** x_0, and the **coefficients** $a_0, a_1, \cdots$ to be real. The **nth partial sum** of (1) is

(2) $$s_n(x) = a_0 + a_1(x - x_0) + a_2(x - x_0)^2 + \cdots + a_n(x - x_0)^n$$

where $n = 0, 1, \cdots$. Clearly, if we omit the terms of s_n from (1), the remaining expression is

(3) $$R_n(x) = a_{n+1}(x - x_0)^{n+1} + a_{n+2}(x - x_0)^{n+2} + \cdots.$$

This expression is called the **remainder** *of* (1) *after the term* $a_n(x - x_0)^n$.

For example, in the case of the geometric series

$$1 + x + x^2 + \cdots + x^n + \cdots$$

we have

$$s_0 = 1, \qquad\qquad R_0 = x + x^2 + x^3 + \cdots,$$
$$s_1 = 1 + x, \qquad\quad R_1 = x^2 + x^3 + x^4 + \cdots,$$
$$s_2 = 1 + x + x^2, \quad R_2 = x^3 + x^4 + x^5 + \cdots, \qquad \text{etc.}$$

In this way we have now associated with (1) the sequence of the partial sums $s_0(x), s_1(x), s_2(x), \cdots$. If for some $x = x_1$ this sequence converges, say,

$$\lim_{n \to \infty} s_n(x_1) = s(x_1),$$

then the series (1) is called **convergent** *at* $x = x_1$, the number $s(x_1)$ is called the **value** or *sum* of (1) at x_1, and we write

$$s(x_1) = \sum_{m=0}^{\infty} a_m(x_1 - x_0)^m.$$

Then we have for every n,

(4) $$s(x_1) = s_n(x_1) + R_n(x_1).$$

If that sequence diverges at $x = x_1$, the series (1) is called **divergent** at $x = x_1$.

In the case of convergence, for any positive ϵ there is an N (depending on ϵ) such that, by (4),

(5) $$|R_n(x_1)| = |s(x_1) - s_n(x_1)| < \epsilon \qquad\qquad \text{for all } n > N.$$

Geometrically, this means that all $s_n(x_1)$ with $n > N$ lie between $s(x_1) - \epsilon$ and $s(x_1) + \epsilon$ (Fig. 99). Practically, this means that in the case of convergence we can approximate the sum $s(x_1)$ of (1) at x_1 by $s_n(x_1)$ as accurately as we please, by taking n large enough.

Fig. 99. Inequality (5), where s denotes $s(x_1)$

Convergence Interval. Radius of Convergence

1. The series (1) always converges at $x = x_0$, because then all its terms except for the first, a_0, are zero. In exceptional cases this may be the only x for which (1) converges. Such a series is of no practical interest.

2. If there are further values of x for which the series converges, these values form an interval, called the **convergence interval.** If this interval is finite, it has the midpoint x_0, so that it is of the form

(6) $$|x - x_0| < R$$ (Fig. 100)

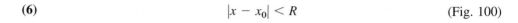
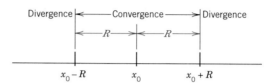

Fig. 100. Convergence interval (6) of a power series with center x_0

and the series (1) converges for all x such that $|x - x_0| < R$ and diverges for all x such that $|x - x_0| > R$. The number R is called the **radius^2 of convergence** of (1). It can be obtained from either of the formulas

(7)   (a) $R = 1 \Big/ \lim_{m \to \infty} \sqrt[m]{|a_m|}$ (b) $R = 1 \Big/ \lim_{m \to \infty} \left| \dfrac{a_{m+1}}{a_m} \right|$

provided these limits exist and are not zero. [If they are infinite, then (1) converges only at the center x_0.]

3. The convergence interval may sometimes be infinite, that is, (1) converges for all x. For instance, if the limit in (7a) or (7b) is zero, this case occurs. One then writes $R = \infty$, for convenience. (Proofs of all these facts can be found in Sec. 14.2.)

For each x for which (1) converges, it has a certain value $s(x)$. We say that (1) **represents** the function $s(x)$ in the convergence interval and write

$$s(x) = \sum_{m=0}^{\infty} a_m (x - x_0)^m \qquad (|x - x_0| < R).$$

Let us illustrate the three possibilities with typical examples.

EXAMPLE 1   **The useless Case 1 of convergence only at the center**

In the case of the series

$$\sum_{m=0}^{\infty} m! x^m = 1 + x + 2x^2 + 6x^3 + \cdots$$

---

^{2}Because for a *complex* power series, one has convergence in an open disk of *radius R.* No general statement about convergence or divergence can be made for $x - x_0 = R$ or $-R$.

we have $a_m = m!$, and in (7b),

$$\frac{a_{m+1}}{a_m} = \frac{(m+1)!}{m!} = m + 1 \;\;\to\;\; \infty \qquad\qquad \text{as } m \to \infty.$$

Thus this series converges only at the center $x = 0$. Such a series is useless. ◀

EXAMPLE 2   **The usual Case 2 of convergence in a finite interval. Geometric series**

For the **geometric series** we have

$$\frac{1}{1-x} = \sum_{m=0}^{\infty} x^m = 1 + x + x^2 + \cdots \qquad\qquad (|x| < 1).$$

In fact, $a_m = 1$ for all m, and from (7) we obtain $R = 1$, that is, the geometric series converges and represents $1/(1 - x)$ when $|x| < 1$. ◀

EXAMPLE 3   **The best Case 3 of convergence for all x**

In the case of the series

$$e^x = \sum_{m=0}^{\infty} \frac{x^m}{m!} = 1 + x + \frac{x^2}{2!} + \cdots$$

we have $a_m = 1/m!$. Hence in (7b),

$$\frac{a_{m+1}}{a_m} = \frac{1/(m+1)!}{1/m!} = \frac{1}{m+1} \;\;\to\;\; 0 \qquad\qquad \text{as } m \to \infty,$$

so that the series converges for all x. ◀

EXAMPLE 4   **A hint for some of the problems**

Find the radius of convergence of the series

$$\sum_{m=0}^{\infty} \frac{(-1)^m}{8^m} x^{3m} = 1 - \frac{x^3}{8} + \frac{x^6}{64} - \frac{x^9}{512} + - \cdots.$$

Solution.   This is a series in powers of $t = x^3$ with coefficients $a_m = (-1)^m/8^m$, so that in (7b),

$$\left| \frac{a_{m+1}}{a_m} \right| = \frac{8^m}{8^{m+1}} = \frac{1}{8}.$$

Thus $R = 8$. Hence the series converges for $|t| < 8$, that is, $|x| < 2$. ◀

Operations on Power Series

In the power series method we differentiate, add, and multiply power series. These three operations are permissible, in the sense explained in what follows. We also list a condition about the vanishing of all coefficients of a power series, which is a basic tool of the power series method. (Proofs can be found in Sec. 14.3.)

Termwise Differentiation

A power series may be differentiated term by term. More precisely: if

$$y(x) = \sum_{m=0}^{\infty} a_m(x - x_0)^m$$

converges for $|x - x_0| < R$, where $R > 0$, then the series obtained by differentiating term by term also converges for those x and represents the derivative y' of y for those x, that is,

$$y'(x) = \sum_{m=1}^{\infty} m a_m (x - x_0)^{m-1} \qquad (|x - x_0| < R).$$

Similarly,

$$y''(x) = \sum_{m=2}^{\infty} m(m - 1) a_m (x - x_0)^{m-2} \qquad (|x - x_0| < R), \text{ etc.}$$

Termwise Addition

Two power series may be added term by term. More precisely: if the series

(8)
$$\sum_{m=0}^{\infty} a_m (x - x_0)^m \qquad \text{and} \qquad \sum_{m=0}^{\infty} b_m (x - x_0)^m$$

have positive radii of convergence and their sums are $f(x)$ and $g(x)$, then the series

$$\sum_{m=0}^{\infty} (a_m + b_m)(x - x_0)^m$$

converges and represents $f(x) + g(x)$ for each x that lies in the interior of the convergence interval of each of the given series.

Termwise Multiplication

Two power series may be multiplied term by term. More precisely: Suppose that the series (8) have positive radii of convergence and let $f(x)$ and $g(x)$ be their sums. Then the series obtained by multiplying each term of the first series by each term of the second series and collecting like powers of $x - x_0$, that is,

$$\sum_{m=0}^{\infty} (a_0 b_m + a_1 b_{m-1} + \cdots + a_m b_0)(x - x_0)^m$$

$$= a_0 b_0 + (a_0 b_1 + a_1 b_0)(x - x_0) + (a_0 b_2 + a_1 b_1 + a_2 b_0)(x - x_0)^2 + \cdots$$

converges and represents $f(x)g(x)$ for each x in the interior of the convergence interval of each of the given series.

Vanishing of All Coefficients

If a power series has a positive radius of convergence and a sum that is identically zero throughout its interval of convergence, then each coefficient of the series must be zero.

Shifting Summation Indices

This is a technicality best explained in terms of a typical example. Suppose we are given

$$x^2 \sum_{m=2}^{\infty} m(m - 1) a_m x^{m-2} + 2 \sum_{m=1}^{\infty} m a_m x^{m-1}$$

$$= x^2 (2a_2 + 6a_3 x + 12a_4 x^2 + \cdots) + 2(a_1 + 2a_2 x + 3a_3 x^2 + \cdots)$$

and we want to write this as a single series. We first take x^2 inside the summation, obtaining

$$\sum_{m=2}^{\infty} m(m-1)a_m x^m + \sum_{m=1}^{\infty} 2m a_m x^{m-1}.$$

Suppose that we want to use s as the summation letter of the series to be obtained. Then in the first series we simply write s for m. Such a change of notation can always be made because a summation letter is just a dummy index and any letter that has not been used previously is acceptable. The shift takes place in the second series. We shift by one unit, setting $m - 1 = s$. (We can recognize this directly from the exponent in the series—can you see it?) Then $m = s + 1$. The summation now begins with $s = 0$ because then $m = 0 + 1 = 1$, the old beginning. Together,

$$\sum_{s=2}^{\infty} s(s-1)a_s x^s + \sum_{s=0}^{\infty} 2(s+1)a_{s+1} x^s.$$

In the first series we can replace $s = 2$ by $s = 0$ (why?). We then obtain

$$\sum_{s=0}^{\infty} [s(s-1)a_s + 2(s+1)a_{s+1}]x^s = 2a_1 + 4a_2 x + (2a_2 + 6a_3)x^2 + (6a_3 + 8a_4)x^3 + \cdots.$$

Existence of Power Series Solutions. Real Analytic Functions

The properties of power series just discussed form the foundation of the power series method. The remaining question is whether an equation has power series solutions at all. An answer is simple: If the coefficients p and q and the function r on the right side of

$$(9) \qquad\qquad y'' + p(x)y' + q(x)y = r(x)$$

have power series representations, then (9) has power series solutions. The same is true if $\tilde{h}$, $\tilde{p}$, $\tilde{q}$, and $\tilde{r}$ in

$$(10) \qquad\qquad \tilde{h}(x)y'' + \tilde{p}(x)y' + \tilde{q}(x)y = \tilde{r}(x)$$

have power series representations and $\tilde{h}(x_0) \neq 0$ (x_0 the center of the series). Almost all equations in practice have polynomials as coefficients (thus terminating power series), so that (when $r(x) \equiv 0$ or is a power series, too) those conditions are satisfied, except perhaps the condition $\tilde{h}(x_0) \neq 0$, a fact that will keep us busy in later sections.

To formulate all of this in a precise and simple way, we use the following concept (which is of general interest).

Definition of real analytic function

A real function $f(x)$ is called **analytic** *at a point* $x = x_0$ if it can be represented by a power series in powers of $x - x_0$ with radius of convergence $R > 0$.

Using this concept, we can state the following basic theorem.

THEOREM 1   **(Existence of power series solutions)**

If p, q, and r in (9) are analytic at $x = x_0$, then every solution of (9) is analytic at $x = x_0$ and can thus be represented by a power series in powers of $x - x_0$ with radius of convergence $R > 0$. Hence the same is true if $\tilde{h}$, $\tilde{p}$, $\tilde{q}$, and $\tilde{r}$ in (10) are analytic *at $x = x_0$ and $\tilde{h}(x_0) \neq 0$.[3]*

The proof of this theorem requires advanced methods of complex analysis and can be found in Ref. [A5] listed in Appendix 1.

PROBLEM SET 4.2

Power Series Solutions. Find a power series solution in powers of x of the following differential equations. (Show the details of your work.)

1. $y' = -2xy$ **2.** $(x - 2)y' = xy$
3. $xy' - 3y = k \; (= const)$ **4.** $(1 - x^2)y' = 2xy$
5. $y'' - 3y' + 2y = 0$ **6.** $y'' - 4xy' + (4x^2 - 2)y = 0$
7. $y'' + 4y = 0$ **8.** $(1 - x^2)y'' - 2xy' + 2y = 0$

Series Solutions in Powers of $x - x_0$ can be obtained if you introduce $t = x - x_0$ as a new independent variable and solve the resulting equation for y as a power series in t. Do this for the following equations, with $x_0 = 1$ (for simplicity). Show the details of your work.

9. $y' = ky$ **10.** $y'' - y = 0$ **11.** $y' = (y/x) + 1$

12. Why can the equation in Prob. 11 not be solved by a power series in powers of x?

Radius of Convergence. Determine the radius of convergence of the following series.

13. $\displaystyle\sum_{m=0}^{\infty} \frac{x^{2m}}{m!}$ **14.** $\displaystyle\sum_{m=0}^{\infty} (m + 1)mx^m$ **15.** $\displaystyle\sum_{m=0}^{\infty} \frac{1}{3^m} (x - 3)^{2m}$

16. $\displaystyle\sum_{m=0}^{\infty} (-1)^m x^{4m}$ **17.** $\displaystyle\sum_{m=0}^{\infty} \frac{(-1)^m}{k^m} x^{2m}$ **18.** $\displaystyle\sum_{m=0}^{\infty} \frac{x^{2m+1}}{(2m + 1)!}$

19. $\displaystyle\sum_{m=0}^{\infty} \left(\tfrac{2}{3}\right)^m x^{2m}$ **20.** $\displaystyle\sum_{m=0}^{\infty} m^m x^m$ **21.** $\displaystyle\sum_{m=0}^{\infty} \frac{1}{2^m} (x - x_0)^{2m}$

22. TEAM PROJECT. Properties from Power Series. In the next sections we shall define new functions (Legendre functions, etc.) by power series of importance to the engineer. And we shall see that various properties of these functions can be derived directly from their series. To understand this idea, explore it for functions familiar from calculus. Using power series, do:

 (a) Derive the differentiation formulas of e^x, $\cos x$, $1/(1 - x)$, and other functions of your own choice. Also show that $(\sin x)'' = -\sin x$.

 (b) Show that $\cosh x + \sinh x = e^x$.

 (c) What can you conclude if you notice that the series of $\cot x$ and $\sinh x$ contain only odd powers of x? That a certain series contains only even powers of x? That a certain series begins with x^{10} as the smallest power? That all the coefficients of a series are positive? (Give examples.)

 (d) Find further properties from series of your own choice. Can you see from the series that $\cos x$ and $\sin x$ are periodic? That they are of absolute value not exceeding 1?

Shift of Summation Index. This operation occurs often in the power series method. Shift the index so that the power under the summation sign is x^m. Check your result by writing the first few terms explicitly. Also determine the radius of convergence.

---

[3]R is at least equal to the distance between the point $x = x_0$ and that point (or those points) closest to $x = x_0$ at which one of the functions p, q, r, *as functions of a complex variable*, is not analytic. (Note that that point may not lie on the x-axis, but somewhere in the complex plane.)

23. $\displaystyle\sum_{s=2}^{\infty} \frac{s(s+1)}{s^2+1} x^{s-1}$ **24.** $\displaystyle\sum_{p=2}^{\infty} p(p-1)x^{p-2}$ **25.** $\displaystyle\sum_{n=0}^{\infty} \frac{(n+1)^3}{(n+2)!} x^{n+1}$

 26. CAS PROJECT. Information from Partial Sums. Partial sums (2) of power series (1) are often used in computing approximate values of the function given by (1), with the choice of n in (2) depending on the accuracy desired and the location of x relative to x_0. Let $x_0 = 0$ for simplicity. A very rough first impression of accuracy can often be obtained by experimenting with plots of (2) for increasing n.

(a) Do this for $\cos x$ and $n = 2, 4, 6, \cdots, 20$.

(b) Do this for the solution of the initial value problem $(1 - x^2)y'' - 2xy' + 2y = 0$, $y(0) = 1$, $y'(0) = 0$ and n of your choice. State what the plot suggests with respect to convergence.

(c) Experiment with two initial value problems of your own choice, one in which the solution varies only slowly with x and one in which it varies rapidly.

4.3 Legendre's Equation Legendre Polynomials $P_n(x)$

So far we have applied the power series method to equations that can also be solved by other methods, in order to gain skill. We now turn to the first "big" equation of physics, for which we do need the power series method. This is **Legendre's equation**[4]

(1)
$$(1 - x^2)y'' - 2xy' + n(n+1)y = 0.$$

It arises in numerous problems, particularly in boundary value problems for spheres (take a quick look at Example 1 in Sec. 11.11). The **parameter** n in (1) is a given real number. Any solution of (1) is called a **Legendre function.** The study of these and other "higher" functions not occurring in calculus is called the **theory of special functions.** (Further special functions will occur in the next sections.)

Dividing (1) by the coefficient $1 - x^2$ of y'', we see that the coefficients $-2x/(1 - x^2)$ and $n(n+1)/(1 - x^2)$ of the new equation are analytic at $x = 0$. Hence by Theorem 1, Sec. 4.2, we may apply the power series method. We substitute

(2)
$$y = \sum_{m=0}^{\infty} a_m x^m$$

and its derivatives into (1). Denoting the constant $n(n+1)$ simply by k, we obtain

$$(1 - x^2)\sum_{m=2}^{\infty} m(m-1)a_m x^{m-2} - 2x\sum_{m=1}^{\infty} ma_m x^{m-1} + k\sum_{m=0}^{\infty} a_m x^m = 0.$$

---

[4]ADRIEN MARIE LEGENDRE (1752—1833), French mathematician, who became a professor in Paris in 1775 and made important contributions to special functions, elliptic integrals, number theory, and the calculus of variations. His book *Éléments de géométrie* (1794) became very famous and had 12 editions in less than 30 years.

Formulas on Legendre functions may be found in Refs. [1], [6], and [11].

By writing the first expression as two separate series we have the equation

$$(1^*) \quad \sum_{m=2}^{\infty} m(m-1)a_m x^{m-2} - \sum_{m=2}^{\infty} m(m-1)a_m x^m - 2\sum_{m=1}^{\infty} ma_m x^m + k\sum_{m=0}^{\infty} a_m x^m = 0.$$

By writing out each series and arranging each power in a column we obtain

$$2\cdot 1a_2 + 3\cdot 2a_3 x + 4\cdot 3a_4 x^2 + \cdots \qquad + (s+2)(s+1)a_{s+2}x^s \cdots$$
$$-2\cdot 1a_2 x^2 - \cdots \qquad\qquad + \cdots$$
$$-2\cdot 1a_1 x - 2\cdot 2a_2 x^2 - \cdots \qquad -s(s-1)a_s x^s - \cdots$$
$$+ ka_0 + ka_1 x + ka_2 x^2 + \cdots \qquad -2sa_s x^s - \cdots = 0.$$

Since this must be an identity in x if (2) is to be a solution of (1), the sum of the coefficients of each power of x must be zero. Remembering that $k = n(n+1)$, we thus have

$$(3a) \qquad 2a_2 + n(n+1)a_0 = 0 \qquad\qquad \text{coefficients of } x^0$$
$$6a_3 + [-2 + n(n+1)]a_1 = 0 \qquad\qquad \text{coefficients of } x^1$$

and in general, when $s = 2, 3, \cdots$,

$$(3b) \qquad (s+2)(s+1)a_{s+2} + [-s(s-1) - 2s + n(n+1)]a_s = 0.$$

Now the expression in brackets $[\cdots]$ can be written $(n-s)(n+s+1)$, as you may readily verify. We thus obtain from (3)

$$(4) \qquad a_{s+2} = -\frac{(n-s)(n+s+1)}{(s+2)(s+1)} a_s \qquad (s = 0, 1, \cdots).$$

This is called a **recurrence relation** or **recursion formula.** It gives each coefficient in terms of the second one preceding it, except for a_0 and a_1, which are left as arbitrary constants. We find successively

$$a_2 = -\frac{n(n+1)}{2!} a_0 \qquad\qquad a_3 = -\frac{(n-1)(n+2)}{3!} a_1$$

$$a_4 = -\frac{(n-2)(n+3)}{4\cdot 3} a_2 \qquad\qquad a_5 = -\frac{(n-3)(n+4)}{5\cdot 4} a_3$$

$$= \frac{(n-2)n(n+1)(n+3)}{4!} a_0 \qquad\qquad = \frac{(n-3)(n-1)(n+2)(n+4)}{5!} a_1$$

etc. By inserting these values for the coefficients into (2) we obtain

$$(5) \qquad y(x) = a_0 y_1(x) + a_1 y_2(x)$$

where

$$(6) \qquad y_1(x) = 1 - \frac{n(n+1)}{2!} x^2 + \frac{(n-2)n(n+1)(n+3)}{4!} x^4 - + \cdots$$

and

$$(7) \qquad y_2(x) = x - \frac{(n-1)(n+2)}{3!} x^3 + \frac{(n-3)(n-1)(n+2)(n+4)}{5!} x^5 - + \cdots.$$

These series converge for $|x| < 1$ (see Prob. 9; or they may terminate, see below). Since (6) contains even powers of x only, while (7) contains odd powers of x only, the ratio y_1/y_2 is not a constant, so that y_1 and y_2 are not proportional and are thus linearly independent solutions. Hence (5) is a general solution of (1) on the interval $-1 < x < 1$.

Legendre Polynomials

In many applications the parameter n in Legendre's equation will be a nonnegative integer. Then the right side of (4) is zero when $s = n$, and, therefore, $a_{n+2} = 0$, $a_{n+4} = 0$, $a_{n+6} = 0, \cdots$. Hence, if n is even, $y_1(x)$ reduces to a polynomial of degree n. If n is odd, the same is true for $y_2(x)$. These polynomials, multiplied by some constants, are called **Legendre polynomials.** Since they are of great practical importance, let us consider them in more detail. For this purpose we solve (4) for a_s, obtaining

$$(8) \qquad a_s = -\frac{(s+2)(s+1)}{(n-s)(n+s+1)} a_{s+2} \qquad (s \leqq n-2).$$

We may then express all the nonvanishing coefficients in terms of the coefficient a_n of the highest power of x of the polynomial. The coefficient a_n is at first still arbitrary. It is standard to choose $a_n = 1$ when $n = 0$ and

$$(9) \qquad a_n = \frac{(2n)!}{2^n (n!)^2} = \frac{1 \cdot 3 \cdot 5 \cdots (2n-1)}{n!}, \qquad n = 1, 2, \cdots.$$

The reason is that for this choice of a_n all those polynomials will have the value 1 when $x = 1$; this follows from (13) in Team Project 10. We then obtain from (8) and (9)

$$(9^*) \qquad
\begin{aligned}
a_{n-2} &= -\frac{n(n-1)}{2(2n-1)} a_n \\[1ex]
&= -\frac{n(n-1)(2n)!}{2(2n-1)2^n(n!)^2} \\[1ex]
&= -\frac{n(n-1)2n(2n-1)(2n-2)!}{2(2n-1)2^n n(n-1)!\, n(n-1)(n-2)!},
\end{aligned}$$

that is,

$$a_{n-2} = -\frac{(2n-2)!}{2^n(n-1)!\,(n-2)!}.$$

Similarly,

$$\begin{aligned}
a_{n-4} &= -\frac{(n-2)(n-3)}{4(2n-3)} a_{n-2} \\[1ex]
&= \frac{(2n-4)!}{2^n 2!\,(n-2)!\,(n-4)!}
\end{aligned}$$

etc., and in general, when $n - 2m \geqq 0$,

(10) $$a_{n-2m} = (-1)^m \frac{(2n - 2m)!}{2^n m! \, (n - m)! \, (n - 2m)!} .$$

The resulting solution of Legendre's differential equation (1) is called the **Legendre polynomial** *of degree n* and is denoted by $P_n(x)$.

From (10) we obtain

(11)
$$P_n(x) = \sum_{m=0}^{M} (-1)^m \frac{(2n - 2m)!}{2^n m! \, (n - m)! \, (n - 2m)!} x^{n-2m}$$
$$= \frac{(2n)!}{2^n (n!)^2} x^n - \frac{(2n - 2)!}{2^n 1! \, (n - 1)! \, (n - 2)!} x^{n-2} + - \cdots$$

where $M = n/2$ or $(n - 1)/2$, whichever is an integer. The first few of these functions are (Fig. 101)

(11′)
$$P_0(x) = 1,$$
$$P_1(x) = x,$$
$$P_2(x) = \tfrac{1}{2}(3x^2 - 1),$$
$$P_3(x) = \tfrac{1}{2}(5x^3 - 3x),$$
$$P_4(x) = \tfrac{1}{8}(35x^4 - 30x^2 + 3),$$
$$P_5(x) = \tfrac{1}{8}(63x^5 - 70x^3 + 15x),$$

etc.

The so-called ***orthogonality*** of the Legendre polynomials will be considered in Secs. 4.7 and 4.8.

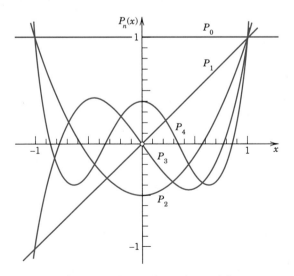

Fig. 101. Legendre polynomials

PROBLEM SET 4.3

1. **(Legendre functions for $n = 0$)** Show that (6) with $n = 0$ gives $y_1(x) = P_0(x) = 1$ and (7) gives

$$y_2(x) = x + \frac{2}{3!}x^3 + \frac{(-3)(-1) \cdot 2 \cdot 4}{5!}x^5 + \cdots = x + \frac{x^3}{3} + \frac{x^5}{5} + \cdots = \frac{1}{2}\ln\frac{1+x}{1-x}.$$

   Verify this by solving (1) with $n = 0$, setting $z = y'$ and separating variables.

2. **(Legendre functions for $n = 1$)** Show that (7) with $n = 1$ gives $y_2(x) = P_1(x) = x$ and (6) gives

$$y_1(x) = 1 - \frac{x^2}{1} - \frac{x^4}{3} - \frac{x^6}{5} - \cdots = 1 - x\left(x + \frac{x^3}{3} + \frac{x^5}{5} + \cdots\right) = 1 - \frac{1}{2}x\ln\frac{1+x}{1-x}.$$

3. **(Special n)** Derive (11$'$) from (11).

4. **(Verification)** Verify that the functions in (11$'$) satisfy Legendre's equation.

5. **(Shortcut)** Get (3) from (1*) more quickly by writing $m - 2 = s$ in the first sum in (1*) and $m = s$ in the other sums, obtaining

$$\sum_{s=0}^{\infty} \{(s + 2)(s + 1)a_{s+2} - [s(s - 1) + 2s - k]a_s\}x^s = 0.$$

6. **(Rodrigues's formula[5])** Applying the binomial theorem to $(x^2 - 1)^n$, differentiating n times term by term, and comparing with (11), show that

   (12)
   $$\boxed{P_n(x) = \frac{1}{2^n n!}\frac{d^n}{dx^n}[(x^2 - 1)^n]}$$
   (Rodrigues's formula).

7. **(Differential equation)** Find a solution of $(a^2 - x^2)y'' - 2xy' + 12y = 0$, $a \neq 0$.

8. **(Rodrigues's formula)** Obtain (11$'$) from (12).

9. **(Convergence)** Show that for any n for which (6) and (7) do not reduce to a polynomial, these series have radius of convergence 1.

10. **TEAM PROJECT. Generating Functions.** Generating functions play a significant role in modern applied mathematics (see [7]). The idea is simple. If we want to study a certain sequence $(f_n(x))$ and can find a function

$$G(u, x) = \sum_{n=0}^{\infty} f_n(x)u^n,$$

   we may obtain properties of $(f_n(x))$ from those of G, which "generates" this sequence and is called a **generating function** of it.

   (a) Legendre Polynomials. Show that

   (13)
   $$G(u, x) = \frac{1}{\sqrt{1 - 2xu + u^2}} = \sum_{n=0}^{\infty} P_n(x)u^n$$

   is a generating function of the Legendre polynomials. *Hint.* Start from the binomial expansion of $1/\sqrt{1 - v}$, then set $v = 2xu - u^2$, multiply the powers of $2xu - u^2$ out, collect all the terms involving u^n, and verify that the sum of these terms is $P_n(x)u^n$.

---

[5]OLINDE RODRIGUES (1794—1851), French mathematician and economist.

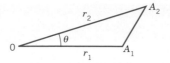

Fig. 102.   Team Project 10

(b) Potential Theory. Let A_1 and A_2 be two points in space (Fig. 102, $r_2 > 0$). Using (13), show that

$$\frac{1}{r} = \frac{1}{\sqrt{r_1{}^2 + r_2{}^2 - 2r_1 r_2 \cos \theta}} = \frac{1}{r_2} \sum_{m=0}^{\infty} P_m(\cos \theta) \left(\frac{r_1}{r_2}\right)^m.$$

This formula has applications in potential theory.

(c) Further Applications of (13). Show that $P_n(1) = 1$, $P_n(-1) = (-1)^n$, and

$$P_{2n+1}(0) = 0, \qquad P_{2n}(0) = (-1)^n \cdot 1 \cdot 3 \cdots (2n-1)/[2 \cdot 4 \cdots (2n)].$$

(d) Bonnet's Recursion.[6] Differentiating (13) with respect to u, using (13) in the resulting formula, and comparing coefficients of u^n, obtain the *Bonnet recursion*

(14) $$(n+1)P_{n+1}(x) = (2n+1)xP_n(x) - nP_{n-1}(x), \qquad n = 1, 2, \cdots.$$

This formula is useful for computations, the loss of significant figures being small (except near zeros). Try (14) out for a few computations of your own choice.

11. (Associated Legendre functions) Consider

$$(1 - x^2)y'' - 2xy' + \left[n(n+1) - \frac{m^2}{1 - x^2}\right]y = 0.$$

Substituting $y(x) = (1 - x^2)^{m/2}v(x)$, show that v satisfies

(15) $$(1 - x^2)v'' - 2(m+1)xv' + [n(n+1) - m(m+1)]v = 0.$$

Starting from (1) and differentiating it m times, show that a solution of (15) is

$$v = \frac{d^m P_n}{dx^m}.$$

The corresponding $y(x)$ is denoted by $P_n{}^m(x)$. It is called an *associated Legendre function* and plays a role in quantum physics. Thus

(16) $$\boxed{P_n{}^m(x) = (1 - x^2)^{m/2} \frac{d^m P_n}{dx^m}.}$$

12. (Associated Legendre functions) Find $P_1{}^1(x)$, $P_2{}^1(x)$, $P_2{}^2(x)$, $P_4{}^2(x)$.

---

[6]OSSIAN BONNET (1819—1892), French mathematician, whose main work was in the differential geometry of surfaces.

4.4 Frobenius Method

Several second-order differential equations of great practical importance—the famous Bessel equation among them—have coefficients that are not analytic (definition in Sec. 4.2), but are such that the following theorem holds, providing an extension of the power series method, called the **Frobenius method.**[7]

THEOREM 1   **(Frobenius method)**

Any differential equation of the form

(1)
$$y'' + \frac{b(x)}{x} y' + \frac{c(x)}{x^2} y = 0,$$

where the functions $b(x)$ and $c(x)$ are analytic at $x = 0$, has at least one solution that can be represented in the form

(2)
$$y(x) = x^r \sum_{m=0}^{\infty} a_m x^m = x^r(a_0 + a_1 x + a_2 x^2 + \cdots) \qquad (a_0 \neq 0)$$

where the exponent r may be any (real or complex) number (and r is chosen so that $a_0 \neq 0$).[8]

The equation also has a second solution (such that these two solutions are linearly independent) that may be similar to (2) (with a different r and different coefficients) or may contain a logarithmic term (details in Theorem 2, below).

For example, Bessel's equation (to be discussed in the next section)

$$y'' + \frac{1}{x} y' + \left(\frac{x^2 - \nu^2}{x^2}\right) y = 0$$

is of the form (1) with $b(x) = 1$ and $c(x) = x^2 - \nu^2$ analytic at $x = 0$, so that the theorem applies. This equation could not be handled in full generally by the power series method.

Similarly, the so-called hypergeometric differential equation (see Problem Set 4.4) also requires the Frobenius method.

The point is that in (2) we have a power series times a single power of x whose exponent r is not restricted to be a nonnegative integer. (The latter restriction would make the whole expression a power series, by definition; see footnote 1 in Sec. 4.1.)

The proof of the theorem requires advanced methods of complex analysis and can be found in Ref. [A5] listed in Appendix 1.

---

[7]GEORG FROBENIUS (1849—1917), German mathematician, who also made important contributions to the theory of matrices and groups.

[8]In this theorem, we may replace x by $x - x_0$, where x_0 is any number. Note that the condition $a_0 \neq 0$ is no restriction of generality; it simply means that we factor out the highest possible power of x.

The singular point of (1) at $x = 0$ is sometimes called a **regular singular point,** a term, confusing to the student, which we shall not use.

Regular and Singular Points

The following terms are practical. A **regular point** of

$$y'' + p(x)y' + q(x)y = 0$$

is a point x_0 at which the coefficients p and q are analytic. Then the power series method can be applied. If x_0 is not regular, it is called **singular.** Similarly, a regular point of

$$\tilde{h}(x)y'' + \tilde{p}(x)y'(x) + \tilde{q}(x)y = 0$$

is an x_0 at which $\tilde{h}, \tilde{p}, \tilde{q}$ are analytic and $\tilde{h}(x_0) \neq 0$ (so that we can divide by $\tilde{h}$ and get the previous standard form). If x_0 is not regular, it is called **singular.**

Indicial Equation, Indicating the Form of Solutions

We shall now explain the Frobenius method for solving (1). Multiplication of (1) by x^2 gives the more convenient form

$$\textbf{(1}')\qquad\qquad x^2 y'' + xb(x)y' + c(x)y = 0.$$

We first expand $b(x)$ and $c(x)$ in power series,

$$b(x) = b_0 + b_1 x + b_2 x^2 + \cdots, \qquad c(x) = c_0 + c_1 x + c_2 x^2 + \cdots.$$

Then we differentiate (2) term by term, finding

$$y'(x) = \sum_{m=0}^{\infty} (m + r)a_m x^{m+r-1} = x^{r-1}[ra_0 + (r + 1)a_1 x + \cdots],$$

$$\textbf{(2*)}\qquad y''(x) = \sum_{m=0}^{\infty} (m + r)(m + r - 1)a_m x^{m+r-2}$$

$$= x^{r-2}[r(r - 1)a_0 + (r + 1)ra_1 x + \cdots].$$

By inserting all these series into $(1')$ we readily obtain

$$\textbf{(3)}\qquad \begin{aligned} x^r[r(r - 1)a_0 + \cdots] &+ (b_0 + b_1 x + \cdots)x^r(ra_0 + \cdots) \\ &+ (c_0 + c_1 x + \cdots)x^r(a_0 + a_1 x + \cdots) = 0. \end{aligned}$$

We now equate the sum of the coefficients of each power x^r, x^{r+1}, x^{r+2}, $\cdots$ to zero. This yields a system of equations involving the unknown coefficients a_m. The equation corresponding to the power x^r is

$$[r(r - 1) + b_0 r + c_0]a_0 = 0.$$

Since by assumption $a_0 \neq 0$, the expression in the brackets $[\cdots]$ must be zero. This gives

(4) $$r(r - 1) + b_0 r + c_0 = 0.$$

This important quadratic equation is called the **indicial equation** of the differential equation (1). Its role is as follows.

Our method will yield a basis of solutions. One of the two solutions will always be of the form (2), where r is a root of (4). The form of the other solution will be indicated by the indicial equation. There are three cases:

Case 1. Distinct roots not differing by an integer 1, 2, 3, $\cdots$.[9]
Case 2. A double root.
Case 3. Roots differing by an integer 1, 2, 3, $\cdots$.

Cases 1 and 2 are not unexpected, because of the Euler–Cauchy equation (Sec. 2.6), the simplest equation of the form (1). Case 3 perhaps is unexpected. In each case the general form of a basis can be indicated as follows.[10] Proofs are given in Appendix 4.

Note that in Case 2 we *must* have a logarithm, whereas in Case 3 we may or may not.

THEOREM 2   **(Frobenius method. Basis of solutions. Three cases)**

Suppose that the differential equation (1) satisfies the assumptions in Theorem 1. Let r_1 and r_2 be the roots of the indicial equation (4). Then we have the following three cases.

Case 1. Distinct roots not differing by an integer. *A basis is*

(5) $$y_1(x) = x^{r_1}(a_0 + a_1 x + a_2 x^2 + \cdots)$$

and

(6) $$y_2(x) = x^{r_2}(A_0 + A_1 x + A_2 x^2 + \cdots)$$

with coefficients obtained successively from (3) with $r = r_1$ and $r = r_2$, respectively.

Case 2. Double root $r_1 = r_2 = r$. *A basis is*

(7) $$y_1(x) = x^r(a_0 + a_1 x + a_2 x^2 + \cdots) \qquad \left[r = \tfrac{1}{2}(1 - b_0)\right]$$

(of the same general form as before) and

(8) $$y_2(x) = y_1(x) \ln x + x^r(A_1 x + A_2 x^2 + \cdots) \qquad (x > 0).$$

Case 3. Roots differing by an integer. *A basis is*

(9) $$y_1(x) = x^{r_1}(a_0 + a_1 x + a_2 x^2 + \cdots)$$

(of the same general form as before) and

(10) $$y_2(x) = k y_1(x) \ln x + x^{r_2}(A_0 + A_1 x + A_2 x^2 + \cdots),$$

where the roots are so denoted that $r_1 - r_2 > 0$ and k may turn out to be zero.

---

[9]Note that this case includes complex conjugate roots r_1 and $r_2 = \bar{r}_1$, because we have $r_1 - r_2 = r_1 - \bar{r}_1 = 2i \operatorname{Im} r_1$, which is imaginary and hence cannot be a *real* integer.

[10]A general theory of convergence of the occurring series will not be presented here, but in each individual case convergence may be tested in the usual way.

Typical Applications

Technically, the Frobenius method is similar to the power series method, once the roots of the indicial equation have been determined. However, (5)−(10) merely indicate the general form of a basis, and a second solution can often be obtained more rapidly by reduction of order (Sec. 2.1).

EXAMPLE 1 **Euler–Cauchy equation, illustrating Cases 1 and 2 and Case 3 without a logarithm**

For the Euler–Cauchy equation (Sec. 2.6)

$$x^2 y'' + b_0 xy' + c_0 y = 0 \qquad (b_0, c_0 \text{ constant})$$

substitution of $y = x^r$ gives the auxiliary equation

$$\boxed{r(r - 1) + b_0 r + c_0 = 0,}$$

which is the indicial equation [and $y = x^r$ is a very special form of (2)!]. For different roots r_1, r_2 we get a basis $y_1 = x^{r_1}$, $y_2 = x^{r_2}$, and for a double root r we get a basis x^r, $x^r \ln x$. Accordingly, for this simple equation, Case 3 without a logarithmic term plays no extra role, and Case 3 with a logarithmic term is not possible. ◄

EXAMPLE 2 **Illustration of Case 2 (Double root)**

Solve the differential equation

$$(11) \qquad x(x - 1)y'' + (3x - 1)y' + y = 0.$$

(This is a special hypergeometric differential equation, as we shall see in the problem set.)

Solution. Writing this equation in the standard form (1), we see that it satisfies the assumptions in Theorem 1. By inserting (2) and its derivatives (2*) into (11) we obtain

$$
\begin{aligned}
(12) \qquad & \sum_{m=0}^{\infty} (m + r)(m + r - 1)a_m x^{m+r} - \sum_{m=0}^{\infty} (m + r)(m + r - 1)a_m x^{m+r-1} \\
& + 3 \sum_{m=0}^{\infty} (m + r)a_m x^{m+r} - \sum_{m=0}^{\infty} (m + r)a_m x^{m+r-1} + \sum_{m=0}^{\infty} a_m x^{m+r} = 0.
\end{aligned}
$$

The smallest power is x^{r-1}; by equating the sum of its coefficients to zero we have

$$[-r(r - 1) - r]a_0 = 0, \qquad \text{thus} \qquad r^2 = 0.$$

Hence this indicial equation has the double root $r = 0$.

First solution. We insert this value into (12) and equate the sum of the coefficients of the power x^s to zero, finding

$$s(s - 1)a_s - (s + 1)sa_{s+1} + 3sa_s - (s + 1)a_{s+1} + a_s = 0,$$

thus $a_{s+1} = a_s$. Hence $a_0 = a_1 = a_2 = \cdots$, and by choosing $a_0 = 1$ we obtain the solution

$$y_1(x) = \sum_{m=0}^{\infty} x^m = \frac{1}{1 - x} \qquad (|x| < 1).$$

Second solution. We get a second independent solution y_2 by the method of reduction of order (Sec. 2.1), substituting $y_2 = uy_1$ and its derivatives into the equation. This leads to (9), Sec. 2.1, which we shall use in this example, instead of starting from scratch (as we shall do in the next example). In (9) of Sec. 2.1, we have

$p = (3x - 1)/(x^2 - x)$, the coefficient of y' in (11) *in standard form.* By partial fractions

$$-\int p\, dx = -\int \frac{3x - 1}{x(x - 1)}\, dx = -\int \left(\frac{2}{x - 1} + \frac{1}{x}\right) dx = -2 \ln (x - 1) - \ln x.$$

Hence (9), Sec. 2.1, becomes

$$u' = U = y_1^{-2}\, e^{-\int p\, dx} = \frac{(x - 1)^2}{(x - 1)^2 x}, \qquad u = \ln x, \qquad y_2 = u y_1 = \frac{\ln x}{1 - x}.$$

y_1 and y_2 are linearly independent and thus form a basis on the interval $0 < x < 1$ (as well as on $1 < x < \infty$). ◀

EXAMPLE 3 **Case 3, second solution with logarithmic term**

Solve

$$(13) \qquad\qquad (x^2 - x)y'' - xy' + y = 0.$$

Solution. Substituting (2) and (2*) into (13), we have

$$(x^2 - x) \sum_{m=0}^{\infty} (m + r)(m + r - 1)a_m x^{m+r-2} - x \sum_{m=0}^{\infty} (m + r)a_m x^{m+r-1} + \sum_{m=0}^{\infty} a_m x^{m+r} = 0.$$

We now take x^2, x and x inside the summations and collect all terms with power x^{m+r} and simplify algebraically, obtaining

$$\sum_{m=0}^{\infty} (m + r - 1)^2 a_m x^{m+r} - \sum_{m=0}^{\infty} (m + r)(m + r - 1)a_m x^{m+r-1} = 0.$$

In the first series we set $m = s$ and in the second $m = s + 1$, thus $s = m - 1$. Then

$$(14) \qquad \sum_{s=0}^{\infty} (s + r - 1)^2 a_s x^{s+r} - \sum_{s=-1}^{\infty} (s + r + 1)(s + r)a_{s+1} x^{s+r} = 0.$$

The lowest power is x^{r-1} (take $s = -1$ in the second series) and gives the indicial equation

$$r(r - 1) = 0.$$

The roots are $r_1 = 1$ and $r_2 = 0$. They differ by an integer. This is Case 3.

First solution. From (14) with $r = r_1 = 1$ we have

$$\sum_{s=0}^{\infty} \left[s^2 a_s - (s + 2)(s + 1)a_{s+1}\right] x^{s+1} = 0.$$

This gives the recurrence relation

$$a_{s+1} = \frac{s^2}{(s + 2)(s + 1)} a_s \qquad\qquad (s = 0, 1, \cdots).$$

Hence $a_1 = 0$, $a_2 = 0$, $\cdots$ successively. Taking $a_0 = 1$, we get as a first solution $y_1 = x^{r_1} a_0 = x$.

Second solution. Applying reduction of order (Sec. 2.1), we substitute $y_2 = y_1 u = xu$, $y_2' = xu' + u$ and $y_2'' = xu'' + 2u'$ into the equation, obtaining

$$(x^2 - x)(xu'' + 2u') - x(xu' + u) + xu = 0.$$

xu drops out. Division by x and simplification give

$$(x^2 - x)u'' + (x - 2)u' = 0.$$

From this, using partial fractions and integrating, we get

$$\frac{u''}{u'} = -\frac{x-2}{x^2-x} = -\frac{2}{x} + \frac{1}{x-1}, \qquad \ln u' = \ln \frac{x-1}{x^2}.$$

Taking exponentials and integrating (again taking the integration constant zero), we obtain

$$u' = \frac{x-1}{x^2} = \frac{1}{x} - \frac{1}{x^2}, \qquad u = \ln x + \frac{1}{x}, \qquad y_2 = xu = x \ln x + 1.$$

y_1 and y_2 are linearly independent, and y_2 has a logarithmic term. ◀

The Frobenius method solves the **hypergeometric equation,** whose solutions include many known functions as special cases (see the problem set). In the next section we use the method for solving Bessel's equation.

PROBLEM SET 4.4

1. **WRITING PROJECT. Power Series Method and Frobenius Method.** Write a short essay ($2-3$ pages) explaining the difference between the two methods and their range of applicability (in your own words and illustrated with simple examples of your own). Also think of answers to some practical questions. For instance, can you apply the power series method with $x_0 = 2$ to an equation with singular points at ± 1? What radius of convergence can you expect in general?

Basis of Solutions by the Frobenius Method. Find a basis of solutions of the following differential equations. Show the details of your work. Try to identify the series as expansions of known functions.

2. $(x + 1)^2 y'' + (x + 1)y' - y = 0$
3. $x(1 - x)y'' + 2(1 - 2x)y' - 2y = 0$
4. $4xy'' + 2y' + y = 0$
5. $xy'' + 2y' + xy = 0$
6. $xy'' + 2y' + 4xy = 0$
7. $xy'' + (1 - 2x)y' + (x - 1)y = 0$
8. $xy'' + 2(1 - x)y' + (x - 2)y = 0$
9. $(x + 2)^2 y'' + (x + 2)y' - y = 0$
10. $(x - 1)^2 y'' + (x - 1)y' - 4y = 0$
11. $2x(x - 1)y'' - (x + 1)y' + y = 0$
12. $xy'' + y' - xy = 0$
13. $x^2 y'' + 2xy' - 6y = 0$
14. $x^2 y'' + x^3 y' + (x^2 - 2)y = 0$
15. $xy'' + 3y' + 4x^3 y = 0$

16. **TEAM PROJECT. Hypergeometric Equation, Hypergeometric Series, Hypergeometric Functions.** Gauss's **hypergeometric differential equation**[11] is

(15)
$$x(1 - x)y'' + [c - (a + b + 1)x]y' - aby = 0.$$

In this equation a, b, and c are constants, and by choosing specific values we can obtain an

---

[11]CARL FRIEDRICH GAUSS (1777—1855), great German mathematician. He already made the first of his great discoveries as a student at Helmstedt and Göttingen. In 1807 he became a professor and director of the Observatory at Göttingen. His work was of basic importance in algebra, number theory, differential equations, differential geometry, non-Euclidean geometry, complex analysis, numerical analysis, astronomy, geodesy, electromagnetism, and theoretical mechanics. He also paved the way for a general and systematic use of complex numbers.

incredibly large number of elementary and higher special functions as solutions of (15). (See part (c) for a small sample.) This accounts for the practical importance of (15).

(a) Hypergeometric Series. Show that the indicial equation of (15) has the roots $r_1 = 0$ and $r_2 = 1 - c$. Show that for $r_1 = 0$ the Frobenius method gives

(16) $y_1(x) = 1 + \dfrac{ab}{1!\,c}\,x + \dfrac{a(a+1)b(b+1)}{2!\,c(c+1)}\,x^2 + \dfrac{a(a+1)(a+2)b(b+1)(b+2)}{3!\,c(c+1)(c+2)}\,x^3 + \cdots$

where $c \neq 0, -1, -2, \cdots$. This series is called the **hypergeometric series.** Its sum $y_1(x)$ is denoted by $F(a, b, c; x)$ and is called the **hypergeometric function.** Motivate the name for (16) by showing that

$$F(1, 1, 1; x) = F(1, b, b; x) = F(a, 1, a; x) = \frac{1}{1 - x}.$$

(b) Convergence. Show that if a or b is a negative integer, the series (16) reduces to a polynomial, and for any other values of a, b, c ($c \neq 0, -1, -2, \cdots$) it converges when $|x| < 1$.

(c) Special Cases. Show that

$$(1 + x)^n = F(-n, b, b; -x), \qquad (1 - x)^n = 1 - nxF(1 - n, 1, 2; x),$$

$$\text{arc tan } x = xF(\tfrac{1}{2}, 1, \tfrac{3}{2}; -x^2), \qquad \text{arc sin } x = xF(\tfrac{1}{2}, \tfrac{1}{2}, \tfrac{3}{2}; x^2),$$

$$\ln(1 + x) = xF(1, 1, 2; -x), \qquad \ln\frac{1 + x}{1 - x} = 2xF(\tfrac{1}{2}, 1, \tfrac{3}{2}; x^2).$$

Find more such relations from the literature on special functions.

(d) Second Solution. Show that for $r_2 = 1 - c$ the Frobenius method yields the following solution (where $c \neq 2, 3, 4, \cdots$):

(17)

$$y_2(x) = x^{1-c}\left(1 + \frac{(a - c + 1)(b - c + 1)}{1!(-c + 2)}\,x \right.$$
$$\left. + \frac{(a - c + 1)(a - c + 2)(b - c + 1)(b - c + 2)}{2!(-c + 2)(-c + 3)}\,x^2 + \cdots \right).$$

Show that

$$y_2(x) = x^{1-c}F(a - c + 1, b - c + 1, 2 - c; x).$$

(e) Other Differential Equations. Show that

(18) $(t^2 + At + B)\ddot{y} + (Ct + D)\dot{y} + Ky = 0$

with $\dot{y} = dy/dt$, etc., constant A, B, C, D, K, and $t^2 + At + B = (t - t_1)(t - t_2)$, $t_1 \neq t_2$, can be reduced to the hypergeometric equation with independent variable

$$x = \frac{t - t_1}{t_2 - t_1}$$

and parameters related by $Ct_1 + D = -c(t_2 - t_1)$, $C = a + b + 1$, $K = ab$.

From this you see that (15) is a "normalized form" of the more general (18) and that various cases of (18) can thus be solved in terms of hypergeometric functions.

Hypergeometric Differential Equations. Find a general solution in terms of hypergeometric functions:

17. $8x(1 - x)y'' + (4 - 14x)y' - y = 0$ **18.** $x(1 - x)y'' + (3 - 5x)y' - 4y = 0$

19. $4x(1 - x)y'' + y' + 8y = 0$ **20.** $x(1 - x)y'' + (\tfrac{1}{2} + 2x)y' - 2y = 0$

4.5 Bessel's Equation. Bessel Functions $J_\nu(x)$

One of the most important differential equations in applied mathematics is **Bessel's differential equation**[12]

(1)
$$x^2 y'' + xy' + (x^2 - \nu^2)y = 0.$$

It appears in connection with electrical fields, vibrations (see Sec. 11.10), heat conduction, etc., very often when a problem shows cylindrical symmetry (just as Legendre's equation may appear in cases of *spherical* symmetry). The parameter ν in (1) is a given number. We assume that ν is real and nonnegative.

Bessel's equation can be solved by the Frobenius method, as we mentioned at the beginning of the preceding section, where the equation is written in standard form (obtained by dividing (1) by x^2). Accordingly, we substitute the series

(2)
$$y(x) = \sum_{m=0}^{\infty} a_m x^{m+r} \qquad (a_0 \neq 0)$$

with undetermined coefficients and its derivatives into (1). This gives

$$\sum_{m=0}^{\infty} (m+r)(m+r-1)a_m x^{m+r} + \sum_{m=0}^{\infty} (m+r)a_m x^{m+r}$$
$$+ \sum_{m=0}^{\infty} a_m x^{m+r+2} - \nu^2 \sum_{m=0}^{\infty} a_m x^{m+r} = 0.$$

We equate the sum of the coefficients of x^{s+r} to zero. Note that this power x^{s+r} corresponds to $m = s$ in the first, second, and fourth series, and to $m = s - 2$ in the third series. Hence for $s = 0$ and $s = 1$, the third series does not contribute since $m \geq 0$. For $s = 2, 3, \cdots$ all four series contribute, so that we get a general formula for all these s. We find

(a) $\qquad\qquad r(r-1)a_0 + ra_0 - \nu^2 a_0 = 0$ $\qquad\qquad$ ($s = 0$)

(3) (b) $\qquad\qquad (r+1)ra_1 + (r+1)a_1 - \nu^2 a_1 = 0$ $\qquad\qquad$ ($s = 1$)

(c) $\quad (s+r)(s+r-1)a_s + (s+r)a_s + a_{s-2} - \nu^2 a_s = 0$ $\qquad$ ($s = 2, 3, \cdots$).

From (3a) we obtain the **indicial equation**

(4)
$$(r + \nu)(r - \nu) = 0.$$

The roots are $r_1 = \nu \ (\geq 0)$ and $r_2 = -\nu$.

---

[12]FRIEDRICH WILHELM BESSEL (1784—1846), German astronomer and mathematician, started out as an apprentice of a trade company, studying astronomy on his own in his spare time, later became an assistant at a small private observatory, and finally director of the new Königsberg observatory. His paper on the Bessel functions (dated 1824) appeared in 1826.

Formulas are contained in Refs. [1], [6], [11], and the standard treatise [A7].

Coefficient Recurrence in the Case of $r = r_1 = \nu$

For $r = r_1 = \nu$, equation (3b) yields $a_1 = 0$. Equation (3c) may be written

$$(s + r + \nu)(s + r - \nu)a_s + a_{s-2} = 0.$$

For $r = \nu$ this takes the form

(5) $$(s + 2\nu)sa_s + a_{s-2} = 0.$$

Since $a_1 = 0$ and $\nu \geqq 0$, it follows from (5) that

$$a_3 = 0, a_5 = 0, \cdots$$

successively. Hence we have to deal only with coefficients of *even* numbers $s = 2m$. Equation (5) with $s = 2m$ becomes

$$(2m + 2\nu)2ma_{2m} + a_{2m-2} = 0.$$

We solve this for a_{2m},

(**6**) $$a_{2m} = -\frac{1}{2^2 m(\nu + m)} a_{2m-2}, \qquad m = 1, 2, \cdots.$$

From (6) we may now determine the even-numbered coefficients $a_2, a_4, \cdots$ successively. This gives

$$a_2 = -\frac{a_0}{2^2(\nu + 1)}$$

$$a_4 = -\frac{a_2}{2^2 2(\nu + 2)}$$

$$= \frac{a_0}{2^4 2! \, (\nu + 1)(\nu + 2)}$$

and so on, and in general

(7) $$a_{2m} = \frac{(-1)^m a_0}{2^{2m} m! \, (\nu + 1)(\nu + 2) \cdots (\nu + m)}, \qquad m = 1, 2, \cdots.$$

Bessel Function $J_n(x)$ for Integer $\nu = n$

Integer values of ν are denoted by n. This is standard. For $\nu = n$ the relation (7) becomes

(8) $$a_{2m} = \frac{(-1)^m a_0}{2^{2m} m! \, (n + 1)(n + 2) \cdots (n + m)}, \qquad m = 1, 2, \cdots.$$

a_0 is still arbitrary, so that the series (2) with these coefficients would contain this arbitrary factor a_0, a highly impractical situation for developing formulas or computing values of this new function. Accordingly, we have to make a choice: $a_0 = 1$ would be possible, but more practical is

$$(9) \qquad a_0 = \frac{1}{2^n n!}$$

because then $n!(n + 1) \cdots (n + m) = (m + n)!$ in (8), so that

$$(10) \qquad a_{2m} = \frac{(-1)^m}{2^{2m+n} m!(n + m)!}, \qquad m = 1, 2, \cdots.$$

With these coefficients and $r_1 = \nu = n$ we get from (2) a particular solution of (1), denoted by $J_n(x)$ and given by

$$(11) \qquad \boxed{J_n(x) = x^n \sum_{m=0}^{\infty} \frac{(-1)^m x^{2m}}{2^{2m+n} m! \, (n + m)!}.}$$

It is called the **Bessel function of the first kind** *of order n.* This series converges for all *x,* as the ratio test shows. In fact, it converges very rapidly because of the factorials in the denominator.

EXAMPLE 1   **Bessel functions $J_0(x)$ and $J_1(x)$**

For $n = 0$ we obtain from (11) the Bessel function of order 0

$$(12) \qquad J_0(x) = \sum_{m=0}^{\infty} \frac{(-1)^m x^{2m}}{2^{2m}(m!)^2} = 1 - \frac{x^2}{2^2(1!)^2} + \frac{x^4}{2^4(2!)^2} - \frac{x^6}{2^6(3!)^2} + - \cdots,$$

which looks similar to a cosine (Fig. 103). For $n = 1$ we obtain the Bessel function of order 1

$$(13) \qquad J_1(x) = \sum_{m=0}^{\infty} \frac{(-1)^m x^{2m+1}}{2^{2m+1} m! \, (m + 1)!} = \frac{x}{2} - \frac{x^3}{2^3 1!2!} + \frac{x^5}{2^5 2!3!} - \frac{x^7}{2^7 3!4!} + - \cdots,$$

which looks similar to a sine (Fig. 103). But the zeros of these functions are not completely regularly spaced (see also Table A1 in Appendix 5) and the height of the "waves" decreases with increasing *x.* Heuristically, n^2/x^2 in (1) in standard form $[(1)$ divided by $x^2]$ is zero (if $n = 0$) or small in absolute value for large *x,* and so is $y'/x,$ so that then Bessel's equation comes close to $y'' + y = 0,$ the equation of $\cos x$ and $\sin x$; also, y'/x acts as a "damping term," in part responsible for the decrease in height. One can show that for large *x,*

$$(14) \qquad J_n(x) \approx \sqrt{\frac{2}{\pi x}} \cos\left(x - \frac{n\pi}{2} - \frac{\pi}{4}\right). \qquad \blacktriangleleft$$

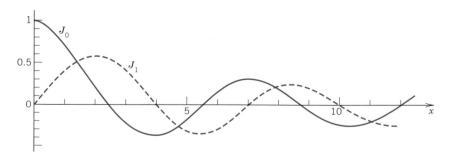

Fig. 103.   Bessel functions of the first kind

Bessel Functions $J_\nu(x)$ for Any $\nu \geq 0$. Gamma Function

We now extend our discussion from integer $\nu = n$ to any $\nu \geq 0$. All we need is an extension of the factorials in (9) and (11) to any ν. This is done by the **gamma function** $\Gamma(\nu)$ defined by the integral

(15)
$$\Gamma(\nu) = \int_0^\infty e^{-t} t^{\nu-1} \, dt \qquad\qquad (\nu > 0).$$

By integration by parts we obtain

$$\Gamma(\nu + 1) = \int_0^\infty e^{-t} t^\nu \, dt = -e^{-t} t^\nu \Big|_0^\infty + \nu \int_0^\infty e^{-t} t^{\nu-1} \, dt.$$

The first expression on the right is zero. The integral on the right is $\Gamma(\nu)$. This yields the basic relation

(16)
$$\Gamma(\nu + 1) = \nu \Gamma(\nu).$$

Now by (15)

$$\Gamma(1) = \int_0^\infty e^{-t} \, dt = -e^{-t} \Big|_0^\infty = 0 - (-1) = 1.$$

From this and (16) we obtain successively $\Gamma(2) = \Gamma(1) = 1!$, $\Gamma(3) = 2\Gamma(2) = 2!$, $\cdots$ and in general

(17)
$$\Gamma(n + 1) = n! \qquad\qquad (n = 0, 1, \cdots).$$

This shows that *the gamma function does in fact generalize the factorial function.*

Now in (9) we had $a_0 = 1/(2^n n!)$. This is $1/(2^n \Gamma(n + 1))$ by (17). It suggests to choose, for any ν.

(18)
$$a_0 = \frac{1}{2^\nu \Gamma(\nu + 1)}.$$

Then (7) becomes

$$a_{2m} = \frac{(-1)^m}{2^{2m+\nu} m! \, (\nu + 1)(\nu + 2) \cdots (\nu + m) \Gamma(\nu + 1)}.$$

But (16) gives in the denominator

$$(\nu + 1)\Gamma(\nu + 1) = \Gamma(\nu + 2), \qquad (\nu + 2)\Gamma(\nu + 2) = \Gamma(\nu + 3)$$

and so on, so that

$$(\nu + 1)(\nu + 2) \cdots (\nu + m) \Gamma(\nu + 1) = \Gamma(\nu + m + 1).$$

Hence the coefficients are

(19)
$$a_{2m} = \frac{(-1)^m}{2^{2m+\nu}m!\,\Gamma(\nu + m + 1)}.$$

With these coefficients and $r = r_1 = \nu$ we get from (2) a particular solution of (1), denoted by $J_\nu(x)$ and given by

(20)
$$J_\nu(x) = x^\nu \sum_{m=0}^{\infty} \frac{(-1)^m x^{2m}}{2^{2m+\nu}m!\,\Gamma(\nu + m + 1)}.$$

It is called the **Bessel function of the first kind of order ν.** This series converges for all x, as one can verify by the ratio test.

Solution $J_{-\nu}$ of the Bessel Equation

For a general solution, in addition to J_ν we need a second linearly independent solution. For ν not an integer this is easy. Replacing ν by $-\nu$ in (20), we have

(21)
$$J_{-\nu}(x) = x^{-\nu} \sum_{m=0}^{\infty} \frac{(-1)^m x^{2m}}{2^{2m-\nu}m!\,\Gamma(m - \nu + 1)}.$$

Since Bessel's equation involves ν^2, the functions J_ν and $J_{-\nu}$ are solutions of the equation for the same ν. If ν is not an integer, they are linearly independent, because the first term in (20) and the first term in (21) are finite nonzero multiples of x^ν and $x^{-\nu}$, respectively. $x = 0$ must be excluded in (21) because of the factor $x^{-\nu}$ (with $\nu > 0$). This gives

THEOREM 1   **(General solution of Bessel's equation)**

If ν is not an integer, a general solution of Bessel's equation for all $x \neq 0$ is

(22)
$$y(x) = c_1 J_\nu(x) + c_2 J_{-\nu}(x).$$

But if ν is an integer, then (22) is not a general solution because of linear dependence.

THEOREM 2   **(Linear dependence of Bessel functions J_n and J_{-n})**

For integer $\nu = n$ the Bessel functions $J_n(x)$ and $J_{-n}(x)$ are linearly dependent, because

(23)
$$J_{-n}(x) = (-1)^n J_n(x) \qquad\qquad (n = 1, 2, \cdots).$$

PROOF.   We use (21) and let ν approach a positive integer n. Then the gamma functions in the coefficients of the first n terms become infinite (see Fig. 517 in Appendix A3.1), the coefficients become zero, and the summation starts with $m = n$. Since in this case $\Gamma(m - n + 1) = (m - n)!$ by (17), we obtain

$$J_{-n}(x) = \sum_{m=n}^{\infty} \frac{(-1)^m x^{2m-n}}{2^{2m-n}m!\,(m - n)!} = \sum_{s=0}^{\infty} \frac{(-1)^{n+s} x^{2s+n}}{2^{2s+n}(n + s)!\,s!}$$

where $m = n + s$ and $s = m - n$. From (11) we see that the last series represents $(-1)^n J_n(x)$. This completes the proof.   ◀

A general solution of the Bessel equation with integer $\nu = n$ will be given in Sec. 4.6. This will require some further interesting ideas and work.

Backbones (24)–(27) of Bessel's Theory

Bessel functions satisfy an incredibly large number of relationships—just look at Ref. [A7] in Appendix 1. We discuss four real backbones (24)–(27) of Bessel's theory, not only because of their great practical importance but also as a model case for showing how properties of functions can be discovered from their series.

Multiplying (20) by x^ν and pulling x^2 under the summation sign, we have

$$x^\nu J_\nu(x) = \sum_{m=0}^{\infty} \frac{(-1)^m x^{2m+2\nu}}{2^{2m+\nu} m!\, \Gamma(\nu + m + 1)}.$$

We differentiate this, cancel a factor 2, pull $x^{2\nu-1}$ out, and use the functional relationship $\Gamma(\nu + m + 1) = (\nu + m)\Gamma(\nu + m)$ [see (16)]. Then

$$(x^\nu J_\nu)' = \sum_{m=0}^{\infty} \frac{(-1)^m 2(m + \nu)x^{2m+2\nu-1}}{2^{2m+\nu} m!\, \Gamma(\nu + m + 1)} = x^\nu x^{\nu-1} \sum_{m=0}^{\infty} \frac{(-1)^m x^{2m}}{2^{2m+\nu-1} m!\, \Gamma(\nu + m)}.$$

Equation (20) shows that the right side is $x^\nu J_{\nu-1}(x)$. This proves our first formula

(24)
$$\boxed{\frac{d}{dx}\left[x^\nu J_\nu(x)\right] = x^\nu J_{\nu-1}(x).}$$

Similarly, we multiply (20) by $x^{-\nu}$, differentiate, cancel $2m$, using $m! = m(m - 1)!$, and set $m = s + 1$. Then

$$(x^{-\nu} J_\nu)' = \sum_{m=1}^{\infty} \frac{(-1)^m x^{2m-1}}{2^{2m+\nu-1}(m - 1)!\Gamma(\nu + m + 1)} = \sum_{s=0}^{\infty} \frac{(-1)^{s+1} x^{2s+1}}{2^{2s+\nu+1} s!\Gamma(\nu + s + 2)}.$$

Equation (20) with $\nu + 1$ instead of ν and s instead of m shows that the expression on the right is $-x^{-\nu} J_{\nu+1}(x)$. This proves our second formula

(25)
$$\boxed{\frac{d}{dx}\left[x^{-\nu} J_\nu(x)\right] = -x^{-\nu} J_{\nu+1}(x).}$$

Next we perform the differentiation in (24),

(24*) $$\nu x^{\nu-1} J_\nu + x^\nu J_\nu' = x^\nu J_{\nu-1}.$$

Then we do the same in (25) and multiply by $x^{2\nu}$. This gives

(25*) $$-\nu x^{\nu-1} J_\nu + x^\nu J_\nu' = -x^\nu J_{\nu+1}.$$

Subtracting (25*) from (24*) and dividing by x^ν, we obtain the first recurrence relation

(26)

$$J_{\nu-1}(x) + J_{\nu+1}(x) = \frac{2\nu}{x} J_\nu(x).$$

Adding (24*) and (25*) and dividing the result by x^ν, we obtain the second recurrence relation

(27)

$$J_{\nu-1}(x) - J_{\nu+1}(x) = 2J_\nu'(x).$$

Formulas (24) and (25) are useful for integrals involving Bessel functions. Formulas (26) and (27) are of practical interest, for instance, in numerical work, where (26) can be used to express Bessel functions of high orders in terms of Bessel functions of low orders for computing tables.

EXAMPLE 2   **Integral involving a Bessel function**

Using the table of J_0 and J_1 (Table A1 in Appendix 5), integrate

$$I = \int_1^2 x^{-3} J_4(x)\, dx.$$

Solution.   From (25) with $\nu = 3$ we obtain by integration on both sides

$$x^{-3} J_3(x) \Big|_{x=1}^{x=2} = -\int_1^2 x^{-3} J_4(x)\, dx.$$

Now by (26) with $\nu = 2$ we have

$$J_1(x) + J_3(x) = \frac{4}{x} J_2(x), \qquad \text{thus} \qquad J_3(x) = \frac{4}{x} J_2(x) - J_1(x)$$

and by (26) with $\nu = 1$

$$J_0(x) + J_2(x) = \frac{2}{x} J_1(x), \qquad \text{thus} \qquad J_2(x) = \frac{2}{x} J_1(x) - J_0(x).$$

Together,

$$J_3(x) = \frac{4}{x}\left(\frac{2}{x} J_1(x) - J_0(x)\right) - J_1(x) = \left(\frac{8}{x^2} - 1\right) J_1(x) - \frac{4}{x} J_0(x).$$

From Table A1, $J_1(2) = 0.5767$, $J_0(2) = 0.2239$, $J_1(1) = 0.4401$, and $J_0(1) = 0.7652$. With these values we obtain

$$I = -\tfrac{1}{8} J_3(2) + J_3(1) = -\tfrac{1}{8} \cdot 0.1289 + 0.0199 = 0.0038.$$

(0.1289 is correct to 4 digits; 0.0199 deviates from the correct 0.0196 due to round-off. Actually, $I = 0.003445$ is correct to 6 digits.)   ◀

$J_\nu(x)$ with $\nu = \pm\frac{1}{2}, \pm\frac{3}{2}, \pm\frac{5}{2}, \cdots$ Are Elementary

It often happens that in special cases, higher functions become functions known from calculus. We study this for $J_\nu(x)$.

When $\nu = \frac{1}{2}$, then (20) is

$$J_{1/2}(x) = \sqrt{x} \sum_{m=0}^{\infty} \frac{(-1)^m x^{2m}}{2^{2m+1/2} m!\, \Gamma(m + \frac{3}{2})} = \sqrt{\frac{2}{x}} \sum_{m=0}^{\infty} \frac{(-1)^m x^{2m+1}}{2^{2m+1} m!\, \Gamma(m + \frac{3}{2})}.$$

Now we use without proof that

(28)
$$\Gamma(\tfrac{1}{2}) = \sqrt{\pi}.$$

From this and (16) we get in the denominator

$$\Gamma(m + \tfrac{3}{2}) = (m + \tfrac{1}{2})(m - \tfrac{1}{2}) \cdots \tfrac{3}{2} \cdot \tfrac{1}{2}\Gamma(\tfrac{1}{2}) = 2^{-(m+1)}(2m + 1)(2m - 1) \cdots 3 \cdot 1\sqrt{\pi}.$$

In the denominator we further have

$$2^{2m+1}m! = 2^{2m+1}m(m - 1) \cdots 2 \cdot 1 = 2^{m+1}2m(2m - 2) \cdots 4 \cdot 2.$$

Together the denominator becomes $(2m + 1)!\sqrt{\pi}$, so that

$$J_{1/2}(x) = \sqrt{\frac{2}{\pi x}} \sum_{m=0}^{\infty} \frac{(-1)^m x^{2m+1}}{(2m + 1)!}.$$

This series is the familiar Maclaurin series of $\sin x$. Thus

(29)
$$J_{1/2}(x) = \sqrt{\frac{2}{\pi x}} \sin x.$$

By differentiation and by (24) with $\nu = \tfrac{1}{2}$ we get from this

$$\left[\sqrt{x}J_{1/2}(x)\right]' = \sqrt{\frac{2}{\pi}} \cos x = x^{1/2}J_{-1/2}(x).$$

Hence our next result is

(30)
$$J_{-1/2}(x) = \sqrt{\frac{2}{\pi x}} \cos x.$$

From this and (26) we can draw the following interesting conclusion.

THEOREM 3   **(Elementary Bessel functions)**
Bessel functions J_ν of orders $\nu = \pm\tfrac{1}{2}, \pm\tfrac{3}{2}, \pm\tfrac{5}{2}, \cdots$ are elementary; they can be expressed by finitely many cosines and sines and powers of x.

EXAMPLE 3   **Further elementary Bessel functions**
From (26), (29), and (30) we get

$$J_{3/2}(x) = \frac{1}{x} J_{1/2}(x) - J_{-1/2}(x) = \sqrt{\frac{2}{\pi x}} \left(\frac{\sin x}{x} - \cos x\right)$$

$$J_{-3/2}(x) = -\frac{1}{x} J_{-1/2}(x) - J_{1/2}(x) = -\sqrt{\frac{2}{\pi x}} \left(\frac{\cos x}{x} + \sin x\right)$$

and so on.   ◀

We hope that our study has not only helped you to become acquainted with Bessel functions, but has also convinced you that series can be quite useful in obtaining various properties of the corresponding functions.

PROBLEM SET 4.5

Differential Equations Reducible to Bessel's Equation

Various differential equations can be reduced to Bessel's equation. To see this, use the indicated substitutions and find a general solution in terms of J_ν and $J_{-\nu}$, or indicate why these functions do not give a general solution. Show the details of your work.

    (*More such equations follow in Problem Set* **4.6.**)

1. $x^2 y'' + xy' + (x^2 - \frac{1}{9})y = 0$
2. $x^2 y'' + xy' + (\lambda^2 x^2 - \nu^2)y = 0$ (Set $\lambda x = z$.)
3. $x^2 y'' + xy' + (4x^4 - \frac{1}{4})y = 0$ ($x^2 = z$)
4. $4x^2 y'' + 4xy' + (x - \frac{1}{36})y = 0$ ($\sqrt{x} = z$)
5. $9x^2 y'' + 9xy' + (36x^4 - 16)y = 0$ ($x^2 = z$)
6. $xy'' + y' + \frac{1}{4}y = 0$ ($\sqrt{x} = z$)
7. $xy'' + 5y' + xy = 0$ ($y = u/x^2$)
8. $x^2 y'' + \frac{1}{4}(x + \frac{3}{4})y = 0$ ($y = u\sqrt{x}, \quad \sqrt{x} = z$)
9. $81x^2 y'' + 27xy' + (9x^{2/3} + 8)y = 0$ ($y = x^{1/3}u, \quad x^{1/3} = z$)
10. $xy'' - 5y' + xy = 0$ ($y = x^3 u$)

11. (**Derivatives**) Show that $J_0'(x) = -J_1(x)$, $J_2'(x) = \frac{1}{2}[J_1(x) - J_3(x)]$. [Use (24), (25), (26), or (27).]

12. (**Derivative**) Derive $J_0'(x) = -J_1(x)$ from (12) and (13).

13. (**Derivative**) Using (24), show that $J_1'(x) = J_0(x) - x^{-1}J_1(x)$.

14. (**Convergence**) Show that the series in (11) converges for all x.

15. (**Tabulation**) Using (26) and Table A1 in Appendix 5, compute $J_2(x)$ for $x = 0, 0.1, 0.2, \cdots,$ 1.0.

16. (**Approximation**) Show that for small $|x|$ we have $J_0(x) \approx 1 - 0.25x^2$. From this compute $J_0(x)$ for $x = 0, 0.1, 0.2, \cdots, 1.0$ and determine the relative error by comparing with Table A1 in Appendix 5.

17. (**Tabulation**) Compute $J_3(x)$ for $x = 2.0, 2.2, 2.4, 2.6, 2.8$ from (26) and Table A1 in Appendix 5.

18. (**Zero of J_0**) Using (12) and the Leibniz test in Appendix A3.3, can you think of an argument why $2 < x_0 < \sqrt{8}$, where $x_0 \approx 2.405$ is the smallest positive zero of $J_0(x)$?

19. (**Bessel's equation**) To understand that (24) and (25) are really fundamental, derive Bessel's differential equation from these formulas.

20. (**Interlacing of zeros**) Using (24), (25), and Rolle's theorem, show that between two consecutive zeros of $J_0(x)$ there is precisely one zero of $J_1(x)$.

21. (**Interlacing of zeros**) Show that between any two consecutive positive zeros of $J_n(x)$ there is precisely one zero of $J_{n+1}(x)$.

Integrals involving Bessel functions can often be evaluated or at least simplified by the use of (24)−(27). Show that

22. $\int x^\nu J_{\nu-1}(x)\, dx = x^\nu J_\nu(x) + c$

23. $\int x^{-\nu} J_{\nu+1}(x)\, dx = -x^{-\nu} J_\nu(x) + c$

24. $\int J_{\nu+1}(x)\, dx = \int J_{\nu-1}(x)\, dx - 2J_\nu(x)$

Using the formulas in Probs. 22−24 and, if necessary, integration by parts, evaluate

25. $\int J_3(x)\, dx$ 26. $\int x^3 J_0(x)\, dx$ 27. $\int J_5(x)\, dx$

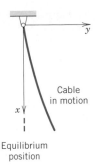

Fig. 104.   Vibrating cable in Team Project 28

28. **TEAM PROJECT. Vibrating Cable or Chain (Fig. 104).** A flexible cable, chain, or rope of length L and density (mass per unit length) ρ is fixed at the upper end ($x = 0$) and allowed to make *small* vibrations [small angles α in the horizontal displacement $u(x, t)$, $t =$ time] in a vertical plane.

(a) Show that the weight of the cable below a point x is $W(x) = \rho g(L - x)$, so that the restoring force is $F(x) = W \sin \alpha \approx W u_x$ ($u_x = \partial u/\partial x$) and the difference in force between x and $x + \Delta x$ is

$$F(x + \Delta x) - F(x) = \Delta x\,(W u_x)_x.$$

Conclude by Newton's second law that

$$\rho\,\Delta x\,u_{tt} = \Delta x\,\rho g[(L - x)u_x]_x.$$

Show that for the expected periodic motions $u(x, t) = y(x) \cos(\omega t + \delta)$ we obtain as the model of our problem

$$(L - x)y'' - y' + \lambda^2 y = 0, \qquad\qquad \lambda^2 = \omega^2/g.$$

(b) Transform this to Bessel's equation with parameter $\nu = 0$ by setting $L - x = z$ and then $s = 2\lambda z^{1/2}$, obtaining

$$\frac{d^2 y}{ds^2} + \frac{1}{s}\frac{dy}{ds} + y = 0,$$

so that the solution is

$$y(x) = J_0(2\omega\sqrt{(L - x)/g}).$$

(c) Conclude that possible frequencies $\omega/2\pi$ are those for which $s = 2\omega\sqrt{L/g}$ is a zero of J_0. The corresponding solutions are called *normal modes*. Figure 103 shows the first of them. Note that because of the root in $y(x)$ it does not look exactly like the first portion of $J_0(x)$ in Fig. 103. Can you imagine what the second normal mode looks like? The third? What is the frequency (cycles/min) of a cable of length 2 meters? Of length 10 meters?

29. **(Elimination of first derivative)** Substitute $y(x) = u(x)v(x)$ into $y'' + p(x)y' + q(x)y = 0$ and show that for obtaining a second-order differential equation for u not containing u', we must take

$$v(x) = \exp\left(-\tfrac{1}{2}\int p(x)\,dx\right).$$

Show that for the Bessel equation the substitution is $y = ux^{-1/2}$ and gives

(31) $$x^2 u'' + (x^2 + \tfrac{1}{4} - \nu^2)u = 0.$$

Solve this equation with $\nu = \tfrac{1}{2}$ and compare the result with (29) and (30). Comment.

30. CAS PROJECT. Bessel Functions J_n for Large x. If your CAS can handle Bessel functions, do the following.

    (a) Graph $J_n(x)$ for $n = 0, \cdots, 4$ on common axes.

    (b) Experiment with (14) for integer n. Using plots, find out from which $x = x_n$ on the curves of (11) and (14) practically coincide. How does x_n change with n?

    (c) What happens in (b) if $n = \pm\frac{1}{2}$? (Our usual notation in this case would be ν.)

    (d) How does the error of (14) behave as a function of x for fixed n?

    (e) Show from the graph that the extrema of J_0 are at x's where $J_1(x) = 0$. By which formula in the text can you prove this?

    (f) Raise and answer further questions of your own.

4.6 Bessel Functions of the Second Kind $Y_\nu(x)$

What is left is the task of getting a general solution of Bessel's equation for integer $\nu = n$. Remember from Sec. 4.5 that for noninteger ν we already have a basis J_ν, $J_{-\nu}$, but for $\nu = n$ these two solutions become linearly dependent, so that we need a second independent solution. This solution will be denoted by Y_n.

$n = 0$: Bessel Function of the Second Kind $Y_0(x)$

We first consider the case $n = 0$. Then Bessel's equation can be written

(1)
$$xy'' + y' + xy = 0,$$

and the indicial equation (4), Sec. 4.5, has a double root $r = 0$. This is Case 2 in Sec. 4.4. In this case we first have only one solution, $J_0(x)$. From (8) in Sec. 4.4 we see that the desired second solution must be of the form

(2)
$$y_2(x) = J_0(x) \ln x + \sum_{m=1}^{\infty} A_m x^m.$$

We substitute y_2 and its derivatives

$$y_2' = J_0' \ln x + \frac{J_0}{x} + \sum_{m=1}^{\infty} m A_m x^{m-1}$$

$$y_2'' = J_0'' \ln x + \frac{2J_0'}{x} - \frac{J_0}{x^2} + \sum_{m=1}^{\infty} m(m-1) A_m x^{m-2}$$

into (1). Then the logarithmic terms disappear because J_0 is a solution of (1). The other two terms containing J_0 cancel. Hence we are left with

$$2J_0' + \sum_{m=1}^{\infty} m(m-1) A_m x^{m-1} + \sum_{m=1}^{\infty} m A_m x^{m-1} + \sum_{m=1}^{\infty} A_m x^{m+1} = 0.$$

From (12) in Sec. 4.5 we obtain the power series of J_0' in the form

$$J_0'(x) = \sum_{m=1}^{\infty} \frac{(-1)^m 2m x^{2m-1}}{2^{2m}(m!)^2} = \sum_{m=1}^{\infty} \frac{(-1)^m x^{2m-1}}{2^{2m-1} m! \, (m-1)!} \, .$$

We insert this series and note that $\Sigma \, m A_m x^{m-1}$ and part of the series resulting from xy'', namely, $\Sigma \, m(m-1) A_m x^{m-1} = \Sigma \, (m^2 - m) A_m x^{m-1}$, cancel. This gives

$$\sum_{m=1}^{\infty} \frac{(-1)^m x^{2m-1}}{2^{2m-2} m! \, (m-1)!} + \sum_{m=1}^{\infty} m^2 A_m x^{m-1} + \sum_{m=1}^{\infty} A_m x^{m+1} = 0.$$

First, we show that the A_m with odd subscripts are all zero. The power x^0 occurs only in the second series, with coefficient A_1. Hence $A_1 = 0$. Next, we consider the even powers x^{2s}. The first series contains none. In the second series, $m - 1 = 2s$ gives the term $(2s + 1)^2 A_{2s+1} x^{2s}$. In the third series, $m + 1 = 2s$. Hence by equating the sum of the coefficients of x^{2s} to zero we have

$$(2s + 1)^2 A_{2s+1} + A_{2s-1} = 0, \qquad\qquad s = 1, 2, \cdots.$$

Since $A_1 = 0$, we thus obtain $A_3 = 0$, $A_5 = 0, \cdots$, successively.

We now equate the sum of the coefficients of x^{2s+1} to zero. For $s = 0$ this gives

$$-1 + 4A_2 = 0 \qquad \text{or} \qquad A_2 = \tfrac{1}{4}.$$

For the other values of s we have in the first series $2m - 1 = 2s + 1$, hence $m = s + 1$, in the second $m - 1 = 2s + 1$, and in the third $m + 1 = 2s + 1$. Thus we obtain

$$\frac{(-1)^{s+1}}{2^{2s}(s + 1)! s!} + (2s + 2)^2 A_{2s+2} + A_{2s} = 0.$$

For $s = 1$ this yields

$$\tfrac{1}{8} + 16A_4 + A_2 = 0 \qquad \text{or} \qquad A_4 = -\tfrac{3}{128}$$

and in general

$$(3) \qquad A_{2m} = \frac{(-1)^{m-1}}{2^{2m}(m!)^2} \left(1 + \frac{1}{2} + \frac{1}{3} + \cdots + \frac{1}{m} \right), \qquad m = 1, 2, \cdots.$$

Using the short notation

$$(4) \qquad\qquad h_m = 1 + \frac{1}{2} + \cdots + \frac{1}{m}$$

and inserting (3) and $A_1 = A_3 = \cdots = 0$ into (2), we obtain the result

$$y_2(x) = J_0(x) \ln x + \sum_{m=1}^{\infty} \frac{(-1)^{m-1} h_m}{2^{2m}(m!)^2} x^{2m}$$

$$(5)$$

$$= J_0(x) \ln x + \frac{1}{4}x^2 - \frac{3}{128}x^4 + \frac{11}{13824}x^6 - + \cdots.$$

Since J_0 and y_2 are linearly independent functions, they form a basis of (1). Of course, another basis is obtained if we replace y_2 by an independent particular solution of the form $a(y_2 + bJ_0)$, where $a \, (\neq 0)$ and b are constants. It is customary to choose $a = 2/\pi$

and $b = \gamma - \ln 2$, where the number $\gamma = 0.577\ 215\ 664\ 90 \cdots$ is the so-called **Euler constant,** which is defined as the limit of

$$1 + \frac{1}{2} + \cdots + \frac{1}{s} - \ln s$$

as s approaches infinity. The standard particular solution thus obtained is called the **Bessel function of the second kind** *of order zero* (Fig. 105) or **Neumann's function**[13] *of order zero* and is denoted by $Y_0(x)$. Thus [see (4)]

$$(6) \qquad Y_0(x) = \frac{2}{\pi} \left[J_0(x) \left(\ln \frac{x}{2} + \gamma \right) + \sum_{m=1}^{\infty} \frac{(-1)^{m-1} h_m}{2^{2m}(m!)^2} x^{2m} \right].$$

For small $x > 0$ the function $Y_0(x)$ behaves about like $\ln x$ (see Fig. 105; why?), and $Y_0(x) \to -\infty$ as $x \to 0$.

Bessel Functions of the Second Kind $Y_n(x)$

For $\nu = n = 1, 2, \cdots$ a second solution can be obtained by manipulations similar to those for $n = 0$, starting from (10), Sec. 4.4. It turns out that in these cases the solution also contains a logarithmic term.

The situation is not yet completely satisfactory, because the second solution is defined differently, depending on whether the order ν is an integer or not. To provide uniformity of formalism and numerical tabulation, it is desirable to adopt a form of the second solution that is valid for all values of the order. This is the reason for introducing a standard second solution $Y_\nu(x)$ defined for all ν by the formula

$$(7) \qquad \text{(a)} \quad Y_\nu(x) = \frac{1}{\sin \nu\pi} [J_\nu(x) \cos \nu\pi - J_{-\nu}(x)]$$

$$\text{(b)} \qquad Y_n(x) = \lim_{\nu \to n} Y_\nu(x).$$

This function is called the **Bessel function of the second kind** *of order ν* or **Neumann's function**[14] *of order ν.* Figure 105 shows $Y_0(x)$ and $Y_1(x)$.

We now discuss the linear independence of J_ν and Y_ν.

For noninteger order ν, the function $Y_\nu(x)$ is evidently a solution of Bessel's equation because $J_\nu(x)$ and $J_{-\nu}(x)$ are solutions of that equation. Since for those ν the solutions J_ν and $J_{-\nu}$ are linearly independent and Y_ν involves $J_{-\nu}$, the functions J_ν and Y_ν are linearly

---

[13]CARL NEUMANN (1832—1925), German mathematician and physicist, became a professor at Leipzig in 1868. His work on potential theory sparked the development in the field of integral equations by VITO VOLTERRA (1860—1940) of Rome, ERIC IVAR FREDHOLM (1866—1927) of Stockholm, and DAVID HILBERT (1862—1943) of Göttingen (see the footnote in Sec. 6.8).

[14]See footnote 13. The solutions $Y_\nu(x)$ are sometimes denoted by $N_\nu(x)$; in Ref. [A7] they are called **Weber's functions;** Euler's constant in (6) is often denoted by C or $\ln \gamma$.

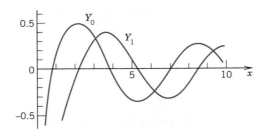

Fig. 105. Bessel functions of the second kind.
(For a small table, see Appendix 5.)

independent. Furthermore, it can be shown that the limit in (7b) exists and Y_n is a solution of Bessel's equation for integer order; see Ref. [A7] in Appendix 1. We shall see that the series development of $Y_n(x)$ contains a logarithmic term. Hence $J_n(x)$ and $Y_n(x)$ are linearly independent solutions of Bessel's equation. The series development of $Y_n(x)$ can be obtained if we insert the series (20) and (21), Sec. 4.5, for $J_\nu(x)$ and $J_{-\nu}(x)$ into (7a) and then let ν approach n; for details see Ref. [A7]. The result is

(8)

$$Y_n(x) = \frac{2}{\pi} J_n(x) \left(\ln \frac{x}{2} + \gamma \right) + \frac{x^n}{\pi} \sum_{m=0}^{\infty} \frac{(-1)^{m-1}(h_m + h_{m+n})}{2^{2m+n} m! \, (m + n)!} x^{2m}$$
$$- \frac{x^{-n}}{\pi} \sum_{m=0}^{n-1} \frac{(n - m - 1)!}{2^{2m-n} m!} x^{2m}$$

where $x > 0$, $n = 0, 1, \cdots$, and

$$h_0 = 0, \qquad h_s = 1 + \frac{1}{2} + \frac{1}{3} + \cdots + \frac{1}{s} \qquad (s = 1, 2, \cdots),$$

and when $n = 0$ the last sum in (8) is to be replaced by 0. For $n = 0$ the representation (8) takes the form (6). Furthermore, it can be shown that

$$Y_{-n}(x) = (-1)^n Y_n(x).$$

We may formulate our main result as follows.

THEOREM 1   **(General solution of Bessel's equation)**

A general solution of Bessel's equation for all values of ν is

(9)

$$y(x) = C_1 J_\nu(x) + C_2 Y_\nu(x).$$

We finally mention that there is a practical need for solutions of Bessel's equation that are complex for real values of x. For this purpose the solutions

(10)

$$H_\nu^{(1)}(x) = J_\nu(x) + i Y_\nu(x)$$
$$H_\nu^{(2)}(x) = J_\nu(x) - i Y_\nu(x)$$

are frequently used. These linearly independent functions are called **Bessel functions of the third kind** *of order* ν or *first and second* **Hankel functions**[15] *of order* ν.

This finishes our discussion on Bessel functions, except for their "orthogonality," which we explain in Sec. 4.7. Applications to vibrations follow in Sec. 11.10.

PROBLEM SET 4.6

1. **WRITING PROJECT. Bessel Functions of First and Second Kind.** Summarize the most important ideas, definitions, and results in this section and the preceding one, answering concrete questions one might raise. For instance, how is J_ν defined? What can one do with the definition? Why did we sometimes write ν, sometimes n? Why did we introduce Y_ν? Wouldn't J_ν have been enough? What was the role of the gamma function? What is the most important property by which J_ν and Y_ν differ? What is Euler's constant and its role here? And so on.

Some Further Differential Equations Reducible to Bessel's Equation
(See also Sec. 4.5.)

Using the indicated substitutions, reduce the following equations to Bessel's differential equation and find a general solution in terms of Bessel functions. (Show the details of your work.)

2. $x^2 y'' + xy' + (x^2 - 25)y = 0$
3. $4x^2 y'' + 4xy' + (100x^2 - 9)y = 0$ $(5x = z)$
4. $y'' + xy = 0$ $(y = u\sqrt{x}, \;\; \frac{2}{3}x^{3/2} = z)$
5. $y'' + x^2 y = 0$ $(y = u\sqrt{x}, \;\; \frac{1}{2}x^2 = z)$
6. $y'' + k^2 xy = 0$ $(y = u\sqrt{x}, \;\; \frac{2}{3}kx^{3/2} = z)$
7. $y'' + k^2 x^2 y = 0$ $(y = u\sqrt{x}, \;\; \frac{1}{2}kx^2 = z)$
8. $y'' + k^2 x^4 y = 0$ $(y = u\sqrt{x}, \;\; \frac{1}{3}kx^3 = z)$
9. $x^2 y'' + \frac{1}{2}xy' + \frac{1}{16}(x^{1/2} + \frac{15}{16})y = 0$ $(y = x^{1/4}u, \;\; x^{1/4} = z)$
10. $x^2 y'' + (1 - 2\nu)xy' + \nu^2(x^{2\nu} + 1 - \nu^2)y = 0$ $(y = x^\nu u, \;\; x^\nu = z)$

11. **(Y_0 for small x)** Show that for small $x > 0$ we have $Y_0(x) \approx 2(\ln \frac{1}{2}x + \gamma)/\pi$. Using this formula, compute an approximate value of the smallest positive zero of $Y_0(x)$ and compare it with the more accurate value 0.9.

12. **(Large x)** It can be shown that for large x,

    (11) $$Y_n(x) \approx \sqrt{2/(\pi x)} \sin (x - \tfrac{1}{2}n\pi - \tfrac{1}{4}\pi).$$

    Using (11), sketch $Y_0(x)$ and $Y_1(x)$ for $0 < x \leq 15$. Using (11), compute approximate values of the first three positive zeros of $Y_0(x)$ and compare these values with the more accurate values 0.89, 3.96, and 7.09.

13. **(Hankel functions)** Show that the Hankel functions (10) constitute a basis of solutions of Bessel's equation for any ν.

Modified Bessel Functions

14. The function $I_\nu(x) = i^{-\nu} J_\nu(ix)$, $i = \sqrt{-1}$, is called the *modified Bessel function of the first kind of order* ν. Show that $I_\nu(x)$ is a solution of the differential equation

    (12) $$x^2 y'' + xy' - (x^2 + \nu^2)y = 0$$

    and has the representation

    (13) $$I_\nu(x) = \sum_{m=0}^{\infty} \frac{x^{2m+\nu}}{2^{2m+\nu} m! \, \Gamma(m + \nu + 1)} \, .$$

---

[15] HERMANN HANKEL (1839—1873), German mathematician.

15. Show that $I_\nu(x)$ is real for all real x (and real ν), $I_\nu(x) \neq 0$ for all real $x \neq 0$, and $I_{-n}(x) = I_n(x)$, where n is any integer.

16. Show that another solution of the differential equation (12) is the so-called *modified Bessel function of the third kind* (sometimes called *of the second kind*)

$$(14) \qquad K_\nu(x) = \frac{\pi}{2 \sin \nu\pi} [I_{-\nu}(x) - I_\nu(x)].$$

4.7 Sturm–Liouville Problems. Orthogonal Functions

Legendre's, Bessel's, and other differential equations of importance in engineering science can be written in the form

$$(1) \qquad \boxed{[r(x)y']' + [q(x) + \lambda p(x)]y = 0.}$$

This is called a **Sturm–Liouville equation.**[16]

Indeed, for Legendre's equation

$$(1 - x^2)y'' - 2xy' + n(n + 1)y = 0$$

we can write

$$[(1 - x^2)y']' + \lambda y = 0, \qquad\qquad \lambda = n(n + 1).$$

This is (1) with $r = 1 - x^2$, $q = 0$, and $p = 1$. For Bessel's equation

$$\tilde{x}^2 \ddot{y} + \tilde{x}\dot{y} + (\tilde{x}^2 - n^2)y = 0, \qquad\qquad \dot{y} = dy/d\tilde{x}, \text{ etc.,}$$

we need a simple transformation $\tilde{x} = kx$, so that $\dot{y} = y'/k$, $\ddot{y} = y''/k^2$ and we get (k^2 and k drop out in the first two terms)

$$x^2 y'' + xy' + (k^2 x^2 - n^2)y = 0.$$

Dividing by x we get

$$[xy']' + \left(-\frac{n^2}{x} + \lambda x\right)y = 0, \qquad\qquad \lambda = k^2.$$

This is (1) with $r = x$, $q = -n^2/x$, and $p = x$.

Sturm and Liouville have developed a theory of equations (1) that leads to practically useful series developments in terms of particular solutions of (1) on a given interval $a \leqq x \leqq b$ satisfying boundary conditions at the two endpoints a and b,

---

[16]JACQUES CHARLES FRANÇOIS STURM (1803—1855), was born and studied in Switzerland and then moved to Paris, where he later became the successor of Poisson in the chair of mechanics at the Sorbonne (the University of Paris).

JOSEPH LIOUVILLE (1809—1882), French mathematician and professor in Paris, contributed to various fields in mathematics and is particularly known by his important work in complex analysis (Liouville's theorem; Sec. 13.4), special functions, differential geometry, and number theory.

$$\text{(a)} \quad \boxed{k_1 y(a) + k_2 y'(a) = 0}$$

(2)

$$\text{(b)} \quad \boxed{l_1 y(b) + l_2 y'(b) = 0.}$$

k_1, k_2 in (2) are given constants, not both zero, and so are l_1, l_2. The **boundary value problem** consisting of (1) and (2) is called a **Sturm–Liouville problem.**

On that interval we assume continuity of p, q, r, r' as well as

$$\boxed{p(x) > 0.}$$

Clearly, $y \equiv 0$ is always a solution of the problem, but is of no practical use. What we want to find is solutions of (1) satisfying (2) without being identically zero. We call such a solution $y(x)$—if it exists—an **eigenfunction** and a number λ for which an eigenfunction exists, an **eigenvalue** of the problem.

EXAMPLE 1   **Vibrating elastic string**

Find the eigenvalues and eigenfunctions of the Sturm–Liouville problem

(3)   (a) $y'' + \lambda y = 0$ (b) $y(0) = 0$, $y(\pi) = 0$.

This problem arises, for instance, if an elastic string (a violin string, for example) is stretched a little and then fixed at its ends $x = 0$ and $x = \pi$ and allowed to vibrate. Then $y(x)$ is the "space function" of the deflection $u(x, t)$ of the string, assumed in the form $u(x, t) = y(x)w(t)$, where t is time. (This model will be discussed in Secs. 11.2–11.4.)

Solution. $r = 1$, $q = 0$, $p = 1$ in (1), $a = 0$, $b = \pi$, $k_1 = l_1 = 1$, $k_2 = l_2 = 0$ in (2). For negative $\lambda = -\nu^2$ a general solution of the equation is $y(x) = c_1 e^{\nu x} + c_2 e^{-\nu x}$. From (3b) we obtain $c_1 = c_2 = 0$, so that $y \equiv 0$, which is not an eigenfunction. For $\lambda = 0$ the situation is similar. For positive $\lambda = \nu^2$ a general solution is

$$y(x) = A \cos \nu x + B \sin \nu x.$$

From the first boundary condition we obtain $y(0) = A = 0$. The second boundary condition then yields

$$y(\pi) = B \sin \nu \pi = 0, \qquad \text{thus} \qquad \nu = 0, \pm 1, \pm 2, \cdots.$$

For $\nu = 0$ we have $y \equiv 0$. For $\lambda = \nu^2 = 1, 4, 9, 16, \cdots$, taking $B = 1$, we obtain

$$y(x) = \sin \nu x \qquad\qquad (\nu = 1, 2, \cdots).$$

Hence the eigenvalues of the problem are $\lambda = \nu^2$, where $\nu = 1, 2, \cdots$; and corresponding eigenfunctions are $y(x) = \sin \nu x$, where $\nu = 1, 2, \cdots$.   ◀

Existence of eigenvalues

Eigenvalues of a Sturm–Liouville problem (1), (2), even infinitely many, exist under rather general conditions on p, q, r in (1). (Sufficient are the conditions in Theorem 1, below, together with $p(x) > 0$ and $r(x) > 0$ on $a < x < b$. Proofs are complicated; see Ref. [A1] or [A5] listed in Appendix 1.)

Reality of eigenvalues

Furthermore, if p, q, r, and r' in (1) are real-valued and continuous on the interval $a \leqq x \leqq b$ and p is positive throughout that interval (or negative throughout that interval), then all the eigenvalues of the Sturm–Liouville problem (1), (2) are real. (A proof is included in Appendix 4.)

This is what the engineer would expect because eigenvalues are often related to frequencies, energies, or other physical quantities that must be real.

Orthogonality

Eigenfunctions of Sturm–Liouville problems have remarkable general properties—above all, orthogonality, which is defined as follows.

Definition of orthogonality

Functions y_1, y_2, $\cdots$ defined on some interval $a \leqq x \leqq b$ are called **orthogonal** on $a \leqq x \leqq b$ with respect to a **weight function** $p(x) > 0$ if

$$(4) \qquad \boxed{\int_a^b p(x) y_m(x) y_n(x)\, dx = 0} \qquad \text{for } m \neq n.$$

The **norm** $\|y_m\|$ of y_m is defined by

$$(5) \qquad \boxed{\|y_m\| = \sqrt{\int_a^b p(x) y_m^2(x)\, dx}.}$$

The functions are called **orthonormal** on $a \leqq x \leqq b$ if they are orthogonal on $a \leqq x \leqq b$ and all have norm 1.

For "orthogonal with respect to $p(x) = 1$" we simply say "orthogonal." Thus, functions y_1, y_2, $\cdots$ are **orthogonal** on some interval $a \leqq x \leqq b$ if

$$(4') \qquad \boxed{\int_a^b y_m(x) y_n(x)\, dx = 0} \qquad \text{for } m \neq n.$$

The **norm** $\|y_m\|$ of y_m is then simply defined by

$$(5') \qquad \boxed{\|y_m\| = \sqrt{\int_a^b y_m^2(x)\, dx}.}$$

And the functions are called **orthonormal** on $a \leqq x \leqq b$ if they are orthogonal there and all have norm (5') equal to 1. ◀

EXAMPLE 2   **Orthogonal set, orthonormal set**

The functions $y_m(x) = \sin mx$, $m = 1, 2, \cdots$ form an orthogonal set on the interval $-\pi \leqq x \leqq \pi$, because for $m \neq n$ we obtain [see (11) in Appendix A3.1]

$$\int_{-\pi}^{\pi} y_m(x) y_n(x)\, dx = \int_{-\pi}^{\pi} \sin mx \sin nx\, dx = \frac{1}{2} \int_{-\pi}^{\pi} \cos (m-n)x\, dx - \frac{1}{2} \int_{-\pi}^{\pi} \cos (m+n)x\, dx = 0.$$

The norm $\|y_m\|$ equals $\sqrt{\pi}$, because

$$\|y_m\|^2 = \int_{-\pi}^{\pi} \sin^2 mx\, dx = \pi \qquad\qquad (m = 1, 2, \cdots).$$

Hence the corresponding orthonormal set, obtained by division by the norm, is

$$\frac{\sin x}{\sqrt{\pi}}, \qquad \frac{\sin 2x}{\sqrt{\pi}}, \qquad \frac{\sin 3x}{\sqrt{\pi}}, \qquad \cdots \qquad \blacktriangleleft$$

Orthogonality of Eigenfunctions

THEOREM 1   **(Orthogonality of eigenfunctions)**

Suppose that the functions p, q, r, and r' in the Sturm–Liouville equation (1) are real-valued and continuous and $p(x) > 0$ on the interval $a \leqq x \leqq b$. Let $y_m(x)$ and $y_n(x)$ be eigenfunctions of the Sturm–Liouville problem (1), (2) that correspond to different eigenvalues λ_m and λ_n, respectively. Then y_m, y_n are orthogonal on that interval with respect to the weight function p.

If $r(a) = 0$, then (2a) can be dropped from the problem. If $r(b) = 0$, then (2b) can be dropped. [It is then required that y and y' remain bounded at such a point, and the problem is called **singular,** *as opposed to a* **regular problem** *in which (2) is used.]*

If $r(a) = r(b)$, then (2) can be replaced by the **"periodic boundary conditions"**

(6) $$y(a) = y(b), \qquad y'(a) = y'(b).$$

Remark. The boundary value problem consisting of the Sturm–Liouville equation (1) and the periodic boundary conditions (6) is called a **periodic Sturm–Liouville problem.**

PROOF OF THEOREM 1. By assumption, y_m satisfies

$$(ry'_m)' + (q + \lambda_m p)y_m = 0,$$

and y_n satisfies

$$(ry'_n)' + (q + \lambda_n p)y_n = 0.$$

Multiplying the first equation by y_n, the second by $-y_m$ and adding, we get

$$(\lambda_m - \lambda_n)py_m y_n = y_m(ry'_n)' - y_n(ry'_m)' = [(ry'_n)y_m - (ry'_m)y_n]'$$

where the last equality can be readily verified by performing the indicated differentiation of the last expression in brackets. This expression is continuous on $a \leqq x \leqq b$ since r and r' are continuous by assumption and y_m, y_n are solutions of (1). Integrating over x from a to b, we thus obtain

(7) $$(\lambda_m - \lambda_n) \int_a^b py_m y_n \, dx = \left[r(y'_n y_m - y'_m y_n) \right]_a^b.$$

The expression on the right equals

(8)
$$r(b)\left[y'_n(b)y_m(b) - y'_m(b)y_n(b)\right]$$
$$-r(a)\left[y'_n(a)y_m(a) - y'_m(a)y_n(a)\right].$$

We now have to consider several cases depending on whether r vanishes or does not vanish at a or b.

Case 1. If $r(a) = 0$ and $r(b) = 0$, then the expression in (8) is zero. Hence the expression on the left side of (7) must be zero, because y_m, y'_m, y_n, y'_n on the right remain continuous at a and b (by assumption) and λ_m and λ_n are distinct. We thus obtain the desired orthogonality

$$(9) \qquad \int_a^b p(x)y_m(x)y_n(x)\,dx = 0 \qquad\qquad (m \neq n)$$

without the use of the boundary conditions (2).

Case 2. Let $r(b) = 0$, but $r(a) \neq 0$. Then the first line in (8) is zero. We consider the remaining expression in (8). From (2a) we have

$$k_1 y_n(a) + k_2 y'_n(a) = 0,$$
$$k_1 y_m(a) + k_2 y'_m(a) = 0.$$

Let $k_2 \neq 0$. We multiply the first equation by $y_m(a)$, the last by $-y_n(a)$ and add,

$$k_2 \big[y'_n(a)y_m(a) - y'_m(a)y_n(a) \big] = 0.$$

Since $k_2 \neq 0$, the expression in brackets must be zero. This expression is identical with that in the last line of (8). Hence (8) is zero, and from (7) we obtain (9) as before. If $k_2 = 0$, then by assumption $k_1 \neq 0$, and the argument of proof is similar.

Case 3. If $r(a) = 0$, but $r(b) \neq 0$, the proof is similar to that in Case 2, but instead of (2a) we now have to use (2b).

Case 4. If $r(a) \neq 0$ and $r(b) \neq 0$, we have to use both boundary conditions (2) and proceed as in Cases 2 and 3.

Case 5. Let $r(a) = r(b)$. Then (8) takes the form

$$r(b)\big[y'_n(b)y_m(b) - y'_m(b)y_n(b) - y'_n(a)y_m(a) + y'_m(a)y_n(a) \big].$$

We may use (2) as before and conclude that the expression in brackets is zero. However, we immediately see that this also follows from (6), so that we may replace (2) by (6). Hence (7) yields (9), as before. This completes the proof of Theorem 1. ◀

EXAMPLE 3   **Vibrating elastic string**

The differential equation in Example 1 is of the form (1) where $r = 1$, $q = 0$, and $p = 1$. From Theorem 1 it follows that the eigenfunctions are orthogonal on the interval $0 \leqq x \leqq \pi$. ◀

EXAMPLE 4   **Orthogonality of Legendre polynomials**

Legendre's equation is a Sturm–Liouville equation (see the beginning of this section)

$$[(1 - x^2)y']' + \lambda y = 0, \qquad\qquad \lambda = n(n + 1),$$

with $r = 1 - x^2$, $q = 0$, and $p = 1$. Since $r(-1) = r(1) = 0$, we need no boundary conditions, but have a *singular Sturm–Liouville problem* on the interval $-1 \leqq x \leqq 1$. We know that for $n = 0, 1, \cdots$, hence $\lambda = 0, 1 \cdot 2, 2 \cdot 3, \cdots$, the Legendre polynomials $P_n(x)$ are solutions of the problem. Hence these are the eigenfunctions. From Theorem 1 it follows that they are orthogonal on that interval, that is,

$$(10) \qquad \int_{-1}^1 P_m(x)P_n(x)\,dx = 0 \qquad\qquad (m \neq n). \quad ◀$$

EXAMPLE 5   **Orthogonality of Bessel functions $J_n(x)$**

The Bessel function $J_n(\tilde{x})$ with fixed integer $n \geqq 0$ satisfies Bessel's equation (Sec. 4.5)

$$\tilde{x}^2 \ddot{J}_n(\tilde{x}) + \tilde{x}\dot{J}_n(\tilde{x}) + (\tilde{x}^2 - n^2)J_n(\tilde{x}) = 0,$$

where $\dot{J}_n = dJ_n/d\tilde{x}$, $\ddot{J}_n = d^2J_n/d\tilde{x}^2$. At the beginning of this section we transformed this equation, by setting $\tilde{x} = kx$, into a Sturm–Liouville equation

$$\left[xJ_n'(kx)\right]' + \left(-\frac{n^2}{x} + k^2x\right)J_n(kx) = 0$$

with $p(x) = x$, $q(x) = -n^2/x$, $r(x) = x$, and parameter $\lambda = k^2$. Since $r(0) = 0$, Theorem 1 implies orthogonality on an interval $0 \leqq x \leqq R$ of those solutions $J_n(kx)$ that are zero at $x = R$, that is,

(11) $$J_n(kR) = 0 \qquad\qquad (n \text{ fixed}).$$

[Note that $q(x) = -n^2/x$ is discontinuous at 0, but this does not affect the proof of Theorem 1.] It can be shown (see Ref. [A7]) that $J_n(\tilde{x})$ has infinitely many real zeros, say, $\tilde{x} = \alpha_{1n} < \alpha_{2n} < \cdots$ (see Fig. 103 in Sec. 4.5 for $n = 0$ and 1). Hence we must have

(12) $$kR = \alpha_{mn}, \qquad \text{thus} \qquad k = k_{mn} = \alpha_{mn}/R \qquad (m = 1, 2, \cdots).$$

This proves the following orthogonality property.

THEOREM 2   **(Orthogonality of Bessel functions)**

*For each **fixed** nonnegative integer n the sequence of the Bessel functions of the first kind $J_n(k_{1n}x)$, $J_n(k_{2n}x)$, $J_n(k_{3n}x)$, $\cdots$, with k_{mn} as in (12), forms an orthogonal set on the interval $0 \leqq x \leqq R$ with respect to the weight $p(x) = x$, that is,*

(13) $$\int_0^R xJ_n(k_{mn}x)J_n(k_{jn}x)\,dx = 0 \qquad\qquad (j \neq m).$$

Hence we have obtained infinitely many orthogonal sets, each corresponding to one of the fixed values n. This also illustrates the importance of the zeros of the Bessel functions. ◀

PROBLEM SET 4.7

1. **(Proof of Theorem 1)** Carry out the details in Cases 3 and 4.

2. **Normalization of eigenfunctions** y_m of (1), (2) means that we multiply y_m by a nonzero constant c_m such that $c_m y_m$ has norm 1. Show that $z_m = cy_m$ with *any* $c \neq 0$ is an eigenfunction for the eigenvalue corresponding to y_m.

Sturm–Liouville Problems. Find the eigenvalues and eigenfunctions of the following problems. In Probs. 3–6 also verify orthogonality by direct calculation.

3. $y'' + \lambda y = 0$, $y(0) = 0$, $y'(1) = 0$
4. $y'' + \lambda y = 0$, $y(0) = 0$, $y(L) = 0$
5. $y'' + \lambda y = 0$, $y(0) = 0$, $y'(L) = 0$
6. $y'' + \lambda y = 0$, $y(0) = y(2\pi)$, $y'(0) = y'(2\pi)$
7. $(xy')' + \lambda x^{-1}y = 0$, $y(1) = 0$, $y'(e) = 0$ *Hint.* You may set $x = e^t$.
8. $(e^{2x}y')' + e^{2x}(\lambda + 1)y = 0$, $y(0) = 0$, $y(\pi) = 0$ *Hint.* Set $y = e^{-x}u$.
9. $(x^{-1}y')' + (\lambda + 1)x^{-3}y = 0$, $y(1) = 0$, $y(e) = 0$

10. (Sturm–Liouville problem) Find a problem whose eigenfunctions are 1, $\cos x$, $\cos 2x$, $\cdots$.

11. (Sturm–Liouville problem) Find a problem whose eigenfunctions are 1, $\cos (m\pi x/L)$, $\sin (m\pi x/L)$, $m = 1, 2, \cdots$.

12. (Transcendental equation) Show that the eigenvalues of the problem $y'' + \lambda y = 0$, $y(0) = 0$, $y(1) + y'(1) = 0$ are obtained as solutions of $\tan k = -k$, where $k = \sqrt{\lambda}$. Conclude from a plot that this equation must have infinitely many solutions $k = k_m$ and that k_m is of the form $k_m = \frac{1}{2}(2m + 1)\pi + \delta_m$ with small positive δ_m such that $\delta_m \to 0$ as $m \to \infty$. Show that the eigenfunctions are $y_m = \sin k_m x$ ($k_m \neq 0$). Compute k_0 and k_1 (by Newton's method; see Sec. 17.2).

13. (Change of x) Show that if the functions $y_0(x)$, $y_1(x)$, $\cdots$ form an orthogonal set on an interval $a \leqq x \leqq b$ (with $p(x) = 1$), then the functions $y_0(ct + k)$, $y_1(ct + k)$, $\cdots$, $c > 0$, form an orthogonal set on the interval $(a - k)/c \leqq t \leqq (b - k)/c$.

14. (Change of x) Using Prob. 13, derive the orthogonality of 1, $\cos \pi x$, $\sin \pi x$, $\cos 2\pi x$, $\sin 2\pi x$, $\cdots$ on $-1 \leqq x \leqq 1$ ($p(x) = 1$) from that of 1, $\cos x$, $\sin x$, $\cos 2x$, $\sin 2x$, $\cdots$ on $-\pi \leqq x \leqq \pi$.

15. (Legendre polynomials) Show that the functions $P_n(\cos \theta)$, $n = 0, 1, \cdots$, form an orthogonal set on the interval $0 \leqq \theta \leqq \pi$ with respect to the weight function $\sin \theta$.

16. TEAM PROJECT. Special Functions. Orthogonal polynomials play a great role in applications. For this reason, besides Legendre polynomials various other orthogonal polynomials have been extensively studied; see Refs. [1], [11], [12] in Appendix 1. Consider some of the most important ones as follows.

(a) Chebyshev polynomials[17] *of the first and second kind* are defined by

$$T_n(x) = \cos (n \text{ arc cos } x), \quad U_n(x) = \frac{\sin [(n + 1) \text{ arc cos } x]}{\sqrt{1 - x^2}} \quad (n = 0, 1, \cdots)$$

respectively. Show that

$$T_0 = 1, \qquad T_1(x) = x, \qquad T_2(x) = 2x^2 - 1, \qquad T_3(x) = 4x^3 - 3x,$$

$$U_0 = 1, \qquad U_1(x) = 2x, \qquad U_2(x) = 4x^2 - 1, \qquad U_3(x) = 8x^3 - 4x.$$

Show that the Chebyshev polynomials $T_n(x)$ are orthogonal on the interval $-1 \leqq x \leqq 1$ with respect to the weight function $p(x) = 1/\sqrt{1 - x^2}$. *Hint.* To evaluate the integral, set arc cos $x = \theta$.

(b) Orthogonality on an infinite interval: Laguerre polynomials[18] are defined by

$$L_0 = 1, \qquad L_n(x) = \frac{e^x}{n!} \frac{d^n(x^n e^{-x})}{dx^n}, \qquad n = 1, 2, \cdots.$$

Show that

$$L_1(x) = 1 - x, \qquad L_2(x) = 1 - 2x + x^2/2, \qquad L_3(x) = 1 - 3x + 3x^2/2 - x^3/6.$$

Prove that the Laguerre polynomials are orthogonal on the positive axis $0 \leqq x < \infty$ with respect to the weight function $p(x) = e^{-x}$. *Hint.* Since the highest power in L_m is x^m, it suffices to show that $\int e^{-x} x^k L_n \, dx = 0$ for $k < n$. Do this by k integrations by parts.

---

[17]PAFNUTI CHEBYSHEV (1821—1894), Russian mathematician, is known for his work in approximation theory and the theory of numbers. Another transliteration of the name is TCHEBICHEF.

[18]EDMOND LAGUERRE (1834—1886), French mathematician, who did research work in geometry and in the theory of infinite series.

4.8 Orthogonal Eigenfunction Expansions

What are orthogonal functions (from Sturm–Liouville problems or elsewhere) good for? We show that they yield series developments of given functions in a simple fashion. This includes the famous Fourier series, the daily bread of the physicist and engineer in heat conduction, vibrations, fluid flow (to which we devote Chaps. 10 and 11). Indeed, orthogonality is one of the most useful ideas ever introduced in applied mathematics.

To explain this, we first introduce a practical standard notation. We denote the integral (4), Sec. 4.7, defining orthogonality and orthonormality, simply by (y_m, y_n). Thus for orthonormal functions $y_0, y_1, y_2, \cdots$ with respect to weight $p(x) > 0$ on $a \leqq x \leqq b$,

$$(1) \qquad (y_m, y_n) = \int_a^b p(x) y_m(x) y_n(x)\, dx = \begin{cases} 0 & \text{if} \quad m \neq n \\ 1 & \text{if} \quad m = n \end{cases}$$

where $m = 0, 1, 2, \cdots$ and $n = 0, 1, 2, \cdots$. Even more briefly,

$$(1^*) \qquad (y_m, y_n) = \delta_{mn} \qquad \text{where} \qquad \delta_{mn} = \begin{cases} 0 & \text{if} \quad m \neq n \\ 1 & \text{if} \quad m = n \end{cases}$$

is called the **Kronecker delta.**[19] For the norm $\|y_m\|$ of y_m we can now write

$$(2) \qquad \|y_m\| = \sqrt{(y_m, y_m)} = \sqrt{\int_a^b p(x) y_m^2(x)\, dx}.$$

Now let $y_0, y_1, \cdots$ be an orthogonal set with respect to weight $p(x)$ on an interval $a \leqq x \leqq b$ and $f(x)$ a function that can be represented by a convergent series

$$(3) \qquad f(x) = \sum_{m=0}^{\infty} a_m y_m(x) = a_0 y_0(x) + a_1 y_1(x) + \cdots.$$

This is called an **orthogonal expansion** or **generalized Fourier series** or, if the y_m are eigenfunctions of a Sturm–Liouville problem, an **eigenfunction expansion.** [We use m in (3) because we need n later as a fixed index of Bessel functions.]

The point is that because of the orthogonality we can obtain the unknown coefficients $a_0, a_1, \cdots$ in a simple fashion; these are called the **Fourier constants** of $f(x)$ with respect to $y_0, y_1, \cdots$. In fact, if we multiply both sides of (3) by $p(x) y_n(x)$ (n fixed) and then integrate over $a \leqq x \leqq b$, we find, assuming that term-by-term integration is permissible,[20]

$$(f, y_n) = \int_a^b p f y_n\, dx = \int_a^b p \left(\sum_{m=0}^{\infty} a_m y_m \right) y_n\, dx = \sum_{m=0}^{\infty} a_m (y_m, y_n).$$

Crucial now is that because of the orthogonality, all the integrals (y_m, y_n) on the right are zero, except when $m = n$; then $(y_n, y_n) = \|y_n\|^2$. Hence the whole formula reduces to

$$(3^*) \qquad (f, y_n) = a_n \|y_n\|^2.$$

---

[19]LEOPOLD KRONECKER (1823—1891), German mathematician at Berlin University, who made important contributions to algebra, group theory, and number theory.

[20]This is justified, for instance, in the case of uniform convergence (Theorem 3, Sec. 14.5).

Writing m for n, to be in agreement with the notation in (3), we get the desired formula for the Fourier constants

(4)
$$a_m = \frac{(f, y_m)}{\|y_m\|^2} = \frac{1}{\|y_m\|^2} \int_a^b p(x)f(x)y_m(x)\,dx \qquad (m = 0, 1, \cdots).$$

EXAMPLE 1   **Fourier series**

For the periodic Sturm–Liouville problem

$$y'' + \lambda y = 0, \qquad y(\pi) = y(-\pi), \qquad y'(\pi) = y'(-\pi)$$

we obtain from the general solution $y = A \cos kx + B \sin kx$, where $k = \sqrt{\lambda}$, and the boundary conditions the two equations

$$A \cos k\pi + B \sin k\pi = A \cos (-k\pi) + B \sin (-k\pi)$$

$$-kA \sin k\pi + kB \cos k\pi = -kA \sin (-k\pi) + kB \cos (-k\pi).$$

Since $\cos (-\alpha) = \cos \alpha$ and $\sin (-\alpha) = -\sin \alpha$, this gives

$$\sin k\pi = 0, \qquad \lambda = k^2 = n^2 = 0, 1, 4, 9, \cdots .$$

Hence the eigenfunctions are

$$1, \qquad \cos x, \qquad \sin x, \qquad \cos 2x, \qquad \sin 2x, \qquad \cdots .$$

By Theorem 1 in Sec. 4.7, any two of these belonging to different eigenvalues are orthogonal on the interval $-\pi \le x \le \pi$ (note that $p(x) = 1$ for the present equation). The orthogonality of $\cos mx$ and $\sin mx$ for the same m follows by integration,

$$\int_{-\pi}^{\pi} \cos mx \sin mx\,dx = \frac{1}{2} \int_{-\pi}^{\pi} \sin 2mx\,dx = 0.$$

For the norms we get $\|1\| = \sqrt{2\pi}$ and $\sqrt{\pi}$ for all the others, as the student may verify by integrating 1, $\cos^2 x$, $\sin^2 x$, etc., from $-\pi$ to π. The corresponding series (3) is

(5)
$$f(x) = a_0 + \sum_{m=1}^{\infty} (a_m \cos mx + b_m \sin mx).$$

This is called the **Fourier series** of $f(x)$. The a_m, b_m are called the **Fourier coefficients** of $f(x)$. From (4) with $p(x) = 1$ and those values of the norm we see that they are given by the so-called **Euler formulas**

(6)
$$a_0 = \frac{1}{2\pi} \int_{-\pi}^{\pi} f(x)\,dx$$

$$a_m = \frac{1}{\pi} \int_{-\pi}^{\pi} f(x) \cos mx\,dx \qquad (m = 1, 2, \cdots)$$

$$b_m = \frac{1}{\pi} \int_{-\pi}^{\pi} f(x) \sin mx\,dx \qquad (m = 1, 2, \cdots).$$

For instance, for the **"periodic square wave"** in Fig. 106 on the next page, given by

$$f(x) = \begin{cases} -1 & \text{if} \quad -\pi < x < 0 \\ 1 & \text{if} \quad 0 < x < \pi \end{cases} \qquad \text{and} \qquad f(x + 2\pi) = f(x),$$

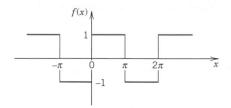

Fig. 106.   Periodic square wave in Example 1

we get from (6) the values $a_0 = 0$ and

$$a_m = \frac{1}{\pi} \left[\int_{-\pi}^{0} (-1) \cos mx \, dx + \int_{0}^{\pi} 1 \cdot \cos mx \, dx \right] = 0,$$

$$b_m = \frac{1}{\pi} \left[\int_{-\pi}^{0} (-1) \sin mx \, dx + \int_{0}^{\pi} 1 \cdot \sin mx \, dx \right]$$

$$= \frac{1}{\pi} \left[\frac{\cos mx}{m} \bigg|_{-\pi}^{0} - \frac{\cos mx}{m} \bigg|_{0}^{\pi} \right]$$

$$= \frac{1}{\pi m} [1 - 2 \cos m\pi + 1] = \begin{cases} 4/(\pi m) & \text{if } m = 1, 3, \cdots, \\ 0 & \text{if } m = 2, 4, \cdots. \end{cases}$$

Hence the Fourier series of the periodic square wave is

$$f(x) = \frac{4}{\pi} \left(\sin x + \frac{1}{3} \sin 3x + \frac{1}{5} \sin 5x + \cdots \right).$$ ◀

Fourier series are by far the most important eigenfunction expansions, so important to the engineer that we shall devote two chapters (10 and 11) to them and their applications, and discuss numerous examples.

EXAMPLE 2   **Fourier–Legendre series**

This is an eigenfunction expansion

$$f(x) = \sum_{m=0}^{\infty} a_m P_m(x) = a_0 P_0 + a_1 P_1(x) + \cdots = a_0 + a_1 x + a_2 \left(\frac{3}{2} x^2 - \frac{1}{2} \right) + \cdots$$

in terms of Legendre polynomials (Sec. 4.3), the eigenfunctions of the Sturm–Liouville problem in Example 4 of Sec. 4.7 on the interval $-1 \leqq x \leqq 1$. We have $p(x) = 1$ for Legendre's equation, and (4) gives

(7) $$a_m = \frac{2m + 1}{2} \int_{-1}^{1} f(x) P_m(x) \, dx,$$ $m = 0, 1, \cdots,$

because the norm is

(8) $$\|P_m\| = \sqrt{\int_{-1}^{1} P_m(x)^2 \, dx} = \sqrt{\frac{2}{2m + 1}}$$ $(m = 0, 1, \cdots),$

as we state without proof (which is tricky; it uses Rodrigues's formula in Sec. 4.3 and a reduction of the resulting integral to a quotient of gamma functions).

For instance, if $f(x) = \sin \pi x$, we obtain the coefficients

$$a_m = \frac{2m + 1}{2} \int_{-1}^{1} (\sin \pi x) \, P_m(x) \, dx$$

and the Fourier–Legendre series

$$\sin \pi x = 0.95493 P_1(x) - 1.15824 P_3(x) + 0.21429 P_5(x) - 0.01664 P_7(x) + 0.00068 P_9(x) - 0.00002 P_{11}(x) + \cdots$$

The coefficient of P_{13} is about $3 \cdot 10^{-7}$. The sum of the first three nonzero terms gives a curve that practically coincides with the sine curve. Can you see why the even-numbered coefficients are zero? Why a_3 is the absolutely biggest coefficient? ◀

EXAMPLE 3   **Fourier–Bessel series**

In Example 5 of Sec. 4.7 we obtained infinitely many orthogonal sets of Bessel functions, one for each of $J_0, J_1, J_2, \cdots$. The orthogonality is on an interval $0 \leqq x \leqq R$ with any fixed positive R and with respect to weight x. The set for J_n is $J_n(k_{1n}x)$, $J_n(k_{2n}x)$, $J_n(k_{3n}x)$, $\cdots$, where n is fixed, and k_{mn} is given in (12), Sec. 4.7. The corresponding Fourier–Bessel series is

$$(9) \qquad f(x) = \sum_{m=1}^{\infty} a_m J_n(k_{mn}x) = a_1 J_n(k_{1n}x) + a_2 J_n(k_{2n}x) + a_3 J_n(k_{3n}x) + \cdots.$$

The coefficients are (with $\alpha_{mn} = k_{mn}R$)

$$(10) \qquad \boxed{a_m = \frac{2}{R^2 J_{n+1}^2(\alpha_{mn})} \int_0^R x f(x) J_n(k_{mn}x)\, dx,} \qquad m = 1, 2, \cdots$$

because the norm is

$$(11) \qquad \| J_n(k_{mn}x) \|^2 = \int_0^R x J_n^{\,2}(k_{mn}x)\, dx = \frac{R^2}{2} J_{n+1}^2(k_{mn}R),$$

as we state without a proof (which is tricky; see the discussion beginning on p. 576 of [A7]).

For instance, we consider $f(x) = 1 - x^2$ and take $R = 1$ and $n = 0$ in the series (9), simply writing λ for α_{m0}. Then $k_{mn} = \alpha_{m0} = \lambda = 2.405, 5.520, 8.654, 11.792$, etc. (see Table A1 in Appendix 5). Next we calculate the coefficients (10). We first use (24), Sec. 4.5, with $\nu = 1$, that is,

$$[x J_1(\lambda x)]' = \lambda x J_0(\lambda x),$$

and an integration by parts (in which the integral-free part vanishes). Then we use (24), Sec. 4.5, with $\nu = 2$,

$$[x^2 J_2(\lambda x)]' = \lambda x^2 J_1(\lambda x).$$

This gives

$$a_m = \frac{2}{J_1^{\,2}(\lambda)} \int_0^1 x(1 - x^2) J_0(\lambda x)\, dx = \frac{2}{J_1^{\,2}(\lambda)} \left[\frac{1}{\lambda}(1 - x^2) x J_1(\lambda x) \Big|_0^1 - \frac{1}{\lambda} \int_0^1 x J_1(\lambda x)(-2x)\, dx \right]$$

$$= \frac{4 J_2(\lambda)}{\lambda^2 J_1^{\,2}(\lambda)} \qquad\qquad (\lambda = \alpha_{m0}).$$

This gives the orthogonal eigenfunction expansion (9),

$$1 - x^2 = 1.1081 J_0(2.405x) - 0.1398 J_0(5.520x) + 0.0455 J_0(8.654x) - 0.0210 J_0(11.792x) + \cdots.$$

(Recall from Sec. 4.5 that $J_2 = 2x^{-1}J_1 - J_0$ or use tables in Ref. [1] in Appendix 1 or a CAS.) A plot shows that the curve of $1 - x^2$ and that of the sum of the first three terms on the right practically coincide. ◀

Completeness of Orthonormal Sets

In practice, one uses only orthonormal sets that consist of "sufficiently many" functions, so that one can represent large classes of functions—certainly all continuous functions on an interval $a \leqq x \leqq b$—by a generalized Fourier series (3). These orthonormal sets are called *"complete"* (in the set of functions considered; definition below). For instance, the orthonormal set in Example 1 is complete in the set of continuous[21] functions on the interval $-\pi \leqq x \leqq \pi$, and so are the sets of Legendre polynomials and Bessel functions in Examples 2 and 3 on their respective intervals.

In this connection, convergence is **convergence in the norm** (also called **mean-square convergence** or **mean convergence**); that is, a sequence of functions f_n is called **convergent** with the limit f if

$$(12^*) \qquad \lim_{k \to \infty} \|f_k - f\| = 0;$$

written out by (2) (where we can drop the square root),

$$(12) \qquad \lim_{k \to \infty} \int_a^b p(x)[f_k(x) - f(x)]^2 \, dx = 0.$$

Accordingly, (3) converges and represents f if

$$(13) \qquad \lim_{k \to \infty} \int_a^b p(x)[s_k(x) - f(x)]^2 \, dx = 0$$

where s_k is the kth partial sum of (3),

$$(14) \qquad s_k(x) = \sum_{m=0}^{k} a_m y_m(x).$$

By definition, an orthonormal set $y_0, y_1, \cdots$ on an interval $a \leqq x \leqq b$ is **complete in a set of functions** S defined on $a \leqq x \leqq b$ if we can approximate every f belonging to S arbitrarily closely by a linear combination $a_0 y_0 + a_1 y_1 + \cdots + a_k y_k$, that is, technically, if for every $\epsilon > 0$ we can find constants $a_0, \cdots, a_k$ (with k large enough) such that

$$(15) \qquad \|f - (a_0 y_0 + \cdots + a_k y_k)\| < \epsilon.$$

An interesting and basic consequence of the integral in (13) is obtained as follows. Performing the square and using (14), we first have

$$\int_a^b p(x)[s_k(x) - f(x)]^2 \, dx = \int_a^b p s_k^2 \, dx - 2 \int_a^b p f s_k \, dx + \int_a^b p f^2 \, dx$$

$$= \int_a^b p \left[\sum_{m=0}^{k} a_m y_m \right]^2 dx - 2 \sum_{m=0}^{k} a_m \int_a^b p f y_m \, dx + \int_a^b p f^2 \, dx.$$

The first integral on the right equals Σa_m^2 because $\int p y_m y_l \, dx = 0$ for $m \neq l$, and $\int p y_m^2 \, dx = 1$. In the second sum on the right, the integral equals a_m, by (4) with $\|y_m\|^2 = 1$. Hence the right side reduces to

$$-\sum_{m=0}^{k} a_m^2 + \int_a^b p f^2 \, dx.$$

---

[21]Actually, piecewise continuous and much more general functions, but a full discussion of completeness would need prerequisites not required in this book. See Ref. [9], Secs. 3.4–3.7, listed in Appendix 1. (There "complete sets" are called "total sets," a more modern term.)

This is nonnegative because in the previous formula the integrand on the left and thus the integral on the left are nonnegative. This proves the important **Bessel's inequality**

(16)
$$\sum_{m=0}^{k} a_m{}^2 \leqq \|f\|^2 = \int_a^b p(x) f(x)^2 \, dx \qquad (k = 1, 2, \cdots).$$

Here we can let $k \to \infty$, because the left sides form a monotone increasing sequence that is bounded by the right side, so that we have convergence by the familiar Theorem 1 in Appendix A3.3. Hence

(17)
$$\sum_{m=0}^{\infty} a_m{}^2 \leqq \|f\|^2.$$

Furthermore, if $y_0, y_1, \cdots$ is complete in a set of functions S, then (13) holds for every f belonging to S. By (15) this implies equality in (16) with $k \to \infty$. Hence in the case of completeness every f in S satisfies the so-called **Parseval's equality**

(18)
$$\sum_{m=0}^{\infty} a_m{}^2 = \|f\|^2 = \int_a^b p(x) f(x)^2 \, dx.$$

As a consequence of (18) we prove that in the case of *completeness* there is no function orthogonal to *every* function of the orthonormal set, with the trivial exception of a function of zero norm:

THEOREM 1 **(Completeness)**

Let $y_0, y_1, \cdots$ be a complete orthonormal set on $a \leqq x \leqq b$ in a set of functions S. Then if a function f belongs to S and is orthogonal to every y_m, it must have norm zero. In particular, if f is continuous, it must be identically zero.

PROOF. From the orthogonality assumption we see that the left side of (18) must be zero. This proves the first statement. If f is continuous, then $\|f\| = 0$ implies $f(x) \equiv 0$, as can be seen directly from (2), with f instead of y_m. ◀

EXAMPLE 4 **Fourier series**

The orthonormal set in Example 1 is complete in the set of continuous functions on $-\pi \leqq x \leqq \pi$. Verify directly that $f(x) \equiv 0$ is the only continuous function orthogonal to all the functions of that set.

Solution. Let f be any continuous function. By the orthogonality (we can omit $\sqrt{2\pi}$ and $\sqrt{\pi}$),

$$\int_{-\pi}^{\pi} 1 \cdot f(x) \, dx = 0, \qquad \int_{-\pi}^{\pi} f(x) \cos mx \, dx = 0, \qquad \int_{-\pi}^{\pi} f(x) \sin mx \, dx = 0.$$

Hence $a_m = 0$ and $b_m = 0$ in (6) for all m, so that (3) reduces to $f(x) \equiv 0$. ◀

This is the end of Chap. 4 on the power series and Frobenius methods, which are indispensable in solving linear differential equations with variable coefficients, some of the most important of which we have discussed and solved. We have also seen that the latter equations are important sources of special functions having orthogonality properties that make them suitable for orthogonal series representations of given functions.

PROBLEM SET 4.8

Fourier–Legendre Series. Showing the details of your calculations, develop:

1. $70x^4 - 84x^2 + 30$ **2.** $20x^3 - 6x^2 - 10x - 2$

3. $1 - x^4$ **4.** $1, x, x^2, x^3, x^4$

Find the first few terms of the Fourier–Legendre series and plot the given function and the first few partial sums. (Show the details of your work.)

5. $\cos \frac{1}{2}\pi x$ **6.** e^x

7. $f(x) = \begin{cases} 0 & \text{if} \quad -1 < x < 0 \\ x & \text{if} \quad\ \ 0 < x < 1 \end{cases}$ **8.** $f(x) = \begin{cases} 1 & \text{if } -\frac{1}{2} < x < \frac{1}{2} \\ 0 & \text{elsewhere} \end{cases}$

9. CAS PROJECT. Fourier–Bessel Series. This refers to Example 3 in the text. We again take $n = 0$ and $R = 1$, so that we get series

$$(19) \qquad f(x) = a_1 J_0(\alpha_{10}x) + a_2 J_0(\alpha_{20}x) + a_3 J_0(\alpha_{30}x) + \cdots$$

with $\alpha_{10}, \cdots, \alpha_{50}$ from Table A1 in Appendix 5 and the further zeros $\alpha_{60}, \cdots, \alpha_{10,0}$ given by 14.931, 18.071, 21.212, 24.352, 27.493, 30.635.

(a) Plot the terms $J_0(\alpha_{10}x), \cdots, J_0(\alpha_{10,0}x)$ for $0 \leq x \leq 1$ on common axes.

(b) Write a program for calculating partial sums of (19) of any number of terms. Find out for what functions $f(x)$ your CAS can evaluate the integrals. Take two such $f(x)$ and comment empirically on the speed of convergence by observing how fast the coefficients decrease.

(c) Take $f(x) = 1$ in (19) and evaluate the integrals for the coefficients analytically by using (24), Sec. 4.5, with $\nu = 1$. Plot the first few partial sums on common axes.

10. TEAM PROJECT. Orthogonality on the Entire Real Axis: Hermite Polynomials.[22] These orthogonal polynomials are defined by

$$He_0 = 1, \qquad He_n(x) = (-1)^n e^{x^2/2} \frac{d^n}{dx^n}(e^{-x^2/2}), \qquad n = 1, 2, \cdots.$$

Remark. As is true for many special functions, the literature contains more than one notation, and one sometimes defines as Hermite polynomials the functions

$$H_0^* = 1, \qquad H_n^*(x) = (-1)^n e^{x^2} \frac{d^n e^{-x^2}}{dx^n}.$$

This differs from our definition, which is preferably used in applications.

(a) Small values of n. Show that

$$He_1(x) = x, \quad He_2(x) = x^2 - 1, \quad He_3(x) = x^3 - 3x, \quad He_4(x) = x^4 - 6x^2 + 3.$$

(b) Generating function. A generating function of the Hermite polynomials is

$$(20) \qquad\qquad e^{tx - t^2/2} = \sum_{n=0}^{\infty} a_n(x)t^n$$

---

[22]CHARLES HERMITE (1822—1901), French mathematician, is known for his work in algebra and number theory. The great HENRI POINCARÉ (1854—1912) was one of his students.

because $He_n(x) = n!a_n(x)$. Prove this. *Hint.* Use the formula for the coefficients of a Maclaurin series and note that $tx - \frac{1}{2}t^2 = \frac{1}{2}x^2 - \frac{1}{2}(x - t)^2$.

(c) Derivative. Differentiating the generating function with respect to x, show that

(21) $$He_n'(x) = nHe_{n-1}(x).$$

(d) Orthogonality on the x-axis needs a weight function that goes to zero sufficiently fast as $x \to \pm\infty$. (Why?) Show that the Hermite polynomials are orthogonal on $-\infty < x < \infty$ with respect to the weight function $p(x) = e^{-x^2/2}$. *Hint.* Use integration by parts and (21).

(e) Differential equations. Show that

(22) $$He_n'(x) = xHe_n(x) - He_{n+1}(x).$$

Using this with $n - 1$ instead of n and (21), show that $y = He_n(x)$ satisfies the differential equation

(23) $$y'' - xy' + ny = 0.$$

Show that $w = e^{-x^2/4}y$ is a solution of **Weber's equation**[23]

(24) $$w'' + (n + \tfrac{1}{2} - \tfrac{1}{4}x^2)w = 0 \qquad\qquad (n = 0, 1, \cdots).$$

11. **WRITING PROJECT. Orthogonality.** Write a short essay $(2-3$ pages) about the most important ideas and facts related to orthogonality and orthogonal series and their applications.

CHAPTER 4 REVIEW

1. What is a power series? Its center? Can it contain fractional powers? Negative powers?
2. What does convergence of a power series mean? Why is it important? How would you test it?
3. What is the power series method? Why do we need it? Can we discover properties of a solution from its power series? Give examples (without looking in the text).
4. What were the most important equations considered by the power series method or the Frobenius method?
5. What is the difference between the two methods in Prob. 4? Why were both of them necessary?
6. Why did we introduce two kinds of Bessel functions J and Y?
7. Can a power series solution reduce to a polynomial? Give three important kinds of solutions for which that can happen.
8. What is the hypergeometric equation? Where does the name come from? Why do its solutions include so many special functions?
9. List the three cases in the Frobenius method, giving examples of your own.
10. Why is orthogonality of functions important? How is it defined?
11. What is a Sturm–Liouville problem? What does it have to do with orthogonality?
12. What do you remember about Legendre polynomials? About Fourier–Legendre series?
13. How many orthogonal sets of Bessel functions did we derive? What role did zeros of Bessel functions play in this context?
14. What is completeness of orthogonal sets? Why is it important?
15. What is the indicial equation? Can it have complex roots?

---

[23]HEINRICH WEBER (1842—1913), German mathematician.

Power Series Method or Frobenius Method. Find a basis of solutions. Try to identify the series obtained as expansions of known functions. (Show the details of your work.)

16. $(x - 2)^2 y'' + 2(x - 2)y' - 6y = 0$

17. $4y'' + y = 0$

18. $y'' + 4xy' + (4x^2 + 2)y = 0$

19. $16(x + 1)^2 y'' + 3y = 0$

20. $xy'' + 3y' + 4x^3 y = 0$

21. $x^2 y'' + xy' + (x^2 - 3)y = 0$

22. $xy'' + (1 - 2x)y' + (x - 1)y = 0$

23. $xy'' + 2y' + 4xy = 0$

24. $(x + 1)x^2 y'' - (2x + 1)xy' + (2x + 1)y = 0$

25. $(x^2 + 2x)y'' + (x^2 - 2)y' - (2x + 2)y = 0$

Orthogonality. Show orthogonality on the given interval. (Show the details.) Determine the corresponding orthonormal set of functions.

26. $1, \cos x, \cos 2x, \cos 3x, \cdots$ $(0 \leq x \leq \pi)$

27. $\sin \omega x, \sin 2\omega x, \sin 3\omega x, \cdots$ $(-\pi/\omega \leq x \leq \pi/\omega)$

28. $1, x, x^2 - \frac{1}{3}, x^3 - \frac{3}{5}x$ $(-1 \leq x \leq 1)$

29. $1, \cos 4nx, \sin 4nx,$ $n = 1, 2, \cdots$ $(0 \leq x \leq \frac{1}{2}\pi)$

Eigenvalues and Eigenfunctions. Find the eigenvalues and eigenfunctions of the following problems.

30. $y'' + \lambda y = 0,$ $y(0) = 0,$ $y(\pi) = 0$

31. $y'' + \lambda y = 0,$ $y(0) = y(2L),$ $y'(0) = y'(2L)$

32. $x^2 y'' + xy' + (\lambda^2 x^2 - 1)y = 0,$ $y(0) = 0,$ $y(1) = 0$

33. $y'' + \lambda y = 0,$ $y(-\pi/2) = 0,$ $y(\pi/2) = 0$

Fourier–Legendre Series. Develop in terms of Legendre polynomials:

34. x^5, x^6

35. $15 - 42x^2 + 35x^4$

36. $5 - 105x^2 + 315x^4 - 231x^6$

Bessel Functions. Solve each of the following in terms of Bessel functions. (Use the indicated transformations. Show the details of your work.)

37. $y'' + 4xy = 0$ $(y = u\sqrt{x}, \; \frac{4}{3}x^{3/2} = z)$

38. $x^2 y'' + xy' + (4x^2 - \nu^2)y = 0$ $(2x = z)$

39. $x^2 y'' - 3xy' + 4(x^4 - 3)y = 0$ $(y = x^2 u, \; x^2 = z)$

40. $x^2 y'' + xy' + (x^2 - \frac{1}{16})y = 0$

SUMMARY OF CHAPTER 4
SERIES SOLUTIONS. SPECIAL FUNCTIONS

The **power series method** is a general method for solving linear differential equations

$$(1) \qquad\qquad y'' + p(x)y' + q(x)y = r(x)$$

with variable $p(x)$, $q(x)$, and $r(x)$; it also applies to higher order equations. It gives solutions in the form of power series; this motivates the name. In this method, one substitutes a power series (with any center x_0, e.g., $x_0 = 0$)

(2) $$y(x) = a_0 + a_1(x - x_0) + a_2(x - x_0)^2 + \cdots$$

and its derivatives $y'(x) = a_1 + 2a_2(x - x_0) + \cdots$ and $y''(x)$ into (1). In this way, one determines the undetermined coefficients a_m in (2), as explained in Sec. 4.1 and other sections. This gives solutions y represented by power series. If $p(x)$, $q(x)$, and $r(x)$ are **analytic** at $x = x_0$ (as defined in Sec. 4.2), then (1) has solutions of this form. The same holds if $\tilde{h}(x)$, $\tilde{p}(x)$, $\tilde{q}(x)$, and $\tilde{r}(x)$ in

$$\tilde{h}(x)y'' + \tilde{p}(x)y' + \tilde{q}(x)y = \tilde{r}(x)$$

are analytic at $x = x_0$, and $\tilde{h}(x_0) \neq 0$ [so that we can divide by $\tilde{h}$ and obtain the standard form (1)].

The **Frobenius method** (Sec. 4.4) extends the power series method to equations

(3) $$y'' + \frac{a(x)}{x - x_0} y' + \frac{b(x)}{(x - x_0)^2} y = 0$$

whose coefficients are **singular** (i.e., not analytic) at $x = x_0$, but are not "too bad," namely, such that $a(x)$ and $b(x)$ are analytic at $x = x_0$. Then (3) has at least one solution of the form

(4) $$y(x) = x^r[a_0 + a_1(x - x_0) + a_2(x - x_0)^2 + \cdots],$$

where r can be any real number or even a complex number and is determined by substituting (4) into (3); this also gives the a_m's. A second independent solution may be of a similar form (with different r and a_m's) or may involve a logarithmic term.

"Special functions" is a common name for higher functions, as opposed to the usual functions of calculus. They arise from (1) or (3) and get a special name and notation if they are important in applications. Of this kind, and particularly useful to the engineer and physicist, are **Legendre's equation** and the Legendre polynomials $P_0(x)$, $P_1(x)$, $P_2(x)$, $\cdots$ (Sec. 4.3), the **hypergeometric equation** and the hypergeometric functions $F(a, b, c; x)$ (Sec. 4.4), and **Bessel's equation** and the Bessel functions J_ν and Y_ν (Secs. 4.5, 4.6). Indeed, second-order linear differential equations are one of the two main sources of such "higher functions." [Other special functions arise from nonelementary integrals, such as those listed in Appendix A3.1. The gamma function (Sec. 4.5) is of that type.]

Modeling involving differential equations in most cases leads to initial value problems (Chap. 2) or boundary value problems. Many of the latter can be written in the form of **Sturm–Liouville problems** (Sec. 4.7). These are **eigenvalue problems** involving a parameter λ, which in applications may be related to frequencies, energies, or other physical quantities. Solutions of these problems, called *eigenfunctions,* have many general properties in common, notably the highly important **orthogonality** (Sec. 4.8). This leads to **eigenfunction expansions** (Sec. 4.8), such as those involving cosine and sine ("Fourier series," to be discussed in great detail in Chap. 10), Legendre polynomials, or Bessel functions.

CHAPTER 5

Laplace Transforms

The Laplace transform method solves differential equations and corresponding initial and boundary value problems. The process of solution consists of three main steps:

1st step. The given "hard" problem is transformed into a "simple" equation **(subsidiary equation).**

2nd step. The subsidiary equation is solved by purely algebraic manipulations.

3rd step. The solution of the subsidiary equation is transformed back to obtain the solution of the given problem.

In this way Laplace transforms reduce the problem of solving a differential equation to an algebraic problem. This process is made easier by tables of functions and their transforms, whose role is similar to that of integral tables in calculus. Such a table is included at the end of the chapter.

This switching from operations of calculus to *algebraic* operations on transforms is called **operational calculus,** a very important area of applied mathematics, and for the engineer, the Laplace transform method is practically the most important operational method. It is particularly useful in problems where the mechanical or electrical driving force has discontinuities, is impulsive or is a complicated periodic function, not merely a sine or cosine. (For another operational method, the Fourier transform, see Sec. 10.10.)

The Laplace transform also has the advantage that it solves problems directly, initial value problems without first determining a general solution, and nonhomogeneous differential equations without first solving the corresponding homogeneous equation.

In this chapter we consider Laplace transforms from a practical point of view and illustrate their use by important engineering problems, many of them related to *ordinary* differential equations.

Partial differential equations can also be treated by Laplace transforms, as we show in Sec. 11.12.

Section 5.8 contains a list of general formulas and Sec. 5.9 a list of transforms F(s) and corresponding functions f(t).

Prerequisite for this chapter: Chap. 2.
Sections that may be omitted in a very short course: 5.4–5.6.
References: Appendix 1, Part A.
Answers to problems: Appendix 2.

5.1 Laplace Transform. Inverse Transform Linearity. Shifting

Let $f(t)$ be a given function that is defined for all $t \geq 0$. We multiply $f(t)$ by e^{-st} and integrate with respect to t from zero to infinity. Then, if the resulting integral exists (that is, has some finite value), it is a function of s, say, $F(s)$:

$$F(s) = \int_0^\infty e^{-st} f(t)\, dt.$$

This function $F(s)$ of the variable s is called the **Laplace transform**[1] of the original function $f(t)$, and will be denoted by $\mathcal{L}(f)$. Thus

(1)
$$F(s) = \mathcal{L}(f) = \int_0^\infty e^{-st} f(t)\, dt.$$

So remember: the original function f depends on t and the new function F (its transform) depends on s.

The operation just described, which yields $F(s)$ from a given $f(t)$, is also called the **Laplace transform.**

Furthermore, the original function $f(t)$ in (1) is called the *inverse transform* or **inverse** of $F(s)$ and will be denoted by $\mathcal{L}^{-1}(F)$; that is, we shall write

$$f(t) = \mathcal{L}^{-1}(F).$$

Notation

*Original functions are denoted by **lowercase letters** and their transforms by the same letters in **capitals**,* so that $F(s)$ denotes the transform of $f(t)$, and $Y(s)$ denotes the transform of $y(t)$, and so on.

EXAMPLE 1   **Laplace transform**

Let $f(t) = 1$ when $t \geq 0$. Find $F(s)$.

Solution.   From (1) we obtain by integration

$$\mathcal{L}(f) = \mathcal{L}(1) = \int_0^\infty e^{-st}\, dt = -\frac{1}{s} e^{-st} \Big|_0^\infty = \frac{1}{s} \qquad\qquad (s > 0).$$

---

[1]PIERRE SIMON MARQUIS DE LAPLACE (1749—1827), great French mathematician, was a professor in Paris. He developed the foundation of potential theory and made important contributions to celestial mechanics, astronomy in general, special functions, and probability theory. Napoléon Bonaparte was his student for a year. For Laplace's interesting political involvements, see Ref. [2], p. 260, listed in Appendix 1.

The powerful practical Laplace transform techniques were developed over a century later by the English electrical engineer OLIVER HEAVISIDE (1850—1925) and were often called "Heaviside calculus."

We shall drop variables when this simplifies formulas without causing confusion. For instance, in (1) we wrote $\mathcal{L}(f)$ instead of $\mathcal{L}(f)(s)$.

Our notation is convenient, but we should say a word about it. The interval of integration in (1) is infinite. Such an integral is called an **improper integral** and, by definition, is evaluated according to the rule

$$\int_0^\infty e^{-st} f(t)\, dt = \lim_{T \to \infty} \int_0^T e^{-st} f(t)\, dt.$$

Hence our convenient notation means

$$\int_0^\infty e^{-st}\, dt = \lim_{T \to \infty}\left[-\frac{1}{s}\, e^{-st} \right]_0^T = \lim_{T \to \infty}\left[-\frac{1}{s}\, e^{-sT} + \frac{1}{s}\, e^0 \right] = \frac{1}{s} \qquad (s > 0).$$

We shall use that notation throughout this chapter.   ◄

EXAMPLE 2 **Laplace transform of the exponential function**

Let $f(t) = e^{at}$ when $t \geqq 0$, where a is a constant. Find $\mathscr{L}(f)$.

Solution. Again by (1),

$$\mathscr{L}(e^{at}) = \int_0^\infty e^{-st} e^{at}\, dt = \frac{1}{a - s}\, e^{-(s-a)t}\, \Big|_0^\infty ;$$

hence, when $s - a > 0$,

$$\mathscr{L}(e^{at}) = \frac{1}{s - a}.$$ ◄

Must we go on in this fashion and obtain the transform of one function after another directly from the definition? The answer is no. And the reason is that the Laplace transform has many general properties that are helpful for that purpose. Above all, the Laplace transform is a "linear operation," just as differentiation and integration are. By this we mean the following.

THEOREM 1 **(Linearity of the Laplace transform)**

The Laplace transform is a linear operation; that is, for any functions $f(t)$ and $g(t)$ whose Laplace transforms exist and any constants a and b,

$$\boxed{\mathscr{L}\{af(t) + bg(t)\} = a\mathscr{L}\{f(t)\} + b\mathscr{L}\{g(t)\}.}$$

PROOF. By the definition,

$$\mathscr{L}\{af(t) + bg(t)\} = \int_0^\infty e^{-st}[af(t) + bg(t)]\, dt$$

$$= a\int_0^\infty e^{-st} f(t)\, dt + b\int_0^\infty e^{-st} g(t)\, dt$$

$$= a\mathscr{L}\{f(t)\} + b\mathscr{L}\{g(t)\}.$$ ◄

EXAMPLE 3 **An application of Theorem 1: Hyperbolic functions**

Let $f(t) = \cosh at = \frac{1}{2}(e^{at} + e^{-at})$. Find $\mathscr{L}(f)$.

Solution. From Theorem 1 and Example 2 we obtain

$$\mathscr{L}(\cosh at) = \frac{1}{2}\,\mathscr{L}(e^{at}) + \frac{1}{2}\,\mathscr{L}(e^{-at}) = \frac{1}{2}\left(\frac{1}{s - a} + \frac{1}{s + a} \right);$$

and by taking the common denominator, when $s > a \ (\geqq 0)$,

$$\mathscr{L}(\cosh at) = \frac{s}{s^2 - a^2} .$$

Similarly, for the transform of the **hyperbolic sine** we obtain

$$\mathscr{L}(\sinh at) = \frac{1}{2} \mathscr{L}(e^{at}) - \frac{1}{2} \mathscr{L}(e^{-at}) = \frac{1}{2} \left(\frac{1}{s - a} - \frac{1}{s + a} \right) = \frac{a}{s^2 - a^2} \qquad (s > a \geqq 0). \qquad \blacktriangleleft$$

EXAMPLE 4 **Cosine and Sine**

In Example 2, if we set $a = i\omega$ with $i = \sqrt{-1}$, we obtain

$$\mathscr{L}(e^{i\omega t}) = \frac{1}{s - i\omega} = \frac{s + i\omega}{(s - i\omega)(s + i\omega)} = \frac{s + i\omega}{s^2 + \omega^2} = \frac{s}{s^2 + \omega^2} + i \frac{\omega}{s^2 + \omega^2} .$$

On the other hand, by Theorem 1 and $e^{i\omega t} = \cos \omega t + i \sin \omega t$ (Sec. 2.3),

$$\mathscr{L}(e^{i\omega t}) = \mathscr{L}(\cos \omega t + i \sin \omega t) = \mathscr{L}(\cos \omega t) + i \, \mathscr{L}(\sin \omega t).$$

Equating the real and imaginary parts of these two equations, we obtain the transforms of cosine and sine,

$$\mathscr{L}(\cos \omega t) = \frac{s}{s^2 + \omega^2} , \qquad \mathscr{L}(\sin \omega t) = \frac{\omega}{s^2 + \omega^2} .$$

We shall use these formulas frequently. You may derive them without going into complex by using the definition (1) and integration by parts. Another derivation follows in the next section. $\qquad \blacktriangleleft$

First Shifting Theorem:
Replacement of *s* by *s* − *a* in the Transform

The Laplace transform has the following very useful general property.

THEOREM 2 **(First Shifting theorem)**

If $f(t)$ has the transform $F(s)$ (where $s > k$), then $e^{at} f(t)$ has the transform $F(s - a)$ (where $s - a > k$). In formulas,

$$\boxed{\mathscr{L}\{e^{at} f(t)\} = F(s - a)}$$

or, if we take the inverse on both sides,

$$\boxed{e^{at} f(t) = \mathscr{L}^{-1}\{F(s - a)\}.}$$

PROOF. We obtain $F(s - a)$ by replacing s by $s - a$ in the integral in (1), so that we get

$$F(s - a) = \int_0^\infty e^{-(s-a)t} f(t) \, dt = \int_0^\infty e^{-st} \left[e^{at} f(t) \right] dt = \mathscr{L}\{e^{at} f(t)\}.$$

If $F(s)$ exists (i.e., is finite) for s greater than some k, then our first integral exists for $s - a > k$. Now take the inverse on both sides to obtain the second formula in the theorem. $\qquad \blacktriangleleft$

EXAMPLE 5   **Damped vibrations**

From Example 4 and the first shifting theorem we immediately obtain the very useful formulas

$$\mathscr{L}(e^{at} \cos \omega t) = \frac{s - a}{(s - a)^2 + \omega^2}$$

$$\mathscr{L}(e^{at} \sin \omega t) = \frac{\omega}{(s - a)^2 + \omega^2}.$$

For negative a these $f(t)$ are damped vibrations.   ◀

A Short List of Important Transforms

Table 5.1 gives a short list of transforms that are basic. From these transforms we can obtain nearly all the other transforms that we shall need in this chapter, through the use of some simple general theorems that we shall consider in the next sections.

A more extensive list of functions and transforms follows in Sec. 5.9. See also Refs. [A4] and [A6] in Appendix 1.

PROOFS.   Formulas 1, 2, and 3 in Table 5.1 are special cases of formula 4. We prove formula 4 by induction. It is true for $n = 0$ because of Example 1 and $0! = 1$. We now make the induction hypothesis that it holds for any positive integer n. From (1) we get by integration by parts

$$\mathscr{L}(t^{n+1}) = \int_0^\infty e^{-st} t^{n+1}\, dt = -\frac{1}{s} e^{-st} t^{n+1} \Big|_0^\infty + \frac{(n + 1)}{s} \int_0^\infty e^{-st} t^n\, dt.$$

Table 5.1
Some Functions $f(t)$ and Their Laplace Transforms $\mathscr{L}(f)$

	$f(t)$	$\mathscr{L}(f)$		$f(t)$	$\mathscr{L}(f)$
1	1	$1/s$	7	$\cos \omega t$	$\dfrac{s}{s^2 + \omega^2}$
2	t	$1/s^2$	8	$\sin \omega t$	$\dfrac{\omega}{s^2 + \omega^2}$
3	t^2	$2!/s^3$	9	$\cosh at$	$\dfrac{s}{s^2 - a^2}$
4	t^n ($n = 0, 1, \cdots$)	$\dfrac{n!}{s^{n+1}}$	10	$\sinh at$	$\dfrac{a}{s^2 - a^2}$
5	t^a (a positive)	$\dfrac{\Gamma(a + 1)}{s^{a+1}}$	11	$e^{at} \cos \omega t$	$\dfrac{s - a}{(s - a)^2 + \omega^2}$
6	e^{at}	$\dfrac{1}{s - a}$	12	$e^{at} \sin \omega t$	$\dfrac{\omega}{(s - a)^2 + \omega^2}$

The integral-free part is zero at $t = 0$ and for $t \to \infty$. The right side equals $(n + 1)\mathcal{L}(t^n)/s$. From this and the induction hypothesis we obtain

$$\mathcal{L}(t^{n+1}) = \frac{n + 1}{s} \mathcal{L}(t^n) = \frac{(n + 1)n!}{s \cdot s^{n+1}} = \frac{(n + 1)!}{s^{n+2}} .$$

This proves formula 4.

$\Gamma(a + 1)$ in formula 5 is the so-called *gamma function* [(15) in Sec. 4.5 or (24) in Appendix A3.1]. We get formula 5 from (1), setting $st = x$:

$$\mathcal{L}(t^a) = \int_0^\infty e^{-st} t^a \, dt = \int_0^\infty e^{-x} \left(\frac{x}{s}\right)^a \frac{dx}{s} = \frac{1}{s^{a+1}} \int_0^\infty e^{-x} x^a \, dx$$

where $s > 0$. The last integral is precisely that defining $\Gamma(a + 1)$, so we have $\Gamma(a + 1)/s^{a+1}$, as claimed.

Note that $\Gamma(n + 1) = n!$ for nonnegative integer n, so that formula 4 also follows from 5.

Formulas 6–12 were derived in Examples 2–5. ◀

EXAMPLE 6   **First shifting theorem**

From Table 5.1 and the first shifting theorem we immediately obtain another useful formula,

$$\mathcal{L}(t^n e^{at}) = \frac{n!}{(s - a)^{n+1}} .$$

For instance, $\mathcal{L}(te^{at}) = 1/(s - a)^2$. ◀

Existence of Laplace Transforms

This is not a big *practical* problem, because in most cases we can check a solution of a differential equation by substitution without too much trouble. Of course, the existence of a transform is not always guaranteed because in (1) we integrate over an infinite interval. For a fixed s the integral in (1) will exist if the whole integrand $e^{-st} f(t)$ goes to zero fast enough as $t \to \infty$, say, at least like an exponential function with a negative exponent. This implies that $f(t)$ itself should not grow faster than, say, e^{kt}, for instance not like e^{t^2}. This motivates the inequality (2) below.

The function $f(t)$ need not be continuous. Sufficient is so-called piecewise continuity (a concept of general mathematical interest). This is of practical importance because discontinuous inputs (driving forces, electromotive forces) are just those for which the Laplace transform method becomes particularly useful.

By definition, a function $f(t)$ is **piecewise continuous** on a finite interval $a \leqq t \leqq b$ if $f(t)$ is defined on that interval and is such that the interval can be subdivided into finitely many intervals in each of which $f(t)$ is continuous and has finite limits as t approaches either endpoint of the interval of subdivision from the interior.

It follows from this definition that finite jumps are the only discontinuities that a piecewise continuous function may have; these are known as *ordinary discontinuities*. Figure 107 on the next page shows an example. Clearly, the class of piecewise continuous functions includes every continuous function.

Fig. 107. Example of a piecewise continuous function $f(t)$
(The dots mark the function values at the jumps.)

THEOREM 3   **(Existence theorem for Laplace transforms)**

Let $f(t)$ be a function that is piecewise continuous on every finite interval in the range $t \geqq 0$ and satisfies

(2) $$\boxed{|f(t)| \leqq Me^{-kt}}$$ *for all $t \geqq 0$*

and for some constants k and M. Then the Laplace transform of $f(t)$ exists for all $s > k$.

PROOF.   Since $f(t)$ is piecewise continuous, $e^{-st}f(t)$ is integrable over any finite interval on the t-axis. From (2), assuming that $s > k$, we obtain

$$|\mathscr{L}(f)| = \left| \int_0^\infty e^{-st}f(t)\, dt \right| \leqq \int_0^\infty |f(t)|e^{-st}\, dt \leqq \int_0^\infty Me^{kt}e^{-st}\, dt = \frac{M}{s-k}$$

where the condition $s > k$ was needed for the existence of the last integral. This completes the proof.   ◀

The conditions in Theorem 3 are sufficient for most applications, and it is easy to find out whether a given function satisfies an inequality of the form (2). For example,

(3) $\cosh t < e^t,$ $t^n < n!\, e^t$ $(n = 0, 1, \cdots)$ for all $t > 0$,

and any function that is bounded in absolute value for all $t \geqq 0$, such as the sine and cosine functions of a real variable, satisfies that condition. An example of a function that does not satisfy a relation of the form (2) is the exponential function e^{t^2}, because, no matter how large we choose M and k in (2),

$$e^{t^2} > Me^{kt}$$ for all $t > t_0$

where t_0 is a sufficiently large number, depending on M and k.

It should be noted that the conditions in Theorem 3 are sufficient rather than necessary. For example, the function $1/\sqrt{t}$ is infinite at $t = 0$, but its transform exists; in fact, from the definition and $\Gamma(\tfrac{1}{2}) = \sqrt{\pi}$ [see (30) in Appendix A3.1] we obtain, setting $st = x$,

$$\mathscr{L}(t^{-1/2}) = \int_0^\infty e^{-st}t^{-1/2}\, dt = \frac{1}{\sqrt{s}} \int_0^\infty e^{-x}x^{-1/2}\, dx = \frac{1}{\sqrt{s}}\,\Gamma(\tfrac{1}{2}) = \sqrt{\frac{\pi}{s}}.$$

Uniqueness. If the Laplace transform of a given function exists, it is uniquely determined. Conversely, it can be shown that if two functions (both defined on the positive real axis) have the same transform, these functions cannot differ over an interval of positive length, although they may differ at various isolated points (see Ref. [A8] in Appendix 1). Since this is of no importance in applications, we may say that the inverse of a given transform is essentially unique. In particular, if two *continuous* functions have the same transform, they are completely identical. Of course, this *is* of practical importance. Why? (Remember the introduction to the chapter.)

PROBLEM SET 5.1

Laplace Transforms. Find the Laplace transforms of the following functions. Show the details of your work. (a, b, c, ω, δ are constant.)

1. $2t + 6$

2. $a + bt + ct^2$

3. $\sin \pi t$

4. $\cos^2 \omega t$

5. e^{a-bt}

6. $e^t \cosh 3t$

7. $\sin(\omega t + \delta)$

8. $\sin 2t \cos 2t$

9.

10.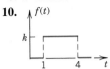

11.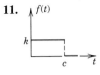

12.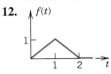

13.

14.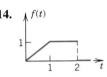

15.

16.

Inverse Laplace Transforms. Given $F(s) = \mathcal{L}(f)$, find $f(t)$. Show the details. (L, n, etc. are constant.)

17. $\dfrac{0.1s + 0.9}{s^2 + 3.24}$

18. $\dfrac{5s}{s^2 - 25}$

19. $\dfrac{-s - 10}{s^2 - s - 2}$

20. $\dfrac{s - 4}{s^2 - 4}$

21. $\dfrac{2.4}{s^4} - \dfrac{228}{s^6}$

22. $\dfrac{60 + 6s^2 + s^4}{s^7}$

23. $\dfrac{s}{L^2 s^2 + n^2 \pi^2}$

24. $\dfrac{1 - 7s}{(s - 3)(s - 1)(s + 2)}$

25. $\displaystyle\sum_{k=1}^{5} \dfrac{a_k}{s + k^2}$

26. $\dfrac{s^4 + 6s - 18}{s^5 - 3s^4}$

27. $\dfrac{1}{(s + \sqrt{2})(s - \sqrt{3})}$

28. $\dfrac{2s^3}{s^4 - 1}$

Applications of the First Shifting Theorem

Find the Laplace transform. (Show the details.)

29. $t^2 e^{-3t}$

30. $e^{-\alpha t} \cos \beta t$

31. $5e^{2t} \sinh 2t$

32. $2e^{-t} \cos^2 \tfrac{1}{2} t$

33. $\sinh t \cos t$

34. $(t + 1)^2 e^t$

Find the inverse transform. (Show the details.)

35. $\dfrac{1}{(s + 1)^2}$

36. $\dfrac{12}{(s - 3)^4}$

37. $\dfrac{3}{s^2 + 6s + 18}$

38. $\dfrac{4}{s^2 - 2s - 3}$

39. $\dfrac{s}{(s + \tfrac{1}{2})^2 + 1}$

40. $\dfrac{2}{s^2 + s + \tfrac{1}{2}}$

41. (Growth) Prove (3).

42. (Inverse transform) Prove that $\mathscr{L}^{-1}$ is linear. *Hint.* Use the fact that $\mathscr{L}$ is linear.

43. (Inverse transform) Rewrite Table 5.1, using $\mathscr{L}^{-1}$ (e.g., $\mathscr{L}^{-1}(1/s^3) = t^2/2$).

44. (Replacement of t by ct) If $\mathscr{L}(f(t)) = F(t)$ and c is any positive constant, show that $\mathscr{L}(f(ct)) = F(s/c)/c$. [*Hint.* Use (1).] Use this to obtain $\mathscr{L}(\cos \omega t)$ from $\mathscr{L}(\cos t)$.

45. (Nonexistence) Give simple examples of functions that have no Laplace transform. Indicate the reason.

5.2 Transforms of Derivatives and Integrals Differential Equations

The Laplace transform is a method of solving differential equations. The crucial idea is that ***the Laplace transform replaces operations of calculus by operations of algebra on transforms.*** Roughly, *differentiation* of $f(t)$ is replaced by *multiplication* of $\mathscr{L}(s)$ by s (see Theorems 1 and 2). *Integration* of $f(t)$ is replaced by *division* of $\mathscr{L}(f)$ by s (see Theorem 3).

THEOREM 1

[Laplace transform of the derivative of $f(t)$]

Suppose that $f(t)$ is continuous for all $t \geqq 0$, satisfies (2), Sec. 5.1, for some k and M, and has a derivative $f'(t)$ that is piecewise continuous on every finite interval in the range $t \geqq 0$. Then the Laplace transform of the derivative $f'(t)$ exists when $s > k$, and

(1)
$$\boxed{\mathscr{L}(f') = s\mathscr{L}(f) - f(0)}$$
$$(s > k).$$

PROOF. We first consider the case when $f'(t)$ is continuous for all $t \geqq 0$. Then, by the definition and by integration by parts,

$$\mathscr{L}(f') = \int_0^\infty e^{-st} f'(t) \, dt = \left[e^{-st} f(t) \right] \Big|_0^\infty + s \int_0^\infty e^{-st} f(t) \, dt.$$

Since f satisfies (2), Sec. 5.1, the integrated portion on the right is zero at the upper limit when $s > k$, and at the lower limit it contributes $-f(0)$. The last integral is $\mathscr{L}(f)$, the existence for $s > k$ being a consequence of Theorem 3 in Sec. 5.1. This proves that the expression on the right exists when $s > k$ and is equal to $-f(0) + s\mathscr{L}(f)$. Consequently, $\mathscr{L}(f')$ exists when $s > k$, and (1) holds.

If the derivative $f'(t)$ is merely piecewise continuous, the proof is quite similar. In this case, the range of integration in the original integral must be broken up into parts such that f' is continuous in each such part. ◄

REMARK This theorem may be extended to piecewise continuous functions $f(t)$, but in place of (1) we then obtain the formula (1*) in Project 10 in the problem set of this section.

By applying (1) to the second derivative $f''(t)$ we obtain

$$\mathscr{L}(f'') = s\mathscr{L}(f') - f'(0)$$

$$= s[s\mathscr{L}(f) - f(0)] - f'(0);$$

that is,

(2) $$\mathcal{L}(f'') = s^2\mathcal{L}(f) - sf(0) - f'(0).$$

Similarly,

(3) $$\mathcal{L}(f''') = s^3\mathcal{L}(f) - s^2 f(0) - sf'(0) - f''(0),$$

etc. By induction we thus obtain the following extension of Theorem 1.

THEOREM 2   **(Laplace transform of the derivative of any order *n*)**

Let $f(t)$ and its derivatives $f'(t)$, $f''(t)$, $\cdots$, $f^{(n-1)}(t)$ be continuous functions for all $t \geqq 0$, satisfying (2), Sec. 5.1, for some k and M, and let the derivative $f^{(n)}(t)$ be piecewise continuous on every finite interval in the range $t \geqq 0$. Then the Laplace transform of $f^{(n)}(t)$ exists when $s > k$ and is given by

(4) $$\mathcal{L}(f^{(n)}) = s^n\mathcal{L}(f) - s^{n-1}f(0) - s^{n-2}f'(0) - \cdots - f^{(n-1)}(0).$$

EXAMPLE 1   Let $f(t) = t^2$. Derive $\mathcal{L}(f)$ from $\mathcal{L}(1)$.

Solution.   Since $f(0) = 0$, $f'(0) = 0$, $f''(t) = 2$, and $\mathcal{L}(2) = 2\mathcal{L}(1) = 2/s$, we obtain from (2)

$$\mathcal{L}(f'') = \mathcal{L}(2) = \frac{2}{s} = s^2\mathcal{L}(f), \qquad \text{hence} \qquad \mathcal{L}(t^2) = \frac{2}{s^3},$$

in agreement with Table 5.1. The example is typical: it illustrates that in general there are several ways of obtaining the transforms of given functions.   ◀

EXAMPLE 2   Derive the Laplace transform of cos ωt.

Solution.   Let $f(t) = \cos \omega t$. Then $f''(t) = -\omega^2 \cos \omega t = -\omega^2 f(t)$. Also $f(0) = 1$, $f'(0) = 0$. Now we take the transform, $\mathcal{L}(f'') = -\omega^2\mathcal{L}(f)$. From this and (2),

$$-\omega^2\mathcal{L}(f) = \mathcal{L}(f'') = s^2\mathcal{L}(f) - s, \qquad \text{hence} \qquad \mathcal{L}(f) = \mathcal{L}(\cos \omega t) = \frac{s}{s^2 + \omega^2}.$$

Can you obtain $\mathcal{L}(\sin \omega t)$ by the present method?   ◀

EXAMPLE 3   Let $f(t) = \sin^2 t$. Find $\mathcal{L}(f)$.

Solution.   We have $f(0) = 0$, $f'(t) = 2\sin t \cos t = \sin 2t$, and (1) gives

$$\mathcal{L}(\sin 2t) = \frac{2}{s^2 + 4} = s\mathcal{L}(f) \qquad \text{or} \qquad \mathcal{L}(\sin^2 t) = \frac{2}{s(s^2 + 4)}.$$ ◀

EXAMPLE 4   Let $f(t) = t \sin \omega t$. Find $\mathcal{L}(f)$.

Solution.   We have $f(0) = 0$ and

$$f'(t) = \sin \omega t + \omega t \cos \omega t, \qquad f'(0) = 0,$$
$$f''(t) = 2\omega \cos \omega t - \omega^2 t \sin \omega t$$
$$= 2\omega \cos \omega t - \omega^2 f(t),$$

so that by (2),

$$\mathcal{L}(f'') = 2\omega\mathcal{L}(\cos \omega t) - \omega^2\mathcal{L}(f) = s^2\mathcal{L}(f).$$

Using the formula for the Laplace transform of cos ωt, we thus obtain

$$(s^2 + \omega^2)\mathscr{L}(f) = 2\omega \mathscr{L}(\cos \omega t) = \frac{2\omega s}{s^2 + \omega^2}.$$

Hence the result is

$$\mathscr{L}(t \sin \omega t) = \frac{2\omega s}{(s^2 + \omega^2)^2}.$$ ◀

Differential Equations, Initial Value Problems

We shall now discuss how the Laplace transform method solves differential equations. We begin with an initial value problem

(5) $\qquad y'' + ay' + by = r(t), \qquad y(0) = K_0, \qquad y'(0) = K_1$

with constant a and b. Here $r(t)$ is the **input** (driving force) applied to the mechanical system and $y(t)$ is the **output** (response of the system). In Laplace's method we do three steps:

1st Step. We transform (5) by means of (1) and (2), writing $Y = \mathscr{L}(y)$ and $R = \mathscr{L}(r)$. This gives

$$\left[s^2 Y - sy(0) - y'(0)\right] + a[sY - y(0)] + bY = R(s).$$

This is called the **subsidiary equation.** Collecting Y-terms, we have

$$(s^2 + as + b)Y = (s + a)y(0) + y'(0) + R(s).$$

2nd Step. We solve the subsidiary equation *algebraically* for Y. Division by $s^2 + as + b$ and use of the so-called **transfer function**[2]

(6) $\qquad\boxed{Q(s) = \dfrac{1}{s^2 + as + b}}$

gives the solution

(7) $\qquad\boxed{Y(s) = \left[(s + a)y(0) + y'(0)\right]Q(s) + R(s)Q(s).}$

If $y(0) = y'(0) = 0$, this is simply $Y = RQ$; thus Q is the quotient

$$\boxed{Q = \frac{Y}{R} = \frac{\mathscr{L}(\text{output})}{\mathscr{L}(\text{input})}}$$

and this explains the name of Q. Note that Q depends only on a and b, but neither on $r(t)$ nor on the initial conditions.

3rd Step. We reduce (7) (usually by partial fractions, as in calculus) to a sum of terms whose inverses can be found from the table, so that the solution $y(t) = \mathscr{L}^{-1}(Y)$ of (5) is obtained.

---

[2]Often denoted by H, but we need H much more frequently for other purposes.

EXAMPLE 5   **Initial value problem: Explanation of the basic steps**

Solve

$$y'' - y = t, \qquad y(0) = 1, \qquad y'(0) = 1.$$

Solution. 1st Step.   From (2) and Table 5.1 we get the subsidiary equation

$$s^2 Y - sy(0) - y'(0) - Y = 1/s^2, \qquad \text{thus} \qquad (s^2 - 1)Y = s + 1 + 1/s^2.$$

2nd Step.   The transfer function is $Q = 1/(s^2 - 1)$, and (7) becomes

$$Y = (s + 1)Q + \frac{1}{s^2} Q = \frac{s + 1}{s^2 - 1} + \frac{1}{s^2(s^2 - 1)}$$

$$= \frac{1}{s - 1} + \left(\frac{1}{s^2 - 1} - \frac{1}{s^2} \right).$$

3rd Step.   From this expression for Y and Table 5.1 we obtain the solution

$$y(t) = \mathcal{L}^{-1}(Y) = \mathcal{L}^{-1} \left\{ \frac{1}{s - 1} \right\} + \mathcal{L}^{-1} \left\{ \frac{1}{s^2 - 1} \right\} - \mathcal{L}^{-1} \left\{ \frac{1}{s^2} \right\} = e^t + \sinh t - t.$$

The diagram in Fig. 108 summarizes our approach.   ◀

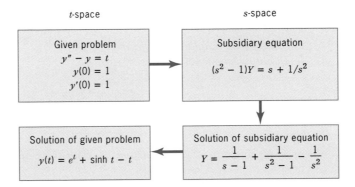

Fig. 108.   Laplace transform method

EXAMPLE 6   **Comparison with the usual method**

Solve the initial value problem

$$y'' + 2y' + y = e^{-t}, \qquad y(0) = -1, \qquad y'(0) = 1.$$

Solution.   From the formulas (1) and (2) for the transforms of the derivatives and from the initial conditions we obtain the subsidiary equation

$$(s^2 Y + s - 1) + 2(sY + 1) + Y = \frac{1}{s + 1}.$$

Collecting Y-terms gives

$$(s^2 + 2s + 1)Y = (s + 1)^2 Y = -s - 1 + \frac{1}{s + 1}.$$

Dividing by $(s + 1)^2$, we solve algebraically for Y and simplify:

$$Y = \frac{-s - 1}{(s + 1)^2} + \frac{1}{(s + 1)^3} = -\frac{1}{s + 1} + \frac{1}{(s + 1)^3}.$$

The first term on the right has the inverse $-e^{-t}$. For the other term on the right the first shifting theorem gives the inverse $\frac{1}{2}t^2 e^{-t}$. Hence the solution of our initial value problem is

$$y = (\tfrac{1}{2}t^2 - 1)e^{-t}.$$

This agrees with the result in Example 3 of Sec. 2.9 and was obtained with much less work—take a look and compare. ◀

In practice, instead of justifying the use of formulas and theorems in this method, one simply checks at the end whether $y(t)$ satisfies the given equation and initial conditions. Example 6 illustrates the **advantages of the method** compared to that in Chap. 2:

1. No determination of a general solution of the homogeneous equation.
2. No determination of values for arbitrary constants in a general solution.

Laplace Transform of the Integral of a Function

Differentiation and integration are inverse processes. Accordingly, since (roughly speaking) differentiation of a function corresponds to the multiplication of its transform by s, we expect integration of a function to correspond to division of its transform by s, because division is the inverse operation of multiplication:

THEOREM 3   **[Integration of $f(t)$]**

Let $F(s)$ be the Laplace transform of $f(t)$. If $f(t)$ is piecewise continuous and satisfies an inequality of the form (2), Sec. 5.1, then

(8)
$$\mathcal{L}\left\{ \int_0^t f(\tau)\, d\tau \right\} = \frac{1}{s} F(s) \qquad (s > 0, \, s > k)$$

or, if we take the inverse transform on both sides of (8),

(9)
$$\int_0^t f(\tau)\, d\tau = \mathcal{L}^{-1}\left\{ \frac{1}{s} F(s) \right\}.$$

PROOF.   Suppose that $f(t)$ is piecewise continuous and satisfies (2), Sec. 5.1, for some k and M. Clearly, if (2) holds for some negative k, it also holds for positive k, and we may assume that k is positive. Then the integral

$$g(t) = \int_0^t f(\tau)\, d\tau$$

is continuous, and by using (2) in Sec. 5.1 we obtain for any positive t

$$|g(t)| \leq \int_0^t |f(\tau)|\, d\tau \leq M \int_0^t e^{k\tau}\, d\tau = \frac{M}{k}(e^{kt} - 1) \leq \frac{M}{k} e^{kt} \qquad (k > 0).$$

This shows that $g(t)$ also satisfies an inequality of the form (2), Sec. 5.1. Also, $g'(t) = f(t)$, except for points at which $f(t)$ is discontinuous. Hence $g'(t)$ is piecewise continuous on each finite interval, and, by Thereom 1,

$$\mathcal{L}\{f(t)\} = \mathcal{L}\{g'(t)\} = s\mathcal{L}\{g(t)\} - g(0) \qquad (s > k).$$

Here, clearly, $g(0) = 0$, so that $\mathcal{L}(f) = s\mathcal{L}(g)$. This implies (8), and (9) follows as indicated in the theorem.   ◄

EXAMPLE 7   **An application of Theorem 3**

Let $\mathcal{L}(f) = \dfrac{1}{s(s^2 + \omega^2)}$. Find $f(t)$.

Solution.   From Table 5.1 in Sec. 5.1 we have

$$\mathcal{L}^{-1}\left(\frac{1}{s^2 + \omega^2}\right) = \frac{1}{\omega}\sin \omega t.$$

From this and Theorem 3 we obtain the answer

$$\mathcal{L}^{-1}\left\{\frac{1}{s}\left(\frac{1}{s^2 + \omega^2}\right)\right\} = \frac{1}{\omega}\int_0^t \sin \omega \tau \, d\tau = \frac{1}{\omega^2}(1 - \cos \omega t).$$

This proves formula 19 in the table in Sec. 5.9.   ◄

EXAMPLE 8   **Another application of Theorem 3**

Derive formula 20 in the table in Sec. 5.9.

Solution.   Applying Theorem 3 to the answer in Example 7, we obtain the desired formula

$$\mathcal{L}^{-1}\left\{\frac{1}{s^2}\left(\frac{1}{s^2 + \omega^2}\right)\right\} = \frac{1}{\omega^2}\int_0^t (1 - \cos \omega \tau) \, d\tau = \frac{1}{\omega^2}\left(t - \frac{\sin \omega t}{\omega}\right).$$

This proves formula 20 in the table in Sec. 5.9.   ◄

EXAMPLE 9   **Shifted data problems**

This is a short name for initial value problems with initial conditions referring to some later instant $t = t_0$ instead of $t = 0$. In this case, the conditions $y(0)$ and $y'(0)$ occurring in the Laplace transform approach cannot be used immediately. We may now proceed in two ways.

First Solution Method.   Find the general solution by Laplace transform and from it find the solution of the problem as in the classical method.

Second Solution Method.   Set $t = \tilde{t} + t_0$, so that $t = t_0$ gives $\tilde{t} = 0$ and the Laplace transform becomes applicable throughout. We show this method for the problem

$$y'' + y = 2t, \qquad y(\tfrac{1}{4}\pi) = \tfrac{1}{2}\pi, \qquad y'(\tfrac{1}{4}\pi) = 2 - \sqrt{2}.$$

Solution.   We have $t_0 = \tfrac{1}{4}\pi$ and we set $t = \tilde{t} + \tfrac{1}{4}\pi$. Then the problem is

$$\tilde{y}'' + \tilde{y} = 2(\tilde{t} + \tfrac{1}{4}\pi), \qquad \tilde{y}(0) = \tfrac{1}{2}\pi, \qquad \tilde{y}'(0) = 2 - \sqrt{2}$$

where $\tilde{y}(\tilde{t}) = y(t)$.

1st Step. Setting up the subsidiary equation.   From (2) and Table 5.1 in Sec. 5.1 we obtain

$$s^2\tilde{Y} - s\tilde{y}(0) - \tilde{y}'(0) + \tilde{Y} = \frac{2}{s^2} + \frac{\pi/2}{s}$$

where $\tilde{Y}$ is the transform of $\tilde{y}$.

2nd Step. Solution of the subsidiary equation. Solving algebraically, we have

$$\tilde{Y} = \frac{2}{(s^2 + 1)s^2} + \frac{\pi/2}{s(s^2 + 1)} + \tilde{y}(0)\frac{s}{s^2 + 1} + \tilde{y}'(0)\frac{1}{s^2 + 1}.$$

The first two terms on the right are those in Examples 7 and 8, so that

$$\tilde{y} = 2(\tilde{t} - \sin \tilde{t}) + \tfrac{1}{2}\pi(1 - \cos \tilde{t}) + \tfrac{1}{2}\pi \cos \tilde{t} + (2 - \sqrt{2})\sin \tilde{t}.$$

Substituting $\tilde{t} = t - \tfrac{1}{4}\pi$, canceling terms, and using $\cos \tfrac{1}{4}\pi = \sin \tfrac{1}{4}\pi = 1/\sqrt{2}$, we obtain the solution

$$y = 2t - \sin t + \cos t.$$ ◀

PROBLEM SET 5.2

Initial Value Problems. Solve the following initial value problems by the Laplace transform. (Show the details of your work.)

1. $y' + 3y = 10 \sin t$, $y(0) = 0$
2. $y' - 5y = 1.5e^{-4t}$, $y(0) = 1$
3. $y' + 0.2y = 0.01t$, $y(0) = -0.25$
4. $y'' - y' - 2y = 0$, $y(0) = 8$, $y'(0) = 7$
5. $y'' + ay' - 2a^2y = 0$, $y(0) = 6$, $y'(0) = 0$
6. $y'' + y = 2 \cos t$, $y(0) = 3$, $y'(0) = 4$
7. $y'' - 4y' + 3y = 6t - 8$, $y(0) = 0$, $y'(0) = 0$
8. $y'' + 0.04y = 0.02t^2$, $y(0) = -25$, $y'(0) = 0$
9. $y'' + 2y' - 3y = 6e^{-2t}$, $y(0) = 2$, $y'(0) = -14$

10. **PROJECT. Summary of Sec. 5.2. (a)** Compare the Laplace transform with the classical method of solving differential equations, explaining the advantages and illustrating the comparison with examples of your own.

    (b) Theorems 1 and 2 play a role different from that of Theorem 3. Explain the difference.

    (c) Extension of Theorem 1. Show that if $f(t)$ is continuous, except for an ordinary discontinuity (finite jump) at $t = a$ (> 0), the other conditions remaining the same as in Theorem 1, then (see Fig. 109)

    (1*) $\mathcal{L}(f') = s\mathcal{L}(f) - f(0) - [f(a + 0) - f(a - 0)]e^{-as}.$

    (d) Using (1*), find the Laplace transform of $f(t) = t$ if $0 < t < 1$, $f(t) = 1$ if $1 < t < 2$, $f(t) = 0$ otherwise.

11. **Derivation by different methods** is possible for various formulas and is typical of Laplace transforms. Find $\mathcal{L}(\cos^2 t)$ (a) by using the result in Example 3, (b) by the method used in that example, (c) by expressing $\cos^2 t$ in terms of $\cos 2t$.

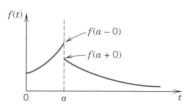

Fig. 109. Formula (1*)

12. PROJECT. Extension of Example 4. Extend the method of differentiation in Example 4 to obtain

(a) $\mathcal{L}(t \cos \omega t) = \dfrac{s^2 - \omega^2}{(s^2 + \omega^2)^2}$.

From this and Example 4 derive

(b) $\mathcal{L}^{-1}\left(\dfrac{1}{(s^2 + \omega^2)^2}\right) = \dfrac{1}{2\omega^3}(\sin \omega t - \omega t \cos \omega t)$

(c) $\mathcal{L}^{-1}\left(\dfrac{s}{(s^2 + \omega^2)^2}\right) = \dfrac{1}{2\omega} t \sin \omega t$

(d) $\mathcal{L}^{-1}\left(\dfrac{s^2}{(s^2 + \omega^2)^2}\right) = \dfrac{1}{2\omega}(\sin \omega t + \omega t \cos \omega t)$.

Obtain similar formulas for hyperbolic functions, namely,

(e) $\mathcal{L}(t \cosh at) = \dfrac{s^2 + a^2}{(s^2 - a^2)^2}$

(f) $\mathcal{L}(t \sinh at) = \dfrac{2as}{(s^2 - a^2)^2}$.

New Inverse Transforms by Integration (Theorem 3). Given $\mathcal{L}(f)$, find $f(t)$. (Show the details of your work.)

13. $\dfrac{1}{s^2 + 4s}$

14. $\dfrac{4}{s^3 - 2s^2}$

15. $\dfrac{1}{s(s^2 + \omega^2)}$

16. $\dfrac{1}{s^5 + s^3}$

17. $\dfrac{1}{s^3 - s}$

18. $\dfrac{1}{s^2}\left(\dfrac{s - 1}{s + 1}\right)$

19. $\dfrac{9}{s^2}\left(\dfrac{s + 1}{s^2 + 9}\right)$

20. $\dfrac{\pi^5}{s^4(s^2 + \pi^2)}$

5.3 Unit Step Function
Second Shifting Theorem
Dirac's Delta Function

What state have we reached and what is our next goal? We know that differentiation of $f(t)$ roughly corresponds to multiplication of the transform $\mathcal{L}(f)$ by s (Theorems 1 and 2, Sec. 5.2) and that this property is essential in solving differential equations. Example 5 in Sec. 5.2 explains the three steps of this Laplace technique. But that equation can be solved easily by the usual methods, and our goal is to derive further properties of the Laplace transform in order to show the real power of this method in applications. We accomplish this essentially by defining two important functions, the unit step function and Dirac's delta function.

Unit Step Function $u(t - a)$

By definition, $u(t - a)$ is 0 for $t < a$, has a jump of size 1 at $t = a$ (where we can leave it undefined) and is 1 for $t > a$:

(1)
$$u(t - a) = \begin{cases} 0 & \text{if } t < a \\ 1 & \text{if } t > a \end{cases} \qquad (a \geqq 0).$$

Figure 110 shows the special case $u(t)$, which has the jump at zero, and Fig. 111 the general case $u(t - a)$ for an arbitrary positive a. The unit step function is also called the **Heaviside function.**[3]

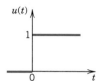

Fig. 110.   Unit step function $u(t)$

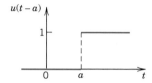

Fig. 111.   Unit step function $u(t - a)$

The unit step function is a typical "engineering function" made to measure for engineering applications, which often involve functions (mechanical or electrical driving forces) that are either "off" or "on." Multiplying functions $f(t)$ with $u(t - a)$, we can produce all sorts of effects. The simple basic idea is illustrated in Figs. 112 and 113. In Fig. 112 the given function is shown in (a). In (b) it is switched off between $t = 0$ and $t = 2$ (because $u(t - 2) = 0$ when $t < 2$) and is switched on beginning at $t = 2$. In (c) it is shifted to the right by 2 seconds, so that it begins 2 seconds later in the same fashion as before. Figure 113 shows the effect of many unit step functions, three of them in (A) and infinitely many in (B); this is the effect of a rectifier that clips off the negative half-waves of a sinusoidal voltage. Before going on, make sure that you fully understand these figures, in particular the difference between parts (b) and (c) of Figure 112. Figure 112(c) will be applied next.

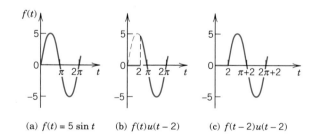

(a) $f(t) = 5 \sin t$ (b) $f(t)u(t - 2)$ (c) $f(t - 2)u(t - 2)$

Fig. 112.   Effects of the unit step function: (a) Given function.
(b) Switching off and on. (c) Shift.

---

[3]See footnote 1 in Sec. 5.1.

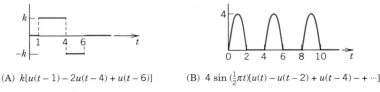

(A) $k[u(t-1) - 2u(t-4) + u(t-6)]$ (B) $4 \sin (\frac{1}{2}\pi t)[u(t) - u(t-2) + u(t-4) - + \cdots]$

Fig. 113. Use of many unit step functions

t-Shifting: Replacing *t* by *t* − *a* in *f(t)*

If $f(t)$ has the transform $F(s)$, then $e^{at} f(t)$ has the transform $F(s - a)$. This is *s*-shifting (see Sec. 5.1). Now comes *t*-**shifting:**

THEOREM 1   **(Second shifting theorem; *t*-shifting)**

*If $f(t)$ has the transform $F(s)$, then the "**shifted function**"*

$$(2) \qquad \tilde{f}(t) = f(t - a)u(t - a) = \begin{cases} 0 & \text{if } t < a \\ f(t - a) & \text{if } t > a \end{cases}$$

has the transform $e^{-as}F(s)$. That is,

$$(3) \qquad \boxed{\mathcal{L}\{f(t - a)u(t - a)\} = e^{-as}F(s).}$$

Or, if we take the inverse on both sides, we can write

$$(3^*) \qquad \boxed{f(t - a)u(t - a) = \mathcal{L}^{-1}\{e^{-as}F(s)\}.}$$

Practically speaking, if we know $F(s)$, we can obtain the transform of (2) by multiplying $F(s)$ by e^{-as}. In Fig. 112, the transform of $5 \sin t$ is $F(s) = 5/(s^2 + 1)$, hence $5 \sin (t - 2) u(t - 2)$ shown in Fig. 112(c) has the transform $e^{-2s}F(s) = 5e^{-2s}/(s^2 + 1)$.

PROOF.   We prove Theorem 1. From the definition of the Laplace transform we have

$$e^{-as}F(s) = e^{-as} \int_0^\infty e^{-s\tau}f(\tau)\, d\tau = \int_0^\infty e^{-s(\tau+a)}f(\tau)\, d\tau.$$

Substituting $\tau + a = t$ in the integral, we obtain (note the lower limit of integration!)

$$e^{-as}F(s) = \int_a^\infty e^{-st}f(t - a)\, dt.$$

We can write this as an integral from 0 to ∞ (as required for a Laplace transform!) if we make sure that the integrand is zero for all t from 0 to a. This is easy. We just multiply the integrand by the unit step function $u(t - a)$—do you now see why and how it comes in? This gives (3) and completes the proof:

$$e^{-as}F(s) = \int_0^\infty e^{-st}f(t-a)u(t-a)\,dt = \mathcal{L}\{f(t-a)u(t-a)\}. \qquad \blacktriangleleft$$

It is fair to say that we are already approaching the stage where we can attack problems for which the Laplace transform method is preferable to the usual method, as the examples in the next section will illustrate. In this connection we need the transform of the unit step function $u(t-a)$,

(4)
$$\mathcal{L}\{u(t-a)\} = \frac{e^{-as}}{s} \qquad\qquad (s > 0).$$

This formula follows directly from the definition because

$$\mathcal{L}\{u(t-a)\} = \int_0^\infty e^{-st}u(t-a)\,dt$$

$$= \int_0^a e^{-st}0\,dt + \int_a^\infty e^{-st}1\,dt = -\frac{1}{s}e^{-st}\Big|_a^\infty.$$

Let us consider two further examples. More applications follow in the problem set and in the next sections.

EXAMPLE 1   **Application of Theorem 1. Use of unit step functions**

Find the transform of the function (Fig. 114)

$$f(t) = \begin{cases} 2 & \text{if } 0 < t < \pi \\ 0 & \text{if } \pi < t < 2\pi \\ \sin t & \text{if } \quad t > 2\pi. \end{cases}$$

Solution. 1st Step.   We write $f(t)$ in terms of unit step functions. For $0 < t < \pi$, we take $2u(t)$. For $t > \pi$ we want 0, so we must subtract the step function $2u(t-\pi)$ with step at π. Then we have $2u(t) - 2u(t-\pi) = 0$ when $t > \pi$. This is fine until we reach 2π where we want $\sin t$ to come in; so we add $u(t-2\pi)\sin t$. Together,

$$f(t) = 2u(t) - 2u(t-\pi) + u(t-2\pi)\sin t.$$

2nd Step.   The last term equals $u(t-2\pi)\sin(t-2\pi)$ because of the periodicity, so that (4), (3), and Table 5.1 (Sec. 5.1) give

$$\mathcal{L}(f) = \frac{2}{s} - \frac{2e^{-\pi s}}{s} + \frac{e^{-2\pi s}}{s^2+1}. \qquad \blacktriangleleft$$

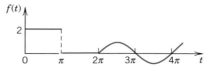

Fig. 114.   Example 1

EXAMPLE 2   Application of Theorem 1. Inverse transform

Find the inverse Laplace transform $f(t)$ of

$$F(s) = \frac{2}{s^2} - \frac{2e^{-2s}}{s^2} - \frac{4e^{-2s}}{s} + \frac{se^{-\pi s}}{s^2 + 1}.$$

Solution.   Without the exponential functions the four terms of $F(s)$ would have the inverses $2t$, $-2t$, -4, $\cos t$ (see Table 5.1 in Sec. 5.1). Hence by Theorem 1,

$$f(t) = 2t - 2(t - 2)u(t - 2) - 4u(t - 2) + \cos(t - \pi)\,u(t - \pi)$$

$$= 2t - 2tu(t - 2) - \cos t\,u(t - \pi)$$

because $4u(t - 2)$ and $-4u(t - 2)$ cancel each other. This gives $2t$ if $0 < t < 2$, $2t - 2t = 0$ if $2 < t < \pi$, and $2t - 2t - \cos t = -\cos t$ if $t > \pi$.   ◀

EXAMPLE 3   Response of an *RC*-circuit to a single square wave

Find the current $i(t)$ in the circuit in Fig. 115 if a single square wave with voltage V_0 is applied. The circuit is assumed to be quiescent before the square wave is applied.

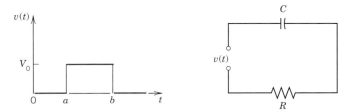

Fig. 115.   Example 3

Solution.   The equation of the circuit is (see Sec. 1.7)

$$Ri(t) + \frac{q(t)}{C} = Ri(t) + \frac{1}{C}\int_0^t i(\tau)\,d\tau = v(t)$$

where $v(t)$ can be represented in terms of two unit step functions:

$$v(t) = V_0[u(t - a) - u(t - b)].$$

Using Theorem 3 in Sec. 5.2 and formula (1) in this section, we obtain the subsidiary equation

$$RI(s) + \frac{I(s)}{sC} = \frac{V_0}{s}\left[e^{-as} - e^{-bs}\right].$$

Solving this equation algebraically for $I(s)$, we get

$$I(s) = F(s)(e^{-as} - e^{-bs}) \qquad \text{where} \qquad F(s) = \frac{V_0/R}{s + 1/(RC)}.$$

From Table 5.1 in Sec. 5.1 we have

$$\mathscr{L}^{-1}(F) = \frac{V_0}{R}e^{-t/(RC)}.$$

Hence Theorem 1 yields the solution (Fig. 116)

$$i(t) = \mathcal{L}^{-1}(I) = \mathcal{L}^{-1}\{e^{-as}F(s)\} - \mathcal{L}^{-1}\{e^{-bs}F(s)\} = \frac{V_0}{R}\left[e^{-(t-a)/(RC)}u(t-a) - e^{-(t-b)/(RC)}u(t-b)\right];$$

that is, $i = 0$ if $t < a$, and

$$i(t) = \begin{cases} K_1 e^{-t/(RC)} & \text{if } a < t < b \\ (K_1 - K_2)e^{-t/(RC)} & \text{if } a > b \end{cases}$$

where $K_1 = V_0 e^{a/(RC)}/R$ and $K_2 = V_0 e^{b/(RC)}/R$.   ◀

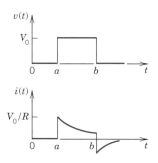

Fig. 116.   Voltage and current in Example 3

Short Impulses. Dirac's Delta Function

Phenomena of an impulsive nature, such as the action of very large forces (or voltages) over very short intervals of time, are of great practical interest, since they arise in various applications. This situation occurs, for instance, when a tennis ball is hit, a system is given a blow by a hammer, an airplane makes a "hard" landing, a ship is hit by a high single wave, and so on. Our present goal is to show how to solve problems involving short impulses by Laplace transforms.

In mechanics, the **impulse** of a force $f(t)$ over a time interval, say, $a \leqq t \leqq a + k$, is defined to be the integral of $f(t)$ from a to $a + k$. The analog for an electric circuit is the integral of the electromotive force applied to the circuit, integrated from a to $a + k$. Of particular practical interest is the case of a very short k (and its limit $k \to 0$), that is, the impulse of a force acting only for an instant. To handle the case, we consider the function

(5)
$$f_k(t - a) = \begin{cases} 1/k & \text{if } a \leqq t \leqq a + k \\ 0 & \text{otherwise} \end{cases}$$
(Fig. 117).

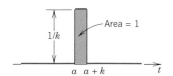

Fig. 117.   The function $f_k(t - a)$ in (5)

Its impulse I_k is 1, since the integral evidently gives the area of the rectangle in Fig. 117:

(6)
$$I_k = \int_0^\infty f_k(t - a)\, dt = \int_a^{a+k} \frac{1}{k}\, dt = 1.$$

We can represent $f_k(t - a)$ in terms of two unit step functions, namely,

$$f_k(t - a) = \frac{1}{k}\left[u(t - a) - u(t - (a + k))\right].$$

From (4) we obtain the Laplace transform

(7)
$$\mathcal{L}\{f_k(t - a)\} = \frac{1}{ks}\left[e^{-as} - e^{-(a+k)s}\right] = e^{-as}\, \frac{1 - e^{-ks}}{ks}.$$

The limit of $f_k(t - a)$ as $k \to 0$ $(k > 0)$ is denoted by $\delta(t - a)$, that is,

$$\delta(t - a) = \lim_{k \to 0} f_k(t - a).$$

$\delta(t - a)$ is called the **Dirac delta function**[4] (sometimes the **unit impulse function**). The quotient in (7) has the limit 1 as $k \to 0$, as follows by l'Hôpital's rule (differentiate the numerator and also the denominator with respect to k). Hence the right side of (7) has the limit e^{-as}. This suggests defining the Laplace transform of $\delta(t - a)$ by this limit in (7), that is,

(8)
$$\boxed{\mathcal{L}\{\delta(t - a)\} = e^{-as}.}$$

We note that $\delta(t - a)$ is not a function in the ordinary sense as used in calculus, but a so-called *"generalized function,"*[5] because (5) and (6) with $k \to 0$ imply

$$\delta(t - a) = \begin{cases} \infty & \text{if } t = a \\ 0 & \text{otherwise} \end{cases} \quad \text{and} \quad \int_0^\infty \delta(t - a)\, dt = 1,$$

but an ordinary function that is everywhere 0 except at a single point must have the integral 0. Nevertheless, in impulse problems it is convenient to operate on $\delta(t - a)$ as though it were an ordinary function.

EXAMPLE 4   **Response of a damped vibrating system to a single square wave and to a unit impulse**

Determine the response of the damped mass–spring system (see Sec. 2.11) governed by

$$y'' + 3y' + 2y = r(t), \qquad y(0) = 0, \qquad y'(0) = 0$$

---

[4]PAUL DIRAC (1902—1984), English physicist, was awarded the Nobel Prize [jointly with ERWIN SCHRÖDINGER (1887—1961)] in 1933 for his work in quantum mechanics.

[5]Or *"distribution."* A systematic theory of generalized functions was created in 1936 by the Russian mathematician, SERGEI L'VOVICH SOBOLEV (1908—1989) and in 1945, under wider aspects, by the French mathematician, LAURENT SCHWARTZ (born 1915).

where $r(t)$ is (A) the square wave

$$r(t) = u(t - 1) - u(t - 2) \qquad \text{(see Fig. 118)}$$

and (B) the unit impulse at time $t = 1$,

$$r(t) = \delta(t - 1).$$

Solution (A). From (1) and (2) in Sec. 5.2 and (1) and (4) in this section we obtain the subsidiary equation

$$s^2 Y + 3sY + 2Y = \frac{1}{s}(e^{-s} - e^{-2s}).$$

Algebraically solving for Y, we have

$$Y(s) = F(s)(e^{-s} - e^{-2s}) \qquad \text{where} \qquad F(s) = \frac{1}{s(s + 1)(s + 2)}.$$

In terms of partial fractions,

$$F(s) = \frac{1/2}{s} - \frac{1}{s + 1} + \frac{1/2}{s + 2}.$$

Hence by Table 5.1 in Sec. 5.1,

$$f(t) = \mathcal{L}^{-1}(F) = \tfrac{1}{2} - e^{-t} + \tfrac{1}{2}e^{-2t}.$$

Therefore, by Theorem 1 we have

$$y = \mathcal{L}^{-1}(F(s)e^{-s} - F(s)e^{-2s})$$
$$= f(t - 1)u(t - 1) - f(t - 2)u(t - 2)$$
$$= \begin{cases} 0 & (0 < t < 1) \\ \tfrac{1}{2} - e^{-(t-1)} + \tfrac{1}{2}e^{-2(t-1)} & (1 < t < 2) \\ -e^{-(t-1)} + e^{-(t-2)} + \tfrac{1}{2}e^{-2(t-1)} - \tfrac{1}{2}e^{-2(t-2)} & (t > 2) \end{cases}$$

(with $\tfrac{1}{2} - \tfrac{1}{2} = 0$ in the third line). This solution is shown in Fig. 118.

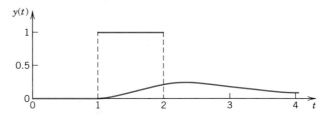

Fig. 118. Square wave and response in Example 4

Solution (B). With $r(t)$ as given in (B) we now obtain the subsidiary equation [see (8)]

$$s^2 Y + 3sY + 2Y = e^{-s}.$$

Solving for Y, we have

$$Y = F(s)e^{-s} \qquad \text{where} \qquad F(s) = \frac{1}{(s + 1)(s + 2)} = \frac{1}{s + 1} - \frac{1}{s + 2}.$$

Taking the inverse transform of F, we obtain

$$f(t) = \mathcal{L}^{-1}(F) = e^{-t} - e^{-2t}.$$

Hence by Theorem 1 we have

$$y(t) = \mathcal{L}^{-1}\{e^{-s}F(s)\} = f(t-1)u(t-1) = \begin{cases} 0 & (0 \leq t < 1) \\ e^{-(t-1)} - e^{-2(t-1)} & (t > 1). \end{cases}$$

Figure 119 shows the solution. Can you explain physically why this curve is steeper near $t = 1$ than the curve in Fig. 118?   ◀

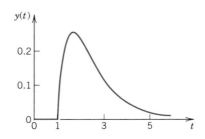

Fig. 119.   Response to a hammer blow at $t = 1$ in Example 4

PROBLEM SET 5.3

1. **WRITING PROJECT. Shifting Theorems.** Explain and compare the different roles of the two shifting theorems. Use your own formulations and examples; do not copy phrases from the text.

Applications of the Second Shifting Theorem
Laplace transform. Sketch the following functions and find their Laplace transforms. (Show the details of your work.)

2. $tu(t-1)$
3. $(t-1)u(t-1)$
4. $(t-1)^2 u(t-1)$
5. $t^2 u(t-1)$
6. $e^{-2t}u(t-3)$
7. $4u(t-\pi)\cos t$

Laplace Transform. Sketch the given function, which is assumed to be zero outside the given interval. Find its Laplace transform. (Show the details of your work.)

8. t^2 $(0 < t < 1)$
9. $\sin \omega t$ $(0 < t < \pi/\omega)$
10. $1 - e^{-t}$ $(0 < t < 2)$
11. e^t $(0 < t < 1)$
12. $\sin t$ $(2\pi < t < 4\pi)$
13. $10 \cos \pi t$ $(1 < t < 2)$

Inverse Transform. Find and sketch the inverse Laplace transform. (Show the details of your work.)

14. $4(e^{-2s} - 2e^{-5s})/s$
15. e^{-3s}/s^3
16. $e^{-3s}/(s-1)^3$
17. $3(1 - e^{-\pi s})/(s^2 + 9)$
18. $e^{-2\pi s}/(s^2 + 2s + 2)$
19. $se^{-2s}/(s^2 + \pi^2)$

Initial Value Problems. Some with Discontinuous or Impulse Inputs. Using the Laplace transform, solve the following problems. (Show the details.)

20. $4y'' - 4y' + 37y = 0$, $y(0) = 3$, $y'(0) = 10.5$
21. $y'' + 6y' + 8y = e^{-3t} - e^{-5t}$, $y(0) = 0$, $y'(0) = 0$
22. $y'' + 3y' + 2y = 4t$ if $0 < t < 1$ and 8 if $t > 1$; $y(0) = 0$, $y'(0) = 0$
23. $y'' + 9y = 8 \sin t$ if $0 < t < \pi$ and 0 if $t > \pi$; $y(0) = 0$, $y'(0) = 4$

24. $y'' - 5y' + 6y = 4e^t$ if $0 < t < 2$ and 0 if $t > 2$; $\quad y(0) = 1, \quad y'(0) = -2$

25. $y'' + y' - 2y = 3 \sin t - \cos t$ if $0 < t < 2\pi$ and $3 \sin 2t - \cos 2t$ if $t > 2\pi$;
$\quad y(0) = 1, \qquad y'(0) = 0$

26. $y'' + 16y = 4\delta(t - \pi), \qquad\qquad y(0) = 2, \qquad y'(0) = 0$

27. $y'' + y = \delta(t - \pi) - \delta(t - 2\pi), \qquad\qquad y(0) = 0, \qquad y'(0) = 1$

28. $y'' + 4y' + 5y = \delta(t - 1), \qquad\qquad y(0) = 0, \qquad y'(0) = 3$

29. $y'' + 2y' - 3y = 8e^{-t} + \delta(t - \frac{1}{2}), \qquad\qquad y(0) = 3, \qquad y'(0) = -5$

30. $y'' + 5y' + 6y = u(t - 1) + \delta(t - 2), \qquad\qquad y(0) = 0, \qquad y'(0) = 1$

Models of Electric Circuits

RL-Circuit. Using the Laplace transform (and showing the details of your work), find the current $i(t)$ in the circuit in Fig. 120, assuming $i(0) = 0$ and

31. $v(t) = t$ if $0 < t < 4\pi$ and 0 if $t > 4\pi$

32. $v(t) = \sin t$ if $0 < t < 2\pi$ and 0 otherwise

LC-Circuit. Using the Laplace transform (and showing the details of your work), find the current $i(t)$ in the circuit in Fig. 121, assuming $L = 1$ henry, $C = 1$ farad, zero initial current and charge on the capacitor, and

33. $v(t) = t$ if $0 < t < 1$ and $v(t) = 1$ if $t > 1$

34. $v(t) = 1$ if $0 < t < a$ and 0 otherwise

35. $v(t) = 1 - e^{-t}$ if $0 < t < \pi$ and 0 otherwise

RC-Circuit. Using the Laplace transform (and showing your work), find the current $i(t)$ in the circuit in Fig. 122 with $R = 100$ ohms, $C = 0.1$ farad, and $v(t)$ volts as follows. Assume that current and charge are zero at $t = 0$.

36. $v(t) = 10000$ if $1 < t < 1.01$ and 0 otherwise

37. $v(t) = 100$ if $1 < t < 2$ and 0 otherwise. Compare with Prob. 36.

38. $v(t) = 0$ if $t < 3$ and $50(t - 3)$ if $t > 3$

39. $v(t) = 0$ if $t < 2$ and e^{-t} if $t > 2$

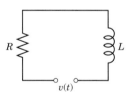 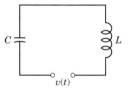 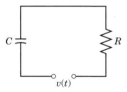

Fig. 120. Problems 31, 32   **Fig. 121.** Problems 33–35   **Fig. 122.** Problems 36–39

40. CAS PROJECT. Limit of Square Wave. Effects of Impulse.

    (a) In Example 4, take a square wave of area 1 from 1 to $1 + k$. Plot the responses for a sequence of values of k approaching zero, illustrating that for smaller and smaller k those curves approach the curve shown in Fig. 119. *Hint.* If your CAS gives no solution for the differential equation involving k, take specific k's from the beginning.

    (b) Experiment on the response of the equation in Example 4 (or of another equation of your choice) to an impulse $\delta(t - a)$ for various systematically chosen a (> 0); choose initial conditions $y(0) \neq 0$, $y'(0) = 0$. Also consider the solution if no impulse is applied. Is there a dependence of the response on a? On b if you choose $b\delta(t - a)$? Would $-\delta(t - \tilde{a})$ with $\tilde{a} > a$ annihilate the effect of $\delta(t - a)$? Can you think of other questions that one could consider experimentally by inspecting plots?

5.4 Differentiation and Integration of Transforms

The variety of methods for obtaining transforms or inverse transforms and their application in solving differential equations is surprisingly large. They include direct integration (Sec. 5.1), the use of linearity (Sec. 5.1), shifting (Secs. 5.1, 5.3), and differentiation or integration of original functions $f(t)$ (Sec. 5.2). But this is not all: in this section we consider differentiation and integration of *transforms* $F(s)$ and find out the corresponding operations for original functions $f(t)$.

Differentiation of Transforms

It can be shown that if $f(t)$ satisfies the conditions of the existence theorem in Sec. 5.1, then the derivative $F'(s)$ of its transform

$$F(s) = \mathcal{L}(f) = \int_0^\infty e^{-st} f(t)\, dt$$

with respect to s can be obtained by differentiating under the integral sign with respect to s (proof in Ref. [5] listed in Appendix 1); thus

$$F'(s) = -\int_0^\infty e^{-st} [t f(t)]\, dt.$$

Consequently, if $\mathcal{L}(f) = F(s)$, then

(1) $$\boxed{\mathcal{L}\{t f(t)\} = -F'(s);}$$

differentiation of the transform of a function corresponds to the multiplication of the function by $-t$. Equivalently,

(1*) $$\boxed{\mathcal{L}^{-1}\{F'(s)\} = -t f(t).}$$

This property enables us to get new transforms from given ones, as we show next.

EXAMPLE 1 **Differentiation of transforms**

We shall derive the following three formulas (formulas 21–23 in the table in Sec. 5.9):

	$\mathcal{L}(f)$	$f(t)$
(2)	$\dfrac{1}{(s^2 + \beta^2)^2}$	$\dfrac{1}{2\beta^3}(\sin \beta t - \beta t \cos \beta t)$
(3)	$\dfrac{s}{(s^2 + \beta^2)^2}$	$\dfrac{t}{2\beta} \sin \beta t$
(4)	$\dfrac{s^2}{(s^2 + \beta^2)^2}$	$\dfrac{1}{2\beta}(\sin \beta t + \beta t \cos \beta t)$

Solution. From (1) and formula 8 (with $\omega = \beta$) in Table 5.1, Sec. 5.1, we obtain by differentiation (don't forget to apply the chain rule!)

$$\mathcal{L}(t \sin \beta t) = \frac{2\beta s}{(s^2 + \beta^2)^2} .$$

By dividing by 2β we obtain (3).

Formulas (2) and (4) are obtained as follows. From (1) and formula 7 (with $\omega = \beta$) in Table 5.1 we find

(5)
$$\mathcal{L}(t \cos \beta t) = -\frac{(s^2 + \beta^2) - 2s^2}{(s^2 + \beta^2)^2} = \frac{s^2 - \beta^2}{(s^2 + \beta^2)^2} .$$

From this and formula 8 (with $\omega = \beta$) in Table 5.1 we have

$$\mathcal{L}\left(t \cos \beta t \pm \frac{1}{\beta} \sin \beta t\right) = \frac{s^2 - \beta^2}{(s^2 + \beta^2)^2} \pm \frac{1}{s^2 + \beta^2} .$$

On the right we now take the common denominator. Then we see that for the plus sign the numerator becomes $s^2 - \beta^2 + s^2 + \beta^2 = 2s^2$, so that (4) follows by division by 2. Similarly, for the minus sign the numerator takes the form $s^2 - \beta^2 - s^2 - \beta^2 = -2\beta^2$, and we obtain (2). ◀

Integration of Transforms

Similarly, if $f(t)$ satisfies the conditions of the existence theorem in Sec. 5.1 and the limit of $f(t)/t$, as t approaches 0 from the right, exists, then

(6)
$$\mathcal{L}\left\{\frac{f(t)}{t}\right\} = \int_s^\infty F(\tilde{s}) \, d\tilde{s} \qquad (s > k);$$

in this manner, *integration of the transform of a function $f(t)$ corresponds to the division of $f(t)$ by t.* Equivalently,

(6*)
$$\mathcal{L}^{-1}\left\{\int_s^\infty F(\tilde{s}) \, d\tilde{s}\right\} = \frac{f(t)}{t} .$$

In fact, from the definition it follows that

$$\int_s^\infty F(\tilde{s}) \, d\tilde{s} = \int_s^\infty \left[\int_0^\infty e^{-\tilde{s}t} f(t) \, dt\right] d\tilde{s},$$

and it can be shown (see Ref. [5] in Appendix 1) that under the above assumptions we may reverse the order of integration, that is,

$$\int_s^\infty F(\tilde{s}) \, d\tilde{s} = \int_0^\infty \left[\int_s^\infty e^{-\tilde{s}t} f(t) \, d\tilde{s}\right] dt = \int_0^\infty f(t) \left[\int_s^\infty e^{-\tilde{s}t} \, d\tilde{s}\right] dt.$$

Integration of $e^{-\tilde{s}t}$ with respect to $\tilde{s}$ gives $e^{-\tilde{s}t}/(-t)$. Here the integral over $\tilde{s}$ on the right equals e^{-st}/t. Therefore,

$$\int_s^\infty F(\tilde{s}) \, d\tilde{s} = \int_0^\infty e^{-st} \frac{f(t)}{t} \, dt = \mathcal{L}\left\{\frac{f(t)}{t}\right\} \qquad (s > k). ◀$$

EXAMPLE 2　**Integration of transforms**

Find the inverse transform of the function $\ln\left(1 + \dfrac{\omega^2}{s^2}\right)$.

Solution.　By differentiation,

$$-\frac{d}{ds}\ln\left(1 + \frac{\omega^2}{s^2}\right) = -\frac{1}{1 + \dfrac{\omega^2}{s^2}}\cdot(-2)\frac{\omega^2}{s^3} = \frac{2\omega^2}{s(s^2 + \omega^2)} = \frac{2}{s} - 2\frac{s}{s^2 + \omega^2},$$

where the last equality can be readily verified by direct calculation. This is our present $F(s)$. It is the derivative of the given function (times -1), so that the latter is the integral of $F(s)$ from s to ∞. From Table 5.1 in Sec. 5.1 we obtain

$$f(t) = \mathcal{L}^{-1}(F) = \mathcal{L}^{-1}\left\{\frac{2}{s} - 2\frac{s}{s^2 + \omega^2}\right\} = 2 - 2\cos\omega t.$$

This function satisfies the conditions under which (6) holds. Therefore,

$$\mathcal{L}^{-1}\left\{\ln\left(1 + \frac{\omega^2}{s^2}\right)\right\} = \mathcal{L}^{-1}\left\{\int_s^\infty F(\tilde{s})\,d\tilde{s}\right\} = \frac{f(t)}{t}.$$

Our result is

$$\mathcal{L}^{-1}\left\{\ln\left(1 + \frac{\omega^2}{s^2}\right)\right\} = \frac{2}{t}(1 - \cos\omega t).$$

This proves formula 42 in the table in Sec. 5.9. ◀

EXAMPLE 3　**Integration of transforms**

Reasoning as in Example 2, we obtain (see formula 43 in the table in Sec. 5.9)

$$\mathcal{L}^{-1}\left\{\ln\left(1 - \frac{a^2}{s^2}\right)\right\} = \frac{2}{t}(1 - \cosh at). \quad ◀$$

Differential Equations with Variable Coefficients

From (1) with $f = y'\ dy/dt$ and $\mathcal{L}(y') = sY - y(0)$ (see Sec. 5.2) and subsequent product differentiation we obtain

(7)　　　　　　　　$\mathcal{L}(ty') = -\dfrac{d}{ds}[sY - y(0)] = -Y - s\dfrac{dY}{ds},$

Similarly, by (1) with $f = y''$ and (2), Sec. 5.2, we find

(8)　　　$\mathcal{L}(ty'') = -\dfrac{d}{ds}[s^2Y - sy(0) - y'(0)] = -2sY - s^2\dfrac{dY}{ds} + y(0).$

Hence if a differential equation has coefficients such as $at + b$, we get a first-order differential equation for Y, which is sometimes simpler than the given equation. But if the latter has coefficients $at^2 + bt + c$, we get, by two applications of (1), a second-order differential equation for Y, and this shows that the Laplace transform method works well only for very special equations with *variable* coefficients. We illustrate it for an important equation in the following example.

EXAMPLE 4   **Laguerre's differential equation, Laguerre polynomials**

Laguerre's differential equation is

(9) $$ty'' + (1 - t)y' + ny = 0.$$

We determine a solution of (9) with $n = 0, 1, 2, \cdots$. From (7)–(9) we get

$$\left[-2sY - s^2 \frac{dY}{ds} + y(0)\right] + sY - y(0) - \left(-Y - s \frac{dY}{ds}\right) + nY = 0.$$

Simplification gives

$$(s - s^2) \frac{dY}{ds} + (n + 1 - s)Y = 0.$$

Separating variables, using partial fractions, integrating (with the constant of integration taken zero), and taking exponentials, we get

(10*) $$\frac{dY}{Y} = -\frac{n + 1 - s}{s - s^2} \, ds = \left(\frac{n}{s - 1} - \frac{n + 1}{s}\right) ds \qquad \text{and} \qquad Y = \frac{(s - 1)^n}{s^{n+1}}.$$

We write $l_n = \mathcal{L}^{-1}(Y)$ and show that

(10) $$l_0 = 1, \qquad l_n(t) = \frac{e^t}{n!} \frac{d^n}{dt^n} (t^n e^{-t}), \qquad\qquad n = 1, 2, \cdots.$$

These are polynomials because the exponential terms cancel if we perform the indicated differentiations. They are called **Laguerre polynomials** and are usually denoted by L_n (see Problem Set 4.7; but we conform to our convention of reserving capital letters for transforms). We prove (10). By Table 5.1 and the first shifting theorem,

$$\mathcal{L}(t^n e^{-t}) = \frac{n!}{(s + 1)^{n+1}}.$$

By (4) in Sec. 5.2, since the derivatives are zero at 0,

$$\mathcal{L}\{(t^n e^{-t})^{(n)}\} = \frac{n! s^n}{(s + 1)^{n+1}}.$$

Now make another shift and divide by $n!$ to get [see (10) and (10*)]

$$\mathcal{L}(l_n) = \frac{(s - 1)^n}{s^{n+1}} = Y. \qquad\qquad \blacktriangleleft$$

PROBLEM SET 5.4

Transforms by Differentiation. Find the Laplace transform. (Show the details of your work.)

1. te^t **2.** $3t \sinh 4t$ **3.** $t^2 \cosh \pi t$ **4.** $te^{-t} \cos t$
5. $t \cos \omega t$ **6.** $t^2 \sin 2t$ **7.** $te^{-t} \sin t$ **8.** $t^2 \cos \omega t$

Inverse Transforms by Differentiation or Integration. Using (6) or (1), find the inverse transform. (Show your work.)

9. $\dfrac{1}{(s - 3)^3}$ **10.** $\dfrac{s}{(s^2 - 9)^2}$ **11.** $\dfrac{s^2 - \pi^2}{(s^2 + \pi^2)^2}$ **12.** $\dfrac{2s + 6}{(s^2 + 6s + 10)^2}$

13. $\ln \dfrac{s^2 + 1}{(s - 1)^2}$ **14.** $\ln \dfrac{s + a}{s + b}$ **15.** $\dfrac{s}{(s^2 + 4)^2}$ **16.** $\operatorname{arc\,cot} \dfrac{s}{\pi}$

17. (Shifting) Can you solve Probs. 1 and 3 by the first shifting theorem?

18. (Differentiation) Find $\mathcal{L}(t^n e^{at})$ by repeated application of (1), choosing $f(t) = e^{at}$.

19. WRITING PROJECT. Differentiation and Integration of Functions and Transforms. Make a short draft on these four operations from memory. Then compare your notes with the text and write an essay of 2–3 pages on these operations and their significance in applications.

 20. CAS PROJECT. Laguerre Polynomials. (a) Write a CAS program for finding $l_n(t)$ in explicit form from (10). Apply it to calculate $l_0, \cdots, l_{10}$. Verify that $l_0, \cdots, l_{10}$ satisfy Laguerre's differential equation (9).

(b) Show that

$$l_n(t) = \sum_{m=0}^{n} \frac{(-1)^m}{m!} \binom{n}{m} t^m$$

and calculate $l_0, \cdots, l_{10}$ from this formula.

(c) Calculate $l_0, \cdots, l_{10}$ recursively from $l_0 = 1$, $l_1 = 1 - t$ by

$$(n + 1)l_{n+1} = (2n + 1 - t)l_n - nl_{n-1}.$$

CAUTION! Sometimes the functions $\tilde{l}_n = n! l_n$ are also called *Laguerre polynomials.* Their recursion is different! (Such differences in normalization are typical of several special functions— and a possible source of errors.)

5.5 Convolution. Integral Equations

Another important general property of the Laplace transform has to do with products of transforms. It often happens that we are given two transforms $F(s)$ and $G(s)$ whose inverses $f(t)$ and $g(t)$ we know, and we would like to calculate the inverse of the product $H(s) = F(s)G(s)$ from those known inverses $f(t)$ and $g(t)$. This inverse $h(t)$ is written $(f * g)(t)$, which is a standard notation, and is called the **convolution** of f and g. How can we find h from f and g? This is stated in the following theorem. Since the situation and task just described arise quite often in applications, this theorem is of considerable practical importance.

THEOREM 1   **(Convolution theorem)**

Let $f(t)$ and $g(t)$ satisfy the hypothesis of the existence theorem (Sec. 5.1). Then the product of their transforms $F(s) = \mathcal{L}(f)$ and $G(s) = \mathcal{L}(g)$ is the transform $H(s) = \mathcal{L}(h)$ of the **convolution** *$h(t)$ of $f(t)$ and $g(t)$, which is denoted by $(f * g)(t)$ and defined by*

(1)
$$h(t) = (f * g)(t) = \int_0^t f(\tau)g(t - \tau)\,d\tau.$$

(The proof follows after Example 3.)

EXAMPLE 1   **Convolution**

Using convolution, find the inverse $h(t)$ of

$$H(s) = \frac{1}{(s^2 + 1)^2} = \frac{1}{s^2 + 1} \cdot \frac{1}{s^2 + 1}.$$

Solution. We know that each factor on the right has the inverse sin t. Hence by the convolution theorem and by (11) in Appendix A3.1, we get

$$h(t) = \mathcal{L}^{-1}(H) = \sin t * \sin t$$

$$= \int_0^t \sin \tau \sin (t - \tau) \, d\tau$$

$$= \frac{1}{2} \int_0^t -\cos t \, d\tau + \frac{1}{2} \int_0^t \cos (2\tau - t) \, d\tau$$

$$= -\tfrac{1}{2} t \cos t + \tfrac{1}{2} \sin t. \qquad \blacktriangleleft$$

EXAMPLE 2   **Convolution**

$1/s^2$ has the inverse t and $1/s$ has the inverse 1, and the convolution theorem confirms that $1/s^3 = (1/s^2)(1/s)$ has the inverse

$$t * 1 = \int_0^t \tau \cdot 1 \, d\tau = \frac{t^2}{2} \, . \qquad \blacktriangleleft$$

EXAMPLE 3   **Convolution**

Let $H(s) = 1/[s^2(s - a)]$. Find $h(t)$.

Solution. From Table 5.1 in Sec. 5.1 we know that

$$\mathcal{L}^{-1} \left\{ \frac{1}{s^2} \right\} = t, \qquad \mathcal{L}^{-1} \left\{ \frac{1}{s - a} \right\} = e^{at}.$$

Using the convolution theorem and integrating by parts, we get the answer

$$h(t) = t * e^{at} = \int_0^t \tau e^{a(t - \tau)} \, d\tau = e^{at} \int_0^t \tau e^{-a\tau} \, d\tau$$

$$= \frac{1}{a^2} (e^{at} - at - 1). \qquad \blacktriangleleft$$

PROOF.   Theorem 1 can be proved as follows. By the definition of $G(s)$ and the second shifting theorem, for each fixed τ $(\tau \geqq 0)$ we have

$$e^{-s\tau} G(s) = \mathcal{L}\{g(t - \tau) u(t - \tau)\}$$

$$= \int_0^\infty e^{-st} g(t - \tau) u(t - \tau) \, dt$$

$$= \int_\tau^\infty e^{-st} g(t - \tau) \, dt$$

where $s > k$. From this and the definition of $F(s)$ we obtain

$$F(s)G(s) = \int_0^\infty e^{-s\tau} f(\tau) G(s) \, d\tau = \int_0^\infty f(\tau) \int_\tau^\infty e^{-st} g(t - \tau) \, dt \, d\tau$$

where $s > k$. Here we integrate over t from τ to ∞ and then over τ from 0 to ∞; this corresponds to the colored wedge-shaped region extending to infinity in the $t\tau$-plane shown in Fig. 123 on p. 281. Our assumptions on f and g are such that the order of integration can be reversed. (A proof requiring the knowledge of uniform convergence is included in Ref. [A2] listed in Appendix 1.) We then integrate first over τ from 0 to t (see Fig. 123)

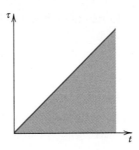

Fig. 123.   Region of integration in
the $t\tau$-plane in the proof of Theorem 1

and then over t from 0 to ∞; thus

$$F(s)G(s) = \int_0^\infty e^{-st} \int_0^t f(\tau)g(t - \tau)\,d\tau\,dt$$

$$= \int_0^\infty e^{-st}h(t)\,dt = \mathcal{L}(h)$$

where h is given by (1). This completes the proof.   ◄

Using the definition, the reader may show that the convolution $f * g$ has the properties

$$f * g = g * f \qquad\qquad \text{(commutative law)}$$

$$f * (g_1 + g_2) = f * g_1 + f * g_2 \qquad\qquad \text{(distributive law)}$$

$$(f * g) * v = f * (g * v) \qquad\qquad \text{(associative law)}$$

$$f * 0 = 0 * f = 0,$$

just as for numbers. But $f * 1 \neq f$ in general, as Example 2 shows. Another unusual property is that $(f * f)(t) \geq 0$ may not hold, as we can see from Example 1.

Very useful applications of convolution occur in a natural way in the solution of differential equations, as we shall now discuss.

Differential Equations

From Sec. 5.2 we recall that the differential equation

$$(2) \qquad\qquad y'' + ay' + by = r(t)$$

has the subsidiary equation $(s^2 + as + b)Y = (s + a)y(0) + y'(0) + \mathcal{L}(r)$. The solution of the latter is

$$(3) \qquad\qquad Y(s) = [(s + a)y(0) + y'(0)]Q(s) + R(s)Q(s)$$

with $R(s) = \mathcal{L}(r)$ and $Q(s) = 1/(s^2 + as + b)$ the transfer function. Hence for the solution $y(t)$ of (2) satisfying $y(0) = y'(0) = 0$ we have $Y = RQ$ in (3) and obtain from the convolution theorem the integral representation

(4)
$$y(t) = \int_0^t q(t - \tau)r(\tau)\, d\tau, \qquad\qquad q(t) = \mathcal{L}^{-1}(Q).$$

EXAMPLE 4   **Response of a damped system to a single square wave**

We reconsider the model in Sec. 5.3, Example 4 (Fig. 118):

$$y'' + 3y' + 2y = r(t), \qquad r(t) = 1 \text{ if } 1 < t < 2 \text{ and } 0 \text{ otherwise}, \qquad y(0) = y'(0) = 0.$$

We solve it by the convolution technique in order to see how it works for *inputs that act for some time only.*

Solution.   We have

$$Q(s) = \frac{1}{s^2 + 3s + 2} = \frac{1}{s + 1} - \frac{1}{s + 2}, \qquad \text{hence} \qquad q(t) = e^{-t} - e^{-2t}.$$

Hence the integral (4) is (except for the limits of integration)

$$\int \left[e^{-(t-\tau)} - e^{-2(t-\tau)}\right] d\tau = e^{-(t-\tau)} - \tfrac{1}{2}e^{-2(t-\tau)}.$$

Now comes an important point. $r(t) = 1$ if $1 < t < 2$. Hence if $t < 1$, the integral is 0. If $1 < t < 2$, we have to integrate from 1 to t. This gives

$$e^{-0} - e^{-(t-1)} - \tfrac{1}{2}(e^{-0} - e^{-2(t-1)}) = \tfrac{1}{2} - e^{-(t-1)} + \tfrac{1}{2}e^{-2(t-1)}.$$

If $t > 2$, we have to integrate from 1 to 2 only (not to t). This gives

$$e^{-(t-2)} - e^{-(t-1)} - \tfrac{1}{2}\left[e^{-2(t-2)} - e^{-2(t-1)}\right].$$

All three results agree with those in Example 4(A) in Sec. 5.3.   ◀

Integral Equations

Convolution also helps in solving certain **integral equations,** that is, equations in which the unknown function $y(t)$ appears under the integral (and perhaps also outside of it). This concerns only very special ones (those whose integral is of the form of a convolution), so that it suffices to consider a typical example and a handful of problems, but we do this because integral equations are practically important and often difficult to solve.

EXAMPLE 5   **Integral equation**

Solve the integral equation

$$y(t) = t + \int_0^t y(\tau) \sin (t - \tau)\, d\tau.$$

Solution. 1st Step. Equation in terms of convolution.   We see that the given equation can be written

$$y = t + y * \sin t.$$

2nd Step. Application of the convolution theorem.   We write $Y = \mathcal{L}(y)$. By the convolution theorem,

$$Y(s) = \frac{1}{s^2} + Y(s)\, \frac{1}{s^2 + 1}.$$

Solving for $Y(s)$, we obtain

$$Y(s) = \frac{s^2 + 1}{s^4} = \frac{1}{s^2} + \frac{1}{s^4} .$$

3rd Step. Taking the inverse transform. This gives the solution

$$y(t) = t + \tfrac{1}{6}t^3.$$

The reader may check this by substitution and evaluating the integral by repeated integration by parts (which will need patience).   ◀

PROBLEM SET 5.5

Calculation of Convolutions by Integrating (1). Do this for the following. (Show the details of your work.)

1. $1 * 1$

2. $1 * \sin \omega t$

3. $e^t * e^{-t}$

4. $\cos \omega t * \cos \omega t$

5. $\sin \omega t * \cos \omega t$

6. $e^{at} * e^{bt}$ $(a \neq b)$

7. $t * e^t$

8. $u(t - 1) * t^2$

9. $u(t - 3) * e^{-2t}$

Inverse Transforms by Convolution. Find $h(t)$ by the convolution theorem from the given $H(s) = \mathcal{L}(h)$. (Show the details of your work.)

10. $\dfrac{6}{s(s + 3)}$

11. $\dfrac{1}{s^2(s - 1)}$

12. $\dfrac{1}{(s - a)^2}$

13. $\dfrac{1}{s(s^2 + 4)}$

14. $\dfrac{s^2}{(s^2 + \omega^2)^2}$

15. $\dfrac{s}{(s^2 + \pi^2)^2}$

16. $\dfrac{e^{-as}}{s(s - 2)}$

17. $\dfrac{\omega}{s^2(s^2 + \omega^2)}$

18. $\dfrac{1}{(s + 3)(s - 2)}$

Initial Value Problems. Applying convolution, find and sketch or plot the solution. First guess what the solution might look like. (Show the details of your work.)

19. $y'' + y = 3 \cos 2t;$ $y(0) = 0,$ $y'(0) = 0$

20. $y'' + y = t;$ $y(0) = 0,$ $y'(0) = 0$

21. $y'' + 4y = r(t), r(t) = 1$ if $0 < t < 1$ and 0 if $t > 1;$ $y(0) = 1,$ $y'(0) = 0$

22. $y'' + 3y' + 2y = r(t), r(t)$ as in Prob. 21; $y(0) = 0,$ $y'(0) = 1$

23. $y'' + 2y' + 2y = r(t), r(t) = 5u(t - 2\pi) \sin t;$ $y(0) = 1,$ $y'(0) = 0$

24. $y'' - 5y' + 6y = r(t), r(t) = 4e^t$ if $0 < t < 2$ and 0 if $t > 2;$ $y(0) = 1,$ $y'(0) = -2$

25. $y'' + y = r(t), r(t) = t$ if $1 < t < 2$ and 0 otherwise; $y(0) = 0,$ $y'(0) = 0$

26. $y'' + 3y' + 2y = r(t), r(t) = 4t$ if $0 < t < 1$ and 8 if $t > 1;$ $y(0) = 0;$ $y'(0) = 0$

Integral Equations. Using Laplace transforms, solve the integral equations in Probs. 27–33. (Show the details of your work.)

27. $y(t) = 1 + \displaystyle\int_0^t y(\tau) \, d\tau$

28. $y = 2t - 4 \displaystyle\int_0^t y(\tau)(t - \tau) \, d\tau$

29. $y(t) = 1 - \displaystyle\int_0^t (t - \tau)y(\tau) \, d\tau$

30. $y(t) = \sin 2t + \displaystyle\int_0^t y(\tau) \sin 2(t - \tau) \, d\tau$

31. $y(t) = te^t - 2e^t \int_0^t e^{-\tau} y(\tau) \, d\tau$ **32.** $y(t) = \sin t + \int_0^t y(\tau) \sin (t - \tau) \, d\tau$

33. $y(t) = 1 - \sinh t + \int_0^t (1 + \tau) y(t - \tau) \, d\tau$

34. TEAM PROJECT. Properties of Convolution. Prove:

    (a) Commutativity, $f * g = g * f$

    (b) Associativity, $(f * g) * v = f * (g * v)$

    (c) Distributivity, $f * (g_1 + g_2) = f * g_1 + f * g_2$

    (d) Dirac's delta. Using the convolution theorem and treating $\delta(t)$ (Sec. 5.3) as though it were an ordinary function, show that

$$(\delta * f)(t) = f(t).$$

    (e) Derive the formula in (d) by using f_k with $a = 0$ (Sec. 5.3) and applying the mean value theorem for integrals.

    (f) Unspecified driving force. Show that forced vibrations governed by

$$y'' + \omega^2 y = r(t), \qquad y(0) = K_1, \qquad y'(0) = K_2$$

with unspecified driving force $r(t)$ can be written in convolution form,

$$y = \frac{1}{\omega} \sin \omega t * r(t) + K_1 \cos \omega t + \frac{K_2}{\omega} \sin \omega t \qquad\qquad (\omega \neq 0).$$

5.6 Partial Fractions
Differential Equations

The solution $Y(s)$ of a subsidiary equation of a differential equation (see Sec. 5.2) usually comes out as a quotient of two polynomials,

$$Y(s) = \frac{F(s)}{G(s)}.$$

Hence we can often determine its inverse by writing $Y(s)$ as a sum of **partial fractions** (as in calculus) and obtain the inverse of the latter from a table and the first shifting theorem (Sec. 5.1). This is an easy matter, which we shall now explain by typical examples in the context of Laplace transforms, where partial fraction representations and their inverse transforms are sometimes called **Heaviside expansions.** The form of the partial fractions depends on the kind of factors in the product form of $G(s)$. Of practical interest are:

    (Case 1)   Unrepeated factors $s - a$

    (Case 2)   Repeated factors $(s - a)^m$

    (Case 3)   Complex factors $(s - a)(s - \bar{a})$

    (Case 4)   Repeated complex factors $[(s - a)(s - \bar{a})]^2$.

For determining coefficients use your own favorite method from calculus and any shortcut you can discover.

Case 1. Unrepeated Factor $s - a$

EXAMPLE 1 **Unrepeated factors. An initial value problem**

Solve the initial value problem

$$y'' + y' - 6y = 1, \qquad y(0) = 0, \qquad y'(0) = 1.$$

Solution. From $\mathcal{L}(1) = 1/s$ and the formulas (1), (2) in Sec. 5.2 for the derivatives, we obtain the subsidiary equation

$$(s^2 + s - 6)Y = 1 + \frac{1}{s} = \frac{s+1}{s}.$$

Now $s^2 + s - 6 = (s - 2)(s + 3)$. These are unrepeated factors. Hence the solution $Y(s)$ and its partial fraction representation are

$$Y(s) = \frac{s+1}{s(s-2)(s+3)} = \frac{A_1}{s} + \frac{A_2}{s-2} + \frac{A_3}{s+3}.$$

Next we determine A_1, A_2, and A_3. Multiplication by the common denominator $s(s - 2)(s + 3)$ gives

$$s + 1 = (s - 2)(s + 3)A_1 + s(s + 3)A_2 + s(s - 2)A_3.$$

Taking $s = 0$, $s = 2$, $s = -3$, we obtain

$$1 = -2 \cdot 3A_1$$

$$3 = 2 \cdot 5A_2$$

$$-2 = -3(-5)A_3.$$

Hence $A_1 = -1/6$, $A_2 = 3/10$, and $A_3 = -2/15$. The answer is

$$\mathcal{L}^{-1}(Y) = -\tfrac{1}{6} + \tfrac{3}{10}e^{2t} - \tfrac{2}{15}e^{-3t}. \qquad\qquad \blacktriangleleft$$

Case 2. Repeated Factor $(s - a)^m$

Repeated factors $(s - a)^2$, $(s - a)^3$, etc., require partial fractions

$$(1) \qquad \frac{A_2}{(s-a)^2} + \frac{A_1}{s-a}, \qquad\qquad \frac{A_3}{(s-a)^3} + \frac{A_2}{(s-a)^2} + \frac{A_1}{s-a}, \text{ etc.,}$$

respectively.

EXAMPLE 2 **Repeated factor. An initial value problem**

Solve the initial value problem

$$y'' - 3y' + 2y = 4t \qquad y(0) = 1, \qquad y'(0) = -1.$$

Solution. From Table 5.1, Sec. 5.1, and (1) and (2) in Sec. 5.2 we get the subsidiary equation

$$s^2 Y - s + 1 - 3(sY - 1) + 2Y = \frac{4}{s^2}.$$

Collecting the terms in Y on the left and the others on the right, we have

$$(s^2 - 3s + 2)Y = \frac{4}{s^2} + s - 4 = \frac{4 + s^3 - 4s^2}{s^2}.$$

Since s^2 is a double factor and $s^2 - 3s + 2 = (s - 2)(s - 1)$ has simple factors, we obtain the partial fraction representation

(2*) $$Y(s) = \frac{s^3 - 4s^2 + 4}{s^2(s - 2)(s - 1)} = \frac{A_2}{s^2} + \frac{A_1}{s} + \frac{B}{s - 2} + \frac{C}{s - 1}.$$

Multiplication with $s^2(s - 2)(s - 1)$ gives

(2) $$s^3 - 4s^2 + 4 = A_2(s - 2)(s - 1) + A_1 s(s - 2)(s - 1) + Bs^2(s - 1) + Cs^2(s - 2).$$

For $s = 1$ this is $1 = C(-1)$, hence $C = -1$. For $s = 2$ it is $-4 = 4B$, hence $B = -1$. This was nothing new. New are the double root $s = 0$ and the two fractions A_2/s^2 and A_1/s. For $s = 0$ we get $4 = 2A_2$, hence $A_2 = 2$. Differentiation of (2) gives

$$3s^2 - 8s = A_2(2s - 3) + A_1(s - 2)(s - 1) + \text{further terms all containing a factor } s.$$

For $s = 0$ this is $0 = -3A_2 + 2A_1$, hence $A_1 = 3A_2/2 = 3$. Alternatively, you could get A_1 by taking the terms in s in (2), $0 = -3A_2 + 2A_1$. Substituting all these constants into (2*), we obtain from Table 5.1 in Sec. 5.1 the answer

$$y(t) = \mathcal{L}^{-1}(Y) = \mathcal{L}^{-1}\left\{ \frac{2}{s^2} + \frac{3}{s} - \frac{1}{s - 2} - \frac{1}{s - 1} \right\}$$

$$= 2t + 3 - e^{2t} - e^t.$$

The first two terms result from the "driving force" $4t$ and the exponential terms from the general solution of the homogeneous equation. ◀

For a triple root we would proceed similarly, obtaining A_3, A_2, A_1 in (1) from the analog of (2) and its first *and second* derivatives, respectively.

Case 3. Unrepeated Complex Factors $(s - a)(s - \bar{a})$

Such factors occur, for instance, in connection with vibrations. If $s - a$ with complex $a = \alpha + i\beta$ is a factor of $G(s)$, so is $s - \bar{a}$ with $\bar{a} = \alpha - i\beta$ the conjugate. To $(s - a)(s - \bar{a}) = (s - \alpha)^2 + \beta^2$ there corresponds the partial fraction

(3) $$\frac{As + B}{(s - a)(s - \bar{a})} \quad \text{or} \quad \frac{As + B}{(s - \alpha)^2 + \beta^2}.$$

EXAMPLE 3 **Unrepeated complex factors. Damped forced vibrations**

Solve the initial value problem

$$y'' + 2y' + 2y = r(t), \ r(t) = 10 \sin 2t \text{ if } 0 < t < \pi \text{ and } 0 \text{ if } t > \pi; \quad y(0) = 1, \quad y'(0) = -5.$$

This is a damped mass–spring system (Fig. 124) with a sinusoidal driving force acting during the interval $0 < t < \pi$ only.

Solution. From Table 5.1 in Sec. 5.1, formulas (1), (2) in Sec. 5.2, and the second shifting theorem in Sec. 5.3, we obtain the subsidiary equation

$$(s^2 Y - s + 5) + 2(sY - 1) + 2Y = 10 \frac{2}{s^2 + 4}(1 - e^{-\pi s}).$$

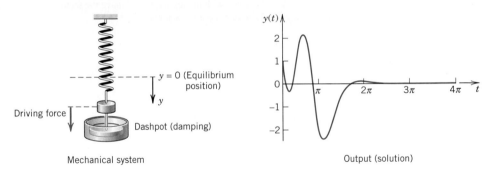

Fig. 124.   Example 3

We take $-s + 5 - 2 = -s + 3$ to the right, collect the Y-terms to get $(s^2 + 2s + 2)Y$, and solve the equation for Y,

(4)
$$Y = \frac{20}{(s^2 + 4)(s^2 + 2s + 2)} - \frac{20e^{-\pi s}}{(s^2 + 4)(s^2 + 2s + 2)} + \frac{s - 3}{s^2 + 2s + 2}.$$

For the last fraction we get from Table 5.1 and the first shifting theorem

(5)
$$\mathcal{L}^{-1}\left\{\frac{s + 1 - 4}{(s + 1)^2 + 1}\right\} = e^{-t}(\cos t - 4 \sin t).$$

In the first fraction in (4) we have unrepeated complex roots, hence a partial fraction representation

$$\frac{20}{(s^2 + 4)(s^2 + 2s + 2)} = \frac{As + B}{s^2 + 4} + \frac{Ms + N}{s^2 + 2s + 2}.$$

Multiplication by the common denominator gives

$$20 = (As + B)(s^2 + 2s + 2) + (Ms + N)(s^2 + 4).$$

We determine A, B, M, N. Equating the coefficients of each power of s on both sides gives the four equations

(a) $[s^3]$: $0 = A + M$

(b) $[s^2]$: $0 = 2A + B + N$

(c) $[s]$: $0 = 2A + 2B + 4M$

(d) $[s^0]$: $20 = 2B + 4N.$

We can solve this, for instance, obtaining $M = -A$ from (a), then $A = B$ from (c), then $N = -3A$ from (b), and finally $A = -2$ from (d). Hence $A = -2$, $B = -2$, $M = 2$, $N = 6$, and the first fraction in (4) has the representation

$$\frac{-2s - 2}{s^2 + 4} + \frac{2(s + 1) + 6 - 2}{(s + 1)^2 + 1}.$$

By Table 5.1 and the first shifting (Sec. 5.1), the inverse transform is

(6)
$$-2 \cos 2t - \sin 2t + e^{-t}(2 \cos t + 4 \sin t).$$

The sum of (5) and (6) is the solution of the problem for $0 < t < \pi$, namely (the sines cancel),

(7)
$$y(t) = 3e^{-t} \cos t - 2 \cos 2t - \sin 2t \qquad\qquad \text{if } 0 < t < \pi.$$

In the second fraction in (4) taken with the minus sign we have the factor $e^{-\pi s}$, so that from (6) and the second shifting theorem (Sec. 5.3) we get the inverse transform

$$+2 \cos (2t - 2\pi) + \sin (2t - 2\pi) - e^{-(t-\pi)} [2 \cos (t - \pi) + 4 \sin (t - \pi)]$$

$$= 2 \cos 2t + \sin 2t + e^{-(t-\pi)} (2 \cos t + 4 \sin t).$$

The sum of this and (7) is the solution for $t > \pi$,

(8) $$y(t) = e^{-t}[(3 + 2e^{\pi}) \cos t + 4e^{\pi} \sin t]$$ if $t > \pi$.

Figure 124 shows (7) (for $0 < t < \pi$) and (8) (for $t > 0$), a beginning vibration, which goes to zero rapidly because of the damping and the absence of a driving force after $t = \pi$. ◀

Case 4. Repeated Complex Factors $[(s - a)(s - \bar{a})]^2$

In this case the partial fractions are of the form

(9) $$\frac{As + B}{[(s - a)(s - \bar{a})]^2} + \frac{Ms + N}{(s - a)(s - \bar{a})}.$$

This case is important, for instance, in connection with resonance.

EXAMPLE 4   **Repeated complex factors. Resonance**

In an undamped mass–spring system, resonance occurs if the frequency of the driving force equals the natural frequency of the system. Then the model is (see Sec. 2.11)

$$y'' + \omega_0^2 y = K \sin \omega_0 t$$

where $\omega_0^2 = k/m$, k is the spring constant, and m is the mass of the body attached to the spring. We assume $y(0) = 0$ and $y'(0) = 0$, for simplicity. Then the subsidiary equation is

$$s^2 Y + \omega_0^2 Y = \frac{K\omega_0}{s^2 + \omega_0^2}.$$

Hence

$$Y = \frac{K\omega_0}{(s^2 + \omega_0^2)^2}.$$

The denominator is a repeated complex factor with double roots $s = i\omega_0$ and $-i\omega_0$. Hence our Y consists of this single partial fraction. Its inverse can be obtained by convolution. This is similar to Example 1 in Sec. 5.5. Using the convolution theorem and (11) in Appendix A3.1, we obtain

$$y(t) = \mathcal{L}^{-1}(Y) = \frac{K}{\omega_0} \sin \omega_0 t * \sin \omega_0 t$$

$$= \frac{K}{\omega_0} \int_0^t \sin \omega_0 \tau \sin (\omega_0 t - \omega_0 \tau) \, d\tau$$

$$= \frac{K}{2\omega_0} \left[\int_0^t -\cos \omega_0 t \, d\tau + \int_0^t \cos (2\omega_0 \tau - \omega_0 t) \, d\tau \right]$$

$$= \frac{K}{2\omega_0^2} (-\omega_0 t \cos \omega_0 t + \sin \omega_0 t).$$

This is the solution of our problem. The first term grows without bound. It is clear that in the case of resonance such a term must occur. (See also Fig. 58 in Sec. 2.11, which shows a similar kind of solution.) ◀

PROBLEM SET 5.6

Inverse Transforms. Find the function $f(t)$ for a given transform $\mathcal{L}(f)$ by partial fraction reduction or any other method that you think is simplest or fastest. Indicate the method used and show the details of your work.

1. $\dfrac{6}{(s + 2)(s - 4)}$

2. $\dfrac{s^3 + 2s^2 + 2}{s^3(s^2 + 1)}$

3. $\dfrac{s^2 + 9s - 9}{s^3 - 9s}$

4. $\dfrac{s}{(s + 1)^2}$

5. $\dfrac{2s^3}{s^4 - 81}$

6. $\dfrac{s^3 - 3s^2 + 6s - 4}{(s^2 - 2s + 2)^2}$

7. $\dfrac{s^4 + 3(s + 1)^3}{s^4(s + 1)^3}$

8. $\dfrac{s^3 - 7s^2 + 14s - 9}{(s - 1)^2(s - 2)^3}$

9. $\dfrac{s^3 + 6s^2 + 14s}{(s + 2)^4}$

Inverse Transforms. Derive the following formulas, showing the details of your work.

10. $\mathcal{L}^{-1}\left\{\dfrac{1}{s^4 + 4a^4}\right\} = \dfrac{1}{4a^3}(\cosh at \sin at - \sinh at \cos at)$

11. $\mathcal{L}^{-1}\left\{\dfrac{s}{s^4 + 4a^4}\right\} = \dfrac{1}{2a^2}\sinh at \sin at$

12. $\mathcal{L}^{-1}\left\{\dfrac{s^2}{s^4 + 4a^4}\right\} = \dfrac{1}{2a}(\cosh at \sin at + \sinh at \cos at)$

13. $\mathcal{L}^{-1}\left\{\dfrac{s^3}{s^4 + 4a^4}\right\} = \cosh at \cos at$

14. (Undamped system, no resonance) What happens in Example 4 if you replace the driving force by $K \sin pt$ with $p^2 \neq \omega_0^2$? Find the solution by the Laplace transform. (Show the details of your work.)

15. PROJECT. Heaviside Formulas. (a) Show that for a simple root a and fraction $A/(s - a)$ in $F(s)/G(s)$ we have the *Heaviside formula*

$$A = \lim_{s \to a} \frac{(s - a)F(s)}{G(s)} .$$

(b) Similarly, show that for a root a of order m and fractions in

$$\frac{F(s)}{G(s)} = \frac{A_m}{(s - a)^m} + \frac{A_{m-1}}{(s - a)^{m-1}} + \cdots + \frac{A_1}{s - a} + \text{further fractions}$$

we have the *Heaviside formulas* for the first coefficient

$$A_m = \lim_{s \to a} \frac{(s - a)^m F(s)}{G(s)}$$

and for the other coefficients

$$A_k = \frac{1}{(m - k)!} \lim_{s \to a} \frac{d^{m-k}}{ds^{m-k}}\left[\frac{(s - a)^m F(s)}{G(s)}\right], \qquad k = 1, \cdots, m - 1.$$

16. TEAM PROJECT. Laplace Transform of Periodic Functions

    (a) Theorem. *The Laplace transform of a piecewise continuous function* $f(t)$ *with period* p *is*

(10)
$$\mathcal{L}(f) = \frac{1}{1 - e^{-ps}} \int_0^p e^{-st} f(t)\, dt \qquad (s > 0).$$

Prove this theorem. *Hint.* Write $\int_0^\infty = \int_0^p + \int_p^{2p} + \cdots$. Set $t = (n - 1)p$ in the nth integral. Take out $e^{-(n-1)p}$ from under the integral sign. Use the sum formula for the geometric series.

    (b) Half-wave rectifier. Using (10), show that the half-wave rectification of $\sin \omega t$ in Fig. 125 has the Laplace transform

$$\mathcal{L}(f) = \frac{\omega(1 + e^{-\pi s/\omega})}{(s^2 + \omega^2)(1 - e^{-2\pi s/\omega})} = \frac{\omega}{(s^2 + \omega^2)(1 - e^{-\pi s/\omega})}.$$

(A *half-wave rectifier* clips the negative portions of the curve. A *full-wave rectifier* converts them to positive; see Fig. 126.)

       Fig. 125.   Half-wave rectification

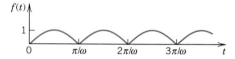

       Fig. 126.   Full-wave rectification

    (c) Full-wave rectifier. Show that the Laplace transform of the full-wave rectification of $\sin \omega t$ is

$$\frac{\omega}{s^2 + \omega^2} \coth \frac{\pi s}{2\omega}.$$

    (d) Saw-tooth wave. Find the Laplace transform of the saw-tooth wave in Fig. 127.

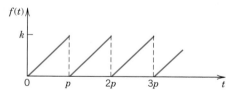

       Fig. 127.   Saw-tooth wave

    (e) Staircase function. Find the Laplace transform of the staircase function in Fig. 128 by noting that it is the difference of kt/p and the function in (d).

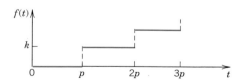

       Fig. 128.   Staircase function

5.7 Systems of Differential Equations

The Laplace transform method may also be used for solving *systems* of differential equations. We shall explain this in terms of typical applications.

For a first-order linear system (as discussed in Sec. 3.1),

(1)
$$y_1' = a_{11}y_1 + a_{12}y_2 + g_1(t)$$
$$y_2' = a_{21}y_1 + a_{22}y_2 + g_2(t),$$

writing $Y_1 = \mathcal{L}(y_1)$, $Y_2 = \mathcal{L}(y_2)$, $G_1 = \mathcal{L}(g_1)$, $G_2 = \mathcal{L}(g_2)$, we obtain from (1) in Sec. 5.2 the subsidiary equations

$$sY_1 - y_1(0) = a_{11}Y_1 + a_{12}Y_2 + G_1(s)$$
$$sY_2 - y_2(0) = a_{21}Y_1 + a_{22}Y_2 + G_2(s)$$

or, by collecting the Y_1- and Y_2-terms,

(2)
$$(a_{11} - s)Y_1 + a_{12}Y_2 = -y_1(0) - G_1(s)$$
$$a_{21}Y_1 + (a_{22} - s)Y_2 = -y_2(0) - G_2(s).$$

This must be solved algebraically for $Y_1(s)$ and $Y_2(s)$. The solution of the given system is then obtained if we take the inverse $y_1 = \mathcal{L}^{-1}(Y_1)$, $y_2 = \mathcal{L}^{-1}(Y_2)$.

EXAMPLE 1   **Mixing problem involving two tanks**

Tank T_1 in Fig. 129 contains initially 100 gal of pure water. Tank T_2 contains initially 100 gal of water in which 150 lb of salt are dissolved. The inflow into T_1 is 2 gal/min from T_2 and 6 gal/min containing 6 lb of salt from the outside. The inflow into T_2 is 8 gal/min from T_1. The outflow from T_2 is 2 + 6 = 8 gal/min, as shown in the figure. The mixtures are kept uniform by stirring. Find and plot the salt contents $y_1(t)$ and $y_2(t)$ in T_1 and T_2, respectively.

Solution.   The model is obtained in the form of two equations

$$\text{Time rate of change} = \text{Inflow/min} - \text{Outflow/min}$$

for the two tanks (see Sec. 3.1). Thus,

$$y_1' = -\frac{8}{100}y_1 + \frac{2}{100}y_2 + 6, \qquad y_2' = \frac{8}{100}y_1 - \frac{8}{100}y_2.$$

The initial conditions are $y_1(0) = 0$, $y_2(0) = 150$. From this we see that the subsidiary equations (2) are

$$(-0.08 - s)Y_1 + 0.02Y_2 = -\frac{6}{s}$$
$$0.08Y_1 + (-0.08 - s)Y_2 = -150.$$

We solve these algebraically for Y_1 and Y_2 by Cramer's rule (Sec. 6.6) or by elimination, and we write the solutions in terms of partial fractions,

$$Y_1 = \frac{9s + 0.48}{s(s + 0.12)(s + 0.04)} = \frac{100}{s} - \frac{62.5}{s + 0.12} - \frac{37.5}{s + 0.04}$$

$$Y_2 = \frac{150s^2 + 12s + 0.48}{s(s + 0.12)(s + 0.04)} = \frac{100}{s} + \frac{125}{s + 0.12} - \frac{75}{s + 0.04}.$$

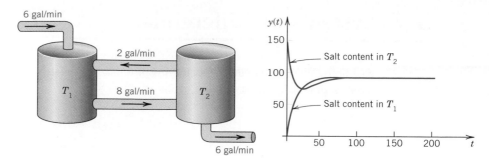

Fig. 129. Mixing problem in Example 1

By taking the inverse transform we arrive at the solution

$$y_1 = 100 - 62.5e^{-0.12t} - 37.5e^{-0.04t}$$

$$y_2 = 100 + 125e^{-0.12t} - 75e^{-0.04t}.$$

Figure 129 shows the interesting plot of these functions. Can you give physical explanations for their main features? Why do they have the limit 100? Why is y_2 not monotone, whereas y_1 is? Why is y_1 from some time on all of a sudden larger than y_2? Etc. ◀

Other systems of differential equations of practical importance can be solved by the Laplace transform method in a similar way, and eigenvalues and eigenvectors as we had to determine them in Chap. 3 will come out automatically, as we have seen in Example 1.

EXAMPLE 2   **Electrical network**

Find the currents $i_1(t)$ and $i_2(t)$ in the network in Fig. 130 with L and R measured in terms of the usual units (see Sec. 1.7), $v(t) = 100$ volts if $0 \leqq t \leqq 0.5$ sec and 0 thereafter, and $i(0) = 0$, $i'(0) = 0$.

Solution.   The model of the network is obtained from Kirchhoff's voltage law as in Secs. 1.7 and 2.12,

$$0.8\,i_1' + 1(i_1 - i_2) + 1.4\,i_1 = 100\left[1 - u(t - \tfrac{1}{2})\right]$$

$$1 \cdot i_2' + 1(i_2 - i_1) = 0.$$

Division by 0.8 and ordering gives

$$i_1' + 3\,i_1 - 1.25\,i_2 = 125\left[1 - u(t - \tfrac{1}{2})\right]$$

$$i_2' - i_1 + i_2 = 0.$$

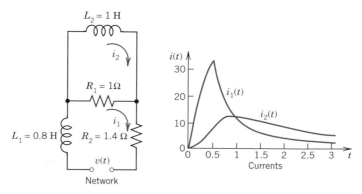

Fig. 130.   Electrical network in Example 2

With $i_1(0) = 0$, $i_2(0) = 0$ we obtain from (1) in Sec. 5.2 and the second shifting theorem the subsidiary equations

$$(s + 3)I_1 - 1.25I_2 = 125 \left(\frac{1}{s} - \frac{e^{-s/2}}{s} \right)$$
$$-I_1 + (s + 1)I_2 = 0.$$

Solving algebraically for I_1 and I_2 gives

$$I_1 = \frac{125(s + 1)}{s(s + \frac{1}{2})(s + \frac{7}{2})} (1 - e^{-s/2})$$

$$I_2 = \frac{125}{s(s + \frac{1}{2})(s + \frac{7}{2})} (1 - e^{-s/2}),$$

The right sides without the factor $1 - e^{-s/2}$ have the partial fraction expansions

$$\frac{500}{7s} - \frac{125}{3(s + \frac{1}{2})} - \frac{625}{21(s + \frac{7}{2})}, \qquad \frac{500}{7s} - \frac{250}{3(s + \frac{1}{2})} + \frac{250}{21(s + \frac{7}{2})},$$

respectively. The inverse transform of this gives the solution for $0 \leqq t \leqq \frac{1}{2}$,

$$i_1(t) = -\frac{125}{3} e^{-t/2} - \frac{625}{21} e^{-7t/2} + \frac{500}{7}$$

$$i_2(t) = -\frac{250}{3} e^{-t/2} + \frac{250}{21} e^{-7t/2} + \frac{500}{7}$$

$$(0 \leqq t \leqq \tfrac{1}{2}).$$

According to the second shifting theorem, the solution for $t > \frac{1}{2}$ is obtained if we subtract from this $i_1(t - \frac{1}{2})$ and $i_2(t - \frac{1}{2})$, respectively. This gives

$$i_1(t) = -\frac{125}{3} (1 - e^{1/4})e^{-t/2} - \frac{625}{21} (1 - e^{7/4})e^{-7t/2}$$

$$i_2(t) = -\frac{250}{3} (1 - e^{1/4})e^{-t/2} + \frac{250}{21} (1 - e^{7/4})e^{-7t/2}$$

$$(t > \tfrac{1}{2}).$$

Can you explain physically why both currents eventually go to zero, and why $i_1(t)$ has a sharp cusp whereas $i_2(t)$ has a continuous tangent direction at $t = \frac{1}{2}$? ◀

Systems of differential equations of higher order can be solved by the Laplace transform method in a similar fashion. As an important application, typical of many similar mechanical systems, we consider coupled vibrating masses on springs.

EXAMPLE 3   **Model of two masses on springs (Fig. 131)**

The mechanical system in Fig. 131 consists of two bodies of mass 1 on three springs and is governed by the differential equations

(3)
$$y_1'' = -ky_1 + k(y_2 - y_1)$$
$$y_2'' = -k(y_2 - y_1) - ky_2,$$

where k is the spring constant of each of the three springs, and y_1 and y_2 are the displacements of the bodies from their positions of static equilibrium; the masses of the springs and damping are neglected. The equations follow from *Newton's second law, Mass $\times$ Acceleration = Force*, as in Sec. 2.5 for a single body. In the equations, $-ky_1$ is the force of the upper spring on the first body and $k(y_2 - y_1)$ that of the middle spring, where $y_2 - y_1$ is the net change in spring length—think this over before going on. Also, $-k(y_2 - y_1)$ is the force of the middle spring on the lower body and $-ky_2$ that of the lower spring.

We shall determine the solution corresponding to the initial conditions $y_1(0) = 1$, $y_2(0) = 1$, $y_1'(0) = \sqrt{3k}$, $y_2'(0) = -\sqrt{3k}$. Let $Y_1 = \mathscr{L}(y_1)$ and $Y_2 = \mathscr{L}(y_2)$. Then from (2) in Sec. 5.2 and the initial conditions we obtain the subsidiary equations

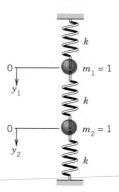

Fig. 131. Example 3

$$s^2Y_1 - s - \sqrt{3k} = -kY_1 + k(Y_2 - Y_1)$$

$$s^2Y_2 - s + \sqrt{3k} = -k(Y_2 - Y_1) - kY_2.$$

This system of linear algebraic equations in the unknowns Y_1 and Y_2 may be written

$$(s^2 + 2k)Y_1 - \quad kY_2 \quad = s + \sqrt{3k}$$

$$-kY_1 \quad + (s^2 + 2k)Y_2 = s - \sqrt{3k}.$$

Cramer's rule (Sec. 6.6) or elimination yields the solution, which we can expand in terms of partial fractions,

$$Y_1 = \frac{(s + \sqrt{3k})(s^2 + 2k) + k(s - \sqrt{3k})}{(s^2 + 2k)^2 - k^2} = \frac{s}{s^2 + k} + \frac{\sqrt{3k}}{s^2 + 3k}$$

$$Y_2 = \frac{(s^2 + 2k)(s - \sqrt{3k}) + k(s + \sqrt{3k})}{(s^2 + 2k)^2 - k^2} = \frac{s}{s^2 + k} - \frac{\sqrt{3k}}{s^2 + 3k}.$$

Hence the solution of our initial value problem is

$$y_1(t) = \mathcal{L}^{-1}(Y_1) = \cos\sqrt{k}t + \sin\sqrt{3k}t$$

$$y_2(t) = \mathcal{L}^{-1}(Y_2) = \cos\sqrt{k}t - \sin\sqrt{3k}t.$$

We see that the motion of each mass is harmonic (the system is undamped!), being the superposition of a "slow" and a "rapid" oscillation. ◀

PROBLEM SET 5.7

Systems of Differential Equations. In Probs. 1–14 solve the given initial value problem by means of Laplace transforms. (Show the details of your work.)

1. $y_1' = -y_1 + y_2$, $y_2' = -y_1 - y_2$, $\quad y_1(0) = 1$, $y_2(0) = 0$
2. $y_1' = 6y_1 + 9y_2$, $y_2' = y_1 + 6y_2$, $\quad y_1(0) = -3$, $y_2(0) = -3$
3. $y_1' = -y_1 + 4y_2$, $y_2' = 3y_1 - 2y_2$, $\quad y_1(0) = 3$, $y_2(0) = 4$
4. $y_1' = 5y_1 + y_2$, $y_2' = y_1 + 5y_2$, $\quad y_1(0) = -3$, $y_2(0) = 7$
5. $y_1' + y_2 = 2\cos t$, $\quad y_1 + y_2' = 0$, $\quad y_1(0) = 0$, $y_2(0) = 1$

6. $y_1'' + y_2 = -5 \cos 2t, \quad y_2'' + y_1 = 5 \cos 2t,$
$\quad y_1(0) = 1, \quad y_1'(0) = 1, \quad y_2(0) = -1, \quad y_2'(0) = 1$

7. $y_1'' = y_1 + 3y_2, \quad y_2'' = 4y_1 - 4e^t,$
$\quad y_1(0) = 2, \quad y_1'(0) = 3, \quad y_2(0) = 1, \quad y_2'(0) = 2$

8. $y_1'' = -5y_1 + 2y_2, \quad y_2'' = 2y_1 - 2y_2,$
$\quad y_1(0) = 3, \quad y_1'(0) = 0, \quad y_2(0) = 1, \quad y_2'(0) = 0$

9. $y_1' + y_2' = 2 \sinh t, \quad y_2' + y_3' = e^t, \quad y_3' + y_1' = 2e^t + e^{-t},$
$\quad y_1(0) = 1, \quad y_2(0) = 1, \quad y_3(0) = 0$

10. $2y_1' - y_2' - y_3' = 0, \quad y_1' + y_2' = 4t + 2, \quad y_2' + y_3 = t^2 + 2,$
$\quad y_1(0) = y_2(0) = y_3(0) = 0$

11. $y_1' = -y_2 + 1 - u(t-1), \quad y_2' = y_1 + 1 - u(t-1), \quad y_1(0) = 0, \quad y_2(0) = 0$

12. $y_1' + y_2 = 2[1 - u(t - 2\pi)] \cos t, \quad y_1 + y_2' = 0, \quad y_1(0) = 0, \quad y_2(0) = 1$

13. $y_1' = 2y_1 - 4y_2 + u(t-1)e^t, \quad y_2' = y_1 - 3y_2 + u(t-1)e^t, \quad y_1(0) = 3, \quad y_2(0) = 0$

14. $y_1' = 2y_1 + 4y_2 + 64tu(t-1), \quad y_2' = y_1 + 2y_2, \quad y_1(0) = -4, \quad y_2(0) = -4$

15. TEAM PROJECT. First-Order Linear Systems of Differential Equations.
    (a) Models. Solve the models in Examples 1 and 2 of Sec. 3.1 by the Laplace transform method and compare the work with that in Sec. 3.1.
    (b) Homogeneous Systems. Solve the systems (8), (11)–(13) in Sec. 3.3 by the Laplace transform method.
    (c) Nonhomogeneous Systems. Solve the systems (3) and (4) in Sec. 3.6 by the Laplace transform method.
    In (a)–(c) always show the details of your work.

Further Applications

16. (Mixing problem) What will happen in Example 1 if you double all flows, leaving the size of the tanks and the initial conditions as before. First guess, then calculate. Can you relate the new solution to the old one?

17. (Sinusoidal inflow from the outside) What will happen in Example 1 if you assume that the total salt content of the inflowing 6 gal/min varies between 0 and 6 according to $6 \sin^2 t$? Why is $y_2(t)$ much less wavy than $y_1(t)$?

18. (Forced vibrations of two masses) Solve the model in Example 3 with $k = 3$ and initial conditions $y_1(0) = 1, y_2(0) = 1, y_1'(0) = 3, y_2'(0) = -3$ under the assumption that the force $8 \sin t$ acts on the first body and $-8 \sin t$ acts on the second.

19. (Electrical network) Using the Laplace transform method, find the currents $i_1(t)$ and $i_2(t)$ in Fig. 132, where $v(t) = 195 \sin t$ and $i_1(0) = 0, i_2(0) = 0$. How soon will the currents practically reach their steady state? Guess what the little curve in Fig. 132 is.

20. (Single sine wave) Solve Prob. 19 when the electromotive force acts from 0 to 2π only. Can you obtain the solution from that in Prob. 19 practically without calculation?

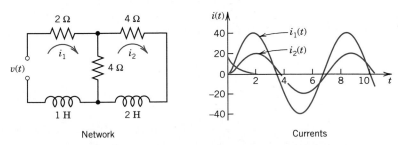

Network   Currents

Fig. 132. Electrical network and currents in Problem 19

5.8 Laplace Transform: General Formulas

Formula	Name, Comments	Sec.
$F(s) = \mathcal{L}\{f(t)\} = \int_0^\infty e^{-st} f(t)\, dt$ $f(t) = \mathcal{L}^{-1}\{F(s)\}$	Definition of Transform Inverse Transform	5.1
$\mathcal{L}\{af(t) + bg(t)\} = a\mathcal{L}\{f(t)\} + b\mathcal{L}\{g(t)\}$	Linearity	5.1
$\mathcal{L}\{e^{at} f(t)\} = F(s - a)$ $\mathcal{L}^{-1}\{F(s - a)\} = e^{at} f(t)$	s-Shifting (First Shifting Theorem)	5.1
$\mathcal{L}(f') = s\mathcal{L}(f) - f(0)$ $\mathcal{L}(f'') = s^2 \mathcal{L}(f) - sf(0) - f'(0)$ $\mathcal{L}(f^{(n)}) = s^n \mathcal{L}(f) - s^{n-1} f(0) - \cdots$ $\cdots - f^{(n-1)}(0)$ $\mathcal{L}\left\{ \int_0^t f(\tau)\, d\tau \right\} = \dfrac{1}{s}\, \mathcal{L}(f)$	Differentiation of Function Integration of Function	5.2
$\mathcal{L}\{f(t - a)\,u(t - a)\} = e^{-as} F(s)$ $\mathcal{L}^{-1}\{e^{-as} F(s)\} = f(t - a)\,u(t - a)$	t-Shifting (Second Shifting Theorem)	5.3
$\mathcal{L}\{tf(t)\} = -F'(s)$ $\mathcal{L}\left\{ \dfrac{f(t)}{t} \right\} = \int_s^\infty F(\tilde{s})\, d\tilde{s}$	Differentiation of Transform Integration of Transform	5.4
$(f * g)(t) = \int_0^t f(\tau) g(t - \tau)\, d\tau$ $= \int_0^t f(t - \tau) g(\tau)\, d\tau$ $\mathcal{L}(f * g) = \mathcal{L}(f)\mathcal{L}(g)$	Convolution	5.5
$\mathcal{L}(f) = \dfrac{1}{1 - e^{-ps}} \int_0^p e^{-st} f(t)\, dt$	f Periodic with Period p	5.6 Project 16

5.9   Table of Laplace Transforms

For more extensive tables, see Refs. [A4] and [A6] in Appendix 1.

	$F(s) = \mathcal{L}\{f(t)\}$	$f(t)$	Sec.
1	$1/s$	1	
2	$1/s^2$	t	
3	$1/s^n \quad (n = 1, 2, \cdots)$	$t^{n-1}/(n-1)!$	
4	$1/\sqrt{s}$	$1/\sqrt{\pi t}$	5.1
5	$1/s^{3/2}$	$2\sqrt{t/\pi}$	
6	$1/s^a \quad (a > 0)$	$t^{a-1}/\Gamma(a)$	
7	$\dfrac{1}{s-a}$	e^{at}	
8	$\dfrac{1}{(s-a)^2}$	te^{at}	
9	$\dfrac{1}{(s-a)^n} \quad (n = 1, 2, \cdots)$	$\dfrac{1}{(n-1)!}\, t^{n-1} e^{at}$	5.1
10	$\dfrac{1}{(s-a)^k} \quad (k > 0)$	$\dfrac{1}{\Gamma(k)}\, t^{k-1} e^{at}$	
11	$\dfrac{1}{(s-a)(s-b)} \quad (a \neq b)$	$\dfrac{1}{(a-b)}(e^{at} - e^{bt})$	
12	$\dfrac{s}{(s-a)(s-b)} \quad (a \neq b)$	$\dfrac{1}{(a-b)}(ae^{at} - be^{bt})$	
13	$\dfrac{1}{s^2 + \omega^2}$	$\dfrac{1}{\omega}\sin \omega t$	
14	$\dfrac{s}{s^2 + \omega^2}$	$\cos \omega t$	
15	$\dfrac{1}{s^2 - a^2}$	$\dfrac{1}{a}\sinh at$	
16	$\dfrac{s}{s^2 - a^2}$	$\cosh at$	5.1
17	$\dfrac{1}{(s-a)^2 + \omega^2}$	$\dfrac{1}{\omega} e^{at} \sin \omega t$	
18	$\dfrac{s-a}{(s-a)^2 + \omega^2}$	$e^{at} \cos \omega t$	
19	$\dfrac{1}{s(s^2 + \omega^2)}$	$\dfrac{1}{\omega^2}(1 - \cos \omega t)$	
20	$\dfrac{1}{s^2(s^2 + \omega^2)}$	$\dfrac{1}{\omega^3}(\omega t - \sin \omega t)$	5.2
21	$\dfrac{1}{(s^2 + \omega^2)^2}$	$\dfrac{1}{2\omega^3}(\sin \omega t - \omega t \cos \omega t)$	5.4

(continued)

Table of Laplace Transforms (*continued*)

	$F(s) = \mathcal{L}\{f(t)\}$	$f(t)$	Sec.
22	$\dfrac{s}{(s^2 + \omega^2)^2}$	$\dfrac{t}{2\omega} \sin \omega t$	
23	$\dfrac{s^2}{(s^2 + \omega^2)^2}$	$\dfrac{1}{2\omega} (\sin \omega t + \omega t \cos \omega t)$	5.4
24	$\dfrac{s}{(s^2 + a^2)(s^2 + b^2)}$ $(a^2 \neq b^2)$	$\dfrac{1}{b^2 - a^2} (\cos at - \cos bt)$	
25	$\dfrac{1}{s^4 + 4k^4}$	$\dfrac{1}{4k^3} (\sin kt \cos kt - \cos kt \sinh kt)$	
26	$\dfrac{s}{s^4 + 4k^4}$	$\dfrac{1}{2k^2} \sin kt \sinh kt$	5.6
27	$\dfrac{1}{s^4 - k^4}$	$\dfrac{1}{2k^3} (\sinh kt - \sin kt)$	
28	$\dfrac{s}{s^4 - k^4}$	$\dfrac{1}{2k^2} (\cosh kt - \cos kt)$	
29	$\sqrt{s - a} - \sqrt{s - b}$	$\dfrac{1}{2\sqrt{\pi t^3}} (e^{bt} - e^{at})$	
30	$\dfrac{1}{\sqrt{s + a} \sqrt{s + b}}$	$e^{-(a+b)t/2} I_0 \left(\dfrac{a - b}{2} t \right)$	4.6
31	$\dfrac{1}{\sqrt{s^2 + a^2}}$	$J_0(at)$	4.5
32	$\dfrac{s}{(s - a)^{3/2}}$	$\dfrac{1}{\sqrt{\pi t}} e^{at}(1 + 2at)$	
33	$\dfrac{1}{(s^2 - a^2)^k}$ $(k > 0)$	$\dfrac{\sqrt{\pi}}{\Gamma(k)} \left(\dfrac{t}{2a} \right)^{k-1/2} I_{k-1/2}(at)$	4.6
34	e^{-as}/s	$u(t - a)$	5.3
35	e^{-as}	$\delta(t - a)$	
36	$\dfrac{1}{s} e^{-k/s}$	$J_0(2\sqrt{kt})$	4.5
37	$\dfrac{1}{\sqrt{s}} e^{-k/s}$	$\dfrac{1}{\sqrt{\pi t}} \cos 2\sqrt{kt}$	
38	$\dfrac{1}{s^{3/2}} e^{k/s}$	$\dfrac{1}{\sqrt{\pi k}} \sinh 2\sqrt{kt}$	
39	$e^{-k\sqrt{s}}$ $(k > 0)$	$\dfrac{k}{2\sqrt{\pi t^3}} e^{-k^2/4t}$	
40	$\dfrac{1}{s} \ln s$	$-\ln t - \gamma$ $(\gamma \approx 0.5772)$	4.6
41	$\ln \dfrac{s - a}{s - b}$	$\dfrac{1}{t} (e^{bt} - e^{at})$	5.4

(*continued*)

Table of Laplace Transforms (*continued*)

	$F(s) = \mathcal{L}\{f(t)\}$	$f(t)$	Sec.
42	$\ln \dfrac{s^2 + \omega^2}{s^2}$	$\dfrac{2}{t}(1 - \cos \omega t)$	5.4
43	$\ln \dfrac{s^2 - a^2}{s^2}$	$\dfrac{2}{t}(1 - \cosh at)$	
44	$\arctan \dfrac{\omega}{s}$	$\dfrac{1}{t}\sin \omega t$	
45	$\dfrac{1}{s} \operatorname{arc\,cot} s$	$\mathrm{Si}(t)$	App. A3.1

CHAPTER 5 REVIEW

1. Compared to the usual method, what are the advantages of the Laplace transform in solving differential equations?
2. What is the crucial property of the Laplace transform that makes it suitable for solving differential equations?
3. What do we mean by saying that the Laplace transform is a *linear* operation? Why is this practically important?
4. For what problems would you prefer the Laplace transform over the usual method? Give a reason.
5. What is the subsidiary equation? How is it used?
6. Does every continuous function have a Laplace transform? Give a reason or a counterexample.
7. What is the unit step function? Why is it important?
8. What is Dirac's delta function? How did we use it?
9. State the Laplace transforms of a few simple functions from memory.
10. State the formula for the Laplace transform of the nth derivative of a function $f(t)$ from memory.
11. Can a discontinuous function have a Laplace transform? (Give a reason for your answer.)
12. Does $\tan t$ have a Laplace transform? Is it piecewise continuous?
13. If you know $f(t) = \mathcal{L}^{-1}\{F(s)\}$, how would you find $\mathcal{L}^{-1}\{F(s)/s^2\}$?
14. Is $\mathcal{L}\{f(t)g(t)\} = \mathcal{L}\{f(t)\}\mathcal{L}\{g(t)\}$? Or what?
15. What is the difference in the shifting by the first shifting theorem and by the second shifting theorem?

Laplace Transform. Find the Laplace transform of the given function. (Show the details of your work.)

16. $e^{-t} \sin \pi t$ 17. $\cos^2 t$ 18. $\sin^2 (\pi t/2)$

19. $e^t u(t - 2)$ 20. $t^2 u(t - \tfrac{1}{4})$ 21. $t * e^{-3t}$

22. $e^{2t} * \cos 4t$ 23. $\cosh \tfrac{1}{10} t$ 24. $t \cos t + \sin t$

Inverse Laplace Transform. In Probs. 25–33 find the inverse Laplace transform of the given function. (Show your work.)

25. $\dfrac{s + 3}{s^2 + 9}$ 26. $\dfrac{1}{s^2 - 2s - 8}$ 27. $\dfrac{s + 1}{s^2} e^{-s}$

28. $\dfrac{6(s + 1)}{s^4}$ **29.** $\dfrac{s^2 - 6s + 4}{s^3 - 3s^2 + 2s}$ **30.** $\dfrac{2s^2 - 3s + 4}{(s^2 + 4)(s - 3)}$

31. $\dfrac{s^3 - s^2 - s + 4}{s^4 - 5s^2 + 4}$ **32.** $\dfrac{\omega \cos \theta + s \sin \theta}{s^2 + \omega^2}$ **33.** $\dfrac{3s + 4}{s^2 + 4s + 5}$

Differential Equations. Systems of Differential Equations. Solve the given initial value problems by Laplace transforms. Show the details of your work.

34. $y'' + y = \delta(t - 2)$, $y(0) = 2.5$, $y'(0) = 0$

35. $y'' + 4y = u(t - 3)$, $y(0) = 1$, $y'(0) = 0$

36. $y'' - 2y' + 2y = 8e^{-t} \cos t$, $y(0) = 16$, $y'(0) = -16$

37. $y'' + 9y = 6 \sin t$ if $0 \leq t \leq \pi$ and 0 if $t > \pi$, $y(0) = 0$, $y'(0) = 0$

38. $y'' + 3y' + 2y = 2u(t - 2)$, $y(0) = 0$, $y'(0) = 0$

39. $y_1' = -y_2$, $y_2' = y_1$, $y_1(0) = 1$, $y_2(0) = 0$

40. $y_1' = 2y_1 + 4y_2$, $y_2' = y_1 + 2y_2$, $y_1(0) = -4$, $y_2(0) = -4$

41. $y_1' = 2y_1 - 4y_2$, $y_2' = y_1 - 3y_2$, $y_1(0) = 3$, $y_2(0) = 0$

42. $y_1' = -2y_1 + 3y_2$, $y_2' = 4y_1 - y_2$, $y_1(0) = 4$, $y_2(0) = 3$

Models of Mass–Spring Systems, Circuits, Networks

Solve the following problems by Laplace transforms. (Show your work.)

43. Show that the model of the mechanical system in Fig. 133 (no friction, no damping) is

$$m_1 y_1'' = -k_1 y_1 + k_2(y_2 - y_1)$$

$$m_2 y_2'' = -k_2(y_2 - y_1) - k_3 y_2.$$

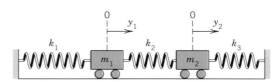

Fig. 133. System in Problems 43 and 44

44. In Prob. 43, let $m_1 = m_2 = 10$ kg, $k_1 = k_3 = 20$ kg/sec^2, $k_2 = 40$ kg/sec^2. Find the solution satisfying the initial conditions $y_1(0) = y_2(0) = 0$, $y_1'(0) = 1$ meter/sec, $y_2'(0) = -1$ meter/sec.

45. Find the current $i(t)$ in the *RC*-circuit in Fig. 134, where $R = 10$ ohms, $C = 0.1$ farad, $e(t) = 10t$ volts if $0 < t < 4$, $v(t) = 40$ volts if $t > 4$, and the initial charge on the capacitor is 0.

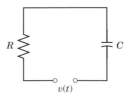

Fig. 134. *RC*-circuit

46. Find the charge $q(t)$ and the current $i(t)$ in the LC-circuit in Fig. 135, assuming $L = 1$ henry, $C = 1$ farad, $v(t) = 1 - e^{-t}$ if $0 < t < \pi$, $e(t) = 0$ if $t > \pi$, and zero initial current and charge.

47. Find the current $i(t)$ in the RLC-circuit in Fig. 136, where $R = 160$ ohms, $L = 20$ henrys, $C = 0.002$ farad, $v(t) = 37 \sin 10t$ volts, assuming zero initial current and charge.

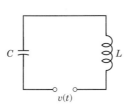

Fig. 135. LC-circuit

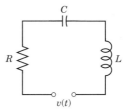

Fig. 136. RLC-circuit

48. Show that, by Kirchhoff's voltage law (Sec. 1.7), the currents in the network in Fig. 137 are obtained from the system

$$Li_1' + R(i_1 - i_2) = v(t)$$
$$R(i_2' - i_1') + \frac{1}{C} i_2 = 0.$$

Solve the system, assuming that $R = 10$ ohms, $L = 20$ henrys, $C = 0.05$ farad, $v = 20$ volts, $i_1(0) = 0$, $i_2(0) = 2$ amperes.

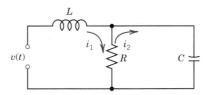

Fig. 137.   Network in Problems 48 and 49

49. Solve the system in Prob. 48 when $R = 0.8$ ohm, $L = 1$ henry, $C = 0.25$ farad, $v = \frac{4}{5}t + \frac{21}{25}$ volt, $i_1(0) = 1$ ampere, and $i_2(0) = -3.8$ ampere. What do the initial conditions mean physically?

50. Set up the model of the network in Fig. 138 and find the solution, assuming that all charges and currents are 0 when the switch is closed at $t = 0$. Find the limits of $i_1(t)$ and $i_2(t)$ as $t \to \infty$, (i) from the solution, (ii) directly from the given network.

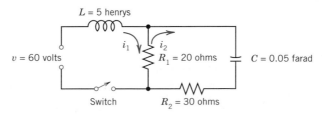

Fig. 138.   Network in Problem 50

SUMMARY OF CHAPTER 5. LAPLACE TRANSFORMS

The main purpose of Laplace transforms is the solution of differential equations and systems of such equations, as well as corresponding initial value problems. The **Laplace transform** $F(s) = \mathcal{L}(f)$ of a function $f(t)$ is defined by

$$(1) \qquad\qquad F(s) = \mathcal{L}(f) = \int_0^\infty e^{-st} f(t)\, dt. \qquad\qquad \text{(Sec. 5.1)}$$

This definition is motivated by the property that the differentiation of f with respect to t corresponds to the multiplication of the transform F by s; more precisely,

$$(2) \qquad\qquad \begin{aligned} \mathcal{L}(f') &= s\mathcal{L}(f) - f(0), \\ \mathcal{L}(f'') &= s^2\mathcal{L}(f) - sf(0) - f'(0), \end{aligned} \qquad\qquad \text{(Sec. 5.2)}$$

etc. Hence by taking the transform of a given differential equation

$$(3) \qquad\qquad y'' + ay' + by = r(t)$$

and writing $\mathcal{L}(y) = Y(s)$, we obtain the **subsidiary equation**

$$(4) \qquad\qquad (s^2 + as + b)Y = \mathcal{L}(r) + sf(0) + f'(0) + af(0).$$

Here, in obtaining the transform $\mathcal{L}(r)$ we can get help from the small table in Sec. 5.1 or the larger table in Sec. 5.9. This is the first step. In the second step we solve the subsidiary equation *algebraically* for $Y(s)$. In the third step we determine the inverse transform $y(t) = \mathcal{L}^{-1}(Y)$, that is, the solution of the problem. This is generally the hardest step, and in it we may again use one of those two tables. $Y(s)$ will often be a rational function, so that we can obtain the inverse $\mathcal{L}^{-1}(Y)$ by partial fraction reduction (Sec. 5.6) if we see no simpler way.

   The Laplace method avoids the determination of a general solution of the homogeneous equation, and we also need not determine values of arbitrary constants in a general solution from initial conditions; instead, we can insert the latter directly into (4). Two further facts account for the practical importance of the Laplace transform. First, it has some basic properties and resulting techniques that simplify the determination of transforms and inverses. The most important of these properties are listed in Sec. 5.8, together with references to the corresponding sections. More on the use of unit step functions and Dirac's delta can be found in Sec. 5.3, and more on convolution in Sec. 5.5. Second, due to these properties, the present method is particularly suitable for handling right sides $r(t)$ given by different expressions over different intervals of time, for instance, when $r(t)$ is a square wave or an impulse or of a form such as $r(t) = \cos t$ if $0 \leqq t \leqq 4\pi$ and 0 elsewhere.

   The application of the Laplace transform to systems of differential equations is shown in Sec. 5.6. (The application to partial differential equations follows in Sec. 11.12.)

P A R T   B

Linear Algebra, Vector Calculus

Chapter 6   **Linear Algebra: Matrices, Vectors, Determinants. Linear Systems of Equations**

Chapter 7   **Linear Algebra: Matrix Eigenvalue Problems**

Chapter 8   **Vector Differential Calculus. Grad, Div, Curl**

Chapter 9   **Vector Integral Calculus. Integral Theorems**

Two main factors have affected the development of engineering mathematics during the past decades, namely, the extensive application of computers to engineering problems and the increased use of linear algebra and linear analysis, for instance, in handling large-scale problems in systems analysis.

The first two chapters of Part B are devoted to **linear algebra,** consisting of the theory and application of vectors and matrices, mainly in connection with the solution of linear systems of equations, with eigenvalue problems, and with linear transformations.

Numerical methods in linear algebra are presented in Chap. 18, which is independent of the other chapters in Part E on numerical methods. Thus Secs. 18.1–18.5 can be studied immediately after Chap. 6 and Secs. 18.6–18.9 after Chap. 7.

The last two chapters of Part B are devoted to **linear analysis,** usually called **vector calculus.** Chapter 8 concerns vector *differential* calculus and includes vector fields, curves, velocity, directional derivative, gradient, divergence, and curl. Chapter 9 covers vector *integral* calculus first discussing line, surface, and triple integrals and then their transformation by the integral theorems of Green, Gauss, and Stokes.

CHAPTER 6

Linear Algebra: Matrices, Vectors, Determinants Linear Systems of Equations

Linear algebra includes the theory and application of linear systems of equations (briefly called linear systems), linear transformations, and eigenvalue problems, as they arise, for instance, from electrical networks, frameworks in mechanics, curve fitting and other optimization problems, systems of differential equations, and processes in statistics.

Linear algebra makes systematic use of **vectors** and **matrices** (Sec. 6.1) and, to a lesser extent, **determinants** (Sec. 6.6). This requires the study of properties of matrices as a central task by itself.

A matrix is a rectangular array of numbers. Matrices occur in various problems, for instance, as arrays of coefficients of equations. Matrices (and vectors) are useful because they enable us to consider an array of many numbers as a single object, denote it by a single symbol, and perform calculations with these symbols in a very compact form. The "mathematical shorthand" thus obtained is very elegant and powerful and is suitable for various practical problems.

The definitions of matrices and vectors and related concepts are given in Sec. 6.1, together with a discussion of the two basic algebraic operations for matrices, addition and scalar multiplication. Matrix multiplication follows in Sec. 6.2, where we also define special matrices and the operation of transposition of a matrix or vector.

The remaining sections center around **linear systems of equations.** This includes the Gauss elimination (Sec. 6.3), the important role of rank (Secs. 6.4, 6.5, 6.7), special vector spaces (Sec. 6.4), the basic existence and uniqueness problem for solutions (Sec. 6.5), determinants and their use in linear systems (Cramer's rule, Sec. 6.6), and the inverse of a matrix and its calculation by the Gauss–Jordan method (Sec. 6.7).

The last section (6.8) on vector spaces, inner product spaces, and linear transformations is more abstract.

Applications to practical problems are shown throughout the chapter.

NUMERICAL METHODS in Secs. 18.1–18.5 can be studied immediately after the corresponding material in this chapter.

Eigenvalue problems for matrices follow in Chap. 7.

Prerequisite for this chapter: None.
Sections that may be omitted in a shorter course: 6.5, 6.8.
References: Appendix 1, Part B.
Answers to problems: Appendix 2.

6.1 Basic Concepts
Matrix Addition, Scalar Multiplication

The first two sections of this chapter introduce the basic concepts and rules of matrix and vector algebra. The main application to linear systems of equations begins in Sec. 6.3.

A **matrix** is a rectangular array of numbers (or functions) enclosed in brackets. These numbers (or functions) are called *entries* or *elements* of the matrix. For example,

$$(1) \qquad \begin{bmatrix} 2 & 0.4 & 8 \\ 5 & -32 & 0 \end{bmatrix}, \quad \begin{bmatrix} 6 \\ 1 \end{bmatrix}, \quad [a_1 \ a_2 \ a_3], \quad \begin{bmatrix} a & b \\ c & d \end{bmatrix}, \quad \begin{bmatrix} e^x & 3x \\ e^{2x} & x^2 \end{bmatrix}$$

are matrices. The first has two *"rows"* (horizontal lines) and three *"columns"* (vertical lines). The second consists of a single column, and we call it a *column vector*. The third consists of a single row, and we call it a *row vector*. The last two are *square matrices,* that is, each has as many rows as columns (two in this case).

Matrices are practical in connection with various applications, as we shall see throughout this chapter. As a first illustration, let us consider two examples that are typical.

EXAMPLE 1   **Coefficient matrix of a linear system of equations**

In a system of equations, such as

$$5x - 2y + \ z = 0$$
$$3x + \qquad 4z = 0$$

the coefficients of the unknowns x, y, z are the entries of the *coefficient matrix,* call it **A,**

$$\mathbf{A} = \begin{bmatrix} 5 & -2 & 1 \\ 3 & 0 & 4 \end{bmatrix},$$

which displays these coefficients in the pattern of the equations. That is, their position in **A** corresponds to that in the system when written as shown.

The notation for the unknowns is not very essential. We could denote them by x_1, x_2, x_3 or by some other letters. Essential is the coefficient matrix **A** because the coefficients of the system (and the numbers on the right, 0, 0 in the present case) contain all the information on solutions. We shall discuss this in great detail, beginning in Sec. 6.3.   ◀

EXAMPLE 2   **Sales figures in matrix form**

Sales figures for three products I, II, III in a store on Monday (M), Tuesday (T), · · · may for each week be arranged in a matrix

$$\mathbf{A} = \begin{bmatrix} 40 & 33 & 81 & 0 & 21 & 47 \\ 0 & 12 & 78 & 50 & 50 & 96 \\ 10 & 0 & 0 & 27 & 43 & 78 \end{bmatrix} \begin{matrix} \text{I} \\ \text{II} \\ \text{III} \end{matrix}$$

	M	T	W	Th	F	S

and if the company has ten stores, we can set up ten such matrices, one for each store; then by adding corresponding entries of these matrices we can get a matrix showing the total sales of each product on each day. Can you think of other data for which matrices are feasible? For instance, in transportation or storage problems? Or in recording phone calls, or in listing distances in a network of roads?   ◀

General Notations and Concepts

Our discussion suggests the following. We denote matrices by capital boldface letters **A, B, C,** $\cdots$, or by writing the general entry in brackets; thus, $\mathbf{A} = [a_{jk}]$, and so on. By an **$m \times n$ matrix** (read "m by n matrix") we mean a matrix with m rows and n columns. Thus an $m \times n$ matrix is of the form

$$(2) \qquad \mathbf{A} = [a_{jk}] = \begin{bmatrix} a_{11} & a_{12} & \cdots & a_{1n} \\ a_{21} & a_{22} & \cdots & a_{2n} \\ . & . & \cdots & . \\ a_{m1} & a_{m2} & \cdots & a_{mn} \end{bmatrix}.$$

Hence the matrices in (1) are 2×3, 2×1, 1×3, 2×2, and 2×2.

In the **double-subscript notation** *for the entries, the first subscript always denotes the* **row** *and the second the* **column** *in which the given entry stands.* Thus a_{23} is the entry in the second row and third column.

If $m = n$, we call **A** an $n \times n$ **square matrix.** Then its diagonal containing the entries $a_{11}, a_{22}, \cdots, a_{nn}$ is called the **main diagonal** or *principal diagonal* of **A**. Thus the last two matrices in (1) are square. Their main diagonals are a, d and e^x, x^2, respectively. Square matrices are particularly important, as we shall see.

A matrix that is not square is called a **rectangular matrix.**

Vectors

A **vector** is a matrix that has only one row—then we call the matrix a **row vector**—or only one column—then we call it a **column vector.** In both cases we call its entries **components** and denote the vector by a *lowercase* boldface letter such as **a, b,** $\cdots$, or by its general component in brackets, $\mathbf{a} = [a_j]$, and so on.

EXAMPLE 3   **Row vectors. Column vectors. Transposition of vectors**

A **row vector** is of the form

$$\mathbf{a} = \begin{bmatrix} a_1 & a_2 & \cdots & a_n \end{bmatrix}. \qquad \text{For instance,} \qquad \mathbf{a} = \begin{bmatrix} 5 & 3 & \frac{1}{2} \end{bmatrix}.$$

A **column vector** is of the form

$$\mathbf{b} = \begin{bmatrix} b_1 \\ b_2 \\ \vdots \\ b_m \end{bmatrix}. \qquad \text{For instance,} \qquad \mathbf{b} = \begin{bmatrix} 4 \\ 0 \\ -7 \end{bmatrix}.$$

Row vectors can be converted to column vectors (and conversely) by an operation that is called **transposition** and is indicated by ^{T}. Thus $\mathbf{a}^{\mathsf{T}}$ is a column vector and $\mathbf{b}^{\mathsf{T}}$ a row vector:

$$\mathbf{a}^{\mathsf{T}} = \begin{bmatrix} 5 \\ 3 \\ \frac{1}{2} \end{bmatrix}, \qquad \mathbf{b}^{\mathsf{T}} = \begin{bmatrix} 4 & 0 & -7 \end{bmatrix}.$$

It will depend on the purpose as to which of the two forms of vector is more practical. We shall preferably use column vectors. ◀

The rows and columns of an $m \times n$ matrix **A** are sometimes also called the *row vectors* and *column vectors of* **A.** This should not cause any confusion.

Transposition

It is practical to define **transposition** for any matrix. The **transpose** $\mathbf{A}^\mathsf{T}$ of an $m \times n$ matrix $\mathbf{A} = [a_{jk}]$ as given in (2) is the $n \times m$ matrix that has the first *row* of **A** as its first *column,* the second *row* of **A** as its second *column,* and so on. Thus the transpose of **A** in (2) is

$$
(3) \qquad \mathbf{A}^\mathsf{T} = [a_{kj}] = \begin{bmatrix} a_{11} & a_{21} & \cdots & a_{m1} \\ a_{12} & a_{22} & \cdots & a_{m2} \\ \cdot & \cdot & \cdots & \cdot \\ a_{1n} & a_{2n} & \cdots & a_{mn} \end{bmatrix}.
$$

EXAMPLE 4   **Transposition of a matrix**

If
$$
\mathbf{A} = \begin{bmatrix} 5 & -8 & 1 \\ 4 & 0 & 0 \end{bmatrix}, \qquad \text{then} \qquad \mathbf{A}^\mathsf{T} = \begin{bmatrix} 5 & 4 \\ -8 & 0 \\ 1 & 0 \end{bmatrix}.
$$
◀

Symmetric matrices and **skew-symmetric matrices** are *square* matrices whose transpose equals the matrix or minus the matrix, respectively:

$$\boxed{\mathbf{A}^\mathsf{T} = \mathbf{A}} \qquad\qquad\qquad \boxed{\mathbf{A}^\mathsf{T} = -\mathbf{A}}$$

Symmetric matrix   Skew-symmetric matrix

Here, $-\mathbf{A} = [-a_{jk}]$ is obtained from **A** by replacing each a_{jk} with the corresponding $-a_{jk}$ (e.g., 4 by -4, -3 by 3, etc.). Symmetric and skew-symmetric matrices are quite important and will occur frequently in this chapter.

Equality of Matrices

What makes matrices and vectors really useful is that we can calculate with them almost as easily as with numbers. Indeed, practical applications suggested the rules of addition and multiplication by scalars (numbers), which we now introduce. (Multiplication of matrices by matrices follows in the next section.)

We say briefly that two matrices have the **same size** if they are both $m \times n$, for instance, both 3×4. We begin by defining equality.

Definition. Equality of matrices

Two matrices $\mathbf{A} = [a_{jk}]$ and $\mathbf{B} = [b_{jk}]$ are equal, written $\mathbf{A} = \mathbf{B}$, if and only if they have the same size and the corresponding entries are equal, that is, $a_{11} = b_{11}$, $a_{12} = b_{12}$, and so on.

EXAMPLE 5   Equality of matrices

$$\mathbf{A} = \begin{bmatrix} a_{11} & a_{12} \\ a_{21} & a_{22} \end{bmatrix} = \mathbf{B} = \begin{bmatrix} 4 & 0 \\ 3 & -1 \end{bmatrix} \quad \text{if and only if} \quad \begin{aligned} a_{11} = 4, & \quad a_{12} = 0, \\ a_{21} = 3, & \quad a_{22} = -1. \end{aligned}$$

Matrices of the same size that are all different (not equal) are, for instance,

$$\begin{bmatrix} 3 & \frac{1}{4} \\ 0 & -7 \end{bmatrix}, \qquad \begin{bmatrix} 3 & -7 \\ 0 & \frac{1}{4} \end{bmatrix}, \qquad \begin{bmatrix} 3 & \frac{1}{4} \\ 0 & 7 \end{bmatrix}, \qquad \begin{bmatrix} 3 & 1 \\ 0 & -7 \end{bmatrix}.$$

$\mathbf{A}$ and $\mathbf{A}^\mathsf{T}$ in Example 4 are not equal because they are not even of the same size. (What are their sizes?) Similarly for $\mathbf{a}$ and $\mathbf{a}^\mathsf{T}$ in Example 3, as well as for $\mathbf{b}$ and $\mathbf{b}^\mathsf{T}$.   ◀

Matrix Addition

Definition. Addition of matrices

Addition is defined only for matrices $\mathbf{A} = [a_{jk}]$ and $\mathbf{B} = [b_{jk}]$ of the same size; their **sum,** written $\mathbf{A} + \mathbf{B},$ is then obtained by adding the corresponding entries. Matrices of different sizes cannot be added.

As a special case, the **sum a** $+$ **b** of two row vectors or two column vectors, which must have the same number of components, is obtained by adding the corresponding components.

EXAMPLE 6   Addition of matrices and vectors

If $\quad \mathbf{A} = \begin{bmatrix} -4 & 6 & 3 \\ 0 & 1 & 2 \end{bmatrix} \quad$ and $\quad \mathbf{B} = \begin{bmatrix} 5 & -1 & 0 \\ 3 & 1 & 0 \end{bmatrix}, \quad$ then $\quad \mathbf{A} + \mathbf{B} = \begin{bmatrix} 1 & 5 & 3 \\ 3 & 2 & 2 \end{bmatrix}.$

Our present $\mathbf{A}$ and $\mathbf{A}^\mathsf{T}$ cannot be added. $\mathbf{B}$ in Example 5 and the present $\mathbf{B}$ cannot be added. If $\mathbf{a} = \begin{bmatrix} 5 & 7 & 2 \end{bmatrix}$ and $\mathbf{b} = \begin{bmatrix} -6 & 2 & 0 \end{bmatrix}$, then $\mathbf{a} + \mathbf{b} = \begin{bmatrix} -1 & 9 & 2 \end{bmatrix}$.

An application of matrix addition was suggested in Example 2. Many others will follow.   ◀

Scalar Multiplication

Definition. Scalar multiplication (Multiplication by a number)

The **product** of any $m \times n$ matrix $\mathbf{A} = [a_{jk}]$ and any scalar c (number c), written $c\mathbf{A}$, is the $m \times n$ matrix $c\mathbf{A} = [ca_{jk}]$ obtained by multiplying each entry in $\mathbf{A}$ by c.

Here $(-1)\mathbf{A}$ is simply written $-\mathbf{A}$ and is called the **negative** of $\mathbf{A}$. Similarly, $(-k)\mathbf{A}$ is written $-k\mathbf{A}.$ Also, $\mathbf{A} + (-\mathbf{B})$ is written $\mathbf{A} - \mathbf{B}$ and is called the **difference** of $\mathbf{A}$ and $\mathbf{B}$ (which must have the same size!).

EXAMPLE 7   Scalar multiplication

If $\quad \mathbf{A} = \begin{bmatrix} 2.7 & -1.8 \\ 0 & 0.9 \\ 9.0 & -4.5 \end{bmatrix}, \quad$ then $\quad -\mathbf{A} = \begin{bmatrix} -2.7 & 1.8 \\ 0 & -0.9 \\ -9.0 & 4.5 \end{bmatrix}, \quad \frac{10}{9}\mathbf{A} = \begin{bmatrix} 3 & -2 \\ 0 & 1 \\ 10 & -5 \end{bmatrix}, \quad 0\mathbf{A} = \begin{bmatrix} 0 & 0 \\ 0 & 0 \\ 0 & 0 \end{bmatrix}.$ ◀

From the familiar laws for numbers we obtain similar laws for matrix addition and scalar multiplication. Indeed, for matrices of the same size $m \times n$ we obtain for addition

(4)

　(a) $\qquad \mathbf{A} + \mathbf{B} = \mathbf{B} + \mathbf{A}$

　(b) $\qquad (\mathbf{U} + \mathbf{V}) + \mathbf{W} = \mathbf{U} + (\mathbf{V} + \mathbf{W}) \qquad$ (written $\mathbf{U} + \mathbf{V} + \mathbf{W}$)

　(c) $\qquad \mathbf{A} + \mathbf{0} = \mathbf{A}$

　(d) $\qquad \mathbf{A} + (-\mathbf{A}) = \mathbf{0}$

and for scalar multiplication

(5)

　(a) $\qquad c(\mathbf{A} + \mathbf{B}) = c\mathbf{A} + c\mathbf{B}$

　(b) $\qquad (c + k)\mathbf{A} = c\mathbf{A} + k\mathbf{A}$

　(c) $\qquad c(k\mathbf{A}) = (ck)\mathbf{A} \qquad$ (written $ck\mathbf{A}$)

　(d) $\qquad 1\mathbf{A} = \mathbf{A}.$

Here, $\mathbf{0}$ in (4) denotes the **zero matrix** (of size $m \times n$), that is, the matrix with all entries zero.

Transposition of a sum can be done term by term,

(6) $$(\mathbf{A} + \mathbf{B})^{\mathsf{T}} = \mathbf{A}^{\mathsf{T}} + \mathbf{B}^{\mathsf{T}},$$

as you may prove, and for scalar multiplication we have

(7) $$(c\mathbf{A})^{\mathsf{T}} = c\mathbf{A}^{\mathsf{T}}.$$

One more algebraic operation, the multiplication of matrices by matrices, follows in the next section. Then we shall be ready for applications.

PROBLEM SET 6.1

Matrix Addition and Scalar Multiplication. Let

$$\mathbf{A} = \begin{bmatrix} 2 & 1 \\ 1 & 7 \end{bmatrix}, \quad \mathbf{B} = \begin{bmatrix} -2 & 5 \\ 0 & 8 \end{bmatrix}, \quad \mathbf{C} = \begin{bmatrix} 6 & 0 & 3 \\ 1 & 0 & -5 \end{bmatrix}, \quad \mathbf{D} = \begin{bmatrix} 4 & 0 & -4 \\ -3 & 4 & 9 \end{bmatrix}.$$

Find the following expressions or give reasons why they are undefined.

1. $\mathbf{A} + \mathbf{B}, \mathbf{B} + \mathbf{A}, \mathbf{A} + \mathbf{B} + \mathbf{C}$　　　　　2. $4\mathbf{A} - 8\mathbf{B}, 4(2\mathbf{B} - \mathbf{A}), 8\mathbf{B} - 4\mathbf{A}$

3. $\mathbf{A} + \mathbf{C}, (\mathbf{C}^{\mathsf{T}})^{\mathsf{T}}, \mathbf{C} + \mathbf{C}^{\mathsf{T}}$　　　　　4. $6\mathbf{C} - 5\mathbf{C}, 3\mathbf{C}^{\mathsf{T}} + 2\mathbf{D}^{\mathsf{T}}, \mathbf{C} - 2\mathbf{C}^{\mathsf{T}}$

5. $5\mathbf{D} - 3\mathbf{C}, 5\mathbf{D}^{\mathsf{T}} - 3\mathbf{C}^{\mathsf{T}}$　　　　　6. $\mathbf{A} + 0\mathbf{C}, \mathbf{C} - 0\mathbf{A}, 0\mathbf{B}$

7. $(\mathbf{B}^{\mathsf{T}} - \mathbf{A}^{\mathsf{T}})^{\mathsf{T}}, \mathbf{B} - \mathbf{A}, \mathbf{A} - \mathbf{A}^{\mathsf{T}}$　　　　　8. $6(\mathbf{C}^{\mathsf{T}} + 3\mathbf{D}^{\mathsf{T}})^{\mathsf{T}}, 6\mathbf{C} + 18\mathbf{D}$

Addition and Scalar Multiplication of Vectors. Let

$$\mathbf{a} = [3 \quad 0 \quad 4], \quad \mathbf{b} = [-1 \quad 8 \quad 2], \quad \mathbf{c} = \begin{bmatrix} 9 \\ 5 \\ 7 \end{bmatrix}, \quad \mathbf{d} = \begin{bmatrix} 2 \\ -2 \\ 6 \end{bmatrix}.$$

Find the following expressions or give reasons why they are undefined.

9. $7\mathbf{a} - 5\mathbf{b}, 7\mathbf{a}^\mathsf{T} - 5\mathbf{b}^\mathsf{T}$

10. $\mathbf{a} - \mathbf{c}^\mathsf{T}, \mathbf{c} - \mathbf{a}^\mathsf{T}, \mathbf{a} + \mathbf{b} + \mathbf{c}$

11. $3(\mathbf{c} - 4\mathbf{d}), 3\mathbf{c} - 12\mathbf{d}$

12. $\mathbf{a} - \mathbf{b} + \mathbf{c}^\mathsf{T} - \mathbf{d}^\mathsf{T}, 12(\mathbf{b} - \mathbf{d}^\mathsf{T})$

13. $5(\mathbf{c} - 2\mathbf{d}), 10\mathbf{d} - 5\mathbf{c}$

14. $\mathbf{b} + \mathbf{c}, \mathbf{c} - \mathbf{a}^\mathsf{T}, \mathbf{c} + \mathbf{c}^\mathsf{T}$

15. $6\mathbf{a} - 5\mathbf{a}, (\mathbf{a}^\mathsf{T})^\mathsf{T} - \mathbf{a}, \mathbf{a}^\mathsf{T} - \mathbf{a}$

16. $(3\mathbf{c})^\mathsf{T} - 3\mathbf{c}^\mathsf{T}, 4\mathbf{c} - 4\mathbf{c}^\mathsf{T}, \mathbf{b} + 4\mathbf{d}^\mathsf{T}$

17. (Matrix addition, scalar multiplication) Prove (4) and (5) for general 2×3 matrices and scalars c and k.

18. (Transposition) Prove (6) and (7) for general 4×3 matrices and scalar c.

19. TEAM PROJECT. Symmetric and skew-symmetric matrices are square matrices defined by $\mathbf{A}^\mathsf{T} = \mathbf{A}$ and $\mathbf{B}^\mathsf{T} = -\mathbf{B}$, respectively (see before). They occur quite frequently in applications, so it is worthwhile to study some of their most important properties.

(a) Show that $\mathbf{A} = \begin{bmatrix} a_{jk} \end{bmatrix}$ is symmetric if and only if $a_{kj} = a_{jk}$ for all j and k, and $\mathbf{B} = \begin{bmatrix} b_{jk} \end{bmatrix}$ is skew-symmetric if and only if $b_{kj} = -b_{jk}$, in particular, $b_{jj} = 0$.

(b) Show that for every square matrix $\mathbf{C}$ the matrix $\mathbf{C} + \mathbf{C}^\mathsf{T}$ is symmetric and that $\mathbf{C} - \mathbf{C}^\mathsf{T}$ is skew-symmetric. Hence $\mathbf{C}$ can be written $\mathbf{C} = \mathbf{S} + \mathbf{T}$, where $\mathbf{S}$ is symmetric and $\mathbf{T}$ is skew-symmetric. Find $\mathbf{S}$ and $\mathbf{T}$. Represent $\mathbf{A}$ and $\mathbf{B}$ in Probs. 1–8 in this form.

(c) A **linear combination** of matrices $\mathbf{A}, \mathbf{B}, \mathbf{C}, \cdots, \mathbf{M}$ of the same size is an expression of the form

$$(8) \qquad\qquad a\mathbf{A} + b\mathbf{B} + c\mathbf{C} + \cdots + m\mathbf{M},$$

where $a, \cdots, m$ are any scalars. Show that if these matrices are symmetric, so is (8). Similarly, show that if they are skew-symmetric, so is (8).

20. TEAM PROJECT. Matrices for Networks. Matrices have various engineering applications, as we shall see. For instance, they can be used to characterize connections in electrical networks, in nets of roads, in production processes, etc., as follows.

(a) Nodal Incidence Matrix. The network in Fig. 139 consists of 6 *branches* (connections) and 4 *nodes* (points where two or more branches come together). One node is the *reference node* (grounded node, whose voltage is zero). We number the other nodes and number and direct the branches. This we do arbitrarily. The network can now be described by a matrix $\mathbf{A} = \begin{bmatrix} a_{jk} \end{bmatrix}$, where

$$a_{jk} = \begin{cases} +1 & \text{if branch } k \text{ leaves node } \textcircled{j} \\ -1 & \text{if branch } k \text{ enters node } \textcircled{j} \\ 0 & \text{if branch } k \text{ does not touch node } \textcircled{j}. \end{cases}$$

$\mathbf{A}$ is called the *nodal incidence matrix* of the network. Show that for the network in Fig. 139 the matrix $\mathbf{A}$ has the given form.

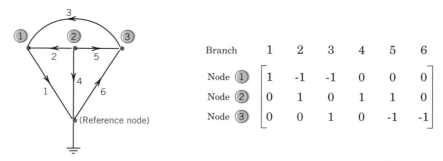

Branch	1	2	3	4	5	6
Node ①	1	-1	-1	0	0	0
Node ②	0	1	0	1	1	0
Node ③	0	0	1	0	-1	-1

Fig. 139. Network and nodal incidence matrix in Team Project 20(a)

(b) Find the nodal incidence matrices in the networks in Fig. 140.

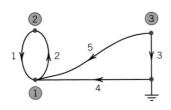 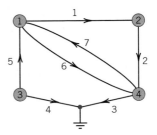

Fig. 140. Electrical networks in Team Project 20(b)

(c) Sketch the three networks corresponding to the nodal incidence matrices

$$
\begin{bmatrix} 1 & 0 & 0 & 1 \\ -1 & 1 & 0 & 0 \\ 0 & -1 & 1 & 0 \end{bmatrix}, \quad
\begin{bmatrix} 1 & -1 & 0 & 0 & 1 \\ -1 & 1 & -1 & 1 & 0 \\ 0 & 0 & 1 & -1 & 0 \end{bmatrix}, \quad
\begin{bmatrix} 1 & 0 & 1 & 0 & 0 \\ -1 & 1 & 0 & 1 & 0 \\ 0 & -1 & -1 & 0 & 1 \end{bmatrix}.
$$

(d) Mesh Incidence Matrix. A network can also be characterized by the *mesh incidence matrix* $\mathbf{M} = [m_{jk}]$, where

$$
m_{jk} = \begin{cases} +1 & \text{if branch } k \text{ is in mesh } \boxed{j} \text{ and has the same orientation} \\ -1 & \text{if branch } k \text{ is in mesh } \boxed{j} \text{ and has the opposite orientation} \\ 0 & \text{if branch } k \text{ is not in mesh } \boxed{j} \end{cases}
$$

and a mesh is a loop with no branch in its interior (or in its exterior). Here, the meshes are numbered and directed (oriented) in an arbitrary fashion. Show that for the network in Fig. 141, the matrix $\mathbf{M}$ has the given form, where row 1 corresponds to mesh 1, etc.

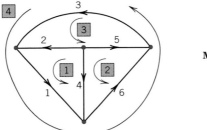
$$
\mathbf{M} = \begin{bmatrix} 1 & 1 & 0 & -1 & 0 & 0 \\ 0 & 0 & 0 & 1 & -1 & 1 \\ 0 & -1 & 1 & 0 & 1 & 0 \\ 1 & 0 & 1 & 0 & 0 & 1 \end{bmatrix}
$$

Fig. 141. Network and matrix $\mathbf{M}$ in Team Project 20(d)

6.2 Matrix Multiplication

Matrix multiplication means multiplication of matrices by matrices. This is the last algebraic operation to be defined. Now matrices are *added* by *adding* corresponding entries. Is the same true for multiplication? The answer is no. The definition of multiplication looks artificial. But please be patient, it will be fully motivated by the use of matrices in "linear transformations," by which this multiplication is suggested.

Definition. Multiplication of a matrix by a matrix

The product $C = AB$ (in this order) of an $m \times n$ matrix $A = [a_{jk}]$ and an $r \times p$ matrix $B = [b_{jk}]$ is defined if and only if $r = n$, that is,

Number of rows of 2nd factor B = *Number of columns of 1st factor* A,

and is then defined as the $m \times p$ matrix $C = [c_{jk}]$ with entries

(1)
$$c_{jk} = \sum_{l=1}^{n} a_{jl}b_{lk} = a_{j1}b_{1k} + a_{j2}b_{2k} + \cdots + a_{jn}b_{nk}$$
$$j = 1, \cdots, m$$
$$k = 1, \cdots, p.$$

That is, to get c_{jk}, multiply each entry in the jth **row** of A by the corresponding entry in the kth **column** of B and then add these n products. One says briefly that this is a *"multiplication of rows into columns."* Figure 142 illustrates this.

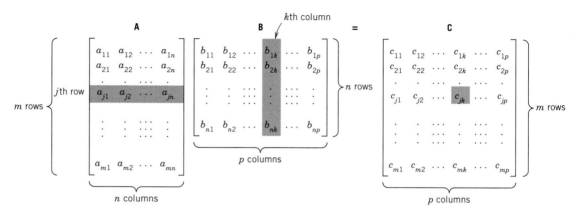

Fig. 142. Matrix multiplication $AB = C$

EXAMPLE 1   **Matrix multiplication**

$$AB = \begin{bmatrix} 4 & 3 \\ 7 & 2 \\ 9 & 0 \end{bmatrix} \begin{bmatrix} 2 & 5 \\ 1 & 6 \end{bmatrix} = \begin{bmatrix} 4 \cdot 2 + 3 \cdot 1 & 4 \cdot 5 + 3 \cdot 6 \\ 7 \cdot 2 + 2 \cdot 1 & 7 \cdot 5 + 2 \cdot 6 \\ 9 \cdot 2 + 0 \cdot 1 & 9 \cdot 5 + 0 \cdot 6 \end{bmatrix} = \begin{bmatrix} 11 & 38 \\ 16 & 47 \\ 18 & 45 \end{bmatrix}.$$

Here A is 3×2 and B is 2×2, so that AB comes out 3×2, whereas BA is not defined. ◀

EXAMPLE 2   **Multiplication of a matrix and a vector**

$$\begin{bmatrix} 4 & 2 \\ 1 & 8 \end{bmatrix} \begin{bmatrix} 3 \\ 5 \end{bmatrix} = \begin{bmatrix} 12 + 10 \\ 3 + 40 \end{bmatrix} = \begin{bmatrix} 22 \\ 43 \end{bmatrix} \qquad \text{whereas} \qquad \begin{bmatrix} 3 \\ 5 \end{bmatrix} \begin{bmatrix} 4 & 2 \\ 1 & 8 \end{bmatrix} \quad \text{is undefined.} \quad ◀$$

EXAMPLE 3   **Products of row and column vectors**

$$[3 \quad 6 \quad 1] \begin{bmatrix} 1 \\ 2 \\ 4 \end{bmatrix} = [19], \qquad \begin{bmatrix} 1 \\ 2 \\ 4 \end{bmatrix} [3 \quad 6 \quad 1] = \begin{bmatrix} 3 & 6 & 1 \\ 6 & 12 & 2 \\ 12 & 24 & 4 \end{bmatrix}. \qquad ◀$$

Differences from Multiplication of Numbers

Matrix multiplication differs from the multiplication of numbers by some unusual properties. Of course, this is of great practical importance.

EXAMPLE 4 **CAUTION! Matrix multiplication is not commutative, AB ≠ BA in general**

This is illustrated by Examples 1, 2, and 3, but also holds for square matrices; for instance,

$$
\begin{bmatrix} 9 & 3 \\ -2 & 0 \end{bmatrix} \begin{bmatrix} 1 & -4 \\ 2 & 5 \end{bmatrix} = \begin{bmatrix} 9 \cdot 1 + 3 \cdot 2 & 9 \cdot (-4) + 3 \cdot 5 \\ -2 \cdot 1 + 0 \cdot 2 & (-2) \cdot (-4) + 0 \cdot 5 \end{bmatrix} = \begin{bmatrix} 15 & -21 \\ -2 & 8 \end{bmatrix}
$$

whereas

$$
\begin{bmatrix} 1 & -4 \\ 2 & 5 \end{bmatrix} \begin{bmatrix} 9 & 3 \\ -2 & 0 \end{bmatrix} = \begin{bmatrix} 1 \cdot 9 + (-4) \cdot (-2) & 1 \cdot 3 + (-4) \cdot 0 \\ 2 \cdot 9 + 5 \cdot (-2) & 2 \cdot 3 + 5 \cdot 0 \end{bmatrix} = \begin{bmatrix} 17 & 3 \\ 8 & 6 \end{bmatrix}.
$$
◀

EXAMPLE 5 **CAUTION! AB = 0 does not necessarily imply A = 0 or B = 0 or BA = 0**

$$
\begin{bmatrix} 1 & 1 \\ 2 & 2 \end{bmatrix} \begin{bmatrix} -1 & 1 \\ 1 & -1 \end{bmatrix} = \begin{bmatrix} 0 & 0 \\ 0 & 0 \end{bmatrix}, \qquad \begin{bmatrix} -1 & 1 \\ 1 & -1 \end{bmatrix} \begin{bmatrix} 1 & 1 \\ 2 & 2 \end{bmatrix} = \begin{bmatrix} 1 & 1 \\ -1 & -1 \end{bmatrix}.
$$
◀

EXAMPLE 6 **CAUTION! AC = AD does not necessarily imply C = D (even when A ≠ 0)**

$$
\begin{bmatrix} 1 & 1 \\ 2 & 2 \end{bmatrix} \begin{bmatrix} 2 & 1 \\ 2 & 2 \end{bmatrix} = \begin{bmatrix} 4 & 3 \\ 8 & 6 \end{bmatrix} = \begin{bmatrix} 1 & 1 \\ 2 & 2 \end{bmatrix} \begin{bmatrix} 3 & 0 \\ 1 & 3 \end{bmatrix}, \qquad \text{but} \quad \begin{bmatrix} 2 & 1 \\ 2 & 2 \end{bmatrix} \neq \begin{bmatrix} 3 & 0 \\ 1 & 3 \end{bmatrix}.
$$

That is, the cancellation law for numbers does not extend to matrices.

The illegitimate use of the cancellation law is a major source of errors among beginners! ◀

We have thus discovered three properties by which matrix multiplication differs from that for numbers,

(2a) $$\mathbf{AB} \neq \mathbf{BA} \qquad \text{in general;}$$

(2b) $$\mathbf{AB} = \mathbf{0} \quad \text{does not necessarily imply} \quad \mathbf{A} = \mathbf{0} \text{ or } \mathbf{B} = \mathbf{0} \quad \text{or} \quad \mathbf{BA} = \mathbf{0},$$

(2c) $$\mathbf{AC} = \mathbf{AD} \quad \text{does not necessarily imply} \quad \mathbf{C} = \mathbf{D}.$$

[(2c) follows from (2b) with $\mathbf{B} = \mathbf{C} - \mathbf{D}$; thus $\mathbf{A}(\mathbf{C} - \mathbf{D}) = \mathbf{0}$ or $\mathbf{AC} = \mathbf{AD}$.] Hence, always observe the order of factors very carefully! To emphasize this, we say that in **AB,** the matrix **B** is *premultiplied,* or *multiplied from the left,* by **A,** and **A** is *postmultiplied,* or *multiplied from the right,* by **B.** More about (2b) will be said in Sec. 6.7. The other properties of matrix multiplication are similar to those for numbers, namely,

(2)

      (d) $(k\mathbf{A})\mathbf{B} = k(\mathbf{AB}) = \mathbf{A}(k\mathbf{B})$ *written k**AB** or **A**k**B***

      (e) $\mathbf{A}(\mathbf{BC}) = (\mathbf{AB})\mathbf{C}$ *written **ABC***

      (f) $(\mathbf{A} + \mathbf{B})\mathbf{C} = \mathbf{AC} + \mathbf{BC}$

      (g) $\mathbf{C}(\mathbf{A} + \mathbf{B}) = \mathbf{CA} + \mathbf{CB}$

provided **A, B,** and **C** are such that the expressions on the left are defined; here, k is any scalar. (2e) is called the *associative law*. (2f) and (2g) are called the *distributive laws*.

Special Matrices

Certain kinds of matrices will occur quite frequently in our work and we now list the most important ones of them.

Triangular Matrices

Upper triangular matrices are square matrices that can have nonzero entries only on and ***above*** the main diagonal, whereas any entry below the diagonal must be zero. Similarly, **lower triangular matrices** can have nonzero entries only on and ***below*** the main diagonal. Any entry *on* the main diagonal of a triangular matrix may be zero or not.

EXAMPLE 7   **Upper and lower triangular matrices**

$$\begin{bmatrix} 1 & 3 \\ 0 & 2 \end{bmatrix}, \quad \begin{bmatrix} 1 & 4 & 2 \\ 0 & 3 & 2 \\ 0 & 0 & 6 \end{bmatrix}, \quad \begin{bmatrix} 4 & 2 & 2 & 0 \\ 0 & -3 & 5 & 1 \\ 0 & 0 & 0 & -6 \\ 0 & 0 & 0 & 5 \end{bmatrix}$$

Upper triangular

$$\begin{bmatrix} 5 & 0 \\ 2 & 3 \end{bmatrix}, \quad \begin{bmatrix} 2 & 0 & 0 \\ 8 & -1 & 0 \\ 7 & 6 & 8 \end{bmatrix}, \quad \begin{bmatrix} 3 & 0 & 0 & 0 \\ 9 & -3 & 0 & 0 \\ 1 & 0 & 2 & 0 \\ 1 & 9 & 3 & 6 \end{bmatrix} \qquad \blacktriangleleft$$

Lower triangular

Diagonal Matrices

These are square matrices that can have nonzero entries only on the main diagonal. Any entry above or below the main diagonal must be zero.

If all the diagonal entries of a diagonal matrix **S** are equal, say, *c*, we call **S** a **scalar matrix** because multiplication of any square matrix **A** of the same size by **S** has the same effect as the multiplication by a scalar, that is,

$$(3) \qquad\qquad\qquad \mathbf{AS} = \mathbf{SA} = c\mathbf{A}.$$

In particular, a scalar matrix whose entries on the main diagonal are all 1 is called a **unit matrix** (or **identity matrix**) and is denoted by $\mathbf{I}_n$ or simply by **I.** For **I,** formula (3) becomes

$$(4) \qquad\qquad\qquad \mathbf{AI} = \mathbf{IA} = \mathbf{A}.$$

EXAMPLE 8   **Diagonal matrix D. Scalar matrix S. Unit matrix I**

$$\mathbf{D} = \begin{bmatrix} 2 & 0 & 0 \\ 0 & -3 & 0 \\ 0 & 0 & 0 \end{bmatrix}, \quad \mathbf{S} = \begin{bmatrix} c & 0 & 0 \\ 0 & c & 0 \\ 0 & 0 & c \end{bmatrix}, \quad \mathbf{I} = \begin{bmatrix} 1 & 0 & 0 \\ 0 & 1 & 0 \\ 0 & 0 & 1 \end{bmatrix} \qquad \blacktriangleleft$$

Transpose of a Product

The transpose (see Sec. 6.1) *of a product equals the product of the transposed factors,* ***taken in reverse order,***

(5)
$$(\mathbf{AB})^\mathsf{T} = \mathbf{B}^\mathsf{T}\mathbf{A}^\mathsf{T}.$$

The proof of the useful formula (5) follows from the definition of matrix multiplication and is left to the student. (See Team Project 10.)

EXAMPLE 9   **Transposition of a product**

Formula (5) is illustrated by

$$(\mathbf{AB})^\mathsf{T} = \left(\begin{bmatrix} 4 & 9 \\ 0 & 2 \\ 1 & 6 \end{bmatrix} \begin{bmatrix} 3 & 7 \\ 2 & 8 \end{bmatrix} \right)^\mathsf{T} = \begin{bmatrix} 30 & 100 \\ 4 & 16 \\ 15 & 55 \end{bmatrix}^\mathsf{T} = \begin{bmatrix} 30 & 4 & 15 \\ 100 & 16 & 55 \end{bmatrix},$$

$$\mathbf{B}^\mathsf{T}\mathbf{A}^\mathsf{T} = \begin{bmatrix} 3 & 2 \\ 7 & 8 \end{bmatrix} \begin{bmatrix} 4 & 0 & 1 \\ 9 & 2 & 6 \end{bmatrix} = \begin{bmatrix} 30 & 4 & 15 \\ 100 & 16 & 55 \end{bmatrix}. \qquad \blacktriangleleft$$

Inner Product of Vectors

This is just a special case of our definition of matrix multiplication, which occurs frequently, so that it pays to give it a special name and notation, as follows.

If $\mathbf{a}$ is a row vector and $\mathbf{b}$ a column vector, both with n components, then matrix multiplication (row times column) gives a 1×1 matrix, which we can identify with its single entry. This product is called the **inner product** or **dot product** of $\mathbf{a}$ and $\mathbf{b}$ and is denoted by $\mathbf{a} \cdot \mathbf{b}$; thus

(6)
$$\mathbf{a} \cdot \mathbf{b} = [a_1 \cdots a_n] \begin{bmatrix} b_1 \\ \vdots \\ b_n \end{bmatrix} = \sum_{l=1}^{n} a_l b_l = a_1 b_1 + \cdots + a_n b_n.$$

EXAMPLE 10   **Inner product**

Let $\mathbf{a} = [4 \quad -1 \quad 5]$ and $\mathbf{b} = [2 \quad 5 \quad 8]^\mathsf{T}$, so that $\mathbf{b}$ is a column vector. The inner product of these vectors is given by

$$\mathbf{a} \cdot \mathbf{b} = [4 \quad -1 \quad 5] \begin{bmatrix} 2 \\ 5 \\ 8 \end{bmatrix} = 4 \cdot 2 + (-1) \cdot 5 + 5 \cdot 8 = 43. \qquad \blacktriangleleft$$

Inner products have interesting applications in mechanics and geometry, as we shall see in Sec. 8.2. At present we shall use them to express matrix products in a condensed form, which is often quite useful.

Product in Terms of Row and Column Vectors

Matrix multiplication is a multiplication of rows into columns, as we know, and we can thus write (1) in terms of inner products. Indeed, every entry of $\mathbf{C} = \mathbf{AB}$ is an inner product,

$$c_{11} = \mathbf{a}_1 \bullet \mathbf{b}_1 = (\text{first row of } \mathbf{A}) \bullet (\text{first column of } \mathbf{B})$$

$$c_{12} = \mathbf{a}_1 \bullet \mathbf{b}_2 = (\text{first row of } \mathbf{A}) \bullet (\text{second column of } \mathbf{B})$$

and so on, the general term being

(7)
$$c_{jk} = \mathbf{a}_j \bullet \mathbf{b}_k = (j\text{th row of } \mathbf{A}) \bullet (k\text{th column of } \mathbf{B}).$$

EXAMPLE 11   **Matrix product in terms of row and column vectors**

The row vectors $\mathbf{a}_1$, $\mathbf{a}_2$, $\mathbf{a}_3$ of the matrix

$$\mathbf{A} = \begin{bmatrix} \mathbf{a}_1 \\ \mathbf{a}_2 \\ \mathbf{a}_3 \end{bmatrix} = \begin{bmatrix} 4 & 3 \\ 7 & 2 \\ 9 & 0 \end{bmatrix} \quad \text{are} \quad \begin{array}{l} \mathbf{a}_1 = [4 \quad 3] \\ \mathbf{a}_2 = [7 \quad 2] \,. \\ \mathbf{a}_3 = [9 \quad 0] \end{array}$$

The column vectors $\mathbf{b}_1$, $\mathbf{b}_2$ of the matrix

$$\mathbf{B} = [\mathbf{b}_1 \quad \mathbf{b}_2] = \begin{bmatrix} 2 & 5 \\ 1 & 6 \end{bmatrix} \quad \text{are} \quad \mathbf{b}_1 = \begin{bmatrix} 2 \\ 1 \end{bmatrix}, \quad \mathbf{b}_2 = \begin{bmatrix} 5 \\ 6 \end{bmatrix}.$$

Hence from (6) and (7) we obtain for the product

$$\mathbf{AB} = \begin{bmatrix} \mathbf{a}_1 \bullet \mathbf{b}_1 & \mathbf{a}_1 \bullet \mathbf{b}_2 \\ \mathbf{a}_2 \bullet \mathbf{b}_1 & \mathbf{a}_2 \bullet \mathbf{b}_2 \\ \mathbf{a}_3 \bullet \mathbf{b}_1 & \mathbf{a}_3 \bullet \mathbf{b}_2 \end{bmatrix} = \begin{bmatrix} 4 \cdot 2 + 3 \cdot 1 & 4 \cdot 5 + 3 \cdot 6 \\ 7 \cdot 2 + 2 \cdot 1 & 7 \cdot 5 + 2 \cdot 6 \\ 9 \cdot 2 + 0 \cdot 1 & 9 \cdot 5 + 0 \cdot 6 \end{bmatrix} = \begin{bmatrix} 11 & 38 \\ 16 & 47 \\ 18 & 45 \end{bmatrix}.$$

This agrees with Example 1.   ◀

In the general case, if $\mathbf{A}$ is $m \times n$ and $\mathbf{B}$ is $n \times p$, then $\mathbf{AB}$ is $m \times p$, and from (7) we see that it takes the form

(8)
$$\mathbf{AB} = \begin{bmatrix} \mathbf{a}_1 \bullet \mathbf{b}_1 & \mathbf{a}_1 \bullet \mathbf{b}_2 & \cdots & \mathbf{a}_1 \bullet \mathbf{b}_p \\ \mathbf{a}_2 \bullet \mathbf{b}_1 & \mathbf{a}_2 \bullet \mathbf{b}_2 & \cdots & \mathbf{a}_2 \bullet \mathbf{b}_p \\ \cdot & \cdot & \cdots & \cdot \\ \mathbf{a}_m \bullet \mathbf{b}_1 & \mathbf{a}_m \bullet \mathbf{b}_2 & \cdots & \mathbf{a}_m \bullet \mathbf{b}_p \end{bmatrix}.$$

Motivation of Matrix Multiplication by Linear Transformations

We now motivate the "unnatural" matrix multiplication, as promised. The motivation comes from the use of matrices in **linear transformations,** which we consider in their general form in Sec. 6.8. For two variables these transformations have the form

(9)
$$y_1 = a_{11}x_1 + a_{12}x_2$$
$$y_2 = a_{21}x_1 + a_{22}x_2.$$

This may be a relationship between two coordinate systems in the plane: the x_1x_2-system and the y_1y_2-system. Assume that the x_1x_2-system is related to a third system, the

w_1w_2-system, by another linear transformation

(10)
$$x_1 = b_{11}w_1 + b_{12}w_2$$
$$x_2 = b_{21}w_1 + b_{22}w_2.$$

By substituting (10) into (9) we see that the y_1y_2-system is related directly to the w_1w_2-system by a linear transformation

(11)
$$y_1 = c_{11}w_1 + c_{12}w_2$$
$$y_2 = c_{21}w_1 + c_{22}w_2.$$

Now this substitution gives

$$y_1 = a_{11}(b_{11}w_1 + b_{12}w_2) + a_{12}(b_{21}w_1 + b_{22}w_2)$$
$$y_2 = a_{21}(b_{11}w_1 + b_{12}w_2) + a_{22}(b_{21}w_1 + b_{22}w_2).$$

Comparing this with (11), we see that we must have

$$c_{11} = a_{11}b_{11} + a_{12}b_{21} \qquad c_{12} = a_{11}b_{12} + a_{12}b_{22}$$
$$c_{21} = a_{21}b_{11} + a_{22}b_{21} \qquad c_{22} = a_{21}b_{12} + a_{22}b_{22}$$

or briefly

(12)
$$c_{jk} = a_{j1}b_{1k} + a_{j2}b_{2k} = \sum_{i=1}^{2} a_{ji}b_{ik} \qquad\qquad j, k = 1, 2.$$

This is (1) with $m = n = p = 2$.

What does our calculation show? Essentially two things. First, matrix multiplication is defined in such a way that linear transformations can be written in compact form using matrices; in our case, (9) becomes $\mathbf{y} = \mathbf{Ax}$, written out,

(9*)
$$\mathbf{y} = \begin{bmatrix} y_1 \\ y_2 \end{bmatrix} = \mathbf{Ax} = \begin{bmatrix} a_{11} & a_{12} \\ a_{21} & a_{22} \end{bmatrix} \begin{bmatrix} x_1 \\ x_2 \end{bmatrix}$$

and similarly for (10),

(10*) $\qquad \mathbf{x} = \mathbf{Bw} \quad$ where $\quad \mathbf{x} = \begin{bmatrix} x_1 \\ x_2 \end{bmatrix}, \quad \mathbf{B} = \begin{bmatrix} b_{11} & b_{12} \\ b_{21} & b_{22} \end{bmatrix}, \quad \mathbf{w} = \begin{bmatrix} w_1 \\ w_2 \end{bmatrix}.$

Second, if we substitute linear transformations into each other, we can obtain the coefficient matrix $\mathbf{C}$ of the **composite transformation** (the transformation obtained by the substitution) simply by multiplying the coefficient matrices $\mathbf{A}$ and $\mathbf{B}$ of the given transformations, in the right order suggested by the substitution; from (9*), (10*), and (11) we get

$$\mathbf{y} = \mathbf{Ax} = \mathbf{A(Bw)} = \mathbf{ABw} = \mathbf{Cw}, \qquad \text{where} \qquad \mathbf{C = AB}.$$

For higher dimensions the idea and the result are exactly the same; only the number of variables changes. We then have m variables $y_1, \cdots, y_m$ and n variables $x_1, \cdots, x_n$ and p variables $w_1, \cdots, w_p$. The matrix $\mathbf{A}$ is $m \times n$, the matrix $\mathbf{B}$ is $n \times p$, and $\mathbf{C}$ is $m \times p$, as in Fig. 142. And the requirement that $\mathbf{C}$ be the product $\mathbf{AB}$ leads to formula (1) in its general form. This motivates matrix multiplication completely. ◀

Applications of Matrix Multiplication

EXAMPLE 12   **Weight watching. Matrix times vector**

Suppose that in a weight-watching program, walking, bicycling, and jogging is recommended with the following effects for a person weighing 180 lb: 300 cal/hr burnt in walking (3 mph), 400 in bicycling (12 mph), and 900 in jogging (5 mph). Jim, weighing 180 lb, plans to exercise according to the matrix shown. Verify the calculations. (w = walking, b = bicycling, j = jogging.)

$$
\begin{array}{c}
 \\
\text{Monday} \\
\text{Wednesday} \\
\text{Friday} \\
\text{Saturday}
\end{array}
\begin{array}{ccc}
w & b & j \\
\end{array}
\left[\begin{array}{ccc}
1.0 & 0 & 0.5 \\
1.0 & 0.75 & 0 \\
1.0 & 0 & 0.5 \\
1.5 & 1.0 & 0.5
\end{array}\right]
\left[\begin{array}{c}
300 \\
400 \\
900
\end{array}\right]
=
\left[\begin{array}{c}
750 \\
600 \\
750 \\
1300
\end{array}\right]
\begin{array}{l}
\text{Monday} \\
\text{Wednesday} \\
\text{Friday} \\
\text{Saturday}
\end{array}
$$

◀

EXAMPLE 13   **Computer production. Matrix times matrix**

Fastcomp Ltd produces computer models PC786 and PC886. The matrix $\mathbf{A}$ shows the cost per computer (in thousands of dollars) and $\mathbf{B}$ the production figures for the year 2000 (in multiples of 1000 units). Find a matrix $\mathbf{C}$ that shows the stockholders the costs per quarter (in millions of dollars) for raw material, labor, and miscellaneous.

$$
\mathbf{A} = \begin{array}{c}
\begin{array}{cc} \text{PC786} & \text{PC886} \end{array} \\
\left[\begin{array}{cc}
0.3 & 0.5 \\
0.8 & 1.3 \\
0.7 & 0.9
\end{array}\right]
\begin{array}{l}
\text{Raw material} \\
\text{Labor} \\
\text{Miscellaneous}
\end{array}
\end{array}
\qquad
\mathbf{B} = \begin{array}{c}
\text{Quarter} \\
\begin{array}{cccc} 1 & 2 & 3 & 4 \end{array} \\
\left[\begin{array}{cccc}
4 & 1 & 2 & 3 \\
2 & 3 & 2 & 2
\end{array}\right]
\begin{array}{l}
\text{PC786} \\
\text{PC886}
\end{array}
\end{array}
$$

Solution.

$$
\mathbf{C} = \mathbf{AB} = \begin{array}{c}
\text{Quarter} \\
\begin{array}{cccc} 1 & 2 & 3 & 4 \end{array} \\
\left[\begin{array}{cccc}
2.2 & 1.8 & 1.6 & 1.9 \\
5.8 & 4.7 & 4.2 & 5.0 \\
4.6 & 3.4 & 3.2 & 3.9
\end{array}\right]
\begin{array}{l}
\text{Raw material} \\
\text{Labor} \\
\text{Miscellaneous}
\end{array}
\end{array}
$$

Indeed, for the first quarter the cost for raw material is $c_{11} = 0.3 \cdot 4 + 0.5 \cdot 2 = 2.2$. For labor it is $c_{21} = 5.8$. For miscellaneous it is $c_{31} = 4.6$. Similarly for the other quarters. ◀

EXAMPLE 14   **Stochastic matrix. Markov process. Powers of a matrix**

Suppose that the 1998 state of land use in a city of 50 square miles of (nonvacant) area is

I	(Residentially used)	30%
II	(Commercially used)	20%
III	(Industrially used)	50%.

Find the states in 2003, 2008, and 2013, assuming that the transition probabilities for 5-year intervals are given by the matrix

$$\begin{array}{ccc} \text{To I} & \text{To II} & \text{To III} \end{array}$$

$$\mathbf{A} = \begin{bmatrix} 0.8 & 0.1 & 0.1 \\ 0.1 & 0.7 & 0.2 \\ 0 & 0.1 & 0.9 \end{bmatrix} \begin{array}{l} \text{From I} \\ \text{From II} \\ \text{From III} \end{array}$$

Remark. A square matrix with nonnegative entries and row sums all equal to 1 is called a **stochastic matrix.** Thus $\mathbf{A}$ is a stochastic matrix. A stochastic process for which the probability of entering a certain state depends only on the *last* state occupied (and on the matrix governing the process) is called a **Markov process.**[1] Thus our example concerns a Markov process.

Solution. From the matrix $\mathbf{A}$ and the 1998 state we can compute the 2003 state:

$$\begin{array}{lll} \text{I} & \text{(Residential)} & 0.8 \cdot 30 + 0.1 \cdot 20 + \quad 0 \cdot 50 = 26 \ [\%] \\ \text{II} & \text{(Commercial)} & 0.1 \cdot 30 + 0.7 \cdot 20 + 0.1 \cdot 50 = 22 \ [\%] \\ \text{III} & \text{(Industrial)} & 0.1 \cdot 30 + 0.2 \cdot 20 + 0.9 \cdot 50 = 52 \ [\%]. \end{array}$$

The sum is 100%, as it should be. We write this in matrix form. Let the column vector $\mathbf{x}$ denote the 1998 state; thus, $\mathbf{x}^\mathsf{T} = [30 \quad 20 \quad 50]$. Let $\mathbf{y}$ denote the 2003 state. Then

$$\mathbf{y}^\mathsf{T} = \mathbf{x}^\mathsf{T}\mathbf{A} = [30 \quad 20 \quad 50] \begin{bmatrix} 0.8 & 0.1 & 0.1 \\ 0.1 & 0.7 & 0.2 \\ 0 & 0.1 & 0.9 \end{bmatrix} = [26 \quad 22 \quad 52].$$

Similarly, for 2008 and 2013 we get the state vectors, as you may verify,

$$\mathbf{z}^\mathsf{T} = \mathbf{y}^\mathsf{T}\mathbf{A} = (\mathbf{x}^\mathsf{T}\mathbf{A})\mathbf{A} = \mathbf{x}^\mathsf{T}\mathbf{A}^2 = [23.0 \quad 23.2 \quad 53.8]$$

$$\mathbf{u}^\mathsf{T} = \mathbf{z}^\mathsf{T}\mathbf{A} = (\mathbf{x}^\mathsf{T}\mathbf{A}^2)\mathbf{A} = \mathbf{x}^\mathsf{T}\mathbf{A}^3 = [20.72 \quad 23.92 \quad 55.36].$$

Answer. In 2003, the residential area will be 26% (13 square miles), the commercial 22% (11 square miles) and the industrial 52% (26 square miles). For 2008 the corresponding figures are 23%, 23.2%, 53.8%. For 2013 they are 20.72%, 22.92%, 55.36%. ◀

PROBLEM SET 6.2

Multiplication of Matrices by Matrices and by Vectors. Let

$$\mathbf{a} = \begin{bmatrix} 1 \\ 4 \\ 3 \end{bmatrix}, \quad \mathbf{B} = \begin{bmatrix} 2 & -3 \\ 0 & 2 \\ 0 & 1 \end{bmatrix}, \quad \mathbf{C} = \begin{bmatrix} 4 & 6 & 2 \\ 6 & 0 & 3 \\ 2 & 3 & -1 \end{bmatrix}, \quad \mathbf{d} = [4 \quad 3 \quad 0].$$

Calculate the following products or give reasons why they are not defined. (Show intermediate products.)

1. $\mathbf{Ba}$, $\mathbf{a}^\mathsf{T}\mathbf{B}$, $\mathbf{aB}$
2. $\mathbf{Ca}$, $\mathbf{C}^2\mathbf{a}$, $\mathbf{C}^3\mathbf{a}$
3. $\mathbf{C}^2$, $\mathbf{C}^\mathsf{T}\mathbf{C}$, $\mathbf{CC}^\mathsf{T}$
4. $\mathbf{Ca}$, $\mathbf{Cd}$, $\mathbf{dC}$
5. $\mathbf{a}^\mathsf{T}\mathbf{d}$, $\mathbf{a}^\mathsf{T}\mathbf{d}^\mathsf{T}$, $\mathbf{da}$, $\mathbf{ad}$
6. $5\mathbf{aa}^\mathsf{T}$, $5\mathbf{a}^\mathsf{T}\mathbf{a}$, $(\mathbf{a}^\mathsf{T} - \mathbf{d})\mathbf{B}$
7. $\mathbf{BB}^\mathsf{T}$, $\mathbf{B}^\mathsf{T}\mathbf{B}$, $\mathbf{BB}^\mathsf{T}\mathbf{B}$
8. $\mathbf{a}^\mathsf{T}\mathbf{BB}^\mathsf{T}\mathbf{d}^\mathsf{T}$, $\mathbf{dBB}^\mathsf{T}\mathbf{a}$, $\mathbf{CB}$, $\mathbf{B}^\mathsf{T}\mathbf{C}^\mathsf{T}$
9. $\frac{1}{2}\mathbf{C}^2 - \mathbf{C}$, $\mathbf{C}(\frac{1}{2}\mathbf{C} - \mathbf{I})$, $\mathbf{C} - \mathbf{C}^\mathsf{T}$

---

[1]ANDREI ANDRĖJEVITCH MARKOV (1856—1922), Russian mathematician, known for his work in probability theory.

10. **TEAM PROJECT. Special Matrices.** This project should familiarize you with practically important features of matrix algebra, some of them without analog for numbers.

(a) **Symmetric matrices.** (See Sec. 6.1.) Show that for any $\mathbf{A}$ the matrix $\mathbf{B} = \mathbf{A}\mathbf{A}^{\mathsf{T}}$ is square and symmetric. Show that $\mathbf{A}\mathbf{B}$ with symmetric $\mathbf{A}$ and $\mathbf{B}$ is symmetric if and only if $\mathbf{A}$ and $\mathbf{B}$ commute, $\mathbf{A}\mathbf{B} = \mathbf{B}\mathbf{A}.$ Find all real square matrices that are both symmetric and skew-symmetric.

(b) **Idempotent and nilpotent matrices.** By definition, $\mathbf{A}$ is *idempotent* if $\mathbf{A}^2 = \mathbf{A},$ and $\mathbf{B}$ is *nilpotent* if $\mathbf{B}^m = \mathbf{0}$ for some integer $m.$ Give examples (different from $\mathbf{0}$ or $\mathbf{I}$); also give examples of matrices such that $\mathbf{A}^2 = \mathbf{I}$ (the unit matrix).

(c) **Triangular matrices.** Let $\mathbf{U}_1, \mathbf{U}_2$ be upper triangular and $\mathbf{L}_1, \mathbf{L}_2$ lower triangular. Find out which of the following expressions are triangular. Give examples.

$$\mathbf{U}_1 + \mathbf{U}_2, \quad \mathbf{U}_1\mathbf{U}_2, \quad \mathbf{U}_1^2, \quad \mathbf{U}_1 + \mathbf{L}_1, \quad \mathbf{U}_1\mathbf{L}_1, \quad \mathbf{L}_1 + \mathbf{L}_2, \quad \mathbf{L}_1\mathbf{L}_2, \quad \mathbf{L}_1^2$$

How can you save half of your work by transposition?

(d) **Transposition of products.** Give examples of your own illustrating the basic formula (5). Then prove (5).

(e) **Differences from the multiplication of numbers** are illustrated in Examples 3–6 of the text. Find similar examples in terms of 3×3 matrices.

Applications

11. **(Profit vector)** Two factory outlets F_1 and F_2 in New York and Chicago sell bedsteads $(B),$ tables $(T),$ and chairs (C) with a profit of \$85, \$62, and \$30, respectively. Let the sales in a certain week be given by the matrix

$$\mathbf{A} = \begin{matrix} & B & T & C \\ \begin{bmatrix} 400 & 60 & 240 \\ 100 & 120 & 500 \end{bmatrix} & & \end{matrix} \begin{matrix} F_1 \\ F_2 \end{matrix}$$

Introduce a "profit vector" $\mathbf{p}$ such that the components of the vector $\mathbf{v} = \mathbf{A}\mathbf{p}$ give the total profits of F_1 and $F_2.$

12. **(Production)** In a production process, let N mean "no trouble" and T "trouble." Let the transition probabilities from one day to the next be 0.8 for $N \to N,$ hence 0.2 for $N \to T,$ and 0.5 for $T \to N,$ hence 0.5 for $T \to T.$ If today there is no trouble, what is the probability of trouble 2 days after today? 3 days after today?

13. **(Markov process)** For the Markov process with transition matrix $\mathbf{A} = [a_{jk}],$ whose entries are $a_{11} = a_{12} = 0.5, a_{21} = 0.2, a_{22} = 0.8,$ and initial state $[0.7 \quad 0.7]^{\mathsf{T}},$ compute the next 3 states.

14. **(Concert subscription)** In a community of 100 000 adults, subscribers to a concert series tend to renew their subscription with probability 90% and persons not subscribing will subscribe for the next season with probability 0.2%. If the present number of subscribers is 1200, can one predict an increase or decrease or stability over each of the next three seasons?

15. **(Markov process)** Reformulate Example 14 by using $\mathbf{A}^{\mathsf{T}},$ so that you can work with column vectors instead of row vectors.

16. **(Markov process)** What does it mean in Example 14 if you replace $\mathbf{A}$ by $\mathbf{A}^2$?

17. **CAS PROJECT. Markov Process.** Write a program for a Markov process involving a 3×3 matrix. Use it to calculate further steps in Example 14. Do the values seem to converge to 12.5%, 25%, and 62.5%? Experiment with other stochastic 3×3 matrices, also using different starting vectors.

18. **TEAM PROJECT. Special Linear Transformations. Rotations** have various applications. We show in this project how they can be handled by matrices.

(a) **Rotation in the plane.** Show that the linear transformation $\mathbf{y} = \mathbf{A}\mathbf{x}$ with matrix

$$\mathbf{A} = \begin{bmatrix} \cos\theta & -\sin\theta \\ \sin\theta & \cos\theta \end{bmatrix} \quad \text{and} \quad \mathbf{x} = \begin{bmatrix} x_1 \\ x_2 \end{bmatrix}, \quad \mathbf{y} = \begin{bmatrix} y_1 \\ y_2 \end{bmatrix}$$

is a counterclockwise rotation of the Cartesian x_1x_2-coordinate system in the plane about the origin, where θ is the angle of rotation.

(b) **Rotation through $n\theta$.** Show that in (a)

$$\mathbf{A}^n = \begin{bmatrix} \cos n\theta & -\sin n\theta \\ \sin n\theta & \cos n\theta \end{bmatrix}.$$

Is this plausible? Explain this in words.

(c) **Addition formulas for cosine and sine.** By geometry we should have

$$\begin{bmatrix} \cos \alpha & -\sin \alpha \\ \sin \alpha & \cos \alpha \end{bmatrix} \begin{bmatrix} \cos \beta & -\sin \beta \\ \sin \beta & \cos \beta \end{bmatrix} = \begin{bmatrix} \cos (\alpha + \beta) & -\sin (\alpha + \beta) \\ \sin (\alpha + \beta) & \cos (\alpha + \beta) \end{bmatrix}.$$

Derive from this the addition formulas (6) in Appendix A3.1.

(d) **Computer graphics.** To visualize a three-dimensional object with plane faces (e.g., a cube), we may store the position vectors of the vertices with respect to a suitable $x_1x_2x_3$-coordinate system (and a list of the connecting edges) and then obtain a two-dimensional image on a video screen by projecting the object onto a coordinate plane, for instance, onto the x_1x_2-plane by setting $x_3 = 0$. To change the appearance of the image, we can impose a linear transformation on the position vectors stored. Show that a diagonal matrix $\mathbf{D}$ with main diagonal entries $3, 1, \frac{1}{2}$ gives from an $\mathbf{x} = [x_j]$ the new position vector $\mathbf{y} = \mathbf{D}\mathbf{x}$, where $y_1 = 3x_1$ (stretch in the x_1-direction by a factor 3), $y_2 = x_2$ (unchanged), $y_3 = \frac{1}{2}x_3$ (contraction in the x_3-direction). What effect would a scalar matrix have?

(e) **Rotations in space.** Explain $\mathbf{y} = \mathbf{A}\mathbf{x}$ geometrically when $\mathbf{A}$ is one of the three matrices

$$\begin{bmatrix} 1 & 0 & 0 \\ 0 & \cos \theta & -\sin \theta \\ 0 & \sin \theta & \cos \theta \end{bmatrix}, \quad \begin{bmatrix} \cos \varphi & 0 & -\sin \varphi \\ 0 & 1 & 0 \\ \sin \varphi & 0 & \cos \varphi \end{bmatrix}, \quad \begin{bmatrix} \cos \psi & -\sin \psi & 0 \\ \sin \psi & \cos \psi & 0 \\ 0 & 0 & 1 \end{bmatrix}.$$

What effect would these transformations have in situations such as that described in (d)?

6.3 Linear Systems of Equations Gauss Elimination

The most important practical use of matrices is in the solution of linear systems of equations, which appear frequently as models of various problems, for instance, in frameworks, electrical networks, traffic flow, production and consumption, assignment of jobs to workers, population growth, statistics, numerical methods for differential equations (Chap. 19), and many others. We begin in this section with an important solution method, the Gauss elimination. General properties of solutions will be discussed in the next sections.

Linear System, Coefficient Matrix, Augmented Matrix

A **linear system of m equations in n unknowns** $x_1, \cdots, x_n$ is a set of equations of the form

$$\begin{aligned}
a_{11}x_1 + \cdots + a_{1n}x_n &= b_1 \\
a_{21}x_1 + \cdots + a_{2n}x_n &= b_2 \\
\cdots\cdots\cdots\cdots\cdots\cdots\cdots \\
a_{m1}x_1 + \cdots + a_{mn}x_n &= b_m
\end{aligned}$$

(1)

Thus, a system of two equations in three unknowns is

$$\begin{aligned}
a_{11}x_1 + a_{12}x_2 + a_{13}x_3 &= b_1 \\
a_{21}x_1 + a_{22}x_2 + a_{23}x_3 &= b_2,
\end{aligned}
\qquad \text{for example,} \qquad
\begin{aligned}
5x_1 + 2x_2 - x_3 &= 4 \\
x_1 - 4x_2 + 3x_3 &= 6.
\end{aligned}$$

The a_{jk} are given numbers, which are called the **coefficients** of the system. The b_i are also given numbers. If the b_i are all zero, then (1) is called a **homogeneous system.** If at least one b_i is not zero, then (1) is called a **nonhomogeneous system.**

A **solution** of (1) is a set of numbers $x_1, \cdots, x_n$ that satisfies all the m equations. A **solution vector** of (1) is a vector $\mathbf{x}$ whose components constitute a solution of (1). If the system (1) is homogeneous, it has at least the **trivial solution** $x_1 = 0, \cdots, x_n = 0$.

Matrix Form of the Linear System (1). From the definition of matrix multiplication we see that the m equations of (1) may be written as a single vector equation

(2)
$$\mathbf{Ax} = \mathbf{b}$$

where the **coefficient matrix** $\mathbf{A} = [a_{jk}]$ is the $m \times n$ matrix

$$\mathbf{A} = \begin{bmatrix} a_{11} & a_{12} & \cdots & a_{1n} \\ a_{21} & a_{22} & \cdots & a_{2n} \\ \cdot & \cdot & \cdots & \cdot \\ a_{m1} & a_{m2} & \cdots & a_{mn} \end{bmatrix}, \quad \text{and} \quad \mathbf{x} = \begin{bmatrix} x_1 \\ \cdot \\ \cdot \\ \cdot \\ x_n \end{bmatrix} \quad \text{and} \quad \mathbf{b} = \begin{bmatrix} b_1 \\ \cdot \\ \cdot \\ \cdot \\ b_m \end{bmatrix}$$

are column vectors. We assume that the coefficients a_{jk} are not all zero, so that $\mathbf{A}$ is not a zero matrix. Note that $\mathbf{x}$ has n components, whereas $\mathbf{b}$ has m components. The matrix

$$\widetilde{\mathbf{A}} = \begin{bmatrix} a_{11} & \cdots & a_{1n} & b_1 \\ \cdot & \cdots & \cdot & \cdot \\ \cdot & \cdots & \cdot & \cdot \\ a_{m1} & \cdots & a_{mn} & b_m \end{bmatrix}$$

is called the **augmented matrix** of the system (1). We see that $\widetilde{\mathbf{A}}$ is obtained by augmenting

A by the column **b.** The augmented matrix is sometimes written

$$\widetilde{\mathbf{A}} = \begin{bmatrix} a_{11} & \cdots & a_{1n} & \vline & b_1 \\ \cdot & \cdots & \cdot & \vline & \cdot \\ \cdot & \cdots & \cdot & \vline & \cdot \\ a_{m1} & \cdots & a_{mn} & \vline & b_m \end{bmatrix}$$

where the vertical line is merely a reminder that the last column of $\widetilde{\mathbf{A}}$ is the right side of the system.

The augmented matrix $\widetilde{\mathbf{A}}$ determines the system (1) completely because it contains all the given numbers appearing in (1).

EXAMPLE 1   **Geometric interpretation. Existence of solutions**

If $m = n = 2$, we have two equations in two unknowns x_1, x_2

$$a_{11}x_1 + a_{12}x_2 = b_1$$
$$a_{21}x_1 + a_{22}x_2 = b_2.$$

If we interpret x_1, x_2 as coordinates in the x_1x_2-plane, then each of the two equations represents a straight line, and (x_1, x_2) is a solution if and only if the point P with coordinates x_1, x_2 lies on both lines. Hence there are three possible cases:

   (**a**) No solution if the lines are parallel.

   (**b**) Precisely one solution if they intersect.

   (**c**) Infinitely many solutions if they coincide.

For instance,

$x + y = 1$	$x + y = 1$	$x + y = 1$
$x + y = 0$	$x - y = 0$	$2x + 2y = 2$
Case (*a*)	Case (*b*)	Case (*c*)

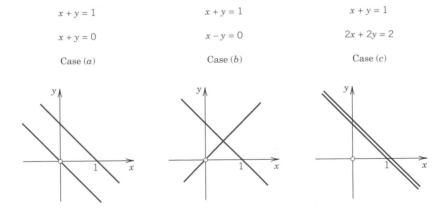

If the system is homogeneous, Case (*a*) cannot happen, because then those two straight lines pass through the origin, whose coordinates 0, 0 constitute the trivial solution. The reader may consider three equations in three unknowns as representations of three planes in space and discuss the various possible cases in a similar fashion.   ◄

Our simple example illustrates that a system (1) may perhaps have no solution. This poses the following problems. Does a given system (1) have a solution? Under what conditions does it have precisely one solution? If it has more than one solution, how can we characterize the set of all solutions? How can we actually *obtain* the solutions? Perhaps this is the most immediate question from a practical viewpoint. We discuss it first and the other problems in Sec. 6.5.

Gauss Elimination

The Gauss elimination is a standard method for solving linear systems. It is a systematic elimination process, a method of great importance that works in practice and is reasonable with respect to computing time and storage demand (two aspects we shall consider in Sec. 18.1 on numerical methods). For instance, to solve the system

$$2x_1 + 5x_2 = 2$$
$$4x_1 + 3x_2 = 18$$

we multiply the first equation by 2 and subtract it from the second, obtaining

$$2x_1 + 5x_2 = 2$$
$$-7x_2 = 14.$$

This is "Gauss elimination" for 2 equations. The solution now follows by "back substitution," $x_2 = 14/(-7) = -2$, $x_1 = (2 - 5x_2)/2 = (2 + 10)/2 = 6$.

Since a linear system is completely determined by its augmented matrix, the elimination process can be done by merely considering the matrices. To see this correspondence we shall write systems of equations and augmented matrices side by side.

EXAMPLE 2   **Gauss elimination. Electrical network**

Solve the linear system

$$x_1 - x_2 + x_3 = 0$$
$$-x_1 + x_2 - x_3 = 0$$
$$10x_2 + 25x_3 = 90$$
$$20x_1 + 10x_2 = 80.$$

Derivation from the circuit in Fig. 143 (Optional). This is the system for the unknown currents $x_1 = i_1$, $x_2 = i_2$, $x_3 = i_3$ in the electrical network in Fig. 143. To obtain it, we label the currents as shown, choosing directions arbitrarily; if a current will come out negative, this will simply mean that the current flows against the direction of our arrow. The current entering each battery will be the same as the current leaving it. The equations for the currents result from Kirchhoff's laws:

> **Kirchhoff's current law (KCL).** *At any point of a circuit, the sum of the inflowing currents equals the sum of the outflowing currents.*

> **Kirchhoff's voltage law (KVL).** *In any closed loop, the sum of all voltage drops equals the impressed electromotive force.*

Node P gives the first equation, node Q the second, the right loop the third, and the left loop the fourth, as indicated in the figure.

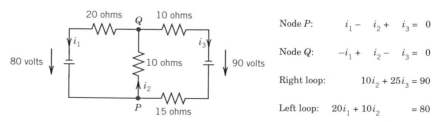

Node P:	$i_1 - i_2 + i_3 = 0$
Node Q:	$-i_1 + i_2 - i_3 = 0$
Right loop:	$10i_2 + 25i_3 = 90$
Left loop:	$20i_1 + 10i_2 = 80$

Fig. 143. Network in Example 2 and equations relating the currents

Solution by Gauss's method.   This system is so simple that we could almost solve it by inspection. This is not the point. The point is to perform a systematic elimination—the Gauss elimination—which will work in general, also for large systems. It is a reduction to *"triangular form"* from which we shall then readily obtain the values of the unknowns by *"back substitution."*

We write the system and its augmented matrix side by side:

Equations   Augmented Matrix $\tilde{\mathbf{A}}$

Pivot 1 $\longrightarrow$ $\left(x_1\right)$ $-$ $x_2 +$ $x_3 = 0$

$\boxed{-x_1}$ $+$ $x_2 -$ $x_3 = 0$

Eliminate $\longrightarrow$ $10x_2 + 25x_3 = 90$

$\boxed{20x_1}$ $+ 10x_2$ $= 80$

$$\left[\begin{array}{rrr|r} 1 & -1 & 1 & 0 \\ -1 & 1 & -1 & 0 \\ 0 & 10 & 25 & 90 \\ 20 & 10 & 0 & 80 \end{array}\right]$$

First Step. Elimination of x_1

Call the first equation the **pivot equation** and the coefficient 1 of its x_1-term the **pivot** in this step. Use this equation to eliminate x_1 (get rid of x_1) in the other equations. For this, do these operations:

Add 1 times the pivot equation to the second equation.[2]

Add -20 times the pivot equation to the fourth equation.[2]

This corresponds to row operations on the augmented matrix, which we indicate behind the *new* matrix in (3). The result is

(3)

$x_1 -$ $x_2 +$ $x_3 = 0$

$0 = 0$

$10x_2 + 25x_3 = 90$

$30x_2 - 20x_3 = 80$

$$\left[\begin{array}{rrr|r} 1 & -1 & 1 & 0 \\ 0 & 0 & 0 & 0 \\ 0 & 10 & 25 & 90 \\ 0 & 30 & -20 & 80 \end{array}\right] \quad \begin{array}{l} \\ \text{Row 2 + Row 1} \\ \\ \text{Row 4} - 20 \text{ Row 1} \end{array}$$

Second Step. Elimination of x_2

The first equation, which has just served as the pivot equation, remains ***untouched.*** We want to take the (new!) second equation as the next pivot equation. Since it contains no x_2-term (needed as the next pivot)—in fact, it is $0 = 0$—we first have to change the order of equations (and corresponding rows of the new matrix) to get a nonzero pivot. We put the second equation ($0 = 0$) at the end and move the third and fourth equations one place up; this is called **partial pivoting.**[3] We get

$x_1 -$ x_2 $+$ $x_3 = 0$

Pivot 10 $\longrightarrow$ $\left(10x_2\right)$ $+ 25x_3 = 90$

Eliminate $\longrightarrow$ $\boxed{30x_2}$ $- 20x_3 = 80$

$0 = 0$

$$\left[\begin{array}{rrr|r} 1 & -1 & 1 & 0 \\ 0 & 10 & 25 & 90 \\ 0 & 30 & -20 & 80 \\ 0 & 0 & 0 & 0 \end{array}\right]$$

To eliminate x_2, do:

Add -3 times the pivot equation to the third equation.

---

[2]To call all the operations "additions" rather than "additions" and "subtractions" is preferable from the viewpoint of uniformity of numerical algorithms. See also Sec. 18.1.

[3]As opposed to **total pivoting,** in which the order of the unknowns is also changed. Total pivoting is hardly used in practice.

The result is

(4)

$$
\begin{aligned}
x_1 - x_2 + x_3 &= 0 \\
10x_2 + 25x_3 &= 90 \\
- 95x_3 &= -190 \\
0 &= 0
\end{aligned}
\qquad
\begin{bmatrix}
1 & -1 & 1 & \vdots & 0 \\
0 & 10 & 25 & \vdots & 90 \\
0 & 0 & -95 & \vdots & -190 \\
0 & 0 & 0 & \vdots & 0
\end{bmatrix}
\quad \text{Row } 3 - 3 \text{ Row } 2
$$

Back Substitution. Determination of x_3, x_2, x_1 (in this order)

Working backward from the last to the first equation of this "triangular" system (4), we can now readily find x_3, then x_2, and then x_1:

$$
\begin{aligned}
-95x_3 &= -190, & x_3 &= i_3 = 2 \text{ [amperes]}, \\
10x_2 + 25x_3 &= 90, & x_2 &= \tfrac{1}{10}(90 - 25x_3) = i_2 = 4 \text{ [amperes]}, \\
x_1 - x_2 + x_3 &= 0, & x_1 &= x_2 - x_3 = i_1 = 2 \text{ [amperes]}.
\end{aligned}
$$

This is the answer to our problem. The solution is unique.   ◀

Elementary Row Operations. Row-Equivalent Systems

Example 2 illustrates the operations of the Gauss elimination. These are the first two of three operations, which are called

Elementary Operations for Equations

Interchange of two equations
Addition of a constant multiple of one equation to another equation
*Multiplication of an equation by a **nonzero** constant c.*

To these correspond the following

Elementary Row Operations for Matrices

Interchange of two rows
Addition of a constant multiple of one row to another row
*Multiplication of a row by a **nonzero** constant c.*

Clearly, the interchange of two equations does not alter the solution set. Neither does the addition because we can undo it by a corresponding subtraction. Similarly for the multiplication, which we can undo by multiplying the new equation by $1/c$, producing the original equation.

We now call a linear system S_1 **row-equivalent** to a linear system S_2 if S_1 can be obtained from S_2 by (finitely many!) elementary row operations. Using this concept, we can formulate our result that justifies the Gauss elimination.

THEOREM 1   **(Row-equivalent systems)**

Row-equivalent linear systems have the same sets of solutions.

A linear system (1) is called **overdetermined** if it has more equations than unknowns, as in Example 2, **determined** if $m = n$, as in Example 1, and **underdetermined** if (1) has fewer equations than unknowns.

Furthermore, a system (1) is called **consistent** if it has at least one solution, but **inconsistent** if it has no solutions at all.

Gauss Elimination: The Three Possible Cases of Systems

We show next that the Gauss elimination can take care of all three possible cases that a system has infinitely many solutions (Example 3, below), a unique solution (Example 4) or no solutions (Example 5).

EXAMPLE 3   **Gauss elimination if infinitely many solutions exist**

Solve the linear system of three equations in four unknowns

(5)
$$
\begin{aligned}
3.0x_1 + 2.0x_2 + 2.0x_3 - 5.0x_4 &= 8.0 \\
0.6x_1 + 1.5x_2 + 1.5x_3 - 5.4x_4 &= 2.7 \\
1.2x_1 - 0.3x_2 - 0.3x_3 + 2.4x_4 &= 2.1
\end{aligned}
\qquad
\left[\begin{array}{cccc|c}
3.0 & 2.0 & 2.0 & -5.0 & 8.0 \\
0.6 & 1.5 & 1.5 & -5.4 & 2.7 \\
1.2 & -0.3 & -0.3 & 2.4 & 2.1
\end{array}\right]
$$

Solution.   As in the previous example, we circle pivots and box terms to be eliminated.

First Step. Elimination of x_1 from the second and third equations by adding

$$- 0.6/3.0 = -0.2 \text{ times the first equation to the second equation,}$$

$$- 1.2/3.0 = -0.4 \text{ times the first equation to the third equation.}$$

This gives a new system of equations

(6)
$$
\begin{aligned}
3.0x_1 + 2.0x_2 + 2.0x_3 - 5.0x_4 &= 8.0 \\
1.1x_2 + 1.1x_3 - 4.4x_4 &= 1.1 \\
-1.1x_2 - 1.1x_3 + 4.4x_4 &= -1.1
\end{aligned}
\qquad
\left[\begin{array}{cccc|c}
3.0 & 2.0 & 2.0 & -5.0 & 8.0 \\
0 & 1.1 & 1.1 & -4.4 & 1.1 \\
0 & -1.1 & -1.1 & 4.4 & -1.1
\end{array}\right]
$$

and we circle the pivot to be used in the next step.

Second Step. Elimination of x_2 from the third equation of (6) by adding

$$1.1/1.1 = 1 \text{ times the second equation to the third equation.}$$

This gives

(7)
$$
\begin{aligned}
3.0x_1 + 2.0x_2 + 2.0x_3 - 5.0x_4 &= 8.0 \\
1.1x_2 + 1.1x_3 - 4.4x_4 &= 1.1 \\
0 &= 0
\end{aligned}
\qquad
\left[\begin{array}{cccc|c}
3.0 & 2.0 & 2.0 & -5.0 & 8.0 \\
0 & 1.1 & 1.1 & -4.4 & 1.1 \\
0 & 0 & 0 & 0 & 0
\end{array}\right]
$$

Back Substitution.   From the second equation, $x_2 = 1 - x_3 + 4x_4$. From this and the first equation, $x_1 = 2 - x_4$. Since x_3 and x_4 remain arbitrary, we have infinitely many solutions. If we choose a value of x_3 and a value of x_4, then the corresponding values of x_1 and x_2 are uniquely determined.   ◀

EXAMPLE 4   **Gauss elimination if a unique solution exists**

Solve the linear system

$$
\begin{aligned}
-x_1 + x_2 + 2x_3 &= 2 \\
3x_1 - x_2 + x_3 &= 6 \\
-x_1 + 3x_2 + 4x_3 &= 4
\end{aligned}
\qquad
\left[\begin{array}{ccc|c}
-1 & 1 & 2 & 2 \\
3 & -1 & 1 & 6 \\
-1 & 3 & 4 & 4
\end{array}\right]
$$

First Step. Elimination of x_1 from the second and third equations gives

$$
\begin{aligned}
-x_1 + x_2 + 2x_3 &= 2 \\
2x_2 + 7x_3 &= 12 \\
2x_2 + 2x_3 &= 2
\end{aligned}
\qquad
\left[\begin{array}{ccc|c}
-1 & 1 & 2 & 2 \\
0 & 2 & 7 & 12 \\
0 & 2 & 2 & 2
\end{array}\right]
\begin{array}{l}
\\
\text{Row 2 + 3 Row 1} \\
\text{Row 3 - Row 1}
\end{array}
$$

Second Step. Elimination of x_2 from the third equation gives

$$
\begin{aligned}
-x_1 + x_2 + 2x_3 &= 2 \\
2x_2 + 7x_3 &= 12 \\
- 5x_3 &= -10
\end{aligned}
\qquad
\begin{bmatrix}
-1 & 1 & 2 & \vdots & 2 \\
0 & 2 & 7 & \vdots & 12 \\
0 & 0 & -5 & \vdots & -10
\end{bmatrix}
\quad \text{Row } 3 - \text{Row } 2
$$

Back Substitution. Beginning with the last equation, we obtain successively $x_3 = 2$, $x_2 = -1$, $x_1 = 1$. We see that the system has a unique solution.   ◀

EXAMPLE 5   **Gauss elimination if no solution exists**

What will happen if we apply the Gauss elimination to a linear system that has no solution? The answer is that in this case the method will show this fact by producing a contradiction. For instance, consider

$$
\begin{aligned}
3x_1 + 2x_2 + x_3 &= 3 \\
2x_1 + x_2 + x_3 &= 0 \\
6x_1 + 2x_2 + 4x_3 &= 6
\end{aligned}
\qquad
\begin{bmatrix}
3 & 2 & 1 & \vdots & 3 \\
2 & 1 & 1 & \vdots & 0 \\
6 & 2 & 4 & \vdots & 6
\end{bmatrix}
$$

First Step. Elimination of x_1 from the second and third equations by adding

$-2/3$ times the first equation to the second equation,

$-6/3 = -2$ times the first equation to the third equation.

This gives

$$
\begin{aligned}
3x_1 + 2x_2 + x_3 &= 3 \\
-\tfrac{1}{3}x_2 + \tfrac{1}{3}x_3 &= -2 \\
-2x_2 + 2x_3 &= 0
\end{aligned}
\qquad
\begin{bmatrix}
3 & 2 & 1 & \vdots & 3 \\
0 & -\tfrac{1}{3} & \tfrac{1}{3} & \vdots & -2 \\
0 & -2 & 2 & \vdots & 0
\end{bmatrix}
$$

Second Step. Elimination of x_2 from the third equation gives

$$
\begin{aligned}
3x_1 + 2x_2 + x_3 &= 3 \\
-\tfrac{1}{3}x_2 + \tfrac{1}{3}x_3 &= -2 \\
0 &= 12
\end{aligned}
\qquad
\begin{bmatrix}
3 & 2 & 1 & \vdots & 3 \\
0 & -\tfrac{1}{3} & \tfrac{1}{3} & \vdots & -2 \\
0 & 0 & 0 & \vdots & 12
\end{bmatrix}
$$

This shows that the system has no solution.   ◀

Echelon Form. Information Resulting from It

The form of the system and of the matrix in the last step of the Gauss elimination is called the **echelon form.** Thus in Example 5 the echelon forms of the coefficient matrix and the augmented matrix are

$$
\begin{bmatrix}
3 & 2 & 1 \\
0 & -\tfrac{1}{3} & \tfrac{1}{3} \\
0 & 0 & 0
\end{bmatrix}
\qquad \text{and} \qquad
\begin{bmatrix}
3 & 2 & 1 & \vdots & 3 \\
0 & -\tfrac{1}{3} & \tfrac{1}{3} & \vdots & -2 \\
0 & 0 & 0 & \vdots & 12
\end{bmatrix}.
$$

At the end of the Gauss elimination (before the back substitution) the reduced system will have the form

$$a_{11}x_1 + a_{12}x_2 + \cdots\cdots + a_{1n}x_n = b_1$$

$$c_{22}x_2 + \cdots\cdots + c_{2n}x_n = b_2{}^*$$

$$\vdots$$

(8) $$k_{rr}x_r + \cdots + k_{rn}x_n = \tilde{b}_r$$

$$0 = \tilde{b}_{r+1}$$

$$\vdots$$

$$0 = \tilde{b}_m$$

where $r \le m$ (and $a_{11} \ne 0$, $c_{22} \ne 0, \cdots, k_{rr} \ne 0$). From this we see that with respect to solutions of this system (8), there are three possible cases:

(a) No solution if $r < m$ and one of the numbers $\tilde{b}_{r+1}, \cdots, \tilde{b}_m$ is not zero. This is illustrated in Example 5, where $r = 2 < m = 3$ and $\tilde{b}_{r+1} = \tilde{b}_3 = 12$.

(b) Precisely one solution if $r = n$ and $\tilde{b}_{r+1}, \cdots, \tilde{b}_m$, if present, are zero. This solution is obtained by solving the nth equation of (8) for x_n, then the $(n-1)$th equation for x_{n-1}, and so on up the line. See Example 2, where $r = n = 3$ and $m = 4$.

(c) Infinitely many solutions if $r < n$ and $\tilde{b}_{r+1}, \cdots, \tilde{b}_m$, if present, are zero. Then any of these solutions is obtained by choosing values at pleasure for the unknowns $x_{r+1}, \cdots, x_n$, solving the rth equation for x_r, then the $(r-1)$th equation for x_{r-1}, and so on up the line. Example 3 illustrates this case.

Remark. For the so-called **Gauss–Jordan elimination** see Sec. 6.7. This method is inferior to the Gauss elimination because it requires more computational operations.
    More on **pivoting** will be said in Sec. 18.1

PROBLEM SET 6.3

Gauss Elimination. Solve the following systems by the Gauss elimination or indicate the nonexistence of solutions. (Show the details of your work.)

1. $6x + 4y = 2$
$3x - 5y = -34$

2. $0.4x + 1.2y = -2.0$
$1.7x - 3.2y = 8.1$

3. $3.0x - 0.5y = 0.6$
$1.5x + 4.5y = 6.0$

4. $ 7y + 3z = -12$
$2x + 8y + z = 0$
$-5x + 2y - 9z = 26$

5. $x + y - z = 9$
$8y + 6z = -6$
$-2x + 4y - 6z = 40$

6. $4x + y = 4$
$5x - 3y + z = 2$
$-9x + 2y - z = 5$

7. $13x + 12y = -6$
$-4x + 7y = -73$
$11x - 13y = 157$

8. $4x - 8y + 3z = 16$
$-x + 2y - 5z = -21$
$3x - 6y + z = 7$

9. $4y + 3z = 8$
$2x - z = 2$
$3x + 2y = 5$

10. $1.3x - 9.1y + 11.7z = 0$
$-0.9x + 6.3y - 8.1z = 0$

11. $7x - 4y - 2z = -6$
$16x + 2y + z = 3$

12. $12x - 26y + 34z = 18$
$-30x + 65y - 85z = -46$

13. $5x + 5y - 10z = 0$
$2w - 3x - 3y + 6z = 2$
$4w + x + y - 2z = 4$

14. $2w - 2x + 4y = 0$
$-3w + 3x - 6y + 5z = 15$
$w - x + 2y = 0$

15. $10x + 4y - 2z = -4$
$-3w - 17x + y + 2z = 2$
$w + x + y = 6$
$8w - 34x + 16y - 10z = 4$

16. $2w + 3x + y - 11z = 1$
$5w - 2x + 5y - 4z = 5$
$w - x + 3y - 3z = 3$
$3w + 4x - 7y + 2z = -7$

Models of Electrical Networks. Using Kirchhoff's laws (see Example 2), find the currents in the following networks. (Show the details of your work.)

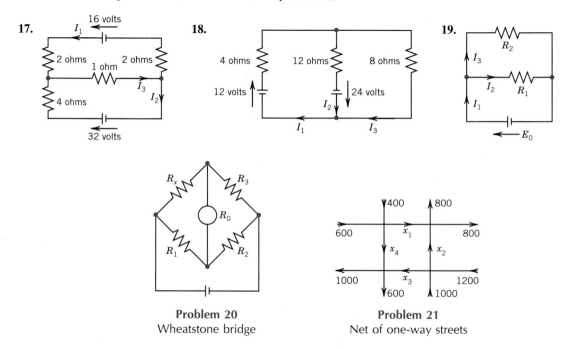

Problem 20
Wheatstone bridge

Problem 21
Net of one-way streets

20. (Wheatstone bridge) Show that if $R_x/R_3 = R_1/R_2$ in the figure, then $I = 0$. (R_0 is the resistance of the instrument by which I is measured.) This bridge is a method for determining R_x. R_1, R_2, R_3 are known. R_3 is variable. To get R_x, make $I = 0$ by varying R_3. Then calculate $R_x = R_3R_1/R_2$.

21. (Traffic flow) Methods of electrical circuit analysis have applications to other fields. For instance, applying the analog of Kirchhoff's current law, find the traffic flow (cars per hour) in the net of one-way streets (in the directions indicated by the arrows) shown in the figure. Is the solution unique?

22. (Models of markets) Determine the equilibrium solution ($D_1 = S_1$, $D_2 = S_2$) of the two-commodity market with linear model (D, S, P = demand, supply, price; index 1 = first commodity, index 2 = second commodity)

$$D_1 = 40 - 2P_1 - P_2, \qquad S_1 = 4P_1 - P_2 + 4$$
$$D_2 = 5P_1 - 2P_2 + 16, \qquad S_2 = 3P_2 - 4.$$

23. (Equivalence relation) By definition, an *equivalence relation* on a set is a relation satisfying three conditions (named as indicated):

(i) Each element A of the set is equivalent to itself (*"Reflexivity"*).

(ii) If A is equivalent to B, then B is equivalent to A (*"Symmetry"*).

(iii) If A is equivalent to B and B is equivalent to C, then A is equivalent to C (*"Transitivity"*).

Show that row equivalence of matrices satisfies these three conditions. *Hint.* Show that for each of the three elementary row operations these conditions hold.

24. PROJECT. Elementary Matrices. The idea is that elementary operations can be accomplished by matrix multiplication. If $\mathbf{A}$ is an $m \times n$ matrix on which we want to do an elementary operation, then there is a matrix $\mathbf{E}$ such that $\mathbf{EA}$ is the new matrix after the operation. Such an $\mathbf{E}$ is called an **elementary matrix.** (This idea is primarily of *theoretical* interest, for instance, in the design of algorithms. *Computationally,* it is preferable to do row operations directly rather than by matrix multiplication by $\mathbf{E}$.)

(a) Show that the following are elementary matrices, for interchanging rows 2 and 3, for adding -5 times the first row to the third, and for multiplying the fourth row by 8.

$$\mathbf{E}_1 = \begin{bmatrix} 1 & 0 & 0 & 0 \\ 0 & 0 & 1 & 0 \\ 0 & 1 & 0 & 0 \\ 0 & 0 & 0 & 1 \end{bmatrix}, \quad \mathbf{E}_2 = \begin{bmatrix} 1 & 0 & 0 & 0 \\ 0 & 1 & 0 & 0 \\ -5 & 0 & 1 & 0 \\ 0 & 0 & 0 & 1 \end{bmatrix}, \quad \mathbf{E}_3 = \begin{bmatrix} 1 & 0 & 0 & 0 \\ 0 & 1 & 0 & 0 \\ 0 & 0 & 1 & 0 \\ 0 & 0 & 0 & 8 \end{bmatrix}.$$

Apply $\mathbf{E}_1$, $\mathbf{E}_2$, $\mathbf{E}_3$ to a vector and to a 4×3 matrix of your choice. Find $\mathbf{B} = \mathbf{E}_3\mathbf{E}_2\mathbf{E}_1\mathbf{A}$, where $\mathbf{A} = [a_{jk}]$ is the general 4×2 matrix. Is $\mathbf{B}$ equal to $\mathbf{C} = \mathbf{E}_1\mathbf{E}_2\mathbf{E}_3\mathbf{A}$?

(b) Conclude that E_1, E_2, E_3 are obtained by doing the corresponding elementary operations on the 4×4 unit matrix. Prove that *if* $\mathbf{M}$ *is obtained from* $\mathbf{A}$ *by an elementary row operation, then*

$$\mathbf{M} = \mathbf{EA},$$

where $\mathbf{E}$ *is obtained from the* $n \times n$ *unit matrix* $\mathbf{I}_n$ *by the same row operation.*

(c) Find the elementary matrices that do the Gauss elimination (not the back substitution) of the general 3×3 matrix $\mathbf{A}$. Find their product (in the proper order!).

 25. CAS PROJECT. Gauss Elimination. Write a program for the Gauss elimination (including back substitution) without pivoting. Apply it to Probs. 1–10. Find out what happens when pivoting would be needed.

6.4 Rank of a Matrix. Linear Independence. Vector Space

In the last section we explained the Gauss algorithm, the most important solution method for linear systems of equations. We also saw that such a system may have no solution at all, or one solution, or more than one (and then infinitely many solutions). So we are confronted with the questions of *existence and uniqueness of solutions.* We shall answer these questions in the next section. For this we introduce as the key concept the *rank of a matrix.* To define it, we first need the following concepts, which are of general importance.

Linear Independence and Dependence of Vectors

Given any set of m vectors[4] $\mathbf{a}_{(1)}, \cdots, \mathbf{a}_{(m)}$ (with the same number of components), a **linear combination** of these vectors is an expression of the form

$$c_1 \mathbf{a}_{(1)} + \cdots + c_m \mathbf{a}_{(m)}$$

where $c_1, \cdots, c_m$ are any scalars.[5] Now consider the equation

(1)
$$c_1 \mathbf{a}_{(1)} + c_2 \mathbf{a}_{(2)} + \cdots + c_m \mathbf{a}_{(m)} = \mathbf{0}.$$

Clearly, this holds if we choose all c_j's zero, because then it becomes $\mathbf{0} = \mathbf{0}$. If this is the only m-tuple of scalars for which (1) holds, then our vectors $\mathbf{a}_{(1)}, \cdots, \mathbf{a}_{(m)}$ are said to form a *linearly independent set* or, more briefly, we call them **linearly independent.** Otherwise, if (1) also holds with scalars not all zero, we call these vectors **linearly dependent,** because then we can express (at least) one of them as a linear combination of the others. For instance, if (1) holds with, say, $c_1 \neq 0$, we can solve (1) for $\mathbf{a}_{(1)}$:

$$\mathbf{a}_{(1)} = k_2 \mathbf{a}_{(2)} + \cdots + k_m \mathbf{a}_{(m)} \qquad \text{where } k_j = -c_j / c_1.$$

(Some k_j's may be zero. Or even all of them, namely, if $\mathbf{a}_{(1)} = \mathbf{0}$.)

EXAMPLE 1 **Linear independence and dependence**

The three vectors

$$\mathbf{a}_{(1)} = [\quad 3 \qquad 0 \qquad 2 \qquad 2]$$
$$\mathbf{a}_{(2)} = [-6 \qquad 42 \qquad 24 \qquad 54]$$
$$\mathbf{a}_{(3)} = [\quad 21 \quad -21 \qquad 0 \quad -15]$$

are linearly dependent because

$$6\,\mathbf{a}_{(1)} - \tfrac{1}{2}\mathbf{a}_{(2)} - \mathbf{a}_{(3)} = \mathbf{0}.$$

Although this is easily checked (do it!), it is not so easy to discover. However, a method for finding out about linear independence and dependence follows below.

The first two of the three vectors are linearly independent because $c_1 \mathbf{a}_{(1)} + c_2 \mathbf{a}_{(2)} = \mathbf{0}$ implies $c_2 = 0$ (from the second components) and then $c_1 = 0$ (from any other component of $\mathbf{a}_{(1)}$). ◀

EXAMPLE 2 **Linear independence and dependence**

If you have studied Chap. 2, recall that linear independence and dependence are basic in linear differential equations, namely, in connection with bases (fundamental systems) of solutions. ◀

What is the point of linear independence and dependence? Well, from a linearly dependent set we may often omit vectors that are linear combinations of others until we are finally left with a *linearly independent* subset of the "really essential" vectors, which can no longer be expressed linearly in terms of each other.

---

[4]Write simply $\mathbf{a}_1, \cdots, \mathbf{a}_m$ if you wish, but keep in mind that these are *vectors,* not vector *components.*

[5]In this section, scalars will be *real* numbers.

Rank of a Matrix

The maximum number of linearly independent row vectors of a matrix $\mathbf{A} = [a_{jk}]$ is called the **rank** of $\mathbf{A}$ and is denoted by

$$\text{rank } \mathbf{A}.$$

EXAMPLE 3   Rank

The matrix

$$(2) \qquad \mathbf{A} = \begin{bmatrix} 3 & 0 & 2 & 2 \\ -6 & 42 & 24 & 54 \\ 21 & -21 & 0 & -15 \end{bmatrix}$$

has rank 2, because Example 1 shows that the first two row vectors are linearly independent, whereas all three row vectors are linearly dependent. ◀

Note further that rank $\mathbf{A} = 0$ if and only if $\mathbf{A} = \mathbf{0}$. This follows directly from the definition.

In our proposed discussion of the existence and uniqueness of solutions of systems of linear equations we shall need the following very important theorem.

THEOREM 1   **(Rank in terms of column vectors)**

The rank of a matrix $\mathbf{A}$ equals the maximum number of linearly independent ***column*** *vectors of $\mathbf{A}$.*

Hence $\mathbf{A}$ and its transpose $\mathbf{A}^\mathsf{T}$ have the same rank.

PROOF.   Let $\mathbf{A} = [a_{jk}]$ be an $m \times n$ matrix, and let rank $\mathbf{A} = r$. Then, by definition, $\mathbf{A}$ has a linearly independent set of r row vectors, call them $\mathbf{v}_{(1)}, \cdots, \mathbf{v}_{(r)}$, and all row vectors $\mathbf{a}_{(1)}, \cdots, \mathbf{a}_{(m)}$ of $\mathbf{A}$ are linear combinations of those independent ones, say,

$$\begin{aligned}
\mathbf{a}_{(1)} &= c_{11}\mathbf{v}_{(1)} + c_{12}\mathbf{v}_{(2)} + \cdots + c_{1r}\mathbf{v}_{(r)} \\
\mathbf{a}_{(2)} &= c_{21}\mathbf{v}_{(1)} + c_{22}\mathbf{v}_{(2)} + \cdots + c_{2r}\mathbf{v}_{(r)} \\
&\;\;\vdots \qquad\quad \vdots \qquad\quad \vdots \qquad\qquad \vdots \\
\mathbf{a}_{(m)} &= c_{m1}\mathbf{v}_{(1)} + c_{m2}\mathbf{v}_{(2)} + \cdots + c_{mr}\mathbf{v}_{(r)}.
\end{aligned}$$

These are vector equations. Each of them is equivalent to n equations for corresponding components. Denoting the components of $\mathbf{v}_{(1)}$ by $v_{11}, \cdots, v_{1n}$, the components of $\mathbf{v}_{(2)}$ by $v_{21}, \cdots, v_{2n}$, etc., and similarly for the vectors on the left side, we thus have, for $k = 1, \cdots, n$,

$$\begin{aligned}
a_{1k} &= c_{11}v_{1k} + c_{12}v_{2k} + \cdots + c_{1r}v_{rk} \\
a_{2k} &= c_{21}v_{1k} + c_{22}v_{2k} + \cdots + c_{2r}v_{rk} \\
&\;\;\vdots \qquad\quad \vdots \qquad\quad \vdots \qquad\qquad \vdots \\
a_{mk} &= c_{m1}v_{1k} + c_{m2}v_{2k} + \cdots + c_{mr}v_{rk}.
\end{aligned}$$

This can be written

$$
\begin{bmatrix} a_{1k} \\ a_{2k} \\ \cdot \\ \cdot \\ \cdot \\ a_{mk} \end{bmatrix} = v_{1k} \begin{bmatrix} c_{11} \\ c_{21} \\ \cdot \\ \cdot \\ \cdot \\ c_{m1} \end{bmatrix} + v_{2k} \begin{bmatrix} c_{12} \\ c_{22} \\ \cdot \\ \cdot \\ \cdot \\ c_{m2} \end{bmatrix} + \cdots + v_{rk} \begin{bmatrix} c_{1r} \\ c_{2r} \\ \cdot \\ \cdot \\ \cdot \\ c_{mr} \end{bmatrix}
$$

where $k = 1, \cdots, n$. The vector on the left is the kth column vector of $\mathbf{A}$. Hence the equation shows that each column vector of $\mathbf{A}$ is a linear combination of the r vectors on the right. Hence the maximum number of linearly independent column vectors of $\mathbf{A}$ cannot exceed r, which is the maximum number of linearly independent row vectors of $\mathbf{A}$, by the definition of rank.

Now the same conclusion applies to the transpose $\mathbf{A}^\mathsf{T}$ of $\mathbf{A}$. Since the row vectors of $\mathbf{A}^\mathsf{T}$ are the column vectors of $\mathbf{A}$, and the column vectors of $\mathbf{A}^\mathsf{T}$ are the row vectors of $\mathbf{A}$, that conclusion means that the maximum number of linearly independent row vectors of $\mathbf{A}$ (which is r) cannot exceed the maximum number of linearly independent column vectors of $\mathbf{A}$. Hence that number must equal r, and the proof is complete. ◄

EXAMPLE 4   **Illustration of Theorem 1**

What does Theorem 1 mean with respect to our matrix $\mathbf{A}$ in (2)? Since we have rank $\mathbf{A} = 2$, the column vectors should contain two linearly independent ones, and the other two should be linear combinations of them. Indeed, the first two column vectors are linearly independent, and

$$
\begin{bmatrix} 2 \\ 24 \\ 0 \end{bmatrix} = \frac{2}{3} \begin{bmatrix} 3 \\ -6 \\ 21 \end{bmatrix} + \frac{2}{3} \begin{bmatrix} 0 \\ 42 \\ -21 \end{bmatrix} \quad \text{and} \quad \begin{bmatrix} 2 \\ 54 \\ -15 \end{bmatrix} = \frac{2}{3} \begin{bmatrix} 3 \\ -6 \\ 21 \end{bmatrix} + \frac{29}{21} \begin{bmatrix} 0 \\ 42 \\ -21 \end{bmatrix}.
$$

This is easy to verify but not so easy to see. Indeed, we realize that the determination of the rank of a matrix by a direct application of the definition is not the proper way, unless the matrix is sufficiently simple. This suggests asking whether we can "simplify" (transform) a matrix without altering its rank. The answer is yes, as we show next. ◄

In discussing our further ideas we shall use the following related concepts, which are of general importance in linear algebra.

Vector Space, Dimension, Basis

A **vector space** is a (nonempty) set V of vectors such that with any two vectors $\mathbf{a}$ and $\mathbf{b}$ in V all their linear combinations $\alpha\mathbf{a} + \beta\mathbf{b}$ (α, β any real numbers) are elements of V, and these vectors satisfy the familiar laws (4) and (5) in Sec. 6.1 (written in lowercase letters $\mathbf{a}, \mathbf{b}, \mathbf{u}, \cdots$, which is our notation for vectors).[6]

The maximum number of linearly independent vectors in V is called the **dimension** of V and is denoted by dim V.

A linearly independent set in V consisting of a maximum possible number of vectors in V is called a **basis** for V. Thus the number of vectors of a basis for V equals dim V.

---

[6]This suffices presently. General vector spaces will be discussed in Sec. 6.8.

The set of all linear combinations of given vectors $\mathbf{a}_{(1)}, \cdots, \mathbf{a}_{(p)}$ with the same number of components is called the **span** of these vectors. Obviously, a span is a vector space.

By a **subspace** of a vector space V we mean a nonempty subset of V that itself forms a vector space with respect to the two algebraic operations defined for the vectors of V.

EXAMPLE 5   **Vector space, Dimension, Basis**

The span of the three vectors in Example 1 is a vector space of dimension 2, and a basis is $\mathbf{a}_{(1)}, \mathbf{a}_{(2)}$, for instance, or $\mathbf{a}_{(1)}, \mathbf{a}_{(3)}$, etc.   ◀

The span of the row vectors of a matrix $\mathbf{A}$ is called the **row space** of $\mathbf{A}$ and the span of the columns the **column space** of $\mathbf{A}$.

Now, by Theorem 1, $\mathbf{A}$ has as many linearly independent rows as it has linearly independent columns. By the definition of dimension, their number is the dimension of the row or column space of $\mathbf{A}$. This proves

THEOREM 2   **(Row space and column space)**

The row space and the column space of a matrix $\mathbf{A}$ *have the same dimension, equal to* rank $\mathbf{A}$.

Invariance of Rank Under Elementary Row Operations

Now rank $\mathbf{A}$ is the maximum number of linearly independent row vectors of $\mathbf{A}$. This number does not change if we change the order of these rows, or multiply one of them by a nonzero c, or take a linear combination by adding a multiple of a row to another row. Hence we have

THEOREM 3   **(Row-equivalent matrices)**

Row-equivalent matrices have the same rank.

This theorem tells us what we can do to determine the rank of a matrix $\mathbf{A}$. Namely, we can reduce $\mathbf{A}$ to echelon form (Sec. 6.3) using the technique of the Gauss elimination. This leaves the rank unchanged, by Theorem 3, and from the echelon form we can recognize the rank directly.

EXAMPLE 6   **Determination of rank**

For the matrix in Example 3 we obtain successively

$$
\mathbf{A} = \begin{bmatrix} 3 & 0 & 2 & 2 \\ -6 & 42 & 24 & 54 \\ 21 & -21 & 0 & -15 \end{bmatrix} \qquad \text{(given)}
$$

$$
\begin{bmatrix} 3 & 0 & 2 & 2 \\ 0 & 42 & 28 & 58 \\ 0 & -21 & -14 & -29 \end{bmatrix} \qquad \begin{array}{l} \text{Row 2 + 2 Row 1} \\ \text{Row 3} - 7 \text{ Row 1} \end{array}
$$

$$
\begin{bmatrix} 3 & 0 & 2 & 2 \\ 0 & 42 & 28 & 58 \\ 0 & 0 & 0 & 0 \end{bmatrix} \qquad \text{Row 3} + \tfrac{1}{2} \text{ Row 2}
$$

The last matrix is in echelon form. From the row vectors and Theorem 3 we see immediately that rank $\mathbf{A} \leqq 2$, and rank $\mathbf{A} = 2$ by Theorem 1, since the first two column vectors are certainly linearly independent. ◀

This method of determining rank has practical application in connection with the determination of linear dependence and independence of vectors. The key to this is the following theorem, which results immediately from the definition of rank.

THEOREM 4   **(Linear dependence and independence)**

p vectors $\mathbf{x}_{(1)}, \cdots, \mathbf{x}_{(p)}$ (with n components each) are linearly independent if the matrix with row vectors $\mathbf{x}_{(1)}, \cdots, \mathbf{x}_{(p)}$ has rank p; they are linearly dependent if that rank is less than p.

Since each of those p vectors has n components, that matrix, call it $\mathbf{A}$, has p rows and n columns; and if $n < p$, then by Theorem 1 we must have rank $\mathbf{A} \leqq n < p$, so that Theorem 4 yields the following result, which one should keep in mind.

THEOREM 5   *p vectors with $n < p$ components are always linearly dependent.*

For instance, three or more vectors in the plane are linearly dependent. Similarly, four or more vectors in space are linearly dependent.

By the definition of dimension, we also have

THEOREM 6   *The vector space R^n consisting of all vectors with n components has dimension n.*

In the next section we apply all these results to linear systems, with rank playing the basic role.

PROBLEM SET 6.4

Linear Independence

Are the following sets of vectors linearly independent or dependent? (Show the details of your work.)

1. $[1 \quad 0 \quad 0]$, $[1 \quad 1 \quad 0]$, $[1 \quad 1 \quad 1]$
2. $[7 \quad -3 \quad 11 \quad -6]$, $[-56 \quad 24 \quad -88 \quad 48]$
3. $[-1 \quad 5 \quad 0]$, $[16 \quad 8 \quad -3]$, $[-64 \quad 56 \quad 9]$
4. $[1 \quad -1 \quad 1]$, $[1 \quad 1 \quad -1]$, $[-1 \quad 1 \quad 1]$, $[0 \quad 1 \quad 0]$
5. $[2 \quad -4]$, $[1 \quad 9]$, $[3 \quad 5]$
6. $[1 \quad 2 \quad 3]$, $[0 \quad 0 \quad 0]$, $[5 \quad 5 \quad 1]$
7. $[1 \quad 9 \quad 9 \quad 8]$, $[2 \quad 0 \quad 0 \quad 3]$, $[2 \quad 0 \quad 0 \quad 8]$
8. $[\frac{1}{4} \quad 0 \quad -\frac{1}{4}]$, $[0 \quad \frac{1}{2} \quad -\frac{1}{2}]$, $[\frac{1}{3} \quad -\frac{1}{3} \quad 0]$

Rank

In Probs. 9–17 find the rank by the method in the text or by inspection. (Show the details.)

9. $\begin{bmatrix} 8 & -4 \\ -2 & 1 \\ 6 & -3 \end{bmatrix}$

10. $\begin{bmatrix} m & n & p \\ n & m & p \end{bmatrix}$
$m^2 \neq n^2$

11. $\begin{bmatrix} 8 & -3 & 7 \\ -20 & -17 & -15 \\ 11 & 2 & 9 \end{bmatrix}$

12. $\begin{bmatrix} 0 & 1 & 1 \\ 1 & 0 & 1 \\ 1 & 1 & 0 \end{bmatrix}$　　**13.** $\begin{bmatrix} 3 & -1 & 5 \\ 2 & -4 & 6 \\ 10 & 0 & 14 \end{bmatrix}$　　**14.** $\begin{bmatrix} 4 & 0 & 2 & 8 \\ 5 & 7 & 3 & 1 \\ 0 & 6 & 9 & 0 \end{bmatrix}$

15. $\begin{bmatrix} 3 & 1 & 4 \\ 0 & 5 & 8 \\ -3 & 4 & 4 \\ 1 & 2 & 4 \end{bmatrix}$　　**16.** $\begin{bmatrix} 0 & 8 & -1 \\ 1 & 2 & 0 \\ 0 & 0 & 3 \\ 0 & 4 & 5 \end{bmatrix}$　　**17.** $\begin{bmatrix} 9 & 3 & 1 & 0 \\ 3 & 0 & 1 & -6 \\ 1 & 1 & 1 & 1 \\ 0 & -6 & 1 & 9 \end{bmatrix}$

Vector Spaces

Is the given set of vectors a vector space? (Give a reason.) If your answer is yes, determine the dimension and find a basis. ($v_1, v_2, \cdots$ denote components.)

18. All vectors in R^3 such that $4v_2 + v_3 = k$

19. All vectors in R^3 such that $2v_1 + 3v_3 = 0$

20. All vectors in R^5 such that $v_1 = 0$, $v_3 = 0$, $v_2 + v_4 \geq 0$

21. All ordered quadruples of positive real numbers

22. All vectors in R^n with the first $n - 2$ components zero

23. All real numbers

24. All vectors in R^3 such that $3v_1 - 2v_2 + v_3 = 0$, $4v_1 + 5v_2 = 0$

25. All vectors in R^2 with components less than 1 in absolute value

26. TEAM PROJECT. Rank of a Matrix. The purpose of this project is to familiarize you more thoroughly with the concept of rank, which is central to the study of matrices in connection with linear systems of equations, linear transformations, etc.

　(a) On the definition. Illustrate the definition of rank and its characterization by column vectors with a 3×4 matrix of rank 2 of your choice.

　(b) Simple properties. Show that rank $\mathbf{B}^\mathsf{T}\mathbf{A}^\mathsf{T} = $ rank $\mathbf{AB}$. Show that for a square matrix the linear independence of the row vectors implies that of the column vectors and conversely. Show that for a nonsquare matrix either the row vectors or the column vectors must always be linearly dependent.

　(c) Product. *The rank of the product of two matrices cannot exceed the rank of either factor.* Illustrate this theorem with examples.

　(d) Sylvester's inequality. The rank r of the product of an $m \times n$ matrix $\mathbf{A}$ of rank r_A and an $n \times p$ matrix $\mathbf{B}$ of rank r_B satisfies

$$r_A + r_B - n \leq r \leq \min(r_A, r_B) \; (= \text{the smaller of } r_A \text{ and } r_B).$$

　Illustrate this with examples. Find an example in which both equality signs hold.

　(e) Square of a matrix. Illustrate with an example that rank $\mathbf{A} = $ rank $\mathbf{B}$ *does not imply* rank $\mathbf{A}^2 = $ rank $\mathbf{B}^2$.

27. (Row space) Prove that row-equivalent matrices have the same row space.

28. (Row and column spaces) Find a basis of the row space and of the column space of the matrix in Prob. 9.

29. (Row and column spaces) Do the same tasks as in Prob. 28 for the matrix in Prob. 15.

30. (Subspace) Give examples of one- and two-dimensional subspaces of R^3.

6.5 Solutions of Linear Systems: Existence, Uniqueness, General Form

Using rank (see Sec. 6.4), we can now give necessary and sufficient conditions for the existence of solutions and for the uniqueness, and we can characterize their general form in Theorem 1 (which you should memorize!). (For typical examples, see Sec. 6.3.)

Here a **submatrix** of a matrix **A** (a concept of general importance occurring in the theorem) is a matrix obtained from **A** by omitting some rows or columns (or both). This includes **A** itself (as the matrix obtained by omitting no rows or columns).

THEOREM 1   **Fundamental Theorem for linear systems**

(a) **Existence.** *A linear system of m equations in n unknowns* $x_1, \cdots, x_n$

(1)
$$
\begin{aligned}
a_{11}x_1 + a_{12}x_2 + \cdots + a_{1n}x_n &= b_1 \\
a_{21}x_1 + a_{22}x_2 + \cdots + a_{2n}x_n &= b_2 \\
&\cdots \cdots \\
a_{m1}x_1 + a_{m2}x_2 + \cdots + a_{mn}x_n &= b_m
\end{aligned}
$$

has solutions if and only if the coefficient matrix **A** *and the augmented matrix* $\widetilde{\mathbf{A}}$ *have the same rank. Here,*

$$
\mathbf{A} = \begin{bmatrix} a_{11} & \cdots & a_{1n} \\ \cdot & \cdots & \cdot \\ \cdot & \cdots & \cdot \\ a_{m1} & \cdots & a_{mn} \end{bmatrix} \quad and \quad \widetilde{\mathbf{A}} = \begin{bmatrix} a_{11} & \cdots & a_{1n} & \vdots & b_1 \\ \cdot & \cdots & \cdot & \vdots & \cdot \\ \cdot & \cdots & \cdot & \vdots & \cdot \\ a_{m1} & \cdots & a_{mn} & \vdots & b_m \end{bmatrix}.
$$

(b) **Uniqueness.** *The system* (1) *has precisely one solution if and only if this common rank r of* **A** *and* $\widetilde{\mathbf{A}}$ *equals n.*

(c) **Infinitely many solutions.** *If this rank r is less than n, the system* (1) *has infinitely many solutions. All of these are obtained by determining r suitable unknowns (whose submatrix of coefficients must have rank r) in terms of the remaining n − r unknowns, to which arbitrary values can be assigned.* (See Example 3 in Sec. 6.3.)

(d) **Gauss elimination (Sec. 6.3).** *If solutions exist, they can all be obtained by the Gauss elimination.* (This elimination may be started without first looking at the ranks of **A** and $\widetilde{\mathbf{A}}$ because it will automatically reveal whether or not solutions exist; see Sec. 6.3.)

PROOF.   (a) We can write the system (1) in the form

(1)
$$
\mathbf{Ax = b}
$$

or in terms of the column vectors $\mathbf{c}_{(1)}, \cdots, \mathbf{c}_{(n)}$ of **A**:

(2)
$$
\mathbf{c}_{(1)}x_1 + \mathbf{c}_{(2)}x_2 + \cdots + \mathbf{c}_{(n)}x_n = \mathbf{b}.
$$

Since $\tilde{\mathbf{A}}$ is obtained by attaching to $\mathbf{A}$ the additional column $\mathbf{b}$, Theorem 1 in Sec. 6.4 implies that rank $\tilde{\mathbf{A}}$ equals rank $\mathbf{A}$ or rank $\mathbf{A} + 1$. Now if (1) has a solution $\mathbf{x}$, then (2) shows that $\mathbf{b}$ must be a linear combination of those column vectors. Hence rank $\tilde{\mathbf{A}}$ cannot exceed rank $\mathbf{A}$, so that we must have rank $\tilde{\mathbf{A}} = $ rank $\mathbf{A}$.

Conversely, if rank $\tilde{\mathbf{A}} = $ rank $\mathbf{A}$, then $\mathbf{b}$ must be a linear combination of the column vectors of $\mathbf{A}$, say,

$$(2^*) \qquad\qquad \mathbf{b} = \alpha_1 \mathbf{c}_{(1)} + \cdots + \alpha_n \mathbf{c}_{(n)}$$

since otherwise rank $\tilde{\mathbf{A}} = $ rank $\mathbf{A} + 1$. But this means that (1) has a solution, namely, $x_1 = \alpha_1, \cdots, x_n = \alpha_n$, as can be seen by comparing (2^*) and (2).

(b) If rank $\mathbf{A} = r = n$, then the set $C = \{\mathbf{c}_{(1)}, \cdots, \mathbf{c}_{(n)}\}$ is linearly independent, by Theorem 1 in Sec. 6.4. It follows that then the representation (2) of $\mathbf{b}$ is unique because nonuniqueness, say,

$$\mathbf{c}_{(1)} x_1 + \cdots + \mathbf{c}_{(n)} x_n = \mathbf{c}_{(1)} \tilde{x}_1 + \cdots + \mathbf{c}_{(n)} \tilde{x}_n$$

would imply

$$(x_1 - \tilde{x}_1)\mathbf{c}_{(1)} + \cdots + (x_n - \tilde{x}_n)\mathbf{c}_{(n)} = \mathbf{0}$$

and $x_1 - \tilde{x}_1 = 0, \cdots, x_n - \tilde{x}_n = 0$ by the linear independence. Hence the scalars $x_1, \cdots, x_n$ in (2) are uniquely determined, that is, the solution of (1) is unique.

(c) If rank $\mathbf{A} = $ rank $\tilde{\mathbf{A}} = r < n$, then by Theorem 1 in Sec. 6.4 there is a linearly independent set K of r column vectors of $\mathbf{A}$ such that the other $n - r$ column vectors of $\mathbf{A}$ are linear combinations of those vectors. We renumber the columns and unknowns, denoting the renumbered quantities by $\hat{\ }$, so that $\{\hat{\mathbf{c}}_{(1)}, \cdots, \hat{\mathbf{c}}_{(r)}\}$ is that linearly independent set K. Then (2) becomes

$$\hat{\mathbf{c}}_{(1)} \hat{x}_1 + \cdots + \hat{\mathbf{c}}_{(n)} \hat{x}_n = \mathbf{b},$$

$\hat{\mathbf{c}}_{(r+1)}, \cdots, \hat{\mathbf{c}}_{(n)}$ are linear combinations of the vectors of K, and so are the vectors $\hat{x}_{r+1}\hat{\mathbf{c}}_{(r+1)}, \cdots, \hat{x}_n\hat{\mathbf{c}}_{(n)}$. Expressing these vectors in terms of the vectors of K and collecting terms, we can thus write the system in the form

$$(3) \qquad\qquad \hat{\mathbf{c}}_{(1)} y_1 + \cdots + \hat{\mathbf{c}}_{(r)} y_r = \mathbf{b}$$

with $y_j = \hat{x}_j + \beta_j$, where β_j results from the terms $\hat{\mathbf{c}}_{(r+1)}\hat{x}_{r+1}, \cdots, \hat{\mathbf{c}}_{(n)}\hat{x}_n$; here, $j = 1, \cdots, r$. Since the system has a solution, there are $y_1, \cdots, y_r$ satisfying (3). These scalars are unique since K is linearly independent. Choosing $\hat{x}_{r+1}, \cdots, \hat{x}_n$ fixes the β_j and corresponding $\hat{x}_j = y_j - \beta_j$, where $j = 1, \cdots, r$.

(d) This was proved in Sec. 6.3 and is restated here as a reminder. ◀

The theorem is illustrated by the examples in Sec. 6.3: in Example 3 we have rank $\mathbf{A} = $ rank $\hat{\mathbf{A}} = 2 < n = 4$ and can choose x_3 and x_4 arbitrarily; in Example 4 there is a unique solution since rank $\mathbf{A} = $ rank $\tilde{\mathbf{A}} = n = 3$; and in Example 5 there is no solution, since rank $\mathbf{A} = 2 < $ rank $\tilde{\mathbf{A}} = 3$.

The Homogeneous Linear System

The system (1) is called **homogeneous** if all the b_j's on the right side are zero. Otherwise it is called **nonhomogeneous.** (See also Sec. 6.3.) From the Fundamental Theorem we readily obtain the following results.

THEOREM 2   **(Homogeneous system)**

A homogeneous linear system

(4)

$$\begin{aligned}
a_{11}x_1 + a_{12}x_2 + \cdots + a_{1n}x_n &= 0 \\
a_{21}x_1 + a_{22}x_2 + \cdots + a_{2n}x_n &= 0 \\
\cdots\cdots\cdots\cdots\cdots\cdots\cdots\cdots\cdots\cdots& \\
a_{m1}x_1 + a_{m2}x_2 + \cdots + a_{mn}x_n &= 0
\end{aligned}$$

always has the **trivial solution** $x_1 = 0, \cdots, x_n = 0$. *Nontrivial solutions exist if and only if rank* $A < n$. *If rank* $A = r < n$, *these solutions, together with* $x = 0$, *form a vector space* (see Sec. 6.4) *of dimension* $n - r$, *called the* **solution space** *of* (4).

In particular, if $x_{(1)}$ *and* $x_{(2)}$ *are solution vectors of* (4), *then* $x = c_1 x_{(1)} + c_2 x_{(2)}$ *with any scalars* c_1 *and* c_2 *is a solution vector of* (4). (This ***does not hold*** for nonhomogeneous systems. Also, the term *solution space* is used for *homogeneous* systems only.)

PROOF.   The first proposition is obvious. It agrees with the fact that for a homogeneous system the matrix of the coefficients and the augmented matrix have the same rank. The solution vectors form a vector space because if $x_{(1)}$ and $x_{(2)}$ are any of them, then $Ax_{(1)} = 0$, $Ax_{(2)} = 0$, and this implies $A(x_{(1)} + x_{(2)}) = Ax_{(1)} + Ax_{(2)} = 0$ as well as $A(cx_{(1)}) = cAx_{(1)} = 0$, where c is arbitrary. If rank $A = r < n$, the Fundamental Theorem implies that we can choose $n - r$ suitable unknowns, call them $x_{r+1}, \cdots, x_n$, in an arbitrary fashion, and every solution is obtained in this way. Hence a basis for the solution space, briefly called a **basis of solutions** of (4), is $y_{(1)}, \cdots, y_{(n-r)}$, where the solution vector $y_{(j)}$ is obtained by choosing $x_{r+j} = 1$ and the other $x_{r+1}, \cdots, x_n$ zero; the corresponding $x_1, \cdots, x_r$ are then determined. Thus the solution space of (4) has dimension $n - r$. This completes the proof. ◀

The solution space of (4) is also called the **null space** of A because $Ax = 0$ for every x in the solution space. Its dimension is called the **nullity** of A. In terms of these concepts, Theorem 2 states that

(5)
$$\boxed{\text{rank } A + \text{nullity } A = n}$$

where n is the number of unknowns (number of columns of A).

Since in (4), rank $A = r \leqq m$ by the definition of rank, we see that if $m < n$, then $r < n$. By Theorem 2 this gives the practically important

THEOREM 3   **(Homogeneous linear system with fewer equations than unknowns)**

A homogeneous linear system with fewer equations than unknowns always has nontrivial solutions.

The Nonhomogeneous Linear System

The totality of solutions of a nonhomogeneous linear system can now be characterized as follows.

THEOREM 4   **(Nonhomogeneous system)**

If a nonhomogeneous linear system of equations of the form (1) *has solutions, then all these solutions are of the form*

$$\mathbf{x} = \mathbf{x}_0 + \mathbf{x}_h$$

where $\mathbf{x}_0$ *is any fixed solution of* (1) *and* $\mathbf{x}_h$ *runs through all the solutions of the corresponding homogeneous system* (4).

PROOF.   Let $\mathbf{x}$ be any given solution of (1) and $\mathbf{x}_0$ an arbitrarily chosen solution of (1). Then $\mathbf{Ax} = \mathbf{b}$, $\mathbf{Ax}_0 = \mathbf{b}$ and, therefore,

$$\mathbf{A}(\mathbf{x} - \mathbf{x}_0) = \mathbf{Ax} - \mathbf{Ax}_0 = \mathbf{0}.$$

This shows that the difference $\mathbf{x} - \mathbf{x}_0$ of any solution $\mathbf{x}$ of (1) and any fixed solution $\mathbf{x}_0$ of (1) is a solution of (4), say, $\mathbf{x}_h$. Hence all solutions of (1) are obtained by letting $\mathbf{x}_h$ run through all the solutions of the homogeneous system (4), and the proof is complete. ◀

6.6 Determinants
Cramer's Rule

Determinants were originally introduced for solving linear systems. Although ***impractical in computations,***[7] they have important engineering applications in eigenvalue problems (Sec. 7.1), differential equations (Chaps. 2, 3), vector algebra (Sec. 8.3), and so on. They can be introduced in several equivalent ways. Our definition is particularly practical in connection with linear systems.

An *nth-order determinant* is an expression associated with an $n \times n$ (hence ***square!***) matrix $\mathbf{A} = [a_{jk}]$, as we now explain, beginning with $n = 2$.

Second-Order Determinants

A **determinant of second order** is denoted and defined by

(1) $$D = \det \mathbf{A} = \begin{vmatrix} a_{11} & a_{12} \\ a_{21} & a_{22} \end{vmatrix} = a_{11}a_{22} - a_{12}a_{21}.$$

So here we have ***bars*** (whereas a matrix has ***brackets***).

---

[7]In numerical work, use a method from Secs. 6.3, 18.1–18.3; ***do not use*** **Cramer's rule.**

EXAMPLE 1

$$\begin{vmatrix} 4 & 3 \\ 2 & 5 \end{vmatrix} = 4 \cdot 5 - 3 \cdot 2 = 14$$ ◀

EXAMPLE 2 **Cramer's rule[8] for linear systems of 2 equations**

Derive a solution formula for the system

(2)
$$\text{(a)} \quad a_{11}x_1 + a_{12}x_2 = b_1$$
$$\text{(b)} \quad a_{21}x_1 + a_{22}x_2 = b_2.$$

Solution. Eliminate x_2 by multiplying (2a) by a_{22} and (2b) by $-a_{12}$ and adding,

$$(a_{11}a_{22} - a_{12}a_{21})x_1 = b_1a_{22} - a_{12}b_2.$$

Now eliminate x_1 by multiplying (2a) by $-a_{21}$ and (2b) by a_{11} and adding,

$$(a_{11}a_{22} - a_{12}a_{21})x_2 = a_{11}b_2 - b_1a_{21}.$$

Assuming that $D = a_{11}a_{22} - a_{12}a_{21} \neq 0$, we divide these two equations by D. We can write also the right sides as determinants. This gives **Cramer's rule for $n = 2$ equations (2)**

(3)
$$x_1 = \frac{\begin{vmatrix} b_1 & a_{12} \\ b_2 & a_{22} \end{vmatrix}}{D} = \frac{b_1a_{22} - a_{12}b_2}{D}, \quad x_2 = \frac{\begin{vmatrix} a_{11} & b_1 \\ a_{21} & b_2 \end{vmatrix}}{D} = \frac{a_{11}b_2 - b_1a_{21}}{D}.$$

For instance, if

$$4x_1 + 3x_2 = 12$$
$$2x_1 + 5x_2 = -8$$

then

$$x_1 = \frac{\begin{vmatrix} 12 & 3 \\ -8 & 5 \end{vmatrix}}{\begin{vmatrix} 4 & 3 \\ 2 & 5 \end{vmatrix}} = \frac{84}{14} = 6, \quad x_2 = \frac{\begin{vmatrix} 4 & 12 \\ 2 & -8 \end{vmatrix}}{\begin{vmatrix} 4 & 3 \\ 2 & 5 \end{vmatrix}} = \frac{-56}{14} = -4.$$

If the system (2) is homogeneous ($b_1 = b_2 = 0$) and $D \neq 0$, it has only the trivial solution $x_1 = x_2 = 0$, and if $D = 0$, it also has nontrivial solutions. ◀

Third-Order Determinants

A **determinant of third order** can be defined by

(4) $$D = \begin{vmatrix} a_{11} & a_{12} & a_{13} \\ a_{21} & a_{22} & a_{23} \\ a_{31} & a_{32} & a_{33} \end{vmatrix} = a_{11} \begin{vmatrix} a_{22} & a_{23} \\ a_{32} & a_{33} \end{vmatrix} - a_{21} \begin{vmatrix} a_{12} & a_{13} \\ a_{32} & a_{33} \end{vmatrix} + a_{31} \begin{vmatrix} a_{12} & a_{13} \\ a_{22} & a_{23} \end{vmatrix}.$$

Note the following. The signs on the right are $+ - +$. Each of the three terms on the right is an entry in the first column of D times its **"minor,"** that is, the second-order determinant obtained by deleting from D the row and column of that entry (thus for a_{11} delete the first row and first column, etc.).

[8]GABRIEL CRAMER (1704—1752). Swiss mathematician.

If we write out the minors, we get

(4) $D = a_{11}a_{22}a_{33} - a_{11}a_{23}a_{32} + a_{21}a_{13}a_{32} - a_{21}a_{12}a_{33} + a_{31}a_{12}a_{23} - a_{31}a_{13}a_{22}.$

EXAMPLE 3   **Cramer's rule for linear systems of 3 equations**

For linear systems of three equations in three unknowns

$$a_{11}x_1 + a_{12}x_2 + a_{13}x_3 = b_1$$

(5) $a_{21}x_1 + a_{22}x_2 + a_{23}x_3 = b_2$

$$a_{31}x_1 + a_{32}x_2 + a_{33}x_3 = b_3$$

Cramer's rule is

(6) $x_1 = \dfrac{D_1}{D}, \qquad x_2 = \dfrac{D_2}{D}, \qquad x_3 = \dfrac{D_3}{D}$ $(D \neq 0)$

with the *"determinant of the system"* D given by (4) and

$$D_1 = \begin{vmatrix} b_1 & a_{12} & a_{13} \\ b_2 & a_{22} & a_{23} \\ b_3 & a_{32} & a_{33} \end{vmatrix}, \qquad D_2 = \begin{vmatrix} a_{11} & b_1 & a_{13} \\ a_{21} & b_2 & a_{23} \\ a_{31} & b_3 & a_{33} \end{vmatrix}, \qquad D_3 = \begin{vmatrix} a_{11} & a_{12} & b_1 \\ a_{21} & a_{22} & b_2 \\ a_{31} & a_{32} & b_3 \end{vmatrix}.$$

This could be derived by elimination similar to that in Example 2. Instead, we shall obtain Cramer's rule for general n below.   ◀

Determinant of Any Order n

A **determinant of order n** is a scalar associated with an $n \times n$ matrix $\mathbf{A} = [a_{jk}]$, which is written

(7) $D = \det \mathbf{A} = \begin{vmatrix} a_{11} & a_{12} & \cdots & a_{1n} \\ a_{21} & a_{22} & \cdots & a_{2n} \\ \cdot & \cdot & \cdots & \cdot \\ \cdot & \cdot & \cdots & \cdot \\ a_{n1} & a_{n2} & \cdots & a_{nn} \end{vmatrix}$

and is defined for $n = 1$ by

(8) $D = a_{11}$

and for $n \geqq 2$ by

(9a) $\boxed{D = a_{j1}C_{j1} + a_{j2}C_{j2} + \cdots + a_{jn}C_{jn}}$ $(j = 1, 2, \cdots, \text{ or } n)$

or

(9b) $\boxed{D = a_{1k}C_{1k} + a_{2k}C_{2k} + \cdots + a_{nk}C_{nk}}$ $(k = 1, 2, \cdots, \text{ or } n)$

where

$$C_{jk} = (-1)^{j+k} M_{jk}$$

and M_{jk} is a determinant of order $n - 1$, namely, the determinant of the submatrix of **A** obtained from **A** by deleting the row and column of the entry a_{jk} (the jth row and the kth column). ◄

In this way, D is defined in terms of n determinants of order $n - 1$, each of which is, in turn, defined in terms of $n - 1$ determinants of order $n - 2$, and so on; we finally arrive at second-order determinants, in which those submatrices consist of single entries whose determinant is defined to be the entry itself.

From the definition it follows that *we may* **expand** D *by any row or column,* that is, choose in (9) the entries in any row or column, similarly when expanding the C_{jk}'s in (9), and so on.

This definition is unambiguous, that is, yields the same value for D no matter which columns or rows we choose in expanding. A proof is given in Appendix 4.

Terms used in connection with determinants are taken from matrices. In D we have n^2 **entries** or *elements* a_{jk}, also n **rows** and n **columns,** a **main diagonal** or *principal diagonal* on which $a_{11}, a_{22}, \cdots, a_{nn}$ stand. Two terms are new:

M_{jk} is called the **minor** *of* a_{jk} *in* D, and C_{jk} the **cofactor** *of* a_{jk} *in* D.

For later use we note that (9) may also be written in terms of minors

(10a) $$D = \sum_{k=1}^{n} (-1)^{j+k} a_{jk} M_{jk} \qquad (j = 1, 2, \cdots, \text{or } n)$$

(10b) $$D = \sum_{j=1}^{n} (-1)^{j+k} a_{jk} M_{jk} \qquad (k = 1, 2, \cdots, \text{or } n).$$

EXAMPLE 4   **Minors and cofactors of a third-order determinant**

In (4) the minors and cofactors of the entries in the first row of D can be seen directly. For the entries in the second row the minors are

$$M_{21} = \begin{vmatrix} a_{12} & a_{13} \\ a_{32} & a_{33} \end{vmatrix}, \qquad M_{22} = \begin{vmatrix} a_{11} & a_{13} \\ a_{31} & a_{33} \end{vmatrix}, \qquad M_{23} = \begin{vmatrix} a_{11} & a_{12} \\ a_{31} & a_{32} \end{vmatrix}$$

and the cofactors are $C_{21} = -M_{21}$, $C_{22} = +M_{22}$, and $C_{23} = -M_{23}$. Similarly for the third row—write these down yourself. And verify that the signs in C_{jk} form a **checkerboard pattern**

$$\begin{matrix} + & - & + \\ - & + & - \\ + & - & + \end{matrix}$$

◄

EXAMPLE 5   **Expansions of a third-order determinant**

$$D = \begin{vmatrix} 1 & 3 & 0 \\ 2 & 6 & 4 \\ -1 & 0 & 2 \end{vmatrix} = 1 \begin{vmatrix} 6 & 4 \\ 0 & 2 \end{vmatrix} - 3 \begin{vmatrix} 2 & 4 \\ -1 & 2 \end{vmatrix} + 0 \begin{vmatrix} 2 & 6 \\ -1 & 0 \end{vmatrix}$$

$$= 1(12 - 0) - 3(4 + 4) + 0(0 + 6) = -12.$$

This is the expansion by the first row. The expansion by the third column is

$$D = 0 \begin{vmatrix} 2 & 6 \\ -1 & 0 \end{vmatrix} - 4 \begin{vmatrix} 1 & 3 \\ -1 & 0 \end{vmatrix} + 2 \begin{vmatrix} 1 & 3 \\ 2 & 6 \end{vmatrix} = 0 - 12 + 0 = -12,$$

Verify that the other four expansions also give the value -12.   ◀

EXAMPLE 6   **Determinant of a triangular matrix**

$$\begin{vmatrix} -3 & 0 & 0 \\ 6 & 4 & 0 \\ -1 & 2 & 5 \end{vmatrix} = -3 \begin{vmatrix} 4 & 0 \\ 2 & 5 \end{vmatrix} = -3 \cdot 4 \cdot 5 = -60.$$

Inspired by this, can you formulate a little theorem on determinants of triangular matrices? Of diagonal matrices?   ◀

General Properties of Determinants

THEOREM 1   **(Behavior of an nth-order determinant under elementary row operations)**

(**a**) *Interchange of two rows multiplies the value of the determinant by* -1.

(**b**) *Addition of a multiple of a row to another row does not alter the value of the determinant.*

(**c**) *Multiplication of a row by c multiplies the value of the determinant by c.*

PROOF.   (**a**) By induction. The statement holds for $n = 2$ because

$$\begin{vmatrix} a & b \\ c & d \end{vmatrix} = ad - bc, \qquad \text{but} \qquad \begin{vmatrix} c & d \\ a & b \end{vmatrix} = bc - ad.$$

We now make the induction hypothesis that (a) holds for determinants of order $n - 1$ and show that it then holds for determinants of order n. Let D be of order n. Let E be obtained from D by the interchange of two rows. Expand D and E by a row that is **not** one of those interchanged, call it the jth row. Then by (10a),

$$(11) \qquad D = \sum_{k=1}^{n} (-1)^{j+k} a_{jk} M_{jk}, \qquad E = \sum_{k=1}^{n} (-1)^{j+k} a_{jk} N_{jk}$$

where N_{jk} is obtained from the minor M_{jk} of a_{jk} in D by the interchange of those two rows which have been interchanged in D (and which N_{jk} must both contain because we expand by another row!). Now these minors are of order $n - 1$. Hence the induction hypothesis applies and gives $N_{jk} = -M_{jk}$. Thus $E = -D$ by (11).

(**b**) Add c times row i to row j. Let $\tilde{D}$ be the new determinant. Its entries in row j are $a_{jk} + ca_{ik}$. If we expand $\tilde{D}$ by this row j, we see that we can write it as $\tilde{D} = D_1 + cD_2$, where $D_1 = D$ has the a_{jk} in row j and D_2 has the a_{ik} in row j. Hence row j = row i in D_2. Interchanging these gives D_2 back, but on the other hand $-D_2$ by (a). Together $D_2 = -D_2 = 0$, and $\tilde{D} = D_1 = D$.

(**c**) Expand the determinant by the row that has been multiplied.   ◀

Caution! $\det (k\mathbf{A}) = k^n \det \mathbf{A}$ (not $k \det \mathbf{A}$). Explain why.

EXAMPLE 7　　**Evaluation of determinants by reduction to triangular form**

Because of Theorem 1 we may evaluate a determinant by reduction to triangular form, as in the Gauss elimination for a matrix. For instance (with blue explanations always referring to the preceding determinant),

$$
D = \begin{vmatrix}
2 & 0 & -4 & 6 \\
4 & 5 & 1 & 0 \\
0 & 2 & 6 & -1 \\
-3 & 8 & 9 & 1
\end{vmatrix}
$$

$$
= \begin{vmatrix}
2 & 0 & -4 & 6 \\
0 & 5 & 9 & -12 \\
0 & 2 & 6 & -1 \\
0 & 8 & 3 & 10
\end{vmatrix}
\quad
\begin{array}{l}
\text{Row } 2 - 2 \text{ Row } 1 \\[1em]
\\[0.5em]
\text{Row } 4 + 1.5 \text{ Row } 1
\end{array}
$$

$$
= \begin{vmatrix}
2 & 0 & -4 & 6 \\
0 & 5 & 9 & -12 \\
0 & 0 & 2.4 & 3.8 \\
0 & 0 & -11.4 & 29.2
\end{vmatrix}
\quad
\begin{array}{l}
\\[1em]
\text{Row } 3 - 0.4 \text{ Row } 2 \\[0.5em]
\text{Row } 4 - 1.6 \text{ Row } 2
\end{array}
$$

$$
= \begin{vmatrix}
2 & 0 & -4 & 6 \\
0 & 5 & 9 & -12 \\
0 & 0 & 2.4 & 3.8 \\
0 & 0 & 0 & 47.25
\end{vmatrix}
\quad
\begin{array}{l}
\\[2em]
\text{Row } 4 + 4.75 \text{ Row } 3
\end{array}
$$

$$
= 2 \times 5 \times 2.4 \times 47.25 = 1134. \blacktriangleleft
$$

THEOREM 2　　**(Further properties of nth-order determinants)**

　　(a)–(c) *in Theorem* 1 *hold also for columns.*

　　(d) Transposition *leaves the value of a determinant unaltered.*

　　(e) A zero row or column *renders the value of a determinant zero.*

　　(f) Proportional rows or columns *render the value of a determinant zero. In particular, a determinant with two identical rows or columns has the value zero.*

PROOF.　　**(a)–(e)** follow directly from the fact that a determinant can be expanded by any row or column. In (d), transposition is defined as for matrices, that is, the jth row becomes the jth column of the transpose.

　　(f) If row $j = c$ times row i, then $D = cD_1$, where D_1 has row $j =$ row i. Hence an interchange of these rows reproduces D_1 but gives $-D_1$ by Theorem 1(a). Hence $D_1 = 0$ and $D = cD_1 = 0$. Similarly for columns. $\blacktriangleleft$

Rank in Terms of Determinants

It is most remarkable that the important concept of the rank of an $m \times n$ matrix $\mathbf{A}$ (the maximum number of linearly independent row or column vectors of $\mathbf{A}$; see Sec. 6.4) can be related to determinants. (This is sometimes used for *defining* rank.) Here we may assume that rank $\mathbf{A} > 0$ (because rank $\mathbf{A} = 0$ if and only if $\mathbf{A} = \mathbf{0}$; see Sec. 6.4).

THEOREM 3   **(Rank in terms of determinants)**

An $m \times n$ matrix $\mathbf{A} = [a_{jk}]$ has rank $r \geqq 1$ if and only if $\mathbf{A}$ has an $r \times r$ submatrix with nonzero determinant, whereas the determinant of every square submatrix with $r + 1$ or more rows that $\mathbf{A}$ has (or does not have!) is zero.

In particular, if $\mathbf{A}$ is square, $n \times n$, it has rank n if and only if

$$\det \mathbf{A} \neq 0.$$

PROOF.   The key idea is that elementary row operations (Sec. 6.3) alter neither rank (by Theorem 3, Sec. 6.4) nor the property of a determinant being nonzero (by Theorem 1, this section). The echelon form $\hat{\mathbf{A}}$ of $\mathbf{A}$ (see Sec. 6.3) has r nonzero row vectors (which are the first r row vectors) if and only if rank $\mathbf{A} = r$. Let $\hat{\mathbf{R}}$ be the $r \times r$ submatrix in the left upper corner of $\hat{\mathbf{A}}$ (so that the entries of $\hat{\mathbf{R}}$ are in both the first r rows and columns of $\hat{\mathbf{A}}$). Now $\hat{\mathbf{R}}$ is triangular, with all diagonal entries r_{jj} nonzero. Thus, $\det \hat{\mathbf{R}} = r_{11} \cdots r_{rr} \neq 0$. Also $\det \mathbf{R} \neq 0$ for the corresponding $r \times r$ submatrix $\mathbf{R}$ of $\mathbf{A}$ because $\hat{\mathbf{R}}$ results from $\mathbf{R}$ by elementary row operations. Similarly, $\det \mathbf{S} = 0$ for any square submatrix $\mathbf{S}$ of $r + 1$ or more rows perhaps contained in $\mathbf{A}$ because the corresponding submatrix $\hat{\mathbf{S}}$ of $\hat{\mathbf{A}}$ must contain a row of zeros (or else rank $\mathbf{A} \geqq r + 1$), so that $\det \hat{\mathbf{S}} = 0$ by Theorem 2. This proves the theorem for an $m \times n$ matrix.

In particular, if $\mathbf{A}$ is square, $n \times n$, then rank $\mathbf{A} = n$ if and only if $\mathbf{A}$ contains an $n \times n$ submatrix with nonzero determinant. But the only such submatrix can be $\mathbf{A}$ itself, hence $\det \mathbf{A} \neq 0$.   ◀

Cramer's Rule

Theorem 3 opens the way to the classical solution formula for linear systems, Cramer's rule, which gives solutions of linear systems as quotients of determinants. ***Cramer's rule is not practical in computations*** (for which the methods in Secs. 6.3 and 18.1–18.3 are suitable), but is of ***theoretical interest*** in differential equations (Secs. 2.10, 2.15) and other theories that have engineering applications.

THEOREM 4   **Cramer's Theorem (Solution of linear systems by determinants)**

(a) *If a linear system of n equations in the same number of unknowns $x_1, \cdots, x_n$*

(12)
$$\begin{aligned}
a_{11}x_1 + a_{12}x_2 + \cdots + a_{1n}x_n &= b_1 \\
a_{21}x_1 + a_{22}x_2 + \cdots + a_{2n}x_n &= b_2 \\
\cdots \cdots \cdots \cdots \cdots \cdots \cdots \cdots \cdots \\
a_{n1}x_1 + a_{n2}x_2 + \cdots + a_{nn}x_n &= b_n
\end{aligned}$$

has a nonzero coefficient determinant $D = \det \mathbf{A}$, the system has precisely one solution. This solution is given by the formulas

(13)
$$x_1 = \frac{D_1}{D}, \quad x_2 = \frac{D_2}{D}, \cdots, \quad x_n = \frac{D_n}{D}$$
          (Cramer's rule)

where D_k is the determinant obtained from D by replacing in D the kth column by the column with the entries $b_1, \cdots, b_n$.

(b) *Hence if the system* (12) *is* **homogeneous** *and $D \neq 0$, it has only the trivial solution $x_1 = 0$, $x_2 = 0, \cdots, x_n = 0$. If $D = 0$, the homogeneous system also has nontrivial solutions.*

PROOF.　The augmented matrix $\widetilde{\mathbf{A}}$ of the system (12) is of size $n \times (n + 1)$. Hence its rank can be at most n. Now if

$$(14) \qquad D = \det \mathbf{A} = \begin{vmatrix} a_{11} & \cdots & a_{1n} \\ \cdot & \cdots & \cdot \\ \cdot & \cdots & \cdot \\ a_{n1} & \cdots & a_{nn} \end{vmatrix} \neq 0,$$

then rank $\mathbf{A} = n$ by Theorem 3. Thus rank $\widetilde{\mathbf{A}} = $ rank $\mathbf{A}$. Hence, by the Fundamental Theorem in Sec. 6.5, the system (12) has a unique solution.

Let us now prove (13). Expanding D by its kth column, we obtain

$$(15) \qquad D = a_{1k}C_{1k} + a_{2k}C_{2k} + \cdots + a_{nk}C_{nk},$$

where C_{ik} is the cofactor of the entry a_{ik} in D. If we replace the entries in the kth column of D by any other numbers, we obtain a new determinant, say, $\hat{D}$. Clearly, its expansion by the kth column will be of the form (15), with $a_{1k}, \cdots, a_{nk}$ replaced by those new numbers and the cofactors C_{ik} as before. In particular, if we choose as new numbers the entries $a_{1l}, \cdots, a_{nl}$ in the lth column of D (where $l \neq k$), then the expansion of the resulting determinant $\hat{D}$ becomes

$$(16) \qquad a_{1l}C_{1k} + a_{2l}C_{2k} + \cdots + a_{nl}C_{nk} = 0 \qquad\qquad (l \neq k)$$

because $\hat{D}$ has two identical columns and is zero by Theorem 2(f). We now multiply the first equation in (12) by C_{1k} on both sides, the second by $C_{2k}, \cdots$, the last by C_{nk}, and add the resulting equations. This gives

$$(17) \quad C_{1k}(a_{11}x_1 + \cdots + a_{1n}x_n) + \cdots + C_{nk}(a_{n1}x_1 + \cdots + a_{nn}x_n)$$
$$= b_1 C_{1k} + \cdots + b_n C_{nk}.$$

Collecting terms with the same x_j, we can write the left side as

$$x_1(a_{11}C_{1k} + a_{21}C_{2k} + \cdots + a_{n1}C_{nk}) + \cdots + x_n(a_{1n}C_{1k} + a_{2n}C_{2k} + \cdots + a_{nn}C_{nk}).$$

From this we see that x_k is multiplied by

$$a_{1k}C_{1k} + a_{2k}C_{2k} + \cdots + a_{nk}C_{nk}.$$

Equation (15) shows that this equals D. Similarly, x_l is multiplied by

$$a_{1l}C_{1k} + a_{2l}C_{2k} + \cdots + a_{nl}C_{nk}.$$

Equation (16) shows that this is zero when $l \neq k$. Accordingly, the left side of (17) equals simply $x_k D$, so that (17) becomes

$$x_k D = b_1 C_{1k} + b_2 C_{2k} + \cdots + b_n C_{nk}.$$

Now the right side of this is D_k as defined in the theorem, expanded by its kth column, so that division by D gives (13). This proves Cramer's rule.

If (12) is homogeneous and $D \neq 0$, then each D_k has a column of zeros, so that $D_k = 0$ by Theorem 2(e), and (13) gives the trivial solution.

Finally, if (12) is homogeneous and $D = 0$, then rank $\mathbf{A} < n$ by Theorem 3, so that nontrivial solutions exist by Theorem 2 in Sec. 6.5.   ◄

Illustrations of Theorem 4 are given in Examples 2 and 3, and an important application follows in the next section.

PROBLEM SET 6.6

1. **(Second-order determinant)** Expand (1) in four possible ways and show that the four results agree.
2. **(Third-order determinant)** Do the task indicated in Example 5. Also evaluate D by reduction to triangular form.
3. **(Minors, cofactors)** Complete the list of minors and cofactors in Example 4.
4. **WRITING PROJECT. General Properties of Determinants.** Illustrate each statement in Theorems 1 and 2 with an example of your choice.

Evaluation of Determinants. Evaluate

5. $\begin{vmatrix} 3.8 & 0.6 \\ -1.4 & 9.3 \end{vmatrix}$

6. $\begin{vmatrix} \cos n\theta & \sin n\theta \\ -\sin n\theta & \cos n\theta \end{vmatrix}$

7. $\begin{vmatrix} 104 & 624 \\ 102 & 612 \end{vmatrix}$

8. $\begin{vmatrix} 205 & 16 & 81 \\ 0 & -13 & 2 \\ 0 & 0 & 16 \end{vmatrix}$

9. $\begin{vmatrix} m & n & p \\ p & m & n \\ n & p & m \end{vmatrix}$

10. $\begin{vmatrix} 7 & 14 & 21 \\ 36 & 18 & 6 \\ 87 & 12 & -45 \end{vmatrix}$

11. $\begin{vmatrix} 1 & 0 & 3 & 7 \\ 4 & 2 & 0 & 1 \\ 7 & 7 & 3 & 0 \\ 5 & 0 & 6 & 8 \end{vmatrix}$

12. $\begin{vmatrix} 1 & 2 & 0 & 0 \\ 2 & 4 & 2 & 0 \\ 0 & 2 & 9 & 2 \\ 0 & 0 & 2 & 16 \end{vmatrix}$

13. $\begin{vmatrix} 3 & 2 & 0 & 0 \\ 6 & 8 & 0 & 0 \\ 0 & 0 & 4 & 7 \\ 0 & 0 & 2 & 5 \end{vmatrix}$

Rank by Determinants. Find the rank by Theorem 3. Check by row reduction. (Show the details of your work.)

14. $\begin{bmatrix} 4 & 3 \\ -8 & -6 \\ 16 & 12 \end{bmatrix}$

15. $\begin{bmatrix} 0 & 2 & -3 \\ 2 & 0 & 5 \\ -3 & 5 & 0 \end{bmatrix}$

16. $\begin{bmatrix} 21 & -3 & 17 & 13 \\ 46 & 11 & 52 & 14 \\ 33 & 48 & 71 & -23 \end{bmatrix}$

Cramer's Rule. Solve by Cramer's rule and check by Gauss elimination:

17. $5x - 3y = 37$
$-2x + 7y = -38$

18. $x + 2y + 3z = 20$
$7x + 3y + z = 13$
$x + 6y + 2z = 0$

19. $3x + 7y + 8z = -13$
$2x + 9z = -5$
$-4x + y - 26z = 2$

20. TEAM PROJECT. Geometrical Applications: Curves and Surfaces Through Given Points. The idea is to get an equation from the vanishing of the determinant of a homogeneous linear system as the condition for a nontrivial solution in Cramer's theorem. We explain the trick for obtaining such a system for the case of a line L through two given points P_1: (x_1, y_1) and P_2: (x_2, y_2). The unknown line is $ax + by = -c$, say. We write it as $ax + by + c \cdot 1 = 0$. To get a nontrivial solution a, b, c, the determinant of the "coefficients" x, y, 1 must be zero. The system is

$$ax + by + c \cdot 1 = 0 \quad \text{(Line } L)$$

(18)
$$ax_1 + by_1 + c \cdot 1 = 0 \quad (P_1 \text{ on } L)$$

$$ax_2 + by_2 + c \cdot 1 = 0 \quad (P_2 \text{ on } L).$$

(a) Line through two points. Derive from $D = 0$ in (18) the familiar formula

$$\frac{x - x_1}{x_1 - x_2} = \frac{y - y_1}{y_1 - y_2}.$$

(b) Plane. Find the analog of (18) for a plane through three given points. Apply it when the points are $(1, 1, 1)$, $(3, 2, 6)$, $(5, 0, 5)$.

(c) Circle. Find a similar formula for a circle in the plane through three given points. Find and sketch the circle through $(2, 6)$, $(6, 4)$, $(7, 1)$.

(d) Sphere. Find the analog of the formula in (c) for a sphere through four given points. Find the sphere through $(0, 0, 5)$, $(4, 0, 1)$, $(0, 4, 1)$, $(0, 0, -3)$ by this formula or by inspection.

(e) General conic section. Find a formula for a general conic section (the vanishing of a determinant of 6th order). Try it out for a quadratic parabola and for a more general conic section of your own choice.

6.7 Inverse of a Matrix
Gauss–Jordan Elimination

In this section we consider square *matrices exclusively.*

The **inverse** of an $n \times n$ matrix $\mathbf{A} = [a_{jk}]$ is denoted by $\mathbf{A}^{-1}$ and is an $n \times n$ matrix such that

(1)
$$\mathbf{A}\mathbf{A}^{-1} = \mathbf{A}^{-1}\mathbf{A} = \mathbf{I},$$

where $\mathbf{I}$ is the $n \times n$ unit matrix (see Sec. 6.2).

If $\mathbf{A}$ has an inverse, then $\mathbf{A}$ is called a **nonsingular matrix.** If $\mathbf{A}$ has no inverse, then $\mathbf{A}$ is called a **singular matrix.**

If $\mathbf{A}$ *has an inverse, the inverse is unique.*

Indeed, if both **B** and **C** are inverses of **A**, then $\mathbf{AB} = \mathbf{I}$ and $\mathbf{CA} = \mathbf{I}$, so that we obtain the uniqueness from

$$\mathbf{B} = \mathbf{IB} = (\mathbf{CA})\mathbf{B} = \mathbf{C}(\mathbf{AB}) = \mathbf{CI} = \mathbf{C}.$$

We prove next that **A** has an inverse (is nonsingular) if and only if it has maximum possible rank n. The proof will also show that $\mathbf{Ax} = \mathbf{b}$ implies $\mathbf{x} = \mathbf{A}^{-1}\mathbf{b}$ provided $\mathbf{A}^{-1}$ exists, and thus give a motivation for the inverse as well as a relation to linear systems.[9]

THEOREM 1　　**(Existence of the inverse)**

*The inverse $\mathbf{A}^{-1}$ of an $n \times n$ matrix **A** exists if and only if* rank $\mathbf{A} = n$, *hence* (by Theorem 3, Sec. 6.6) *if and only if* det $\mathbf{A} \neq 0$. *Hence **A** is nonsingular if* rank $\mathbf{A} = n$, *and is singular if* rank $\mathbf{A} < n$.

PROOF.　　Consider the linear system

(2)　　　　　　　　　　　　　　　$\mathbf{Ax} = \mathbf{b}$

with the given matrix **A** as coefficient matrix. If the inverse exists, then multiplication from the left on both sides gives by (1)

$$\mathbf{A}^{-1}\mathbf{Ax} = \mathbf{x} = \mathbf{A}^{-1}\mathbf{b}.$$

This shows that (2) has a unique solution **x,** so that **A** must have rank n by the Fundamental Theorem in Sec. 6.5.

Conversely, let rank $\mathbf{A} = n$. Then by the same theorem, the system (2) has a unique solution **x** for any **b,** and the back substitution following the Gauss elimination (in Sec. 6.3) shows that its components x_j are linear combinations of those of **b,** so that we can write

(3)　　　　　　　　　　　　　　　$\mathbf{x} = \mathbf{Bb}.$

Substitution into (2) gives

$$\mathbf{Ax} = \mathbf{A}(\mathbf{Bb}) = (\mathbf{AB})\mathbf{b} = \mathbf{Cb} = \mathbf{b} \qquad (\mathbf{C} = \mathbf{AB})$$

for any **b.** Hence $\mathbf{C} = \mathbf{AB} = \mathbf{I}$, the unit matrix. Similarly, if we substitute (2) into (3) we get

$$\mathbf{x} = \mathbf{Bb} = \mathbf{B}(\mathbf{Ax}) = (\mathbf{BA})\mathbf{x}$$

for any **x** (and $\mathbf{b} = \mathbf{Ax}$). Hence $\mathbf{BA} = \mathbf{I}$. Together, $\mathbf{B} = \mathbf{A}^{-1}$ exists.　　　　◀

---

[9]But **not** a method of solving $\mathbf{Ax} = \mathbf{b}$ **numerically,** because the Gauss elimination (Sec. 6.3) requires fewer computations.

Determination of the Inverse

For determining the inverse $\mathbf{A}^{-1}$ of a nonsingular $n \times n$ matrix $\mathbf{A}$ practically, we can use the Gauss elimination (Sec. 6.3), actually a variant of it, called the **Gauss–Jordan elimination**.[10] The idea of the method is as follows.

Using $\mathbf{A}$, we form n linear systems $\mathbf{A}\mathbf{x}_{(1)} = \mathbf{e}_{(1)}, \cdots, \mathbf{A}\mathbf{x}_{(n)} = \mathbf{e}_{(n)}$, where $\mathbf{e}_{(1)}, \cdots, \mathbf{e}_{(n)}$ are the columns of the $n \times n$ unit matrix $\mathbf{I}$. These are n vector equations in the unknown vectors $\mathbf{x}_{(1)}, \cdots, \mathbf{x}_{(n)}$. We combine them into a single matrix equation $\mathbf{A}\mathbf{X} = \mathbf{I}$, with the unknown matrix $\mathbf{X}$ having the columns $\mathbf{x}_{(1)}, \cdots, \mathbf{x}_{(n)}$. Correspondingly, we combine the n augmented matrices $[\mathbf{A} \quad \mathbf{e}_{(1)}], \cdots, [\mathbf{A} \quad \mathbf{e}_{(n)}]$ into the "augmented matrix" $\widetilde{\mathbf{A}} = [\mathbf{A} \quad \mathbf{I}]$. Now multiplication of $\mathbf{A}\mathbf{X} = \mathbf{I}$ by $\mathbf{A}^{-1}$ from the left gives $\mathbf{X} = \mathbf{A}^{-1}\mathbf{I} = \mathbf{A}^{-1}$. Hence, to solve $\mathbf{A}\mathbf{X} = \mathbf{I}$ for $\mathbf{X}$, we can apply the Gauss elimination to $\widetilde{\mathbf{A}} = [\mathbf{A} \quad \mathbf{I}]$. This gives a matrix of the form $[\mathbf{U} \quad \mathbf{H}]$ with upper triangular $\mathbf{U}$ because the Gauss elimination triangularizes systems. The Gauss–Jordan method reduces $\mathbf{U}$ by further elementary row operations to diagonal form, in fact to the unit matrix $\mathbf{I}$. This is done by eliminating the entries of $\mathbf{U}$ above the main diagonal and making the diagonal entries all 1 by multiplication (see the example below). Of course, the method operates on the entire matrix $[\mathbf{U} \quad \mathbf{H}]$, transforming $\mathbf{H}$ into some matrix $\mathbf{K}$, hence the entire $[\mathbf{U} \quad \mathbf{H}]$ to $[\mathbf{I} \quad \mathbf{K}]$. This is the "augmented matrix" of $\mathbf{I}\mathbf{X} = \mathbf{K}$. Now $\mathbf{I}\mathbf{X} = \mathbf{X} = \mathbf{A}^{-1}$, as shown before. By comparison, $\mathbf{K} = \mathbf{A}^{-1}$, so that we can read $\mathbf{A}^{-1}$ directly from $[\mathbf{I} \quad \mathbf{K}]$.

The following example illustrates the practical details of the method.

EXAMPLE 1   **Inverse of a matrix. Gauss–Jordan elimination**

Find the inverse $\mathbf{A}^{-1}$ of

$$\mathbf{A} = \begin{bmatrix} -1 & 1 & 2 \\ 3 & -1 & 1 \\ -1 & 3 & 4 \end{bmatrix}.$$

Solution.   We apply the Gauss elimination (Sec. 6.3) to

$$[\mathbf{A} \quad \mathbf{I}] = \left[\begin{array}{ccc|ccc} -1 & 1 & 2 & 1 & 0 & 0 \\ 3 & -1 & 1 & 0 & 1 & 0 \\ -1 & 3 & 4 & 0 & 0 & 1 \end{array}\right]$$

$$\left[\begin{array}{ccc|ccc} -1 & 1 & 2 & 1 & 0 & 0 \\ 0 & 2 & 7 & 3 & 1 & 0 \\ 0 & 2 & 2 & -1 & 0 & 1 \end{array}\right] \quad \begin{array}{l} \text{Row 2 + 3 Row 1} \\ \text{Row 3 − Row 1} \end{array}$$

$$\left[\begin{array}{ccc|ccc} -1 & 1 & 2 & 1 & 0 & 0 \\ 0 & 2 & 7 & 3 & 1 & 0 \\ 0 & 0 & -5 & -4 & -1 & 1 \end{array}\right] \quad \text{Row 3 − Row 2}$$

---

[10]WILHELM JORDAN (1842—1899), German mathematician and geodesist. [See *American Mathematical Monthly* **94** (1987), 130–142.]

We do **not recommend** it as a method for solving systems of linear equations, since the number of operations in addition to those of the Gauss elimination is larger than that for back substitution, which the Gauss–Jordan elimination avoids. See also Sec. 18.1.

This is [**U   H**] as produced by the Gauss elimination, and **U** agrees with Example 4 in Sec. 6.3. Now follow the additional Gauss–Jordan steps, reducing **U** to **I**, that is, to diagonal form with entries 1 on the main diagonal.

$$
\left[\begin{array}{ccc|ccc}
1 & -1 & -2 & -1 & 0 & 0 \\
0 & 1 & 3.5 & 1.5 & 0.5 & 0 \\
0 & 0 & 1 & 0.8 & 0.2 & -0.2
\end{array}\right]
\begin{array}{l}
-\text{ Row 1} \\
0.5\ \text{Row 2} \\
-0.2\ \text{Row 3}
\end{array}
$$

$$
\left[\begin{array}{ccc|ccc}
1 & -1 & 0 & 0.6 & 0.4 & -0.4 \\
0 & 1 & 0 & -1.3 & -0.2 & 0.7 \\
0 & 0 & 1 & 0.8 & 0.2 & -0.2
\end{array}\right]
\begin{array}{l}
\text{Row 1 + 2 Row 3} \\
\text{Row 2 - 3.5 Row 3}
\end{array}
$$

$$
\left[\begin{array}{ccc|ccc}
1 & 0 & 0 & -0.7 & 0.2 & 0.3 \\
0 & 1 & 0 & -1.3 & -0.2 & 0.7 \\
0 & 0 & 1 & 0.8 & 0.2 & -0.2
\end{array}\right]
\begin{array}{l}
\text{Row 1 + Row 2}
\end{array}
$$

The last three columns constitute $\mathbf{A}^{-1}$. Check:

$$
\left[\begin{array}{ccc}
-1 & 1 & 2 \\
3 & -1 & 1 \\
-1 & 3 & 4
\end{array}\right]
\left[\begin{array}{ccc}
-0.7 & 0.2 & 0.3 \\
-1.3 & -0.2 & 0.7 \\
0.8 & 0.2 & -0.2
\end{array}\right]
=
\left[\begin{array}{ccc}
1 & 0 & 0 \\
0 & 1 & 0 \\
0 & 0 & 1
\end{array}\right].
$$

Hence $\mathbf{A}\mathbf{A}^{-1} = \mathbf{I}$. Similarly, $\mathbf{A}^{-1}\mathbf{A} = \mathbf{I}$.   ◀

Some Useful Formulas for Inverses

The explicit formula (4) in the following theorem is often useful in theoretical studies (as opposed to *computing* inverses). Indeed, the special case $n = 2$ occurs quite frequently in geometrical and other applications.

THEOREM 2   **(Inverse of a matrix)**

The inverse of a nonsingular $n \times n$ matrix $\mathbf{A} = [a_{jk}]$ is given by

$$
(4) \qquad \mathbf{A}^{-1} = \frac{1}{\det \mathbf{A}} [A_{jk}]^{\mathsf{T}} = \frac{1}{\det \mathbf{A}}
\left[\begin{array}{cccc}
A_{11} & A_{21} & \cdots & A_{n1} \\
A_{12} & A_{22} & \cdots & A_{n2} \\
. & . & \cdots & . \\
A_{1n} & A_{2n} & \cdots & A_{nn}
\end{array}\right]
$$

where A_{jk} is the cofactor of a_{jk} in $\det \mathbf{A}$ (see Sec. 6.6). Note well that in $\mathbf{A}^{-1}$, the cofactor A_{jk} occupies the same place as a_{kj} (not a_{jk}) does in $\mathbf{A}$.

In particular, the inverse of

$$
(4^*) \qquad \mathbf{A} =
\left[\begin{array}{cc}
a_{11} & a_{12} \\
a_{21} & a_{22}
\end{array}\right]
\qquad \text{is} \qquad
\mathbf{A}^{-1} = \frac{1}{\det \mathbf{A}}
\left[\begin{array}{cc}
a_{22} & -a_{12} \\
-a_{21} & a_{11}
\end{array}\right].
$$

PROOF.   We denote the right side of (4) by **B** and show that $\mathbf{BA} = \mathbf{I}$. We write

$$
(5) \qquad\qquad\qquad \mathbf{BA} = \mathbf{G} = [g_{kl}].
$$

Here, by the definition of matrix multiplication, and by the form of the entries of **B** as given in (4),

(6) $$g_{kl} = \sum_{s=1}^{n} \frac{A_{sk}}{\det \mathbf{A}} \, a_{sl} = \frac{1}{\det \mathbf{A}} (a_{1l}A_{1k} + \cdots + a_{nl}A_{nk}).$$

Now (15) and (16) in Sec. 6.6 (with C_{jk} written in our present notation A_{jk}) show that the sum $(\cdots)$ on the right is $D = \det \mathbf{A}$ when $l = k$ and zero when $l \neq k$. Hence

$$g_{kk} = \frac{1}{\det \mathbf{A}} \det \mathbf{A} = 1, \qquad g_{kl} = 0 \quad (l \neq k),$$

so that $\mathbf{G} = [g_{kl}] = \mathbf{BA} = \mathbf{I}$ in (5). Similarly $\mathbf{AB} = \mathbf{I}$. Hence $\mathbf{B} = \mathbf{A}^{-1}$.

In particular, when $n = 2$, in the first row of (4) we have $A_{11} = a_{22}$, $A_{21} = -a_{12}$ and in the second row $A_{12} = -a_{21}$, $A_{22} = a_{11}$. This gives (4*). ◀

EXAMPLE 2   Inverse of a 2 × 2 matrix

$$\mathbf{A} = \begin{bmatrix} 3 & 1 \\ 2 & 4 \end{bmatrix}, \qquad \mathbf{A}^{-1} = \frac{1}{10} \begin{bmatrix} 4 & -1 \\ -2 & 3 \end{bmatrix} = \begin{bmatrix} 0.4 & -0.1 \\ -0.2 & 0.3 \end{bmatrix}$$ ◀

EXAMPLE 3   Further Illustration of Theorem 2

Using (4), find the inverse of

$$\mathbf{A} = \begin{bmatrix} -1 & 1 & 2 \\ 3 & -1 & 1 \\ -1 & 3 & 4 \end{bmatrix}.$$

Solution.   We get $\det \mathbf{A} = -1(-7) - 13 + 2 \cdot 8 = 10$, and in (4),   ◀

$$A_{11} = \begin{vmatrix} -1 & 1 \\ 3 & 4 \end{vmatrix} = -7, \qquad A_{21} = -\begin{vmatrix} 1 & 2 \\ 3 & 4 \end{vmatrix} = 2, \qquad A_{31} = \begin{vmatrix} 1 & 2 \\ -1 & 1 \end{vmatrix} = 3,$$

$$A_{12} = -\begin{vmatrix} 3 & 1 \\ -1 & 4 \end{vmatrix} = -13, \qquad A_{22} = \begin{vmatrix} -1 & 2 \\ -1 & 4 \end{vmatrix} = -2, \qquad A_{32} = -\begin{vmatrix} -1 & 2 \\ 3 & 1 \end{vmatrix} = 7,$$

$$A_{13} = \begin{vmatrix} 3 & -1 \\ -1 & 3 \end{vmatrix} = 8, \qquad A_{23} = -\begin{vmatrix} -1 & 1 \\ -1 & 3 \end{vmatrix} = 2, \qquad A_{33} = \begin{vmatrix} -1 & 1 \\ 3 & -1 \end{vmatrix} = -2,$$

so that by (4), in agreement with Example 1,

$$\mathbf{A}^{-1} = \begin{bmatrix} -0.7 & 0.2 & 0.3 \\ -1.3 & -0.2 & 0.7 \\ 0.8 & 0.2 & -0.2 \end{bmatrix}.$$ ◀

Diagonal matrices $\mathbf{A} = [a_{jk}]$, $a_{jk} = 0$ when $j \neq k$, have an inverse if and only if all $a_{jj} \neq 0$. Then $\mathbf{A}^{-1}$ is diagonal with entries $1/a_{11}, \cdots, 1/a_{nn}$. *Proof.* In (4), for a diagonal matrix,

$$\frac{A_{11}}{D} = \frac{a_{22} \cdots a_{nn}}{a_{11}a_{22} \cdots a_{nn}} = \frac{1}{a_{11}}, \qquad \text{etc.}$$ ◀

EXAMPLE 4   **Inverse of a diagonal matrix**

$$A = \begin{bmatrix} -0.5 & 0 & 0 \\ 0 & 4 & 0 \\ 0 & 0 & 1 \end{bmatrix}, \qquad A^{-1} = \begin{bmatrix} -2 & 0 & 0 \\ 0 & 0.25 & 0 \\ 0 & 0 & 1 \end{bmatrix} \qquad \blacktriangleleft$$

Products can be inverted by taking the inverse of each factor and multiplying these inverses *in reverse order,*

(7)
$$(AC)^{-1} = C^{-1}A^{-1}.$$

Hence for more than two factors,

(8)
$$(AC \cdots PQ)^{-1} = Q^{-1}P^{-1} \cdots C^{-1}A^{-1}.$$

PROOF.   Equation (1) with A replaced by AC is

$$AC(AC)^{-1} = I.$$

Multiplying this by A^{-1} from the left and using $A^{-1}A = I$, we obtain

$$C(AC)^{-1} = A^{-1}I = A^{-1}.$$

Multiplying this by C^{-1} from the left we obtain (7). Equation (8) then follows from (7) by induction. $\blacktriangleleft$

We also note that *the inverse of the inverse is the given matrix,* as you may prove,

(9)
$$(A^{-1})^{-1} = A.$$

Vanishing of Matrix Products. Cancellation Law

Matrix multiplication differs from the multiplication of numbers. The following deviations from familiar rules for numbers must be carefully observed.

[1.] It is not commutative; that is, in general we have

$$AB \neq BA.$$

[2.] $AB = 0$ does not generally imply $A = 0$ or $B = 0$ (or $BA = 0$); for example,

$$\begin{bmatrix} 1 & 1 \\ 2 & 2 \end{bmatrix} \begin{bmatrix} -1 & 1 \\ 1 & -1 \end{bmatrix} = \begin{bmatrix} 0 & 0 \\ 0 & 0 \end{bmatrix}.$$

[3.] $AC = AD$ does not generally imply $C = D$ (even when $A \neq 0$).

These facts of great practical importance were noted in Sec. 6.2, and we are now able to explain the restricted validity of the cancellation law ([2.] and [3.]) using rank and inverse, concepts that were not yet available in Sec. 6.2.

THEOREM 3   **(Cancellation law)**

Let **A, B, C** *be* $n \times n$ *matrices. Then:*

   (a) *If* rank **A** $= n$ *and* **AB** $=$ **AC,** *then* **B** $=$ **C.**

   (b) *If* rank **A** $= n$, *then* **AB** $=$ **0** *implies* **B** $=$ **0.** *Hence if* **AB** $=$ **0,** *but* **A** $\neq$ **0** *as well as* **B** $\neq$ **0,** *then* rank **A** $< n$ *and* rank **B** $< n$.

   (c) *If* **A** *is singular, so are* **BA** *and* **AB.**

PROOF.   **(a)** Premultiply **AB** $=$ **AC** on both sides by $\mathbf{A}^{-1}$, which exists by Theorem 1.

   (b) Premultiply **AB** $=$ **0** on both sides by $\mathbf{A}^{-1}$.

   (c$_1$) Rank **A** $< n$ by Theorem 1. Hence **Ax** $=$ **0** has nontrivial solutions, by Theorem 2 in Sec. 6.5. Multiplication gives **BAx** $=$ **0.** Hence those solutions also satisfy **BAx** $=$ **0.** Hence rank (**BA**) $< n$ by Theorem 2 in Sec. 6.5, and **BA** is singular by Theorem 1.

   (c$_2$) $\mathbf{A}^{\mathsf{T}}$ is singular by Theorem 2(d) in Sec. 6.6. Hence $\mathbf{B}^{\mathsf{T}}\mathbf{A}^{\mathsf{T}}$ is singular by part (c$_1$). But $\mathbf{B}^{\mathsf{T}}\mathbf{A}^{\mathsf{T}} = (\mathbf{AB})^{\mathsf{T}}$; see (5) in Sec. 6.2. Hence **AB** is singular by Theorem 2(d) in Sec. 6.6.   ◀

Determinants of Matrix Products

Determinants of matrix products can be written as products of the determinants of the factors. This interesting formula is needed occasionally. We obtain it here with the help of the Gauss–Jordan diagonalization (see Example 1) and Theorem 3 just proved.

THEOREM 4   **(Determinant of a product of matrices)**

For any $n \times n$ *matrices* **A** *and* **B,**

(10)
$$\det (\mathbf{AB}) = \det (\mathbf{BA}) = \det \mathbf{A} \det \mathbf{B}.$$

PROOF.   If **A** is singular, so is **AB** by Theorem 3(c). Hence we have det **A** $= 0$, det (**AB**) $= 0$ by Theorem 3 in Sec. 6.6, and (10) is $0 = 0$, which holds.

Let **A** be nonsingular. Then we can reduce **A** to a diagonal matrix $\hat{\mathbf{A}} = [\hat{a}_{jk}]$ by Gauss–Jordan steps. Under these operations, det **A** retains its value, by Theorem 1 in Sec. 6.6, (a) and (b) [not (c)!] except perhaps for a sign reversal if we have to interchange two rows to get a nonzero pivot. But the same operations reduce **AB** to $\hat{\mathbf{A}}\mathbf{B}$ with the same effect on det (**AB**). Hence it remains to prove (10) for $\hat{\mathbf{A}}\mathbf{B}$; written out,

$$\hat{\mathbf{A}}\mathbf{B} = \begin{bmatrix} \hat{a}_{11} & 0 & \cdots & 0 \\ 0 & \hat{a}_{22} & \cdots & 0 \\ & & \ddots & \\ 0 & 0 & \cdots & \hat{a}_{nn} \end{bmatrix} \begin{bmatrix} b_{11} & b_{12} & \cdots & b_{1n} \\ b_{21} & b_{22} & \cdots & b_{2n} \\ & & \vdots & \\ b_{n1} & b_{n2} & \cdots & b_{nn} \end{bmatrix}$$

$$= \begin{bmatrix} \hat{a}_{11}b_{11} & \hat{a}_{11}b_{12} & \cdots & \hat{a}_{11}b_{1n} \\ \hat{a}_{22}b_{21} & \hat{a}_{22}b_{22} & \cdots & \hat{a}_{22}b_{2n} \\ & & \vdots & \\ \hat{a}_{nn}b_{n1} & \hat{a}_{nn}b_{n2} & \cdots & \hat{a}_{nn}b_{nn} \end{bmatrix}.$$

We now take the determinant det $(\hat{A}B)$. On the right we can take out a factor $\hat{a}_{11}$ from the first row, $\hat{a}_{22}$ from the second, $\cdots$, $\hat{a}_{nn}$ from the nth. But this product $\hat{a}_{11}\hat{a}_{22} \cdots \hat{a}_{nn}$ equals det $\hat{A}$, since $\hat{A}$ is diagonal. The remaining determinant is det B, and (10) is proved. ◀

This completes our discussion of linear systems (Secs. 6.3–6.7). (For **numerical methods,** see Secs. 18.1–18.4, which are independent of other sections on numerical methods.) Section 6.8 on vector spaces and linear transformations is optional.

PROBLEM SET 6.7

Inverse. Calculate the inverse by the Gauss–Jordan elimination or state that it does not exist. Check by using (1). (Show the details of your work.)

1. $\begin{bmatrix} 2 & 0 & -1 \\ 5 & 1 & 0 \\ 0 & 1 & 3 \end{bmatrix}$

2. $\begin{bmatrix} 1 & 2 & 5 \\ 0 & -1 & 2 \\ 2 & 4 & 11 \end{bmatrix}$

3. $\begin{bmatrix} 3 & -1 & 5 \\ 2 & 6 & 4 \\ 5 & 5 & 9 \end{bmatrix}$

4. $\begin{bmatrix} -7 & 0 & 0 \\ 0 & 8 & 13 \\ 0 & 3 & 5 \end{bmatrix}$

5. $\begin{bmatrix} 4 & -1 & -5 \\ 15 & 1 & -5 \\ 5 & 4 & 9 \end{bmatrix}$

6. $\begin{bmatrix} 0 & 0 & 0.3 \\ 0 & -0.2 & 0 \\ 0.4 & 0 & 0 \end{bmatrix}$

7. $\begin{bmatrix} -1 & 2 & 2 \\ 2 & -1 & 2 \\ 2 & 2 & -1 \end{bmatrix}$

8. $\begin{bmatrix} 1 & 2 & 3 \\ 4 & 5 & 6 \\ 7 & 8 & 9 \end{bmatrix}$

9. $\begin{bmatrix} 1 & 8 & -7 \\ 0 & 1 & 3 \\ 0 & 0 & 1 \end{bmatrix}$

10. (Inverse of the square) Show that $(A^2)^{-1} = (A^{-1})^2$.

11. (Inverse of the square) Verify $(A^2)^{-1} = (A^{-1})^2$ for A in Prob. 1.

12. (Inverse of the transpose) Show that $(A^{-1})^\mathsf{T} = (A^\mathsf{T})^{-1}$.

13. (Inverse of a symmetric matrix) Show that the inverse of a (nonsingular) symmetric matrix is symmetric.

14. (Inverse of the inverse) Show that $(A^{-1})^{-1} = A$.

Explicit Formula for the Inverse. Calculate the inverse from (4). Check the result by (1). (Show the details of your work.)

15. $\begin{bmatrix} 2 & 0 & -1 \\ 5 & 1 & 0 \\ 0 & 1 & 3 \end{bmatrix}$

16. $\begin{bmatrix} 4 & 0 & 4 \\ 0 & 2 & 0 \\ 6 & 0 & 1 \end{bmatrix}$

17. $\begin{bmatrix} 0.5 & 0 & -0.5 \\ -0.1 & 0.2 & 0.3 \\ 0.5 & 0 & -1.5 \end{bmatrix}$

18. $\begin{bmatrix} 2 & 0 & 0 \\ 6 & -1 & 0 \\ 5 & 3 & 4 \end{bmatrix}$

19. $\begin{bmatrix} 1.5 & -1.5 & 0.5 \\ -1.5 & 0.5 & 0.5 \\ 0.5 & 0.5 & -0.5 \end{bmatrix}$

20. $\begin{bmatrix} 1 & -3 & 0 \\ 0 & 2 & 4 \\ 3 & 3 & 0 \end{bmatrix}$

21. CAS PROJECT. Experiments with Hilbert Matrices. The $n \times n$ Hilbert matrix $\mathbf{H}_n = [h_{jk}]$ is the matrix with entries $h_{jk} = 1/(j + k - 1)$. Thus

$$\mathbf{H}_2 = \begin{bmatrix} 1 & \frac{1}{2} \\ \frac{1}{2} & 1 \end{bmatrix}, \qquad \mathbf{H}_3 = \begin{bmatrix} 1 & \frac{1}{2} & \frac{1}{3} \\ \frac{1}{2} & \frac{1}{3} & \frac{1}{4} \\ \frac{1}{3} & \frac{1}{4} & \frac{1}{5} \end{bmatrix}, \qquad \text{etc.}$$

Experiment with these matrices as follows.

(a) Compute det $\mathbf{H}_n$ and $\mathbf{H}_n^{-1}$ for $n = 2, \cdots, 8$. What can you observe about the size of the determinants? Of the entries of $\mathbf{H}_n^{-1}$?

(b) Find out experimentally how the decrease of the values of the determinants and the increase of the entries of the inverse change if you change the formula for the entries, for instance, to $1/(2j + 2k - 3)$, etc.

(c) The entries of $\mathbf{H}_n^{-1}$ are integers. Can you find other formulas for h_{jk} for which this continues to hold?

6.8 Vector Spaces, Inner Product Spaces, Linear Transformations
Optional

Special vector spaces were considered in Sec. 6.4, at which you may wish to take a quick look before going on. Recall that we can obtain a vector space by taking a set of vectors with n (real) components and forming their *span*, that is, the set of all their linear combinations (which are again vectors with n components).

Now if we take *all* vectors with n real numbers as components (*"real vectors"*) and real numbers as scalars, we obtain the very important **real n-dimensional vector space** R^n. This is a standard name and notation. Thus each vector in R^n is an ordered n-tuple of real numbers.

In particular, for $n = 3$ we get R^3 consisting of ordered triples (**"vectors in 3-space"**). For $n = 2$ we get R^2 consisting of ordered pairs (**"vectors in the plane"**). In Chaps. 8 and 9 we shall see that vectors in R^2 and R^3 have wide applications in mechanics, geometry, and calculus that are basic to the engineer and physicist.

Similarly, taking all ordered n-tuples of *complex* numbers as vectors and *complex* numbers as scalars, we obtain the **complex vector space** C^n, which we shall apply in Sec. 7.4.

There are other sets of practical interest (sets of matrices, functions, transformations, etc.) for which an addition and a scalar multiplication can be defined in a natural way. The desire to treat such sets as "vector spaces" suggests that we create from the **"concrete model"** R^n the **"abstract concept"** of a "real vector space" V by taking the most basic properties of R^n as axioms by which V is defined, properties without which one would not be able to create a useful and applicable theory of those more general situations. Selecting good axioms is not easy, but needs experience, sometimes gained only over a long period of time. In the present case, the following system of axioms turned out to be useful; note that each axiom expresses a simple property of R^n, or of R^3, as a matter of fact. Indeed, these axioms are suggested *literally* by (4) and (5) in Sec. 6.1—take a look there and compare.

Definition of a real vector space

A nonempty set V of elements $\mathbf{a}, \mathbf{b}, \cdots$ is called a **real vector space** (or *real linear space*), and these elements are called **vectors,**[11] if in V there are defined two algebraic operations (called *vector addition* and *scalar multiplication*) as follows.

I. Vector addition associates with every pair of vectors $\mathbf{a}$ and $\mathbf{b}$ of V a unique vector of V, called the *sum* of $\mathbf{a}$ and $\mathbf{b}$ and denoted by $\mathbf{a} + \mathbf{b}$, such that the following axioms are satisfied.

I.1 *Commutativity.* For any two vectors $\mathbf{a}$ and $\mathbf{b}$ of V,

$$\mathbf{a} + \mathbf{b} = \mathbf{b} + \mathbf{a}.$$

I.2 *Associativity.* For any three vectors $\mathbf{u}, \mathbf{v}, \mathbf{w}$ of V,

$$(\mathbf{u} + \mathbf{v}) + \mathbf{w} = \mathbf{u} + (\mathbf{v} + \mathbf{w}) \qquad \text{(written } \mathbf{u} + \mathbf{v} + \mathbf{w}).$$

I.3 There is a unique vector in V, called the *zero vector* and denoted by $\mathbf{0}$, such that for every $\mathbf{a}$ in V,

$$\mathbf{a} + \mathbf{0} = \mathbf{a}.$$

I.4 For every $\mathbf{a}$ in V there is a unique vector in V that is denoted by $-\mathbf{a}$ and is such that

$$\mathbf{a} + (-\mathbf{a}) = \mathbf{0}.$$

II. Scalar multiplication. The real numbers are called **scalars.** Scalar multiplication associates with every $\mathbf{a}$ in V and every scalar c a unique vector of V, called the *product* of c and $\mathbf{a}$ and denoted by $c\mathbf{a}$ (or $\mathbf{a}c$) such that the following axioms are satisfied.

II.1 *Distributivity.* For every scalar c and vectors $\mathbf{a}$ and $\mathbf{b}$ in V,

$$c(\mathbf{a} + \mathbf{b}) = c\mathbf{a} + c\mathbf{b}.$$

II.2 *Distributivity.* For all scalars c and k and every $\mathbf{a}$ in V,

$$(c + k)\mathbf{a} = c\mathbf{a} + k\mathbf{a}.$$

II.3 *Associativity.* For all scalars c and k and every $\mathbf{a}$ in V,

$$c(k\mathbf{a}) = (ck)\mathbf{a} \qquad \text{(written } ck\mathbf{a}).$$

II.4 For every $\mathbf{a}$ in V,

$$1\mathbf{a} = \mathbf{a}. \qquad \blacktriangleleft$$

A **complex vector space** is obtained if, instead of real numbers, we take complex numbers as scalars.

---

[11]Regardless of what they actually are; this convention causes no confusion because in any specific case the nature of those elements is clear from the context.

Basic Concepts Related to Vector Space

These concepts are defined as in Sec. 6.4.

A **linear combination** of vectors $\mathbf{a}_{(1)}, \cdots, \mathbf{a}_{(m)}$ in a vector space V is an expression

$$c_1\mathbf{a}_{(1)} + \cdots + c_m\mathbf{a}_{(m)} \qquad (c_1, \cdots, c_m \text{ any scalars}).$$

These vectors form a **linearly independent set** (briefly, they are called **linearly independent**) if

$$(1) \qquad\qquad\qquad c_1\mathbf{a}_{(1)} + \cdots + c_m\mathbf{a}_{(m)} = \mathbf{0}$$

implies that $c_1 = 0, \cdots, c_m = 0$. Otherwise they are called **linearly dependent.** Note that (1) with $m = 1$ is $c\mathbf{a} = \mathbf{0}$ and shows that a single vector $\mathbf{a}$ is linearly independent if and only if $\mathbf{a} \neq \mathbf{0}$.

V has **dimension n,** or is **n-dimensional,** if it contains a linearly independent set of n vectors, called a **basis** for V, whereas any set of more than n vectors in V is linearly dependent. Then every vector in V can be uniquely written as a linear combination of the basis vectors.

EXAMPLE 1 **Vector space of matrices**

The real 2×2 matrices form a four-dimensional real vector space. A basis is

$$\mathbf{B}_{11} = \begin{bmatrix} 1 & 0 \\ 0 & 0 \end{bmatrix}, \qquad \mathbf{B}_{12} = \begin{bmatrix} 0 & 1 \\ 0 & 0 \end{bmatrix}, \qquad \mathbf{B}_{21} = \begin{bmatrix} 0 & 0 \\ 1 & 0 \end{bmatrix}, \qquad \mathbf{B}_{22} = \begin{bmatrix} 0 & 0 \\ 0 & 1 \end{bmatrix}$$

because any $\mathbf{A} = [a_{jk}] = a_{11}\mathbf{B}_{11} + a_{12}\mathbf{B}_{12} + a_{21}\mathbf{B}_{21} + a_{22}\mathbf{B}_{22}$ in a unique fashion. Similarly, the real $m \times n$ matrices with fixed m and n form an mn-dimensional vector space. What is the dimension of the vector space of all skew-symmetric 3×3 matrices? Can you find a basis? ◄

EXAMPLE 2 **Polynomials**

The set of all constant, linear and quadratic polynomials in x together forms a vector space under the usual addition and multiplication by real numbers, since these two operations give polynomials of degree not exceeding 2, and the axioms in our definition follow by direct calculation. This space has dimension 3. A basis is $\{1, \ x, \ x^2\}$. ◄

EXAMPLE 3 **Second-order homogeneous linear differential equations**

The solutions of such an equation on a fixed interval $a < x < b$ form a vector space under the usual addition and multiplication by numbers since these two operations give again such a solution, by Fundamental Theorem 1 in Sec. 2.1, and I.1 to II.4 follow by direct calculation. Do the solutions of a nonhomogeneous linear differential equation form a vector space? ◄

If a vector space V contains a linearly independent set of n vectors for every n, no matter how large, then V is called **infinite dimensional,** as opposed to a *finite dimensional* (n-dimensional) vector space as defined before. An example is the space of all continuous functions on some interval $[a, \ b]$ of the x-axis, as we mention without proof.

Inner Product Spaces

From Sec. 6.2 we know that for vectors $\mathbf{a}$ and $\mathbf{b}$ in R^n we can define an inner product $\mathbf{a} \cdot \mathbf{b} = \mathbf{a}^\mathsf{T}\mathbf{b}$. This definition can be extended to general real vector spaces by taking basic properties of $\mathbf{a} \cdot \mathbf{b}$ as axioms for an "abstract" inner product, denoted by $(\mathbf{a}, \mathbf{b})$.

Definition of a real inner product space

A real vector space V is called a *real inner product space* (or *real pre-Hilbert*[12] *space*) if it has the following property. With every pair of vectors $\mathbf{a}$ and $\mathbf{b}$ in V there is associated a real number, which is denoted by $(\mathbf{a}, \mathbf{b})$ and is called the **inner product** of $\mathbf{a}$ and $\mathbf{b}$, such that the following axioms are satisfied.

I. For all scalars q_1 and q_2 and all vectors $\mathbf{a}, \mathbf{b}, \mathbf{c}$ in V,

$$(q_1\mathbf{a} + q_2\mathbf{b}, \mathbf{c}) = q_1(\mathbf{a}, \mathbf{c}) + q_2(\mathbf{b}, \mathbf{c}) \qquad (Linearity).$$

II. For all vectors $\mathbf{a}$ and $\mathbf{b}$ in V,

$$(\mathbf{a}, \mathbf{b}) = (\mathbf{b}, \mathbf{a}) \qquad (Symmetry).$$

III. For every $\mathbf{a}$ in V,

$$(\mathbf{a}, \mathbf{a}) \geqq 0,$$
$$(\mathbf{a}, \mathbf{a}) = 0 \quad \text{if and only if} \quad \mathbf{a} = \mathbf{0} \qquad \left.\right\} \quad (Positive\text{-}definiteness). \quad \blacktriangleleft$$

Vectors whose inner product is zero are called **orthogonal.**
The *length* or **norm** of a vector in V is defined by

$$(2) \qquad \|\mathbf{a}\| = \sqrt{(\mathbf{a}, \mathbf{a})} \qquad (\geqq 0).$$

A vector of norm 1 is called a **unit vector.**
From these axioms and from (2) one can derive the basic inequality

$$(3) \qquad |(\mathbf{a}, \mathbf{b})| \leqq \|\mathbf{a}\| \, \|\mathbf{b}\| \qquad (Schwarz^{13}\ inequality).$$

From this follows

$$(4) \qquad \|\mathbf{a} + \mathbf{b}\| \leqq \|\mathbf{a}\| + \|\mathbf{b}\| \qquad (Triangle\ inequality).$$

A simple direct calculation gives

$$(5) \qquad \|\mathbf{a} + \mathbf{b}\|^2 + \|\mathbf{a} - \mathbf{b}\|^2 = 2(\|\mathbf{a}\|^2 + \|\mathbf{b}\|^2) \qquad (Parallelogram\ equality).$$

---

[12]DAVID HILBERT (1862—1943), great German mathematician, taught at Königsberg and Göttingen and was the creator of the famous Göttingen mathematical school. He is known for his basic work in algebra, the calculus of variations, integral equations, functional analysis, and mathematical logic. His "Foundations of Geometry" helped the axiomatic method to gain general recognition. His famous 23 problems (presented in 1900 at the International Congress of Mathematicians in Paris) considerably influenced the development of modern mathematics.

If V is finite dimensional, it is actually a so-called *Hilbert space;* see Ref. [9], p. 73, listed in Appendix I.

[13]HERMANN AMANDUS SCHWARZ (1843—1921). German mathematician, successor of Weierstrass at Berlin, known by his work in complex analysis (conformal mapping), differential geometry, and the calculus of variations (minimal surfaces).

EXAMPLE 4   ***n*-dimensional Euclidean space**

R^n with the inner product of Sec. 6.2,

$$(6) \qquad (\mathbf{a}, \mathbf{b}) = \mathbf{a}^\mathsf{T}\mathbf{b} = a_1 b_1 + \cdots + a_n b_n,$$

(where both **a** and **b** are *column* vectors) is called the ***n*-dimensional Euclidean space** and is denoted by E^n or again simply by R^n. Axioms I–III hold, as direct calculation shows. Equation (2) gives the **"Euclidean norm"**

$$(7) \qquad \|\mathbf{a}\| = \sqrt{(\mathbf{a}, \mathbf{a})} = \sqrt{\mathbf{a}^\mathsf{T}\mathbf{a}} = \sqrt{a_1{}^2 + \cdots + a_n{}^2}. \qquad \blacktriangleleft$$

EXAMPLE 5   **An inner product for functions**

The set of all real-valued continuous functions $f(x), g(x), \cdots$ on a given interval $\alpha \leqq x \leqq \beta$ forms a real vector space under the usual addition of functions and multiplication by scalars (real numbers). On this space we can define an inner product by the integral

$$(8) \qquad (f, g) = \int_\alpha^\beta f(x)g(x)\,dx.$$

Axioms I–III can be verified by direct calculation. Equation (2) gives the norm

$$(9) \qquad \|f\| = \sqrt{(f, f)} = \sqrt{\int_\alpha^\beta f(x)^2\,dx}. \qquad \blacktriangleleft$$

Our examples give a first impression of the great generality of the abstract concepts of vector spaces and inner product spaces. Further details belong to more advanced courses (on functional analysis, meaning abstract modern analysis; see Ref. [9] listed in Appendix 1) and cannot be discussed here. Instead we now take up a related topic where matrices play a central role.

Linear Transformations

Let X and Y be any vector spaces. To each vector **x** in X we assign a unique vector **y** in Y. Then we say that a **mapping** (or **transformation** or **operator**) of X into Y is given. Such a mapping is denoted by a capital letter, say F. The vector **y** in Y assigned to a vector **x** in X is called the **image** of **x** and is denoted by $F(\mathbf{x})$ [or $F\mathbf{x}$, without parentheses].

F is called a **linear mapping** or **linear transformation** if for all vectors **v** and **x** in X and scalars c,

$$(10) \qquad \begin{aligned} F(\mathbf{v} + \mathbf{x}) &= F(\mathbf{v}) + F(\mathbf{x}) \\ F(c\mathbf{x}) &= cF(\mathbf{x}). \end{aligned}$$

Linear transformation of space R^n into space R^m

From now on we let $X = R^n$ and $Y = R^m$. Then any real $m \times n$ matrix $\mathbf{A} = [a_{jk}]$ gives a transformation of R^n into R^m,

$$(11) \qquad \mathbf{y} = \mathbf{Ax}.$$

Since $\mathbf{A}(\mathbf{u} + \mathbf{x}) = \mathbf{Au} + \mathbf{Ax}$ and $\mathbf{A}(c\mathbf{x}) = c\mathbf{Ax}$, this transformation is linear.

We show that, conversely, every linear transformation F of R^n into R^m can be given in terms of an $m \times n$ matrix $\mathbf{A}$, after a basis for R^n and a basis for R^m have been chosen. This can be proved as follows.

Let $\mathbf{e}_{(1)}, \cdots, \mathbf{e}_{(n)}$ be any basis for R^n. Then every $\mathbf{x}$ in R^n has a unique representation

$$\mathbf{x} = x_1 \mathbf{e}_{(1)} + \cdots + x_n \mathbf{e}_{(n)}.$$

Since F is linear, this representation implies for the image $F(\mathbf{x})$:

$$F(\mathbf{x}) = F(x_1 \mathbf{e}_{(1)} + \cdots + x_n \mathbf{e}_{(n)}) = x_1 F(\mathbf{e}_{(1)}) + \cdots + x_n F(\mathbf{e}_{(n)}).$$

Hence F is uniquely determined by the images of the vectors of a basis for R^n. We now choose for R^n the **"standard basis"**

$$(12) \qquad \mathbf{e}_{(1)} = \begin{bmatrix} 1 \\ 0 \\ \vdots \\ 0 \end{bmatrix}, \qquad \mathbf{e}_{(2)} = \begin{bmatrix} 0 \\ 1 \\ \vdots \\ 0 \end{bmatrix}, \qquad \cdots, \qquad \mathbf{e}_{(n)} = \begin{bmatrix} 0 \\ 0 \\ \vdots \\ 1 \end{bmatrix}$$

where $\mathbf{e}_{(j)}$ has its jth component equal to 1 and all others 0. We show that we can now determine an $m \times n$ matrix $\mathbf{A} = [a_{jk}]$ such that for every $\mathbf{x}$ in R^n and image $\mathbf{y} = F(\mathbf{x})$ in R^m,

$$\mathbf{y} = F(\mathbf{x}) = \mathbf{A}\mathbf{x}.$$

Indeed, from the image $\mathbf{y}^{(1)} = F(\mathbf{e}_{(1)})$ of $\mathbf{e}_{(1)}$ we get the condition

$$\mathbf{y}^{(1)} = \begin{bmatrix} y_1^{(1)} \\ y_2^{(1)} \\ \vdots \\ y_m^{(1)} \end{bmatrix} = \begin{bmatrix} a_{11} & \cdots & a_{1n} \\ a_{21} & \cdots & a_{2n} \\ \vdots & & \vdots \\ a_{m1} & \cdots & a_{mn} \end{bmatrix} \begin{bmatrix} 1 \\ 0 \\ \vdots \\ 0 \end{bmatrix}$$

from which we can determine the first column of $\mathbf{A}$, namely $a_{11} = y_1^{(1)}$, $a_{21} = y_2^{(1)}, \cdots$, $a_{m1} = y_m^{(1)}$. Similarly, from the image of $\mathbf{e}_{(2)}$ we get the second column of $\mathbf{A}$, and so on. This completes the proof. ◀

We say that **A represents** F, or *is a representation of F,* with respect to the bases for R^n and R^m. Quite generally, the purpose of a *"representation"* is the replacement of one object of study by another object whose properties are more readily apparent.

In three-dimensional space R^3 the standard basis is usually written $\mathbf{e}_{(1)} = \mathbf{i}$, $\mathbf{e}_{(2)} = \mathbf{j}$, $\mathbf{e}_{(3)} = \mathbf{k}$. Thus,

$$(13) \qquad \mathbf{i} = \begin{bmatrix} 1 \\ 0 \\ 0 \end{bmatrix}, \qquad \mathbf{j} = \begin{bmatrix} 0 \\ 1 \\ 0 \end{bmatrix}, \qquad \mathbf{k} = \begin{bmatrix} 0 \\ 0 \\ 1 \end{bmatrix}.$$

These are the three unit vectors in the positive directions of the axes of the Cartesian coordinate system in space, that is, the usual coordinate system with the same scale of measurement on the three mutually perpendicular coordinate axes.

EXAMPLE 6   **Linear transformations**

Interpreted as transformations of Cartesian coordinates in the plane, the matrices

$$\begin{bmatrix} 0 & 1 \\ 1 & 0 \end{bmatrix}, \quad \begin{bmatrix} 1 & 0 \\ 0 & -1 \end{bmatrix}, \quad \begin{bmatrix} -1 & 0 \\ 0 & -1 \end{bmatrix}, \quad \begin{bmatrix} a & 0 \\ 0 & 1 \end{bmatrix}$$

represent a reflection in the line $x_2 = x_1$, a reflection in the x_1-axis, a reflection in the origin, and a stretch (when $a > 1$, or a contraction when $0 < a < 1$) in the x_1-direction, respectively. ◀

EXAMPLE 7   **Linear transformations**

Our discussion preceding Example 6 is simpler than it may look at first sight. To see this, find **A** representing the linear transformation that maps (x_1, x_2) onto $(2x_1 - 5x_2, 3x_1 + 4x_2)$.

Solution.   Obviously, the transformation is

$$y_1 = 2x_1 - 5x_2$$

$$y_2 = 3x_1 + 4x_2.$$

From this we can directly see that the matrix is

$$\mathbf{A} = \begin{bmatrix} 2 & -5 \\ 3 & 4 \end{bmatrix}. \quad \text{Check:} \quad \begin{bmatrix} y_1 \\ y_2 \end{bmatrix} = \begin{bmatrix} 2 & -5 \\ 3 & 4 \end{bmatrix} \begin{bmatrix} x_1 \\ x_2 \end{bmatrix} = \begin{bmatrix} 2x_1 - 5x_2 \\ 3x_1 + 4x_2 \end{bmatrix}. \quad ◀$$

If **A** in (11) is square, $n \times n$, then (11) maps R^n into R^n. If this **A** is nonsingular, so that $\mathbf{A}^{-1}$ exists (see Sec. 6.7), then multiplication of (11) by $\mathbf{A}^{-1}$ from the left and use of $\mathbf{A}^{-1}\mathbf{A} = \mathbf{I}$ gives the **inverse transformation**

$$(14) \qquad\qquad\qquad \mathbf{x} = \mathbf{A}^{-1}\mathbf{y}.$$

It maps every $\mathbf{y} = \mathbf{y}_0$ onto that **x,** which by (11) is mapped onto $\mathbf{y}_0$. *The inverse of a linear transformation is itself linear,* because it is given by a matrix, as (14) shows.

PROBLEM SET 6.8

Vector Spaces, Bases. (See Sec. 6.4 for additional problems.)

Is the given set (taken with the usual addition and scalar multiplication) a vector space or not? (Give a reason.) If your answer is yes, determine the dimension and find a basis.

1. All vectors in R^3 satisfying $v_1 - 3v_2 + 2v_3 = 0$
2. All ordered quadruples of nonnegative real numbers
3. All polynomials in x, of degree not exceeding 3
4. All functions $y(x) = a \cos x + b \sin x$ with arbitrary constant b and c
5. All skew-symmetric 2×2 matrices
6. All symmetric 3×3 matrices
7. All 2×2 matrices such that $a_{11} + a_{22} = 0$
8. All 2×3 matrices with first row any multiple of $[2 \quad -1 \quad 3]$
9. All $m \times n$ matrices with positive entries
10. All 3×3 matrices with main diagonal 1, 1, 1.

11. Find three different bases for R^2.

12. (Uniqueness) Show that the representation $\mathbf{v} = c_1 \mathbf{a}_{(1)} + \cdots + c_n \mathbf{a}_{(n)}$ of any given vector $\mathbf{v}$ in an n-dimensional vector space V in terms of a basis $\mathbf{a}_{(1)}, \cdots, \mathbf{a}_{(n)}$ for V is unique.

Inner Product, Orthogonality

Find the Euclidean norm of the vectors:

13. $[0.4 \quad 1.3 \quad -2.2]^{\mathsf{T}}$ **14.** $[2 \quad 0 \quad 3 \quad 0 \quad 8]^{\mathsf{T}}$ **15.** $[\frac{1}{2} \quad \frac{1}{3} \quad -\frac{1}{2} \quad -\frac{1}{3}]^{\mathsf{T}}$

16. $[2 \quad 3 \quad -5]^{\mathsf{T}}$ **17.** $[3 \quad 2 \quad -2 \quad 4 \quad 0]^{\mathsf{T}}$ **18.** $[6 \quad 1 \quad 0 \quad 5]^{\mathsf{T}}$

19. Show that the vectors in Probs. 14 and 17 are orthogonal.

20. Find all unit vectors $\mathbf{v} = [v_1 \quad v_2]$ orthogonal to $[3 \quad -4]$.

21. Find all vectors $\mathbf{v}$ orthogonal to $\mathbf{a} = [1 \quad 2 \quad 0]^{\mathsf{T}}$. Do they form a vector space?

Using the Euclidean norm, verify:

22. The Schwarz inequality for the vectors in Probs. 13 and 16.

23. The triangle inequality for the vectors in Probs. 13 and 16.

24. The parallelogram equality for the vectors $[2 \quad 1]^{\mathsf{T}}$ and $[1 \quad 3]^{\mathsf{T}}$. Graph the parallelogram with these sides and explain the geometric meaning of this equality.

Linear Transformations

Find the inverse transformation. (Show the details of your work.)

25. $\begin{aligned} y_1 &= 3x_1 + 2x_2 \\ y_2 &= 4x_1 + x_2 \end{aligned}$ **26.** $\begin{aligned} y_1 &= 0.5x_1 - 0.5x_2 \\ y_2 &= 1.5x_1 - 2.5x_2 \end{aligned}$

27. $\begin{aligned} y_1 &= 5x_1 + 3x_2 - 3x_3 \\ y_2 &= 3x_1 + 2x_2 - 2x_3 \\ y_3 &= 2x_1 - x_2 + 2x_3 \end{aligned}$ **28.** $\begin{aligned} y_1 &= 4x_1 \quad + 2x_3 \\ y_2 &= \quad x_2 + 4x_3 \\ y_3 &= \quad 5x_3 \end{aligned}$

29. $\begin{aligned} y_1 &= 0.2x_1 - 0.1x_2 \\ y_2 &= \quad -0.2x_2 + 0.1x_3 \\ y_3 &= 0.1x_1 \quad + 0.1x_3 \end{aligned}$ **30.** $\begin{aligned} y_1 &= x_1 \\ y_2 &= \quad x_2 \cos \theta + x_3 \sin \theta \\ y_3 &= \quad -x_2 \sin \theta + x_3 \cos \theta \end{aligned}$

CHAPTER 6 REVIEW

1. Let $\mathbf{A}$ be a 10×5 matrix and $\mathbf{B}$ a 10×10 matrix. Indicate whether or not the following expressions are defined: $5\mathbf{A}$, $\mathbf{A} + \mathbf{B}$, $\mathbf{AB}$, $\mathbf{BA}$, $\mathbf{A}^2$, $\mathbf{B}^2$, $\mathbf{A}^{\mathsf{T}}\mathbf{B}$, $\mathbf{AB}^{\mathsf{T}}$, $\mathbf{A}^{\mathsf{T}}\mathbf{A}$, $\mathbf{BB}^{\mathsf{T}}$, $\mathbf{A}^{\mathsf{T}}\mathbf{BA}$, $\mathbf{AA}^{\mathsf{T}} - \mathbf{B}$.

2. How is matrix multiplication motivated?

3. Write down from memory the unusual properties of matrix multiplication by which it differs from the multiplication of numbers. Give examples.

4. What is the inverse of a matrix? When does it exist? How would you calculate it?

5. What is the rank of a matrix? Why is it of basic importance?

6. Do there exist linear systems of equations without any solution?

7. What is the Gauss elimination? The Gauss–Jordan elimination?

8. What do you know about the number of solutions of a nonhomogeneous linear system? Of a homogeneous system?

9. What is Cramer's rule? When would you apply it?

10. Can the row vectors of a 20×12 matrix be linearly independent?

Addition, Scalar Multiplication, Multiplication. Calculate the following expressions (showing the details of your work) when

$$\mathbf{A} = \begin{bmatrix} 3 & 1 & -3 \\ 1 & 4 & 2 \\ -3 & 2 & 5 \end{bmatrix}, \quad \mathbf{B} = \begin{bmatrix} 0 & 4 & 1 \\ -4 & 0 & -2 \\ -1 & 2 & 0 \end{bmatrix}, \quad \mathbf{c} = \begin{bmatrix} 2 \\ 0 \\ -5 \end{bmatrix}, \quad \mathbf{d} = \begin{bmatrix} 7 \\ -3 \\ 3 \end{bmatrix}.$$

11. AB	**12. BA**	**13.** $\frac{1}{5}$**ABA**	**14.** $\mathbf{A}^2$, $\mathbf{AA}^\mathsf{T}$	**15.** $\mathbf{B}^2$
16. AB − BA	**17.** $\mathbf{BB}^\mathsf{T}$, $\mathbf{B}^\mathsf{T}\mathbf{B}$	**18.** $\frac{1}{21}$$\mathbf{BB}^\mathsf{T}\mathbf{B}$	**19. Ac**	**20. Bd**
21. $\mathbf{c}^\mathsf{T}\mathbf{d}$	**22.** det **A**	**23.** $0.1\mathbf{c}^\mathsf{T}\mathbf{A}$	**24.** $\mathbf{d}^\mathsf{T}\mathbf{B}$	**25.** $\mathbf{d}^\mathsf{T}\mathbf{Bd}$

Linear Systems of Equations. Solve or indicate that no solution exists. (Show the details of your work.)

26.
$$\begin{aligned} 4y + z &= 0 \\ 12x - 5y - 3z &= 17 \\ -6x + 4z &= 4 \end{aligned}$$

27.
$$\begin{aligned} 9x + 3y - 6z &= 30 \\ 2x - 4y + 8z &= 2 \end{aligned}$$

28.
$$\begin{aligned} 2x + 3y - 7z &= 3 \\ x - 4y + z &= 0 \\ -14x - 21y + 49z &= -20 \end{aligned}$$

29.
$$\begin{aligned} 3x - 7y + 13z &= 32.4 \\ 9y - 8z &= -25.3 \\ 7z &= 11.9 \end{aligned}$$

30.
$$\begin{aligned} 5x - 3y + 7z &= -27 \\ 10x - 9z &= 27 \\ -x + 6y + 4z &= 0 \end{aligned}$$

31.
$$\begin{aligned} 22x - 13y + z &= 0 \\ -13x - 4z &= 0 \\ x - 4y - 11z &= 0 \end{aligned}$$

32.
$$\begin{aligned} 2x + 3y - z &= 0 \\ 5x - 3y + z &= 7 \\ 8x + 9y - 3z &= 2 \end{aligned}$$

33.
$$\begin{aligned} -3y + 9z &= 36 \\ 2x + 17z &= 35 \\ -x - y + 34z &= 34 \end{aligned}$$

34.
$$\begin{aligned} 2x - 13y + 3z &= 4 \\ -6x + 39y - 9z &= -12 \end{aligned}$$

Rank. Determine the ranks of the coefficient matrix and of the augmented matrix in the following problems. (Show the details of your work.) What does your answer imply regarding solutions?

35. Prob. 27	**36.** Prob. 28	**37.** Prob. 29
38. Prob. 30	**39.** Prob. 31	**40.** Prob. 32

Inverse of a Matrix. Let **A** and **B** be as in Probs. 11–25 and **I** the 3×3 unit matrix. Find the inverse of the following matrices or indicate nonexistence. Check your result. (Show the details.)

41. $\frac{1}{5}\mathbf{A}$	**42. AB, BA**	**43. B**	**44.** $\frac{1}{11}(\mathbf{B} + \mathbf{I})$	**45.** $\frac{1}{9}(\mathbf{A} - 4\mathbf{I})$

Networks

Find the currents in the following networks.

46.

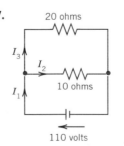

47.

48.

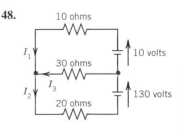

Four-Terminal Networks. Assume that the input current i_1 and voltage u_1 of the four-terminal network in Fig. 144 are related to the output current i_2 and voltage u_2 according to

$$\mathbf{v}_1 = \mathbf{T}\mathbf{v}_2, \qquad \text{where} \qquad \mathbf{v}_1 = \begin{bmatrix} u_1 \\ i_1 \end{bmatrix}, \quad \mathbf{T} = \begin{bmatrix} t_{11} & t_{12} \\ t_{21} & t_{22} \end{bmatrix}, \quad \mathbf{v}_2 = \begin{bmatrix} u_2 \\ i_2 \end{bmatrix}$$

Fig. 144.   Four-terminal network

and where $\mathbf{T}$ is called the transmission matrix of the network. Verify the form of $\mathbf{T}$:

49.

$$\mathbf{T} = \begin{bmatrix} 1 & Z \\ 0 & 1 \end{bmatrix}$$

50.

$$\mathbf{T} = \begin{bmatrix} 1 + Z_1/Z_2 & Z_1 \\ 1/Z_2 & 1 \end{bmatrix}$$

51.

$$\mathbf{T} = \begin{bmatrix} 1 & 0 \\ 1/Z & 1 \end{bmatrix}$$

SUMMARY OF CHAPTER 6
LINEAR ALGEBRA: MATRICES, VECTORS, DETERMINANTS
LINEAR SYSTEMS OF EQUATIONS

An $m \times n$ **matrix** $\mathbf{A} = [a_{jk}]$ is a rectangular array of numbers ("entries" or "elements") arranged in m horizontal rows and n vertical columns. If $m = n$, the matrix is called **square**. A $1 \times n$ matrix is called a **row vector** and an $m \times 1$ matrix a **column vector** (Sec. 6.1).

The **sum** $\mathbf{A} + \mathbf{B}$ of matrices of the same **size** (i.e., both $m \times n$) is obtained by adding corresponding entries. The **product** of $\mathbf{A}$ by a scalar c is obtained by multiplying each a_{jk} by c (Sec. 6.1).

The **product** $\mathbf{C} = \mathbf{AB}$ of an $m \times n$ matrix $\mathbf{A}$ by an $r \times p$ matrix $\mathbf{B} = [b_{jk}]$ is defined only when $r = n$, and is the $m \times p$ matrix $\mathbf{C} = [c_{jk}]$ with entries

$$(1) \qquad\qquad c_{jk} = a_{j1}b_{1k} + a_{j2}b_{2k} + \cdots + a_{jn}b_{nk} \qquad \begin{array}{l} \text{(row } j \text{ of } \mathbf{A} \text{ times} \\ \text{column } k \text{ of } \mathbf{B}\text{).} \end{array}$$

This multiplication is motivated by the composition of **linear transformations** (Secs. 6.2, 6.8). It is associative, but is *not commutative:* if **AB** is defined, **BA** may not be defined, but even if **BA** is defined, **AB** ≠ **BA** in general. Also **AB** = **0** may not imply **A** = **0** or **B** = **0** or **BA** = **0** (Secs. 6.2, 6.7). Illustrations:

$$\begin{bmatrix} 1 & 1 \\ 2 & 2 \end{bmatrix} \begin{bmatrix} -1 & 1 \\ 1 & -1 \end{bmatrix} = \begin{bmatrix} 0 & 0 \\ 0 & 0 \end{bmatrix}$$

$$\begin{bmatrix} -1 & 1 \\ 1 & -1 \end{bmatrix} \begin{bmatrix} 1 & 1 \\ 2 & 2 \end{bmatrix} = \begin{bmatrix} 1 & 1 \\ -1 & -1 \end{bmatrix}$$

$$\begin{bmatrix} 1 & 2 \end{bmatrix} \begin{bmatrix} 3 \\ 4 \end{bmatrix} = \begin{bmatrix} 11 \end{bmatrix}, \qquad \begin{bmatrix} 3 \\ 4 \end{bmatrix} \begin{bmatrix} 1 & 2 \end{bmatrix} = \begin{bmatrix} 3 & 6 \\ 4 & 8 \end{bmatrix}.$$

The *transpose* $\mathbf{A}^\mathsf{T}$ of a matrix $\mathbf{A} = [a_{jk}]$ is $\mathbf{A}^\mathsf{T} = [a_{kj}]$; rows become columns and conversely (Sec. 6.1). Here, **A** need not be square. If it is and $\mathbf{A} = \mathbf{A}^\mathsf{T}$, then **A** is called **symmetric;** if $\mathbf{A} = -\mathbf{A}^\mathsf{T}$, it is called **skew-symmetric.** For a product, $(\mathbf{AB})^\mathsf{T} = \mathbf{B}^\mathsf{T}\mathbf{A}^\mathsf{T}$ (Sec. 6.2).

A main application of matrices concerns **linear systems of equations**

$$(2) \qquad\qquad\qquad \mathbf{Ax} = \mathbf{b} \qquad\qquad\qquad \text{(Sec. 6.3)}$$

(*m* equations in *n* unknowns $x_1, \cdots, x_n$; **b** given). The most important method of solution is the **Gauss elimination** (Sec. 6.3), which reduces the system to "triangular" form by *elementary row operations,* which leave the set of solutions unchanged. (Numerical aspects and variants, such as *Doolittle's* and *Cholesky's methods,* are discussed in Secs. 18.1 and 18.2.)

Cramer's rule (Sec. 6.6) represents the unknowns in a system (2) of *n* equations in *n* unknowns as quotients of determinants; for numerical work it is impractical. **Determinants** (Sec. 6.6) have decreased in importance, but will retain their place in eigenvalue problems, elementary geometry, etc.

The **inverse** $\mathbf{A}^{-1}$ of a square matrix satisfies $\mathbf{AA}^{-1} = \mathbf{A}^{-1}\mathbf{A} = \mathbf{I}$. It exists if and only if det **A** ≠ 0. It can be computed by the *Gauss–Jordan elimination* (Sec. 6.7).

The **rank** *r* of a matrix **A** is the maximum number of linearly independent rows or columns of **A** or, equivalently, the number of rows of the largest square submatrix of **A** with nonzero determinant (Secs. 6.4, 6.6).

The system (2) has solutions if and only if rank **A** = rank [**A   b**], where [**A   b**] is the *augmented matrix* (Fundamental Theorem, Sec. 6.5).

The *homogeneous system*

$$(3) \qquad\qquad\qquad \mathbf{Ax} = \mathbf{0}$$

has solutions $\mathbf{x} \neq \mathbf{0}$ ("nontrivial solutions") if and only if rank $\mathbf{A} < n$, in the case $m = n$ equivalently if and only if det **A** = 0 (Secs. 6.5, 6.6).

Vector spaces, inner product spaces, and linear transformations are discussed in Sec. 6.8. See also Sec. 6.4.

CHAPTER 7

Linear Algebra: Matrix Eigenvalue Problems

Matrix eigenvalue problems concern vector equations of the form

$$(1) \qquad\qquad \mathbf{A} = \lambda\mathbf{x}$$

where $\mathbf{A}$ is a given square matrix. *All matrices in this chapter are square:* 2×2, 3×3, or $n \times n$. Furthermore, $\mathbf{x}$ is an unknown vector and λ an unknown scalar. Our goal is to solve (1). Obviously, $\mathbf{x} = \mathbf{0}$ is a solution, giving $\mathbf{0} = \mathbf{0}$, but this is of no practical interest. We want solutions $\mathbf{x} \neq \mathbf{0}$. These are called **eigenvectors** of $\mathbf{A}$. We shall see that such $\mathbf{x} \neq \mathbf{0}$ exist only for certain values of λ. These values are called **eigenvalues**[1] (or *characteristic values*) of $\mathbf{A}$. Geometrically, solving (1) means we are looking for $\mathbf{x}$ for which the multiplication by $\mathbf{A}$ has the same effect as the multiplication of $\mathbf{x}$ by a scalar, λ, giving a vector $\lambda\mathbf{x}$, with components proportional to those of $\mathbf{x}$, and λ as the factor of proportionality.

We shall see that eigenvalue problems are of greatest importance to the engineer and physicist, and they make up a beautiful chapter in linear algebra. Of course, this is not obvious from (1) and thus needs further explanation.

In Sec. 7.1 we explain the basic concepts and show how to systematically find eigenvalues and eigenvectors. Typical applications follow in Sec. 7.2. Sections 7.3 and 7.4 concern properties and eigenvalue problems of symmetric, skew-symmetric, and orthogonal matrices and their complex counterparts (Hermitian, skew-Hermitian, and unitary matrices). In Sec. 7.5 we show that diagonalization of matrices also leads to eigenvalues.

Numerical methods in Secs. 18.6–18.9 can be studied immediately after the corresponding material in the present chapter.

Prerequisite for this chapter: Chap. 6.
Sections that may be omitted in a shorter course: 7.4, 7.5
References: Appendix 1, Part B.
Answers to problems: Appendix 2.

7.1 Eigenvalues, Eigenvectors

From the standpoint of engineering applications, eigenvalue problems are among the most important problems in connection with matrices, and the student should follow our present discussion with particular attention. We first define the basic concepts and explain them in terms of typical examples. Then we shall turn to practical applications.

Let $\mathbf{A} = [a_{jk}]$ be a given $n \times n$ matrix and consider the vector equation

(1)
$$\mathbf{Ax} = \lambda\mathbf{x}.$$

Here, $\mathbf{x}$ is an unknown vector and λ an unknown scalar, and we want to determine both.

Clearly, the zero vector $\mathbf{x} = \mathbf{0}$ is a solution of (1) for any value of λ. This is of no practical interest. A value of λ for which (1) has a solution $\mathbf{x} \neq \mathbf{0}$ is called an **eigenvalue**[1] or **characteristic value** (or *latent root*) of the matrix $\mathbf{A}$. The corresponding solutions $\mathbf{x} \neq \mathbf{0}$ of (1) are called **eigenvectors** or **characteristic vectors** of $\mathbf{A}$ corresponding to that eigenvalue λ. The set of the eigenvalues is called the **spectrum** of $\mathbf{A}$. The largest of the absolute values of the eigenvalues of $\mathbf{A}$ is called the **spectral radius** of $\mathbf{A}$.

The set of all eigenvectors corresponding to an eigenvalue of $\mathbf{A}$, together with $\mathbf{0}$, forms a vector space (Sec. 6.4), called the **eigenspace** of $\mathbf{A}$ corresponding to this eigenvalue.

The problem of determining the eigenvalues and eigenvectors of a matrix is called an *eigenvalue problem*.[2] Problems of this type occur in connection with physical, technical, geometrical, and other applications, as we shall see.

How to Find Eigenvalues and Eigenvectors

EXAMPLE 1 **Determination of eigenvalues and eigenvectors**

We illustrate all the steps in terms of the matrix

$$\mathbf{A} = \begin{bmatrix} -5 & 2 \\ 2 & -2 \end{bmatrix}.$$

Solution. (a) *Eigenvalues.* These must be determined ***first.*** Equation (1) is

$$\mathbf{Ax} = \begin{bmatrix} -5 & 2 \\ 2 & -2 \end{bmatrix} \begin{bmatrix} x_1 \\ x_2 \end{bmatrix} = \lambda \begin{bmatrix} x_1 \\ x_2 \end{bmatrix};$$

written out in components,

$$-5x_1 + 2x_2 = \lambda x_1$$
$$2x_1 - 2x_2 = \lambda x_2.$$

Transferring the terms on the right to the left, we get

(2*)
$$(-5 - \lambda)x_1 + \quad 2x_2 \quad = 0$$
$$2x_1 \quad + (-2 - \lambda)x_2 = 0.$$

This can be written in matrix notation

(3*)
$$(\mathbf{A} - \lambda\mathbf{I})\mathbf{x} = \mathbf{0}.$$

---

[1] *"Eigen"* is German and means "proper" or "characteristic."

[2] More precisely: an *algebraic* eigenvalue problem, because there are other eigenvalue problems involving a differential equation (see Secs. 4.7 and 11.3) or an integral equation.

[Indeed, (1) is $\mathbf{Ax} - \lambda\mathbf{x} = \mathbf{0}$ or $\mathbf{Ax} - \lambda\mathbf{Ix} = \mathbf{0}$, which gives (3*).] We see that this is a ***homogeneous*** linear system. By Cramer's Theorem in Sec. 6.6 it has a nontrivial solution $\mathbf{x} \neq \mathbf{0}$ (an eigenvector of $\mathbf{A}$ we are looking for) if and only if its coefficient determinant is zero,

$$(4^*) \qquad D(\lambda) = \det(\mathbf{A} - \lambda\mathbf{I}) = \begin{vmatrix} -5 - \lambda & 2 \\ 2 & -2 - \lambda \end{vmatrix}$$

$$= (-5 - \lambda)(-2 - \lambda) - 4 = \lambda^2 + 7\lambda + 6 = 0.$$

We call $D(\lambda)$ the **characteristic determinant** or, if expanded, the **characteristic polynomial,** and $D(\lambda) = 0$ the **characteristic equation** of $\mathbf{A}$. The solutions of this quadratic equation are $\lambda_1 = -1$ and $\lambda_2 = -6$. These are the eigenvalues of $\mathbf{A}$.

(b_1) ***Eigenvector of*** $\mathbf{A}$ ***corresponding to*** $\boldsymbol{\lambda_1}$. This vector is obtained from (2*) with $\lambda = \lambda_1 = -1$, that is,

$$-4x_1 + 2x_2 = 0$$

$$2x_1 - x_2 = 0.$$

A solution is $x_2 = 2x_1$. This determines an eigenvector corresponding to $\lambda_1 = -1$ up to a scalar multiple. If we choose $x_1 = 1$, we obtain the eigenvector

$$\mathbf{x}_1 = \begin{bmatrix} 1 \\ 2 \end{bmatrix}.$$

We can easily check this:

$$\mathbf{Ax}_1 = \begin{bmatrix} -5 & 2 \\ 2 & -2 \end{bmatrix} \begin{bmatrix} 1 \\ 2 \end{bmatrix} = \begin{bmatrix} -1 \\ -2 \end{bmatrix} = (-1)\mathbf{x}_1 = \lambda_1\mathbf{x}_1.$$

(b_2) ***Eigenvector of*** $\mathbf{A}$ ***corresponding to*** $\boldsymbol{\lambda_2}$. For $\lambda = \lambda_2 = -6$, equation (2*) becomes

$$x_1 + 2x_2 = 0$$

$$2x_1 + 4x_2 = 0.$$

A solution is $x_2 = -x_1/2$ with arbitrary x_1. If we choose $x_1 = 2$, we get $x_2 = -1$. Thus an eigenvector of $\mathbf{A}$ corresponding to $\lambda_2 = -6$ is (check this!)

$$\mathbf{x}_2 = \begin{bmatrix} 2 \\ -1 \end{bmatrix}. \qquad \blacktriangleleft$$

This example illustrates the general case as follows. Equation (1) written in components is

$$a_{11}x_1 + \cdots + a_{1n}x_n = \lambda x_1$$

$$a_{21}x_1 + \cdots + a_{2n}x_n = \lambda x_2$$

$$\cdots\cdots\cdots\cdots\cdots\cdots\cdots$$

$$a_{n1}x_1 + \cdots + a_{nn}x_n = \lambda x_n.$$

Transferring the terms on the right side to the left side, we have

$$(2) \qquad \begin{array}{cccccc} (a_{11} - \lambda)x_1 & + & a_{12}x_2 & + \cdots + & a_{1n}x_n & = 0 \\ a_{21}x_1 & + & (a_{22} - \lambda)x_2 & + \cdots + & a_{2n}x_n & = 0 \\ \multicolumn{6}{c}{\cdots\cdots\cdots\cdots\cdots\cdots\cdots\cdots\cdots\cdots\cdots} \\ a_{n1}x_1 & + & a_{n2}x_2 & + \cdots + & (a_{nn} - \lambda)x_n & = 0. \end{array}$$

In matrix notation,

$$\boxed{(\mathbf{A} - \lambda \mathbf{I})\mathbf{x} = \mathbf{0}.}$$

(3)

By Cramer's Theorem in Sec. 6.6, this homogeneous linear system of equations has a nontrivial solution if and only if the corresponding determinant of the coefficients is zero:

(4) $$D(\lambda) = \det(\mathbf{A} - \lambda \mathbf{I}) = \begin{vmatrix} a_{11} - \lambda & a_{12} & \cdots & a_{1n} \\ a_{21} & a_{22} - \lambda & \cdots & a_{2n} \\ \cdot & \cdot & \cdots & \cdot \\ a_{n1} & a_{n2} & \cdots & a_{nn} - \lambda \end{vmatrix} = 0.$$

$D(\lambda)$ is called the **characteristic determinant.** Equation (4) is called the **characteristic equation** of the matrix **A.** By developing $D(\lambda)$ we obtain a polynomial of nth degree in λ. This is called the **characteristic polynomial** of **A.**

This proves the following important theorem.

THEOREM 1   **(Eigenvalues)**

The eigenvalues of a square matrix **A** *are the roots of the characteristic equation* (4) *of* **A.**

Hence an $n \times n$ matrix has at least one eigenvalue and at most n numerically different eigenvalues.

For larger n, the actual computation of eigenvalues will in general require the use of Newton's method (Sec. 17.2) or another numerical approximation method in Secs. 18.7–18.9.

*The eigen***values *must be determined first.*** Once these are known, corresponding eigen*vectors* are obtained from the system (2), for instance, by the Gauss elimination, where λ is the eigenvalue for which an eigenvector is wanted. This is what we did in Example 1 and shall do again in the examples below.[3]

Eigenvectors are determined only up to a (nonzero) scalar factor:

THEOREM 2   **(Eigenvectors)**

If **x** *is an eigenvector of a matrix* **A** *corresponding to an eigenvalue λ, so is* **kx** *with any $k \neq 0$.*

PROOF. $\mathbf{Ax} = \lambda \mathbf{x}$ implies $k(\mathbf{Ax}) = \mathbf{A}(k\mathbf{x}) = \lambda(k\mathbf{x})$.   ◀

Examples 2 and 3 will illustrate that an $n \times n$ matrix may have n linearly independent eigenvectors, or it may have fewer than n. In Example 4 we shall see that a *real* matrix may have *complex* eigenvalues and eigenvectors.

---

[3] To avoid misunderstandings: In Sec. 18.8 we shall see that there are numerical methods that give (approximations of) eigen*vectors* first.

EXAMPLE 2   **Multiple eigenvalues**

Find the eigenvalues and eigenvectors of the matrix

$$\mathbf{A} = \begin{bmatrix} -2 & 2 & -3 \\ 2 & 1 & -6 \\ -1 & -2 & 0 \end{bmatrix}.$$

Solution.   For our matrix, the characteristic determinant gives the characteristic equation

$$-\lambda^3 - \lambda^2 + 21\lambda + 45 = 0.$$

The roots (eigenvalues of $\mathbf{A}$) are $\lambda_1 = 5$, $\lambda_2 = \lambda_3 = -3$. To find eigenvectors, we apply the Gauss elimination (Sec. 6.3) to the system $(\mathbf{A} - \lambda\mathbf{I})\mathbf{x} = \mathbf{0}$, first with $\lambda = 5$ and then with $\lambda = -3$. We find that an eigenvector of $\mathbf{A}$ corresponding to the eigenvalue 5 is

$$\mathbf{x}_1 = \begin{bmatrix} 1 \\ 2 \\ -1 \end{bmatrix}.$$

For $\lambda = -3$ the characteristic matrix

$$\mathbf{A} - \lambda\mathbf{I} = \mathbf{A} + 3\mathbf{I} = \begin{bmatrix} 1 & 2 & -3 \\ 2 & 4 & -6 \\ -1 & -2 & 3 \end{bmatrix} \quad \text{row-reduces to} \quad \begin{bmatrix} 1 & 2 & -3 \\ 0 & 0 & 0 \\ 0 & 0 & 0 \end{bmatrix}.$$

Hence it has rank 1. From $x_1 + 2x_2 - 3x_3 = 0$ we have $x_1 = -2x_2 + 3x_3$. Choosing $x_2 = 1$, $x_3 = 0$ and $x_2 = 0$, $x_3 = 1$, we obtain two linearly independent eigenvectors of $\mathbf{A}$ corresponding to $\lambda = -3$ [as they must exist by (5), Sec. 6.5, with rank = 1 and $n = 3$],

$$\mathbf{x}_2 = \begin{bmatrix} -2 \\ 1 \\ 0 \end{bmatrix} \quad \text{and} \quad \mathbf{x}_3 = \begin{bmatrix} 3 \\ 0 \\ 1 \end{bmatrix}. \qquad \blacktriangleleft$$

The order M_λ of an eigenvalue λ as a root of the characteristic polynomial is called the **algebraic multiplicity** of λ. The number m_λ of linear independent eigenvectors corresponding to λ is called the **geometric multiplicity** of λ. Thus m_λ is the dimension of the eigenspace corresponding to this λ. Since the characteristic polynomial has degree n, the sum of all the algebraic multiplicities must equal n. In Example 2 for $\lambda = -3$ we have $m_\lambda = M_\lambda = 2$. In general, $m_\lambda \leqq M_\lambda$, as can be shown. The difference $\Delta_\lambda = M_\lambda - m_\lambda$ is called the **defect** of λ. Thus $\Delta_{-3} = 0$ in Example 2, but positive Δ_λ can easily occur:

EXAMPLE 3   **Algebraic and geometric multiplicity. Positive defect**

The characteristic equation of the matrix

$$\mathbf{A} = \begin{bmatrix} 0 & 1 \\ 0 & 0 \end{bmatrix} \quad \text{is} \quad \det (\mathbf{A} - \lambda\mathbf{I}) = \begin{vmatrix} -\lambda & 1 \\ 0 & -\lambda \end{vmatrix} = \lambda^2 = 0.$$

Hence $\lambda = 0$ is an eigenvalue of algebraic multiplicity 2. But its geometric multiplicity is only 1, since eigenvectors result from $-0x_1 + x_2 = 0$, hence $x_2 = 0$, in the form $[x_1 \quad 0]^\mathsf{T}$. Hence $\Delta_0 = 1$. $\blacktriangleleft$

EXAMPLE 4 **Real matrices with complex eigenvalues and eigenvectors**

Since real polynomials may have complex roots (which then occur in conjugate pairs), a real matrix may have complex eigenvalues and eigenvectors. For instance, the characteristic equation of the skew-symmetric matrix

$$\mathbf{A} = \begin{bmatrix} 0 & 1 \\ -1 & 0 \end{bmatrix} \qquad \text{is} \qquad \det(\mathbf{A} - \lambda\mathbf{I}) = \begin{vmatrix} -\lambda & 1 \\ -1 & -\lambda \end{vmatrix} = \lambda^2 + 1 = 0.$$

It gives the eigenvalues $\lambda_1 = i \; (=\sqrt{-1})$, $\lambda_2 = -i$. Eigenvectors are obtained from $-ix_1 + x_2 = 0$ and $ix_1 + x_2 = 0$, respectively, and we can choose $x_1 = 1$ to get

$$\begin{bmatrix} 1 \\ i \end{bmatrix} \quad \text{and} \quad \begin{bmatrix} 1 \\ -i \end{bmatrix}.$$

The reader may show that, more generally, these vectors are eigenvectors of the matrix

$$\mathbf{A} = \begin{bmatrix} a & b \\ -b & a \end{bmatrix} \qquad\qquad (a, b \text{ real})$$

and that $\mathbf{A}$ has the eigenvalues $a + ib$ and $a - ib$.   ◀

Having gained a first impression of matrix eigenvalue problems, in the next section we illustrate their importance with some typical applications.

PROBLEM SET 7.1

Eigenvalues and Eigenvectors. Find the spectrum and eigenvectors of the following matrices. (Show the details of your work.)

1. $\begin{bmatrix} 4 & 0 \\ 0 & -6 \end{bmatrix}$ 2. $\begin{bmatrix} 0 & 0 \\ 0 & 0 \end{bmatrix}$ 3. $\begin{bmatrix} 10 & -4 \\ 18 & -12 \end{bmatrix}$ 4. $\begin{bmatrix} 0 & a \\ a & 0 \end{bmatrix}$

5. $\begin{bmatrix} 3 & 4 \\ 4 & -3 \end{bmatrix}$ 6. $\begin{bmatrix} 0 & 3 \\ -3 & 0 \end{bmatrix}$ 7. $\begin{bmatrix} 1 & 2 \\ 0 & 3 \end{bmatrix}$ 8. $\begin{bmatrix} 0.8 & -0.6 \\ 0.6 & 0.8 \end{bmatrix}$

9. $\begin{bmatrix} 3 & 0 & 0 \\ 0 & -8 & 0 \\ 0 & 0 & 4 \end{bmatrix}$ 10. $\begin{bmatrix} -10 & 10 & -15 \\ 10 & 5 & -30 \\ -5 & -10 & 0 \end{bmatrix}$ 11. $\begin{bmatrix} 3 & 5 & 3 \\ 0 & 4 & 6 \\ 0 & 0 & 1 \end{bmatrix}$

12. $\begin{bmatrix} a & 1 & 0 \\ 1 & a & 1 \\ 0 & 1 & a \end{bmatrix}$ 13. $\begin{bmatrix} 2 & 0 & -1 \\ 0 & \frac{1}{2} & 0 \\ 1 & 0 & 4 \end{bmatrix}$ 14. $\begin{bmatrix} 0 & 7 & 0 \\ 0 & 0 & 0 \\ 0 & 0 & -2 \end{bmatrix}$

Significance of Eigenvalues and Eigenvectors in Linear Transformations y = Ax

In Probs. 15-20 find the matrix $\mathbf{A}$ in the indicated linear transformations $\mathbf{y} = \mathbf{Ax}$. Find its eigenvalues and eigenvectors and explain their geometric significance. This should help you to better comprehend the nature of eigenvalues and eigenvectors. ($\mathbf{x} = [x \quad y]$ in R^2 or $\mathbf{x} = [x \quad y \quad z]$ in R^3 are Cartesian coordinates.) Show the details of your work.

15. Reflection about the x-axis in R^2

16. Counterclockwise rotation through the angle $\pi/2$ about the origin in R^2

17. Dilatation in R^2 by a factor 4

18. Reflection about the xy-plane in R^3

19. Orthogonal projection (perpendicular projection) of R^2 onto the y-axis

20. Orthogonal projection of R^3 onto the plane $y = x$

7.2 Some Applications of Eigenvalue Problems

In this section we discuss a few typical examples from the range of applications of matrix eigenvalue problems, which is incredibly large. Chapter 3 shows matrix eigenvalue problems related to differential equations governing mechanical systems and electrical networks. To keep our present discussion independent of Chap. 3, we include a typical application of that kind as our last example.

EXAMPLE 1   **Stretching of an elastic membrane**

An elastic membrane in the x_1x_2-plane with boundary circle $x_1{}^2 + x_2{}^2 = 1$ (Fig. 145) is stretched so that a point $P\colon (x_1,\ x_2)$ goes over into the point $Q\colon (y_1,\ y_2)$ given by

$$(1) \qquad \mathbf{y} = \begin{bmatrix} y_1 \\ y_2 \end{bmatrix} = \mathbf{A}\mathbf{x} = \begin{bmatrix} 5 & 3 \\ 3 & 5 \end{bmatrix}\begin{bmatrix} x_1 \\ x_2 \end{bmatrix}; \qquad \text{in components,} \qquad \begin{aligned} y_1 &= 5x_1 + 3x_2 \\ y_2 &= 3x_1 + 5x_2. \end{aligned}$$

Find the **principal directions,** that is, the directions of the position vector $\mathbf{x}$ of P for which the direction of the position vector $\mathbf{y}$ of Q is the same or exactly opposite. What shape does the boundary circle take under this deformation?

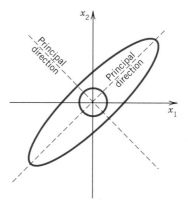

Fig. 145.   Undeformed and deformed membrane in Example 1

Solution.   We are looking for vectors $\mathbf{x}$ such that $\mathbf{y} = \lambda\mathbf{x}.$ Since $\mathbf{y} = \mathbf{A}\mathbf{x},$ this gives $\mathbf{A}\mathbf{x} = \lambda\mathbf{x},$ an equation of the form (1), an eigenvalue problem. In components, $\mathbf{A}\mathbf{x} = \lambda\mathbf{x}$ is

$$(2) \qquad \begin{aligned} 5x_1 + 3x_2 &= \lambda x_1 \\ 3x_1 + 5x_2 &= \lambda x_2 \end{aligned} \qquad \text{or} \qquad \begin{aligned} (5 - \lambda)x_1 + \quad 3x_2 &= 0 \\ 3x_1 + (5 - \lambda)\,x_2 &= 0. \end{aligned}$$

The characteristic equation is

(3)
$$\begin{vmatrix} 5 - \lambda & 3 \\ 3 & 5 - \lambda \end{vmatrix} = (5 - \lambda)^2 - 9 = 0.$$

Its solutions are $\lambda_1 = 8$ and $\lambda_2 = 2$. These are the eigenvalues of our problem. For $\lambda = \lambda_1 = 8$, our system (2) becomes

$$-3x_1 + 3x_2 = 0, \quad \left| \quad \text{Solution } x_2 = x_1, \quad x_1 \text{ arbitrary,} \right.$$
$$3x_1 - 3x_2 = 0. \quad \left| \quad \text{for instance, } x_1 = x_2 = 1. \right.$$

For $\lambda_2 = 2$, our system (2) becomes

$$3x_1 + 3x_2 = 0, \quad \left| \quad \text{Solution } x_2 = -x_1, \quad x_1 \text{ arbitrary,} \right.$$
$$3x_1 + 3x_2 = 0. \quad \left| \quad \text{for instance, } x_1 = 1, x_2 = -1. \right.$$

We thus obtain as eigenvectors of **A,** for instance,

$$\begin{bmatrix} 1 \\ 1 \end{bmatrix} \text{ corresponding to } \lambda_1; \qquad \begin{bmatrix} 1 \\ -1 \end{bmatrix} \text{ corresponding to } \lambda_2;$$

(or a nonzero scalar multiple of these). These vectors make $45°$ and $135°$ angles with the positive x_1-direction. They give the principal directions, the answer to our problem. The eigenvalues show that in the principal directions the membrane is stretched by factors 8 and 2, respectively; see Fig. 145.

Accordingly, if we choose the principal directions as directions of a new Cartesian u_1u_2-coordinate system, say, with the positive u_1-semiaxis in the first quadrant and the positive u_2-semiaxis in the second quadrant of the x_1x_2-system, and if we set

$$u_1 = r \cos \phi, \qquad u_2 = r \sin \phi,$$

then a boundary point of the unstretched circular membrane has coordinates $\cos \phi$, $\sin \phi$. Hence, after the stretch we have

$$z_1 = 8 \cos \phi, \qquad z_2 = 2 \sin \phi.$$

Since $\cos^2 \phi + \sin^2 \phi = 1$, this shows that the deformed boundary is an ellipse (Fig. 145)

$$\frac{z_1^2}{8^2} + \frac{z_2^2}{2^2} = 1$$

with principal semiaxes 8 and 2 in the principal directions.   ◄

EXAMPLE 2   Eigenvalue problems arising from Markov processes

As another application, let us show that Markov processes also lead to eigenvalue problems. To see this, let us determine the limit state of the land-use succession in Example 14, Sec. 6.2.

Solution.   We recall that Example 14 in Sec. 6.2 concerns a *Markov process* and that such a transition process is governed by a **stochastic matrix** $\mathbf{A} = [a_{jk}]$, that is, a square matrix with nonnegative entries a_{jk} (giving transition probabilities) and all row sums equal to 1. Furthermore, state **y** (a column vector) is obtained from state **x** according to $\mathbf{y}^\mathsf{T} = \mathbf{x}^\mathsf{T}\mathbf{A}$. Taking the transpose, we get the equivalent relationship $\mathbf{y} = \mathbf{A}^\mathsf{T}\mathbf{x}$. A limit is reached if states remain unchanged, if $\mathbf{x}^\mathsf{T} = \mathbf{x}^\mathsf{T}\mathbf{A}$ or

(4) $$\mathbf{A}^\mathsf{T}\mathbf{x} = \mathbf{x}.$$

This means that $\mathbf{A}^\mathsf{T}$ should have the eigenvalue 1. But $\mathbf{A}^\mathsf{T}$ has the same eigenvalues as **A**, by Theorem 2(d) in Sec. 6.6. And **A** indeed has the eigenvalue 1, with eigenvector $\mathbf{v}^\mathsf{T} = [1 \quad \cdots \quad 1]^\mathsf{T}$ because the row sums of **A** equal 1. In our example,

$$\mathbf{Av} = \begin{bmatrix} 0.8 & 0.1 & 0.1 \\ 0.1 & 0.7 & 0.2 \\ 0 & 0.1 & 0.9 \end{bmatrix} \begin{bmatrix} 1 \\ 1 \\ 1 \end{bmatrix} = \begin{bmatrix} 1 \\ 1 \\ 1 \end{bmatrix}$$

as claimed. Hence (4) has a nontrivial solution $\mathbf{x} \neq \mathbf{0}$, which is an eigenvector of $\mathbf{A}^T$ corresponding to $\lambda = 1$. Now (4) is $(\mathbf{A}^T - \mathbf{I})\mathbf{x} = \mathbf{0}$; in components,

$$-0.2x_1 + 0.1x_2 \qquad\qquad = 0$$
$$0.1x_1 - 0.3x_2 + 0.1x_3 = 0$$
$$0.1x_1 + 0.2x_2 - 0.1x_3 = 0.$$

The solution with the sum of the components equal to 100% is $\mathbf{x}^T = [12.5 \quad 25 \quad 62.5]$. *Answer.* Assuming that the probabilities remain the same as time progresses, we see that the states tend to 12.5% residentially, 25% commercially, and 62.5% industrially used area. ◀

EXAMPLE 3   Eigenvalue problems arising from population models. Leslie model

The Leslie model describes age-specified population growth, as follows. Let the oldest age attained by the females in some animal population be 6 years. Divide the population into three age classes of 2 years each. Let the *"Leslie matrix"* be

$$(5) \qquad\qquad \mathbf{L} = [l_{jk}] = \begin{bmatrix} 0 & 2.3 & 0.4 \\ 0.6 & 0 & 0 \\ 0 & 0.3 & 0 \end{bmatrix}$$

where l_{1k} is the average number of daughters born to a single female during the time she is in age class k, and $l_{j,j-1}$ ($j = 2, 3$) is the fraction of females in age class $j - 1$ that will survive and pass into class j. (a) What is the number of females in each class after 2, 4, 6 years if each class initially consists of 500 females? (b) For what initial distribution will the number of females in each class change by the same proportion? What is this rate of change?

Solution. (a) Initially, $\mathbf{x}_{(0)}^T = [500 \quad 500 \quad 500]$. After 2 years,

$$\mathbf{x}_{(2)} = \mathbf{L}\mathbf{x}_{(0)} = \begin{bmatrix} 0 & 2.3 & 0.4 \\ 0.6 & 0 & 0 \\ 0 & 0.3 & 0 \end{bmatrix} \begin{bmatrix} 500 \\ 500 \\ 500 \end{bmatrix} = \begin{bmatrix} 1350 \\ 300 \\ 150 \end{bmatrix}.$$

Similarly, after 4 years the number of females in each class is given by $\mathbf{x}_{(4)}^T = (\mathbf{L}\mathbf{x}_{(2)})^T = [750 \quad 810 \quad 90]$, and after 6 years we have $\mathbf{x}_{(6)}^T = (\mathbf{L}\mathbf{x}_{(4)})^T = [1899 \quad 450 \quad 243]$.

(b) Proportional change means that we are looking for a distribution vector $\mathbf{x}$ such that $\mathbf{L}\mathbf{x} = \lambda\mathbf{x}$, where λ is the rate of change (growth if $\lambda > 1$, decrease if $\lambda < 1$). The characteristic equation is

$$\det(\mathbf{L} - \lambda\mathbf{I}) = -\lambda^3 - 0.6(-2.3\lambda - 0.3 \cdot 0.4) = -\lambda^3 + 1.38\lambda + 0.072 = 0.$$

A positive root is found to be (for instance, by Newton's method, Sec. 17.2) $\lambda = 1.2$. A corresponding eigenvector can be determined from $0.6x_1 - 1.2x_2 = 0$, $0.3x_2 - 1.2x_3 = 0$, resulting from the second and third components of the vector equation $(\mathbf{L} - 1.2\mathbf{I})\mathbf{x} = \mathbf{0}$. Thus, $\mathbf{x}^T = [1 \quad 0.5 \quad 0.125]$. To get an initial population of 1500, as before, we multiply $\mathbf{x}^T$ by 923. *Answer.* 923 females in class 1, 462 in class 2, 115 in class 3. Growth rate 1.2. ◀

EXAMPLE 4   Vibrating system of two masses on two springs (Fig. 77 in Problem Set 3.1)

Mass–spring systems involving several masses and springs can be treated as eigenvalue problems. For instance, the mechanical system in Fig. 77 in Problem Set 3.1 is governed by the differential equations

$$(6) \qquad\qquad \begin{aligned} y_1'' &= -5y_1 + 2y_2 \\ y_2'' &= 2y_1 - 2y_2 \end{aligned}$$

where y_1 and y_2 are the displacements of the masses from rest, as shown in the figure, and primes denote derivatives with respect to time t. In vector form, this becomes

$$(7) \qquad\qquad \mathbf{y}'' = \begin{bmatrix} y_1'' \\ y_2'' \end{bmatrix} = \mathbf{A}\mathbf{y} = \begin{bmatrix} -5 & 2 \\ 2 & -2 \end{bmatrix} \begin{bmatrix} y_1 \\ y_2 \end{bmatrix}.$$

We try a vector solution of the form

(8) $$\mathbf{y} = \mathbf{x}e^{\omega t}.$$

This is suggested by a mechanical system of a single mass on a spring (Sec. 2.5), whose motion is given by exponential functions (and sines and cosines). Substitution into (7) gives

$$\omega^2 \mathbf{x}e^{\omega t} = \mathbf{A}\mathbf{x}e^{\omega t}.$$

Dividing by $e^{\omega t}$ and writing $\omega^2 = \lambda$, we see that our mechanical system leads to the eigenvalue problem

(9) $$\mathbf{A}\mathbf{x} = \lambda\mathbf{x}$$ where $\lambda = \omega^2$.

From Example 1 in Sec. 7.1 we see that $\mathbf{A}$ has the eigenvalues $\lambda_1 = -1$ and $\lambda_2 = -6$. Consequently, $\omega = \sqrt{-1} = \pm i$ and $\sqrt{-6} = \pm i\sqrt{6}$, respectively. Corresponding eigenvectors are

(10) $$\mathbf{x}_1 = \begin{bmatrix} 1 \\ 2 \end{bmatrix} \quad \text{and} \quad \mathbf{x}_2 = \begin{bmatrix} 2 \\ -1 \end{bmatrix}.$$

From (8) we thus obtain the four complex solutions [see (7), Sec. 2.3]

$$\mathbf{x}_1 e^{\pm it} = \mathbf{x}_1(\cos t \pm i \sin t),$$

$$\mathbf{x}_2 e^{\pm i\sqrt{6}t} = \mathbf{x}_2(\cos \sqrt{6}\, t \pm i \sin \sqrt{6}\, t).$$

By addition and subtraction (see Sec. 2.3) we get the four real solutions

$$\mathbf{x}_1 \cos t, \quad \mathbf{x}_1 \sin t, \quad \mathbf{x}_2 \cos \sqrt{6}\, t, \quad \mathbf{x}_2 \sin \sqrt{6}\, t.$$

A general solution is obtained by taking a linear combination of these,

$$\mathbf{y} = \mathbf{x}_1(a_1 \cos t + b_1 \sin t) + \mathbf{x}_2(a_2 \cos \sqrt{6}\, t + b_2 \sin \sqrt{6}\, t)$$

with arbitrary constants a_1, b_1, a_2, b_2 (to which values can be assigned by prescribing initial displacement and initial velocity of each of the two masses). By (10), the components of $\mathbf{y}$ are

$$y_1 = a_1 \cos t + b_1 \sin t + 2a_2 \cos \sqrt{6}\, t + 2b_2 \sin \sqrt{6}\, t$$

$$y_2 = 2a_1 \cos t + 2b_1 \sin t - a_2 \cos \sqrt{6}\, t - b_2 \sin \sqrt{6}\, t.$$

These functions describe harmonic oscillations of the two masses. Physically, this had to be expected because we have neglected damping.   ◀

PROBLEM SET 7.2

Elastic Deformations. Find the principal directions and corresponding factors of extension or contraction of the elastic deformation $\mathbf{y} = \mathbf{A}\mathbf{x}$ with given $\mathbf{A}$. (Show the details of your work.)

1. $\begin{bmatrix} 4 & \sqrt{8} \\ \sqrt{8} & 6 \end{bmatrix}$

2. $\begin{bmatrix} 2.0 & 0.4 \\ 0.4 & 2.0 \end{bmatrix}$

3. $\begin{bmatrix} 3.0 & 1.5 \\ 1.5 & 3.0 \end{bmatrix}$

4. $\begin{bmatrix} 1 & \frac{1}{2} \\ \frac{1}{2} & 1 \end{bmatrix}$

5. $\begin{bmatrix} 3/2 & 1/\sqrt{2} \\ 1/\sqrt{2} & 1 \end{bmatrix}$

6. $\begin{bmatrix} 1.25 & 0.75 \\ 0.75 & 1.25 \end{bmatrix}$

Markov Processes. Find limit states of the Markov processes governed by the following stochastic matrices. (Show details.)

7.
$$\begin{bmatrix} 0.2 & 0.8 \\ 0.5 & 0.5 \end{bmatrix}$$

8.
$$\begin{bmatrix} 0.4 & 0.3 & 0.3 \\ 0.3 & 0.6 & 0.1 \\ 0.3 & 0.1 & 0.6 \end{bmatrix}$$

9.
$$\begin{bmatrix} 0.4 & 0.3 & 0.3 \\ 0.2 & 0.6 & 0.2 \\ 0.1 & 0.1 & 0.8 \end{bmatrix}$$

Population Model with Age Specification. Find the growth rate in the Leslie model with the matrix as given. (Show details.)

10.
$$\begin{bmatrix} 0 & 9 & 5 \\ 0.4 & 0 & 0 \\ 0 & 0.4 & 0 \end{bmatrix}$$

11.
$$\begin{bmatrix} 0 & 6 & 0 \\ 0.375 & 0 & 0 \\ 0 & 0.15 & 0 \end{bmatrix}$$

12.
$$\begin{bmatrix} 0 & 10.4 & 4.25 \\ 0.8 & 0 & 0 \\ 0 & 0.6 & 0 \end{bmatrix}$$

13. (Leontief[4] input–output model) Suppose that three industries are interrelated so that their outputs are used as inputs by themselves, according to the 3×3 **consumption matrix**

$$\mathbf{A} = [a_{jk}] = \begin{bmatrix} 0.1 & 0.5 & 0 \\ 0.8 & 0 & 0.4 \\ 0.1 & 0.5 & 0.6 \end{bmatrix}$$

where a_{jk} is the fraction of the output of industry k consumed (purchased) by industry j. Let p_j be the price charged by industry j for its total output. A problem is to find prices so that for each industry, total expenditures equal total income. Show that this leads to $\mathbf{Ap} = \mathbf{p}$, where $\mathbf{p} = [p_1 \quad p_2 \quad p_3]^T$, and find a solution $\mathbf{p}$ with nonnegative p_1, p_2, p_3.

14. Show that a consumption matrix as considered in Prob. 13 must have column sums 1 and always has the eigenvalue 1.

15. (Open Leontief input–output model) If not the whole output but only a portion of it is consumed by the industries themselves, then instead of $\mathbf{Ax} = \mathbf{x}$ (as in Prob. 13), we have $\mathbf{x} - \mathbf{Ax} = \mathbf{y}$, where $\mathbf{x} = [x_1 \quad x_2 \quad x_3]^T$ is produced, $\mathbf{Ax}$ is consumed by the industries, and, thus, $\mathbf{y}$ is the net production available for other consumers. Find for what production $\mathbf{x}$ a given $\mathbf{y} = [0.1 \quad 0.3 \quad 0.1]^T$ can be achieved if the consumption matrix is

$$\mathbf{A} = \begin{bmatrix} 0.1 & 0.4 & 0.2 \\ 0.5 & 0 & 0.1 \\ 0.1 & 0.4 & 0.4 \end{bmatrix}.$$

16. TEAM PROJECT. General Properties of Eigenvalues and Eigenvectors. Prove the following statements and illustrate them with examples of your own choice. Here, $\lambda_1, \cdots, \lambda_n$ are the (not necessarily distinct) eigenvalues of a given matrix $\mathbf{A} = [a_{jk}]$.

(a) **Real and complex eigenvalues.** If $\mathbf{A}$ is real, its eigenvalues are real or complex conjugates in pairs.

(b) **Inverse.** $\mathbf{A}^{-1}$ exists if and only if 0 is not an eigenvalue of $\mathbf{A}$. It has the eigenvalues $1/\lambda_1, \cdots, 1/\lambda_n$.

(c) **Trace.** The sum of the main diagonal entries is called the *trace* of $\mathbf{A}$. It equals the sum of the eigenvalues.

---

[4]WASSILY LEONTIEF (1906–1999). American economist at New York University. For his input–output analysis he was awarded the Nobel Prize in 1973.

(d) **"Spectral shift."** $\mathbf{A} - k\mathbf{I}$ has the eigenvalues $\lambda_1 - k, \cdots, \lambda_n - k$ and the same eigenvectors as $\mathbf{A}$.

(e) **Scalar multiples, powers.** $k\mathbf{A}$ has the eigenvalues $k\lambda_1, \cdots, k\lambda_n$. $\mathbf{A}^m$ $(m = 1, 2, \cdots)$ has the eigenvalues $\lambda_1^m, \cdots, \lambda_n^m$. The eigenvectors are those of $\mathbf{A}$.

(f) **Spectral Mapping Theorem.** The **"polynomial matrix"**

$$p(\mathbf{A}) = k_m \mathbf{A}^m + k_{m-1}\mathbf{A}^{m-1} + \cdots + k_1 \mathbf{A} + k_0 \mathbf{I}$$

has the eigenvalues

$$p(\lambda_j) = k_m \lambda_j^m + k_{m-1}\lambda_j^{m-1} + \cdots + k_1 \lambda_j + k_0,$$

where $j = 1, \cdots, n$, and the same eigenvectors as $\mathbf{A}$.

(g) **Perron's theorem.** Show that a Leslie matrix $\mathbf{L}$ with positive $l_{12}, l_{13}, l_{31}, l_{32}$ has a positive eigenvalue. (This is a special case of the famous Perron–Frobenius theorem in Sec. 18.7, which is difficult to prove in its general form.)

7.3 Symmetric, Skew-Symmetric, and Orthogonal Matrices

We consider three classes of real square matrices that occur quite frequently in applications. These are defined as follows.

Definitions of symmetric, skew-symmetric, and orthogonal matrices

A *real* square matrix $\mathbf{A} = [a_{jk}]$ is called

symmetric if transposition leaves it unchanged,

(1)
$$\boxed{\mathbf{A}^\mathsf{T} = \mathbf{A},} \qquad \text{thus} \qquad a_{kj} = a_{jk},$$

skew-symmetric if transposition gives the negative of $\mathbf{A}$,

(2)
$$\boxed{\mathbf{A}^\mathsf{T} = -\mathbf{A},} \qquad \text{thus} \qquad a_{kj} = -a_{jk},$$

orthogonal if transposition gives the inverse of $\mathbf{A}$,

(3)
$$\boxed{\mathbf{A}^\mathsf{T} = \mathbf{A}^{-1}.}$$

EXAMPLE 1 **Symmetric, skew-symmetric, and orthogonal matrices**

The matrices

$$\begin{bmatrix} -3 & 1 & 5 \\ 1 & 0 & -2 \\ 5 & -2 & 4 \end{bmatrix}, \quad \begin{bmatrix} 0 & 9 & -12 \\ -9 & 0 & 20 \\ 12 & -20 & 0 \end{bmatrix}, \quad \begin{bmatrix} \frac{2}{3} & \frac{1}{3} & \frac{2}{3} \\ -\frac{2}{3} & \frac{2}{3} & \frac{1}{3} \\ \frac{1}{3} & \frac{2}{3} & -\frac{2}{3} \end{bmatrix}$$

are symmetric, skew-symmetric, and orthogonal, respectively, as you should verify. Every skew-symmetric matrix has all main diagonal entries zero. (Can you prove this?) ◀

Any real square matrix $\mathbf{A}$ may be written as the sum of a symmetric matrix $\mathbf{R}$ and a skew-symmetric matrix $\mathbf{S}$, where

(4) $$\mathbf{R} = \tfrac{1}{2}(\mathbf{A} + \mathbf{A}^\mathsf{T}) \qquad \text{and} \qquad \mathbf{S} = \tfrac{1}{2}(\mathbf{A} - \mathbf{A}^\mathsf{T}).$$

EXAMPLE 2   **Illustration of formula (4)**

$$\mathbf{A} = \begin{bmatrix} 3 & -4 & -1 \\ 6 & 0 & -1 \\ -3 & 13 & -4 \end{bmatrix} = \mathbf{R} + \mathbf{S} = \begin{bmatrix} 3 & 1 & -2 \\ 1 & 0 & 6 \\ -2 & 6 & -4 \end{bmatrix} + \begin{bmatrix} 0 & -5 & 1 \\ 5 & 0 & -7 \\ -1 & 7 & 0 \end{bmatrix} \qquad \blacktriangleleft$$

THEOREM 1   **(Eigenvalues of symmetric and skew-symmetric matrices) (Proof in Sec. 7.4)**

    (a) *The eigenvalues of a symmetric matrix are real.*

    (b) *The eigenvalues of a skew-symmetric matrix are pure imaginary or zero.*

EXAMPLE 3   **Eigenvalues of symmetric and skew-symmetric matrices**

The matrices in (1) and (7) of Sec. 7.2 are symmetric and have real eigenvalues. The skew-symmetric matrix in Example 1 has the eigenvalues 0, $-25i$, and $25i$. (Verify this.) The following matrix has the real eigenvalues 1 and 5 and is not symmetric. Does this contradict Theorem 1?

$$\begin{bmatrix} 3 & 4 \\ 1 & 3 \end{bmatrix} \qquad \blacktriangleleft$$

Orthogonal Transformations and Matrices

Orthogonal transformations are transformations

(5) $$\mathbf{y} = \mathbf{A}\mathbf{x} \qquad\qquad \text{with } \mathbf{A} \text{ an orthogonal matrix.}$$

With each vector $\mathbf{x}$ in R^n such a transformation assigns a vector $\mathbf{y}$ in R^n. For instance, the plane rotation through an angle θ

(6) $$\mathbf{y} = \begin{bmatrix} y_1 \\ y_2 \end{bmatrix} = \begin{bmatrix} \cos\theta & -\sin\theta \\ \sin\theta & \cos\theta \end{bmatrix} \begin{bmatrix} x_1 \\ x_2 \end{bmatrix}$$

is an orthogonal transformation. It can be shown that any orthogonal transformation in the plane or in three-dimensional space is a **rotation** (possibly combined with a reflection in a straight line or a plane, respectively).

The main reason for the importance of orthogonal matrices is as follows.

THEOREM 2   **(Invariance of inner product)**

An orthogonal transformation preserves the value of the inner product of vectors (Sec. 6.2)

(7) $$\mathbf{a} \cdot \mathbf{b} = \mathbf{a}^\mathsf{T}\mathbf{b}.$$

(**a** and **b** are column vectors.) *Hence it preserves also the **length** or **norm** of a vector in R^n given by*

(8) $$\|\mathbf{a}\| = \sqrt{\mathbf{a} \cdot \mathbf{a}} = \sqrt{\mathbf{a}^\mathsf{T}\mathbf{a}}.$$

PROOF. Let $\mathbf{u} = \mathbf{Aa}$ and $\mathbf{v} = \mathbf{Ab}$, where $\mathbf{A}$ is orthogonal. We must show that $\mathbf{u} \cdot \mathbf{v} = \mathbf{a} \cdot \mathbf{b}$. Now (5) in Sec. 6.2 gives $\mathbf{u}^\mathsf{T} = (\mathbf{Aa})^\mathsf{T} = \mathbf{a}^\mathsf{T}\mathbf{A}^\mathsf{T}$. Also, $\mathbf{A}^\mathsf{T}\mathbf{A} = \mathbf{A}^{-1}\mathbf{A} = \mathbf{I}$ by (3). Hence

$$\text{(9)} \qquad \mathbf{u} \cdot \mathbf{v} = \mathbf{u}^\mathsf{T}\mathbf{v} = (\mathbf{Aa})^\mathsf{T}\mathbf{Ab} = \mathbf{a}^\mathsf{T}\mathbf{A}^\mathsf{T}\mathbf{Ab} = \mathbf{a}^\mathsf{T}\mathbf{I}\,\mathbf{b} = \mathbf{a}^\mathsf{T}\mathbf{b} = \mathbf{a} \cdot \mathbf{b}.$$

This also implies (8) because $\|\mathbf{a}\|$ is given in terms of an inner product.   ◀

Orthogonal matrices have further interesting properties, as follows.

THEOREM 3   **(Orthonormality of column and row vectors)**

A real square matrix is orthogonal if and only if its column vectors $\mathbf{a}_1, \cdots, \mathbf{a}_n$ (and also its row vectors) form an **orthonormal system,** *that is,*

$$\text{(10)} \qquad \mathbf{a}_j \cdot \mathbf{a}_k = \mathbf{a}_j^\mathsf{T}\mathbf{a}_k = \begin{cases} 0 & \text{if } j \neq k \\ 1 & \text{if } j = k. \end{cases}$$

PROOF. **(a)** Let $\mathbf{A}$ be orthogonal. Then $\mathbf{A}^{-1}\mathbf{A} = \mathbf{A}^\mathsf{T}\mathbf{A} = \mathbf{I}$. In terms of column vectors $\mathbf{a}_1, \cdots, \mathbf{a}_n$,

$$\mathbf{A}^{-1}\mathbf{A} = \mathbf{A}^\mathsf{T}\mathbf{A} = \begin{bmatrix} \mathbf{a}_1^\mathsf{T} \\ \vdots \\ \mathbf{a}_n^\mathsf{T} \end{bmatrix} [\mathbf{a}_1 \cdots \mathbf{a}_n]$$

$$\text{(11)}$$

$$= \begin{bmatrix} \mathbf{a}_1^\mathsf{T}\mathbf{a}_1 & \mathbf{a}_1^\mathsf{T}\mathbf{a}_2 & \cdots & \mathbf{a}_1^\mathsf{T}\mathbf{a}_n \\ \mathbf{a}_2^\mathsf{T}\mathbf{a}_1 & \mathbf{a}_2^\mathsf{T}\mathbf{a}_2 & \cdots & \mathbf{a}_2^\mathsf{T}\mathbf{a}_n \\ \cdot & \cdot & \cdots & \cdot \\ \mathbf{a}_n^\mathsf{T}\mathbf{a}_1 & \mathbf{a}_n^\mathsf{T}\mathbf{a}_2 & \cdots & \mathbf{a}_n^\mathsf{T}\mathbf{a}_n \end{bmatrix} = \mathbf{I}.$$

The last equality implies (10), by the definition of the $n \times n$ unit matrix $\mathbf{I}$. From (3) it follows that the inverse of an orthogonal matrix is orthogonal (see CAS Project 14). Now the column vectors of $\mathbf{A}^{-1} (= \mathbf{A}^\mathsf{T})$ are the row vectors of $\mathbf{A}$. Hence the row vectors of $\mathbf{A}$ also form an orthonormal system.

(b) Conversely, if the column vectors of $\mathbf{A}$ satisfy (10), the off-diagonal entries in (11) must be 0 and the diagonal entries 1. Hence $\mathbf{A}^\mathsf{T}\mathbf{A} = \mathbf{I}$, as (11) shows. Similarly, $\mathbf{A}\mathbf{A}^\mathsf{T} = \mathbf{I}$. This implies $\mathbf{A}^\mathsf{T} = \mathbf{A}^{-1}$ because also $\mathbf{A}^{-1}\mathbf{A} = \mathbf{A}\mathbf{A}^{-1} = \mathbf{I}$ and the inverse is unique. Hence $\mathbf{A}$ is orthogonal. Similarly when the row vectors of $\mathbf{A}$ form an orthonormal system, by what has been said at the end of part **(a)**.   ◀

THEOREM 4   **(Determinant of an orthogonal matrix)**

The determinant of an orthogonal matrix has the value $+1$ or -1.

PROOF. From $\det \mathbf{AB} = \det \mathbf{A} \det \mathbf{B}$ (Sec. 6.7, Theorem 4) and $\det \mathbf{A}^\mathsf{T} = \det \mathbf{A}$ (Sec. 6.6, Theorem 2), we get for an orthogonal matrix

$$1 = \det \mathbf{I} = \det (\mathbf{A}\mathbf{A}^{-1}) = \det (\mathbf{A}\mathbf{A}^\mathsf{T}) = \det \mathbf{A} \det \mathbf{A}^\mathsf{T} = (\det \mathbf{A})^2.$$ ◀

EXAMPLE 4   **Illustration of Theorems 3 and 4**

The last matrix in Example 1 and the matrix in (6) illustrate Theorems 3 and 4, their determinants being -1 and $+1$, as you should verify.   ◀

THEOREM 5   **(Eigenvalues of an orthogonal matrix)**

The eigenvalues of an orthogonal matrix $\mathbf{A}$ are real or complex conjugates in pairs and have absolute value 1.

PROOF.   The first part of the statement holds for any real matrix $\mathbf{A}$ because its characteristic polynomial has real coefficients, so that its zeros (the eigenvalues of $\mathbf{A}$) must be as indicated. The claim that $|\lambda| = 1$ will be proved in the next section.   ◀

EXAMPLE 5   **Eigenvalues of an orthogonal matrix**

The orthogonal matrix in Example 1 has the characteristic equation

$$-\lambda^3 + \tfrac{2}{3}\lambda^2 + \tfrac{2}{3}\lambda - 1 = 0.$$

Now one of the eigenvalues must be real (why?), hence $+1$ or -1. Trying, we find -1. Division by $\lambda + 1$ gives $\lambda^2 - 5\lambda/3 + 1 = 0$ and the two eigenvalues $(5 + i\sqrt{11})/6$ and $(5 - i\sqrt{11})/6$. Verify all of this.   ◀

PROBLEM SET 7.3

Symmetric, Skew-Symmetric, and Orthogonal Matrices and Their Spectrum. Are the following matrices symmetric, skew-symmetric, or orthogonal? Find their eigenvalues (thereby illustrating Theorems 1 and 5). (Show the details of your work.)

1. $\begin{bmatrix} 0.96 & -0.28 \\ 0.28 & 0.96 \end{bmatrix}$

2. $\begin{bmatrix} a & b \\ -b & a \end{bmatrix}$

3. $\begin{bmatrix} 1 & 4 \\ -4 & 1 \end{bmatrix}$

4. $\begin{bmatrix} 1 & 0 & 0 \\ 0 & \cos\theta & -\sin\theta \\ 0 & \sin\theta & \cos\theta \end{bmatrix}$

5. $\begin{bmatrix} 0 & 9 & -12 \\ -9 & 0 & 20 \\ 12 & -20 & 0 \end{bmatrix}$

6. $\begin{bmatrix} a & k & k \\ k & a & k \\ k & k & a \end{bmatrix}$

7. **(Skew-symmetric matrix)** Show that the main diagonal entries of a skew-symmetric matrix must be zero.

8. **(Symmetric matrix)** Prove that eigenvectors of a symmetric matrix corresponding to different eigenvalues are orthogonal. Give an example.

9. **(Rotation in space)** Give a geometrical interpretation of the transformation $\mathbf{y} = \mathbf{A}\mathbf{x}$ with $\mathbf{A}$ as in Prob. 4 and $\mathbf{x}$ and $\mathbf{y}$ referred to a Cartesian coordinate system.

10. **(Symmetric matrix)** Do there exist symmetric 3×3 matrices that are orthogonal (except for the unit matrix)?

11. **(Skew-symmetric matrix)** Show that the inverse of a skew-symmetric matrix is skew-symmetric.

12. **(Skew-symmetric matrix)** Do there exist nonsingular skew-symmetric 3×3 matrices? 4×4? 5×5?

13. **(Orthogonal matrix)** Do there exist skew-symmetric orthogonal 3×3 matrices?

14. CAS PROJECT. Orthogonal Matrices.

   (a) Products. Inverse. Prove that the product of two orthogonal matrices is orthogonal, and so is the inverse of an orthogonal matrix. What does this mean in terms of rotations?

   (b) Rotation. Show that (6) is an orthogonal transformation. Verify that it satisfies Theorem 3. Find the inverse transformation.

   (c) Powers. Write a program for computing powers $\mathbf{A}^m$ ($m = 1, 2, \cdots$) of a 2×2 matrix $\mathbf{A}$ and their spectra. Apply it to the matrix in Prob. 1 (call it $\mathbf{A}$). To what rotation does $\mathbf{A}$ correspond? Do the eigenvalues of $\mathbf{A}^m$ have a limit as $m \to \infty$?

   (d) Compute the eigenvalues of $(0.9\mathbf{A})^m$, where $\mathbf{A}$ is the matrix in Prob. 1. Plot them as points. What is their limit? Along what kind of curve do these points approach it?

   (e) Find $\mathbf{A}$ such that $\mathbf{y} = \mathbf{A}\mathbf{x}$ is a counterclockwise rotation through $30°$ in the plane.

7.4 Complex Matrices: Hermitian, Skew-Hermitian, Unitary

We shall now introduce three classes of *complex square matrices*. These generalize the three classes of real matrices just considered. They are important in certain applications, mainly because of their spectra (see Theorem 1, below), for instance, in quantum mechanics.

In this connection we use the standard notation

$$\overline{\mathbf{A}} = [\bar{a}_{jk}]$$

for the matrix obtained from $\mathbf{A} = [a_{jk}]$ by replacing each entry by its complex conjugate. We also use the notation

$$\boxed{\overline{\mathbf{A}}^\mathsf{T} = [\bar{a}_{kj}]}$$

for the conjugate transpose. For example, if

$$\mathbf{A} = \begin{bmatrix} 3 + 4i & -5i \\ -7 & 6 - 2i \end{bmatrix}, \quad \text{then} \quad \overline{\mathbf{A}}^\mathsf{T} = \begin{bmatrix} 3 - 4i & -7 \\ 5i & 6 + 2i \end{bmatrix}.$$

Definitions of Hermitian,[5] Skew-Hermitian, and unitary matrices

A square matrix $\mathbf{A} = [a_{kj}]$ is called

        Hermitian if $\overline{\mathbf{A}}^\mathsf{T} = \mathbf{A},$ that is, $\bar{a}_{kj} = a_{jk}$

        skew-Hermitian if $\overline{\mathbf{A}}^\mathsf{T} = -\mathbf{A},$ that is, $\bar{a}_{kj} = -a_{jk}$

        unitary if $\overline{\mathbf{A}}^\mathsf{T} = \mathbf{A}^{-1}.$

From these definitions we see the following. If $\mathbf{A}$ is Hermitian, the entries on the main diagonal must satisfy $\bar{a}_{jj} = a_{jj}$, that is, they are real. Similarly, if $\mathbf{A}$ is skew-Hermitian, then

---

[5]See footnote 22 in Problem Set 4.8.

$\bar{a}_{jj} = -a_{jj}$. If we set $a_{jj} = \alpha + i\beta$, this becomes $\alpha - i\beta = -(\alpha + i\beta)$. Hence $\alpha = 0$, so that a_{jj} must be pure imaginary or 0.

EXAMPLE 1   **Hermitian, skew-Hermitian, and unitary matrices**

$$\mathbf{A} = \begin{bmatrix} 4 & 1 - 3i \\ 1 + 3i & 7 \end{bmatrix}, \quad \mathbf{B} = \begin{bmatrix} 3i & 2 + i \\ -2 + i & -i \end{bmatrix}, \quad \mathbf{C} = \begin{bmatrix} \tfrac{1}{2}i & \tfrac{1}{2}\sqrt{3} \\ \tfrac{1}{2}\sqrt{3} & \tfrac{1}{2}i \end{bmatrix}$$

are Hermitian, skew-Hermitian, and unitary matrices, respectively, as you may verify by using the definitions. ◀

If a Hermitian matrix is real, then $\overline{\mathbf{A}}^\mathsf{T} = \mathbf{A}^\mathsf{T} = \mathbf{A}$. Hence a real Hermitian matrix is a symmetric matrix (Sec. 7.3).

Similarly, if a skew-Hermitian matrix is real, then $\overline{\mathbf{A}}^\mathsf{T} = \mathbf{A}^\mathsf{T} = -\mathbf{A}$. Hence a real skew-Hermitian matrix is a skew-symmetric matrix.

Finally, if a unitary matrix is real, then $\overline{\mathbf{A}}^\mathsf{T} = \mathbf{A}^\mathsf{T} = \mathbf{A}^{-1}$. Hence a real unitary matrix is an orthogonal matrix.

This shows that *Hermitian, skew-Hermitian, and unitary matrices generalize symmetric, skew-symmetric, and orthogonal matrices, respectively.*

Eigenvalues

It is quite remarkable and in part accounts for the importance of the matrices under consideration that their spectra (their sets of eigenvalues; see Sec. 7.1) can be characterized in a general way as follows (see Fig. 146).

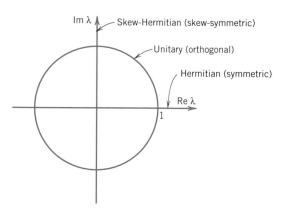

Fig. 146. Location of the eigenvalues of Hermitian, skew-Hermitian, and unitary matrices in the complex λ-plane

THEOREM 1   **(Eigenvalues)**

    (a) *The eigenvalues of a Hermitian matrix (and thus of a symmetric matrix) are real.*

    (b) *The eigenvalues of a skew-Hermitian matrix (and thus of a skew-symmetric matrix) are pure imaginary or zero.*

    (c) *The eigenvalues of a unitary matrix (and thus of an orthogonal matrix) have absolute value 1.*

EXAMPLE 2   Illustration of Theorem 1

For the matrices in Example 1 we find by direct calculation

	Matrix	Characteristic Equation	Eigenvalues
A	Hermitian	$\lambda^2 - 11\lambda + 18 = 0$	$9, \quad 2$
B	Skew-Hermitian	$\lambda^2 - 2i\lambda + 8 = 0$	$4i, \quad -2i$
C	Unitary	$\lambda^2 - i\lambda - 1 = 0$	$\frac{1}{2}\sqrt{3} + \frac{1}{2}i, \quad -\frac{1}{2}\sqrt{3} + \frac{1}{2}i$

and $\left|\pm\frac{1}{2}\sqrt{3} + \frac{1}{2}i\right|^2 = \frac{3}{4} + \frac{1}{4} = 1$.   ◄

PROOF OF THEOREM 1.   Let λ be an eigenvalue of **A** and **x** a corresponding eigenvector. Then

$$(1) \qquad\qquad \mathbf{Ax} = \lambda\mathbf{x}.$$

(**a**) Let **A** be Hermitian. Multiplying (1) by $\bar{\mathbf{x}}^\mathsf{T}$ from the left, we obtain

$$\bar{\mathbf{x}}^\mathsf{T}\mathbf{Ax} = \bar{\mathbf{x}}^\mathsf{T}\lambda\mathbf{x} = \lambda\bar{\mathbf{x}}^\mathsf{T}\mathbf{x}.$$

Now $\bar{\mathbf{x}}^\mathsf{T}\mathbf{x} = \bar{x}_1 x_1 + \cdots + \bar{x}_n x_n = |x_1|^2 + \cdots + |x_n|^2$ is real, and is not 0 since $\mathbf{x} \neq \mathbf{0}.$ Hence we may divide to get

$$(2) \qquad\qquad \lambda = \frac{\bar{\mathbf{x}}^\mathsf{T}\mathbf{Ax}}{\bar{\mathbf{x}}^\mathsf{T}\mathbf{x}}.$$

We see that λ is real if the numerator is real. We prove that the numerator is real by showing that it is equal to its complex conjugate, using $\overline{\mathbf{A}}^\mathsf{T} = \mathbf{A}$ or $\overline{\mathbf{A}} = \mathbf{A}^\mathsf{T}$ and (5) in Sec. 6.2. Indeed, beginning with the application of a transposition, which has no effect on a number (the numerator), we get

$$(3) \qquad\qquad \bar{\mathbf{x}}^\mathsf{T}\mathbf{Ax} = (\bar{\mathbf{x}}^\mathsf{T}\mathbf{Ax})^\mathsf{T} = \mathbf{x}^\mathsf{T}\mathbf{A}^\mathsf{T}\bar{\mathbf{x}} = \mathbf{x}^\mathsf{T}\overline{\mathbf{A}}\bar{\mathbf{x}} = (\overline{\bar{\mathbf{x}}^\mathsf{T}\mathbf{Ax}}).$$

Hence $\bar{\mathbf{x}}^\mathsf{T}\mathbf{Ax}$ is real. From this and (2), whose denominator is real, we see that λ is real.

(**b**) If **A** is skew-Hermitian, then $\overline{\mathbf{A}}^\mathsf{T} = -\mathbf{A},$ thus $\overline{\mathbf{A}} = -\mathbf{A}^\mathsf{T}$, so that we get a minus sign in (3),

$$(4) \qquad\qquad \bar{\mathbf{x}}^\mathsf{T}\mathbf{Ax} = -(\overline{\bar{\mathbf{x}}^\mathsf{T}\mathbf{Ax}}).$$

So this is a complex number $c = a + ib$ that equals minus its conjugate $\bar{c} = a - ib$, that is, $a + ib = -(a - ib)$. Hence $a = -a = 0$, so that c is pure imaginary or zero. Division by the real $\bar{\mathbf{x}}^\mathsf{T}\mathbf{x}$ in (2) now gives a pure imaginary λ or $\lambda = 0$.

(**c**) Let **A** be unitary. We take (1) and its conjugate transpose,

$$\mathbf{Ax} = \lambda\mathbf{x} \qquad \text{and} \qquad (\overline{\mathbf{Ax}})^\mathsf{T} = (\overline{\lambda}\overline{\mathbf{x}})^\mathsf{T} = \overline{\lambda}\bar{\mathbf{x}}^\mathsf{T}$$

and multiply the two left sides and the two right sides,

$$(\overline{\mathbf{Ax}})^\mathsf{T}\mathbf{Ax} = \overline{\lambda}\lambda\bar{\mathbf{x}}^\mathsf{T}\mathbf{x} = |\lambda|^2\,\bar{\mathbf{x}}^\mathsf{T}\mathbf{x}.$$

But **A** is unitary, $\overline{\mathbf{A}}^{\mathsf{T}} = \mathbf{A}^{-1}$, so that on the left we obtain

$$(\overline{\mathbf{A}\mathbf{x}})^{\mathsf{T}}\mathbf{A}\mathbf{x} = \overline{\mathbf{x}}^{\mathsf{T}}\overline{\mathbf{A}}^{\mathsf{T}}\mathbf{A}\mathbf{x} = \overline{\mathbf{x}}^{\mathsf{T}}\mathbf{A}^{-1}\mathbf{A}\mathbf{x} = \overline{\mathbf{x}}^{\mathsf{T}}\mathbf{I}\,\mathbf{x} = \overline{\mathbf{x}}^{\mathsf{T}}\mathbf{x}.$$

Together, $\overline{\mathbf{x}}^{\mathsf{T}}\mathbf{x} = |\lambda|^2\,\overline{\mathbf{x}}^{\mathsf{T}}\mathbf{x}$. We now divide by $\overline{\mathbf{x}}^{\mathsf{T}}\mathbf{x}$ ($\neq 0$) to get $|\lambda|^2 = 1$. Hence $|\lambda| = 1$. This proves Theorem 1 as well as Theorems 1 and 5 in Sec. 7.3.   ◀

Forms

The numerator $\overline{\mathbf{x}}^{\mathsf{T}}\mathbf{A}\mathbf{x}$ in (2) is called a **form** in the components $x_1, \cdots, x_n$ of **x,** and **A** is called its *coefficient matrix*. When $n = 2$, we get

$$\overline{\mathbf{x}}^{\mathsf{T}}\mathbf{A}\mathbf{x} = \begin{bmatrix} \overline{x}_1 & \overline{x}_2 \end{bmatrix} \begin{bmatrix} a_{11} & a_{12} \\ a_{21} & a_{22} \end{bmatrix} \begin{bmatrix} x_1 \\ x_2 \end{bmatrix} = \begin{bmatrix} \overline{x}_1 & \overline{x}_2 \end{bmatrix} \begin{bmatrix} a_{11}x_1 + a_{12}x_2 \\ a_{21}x_1 + a_{22}x_2 \end{bmatrix}$$

$$= a_{11}\overline{x}_1 x_1 + a_{12}\overline{x}_1 x_2 + a_{21}\overline{x}_2 x_1 + a_{22}\overline{x}_2 x_2.$$

Similarly for general n,

$$(5) \qquad \begin{aligned} \overline{\mathbf{x}}^{\mathsf{T}}\mathbf{A}\mathbf{x} = \sum_{j=1}^{n}\sum_{k=1}^{n} a_{jk}\overline{x}_j x_k = \quad & a_{11}\overline{x}_1 x_1 + \cdots + a_{1n}\overline{x}_1 x_n \\ + \; & a_{21}\overline{x}_2 x_1 + \cdots + a_{2n}\overline{x}_2 x_n \\ + \; & \cdots\cdots\cdots\cdots\cdots\cdots \\ + \; & a_{n1}\overline{x}_n x_1 + \cdots + a_{nn}\overline{x}_n x_n. \end{aligned}$$

So this form (5) is a sum of n^2 terms. If **x** and **A** are real, it becomes

$$(6) \qquad \begin{aligned} \mathbf{x}^{\mathsf{T}}\mathbf{A}\mathbf{x} = \sum_{j=1}^{n}\sum_{k=1}^{n} a_{jk} x_j x_k = \quad & a_{11}x_1{}^2 + a_{12}x_1 x_2 + \cdots + a_{1n}x_1 x_n \\ + \; & a_{21}x_2 x_1 + a_{22}x_2{}^2 + \cdots + a_{2n}x_2 x_n \\ + \; & \cdots\cdots\cdots\cdots\cdots\cdots\cdots\cdots \\ + \; & a_{n1}x_n x_1 + a_{n2}x_n x_2 + \cdots + a_{nn}x_n{}^2. \end{aligned}$$

This is called a **quadratic form.** Without restriction we may then assume the coefficient matrix to be **symmetric,** because we can take off-diagonal terms together in pairs and then write the result as a sum of two equal terms, as the next example (Example 3) illustrates. Quadratic forms occur in physics and geometry, for instance, in connection with conic sections (ellipses $x_1{}^2/a^2 + x_2{}^2/b^2 = 1$, etc.) and quadratic surfaces. (Their "transformation to principal axes" will be discussed in the next section.)

EXAMPLE 3   **Quadratic form. Symmetric coefficient matrix C**

Let

$$\mathbf{x}^{\mathsf{T}}\mathbf{A}\mathbf{x} = \begin{bmatrix} x_1 & x_2 \end{bmatrix} \begin{bmatrix} 3 & 4 \\ 6 & 2 \end{bmatrix} \begin{bmatrix} x_1 \\ x_2 \end{bmatrix} = 3x_1{}^2 + 4x_1 x_2 + 6x_2 x_1 + 2x_2{}^2 = 3x_1{}^2 + 10x_1 x_2 + 2x_2{}^2.$$

Here $4 + 6 = 10 = 5 + 5$. From the corresponding *symmetric* matrix $\mathbf{C} = [c_{jk}]$, where $c_{jk} = \frac{1}{2}(a_{jk} + a_{kj})$, thus $c_{11} = 3$, $c_{12} = c_{21} = 5$, $c_{22} = 2$, we get the same result; indeed,

$$\mathbf{x}^T\mathbf{C}\mathbf{x} = [x_1 \quad x_2]\begin{bmatrix} 3 & 5 \\ 5 & 2 \end{bmatrix}\begin{bmatrix} x_1 \\ x_2 \end{bmatrix} = 3x_1{}^2 + 5x_1x_2 + 5x_2x_1 + 2x_2{}^2 = 3x_1{}^2 + 10x_1x_2 + 2x_2{}^2. \quad \blacktriangleleft$$

If the matrix $\mathbf{A}$ in (5) is Hermitian or skew-Hermitian, the form (5) is called a **Hermitian form** or **skew-Hermitian form,** respectively. These forms have the following property, which accounts for their importance in physics.

THEOREM 1*   **(Hermitian and skew-Hermitian forms)**

For every choice of the vector $\mathbf{x}$ *the value of a Hermitian form is real, and the value of a skew-Hermitian form is pure imaginary or* 0.

PROOF.   In proving (3) and (4), we made no use of the fact that $\mathbf{x}$ was an eigenvector, and the proofs remain valid for any vectors (and Hermitian or skew-Hermitian matrices). From this, our present theorem follows. $\quad \blacktriangleleft$

EXAMPLE 4   **Hermitian form**

If

$$\mathbf{A} = \begin{bmatrix} 3 & 2 - i \\ 2 + i & 4 \end{bmatrix} \quad \text{and} \quad \mathbf{x} = \begin{bmatrix} 1 + i \\ 2i \end{bmatrix},$$

then

$$\bar{\mathbf{x}}^T\mathbf{A}\mathbf{x} = [1 - i \quad -2i]\begin{bmatrix} 3 & 2 - i \\ 2 + i & 4 \end{bmatrix}\begin{bmatrix} 1 + i \\ 2i \end{bmatrix} = [1 - i \quad -2i]\begin{bmatrix} 3(1 + i) + (2 - i)2i \\ (2 + i)(1 + i) + 4 \cdot 2i \end{bmatrix} = 34. \quad \blacktriangleleft$$

Properties of Unitary Matrices. Complex Vector Space C^n

We now extend our discussion of orthogonal matrices in Sec. 7.3 to unitary matrices. Instead of the real vector space R^n of all real vectors with n components and real numbers as scalars we now use the **complex vector space C^n** of all complex vectors with n complex numbers as components and complex numbers as scalars. For such complex vectors, the **inner product** is defined by

(7)
$$\boxed{\mathbf{a} \cdot \mathbf{b} = \bar{\mathbf{a}}^T\mathbf{b}.}$$

The **length** or **norm** of such a complex vector is defined by

(8) $$\|\mathbf{a}\| = \sqrt{\mathbf{a} \cdot \mathbf{a}} = \sqrt{\bar{\mathbf{a}}^T\mathbf{a}} = \sqrt{\bar{a}_1 a_1 + \cdots + \bar{a}_n a_n} = \sqrt{|a_1|^2 + \cdots + |a_n|^2}.$$

Note that for *real* vectors, (7) reduces to the inner product as defined in Sec. 6.2.

THEOREM 2   **(Invariance of inner product)**

A **unitary transformation,** *that is,* $\mathbf{y} = \mathbf{A}\mathbf{x}$ *with a unitary matrix* $\mathbf{A}$, *preserves the value of the inner product* (7), *hence also the norm* (8).

PROOF.   The proof is the same as that of Theorem 2 in Sec. 7.3, which the theorem generalizes. In the analog of (9), Sec. 7.3, we now have bars,

$$\mathbf{u} \cdot \mathbf{v} = \bar{\mathbf{u}}^T\mathbf{v} = (\overline{\mathbf{A}\mathbf{a}})^T\mathbf{A}\mathbf{b} = \bar{\mathbf{a}}^T\bar{\mathbf{A}}^T\mathbf{A}\mathbf{b} = \bar{\mathbf{a}}^T\mathbf{I}\,\mathbf{b} = \bar{\mathbf{a}}^T\mathbf{b} = \mathbf{a} \cdot \mathbf{b}. \quad \blacktriangleleft$$

The complex analog of an *orthonormal system* of real vectors (see Sec. 7.3) is a **unitary system,** defined by

$$(9) \qquad \mathbf{a}_j \cdot \mathbf{a}_k = \bar{\mathbf{a}}_j^{\mathsf{T}} \mathbf{a}_k = \begin{cases} 0 & \text{if} \quad j \neq k \\ 1 & \text{if} \quad j = k. \end{cases}$$

The extension of Theorem 3, Sec. 7.3, to complex is as follows.

THEOREM 3   **(Unitary systems of column and row vectors)**

A square matrix is unitary if and only if its column vectors (and also its row vectors) form a unitary system.

PROOF. The proof is the same as that of Theorem 3 in Sec. 7.3, except for the bars required in $\bar{\mathbf{A}}^{\mathsf{T}} = \mathbf{A}^{-1}$ and in (7) and (9) in this section. ◀

THEOREM 4   **(Determinant of a unitary matrix)**

The determinant of a unitary matrix has absolute value 1.

PROOF. Similarly as in Sec. 7.3 we get

$$1 = \det (\mathbf{A}\,\mathbf{A}^{-1}) = \det (\mathbf{A}\,\bar{\mathbf{A}}^{\mathsf{T}}) = \det \mathbf{A} \det \bar{\mathbf{A}}^{\mathsf{T}} = \det \mathbf{A} \det \bar{\mathbf{A}}$$

$$= \det \mathbf{A} \,\overline{\det \mathbf{A}} = |\det \mathbf{A}|^2.$$

Hence $|\det \mathbf{A}| = 1$ (where $\det \mathbf{A}$ may now be complex). ◀

EXAMPLE 5   **Unitary matrix illustrating Theorems 2–4**

For the vectors $\mathbf{a}^{\mathsf{T}} = [1 \quad i]$ and $\mathbf{b}^{\mathsf{T}} = [3i \quad 2 + i]$ we get $\bar{\mathbf{a}}^{\mathsf{T}}\mathbf{b} = 3i - i(2 + i) = 1 + i$, and with

$$\mathbf{A} = \begin{bmatrix} 0.6i & 0.8 \\ 0.8 & 0.6i \end{bmatrix} \quad \text{also} \quad \mathbf{Aa} = \begin{bmatrix} 1.4i \\ 0.2 \end{bmatrix} \quad \text{and} \quad \mathbf{Ab} = \begin{bmatrix} -0.2 + 0.8i \\ -0.6 + 3.6i \end{bmatrix},$$

as one can readily verify. This gives $(\overline{\mathbf{Aa}})^{\mathsf{T}}\mathbf{Ab} = 1 + i$, illustrating Theorem 2. The matrix is unitary. Its columns form a unitary system,

$$\bar{\mathbf{a}}_1^{\mathsf{T}} \mathbf{a}_1 = -0.6i \cdot 0.6i + 0.8^2 = 1,$$

$$\bar{\mathbf{a}}_1^{\mathsf{T}} \mathbf{a}_2 = -0.6i \cdot 0.8 + 0.8 \cdot 0.6i = 0, \qquad \bar{\mathbf{a}}_2^{\mathsf{T}} \mathbf{a}_2 = 1,$$

and so do the rows. Also, $\det \mathbf{A} = -1$. ◀

PROBLEM SET 7.4

1. Verify the eigenvalues in Example 2.

In Examples 1 and 2, find eigenvectors of

2. The matrix $\mathbf{A}$ **3.** The matrix $\mathbf{B}$ **4.** The matrix $\mathbf{C}$

Hermitian, Skew-Hermitian, and Unitary Matrices and Their Spectrum

Are the matrices in Probs. 5-10 Hermitian, skew-Hermitian or unitary? Find their eigenvalues (thereby verifying Theorem 1) and eigenvectors.

5. $\begin{bmatrix} 2 & 3 + 4i \\ 3 - 4i & 2 \end{bmatrix}$ **6.** $\begin{bmatrix} i & 1 + i \\ -1 + i & 0 \end{bmatrix}$ **7.** $\begin{bmatrix} \frac{1}{2} & i\sqrt{\frac{3}{4}} \\ i\sqrt{\frac{3}{4}} & \frac{1}{2} \end{bmatrix}$

8. $\begin{bmatrix} 4i & 0 & i \\ 0 & i & 0 \\ i & 0 & 4i \end{bmatrix}$ **9.** $\begin{bmatrix} i & 0 & 0 \\ 0 & 0 & i \\ 0 & i & 0 \end{bmatrix}$ **10.** $\begin{bmatrix} 0 & 1+i & 0 \\ 1-i & 0 & 1+i \\ 0 & 1-i & 0 \end{bmatrix}$

11. (Decomposition) Show that any square matrix may be written as the sum of a Hermitian matrix and a skew-Hermitian matrix. Give examples.

12. PROJECT. Unitary Matrices.

    (a) Products, Inverse. Prove that the product of two unitary matrices (of the same size) and the inverse of a unitary matrix are unitary. Give examples.

    (b) Powers of unitary matrices occurring in applications may sometimes be familiar real matrices. Show that $\mathbf{A}^3 = -\mathbf{I}$ for $\mathbf{A}$ in Prob. 7.

    (c) Normal matrix. This important concept denotes a matrix that commutes with its conjugate transpose, $\mathbf{A}\overline{\mathbf{A}}^\mathsf{T} = \overline{\mathbf{A}}^\mathsf{T}\mathbf{A}$. Prove that Hermitian, skew-Hermitian, and unitary matrices are normal. Illustrate this with examples of your own.

    (d) Normality Criterion. Prove that $\mathbf{A}$ is normal if and only if the Hermitian and skew-Hermitian matrices in Prob. 11 commute.

    (e) Find a simple matrix that is not normal. Find a normal matrix that is not Hermitian, skew-Hermitian, or unitary.

Quadratic Forms

Find the symmetric coefficient matrix $\mathbf{C}$ of the quadratic form $Q = \mathbf{x}^\mathsf{T}\mathbf{C}\mathbf{x}$ given by

13. $4x_1^2 - 8x_1x_2 + 5x_2^2$ **14.** $-2x_1^2 + 2x_1x_3 + 4x_2x_3 - 9x_3^2$

15. $0.5x_1^2 + 0.8x_1x_2 - 1.4x_2x_3$ **16.** $(x_1 - 2x_2 + 3x_3 - x_4)^2$

17. $(x_1 - x_2)^2 - 4x_3^2$ **18.** $(x_1 - x_2 + 4x_3)^2 - 4(x_2 - x_4)^2$

19. (Definiteness) A real quadratic form $Q = \mathbf{x}^\mathsf{T}\mathbf{C}\mathbf{x}$ and its symmetric matrix $\mathbf{C} = [c_{jk}]$ are said to be **positive definite** if $Q > 0$ for all $[x_1 \cdots x_n] \neq [0 \cdots 0]$. A necessary and sufficient condition for positive definiteness is that all the determinants

$$C_1 = c_{11}, \quad C_2 = \begin{vmatrix} c_{11} & c_{12} \\ c_{21} & c_{22} \end{vmatrix}, \quad C_3 = \begin{vmatrix} c_{11} & c_{12} & c_{13} \\ c_{21} & c_{22} & c_{23} \\ c_{31} & c_{32} & c_{33} \end{vmatrix}, \cdots, C_n = \det \mathbf{C}$$

are positive (see Ref. [B2], vol. 1, p. 306). Show that the form in Prob. 13 is positive definite, whereas that in Example 3 is not positive definite.

Hermitian and Skew-Hermitian Forms

Is $\mathbf{A}$ Hermitian or skew-Hermitian? Find $\overline{\mathbf{x}}^\mathsf{T}\mathbf{A}\mathbf{x}$.

20. $\mathbf{A} = \begin{bmatrix} i & 1+i \\ -1+i & -2i \end{bmatrix}$, $\mathbf{x} = \begin{bmatrix} 2 \\ i \end{bmatrix}$ **21.** $\mathbf{A} = \begin{bmatrix} 4 & 3-2i \\ 3+2i & -4 \end{bmatrix}$, $\mathbf{x} = \begin{bmatrix} -2i \\ 1+i \end{bmatrix}$

22. $\mathbf{A} = \begin{bmatrix} a & b+ic \\ b-ic & k \end{bmatrix}$, $\mathbf{x} = \begin{bmatrix} x_1 \\ x_2 \end{bmatrix}$ **23.** $\mathbf{A} = \begin{bmatrix} i & -2+3i \\ 2+3i & 0 \end{bmatrix}$, $\mathbf{x} = \begin{bmatrix} i \\ 4 \end{bmatrix}$

    (a, b, c, k real)

24. $\mathbf{A} = \begin{bmatrix} 1 & i & 4 \\ -i & 3 & 0 \\ 4 & 0 & 2 \end{bmatrix}$, $\mathbf{x} = \begin{bmatrix} 1 \\ i \\ -i \end{bmatrix}$ **25.** $\mathbf{A} = \begin{bmatrix} -i & 1 & 2+i \\ -1 & 0 & 3i \\ -2+i & 3i & i \end{bmatrix}$, $\mathbf{x} = \begin{bmatrix} 0 \\ 1 \\ 2 \end{bmatrix}$

7.5 Similarity of Matrices
Basis of Eigenvectors
Diagonalization

So far we have emphasized properties of eigen*values.* We now turn to eigen*vectors* and their properties. Eigenvectors of an $n \times n$ matrix $\mathbf{A}$ may (or may not!) form a basis for R^n or C^n (defined in Sec. 7.4). If they do, we can use them for *"diagonalizing"* $\mathbf{A},$ that is, for transforming it into diagonal form with the eigenvalues on the main diagonal. These are the key issues in this section.

We begin with a concept of central interest in eigenvalue problems:

Similarity of Matrices

An $n \times n$ matrix $\hat{\mathbf{A}}$ is called **similar** to an $n \times n$ matrix $\mathbf{A}$ if

(1)
$$\hat{\mathbf{A}} = \mathbf{P}^{-1}\mathbf{A}\mathbf{P}$$

for some (nonsingular!) $n \times n$ matrix $\mathbf{P}.$ This transformation, which gives $\hat{\mathbf{A}}$ from $\mathbf{A},$ is called a **similarity transformation.**

Similarity transformations are important since they preserve eigenvalues:

THEOREM 1   **(Eigenvalues and eigenvectors of similar matrices)**

If $\hat{\mathbf{A}}$ is similar to $\mathbf{A},$ then $\hat{\mathbf{A}}$ has the same eigenvalues as $\mathbf{A}.$

Furthermore, if $\mathbf{x}$ is an eigenvector of $\mathbf{A},$ then $\mathbf{y} = \mathbf{P}^{-1}\mathbf{x}$ is an eigenvector of $\hat{\mathbf{A}}$ corresponding to the same eigenvalue.

PROOF.   From $\mathbf{A}\mathbf{x} = \lambda\mathbf{x}$ (λ an eigenvalue, $\mathbf{x} \neq \mathbf{0}$) we get $\mathbf{P}^{-1}\mathbf{A}\mathbf{x} = \lambda\mathbf{P}^{-1}\mathbf{x}.$ Now $\mathbf{I} = \mathbf{P}\mathbf{P}^{-1}.$ By this *"identity trick"* the previous equation gives

$$\mathbf{P}^{-1}\mathbf{A}\mathbf{x} = \mathbf{P}^{-1}\mathbf{A}\mathbf{I}\mathbf{x} = \mathbf{P}^{-1}\mathbf{A}\mathbf{P}\mathbf{P}^{-1}\mathbf{x} = \hat{\mathbf{A}}(\mathbf{P}^{-1}\mathbf{x}) = \lambda\mathbf{P}^{-1}\mathbf{x}.$$

Hence λ is an eigenvalue of $\hat{\mathbf{A}}$ and $\mathbf{P}^{-1}\mathbf{x}$ a corresponding eigenvector. Indeed, $\mathbf{P}^{-1}\mathbf{x} = \mathbf{0}$ would give $\mathbf{x} = \mathbf{I}\mathbf{x} = \mathbf{P}\mathbf{P}^{-1}\mathbf{x} = \mathbf{P}\mathbf{0} = \mathbf{0},$ contradicting $\mathbf{x} \neq \mathbf{0}.$ ◀

Properties of Eigenvectors

The next theorem is of interest in itself and of help in connection with bases of eigenvectors.

THEOREM 2   **(Linear independence of eigenvectors)**

*Let $\lambda_1, \lambda_2, \cdots, \lambda_k$ be **distinct** eigenvalues of an $n \times n$ matrix. Then corresponding eigenvectors $\mathbf{x}_1, \mathbf{x}_2, \cdots, \mathbf{x}_k$ form a linearly independent set.*

PROOF.   Suppose that the conclusion is false. Let r be the largest integer such that $\{\mathbf{x}_1, \cdots, \mathbf{x}_r\}$ is a linearly independent set. Then $r < k$ and the set $\{\mathbf{x}_1, \cdots, \mathbf{x}_{r+1}\}$ is linearly dependent. Thus there are scalars $c_1, \cdots, c_{r+1},$ not all zero, such that

(2) $$c_1\mathbf{x}_1 + \cdots + c_{r+1}\mathbf{x}_{r+1} = \mathbf{0}$$

(see Sec. 6.4). Multiplying both sides by $\mathbf{A}$ and using $\mathbf{A}\mathbf{x}_j = \lambda_j\mathbf{x}_j$, we obtain

(3) $$c_1\lambda_1\mathbf{x}_1 + \cdots + c_{r+1}\lambda_{r+1}\mathbf{x}_{r+1} = \mathbf{0}.$$

To get rid of the last term, we subtract λ_{r+1} times (2) from this, obtaining

$$c_1(\lambda_1 - \lambda_{r+1})\mathbf{x}_1 + \cdots + c_r(\lambda_r - \lambda_{r+1})\mathbf{x}_r = \mathbf{0}.$$

Here $c_1(\lambda_1 - \lambda_{r+1}) = 0, \cdots, c_r(\lambda_r - \lambda_{r+1}) = 0$ since $\{x_1, \cdots, x_r\}$ is linearly independent. Hence $c_1 = \cdots = c_r = 0$, since all the eigenvalues are distinct. But with this, (2) reduces to $c_{r+1}\mathbf{x}_{r+1} = \mathbf{0}$, hence $c_{r+1} = 0$, since $\mathbf{x}_{r+1} \neq \mathbf{0}$ (an eigenvector!). This contradicts the fact that not all scalars in (2) are zero. Hence the conclusion of the theorem must hold.   ◄

This theorem immediately implies the following.

THEOREM 3   **(Basis of eigenvectors)**

*If an $n \times n$ matrix $\mathbf{A}$ has n **distinct** eigenvalues, then $\mathbf{A}$ has a basis of eigenvectors for C^n (or R^n).*

EXAMPLE 1   **Basis of eigenvectors**

The matrix

$$\mathbf{A} = \begin{bmatrix} 5 & 3 \\ 3 & 5 \end{bmatrix} \quad \text{has a basis of eigenvectors} \quad \begin{bmatrix} 1 \\ 1 \end{bmatrix}, \quad \begin{bmatrix} 1 \\ -1 \end{bmatrix}$$

corresponding to the eigenvalues $\lambda_1 = 8$, $\lambda_2 = 2$. (See Example 1 in Sec. 7.2.)   ◄

EXAMPLE 2   **Basis when not all eigenvalues are distinct. Nonexistence of basis**

Even if not all n eigenvalues are different, a matrix $\mathbf{A}$ may still provide a basis of eigenvectors for C^n or R^n. This is illustrated by Example 2 in Sec. 7.1, where $n = 3$. On the other hand, $\mathbf{A}$ may not have enough linearly independent eigenvectors to make up a basis. For instance, the matrix in Example 3, Sec. 7.1,

$$\mathbf{A} = \begin{bmatrix} 0 & 1 \\ 0 & 0 \end{bmatrix} \quad \text{has only one eigenvector} \quad \begin{bmatrix} k \\ 0 \end{bmatrix},$$

where k is arbitrary, not zero. Hence $\mathbf{A}$ does not provide a basis of eigenvectors for R^2.   ◄

Actually, bases of eigenvectors exist under much more general conditions than those given in Theorem 3, and for the matrices in the previous section we can even choose a unitary system of eigenvectors, as follows.

THEOREM 4   **(Basis of eigenvectors)**

A Hermitian, skew-Hermitian, or unitary matrix has a basis of eigenvectors for C^n that is a unitary system (see Sec. 7.4). A symmetric matrix has an orthonormal basis of eigenvectors for R^n. (Proof: see Ref. [B2], vol. 1, pp. 270–272.)

EXAMPLE 3   **Orthonormal basis of eigenvectors**

The matrix in Example 1 is symmetric, and an orthonormal basis of eigenvectors is $[1/\sqrt{2} \quad 1/\sqrt{2}]^{\mathsf{T}}$, $[1/\sqrt{2} \quad -1/\sqrt{2}]^{\mathsf{T}}$.   ◄

A basis of eigenvectors of a matrix $\mathbf{A}$ is of great advantage if we are interested in a transformation $\mathbf{y} = \mathbf{A}\mathbf{x}$ because then we can represent any $\mathbf{x}$ uniquely as

$$\mathbf{x} = c_1\mathbf{x}_1 + c_2\mathbf{x}_2 + \cdots + c_n\mathbf{x}_n$$

in terms of such a basis $\mathbf{x}_1, \cdots, \mathbf{x}_n$; if these eigenvectors of $\mathbf{A}$ correspond to (not necessarily distinct) eigenvalues $\lambda_1, \cdots, \lambda_n$ of $\mathbf{A}$, then we get

$$\mathbf{y} = \mathbf{A}\mathbf{x} = \mathbf{A}(c_1\mathbf{x}_1 + \cdots + c_n\mathbf{x}_n)$$

(4)
$$= c_1\mathbf{A}\mathbf{x}_1 + \cdots + c_n\mathbf{A}\mathbf{x}_n$$

$$= c_1\lambda_1\mathbf{x}_1 + \cdots + c_n\lambda_n\mathbf{x}_n.$$

This shows the advantage: we have decomposed the complicated action of $\mathbf{A}$ on arbitrary vectors $\mathbf{x}$ into a sum of simple actions (multiplication by scalars) on the eigenvectors of $\mathbf{A}$.

Diagonalization

Bases of eigenvectors also play a central role in the diagonalization of an $n \times n$ matrix $\mathbf{A}$, as the following theorem explains.

THEOREM 5   **(Diagonalization of a matrix)**

If an $n \times n$ matrix $\mathbf{A}$ has a basis of eigenvectors, then

(5)
$$\boxed{\mathbf{D} = \mathbf{X}^{-1}\mathbf{A}\mathbf{X}}$$

is diagonal, with the eigenvalues of $\mathbf{A}$ as the entries on the main diagonal. Here $\mathbf{X}$ is the matrix with these eigenvectors as column vectors. Also,

(5*)
$$\mathbf{D}^m = \mathbf{X}^{-1}\mathbf{A}^m\mathbf{X} \qquad\qquad (m = 2, 3, \cdots).$$

EXAMPLE 4   **Diagonalization of a matrix. Illustration of Theorem 5**

Calculation as in the examples in Sec. 7.1, etc. shows that the matrix

$$\mathbf{A} = \begin{bmatrix} 5 & 4 \\ 1 & 2 \end{bmatrix} \quad \text{has eigenvectors} \quad \begin{bmatrix} 4 \\ 1 \end{bmatrix} \quad \text{and} \quad \begin{bmatrix} 1 \\ -1 \end{bmatrix}. \quad \text{Hence } \mathbf{X} = \begin{bmatrix} 4 & 1 \\ 1 & -1 \end{bmatrix}$$

and, using (4*) in Sec. 6.7, we obtain

$$\mathbf{X}^{-1}\mathbf{A}\mathbf{X} = \frac{1}{-5}\begin{bmatrix} -1 & -1 \\ -1 & 4 \end{bmatrix}\begin{bmatrix} 5 & 4 \\ 1 & 2 \end{bmatrix}\begin{bmatrix} 4 & 1 \\ 1 & -1 \end{bmatrix} = \begin{bmatrix} 0.2 & 0.2 \\ 0.2 & -0.8 \end{bmatrix}\begin{bmatrix} 24 & 1 \\ 6 & -1 \end{bmatrix} = \begin{bmatrix} 6 & 0 \\ 0 & 1 \end{bmatrix}.$$

What happens to $\mathbf{D}$ if you interchange the columns of $\mathbf{X}$?   ◀

PROOF OF THEOREM 5.   Let $\mathbf{x}_1, \cdots, \mathbf{x}_n$ form a basis of eigenvectors of $\mathbf{A}$ for C^n (or R^n) corresponding to the eigenvalues $\lambda_1, \cdots, \lambda_n$, respectively, of $\mathbf{A}$. Then $\mathbf{X} = [\mathbf{x}_1 \cdots \mathbf{x}_n]$ has rank n, by Theorem 1 in Sec. 6.4. Hence $\mathbf{X}^{-1}$ exists by Theorem 1 in Sec. 6.7. We now claim that

(6) $$\mathbf{A}\mathbf{X} = \mathbf{A}[\mathbf{x}_1 \quad \cdots \quad \mathbf{x}_n] = [\mathbf{A}\mathbf{x}_1 \quad \cdots \quad \mathbf{A}\mathbf{x}_n] = [\lambda_1\mathbf{x}_1 \quad \cdots \quad \lambda_n\mathbf{x}_n].$$

Indeed, the last equality follows from $\mathbf{A}\mathbf{x}_j = \lambda_j\mathbf{x}_j$. This is simple. The second equality is perhaps best understood if we write $\mathbf{A}$ in terms of row vectors $\mathbf{a}_1, \cdots, \mathbf{a}_n$ and note that

$$
\mathbf{A}\mathbf{x}_1 = \begin{bmatrix} a_{11} & \cdots & a_{1n} \\ a_{21} & \cdots & a_{2n} \\ \cdot & \cdots & \cdot \\ \cdot & \cdots & \cdot \\ a_{n1} & \cdots & a_{nn} \end{bmatrix} \mathbf{x}_1 = \begin{bmatrix} \mathbf{a}_1 \bullet \mathbf{x}_1 \\ \mathbf{a}_2 \bullet \mathbf{x}_1 \\ \cdot \\ \cdot \\ \mathbf{a}_n \bullet \mathbf{x}_1 \end{bmatrix}
$$

is the first column of $\mathbf{AX}$, as we can see by comparing (8) in Sec. 6.2 with $\mathbf{B} = \mathbf{X}$. Similarly for the other $n - 1$ columns of $\mathbf{AX}$. Now direct calculation shows that $[\lambda_1\mathbf{x}_1 \ \cdots \ \lambda_n\mathbf{x}_n] = \mathbf{XD}$ with $\mathbf{D}$ as defined in (5). (Try this out for $n = 2$ and then for general n.) From this and (6) we have $\mathbf{AX} = \mathbf{XD}$. Multiplication by $\mathbf{X}^{-1}$ from the left gives $\mathbf{X}^{-1}\mathbf{AX} = \mathbf{X}^{-1}\mathbf{XD} = \mathbf{D}$, which is (5). Equation (5*) now follows by noting that

$$
\mathbf{D}^2 = \mathbf{DD} = \mathbf{X}^{-1}\mathbf{AXX}^{-1}\mathbf{AX} = \mathbf{X}^{-1}\mathbf{AAX} = \mathbf{X}^{-1}\mathbf{A}^2\mathbf{X}, \qquad \text{etc.} \qquad \blacktriangleleft
$$

EXAMPLE 5   **Diagonalization**

Diagonalize

$$
\mathbf{A} = \begin{bmatrix} 7.3 & 0.2 & -3.7 \\ -11.5 & 1.0 & 5.5 \\ 17.7 & 1.8 & -9.3 \end{bmatrix}.
$$

Solution.   The characteristic determinant gives the characteristic equation $-\lambda^3 - \lambda^2 + 12\lambda = 0$. The roots (eigenvalues of $\mathbf{A}$) are $\lambda_1 = 3$, $\lambda_2 = -4$, $\lambda_3 = 0$. By the Gauss elimination applied to $(\mathbf{A} - \lambda\mathbf{I})\mathbf{x} = \mathbf{0}$ with $\lambda = \lambda_1, \lambda_2, \lambda_3$ we find eigenvectors and then $\mathbf{X}^{-1}$ by the Gauss–Jordan elimination (Sec. 6.7, Example 1). The results are

$$
\begin{bmatrix} -1 \\ 3 \\ -1 \end{bmatrix}, \quad \begin{bmatrix} 1 \\ -1 \\ 3 \end{bmatrix}, \quad \begin{bmatrix} 2 \\ 1 \\ 4 \end{bmatrix}, \quad \mathbf{X} = \begin{bmatrix} -1 & 1 & 2 \\ 3 & -1 & 1 \\ -1 & 3 & 4 \end{bmatrix}, \quad \mathbf{X}^{-1} = \begin{bmatrix} -0.7 & 0.2 & 0.3 \\ -1.3 & -0.2 & 0.7 \\ 0.8 & 0.2 & -0.2 \end{bmatrix}.
$$

Calculating $\mathbf{AX}$ and multiplying by $\mathbf{X}^{-1}$ from the left, we thus obtain

$$
\mathbf{D} = \mathbf{X}^{-1}\mathbf{AX} = \begin{bmatrix} -0.7 & 0.2 & 0.3 \\ -1.3 & -0.2 & 0.7 \\ 0.8 & 0.2 & -0.2 \end{bmatrix} \begin{bmatrix} -3 & -4 & 0 \\ 9 & 4 & 0 \\ -3 & -12 & 0 \end{bmatrix} = \begin{bmatrix} 3 & 0 & 0 \\ 0 & -4 & 0 \\ 0 & 0 & 0 \end{bmatrix}. \qquad \blacktriangleleft
$$

Transformation of Forms to Principal Axes

This is an important practical task related to the diagonalization of matrices. We explain the idea for quadratic forms (see Sec. 7.4)

$$
(7) \qquad \boxed{Q = \mathbf{x}^{\mathsf{T}}\mathbf{A}\mathbf{x}.}
$$

Without restriction we can assume that $\mathbf{A}$ is real *symmetric* (see Sec. 7.4). Then $\mathbf{A}$ has an *orthonormal basis* of n eigenvectors, by Theorem 4. Hence the matrix $\mathbf{X}$ with these vectors as column vectors is orthogonal, so that $\mathbf{X}^{-1} = \mathbf{X}^{\mathsf{T}}$. From (5) we thus have $\mathbf{A} = \mathbf{XDX}^{-1} = \mathbf{XDX}^{\mathsf{T}}$. Substitution into (7) gives

(8) $Q = \mathbf{x}^T \mathbf{X} \mathbf{D} \mathbf{X}^T \mathbf{x}.$

If we set $\mathbf{X}^T \mathbf{x} = \mathbf{y}$, then, since $\mathbf{X}^T = \mathbf{X}^{-1}$, we get

(9) $\mathbf{x} = \mathbf{X} \mathbf{y}.$

Furthermore, in (8) we have $\mathbf{x}^T \mathbf{X} = (\mathbf{X}^T \mathbf{x})^T = \mathbf{y}^T$ and $\mathbf{X}^T \mathbf{x} = \mathbf{y}$, so that Q becomes simply

(10) $Q = \mathbf{y}^T \mathbf{D} \mathbf{y} = \lambda_1 y_1^2 + \lambda_2 y_2^2 + \cdots + \lambda_n y_n^2.$

This proves

THEOREM 6   **(Principal axes theorem)**

The substitution (9) *transforms a quadratic form*

$$Q = \mathbf{x}^T \mathbf{A} \mathbf{x} = \sum_{j=1}^{n} \sum_{k=1}^{n} a_{jk} x_j x_k$$

to the principal axes form (10), *where* $\lambda_1, \cdots, \lambda_n$ *are the* (*not necessarily distinct*) *eigenvalues of the* (*symmetric!*) *matrix* **A,** *and* **X** *is an orthogonal matrix with corresponding eigenvectors* $\mathbf{x}_1, \cdots, \mathbf{x}_n$, *respectively, as column vectors.*

EXAMPLE 6   **Transformation to principal axes. Conic sections**

Find out what type of conic section the following quadratic form represents and transform it to principal axes:

$$Q = 17 x_1^2 - 30 x_1 x_2 + 17 x_2^2 = 128.$$

Solution.   We have $Q = \mathbf{x}^T \mathbf{A} \mathbf{x}$, where

$$\mathbf{A} = \begin{bmatrix} 17 & -15 \\ -15 & 17 \end{bmatrix}, \qquad \mathbf{x} = \begin{bmatrix} x_1 \\ x_2 \end{bmatrix}.$$

This gives the characteristic equation $(17 - \lambda)^2 - 15^2 = 0$. It has the roots $\lambda_1 = 2$, $\lambda_2 = 32$. Hence (10) becomes

$$Q = 2 y_1^2 + 32 y_2^2.$$

We see that $Q = 128$ represents the ellipse $2 y_1^2 + 32 y_2^2 = 128$, that is,

$$\frac{y_1^2}{8^2} + \frac{y_2^2}{2^2} = 1.$$

If we want to know the direction of the principal axes in the $x_1 x_2$-coordinates, we have to determine normalized eigenvectors from $(\mathbf{A} - \lambda \mathbf{I}) \mathbf{x} = \mathbf{0}$ with $\lambda = \lambda_1 = 2$ and $\lambda = \lambda_2 = 32$ and then use (9). We get

$$\begin{bmatrix} 1/\sqrt{2} \\ 1/\sqrt{2} \end{bmatrix} \quad \text{and} \quad \begin{bmatrix} -1/\sqrt{2} \\ 1/\sqrt{2} \end{bmatrix},$$

hence

$$\mathbf{x} = \mathbf{X} \mathbf{y} = \begin{bmatrix} 1/\sqrt{2} & -1/\sqrt{2} \\ 1/\sqrt{2} & 1/\sqrt{2} \end{bmatrix} \begin{bmatrix} y_1 \\ y_2 \end{bmatrix}, \qquad \begin{aligned} x_1 &= y_1/\sqrt{2} - y_2/\sqrt{2} \\ x_2 &= y_1/\sqrt{2} + y_2/\sqrt{2}. \end{aligned}$$

This is a 45° rotation. Our results agree with those in Sec. 7.2, Example 1, except for the notations. See also Fig. 145 in that example.   ◀

PROBLEM SET 7.5

Similar Matrices Have Equal Spectra

Verify this for **A** and $\hat{\mathbf{A}} = \mathbf{P}^{-1}\mathbf{AP}$. Find eigenvectors **y** of $\hat{\mathbf{A}}$. Show that $\mathbf{x} = \mathbf{Py}$ are eigenvectors of **A**. (Show the details of your work.)

1. $\mathbf{A} = \begin{bmatrix} 1 & 2 \\ 2 & 4 \end{bmatrix}$, $\mathbf{P} = \begin{bmatrix} 1 & 3 \\ 3 & 6 \end{bmatrix}$ **2.** $\mathbf{A} = \begin{bmatrix} 8 & -4 \\ 2 & 2 \end{bmatrix}$, $\mathbf{P} = \begin{bmatrix} 0.28 & 0.96 \\ -0.96 & 0.28 \end{bmatrix}$

3. $\mathbf{A} = \begin{bmatrix} 3 & 4 \\ 4 & -3 \end{bmatrix}$, $\mathbf{P} = \begin{bmatrix} -4 & 2 \\ 3 & -1 \end{bmatrix}$ **4.** $\mathbf{A} = \begin{bmatrix} 1 & 0 \\ 2 & -1 \end{bmatrix}$, $\mathbf{P} = \begin{bmatrix} 7 & -5 \\ 10 & -7 \end{bmatrix}$

5. $\mathbf{A} = \begin{bmatrix} 10 & -3 & 5 \\ 0 & 1 & 0 \\ -15 & 9 & -10 \end{bmatrix}$, $\mathbf{P} = \begin{bmatrix} 2 & 0 & 3 \\ 0 & 1 & 0 \\ 3 & 0 & 5 \end{bmatrix}$

6. $\mathbf{A} = \begin{bmatrix} 7 & 0 & 3 \\ 2 & 1 & 1 \\ 2 & 0 & 2 \end{bmatrix}$, $\mathbf{P} = \begin{bmatrix} 0 & 1 & 0 \\ 1 & 0 & 0 \\ 0 & 0 & 1 \end{bmatrix}$

Unitary Eigenbasis

Find a basis of eigenvectors that form a unitary system. (Show the details.)

7. $\begin{bmatrix} 0 & 3i \\ -3i & 0 \end{bmatrix}$ **8.** $\begin{bmatrix} 4 & 1+i \\ 1-i & 4 \end{bmatrix}$ **9.** $\begin{bmatrix} i & 1 \\ -1 & i \end{bmatrix}$

Diagonalization

Find a basis of eigenvectors and diagonalize. (Show the details.)

10. $\begin{bmatrix} -19 & 7 \\ -42 & 16 \end{bmatrix}$ **11.** $\begin{bmatrix} 2 & 1 \\ 2 & 1 \end{bmatrix}$ **12.** $\begin{bmatrix} -43 & 77 \\ 13 & 93 \end{bmatrix}$

13. $\begin{bmatrix} 16 & 0 & 0 \\ 48 & -8 & 0 \\ 84 & -24 & 4 \end{bmatrix}$ **14.** $\begin{bmatrix} -2.5 & -3 & 3 \\ -4.5 & -4 & 6 \\ -6 & -6 & 8 \end{bmatrix}$, $\lambda_1 = 2$ **15.** $\begin{bmatrix} 5 & 10 & -10 \\ 10 & 5 & -20 \\ 5 & -5 & -10 \end{bmatrix}$

Transformation of Quadratic Forms to Principal Axes. Conic Sections

Find out what type of conic section (or pair of straight lines) is represented by the given quadratic form. Transform it to principal axes. Express $\mathbf{x}^\mathsf{T} = [x_1 \ \ x_2]$ in terms of the new coordinate vector $\mathbf{y}^\mathsf{T} = [y_1 \ \ y_2]$, as in Example 6.

16. $-11x_1{}^2 + 84x_1x_2 + 24x_2{}^2 = 156$ **17.** $7x_1{}^2 + 6x_1x_2 + 7x_2{}^2 = 200$

18. $41x_1{}^2 - 24x_1x_2 + 34x_2{}^2 = 0$ **19.** $9x_1{}^2 - 6x_1x_2 + x_2{}^2 = 40$

20. $4x_1{}^2 + 12x_1x_2 + 13x_2{}^2 = 16$ **21.** $32x_1{}^2 - 60x_1x_2 + 7x_2{}^2 = -52$

22. **PROJECT. Similarity of matrices** is basic, for instance, in designing numerical methods.
    (a) Trace. Let $\mathbf{A} = [a_{jk}]$ have the eigenvalues $\lambda_1, \cdots, \lambda_n$. By definition, its trace is

$$\text{trace } \mathbf{A} = a_{11} + a_{22} + \cdots + a_{nn}.$$

Show that

$$\text{trace } \mathbf{A} = \lambda_1 + \lambda_2 + \cdots + \lambda_n,$$

each eigenvalue counted as often as its algebraic multiplicity indicates.

(b) Trace of product. Let $\mathbf{B} = [b_{jk}]$ be $n \times n$. Show that similar matrices have equal traces, by first proving

$$\text{trace } \mathbf{AB} = \sum_{i=1}^{n} \sum_{l=1}^{n} a_{il} b_{li} = \text{trace } \mathbf{BA}.$$

(c) Find a relationship between $\hat{\mathbf{A}}$ in (1) and $\widetilde{\mathbf{A}} = \mathbf{PAP}^{-1}$.

(d) Diagonalization. What can you do in (5) if you want to change the order of the eigenvalues in $\mathbf{D}$, for instance, interchange $d_{11} = \lambda_1$ and $d_{22} = \lambda_2$?

CHAPTER 7 REVIEW

1. State the definitions of "eigenvalue" and "eigenvector" from memory.
2. Do there exist square matrices without eigenvalues? Can 0 be an eigenvalue? Can a real matrix have complex eigenvalues?
3. Can a complex matrix have real eigenvalues? Real eigenvectors?
4. What is the algebraic multiplicity and the geometric multiplicity of an eigenvalue? Why are these concepts important?
5. What is a basis of eigenvectors? When does it exist?
6. We have discussed three classes of real matrices and three classes of complex matrices. State their definitions and some of the major properties (from memory!).
7. In which cases can we expect orthogonal eigenvectors?
8. What is the spectral mapping theorem? Give examples illustrating it.
9. Give a few typical applications in which eigenvalue problems occur.
10. What do you know about the eigenvalues of the inverse of a matrix?

Eigenvalues and Eigenvectors. Find the spectrum and eigenvectors. (Show the details of your work.)

11. $\begin{bmatrix} 2.5 & 0.5 \\ 0.5 & 2.5 \end{bmatrix}$ 12. $\begin{bmatrix} -28 & 16 \\ -48 & 28 \end{bmatrix}$ 13. $\begin{bmatrix} 0.35 & 0.30 \\ 0.30 & -0.10 \end{bmatrix}$

14. $\begin{bmatrix} 1 & -1 & 0 \\ 0 & 0 & 0 \\ 0 & -3 & 3 \end{bmatrix}$ 15. $\begin{bmatrix} 4 & -6 & -6 \\ 0 & -2 & 0 \\ 1 & -1 & -1 \end{bmatrix}$ 16. $\begin{bmatrix} 9 & -10 & 2 \\ -6 & 5 & 2 \\ 6 & -2 & 1 \end{bmatrix}$

Diagonalization. Find a basis of eigenvectors. Diagonalize. (Show the details.)

17. $\begin{bmatrix} 8 & -1 \\ 5 & 2 \end{bmatrix}$ 18. $\begin{bmatrix} 15 & 6 & -12 \\ 4 & 10 & -2 \\ -4 & 8 & -7 \end{bmatrix}$ 19. $\begin{bmatrix} -8 & 11 & 3 \\ 4 & -1 & 3 \\ -4 & 10 & 6 \end{bmatrix}$

Conic Sections. Reduction to Principal Axes. Reduce to principal axes. Express $[x_1 \quad x_2]^T$ in terms of the new variables. (Show the details.)

20. $2x_1{}^2 + 12x_1x_2 - 7x_2{}^2 = 10$ **21.** $9x_1{}^2 - 6x_1x_2 + 17x_2{}^2 = 72$

22. $5x_1{}^2 + 24x_1x_2 - 5x_2{}^2 = 0$ **23.** $7.4x_1{}^2 + 6.4x_1x_2 + 2.6x_2{}^2 = 9$

24. (Pauli spin matrices) Find the eigenvalues and eigenvectors of the so-called *Pauli spin matrices* and show that $\mathbf{S}_x\mathbf{S}_y = i\mathbf{S}_z$, $\mathbf{S}_y\mathbf{S}_x = -i\mathbf{S}_z$, $\mathbf{S}_x{}^2 = \mathbf{S}_y{}^2 = \mathbf{S}_z{}^2 = \mathbf{I}$, where

$$\mathbf{S}_x = \begin{bmatrix} 0 & 1 \\ 1 & 0 \end{bmatrix}, \qquad \mathbf{S}_y = \begin{bmatrix} 0 & -i \\ i & 0 \end{bmatrix}, \qquad \mathbf{S}_z = \begin{bmatrix} 1 & 0 \\ 0 & -1 \end{bmatrix}.$$

SUMMARY OF CHAPTER 7
LINEAR ALGEBRA: MATRIX EIGENVALUE PROBLEMS

The practical importance of these problems cannot be overrated. They are defined by the equation

(1) $$\mathbf{A}\mathbf{x} = \lambda\mathbf{x}.$$

$\mathbf{A}$ is a given square matrix. All matrices in this chapter are *square*. λ is a scalar. To *solve* the problem (1) means to determine values of λ, called **eigenvalues** (or **characteristic values**) of $\mathbf{A}$, such that (1) has a nontrivial solution $\mathbf{x}$, called an **eigenvector** of $\mathbf{A}$ corresponding to that λ. An $n \times n$ matrix has at least one and at most n numerically different eigenvalues. These are the solutions of the **characteristic equation** (Sec. 7.1)

(2) $$D(\lambda) = \det (\mathbf{A} - \lambda\mathbf{I}) = \begin{vmatrix} a_{11} - \lambda & a_{12} & \cdots & a_{1n} \\ a_{21} & a_{22} - \lambda & \cdots & a_{2n} \\ . & . & \cdots & . \\ a_{n1} & a_{n2} & \cdots & a_{nn} - \lambda \end{vmatrix} = 0.$$

$D(\lambda)$ is called the **characteristic determinant** of $\mathbf{A}$. By expanding it we get the **characteristic polynomial** of $\mathbf{A}$, which is of degree n in λ. Some typical applications are shown in Sec. 7.2.

Section 7.3 is concerned with eigenvalue problems for **symmetric** ($\mathbf{A}^T = \mathbf{A}$), **skew-symmetric** ($\mathbf{A}^T = -\mathbf{A}$), and **orthogonal matrices** ($\mathbf{A}^T = \mathbf{A}^{-1}$). Section 7.4 presents eigenvalue problems for the complex analogs of these matrices, called **Hermitian** ($\overline{\mathbf{A}}^T = \mathbf{A}$), **skew-Hermitian** ($\overline{\mathbf{A}}^T = -\mathbf{A}$), and **unitary matrices** ($\overline{\mathbf{A}}^T = \mathbf{A}^{-1}$). All the eigenvalues of a Hermitian matrix (and a symmetric one) are real. For a skew-Hermitian (and a skew-symmetric) matrix they are pure imaginary or zero. For a unitary (and an orthogonal) matrix they have absolute value 1.

The diagonalization of matrices and the transformation of quadratic forms to principal axes are related to eigenvalues, as explained in Sec. 7.5.

Vector Differential Calculus. Grad, Div, Curl

This chapter deals with vectors and vector functions in 3-space and extends the differential calculus to these vector functions. Forces, velocities and various other quantities are vectors. This makes the algebra and calculus of these vector functions the natural instrument for the engineer and physicist in solid mechanics, fluid flow, heat flow, electrostatics, and so on. The engineer must understand these fields as the basis of the design and construction of systems, such as airplanes, laser generators, thermodynamical systems, or robots. In three dimensions (as opposed to higher dimensions), geometrical ideas become influential, enriching the theory, and many geometrical quantities (tangents and normals, for example) can be given by vectors.

As a preparation, in Secs. 8.1−8.3 we explain the basic *algebraic* operations with vectors in 3-space. Vector differential calculus begins in Sec. 8.4 with a discussion of vector functions, which represent vector fields and have various physical and geometrical applications. Then the basic concepts of differential calculus are extended to vector functions in a simple and natural fashion. In Secs. 8.5−8.7 we shall see that vector functions are useful in studying curves and their applications as paths of moving bodies in mechanics.

We finally discuss three physically and geometrically important concepts related to scalar and vector fields, namely, the gradient (Sec. 8.9), divergence (Sec. 8.10), and curl (Sec. 8.11). (Integral theorems involving these concepts follow in Chap. 9 on **vector integral calculus.** The form of these quantities in curvilinear coordinates is given in Appendix A3.4.)

We shall keep this chapter independent[1] of Chaps. 6 and 7.

Prerequisites for this chapter: In Sec. 8.3 we shall make elementary use of second- and third-order determinants.
Sections that may be omitted in a shorter course: 8.6−8.8, 8.12.
References: Appendix 1, Part B.
Answers to problems: Appendix 2.

---

[1]Readers familiar with Chap. 6 will notice that our present approach is in harmony with that in Chap. 6. The restriction to two and three dimensions will provide for a richer theory with basic physical, engineering, and geometrical applications.

8.1 Vector Algebra in 2-Space and 3-Space

In geometry and physics and its engineering applications we use two kinds of quantities, scalars and vectors. A **scalar** is a quantity that is determined by its magnitude, its number of units measured on a suitable scale. For instance, length, temperature, and voltage are scalars.

A **vector** is a quantity that is determined by both its magnitude and its direction; thus it is an **arrow** or **directed line segment.** For instance, a force is a vector, and so is a velocity, giving the speed and direction of motion (Fig. 147).

We denote vectors by lowercase boldface letters[2] **a, b, v,** etc.

A vector (arrow) has a tail, called its **initial point,** and a tip, called its **terminal point.** For instance, in Fig. 148, the triangle is translated (displaced without rotation); the initial point P of the vector **a** is the original position of a point and the terminal point Q is its position after the translation.

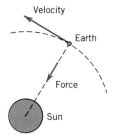

Fig. 147.   Force and velocity

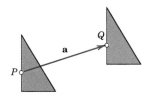

Fig. 148.   Translation

The **length** (or *magnitude*) of a vector **a** (length of the arrow) is also called the **norm** (or Euclidean norm) of **a** and is denoted by $|\mathbf{a}|$.

A vector of length 1 is called a **unit vector.**

Of course, we want to *calculate* with vectors. For instance, we want to find the resultant of forces or compare parallel forces of different magnitude. This motivates our next ideas: to define *equality of vectors,* the notion of *components,* and then the two basic algebraic operations of *vector addition* and *scalar multiplication.*

Equality of Vectors

By definition, two vectors **a** and **b** are **equal,** written, **a = b,** if they have the same length and the same direction (Fig. 149). Hence a vector can be arbitrarily translated, that is, its

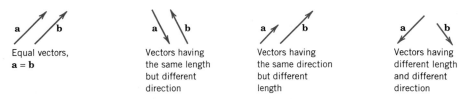

Fig. 149.   Vectors

---

[2]This is customary in printed work; in handwritten work one may characterize vectors by arrows, for example, $\vec{a}$ (in place of **a**), $\vec{b}$, etc.

initial point can be chosen arbitrarily. This definition is practical in connection with forces and other applications.

Components of a Vector

We choose an *xyz* **Cartesian coordinate system**[3] in space (Fig. 150), that is, a usual rectangular coordinate system with the same scale of measurement on the three mutually perpendicular coordinate axes. Then if a given vector **a** has initial point P: (x_1, y_1, z_1) and terminal point Q: (x_2, y_2, z_2), the three numbers (Fig. 151)

(1)
$$a_1 = x_2 - x_1, \quad a_2 = y_2 - y_1, \quad a_3 = z_2 - z_1$$

are called the **components** of the vector **a** with respect to that coordinate system, and we write simply

$$\mathbf{a} = [a_1, a_2, a_3].$$

Length in Terms of Components

By definition, the **length** $|\mathbf{a}|$ of a vector **a** is the distance between its initial point P and terminal point Q. From the Pythagorean theorem and (1) we see that

(2)
$$|\mathbf{a}| = \sqrt{a_1{}^2 + a_2{}^2 + a_3{}^2}.$$

EXAMPLE 1   **Components and length of a vector**

The vector **a** with initial point P: (4, 0, 2) and terminal point Q: (6, −1, 2) has the components

$$a_1 = 6 - 4 = 2, \quad a_2 = -1 - 0 = -1, \quad a_3 = 2 - 2 = 0.$$

Hence **a** = [2, −1, 0]. (Can you sketch **a**, as in Fig. 151?) Equation (2) gives the length

$$|\mathbf{a}| = \sqrt{2^2 + (-1)^2 + 0^2} = \sqrt{5}.$$

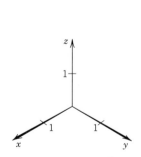

Fig. 150.   Cartesian coordinate system

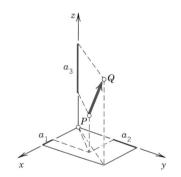

Fig. 151.   Components of a vector

---

[3]Named after the French philosopher and mathematician RENATUS CARTESIUS, latinized for RENÉ DESCARTES (1596—1650), who invented analytic geometry. His basic work *Géométrie* appeared in 1637, as an appendix to his *Discours de la méthode*.

If we choose $(-1, 5, 8)$ as the initial point of **a**, the corresponding terminal point is $(1, 4, 8)$.

If we choose the origin $(0, 0, 0)$ as the initial point of **a**, the corresponding terminal point is $(2, -1, 0)$; its coordinates equal the components of **a**. This suggests that we can determine each point in space by a vector, as follows. ◄

Position Vector

A Cartesian coordinate system being given, the **position vector r** of a point $A: (x, y, z)$ is the vector with the origin $(0, 0, 0)$ as the initial point and A as the terminal point (see Fig. 152). Thus $\mathbf{r} = [x, y, z]$. This can be seen directly from (1).

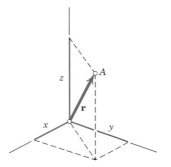

Fig. 152.   Position vector **r** of a point $A: (x, y, z)$

Furthermore, if we translate a vector **a**, with initial point P and terminal point Q, then corresponding coordinates of P and Q change by the same amount, so that the differences in (1) remain unchanged. This proves

THEOREM 1   **(Vectors as ordered triples of real numbers)**

*A fixed Cartesian coordinate system being given, each vector is uniquely determined by its ordered triple of corresponding components. Conversely, to each ordered triple of real numbers (a_1, a_2, a_3) there corresponds precisely one vector $\mathbf{a} = [a_1, a_2, a_3]$, with $(0, 0, 0)$ corresponding to the **zero vector 0**, which has length 0 and no direction.*

Hence a vector equation $\mathbf{a} = \mathbf{b}$ is equivalent to the three equations $a_1 = b_1$, $a_2 = b_2$, $a_3 = b_3$ for the components.

We see that from our "geometrical" definition of vectors as arrows we have arrived at an "algebraic" characterization by Theorem 1. We could have started from the latter[4] and reversed our process. This shows that the two approaches are equivalent.

Vector Addition, Scalar Multiplication

Applications have suggested algebraic calculations with vectors that are practically useful and almost as simple as calculations with numbers.

Definition 1. Addition of Vectors

The **sum a + b** of two vectors $\mathbf{a} = [a_1, a_2, a_3]$ and $\mathbf{b} = [b_1, b_2, b_3]$ is obtained by adding the corresponding components,

$$(3) \qquad \mathbf{a} + \mathbf{b} = [a_1 + b_1, \quad a_2 + b_2, \quad a_3 + b_3].$$

---

[4] This is in agreement with Chap. 6 (which we shall not use here).

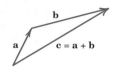

Fig. 153. Vector addition

Geometrically, place the vectors as in Fig. 153 (the initial point of **b** at the terminal point of **a**); then **a** + **b** is the vector drawn from the initial point of **a** to the terminal point of **b**.   ◄

Figure 154 shows that for forces, this addition is the parallelogram law by which we obtain the **resultant** of two forces in mechanics.

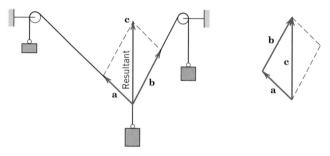

Fig. 154. Resultant of two forces (parallelogram law)

Figure 155 illustrates (for the plane) that the "algebraic" way and the "geometric" way of vector addition amount to the same thing.

Basic properties of vector addition follow immediately from the familiar laws for real numbers (see also Figs. 156 and 157):

(4)

(a) $\mathbf{a} + \mathbf{b} = \mathbf{b} + \mathbf{a}$ (*commutativity*)

(b) $(\mathbf{u} + \mathbf{v}) + \mathbf{w} = \mathbf{u} + (\mathbf{v} + \mathbf{w})$ (*associativity*)

(c) $\mathbf{a} + \mathbf{0} = \mathbf{0} + \mathbf{a} = \mathbf{a}$

(d) $\mathbf{a} + (-\mathbf{a}) = \mathbf{0}$

where $-\mathbf{a}$ denotes the vector having the length $|\mathbf{a}|$ and the direction opposite to that of **a.**

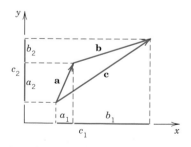

Fig. 155.   Vector addition

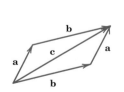

Fig. 156.   Cummutativity of vector addition

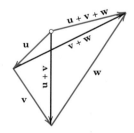

Fig. 157.   Associativity of vector addition

In (4b) we may simply write $\mathbf{u} + \mathbf{v} + \mathbf{w},$ and similarly for sums of more than three vectors. Instead of $\mathbf{a} + \mathbf{a}$ we also write $2\mathbf{a},$ and so on. This (and the notation $-\mathbf{a}$ used before) suggests that we define the second algebraic operation for vectors, the multiplication of a vector by a scalar, as follows.

Definition 2. Scalar Multiplication (Multiplication by a Number)

The product $c\mathbf{a}$ of any vector $\mathbf{a} = [a_1, a_2, a_3]$ and any scalar c (real number c) is the vector obtained by multiplying each component of $\mathbf{a}$ by $c,$

(5)
$$c\mathbf{a} = [ca_1, ca_2, ca_3].$$

Geometrically, if $\mathbf{a} \neq \mathbf{0},$ then $c\mathbf{a}$ with $c > 0$ has the direction of $\mathbf{a}$ and with $c < 0$ the direction opposite to $\mathbf{a}.$ In any case, the length of $c\mathbf{a}$ is $|c\mathbf{a}| = |c||\mathbf{a}|,$ and $c\mathbf{a} = \mathbf{0}$ if $\mathbf{a} = \mathbf{0}$ or $c = 0$ (or both). (See Fig. 158.) ◀

Basic properties of scalar multiplication are obtained from Definitions 1 and 2:

(6)

     (a) $c(\mathbf{a} + \mathbf{b}) = c\mathbf{a} + c\mathbf{b}$

     (b) $(c + k)\mathbf{a} = c\mathbf{a} + k\mathbf{a}$

     (c) $c(k\mathbf{a}) = (ck)\mathbf{a}$ (written $ck\mathbf{a}$)

     (d) $1\mathbf{a} = \mathbf{a}.$

You may prove that (4) and (6) imply for any vector $\mathbf{a}$

(7)

     (a) $0\mathbf{a} = \mathbf{0}$

     (b) $(-1)\mathbf{a} = -\mathbf{a}.$

Instead of $\mathbf{b} + (-\mathbf{a})$ we simply write $\mathbf{b} - \mathbf{a}$ (Fig. 159).

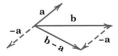

Fig. 158. Scalar multiplication [multiplication of vectors by scalars (numbers)]

Fig. 159. Difference of vectors

EXAMPLE 2   **Vector addition. Multiplication by scalars**

With respect to a given coordinate system, let

$$\mathbf{a} = [4, 0, 1] \qquad \text{and} \qquad \mathbf{b} = [2, -5, \tfrac{1}{3}].$$

Then $-\mathbf{a} = [-4, 0, -1],$ $7\mathbf{a} = [28, 0, 7],$ $\mathbf{a} + \mathbf{b} = [6, -5, \tfrac{4}{3}],$ and

$$2(\mathbf{a} - \mathbf{b}) = 2[2, 5, \tfrac{2}{3}] = [4, 10, \tfrac{4}{3}] = 2\mathbf{a} - 2\mathbf{b}. \qquad ◀$$

Unit Vectors i, j, k

Another popular representation of vectors is

(8)
$$\mathbf{a} = [a_1, a_2, a_3] = a_1\mathbf{i} + a_2\mathbf{j} + a_3\mathbf{k}.$$

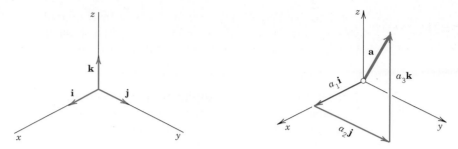

Fig. 160. The unit vectors **i**, **j**, **k** and the representation (8)

In this representation, **i**, **j**, **k** are the **unit vectors** in the positive directions of the axes of a Cartesian coordinate system (Fig. 160). Hence

(9) $\mathbf{i} = [1, 0, 0]$, $\mathbf{j} = [0, 1, 0]$, $\mathbf{k} = [0, 0, 1]$

and the right side of (8) is a sum of three vectors parallel to the three axes.

EXAMPLE 3   **i j k notation for vectors**

In Example 2 we have $\mathbf{a} = 4\mathbf{i} + \mathbf{k}$, $\mathbf{b} = 2\mathbf{i} - 5\mathbf{j} + \frac{1}{3}\mathbf{k}$, and so on.   ◀

Vector Space R^3 (*Optional*)

We mention that the set of all vectors considered are said to form the *real three-dimensional vector space* R^3 with the two *algebraic operations* of vector addition and scalar multiplication. To explain "dimension," we need the concept of linear independence. By a **linear combination** of given vectors $\mathbf{a}_{(1)}, \mathbf{a}_{(2)}, \cdots, \mathbf{a}_{(m)}$ we mean an expression of the form

$$c_1\mathbf{a}_{(1)} + c_2\mathbf{a}_{(2)} + \cdots + c_m\mathbf{a}_{(m)}$$

where $c_1, \cdots, c_m$ are any scalars; clearly, this is again a vector. The given vectors are called a *linearly independent set* or, briefly, **linearly independent** if and only if the only solution of

(10) $$c_1\mathbf{a}_{(1)} + \cdots + c_m\mathbf{a}_{(m)} = \mathbf{0}$$ $(c_1, \cdots, c_m$ any scalars)

is $c_1 = c_2 = \cdots = c_m = 0$. If (10) also holds with scalars not all zero, then those m vectors are called **linearly dependent.** Now it can be shown that 4 or more vectors in R^3 are always linearly dependent, but R^3 contains linearly independent sets of 3 vectors, so that the maximum possible number of vectors in a linearly independent set in R^3 is 3. This number 3 is called the **dimension** of R^3, and any such set is called a **basis** for R^3. Particularly useful is the **"standard basis"** (9). Any basis being given, every vector in R^3 can be written as a linear combination of the basis vectors in a unique fashion. Representation (8) is an example of this.

Vector space R^3 is a model of a *general vector space,* as discussed in Sec. 6.8 but not needed in this chapter.

PROBLEM SET 8.1

Components and Length

Find the components of the vector **v** with given initial point $P: (x_1, y_1, z_1)$ and terminal point $Q: (x_2, y_2, z_2)$. Find $|\mathbf{v}|$. Sketch **v**.

1. $P: (1, 1, 0), \quad Q: (4, 5, 0)$
2. $P: (2, 3, 0), \quad Q: (1, 7, 3)$
3. $P: (1, 2, 3), \quad Q: (2, 4, 6)$
4. $P: (5, -1, 3), \quad Q: (-5, 1, -3)$
5. $P: (4, -4, 0), \quad Q: (0, 0, 0)$
6. $P: (0, 0, 0), \quad Q: (a, b, c)$
7. $P: (3, 9, 1), \quad Q: (3, -9, 1)$
8. $P: (4, 0, -2), \quad Q: (2, 0, -4)$

Given the components v_1, v_2, v_3 of a vector **v** and a particular initial point P, find the corresponding terminal point and the length (norm) of **v**.

9. $2, 3, 0; \quad P: (1, 0, 0)$
10. $0, 2, -3; \quad P: (0, -2, 4)$
11. $3, 8, -1; \quad P: (-3, -8, 1)$
12. $0, 0, 0; \quad P: (3, -1, -2)$
13. $\frac{1}{4}, \frac{1}{2}, -\frac{1}{2}; \quad P: (1, \frac{1}{2}, -4)$
14. $3, -1, 6; \quad P: (0, 0, 0)$

Vector Addition, Scalar Multiplication

Let $\mathbf{a} = [3, -2, 1] = 3\mathbf{i} - 2\mathbf{j} + \mathbf{k}, \mathbf{b} = [0, 3, 0] = 3\mathbf{j}, \mathbf{c} = [4, 1, -1] = 4\mathbf{i} + \mathbf{j} - \mathbf{k}$. Find

15. $\mathbf{a} + \mathbf{c}, \mathbf{c} + \mathbf{a}$
16. $-\mathbf{a}, 3\mathbf{a}, -\frac{1}{2}\mathbf{a}$
17. $(\mathbf{a} + \mathbf{b}) + \mathbf{c}, \mathbf{a} + (\mathbf{b} + \mathbf{c})$
18. $|\mathbf{a} + \mathbf{b}|, |\mathbf{a}| + |\mathbf{b}|$
19. $4\mathbf{a} + 8\mathbf{c}, 4(\mathbf{a} + 2\mathbf{c})$
20. $(1/|\mathbf{a}|)\mathbf{a}, (1/|\mathbf{b}|)\mathbf{b}$
21. $6\mathbf{a} - 12\mathbf{b}, 3(2\mathbf{a} - 4\mathbf{b})$
22. $5\mathbf{a} - 4\mathbf{b} + 3\mathbf{c}$

23. What laws do Probs. 15, 17, and 19 illustrate?

Forces

Find the resultant (in components) and its magnitude.

24. $\mathbf{p} = [0, 1, 2], \mathbf{q} = [3, -1, 4], \mathbf{u} = [2, 0, 2], \mathbf{v} = [3, -3, 0]$
25. $\mathbf{p} = [4, -2, -3], \mathbf{q} = [8, 8, 1], \mathbf{u} = [-12, -6, 2]$
26. $\mathbf{p} = [1, 0, -3], \mathbf{q} = [5, -5, 2], \mathbf{u} = [-6, -5, 1]$
27. $\mathbf{p} = [3, 5, 7], \mathbf{q} = 3\mathbf{p}, \mathbf{u} = -6\mathbf{p}$
28. $\mathbf{p} = [8, 0, 1], \mathbf{q} = [4, 0, 6], \mathbf{u} = [0, 0, 9]$

29. Find **p** such that **p**, $\mathbf{q} = [3, 2, 0]$, and $\mathbf{u} = [-2, 4, 0]$ are in equilibrium.
30. If $|\mathbf{p}| = 6$ and $|\mathbf{q}| = 4$, what can you say about the magnitude and direction of the resultant? Of the resultant of $4\mathbf{p} - 3\mathbf{q}$?
31. Find $\mathbf{p} = [0, 0, p_3]$ such that the resultant of **p**, $\mathbf{q} = [4, 0, 8], \mathbf{u} = [2, -2, 6], \mathbf{v} = [-1, 1, 3]$ is parallel to the xy-plane.
32. Find forces **p, q, u** in the directions of the coordinate axes such that **p, q, u**, $\mathbf{v} = [3, 4, -1]$, and $\mathbf{w} = [0, 5, -2]$ are in equilibrium. Are **p, q, u** uniquely determined?
33. Find the magnitude of the force in each rope in the figure for any weight **w** and angle α.
34. **TEAM PROJECT. Geometrical applications** can help you develop skill with vector methods. Using vectors, prove the following.
    (a) The diagonals of a parallelogram bisect each other (see the figure).
    (b) The line through the midpoints of adjacent sides of a parallelogram bisect one of the diagonals in the ratio $1:3$.

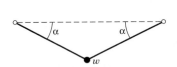

Problem 33

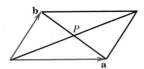

Team Project 34(a)

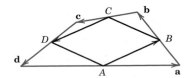

Team Project 34(d)

(c) Obtain (b) from (a).

(d) The quadrilateral whose vertices are the midpoints of the sides of an arbitrary quadrilateral is a parallelogram (see the figure).

(e) The sum of the vectors drawn from the center of a regular polygon to its vertices is the zero vector.

(f) The four space diagonals of a parallelepiped meet and bisect each other.

8.2 Inner Product (Dot Product)

We shall now define a multiplication of two vectors that gives a scalar as the product and is suggested by various applications.

Definition. Inner Product (Dot Product) of Vectors

The **inner product** or **dot product a·b** (read "**a** dot **b**") of two vectors **a** and **b** is the product of their lengths times the cosine of their angle (see Fig. 161),

(1)

$$\mathbf{a} \cdot \mathbf{b} = |\mathbf{a}||\mathbf{b}| \cos \gamma \qquad \text{if} \quad \mathbf{a} \neq \mathbf{0}, \mathbf{b} \neq \mathbf{0}$$

$$\mathbf{a} \cdot \mathbf{b} = 0 \qquad \text{if} \quad \mathbf{a} = \mathbf{0} \text{ or } \mathbf{b} = \mathbf{0}.$$

The angle γ, $0 \leq \gamma \leq \pi$, between **a** and **b** is measured when the vectors have their initial points coinciding, as in Fig. 161. In components, $\mathbf{a} = [a_1, a_2, a_3]$, $\mathbf{b} = [b_1, b_2, b_3]$, and

(2)

$$\mathbf{a} \cdot \mathbf{b} = a_1 b_1 + a_2 b_2 + a_3 b_3.$$

For the derivation of (2) from (1), see below. ◀

Since the cosine in (1) may be positive, zero, or negative, so may be the inner product (Fig. 161). The case that the inner product is zero is of great practical interest and suggests the following concept.

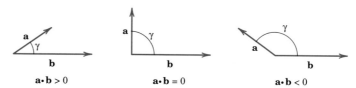

Fig. 161.   Angle between vectors and value of inner product

A vector **a** is called **orthogonal** *to a vector* **b** if **a·b** = 0. Then **b** is also orthogonal to **a** and we call these vectors **orthogonal vectors.** Clearly, the zero vector is orthogonal to every vector. For nonzero vectors we have **a·b** = 0 if and only if cos γ = 0; thus $\gamma = \pi/2$ (90°). This proves the important

THEOREM 1   **(Orthogonality)**

The inner product of two nonzero vectors is zero if and only if these vectors are perpendicular.

Length and Angle in Terms of Inner Product

Equation (1) with $\mathbf{b} = \mathbf{a}$ gives $\mathbf{a} \cdot \mathbf{a} = |\mathbf{a}|^2$. Hence

(3)
$$|\mathbf{a}| = \sqrt{\mathbf{a} \cdot \mathbf{a}} \ .$$

From (3) and (1) we obtain for the angle γ between two nonzero vectors

(4)
$$\cos \gamma = \frac{\mathbf{a} \cdot \mathbf{b}}{|\mathbf{a}||\mathbf{b}|} = \frac{\mathbf{a} \cdot \mathbf{b}}{\sqrt{\mathbf{a} \cdot \mathbf{a}} \sqrt{\mathbf{b} \cdot \mathbf{b}}} \ .$$

EXAMPLE 1   **Inner product. Angle between vectors**

Find the inner product and the lengths of $\mathbf{a} = [1, 2, 0]$ and $\mathbf{b} = [3, -2, 1]$ as well as the angle between these vectors.

Solution. $\mathbf{a} \cdot \mathbf{b} = 1 \cdot 3 + 2 \cdot (-2) + 0 \cdot 1 = -1$, $|\mathbf{a}| = \sqrt{\mathbf{a} \cdot \mathbf{a}} = \sqrt{5}$, $|\mathbf{b}| = \sqrt{\mathbf{b} \cdot \mathbf{b}} = \sqrt{14}$, and (4) gives the angle

$$\gamma = \text{arc cos} \frac{\mathbf{a} \cdot \mathbf{b}}{|\mathbf{a}||\mathbf{b}|} = \text{arc cos} \, (-0.11952) = 1.69061 = 96.865°.$$

Can you sketch these vectors and convince yourself that they make an angle greater than 90°, so that the inner product comes out negative? ◀

General Properties of Inner Products

From the definition we see that the inner product has the following properties. For any vectors $\mathbf{a}, \mathbf{b}, \mathbf{c}$ and scalars q_1, q_2,

(5)

(a) $\qquad [q_1\mathbf{a} + q_2\mathbf{b}] \cdot \mathbf{c} = q_1\mathbf{a} \cdot \mathbf{c} + q_2\mathbf{b} \cdot \mathbf{c}$ $\qquad$ (*Linearity*)

(b) $\qquad \mathbf{a} \cdot \mathbf{b} = \mathbf{b} \cdot \mathbf{a}$ $\qquad$ (*Symmetry*)

(c) $\qquad \left. \begin{array}{l} \mathbf{a} \cdot \mathbf{a} \geq 0 \\ \mathbf{a} \cdot \mathbf{a} = 0 \quad \text{if and only if} \quad \mathbf{a} = \mathbf{0} \end{array} \right\}$ $\qquad$ (*Positive-definiteness*).

Hence *dot multiplication is commutative* [see (5b)] *and is distributive with respect to vector addition;* in fact, from (5a) with $q_1 = 1$ and $q_2 = 1$ we have

(5a*) $\qquad (\mathbf{a} + \mathbf{b}) \cdot \mathbf{c} = \mathbf{a} \cdot \mathbf{c} + \mathbf{b} \cdot \mathbf{c}$ $\qquad$ (*Distributivity*).

Furthermore, from (1) and $|\cos \gamma| \leq 1$ we see that

(6) $\qquad |\mathbf{a} \cdot \mathbf{b}| \leq |\mathbf{a}||\mathbf{b}|$ $\qquad$ (*Schwarz*[5] *inequality*).

Using this and (3), you may prove

(7) $\qquad |\mathbf{a} + \mathbf{b}| \leq |\mathbf{a}| + |\mathbf{b}|$ $\qquad$ **(Triangle inequality).**

---

[5]See the footnote in Sec. 6.8.

A simple direct calculation with inner products shows that

(8) $$|\mathbf{a} + \mathbf{b}|^2 + |\mathbf{a} - \mathbf{b}|^2 = 2(|\mathbf{a}|^2 + |\mathbf{b}|^2)$$ (*Parallelogram equality*).

Equations (6)−(8) play a basic role in so-called *Hilbert spaces* (abstract inner product spaces), which form the basis of quantum mechanics (see Ref. [9] listed in Appendix 1).

Derivation of (2) from (1)
Using (8) in Sec. 8.1, we can write

$$\mathbf{a} = a_1\mathbf{i} + a_2\mathbf{j} + a_3\mathbf{k} \qquad \text{and} \qquad \mathbf{b} = b_1\mathbf{i} + b_2\mathbf{j} + b_3\mathbf{k}.$$

Since $\mathbf{i}$, $\mathbf{j}$, and $\mathbf{k}$ are unit vectors, we have from (3)

$$\mathbf{i}\cdot\mathbf{i} = 1, \qquad \mathbf{j}\cdot\mathbf{j} = 1, \qquad \mathbf{k}\cdot\mathbf{k} = 1.$$

Since they are orthogonal (because the coordinate axes are perpendicular), Theorem 1 gives

$$\mathbf{i}\cdot\mathbf{j} = 0, \qquad \mathbf{j}\cdot\mathbf{k} = 0, \qquad \mathbf{k}\cdot\mathbf{i} = 0.$$

Hence if we substitute those representations of $\mathbf{a}$ and $\mathbf{b}$ into $\mathbf{a}\cdot\mathbf{b}$ and use (5a*) and (5b), we first have a sum of nine inner products,

$$\mathbf{a}\cdot\mathbf{b} = a_1 b_1\mathbf{i}\cdot\mathbf{i} + a_1 b_2\mathbf{i}\cdot\mathbf{j} + \cdots + a_3 b_3\mathbf{k}\cdot\mathbf{k}.$$

Since six of these products are zero, we obtain (2).   ◀

Applications of Inner Products

Typical applications of inner products are shown in the following examples and in Problem Set 8.2.

EXAMPLE 2   **Work done by a force as inner product**

Consider a body on which a constant force $\mathbf{p}$ acts. Let the body be given a displacement $\mathbf{d}$. Then the work done by $\mathbf{p}$ in the displacement is defined as

(9) $$W = |\mathbf{p}||\mathbf{d}| \cos \alpha = \mathbf{p}\cdot\mathbf{d},$$

that is, magnitude $|\mathbf{p}|$ of the force times length $|\mathbf{d}|$ of the displacement times the cosine of the angle α between $\mathbf{p}$ and $\mathbf{d}$ (Fig. 162). If $\alpha < 90°$, as in Fig. 162, then $W > 0$. If $\mathbf{p}$ and $\mathbf{d}$ are orthogonal, then the work is zero (why?). If $\alpha > 90°$, then $W < 0$, which means that in the displacement one has to do work against the force.   ◀

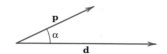

Fig. 162.   Work done by a force

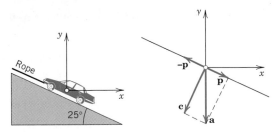

Fig. 163.   Example 3

EXAMPLE 3   **Component of a force in a given direction**

What force in the rope in Fig. 163 will hold a car of 5000 lb in equilibrium if the ramp makes an angle of 25°
with the horizontal?

Solution.   Introducing coordinates as shown, the weight is $\mathbf{a} = [0, -5000]$ because this force points downward,
in the negative y-direction. We have to represent $\mathbf{a}$ as a sum (resultant) of two forces, $\mathbf{a} = \mathbf{c} + \mathbf{p}$, where $\mathbf{c}$ is
the force the car exerts on the ramp, which is of no interest to us, and $\mathbf{p}$ is parallel to the rope, of magnitude
(see Fig. 163)

$$|\mathbf{p}| = |\mathbf{a}| \cos \gamma = 5000 \cos 65° = 2113 \; [\mathrm{lb}]$$

and direction of the unit vector $\mathbf{u}$ opposite to the direction of the rope; here $\gamma = 90° - 25° = 65°$ is the angle
between $\mathbf{a}$ and $\mathbf{p}.$ Now a vector in the direction of the rope is

$$\mathbf{b} = [-1, \tan 25°] = [-1, 0.46631], \qquad \text{thus} \qquad |\mathbf{b}| = 1.10338,$$

so that

$$\mathbf{u} = -\frac{1}{|\mathbf{b}|} \, \mathbf{b} = [0.90631, -0.42262].$$

Since $|\mathbf{u}| = 1$ and $\cos \gamma > 0$, we see that we can also write our result as

$$|\mathbf{p}| = (|\mathbf{a}| \cos \gamma)|\mathbf{u}| = \mathbf{a} \cdot \mathbf{u} = -\frac{\mathbf{a} \cdot \mathbf{b}}{|\mathbf{b}|} = \frac{5000 \cdot 0.46631}{1.10338} = 2113 \; [\mathrm{lb}].$$

Answer. About 2100 lb.   ◀

Example 3 is typical of applications in which one uses the concept of the **component**
or **projection** *of a vector* $\mathbf{a}$ *in the direction of a vector* $\mathbf{b}$ ($\neq \mathbf{0}$), defined by (see Fig. 164)

(10) $$p = |\mathbf{a}| \cos \gamma.$$

Thus p is the length of the orthogonal projection of $\mathbf{a}$ on a straight line l parallel to $\mathbf{b}$,
taken with the plus sign if $p\mathbf{b}$ has the direction of $\mathbf{b}$ and with the minus sign if $p\mathbf{b}$ has the
direction opposite to $\mathbf{b}$; see Fig. 164.

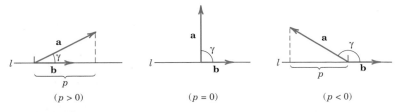

Fig. 164.   Component of a vector $\mathbf{a}$ in the direction of a vector $\mathbf{b}$

Multiplying (10) by $|\mathbf{b}|/|\mathbf{b}| = 1$, we have $\mathbf{a} \cdot \mathbf{b}$ in the numerator and thus

(11)
$$p = \frac{\mathbf{a} \cdot \mathbf{b}}{|\mathbf{b}|}$$
$(\mathbf{b} \neq \mathbf{0}).$

If $\mathbf{b}$ is a unit vector, as it is often used for fixing a direction, then (11) simply gives

(12) $p = \mathbf{a} \cdot \mathbf{b}$ $(|\mathbf{b}| = 1).$

EXAMPLE 4   **Orthonormal basis**

By definition, an *orthonormal basis* for 3-space is a basis $\{\mathbf{a}, \mathbf{b}, \mathbf{c}\}$ consisting of orthogonal unit vectors. It has the great advantage that the determination of the coefficients in representations

$$\mathbf{v} = l_1\mathbf{a} + l_2\mathbf{b} + l_3\mathbf{c}$$
                                    (**v** a given vector)

is very simple. We claim that $l_1 = \mathbf{a} \cdot \mathbf{v}$, $l_2 = \mathbf{b} \cdot \mathbf{v}$, $l_3 = \mathbf{c} \cdot \mathbf{v}$. Indeed, this follows simply by taking the inner products of the representation with $\mathbf{a}, \mathbf{b}, \mathbf{c}$, respectively, and using the orthonormality of the basis, $\mathbf{a} \cdot \mathbf{v} = l_1\mathbf{a} \cdot \mathbf{a} + l_2\mathbf{a} \cdot \mathbf{b} + l_3\mathbf{a} \cdot \mathbf{c} = l_1$, etc.

For example, the unit vectors $\mathbf{i}, \mathbf{j}, \mathbf{k}$ in (8), Sec. 8.1, associated with a Cartesian coordinate system form an orthonormal basis, called the **standard basis.**   ◀

EXAMPLE 5   **Orthogonal straight lines in the plane**

Find the straight line L_1 through the point P: (1, 3) in the xy-plane and perpendicular to the straight line L_2: $x - 2y + 2 = 0$; see Fig. 165.

Solution.   The idea is to write a general straight line L_1: $a_1x + a_2y = c$ as $\mathbf{a} \cdot \mathbf{r} = c$ with $\mathbf{a} = [a_1, a_2] \neq \mathbf{0}$ and $\mathbf{r} = [x, y]$, according to (2). Now the line L_1^* through the origin and parallel to L_1 is $\mathbf{a} \cdot \mathbf{r} = 0$. Hence, by Theorem 1, the vector $\mathbf{a}$ is perpendicular to $\mathbf{r}$. Hence it is perpendicular to L_1^* and also to L_1 because L_1 and L_1^* are parallel. $\mathbf{a}$ is called a **normal vector** to L_1 (and to L_1^*).

Now a normal vector to the given line $x - 2y + 2 = 0$ is $\mathbf{b} = [1, -2]$. Thus L_1 is perpendicular to L_2 if $\mathbf{b} \cdot \mathbf{a} = a_1 - 2a_2 = 0$, for instance, if $\mathbf{a} = [2, 1]$. Hence L_1 is given by $2x + y = c$. It passes through P: (1, 3) when $2 \cdot 1 + 3 = c = 5$.

Answer: $y = -2x + 5$. Show that the point of intersection is $(x, y) = (1.6, 1.8)$.   ◀

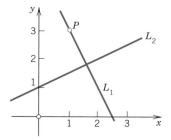

Fig. 165.   Example 5

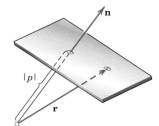

Fig. 166.   Normal vector to a plane

EXAMPLE 6   **Normal vector to a plane**

Find a unit vector perpendicular to the plane $4x + 2y + 4z = -7$.

Solution.   Using (2), we may write any plane in space as

(13) $\mathbf{a} \cdot \mathbf{r} = a_1x + a_2y + a_3z = c$

where $\mathbf{a} = [a_1, a_2, a_3] \neq \mathbf{0}$ and $\mathbf{r} = [x, y, z]$. The unit vector in the direction of $\mathbf{a}$ is (Fig. 166)

$$\mathbf{n} = \frac{1}{|\mathbf{a}|}\mathbf{a}.$$

Dividing by $|\mathbf{a}|$, we obtain from (13)

(14) $\mathbf{n} \cdot \mathbf{r} = p$ where $p = \dfrac{c}{|\mathbf{a}|}$.

From (12) we see that p is the projection of $\mathbf{r}$ in the direction of $\mathbf{n}$. This projection has the same constant value $c/|\mathbf{a}|$ for the position vector $\mathbf{r}$ of any point in the plane. Clearly this holds if and only if $\mathbf{n}$ is perpendicular to the plane. $\mathbf{n}$ is called a **unit normal vector** to the plane (the other being $-\mathbf{n}$).

Furthermore, from this and the definition of projection it follows that $|p|$ is the distance of the plane from the origin. Representation (14) is called **Hesse's**[6] **normal form** of a plane. In our case, $\mathbf{a} = [4, 2, 4]$, $c = -7$, $|\mathbf{a}| = 6$, $\mathbf{n} = \frac{1}{6}\mathbf{a} = [\frac{2}{3}, \frac{1}{3}, \frac{2}{3}]$, and the plane has the distance 7/6 from the origin. ◀

PROBLEM SET 8.2

Inner Product. Work
Let $\mathbf{a} = [1, 3, 2]$, $\mathbf{b} = [2, 0, -5]$, $\mathbf{c} = [4, -2, 1]$. Find

1. $\mathbf{a} \cdot \mathbf{b}$, $\mathbf{b} \cdot \mathbf{a}$

2. $|\mathbf{a}|$, $|2\mathbf{a}|$, $|\mathbf{c}|$

3. $|4\mathbf{a}|$, $4|\mathbf{a}|$, $|\mathbf{a} - \mathbf{b}|$

4. $\mathbf{a} \cdot (2\mathbf{b} + 3\mathbf{c})$, $2\mathbf{a} \cdot \mathbf{b} + 3\mathbf{a} \cdot \mathbf{c}$

5. $2\mathbf{b} \cdot 5\mathbf{c}$, $10\mathbf{b} \cdot \mathbf{c}$

6. $(\mathbf{a} - \mathbf{b}) \cdot \mathbf{b}$, $\mathbf{a} \cdot \mathbf{b} - \mathbf{b} \cdot \mathbf{b}$

7. $\mathbf{a} \cdot (\mathbf{b} - \mathbf{a})$, $\mathbf{a} \cdot (\mathbf{a} - \mathbf{b})$

8. $|\mathbf{a} + \mathbf{b}|$, $|\mathbf{a}| + |\mathbf{b}|$

9. $\mathbf{a} \cdot (\mathbf{b} + \mathbf{c})$, $(\mathbf{a} \cdot \mathbf{b})\mathbf{c}$

10. What does $\mathbf{u} \cdot \mathbf{v} = \mathbf{u} \cdot \mathbf{w}$ with $\mathbf{u} \neq \mathbf{0}$ imply?

11. What laws do Probs. 1, 4, 5, 6 illustrate?

12. Make a sketch similar to Fig. 164 when $\mathbf{a} = 3\mathbf{i}$, $\mathbf{b} = \mathbf{i} + \mathbf{j}$, $\mathbf{b} = \mathbf{j}$, $\mathbf{b} = -\mathbf{i} + \mathbf{j}$.

Work. Find the work done by a force $\mathbf{p}$ acting on a body if the body is displaced from a point A to a point B along the straight segment AB. Sketch $\mathbf{p}$ and AB. Show the details of your work.

13. $\mathbf{p} = [2, 6, 6]$, A: $(3, 4, 0)$, B: $(5, 8, 0)$

14. $\mathbf{p} = [3, 2, 0]$, A: $(1, 5, 0)$, B: $(7, -4, 0)$

15. $\mathbf{p} = [1, 1, 1]$, A: $(2, 2, 2)$, B: $(4, 0, 2)$

16. $\mathbf{p} = [5, 3, 8]$, A: $(2, 9, 6)$, B: $(0, 1, 0)$

17. Can work be zero or negative? In what cases?

18. **(Resultant)** Is the work done by the resultant of two forces in a displacement the sum of the works done by each of the two forces separately? Give a proof or counterexample.

Angle Between Vectors. Component in the Direction of a Vector
Angle. Let $\mathbf{a} = [1, 1, 0]$, $\mathbf{b} = [3, 2, 1]$, $\mathbf{c} = [1, 0, 2]$. Find the angle between

19. $\mathbf{a}$, $\mathbf{b}$ **20.** $\mathbf{b} - \mathbf{a}$, $\mathbf{c} - \mathbf{a}$ **21.** $\mathbf{a}$, $\mathbf{b} + \mathbf{c}$ **22.** $\mathbf{a} + \mathbf{b}$, $\mathbf{c}$

23. **(Straight lines)** Find the angle between $x - y = 1$ and $x - 2y = -1$.

24. **(Straight lines)** Find the angle between $3x + 5y = 0$ and $4x - 2y = 1$.

25. **(Planes)** Find the angle between $x + y + z = 1$ and $x + 2y + 3z = 6$. (By definition, this is the angle between the normals of these planes.)

26. **(Planes)** Find the angle between $x - y = 0$ and $x - z = 1$.

27. **(Triangle)** Find the angles if the vertices are A: $(0, 0, 0)$, B: $(4, 2, 1)$, C: $(1, 2, 4)$.

28. **(Triangle)** Find the angles if the vertices are A: $(1, 1, 0)$, B: $(5, 3, 0)$, C: $(2, 8, 0)$.

29. **(Parallelogram)** Find the angles if the vertices are $(1, 2, 3)$, $(3, 5, 7)$, $(2, 0, 9)$, $(4, 3, 13)$.

Components. Find the component of $\mathbf{a}$ in the direction of $\mathbf{b}$:

30. $\mathbf{a} = [3, 5, 1]$, $\mathbf{b} = [1, 0, 0]$

31. $\mathbf{a} = [4, 0, -3]$, $\mathbf{b} = [1, 1, 1]$

32. $\mathbf{a} = [2.0, 0.5, 0.3]$, $\mathbf{b} = [0.2, 1:4, 3.0]$

33. $\mathbf{a} = [-2, 3, -1]$, $\mathbf{b} = [4, -2, 0]$

34. $\mathbf{a} = [3, -4, 7]$, $\mathbf{b} = [2, 5, 2]$

35. $\mathbf{a} = [8, 2, -3]$, $\mathbf{b} = [0.8, 0, 0.6]$

---

[6]LUDWIG OTTO HESSE (1811—1874), German mathematician, who contributed to the theory of curves and surfaces.

36. (Triangle inequality) Prove (7).

37. (Parallelogram equality) Prove (8).

38. TEAM PROJECT. Orthogonality is perhaps the most important concept in this section, mainly because of the importance of orthogonal coordinates, such as **Cartesian coordinates,** whose **"natural basis"** (9), Sec. 8.1, consists of three orthogonal unit vectors.

(a) For what a_1 are $\mathbf{a} = [a_1, 4, 3]$ and $\mathbf{b} = [5, -2, 1]$ orthogonal?

(b) Find all unit vectors $\mathbf{a} = [a_1, a_2]$ orthogonal to $[5, -2]$.

(c) Find all vectors orthogonal to $\mathbf{a} = [2, 1, 0]$. Do these form a vector space?

(d) For what c are the straight lines $x - 4y = 3$ and $3x + cy = 8$ orthogonal? Sketch them.

(e) For what c are the planes $x + 2y + 3z = 6$ and $x + cy + z = 0$ orthogonal?

(f) Find an orthonormal basis $\{\mathbf{a}, \mathbf{b}, \mathbf{c}\}$ in three-dimensional space, where $\mathbf{b} = q_1[4, -3, 0]$ and $\mathbf{c} = q_2[3, 4, 0]$ and q_1, q_2 are suitable scalars.

(g) If the diagonals of a rectangle are orthogonal, what obvious conclusion can you draw and how can you prove it by vectors?

(h) Discuss further mechanical and geometrical applications in which orthogonality plays a role.

8.3 Vector Product (Cross Product)

The dot product is a *scalar* (Sec. 8.2). We shall see that some applications, for instance, in connection with rotations, require a product that is again a *vector:*

Definition. Vector product (Cross product)

The **vector product (cross product)** $\mathbf{a} \times \mathbf{b}$ of two vectors $\mathbf{a} = [a_1, a_2, a_3]$ and $\mathbf{b} = [b_1, b_2, b_3]$ is a vector

$$\mathbf{v} = \mathbf{a} \times \mathbf{b}$$

as follows. If $\mathbf{a}$ and $\mathbf{b}$ have the same or opposite direction or if one of these vectors is the zero vector, then $\mathbf{v} = \mathbf{a} \times \mathbf{b} = \mathbf{0}$. In any other case, $\mathbf{v} = \mathbf{a} \times \mathbf{b}$ has the length

(1) $$|\mathbf{v}| = |\mathbf{a}||\mathbf{b}| \sin \gamma.$$

This is the area of the parallelogram in Fig. 167 with $\mathbf{a}$ and $\mathbf{b}$ as adjacent sides. (γ is the angle between $\mathbf{a}$ and $\mathbf{b}$, as in the last section.) The direction of $\mathbf{v} = \mathbf{a} \times \mathbf{b}$ is **perpendicular to both $\mathbf{a}$ and $\mathbf{b}$** and such that $\mathbf{a}, \mathbf{b}, \mathbf{v}$, in this order, form a *right-handed triple* as in Fig. 167 (explanation below).   ◀

In components, $\mathbf{v} = [v_1, v_2, v_3] = \mathbf{a} \times \mathbf{b}$ is

(2) $$\boxed{v_1 = a_2 b_3 - a_3 b_2, \quad v_2 = a_3 b_1 - a_1 b_3, \quad v_3 = a_1 b_2 - a_2 b_1.}$$

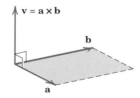

Fig. 167. Vector product

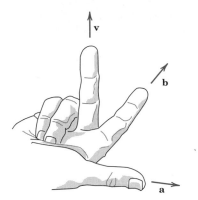

Fig. 168.   Right-handed triple of vectors
a, b, v

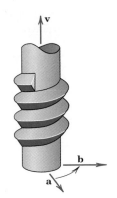

Fig. 169.   Right-handed screw

Here we assume that the Cartesian coordinate system is *right-handed* (explanation below). For a *left-handed* system each of these three components must be multiplied by -1. Formula (2) is derived in Appendix 4.

Explanations. First, a **right-handed triple** of vectors **a, b, v** is one in which the vectors, in the order given, assume the same sort of orientation as the thumb, index finger, and middle finger of the right hand when these are held as shown in Fig. 168. We may also say that if **a** is rotated into the direction of **b** through the angle α $(< \pi)$, then **v** advances in the same direction as a right-handed screw would if turned in the same way (Fig. 169).

Second, a Cartesian coordinate system is called **right-handed** if the corresponding unit vectors **i, j, k** in the positive directions of the axes (see Sec. 8.1) form a right-handed triple as in Fig. 170a. The system is called *left-handed* if the sense of **k** is reversed, as in Fig. 170b. In applications, we prefer right-handed systems.

EXAMPLE 1   **Vector product**

For the vector product $\mathbf{v} = \mathbf{a} \times \mathbf{b}$ of $\mathbf{a} = [1, \quad 1, \quad 0]$ and $\mathbf{b} = [3, \quad 0, \quad 0]$ in right-handed coordinates we obtain from (2)

$$v_1 = 0, \qquad v_2 = 0, \qquad v_3 = 1 \cdot 0 - 1 \cdot 3 = -3.$$

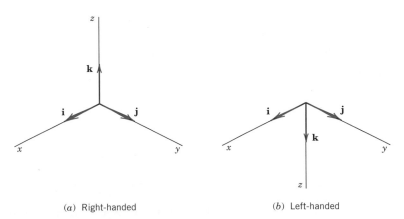

(*a*) Right-handed   (*b*) Left-handed

Fig. 170.   The two types of Cartesian coordinate systems

To check the result in this simple case, sketch **a, b,** and **v.** Can you see that two vectors in the xy-plane must always have their vector product parallel to the z-axis (or equal to the zero vector)? ◀

How to memorize formula (2). Students familiar with second-order and third-order determinants will notice that in (2),

$$(2^*) \qquad v_1 = \begin{vmatrix} a_2 & a_3 \\ b_2 & b_3 \end{vmatrix}, \qquad v_2 = \begin{vmatrix} a_3 & a_1 \\ b_3 & b_1 \end{vmatrix}, \qquad v_3 = \begin{vmatrix} a_1 & a_2 \\ b_1 & b_2 \end{vmatrix}.$$

Hence $\mathbf{v} = [v_1, v_2, v_3] = v_1\mathbf{i} + v_2\mathbf{j} + v_3\mathbf{k}$ is the expansion of the symbolical third-order determinant

$$(2^{**}) \qquad \mathbf{a} \times \mathbf{b} = \begin{vmatrix} \mathbf{i} & \mathbf{j} & \mathbf{k} \\ a_1 & a_2 & a_3 \\ b_1 & b_2 & b_3 \end{vmatrix}$$

by the first row. (We call it "symbolical" because the first row consists of vectors rather than numbers.) In left-handed coordinates these determinants (2^*) and (2^{**}) must be multiplied by -1.

EXAMPLE 2   **An application of (2^{**})**

With respect to a right-handed Cartesian coordinate system, let $\mathbf{a} = [4, 0, -1]$ and $\mathbf{b} = [-2, 1, 3]$. Then

$$\mathbf{a} \times \mathbf{b} = \begin{vmatrix} \mathbf{i} & \mathbf{j} & \mathbf{k} \\ 4 & 0 & -1 \\ -2 & 1 & 3 \end{vmatrix} = \mathbf{i} - 10\mathbf{j} + 4\mathbf{k} = [1, -10, \quad 4].$$ ◀

EXAMPLE 3   **Vector products of the standard basis vectors**

Since **i, j, k** are orthogonal (mutually perpendicular) unit vectors, the definition of vector product gives some useful formulas for simplifying vector products; in right-handed coordinates these are

$$(3) \qquad \begin{matrix} \mathbf{i} \times \mathbf{j} = \mathbf{k}, & \mathbf{j} \times \mathbf{k} = \mathbf{i}, & \mathbf{k} \times \mathbf{i} = \mathbf{j} \\ \mathbf{j} \times \mathbf{i} = -\mathbf{k}, & \mathbf{k} \times \mathbf{j} = -\mathbf{i}, & \mathbf{i} \times \mathbf{k} = -\mathbf{j}. \end{matrix}$$

For left-handed coordinates, replace **k** by $-\mathbf{k}$. Thus $\mathbf{i} \times \mathbf{j} = -\mathbf{k}, \mathbf{j} \times (-\mathbf{k}) = \mathbf{i}$ or $\mathbf{k} \times \mathbf{j} = \mathbf{i}$, etc. ◀

General Properties of Vector Products

Cross multiplication has the property that for every scalar l,

$$(4) \qquad (l\mathbf{a}) \times \mathbf{b} = l(\mathbf{a} \times \mathbf{b}) = \mathbf{a} \times (l\mathbf{b}).$$

It is distributive with respect to vector addition, that is,

$$(5) \qquad \begin{matrix} \text{(a)} & \mathbf{a} \times (\mathbf{b} + \mathbf{c}) = (\mathbf{a} \times \mathbf{b}) + (\mathbf{a} \times \mathbf{c}), \\ \text{(b)} & (\mathbf{a} + \mathbf{b}) \times \mathbf{c} = (\mathbf{a} \times \mathbf{c}) + (\mathbf{b} \times \mathbf{c}). \end{matrix}$$

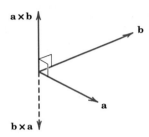

Fig. 171.   Anticommutativity of cross multiplication

It is **not commutative** but **anticommutative,** that is,

(6) $$\mathbf{b} \times \mathbf{a} = -(\mathbf{a} \times \mathbf{b})$$ (Fig. 171).

It is **not associative,** that is,

(7) $$\mathbf{a} \times (\mathbf{b} \times \mathbf{c}) \neq (\mathbf{a} \times \mathbf{b}) \times \mathbf{c}$$ in general,

so that the parentheses cannot be omitted.

PROOFS.   (4) follows directly from the definition. In (5a), formula (2*) gives for the first component on the left

$$\begin{vmatrix} a_2 & a_3 \\ b_2 + c_2 & b_3 + c_3 \end{vmatrix} = a_2(b_3 + c_3) - a_3(b_2 + c_2)$$

$$= (a_2 b_3 - a_3 b_2) + (a_2 c_3 - a_3 c_2)$$

$$= \begin{vmatrix} a_2 & a_3 \\ b_2 & b_3 \end{vmatrix} + \begin{vmatrix} a_2 & a_3 \\ c_2 & c_3 \end{vmatrix}.$$

By (2*) the sum of the two determinants is the first component of $(\mathbf{a} \times \mathbf{b}) + (\mathbf{a} \times \mathbf{c})$, the right side of (5a). For the other components in (5a) and in (5b), equality follows by the same idea.

Anticommutativity (6) follows from (2**) by noting that the interchange of rows 2 and 3 multiplies the determinant by -1. We can confirm this geometrically if we set $\mathbf{a} \times \mathbf{b} = \mathbf{v}$ and $\mathbf{b} \times \mathbf{a} = \mathbf{w}$; then $|\mathbf{v}| = |\mathbf{w}|$ by (1), and for $\mathbf{b}$, $\mathbf{a}$, $\mathbf{w}$ to form a *right-handed* triple, we must have $\mathbf{w} = -\mathbf{v}$.

Finally, $\mathbf{i} \times (\mathbf{i} \times \mathbf{j}) = \mathbf{i} \times \mathbf{k} = -\mathbf{j}$, whereas $(\mathbf{i} \times \mathbf{i}) \times \mathbf{j} = \mathbf{0} \times \mathbf{j} = \mathbf{0}$ (see Example 3). This proves (7).   ◄

Typical Applications of Vector Products

EXAMPLE 4   **Moment of a force**

In mechanics the moment m of a force $\mathbf{p}$ about a point Q is defined as the product $m = |\mathbf{p}| d$, where d is the (perpendicular) distance between Q and the line of action L of $\mathbf{p}$ (Fig. 172 on the next page). If $\mathbf{r}$ is the vector from Q to any point A on L, then $d = |\mathbf{r}| \sin \gamma$ (Fig. 172) and

$$m = |\mathbf{r}|\,|\mathbf{p}| \sin \gamma.$$

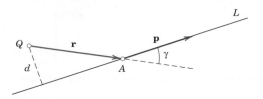

Fig. 172. Moment of a force

Since γ is the angle between $\mathbf{r}$ and $\mathbf{p}$,

$$m = |\mathbf{r} \times \mathbf{p}|,$$

as follows from (1). The vector,

(8)
$$\boxed{\mathbf{m} = \mathbf{r} \times \mathbf{p}}$$

is called the **moment vector** or **vector moment** of $\mathbf{p}$ about Q. Its magnitude is m. If $\mathbf{m} \neq \mathbf{0}$, its direction is that of the axis of the rotation about Q that $\mathbf{p}$ has the tendency to produce. This axis is perpendicular to both $\mathbf{r}$ and $\mathbf{p}$. ◀

EXAMPLE 5   **Moment of a force**

Find the moment of the force $\mathbf{p}$ in Fig. 173 about the center of the wheel.

Solution.   Introducing coordinates as shown in Fig. 173, we have

$$\mathbf{p} = [1000 \cos 30°, \quad 1000 \sin 30°, \quad 0] = [866, \quad 500, \quad 0], \qquad \mathbf{r} = [0, \quad 1.5, \quad 0].$$

(Note that the center of the wheel is at $y = -1.5$ on the y-axis.) Hence (8) and (2**) give

$$\mathbf{m} = \mathbf{r} \times \mathbf{p} = \begin{vmatrix} \mathbf{i} & \mathbf{j} & \mathbf{k} \\ 0 & 1.5 & 0 \\ 866 & 500 & 0 \end{vmatrix} = 0\mathbf{i} - 0\mathbf{j} + \begin{vmatrix} 0 & 1.5 \\ 866 & 500 \end{vmatrix} \mathbf{k} = [0, \quad 0, \quad -1299].$$

This moment vector is normal (perpendicular) to the plane of the wheel; hence it has the direction of the axis of rotation about the center of the wheel that the force has the tendency to produce. $\mathbf{m}$ points in the negative z-direction, the direction in which a right-handed screw would advance if turned in that way. ◀

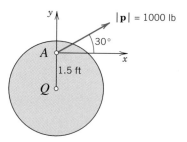

Fig. 173. Moment of a force $\mathbf{p}$

EXAMPLE 6   **Velocity of a rotating body**

A rotation of a rigid body B in space can be simply and uniquely described by a vector $\mathbf{w}$ as follows. The direction of $\mathbf{w}$ is that of the axis of rotation and such that the rotation appears clockwise if one looks from the initial point of $\mathbf{w}$ to its terminal point. The length of $\mathbf{w}$ is equal to the **angular speed** ω (> 0) of the rotation, that is, the linear (or tangential) speed of a point of B divided by its distance from the axis of rotation.

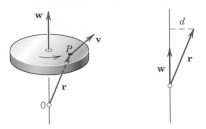

Fig. 174.   Rotation of a rigid body

Let P be any point of B and d its distance from the axis. Then P has the speed ωd. Let $\mathbf{r}$ be the position vector of P referred to a coordinate system with origin 0 on the axis of rotation. Then $d = |\mathbf{r}| \sin \gamma$, where γ is the angle between $\mathbf{w}$ and $\mathbf{r}$. Therefore,

$$\omega d = |\mathbf{w}|\, |\mathbf{r}| \sin \gamma = |\mathbf{w} \times \mathbf{r}|.$$

From this and the definition of vector product we see that the velocity vector $\mathbf{v}$ of P can be represented in the form (Fig. 174)

(9)
$$\boxed{\mathbf{v} = \mathbf{w} \times \mathbf{r}.}$$

This simple formula is useful for determining $\mathbf{v}$ at any point of B.   ◀

Scalar Triple Product

The **scalar triple product** or **mixed triple product** of three vectors

$$\mathbf{a} = [a_1, a_2, a_3], \qquad \mathbf{b} = [b_1, b_2, b_3], \qquad \mathbf{c} = [c_1, c_2, c_3]$$

is denoted by $(\mathbf{a}\ \ \mathbf{b}\ \ \mathbf{c})$ and is defined by

$$(\mathbf{a}\ \ \mathbf{b}\ \ \mathbf{c}) = \mathbf{a} \cdot (\mathbf{b} \times \mathbf{c}).$$

We can write this as a third-order determinant. For this we set $\mathbf{b} \times \mathbf{c} = \mathbf{v} = [v_1, v_2, v_3]$. Then from the dot product in components [formula (2) in Sec. 8.2] and from (2*) (with $\mathbf{b}$ and $\mathbf{c}$ instead of $\mathbf{a}$ and $\mathbf{b}$) we first obtain

$$\mathbf{a} \cdot (\mathbf{b} \times \mathbf{c}) = \mathbf{a} \cdot \mathbf{v} = a_1 v_1 + a_2 v_2 + a_3 v_3$$

$$= a_1 \begin{vmatrix} b_2 & b_3 \\ c_2 & c_3 \end{vmatrix} - a_2 \left(- \begin{vmatrix} b_3 & b_1 \\ c_3 & c_1 \end{vmatrix} \right) + a_3 \begin{vmatrix} b_1 & b_2 \\ c_1 & c_2 \end{vmatrix}.$$

The expression on the right is the expansion of a third-order determinant by its first row. Thus

(10)
$$(\mathbf{a}\ \ \mathbf{b}\ \ \mathbf{c}) = \mathbf{a} \cdot (\mathbf{b} \times \mathbf{c}) = \begin{vmatrix} a_1 & a_2 & a_3 \\ b_1 & b_2 & b_3 \\ c_1 & c_2 & c_3 \end{vmatrix}.$$

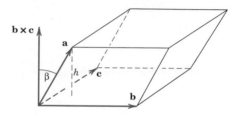

Fig. 175. Geometrical interpretation of a scalar triple product

We also have for any scalar k

(11)
$$(k\mathbf{a} \quad \mathbf{b} \quad \mathbf{c}) = k(\mathbf{a} \quad \mathbf{b} \quad \mathbf{c})$$

because the multiplication of a row of a determinant by k multiplies the value of the determinant by k. Furthermore, we prove that

(12)
$$\mathbf{a} \cdot (\mathbf{b} \times \mathbf{c}) = (\mathbf{a} \times \mathbf{b}) \cdot \mathbf{c}.$$

Dot multiplication is commutative, so that on the right in (12),

$$(\mathbf{a} \times \mathbf{b}) \cdot \mathbf{c} = \mathbf{c} \cdot (\mathbf{a} \times \mathbf{b}) = \begin{vmatrix} c_1 & c_2 & c_3 \\ a_1 & a_2 & a_3 \\ b_1 & b_2 & b_3 \end{vmatrix}.$$

Now interchange in this determinant rows 1 and 2 and in the result rows 2 and 3. This produces the determinant in (10) but leaves the value of the determinant unchanged because each interchange gives a factor -1, and $(-1)(-1) = 1$. This proves (12).

Geometric Interpretation of Scalar Triple Products

The absolute value of the scalar triple product (10) is the volume of the parallelepiped with $\mathbf{a}, \mathbf{b}, \mathbf{c}$ as edge vectors (Fig. 175). Indeed, by (1) in Sec. 8.2,

$$|\mathbf{a} \cdot (\mathbf{b} \times \mathbf{c})| = |\mathbf{a}||\mathbf{b} \times \mathbf{c}| \cos \beta \qquad \text{(Fig. 175)}$$

where $|\mathbf{a}| |\cos \beta|$ is the height h and, by (1), the base, the parallelogram with sides $\mathbf{b}$ and $\mathbf{c}$, has area $|\mathbf{b} \times \mathbf{c}|$.

EXAMPLE 7   **Tetrahedron**

A tetrahedron is determined by three edge vectors $\mathbf{a}, \mathbf{b}, \mathbf{c}$ as indicated in Fig. 176.
   Find its volume if with respect to right-handed Cartesian coordinates, $\mathbf{a} = [2, 0, 3]$, $\mathbf{b} = [0, 6, 2]$, $\mathbf{c} = [3, 3, 0]$.

Fig. 176.   Tetrahedron

Solution. The volume V of the parallelepiped with these vectors as edge vectors is the absolute value of the scalar triple product

$$(\mathbf{a} \quad \mathbf{b} \quad \mathbf{c}) = \begin{vmatrix} 2 & 0 & 3 \\ 0 & 6 & 2 \\ 3 & 3 & 0 \end{vmatrix} = 2 \begin{vmatrix} 6 & 2 \\ 3 & 0 \end{vmatrix} + 3 \begin{vmatrix} 0 & 6 \\ 3 & 3 \end{vmatrix} = -12 - 54 = -66,$$

that is, $V = 66$. The minus sign indicates that $\mathbf{a}, \mathbf{b}, \mathbf{c}$, in this order, form a *left-handed* triple. The volume of the tetrahedron is $\frac{1}{6}$ of that of the parallelepiped, hence 11.

Can you sketch the tetrahedron, choosing the origin as the common initial point of the three vectors? What are the coordinates of the four vertices? ◀

Linear independence of three vectors can be tested by scalar triple products, as follows. We call a given set of vectors $\mathbf{a}_{(1)}, \cdots, \mathbf{a}_{(m)}$ **linearly independent** if the only scalars $c_1, \cdots, c_m$ for which the vector equation

$$\boxed{c_1 \mathbf{a}_{(1)} + c_2 \mathbf{a}_{(2)} + \cdots + c_m \mathbf{a}_{(m)} = \mathbf{0}}$$

is satisfied are $c_1 = 0, c_2 = 0, \cdots, c_m = 0$. Otherwise, that is, if that equation also holds for an m-tuple of scalars not all zero, we call that set of vectors **linearly dependent.**

Now three vectors, if we let their initial point coincide, form a linearly independent set if and only if they do not lie in the same plane (or on the same line). The interpretation of a scalar triple product as a volume thus gives the following criterion.

THEOREM 1   **(Linear independence of three vectors)**

Three vectors form a linearly independent set if and only if their scalar triple product is not zero.

The scalar triple product is the most important "repeated product." Several others that one needs occasionally are included in the problem set.

This is the end of vector *algebra* (in 3-space and in the plane). Vector *calculus* (differentiation) begins in the next section.

PROBLEM SET 8.3

Vector Product, Scalar Triple Product

In Probs. 1–18, with respect to a right-handed Cartesian coordinate system, let $\mathbf{a} = [1, 2, 0]$, $\mathbf{b} = [-3, 2, 0]$, $\mathbf{c} = [2, 3, 4]$, $\mathbf{d} = [6, -7, 2]$. Find the following expressions. (Show the details of your work.)

1. $\mathbf{a} \times \mathbf{b}, \mathbf{b} \times \mathbf{a}, \mathbf{a} \cdot \mathbf{b}, \mathbf{b} \cdot \mathbf{a}$
2. $\mathbf{b} \times \mathbf{c}, \mathbf{c} \times \mathbf{b}, \mathbf{b} \cdot \mathbf{c}, \mathbf{c} \cdot \mathbf{b}$
3. $\mathbf{a} \times \mathbf{c}, |\mathbf{a} \times \mathbf{c}|, |\mathbf{c} \times \mathbf{a}|, \mathbf{a} \cdot \mathbf{c}$
4. $\mathbf{b} \times \mathbf{d} - \mathbf{d} \times \mathbf{b}$
5. $3\mathbf{a} \times 5\mathbf{b}, 5\mathbf{a} \times 3\mathbf{b}, 15\mathbf{a} \times \mathbf{b}$
6. $\mathbf{b} \times \mathbf{b}, (\mathbf{b} - \mathbf{c}) \times (\mathbf{c} - \mathbf{b}), \mathbf{b} \cdot \mathbf{b}$
7. $(\mathbf{a} + \mathbf{b}) \times \mathbf{c}, \mathbf{a} \times \mathbf{c} + \mathbf{b} \times \mathbf{c}$
8. $(\mathbf{b} - \mathbf{d}) \times \mathbf{a}, \mathbf{b} \times \mathbf{a} - \mathbf{d} \times \mathbf{a}$
9. $(\mathbf{a} \times \mathbf{b}) \times \mathbf{c}, \mathbf{a} \times (\mathbf{b} \times \mathbf{c})$
10. $(\mathbf{a} + \mathbf{b}) \times (\mathbf{b} + \mathbf{a})$
11. $(\mathbf{a} \cdot \mathbf{b})\mathbf{c}, (\mathbf{a} \times \mathbf{b}) \cdot \mathbf{c}$
12. $(\mathbf{a} \times \mathbf{b}) \times \mathbf{b}, \mathbf{a} \times (\mathbf{b} \times \mathbf{b})$
13. $(\mathbf{i} \quad \mathbf{j} \quad \mathbf{k}), (\mathbf{i} \quad \mathbf{k} \quad \mathbf{j})$
14. $(\mathbf{b} \times \mathbf{c}) \cdot \mathbf{d}, \mathbf{b} \cdot (\mathbf{c} \times \mathbf{d})$

15. $(\mathbf{a} \times \mathbf{b}) \cdot (\mathbf{c} \times \mathbf{d})$, $(\mathbf{b} \times \mathbf{a}) \cdot (\mathbf{d} \times \mathbf{c})$

16. $(\mathbf{a} \ \ \mathbf{b} \ \ \mathbf{c})$, $(\mathbf{a} - \mathbf{b} \ \ \mathbf{b} - \mathbf{c} \ \ \mathbf{c})$

17. $(\mathbf{b} \ \ \mathbf{a} \ \ \mathbf{d})$, $(\mathbf{a} \ \ \mathbf{b} \ \ \mathbf{d})$

18. $(3\mathbf{a} \ \ 2\mathbf{c} \ \ 4\mathbf{d})$, $24(\mathbf{a} \ \ \mathbf{c} \ \ \mathbf{d})$

19. What properties of cross multiplication do Probs. 1, 5, 7, 9, and 12 illustrate?

Applications of Vector Products and Scalar Triple Products

20. (Rotation) A wheel is rotating about the x-axis with angular speed $3 \ \sec^{-1}$. The rotation appears clockwise if one looks from the origin in the positive x-direction. Find the velocity and the speed at the point $(2, 2, 2)$.

21. (Rotation) What are the velocity and speed in Prob. 20 at the point $(3, 4, 8)$ if the wheel rotates about the y-axis and $\omega = 9 \ \sec^{-1}$?

Moment of a force. A force $\mathbf{p}$ acts on a line through a point A. Find the moment vector $\mathbf{m}$ of $\mathbf{p}$ about a point Q, where the force, the point A, and the point Q are

22. $\mathbf{p} = [2, 1, 0]$, A: $(1, 3, 0)$, Q: $(4, -1, 0)$

23. $\mathbf{p} = [0, 0, 10]$, A: $(0, 0, 0)$, Q: $(2, 2, 0)$

24. $\mathbf{p} = [3, -1, 2]$, A: $(0, -1, 4)$, Q: $(3, 0, 2)$

25. $\mathbf{p} = [3, 0, -6]$, A: $(0, -1, 4)$, Q: $(4, 6, -1)$

26. (Area of a parallelogram) Find the area if the vertices are $(1, 1)$, $(4, -2)$, $(9, 3)$, $(12, 0)$.

27. (Area of a parallelogram) Find the area if the vertices are $(1, 1, 1)$, $(4, 4, 4)$, $(8, -3, 14)$, $(11, 0, 17)$.

28. (Area of a triangle) Find the area if the vertices are $(2, 1)$, $(4, -1)$, $(6, 3)$. Sketch the triangle.

29. (Triangle in space) Find the area if the vertices are $(1, 3, 2)$, $(3, -4, 2)$, $(5, 0, -5)$.

30. (Plane) Find a normal vector to the plane through the points $(1, 3, 0)$, $(2, 0, 8)$, $(0, 2, 2)$ and from it find an equation for the plane.

31. (Plane) Find the plane through $(1, 2, \frac{1}{4})$, $(4, 2, -2)$, $(0, 8, 4)$.

32. (Parallelepiped) Find the volume if the edge vectors are $\mathbf{i} + \mathbf{j}$, $-2\mathbf{i} + 2\mathbf{k}$, $-2\mathbf{i} - 3\mathbf{k}$. Make a sketch.

33. (Parallelepiped) Find the volume if the edge vectors are $[4, 9, -1]$, $[2, 6, 0]$, $[5, -4, 2]$.

34. (Tetrahedron) Find the volume if the vertices are $(1, 3, 6)$, $(3, 7, 12)$, $(8, 8, 9)$, $(2, 2, 8)$.

35. (Tetrahedron) Find the volume if the vertices are $(1, 1, 1)$, $(5, -7, 3)$, $(7, 4, 8)$, $(10, 7, 4)$.

36. (Linear independence) Are the vectors $[4, 2, 9]$, $[3, 2, 1]$, $[-4, 6, 9]$ linearly independent?

37. (Linear independence) Are the vectors $[3, 5, 9]$, $[73, -56, 76]$, $[-4, 7, -1]$ linearly independent?

38. TEAM PROJECT. Useful Formulas for Repeated Products. The following formulas are often useful in practical work. In each case give a proof and an illustration with two or three typical examples of your own choice.

(a)
$$|\mathbf{a} \times \mathbf{b}| = \sqrt{(\mathbf{a} \cdot \mathbf{a})(\mathbf{b} \cdot \mathbf{b}) - (\mathbf{a} \cdot \mathbf{b})^2}.$$

Could you explain this formula geometrically? What do you obtain for orthogonal vectors?

(b)
$$\mathbf{b} \times (\mathbf{c} \times \mathbf{d}) = (\mathbf{b} \cdot \mathbf{d})\mathbf{c} - (\mathbf{b} \cdot \mathbf{c})\mathbf{d}.$$

Hint. Choose special right-handed Cartesian coordinates such that $\mathbf{d} = [d_1, 0, 0]$ and $\mathbf{c} = [c_1, c_2, 0]$. Then verify that each side equals $[-b_2 c_2 d_1, b_1 c_2 d_1, 0]$. Then give a reason why the two sides must thus be equal in any Cartesian coordinate system.

(c)
$$(\mathbf{a} \times \mathbf{b}) \times (\mathbf{c} \times \mathbf{d}) = (\mathbf{a} \ \ \mathbf{b} \ \ \mathbf{d})\mathbf{c} - (\mathbf{a} \ \ \mathbf{b} \ \ \mathbf{c})\mathbf{d}. \qquad \text{[Use (b).]}$$

(d) **Lagrange's identity** [use (b)]:

$$(\mathbf{a} \times \mathbf{b}) \cdot (\mathbf{c} \times \mathbf{d}) = (\mathbf{a} \cdot \mathbf{c})(\mathbf{b} \cdot \mathbf{d}) - (\mathbf{a} \cdot \mathbf{d})(\mathbf{b} \cdot \mathbf{c}).$$

(e)
$$(\mathbf{a} \ \ \mathbf{b} \ \ \mathbf{c}) = -(\mathbf{b} \ \ \mathbf{a} \ \ \mathbf{c}) = -(\mathbf{a} \ \ \mathbf{c} \ \ \mathbf{b})$$

$$(\mathbf{a} \ \ \mathbf{b} \ \ \mathbf{c}) = (\mathbf{b} \ \ \mathbf{c} \ \ \mathbf{a}) = (\mathbf{c} \ \ \mathbf{a} \ \ \mathbf{b})$$

8.4 Vector and Scalar Functions and Fields. Derivatives

This is the beginning of vector *calculus,* which involves two kinds of functions, **vector functions,** whose values are vectors

$$\mathbf{v} = \mathbf{v}(P) = \big[v_1(P),\ v_2(P),\ v_3(P)\big]$$

depending on the points P in space, and **scalar functions,** whose values are scalars

$$f = f(P)$$

depending on P. In applications, the domain of definition for such a function is a region of space or a surface in space or a curve in space. We say that a vector function defines a **vector field** in that region (or on that surface or curve). Examples are shown in Figs. 177–180. Similarly, a scalar function defines a **scalar field** in a region or on a surface or a curve. Examples are the temperature field in a body and the pressure field of the air in the earth's atmosphere. Vector and scalar functions may also depend on time t or on further parameters.

Fig. 177. Field of tangent vectors
of a curve

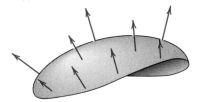

Fig. 178. Field of normal vectors
of a surface

Comment on Notation. If we introduce Cartesian coordinates x, y, z, then instead of $\mathbf{v}(P)$ and $f(P)$ we can also write

$$\mathbf{v}(x,\ y,\ z) = \big[v_1(x,\ y,\ z),\quad v_2(x,\ y,\ z),\quad v_3(x,\ y,\ z)\big]$$

and $f(x,\ y,\ z)$, but we keep in mind that a vector or scalar field that has a geometrical or physical meaning should depend only on the points P where it is defined but not on the particular choice of Cartesian coordinates.

EXAMPLE 1   **Scalar function (Euclidean distance in space)**

The distance $f(P)$ of any point P from a fixed point P_0 in space is a scalar function whose domain of definition is the whole space. $f(P)$ defines a scalar field in space. If we introduce a Cartesian coordinate system and P_0 has the coordinates x_0, y_0, z_0, then f is given by the well-known formula

$$f(P) = f(x,\ y,\ z) = \sqrt{(x - x_0)^2 + (y - y_0)^2 + (z - z_0)^2}$$

where x, y, z are the coordinates of P. If we replace the given Cartesian coordinate system by another such system by translating and rotating the given system, then the values of the coordinates of P and P_0 will in general change, but $f(P)$ will have the same value as before. Hence $f(P)$ is a scalar function. The direction cosines of the line through P and P_0 are not scalars because their values will depend on the choice of the coordinate system. ◀

EXAMPLE 2   Vector field (Velocity field)

At any instant the velocity vectors $\mathbf{v}(P)$ of a rotating body B constitute a vector field, the so-called **velocity field** of the rotation. If we introduce a Cartesian coordinate system having the origin on the axis of rotation, then (see Example 6 in Sec. 8.3)

(1) $$\mathbf{v}(x, y, z) = \mathbf{w} \times \mathbf{r} = \mathbf{w} \times [x, y, z] = \mathbf{w} \times (x\mathbf{i} + y\mathbf{j} + z\mathbf{k})$$

where x, y, z are the coordinates of any point P of B at the instant under consideration. If the coordinates are such that the z-axis is the axis of rotation and $\mathbf{w}$ points in the positive z-direction, then $\mathbf{w} = \omega\mathbf{k}$ and

$$\mathbf{v} = \begin{vmatrix} \mathbf{i} & \mathbf{j} & \mathbf{k} \\ 0 & 0 & \omega \\ x & y & z \end{vmatrix} = \omega[-y, x, 0] = \omega(-y\mathbf{i} + x\mathbf{j}).$$

An example of a rotating body and the corresponding velocity field are shown in Fig. 179. ◀

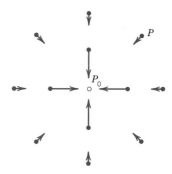

Fig. 179.   Velocity field
of a rotating body

Fig. 180.   Gravitational field
in Example 3

EXAMPLE 3   Vector field (Field of force)

Let a particle A of mass M be fixed at a point P_0 and let a particle B of mass m be free to take up various positions P in space. Then A attracts B. According to **Newton's law of gravitation** the corresponding gravitational force $\mathbf{p}$ is directed from P to P_0, and its magnitude is proportional to $1/r^2$, where r is the distance between P and P_0, say,

(2) $$|\mathbf{p}| = \frac{c}{r^2}, \qquad\qquad c = GMm,$$

where G ($= 6.67 \cdot 10^{-8}$ cm^3/gm $\cdot$ sec^2) is the gravitational constant. Hence $\mathbf{p}$ defines a vector field in space. If we introduce Cartesian coordinates such that P_0 has the coordinates x_0, y_0, z_0 and P has the coordinates x, y, z, then by the Pythagorean theorem,

$$r = \sqrt{(x - x_0)^2 + (y - y_0)^2 + (z - z_0)^2} \qquad (\geqq 0).$$

Assuming that $r > 0$ and introducing the vector

$$\mathbf{r} = [x - x_0, \quad y - y_0, \quad z - z_0] = (x - x_0)\mathbf{i} + (y - y_0)\mathbf{j} + (z - z_0)\mathbf{k},$$

we have $|\mathbf{r}| = r$, and $(-1/r)\mathbf{r}$ is a unit vector in the direction of $\mathbf{p}$; the minus sign indicates that $\mathbf{p}$ is directed from P to P_0 (Fig. 180). From this and (2) we obtain

(3) $$\mathbf{p} = |\mathbf{p}| \left(-\frac{1}{r}\mathbf{r} \right) = -\frac{c}{r^3}\mathbf{r} = -c\frac{x - x_0}{r^3}\mathbf{i} - c\frac{y - y_0}{r^3}\mathbf{j} - c\frac{z - z_0}{r^3}\mathbf{k.}$$

This vector function describes the gravitational force acting on B. ◀

Vector Calculus

We show next that the basic concepts of calculus, such as convergence, continuity, and differentiability, can be defined for vector functions in a simple and natural way. Most important here is the derivative.

Convergence. An infinite sequence of vectors $\mathbf{a}_{(n)}$, $n = 1, 2, \cdots$, is said to **converge** if there is a vector $\mathbf{a}$ such that

$$(4) \qquad \lim_{n \to \infty} |\mathbf{a}_{(n)} - \mathbf{a}| = 0.$$

$\mathbf{a}$ is called the **limit vector** of that sequence, and we write

$$(5) \qquad \lim_{n \to \infty} \mathbf{a}_{(n)} = \mathbf{a}.$$

Cartesian coordinates being given, this sequence of vectors converges to $\mathbf{a}$ if and only if the three sequences of components of the vectors converge to the corresponding components of $\mathbf{a}$. We leave the simple proof to the student.

Similarly, a vector function $\mathbf{v}(t)$ of a real variable t is said to have the **limit** l as t approaches t_0, if $\mathbf{v}(t)$ is defined in some *neighborhood*[7] of t_0 (possibly except at t_0) and

$$(6) \qquad \lim_{t \to t_0} |\mathbf{v}(t) - l| = 0.$$

Then we write

$$(7) \qquad \lim_{t \to t_0} \mathbf{v}(t) = l.$$

Continuity. A vector function $\mathbf{v}(t)$ is said to be **continuous** at $t = t_0$ if it is defined in some neighborhood of t_0 and

$$(8) \qquad \lim_{t \to t_0} \mathbf{v}(t) = \mathbf{v}(t_0).$$

If we introduce a Cartesian coordinate system, we may write

$$\mathbf{v}(t) = [v_1(t), v_2(t), v_3(t)] = v_1(t)\mathbf{i} + v_2(t)\mathbf{j} + v_3(t)\mathbf{k}.$$

Then $\mathbf{v}(t)$ is continuous at t_0 if and only if its three components are continuous at t_0.

We now state the most important of these definitions.

Definition. Derivative of a vector function

A vector function $\mathbf{v}(t)$ is said to be **differentiable** at a point t if the following limit exists:

$$(9) \qquad \boxed{\mathbf{v}'(t) = \lim_{\Delta t \to 0} \frac{\mathbf{v}(t + \Delta t) - \mathbf{v}(t)}{\Delta t}.}$$

---

[7]That is, in some interval (segment) on the t-axis containing t_0 as an interior point (not as an endpoint). In mathematics (as opposed to everyday language), a point (here, t_0) is always regarded as a point of any of its neighborhoods, by definition. This turns out to be practical.

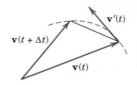

Fig. 181.   Derivative of a vector function

exists. The vector $\mathbf{v}'(t)$ is called the **derivative** of $\mathbf{v}(t)$. See Fig. 181. (The curve in this figure is the locus of the heads of the arrows representing $\mathbf{v}$ for values of the independent variable in some interval containing t and $t + \Delta t$.)   ◀

In terms of components with respect to a given Cartesian coordinate system, $\mathbf{v}(t)$ is differentiable at a point t if and only if its three components are differentiable at t, and then *the derivative $\mathbf{v}'(t)$ is obtained by differentiating each component separately,*

(10)
$$\mathbf{v}'(t) = [v_1'(t), \quad v_2'(t), \quad v_3'(t)].$$

It follows that the familiar rules of differentiation yield corresponding rules for differentiating vector functions, for example,

$$(c\mathbf{v})' = c\mathbf{v}' \qquad\qquad (c \text{ constant}),$$

$$(\mathbf{u} + \mathbf{v})' = \mathbf{u}' + \mathbf{v}'$$

and in particular

(11)
$$(\mathbf{u} \cdot \mathbf{v})' = \mathbf{u}' \cdot \mathbf{v} + \mathbf{u} \cdot \mathbf{v}'$$

(12)
$$(\mathbf{u} \times \mathbf{v})' = \mathbf{u}' \times \mathbf{v} + \mathbf{u} \times \mathbf{v}'$$

(13)
$$(\mathbf{u}\ \ \mathbf{v}\ \ \mathbf{w})' = (\mathbf{u}'\ \ \mathbf{v}\ \ \mathbf{w}) + (\mathbf{u}\ \ \mathbf{v}'\ \ \mathbf{w}) + (\mathbf{u}\ \ \mathbf{v}\ \ \mathbf{w}').$$

The simple proofs are left to the reader. In (12), the order of the vectors must be carefully observed because cross multiplication is not commutative.

EXAMPLE 4   **Derivative of a vector function of constant length**

Let $\mathbf{v}(t)$ be a vector function whose length is constant, say, $|\mathbf{v}(t)| = c$. Then $|\mathbf{v}|^2 = \mathbf{v} \cdot \mathbf{v} = c^2$, and $(\mathbf{v} \cdot \mathbf{v})' = 2\mathbf{v} \cdot \mathbf{v}' = 0$, by differentiation [see (11)]. This yields the following result. *The derivative of a vector function $\mathbf{v}(t)$ of constant length is either the zero vector or is perpendicular to $\mathbf{v}(t)$.*   ◀

Partial Derivatives of a Vector Function

From our present discussion we see that partial differentiation of vector functions depending on two or more variables can be introduced as follows. Suppose that the components of a vector function

$$\mathbf{v} = [v_1, \quad v_2, \quad v_3] = v_1\mathbf{i} + v_2\mathbf{j} + v_3\mathbf{k}$$

are differentiable functions of n variables $t_1, \cdots, t_n$. Then the **partial derivative** of $\mathbf{v}$

with respect to t_l is denoted by $\partial \mathbf{v}/\partial t_l$ and is defined as the vector function

$$\frac{\partial \mathbf{v}}{\partial t_l} = \frac{\partial v_1}{\partial t_l}\mathbf{i} + \frac{\partial v_2}{\partial t_l}\mathbf{j} + \frac{\partial v_3}{\partial t_l}\mathbf{k}.$$

Similarly,

$$\frac{\partial^2 \mathbf{v}}{\partial t_l\, \partial t_m} = \frac{\partial^2 v_1}{\partial t_l\, \partial t_m}\mathbf{i} + \frac{\partial^2 v_2}{\partial t_l\, \partial t_m}\mathbf{j} + \frac{\partial^2 v_3}{\partial t_l\, \partial t_m}\mathbf{k},$$

and so on.

EXAMPLE 5   **Partial derivatives**

Let $\mathbf{r}(t_1, t_2) = a \cos t_1\, \mathbf{i} + a \sin t_1\, \mathbf{j} + t_2 \mathbf{k}$. Then

$$\frac{\partial \mathbf{r}}{\partial t_1} = -a \sin t_1\, \mathbf{i} + a \cos t_1\, \mathbf{j}, \qquad \frac{\partial \mathbf{r}}{\partial t_2} = \mathbf{k}. \qquad \blacktriangleleft$$

Various physical and geometrical applications of derivatives of vector functions will be discussed in the next sections and in Chap. 9.

PROBLEM SET 8.4

Scalar Fields
Scalar Field in the Plane. Consider the scalar field (pressure field) given by $f(x, y) = 9x^2 + 4y^2$. Find:

1. The pressure at the points $(2, 4)$, $(0.5, -3.25)$, $(\sqrt{17}, 1/\sqrt{6})$.

2. The **isobars** (curves of constant pressure). Sketch some of them.

3. The region in which the pressure varies between 36 and 144.

Level Curves. Determine the **isotherms** (curves of constant temperature) of the temperature fields given by the following functions. Sketch some of them.

4. $T = \ln(x^2 + y^2)$	**5.** $T = \arctan(y/x)$	**6.** $T = x^2 - y^2$
7. $T = xy$	**8.** $T = x/(x^2 + y^2)$	**9.** $T = y/(x^2 + y^2)$

 10. CAS PROJECT. Plane Scalar Fields. Plot isotherms of the following fields and describe what they look like.

(a) $x^3 - 3xy^2$	(b) $3x^2y - y^3$	(c) $\cos x \cosh y$
(d) $\sin x \cosh y$	(e) $\ln[(x-1)^2 + y^2]$	(f) $\cos^2 x + \sinh^2 y$

Scalar Fields in Space. What kind of surfaces are the level surfaces of the following scalar fields?

11. $f = 4x + 3y - z$	**12.** $f = x^2 + 3y^2$	**13.** $f = 4x^2 + y^2 + 9z^2$
14. $f = x^2 + y^2 - z$	**15.** $f = z - \sqrt{x^2 + y^2}$	**16.** $f = y^2 - z$

Vector Fields
Sketch figures similar to Fig. 180 of vector fields given by the vector functions

17. $\mathbf{v} = \mathbf{i} + \mathbf{j}$	**18.** $\mathbf{v} = x\mathbf{i} + y\mathbf{j}$	**19.** $\mathbf{v} = y^2\mathbf{i} + \mathbf{j}$
20. $\mathbf{v} = y\mathbf{i} - x\mathbf{j}$	**21.** $\mathbf{v} = -(x^2 + y^2)\mathbf{i}$	**22.** $\mathbf{v} = (1/x)\mathbf{i} + (1/y)\mathbf{j}$

 23. CAS PROJECT. Vector Fields. Plot the following vector fields (by arrows).

(a) $\mathbf{v} = [x, x^3]$	(b) $\mathbf{v} = [1/y, 1/x]$	(c) $\mathbf{v} = (x^2 + y^2)^{-1}[x, -y]$
(d) $\mathbf{v} = e^{(x^2+y^2)}[x, -y]$	(e) $\mathbf{v} = [\cos x, \sin x]$	

24. WRITING PROJECT. Differentiation of Vector Functions. Write a short essay on the essential ideas, including examples of your own and proofs of (11)−(13).

Differentiation of Vector Functions. Find the first partial derivatives.

25. $[y^2, \quad z^2, \quad x^2]$

26. $\cos xyz \, (\mathbf{i} + \mathbf{j})$

27. $[xy, \quad yz, \quad zx]$

28. $[e^x \cos y, \quad e^x \sin y, \quad 0]$

29. $[\cos x \cosh y, \quad -\sin x \sinh y]$

30. $\left[\dfrac{1}{2} \ln (x^2 + y^2), \quad \arctan \dfrac{y}{x}, \quad 0 \right]$

8.5 Curves. Tangents. Arc Length

Curves in space are important in calculus and in physics (for instance, as paths of moving bodies), and we show in this section that they constitute a major field of application of vector calculus. Their theory, together with that of surfaces in space (Sec. 9.5) is called **differential geometry.** It plays a significant role in engineering design, geodesy, geography, space travel, and (in extended form) relativity theory (see Ref. [10] in Appendix 1).

A curve C in space can be represented by a vector function (see Fig. 182)

(1)
$$\mathbf{r}(t) = [x(t), \quad y(t), \quad z(t)] = x(t)\mathbf{i} + y(t)\mathbf{j} + z(t)\mathbf{k}$$

where x, y, z are Cartesian coordinates (the usual rectangular coordinates, see Sec. 8.1). This is called a **parametric representation** of the curve. t is called the **parameter** of the representation. To each value t_0 of t there corresponds a point of C with position vector $\mathbf{r}(t_0)$, that is, with coordinates $x(t_0)$, $y(t_0)$, $z(t_0)$. The parameter t may be time or something else.

In (1) the coordinates x, y, z play the same role—all three are *dependent* variables. We shall see that very often this is a great practical advantage, the key property of a parametric representation. Also, (1) gives an **orientation** of C, a direction of traveling along C so that t increases. This is called the **positive sense** on C given by (1). That of decreasing t is the *negative sense*. Obviously, there are two ways of orienting a curve.

Two other kinds of representations suitable for most space curves C are

(2)
$$y = f(x), \qquad z = g(x)$$

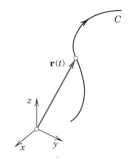

Fig. 182.   Parametric representation
of a curve

(or a pair of equations with y or with z as the independent variable) and

(3) $$F(x, y, z) = 0,\qquad G(x, y, z) = 0.$$

Geometrically, $y = f(x)$ is the projection of C into the xy-plane and $z = g(x)$ the projection into the xz-plane. Equation (3) gives C as the intersection of two surfaces $F(x, y, z) = 0$ and $G(x, y, z) = 0$.

Typical Examples. Kinds of Curves

EXAMPLE 1   **Straight line**

A straight line L through a point A with position vector $\mathbf{a}$ in the direction of a constant vector $\mathbf{b}$ (see Fig. 183) can be represented in the form

(4) $$\mathbf{r}(t) = \mathbf{a} + t\mathbf{b} = [a_1 + tb_1,\ \ a_2 + tb_2,\ \ a_3 + tb_3].$$

If $\mathbf{b}$ is a unit vector, its components are the **direction cosines** of L. In this case, $|t|$ measures the distance of the points of L from A. For instance, the straight line in the xy-plane through A: $(3, 2)$ having slope 1 is (sketch it)

$$\mathbf{r}(t) = [3,\quad 2,\quad 0] + t[1,\quad 1,\quad 0] = [3 + t,\quad 2 + t,\quad 0]. \qquad \blacktriangleleft$$

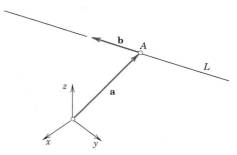

Fig. 183.   Parametric representation
of a straight line

EXAMPLE 2   **Ellipse, circle**

The vector function

(5) $$\mathbf{r}(t) = [a \cos t,\quad b \sin t,\quad 0] = a \cos t\,\mathbf{i} + b \sin t\,\mathbf{j}$$

represents an ellipse in the xy-plane with center at the origin and principal axes in the direction of the x and y axes. In fact, since $\cos^2 t + \sin^2 t = 1$, we obtain from (5)

$$\frac{x^2}{a^2} + \frac{y^2}{b^2} = 1, \qquad z = 0.$$

If $b = a$, then (5) represents a *circle* of radius a. $\blacktriangleleft$

A **plane curve** is a curve that lies in a plane in space. A curve that is not plane is called a **twisted curve.** A standard example is the following.

EXAMPLE 3   **Circular helix**

The twisted curve C represented by the vector function

(6) $$\mathbf{r}(t) = [a \cos t,\quad a \sin t,\quad ct] = a \cos t\,\mathbf{i} + a \sin t\,\mathbf{j} + ct\mathbf{k} \qquad\qquad (c \neq 0)$$

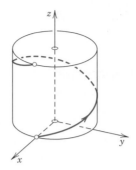

Fig. 184.   Right-handed circular helix

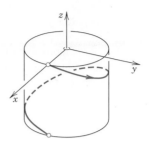

Fig. 185.   Left-handed circular helix

is called a *circular helix*. It lies on the cylinder $x^2 + y^2 = a^2$. If $c > 0$, the helix is shaped like a right-handed screw (Fig. 184). If $c < 0$, it looks like a left-handed screw (Fig. 185). If $c = 0$, then (6) is a circle. ◀

A **simple curve** is a curve without **multiple points,** that is, without points at which the curve intersects or touches itself. Circle and helix are simple. Figure 186 shows curves that are not simple. An example is $[\sin 2t, \quad \cos t, \quad 0]$. Can you sketch it?

An **arc** of a curve is the portion between any two points of the curve. For simplicity, we say "curve" for curves as well as for arcs.

Fig. 186.   Curves with multiple points

Comment on parameter. A curve C may be given by various vector functions. If C is given by (1) and we set $t = h(t^*)$, we obtain a new vector function $\tilde{\mathbf{r}}(t^*)$ representing C. In mechanics, when t is time, this means that we change the motion of a body in time without changing its path. *Example.* For the parabola $\mathbf{r}(t) = [t, \quad t^2]$, by setting $t = -2t^*$, we get the new representation $\tilde{\mathbf{r}}(t^*) = \mathbf{r}(-2t^*) = [-2t^*, \quad 4t^{*2}]$. Do you see that $t = -2t^*$ reverses the orientation? Why? (*Answer:* It has a negative derivative -2.) ◀

The next idea is the approximation of a curve by straight lines, leading to tangents and to a definition of length. Tangents are straight lines touching a curve, as follows.

Tangent to a Curve

The **tangent** to a curve C at a point P of C is the limiting position of a straight line L through P and a point Q of C as Q approaches P along C. See Fig. 187.

If C is given by $\mathbf{r}(t)$, with P and Q corresponding to t and $t + \Delta t$, respectively, then the following vector has the direction of L:

$$\frac{1}{\Delta t} \, [\mathbf{r}(t + \Delta t) - \mathbf{r}(t)].$$

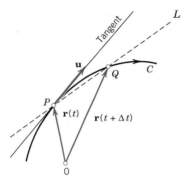

Fig. 187.   Tangent to a curve

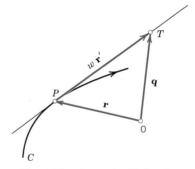

Fig. 188.   Formula (9) for the
tangent to a curve

In the limit this vector becomes the derivative

(7)
$$\mathbf{r}'(t) = \lim_{\Delta t \to 0} \frac{1}{\Delta t} [\mathbf{r}(t + \Delta t) - \mathbf{r}(t)],$$

provided $\mathbf{r}(t)$ is differentiable, as we shall assume from now on. If $\mathbf{r}'(t) \neq \mathbf{0}$, we call $\mathbf{r}'(t)$ a **tangent vector** of C at P because it has the direction of the tangent. The corresponding unit vector is the **unit tangent vector** (see Fig. 187)

(8)
$$\mathbf{u} = \frac{1}{|\mathbf{r}'|} \mathbf{r}'.$$

Note that both $\mathbf{r}'$ and $\mathbf{u}$ point in the direction of increasing t. Hence their sense depends on the orientation of C. It is reversed if we reverse the orientation.

It is now easy to see that the **tangent** to C at P is given by

(9) $\mathbf{q}(w) = \mathbf{r} + w\mathbf{r}'$ (Fig. 188).

This is the sum of the position vector $\mathbf{r}$ of P and a multiple of the tangent vector $\mathbf{r}'$ of C at P. Both vectors depend on P. The variable w is the parameter in (9).

EXAMPLE 4   **Tangent to an ellipse**

Find the tangent to the ellipse $\frac{1}{4}x^2 + y^2 = 1$ at P: $(\sqrt{2}, 1/\sqrt{2})$.

Solution. $\mathbf{r}(t) = 2 \cos t\,\mathbf{i} + \sin t\,\mathbf{j}$, hence $\mathbf{r}'(t) = -2 \sin t\,\mathbf{i} + \cos t\,\mathbf{j}$, and P corresponds to $t = \pi/4$, since $2 \cos (\pi/4) = \sqrt{2}$ and $\sin (\pi/4) = 1/\sqrt{2}$. Thus $\mathbf{r}'(\pi/4) = [-\sqrt{2}, 1/\sqrt{2}]$. *Answer:*

$$\mathbf{q}(w) = [\sqrt{2}, \quad 1/\sqrt{2}] + w[-\sqrt{2}, \quad 1/\sqrt{2}] = \sqrt{2}(1 - w)\mathbf{i} + (1/\sqrt{2})(1 + w)\mathbf{j}.$$

To check the result, sketch the ellipse and the tangent.   ◄

Length of a Curve

We are now ready to define the length l of a curve. l will be the limit of the lengths of broken lines of n chords (see Fig. 189, where $n = 5$) with larger and larger n. For this,

Fig. 189.   Length of a curve

let $\mathbf{r}(t)$, $a \leqq t \leqq b$, represent C. For each $n = 1, 2, \cdots$ we subdivide ("partition") the interval $a \leqq t \leqq b$ by points

$$t_0 \, (= a), \quad t_1, \quad \cdots, \quad t_{n-1}, \quad t_n \, (= b),$$

where $t_0 < t_1 < \cdots < t_n$. This gives a broken line of chords with endpoints $\mathbf{r}(t_0), \cdots,$ $\mathbf{r}(t_n)$. We do this arbitrarily but so that the greatest $|\Delta t_m| = |t_m - t_{m-1}|$ approaches 0 as $n \to \infty$. The lengths $l_1, l_2, \cdots$ of these lines of chords can be obtained from the Pythagorean theorem. If $\mathbf{r}(t)$ has a continuous derivative $\mathbf{r}'(t)$, it can be shown that the sequence $l_1, l_2, \cdots$ has a limit, which is independent of the particular choice of the representation of C and of the choice of subdivisions. This limit is given by the integral

$$\textbf{(10)} \qquad l = \int_a^b \sqrt{\mathbf{r}' \boldsymbol{\cdot} \mathbf{r}'} \; dt \qquad\qquad \left(\mathbf{r}' = \frac{d\mathbf{r}}{dt} \right).$$

l is called the **length** of C, and C is called **rectifiable.** The proof is similar to that for plane curves given in calculus (see Ref. [13]) and can be found in Ref. [10] listed in Appendix 1. The practical evaluation of the integral (10) will be difficult in general. Some simple cases are given in the problem set.

Arc Length s of a Curve

The length L of a curve C is a constant, a positive number. But if we replace the fixed upper limit b in (10) with a variable upper limit t, the integral becomes a function of t, commonly denoted by $s(t)$ and called the *arc length function* or, simply, the **arc length** of C. Thus

$$\textbf{(11)} \qquad s(t) = \int_a^t \sqrt{\mathbf{r}' \boldsymbol{\cdot} \mathbf{r}'} \; d\tilde{t} \qquad\qquad \left(\mathbf{r}' = \frac{d\mathbf{r}}{d\tilde{t}} \right).$$

Here the variable of integration is denoted by $\tilde{t}$ because t is used in the upper limit.

Geometrically, $s(t_0)$ with some $t_0 > a$ is the length of the arc of C between the points with parametric values a and t_0. The choice of a (the point $s = 0$) is arbitrary; changing a means changing s by a constant.

Linear element ds. If we differentiate (11) and square, we have

$$\textbf{(12)} \qquad \left(\frac{ds}{dt} \right)^2 = \frac{d\mathbf{r}}{dt} \boldsymbol{\cdot} \frac{d\mathbf{r}}{dt} = |\mathbf{r}'(t)|^2 = \left(\frac{dx}{dt} \right)^2 + \left(\frac{dy}{dt} \right)^2 + \left(\frac{dz}{dt} \right)^2.$$

It is customary to write

$$\textbf{(13*)} \qquad d\mathbf{r} = [dx, \, dy, \, dz] = dx \, \mathbf{i} + dy \, \mathbf{j} + dz \, \mathbf{k}$$

and

(13)
$$ds^2 = d\mathbf{r} \cdot d\mathbf{r} = dx^2 + dy^2 + dz^2.$$

ds is called the **linear element** of C.

Arc length s as parameter in representations of curves. The use of the special $t = s$ (the arc length) in representations (1) of curves simplifies various formulas. A first instance of this arises for the unit tangent vector, namely,

(14)
$$\mathbf{u}(s) = \mathbf{r}'(s).$$

This follows from (12) with $t = s$ because $ds/ds = 1$. Even greater simplifications occur in curvature and torsion (Sec. 8.7).

EXAMPLE 5   **Circular helix. Circle. Arc length as parameter**

For the helix in Example 3,

$$\mathbf{r}(t) = [a \cos t, \quad a \sin t, \quad ct] = a \cos t\, \mathbf{i} + a \sin t\, \mathbf{j} + ct\mathbf{k},$$

we get the derivative

$$\mathbf{r}'(t) = [-a \sin t, \quad a \cos t, \quad c] = -a \sin t\, \mathbf{i} + a \cos t\, \mathbf{j} + c\mathbf{k},$$

From this, $\mathbf{r}' \cdot \mathbf{r}' = a^2 + c^2$, so that (11) gives

$$s = \int_0^t \sqrt{a^2 + c^2}\, d\tilde{t} = t\sqrt{a^2 + c^2}.$$

Hence $t = s/\sqrt{a^2 + c^2}$, and a formula for the helix with the arc length s as parameter is

$$\mathbf{r}^*(s) = \mathbf{r}\left(\frac{s}{\sqrt{a^2 + c^2}} \right) = a \cos \frac{s}{\sqrt{a^2 + c^2}}\, \mathbf{i} + a \sin \frac{s}{\sqrt{a^2 + c^2}}\, \mathbf{j} + \frac{cs}{\sqrt{a^2 + c^2}}\, \mathbf{k}.$$

Setting $c = 0$, we have $t = s/a$ and obtain for a circle of radius a the representation

$$\mathbf{r}\left(\frac{s}{a} \right) = \left[a \cos \frac{s}{a}, \quad a \sin \frac{s}{a} \right] = a \cos \frac{s}{a}\, \mathbf{i} + a \sin \frac{s}{a}\, \mathbf{j}.$$

The circle is oriented in the counterclockwise sense, which corresponds to increasing values of s. Setting $s = -s^*$ and using $\cos(-\alpha) = \cos \alpha$ and $\sin(-\alpha) = -\sin \alpha$, we obtain

$$\mathbf{r}\left(-\frac{s^*}{a} \right) = \left[a \cos \frac{s^*}{a}, \quad -a \sin \frac{s^*}{a} \right] = a \cos \frac{s^*}{a}\, \mathbf{i} - a \sin \frac{s^*}{a}\, \mathbf{j};$$

we have $ds/ds^* = -1 < 0$, and the circle is now oriented clockwise. ◀

PROBLEM SET 8.5

Straight Lines

Find a parametric representation of the straight line through a point A in the direction of a vector **b** and sketch the line.

1. A: $(4, 2, 0)$, $\mathbf{b} = \mathbf{i} + \mathbf{j}$

2. A: $(-1, 3, 8)$, $\mathbf{b} = [3, 1, 0]$

3. A: $(3, 1, 5)$, $\mathbf{b} = [4, 7, -1]$

4. A: $(1, 1, 1)$, $\mathbf{b} = [-1, 1, -1]$

Find a parametric representation of the straight line through the points A and B. Sketch the line.

5. A: $(2, 3, 0)$, B: $(5, -1, 0)$ **6.** A: $(0, 0, 0)$, B: $(4, 4, 1)$

7. A: $(1, 2, 3)$, B: $(3, 2, 0)$ **8.** A: (a, b, c), B: $(a + 4, 2 - b, c - 1)$

General Curves

What curves are given by the following parametric representations?

9. $[t,\quad t^3 + 2,\quad 0]$ **10.** $[3 \cos t,\quad 4 \sin t,\quad t]$

11. $[0,\quad 5 \cos t,\quad 5 \sin t]$ **12.** $[a + 2 \cos 2t,\quad b - 2 \sin 2t,\quad 0]$

13. $[\cosh t,\quad \sinh t,\quad 0]$ **14.** $[3 + 6 \cos t,\quad -2 + \sin t,\quad 4]$

15. **(Orientation)** What orientation does the curve in Prob. 14 have? What is the simplest transformation to reverse it?

16. **(Parametric transformation)** If we set $t = e^{\tilde{t}}$ in Prob. 9, do we get a representation of the entire curve? (Give a reason.)

Represent the following curves parametrically and sketch them.

17. $y^2 + (z - 3)^2 = 9$, $x = 0$ **18.** $4x^2 - 3y^2 = 12$, $z = 1$

19. $x^2 + y^2 = 1$, $y = z$ **20.** $x^2 + y^2 = 9$, $z = 5 \text{ arc tan } (y/x)$

Tangent Vectors, Tangents

Given a curve C: $\mathbf{r}(t)$, find (a) a tangent vector $\mathbf{r}'(t)$ and the corresponding unit tangent vector $\mathbf{u}(t)$, (b) $\mathbf{r}'$ and $\mathbf{u}$ at the given point P, and (c) the tangent at P. Sketch the curve and the tangent.

21. $\mathbf{r}(t) = t\mathbf{i} + t^3\mathbf{j}$, P: $(1, 1, 0)$ **22.** $\mathbf{r}(t) = 2 \cos t\,\mathbf{i} + 2 \sin t\,\mathbf{j}$, P: $(\sqrt{2}, \sqrt{2}, 0)$

23. $\mathbf{r}(t) = \cos t\,\mathbf{i} + 2 \sin t\,\mathbf{j}$, P: $(\frac{1}{2}, \sqrt{3}, 0)$ **24.** $\mathbf{r}(t) = 2 \cos t\,\mathbf{i} + 2 \sin t\,\mathbf{j} + t\mathbf{k}$, P: $(2, 0, 0)$

25. $\mathbf{r}(t) = \cosh t\,\mathbf{i} + \sinh t\,\mathbf{j}$, P: $(\frac{5}{3}, \frac{4}{3}, 0)$ **26.** $\mathbf{r}(t) = t\mathbf{i} + t^2\mathbf{j} + t^3\mathbf{k}$, P: $(1, 1, 1)$

Length of a Curve

Find the lengths of the following curves. Sketch the curves.

27. **Catenary** $\mathbf{r}(t) = t\mathbf{i} + \cosh t\,\mathbf{j}$ from $t = 0$ to $t = 1$

28. **Circular helix** $\mathbf{r}(t) = a \cos t\,\mathbf{i} + a \sin t\,\mathbf{j} + ct\mathbf{k}$ from $(a, 0, 0)$ to $(a, 0, 2\pi c)$

29. **Semicubical parabola** $\mathbf{r}(t) = t\mathbf{i} + t^{3/2}\mathbf{j}$ from $(0, 0, 0)$ to $(4, 8, 0)$

30. Four-cusped **hypocycloid** $\mathbf{r}(t) = a \cos^3 t\,\mathbf{i} + a \sin^3 t\,\mathbf{j}$, total length

31. **(Representation $y = f(x)$)** From (10) obtain the following formula for the length between $x = a$ and $x = b$:

$$l = \int_a^b \sqrt{1 + y'^2}\, dx \qquad\qquad y' = \frac{dy}{dx}.$$

32. **(Polar representation $\rho = \rho(\theta)$)** If $\rho^2 = x^2 + y^2$ and $\theta \doteq \text{arc tan } (y/x)$, show that $ds^2 = \rho^2\, d\theta^2 + d\rho^2$ and thus

$$l = \int_\alpha^\beta \sqrt{\rho^2 + \rho'^2}\, d\theta \qquad\qquad (\rho' = d\rho/d\theta).$$

Find the total length of the **cardioid** $\rho = a(1 - \cos \theta)$. Sketch this curve.

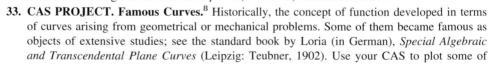 **33.** **CAS PROJECT. Famous Curves.**[8] Historically, the concept of function developed in terms of curves arising from geometrical or mechanical problems. Some of them became famous as objects of extensive studies; see the standard book by Loria (in German), *Special Algebraic and Transcendental Plane Curves* (Leipzig: Teubner, 1902). Use your CAS to plot some of

---

[8]Named after ARCHIMEDES (c. 287—212 B.C.), DESCARTES (Sec. 8.1), DIOCLES (200 B.C.), GABRIEL LAMÉ (1795—1870), MACLAURIN (Sec. 14.4), NICOMEDES (250? B.C.), PASCAL (Sec. 9.4), and STEINER (Sec. 9.6).

them and investigate the dependence of their form on parameters a and b. (r and θ are polar coordinates.)

Archimedes spiral	$r = a\theta$
Hyperbolic spiral	$r = a/\theta,$
Logarithmic spiral	$r = ae^{b\theta}$
Astroid (four-cusped hypocycloid)	$x = 4\cos^3 t,\ y = 4\sin^3 t$
Cardioid	$r = a(1 - \cos\theta)$
Descartes folium	$r = \dfrac{3a\sin 2\theta}{\cos^3\theta + \sin^3\theta}$
Diocles cissoid	$r = \dfrac{2a\sin^2\theta}{\cos\theta}$
Lamé curve	$x^4 + y^4 = 1$
Maclaurin trisectrix	$r = 2a\,\dfrac{\sin 3\theta}{\sin 2\theta}$
Nicomedes conchoid	$r = \dfrac{a}{\cos\theta} + b$
Pascal limaçon (snail)	$r = 2a\cos\theta + b$
Steiner hypocycloid	$x = 2\cos t + \cos 2t,\quad y = 2\sin t - \sin 2t$

8.6 Curves in Mechanics Velocity and Acceleration

Curves as just discussed from the viewpoint of geometry also play an important role in mechanics as paths of moving bodies. To see this, we consider a path C (a curve) given by $\mathbf{r}(t)$, where t now is *time,* and we show that the derivative $\mathbf{r}'(t)$, interpreted geometrically in the last section, is also basic in mechanics, and so is the second derivative $\mathbf{r}''(t)$.

From Sec. 8.5 we know that the vector

(1)
$$\mathbf{v} = \mathbf{r}' = \frac{d\mathbf{r}}{dt}$$

is tangent to C and, therefore, points in the instantaneous direction of motion of the moving body P. From (12) in Sec. 8.5 we see that this vector has the length

$$|\mathbf{v}| = \sqrt{\mathbf{r}' \cdot \mathbf{r}'} = \frac{ds}{dt}.$$

Here s is the arc length, which measures the distance of P from a fixed point ($s = 0$) on C along the curve. Hence ds/dt is the **speed** of P. The vector $\mathbf{v}$ is therefore called the **velocity vector**[9] of the motion.

---

[9]When no confusion is likely to arise, the speed $|\mathbf{v}|$ is also called the **velocity.**

The derivative of the velocity vector is called the **acceleration vector** and will be denoted by **a.** Thus

(2)
$$\mathbf{a}(t) = \mathbf{v}'(t) = \mathbf{r}''(t).$$

EXAMPLE 1   **Centripetal acceleration. Centrifugal force**

The vector function

$$\mathbf{r}(t) = R \cos \omega t\, \mathbf{i} + R \sin \omega t\, \mathbf{j} \qquad\qquad (\omega > 0)$$

represents a circle C of radius R with center at the origin of the xy-plane and describes a motion of a particle P in the counterclockwise sense. The velocity vector

$$\mathbf{v} = \mathbf{r}' = -R\omega \sin \omega t\, \mathbf{i} + R\omega \cos \omega t\, \mathbf{j}$$

(see Fig. 190) is tangent to C. Its magnitude, the speed

$$|\mathbf{v}| = \sqrt{\mathbf{r}' \cdot \mathbf{r}'} = R\omega$$

is constant. The **angular speed** (speed divided by the distance R from the center) is equal to ω. The acceleration vector is

(3)
$$\mathbf{a} = \mathbf{v}' = -R\omega^2 \cos \omega t\, \mathbf{i} - R\omega^2 \sin \omega t\, \mathbf{j} = -\omega^2 \mathbf{r}.$$

We see that there is an acceleration of constant magnitude $|\mathbf{a}| = \omega^2 |\mathbf{r}| = \omega^2 R$ toward the origin. **a** is called the **centripetal acceleration.** It results from the fact that the velocity vector is changing direction at a constant rate. The **centripetal force** is $m\mathbf{a}$, where m is the mass of P. The opposite vector $-m\mathbf{a}$ is called the **centrifugal force,** and the two forces are in equilibrium at each instant of the motion.   ◄

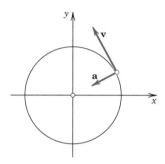

Fig. 190. Centripetal acceleration

Tangential Acceleration and Normal Acceleration

The next idea is the decomposition of acceleration into a component *in* the direction of motion and a component perpendicular ("normal") to it. The practical importance of this idea is almost obvious—think of a moving car, airplane, or rocket.

It is clear that **a** is the time rate of change of **v.** In Example 1 we have $|\mathbf{v}| = const$, but $|\mathbf{a}| \neq 0$. This illustrates that the magnitude of **a** is not in general the rate of change of $|\mathbf{v}|$. The reason is that, in general, **a** is not tangent to the path C. In fact, by applying the chain rule to (1) we have

$$\mathbf{v}(t) = \frac{d\mathbf{r}}{dt} = \frac{d\mathbf{r}}{ds}\frac{ds}{dt} = \mathbf{u}(s)\frac{ds}{dt}$$

where $\mathbf{u}(s) = d\mathbf{r}/ds$ is the unit tangent vector of C (Sec. 8.5), and, by differentiating this again,

(4) $$\mathbf{a}(t) = \frac{d\mathbf{v}}{dt} = \frac{d}{dt}\left(\mathbf{u}(s)\frac{ds}{dt}\right) = \frac{d\mathbf{u}}{ds}\left(\frac{ds}{dt}\right)^2 + \mathbf{u}(s)\frac{d^2s}{dt^2} .$$

Now $\mathbf{u}(s)$ is tangent to C and of constant length (one), so that $d\mathbf{u}/ds$ is perpendicular to $\mathbf{u}(s)$ (recall Example 4 in Sec. 8.4). Hence (4) is a decomposition of the acceleration vector into its normal component $(d\mathbf{u}/ds)(ds/dt)^2$, called the **normal acceleration,** and its tangential component $\mathbf{u}(s)(d^2s/dt^2)$, called the **tangential acceleration.** From this we see that if and only if the normal acceleration is zero, $|\mathbf{a}|$ equals the time rate of change of $|\mathbf{v}| = ds/dt$ (except for the sign), because then we have $|\mathbf{a}| = |\mathbf{u}(s)|\,|d^2s/dt^2| = |d^2s/dt^2|$ from (4).

We denote the normal and tangential accelerations by $\mathbf{a}_{norm}$ and $\mathbf{a}_{tan}$, so that we can write (4) as

$$\mathbf{a} = \mathbf{a}_{norm} + \mathbf{a}_{tan}.$$

The length of $\mathbf{a}_{tan}$ is the projection of $\mathbf{a}$ in the direction of $\mathbf{v}$, given by (11) in Sec. 8.2 as $|\mathbf{a}_{tan}| = \mathbf{a}\cdot\mathbf{v}/|\mathbf{v}|$. Hence $\mathbf{a}_{tan}$ is this expression times the unit vector $(1/|\mathbf{v}|)\mathbf{v}$ in the direction of $\mathbf{v}$, that is,

(4*) $$\mathbf{a}_{tan} = \frac{\mathbf{a}\cdot\mathbf{v}}{\mathbf{v}\cdot\mathbf{v}}\,\mathbf{v}.$$

In Example 1 we can see from (3) and Fig. 190 that the tangential acceleration is **0**. Of course, this will not be the case in general, as the following interesting example illustrates.

EXAMPLE 2   **Tangential and normal accelerations. Coriolis acceleration.**[10]

A small body B moves on a disk toward the edge, its position vector being

(5) $$\mathbf{r}(t) = t\mathbf{b}.$$

The disk is rotating counterclockwise with constant angular speed $\omega = 1$, and $\mathbf{b}$ is a unit vector rotating with the disk (Fig. 191 on the next page). Find the acceleration $\mathbf{a}$ of B.

Solution.   Because of the rotation, $\mathbf{b}$ is of the form

(5′) $$\mathbf{b} = \cos t\,\mathbf{i} + \sin t\,\mathbf{j}.$$

Differentiation of (5) gives the velocity

(6) $$\mathbf{v} = \mathbf{r}' = (t\mathbf{b})' = \mathbf{b} + t\mathbf{b}'.$$

Obviously, $\mathbf{b}$ is the velocity of B relative to the disk. $t\mathbf{b}'$ is the additional velocity due to the rotation. Differentiating once more, we obtain the acceleration

(7) $$\mathbf{a} = \mathbf{v}' = \mathbf{r}'' = (\mathbf{b} + t\mathbf{b}')' = 2\mathbf{b}' + t\mathbf{b}''.$$

---

[10]GUSTAVE GASPARD CORIOLIS (1792—1843), French engineer, who did research in mechanics.

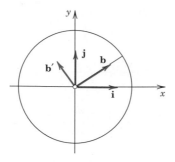

Fig. 191.   Vectors in Example 2

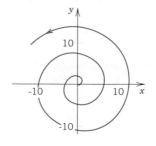

Fig. 192.   Motion in Example 2

Since $\mathbf{b}'' = -\mathbf{b}$, as two differentiations of (5') show, we have $t\mathbf{b}'' = -t\mathbf{b}$. From Example 1 we see that this is the centripetal acceleration due to the rotation. In fact, the instant distance of B from the center is t (because $\mathbf{b}$ is a unit vector), which therefore plays the role of R in Example 1.

The most interesting and probably unexpected term in (7) is $2\mathbf{b}'$, the so-called **Coriolis acceleration,** which results from the interaction of the rotation of the disk and the motion of B on the disk. It has the direction of $\mathbf{b}'$; that is, it is tangential to the edge of the disk and, referred to the fixed xy-coordinate system, it points in the direction of the rotation. If B is a person of mass m_0 walking on the disk according to (5), then B will feel a force $-2m_0\mathbf{b}'$ in the opposite direction, that is, against the sense of the rotation.

In the fixed xy-plane, B describes a spiraling motion (Fig. 192). The tangential acceleration $\mathbf{a}_{\text{tang}}$ is the component of $\mathbf{a}$ in the direction of $\mathbf{v}$ times the unit tangent vector $(1/|\mathbf{v}|)\mathbf{v}$. From (12) in Sec. 8.2 (with $\mathbf{v}$ instead of $\mathbf{b}$), using $|\mathbf{b}| = |\mathbf{b}'| = 1$, $\mathbf{b} \cdot \mathbf{b}' = 0$, $\mathbf{b}'' = -\mathbf{b}$, we obtain

$$\mathbf{a}_{\text{tang}} = \frac{\mathbf{a} \cdot \mathbf{v}}{|\mathbf{v}|^2}\mathbf{v} = \frac{(2\mathbf{b}' + t\mathbf{b}'') \cdot (\mathbf{b} + t\mathbf{b}')}{|\mathbf{b} + t\mathbf{b}'|^2}(\mathbf{b} + t\mathbf{b}')$$

$$= \frac{t}{1 + t^2}(\mathbf{b} + t\mathbf{b}').$$

Hence the normal acceleration is

$$\mathbf{a}_{\text{norm}} = \mathbf{a} - \mathbf{a}_{\text{tang}} = \frac{2 + t^2}{1 + t^2}(\mathbf{b}' - t\mathbf{b}).$$

Can you see that these two vectors are orthogonal?   ◀

EXAMPLE 3   **Superposition of two rotations, Coriolis acceleration**

A uniform motion of a projectile B along a meridian M of the rotating earth (of radius R) can be given by

$$(8) \qquad\qquad \mathbf{r}(t) = R \cos \gamma t \, \mathbf{b} + R \sin \gamma t \, \mathbf{k} \qquad\qquad (\gamma > 0)$$

with

$$\mathbf{b}(t) = \cos \omega t \, \mathbf{i} + \sin \omega t \, \mathbf{j} \qquad\qquad \text{(Fig. 193)}$$

where $\omega \, (> 0)$ is the angular speed of the earth and $\mathbf{i}$ and $\mathbf{j}$ are the unit vectors in the positive x- and y-directions, which are fixed in space. By differentiating (8) we obtain the **velocity**

$$\mathbf{v} = \mathbf{r}' = R \cos \gamma t \, \mathbf{b}' - \gamma R \sin \gamma t \, \mathbf{b} + \gamma R \cos \gamma t \, \mathbf{k}.$$

By differentiating this again with respect to t we obtain the **acceleration**

$$(9) \qquad\qquad \mathbf{a} = \mathbf{v}' = R \cos \gamma t \, \mathbf{b}'' - 2\gamma R \sin \gamma t \, \mathbf{b}' - \gamma^2 R \cos \gamma t \, \mathbf{b} - \gamma^2 R \sin \gamma t \, \mathbf{k}.$$

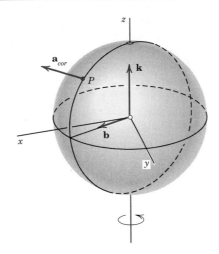

Fig. 193.   Superposition of two rotations

The first term is the centripetal acceleration due to the rotation of the earth. The second term is the **Coriolis acceleration,** call it $\mathbf{a}_{cor}$ (Fig. 193). The third term is the centripetal acceleration due to the motion of B on M.

On the Northern Hemisphere, $\sin \gamma t > 0$ [see (8)], so that $\mathbf{a}_{cor}$ has the direction of $-\mathbf{b}'$, that is, opposite to the rotation of the earth. $|\mathbf{a}_{cor}|$ is maximum at the North Pole and zero at the equator. The projectile B of mass m_0 experiences a force $-m_0\mathbf{a}_{cor}$ opposite to $m_0\mathbf{a}_{cor}$, which tends to let B deviate from M to the right (and on the Southern Hemisphere, where $\sin \gamma t < 0$, to the left). This has been observed for missiles, rockets, shells, and atmospheric air flow. ◀

PROBLEM SET 8.6

Forces on moving objects (cars, planes, etc.) make it necessary that the engineer knows corresponding tangential and normal accelerations. Find them for the following motions. Sketch the path.

1. $\mathbf{r}(t) = 3t\mathbf{i} - 3t\mathbf{j} + 2t\mathbf{k}$

2. $\mathbf{r}(t) = \sin t\,\mathbf{i}$

3. $\mathbf{r}(t) = 5t^2\mathbf{k}$

4. $\mathbf{r}(t) = 2\cos 2t\,\mathbf{i} - 2\sin 2t\,\mathbf{j}$

5. $\mathbf{r}(t) = b\cos t\,\mathbf{i} + b\sin t\,\mathbf{j} + c\mathbf{k}$

6. $\mathbf{r}(t) = \cos t\,\mathbf{i} + \sin 2t\,\mathbf{j}$

7. $\mathbf{r}(t) = e^t\mathbf{i} + e^{-t}\mathbf{j}$

8. CAS PROJECT. Paths of motions more complicated than those in Probs. 1–7 occur in gear transmissions and other engineering constructions. These can be plotted and studied more easily by using a CAS. To grasp the idea, plot the following paths and find the corresponding velocity, speed, and tangential and normal accelerations.

  (a) $\mathbf{r}(t) = [2\cos t + \cos 2t, \quad 2\sin t - \sin 2t]$ (*Steiner's hypocycloid*)

  (b) $\mathbf{r}(t) = [\cos t + \cos 2t, \quad \sin t - \sin 2t]$

  (c) $\mathbf{r}(t) = [\cos t, \quad \sin 2t, \quad \cos 2t]$

  (d) $\mathbf{r}(t) = [ct\cos t, \quad ct\sin t, \quad ct]$ $(c \neq 0)$

9. (Cycloid) Sketch $\mathbf{r}(t) = (R\sin \omega t + \omega Rt)\mathbf{i} + (R\cos \omega t + R)\mathbf{j}$, taking $R = 1$ and $\omega = 1$. This so-called *cycloid* is the path of a point on the rim of a wheel of radius R that rolls without slipping along the x-axis. Find $\mathbf{v}$ and $\mathbf{a}$ at the maximum and minimum y-values of the curve.

10. (Elliptical orbit) Consider the motion $\mathbf{r}(t) = \cos t\,\mathbf{i} + 2\sin t\,\mathbf{j}$. Find the points of maximum speed and acceleration. (Guess first.) Find the tangential and normal accelerations.

11. (Rotation) Obtain (3) from $\mathbf{v} = \mathbf{w} \times \mathbf{r}$ [(9), Sec. 8.3] by differentiation.

12. (Coriolis acceleration) Find the Coriolis acceleration in Example 2 with (5) replaced by $\mathbf{r} = t^2\mathbf{b}$.

13. **(Sun and earth)** Find the acceleration of the earth toward the sun from (3) and the fact that the earth revolves about the sun in a nearly circular orbit with an almost constant speed of 30 km/sec.

14. **(Earth and moon)** Find the centripetal acceleration of the moon toward the earth, assuming that the orbit of the moon is a circle of radius 239,000 miles $= 3.85 \cdot 10^8$ meters and the time for one complete revolution is 27.3 days $= 2.36 \cdot 10^6$ sec.

15. **(Satellite)** Find the speed of an artificial earth satellite traveling at an altitude of 80 miles above the earth's surface where $g = 31$ ft/sec^2. (The radius of the earth is 3960 miles.) *Hint.* See Example 1.

16. **(Satellite)** A satellite moves in a circular orbit 450 miles above the earth's surface and completes 1 revolution in 100 min. Find the acceleration of gravity at the orbit from these data and from the radius of the earth (3960 miles).

8.7 Curvature and Torsion of a Curve
Optional

Together with Sec. 8.5, this section gives the foundations of the theory of space curves, but we shall not need the material included here, so that we can leave this section optional.

The **curvature** $\kappa(s)$ of a curve C, represented by $\mathbf{r}(s)$ with arc length s as parameter, is defined by

(1)
$$\kappa(s) = |\mathbf{u}'(s)| = |\mathbf{r}''(s)| \qquad (' = d/ds).$$

Here $\mathbf{u}(s) = \mathbf{r}'(s)$ is the unit tangent vector of C (Sec. 8.5), and we have to assume that $\mathbf{r}(s)$ is twice differentiable, so that $\mathbf{r}''(s)$ exists.

We see from (1) that κ is the length of the rate of change of the unit tangent vector with s. Hence at each point, κ measures the deviation of C from the tangent. Examples are given below and in the problem set. For general parameters t, the formula for κ is complicated (Prob. 2).

Figure 194 shows the **trihedron** of a curve C with $\kappa \neq 0$ (hence $\kappa > 0$), which is a right-handed triple (Sec. 8.3) consisting of three mutually perpendicular unit vectors,

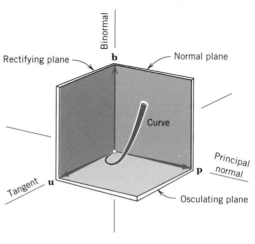

Fig. 194. Trihedron

namely, the **unit tangent vector**

$$\mathbf{u} = \mathbf{r}'$$

the **unit principal normal vector**

(2) $$\mathbf{p} = \frac{1}{|\mathbf{u}'|}\mathbf{u}' = \frac{1}{\kappa}\mathbf{u}' \qquad (\kappa > 0),$$

and the **unit binormal vector**

(3) $$\mathbf{b} = \mathbf{u} \times \mathbf{p} \qquad (\kappa > 0).$$

$\mathbf{p}$ is a unit vector because $|\mathbf{u}'| = \kappa$ by (1). It is perpendicular to $\mathbf{u}$ because $|\mathbf{u}|^2 = \mathbf{u} \cdot \mathbf{u} = 1$ implies $\mathbf{u} \cdot \mathbf{u}' = 0$. The vector $\mathbf{b}$ is perpendicular to $\mathbf{u}$ and $\mathbf{p}$ by the definition of vector product, which also gives $|\mathbf{b}| = 1$ and right-handedness of $\mathbf{u}, \mathbf{p}, \mathbf{b}$.

The straight lines in the directions of $\mathbf{u}, \mathbf{p}, \mathbf{b}$ are called the **tangent,** the **principal normal,** and the **binormal** of C. Figure 194 also shows the names of the three planes spanned by each pair of those vectors.

Torsion of a Curve

Whereas at each point of a curve C the curvature $\kappa(s)$ measures the deviation of C from the tangent, the torsion $\tau(s)$ measures the deviation of C from the osculating plane in Fig. 194. The **torsion** is defined by

(4) $$\tau(s) = -\mathbf{p}(s) \cdot \mathbf{b}'(s).$$

This involves $\mathbf{b}'$, the rate of change of $\mathbf{b}$, hence of the osculating plane because $\mathbf{b}$ is the normal vector of this plane (see Fig. 194). The minus sign is conventional to make τ positive for a right-handed helix (Sec. 8.5). For a plane curve, $\mathbf{b}$ is constant, hence $\tau \equiv 0$.

In (4) we assume that $\kappa(s) > 0$ as before, so that $\mathbf{p}$ exists, and that $\mathbf{r}(s)$ representing C is three times differentiable, so that $\mathbf{b}'$ exists. We show that if $\mathbf{b}' \neq \mathbf{0}$, it has the direction of $\mathbf{p}$ and we get (4) from this. $\mathbf{b}'$ is perpendicular to $\mathbf{b}$ because $|\mathbf{b}| = 1$. It is also perpendicular to $\mathbf{u}$. Indeed, $(\mathbf{b} \cdot \mathbf{u})' = \mathbf{b}' \cdot \mathbf{u} + \mathbf{b} \cdot \mathbf{u}' = \mathbf{b}' \cdot \mathbf{u} = 0$ because $\mathbf{b} \cdot \mathbf{u}' = 0$, as follows from (2), namely, $\mathbf{b} \cdot \mathbf{u}' = \mathbf{b} \cdot (\kappa \mathbf{p}) = \kappa \mathbf{b} \cdot \mathbf{p} = 0$. Hence $\mathbf{b}'$ has the direction of $\mathbf{p}$, say,

(5) $$\mathbf{b}' = -\tau \mathbf{p} \qquad (\kappa > 0).$$

Taking the dot product by $\mathbf{p}$ gives (4).

The concepts just introduced are basic in the theory and application of curves. Let us illustrate them with a typical example. Further applications are given in the problem set.

EXAMPLE 1   **Circular helix. Circle**

In the case of the circular helix (6) in Sec. 8.5 we obtain the arc length $s = t\sqrt{a^2 + c^2}$; see Example 5 in Sec. 8.5. Hence we may represent the helix in the form

$$\mathbf{r}(s) = a\cos\frac{s}{K}\mathbf{i} + a\sin\frac{s}{K}\mathbf{j} + c\frac{s}{K}\mathbf{k} \qquad \text{where} \qquad K = \sqrt{a^2 + c^2}.$$

It follows that

$$\mathbf{u}(s) = \mathbf{r}'(s) = -\frac{a}{K} \sin \frac{s}{K} \mathbf{i} + \frac{a}{K} \cos \frac{s}{K} \mathbf{j} + \frac{c}{K} \mathbf{k}$$

$$\mathbf{r}''(s) = -\frac{a}{K^2} \cos \frac{s}{K} \mathbf{i} - \frac{a}{K^2} \sin \frac{s}{K} \mathbf{j}$$

$$\kappa(s) = |\mathbf{r}''| = \sqrt{\mathbf{r}'' \cdot \mathbf{r}''} = \frac{a}{K^2} = \frac{a}{a^2 + c^2}$$

$$\mathbf{p}(s) = \frac{1}{\kappa(s)} \mathbf{r}''(s) = -\cos \frac{s}{K} \mathbf{i} - \sin \frac{s}{K} \mathbf{j}$$

$$\mathbf{b}(s) = \mathbf{u}(s) \times \mathbf{p}(s) = \frac{c}{K} \sin \frac{s}{K} \mathbf{i} - \frac{c}{K} \cos \frac{s}{K} \mathbf{j} + \frac{a}{K} \mathbf{k}$$

$$\mathbf{b}'(s) = \frac{c}{K^2} \cos \frac{s}{K} \mathbf{i} + \frac{c}{K^2} \sin \frac{s}{K} \mathbf{j}$$

$$\tau(s) = -\mathbf{p}(s) \cdot \mathbf{b}'(s) = \frac{c}{K^2} = \frac{c}{a^2 + c^2}.$$

Hence the circular helix has constant curvature and torsion. If $c > 0$ (right-handed helix, see Fig. 184 in Sec. 8.5), then $\tau > 0$, and if $c < 0$ (left-handed helix, Fig. 185), then $\tau < 0$.

If $c = 0$, we get a circle of radius a, and our formulas then yield $\kappa = 1/a$ (thus the curvature is the reciprocal of the radius) and $\tau = 0$; also, $\mathbf{b}(s) = \mathbf{k}$ is constant, namely, perpendicular to the plane of the circle (the xy-plane).   ◀

Since $\mathbf{u}$, $\mathbf{p}$, and $\mathbf{b}$ are linearly independent vectors, we may represent any vector in space as a linear combination of these vectors. Hence if the derivatives $\mathbf{u}'$, $\mathbf{p}'$, and $\mathbf{b}'$ exist, they may be represented in this fashion. The corresponding formulas are the famous **Frenet formulas**[11]

(6)

(a) $\mathbf{u}' = \qquad\qquad \kappa\mathbf{p}$

(b) $\mathbf{p}' = -\kappa\mathbf{u} \qquad\quad + \tau\mathbf{b}$

(c) $\mathbf{b}' = \qquad\qquad - \tau\mathbf{p}$

Formula (6a) follows from (2), and (6c) is identical to (5). The derivation of (6b) is not very difficult either and is left to the reader (Prob. 16).

A curve is uniquely determined (except for its position in space) if we prescribe its curvature κ (> 0) and torsion τ as continuous functions of arc length s. (Proof in Ref. [10] listed in Appendix 1.) For this reason, one calls $\kappa = \kappa(s)$ and $\tau = \tau(s)$ the **natural equations** of a curve. This also shows why curvature and torsion are basic in the differential geometry of space curves.

This is the end of our discussion of vector functions of a single variable and their application in geometry and mechanics.

---

[11]JEAN-FRÉDÉRIC FRENET (1816—1900), French mathematician.

PROBLEM SET 8.7

Curvature of a Curve

1. Show that the curvature of a circle of radius a equals $1/a$.

2. Using (1), show that if a curve is represented by $\mathbf{r}(t)$, where t is any parameter, then its curvature is

(1′)
$$\kappa(t) = \frac{\sqrt{(\mathbf{r}' \cdot \mathbf{r}')(\mathbf{r}'' \cdot \mathbf{r}'') - (\mathbf{r}' \cdot \mathbf{r}'')^2}}{(\mathbf{r}' \cdot \mathbf{r}')^{3/2}}.$$

3. Using (1′), show that for a curve $y = y(x)$ in the xy-plane,

(1″)
$$\kappa(x) = \frac{|y''|}{(1 + y'^2)^{3/2}} \qquad \left(y' = \frac{dy}{dx}, \text{ etc.}\right).$$

Indicate what kind of curve is represented and find its curvature, using (1′) or (1″).

4. $\mathbf{r} = a \cos t\, \mathbf{i} + b \sin t\, \mathbf{j}$

5. $y = x^2$

6. $xy = c$

7. $\mathbf{r} = a \cos t\, \mathbf{i} + a \sin t\, \mathbf{j} + ct\mathbf{k}$

8. $\mathbf{r} = \cosh t\, \mathbf{i} + \sinh t\, \mathbf{j}$

9. $\mathbf{r} = t\mathbf{i} + t^{3/2}\mathbf{j} \quad (t \geqq 0)$

Torsion of a Curve

10. Using (3) and (4), show that the torsion $\tau(s)$ of a curve $C: \mathbf{r}(s)$ is

(4′)
$$\tau(s) = (\mathbf{u} \quad \mathbf{p} \quad \mathbf{p}') \qquad\qquad (\kappa > 0).$$

11. Using (2), show that (4′) may be written ($' = d/ds$, etc.)

(4″)
$$\tau(s) = (\mathbf{r}' \quad \mathbf{r}'' \quad \mathbf{r}''')/\kappa^2 \qquad\qquad (\kappa > 0).$$

12. Show that if a curve C is represented by $\mathbf{r}(t)$, where t is any parameter, then (4″) becomes ($' = d/dt$, etc.)

(4‴)
$$\tau(t) = \frac{(\mathbf{r}' \quad \mathbf{r}'' \quad \mathbf{r}''')}{(\mathbf{r}' \cdot \mathbf{r}')(\mathbf{r}'' \cdot \mathbf{r}'') - (\mathbf{r}' \cdot \mathbf{r}'')^2} \qquad\qquad (\kappa > 0).$$

13. Show that the torsion of a plane curve (with $\kappa > 0$) is identically zero.

14. Find the torsion of the helix $\mathbf{r}(t) = a \cos t\, \mathbf{i} + a \sin t\, \mathbf{j} + ct\mathbf{k}$. Use (4‴). Compare the result with Example 1.

15. Find the torsion of the curve $C: \mathbf{r}(t) = t\mathbf{i} + t^2\mathbf{j} + t^3\mathbf{k}$ (which looks similar to the curve in Fig. 194).

16. Prove the Frenet formula (6b). *Hint.* Start by differentiating $\mathbf{p} = \mathbf{b} \times \mathbf{u}$.

8.8 Review from Calculus in Several Variables
Optional

From vector functions of a single variable we now proceed to vector functions of several variables, beginning with a review from calculus. ***You should go on to the next section, consulting this material only when needed.*** (We include this short section to keep the book reasonably self-contained. For partial derivatives see Appendix A3.2.)

Chain Rules

THEOREM 1   **(Chain rule)**

Let $w = f(x, y, z)$ be continuous and have continuous first partial derivatives in a domain[12] D in xyz-space. Let $x = x(u, v)$, $y = y(u, v)$, $z = z(u, v)$ be functions that are continuous and have first partial derivatives in a domain B in the uv-plane, where B is such that for every point (u, v) in B, the corresponding point $[x(u, v), y(u, v), z(u, v)]$ lies in D. Then the function

$$w = f(x(u, v), y(u, v), z(u, v))$$

is defined in B, has first partial derivatives with respect to u and v in B, and

(1)
$$\frac{\partial w}{\partial u} = \frac{\partial w}{\partial x}\frac{\partial x}{\partial u} + \frac{\partial w}{\partial y}\frac{\partial y}{\partial u} + \frac{\partial w}{\partial z}\frac{\partial z}{\partial u},$$

$$\frac{\partial w}{\partial v} = \frac{\partial w}{\partial x}\frac{\partial x}{\partial v} + \frac{\partial w}{\partial y}\frac{\partial y}{\partial v} + \frac{\partial w}{\partial z}\frac{\partial z}{\partial v}.$$

Special Cases of Practical Interest

If $w = f(x, y)$ and $x = x(u, v)$, $y = y(u, v)$ as before, then (1) becomes

(2)
$$\frac{\partial w}{\partial u} = \frac{\partial w}{\partial x}\frac{\partial x}{\partial u} + \frac{\partial w}{\partial y}\frac{\partial y}{\partial u},$$

$$\frac{\partial w}{\partial v} = \frac{\partial w}{\partial x}\frac{\partial x}{\partial v} + \frac{\partial w}{\partial y}\frac{\partial y}{\partial v}.$$

If $w = f(x, y, z)$ and $x = x(t)$, $y = y(t)$, $z = z(t)$, then (1) gives

(3)
$$\frac{dw}{dt} = \frac{\partial w}{\partial x}\frac{dx}{dt} + \frac{\partial w}{\partial y}\frac{dy}{dt} + \frac{\partial w}{\partial z}\frac{dz}{dt}.$$

If $w = f(x, y)$ and $x = x(t)$, $y = y(t)$, then (3) reduces to

(4)
$$\frac{dw}{dt} = \frac{\partial w}{\partial x}\frac{dx}{dt} + \frac{\partial w}{\partial y}\frac{dy}{dt}.$$

---

[12]A **domain** D is an open connected point set, where "connected" means that any two points of D can be joined by a broken line of finitely many linear segments all of whose points belong to D. "Open" means that every point P of D has a neighborhood (a little ball with center P) all of whose points belong to D. For example, the interior of a cube or of an ellipsoid (the solid without the boundary surface) is a domain.

Finally, the simplest case $w = f(x)$, $x = x(t)$ gives

(5)
$$\frac{dw}{dt} = \frac{dw}{dx}\frac{dx}{dt} \cdot$$

EXAMPLE 1   **Chain rule**

If $w = x^2 - y^2$ and we define polar coordinates r, θ by $x = r\cos\theta$, $y = r\sin\theta$, then (2) gives

$$\frac{\partial w}{\partial r} = 2x\cos\theta - 2y\sin\theta = 2r\cos^2\theta - 2r\sin^2\theta = 2r\cos 2\theta$$

$$\frac{\partial w}{\partial\theta} = 2x(-r\sin\theta) - 2y(r\cos\theta) = -2r^2\cos\theta\sin\theta - 2r^2\sin\theta\cos\theta = -2r^2\sin 2\theta. \qquad \blacktriangleleft$$

Mean Value Theorems

THEOREM 2   **(Mean value theorem)**

Let $f(x, y, z)$ be continuous and have continuous first partial derivatives in a domain D in xyz-space. Let P_0: (x_0, y_0, z_0) and P: $(x_0 + h, y_0 + k, z_0 + l)$ be points in D such that the straight line segment P_0P joining these points lies entirely in D. Then

(6)
$$f(x_0 + h, y_0 + k, z_0 + l) - f(x_0, y_0, z_0) = h\frac{\partial f}{\partial x} + k\frac{\partial f}{\partial y} + l\frac{\partial f}{\partial z},$$

the partial derivatives being evaluated at a suitable point of that segment.

Special Cases

For a function $f(x, y)$ of two variables (satisfying assumptions as in the theorem), formula (6) reduces to (Fig. 195)

(7)
$$f(x_0 + h, y_0 + k) - f(x_0, y_0) = h\frac{\partial f}{\partial x} + k\frac{\partial f}{\partial y},$$

and for a function $f(x)$ of a single variable, (6) becomes

(8)
$$f(x_0 + h) - f(x_0) = h\frac{df}{dx},$$

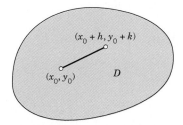

Fig. 195.   Mean value theorem for a function of two variables [Formula (7)]

where in (8), the domain D is a segment of the x-axis and the derivative is taken at a suitable point between x_0 and $x_0 + h$.

PROBLEM SET 8.8

Find dw/dt by (3) or (4) and check the result by substitution and differentiation. (Show the details of your work.)

1. $w = \sqrt{x^2 + y^2}, \quad x = e^{4t}, \quad y = e^{-4t}$

2. $w = x/y, \quad x = g(t), \quad y = h(t)$

3. $w = x^y, \quad x = \cos t, \quad y = \sin t$

4. $w = xy + yz + zx, \quad x = t, \quad y = \cos t, \quad z = \sin t$

5. $w = (x^2 + y^2 + z^2)^{-1/2}, \quad x = \cos t, \quad y = \sin t, \quad z = t$

Find $\partial w/\partial u$ and $\partial w/\partial v$ and check the result by substitution and differentiation, where

6. $w = x^2 + y^2, \quad x = u + v, \quad y = u - v$

7. $w = xy, \quad x = e^u \cos v, \quad y = e^u \sin v$

8. $w = x^4 - 4x^2y^2 + y^4, \quad x = uv, \quad y = u/v$

9. $w = \frac{1}{2}(x^2 + y^2 + z^2)^{-1}, \quad x = u^2 + v^2, \quad y = u^2 - v^2, \quad z = 2uv$

10. **(Partial derivatives on a surface)** Let $w = f(x, y, z)$, and let $z = g(x, y)$ represent a surface S in space. Then on S, the function becomes

$$\widetilde{w}(x, y) = f[x, y, g(x, y)].$$

Show that its partial derivatives are obtained from

$$\frac{\partial \widetilde{w}}{\partial x} = \frac{\partial f}{\partial x} + \frac{\partial f}{\partial z}\frac{\partial g}{\partial x}, \qquad \frac{\partial \widetilde{w}}{\partial y} = \frac{\partial f}{\partial y} + \frac{\partial f}{\partial z}\frac{\partial g}{\partial y} \qquad [z = g(x, y)].$$

Apply this to $f = x^3 + y^3 + z^2$, $g = x^2 + y^2$ and check by substitution and direct differentiation. (The general formula will be needed in Sec. 9.9.)

8.9 Gradient of a Scalar Field Directional Derivative

We shall see that some of the vector fields in applications—not all of them!—can be obtained from scalar fields. This is a considerable advantage because scalar fields can be handled more easily. The relation between the two types of fields is accomplished by the "gradient." Hence the gradient is of great practical importance.

Definition of Gradient

The **gradient** grad f of a given scalar function $f(x, y, z)$ is the vector function defined by

(1*)
$$\text{grad } f = \frac{\partial f}{\partial x}\mathbf{i} + \frac{\partial f}{\partial y}\mathbf{j} + \frac{\partial f}{\partial z}\mathbf{k}.$$

Here we must assume that f is differentiable. It has become popular, particularly with physicists and engineers, to introduce the differential operator

$$(2) \qquad \nabla = \frac{\partial}{\partial x}\mathbf{i} + \frac{\partial}{\partial y}\mathbf{j} + \frac{\partial}{\partial z}\mathbf{k}$$

(read **nabla** or *del*) and to write

$$(1) \qquad \text{grad } f = \nabla f = \frac{\partial f}{\partial x}\mathbf{i} + \frac{\partial f}{\partial y}\mathbf{j} + \frac{\partial f}{\partial z}\mathbf{k}.$$

For instance, if $f(x, y, z) = 2x + yz - 3y^2$, then grad $f = \nabla f = 2\mathbf{i} + (z - 6y)\mathbf{j} + y\mathbf{k}.$

We show later that ***grad f is a vector;*** that is, although it is defined in (1) in terms of components, it has a length and direction that is independent of the particular choice of Cartesian coordinates. (Read this sentence again, to grasp the idea.) But first we explore how the gradient is related to the rate of change of f in various directions. In the directions of the three coordinate axes, this rate is given by the partial derivatives, as we know from calculus. The idea of extending this to arbitrary directions seems natural and leads to the concept of directional derivative.

Directional Derivative

The rate of change of f at any point P in any fixed direction given by a vector **b** is defined as in calculus. We denote it by $D_\mathbf{b}f$ or df/ds, call it the **directional derivative** of f at P *in the direction of* **b,** and define it by (see Fig. 196)

$$(3) \qquad D_\mathbf{b}f = \frac{df}{ds} = \lim_{s \to 0} \frac{f(Q) - f(P)}{s}.$$

Here Q is a variable point on the line L in the direction of **b** and $|s|$ is the distance between P and Q. Also, $s > 0$ if Q lies in the direction of **b** (as in Fig. 196), $s < 0$ if Q lies in the direction of $-\mathbf{b}$, and $s = 0$ if $Q = P$.

The next idea is to use Cartesian xyz-coordinates and for **b** a unit vector. Then the line L is given by

$$(4) \qquad \mathbf{r}(s) = x(s)\mathbf{i} + y(s)\mathbf{j} + z(s)\mathbf{k} = \mathbf{p}_0 + s\mathbf{b} \qquad (|\mathbf{b}| = 1)$$

($\mathbf{p}_0$ the position vector of P). Equation (3) now shows that $D_\mathbf{b}f = df/ds$ is the derivative of the function $f(x(s), y(s), z(s))$ with respect to the arc length s of L. Hence, assuming that f has continuous partial derivatives and applying the chain rule [formula (3) in the

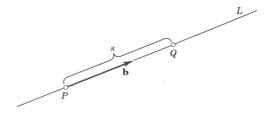

Fig. 196.   Directional derivative

previous section], we obtain

$$(5) \qquad D_{\mathbf{b}}f = \frac{df}{ds} = \frac{\partial f}{\partial x}x' + \frac{\partial f}{\partial y}y' + \frac{\partial f}{\partial z}z'$$

where primes denote derivatives with respect to s (which are taken at $s = 0$). But here, $\mathbf{r}' = x'\mathbf{i} + y'\mathbf{j} + z'\mathbf{k} = \mathbf{b}$ by (4). Hence (5) is simply the inner product of $\mathbf{b}$ and grad f [see (2), Sec. 8.2],

$$(6) \qquad \boxed{D_{\mathbf{b}}f = \frac{df}{ds} = \mathbf{b} \cdot \operatorname{grad} f} \qquad\qquad (|\mathbf{b}| = 1).$$

Attention! If the direction is given by a vector $\mathbf{a}$ of any length $(\neq 0)$, then

$$(6') \qquad \boxed{D_{\mathbf{a}}f = \frac{df}{ds} = \frac{1}{|\mathbf{a}|}\,\mathbf{a} \cdot \operatorname{grad} f.}$$

EXAMPLE 1   **Gradient. Directional derivative**

Find the directional derivative of $f(x, y, z) = 2x^2 + 3y^2 + z^2$ at the point P: $(2, 1, 3)$ in the direction of the vector $\mathbf{a} = \mathbf{i} - 2\mathbf{k}.$

Solution.   We obtain

$$\operatorname{grad} f = 4x\mathbf{i} + 6y\mathbf{j} + 2z\mathbf{k}, \qquad \text{and at } P, \qquad \operatorname{grad} f = 8\mathbf{i} + 6\mathbf{j} + 6\mathbf{k}.$$

From this and $(6')$,

$$D_{\mathbf{a}}f = \frac{1}{\sqrt{5}}(\mathbf{i} - 2\mathbf{k}) \cdot (8\mathbf{i} + 6\mathbf{j} + 6\mathbf{k}) = \frac{1 \cdot 8 - 2 \cdot 6}{\sqrt{5}} = -\frac{4}{\sqrt{5}} \approx -1.789.$$

The minus sign indicates that f decreases at P in the direction of $\mathbf{a}$.   ◄

Gradient Characterizes Maximum Increase

THEOREM 1   **(Gradient, maximum increase)**

Let $f(P) = f(x, y, z)$ be a scalar function having continuous first partial derivatives. Then grad f exists and its length and direction are independent of the particular choice of Cartesian coordinates in space. If at a point P the gradient of f is not the zero vector, it has the direction of maximum increase of f at P.

PROOF.   From (6) and the definition of inner product [(1) in Sec. 8.2] we have

$$(7) \qquad D_{\mathbf{b}}f = |\mathbf{b}|\,|\operatorname{grad} f|\cos\gamma = |\operatorname{grad} f|\cos\gamma$$

where γ is the angle between $\mathbf{b}$ and grad f. Now f is a scalar function. Hence its value at a point P depends on P but not on the particular choice of coordinates. The same holds for the arc length s of the ray C (see Fig. 196), hence also for $D_{\mathbf{b}}f$. Now (7) shows that $D_{\mathbf{b}}f$ is maximum when $\cos\gamma = 1$, $\gamma = 0$, and then $D_{\mathbf{b}}f = |\operatorname{grad} f|$. It follows that the length and direction of grad f are independent of the coordinates. Since $\gamma = 0$ if and only if $\mathbf{b}$ has the direction of grad f, the latter is the direction of maximum increase of f at P, provided grad $f \neq \mathbf{0}$ at P.   ◄

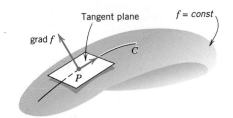

Fig. 197.   Gradient as surface normal vector

Gradient as Surface Normal Vector

Another basic use of the gradient results in connection with surfaces S in space given by

$$(8) \qquad f(x, y, z) = c = const,$$

as follows. We recall that a curve C in space can be given by

$$(9) \qquad \mathbf{r}(t) = x(t)\mathbf{i} + y(t)\mathbf{j} + z(t)\mathbf{k} \qquad \text{[see (1), Sec. 8.5].}$$

Now if we want C to lie on S, its components must satisfy (8); thus

$$(10) \qquad f(x(t), y(t), z(t)) = c.$$

A tangent vector of C is [see (7), Sec. 8.5]

$$\mathbf{r}'(t) = x'(t)\mathbf{i} + y'(t)\mathbf{j} + z'(t)\mathbf{k}.$$

If C lies on S, this vector is tangent to S. At a fixed point P on S, these tangent vectors of all curves on S through P will generally form a plane, called the **tangent plane** of S at P (Fig. 197). Its normal (the straight line through P and perpendicular to the tangent plane) is called the **surface normal** of S at P. A vector parallel to it is called a **surface normal vector** of S at P. Now if we differentiate (10) with respect to t, we get by the chain rule

$$(11) \qquad \frac{\partial f}{\partial x} x' + \frac{\partial f}{\partial y} y' + \frac{\partial f}{\partial z} z' = (\text{grad } f) \cdot \mathbf{r}' = 0.$$

This means orthogonality of grad f and all the vectors $\mathbf{r}'$ in the tangent plane. Our result is (see Fig. 197)

THEOREM 2   **(Gradient as surface normal vector)**

Let f be a differentiable scalar function that represents a surface S: $f(x, y, z) = c = const$. Then if the gradient of f at a point P of S is not the zero vector, it is a normal vector of S at P.

Comment. The surfaces given by (8) with various values of c are called the **level surfaces** of the scalar function f.

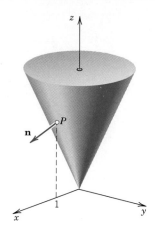

Fig. 198.   Cone and unit normal vector **n**

EXAMPLE 2   **Gradient as surface normal vector**

Find a unit normal vector **n** of the cone of revolution $z^2 = 4(x^2 + y^2)$ at the point P: (1, 0, 2).

Solution.   The cone is the level surface $f = 0$ of $f(x, y, z) = 4(x^2 + y^2) - z^2$. Thus

$$\text{grad } f = 8x\mathbf{i} + 8y\mathbf{j} - 2z\mathbf{k} \qquad \text{and at } P, \qquad \text{grad } f = 8\mathbf{i} - 4\mathbf{k}.$$

Hence, by Theorem 2, a unit normal vector of the cone at P is

$$\mathbf{n} = \frac{1}{|\text{grad } f|} \text{grad } f = \frac{2}{\sqrt{5}}\mathbf{i} - \frac{1}{\sqrt{5}}\mathbf{k} \qquad\qquad \text{(Fig. 198)}$$

and the other one is $-\mathbf{n}$.   ◀

Vector Fields That Are Gradients of a Scalar Field ("Potential")

At the beginning of this section we mentioned that some vector fields have the advantage that they can be obtained from scalar fields, which can be handled more easily. Such a vector field is given by a vector function $\mathbf{v}(P)$, which is obtained as the gradient of a scalar function, say, $\mathbf{v}(P) = \text{grad } f(P)$. The function $f(P)$ is called a *potential function* or a **potential** of $\mathbf{v}(P)$. Such a $\mathbf{v}(P)$ and the corresponding vector field are called **conservative** because in such a vector field, energy is conserved; that is, no energy is lost (or gained) in displacing a body (or a charge in the case of an electrical field) from a point P to another point in the field and back to P. We show this in Sec. 9.2. To obtain a first impression of this method of handling vector fields, let us consider an important application.

EXAMPLE 3   **Gravitational field. Laplace's equation**

In Example 3 of Sec. 8.4 we saw that, by Newton's law of gravitation, the force of attraction between two particles is

(12) $$\mathbf{p} = -\frac{c}{r^3}\mathbf{r} = -c\left(\frac{x - x_0}{r^3}\mathbf{i} + \frac{y - y_0}{r^3}\mathbf{j} + \frac{z - z_0}{r^3}\mathbf{k}\right).$$

Here r is the distance between the two particles at P_0: (x_0, y_0, z_0) and P: (x, y, z); thus

$$r = \sqrt{(x - x_0)^2 + (y - y_0)^2 + (z - z_0)^2}.$$

The key observation now is

(13a)
$$\frac{\partial}{\partial x}\left(\frac{1}{r}\right) = \frac{-2(x - x_0)}{2[(x - x_0)^2 + (y - y_0)^2 + (z - z_0)^2]^{3/2}} = -\frac{x - x_0}{r^3}$$

and similarly

(13b)
$$\frac{\partial}{\partial y}\left(\frac{1}{r}\right) = -\frac{y - y_0}{r^3}, \qquad \frac{\partial}{\partial z}\left(\frac{1}{r}\right) = -\frac{z - z_0}{r^3}.$$

From this we see that **p** is the gradient of the scalar function

$$f(x,\ y,\ z) = \frac{c}{r} \qquad\qquad (r > 0).$$

Thus, f is a potential of that gravitational field (and further potentials are $f + k$ with constant k).

Furthermore, we show next that f satisfies the most important partial differential equation of physics, the so-called **Laplace's equation**

(14)
$$\boxed{\frac{\partial^2 f}{\partial x^2} + \frac{\partial^2 f}{\partial y^2} + \frac{\partial^2 f}{\partial z^2} = 0.}$$

Indeed, to see this, all we have to do is differentiate (13),

$$\frac{\partial^2}{\partial x^2}\left(\frac{1}{r}\right) = -\frac{1}{r^3} + \frac{3(x - x_0)^2}{r^5}, \qquad \frac{\partial^2}{\partial y^2}\left(\frac{1}{r}\right) = -\frac{1}{r^3} + \frac{3(y - y_0)^2}{r^5},$$

$$\frac{\partial^2}{\partial z^2}\left(\frac{1}{r}\right) = -\frac{1}{r^3} + \frac{3(z - z_0)^2}{r^5},$$

and then add these three expressions. Their common denominator is r^5. Hence the three terms $-1/r^3$ contribute $-3r^2$ to the numerator, and the three other terms give $3(x - x_0)^2 + 3(y - y_0)^2 + 3(z - z_0)^2 = 3r^2$, so that the numerator is 0 and we obtain (14), as claimed.

In (14), the expression on the left is called the **Laplacian** of f and is denoted by $\nabla^2 f$ or Δf. The differential operator

(15)
$$\boxed{\nabla^2 = \Delta = \frac{\partial^2}{\partial x^2} + \frac{\partial^2}{\partial y^2} + \frac{\partial^2}{\partial z^2}}$$

(read "nabla squared" or "delta") is called the **Laplace operator.** Using this operator, we may write (14) in the form

$$\boxed{\nabla^2 f = 0.}$$

It can be shown that the field of force produced by any distribution of masses is given by a vector function that is the gradient of a scalar function f, and f satisfies (14) in any region of space that is free of matter.

There are other laws in physics that are of the same form as Newton's law of gravitation. For example, in electrostatics the force of attraction (or repulsion) between two particles of opposite (or like) charges Q_1 and Q_2 is

$$\boxed{\mathbf{p} = \frac{k}{r^3}\mathbf{r}} \qquad\qquad \textbf{(Coulomb's law}[13])$$

---

[13]CHARLES AUGUSTIN DE COULOMB (1736—1806), French physicist and engineer. Coulomb's law was derived by him from his own very precise measurements.

where $k = Q_1 Q_2 / 4\pi\epsilon$, and ϵ is the dielectric constant. Hence **p** is the gradient of the potential $f = -k/r$, and f satisfies (14) when $r > 0$.

Laplace's equation will be discussed in detail in Chaps. 11 and 16.   ◀

PROBLEM SET 8.9

Use of Gradients

Electric Force. The force in an electrostatic field $f(x, y, z)$ has the direction of the gradient grad f. Find grad f and its value at P. Make a sketch. Here, f and P are as follows.

1. $x^2 - y^2$, P: $(-1, 3)$
2. xy, P: $(1, 1)$
3. $\ln (x^2 + y^2)$, P: $(2, 0)$
4. $x^2 + 9y^2$, P: $(-2, 2)$
5. $(x^2 + y^2 + z^2)^{-1/2}$, P: $(2, 1, 3)$
6. $e^x \sin y$, P: $(\ln 2, \frac{1}{4}\pi)$

Heat Flow. The flow of heat in a temperature field takes place in the direction of maximum decrease of temperature T. Find this direction in general and at the given point.

7. $z/(x^2 + y^2)$, P: $(0, 1, 2)$
8. $\sin (x + z)$, P: $\left(\dfrac{\pi}{8}, 1, \dfrac{\pi}{8}\right)$
9. $\cos x \cosh y$, P: $\left(\dfrac{\pi}{2}, 1\right)$
10. $\arctan \dfrac{y}{x}$, P: $(3, 4)$
11. x/y, P: $(8, -1)$
12. $e^{x^2 - y^2} \sin 2xy$, P: $(1, 1)$

13. **(Gradient)** What does it mean if $|\text{grad } f(P)| > |\text{grad } f(Q)| > 0$ at two points P and Q in a scalar field?

14. **(Landscape)** If on a mountain the elevation above sea level is $z(x, y) = 1500 - 3x^2 - 5y^2$ [meters], what is the direction of steepest ascent at P: $(-0.2, 0.1)$?

Curve and Surface Normals. Using gradients, find unit normal vectors for the given curves and surfaces at the given points.

15. $y = \frac{4}{3}x - \frac{2}{3}$, P: $(2, 2)$
16. $y = 1 - x^2$, P: $(1, 0)$
17. $x^2 + y^2 = 25$, P: $(3, 4)$
18. $ax + by + cz + d = 0$, any P
19. $z = \sqrt{x^2 + y^2}$, P: $(6, 8, 10)$
20. $x^2 + y^2 + 2z^2 = 26$, P: $(2, 2, 3)$

Potential Fields. We can handle potential fields f more easily than vector fields $\mathbf{v} = \text{grad } f$. Find f for given **v** or state that **v** has no potential.

21. $[2x, 4y, 8z]$
22. $[yz, xz, xy]$
23. $[xy, 2xy, 0]$
24. $[ye^x, e^x, 1]$
25. $\left[\dfrac{y}{z}, \dfrac{x}{z}, -\dfrac{xy}{z^2}\right]$
26. $(x^2 + y^2)^{-1}[x, y]$

27. **CAS PROJECT. Equipotential Curves.** Plot some **isotherms** (curves of constant temperature) and indicate directions of heat flow by arrows, when the temperature $T(x, y)$ equals

     (a) $x^3 - 3xy^2$,   (b) $\sin x \sinh y$,   (c) $e^x \cos y$.

28. PROJECT. Useful Formulas for Gradients and Laplacians. Prove the following formulas and illustrate each of them with two examples showing where they are advantageous.

$$\nabla(fg) = f\nabla g + g\nabla f, \qquad\qquad \nabla(f^n) = nf^{n-1}\nabla f$$

$$\nabla(f/g) = (1/g^2)(g\nabla f - f\nabla g), \qquad\qquad \nabla^2(fg) = g\nabla^2 f + 2\nabla f \cdot \nabla g + f\nabla^2 g$$

Directional Derivatives

Find the directional derivative of f at P in the direction of $\mathbf{a}$, where

29. $f = x^2 + y^2$, P: $(1, 1)$, $\mathbf{a} = 2\mathbf{i} - 4\mathbf{j}$

30. $f = x - y$, P: $(4, 5)$, $\mathbf{a} = 2\mathbf{i} + \mathbf{j}$

31. $f = 1/\sqrt{x^2 + y^2 + z^2}$, P: $(3, 0, 4)$, $\mathbf{a} = \mathbf{i} + \mathbf{j} + \mathbf{k}$

32. $f = \ln(x^2 + y^2)$, P: $(4, 0)$, $\mathbf{a} = \mathbf{i} - \mathbf{j}$

33. $f = xyz$, P: $(-1, 1, 3)$, $\mathbf{a} = \mathbf{i} - 2\mathbf{j} + 2\mathbf{k}$

34. $f = x^2 + 3y^2 + 4z^2$, P: $(1, 0, 1)$, $\mathbf{a} = -\mathbf{i} - \mathbf{j} + \mathbf{k}$

35. $f = e^x \cos y$, P: $(2, \pi, 0)$, $\mathbf{a} = 2\mathbf{i} + 3\mathbf{j}$

8.10   Divergence of a Vector Field

Vector calculus owes much of its importance in engineering and physics to the gradient, divergence, and curl. Having discussed the gradient, we turn next to the divergence. The curl follows in Sec. 8.11.

Let $\mathbf{v}(x, y, z)$ be a differentiable vector function, where x, y, z are Cartesian coordinates, and let v_1, v_2, v_3 be the components of $\mathbf{v}$. Then the function

(1)
$$\operatorname{div} \mathbf{v} = \frac{\partial v_1}{\partial x} + \frac{\partial v_2}{\partial y} + \frac{\partial v_3}{\partial z}$$

is called the **divergence** *of* $\mathbf{v}$ or the *divergence of the vector field defined by* $\mathbf{v}$. Another common notation for the divergence of $\mathbf{v}$ is $\nabla \cdot \mathbf{v}$,

$$\operatorname{div} \mathbf{v} = \nabla \cdot \mathbf{v} = \left(\frac{\partial}{\partial x}\mathbf{i} + \frac{\partial}{\partial y}\mathbf{j} + \frac{\partial}{\partial z}\mathbf{k}\right) \cdot (v_1\mathbf{i} + v_2\mathbf{j} + v_3\mathbf{k})$$

$$= \frac{\partial v_1}{\partial x} + \frac{\partial v_2}{\partial y} + \frac{\partial v_3}{\partial z},$$

with the understanding that the "product" $(\partial/\partial x)v_1$ in the dot product means the partial derivative $\partial v_1/\partial x$, etc. This is a convenient notation, but nothing more. Note that $\nabla \cdot \mathbf{v}$ means the scalar div $\mathbf{v}$, whereas ∇f means the vector grad f defined in Sec. 8.9.

For example, if

$$\mathbf{v} = 3xz\mathbf{i} + 2xy\mathbf{j} - yz^2\mathbf{k}, \qquad \text{then} \qquad \operatorname{div} \mathbf{v} = 3z + 2x - 2yz.$$

We shall see below that the divergence has an important physical meaning. Clearly the values of a function that characterize a physical or geometrical property must be independent of the particular choice of coordinates; that is, those values must be invariant with respect to coordinate transformations.

THEOREM 1   **(Invariance of the divergence)**

The values of div **v** *depend only on the points in space (and, of course, on* **v***) but not on the particular choice of the coordinates in* (1), *so that with respect to other Cartesian coordinates* x^*, y^*, z^* *and corresponding components* v_1^*, v_2^*, v_3^* *of* **v** *the function* div **v** *is given by*

$$(2) \qquad \text{div } \mathbf{v} = \frac{\partial v_1^*}{\partial x^*} + \frac{\partial v_2^*}{\partial y^*} + \frac{\partial v_3^*}{\partial z^*} .$$

A proof (based on integrals) will be given in Sec. 9.8. Presently, let us turn to the more immediate practical task of getting a feel for the significance of the divergence.

If $f(x, y, z)$ is a twice differentiable scalar function, then

$$\text{grad } f = \frac{\partial f}{\partial x} \mathbf{i} + \frac{\partial f}{\partial y} \mathbf{j} + \frac{\partial f}{\partial z} \mathbf{k}$$

and by (1),

$$\text{div (grad } f) = \frac{\partial^2 f}{\partial x^2} + \frac{\partial^2 f}{\partial y^2} + \frac{\partial^2 f}{\partial z^2} .$$

The expression on the right is the **Laplacian** of f (see Sec. 8.9). Thus

$$(3) \qquad \boxed{\text{div (grad } f) = \nabla^2 f.}$$

EXAMPLE 1   **Gravitational force**

The gravitational force **p** in Example 3, Sec. 8.9, is the gradient of the scalar function $f(x, y, z) = c/r$, which satisfies Laplace's equation $\nabla^2 f = 0$. According to (3), this means that div **p** $= 0$ ($r > 0$).   ◀

The following example, taken from hydrodynamics, shows the physical significance of the divergence of a vector field (and more will be added in Sec. 9.8 when the so-called divergence theorem of Gauss will be available).

EXAMPLE 2   **Motion of a compressible fluid. Physical meaning of the divergence**

We consider the motion of a fluid in a region R having no **sources** or **sinks** in R, that is, no points at which fluid is produced or disappears. The concept of **fluid state** is meant to cover also gases and vapors. Fluids in the restricted **sense,** or **liquids** (water or oil, for instance), have very small compressibility, which can be neglected in many problems. Gases and vapors have large compressibility; that is, their density ρ ($=$ mass per unit volume) depends on the coordinates x, y, z in space (and may depend on time t). We assume that our fluid is compressible.

We consider the flow through a small rectangular box W of dimensions[14] Δx, Δy, Δz with edges parallel to the coordinate axes (Fig. 199). W has the volume $\Delta V = \Delta x \, \Delta y \, \Delta z$. Let **v** $= [v_1, v_2, v_3] = v_1 \mathbf{i} + v_2 \mathbf{j} + v_3 \mathbf{k}$ be the velocity vector of the motion. We set

$$(4) \qquad \mathbf{u} = \rho \mathbf{v} = [u_1, u_2, u_3] = u_1 \mathbf{i} + u_2 \mathbf{j} + u_3 \mathbf{k}$$

and assume that **u** and **v** are continuously differentiable vector functions of x, y, z, and t (that is, they have first partial derivatives, which are continuous). Let us calculate the change in the mass included in W by considering the **flux** across the boundary, that is, the total loss of mass leaving W per unit time. Consider the flow through

---

[14]It is a standard usage to indicate small quantities by Δ; this has, of course, nothing to do with the notation Δ for the Laplacian in (15), Sec. 8.9.

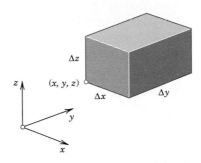

Fig. 199. Physical interpretation of the divergence

the left face of W, whose area is $\Delta x\,\Delta z$. The components v_1 and v_3 of $\mathbf{v}$ are parallel to that face and contribute nothing to this flow. Hence the mass of fluid entering through that face during a short time interval Δt is given approximately by

$$(\rho v_2)_y\,\Delta x\,\Delta z\,\Delta t = (u_2)_y\,\Delta x\,\Delta z\,\Delta t,$$

where the subscript y indicates that this expression refers to the left face. The mass of fluid leaving the box W through the opposite face during the same time interval is approximately $(u_2)_{y+\Delta y}\,\Delta x\,\Delta z\,\Delta t$, where the subscript $y + \Delta y$ indicates that this expression refers to the right face (which is not visible in Fig. 199). The difference

$$\Delta u_2\,\Delta x\,\Delta z\,\Delta t = \frac{\Delta u_2}{\Delta y}\,\Delta V\,\Delta t \qquad \left[\Delta u_2 = (u_2)_{y+\Delta y} - (u_2)_y\right]$$

is the approximate loss of mass. Two similar expressions are obtained by considering the other two pairs of parallel faces of W. If we add these three expressions, we find that the total loss of mass in W during the time interval Δt is approximately

$$\left(\frac{\Delta u_1}{\Delta x} + \frac{\Delta u_2}{\Delta y} + \frac{\Delta u_3}{\Delta z}\right)\Delta V\,\Delta t,$$

where

$$\Delta u_1 = (u_1)_{x+\Delta x} - (u_1)_x$$

and

$$\Delta u_3 = (u_3)_{z+\Delta z} - (u_3)_z.$$

This loss of mass in W is caused by the time rate of change of the density and is thus equal to

$$-\frac{\partial \rho}{\partial t}\,\Delta V\,\Delta t.$$

If we equate both expressions, divide the resulting equation by $\Delta V\,\Delta t$, and let Δx, Δy, Δz and Δt approach zero, then we obtain

$$\operatorname{div}\mathbf{u} = \operatorname{div}(\rho\mathbf{v}) = -\frac{\partial \rho}{\partial t}$$

or

(5)
$$\boxed{\frac{\partial \rho}{\partial t} + \operatorname{div}(\rho\mathbf{v}) = 0.}$$

This important relation is called the *condition for the conservation of mass* or the **continuity equation** *of a compressible fluid flow.*

If the flow is **steady,** that is, independent of time, then $\partial\rho/\partial t = 0$ and the continuity equation is

(6) $$\text{div}\,(\rho\mathbf{v}) = 0.$$

If the density ρ is constant, so that the fluid is incompressible, then equation (6) becomes

(7) $$\boxed{\text{div}\,\mathbf{v} = 0.}$$

This relation is known as the **condition of incompressibility.** It expresses the fact that the balance of outflow and inflow for a given volume element is zero at any time. Clearly, the assumption that the flow has no sources or sinks in R is essential to our argument.

From this discussion you should conclude and remember that, roughly speaking, *the divergence measures outflow minus inflow.* ◀

Comment. The **divergence theorem** of Gauss, an integral theorem involving the divergence, follows in the next chapter (Sec. 9.7).

PROBLEM SET 8.10

Calculation of the Divergence. Find div $\mathbf{v}$ if $\mathbf{v}$ equals

1. $x\mathbf{i} + y\mathbf{j} + z\mathbf{k}$

2. $x^2\mathbf{i} + y^2\mathbf{j} + z^2\mathbf{k}$

3. $e^x(\cos y\,\mathbf{i} + \sin y\,\mathbf{j})$

4. $v_1(y, z)\mathbf{i} + v_2(z, x)\mathbf{j} + v_3(x, y)\mathbf{k}$

5. $(x^2 + y^2)^{-1}(-y\mathbf{i} + x\mathbf{j})$

6. $(x^2 + y^2 + z^2)^{-3/2}(x\mathbf{i} + y\mathbf{j} + z\mathbf{k})$

7. $e^x\mathbf{i} + ye^{-x}\mathbf{j} + 2z\sinh x\,\mathbf{k}$

8. $xyz(x\mathbf{i} + y\mathbf{j} + z\mathbf{k})$

 9. CAS PROJECT. Visualizing the Divergence. Plot the given velocity field $\mathbf{v}$ of a fluid flow in a square centered at the origin. Recall that the divergence measures outflow minus inflow. By looking at the flow near the sides of the square, can you see whether div $\mathbf{v}$ must be positive or negative or may perhaps be zero? Then calculate div $\mathbf{v}$. First do the given flows and then do some of your own. Enjoy it.

(a) $\mathbf{v} = \mathbf{i}$

(b) $\mathbf{v} = x\mathbf{i}$

(c) $\mathbf{v} = x\mathbf{i} - y\mathbf{j}$

(d) $\mathbf{v} = x\mathbf{i} + y\mathbf{j}$

(e) $\mathbf{v} = -x\mathbf{i} - y\mathbf{j}$

(f) $\mathbf{v} = (x^2 + y^2)^{-1}(-y\mathbf{i} + x\mathbf{j})$

10. (Rotational flow) The velocity vector $\mathbf{v}(x, y, z)$ of an incompressible fluid rotating in a cylindrical vessel is of the form $\mathbf{v} = \mathbf{w} \times \mathbf{r}$, where $\mathbf{w}$ is the (constant) rotation vector; see Example 6 in Sec. 8.3. Show that div $\mathbf{v} = 0$. Is this plausible from our present Example 2?

11. (Flow) Consider the flow with velocity vector $\mathbf{v} = y\mathbf{i}$. Show that this flow is incompressible. Show that the particles that at time $t = 0$ are in the cube bounded by the planes $x = 0$, $x = 1$, $y = 0$, $y = 1$, $z = 0$, $z = 1$ occupy at $t = 1$ the volume 1.

12. (Flow) Consider the flow having the velocity $\mathbf{v} = x\mathbf{i}$. Show that the individual particles have the position vectors $\mathbf{r}(t) = c_1 e^t\mathbf{i} + c_2\mathbf{j} + c_3\mathbf{k}$, where c_1, c_2, c_3 are constants, the flow is compressible, and e is the volume occupied at $t = 1$ by the particles that at $t = 0$ fill the cube in Prob. 11.

13. PROJECT. Useful Formulas for the Divergence. Prove

(a) div $(k\mathbf{v}) = k$ div $\mathbf{v}$ (k constant)

(b) div $(f\mathbf{v}) = f$ div $\mathbf{v} + \mathbf{v} \cdot \nabla f$

(c) div $(f \nabla g) = f\nabla^2 g + \nabla f \cdot \nabla g$

(d) div $(f \nabla g) -$ div $(g\nabla f) = f\nabla^2 g - g\nabla^2 f$

Verify (b) for $f = e^{xyz}$ and $\mathbf{v} = ax\mathbf{i} + by\mathbf{j} + cz\mathbf{k}$. Obtain the answer to Prob. 6 from (b). Verify (c) for $f = x^2 - y^2$ and $g = e^{x+y}$. Give examples of your own for which (a)−(d) are advantageous.

Calculation of the Laplacian by (3). Find $\nabla^2 f$ by (3). Check by direct differentiation. Indicate where (3) is simpler. (Show the details of your work.)

14. $f = (x - y)/(x + y)$ **15.** $f = 4x^2 + 9y^2 + z^2$

16. $f = e^{2x} \sin 2y$ **17.** $f = xy/z$

18. $f = \cosh^2 x - \sinh^2 y$ **19.** $f = \text{arc tan } (y/x)$

20. $f = z - \sqrt{x^2 + y^2}$

8.11   Curl of a Vector Field

Gradient (Sec. 8.9), divergence (Sec. 8.10), and curl are basic in connection with fields. We now define and discuss the curl.

Let x, y, z be right-handed Cartesian coordinates (Sec. 8.3), and let

$$\mathbf{v}(x, y, z) = v_1\mathbf{i} + v_2\mathbf{j} + v_3\mathbf{k}$$

be a differentiable vector function. Then the function

$$
\text{curl } \mathbf{v} = \nabla \times \mathbf{v} =
\begin{vmatrix}
\mathbf{i} & \mathbf{j} & \mathbf{k} \\
\dfrac{\partial}{\partial x} & \dfrac{\partial}{\partial y} & \dfrac{\partial}{\partial z} \\
v_1 & v_2 & v_3
\end{vmatrix}
$$

(1)

$$
= \left(\frac{\partial v_3}{\partial y} - \frac{\partial v_2}{\partial z} \right)\mathbf{i} + \left(\frac{\partial v_1}{\partial z} - \frac{\partial v_3}{\partial x} \right)\mathbf{j} + \left(\frac{\partial v_2}{\partial x} - \frac{\partial v_1}{\partial y} \right)\mathbf{k}
$$

is called the **curl** *of the vector function* $\mathbf{v}$ or the *curl of the vector field defined by* $\mathbf{v}$. For a left-handed Cartesian coordinate system, the determinant in (1) has a minus sign in front, in agreement with the remark following (2**) in Sec. 8.3.

Instead of curl $\mathbf{v}$ the notation

$$\text{rot } \mathbf{v}$$

(suggested by "rotation"; see Example 2, below) is also used.

EXAMPLE 1   **Curl of a vector function**

With respect to right-handed Cartesian coordinates, let

$$\mathbf{v} = yz\mathbf{i} + 3zx\mathbf{j} + z\mathbf{k}.$$

Then (1) gives

$$
\text{curl } \mathbf{v} =
\begin{vmatrix}
\mathbf{i} & \mathbf{j} & \mathbf{k} \\
\partial/\partial x & \partial/\partial y & \partial/\partial z \\
yz & 3zx & z
\end{vmatrix}
= -3x\mathbf{i} + y\mathbf{j} + (3z - z)\mathbf{k} = -3x\mathbf{i} + y\mathbf{j} + 2z\mathbf{k}. \qquad \blacktriangleleft
$$

The curl plays an important role in many applications. Let us illustrate this with a typical basic example. (We shall say more about the role and nature of the curl in Sec. 9.9.)

EXAMPLE 2   **Rotation of a rigid body. Relation to the curl**

We have seen in Example 6, Sec. 8.3, that a rotation of a rigid body B about a fixed axis in space can be described by a vector $\mathbf{w}$ of magnitude ω in the direction of the axis of rotation, where ω (> 0) is the angular speed of the rotation, and $\mathbf{w}$ is directed so that the rotation appears clockwise if we look in the direction of $\mathbf{w}$. According to (9), Sec. 8.3, the velocity field of the rotation can be represented in the form

$$\mathbf{v} = \mathbf{w} \times \mathbf{r}$$

where $\mathbf{r}$ is the position vector of a moving point with respect to a Cartesian coordinate system having the origin on the axis of rotation. Let us choose right-handed Cartesian coordinates such that

$$\mathbf{w} = \omega\mathbf{k};$$

that is, the axis of rotation is the z-axis. Then (see Example 2 in Sec. 8.4)

$$\mathbf{v} = \mathbf{w} \times \mathbf{r} = -\omega y\mathbf{i} + \omega x\mathbf{j}$$

and, therefore,

$$\text{curl } \mathbf{v} = \begin{vmatrix} \mathbf{i} & \mathbf{j} & \mathbf{k} \\ \dfrac{\partial}{\partial x} & \dfrac{\partial}{\partial y} & \dfrac{\partial}{\partial z} \\ -\omega y & \omega x & 0 \end{vmatrix} = 2\omega\mathbf{k},$$

that is,

(2) $\text{curl } \mathbf{v} = 2\mathbf{w}.$

Hence, in the case of a rotation of a rigid body, the curl of the velocity field has the direction of the axis of rotation, and its magnitude equals twice the angular speed ω of the rotation.

Note that our result does not depend on the particular choice of the Cartesian coordinate system in space. ◄

For any twice continuously differentiable scalar function f,

(3) $\boxed{\text{curl (grad } f) = \mathbf{0},}$

as can easily be verified by direct calculation. *Hence if a vector function is the gradient of a scalar function, its curl is the zero vector.* Since the curl characterizes the rotation in a field, we also say more briefly that *gradient fields describing a motion are* **irrotational.** (If such a field occurs in some other connection, not as a velocity field, it is usually called **conservative;** see Sec. 8.9.)

EXAMPLE 3   The gravitational field in Example 3, Sec. 8.9, has curl $\mathbf{p} = \mathbf{0}$. The field in Example 2 of this section is not irrotational. A similar velocity field is obtained by stirring coffee in a cup. ◄

Other than (3), another key formula for any twice continuously differentiable scalar function is

(4) $\boxed{\text{div (curl } \mathbf{v}) = 0.}$

It is plausible because of the interpretation of the curl as a rotation and the divergence as a flux (see Example 2 in Sec. 8.10). A proof of (4) follows readily from the definitions of curl and div; the six terms cancel in pairs.

The curl is defined in (1) in terms of coordinates, but if it is supposed to have a physical or geometrical significance, it should not depend on the choice of these coordinates. This is true, as follows.

THEOREM 1   **(Invariance of the curl)**

The length and direction of curl **v** *are independent of the particular choice of Cartesian coordinate systems in space.* (Proof in Appendix 4.)

PROBLEM SET 8.11

1. **WRITING PROJECT. Summary on grad, div, curl.** Make a list of the definitions and the most important facts and formulas for grad, div, curl, and ∇^2. Use your list to write a corresponding essay of $3-4$ pages. Include typical examples.

Calculation of curl. Find curl **v**, where with respect to right-handed Cartesian coordinates, **v** equals

2. $[2y,\ 5x,\ 0]$

3. $\frac{1}{2}(x^2 + y^2 + z^2)(\mathbf{i} + \mathbf{j} + \mathbf{k})$

4. $[v_1(x),\ v_2(y),\ v_3(z)]$

5. $(x^2 + y^2 + z^2)^{-3/2}\ (x\mathbf{i} + y\mathbf{j} + z\mathbf{k})$

6. $[\sin y,\ \cos z,\ 0]$

7. $xyz(x\mathbf{i} + y\mathbf{j} + z\mathbf{k})$

Fluid Motion. In each case, the velocity vector **v** of a fluid motion is given. Is the flow irrotational? Incompressible? Find the paths of the particles. (Show the details of your work.)

8. $\mathbf{v} = [2y^2,\ 0,\ 0]$

9. $\mathbf{v} = x^3\mathbf{k}$

10. $\mathbf{v} = [\sec x,\ \operatorname{cosec} x,\ 0]$

11. $\mathbf{v} = y\mathbf{i} - x\mathbf{j}$

12. $\mathbf{v} = \left[-\frac{1}{4}y,\ 4x,\ 0\right]$

13. $\mathbf{v} = x\mathbf{i} + y\mathbf{j} - z\mathbf{k}$

14. **PROJECT. Useful Formulas Involving the Curl.** Assuming sufficient differentiability, show that

   (a) curl $(\mathbf{u} + \mathbf{v})$ = curl **u** + curl **v**

   (b) div (curl **v**) = 0

   (c) curl $(f\mathbf{v})$ = (grad f) $\times$ **v** + f curl **v**

   (d) curl (grad f) = **0**

   (e) div $(\mathbf{u} \times \mathbf{v})$ = **v**•curl **u** − **u**•curl **v**.

Calculation of Expressions Involving the Curl. With respect to right-handed coordinates, let $\mathbf{u} = y\mathbf{i} + z\mathbf{j} + x\mathbf{k}$, $\mathbf{v} = yz\mathbf{i} + zx\mathbf{j} + xy\mathbf{k}$, and $f = xyz$. Find the following expressions. If one of the formulas in Project 14 applies, use it to check your result. (Show the details of your work.)

15. curl $(f\mathbf{u})$

16. curl **v**,   curl $(f\mathbf{v})$

17. **u** $\times$ curl **v**,   **v** $\times$ curl **u**

18. curl $(\mathbf{u} \times \mathbf{v})$

19. **v**•curl **u**,   **u**•curl **v**

20. **u**•curl **u**,   **u** $\times$ curl **u**

CHAPTER 8 REVIEW

1. What is a vector? A scalar? A vector function? A vector field? A scalar function? A scalar field? Give examples.
2. What is the inner product? What motivates it? What other name does it have?
3. What is the vector product? What motivates it?

4. What is orthogonality? Why is it important?

5. What are right-handed and left-handed coordinates? When is this distinction important?

6. Can you remember the two unusual properties of vector product?

7. When is an inner product zero? A vector product?

8. What is wrong with $\mathbf{a} \times \mathbf{b} \times \mathbf{c}$, $\mathbf{a} \cdot \mathbf{b} \cdot \mathbf{c}$, $(\mathbf{a} \cdot \mathbf{b}) \times \mathbf{c}$?

9. How is the derivative of a vector function defined? What is its significance in mechanics and in geometry?

10. Can a body have constant speed but variable velocity?

11. What do you know about the directional derivative?

12. Write down the definitions of grad, div, and curl from memory and explain the meaning of these highly important concepts.

13. Granted sufficient differentiability, which of the expressions f curl $\mathbf{v}$, $\mathbf{v}$ curl f, $f\mathbf{v}$, $f \cdot \mathbf{v}$, $f \times \mathbf{v}$, $\mathbf{v} \times \mathbf{v}$, $\mathbf{v} \times$ grad f, $\mathbf{v} \times$ (curl $\mathbf{v}$), div $(f\mathbf{v})$, curl $(f\mathbf{v})$ make sense?

Vector Addition, Scalar Multiplication, Products

Let $\mathbf{a} = [3, 1, -2]$, $\mathbf{b} = [-5, 7, 0]$, $\mathbf{c} = [4, -6, 0]$, $\mathbf{d} = [9, 1, 14]$. Find

14. $(4\mathbf{a} + 5\mathbf{b}) \cdot 6\mathbf{c}$

15. $3\mathbf{b} \times 6\mathbf{c}$, $18\mathbf{c} \times \mathbf{b}$

16. $\mathbf{a} \times \mathbf{d}$, $\mathbf{a} \cdot \mathbf{d}$

17. $(2\mathbf{a} \quad \mathbf{b} \quad 3\mathbf{c})$, $6(\mathbf{b} \quad \mathbf{a} \quad \mathbf{c})$

18. $(\mathbf{a} \times \mathbf{b}) \times \mathbf{c}$, $\mathbf{a} \times (\mathbf{b} \times \mathbf{c})$

19. $(\mathbf{b} - 2\mathbf{c}) \times \mathbf{d}$, $\mathbf{d} \times (2\mathbf{c} - \mathbf{b})$

20. $(1/|\mathbf{a}|)\mathbf{a}$, $(1/|\mathbf{c}|)\mathbf{c}$

21. $(\mathbf{d} \quad 4\mathbf{a} \quad \mathbf{c})$, $(\mathbf{c} \quad 4\mathbf{d} \quad \mathbf{a})$

22. $\mathbf{a} \times \mathbf{a}$, $(\mathbf{a} \times \mathbf{b}) \times \mathbf{a}$

23. $(\mathbf{c} \cdot \mathbf{d})(\mathbf{c} \times \mathbf{d} - \mathbf{d} \times \mathbf{c})$

24. $|\mathbf{a} + \mathbf{b}|$, $|\mathbf{a}| + |\mathbf{b}|$

25. $|\mathbf{c} - \mathbf{d}|$, $||\mathbf{c}| - |\mathbf{d}||$

26. **(Angle)** Find the angle between $\mathbf{a}$ and $\mathbf{b}$. Between $\mathbf{a}$ and $\mathbf{d}$.

27. **(Angle)** Find the angle between $\mathbf{b}$ and $\mathbf{c}$. Make a sketch.

28. **(Resultant)** Find a force $\mathbf{p}$ such that the resultant of $\mathbf{p}$, $\mathbf{q} = [2, -2, 3]$, and $\mathbf{u} = [-3, 0, 4]$ is the zero vector.

29. **(Work)** Find the work done by $\mathbf{q}$ in Prob. 28 in the displacement from $(1, 0, 0)$ to $(3, 8, 0)$. Make a sketch.

30. **(Component)** Find the component of $\mathbf{a} = [4, 0, 2]$ in the direction of $\mathbf{b} = [3, 3, 0]$.

31. **(Component)** In what case is the component of $\mathbf{a}$ in the direction of $\mathbf{b}$ equal to the component of $\mathbf{b}$ in the direction of $\mathbf{a}$?

32. **(Moment)** Find the moment vector $\mathbf{m}$ of $\mathbf{p} = [3, 8, 0]$ about P: $(2, 1, 0)$ if $\mathbf{p}$ acts on a line through $(1, 3, 0)$. Make a sketch.

33. **(Moment)** In what case is the moment of a force zero?

34. **(Tetrahedron)** Find the volume of the tetrahedron with vertices $(2, 1, 8)$, $(3, 2, 9)$, $(2, 1, 4)$, and $(3, 3, 10)$.

35. **(Plane)** Find an equation of the plane through the points $(1, 1, 1)$, $(5, 0, -5)$, and $(3, 2, 0)$.

36. **(Velocity, acceleration)** Find the velocity, speed, and acceleration of the motion given by the vector function $\mathbf{r}(t) = [3 \cos t, -2 \sin t, \frac{1}{2}t]$ at P: $(3/\sqrt{2}, -2/\sqrt{2}, \pi/8)$. Make a sketch.

Grad, Div, Curl, ∇^2, Directional Derivative.

Let $f = xy - yz$, $\mathbf{v} = [2y, 2z, 4x + z]$, $\mathbf{w} = [3z^2, 2x^2 - y^2, y^2]$. Find

37. grad f at $(2, 0, 7)$

38. grad (f^2), (grad f) $\times$ (grad f)

39. div $\mathbf{v}$, div $\mathbf{w}$

40. curl $\mathbf{v}$, curl $\mathbf{w}$

41. div (grad f)

42. div (grad $(x^2 f)$)

43. grad (div $\mathbf{w}$)

44. (grad f) $\cdot \mathbf{v}$

45. $D_{\mathbf{v}}f$ at $(2, 3, 1)$

46. $D_{\mathbf{w}}f$ at $(1, 1, 0)$

47. curl (grad (f^2))

48. div (curl $\mathbf{v}$), div (curl $\mathbf{w}$)

49. [(curl $\mathbf{v}$) $\times \mathbf{w}$] $\cdot \mathbf{w}$

50. [(grad f) $\times \mathbf{v}$] $\cdot \mathbf{v}$

Summary of Chapter 8
Vector Differential Calculus.
Grad, Div, Curl

All vectors of the form

$$(1) \qquad \mathbf{a} = [a_1, a_2, a_3] = a_1\mathbf{i} + a_2\mathbf{j} + a_3\mathbf{k} \qquad \text{(Sec. 8.1)}$$

constitute the real vector space R^3 with *vector addition* defined by

$$(2) \qquad [a_1, a_2, a_3] + [b_1, b_2, b_3] = [a_1 + b_1, a_2 + b_2, a_3 + b_3]$$

and *scalar multiplication* defined by

$$(3) \qquad c[a_1, a_2, a_3] = [ca_1, ca_2, ca_3] \qquad \text{(Sec. 8.1)},$$

c being a scalar (a real number). For instance, the *resultant* of forces $\mathbf{a}$ and $\mathbf{b}$ is the sum $\mathbf{a} + \mathbf{b}$ of the two forces.

The **inner product** or **dot product** of two vectors is defined by

$$(4) \qquad \mathbf{a} \cdot \mathbf{b} = |\mathbf{a}||\mathbf{b}| \cos \gamma = a_1b_1 + a_2b_2 + a_3b_3 \qquad \text{(Sec. 8.2)}$$

where γ is the angle between $\mathbf{a}$ and $\mathbf{b}$. This gives for the **norm** or **length** $|\mathbf{a}|$ of $\mathbf{a}$ the formula

$$(5) \qquad |\mathbf{a}| = \sqrt{\mathbf{a} \cdot \mathbf{a}} = \sqrt{a_1{}^2 + a_2{}^2 + a_3{}^2}$$

as well as a formula for γ. If $\mathbf{a} \cdot \mathbf{b} = 0$, we call $\mathbf{a}$ and $\mathbf{b}$ **orthogonal.** The dot product is suggested by the *work* $W = \mathbf{p} \cdot \mathbf{d}$ done by a force $\mathbf{p}$ in a displacement $\mathbf{d}$.

The **vector product** or **cross product** $\mathbf{v} = \mathbf{a} \times \mathbf{b}$ is a vector of length

$$(6) \qquad |\mathbf{a} \times \mathbf{b}| = |\mathbf{a}| \, |\mathbf{b}| \sin \gamma \qquad \text{(Sec. 8.3)}$$

and perpendicular to both $\mathbf{a}$ and $\mathbf{b}$ such that $\mathbf{a}, \mathbf{b}, \mathbf{v}$ form a *right-handed* triple. In terms of components with respect to right-handed coordinates,

$$(7) \qquad \mathbf{a} \times \mathbf{b} = \begin{vmatrix} \mathbf{i} & \mathbf{j} & \mathbf{k} \\ a_1 & a_2 & a_3 \\ b_1 & b_2 & b_3 \end{vmatrix} \qquad \text{(Sec. 8.3)}.$$

The vector product is suggested, for instance, by moments of forces or by rotations. Caution! This multiplication is *anti*commutative, $\mathbf{a} \times \mathbf{b} = -\mathbf{b} \times \mathbf{a},$ and is *not* associative.

Most important among repeated products is the scalar triple product

$$(8) \qquad (\mathbf{a} \quad \mathbf{b} \quad \mathbf{c}) = \mathbf{a} \cdot (\mathbf{b} \times \mathbf{c}) = (\mathbf{a} \times \mathbf{b}) \cdot \mathbf{c}.$$

Sections 8.4−8.11 concern the extension of differential calculus to **vector functions.** If these functions depend on a single variable, t, they are of the form

$$(9) \qquad \mathbf{v}(t) = [v_1(t),\, v_2(t),\, v_3(t)] = v_1(t)\mathbf{i} + v_2(t)\mathbf{j} + v_3(t)\mathbf{k}.$$

Then the derivative is

$$(10) \qquad \mathbf{v}' = \frac{d\mathbf{v}}{dt} = \lim_{\Delta t \to 0} \frac{\mathbf{v}(t + \Delta t) - \mathbf{v}(t)}{\Delta t} \qquad \text{(Sec. 8.4).}$$

Formula (10) looks quite similar to the familiar formula in calculus. It implies that

$$\mathbf{v}' = [v_1',\, v_2',\, v_3'] = v_1'\mathbf{i} + v_2'\mathbf{j} + v_3'\mathbf{k}.$$

Differentiation rules are as in calculus. They imply (Sec. 8.4)

$$(\mathbf{u} \cdot \mathbf{v})' = \mathbf{u}' \cdot \mathbf{v} + \mathbf{u} \cdot \mathbf{v}', \qquad (\mathbf{u} \times \mathbf{v})' = \mathbf{u}' \times \mathbf{v} + \mathbf{u} \times \mathbf{v}'.$$

A vector function of a single variable t, usually denoted by $\mathbf{r}(t)$, can be used to represent a curve C in space. Then $\mathbf{r}(t)$ associates with each $t = t_0$ in some interval $a < t < b$ the point of C with position vector $\mathbf{r}(t_0)$. The derivative $\mathbf{r}'(t)$ is a **tangent vector** of C (Sec. 8.5). If we choose the **arc length** s (Sec. 8.5) as a parameter, then we get the *unit tangent vector* $\mathbf{r}'(s)$. The length of the derivative of this vector is the **curvature** $\kappa(s) = |\mathbf{r}''(s)|$; see Sec. 8.7.

In mechanics, $\mathbf{r}(t)$ may represent the path of a moving body, where t is time. Then $\mathbf{v}(t) = \mathbf{r}'(t)$ is the **velocity vector,** its length is the *speed,* and its derivative $\mathbf{a}(t) = \mathbf{v}'(t) = \mathbf{r}''(t)$ is the **acceleration vector** (see Sec. 8.6).

A vector function of Cartesian coordinates x, y, z in space, say,

$$(11) \qquad \begin{aligned} \mathbf{v}(x,\, y,\, z) &= [v_1(x,\, y,\, z),\quad v_2(x,\, y,\, z),\quad v_3(x,\, y,\, z)] \\ &= v_1(x,\, y,\, z)\mathbf{i} + v_2(x,\, y,\, z)\mathbf{j} + v_3(x,\, y,\, z)\mathbf{k}, \end{aligned}$$

defines a **vector field** in the region in space in which $\mathbf{v}$ is defined; with every point P_0: $(x_0,\, y_0,\, z_0)$ in that region, $\mathbf{v}$ associates a vector $\mathbf{v}(P_0) = \mathbf{v}(x_0,\, y_0,\, z_0)$. Partial derivatives of $\mathbf{v}$ are obtained by taking partial derivatives of components; for instance,

$$\frac{\partial \mathbf{v}}{\partial x} = \left[\frac{\partial v_1}{\partial x},\, \frac{\partial v_2}{\partial x},\, \frac{\partial v_3}{\partial x} \right] = \frac{\partial v_1}{\partial x}\mathbf{i} + \frac{\partial v_2}{\partial x}\mathbf{j} + \frac{\partial v_3}{\partial x}\mathbf{k}.$$

Section 8.8 on the chain rule and a few other facts about functions of several variables is a preparation for Secs. 8.9−8.11 on the **gradient** of a scalar function f (Sec. 8.9),

(12)
$$\text{grad } f = \nabla f = \left[\frac{\partial f}{\partial x}, \frac{\partial f}{\partial y}, \frac{\partial f}{\partial z} \right],$$

the **directional derivative** of f in the direction of a **unit** vector **b**,

(13)
$$D_{\mathbf{b}}f = \frac{df}{ds} = \mathbf{b} \cdot \nabla f \qquad \text{(Sec. 8.9)},$$

the **divergence** of a vector function **v** (Sec. 8.10),

(14)
$$\text{div } \mathbf{v} = \nabla \cdot \mathbf{v} = \frac{\partial v_1}{\partial x} + \frac{\partial v_2}{\partial y} + \frac{\partial v_3}{\partial z},$$

and the **curl** of **v** (Sec. 8.11),

(15)
$$\text{curl } \mathbf{v} = \nabla \times \mathbf{v} = \begin{vmatrix} \mathbf{i} & \mathbf{j} & \mathbf{k} \\ \dfrac{\partial}{\partial x} & \dfrac{\partial}{\partial y} & \dfrac{\partial}{\partial z} \\ v_1 & v_2 & v_3 \end{vmatrix}.$$

(Formula (15) has a minus sign in front of the determinant for left-handed coordinates).

Some basic formulas for grad, div, curl are (Secs. 8.9−8.11)

(16)
$$\nabla(fg) = f\nabla g + g\nabla f$$
$$\nabla(f/g) = (1/g^2)(g\nabla f - f\nabla g)$$

(17)
$$\text{div } (f\mathbf{v}) = f \text{ div } \mathbf{v} + \mathbf{v} \cdot \nabla f$$
$$\text{div } (f\nabla g) = f\nabla^2 g + \nabla f \cdot \nabla g$$

(18)
$$\nabla^2 f = \text{div } (\nabla f)$$
$$\nabla^2(fg) = g\nabla^2 f + 2\nabla f \cdot \nabla g + f\nabla^2 g$$

(19)
$$\text{curl } (f\mathbf{v}) = \nabla f \times \mathbf{v} + f \text{ curl } \mathbf{v}$$
$$\text{div } (\mathbf{u} \times \mathbf{v}) = \mathbf{v} \cdot \text{curl } \mathbf{u} - \mathbf{u} \cdot \text{curl } \mathbf{v}$$

(20)
$$\text{curl } (\nabla f) = \mathbf{0}$$
$$\text{div } (\text{curl } \mathbf{v}) = 0.$$

The form of grad, div, curl (and the Laplacian) in curvilinear coordinates is shown in Appendix A3.4.

Vector Integral Calculus Integral Theorems

In this chapter we shall define line integrals and surface integrals and consider some of their basic engineering applications in solid mechanics, fluid flow, and heat problems. We shall see that a line integral is a natural generalization of a definite integral, and a surface integral is a generalization of a double integral.

Line integrals can be transformed into double integrals (Sec. 9.4) or into surface integrals (Sec. 9.9), and conversely. Triple integrals can be transformed into surface integrals (Sec. 9.7), and vice versa. These transformations are of great practical importance. The corresponding formulas of Green, Gauss, and Stokes (Secs. 9.4, 9.7, 9.9) serve as powerful tools in many applications as well as in theoretical problems (for instance, in potential theory; see Sec. 9.8). We shall see that they also lead to a better understanding of the physical meaning of the divergence and the curl of a vector function.

Prerequisites for this chapter: elementary integral calculus and Chap. 8.
Sections that may be omitted in a shorter course: 9.3, 9.5, 9.8.
References: Appendix 1, Part B.
Answers to problems: Appendix 2.

9.1 Line Integrals

The concept of a line integral is a simple and natural generalization of a definite integral

$$(1) \qquad \int_a^b f(x)\, dx$$

known from calculus. In (1) we integrate the *integrand* $f(x)$ from $x = a$ along the x-axis to $x = b$. In a line integral we shall integrate a given function, called the *integrand,* along a curve C in space (or in the plane). Hence *curve integral* would be a better term, but *line integral* is standard.

We represent the curve C by a parametric representation (see Sec. 8.5)

$$(2) \qquad \mathbf{r}(t) = [x(t), y(t), z(t)] = x(t)\mathbf{i} + y(t)\mathbf{j} + z(t)\mathbf{k} \qquad (a \leqq t \leqq b).$$

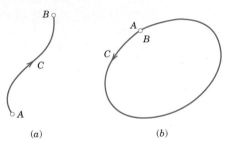

Fig. 200.   Oriented curve

We call C the **path of integration,** $A:$ $\mathbf{r}(a)$ its *initial point,* and $B:$ $\mathbf{r}(b)$ its *terminal point.*
C is now *oriented.* The direction from A to B, in which t increases, is called the *positive direction* on C. We can indicate the direction by an arrow (as in Fig. 200a). The points A and B may coincide (as in Fig. 200b). Then C is called a *closed path.*

We call C a **smooth curve** if C has a unique tangent at each of its points whose direction varies *continuously* as we move along C. Technically: C has a representation (2) such that $\mathbf{r}(t)$ is differentiable and the derivative $\mathbf{r}'(t) = d\mathbf{r}/dt$ is continuous and different from the zero vector at every point of C.

General assumption

In this book, every path of integration of a line integral is assumed to be **piecewise smooth;** *that is, it consists of finitely many smooth curves.*

Definition and Evaluation of Line Integrals

A **line integral** of a vector function $\mathbf{F}(\mathbf{r})$ over a curve C is defined by

(3)
$$\int_C \mathbf{F}(\mathbf{r}) \cdot d\mathbf{r} = \int_a^b \mathbf{F}(\mathbf{r}(t)) \cdot \frac{d\mathbf{r}}{dt}\, dt \qquad \text{[see (2)]}$$

(see Sec. 8.2 for the dot product). In terms of components, with $d\mathbf{r} = [dx,\quad dy,\quad dz]$ as in Sec. 8.5 and $' = d/dt$, formula (3) becomes

(3′)
$$\int_C \mathbf{F}(\mathbf{r}) \cdot d\mathbf{r} = \int_C (F_1\, dx + F_2\, dy + F_3\, dz)$$
$$= \int_a^b (F_1 x' + F_2 y' + F_3 z')\, dt.$$

If the path of integration C in (3) is a *closed* curve, then instead of

$$\int_C \qquad \text{we also write} \qquad \oint_C .$$

Note that the integrand is a scalar, not a vector, because we take the dot product. Indeed, $\mathbf{F} \cdot d\mathbf{r}/dt$ with $t = s$ (the arc length of C) is the tangential component of $\mathbf{F}$.

Such line integrals (3) arise naturally in mechanics, where they give the work done by a force **F** in a displacement along C (details and examples below). We may thus call the line integral (3) the **work integral.** Other forms of the line integral will be discussed later in this section.

We see that the integral in (3) on the right is a definite integral of a function of t taken over the interval $a \leqq t \leqq b$ on the t-axis in the **positive** direction (the direction of increasing t). This definite integral exists for continuous **F** and piecewise smooth C, because this makes $\mathbf{F} \cdot \mathbf{r}'$ piecewise continuous.

EXAMPLE 1   Evaluation of a line integral in the plane

Find the value of the line integral (3) when $\mathbf{F(r)} = [-y, -xy] = -y\mathbf{i} - xy\mathbf{j}$ and C is the circular arc in Fig. 201 from A to B.

Solution.   We may represent C by

$$\mathbf{r}(t) = [\cos t, \sin t] = \cos t\, \mathbf{i} + \sin t\, \mathbf{j} \qquad (0 \leqq t \leqq \pi/2).$$

Thus $x(t) = \cos t$, $y(t) = \sin t$, so that

$$\mathbf{F(r}(t)) = -y(t)\mathbf{i} - x(t)y(t)\mathbf{j} = [-\sin t, \ -\cos t \sin t] = -\sin t\, \mathbf{i} - \cos t \sin t\, \mathbf{j}.$$

By differentiation,

$$\mathbf{r}'(t) = -\sin t\, \mathbf{i} + \cos t\, \mathbf{j},$$

so that by (3)

$$\int_C \mathbf{F(r)} \cdot d\mathbf{r} = \int_0^{\pi/2} (-\sin t\, \mathbf{i} - \cos t \sin t\, \mathbf{j}) \cdot (-\sin t\, \mathbf{i} + \cos t\, \mathbf{j})\, dt$$

$$= \int_0^{\pi/2} (\sin^2 t - \cos^2 t \sin t)\, dt = \frac{\pi}{4} - \frac{1}{3} \approx 0.4521$$

[use (10) in Appendix A3.1; set $\cos t = u$ in the second term].   ◀

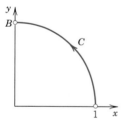

Fig. 201.   Example 1

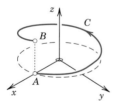

Fig. 202.   Example 2

EXAMPLE 2   Line integral in space

Evaluation of line integrals in space is practically the same as it is in the plane. To see this, find the value of (3) when $\mathbf{F(r)} = [z, x, y] = z\mathbf{i} + x\mathbf{j} + y\mathbf{k}$ and C is the helix (Fig. 202)

(4) $$\mathbf{r}(t) = [\cos t, \sin t, 3t] = \cos t\, \mathbf{i} + \sin t\, \mathbf{j} + 3t\mathbf{k} \qquad (0 \leqq t \leqq 2\pi).$$

Solution.   From (4) we have $x(t) = \cos t$, $y(t) = \sin t$, $z(t) = 3t$. Thus

$$\mathbf{F(r}(t)) \cdot \mathbf{r}'(t) = (3t\mathbf{i} + \cos t\, \mathbf{j} + \sin t\, \mathbf{k}) \cdot (-\sin t\, \mathbf{i} + \cos t\, \mathbf{j} + 3\mathbf{k}).$$

The dot product is $3t(-\sin t) + \cos^2 t + 3 \sin t$. Hence (3) gives

$$\int_C \mathbf{F(r)} \cdot d\mathbf{r} = \int_0^{2\pi} (-3t \sin t + \cos^2 t + 3 \sin t)\, dt = 6\pi + \pi + 0 = 7\pi \approx 21.99.$$ ◀

Two important questions now arise:

1. Choice of representation. Does the value of a line integral (3) with given **F** and C depend on the particular choice of a representation (2) of C? The answer is no; see Theorem 1 below.

2. Choice of path. Does this value change if we integrate from the old point A to the old point B but along another path? The answer is yes, in general; see Example 3.

EXAMPLE 3   **Dependence of a line integral on path (same endpoints)**

Evaluate the line integral (3) with

$$\mathbf{F(r)} = [5z,\ xy,\ x^2z] = 5z\mathbf{i} + xy\mathbf{j} + x^2z\mathbf{k}$$

along two different paths with the same initial point A: $(0, 0, 0)$ and the same terminal point B: $(1, 1, 1)$, namely (Fig. 203),

(a) C_1: the straight-line segment $\mathbf{r}_1(t) = [t,\ \ t,\ \ t] = t\mathbf{i} + t\mathbf{j} + t\mathbf{k}$, $0 \leqq t \leqq 1$, and

(b) C_2: the parabolic arc $\mathbf{r}_2(t) = \left[t,\ \ t,\ \ t^2\right] = t\mathbf{i} + t\mathbf{j} + t^2\mathbf{k}$, $0 \leqq t \leqq 1$.

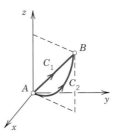

Fig. 203.   Example 3

Solution.   (a) By substituting $\mathbf{r}_1$ into **F** we obtain $\mathbf{F(r}_1(t)) = \left[5t,\ t^2,\ t^3\right] = 5t\mathbf{i} + t^2\mathbf{j} + t^3\mathbf{k}$. We also need $\mathbf{r}'_1 = [1,\ \ 1,\ \ 1] = \mathbf{i} + \mathbf{j} + \mathbf{k}$. Hence the integral over C_1 is

$$\int_{C_1} \mathbf{F(r)} \cdot d\mathbf{r} = \int_0^1 \mathbf{F(r}_1(t)) \cdot \mathbf{r}'_1(t)\, dt$$

$$= \int_0^1 (5t\mathbf{i} + t^2\mathbf{j} + t^3\mathbf{k}) \cdot (\mathbf{i} + \mathbf{j} + \mathbf{k})\, dt$$

$$= \int_0^1 (5t + t^2 + t^3)\, dt$$

$$= \frac{5}{2} + \frac{1}{3} + \frac{1}{4} = \frac{37}{12}\,.$$

(b) Similarly, by substituting $\mathbf{r}_2$ into **F** and calculating $\mathbf{r}'_2$ we obtain for the integral over the path C_2

$$\int_{C_2} \mathbf{F(r)} \cdot d\mathbf{r} = \int_0^1 \mathbf{F(r}_2(t)) \cdot \mathbf{r}'_2(t)\, dt = \int_0^1 (5t^2 + t^2 + 2t^5)\, dt$$

$$= \frac{5}{3} + \frac{1}{3} + \frac{2}{6} = \frac{28}{12}\,.$$

The two results are different, although the endpoints are the same. This shows that **the value of a line integral (3) will in general depend not only on F and on the endpoints *A, B* of the path but also on the path along which we integrate from *A* to *B*.**

Can we find conditions that guarantee independence? This is a basic question in connection with physical applications. The answer is yes, as we show in Sec. 9.2.   ◄

Motivation of the Line Integral (3): Work Done by a Force

The work W done by a *constant* force $\mathbf{F}$ in the displacement along a *straight* segment $\mathbf{d}$ is $W = \mathbf{F} \cdot \mathbf{d}$; see Example 2 in Sec. 8.2. This suggests that we define the work W done by a *variable* force $\mathbf{F}$ in the displacement along a curve C: $\mathbf{r}(t)$ as the limit of sums of works done in displacements along small chords of C. We shall now show that this definition amounts to defining W by the line integral (3).

For this we choose points $t_0 (= a) < t_1 < \cdots < t_n (= b)$. Then the work ΔW_m done by $\mathbf{F}(\mathbf{r}(t_m))$ in the straight displacement from $\mathbf{r}(t_m)$ to $\mathbf{r}(t_{m+1})$ is

$$\Delta W_m = \mathbf{F}(\mathbf{r}(t_m)) \cdot [\mathbf{r}(t_{m+1}) - \mathbf{r}(t_m)] \approx \mathbf{F}(\mathbf{r}(t_m)) \cdot \mathbf{r}'(t_m) \, \Delta t_m \qquad (\Delta t_m = t_{m+1} - t_m).$$

The sum of these n works is $W_n = \Delta W_0 + \cdots + \Delta W_{n-1}$. If we choose points and consider W_n for every n arbitrarily but so that the greatest Δt_m approaches zero as $n \to \infty$, then the limit of W_n as $n \to \infty$ is the line integral (3). This integral exists because of our general assumption that $\mathbf{F}$ is continuous and C is piecewise smooth; this makes $\mathbf{r}'(t)$ continuous, except at finitely many points where C may have corners or cusps.

EXAMPLE 4   **Work done by a variable force**

If $\mathbf{F}$ in Example 1 is a force, the work done by $\mathbf{F}$ in the displacement along the quarter-circle is 0.4521, measured in suitable units, say, newton-meters (nt·m, also called joules, abbreviation J; see also front cover). Similarly in Examples 2 and 3.   ◄

EXAMPLE 5   **Work done equals the gain in kinetic energy**

Let $\mathbf{F}$ be a force, so that (3) is work. Let t be time, so that $d\mathbf{r}/dt = \mathbf{v}$, velocity. Then we can write (3) as

$$(5) \qquad W = \int_C \mathbf{F} \cdot d\mathbf{r} = \int_a^b \mathbf{F}(\mathbf{r}(t)) \cdot \mathbf{v}(t) \, dt.$$

Now by Newton's second law (force = mass × acceleration),

$$\mathbf{F} = m\mathbf{r}''(t) = m\mathbf{v}'(t),$$

where m is the mass of the body displaced. Substitution into (5) gives [see (11), Sec. 8.4]

$$W = \int_a^b m\mathbf{v}' \cdot \mathbf{v} \, dt = \int_a^b m \left(\frac{\mathbf{v} \cdot \mathbf{v}}{2} \right)' \, dt = \frac{m}{2} |\mathbf{v}|^2 \Big|_{t=a}^{t=b}.$$

On the right, $m|\mathbf{v}|^2/2$ is the kinetic energy. Hence *the work done equals the gain in kinetic energy.* This is a basic law in mechanics.   ◄

Other Forms of Line Integrals

The line integrals

$$(6) \qquad \int_C F_1 \, dx, \qquad \int_C F_2 \, dy, \qquad \int_C F_3 \, dz$$

are special cases of (3) when $\mathbf{F} = F_1 \mathbf{i}$ or $F_2 \mathbf{j}$ or $F_3 \mathbf{k}$, respectively. Another form is

$$(7) \qquad \boxed{\int_C f(\mathbf{r}) \, dt = \int_a^b f(\mathbf{r}(t)) \, dt}$$

with C as in (2). But this definition can also be regarded as a special case of (3), with $\mathbf{F} = F_1 \mathbf{i}$ and $F_1 = f/(dx/dt)$, so that $f = F_1 x'$, as in (3'). The evaluation of (7) is similar to that in the previous examples.

EXAMPLE 6   **A line integral of the form (7)**

Find the value of (7) when $f = (x^2 + y^2 + z^2)^2$ and C is the helix (4) in Example 2.

Solution.   From $\mathbf{r}(t) = \cos t \, \mathbf{i} + \sin t \, \mathbf{j} + 3t \mathbf{k}$ we see that on C,

$$(x^2 + y^2 + z^2)^2 = [\cos^2 t + \sin^2 t + (3t)^2]^2 = (1 + 9t^2)^2.$$

This gives

$$\int_C f(\mathbf{r}) \, dt = \int_0^{2\pi} (1 + 9t^2)^2 \, dt = 2\pi + 6(2\pi)^3 + \frac{81}{5}(2\pi)^5 \approx 160\ 135.$$ ◀

General Properties of the Line Integral (3)

From familiar properties of integrals in calculus we obtain corresponding formulas for line integrals (3):

(8a)
$$\int_C k\mathbf{F} \cdot d\mathbf{r} = k \int_C \mathbf{F} \cdot d\mathbf{r}$$
            (k constant)

(8b)
$$\int_C (\mathbf{F} + \mathbf{G}) \cdot d\mathbf{r} = \int_C \mathbf{F} \cdot d\mathbf{r} + \int_C \mathbf{G} \cdot d\mathbf{r}$$

(8c)
$$\int_C \mathbf{F} \cdot d\mathbf{r} = \int_{C_1} \mathbf{F} \cdot d\mathbf{r} + \int_{C_2} \mathbf{F} \cdot d\mathbf{r}$$
            (Fig. 204)

where in (8c) the path C is subdivided into two arcs C_1 and C_2 that have the same orientation as C (Fig. 204). In (8b) the orientation of C is the same in all three integrals. If the sense of integration along C is reversed, the value of the integral is multiplied by -1.

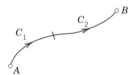

Fig. 204.   Formula (8c)

If a line integral (3) is supposed to represent physical quantities, such as work, the choice of one or another representation of a given curve C should not be essential, as long as the positive directions are the same in both cases. This is true, as follows:

THEOREM 1   **(Direction-preserving transformations of parameter)**

Any representations of C that give the same positive direction on C also yield the same value of the line integral (3).

PROOF. We represent C in (3) using another parameter t^* given by a function $t = \phi(t^*)$ that has a positive derivative and is such that $a^* \leq t^* \leq b^*$ corresponds to $a \leq t \leq b$. Then, writing $\mathbf{r}(\phi(t^*)) = \mathbf{r}^*(t^*)$ and using the chain rule, we have $dt^* = (dt^*/dt)\, dt$ and thus

$$\int_C \mathbf{F}(\mathbf{r}^*) \cdot d\mathbf{r}^* = \int_{a^*}^{b^*} \left[\mathbf{F}(\mathbf{r}^*(t^*)) \cdot \frac{d\mathbf{r}^*}{dt^*} \right] dt^*$$

$$= \int_{a^*}^{b^*} \mathbf{F}(\mathbf{r}(\phi(t^*))) \cdot \frac{d\mathbf{r}}{dt} \frac{dt}{dt^*}\, dt^*$$

$$= \int_a^b \mathbf{F}(\mathbf{r}(t)) \cdot \frac{d\mathbf{r}}{dt}\, dt = \int_C \mathbf{F}(\mathbf{r}) \cdot d\mathbf{r}. \qquad \blacktriangleleft$$

PROBLEM SET 9.1

Line Integral. Work Done by a Force

Calculate $\int_C \mathbf{F}(\mathbf{r}) \cdot d\mathbf{r}$ for the following data. (If $\mathbf{F}$ is a force, this gives the work in the displacement along C.) Show the details of your work.

1. $\mathbf{F} = [y^2, -x^2]$, C the straight-line segment from $(0, 0)$ to $(1, 4)$
2. $\mathbf{F}$ as in Prob. 1, $C: y = 4x^2$ from $(0, 0)$ to $(1, 4)$
3. $\mathbf{F} = [xy, x^2y^2]$, C the quarter-circle from $(2, 0)$ to $(0, 2)$ with center at $(0, 0)$
4. $\mathbf{F}$ as in Prob. 3, C the straight-line segment from $(2, 0)$ to $(0, 2)$
5. $\mathbf{F} = [(x - y)^2, (y - x)^2]$, $C: xy = 1, 1 \leq x \leq 4$
6. $\mathbf{F} = [\exp(y^{2/3}), -\exp(x^{3/2})]$, C the semicubical parabola $y = x^{3/2}$ from $(0, 0)$ to $(1, 1)$
7. $\mathbf{F} = [2z, x, -y]$, $C: \mathbf{r} = [\cos t, \sin t, 2t]$ from $(1, 0, 0)$ to $(1, 0, 4\pi)$
8. $\mathbf{F} = [x - y, y - z, z - x]$, $C: [2\cos t, t, 2\sin t]$ from $(2, 0, 0)$ to $(2, 2\pi, 0)$
9. $\mathbf{F} = [e^x, e^{-y}, e^z]$, $C: \mathbf{r} = [t, t^2, t]$ from $(0, 0, 0)$ to $(1, 1, 1)$
10. $\mathbf{F} = [\cosh x, \sinh y, e^z]$, $C: \mathbf{r} = [t, t^2, t^3]$ from $(0, 0, 0)$ to $(2, 4, 8)$

11. **WRITING PROJECT. Line Integral Generalizes Definite Integral.** Write a short essay on this topic. Geometrically, the definite integral gives the area under the curve of the integrand. Explain the corresponding interpretation for a line integral. Include examples.

12. **PROJECT. Independence of Representation. Dependence on Path.** Consider the integral $\int_C \mathbf{F}(\mathbf{r}) \cdot d\mathbf{r}$, where $\mathbf{F} = [-x^2, xy]$.

   (a) **One path, several representations.** Find the value of the integral when $\mathbf{r} = [\cos t, \sin t]$, $0 \leq t \leq \pi$. Show that the value remains the same if you set $t = -p$ or $t = p^2$ or apply two other parametric transformations of your own choice.

   (b) **Several paths.** Evaluate the integral when $C: y = x^n$, thus $\mathbf{r} = [t, t^n]$, $0 \leq t \leq 1$, where $n = 1, 2, 3, \cdots$. Note that these infinitely many paths have the same endpoints.

   (c) **Limit.** What is the limit in (b) as $n \to \infty$? Can you confirm your result by direct integration without referring to (b)?

   (d) Show path dependence with a simple example of your own choice involving two paths.

Integrals $\int_C f(\mathbf{r})\,ds$ with Arc Length as Parameter

Evaluate this integral with f and C as follows. (Show the details.)

13. $f = x^2 + y^2$, $C: y = 3x$ from $(0, 0)$ to $(2, 6)$

14. $f = x^3 y$, $C: \mathbf{r} = [2\cos t,\ 2\sin t],\ 0 \leq t \leq \pi/2$

15. $f = x^2 + y^2 + z^2$, $C: [\cos t,\ \sin t,\ 2t],\ 0 \leq t \leq 4\pi$

16. $f = \sqrt{2 + x^2 + 3y^2}$, $C: \mathbf{r} = [t,\ t,\ t^2],\ 0 \leq t \leq 3$

17. $f = 1 + y^2 + z^2$, $C: \mathbf{r} = [t,\ \cos t,\ \sin t],\ 0 \leq t \leq \pi$

18. $f = 1 - \sinh^2 x$, C the catenary $\mathbf{r} = [t,\ \cosh t],\ 0 \leq t \leq 2$

19. $f = x^2 + (xy)^{1/3}$, C the hypocycloid $\mathbf{r} = [\cos^3 t,\ \sin^3 t],\ 0 \leq t \leq \pi$

20. $f = \sqrt{16x^2 + 81y^2}$, $C: \mathbf{r} = [3\cos t,\ 2\sin t],\ 0 \leq t \leq \pi$

9.2 Line Integrals Independent of Path

In this section we are concerned with line integrals

$$(1) \qquad \int_C \mathbf{F}(\mathbf{r}) \cdot d\mathbf{r} = \int_C (F_1\,dx + F_2\,dy + F_3\,dz)$$

as before. In (1) we integrate from a point A to a point B over a path C (as in Fig. 200 in Sec. 9.1). The value of such an integral generally depends not only on A and B, but also on the path C along which we integrate. This was shown in Example 3 of the last section. It raises the question of conditions for independence of path, so that we get the same value in integrating from A to B along any path C. This is of great practical importance. For instance, in mechanics, independence of path may mean that we have to do the same amount of work regardless of the path to the mountaintop, be it short and steep or long and gentle, or that we gain back the work done in extending an elastic spring when we release it. Not all forces are of this type—think of swimming in a big whirlpool.

We define a line integral (1) to be **independent of path in a domain D in space** if for every pair of endpoints A, B in D the integral (1) has the same value for all paths in D that begin at A and end at B.

A very practical criterion for path independence is the following. (For the gradient, see Sec. 8.9.)

THEOREM 1   **(Independence of path)**

A line integral (1) with continuous F_1, F_2, F_3 in a domain D in space is independent of path in D if and only if $\mathbf{F} = [F_1, F_2, F_3]$ is the gradient of some function f in D,

$$(2) \qquad \mathbf{F} = \operatorname{grad} f;$$

in components,

$$(2') \qquad F_1 = \frac{\partial f}{\partial x}, \qquad F_2 = \frac{\partial f}{\partial y}, \qquad F_3 = \frac{\partial f}{\partial z}.$$

EXAMPLE 1   **Independence of path**

Show that the integral

$$\int_C \mathbf{F} \cdot d\mathbf{r} = \int_C (2x\,dx + 2y\,dy + 4z\,dz)$$

is independent of path in any domain in space and find its value if C has the initial point A: $(0, 0, 0)$ and terminal point B: $(2, 2, 2)$.

Solution.   By inspection we find that

$$\mathbf{F} = [2x, 2y, 4z] = 2x\mathbf{i} + 2y\mathbf{j} + 4z\mathbf{k} = \text{grad } f, \qquad \text{where} \qquad f = x^2 + y^2 + 2z^2.$$

(If $\mathbf{F}$ is more complicated, proceed by integration, as in Example 2, below.) Theorem 1 now implies independence of path. To find the value of the integral, we can choose the convenient straight path

$$C: \quad \mathbf{r}(t) = [t, \quad t, \quad t] = t(\mathbf{i} + \mathbf{j} + \mathbf{k}), \qquad 0 \leq t \leq 2,$$

and get $\mathbf{r}'(t) = \mathbf{i} + \mathbf{j} + \mathbf{k}$; thus $\mathbf{F} \cdot \mathbf{r}' = 2t + 2t + 4t = 8t$ and from this

$$\int_C (2x\,dx + 2y\,dy + 4z\,dz) = \int_0^2 \mathbf{F} \cdot \mathbf{r}'\,dt = \int_0^2 8t\,dt = 16.$$

Better methods of solution follow below.   ◀

PROOF OF THEOREM 1.   **(a)** Let (2) hold for some function f in D. Let C be any path in D from any point A to any point B, given by

$$\mathbf{r}(t) = x(t)\mathbf{i} + y(t)\mathbf{j} + z(t)\mathbf{k}, \qquad a \leq t \leq b.$$

Then from $(2')$, the chain rule (Sec. 8.8), and $(3')$ in Sec. 9.1 we get

$$\int_A^B (F_1\,dx + F_2\,dy + F_3\,dz) = \int_A^B \left(\frac{\partial f}{\partial x}\,dx + \frac{\partial f}{\partial y}\,dy + \frac{\partial f}{\partial z}\,dz \right)$$

$$= \int_a^b \left(\frac{\partial f}{\partial x}\frac{dx}{dt} + \frac{\partial f}{\partial y}\frac{dy}{dt} + \frac{\partial f}{\partial z}\frac{dz}{dt} \right) dt$$

$$= \int_a^b \frac{df}{dt}\,dt = f[x(t), y(t), z(t)] \Big|_{t=a}^{t=b}$$

$$= f(B) - f(A).$$

This shows that the value of the integral is simply the difference of the values of f at the two endpoints of C and is, therefore, independent of the path C.

(b) The more complicated proof of the converse, that independence of path implies (2) for some f, is given in Appendix 4.   ◀

The last formula in part (a) of the proof,

(3)
$$\int_A^B (F_1\,dx + F_2\,dy + F_3\,dz) = f(B) - f(A) \qquad [\mathbf{F} = \text{grad } f]$$

is the analog of the usual formula for definite integrals in calculus.

$$\int_a^b g(x)\,dx = G(x)\,\Big|_a^b = G(b) - G(a) \qquad [G'(x) = g(x)].$$

Formula (3) should be applied whenever a line integral is independent of path.

Potential theory relates to our present discussion if we remember from Sec. 8.9 that f is called a *potential* of $\mathbf{F} = \text{grad } f$. Thus the integral (1) is independent of path in D if and only if $\mathbf{F}$ is the gradient of a potential in D.

EXAMPLE 2   **Independence of path. Determination of a potential**

Evaluate the integral

$$I = \int_C (3x^2\,dx + 2yz\,dy + y^2\,dz)$$

from A: $(0, 1, 2)$ to B: $(1, -1, 7)$ by showing that $\mathbf{F}$ has a potential and applying (3).

Solution.   If $\mathbf{F}$ has a potential f, we should have

$$f_x = F_1 = 3x^2, \qquad f_y = F_2 = 2yz, \qquad f_z = F_3 = y^2.$$

We show that we can satisfy these conditions. By integration and differentiation,

$$f = x^3 + g(y, z), \qquad f_y = g_y = 2yz, \qquad g = y^2 z + h(z),$$
$$f_z = y^2 + h' = y^2, \qquad h' = 0 \qquad\qquad h = 0, \quad \text{say.}$$

This gives $f(x, y, z) = x^3 + y^2 z$ and by (3),

$$I = f(1, -1, 7) - f(0, 1, 2)$$
$$= 1 + 7 - (0 + 2) = 6. \qquad\qquad \blacktriangleleft$$

Integration Around Closed Curves and Independence of Path

The simple idea that two paths with common endpoints (Fig. 205) make up a single closed curve gives almost immediately

THEOREM 2   **(Independence of path)**

The integral (1) is independent of path in a domain D if and only if its value around every closed path in D is zero.

PROOF.   If we have independence of path, integration from A to B along C_1 and along C_2 in Fig. 205 gives the same value. Now C_1 and C_2 together make up a closed curve C, and if we integrate from A along C_1 to B as before, but then in the opposite sense along C_2 back to A (so that this integral is multiplied by -1), the sum of the two integrals is zero, but this is the integral around the closed curve C.

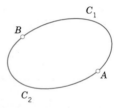

Fig. 205.   Proof of Theorem 2

Conversely, assume that the integral around any closed path C in D is zero. Given any points A and B and any two curves C_1 and C_2 from A to B in D, we see that C_1 with the orientation reversed and C_2 together form a closed path C. By assumption, the integral over C is zero. Hence the integrals over C_1 and C_2, both taken from A to B, must be equal. This proves the theorem.   ◀

Work. Conservative and Nonconservative (Dissipative) Physical Systems

Recall from the last section that in mechanics, the integral (1) represents the work done by a force $\mathbf{F}$ in the displacement of a body along C. Then Theorem 2 states that work is independent of path if and only if it is zero for displacement around any closed path. Furthermore, Theorem 1 tells us that this happens if and only if $\mathbf{F}$ is the gradient of a potential. In this case, $\mathbf{F}$ and the vector field defined by $\mathbf{F}$ are called **conservative,** because in this case mechanical energy is conserved, that is, no work is done in the displacement from a point A and back to A. Similarly for the displacement of an electrical charge (an electron, for instance) in an electrostatic field.

Physically, the kinetic energy of a body can be interpreted as the ability of the body to do work by virtue of its motion, and if the body moves in a conservative field of force, after the completion of a round-trip the body will return to its initial position with the same kinetic energy it had originally. For instance, the gravitational force is conservative; if we throw a ball vertically up, it will (if we assume air resistance to be negligible) return to our hand with the same kinetic energy it had when it left our hand.

Friction, air resistance, and water resistance always act against the direction of motion, tending to diminish the total mechanical energy of a system (usually converting it into heat or mechanical energy of the surrounding medium, or both), and if in the motion of a body these forces are so large that they can no longer be neglected, then the resultant $\mathbf{F}$ of the forces acting on the body is no longer conservative. Quite generally, a physical system is called **conservative** if all the forces acting in it are conservative; otherwise it is called **nonconservative** or **dissipative.**

Exactness and Independence of Path

Theorem 1 relates path independence of the line integral (1) to the gradient and Theorem 2 to integration around closed curves. A third idea and theorem (Theorem 3, below) relate path independence to the exactness of the **differential form**

$$(4) \qquad\qquad F_1 \, dx + F_2 \, dy + F_3 \, dz$$

under the integral sign in (1). This form (4) is called **exact** in a domain D in space if it is the differential

$$df = \frac{\partial f}{\partial x} \, dx + \frac{\partial f}{\partial y} \, dy + \frac{\partial f}{\partial z} \, dz$$

of a differentiable function $f(x, y, z)$ everywhere in D, that is, if we have

$$F_1 \, dx + F_2 \, dy + F_3 \, dz = df.$$

Comparing these two formulas, we see that the form (4) is exact if and only if there is a differentiable function $f(x, y, z)$ in D such that everywhere in D,

$$(5') \qquad\qquad F_1 = \frac{\partial f}{\partial x}, \qquad F_2 = \frac{\partial f}{\partial y}, \qquad F_3 = \frac{\partial f}{\partial z}$$

In vectorial form these three equations (5′) can be written

(5) $$\mathbf{F} = \operatorname{grad} f.$$

Hence, by Theorem 1, *the integral* (1) *is independent of path in D if and only if the differential form* (4) *has continuous components* F_1, F_2, F_3 *and is exact in D.*

This is practically important because there is a useful exactness criterion involving the following concept.

A domain D is called **simply connected** if every closed curve in D can be continuously shrunk to any point in D without leaving D.

For example, the interior of a sphere or a cube, the interior of a sphere with finitely many points removed, and the domain between two concentric spheres are simply connected, while the interior of a torus (a doughnut; see Fig. 229 in Sec. 9.6) and the interior of a cube with one space diagonal removed are not simply connected.

The criterion for path independence based on exactness is then as follows.

THEOREM 3　**(Criterion for exactness and independence of path)**

Let F_1, F_2, F_3 *in the line integral* (1),

$$\int_C \mathbf{F}(\mathbf{r}) \cdot d\mathbf{r} = \int_C (F_1 \, dx + F_2 \, dy + F_3 \, dz),$$

be continuous and have continuous first partial derivatives in a domain D in space. Then:

(a) If (1) *is independent of path in D—and thus the differential form* (4) *under the integral sign is exact—then in D,*

(6) $$\boxed{\operatorname{curl} \mathbf{F} = \mathbf{0};}$$

in components (see Sec. 8.11)

(6′) $$\frac{\partial F_3}{\partial y} = \frac{\partial F_2}{\partial z}, \qquad \frac{\partial F_1}{\partial z} = \frac{\partial F_3}{\partial x}, \qquad \frac{\partial F_2}{\partial x} = \frac{\partial F_1}{\partial y}.$$

(b) If (6′) *holds in D and D is simply connected, then* (1) *is independent of path in D.*

PROOF.　**(a)** If (1) is independent of path in D, then $\mathbf{F} = \operatorname{grad} f$ by (2) and

$$\operatorname{curl} \mathbf{F} = \operatorname{curl} (\operatorname{grad} f) = \mathbf{0}$$

[see (3) in Sec. 8.11], so that (6) holds.

(b) The proof of the converse requires "Stokes's theorem" and is given in Sec. 9.9.　◀

Comment For a line integral in the plane

$$\int_C \mathbf{F}(\mathbf{r}) \cdot d\mathbf{r} = \int_C (F_1 \, dx + F_2 \, dy),$$

curl **F** has just one component and (6′) reduces to the single relation

(6″)
$$\boxed{\frac{\partial F_2}{\partial x} = \frac{\partial F_1}{\partial y}.}$$

EXAMPLE 3   **Exactness and independence of path. Determination of a potential**

Using (6′), show that the differential form under the integral sign of

$$I = \int_C \left[2xyz^2 \, dx + (x^2 z^2 + z \cos yz) \, dy + (2x^2 yz + y \cos yz) \, dz \right]$$

is exact, so that we have independence of path in any domain, and find the value of I from A: $(0, 0, 1)$ to B: $(1, \pi/4, 2)$.

Solution.   Exactness follows from (6′), which gives

$$(F_3)_y = 2x^2 z + \cos yz - yz \sin yz = (F_2)_z$$

$$(F_1)_z = 4xyz = (F_3)_x$$

$$(F_2)_x = 2xz^2 = (F_1)_y.$$

To find f, we integrate F_2 (which is "long," so that we save work) and then differentiate to compare with F_1 and F_3,

$$f = \int F_2 \, dy = \int (x^2 z^2 + z \cos yz) \, dy = x^2 z^2 y + \sin yz + g(x, z)$$

$$f_x = 2xz^2 y + g_x = F_1 = 2xyz^2, \qquad g_x = 0, \qquad g = h(z)$$

$$f_z = 2x^2 zy + y \cos yz + h' = F_3 = 2x^2 zy + y \cos yz, \qquad h' = 0,$$

so that, taking $h = 0$, we have

$$f(x, y, z) = x^2 yz^2 + \sin yz.$$

From this and (3) we get

$$I = f(1, \pi/4, 2) - f(0, 0, 1) = \pi + \sin \tfrac{1}{2}\pi - 0 = \pi + 1. \qquad \blacktriangleleft$$

The assumption in Theorem 3 that D be simply connected is essential and cannot be omitted. This can be seen from the following example.

EXAMPLE 4   **On the assumption of simple connectedness in Theorem 3**

Let

(7)
$$F_1 = -\frac{y}{x^2 + y^2}, \qquad F_2 = \frac{x}{x^2 + y^2}, \qquad F_3 = 0.$$

Differentiation shows that (6′) is satisfied in any domain of the xy-plane not containing the origin, for example, in the domain D: $\frac{1}{2} < \sqrt{x^2 + y^2} < \frac{3}{2}$ shown in Fig. 206 on the next page. Indeed, F_1 and F_2 do not depend on z, and $F_3 = 0$, so that the first two relations in (6′) are trivially true, and the third is verified by differentiation:

$$\frac{\partial F_2}{\partial x} = \frac{x^2 + y^2 - x \cdot 2x}{(x^2 + y^2)^2} = \frac{y^2 - x^2}{(x^2 + y^2)^2},$$

$$\frac{\partial F_1}{\partial y} = -\frac{x^2 + y^2 - y \cdot 2y}{(x^2 + y^2)^2} = \frac{y^2 - x^2}{(x^2 + y^2)^2}.$$

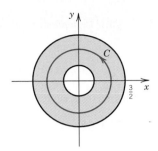

Fig. 206.   Example 4

Clearly, D in Fig. 206 is not simply connected. If the integral

$$I = \int_C (F_1 \, dx + F_2 \, dy) = \int_C \frac{-y \, dx + x \, dy}{x^2 + y^2}$$

were independent of path in D, then $I = 0$ on any closed curve in D, for example, on the circle $x^2 + y^2 = 1$. But setting $x = r \cos \theta$, $y = r \sin \theta$, and noting that the circle is represented by $r = 1$, we have

$$x = \cos \theta, \qquad dx = -\sin \theta \, d\theta, \qquad y = \sin \theta, \qquad dy = \cos \theta \, d\theta,$$

so that $-y \, dx + x \, dy = \sin^2 \theta \, d\theta + \cos^2 \theta \, d\theta = d\theta$ and counterclockwise integration gives

$$I = \int_0^{2\pi} \frac{d\theta}{1} = 2\pi.$$

Since D is not simply connected, we cannot apply Theorem 3 and conclude that I is independent of path in D.

Although $\mathbf{F} = \operatorname{grad} f$, where $f = \arctan (y/x)$ (verify!), we cannot apply Theorem 1 either because the polar angle $\theta = \arctan (y/x)$ is not single-valued, as it is required for a function in calculus. ◀

PROBLEM SET 9.2

1. **WRITING PROJECT. Key Ideas in This Section.** Make a list of the main ideas on path independence and dependence in this section. Then work the list into an essay, including explanations of all definitions, comments on the practical usefulness of the theorems, and examples illustrating each statement. Include no proofs.

Path Independent Integrals

Show that the form under the integral sign is exact in the plane (Probs. 2–5) or in space (Probs. 6–9) and evaluate the integral. (Show the details of your work.)

2. $\displaystyle\int_{(0,\pi)}^{(3,\pi/2)} e^x (\cos y \, dx - \sin y \, dy)$

3. $\displaystyle\int_{(-1,5)}^{(4,3)} (3z^2 \, dx + 6xz \, dz)$

4. $\displaystyle\int_{(\pi/2,-\pi)}^{(\pi/4,0)} (\cos x \cos 2y \, dx - 2 \sin x \sin 2y \, dy)$

5. $\displaystyle\int_{(3,3/2)}^{(4,1/2)} (2x \sin \pi y \, dx + \pi x^2 \cos \pi y \, dy)$

6. $\displaystyle\int_{(0,0,0)}^{(4,1,2)} (3y \, dx + 3x \, dy + 2z \, dz)$

7. $\displaystyle\int_{(0,-1,1)}^{(2,4,0)} e^{x-y+z^2} (dx - dy + 2z \, dz)$

8. $\displaystyle\int_{(\pi,\pi/2,2)}^{(0,\pi,1)} (-z \sin xz \, dx + \cos y \, dy - x \sin xz \, dz)$

9. $\displaystyle\int_{(0,2,3)}^{(1,1,1)} (yz \sinh xz \, dx + \cosh xz \, dy + xy \sinh xz \, dz)$

10. PROJECT. Path Dependence. (a) Show that $I = \int_C (x^2 y\, dx + 2xy^2\, dy)$ is path dependent in the xy-plane.

    (b) Integrate from $(0, 0)$ along the straight-line segment to $(1, b)$, $0 \leq b \leq 1$, and then vertically up to $(1, 1)$; see the figure. For which of these paths is I maximum? What is its maximum value?

    (c) Integrate from $(0, 0)$ along the straight-line segment to $(c, 1)$, $0 \leq c \leq 1$, and then horizontally to $(1, 1)$. For $c = 1$, do you get the same value as for $b = 1$ in (b)? For which c is I maximum? What is its maximum value?

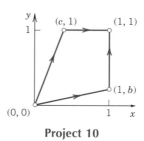

Project 10

Path-Independent and Dependent Integrals

Check for path-independence. In the case of independence integrate from $(0, 0, 0)$ to (a, b, c). (Show the details of your work.)

11. $2xy^2\, dx + 2x^2 y\, dy + dz$

12. $y\, dx - zx\, dy + z\, dz$

13. $\sinh xz\, (z\, dx - x\, dz)$

14. $yz\, dx + xz\, dy + xy\, dz$

15. $ye^{2z}\, dy - ze^y\, dz$

16. $3(x + y)^2 (dx + 2\, dy) + dz$

17. $\cos (x + yz)\, (dx + z\, dy + y\, dz)$

18. $(y\, dx + 2x\, dy)\, e^{xy} + e^{2z}\, dz$

19. $e^z\, dx + 2y\, dy + xe^z\, dz$

20. $-2x\, dx + z \sinh y\, dy + \cosh y\, dz$

9.3 From Calculus: Double Integrals. *Optional*

Students familiar with double integrals from calculus should go on to the next section, skipping the present review (included to make the book reasonably self-contained).

In a definite integral (1), Sec. 9.1, we integrate a function $f(x)$ over an interval (a segment) of the x-axis. In a double integral we integrate a function $f(x, y)$, called the *integrand,* over a closed bounded[1] region R in the xy-plane, whose boundary curve has a unique tangent at each point, but may have finitely many cusps (such as the vertices of a triangle or a rectangle).

The definition of the double integral is quite similar to that of the definite integral. We subdivide the region R by drawing parallels to the x- and y-axes (Fig. 207). We number the rectangles that are within R from 1 to n. In each such rectangle we choose a point, say, (x_k, y_k) in the kth rectangle, and then we form the sum

$$J_n = \sum_{k=1}^{n} f(x_k, y_k)\, \Delta A_k$$

---

[1] "Closed" means that the boundary is part of the region, and "bounded" means that the region can be enclosed in a circle of sufficiently large radius.

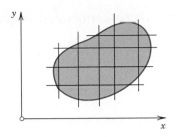

Fig. 207.   Subdivision of *R*

where ΔA_k is the area of the *k*th rectangle. This we do for larger and larger positive integers *n* in a completely independent manner but so that the length of the maximum diagonal of the rectangles approaches zero as *n* approaches infinity. In this fashion we obtain a sequence of real numbers $J_{n_1}, J_{n_2}, \cdots$. Assuming that $f(x, y)$ is continuous in *R* and *R* is bounded by finitely many smooth curves (see Sec. 9.1), one can show[2] that this sequence converges and its limit is independent of the choice of subdivisions and corresponding points (x_k, y_k). This limit is called the **double integral** *of* $f(x, y)$ *over the region R*, and is denoted by

$$\iint\limits_{R} f(x, y) \, dx \, dy \qquad \text{or} \qquad \iint\limits_{R} f(x, y) \, dA.$$

Double integrals have properties quite similar to those of definite integrals. Indeed, for any functions *f* and *g* of (x, y), defined and continuous in a region *R*,

$$\iint\limits_{R} kf \, dx \, dy = k \iint\limits_{R} f \, dx \, dy \qquad \qquad (k \text{ constant})$$

(1)
$$\iint\limits_{R} (f + g) \, dx \, dy = \iint\limits_{R} f \, dx \, dy + \iint\limits_{R} g \, dx \, dy$$

$$\iint\limits_{R} f \, dx \, dy = \iint\limits_{R_1} f \, dx \, dy + \iint\limits_{R_2} f \, dx \, dy \qquad \qquad (\text{Fig. 208}).$$

Furthermore, there exists at least one point (x_0, y_0) in *R* such that we have

(2)
$$\iint\limits_{R} f(x, y) \, dx \, dy = f(x_0, y_0) A,$$

where *A* is the area of *R*. This is called the **mean value theorem** *for double integrals*.

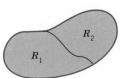

Fig. 208.   Formula (1)

---

[2] See Ref. [5] in Appendix 1.

Evaluation of Double Integrals

Double integrals over a region R may be evaluated by two successive integrations as follows. Suppose that R can be described by inequalities of the form

$$a \leqq x \leqq b, \qquad g(x) \leqq y \leqq h(x) \qquad \text{(Fig. 209)}$$

so that $y = g(x)$ and $y = h(x)$ represent the boundary of R. Then

(3)
$$\iint_R f(x, y) \, dx \, dy = \int_a^b \left[\int_{g(x)}^{h(x)} f(x, y) \, dy \right] dx.$$

We first integrate the inner integral

$$\int_{g(x)}^{h(x)} f(x, y) \, dy.$$

In this integration we keep x fixed, that is, we regard x as a constant. The result of this integration will be a function of x, say, $F(x)$. Integrating $F(x)$ over x from a to b, we then obtain the value of the double integral in (3).

Similarly, if R can be described by inequalities of the form

$$c \leqq y \leqq d, \qquad p(y) \leqq x \leqq q(y) \qquad \text{(Fig. 210)}$$

we obtain

(4)
$$\iint_R f(x, y) \, dx \, dy = \int_c^d \left[\int_{p(y)}^{q(y)} f(x, y) \, dx \right] dy;$$

we now integrate first over x (treating y as a constant) and then the resulting function of y from c to d.

If R cannot be represented by those inequalities, but can be subdivided into finitely many portions that have that property, we may integrate $f(x, y)$ over each portion separately and add the results; this will give us the value of the double integral of $f(x, y)$ over that region R.

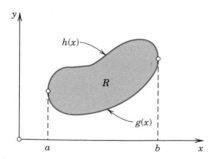

Fig. 209. Evaluation of a double integral

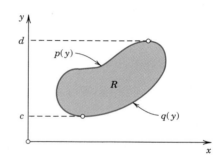

Fig. 210. Evaluation of a double integral

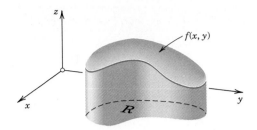

Fig. 211.   Double integral as volume

Applications of Double Integrals

Double integrals have various geometrical and physical applications. For example, the **area** A of a region R in the xy-plane is given by the double integral

$$A = \iint_R dx\,dy.$$

The **volume** V beneath the surface $z = f(x, y)$ (> 0) and above a region R in the xy-plane is (Fig. 211)

$$V = \iint_R f(x, y)\,dx\,dy,$$

because the term $f(x_k, y_k)\,\Delta A_k$ in J_n at the beginning of this section represents the volume of a rectangular box with base ΔA_k and altitude $f(x_k, y_k)$.

Let $f(x, y)$ be the density (= mass per unit area) of a distribution of mass in the xy-plane. Then the total mass M in R is

$$M = \iint_R f(x, y)\,dx\,dy;$$

the **center of gravity** of the mass in R has the coordinates $\bar{x}$, $\bar{y}$, where

$$\bar{x} = \frac{1}{M} \iint_R x f(x, y)\,dx\,dy \qquad \text{and} \qquad \bar{y} = \frac{1}{M} \iint_R y f(x, y)\,dx\,dy;$$

the **moments of inertia** I_x and I_y of the mass in R about the x- and y-axes, respectively, are

$$I_x = \iint_R y^2 f(x, y)\,dx\,dy, \qquad I_y = \iint_R x^2 f(x, y)\,dx\,dy;$$

and the **polar moment of inertia** I_0 about the origin of the mass in R is

$$I_0 = I_x + I_y = \iint_R (x^2 + y^2) f(x, y)\,dx\,dy.$$

EXAMPLE 1   **Center of gravity. Moments of inertia**

Let $f(x, y) = 1$ be the density of mass in the region R: $0 \leq y \leq \sqrt{1 - x^2}$, $0 \leq x \leq 1$ (Fig. 212). Find the center of gravity and the moments of inertia I_x, I_y, and I_0.

Solution.   The total mass in R is obtained as the double integral

Fig. 212.   Example 1

$$M = \iint_R dx\,dy = \int_0^1\left[\int_0^{\sqrt{1-x^2}} dy\right]dx = \int_0^1\sqrt{1-x^2}\,dx = \int_0^{\pi/2}\cos^2\theta\,d\theta = \frac{\pi}{4}$$

$(x = \sin\theta)$, which is the area of R. The coordinates of the center of gravity are

$$\bar{x} = \frac{4}{\pi}\iint_R x\,dx\,dy = \frac{4}{\pi}\int_0^1\left[\int_0^{\sqrt{1-x^2}} x\,dy\right]dx = \frac{4}{\pi}\int_0^1 x\sqrt{1-x^2}\,dx = -\frac{4}{\pi}\int_1^0 z^2\,dz = \frac{4}{3\pi}$$

$(\sqrt{1-x^2} = z)$, and $\bar{y} = \bar{x}$, for reasons of symmetry. Furthermore,

$$I_x = \iint_R y^2\,dx\,dy = \int_0^1\left[\int_0^{\sqrt{1-x^2}} y^2\,dy\right]dx = \frac{1}{3}\int_0^1(\sqrt{1-x^2})^3\,dx$$

$$= \frac{1}{3}\int_0^{\pi/2}\cos^4\theta\,d\theta = \frac{\pi}{16}, \qquad I_y = \frac{\pi}{16}, \qquad I_0 = I_x + I_y = \frac{\pi}{8} \approx 0.3927.$$

These integrations will become much simpler if we first make a suitable change of variables. This is what we show next.   ◀

Change of Variables in Double Integrals. Jacobian

Practical problems often require a change of the variables of integration in double integrals. Recall from calculus that for a definite integral the formula for the change from x to u is

(5)
$$\int_a^b f(x)\,dx = \int_\alpha^\beta f(x(u))\,\frac{dx}{du}\,du.$$

Here we assume that $x = x(u)$ is continuous and has a continuous derivative in some interval $\alpha \leqq u \leqq \beta$ such that $x(\alpha) = a$, $x(\beta) = b$ [or $x(\alpha) = b$, $x(\beta) = a$] and $x(u)$ varies between a and b when u varies between α and β.

The formula for a change of variables in double integrals from x, y to u, v is

(6)
$$\iint_R f(x, y)\,dx\,dy = \iint_{R^*} f(x(u, v), y(u, v))\left|\frac{\partial(x, y)}{\partial(u, v)}\right|\,du\,dv;$$

that is, the integrand is expressed in terms of u and v, and $dx\,dy$ is replaced by $du\,dv$ times the absolute value of the **Jacobian**[3]

---

[3]Named after the German mathematician CARL GUSTAV JACOB JACOBI (1804—1851), known for his contributions to elliptic functions, partial differential equations, and mechanics.

(7)
$$J = \frac{\partial(x, y)}{\partial(u, v)} = \begin{vmatrix} \dfrac{\partial x}{\partial u} & \dfrac{\partial x}{\partial v} \\[2mm] \dfrac{\partial y}{\partial u} & \dfrac{\partial y}{\partial v} \end{vmatrix}.$$

Here we assume the following. The functions

$$x = x(u, v), \qquad y = y(u, v)$$

effecting the change are continuous and have continuous partial derivatives in some region R^* in the uv-plane such that for every (u, v) in R^* the corresponding point (x, y) lies in R and, conversely, to every (x, y) in R there corresponds one and only one (u, v) in R^*; furthermore, the Jacobian J is either positive throughout R^* or negative throughout R^*. For a proof, see Ref. [5] in Appendix 1.

Of particular practical interest are **polar coordinates** r and θ, which can be introduced by setting $x = r \cos \theta$, $\quad y = r \sin \theta$. Then

$$J = \frac{\partial(x, y)}{\partial(r, \theta)} = \begin{vmatrix} \cos \theta & -r \sin \theta \\ \sin \theta & r \cos \theta \end{vmatrix} = r$$

and

(8)
$$\iint_R f(x, y) \, dx \, dy = \iint_{R^*} f(r \cos \theta, r \sin \theta) \, r \, dr \, d\theta$$

where R^* is the region in the $r\theta$-plane corresponding to R in the xy-plane.

EXAMPLE 2   Double integral in polar coordinates

Using (8), we obtain for I_x in Example 1

$$I_x = \iint_R y^2 \, dx \, dy = \int_0^{\pi/2} \int_0^1 r^2 \sin^2 \theta \, r \, dr \, d\theta = \int_0^{\pi/2} \sin^2 \theta \, d\theta \int_0^1 r^3 \, dr = \frac{\pi}{4} \cdot \frac{1}{4} = \frac{\pi}{16}. \quad \blacktriangleleft$$

EXAMPLE 3   Change of variables in a double integral

Evaluate the following double integral over the square R in Fig. 213.

$$\iint_R (x^2 + y^2) \, dx \, dy$$

Solution.   The shape of R suggests the transformation $x + y = u$, $x - y = v$. Then $x = \frac{1}{2}(u + v)$, $y = \frac{1}{2}(u - v)$, the Jacobian is

$$J = \frac{\partial(x, y)}{\partial(u, v)} = \begin{vmatrix} \frac{1}{2} & \frac{1}{2} \\[1mm] \frac{1}{2} & -\frac{1}{2} \end{vmatrix} = -\frac{1}{2},$$

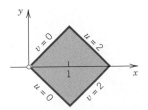

Fig. 213.   Region in Example 3

R corresponds to the square $0 \leqq u \leqq 2$, $0 \leqq v \leqq 2$, and, therefore,

$$\iint\limits_{R} (x^2 + y^2)\, dx\, dy = \int_0^2 \int_0^2 \tfrac{1}{2}(u^2 + v^2)\tfrac{1}{2}\, du\, dv = \tfrac{8}{3}.$$ ◄

This is the end of our review on double integrals. These integrals will be needed in this chapter, beginning in the next section.

PROBLEM SET 9.3

Double Integrals
Describe the region of integration and evaluate. (Show the details of your work.)

1. $\displaystyle\int_0^2 \int_0^4 (x^2 + y^2)\, dx\, dy$

2. As Prob. 1, order reversed

3. $\displaystyle\int_0^3 \int_{-y}^{y} (x^2 + y^2)\, dx\, dy$

4. As Prob. 3, order reversed

5. $\displaystyle\int_0^{\pi/4} \int_0^{y} \frac{\sin y}{y}\, dx\, dy$

6. $\displaystyle\int_0^{\pi/4} \int_{\sin x}^{\cos x} xy\, dy\, dx$

7. $\displaystyle\int_0^2 \int_0^{y} \sinh(x + y)\, dx\, dy$

8. As Prob. 7, order reversed

9. $\displaystyle\int_1^5 \int_0^{x^2} (1 + 2x)\, e^{x+y}\, dy\, dx$

10. $\displaystyle\int_0^{\pi/4} \int_0^{\cos y} x^2 \sin y\, dx\, dy$

Applications
Volume. Find the volume of the following regions in space.
11. The region beneath $z = 4x^2 + 9y^2$ and above the rectangle with vertices $(0, 0)$, $(3, 0)$, $(3, 2)$, $(0, 2)$
12. The first octant region bounded by the coordinate planes and the surfaces $y = 1 - x^2$, $z = 1 - x^2$
13. The first octant section cut from the region inside the cylinder $x^2 + z^2 = a^2$ by the planes $y = 0$, $z = 0$, $x = y$

Center of Gravity. Find the coordinates $\bar{x}$, $\bar{y}$ of the center of gravity of a mass of density $f(x, y) = 1$ in a region R, where R is
14. The triangle with vertices $(0, 0)$, $(b, 0)$, (b, h)
15. The region $x^2 + y^2 \leqq a^2$ in the first quadrant
16. Check the result in Prob. 15 using polar coordinates

Moments of Inertia. In Probs. 17–20 find the moments of inertia I_x, I_y, I_0 of a mass of density $f(x, y) = 1$ in a region R shown in the following figures (which the engineer is likely to need, along with other profiles listed in engineering handbooks).

17.
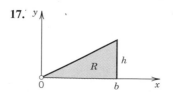

18.
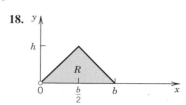

19.

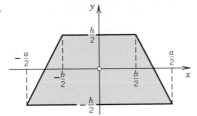

20.

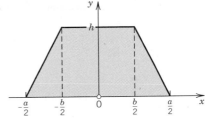

9.4 Green's Theorem in the Plane

Double integrals over a plane region may be transformed into line integrals over the boundary of the region and conversely. This is of practical interest because it may help to make the evaluation of an integral easier. It also helps in the theory whenever one wants to switch from one kind of integral to the other. The transformation can be done by the following theorem.

THEOREM 1   **Green's theorem in the plane[4]**

(Transformation between double integrals and line integrals)

Let R be a closed bounded region (see Sec. 9.3) *in the xy-plane whose boundary C consists of finitely many smooth curves* (see Sec. 9.1). *Let $F_1(x, y)$ and $F_2(x, y)$ be functions that are continuous and have continuous partial derivatives $\partial F_1/\partial y$ and $\partial F_2/\partial x$ everywhere in some domain containing R. Then*

(1)
$$\iint_R \left(\frac{\partial F_2}{\partial x} - \frac{\partial F_1}{\partial y} \right) dx\, dy = \oint_C (F_1\, dx + F_2\, dy);$$

here we integrate along the entire boundary C of R such that R is on the left as we advance in the direction of integration (see Fig. 214). (Proof below.)

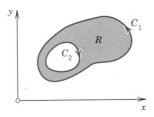

Fig. 214. Region *R* whose boundary *C* consists of two parts: C_1 is traversed counterclockwise, while C_2 is traversed clockwise

---

[4]GEORGE GREEN (1793—1841), English mathematician, who was self-educated, started out as a baker, and at his death was fellow of Caius College, Cambridge. His work concerned potential theory in connection with electricity and magnetism, vibrations, waves, and elasticity theory. It remained almost unknown, even in England, until after his death.

Comment. Formula (1) can be written in vectorial form

(1′)
$$\iint_R (\text{curl } \mathbf{F}) \cdot \mathbf{k} \, dx \, dy = \oint_C \mathbf{F} \cdot d\mathbf{r} \qquad (\mathbf{F} = [F_1, F_2] = F_1 \mathbf{i} + F_2 \mathbf{j}).$$

This follows from (1), Sec. 8.11, which shows that the third component of curl $\mathbf{F}$ is $\partial F_2/\partial x - \partial F_1/\partial y$.

EXAMPLE 1   **Verification of Green's theorem in the plane**

Green's theorem in the plane will be quite important in our further work. Before proving it, let us get used to it by verifying it for $F_1 = y^2 - 7y$, $F_2 = 2xy + 2x$ and C the circle $x^2 + y^2 = 1$.

Solution. In (1) on the left we get

$$\iint_R \left(\frac{\partial F_2}{\partial x} - \frac{\partial F_1}{\partial y} \right) dx \, dy = \iint_R [(2y + 2) - (2y - 7)] \, dx \, dy = 9 \iint_R dx \, dy = 9\pi$$

since the circular disk R has area π. On the right in (1) we represent C (oriented counterclockwise!) by

$$\mathbf{r}(t) = [\cos t, \sin t]. \qquad \text{Then} \qquad \mathbf{r}'(t) = [-\sin t, \cos t].$$

On C we thus obtain

$$F_1 = \sin^2 t - 7 \sin t, \qquad F_2 = 2 \cos t \sin t + 2 \cos t.$$

Hence the integral in (1) on the right becomes

$$\oint_C (F_1 x' + F_2 y') \, dt = \int_0^{2\pi} \left[(\sin^2 t - 7 \sin t)(-\sin t) + 2(\cos t \sin t + \cos t)(\cos t) \right] dt$$

$$= 0 + 7\pi + 0 + 2\pi = 9\pi.$$

This verifies Green's theorem in the plane.   ◀

PROOF OF GREEN'S THEOREM. We first prove Green's theorem for a *special region R* that can be represented in both the forms

$$a \leqq x \leqq b, \qquad u(x) \leqq y \leqq v(x) \qquad \text{(Fig. 215)}$$

and

$$c \leqq y \leqq d, \qquad p(y) \leqq x \leqq q(y) \qquad \text{(Fig. 216)}.$$

Using (3) in Sec. 9.3, we obtain for the second term on the left side of (1) (without the minus sign)

(2)
$$\iint_R \frac{\partial F_1}{\partial y} \, dx \, dy = \int_a^b \left[\int_{u(x)}^{v(x)} \frac{\partial F_1}{\partial y} \, dy \right] dx.$$

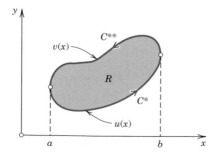

Fig. 215.   Example of a special region

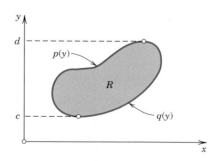

Fig. 216.   Example of a special region

(The first term will be considered later.) We integrate the inner integral:

$$\int_{u(x)}^{v(x)} \frac{\partial F_1}{\partial y}\, dy = F_1(x, y)\, \bigg|_{y=u(x)}^{y=v(x)} = F_1[x, v(x)] - F_1[x, u(x)].$$

By inserting this into (2) we find

$$\iint_R \frac{\partial F_1}{\partial y}\, dx\, dy = \int_a^b F_1[x, v(x)]\, dx - \int_a^b F_1[x, u(x)]\, dx$$

$$= -\int_b^a F_1[x, v(x)]\, dx - \int_a^b F_1[x, u(x)]\, dx.$$

Since $y = v(x)$ represents the curve C^{**} (Fig. 215) and $y = u(x)$ represents C^*, the last two integrals may be written as line integrals over C^{**} and C^* (oriented as in Fig. 215); therefore,

$$\iint_R \frac{\partial F_1}{\partial y}\, dx\, dy = -\int_{C^{**}} F_1(x, y)\, dx - \int_{C^*} F_1(x, y)\, dx$$

(3)

$$= -\oint_C F_1(x, y)\, dx.$$

This proves (1) in Green's theorem if $F_2 = 0$.

The result remains valid if C has portions parallel to the y-axis (such as $\tilde{C}$ and $\tilde{\tilde{C}}$ in Fig. 217). Indeed, the integrals over these portions are zero because in (3) on the right we integrate with respect to x. Hence we may add these integrals to the integrals over C^* and C^{**} to obtain the integral over the whole boundary C in (3).

Similarly, using (4) in Sec. 9.3, we obtain for the first term in (1) on the left by means of the second representation of the special region (see Fig. 216)

$$\iint_R \frac{\partial F_2}{\partial x}\, dx\, dy = \int_c^d \left[\int_{p(y)}^{q(y)} \frac{\partial F_2}{\partial x}\, dx \right] dy$$

$$= \int_c^d F_2(q(y), y)\, dy + \int_d^c F_2(p(y), y)\, dy$$

$$= \oint_C F_2(x, y)\, dy.$$

Together with (3) this gives (1) and proves Green's theorem for special regions.

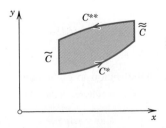

Fig. 217.   Proof of Green's theorem

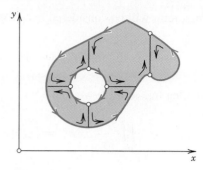

Fig. 218. Proof of Green's theorem

We now prove the theorem for a region R that itself is not a special region but can be subdivided into finitely many special regions (Fig. 218). In this case we apply the theorem to each subregion and then add the results; the left-hand members add up to the integral over R while the right-hand members add up to the line integral over C plus integrals over the curves introduced for subdividing R. Each of the latter integrals occurs twice, taken once in each direction. Hence these two integrals cancel each other, and we are left with the line integral over C.

The proof thus far covers all regions that are of interest in practical problems. To prove the theorem for the most general region R satisfying the conditions in the theorem, we must approximate R by a region of the type just considered and then use a limiting process. For details of this see Ref. [5] in Appendix 1. ◀

Further Applications of Green's Theorem

EXAMPLE 2   **Area of a plane region as a line integral over the boundary**

In (1) we first choose $F_1 = 0$, $F_2 = x$ and then $F_1 = -y$, $F_2 = 0$. This gives

$$\iint_R dx\, dy = \oint_C x\, dy \qquad \text{and} \qquad \iint_R dx\, dy = -\oint_C y\, dx,$$

respectively. The double integral is the area A of R. By addition we have

(4)
$$A = \frac{1}{2} \oint_C (x\, dy - y\, dx),$$

where we integrate as indicated in Green's theorem. This interesting formula expresses the area of R in terms of a line integral over the boundary. It has various applications; for instance, the theory of certain **planimeters** (instruments for measuring area) is based on it.

For an **ellipse** $x^2/a^2 + y^2/b^2 = 1$ or $x = a \cos t$, $y = b \sin t$ we get $x' = -a \sin t$, $y' = b \cos t$; thus from (4) we obtain the familiar result

$$A = \frac{1}{2} \int_0^{2\pi} (xy' - yx')\, dt = \frac{1}{2} \int_0^{2\pi} \left[ab \cos^2 t - (-ab \sin^2 t) \right] dt = \pi ab. \qquad ◀$$

EXAMPLE 3   **Area of a plane region in polar coordinates**

Let r and θ be polar coordinates defined by $x = r \cos \theta$, $y = r \sin \theta$. Then

$$dx = \cos \theta\, dr - r \sin \theta\, d\theta, \qquad dy = \sin \theta\, dr + r \cos \theta\, d\theta,$$

and (4) becomes a formula that is well known from calculus, namely,

$$(5) \qquad\qquad A = \frac{1}{2} \oint_C r^2 \, d\theta.$$

As an application of (5), we consider the **cardioid** $r = a(1 - \cos \theta)$, where $0 \leq \theta \leq 2\pi$ (Fig. 219). We find

$$A = \frac{a^2}{2} \int_0^{2\pi} (1 - \cos \theta)^2 \, d\theta = \frac{3\pi}{2} a^2. \qquad\qquad \blacktriangleleft$$

EXAMPLE 4   **Transformation of a double integral of the Laplacian of a function into a line integral of its normal derivative**

The Laplacian plays an important role in physics and engineering. A first impression of this was obtained in Sec. 8.9, and we shall discuss this further in Chap. 11. At present, let us use Green's theorem for deriving a basic integral formula involving the Laplacian.

   We take a function $w(x, y)$ that is continuous and has continuous first and second partial derivatives in a domain of the xy-plane containing a region R of the type indicated in Green's theorem. We set $F_1 = -\partial w/\partial y$ and $F_2 = \partial w/\partial x$. Then $\partial F_1/\partial y$ and $\partial F_2/\partial x$ are continuous in R, and in (1) on the left we obtain

$$(6) \qquad\qquad \frac{\partial F_2}{\partial x} - \frac{\partial F_1}{\partial y} = \frac{\partial^2 w}{\partial x^2} + \frac{\partial^2 w}{\partial y^2} = \nabla^2 w,$$

the Laplacian of w (see Sec. 8.9). Furthermore, using those expressions for F_1 and F_2, we get in (1) on the right

$$(7) \qquad \oint_C (F_1 \, dx + F_2 \, dy) = \oint_C \left(F_1 \frac{dx}{ds} + F_2 \frac{dy}{ds} \right) ds = \oint_C \left(-\frac{\partial w}{\partial y} \frac{dx}{ds} + \frac{\partial w}{\partial x} \frac{dy}{ds} \right) ds$$

where s is the arc length of C, and C is oriented as shown in Fig. 220. The integrand of the last integral may be written as the dot product of the vectors

$$\operatorname{grad} w = \frac{\partial w}{\partial x} \mathbf{i} + \frac{\partial w}{\partial y} \mathbf{j} \qquad\text{and}\qquad \mathbf{n} = \frac{dy}{ds} \mathbf{i} - \frac{dx}{ds} \mathbf{j};$$

that is,

$$(8) \qquad\qquad -\frac{\partial w}{\partial y} \frac{dx}{ds} + \frac{\partial w}{\partial x} \frac{dy}{ds} = (\operatorname{grad} w) \cdot \mathbf{n}.$$

The vector $\mathbf{n}$ is a unit normal vector to C, because the vector

$$\mathbf{r}'(s) = \frac{d\mathbf{r}}{ds} = \frac{dx}{ds} \mathbf{i} + \frac{dy}{ds} \mathbf{j} \qquad\qquad \text{(Sec. 8.5)}$$

is the unit tangent vector to C, and $\mathbf{r}' \cdot \mathbf{n} = 0$, so that $\mathbf{n}$ is perpendicular to $\mathbf{r}'$. Furthermore, it is not difficult to see that $\mathbf{n}$ is directed to the *exterior* of C. From this and (6) in Sec. 8.9, it follows that the expression on the right side of (8) is the derivative of w in the direction of the outward normal to C. Denoting this directional

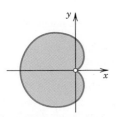

Fig. 219.   Cardioid

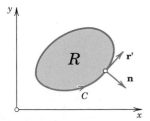

Fig. 220.   Example 4

derivative, the so-called **normal derivative** of R, by $\partial w/\partial n$ and taking (6), (7), and (8) into account, we obtain from Green's theorem the desired integral formula

(9)
$$\iint\limits_{R} \nabla^2 w \, dx \, dy = \oint_{C} \frac{\partial w}{\partial n} \, ds.$$

For instance, $w = x^2 - y^2$ satisfies Laplace's equation $\nabla^2 w = 0$. Hence its normal derivative integrated over a closed curve must give 0. Can you verify this directly by integration, say, for the square $0 \leqq x \leqq 1, 0 \leqq y \leqq 1$? ◀

    Green's theorem in the plane may facilitate the evaluation of integrals and can be used in both directions, depending on the kind of integral that is simpler in a concrete case. This is illustrated further in the problem set. Moreover, and perhaps more fundamentally, Green's theorem will be the essential tool in the proof of a very important integral theorem (Stokes's theorem in Sec. 9.9).

PROBLEM SET 9.4

Evaluation of Line Integrals by Green's Theorem

Using Green's theorem, evaluate the line integral $\oint_{C} \mathbf{F(r)} \cdot d\mathbf{r}$ counterclockwise around the boundary C of the region R, where

1. $\mathbf{F} = [x^2 e^y, y^2 e^x]$, C the rectangle with vertices $(0, 0)$, $(2, 0)$, $(2, 3)$, $(0, 3)$
2. $\mathbf{F} = [3y^2, x - y^4]$, R the square with vertices $(1, 1)$, $(-1, 1)$, $(-1, -1)$, $(1, -1)$
3. $\mathbf{F} = [y, -x]$, C the circle $x^2 + y^2 = 1/4$
4. $\mathbf{F} = [2xy^3, 3x^2y^2]$, C: $x^4 + y^4 = 1$ (sketch it)
5. $\mathbf{F} = \text{grad} (\sin x \cos y)$, C the ellipse $25x^2 + 9y^2 = 225$
6. $\mathbf{F} = [\sin y, \cos x]$, R the triangle with vertices $(0, 0)$, $(\pi, 0)$, $(\pi, 1)$
7. $\mathbf{F} = [\tan 0.2x, x^5 y]$, R: $x^2 + y^2 \leqq 25, y \geqq 0$
8. $\mathbf{F} = [\cosh y, -\sinh x]$, R: $1 \leqq x \leqq 3, x \leqq y \leqq 3x$
9. $\mathbf{F} = [e^y/x, e^y \ln x + 2x]$, R: $1 + x^4 \leqq y \leqq 2$
10. $\mathbf{F} = [x \cosh 2y, 2x^2 \sinh 2y]$, R: $x^2 \leqq y \leqq x$

Further Applications of Green's Theorem

Area. Find the area of the following regions.

11. The region in the first quadrant within the **cardioid** (see Example 3)
12. The region under one arch of the **cycloid** $\mathbf{r} = a(t - \sin t)\mathbf{i} + a(1 - \cos t)\mathbf{j}, 0 \leqq t \leqq 2\pi$. [Sketch it. Use (4).]
13. The region in the first quadrant under the arc of the **limaçon** (snail of Pascal[5]) $r = 1 + 2 \cos \theta$, $0 \leqq \theta \leqq \pi/2$. [Use (5).]

Integral of the normal derivative. In Probs. 14–18, using (9), evaluate $\oint_{C} \dfrac{\partial w}{\partial n} \, ds$

counterclockwise over the boundary curve C of the region R. (Show the details of your work.)

14. $w = e^x + e^y$, R the rectangle $0 \leqq x \leqq 2, 0 \leqq y \leqq 1$
15. $w = \cosh x$, R the triangle with vertices $(0, 0)$, $(4, 2)$, $(0, 2)$
16. $w = e^x \sin y$, R as in Prob. 15

---

[5]ETIENNE PASCAL (1588—1651), father of the famous French mathematician and philosopher BLAISE PASCAL (1623—1662).

17. $w = 3x^2y - y^3 + y^2$, $C\text{: } 25x^2 + y^2 = 25$

18. $w = x^5y + xy^5$, $R\text{: } x^2 + y^2 \leqq 1, x \geqq 0, y \geqq 0$

19. (Laplace's equation) Show that for a solution $w(x, y)$ of Laplace's equation $\nabla^2 w = 0$ in a region R with boundary curve C and outer unit normal vector $\mathbf{n}$,

(10)
$$\iint\limits_{R} \left[\left(\frac{\partial w}{\partial x} \right)^2 + \left(\frac{\partial w}{\partial y} \right)^2 \right] dx\, dy = \oint_{C} w \frac{\partial w}{\partial n}\, ds.$$

20. PROJECT. Other Forms of Green's Theorem in the Plane. Let R and C be as in Green's theorem, $\mathbf{r}'$ a unit tangent vector, and $\mathbf{n}$ an outer unit normal vector of C (Fig. 220 in Example 4). Show that (1) may be written

(11)
$$\iint\limits_{R} \text{div } \mathbf{F}\, dx\, dy = \oint_{C} \mathbf{F} \cdot \mathbf{n}\, ds$$

or

(12)
$$\iint\limits_{R} (\text{curl } \mathbf{F}) \cdot \mathbf{k}\, dx\, dy = \oint_{C} \mathbf{F} \cdot \mathbf{r}'\, ds$$

where $\mathbf{k}$ is a unit vector perpendicular to the xy-plane. Verify (11) and (12) for $\mathbf{F} = [7x, -3y]$ and C the circle $x^2 + y^2 = 4$ as well as for an example of your own choice.

9.5 Surfaces for Surface Integrals

Having introduced line and double integrals over regions in the plane, we turn next to surface integrals, in which we integrate over surfaces in space (a sphere, a portion of a cylinder, etc.). Hence we must first see how we can represent surfaces. We do this now and then discuss surface normals, which will also be needed in surface integrals.

Representations of Surfaces

We say briefly "surface," also for a *portion* of a surface, just as we said "curve" for an *arc* of a curve, for simplicity.

Representations of a surface S in xyz-space are

(1) $z = f(x, y)$ or $g(x, y, z) = 0.$

For example, $z = +\sqrt{a^2 - x^2 - y^2}$ or $x^2 + y^2 + z^2 - a^2 = 0$ ($z \geqq 0$) represents a hemisphere of radius a and center 0.

Now for *curves* C in line integrals, it was more practical and gave greater flexibility to use a *parametric* representation $\mathbf{r} = \mathbf{r}(t)$, where $a \leqq t \leqq b$. This is a mapping of the interval $a \leqq t \leqq b$, located on the t-axis, onto the curve C in xyz-space. It maps every t in that interval onto the point of C with position vector $\mathbf{r}(t)$. See Fig. 221.

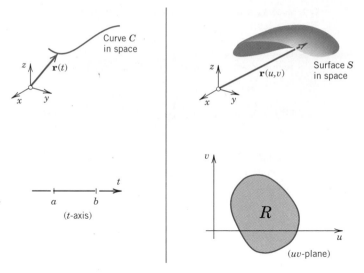

Fig. 221. Parametric representations of a curve and a surface

Similarly, for surfaces S in surface integrals, it will often be more practical to use a *parametric* representation. Surfaces are *two*-dimensional. Hence we need *two* parameters, which we call u and v. Thus a **parametric representation** of a surface S in space is of the form

(2)
$$\mathbf{r}(u, v) = x(u, v)\mathbf{i} + y(u, v)\mathbf{j} + z(u, v)\mathbf{k} \qquad (u, v) \text{ in } R$$

where R is some region in the uv-plane. This mapping (2) maps every point (u, v) in R onto the point of S with position vector $\mathbf{r}(u, v)$. See Fig. 221.

EXAMPLE 1   **Parametric representation of a cylinder**

The circular cylinder $x^2 + y^2 = a^2$, $-1 \leqq z \leqq 1$, has radius a, height 2, and the z-axis as axis. A parametric representation is

$$\mathbf{r}(u, v) = [a \cos u, \quad a \sin u, \quad v] = a \cos u\,\mathbf{i} + a \sin u\,\mathbf{j} + v\mathbf{k} \qquad \text{(Fig. 222)}.$$

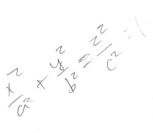

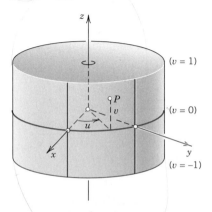

Fig. 222. Parametric representation
of a cylinder

In this representation the parameters u, v vary in the rectangle R: $0 \leq u \leq 2\pi$, $-1 \leq v \leq 1$ in the uv-plane. The components of $\mathbf{r}(u, v)$ are

$$x = a \cos u, \qquad y = a \sin u, \qquad z = v.$$

The curves $v = const$ are parallel circles. The curves $u = const$ are vertical straight lines. The point P in Fig. 222 corresponds to $u = \pi/3 = 60°$, $v = 0.7$. ◄

EXAMPLE 2   **Parametric representation of a sphere**

A sphere $x^2 + y^2 + z^2 = a^2$ can be represented in the form

(3)
$$\mathbf{r}(u, v) = a \cos v \cos u \, \mathbf{i} + a \cos v \sin u \, \mathbf{j} + a \sin v \, \mathbf{k}$$

where the parameters u, v vary in the rectangle R in the uv-plane given by the inequalities $0 \leq u \leq 2\pi$, $-\pi/2 \leq v \leq \pi/2$. The components of $\mathbf{r}$ are

$$x = a \cos v \cos u, \qquad y = a \cos v \sin u, \qquad z = a \sin v.$$

The curves $u = const$ and $v = const$ are the "meridians" and "parallels" on S (see Fig. 223). *This representation is used in **geography** for measuring the latitude and longitude of points on the globe.*

Another parametric representation of the sphere also used in mathematics is

(3*)
$$\mathbf{r}(u, v) = a \cos u \sin v \, \mathbf{i} + a \sin u \sin v \, \mathbf{j} + a \cos v \, \mathbf{k}$$

where $0 \leq u \leq 2\pi$, $0 \leq v \leq \pi$. ◄

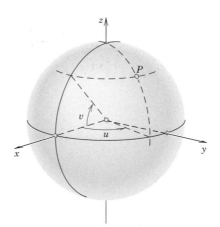

Fig. 223.   Parametric representation
of a sphere

EXAMPLE 3   **Parametric representation of a cone**

A circular cone $z = \sqrt{x^2 + y^2}$, $0 \leq z \leq H$ can be represented by

$$\mathbf{r}(u, v) = [u \cos v, \quad u \sin v, \quad u] = u \cos v \, \mathbf{i} + u \sin v \, \mathbf{j} + u\mathbf{k}$$

where u, v vary in the rectangle R: $0 \leq u \leq H$, $0 \leq v \leq 2\pi$. The components of $\mathbf{r}(u, v)$ are

$$x = u \cos v, \qquad y = u \sin v, \qquad z = u$$

and we can check that $x^2 + y^2 = z^2$, as it should be. What are the curves $u = const$ and $v = const$? Make a sketch. ◄

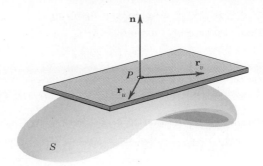

Fig. 224.   Tangent plane and normal vector

Tangent Plane and Surface Normal

Before defining surface integrals we go one step further and introduce surface normal vectors, which we shall need. A **normal vector** of a surface S at a point P is a vector perpendicular to the **tangent plane** of S at P (Fig. 224), the plane containing all the tangent vectors of curves on S through P, as we know from Sec. 8.9. Since S is given by $\mathbf{r} = \mathbf{r}(u, v)$ in (2), the idea is that we get a curve C on S by taking a pair of continuous functions (not both constant)

$$u = u(t), \qquad v = v(t),$$

so that C has the position vector $\tilde{\mathbf{r}}(t) = \mathbf{r}(u(t), v(t))$. Assuming these functions to be differentiable and applying the chain rule (Sec. 8.8), we get a tangent vector of C given by

$$\tilde{\mathbf{r}}'(t) = \frac{d\tilde{\mathbf{r}}}{dt} = \frac{\partial \mathbf{r}}{\partial u} u' + \frac{\partial \mathbf{r}}{\partial v} v'.$$

Hence the partial derivatives $\mathbf{r}_u$ and $\mathbf{r}_v$ at P are tangential to S at P, and we assume that they are linearly independent, so that they span the tangent plane of S at P. Then their vector product gives a normal vector $\mathbf{N}$ of S at P,

(4)
$$\boxed{\mathbf{N} = \mathbf{r}_u \times \mathbf{r}_v \neq \mathbf{0}.}$$

The corresponding **unit normal vector n** of S at P is (Fig. 224)

(5)
$$\boxed{\mathbf{n} = \frac{1}{|\mathbf{N}|} \mathbf{N} = \frac{1}{|\mathbf{r}_u \times \mathbf{r}_v|} \mathbf{r}_u \times \mathbf{r}_v.}$$

Also, if S is represented by $g(x, y, z) = 0$, then, by Theorem 2 in Sec. 8.9,

(5*)
$$\boxed{\mathbf{n} = \frac{1}{|\text{grad } g|} \text{grad } g.}$$

A surface S is called a **smooth surface** if its surface normal depends continuously on the points of S.

S is called **piecewise smooth** if it consists of finitely many smooth portions.

For instance, a sphere is smooth, and the surface of a cube is piecewise smooth (explain!). We can now summarize our discussion as follows.

THEOREM 1   **(Tangent plane and surface normal)**

If a surface S is given by (2) with continuous $\mathbf{r}_u = \partial\mathbf{r}/\partial u$ and $\mathbf{r}_v = \partial\mathbf{r}/\partial v$ satisfying (4) at every point of S, then S has at every point P a unique tangent plane passing through P and spanned by $\mathbf{r}_u$ and $\mathbf{r}_v$, and a unique normal whose direction depends continuously on the points of S.

A unit normal vector **n** of S is given by (5). (See Fig. 224 on the previous page.)

EXAMPLE 4   **Unit normal vector of a sphere**

From (5*) we find that the sphere $g(x,\,y,\,z) = x^2 + y^2 + z^2 - a^2 = 0$ has the unit normal vector

$$\mathbf{n}(x,\,y,\,z) = \left[\frac{x}{a},\ \frac{y}{a},\ \frac{z}{a}\right] = \frac{x}{a}\mathbf{i} + \frac{y}{a}\mathbf{j} + \frac{z}{a}\mathbf{k}. \qquad \blacktriangleleft$$

EXAMPLE 5   **Unit normal vector of a cone**

At the apex of the cone $g(x,\,y,\,z) = -z + \sqrt{x^2 + y^2} = 0$ in Example 3, the unit normal vector **n** becomes undetermined because from (5*) we get

$$\mathbf{n} = \frac{1}{\sqrt{2}}\left(\frac{x}{\sqrt{x^2 + y^2}}\mathbf{i} + \frac{y}{\sqrt{x^2 + y^2}}\mathbf{j} - \mathbf{k}\right). \qquad \blacktriangleleft$$

What we have learned about surfaces in this section will be applied to discuss **surface integrals** in the next section.

PROBLEM SET 9.5

Preparation for Surface Integrals: Parametric Surface Representations, Normals

Familiarize yourself with parametric representations of practically important surfaces by deriving a representation (1), by finding the **parameter curves** (curves $u = const$ and $v = const$) on the surface and a normal vector $\mathbf{N} = \mathbf{r}_u \times \mathbf{r}_v$ of the surface. (Show the details of your work.)

1. xy-plane $\mathbf{r}(u,\,v) = [u,\,v]$ (that is, $u\mathbf{i} + v\mathbf{j}$; similarly in Probs. 2–10)
2. xy-plane in polar coordinates $\mathbf{r}(u,\,v) = [u\cos v,\ u\sin v]$ (thus, $u = r$, $v = \theta$)
3. Cone $\mathbf{r}(u,\,v) = [u\cos v,\ u\sin v,\ cu]$
4. Paraboloid of revolution $\mathbf{r}(u,\,v) = [u\cos v,\ u\sin v,\ u^2]$
5. Elliptic paraboloid $\mathbf{r}(u,\,v) = [au\cos v,\ bu\sin v,\ u^2]$
6. Ellipsoid $\mathbf{r}(u,\,v) = [a\cos v\cos u,\ b\cos v\sin u,\ c\sin v]$
7. Hyperbolic paraboloid $\mathbf{r}(u,\,v) = [au\cosh v,\ bu\sinh v,\ u^2]$
8. Hyperboloid $\mathbf{r}(u,\,v) = [a\sinh u\cos v,\ b\sinh u\sin v,\ c\cosh u]$
9. Elliptic cylinder $\mathbf{r}(u,\,v) = [a\cos v,\ b\sin v,\ u]$
10. Helicoid $\mathbf{r}(u,\,v) = [u\cos v,\ u\sin v,\ v]$ (Explain the name.)

 11. **CAS PROJECT. Plotting Surfaces.** Plot the surfaces in Probs. 1–10. Add $z = 0$ in Probs. 1 and 2. Choose values of a, b, c and describe how a surface changes if you change these values.

Derivation of parametric representations of surfaces. Find a parametric representation of the following surfaces. (The answer gives *one* such representation and there are many others.) Find a normal vector. (Show the details.)

12. Plane $x = z$
13. Plane $3x + 4y + 6z = 24$
14. Elliptic cylinder $9x^2 + 4y^2 = 36$
15. Ellipsoid $x^2 + y^2 + \frac{1}{4}z^2 = 1$
16. Sphere $x^2 + (y - 1)^2 + (z + 2)^2 = 4$
17. Hyperbolic cylinder $x^2 - y^2 = 1$
18. Elliptic cone $z = \sqrt{x^2 + 4y^2}$
19. Paraboloid $z = 9(x^2 + y^2)$

20. (Normal vector) Find the points in Probs. 2–5 at which (4) does not hold. Indicate whether this results from the shape of the surface or from the choice of the representation.

21. (Normal vector) Represent the paraboloid in Prob. 19 so that $N(0, 0) \neq \mathbf{0}$. Find N.

22. (Orthogonal parameters on a surface) Show that the parameter curves $u = const$ and $v = const$ on a surface $\mathbf{r} = \mathbf{r}(u, v)$ intersect at right angles if and only if $\mathbf{r}_u \cdot \mathbf{r}_v = 0$.

23. (Representation $z = f(x, y)$) Show that $z = f(x, y)$ can be written ($f_u = \partial f / \partial u$, etc.)

$$(6) \qquad \mathbf{r}(u, v) = [u, v, f(u, v)] \qquad \text{and} \qquad N = [-f_u, -f_v, 1].$$

Surface Normal and Tangent Plane

Surface Normal. (See also Probs. 1–10.) Find a unit normal vector of the surfaces represented by

24. $4x^2 + y^2 + 9z^2 = 36$ **25.** $4x - 4y + 7z = -3$

26. $z = 5xy$ **27.** $y^2 + z^2 = a^2$

28. $x^2 - y^2 + z^2 = 1$ **29.** $x^2 + y^2 + z^2 = 36$

30. PROJECT. Tangent planes $T(P)$ will be less important in our work, but you should know how to represent them. Show that:

    (a) If S: $\mathbf{r}(u, v)$, then $T(P)$: $(\mathbf{r}^* - \mathbf{r} \quad \mathbf{r}_u \quad \mathbf{r}_v) = 0$ (a scalar triple product) or $\mathbf{r}^*(p, q) = \mathbf{r}(P) + p\mathbf{r}_u(P) + q\mathbf{r}_v(P)$.

    (b) If S: $g(x, y, z) = 0$, then $T(P)$: $(\mathbf{r}^* - \mathbf{r}(P)) \cdot \nabla g = 0$.

    (c) If S: $z = f(x, y)$, then $T(P)$: $z^* - z = (x^* - x)f_x(P) + (y^* - y)f_y(P)$.

    Interpret (a)–(c) geometrically. Give two examples for (a), two for (b), and two for (c).

9.6 Surface Integrals

To define a surface integral, we take a surface S, given by a parametric representation as just discussed,

$$(1) \qquad \mathbf{r}(u, v) = [x(u, v), y(u, v), z(u, v)] = x(u, v)\mathbf{i} + y(u, v)\mathbf{j} + z(u, v)\mathbf{k} \qquad (u, v) \text{ in } R.$$

We assume that S is piecewise smooth, so that S has a normal vector

$$(2) \qquad N = \mathbf{r}_u \times \mathbf{r}_v \qquad \text{and unit normal vector} \qquad \mathbf{n} = \frac{1}{|N|}N$$

(see Sec. 9.5) at every point (except perhaps for some edges or cusps, as for a cube or cone). For a given vector function $\mathbf{F}$ we can now define the surface integral over S by

$$(3) \qquad \iint_S \mathbf{F} \cdot \mathbf{n} \, dA = \iint_R \mathbf{F}[\mathbf{r}(u, v)] \cdot N(u, v) \, du \, dv.$$

Note that the integrand is a scalar, not a vector because we take dot products. Indeed, $\mathbf{F} \cdot \mathbf{n}$ is the normal component of $\mathbf{F}$. This integral arises naturally in flow problems, where it gives the *flux* across S (= mass of fluid crossing S per unit time; see Sec. 8.10) when $\mathbf{F} = \rho\mathbf{v}$. Here, ρ is the density of the fluid and $\mathbf{v}$ the velocity vector of the flow (example below). We may thus call the surface integral (3) the **flux integral.**

The integral on the right in (3) is a double integral (Sec. 9.3) over the region R in the uv-plane corresponding to S. It exists for continuous $\mathbf{F}$ and piecewise smooth S, because this makes $\mathbf{F} \cdot \mathbf{N}$ piecewise continuous.

We motivate (3). By the definition of a vector product, $|\mathbf{N}| = |\mathbf{r}_u \times \mathbf{r}_v|$ is the area of the parallelogram with sides $\mathbf{r}_u$ and $\mathbf{r}_v$. Hence $dA = |\mathbf{N}| \, du \, dv$ is the element of area of S. Multiplication by $\mathbf{n}$ and use of (2) gives

(3*)
$$\mathbf{n} \, dA = \mathbf{n}|\mathbf{N}| \, du \, dv = \mathbf{N} \, du \, dv.$$

Dot multiplication by $\mathbf{F}$ and integration gives (3), which we wanted to motivate.

In terms of components we write

$$\mathbf{F} = [F_1, F_2, F_3] = F_1\mathbf{i} + F_2\mathbf{j} + F_3\mathbf{k}$$

(3**) $$\mathbf{n} = [\cos\alpha, \cos\beta, \cos\gamma] = \cos\alpha\,\mathbf{i} + \cos\beta\,\mathbf{j} + \cos\gamma\,\mathbf{k}$$

$$\mathbf{N} = [N_1, N_2, N_3] = N_1\mathbf{i} + N_2\mathbf{j} + N_3\mathbf{k}.$$

We claim that (3) can now be written

(4)
$$\iint_S \mathbf{F} \cdot \mathbf{n} \, dA = \iint_S (F_1 \cos\alpha + F_2 \cos\beta + F_3 \cos\gamma) \, dA$$
$$= \iint_R (F_1 N_1 + F_2 N_2 + F_3 N_3) \, du \, dv.$$

Indeed, both right sides follow by (2), Sec. 8.2, the first from $\mathbf{F} \cdot \mathbf{n}$ and the second from $\mathbf{F} \cdot \mathbf{N}$. Here, α, β, γ are the angles between $\mathbf{n}$ and the positive directions of the coordinate axes. In fact, we have $\mathbf{n} \cdot \mathbf{i} = \cos\alpha$ from the formula for $\mathbf{n}$, and the same from (4) in Sec. 8.2, namely, $\cos\alpha = \mathbf{n} \cdot \mathbf{i}/(|\mathbf{n}||\mathbf{i}|) = \mathbf{n} \cdot \mathbf{i}$ (because $\mathbf{n}$ and $\mathbf{i}$ are unit vectors). Similarly for β and γ. ◀

Other forms of surface integrals will be discussed later in this section.

EXAMPLE 1   **Flux through a surface**

Compute the flux of water through the parabolic cylinder $S: y = x^2$, $0 \le x \le 2$, $0 \le z \le 3$ (Fig. 225) if the velocity vector is $\mathbf{v} = \mathbf{F} = [3z^2, 6, 6xz]$, speed being measured in meters/sec. (Generally, $\mathbf{F} = \rho\mathbf{v}$, but water has the density $\rho = 1$ gram/cm^3 = 1 ton/meter^3.)

Solution.   Writing $x = u$ and $z = v$, we have $y = x^2 = u^2$. Hence a representation of S is

$$S: \qquad \mathbf{r} = [u, u^2, v] \qquad\qquad (0 \le u \le 2, 0 \le v \le 3).$$

From this,

$$\mathbf{r}_u = [1, 2u, 0]$$
$$\mathbf{r}_v = [0, 0, 1]$$
$$\mathbf{N} = \mathbf{r}_u \times \mathbf{r}_v = [2u, -1, 0].$$

On S, writing simply $\mathbf{F}(S)$ for $\mathbf{F}[\mathbf{r}(u, v)]$, we have $\mathbf{F}(S) = [3v^2, 6, 6uv]$. Hence $\mathbf{F}(S) \cdot \mathbf{N} = 6uv^2 - 6$. By integration we thus get from (3) the flux

$$\iint\limits_{S} \mathbf{F} \cdot \mathbf{n} \, dA = \int_0^3 \int_0^2 (6uv^2 - 6) \, du \, dv = \int_0^3 (3u^2v^2 - 6u)\Big|_{u=0}^2 \, dv$$

$$= \int_0^3 (12v^2 - 12) \, dv = (4v^3 - 12v)\Big|_{v=0}^3 = 108 - 36 = 72 \, [\text{meters}^3/\text{sec}]$$

or 72 000 liters/sec. ◄

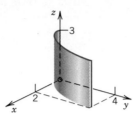

Fig. 225. Surface S in
Example 1

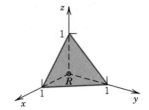

Fig. 226. Portion of a plane
in Example 2

EXAMPLE 2   Surface integral

Evaluate (3) when $\mathbf{F} = [x^2, 0, 3y^2]$ and S is the portion of the plane $x + y + z = 1$ in the first octant (Fig. 226).

Solution. Writing $x = u$ and $y = v$, we have $z = 1 - x - y = 1 - u - v$. Hence we can represent the plane $x + y + z = 1$ in the form $\mathbf{r}(u, v) = [u, v, 1 - u - v]$. We obtain the first-octant portion S of this plane by restricting $x = u$ and $y = v$ to the projection R of S in the xy-plane. R is the triangle bounded by the two coordinate axes and the straight line $x + y = 1$; thus $0 \leqq x \leqq 1 - y$, $0 \leqq y \leqq 1$.

By inspection or by differentiation,

$$\mathbf{N} = \mathbf{r}_u \times \mathbf{r}_v = [1, 0, -1] \times [0, 1, -1] = [1, 1, 1].$$

Hence $\mathbf{F}(S) \cdot \mathbf{N} = [u^2, 0, 3v^2] \cdot [1, 1, 1] = u^2 + 3v^2$. By (3),

$$\iint\limits_{S} \mathbf{F} \cdot \mathbf{n} \, dA = \iint\limits_{R} (u^2 + 3v^2) \, du \, dv = \int_0^1 \int_0^{1-v} (u^2 + 3v^2) \, du \, dv = \int_0^1 \left[\frac{1}{3}(1 - v)^3 + 3v^2(1 - v) \right] dv = \frac{1}{3}.$$

◄

From (3) or (4) we see that the value of the integral depends on the choice of the unit normal vector **n.** (Instead of **n** we could choose $-\mathbf{n}$.) We express this by saying that such an integral is an *integral over an* **oriented surface** S, that is, over a surface S on which we have chosen one of the two possible unit normal vectors in a continuous fashion. (For a piecewise smooth surface, this needs some further discussion, which we give below.) If we change the orientation of S, which means that we replace **n** by $-\mathbf{n}$, then each component of **n** in (4) is multiplied by -1, so that we have

THEOREM 1   (Change of orientation)

The replacement of **n** *by* $-\mathbf{n}$ *(hence of* **N** *by* $-\mathbf{N}$*) corresponds to the multiplication of the integral in* (3) *or* (4) *by* -1.

How to effect such a change of **N** in practice if S is given in the form (1)? The simplest way is to interchange u and v, because then $\mathbf{r}_u$ becomes $\mathbf{r}_v$ and conversely, so that $\mathbf{N} = \mathbf{r}_u \times \mathbf{r}_v$ becomes $\mathbf{r}_v \times \mathbf{r}_u = -\mathbf{r}_u \times \mathbf{r}_v = -\mathbf{N}$, as wanted. Let us illustrate this.

EXAMPLE 3   **Change of orientation in a surface integral**

In Example 1 we now represent S by $\tilde{\mathbf{r}} = [v, v^2, u]$, $0 \leqq v \leqq 2$, $0 \leqq u \leqq 3$. Then
$\tilde{\mathbf{N}} = \tilde{\mathbf{r}}_u \times \tilde{\mathbf{r}}_v = [0, 0, 1] \times [1, 2v, 0] = [-2v, 1, 0]$. For $\mathbf{F} = [3z^2, 6, 6xz]$ we now get $\tilde{\mathbf{F}}(S) = [3u^2, 6, 6uv]$.
Hence $\tilde{\mathbf{F}}(S) \cdot \tilde{\mathbf{N}} = -6u^2v + 6$ and integration gives the old result times -1,

$$\iint\limits_{R} \tilde{\mathbf{F}}(S) \cdot \tilde{\mathbf{N}} \, dv \, du = \int_0^3 \int_0^2 (-6u^2v + 6) \, dv \, du = \int_0^3 (-12u^2 + 12) \, du = -72. \quad \blacktriangleleft$$

More About Orientation

Smooth Surfaces. If a surface S is smooth and P any of its points, we may choose a unit normal vector $\mathbf{n}$ of S at P. The direction of $\mathbf{n}$ is then called the *positive normal direction* of S at P. Obviously there are two possibilities in choosing $\mathbf{n}$.

A smooth surface S is said to be **orientable** if the positive normal direction, when given at an arbitrary point P_0 of S, can be continued in a unique and continuous way to the entire surface.

Essential in practice is the fact that *a sufficiently small portion of a smooth surface is always orientable.* From a theoretical point of view it is interesting that this may not hold in the large. There are nonorientable surfaces. A well-known example of such a surface is the **Möbius strip**[6] shown in Fig. 227. When a normal vector, which is given at P_0, is displaced continuously along the curve C in Fig. 227, the resulting normal vector upon returning to P_0 is opposite to the original vector at P_0. A model of a Möbius strip can be made by taking a long rectangular piece of paper, making a half-twist and sticking the shorter sides together so that the two points A and the two points B in Fig. 227 coincide.

Piecewise Smooth Surfaces. These can be oriented by using the following simple idea. If S is *smooth* and orientable and is bounded by a simple closed curve C, we may associate with each of the two possible orientations of S an orientation of C, as shown in Fig. 228*a*. If S is *piecewise smooth*, we call it **orientable** if we can orient each smooth piece of S in such a manner that along each curve C^* which is a common boundary of two pieces S_1 and S_2 the positive direction of C^* relative to S_1 is opposite to the positive direction of C^* relative to S_2.

Figure 228*b* illustrates the ideas for a surface consisting of two smooth pieces.

Another Way of Writing Surface Integrals (3) or (4)

Our discussion of orientation allows us to explain another way of writing (4). It is also customary to write in (4)

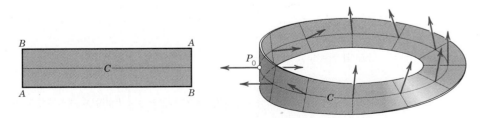

Fig. 227.   Möbius strip

---

[6]AUGUST FERDINAND MÖBIUS (1790—1868), German mathematician, student of Gauss, known for his work in the theory of surfaces, projective geometry, and mechanics.

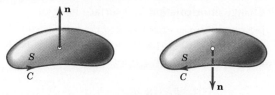

(a) Smooth surface

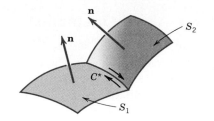

(b) Piecewise smooth surface

Fig. 228. Orientation of a surface

$$\textbf{(a)} \qquad \iint_S F_1 \cos \alpha \, dA = \iint_S F_1 \, dy \, dz$$

(5)

$$\textbf{(b)} \qquad \iint_S F_2 \cos \beta \, dA = \iint_S F_2 \, dz \, dx$$

$$\textbf{(c)} \qquad \iint_S F_3 \cos \gamma \, dA = \iint_S F_3 \, dx \, dy$$

and together

$$\textbf{(6)} \qquad \iint_S \mathbf{F} \cdot \mathbf{n} \, dA = \iint_S (F_1 \, dy \, dz + F_2 \, dz \, dx + F_3 \, dx \, dy).$$

This is an analog of (3′) in Sec. 9.1 for line integrals. We can use these formulas for evaluating surface integrals by converting them to double integrals over plane regions, but must carefully take into account the orientation of S (the choice of $\mathbf{n}$). We explain this for (5c). If the surface S is given by $z = h(x, y)$ with (x, y) varying in a region R in the xy-plane, and if S is oriented so that $\cos \gamma > 0$, then

$$\text{(5c′)} \qquad \iint_S F_3 \cos \gamma \, dA = + \iint_R F_3[x, y, h(x, y)] \, dx \, dy,$$

but if $\cos \gamma < 0$, then

$$\text{(5c″)} \qquad \iint_S F_3 \cos \gamma \, dA = - \iint_R F_3[x, y, h(x, y)] \, dx \, dy.$$

This follows by noting that the element of area $dx \, dy$ in the xy-plane is the projection $|\cos \gamma| \, dA$ of the element of area dA of S, and we have $\cos \gamma = +|\cos \gamma|$ in (5c′), where $\cos \gamma > 0$, but $\cos \gamma = -|\cos \gamma|$ in (5c″), where $\cos \gamma < 0$. Similarly for (5a), (5b) and in (6). ◀

EXAMPLE 4   **An application of (6)**

Verify the result in Example 1 by (6).

Solution. $F_1 = 3z^2$, $F_2 = 6$, $F_3 = 6xz$.   *S:* $y = x^2$ with $0 \leqq x \leqq 2$, thus $0 \leqq y \leqq 4$, and $0 \leqq z \leqq 3$ give the limits of integration in the coordinate planes. Also [see (2) and (3**)],

$$\mathbf{N} = |\mathbf{N}|\mathbf{n} = |\mathbf{N}|[\cos \alpha, \cos \beta, \cos \gamma] = [2u, -1, 0] = [2x, -1, 0]$$

shows that $\cos \alpha > 0$, $\cos \beta < 0$, causing a minus sign in (5b), and $\cos \gamma = 0$, causing (5c) to be absent. From all this we get in (6)

$$\iint_S \mathbf{F} \cdot \mathbf{n}\, dA = \int_0^3 \int_0^4 3z^2\, dy\, dz - \int_0^2 \int_0^3 6\, dz\, dx = 4 \cdot 3^3 - 6 \cdot 3 \cdot 2 = 72.$$

This confirms the result in Example 1.   ◀

Surface Integrals Without Regard to Orientation

Another type of surface integral is

(7)
$$\iint_S G(\mathbf{r})\, dA = \iint_R G(\mathbf{r}(u, v))|\mathbf{N}(u, v)|\, du\, dv.$$

Here $dA = |\mathbf{N}|\, du\, dv = |\mathbf{r}_u \times \mathbf{r}_v|\, du\, dv$ is the element of area of the surface S represented by (1) [see before (3*)] and we disregard the orientation.

As for applications, if $G(\mathbf{r})$ is the mass density of S, then (7) is the total mass of S. If $G = 1$, then (7) gives the **area** $A(S)$ of S,

(8)
$$A(S) = \iint_S dA = \iint_R |\mathbf{r}_u \times \mathbf{r}_v|\, du\, dv.$$

EXAMPLE 5   **Area of a sphere**

A sphere of radius a can be represented by (3), Sec. 9.5, that is,

$$\mathbf{r}(u, v) = [a \cos v \cos u, \quad a \cos v \sin u, \quad a \sin v]$$

where $0 \leqq u \leqq 2\pi$, $-\pi/2 \leqq v \leqq \pi/2$. By direct calculation we obtain (verify!)

$$\mathbf{r}_u \times \mathbf{r}_v = [a^2 \cos^2 v \cos u, \quad a^2 \cos^2 v \sin u, \quad a^2 \cos v \sin v].$$

Hence, using $\cos^2 u + \sin^2 u = 1$, etc., we obtain

$$|\mathbf{r}_u \times \mathbf{r}_v| = a^2(\cos^4 v \cos^2 u + \cos^4 v \sin^2 u + \cos^2 v \sin^2 v)^{1/2} = a^2 |\cos v|.$$

With this, (8) gives the familiar formula

$$A(S) = a^2 \int_{-\pi/2}^{\pi/2} \int_0^{2\pi} |\cos v|\, du\, dv = 2\pi a^2 \int_{-\pi/2}^{\pi/2} \cos v\, dv = 4\pi a^2.$$ ◀

EXAMPLE 6   **Representation and area of the surface of a torus (surface of a doughnut)**

A *torus surface* S is obtained by rotating a circle C about a straight line L in space so that C does not intersect or touch L but its plane always passes through L. If L is the z-axis and C has radius b and its center has distance a ($> b$) from L, as in Fig. 229, then S can be represented by

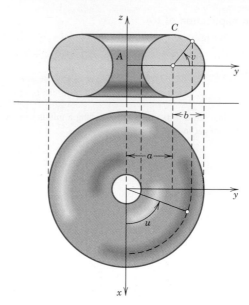

Fig. 229. Torus in Example 6

$$\mathbf{r}(u, v) = (a + b \cos v) \cos u\, \mathbf{i} + (a + b \cos v) \sin u\, \mathbf{j} + b \sin v\, \mathbf{k}.$$

Thus

$$\mathbf{r}_u = -(a + b \cos v) \sin u\, \mathbf{i} + (a + b \cos v) \cos u\, \mathbf{j}$$

$$\mathbf{r}_v = -b \sin v \cos u\, \mathbf{i} - b \sin v \sin u\, \mathbf{j} + b \cos v\, \mathbf{k}$$

$$\mathbf{r}_u \times \mathbf{r}_v = b(a + b \cos v)[\cos u \cos v\, \mathbf{i} + \sin u \cos v\, \mathbf{j} + \sin v\, \mathbf{k}].$$

Hence $|\mathbf{r}_u \times \mathbf{r}_v| = b(a + b \cos v)$, and (8) gives the total area of the torus

$$(9) \qquad A(S) = \int_0^{2\pi} \int_0^{2\pi} b(a + b \cos v)\, du\, dv = 4\pi^2 ab. \qquad \blacktriangleleft$$

EXAMPLE 7   **Moment of inertia**

Find the moment of inertia I of a homogeneous spherical lamina $S: x^2 + y^2 + z^2 = a^2$ of mass M about the z-axis.

Solution.   If a mass is distributed over a surface S and $\mu(x, y, z)$ is the density of the mass (= mass per unit area), then the moment of inertia I of the mass with respect to a given axis L is defined by the surface integral

$$(10) \qquad I = \int\int_S \mu D^2\, dA$$

where $D(x, y, z)$ is the distance of the point (x, y, z) from L. Since, in the present example, μ is constant and S has the area $A = 4\pi a^2$, we have $\mu = M/A = M/(4\pi a^2)$.

For S we use the same representation as in Example 5. Then $D^2 = x^2 + y^2 = a^2 \cos^2 v$. Also, as in that example, $dA = a^2 \cos v\, du\, dv$. This gives the following result. [In the integration, use $\cos^3 v = \cos v\, (1 - \sin^2 v)$.]

$$I = \int\int_S \mu D^2\, dA = \frac{M}{4\pi a^2} \int_{-\pi/2}^{\pi/2} \int_0^{2\pi} a^4 \cos^3 v\, du\, dv = \frac{Ma^2}{2} \int_{-\pi/2}^{\pi/2} \cos^3 v\, dv = \frac{2Ma^2}{3}. \qquad \blacktriangleleft$$

Representations $z = f(x, y)$. If a surface S is given by $z = f(x, y)$, then setting $u = x$, $v = y$, $\mathbf{r} = [u, v, f]$ gives

$$|\mathbf{N}| = |\mathbf{r}_u \times \mathbf{r}_v| = |[1, 0, f_u] \times [0, 1, f_v]| = |[-f_u, -f_v, 1]| = \sqrt{1 + f_u^2 + f_v^2}$$

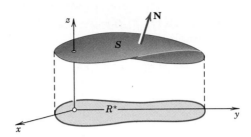

Fig. 230. Formula (11)

and, since $f_u = f_x$, $f_v = f_y$, formula (7) becomes

(11)
$$\iint\limits_{S} G(\mathbf{r})\, dA = \iint\limits_{R^*} G(x, y, f(x, y)) \sqrt{1 + \left(\frac{\partial f}{\partial x}\right)^2 + \left(\frac{\partial f}{\partial y}\right)^2}\; dx\, dy$$

where R^* is the projection of S into the xy-plane (Fig. 230) and the normal vector $\mathbf{N}$ on S points *up*. If it points *down*, the integral on the right is preceded by a minus sign.

From (11) with $G = 1$ we obtain for the **area** $A(S)$ of S: $z = f(x, y)$ the formula

(12)
$$A(S) = \iint\limits_{R^*} \sqrt{1 + \left(\frac{\partial f}{\partial x}\right)^2 + \left(\frac{\partial f}{\partial y}\right)^2}\; dx\, dy$$

where R^* is the projection of S into the xy-plane, as before.

PROBLEM SET 9.6

Surface Integrals

Flux Integrals (3) $\iint\limits_{S} \mathbf{F} \cdot \mathbf{n}\, dA$. Evaluate these integrals for the following data. Indicate the kind of surface. (Show the details of your work.)

1. $\mathbf{F} = [3x^2,\ y^2,\ 0]$, S: $\mathbf{r} = [u,\ v,\ 2u + 3v]$, $0 \le u \le 2$, $-1 \le v \le 1$
2. $\mathbf{F} = [x^2,\ e^y,\ 1]$, S: $x + y + z = 1$, $x \ge 0$, $y \ge 0$, $z \ge 0$
3. $\mathbf{F} = [e^{2y},\ e^{-2z},\ e^{2x}]$, S: $\mathbf{r} = [3 \cos u,\ 3 \sin u,\ v]$, $0 \le u \le \tfrac{1}{2}\pi$, $0 \le v \le 2$
4. $\mathbf{F} = [\sinh yz,\ 0,\ y^4]$, S: $\mathbf{r} = [u,\ \cos v,\ \sin v]$, $-4 \le u \le 4$, $0 \le v \le \pi$
5. $\mathbf{F} = [x - z,\ y - x,\ z - y]$, S: $\mathbf{r} = [u \cos v,\ u \sin v,\ u]$, $0 \le u \le 3$, $0 \le v \le 2\pi$
6. $\mathbf{F} = [y^3,\ x^3,\ z^3]$, S: $x^2 + 4y^2 = 1$, $x \ge 0$, $y \ge 0$, $0 \le z \le h$
7. $\mathbf{F} = [0,\ x,\ 0]$, S: $x^2 + y^2 + z^2 = 1$, $x \ge 0$, $y \ge 0$, $z \ge 0$
8. $\mathbf{F} = [4xy,\ 2x^2,\ 0]$, S: $\mathbf{r} = [\cosh u,\ \sinh u,\ v]$, $0 \le u \le 2$, $-3 \le v \le 3$
9. $\mathbf{F} = [x,\ y,\ z]$, S: $\mathbf{r} = [u \cos v,\ u \sin v,\ u^2]$, $0 \le u \le 4$, $-\pi \le v \le \pi$
10. $\mathbf{F} = [x^2,\ y^2,\ z^2]$, S: $\mathbf{r} = [u \cos v,\ u \sin v,\ 3v]$, $0 \le u \le 1$, $0 \le v \le 2\pi$

 11. **CAS PROJECT. Program for Surface Integrals.** Write a program for evaluating surface integrals (3) that prints (or at least shows) intermediate results ($\mathbf{F}(S)$, $\mathbf{N}$, the integrand, the first integral), so that you get a feeling for possible difficulties in each step. Apply the program to Probs. 5 and 9 and to three examples of your choice, in which S is a sphere, an ellipsoid, and a helicoid.

Surface Integrals (7) $\iint_S G(\mathbf{r})\, dA$. Using (7) or (11), evaluate these integrals for the given data. (Show the details.)

12. $G = \cos x + \sin y$,   S: the portion of $x + y + z = 1$ in the first octant

13. $G = ye^{-xy}$,   S: $z = 3x + 4y$, $x \geq 1,\ y \geq 1$

14. $G = x^4 + y^4$,   S: $\mathbf{r} = [5 \cos u, \ 5 \sin u, \ v]$, $0 \leq u \leq \pi,\ -0.2 \leq v \leq 0.2$

15. $G = (1 + 9xz)^{3/2}$,   S: $\mathbf{r} = [u, \ v, \ u^3]$, $0 \leq u \leq 1,\ -2 \leq v \leq 2$

16. $G = (x^2 + y^2)^2 - z^2$,   S: $\mathbf{r} = [u \cos v, \ u \sin v, \ 2u]$, $0 \leq u \leq 1,\ -\pi \leq v \leq \pi$

17. $G = z$,   S: $x^2 + y^2 + z^2 = 9$, $z \geq 0$

Applications: Center of Gravity, Moments of Inertia

18. (Center of gravity) Justify the following formulas for the mass M and the center of gravity $(\bar{x}, \bar{y}, \bar{z})$ of a lamina S of density (mass per unit area) $\sigma(x, y, z)$ in space:

$$M = \iint_S \sigma\, dA, \quad \bar{x} = \frac{1}{M} \iint_S x\sigma\, dA, \quad \bar{y} = \frac{1}{M} \iint_S y\sigma\, dA, \quad \bar{z} = \frac{1}{M} \iint_S z\sigma\, dA.$$

19. (Moments of inertia) Justify the following formulas for the moments of inertia of the lamina in Prob. 18 about the x-, y- and z-axes, respectively:

$$I_x = \iint_S (y^2 + z^2)\sigma\, dA, \quad I_y = \iint_S (x^2 + z^2)\sigma\, dA, \quad I_z = \iint_S (x^2 + y^2)\sigma\, dA.$$

20. Find a formula for the moment of inertia of the lamina in Prob. 18 about the line $y = x$, $z = 0$.

Find the moment of inertia of a lamina S of density 1 about an axis A, where

21. S: $x^2 + y^2 = 1$, $0 \leq z \leq h$,   A: the z-axis

22. S as in Prob. 21,   A: the line $z = h/2$ in the xz-plane

23. S: $x^2 + y^2 = z^2$, $0 \leq z \leq h$,   A: the z-axis

24. (Steiner's theorem[7]) If I_A is the moment of inertia of a mass distribution of total mass M with respect to an axis A through the center of gravity, show that its moment of inertia I_B with respect to an axis B, which is parallel to A and has the distance k from it, is

$$I_B = I_A + k^2 M.$$

25. Using Steiner's theorem, find the moment of inertia of S in Prob. 22 about the x-axis.

26. TEAM PROJECT. First Fundamental Form of a Surface. Given a surface S: $\mathbf{r}(u, v)$, the corresponding quadratic differential form

(13) $$ds^2 = E\, du^2 + 2F\, du\, dv + G\, dv^2$$

with coefficients[8]

(14) $$E = \mathbf{r}_u \cdot \mathbf{r}_u, \qquad F = \mathbf{r}_u \cdot \mathbf{r}_v, \qquad G = \mathbf{r}_v \cdot \mathbf{r}_v$$

is called the **first fundamental form** of S. It is basic in the theory of surfaces, since with its help we can determine lengths, angles, and areas on S. To show this, prove the following.

---

[7]JACOB STEINER (1796—1863), Swiss geometer, born in a small village, learned to write only at age 14, became a pupil of Pestalozzi at 18, later studied at Heidelberg and Berlin and, finally, because of his outstanding research, was appointed professor at Berlin University.

[8]E, F, G are standard notations; of course, they have nothing to do with the functions F and G occurring at some places in this chapter.

(a) For a curve $C: u = u(t), v = v(t), a \leqq t \leqq b$, on S, formulas (10), Sec. 8.5, and (14) give the length

$$(15) \qquad l = \int_a^b \sqrt{\mathbf{r}'(t) \cdot \mathbf{r}'(t)}\, dt = \int_a^b \sqrt{Eu'^2 + 2Fu'v' + Gv'^2}\, dt.$$

(b) The angle γ between two intersecting curves $C_1: u = g(t), v = h(t)$ and $C_2: u = p(t), v = q(t)$ on $S: \mathbf{r}(u, v)$ is obtained from

$$(16) \qquad\qquad\qquad \cos \gamma = \frac{\mathbf{a} \cdot \mathbf{b}}{|\mathbf{a}|\, |\mathbf{b}|}$$

where $\mathbf{a} = \mathbf{r}_u g' + \mathbf{r}_v h'$ and $\mathbf{b} = \mathbf{r}_u p' + \mathbf{r}_v q'$ are tangent vectors of C_1 and C_2.

(c) The length of the normal vector $\mathbf{N}$ can be written

$$(17) \qquad\qquad\qquad |\mathbf{N}|^2 = |\mathbf{r}_u \times \mathbf{r}_v|^2 = EG - F^2,$$

so that formula (8) for the area $A(S)$ of S becomes

$$(18) \qquad A(S) = \iint_S dA = \iint_R |\mathbf{N}|\, du\, dv = \iint_R \sqrt{EG - F^2}\, du\, dv.$$

(d) For polar coordinates $u\ (= r)$ and $v\ (= \theta)$ defined by $x = u \cos v, y = u \sin v$ we have $E = 1, F = 0, G = u^2$, so that

$$ds^2 = du^2 + u^2\, dv^2 = dr^2 + r^2\, d\theta^2.$$

Calculate from this and (18) the area of a disk of radius a.

(e) Find the first fundamental form of the torus in Example 6. Use it to calculate the area A of the torus. Show that A can also be obtained by the **theorem of Pappus**,[9] which states that the area of a surface of revolution equals the product of the length of a meridian C and the length of the path of the center of gravity of C when C is rotated through the angle 2π.

(f) Calculate the first fundamental form for the usual representations of important surfaces of your own choice (cylinder, cone, etc.) and apply them to the calculation of lengths and areas on these surfaces.

9.7 Triple Integrals
Divergence Theorem of Gauss

In this section we first discuss triple integrals. Then we obtain the first "big" integral theorem, which transforms surface integrals into triple integrals. It is called **Gauss's divergence theorem** because it involves the divergence of a vector function (see Sec. 8.10, which you may wish to review).

The triple integral is a generalization of the double integral introduced in Sec. 9.3. For defining this integral we consider a function $f(x, y, z)$ defined in a bounded closed[10] region T in space. We subdivide this three-dimensional region T by planes parallel to the three coordinate planes. Then those boxes of subdivision (rectangular parallelepipeds) that lie

---

[9] PAPPUS OF ALEXANDRIA (about 300 A.D.), Greek mathematician. The theorem is also called Guldin's theorem. HABAKUK (after his conversion: PAUL) GULDIN (1577—1643) was born in St. Gallen, Switzerland, and later became professor in Graz and Vienna.

[10] Explained in footnote 1, Sec. 9.3 (with "sphere" instead of "circle").

entirely inside T are numbered 1 to n. In each such box we choose an arbitrary point, say, (x_k, y_k, z_k) in box k, and form the sum

$$J_n = \sum_{k=1}^{n} f(x_k, y_k, z_k) \, \Delta V_k$$

where ΔV_k is the volume of box k. This we do for larger and larger positive integers n arbitrarily but so that the maximum length of all the edges of those n boxes approaches zero as n approaches infinity. This gives a sequence of real numbers $J_{n_1}, J_{n_2}, \cdots$. We assume that $f(x, y, z)$ is continuous in a domain containing T and T is bounded by finitely many *smooth surfaces* (see Sec. 9.5). Then it can be shown (see Ref. [5] in Appendix 1) that the sequence converges to a limit that is independent of the choice of subdivisions and corresponding points (x_k, y_k, z_k). This limit is called the **triple integral** *of* $f(x, y, z)$ *over the region* T and is denoted by

$$\iiint_T f(x, y, z) \, dx \, dy \, dz \qquad \text{or} \qquad \iiint_T f(x, y, z) \, dV.$$

Triple integrals can be evaluated by three successive integrations. This is similar to the evaluation of double integrals by two successive integrations, as discussed in Sec. 9.3. An example is shown below (Example 1).

Divergence Theorem of Gauss

Triple integrals can be transformed into surface integrals over the boundary surface of a region in space and conversely. This is of practical interest because one of the two kinds of integral is often simpler than the other. It also helps in establishing fundamental equations in fluid flow, heat conduction, etc., as we shall see. The transformation is done by the *divergence theorem*, which involves the **divergence** of a vector function $\mathbf{F} = [F_1, F_2, F_3] = F_1\mathbf{i} + F_2\mathbf{j} + F_3\mathbf{k}$,

$$(1) \qquad\qquad \operatorname{div} \mathbf{F} = \frac{\partial F_1}{\partial x} + \frac{\partial F_2}{\partial y} + \frac{\partial F_3}{\partial z} \qquad\qquad \text{(Sec. 8.10)}.$$

THEOREM 1   **Divergence theorem of Gauss**

(Transformation between volume integrals and surface integrals)

Let T be a closed[11] bounded region in space whose boundary is a piecewise smooth[12] orientable surface S. Let $\mathbf{F}(x, y, z)$ be a vector function that is continuous and has continuous first partial derivatives in some domain containing T. Then

$$(2) \qquad\qquad \boxed{\iiint_T \operatorname{div} \mathbf{F} \, dV = \iint_S \mathbf{F} \cdot \mathbf{n} \, dA}$$

where $\mathbf{n}$ is the outer unit normal vector of S (pointing to the outside of S, as in Fig. 231). Proof after Example 1.

---

[11]"Closed" means that the boundary surface S is part of the region.

[12]See Sec. 9.5.

Formula (2) in Components. Using (1) and $\mathbf{n} = [\cos \alpha, \cos \beta, \cos \gamma]$, we can write (2)

$$(3^*) \qquad \iiint_T \left(\frac{\partial F_1}{\partial x} + \frac{\partial F_2}{\partial y} + \frac{\partial F_3}{\partial z} \right) dx\, dy\, dz = \iint_S (F_1 \cos \alpha + F_2 \cos \beta + F_3 \cos \gamma)\, dA.$$

Because of (6) in the last section this may also be written

$$(3) \qquad \iiint_T \left(\frac{\partial F_1}{\partial x} + \frac{\partial F_2}{\partial y} + \frac{\partial F_3}{\partial z} \right) dx\, dy\, dz = \iint_S (F_1\, dy\, dz + F_2\, dz\, dx + F_3\, dx\, dy).$$

EXAMPLE 1   **Evaluation of a surface integral by the divergence theorem**

Before we prove the divergence theorem, let us show a typical application. By transforming to a triple integral, evaluate

$$I = \iint_S (x^3\, dy\, dz + x^2 y\, dz\, dx + x^2 z\, dx\, dy)$$

where S is the closed surface consisting of the cylinder $x^2 + y^2 = a^2$ $(0 \le z \le b)$ and the circular disks $z = 0$ and $z = b$ $(x^2 + y^2 \le a^2)$. (Sketch S.)

Solution.   In (3) we now have

$$F_1 = x^3, \quad F_2 = x^2 y, \quad F_3 = x^2 z, \qquad \text{hence} \qquad \operatorname{div} \mathbf{F} = 3x^2 + x^2 + x^2 = 5x^2.$$

Introducing polar coordinates r, θ defined by $x = r \cos \theta$, $y = r \sin \theta$ (thus, cylindrical coordinates r, θ, z), we have $dx\, dy\, dz = r\, dr\, d\theta\, dz$, and we obtain

$$I = \iiint_T 5x^2\, dx\, dy\, dz = 5 \int_{z=0}^{b} \int_{r=0}^{a} \int_{\theta=0}^{2\pi} r^2 \cos^2 \theta\, r\, dr\, d\theta\, dz$$

$$= 5b \int_0^a \int_0^{2\pi} r^3 \cos^2 \theta\, dr\, d\theta = 5b \frac{a^4}{4} \int_0^{2\pi} \cos^2 \theta\, d\theta = \frac{5}{4} \pi b a^4. \qquad \blacktriangleleft$$

PROOF OF THE DIVERGENCE THEOREM.   Clearly, (3^*) is true if the integrals of each component on both sides of (3^*) are equal, that is,

$$(4) \qquad \iiint_T \frac{\partial F_1}{\partial x}\, dx\, dy\, dz = \iint_S F_1 \cos \alpha\, dA,$$

$$(5) \qquad \iiint_T \frac{\partial F_2}{\partial y}\, dx\, dy\, dz = \iint_S F_2 \cos \beta\, dA,$$

$$(6) \qquad \iiint_T \frac{\partial F_3}{\partial z}\, dx\, dy\, dz = \iint_S F_3 \cos \gamma\, dA.$$

We first prove (6) for a *special region* T that is bounded by a piecewise smooth orientable surface S and has the property that any straight line parallel to any one of the coordinate axes and intersecting T has at most *one* segment (or a single point) in common with T. This implies that T can be represented in the form

$$(7) \qquad g(x, y) \le z \le h(x, y)$$

where (x, y) varies in the orthogonal projection R of T in the xy-plane. Clearly,

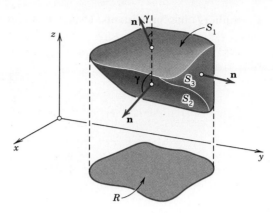

Fig. 231. Example of a special region

$z = g(x, y)$ represents the "bottom" S_2 of S (Fig. 231), whereas $z = h(x, y)$ represents the "top" S_1 of S, and there may be a remaining vertical portion S_3 of S. (The portion S_3 may degenerate into a curve, as for a sphere.)

To prove (6), we use (7). Since **F** is continuously differentiable in some domain containing T, we have

$$(8) \qquad \iiint\limits_T \frac{\partial F_3}{\partial z} \, dx \, dy \, dz = \iint\limits_R \left[\int_{g(x,\,y)}^{h(x,\,y)} \frac{\partial F_3}{\partial z} \, dz \right] dx \, dy.$$

We integrate the inner integral:

$$\int_g^h \frac{\partial F_3}{\partial z} \, dz = F_3[x, y, h(x, y)] - F_3[x, y, g(x, y)].$$

Hence the left side of (8) equals

$$(9) \qquad \iint\limits_R F_3[x, y, h(x, y)] \, dx \, dy - \iint\limits_R F_3[x, y, g(x, y)] \, dx \, dy.$$

But the same result is also obtained by evaluating the right side of (6), that is [see also (3)],

$$\iint\limits_S F_3 \cos \gamma \, dA = \iint\limits_S F_3 \, dx \, dy$$

$$= + \iint\limits_R F_3[x, y, h(x, y)] \, dx \, dy - \iint\limits_R F_3[x, y, g(x, y)] \, dx \, dy,$$

where the first integral over R gets a plus sign because $\cos \gamma > 0$ on S_1 in Fig. 231 [as in (5c$'$), Sec. 9.6], and the second integral gets a minus sign because $\cos \gamma < 0$ on S_2 [as in (5c$''$), Sec. 9.6, with g instead of h]. This proves (6).

The relations (4) and (5) now follow by merely relabeling the variables and using the fact that, by assumption, T has representations similar to (7), namely,

$$\tilde{g}(y, z) \leqq x \leqq \tilde{h}(y, z) \qquad \text{and} \qquad \tilde{\tilde{g}}(z, x) \leqq y \leqq \tilde{\tilde{h}}(z, x).$$

This establishes the divergence theorem for special regions.

For any region T that can be subdivided into *finitely many* special regions by means of auxiliary surfaces, the theorem follows by adding the result for each part separately; this procedure is analogous to that in the proof of Green's theorem in Sec. 9.4. The surface integrals over the auxiliary surfaces cancel in pairs, and the sum of the remaining surface integrals is the surface integral over the whole boundary surface S of T; the volume integrals over the parts of T add up to the volume integral over T.

The divergence theorem is now proved for any bounded region that is of interest in practical problems. The extension to the most general region T of the type characterized in the theorem would require a certain limit process; this is similar to the situation in the case of Green's theorem in Sec. 9.4.   ◄

EXAMPLE 2

Verification of the divergence theorem

Evaluate $\iint\limits_{S} (7x\mathbf{i} - z\mathbf{k}) \cdot \mathbf{n}\, dA$ over the sphere S: $x^2 + y^2 + z^2 = 4$ (a) by (2), (b) directly.

Solution.   (a) div $\mathbf{F}$ = div $[7x, 0, -z]$ = div $[7x\mathbf{i} - z\mathbf{k}]$ = $7 - 1 = 6$. *Answer:* $6 \cdot (4/3)\pi \cdot 2^3 = 64\pi$.
  (b) We can represent S by (3), Sec. 9.5 (with $a = 2$), and we shall use $\mathbf{n}\, dA = \mathbf{N}\, du\, dv$ [see (3^*), Sec. 9.6]. Accordingly,

$$S: \quad \mathbf{r} = [2 \cos v \cos u, \quad 2 \cos v \sin u, \quad 2 \sin v].$$

Then

$$\mathbf{r}_u = [-2 \cos v \sin u, \quad 2 \cos v \cos u, \quad 0]$$

$$\mathbf{r}_v = [-2 \sin v \cos u, \; -2 \sin v \sin u, \quad 2 \cos v]$$

$$\mathbf{N} = \mathbf{r}_u \times \mathbf{r}_v = [4 \cos^2 v \cos u, \quad 4 \cos^2 v \sin u, \quad 4 \cos v \sin v].$$

Now on S we have $x = 2 \cos v \cos u$, $z = 2 \sin v$, so that $\mathbf{F} = [7x, 0, -z]$ becomes on S

$$\mathbf{F}(S) = [14 \cos v \cos u, \quad 0, \quad -2 \sin v]$$

and

$$\mathbf{F}(S) \cdot \mathbf{N} = (14 \cos v \cos u) 4 \cos^2 v \cos u + (-2 \sin v) 4 \cos v \sin v$$

$$= 56 \cos^3 v \cos^2 u - 8 \cos v \sin^2 v.$$

On S we have to integrate over u from 0 to 2π. This gives

$$\pi \cdot 56 \cos^3 v - 2\pi \cdot 8 \cos v \sin^2 v.$$

The integral of $\cos v \sin^2 v$ is $(\sin^3 v)/3$, and that of $\cos^3 v = \cos v\, (1 - \sin^2 v)$ is $\sin v - (\sin^3 v)/3$. On S we have $-\pi/2 \leqq v \leqq \pi/2$, so that by substituting these limits we get

$$56\pi(2 - 2/3) - 16\pi \cdot 2/3 = 64\pi$$

as hoped for. To see the point of Gauss's theorem, compare the amounts of work.   ◄

Further applications of the divergence theorem follow in the problem set and in the next section. The examples in the next section will shed further light on the nature of the divergence and its applications.

PROBLEM SET 9.7

Application of Triple Integrals
Mass Distribution. In Probs. 1–7 find the total mass of a mass distribution of density σ in a region T in space. (Show the details of your work.)

1. $\sigma = x^2 + y^2 + z^2$, T the box $|x| \leqq 1$, $|y| \leqq 3$, $|z| \leqq 2$
2. $\sigma = e^{-x-y-z}$, T: $0 \leqq x \leqq 1 - y$, $0 \leqq y \leqq 1$, $0 \leqq z \leqq 2$

3. $\sigma = \sin \pi x \cos \pi y + 2,$ T the box $0 \leqq x \leqq 1,$ $0 \leqq y \leqq \frac{1}{2},$ $|z| \leqq 2$

4. $\sigma = \frac{1}{3}(x^2 + y^2)^2,$ T the cylinder $x^2 + y^2 \leqq 9,$ $-3 \leqq z \leqq 3$

5. $\sigma = 12xy,$ T the tetrahedron with vertices $(0, 0, 0),$ $(1, 0, 0),$ $(0, 1, 0),$ $(0, 0, 1)$

6. $\sigma = 4z,$ T the region in the first octant bounded by $y = 1 - x^2$ and $z = x$

7. $\sigma = 2/yz,$ $T: 0 \leqq x \leqq 2,$ $e^{-x} \leqq y \leqq 1,$ $e^{-x} \leqq z \leqq 1$

Moment of Inertia. Find the moment of inertia $I_x = \iiint\limits_{T} (y^2 + z^2)\, dx\, dy\, dz$ of a mass of density 1 in T about the x-axis, where T is

8. The box $0 \leqq x \leqq a,$ $-b/2 \leqq y \leqq b/2,$ $-c/2 \leqq z \leqq c/2$

9. The cube $0 \leqq x \leqq a,$ $0 \leqq y \leqq a,$ $0 \leqq z \leqq a$

10. The cone $y^2 + z^2 \leqq x^2,$ $0 \leqq x \leqq h$

11. The cylinder $y^2 + z^2 \leqq a^2,$ $0 \leqq x \leqq h$

12. The ball $x^2 + y^2 + z^2 \leqq a^2$

Application of the Divergence Theorem

Surface Integral $\iint\limits_{S} \mathbf{F} \cdot \mathbf{n}\, dA.$ Evaluate this integral by the divergence theorem for the following data. (Show the details. More such problems in the next problem set.)

13. $\mathbf{F} = [x^2,\ \ 0,\ \ z^2],$ S the surface of the box in Prob. 1

14. $\mathbf{F} = [e^x,\ \ e^y,\ \ e^z],$ S the surface of the cube $|x| \leqq 1,$ $|y| \leqq 1,$ $|z| \leqq 1$

15. $\mathbf{F} = [\cos y,\ \ \sin x,\ \ \cos z],$ S the surface of $x^2 + y^2 \leqq 4,$ $|z| \leqq 2$

16. $\mathbf{F}$ as in Prob. 15, S the surface of $x^2 + y^2 \leqq 9,$ $0 \leqq z \leqq 2$

17. $\mathbf{F} = [4x,\ \ x^2y,\ \ -x^2z],$ S the surface of the tetrahedron in Prob. 5

18. $\mathbf{F} = [2x^2,\ \ \frac{1}{2}y^2,\ \ -\cos \pi z],$ S as in Prob. 5

19. $\mathbf{F} = [x^3,\ \ y^3,\ \ z^3],$ S the sphere $x^2 + y^2 + z^2 = 9$

20. WRITING PROJECT. Simplifications occur in the use of the divergence theorem if div $\mathbf{F}$ is constant. Find other cases of simplifications (div $\mathbf{F}$ depending on one variable only, symmetry, the use of cylindrical or spherical coordinates—in Prob. 19, e.g.—etc.). Invent examples. Write down your results systematically in a short essay.

9.8 Divergence Theorem: Further Applications

The divergence theorem has various applications and consequences. We illustrate some of them in the following examples. Here, the occurring regions and functions are assumed to satisfy the conditions under which the divergence theorem is valid, and in each case, **n** is the *outer* unit normal vector of the boundary surface of the region, as before.

EXAMPLE 1   **Representation of the divergence independent of coordinates.**
Invariance of the divergence

We have defined the divergence of a vector function $\mathbf{F}$ in terms of coordinates [see (1) in the last section], but we want to use the divergence theorem for showing that div $\mathbf{F}$ has a meaning independent of the particular choice of coordinates.

For this purpose we first note that the basic properties of triple integrals are essentially the same as those of double integrals considered in Sec. 9.3. In particular, the **mean value theorem for triple integrals** asserts that for any *continuous* function $f(x, y, z)$ in a region T there is a point Q: (x_0, y_0, z_0) in T such that

(1)
$$\iiint\limits_{T} f(x, y, z)\, dV = f(x_0, y_0, z_0)\, V(T) \qquad (V(T) = \text{volume of } T).$$

We now interchange the two sides, divide by $V(T)$, and set $f - \operatorname{div} \mathbf{F}$. Then by the divergence theorem we obtain for the divergence the formula

$$
(2) \qquad \operatorname{div} \mathbf{F}(x_0, y_0, z_0) = \frac{1}{V(T)} \iiint\limits_{T} \operatorname{div} \mathbf{F} \, dV = \frac{1}{V(T)} \iint\limits_{S(T)} \mathbf{F} \cdot \mathbf{n} \, dA
$$

where $S(T)$ is the boundary surface of T.

We now choose any fixed point P: (x_1, y_1, z_1) in T and let T shrink down onto P, so that the maximum distance $d(T)$ of the points of T from P goes to zero. Then Q: (x_0, y_0, z_0) must approach P. Hence (2) becomes

$$
(3) \qquad \boxed{\operatorname{div} \mathbf{F}(x_1, y_1, z_1) = \lim_{d(T) \to 0} \frac{1}{V(T)} \iint\limits_{S(T)} \mathbf{F} \cdot \mathbf{n} \, dA.}
$$

This formula is sometimes used as a ***definition*** of the divergence. While the definition of the divergence in Sec. 8.10 involves coordinates, formula (3) is independent of coordinates. Hence from (3) it follows immediately that *the divergence is independent of the particular choice of Cartesian coordinates.* ◀

EXAMPLE 2   Physical interpretation of the divergence

From the divergence theorem we may obtain an intuitive interpretation of the divergence of a vector. For this purpose we consider the flow of an incompressible fluid (see Sec. 8.10) of constant density $\rho = 1$ which is **steady,** that is, does not vary with time. Such a flow is determined by the field of its velocity vector $\mathbf{v}(P)$ at any point P.

Let S be the boundary surface of a region T in space, and let $\mathbf{n}$ be the outer unit normal vector of S. The mass of fluid that flows through a small portion ΔS of S of area ΔA per unit time from the interior of S to the exterior is equal to $\mathbf{v} \cdot \mathbf{n} \, \Delta A$, where[13] $\mathbf{v} \cdot \mathbf{n}$ is the normal component of $\mathbf{v}$ in the direction of $\mathbf{n}$, taken at a suitable point of ΔS. Consequently, the total mass of fluid that flows across S from T to the outside per unit of time is given by the surface integral

$$
\iint\limits_{S} \mathbf{v} \cdot \mathbf{n} \, dA.
$$

Hence this integral represents the total flow out of T, and the integral

$$
(4) \qquad \frac{1}{V} \iint\limits_{S} \mathbf{v} \cdot \mathbf{n} \, dA
$$

where V is the volume of T, represents the average flow out of T. Since the flow is steady and the fluid is incompressible, the amount of fluid flowing outward must be continuously supplied. Hence, if the value of the integral (4) is different from zero, there must be **sources** (*positive sources and negative sources, called* **sinks**) in T, that is, points where fluid is produced or disappears.

If we let T shrink down to a fixed point P in T, we obtain from (4) the **source intensity** at P given by the right side of (3) with $\mathbf{F} \cdot \mathbf{n}$ replaced by $\mathbf{v} \cdot \mathbf{n}$. From this and (3) it follows that *the divergence of the velocity vector $\mathbf{v}$ of a steady incompressible flow is the source intensity of the flow at the corresponding point.* There are no sources in T if and only if $\operatorname{div} \mathbf{v} \equiv 0$; in this case,

$$
\iint\limits_{S} \mathbf{v} \cdot \mathbf{n} \, dA = 0
$$

for any closed surface S in T. ◀

EXAMPLE 3   Modeling of heat flow. Heat equation

We know that in a body heat will flow in the direction of decreasing temperature. Physical experiments show that the rate of flow is proportional to the gradient of the temperature. This means that the velocity $\mathbf{v}$ of the heat flow in a body is of the form

---

[13]$\mathbf{v} \cdot \mathbf{n}$ may be negative at a certain point, which means that fluid *enters* the interior of S at such a point.

(5) $\mathbf{v} = -K \operatorname{grad} U$

where $U(x, y, z, t)$ is temperature, t is time, and K is called the *thermal conductivity* of the body; in ordinary physical circumstances K is a constant. Using this information, set up the mathematical model of heat flow, the so-called **heat equation.**

Solution.   Let T be a region in the body and let S be its boundary surface. Then the amount of heat leaving T per unit of time is

$$\iint_S \mathbf{v} \cdot \mathbf{n} \, dA,$$

where $\mathbf{v} \cdot \mathbf{n}$ is the component of $\mathbf{v}$ in the direction of the outer unit normal vector $\mathbf{n}$ of S. This expression is obtained in a fashion similar to that in the preceding example. From (5) and the divergence theorem we obtain [see (3), Sec. 8.10]

(6) $$\iint_S \mathbf{v} \cdot \mathbf{n} \, dA = -K \iiint_T \operatorname{div}\,(\operatorname{grad} U) \, dx \, dy \, dz = -K \iiint_T \nabla^2 U \, dx \, dy \, dz$$

where $\nabla^2 U = U_{xx} + U_{yy} + U_{zz}$ is the Laplacian of U.
   On the other hand, the total amount of heat H in T is

$$H = \iiint_T \sigma \rho U \, dx \, dy \, dz$$

where the constant σ is the specific heat of the material of the body and ρ is the density (= mass per unit volume) of the material. Hence the time rate of decrease of H is

$$-\frac{\partial H}{\partial t} = -\iiint_T \sigma \rho \frac{\partial U}{\partial t} \, dx \, dy \, dz$$

and this must be equal to the above amount of heat leaving T. From (6) we thus have

$$-\iiint_T \sigma \rho \frac{\partial U}{\partial t} \, dx \, dy \, dz = -K \iiint_T \nabla^2 U \, dx \, dy \, dz$$

or

$$\iiint_T \left(\sigma \rho \frac{\partial U}{\partial t} - K \nabla^2 U \right) dx \, dy \, dz = 0.$$

Since this holds for any region T in the body, the integrand (if continuous) must be zero everywhere; that is,

(7) $$\boxed{\frac{\partial U}{\partial t} = c^2 \nabla^2 U}$$ $$c^2 = \frac{K}{\sigma \rho}$$

where c^2 is called the *thermal diffusivity* of the material. This partial differential equation is called the **heat equation.** It is fundamental for heat conduction. And our derivation is another impressive demonstration of the great importance of the divergence theorem.
   Methods for solving heat problems will be shown in Chap. 11. If heat flow does not depend on time (**"steady-state flow"**), then $\partial U / \partial t = 0$ and the heat equation (7) reduces to Laplace's equation $\nabla^2 U = 0$, which we shall now consider.   ◀

Potential Theory. Harmonic Functions

The theory of solutions of **Laplace's equation**

(8) $$\boxed{\nabla^2 f = \frac{\partial^2 f}{\partial x^2} + \frac{\partial^2 f}{\partial y^2} + \frac{\partial^2 f}{\partial z^2} = 0}$$

is called **potential theory.** A solution of (8) that has ***continuous*** second-order partial derivatives is called a **harmonic function.**[14] We shall consider details of potential theory in Chaps. 11 and 16. At present let us show that the divergence theorem plays a key role in potential theory.

EXAMPLE 4   **A basic property of solutions of Laplace's equation**

The integrands in the divergence theorem are div $\mathbf{F}$ and $\mathbf{F} \cdot \mathbf{n}$ (Sec. 9.7). If $\mathbf{F}$ is the gradient of a scalar function, say, $\mathbf{F} = \operatorname{grad} f$, then div $\mathbf{F} = \operatorname{div}(\operatorname{grad} f) = \nabla^2 f$; see (3), Sec. 8.10. Also, $\mathbf{F} \cdot \mathbf{n} = \mathbf{n} \cdot \mathbf{F} = \mathbf{n} \cdot \operatorname{grad} f$. This is the directional derivative of f in the outer normal direction of S, the boundary surface of the region T in the theorem. This derivative is commonly denoted by $\partial f / \partial n$. Thus the formula in the divergence theorem becomes

(9)
$$\iiint_T \nabla^2 f \, dV = \iint_S \frac{\partial f}{\partial n} \, dA.$$

Obviously this is the three-dimensional analog of the formula (9) in Sec. 9.4.

Taking into account the assumptions under which the divergence theorem holds, we immediately obtain from (9) the following result.

THEOREM 1   **(A basic property of harmonic functions)**

Let $f(x, y, z)$ be a harmonic function in some domain D. Then the integral of the normal derivative of the function f over any piecewise smooth[15] closed orientable surface S in D whose entire interior belongs to D is zero. ◄

EXAMPLE 5   **Green's theorems**

Let f and g be scalar functions such that $\mathbf{F} = f \operatorname{grad} g$ satisfies the assumptions of the divergence theorem in some region T. Then, by the formula in Project 13(c) of Problem Set 8.10,

$$\operatorname{div} \mathbf{F} = \operatorname{div}(f \operatorname{grad} g) = f \nabla^2 g + \operatorname{grad} f \cdot \operatorname{grad} g.$$

Furthermore, since f is a scalar function,

$$\mathbf{F} \cdot \mathbf{n} = \mathbf{n} \cdot \mathbf{F} = \mathbf{n} \cdot (f \operatorname{grad} g) = (\mathbf{n} \cdot \operatorname{grad} g) f.$$

$\mathbf{n} \cdot \operatorname{grad} g$ is the directional derivative $\partial g / \partial n$ of g in the outer normal direction of S. Hence the formula in the divergence theorem becomes **"Green's first formula"**

(10)
$$\iiint_T (f \nabla^2 g + \operatorname{grad} f \cdot \operatorname{grad} g) \, dV = \iint_S f \frac{\partial g}{\partial n} \, dA.$$

Formula (10) together with the assumptions is known as the *first form of Green's theorem.*

Interchanging f and g we obtain a similar formula. Subtracting this formula from (10) we find

(11)
$$\iiint_T (f \nabla^2 g - g \nabla^2 f) \, dV = \iint_S \left(f \frac{\partial g}{\partial n} - g \frac{\partial f}{\partial n} \right) dA.$$

This formula is called **Green's second formula** or (together with the assumptions) the *second form of Green's theorem.* ◄

---

[14] This continuity requirement should not be deleted from the definition of a harmonic function, as some older books do.

[15] See Sec. 9.6.

EXAMPLE 6   **Uniqueness of solutions of Laplace's equation**

If f is harmonic in a domain D, so that $\nabla^2 f = 0$, and if f is zero everywhere on a piecewise smooth closed orientable surface S in D whose entire interior T belongs to D, then the surface integral in (10) is zero, and (10) with $g = f$ gives

$$\iiint_T \text{grad } f \cdot \text{grad } f \, dV = \iiint_T |\text{grad } f|^2 \, dV = 0.$$

Since by assumption $|\text{grad } f|$ is continuous in T and on S and is nonnegative, it must be zero everywhere in T. Hence $f_x = f_y = f_z = 0$, and f is constant in T and, because of continuity, equal to its value 0 on S. This proves

THEOREM 2   **(Harmonic functions)**

If a function $f(x, y, z)$ is harmonic in some domain D and is zero at every point of a piecewise smooth closed orientable surface S in D whose entire interior T belongs to D, then f is identically zero in T.

This theorem has an important consequence. Let f_1 and f_2 be functions that satisfy the assumptions of Theorem 1 and take on the same values on S. Then their difference $f_1 - f_2$ satisfies those assumptions and has the value 0 everywhere on S. Hence, Theorem 2 implies that $f_1 - f_2 = 0$ throughout T, and we have the following fundamental result.

THEOREM 3   **(Uniqueness theorem for Laplace's equation)**

Let T be a region that satisfies the assumptions of the divergence theorem, and let $f(x, y, z)$ be a harmonic function in a domain D that contains T and its boundary surface S. Then f is uniquely determined in T by its values on S.

The problem of determining a solution u of a partial differential equation in a region T such that u assumes given values on the boundary surface S of T is called the **Dirichlet problem.**[16] We may thus reformulate Theorem 3 as follows.

THEOREM 3*   **(Uniqueness theorem for the Dirichlet problem)**

If the assumptions in Theorem 3 are satisfied and the Dirichlet problem for the Laplace equation has a solution in T, then this solution is unique.

These theorems demonstrate the extreme importance of Gauss's theorem in potential theory.   ◀

PROBLEM SET 9.8

1. **(Verification)** Verify Theorem 1 for $f = 2z^2 - x^2 - y^2$ and S the surface of the box $0 \leqq x \leqq 1, \, 0 \leqq y \leqq 2, \, 0 \leqq z \leqq 4$.

2. **(Verification)** Verify Theorem 1 for $f = x^2 - y^2$ and the surface of the cylinder $x^2 + y^2 \leqq 4$, $0 \leqq z \leqq 1$.

Evaluation of Surface Integrals by the Divergence Theorem

In Probs. 3–8, using the divergence theorem, evaluate $\iint_S \mathbf{F} \cdot \mathbf{n} \, dA$ for the given data. (Show the details of your work. For similar problems, see Sec. 9.7.)

3. $\mathbf{F} = [x, \, z, \, y]$, S the hemisphere $x^2 + y^2 + z^2 = 4$, $z \geqq 0$

---

[16]PETER GUSTAV LEJEUNE DIRICHLET (1805—1859), German mathematician, studied in Paris under Cauchy and others and succeeded Gauss at Göttingen in 1855. He became known by his important research on Fourier series (he knew Fourier personally) and in number theory.

4. $\mathbf{F} = [x, \quad 3y, \quad 6z],$ $\quad$ S the surface of the cone $\sqrt{x^2 + y^2} \leq z, \quad 0 \leq z \leq 3$

5. $\mathbf{F} = [9x, \quad y \cosh^2 x, \quad -z \sinh^2 x],$ $\quad$ S the ellipsoid $4x^2 + y^2 + 9z^2 = 36$

6. $\mathbf{F} = [5, \quad 10y, \quad z^3],$ $\quad$ S the surface of $-1 \leq x \leq 1, \quad \frac{1}{2}x \leq y \leq x, \quad 0 \leq z \leq y$

7. $\mathbf{F} = [\sin x, \quad y, \quad z],$ $\quad$ S the surface of $0 \leq x \leq \pi/2, \quad x \leq y \leq z, \quad 0 \leq z \leq 1$

8. $\mathbf{F} = [1, \quad xy, \quad yz],$ $\quad$ S the surface of $x^2 + y^2 \leq z, \quad y \geq 0, \quad z \leq 4$

9. (**Volume as a surface integral**) Show that a region T with boundary surface S has the volume

$$V = \frac{1}{3} \iint_S r \cos \phi \, dA$$

where r is the distance of a variable point $P: (x, y, z)$ on S from the origin O and ϕ is the angle between the directed line OP and the outer normal of S at P.

10. Find the volume of a ball of radius a by means of the formula in Prob. 9.

11. Show that a region T with boundary surface S has the volume

$$V = \iint_S x \, dy \, dz = \iint_S y \, dz \, dx = \iint_S z \, dx \, dy = \frac{1}{3} \iint_S (x \, dy \, dz + y \, dz \, dx + z \, dx \, dy).$$

12. **TEAM PROJECT. Divergence Theorem and Potential Theory.** The importance of the divergence theorem in potential theory is obvious from (9)–(11) and Theorems 1–3. To emphasize it further, consider functions f and g that are harmonic in some domain D containing a region T with boundary surface S such that T satisfies the assumptions in the divergence theorem. Prove and illustrate by examples that then:

(a) $$\iint_S g \frac{\partial g}{\partial n} \, dA = \iiint_T |\text{grad } g|^2 \, dV.$$

(b) If $\partial g/\partial n = 0$ on S, then g is constant in T.

(c) $$\iint_S \left(f \frac{\partial g}{\partial n} - g \frac{\partial f}{\partial n} \right) dA = 0.$$

(d) If $\partial f/\partial n = \partial g/\partial n$ on S, then $f = g + c$ in T, where c is a constant.

(e) The **Laplacian** can be represented *independently of coordinate systems* in the form

$$\nabla^2 f = \lim_{d(T) \to 0} \frac{1}{V(T)} \iint_{S(T)} \frac{\partial f}{\partial n} \, dA$$

where $d(T)$ is the maximum distance of the points of a region T bounded by $S(T)$ from the point at which the Laplacian is evaluated and $V(T)$ is the volume of T.

9.9 Stokes's Theorem

Having seen the great usefulness of Gauss's theorem (Secs. 9.7, 9.8), we now turn to the second "big" theorem in this chapter, *Stokes's theorem*, which transforms line integrals into surface integrals and conversely. Hence this theorem generalizes Green's theorem of Sec. 9.4. It involves the curl,

(1) $$\text{curl } \mathbf{F} = \begin{vmatrix} \mathbf{i} & \mathbf{j} & \mathbf{k} \\ \partial/\partial x & \partial/\partial y & \partial/\partial z \\ F_1 & F_2 & F_3 \end{vmatrix} \qquad \text{(see Sec. 8.11)}.$$

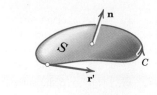

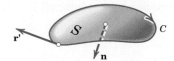

Fig. 232.   Stokes's theorem

THEOREM 1   **Stokes's theorem[17]**

(Transformation between surface integrals and line integrals)
Let S be a piecewise smooth[18] oriented surface in space and let the boundary of S be a piecewise smooth[18] simple closed curve C. Let $\mathbf{F}(x, y, z)$ *be a continuous vector function that has continuous first partial derivatives in a domain in space containing S. Then*

(2)
$$\iint_S (\text{curl } \mathbf{F}) \cdot \mathbf{n} \, dA = \oint_C \mathbf{F} \cdot \mathbf{r}'(s) \, ds$$

where $\mathbf{n}$ *is a unit normal vector of S and, depending on* $\mathbf{n}$, *the integration around C is taken in the sense shown in Fig. 232. Furthermore,* $\mathbf{r}' = d\mathbf{r}/ds$ *is the unit tangent vector and s the arc length of C.* (The proof follows after Example 1.)

Formula (2) in Components. From (3) in Sec. 9.6 with curl $\mathbf{F}$ instead of $\mathbf{F}$ we get (2) in terms of components [see (1)]

(3)
$$\iint_R \left[\left(\frac{\partial F_3}{\partial y} - \frac{\partial F_2}{\partial z} \right) N_1 + \left(\frac{\partial F_1}{\partial z} - \frac{\partial F_3}{\partial x} \right) N_2 + \left(\frac{\partial F_2}{\partial x} - \frac{\partial F_1}{\partial y} \right) N_3 \right] du \, dv$$
$$= \oint_{\overline{C}} (F_1 \, dx + F_2 \, dy + F_3 \, dz)$$

where R is the region with boundary curve $\overline{C}$ in the uv-plane corresponding to S represented by $\mathbf{r}(u, v)$, and $\mathbf{N} = [N_1, N_2, N_3] = \mathbf{r}_u \times \mathbf{r}_v$.

EXAMPLE 1   **Verification of Stokes's theorem**

Before proving Stokes's theorem, let us get used to it by verifying it for $\mathbf{F} = [y, z, x] = y\mathbf{i} + z\mathbf{j} + x\mathbf{k}$ and S the paraboloid (Fig. 233)

$$z = f(x, y) = 1 - (x^2 + y^2), \qquad z \geqq 0.$$

Solution.   The curve C is the circle $\mathbf{r}(s) = [\cos s, \sin s, 0] = \cos s \, \mathbf{i} + \sin s \, \mathbf{j}$. It has the unit tangent vector $\mathbf{r}'(s) = [-\sin s, \cos s, 0] = -\sin s \, \mathbf{i} + \cos s \, \mathbf{j}$. Consequently, the line integral in (2) on the right is simply

---

[17]Sir GEORGE GABRIEL STOKES (1819—1903), Irish mathematician and physicist, who became a professor in Cambridge in 1849. He is also known for his important contribution to the theory of infinite series and to viscous flow (Navier–Stokes equations), geodesy, and optics.

[18]"Piecewise smooth" is defined in Secs. 9.1 and 9.6.

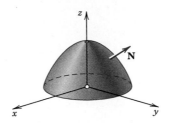

Fig. 233.   Surface S in Example 1

$$\oint_C \mathbf{F} \cdot d\mathbf{r} = \int_0^{2\pi} [(\sin s)(-\sin s) + 0 + 0]\, ds = -\pi.$$

On the other hand, in (2) on the left we need (verify this)

$$\operatorname{curl} \mathbf{F} = [-1, -1, -1] \quad \text{and} \quad \mathbf{N} = \operatorname{grad}\,(z - f(x, y)) = [2x, 2y, 1]$$

so that $(\operatorname{curl} \mathbf{F}) \cdot \mathbf{N} = -2x - 2y - 1$. From (3) in Sec. 9.6 with curl $\mathbf{F}$ instead of $\mathbf{F}$ and x, y instead of u, v we thus obtain

$$\iint_S (\operatorname{curl} \mathbf{F}) \cdot \mathbf{n}\, dA = \iint_R (-2x - 2y - 1)\, dx\, dy$$

$$= \iint_{\tilde{R}} (-2r\cos\theta - 2r\sin\theta - 1) r\, dr\, d\theta$$

where $x = r\cos\theta$, $y = r\sin\theta$, and $dx\,dy = r\,dr\,d\theta$. Now the projection R of S in the xy-plane is given in polar coordinates by $\tilde{R}$: $r \leqq 1$, $0 \leqq \theta \leqq 2\pi$. The integration of the cosine and sine terms over θ from 0 to 2π gives zero. The remaining term $-1 \cdot r$ has the integral $(-1/2)2\pi = -\pi$, in agreement with the previous result.

Note well that $\mathbf{N}$ is an *upper* normal vector of S (why?), and $\mathbf{r}(s)$ orients C counterclockwise, as required in Stokes's theorem.   ◀

PROOF OF STOKES'S THEOREM.   Obviously, (2) holds if the integrals of each component on both sides of (3) are equal, that is,

$$\text{(4)} \qquad \iint_R \left(\frac{\partial F_1}{\partial z} N_2 - \frac{\partial F_1}{\partial y} N_3 \right) du\, dv = \oint_{\overline{C}} F_1\, dx$$

$$\text{(5)} \qquad \iint_R \left(-\frac{\partial F_2}{\partial z} N_1 + \frac{\partial F_2}{\partial x} N_3 \right) du\, dv = \oint_{\overline{C}} F_2\, dy$$

$$\text{(6)} \qquad \iint_R \left(\frac{\partial F_3}{\partial y} N_1 - \frac{\partial F_3}{\partial x} N_2 \right) du\, dv = \oint_{\overline{C}} F_3\, dz.$$

We prove this first for a surface S that can be represented simultaneously in the forms

$$\text{(7)} \qquad \text{(a)} \quad z = f(x, y), \qquad \text{(b)} \quad y = g(x, z), \qquad \text{(c)} \quad x = h(y, z).$$

We prove (4), using (7a). Setting $u = x$, $v = y$, we have from (7a)

$$\mathbf{r}(u, v) = \mathbf{r}(x, y) = [x, y, f(x, y)] = x\mathbf{i} + y\mathbf{j} + f\mathbf{k}$$

and in (3), Sec. 9.6, by direct calculation

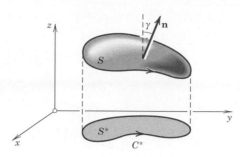

Fig. 234.   Proof of Stokes's theorem

$$\mathbf{N} = \mathbf{r}_u \times \mathbf{r}_v = \mathbf{r}_x \times \mathbf{r}_y = [-f_x, -f_y, 1] = -f_x \mathbf{i} - f_y \mathbf{j} + \mathbf{k}.$$

Note that $\mathbf{N}$ is an *upper* normal vector of S, since it has a *positive* z-component. Also, $R = S^*$, the projection of S into the xy-plane, with boundary curve $\overline{C} = C^*$ (Fig. 234). Hence the left side of (4) is

(8)
$$\iint\limits_{S^*} \left[\frac{\partial F_1}{\partial z} (-f_y) - \frac{\partial F_1}{\partial y} \right] dx\, dy.$$

We now consider the right side of (4). We transform this line integral over $\overline{C} = C^*$ into a double integral over S^* by applying Green's theorem [formula (1) in Sec. 9.4 with $F_2 = 0$]. This gives

$$\oint\limits_{C^*} F_1 \, dx = \iint\limits_{S^*} - \frac{\partial F_1}{\partial y} \, dx\, dy.$$

Here, $F_1 = F_1(x, y, f(x, y))$. Hence by the chain rule (see also Prob. 10 in Problem Set 8.8),

$$-\frac{\partial F_1(x, y, f(x, y))}{\partial y} = -\frac{\partial F_1(x, y, z)}{\partial y} - \frac{\partial F_1(x, y, z)}{\partial z} \frac{\partial f}{\partial y} \qquad [z = f(x, y)].$$

We see that the right side of this equals the integrand in (8). This proves (4). Relations (5) and (6) follow in the same way if we use (7b) and (7c), respectively. By addition we obtain (3). This proves Stokes's theorem for a surface S that can be represented simultaneously in the forms (7a), (7b), (7c).

   As in the proof of the divergence theorem, our result may be immediately extended to a surface S that can be decomposed into finitely many pieces, each of which is of the kind just considered. This covers most of the cases of practical interest. The proof in the case of a most general surface S satisfying the assumptions of the theorem would require a limit process; this is similar to the situation in the case of Green's theorem in Sec. 9.4.   ◀

EXAMPLE 2   **Green's theorem in the plane as a special case of Stokes's theorem**

Let $\mathbf{F} = [F_1, F_2] = F_1 \mathbf{i} + F_2 \mathbf{j}$ be a vector function that is continuously differentiable in a domain in the xy-plane containing a simply connected bounded closed region S whose boundary C is a piecewise smooth simple closed curve. Then, according to (1),

$$(\text{curl } \mathbf{F}) \boldsymbol{\cdot} \mathbf{n} = (\text{curl } \mathbf{F}) \boldsymbol{\cdot} \mathbf{k} = \frac{\partial F_2}{\partial x} - \frac{\partial F_1}{\partial y}.$$

Hence the formula in Stokes's theorem now takes the form

$$\iint\limits_{S} \left(\frac{\partial F_2}{\partial x} - \frac{\partial F_1}{\partial y} \right) dA = \oint_C (F_1\,dx + F_2\,dy).$$

This shows that Green's theorem in the plane (Sec. 9.4) is a special case of Stokes's theorem (which we needed in the proof of the latter!).   ◀

EXAMPLE 3   **Evaluation of a line integral by Stokes's theorem**

Evaluate $\int_C \mathbf{F} \cdot \mathbf{r}'\,ds$, where C is the circle $x^2 + y^2 = 4$, $z = -3$, oriented counterclockwise as seen by a person standing at the origin, and, with respect to right-handed Cartesian coordinates,

$$\mathbf{F} = [y, \quad xz^3, \quad -zy^3] = y\mathbf{i} + xz^3\mathbf{j} - zy^3\mathbf{k}.$$

Solution.   As a surface S bounded by C we can take the plane circular disk $x^2 + y^2 \leqq 4$ in the plane $z = -3$. Then $\mathbf{n}$ in Stokes's theorem points in the positive z-direction; thus $\mathbf{n} = \mathbf{k}$. Hence (curl $\mathbf{F}$) $\cdot \mathbf{n}$ is simply the component of curl $\mathbf{F}$ in the positive z-direction. Since $\mathbf{F}$ with $z = -3$ has the components $F_1 = y$, $F_2 = -27x$, $F_3 = 3y^3$, we thus obtain

$$(\text{curl }\mathbf{F}) \cdot \mathbf{n} = \frac{\partial F_2}{\partial x} - \frac{\partial F_1}{\partial y} = -27 - 1 = -28.$$

Hence the integral over S in Stokes's theorem equals -28 times the area 4π of the disk S. This yields the answer $-28 \cdot 4\pi = -112\pi \approx -352$. Confirm this result by direct calculation, which involves somewhat more work.   ◀

EXAMPLE 4   **Physical interpretation of the curl. Circulation**

Let S_r be a circular disk of radius r and center P bounded by the circle C_r (Fig. 235), and let $\mathbf{F}(Q) \equiv \mathbf{F}(x, y, z)$ be a continuously differentiable vector function in a domain containing S_r. Then by Stokes's theorem and the mean value theorem for surface integrals,

$$\oint_{C_r} \mathbf{F} \cdot \mathbf{r}'\,ds = \iint\limits_{S_r} (\text{curl }\mathbf{F}) \cdot \mathbf{n}\,dA = (\text{curl }\mathbf{F}) \cdot \mathbf{n}(P^*) A_r$$

where A_r is the area of S_r and P^* is a suitable point of S_r. This may be written in the form

$$(\text{curl }\mathbf{F}) \cdot \mathbf{n}(P^*) = \frac{1}{A_r} \oint_{C_r} \mathbf{F} \cdot \mathbf{r}'\,ds.$$

In the case of a fluid motion with velocity vector $\mathbf{F} = \mathbf{v}$, the integral

$$\oint_{C_r} \mathbf{v} \cdot \mathbf{r}'\,ds$$

is called the **circulation** of the flow around C_r. It measures the extent to which the corresponding fluid motion is a rotation around the circle C_r. If we now let r approach zero, we find

(9) $$(\text{curl }\mathbf{v}) \cdot \mathbf{n}(P) = \lim_{r \to 0} \frac{1}{A_r} \oint_C \mathbf{v} \cdot \mathbf{r}'\,ds;$$

that is, the component of the curl in the positive normal direction can be regarded as the **specific circulation** (circulation per unit area) of the flow in the surface at the corresponding point.   ◀

Fig. 235.   Example 4

EXAMPLE 5   **Work done in the displacement around a closed curve**

Find the work done by the force $\mathbf{F} = 2xy^3 \sin z\, \mathbf{i} + 3x^2y^2 \sin z\, \mathbf{j} + x^2y^3 \cos z\, \mathbf{k}$ in the displacement around the curve of intersection of the paraboloid $z = x^2 + y^2$ and the cylinder $(x - 1)^2 + y^2 = 1$.

Solution.   This work is given by the line integral in Stokes's theorem. Now $\mathbf{F} = \operatorname{grad} f$, where $f = x^2y^3 \sin z$ and curl grad $f = \mathbf{0}$ [see (3) in Sec. 8.11], so that (curl $\mathbf{F}$)$\cdot\mathbf{n} = 0$ and the work is 0 by Stokes's theorem. ◀

Stokes's Theorem Applied to Path Independence

We have seen and emphasized in Sec. 9.2 that the value of a line integral generally depends not only on the function to be integrated and on the two endpoints A and B of the path of integration C, but also on the particular choice of a path from A to B. In Theorem 3 of Sec. 9.2 we proved that if a line integral

$$(10) \qquad \int_C \mathbf{F}(\mathbf{r})\cdot d\mathbf{r} = \int_C (F_1\, dx + F_2\, dy + F_3\, dz)$$

(involving continuous F_1, F_2, F_3 that have continuous first partial derivatives) is independent of path in a domain D, then

$$(11) \qquad\qquad\qquad\qquad \operatorname{curl} \mathbf{F} = \mathbf{0}.$$

We claimed that the converse is also true, that is, if (11) holds, then (10) is independent of path in D, provided D is simply connected (see Sec. 9.2).

The proof needs Stokes's theorem and can now be given as follows. Let C be any simple closed path in D. Since D is simply connected, we can find a surface S in D bounded by C. Stokes's theorem is applicable and gives

$$\oint_C (F_1\, dx + F_2\, dy + F_3\, dz) = \oint_C \mathbf{F}\cdot\mathbf{r}'\, ds = \iint_S (\operatorname{curl} \mathbf{F})\cdot\mathbf{n}\, dA$$

for proper direction on C and normal vector $\mathbf{n}$ on S. Now, regardless of the choice of C, the integral over S is zero because, by (11), its integrand is identically zero. From this and Theorem 2 in Sec. 9.2 it follows that the integral (10) is independent of path in D. This completes the proof. ◀

PROBLEM SET 9.9

Direct Integration of the Surface Integral $\iint\limits_S (\operatorname{curl} \mathbf{F})\cdot\mathbf{n}\, dA$. **Verification by Stokes's Theorem.**

Integrate the surface integral directly. Then check the result by integrating the corresponding line integral in (2). (Show the details of your work.)

1. $\mathbf{F} = [z^2, \quad 5x, \quad 0]$, S the square $0 \leqq x \leqq 1$, $0 \leqq y \leqq 1$, $z = 1$
2. $\mathbf{F} = [y^2, \quad -x^2, \quad 0]$, S the circular semidisk $x^2 + y^2 \leqq 4$, $y \geqq 0$, $z = 0$
3. $\mathbf{F} = [e^z, \quad e^z \sin y, \quad e^z \cos y]$, S: $z = y^2$, $0 \leqq x \leqq 4$, $0 \leqq y \leqq 2$
4. $\mathbf{F} = [0, \quad 0, \quad x \cos 2z]$, S: $x^2 + y^2 = 1$, $y \geqq 0$, $0 \leqq z \leqq \pi/4$
5. $\mathbf{F} = [y^2, \quad z^2, \quad x^2]$, S the portion of the paraboloid $x^2 + y^2 = z$, $y \geqq 0$, $z \leqq 1$
6. $\mathbf{F}$ as in Prob. 5, S the portion of the sphere $x^2 + y^2 + (z - 1)^2 = 1$, $y \geqq 0$, $z \leqq 1$

Evaluation of the Line Integral $\int_C \mathbf{F} \cdot \mathbf{r}'(s)\, ds$ **by Stokes's Theorem** (clockwise as seen by a person standing at the origin). Calculate this integral by Stokes's theorem for the following C and $\mathbf{F}$ (referred to right-handed Cartesian coordinates). (Show the details.)

   7. $\mathbf{F} = [-5y, \quad 4x, \quad z]$, C the circle $x^2 + y^2 = 4$, $z = 1$
   8. $\mathbf{F} = [2y^2, \quad x, \quad -z^3]$, C the circle $x^2 + y^2 = a^2$, $z = b\ (> 0)$
   9. $\mathbf{F} = [4z, \quad -2x, \quad 2x]$, C the ellipse $x^2 + y^2 = 1$, $z = y + 1$
  10. $\mathbf{F} = [y, \quad \frac{1}{2}z, \quad \frac{3}{2}y]$, C the circle $x^2 + y^2 + z^2 = 6z$, $z = x + 3$
  11. $\mathbf{F} = [0, \quad xyz, \quad 0]$, C the boundary of the triangle with vertices $(1, 0, 0)$, $(0, 1, 0)$, $(0, 0, 1)$
  12. $\mathbf{F} = [x^4, \quad y^4, \quad z^4]$, C the intersection of $x^2 + y^2 + z^2 = a^2$ and $z = y^2$
  13. $\mathbf{F} = [y^3, \quad 0, \quad x^3]$, C as in Prob. 11
  14. **(Stokes's theorem not applicable)** Evaluate $\int_C \mathbf{F} \cdot \mathbf{r}'\, ds$, $\mathbf{F} = (x^2 + y^2)^{-1}[-y, x, 0]$, C the circle $x^2 + y^2 = 1$, $z = 0$, oriented clockwise. Why can Stokes's theorem not be applied?

  15. **WRITING PROJECT. Grad, Div, Curl in Connection with Integrals.** Make a list of ideas and results on this subject by carefully searching through this chapter. See whether you can improve your list by rearranging or combining certain topics. See how you can subdivide your material into $3-5$ large sections. Then work out each point, so that you obtain a well-balanced essay. Include no proofs, but use typical examples whenever you feel that they lead to a better understanding.

CHAPTER 9 REVIEW

   1. State from memory how you can evaluate a line integral.
   2. Explain how a line integral generalizes a definite integral known from calculus.
   3. What do we mean by path independence of a line integral? Why is this practically important? What does it mean physically?
   4. List the kinds of integral we have discussed in this chapter.
   5. How can we convert line integrals to surface integrals? State the conditions in the corresponding theorem.
   6. How can we convert surface integrals to volume integrals?
   7. What role did the gradient play in this chapter?
   8. How is the divergence defined and how did we use it in this chapter?
   9. Why was it essential in Stokes's theorem whether we used right-handed or left-handed coordinates?
  10. What are typical applications of line integrals? Of surface integrals?
  11. Orientation played a role in connection with surface integrals. Explain.
  12. What is a smooth surface? A piecewise smooth surface? Give examples. Why did we need these concepts?
  13. State Laplace's equation. Where in physics is it important?
  14. Summarize our discussions on harmonic functions.
  15. In line and surface integrals we used *vector* functions $\mathbf{F}$, but the integrands were actually *scalar* functions. Why did we proceed in this way (which at first sight looks somewhat like a detour)?

Line Integrals $\int_C \mathbf{F}(\mathbf{r}) \cdot d\mathbf{r}$. **Work Done by a Force.** In Probs. 16–30, with $\mathbf{F}$ and C as given, evaluate this integral by the method that seems most suitable (direct integration, use of exactness or Green's theorem or Stokes's theorem). Recall that if $\mathbf{F}$ is a force, the integral gives the work done in a displacement. (Show the details of your work.)

16. $\mathbf{F} = [x^2, \quad -2y^2]$, $\quad C$ the straight-line segment from $(4, 2)$ to $(-3, 5)$

17. $\mathbf{F} = [y \cos xy, \quad x \cos xy, \quad e^z]$, $\quad C$ the straight-line segment from $(\pi, 1, 0)$ to $(\frac{1}{2}, \pi, 1)$

18. $\mathbf{F} = [xy, \quad z, \quad 0]$, $\quad C: y = 2x^2, \quad z = x$ from $(1, 2, 1)$ to $(2, 8, 2)$

19. $\mathbf{F} = [y^2, \quad 2xy + \sin x, \quad 0]$, $\quad C$ the boundary of $0 \le x \le \pi/2, 0 \le y \le 2, z = 0$

20. $\mathbf{F} = [-y^3, \quad x^3 + e^y, \quad 0]$, $\quad C$ the circle $x^2 + y^2 = 4, z = 0$

21. $\mathbf{F} = [\cos \pi y, \quad \sin \pi x, \quad \cos \pi x]$, $\quad C$ the boundary of $0 \le x \le 1/2, 0 \le y \le 4, z = x$

22. $\mathbf{F} = [-z, \quad 5x, \quad -y]$, $\quad C$ the ellipse $x^2 + y^2 = 4, z = x + 2$

23. $\mathbf{F} = [8xy, \quad 4x^2, \quad 2 \cos 2z]$, $\quad C$ the helix $\mathbf{r} = [\cos t, \quad \sin t, \quad t], 0 \le t \le \pi/4$

24. $\mathbf{F} = [ze^{xz}, \quad 2 \sinh 2y, \quad xe^{xz}]$, $\quad C$ the parabola $y = x, z = x^2, -1 \le x \le 1$

25. $\mathbf{F} = [e^x, \quad e^y, \quad e^z]$, $\quad C: x = \ln y, z = \ln y, 1 \le y \le 2$

26. $\mathbf{F} = [x \ln y, \quad 2ye^x, \quad 0]$, $\quad C$ the boundary of the rectangle $0 \le x \le 2, 1 \le y \le 2, z = 0$

27. $\mathbf{F} = [x^3, \quad e^{3y}, \quad e^{-3z}]$, $\quad C: x^2 + 9y^2 = 9, z = x^2$

28. $\mathbf{F} = [\cosh y, \quad xy^2, \quad z]$, $\quad C: \mathbf{r} = [t, \quad 3t, \quad t^2], 0 \le t \le 1$

29. $\mathbf{F} = [\sin \pi x, \quad z, \quad 0]$, $\quad C$ the boundary of the triangle with vertices $(0, 0, 0)$, $(1, 0, 0)$, $(1, 1, 0)$

30. $\mathbf{F} = [x^2, \quad y^2, \quad y^2 x]$, $\quad C$ the helix $\mathbf{r} = [\cos t, \quad \sin t, \quad 3t], 0 \le t \le \pi/2$

Double Integrals. Center of Gravity. Find the coordinates $\bar{x}, \bar{y}$ of the center of gravity of a mass of density $f(x, y)$ in the region R. (This amounts to the evaluation of double integrals. Show the details. Sketch R.)

31. $f = xy$, $\quad R: 0 \le x \le 1, 0 \le y \le x$ **32.** $f = x^2 y$, $\quad R: -1 \le x \le 1, x^2 \le y \le 1$

33. $f = 1$, $\quad R: 1 \le x \le 2, 0 \le y \le \ln x$ **34.** $f = 1$, $\quad R: x^2 + y^2 \le a^2, y \ge 0$

35. $f = x^2$, $\quad R: -1 \le x \le 2, x^2 \le y \le x + 2$

36. $f = x^2 + y^2$, $\quad R: x^2 + y^2 \le a^2, x \ge 0, y \ge 0$

Surface Integrals $\iint\limits_{S} \mathbf{F} \cdot \mathbf{n} \, dA$. **Divergence Theorem.** Evaluate this integral directly or, if possible, by the divergence theorem. (Show the details.)

37. $\mathbf{F} = [x, \quad y]$, $\quad S: z = 2x + 5y, \quad 0 \le x \le 2, \quad -1 \le y \le 1$

38. $\mathbf{F} = [\sin x, \quad z, \quad y]$, $\quad S: y^2 + z^2 = 4, \quad -1/2 \le x \le 1/2, \quad y \ge 0, \quad z \ge 0$

39. $\mathbf{F} = [0, \quad 20y, \quad 2z^3]$, $\quad S$ the surface of $0 \le x \le 6, \quad 0 \le y \le 1, \quad 0 \le z \le y$

40. $\mathbf{F} = [x^3, \quad y^3, \quad z^3]$, $\quad S$ the sphere $x^2 + y^2 + z^2 = 4$

41. $\mathbf{F} = [0, \quad x^2, \quad -xz]$, $\quad S: \mathbf{r} = [u, \quad u^2, \quad v], \quad 0 \le u \le 1, \quad -2 \le v \le 2$

42. $\mathbf{F} = [e^y, \quad 0, \quad ze^x]$, $\quad S: \mathbf{r} = [u, \quad 2u, \quad v], -1 \le u \le 1, \quad 0 \le v \le 3$

43. $\mathbf{F} = [1, \quad 1, \quad 1]$, $\quad S: x^2 + y^2 + 4z^2 = 4, \quad z \ge 0$

44. $\mathbf{F} = [y^2, \quad x^2, \quad z^2]$, $\quad S$ the surface of the cylinder $x^2 + y^2 \le 4, \quad 0 \le z \le 2$

45. $\mathbf{F} = [x + z, \quad y + z, \quad x + y]$, $\quad S$ the sphere $x^2 + y^2 + z^2 = 9$

Summary of Chapter 9
Vector Integral Calculus.
Integral Theorems

Chapter 8 extended *differential* calculus to vectors, that is, to vector functions $\mathbf{v}(x, y, z)$ or $\mathbf{v}(t)$. Chapter 9 extends *integral* calculus to vector functions. This involves *line integrals* (Sec. 9.1), *double integrals* (Sec. 9.3), *surface integrals* (Sec. 9.6), and *triple integrals* (Sec. 9.7) and the three "big" theorems for transforming these integrals into one another, the theorems by Green (Sec. 9.4), Gauss (Sec. 9.7), and Stokes (Sec. 9.9).

The analog of the definite integral of calculus is the **line integral**

$$(1) \qquad \int_C \mathbf{F}(\mathbf{r}) \cdot d\mathbf{r} = \int_C (F_1\, dx + F_2\, dy + F_3\, dz) = \int_a^b \mathbf{F}(\mathbf{r}(t)) \cdot \frac{d\mathbf{r}}{dt}\, dt$$

where $C:\ \mathbf{r}(t) = [x(t), y(t), z(t)] = x(t)\mathbf{i} + y(t)\mathbf{j} + z(t)\mathbf{k}$ $(a \leqq t \leqq b)$ is a curve in space (or in the plane). See Sec. 9.1. Physically, (1) may represent the work done by a (variable) force in a displacement. Other kinds of line integrals and their applications are also discussed in Sec. 9.1.

Independence of path of a line integral in a region D means that the integral of a given function over any path C with endpoints P and Q has the same value for all paths from P to Q that lie in D; here P and Q are fixed. An integral (1) is independent of path in D if and only if the differential form $F_1\, dx + F_2\, dy + F_3\, dz$ is exact in D (Sec. 9.2). Also, if curl $\mathbf{F} = \mathbf{0}$, where $\mathbf{F} = [F_1, F_2, F_3]$ has continuous first partial derivatives in a *simply connected* domain D, then the integral (1) is independent of path in D (Sec. 9.2).

Double integrals over a region R in the plane (Sec. 9.3) can be transformed into line integrals over the boundary C of R (and conversely) by **Green's theorem in the plane** (Sec. 9.4)

$$(2) \qquad \iint_R \left(\frac{\partial F_2}{\partial x} - \frac{\partial F_1}{\partial y} \right) dx\, dy = \oint_C (F_1\, dx + F_2\, dy).$$

Other forms of this formula are given in Sec. 9.4. Green's theorem is also needed as the essential tool in proving Stokes's theorem (see below), and is a special case of the latter.

Triple integrals (Sec. 9.7) taken over a region T in space can be transformed into surface integrals over the boundary surface S of T (and conversely) by the **divergence theorem of Gauss** (Sec. 9.7)

$$(3) \qquad \iiint_T \text{div } \mathbf{F}\, dV = \iint_S \mathbf{F} \cdot \mathbf{n}\, dA.$$

Among other things, this theorem implies **Green's formulas** (Sec. 9.8)

$$(4) \qquad \iiint_T (f\nabla^2 g + \nabla f \cdot \nabla g)\, dV = \iint_S f\, \frac{\partial g}{\partial n}\, dA,$$

$$(5) \qquad \iiint_T (f\nabla^2 g - g\nabla^2 f)\, dV = \iint_S \left(f\, \frac{\partial g}{\partial n} - g\, \frac{\partial f}{\partial n} \right) dA.$$

Surface integrals over a surface S with boundary curve C can be transformed into line integrals over C (and conversely) by **Stokes's theorem** (Sec. 9.9)

$$(6) \qquad \iint_S (\text{curl } \mathbf{F}) \cdot \mathbf{n}\, dA = \oint_C \mathbf{F} \cdot \mathbf{r}'(s)\, ds.$$

PART C

Fourier Analysis and Partial Differential Equations

Chapter 10 **Fourier Series, Integrals, and Transforms**

Chapter 11 **Partial Differential Equations**

Periodic phenomena occur quite frequently—think of motors, rotating machines, sound waves, the motion of the earth, and the heart under normal conditions. In such a case it is an important practical problem to represent the corresponding periodic functions in terms of simple periodic functions, namely, cosine and sine. These representations will be series, called **Fourier series.** Their introduction by Fourier (after work by Euler and Daniel Bernoulli) was one of the most path-breaking events in applied mathematics.

Chapter 10 is primarily concerned with Fourier series. The corresponding ideas and techniques can also be extended to nonperiodic phenomena. This leads to **Fourier integrals** and **Fourier transforms** (Secs. 10.8−10.10). A common name for the whole area is **Fourier analysis.**

Chapter 11 is concerned with the most important **partial differential equations** of physics and engineering. In this area, Fourier analysis has its most basic applications, namely, for solving boundary and initial value problems in mechanics, heat flow, electrostatics, and other fields.

525

Fourier Series, Integrals, and Transforms

Fourier series[1] (Sec. 10.2) are series of cosine and sine terms and arise in the important practical task of representing general periodic functions. They constitute a very important tool in solving problems that involve ordinary and partial differential equations. In the present chapter we discuss these series and their engineering use from a practical viewpoint. Further applications follow in the next chapter on partial differential equations.

The *theory* of Fourier series is rather complicated, but the *application* of these series is simple. Fourier series are, in a certain sense, more universal than Taylor series, because many *discontinuous* periodic functions of practical interest can be developed in Fourier series, but, of course, do not have Taylor series representations.

The last four sections (10.8–10.11) concern **Fourier integrals** and **Fourier transforms,** which extend the ideas and techniques of Fourier series to nonperiodic functions defined for all *x*. (Corresponding applications to partial differential equations will be considered in the next chapter, in Sec. 11.6.

Prerequisite for this chapter: Elementary integral calculus.
Sections that may be omitted in a shorter course: 10.5 – 10.10.
References: Appendix 1, Part C.
Answers to problems: Appendix 2.

---

[1]JEAN-BAPTISTE JOSEPH FOURIER (1768—1830), French physicist and mathematician, lived and taught in Paris, accompanied Napoleon to Egypt, and was later made prefect of Grenoble. He utilized Fourier series in his main work *Théorie analytique de la chaleur* (*Analytic Theory of Heat,* Paris 1822), in which he developed the theory of heat conduction (heat equation, see Sec. 11.5). These new series became a most important tool in mathematical physics and also had considerable influence on the further development of mathematics itself; see Ref. [9] in Appendix 1.

10.1   Periodic Functions. Trigonometric Series

A function $f(x)$ is called **periodic** if it is defined for all[2] real x and if there is some positive number p such that

$$(1) \qquad\qquad f(x + p) = f(x) \qquad\qquad \text{for all } x.$$

This number p is called a **period** of $f(x)$. The graph of such a function is obtained by periodic repetition of its graph in any interval of length p (Fig. 236). Periodic phenomena and functions have many applications, as was mentioned before.

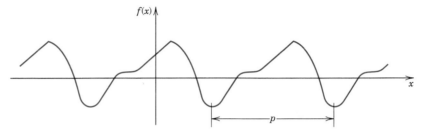

Fig. 236.   Periodic function

Familiar periodic functions are the sine and cosine functions. We note that the function $f = c = const$ is also a periodic function in the sense of the definition, because it satisfies (1) for every positive p. Examples of functions that are *not* periodic are x, x^2, x^3, e^x, $\cosh x$, and $\ln x$, to mention just a few.

From (1) we have $f(x + 2p) = f[(x + p) + p] = f(x + p) = f(x)$, etc., and for any integer n,

$$(2) \qquad\qquad f(x + np) = f(x) \qquad\qquad \text{for all } x.$$

Hence $2p$, $3p$, $4p$, $\cdots$ are also periods of $f(x)$. Furthermore, if $f(x)$ and $g(x)$ have period p, then the function

$$h(x) = af(x) + bg(x) \qquad\qquad (a, b \text{ constant})$$

also has the period p.

If a periodic function $f(x)$ has a smallest period p (> 0), this is often called the **fundamental period** of $f(x)$. For $\cos x$ and $\sin x$ the fundamental period is 2π, for $\cos 2x$ and $\sin 2x$ it is π, and so on. A function without fundamental period is $f = const$.

Trigonometric Series

Our problem in the first few sections of this chapter will be the representation of various functions of period $p = 2\pi$ in terms of the simple functions

$$(3) \qquad 1, \qquad \cos x, \qquad \sin x, \qquad \cos 2x, \qquad \sin 2x, \cdots, \qquad \cos nx, \qquad \sin nx, \cdots.$$

---

[2]Except perhaps for certain isolated x, such as $\pm\pi/2$, $\pm 3\pi/2$, $\cdots$ in the case of $\tan x$ (which is periodic with period π).

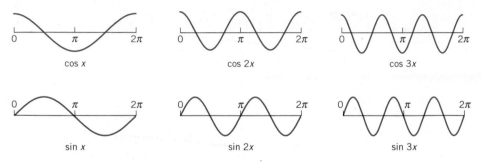

Fig. 237.   Cosine and sine functions having the period 2π

These functions have the period 2π. Figure 237 shows the first few of them.
  The series that will arise in this connection will be of the form

(4) $a_0 + a_1 \cos x + b_1 \sin x + a_2 \cos 2x + b_2 \sin 2x + \cdots,$

where $a_0, a_1, a_2, \cdots, b_1, b_2, \cdots$ are real constants. Such a series is called a **trigonometric series,** and the a_n and b_n are called the **coefficients** of the series. Using the summation sign,[3] we may write this series

(4)
$$a_0 + \sum_{n=1}^{\infty} (a_n \cos nx + b_n \sin nx).$$

The set of functions (3) from which we have made up the series (4) is often called the **trigonometric system,** to have a short name for it.
  We see that each term of the series (4) has the period 2π. Hence *if the series (4) converges, its sum will be a function of period 2π.*
  The point is that trigonometric series can be used for representing any practically important periodic function f, simple or complicated, of any period p. (This series will then be called the *Fourier series* of f.)

PROBLEM SET 10.1

Fundamental Period. Find the smallest positive period p of

1. $\cos x,$ $\sin x,$ $\cos 2x,$ $\sin 2x,$ $\cos \pi x,$ $\sin \pi x,$ $\cos 2\pi x,$ $\sin 2\pi x$

2. $\cos nx,$ $\sin nx,$ $\cos \dfrac{2\pi x}{k},$ $\sin \dfrac{2\pi x}{k},$ $\cos \dfrac{2\pi n x}{k},$ $\sin \dfrac{2\pi n x}{k}$

3. **(Vector space)** If $f(x)$ and $g(x)$ have period p, show that $h = af + bg$ (a, b constant) has the period p. Thus all functions of period p form a vector space.

4. **(Integer multiples of period)** If p is a period of $f(x)$, show that $np, n = 2, 3, \cdots$, is a period of $f(x)$.

---

[3]And inserting parentheses; from a convergent series this gives again a convergent series with the same sum, as can be proved.

5. **(Constant)** Show that the function $f(x) = const$ is a periodic function of period p for every positive p.

6. **(Change of scale)** If $f(x)$ is a periodic function of x of period p, show that $f(ax)$, $a \neq 0$, is a periodic function of x of period p/a, and $f(x/b)$, $b \neq 0$, is a periodic function of x of period bp. Verify these results for $f(x) = \cos x$, $a = b = 2$.

Graphs of 2π-Periodic Functions

Sketch or plot the following functions $f(x)$, which are assumed to be periodic with period 2π and, for $-\pi < x < \pi$, are given by the formulas

7. $f(x) = x$ 8. $f(x) = x^2$ 9. $f(x) = |x|$

10. $f(x) = \pi - |x|$ 11. $f(x) = |\sin x|$ 12. $f(x) = e^{-|x|}$

13. $f(x) = \begin{cases} x & \text{if } -\pi \leq x \leq 0 \\ 0 & \text{if } 0 \leq x \leq \pi \end{cases}$ 14. $f(x) = \begin{cases} 0 & \text{if } -\pi \leq x \leq 0 \\ x^2 & \text{if } 0 \leq x \leq \pi \end{cases}$

15. $f(x) = \begin{cases} -1 & \text{if } -\pi < x < 0 \\ 1 & \text{if } 0 < x < \pi \end{cases}$ 16. $f(x) = \begin{cases} x & \text{if } -\pi < x < 0 \\ \pi - x & \text{if } 0 < x < \pi \end{cases}$

17. $f(x) = \begin{cases} 0 & \text{if } -\pi < x < 0 \\ e^{-x} & \text{if } 0 < x < \pi \end{cases}$ 18. $f(x) = \begin{cases} x^2 & \text{if } -\pi < x < 0 \\ -x^2 & \text{if } 0 < x < \pi \end{cases}$

 19. **CAS PROJECT. Plotting Periodic Functions. (a)** Write a program for plotting periodic functions $f(x)$ of period 2π given for $-\pi < x \leq \pi$. Using your program, plot the functions in Probs. 7–12 for $-10\pi \leq x \leq 10\pi$. Also plot some functions of your own choice.

   (b) Extend your program to 2π-periodic functions given on two subintervals of the same length, as in Probs. 13–18. Apply your program to those problems with $-10\pi \leq x \leq 10\pi$.

 20. **CAS PROJECT. Partial Sums of Trigonometric Series. (a)** Write a program that prints a partial sum[4] of a trigonometric series (4). Applying it, list all partial sums of up to five nonzero terms of each of the series

$$\frac{1}{3}\pi^2 - 4\left(\cos x - \frac{1}{4}\cos 2x + \frac{1}{9}\cos 3x - \frac{1}{16}\cos 4x + - \cdots\right)$$

$$\frac{4}{\pi}\left(\sin x + \frac{1}{3}\sin 3x + \frac{1}{5}\sin 5x + \frac{1}{7}\sin 7x + \cdots\right)$$

$$2\left(\sin x - \frac{1}{2}\sin 2x + \frac{1}{3}\sin 3x - \frac{1}{4}\sin 4x + - \cdots\right).$$

   (b) Plot the partial sums in (a) (for each series on common axes). Guess what periodic function the series might represent.

10.2 Fourier Series

Fourier series arise from the practical task of representing a given periodic function $f(x)$ in terms of cosine and sine functions. These series are trigonometric series (Sec. 10.1) whose coefficients are determined from $f(x)$ by the "Euler formulas" [(6), below], which we shall derive first. Afterwards we shall take a look at the theory of Fourier series.

---

[4]That is, $a_0 + \sum_{n=1}^{N} (a_n \cos nx + b_n \sin nx)$ for $N = 1, 2, 3, \cdots$.

Euler Formulas for the Fourier Coefficients

Let us assume that $f(x)$ is a periodic function of period 2π and is integrable over a period. Let us further assume that $f(x)$ can be **represented** by a trigonometric series,

$$(1) \qquad f(x) = a_0 + \sum_{n=1}^{\infty} (a_n \cos nx + b_n \sin nx);$$

that is, we assume that this series converges and has $f(x)$ as its sum. Given such a function $f(x)$, we want to determine the coefficients a_n and b_n of the corresponding series (1).

Determination of the constant term a_0. Integrating on both sides of (1) from $-\pi$ to π, we get

$$\int_{-\pi}^{\pi} f(x)\, dx = \int_{-\pi}^{\pi} \left[a_0 + \sum_{n=1}^{\infty} (a_n \cos nx + b_n \sin nx) \right] dx.$$

If term-by-term integration of the series is allowed,[5] we obtain

$$\int_{-\pi}^{\pi} f(x)\, dx = a_0 \int_{-\pi}^{\pi} dx + \sum_{n=1}^{\infty} \left(a_n \int_{-\pi}^{\pi} \cos nx\, dx + b_n \int_{-\pi}^{\pi} \sin nx\, dx \right).$$

The first term on the right equals $2\pi a_0$. All the other integrals on the right are zero, as can be readily seen by integration. Hence our first result is

$$(2) \qquad a_0 = \frac{1}{2\pi} \int_{-\pi}^{\pi} f(x)\, dx.$$

Determination of the coefficients a_n of the cosine terms. Similarly, we multiply (1) by $\cos mx$, where m is any *fixed* positive integer, and integrate from $-\pi$ to π:

$$(3) \qquad \int_{-\pi}^{\pi} f(x) \cos mx\, dx = \int_{-\pi}^{\pi} \left[a_0 + \sum_{n=1}^{\infty} (a_n \cos nx + b_n \sin nx) \right] \cos mx\, dx.$$

Integrating term by term, we see that the right side becomes

$$a_0 \int_{-\pi}^{\pi} \cos mx\, dx + \sum_{n=1}^{\infty} \left[a_n \int_{-\pi}^{\pi} \cos nx \cos mx\, dx + b_n \int_{-\pi}^{\pi} \sin nx \cos mx\, dx \right].$$

The first integral is zero. By applying (11) in Appendix A3.1 we obtain

$$\int_{-\pi}^{\pi} \cos nx \cos mx\, dx = \frac{1}{2} \int_{-\pi}^{\pi} \cos (n + m)x\, dx + \frac{1}{2} \int_{-\pi}^{\pi} \cos (n - m)x\, dx,$$

$$\int_{-\pi}^{\pi} \sin nx \cos mx\, dx = \frac{1}{2} \int_{-\pi}^{\pi} \sin (n + m)x\, dx + \frac{1}{2} \int_{-\pi}^{\pi} \sin (n - m)x\, dx.$$

---

[5]This is justified, for instance, in the case of uniform convergence (see Theorem 3 in Sec. 14.5).

Integration shows that the four terms on the right are zero, except for the last term in the first line, which equals π when $n = m$. Since in (3) this term is multiplied by a_m, the right side in (3) equals $a_m\pi$. Our second result is

$$(4) \qquad\qquad a_m = \frac{1}{\pi} \int_{-\pi}^{\pi} f(x) \cos mx\, dx, \qquad\qquad m = 1, 2, \cdots.$$

Determination of the coefficients b_n of the sine terms. We finally multiply (1) by $\sin mx$, where m is any *fixed* positive integer, and then integrate from $-\pi$ to π:

$$(5) \qquad \int_{-\pi}^{\pi} f(x) \sin mx\, dx = \int_{-\pi}^{\pi} \left[a_0 + \sum_{n=1}^{\infty} (a_n \cos nx + b_n \sin nx) \right] \sin mx\, dx.$$

Integrating term by term, we see that the right side becomes

$$a_0 \int_{-\pi}^{\pi} \sin mx\, dx + \sum_{n=1}^{\infty} \left[a_n \int_{-\pi}^{\pi} \cos nx \sin mx\, dx + b_n \int_{-\pi}^{\pi} \sin nx \sin mx\, dx \right].$$

The first integral is zero. The next integral is of the kind considered before, and is zero for all $n = 1, 2, \cdots$. For the last integral we obtain

$$\int_{-\pi}^{\pi} \sin nx \sin mx\, dx = \frac{1}{2} \int_{-\pi}^{\pi} \cos (n - m)x\, dx - \frac{1}{2} \int_{-\pi}^{\pi} \cos (n + m)x\, dx.$$

The last term is zero. The first term on the right is zero when $n \neq m$ and is π when $n = m$. Since in (5) this term is multiplied by b_m, the right side in (5) is equal to $b_m\pi$, and our last result is

$$b_m = \frac{1}{\pi} \int_{-\pi}^{\pi} f(x) \sin mx\, dx, \qquad\qquad m = 1, 2, \cdots.$$

Summary of These Calculations: Fourier Coefficients, Fourier Series

From (2), (4), and the formula just obtained, writing n in place of m, we have the so-called **Euler**[6] **formulas**

$$(6)$$

$$\textbf{(a)} \qquad a_0 = \frac{1}{2\pi} \int_{-\pi}^{\pi} f(x)\, dx$$

$$\textbf{(b)} \qquad a_n = \frac{1}{\pi} \int_{-\pi}^{\pi} f(x) \cos nx\, dx \qquad n = 1, 2, \cdots,$$

$$\textbf{(c)} \qquad b_n = \frac{1}{\pi} \int_{-\pi}^{\pi} f(x) \sin nx\, dx \qquad n = 1, 2, \cdots.$$

[6]See footnote 9 in Sec. 2.6.

These numbers given by (6) are called the **Fourier coefficients** of $f(x)$. The trigonometric series

(7)

$$a_0 + \sum_{n=1}^{\infty} (a_n \cos nx + b_n \sin nx)$$

with coefficients given in (6) is called the **Fourier series** of $f(x)$ (regardless of convergence—we shall discuss this later in this section).

EXAMPLE 1 **Rectangular wave**

Find the Fourier coefficients of the periodic function $f(x)$ in Fig. 238a. The formula is

$$f(x) = \begin{cases} -k & \text{if} & -\pi < x < 0 \\ k & \text{if} & 0 < x < \pi \end{cases} \quad \text{and} \quad f(x + 2\pi) = f(x).$$

Functions of this kind occur as external forces acting on mechanical systems, electromotive forces in electric circuits, etc. (The value of $f(x)$ at a single point does not affect the integral; hence we can leave $f(x)$ undefined at $x = 0$ and $x = \pm\pi$.)

Solution. From (6a) we obtain $a_0 = 0$. This can also be seen without integration, since the area under the curve of $f(x)$ between $-\pi$ and π is zero. From (6b),

$$a_n = \frac{1}{\pi} \int_{-\pi}^{\pi} f(x) \cos nx \, dx = \frac{1}{\pi} \left[\int_{-\pi}^{0} (-k) \cos nx \, dx + \int_{0}^{\pi} k \cos nx \, dx \right]$$

$$= \frac{1}{\pi} \left[-k \frac{\sin nx}{n} \Big|_{-\pi}^{0} + k \frac{\sin nx}{n} \Big|_{0}^{\pi} \right] = 0$$

because $\sin nx = 0$ at $-\pi$, 0, and π for all $n = 1, 2, \cdots$. Similarly, from (6c) we obtain

$$b_n = \frac{1}{\pi} \int_{-\pi}^{\pi} f(x) \sin nx \, dx = \frac{1}{\pi} \left[\int_{-\pi}^{0} (-k) \sin nx \, dx + \int_{0}^{\pi} k \sin nx \, dx \right]$$

$$= \frac{1}{\pi} \left[k \frac{\cos nx}{n} \Big|_{-\pi}^{0} - k \frac{\cos nx}{n} \Big|_{0}^{\pi} \right].$$

Since $\cos (-\alpha) = \cos \alpha$ and $\cos 0 = 1$, this yields

$$b_n = \frac{k}{n\pi} [\cos 0 - \cos (-n\pi) - \cos n\pi + \cos 0] = \frac{2k}{n\pi} (1 - \cos n\pi).$$

Now, $\cos \pi = -1$, $\cos 2\pi = 1$, $\cos 3\pi = -1$, etc.; in general,

$$\cos n\pi = \begin{cases} -1 & \text{for odd } n, \\ 1 & \text{for even } n, \end{cases} \quad \text{and thus} \quad 1 - \cos n\pi = \begin{cases} 2 & \text{for odd } n, \\ 0 & \text{for even } n. \end{cases}$$

Hence the Fourier coefficients b_n of our function are

$$b_1 = \frac{4k}{\pi}, \qquad b_2 = 0, \qquad b_3 = \frac{4k}{3\pi}, \qquad b_4 = 0, \qquad b_5 = \frac{4k}{5\pi}, \cdots.$$

Since the a_n are zero, the Fourier series of $f(x)$ is

(8)

$$\frac{4k}{\pi} \left(\sin x + \frac{1}{3} \sin 3x + \frac{1}{5} \sin 5x + \cdots \right).$$

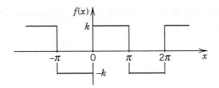

(a) The given function $f(x)$ (Periodic square wave)

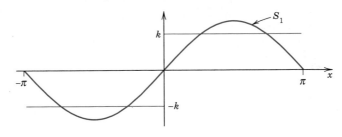

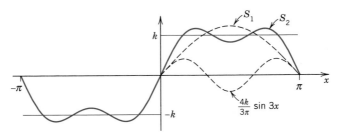

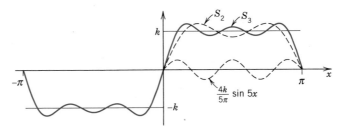

(b) The first three partial sums of the corresponding Fourier series

Fig. 238.   Example 1

The partial sums are

$$S_1 = \frac{4k}{\pi} \sin x, \qquad S_2 = \frac{4k}{\pi}\left(\sin x + \frac{1}{3}\sin 3x\right), \qquad \text{etc.,}$$

Their graphs in Fig. 238 seem to indicate that the series is convergent and has the sum $f(x)$, the given function. We notice that at $x = 0$ and $x = \pi$, the points of discontinuity of $f(x)$, all partial sums have the value zero, the arithmetic mean of the values $-k$ and k of our function.

Furthermore, assuming that $f(x)$ is the sum of the series and setting $x = \pi/2$, we have

$$f\left(\frac{\pi}{2}\right) = k = \frac{4k}{\pi}\left(1 - \frac{1}{3} + \frac{1}{5} - + \cdots\right),$$

thus

$$1 - \frac{1}{3} + \frac{1}{5} - \frac{1}{7} + - \cdots = \frac{\pi}{4}.$$

This is a famous result by Leibniz (obtained in 1673 from geometrical considerations). It illustrates that the values of various series with constant terms can be obtained by evaluating Fourier series at specific points.   ◄

Orthogonality of the Trigonometric System

The **trigonometric system** (3), Sec. 10.1,

$$1, \quad \cos x, \quad \sin x, \quad \cos 2x, \quad \sin 2x, \quad \cdots, \quad \cos nx, \quad \sin nx, \quad \cdots$$

is **orthogonal** *on the interval* $-\pi \leqq x \leqq \pi$ (hence on any interval of length 2π, because of periodicity). By definition, this means that the integral of the product of any two different of these functions over that interval is zero; in formulas, for any integers m and $n \neq m$ we have

$$\int_{-\pi}^{\pi} \cos mx \cos nx \, dx = 0 \qquad (m \neq n)$$

and

$$\int_{-\pi}^{\pi} \sin mx \sin nx \, dx = 0 \qquad (m \neq n),$$

and for any integers m and n (including $m = n$) we have

$$\int_{-\pi}^{\pi} \cos mx \sin nx \, dx = 0.$$

This is the most important property of the trigonometric system, the key in deriving the Euler formulas (where we proved this orthogonality).

Convergence and Sum of Fourier Series

Throughout this chapter we consider Fourier series from a practical point of view. We shall see that the application of these series is rather simple. In contrast to this, their theory is complicated, and we shall not go into any details of it. However, we present a theorem on the convergence and the sum of Fourier series that takes care of most applications.

Suppose that $f(x)$ is any given periodic function of period 2π for which the integrals in (6) exist; for instance, $f(x)$ is continuous or merely piecewise continuous (continuous except for finitely many finite jumps in the interval of integration). Then we can compute the Fourier coefficients (6) of $f(x)$ and use them to form the Fourier series (7) of $f(x)$. It would be nice if the series thus obtained converged and had the sum $f(x)$. Most functions appearing in applications are such that this is true (except at jumps of $f(x)$, which we discuss below). In this case, in which the Fourier series of $f(x)$ does represent $f(x)$, we write

$$f(x) = a_0 + \sum_{n=1}^{\infty} (a_n \cos nx + b_n \sin nx)$$

with an equality sign. If the Fourier series of $f(x)$ does not have the sum $f(x)$ or does not converge, one still writes

$$f(x) \sim a_0 + \sum_{n=1}^{\infty} (a_n \cos nx + b_n \sin nx)$$

with a tilde $\sim$, which indicates that the trigonometric series on the right has the Fourier coefficients of $f(x)$ as its coefficients, so it is the Fourier series of $f(x)$.

The class of functions that can be represented by Fourier series is surprisingly large and general. Corresponding sufficient conditions covering almost any conceivable application are as follows.

THEOREM 1   **(Representation by a Fourier series)**

If a periodic function $f(x)$ with period 2π is piecewise continuous[7] in the interval $-\pi \leq x \leq \pi$ and has a left-hand derivative and right-hand derivative[8] at each point of that interval, then the Fourier series (7) of $f(x)$ [with coefficients (6)] is convergent. Its sum is $f(x)$, except at a point x_0 at which $f(x)$ is discontinuous and the sum of the series is the average of the left- and right-hand limits[8] of $f(x)$ at x_0.

PROOF OF CONVERGENCE IN THEOREM 1.   We prove convergence for a continuous function $f(x)$ having continuous first and second derivatives. Integrating (6b) by parts, we get

$$a_n = \frac{1}{\pi} \int_{-\pi}^{\pi} f(x) \cos nx \, dx = \frac{f(x) \sin nx}{n\pi} \bigg|_{-\pi}^{\pi} - \frac{1}{n\pi} \int_{-\pi}^{\pi} f'(x) \sin nx \, dx.$$

The first term on the right is zero. Another integration by parts gives

$$a_n = \frac{f'(x) \cos nx}{n^2 \pi} \bigg|_{-\pi}^{\pi} - \frac{1}{n^2 \pi} \int_{-\pi}^{\pi} f''(x) \cos nx \, dx.$$

The first term on the right is zero because of the periodicity and continuity of $f'(x)$. Since f'' is continuous in the interval of integration, we have

$$|f''(x)| < M$$

for an appropriate constant M. Furthermore, $|\cos nx| \leq 1$. It follows that

$$|a_n| = \frac{1}{n^2 \pi} \left| \int_{-\pi}^{\pi} f''(x) \cos nx \, dx \right| < \frac{1}{n^2 \pi} \int_{-\pi}^{\pi} M \, dx = \frac{2M}{n^2}.$$

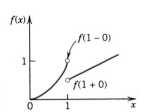

Fig. 239.   Left- and right-hand limits

$$f(1 - 0) = 1,$$
$$f(1 + 0) = \tfrac{1}{2}$$

of the function

$$f(x) = \begin{cases} x^2 & \text{if } x < 1 \\ x/2 \end{cases}$$

[7]Definition in Sec. 5.1.

[8]The **left-hand limit** of $f(x)$ at x_0 is defined as the limit of $f(x)$ as x approaches x_0 from the left and is frequently denoted by $f(x_0 - 0)$. Thus

$$f(x_0 - 0) = \lim_{h \to 0} f(x_0 - h) \text{ as } h \to 0 \text{ through positive values.}$$

The **right-hand limit** is denoted by $f(x_0 + 0)$ and

$$f(x_0 + 0) = \lim_{h \to 0} f(x_0 + h) \text{ as } h \to 0 \text{ through positive values.}$$

The **left-** and **right-hand derivatives** of $f(x)$ at x_0 are defined as the limits of

$$\frac{f(x_0 - h) - f(x_0 - 0)}{-h} \quad \text{and} \quad \frac{f(x_0 + h) - f(x_0 + 0)}{h},$$

respectively, as $h \to 0$ through positive values. Of course if $f(x)$ is continuous at x_0, the last term in both numerators is simply $f(x_0)$.

Similarly, $|b_n| < 2\ M/n^2$ for all n. Hence the absolute value of each term of the Fourier series of $f(x)$ is at most equal to the corresponding term of the series

$$|a_0| + 2M \left(1 + 1 + \frac{1}{2^2} + \frac{1}{2^2} + \frac{1}{3^2} + \frac{1}{3^2} + \cdots \right)$$

which is convergent. Hence that Fourier series converges and the proof is complete. (Readers already familiar with uniform convergence will see that, by the Weierstrass test in Sec. 14.5, under our present assumptions the Fourier series converges uniformly, and our derivation of (6) by integrating term by term is then justified by Theorem 3 of Sec. 14.5.)

The proof of convergence in the case of a piecewise continuous function $f(x)$ and the proof that under the assumptions in the theorem the Fourier series (7) with coefficients (6) represents $f(x)$ are substantially more complicated; see, for instance, Ref. [C9]. ◀

EXAMPLE 2 **Convergence at a jump as indicated in Theorem 1**

The square wave in Example 1 has a jump at $x = 0$. Its left-hand limit there is $-k$ and its right-hand limit is k (Fig. 238). Hence the average of these limits is 0. The Fourier series (8) of the square wave does indeed converge to this value when $x = 0$ because then all its terms are 0. Similarly for the other jumps. This is in agreement with Theorem 1. ◀

Summary. A Fourier series of a given function $f(x)$ of period 2π is a series of the form (7) with coefficients given by the Euler formulas (6). Theorem 1 gives conditions that are sufficient for this series to converge and at each x to have the value $f(x)$, except at discontinuities of $f(x)$, where the series equals the arithmetic mean of the left-hand and right-hand limits of $f(x)$ at that point.

PROBLEM SET 10.2

Fourier Series

Showing the details of your work, find the Fourier series of the function $f(x)$, which is assumed to have the period 2π, and plot accurate graphs of the first three partial sums, where $f(x)$ equals

1.

2.

3.
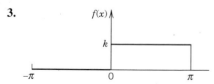

4.
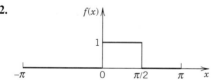

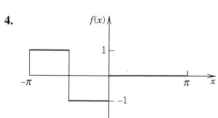

5. $f(x) = x \quad (-\pi < x < \pi)$
7. $f(x) = x^2 \quad (-\pi < x < \pi)$
9. $f(x) = x^3 \quad (-\pi < x < \pi)$

6. $f(x) = x \quad (0 < x < 2\pi)$
8. $f(x) = x^2 \quad (0 < x < 2\pi)$
10. $f(x) = x + |x| \quad (-\pi < x < \pi)$

11. $f(x) = \begin{cases} 1 & \text{if } -\pi < x < 0 \\ -1 & \text{if } \quad 0 < x < \pi \end{cases}$

12. $f(x) = \begin{cases} -1 & \text{if } \quad 0 < x < \pi/2 \\ 0 & \text{if } \pi/2 < x < 2\pi \end{cases}$

13. $f(x) = \begin{cases} 1 & \text{if } -\pi/2 < x < \pi/2 \\ -1 & \text{if } \quad \pi/2 < x < 3\pi/2 \end{cases}$

14. $f(x) = \begin{cases} x & \text{if } -\pi/2 < x < \pi/2 \\ \pi - x & \text{if } \quad \pi/2 < x < 3\pi/2 \end{cases}$

15. $f(x) = \begin{cases} x & \text{if } -\pi/2 < x < \pi/2 \\ 0 & \text{if } \quad \pi/2 < x < 3\pi/2 \end{cases}$

16. $f(x) = \begin{cases} x^2 & \text{if } -\pi/2 < x < \pi/2 \\ \pi^2/4 & \text{if } \quad \pi/2 < x < 3\pi/2 \end{cases}$

17. (Discontinuity) Verify the last statement in Theorem 1 regarding discontinuities for the function in Prob. 1.

18. CAS (Orthogonality). Integrate and plot a typical integral, for instance, that of $\sin 3x \sin 4x$, from $-a$ to a, as a function of a, and conclude orthogonality of $\sin 3x$ and $\sin 4x$ for $a = \pi$ from the plot.

19. CAS PROJECT. Fourier Series. (a) Write a program for obtaining any partial sum of a Fourier series (7).

(b) Using the program, list all partial sums of up to five nonzero terms of the Fourier series in Probs. 5, 11, and 15, and make three corresponding plots. Comment on the accuracy.

20. (Calculus review) Review integration techniques for integrals as they may arise from the Euler formulas, for instance, definite integrals of $x \sin nx$, $x^2 \cos nx$, $e^{-x} \sin nx$, etc.

10.3 Functions of Any Period $p = 2L$

The functions considered so far had period 2π, for simplicity. Of course, in applications, periodic functions will generally have other periods. But we show that the transition from period $p = 2\pi$ to period[9] $p = 2L$ is quite simple. It amounts to a stretch (or contraction) of scale on the axis.

If a function $f(x)$ of period $p = 2L$ has a **Fourier series,** we claim that this series is

(1)
$$f(x) = a_0 + \sum_{n=1}^{\infty} \left(a_n \cos \frac{n\pi}{L} x + b_n \sin \frac{n\pi}{L} x \right)$$

with the **Fourier coefficients** of $f(x)$ given by the **Euler formulas**

(2)

(a) $\quad a_0 = \dfrac{1}{2L} \displaystyle\int_{-L}^{L} f(x)\, dx$

(b) $\quad a_n = \dfrac{1}{L} \displaystyle\int_{-L}^{L} f(x) \cos \frac{n\pi x}{L}\, dx \qquad n = 1, 2, \cdots$

(c) $\quad b_n = \dfrac{1}{L} \displaystyle\int_{-L}^{L} f(x) \sin \frac{n\pi x}{L}\, dx \qquad n = 1, 2, \cdots$.

[9]This notation is practical since in applications, L will be the length of a vibrating string (Sec. 11.2), of a rod in heat conduction (Sec. 11.5), etc.

(The series in (1) with *arbitrary* coefficients is called a **trigonometric series,** and Theorem 1 in Sec. 10.2 extends to any period p.)

PROOF. Equations (1) and (2) follow from Sec. 10.2 by a change of scale, say, by setting $v = \pi x/L$. Then $x = Lv/\pi$. Also, $x = \pm L$ corresponds to $v = \pm \pi$. Thus f, regarded as a function of v that we call $g(v)$,

$$f(x) = g(v),$$

has period 2π. Accordingly, by (7) and (6), Sec. 10.2, with v instead of x, this 2π-periodic function $g(v)$ has the Fourier series

$$(3) \qquad g(v) = a_0 + \sum_{n=1}^{\infty} (a_n \cos nv + b_n \sin nv)$$

with coefficients

$$a_0 = \frac{1}{2\pi} \int_{-\pi}^{\pi} g(v)\, dv$$

$$(4) \qquad a_n = \frac{1}{\pi} \int_{-\pi}^{\pi} g(v) \cos nv\, dv$$

$$b_n = \frac{1}{\pi} \int_{-\pi}^{\pi} g(v) \sin nv\, dv.$$

Since $v = \pi x/L$ and $g(v) = f(x)$, formula (3) gives (1). In (4) we introduce $x = Lv/\pi$ as variable of integration. Then the limits of integration $v = \pm \pi$ become $x = \pm L$. Also, $v = \pi x/L$ implies $dv = \pi\, dx/L$. Thus $dv/2\pi = dx/2L$ in a_0. Similarly, $dv/\pi = dx/L$ in a_n and b_n. Hence (4) gives (2). ◀

Interval of integration. In (2) we may replace the interval of integration by any interval of length $p = 2L$, for example, by the interval $0 \leqq x \leqq 2L$.

EXAMPLE 1   **Periodic square wave**

Find the Fourier series of the function (see Fig. 240)

$$f(x) = \begin{cases} 0 & \text{if} \quad -2 < x < -1 \\ k & \text{if} \quad -1 < x < 1 \qquad p = 2L = 4, \quad L = 2. \\ 0 & \text{if} \quad 1 < x < 2 \end{cases}$$

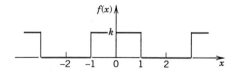

Fig. 240.   Example 1

Solution. From (2a) and (2b) we obtain

$$a_0 = \frac{1}{4} \int_{-2}^{2} f(x)\, dx = \frac{1}{4} \int_{-1}^{1} k\, dx = \frac{k}{2},$$

$$a_n = \frac{1}{2} \int_{-2}^{2} f(x) \cos \frac{n\pi x}{2}\, dx = \frac{1}{2} \int_{-1}^{1} k \cos \frac{n\pi x}{2}\, dx = \frac{2k}{n\pi} \sin \frac{n\pi}{2}.$$

Thus $a_n = 0$ if n is even and

$$a_n = 2k/n\pi \quad \text{if} \quad n = 1, 5, 9, \cdots, \qquad a_n = -2k/n\pi \quad \text{if} \quad n = 3, 7, 11, \cdots.$$

From (2c) we find that $b_n = 0$ for $n = 1, 2, \cdots$. Hence the result is

$$f(x) = \frac{k}{2} + \frac{2k}{\pi} \left(\cos \frac{\pi}{2} x - \frac{1}{3} \cos \frac{3\pi}{2} x + \frac{1}{5} \cos \frac{5\pi}{2} x - + \cdots \right).$$

Could you obtain this from (8) in Sec. 10.2?   ◀

EXAMPLE 2   **Half-wave rectifier**

A sinusoidal voltage $E \sin \omega t$, where t is time, is passed through a half-wave rectifier that clips the negative portion of the wave (Fig. 241). Find the Fourier series of the resulting periodic function

$$u(t) = \begin{cases} 0 & \text{if} \quad -L < t < 0, \\ E \sin \omega t & \text{if} \quad 0 < t < L \end{cases} \qquad p = 2L = \frac{2\pi}{\omega}, \qquad L = \frac{\pi}{\omega}.$$

Solution. Since $u = 0$ when $-L < t < 0$, we obtain from (2a), with t instead of x,

$$a_0 = \frac{\omega}{2\pi} \int_0^{\pi/\omega} E \sin \omega t\, dt = \frac{E}{\pi}$$

and from (2b), by using formula (11) in Appendix A3.1 with $x = \omega t$ and $y = n\omega t$,

$$a_n = \frac{\omega}{\pi} \int_0^{\pi/\omega} E \sin \omega t \cos n\omega t\, dt = \frac{\omega E}{2\pi} \int_0^{\pi/\omega} [\sin (1 + n)\omega t + \sin (1 - n)\omega t]\, dt.$$

If $n = 1$, the integral on the right is zero, and if $n = 2, 3, \cdots$, we readily obtain

$$a_n = \frac{\omega E}{2\pi} \left[-\frac{\cos (1 + n)\omega t}{(1 + n)\omega} - \frac{\cos (1 - n)\omega t}{(1 - n)\omega} \right]_0^{\pi/\omega}$$

$$= \frac{E}{2\pi} \left(\frac{-\cos (1 + n)\pi + 1}{1 + n} + \frac{-\cos (1 - n)\pi + 1}{1 - n} \right).$$

If n is odd, this is equal to zero, and for even n we have

$$a_n = \frac{E}{2\pi} \left(\frac{2}{1 + n} + \frac{2}{1 - n} \right) = -\frac{2E}{(n - 1)(n + 1)\pi} \qquad (n = 2, 4, \cdots).$$

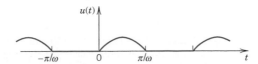

$$-\pi/\omega \qquad\qquad 0 \qquad\qquad \pi/\omega \qquad\qquad\qquad\qquad t$$

Fig. 241.   Half-wave rectifier

In a similar fashion we find from (2c) that $b_1 = E/2$ and $b_n = 0$ for $n = 2, 3, \cdots$. Consequently,

$$u(t) = \frac{E}{\pi} + \frac{E}{2} \sin \omega t - \frac{2E}{\pi} \left(\frac{1}{1 \cdot 3} \cos 2\omega t + \frac{1}{3 \cdot 5} \cos 4\omega t + \cdots \right).$$ ◀

PROBLEM SET 10.3

Fourier Series for Period $p = 2L$

Find the Fourier series of the periodic function $f(x)$, of period $p = 2L$, and sketch $f(x)$ and the first three partial sums. (Show the details of your work.)

1. $f(x) = -1$ $(-1 < x < 0)$, $f(x) = 1$ $(0 < x < 1)$, $p = 2L = 2$
2. $f(x) = 1$ $(-1 < x < 0)$, $f(x) = -1$ $(0 < x < 1)$, $p = 2L = 2$
3. $f(x) = 0$ $(-2 < x < 0)$, $f(x) = 2$ $(0 < x < 2)$, $p = 2L = 4$
4. $f(x) = |x|$ $(-2 < x < 2)$, $p = 2L = 4$
5. $f(x) = 2x$ $(-1 < x < 1)$, $p = 2L = 2$
6. $f(x) = 1 - x^2$ $(-1 < x < 1)$, $p = 2L = 2$
7. $f(x) = 3x^2$ $(-1 < x < 1)$, $p = 2L = 2$
8. $f(x) = \frac{1}{2} + x$ $(-\frac{1}{2} < x < 0)$, $f(x) = \frac{1}{2} - x$ $(0 < x < \frac{1}{2})$, $p = 2L = 1$
9. $f(x) = 0$, $(-1 < x < 0)$, $f(x) = x$ $(0 < x < 1)$, $p = 2L = 2$
10. $f(x) = x$ $(0 < x < 1)$, $f(x) = 1 - x$ $(1 < x < 2)$, $p = 2L = 2$
11. $f(x) = \pi \sin \pi x$ $(0 < x < 1)$, $p = 2L = 1$
12. $f(x) = \pi x^3/2$ $(-1 < x < 1)$, $p = 2L = 2$

13. **(Periodicity)** Show that each term in (1) has the period $p = 2L$.

14. **(Rectifier)** Find the Fourier series of the periodic function that is obtained by passing the voltage $v(t) = V_0 \cos 100\pi t$ through a half-wave rectifier.

15. **(Transformation)** Obtain the Fourier series in Prob. 1 from that in Example 1, Sec. 10.2.

16. **(Transformation)** Obtain the Fourier series in Prob. 7 from that in Prob. 7, Sec. 10.2.

17. **(Transformation)** Obtain the Fourier series in Prob. 3 from that in Example 1, Sec. 10.2.

18. **(Interval of Integration)** Show that in (2) the interval of integration may be replaced by any other interval of length $p = 2L$.

19. **CAS PROJECT. Fourier Series of 2L-Periodic Functions.** **(a)** Write a program for obtaining any partial sum of a Fourier series (1).
    (b) Apply the program to Probs. 5–7, plotting the first few partial sums of each of the three series on common axes. Choose the first five or more partial sums until they approximate the given function reasonably well.

20. **CAS PROJECT. Gibbs Phenomenon.** The partial sums $s_n(x)$ of a Fourier series show oscillations near a discontinuity point. These do not disappear as n increases but instead become sharp "spikes." They were explained mathematically by J. W. Gibbs.[10] Plot $s_n(x)$ in Prob. 5. When $n = 20$, say, you will see those oscillations quite distinctly. Consider two other Fourier series of your choice in a similar way.

---

[10]JOSIAH WILLARD GIBBS (1839—1903), American mathematician, professor of mathematical physics at Yale from 1871, one of the founders of vector calculus [another being O. Heaviside (see Sec. 5.1)], mathematical thermodynamics, and statistical mechanics. His work was of great importance to the development of mathematical physics.

10.4 Even and Odd Functions Half-Range Expansions

The function in Example 1 of the last section was even and had only cosine terms in its Fourier series, no sine terms. This is typical. In fact, unnecessary work (and corresponding sources of errors) in determining Fourier coefficients can be avoided if a function is even or odd.

Even and Odd Functions

We first remember that a function $y = g(x)$ is **even** if

$$g(-x) = g(x) \qquad \text{for all } x.$$

The graph of such a function is symmetric with respect to the y-axis (Fig. 242). A function $h(x)$ is **odd** if

$$h(-x) = -h(x) \qquad \text{for all } x.$$

(See Fig. 243.) The function $\cos nx$ is even, while $\sin nx$ is odd.

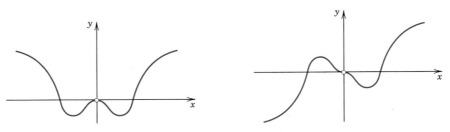

Fig. 242.   Even function   **Fig. 243.**   Odd function

Three Key Facts for the Present Discussion

1. If $g(x)$ is an **even** function, then

$$(1) \qquad \int_{-L}^{L} g(x)\, dx = 2 \int_{0}^{L} g(x)\, dx \qquad (g \text{ even}).$$

2. If $h(x)$ is an **odd** function, then

$$(2) \qquad \int_{-L}^{L} h(x)\, dx = 0 \qquad (h \text{ odd}).$$

3. The product of an even and an odd function is odd.

PROOF.   Equations (1) and (2) are obvious from the graphs of g and h. (Give a formal proof.) $q = gh$ with even g and odd h is odd because

$$q(-x) = g(-x)h(-x) = g(x)(-h(x)) = -g(x)h(x) = -q(x). \qquad \blacktriangleleft$$

Hence if $f(x)$ is even, then $f(x) \sin (n\pi x/L)$ is odd, so (2) implies that $b_n = 0$ in (2c), Sec. 10.3. Similarly, if $f(x)$ is odd, so is $f(x) \cos (n\pi x/L)$ and $a_0 = 0$, $a_n = 0$ in (2a), (2b), Sec. 10.3. From this and (1) we have

THEOREM 1   **(Fourier cosine series, Fourier sine series)**

*The Fourier series of an **even** function of period 2L is a "**Fourier cosine series**"*

(3)
$$f(x) = a_0 + \sum_{n=1}^{\infty} a_n \cos \frac{n\pi}{L} x \qquad (f \text{ even})$$

with coefficients

(4) $a_0 = \dfrac{1}{L} \displaystyle\int_0^L f(x)\, dx, \qquad a_n = \dfrac{2}{L} \int_0^L f(x) \cos \dfrac{n\pi x}{L}\, dx, \qquad n = 1, 2, \cdots.$

*The Fourier series of an **odd** function of period 2L is a "**Fourier sine series**"*

(5)
$$f(x) = \sum_{n=1}^{\infty} b_n \sin \frac{n\pi}{L} x \qquad (f \text{ odd})$$

with coefficients

(6) $\qquad\qquad b_n = \dfrac{2}{L} \displaystyle\int_0^L f(x) \sin \dfrac{n\pi x}{L}\, dx.$ ◀

The case of period 2π. In this case Theorem 1 gives for an even function simply

(3*)
$$f(x) = a_0 + \sum_{n=1}^{\infty} a_n \cos nx \qquad (f \text{ even})$$

with coefficients

(4*) $a_0 = \dfrac{1}{\pi} \displaystyle\int_0^{\pi} f(x)\, dx, \qquad a_n = \dfrac{2}{\pi} \int_0^{\pi} f(x) \cos nx\, dx, \qquad n = 1, 2, \cdots.$

Similarly, for an odd 2π-periodic function we simply have

(5*)
$$f(x) = \sum_{n=1}^{\infty} b_n \sin nx \qquad (f \text{ odd})$$

with coefficients

(6*) $\qquad\qquad b_n = \dfrac{2}{\pi} \displaystyle\int_0^{\pi} f(x) \sin nx\, dx, \qquad n = 1, 2, \cdots.$

For instance, $f(x)$ in Example 1, Sec. 10.2, is odd and, therefore, is represented by a Fourier sine series.

Further simplifications result from the following property.

THEOREM 2   **(Sum of functions)**

The Fourier coefficients of a sum $f_1 + f_2$ are the sums of the corresponding Fourier coefficients of f_1 and f_2.

The Fourier coefficients of cf are c times the corresponding Fourier coefficients of f.

EXAMPLE 1   **Rectangular pulse**

The function $f^*(x)$ in Fig. 244 is the sum of the function $f(x)$ in Example 1 of Sec. 10.2 and the constant k. Hence, from that example and Theorem 2 we conclude that

$$f^*(x) = k + \frac{4k}{\pi} \left(\sin x + \frac{1}{3} \sin 3x + \frac{1}{5} \sin 5x + \cdots \right).$$ ◄

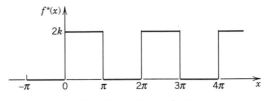

Fig. 244.   Example 1

EXAMPLE 2   **Sawtooth wave**

Find the Fourier series of the function (Fig. 245a)

$$f(x) = x + \pi \quad \text{if} \quad -\pi < x < \pi \quad \text{and} \quad f(x + 2\pi) = f(x).$$

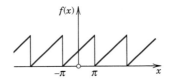

(*a*) The function $f(x)$

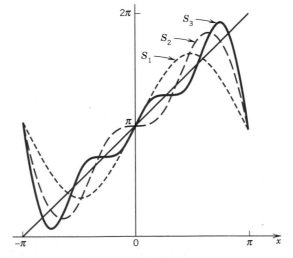

(*b*) Partial sums $S_n(x)$

Fig. 245.   Example 2

Solution. We may write

$$f = f_1 + f_2, \qquad \text{where} \qquad f_1 = x \qquad \text{and} \qquad f_2 = \pi.$$

The Fourier coefficients of f_2 are zero, except for the first one (the constant term), which is π. Hence, by Theorem 2, the Fourier coefficients a_n, b_n are those of f_1, except for a_0, which is π. Since f_1 is odd, $a_n = 0$ for $n = 1, 2, \cdots$, and

$$b_n = \frac{2}{\pi} \int_0^{\pi} f_1(x) \sin nx \, dx = \frac{2}{\pi} \int_0^{\pi} x \sin nx \, dx.$$

Integrating by parts we obtain

$$b_n = \frac{2}{\pi} \left[\frac{-x \cos nx}{n} \Big|_0^{\pi} + \frac{1}{n} \int_0^{\pi} \cos nx \, dx \right] = -\frac{2}{n} \cos n\pi.$$

Hence $b_1 = 2$, $b_2 = -2/2$, $b_3 = 2/3$, $b_4 = -2/4$, $\cdots$, and the Fourier series of $f(x)$ is

$$f(x) = \pi + 2 \left(\sin x - \frac{1}{2} \sin 2x + \frac{1}{3} \sin 3x - + \cdots \right). \qquad \blacktriangleleft$$

Half-Range Expansions

This concerns a practically useful simple idea. In applications we often want to employ a Fourier series for a function f that is given only on some interval, say, $0 \leqq x \leqq L$, as in Fig. 246(a). This function f can be the displacement of a violin string of (undistorted) length L or the temperature in a metal bar of length L, and so on (as we shall discuss in Secs. 11.3 and 11.5). Now the key idea is as follows. For our function f we can calculate Fourier coefficients from (4) or (6) in Theorem 1. And we have a choice. If we use (4), we get a Fourier cosine series (3). This series represents the **even periodic extension** f_1

(a) The given function $f(x)$

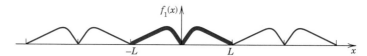

(b) $f(x)$ extended as an even periodic function of period $2L$

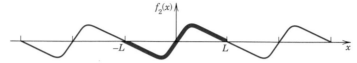

(c) $f(x)$ extended as an odd periodic function of period $2L$

Fig. 246. (a) Function $f(x)$ given on an interval $0 \leqq x \leqq L$

(b) Even extension to the full "range" (interval) $-L \leqq x \leqq L$ (heavy curve) and the periodic extension of period $2L$ to the x-axis

(c) Odd extension to $-L \leqq x \leqq L$ (heavy curve) and the periodic extension of period $2L$ to the x-axis of f in Fig. 246(b). If in a practical problem we think that using (6) is better, we get a Fourier sine series (5). This series represents the **odd periodic extention** f_2 of f in Fig. 246(c). Both extensions have period $2L$. This motivates the name **half-range expansions:** f is given (and of physical interest) only on half the range, half the interval of periodicity of length $2L$. Let us illustrate these ideas with an example that we shall also use in Chap. 11.

EXAMPLE 3   **"Triangle" and its half-range expansions**

Find the two half-range expansions of the function (Fig. 247)

$$f(x) = \begin{cases} \dfrac{2k}{L}\, x & \text{if} \quad 0 < x < \dfrac{L}{2} \\[2mm] \dfrac{2k}{L}\, (L - x) & \text{if} \quad \dfrac{L}{2} < x < L. \end{cases}$$

Solution. (a) **Even periodic extension.**   From (4) we obtain

$$a_0 = \frac{1}{L}\left[\frac{2k}{L} \int_0^{L/2} x\, dx + \frac{2k}{L}\int_{L/2}^{L} (L - x)\, dx \right] = \frac{k}{2},$$

$$a_n = \frac{2}{L}\left[\frac{2k}{L} \int_0^{L/2} x \cos \frac{n\pi}{L} x\, dx + \frac{2k}{L}\int_{L/2}^{L} (L - x) \cos \frac{n\pi}{L} x\, dx \right].$$

We consider a_n. For the first integral we obtain by integration by parts

$$\int_0^{L/2} x \cos \frac{n\pi}{L} x\, dx = \frac{Lx}{n\pi} \sin \frac{n\pi}{L} x \Big|_0^{L/2} - \frac{L}{n\pi}\int_0^{L/2} \sin \frac{n\pi}{L} x\, dx$$

$$= \frac{L^2}{2n\pi} \sin \frac{n\pi}{2} + \frac{L^2}{n^2\pi^2}\left(\cos \frac{n\pi}{2} - 1 \right).$$

Similarly, for the second integral we obtain

$$\int_{L/2}^{L} (L - x) \cos \frac{n\pi}{L} x\, dx = \frac{L}{n\pi} (L - x) \sin \frac{n\pi}{L} x \Big|_{L/2}^{L} + \frac{L}{n\pi}\int_{L/2}^{L} \sin \frac{n\pi}{L} x\, dx$$

$$= 0 - \frac{L}{n\pi}\left(L - \frac{L}{2} \right) \sin \frac{n\pi}{2} - \frac{L^2}{n^2\pi^2}\left(\cos n\pi - \cos \frac{n\pi}{2} \right).$$

We insert these two results into the formula for a_n. The sine terms cancel and so does a factor L^2. This gives

$$a_n = \frac{4k}{n^2\pi^2}\left(2 \cos \frac{n\pi}{2} - \cos n\pi - 1 \right).$$

Thus,

$$a_2 = -16k/2^2\pi^2, \qquad a_6 = -16k/6^2\pi^2, \qquad a_{10} = -16k/10^2\pi^2, \cdots,$$

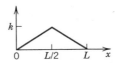

Fig. 247.   The given function in Example 3

(a) Even extension

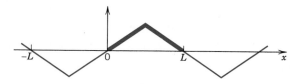

(b) Odd extension

Fig. 248. Periodic extensions of $f(x)$ in Example 3

and $a_n = 0$ if $n \neq 2, 6, 10, 14, \cdots$. Hence the first half-range expansion of $f(x)$ is

$$f(x) = \frac{k}{2} - \frac{16k}{\pi^2} \left(\frac{1}{2^2} \cos \frac{2\pi}{L} x + \frac{1}{6^2} \cos \frac{6\pi}{L} x + \cdots \right).$$

This Fourier cosine series represents the even periodic extension of the given function $f(x)$, of period $2L$, shown in Fig. 248a.

(b) Odd periodic extension. Similarly, from (6) we obtain

(7)
$$b_n = \frac{8k}{n^2 \pi^2} \sin \frac{n\pi}{2}.$$

Hence the other half-range expansion of $f(x)$ is

$$f(x) = \frac{8k}{\pi^2} \left(\frac{1}{1^2} \sin \frac{\pi}{L} x - \frac{1}{3^2} \sin \frac{3\pi}{L} x + \frac{1}{5^2} \sin \frac{5\pi}{L} x - + \cdots \right).$$

This series represents the odd periodic extension of $f(x)$, of period $2L$, shown in Fig. 248b.
  Basic applications of these results will be shown in Secs. 11.3 and 11.5. ◀

PROBLEM SET 10.4

Even and Odd Functions
Are the following functions odd, even, or neither odd nor even?
1. $|x^3|$, $x \cos nx$, $x^2 \cos nx$, $\cosh x$, $\sinh x$, $\sin x + \cos x$, $x|x|$
2. $x + x^2$, $|x|$, e^x, e^{x^2}, $\sin^2 x$, $x \sin x$, $\ln x$, $x \cos x$, $e^{-|x|}$

Are the following functions $f(x)$, which are assumed to be periodic, of period 2π, even, odd or neither even nor odd?

3. $f(x) = x^2$ $(0 < x < 2\pi)$

4. $f(x) = x^4$ $(0 < x < 2\pi)$

5. $f(x) = e^{-|x|}$ $(-\pi < x < \pi)$

6. $f(x) = |\sin 5x|$ $(-\pi < x < \pi)$

7. $f(x) = \begin{cases} 0 & \text{if } 2 < x < 2\pi - 2 \\ x & \text{if } -2 < x < 2 \end{cases}$

8. $f(x) = \begin{cases} \cos^2 x & \text{if } -\pi < x < 0 \\ \sin^2 x & \text{if } 0 < x < \pi \end{cases}$

9. $f(x) = x^3$ $(-\pi/2 < x < 3\pi/2)$

10. PROJECT. Even and Odd Functions. (a) Are the following expressions even or odd? Sums and products of even functions and of odd functions. Products of even times odd functions. Absolute values of odd functions. $f(x) + f(-x)$ and $f(x) - f(-x)$ for arbitrary $f(x)$.

(b) Write e^{kx}, $1/(1-x)$, $\sin(x+k)$, $\cosh(x+k)$ as sums of an even and an odd function.

(c) Find all functions that are both even and odd.

(d) Is $\cos^3 x$ even or odd? $\sin^3 x$? Find the Fourier series of these two functions. Do you recognize familiar identities?

Fourier Series of Even and Odd Functions

State whether the given function is even or odd. Find its Fourier series. Sketch the function and some partial sums. (Show the details of your work.)

11. $f(x) = \begin{cases} k & \text{if } -\pi/2 < x < \pi/2 \\ 0 & \text{if } \pi/2 < x < 3\pi/2 \end{cases}$

12. $f(x) = \begin{cases} -2x & \text{if } -\pi < x < 0 \\ 2x & \text{if } 0 < x < \pi \end{cases}$

13. $f(x) = \begin{cases} x & \text{if } -\pi/2 < x < \pi/2 \\ \pi - x & \text{if } \pi/2 < x < 3\pi/2 \end{cases}$

14. $f(x) = \begin{cases} x & \text{if } 0 < x < \pi \\ \pi - x & \text{if } \pi < x < 2\pi \end{cases}$

15. $f(x) = x^2/2 \quad (-\pi < x < \pi)$

16. $f(x) = 3x(\pi^2 - x^2) \quad (-\pi < x < \pi)$

Show that

17. $1 - \dfrac{1}{3} + \dfrac{1}{5} - \dfrac{1}{7} + - \cdots = \dfrac{\pi}{4}$ (Use Prob. 11.)

18. $1 + \dfrac{1}{4} + \dfrac{1}{9} + \dfrac{1}{16} + \dfrac{1}{25} + \cdots = \dfrac{\pi^2}{6}$ (Use Prob. 15.)

19. $1 - \dfrac{1}{4} + \dfrac{1}{9} - \dfrac{1}{16} + - \cdots = \dfrac{\pi^2}{12}$ (Use Prob. 15.)

Half-Range Expansions

Find the Fourier cosine series as well as the Fourier sine series. Sketch $f(x)$ and its two periodic extensions. (Show the details.)

20. $f(x) = 1 \quad (0 < x < L)$ **21.** $f(x) = x \quad (0 < x < L)$ **22.** $f(x) = x^2 \quad (0 < x < L)$

23. $f(x) = \pi - x \quad (0 < x < \pi)$ **24.** $f(x) = x^3 \quad (0 < x < L)$ **25.** $f(x) = e^x \quad (0 < x < L)$

10.5 Complex Fourier Series. *Optional*

In this optional section we show that the Fourier series

(1) $$f(x) = a_0 + \sum_{n=1}^{\infty} (a_n \cos nx + b_n \sin nx)$$

can be written in complex form, which sometimes simplifies calculations (see Example 1, below). This is done by the Euler formula (5), Sec. 2.3, with nx instead of x, that is,

(2) $$e^{inx} = \cos nx + i \sin nx,$$

(3) $$e^{-inx} = \cos nx - i \sin nx.$$

By addition of these two formulas (2) and (3) and division by 2 we get

(4)
$$\cos nx = \frac{1}{2}(e^{inx} + e^{-inx}).$$

Subtraction and division by $2i$ gives

(5)
$$\sin nx = \frac{1}{2i}(e^{inx} - e^{-inx}).$$

From this, using $1/i = -i$, we have in (1)

$$a_n \cos nx + b_n \sin nx = \frac{1}{2} a_n(e^{inx} + e^{-inx}) + \frac{1}{2i} b_n(e^{inx} - e^{-inx})$$

$$= \tfrac{1}{2}(a_n - ib_n)e^{inx} + \tfrac{1}{2}(a_n + ib_n)e^{-inx}.$$

We insert this into (1), writing $a_0 = c_0$, $a_n - ib_n = c_n$, and $a_n + ib_n = k_n$. Then (1) becomes

(6)
$$f(x) = c_0 + \sum_{n=1}^{\infty}(c_n e^{inx} + k_n e^{-inx}).$$

For the coefficients c_1, c_2, $\cdots$ and k_1, k_2, $\cdots$ we obtain from (2), (3) and the Euler formulas (6), Sec. 10.2,

(7)
$$c_n = \frac{1}{2}(a_n - ib_n) = \frac{1}{2\pi}\int_{-\pi}^{\pi} f(x)(\cos nx - i \sin nx)\,dx = \frac{1}{2\pi}\int_{-\pi}^{\pi} f(x)e^{-inx}\,dx$$

$$k_n = \frac{1}{2}(a_n + ib_n) = \frac{1}{2\pi}\int_{-\pi}^{\pi} f(x)(\cos nx + i \sin nx)\,dx = \frac{1}{2\pi}\int_{-\pi}^{\pi} f(x)e^{inx}\,dx.$$

Finally, we can combine the two formulas (7) into one by the trick of writing $k_n = c_{-n}$. Then (6), (7), together with (6a) in Sec. 10.2, give

(8)
$$f(x) = \sum_{n=-\infty}^{\infty} c_n e^{inx},$$

$$c_n = \frac{1}{2\pi}\int_{-\pi}^{\pi} f(x)e^{-inx}\,dx, \qquad n = 0, \pm 1, \pm 2, \cdots.$$

This is the so-called **complex form of the Fourier series** or, more briefly, the **complex Fourier series,** of $f(x)$. The c_n are called the **complex Fourier coefficients** of $f(x)$.

For a function of period $2L$ our reasoning gives the **complex Fourier series**

(9)
$$f(x) = \sum_{n=-\infty}^{\infty} c_n e^{in\pi x/L}, \qquad c_n = \frac{1}{2L}\int_{-L}^{L} f(x)e^{-in\pi x/L}\,dx.$$

EXAMPLE 1   **Complex Fourier series**

Find the complex Fourier series of $f(x) = e^x$ if $-\pi < x < \pi$ and $f(x + 2\pi) = f(x)$ and obtain from it the usual Fourier series.

Solution.   Since $\sin n\pi = 0$ for integer n, we have

$$e^{\pm in\pi} = \cos n\pi \pm i \sin n\pi = \cos n\pi = (-1)^n.$$

With this we obtain from (8) by integration

$$c_n = \frac{1}{2\pi} \int_{-\pi}^{\pi} e^x e^{-inx}\, dx = \frac{1}{2\pi} \frac{1}{1 - in} e^{x-inx}\Big|_{x=-\pi}^{\pi} = \frac{1}{2\pi} \frac{1}{1 - in}(e^\pi - e^{-\pi})(-1)^n.$$

On the right,

$$\frac{1}{1 - in} = \frac{1 + in}{(1 - in)(1 + in)} = \frac{1 + in}{1 + n^2} \qquad \text{and} \qquad e^\pi - e^{-\pi} = 2\sinh \pi.$$

Hence the complex Fourier series is

(10) $$e^x = \frac{\sinh \pi}{\pi} \sum_{n=-\infty}^{\infty} (-1)^n \frac{1 + in}{1 + n^2} e^{inx} \qquad (-\pi < x < \pi).$$

From this let us derive the real Fourier series. Using (2) and $i^2 = -1$ we have in (10)

$$(1 + in)e^{inx} = (1 + in)(\cos nx + i \sin nx) = (\cos nx - n \sin nx) + i(n \cos nx + \sin nx).$$

Now (10) also has a corresponding term with $-n$ instead of n. Since $\cos(-nx) = \cos nx$ and $\sin(-nx) = -\sin nx$, we obtain in this term

$$(1 - in)e^{-inx} = (1 - in)(\cos nx - i \sin nx) = (\cos nx - n \sin nx) - i(n \cos nx + \sin nx).$$

If we add these two expressions, the imaginary parts cancel. Hence their sum is

$$2(\cos nx - n \sin nx), \qquad\qquad n = 1, 2, \cdots.$$

For $n = 0$ we get 1 (not 2) because there is only one term. Hence the real Fourier series is

(11) $$e^x = \frac{2\sinh \pi}{\pi}\left[\frac{1}{2} - \frac{1}{1 + 1^2}(\cos x - \sin x) + \frac{1}{1 + 2^2}(\cos 2x - 2\sin 2x) - + \cdots\right]$$

where $-\pi < x < \pi$. ◀

PROBLEM SET 10.5

1. **(Calculus review)** Review complex numbers.

Complex Fourier Series. Find the complex Fourier series of the following functions. (Show the details of your work.)

2. $f(x) = -1$ if $-\pi < x < 0$, $f(x) = 1$ if $0 < x < \pi$
3. $f(x) = x$ $(-\pi < x < \pi)$
4. $f(x) = 0$ if $-\pi < x < 0$, $f(x) = 1$ if $0 < x < \pi$
5. $f(x) = x$ $(0 < x < 2\pi)$
6. $f(x) = x^2$ $(-\pi < x < \pi)$

7. **(Even and odd functions)** Show that the complex Fourier coefficients of an even function are real and those of an odd function are pure imaginary.

8. **(Conversion)** Convert the Fourier series in Prob. 5 to real form.

9. **(Fourier coefficients)** Show that $a_0 = c_0$, $a_n = c_n + c_{-n}$, $b_n = i(c_n - c_{-n})$, $n = 1, 2, \cdots$.

10. **PROJECT. Complex Fourier Coefficients.** It is very interesting that the c_n in (8) can be derived directly by a method similar to that for the a_n and b_n in Sec. 10.2. For this, multiply the series in (8) by e^{-imx} with fixed integer m and integrate termwise from $-\pi$ to π on both sides (allowed, for instance, in the case of uniform convergence), to get

$$\int_{-\pi}^{\pi} f(x)e^{-imx}\,dx = \sum_{n=-\infty}^{\infty} c_n \int_{-\pi}^{\pi} e^{i(n-m)x}\,dx.$$

Show that the integral on the right equals 2π when $n = m$ and 0 when $n \neq m$ [use (5)], so that you get the coefficient formula in (8).

10.6 Forced Oscillations

Fourier series have important applications in differential equations. We show this for a basic problem involving an ordinary differential equation. Numerous applications to partial differential equations will follow in Chap. 11. All this will justify Euler's and Fourier's idea of splitting up a periodic function in a series of (simpler) such functions, an idea whose enormous usefulness was far from obvious.

From Sec. 2.11 we know that forced oscillations of a body of mass m on a spring of modulus k are governed by the equation

(1)
$$my'' + cy' + ky = r(t),$$

where $y = y(t)$ is the displacement from rest, c the damping constant, and $r(t)$ the external force depending on time t. Figure 249 shows the model and Fig. 250 its electrical analog, an *RLC*-circuit governed by

(1*)
$$LI'' + RI' + \frac{1}{C}I = E'(t)$$
(Sec. 2.12).

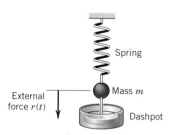

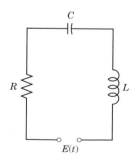

Fig. 249. Vibrating system under consideration

Fig. 250. Electrical analog of the system in Fig. 249 (*RLC*-circuit)

We consider (1). If $r(t)$ is a sine or cosine function and if there is damping ($c > 0$), then the steady-state solution is a harmonic oscillation with frequency equal to that of $r(t)$. However, if $r(t)$ is not a pure sine or cosine function but is any other periodic function, then the steady-state solution will be a superposition of harmonic oscillations with frequencies equal to that of $r(t)$ and integer multiples of the latter. And if one of these frequencies is close to the (practical) resonant frequency of the vibrating system (see Sec. 2.11), then the corresponding oscillation may be the dominant part of the response of the system to the external force. This is what the use of Fourier series will show us. Of course, this is quite surprising to an observer unfamiliar with Fourier series, which are highly important in the study of vibrating systems and resonance. Let us discuss the entire situation in terms of a typical example.

EXAMPLE 1   **Forced oscillations under a nonsinusoidal periodic driving force**

In (1), let $m = 1$ (gm), $c = 0.02$ (gm/sec), and $k = 25$ (gm/sec^2), so that (1) becomes

(2) $$y'' + 0.02y' + 25y = r(t)$$

where $r(t)$ is measured in gm $\cdot$ cm/sec^2. Let (Fig. 251)

$$r(t) = \begin{cases} t + \dfrac{\pi}{2} & \text{if} \quad -\pi < t < 0, \\[2mm] -t + \dfrac{\pi}{2} & \text{if} \quad 0 < t < \pi, \end{cases} \qquad r(t + 2\pi) = r(t).$$

Find the steady-state solution $y(t)$.

Solution.   We represent $r(t)$ by a Fourier series, finding

(3) $$r(t) = \frac{4}{\pi}\left(\cos t + \frac{1}{3^2}\cos 3t + \frac{1}{5^2}\cos 5t + \cdots\right)$$

(take $\pi/2$ minus the answer to Prob. 21 in Problem Set 10.4 with $L = \pi$ and $x = t$). Then we consider the differential equation

(4) $$y'' + 0.02y' + 25y = \frac{4}{n^2\pi}\cos nt \qquad (n = 1, 3, \cdots)$$

whose right side is a single term of the series (3). From Sec. 2.11 we know that the steady-state solution $y_n(t)$ of (4) is of the form

(5) $$y_n = A_n \cos nt + B_n \sin nt.$$

By substituting this into (4) we find that

(6) $$A_n = \frac{4(25 - n^2)}{n^2\pi D}, \qquad B_n = \frac{0.08}{n\pi D}, \qquad \text{where} \qquad D = (25 - n^2)^2 + (0.02n)^2.$$

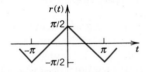

Fig. 251.   Force in Example 1

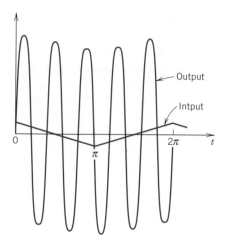

Fig. 252.   Input and steady-state output
in Example 1

Since the differential equation (2) is linear, we may expect the steady-state solution to be

$$(7) \qquad\qquad y = y_1 + y_3 + y_5 + \cdots$$

where y_n is given by (5) and (6). In fact, this follows readily by substituting (7) into (2) and using the Fourier series of $r(t)$, provided that termwise differentiation of (7) is permissible. (Readers already familiar with the notion of uniform convergence [Sec. 14.5] may prove that (7) may be differentiated term by term.)
  From (6) we find that the amplitude of (5) is

$$C_n = \sqrt{A_n{}^2 + B_n{}^2} = \frac{4}{n^2 \pi \sqrt{D}}.$$

Numerical values are

$$C_1 = 0.0530$$
$$C_3 = 0.0088$$
$$\boxed{C_5 = 0.5100}$$
$$C_7 = 0.0011$$
$$C_9 = 0.0003.$$

For $n = 5$ the quantity D is very small, the denominator of C_5 is small, and C_5 is so large that y_5 is the dominating term in (7). This implies that the steady-state motion is almost a harmonic oscillation whose frequency equals five times that of the exciting force (Fig. 252).   ◀

  The application of Fourier series to more general vibrating systems, heat conduction, and other problems follows in Chap. 11.

PROBLEM SET 10.6

1. **(Change of spring and damping)** What would happen to the amplitudes C_n in Example 1 (and, accordingly, to the form of the vibrations) if we changed the spring constant to the value 9? If we took a stiffer spring with $k = 49$? If we increased the damping?
2. **(Change of the input)** What would happen in Example 1 if we replaced $r(t)$ by its derivative (the square wave)? What is the ratio of the new C_n to the old?

General Solution

Find a general solution of the differential equation $y'' + \omega^2 y = r(t)$ with $r(t)$ as given. (Show the details of your work.)

3. $r(t) = \sin t,\quad \omega = 0.5, 0.7, 0.9, 1.1, 1.5, 2.0, 10.0$

4. $r(t) = \cos \alpha t + \cos \beta t \qquad (\omega^2 \neq \alpha^2, \beta^2)$

5. $r(t) = \displaystyle\sum_{n=1}^{N} b_n \sin nt,\qquad |\omega| \neq 1, 2, \cdots, N$

6. $r(t) = \sin t + \frac{1}{9} \sin 3t + \frac{1}{25} \sin 5t,\quad \omega = 0.5, 0.9, 1.1, 2, 2.9, 3.1, 4, 4.9, 5.1, 6, 8$

7. $r(t) = \begin{cases} t + \pi & \text{if} \quad -\pi < t < 0 \\ -t + \pi & \text{if} \quad\ \ 0 < t < \pi \end{cases}$ and $r(t + 2\pi) = r(t), |\omega| \neq 0, 1, 3, \cdots$

8. $r(t) = \dfrac{\pi}{4} |\cos t|$ if $-\pi < t < \pi$ and $r(t + 2\pi) = r(t), |\omega| \neq 0, 2, 4, \cdots$

 9. **(CAS Program)** Write a program for solving the differential equation just considered and for plotting solutions of corresponding initial value problems.

10. **(Sign of coefficients)** Some of the $A(\omega)$ in Prob. 3 are positive and some negative. Is this physically understandable?

Steady-State Oscillations

Find the steady-state oscillation of $y'' + cy' + y = r(t)$ with $c > 0$ and $r(t)$ as given. (Show the details of your work.)

11. $r(t) = \displaystyle\sum_{n=1}^{N} (a_n \cos nt + b_n \sin nt)$

12. $r(t) = \begin{cases} \pi t/4 & \text{if} \quad -\pi/2 < t < \pi/2 \\ \pi(\pi - t)/4 & \text{if} \quad\ \ \pi/2 < t < 3\pi/2 \end{cases}$ and $r(t + 2\pi) = r(t)$

13. $r(t) = \dfrac{t}{12} (\pi^2 - t^2)$ if $-\pi < t < \pi$ and $r(t + 2\pi) = r(t)$

14. **(RLC-circuit)** Find the steady-state current $I(t)$ in the RLC-circuit in Fig. 250, where $R = 100$ ohms, $L = 10$ henrys, $C = 10^{-2}$ farad,

$$E(t) = \begin{cases} 100(\pi t + t^2) & \text{if} \quad -\pi < t < 0 \\ 100(\pi t - t^2) & \text{if} \quad\ \ 0 < t < \pi \end{cases} \qquad \text{and} \qquad E(t + 2\pi) = E(t).$$

Sketch or plot the first four partial sums. Note that the coefficients of the Fourier series of the solution decrease rapidly.

15. **(RLC-circuit)** Perform the same task as in Prob. 14 with R, L, C as before and $E(t) = 200t(\pi^2 - t^2)$ volts if $-\pi < t < \pi$ and $E(t + 2\pi) = E(t)$.

10.7 Approximation by Trigonometric Polynomials

Fourier series play a prominent role in differential equations. Another field in which they have major applications is **approximation theory,** that is, the approximation of functions by simpler functions. In connection with Fourier series, the idea can be explained as follows.

Let $f(x)$ be a periodic function, of period 2π for simplicity, that can be represented by a Fourier series. Then the Nth partial sum of this series is an approximation to $f(x)$:

$$(1) \qquad f(x) \approx a_0 + \sum_{n=1}^{N} (a_n \cos nx + b_n \sin nx).$$

It is natural to ask whether (1) is the "best" approximation to f by a **trigonometric polynomial** *of degree N* (N fixed), that is, by a function of the form

$$(2) \qquad F(x) = A_0 + \sum_{n=1}^{N} (A_n \cos nx + B_n \sin nx),$$

where "best" means that the "error" of the approximation is minimum.

Of course, we must first define what we mean by the error E of such an approximation. We want to choose a definition that measures the goodness of agreement between f and F *on the whole interval* $-\pi \leqq x \leqq \pi$. Obviously, the maximum of $|f - F|$ is not suitable for that purpose: in Fig. 253 the function F is a good approximation to f, but $|f - F|$ is large near x_0. We choose

$$(3) \qquad \boxed{E = \int_{-\pi}^{\pi} (f - F)^2 \, dx.}$$

This is called the **total square error** of F relative to the function f on the interval $-\pi \leqq x \leqq \pi$. Clearly, $E \geqq 0$.

N being fixed, we want to determine the coefficients in (2) such that E is minimum. Since $(f - F)^2 = f^2 - 2fF + F^2$, we have

$$(4) \qquad E = \int_{-\pi}^{\pi} f^2 \, dx - 2 \int_{-\pi}^{\pi} fF \, dx + \int_{-\pi}^{\pi} F^2 \, dx.$$

We square (2), insert it into the last integral, and evaluate the occurring integrals. This gives integrals of $\cos^2 nx$ and $\sin^2 nx$ ($n \geqq 1$), which equal π, and integrals of $(\cos nx)(\sin mx)$, which are zero (just as in Sec. 10.2). Thus

$$\int_{-\pi}^{\pi} F^2 \, dx = \pi (2A_0{}^2 + A_1{}^2 + \cdots + A_N{}^2 + B_1{}^2 + \cdots + B_N{}^2).$$

We now insert (2) into the integral of fF in (4). This gives integrals of $f \cos nx$ as well as $f \sin nx$, just as in Euler's formulas, Sec. 10.2, for a_n and b_n (each multiplied by A_n or B_n). Hence

$$\int_{-\pi}^{\pi} fF \, dx = \pi (2A_0 a_0 + A_1 a_1 + \cdots + A_N a_N + B_1 b_1 + \cdots + B_N b_N).$$

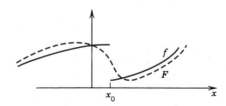

Fig. 253. Error of approximation

With these expressions, (4) becomes

$$E = \int_{-\pi}^{\pi} f^2 \, dx - 2\pi \left[2A_0 a_0 + \sum_{n=1}^{N} (A_n a_n + B_n b_n) \right]$$

(5)

$$+ \pi \left[2A_0^2 + \sum_{n=1}^{N} (A_n^2 + B_n^2) \right].$$

If we take $A_n = a_n$ and $B_n = b_n$ in (2), then in (5) the second line cancels half of the integral-free expression in the first line. Hence for this choice of the coefficients of F, the square error, call it E^*, is

(6)
$$E^* = \int_{-\pi}^{\pi} f^2 \, dx - \pi \left[2a_0^2 + \sum_{n=1}^{N} (a_n^2 + b_n^2) \right].$$

We finally subtract (6) from (5). Then the integrals drop out and we get terms $A_n^2 - 2A_n a_n + a_n^2 = (A_n - a_n)^2$ and similar terms $(B_n - b_n)^2$:

$$E - E^* = \pi \left\{ 2(A_0 - a_0)^2 + \sum_{n=1}^{N} [(A_n - a_n)^2 + (B_n - b_n)^2] \right\}.$$

Since the sum of squares of real numbers on the right cannot be negative,

$$E - E^* \geqq 0, \qquad \text{thus} \qquad E \geqq E^*,$$

and $E = E^*$ if and only if $A_0 = a_0, \cdots, B_N = b_N$. This proves the following fundamental minimum property of the partial sums of Fourier series.

THEOREM 1   **(Minimum square error)**

The total square error of F in (2) (with fixed N) relative to f on the interval $-\pi \leqq x \leqq \pi$ is minimum if and only if the coefficients of F in (2) are the Fourier coefficients of f. This minimum value E^* is given by (6).

From (6) we see that E^* cannot increase as N increases, but may decrease. Hence *with increasing N the partial sums of the Fourier series of f yield better and better approximations to f,* considered from the viewpoint of the square error.

Since $E^* \geqq 0$ and (6) holds for every N, we obtain from (6) the important **Bessel inequality**[11]

(7)
$$2a_0^2 + \sum_{n=1}^{\infty} (a_n^2 + b_n^2) \leqq \frac{1}{\pi} \int_{-\pi}^{\pi} f(x)^2 \, dx$$

for the Fourier coefficients of any function f for which the integral on the right exists.

It can be shown (see [C9] in Appendix 1) that for such a function f, **Parseval's theorem**

---

[11] See footnote 12 in Sec. 4.5.

holds, that is, formula (7) holds with the equality sign, so that it becomes **"Parseval's identity"**[12]

$$(8) \qquad 2a_0{}^2 + \sum_{n=1}^{\infty} (a_n{}^2 + b_n{}^2) = \frac{1}{\pi} \int_{-\pi}^{\pi} f(x)^2 \, dx.$$

EXAMPLE 1   **Square error for the sawtooth wave**

Compute the total square error of F with $N = 3$ relative to

$$f(x) = x + \pi \quad (-\pi < x < \pi) \qquad \text{(Fig. 245a, Sec. 10.4)}$$

on the interval $-\pi \le x \le \pi$.

Solution. $F(x) = \pi + 2 \sin x - \sin 2x + \frac{2}{3} \sin 3x$ by Example 2, Sec. 10.4. From this and (6),

$$E^* = \int_{-\pi}^{\pi} (x + \pi)^2 \, dx - \pi \left[2\pi^2 + 2^2 + 1^2 + (\tfrac{2}{3})^2 \right],$$

hence

$$E^* = \tfrac{8}{3}\pi^3 - \pi(2\pi^2 + \tfrac{49}{9}) \approx 3.567.$$

$F = S_3$ is shown in Fig. 245b, Sec. 10.4, and although $|f(x) - F(x)|$ is large at $x = \pm\pi$ (how large?), where f is discontinuous, F approximates f quite well on the whole interval.   ◀

This brings to an end our discussion of Fourier series, which has emphasized the practical aspects of these series, as needed in applications. In the last four sections of this chapter we show how ideas and techniques in Fourier series can be extended to *nonperiodic* functions.

PROBLEM SET 10.7

Minimum Square Error

In each case find the function $F(x)$ of the form (2) for which the total square error E on the interval $-\pi \le x \le \pi$ is minimum and compute this minimum value for $N = 1, 2, \cdots, 5$, where, for $-\pi < x < \pi$,

1. $f(x) = -1$ if $-\pi < x < 0$, $f(x) = 1$ if $0 < x < \pi$
2. $f(x) = |x|$ 3. $f(x) = x$
3. $f(x) = x^2$ 5. $f(x) = x^3$
6. $f(x) = x$ if $-\pi/2 < x < \pi/2$, $f(x) = \pi - x$ if $\pi/2 < x < 3\pi/2$
7. $f(x) = x$ if $-\pi/2 < x < \pi/2$, $f(x) = 0$ elsewhere in $-\pi < x < \pi$
8. $f(x) = x(\pi^2 - x^2)/12$

9. **(Monotonicity)** Show that the minimum square error (6) is a monotone decreasing function of N. How can you use this in practice? What is the smallest N in Prob. 1 for which $E^* \le 0.2$?

10. **CAS PROJECT. Square Error for Continuous and Discontinuous Functions. (a)** Why can you expect the decrease of the minimum square error to be more rapid for a continuous function than for a discontinuous one?

     (b) Illustrate the claim in (a) by more extensive computations for Probs. 4 and 5 and Example 1, say, for $N = 1, \cdots, 1000$.

---

[12]MARC ANTOINE PARSEVAL (1755—1836), French mathematician. A physical interpretation of the identity follows in the next section.

Applications of Parseval's Identity

Using Parseval's identity, prove the following. In Probs. 11−13 compute the first few partial sums to see that the convergence is rather rapid.

11. $1 + \dfrac{1}{2^4} + \dfrac{1}{3^4} + \dfrac{1}{4^4} + \cdots = \dfrac{\pi^4}{90}$ (Use Prob. 7 in Sec. 10.2.)

12. $1 + \dfrac{1}{9} + \dfrac{1}{25} + \cdots = \dfrac{\pi^2}{8}$ (Use Prob. 13 in Sec. 10.2.)

13. $1 + \dfrac{1}{3^4} + \dfrac{1}{5^4} + \dfrac{1}{7^4} + \cdots = \dfrac{\pi^4}{96}$ (Use Prob. 13 in Sec. 10.4.)

14. $\displaystyle\int_{-\pi}^{\pi} \cos^4 x \, dx = \dfrac{3\pi}{4}$ **15.** $\displaystyle\int_{-\pi}^{\pi} \cos^6 x \, dx = \dfrac{5\pi}{8}$

10.8 Fourier Integrals

Fourier series are powerful tools in treating various problems involving *periodic* functions. Section 10.6 contained a first illustration of this, and various further applications follow in Chap. 11. Since, of course, many practical problems involve **nonperiodic functions,** we ask what can be done to extend the method of Fourier series to such functions. This is our goal in this section. In Example 1 we begin with a special function $f_L(x)$ of period $2L$ and see what happens to its Fourier series if we let $L \to \infty$. Then we consider the Fourier series of an arbitrary function f_L of period $2L$ and again let $L \to \infty$. This will motivate and suggest the main result of this section, which is an integral representation given in Theorem 1 (below).

EXAMPLE 1   **Square wave**

Consider the periodic square wave $f_L(x)$ of period $2L > 2$ given by

$$
f_L(x) = \begin{cases} 0 & \text{if} \quad -L < x < -1 \\ 1 & \text{if} \quad -1 < x < \ \ 1 \\ 0 & \text{if} \quad \ \ 1 < x < \ \ L. \end{cases}
$$

The left part of Fig. 254 on the next page shows this function for $2L = 4, 8, 16$ as well as the nonperiodic function $f(x)$, which we obtain from f_L if we let $L \to \infty$,

$$
f(x) = \lim_{L \to \infty} f_L(x) = \begin{cases} 1 & \text{if } -1 < x < 1 \\ 0 & \text{otherwise.} \end{cases}
$$

We now explore what happens to the Fourier coefficients of f_L as L increases. Since f_L is even, $b_n = 0$ for all n. For a_n the Euler formulas (2), Sec. 10.3, give

$$
a_0 = \frac{1}{2L} \int_{-1}^{1} dx = \frac{1}{L}, \qquad a_n = \frac{1}{L} \int_{-1}^{1} \cos \frac{n\pi x}{L} \, dx = \frac{2}{L} \int_{0}^{1} \cos \frac{n\pi x}{L} \, dx = \frac{2}{L} \frac{\sin (n\pi/L)}{n\pi/L} .
$$

This sequence of Fourier coefficients is called the **amplitude spectrum** of f_L because $|a_n|$ is the maximum amplitude of the wave $a_n \cos (n\pi x/L)$. Figure 254 shows this spectrum for the periods $2L = 4, 8, 16$. We see that for increasing L the amplitudes become more and more dense on the positive w_n-axis, where $w_n = n\pi/L$. Indeed, for $2L = 4, 8, 16$ we have 1, 3, 7 amplitudes per "half-wave" of the function $(2 \sin w_n)/(Lw_n)$ (dashed in the figure). Hence for $2L = 2^k$ we have $2^{k-1} - 1$ amplitudes per half-wave, so that these amplitudes will eventually be everywhere dense on the positive w_n-axis (and will decrease to zero). ◀

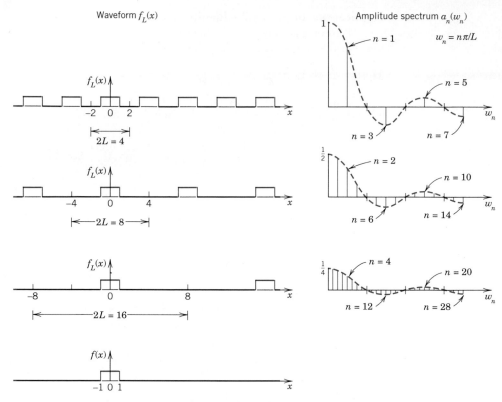

Fig. 254. Waveforms and amplitude spectra in Example 1

From Fourier Series to the Fourier Integral

We now consider any periodic function $f_L(x)$ of period $2L$ that can be represented by a Fourier series

$$f_L(x) = a_0 + \sum_{n=1}^{\infty} (a_n \cos w_n x + b_n \sin w_n x), \qquad w_n = \frac{n\pi}{L},$$

and find out what happens if we let $L \to \infty$. Together with our example, the present calculation will suggest that we should expect an integral (instead of a series) involving $\cos wx$ and $\sin wx$ with w no longer restricted to integer multiples $w = w_n = n\pi/L$ of π/L but taking *all* values. We shall also see what form such an integral might have.

If we insert a_n and b_n from the Euler formulas (2), Sec. 10.3, and denote the variable of integration by v, the Fourier series of $f_L(x)$ becomes

$$f_L(x) = \frac{1}{2L} \int_{-L}^{L} f_L(v)\, dv + \frac{1}{L} \sum_{n=1}^{\infty} \left[\cos w_n x \int_{-L}^{L} f_L(v) \cos w_n v\, dv \right.$$

$$\left. + \sin w_n x \int_{-L}^{L} f_L(v) \sin w_n v\, dv \right].$$

We now set

$$\Delta w = w_{n+1} - w_n = \frac{(n+1)\pi}{L} - \frac{n\pi}{L} = \frac{\pi}{L}.$$

Then $1/L = \Delta w/\pi$, and we may write the Fourier series in the form

(1)
$$f_L(x) = \frac{1}{2L} \int_{-L}^{L} f_L(v)\, dv + \frac{1}{\pi} \sum_{n=1}^{\infty} \left[(\cos w_n x)\, \Delta w \int_{-L}^{L} f_L(v) \cos w_n v\, dv \right.$$

$$\left. + (\sin w_n x)\, \Delta w \int_{-L}^{L} f_L(v) \sin w_n v\, dv \right].$$

This representation is valid for any fixed L, arbitrarily large, but finite.

We now let $L \to \infty$ and assume that the resulting nonperiodic function

$$f(x) = \lim_{L \to \infty} f_L(x)$$

is **absolutely integrable** on the x-axis; that is, the following (finite!) limits exist:

(2)
$$\lim_{a \to -\infty} \int_{a}^{0} |f(x)|\, dx + \lim_{b \to \infty} \int_{0}^{b} |f(x)|\, dx \qquad \left(\text{written } \int_{-\infty}^{\infty} |f(x)|\, dx \right).$$

Then $1/L \to 0$, and the value of the first term on the right side of (1) approaches zero. Also $\Delta w = \pi/L \to 0$ and it seems **plausible** that the infinite series in (1) becomes an integral from 0 to ∞, which represents $f(x)$, namely,

(3)
$$f(x) = \frac{1}{\pi} \int_{0}^{\infty} \left[\cos wx \int_{-\infty}^{\infty} f(v) \cos wv\, dv + \sin wx \int_{-\infty}^{\infty} f(v) \sin wv\, dv \right] dw.$$

If we introduce the notations

(4)
$$A(w) = \frac{1}{\pi} \int_{-\infty}^{\infty} f(v) \cos wv\, dv, \qquad B(w) = \frac{1}{\pi} \int_{-\infty}^{\infty} f(v) \sin wv\, dv,$$

we can write this in the form

(5)
$$f(x) = \int_{0}^{\infty} [A(w) \cos wx + B(w) \sin wx]\, dw.$$

This is called a representation of $f(x)$ by a **Fourier integral.**

It is clear that our naive approach merely **suggests** the representation (5), but by no means establishes it; in fact, the limit of the series in (1) as Δw approaches zero is not the definition of the integral (3). Sufficient conditions for the validity of (5) are as follows.

THEOREM 1   **(Fourier integral)**

If $f(x)$ is piecewise continuous (see Sec. 5.1) in every finite interval and has a right-hand derivative and a left-hand derivative at every point (see Sec. 10.2) and if the integral (2) exists, then $f(x)$ can be represented by a Fourier integral (5). At a point where $f(x)$ is discontinuous the value of the Fourier integral equals the average of the left- and right-hand limits of $f(x)$ at that point (see Sec. 10.2). (Proof is in Ref. [C9]; see Appendix 1.)

Applications of the Fourier Integral

The main use of the Fourier integral is in solving differential equations, as we shall see in Sec. 11.6. However, we can also use the Fourier integral in integration and in discussing functions defined by integrals, as the next examples illustrate.

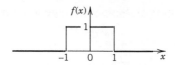

Fig. 255. Example 2

EXAMPLE 2 **Single pulse, sine integral**

Find the Fourier integral representation of the function (Fig. 255)

$$f(x) = \begin{cases} 1 & \text{if} & |x| < 1, \\ 0 & \text{if} & |x| > 1. \end{cases}$$

Solution. From (4) we obtain

$$A(w) = \frac{1}{\pi} \int_{-\infty}^{\infty} f(v) \cos wv \, dv = \frac{1}{\pi} \int_{-1}^{1} \cos wv \, dv = \frac{\sin wv}{\pi w} \Big|_{-1}^{1} = \frac{2 \sin w}{\pi w}, \quad B(w) = \frac{1}{\pi} \int_{-1}^{1} \sin wv \, dv = 0,$$

and (5) gives the answer

(6)
$$f(x) = \frac{2}{\pi} \int_{0}^{\infty} \frac{\cos wx \sin w}{w} \, dw.$$

The average of the left- and right-hand limits of $f(x)$ at $x = 1$ is equal to $(1 + 0)/2$, that is, $1/2$.

Furthermore, from (6) and Theorem 1 we obtain

(7*)
$$\int_{0}^{\infty} \frac{\cos wx \sin w}{w} \, dw = \begin{cases} \pi/2 & \text{if} & 0 \leqq x < 1, \\ \pi/4 & \text{if} & x = 1, \\ 0 & \text{if} & x > 1. \end{cases}$$

We mention that this integral is called **Dirichlet's discontinuous factor.**[13] Let us consider the case $x = 0$, which is of particular interest. If $x = 0$, then

(7)
$$\int_{0}^{\infty} \frac{\sin w}{w} \, dw = \frac{\pi}{2}.$$

We see that this integral is the limit of the so-called **sine integral**

(8)
$$\text{Si}(u) = \int_{0}^{u} \frac{\sin w}{w} \, dw$$

as $u \to \infty$. The graph of $\text{Si}(u)$ is shown in Fig. 256.

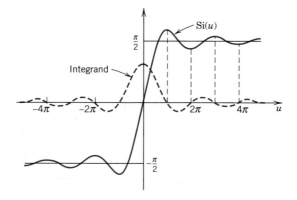

Fig. 256. Sine integral $\text{Si}(u)$

[13] See footnote 16 in Sec. 9.8.

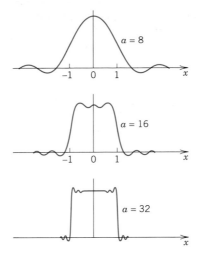

Fig. 257. The integral (9) for $a = 8$, 16, and 32

In the case of a Fourier series the graphs of the partial sums are approximation curves of the curve of the periodic function represented by the series. Similarly, in the case of the Fourier integral (5), approximations are obtained by replacing ∞ by numbers a. Hence the integral

$$(9) \qquad \int_0^a \frac{\cos wx \sin w}{w}\, dw$$

approximates the integral in (6) and therefore $f(x)$; see Fig. 257.

Figure 257 shows oscillations near the points of discontinuity of $f(x)$. We might expect that these oscillations disappear as a approaches infinity. But this is not true; with increasing a, they are shifted closer to the points $x = \pm 1$. This unexpected behavior, which also occurs in connection with Fourier series, is known as the **Gibbs phenomenon.**[14] It can be explained by representing (9) in terms of the sine integral as follows. Using (11) in Appendix A3.1, we have

$$\frac{2}{\pi} \int_0^a \frac{\cos wx \sin w}{w}\, dw = \frac{1}{\pi} \int_0^a \frac{\sin (w + wx)}{w}\, dw + \frac{1}{\pi} \int_0^a \frac{\sin (w - wx)}{w}\, dw.$$

In the first integral on the right we set $w + wx = t$. Then $dw/w = dt/t$, and $0 \leqq w \leqq a$ corresponds to $0 \leqq t \leqq (x + 1)a$. In the last integral we set $w - wx = -t$. Then $dw/w = dt/t$, and $0 \leqq w \leqq a$ corresponds to $0 \leqq t \leqq (x - 1)a$. Since $\sin (-t) = -\sin t$, we thus obtain

$$\frac{2}{\pi} \int_0^a \frac{\cos wx \sin w}{w}\, dw = \frac{1}{\pi} \int_0^{(x+1)a} \frac{\sin t}{t}\, dt - \frac{1}{\pi} \int_0^{(x-1)a} \frac{\sin t}{t}\, dt.$$

From this and (8) we see that our integral equals

$$\frac{1}{\pi} \operatorname{Si}(a[x + 1]) - \frac{1}{\pi} \operatorname{Si}(a[x - 1]),$$

and the oscillations in Fig. 257 result from those in Fig. 256. The increase of a amounts to a transformation of the scale on the axis and causes the shift of the oscillations.   ◀

Fourier Cosine and Sine Integrals

For an even or odd function the Fourier integral becomes simpler. Just as in the case of Fourier series (Sec. 10.4), this is of practical interest in saving work and avoiding errors. The simplifications follow immediately from our formulas just obtained.

---

[14]See Problem Set 10.3.

Indeed, if $f(x)$ is an even function, then $B(w) = 0$ in (4) and

(10)
$$A(w) = \frac{2}{\pi} \int_0^\infty f(v) \cos wv \, dv.$$

The Fourier integral (5) then reduces to the **Fourier cosine integral**

(11)
$$f(x) = \int_0^\infty A(w) \cos wx \, dw \qquad\qquad (f \text{ even}).$$

Similarly, if $f(x)$ is odd, then in (4) we have $A(w) = 0$ and

(12)
$$B(w) = \frac{2}{\pi} \int_0^\infty f(v) \sin wv \, dv.$$

The Fourier integral (5) then reduces to the **Fourier sine integral**

(13)
$$f(x) = \int_0^\infty B(w) \sin wx \, dw \qquad\qquad (f \text{ odd}).$$

Evaluation of Integrals

Fourier integral representations may also be used for evaluating integrals. We illustrate this method with a typical example.

EXAMPLE 3   **Laplace integrals**

Find the Fourier cosine and sine integrals of

$$f(x) = e^{-kx} \qquad\qquad (x > 0, k > 0).$$

Solution.   **(a)** From (10) we have

$$A(w) = \frac{2}{\pi} \int_0^\infty e^{-kv} \cos wv \, dv.$$

Now, by integration by parts,

$$\int e^{-kv} \cos wv \, dv = -\frac{k}{k^2 + w^2} e^{-kv} \left(-\frac{w}{k} \sin wv + \cos wv \right).$$

If $v = 0$, the expression on the right equals $-k/(k^2 + w^2)$; if v approaches infinity, it approaches zero because of the exponential factor. Thus

(14)
$$A(w) = \frac{2k/\pi}{k^2 + w^2}.$$

By substituting this into (11) we thus obtain the Fourier cosine integral representation

$$f(x) = e^{-kx} = \frac{2k}{\pi} \int_0^\infty \frac{\cos wx}{k^2 + w^2} \, dw \qquad\qquad (x > 0, k > 0).$$

From this representation we see that

(15)
$$\int_0^\infty \frac{\cos wx}{k^2 + w^2} \, dw = \frac{\pi}{2k} e^{-kx} \qquad\qquad (x > 0, k > 0).$$

(b) Similarly, from (12) we have

$$B(w) = \frac{2}{\pi} \int_0^\infty e^{-kv} \sin wv \, dv.$$

By integration by parts,

$$\int e^{-kv} \sin wv \, dv = -\frac{w}{k^2 + w^2} e^{-kv} \left(\frac{k}{w} \sin wv + \cos wv \right).$$

This equals $-w/(k^2 + w^2)$ if $v = 0$, and approaches 0 as $v \to \infty$. Thus

(16)
$$B(w) = \frac{2w/\pi}{k^2 + w^2}.$$

From (13) we thus obtain the Fourier sine integral representation

$$f(x) = e^{-kx} = \frac{2}{\pi} \int_0^\infty \frac{w \sin wx}{k^2 + w^2} \, dw.$$

From this we see that

(17)
$$\int_0^\infty \frac{w \sin wx}{k^2 + w^2} \, dw = \frac{\pi}{2} e^{-kx} \qquad\qquad (x > 0, k > 0).$$

The integrals (15) and (17) are called the **Laplace integrals.** ◀

PROBLEM SET 10.8

Evaluation of Integrals

Using (5), (11), or (13), show that the given integrals represent the indicated functions. (Can you see that the integral tells you which formula to use? Show the details of your work.)

1. $\displaystyle\int_0^\infty \frac{\cos xw + w \sin xw}{1 + w^2} \, dw = \begin{cases} 0 & \text{if } x < 0 \\ \pi/2 & \text{if } x = 0 \\ \pi e^{-x} & \text{if } x > 0 \end{cases}$

2. $\displaystyle\int_0^\infty \frac{\sin w \cos xw}{w} \, dw = \begin{cases} \pi/2 & \text{if } 0 \leqq x < 1 \\ \pi/4 & \text{if } x = 1 \\ 0 & \text{if } x > 1 \end{cases}$

3. $\displaystyle\int_0^\infty \frac{1 - \cos \pi w}{w} \sin xw \, dw = \begin{cases} \pi/2 & \text{if } 0 < x < \pi \\ 0 & \text{if } x > \pi \end{cases}$

4. $\displaystyle\int_0^\infty \frac{\cos (\pi w/2) \cos xw}{1 - w^2} \, dw = \begin{cases} (\pi/2) \cos x & \text{if } |x| < \pi/2 \\ 0 & \text{if } |x| > \pi/2 \end{cases}$

5. $\displaystyle\int_0^\infty \frac{\cos xw}{1 + w^2} \, dw = \frac{\pi}{2} e^{-x} \;\; \text{if } x > 0$　　　　**6.** $\displaystyle\int_0^\infty \frac{w^3 \sin xw}{w^4 + 4} \, dw = \frac{\pi}{2} e^{-x} \cos x \;\; \text{if } x > 0$

Fourier Cosine Integral Representation

Represent the following functions $f(x)$ in the form (11).

7. $f(x) = \begin{cases} 1 & \text{if} \quad 0 < x < 1 \\ 0 & \text{if} \quad x > 1 \end{cases}$

8. $f(x) = \begin{cases} x^2 & \text{if} \quad 0 < x < 1 \\ 0 & \text{if} \quad x > 1 \end{cases}$

9. $f(x) = \begin{cases} x & \text{if} \quad 0 < x < a \\ 0 & \text{if} \quad x > a \end{cases}$

10. $f(x) = \begin{cases} a^2 - x^2 & \text{if} \quad 0 < x < a \\ 0 & \text{if} \quad x > a \end{cases}$

11. $f(x) = 1/(1 + x^2)$ $[x > 0,\ \text{see (15)}]$

12. $f(x) = e^{-x} + e^{-2x}$ $(x > 0)$

Fourier Sine Integral Representation

Represent the following functions $f(x)$ in the form (13).

13. $f(x) = \begin{cases} 1 & \text{if} \quad 0 < x < a \\ 0 & \text{if} \quad\quad x > a \end{cases}$

14. $f(x) = \begin{cases} x & \text{if} \quad 0 < x < a \\ 0 & \text{if} \quad\quad x > a \end{cases}$

15. $f(x) = \begin{cases} \sin x & \text{if } 0 < x < \pi \\ 0 & \text{if} \quad\quad x > \pi \end{cases}$

16. $f(x) = \begin{cases} \pi - x & \text{if} \quad 0 < x < \pi \\ 0 & \text{if} \quad\quad x > \pi \end{cases}$

17. $f(x) = \begin{cases} e^x & \text{if} \quad 0 < x < 1 \\ 0 & \text{if} \quad\quad x > 1 \end{cases}$

18. $f(x) = \begin{cases} e^{-x} & \text{if} \quad 0 < x < 1 \\ 0 & \text{if} \quad\quad x > 1 \end{cases}$

 19. **(CAS. Sine integral)** Plot $\text{Si}(u)$ for positive u. Does the sequence of the maximum and minimum values make the impression that it converges and has the limit $\pi/2$? Investigate the Gibbs phenomenon graphically.

20. **PROJECT. Properties of Fourier Integrals. (a) Fourier Cosine Integral.** Show that (11) implies

$$(a1) \quad f(ax) = \frac{1}{a} \int_0^\infty A\left(\frac{w}{a}\right) \cos xw\, dw \quad (a > 0)$$

$$(a2) \quad xf(x) = \int_0^\infty B^*(w) \sin xw\, dw, \quad B^* = -\frac{dA}{dw}, \quad A \text{ as in (10)}$$

$$(a3) \quad x^2 f(x) = \int_0^\infty A^*(w) \cos xw\, dw, \quad A^* = -\frac{d^2A}{dw^2}.$$

(b) Solve Prob. 8 by applying (a3) to the result of Prob. 7.

(c) Verify (a2) for $f(x) = 1$ if $0 < x < a$ and $f(x) = 0$ if $x > a$.

(d) Fourier Sine Integral. Find formulas for the Fourier sine integral similar to those in (a).

10.9 Fourier Cosine and Sine Transforms

An **integral transform** is a transformation that produces from given functions new functions that depend on a different variable and appear in the form of an integral. These transformations are of interest mainly as tools in solving ordinary differential equations, partial differential equations, and integral equations, and they often also help in handling and applying special functions. The **Laplace transform** (Chap. 5) is of this kind and is by far the most important integral transform in engineering. From the viewpoint of applications, the next in order of importance are perhaps the **Fourier transforms,** although these are somewhat more difficult to handle than the Laplace transform. We shall see that they can be obtained from the Fourier integral representations in Sec. 10.8. In this section we consider two of them, called the *Fourier cosine* and *sine transforms,* which are real, and in the next section a third one, which is complex.

Fourier Cosine Transforms

For an **even** function $f(x)$, the Fourier integral is the Fourier cosine integral

(1) (a) $\quad f(x) = \int_0^\infty A(w) \cos wx \, dw,$ where (b) $\quad A(w) = \dfrac{2}{\pi} \int_0^\infty f(v) \cos wv \, dv$

[see (10), (11), Sec. 10.8]. We now set $A(w) = \sqrt{2/\pi}\, \hat{f}_c(w)$, where c suggests "cosine." Then from (1b), writing $v = x$, we have

(2)
$$\hat{f}_c(w) = \sqrt{\dfrac{2}{\pi}} \int_0^\infty f(x) \cos wx \, dx$$

and from (1a),

(3)
$$f(x) = \sqrt{\dfrac{2}{\pi}} \int_0^\infty \hat{f}_c(w) \cos wx \, dw.$$

Attention! In (2) we integrate with respect to x and in (3) with respect to w. Formula (2) gives from $f(x)$ a new function $\hat{f}_c(w)$, called the **Fourier cosine transform** of $f(x)$. Formula (3) gives us back $f(x)$ from $\hat{f}_c(w)$, and we therefore call $f(x)$ the **inverse Fourier cosine transform** of $\hat{f}_c(w)$.

The process of obtaining the transform $\hat{f}_c$ from a given f is also called the **Fourier cosine transform** or the *Fourier cosine transform method.*

Fourier Sine Transforms

Similarly, for an **odd** function $f(x)$, the Fourier integral is the Fourier sine integral [see (12), (13), Sec. 10.8]

(4) (a) $\quad f(x) = \int_0^\infty B(w) \sin wx \, dw,$ where (b) $\quad B(w) = \dfrac{2}{\pi} \int_0^\infty f(v) \sin wv \, dv.$

We now set $B(w) = \sqrt{2/\pi}\, \hat{f}_s(w)$, where s suggests "sine." Then from (4b), writing $v = x$, we have

(5)
$$\hat{f}_s(w) = \sqrt{\dfrac{2}{\pi}} \int_0^\infty f(x) \sin wx \, dx.$$

This is called the **Fourier sine transform** of $f(x)$. Similarly, from (4a) we have

(6)
$$f(x) = \sqrt{\dfrac{2}{\pi}} \int_0^\infty \hat{f}_s(w) \sin wx \, dw.$$

This is called the **inverse Fourier sine transform** of $\hat{f}_s(w)$. The process of obtaining $\hat{f}_s(w)$ from $f(x)$ is also called the **Fourier sine transform** or the *Fourier sine transform method.*

Other notations are

$$\mathscr{F}_c(f) = \hat{f}_c, \qquad \mathscr{F}_s(f) = \hat{f}_s$$

and $\mathscr{F}_c^{-1}$ and $\mathscr{F}_s^{-1}$ for the inverses of $\mathscr{F}_c$ and $\mathscr{F}_s$, respectively.

EXAMPLE 1   **Fourier cosine and Fourier sine transforms**

Find the Fourier cosine and sine transforms of the function

$$f(x) = \begin{cases} k & \text{if } 0 < x < a \\ 0 & \text{if } x > a. \end{cases}$$

Solution.   From the definitions (2) and (5) we obtain by integration

$$\hat{f}_c(w) = \sqrt{\frac{2}{\pi}}\, k \int_0^a \cos wx\, dx = \sqrt{\frac{2}{\pi}}\, k \left(\frac{\sin aw}{w} \right)$$

$$\hat{f}_s(w) = \sqrt{\frac{2}{\pi}}\, k \int_0^a \sin wx\, dx = \sqrt{\frac{2}{\pi}}\, k \left(\frac{1 - \cos aw}{w} \right).$$

This agrees with formulas 1 in the first two tables in Sec. 10.11 (where $k = 1$).

Note that for $f(x) = k = const$ $(0 < x < \infty)$, these transforms do not exist. (Why?)   ◀

EXAMPLE 2   **Fourier cosine transform of the exponential function**

Find $\mathscr{F}_c(e^{-x})$.

Solution.   By integration by parts and recursion,

$$\mathscr{F}_c(e^{-x}) = \sqrt{\frac{2}{\pi}} \int_0^{\infty} e^{-x} \cos wx\, dx = \sqrt{\frac{2}{\pi}}\, \frac{e^{-x}}{1 + w^2} (-\cos wx + w \sin wx) \Big|_0^{\infty} = \frac{\sqrt{2/\pi}}{1 + w^2}.$$

This agrees with formula 3 in Table I, Sec. 10.11, with $a = 1$.   ◀

What have we done in order to introduce the two integral transforms under consideration? Actually not much: We have changed the notations A and B to get a "symmetric" distribution of the constant $2/\pi$ in the original formulas (10)–(13), Sec. 10.8. This redistribution is a standard convenience, but is not essential. One could do without it.

What have we gained? We show next that these transforms have operational properties that permit them to convert differentiations into algebraic operations (just as the Laplace transform does). This is the key to their application in solving differential equations.

Linearity, Transforms of Derivatives

If $f(x)$ is absolutely integrable (see Sec. 10.8) on the positive x-axis and piecewise continuous (see Sec. 5.1) on every finite interval, then the Fourier cosine and sine transforms of f exist.

Furthermore, for a function $af(x) + bg(x)$ we have from (2)

$$\mathscr{F}_c(af + bg) = \sqrt{\frac{2}{\pi}} \int_0^{\infty} [af(x) + bg(x)] \cos wx\, dx$$

$$= a \sqrt{\frac{2}{\pi}} \int_0^{\infty} f(x) \cos wx\, dx + b \sqrt{\frac{2}{\pi}} \int_0^{\infty} g(x) \cos wx\, dx.$$

The right side is $a\mathscr{F}_c(f) + b\mathscr{F}_c(g)$. Similarly for $\mathscr{F}_s$, by (5). This shows that the Fourier cosine and sine transforms are **linear operations,**

$$\text{(a)} \qquad \mathscr{F}_c(af + bg) = a\mathscr{F}_c(f) + b\mathscr{F}_c(g),$$

(7)

$$\text{(b)} \qquad \mathscr{F}_s(af + bg) = a\mathscr{F}_s(f) + b\mathscr{F}_s(g).$$

THEOREM 1　**(Cosine and sine transforms of derivatives)**

Let $f(x)$ be continuous and absolutely integrable on the x-axis, let $f'(x)$ be piecewise continuous on each finite interval, and let $f(x) \to 0$ as $x \to \infty$. Then

$$\text{(a)} \qquad \mathscr{F}_c\{f'(x)\} = w\mathscr{F}_s\{f(x)\} - \sqrt{\frac{2}{\pi}}\, f(0),$$

(8)

$$\text{(b)} \qquad \mathscr{F}_s\{f'(x)\} = -w\mathscr{F}_c\{f(x)\}.$$

PROOF.　This follows from the definitions by integration by parts, namely,

$$\mathscr{F}_c\{f'(x)\} = \sqrt{\frac{2}{\pi}} \int_0^\infty f'(x) \cos wx\, dx$$

$$= \sqrt{\frac{2}{\pi}} \left[f(x) \cos wx \,\bigg|_0^\infty + w \int_0^\infty f(x) \sin wx\, dx \right]$$

$$= -\sqrt{\frac{2}{\pi}}\, f(0) + w\mathscr{F}_s\{f(x)\};$$

similarly,

$$\mathscr{F}_s\{f'(x)\} = \sqrt{\frac{2}{\pi}} \int_0^\infty f'(x) \sin wx\, dx$$

$$= \sqrt{\frac{2}{\pi}} \left[f(x) \sin wx \,\bigg|_0^\infty - w \int_0^\infty f(x) \cos wx\, dx \right]$$

$$= 0 - w\mathscr{F}_c\{f(x)\}. \qquad \blacktriangleleft$$

Formula (8a) with f' instead of f gives

$$\mathscr{F}_c\{f''(x)\} = w\mathscr{F}_s\{f'(x)\} - \sqrt{\frac{2}{\pi}}\, f'(0);$$

hence by (8b),

(9a)

$$\mathscr{F}_c\{f''(x)\} = -w^2 \mathscr{F}_c\{f(x)\} - \sqrt{\frac{2}{\pi}}\, f'(0).$$

Similarly,

(9b)

$$\mathscr{F}_s\{f''(x)\} = -w^2 \mathscr{F}_s\{f(x)\} + \sqrt{\frac{2}{\pi}}\, wf(0).$$

An application of (9) to differential equations will be given in Sec. 11.6. For the time being, we show how (9) can be used to derive transforms.

EXAMPLE 3 **An application of the operational formula (9)**

Find the Fourier cosine transform of $f(x) = e^{-ax}$, where $a > 0$.

Solution. By differentiation, $(e^{-ax})'' = a^2 e^{-ax}$; thus $a^2 f(x) = f''(x)$. From this and (9a),

$$a^2 \mathcal{F}_c(f) = \mathcal{F}_c(f'') = -w^2 \mathcal{F}_c(f) - \sqrt{\frac{2}{\pi}} f'(0) = -w^2 \mathcal{F}_c(f) + a\sqrt{\frac{2}{\pi}} \,.$$

Hence $(a^2 + w^2)\mathcal{F}_c(f) = a\sqrt{2/\pi}$. The answer is (see Table I, Sec. 10.11)

$$\mathcal{F}_c(e^{-ax}) = \sqrt{\frac{2}{\pi}} \left(\frac{a}{a^2 + w^2} \right) \qquad (a > 0). \qquad \blacktriangleleft$$

Tables of Fourier cosine and sine transforms are included in Sec. 10.11. For more extensive tables, see Ref. [C3] in Appendix 1.

PROBLEM SET 10.9

Fourier Cosine Transform

1. Find the cosine transform $\hat{f}_c(w)$ of $f(x) = 1$ if $0 < x < 1$, $f(x) = -1$ if $1 < x < 2$, $f(x) = 0$ if $x > 2$.
2. Obtain $f(x)$ in Prob. 1 from the answer to Prob. 1. (Use Prob. 2 in Problem Set 10.8.)
3. Find $\hat{f}_c(w)$ for $f(x) = x$ if $0 < x < a$, $f(x) = 0$ if $x > a$.
4. Derive formula 3 in Table I, Sec. 10.11, by integration.
5. Obtain $\mathcal{F}_c(1/(1 + x^2))$. (Use Prob. 5 in Problem Set 10.8.)
6. Obtain the inverse cosine transform of e^{-x}.
7. Find $\hat{f}_c(w)$ for $f(x) = x^2$ if $0 < x < 1$, $f(x) = 0$ if $x > 1$.
8. Find the cosine transform of $(\cos \frac{1}{2}\pi x)/(1 - x^2)$. (Use Prob. 4 in Problem Set 10.8.)
9. Obtain formula 10 in Table I, Sec. 10.11, with $a = 1$ from Example 2 in Sec. 10.8.
10. **(Nonexistence)** Give reasons why $f(x) = 1$ has neither a Fourier cosine transform nor a Fourier sine transform.

Fourier Sine Transform

11. Find $\mathcal{F}_s(e^{-ax})$, $a > 0$, by integration.
12. Obtain the answer to Prob. 11 from (9b).
13. Find the Fourier sine transform of $f(x) = x^2$ if $0 < x < 1$, $f(x) = 0$ if $x > 1$.
14. Find $\mathcal{F}_s(x^{-1} - x^{-1} \cos \pi x)$. *Hint.* Use Prob. 3 in Problem Set 10.8 with x and w interchanged.
15. Find $\mathcal{F}_s(e^{-x})$ from (8a) and formula 3 of Table I, Sec. 10.11.
16. Find $\mathcal{F}_s(xe^{-x^2/2})$ from (8b) and a suitable formula in Table I, Sec. 10.11.
17. Let $f(x) = x^3/(x^4 + 4)$. Find $\hat{f}_s(w)$ for $w > 0$. *Hint.* Use Prob. 6 in Sec. 10.8.
18. Using $\Gamma(\frac{1}{2}) = \sqrt{\pi}$, obtain formula 2 in Table II, Sec. 10.11, from formula 4 in that table.
19. Do the Fourier sine and cosine transforms of e^x exist?

20. **WRITING PROJECT. Finding Fourier Cosine and Sine Transforms.** Write a short essay on ways of obtaining these transforms, with illustration by examples of your own.

10.10 Fourier Transform

The preceding section concerned two transforms obtained from the Fourier cosine and sine integrals in Sec. 10.8. We now consider a third transform, the *Fourier transform,* which is obtained from the Fourier integral in complex form. (For a motivation of this transform, see the beginning of Sec. 10.9.) We therefore consider first the complex form of the Fourier integral.

Complex Form of the Fourier Integral

The (real) Fourier integral is [see (4), (5), Sec. 10.8]

$$f(x) = \int_0^\infty [A(w) \cos wx + B(w) \sin wx] \, dw$$

where

$$A(w) = \frac{1}{\pi} \int_{-\infty}^\infty f(v) \cos wv \, dv, \qquad B(w) = \frac{1}{\pi} \int_{-\infty}^\infty f(v) \sin wv \, dv.$$

Substituting A and B into the integral for f, we have

$$f(x) = \frac{1}{\pi} \int_0^\infty \int_{-\infty}^\infty f(v)[\cos wv \cos wx + \sin wv \sin wx] \, dv \, dw.$$

By the addition formula for the cosine [(6) in Appendix A3.1] the expression in the brackets [· · ·] equals $\cos (wv - wx)$ or, since the cosine is even, $\cos (wx - wv)$. We thus obtain

$$(1^*) \qquad f(x) = \frac{1}{\pi} \int_0^\infty \left[\int_{-\infty}^\infty f(v) \cos (wx - wv) \, dv \right] dw.$$

The integral in brackets is an *even* function of w, call it $F(w)$, because $\cos (wx - wv)$ is an even function of w, the function f does not depend on w, and we integrate with respect to v (not w). Hence the integral of $F(w)$ from $w = 0$ to ∞ is 1/2 times the integral of $F(w)$ from $-\infty$ to ∞. Thus

$$(1) \qquad f(x) = \frac{1}{2\pi} \int_{-\infty}^\infty \left[\int_{-\infty}^\infty f(v) \cos (wx - wv) \, dv \right] dw.$$

We claim that the integral of the form (1) with sin instead of cos is zero:

$$(2) \qquad \frac{1}{2\pi} \int_{-\infty}^\infty \left[\int_{-\infty}^\infty f(v) \sin (wx - wv) \, dv \right] dw = 0.$$

This is true since $\sin (wx - wv)$ is an odd function of w, which makes the integral in brackets an odd function of w, call it $G(w)$. Hence the integral of $G(w)$ from $-\infty$ to ∞ is zero, as claimed.

We now take the integrand of (1) plus $i (= \sqrt{-1})$ times the integrand of (2) and use the **Euler formula** (Sec. 2.3)

$$(3) \qquad e^{ix} = \cos x + i \sin x.$$

Taking $wx - wv$ instead of x in (3) gives

$$f(v) \cos (wx - wv) + if(v) \sin (wx - wv) = f(v)e^{i(wx-wv)}.$$

Hence the result of adding (1) plus i times (2), called the **complex Fourier integral,** is

(4)
$$f(x) = \frac{1}{2\pi} \int_{-\infty}^{\infty} \int_{-\infty}^{\infty} f(v)e^{iw(x-v)} \, dv \, dw \qquad (i = \sqrt{-1}).$$

It is now only a very short step to our present goal, the Fourier transform.

Fourier Transform and Its Inverse

Writing the exponential function in (4) as a product of exponential functions, we have

(5)
$$f(x) = \frac{1}{\sqrt{2\pi}} \int_{-\infty}^{\infty} \left[\frac{1}{\sqrt{2\pi}} \int_{-\infty}^{\infty} f(v)e^{-iwv} \, dv \right] e^{iwx} \, dw.$$

The expression in brackets is a function of w, is denoted by $\hat{f}(w)$, and is called the **Fourier transform** of f; writing $v = x$, we have

(6)
$$\hat{f}(w) = \frac{1}{\sqrt{2\pi}} \int_{-\infty}^{\infty} f(x)e^{-iwx} \, dx.$$

With this, (5) becomes

(7)
$$f(x) = \frac{1}{\sqrt{2\pi}} \int_{-\infty}^{\infty} \hat{f}(w)e^{iwx} \, dw$$

and is called the **inverse Fourier transform** of $\hat{f}(w)$.

Another notation is $\mathcal{F}(f) = \hat{f}(w)$ and $\mathcal{F}^{-1}$ for the inverse.

The process of obtaining the Fourier transform $\mathcal{F}(f) = \hat{f}$ from a given f is also called the **Fourier transform** or the *Fourier transform method.*

Existence of the Fourier Transform (6)

Sufficient for the existence of the Fourier transform (6) are the following two conditions (involving concepts defined in Secs. 5.1 and 10.8), as we mention without proof.

1. $f(x)$ is piecewise continuous on every finite interval.

2. $f(x)$ is absolutely integrable on the x-axis.

EXAMPLE 1   **Fourier transform**

Find the Fourier transform of $f(x) = k$ if $0 < x < a$ and $f(x) = 0$ otherwise.

Solution.   From (6) by integration,

$$\hat{f}(w) = \frac{1}{\sqrt{2\pi}} \int_0^a ke^{-iwx} \, dx = \frac{k}{\sqrt{2\pi}} \left(\frac{e^{-iwa} - 1}{-iw} \right) = \frac{k(1 - e^{-iaw})}{iw\sqrt{2\pi}}.$$

This shows that the Fourier transform will in general be a complex-valued function.   ◀

EXAMPLE 2   **Fourier transform**

Find the Fourier transform of e^{-ax^2}, where $a > 0$.

Solution.   We use the definition, complete the square in the exponent, and pull out the exponential factor that contains no x:

$$\mathscr{F}(e^{-ax^2}) = \frac{1}{\sqrt{2\pi}} \int_{-\infty}^{\infty} \exp\left[-ax^2 - iwx\right] dx$$

$$= \frac{1}{\sqrt{2\pi}} \int_{-\infty}^{\infty} \exp\left[-\left(\sqrt{a}\,x + \frac{iw}{2\sqrt{a}}\right)^2 + \left(\frac{iw}{2\sqrt{a}}\right)^2\right] dx$$

$$= \frac{1}{\sqrt{2\pi}} \exp\left(-\frac{w^2}{4a}\right) \int_{-\infty}^{\infty} \exp\left[-\left(\sqrt{a}\,x + \frac{iw}{2\sqrt{a}}\right)^2\right] dx.$$

We denote the integral by I and show that it equals $\sqrt{\pi/a}$. For this we use $\sqrt{a}\,x + iw/2\sqrt{a} = v$ as a new variable of integration. Then $dx = dv/\sqrt{a}$, so that

$$I = \frac{1}{\sqrt{a}} \int_{-\infty}^{\infty} e^{-v^2} dv.$$

We now get the result by the following trick. We square the integral, convert it to a double integral, and use polar coordinates $r = \sqrt{u^2 + v^2}$ and θ. Since $du\,dv = r\,dr\,d\theta$, we get

$$I^2 = \frac{1}{a} \int_{-\infty}^{\infty} e^{-u^2} du \int_{-\infty}^{\infty} e^{-v^2} dv = \frac{1}{a} \int_{-\infty}^{\infty} \int_{-\infty}^{\infty} e^{-(u^2 + v^2)} du\,dv$$

$$= \frac{1}{a} \int_{0}^{2\pi} \int_{0}^{\infty} e^{-r^2} r\,dr\,d\theta = \frac{2\pi}{a} \left(-\frac{1}{2} e^{-r^2}\right)\Bigg|_{0}^{\infty} = \frac{\pi}{a}.$$

Hence $I = \sqrt{\pi/a}$. From this and the first formula in this solution,

$$\mathscr{F}(e^{-ax^2}) = \frac{1}{\sqrt{2\pi}} \exp\left(-\frac{w^2}{4a}\right) \sqrt{\frac{\pi}{a}} = \frac{1}{\sqrt{2a}} e^{-w^2/4a}.$$

This agrees with formula 9 in Table III, Sec. 10.11.   ◀

Physical Interpretation: Spectrum

The nature of the representation (7) of $f(x)$ becomes clear if we think of it as a superposition of sinusoidal oscillations of all possible frequencies, called a **spectral representation.** This name is suggested by optics, where light is such a superposition of colors (frequencies). In (7), the **"spectral density"** $\hat{f}(w)$ measures the intensity of $f(x)$ in the frequency interval between w and $w + \Delta w$ (Δw small, fixed). We claim that in connection with vibrations, the integral

$$\int_{-\infty}^{\infty} |\hat{f}(w)|^2 \, dw$$

can be interpreted as the **total energy** of the physical system. Hence an integral of $|\hat{f}(w)|^2$ from a to b gives the contribution of the frequencies w between a and b to the total energy.

To make this plausible, we begin with a mechanical system giving a single frequency, namely, the harmonic oscillator (mass on a spring, Sec. 2.5)

$$my'' + ky = 0.$$

Here we denote time t by x. Multiplication by y' gives $my'y'' + ky'y = 0$. By integration,

$$\tfrac{1}{2}mv^2 + \tfrac{1}{2}ky^2 = E_0 = const$$

where $v = y'$ is the velocity. The first term is the kinetic energy, the second the potential energy, and E_0 the total energy of the system. Now a general solution is [use (4), (5), Sec. 10.5]

$$y = a_1 \cos w_0 x + b_1 \sin w_0 x = c_1 e^{iw_0 x} + c_{-1} e^{-iw_0 x}, \quad w_0^2 = k/m,$$

where $c_1 = (a_1 - ib_1)/2$, $c_{-1} = \bar{c}_1 = (a_1 + ib_1)/2$. We write simply $A = c_1 e^{iw_0 x}$, $B = c_{-1} e^{-iw_0 x}$. Then $y = A + B$. By differentiation, $v = y' = A' + B' = iw_0(A - B)$. Substitution of v and y on the left side of the equation for E_0 gives

$$E_0 = \tfrac{1}{2}mv^2 + \tfrac{1}{2}ky^2 = \tfrac{1}{2}m(iw_0)^2(A - B)^2 + \tfrac{1}{2}k(A + B)^2.$$

Here $w_0^2 = k/m$, as just stated; hence $mw_0^2 = k$. Also $i^2 = -1$, so that

$$E_0 = \tfrac{1}{2}k[-(A - B)^2 + (A + B)^2] = 2kAB = 2kc_1 e^{iw_0 x} c_{-1} e^{-iw_0 x} = 2kc_1 c_{-1} = 2k|c_1|^2.$$

Hence *the energy is proportional to the square of the amplitude* $|c_1|$.

As the next step, if a more complicated system leads to a periodic solution $y = f(x)$ that can be represented by a Fourier series, then instead of the single energy term $|c_1|^2$ we get a series of squares $|c_n|^2$ of Fourier coefficients c_n given by (8), Sec. 10.5. In this case we have a **"discrete spectrum"** (or **"point spectrum"**) consisting of countably many isolated frequencies (infinitely many, in general), the corresponding $|c_n|^2$ being the contributions to the total energy.

Finally, a system whose solution can be represented by a Fourier integral (7) leads to the above integral for the energy, as is plausible from the cases just discussed.

Linearity. Fourier Transform of Derivatives

New transforms can be obtained from given ones by

THEOREM 1   **(Linearity of the Fourier transform)**

The Fourier transform is a linear operation; that is, for any functions $f(x)$ and $g(x)$ whose Fourier transforms exist and any constants a and b,

(8)
$$\boxed{\mathscr{F}(af + bg) = a\mathscr{F}(f) + b\mathscr{F}(g).}$$

PROOF.   This is true because integration is a linear operation, so that (6) gives

$$\mathscr{F}\{af(x) + bg(x)\} = \frac{1}{\sqrt{2\pi}} \int_{-\infty}^{\infty} [af(x) + bg(x)]e^{-iwx}\, dx$$

$$= a\frac{1}{\sqrt{2\pi}} \int_{-\infty}^{\infty} f(x)e^{-iwx}\, dx + b\frac{1}{\sqrt{2\pi}} \int_{-\infty}^{\infty} g(x)e^{-iwx}\, dx$$

$$= a\mathscr{F}\{f(x)\} + b\mathscr{F}\{g(x)\}. \qquad \blacktriangleleft$$

In the application of the Fourier transform to differential equations, the key property is that differentiation of functions corresponds to multiplication of transforms by iw:

THEOREM 2   **[Fourier transform of the derivative of $f(x)$]**

Let $f(x)$ be continuous on the x-axis and $f(x) \to 0$ as $|x| \to \infty$. Furthermore, let $f'(x)$ be absolutely integrable on the x-axis. Then

(9)
$$\mathscr{F}\{f'(x)\} = iw\mathscr{F}\{f(x)\}.$$

PROOF.   From the definition of the Fourier transform we have

$$\mathscr{F}\{f'(x)\} = \frac{1}{\sqrt{2\pi}} \int_{-\infty}^{\infty} f'(x)\, e^{-iwx}\, dx.$$

Integrating by parts, we obtain

$$\mathscr{F}\{f'(x)\} = \frac{1}{\sqrt{2\pi}} \left[f(x)\, e^{-iwx} \Big|_{-\infty}^{\infty} - (-iw) \int_{-\infty}^{\infty} f(x)\, e^{-iwx}\, dx \right].$$

Since $f(x) \to 0$ as $|x| \to \infty$, the desired result follows, namely,

$$\mathscr{F}\{f'(x)\} = 0 + iw\mathscr{F}\{f(x)\}. \qquad \blacktriangleleft$$

Two successive applications of (9) give

$$\mathscr{F}(f'') = iw\mathscr{F}(f') = (iw)^2\mathscr{F}(f).$$

Since $(iw)^2 = -w^2$, we have for the transform of the second derivative of f

(10)
$$\mathscr{F}\{f''(x)\} = -w^2\, \mathscr{F}\{f(x)\}.$$

Similarly for higher derivatives.

An application of (10) to differential equations will be given in Sec. 11.6. For the time being we show how (9) can be used to derive transforms.

EXAMPLE 3   **An application of the operational formula (9)**

Find the Fourier transform of xe^{-x^2} from Table III, Sec. 10.11.

Solution.   We use (9). By formula 9 in Table III,

$$\mathscr{F}(xe^{-x^2}) = \mathscr{F}\left\{ -\frac{1}{2}\, (e^{-x^2})' \right\} = -\frac{1}{2}\, \mathscr{F}\left\{ (e^{-x^2})' \right\}$$

$$= -\frac{1}{2}\, iw\, \mathscr{F}(e^{-x^2}) = -\frac{1}{2}\, iw\, \frac{1}{\sqrt{2}}\, e^{-w^2/4} = -\frac{iw}{2\sqrt{2}}\, e^{-w^2/4}. \qquad \blacktriangleleft$$

Convolution

The **convolution** $f * g$ of functions f and g is defined by

$$(11) \qquad h(x) = (f * g)(x) = \int_{-\infty}^{\infty} f(p)g(x - p)\, dp = \int_{-\infty}^{\infty} f(x - p)g(p)\, dp.$$

The purpose is the same as in the case of Laplace transforms (Sec. 5.5): the convolution of functions corresponds to the multiplication of their Fourier transforms (except for a factor $\sqrt{2\pi}$):

THEOREM 3   **(Convolution theorem)**

Suppose that $f(x)$ and $g(x)$ are piecewise continuous, bounded, and absolutely integrable on the x-axis. Then

$$(12) \qquad \boxed{\mathscr{F}(f * g) = \sqrt{2\pi}\, \mathscr{F}(f)\mathscr{F}(g).}$$

PROOF.   By the definition,

$$\mathscr{F}(f * g) = \frac{1}{\sqrt{2\pi}} \int_{-\infty}^{\infty} \int_{-\infty}^{\infty} f(p)g(x - p)e^{-iwx}\, dp\, dx.$$

An interchange of the order of integration gives

$$\mathscr{F}(f * g) = \frac{1}{\sqrt{2\pi}} \int_{-\infty}^{\infty} \int_{-\infty}^{\infty} f(p)g(x - p)e^{-iwx}\, dx\, dp.$$

Instead of x we now take $x - p = q$ as a new variable of integration. Then $x = p + q$ and

$$\mathscr{F}(f * g) = \frac{1}{\sqrt{2\pi}} \int_{-\infty}^{\infty} \int_{-\infty}^{\infty} f(p)g(q)e^{-iw(p+q)}\, dq\, dp.$$

This double integral can be written as a product of two integrals and gives the desired result

$$\mathscr{F}(f * g) = \frac{1}{\sqrt{2\pi}} \int_{-\infty}^{\infty} f(p)e^{-iwp}\, dp \int_{-\infty}^{\infty} g(q)e^{-iwq}\, dq$$

$$= \sqrt{2\pi}\, \mathscr{F}(f)\mathscr{F}(g). \qquad \blacktriangleleft$$

By taking the inverse Fourier transform on both sides of (12), writing $\hat{f} = \mathscr{F}(f)$ and $\hat{g} = \mathscr{F}(g)$ as before, and noting that $\sqrt{2\pi}$ and $1/\sqrt{2\pi}$ in (12) and (7) cancel each other, we obtain

$$(13) \qquad (f * g)(x) = \int_{-\infty}^{\infty} \hat{f}(w)\hat{g}(w)\, e^{iwx}\, dw,$$

a formula that will help us in solving partial differential equations (Sec. 11.6).

A **table** of Fourier transforms is included in the next section. For more extensive tables, see Ref. [C3] in Appendix 1.

We finally mention that the so-called **discrete Fourier transform** gives approximations to the Fourier transform of a function $f(x)$ from (equally spaced) recorded values of $f(x)$. An economical way of computing the discrete Fourier transform is the so-called **fast Fourier transform.** See, for instance, the explanations and literature given in [E13].

This is the end of Chap. 10 on Fourier series, Fourier integrals, and Fourier transforms. The introduction of Fourier series (and Fourier integrals) was one of the greatest advances ever made in mathematical physics and its engineering applications, because Fourier series (and Fourier integrals) are probably the most important tools in solving boundary value problems. This will be explained in the next chapter.

PROBLEM SET 10.10

Calculation of Fourier Transforms

Find the Fourier transforms of the following functions $f(x)$ (without using Table III, Sec. 10.11). Show the details of your work.

1. $f(x) = \begin{cases} 1 & \text{if } a < x < b \\ 0 & \text{otherwise} \end{cases}$

2. $f(x) = \begin{cases} e^{-kx} & \text{if } x > 0 \quad (k > 0) \\ 0 & \text{if } x < 0 \end{cases}$

3. $f(x) = \begin{cases} e^x & \text{if } -a < x < a \\ 0 & \text{otherwise} \end{cases}$

4. $f(x) = \begin{cases} e^{kx} & \text{if } x < 0 \quad (k > 0) \\ 0 & \text{if } x > 0 \end{cases}$

5. $f(x) = \begin{cases} x & \text{if } 0 < x < a \\ 0 & \text{otherwise} \end{cases}$

6. $f(x) = \begin{cases} x^2 & \text{if } 0 < x < 1 \\ 0 & \text{otherwise} \end{cases}$

7. $f(x) = \begin{cases} xe^{-x} & \text{if } x > 0 \\ 0 & \text{if } x < 0 \end{cases}$

8. $f(x) = \begin{cases} e^x & \text{if } x < 0 \\ e^{-x} & \text{if } x > 0 \end{cases}$

9. $f(x) = \begin{cases} -1 & \text{if } -a < x < 0 \\ 1 & \text{if } 0 < x < a \\ 0 & \text{otherwise} \end{cases}$

10. $f(x) = \begin{cases} |x| & \text{if } -1 < x < 1 \\ 0 & \text{otherwise} \end{cases}$

Use of Table III in Sec. 10.11

11. Obtain $\mathscr{F}(e^{-x^2/2})$ from formula 9 in Table III.

12. Solve Prob. 7 by (9) in the text and formula 5 in Table III.

13. In Table III obtain formula 7 from formula 8.

14. Solve Prob. 8 by formula 5 in Table III.

15. **(Convolution)** Solve Prob. 7 by convolution.

16. **TEAM PROJECT. Shifting. (a)** Show that if $f(x)$ has a Fourier transform, so does $f(x - a)$, and $\mathscr{F}\{f(x - a)\} = e^{-iwa}\,\mathscr{F}\{f(x)\}$.

   (b) Using (a), obtain formula 1 in Table III, Sec. 10.11, from formula 2.

   (c) Shifting on the w-axis. Show that if $\hat{f}(w)$ is the Fourier transform of $f(x)$, then $\hat{f}(w - a)$ is the Fourier transform of $e^{iax}f(x)$.

   (d) Using (c), obtain formula 7 in Table III from formula 1, and formula 8 from formula 2.

10.11 Tables of Transforms

For more extensive tables, see Ref. [C3] in Appendix 1.

Table I. Fourier Cosine Transforms

See (2) in Sec. 10.9.

	$f(x)$	$\hat{f}_c(w) = \mathscr{F}_c(f)$	
1	$\begin{cases} 1 & \text{if } 0 < x < a \\ 0 & \text{otherwise} \end{cases}$	$\sqrt{\dfrac{2}{\pi}}\,\dfrac{\sin aw}{w}$	
2	x^{a-1} $(0 < a < 1)$	$\sqrt{\dfrac{2}{\pi}}\,\dfrac{\Gamma(a)}{w^a}\cos\dfrac{aw}{2}$	($\Gamma(a)$ see Appendix A3.1.)
3	e^{-ax} $(a > 0)$	$\sqrt{\dfrac{2}{\pi}}\left(\dfrac{a}{a^2 + w^2}\right)$	
4	$e^{-x^2/2}$	$e^{-w^2/2}$	
5	e^{-ax^2} $(a > 0)$	$\dfrac{1}{\sqrt{2a}}\,e^{-w^2/4a}$	
6	$x^n e^{-ax}$ $(a > 0)$	$\sqrt{\dfrac{2}{\pi}}\,\dfrac{n!}{(a^2 + w^2)^{n+1}}\,\text{Re}\,(a + iw)^{n+1}$	$\text{Re} = $ Real part
7	$\begin{cases} \cos x & \text{if } 0 < x < a \\ 0 & \text{otherwise} \end{cases}$	$\dfrac{1}{\sqrt{2\pi}}\left[\dfrac{\sin a(1 - w)}{1 - w} + \dfrac{\sin a(1 + w)}{1 + w}\right]$	
8	$\cos ax^2$ $(a > 0)$	$\dfrac{1}{\sqrt{2a}}\cos\left(\dfrac{w^2}{4a} - \dfrac{\pi}{4}\right)$	
9	$\sin ax^2$ $(a > 0)$	$\dfrac{1}{\sqrt{2a}}\cos\left(\dfrac{w^2}{4a} + \dfrac{\pi}{4}\right)$	
10	$\dfrac{\sin ax}{x}$ $(a > 0)$	$\sqrt{\dfrac{\pi}{2}}\,u(a - w)$	(See Sec. 5.3.)[15]
11	$\dfrac{e^{-x}\sin x}{x}$	$\dfrac{1}{\sqrt{2\pi}}\arctan\dfrac{2}{w^2}$	
12	$J_0(ax)$ $(a > 0)$	$\sqrt{\dfrac{2}{\pi}}\,\dfrac{u(a - w)}{\sqrt{a^2 - w^2}}$	(See Secs. 4.5, 5.3.)[15]

---

[15] $u(a - w) = 1 - u(w - a)$

Table II. Fourier Sine Transforms

See (5) in Sec. 10.9.

	$f(x)$	$\hat{f}_s(w) = \mathscr{F}_s(f)$	
1	$\begin{cases} 1 & \text{if } 0 < x < a \\ 0 & \text{otherwise} \end{cases}$	$\sqrt{\dfrac{2}{\pi}} \left[\dfrac{1 - \cos aw}{w} \right]$	
2	$1/\sqrt{x}$	$1/\sqrt{w}$	
3	$1/x^{3/2}$	$2\sqrt{w}$	
4	x^{a-1} $(0 < a < 1)$	$\sqrt{\dfrac{2}{\pi}}\, \dfrac{\Gamma(a)}{w^a} \sin \dfrac{a\pi}{2}$	($\Gamma(a)$ see Appendix A3.1.)
5	e^{-x}	$\sqrt{\dfrac{2}{\pi}} \left(\dfrac{w}{1 + w^2} \right)$	
6	$\dfrac{e^{-ax}}{x}$ $(a > 0)$	$\sqrt{\dfrac{2}{\pi}} \arctan \dfrac{w}{a}$	
7	$x^n e^{-ax}$ $(a > 0)$	$\sqrt{\dfrac{2}{\pi}}\, \dfrac{n!}{(a^2 + w^2)^{n+1}} \operatorname{Im}(a + iw)^{n+1}$	$\operatorname{Im} = $ Imaginary part
8	$x e^{-x^2/2}$	$w e^{-w^2/2}$	
9	$x e^{-ax^2}$ $(a > 0)$	$\dfrac{w}{(2a)^{3/2}} e^{-w^2/4a}$	
10	$\begin{cases} \sin x & \text{if } 0 < x < a \\ 0 & \text{otherwise} \end{cases}$	$\dfrac{1}{\sqrt{2\pi}} \left[\dfrac{\sin a(1 - w)}{1 - w} - \dfrac{\sin a(1 + w)}{1 + w} \right]$	
11	$\dfrac{\cos ax}{x}$ $(a > 0)$	$\sqrt{\dfrac{\pi}{2}}\, u(w - a)$	(See Sec. 5.3.)
12	$\arctan \dfrac{2a}{x}$ $(a > 0)$	$\sqrt{2\pi}\, \dfrac{\sinh aw}{w} e^{-aw}$	

Table III. Fourier Transforms

See (6) in Sec. 10.10.

	$f(x)$	$\hat{f}(w) = \mathcal{F}(f)$
1	$\begin{cases} 1 & \text{if } -b < x < b \\ 0 & \text{otherwise} \end{cases}$	$\sqrt{\dfrac{2}{\pi}} \dfrac{\sin bw}{w}$
2	$\begin{cases} 1 & \text{if } b < x < c \\ 0 & \text{otherwise} \end{cases}$	$\dfrac{e^{-ibw} - e^{-icw}}{iw\sqrt{2\pi}}$
3	$\dfrac{1}{x^2 + a^2} \quad (a > 0)$	$\sqrt{\dfrac{\pi}{2}} \dfrac{e^{-a|w|}}{a}$
4	$\begin{cases} x & \text{if } 0 < x < b \\ 2x - b & \text{if } b < x < 2b \\ 0 & \text{otherwise} \end{cases}$	$\dfrac{-1 + 2e^{ibw} - e^{-2ibw}}{\sqrt{2\pi}\, w^2}$
5	$\begin{cases} e^{-ax} & \text{if } x > 0 \\ 0 & \text{otherwise} \end{cases} \quad (a > 0)$	$\dfrac{1}{\sqrt{2\pi}(a + iw)}$
6	$\begin{cases} e^{ax} & \text{if } b < x < c \\ 0 & \text{otherwise} \end{cases}$	$\dfrac{e^{(a-iw)c} - e^{(a-iw)b}}{\sqrt{2\pi}(a - iw)}$
7	$\begin{cases} e^{iax} & \text{if } -b < x < b \\ 0 & \text{otherwise} \end{cases}$	$\sqrt{\dfrac{2}{\pi}} \dfrac{\sin b(w - a)}{w - a}$
8	$\begin{cases} e^{iax} & \text{if } b < x < c \\ 0 & \text{otherwise} \end{cases}$	$\dfrac{i}{\sqrt{2\pi}} \dfrac{e^{ib(a-w)} - e^{ic(a-w)}}{a - w}$
9	$e^{-ax^2} \quad (a > 0)$	$\dfrac{1}{\sqrt{2a}} e^{-w^2/4a}$
10	$\dfrac{\sin ax}{x} \quad (a > 0)$	$\sqrt{\dfrac{\pi}{2}} \text{ if } |w| < a; \quad 0 \text{ if } |w| > a$

CHAPTER 10 REVIEW

1. What is a trigonometric series? A Fourier series? (Answer from memory.)
2. What are the Euler formulas for the Fourier coefficients? By what idea did we get them?
3. How did we accomplish the transition from a 2π-periodic function to one of arbitrary period?
4. Why did we start with 2π-periodicity? What simplification did this entail?
5. Can a discontinuous function be developed in a Fourier series? In a Taylor series?
6. Isn't it mysterious that a series of continuous terms can have a discontinuous function as its sum? Explain.
7. State from memory what you know about approximation by trigonometric polynomials.
8. We gave an example of a differential equation for which the output was oscillating five times as fast as the input. Why?
9. What are half-range expansions? What are they good for?
10. If $f(x)$ has a Fourier series of cosine and sine terms, what function does the series of the cosine terms represent? The series of the sine terms?
11. What is piecewise continuity? Why did this concept occur?
12. What is the Fourier integral representation? The Fourier integral transform?
13. What is the Gibbs phenomenon?
14. What are the Fourier cosine and sine transforms?
15. What are even and odd periodic extensions of a function? Why did they occur in this chapter?

Fourier Series. Find the Fourier series of the following periodic functions which for a period are given by the following formulas. Sketch them. (Show the details of your work.)

16. $f(x) = x \quad (-\pi < x < \pi)$

17. $f(x) = x^2 \quad (-\pi < x < \pi)$

18. $f(x) = \begin{cases} -k & \text{if} \quad -1 < x < 0 \\ k & \text{if} \quad 0 < x < 1 \end{cases}$

19. $f(x) = \begin{cases} -x & \text{if} \quad -2 < x < 0 \\ x & \text{if} \quad 0 < x < 2 \end{cases}$

20. $f(x) = \begin{cases} 4 + x & \text{if} \quad -4 < x < 0 \\ 4 - x & \text{if} \quad 0 < x < 4 \end{cases}$

21. $f(x) = \begin{cases} 1 & \text{if} \quad -1 < x < 0 \\ 0 & \text{if} \quad 0 < x < 1 \end{cases}$

22. $f(x) = |\sin x| \quad (-\pi < x < \pi)$

23. $f(x) = |\cos x| \quad (-\pi < x < \pi)$

24. $f(x) = \pi - x \quad (0 < x < 2\pi)$

25. $f(x) = |x| \quad (-\pi < x < \pi)$

26. $f(x) = x \quad (0 < x < 2\pi)$

27. $f(x) = \pi - 2|x| \quad (-\pi < x < \pi)$

28. $f(x) = x^3 \quad (-1 < x < 1)$

29. $f(x) = x^2 \quad (0 < x < 2)$

Using the answers to suitable odd-numbered review problems, show that

30. $\dfrac{1}{1 \cdot 3} - \dfrac{1}{3 \cdot 5} + \dfrac{1}{5 \cdot 7} - \dfrac{1}{7 \cdot 9} + - \cdots = \dfrac{\pi}{4} - \dfrac{1}{2}$

31. $1 + \dfrac{1}{9} + \dfrac{1}{25} + \dfrac{1}{49} + \dfrac{1}{81} + \cdots = \dfrac{\pi^2}{8}$

32. $1 - \dfrac{1}{4} + \dfrac{1}{9} - \dfrac{1}{16} + \dfrac{1}{25} - + \cdots = \dfrac{\pi^2}{12}$

33. **(Minimum square error)** Compute the minimum square error for the first six partial sums in Prob. 25.
34. **(Minimum square error)** Compute the minimum square error for the first ten partial sums in Prob. 17.
35. **(General solution)** Solve $y'' + \omega^2 y = r(t)$, where $r(t) = t^2/4 \quad (-\pi < t < \pi)$, $r(t + 2\pi) = r(t)$, $|\omega| \neq 0, 1, 2, \cdots$.

36. Perform the same task as in Prob. 35 when $r(t) = t(\pi^2 - t^2)/12 \ (-\pi < t < \pi), \ r(t + 2\pi) = r(t),$ $|\omega| \neq 1, 2, \cdots$.

37. **(Fourier cosine transform)** Find the Fourier cosine transform of $f(x) = x$ if $1 < x < a$ and $f(x) = 0$ otherwise.

38. **(Fourier sine transform)** Find the Fourier sine transform of $f(x) = x + 1$ if $0 < x < 1$ and $f(x) = 0$ otherwise.

39. **(Fourier transform)** Find the Fourier transform of $f(x) = e^{-2x}$ if $x > 0$ and $f(x) = 0$ if $x < 0$.

40. **(Fourier transform)** Find the Fourier transform of $f(x) = kx$ if $a < x < b$ and $f(x) = 0$ otherwise.

SUMMARY OF CHAPTER 10
FOURIER SERIES, INTEGRALS, AND TRANSFORMS

A **trigonometric series** (Secs. 10.1, 10.3) is a series of the form

$$
(1) \qquad a_0 + \sum_{n=1}^{\infty} \left(a_n \cos \frac{n\pi}{L} x + b_n \sin \frac{n\pi}{L} x \right).
$$

If it converges, its sum is a function of period $p = 2L$. The **Fourier series** of a given periodic function $f(x)$ of period $p = 2L$ is a trigonometric series (1) whose coefficients are the **Fourier coefficients** of $f(x)$, given by the **Euler formulas** (Sec. 10.3)

$$
a_0 = \frac{1}{2L} \int_{-L}^{L} f(x)\, dx, \qquad a_n = \frac{1}{L} \int_{-L}^{L} f(x) \cos \frac{n\pi x}{L}\, dx,
$$

$$
(2)
$$

$$
b_n = \frac{1}{L} \int_{-L}^{L} f(x) \sin \frac{n\pi x}{L}\, dx.
$$

For period 2π we simply have (Sec. 10.2)

$$
(1^*) \qquad f(x) = a_0 + \sum_{n=1}^{\infty} (a_n \cos nx + b_n \sin nx)
$$

with Fourier coefficients (Sec. 10.2)

$$
a_0 = \frac{1}{2\pi} \int_{-\pi}^{\pi} f(x)\, dx, \qquad a_n = \frac{1}{\pi} \int_{-\pi}^{\pi} f(x) \cos nx\, dx,
$$

$$
(2^*)
$$

$$
b_n = \frac{1}{\pi} \int_{-\pi}^{\pi} f(x) \sin nx\, dx.
$$

Fourier series are fundamental in connection with periodic phenomena, particularly in models involving differential equations (Sec. 10.6, Chap. 11). If $f(x)$ is even [$f(-x) = f(x)$] or odd [$f(-x) = -f(x)$], they reduce to **Fourier cosine** or **Fourier sine series,** respectively (Sec. 10.4). If $f(x)$ is given for $0 \leq x \leq L$ only, it has two **half-range expansions** of period 2L, namely, a cosine and a sine series (Sec. 10.4).

The set of cosine and sine functions in (2) is called the **trigonometric system.** Its most basic property is its **orthogonality** on an interval of length 2L; that is, for all integers m and $n \neq m$ we have

$$\int_{-L}^{L} \cos\frac{m\pi x}{L} \cos\frac{n\pi x}{L} \, dx = 0, \qquad \int_{-L}^{L} \sin\frac{m\pi x}{L} \sin\frac{n\pi x}{L} \, dx = 0$$

and for all integers m and n,

$$\int_{-L}^{L} \cos\frac{m\pi x}{L} \sin\frac{n\pi x}{L} \, dx = 0.$$

This orthogonality was crucial in deriving the Euler formulas for the Fourier coefficients.

Partial sums of Fourier series minimize the square error (Sec. 10.7).

Ideas and techniques of Fourier series extend to nonperiodic functions $f(x)$ defined on the entire real line; this leads to the **Fourier integral**

$$(3) \qquad f(x) = \int_{0}^{\infty} [A(w) \cos wx + B(w) \sin wx] \, dw$$

(Sec. 10.8), where

$$(4) \qquad A(w) = \frac{1}{\pi} \int_{-\infty}^{\infty} f(v) \cos wv \, dv, \qquad B(w) = \frac{1}{\pi} \int_{-\infty}^{\infty} f(v) \sin wv \, dv$$

or, in complex form (Sec. 10.10),

$$(5) \qquad f(x) = \frac{1}{\sqrt{2\pi}} \int_{-\infty}^{\infty} \hat{f}(w) e^{iwx} \, dw \qquad\qquad (i = \sqrt{-1}),$$

where

$$(6) \qquad \hat{f}(w) = \frac{1}{\sqrt{2\pi}} \int_{-\infty}^{\infty} f(x) e^{-iwx} \, dx.$$

Formula (6) transforms $f(x)$ into its **Fourier transform** $\hat{f}(w)$.

Related to this are the **Fourier cosine transform**

$$(7) \qquad \hat{f}_c(w) = \sqrt{\frac{2}{\pi}} \int_{0}^{\infty} f(x) \cos wx \, dx$$

and the **Fourier sine transform** (Sec. 10.9)

$$(8) \qquad \hat{f}_s(w) = \sqrt{\frac{2}{\pi}} \int_{0}^{\infty} f(x) \sin wx \, dx.$$

CHAPTER 11

Partial Differential Equations

Partial differential equations arise in connection with various physical and geometrical problems when the functions involved depend on two or more independent variables, usually on time t and on one or several space variables. It is fair to say that only the simplest physical systems can be modeled by *ordinary* differential equations, whereas most problems in fluid mechanics elasticity, heat transfer, electromagnetic theory, quantum mechanics, and other areas of physics lead to *partial* differential equations. Indeed, the range of application of the latter is enormous, compared to that of ordinary differential equations.

In this chapter we consider some of the most important partial differential equations occurring in engineering applications. We derive these equations as models of physical systems and develop methods for solving **initial** and **boundary value problems,** consisting of such an equation and additional physical conditions.

In Sec. 11.1 we define the notion of a solution of a partial differential equation. Sections 11.2–11.4 concern the one-dimensional wave equation, governing the motion of a vibrating string. The heat equation is considered in Secs. 11.5 and 11.6, the two-dimensional wave equation modeling vibrating membranes, in Secs. 11.7–11.10, and Laplace's equation in Sec. 11.11.

In Secs. 11.6 and 11.12 we see that partial differential equations can also be solved by Fourier transform or Laplace transform methods.

Numerical methods for partial differential equations are presented in Secs. 19.4–19.7.

> *Prerequisites for this chapter:* Ordinary linear differential equations (Chap. 2) and Fourier series (Chap. 10).
> *Sections that may be omitted in a shorter course:* 11.6, 11.9–11.12.
> *References:* Appendix 1, Part C.
> *Answers to problems:* Appendix 2.

11.1 Basic Concepts

An equation involving one or more partial derivatives of an (unknown) function of two or more independent variables is called a **partial differential equation.** The order of the highest derivative is called the **order** of the equation.

Just as in the case of an ordinary differential equation, we say that a partial differential equation is **linear** if it is of the first degree in the dependent variable (the unknown function) and its partial derivatives. If each term of such an equation contains either the dependent variable or one of its derivatives, the equation is said to be **homogeneous;** otherwise it is said to be **nonhomogeneous.**

EXAMPLE 1   **Important linear partial differential equations of the second order**

$$(1) \qquad \frac{\partial^2 u}{\partial t^2} = c^2 \frac{\partial^2 u}{\partial x^2} \qquad\qquad \textit{One-dimensional wave equation}$$

$$(2) \qquad \frac{\partial u}{\partial t} = c^2 \frac{\partial^2 u}{\partial x^2} \qquad\qquad \textit{One-dimensional heat equation}$$

$$(3) \qquad \frac{\partial^2 u}{\partial x^2} + \frac{\partial^2 u}{\partial y^2} = 0 \qquad\qquad \textit{Two-dimensional Laplace equation}$$

$$(4) \qquad \frac{\partial^2 u}{\partial x^2} + \frac{\partial^2 u}{\partial y^2} = f(x,\, y) \qquad\qquad \textit{Two-dimensional Poisson equation}$$

$$(5) \qquad \frac{\partial^2 u}{\partial t^2} = c^2 \left(\frac{\partial^2 u}{\partial x^2} + \frac{\partial^2 u}{\partial y^2} \right) \qquad\qquad \textit{Two-dimensional wave equation}$$

$$(6) \qquad \frac{\partial^2 u}{\partial x^2} + \frac{\partial^2 u}{\partial y^2} + \frac{\partial^2 u}{\partial z^2} = 0 \qquad\qquad \textit{Three-dimensional Laplace equation}$$

Here c is a constant, t is time, x, y, z are Cartesian coordinates and *dimension* is the number of these coordinates in the equation. Equation (4) (with $f(x,\, y) \neq 0$) is nonhomogeneous, while the other equations are homogeneous. ◀

A **solution** *of a partial differential equation in some region R of the space of the independent variables* is a function that has all the partial derivatives appearing in the equation in some domain containing R and satisfies the equation everywhere in R. (Often one merely requires that that function is continuous on the boundary of R, has those derivatives in the interior of R, and satisfies the equation in the interior of R.)

In general, the totality of solutions of a partial differential equation is very large. For example, the functions

$$(7) \qquad u = x^2 - y^2, \qquad u = e^x \cos y, \qquad u = \ln (x^2 + y^2),$$

which are entirely different from each other, are solutions of (3), as you may verify. We shall see later that the unique solution of a partial differential equation corresponding to a given physical problem will be obtained by the use of additional conditions arising from the problem, for instance, the condition that the solution u assume given values on the boundary of the region considered (**"boundary conditions"**), or, when time t is one of the variables, that u (or $u_t = \partial u/\partial t$ or both) are prescribed at $t = 0$ (**"initial conditions"**).

We know that if an *ordinary* differential equation is linear and homogeneous, then from known solutions we can obtain further solutions by superposition. For a homogeneous linear *partial* differential equation the situation is quite similar:

THEOREM 1   **Fundamental Theorem (Superposition or linearity principle)**

If u_1 and u_2 are any solutions of a linear homogeneous partial differential equation in some region R, then

$$u = c_1 u_1 + c_2 u_2,$$

with any constants c_1 and c_2, is also a solution of that equation in R.

The proof of this important theorem is simple and quite similar to that of Theorem 1 in Sec. 2.1 and is left to the student.

Verification of solutions in Probs. 2–14 proceeds as for ordinary differential equations. Problems 15–22 concern partial differential equations that can be solved like ordinary ones. To help the student with them, we consider two typical examples.

EXAMPLE 2   **Find a solution $u(x, y)$ of the partial differential equation $u_{xx} - u = 0$.**

Solution. Since no y-derivatives occur, we can solve this like $u'' - u = 0$. In Sec. 2.2 we would have obtained $u = Ae^x + Be^{-x}$ with constant A and B. Here A and B may be functions of y, so that the answer is

$$u(x, y) = A(y)e^x + B(y)e^{-x}$$

with arbitrary functions A and B. We thus have a great variety of solutions. Check the result by differentiation. ◀

EXAMPLE 3   **Solve the partial differential equation $u_{xy} = -u_x$.**

Solution. Setting $u_x = p$, we have $p_y = -p$, $p_y/p = -1$, $\ln p = -y + \tilde{c}(x)$, $p = c(x)e^{-y}$ and by integration with respect to x,

$$u(x, y) = f(x)e^{-y} + g(y) \qquad \text{where} \qquad f(x) = \int c(x)\, dx;$$

here, $f(x)$ and $g(y)$ are arbitrary. ◀

PROBLEM SET 11.1

1. **(Fundamental Theorem)** Prove Fundamental Theorem 1 for second-order differential equations in two and three independent variables.

Verification of Solutions

In each case verify that the given function is a solution of the indicated equation and sketch or plot a figure of the solution as a surface in space.

Wave Equation (1) (with suitable c)

  2. $u = x^2 + t^2$ **3.** $u = \sin 9t \sin \frac{1}{4}x$ **4.** $u = \cos 4t \sin 2x$ **5.** $u = \sin ct \sin x$

Heat Equation (2) (with suitable c)

  6. $u = e^{-t} \sin x$ **7.** $u = e^{-4t} \cos 3x$ **8.** $u = e^{-9t} \cos \omega x$ **9.** $u = e^{-\omega^2 c^2 t} \sin \omega x$

Laplace Equation (3)

  10. $u = 2xy$ **11.** $u = e^x \sin y$ **12.** $u = \cos x \sinh y$ **13.** $u = \arctan (y/x)$

14. **TEAM PROJECT. Verification of Solutions (a) Poisson Equation.** Verify that u satisfies (4) with $f(x, y)$ as indicated.

$$u = x^2 + y^2, \qquad f = 4$$
$$u = \cos (xy), \qquad f = -(x^2 + y^2) \cos (xy)$$
$$u = y/x, \qquad f = 2y/x^3$$

(b) **Laplace Equation.** Verify that $u = 1/\sqrt{x^2 + y^2 + z^2}$ satisfies (6).

(c) Verify that u with arbitrary (sufficiently often differentiable) v and w satisfies the given equation.

$$u = v(x) + w(y), \qquad u_{xy} = 0$$

$$u = v(x)w(y), \qquad uu_{xy} = u_x u_y$$

$$u = v(x + 2t) + w(x - 2t), \qquad u_{tt} = 4u_{xx}$$

Partial Differential Equations Solvable as Ordinary Differential Equations

If an equation involves derivatives with respect to one variable only, we can solve it like an ordinary differential equation, treating the other variable (or variables) as parameters. Find solutions $u(x, y)$ of

15. $u_y = u$ **16.** $u_{xx} + 9u = 0$ **17.** $u_{yy} = 0$ **18.** $u_y + 2yu = 0$

19. $u_{xy} = u_x$ **20.** $u_{yy} = u$ **21.** $u_y = 2xyu$ **22.** $u_{yy} = u_y$

23. **(Boundary value problem)** Verify that $u(x, y) = a \ln (x^2 + y^2) + b$ satisfies Laplace's equation (3) and determine a and b so that u satisfies the boundary conditions $u = 0$ on the circle $x^2 + y^2 = 1$ and $u = 3$ on the circle $x^2 + y^2 = 4$. Sketch a figure of the surface represented by this function.

24. **(Surface of revolution)** Show that the solutions $z = z(x, y)$ of $yz_x = xz_y$ represent surfaces of revolution. Give examples. *Hint.* Use polar coordinates r, θ and show that the equation becomes $z_\theta = 0$.

25. **(System)** Solve the system $u_{xx} = 0$, $u_{yy} = 0$.

11.2 Modeling: Vibrating String, Wave Equation

As a first important partial differential equation, let us derive the equation governing small transverse vibrations of an elastic string, such as a violin string. We place the string along the x-axis, stretch it to length L, and fix it at the ends $x = 0$ and $x = L$. We then distort it and at some instant, say, $t = 0$, we release it and allow it to vibrate. The problem is to determine the vibrations of the string, that is, to find its deflection $u(x, t)$ at any point x and at any time $t > 0$; see Fig. 258.

We shall obtain $u(x, t)$ as the solution of a partial differential equation, which will be the model of our physical system. This equation should not be too complicated, so that we can solve it. Hence we have to make simplifying assumptions (just as for an ordinary differential equation in Chaps. 1–5), as follows.

Physical Assumptions

1. *The mass of the string per unit length is constant ("homogeneous string"). The string is perfectly elastic and does not offer any resistance to bending.*

2. *The tension caused by stretching the string before fixing it at the ends is so large that the action of the gravitational force on the string can be neglected.*

3. *The string performs small transverse motions in a vertical plane; that is, every particle of the string moves strictly vertically and so that the deflection and the slope at every point of the string always remain small in absolute value.*

Under these assumptions we may expect solutions $u(x, t)$ that describe the physical reality sufficiently well.

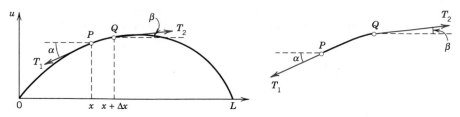

Fig. 258.   Deflected string at fixed time t

Derivation of the Differential Equation from Forces

To obtain the differential equation we consider the forces acting on a small portion of the string (Fig. 258). Since the string does not offer resistance to bending, the tension is tangential to the curve of the string at each point. Let T_1 and T_2 be the tension at the endpoints P and Q of that portion. Since the points of the string move vertically, there is no motion in the horizontal direction. Hence the horizontal components of the tension must be constant. Using the notation shown in Fig. 258, we thus obtain

$$(1) \qquad\qquad T_1 \cos \alpha = T_2 \cos \beta = T = const.$$

In the vertical direction we have two forces, namely, the vertical components $-T_1 \sin \alpha$ and $T_2 \sin \beta$ of T_1 and T_2; here the minus sign appears because the component at P is directed downward. By Newton's second law the resultant of these two forces is equal to the mass $\rho \, \Delta x$ of the portion times the acceleration $\partial^2 u/\partial t^2$, evaluated at some point between x and $x + \Delta x$; here ρ is the mass of the undeflected string per unit length, and Δx is the length of the portion of the undeflected string. Hence

$$T_2 \sin \beta - T_1 \sin \alpha = \rho \, \Delta x \, \frac{\partial^2 u}{\partial t^2} .$$

Using (1), we can divide this by $T_2 \cos \beta = T_1 \cos \alpha = T$, obtaining

$$(2) \qquad \frac{T_2 \sin \beta}{T_2 \cos \beta} - \frac{T_1 \sin \alpha}{T_1 \cos \alpha} = \tan \beta - \tan \alpha = \frac{\rho \, \Delta x}{T} \frac{\partial^2 u}{\partial t^2} .$$

Now $\tan \alpha$ and $\tan \beta$ are the slopes of the string at x and $x + \Delta x$:

$$\tan \alpha = \left(\frac{\partial u}{\partial x} \right)\bigg|_{x} \qquad \text{and} \qquad \tan \beta = \left(\frac{\partial u}{\partial x} \right)\bigg|_{x+\Delta x} .$$

Here we have to write *partial* derivatives because u also depends on time t. Dividing (2) by Δx, we thus have

$$\frac{1}{\Delta x} \left[\left(\frac{\partial u}{\partial x} \right)\bigg|_{x+\Delta x} - \left(\frac{\partial u}{\partial x} \right)\bigg|_{x} \right] = \frac{\rho}{T} \frac{\partial^2 u}{\partial t^2} .$$

If we let Δx approach zero, we obtain the linear partial differential equation

$$(3) \qquad\qquad \boxed{\frac{\partial^2 u}{\partial t^2} = c^2 \frac{\partial^2 u}{\partial x^2} ,} \qquad\qquad c^2 = \frac{T}{\rho} .$$

This is the so-called **one-dimensional wave equation,** which governs our problem. We see that it is homogeneous and of the second order. For the physical constant T/ρ we have written c^2 (instead of c) to indicate that this constant is positive. "One-dimensional" indicates that the equation involves only one space variable, x. Solutions will be obtained in the next section.

11.3 Separation of Variables
Use of Fourier Series

In the last section we showed that the vibrations of an elastic string, such as a violin string, are governed by the **one-dimensional wave equation**

(1)
$$\frac{\partial^2 u}{\partial t^2} = c^2 \frac{\partial^2 u}{\partial x^2},$$

where $u(x,\ t)$ is the deflection of the string. To find out how the string moves, we solve this equation; more precisely, we determine a solution u of (1) that also satisfies the conditions imposed by the physical system. Since the string is fixed at the ends $x = 0$ and $x = L$, we have the two **boundary conditions**

(2)
$$u(0,\ t) = 0, \qquad u(L,\ t) = 0 \qquad \text{for all } t.$$

The form of the motion of the string will depend on the initial deflection (deflection at time $t = 0$) and on the initial velocity (velocity at $t = 0$). Denoting the initial deflection by $f(x)$ and the initial velocity by $g(x)$, we thus obtain the two **initial conditions**

(3)
$$u(x,\ 0) = f(x)$$

and

(4)
$$\left. \frac{\partial u}{\partial t} \right|_{t=0} = g(x).$$

Our problem is now to find a solution of (1) satisfying the conditions (2)–(4). We shall proceed step by step, as follows.

First Step. By applying the so-called **method of separating variables** or *product method,* we shall obtain two *ordinary* differential equations.

Second Step. We shall determine solutions of those two equations that satisfy the boundary conditions (2).

Third Step. Using **Fourier series,** we shall compose those solutions, in order to get a solution of the wave equation (1) that also satisfies the initial conditions (3) and (4).

First Step. Two Ordinary Differential Equations

In the **method of separating variables,** or *product method,* we determine solutions of the wave equation (1) of the form

(5)
$$u(x,\ t) = F(x)G(t)$$

which are a product of two functions, each depending only on one of the variables x and t. We shall see later that this method has various other applications in engineering mathematics. By differentiating (5) we obtain

$$\frac{\partial^2 u}{\partial t^2} = F\ddot{G} \qquad \text{and} \qquad \frac{\partial^2 u}{\partial x^2} = F''G,$$

where dots denote derivatives with respect to t and primes derivatives with respect to x. By inserting this into our differential equation (1) we have

$$F\ddot{G} = c^2 F''G.$$

Dividing by $c^2 FG$, we find

$$\frac{\ddot{G}}{c^2 G} = \frac{F''}{F}\ .$$

We now claim that both sides must be constant. Indeed, the expression on the left depends only on t and that on the right only on x, and if they were variable, then changing t or x would affect only the left or the right side, respectively, leaving the other side unaltered. Thus,

$$\frac{\ddot{G}}{c^2 G} = \frac{F''}{F} = k.$$

This yields immediately two ordinary linear differential equations, namely,

(6)
$$F'' - kF = 0$$

and

(7)
$$\ddot{G} - c^2 kG = 0.$$

Here the constant k is still arbitrary.

Second Step. Satisfying the Boundary Conditions (2)

We shall now determine solutions F and G of (6) and (7) so that $u = FG$ satisfies the boundary conditions (2), that is,

$$u(0,\ t) = F(0)G(t) = 0, \qquad u(L,\ t) = F(L)G(t) = 0 \qquad \text{for all } t.$$

Solving (6). If $G \equiv 0$, then $u \equiv 0$, which is of no interest. Thus $G \not\equiv 0$ and then

$$
\text{(8)} \qquad\qquad \text{(a)} \quad F(0) = 0, \qquad \text{(b)} \quad F(L) = 0.
$$

For $k = 0$ the general solution of (6) is $F = ax + b$, and from (8) we obtain $a = b = 0$. Hence $F \equiv 0$, which is of no interest because then $u \equiv 0$. For positive $k = \mu^2$ the general solution of (6) is

$$
F = Ae^{\mu x} + Be^{-\mu x},
$$

and from (8) we obtain $F \equiv 0$, as before. Hence we are left with the possibility of choosing k negative, say, $k = -p^2$. Then (6) takes the form

$$
F'' + p^2 F = 0.
$$

Its general solution is

$$
F(x) = A \cos px + B \sin px.
$$

From this and (8) we have

$$
F(0) = A = 0 \qquad \text{and then} \qquad F(L) = B \sin pL = 0.
$$

We must take $B \neq 0$ since otherwise $F \equiv 0$. Hence $\sin pL = 0$. Thus

$$
\text{(9)} \qquad\qquad pL = n\pi, \qquad \text{so that} \qquad p = \frac{n\pi}{L} \qquad\qquad (n \text{ integer}).
$$

Setting $B = 1$, we thus obtain infinitely many solutions $F(x) = F_n(x)$, where

$$
\text{(10)} \qquad\qquad F_n(x) = \sin \frac{n\pi}{L} x \qquad\qquad (n = 1, 2, \cdots).
$$

These solutions satisfy (8). [For negative integer n we obtain essentially the same solutions, except for a minus sign, because $\sin(-\alpha) = -\sin \alpha$.]

Solving (7). The constant k is now restricted to the values $k = -p^2 = -(n\pi/L)^2$, resulting from (9). For these k, equation (7) becomes

$$
\text{(11*)} \qquad\qquad \ddot{G} + \lambda_n^2 G = 0 \qquad \text{where} \qquad \lambda_n = \frac{cn\pi}{L}.
$$

A general solution is

$$
G_n(t) = B_n \cos \lambda_n t + B_n{}^* \sin \lambda_n t.
$$

Hence solutions of (1) satisfying (2) are $u_n(x, t) = F_n(x)G_n(t)$, written out

$$
\boxed{
\text{(11)} \qquad u_n(x, t) = (B_n \cos \lambda_n t + B_n{}^* \sin \lambda_n t) \sin \frac{n\pi}{L} x
} \qquad (n = 1, 2, \cdots).
$$

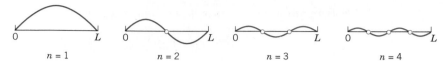

Fig. 259. Normal modes of the vibrating string

These functions are called the **eigenfunctions,** or *characteristic functions,* and the values $\lambda_n = cn\pi/L$ are called the **eigenvalues,** or *characteristic values,* of the vibrating string. The set $\{\lambda_1, \lambda_2, \cdots\}$ is called the **spectrum.**

Discussion of Eigenfunctions. We see that each u_n represents a harmonic motion having the frequency $\lambda_n/2\pi = cn/2L$ cycles per unit time. This motion is called the nth **normal mode** of the string. The first normal mode is known as the *fundamental mode* ($n = 1$), and the others are known as *overtones;* musically they give the octave, octave plus fifth, etc. Since in (11)

$$\sin \frac{n\pi x}{L} = 0 \qquad \text{at} \qquad x = \frac{L}{n}, \frac{2L}{n}, \cdots, \frac{n-1}{n} L,$$

the nth normal mode has $n - 1$ so-called **nodes,** that is, points of the string that do not move (in addition to the fixed endpoints; see Fig. 259).

   Figure 260 shows the second normal mode for various values of t. At any instant the string has the form of a sine wave. When the left part of the string is moving down the other half is moving up, and conversely. For the other modes the situation is similar.

Tuning is done by changing the tension T. Our above formula for the frequency $\lambda_n/2\pi = cn/2L$ of u_n with $c = T/\rho$ [see (3), Sec. 11.2] confirms that effect because it shows that the frequency is proportional to the tension. T cannot be increased indefinitely, but can you see what to do to get a string with a high fundamental mode? (Think of both L and ρ.) Why is a violin smaller than a double-bass?

Third Step. Solution of the Entire Problem. Fourier Series

Clearly, a single solution $u_n(x, t)$ will in general not satisfy the initial conditions (3) and (4). Now, since the equation (1) is linear and homogeneous, it follows from Fundamental Theorem 1 in Sec. 11.1 that the sum of finitely many solutions u_n is a solution of (1). To obtain a solution that satisfies (3) and (4), we consider the infinite series (with $\lambda_n = cn\pi/L$

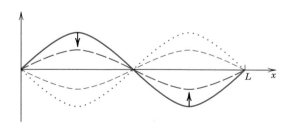

Fig. 260. Second normal mode for various values of t

as before)

$$(12) \qquad u(x,\ t) = \sum_{n=1}^{\infty} u_n(x,\ t) = \sum_{n=1}^{\infty} (B_n \cos \lambda_n t + B_n{}^* \sin \lambda_n t) \sin \frac{n\pi}{L} x.$$

Satisfying Initial Condition (3) (Given Initial Displacement). From (12) and (3) we obtain

$$(13) \qquad u(x,\ 0) = \sum_{n=1}^{\infty} B_n \sin \frac{n\pi}{L} x = f(x).$$

Hence we must choose the B_n's so that $u(x,\ 0)$ becomes the Fourier sine series of $f(x)$. Thus, by (6) in Sec. 10.4,

$$(14) \qquad B_n = \frac{2}{L} \int_0^L f(x) \sin \frac{n\pi x}{L}\ dx, \qquad\qquad n = 1,\ 2,\ \cdots.$$

Satisfying Initial Condition (4) (Given Initial Velocity). Similarly, by differentiating (12) with respect to t and using (4), we obtain

$$\frac{\partial u}{\partial t}\bigg|_{t=0} = \left[\sum_{n=1}^{\infty} (-B_n \lambda_n \sin \lambda_n t + B_n{}^* \lambda_n \cos \lambda_n t) \sin \frac{n\pi x}{L} \right]_{t=0}$$

$$= \sum_{n=1}^{\infty} B_n{}^* \lambda_n \sin \frac{n\pi x}{L} = g(x).$$

Hence we must choose the $B_n{}^*$'s so that for $t = 0$ the derivative $\partial u / \partial t$ becomes the Fourier sine series of $g(x)$. Thus, by (6) in Sec. 10.4,

$$B_n{}^* \lambda_n = \frac{2}{L} \int_0^L g(x) \sin \frac{n\pi x}{L}\ dx.$$

Since $\lambda_n = cn\pi/L$, we obtain by division

$$(15) \qquad B_n{}^* = \frac{2}{cn\pi} \int_0^L g(x) \sin \frac{n\pi x}{L}\ dx, \qquad\qquad n = 1,\ 2,\ \cdots.$$

Result. Our discussion shows that $u(x,\ t)$ given by (12) with coefficients (14) and (15) is a solution of (1) that satisfies all the conditions (2), (3), (4) of our problem, provided the series (12) converges and so do the series obtained by differentiating (12) twice termwise with respect to x and t and have the sums $\partial^2 u/\partial x^2$ and $\partial^2 u/\partial t^2$, respectively, which are continuous.

Solution (12) established. According to our discussion, the solution (12) is at first a purely formal expression, but we shall now establish it. For the sake of simplicity we consider only the case when the initial velocity $g(x)$ is identically zero. Then the $B_n{}^*$ are zero, and (12) reduces to

$$(16) \qquad u(x,\ t) = \sum_{n=1}^{\infty} B_n \cos \lambda_n t \sin \frac{n\pi x}{L}, \qquad \lambda_n = \frac{cn\pi}{L}.$$

Fig. 261.   Odd periodic extension of $f(x)$

It is possible to **sum this series,** that is, to write the result in a closed or finite form. For this purpose we use the formula [see (11), Appendix A3.1]

$$\cos \frac{cn\pi}{L} t \sin \frac{n\pi}{L} x = \frac{1}{2} \left[\sin \left\{ \frac{n\pi}{L} (x - ct) \right\} + \sin \left\{ \frac{n\pi}{L} (x + ct) \right\} \right].$$

Consequently, we may write (16) in the form

$$u(x, t) = \frac{1}{2} \sum_{n=1}^{\infty} B_n \sin \left\{ \frac{n\pi}{L} (x - ct) \right\} + \frac{1}{2} \sum_{n=1}^{\infty} B_n \sin \left\{ \frac{n\pi}{L} (x + ct) \right\}.$$

These two series are those obtained by substituting $x - ct$ and $x + ct$, respectively, for the variable x in the Fourier sine series (13) for $f(x)$. Thus

(17)
$$u(x, t) = \tfrac{1}{2}[f^*(x - ct) + f^*(x + ct)]$$

where f^* is the odd periodic extension of f with the period $2L$ (Fig. 261). Since the initial deflection $f(x)$ is continuous on the interval $0 \le x \le L$ and zero at the endpoints, it follows from (17) that $u(x, t)$ is a continuous function of both variables x and t for all values of the variables. By differentiating (17) we see that $u(x, t)$ is a solution of (1), provided $f(x)$ is twice differentiable on the interval $0 < x < L$, and has one-sided second derivatives at $x = 0$ and $x = L$, which are zero. Under these conditions $u(x, t)$ is established as a solution of (1), satisfying (2)–(4).   ◀

Generalized Solution. If $f'(x)$ and $f''(x)$ are merely piecewise continuous (see Sec. 5.1), or if those one-sided derivatives are not zero, then for each t there will be finitely many values of x at which the second derivatives of u appearing in (1) do not exist. Except at these points the wave equation will still be satisfied. We may then regard $u(x, t)$ as a **"generalized solution,"** as it is called, that is, as a solution in a broader sense. For instance, a triangular initial deflection as in Example 1 (below) leads to a generalized solution.

Physical Interpretation of the Solution (17). The graph of $f^*(x - ct)$ is obtained from the graph of $f^*(x)$ by shifting the latter ct units to the right (Fig. 262). This means that $f^*(x - ct)$ $(c > 0)$ represents a wave that is traveling to the right as t increases. Similarly, $f^*(x + ct)$ represents a wave that is traveling to the left, and $u(x, t)$ is the superposition of these two waves.

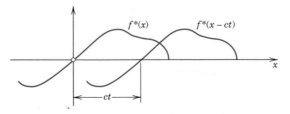

Fig. 262.   Interpretation of (17)

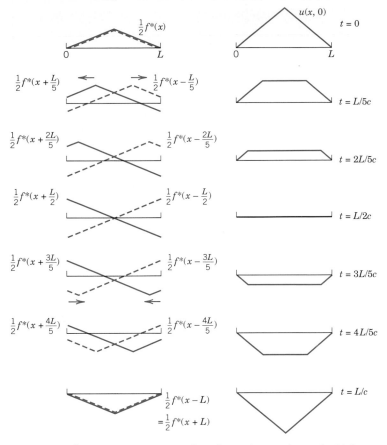

Fig. 263. Solution $u(x, t)$ in Example 1 for various values of t (right part of the figure) obtained as the superposition of a wave traveling to the right (dashed) and a wave traveling to the left (left part of the figure)

EXAMPLE 1   **Vibrating string if the initial deflection is triangular**

Find the solution of the wave equation (1) corresponding to the triangular initial deflection

$$
f(x) = \begin{cases} \dfrac{2k}{L}\,x & \text{if} \quad 0 < x < \dfrac{L}{2} \\[2mm] \dfrac{2k}{L}\,(L - x) & \text{if} \quad \dfrac{L}{2} < x < L \end{cases}
$$

and initial velocity zero. (Figure 263 shows $f(x) = u(x, 0)$ at the top.)

Solution.   Since $g(x) \equiv 0$, we have $B_n{}^* = 0$ in (12), and from Example 3 in Sec. 10.4 we see that the B_n are given by (7), Sec. 10.4. Thus (12) takes the form

$$
u(x, t) = \frac{8k}{\pi^2}\left[\frac{1}{1^2}\sin\frac{\pi}{L}x\cos\frac{\pi c}{L}t - \frac{1}{3^2}\sin\frac{3\pi}{L}x\cos\frac{3\pi c}{L}t + - \cdots\right].
$$

For plotting the graph of the solution we may use $u(x, 0) = f(x)$ and the above interpretation of the two functions in the representation (17). This leads to the graph shown in Fig. 263.   ◀

PROBLEM SET 11.3

1. **(Frequency)** How does the frequency of the fundamental mode of the vibrating string depend on the length of the string? On the mass per unit length? On the tension? What happens to that frequency if we double the tension?

Deflection $u(x, t)$ of the String

Find $u(x, t)$ of the string of length $L = \pi$ when $c^2 = 1$, the initial velocity is zero, and the initial deflection is

2. $0.01 \sin 3x$
3. $k(\sin x - \frac{1}{2} \sin 2x)$
4. $0.1x(\pi - x)$
5. $0.1x(\pi^2 - x^2)$

6.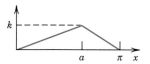

7.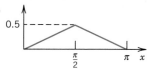

8.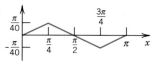

9.

10. **(Nonzero initial velocity)** Find the deflection $u(x, t)$ of the string of length $L = \pi$ and $c^2 = 1$ for zero initial displacement and "triangular" initial velocity $u_t(x, 0) = 0.01x$ if $0 \leqq x \leqq \frac{1}{2}\pi$, $u_t(x, 0) = 0.01(\pi - x)$ if $\frac{1}{2}\pi \leqq x \leqq \pi$. (Initial conditions with $u_t(x, 0) \neq 0$ are hard to realize experimentally.)

11. **CAS PROJECT. Plots of Normal Modes.** Write a program for plotting a figure of u_n with $L = \pi$ similar to Fig. 260. Apply the program to u_2, u_3, u_4. Plot these functions as surfaces over the xt-plane. Explain the connection between these two kinds of plot.

Separation of Variables

Find solutions $u(x, y)$ of the following equations by separating variables.

12. $u_x + u_y = 0$
13. $u_x - u_y = 0$
14. $y^2 u_x - x^2 u_y = 0$
15. $u_x + u_y = (x + y)u$
16. $u_{xx} + u_{yy} = 0$
17. $u_{xy} - u = 0$
18. $u_{xx} - u_{yy} = 0$
19. $xu_{xy} + 2yu = 0$

20. **TEAM PROJECT. Forced Vibrations of an Elastic String.** Show the following.

(a) Substitution of

$$(18) \qquad u(x, t) = \sum_{n=1}^{\infty} G_n(t) \sin \frac{n\pi x}{L} \qquad (L = \text{length of the string})$$

into the wave equation (1) governing free vibrations leads to

$$\ddot{G}_n + \lambda_n^2 G = 0, \qquad \lambda_n = \frac{cn\pi}{L} \qquad \text{[see (11*)].}$$

(b) Forced vibrations of the string under an external force $P(x, t)$ per unit length acting normal to the string are governed by the equation

$$(19) \qquad u_{tt} = c^2 u_{xx} + \frac{P}{\rho}.$$

(c) For a sinusoidal force $P = A\rho \sin \omega t$ we obtain

$$(20) \qquad \frac{P}{\rho} = A \sin \omega t = \sum_{n=1}^{\infty} k_n(t) \sin \frac{n\pi x}{L}, \qquad k_n(t) = \begin{cases} (4A/n\pi) \sin \omega t & (n \text{ odd}) \\ 0 & (n \text{ even}) \end{cases}$$

Substituting (18) and (20) into (19) gives

$$\ddot{G}_n + \lambda_n^2 G_n = \frac{2A}{n\pi} (1 - \cos n\pi) \sin \omega t.$$

If $\lambda_n{}^2 \neq \omega^2$, the solution is

$$G_n(t) = B_n \cos \lambda_n t + B_n{}^* \sin \lambda_n t + \frac{2A(1 - \cos n\pi)}{n\pi(\lambda_n{}^2 - \omega^2)} \sin \omega t.$$

Determine B_n and $B_n{}^*$ so that u satisfies the initial conditions $u(x, 0) = f(x)$, $u_t(x, 0) = 0$.

(d) (Resonance) Show that if $\lambda_n = \omega$, then

$$G_n(t) = B_n \cos \omega t + B_n{}^* \sin \omega t - \frac{A}{n\pi\omega} (1 - \cos n\pi)t \cos \omega t.$$

(e) (Reduction of boundary conditions) Show that a problem (1)–(4) with more complicated boundary conditions $u(0, t) = 0$, $u(L, t) = h(t)$, can be reduced to a problem for a new function v satisfying conditions $v(0, t) = v(L, t) = 0$, $v(x, 0) = f_1(x)$, $v_t(x, 0) = g_1(x)$ but a nonhomogeneous wave equation. *Hint.* Set $u = v + w$ and determine w suitably.

11.4 D'Alembert's Solution of the Wave Equation

It is interesting that the solution (17), Sec. 11.3, of the wave equation

(1)
$$\frac{\partial^2 u}{\partial t^2} = c^2 \frac{\partial^2 u}{\partial x^2}, \qquad\qquad c^2 = \frac{T}{\rho},$$

can be immediately obtained by transforming (1) in a suitable way, namely, by introducing the new independent variables[1]

(2)
$$v = x + ct, \qquad z = x - ct.$$

Then u becomes a function of v and z. The derivatives in (1) can now be expressed in terms of derivatives with respect to v and z by the use of the chain rule in Sec. 8.8. Denoting partial derivatives by subscripts, we see from (2) that $v_x = 1$ and $z_x = 1$. For simplicity let us denote $u(x, t)$, as a function of v and z, by the same letter u. Then

$$u_x = u_v v_x + u_z z_x = u_v + u_z.$$

We now apply the chain rule to the right side. We assume that all the partial derivatives involved are continuous, so that $u_{zv} = u_{vz}$. Since $v_x = 1$ and $z_x = 1$, we obtain

$$u_{xx} = (u_v + u_z)_x = (u_v + u_z)_v v_x + (u_v + u_z)_z z_x = u_{vv} + 2u_{vz} + u_{zz}.$$

We transform the other derivative in (1) by the same procedure, finding

$$u_{tt} = c^2(u_{vv} - 2u_{vz} + u_{zz}).$$

---

[1]We mention that the general theory of partial differential equations provides a systematic way for finding this transformation that will simplify the equation. See Ref. [C5] in Appendix 1.

By inserting these two results in (1) we get (see footnote 2 in Appendix A3.2)

(3)
$$u_{vz} \equiv \frac{\partial^2 u}{\partial z\, \partial v} = 0.$$

Obviously, the point of the present method is that the resulting equation (3) can be readily solved by two successive integrations. In fact, integrating (3) with respect to z, we find

$$\frac{\partial u}{\partial v} = h(v)$$

where $h(v)$ is an arbitrary function of v. Integrating this with respect to v gives

$$u = \int h(v)\, dv + \psi(z)$$

where $\psi(z)$ is an arbitrary function of z. Since the integral is a function of v, say, $\phi(v)$, the solution u is of the form $u = \phi(v) + \psi(z)$. Because of (2),

(4)
$$u(x,\, t) = \phi(x + ct) + \psi(x - ct).$$

This is known as **d'Alembert's solution**[2] of the wave equation (1).

This derivation was much more elegant than the method in Sec. 11.3, but d'Alembert's method is special, whereas the use of Fourier series applies to various equations, as we shall see.

D'Alembert's Solution Satisfying the Initial Conditions

(5)
$$u(x,\, 0) = f(x)$$

(6)
$$u_t(x,\, 0) = g(x).$$

These are the same as in Sec. 11.3. By differentiating (4) we have

(7)
$$u_t(x,\, t) = c\phi'(x + ct) - c\psi'(x - ct)$$

where primes denote derivatives with respect to the *entire* arguments $x + ct$ and $x - ct$, respectively. From (4)–(7) we have

(8)
$$u(x,\, 0) = \phi(x) + \psi(x) = f(x),$$

(9)
$$u_t(x,\, 0) = c\phi'(x) - c\psi'(x) = g(x).$$

Dividing (9) by c and integrating with respect to x, we obtain

(10)
$$\phi(x) - \psi(x) = k(x_0) + \frac{1}{c}\int_{x_0}^{x} g(s)\, ds, \qquad k(x_0) = \phi(x_0) - \psi(x_0).$$

---

[2]JEAN LE ROND D'ALEMBERT (1717—1783), French mathematician, who is also known for his important work in mechanics.

If we add this to (8), then ψ drops out and division by 2 gives

(11)
$$\phi(x) = \frac{1}{2} f(x) + \frac{1}{2c} \int_{x_0}^{x} g(s)\, ds + \frac{1}{2} k(x_0).$$

Similarly, subtraction of (10) from (8) and division by 2 gives

(12)
$$\psi(x) = \frac{1}{2} f(x) - \frac{1}{2c} \int_{x_0}^{x} g(s)\, ds - \frac{1}{2} k(x_0).$$

In (11) we replace x by $x + ct$; we then get an integral from x_0 to $x + ct$. In (12) we replace x by $x - ct$ and get minus an integral from x_0 to $x - ct$ or plus an integral from $x - ct$ to x_0. Hence addition of $\phi(x + ct)$ and $\psi(x - ct)$ gives $u(x, t)$ [see (4)] in the form

(13)
$$u(x, t) = \frac{1}{2} [f(x + ct) + f(x - ct)] + \frac{1}{2c} \int_{x-ct}^{x+ct} g(s)\, ds.$$

If the initial velocity is zero, we see that this reduces to

(14)
$$u(x, t) = \tfrac{1}{2}[f(x + ct) + f(x - ct)],$$

in agreement with (17) in Sec. 11.3. You may show that because of the boundary conditions (2) in that section the function f must be odd and must have period $2L$.

Our result shows that the two initial conditions [the functions $f(x)$ and $g(x)$ in (5) and (6)] determine the solution uniquely.

The solution of the wave equation by the Laplace transform method will be shown in Sec. 11.12.

PROBLEM SET 11.4

1. **(Speed)** Show that c is the speed of each of the two waves given by (4).
2. **(Speed)** If a steel wire 2 meters in length weighs 0.8 nt (about 0.18 lb) and is stretched by a tensile force of 200 nt (about 45 lb), what is the corresponding speed c of transverse waves?
3. **(Frequencies)** What are the frequencies of the eigenfunctions in Prob. 2?
4. **(Periodicity)** Show that because of the boundary condition (2) in Sec. 11.3 the function f in (14) of this section must be odd and of period $2L$.

Vibrations for Different Initial Deflections
Using (14), sketch or plot a figure (similar to Fig. 263 in Sec. 11.3) of the deflection $u(x, t)$ of a vibrating string (length $L = 1$, ends fixed, $c = 1$) starting with initial velocity zero and the following initial deflection $f(x)$, where k is small, say, $k = 0.01$.

5. $f(x) = k \sin \pi x$ 6. $f(x) = kx(1 - x)$ 7. $f(x) = k(x - x^3)$ 8. $f(x) = k(1 - \cos 2\pi x)$

Types and Normal Forms of Linear Partial Differential Equations
An equation of the form

(15)
$$Au_{xx} + 2Bu_{xy} + Cu_{yy} = F(x, y, u, u_x, u_y)$$

is said to be **elliptic** if $AC - B^2 > 0$, **parabolic** if $AC - B^2 = 0$, and **hyperbolic** if $AC - B^2 < 0$.

[Here A, B, C may be functions of x and y, and the type of (15) may be different in different parts of the xy-plane.]

9. Show that

> **Laplace's equation** $u_{xx} + u_{yy} = 0$ is elliptic,
> the **heat equation** $u_t = c^2 u_{xx}$ is parabolic,
> the **wave equation** $u_{tt} = c^2 u_{xx}$ is hyperbolic,
> the **Tricomi equation** $yu_{xx} + u_{yy} = 0$ is of mixed type (elliptic in the upper half-plane and hyperbolic in the lower half-plane).

10. **(Tricomi and Airy equations)** Show that by separating variables we can obtain from the *Tricomi equation* in Prob. 9 the *Airy equation* $G'' - yG = 0$. (For solutions, see p. 446 of Ref. [1] listed in Appendix 1.)

11. **(Hyperbolic equation)** If the equation (15) is *hyperbolic*, it can be transformed to the *normal form* $u_{vz} = F^*(v, z, u, u_v, u_z)$ by setting $v = \Phi(x, y)$, $z = \Psi(x, y)$, where $\Phi = const$ and $\Psi = const$ are the solutions $y = y(x)$ of the equation $Ay'^2 - 2By' + C = 0$ (see Ref. [C5]). Show that in the case of the wave equation (1),

$$\Phi = x + ct, \qquad \Psi = x - ct.$$

12. **(Parabolic equation)** If (15) is *parabolic*, the substitution $v = x$, $z = \Psi(x, y)$ with Ψ defined as in Prob. 11 reduces it to the *normal form* $u_{vv} = F^*(v, z, u, u_v, u_z)$. Verify this result for the equation $u_{xx} + 2u_{xy} + u_{yy} = 0$.

Normal Form. Transform the following equation to normal form and solve them. (Use Probs. 11 and 12.)

13. $u_{xx} + 4u_{xy} + 4u_{yy} = 0$ **14.** $u_{xx} + u_{xy} - 2u_{yy} = 0$ **15.** $u_{xx} - 4u_{xy} + 3u_{yy} = 0$

16. $u_{xx} - 2u_{xy} + u_{yy} = 0$ **17.** $4u_{xx} - u_{yy} = 0$

18. A string moving in an elastic medium is modeled by

$$u_{tt} = c^2 u_{xx} - \gamma^2 u$$

where $\gamma^2 = const$ is proportional to the elasticity coefficient of the medium. Solve this equation for a string of length L, fixed at the ends, subject to initial displacement $f(x)$ and initial velocity zero.

19. **Longitudinal vibrations of an elastic bar or rod** in the direction of the x-axis are governed by the wave equation $u_{tt} = c^2 u_{xx}$, $c^2 = E/\rho$ (see Tolstov [C6], p. 275). If the rod is fastened at one end, $x = 0$, and free at the other, $x = L$, we have $u(0, t) = 0$ and $u_x(L, t) = 0$ (because the force at the free end is zero). Show that the motion corresponding to initial displacement $u(x, 0) = f(x)$ and initial velocity zero is

$$u = \sum_{n=0}^{\infty} A_n \sin p_n x \cos p_n ct, \quad A_n = \frac{2}{L} \int_0^L f(x) \sin p_n x \, dx, \quad p_n = \frac{(2n + 1)\pi}{2L}.$$

20. **TEAM PROJECT. Vibration of a Beam.** It can be shown that small free vertical vibrations of a uniform elastic beam (Fig. 264) are governed by the fourth-order equation

$$(16) \qquad \frac{\partial^2 u}{\partial t^2} + c^2 \frac{\partial^4 u}{\partial x^4} = 0 \qquad \text{(Ref. [C8].)}$$

where $c^2 = EI/\rho A$ (E = Young's modulus of elasticity, I = moment of inertia of the cross section with respect to the y-axis in the figure, ρ = density, A = cross-sectional area). (For the bending of a beam under a load see Example 6, Sec. 2.13.)

(a) Separating variables. Substituting $u = F(x)G(t)$ into (16) and separating variables, show that

$$F^{(4)}/F = -\ddot{G}/c^2 G = \beta^4 = const,$$

$$F(x) = A \cos \beta x + B \sin \beta x + C \cosh \beta x + D \sinh \beta x,$$

$$G(t) = a \cos c\beta^2 t + b \sin c\beta^2 t.$$

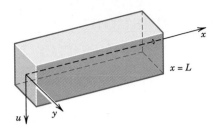

Fig. 264.   Undeformed beam in Team Project 20

(b) Simply supported beam (Fig. 265A). Find solutions $u_n = F_n(x)\,G_n(t)$ of (16) corresponding to zero initial velocity and satisfying the boundary conditions (see Fig. 265A)

$u(0, t) = 0$, $u(L, t) = 0$ (ends simply supported for all times t),

$u_{xx}(0, t) = 0$, $u_{xx}(L, t) = 0$ (zero moments, hence zero curvature, at the ends).

Find the solution of (16) that satisfies these conditions as well as the initial condition $u(x, 0) = f(x) = x(L - x)$. Compare this solution with that in Prob. 4, Sec. 11.3. What is the basic difference between the frequencies of the normal modes of the vibrating string and the vibrating beam?

(c) Clamped beam (Fig. 265B). What are the boundary conditions for the clamped beam in Fig. 265B? Show that F in (a) satisfies these conditions if βL is a solution of the equation

(17) $\cosh \beta L \cos \beta L = 1.$

Determine approximate solutions of (17).

(d) Clamped-free beam (Fig. 265C). If the beam is clamped at the left and free at the right (Fig. 265C), the boundary conditions are

$u(0, t) = 0,$ $u_x(0, t) = 0,$ $u_{xx}(L, t) = 0,$ $u_{xxx}(L, t) = 0.$

Show that F in (a) satisfies these conditions if βL is a solution of the equation

(18) $\cosh \beta L \cos \beta L = -1.$

Find approximate solutions of (18).

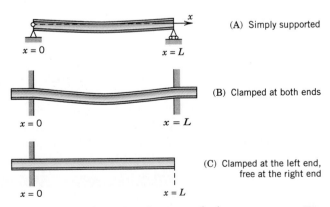

(A) Simply supported

$x = 0$ $x = L$

(B) Clamped at both ends

$x = 0$ $x = L$

(C) Clamped at the left end,
free at the right end

$x = 0$ $x = L$

Fig. 265.   Supports of a beam

11.5 Heat Equation: Solution by Fourier Series

From the wave equation we now turn to the next "big" equation, the **heat equation**

$$\frac{\partial u}{\partial t} = c^2 \nabla^2 u, \qquad\qquad c^2 = \frac{K}{\sigma\rho},$$

which gives the temperature $u(x, y, z, t)$ in a body of homogeneous material. Here c^2 is the thermal diffusivity, K the thermal conductivity, σ the specific heat, and ρ the density of the material of the body. $\nabla^2 u$ is the Laplacian of u, and with respect to Cartesian coordinates $x, y, z,$

$$\nabla^2 u = \frac{\partial^2 u}{\partial x^2} + \frac{\partial^2 u}{\partial y^2} + \frac{\partial^2 u}{\partial z^2}.$$

(The heat equation was derived in Sec. 9.8.)

As an important application, let us first consider the temperature in a long thin bar or wire of constant cross section and homogeneous material, which is oriented along the x-axis (Fig. 266) and is perfectly insulated laterally, so that heat flows in the x-direction only. Then u depends only on x and time t, and the heat equation becomes the **one-dimensional heat equation**

(1)
$$\frac{\partial u}{\partial t} = c^2 \frac{\partial^2 u}{\partial x^2}.$$

This seems to differ only very little from the wave equation, which has a term u_{tt} instead of u_t, but we shall see that this will make the solutions of (1) behave quite differently from those of the wave equation.

We shall solve (1) for some important types of boundary and initial conditions. We begin with the case in which the ends $x = 0$ and $x = L$ of the bar are kept at temperature zero, so that we have the **boundary conditions**

(2)
$$u(0, t) = 0, \qquad u(L, t) = 0 \qquad \text{for all } t,$$

and the initial temperature in the bar at time $t = 0$ is $f(x)$, so that we have the **initial condition**

(3)
$$u(x, 0) = f(x) \qquad\qquad\qquad [f(x) \text{ given}].$$

Here we must have $f(0) = 0$ and $f(L) = 0$ because of (2).

We shall determine a solution $u(x, t)$ of (1) satisfying (2) and (3)—one initial condition will be enough, as opposed to two initial conditions for the wave equation. Technically,

0 $x = L$

Fig. 266.   Bar under consideration

our method will parallel that for the wave equation in Sec. 11.3: an application of **separation of variables,** followed by the use of Fourier series. You may find a step-by-step comparison worthwhile.

First Step. Two Ordinary Differential Equations. Substitution of

(4) $$u(x, t) = F(x)G(t)$$

into (1) gives $F\dot{G} = c^2F''G$ with $\dot{G} = dG/dt$ and $F'' = d^2F/dx^2$. To separate the variables, we divide by c^2FG, obtaining

(5) $$\frac{\dot{G}}{c^2G} = \frac{F''}{F}.$$

The left side depends only on t and the right side only on x, so that both sides must be equal to a constant k (as in Sec. 11.3). You may show that for $k \geq 0$ the only solution $u = FG$ that satisfies (2) is $u \equiv 0$. For negative $k = -p^2$ we have from (5)

$$\frac{\dot{G}}{c^2G} = \frac{F''}{F} = -p^2.$$

We see that this yields the two linear ordinary differential equations

(6) $$\boxed{F'' + p^2F = 0}$$

and

(7) $$\boxed{\dot{G} + c^2p^2G = 0.}$$

Second Step. Satisfying the Boundary Conditions. We solve (6). A general solution is

(8) $$F(x) = A \cos px + B \sin px.$$

From the boundary conditions (2) it follows that

$$u(0, t) = F(0)G(t) = 0 \quad \text{and} \quad u(L, t) = F(L)G(t) = 0.$$

Since $G \equiv 0$ would give $u \equiv 0$, we require $F(0) = 0$, $F(L) = 0$ and get $F(0) = A = 0$ by (8) and then $F(L) = B \sin pL = 0$, with $B \neq 0$ (to avoid $F \equiv 0$); thus,

$$\sin pL = 0, \quad \text{hence} \quad p = \frac{n\pi}{L}, \quad n = 1, 2, \cdots.$$

Setting $B = 1$, we thus obtain the following solutions of (6) satisfying (2):

$$F_n(x) = \sin \frac{n\pi x}{L}, \quad n = 1, 2, \cdots.$$

(As in Sec. 11.3, we need not consider negative integral values of n.)

All this was literally the same as in Sec. 11.3. From now on it differs since (7) differs from (7) in Sec. 11.3. We solve (7). For $p = n\pi/L$, as just obtained, (7) is

$$\dot{G} + \lambda_n^2 G = 0 \qquad \text{where} \qquad \lambda_n = \frac{cn\pi}{L}.$$

It has the general solution

$$G_n(t) = B_n e^{-\lambda_n^2 t}, \qquad\qquad n = 1, 2, \cdots$$

where B_n is a constant. Hence the functions

(9)
$$u_n(x, t) = F_n(x)G_n(t) = B_n \sin \frac{n\pi x}{L} e^{-\lambda_n^2 t} \qquad (n = 1, 2, \cdots)$$

are solutions of the heat equation (1), satisfying (2). These are the **eigenfunctions** of the problem, corresponding to the **eigenvalues** $\lambda_n = cn\pi/L$.

Third Step. Solution of the Entire Problem. So far we have solutions (9) of (1) satisfying the boundary conditions (2). To obtain a solution also satisfying the initial condition (3), we consider a series of these eigenfunctions,

(10)
$$u(x, t) = \sum_{n=1}^{\infty} u_n(x, t) = \sum_{n=1}^{\infty} B_n \sin \frac{n\pi x}{L} e^{-\lambda_n^2 t} \qquad \left(\lambda_n = \frac{cn\pi}{L}\right).$$

From this and (3) we have

$$u(x, 0) = \sum_{n=1}^{\infty} B_n \sin \frac{n\pi x}{L} = f(x).$$

Hence for (10) to satisfy (3), the B_n's must be the coefficients of the Fourier sine series, as given by (6) in Sec. 10.4; thus

(11)
$$B_n = \frac{2}{L} \int_0^L f(x) \sin \frac{n\pi x}{L} dx \qquad (n = 1, 2, \cdots).$$

The solution of our problem can be established, assuming that $f(x)$ is piecewise continuous on the interval $0 \leq x \leq L$ (see Sec. 5.1), and has one-sided derivatives[3] at all interior points of that interval; that is, under these assumptions the series (10) with coefficients (11) is the solution of our physical problem. The proof, which requires the knowledge of uniform convergence of series, will be given at a later occasion (Probs. 15, 16 in Problem Set 14.5).

Because of the exponential factor, all the terms in (10) approach zero as t approaches infinity. The rate of decay increases with n.

---

[3]See footnote 8 in Sec. 10.2.

EXAMPLE 1   Sinusoidal initial temperature

Find the temperature $u(x, t)$ in a laterally insulated copper bar 80 cm long if the initial temperature is $100 \sin (\pi x/80)$ °C and the ends are kept at 0°C. How long will it take for the maximum temperature in the bar to drop to 50°C? First guess, then calculate. Physical data for copper: density 8.92 gm/cm³, specific heat 0.092 cal/(gm °C), thermal conductivity 0.95 cal/(cm sec °C).

Solution.   The initial condition gives

$$u(x, 0) = \sum_{n=1}^{\infty} B_n \sin \frac{n\pi x}{80} = f(x) = 100 \sin \frac{\pi x}{80} .$$

Hence, by inspection or from (10) we get $B_1 = 100$, $B_2 = B_3 = \cdots = 0$. In (10) we need $\lambda_1{}^2 = c^2\pi^2/L^2$, where $c^2 = K/(\sigma\rho) = 0.95/(0.092 \cdot 8.92) = 1.158$ [cm²/sec]. Hence we obtain $\lambda_1{}^2 = 1.158 \cdot 9.870/6400 = 0.001785$ [sec⁻¹]. The solution (10) is

$$u(x, t) = 100 \sin \frac{\pi x}{80} e^{-0.001785t} .$$

Also, $100e^{-0.001785t} = 50$ when $t = (\ln 0.5)/(-0.001785) = 388$ [seconds] ≈ 6.5 [minutes].   ◀

EXAMPLE 2   Speed of decay

Solve the problem in Example 1 when the initial temperature is $100 \sin (3\pi x/80)$ °C and the other data are as before.

Solution.   In (10), instead of $n = 1$ we now have $n = 3$, and $\lambda_3{}^2 = 3^2\lambda_1{}^2 = 9 \cdot 0.001785 = 0.01607$, so that the solution now is

$$u(x, t) = 100 \sin \frac{3\pi x}{80} e^{-0.01607t} .$$

Hence the maximum temperature drops to 50°C in $t = (\ln 0.5)/(-0.01607) \approx 43$ [seconds], which is much faster (9 times as fast as in Example 1).

   Had we chosen a bigger n, the decay would have been still faster, and in a sum or series of such terms, each term has its own rate of decay, and terms with large n are practically 0 after a very short time. Our next example is of this type, and the curve in Fig. 267 on the next page corresponding to $t = 0.5$ looks almost like a sine curve; that is, it is practically the graph of the first term of the solution.   ◀

EXAMPLE 3   "Triangular" initial temperature in a bar

Find the temperature in a laterally insulated bar of length L whose ends are kept at temperature 0, assuming that the initial temperature is

$$f(x) = \begin{cases} x & \text{if} & 0 < x < L/2, \\ L - x & \text{if} & L/2 < x < L. \end{cases}$$

(The uppermost part of Fig. 267 on the next page shows this function for the special $L = \pi$.)

Solution.   From (11) we get

(11*) $$B_n = \frac{2}{L} \left(\int_0^{L/2} x \sin \frac{n\pi x}{L} \, dx + \int_{L/2}^{L} (L - x) \sin \frac{n\pi x}{L} \, dx \right) .$$

Integration gives $B_n = 0$ if n is even,

$$B_n = \frac{4L}{n^2\pi^2} \quad (n = 1, 5, 9, \cdots) \qquad \text{and} \qquad B_n = -\frac{4L}{n^2\pi^2} \quad (n = 3, 7, 11, \cdots).$$

(see also Example 3 in Sec. 10.4 with $k = L/2$). Hence the solution is

$$u(x, t) = \frac{4L}{\pi^2} \left[\sin \frac{\pi x}{L} \exp\left[-\left(\frac{c\pi}{L}\right)^2 t \right] - \frac{1}{9} \sin \frac{3\pi x}{L} \exp\left[-\left(\frac{3c\pi}{L}\right)^2 t \right] + - \cdots \right] .$$

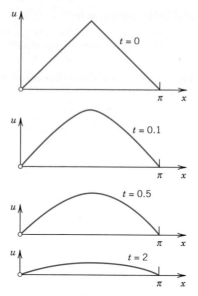

Fig. 267. Solution of Example 3
for $L = \pi$, $c = 1$, and various values of t

Figure 267 shows that the temperature decreases with increasing t, because of the heat loss due to the cooling of the ends.

Compare Fig. 267 and Fig. 263 in Sec. 11.3 and comment.   ◀

EXAMPLE 4   **Bar with insulated ends. Eigenvalue 0**

Find a solution formula of (1), (3) with (2) replaced by the condition that both ends of the bar are insulated.

Solution.   Physical experiments show that the rate of heat flow is proportional to the gradient of the temperature. Hence if the ends $x = 0$ and $x = L$ of the bar are insulated, so that no heat can flow through the ends, we have the boundary conditions

(2*) $$u_x(0, t) = 0, \qquad u_x(L, t) = 0 \qquad\qquad \text{for all } t.$$

Since $u(x, t) = F(x)G(t)$, this gives

$$u_x(0, t) = F'(0)G(t) = 0, \qquad u_x(L, t) = F'(L)G(t) = 0.$$

Differentiating (8), we have $F'(x) = -Ap \sin px + Bp \cos px$, so that

$$F'(0) = Bp = 0 \qquad \text{and then} \qquad F'(L) = -Ap \sin pL = 0.$$

The second of these conditions gives

$$p = p_n = \frac{n\pi}{L}, \qquad\qquad n = 0, 1, 2, \cdots.$$

From this and (8) with $A = 1$ and $B = 0$ we get

$$F_n(x) = \cos \frac{n\pi x}{L}, \qquad\qquad n = 0, 1, 2, \cdots.$$

With G_n as before, this yields the eigenfunctions

(12) $$u_n(x, t) = F_n(x)G_n(t) = A_n \cos \frac{n\pi x}{L} e^{-\lambda_n^2 t} \qquad\qquad (n = 0, 1, \cdots)$$

corresponding to the eigenvalues $\lambda_n = cn\pi/L$. The latter are as before, but we now have the additional eigenvalue $\lambda_0 = 0$ and eigenfunction $u_0 = const$, which is the solution of the problem if the initial temperature $f(x)$ is constant. This shows the remarkable fact that *a separation constant can very well be zero, and zero can be an eigenvalue.*

Furthermore, whereas (9) gave a Fourier sine series, we now get from (12) a Fourier cosine series

(13)
$$u(x, t) = \sum_{n=0}^{\infty} u_n(x, t) = \sum_{n=0}^{\infty} A_n \cos \frac{n\pi x}{L} e^{-\lambda_n^2 t}$$

$$\left(\lambda_n = \frac{cn\pi}{L} \right)$$

insulated

with coefficients resulting from the initial condition (3),

$$u(x, 0) = \sum_{n=0}^{\infty} A_n \cos \frac{n\pi x}{L} = f(x),$$

in the form [see (4), Sec. 10.4]

(14)
$$A_0 = \frac{1}{L} \int_0^L f(x) \, dx, \qquad A_n = \frac{2}{L} \int_0^L f(x) \cos \frac{n\pi x}{L} \, dx, \qquad n = 1, 2, \cdots. \qquad ◀$$

EXAMPLE 5   **"Triangular" initial temperature in a bar with insulated ends**

Find the temperature in the bar in Example 3, assuming that the ends are insulated (instead of being kept at temperature 0).

Solution.   For the triangular initial temperature, (14) gives $A_0 = L/4$ and (see also Example 3 in Sec. 10.4 with $k = L/2$)

$$A_n = \frac{2}{L} \left[\int_0^{L/2} x \cos \frac{n\pi x}{L} \, dx + \int_{L/2}^{L} (L - x) \cos \frac{n\pi x}{L} \, dx \right]$$

$$= \frac{2L}{n^2\pi^2} \left(2 \cos \frac{n\pi}{2} - \cos n\pi - 1 \right).$$

Hence the solution (13) is

$$u(x, t) = \frac{L}{4} - \frac{8L}{\pi^2} \left\{ \frac{1}{2^2} \cos \frac{2\pi x}{L} \exp \left[-\left(\frac{2c\pi}{L} \right)^2 t \right] + \frac{1}{6^2} \cos \frac{6\pi x}{L} \exp \left[-\left(\frac{6c\pi}{L} \right)^2 t \right] + \cdots \right\}.$$

We see that the terms decrease with increasing t, and $u \to L/4$, the mean value of the initial temperature. This is plausible because no heat can escape from this totally insulated bar. In contrast, the cooling of the ends in Example 3 led to heat loss and $u \to 0$, the temperature at which the ends were kept. ◀

Steady-State Two-Dimensional Heat Flow

The two-dimensional heat equation is (see the beginning of this section)

$$\frac{\partial u}{\partial t} = c^2 \nabla^2 u = c^2 \left(\frac{\partial^2 u}{\partial x^2} + \frac{\partial^2 u}{\partial y^2} \right).$$

If the heat flow is **steady** (that is, time independent), then $\partial u/\partial t = 0$, and the heat equation reduces to **Laplace's equation**[4]

(15)
$$\nabla^2 u = \frac{\partial^2 u}{\partial x^2} + \frac{\partial^2 u}{\partial y^2} = 0.$$

---

[4]This very important equation appeared in Sec. 9.8 and will be discussed further in Secs. 11.9, 11.11, 12.4, and in Chap. 16.

A heat problem then consists of this equation to be considered in some region R of the xy-plane and a given boundary condition on the boundary curve of R. This is called a **boundary value problem.** One calls it:

> **Dirichlet problem** if u is prescribed on C,
>
> **Neumann problem** if the normal derivative $u_n = \partial u/\partial n$ is prescribed on C,
>
> **Mixed problem** if u is prescribed on a portion of C and u_n on the rest of C.

Dirichlet Problem in a Rectangle R (Fig. 268). We consider a Dirichlet problem for (15) in a rectangle R, assuming that the temperature $u(x, y)$ equals a given function $f(x)$ on the upper side and 0 on the other three sides of the rectangle.

We solve this problem by separating variables. Substitution of

$$u(x, y) = F(x)G(y)$$

into (15) and division by FG gives

$$\frac{1}{F} \cdot \frac{d^2F}{dx^2} = -\frac{1}{G} \cdot \frac{d^2G}{dy^2} = -k.$$

From this and the left and right boundary conditions,

$$\frac{d^2F}{dx^2} + kF = 0, \qquad F(0) = 0, \quad F(a) = 0.$$

This gives $k = (n\pi/a)^2$ and corresponding nonzero solutions

$$(16) \qquad\qquad F(x) = F_n(x) = \sin \frac{n\pi}{a} x, \qquad\qquad n = 1, 2, \cdots.$$

The differential equation for G then becomes

$$\frac{d^2G}{dy^2} - \left(\frac{n\pi}{a}\right)^2 G = 0.$$

Solutions are

$$G(y) = G_n(y) = A_n e^{n\pi y/a} + B_n e^{-n\pi y/a}.$$

Now the boundary condition $u = 0$ on the lower side of R implies that $G_n(0) = 0$; that is, $G_n(0) = A_n + B_n = 0$ or $B_n = -A_n$. This gives

$$G_n(y) = A_n(e^{n\pi y/a} - e^{-n\pi y/a}) = 2A_n \sinh \frac{n\pi y}{a}.$$

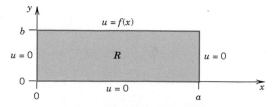

Fig. 268. Rectangle R and given boundary values

From this and (16), writing $2A_n = A_n^*$, we obtain as the **eigenfunctions** of our problem

(17) $$u_n(x, y) = F_n(x)G_n(y) = A_n^* \sin \frac{n\pi x}{a} \sinh \frac{n\pi y}{a}.$$

These satisfy the boundary condition $u = 0$ on the left, right, and lower sides.

　To get a solution also satisfying the boundary condition

(18) $$u(x, b) = f(x)$$

on the upper side, we consider the infinite series

$$u(x, y) = \sum_{n=1}^{\infty} u_n(x, y).$$

From this, (18), and (17) with $y = b$ we obtain

$$u(x, b) = f(x) = \sum_{n=1}^{\infty} A_n^* \sin \frac{n\pi x}{a} \sinh \frac{n\pi b}{a}.$$

We can write this in the form

$$u(x, b) = \sum_{n=1}^{\infty} \left(A_n^* \sinh \frac{n\pi b}{a} \right) \sin \frac{n\pi x}{a}.$$

This shows that the expressions in the parentheses must be the Fourier coefficients b_n of $f(x)$; that is, by (6) in Sec. 10.4,

$$b_n = A_n^* \sinh \frac{n\pi b}{a} = \frac{2}{a} \int_0^a f(x) \sin \frac{n\pi x}{a} \, dx.$$

From this and (17) we see that the solution of our problem is

(19) $$u(x, y) = \sum_{n=1}^{\infty} A_n^* \sin \frac{n\pi x}{a} \sinh \frac{n\pi y}{a}$$

where

(20) $$A_n^* = \frac{2}{a \sinh (n\pi b/a)} \int_0^a f(x) \sin \frac{n\pi x}{a} \, dx.$$

This solution, obtained formally without regard to the convergence and the sums of the series for u, u_{xx}, and u_{yy}, can be established when f and f' are continuous and f'' is piecewise continuous on the interval $0 \leq x \leq a$. The proof is somewhat involved and relies on uniform convergence; it can be found in Ref. [C2] listed in Appendix 1.

Electrostatic Potential. Elastic Membrane

The Laplace equation (15) also governs the electrostatic potential of electrical charges in any region that is free of these charges. Thus our steady-state heat problem can also be interpreted as an electrostatic potential problem. Then (19), (20) is the potential in the rectangle R when the upper side of R is at potential $f(x)$ and the other three sides are grounded.

Actually, in the steady-state case, the two-dimensional wave equation (to be considered in Secs. 11.7, 11.8) also reduces to (15). Then (19), (20) is the displacement of a rectangular elastic membrane (rubber sheet, drumhead) that is fixed along its boundary, with three sides lying in the xy-plane and the fourth side given the displacement $f(x)$.

This is another impressive demonstration of the **unifying power** of mathematics. It illustrates that entirely different physical systems may have the same mathematical model and can thus be treated by the same mathematical methods.

PROBLEM SET 11.5

1. **WRITING PROJECT. Wave and Heat Equations.** Write a short essay on the different general behavior of the solutions of these two equations. Begin with a comparison of Figs. 263 and 267. Then explain what you can say in more general situations.

2. **(Decay)** If the first eigenfunction (9) of the bar decreases to half its value within 20 seconds, what is the value of the diffusivity c^2? How does the rate of decay of (9) for fixed n depend on the specific heat, the density, and the thermal conductivity of the material?

Bar Insulated as in the Text

Find the temperature $u(x, t)$ in a bar of silver (length 10 cm, constant cross section of area 1 cm^2, density 10.6 gm/cm^3, thermal conductivity 1.04 cal/(cm sec °C), specific heat 0.056 cal/(gm °C)) that is perfectly insulated laterally, whose ends are kept at temperature 0°C and whose initial temperature (in °C) is $f(x)$, where

3. $f(x) = \sin 0.1\pi x$ 4. $f(x) = k \sin 0.2\pi x$ 5. $f(x) = x(10 - x)$ 6. $f(x) = 2 - 0.4|x - 5|$

7. **(Different temperatures at the ends)** What is the limit $u_I(x)$ of the temperature of the bar in the text as $t \to \infty$ if the ends are kept at $u(0, t) = U_1 = const$ and $u(L, t) = U_2 = const$?

8. **(Different temperatures)** What is the temperature in the bar in Prob. 7 at any time?

Adiabatic and Other Conditions

9. **(Insulated ends, adiabatic boundary conditions)** The heat flux through the faces at the ends of a bar is found to be proportional to $u_n = \partial u/\partial n$ at the ends. Show that if the bar is perfectly insulated, also at the ends $x = 0$, $x = L$ (**"adiabatic conditions"**), and the initial temperature is $f(x)$, then

$$u_x(0, t) = 0, \qquad u_x(L, t) = 0, \qquad u(x, 0) = f(x)$$

and separating variables gives the solution

$$u(x, t) = A_0 + \sum_{n=1}^{\infty} A_n \cos \frac{n\pi x}{L} \exp\left[-\left(\frac{cn\pi}{L}\right)^2 t\right]$$

with $A_n = a_n$ given by (4), Sec. 10.4. Note that $u \to A_0$ as $t \to \infty$. Does this agree with your physical intuition?

Adiabatic Conditions. Find the temperature in the bar in Prob. 9 when $L = \pi$, $c = 1$, and

10. $f(x) = x$ **11.** $f(x) = k = const$ **12.** $f(x) = \cos 2x$ **13.** $f(x) = 1 - x/\pi$

14. (**Nonhomogeneous heat equation**) Show that the problem consisting of

$$u_t - c^2 u_{xx} = Ne^{-\alpha x}$$

and (2), (3) can be reduced to a problem for the homogeneous heat equation by setting $u(x, t) = v(x, t) + w(x)$ and determining w so that v satisfies the homogeneous equation and the conditions $v(0, t) = v(L, t) = 0$, $v(x, 0) = f(x) - w(x)$. (The term $Ne^{-\alpha x}$ may represent heat loss due to radioactive decay in the bar.)

15. The **boundary condition of heat transfer**

(21) $-u_x(\pi, t) = k[u(\pi, t) - u_0]$

applies when a bar of length π with $c = 1$ is laterally insulated, the left end $x = 0$ is kept at 0°C, and at the right end heat is flowing into air of constant temperature u_0. Let $k = 1$ for simplicity, and $u_0 = 0$. Show that a solution is $u(x, t) = \sin px \, e^{-p^2 t}$, where p is a solution of $\tan p\pi = -p$. Show graphically that this equation has infinitely many positive solutions $p_1, p_2, p_3, \cdots$, where $p_n > n - \frac{1}{2}$ and $\lim_{n\to\infty} (p_n - n + \frac{1}{2}) = 0$. (Formula (21) is also known as radiation boundary condition, but this is misleading; see Ref. [C1], p. 19.)

16. The **heat** of a solution $u(x, t)$ across $x = 0$ is defined by $\phi(t) = -Ku_x(0, t)$. Find the heat flux for the solution (10) and note that it goes to zero as $t \to \infty$. Is this physically understandable? Explain the name.

Two-Dimensional Problems

17. (**Heat flow in a plate**) The faces of a thin square copper plate (Fig. 269, where $a = 24$) are perfectly insulated. The upper side is kept at 20°C and the other sides are kept at 0°C. Find the steady-state temperature $u(x, y)$ in the plate. (Solve the Laplace equation.)

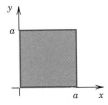

Fig. 269.   Square plate

 18. CAS PROJECT. Isotherms. Find steady-state solutions (temperatures) in the square plate in Fig. 269 with $a = 2$ satisfying the following boundary conditions. Plot isotherms.
(**a**) $u = \sin \pi x$ on the upper side, 0 on the others.
(**b**) $u = 0$ on the vertical sides, assuming that the other sides are perfectly insulated.
(**c**) Boundary conditions of your choice (such that the solution is not identically zero).

19. (**Mixed boundary value problem**) Find the steady-state temperature in the plate in Prob. 17 with the upper and lower sides perfectly insulated, the left side kept at 0°C, and the right side at $f(y)$ °C.

20. Find steady-state temperatures in the rectangle in Fig. 268 in the text if the upper and left sides are perfectly insulated and at the right side heat is flowing into a medium at zero temperature according to $u_x(a, y) + hu(a, y) = 0$, $h > 0$ constant. (You will get many solutions because no condition on the lower side is given.)

11.6 Heat Equation: Solution by Fourier Integrals and Transforms

Our discussion in the last section extends to infinite bars, with Fourier series replaced by Fourier integrals (Sec. 10.8). Such bars are good models of very long bars or wires (such as a wire 300 ft long). Specifically, we shall find solutions of the heat equation

$$(1) \qquad \frac{\partial u}{\partial t} = c^2 \frac{\partial^2 u}{\partial x^2}$$

in a bar that extends to infinity on both sides (and is laterally insulated, as before). Then we do not have boundary conditions, but only the initial condition

$$(2) \qquad u(x, 0) = f(x) \qquad\qquad (-\infty < x < \infty)$$

where $f(x)$ is the given initial temperature of the bar.

To solve this problem, we start as in the last section, substituting $u(x, t) = F(x)G(t)$ into (1). This gives the two ordinary differential equations

$$(3) \qquad\qquad F'' + p^2 F = 0 \qquad\qquad \text{[see (6), Sec. 11.5]}$$

and

$$(4) \qquad\qquad \dot{G} + c^2 p^2 G = 0 \qquad\qquad \text{[see (7), Sec. 11.5].}$$

Solutions are

$$F(x) = A \cos px + B \sin px \qquad \text{and} \qquad G(t) = e^{-c^2 p^2 t},$$

respectively; here A and B are any constants. Hence a solution of (1) is

$$(5) \qquad u(x, t; p) = FG = (A \cos px + B \sin px)e^{-c^2 p^2 t}.$$

[As in the last section, we had to choose the separation constant k negative, $k = -p^2$, because positive values of k would lead to an increasing exponential function in (5), which has no physical meaning.]

Use of Fourier Integrals

Any series of functions (5), found in the usual manner by taking p as multiples of a fixed number, would lead to a function that is periodic in x when $t = 0$. However, since $f(x)$ in (2) is not assumed to be periodic, it is natural to use **Fourier integrals** instead of Fourier series. Also, A and B in (5) are arbitrary and we may regard them as functions of p, writing $A = A(p)$ and $B = B(p)$. Now, since the heat equation (1) is linear and homogeneous, the function

$$(6) \qquad u(x, t) = \int_0^\infty u(x, t; p)\, dp = \int_0^\infty [A(p) \cos px + B(p) \sin px]e^{-c^2 p^2 t}\, dp$$

is then a solution of (1), provided this integral exists and can be differentiated twice with respect to x and once with respect to t.

Determination of $A(p)$ and $B(p)$ from the Initial Condition. From (6) and (2) we get

$$(7) \qquad u(x, 0) = \int_0^\infty [A(p) \cos px + B(p) \sin px] \, dp = f(x).$$

This gives $A(p)$ and $B(p)$ in terms of $f(x)$; indeed, from (4) and (5) in Sec. 10.8 we have

$$(8) \qquad A(p) = \frac{1}{\pi} \int_{-\infty}^\infty f(v) \cos pv \, dv, \qquad B(p) = \frac{1}{\pi} \int_{-\infty}^\infty f(v) \sin pv \, dv.$$

According to (1*), Sec. 10.10, our Fourier integral (7) with these $A(p)$ and $B(p)$ can be written

$$u(x, 0) = \frac{1}{\pi} \int_0^\infty \left[\int_{-\infty}^\infty f(v) \cos (px - pv) \, dv \right] dp$$

and, similarly, (6) in this section becomes

$$u(x, t) = \frac{1}{\pi} \int_0^\infty \left[\int_{-\infty}^\infty f(v) \cos (px - pv) \, e^{-c^2p^2t} \, dv \right] dp.$$

Assuming that we may invert the order of integration, we obtain

$$(9) \qquad u(x, t) = \frac{1}{\pi} \int_{-\infty}^\infty f(v) \left[\int_0^\infty e^{-c^2p^2t} \cos (px - pv) \, dp \right] dv.$$

Then we can evaluate the inner integral by the formula

$$(10) \qquad \int_0^\infty e^{-s^2} \cos 2bs \, ds = \frac{\sqrt{\pi}}{2} e^{-b^2}.$$

[A derivation of (10) is given in Problem Set 15.4 (Team Project 30).] This takes the form of our inner integral if we choose $p = s/c\sqrt{t})$ as a new variable of integration and set

$$b = \frac{x - v}{2c\sqrt{t}}.$$

Then $2bs = (x - v)p$ and $ds = c\sqrt{t} \, dp$, so that (10) becomes

$$\int_0^\infty e^{-c^2p^2t} \cos (px - pv) \, dp = \frac{\sqrt{\pi}}{2c\sqrt{t}} \exp \left\{ -\frac{(x - v)^2}{4c^2t} \right\}.$$

By inserting this result into (9) we obtain the representation

$$(11) \qquad u(x, t) = \frac{1}{2c\sqrt{\pi t}} \int_{-\infty}^\infty f(v) \exp \left\{ -\frac{(x - v)^2}{4c^2t} \right\} dv.$$

Taking $z = (v - x)/(2c\sqrt{t})$ as a variable of integration, we get the alternative form

(12)
$$u(x, t) = \frac{1}{\sqrt{\pi}} \int_{-\infty}^{\infty} f(x + 2cz\sqrt{t})\, e^{-z^2}\, dz.$$

If $f(x)$ is bounded for all values of x and integrable in every finite interval, it can be shown (see Ref. [C7]) that the function (11) or (12) satisfies (1) and (2). Hence this function is the required solution in the present case.

EXAMPLE 1 **Temperature in an infinite bar**

Find the temperature in the infinite bar if the initial temperature is (Fig. 270)

$$f(x) = \begin{cases} U_0 = const & \text{if } |x| < 1, \\ 0 & \text{if } |x| > 1. \end{cases}$$

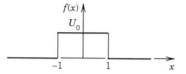

Fig. 270.   Initial temperature in Example 1

Solution.   From (11) we have

$$u(x, t) = \frac{U_0}{2c\sqrt{\pi t}} \int_{-1}^{1} \exp\left\{ -\frac{(x - v)^2}{4c^2 t} \right\} dv.$$

If we introduce the above variable of integration z, then the integration over v from -1 to 1 corresponds to the integration over z from $(-1 - x)/(2c\sqrt{t})$ to $(1 - x)/(2c\sqrt{t})$, and

(13)
$$u(x, t) = \frac{U_0}{\sqrt{\pi}} \int_{-(1+x)/2c\sqrt{t}}^{(1-x)/2c\sqrt{t}} e^{-z^2}\, dz \qquad\qquad (t > 0).$$

We mention that this integral is not an elementary function, but can easily be expressed in terms of the error function, whose values have been tabulated. (Table A4 in Appendix 5 contains a few values; larger tables are listed in Ref. [1] in Appendix 1. See also CAS Project 10, below.) Figure 271 shows $u(x, t)$ for $U_0 = 100°C$, $c^2 = 1$ cm^2/sec, and several values of t. ◀

Use of Fourier Transforms

The Fourier transform is closely related to the Fourier integral, from which we obtained it in Sec. 10.10. And the transition to the Fourier cosine and sine transform in Sec. 10.9 was even simpler. (You may perhaps wish to review this before going on.) Hence it should not surprise you that we can use these transforms for solving our present or similar problems. The Fourier transform applies to problems concerning the entire axis, and the Fourier cosine and sine transforms to problems involving the positive half-axis. Let us explain these transform methods by typical applications that fit our present discussion.

EXAMPLE 2 **Temperature in the infinite bar in Example 1**

Solve Example 1 using the Fourier transform.

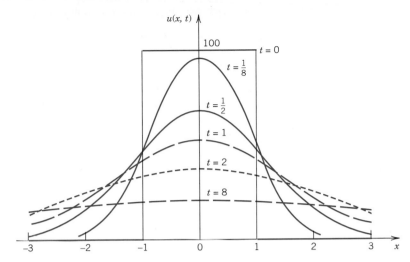

Fig. 271. Solution $u(x, t)$ to Example 1 for $U_0 = 100°C$, $c^2 = 1 \text{ cm}^2/\text{sec}$, and several values of t

Solution. The problem consists of the heat equation (1) and the initial condition (2), which in this example is

$$f(x) = U_0 = const \quad \text{if } |x| < 1 \qquad \text{and 0 otherwise.}$$

Our strategy is to take the Fourier transform with respect to x and then to solve the resulting ordinary differential equation in t. The details are as follows.

Let $\hat{u} = \mathcal{F}(u)$ denote the Fourier transform of u, ***regarded as a function of x.*** From (10) in Sec. 10.10 we see that the heat equation (1) gives

$$\mathcal{F}(u_t) = c^2 \mathcal{F}(u_{xx}) = c^2(-w^2)\mathcal{F}(u) = -c^2 w^2 \hat{u}.$$

On the left, assuming that we may interchange the order of differentiation and integration, we have

$$\mathcal{F}(u_t) = \frac{1}{\sqrt{2\pi}} \int_{-\infty}^{\infty} u_t e^{-iwx} \, dx = \frac{1}{\sqrt{2\pi}} \frac{\partial}{\partial t} \int_{-\infty}^{\infty} u e^{-iwx} \, dx = \frac{\partial \hat{u}}{\partial t}.$$

Thus

$$\frac{\partial \hat{u}}{\partial t} = -c^2 w^2 \hat{u}.$$

Since this equation involves only a derivative with respect to t but none with respect to w, this is a first-order ***ordinary*** differential equation with t as the independent variable and w as a parameter. By separating variables (Sec. 1.3) we get the general solution

$$\hat{u}(w, t) = C(w)e^{-c^2 w^2 t},$$

with the arbitrary "constant" $C(w)$ depending on the parameter w. The initial condition (1) gives $\hat{u}(w, 0) = C(w) = \hat{f}(w) = \mathcal{F}(f)$. Our intermediate result is

$$\hat{u}(w, t) = \hat{f}(w)e^{-c^2 w^2 t}.$$

The inversion formula (7), Sec. 10.10, now gives the solution

$$(14) \qquad\qquad u(x, t) = \frac{1}{\sqrt{2\pi}} \int_{-\infty}^{\infty} \hat{f}(w)e^{-c^2 w^2 t} e^{iwx} \, dw.$$

In this we may insert the Fourier transform

$$\hat{f}(w) = \frac{1}{\sqrt{2\pi}} \int_{-\infty}^{\infty} f(v)e^{-ivw} \, dv.$$

Assuming that we may invert the order of integration, we then obtain

$$u(x, t) = \frac{1}{2\pi} \int_{-\infty}^{\infty} f(v) \left[\int_{-\infty}^{\infty} e^{-c^2 w^2 t} e^{i(wx - wv)} \, dw \right] dv.$$

By the Euler formula (3), Sec. 10.10, the integrand of the inner integral equals

$$e^{-c^2 w^2 t} \cos (wx - wv) + i e^{-c^2 w^2 t} \sin (wx - wv).$$

This shows that its imaginary part is an odd function of w, so that the integral[5] of this part is 0, and the real part is even, so that its integral is twice the integral from 0 to ∞:

$$u(x, t) = \frac{1}{\pi} \int_{-\infty}^{\infty} f(v) \left[\int_{0}^{\infty} e^{-c^2 w^2 t} \cos (wx - wv) \, dw \right] dv.$$

This agrees with (9) and leads to the further formulas (11) and (13).   ◀

EXAMPLE 3   **Solution in Example 1 by the method of convolution**

Solve the heat problem in Example 1 by the method of convolution.

Solution.   The beginning is as in Example 2 and leads to (14), that is,

$$(15) \qquad\qquad u(x, t) = \frac{1}{\sqrt{2\pi}} \int_{-\infty}^{\infty} \hat{f}(w) e^{-c^2 w^2 t} e^{iwx} \, dw.$$

Now comes the crucial idea. We recognize that this is of the form (13) in Sec. 10.10, that is,

$$(16) \qquad\qquad u(x, t) = (f * g)(x) = \int_{-\infty}^{\infty} \hat{f}(w) \hat{g}(w) e^{iwx} \, dw$$

where

$$(17) \qquad\qquad \hat{g}(w) = \frac{1}{\sqrt{2\pi}} e^{-c^2 w^2 t}.$$

Since, by the definition of convolution [(11), Sec. 10.10],

$$(18) \qquad\qquad (f * g)(x) = \int_{-\infty}^{\infty} f(p) g(x - p) \, dp,$$

as our next and last step we must determine the inverse Fourier transform g of $\hat{g}$. For this we can use formula 9 in Table III of Sec. 10.11 (which was derived in Example 2 of Sec. 10.10),

$$\mathcal{F}(e^{-ax^2}) = \frac{1}{\sqrt{2a}} e^{-w^2/4a}$$

with a suitable a. With $c^2 t = 1/4a$ or $a = 1/4c^2 t$, using (17) we obtain

$$\mathcal{F}(e^{-x^2/4c^2 t}) = \sqrt{2c^2 t} \, e^{-c^2 w^2 t} = \sqrt{2c^2 t} \, \sqrt{2\pi} \hat{g}(w).$$

Hence $\hat{g}$ has the inverse

$$\frac{1}{\sqrt{2c^2 t} \sqrt{2\pi}} e^{-x^2/4c^2 t}.$$

---

[5]Actually, the principal part of the integral; see Sec. 15.4.

Replacing x with $x - p$ and substituting this into (18) we finally have

(19)
$$u(x, t) = (f * g)(x) = \frac{1}{2c\sqrt{\pi t}} \int_{-\infty}^{\infty} f(p) \exp\left\{-\frac{(x - p)^2}{4c^2 t}\right\} dp.$$

This solution formula of our problem agrees with (11). We wrote $(f * g)(x)$, without indicating the parameter t with respect to which we did not integrate.   ◄

EXAMPLE 4   **Fourier sine transform applied to the heat equation**

If a laterally insulated bar extends from $x = 0$ to infinity, we can use the Fourier sine transform. We let the initial temperature be $u(x, 0) = f(x)$ and impose the boundary condition $u(0, t) = 0$. Then from the heat equation and (9b) in Sec. 10.9, since $f(0) = u(0, 0) = 0$ we obtain

$$\mathcal{F}_s(u_t) = \frac{\partial \hat{u}_s}{\partial t} = c^2 \mathcal{F}_s(u_{xx}) = -c^2 w^2 \mathcal{F}_s(u) = -c^2 w^2 \hat{u}_s(w, t).$$

This is a first-order ordinary differential equation. Its solution is

$$\hat{u}_s(w, t) = C(w)e^{-c^2 w^2 t}.$$

From the initial condition $u(x, 0) = f(x)$ we have $\hat{u}_s(w, 0) = \hat{f}_s(w) = C(w)$. Hence

$$\hat{u}_s(w, t) = \hat{f}_s(w)e^{-c^2 w^2 t}.$$

Taking the inverse Fourier sine transform and substituting

$$\hat{f}_s(w) = \sqrt{\frac{2}{\pi}} \int_0^{\infty} f(p) \sin wp \, dp,$$

we obtain the solution formula

(20)
$$u(x, t) = \frac{2}{\pi} \int_0^{\infty} \int_0^{\infty} f(p) \sin wp \, e^{-c^2 w^2 t} \sin wx \, dp \, dw.$$ ◄

PROBLEM SET 11.6

 1. CAS PROJECT. Heat Flow. (a) Plot the basic Fig. 271 on the computer.
   (b) Plot $u(x, t)$ shown in Fig. 271 as a surface over the upper xt-half-plane.

Solutions in Integral Form

Using (6), obtain the solution of (1) in integral form subject to the initial condition $u(x, 0) = f(x)$, where

2. $f(x) = 1/(1 + x^2)$. [Use (15), Sec. 10.8.]
3. $f(x) = 1$ if $|x| < 1$ and 0 otherwise
4. $f(x) = e^{-|x|}$. [Use Example 3, Sec. 10.8.]
5. $f(x) = (\sin x)/x$. [Use Prob. 2, Sec. 10.8.]
6. Verify by integration that u in Prob. 5 satisfies the initial condition.

7. (Formula (12)) If $f(x) = 1$ when $x > 0$ and $f(x) = 0$ when $x < 0$, show that

$$u(x, t) = \frac{1}{\sqrt{\pi}} \int_{-x/2c\sqrt{t}}^{\infty} e^{-z^2} dz \qquad (t > 0).$$

8. (Normal distribution) Introducing $w = z\sqrt{2}$ as a new variable of integration, show that (12) becomes

$$u(x, t) = \frac{1}{\sqrt{2\pi}} \int_{-\infty}^{\infty} f(x + cw\sqrt{2t})\, e^{-w^2/2}\, dw.$$

Students familiar with probability theory will notice that this involves the density $e^{-w^2/2}/\sqrt{2\pi}$ of the normal distribution (see Sec. 22.8).

9. (Normal distribution) Show that the distribution function of the normal distribution (Sec. 22.8) can be written in terms of the error function,

$$\Phi(x) = \frac{1}{\sqrt{2\pi}} \int_{-\infty}^{x} e^{-s^2/2}\, ds = \frac{1}{2} + \frac{1}{2}\, \mathrm{erf}\left(\frac{x}{\sqrt{2}}\right).$$

 **10. CAS PROJECT. Error Function**

(21)

$$\mathrm{erf}\, x = \frac{2}{\sqrt{\pi}} \int_{0}^{x} e^{-w^2}\, dw$$

This function is important in applied mathematics and physics (probability theory, thermodynamics, etc.) and fits our present discussion. As a model case on special functions defined by integrals (that cannot be evaluated by the usual methods of calculus), do the following.

(a) Sketch or plot the **bell-shaped curve** [the curve of the integrand in (21)]. Show that $\mathrm{erf}\, x$ is odd. Show that

$$\int_{a}^{b} e^{-w^2}\, dw = \frac{\sqrt{\pi}}{2}\, (\mathrm{erf}\, b - \mathrm{erf}\, a), \qquad \int_{-b}^{b} e^{-w^2}\, dw = \sqrt{\pi}\, \mathrm{erf}\, b.$$

(b) Obtain the Maclaurin series of $\mathrm{erf}\, x$ by integrating that of the integrand. Use this series to compute a table of $\mathrm{erf}\, x$ for $x = 0, 0.01, 0.02, \cdots, 3$.

(c) Obtain the values in (b) by an integration command of your CAS. Compare accuracy.

(d) It can be shown that $\mathrm{erf}\, (\infty) = 1$. Confirm this experimentally by computing $\mathrm{erf}\, x$ for large x.

(e) Express the temperature (13) in terms of the error function.

(f) Using $\mathrm{erf}\, (\infty) = 1$, express the formula in Prob. 7 in terms of the error function.

11.7 Modeling: Membrane, Two-Dimensional Wave Equation

As another basic vibrational problem, let us consider the motion of a stretched elastic membrane, such as a drumhead. This is the two-dimensional analog of the vibrating string problem; indeed, the modeling will be similar to that in Sec. 11.2.

Physical Assumptions. We assume the following:

1. *The mass of the membrane per unit area is constant ("homogeneous membrane"). The membrane is perfectly flexible and offers no resistance to bending.*

2. *The membrane is stretched and then fixed along its entire boundary in the xy-plane. The tension per unit length T caused by stretching the membrane is the same at all points and in all directions and does not change during the motion.*

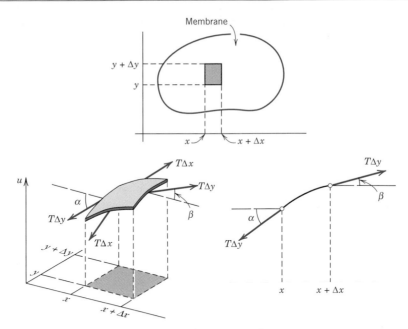

Fig. 272. Vibrating membrane

3. *The deflection u(x, y, t) of the membrane during the motion is small compared to the size of the membrane, and all angles of inclination are small.*

Although one cannot realize these assumptions exactly, they hold relatively accurately for small transverse vibrations of a thin elastic membrane, so that we obtain a good model, for instance, of a drumhead, as we shall see.

Derivation of Differential Equation. We obtain the differential equation that governs the motion of the membrane by considering the forces acting on a small portion of the membrane in Fig. 272. Since the deflections of the membrane and the angles of inclination are small, the sides of the portion are approximately equal to Δx and Δy. The tension T is the force per unit length. Hence the forces acting on the sides of the portion are approximately $T \Delta x$ and $T \Delta y$. Since the membrane is perfectly flexible, these forces are tangent to the membrane.

Horizontal Components of the Forces. We first consider the horizontal components of the forces. These are obtained by multiplying the forces by the cosines of the angles of inclination. Since these angles are small, their cosines are close to 1. Hence the horizontal components of the forces at opposite sides are approximately equal. Therefore, the motion of the particles of the membrane in a horizontal direction will be negligibly small. From this we conclude that we may regard the motion of the membrane as transversal; that is, each particle moves vertically.

Vertical Components of the Forces. These components along the right side and the left side are[6] (Fig. 272), respectively,

$$T \, \Delta y \sin \beta \qquad \text{and} \qquad -T \, \Delta y \sin \alpha.$$

---

[6]Note that the angle of inclination varies along the sides, and α and β represent values of that angle at a suitable point of the sides under consideration.

Here the minus sign appears because the force on the left side is directed downward. Since the angles are small, we may replace their sines by their tangents. Hence the resultant of those two vertical components is

(1) $$T \, \Delta y \, (\sin \beta - \sin \alpha) \approx T \, \Delta y \, (\tan \beta - \tan \alpha)$$

$$= T \, \Delta y \, [u_x(x + \Delta x, y_1) - u_x(x, y_2)]$$

where subscripts x denote partial derivatives and y_1 and y_2 are values between y and $y + \Delta y$. Similarly, the resultant of the vertical components of the forces acting on the other two sides of the portion is

(2) $$T \, \Delta x \, [u_y(x_1, y + \Delta y) - u_y(x_2, y)]$$

where x_1 and x_2 are values between x and $x + \Delta x$.

Newton's Second Law Gives the Differential Equation. By Newton's second law (see Sec. 2.5), the sum of the forces given by (1) and (2) is equal to the mass $\rho \, \Delta A$ of that small portion times the acceleration $\partial^2 u / \partial t^2$; here ρ is the mass of the undeflected membrane per unit area, and $\Delta A = \Delta x \, \Delta y$ is the area of that portion when it is undeflected. Thus

$$\rho \, \Delta x \, \Delta y \, \frac{\partial^2 u}{\partial t^2} = T \, \Delta y \, [u_x(x + \Delta x, y_1) - u_x(x, y_2)]$$

$$+ \, T \, \Delta x \, [u_y(x_1, y + \Delta y) - u_y(x_2, y)]$$

where the derivative on the left is evaluated at some suitable point $(\tilde{x}, \tilde{y})$ corresponding to that portion. Division by $\rho \, \Delta x \, \Delta y$ gives

$$\frac{\partial^2 u}{\partial t^2} = \frac{T}{\rho} \left[\frac{u_x(x + \Delta x, y_1) - u_x(x, y_2)}{\Delta x} + \frac{u_y(x_1, y + \Delta y) - u_y(x_2, y)}{\Delta y} \right].$$

If we let Δx and Δy approach zero, we obtain the partial differential equation

(3) $$\frac{\partial^2 u}{\partial t^2} = c^2 \left(\frac{\partial^2 u}{\partial x^2} + \frac{\partial^2 u}{\partial y^2} \right) \qquad\qquad c^2 = \frac{T}{\rho}.$$

This equation is called the **two-dimensional wave equation.** The expression in parentheses is the Laplacian $\nabla^2 u$ of u (Sec. 11.5). Hence (3) can be written

(3') $$\frac{\partial^2 u}{\partial t^2} = c^2 \nabla^2 u.$$

Solutions will be obtained and discussed in the next section.

11.8 Rectangular Membrane Use of Double Fourier Series

To solve the problem of a vibrating membrane, we have to determine a solution $u(x, y, t)$ of the **two-dimensional wave equation**

(1)
$$\frac{\partial^2 u}{\partial t^2} = c^2 \left(\frac{\partial^2 u}{\partial x^2} + \frac{\partial^2 u}{\partial y^2} \right),$$
$$c^2 = \frac{T}{\rho},$$

that satisfies the **boundary condition**

(2)
$$u = 0$$
on the boundary of the membrane for all $t \geqq 0$

and the two **initial conditions**

(3)
$$u(x, y, 0) = f(x, y)$$
[given initial displacement $f(x, y)$]

and

(4)
$$\left. \frac{\partial u}{\partial t} \right|_{t=0} = g(x, y)$$
[given initial velocity $g(x, y)$].

$u(x, y, t)$ gives the displacement of the point (x, y) of the membrane from rest ($u = 0$) at time t. We see that the conditions (2)–(4) are similar to those for the vibrating string.

As a first important case, let us consider the rectangular membrane R shown in Fig. 273.

First Step. Three Ordinary Differential Equations

By applying the method of separation of variables we first determine solutions of (1) that satisfy the boundary condition (2). We start from

(5)
$$u(x, y, t) = F(x, y)G(t).$$

By substituting this into the wave equation (1) we have

$$F\ddot{G} = c^2(F_{xx}G + F_{yy}G)$$

where subscripts denote partial derivatives and dots denote derivatives with respect to t.

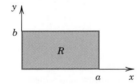

Fig. 273. Rectangular membrane

To separate the variables, we divide both sides by c^2FG:

$$\frac{\ddot{G}}{c^2 G} = \frac{1}{F}(F_{xx} + F_{yy}).$$

Since the left side depends only on t, whereas the right side is independent of t, both sides must equal a constant. By a little investigation we see that only negative values of that constant will lead to solutions that satisfy (2) without being identically zero; this is similar to Sec. 11.3. Denoting that negative constant by $-\nu^2$, we have

$$\frac{\ddot{G}}{c^2 G} = \frac{1}{F}(F_{xx} + F_{yy}) = -\nu^2.$$

This gives two equations: for the **"time function"** $G(t)$ we have the ordinary differential equation

(6) $$\boxed{\ddot{G} + \lambda^2 G = 0}$$ where $\lambda = c\nu$,

and for the **"amplitude function"** $F(x, y)$ the partial differential equation

(7) $$F_{xx} + F_{yy} + \nu^2 F = 0,$$

known as the two-dimensional **Helmholtz[7] equation.**

Separation of the Helmholtz equation is achieved if we set

(8) $$F(x, y) = H(x)Q(y).$$

Substitution of this into (7) gives

$$\frac{d^2 H}{dx^2} Q = -\left(H\frac{d^2 Q}{dy^2} + \nu^2 HQ \right).$$

To separate the variables, we divide both sides by HQ, finding

$$\frac{1}{H}\frac{d^2 H}{dx^2} = -\frac{1}{Q}\left(\frac{d^2 Q}{dy^2} + \nu^2 Q \right).$$

Both sides must equal a constant, by the usual argument. This constant must be negative, say, $-k^2$, because only negative values will lead to solutions that satisfy (2) without being identically zero. Thus

$$\frac{1}{H}\frac{d^2 H}{dx^2} = -\frac{1}{Q}\left(\frac{d^2 Q}{dy^2} + \nu^2 Q \right) = -k^2.$$

---

[7]HERMANN VON HELMHOLTZ (1821—1894), German physicist, known for his basic work in thermodynamics, fluid flow, and acoustics.

This yields two ordinary linear differential equations for H and Q:

(9)
$$\frac{d^2H}{dx^2} + k^2H = 0$$

and

(10)
$$\frac{d^2Q}{dy^2} + p^2Q = 0 \qquad \text{where} \quad p^2 = \nu^2 - k^2.$$

Second Step. Satisfying the Boundary Conditions

The general solutions of (9) and (10) are

$$H(x) = A \cos kx + B \sin kx \qquad \text{and} \qquad Q(y) = C \cos py + D \sin py$$

where A, B, C, and D are constants. From (5) and (2) it follows that the function $F = HQ$ must be zero on the boundary, which corresponds to $x = 0$, $x = a$, $y = 0$, and $y = b$; see Fig. 273. This yields the conditions

$$H(0) = 0, \qquad H(a) = 0, \qquad Q(0) = 0, \qquad Q(b) = 0.$$

Therefore, $H(0) = A = 0$, and then

$$H(a) = B \sin ka = 0.$$

We must take $B \neq 0$, since otherwise $H \equiv 0$ and $F \equiv 0$. Hence $\sin ka = 0$ or $ka = m\pi$, that is,

$$k = \frac{m\pi}{a} \qquad\qquad (m \text{ integer}).$$

In precisely the same fashion we conclude that $C = 0$ and p must be restricted to the values $p = n\pi/b$ where n is an integer. We thus obtain the solutions

$$H_m(x) = \sin \frac{m\pi x}{a} \qquad \text{and} \qquad Q_n(y) = \sin \frac{n\pi y}{b}, \qquad \begin{array}{l} m = 1, 2, \cdots, \\ n = 1, 2, \cdots. \end{array}$$

(As in the case of the vibrating string, it is not necessary to consider m, $n = -1, -2, \cdots$ since the corresponding solutions are essentially the same as for positive m and n, except for a factor -1.) Hence the functions

(11)
$$F_{mn}(x, y) = H_m(x) Q_n(y) = \sin \frac{m\pi x}{a} \sin \frac{n\pi y}{b}, \qquad \begin{array}{l} m = 1, 2, \cdots, \\ n = 1, 2, \cdots, \end{array}$$

are solutions of (7) that are zero on the boundary of our membrane.

Eigenfunctions and Eigenvalues. Having taken care of (7), we turn to (6). Since $p^2 = \nu^2 - k^2$ in (10) and $\lambda = c\nu$ in (6), we have

$$\lambda = c\sqrt{k^2 + p^2}.$$

Hence to $k = m\pi/a$ and $p = n\pi/b$ there corresponds the value

(12) $$\lambda = \lambda_{mn} = c\pi\sqrt{\frac{m^2}{a^2} + \frac{n^2}{b^2}}, \qquad \begin{aligned} m &= 1, 2, \cdots, \\ n &= 1, 2, \cdots, \end{aligned}$$

in the differential equation (6). The corresponding general solution of (6) is

$$G_{mn}(t) = B_{mn}\cos\lambda_{mn}t + B_{mn}^*\sin\lambda_{mn}t.$$

It follows that the functions $u_{mn}(x, y, t) = F_{mn}(x, y)G_{mn}(t)$, written out

(13) $$u_{mn}(x, y, t) = (B_{mn}\cos\lambda_{mn}t + B_{mn}^*\sin\lambda_{mn}t)\sin\frac{m\pi x}{a}\sin\frac{n\pi y}{b}$$

with λ_{mn} according to (12), are solutions of the wave equation (1) that are zero on the boundary of the rectangular membrane in Fig. 273. These functions are called the **eigenfunctions** or *characteristic functions,* and the numbers λ_{mn} are called the **eigenvalues** or *characteristic values* of the vibrating membrane. The frequency of u_{mn} is $\lambda_{mn}/2\pi$.

Discussion of Eigenfunctions. It is very interesting that, depending on a and b, several functions F_{mn} may correspond to the same eigenvalue. Physically this means that there may exist vibrations having the same frequency but entirely different **nodal lines** (curves of points on the membrane that do not move). Let us illustrate this with the following example.

EXAMPLE 1   **Eigenvalues and eigenfunctions of the square membrane**

Consider the square membrane for which $a = b = 1$. From (12) we obtain the eigenvalues

(14) $$\lambda_{mn} = c\pi\sqrt{m^2 + n^2}.$$

Hence

$$\lambda_{mn} = \lambda_{nm}$$

but for $m \neq n$ the corresponding functions

$$F_{mn} = \sin m\pi x \sin n\pi y \qquad \text{and} \qquad F_{nm} = \sin n\pi x \sin m\pi y$$

are certainly different. For example, to $\lambda_{12} = \lambda_{21} = c\pi\sqrt{5}$ there correspond the two functions

$$F_{12} = \sin \pi x \sin 2\pi y \qquad \text{and} \qquad F_{21} = \sin 2\pi x \sin \pi y.$$

Hence the corresponding solutions

$$u_{12} = (B_{12}\cos c\pi\sqrt{5}t + B_{12}^*\sin c\pi\sqrt{5}t)F_{12} \qquad ($$

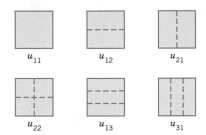

Fig. 274.   Nodal lines of the solutions u_{11}, u_{12}, u_{21}, u_{22}, u_{13}, u_{31} in the case of the square membrane

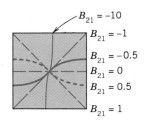

Fig. 275.   Nodal lines of the solution (15) for some values of B_{21}

and

$$u_{21} = (B_{21} \cos c\pi\sqrt{5}t + B_{21}^* \sin c\pi\sqrt{5}t)F_{21}$$

have the nodal lines $y = \frac{1}{2}$ and $x = \frac{1}{2}$, respectively (see Fig. 274). Taking $B_{12} = 1$ and $B_{12}^* = B_{21}^* = 0$, we obtain

$$(15) \qquad u_{12} + u_{21} = \cos c\pi\sqrt{5}t \, (F_{12} + B_{21}F_{21})$$

which represents another vibration corresponding to the eigenvalue $c\pi\sqrt{5}$. The nodal line of this function is the solution of the equation

$$F_{12} + B_{21}F_{21} = \sin \pi x \sin 2\pi y + B_{21} \sin 2\pi x \sin \pi y = 0$$

or, since $\sin 2\alpha = 2 \sin \alpha \cos \alpha$,

$$(16) \qquad \sin \pi x \sin \pi y \, (\cos \pi y + B_{21} \cos \pi x) = 0.$$

This solution depends on the value of B_{21} (see Fig. 275).

From (14) we see that even more than two functions may correspond to the same numerical value of λ_{mn}. For example, the four functions F_{18}, F_{81}, F_{47}, and F_{74} correspond to the value $\lambda_{18} = \lambda_{81} = \lambda_{47} = \lambda_{74} = c\pi\sqrt{65}$, because

$$1^1 + 8^2 = 4^2 + 7^2 = 65.$$

This happens because 65 can be expressed as the sum of two squares of positive integers in several ways. According to a theorem by Gauss, this is the case for every sum of two squares among whose prime factors there are at least two different ones of the form $4n + 1$ where n is a positive integer. In our case, $65 = 5 \cdot 13 = (4 + 1)(12 + 1)$. ◀

Third Step. Solution of the Entire Problem. Double Fourier Series

To obtain the solution that also satisfies the initial conditions (3) and (4), we proceed similarly as in Sec. 11.3. We consider the double series[8]

$$(17) \qquad \begin{aligned} u(x, y, t) &= \sum_{m=1}^{\infty} \sum_{n=1}^{\infty} u_{mn}(x, y, t) \\ &= \sum_{m=1}^{\infty} \sum_{n=1}^{\infty} (B_{mn} \cos \lambda_{mn}t + B_{mn}^* \sin \lambda_{mn}t) \sin \frac{m\pi x}{a} \sin \frac{n\pi y}{b} . \end{aligned}$$

---

[8]We shall not consider the problems of convergence and uniqueness.

From this and (3) we obtain

(18)
$$u(x, y, 0) = \sum_{m=1}^{\infty} \sum_{n=1}^{\infty} B_{mn} \sin \frac{m\pi x}{a} \sin \frac{n\pi y}{b} = f(x, y).$$

This series is called a **double Fourier series.** Suppose that $f(x, y)$ can be developed in such a series.[9] Then the Fourier coefficients B_{mn} of $f(x, y)$ in (18) may be determined as follows. Setting

(19)
$$K_m(y) = \sum_{n=1}^{\infty} B_{mn} \sin \frac{n\pi y}{b}$$

we can write (18) in the form

$$f(x, y) = \sum_{m=1}^{\infty} K_m(y) \sin \frac{m\pi x}{a}.$$

For fixed y this is the Fourier sine series of $f(x, y)$, considered as a function of x. From (6) in Sec. 10.4 we see that the coefficients of this expansion are

(20)
$$K_m(y) = \frac{2}{a} \int_0^a f(x, y) \sin \frac{m\pi x}{a} \, dx.$$

Furthermore, (19) is the Fourier sine series of $K_m(y)$, and from (6) in Sec. 10.4 it follows that the coefficients are

$$B_{mn} = \frac{2}{b} \int_0^b K_m(y) \sin \frac{n\pi y}{b} \, dy.$$

From this and (20) we obtain the **generalized Euler formula**

(21)
$$B_{mn} = \frac{4}{ab} \int_0^b \int_0^a f(x, y) \sin \frac{m\pi x}{a} \sin \frac{n\pi y}{b} \, dx \, dy \qquad \begin{aligned} m &= 1, 2, \cdots \\ n &= 1, 2, \cdots \end{aligned}$$

for the Fourier coefficients of $f(x, y)$ in the double Fourier series (18).

The B_{mn} in (17) are now determined in terms of $f(x, y)$. To determine the B_{mn}^*, we differentiate (17) termwise with respect to t; using (4), we obtain

$$\left. \frac{\partial u}{\partial t} \right|_{t=0} = \sum_{m=1}^{\infty} \sum_{n=1}^{\infty} B_{mn}^* \lambda_{mn} \sin \frac{m\pi x}{a} \sin \frac{n\pi y}{b} = g(x, y).$$

Suppose that $g(x, y)$ can be developed in this double Fourier series. Then, proceeding as before, we find that the coefficients are

(22)
$$B_{mn}^* = \frac{4}{ab\lambda_{mn}} \int_0^b \int_0^a g(x, y) \sin \frac{m\pi x}{a} \sin \frac{n\pi y}{b} \, dx \, dy \qquad \begin{aligned} m &= 1, 2, \cdots \\ n &= 1, 2, \cdots. \end{aligned}$$

---

[9]Sufficient conditions: f, $\partial f/\partial x$, $\partial f/\partial y$, $\partial^2 f/\partial x \, \partial y$ continuous in the rectangle R under consideration.

The result is that, for (17) to satisfy the initial conditions, the coefficients B_{mn} and B_{mn}^* must be chosen according to (21) and (22).

EXAMPLE 2 **Vibrations of a rectangular membrane**

Find the vibrations of a rectangular membrane of sides $a = 4$ ft and $b = 2$ ft (Fig. 276) if the tension is 12.5 lb/ft, the density is 2.5 slugs/ft^2 (as for light rubber), the initial velocity is 0, and the initial displacement is

$$(23) \qquad f(x, y) = 0.1(4x - x^2)(2y - y^2) \text{ ft.}$$

Solution. $c^2 = T/\rho = 12.5/2.5 = 5 \; [\text{ft}^2/\text{sec}^2]$. Also, $B_{mn}^* = 0$ from (22). From (21) and (23),

$$B_{mn} = \frac{4}{4 \cdot 2} \int_0^2 \int_0^4 0.1(4x - x^2)(2y - y^2) \sin \frac{m\pi x}{4} \sin \frac{n\pi y}{2} \, dx \, dy$$

$$= \frac{1}{20} \int_0^4 (4x - x^2) \sin \frac{m\pi x}{4} \, dx \int_0^2 (2y - y^2) \sin \frac{n\pi y}{2} \, dy.$$

Two integrations by parts give for the first integral on the right

$$\frac{128}{m^3 \pi^3} [1 - (-1)^m] = \frac{256}{m^3 \pi^3} \qquad (m \text{ odd})$$

and for the second integral

$$\frac{16}{n^3 \pi^3} [1 - (-1)^n] = \frac{32}{n^3 \pi^3} \qquad (n \text{ odd}).$$

For even m or n we get 0. Together with the factor 1/20 we thus have $B_{mn} = 0$ if m or n is even and

$$B_{mn} = \frac{256 \cdot 32}{20 m^3 n^3 \pi^6} \approx \frac{0.426 \, 050}{m^3 n^3} \qquad (m \text{ and } n \text{ both odd}).$$

From this and (17) we obtain the answer

$$u(x, y, t) = 0.426 \, 050 \sum_{m,n \text{ odd}} \sum \frac{1}{m^3 n^3} \cos \left(\frac{5\pi}{4} \sqrt{m^2 + 4n^2} \right) t \sin \frac{m\pi x}{4} \sin \frac{n\pi y}{2}$$

$$(24) \qquad = 0.426 \, 050 \left(\cos \frac{5\pi \sqrt{5}}{4} t \sin \frac{\pi x}{4} \sin \frac{\pi y}{2} + \frac{1}{27} \cos \frac{5\pi \sqrt{37}}{4} t \sin \frac{\pi x}{4} \sin \frac{3\pi y}{2} \right.$$

$$\left. + \frac{1}{27} \cos \frac{5\pi \sqrt{13}}{4} t \sin \frac{3\pi x}{4} \sin \frac{\pi y}{2} + \frac{1}{729} \cos \frac{5\pi \sqrt{45}}{4} t \sin \frac{3\pi x}{4} \sin \frac{3\pi y}{2} + \cdots \right).$$

To discuss this solution, we note that the first term is very similar to the initial shape of the membrane, has no nodal lines, and is by far the dominating term because the coefficients of the next terms are much smaller. The second term has two horizontal nodal lines ($y = 2/3, 4/3$), the third term two vertical ones ($x = 4/3, 8/3$), the fourth term two horizontal and two vertical ones, and so on. ◀

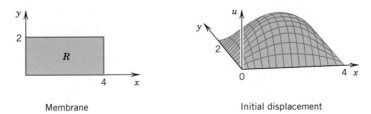

Membrane — Initial displacement

Fig. 276. Example 2

PROBLEM SET 11.8

1. **(Frequency)** How does the frequency of a solution (13) change if the tension of the membrane increased?
2. **(Fourier coefficients)** Verify B_{mn} in Example 2. (Integrate by parts.)
3. **(Solution (24))** Verify the discussion of the terms of (24).

Square Membrane

4. **(Nodal lines)** Determine and sketch the nodal lines of the solutions (13) with $a = b = 1$ for $m = 1, 2, 3, 4$ and $n = 1, 2, 3, 4$.
5. **(Eigenvalues)** Find further eigenvalues of the square membrane with side 1 to each of which correspond four (nonproportional) different eigenfunctions.

Double Fourier Series. Represent $f(x, y)$ by a series (18), where $0 < x < \pi$, $0 < y < \pi$. (f in Prob. 9 satisfies the usual boundary conditions. The other f's do not; we have kept them simpler.)

6. $f(x, y) = 1$ 7. $f(x, y) = y$ 8. $f(x, y) = xy$ 9. $f(x, y) = xy(\pi - x)(\pi - y)$

 10. **CAS PROJECT. Double Fourier Series.** **(a)** Write a program that gives partial sums of (18). Apply it to the initial condition in Prob. 9 and plot the first few partial sums as surfaces over the xy-plane. Also plot single terms of (18). Why is the convergence rapid?
   (b) Do the tasks in (a) for the initial condition in Prob. 6. For several partial sums plot a portion (e.g., $0 < x < \frac{1}{2}\pi$, $0 < y < \frac{1}{2}\pi$) on common axes, so that you can see how they differ.
   (c) Do the tasks in (b) for an example of your own choice.

Deflection. Find the deflection $u(x, y, t)$ of the square membrane with $a = b = 1$ and $c = 1$ if the initial velocity is zero and the initial deflection is

11. $0.1 \sin 3\pi x \sin 4\pi y$ 12. $k \sin \pi x \sin \pi y$ 13. $kxy(1 - x)(1 - y)$ 14. $k \sin^2 \pi x \sin^2 \pi y$

Rectangular Membrane

15. **(Nodal lines)** Repeat the task in Prob. 4 when $a = 3$ and $b = 1$.
16. **(Minimum property)** Show that among all rectangular membranes of the same area $A = ab$ and the same c the square membrane is that for which u_{11} [see (13)] has the lowest frequency.
17. **(Minimum property)** Find a similar result as in Prob. 16 for the frequency of a solution (13) with arbitrary fixed m and n.
18. **(Eigenvalues)** Find eigenvalues of the rectangular membrane of sides $a = 2$, $b = 1$ such that two or more different (independent) eigenfunctions correspond to each such eigenvalue.

Deflection. Find $u(x, y, t)$ for the rectangular membrane with sides a and b and $c = 1$ if the initial velocity is zero and the initial deflection is

19. $\sin \dfrac{2\pi x}{a} \sin \dfrac{3\pi y}{b}$ 20. $xy(a - x)(b - y)$

11.9 Laplacian in Polar Coordinates

In connection with boundary value problems for partial differential equations, it is a general principle to use coordinates with respect to which the boundary of the region under consideration is given by simple formulas. In the next section we shall discuss circular membranes (drumheads). Then the usual polar coordinates r and θ, defined by

$$x = r \cos \theta, \qquad y = r \sin \theta,$$

will be appropriate, because they give the boundary of the membrane by the simple equation $r = const$. Their use requires the transformation of the Laplacian

$$\nabla^2 u = \frac{\partial^2 u}{\partial x^2} + \frac{\partial^2 u}{\partial y^2}$$

in the wave equation into these new coordinates. This we do now once and for all.

    Transformations of differential expressions from one coordinate system into another are frequently needed in practice, and the student should follow our present discussion with great attention.

    We denote partial derivatives by subscripts and $u(x, y, t)$ as a function of r, θ, t by the same letter u, for simplicity. As in Sec. 11.4 we use the chain rule (Sec. 8.8), obtaining

$$u_x = u_r r_x + u_\theta \theta_x.$$

Differentiating again with respect to x and applying the product rule and then the chain rule gives

$$u_{xx} = (u_r r_x)_x + (u_\theta \theta_x)_x$$

(1) $$= (u_r)_x r_x + u_r r_{xx} + (u_{\theta x}\theta_x + u_\theta \theta_{xx}$$

$$= (u_{rr}r_x + u_{r\theta}\theta_x)r_x + u_r r_{xx} + (u_{\theta r}r_x + u_{\theta\theta}\theta_x)\theta_x + u_\theta \theta_{xx}.$$

To determine the partial derivatives r_x and θ_x, we have to differentiate

$$r = \sqrt{x^2 + y^2} \qquad \text{and} \qquad \theta = \text{arc tan}\,\frac{y}{x},$$

finding

$$r_x = \frac{x}{\sqrt{x^2 + y^2}} = \frac{x}{r}, \qquad \theta_x = \frac{1}{1 + (y/x)^2}\left(-\frac{y}{x^2}\right) = -\frac{y}{r^2}.$$

Differentiating these two formulas again, we obtain

$$r_{xx} = \frac{r - xr_x}{r^2} = \frac{1}{r} - \frac{x^2}{r^3} = \frac{y^2}{r^3}, \qquad \theta_{xx} = -y\left(-\frac{2}{r^3}\right)r_x = \frac{2xy}{r^4}.$$

We substitute all these expressions into (1). Assuming continuity of the first and second partial derivatives, we have $u_{r\theta} = u_{\theta r}$, and by simplifying,

(2) $$u_{xx} = \frac{x^2}{r^2}u_{rr} - 2\frac{xy}{r^3}u_{r\theta} + \frac{y^2}{r^4}u_{\theta\theta} + \frac{y^2}{r^3}u_r + 2\frac{xy}{r^4}u_\theta.$$

In a similar fashion it follows that

(3) $$u_{yy} = \frac{y^2}{r^2}u_{rr} + 2\frac{xy}{r^3}u_{r\theta} + \frac{x^2}{r^4}u_{\theta\theta} + \frac{x^2}{r^3}u_r - 2\frac{xy}{r^4}u_\theta.$$

By adding (2) and (3) we see that the **Laplacian of u in polar coordinates is**

(4)
$$\nabla^2 u = \frac{\partial^2 u}{\partial r^2} + \frac{1}{r}\frac{\partial u}{\partial r} + \frac{1}{r^2}\frac{\partial^2 u}{\partial \theta^2}.$$

In the next section we shall apply this formula to the study of the vibrations of a drumhead (circular membrane).

From polar coordinates we obtain cylindrical coordinates in space by taking z as a third coordinate. Accordingly, by (4), the **Laplacian of u in cylindrical coordinates is**

(5)
$$\nabla^2 u = u_{rr} + \frac{1}{r} u_r + \frac{1}{r^2} u_{\theta\theta} + u_{zz}.$$

PROBLEM SET 11.9

1. **(Derivation of (4))** Complete the details of the calculations that lead to (2) and (3).
2. Transform (4) back into Cartesian coordinates.
3. Show that an **alternative form of (4)** is

$$\nabla^2 u = \frac{1}{r}\frac{\partial}{\partial r}\left(r\frac{\partial u}{\partial r}\right) + \frac{1}{r^2}\frac{\partial^2 u}{\partial \theta^2}.$$

4. **(Independence of θ)** If u is independent of θ, then (4) reduces to $\nabla^2 u = u_{rr} + u_r/r$. Derive this directly from the Laplacian in Cartesian coordinates by assuming that u is independent of θ.
5. **(Radial solution)** Show that the only solution of $\nabla^2 u = 0$ depending only on $r = \sqrt{x^2 + y^2}$ is $u = a \ln r + b$ with constant a and b.
6. **TEAM PROJECT. Series for Dirichlet and Neumann Problems.**
   (a) Show that $u_n = r^n \cos n\theta$, $u_n = r^n \sin n\theta$, $n = 0, 1, \cdots$, are solutions of Laplace's equation $\nabla^2 u = 0$ with $\nabla^2 u$ given by (4). (What would u_n be in Cartesian coordinates? Experiment with small n.)
   (b) **(Dirichlet problem)** (See Sec. 11.5) Assuming that termwise differentiation is permissible, show that a solution of the Laplace equation in the disk $r < R$ satisfying the boundary condition $u(R, \theta) = f(\theta)$ (f given) is

(6)
$$u(r, \theta) = a_0 + \sum_{n=1}^{\infty}\left[a_n\left(\frac{r}{R}\right)^n \cos n\theta + b_n\left(\frac{r}{R}\right)^n \sin n\theta\right]$$

   where a_n, b_n are the Fourier coefficients of f (see Sec. 10.2).
   (c) **(Dirichlet problem)** Solve the Dirichlet problem using (6) if $R = 1$ and the boundary values are $u(\theta) = -100$ volts if $-\pi < \theta < 0$, $u(\theta) = 100$ volts if $0 < \theta < \pi$. (Sketch this disk, indicate the boundary values.)
   (d) **(Neumann problem)** Show that the solution of the Neumann problem $\nabla^2 u = 0$ if $r < R$, $u_n(R, \theta) = f(\theta)$ (n the outer normal) is

$$u(r, \theta) = A_0 + \sum_{n=1}^{\infty} r^n (A_n \cos n\theta + B_n \sin n\theta)$$

with arbitrary A_0 and

$$A_n = \frac{1}{\pi n R^{n-1}} \int_{-\pi}^{\pi} f(\theta) \cos n\theta \, d\theta, \qquad B_n = \frac{1}{\pi n R^{n-1}} \int_{-\pi}^{\pi} f(\theta) \sin n\theta \, d\theta.$$

(e) **(Compatibility condition)** Show that (9), Sec. 9.4, imposes on $f(\theta)$ in (d) the *"compatibility condition"*

$$\int_{-\pi}^{\pi} f(\theta) \, d\theta = 0.$$

Electrostatic Potential. Steady-State Heat Problems.
The electrostatic potential u satisfies Laplace's equation $\nabla^2 u = 0$ in any region free of charges. Also, the heat equation $u_t = c^2 \nabla^2 u$ (see Sec. 11.5) reduces to Laplace's equation if the temperature u is independent of time t (**"steady-state case"**). Using (6), find the electrostatic potential (equivalently: the steady-state temperature distribution) in the disk $r < 1$ corresponding to the following boundary values.

7. $u(\theta) = \sin^3 \theta$

8. $u(\theta) = 10 \cos^2 \theta$

9. $u(\theta) = \begin{cases} \theta & \text{if } -\pi/2 < \theta < \pi/2 \\ 0 & \text{if } \pi/2 < \theta < 3\pi/2 \end{cases}$

10. $u(\theta) = \begin{cases} -\theta & \text{if } -\pi < \theta < 0 \\ \theta & \text{if } 0 < \theta < \pi \end{cases}$

11. **(Potential)** Find a formula for the potential u on the x-axis in Prob. 10. Sketch or plot u.

Temperature in a Thin Semicircular Plate. Find the temperature u in a plate $r < 1$, $y > 0$ if the segment $-1 < x < 1$ is kept at $0\,°C$ and the semicircular boundary is kept at temperature

12. $100 \sin^3 \theta$

13. $u_0 = const$

14. $10\theta(\pi - \theta)$

15. **(Invariance)** Show that $\nabla^2 u$ is invariant under translations $x^* = x + a$, $y^* = y + b$ and under rotations $x^* = x \cos \alpha - y \sin \alpha$, $y^* = x \sin \alpha + y \cos \alpha$.

11.10 Circular Membrane
Use of Fourier–Bessel Series

Circular membranes occur in drums, pumps, microphones, telephones, and so on. This accounts for their great importance in engineering. Whenever a circular membrane is plane and its material is elastic, but offers no resistance to bending (this excludes thin metallic membranes!), its vibrations are governed by the two-dimensional wave equation $(3')$, Sec. 11.7. We now write this equation in polar coordinates defined by $x = r \cos \theta$, $y = r \sin \theta$ in the form [see (4) in the last section]

$$\frac{\partial^2 u}{\partial t^2} = c^2 \left(\frac{\partial^2 u}{\partial r^2} + \frac{1}{r} \frac{\partial u}{\partial r} + \frac{1}{r^2} \frac{\partial^2 u}{\partial \theta^2} \right).$$

Figure 277 shows our membrane of radius R, for which we shall determine solutions $u(r, t)$ that are radially symmetric,[10] that is, do not depend on θ. Then the wave equation

---

[10] For solutions depending on θ, see the problem set.

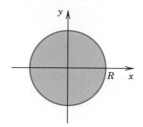

Fig. 277.   Circular membrane

reduces to

(1)
$$\frac{\partial^2 u}{\partial t^2} = c^2 \left(\frac{\partial^2 u}{\partial r^2} + \frac{1}{r} \frac{\partial u}{\partial r} \right).$$

Boundary and Initial Conditions. Since the membrane is fixed along its boundary $r = R$, we have the boundary condition

(2)
$$u(R, t) = 0$$
for all $t \geqq 0$.

Solutions not depending on θ will occur if the initial conditions do not depend on θ, that is, if they are of the form

(3)
$$u(r, 0) = f(r)$$
[initial deflection $f(r)$]

and

(4)
$$\left. \frac{\partial u}{\partial t} \right|_{t=0} = g(r)$$
[initial velocity $g(r)$].

First Step. Ordinary Differential Equations. Bessel's Equation

Separating variables, we first determine solutions

(5)
$$u(r, t) = W(r) G(t)$$

satisfying the boundary condition (2). Substituting (5) and its derivatives into (1) and dividing the result by $c^2 WG$, we get

$$\frac{\ddot{G}}{c^2 G} = \frac{1}{W} \left(W'' + \frac{1}{r} W' \right)$$

where dots denote derivatives with respect to t and primes denote derivatives with respect to r. The expressions on both sides must equal a constant. This constant must be negative, say, $-k^2$, in order to obtain solutions that satisfy the boundary condition without being identically zero. Thus,

$$\frac{\ddot{G}}{c^2 G} = \frac{1}{W} \left(W'' + \frac{1}{r} W' \right) = -k^2.$$

This yields the two ordinary linear differential equations

(6)
$$\ddot{G} + \lambda^2 G = 0$$
where $\lambda = ck$

and

(7)
$$W'' + \frac{1}{r} W' + k^2 W = 0.$$

We can reduce (7) to Bessel's equation (Sec. 4.5) if we set $s = kr$. Then $1/r = k/s$ and the chain rule gives

$$W' = \frac{dW}{dr} = \frac{dW}{ds} \frac{ds}{dr} = \frac{dW}{ds} k \qquad \text{and} \qquad W'' = \frac{d^2 W}{ds^2} k^2.$$

By substituting this into (7) and omitting the common factor k^2 we obtain

(7*)
$$\frac{d^2 W}{ds^2} + \frac{1}{s} \frac{dW}{ds} + W = 0.$$

This is **Bessel's equation** (1), Sec. 4.5, with parameter $\nu = 0$.

Second Step. Satisfying the Boundary Condition

Solutions of (7*) are the Bessel functions J_0 and Y_0 of the first and second kind (see Secs. 4.5, 4.6). Now Y_0 becomes infinite at 0, so that we cannot use it because the deflection of the membrane must always remain finite. This leaves us with

(8)
$$W(r) = J_0(s) = J_0(kr) \qquad (s = kr).$$

On the boundary $r = R$ we get $W(R) = J_0(kR) = 0$ from (2) (because $G \equiv 0$ would imply $u \equiv 0$). We can satisfy this condition because J_0 has (infinitely many) positive zeros, $s = \alpha_1, \alpha_2, \cdots$ (see Fig. 278), with numerical values

$$\alpha_1 = 2.4048, \quad \alpha_2 = 5.5201, \quad \alpha_3 = 8.6537, \quad \alpha_4 = 11.7915, \quad \alpha_5 = 14.9309,$$

and so on (see Ref. [1], Appendix 1, for more extensive tables). These are irregularly spaced, as we see. Equation (8) now implies

(9)
$$kR = \alpha_m \qquad \text{thus} \qquad k = k_m = \frac{\alpha_m}{R}, \qquad m = 1, 2, \cdots.$$

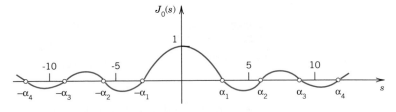

Fig. 278.   Bessel function $J_0(s)$

Hence the functions

$$(10) \qquad W_m(r) = J_0(k_m r) = J_0\left(\frac{\alpha_m}{R} r\right), \qquad m = 1, 2, \cdots,$$

are solutions of (7) that vanish at $r = R$.

Eigenfunctions and Eigenvalues. For W_m in (10), a corresponding general solution of (6) with $\lambda = \lambda_m = ck_m = c\alpha_m/R$ is

$$G_m(t) = a_m \cos \lambda_m t + b_m \sin \lambda_m t.$$

Hence the functions

$$(11) \qquad \boxed{u_m(r, t) = W_m(r)G_m(t) = (a_m \cos \lambda_m t + b_m \sin \lambda_m t)J_0(k_m r)}$$

with $m = 1, 2, \cdots$ are solutions of the wave equation (1) satisfying the boundary condition (2). These are the **eigenfunctions** of our problem. The corresponding **eigenvalues** are λ_m.

The vibration of the membrane corresponding to u_m is called the mth **normal mode;** it has the frequency $\lambda_m/2\pi$ cycles per unit time. Since the zeros of the Bessel function J_0 are not regularly spaced on the axis (in contrast to the zeros of the sine functions appearing in the case of the vibrating string), the sound of a drum is entirely different from that of a violin. The forms of the normal modes can easily be obtained from Fig. 278 and are shown in Fig. 279. For $m = 1$, all the points of the membrane move up (or down) at the same time. For $m = 2$, the situation is as follows. The function

$$W_2(r) = J_0\left(\frac{\alpha_2}{R} r\right)$$

is zero for $\alpha_2 r/R = \alpha_1$, thus $r = \alpha_1 R/\alpha_2$. The circle $r = \alpha_1 R/\alpha_2$ is, therefore, a nodal line, and when at some instant the central part of the membrane moves up, the outer part

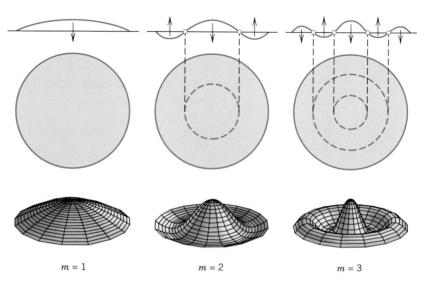

$m = 1$ $m = 2$ $m = 3$

Fig. 279. Normal modes of the circular membrane in the case of vibrations independent of the angle

$(r > \alpha_1 R/\alpha_2)$ moves down, and conversely. The solution $u_m(r, t)$ has $m - 1$ nodal lines, which are circles (Fig. 279).

Third Step. Solution of the Entire Problem

To obtain a solution that also satisfies the initial conditions (3) and (4), we may proceed as in the case of the string. That is, we consider the series[11]

$$(12) \quad u(r, t) = \sum_{m=1}^{\infty} W_m(r)G_m(t) = \sum_{m=1}^{\infty} (a_m \cos \lambda_m t + b_m \sin \lambda_m t) J_0\left(\frac{\alpha_m}{R} r\right).$$

Setting $t = 0$ and using (3), we obtain

$$(13) \qquad\qquad u(r, 0) = \sum_{m=1}^{\infty} a_m J_0\left(\frac{\alpha_m}{R} r\right) = f(r).$$

Thus for the series (12) to satisfy the condition (3), the constants a_m must be the coefficients of the **Fourier–Bessel series** (13) that represents $f(r)$ in terms of $J_0(\alpha_m r/R)$; that is [see (10) in Sec. 4.8 with $n = 0$],

$$(\mathbf{14}) \qquad\qquad \boxed{a_m = \frac{2}{R^2 J_1{}^2(\alpha_m)} \int_0^R r f(r) J_0\left(\frac{\alpha_m}{R} r\right) dr} \qquad (m = 1, 2, \cdots).$$

Differentiability of $f(r)$ in the interval $0 \leqq r \leqq R$ is sufficient for the existence of the development (13); see Ref. [A7]. The coefficients b_m in (12) can be determined from (4) in a similar fashion. To obtain numerical values of a_m and b_m, we may apply one of the usual methods of approximate integration, using tables of J_0 and J_1. Sometimes numerical integration can be avoided, as the following example illustrates.

EXAMPLE 1   **Vibrations of a circular membrane**

Find the vibrations of a circular drumhead of radius 1 ft and density 2 slugs/ft^2 if the tension is 8 lb/ft, the initial velocity is 0, and the initial displacement is

$$f(r) = 1 - r^2 \text{ [ft]}.$$

Solution. $c^2 = T/\rho = 8/2 = 4 \text{ [ft}^2/\text{sec}^2]$. Also $b_m = 0$, since the initial velocity is 0. From (14) and Example 3 in Sec. 4.8, since $R = 1$, we obtain

$$a_m = \frac{2}{J_1{}^2(\alpha_m)} \int_0^1 r(1 - r^2) J_0(\alpha_m r) \, dr = \frac{4 J_2(\alpha_m)}{\alpha_m{}^2 J_1{}^2(\alpha_m)} = \frac{8}{\alpha_m{}^3 J_1(\alpha_m)},$$

where the last equality follows from (26), Sec. 4.5, with $\nu = 1$, that is,

$$J_2(\alpha_m) = \frac{2}{\alpha_m} J_1(\alpha_m) - J_0(\alpha_m) = \frac{2}{\alpha_m} J_1(\alpha_m).$$

---

[11]We shall not consider the problems of convergence and uniqueness.

Table 9.5 on p. 409 of [1] gives α_m and $J_0'(\alpha_m)$. From this we get $J_1(\alpha_m) = -J_0'(\alpha_m)$ by (25), Sec. 4.5, with $\nu = 0$, and compute the a_m's:

m	α_m	$J_1(\alpha_m)$	$J_2(\alpha_m)$	a_m
1	2.40483	0.51915	0.43176	1.10801
2	5.52008	−0.34026	−0.12328	−0.13978
3	8.65373	0.27145	0.06274	0.04548
4	11.79153	−0.23246	−0.03943	−0.02099
5	14.93092	0.20655	0.02767	0.01164
6	18.07106	−0.18773	−0.02078	−0.00722
7	21.21164	0.17327	0.01634	0.00484
8	24.35247	−0.16170	−0.01328	−0.00343
9	27.49348	0.15218	0.01107	0.00253
10	30.63461	−0.14417	−0.00941	−0.00193

Thus

$$f(r) = 1.108 \, J_0(2.4048r) - 0.140 \, J_0(5.5201r) + 0.045 \, J_0(8.6537r) - \cdots .$$

We see that the coefficients decrease relatively slowly. The sum of the explicitly given coefficients in the table is 0.99915. The sum of *all* the coefficients should be 1. (Why?) Hence by the Leibniz test in Appendix A3.3 the partial sum of those terms gives about three correct decimals of the amplitude $f(r)$.
Since

$$\lambda_m = ck_m = c\alpha_m/R = 2\alpha_m,$$

from (12) we thus obtain the solution (with r measured in feet and t in seconds)

$$u(r, t) = 1.108J_0(2.4048r) \cos 4.8097t - 0.140J_0(5.5201r) \cos 11.0402t + 0.045J_0(8.6537r) \cos 17.3075t - \cdots .$$

In Fig. 279, $m = 1$ gives an idea of the motion of the first term of our series, $m = 2$ of the second term, and $m = 3$ of the third term, so that we can "see" our result about as well as for a violin string in Sec. 11.3. ◀

Problem Set 11.10

1. **(Size of drum)** A small drum should have a higher fundamental frequency than a large one, tension and density being the same. How can you conclude this from our formulas?

2. **(Fundamental frequency)** Find a formula for the fundamental frequency of a circular drum. (Use the numerical value of α_1.)

3. **(Tension)** Find a formula for the tension T required to produce a desired fundamental frequency of a circular drum.

4. **(Polar coordinates)** Why did we use polar coordinates?

5. **(Frequency)** What happens to the frequency of a drum if you increase the tension of the membrane?

6. **CAS PROJECT. Normal Modes.** (a) Plot the normal modes u_{mn}, $m = 1, \cdots, 4$, $n = 1, \cdots, 4$, for $R = 1$ as surfaces over the $r\theta$-plane.
    (b) Write a program for calculating the a_m in Example 1 and extend the table to $m = 15$. Verify numerically that $\alpha_m \approx (m - \frac{1}{4})\pi$ and compute the error.
    (c) Plot the initial deflection in Example 1 as a surface.

7. **(Nonzero initial velocity)** Show that for (12) to satisfy (4) we must have

$$(15) \qquad \boxed{b_m = \frac{2}{c\alpha_m R J_1^{\,2}(\alpha_m)} \int_0^R rg(r)J_0(\alpha_m r/R)\, dr,} \qquad m = 1, 2, \cdots .$$

8. (Nonzero initial velocity) Could you think of a situation when a drum has initial displacement zero and initial velocity not zero? As an application of (15) calculate this case with $R = 1$, $c = 1$, $g(r) = 1$. [Use (24) in Sec. 4.5 with $\nu = 1$.]

9. (Nodal lines) Can, for fixed c and R, two or more u_m [see (11)] with different nodal lines correspond to the same eigenvalue? (Give a reason.)

10. (Sum of coefficients) Why is $a_1 + a_2 + \cdots = 1$ in Example 1? Compute the first few partial sums until you get 3-digit accuracy.

Vibrations of a Circular Membrane Depending on r and θ

11. (First separation) Show that substitution of $u = F(r, \theta)G(t)$ into the wave equation

$$(16) \qquad u_{tt} = c^2 \left(u_{rr} + \frac{1}{r} u_r + \frac{1}{r^2} u_{\theta\theta} \right)$$

leads to

$$(17) \qquad \ddot{G} + \lambda^2 G = 0, \qquad \text{where } \lambda = ck,$$

$$(18) \qquad F_{rr} + \frac{1}{r} F_r + \frac{1}{r^2} F_{\theta\theta} + k^2 F = 0.$$

12. (Second separation) Show that substitution of $F = W(r)Q(\theta)$ into (18) gives

$$(19) \qquad Q'' + n^2 Q = 0,$$

$$(20) \qquad r^2 W'' + rW' + (k^2 r^2 - n^2)W = 0.$$

13. (Periodicity) Show that $Q(\theta)$ must be periodic with period 2π, and, therefore, $n = 0, 1, 2, \cdots$ in (19) and (20). Show that this yields the solutions $Q_n = \cos n\theta$, $Q_n^* = \sin n\theta$, $W_n = J_n(kr)$, $n = 0, 1, \cdots$.

14. (Boundary condition) Show that the boundary condition

$$(21) \qquad u(R, \theta, t) = 0$$

leads to $k = k_{mn} = \alpha_{mn}/R$, where $s = \alpha_{mn}$ is the mth positive zero of $J_n(s)$.

15. (Solutions depending on both r and θ) Show that solutions of (16) that satisfy (21) are

$$(22) \qquad \begin{aligned} u_{mn} &= (A_{mn} \cos ck_{mn}t + B_{mn} \sin ck_{mn}t)J_n(k_{mn}r) \cos n\theta \\ u_{mn}^* &= (A_{mn}^* \cos ck_{mn}t + B_{mn}^* \sin ck_{mn}t)J_n(k_{mn}r) \sin n\theta \end{aligned} \qquad \text{(Fig. 280).}$$

16. (Initial condition) Show that $u_t(r, \theta, 0) = 0$ leads to $B_{mn} = 0$, $B_{mn}^* = 0$ in (22).

17. Show that $u_{m0}^* \equiv 0$ and u_{m0} is identical with (11) in the current section.

18. (Semicircular membrane) Show that u_{11} represents the fundamental mode of a semicircular membrane and find the corresponding frequency when $c^2 = 1$ and $R = 1$.

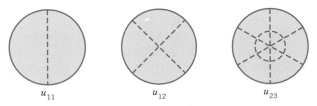

u_{11} u_{12} u_{23}

Fig. 280.   Nodal lines of some of the solutions (22)

11.11 Laplace's Equation in Cylindrical and Spherical Coordinates.   Potential

Laplace's equation

(1)

$$\nabla^2 u = u_{xx} + u_{yy} + u_{zz} = 0$$

is one of the most important partial differential equations in physics and its engineering applications. x, y, z are Cartesian coordinates in space (Fig. 150, Sec. 8.1), $u_{xx} = \partial^2 u/\partial x^2$, etc. The expression $\nabla^2 u$ is called the **Laplacian** of u. The theory of the solutions of (1) is called **potential theory.** Solutions of (1) that have *continuous* second partial derivatives are called **harmonic functions.**

Laplace's equation occurs mainly in **gravitation, electrostatics** (see Example 3, Sec. 8.9), steady-state **heat flow** (Sec. 11.5), and **fluid flow** (to be discussed in Chap. 16). Recall from Sec. 8.9 that the gravitational **potential** $u(x, y, z)$ at a point (x, y, z) resulting from a single mass located at a point (X, Y, Z) is

(2)
$$u(x, y, z) = \frac{c}{r} = \frac{c}{\sqrt{(x - X)^2 + (y - Y)^2 + (z - Z)^2}} \qquad (r > 0)$$

and u satisfies (1). Similarly, if mass is distributed in a region T in space with density $\rho(X, Y, Z)$, its potential at a point (x, y, z) not occupied by mass is

(3)
$$u(x, y, z) = k \iiint_T \frac{\rho(X, Y, Z)}{r} \, dX \, dY \, dZ.$$

It satisfies (1) because $\nabla^2(1/r) = 0$ (Sec. 8.9) and ρ is not a function of x, y, z.

Practical problems involving Laplace's equation are boundary value problems in a region T with boundary surface S (or in a region in the plane bounded by some curve). Such a problem is called (see also Sec. 11.5):

> **(I) First boundary value problem** or **Dirichlet problem** if u is prescribed on S.
>
> **(II) Second boundary value problem** or **Neumann problem** if the normal derivative $u_n = \partial u/\partial n$ is prescribed on S.
>
> **(III) Third** or **mixed boundary value problem** if u is prescribed on a portion of S and u_n on the remaining portion of S.

Laplacian in Cylindrical Coordinates

The first step in solving a boundary value problem is generally the introduction of coordinates in which the boundary surface S has a simple representation. Cylindrical symmetry (a cylinder as region T) calls for cylindrical coordinates r, θ, z related to x, y, z by

(4) $x = r \cos \theta, \qquad y = r \sin \theta, \qquad z = z$ (Fig. 281).

For these we get $\nabla^2 u$ immediately by adding u_{zz} to (4) in Sec. 11.9; thus,

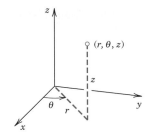

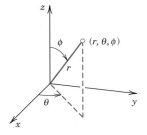

Fig. 281. Cylindrical coordinates   **Fig. 282.** Spherical coordinates

(5)
$$\nabla^2 u = \frac{\partial^2 u}{\partial r^2} + \frac{1}{r}\frac{\partial u}{\partial r} + \frac{1}{r^2}\frac{\partial^2 u}{\partial \theta^2} + \frac{\partial^2 u}{\partial z^2}.$$

Laplacian in Spherical Coordinates

Spherical symmetry (a ball as region T bounded by a sphere S) requires **spherical coordinates** r, θ, ϕ related to x, y, z by[12]

(6) $x = r \cos \theta \sin \phi,$ $y = r \sin \theta \sin \phi,$ $z = r \cos \phi$ (Fig. 282).

Using the chain rule (as in Sec. 11.9), we obtain $\nabla^2 u$ in spherical coordinates

(7)
$$\nabla^2 u = \frac{\partial^2 u}{\partial r^2} + \frac{2}{r}\frac{\partial u}{\partial r} + \frac{1}{r^2}\frac{\partial^2 u}{\partial \phi^2} + \frac{\cot \phi}{r^2}\frac{\partial u}{\partial \phi} + \frac{1}{r^2 \sin^2 \phi}\frac{\partial^2 u}{\partial \theta^2}.$$

We leave the details as an exercise. It is sometimes practical to write (7) in the form

(7′) $$\nabla^2 u = \frac{1}{r^2}\left[\frac{\partial}{\partial r}\left(r^2 \frac{\partial u}{\partial r} \right) + \frac{1}{\sin \phi}\frac{\partial}{\partial \phi}\left(\sin \phi \frac{\partial u}{\partial \phi} \right) + \frac{1}{\sin^2 \phi}\frac{\partial^2 u}{\partial \theta^2} \right].$$

Boundary Value Problem in Spherical Coordinates

If the boundary condition on a sphere S of radius R is independent of θ, say,

(8) $u(R, \theta, \phi) = f(\phi),$

for instance, representing an electrostatic potential (or a temperature) at which the sphere S is kept, then we can expect a solution $u(r, \phi)$ of Laplace's equation that is independent of θ. Thus $u_{\theta\theta} = 0$ in (7′), so that the Laplace's equation becomes

(9)
$$\nabla^2 u = \frac{\partial}{\partial r}\left(r^2 \frac{\partial u}{\partial r} \right) + \frac{1}{\sin \phi}\frac{\partial}{\partial \phi}\left(\sin \phi \frac{\partial u}{\partial \phi} \right) = 0.$$

---

[12]Equation (6) is used in calculus and extends the familiar notation for polar coordinates. Unfortunately, some books use θ and ϕ interchanged, an extension of the notation $x = r \cos \phi$, $y = r \sin \phi$ (used in some European countries).

At infinity the potential will be zero, that is, we must have

(10) $$\lim_{r \to \infty} u(r, \phi) = 0.$$

Solving the Dirichlet Problem (8), (9), (10)

We solve this problem by separating variables, setting $u(r, \phi) = G(r)H(\phi)$. Substitution into (9) and division by GH gives

$$\frac{1}{G} \frac{d}{dr} \left(r^2 \frac{dG}{dr} \right) = -\frac{1}{H \sin \phi} \frac{d}{d\phi} \left(\sin \phi \frac{dH}{d\phi} \right).$$

The variables are now separated. By the usual argument the two sides of this equation must equal a constant, say, k. Thus,

(11) $$\frac{1}{\sin \phi} \frac{d}{d\phi} \left(\sin \phi \frac{dH}{d\phi} \right) + kH = 0$$

and

(12) $$\frac{1}{G} \frac{d}{dr} \left(r^2 \frac{dG}{dr} \right) = k.$$

We first solve (12), which is (multiply by G) $(r^2 G')' = kG$. Its solutions will take a simple form if we write $k = n(n + 1)$. Since $(r^2 G')' = r^2 G'' + 2rG'$ in (12), we have

(13) $$\boxed{r^2 G'' + 2rG' - n(n + 1)G = 0.}$$

This is an **Euler–Cauchy equation** (see Sec. 2.6). Substituting $G = r^a$ and dropping the common factor r^a gives $a(a - 1) + 2a - n(n + 1) = 0$. The roots are $a = n$ and $-n - 1$, where n is arbitrary. This gives the solutions

(14) $$G_n(r) = r^n \qquad \text{and} \qquad G_n^*(r) = \frac{1}{r^{n+1}}.$$

We now solve (11). Setting $\cos \phi = w$, we have $\sin^2 \phi = 1 - w^2$ and

$$\frac{d}{d\phi} = \frac{d}{dw} \frac{dw}{d\phi} = -\sin \phi \frac{d}{dw}.$$

Consequently, (11) with $k = n(n + 1)$ takes the form

(15) $$\frac{d}{dw} \left[(1 - w^2) \frac{dH}{dw} \right] + n(n + 1)H = 0.$$

This is **Legendre's equation** (see Sec. 4.3), written out

(15′) $$\boxed{(1 - w^2) \frac{d^2 H}{dw^2} - 2w \frac{dH}{dw} + n(n + 1)H = 0.}$$

Solution Using a Fourier–Legendre Series

For integer[13] $n = 0, 1, \cdots$, the Legendre polynomials

$$H = P_n(w) = P_n(\cos \phi), \qquad\qquad n = 0, 1, \cdots,$$

are solutions of Legendre's equation (15). We thus obtain the following two sequences of solutions $u = GH$ of Laplace's equation (9), with constant A_n and B_n,

$$(16^*) \qquad u_n(r, \phi) = A_n r^n P_n(\cos \phi), \qquad u_n^*(r, \phi) = \frac{B_n}{r^{n+1}} P_n(\cos \phi) \qquad n = 0, 1, \cdots,$$

Solution of the Interior Problem. This means a solution of (9) *inside* the sphere and satisfying (8). For this we consider the series[14]

$$(16) \qquad\qquad \boxed{u(r, \phi) = \sum_{n=0}^{\infty} A_n r^n P_n(\cos \phi).}$$

For (16) to satisfy (8) we must have

$$(17) \qquad\qquad u(R, \phi) = \sum_{n=0}^{\infty} A_n R^n P_n(\cos \phi) = f(\phi);$$

that is, (17) must be the **Fourier–Legendre series** of $f(\phi)$. From (7) in Sec. 4.8 we get the coefficients

$$(18^*) \qquad\qquad A_n R^n = \frac{2n + 1}{2} \int_{-1}^{1} \tilde{f}(w) P_n(w) \, dw$$

where $\tilde{f}(w)$ denotes $f(\phi)$ as a function of $w = \cos \phi$. Since $dw = -\sin \phi \, d\phi$, and the limits of integration -1 and 1 correspond to $\phi = \pi$ and $\phi = 0$, respectively, we also obtain

$$(18) \qquad\qquad \boxed{A_n = \frac{2n + 1}{2R^n} \int_0^{\pi} f(\phi) P_n(\cos \phi) \sin \phi \, d\phi,} \qquad n = 0, 1, \cdots.$$

Thus the series (16) with coefficients (18) is the solution of our problem for points inside the sphere.

Solution of the Exterior Problem. Outside the sphere we cannot use the functions $u_n(r, \phi)$ because these functions do not satisfy (10), but we may use the functions $u_n^*(r, \phi)$, which do satisfy (10), and proceed as before. This leads to the solution

---

[13]So far, n was arbitrary since k was arbitrary. It can be shown that the restriction of n to integers is necessary to make the solution of (7) continuous, together with its derivative of the first order, in the interval $-1 \leqq w \leqq 1$ or $0 \leqq \phi \leqq \pi$.

[14]Convergence will not be considered. It can be shown that if $f(\phi)$ and $f'(\phi)$ are piecewise continuous in the interval $0 \leqq \phi \leqq \pi$, the series (16) with coefficients (18) can be differentiated termwise twice with respect to r and with respect to ϕ and the resulting series converge and represent $\partial^2 u / \partial r^2$ and $\partial^2 u / \partial \phi^2$, respectively. Hence the series (16) with coefficients (18) is then the solution of our problem inside the sphere.

(19)

$$u(r, \phi) = \sum_{n=0}^{\infty} \frac{B_n}{r^{n+1}} P_n(\cos \phi) \qquad (r \geqq R)$$

with coefficients

(20)

$$B_n = \frac{2n + 1}{2} R^{n+1} \int_0^{\pi} f(\phi) P_n(\cos \phi) \sin \phi \, d\phi.$$

EXAMPLE 1   **Spherical capacitor**

Find the potential inside and outside a spherical capacitor consisting of two metallic hemispheres of radius 1 ft separated by a small slit for reasons of insulation, if the upper hemisphere is kept at 110 volts and the lower is grounded (Fig. 283).

Solution.   The given boundary condition is (recall Fig. 282).

$$f(\phi) = \begin{cases} 110 & \text{if} \quad 0 \leqq \phi < \pi/2 \\ 0 & \text{if} \quad \pi/2 < \phi \leqq \pi. \end{cases}$$

Since $R = 1$, we thus obtain from (18)

$$A_n = \frac{2n + 1}{2} \cdot 110 \int_0^{\pi/2} P_n(\cos \phi) \sin \phi \, d\phi.$$

We set $w = \cos \phi$. Then $P_n(\cos \phi) \sin \phi \, d\phi = -P_n(w) \, dw$ and we integrate from 1 to 0. We get rid of the minus by integrating from 0 to 1. Then from (11) in Sec. 4.3,

$$A_n = 55(2n + 1) \sum_{m=0}^{M} (-1)^m \frac{(2n - 2m)!}{2^n m!(n - m)!(n - 2m)!} \int_0^1 w^{n-2m} \, dw$$

where $M = n/2$ for even n and $M = (n - 1)/2$ for odd n. The integral equals $1/(n - 2m + 1)$. Thus

(21)

$$A_n = \frac{55(2n + 1)}{2^n} \sum_{m=0}^{M} (-1)^m \frac{(2n - 2m)!}{m!(n - m)!(n - 2m + 1)!}.$$

Taking $n = 0$, we get $A_0 = 55$ (since $0! = 1$). For $n = 1, 2, 3, \cdots$ we get

$$A_1 = \frac{165}{2} \cdot \frac{2!}{0!1!2!} = \frac{165}{2},$$

$$A_2 = \frac{275}{4} \left(\frac{4!}{0!2!3!} - \frac{2!}{1!1!1!} \right) = 0,$$

$$A_3 = \frac{385}{8} \left(\frac{6!}{0!3!4!} - \frac{4!}{1!2!2!} \right) = -\frac{385}{8}, \qquad \text{etc.}$$

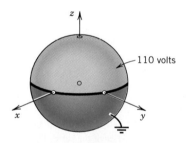

Fig. 283.   Spherical capacitor in Example 1

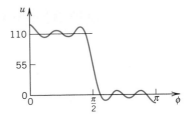

Fig. 284. Partial sum of the first six
nonzero terms of (22) for $r = R = 1$

Hence the potential (16) inside the sphere is (since $P_0 = 1$)

(22)
$$u(r, \phi) = 55 + \frac{165}{2} rP_1(\cos \phi) - \frac{385}{8} r^3 P_3(\cos \phi) + \cdots$$

with P_1, P_3, $\cdots$ given by (11'), Sec. 4.3. Since $R = 1$, we see from (18) and (20) in this section that $B_n = A_n$, and (19) thus gives the potential outside the sphere

(23)
$$u(r, \phi) = \frac{55}{r} + \frac{165}{2r^2} P_1(\cos \phi) - \frac{385}{8r^4} P_3(\cos \phi) + \cdots.$$

Partial sums of these series can now be used for computing approximate values of the potential. Also, it is interesting to see that far away from the sphere the potential is approximately that of a point charge, namely, $55/r$. (Compare with Example 3 in Sec. 8.9.) ◀

PROBLEM SET 11.11

1. **(Cylindrical coordinates)** Verify (5) by transforming $\nabla^2 u$ back into Cartesian coordinates.
2. **(Spherical coordinates)** Derive (7) from $\nabla^2 u$ in Cartesian coordinates.

Potentials Depending only on r

3. **(Dimension 2)** Show that the only solution of Laplace's equation depending only on $r = \sqrt{x^2 + y^2}$ is $u = c \ln r + k$ with constant c and k.
4. **(Dirichlet problem)** Find the electrostatic potential between two coaxial cylinders of radii $r_1 = 2$ cm and $r_2 = 4$ cm kept at the potentials $U_1 = 220$ volts and $U_2 = 140$ volts, respectively.
5. **(Dimension 3)** Verify that $u = c/r$, $r = \sqrt{x^2 + y^2 + z^2}$, satisfies Laplace's equation in spherical coordinates.
6. **(Dimension 3)** Show that the only solution of Laplace's equation depending only on r, $r = \sqrt{x^2 + y^2 + z^2}$, is $u = c/r + k$.
7. **(Dirichlet problem)** Find the electrostatic potential between two concentric spheres of radii $r_1 = 2$ cm and $r_2 = 4$ cm kept at the potentials $U_1 = 220$ volts and $U_2 = 140$ volts, respectively. Sketch and compare the equipotential lines in Probs. 4 and 7. Comment.
8. **(Verification)** Substituting $u(r)$ with r as in Prob. 5 into $u_{xx} + u_{yy} + u_{zz} = 0$, verify that $u'' + 2u'/r = 0$, in agreement with (7).
9. **(Heat problem)** If the surface of the ball $r^2 = x^2 + y^2 + z^2 \leq R^2$ is kept at temperature zero and the initial temperature in the ball is $f(r)$, show that the temperature $u(r, t)$ in the ball is a solution of $u_t = c^2(u_{rr} + 2u_r/r)$ satisfying the conditions $u(R, t) = 0$, $u(r, 0) = f(r)$. Show that setting $v = ru$ gives $v_t = c^2 v_{rr}$, $v(R, t) = 0$, $v(r, 0) = rf(r)$. Include the condition $v(0, t) = 0$ (which holds because u must be bounded at $r = 0$), and solve the resulting problem by separating variables.

Boundary Value Problems in Spherical Coordinates

Let r, θ, ϕ be the spherical coordinates used in the text. Find the potential in the interior of the sphere $R = 1$, assuming that there are no charges in the interior and the potential on the surface is:

10. $f(\phi) = \cos \phi$ **11.** $f(\phi) = 1$ **12.** $f(\phi) = 1 - \cos^2 \phi$ **13.** $f(\phi) = \cos 2\phi$

14. $f(\phi) = 10 \cos^3 \phi - 3 \cos^2 \phi - 5 \cos \phi - 1$

15. Show that in Prob. 11 the potential exterior to the sphere is the same as that of a point charge at the origin.

16. Find the potential exterior to the sphere in Probs. 10 and 13.

17. Sketch the intersections of the equipotential surfaces in Prob. 10 with the xz-plane.

18. In Example 1 in the text verify the values of A_0, A_1, A_2, A_3 and compute $A_4, \cdots, A_{10}$. Try to find out graphically how well the corresponding partial sums of (22) approximate the given boundary function.

19. Find the temperature in a homogeneous ball of radius 1 if its lower boundary hemisphere is kept at 0°C and its upper at 20°C.

20. **(Transformation)** Let r, θ, ϕ be spherical coordinates. If $u(r, \theta, \phi)$ satisfies $\nabla^2 u = 0$, show that $v(r, \theta, \phi) = r^{-1}u(r^{-1}, \theta, \phi)$ satisfies $\nabla^2 v = 0$.

21. **(Transformation)** If $u(r, \theta)$ satisfies $\nabla^2 u = 0$, show that $v(r, \theta) = u(r^{-1}, \theta)$ satisfies $\nabla^2 v = 0$. (r and θ are polar coordinates.)

22. **(Transformation)** What do you get from Prob. 21 if $u = r^2 \cos \theta \sin \theta$? What is v in terms of x and y?

23. **(Fourier–Legendre series)** What do you get by applying Prob. 20 to the functions in (16*)? Using the result, interpret the transformation in Prob. 20 geometrically.

24. **TEAM PROJECT. Transmission Line and Related Equations.** Consider a long cable or telephone wire (Fig. 285) that is imperfectly insulated so that leaks occur along the entire length of the cable. The source S of the current $i(x, t)$ in the cable is at $x = 0$, the receiving end T at $x = l$. The current flows from S to T, through the load, and returns to the ground. Let the constants R, L, C, and G denote the resistance, inductance, capacitance to ground, and conductance to ground, respectively, of the cable per unit length.

(a) Show that

$$-\frac{\partial u}{\partial x} = Ri + L\frac{\partial i}{\partial t} \qquad \textbf{(First transmission line equation)}$$

where $u(x, t)$ is the potential in the cable. *Hint.* Apply Kirchhoff's voltage law to a small portion of the cable between x and $x + \Delta x$ (difference of the potentials at x and $x + \Delta x$ = resistive drop + inductive drop).

(b) Show that for the cable in (a),

$$-\frac{\partial i}{\partial x} = Gu + C\frac{\partial u}{\partial t} \qquad \textbf{(Second transmission line equation).}$$

Hint. Use Kirchhoff's current law (difference of the currents at x and $x + \Delta x$ = loss due to leakage to ground + capacitive loss).

(c) **(Second-order equations)** Show that elimination of i or u from the transmission line equations leads to

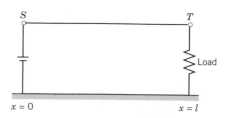

Fig. 285. Transmission line

$$u_{xx} = LCu_{tt} + (RC + GL)u_t + RGu,$$

$$i_{xx} = LCi_{tt} + (RC + GL)i_t + RGi.$$

(d) **(Telegraph equations)** For a submarine cable, G is negligible and the frequencies are low. Show that this leads to the so-called *submarine cable equations* or **telegraph equations**

$$u_{xx} = RCu_t, \qquad i_{xx} = RCi_t.$$

Find the potential in a submarine cable with ends ($x = 0$, $x = l$) grounded and initial voltage distribution $U_0 = const.$

(e) **(High-frequency line equations)** Show that in the case of alternating currents of high frequencies the equations in (c) can be approximated by the so-called **high-frequency line equations**

$$u_{xx} = LCu_{tt}, \qquad i_{xx} = LCi_{tt}.$$

Solve the first of them, assuming that the initial potential is $U_0 \sin(\pi x/l)$, $u_t(x, 0) = 0$ and $u = 0$ at the ends $x = 0$ and $x = l$ for all t.

25. TEAM PROJECT. Two-Dimensional Potential Problems. Such problems in which the functions depend only on two of the three space coordinates can best be solved by methods of **complex analysis** (see Sec. 12.5 and Chap. 16).

(a) For this reason we merely invite you to verify that the given functions $u = f(x, y)$ satisfy Laplace's equation and to sketch or plot some of the equipotential lines $u = const.$
Do this for the functions xy, $x^2 - y^2$, $x/(x^2 + y^2)$, $e^x \cos y$, $e^x \sin y$, $\cos x \cosh y$, $\ln(x^2 + y^2)$, $\arctan(y/x)$.

(b) Determine the most general functions $ax^3 + bx^2y + cxy^2 + ky^3$ that is harmonic.

11.12   Solution by Laplace Transforms

Readers familiar with Chap. 5 may wonder whether Laplace transforms can also be used for solving *partial* differential equations. The answer is yes, particularly if one of the independent variables ranges over the positive axis. The steps of solution are similar to those in Chap. 5. For an equation in two variables they are as follows.

1. Take the Laplace transform with respect to one of the two variables, usually t. This gives an *ordinary differential equation* for the transform of the unknown function. This is so since the derivatives of this function with respect to the other variable slip into the transformed equation. The latter also incorporates the given boundary and initial conditions.

2. Solving that ordinary differential equation, obtain the transform of the unknown function.

3. Taking the inverse transform, obtain the solution of the given problem.

If the coefficients of the given equation do not depend on t, the use of Laplace transforms will simplify the problem.

We explain the method in terms of a typical example.

EXAMPLE 1   **Semi-infinite string**

Find the displacement $w(x, t)$ of an elastic string subject to the following conditions.

(*i*) The string is initially at rest on the x-axis from $x = 0$ to ∞ ("*semi-infinite string*").

(*ii*) For time $t > 0$ the left end of the string is moved in a given fashion, namely,

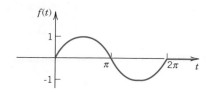

Fig. 286. Motion of the left end of the string in
Example 1 as a function of time t

$$w(0, t) = f(t) = \begin{cases} \sin t & \text{if } 0 \leqq t \leqq 2\pi \\ 0 & \text{otherwise} \end{cases} \qquad \text{(Fig, 286).}$$

(*iii*) Furthermore

$$\lim_{x \to \infty} w(x, t) = 0 \quad \text{for } t \geqq 0.$$

Of course there is no infinite string, but our model describes a long string or rope (of negligible weight) with its right end fixed far out on the x-axis. (We write w since we need u to denote the unit step function.)

Solution. We have to solve the wave equation (Sec. 11.2)

$$(1) \qquad \frac{\partial^2 w}{\partial t^2} = c^2 \frac{\partial^2 w}{\partial x^2}, \qquad\qquad c^2 = \frac{T}{\rho},$$

for positive x and t, subject to the "boundary conditions"

$$(2) \qquad w(0, t) = f(t), \qquad \lim_{x \to \infty} w(x, t) = 0 \qquad (t \geqq 0)$$

with f as given above, and the initial conditions

$$(3) \qquad w(x, 0) = 0,$$

$$(4) \qquad \left.\frac{\partial w}{\partial t}\right|_{t=0} = 0.$$

We take the Laplace transform **with respect to t.** By (2) in Sec. 5.2,

$$\mathcal{L}\left\{\frac{\partial^2 w}{\partial t^2}\right\} = s^2 \mathcal{L}\{w\} - s w(x, 0) - \left.\frac{\partial w}{\partial t}\right|_{t=0} = c^2 \mathcal{L}\left\{\frac{\partial^2 w}{\partial x^2}\right\}.$$

Two terms drop out, by (3) and (4). On the right we assume that we may interchange integration and differentiation:

$$\mathcal{L}\left\{\frac{\partial^2 w}{\partial x^2}\right\} = \int_0^\infty e^{-st} \frac{\partial^2 w}{\partial x^2} \, dt = \frac{\partial^2}{\partial x^2} \int_0^\infty e^{-st} w(x, t) \, dt = \frac{\partial^2}{\partial x^2} \mathcal{L}\{w(x, t)\}.$$

Writing $W(x, s) = \mathcal{L}\{w(x, t)\}$, we thus obtain

$$s^2 W = c^2 \frac{\partial^2 W}{\partial x^2}, \qquad\qquad \text{thus} \qquad\qquad \frac{\partial^2 W}{\partial x^2} - \frac{s^2}{c^2} W = 0.$$

Since this equation contains only a derivative with respect to x, it may be regarded as an ordinary differential equation for $W(x, s)$ considered as a function of x. A general solution is

$$(5) \qquad W(x, s) = A(s) e^{sx/c} + B(s) e^{-sx/c}.$$

From (2) we obtain, writing $F(s) = \mathcal{L}\{f(t)\}$,

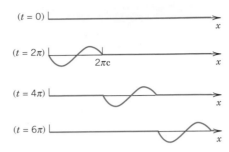

Fig. 287. Traveling wave in Example 1

$$W(0, s) = \mathcal{L}\{w(0, t)\} = \mathcal{L}\{f(t)\} = F(s).$$

Assuming that we can interchange integration and taking the limit, we have

$$\lim_{x \to \infty} W(x, s) = \lim_{x \to \infty} \int_0^\infty e^{-st} w(x, t)\, dt = \int_0^\infty e^{-st} \lim_{x \to \infty} w(x, t)\, dt = 0.$$

This implies $A(s) = 0$ in (5) because $c > 0$, so that for every fixed positive s the function $e^{sx/c}$ increases as x increases. Note that we may assume $s > 0$ since a Laplace transform generally exists for *all* s greater than some fixed k (Sec. 5.2). Hence we have

$$W(0, s) = B(s) = F(s),$$

so that (5) becomes

$$W(x, s) = F(s) e^{-sx/c}.$$

From the second shifting theorem (Sec. 5.3) with $a = x/c$ we obtain the inverse transform

(6) $$w(x, t) = f\left(t - \frac{x}{c}\right) u\left(t - \frac{x}{c}\right)$$ (Fig. 287)

that is,

$$w(x, t) = \sin\left(t - \frac{x}{c}\right) \qquad \text{if} \qquad \frac{x}{c} < t < \frac{x}{c} + 2\pi \qquad \text{or} \qquad ct > x > (t - 2\pi)c$$

and zero otherwise. This is a single sine wave traveling to the right with speed c. Note that a point x remains at rest until $t = x/c$, the time needed to reach that x if one starts at $t = 0$ (start of the motion of the left end) and travels with speed c. The result agrees with our physical intuition. Since we proceeded formally, we must verify that (6) satisfies the given conditions. We leave this to the student.   ◀

This is the end of Chap. 11, in which we concentrated on the most important partial differential equations in physics and engineering. This is also the end of Part C on Fourier analysis and partial differential equations.

We have seen that these equations have various basic engineering applications. For this reason they are the subject of many ongoing research projects.

Numerical methods for partial differential equations follow in Secs. 19.4–19.7, which are independent of the other sections in Part E on numerical methods.

In the next part (Part D, Chaps. 12–16) we again turn to an area of different nature, **complex analysis,** which is also highly important to the engineer, as our examples and problems will show. This will include another approach to the (two-dimensional) Laplace equation in Chap. 16.

PROBLEM SET 11.12

1. Verify the solution in Example 1. What traveling wave do we obtain in Example 1 if we impose a (nonterminating) sinusoidal motion of the left end starting at $t = 0$?
2. How does the speed of the wave in Example 1 depend on the tension and the mass of the string?
3. Sketch a figure similar to Fig. 287 if $c = 1$ and f is "triangular" as in Example 1, Sec. 11.3, with $k = L/2 = 1$.

Solve by Laplace transforms:

4. $\dfrac{\partial u}{\partial x} + 2x\dfrac{\partial u}{\partial t} = 2x$, $u(x, 0) = 1$, $u(0, t) = 1$

5. $x\dfrac{\partial u}{\partial x} + \dfrac{\partial u}{\partial t} = xt$, $u(x, 0) = 0$ if $x \geqq 0$, $u(0, t) = 0$ if $t \geqq 0$.

6. Solve Prob. 5 by another method.

Heat Problem

Find the temperature $w(x, t)$ in a semi-infinite laterally insulated bar extending from $x = 0$ along the x-axis to ∞, assuming that the initial temperature is 0, $w(x, t) \to 0$ as $x \to \infty$ for every fixed $t \geqq 0$, and $w(0, t) = f(t)$. Proceed as follows.

7. Set up the model and show that the Laplace transform leads to

$$sW(x, s) = c^2\frac{\partial^2 W}{\partial x^2}, \qquad\qquad W = \mathcal{L}\{w\},$$

and

$$W(x, s) = F(s)e^{-\sqrt{s}\,x/c}, \qquad\qquad F = \mathcal{L}\{f\}.$$

8. Applying the convolution theorem in Prob. 7, show that

$$w(x, t) = \frac{x}{2c\sqrt{\pi}}\int_0^t f(t - \tau)\tau^{-3/2}e^{-x^2/4c^2\tau}\,d\tau.$$

9. Let $w(0, t) = f(t) = u(t)$ (Sec. 5.3). Denote the corresponding w, W, and F by w_0, W_0, and F_0. Show that then in Prob. 8,

$$w_0(x, t) = \frac{x}{2c\sqrt{\pi}}\int_0^t \tau^{-3/2}e^{-x^2/4c^2\tau}\,d\tau = 1 - \mathrm{erf}\left(\frac{x}{2c\sqrt{t}}\right)$$

with the error function erf as defined in Problem Set 11.6.

10. **(Duhamel's formula[15])** Show that in Prob. 9,

$$W_0(x, s) = \frac{1}{s}e^{-\sqrt{s}\,x/c}$$

and the convolution theorem gives *Duhamel's formula*

$$w(x, t) = \int_0^t f(t - \tau)\frac{\partial w_0}{\partial \tau}\,d\tau.$$

---

[15]JEAN MARIE CONSTANT DUHAMEL (1797—1872), French mathematician.

CHAPTER 11 REVIEW

1. What kinds of problems lead to ordinary differential equations? To partial differential equations?
2. Name three or four most important partial differential equations, write down their form, and explain their main applications.
3. By what physical law did we obtain the equation of the vibrating string? What is the name of this equation?
4. What are the eigenfunctions and their frequencies of the vibrating string?
5. Why did we sometimes use polar or spherical coordinates?
6. Why and where did Bessel's equation occur? Legendre's equation?
7. What is the method of separating variables? Give an example.
8. In what case did we need two subsequent separations of variables?
9. What is d'Alembert's method? To what equation does it apply?
10. Name and explain the three kinds of boundary value problems.
11. What role did Fourier series play in this chapter? What could we accomplish without their use?
12. What kind of additional conditions did we consider for the wave equation? For the heat equation?
13. To what kind of problem does the Fourier integral apply? The Fourier transform?
14. What kind of partial differential equations can be solved by methods for ordinary differential equations? Give examples.
15. What is the idea involved in applying the Laplace transform method to *partial* differential equations?
16. What is the superposition principle? Why is it useful?
17. What are elliptic, parabolic, and hyperbolic equations? Give examples.
18. What are the eigenfunctions of the circular membrane? How do their frequencies differ in principle from those of the vibrating string?
19. In separating the heat equation we got exponential functions. Why? Why not in the case of the wave equation? (Give a *mathematical* reason, not a physical one.)
20. What is the error function? Why did it occur and where?

Solve:

21. $u_{xx} + 9u = 0$
22. $u_{xy} + u_x + x = 0$
23. $u_{yy} + 3u_y - 4u = 12$
24. $u_{xx} + u_x = 0$, $u(0, y) = f(y)$, $u_x(0, y) = g(y)$

25. Find solutions of $u_x = yu_y$ by separating variables.
26. Find all solutions $u(x, y) = F(x)G(y)$ of Laplace's equation in two variables.

Transform to normal form and solve:

27. $u_{xx} - 4u_{yy} = 0$
28. $u_{xx} + 6u_{xy} + 9u_{yy} = 0$
29. $u_{xy} = u_{yy}$

Find and sketch (as in Fig. 263) the deflection $u(x, t)$ of a vibrating string of length π and $c^2 = T/\rho = 4$ for zero initial velocity and initial deflection given by

30. $\sin^3 x$
31. $\sin 5x$
32. $\frac{1}{2}\pi - |x - \frac{1}{2}\pi|$

Find the temperature distribution in a laterally insulated thin copper bar ($c^2 = K/\sigma\rho = 1.158$ cm²/sec), 100 cm long and of constant cross section whose endpoints at $x = 0$ and $x = 100$ are kept at 0°C and whose initial temperature is

33. $\sin 0.01\pi x$
34. $50 - |50 - x|$
35. $\sin^3 0.01\pi x$

Find the temperature $u(x, t)$ in a laterally insulated bar of length π with $c^2 = 1$ for adiabatic boundary conditions (see Problem Set 11.5) and initial temperature

36. $15x^2$
37. $250 \cos 2x$
38. $2\pi - 4|x - \frac{1}{2}\pi|$

Find the temperature $u(x, y)$ in a thin metallic square plate of side $a = 12$ with insulated faces if the left, lower, and right sides are kept at 0°C and the upper side at the temperature (in °C)

39. $\sin(\pi x/12)$ **40.** 100   **41.** $\sin(\pi x/4)$

42. Sketch or plot some of the isotherms in Prob. 39. Give physical reasons for the general shape of these curves.

Show that the following membranes of area 1 with $c^2 = 1$ have the frequencies of the fundamental mode as given (4-decimal values). Compare.

43. Circle: $\alpha_1/2\sqrt{\pi} = 0.6784$ **44.** Square: $1/\sqrt{2} = 0.7071$

45. Rectangle (sides $1:2$): $\sqrt{5/8} = 0.7906$ **46.** Semicircle: $3.832/\sqrt{8\pi} = 0.7644$

47. Quadrant of circle: $\alpha_{12}/4\sqrt{\pi} = 0.7244$ ($\alpha_{12} = 5.13562 =$ first positive zero of J_2)

Find the electrostatic potential:

48. Between two concentric spheres of radii r_0 and r_1 kept at the potentials u_0 and u_1, respectively.

49. Between two coaxial circular cylinders of radii r_0 and r_1 kept at the potential u_0 and u_1, respectively. (Compare with Prob. 48.)

50. In the (charge-free) interior of a sphere of radius 1 kept at the potential $f(\phi) = \cos 3\phi + 3 \cos \phi$ (referred to our usual spherical coordinates).

SUMMARY OF CHAPTER 11
PARTIAL DIFFERENTIAL EQUATIONS

Whereas *ordinary* differential equations (Chaps. 1–5) appear as models of simpler problems involving a single independent variable, problems involving two or more independent variables (space variables, or time t and one or several space variables) lead to *partial* differential equations. Hence the importance of these equations to the engineer and physicist can hardly be overestimated.

In this chapter we were mainly concerned with the most important partial differential equations of physics and engineering, namely:

(1) $u_{tt} = c^2 u_{xx}$ **One-dimensional wave equation**

(Secs. 11.2–11.4)

(2) $u_{tt} = c^2(u_{xx} + u_{yy})$ **Two-dimensional wave equation**

(Secs. 11.7–11.10)

(3) $u_t = c^2 u_{xx}$ **One-dimensional heat equation**

(Secs. 11.5, 11.6)

(4) $\nabla^2 u = u_{xx} + u_{yy} = 0$ **Two-dimensional Laplace equation**

(Secs. 11.5, 11.9)

(5) $\nabla^2 u = u_{xx} + u_{yy} + u_{zz} = 0$ **Three-dimensional Laplace equation**

(Sec. 11.11).

Equations (1) and (2) are hyperbolic, (3) is parabolic, (4) and (5) are elliptic. (See Problem Set 11.4.)

In practice, one is interested in obtaining the solution of such an equation in a given region satisfying given additional conditions, such as **initial conditions** (conditions at time $t = 0$) or **boundary conditions** (prescribed values of the solution u or some of its derivatives on the boundary surface S, or boundary curve C, of the region) or both. For (1) and (2) one prescribes two initial conditions (initial displacement and initial velocity). For (3) one prescribes the initial temperature distribution. For (4) or (5) one prescribes a boundary condition and calls the resulting problem a (see Sec. 11.5)

> **Dirichlet problem** if u is prescribed on S,
> **Neumann problem** if $u_n = \partial u / \partial n$ is prescribed on S,
> **Mixed problem** if u is prescribed on one part of S and u_n on the other.

A general method for solving such problems is the method of **separating variables** or **product method,** in which one assumes solutions in the form of products of functions each depending on one variable only. Thus equation (1) is solved by setting (Sec. 11.3)

$$u(x,\ t) = F(x)G(t);$$

similarly for (3) in Sec. 11.5. Substitution into the given equation yields *ordinary* differential equations for F and G, and from these one gets infinitely many solutions $F = F_n$ and $G = G_n$ such that the corresponding functions

$$u_n(x,\ t) = F_n(x)G_n(t)$$

are solutions of the partial differential equations satisfying the given boundary conditions. These are the **eigenfunctions** of the problem, and the corresponding **eigenvalues** determine the frequency of the vibration (or the rapidity of the decrease of temperature in the case of the heat equation, etc.). To satisfy also the initial condition (or conditions), one must consider infinite series of the u_n, whose coefficients turn out to be the Fourier coefficients of the functions f and g representing the given initial conditions (Secs. 11.3, 11.5). Hence **Fourier series** (and *Fourier integrals*) are of basic importance here (Secs. 11.3, 11.5, 11.6, 11.8).

Steady-state problems are problems in which the solution does not depend on time t. For these, the heat equation $u_t = c^2 \nabla^2 u$ becomes the Laplace equation.

Before solving an initial or boundary value problem, one often transforms the equation into coordinates in which the boundary of the region considered is given by simple formulas. Thus in polar coordinates given by $x = r \cos \theta$, $y = r \sin \theta$, the **Laplacian** becomes (Sec. 11.9)

$$(6) \qquad\qquad \nabla^2 u = u_{rr} + \frac{1}{r} u_r + \frac{1}{r^2} u_{\theta\theta};$$

for spherical coordinates see Sec. 11.11. If one now separates the variables, one gets **Bessel's equation** from (2) and (6) (vibrating circular membrane, Sec. 11.10) and **Legendre's equation** from (5) transformed into spherical coordinates (Sec. 11.11).

Operational methods (Laplace transforms, Fourier transforms) are helpful in solving partial differential equations in infinite regions (Secs. 11.12, 11.16).

PART D

Complex Analysis

Chapter 12 **Complex Numbers and Functions. Conformal Mapping**

Chapter 13 **Complex Integration**

Chapter 14 **Power Series, Taylor Series**

Chapter 15 **Laurent Series, Residue Integration**

Chapter 16 **Complex Analysis Applied to Potential Theory**

Many engineering problems may be treated and solved by methods involving complex numbers and complex functions. There are two kinds of such problems. The first of them consists of "elementary problems" for which some acquaintance with complex numbers is sufficient. This includes many applications to electric circuits or mechanical vibrating systems.

The second kind consists of more advanced problems for which we must be familiar with the theory of complex analytic functions—**"complex function theory"** or **"complex analysis,"** for short—and with its powerful and elegant methods. Interesting problems in heat conduction, fluid flow, and electrostatics belong to this category.

We devote the next five chapters (Chaps. 12–16) to complex analysis and its applications. We shall see that the importance of complex analytic functions in engineering mathematics has the following two main roots.

1. *The real and imaginary parts of an analytic function are solutions of Laplace's equation in two independent variables. Consequently, two-dimensional potential problems can be treated by methods developed for analytic functions.* See Chap. 16.
2. *Most higher functions in engineering mathematics are analytic functions, and their study for complex values of the independent variable leads to a much deeper understanding of their properties. Furthermore, complex integration can help evaluating complicated complex **and real** integrals of practical interest.* See Secs. 15.3, 15.4.

Complex Numbers and Functions. Conformal Mapping

Complex numbers and the complex plane are discussed in Secs. 12.1–12.2. Complex analysis is concerned with complex analytic functions, as defined in Sec. 12.3. In Sec. 12.4 we explain a check for analyticity based on the so-called Cauchy–Riemann equations. The latter are of basic importance. They are related to Laplace's equation (Sec. 12.4). In the remaining sections of Chap. 12 we study the most important elementary complex functions (exponential function, trigonometric functions, etc.), which generalize familiar real functions known from calculus. This includes discussions of geometric properties of these functions in connection with conformal mapping (defined in Sec. 12.5).

Prerequisites for this chapter: Elementary calculus.
References: Appendix 1, Part D.
Answers to problems: Appendix 2.

12.1 Complex Numbers. Complex Plane

Equations without *real* solutions, such as $x^2 = -1$ or $x^2 - 10x + 40 = 0$, were observed early in history and led to the introduction of complex numbers.[1] By definition, a **complex number** z is an ordered pair (x, y) of real numbers x and y, written

$$z = (x, y).$$

x is called the **real part** and y the **imaginary part** of z, written

$$x = \text{Re } z, \qquad y = \text{Im } z.$$

By definition, two complex numbers are **equal** if and only if their real parts are equal and their imaginary parts are equal.

---

[1]First to use complex numbers for this purpose was the Italian mathematician GIROLAMO CARDANO (1501—1576), who found the formula for solving cubic equations. The term "complex number" was introduced by the great German mathematician CARL FRIEDRICH GAUSS (see the footnote in Sec. 4.4), who also paved the way for a general use of complex numbers.

(0, 1) is called the **imaginary unit** and is denoted by i,

(1)
$$i = (0, 1).$$

Addition, Multiplication. Notation $z = x + iy$

Addition of two complex numbers $z_1 = (x_1, y_1)$ and $z_2 = (x_2, y_2)$ is defined by[2]

(2)
$$z_1 + z_2 = (x_1, y_1) + (x_2, y_2) = (x_1 + x_2, \quad y_1 + y_2).$$

Multiplication is defined by

(3)
$$z_1 z_2 = (x_1, y_1)(x_2, y_2) = (x_1 x_2 - y_1 y_2, \quad x_1 y_2 + x_2 y_1).$$

In particular, these two definitions imply that

$$(x_1, 0) + (x_2, 0) = (x_1 + x_2, 0) \qquad \text{and} \qquad (x_1, 0)(x_2, 0) = (x_1 x_2, 0),$$

as for real numbers x_1, x_2. Hence the complex numbers **"extend"** the reals, and we can write

(4*)
$$(x, 0) = x.$$

Similarly, for any real y

(4**)
$$(0, y) = iy$$

because $iy = (0, 1)(y, 0)$ by (1) and (4*) (with y instead of x); and multiplication (3) gives $(0, 1)(y, 0) + (0y - 1 \cdot 0, 0 \cdot 0 + 1y) = (0, y)$, hence (4**). Together with (4) and by addition we thus have $(x, y) = (x, 0) + (0, y) = x + iy$:
In practice, complex numbers $z = (x, y)$ are written[3]

(4)
$$z = x + iy.$$

If $x = 0$, then $z = iy$ and is called **pure imaginary.** Also, (1) and (3) give

(5)
$$i^2 = -1$$

because by the definition of multiplication, $i^2 = ii = (0, 1)(0, 1) = (-1, 0) = -1$.
For **addition** the standard notation (4) gives [see (2)]

$$(x_1 + iy_1) + (x_2 + iy_2) = (x_1 + x_2) + i(y_1 + y_2).$$

---

[2]Students familiar with vectors see that this is *vector addition*, whereas this multiplication has no analog in the usual vector algebra.

[3]Electrical engineers often use j to reserve i for the current.

For **multiplication** it gives the following very simple recipe. Multiply each term by each term and use $i^2 = -1$ when it occurs [see (3)]:

$$(x_1 + iy_1)(x_2 + iy_2) = x_1x_2 + ix_1y_2 + iy_1x_2 + i^2y_1y_2$$

$$= (x_1x_2 - y_1y_2) + i(x_1y_2 + x_2y_1).$$

This agrees with (3).

EXAMPLE 1   **Real part, imaginary part, sum and product of complex numbers**

Let $z_1 = 8 + 3i$ and $z_2 = 9 - 2i$. Then Re $z_1 = 8$, Im $z_1 = 3$, Re $z_2 = 9$, Im $z_2 = -2$ and

$$z_1 + z_2 = (8 + 3i) + (9 - 2i) = 17 + i,$$

$$z_1z_2 = (8 + 3i)(9 - 2i) = 72 + 6 + i(-16 + 27) = 78 + 11i. \quad \blacktriangleleft$$

Subtraction, Division

Subtraction and **division** are defined as the inverse operations of addition and multiplication. Thus the **difference** $z = z_1 - z_2$ is the complex number z for which $z_1 = z + z_2$. Hence by (2),

$$(6) \qquad\qquad z_1 - z_2 = (x_1 - x_2) + i(y_1 - y_2).$$

The **quotient** $z = z_1/z_2$ $(z_2 \neq 0)$ is the complex number z for which $z_1 = zz_2$. If we equate the real and the imaginary parts on both sides of this equation, setting $z = x + iy$, we obtain $x_1 = x_2x - y_2y$, $y_1 = y_2x + x_2y$. The solution is

$$(7^*) \qquad z = \frac{z_1}{z_2} = x + iy, \qquad x = \frac{x_1x_2 + y_1y_2}{x_2{}^2 + y_2{}^2}, \qquad y = \frac{x_2y_1 - x_1y_2}{x_2{}^2 + y_2{}^2}.$$

The *practical rule* used to get this is by multiplying numerator and denominator of z_1/z_2 by $x_2 - iy_2$ and simplifying:

$$(7) \qquad z = \frac{x_1 + iy_1}{x_2 + iy_2} = \frac{(x_1 + iy_1)(x_2 - iy_2)}{(x_2 + iy_2)(x_2 - iy_2)} = \frac{x_1x_2 + y_1y_2}{x_2{}^2 + y_2{}^2} + i\,\frac{x_2y_1 - x_1y_2}{x_2{}^2 + y_2{}^2}.$$

EXAMPLE 2   **Difference and quotient of complex numbers**

For $z_1 = 8 + 3i$ and $z_2 = 9 - 2i$ we get $z_1 - z_2 = (8 + 3i) - (9 - 2i) = -1 + 5i$ and

$$\frac{z_1}{z_2} = \frac{8 + 3i}{9 - 2i} = \frac{(8 + 3i)(9 + 2i)}{(9 - 2i)(9 + 2i)} = \frac{66 + 43i}{81 + 4} = \frac{66}{85} + \frac{43}{85}\,i.$$

Check the division by multiplication to get $8 + 3i$. $\qquad \blacktriangleleft$

Complex numbers satisfy the same commutative, associative, and distributive laws as real numbers (see the problem set).

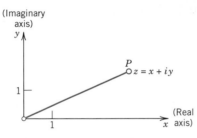

Fig. 288.   The complex plane

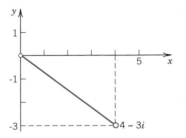

Fig. 289.   The number $4 - 3i$ in the complex plane

Complex Plane

This was algebra. Now comes geometry: the geometrical representation of complex numbers as points in the plane. This is of great practical importance. The idea is quite simple and natural. We choose two perpendicular coordinate axes, the horizontal x-axis, called the **real axis,** and the vertical y-axis, called the **imaginary axis.** On both axes we choose the same unit of length (Fig. 288). This is called a **Cartesian coordinate system.** We now plot a given complex number $z = (x, y) = x + iy$ as the point P with coordinates x, y. The xy-plane in which the complex numbers are represented in this way is called the **complex plane.**[4] Figure 289 shows an example.

Instead of saying "the point represented by z in the complex plane" we say briefly and simply *"the point z in the complex plane."* This will cause no misunderstandings.

Addition and subtraction can now be visualized as illustrated in Figs. 290 and 291.

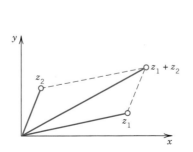

Fig. 290.   Addition of complex numbers

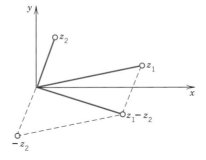

Fig. 291.   Subtraction of complex numbers

Complex Conjugate Numbers

The **complex conjugate** $\bar{z}$ of a complex number $z = x + iy$ is defined by

$$\bar{z} = x - iy.$$

---

[4]Sometimes called the **Argand diagram,** after the French mathematician JEAN ROBERT ARGAND (1768—1822), born in Geneva and later librarian in Paris. His paper on the complex plane appeared in 1806, nine years after a similar memoir by the Norwegian mathematician CASPAR WESSEL (1745—1818), a surveyor of the Danish Academy of Science.

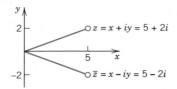

Fig. 292. Complex conjugate numbers

It is obtained geometrically by reflecting the point z in the real axis. Figure 292 shows this for $z = 5 + 2i$ and its conjugate $\bar{z} = 5 - 2i$.

It is important because it permits us to switch from complex to real. Indeed, by multiplication, $z\bar{z} = x^2 + y^2$ (verify!). By addition and subtraction, $z + \bar{z} = 2x$, $z - \bar{z} = 2iy$. We thus obtain for the real part x and the imaginary part y (not iy!) of $z = x + iy$ the important formulas

(8)
$$\text{Re } z = x = \frac{1}{2}(z + \bar{z}), \qquad \text{Im } z = y = \frac{1}{2i}(z - \bar{z}).$$

If z is real, $z = x$, then $\bar{z} = z$ by the definition of $\bar{z}$, and conversely.

Working with conjugates is easy, since we have

(9)
$$\overline{(z_1 + z_2)} = \bar{z}_1 + \bar{z}_2, \qquad \overline{(z_1 - z_2)} = \bar{z}_1 - \bar{z}_2,$$
$$\overline{(z_1 z_2)} = \bar{z}_1 \bar{z}_2, \qquad \overline{\left(\frac{z_1}{z_2}\right)} = \frac{\bar{z}_1}{\bar{z}_2}.$$

EXAMPLE 3   **Illustration of (8) and (9)**

Let $z_1 = 4 + 3i$ and $z_2 = 2 + 5i$. Then by (8),

$$\text{Im } z_1 = \frac{1}{2i}[(4 + 3i) - (4 - 3i)] = \frac{3i + 3i}{2i} = 3.$$

Also, the multiplication formula in (9) is verified by

$$\overline{(z_1 z_2)} = \overline{(4 + 3i)(2 + 5i)} = \overline{-7 + 26i} = -7 - 26i,$$
$$\bar{z}_1 \bar{z}_2 = (4 - 3i)(2 - 5i) = -7 - 26i.$$

◀

PROBLEM SET 12.1

1. (Powers of the imaginary unit) Show that

$$i^2 = -1, \quad i^3 = -i, \quad i^4 = 1, \quad i^5 = i, \cdots$$

(10)
$$\frac{1}{i} = -i, \quad \frac{1}{i^2} = -1, \quad \frac{1}{i^3} = i, \cdots,$$

2. Multiplication by i is geometrically a counterclockwise rotation through $\pi/2$ (90°). Verify this by plotting z and iz and the angle of rotation for $z = 4 + 2i$, $z = -1 + i$, $z = 5 - 2i$.

Arithmetical Operations. Real and Imaginary Parts. Complex Conjugates

Let $z_1 = 4 + 3i$ and $z_2 = 2 - 5i$. Find each of the following in the form $x + iy$, showing the details of your work:

3. $z_1 z_2$ **4.** $(3z_1 - z_2)^2$ **5.** $1/z_1$ **6.** $25z_2/z_1$

7. Re $(z_1{}^3)$, (Re $z_1)^3$ **8.** $(z_1 - z_2)/(z_1 + z_2)$ **9.** $z_1 \bar{z}_2$, $\bar{z}_1 z_2$ **10.** $1/z_1{}^2$, $1/\bar{z}_1{}^2$

11. $\bar{z}_1/\bar{z}_2$, $\overline{(z_1/z_2)}$ **12.** $z_2 \bar{z}_2/(z_1 \bar{z}_1)$

Let $z = x + iy$. Find (showing the details of your work)

13. Im $(1/z)$ **14.** Im z^4, (Im $z^2)^2$ **15.** $(1 + i)^{16}$ **16.** Re $(z/\bar{z})$ **17.** Re $(z^2/\bar{z})$

18. (**Laws of addition and multiplication**) Derive the following laws for complex numbers from the corresponding laws for real numbers.

$$z_1 + z_2 = z_2 + z_1, \qquad z_1 z_2 = z_2 z_1 \qquad \qquad (\textit{Commutative laws})$$

$$(z_1 + z_2) + z_3 = z_1 + (z_2 + z_3), \qquad (z_1 z_2)z_3 = z_1(z_2 z_3) \qquad (\textit{Associative laws})$$

$$z_1(z_2 + z_3) = z_1 z_2 + z_1 z_3 \qquad \qquad (\textit{Distributive law})$$

$$0 + z = z + 0 = z, \qquad z + (-z) = (-z) + z = 0, \qquad z \cdot 1 = z.$$

19. (**Laws for conjugates**) Verify (9) for $z_1 = 38 + 18i$, $z_2 = 3 + 5i$.

20. (**Multiplication**) If the product of two complex numbers is zero, show that at least one factor must be zero.

12.2 Polar Form of Complex Numbers Powers and Roots

We can substantially increase the usefulness of the complex plane and gain further insight into the nature of complex numbers if besides the xy-coordinates we also employ the usual polar coordinates r, θ defined by

(1)
$$x = r \cos \theta, \qquad y = r \sin \theta.$$

Then $z = x + iy$ takes the so-called **polar form**

(2)
$$z = r(\cos \theta + i \sin \theta).$$

r is called the **absolute value** or **modulus** of z and is denoted by $|z|$. Hence

(3)
$$|z| = r = \sqrt{x^2 + y^2} = \sqrt{z\bar{z}}.$$

Geometrically, $|z|$ is the distance of the point z from the origin (Fig. 293). Similarly, $|z_1 - z_2|$ is the distance between z_1 and z_2 (Fig. 294).

θ is called the **argument** of z and is denoted by arg z. Thus (Fig. 293)

(4)
$$\theta = \arg z = \arctan \frac{y}{x} \qquad\qquad (z \neq 0).$$

Geometrically, θ is the directed angle from the positive x-axis to OP in Fig. 293. Here,

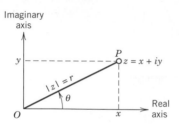

Fig. 293. Complex plane, polar form of a complex number

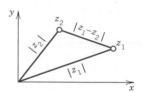

Fig. 294. Distance between two points in the complex plane

as in calculus, *all angles are measured in* **radians** *and positive in the counterclockwise sense.*

For $z = 0$ this angle θ is undefined. (Why?) For given $z \neq 0$ it is determined only up to integer multiples of 2π. The value of θ that lies in the interval $-\pi < \theta \leqq \pi$ is called the **principal value** of the argument of $z \ (\neq 0)$ and is denoted by Arg z, with capital A. Thus, by definition $\theta = \text{Arg } z$ satisfies by definition

$$-\pi < \text{Arg } z \leqq \pi.$$

This principal value is important in connection with roots, the complex logarithm (Sec. 12.8), and certain integrals.

EXAMPLE 1   **Polar form of complex numbers. Principal value**

$z = 1 + i$ (Fig. 295) has the polar form $z = \sqrt{2}(\cos \frac{1}{4}\pi + i \sin \frac{1}{4}\pi)$. Hence $|z| = \sqrt{2}$, arg $z = \frac{1}{4}\pi \pm 2n\pi$ $(n = 0, 1, \cdots)$, and Arg $z = \frac{1}{4}\pi$ (the principal value).

Similarly, $z = 3 + 3\sqrt{3}i = 6(\cos \frac{1}{3}\pi + i \sin \frac{1}{3}\pi)$, $|z| = 6$, and Arg $z = \frac{1}{3}\pi$.   ◀

Caution! In using (4), we must pay attention to the quadrant in which z lies, since $\tan \theta$ has period π, so that the arguments of z and $-z$ have the same tangent. *Example:* for $\theta_1 = \text{arg } (1 + i)$ and $\theta_2 = \text{arg } (-1 - i)$ we have $\tan \theta_1 = \tan \theta_2 = 1$.

Triangle inequality

For any complex numbers we have the important **triangle inequality**

(5) $$|z_1 + z_2| \leqq |z_1| + |z_2|$$ (Fig. 296)

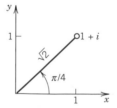

Fig. 295. Example 1

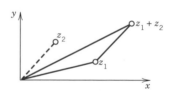

Fig. 296. Triangle inequality

which we shall use quite frequently. This inequality follows by noting that the three points 0, z_1, and $z_1 + z_2$ are the vertices of a triangle (Fig. 296) with sides $|z_1|$, $|z_2|$, and $|z_1 + z_2|$, and one side cannot exceed the sum of the other two sides. A formal proof is left to the reader (Prob. 33). (The triangle degenerates if z_1 and z_2 lie on the same straight line through the origin.)

By induction we obtain from (5) the **generalized triangle inequality**

$$(6) \qquad |z_1 + z_2 + \cdots + z_n| \leq |z_1| + |z_2| + \cdots + |z_n|;$$

that is, *the absolute value of a sum cannot exceed the sum of the absolute values of the terms.*

EXAMPLE 2   **Triangle inequality**

If $z_1 = 1 + i$ and $z_2 = -2 + 3i$, then (sketch a figure!)

$$|z_1 + z_2| = |-1 + 4i| = \sqrt{17} = 4.123 < \sqrt{2} + \sqrt{13} = 5.020. \qquad \blacktriangleleft$$

Multiplication and Division in Polar Form

This will give us a "geometrical" understanding of multiplication and division. Let

$$z_1 = r_1(\cos \theta_1 + i \sin \theta_1) \qquad \text{and} \qquad z_2 = r_2(\cos \theta_2 + i \sin \theta_2).$$

Multiplication. By (3), Sec. 12.1, the product is at first

$$z_1 z_2 = r_1 r_2 [(\cos \theta_1 \cos \theta_2 - \sin \theta_1 \sin \theta_2) + i(\sin \theta_1 \cos \theta_2 + \cos \theta_1 \sin \theta_2)].$$

The addition rules for the sine and cosine [(6) in Appendix A3.1] now yield

$$(7) \qquad z_1 z_2 = r_1 r_2 [\cos (\theta_1 + \theta_2) + i \sin (\theta_1 + \theta_2)].$$

Taking absolute values and arguments on both sides of (7), we thus obtain the important rules

$$(8) \qquad |z_1 z_2| = |z_1||z_2|$$

and

$$(9) \qquad \arg (z_1 z_2) = \arg z_1 + \arg z_2 \qquad \text{(up to multiples of } 2\pi\text{)}.$$

Division. The quotient $z = z_1/z_2$ is the number z satisfying $z z_2 = z_1$. Hence $|z z_2| = |z| |z_2| = |z_1|$, $\arg (z z_2) = \arg z + \arg z_2 = \arg z_1$. This yields

$$(10) \qquad \left| \frac{z_1}{z_2} \right| = \frac{|z_1|}{|z_2|} \qquad (z_2 \neq 0)$$

and

$$(11) \qquad \arg \frac{z_1}{z_2} = \arg z_1 - \arg z_2 \qquad \text{(up to multiples of } 2\pi\text{)}.$$

By combining formulas (10) and (11) we also have

$$(12) \qquad \frac{z_1}{z_2} = \frac{r_1}{r_2} \left[\cos(\theta_1 - \theta_2) + i \sin(\theta_1 - \theta_2)\right].$$

EXAMPLE 3   **Illustration of formulas (8)–(11)**

Let $z_1 = -2 + 2i$ and $z_2 = 3i$. Then $z_1 z_2 = -6 - 6i$, $z_1/z_2 = 2/3 + (2/3)i$. Hence (make a sketch)

$$|z_1 z_2| = 6\sqrt{2} = 3\sqrt{8} = |z_1||z_2|, \qquad |z_1/z_2| = 2\sqrt{2}/3 = |z_1|/|z_2|,$$

and for the arguments we obtain $\operatorname{Arg} z_1 = 3\pi/4$, $\operatorname{Arg} z_2 = \pi/2$,

$$\operatorname{Arg} z_1 z_2 = -\frac{3\pi}{4} = \operatorname{Arg} z_1 + \operatorname{Arg} z_2 - 2\pi, \qquad \operatorname{Arg}(z_1/z_2) = \frac{\pi}{4} = \operatorname{Arg} z_1 - \operatorname{Arg} z_2. \qquad \blacktriangleleft$$

EXAMPLE 4   **Integer powers. De Moivre's formula**

From (8) and (9) with $z_1 = z_2 = z$ we obtain by induction for $n = 0, 1, 2, \cdots$

$$(13) \qquad \boxed{z^n = r^n(\cos n\theta + i \sin n\theta).}$$

Similarly, (12) with $z_1 = 1$ and $z_2 = z^n$ gives (13) for $n = -1, -2, \cdots$. For $|z| = r = 1$, formula (13) becomes **De Moivre's formula**[5]

$$(13^*) \qquad \boxed{(\cos\theta + i\sin\theta)^n = \cos n\theta + i \sin n\theta.}$$

We can use this to express $\cos n\theta$ and $\sin n\theta$ in terms of powers of $\cos\theta$ and $\sin\theta$. For instance, for $n = 2$ we have on the left $\cos^2\theta + 2i\cos\theta\sin\theta - \sin^2\theta$. Taking the real and imaginary parts on both sides of (13^*) with $n = 2$ gives the familiar formulas

$$\cos 2\theta = \cos^2\theta - \sin^2\theta, \qquad \sin 2\theta = 2\cos\theta\sin\theta.$$

This shows that *complex* methods often simplify the derivation of *real* formulas. Try $n = 3$. $\blacktriangleleft$

Roots

If $z = w^n$ $(n = 1, 2, \cdots)$, then to each value of w there corresponds one value of z. We shall immediately see that, conversely, to a given $z \neq 0$ there correspond precisely n distinct values of w. Each of these values is called an **nth root** of z, and we write

$$(14) \qquad w = \sqrt[n]{z}.$$

Hence this symbol is *multivalued*, namely, *n-valued*, in contrast to the usual conventions made in *real* calculus. The n values of $\sqrt[n]{z}$ can easily be obtained as follows. In terms of polar forms

$$z = r(\cos\theta + i\sin\theta) \qquad \text{and} \qquad w = R(\cos\phi + i\sin\phi)$$

the equation $w^n = z$ becomes

$$w^n = R^n(\cos n\phi + i\sin n\phi) = z = r(\cos\theta + i\sin\theta).$$

---

[5]ABRAHAM DE MOIVRE (1667–1754), French mathematician, who introduced imaginary quantities in trigonometry and contributed to probability theory (see Sec. 22.8).

By equating the absolute values on both sides we have

$$R^n = r, \qquad \text{thus} \qquad R = \sqrt[n]{r}$$

where the root is real positive and thus uniquely determined. By equating the arguments we obtain

$$n\phi = \theta + 2k\pi, \qquad \text{thus} \qquad \phi = \frac{\theta}{n} + \frac{2k\pi}{n}$$

where k is an integer. For $k = 0, 1, \cdots, n - 1$ we get n *distinct* values of w. Further integers of k would give values already obtained. For instance, $k = n$ gives $2k\pi/n = 2\pi$, hence the w corresponding to $k = 0$, etc. Consequently, $\sqrt[n]{z}$, for $z \neq 0$, has the n distinct values

(15)
$$\sqrt[n]{z} = \sqrt[n]{r} \left(\cos \frac{\theta + 2k\pi}{n} + i \sin \frac{\theta + 2k\pi}{n} \right)$$

where $k = 0, 1, \cdots, n - 1$. These n values lie on a circle of radius $\sqrt[n]{r}$ with center at the origin and constitute the vertices of a regular polygon of n sides. The value of $\sqrt[n]{z}$ obtained by taking the principal value of arg z and $k = 0$ in (15) is called the **principal value** of $w = \sqrt[n]{z}$.

In particular, taking $z = 1$, we have $|z| = r = 1$ and Arg $z = 0$. Then (15) gives

(16)
$$\sqrt[n]{1} = \cos \frac{2k\pi}{n} + i \sin \frac{2k\pi}{n}, \qquad k = 0, 1, \cdots, n - 1.$$

These n values are called the **nth roots of unity.** They lie on the circle of radius 1 and center 0, briefly called the **unit circle** (and used quite frequently!). Figures 297–299 show $\sqrt[3]{1} = 1, -\frac{1}{2} \pm \frac{1}{2}\sqrt{3}\,i$, $\sqrt[4]{1} = \pm 1, \pm i$, and $\sqrt[5]{1}$. If ω denotes the value corresponding to $k = 1$ in (16), then the n values of $\sqrt[n]{1}$ can be written as

$$1, \omega, \omega^2, \cdots, \omega^{n-1}.$$

Similarly, if w_1 is any nth root of an arbitrary complex number z, then the n values of $\sqrt[n]{z}$ in (15) are

$$w_1, \qquad w_1\omega, \qquad w_1\omega^2, \qquad \cdots, \qquad w_1\omega^{n-1}$$

because multiplying w_1 by ω^k corresponds to increasing the argument of w_1 by $2k\pi/n$.

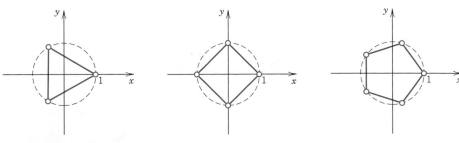

Fig. 297. $\sqrt[3]{1}$ Fig. 298. $\sqrt[4]{1}$ Fig. 299. $\sqrt[5]{1}$

PROBLEM SET 12.2

Polar Form. Principal Value. Conversion to $x + iy$

Polar forms will be needed frequently, so do these problems with great care. Represent each of the following in polar form and plot in the complex plane (showing the details of your work):

1. $1 + i$ **2.** $-2 + 2i$ **3.** $-3 - 4i$ **4.** -10 **5.** $3i, -3i$

6. $\dfrac{1 - i}{1 + i}$ **7.** $\left(\dfrac{6 + 8i}{4 - 3i}\right)^2$ **8.** $\dfrac{i}{3 + 3i}$ **9.** $\dfrac{2 + i}{5 - 3i}$ **10.** $\dfrac{7 - 5i}{4i}$

Determine the principal value of the argument:

11. $1 - i$ **12.** $-10, -10 - i$ **13.** $3 \pm 4i$ **14.** $-5 + 5i$ **15.** $-\pi - \pi i$

Represent each of the following in the form $x + iy$ and plot in the complex plane:

16. $\cos \frac{1}{2}\pi + i \sin \frac{1}{2}\pi$ **17.** $\sqrt{8}(\cos \frac{1}{4}\pi + i \sin \frac{1}{4}\pi)$

18. $6(\cos \frac{1}{3}\pi + i \sin \frac{1}{3}\pi)$ **19.** $\sqrt{18}(\cos \frac{3}{4}\pi + i \sin \frac{3}{4}\pi)$

Roots, Equations

20. TEAM PROJECT. Square Root. (a) Show that $w = \sqrt{z}$ has the values

$$(17) \quad w_1 = \sqrt{r}\left(\cos \frac{\theta}{2} + i \sin \frac{\theta}{2}\right), \quad w_2 = \sqrt{r}\left[\cos\left(\frac{\theta}{2} + \pi\right) + i \sin\left(\frac{\theta}{2} + \pi\right)\right] = -w_1.$$

(b) Obtain from (17) the often more practical formula

$$(18) \quad \sqrt{z} = \pm \left[\sqrt{\tfrac{1}{2}(|z| + x)} + (\text{sign } y)i\sqrt{\tfrac{1}{2}(|z| - x)}\right]$$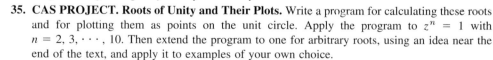

where sign $y = 1$ if $y \geqq 0$, sign $y = -1$ if $y < 0$, and all square roots of positive numbers are taken with the positive sign. *Hint.* Use (10) in Appendix 3 with $x = \theta/2$.

(c) Find the square roots of $4i$, $20 + 48i$, and $23 - 5\sqrt{8}\,i$ by both (17) and (18) and comment on the work involved.

(d) Do some further examples of your own and apply a method of checking your results.

Find and plot all roots:

21. $\sqrt[3]{1 + i}$ **22.** $\sqrt[3]{8i}$ **23.** $\sqrt[3]{216}$ **24.** $\sqrt[4]{-4}$ **25.** $\sqrt{-7 + 24i}$

26. $\sqrt[4]{-7 + 24i}$ **27.** $\sqrt[8]{1}$

Solve the equations:

28. $z^2 - (5 + i)z + 8 + i = 0$ **29.** $z^2 - (7 + i)z + 24 + 7i = 0$

30. $z^4 - (3 + 6i)z^2 - 8 + 6i = 0$ **31.** $z^2 + z + 1 - i = 0$

32. (Triangle inequality) Verify (5) for $z_1 = 4 - 6i$, $z_2 = 2 + 2.5i$.

33. (Triangle inequality) Prove (5).

34. (Parallelogram equality) Show that $|z_1 + z_2|^2 + |z_1 - z_2|^2 = 2(|z_1|^2 + |z_2|^2)$. This is called the *parallelogram equality*. Can you see why?

35. CAS PROJECT. Roots of Unity and Their Plots. Write a program for calculating these roots and for plotting them as points on the unit circle. Apply the program to $z^n = 1$ with $n = 2, 3, \cdots, 10$. Then extend the program to one for arbitrary roots, using an idea near the end of the text, and apply it to examples of your own choice.

36. (Inequalities for Re and Im) Prove the following inequalities, which we shall need occasionally.

$$(19) \quad |\text{Re } z| \leqq |z|, \qquad |\text{Im } z| \leqq |z|.$$

12.3   Derivative. Analytic Function

Our study of complex functions will involve sets in the complex plane. Most important will be the following ones.

Circles and Disks. Half-Planes

The **unit circle** $|z| = 1$ (Fig. 300) has already occurred in Sec. 12.2. Figure 301 shows a general circle of radius ρ and center a. Its equation is

$$|z - a| = \rho$$

because it is the set of all z whose distance $|z - a|$ from the center a equals ρ. Accordingly, its interior (**"open circular disk"**) is given by $|z - a| < \rho$, its interior plus the circle itself (**"closed circular disk"**) by $|z - a| \leqq \rho$, and its exterior by $|z - a| > \rho$. As an example, sketch this for $a = 1 + i$ and $\rho = 2$, to make sure that you understand these inequalities.

An open circular disk $|z - a| < \rho$ is also called a **neighborhood** of a. And a has infinitely many of them, each corresponding to a certain value of $\rho \, (> 0)$, and a is a point of each of them, by definition![6]

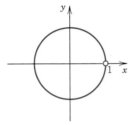

Fig. 300.   Unit circle

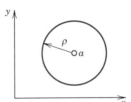

Fig. 301.   Circle in the complex plane

Figure 302 shows an **open annulus** (circular ring) $\rho_1 < |z - a| < \rho_2$, which we shall need later. This is the set of all z whose distance $|z - a|$ from a is greater than ρ_1 but less than ρ_2. Similarly, the **closed annulus** $\rho_1 \leqq |z - a| \leqq \rho_2$ includes the two circles.

Half-planes. By the (open) *upper* **half-plane** we mean the set of all points $z = x + iy$ such that $y > 0$. Similarly, the condition $y < 0$ defines the *lower half-plane, x > 0* the *right half-plane,* and $x < 0$ the *left half-plane.*

For Reference: Concepts on Sets in the Complex Plane

A **point set** in the complex plane means any sort of collection of finitely many or infinitely many points. Examples of sets are the solutions of a quadratic equation, the points on a line, and the points in the interior of a circle.

---

[6]More generally, any set that contains such an open disk is also called a *neighborhood of a.* For distinction, that open disk is often called an *open circular neighborhood of a.*

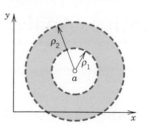

Fig. 302. Annulus
in the complex plane

A set S is called **open** if every point of S has a neighborhood consisting entirely of points that belong to S. For example, the points in the interior of a circle or a square form an open set, and so do the points of the right half-plane Re $z = x > 0$.

An open set S is called **connected** if any two of its points can be joined by a broken line of finitely many straight-line segments all of whose points belong to S. An open connected set is called a **domain.** Thus an open disk and an open annulus are domains. An open square with a diagonal removed is not a domain since this set is not connected. (Why?)

The **complement** of a set S in the complex plane is the set of all points of the complex plane that do *not* belong to S. A set S is called **closed** if its complement is open. For example, the points on and inside the unit circle form a closed set ("closed unit disk") since its complement $|z| > 1$ is open.

A **boundary point** of a set S is a point every neighborhood of which contains both points that belong to S and points that do not belong to S. For example, the boundary points of an annulus are the points on the two bounding circles. Clearly, if a set S is open, then no boundary point belongs to S; if S is closed, then every boundary point belongs to S.

A **region** is a set consisting of a domain plus, perhaps, some or all of its boundary points. (*Warning!* "Domain" is the *modern* term for an open connected set. Nevertheless, some authors still call a domain a "region" and others make no distinction between the two terms.)

Complex Function

Complex analysis is concerned with complex functions that are differentiable in some domain. Hence we should first say what we mean by a complex function and then define the concepts of limit and derivative in complex. This discussion will be similar to that in calculus. Nevertheless it will need great attention because it will show interesting basic differences between real and complex calculus.

Recall from calculus that a *real* function f defined on a set S of real numbers (usually an interval) is a rule that assigns to every x in S a real number $f(x)$, called the *value* of f at x. Now in complex, S is a set of *complex* numbers. And a **function** f defined on S is a rule that assigns to every z in S a complex number w, called the *value* of f at z. We write

$$w = f(z).$$

Here z varies in S and is called a **complex variable.** The set S is called the **domain** *of definition*[7] *of* f.

Example: $w = f(z) = z^2 + 3z$ is a complex function defined for all z; that is, its domain S is the whole complex plane.

The set of all values of a function f is called the **range** *of* f.

w is complex, and we write $w = u + iv$, where u and v are the real and imaginary parts, respectively. Now w depends on $z = x + iy$. Hence u becomes a real function of x and y, and so does v. We may thus write

$$w = f(z) = u(x, y) + i v(x, y).$$

This shows that a *complex* function $f(z)$ is equivalent to a pair of *real* functions $u(x, y)$ and $v(x, y)$, each depending on the two real variables x and y.

EXAMPLE 1   **Function of a complex variable**

Let $w = f(z) = z^2 + 3z$. Find u and v and calculate the value of f at $z = 1 + 3i$.

Solution. $u = \text{Re } f(z) = x^2 - y^2 + 3x$ and $v = 2xy + 3y$. Also,

$$f(1 + 3i) = (1 + 3i)^2 + 3(1 + 3i) = 1 - 9 + 6i + 3 + 9i = -5 + 15i.$$

This shows that $u(1, 3) = -5$ and $v(1, 3) = 15$. Check this by using the expressions for u and v.   ◄

EXAMPLE 2   **Function of a complex variable**

Let $w = f(z) = 2iz + 6\bar{z}$. Find u and v and the value of f at $z = \frac{1}{2} + 4i$.

Solution. $f(z) = 2i(x + iy) + 6(x - iy)$ gives $u(x, y) = 6x - 2y$ and $v(x, y) = 2x - 6y$. Also,

$$f(\tfrac{1}{2} + 4i) = 2i(\tfrac{1}{2} + 4i) + 6(\tfrac{1}{2} - 4i) = i - 8 + 3 - 24i = -5 - 23i.$$

Check this as in Example 1.   ◄

Limit, Continuity

A function $f(z)$ is said to have the **limit** l as z approaches a point z_0, written

(1)
$$\lim_{z \to z_0} f(z) = l,$$

if f is defined in a neighborhood of z_0 (except perhaps at z_0 itself) and if the values of f are "close" to l for all z "close" to z_0; that is, in precise terms, for every positive real ϵ we can find a positive real δ such that for all $z \neq z_0$ in the disk $|z - z_0| < \delta$ (Fig. 303) we have

(2)
$$|f(z) - l| < \epsilon;$$

that is, for every $z \neq z_0$ in that δ-disk the value of f lies in the disk (2).

---

[7]This is a standard term. In most cases, a domain of definition will be an open and connected set (a *domain* as just defined); exceptions rarely occur in applications.

In the literature on complex analysis, one sometimes uses relations such that to a value of z there may correspond more than one value of w, and it is customary to call such a relation a function (a "multivalued function"). We shall not adopt this convention, but assume that all occurring functions are **single-valued** relations, that is, functions in the usual sense: to each z in S there corresponds only *one* value $w = f(z)$. (But, of course, several z may correspond to the same value $w = f(z)$, just as in calculus.)

Strictly speaking, $f(z)$ denotes the value of f at z, but it is a convenient abuse of language to talk about *the function* $f(z)$ (instead of *the function* f), thereby exhibiting the notation for the independent variable.

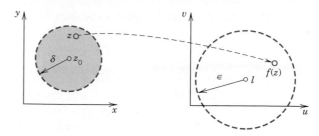

Fig. 303. Limit

Formally, this definition is similar to that in calculus, but there is a big difference. Whereas in the real case, x can approach an x_0 only along the real line, here, by definition, z may approach z_0 *from any direction* in the complex plane. This will be quite essential in what follows.

If a limit exists, it is unique. (See Team Project 24.)

A function $f(z)$ is said to be **continuous** at $z = z_0$ if $f(z_0)$ is defined and

(3) $$\lim_{z \to z_0} f(z) = f(z_0).$$

Note that by the definition of a limit this implies that $f(z)$ is defined in some neighborhood of z_0.

$f(z)$ is said to be *continuous in a domain* if it is continuous at each point of this domain.

Derivative

The **derivative** of a complex function f at a point z_0 is written $f'(z_0)$ and is defined by

(4) $$f'(z_0) = \lim_{\Delta z \to 0} \frac{f(z_0 + \Delta z) - f(z_0)}{\Delta z}$$

provided this limit exists. Then f is said to be **differentiable** at z_0. If we write $\Delta z = z - z_0$, we also have, since $z = z_0 + \Delta z$,

(4') $$f'(z_0) = \lim_{z \to z_0} \frac{f(z) - f(z_0)}{z - z_0}.$$

Now comes an ***important point.*** Remember that, by the definition of limit, $f(z)$ is defined in a neighborhood of z_0 and z in (4') may approach z_0 from any direction in the complex plane. Hence differentiability at z_0 means that, along whatever path z approaches z_0, the quotient in (4') always approaches a certain value and all these values are equal. This is important and should be kept in mind.

EXAMPLE 3   **Differentiability. Derivative**

The function $f(z) = z^2$ is differentiable for all z and has the derivative $f'(z) = 2z$ because

$$f'(z) = \lim_{\Delta z \to 0} \frac{(z + \Delta z)^2 - z^2}{\Delta z} = \lim_{\Delta z \to 0} \frac{z^2 + 2z\Delta z + (\Delta z)^2 - z^2}{\Delta z} = \lim_{\Delta z \to 0} (2z + \Delta z) = 2z. \quad \blacktriangleleft$$

*The **differentiation rules** are the same as in real calculus,* since their proofs are literally the same. Thus,

$$(cf)' = cf', \quad (f + g)' = f' + g', \quad (fg)' = f'g + fg', \quad \left(\frac{f}{g}\right)' = \frac{f'g - fg'}{g^2}$$

as well as the chain rule and the power rule $(z^n)' = nz^{n-1}$ (n integer) hold.

Also, if $f(z)$ is differentiable at z_0, it is continuous at z_0. (See Team Project 24.)

EXAMPLE 4   *z̄* **not differentiable**

It is important to note that there are many simple functions that do not have a derivative at any point. For instance, $f(z) = \bar{z} = x - iy$ is such a function. Indeed, if we write $\Delta z = \Delta x + i\,\Delta y$, we have

(5)
$$\frac{f(z + \Delta z) - f(z)}{\Delta z} = \frac{\overline{(z + \Delta z)} - \bar{z}}{\Delta z} = \frac{\overline{\Delta z}}{\Delta z} = \frac{\Delta x - i\,\Delta y}{\Delta x + i\,\Delta y}.$$

If $\Delta y = 0$, this is $+1$. If $\Delta x = 0$, this is -1. Thus (5) approaches $+1$ along path I in Fig. 304 but -1 along path II. Hence, by definition, the limit of (5) as $\Delta z \to 0$ does not exist at any z.   ◀

The example just discussed may be surprising, but it merely illustrates that differentiability of a complex function is a rather severe requirement.

The idea of proof (approach from different directions) is basic and will be used again in the next section.

Analytic Functions

These are the functions that are differentiable in some domain, so that we can do "calculus in complex." ***They are the main concern of complex analysis.*** Their introduction is our main goal in this section.

Definition (Analyticity)

A function $f(z)$ is said to be *analytic in a domain D* if $f(z)$ is defined and differentiable at all points of D. The function $f(z)$ is said to be *analytic at a point $z = z_0$* in D if $f(z)$ is analytic in a neighborhood of z_0.

Also, by an **analytic function** we mean a function that is analytic in *some* domain.   ◀

Hence analyticity of $f(z)$ at z_0 means that $f(z)$ has a derivative at every point in some neighborhood of z_0 (including z_0 itself since, by definition, z_0 is a point of all its neighborhoods). This concept is *motivated* by the fact that it is of no practical interest if a function is differentiable merely at a single point z_0 but not throughout some neighborhood of z_0. Team Project 24 gives an example.

A more modern term for *analytic in D* is *holomorphic in D*.

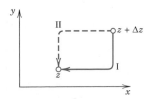

Fig. 304.   Paths in (5)

EXAMPLE 5 **Polynomials, rational functions**

The nonnegative integer powers $1, z, z^2, \cdots$ are analytic in the entire complex plane, and so are **polynomials,** that is, functions of the form

$$f(z) = c_0 + c_1 z + c_2 z^2 + \cdots + c_n z^n$$

where $c_0, \cdots, c_n$ are complex constants.

The quotient of two polynomials $g(z)$ and $h(z)$,

$$f(z) = \frac{g(z)}{h(z)},$$

is called a **rational function.** This f is analytic except at the points where $h(z) = 0$; here we assume that common factors of g and h have been canceled. ◀

The concepts discussed in this section extend familiar concepts of calculus. Most important is the concept of an analytic function. Indeed, complex analysis is concerned exclusively with analytic functions, and although many simple functions are not analytic, the large variety of remaining functions will yield a branch of mathematics that is very useful for practical purposes and most beautiful from a theoretical viewpoint.

PROBLEM SET 12.3

Regions of Practical Interest
Determine and sketch or plot the sets in the complex plane given by

1. $|z + 2 + 5i| \leq \frac{1}{2}$ 2. $\frac{1}{2} < |z - 4 + 2i| < 2$ 3. $0 < |z| < 1$ 4. $0 < |z - 1 - i| < \sqrt{2}$
5. $-\pi < \text{Im } z < \pi$ 6. $\text{Re } (1/z) < 1$ 7. $\text{Re } (z^2) \leq 1$ 8. $|\text{arg } z| < \pi/4$

9. **WRITING PROJECT. Concepts on Sets.** Extend the part of the text on sets in the complex plane by formulating everything in your own words, including examples of your own and comparing with calculus when applicable.

Functions and Their Derivatives
Function Values. Find the values of $\text{Re } f$ and $\text{Im } f$ at the indicated point.

10. $f = z^2 + 2z + 2$ at $1 - i$ 11. $f = 1/(1 - z)$ at $7 + 2i$ 12. $f = (z - 2)/(z + 2)$ at $4i$

Continuity. Find out (and give reason) whether $f(z)$ is continuous at $z = 0$ if $f(0) = 0$ and for $z \neq 0$ the function f is equal to

13. $(\text{Im } z)/|z|$ 14. $(\text{Re } z^2)/|z|$ 15. $(\text{Re } z)/(1 + |z|)$ 16. $(\text{Re } z - \text{Im } z)/|z|^2$

Derivative. Find the value of the derivative of

17. $(z - i)/(z + i)$ at i 18. $(z - 4i)^8$ at $5 + 4i$ 19. $(5 + 3i)/z^3$ at $2 + i$
20. $(3z^2 + iz)^2$ at $1 + i$ 21. $z^4 + 1/z^4$ at $-1 - i$ 22. $(iz^3 + 3z^2)^3$ at $2i$
23. $(iz + 2)/(3z - 6i)$ at any z. (Explain the result.)

24. **TEAM PROJECT. Limit, Continuity, Derivative. (a) Limit.** Prove that (1) is equivalent to the pair of relations

$$\lim_{z \to z_0} \text{Re } f(z) = \text{Re } l, \qquad \lim_{z \to z_0} \text{Im } f(z) = \text{Im } l.$$

(b) **Limit.** If $\lim_{z \to z_0} f(z)$ exists, show that this limit is unique.

(c) **Continuity.** If $z_1, z_2, \cdots$ are complex numbers for which $\lim_{n \to \infty} z_n = a$, and if $f(z)$ is continuous at $z = a$, show that

$$\lim_{n \to \infty} f(z_n) = f(a).$$

(d) **Continuity.** If $f(z)$ is differentiable at z_0, show that $f(z)$ is continuous at z_0.

(e) **Differentiability.** Show that $f(z) = \text{Re } z = x$ is not differentiable at any z. Can you find other such functions?

(f) **Differentiability.** Show that $f(z) = |z|^2$ is differentiable only at $z = 0$; hence it is nowhere analytic.

 25. CAS PROJECT. Plotting Functions. Find and plot Re f, Im f, and $|f|$ as surfaces over the z-plane. Also plot the two families of curves Re $f(z) = const$ and Im $f(z) = const$ in the same figure, and the curves $|f(z)| = const$ in another figure, where   (a) $f(z) = z^2$,   (b) $f(z) = 1/z$, (c) $f(z) = z^4$.

12.4 Cauchy–Riemann Equations Laplace's Equation

The Cauchy–Riemann equations are the most important equations in this chapter and one of the pillars on which complex analysis rests. They provide a criterion (a test) for the analyticity of a complex function

$$w = f(z) = u(x, y) + iv(x, y).$$

Roughly, f is analytic in a domain D if and only if the first partial derivatives of u and v satisfy the two so-called **Cauchy–Riemann equations**[8]

(1)
$$u_x = v_y, \qquad u_y = -v_x$$

everywhere in D; here $u_x = \partial u/\partial x$ and $u_y = \partial u/\partial y$ (and similarly for v) are the usual notations for partial derivatives. The precise formulation of this statement is given in Theorems 1 and 2.

Example: $f(z) = z^2 = x^2 - y^2 + 2ixy$ is analytic for all z, and $u = x^2 - y^2$ and $v = 2xy$ satisfy (1), namely, $u_x = 2x = v_y$ as well as $u_y = -2y = -v_x$. More examples will follow.

THEOREM 1   **(Cauchy–Riemann equations)**

Let $f(z) = u(x, y) + iv(x, y)$ be defined and continuous in some neighborhood of a point $z = x + iy$ and differentiable at z itself. Then at that point, the first-order partial derivatives of u and v exist and satisfy the Cauchy–Riemann equations (1).

Hence if $f(z)$ is analytic in a domain D, those partial derivatives exist and satisfy (1) *at all points of D.*

---

[8]The French mathematician AUGUSTIN-LOUIS CAUCHY (see Sec. 2.6) and the German mathematicians BERNHARD RIEMANN (1826—1866) and KARL WEIERSTRASS (1815—1897; see also Sec. 14.5) are the founders of complex analysis. Riemann received his Ph.D. (in 1851) under Gauss (Sec. 4.4) at Göttingen, where he also taught until he died, when he was only 39 years old. He introduced the concept of the integral as it is used in basic calculus courses, and made important contributions to differential equations, number theory, and mathematical physics. He also developed the so-called Riemannian geometry, which is the mathematical base of Einstein's theory of relativity.

PROOF.   By assumption, the derivative $f'(z)$ at z exists. It is given by

(2)
$$f'(z) = \lim_{\Delta z \to 0} \frac{f(z + \Delta z) - f(z)}{\Delta z}.$$

The idea of the proof is very simple. By the definition of a limit in complex (Sec. 12.3) we can let Δz approach zero along any path in a neighborhood of z. Thus we may choose the two paths I and II in Fig. 305 and equate the results. By comparing the real parts we shall obtain the first Cauchy–Riemann equation and by comparing the imaginary parts the other equation in (1). The technical details are as follows.

We write $\Delta z = \Delta x + i\Delta y$. In terms of u and v, the derivative in (2) becomes

(3) $f'(z) = \lim\limits_{\Delta z \to 0} \dfrac{[u(x + \Delta x, y + \Delta y) + iv(x + \Delta x, y + \Delta y)] - [u(x, y) + iv(x, y)]}{\Delta x + i\Delta y}.$

We first choose path I in Fig. 305. Thus we let $\Delta y \to 0$ first and then $\Delta x \to 0$. After Δy is zero, $\Delta z = \Delta x$. Then (3) becomes, if we first write the two u-terms and then the two v-terms,

$$f'(z) = \lim_{\Delta x \to 0} \frac{u(x + \Delta x,\ y) - u(x, y)}{\Delta x} + i \lim_{\Delta x \to 0} \frac{v(x + \Delta x,\ y) - v(x, y)}{\Delta x}.$$

Since $f'(z)$ exists, the two real limits on the right exist. By definition, they are the partial derivatives of u and v with respect to x. Hence the derivative $f'(z)$ of $f(z)$ can be written

(4)
$$f'(z) = u_x + iv_x.$$

Similarly, if we choose path II in Fig. 305, we let $\Delta x \to 0$ first and then $\Delta y \to 0$. After Δx is zero, $\Delta z = i\Delta y$, so that from (3) we now obtain

$$f'(z) = \lim_{\Delta y \to 0} \frac{u(x,\ y + \Delta y) - u(x, y)}{i\,\Delta y} + i \lim_{\Delta y \to 0} \frac{v(x,\ y + \Delta y) - v(x, y)}{i\,\Delta y}.$$

Since $f'(z)$ exists, the limits on the right exist and give the partial derivatives of u and v with respect to y; noting that $1/i = -i$, we thus obtain

(5)
$$f'(z) = -iu_y + v_y.$$

The existence of the derivative $f'(z)$ thus implies the existence of the four partial derivatives in (4) and (5). By equating the real parts u_x and v_y in (4) and (5) we obtain

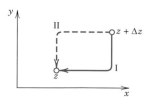

Fig. 305.   Paths in (2)

the first Cauchy–Riemann equation (1). Equating the imaginary parts gives the other. This proves the first statement of the theorem and implies the second because of the definition of analyticity.   ◀

Formulas (4) and (5) are also quite practical for calculating derivatives $f'(z)$, as we shall see.

EXAMPLE 1   **Cauchy–Riemann equations**

$f(z) = z^2$ is analytic for all z. It follows that the Cauchy–Riemann equations must be satisfied (as we have verified above).

For $f(z) = \bar{z} = x - iy$ we have $u = x$, $v = -y$ and see that the second Cauchy–Riemann equation is satisfied, $u_y = -v_x = 0$, but the first is not: $u_x = 1 \neq v_y = -1$. We conclude that $f(z) = \bar{z}$ is not analytic, confirming Example 4 of Sec. 12.3. Note the savings in calculation!   ◀

The Cauchy–Riemann equations are fundamental because they are not only necessary but also sufficient for a function to be analytic. More precisely, the following theorem holds.

THEOREM 2   **(Cauchy–Riemann equations)**

*If two real-valued continuous functions $u(x, y)$ and $v(x, y)$ of two real variables x and y have **continuous** first partial derivatives that satisfy the Cauchy–Riemann equations in some domain D, then the complex function $f(z) = u(x, y) + iv(x, y)$ is analytic in D.*

The proof is more involved than that of Theorem 1 and we leave it optional (see Appendix 4).

Theorems 1 and 2 are of great practical importance, since by using the Cauchy–Riemann equations we can now easily find out whether or not a given complex function is analytic.

EXAMPLE 2   **Cauchy–Riemann equations**

Is $f(z) = z^3$ analytic?

Solution.   We find $u = x^3 - 3xy^2$ and $v = 3x^2y - y^3$. Next we calculate

$$u_x = 3x^2 - 3y^2, \qquad v_y = 3x^2 - 3y^2,$$

$$u_y = -6xy, \qquad v_x = 6xy.$$

We see that the Cauchy–Riemann equations are satisfied for every z. Hence $f(z) = z^3$ is analytic for every z, by Theorem 2.   ◀

EXAMPLE 3   **An analytic function of constant absolute value is constant**

The Cauchy–Riemann equations also help in deriving general properties of analytic functions.

For example, show that if $f(z)$ is analytic in a domain D and $|f(z)| = k = const$ in D, then $f(z) = const$ in D. (We shall make crucial use of this in Sec. 16.6.)

Solution.   By assumption, $u^2 + v^2 = k^2$. By differentiation,

$$uu_x + vv_x = 0, \qquad uu_y + vv_y = 0.$$

Now use $v_x = -u_y$ in the first equation and $v_y = u_x$ in the second, to get

(6)   (a) $uu_x - vu_y = 0,$ (b) $uu_y + vu_x = 0.$

To get rid of u_y, multiply (6a) by u and (6b) by v and add. Similarly, to eliminate u_x, multiply (6a) by $-v$ and (6b) by u and add. This yields

$$(u^2 + v^2)u_x = 0, \qquad (u^2 + v^2)u_y = 0.$$

If $k^2 = u^2 + v^2 = 0$, then $u = v = 0$, hence $f = 0$. If $k \neq 0$, then $u_x = u_y = 0$, hence, by the Cauchy–Riemann equations, also $v_x = v_y = 0$. Together, $u = const$ and $v = const$, hence $f = const$. ◀

We mention that if we use the polar form $z = r(\cos \theta + i \sin \theta)$ and set $f(z) = u(r, \theta) + iv(r, \theta)$, then the **Cauchy–Riemann equations** are

$$(7) \qquad \boxed{u_r = \frac{1}{r} v_\theta, \qquad v_r = -\frac{1}{r} u_\theta} \qquad (r > 0).$$

Laplace's Equation. Harmonic Functions

One of the main reasons for the great practical importance of complex analysis in engineering mathematics results from the fact that both the real part and the imaginary part of an analytic function satisfy the most important differential equation of physics, Laplace's equation, which occurs in gravitation, electrostatics, fluid flow, heat conduction, and so on (see Chaps. 11 and 16).

THEOREM 3

(Laplace's equation)

*If $f(z) = u(x, y) + iv(x, y)$ is analytic in a domain D, then u and v satisfy **Laplace's equation***

$$(8) \qquad \boxed{\nabla^2 u = u_{xx} + u_{yy} = 0}$$

(∇^2 read "nabla squared") and

$$(9) \qquad \boxed{\nabla^2 v = v_{xx} + v_{yy} = 0,}$$

respectively, in D and have continuous second partial derivatives in D.

PROOF. Differentiating $u_x = v_y$ with respect to x and $u_y = -v_x$ with respect to y, we have

$$(10) \qquad\qquad u_{xx} = v_{yx}, \qquad u_{yy} = -v_{xy}.$$

Now the derivative of an analytic function is itself analytic, as we shall prove later (in Sec. 13.4). This implies that u and v have continuous partial derivatives of all orders; in particular, the mixed second derivatives are equal: $v_{yx} = v_{xy}$. By adding (10) we thus obtain (8). Similarly, (9) is obtained by differentiating $u_x = v_y$ with respect to y and $u_y = -v_x$ with respect to x and subtraction, using $u_{xy} = u_{yx}$. ◀

Solutions of Laplace's equation having ***continuous*** second-order partial derivatives are called **harmonic functions** and their theory is called **potential theory** (see also Sec. 11.11). Hence the real and imaginary parts of an analytic function are harmonic functions.

If two harmonic functions u and v satisfy the Cauchy–Riemann equations in a domain D, they are the real and imaginary parts of an analytic function f in D. Then v is said to be a **conjugate harmonic function** of u in D. (Of course, this has absolutely nothing to do with the use of "conjugate" for $\bar{z}$.)

EXAMPLE 4   **How to find a conjugate harmonic function by the Cauchy–Riemann equations**

Verify that $u = x^2 - y^2 - y$ is harmonic in the whole complex plane and find a conjugate harmonic function v of u.

Solution. $\nabla^2 u = 0$ by direct calculation. Now $u_x = 2x$ and $u_y = -2y - 1$. Hence because of the Cauchy–Riemann equations a conjugate v of u must satisfy

$$v_y = u_x = 2x, \qquad v_x = -u_y = 2y + 1.$$

Integrating the first equation with respect to y and differentiating the result with respect to x, we obtain

$$v = 2xy + h(x), \qquad v_x = 2y + \frac{dh}{dx}.$$

A comparison with the second equation shows that $dh/dx = 1$. This gives $h(x) = x + c$. Hence $v = 2xy + x + c$ (c any real constant) is the most general conjugate harmonic of the given u. The corresponding analytic function is

$$f(z) = u + iv = x^2 - y^2 - y + i(2xy + x + c) = z^2 + iz + ic. \qquad \blacktriangleleft$$

Example 4 illustrates that *a conjugate of a given harmonic function is uniquely determined up to an arbitrary real additive constant.*

The curves $u = const$ are called **equipotential lines** or **level curves of u.** They form a **family** of curves. Similarly for v. The two families together form an **orthogonal net.** "Orthogonal" means perpendicular. These are standard terms, which we shall use.

The Cauchy–Riemann equations are the most important equations in this chapter. Their relation to Laplace's equation opens wide ranges of engineering and physical applications, as we shall show in Chap. 16.

PROBLEM SET 12.4

Analytic Functions. Cauchy–Riemann Equations
Check for analyticity by using (1) or (7). (Show the details of your work.)

1. $f(z) = z^6$ **2.** $f(z) = i|z|^3$ **3.** $f(z) = e^x(\cos y + i \sin y)$

4. $f(z) = i/z^5$ **5.** $f(z) = z\bar{z}$ **6.** $f(z) = z + 1/z$

7. $f(z) = \ln|z| + i \operatorname{Arg} z$ **8.** $f(z) = 1/(1 - z^4)$ **9.** $f(z) = \operatorname{Re} z/\operatorname{Im} z$

10. $f(z) = \operatorname{Arg} z$ **11.** $f(z) = \operatorname{Re}(z^3)$ **12.** $f(z) = \operatorname{Re}(z^2) - i \operatorname{Im}(z^2)$

13. (**Cauchy–Riemann equations**) Derive (7) from (1).

14. **TEAM PROJECT. Conditions for $f(z) = const$.** Let $f(z)$ be analytic. Prove that each of the following conditions is sufficient for $f(z) = const.$

   (a) $\operatorname{Re} f(z) = const$ (b) $\operatorname{Im} f(z) = const$ (c) $f'(z) = 0$

   (d) $|f(z)| = const$ (see Example 3)

15. (**Formulas for the derivative**) Show that, in addition to (4) and (5),

(11) $$f'(z) = u_x - iu_y, \qquad f'(z) = v_y + iv_x.$$

16. Formulas (4), (5), and (11) are needed from time to time. Familiarize yourself with them by calculating $(z^3)'$ and verifying that the result is as expected.

Harmonic Functions

Determine whether the following functions are harmonic. If your answer is yes, find a corresponding analytic function $f(z) = u(x, y) + iv(x, y)$.

17. $u = x/(x^2 + y^2)$ **18.** $u = x^2 + y^2$ **19.** $u = \ln |z|$ **20.** $v = \text{Arg } z$

21. $v = -e^{-x} \sin y$ **22.** $u = \sin x \cosh y$ **23.** $v = x^3 - 3x^2$ **24.** $v = (x^2 - y^2)^2$

Determine a and b such that the given functions are harmonic and find a conjugate harmonic.

25. $u = ax^3 + by^3$ **26.** $u = ax^3 + bxy$ **27.** $u = e^{ax} \cos 5y$ **28.** $u = \cos ax \cosh 2y$

29. **(Conjugate harmonic)** Show that if u is harmonic and v is a conjugate harmonic of u, then u is a conjugate harmonic of $-v$.

30. **CAS PROJECT. Equipotential Lines.** Write a program for plotting equipotential lines $u = const$ of a harmonic function u and of its conjugate v on the same axes. Apply the program to (a) $u = x^2 - y^2$, $v = 2xy$, (b) $u = x^3 - 3xy^2$, $v = 3x^2y - y^3$, (c) $u = e^x \cos y$, $v = e^x \sin y$.

12.5 Geometry of Analytic Functions: Conformal Mapping

In calculus, figures of functions $y = f(x)$ in the xy-plane (their graphs) give a visual representation and lead to a better understanding of properties of functions. In complex analysis this is similar. Moreover, it has practical applications in potential problems, as we shall see.

Mapping

For a complex function

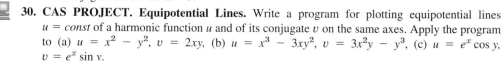

(1) $$w = f(z) = u(x, y) + iv(x, y) \qquad\qquad (z = x + iy)$$

we need two planes, the **z-plane** in which we plot values of z, and the **w-plane** in which we plot the corresponding function values $w = f(z)$. In this way, a given function f assigns to each point z in its domain of definition D the corresponding point $w = f(z)$ in the w-plane. We say that f defines a **mapping** of D into[9] the w-plane. For any point z_0 in D we call the point $w_0 = f(z_0)$ the **image** of z_0 with respect to f. More generally, for the points of a curve C in D, the image points form the **image** of C; similarly for other point sets in D.

Instead of *the mapping by a function* $w = f(z)$ we shall also briefly say *the mapping* $w = f(z)$.

---

[9] The general terminology is as follows. A mapping of a set A into a set B is called **surjective** or a mapping of A **onto** B if every element of B is the image of at least one element of A. It is called **injective** or **one-to-one** if different elements of A have different images in B. Finally, it is called **bijective** if it is both surjective and injective.

Hence (1) maps the domain of definition of f **onto** the range of values of f.

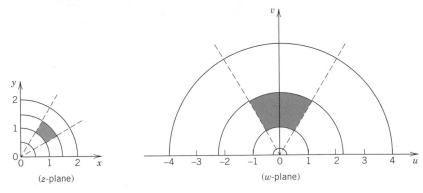

Fig. 306. Mapping $w = z^2$. Lines $|z| = const$,
arg $z = const$ and their images in the w-plane

EXAMPLE 1   **Mapping $w = z^2$**

In polar coordinates we have $z = r(\cos \theta + i \sin \theta)$ and thus [see (13), Sec. 12.2]

$$w = R(\cos \phi + i \sin \phi) = z^2 = r^2 (\cos 2\theta + i \sin 2\theta).$$

Comparing moduli and arguments gives $R = r^2$ and $\phi = 2\theta$. Hence circles $r = r_0$ are mapped onto circles $R = r_0{}^2$ and rays $\theta = \theta_0$ onto rays $\phi = 2\theta_0$. Figure 306 shows this for the region $1 \le |z| \le 3/2$, $\pi/6 \le \theta \le \pi/3$, which is mapped onto the region $1 \le |w| \le 9/4$, $\pi/3 \le \theta \le 2\pi/3$.

In Cartesian coordinates we have $z = x + iy$ and

$$u = \text{Re } (z^2) = x^2 - y^2, \qquad v = \text{Im } (z^2) = 2xy.$$

Hence vertical lines $x = c = const$ are mapped onto $u = c^2 - y^2$, $v = 2cy$. From this we can eliminate y. We obtain $y^2 = c^2 - u$ and $v^2 = 4c^2y^2$. Together,

$$v^2 = 4c^2(c^2 - u) \qquad\qquad \text{(Fig. 307)}.$$

These parabolas open to the left. Similarly, horizontal lines $y = k = const$ are mapped onto parabolas opening to the right,

$$v^2 = 4k^2(k^2 + u) \qquad\qquad \text{(Fig. 307)}. \quad ◀$$

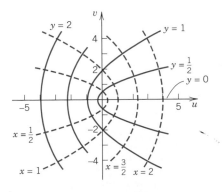

Fig. 307. Images of $x = const$,
$y = const$ under $w = z^2$

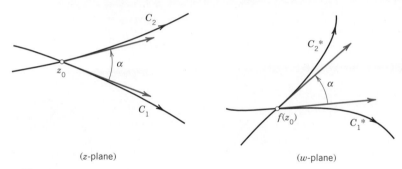

Fig. 308.　Curves C_1 and C_2 and their respective images C_1^* and C_2^* under a conformal mapping.

Conformal Mapping

A **conformal mapping** is a mapping that preserves angles between any oriented curves both in magnitude and in sense.

Here, the concept of an **oriented curve** is defined as follows. Recall from calculus (or from Sec. 8.5) that for a curve C in the xy-plane a **parametric representation** is $x = x(t)$, $y = y(t)$. (Example: $x = \cos t$, $y = \sin t$ represents the circle $x^2 + y^2 = 1$.) In the complex plane we can write this

$$(2) \qquad\qquad C: \qquad \boxed{z(t) = x(t) + iy(t).}$$

(In the example, $z(t) = \cos t + i \sin t$.) We assume that C is **smooth;** that is, $z(t)$ is differentiable and the derivative $\dot{z} = dz/dt$ is continuous and nowhere zero. The sense of increasing values of the *parameter t* is called the **positive sense** on C (in Fig. 308 indicated by the tip of an arrow); in this way $z(t)$ defines an **orientation** of C. Figure 308 also shows the **angle** of intersection α of two curves C_1 and C_2, defined as the angle between the oriented tangents at the intersection point.

THEOREM 1　　**(Conformality of mapping by analytic functions)**

The mapping defined by an analytic function $f(z)$ is conformal, except at **critical points,** *that is, points at which the derivative $f'(z)$ is zero.*

PROOF.　$\dot{z}(t) = dz/dt = \dot{x}(t) + i\dot{y}(t)$ is tangent to C in (2) because this is the limit of $(z_1 - z_0)/\Delta t$ (which has the direction of the secant $z_1 - z_0$ in Fig. 309) as z_1 approaches z_0 along C.

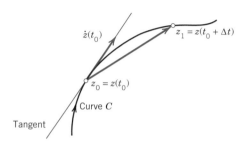

Fig. 309.　Secant and tangent of the curve C

Now the image C^* of C is $w = f(z(t))$. By the chain rule,

$$\dot{w} = f'(z(t))\dot{z}(t).$$

Hence the tangent direction of C^* is given by the argument (use (9), Sec. 12.2)

$$\arg \dot{w} = \arg f' + \arg \dot{z}$$

where $\arg \dot{z}$ gives the tangent direction of C. This shows that the mapping rotates *all* directions at a point z_0 in the domain of analyticity of f through the same angle $\arg f'(z_0)$, which exists as long as $f'(z_0) \neq 0$. But this means conformality, as Fig. 308 illustrates for an angle α between two curves, whose images C_1^* and C_2^* make the same angle (because of the rotation). ◄

EXAMPLE 2   **Conformality of $w = z^n$**

The mapping $w = z^n$, $n = 2, 3, \cdots$, is conformal, except at $z = 0$, where $w' = nz^{n-1} = 0$. For $n = 2$ this is shown in Fig. 306; we see that at 0 the angles are doubled. For general n the angles at 0 are multiplied by a factor n under the mapping. Hence the sector $0 \leq \theta \leq \pi/n$ is mapped by z^n onto the upper half-plane $v \geq 0$ (Fig. 310). ◄

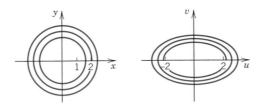

Fig. 310.   Mapping by $w = z^n$

EXAMPLE 3   **Mapping $w = z + 1/z$. Joukowski airfoil**

In terms of polar coordinates this mapping is

$$w = u + iv = r(\cos \theta + i \sin \theta) + \frac{1}{r}(\cos \theta - i \sin \theta).$$

By separating the real and imaginary parts we thus obtain

$$u = a \cos \theta, \qquad v = b \sin \theta \qquad \text{where} \qquad a = r + \frac{1}{r}, \quad b = r - \frac{1}{r}.$$

Hence circles $|z| = r = const \neq 1$ are mapped onto ellipses $x^2/a^2 + y^2/b^2 = 1$. The circle $r = 1$ is mapped onto the segment $-2 \leq u \leq 2$ of the u-axis. See Fig. 311. Now the derivative of w is

 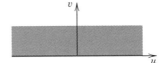

Fig. 311.   Example 3

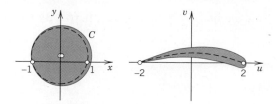

Fig. 312. Joukowski airfoil

$$w' = 1 - \frac{1}{z^2} = \frac{(z+1)(z-1)}{z^2}$$

which is 0 at $z = \pm 1$. These are the points at which the mapping is not conformal. The two circles in Fig. 312 pass through $z = -1$. The larger is mapped onto a *Joukowski airfoil*.[10] The dashed circle passes through both -1 and 1 and is mapped onto a curved segment. ◀

PROBLEM SET 12.5

Mapping of Curves. Find and sketch or plot the images of the given curves under the given mapping $w = u + iv = f(z)$.

1. $x = 0, 1, \cdots, 5;\quad y = 0, 1, \cdots, 5;\quad w = z^2$

2. Curves as in Prob. 1; $w = iz$ (**Rotation**)

3. $|z| = 1/3, 1/2, 1, 2, 3;\quad \text{Arg } z = 0, \pm\pi/4, \pm\pi/2, \pm 3\pi/4, \pi;\quad w = 1/z$

4. Does the mapping $w = \bar{z} = x - iy$ preserve angles in size as well as in sense?

5. Derive the last equation in Example 1.

Mapping of Regions. Find and sketch or plot the image of the given region under the given mapping $w = f(z)$.

6. $x > 0, y < 0, w = z^2$ **7.** $|z| > 1, w = 4z$ **8.** $0 < y < 1, w = z^2$

9. $\pi/2 < \text{Arg } z < 3\pi/4, w = z^2$ **10.** $x \geqq 1, w = 1/z$

Failure of Conformality. Find all points at which the following mappings are not conformal.

11. $w = (z - a)^3$ **12.** $w = (z^3 - a)^2$ **13.** $w = z^2 + 1/z^2$

14. $w = (z^2 + 1)/(z^2 - 1)$ **15.** $w = z^2 + bz + c$

Parametric representation of curves was used in the proof of Theorem 1 and will be needed in integration (Chap. 13), etc. Familiarize yourself with this basic concept by representing the following curves parametrically.

16. $x^2 + 9y^2 = 9$ **17.** $(x - 3)^2 + (y + 1)^2 = 4$

18. $y = kx^2$ **19.** $4x^2 - 16y^2 = 64$

 20. CAS PROJECT. Orthogonal Nets. Plot the orthogonal net consisting of the two families of level curves Re $f(z) = const$ and Im $f(z) = const$, where (a) $f(z) = z^4$, (b) $f(z) = 1/z$, (c) $f(z) = 1/z^2$, (d) $f(z) = (z + i)/(1 + iz)$. Why do these curves intersect at right angles? In your work, experiment to get the best possible figures. Also do the same for other functions of your own choice.

---

[10]NIKOLAI JEGOROVICH JOUKOWSKI (1847—1921), Russian mathematician.

12.6 Exponential Function

In the remaining sections of this chapter we discuss the basic elementary complex functions, the exponential function, trigonometric functions, logarithm, and so on. They will be counterparts of the familiar functions of calculus, to which they reduce when $z = x$ is real. They are indispensable throughout applications, and some of them have interesting properties not apparent in real.

We begin with one of the most important analytic functions, the complex **exponential function**

$$e^z, \qquad \text{also written} \qquad \exp z.$$

The definition of e^z in terms of the real functions e^x, cos y, and sin y is

(1)
$$\boxed{e^z = e^x(\cos y + i \sin y).}$$

This definition[11] is motivated by requirements that make e^z a natural extension of the real exponential function e^x, namely,

(a) e^z should reduce to the latter when $z = x$ is real;

(b) e^z should be an **entire function,** that is, analytic for all z;

(c)   similar to calculus, its derivative should be

(2)
$$(e^z)' = e^z.$$

From (1) we see that (a) holds, since $\cos 0 = 1$ and $\sin 0 = 0$. That e^z is entire is easily verified by the Cauchy–Riemann equations. Formula (2) then follows from (4) in Sec. 12.4:

$$(e^z)' = (e^x \cos y)_x + i(e^x \sin y)_x = e^x \cos y + i e^x \sin y = e^z.$$

Further Properties. e^z has further interesting properties. Let us first show that, as in real, we have the *functional relation*

(3)
$$e^{z_1 + z_2} = e^{z_1} e^{z_2}$$

for any $z_1 = x_1 + iy_1$ and $z_2 = x_2 + iy_2$. Indeed, by (1),

$$e^{z_1} e^{z_2} = e^{x_1}(\cos y_1 + i \sin y_1) e^{x_2}(\cos y_2 + i \sin y_2).$$

Since $e^{x_1} e^{x_2} = e^{x_1 + x_2}$ for these *real* functions, by an application of the addition formulas for the cosine and sine functions (similar to that in Sec. 12.2) we find that this equals

$$e^{z_1} e^{z_2} = e^{x_1 + x_2}[\cos (y_1 + y_2) + i \sin (y_1 + y_2)] = e^{z_1 + z_2},$$

---

[11]This definition provides for a relatively simple discussion. We could also define e^z by the familiar series $e^x = \sum_{n=0}^{\infty} x^n/n!$ with x replaced by z, but then we would first have to discuss complex series at this early stage. We show the connection in Sec. 14.4.

as asserted. An interesting special case is $z_1 = x$, $z_2 = iy$:

(4)
$$e^z = e^x e^{iy}.$$

Furthermore, for $z = iy$ we have from (1) the so-called **Euler formula**

(5)
$$e^{iy} = \cos y + i \sin y.$$

Hence the **polar form** of a complex number, $z = r(\cos \theta + i \sin \theta)$, may now be written

(6)
$$z = re^{i\theta}.$$

From (5) we obtain

(7)
$$e^{2\pi i} = 1$$

as well as the important formulas (verify!)

(8)
$$e^{\pi i/2} = i, \qquad e^{\pi i} = -1, \qquad e^{-\pi i/2} = -i, \qquad e^{-\pi i} = -1.$$

Another consequence of (5) is

(9)
$$|e^{iy}| = |\cos y + i \sin y| = \sqrt{\cos^2 y + \sin^2 y} = 1.$$

That is, for pure imaginary exponents the exponential function has absolute value one, a result you should remember. From (9) and (1),

(10)
$$|e^z| = e^x. \qquad \text{Hence} \qquad \arg e^z = y \pm 2n\pi \qquad (n = 0, 1, 2, \cdots),$$

since $|e^z| = e^x$ shows that (1) is actually e^z in polar form.

From $|e^z| = e^x \neq 0$ in (10) we see that

(11)
$$e^z \neq 0 \qquad\qquad \text{for all } z.$$

So here we have an entire function that never vanishes, in contrast to (nonconstant) polynomials, which are also entire (Example 5 in Sec. 12.3) but always have a zero, as is proved in algebra.

From (10) we also conclude that e^z maps a vertical straight line $x = x_0 = const$ onto the circle $|w| = e^{x_0}$ and a horizontal straight line $y = y_0 = const$ onto the ray $\arg w = y_0$. Hence the rectangle in Fig. 313 is mapped onto a region bounded by circles and rays as shown.

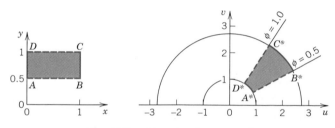

Fig. 313. Mapping by $w = e^z$

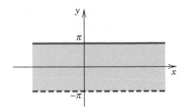

Fig. 314. Fundamental region of the exponential function e^z in the z-plane

Periodicity of e^z with period $2\pi i$,

(12) $$e^{z+2\pi i} = e^z \qquad \text{for all } z$$

is a basic property that follows from (1) and the periodicity of $\cos y$ and $\sin y$. Hence all the values that $w = e^z$ can assume are already assumed in the horizontal strip of width 2π

(13) $$\boxed{-\pi < y \le \pi} \qquad \text{(Fig. 314).}$$

This infinite strip is called a **fundamental region** of e^z.

In geometrical terms: *e^z maps a fundamental region bijectively* (see Sec. 12.5) *onto the entire plane.*

Figure 315 illustrates further that the upper half $0 < y \le \pi$ of the fundamental strip is mapped onto the upper half-plane because this half-plane corresponds to $0 < \arg w \le \pi$. The left half of the strip in Fig. 315 goes inside the unit disk $|w| \le 1$ (because $e^x \le 1$ for $x \le 0$) and the right half goes outside, $|w| > 1$.

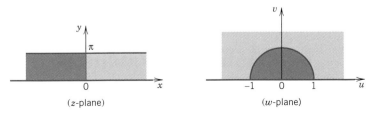

Fig. 315. Mapping by $w = e^z$

EXAMPLE 1   **Function values. Solution of equations**

Computation of values from (1) provides no problem. For instance, verify that

$$e^{1.4-0.6i} = e^{1.4}(\cos 0.6 - i \sin 0.6) = 4.055(0.825 - 0.565i) = 3.347 - 2.290i,$$

$$\left| e^{1.4-0.6i} \right| = e^{1.4} = 4.055, \qquad \text{Arg } e^{1.4-0.6i} = -0.6.$$

To illustrate (3), take the product of

$$e^{2+i} = e^2(\cos 1 + i \sin 1) \qquad \text{and} \qquad e^{4-i} = e^4(\cos 1 - i \sin 1)$$

and verify that it equals $e^2 e^4(\cos^2 1 + \sin^2 1) = e^6 = e^{(2+i)+(4-i)}$.

To solve the equation $e^z = 3 + 4i$, note first that $|e^z| = e^x = 5$, $x = \ln 5 = 1.609$ is the real part of all solutions. Now, since $e^x = 5$,

$$e^x \cos y = 3, \qquad e^x \sin y = 4, \qquad \cos y = 0.6, \qquad \sin y = 0.8, \qquad y = 0.927.$$

Ans. $z = 1.609 + 0.927i \pm 2n\pi i$ ($n = 0, 1, 2, \cdots$). These are infinitely many solutions (due to the periodicity of e^z). They lie on the vertical line $x = 1.609$ at a distance 2π from their neighbors.   ◀

To summarize: many properties of $e^z = \exp z$ parallel those of e^x; an exception is the periodicity of e^z with $2\pi i$, which suggested the concept of a fundamental region. Keep in mind that e^z is an *entire function*. (Do you still remember what that means?)

PROBLEM SET 12.6

Function Values. Find e^z (in the form $u + iv$) and $|e^z|$ if z equals
1. $2 + 3\pi i$ 2. $1 + i$ 3. $2\pi(1 + i)$ 4. $0.95 - 1.6i$ 5. $-\pi i/2$

Polar Form (6). Represent each of the following in the "exponential polar form" (6):
6. $1 + i$ 7. $4 + 3i$ 8. $\sqrt[n]{z}$ 9. -4 10. $\sqrt{i}, \sqrt{-i}$

Equations. Find all solutions and plot some of them in the complex plane.
11. $e^z = 1$ 12. $e^{2z} = 2$ 13. $e^z = -3$ 14. $e^z = 4 + 3i$ 15. $e^z = 0$

Conformal Mapping $w = e^z$. Find and sketch or plot the image of the given region.
16. $-1 \leqq x \leqq 1, -\pi < y < \pi$ 17. $0 < y < \frac{1}{2}\pi$ 18. $\pi < y \leqq 3\pi$ 19. $\ln 3 < x < \ln 5$

20. **TEAM PROJECT. Further Properties of the Exponential Function.** **(a) Analyticity.** Show that e^z is entire. What about $e^{1/z}$? $e^{\bar{z}}$? $e^x(\cos kx + i \sin kx)$? (Use the Cauchy-Riemann equations.)
    (b) Special Values. Find all z such that (i) e^z is real, (ii) $|e^{-z}| < 1$, (iii) $e^{\bar{z}} = \overline{e^z}$.
    (c) Harmonic Function. Show that $u = e^{xy} \cos(x^2/2 - y^2/2)$ is harmonic and find a conjugate.
    (d) Uniqueness. It is interesting that $f(z) = e^z$ is uniquely determined by the two properties $f(x + i0) = e^x$ and $f'(z) = f(z)$, where f is assumed to be entire. Prove this using the Cauchy–Riemann equations.

12.7 Trigonometric Functions Hyperbolic Functions

Just as e^z extends e^x to complex, we want the *complex* trigonometric functions to extend the familiar *real* trigonometric functions. The idea of making the connection is the use of the Euler formulas (Sec. 12.6)

$$e^{ix} = \cos x + i \sin x, \qquad e^{-ix} = \cos x - i \sin x.$$

By addition and subtraction we obtain for the *real* cosine and sine

$$\cos x = \frac{1}{2}(e^{ix} + e^{-ix}), \qquad \sin x = \frac{1}{2i}(e^{ix} - e^{-ix}).$$

This suggests the following definitions for complex values $z = x + iy$:

(1) $$\cos z = \frac{1}{2}(e^{iz} + e^{-iz}), \qquad \sin z = \frac{1}{2i}(e^{iz} - e^{-iz}).$$

It is quite remarkable that here in complex, functions come together that are unrelated in real. This is not an isolated incident but is typical of the general situation and shows the advantage of working in complex.

Furthermore, as in calculus we define

(2)
$$\tan z = \frac{\sin z}{\cos z}, \qquad \cot z = \frac{\cos z}{\sin z}$$

and

(3)
$$\sec z = \frac{1}{\cos z}, \qquad \csc z = \frac{1}{\sin z}.$$

Since e^z is entire, $\cos z$ and $\sin z$ are entire functions. $\tan z$ and $\sec z$ are not entire; they are analytic except at the points where $\cos z$ is zero; and $\cot z$ and $\csc z$ are analytic except where $\sin z$ is zero. Formulas for the derivatives follow readily from $(e^z)' = e^z$ and (1)–(3); as in calculus,

(4)
$$(\cos z)' = -\sin z, \qquad (\sin z)' = \cos z, \qquad (\tan z)' = \sec^2 z,$$

etc. Equation (1) also shows that **Euler's formula** *is valid in complex:*

(5)
$$\boxed{e^{iz} = \cos z + i \sin z} \qquad\qquad \text{for all } z.$$

Real and imaginary parts of $\cos z$ and $\sin z$ are needed in computing values, and they also help in displaying properties of our functions. We illustrate this with a typical example.

EXAMPLE 1   **Real and imaginary parts. Absolute value. Periodicity**

Show that

(6)
    (a) $\cos z = \cos x \cosh y - i \sin x \sinh y$
    (b) $\sin z = \sin x \cosh y + i \cos x \sinh y$

and

(7)
    (a) $|\cos z|^2 = \cos^2 x + \sinh^2 y$
    (b) $|\sin z|^2 = \sin^2 x + \sinh^2 y$

and give some applications of these formulas.

Solution. From (1),

$$\cos z = \tfrac{1}{2}(e^{i(x+iy)} + e^{-i(x+iy)})$$

$$= \tfrac{1}{2}e^{-y}(\cos x + i \sin x) + \tfrac{1}{2}e^{y}(\cos x - i \sin x)$$

$$= \tfrac{1}{2}(e^{y} + e^{-y})\cos x - \tfrac{1}{2}i(e^{y} - e^{-y})\sin x.$$

This yields (6a) since, as is known from calculus,

(8)
$$\cosh y = \tfrac{1}{2}(e^{y} + e^{-y}), \qquad \sinh y = \tfrac{1}{2}(e^{y} - e^{-y});$$

(6b) is obtained similarly. From (6a) and $\cosh^2 y = 1 + \sinh^2 y$ we obtain

$$|\cos z|^2 = \cos^2 x \, (1 + \sinh^2 y) + \sin^2 x \sinh^2 y.$$

Since $\sin^2 x + \cos^2 x = 1$, this gives (7a), and (7b) is obtained similarly.

For instance, $\cos (2 + 3i) = \cos 2 \cosh 3 - i \sin 2 \sinh 3 = -4.190 - 9.109i$.

From (6) we see that $\cos z$ and $\sin z$ are *periodic with period 2π,* just as in real. Periodicity of $\tan z$ and $\cot z$ with period π now follows.

Formula (7) points to an essential difference between the real and the complex cosine and sine: whereas $|\cos x| \leqq 1$ and $|\sin x| \leqq 1$, the complex cosine and sine functions are no longer bounded but approach infinity in absolute value as $y \to \infty$, since then $\sinh y \to \infty$ in (7).   ◀

EXAMPLE 2   **Solution of equations. Zeros of cos z and sin z**

Solve (a) $\cos z = 5$ (which has no real solution!), (b) $\cos z = 0$, (c) $\sin z = 0$.

Solution.   (a) $e^{2iz} - 10e^{iz} + 1 = 0$ from (1) by multiplication by e^{iz}. This is a quadratic equation in e^{iz}, with solutions (rounded off to 3 decimals)

$$e^{iz} = e^{-y+ix} = 5 \pm \sqrt{25 - 1} = 9.899 \quad \text{and} \quad 0.101.$$

Thus $e^{-y} = 9.899$ or 0.101, $e^{ix} = 1$, $y = \pm 2.292$, $x = 2n\pi$. *Ans.* $z = \pm 2n\pi \pm 2.292i$ $(n = 0, 1, 2, \cdots)$. Can you obtain this from (6a)?

(b) $\cos x = 0$, $\sinh y = 0$ by (7a), $y = 0$. *Ans.* $z = \pm\frac{1}{2}(2n + 1)\pi$ $(n = 0, 1, 2, \cdots)$.

(c) $\sin x = 0$, $\sinh y = 0$ by (7b). *Ans.* $z = \pm 2n\pi$ $(n = 0, 1, 2, \cdots)$. Hence the only zeros of $\cos z$ and $\sin z$ are those of the real cosine and sine functions.   ◀

General formulas *for the real trigonometric functions continue to hold for complex values.* This follows immediately from the definitions. We mention in particular the addition rules

(9)
$$\cos (z_1 \pm z_2) = \cos z_1 \cos z_2 \mp \sin z_1 \sin z_2$$

$$\sin (z_1 \pm z_2) = \sin z_1 \cos z_2 \pm \sin z_2 \cos z_1$$

and the formula

(10)
$$\cos^2 z + \sin^2 z = 1.$$

Some further useful formulas are included in the problem set.

Hyperbolic Functions

The complex **hyperbolic cosine** and **sine** are defined by the formulas

(11)
$$\cosh z = \tfrac{1}{2}(e^z + e^{-z}), \qquad \sinh z = \tfrac{1}{2}(e^z - e^{-z}).$$

This is suggested by the familiar definitions for a real variable [see (8)]. These functions are entire, with derivatives

(12)
$$(\cosh z)' = \sinh z, \qquad (\sinh z)' = \cosh z,$$

as in calculus. The other hyperbolic functions are defined by

(13)
$$\tanh z = \frac{\sinh z}{\cosh z}, \qquad \coth z = \frac{\cosh z}{\sinh z},$$

$$\operatorname{sech} z = \frac{1}{\cosh z}, \qquad \operatorname{csch} z = \frac{1}{\sinh z}.$$

Complex trigonometric and hyperbolic functions are related. If in (11), we replace z by iz and use (1), we obtain

(14)
$$\cosh iz = \cos z, \qquad \sinh iz = i \sin z.$$

From this, since cosh is even and sinh is odd, conversely,

(15)
$$\cos iz = \cosh z, \qquad \sin iz = i \sinh z.$$

Here we have another case of *unrelated* real functions that have *related* complex analogs, pointing again to the advantage of working in complex in order to get both a more unified formalism and a deeper understanding of special functions. This is one of the main reasons for the importance of complex analysis, mentioned at the beginning of the chapter.

Conformal Mapping by sin z, cos z, sinh z, cosh z

Sine. Figure 316 shows the mapping by $w = \sin z$. The rectangular net of straight lines $x = const$ and $y = const$ is mapped onto a net in the w-plane consisting of hyperbolas (the images of the vertical lines $x = const$) and ellipses (the images of the horizontal lines $y = const$) intersecting the hyperbolas at right angles. Corresponding calculations are simple. Equation (6b) shows that $u = \mathrm{Re}\,(\sin z) = \sin x \cosh y$ and $v = \mathrm{Im}\,(\sin z) = \cos x \sinh y$. From this and $\cosh^2 y - \sinh^2 y = 1$ and $\sin^2 x + \cos^2 x = 1$ we get

(16)
$$\frac{u^2}{\sin^2 x} - \frac{v^2}{\cos^2 x} = \cosh^2 y - \sinh^2 y = 1 \qquad \text{(Hyperbolas)}$$

(17)
$$\frac{u^2}{\cosh^2 y} + \frac{v^2}{\sinh^2 y} = \sin^2 x + \cos^2 x = 1 \qquad \text{(Ellipses).}$$

Exceptions are the vertical lines $x = \pm\frac{1}{2}\pi$, which are "folded" onto $u \leqq -1$ and $u \geqq 1$ ($v = 0$), respectively.

Figure 317 illustrates this further. The upper and lower sides of the rectangle are mapped onto semi-ellipses and the vertical sides onto $-\cosh 1 \leqq u \leqq -1$ and $1 \leqq u \leqq \cosh 1$ ($v = 0$), respectively. At $z = \pm 1$ the mapping is not conformal. Why?

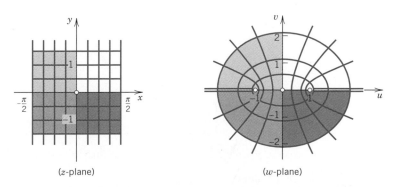

(z-plane)   (w-plane)

Fig. 316.   Mapping $w = u + iv = \sin z$

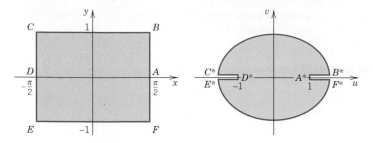

Fig. 317. Mapping by $w = \sin z$

Cosine. The mapping $w = \cos z$ could be discussed independently, but since

$$w = \cos z = \sin (z + \tfrac{1}{2}\pi),$$

we see at once that this is the same mapping as $\sin z$ preceded by a translation to the right through $\tfrac{1}{2}\pi$ units.

Hyperbolic sine. Since

$$w = \sinh z = -i \sin (iz),$$

the mapping is a counterclockwise rotation $Z = iz$ through $\tfrac{1}{2}\pi$ (i.e., 90°), followed by the sine mapping $Z^* = \sin Z$, followed by a clockwise 90°-rotation $w = -iZ^*$.

Hyperbolic cosine. This function

$$w = \cosh z = \cos (iz)$$

defines a mapping that is a rotation $Z = iz$ followed by the mapping $w = \cos Z$.
   The mapping by $w = \tan z$ is more tricky. We shall discuss it in Sec. 12.9.

PROBLEM SET 12.7

Formulas for Hyperbolic Functions. Show that

1.
$$\boxed{\begin{aligned} \cosh z &= \cosh x \cos y + i \sinh x \sin y, \\ \sinh z &= \sinh x \cos y + i \cosh x \sin y. \end{aligned}}$$

2.
$$\boxed{\begin{aligned} \cosh (z_1 + z_2) &= \cosh z_1 \cosh z_2 + \sinh z_1 \sinh z_2, \\ \sinh (z_1 + z_2) &= \sinh z_1 \cosh z_2 + \cosh z_1 \sinh z_2. \end{aligned}}$$

3. $\cosh^2 z - \sinh^2 z = 1$, $\qquad \cosh^2 z + \sinh^2 z = \cosh 2z$

Function Values. Find (in the form $u + iv$)
   4. $\cos (1 + i)$ **5.** $\sin \pi i$ **6.** $\cos (\tfrac{1}{2}\pi - \pi i)$ **7.** $\cosh (-3 - 6i)$ **8.** $\sinh (4 + 5i)$

Equations. Find all solutions of the following equations.

9. $\cos z = 3i$ **10.** $\cosh z = 0$ **11.** $\cosh z = \frac{1}{2}$ **12.** $\sin z = 1000$ **13.** $\sin z = \cosh 3$

Conformal Mapping $w = \sin z$. Find and sketch or plot the image of the region

14. $0 < x < \pi/6$, y arbitrary   **15.** $0 < x < \pi/2, 0 < y < 2$ **16.** $-\pi/4 < x < \pi/4, 0 < y < 3$

Conformal Mapping $w = \cos z$. Find and sketch or plot the image of the region

17. $0 < x < \pi, 0 < y < 1$ **18.** $0 < x < 2\pi, \frac{1}{2} < y < 1$ **19.** $0 < x < \pi, y > 0$

20. CAS PROJECT. Orthogonal Nets of Level Curves. Let $w = u + iv = f(z)$. Plot the orthogonal net of the two families of level curves $u = \text{Re } f(z) = const$ and $v = \text{Im } f(z) = const$ where $f(z)$ equals (a) $\cos z$, (b) $\sin z$, (c) $\cosh z$, (d) $\sinh z$. Then try to find the relations between the four functions experimentally by looking at the four plots. Can you discover these relations completely from the plots?

12.8 Logarithm
General Power

As the last of the functions we introduce the *complex logarithm*, which is more complicated than the real logarithm (which it includes as a special case) and historically puzzled mathematicians for some time (so if you first get puzzled—which need not happen!—be patient and work this section with extra care).

The **natural logarithm** of $z = x + iy$ is denoted by $\ln z$ (sometimes also by $\log z$) and is defined as the inverse of the exponential function; that is, $w = \ln z$ is defined for $z \neq 0$ by the relation

$$e^w = z.$$

(Note that $z = 0$ is impossible, since $e^w \neq 0$ for all w; see Sec. 12.6.) If we set $w = u + iv$ and $z = re^{i\theta}$, this becomes

$$e^w = e^{u+iv} = re^{i\theta}.$$

Now from Sec. 12.6 we know that e^{u+iv} has the absolute value e^u and the argument v. These must be equal to the absolute value and argument on the right:

$$e^u = r, \qquad v = \theta.$$

$e^u = r$ gives $u = \ln r$, where $\ln r$ is the familiar *real* natural logarithm of the positive number $r = |z|$. Hence $w = u + iv = \ln z$ is given by

(1)
$$\boxed{\ln z = \ln r + i\theta} \qquad (r = |z| > 0, \quad \theta = \arg z).$$

Now comes an important point (without analog in real calculus). Since the argument of z is determined only up to integer multiples of 2π, *the complex natural logarithm* $\ln z$ *($z \neq 0$) is infinitely many-valued.*

The value of $\ln z$ corresponding to the principal value $\text{Arg } z$ (see Sec. 12.2) is denoted by $\text{Ln } z$ (Ln with capital L) and is called the **principal value** of $\ln z$. Thus

(2)
$$\boxed{\text{Ln } z = \ln |z| + i \text{ Arg } z} \qquad (z \neq 0).$$

The uniqueness of Arg z for given z ($\neq 0$) implies that Ln z is single-valued, that is, a function in the usual sense. Since the other values of arg z differ by integer multiples of 2π, the other values of ln z are given by

(3)
$$\boxed{\ln z = \text{Ln } z \pm 2n\pi i}$$
$\qquad\qquad (n = 1, 2, \cdots).$

They all have the same real part, and their imaginary parts differ by integer multiples of 2π.

If z is positive real, then Arg $z = 0$, and Ln z becomes identical with the real natural logarithm known from calculus. If z is negative real (so that the natural logarithm of calculus is not defined!), then Arg $z = \pi$ and

$$\text{Ln } z = \ln |z| + \pi i.$$

EXAMPLE 1   **Natural logarithm. Principal value**

$$\ln 1 = 0, \pm 2\pi i, \pm 4\pi i, \cdots \qquad\qquad \text{Ln } 1 = 0$$

$$\ln 4 = 1.386\ 294 \pm 2n\pi i \qquad\qquad \text{Ln } 4 = 1.386\ 294$$

$$\ln(-1) = \pm \pi i, \pm 3\pi i, \pm 5\pi i, \cdots \qquad\qquad \text{Ln }(-1) = \pi i$$

$$\ln(-4) = 1.386\ 294 \pm (2n+1)\pi i \qquad\qquad \text{Ln }(-4) = 1.386\ 294 + \pi i$$

$$\ln i = \pi i/2, -3\pi i/2, 5\pi i/2, \cdots \qquad\qquad \text{Ln } i = \pi i/2$$

$$\ln 4i = 1.386\ 294 + \pi i/2 \pm 2n\pi i \qquad\qquad \text{Ln } 4i = 1.386\ 294 + \pi i/2$$

$$\ln(-4i) = 1.386\ 294 - \pi i/2 \pm 2n\pi i \qquad\qquad \text{Ln }(-4i) = 1.386\ 294 - \pi i/2$$

$$\ln(3 - 4i) = \ln 5 + i \arg(3 - 4i) \qquad\qquad \text{Ln }(3 - 4i) = 1.609\ 438 - 0.927\ 295i$$

$$\qquad\qquad = 1.609\ 438 - 0.927\ 295i \pm 2n\pi i \qquad\qquad \text{(Fig. 318)}$$ ◀

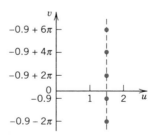

Fig. 318.   Some values of ln $(3 - 4i)$ in Example 1

The familiar relations for the natural logarithm continue to hold for complex values, that is,

(4)   (a) $\ln(z_1 z_2) = \ln z_1 + \ln z_2,$ (b) $\ln(z_1/z_2) = \ln z_1 - \ln z_2$

but these relations are to be understood in the sense that each value of one side is also contained among the values of the other side.

EXAMPLE 2   **Illustration of the functional relations (4) in complex**

Let

$$z_1 = z_2 = e^{\pi i} = -1.$$

If we take the principal values

$$\mathrm{Ln}\, z_1 = \mathrm{Ln}\, z_2 = \pi i,$$

then (4a) holds provided we write $\ln (z_1 z_2) = \ln 1 = 2\pi i$; however, it is not true for the principal value, $\mathrm{Ln}\,(z_1 z_2) = \mathrm{Ln}\, 1 = 0$.   ◀

From (1) and $e^{\ln r} = r$ for positive real r we obtain

(5a)
$$e^{\ln z} = z$$

as expected, but, since $\arg (e^z) = y \pm 2n\pi$ is multivalued, so is

(5b) $$\ln (e^z) = z \pm 2n\pi i, \qquad\qquad n = 0, 1, \cdots.$$

For every fixed nonnegative integer n, formula (3) defines a function. We prove that each of these functions, in particular the principal value $\mathrm{Ln}\, z$, is analytic except at $z = 0$ and except on the negative real axis (where the imaginary part of such a function is not even continuous but has a jump of magnitude 2π). We do this by proving that

(6) $$(\ln z)' = \frac{1}{z} \qquad\qquad [\text{n in (3) fixed, z not negative real or zero}].$$

In (1) we have $\ln z = u + iv$, where

$$u = \ln r = \ln |z| = \tfrac{1}{2} \ln (x^2 + y^2), \qquad v = \arg z = \arctan \frac{y}{x} + c,$$

where c is a constant (a multiple of $2n\pi$). We calculate the partial derivatives of u and v and find that they satisfy the Cauchy–Riemann equations:

$$u_x = \frac{x}{x^2 + y^2} = v_y = \frac{1}{1 + (y/x)^2} \cdot \frac{1}{x},$$

$$u_y = \frac{y}{x^2 + y^2} = -v_x = -\frac{1}{1 + (y/x)^2}\left(-\frac{y}{x^2}\right).$$

Formula (4) in Sec. 12.4 now yields the desired result:

$$(\ln z)' = u_x + iv_x = \frac{x}{x^2 + y^2} + i\,\frac{1}{1 + (y/x)^2}\left(-\frac{y}{x^2}\right) = \frac{x - iy}{x^2 + y^2} = \frac{1}{z}.$$ ◀

Conformal Mapping by ln z

Principle of Inverse Mapping. *The mapping by the inverse $z = f^{-1}(w)$ of $w = f(z)$ is obtained by interchanging the roles of the z-plane and the w-plane in the mapping by $w = f(z)$.* (What would be the analog of this in calculus?)

EXAMPLE 3   **Mapping by $w = z^2$ and by its inverse**

$w = z^2$ restricted to the right half-plane $x > 0$ maps this half-plane onto the w-plane without the negative half $u \leqq 0$ of the real w-axis because at the origin angles are doubled, so that rays $\theta = \text{Arg } z = const$ are mapped onto rays $\phi = \text{Arg } w = 2\theta$. Hence its inverse $w = \sqrt{z}$ (with $\sqrt{z} > 0$ for $z = x > 0$) maps the z-plane without the negative half of the x-axis onto the right half-plane $u = \text{Re } w > 0$.   ◀

Natural logarithm. $\ln z$ is the inverse relation of the exponential function $w = e^z$. The latter maps a fundamental strip (see Sec. 12.6) onto the w-plane without $z = 0$ (because $e^z \neq 0$ for every z). Hence by the principle of inverse mapping, the principal value $w = \text{Ln } z$ maps the z-plane (with $z = 0$ omitted and cut along the negative real axis, where $\theta = \text{Im (Ln } z)$ jumps by 2π) onto the horizontal strip $-\pi < v \leqq \pi$ of the w-plane. Since the mapping

$$w = \text{Ln } z + 2\pi i$$

differs from $w = \text{Ln } z$ by the translation $2\pi i$ (vertically upward), this mapping maps the z-plane (cut as before and 0 omitted) onto the strip $\pi < v \leqq 3\pi$. Similarly for each of the infinitely many mappings

$$w = \ln z = \text{Ln } z \pm 2n\pi i \qquad (n = 0, 1, 2, \cdots).$$

The corresponding horizontal strips of width 2π (images of the z-plane under these mappings) together cover the whole w-plane without overlapping.

General Powers

General powers of a complex number $z = x + iy$ are defined by the formula

(7)
$$z^c = e^{c \ln z} \qquad (c \text{ complex}, z \neq 0).$$

Since $\ln z$ is infinitely many-valued, z^c will, in general, be multivalued. The particular value

$$z^c = e^{c \, \text{Ln } z}$$

is called the **principal value** *of z^c.*

If $c = n = 1, 2, \cdots$, then z^n is single-valued and identical with the usual nth power of z. If $c = -1, -2, \cdots$, the situation is similar.

If $c = 1/n$ where $n = 2, 3, \cdots$, then

$$z^c = \sqrt[n]{z} = e^{(1/n) \ln z} \qquad (z \neq 0),$$

the exponent is determined up to multiples of $2\pi i/n$ and we obtain the n distinct values of the nth root, in agreement with the result in Sec. 12.2. If $c = p/q$, the quotient of two positive integers, the situation is similar, and z^c has only finitely many distinct values. However, if c is real irrational or genuinely complex, then z^c is infinitely many-valued.

EXAMPLE 4   **General power**

$$i^i = e^{i \ln i} = \exp (i \ln i) = \exp \left[i \left(\frac{\pi}{2} i \pm 2n\pi i \right) \right] = e^{-(\pi/2) \mp 2n\pi}.$$

All these values are real, and the principal value ($n = 0$) is $e^{-\pi/2}$.

Similarly, by direct calculation and multiplying out in the exponent,

$$(1 + i)^{2-i} = \exp\left[(2 - i)\ln(1 + i)\right] = \exp\left[(2 - i)\left\{\ln\sqrt{2} + \tfrac{1}{4}\pi i \pm 2n\pi i\right\}\right]$$

$$= 2e^{\pi/4 \pm 2n\pi}\left[\sin\left(\tfrac{1}{2}\ln 2\right) + i\cos\left(\tfrac{1}{2}\ln 2\right)\right]. \qquad \blacktriangleleft$$

It is a *convention* that for real positive $z = x$ the expression z^c means $e^{c\,\ln x}$ where $\ln x$ is the elementary real natural logarithm (that is, the principal value $\operatorname{Ln} z$ ($z = x > 0$) in the sense of our definition). Also, if $z = e$, the base of the natural logarithm, $z^c = e^c$ is *conventionally* regarded as the unique value obtained from (1) in Sec. 12.6.

From (7) we see that for any complex number a,

(8)
$$a^z = e^{z\,\ln a}.$$

We have now introduced the complex functions needed in practical work, some of them (e^z, $\cos z$, $\sin z$, $\cosh z$, $\sinh z$) entire (Sec. 12.6), some of them ($\tan z$, $\cot z$, $\tanh z$, $\coth z$) analytic except at certain points, and one of them ($\ln z$) splitting up into infinitely many functions, each analytic except at 0 and on the negative real axis.

PROBLEM SET 12.8

Verifications in the Text
1. Verify the computations in Example 1.
2. Verify (4) for $z_1 = -i$ and $z_2 = -1$.
3. Prove the analyticity of $\operatorname{Ln} z$ by means of the Cauchy–Riemann equations in polar form (Sec. 12.4).
4. Prove (5a) and (5b).

Complex Natural Logarithm ln z
Principal value. Find $\operatorname{Ln} z$ when z equals
5. -5 6. $-12 - 16i$ 7. $1 \pm i$ 8. $-10 \pm 0.1i$ 9. $\pm3.5 \pm 1.8i$

All Values. Find all values of the given expression and plot some of them in the complex plane.
10. $\ln e$ 11. $\ln 1$ 12. $\ln(-4)$ 13. $\ln(-e^{-i})$ 14. $\ln(4 + 3i)$

Equations. Solve for z:
15. $\ln z = -\pi i/2$ 16. $\ln z = -2 - \tfrac{3}{2}i$ 17. $\ln z = 4 - 3i$ 18. $\ln z = e - \pi i$

19. **(Conformal mapping)** Find and sketch the image of $2 \leqq |z| \leqq 3$, $\pi/4 \leqq \theta \leqq \pi/2$ under the mapping $w = \operatorname{Ln} z$.

General Powers
Find the principal value of the given expression. (Show the details of your work.)
20. $(2i)^{2i}$ 21. 3^{4-i} 22. $(1 + i)^{1-i}$ 23. $(1 + i)^{-1+i}$ 24. $(1 + 3i)^i$
25. $i^{1/2}$ 26. $(-1)^{2-4i}$ 27. $(3 + 4i)^{1/3}$ 28. $(1.3 + 0.4i)^{2\pi i}$ 29. $(-3)^{3-i}$

30. **TEAM PROJECT. Inverse Trigonometric and Hyperbolic Functions.** By definition, the **inverse sine** $w = \sin^{-1} z$ is the relation such that $\sin w = z$. The **inverse cosine** $w = \cos^{-1} z$ is the relation such that $\cos w = z$. The **inverse tangent, inverse cotangent, inverse hyperbolic sine,** etc., are defined and denoted in a similar fashion. (Note that all these relations are

multivalued.) Using $\sin w = (e^{iw} - e^{-iw})/2i$ and similar representations of cos w, etc., show that

(a) $\cos^{-1} z = -i \ln (z + \sqrt{z^2 - 1})$

(b) $\sin^{-1} z = -i \ln (iz + \sqrt{1 - z^2})$

(c) $\cosh^{-1} z = \ln (z + \sqrt{z^2 - 1})$

(d) $\sinh^{-1} z = \ln (z + \sqrt{z^2 + 1})$

(e) $\tan^{-1} z = \dfrac{i}{2} \ln \dfrac{i + z}{i - z}$

(f) $\tanh^{-1} z = \dfrac{1}{2} \ln \dfrac{1 + z}{1 - z}$

(g) Show that $w = \sin^{-1} z$ is infinitely many-valued, and if w_1 is one of these values, the others are of the form $w_1 \pm 2n\pi$ and $\pi - w_1 \pm 2n\pi$, $n = 0, 1, \cdots$. (The *principal value of* $w = u + iv = \sin^{-1} z$ is defined to be the value for which $-\pi/2 \leqq u \leqq \pi/2$ if $v \geqq 0$ and $-\pi/2 < u < \pi/2$ if $v < 0$.)

12.9 Linear Fractional Transformations
Optional

Conformal mappings can help in solving boundary value problems by mapping complicated regions conformally onto standard ones (disks, half-planes, strips). We shall explain this in Sec. 16.2. For this it is useful to know properties of basic mappings, as just discussed, and, perhaps, to get further help from a computer algebra system (CAS). Accordingly, we shall discuss some more conformal mappings of general interest.

Linear fractional transformations (or **Möbius transformations**) are mappings

(1)
$$w = \frac{az + b}{cz + d} \qquad (ad - bc \neq 0)$$

where a, b, c, d are complex or real numbers. Differentiation gives

(2)
$$w' = \frac{a(cz + d) - c(az + b)}{(cz + d)^2} = \frac{ad - bc}{(cz + d)^2}.$$

This motivates our requirement $ad - bc \neq 0$. It implies conformality for all z and excludes the totally uninteresting case $w' \equiv 0$ once and for all.

The family (1) is very interesting. First of all, it includes as special cases

(3)
$$w = z + b \qquad \qquad (Translation)$$
$$w = az \quad \text{with } |a| = 1 \qquad (Rotation)$$
$$w = az + b \qquad \qquad (Linear\ transformation)$$
$$w = 1/z \qquad \qquad (Inversion\ in\ the\ unit\ circle).$$

In polar forms $z = re^{i\theta}$ and $w = Re^{i\phi}$ the inversion $w = 1/z$ is

$$Re^{i\phi} = \frac{1}{re^{i\theta}} = \frac{1}{r} e^{-i\theta} \qquad \text{and gives} \qquad R = \frac{1}{r}, \qquad \phi = -\theta.$$

Hence the unit circle $|z| = r = 1$ is mapped onto the unit circle $|w| = R = 1$; $w = e^{i\phi} = e^{-i\theta}$. For a general z the image $w = 1/z$ can be found geometrically by marking $|w| = R = 1/r$ on the segment from 0 to z and then reflecting the mark in the real axis. (Make a sketch.)

Extended Complex Plane

This concept can be motivated as follows. From (1) we see that each z for which $cz + d \neq 0$ has an image w (which is unique). Let $c \neq 0$. Then $cz + d = 0$ when $z = -d/c$, so that $z = -d/c$ has no image. This suggests that we introduce an additional point, called the **point at infinity** and denoted by the symbol ∞ (*infinity*), and let this point be the image of $z = -d/c$.

Furthermore, when $c = 0$ we must have $a \neq 0$ and $d \neq 0$ (why?), and then we let $w = \infty$ be the image of $z = \infty$.

Finally, the **inverse mapping** of (1) is obtained by solving (1) for z; we find that it is again a linear fractional transformation:

$$(4) \qquad\qquad z = \frac{dw - b}{-cw + a} .$$

When $c \neq 0$, then $-cw + a = 0$ for $w = a/c$, and we let a/c be the image of $z = \infty$.

The complex plane together with the point ∞ is called the **extended complex plane.** The complex plane without the point ∞ is often called the *finite complex plane,* for distinction, or simply the *complex plane,* as before.

With these settings, the linear fractional transformation (1) is now a one-to-one conformal mapping of the extended z-plane onto the extended w-plane. We also say that every linear fractional transformation maps "the extended complex plane in a one-to-one and conformal manner onto itself."

General Remark. If $z = \infty$, then the right side of (1) becomes the meaningless expression $(a \cdot \infty + b)/(c \cdot \infty + d)$. We assign to it the value $w = a/c$ if $c \neq 0$ and $w = \infty$ if $c = 0$.

Fixed Points

Fixed points of a mapping $w = f(z)$ are points that are mapped onto themselves, are "kept fixed" under the mapping. Thus they are obtained from

$$w = f(z) = z.$$

The **identity mapping** $w = z$ has every point as a fixed point. The mapping $w = \bar{z}$ has infinitely many fixed points, $w = 1/z$ has two, a rotation has one, and a translation none in the finite plane. (Find them in each case.) For (1), the fixed-point condition $w = z$ is

$$(5) \qquad z = \frac{az + b}{cz + d} , \qquad\qquad \text{thus} \qquad\qquad cz^2 - (a - d)z - b = 0.$$

This is a quadratic equation in z whose coefficients all vanish if and only if the mapping is the identity mapping $w = z$ (in this case, $a = d \neq 0$, $b = c = 0$). Hence we have the following theorem.

THEOREM 1   **(Fixed points)**

A linear fractional transformation, not the identity, has at most two fixed points. If a linear fractional transformation is known to have three or more fixed points, it must be the identity mapping $w = z$.

How to Find Linear Fractional Transformations

A mapping (1) is determined by a, b, c, d, actually by the ratios of three of these constants to the fourth because we can drop or introduce a common factor. This makes it plausible that three conditions determine a unique mapping (1):

THEOREM 2   **(Three points and their images given)**

Three given distinct points z_1, z_2, z_3 can always be mapped onto three prescribed distinct points w_1, w_2, w_3 by one, and only one, linear fractional transformation $w = f(z)$. This mapping is given implicitly by the equation

(6)
$$\frac{w - w_1}{w - w_3} \cdot \frac{w_2 - w_3}{w_2 - w_1} = \frac{z - z_1}{z - z_3} \cdot \frac{z_2 - z_3}{z_2 - z_1}.$$

(If one of these points is the point ∞, the quotient of the two differences containing this point must be replaced by 1.)

PROOF.   Equation (6) is of the form $F(w) = G(z)$ with linear fractional F and G. Hence $w = F^{-1}(G(z)) = f(z)$, where F^{-1} is the inverse of F and is linear fractional [see (4)], and so is the composite $F^{-1}(G)$ (by Team Project 6), that is, $w = f(z)$ is linear fractional. Now in (6), set $w = w_1$, then $w = w_2$, then $w = w_3$ on the left and $z = z_1, z_2, z_3$ on the right, to see that

$$F(w_1) = 0, \qquad F(w_2) = 1, \qquad F(w_3) = \infty$$

$$G(z_1) = 0, \qquad G(z_2) = 1, \qquad G(z_3) = \infty.$$

From the first column, $F(w_1) = G(z_1)$, thus $w_1 = F^{-1}(G(z_1)) = f(z_1)$. Similarly, $w_2 = f(z_2)$, $w_3 = f(z_3)$. This proves the existence of the desired linear fractional transformation.

To prove uniqueness, let $w = g(z)$ be a linear fractional transformation, which also maps z_j onto w_j, $j = 1, 2, 3$. Thus $w_j = g(z_j)$. Hence $g^{-1}(w_j) = z_j$, where $w_j = f(z_j)$. Together, $g^{-1}(f(z_j)) = z_j$, a mapping with the three fixed points z_1, z_2, z_3. By Theorem 1, this is the identity mapping, $g^{-1}(f(z)) = z$ for all z. Thus $f(z) = g(z)$ for all z, the uniqueness.

The last statement of Theorem 2 follows from the preceding General Remark.   ◀

Mapping of Standard Domains by Theorem 2

Principle. Prescribe three boundary points z_1, z_2, z_3 of the domain D in the z-plane. Choose their images w_1, w_2, w_3 on the boundary of the image D^* of D in the w-plane. Obtain the mapping from (6). Make sure that D is mapped onto D^*, not on its complement. In the latter case, interchange two w-points. (Why does this help?)

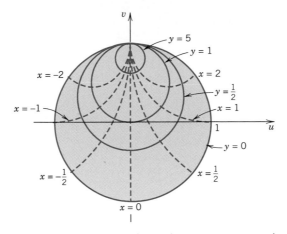

Fig. 319. Linear fractional transformation in Example 1

EXAMPLE 1   Mapping of a half-plane onto a disk

According to the principle, we get such a mapping if we let $z_1 = -1$, $z_2 = 0$, $z_3 = 1$ correspond to, say, $w_1 = -1$, $w_2 = -i$, $w_3 = 1$, respectively. Then (6) gives

$$\frac{w - (-1)}{w - 1} \cdot \frac{-i - 1}{-i - (-1)} = \frac{z - (-1)}{z - 1} \cdot \frac{0 - 1}{0 - (-1)}$$

and by solving for w

$$w = \frac{z - i}{-iz + 1}. \tag{Fig. 319}$$

Figure 319 can be verified without much calculation. $z = x$ gives $|w| = |(x - i)/(-ix + 1)| = 1$. Hence the x-axis maps onto the unit circle. Since $z = i$ gives $w = 0$, the upper half-plane maps onto the interior of the unit circle (and the lower half-plane onto the exterior). $z = 0$, i, ∞ go onto $w = -i$, 0, i, respectively, so that the positive imaginary axis maps onto the vertical diameter (the segment) $-1 \leq v \leq 1$ on the imaginary axis. It can be shown that *every linear fractional transformation maps circles and straight lines onto circles or straight lines.* Hence the lines $x = const$ map onto circles through $w = i$ (the image of $z = \infty$) and perpendicular to $|w| = 1$, by conformality. Similarly, the lines $y = const$ map onto circles through $w = i$ and perpendicular to those circles, again because of conformality. ◀

EXAMPLE 2   Occurrence of ∞

Determine the linear fractional transformation that maps $z_1 = 0$, $z_2 = 1$, $z_3 = \infty$ onto $w_1 = -1$, $w_2 = -i$, $w_3 = 1$, respectively.

Solution. From (6) we obtain the desired mapping

$$w = \frac{z - i}{z + i}.$$

This is sometimes called the *Cayley transformation.*[12] In this case, (6) gave at first the quotient $(1 - \infty)/(z - \infty)$, which we had to replace by 1. ◀

EXAMPLE 3   Mapping of the unit disk onto the right half-plane

Find the linear fractional transformation that maps $z_1 = -1$, $z_2 = i$, $z_3 = 1$ onto $w_1 = 0$, $w_2 = i$, $w_3 = \infty$, respectively. (Make a sketch of the disk and the half-plane. Solution on the next page.)

---

[12]ARTHUR CAYLEY (1821–1895), English mathematician and professor at Cambridge, is known for his important work in algebra, matrix theory, and differential equations.

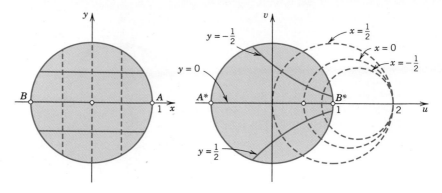

Fig. 320.   Mapping in Example 5

Solution.   From (6) we obtain, after replacing $(i - \infty)/(w - \infty)$ by 1,

$$w = -\frac{z + 1}{z - 1}.$$ ◀

EXAMPLE 4   **Mapping of a half-plane onto a half-plane**

Find the linear fractional transformation that maps the points $z_1 = -2$, $z_2 = 0$, $z_3 = 2$ onto the points $w_1 = \infty$, $w_2 = \frac{1}{2}$, $w_3 = \frac{3}{4}$, respectively.

Solution.   From (6) we obtain

$$w = \frac{z + 1}{z + 2}.$$ ◀

EXAMPLE 5   **Mapping of the unit disk onto the unit disk**

We claim that

(7)
$$w = \frac{z - z_0}{cz - 1}, \qquad c = \bar{z}_0, \qquad |z_0| < 1$$

maps $|z| \leqq 1$ onto $|w| \leqq 1$, with z_0 being mapped onto 0. To see this we take $|z| = 1$ and calculate, writing $c = \bar{z}_0$,

$$|z - z_0| = |\bar{z} - c| = |z| \, |\bar{z} - c| = |z\bar{z} - cz| = |1 - cz| = |cz - 1|.$$

Hence $|w| = |z - z_0|/|cz - 1| = 1$ from (7), so that $|z| = 1$ maps onto $|w| = 1$, as claimed, with z_0 going onto 0, as the numerator in (7) shows. For instance, if we take $z_0 = \frac{1}{2}$, we get (verify!)

$$w = \frac{2z - 1}{z - 2}$$ (Fig. 320). ◀

EXAMPLE 6   **Mapping of an angular region onto the unit disk**

Certain mapping problems can be solved by combining linear fractional transformations with others. For instance, to map the angular region $D: -\pi/6 \leqq \arg z \leqq \pi/6$ (Fig. 321) onto the unit disk $|w| \leqq 1$, we may map D by $Z = z^3$ onto the right Z-half-plane and then the latter onto the disk $|w| \leqq 1$ by

$$w = i\frac{Z - 1}{Z + 1}, \qquad \text{combined} \qquad w = i\frac{z^3 - 1}{z^3 + 1}.$$ ◀

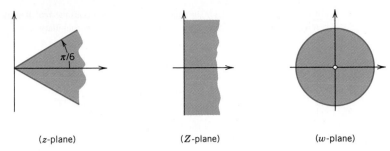

Fig. 321. Mapping in Example 6

EXAMPLE 7   **Mapping of an infinite strip onto a disk by $w = \tan z$**

This is another kind of problem in which linear fractional transformations play an auxiliary role in combination with other mappings. The trick is to write $w = \tan z$ as (see Sec. 12.7)

$$w = \tan z = \frac{\sin z}{\cos z} = \frac{(e^{iz} - e^{-iz})/i}{e^{iz} + e^{-iz}} = \frac{(e^{2iz} - 1)/i}{e^{2iz} + 1}.$$

Hence if we set $Z = e^{2iz}$ and use $1/i = -i$, we have

(8) $$w = \tan z = -iW, \qquad W = \frac{Z - 1}{Z + 1}, \qquad Z = e^{2iz}.$$

We now see that $w = \tan z$ is a linear fractional transformation preceded by an exponential mapping (see Sec. 12.6) and followed by a clockwise rotation through an angle $\frac{1}{2}\pi$ (90°).

For instance, let us map the infinite vertical strip $S: -\pi/4 < x < \pi/4$ in Fig. 322 by $w = \tan z$. We claim that the image will be the unit disk $|w| < 1$ in Fig. 322. Since $Z = e^{2iz} = e^{-2y + 2ix}$, we see from (10), Sec. 12.6, that $|Z| = e^{-2y}$, Arg $Z = 2x$. Hence the vertical lines $x = -\pi/4$, 0, $\pi/4$ are mapped onto the rays Arg $Z = -\pi/2$, 0, $\pi/2$, respectively. Hence S is mapped onto the right Z-half-plane. Also $|Z| = e^{-2y} < 1$ if $y > 0$ and $|Z| > 1$ if $y < 0$. Hence the upper half of S is mapped inside the unit circle $|Z| = 1$ and the lower half of S outside $|Z| = 1$ (Fig. 322).

Now comes the linear fractional transformation in (8), which we denote by $g(Z)$:

(9) $$W = g(Z) = \frac{Z - 1}{Z + 1}.$$

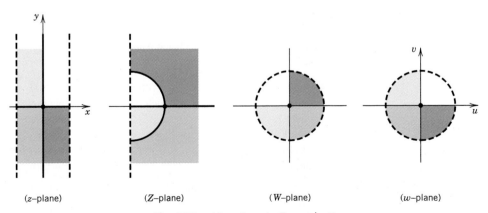

Fig. 322. Mappings in Example 7

For real Z this is real. Hence the real Z-axis is mapped onto the real W-axis. Furthermore, the imaginary Z-axis is mapped onto the unit circle $|W| = 1$ because for pure imaginary $Z = iY$ we get from (9)

$$|W| = |g(iY)| = \left| \frac{iY - 1}{iY + 1} \right| = 1.$$

The right Z-half-plane is mapped inside this unit circle $|W| = 1$, not outside, because $Z = 1$ has its image $g(1) = 0$ inside that circle. Finally, the unit circle $|Z| = 1$ is mapped onto the imaginary W-axis, because this circle is $Z = e^{i\phi}$, so that (9) gives a pure imaginary expression, namely,

$$g(e^{i\phi}) = \frac{e^{i\phi} - 1}{e^{i\phi} + 1} = \frac{e^{i\phi/2} - e^{-i\phi/2}}{e^{i\phi/2} + e^{-i\phi/2}} = \frac{i \sin (\phi/2)}{\cos (\phi/2)} .$$

From the W-plane we get to the w-plane simply by a clockwise rotation through $\pi/2$; see (8).

   Ans. $w = \tan z$ maps S: $-\pi/4 < \operatorname{Re} z < \pi/4$ onto the unit disk $|w| = 1$, with the four quarters of S mapped as indicated in Fig. 322. This mapping is conformal and one-to-one.   ◄

PROBLEM SET 12.9

Derivations in the Text

1. Derive (4) from (1). Check (4) by substitution into (1).
2. Find the inverse of the mapping in Example 1 and show that it maps the lines $x = c = const$ onto circles with centers on the line $v = 1$.
3. Derive the mapping in Example 2 from (6).
4. Derive the mapping in Example 4 from (6). Find its inverse and show that it has the same fixed points as the mapping itself. Is this surprising?
5. Verify the last formula in Example 5.
6. **TEAM PROJECT. Properties of Linear Fractional Transformations (LFTs).**
   (a) Composition. Show that substituting an LFT into an LFT gives an LFT.
   (b) Inversion. Show that $w = 1/z$ maps every straight line or circle onto a circle or straight line.
   (c) Decomposition. Show that (1) can be written as a composite of special cases (3),

$$w = K \frac{1}{cz + d} + \frac{a}{c} \quad \text{where} \quad K = -\frac{ad - bc}{c} .$$

   (d) General Case. Show that the statement in (a) holds for *every* LFT. (Use (c).)
   (e) Composition of LFTs with Other Mappings. Find an analytic function that maps the second quadrant of the z-plane onto the unit disk in the w-plane.

Determination of Linear Fractional Transformations (LFTs)
Find the LFT that maps three given points onto three given points in the respective order.

7. $-1, 0, 1$ onto $1, -1, \infty$ 8. $i, 0, 1$ onto $2 + i, 2, 3$
9. $\infty, 1, 0$ onto $0, 1, \infty$ 10. $0, 1, 2,$ onto $1, \frac{1}{2}, \frac{1}{3}$
11. $0, i, \infty$ onto $0, \frac{1}{2}, \infty$ 12. $0, 1, i$ onto $0, 1, \frac{1}{5}(2 + i)$
13. $i, -i, 0$ onto $0, \infty, -1$ 14. $2i, -2, -2i$ onto $-2, -2i, 2$

15. Find a linear fractional transformation that maps $|z| \leqq 1$ onto $|w| \leqq 1$ such that $z = i/2$ is mapped onto $w = 0$ and sketch the images of the lines $x = const$ and $y = const$.
16. Find all linear fractional transformations $w(z)$ that map the x-axis onto the u-axis.

17. **(Matrices)** Students familiar with 2×2 matrices should prove that the coefficient matrices of (1) and (4) are inverses of each other, provided $ad - bc = 1$, and that the composition of linear fractional transformations by substitution corresponds to the multiplication of the corresponding coefficient matrices.

18. **(Fixed points)** Find all linear fractional transformations whose (only) fixed points are -1 and 1.

19. **(Fixed points)** Find all linear fractional transformations without fixed points in the finite plane.

20. **CAS PROJECT. Linear Fractional Transformations (LFTs).** **(a)** Plot typical regions (squares, disks, etc.) and their images under the LFTs in Examples 1–5.

   (b) Make an experimental study of the continuous dependence of LFTs on the coefficients. For instance, change the LFT in Example 4 continuously and plot the changing image of a fixed region.

12.10 Riemann Surfaces. *Optional*

Riemann surfaces are surfaces on which multivalued relations, such as $w = \sqrt{z}$ or $w = \ln z$, become single-valued, that is, functions in the usual sense. We explain the idea, which is simple—but ingenious!

The mapping given by

(1) $$w = u + iv = z^2$$ (Sec. 12.5)

is conformal, except at $z = 0$, where $w' = 2z = 0$. At $z = 0$, angles are doubled under the mapping. Thus the right z-half-plane (including the positive y-axis) is mapped onto the full w-plane, cut along the negative half of the u-axis; this mapping is one-to-one. Similarly for the left z-half-plane (including the negative y-axis). Hence the image of the full z-plane under $w = z^2$ "covers the w-plane twice" in the sense that every $w \neq 0$ is the image of two z-points; if z_1 is one, the other is $-z_1$. For example, $z = i$ and $-i$ are both mapped onto $w = -1$.

Now comes the crucial idea. We place those two copies of the cut w-plane upon each other so that the upper sheet is the image of the right half z-plane R and the lower sheet is the image of the left half z-plane L. We join the two sheets crosswise along the cuts (along the negative u-axis) so that if z moves from R to L, its image can move from the upper to the lower sheet. The two origins are fastened together because $w = 0$ is the image of just one z-point, $z = 0$. The surface obtained is called a **Riemann surface** (Fig. 323a). $w = 0$ is called a "winding point" or **branch point.** $w = z^2$ maps the full z-plane onto this surface in a one-to-one manner.

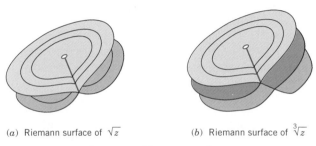

(a) Riemann surface of $\sqrt{z}$ (b) Riemann surface of $\sqrt[3]{z}$

Fig. 323. Riemann surfaces

By interchanging the roles of the variables z and w it follows that the double-valued relation

$$w = \sqrt{z} \tag{2} \quad \text{(Sec. 12.2)}$$

becomes single-valued on the Riemann surface in Fig. 323a, that is, a function in the usual sense. We can let the upper sheet correspond to the principal value of $\sqrt{z}$. Its image is the right w-half-plane. The other sheet is then mapped onto the left w-half-plane.

Similarly, the triple-valued relation $w = \sqrt[3]{z}$ becomes single-valued on the three-sheeted Riemann surface in Fig. 323b, which also has a branch point at $z = 0$.

The infinitely many-valued natural logarithm

$$w = \ln z = \operatorname{Ln} z + 2n\pi i \qquad (n = 0, \pm 1, \pm 2, \cdots)$$

becomes single-valued on a Riemann surface consisting of infinitely many sheets. $w = \operatorname{Ln} z$ corresponds to one of them. This sheet is cut along the negative x-axis and the upper edge of the slit is joined to the lower edge of the next sheet, which corresponds to the argument $\pi < \theta \leqq 3\pi$, that is, to

$$w = \operatorname{Ln} z + 2\pi i.$$

The principal value $\operatorname{Ln} z$ maps its sheet onto the horizontal strip $-\pi < v \leqq \pi$. The function $w = \operatorname{Ln} z + 2\pi i$ maps its sheet onto the neighboring strip $\pi < v \leqq 3\pi$, and so on. The mapping of the points $z \neq 0$ of the Riemann surface onto the points of the w-plane is one-to-one.

PROBLEM SET 12.10

1. Consider $w = \sqrt{z}$. Find the path of the image point w of a point z that moves twice around the unit circle, starting from the initial position $z = 1$.
2. Show that the Riemann surface of $w = \sqrt[n]{z}$ consists of n sheets and has a branch point at $z = 0$.
3. Make a sketch, similar to Fig. 323, of the Riemann surface of $\sqrt[4]{z}$.
4. Consider the Riemann surfaces of $w = \sqrt[4]{z}$ and $w = \sqrt[5]{z}$ in a fashion similar to that in Prob. 1.
5. Determine the path of the image of a point z under the mapping $w = \ln z$ as z moves several times counterclockwise around the unit circle.
6. Show that the Riemann surface of $w = \sqrt{(z-1)(z-2)}$ has branch points at $z = 1$ and $z = 2$ and consists of two sheets that may be cut along the line segment from 1 to 2 and joined crosswise. *Hint.* Introduce polar coordinates $z - 1 = r_1 e^{i\theta_1}$, $z - 2 = r_2 e^{i\theta_2}$.
7. Find the branch points and the number of sheets of the Riemann surface of $w = \sqrt{z^2 - 1}$.
8. Show that the Riemann surface of $w = \sqrt{(1 - z^2)(4 - z^2)}$ has four branch points and two sheets that may be joined crosswise along the segments $-2 \leqq x \leqq -1$ and $1 \leqq x \leqq 2$ of the x-axis.
9. Where does the Riemann surface of $w = 3 + \sqrt[3]{2z + i}$ have a branch point? How many sheets does it have?
10. Answer the questions in Prob. 9 for $\sqrt{z^4 - 1}$.

CHAPTER 12 REVIEW

1. Add, subtract, multiply, and divide $22 + 7i$ and $3 - 2i$.
2. Write the two numbers in Prob. 1 in polar form. Find the principal values of their arguments.
3. The definition of the derivative of a complex function looks similar to that in calculus, but there is a big difference in principle. Explain.
4. State the Cauchy–Riemann equations. What are they good for?
5. What is an analytic function?
6. What does conformal mapping mean? What is its role in connection with analytic functions? How is the angle between curves defined?
7. State the definitions of the functions e^z, $\cos z$, etc. discussed in this chapter. How are these functions related to each other?
8. $\ln z$ is more complicated than $\ln x$. Explain (from memory!).
9. How is a general power z^c defined? Give examples.
10. Why can two-dimensional potential problems be handled by methods of complex analysis?
11. What are linear fractional transformations? State some of their general properties.
12. What is the extended complex plane? How did we motivate it?
13. What is a fixed point of a mapping? Why did fixed points occur in this chapter? Give examples.
14. Does the equation $\sin z = -10$ have a solution? More than one?
15. Interpret the mappings $w = iz$, $w = -z$, $w = az$ (a real), $w = z + c$, $w = 1/z$ geometrically.

Complex Numbers. Find, in the form $x + iy$,
16. $(5 - 7i)^2$ 17. $(1 + i)^8$ 18. $\sqrt{5 - 7i}$ 19. $e^{i\pi/2}$, $e^{-i\pi/2}$ 20. $(32 + 2i)/(17 - 15i)$

Polar Form. Represent in polar form, with the principal argument,
21. $-4 + 4i$ 22. $12 + i$ 23. $-25i$ 24. -7.3 25. $(2.60 + 0.38i)^2$

Roots. Find and plot all values of
26. $\sqrt[4]{81}$ 27. $\sqrt{3 - 4i}$ 28. $\sqrt[4]{-1}$ 29. $\sqrt{-16 - 12i}$ 30. $\sqrt{-32i}$

Analytic Functions. Find $f(z) = u(x, y) + iv(x, y)$ with u or v as given. Check for analyticity by the Cauchy–Riemann equations.
31. $u = x^3 - 3xy$
32. $u = (x^2 - y^2)/(x^2 + y^2)^2$
33. $v = 2y(-1 + x)$
34. $v = \cos 2x \sinh 2y$

Special Functions. Find the value of
35. $\cosh 4\pi i$ 36. e^{4+2i} 37. $\text{Ln}\,(5 - 2i)$ 38. $\sin(4\pi - \pi i/2)$ 39. $\tan(1 + i)$

Conformal Mapping. Find and sketch the image of the given region under $w = u + iv = f(z)$.
40. $|z| < \frac{1}{3}$, $-\frac{1}{4}\pi < \text{Arg}\,z < \frac{1}{4}\pi$; $w = z^2$ 41. $\text{Re}\,z > 0$; $w = 1/z$
42. $0 < \text{Arg}\,z < \frac{1}{2}\pi$; $w = z^3$ 43. $-1 < x < 1$, $-1 < y < 1$; $w = e^z$
44. $1 < x < 2$, $\frac{1}{2}\pi < y < \pi$; $w = e^z$ 45. $2 < |z| < 4$, $0 < \text{Arg}\,z < \frac{1}{2}\pi$; $w = \text{Ln}\,z$
46. At what points is the mapping by $w = \cos(z^2 + 1)$ not conformal?
47. Why do the images of the curves $|z| = const$ and $\text{Arg}\,z = const$ under a mapping by an analytic function intersect at right angles?
48. Find the linear fractional transformation that maps 0, i, $-i$ onto $2i$, ∞, $\frac{1}{2} + i$, respectively.
49. List special linear fractional transformations that have a simple geometrical meaning.
50. Find all linear fractional transformations whose only fixed points are $-i$ and i.

SUMMARY OF CHAPTER 12
COMPLEX NUMBERS AND FUNCTIONS
CONFORMAL MAPPING

For arithmetic operations with **complex numbers**

(1) $$z = x + iy = re^{i\theta} = r(\cos\theta + i\sin\theta),$$

$r = |z| = \sqrt{x^2 + y^2}$, $\theta = $ arc tan (y/x), and for their representation in the complex plane, see Secs. 12.1 and 12.2.

A complex function $f(z) = u(x, y) + iv(x, y)$ is **analytic** in a domain D if it has a **derivative** (Sec. 12.3)

(2) $$f'(z) = \lim_{\Delta z \to 0} \frac{f(z + \Delta z) - f(z)}{\Delta z}$$

everywhere in D. Also, $f(z)$ is *analytic at a point* $z = z_0$ if it has a derivative in a neighborhood of z_0 (not merely at z_0 itself).

If $f(z)$ is analytic in D, then $u(x, y)$ and $v(x, y)$ satisfy the (very important!) **Cauchy–Riemann equations** (Sec. 12.4)

(3) $$\frac{\partial u}{\partial x} = \frac{\partial v}{\partial y}, \qquad \frac{\partial u}{\partial y} = -\frac{\partial v}{\partial x}$$

everywhere in D. Then u and v also satisfy **Laplace's equation**

(4) $$u_{xx} + u_{yy} = 0, \qquad v_{xx} + v_{yy} = 0$$

everywhere in D. If $u(x, y)$ and $v(x, y)$ are continuous and have *continuous* partial derivatives in D that satisfy (3) in D, then $f(z) = u(x, y) + iv(x, y)$ is analytic in D. See Sec. 12.4. (More on Laplace's equation and complex analysis follows in Chap. 16.)

The complex **exponential function** (Sec. 12.6)

(5) $$e^z = \exp z = e^x (\cos y + i\sin y)$$

reduces to e^x if $z = x$ $(y = 0)$. It is periodic with $2\pi i$ and has the derivative e^z.

The **trigonometric functions** are (Sec. 12.7)

(6)
$$\cos z = \frac{1}{2}(e^{iz} + e^{-iz}) = \cos x \cosh y - i\sin x \sinh y$$

$$\sin z = \frac{1}{2i}(e^{iz} - e^{-iz}) = \sin x \cosh y + i\cos x \sinh y$$

$\tan z = (\sin z)/\cos z$, $\cot z = 1/\tan z$, etc.

The **hyperbolic functions** are (Sec. 12.7)

$$\cosh z = \frac{1}{2}(e^z + e^{-z}) = \cos iz,$$

(7)

$$\sinh z = \frac{1}{2}(e^z - e^{-z}) = -i \sin iz,$$

etc. An **entire function** is a function that is analytic everywhere in the complex plane. The functions in (5)–(7) are entire.

The **natural logarithm** is (Sec. 12.8)

(8)
$$\ln z = \ln |z| + i \arg z \qquad\qquad (\arg z = \theta, \, z \neq 0)$$

$$= \ln |z| + i \operatorname{Arg} z \pm 2n\pi i \qquad (n = 0, 1, \cdots),$$

where Arg z is the **principal value** of arg z, that is, $-\pi < \operatorname{Arg} z \leqq \pi$. We see that ln z is infinitely many-valued. Taking $n = 0$ gives the **principal value** Ln z of ln z; thus

(8*)
$$\operatorname{Ln} z = \ln |z| + i \operatorname{Arg} z.$$

General powers are defined by (Sec. 12.8)

(9)
$$z^c = e^{c \ln z} \qquad\qquad (c \text{ complex}, \, z \neq 0).$$

A complex function $w = f(z)$ gives a **mapping** of its domain of definition in the complex z-plane onto its range of values in the complex w-plane. If $f(z)$ is *analytic,* this mapping is **conformal,** that is, angle-preserving: the images of any two intersecting curves make the same angle of intersection, in both magnitude and sense, as the curves themselves (Sec. 12.5). Conformality fails at points where the derivative $f'(z)$ is zero. Mappings by e^z, $\cos z$, etc. are discussed in the corresponding Secs. 12.6–12.8. For tan z and for **linear fractional transformations**

$$w = \frac{az + b}{cz + d} \qquad\qquad (ad - bc \neq 0)$$

see Sec. 12.9.

Riemann surfaces are mentioned in Sec. 12.10.

Complex Integration

Integration in the complex plane is important for two reasons:

1. In applications there occur real integrals that can be evaluated by complex integration, whereas the usual methods of real integral calculus fail.
2. Some basic properties of analytic functions can be established by complex integration, but would be difficult to prove by other methods. The existence of higher derivatives of analytic functions is a striking property of this type.[1]

In this chapter we define and explain complex integrals. The most important result in the whole chapter is **Cauchy's integral theorem** (Sec. 13.2). It implies the useful Cauchy integral formula (Sec. 13.3). In Sec. 13.4 we prove that if a function is analytic, it has derivatives of all orders. Hence in this respect, complex analytic functions behave much more simply than real-valued functions of real variables.

(Integration by means of residues and applications to real integrals will be considered in Chap. 15.)

Prerequisite for this chapter: Chap. 12.
References: Appendix 1, Part D.
Answers to problems: Appendix 2.

13.1 Line Integral in the Complex Plane

As in calculus we distinguish between definite integrals and indefinite integrals or antiderivatives. An **indefinite integral** is a function whose derivative equals a given analytic function in a region. By inverting known differentiation formulas we may find many types of indefinite integrals.

Complex definite integrals are called (complex) **line integrals.** They are written

$$\int_C f(z)\, dz.$$

---

[1]Proved without integration or equivalent methods only relatively recently, in 1961 [by P. Porcelli and E. H. Connell (*Bulletin of the American Mathematical Society,* vol. 67, pp. 177–181), who make use of a topological theorem by G. T. Whyburn].

Here the **integrand** $f(z)$ is integrated over a given curve C in the complex plane,[2] called the **path of integration.** We may represent such a curve C by a parametric representation (see also Sec. 12.5)

$$(1) \qquad \boxed{z(t) = x(t) + iy(t)} \qquad (a \leqq t \leqq b).$$

The sense of increasing t is called the **positive sense** on C, and we say that in this way, (1) **orients** C.

We assume C to be a **smooth curve,** that is, C has a continuous and nonzero derivative $\dot{z} = dz/dt$ (see Fig. 309, Sec. 12.5) at each point. Geometrically this means that C has a unique and continuously turning tangent.

Definition of the Complex Line Integral

This is similar to the method in calculus. Let C be a smooth curve in the complex plane given by (1), and let $f(z)$ be a continuous function given (at least) at each point of C. We now subdivide (we *"partition"*) the interval $a \leqq t \leqq b$ in (1) by points

$$t_0 \, (= a), \quad t_1, \quad \cdots, \quad t_{n-1}, \quad t_n \, (= b)$$

where $t_0 < t_1 \cdots < t_n$. To this subdivision there corresponds a subdivision of C by points

$$z_0, \quad z_1, \quad \cdots, \quad z_{n-1}, \quad z_n \, (= Z) \qquad \text{(Fig. 324),}$$

where $z_j = z(t_j)$. On each portion of subdivision of C we choose an arbitrary point, say, a point ζ_1 between z_0 and z_1 (that is, $\zeta_1 = z(t)$ where t satisfies $t_0 \leqq t \leqq t_1$), a point ζ_2 between z_1 and z_2, etc. Then we form the sum

$$(2) \qquad S_n = \sum_{m=1}^{n} f(\zeta_m)\, \Delta z_m \qquad \text{where} \qquad \Delta z_m = z_m - z_{m-1}.$$

We do this for each $n = 2, 3, \cdots$ in a completely independent manner, but so that the greatest $|\Delta t_m| = |t_m - t_{m-1}|$ approaches zero as $n \to \infty$. This implies that the greatest $|\Delta z_m|$ also approaches zero because it cannot exceed the length of the arc of C from z_{m-1} to z_m and the latter goes to zero since the arc length of the smooth curve C is a continuous function of t. The limit of the sequence of complex numbers $S_2, S_3, \cdots$ thus obtained is called the **line integral** (or simply the *integral*) of $f(z)$ over the oriented curve C. This

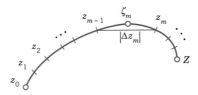

Fig. 324. Complex line integral

---

[2]Actually, often only along a portion (an *arc*) of a curve, but we shall say **"curve"** in either case, for simplicity. Readers who have studied Sec. 9.1 will note that our present discussion is similar.

curve C is called the **path of integration.** The line integral is denoted by

(3) $$\int_C f(z)\,dz, \qquad \text{or by} \qquad \oint_C f(z)\,dz$$

if C is a **closed path** (one whose terminal point Z coincides with its initial point z_0, as for a circle or an 8-shaped curve).

General Assumption. *All paths of integration for complex line integrals are assumed to be* **piecewise smooth,** *that is, they consist of finitely many smooth curves joined end to end.*

Three Basic Properties Directly Implied by the Definition

1. Linearity. Integration is a *linear operation,* that is, we can integrate sums term by term and take out constant factors from under the integral sign,

(4) $$\int_C [k_1 f_1(z) + k_2 f_2(z)]\,dz = k_1 \int_C f_1(z)\,dz + k_2 \int_C f_2(z)\,dz.$$

2. Sense reversal in integrating over the *same* path, from z_0 to Z (left) and from Z to z_0 (right),

(5) $$\int_{z_0}^{Z} f(z)\,dz = -\int_{Z}^{z_0} f(z)\,dz.$$

3. Partitioning of path (see Fig. 325),

(6) $$\int_C f(z)\,dz = \int_{C_1} f(z)\,dz + \int_{C_2} f(z)\,dz.$$

Fig. 325.   Partitioning of path [formula (6)]

Existence of the Complex Line Integral[3]

From our assumptions that $f(z)$ is continuous and C is piecewise smooth, the existence of the line integral (3) follows. In fact, as in the preceding chapter let us write $f(z) = u(x, y) + iv(x, y)$. We also set

$$\zeta_m = \xi_m + i\eta_m \qquad \text{and} \qquad \Delta z_m = \Delta x_m + i\Delta y_m.$$

---

[3]If one does not want to refer to real line integrals, one can take (10) as the *definition* of the complex line integral.

Then (2) may be written

(7)
$$S_n = \sum (u + iv)(\Delta x_m + i\Delta y_m)$$

where $u = u(\xi_m, \eta_m)$, $v = v(\xi_m, \eta_m)$ and we sum over m from 1 to n. Performing the multiplication, we may now split up S_n into four sums:

$$S_n = \sum u \, \Delta x_m - \sum v \, \Delta y_m + i \left[\sum u \, \Delta y_m + \sum v \, \Delta x_m \right].$$

These sums are real. Since f is continuous, u and v are continuous. Hence, if we let n approach infinity in the aforementioned way, then the greatest Δx_m and Δy_m will approach zero and each sum on the right becomes a real line integral:

(8)
$$\lim_{n \to \infty} S_n - \int_C f(z) \, dz = \int_C u \, dx - \int_C v \, dy + i \left[\int_C u \, dy + \int_C v \, dx \right].$$

This shows that under our assumptions on f and C the line integral (3) exists and its value is independent of the choice of subdivisions and intermediate points ζ_m. ◄

Complex integration is rich in methods for evaluating integrals. We discuss the first two of them, and others follow later in this chapter and in Chap. 15.

First Method: Indefinite Integration and Substitution of Limits

This method is simpler than the next one, but is less general. It is restricted to *analytic* functions. Its formula (9) (below) is the analog of the familiar formula from calculus

$$\int_a^b f(x) \, dx = F(b) - F(a) \qquad\qquad [F'(x) = f(x)].$$

THEOREM 1   **(Indefinite integration of analytic functions)**

Let $f(z)$ be analytic in a simply connected[4] domain D. Then there exists an indefinite integral of $f(z)$ in the domain D, that is, an analytic function $F(z)$ such that $F'(z) = f(z)$ in D, and for all paths in D joining two points z_0 and z_1 in D we have

(9)
$$\int_{z_0}^{z_1} f(z) \, dz = F(z_1) - F(z_0) \qquad\qquad [F'(z) = f(z)].$$

(*Note that we can write z_0 and z_1 instead of C, since we get the same value for all those C from z_0 to z_1.*)

This theorem will be proved in the next section.

---

[4]D is called **simply connected** if every simple closed curve (closed curve without self-intersections) in D encloses only points of D.

Simple connectedness is quite essential in Theorem 1, as we shall see in Example 5.

Since analytic functions are our main concern, and since differentiation formulas will often help in finding $F(z)$ for a given $f(z) = F'(z)$, the present method is of great practical interest.

If $f(z)$ is entire (Sec. 12.6), we can take for D the complex plane (which is certainly simply connected).

EXAMPLE 1 $\displaystyle\int_0^{1+i} z^2\, dz = \frac{1}{3} z^3 \Big|_0^{1+i} = \frac{1}{3}(1 + i)^3 = -\frac{2}{3} + \frac{2}{3} i$ ◀

EXAMPLE 2 $\displaystyle\int_{-\pi i}^{\pi i} \cos z\, dz = \sin z \Big|_{-\pi i}^{\pi i} = 2\sin \pi i = 2i \sinh \pi = 23.097i$ ◀

EXAMPLE 3 $\displaystyle\int_{8+\pi i}^{8-3\pi i} e^{z/2}\, dz = 2e^{z/2} \Big|_{8+\pi i}^{8-3\pi i} = 2(e^{4-3\pi i/2} - e^{4+\pi i/2}) = 0$

since e^z is periodic with period $2\pi i$.   ◀

EXAMPLE 4 $\displaystyle\int_{-i}^{i} \frac{dz}{z} = \operatorname{Ln} i - \operatorname{Ln}(-i) = \frac{i\pi}{2} - \left(-\frac{i\pi}{2}\right) = i\pi.$

Here D is the complex plane without 0 and the negative real axis (where $\operatorname{Ln} z$ is not analytic), obviously a simply connected domain.   ◀

Second Method: Use of a Representation of the Path

This method is not restricted to analytic functions but applies to any continuous complex function.

THEOREM 2   **(Integration by the use of the path)**

Let C be a piecewise smooth path, represented by $z = z(t)$, where $a \leqq t \leqq b$. Let $f(z)$ be a continuous function on C. Then

(10)
$$\int_C f(z)\, dz = \int_a^b f[z(t)] \dot{z}(t)\, dt \qquad \left(\dot{z} = \frac{dz}{dt}\right).$$

PROOF.   The left side of (10) is given by (8) in terms of real line integrals, and we show that the right side of (10) also equals (8). We have $z = x + iy$, hence $\dot{z} = \dot{x} + i\dot{y}$. We simply write u for $u[x(t), y(t)]$ and v for $v[x(t), y(t)]$. We also have $dx = \dot{x}\, dt$ and $dy = \dot{y}\, dt$. Consequently, in (10)

$$\int_a^b f[z(t)] \dot{z}(t)\, dt = \int_a^b (u + iv)(\dot{x} + i\dot{y})\, dt$$

$$= \int_C [u\, dx - v\, dy + i(u\, dy + v\, dx)]$$

$$= \int_C (u\, dx - v\, dy) + i \int_C (u\, dy + v\, dx).$$ ◀

Steps in applying Theorem 2

(A) Represent the path C in the form $z(t)$ ($a \le t \le b$).

(B) Calculate the derivative $\dot{z}(t) = dz/dt$.

(C) Substitute $z(t)$ for every z in $f(z)$ (hence $x(t)$ for x and $y(t)$ for y).

(D) Integrate $f[z(t)]\dot{z}(t)$ over t from a to b.

EXAMPLE 5   **A basic result: Integral of $1/z$ around the unit circle**

We show that by integrating $1/z$ counterclockwise around the unit circle (the circle of radius 1 and center 0; see Sec. 12.3) we obtain

(11)
$$\oint_C \frac{dz}{z} = 2\pi i$$
                                                                         (C the unit circle, counterclockwise).

This is a very important result that we shall need quite often.

Solution.   We may represent the unit circle C (see Fig. 300, Sec. 12.3) in the form

$$z(t) = \cos t + i \sin t = e^{it} \qquad (0 \le t \le 2\pi),$$

so that the counterclockwise integration corresponds to an increase of t from 0 to 2π. By differentiation, $\dot{z}(t) = ie^{it}$ (chain rule!) and with $f(z(t)) = 1/z(t) = e^{-it}$ we get from (10) the result

$$\oint_C \frac{dz}{z} = \int_0^{2\pi} e^{-it} i e^{it}\, dt = i \int_0^{2\pi} dt = 2\pi i.$$

Check this result by using $z(t) = \cos t + i \sin t$.

Simple connectedness is essential in Theorem 1. Equation (9) in Theorem 1 gives 0 for any closed path because then $z_1 = z_0$, so that $F(z_1) - F(z_0) = 0$. Now $1/z$ is not analytic at $z = 0$. But any *simply connected* domain containing the unit circle must contain $z = 0$, so that Theorem 1 does not apply—it is not enough that $1/z$ is analytic in an annulus, say, $\frac{1}{2} < |z| < \frac{3}{2}$, because an annulus is not simply connected! ◀

EXAMPLE 6   **Integral of integer powers**

Let $f(z) = (z - z_0)^m$ where m is an integer and z_0 a constant. Integrate counterclockwise around the circle C of radius ρ with center at z_0 (Fig. 326).

Solution.   We may represent C in the form

$$z(t) = z_0 + \rho(\cos t + i \sin t) = z_0 + \rho e^{it} \qquad (0 \le t \le 2\pi).$$

Then we have

$$(z - z_0)^m = \rho^m e^{imt}, \qquad dz = i\rho e^{it}\, dt$$

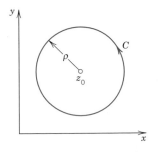

Fig. 326.   Path in Example 6

and obtain

$$\oint_C (z - z_0)^m \, dz = \int_0^{2\pi} \rho^m e^{imt} i\rho e^{it} \, dt = i\rho^{m+1} \int_0^{2\pi} e^{i(m+1)t} \, dt.$$

By the Euler formula (5) in Sec. 12.6 the right side equals

$$i\rho^{m+1} \left[\int_0^{2\pi} \cos(m+1)t \, dt + i \int_0^{2\pi} \sin(m+1)t \, dt \right].$$

If $m = -1$, we have $\rho^{m+1} = 1$, $\cos 0 = 1$, $\sin 0 = 0$. We thus obtain $2\pi i$. For integer $m \neq 1$ each of the two integrals is zero because we integrate over an interval of length 2π, equal to a period of sine and cosine. Hence the result is

(12)
$$\oint_C (z - z_0)^m \, dz = \begin{cases} 2\pi i & (m = -1), \\ 0 & (m \neq -1 \text{ and integer}). \end{cases}$$ ◀

Dependence on path. Now comes a very important fact. If we integrate a given function $f(z)$ from a point z_0 to a point z_1 along different paths, the integrals will in general have different values. In other words, *a complex line integral depends not only on the endpoints of the path but in general also on the path itself.* See the next example.

EXAMPLE 7   **Integral of a nonanalytic function. Dependence on path**

Integrate $f(z) = \text{Re } z = x$ from 0 to $1 + 2i$ (a) along C^* in Fig. 327, (b) along C consisting of C_1 and C_2.

Solution.   (a) C^* can be represented by $z(t) = t + 2it$ $(0 \leq t \leq 1)$. Hence $\dot{z}(t) = 1 + 2i$ and $f[z(t)] = x(t) = t$ on C^*. We now calculate

$$\int_{C^*} \text{Re } z \, dz = \int_0^1 t(1 + 2i) \, dt = \tfrac{1}{2}(1 + 2i) = \tfrac{1}{2} + i.$$

(b) We now have

$$C_1: z(t) = t, \qquad \dot{z}(t) = 1, \qquad f(z(t)) = x(t) = t \qquad (0 \leq t \leq 1)$$

$$C_2: z(t) = 1 + it, \qquad \dot{z}(t) = i, \qquad f(z(t)) = x(t) = 1 \qquad (0 \leq t \leq 2).$$

Using (6) we calculate

$$\int_C \text{Re } z \, dz = \int_{C_1} \text{Re } z \, dz + \int_{C_2} \text{Re } z \, dz = \int_0^1 t \, dt + \int_0^2 1 \cdot i \, dt = \tfrac{1}{2} + 2i.$$

Note that this result differs from the result in (a).   ◀

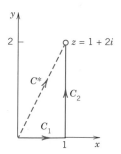

Fig. 327.   Paths in Example 7

Bound for the Absolute Value of Integrals

There will be a frequent need for estimating the absolute value of complex line integrals. The basic formula is

(13)
$$\left| \int_C f(z)\, dz \right| \leq ML$$
(*ML*-inequality);

L is the length of C and M a constant such that $|f(z)| \leq M$ everywhere on C.

PROOF. Taking the absolute value in (2) and applying the generalized triangle inequality (6) in Sec. 12.2, we obtain

$$|S_n| = \left| \sum_{m=1}^{n} f(\zeta_m)\, \Delta z_m \right| \leq \sum_{m=1}^{n} |f(\zeta_m)|\, |\Delta z_m| \leq M \sum_{m=1}^{n} |\Delta z_m|.$$

Now $|\Delta z_m|$ is the length of the chord whose endpoints are z_{m-1} and z_m (see Fig. 324). Hence the sum on the right represents the length L^* of the broken line of chords whose endpoints are $z_0, z_1, \cdots, z_n\ (= Z)$. If n approaches infinity in such a way that the greatest $|\Delta t_m|$ and thus $|\Delta z_m|$ approach zero, then L^* approaches the length L of the curve C, by the definition of the length of a curve. From this the inequality (13) follows. ◀

We cannot see from (13) how close to the bound ML the actual absolute value of the integral is, but this will be no handicap in applying (13), as we shall see. For the time being we explain the practical use of (13) by a simple example.

EXAMPLE 8   **Estimation of an integral**

Find an upper bound for the absolute value of the integral

$$\int_C z^2\, dz,$$
C the straight-line segment from 0 to $1 + i$.

Solution. $L = \sqrt{2}$ and $|f(z)| = |z^2| \leq 2$ on C gives by (13)

$$\left| \int_C z^2\, dz \right| \leq 2\sqrt{2} = 2.8284.$$

The absolute value of the integral is $\left| -\dfrac{2}{3} + \dfrac{2}{3}\, i \right| = \dfrac{2}{3}\sqrt{2} = 0.9428$ (see Example 1).   ◀

PROBLEM SET 13.1

Parametric Representations

Parametric representations are basic in the second integration method, which one must use if Theorem 1 does not apply. Find a parametric representation $z = z(t)$ for

1. The straight-line segment from 0 to $4 - 7i$
2. The straight-line segment from $4 + 3i$ to $-5 - i$
3. The upper half of $|z - 4 + 2i| = 3$
4. $4x^2 + 9y^2 = 36$, counterclockwise
5. $y = 1/x$ from $(1, 1)$ to $(4, \frac{1}{4})$
6. $y = x^3$ from $(-2, -8)$ to $(3, 27)$
7. $|z + 3 - i| = 5$, counterclockwise
8. $x^2 - 4y^2 = 4$, the branch through $(2, 0)$

Find and sketch the curves given by

9. $(3 + 4i)t$ $(0 \leqq t \leqq 2)$

10. $5i + \sqrt{3}e^{it}$ $(0 \leqq t \leqq \pi)$

11. $4 \cos t + (2 + \sin t)i$ $(0 \leqq t \leqq 2\pi)$

12. $t + 3it^4$ $(-1 \leqq t \leqq 1)$

13. $\cosh t + i \sinh t$ $(0 \leqq t \leqq 4)$

14. $t + 4i/t$ $(1 \leqq t \leqq 4)$

Integration

Use the first method or state that it does not apply and use the second method. Integrate (showing the details of your work)

15. $\int_C \operatorname{Re} z \, dz$, C the shortest path from $1 + i$ to $3 + 2i$

16. $\int_C \operatorname{Re} z \, dz$, C vertically from $1 + i$ to $1 + 2i$, then horizontally to $3 + 2i$

17. $\int_C \sin^2 z \, dz$, C from $-\pi i$ along $|z| = \pi$ to πi in the right half-plane

18. $\int_C \bar{z} \, dz$, C from 0 along the parabola $y = x^2$ to $1 + i$

19. $\int_C \operatorname{Re} z^2 \, dz$, C the unit circle, counterclockwise

20. $\int_C \operatorname{Re} z^2 \, dz$, C the boundary of the square with vertices 0, i, $1 + i$, 1, clockwise

21. $\int_C z e^{z^2} \, dz$, C from 1 along the axes to i

22. $\int_C \sinh \pi z \, dz$, C from i along the y-axis to 0

23. $\int_C \cos z \, dz$, C the semicircle $|z| = \pi$, $x \geqq 0$, from $-\pi i$ to πi

24. $\int_C e^{4z} \, dz$, C the shortest path from $8 - 3i$ to $8 - (3 + \pi)i$

25. $\int_C \sec^2 z \, dz$, C any path from $\pi i/4$ to $\pi/4$ in the unit disk

26. $\int_C \left(\dfrac{3}{z - i} - \dfrac{6}{(z - i)^2} \right) dz$, C the circle $|z - i| = 5$, clockwise

27. (**Linearity**) Illustrate (4) using an example of your choice. Prove (4).

28. (***ML*-inequality**) Using (13), find an upper bound for the absolute value of the integral in Prob. 15.

29. CAS PROJECT. Integration. Write programs for the two integration methods. Apply them to problems of your choice. Could you make them into a joint program that also decides which of the two methods to use in a given case?

30. TEAM PROJECT. Integration. (a) **Comparison.** Write a short essay comparing the essential points of the two integration methods.

(b) **Comparison.** Evaluate $\int_C f(z) \, dz$ by Theorem 1 and check the result by Theorem 2, where:

(i) $f(z) = z^4$ and C is the semicircle $|z| = 2$ from $-2i$ to $2i$ in the right half-plane,

(ii) $f(z) = e^{2z}$ and C is the shortest path from 0 to $1 + 2i$.

(c) **Continuous deformation of path.** Experiment with a family of paths with common endpoints, say, $z(t) = t + ia \sin t$, $0 \leqq t \leqq \pi$, with real parameter a. Integrate nonanalytic functions ($\operatorname{Re} z$, $\operatorname{Re} z^2$, etc.) and explore how the result depends on a. Then take analytic functions of your choice. (Show the details of your work.) Compare and comment.

(d) **Continuous deformation of path.** Choose another family, for example, semi-ellipses $z(t) = a \cos t + i \sin t$, $-\pi/2 \leqq t \leqq \pi/2$, and experiment as in (c).

13.2 Cauchy's Integral Theorem

A line integral of a function $f(z)$ depends not merely on the endpoints of the path, but also on the choice of the path itself (see Sec. 13.1). This dependence is awkward and one looks for situations when it does not occur. The answer will be simple. If $f(z)$ is analytic in a domain D and D is simply connected (definition below), then the integral will not depend on path. This result (Theorem 2) follows from Cauchy's integral theorem, along with other basic consequences that make ***Cauchy's theorem the most important theorem in this chapter*** and fundamental throughout complex analysis. To state Cauchy's integral theorem we need the following two concepts.

1. A **simple closed path** is a closed path (Sec. 13.1) that does not intersect or touch itself (Fig. 328). For example, a circle is simple, an 8-shaped curve is not.

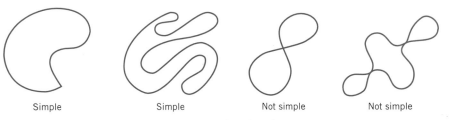

| Simple | Simple | Not simple | Not simple |

Fig. 328. Closed paths

2. A **simply connected domain** D in the complex plane is a domain (Sec. 12.3) such that every simple closed path in D encloses only points of D. *Examples:* The interior of a circle ("open disk"), ellipse, or any simple closed curve. A domain that is not simply connected is called **multiply connected.**[5] *Examples:* An annulus (Sec. 12.3), a disk without the center, for example, $0 < |z| < 1$. See also Fig. 329.

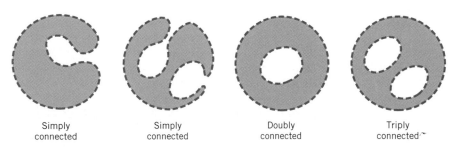

| Simply connected | Simply connected | Doubly connected | Triply connected |

Fig. 329. Simply and multiply connected domains

---

[5]A **bounded domain** D (that is, one that lies entirely in some circle about the origin) is called *p-fold connected* if its boundary consists of p closed connected sets without common points (these sets can be curves, segments, or single points, such as $z = 0$ for $0 < |z| < 1$, for which $p = 2$). Thus, D has $p - 1$ **"holes"** (where "hole" may also mean a segment or even a single point). Thus an annulus is *doubly connected* ($p = 2$).

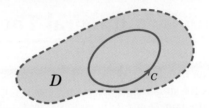

Fig. 330. Cauchy's integral theorem

THEOREM 1   **Cauchy's integral theorem**

If $f(z)$ is analytic in a simply connected domain D, then for every simple closed path C in D,

(1)
$$\oint_C f(z)\, dz = 0.$$
See Fig. 330.

Before we prove the theorem, let us consider some examples in order to really understand what is going on. A simple closed path is sometimes called a *contour* and an integral over such a path a **contour integral.** Thus, (1) and our examples involve contour integrals.

EXAMPLE 1   **No singularities (Entire functions)**

$$\oint_C e^z\, dz = 0, \qquad \oint_C \cos z\, dz = 0, \qquad \oint_C z^n\, dz = 0 \qquad (n = 0, 1, \cdots)$$

for any closed path, since these functions are entire (analytic for all z).   ◄

EXAMPLE 2   **Singularities outside contour**

$$\oint_C \sec z\, dz = 0, \qquad \oint_C \frac{dz}{z^2 + 4} = 0$$

where C is the unit circle, $\sec z = 1/\cos z$ is not analytic at $z = \pm\pi/2, \pm 3\pi/2, \cdots$, but all these points lie outside C; none lies on C or inside C. Similarly for the second integral, whose integrand is not analytic at $z = \pm 2i$ outside C.   ◄

EXAMPLE 3   **Nonanalytic function**

$$\oint_C \bar{z}\, dz = \int_0^{2\pi} e^{-it} i e^{it}\, dt = 2\pi i$$

where $C: z(t) = e^{it}$ is the unit circle. This does not contradict Cauchy's theorem because $f(z) = \bar{z}$ is not analytic.   ◄

EXAMPLE 4   **Analyticity sufficient, not necessary**

$$\oint_C \frac{dz}{z^2} = 0$$

where C is the unit circle. This result does *not* follow from Cauchy's theorem, because $f(z) = 1/z^2$ is not analytic at $z = 0$. Hence *the condition that f be analytic in D is sufficient rather than necessary for* (1) *to be true.*   ◄

EXAMPLE 5   Simple connectedness essential

$$\oint_C \frac{dz}{z} = 2\pi i$$

for counterclockwise integration around the unit circle (see Sec. 13.1). C lies in the annulus $\frac{1}{2} < |z| < \frac{3}{2}$ where $1/z$ is analytic, but this domain is not simply connected, so that Cauchy's theorem cannot be applied. Hence *the condition that the domain D be simply connected is quite essential.*

In other words, by Cauchy's theorem, if $f(z)$ is analytic on a simple closed path C and everywhere inside C, with no exception, not even a single point, then (1) holds. The point that causes trouble here is $z = 0$ where $1/z$ is not analytic.

CAUCHY'S PROOF.   Cauchy proved his integral theorem under the additional assumption that the derivative $f'(z)$ is continuous (which is true, but would need an extra proof). His proof proceeds as follows. From (8) in Sec. 13.1 we have

$$\oint_C f(z)\, dz = \oint_C (u\, dx - v\, dy) + i \oint_C (u\, dy + v\, dx).$$

Since $f(z)$ is analytic in D, its derivative $f'(z)$ exists in D. Since $f'(z)$ is assumed to be continuous, (4) and (5) in Sec. 12.4 imply that u and v have *continuous* partial derivatives in D. Hence Green's theorem (Sec. 9.4) (with u and $-v$ instead of F_1 and F_2) is applicable and gives

$$\oint_C (u\, dx - v\, dy) = \iint_R \left(-\frac{\partial v}{\partial x} - \frac{\partial u}{\partial y} \right) dx\, dy$$

where R is the region bounded by C. The second Cauchy–Riemann equation (Sec. 12.4) shows that the integrand on the right is identically zero. Hence the integral on the left is zero. In the same fashion it follows by the use of the first Cauchy–Riemann equation that the last integral in the above formula is zero. This completes Cauchy's proof.   ◀

Goursat's proof without the condition that $f'(z)$ is continuous[6] is substantially more complicated. We leave it optional and include it in Appendix 4.

Independence of Path

We know from the preceding section that the value of a line integral of a given function $f(z)$ from a point z_1 to a point z_2 will in general depend on the path C over which we integrate, not merely on z_1 and z_2. It is important to characterize situations in which this difficulty of path dependence does not occur. This task suggests the following concept. We call an integral of $f(z)$ **independent of path in a domain D** if for every z_1, z_2 in D its value depends only on the initial point z_1 and the terminal point z_2 but not on the choice of the path C in D (so that any path C_1 in D from z_1 to z_2 gives the same value of the integral as any other path C_2 in D from z_1 to z_2).

---

[6]ÉDOUARD GOURSAT (1858—1936), French mathematician. Cauchy published the theorem in 1825. The removal of that condition by Goursat (see *Transactions Amer. Math. Soc.,* vol. 1, 1900) is quite important, for instance, in connection with the fact that derivatives of analytic functions are also analytic, as we shall prove soon. Goursat also made basic contributions to partial differential equations.

THEOREM 2   **(Independence of path)**

If $f(z)$ is analytic in a simply connected domain D, then the integral of $f(z)$ is independent of path in D.

PROOF.   Let z_1 and z_2 be any points in D. Consider two paths C_1 and C_2 in D from z_1 to z_2 without further common points, as in Fig. 331. Denote by $C_2{}^*$ the path C_2 with the orientation

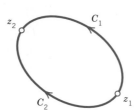

Fig. 331. Formula (2)

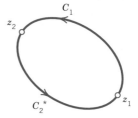

Fig. 332. Formula (2′)

reversed (Fig. 332). Integrate from z_1 over C_1 to z_2 and over $C_2{}^*$ back to z_1. This is a simple closed path, and Cauchy's theorem applies under our assumptions of the present theorem and gives zero:

$$(2')\qquad \int_{C_1} f\, dz + \int_{C_2{}^*} f\, dz = 0, \qquad \text{thus} \qquad \int_{C_1} f\, dz = -\int_{C_2{}^*} f\, dz.$$

But the minus sign on the right disappears if we integrate in the reverse direction, from z_1 to z_2, which shows that the integrals of $f(z)$ over C_1 and C_2 are equal,

$$(2)\qquad\qquad \int_{C_1} f(z)\, dz = \int_{C_2} f(z)\, dz \qquad\qquad \text{(Fig. 331).}$$

This proves the theorem for paths that have only the endpoints in common. For paths that have finitely many further common points, apply the present argument to each "loop" (portions of C_1 and C_2 between consecutive common points; four in Fig. 333). For paths with infinitely many common points we would need additional argumentation not to be presented here.   ◀

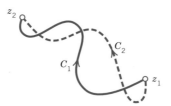

Fig. 333. Paths with more common points

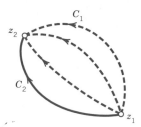

Fig. 334. Continuous
deformation of path

Principle of Deformation of Path

This idea is related to path independence. We may imagine that the path C_2 in (2) was obtained from C_1 by continuously moving C_1 (with ends fixed!) until it coincides with C_2. Figure 334 shows two of the infinitely many intermediate paths for which the integral always retains its value (because of Theorem 2). Hence we may impose a continuous deformation of the path of an integral, keeping the ends fixed. As long as our deforming path always contains only points at which $f(z)$ is analytic, the integral retains the same value. This is called the **principle of deformation of path.**

Existence of Indefinite Integral

This will justify our indefinite integration method in the preceding section [formula (9)]. The proof needs Cauchy's integral theorem.

THEOREM 3   **(Existence of an indefinite integral)**

If $f(z)$ is analytic in a simply connected domain D, then there exists an indefinite integral $F(z)$ of $f(z)$ in D—thus, $F'(z) = f(z)$—which is analytic in D, and for all paths in D joining any two points z_0 and z_1 in D, the integral of $f(z)$ from z_0 to z_1 can be evaluated by formula (9) in Sec. 13.1.

PROOF.   The conditions of Cauchy's integral theorem are satisfied. Hence the line integral of $f(z)$ from any z_0 in D to any z in D is independent of path in D. We keep z_0 fixed. Then this integral becomes a function of z, call it $F(z)$,

$$(3) \qquad\qquad F(z) = \int_{z_0}^{z} f(z^*)\, dz^*,$$

which is uniquely determined. We show that this $F(z)$ is analytic in D and $F'(z) = f(z)$. The idea of doing this is as follows. Using (3) we form the difference quotient

$$(4) \qquad \frac{F(z + \Delta z) - F(z)}{\Delta z} = \frac{1}{\Delta z}\left[\int_{z_0}^{z+\Delta z} f(z^*)\, dz^* - \int_{z_0}^{z} f(z^*)\, dz^*\right] = \frac{1}{\Delta z}\int_{z}^{z+\Delta z} f(z^*)\, dz^*.$$

We now subtract $f(z)$ from (4) and show that the resulting expression approaches zero as $\Delta z \to 0$. The details are as follows.

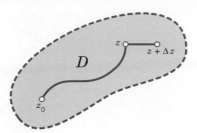

Fig. 335.   Path of integration

We keep z fixed. Then we choose $z + \Delta z$ in D so that the whole segment with endpoints z and $z + \Delta z$ is in D (Fig. 335). This can be done because D is a domain, hence it contains a neighborhood of z. We use this segment as the path of integration in (4). Now we subtract $f(z)$. This is a constant because z is kept fixed. Hence we can write

$$\int_z^{z+\Delta z} f(z)\, dz^* = f(z) \int_z^{z+\Delta z} dz^* = f(z)\, \Delta z. \qquad \text{Thus} \qquad f(z) = \frac{1}{\Delta z} \int_z^{z+\Delta z} f(z)\, dz^*.$$

By this trick and from (4) we get a single integral:

$$\frac{F(z + \Delta z) - F(z)}{\Delta z} - f(z) = \frac{1}{\Delta z} \int_z^{z+\Delta z} \left[f(z^*) - f(z) \right] dz^*.$$

Since $f(z)$ is analytic, it is continuous. An $\epsilon > 0$ being given, we can thus find a $\delta > 0$ such that $|f(z^*) - f(z)| < \epsilon$ when $|z^* - z| < \delta$. Hence, letting $|\Delta z| < \delta$, we see that the ML-inequality (Sec. 13.1) yields

$$\left| \frac{F(z + \Delta z) - F(z)}{\Delta z} - f(z) \right| = \frac{1}{|\Delta z|} \left| \int_z^{z+\Delta z} \left[f(z^*) - f(z) \right] dz^* \right| \leqq \frac{1}{|\Delta z|}\, \epsilon |\Delta z| = \epsilon.$$

By the definition of limit and derivative, this proves that

$$F'(z) = \lim_{\Delta z \to 0} \frac{F(z + \Delta z) - F(z)}{\Delta z} = f(z).$$

Since z is any point in D, this implies that $F(z)$ is analytic in D and is an indefinite integral or antiderivative of $f(z)$ in D, written

$$F(z) = \int f(z)\, dz.$$

Also, if $G'(z) = f(z)$, then $F'(z) - G'(z) \equiv 0$ in D; hence $F(z) - G(z)$ is constant in D (see Team Project 14 in Problem Set 12.4). That is, two indefinite integrals of $f(z)$ can differ only by a constant. The latter drops out in (9) of Sec. 13.1, so that we can use any indefinite integral of $f(z)$. This proves the theorem.   ◀

Cauchy's Theorem for Multiply Connected Domains

Cauchy's theorem applies to multiply connected domains. We first explain this for a **doubly connected domain** D with outer boundary curve C_1 and inner C_2 (Fig. 336). If a function $f(z)$ is analytic in any domain D^* that contains D and its boundary curves, we claim that

$$(5) \qquad \int_{C_1} f(z)\,dz = \int_{C_2} f(z)\,dz \qquad \text{(Fig. 336)},$$

both integrals being taken counterclockwise (or both clockwise, and regardless of whether or not the full interior of C_2 belongs to D^*).

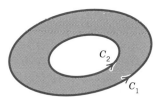

Fig. 336. Paths in (5)

PROOF.   By two cuts $\widetilde{C}_1$ and $\widetilde{C}_2$ (Fig. 337) we cut D into two simply connected domains D_1 and D_2 in which and on whose boundaries $f(z)$ is analytic. By Cauchy's theorem the integral over the entire boundary of D_1 (taken in the sense of the arrows in Fig. 337) is zero, and so is that over the boundary of D_2, and thus their sum. In this sum the integrals over the cuts $\widetilde{C}_1$ and $\widetilde{C}_2$ cancel because we integrate over them in both directions—this is the key— and we are left with the integrals over C_1 (counterclockwise) and C_2 (clockwise; see Fig. 337); hence by reversing the integration over C_2 (to counterclockwise) we have

$$\int_{C_1} f\,dz - \int_{C_2} f\,dz = 0$$

and (5) follows.   ◀

For domains of higher connectivity the idea remains the same. Thus, for a **triply connected domain** we use three cuts $\widetilde{C}_1$, $\widetilde{C}_2$, $\widetilde{C}_3$ (Fig. 338). Adding integrals as before, the integrals over the cuts cancel and the sum of the integrals over C_1 (counterclockwise) and C_2, C_3 (clockwise) is zero. Hence the integral over C_1 equals the sum of the integrals over C_2 and C_3, all three now taken counterclockwise, and so on.

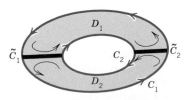

Fig. 337. Doubly connected domain

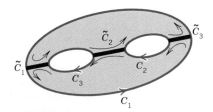

Fig. 338. Triply connected domain

EXAMPLE 6   **A basic result: Integral of integer powers**

From (5) and Sec. 13.1, Example 6, it now follows that

(6)
$$\oint_C (z - z_0)^m \, dz = \begin{cases} 2\pi i & (m = -1) \\ 0 & (m \neq -1 \text{ and integer}) \end{cases}$$

for counterclockwise integration around **any simple closed path containing z_0 in its interior.**   ◄

PROBLEM SET 13.2

Verifications and Comments on the Examples in the Text

1. **(Cauchy's integral theorem)** Verify Theorem 1 for the integral of z^2 over the boundary of the square with vertices $1 + i$, $-1 + i$, $-1 - i$, and $1 - i$ (counterclockwise).

2. **(Singularities)** Can we conclude in Example 2 that the integral of $1/(z^2 + 4)$ taken over (a) $|z - 2| = 2$, (b) $|z - 2| = 3$ is zero? Give reasons.

3. **(Deformation principle)** Can we conclude from Example 4 that the integral is also zero over the contour in Problem 1?

4. **(Cauchy's integral theorem)** For what contours C will it follow from Theorem 1 that

   (a) $\displaystyle\oint_C \frac{dz}{z} = 0$,   (b) $\displaystyle\oint_C \frac{\cos z}{z^6 - z^2} \, dz = 0$,   (c) $\displaystyle\oint_C \frac{e^{1/z}}{z^2 + 9} \, dz = 0$?

5. **(Path independence)** Verify Theorem 2 for the integral of $\cos z$ from 0 to $(1 + i)\pi$ (a) over the shortest path, (b) over the x-axis to π and then straight up to $(1 + i)\pi$.

6. **(Deformation principle)** If the integral of a function $f(z)$ over the unit circle equals 3 and over the circle $|z| = 2$ equals 5, can we conclude that $f(z)$ is analytic everywhere in the annulus $1 < |z| < 2$?

Cauchy's Integral Theorem Applicable?

Integrate $f(z)$ counterclockwise around the unit circle, indicating whether Cauchy's integral theorem applies. (Show the details of your work.)

7. $f(z) = e^{-z^2}$ 8. $f(z) = \tan \frac{1}{2} z$ 9. $f(z) = 1/|z|^2$

10. $f(z) = 1/(\pi z - 1)$ 11. $f(z) = 1/(2z - 1)$ 12. $f(z) = \bar{z}^3$

13. $f(z) = 1/(z^4 + 1.1)$ 14. $f(z) = 1/\bar{z}$ 15. $f(z) = \text{Im } z$

16. **TEAM PROJECT. Cauchy's Integral Theorem. (a) Main aspects.** Each of the five problems in Examples 1–5 explains a basic fact in connection with Cauchy's theorem. Find five examples of your own, more complicated ones if possible, each illustrating one of those facts.

    (b) Partial fractions. Write $f(z)$ in terms of partial fractions and integrate it counterclockwise over the unit circle, where

    (i) $f(z) = \dfrac{2z + 3i}{z^2 + \frac{1}{4}}$ (ii) $f(z) = \dfrac{z + 1}{z^2 + 2z}$.

    (c) Deformation of path. Review (c) and (d) of Team Project 30, Sec. 13.1, in the light of the principle of deformation of path. Then consider another family of paths with common endpoints, say, $z(t) = t + ia(t - t^2)$, $0 \leq t \leq 1$, and experiment with the integration of analytic and nonanalytic functions of your choice over these paths (e.g., z, $\text{Im } z$, z^2, $\text{Re } z^2$, $\text{Im } z^2$, etc.).

Contour Integrals

Evaluate (showing the details and using a partial fraction representation of the integrand if necessary)

17. $\oint_C \dfrac{dz}{z - 3i}$, C the circle $|z| = \pi$, counterclockwise

18. $\oint_C \text{Ln}\,(1 - z)\,dz$, C the boundary of the parallelogram with vertices $\pm i$, $\pm(1 + i)$

19. $\oint_C \dfrac{e^z}{z}\,dz$, C consists of $|z| = 2$ (counterclockwise) and $|z| = 1$ (clockwise)

20. $\oint_C \text{Re}\,z\,dz$, C the contour in Fig. 339

21. $\oint_C \dfrac{dz}{z^2 - 1}$, C the contour in Fig. 340

22. $\oint_C \dfrac{2z - 1}{z^2 - z}\,dz$, C the contour in Fig. 341

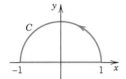

Fig. 339.   Problem 20

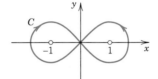

Fig. 340.   Problem 21

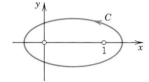

Fig. 341.   Problem 22

23. $\oint_C \dfrac{dz}{z^2 + 1}$, C: (a) $|z + i| = 1$, (b) $|z - i| = 1$, counterclockwise

24. $\oint_C \coth \tfrac{1}{2}z\,dz$, C the circle $|z - \tfrac{1}{2}\pi i| = 1$, clockwise

25. $\oint_C \dfrac{2z^3 + z^2 + 4}{z^4 + 4z^2}\,dz$, C the circle $|z - 2| = 4$, clockwise

13.3 Cauchy's Integral Formula

The most important consequence of Cauchy's integral theorem is Cauchy's integral formula. This formula is useful for evaluating integrals (examples below). Equally important is its key role in proving the surprising fact that analytic functions have derivatives of all orders (Sec. 13.4), in establishing Taylor series representations (Sec. 14.4), and so on. Cauchy's integral formula and its conditions of validity may be stated as follows.

THEOREM 1   **(Cauchy's integral formula)**

Let $f(z)$ be analytic in a simply connected domain D. Then for any point z_0 in D and any simple closed path C in D that encloses z_0 (Fig. 342),

(1)

$$\oint_C \frac{f(z)}{z - z_0} \, dz = 2\pi i f(z_0)$$

           (Cauchy's integral formula),

the integration being taken counterclockwise. Alternatively (for representing $f(z_0)$ by a contour integral, divide (1) by $2\pi i$),

(1*)

$$f(z_0) = \frac{1}{2\pi i} \oint_C \frac{f(z)}{z - z_0} \, dz$$

           (Cauchy's integral formula).

PROOF.   By addition and subtraction, $f(z) = f(z_0) + [f(z) - f(z_0)]$. Inserting this into (1) on the left and taking the constant factor $f(z_0)$ out from under the integral sign, we have

(2)
$$\oint_C \frac{f(z)}{z - z_0} \, dz = f(z_0) \oint_C \frac{dz}{z - z_0} + \oint_C \frac{f(z) - f(z_0)}{z - z_0} \, dz.$$

The first term on the right equals $f(z_0) \cdot 2\pi i$ (see Example 6 in Sec. 13.2 with $m = -1$). This proves the theorem, provided the second integral on the right is zero. This is what we are now going to show. Its integrand is analytic, except at z_0. Hence by the principle of deformation of path (Sec. 13.2) we can replace C by a small circle K of radius ρ and center z_0 (Fig. 343), without altering the value of the integral. Since $f(z)$ is analytic, it is continuous (Team Project 24, Sec. 12.3). Hence an $\epsilon > 0$ being given, we can find a $\delta > 0$ such that $|f(z) - f(z_0)| < \epsilon$ for all z in the disk $|z - z_0| < \delta$. Choosing the radius ρ of K smaller than δ, we thus have the inequality

$$\left| \frac{f(z) - f(z_0)}{z - z_0} \right| < \frac{\epsilon}{\rho}$$

at each point of K. The length of K is $2\pi\rho$. Hence, by the *ML*-inequality in Sec. 13.1,

$$\left| \oint_K \frac{f(z) - f(z_0)}{z - z_0} \, dz \right| < \frac{\epsilon}{\rho} \, 2\pi\rho = 2\pi\epsilon.$$

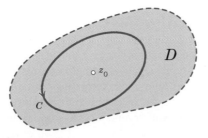

Fig. 342.   Cauchy's integral formula

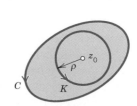

Fig. 343.   Proof of Cauchy's integral formula

Since $\epsilon\,(> 0)$ can be chosen arbitrarily small, it follows that the last integral in (2) must have the value zero, and the theorem is proved. ◀

EXAMPLE 1　Cauchy's integral formula

$$\oint_C \frac{e^z}{z-2}\,dz = 2\pi i e^z\Big|_{z=2} = 2\pi i e^2 \approx 46.4268 i$$

for any contour enclosing $z_0 = 2$ (since e^z is entire), and zero for any contour for which $z_0 = 2$ lies outside (by Cauchy's integral theorem). ◀

EXAMPLE 2　Cauchy's integral formula

$$\oint_C \frac{z^3 - 6}{2z - i}\,dz = \oint_C \frac{\frac{1}{2}z^3 - 3}{z - \frac{1}{2}i}\,dz = 2\pi i\left[\tfrac{1}{2}z^3 - 3\right]\Big|_{z=i/2} = \frac{\pi}{8} - 6\pi i \qquad (z_0 = \tfrac{1}{2}i \text{ inside } C).\ \ ◀$$

EXAMPLE 3　Integration around different contours

Integrate

$$g(z) = \frac{z^2 + 1}{z^2 - 1} = \frac{z^2 + 1}{(z + 1)(z - 1)}$$

counterclockwise around each of the four circles in Fig. 344.

Solution.　$g(z)$ is not analytic at -1 and 1. These are the points we have to watch for.

(a) The circle $|z - 1| = 1$ encloses the point $z_0 = 1$ where $g(z)$ is not analytic. Hence in (1) we have to write

$$g(z) = \frac{z^2 + 1}{z^2 - 1} = \frac{z^2 + 1}{z + 1}\,\frac{1}{z - 1}\,; \qquad \text{thus} \qquad f(z) = \frac{z^2 + 1}{z + 1}\,.$$

and (1) gives

$$\oint_C \frac{z^2 + 1}{z^2 - 1}\,dz = 2\pi i f(1) = 2\pi i\left[\frac{z^2 + 1}{z + 1}\right]_{z=1} = 2\pi i.$$

(b) gives the same as (a) by the principle of deformation of path.

(c) The function $g(z)$ is as before, but $f(z)$ changes because we must take $z_0 = -1$ (instead of 1). This gives a factor $z - z_0 = z + 1$ in (1). Hence we must write

$$g(z) = \frac{z^2 + 1}{z - 1}\,\frac{1}{z + 1}\,; \qquad \text{thus} \qquad f(z) = \frac{z^2 + 1}{z - 1}\,.$$

Compare this for a minute with the previous expression and then go on:

$$\oint_C \frac{z^2 + 1}{z^2 - 1}\,dz = 2\pi i f(-1) = 2\pi i\left[\frac{z^2 + 1}{z - 1}\right]_{z=-1} = -2\pi i.$$

(d) gives 0. Why? ◀

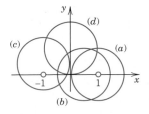

Fig. 344.　Example 3

EXAMPLE 4 **Use of partial fractions**

Integrate $g(z) = (z^2 - 1)^{-1} \tan z$ around the circle C: $|z| = 3/2$ (counterclockwise).

Solution. $\tan z$ is not analytic at $\pm\pi/2$, $\pm3\pi/2, \cdots$, but all these points lie outside the contour. $(z^2 - 1)^{-1} = 1/(z - 1)(z + 1)$ is not analytic at 1 and $- 1$. To get integrals of the form (1), with only a *single* point inside C at which the integrand is not analytic, we use partial fractions:

$$\frac{1}{z^2 - 1} = \frac{1}{2}\left(\frac{1}{z - 1} - \frac{1}{z + 1}\right).$$

From this and (1) we obtain

$$\oint_C \frac{\tan z}{z^2 - 1}\, dz = \frac{1}{2}\left[\oint_C \frac{\tan z}{z - 1}\, dz - \oint_C \frac{\tan z}{z + 1}\, dz\right]$$

$$= \frac{2\pi i}{2}[\tan 1 - \tan(-1)] = 2\pi i \tan 1 \approx 9.785i.$$

(There is a better way for this, as we shall see in Sec. 15.3.) ◀

Multiply connected domains may be handled as in Sec. 13.2. For instance, if $f(z)$ is analytic on C_1 and C_2 and in the ring-shaped domain bounded by C_1 and C_2 (Fig. 345) and z_0 is any point in that domain, then

(3) $$f(z_0) = \frac{1}{2\pi i}\oint_{C_1} \frac{f(z)}{z - z_0}\, dz + \frac{1}{2\pi i}\oint_{C_2} \frac{f(z)}{z - z_0}\, dz,$$

where the outer integral (over C_1) is taken counterclockwise and the inner clockwise, as indicated in Fig. 345.

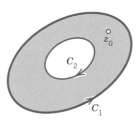

Fig. 345. Formula (3)

PROBLEM SET 13.3

Contour Integration

Integrate $z^2/(z^4 - 1)$ counterclockwise around the circle

  1. $|z + 1| = 1$ **2.** $|z + i| = 1$ **3.** $|z| = 0.9$ **4.** $x^2 + 16y^2 = 4$

Integrate the given function counterclockwise around the unit circle.

  5. $\dfrac{z^3}{2z - i}$ **6.** $\dfrac{e^z}{\pi z - i}$ **7.** $\dfrac{\cosh 3z}{2z}$ **8.** $\dfrac{z^3 \sin z}{3z - 1}$

9. **CAS PROJECT. Contour Integration.** Experiment to find out to what extent your CAS can do contour integration (a) by using the second method in Sec. 13.1, (b) by Cauchy's integral formula.

10. **TEAM PROJECT. Cauchy's Integral Theorem.** Gain additional insight into the proof of Cauchy's integral theorem by producing (2) with a contour enclosing z_0 (as in Fig. 342) and taking the limit as in the text. Choose

$$\text{(a)} \quad \oint_C \frac{z^3 - 6}{z - \frac{1}{2}i}\, dz, \qquad \text{(b)} \quad \oint_C \frac{\sin z}{z - \frac{1}{2}\pi}\, dz,$$

and (c) two other examples of your choice.

Further Contour Integrals

Integrate the given function over the given contour C, counterclockwise or as indicated. (Show the details of your work.)

11. $\dfrac{1}{z^2 + 4}$, C the ellipse $4x^2 + (y - 2)^2 = 4$

12. $\dfrac{4 - \sin z}{z^2 - 2z}$, C the square with vertices ± 1 and $\pm i$

13. $\dfrac{\operatorname{Ln}(z - 1)}{z - 6}$, C the circle $|z - 6| = 4$

14. $\dfrac{e^z}{z e^z - 2iz}$, C the circle $|z| = 0.5$

15. $\dfrac{\cosh(z^2 - \pi i)}{z - \pi i}$, C the rectangle with vertices ± 1 and $\pm 1 + 4i$

16. $\dfrac{\sin z}{4z^2 - 8iz}$, C consists of the boundaries of the squares with vertices ± 3, $\pm 3i$ (counterclockwise) and ± 1, $\pm i$, (clockwise)

17. $\dfrac{e^{z^2}}{z^2(z - 1 - i)}$, C consists of $|z| = 2$ (counterclockwise) and $|z| = 1$ (clockwise)

18. $\dfrac{\operatorname{Ln}(z + 1)}{z^2 + 1}$, C consists of $|z - i| = 1.4$ (counterclockwise) and $|z| = 0.2$ (clockwise)

19. Show that $\oint_C (z - z_1)^{-1}(z - z_2)^{-1}\, dz = 0$ for a simple closed path C enclosing z_1 and z_2, which are arbitrary.

20. Solve Example 4 by applying Cauchy's integral theorem to a suitable triply connected domain, without partial fraction reduction, and then using (1).

13.4 Derivatives of Analytic Functions

In this section we use Cauchy's integral formula to show the basic fact that complex analytic functions have ***derivatives of all orders.*** This is very surprising because it differs strikingly from the situation in real calculus. Indeed, if a *real* function is once differentiable, nothing follows about the existence of second or higher derivatives. Thus, in this respect, complex analytic functions behave much more simply than real functions that are once differentiable.

THEOREM 1 **(Derivatives of an analytic function)**

If $f(z)$ is analytic in a domain D, then it has derivatives of all orders in D, which are then also analytic functions in D. The values of these derivatives at a point z_0 in D are given by the formulas

(1')
$$f'(z_0) = \frac{1}{2\pi i} \oint_C \frac{f(z)}{(z - z_0)^2} \, dz,$$

(1'')
$$f''(z_0) = \frac{2!}{2\pi i} \oint_C \frac{f(z)}{(z - z_0)^3} \, dz,$$

and in general

(1)
$$f^{(n)}(z_0) = \frac{n!}{2\pi i} \oint_C \frac{f(z)}{(z - z_0)^{n+1}} \, dz \qquad\qquad (n = 1, 2, \cdots);$$

here C is any simple closed path in D that encloses z_0 and whose full interior belongs to D; and we integrate counterclockwise around C (Fig. 346).

Comment. For memorizing (1), it is useful to observe that these formulas are obtained formally by differentiating the Cauchy formula (1*), Sec. 13.3, under the integral sign with respect to z_0.

PROOF OF THEOREM 1.   We prove (1'). We start from the definition

$$f'(z_0) = \lim_{\Delta z \to 0} \frac{f(z_0 + \Delta z) - f(z_0)}{\Delta z}.$$

On the right we represent $f(z_0 + \Delta z)$ and $f(z_0)$ by Cauchy's integral formula:

$$\frac{f(z_0 + \Delta z) - f(z_0)}{\Delta z} = \frac{1}{2\pi i \Delta z} \left[\oint_C \frac{f(z)}{z - (z_0 + \Delta z)} \, dz - \oint_C \frac{f(z)}{z - z_0} \, dz \right].$$

We write the two integrals as a single integral. Taking the common denominator gives the numerator $f(z)\{z - z_0 - [z - (z_0 + \Delta z)]\} = f(z)\,\Delta z$, so that Δz drops out and we get

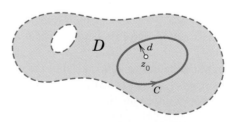

Fig. 346.   Theorem 1 and its proof

$$\frac{f(z_0 + \Delta z) - f(z_0)}{\Delta z} = \frac{1}{2\pi i} \oint_C \frac{f(z)}{(z - z_0 - \Delta z)(z - z_0)} \, dz.$$

Clearly, we can now establish (1′) by showing that, as $\Delta z \to 0$, the integral on the right approaches the integral in (1′). To do this, we consider the difference between these two integrals. We can write this difference as a single integral by taking the common denominator and simplifying the numerator (as just before). This gives

$$\oint_C \frac{f(z)}{(z - z_0 - \Delta z)(z - z_0)} \, dz - \oint_C \frac{f(z)}{(z - z_0)^2} \, dz = \oint_C \frac{f(z)\Delta z}{(z - z_0 - \Delta z)(z - z_0)^2} \, dz.$$

We now show by the *ML*-inequality (Sec. 13.1) that the integral on the right approaches zero as $\Delta z \to 0$.

Being analytic, the function $f(z)$ is continuous on C, hence bounded in absolute value, say, $|f(z)| \le K$. Let d be the smallest distance from z_0 to the points of C (see Fig. 346). Then for all z on C,

$$|z - z_0|^2 \ge d^2, \qquad \text{hence} \qquad \frac{1}{|z - z_0|^2} \le \frac{1}{d^2}.$$

Furthermore, by the triangle inequality for all z on C we then also have

$$d \le |z - z_0| = |z - z_0 - \Delta z + \Delta z| \le |z - z_0 - \Delta z| + |\Delta z|.$$

We now subtract $|\Delta z|$ on both sides and let $|\Delta z| \le d/2$, so that $-|\Delta z| \ge -d/2$. Then

$$\tfrac{1}{2}d \le d - |\Delta z| \le |z - z_0 - \Delta z|. \qquad \text{Hence} \qquad \frac{1}{|z - z_0 - \Delta z|} \le \frac{2}{d}.$$

Let L be the length of C. If $|\Delta z| \le d/2$, then by the *ML*-inequality

$$\left| \oint_C \frac{f(z)\Delta z}{(z - z_0 - \Delta z)(z - z_0)^2} \, dz \right| \le KL \, |\Delta z| \, \frac{2}{d} \cdot \frac{1}{d^2}.$$

This approaches zero as $\Delta z \to 0$. Formula (1′) is proved.

Note that we used Cauchy's integral formula (1*), Sec. 13.3, but if all we had known about $f(z_0)$ is the fact that it can be represented by (1*), Sec. 13.3, our argument would have established the existence of the derivative $f'(z_0)$ of $f(z)$. This is essential to the continuation and completion of this proof, because it implies that (1″) can be proved by a similar argument, with f replaced by f', and that the general formula (1) follows by induction.   ◀

EXAMPLE 1   **Evaluation of line integrals**

From (1′), for any contour enclosing the point πi (counterclockwise)

$$\oint_C \frac{\cos z}{(z - \pi i)^2} \, dz = 2\pi i (\cos z)' \Big|_{z = \pi i} = -2\pi i \sin \pi i = 2\pi \sinh \pi.$$ ◀

EXAMPLE 2   From (1″), for any contour enclosing the point $-i$ (counterclockwise),

$$\oint_C \frac{z^4 - 3z^2 + 6}{(z+i)^3}\, dz = \pi i\, (z^4 - 3z^2 + 6)'' \Big|_{z=-i} = \pi i \big[12z^2 - 6\big]_{z=-i} = -18\pi i. \qquad \blacktriangleleft$$

EXAMPLE 3   By (1′), for any contour for which 1 lies inside and $\pm 2i$ lie outside (counterclockwise),

$$\oint_C \frac{e^z}{(z-1)^2(z^2+4)}\, dz = 2\pi i \left(\frac{e^z}{z^2+4}\right)' \Big|_{z=1} = 2\pi i\, \frac{e^z(z^2+4) - e^z 2z}{(z^2+4)^2} \Big|_{z=1} = \frac{6e\pi}{25}\, i \approx 2.050i. \qquad \blacktriangleleft$$

Cauchy's Inequality. Liouville's and Morera's Theorems

We show that Theorem 1 is also fundamental in deriving general results on analytic functions.

Cauchy's inequality. Theorem 1 yields a basic inequality that has many applications. To get it, all we have to do is to choose for C in (1) a circle of radius r and center z_0 and apply the *ML*-inequality (Sec. 13.1); with $|f(z)| \leq M$ on C we obtain from (1)

$$|f^{(n)}(z_0)| = \frac{n!}{2\pi} \left| \oint_C \frac{f(z)}{(z-z_0)^{n+1}}\, dz \right| \leq \frac{n!}{2\pi}\, M\, \frac{1}{r^{n+1}}\, 2\pi r.$$

This gives **Cauchy's inequality**

(2)
$$\boxed{\; |f^{(n)}(z_0)| \leq \frac{n!M}{r^n}\,. \;}$$

To gain a first impression of the importance of this inequality, let us prove a famous theorem on entire functions (definition in Sec. 12.6).

THEOREM 2   **Liouville's theorem[7]**

If an entire function $f(z)$ is bounded in absolute value for all z, then $f(z)$ must be a constant.

PROOF.   By assumption, $|f(z)|$ is bounded, say, $|f(z)| < K$ for all z. Using (2), we see that $|f'(z_0)| < K/r$. Since $f(z)$ is entire, this is true for every r, so that we can take r as large as we please and conclude that $f'(z_0) = 0$. Since z_0 is arbitrary, $f'(z) = 0$ for all z, and $f(z)$ is constant (see Team Project 14 in Problem Set 12.4). This proves the theorem. $\blacktriangleleft$

Another very interesting consequence of Theorem 1 is

THEOREM 3   **Morera's[8] theorem (Converse of Cauchy's integral theorem)**

If $f(z)$ is continuous in a simply connected domain D and if

(3)
$$\oint_C f(z)\, dz = 0$$

for every closed path in D, then $f(z)$ is analytic in D.

---

[7]See Sec. 4.7, footnote 16.

[8]GIACINTO MORERA (1856—1909), Italian mathematician who worked in Genoa and Turin.

PROOF. In Sec. 13.2 it was shown that if $f(z)$ is analytic in D, then

$$F(z) = \int_{z_0}^{z} f(z^*) \, dz^*$$

is analytic in D and $F'(z) = f(z)$. In the proof we used only the continuity of $f(z)$ and the property that its integral around every closed path in D is zero; from these assumptions we concluded that $F(z)$ is analytic. By Theorem 1, the derivative of $F(z)$ is analytic, that is, $f(z)$ is analytic in D, and Morera's theorem is proved. ◀

PROBLEM SET 13.4

Contour Integration by Theorem 1

Integrate the following functions counterclockwise around the unit circle. (n in Probs. 7 and 8 is a positive integer.) Show the details of your work.

1. $\dfrac{\sinh 2z}{z^4}$

2. $\dfrac{e^{-z} \sin z}{z^2}$

3. $\dfrac{z^2}{(2z - 1)^3}$

4. $\dfrac{z^6}{(2z - 1)^6}$

5. $\dfrac{\tan z}{(z - \frac{1}{4}\pi)^3}$

6. $\dfrac{\sin z}{z^4}$

7. $\dfrac{\cos \pi z}{z^{2n}}$

8. $\dfrac{\cos z}{z^{2n+1}}$

9. $\dfrac{e^{3z}}{(4z - \pi i)^3}$

10. $\dfrac{z^3 e^z}{(z - \frac{1}{2})^3}$

Integrate $f(z)$ around C counterclockwise or as indicated. (Show the details.)

11. $f(z) = z^{-2} \tan \pi z,$ C any contour enclosing 0

12. $f(z) = \dfrac{z^3 + \sin z}{(z - i)^3},$ C the boundary of the square with vertices ± 2 and $\pm 2i$

13. $f(z) = (z - 2)^{-2} \operatorname{Ln} z,$ $C: |z - 3| = 2$

14. $f(z) = (z - \frac{1}{2}\pi)^{-2} \cot z,$ $C: 4x^2 + 9y^2 = 36$

15. $f(z) = \dfrac{2z^3 - 3}{z(z - 1 - i)^2},$ C consists of $|z| = 2$ (counterclockwise) and $|z| = 1$ (clockwise)

16. $f(z) = \dfrac{e^{z^2}}{z(z - 2i)^2},$ C consists of $|z - i| = 3$ (counterclockwise) and $|z| = 1$ (clockwise)

17. $f(z) = \dfrac{(1 + z) \sin z}{(2z - 1)^2},$ $C: |z - i| = 2$

18. $f(z) = \dfrac{\operatorname{Ln}(z + 3)}{(z - 2)(z + 1)^2},$ C the boundary of the square with vertices $\pm 1.5, \pm 1.5i$

19. $f(z) = \dfrac{\cosh 4z}{(z - 4)^3},$ C consists of $|z| = 6$ (counterclockwise) and $|z - 3| = 2$ (clockwise)

20. **TEAM PROJECT. Theory Related to Liouville's Theorem**
    (a) **Growth of entire functions.** If $f(z)$ is not a constant and is analytic for all (finite) z, and R and M are any positive real numbers (no matter how large), show that there exist values of z for which $|z| > R$ and $|f(z)| > M$.
    (b) **Growth of polynomials.** If $f(z)$ is a polynomial of degree $n > 0$ and M is an arbitrary positive real number (no matter how large), show that there exists a positive real number R such that $|f(z)| > M$ for all $|z| > R$.
    (c) **Exponential function.** Show that $f(z) = e^z$ has the property characterized in (a) but does not have that characterized in (b).
    (d) **Fundamental theorem of algebra.** If $f(z)$ is a polynomial in z, not a constant, then $f(z) = 0$ for at least one value of z. Prove this, using (a).

CHAPTER 13 REVIEW

1. State the definition of a complex line integral from memory.
2. Which integration methods in this chapter apply only to analytic functions and which to general continuous complex functions?
3. What is a parametric representation of a curve? What is its advantage over other representations?
4. What did we assume about paths of integration throughout this chapter? What is $\dot{z} = dz/dt$ geometrically if $z(t)$ represents a path?
5. What do you get by integrating $1/z$ counterclockwise around the unit circle? (You should memorize this basic result.) By integrating $1/z^m$, $m = 2, 3, \cdots$?
6. What is independence of path? What is the principle of deformation of path? Why is this important?
7. Don't confuse Cauchy's integral theorem and Cauchy's integral formula. State both from memory. Which follows from which?
8. What is a doubly connected domain? A triply connected domain? How can you extend Cauchy's theorem to such domains?
9. How can you use Cauchy's integral formula in integration? Give examples.
10. If a function $f(z)$ is differentiable in a domain D, is it analytic in D? Does it have higher derivatives in D? How many?
11. What do we mean by saying that integration is a linear operation?
12. Can you remember the relations between complex and real line integrals discussed in this chapter?
13. State an inequality that gives upper bounds for the absolute value of an integral. Does it apply to analytic functions only?
14. What is Liouville's theorem? Give examples. State some of its consequences.
15. Is $\operatorname{Im} \int_C f(z)\, dz = \int_C \operatorname{Im} f(z)\, dz$?
16. Integrate $z^6 - 3z^4$ from 2 along $|z| = 2$, $y \geqq 0$, to -2 by two methods. (Show the details of your work.)
17. Integrate $(1 - z^2)^2$ from $-i$ to i counterclockwise along (a) $|z| = 1$, (b) $x^2 + 4y^2 = 4$. Why are the results equal? (Show details.)
18. Verify Cauchy's integral theorem for counterclockwise integration of $\sin z$ around the boundary of the rectangle with vertices 0, π, $\pi + i$, and i. (Show details.)

Integrate (showing details or giving reasons)

19. $z \cosh (z^2)$ from 0 to πi along any path
20. $1/z + 1/(z - 2)$ clockwise around the ellipse $(x - 1)^2 + 4y^2 = 4$
21. $|z| + z$ counterclockwise around the unit circle
22. e^z/z^4 counterclockwise around $|z| = \frac{1}{2}$
23. $1/\operatorname{Ln} (z + 2i)$ clockwise around the unit circle
24. $\operatorname{Re} z$ from 0 to $3 + 27i$ along $y = x^3$
25. $(z \cosh (z^2))/(z - 2i)^3$ counterclockwise around $|z - i| = 2$
26. $(\tan \pi z)/(z - 1)^2$ counterclockwise around $|z - 1| = 0.1$
27. $4(z + 2i)^{-1} + 2(z + 4i)^{-1}$ clockwise around the circle $|z - 1| = 2.5$
28. $1/|z|$ from 3 counterclockwise along $|z| = 3$ to $3i$
29. $(\operatorname{Ln} z)/(z - 2i)^2$ counterclockwise around $|z - 2i| = 1$
30. $(\cos 4z)/[z^3(4z - \pi)]$ counterclockwise around the circle $|z - 1| = \frac{1}{2}$

SUMMARY OF CHAPTER 13
COMPLEX INTEGRATION

The **complex line integral** of a function $f(z)$ taken over a path C is denoted by

$$(1) \qquad \int_C f(z)\, dz \qquad \text{or, if } C \text{ is closed, also by} \qquad \oint_C f(z)\, dz. \quad \text{(Sec. 13.1)}$$

If $f(z)$ is analytic in a simply connected domain D, then we can evaluate (1) as in calculus by indefinite integration and substitution of limits, that is,

$$(2) \qquad \int_C f(z)\, dz = F(z_1) - F(z_0) \qquad [F'(z) = f(z)]$$

for every path C in D from a point z_0 to a point z_1 (See Sec. 13.1). These assumptions imply **independence of path,** that is, (2) depends only on z_0 and z_1 (and on $f(z)$, of course) but not on the choice of C (Sec. 13.2). The existence of an $F(z)$ such that $F'(z) = f(z)$ is proved in Sec. 13.2 by Cauchy's integral theorem (see below).

A general method of integration, not restricted to analytic functions, uses the equation $z = z(t)$ of C, where $a \leqq t \leqq b$,

$$(3) \qquad \int_C f(z)\, dz = \int_a^b f(z(t)) \dot{z}(t)\, dt \qquad \left(\dot{z} = \frac{dz}{dt} \right).$$

Cauchy's integral theorem is the most important theorem in this chapter. It states that if $f(z)$ is analytic in a simply connected domain D, then for every closed path C in D (Sec. 13.2),

$$(4) \qquad \oint_C f(z)\, dz = 0.$$

Under the same assumptions and for any z_0 in D and closed path C in D containing z_0 in its interior we also have **Cauchy's integral formula**

$$(5) \qquad f(z_0) = \frac{1}{2\pi i} \oint_C \frac{f(z)}{z - z_0}\, dz.$$

Furthermore, then $f(z)$ has derivatives of all orders in D that are themselves analytic functions in D and (Sec. 13.4)

$$(6) \qquad f^{(n)}(z_0) = \frac{n!}{2\pi i} \oint_C \frac{f(z)}{(z - z_0)^{n+1}}\, dz \qquad (n = 1, 2, \cdots).$$

This implies *Morera's theorem* (the converse of Cauchy's integral theorem) and *Cauchy's inequality* (Sec. 13.4)

Power Series, Taylor Series

Complex power series, in particular, Taylor series, are analogs of real power and Taylor series in calculus. However, they are much more fundamental in complex analysis than their counterparts in calculus, because **power series** represent analytic functions (Sec. 14.3) and, conversely, every analytic function can be represented by power series, called **Taylor series** (Sec. 14.4).

Section 14.1 on basic concepts and convergence tests for complex series is similar to the corresponding material for real series. If you are familiar with the latter, use Sec. 14.1 for reference and begin with Sec. 14.2, a thorough discussion of power series. Sections 14.3 and 14.4 have been mentioned, and the concluding Sec. 14.5 concerns uniform convergence of power and other series.

Prerequisites for this chapter: Chaps. 12, 13.
Sections that may be omitted in a shorter course: Secs. 14.1, 14.5.
Reference: Appendix 1, Part D.
Answers to problems: Appendix 2.

14.1 Sequences, Series, Convergence Tests

In this section we define the basic concepts for ***complex*** sequences and series and discuss tests for convergence and divergence. This is very similar to *real* sequences and series in calculus. *If you feel at home with the latter and want to take for granted that the ratio test also holds in complex, skip this section and go to Sec. 14.2.*

Sequences

The basic definitions are as in calculus. An *infinite sequence* or, briefly, a **sequence,** is obtained by assigning to each positive integer n a number z_n, called a **term** of the sequence, and is written

$$z_1, z_2, \cdots \quad \text{or} \quad \{z_1, z_2, \cdots\} \quad \text{or briefly} \quad \{z_n\}.$$

We may also write $z_0, z_1, \cdots$ or $z_2, z_3, \cdots$ or start with some other integer if convenient.

A **real sequence** is one whose terms are real.

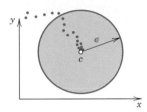

Fig. 347.　Convergent complex sequence

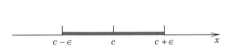

Fig. 348.　Convergent real sequence

Convergence. A **convergent sequence** $z_1, z_2, \cdots$ is one that has a limit c, written

$$\lim_{n \to \infty} z_n = c \qquad \text{or simply} \qquad z_n \to c.$$

By definition of limit this means that for every $\epsilon > 0$ we can find an N such that

$$(1) \qquad\qquad\qquad |z_n - c| < \epsilon \qquad\qquad\qquad \text{for all } n > N;$$

geometrically, all z_n with $n > N$ lie in the open disk of radius ϵ and center c (Fig. 347) and only finitely many do not lie in that disk. [For a *real* sequence, (1) gives an open interval of length 2ϵ and real midpoint c on the real line; see Fig. 348.]

　　A **divergent sequence** is one that does not converge.

EXAMPLE 1　**Convergent and divergent sequences**

The sequence $\{i^n/n\} = \{i, -1/2, -i/3, 1/4, \cdots\}$ is convergent with limit 0.
　　The sequence $\{i^n\} = \{i, -1, -i, 1, \cdots\}$ is divergent, and so is $\{z_n\}$ with $z_n = (1 + i)^n$.　　◄

EXAMPLE 2　**Sequences of the real and imaginary parts**

The sequence $\{z_n\}$ with $z_n = x_n + iy_n = 1 - 1/n^2 + i(2 + 4/n)$ is $6i$, $3/4 + 4i$, $8/9 + 10i/3$, $15/16 + 3i$, $\cdots$. (Sketch it.) It converges with the limit $c = 1 + 2i$. Observe that $\{x_n\}$ has the limit $1 = \text{Re } c$ and $\{y_n\}$ has the limit $2 = \text{Im } c$. This is typical. It illustrates the following theorem by which the convergence of a *complex* sequence can be referred back to that of the two *real* sequences of the real parts and the imaginary parts.　　◄

THEOREM 1　**(Sequences of the real and the imaginary parts)**

A sequence $z_1, z_2, \cdots, z_n, \cdots$ of complex numbers $z_n = x_n + iy_n$ (where $n = 1, 2, \cdots$) converges to $c = a + ib$ if and only if the sequence of the real parts $x_1, x_2, \cdots$ converges to a and the sequence of the imaginary parts $y_1, y_2, \cdots$ converges to b.

PROOF.　If $|z_n - c| < \epsilon$, then $z_n = x_n + iy_n$ is within the circle of radius ϵ about $c = a + ib$ so that necessarily (Fig. 349*a* on the next page)

$$|x_n - a| < \epsilon, \qquad |y_n - b| < \epsilon.$$

Thus convergence $z_n \to c$ implies convergence $x_n \to a$ and $y_n \to b$.

　　Conversely, if $x_n \to a$ and $y_n \to b$ as $n \to \infty$, then for a given $\epsilon > 0$ we can choose N so large that, for every $n > N$,

$$|x_n - a| < \frac{\epsilon}{2}, \qquad |y_n - b| < \frac{\epsilon}{2}.$$

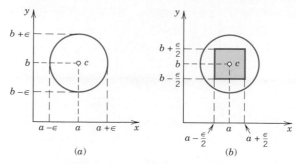

Fig. 349. Proof of Theorem 1

These two inequalities imply that $z_n = x_n + iy_n$ lies in a square with center c and side ϵ. Hence, z_n must lie within a circle of radius ϵ with center c (Fig. 349b). ◀

Series

Given a sequence $z_1, z_2, \cdots, z_m, \cdots$, we may form the sequence of sums

$$s_1 = z_1, \qquad s_2 = z_1 + z_2, \qquad s_3 = z_1 + z_2 + z_3, \quad \cdots$$

and in general

$$(2) \qquad\qquad s_n = z_1 + z_2 + \cdots + z_n \qquad\qquad (n = 1, 2, \cdots).$$

s_n is called the **nth partial sum** of the *infinite series* or **series**

$$(3) \qquad\qquad \sum_{m=1}^{\infty} z_m = z_1 + z_2 + \cdots.$$

The $z_1, z_2, \cdots$ are called the **terms** of the series. (Our usual **summation letter** is n, unless we need n for another purpose, as here.)

A **convergent series** is one whose sequence of partial sums converges, say,

$$\lim_{n \to \infty} s_n = s. \qquad\qquad \text{Then we write} \qquad\qquad s = \sum_{m=1}^{\infty} z_m = z_1 + z_2 + \cdots$$

and call s the **sum** or *value* of the series. A series that is not convergent is called a **divergent series.**

If we omit the terms of s_n from (3), there remains

$$(4) \qquad\qquad R_n = z_{n+1} + z_{n+2} + z_{n+3} + \cdots.$$

This is called the **remainder** *of the series* (3) *after the term* z_n. Clearly, if (3) converges and has the sum s, then

$$s = s_n + R_n, \qquad\qquad \text{thus} \qquad\qquad R_n = s - s_n.$$

Now $s_n \to s$ by the definition of convergence; hence $R_n \to 0$. In applications, when s is unknown and we compute an approximation s_n of s, then $|R_n|$ is the error, and $R_n \to 0$ means that we can make $|R_n|$ as small as we please, by choosing n large enough.

An application of Theorem 1 to the partial sums immediately relates the convergence of a complex series to that of the two series of its real parts and of its imaginary parts:

THEOREM 2 **(Real and imaginary parts)**

A series (3) *with* $z_m = x_m + iy_m$ *converges and has the sum* $s = u + iv$ *if and only if* $x_1 + x_2 + \cdots$ *converges and has the sum* u *and* $y_1 + y_2 + \cdots$ *converges and has the sum* v.

Tests for Convergence and Divergence of Series

Convergence tests in complex are practically the same as in calculus. We apply them before we use a series, to make sure that the series converges.

Divergence can often be shown very simply as follows.

THEOREM 3 **(Divergence)**

If a series $z_1 + z_2 + \cdots$ *converges, then* $\lim\limits_{m \to \infty} z_m = 0$. *Hence if this does not hold, the series diverges.*

PROOF.   If $z_1 + z_2 + \cdots$ converges with the sum s, then, since $z_m = s_m - s_{m-1}$,

$$\lim_{m \to \infty} z_m = \lim_{m \to \infty} (s_m - s_{m-1}) = \lim_{m \to \infty} s_m - \lim_{m \to \infty} s_{m-1} = s - s = 0. \qquad \blacktriangleleft$$

Caution! $z_m \to 0$ is *necessary* for convergence but *not sufficient,* as we see from the harmonic series $1 + \frac{1}{2} + \frac{1}{3} + \frac{1}{4} + \cdots$, which satisfies this condition but diverges, as is shown in calculus (see, for example, Ref. [13] listed in Appendix 1).

The practical difficulty in proving convergence is that in most cases the sum of a series is unknown. Cauchy overcame this by showing that a series converges if and only if its partial sums eventually get close to each other:

THEOREM 4 **(Cauchy's convergence principle for series)**

A series $z_1 + z_2 + \cdots$ *is convergent if and only if for every given* $\epsilon > 0$ (*no matter how small*) *we can find an* N (*which depends on* ϵ, *in general*) *such that*

(5) $\left| z_{n+1} + z_{n+2} + \cdots + z_{n+p} \right| < \epsilon$ *for every* $n > N$ *and* $p = 1, 2, \cdots$.

The somewhat involved proof is left optional (see Appendix 4).

Absolute convergence. A series $z_1 + z_2 + \cdots$ is called **absolutely convergent** if the series of the absolute values of the terms

$$\sum_{m=1}^{\infty} |z_m| = |z_1| + |z_2| + \cdots$$

is convergent.

If $z_1 + z_2 + \cdots$ converges but $|z_1| + |z_2| + \cdots$ diverges, then the series $z_1 + z_2 + \cdots$ is called, more precisely, **conditionally convergent.**

EXAMPLE 3   **A conditionally convergent series**

The series $1 - \frac{1}{2} + \frac{1}{3} - \frac{1}{4} + - \cdots$ converges, but only conditionally since the harmonic series diverges, as mentioned above (after Theorem 3). ◀

> *If a series is absolutely convergent, it is convergent.*

This follows readily from Cauchy's principle (see Team Project 20). This principle also yields the following general convergence test.

THEOREM 5   **(Comparison test)**

If a series $z_1 + z_2 + \cdots$ is given and we can find a converging series $b_1 + b_2 + \cdots$ with nonnegative real terms such that

$$|z_n| \leqq b_n \qquad \text{for } n = 1, 2, \cdots,$$

then the given series converges, even absolutely.

PROOF.   By Cauchy's principle, since $b_1 + b_2 + \cdots$ converges, for any given $\epsilon > 0$ we can find an N such that

$$b_{n+1} + \cdots + b_{n+p} < \epsilon \qquad \text{for every } n > N \text{ and } p = 1, 2, \cdots.$$

From this and $|z_1| \leqq b_1$, $|z_2| \leqq b_2$, $\cdots$ we conclude that for those n and p,

$$|z_{n+1}| + \cdots + |z_{n+p}| \leqq b_{n+1} + \cdots + b_{n+p} < \epsilon.$$

Hence, again by Cauchy's principle, $|z_1| + |z_2| + \cdots$ converges, so that $z_1 + z_2 + \cdots$ is absolutely convergent. ◀

A good comparison series is the geometric series, which behaves as follows.

THEOREM 6   **(Geometric series)**

*The **geometric series***

(6*)
$$\sum_{m=0}^{\infty} q^m = 1 + q + q^2 + \cdots$$

converges with the sum $1/(1 - q)$ if $|q| < 1$ and diverges if $|q| \geqq 1$.

PROOF.   If $|q| \geqq 1$, then $|q^m| \geqq 1$ and Theorem 3 implies divergence.

Now let $|q| < 1$. The nth partial sum is

$$s_n = 1 + q + \cdots + q^n.$$

From this,

$$q s_n = \quad q + \cdots + q^n + q^{n+1}.$$

On subtraction, most terms on the right cancel in pairs, and we are left with

$$s_n - q s_n = (1 - q) s_n = 1 - q^{n+1}.$$

Now $1 - q \neq 0$ since $q \neq 1$, and we may solve for s_n, finding

(6)
$$s_n = \frac{1 - q^{n+1}}{1 - q} = \frac{1}{1 - q} - \frac{q^{n+1}}{1 - q}.$$

Since $|q| < 1$, the last term approaches zero as $n \to \infty$. Hence the series is convergent and has the sum $1/(1 - q)$. This completes the proof. ◀

Ratio Test

This is the most important test in our further work. We get it by taking the geometric series as the comparison series $b_1 + b_2 + \cdots$ in Theorem 5:

THEOREM 7   **(Ratio test)**

If a series $z_1 + z_2 + \cdots$ with $z_n \neq 0$ ($n = 1, 2, \cdots$) has the property that for every n greater than some N,

(7)
$$\left| \frac{z_{n+1}}{z_n} \right| \leq q < 1 \qquad\qquad (n > N)$$

(where $q < 1$ is fixed), this series converges absolutely. If for every $n > N$,

(8)
$$\left| \frac{z_{n+1}}{z_n} \right| \geq 1 \qquad\qquad (n > N),$$

the series diverges.

PROOF.   If (8) holds, then $|z_{n+1}| \geq |z_n|$ for those n, so that divergence of the series follows from Theorem 3.

If (7) holds, then $|z_{n+1}| \leq |z_n| q$ for $n > N$, in particular,

$$|z_{N+2}| \leq |z_{N+1}| q, \qquad |z_{N+3}| \leq |z_{N+2}| q \leq |z_{N+1}| q^2, \qquad \text{etc.,}$$

and in general, $|z_{N+p}| \leq |z_{N+1}| q^{p-1}$. Since $q < 1$, we obtain from this and Theorem 6

$$|z_{N+1}| + |z_{N+2}| + |z_{N+3}| + \cdots \leq |z_{N+1}| (1 + q + q^2 + \cdots) \leq |z_{N+1}| \frac{1}{1 - q}.$$

Absolute convergence of $z_1 + z_2 + \cdots$ now follows from Theorem 5. ◀

Caution! The inequality (7) implies $|z_{n+1}/z_n| < 1$, but this does **not** imply convergence, as we see from the harmonic series, which satisfies $z_{n+1}/z_n = n/(n + 1) < 1$ for all n but diverges.

If the sequence of the ratios in (7) and (8) converges, we get the more convenient

THEOREM 8   **(Ratio test)**

If a series $z_1 + z_2 + \cdots$ with $z_n \neq 0$ ($n = 1, 2, \cdots$) is such that $\lim\limits_{n \to \infty} \left| \dfrac{z_{n+1}}{z_n} \right| = L$, then:

    (a) *The series converges absolutely if $L < 1$.*

    (b) *The series diverges if $L > 1$.*

    (c) *If $L = 1$, the test fails; that is, no conclusion is possible.*

PROOF.   **(a)** Write $k_n = |z_{n+1}/z_n|$. Let $L = 1 - b < 1$. Then, by the definition of a limit, the k_n must eventually get close to $1 - b$, say, $k_n \leqq q = 1 - \frac{1}{2}b < 1$ for all n greater than some N. Convergence of $z_1 + z_2 + \cdots$ now follows from Theorem 7.

(b) Similarly, for $L = 1 + c > 1$ we have $k_n \geqq 1 + \frac{1}{2}c > 1$ for all $n > N^*$ (sufficiently large), which implies divergence of $z_1 + z_2 + \cdots$ by Theorem 7.

(c) The harmonic series has $z_{n+1}/z_n = n/(n + 1)$, hence $L = 1$ and diverges. The series

$$1 + \frac{1}{4} + \frac{1}{9} + \frac{1}{16} + \frac{1}{25} + \cdots \qquad \text{has} \qquad \frac{z_{n+1}}{z_n} = \frac{n^2}{(n + 1)^2},$$

hence also $L = 1$, but converges. Convergence follows from (Fig. 350)

$$s_n = 1 + \frac{1}{4} + \cdots + \frac{1}{n^2} \leqq 1 + \int_1^n \frac{dx}{x^2} = 2 - \frac{1}{n},$$

so that $s_1, s_2, \cdots$ is a bounded sequence and is monotone increasing (since the terms of the series are all positive); both properties together are sufficient for the convergence of the real sequence $s_1, s_2, \cdots$. (In calculus this is proved by the so-called *integral test*, whose idea we have used.)   ◀

EXAMPLE 4   **Ratio test**

Is the following series convergent or divergent? (First guess, then calculate.)

$$\sum_{n=0}^{\infty} \frac{(100 + 75i)^n}{n!} = 1 + (100 + 75i) + \frac{1}{2!}(100 + 75i)^2 + \cdots$$

Solution.   By Theorem 8, the series is convergent, since

$$\left| \frac{z_{n+1}}{z_n} \right| = \frac{|100 + 75i|^{n+1}/(n + 1)!}{|100 + 75i|^n/n!} = \frac{|100 + 75i|}{n + 1} = \frac{125}{n + 1} \;\rightarrow\; L = 0.$$
  ◀

EXAMPLE 5   **Theorem 7 more general than Theorem 8**

Let $a_n = i/2^{3n}$ and $b_n = 1/2^{3n+1}$. Is the following series convergent or divergent?

$$a_0 + b_0 + a_1 + b_1 + \cdots = i + \frac{1}{2} + \frac{i}{8} + \frac{1}{16} + \frac{i}{64} + \frac{1}{128} + \cdots$$

Solution.   The ratios of the absolute values of successive terms are $\frac{1}{2}, \frac{1}{4}, \frac{1}{2}, \frac{1}{4}, \cdots$. Hence convergence follows from Theorem 7. Since the sequence of these ratios has no limit, Theorem 8 is not applicable.   ◀

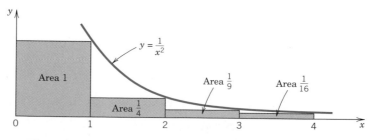

Fig. 350.   Convergence of the series $1 + \frac{1}{4} + \frac{1}{9} + \frac{1}{16} + \cdots$

Root Test

The two practically most important tests are the ratio test and the root test. The ratio test is usually simpler; the root test is somewhat more general.

THEOREM 9   **(Root test)**

If a series $z_1 + z_2 + \cdots$ is such that for every n greater than some N,

(9) $$\sqrt[n]{|z_n|} \leqq q < 1 \qquad (n > N)$$

(where $q < 1$ is fixed), this series converges absolutely. If for infinitely many n,

(10) $$\sqrt[n]{|z_n|} \geqq 1,$$

the series diverges.

PROOF.   If (9) holds, then $|z_n| \leqq q^n < 1$ for all $n > N$. Hence the series $|z_1| + |z_2| + \cdots$ converges by comparison with the geometric series, so that the series $z_1 + z_2 + \cdots$ converges absolutely. If (10) holds, then $|z_n| \geqq 1$ for infinitely many n. Divergence of $z_1 + z_2 + \cdots$ now follows from Theorem 3.   ◀

Caution! Equation (9) implies $\sqrt[n]{|z_n|} < 1$, but <u>this</u> does not imply convergence, as we see from the harmonic series, which satisfies $\sqrt[n]{1/n} < 1$ but diverges.

If the sequence of the roots in (9) and (10) converges, we more conveniently have

THEOREM 10   **(Root test)**

If a series $z_1 + z_2 + \cdots$ is such that $\lim\limits_{n \to \infty} \sqrt[n]{|z_n|} = L$, then:

    (a) *The series converges absolutely if $L < 1$.*

    (b) *The series diverges if $L > 1$.*

    (c) *If $L = 1$, the test fails; that is, no conclusion is possible.*

PROOF.   The proof parallels that of Theorem 8.

    (a) Let $L = 1 - a^* < 1$. Then by the definition of a limit we have $\sqrt[n]{|z_n|} < q = 1 - \frac{1}{2}a^* < 1$ for all n greater than some (sufficiently large) N^*. Hence $|z_n| < q^n < 1$ for all $n > N^*$. Absolute convergence of the series $z_1 + z_2 + \cdots$ now follows by the comparison with the geometric series.

    (b) If $L > 1$, then we also have $\sqrt[n]{|z_n|} > 1$ for all sufficiently large n. Hence $|z_n| > 1$ for those n. Theorem 3 now implies that $z_1 + z_2 + \cdots$ diverges.

    (c) Both the *divergent* harmonic series and the *convergent* series $1 + \frac{1}{4} + \frac{1}{9} + \frac{1}{16} + \frac{1}{25} + \cdots$ give $L = 1$. This can be seen from $(\ln n)/n \to 0$ and

$$\sqrt[n]{\frac{1}{n}} = \frac{1}{n^{1/n}} = \frac{1}{e^{(1/n)\ln n}} \;\to\; \frac{1}{e^0}\,, \qquad \sqrt[n]{\frac{1}{n^2}} = \frac{1}{n^{2/n}} = \frac{1}{e^{(2/n)\ln n}} \;\to\; \frac{1}{e^0}\,. \quad ◀$$

PROBLEM SET 14.1

Sequences

1. **(Uniqueness of limit)** Show that if a sequence converges, its limit is unique.

Are the following sequences $z_1, z_2, \cdots, z_n, \cdots$ bounded? Convergent? Find their limit points. (Show the details of your work.)

2. $z_n = e^{-n\pi i} + (-1)^n i$

3. $z_n = e^{n\pi i/4}/n$

4. $z_n = (1 + i)^{2n}/2^n$

5. $z_n = n\pi i/(n + i)$

6. $z_n = (\cos(n\pi))/n$

7. $z_n = (-1)^n + 100i$

8. $z_n = (12 + 16i)^n/n!$

9. $z_n = n\pi/(1 + 3ni)$

10. **(Addition)** If $z_1, z_2, \cdots$ converges with the limit l and $z_1{}^*, z_2{}^*, \cdots$ converges with the limit l^*, show that $z_1 + z_1{}^*, z_2 + z_2{}^*, \cdots$ converges with the limit $l + l^*$.

11. **(Boundedness)** Show that a complex sequence is bounded if and only if the two corresponding sequences of the real parts and of the imaginary parts are bounded.

Series

Are the following series convergent or divergent?

12. $\displaystyle\sum_{n=0}^{\infty} \frac{(20 + 30i)^n}{n!}$

13. $\displaystyle\sum_{n=1}^{\infty} n^2 \left(\frac{i}{2}\right)^n$

14. $\displaystyle\sum_{n=0}^{\infty} \frac{i^n}{n^2 + i}$

15. $\displaystyle\sum_{n=1}^{\infty} \frac{1}{\sqrt{n}}$

16. $\displaystyle\sum_{n=1}^{\infty} \frac{(n!)^2}{(2n)!}$

17. $\displaystyle\sum_{n=1}^{\infty} \frac{(3i)^n n!}{n^n}$

18. $\displaystyle\sum_{n=2}^{\infty} \frac{1}{\ln n}$

 19. **CAS PROJECT. Sequences and Series. (a)** Write a program for plotting complex **sequences.** Apply it to sequences of your choice that have interesting "geometrical" properties (e.g., lying on an ellipse, spiraling toward its limit, etc.).

    (b) Write a program for computing and plotting numerical values of the first n partial sums of a **series** of complex numbers. Use the program to experiment with the rapidity of convergence of series of your choice.

20. **TEAM PROJECT. Series. (a) Absolute convergence.** Show that if a series converges absolutely, it is convergent.

    (b) Write a short essay on the basic concepts and properties of series of numbers, explaining in each case whether or not they carry over from real series (discussed in calculus) to complex series, with reasons given.

    (c) Estimate of the remainder. Let $|z_{n+1}/z_n| \leqq q < 1$, so that the series $z_1 + z_2 + \cdots$ converges by the ratio test. Show that the remainder $R_n = z_{n+1} + z_{n+2} + \cdots$ satisfies the inequality $|R_n| \leqq |z_{n+1}|/(1 - q)$.

    (d) Using (c), find how many terms suffice for computing the sum s of the series

$$\sum_{n=1}^{\infty} \frac{n + i}{2^n n}$$

with an error not exceeding 0.05 and compute s to this accuracy.

    (e) Find other applications of the estimate in (c).

14.2 Power Series

Power series are the most important series in complex analysis because we shall see that their sums are analytic functions, and every analytic function can be represented by power series (Theorem 5 in Sec. 14.3 and Theorem 1 in Sec. 14.4).

A **power series** *in powers of* $z - z_0$ is a series of the form

(1)
$$\sum_{n=0}^{\infty} a_n(z - z_0)^n = a_0 + a_1(z - z_0) + a_2(z - z_0)^2 + \cdots$$

where z is a complex variable, $a_0, a_1, \cdots$ are complex (or real) constants, called the **coefficients** of the series, and z_0 is a complex (or real) constant, called the **center** of the series.

If $z_0 = 0$, we obtain as a particular case a *power series in powers of z:*

(2)
$$\sum_{n=0}^{\infty} a_n z^n = a_0 + a_1 z + a_2 z^2 + \cdots.$$

Convergence Behavior of Power Series

Power series have variable terms (functions of z), but if we fix z, then all the concepts in the last section apply. Usually a series with variable terms will converge for some z and diverge for others. For a power series the situation is simple. The series (1) may converge in a disk with center z_0 or in the whole z-plane or only at z_0. We illustrate this with typical examples and then prove it.

EXAMPLE 1 **Convergence in a disk. Geometric series**

The *geometric series*
$$\sum_{n=0}^{\infty} z^n = 1 + z + z^2 + \cdots$$

converges absolutely if $|z| < 1$ and diverges if $|z| \geqq 1$ (see Theorem 6 in Sec. 14.1).   ◀

EXAMPLE 2 **Convergence for every z**

The power series (which will be the Maclaurin series of e^z in Sec. 14.4)
$$\sum_{n=0}^{\infty} \frac{z^n}{n!} = 1 + z + \frac{z^2}{2!} + \frac{z^3}{3!} + \cdots$$

is absolutely convergent for every z. In fact, by the ratio test, for any fixed z,
$$\left| \frac{z^{n+1}/(n+1)!}{z^n/n!} \right| = \frac{|z|}{n+1} \; \rightarrow \; 0 \quad \text{as} \quad n \rightarrow \infty.$$ ◀

EXAMPLE 3 **Convergence only at the center. (Useless series)**

The following power series converges only at $z = 0$, but diverges for every $z \neq 0$, as we shall show.
$$\sum_{n=0}^{\infty} n! z^n = 1 + z + 2z^2 + 6z^3 + \cdots$$

In fact, from the ratio test we have

$$\left| \frac{(n + 1)! \, z^{n+1}}{n! \, z^n} \right| = (n + 1)|z| \;\rightarrow\; \infty \quad \text{as} \quad n \to \infty \quad (z \text{ fixed and } \neq 0). \quad \blacktriangleleft$$

THEOREM 1   **(Convergence of a power series)**

(a) *Every power series* (1) *converges at the center* z_0.

(b) *If* (1) *converges at a point* $z = z_1 \neq z_0$, *it converges absolutely for every* z *closer to* z_0 *than* z_1, *that is,* $|z - z_0| < |z_1 - z_0|$. *See Fig. 351.*

(c) *If* (1) *diverges at a* $z = z_2$, *it diverges for every* z *farther away from* z_0 *than* z_2. *See Fig. 351.*

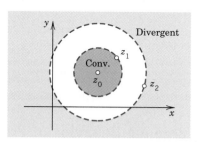

Fig. 351.   Theorem 1

PROOF.   (a) For $z = z_0$ the series (1) reduces to the single term a_0.

(b) Convergence at $z = z_1$ gives by Theorem 3 in Sec. 14.1 $a_n(z_1 - z_0)^n \to 0$ as $n \to \infty$. This implies boundedness in absolute value,

$$\left| a_n(z_1 - z_0)^n \right| < M \qquad \text{for every } n = 0, 1, \cdots.$$

Multiplying and dividing $a_n(z - z_0)^n$ by $(z_1 - z_0)^n$ we obtain from this

$$(3) \qquad \left| a_n(z - z_0)^n \right| = \left| a_n(z_1 - z_0)^n \left(\frac{z - z_0}{z_1 - z_0} \right)^n \right| \leqq M \left| \frac{z - z_0}{z_1 - z_0} \right|^n.$$

Now our assumption $|z - z_0| < |z_1 - z_0|$ implies that

$$\left| \frac{z - z_0}{z_1 - z_0} \right| < 1. \qquad \text{Hence the series} \qquad M \sum_{n=0}^{\infty} \left| \frac{z - z_0}{z_1 - z_0} \right|^n$$

is a converging geometric series (see Theorem 6 in Sec. 14.1). The convergence of (1) when $|z - z_0| < |z_1 - z_0|$ now follows from (3) and the comparison test in Sec. 14.1.

(c) If this were false, we would have convergence at a z_3 farther away from z_0 than z_2. This would imply convergence at z_2, by (b), a contradiction to our assumption of divergence at z_2. $\blacktriangleleft$

Radius of Convergence of a Power Series

Convergence for every z (the nicest case, Example 2) or for no $z \neq z_0$ (the useless case, Example 3) needs no further discussion, and we put these cases aside for a moment. We consider the *smallest* circle with center z_0 that includes all the points at which a given power series (1) converges. Let R denote its radius. The circle

$$\boxed{|z - z_0| = R} \qquad \text{(Fig. 352)}$$

is called the **circle of convergence** and its radius R the **radius of convergence** of (1). Theorem 1 then implies convergence everywhere within that circle, that is, for all z for which

(4) $$|z - z_0| < R$$

(the open disk with center z_0 and radius R). Also, since R is as *small* as possible, the series (1) diverges for all z for which

(5) $$|z - z_0| > R.$$

No general statements can be made about the convergence of a power series (1) *on the circle of convergence* itself. The series (1) may converge at some or all or none of these points. Details will not be essential to us. Hence a simple example may just give us the idea.

EXAMPLE 4   **Behavior on the circle of convergence**

On the circle of convergence (radius $R = 1$ in all three series),

$\sum z^n/n^2$ converges everywhere since $\sum 1/n^2$ converges,

$\sum z^n/n$ converges at -1 (by Leibniz's test) but diverges at 1,

$\sum z^n$ diverges everywhere. ◀

Notations $R = \infty$ and $R = 0$. To include these two excluded cases in the notation, we write

$\quad R = \infty$ if the series (1) converges for all z (as in Example 2),

$\quad R = 0$ if (1) converges only at the center $z = z_0$ (as in Example 3).

These are convenient notations, but nothing else.

Real power series in powers of $x - x_0$ with real coefficients and center. For these, (4) gives the **convergence interval** $|x - x_0| < R$ of length $2R$ on the real line.

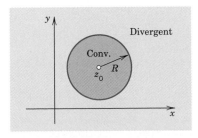

Fig. 352.   Circle of convergence

Determination of the radius of convergence from the coefficients of the series. For this important practical task we can use

THEOREM 2 **(Radius of convergence R)**

Suppose that the sequence $|a_{n+1}/a_n|$, $n = 1, 2, \cdots$, converges with limit L^*. If $L^* = 0$, then $R = \infty$; that is, the power series (1) converges for all z. If $L^* \neq 0$ (hence $L^* > 0$), then

(6)
$$R = \frac{1}{L^*} = \lim_{n \to \infty} \left| \frac{a_n}{a_{n+1}} \right|.$$
(Cauchy–Hadamard formula[1]).

If $|a_{n+1}/a_n| \to \infty$, then $R = 0$ (convergence only at the center z_0).

PROOF. For (1) the ratio of the terms in the ratio test (Sec. 14.1) is

$$\left| \frac{a_{n+1}(z - z_0)^{n+1}}{a_n(z - z_0)^n} \right| = \left| \frac{a_{n+1}}{a_n} \right| |z - z_0|. \qquad \text{The limit is} \qquad L = L^*|z - z_0|.$$

Let $L^* \neq 0$, thus $L^* > 0$. We have convergence if $L = L^*|z - z_0| < 1$, thus $|z - z_0| < 1/L^*$, and divergence if $|z - z_0| > 1/L^*$. By (4) and (5) this shows that $1/L^*$ is the convergence radius and proves (6).

If $L^* = 0$, then $L = 0$ for every z, which gives convergence for all z by the ratio test. If $|a_{n+1}/a_n| \to \infty$, then $|a_{n+1}/a_n||z - z_0| > 1$ for any $z \neq z_0$ and all sufficiently large n. This implies divergence for all $z \neq z_0$ by the ratio test (Theorem 7, Sec. 14.1). ◄

Formula (6) will not help if L^* does not exist, but extensions of Theorem 2 are still possible, as we discuss in Example 6 below.

EXAMPLE 5 **Radius of convergence**

By (6) the radius of convergence of the power series $\sum_{n=0}^{\infty} \frac{(2n)!}{(n!)^2} (z - 3i)^n$ is

$$R = \lim_{n \to \infty} \left[\frac{(2n)!}{(n!)^2} \bigg/ \frac{(2n + 2)!}{((n + 1)!)^2} \right] = \lim_{n \to \infty} \left[\frac{(2n)!}{(2n + 2)!} \cdot \frac{((n + 1)!)^2}{(n!)^2} \right] = \lim_{n \to \infty} \frac{(n + 1)^2}{(2n + 2)(2n + 1)} = \frac{1}{4}.$$

The series converges in the open disk $|z - 3i| < \frac{1}{4}$ of radius $\frac{1}{4}$ and center $3i$. ◄

EXAMPLE 6 **Extension of Theorem 2**

Find the radius of convergence R of the power series

$$\sum_{n=0}^{\infty} \left[1 + (-1)^n + \frac{1}{2^n} \right] z^n = 3 + \frac{1}{2} z + \left(2 + \frac{1}{4} \right) z^2 + \frac{1}{8} z^3 + \left(2 + \frac{1}{16} \right) z^4 + \cdots.$$

Solution. The sequence of the ratios $1/6$, $2(2 + \frac{1}{4})$, $1/(8(2 + \frac{1}{4}))$, $\cdots$ does not converge, so that Theorem 2 is of no help. It can be shown that

(6*)
$$R = 1/\tilde{L}, \qquad \tilde{L} = \lim_{n \to \infty} \sqrt[n]{|a_n|}.$$

---

[1]Named after the French mathematicians A. L. CAUCHY (see the footnote in Sec. 2.6) and JACQUES HADAMARD (1865—1963). Hadamard made basic contributions to the theory of power series and devoted his lifework to partial differential equations.

This still does not help here, since $\left\{\sqrt[n]{|a_n|}\right\}$ does not converge because $\sqrt[n]{|a_n|} = \sqrt[n]{1/2^n} = 1/2$ for odd n and $\sqrt[n]{|a_n|} = \sqrt[n]{2 + 1/2^n} \to 1$ as $n \to \infty$, so that $\sqrt[n]{|a_n|}$ has the two limit points $1/2$ and 1. It can further be shown that

$$(6^{**}) \qquad\qquad R = 1/\tilde{l}, \qquad \tilde{l} \text{ the greatest limit point of the sequence } \left\{\sqrt[n]{|a_n|}\right\}.$$

Here $\tilde{l} = 1$, so that $R = 1$. *Answer.* The series converges for $|z| < 1$.   ◀

Summary. Power series converge in an open circular disk or some even for every z (or some only at the center, but they are useless); for the radius of convergence, see (6) or Example 6.

Except for the useless ones, power series have sums that are analytic functions (as we show in the next section); this accounts for their importance.

PROBLEM SET 14.2

Radius of Convergence

Find the center and the radius of convergence of the following power series. (Show the details of your work.)

1. $\displaystyle\sum_{n=1}^{\infty} n(z + i\sqrt{2})^n$

2. $\displaystyle\sum_{n=0}^{\infty} \frac{2^{20n}}{n!} (z - 3)^n$

3. $\displaystyle\sum_{n=0}^{\infty} \frac{n + 5i}{(2n)!} (z - i)^n$

4. $\displaystyle\sum_{n=1}^{\infty} \frac{1}{n(n + 1)} \left(\frac{z}{4}\right)^{n+1}$

5. $\displaystyle\sum_{n=0}^{\infty} (n + 2i)^n z^n$

6. $\displaystyle\sum_{n=0}^{\infty} \left(\frac{a}{b}\right)^n (z - \pi i)^n$

7. $\displaystyle\sum_{n=0}^{\infty} \frac{1}{(1 + i)^n} (z + 2 - i)^n$

8. $\displaystyle\sum_{n=2}^{\infty} n(n - 1)2^n z^{3n}$

9. $\displaystyle\sum_{n=0}^{\infty} \left(\frac{4 - 2i}{1 + 5i}\right)^n z^n$

10. $\displaystyle\sum_{n=0}^{\infty} \frac{(n!)^2 i^n}{(2n)!} (z + 1)^n$

11. $\displaystyle\sum_{n=0}^{\infty} \frac{(3n)!}{2^n (n!)^3} z^n$

12. $\displaystyle\sum_{n=0}^{\infty} \frac{n^n}{n!} (z - 3i)^n$

13. $\displaystyle\sum_{n=0}^{\infty} \frac{n!}{n^n} (z + \pi)^n$

14. $\displaystyle\sum_{n=0}^{\infty} \frac{(-1)^n}{(2n + 1)!} z^{2n+1}$

15. $\displaystyle\sum_{n=0}^{\infty} \frac{(z - 2i)^n}{n^n}$

16. $\displaystyle\sum_{n=0}^{\infty} \frac{i^n n^3}{2^n} z^{2n}$

17. $3^2 z^2 + z^3 + 3^4 z^4 + z^5 + 3^6 z^6 + z^7 + \cdots$

18. $6^{-2} z^2 + 6^3 z^3 + 6^{-4} z^4 + 6^5 z^5 + 6^{-6} z^6 + \cdots$

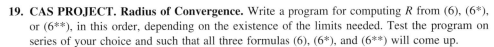 **19. CAS PROJECT. Radius of Convergence.** Write a program for computing R from (6), (6*), or (6**), in this order, depending on the existence of the limits needed. Test the program on series of your choice and such that all three formulas (6), (6*), and (6**) will come up.

20. TEAM PROJECT. Radius of Convergence. (a) Formula (6) for R contains $|a_n/a_{n+1}|$, not $|a_{n+1}/a_n|$. How could you memorize this by using a qualitative argument?

(b) **Change of coefficients.** What happens to R $(0 < R < \infty)$ if you (i) multiply all a_n by $k \neq 0$, (ii) multiply a_n by $k^n \neq 0$, (iii) replace a_n by $1/a_n$?

(c) **Example 6** extends Theorem 2 to nonconvergent cases of a_n/a_{n+1}. Do you understand the principle of "mixing" by which Example 6 was obtained? Use this principle for making up further examples.

(d) **Not all powers present.** Show that if $\Sigma a_n z^n$ has radius of convergence R $(< \infty)$, then $\Sigma a_n z^{2n}$ has radius of convergence $\sqrt{R}$. Give examples.

14.3 Functions Given by Power Series

The main goal of this section is to show that power series represent analytic functions (Theorem 5). On the way we shall see that power series behave nicely under addition, multiplication, differentiation, and integration, which makes them very useful in complex analysis.

To simplify the formulas in this section, we take $z_0 = 0$ and write

$$(1) \qquad \sum_{n=0}^{\infty} a_n z^n.$$

This is no restriction because a series in powers of $\hat{z} - z_0$ with any z_0 can always be reduced to the form (1) if we set $\hat{z} - z_0 = z$.

Terminology and Notation. If any given power series (1) has a nonzero radius of convergence R (thus $R > 0$), its sum is a function of z, say $f(z)$. Then we write

$$(2) \qquad f(z) = \sum_{n=0}^{\infty} a_n z^n = a_0 + a_1 z + a_2 z^2 + \cdots \qquad\qquad (|z| < R).$$

We say that $f(z)$ is **represented** by the power series or that it is **developed** in the power series. For instance, the geometric series *represents* the function $f(z) = 1/(1 - z)$ in the interior of the unit circle $|z| = 1$. (See Theorem 6 in Sec. 14.1.)

Uniqueness of a power series representation. This is our next goal. It means that *a function $f(z)$ cannot be represented by two different power series with the same center.* We claim that if $f(z)$ can at all be developed in a power series with center z_0, the development is unique. This important fact is frequently used in complex analysis (as well as in calculus). We shall prove it in Theorem 2, below. The proof will follow from

THEOREM 1   **(Continuity of the sum of a power series)**

The function $f(z)$ in (2) with $R > 0$ is continuous at $z = 0$.

PROOF.   From (2) with $z = 0$ we have $f(0) = a_0$. Hence by the definition of continuity we must show that $\lim_{z \to 0} f(z) = f(0) = a_0$. That is, we must show that for a given $\epsilon > 0$ there is a $\delta > 0$ such that $|z| < \delta$ implies $|f(z) - a_0| < \epsilon$. Now (2) converges absolutely for $|z| \leqq r$ with any $r < R$, by Theorem 1 in Sec. 14.2. Hence the series

$$\sum_{n=1}^{\infty} |a_n| r^{n-1} = \frac{1}{r} \sum_{n=1}^{\infty} |a_n| r^n$$

with $r > 0$ converges. Let $S \neq 0$ be its sum. ($S = 0$ is trivial.) Then for $0 < |z| \leqq r$,

$$|f(z) - a_0| = \left| \sum_{n=1}^{\infty} a_n z^n \right| \leqq |z| \sum_{n=1}^{\infty} |a_n| \, |z|^{n-1} \leqq |z| \sum_{n=1}^{\infty} |a_n| r^{n-1} = |z| S.$$

This is less than ϵ for $|z| < \delta$, where $\delta > 0$ is less than r (convergence!) as well as ϵ/S. ◀

From this theorem we can now readily obtain the desired uniqueness theorem (again assuming $z_0 = 0$ without loss of generality):

THEOREM 2   **(Identity theorem for power series. Uniqueness of representation)**

Suppose that the power series

$$\sum_{n=0}^{\infty} a_n z^n \qquad and \qquad \sum_{n=0}^{\infty} b_n z^n$$

both converge for $|z| < R$, where R is positive, and have the same sum for all these z. Then these series are identical, that is,

$$a_n = b_n \qquad\qquad\qquad for\ all\ n = 0, 1, \cdots .$$

Hence if a function $f(z)$ has a power series representation with any center z_0, this representation is unique.

PROOF.   We proceed by induction. By assumption,

$$a_0 + a_1 z + a_2 z^2 + \cdots = b_0 + b_1 z + b_2 z^2 + \cdots \qquad (|z| < R).$$

The sums of these two power series are continuous at $z = 0$, by Theorem 1. Hence if we consider $|z| > 0$ and let $z \to 0$ on both sides, we see that $a_0 = b_0$: the assertion is true for $n = 0$. Now assume that $a_n = b_n$ for $n = 0, 1, \cdots, m$. Then on both sides we may omit the terms that are equal and divide the result by z^{m+1} ($\neq 0$); this gives

$$a_{m+1} + a_{m+2} z + a_{m+3} z^2 + \cdots = b_{m+1} + b_{m+2} z + b_{m+3} z^2 + \cdots .$$

Similarly as before by letting $z \to 0$ we conclude from this that $a_{m+1} = b_{m+1}$. This completes the proof.   ◄

Operations on Power Series

Interesting in itself, this discussion will serve to prepare for our main goal, namely, to show that functions represented by power series are analytic.

Termwise addition or subtraction of two power series with radii of convergence R_1 and R_2 yields a power series with radius of convergence at least equal to the smaller of R_1 and R_2. *Proof.* Add (or subtract) the partial sums s_n and s_n^* term by term and use $\lim (s_n \pm s_n^*) = \lim s_n \pm \lim s_n^*$.

Termwise multiplication of two power series

$$f(z) = \sum_{k=0}^{\infty} a_k z^k = a_0 + a_1 z + \cdots$$

and

$$g(z) = \sum_{m=0}^{\infty} b_m z^m = b_0 + b_1 z + \cdots$$

means the multiplication of each term of the first series by each term of the second series and the collection of like powers of z. This gives a power series, which is called the **Cauchy product** of the two series and is given by

$$a_0 b_0 + (a_0 b_1 + a_1 b_0) z + (a_0 b_2 + a_1 b_1 + a_2 b_0) z^2 + \cdots$$

$$= \sum_{n=0}^{\infty} (a_0 b_n + a_1 b_{n-1} + \cdots + a_n b_0) z^n.$$

We mention without proof that this power series converges absolutely for each z within the circle of convergence of each of the two given series and has the sum $s(z) = f(z)g(z)$. For a proof, see [D6] listed in Appendix 1.

Termwise differentiation and integration of power series is permissible, as we show next. We call **derived series** *of the power series* (1) the power series obtained from (1) by termwise differentiation, that is,

$$(3) \qquad \sum_{n=1}^{\infty} n a_n z^{n-1} = a_1 + 2 a_2 z + 3 a_3 z^2 + \cdots.$$

THEOREM 3   **(Termwise differentiation of a power series)**

The derived series of a power series has the same radius of convergence as the original series.

PROOF.   This follows from (6) in Sec. 14.2 because

$$\lim_{n \to \infty} \frac{n|a_n|}{(n+1)|a_{n+1}|} = \lim_{n \to \infty} \frac{n}{n+1} \lim_{n \to \infty} \left| \frac{a_n}{a_{n+1}} \right| = \lim_{n \to \infty} \left| \frac{a_n}{a_{n+1}} \right|$$

or, if the limit does not exist, from (6**) in Sec. 14.2 by noting that $\sqrt[n]{n} \to 1$ as $n \to \infty$.   ◄

EXAMPLE 1   **An application of Theorem 3**

Find the radius of convergence R of the following series by applying Theorem 3.

$$\sum_{n=2}^{\infty} \binom{n}{2} z^n = z^2 + 3 z^3 + 6 z^4 + 10 z^5 + \cdots.$$

Solution.   Differentiate the geometric series twice term by term and multiply the result by $z^2/2$. This yields the given series. Hence $R = 1$ by Theorem 3.   ◄

THEOREM 4   **(Termwise integration of power series)**

The power series

$$\sum_{n=0}^{\infty} \frac{a_n}{n+1} z^{n+1} = a_0 z + \frac{a_1}{2} z^2 + \frac{a_2}{3} z^3 + \cdots$$

obtained by integrating the series $a_0 + a_1 z + a_2 z^2 + \cdots$ term by term has the same radius of convergence as the original series.

The proof is similar to that of Theorem 3.

With Theorem 3 as a tool, we are now ready to establish our main result in this section.

Power Series Represent Analytic Functions

THEOREM 5   **(Analytic functions. Their derivatives)**

A power series with a nonzero radius of convergence R represents an analytic function at every point interior to its circle of convergence. The derivatives of this function are obtained by differentiating the original series term by term. All the series thus obtained have the same radius of convergence as the original series. Hence, by the first statement, each of them represents an analytic function.

PROOF.   **(a)** We consider any power series (1) with positive radius of convergence R. Let $f(z)$ be its sum and $f_1(z)$ the sum of its derived series; thus

$$(4) \qquad f(z) = \sum_{n=0}^{\infty} a_n z^n \quad \text{and} \quad f_1(z) = \sum_{n=1}^{\infty} n a_n z^{n-1}.$$

We show that $f(z)$ is analytic and has the derivative $f_1(z)$ in the interior of the circle of convergence. We do this by proving that for any fixed z with $|z| < R$ and $\Delta z \to 0$ the difference quotient $[f(z + \Delta z) - f(z)]/\Delta z$ approaches $f_1(z)$. By termwise addition we first have from (4)

$$(5) \qquad \frac{f(z + \Delta z) - f(z)}{\Delta z} - f_1(z) = \sum_{n=2}^{\infty} a_n \left[\frac{(z + \Delta z)^n - z^n}{\Delta z} - n z^{n-1} \right].$$

Note that the summation starts with 2, since the constant term drops out in taking the difference $f(z + \Delta z) - f(z)$, and so does the linear term when we subtract $f_1(z)$ from the difference quotient.

(b) We claim that the series in (5) can be written

$$(6) \qquad \sum_{n=2}^{\infty} a_n \Delta z \left[(z + \Delta z)^{n-2} + 2z(z + \Delta z)^{n-3} + \cdots + (n - 1)z^{n-2} \right].$$

The somewhat technical proof of this is given in Appendix 4.

(c) We consider (6). The brackets contain $n - 1$ terms, and the largest coefficient is $n - 1$. Since $(n - 1)^2 \leqq n(n - 1)$, we see that for $|z| \leqq R_0$ and $|z + \Delta z| \leqq R_0$, $R_0 < R$, the absolute value of this series cannot exceed

$$|\Delta z| \sum_{n=2}^{\infty} |a_n| n(n - 1) R_0^{n-2}.$$

This series with a_n instead of $|a_n|$ is the second derived series of (2) at $z = R_0$ and converges absolutely by Theorem 3 of this section and Theorem 1 of Sec. 14.2. Hence our present series converges. Let $K(R_0)$ be its sum. Then we can write our present result

$$\left| \frac{f(z + \Delta z) - f(z)}{\Delta z} - f_1(z) \right| \leqq |\Delta z| \, K(R_0).$$

Letting $\Delta z \to 0$ and noting that $R_0 \, (< R)$ is arbitrary, we conclude that $f(z)$ is analytic at any point interior to the circle of convergence and its derivative is represented by the derived series. From this the statements about the higher derivatives follow by induction. ◄

Summary. The results in this section show that power series are about as nice as we could hope for: we can differentiate and integrate them term by term (Theorems 3 and 4). Theorem 5 accounts for the great importance of power series in complex analysis: the sum of such a series (with a positive radius of convergence) is an analytic function and has derivatives of all orders, which thus are analytic functions. But this is only part of the story. In the next section we show that, conversely, *every* given analytic function $f(z)$ can be represented by power series.

PROBLEM SET 14.3

Calculations in the Text

1. **(Addition and subtraction)** Write out the proof on termwise addition and subtraction of power series.
2. **(On Theorem 3)** In the proof of Theorem 3 we claimed that $\sqrt[n]{n} \to 1$ as $n \to \infty$. Prove this.

Differentiation and Integration (Theorems 3 and 4)

Find the radius of convergence of the following series in two ways: (a) directly by the Cauchy–Hadamard formula (Sec. 14.2), (b) by Theorem 3 or Theorem 4 and a series with simpler coefficients. (Show the details of your work.)

3. $\displaystyle\sum_{n=2}^{\infty} n(n-1)\left(\frac{z}{5}\right)^n$ 4. $\displaystyle\sum_{n=2}^{\infty} \frac{z^n}{n(n-1)}$ 5. $\displaystyle\sum_{n=1}^{\infty} \frac{6^n}{n}(z-i)^n$

6. $\displaystyle\sum_{n=k}^{\infty} \binom{n}{k}\left(\frac{z}{\pi}\right)^n$ 7. $\displaystyle\sum_{n=0}^{\infty} \frac{(-4)^n}{2^n(n+2)(n+1)} z^{2n}$ 8. $\displaystyle\sum_{n=0}^{\infty} \frac{3^n(n+1)n}{5^n} z^{2n}$

9. $\displaystyle\sum_{n=1}^{\infty} \frac{2n}{n!}(z+i)^{2n-1}$ 10. $\displaystyle\sum_{n=0}^{\infty} \frac{(-1)^n z^{2n+1}}{(2n+1)n!}$ 11. $\displaystyle\sum_{n=0}^{\infty} \binom{n+k}{k}(z+2)^n$

12. $\displaystyle\sum_{n=0}^{\infty} \frac{(2n)!}{(n+1)(n!)^2} z^{n+1}$ 13. $\displaystyle\sum_{n=0}^{\infty} \left[\binom{n+k}{n}\right]^{-1} z^{n+k}$ 14. $\displaystyle\sum_{n=1}^{\infty} \frac{3n(3n-1)}{n^n} z^{3n}$

15. **(Cauchy product)** Show that $(1-z)^{-2} = \sum_{n=0}^{\infty} (n+1)z^n$ (a) by using the Cauchy product, (b) by differentiating a suitable series.

Applications of the Identity Theorem (Theorem 2)

16. **(Binomial coefficients)** Using $(1+z)^p(1+z)^q = (1+z)^{p+q}$, obtain the basic relation

$$\sum_{n=0}^{r} \binom{p}{n}\binom{q}{r-n} = \binom{p+q}{r}.$$

17. **(Even function)** If $f(z)$ in (1) is even (i.e., $f(-z) = f(z)$), show that $a_n = 0$ for odd n. (State clearly where and how you are using Theorem 2.)
18. **(Odd function)** State and prove the analog of Prob. 17 for an odd function (i.e., $f(-z) = -f(z)$).

19. Find other applications of Theorem 2 (for instance, in differential equations).

20. **TEAM PROJECT. The Fibonacci numbers**[2] are recursively defined by $a_0 = a_1 = 1$, $a_{n+1} = a_n + a_{n-1}$ if $n \geqq 1$.

   (a) **Fibonacci's rabbit problem.** Compute a list of $a_1, \cdots, a_{12}$. Show that $a_{12} = 233$ is the number of pairs of rabbits after 12 months if initially there is 1 pair and each pair generates 1 pair per month, beginning in the second month of existence (no deaths occurring).

   (b) **Generating function.** Show that the generating function of the Fibonacci numbers is $f(z) = 1/(1 - z - z^2)$; that is, if a power series (1) represents this $f(z)$, its coefficients must be the Fibonacci numbers and conversely. *Hint.* Start from $f(z)(1 - z - z^2) = 1$ and use Theorem 2.

   (c) **Cassini's identity** states that $a_{n+1}a_{n-1} - a_n{}^2 = (-1)^{n+1}$ for $n > 0$. Verify this for the first few n, say, $n \leqq 8$. Use this for $n = 8$ to cut a chessboard into two right triangles of sides 8, 3, $\sqrt{73}$ and two quadrangles of sides 5, 5, 3, $\sqrt{29}$; assemble these pieces (with a little cheating) as a rectangle of sides 13 and 5, hence having 65 squares. (For more on this, see Ref. [7], p. 293.) Explain this "mystery."

14.4 Taylor Series and Maclaurin Series

The **Taylor series**[3] of a function $f(z)$, the complex analog of the real Taylor series (see your calculus book), is

(1)
$$f(z) = \sum_{n=0}^{\infty} a_n(z - z_0)^n \qquad \text{where} \qquad a_n = \frac{1}{n!} f^{(n)}(z_0)$$

or, by (1), Sec. 13.4,

(2)
$$a_n = \frac{1}{2\pi i} \oint_C \frac{f(z^*)}{(z^* - z_0)^{n+1}} \, dz^*$$

with counterclockwise integration around a simple closed path C that contains z_0 in its interior and is such that $f(z)$ is analytic on and everywhere inside C.

The **remainder** of (1) after the term $a_n(z - z_0)^n$ is

(3)
$$R_n(z) = \frac{(z - z_0)^{n+1}}{2\pi i} \oint_C \frac{f(z^*)}{(z^* - z_0)^{n+1}(z^* - z)} \, dz^*$$

(proof below). Writing out the corresponding partial sum of (1), we thus have

---

[2] LEONARDO OF PISA, called FIBONACCI (= son of Bonaccio), about 1180—1250, Italian mathematician, credited with the first renaissance of mathematics on Christian soil.

[3] BROOK TAYLOR (1685—1731), English mathematician, who introduced this formula for functions of a real variable.

$$f(z) = f(z_0) + \frac{z - z_0}{1!} f'(z_0) + \frac{(z - z_0)^2}{2!} f''(z_0) + \cdots$$

(4)

$$+ \frac{(z - z_0)^n}{n!} f^{(n)}(z_0) + R_n(z).$$

This is called **Taylor's formula** *with remainder.*

A **Maclaurin**[4] **series** is a Taylor series with center $z_0 = 0$.

We now show that *every* analytic function can be represented by Taylor series. This accounts for the great importance of these series in complex analysis.

THEOREM 1   **(Taylor's theorem)**

Let $f(z)$ be analytic in a domain D, and let $z = z_0$ be any point in D. Then there exists precisely one Taylor series (1) *with center z_0 that represents $f(z)$. This representation is valid in the largest open disk with center z_0 in which $f(z)$ is analytic. The remainders $R_n(z)$ of* (1) *can be represented in the form* (3). *The coefficients satisfy the inequality*

(5)
$$|a_n| \leqq \frac{M}{r^n}$$

where M is the maximum of $|f(z)|$ on a circle $|z - z_0| = r$ in D whose interior is also in D.

PROOF.   The key tool is Cauchy's integral formula in Sec. 13.3; writing z and z^* instead of z_0 and z, we have

(6)
$$f(z) = \frac{1}{2\pi i} \oint_C \frac{f(z^*)}{z^* - z} \, dz^*.$$

z lies inside C, for which we take a circle of radius r with center z_0 and interior in D (Fig. 353). We develop $1/(z^* - z)$ in (6) in powers of $z - z_0$. By a standard algebraic manipulation (worth remembering!) we first have

(7)
$$\frac{1}{z^* - z} = \frac{1}{z^* - z_0 - (z - z_0)} = \frac{1}{(z^* - z_0)\left(1 - \dfrac{z - z_0}{z^* - z_0}\right)}.$$

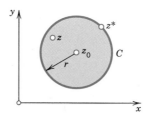

Fig. 353.   Cauchy formula (6)

---

[4]COLIN MACLAURIN (1698—1746), Scots mathematician, professor at Edinburgh.

For later use we note that since z^* is on C while z is inside C, we have

(7*)
$$\left| \frac{z - z_0}{z^* - z_0} \right| < 1$$
(Fig. 353).

To (7) we now apply the sum formula for a finite geometric sum

(8*)
$$1 + q + \cdots + q^n = \frac{1 - q^{n+1}}{1 - q} = \frac{1}{1 - q} - \frac{q^{n+1}}{1 - q} \qquad (q \neq 1),$$

which we use in the form

(8)
$$\frac{1}{1 - q} = 1 + q + \cdots + q^n + \frac{q^{n+1}}{1 - q}.$$

Applying this with $q = (z - z_0)/(z^* - z_0)$ to the right side of (7), we get

$$\frac{1}{z^* - z} = \frac{1}{z^* - z_0} \left[1 + \frac{z - z_0}{z^* - z_0} + \left(\frac{z - z_0}{z^* - z_0} \right)^2 + \cdots + \left(\frac{z - z_0}{z^* - z_0} \right)^n \right]$$

$$+ \frac{1}{z^* - z} \left(\frac{z - z_0}{z^* - z_0} \right)^{n+1}.$$

We insert this into (6). Powers of $z - z_0$ do not depend on the variable of integration z^*, so that we may take them out from under the integral sign. This yields

$$f(z) = \frac{1}{2\pi i} \oint_C \frac{f(z^*)}{z^* - z_0} \, dz^* + \frac{z - z_0}{2\pi i} \oint_C \frac{f(z^*)}{(z^* - z_0)^2} \, dz^* + \cdots$$

$$\cdots + \frac{(z - z_0)^n}{2\pi i} \oint_C \frac{f(z^*)}{(z^* - z_0)^{n+1}} \, dz^* + R_n(z)$$

with $R_n(z)$ given by (3). The integrals are those in (2) related to the derivatives, so that we have proved the Taylor formula (4).

Since analytic functions have derivatives of all orders, we can take n in (4) as large as we please. If we let n approach infinity, we obtain (1). Clearly, (1) will converge and represent $f(z)$ if and only if

(9)
$$\lim_{n \to \infty} R_n(z) = 0.$$

We prove this as follows. Since z^* is on C, whereas z is inside C (Fig. 353), we have $|z^* - z| > 0$. Since $f(z)$ is analytic inside and on C, it is bounded, and so is the function $f(z^*)/(z^* - z)$, say,

$$\left| \frac{f(z^*)}{z^* - z} \right| \leq \tilde{M}$$

for all z^* on C. Also, C has the radius $r = |z^* - z_0|$ and the length $2\pi r$. Hence by the *ML*-inequality (Sec. 13.1) we obtain from (3)

$$
\begin{aligned}
(10) \qquad |R_n| &= \frac{|z - z_0|^{n+1}}{2\pi} \left| \oint_C \frac{f(z^*)}{(z^* - z_0)^{n+1}(z^* - z)} \, dz^* \right| \\[2ex]
&\leq \frac{|z - z_0|^{n+1}}{2\pi} \, \widetilde{M} \, \frac{1}{r^{n+1}} \, 2\pi r = \widetilde{M} r \left| \frac{z - z_0}{r} \right|^{n+1}.
\end{aligned}
$$

Now $|z - z_0| < r$ because z lies *inside* C. Thus $|z - z_0|/r < 1$, so that the right side approaches 0 as $n \to \infty$. This proves the convergence of the Taylor series. Uniqueness follows from Theorem 2 in the last section. Finally, (5) follows from (1) and the Cauchy inequality in Sec. 13.4. This proves Taylor's theorem.   ◀

Accuracy of Approximation. We can achieve any preassigned accuracy in approximating $f(z)$ by a partial sum of (1) by choosing n large enough. This is the practical aspect of formula (9).

Singularity, Radius of Convergence. On the circle of convergence of (1) there is at least one **singular point** of $f(z)$, that is, a point $z = c$ at which $f(z)$ is not differentiable (but such that every disk with center c contains points at which $f(z)$ *is* differentiable). We also say that $f(z)$ **is singular** at c or **has a singularity** at c. Hence the radius of convergence R of (1) is usually equal[5] to the distance from z_0 to the nearest singular point of $f(z)$. *Example.* $1/(1 - z)$ is singular at $z = 1$, and the geometric series has radius of convergence 1.

Power Series as Taylor Series

Taylor series are power series—of course! Conversely, we have

THEOREM 2   **(Relation to the last section)**

A power series with nonzero radius of convergence is the Taylor series of its sum.

PROOF.   Given the power series

$$
f(z) = a_0 + a_1(z - z_0) + a_2(z - z_0)^2 + a_3(z - z_0)^3 + \cdots.
$$

Then $f(z_0) = a_0$. By Theorem 5, Sec. 14.3,

$$
f'(z) = a_1 + 2a_2(z - z_0) + 3a_3(z - z_0)^2 + \cdots, \qquad \text{thus} \qquad f'(z_0) = a_1
$$

$$
f''(z) = 2a_2 + 3 \cdot 2(z - z_0) + \cdots, \qquad\qquad\qquad \text{thus} \qquad f''(z_0) = 2!a_2
$$

and in general $f^{(n)}(z_0) = n!a_n$. With these coefficients the given series becomes the Taylor series of $f(z)$.   ◀

Comparison with Real Functions. One surprising property of complex analytic functions is that they have derivatives of all orders, and now we have discovered the other surprising property that they can always be represented by power series of the form (1). This is not true in general for **real functions**; there are real functions that have derivatives of all

---

[5]Or sometimes larger. *Example.* Ln z is singular at the negative real axis, which has distance 1 from $z_0 = -1 + i$, but the Taylor series of Ln z with center $z_0 = -1 + i$ has radius of convergence $\sqrt{2}$.

orders but cannot be represented by a power series. (Example: $f(x) = \exp(-1/x^2)$ if $x \neq 0$ and $f(0) = 0$; this function cannot be represented by a Maclaurin series since all its derivatives at 0 are zero.)

Important Special Taylor Series

These are as in calculus, with x replaced by complex z. Can you see why? (*Answer.* The coefficient formulas are the same.)

EXAMPLE 1   **Geometric series**

Let $f(z) = 1/(1 - z)$. Then we have $f^{(n)}(z) = n!/(1 - z)^{n+1}$, $f^{(n)}(0) = n!$. Hence the Maclaurin expansion of $1/(1 - z)$ is the geometric series

(11)
$$\frac{1}{1-z} = \sum_{n=0}^{\infty} z^n = 1 + z + z^2 + \cdots \qquad (|z| < 1).$$

$f(z)$ is singular at $z = 1$; this point lies on the circle of convergence.   ◄

EXAMPLE 2   **Exponential function**

We know that the exponential function e^z (Sec. 12.6) is analytic for all z, and $(e^z)' = e^z$. Hence from (1) with $z_0 = 0$ we obtain the Maclaurin series

(12)
$$e^z = \sum_{n=0}^{\infty} \frac{z^n}{n!} = 1 + z + \frac{z^2}{2!} + \cdots.$$

This series is also obtained if we replace x in the familiar Maclaurin series of e^x by z.

Furthermore, by setting $z = iy$ in (12) and separating the series into the real and imaginary parts (see Theorem 2, Sec. 14.1) we obtain

$$e^{iy} = \sum_{n=0}^{\infty} \frac{(iy)^n}{n!} = \sum_{k=0}^{\infty} (-1)^k \frac{y^{2k}}{(2k)!} + i \sum_{k=0}^{\infty} (-1)^k \frac{y^{2k+1}}{(2k+1)!}.$$

Since the series on the right are the familiar Maclaurin series of the real functions $\cos y$ and $\sin y$, this shows that we have rediscovered the **Euler formula**

(13)
$$e^{iy} = \cos y + i \sin y.$$

Indeed, one may use (12) for **defining** e^z and derive from (12) the basic properties of e^z. For instance, the differentiation formula $(e^z)' = e^z$ follows readily from (12) by termwise differentiation.   ◄

EXAMPLE 3   **Trigonometric and hyperbolic functions**

By substituting (12) into (1) of Sec. 12.7 we obtain

(14)
$$\cos z = \sum_{n=0}^{\infty} (-1)^n \frac{z^{2n}}{(2n)!} = 1 - \frac{z^2}{2!} + \frac{z^4}{4!} - + \cdots$$

$$\sin z = \sum_{n=0}^{\infty} (-1)^n \frac{z^{2n+1}}{(2n+1)!} = z - \frac{z^3}{3!} + \frac{z^5}{5!} - + \cdots.$$

When $z = x$ these are the familiar Maclaurin series of the real functions $\cos x$ and $\sin x$. Similarly, by substituting (12) into (11), Sec. 12.7, we obtain

$$(15) \qquad \cosh z = \sum_{n=0}^{\infty} \frac{z^{2n}}{(2n)!} = 1 + \frac{z^2}{2!} + \frac{z^4}{4!} + \cdots$$

$$\sinh z = \sum_{n=0}^{\infty} \frac{z^{2n+1}}{(2n+1)!} = z + \frac{z^3}{3!} + \frac{z^5}{5!} \cdots. \qquad \blacktriangleleft$$

EXAMPLE 4   Logarithm

From (1) it follows that

$$(16) \qquad \boxed{\mathrm{Ln}\,(1 + z) = z - \frac{z^2}{2} + \frac{z^3}{3} - + \cdots} \qquad (|z| < 1).$$

Replacing z by $-z$ and multiplying both sides by -1, we get

$$(17) \qquad -\mathrm{Ln}\,(1 - z) = \mathrm{Ln}\,\frac{1}{1 - z} = z + \frac{z^2}{2} + \frac{z^3}{3} + \cdots \qquad (|z| < 1).$$

By adding both series we obtain

$$(18) \qquad \mathrm{Ln}\,\frac{1 + z}{1 - z} = 2\left(z + \frac{z^3}{3} + \frac{z^5}{5} + \cdots\right) \qquad (|z| < 1). \qquad \blacktriangleleft$$

Practical Methods

The following examples show ways of obtaining Taylor series more quickly than by the use of the coefficient formulas. Regardless of the method used, the result will be the same. This follows from the uniqueness (see Theorem 1).

EXAMPLE 5   Substitution

Find the Maclaurin series of $f(z) = 1/(1 + z^2)$.

Solution. By substituting $-z^2$ for z in (11) we obtain

$$(19) \qquad \frac{1}{1 + z^2} = \frac{1}{1 - (-z^2)} = \sum_{n=0}^{\infty} (-z^2)^n = \sum_{n=0}^{\infty} (-1)^n z^{2n} = 1 - z^2 + z^4 - z^6 + \cdots \quad (|z| < 1). \qquad \blacktriangleleft$$

EXAMPLE 6   Integration

Find the Maclaurin series of $f(z) = \tan^{-1} z$.

Solution.   We have $f'(z) = 1/(1 + z^2)$. Integrating (19) term by term and using $f(0) = 0$ we get

$$\tan^{-1} z = \sum_{n=0}^{\infty} \frac{(-1)^n}{2n + 1} z^{2n+1} = z - \frac{z^3}{3} + \frac{z^5}{5} - + \cdots \qquad (|z| < 1);$$

this series represents the principal value of $w = u + iv = \tan^{-1} z$, defined as that value for which $|u| < \pi/2$. $\qquad \blacktriangleleft$

EXAMPLE 7   Development by using the geometric series

Develop $1/(c - z)$ in powers of $z - z_0$, where $c - z_0 \neq 0$.

Solution.   This was done in the proof of Theorem 1, where $c = z^*$. The beginning was simple algebra and then the use of (11) with z replaced by $(z - z_0)/(c - z_0)$:

$$\frac{1}{c - z} = \frac{1}{c - z_0 - (z - z_0)} = \frac{1}{(c - z_0)\left(1 - \dfrac{z - z_0}{c - z_0}\right)} = \frac{1}{c - z_0} \sum_{n=0}^{\infty} \left(\frac{z - z_0}{c - z_0}\right)^n$$

$$= \frac{1}{c - z_0}\left(1 + \frac{z - z_0}{c - z_0} + \left(\frac{z - z_0}{c - z_0}\right)^2 + \cdots\right).$$

This series converges for

$$\left|\frac{z - z_0}{c - z_0}\right| < 1, \qquad \text{that is,} \qquad |z - z_0| < |c - z_0|. \qquad \blacktriangleleft$$

EXAMPLE 8   **Binomial series, reduction by partial fractions**

Find the Taylor series of the following function with center $z_0 = 1$.

$$f(z) = \frac{2z^2 + 9z + 5}{z^3 + z^2 - 8z - 12}$$

Solution.   We develop $f(z)$ in partial fractions and the first fraction in a **binomial series**

(20)

$$\frac{1}{(1 + z)^m} = (1 + z)^{-m} = \sum_{n=0}^{\infty} \binom{-m}{n} z^n$$

$$= 1 - mz + \frac{m(m + 1)}{2!} z^2 - \frac{m(m + 1)(m + 2)}{3!} z^3 + \cdots$$

with $m = 2$ and the second fraction in a geometric series, and then add the two series term by term. This gives

$$f(z) = \frac{1}{(z + 2)^2} + \frac{2}{z - 3} = \frac{1}{[3 + (z - 1)]^2} - \frac{2}{2 - (z - 1)} = \frac{1}{9}\left(\frac{1}{[1 + \frac{1}{3}(z - 1)]^2}\right) - \frac{1}{1 - \frac{1}{2}(z - 1)}$$

$$= \frac{1}{9} \sum_{n=0}^{\infty} \binom{-2}{n}\left(\frac{z - 1}{3}\right)^n - \sum_{n=0}^{\infty} \left(\frac{z - 1}{2}\right)^n = \sum_{n=0}^{\infty}\left[\frac{(-1)^n(n + 1)}{3^{n+2}} - \frac{1}{2^n}\right](z - 1)^n$$

$$= -\frac{8}{9} - \frac{31}{54}(z - 1) - \frac{23}{108}(z - 1)^2 - \frac{275}{1944}(z - 1)^3 - \cdots.$$

The first series converges for $|z - 1| < 3$ because $z + 2 = 0$ at $z = -2$, the second for $|z - 1| < 2$ because $z - 3 = 0$ at $z = 3$, and thus the sum [the series of $f(z)$] in the disk $|z - 1| < 2$. $\qquad \blacktriangleleft$

PROBLEM SET 14.4

Maclaurin Series

In Probs. 1–8 develop the given function in a Maclaurin series and find the radius of convergence. (Show the details of your work.)

1. $\cos 2z^2$ **2.** $\dfrac{1}{1 - z^4}$ **3.** $\sin^2 z$ **4.** $e^{-z^2/2}$

5. $\dfrac{z+2}{1-z^2}$ **6.** $\dfrac{2-z}{(1-z)^2}$ **7.** $\dfrac{1}{z+3i}$ **8.** $e^{z^2}\displaystyle\int_0^z e^{-t^2}\,dt$

Find the Maclaurin series by termwise integrating the integrand. (The integrals cannot be evaluated by the usual methods of calculus. They define the **error function** erf z, **sine integral** Si(z), and **Fresnel integrals**[6] S(z) and C(z), which occur in statistics, heat conduction, optics, and other applications.)

9. erf $z = \dfrac{2}{\sqrt{\pi}} \displaystyle\int_0^z e^{-t^2}\,dt$ **10.** Si(z) $= \displaystyle\int_0^z \dfrac{\sin t}{t}\,dt$

11. S(z) $= \displaystyle\int_0^z \sin t^2\,dt$ **12.** C(z) $= \displaystyle\int_0^z \cos t^2\,dt$

13. CAS PROJECT. sec, tan, sin^{-1}. (a) Euler numbers. The Maclaurin series

$$\text{(21)} \qquad \sec z = E_0 - \frac{E_2}{2!}\,z^2 + \frac{E_4}{4!}\,z^4 - + \cdots$$

defines the *Euler numbers* E_{2n}. Show that[7] $E_0 = 1$, $E_2 = -1$, $E_4 = 5$, $E_6 = -61$. Write a program that computes the E_{2n} from the coefficient formula in (1) or extracts them as a list from the series.

(b) Bernoulli numbers. The Maclaurin series

$$\text{(22)} \qquad \frac{z}{e^z - 1} = 1 + B_1 z + \frac{B_2}{2!}\,z^2 + \frac{B_3}{3!}\,z^3 + \cdots$$

defines the *Bernoulli numbers* B_n. Using undetermined coefficients, show that[7]

$$\text{(23)} \qquad B_1 = -\frac{1}{2}\,,\ B_2 = \frac{1}{6}\,,\ B_3 = 0,\ B_4 = -\frac{1}{30}\,,\ B_5 = 0,\ B_6 = \frac{1}{42}\,,\ \cdots.$$

Write a program for computing B_n.

(c) Tangent. Using (1), (2), Sec. 12.7, and (22), show that tan z has the following Maclaurin series and calculate from it a table of $B_0, \cdots, B_{20}$:

$$\text{(24)} \qquad \tan z = \frac{2i}{e^{2iz}-1} - \frac{4i}{e^{4iz}-1} - i = \sum_{n=1}^{\infty} (-1)^{n-1}\,\frac{2^{2n}(2^{2n}-1)}{(2n)!}\,B_{2n}z^{2n-1}.$$

14. (Inverse sine) Developing $1/\sqrt{1-z^2}$ and integrating, show that

$$\sin^{-1} z = z + \left(\frac{1}{2}\right)\frac{z^3}{3} + \left(\frac{1\cdot 3}{2\cdot 4}\right)\frac{z^5}{5} + \left(\frac{1\cdot 3\cdot 5}{2\cdot 4\cdot 6}\right)\frac{z^7}{7} + \cdots \qquad (|z| < 1).$$

Show that this series represents the principal value of $\sin^{-1} z$ (defined in Team Project 30, Sec. 12.8).

15. (Undetermined coefficients) Using $\sin z = \tan z \cos z$ and the Maclaurin series of $\sin z$ and $\cos z$, find the first four nonzero terms of the Maclaurin series of tan z. (Show the details.)

---

[6] AUGUSTIN FRESNEL (1788—1827), French physicist, known for his work in optics.

[7] For tables, see Ref. [1], p. 810, in Appendix 1.

Taylor Series

Develop each of the following functions in a Taylor series with the given point as center. Find the radius of convergence. (Show the details.)

16. $1/z$, i **17.** $1/z$, 2 **18.** z^5, -1 **19.** e^z, a

20. $\cos \pi z$, $\frac{1}{2}$ **21.** $\text{Ln } z$, 1 **22.** $\cosh (z - \pi i)$, πi **23.** $\sin z$, $\frac{1}{2}\pi$

24. $1/(z + i)^2$, i **25.** $\sinh (2z - i)$, $\frac{1}{2}i$ **26.** $\cos^2 z$, $\frac{1}{2}\pi$ **27.** $e^{z^2 - 2z}$, 1

28. TEAM PROJECT. Properties from Maclaurin Series. Clearly, from series we can compute function values. In this project we show that properties of functions can often be discovered from their Taylor or Maclaurin series. Using suitable series, prove the following.

(a) The formulas for the derivatives of e^z, $\cos z$, $\sin z$, $\cosh z$, $\sinh z$, and $\text{Ln } (1 + z)$

(b) $\frac{1}{2}(e^{iz} + e^{-iz}) = \cos z$

(c) $\sin (z + \frac{1}{2}\pi) = \cos z$

(d) $\sin z \neq 0$ for all pure imaginary $z = iy \neq 0$

14.5 Uniform Convergence. *Optional*

We know that power series are *absolutely convergent* (Sec. 14.2, Theorem 1) and, as another basic property, we now show that they are *uniformly convergent*. Since uniform convergence is of general importance, for instance, in connection with termwise integration of series, we shall discuss it quite thoroughly.

To define uniform convergence, we consider a series whose terms are functions $f_0(z), f_1(z), \cdots$:

$$(1) \qquad \sum_{m=0}^{\infty} f_m(z) = f_0(z) + f_1(z) + f_2(z) + \cdots .$$

(For the special $f_m(z) = a_m(z - z_0)^m$ this is a power series.) We assume that this series converges for all z in some region G. We call its sum $s(z)$ and its nth partial sum $s_n(z)$; thus

$$s_n(z) = f_0(z) + f_1(z) + \cdots + f_n(z).$$

Convergence in G means the following. If we pick a $z = z_1$ in G, then, by the definition of convergence at z_1, for given $\epsilon > 0$ we can find an $N_1(\epsilon)$ such that

$$|s(z_1) - s_n(z_1)| < \epsilon \qquad\qquad \text{for all } n > N_1(\epsilon).$$

If we pick a z_2 in G, keeping ϵ as before, we can find an $N_2(\epsilon)$ such that

$$|s(z_2) - s_n(z_2)| < \epsilon \qquad\qquad \text{for all } n > N_2(\epsilon),$$

and so on. Hence, given an $\epsilon > 0$, to each z in G there corresponds a number $N_z(\epsilon)$. This number tells us how many terms we need (what s_n we need) at a z to make $|s(z) - s_n(z)|$ smaller than ϵ. It measures the speed of convergence.

Small $N_z(\epsilon)$ means rapid convergence, large $N_z(\epsilon)$ means slow convergence. Now, if we can find an $N(\epsilon)$ larger than *all* these $N_z(\epsilon)$, we say that the convergence of the series (1) in G is *uniform*. This basic concept is defined as follows.

Definition (Uniform convergence)

A series (1) with sum $s(z)$ is called **uniformly convergent** in a region G if for every $\epsilon > 0$ we can find an $N = N(\epsilon)$, *not depending on z*, such that

$$\left| s(z) - s_n(z) \right| < \epsilon \qquad \text{for all } n > N(\epsilon) \text{ and all } z \text{ in } G.$$

Uniformity of convergence is thus a property that always refers to an *infinite set* in the z-plane, that is, a set consisting of infinitely many points.

EXAMPLE 1   **Geometric series**

Show that the geometric series $1 + z + z^2 + \cdots$ is (a) uniformly convergent in any closed disk $|z| \leqq r < 1$, (b) not uniformly convergent in its whole disk of convergence $|z| < 1$.

Solution.   **(a)** For z in that closed disk we have $|1 - z| \geqq 1 - r$ (sketch it). This implies that $1/|1 - z| \leqq 1/(1 - r)$. Hence (remember (8) in Sec. 14.4 with $q = z$)

$$\left| s(z) - s_n(z) \right| = \left| \sum_{m=n+1}^{\infty} z^m \right| = \left| \frac{z^{n+1}}{1 - z} \right| \leqq \frac{r^{n+1}}{1 - r}.$$

Since $r < 1$, we can make the right side as small as we want by choosing n large enough, and since the right side does not depend on z (in the closed disk considered), this means uniform convergence.

(b) For given real K (no matter how large) and n we can always find a z in the disk $|z| < 1$ such that

$$\left| \frac{z^{n+1}}{1 - z} \right| = \frac{|z|^{n+1}}{|1 - z|} > K,$$

simply by taking z close enough to 1. Hence no single $N(\epsilon)$ will suffice to make $|s(z) - s_n(z)|$ smaller than a given $\epsilon > 0$ *throughout the whole disk*. By definition, this shows that the convergence of the geometric series in $|z| < 1$ is not uniform.   ◄

This example suggests that for a power series, the uniformity of convergence may at most be disturbed near the circle of convergence. This is true:

THEOREM 1   **(Uniform convergence of power series)**

A power series

$$(2) \qquad \sum_{m=0}^{\infty} a_m (z - z_0)^m$$

with a nonzero radius of convergence R is uniformly convergent in every circular disk $|z - z_0| \leqq r$ of radius $r < R$.

PROOF.   For $|z - z_0| \leqq r$ and any positive integers n and p we have

$$(3) \quad \left| a_{n+1}(z - z_0)^{n+1} + \cdots + a_{n+p}(z - z_0)^{n+p} \right| \leqq \left| a_{n+1} \right| r^{n+1} + \cdots + \left| a_{n+p} \right| r^{n+p}.$$

Now (2) converges absolutely if $|z - z_0| = r < R$ (by Theorem 1 in Sec. 14.2). Hence it follows from the Cauchy convergence principle (Sec. 14.1) that, an $\epsilon > 0$ being given, we can find an $N(\epsilon)$ such that

$$\left| a_{n+1} \right| r^{n+1} + \cdots + \left| a_{n+p} \right| r^{n+p} < \epsilon \quad \text{for } n > N(\epsilon) \quad \text{and} \quad p = 1, 2, \cdots.$$

From this and (3) we obtain

$$\left| a_{n+1}(z - z_0)^{n+1} + \cdots + a_{n+p}(z - z_0)^{n+p} \right| < \epsilon$$

for all z in the disk $|z - z_0| \leqq r$, every $n > N(\epsilon)$, and every $p = 1, 2, \cdots$. Since $N(\epsilon)$ is independent of z, this shows uniform convergence, and the theorem is proved.   ◀

Theorem 1 meets with our immediate need and concern, which is power series. The remainder of this section should provide a deeper understanding of the concept of uniform convergence as such.

Properties of Uniformly Convergent Series

Uniform convergence derives its main importance from two facts:

1. If a series of *continuous* terms is uniformly convergent, its sum is also continuous (Theorem 2, below).
2. Under the same assumptions, termwise integration is permissible (Theorem 3).

This raises two questions:

1. How can a converging series of continuous terms manage to have a discontinuous sum? (Example 2)
2. How can something go wrong in termwise integration? (Example 3) Another natural question is:
3. What is the relation between absolute convergence and uniform convergence? The surprising answer: none. (Example 4)

These are the ideas we shall discuss.

If we add *finitely many* continuous functions, we get a continuous function as their sum. Example 2 will show that this is no longer true for an infinite series, even if it converges absolutely. However, if it converges *uniformly,* this cannot happen, as follows.

THEOREM 2   **(Continuity of the sum)**

Let the series

$$\sum_{m=0}^{\infty} f_m(z) = f_0(z) + f_1(z) + \cdots$$

be uniformly convergent in a region G. Let $F(z)$ be its sum. Then if each term $f_m(z)$ is continuous at a point z_1 in G, the function $F(z)$ is continuous at z_1.

PROOF.   Let $s_n(z)$ be the nth partial sum of the series and $R_n(z)$ the corresponding remainder:

$$s_n = f_0 + f_1 + \cdots + f_n, \qquad R_n = f_{n+1} + f_{n+2} + \cdots.$$

Since the series converges uniformly, for a given $\epsilon > 0$ we can find an $N = N(\epsilon)$ such that

$$|R_N(z)| < \frac{\epsilon}{3} \qquad\qquad \text{for all } z \text{ in } G.$$

Since $s_N(z)$ is a sum of finitely many functions that are continuous at z_1, this sum is continuous at z_1. Therefore, we can find a $\delta > 0$ such that

$$|s_N(z) - s_N(z_1)| < \frac{\epsilon}{3} \qquad \text{for all } z \text{ in } G \text{ for which } |z - z_1| < \delta.$$

Using $F = s_N + R_N$ and the triangle inequality (Sec. 12.2), for these z we thus obtain

$$|F(z) - F(z_1)| = |s_N(z) + R_N(z) - [s_N(z_1) + R_N(z_1)]|$$

$$\leqq |s_N(z) - s_N(z_1)| + |R_N(z)| + |R_N(z_1)| < \frac{\epsilon}{3} + \frac{\epsilon}{3} + \frac{\epsilon}{3} = \epsilon.$$

This implies that $F(z)$ is continuous at z_1, and the theorem is proved.   ◀

EXAMPLE 2   **Series of continuous terms with a discontinuous sum**
Consider the series

$$x^2 + \frac{x^2}{1 + x^2} + \frac{x^2}{(1 + x^2)^2} + \frac{x^2}{(1 + x^2)^3} + \cdots \qquad (x \text{ real}).$$

This is a geometric series with $q = 1/(1 + x^2)$ times a factor x^2. Its nth partial sum is

$$s_n(x) = x^2 \left[1 + \frac{1}{1 + x^2} + \cdots + \frac{1}{(1 + x^2)^n} \right].$$

By multiplication on both sides,

$$-\frac{1}{1 + x^2} s_n(x) = -x^2 \left[\frac{1}{1 + x^2} + \cdots + \frac{1}{(1 + x^2)^n} + \frac{1}{(1 + x^2)^{n+1}} \right].$$

We now add, simplify on the left and cancel terms on the right, obtaining

$$\frac{x^2}{1 + x^2} s_n(x) = x^2 \left[1 - \frac{1}{(1 + x^2)^{n+1}} \right],$$

thus

$$s_n(x) = 1 + x^2 - \frac{1}{(1 + x^2)^n}.$$

The exciting Fig. 354 "explains" what is going on. We see that if $x \neq 0$, the sum is

$$s(x) = \lim s_n(x) = 1 + x^2,$$

but for $x = 0$ we have $s_n(0) = 1 - 1 = 0$ for all n, hence $s(0) = 0$. So we have the surprising fact that the sum is discontinuous (at $x = 0$), although all the terms are continuous and the series converges even absolutely (its terms are nonnegative, thus equal to their absolute value!).

Theorem 2 now tells us that the convergence cannot be uniform in an interval containing $x = 0$. We can also verify this directly. Indeed, for $x \neq 0$ the remainder has the absolute value

$$|R_n(x)| = |s(x) - s_n(x)| = \frac{1}{(1 + x^2)^n}$$

and we see that for a given ϵ (< 1) we cannot find an N depending only on ϵ such that $|R_n| < \epsilon$ for all $n > N(\epsilon)$ and all x, say, in the interval $0 \leqq x \leqq 1$.   ◀

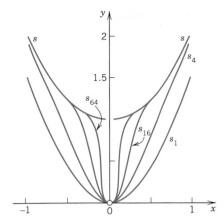

Fig. 354. Partial sums in Example 2

Termwise Integration

This is our second topic in connection with uniform convergence, and we begin with an example to become aware of the danger.

EXAMPLE 3   **A series for which termwise integration is not permissible**

Let

$$u_m(x) = mxe^{-mx^2}$$

and consider the series

$$\sum_{m=1}^{\infty} f_m(x) \qquad \text{where} \qquad f_m(x) = u_m(x) - u_{m-1}(x)$$

in the interval $0 \leqq x \leqq 1$. The nth partial sum is

$$s_n = u_1 - u_0 + u_2 - u_1 + \cdots + u_n - u_{n-1} = u_n - u_0 = u_n.$$

Hence the series has the sum $F(x) = \lim_{n \to \infty} s_n(x) = \lim_{n \to \infty} u_n(x) = 0$ $(0 \leqq x \leqq 1)$. From this we obtain

$$\int_0^1 F(x)\, dx = 0.$$

On the other hand, by integrating term by term and using $f_1 + f_2 + \cdots + f_n = s_n$,

$$\sum_{m=1}^{\infty} \int_0^1 f_m(x)\, dx = \lim_{n \to \infty} \sum_{m=1}^{n} \int_0^1 f_m(x)\, dx = \lim_{n \to \infty} \int_0^1 s_n(x)\, dx.$$

Now $s_n = u_n$ and the expression on the right becomes

$$\lim_{n \to \infty} \int_0^1 u_n(x)\, dx = \lim_{n \to \infty} \int_0^1 nxe^{-nx^2}\, dx = \lim_{n \to \infty} \frac{1}{2}(1 - e^{-n}) = \frac{1}{2},$$

but not 0. This shows that the series under consideration cannot be integrated term by term from $x = 0$ to $x = 1$.   ◄

The series in Example 3 is not uniformly convergent in the interval of integration, and we shall now prove that in the case of a uniformly convergent series of continuous functions we may integrate term by term.

THEOREM 3   **(Termwise integration)**

Let

$$F(z) = \sum_{m=0}^{\infty} f_m(z) = f_0(z) + f_1(z) + \cdots$$

be a uniformly convergent series of continuous functions in a region G. Let C be any path in G. Then the series

(4)
$$\sum_{m=0}^{\infty} \int_C f_m(z)\, dz = \int_C f_0(z)\, dz + \int_C f_1(z)\, dz + \cdots$$

is convergent and has the sum $\int_C F(z)\, dz$.

PROOF.   From Theorem 2 it follows that $F(z)$ is continuous. Let $s_n(z)$ be the nth partial sum of the given series and $R_n(z)$ the corresponding remainder. Then $F = s_n + R_n$ and by integration,

$$\int_C F(z)\, dz = \int_C s_n(z)\, dz + \int_C R_n(z)\, dz.$$

Let L be the length of C. Since the given series converges uniformly, for every given $\epsilon > 0$ we can find a number N such that

$$|R_n(z)| < \frac{\epsilon}{L} \qquad \text{for all } n > N \text{ and all } z \text{ in } G.$$

By applying the ML-inequality (Sec. 13.1) we thus obtain

$$\left| \int_C R_n(z)\, dz \right| < \frac{\epsilon}{L} L = \epsilon \qquad \text{for all } n > N.$$

Since $R_n = F - s_n$, this means that

$$\left| \int_C F(z)\, dz - \int_C s_n(z)\, dz \right| < \epsilon \qquad \text{for all } n > N.$$

Hence, the series (4) converges and has the sum indicated in the theorem.   ◀

Theorems 2 and 3 characterize the two most important properties of uniformly convergent series. Furthermore, since differentiation and integration are inverse processes, Theorem 3 implies

THEOREM 4   **(Termwise differentiation)**

Let the series $f_0(z) + f_1(z) + f_2(z) + \cdots$ be convergent in a region G and let $F(z)$ be its sum. Suppose that the series $f_0'(z) + f_1'(z) + f_2'(z) + \cdots$ converges uniformly in G

and its terms are continuous in G. Then

$$F'(z) = f'_0(z) + f'_1(z) + f'_2(z) + \cdots \qquad \text{for all } z \text{ in } G.$$

Test for Uniform Convergence

Uniform convergence is usually proved by the following comparison test.

THEOREM 5   **(Weierstrass[8] *M*-test for uniform convergence)**

Consider a series of the form (1) *in a region G of the z-plane. Suppose that one can find a convergent series of constant terms,*

$$(5) \qquad\qquad M_0 + M_1 + M_2 + \cdots ,$$

such that $|f_m(z)| \leqq M_m$ *for all z in G and every* $m = 0, 1, \cdots$. *Then* (1) *is uniformly convergent in G.*

The simple proof is left to the student (Team Project 14).

EXAMPLE 4   **Weierstrass *M*-test**

Does the following series converge uniformly in the disk $|z| \leqq 1$?

$$\sum_{m=0}^{\infty} \frac{z^m + 1}{m^2 + \cosh m|z|}$$

Solution.   Uniform convergence follows by the Weierstrass *M*-test and the convergence of $\Sigma 1/m^2$ (see Sec. 14.1, in the proof of Theorem 8) because

$$\left| \frac{z^m + 1}{m^2 + \cosh m|z|} \right| \leqq \frac{|z|^m + 1}{m^2} \leqq \frac{2}{m^2} . \qquad \blacktriangleleft$$

No Relation Between Absolute and Uniform Convergence

We finally show the surprising fact that there are series that converge absolutely but not uniformly, and others that converge uniformly but not absolutely, so that there is no relation between the two concepts.

EXAMPLE 5   **No relation between absolute and uniform convergence**

The series in Example 2 converges absolutely but not uniformly, as we have shown. On the other hand, the series

$$\sum_{m=1}^{\infty} \frac{(-1)^{m-1}}{x^2 + m} = \frac{1}{x^2 + 1} - \frac{1}{x^2 + 2} + \frac{1}{x^2 + 3} - + \cdots \qquad\qquad (x \text{ real})$$

converges uniformly on the whole real line but not absolutely.

Proof. By the familiar Leibniz test of calculus (see Appendix A3.3) the remainder R_n does not exceed its first term in absolute value, since we have a series of alternating terms whose absolute values form a monotone

---

[8]KARL WEIERSTRASS (1815—1897), great German mathematician, whose lifework was the development of complex analysis based on the concept of power series (see the footnote in Sec. 12.4). He also made basic contributions to the calculus, the calculus of variations, approximation theory, and differential geometry. He obtained the concept of uniform convergence in 1841 (published 1894); the first publications on the concept were by G. G. STOKES (see Sec. 9.9) in 1847 and PHILIPP LUDWIG VON SEIDEL (1821—1896) in 1848.

decreasing sequence with limit zero. Hence, given $\epsilon > 0$, for all x we have

$$|R_n(x)| \leqq \frac{1}{x^2 + n + 1} < \frac{1}{n} < \epsilon \qquad \text{if } n > N(\epsilon) \geqq \frac{1}{\epsilon}.$$

This proves uniform convergence, since $N(\epsilon)$ does not depend on x.

The convergence is not absolute because for any fixed x,

$$\left| \frac{(-1)^{m-1}}{x^2 + m} \right| = \frac{1}{x^2 + m} > \frac{k}{m}$$

(k a suitable constant) and $k\Sigma 1/m$ diverges.   ◀

PROBLEM SET 14.5

Uniform Convergence

Prove that the following series converge uniformly in the given regions.

1. $\displaystyle\sum_{n=0}^{\infty} (z - i)^n, \quad |z - i| \leqq 0.99$

2. $\displaystyle\sum_{n=0}^{\infty} \frac{z^{2n}}{(2n)!}, \quad |z| \leqq 10^{20}$

3. $\displaystyle\sum_{n=1}^{\infty} \frac{\cos^n |z|}{n^2}, \quad$ all z

4. $\displaystyle\sum_{n=0}^{\infty} \frac{z^n}{|z|^{2n} + 1}, \quad 2 \leqq |z| \leqq 4$

5. $\displaystyle\sum_{n=0}^{\infty} \frac{(n!)^2}{(2n)!} z^n, \quad |z| \leqq 3$

6. $\displaystyle\sum_{n=0}^{\infty} \frac{\tanh^n |z|}{n(n + 1)}, \quad$ all z

Power Series

Where do the following series converge uniformly?

7. $\displaystyle\sum_{n=0}^{\infty} \frac{(z + i)^{2n}}{5^n}$

8. $\displaystyle\sum_{n=1}^{\infty} \left(\frac{n + 2}{5n - 3} \right)^n z^n$

9. $\displaystyle\sum_{n=1}^{\infty} \frac{(-1)^n}{2^n n^2} (z - 1)^n$

10. $\displaystyle\sum_{n=0}^{\infty} \frac{3^n(1 - i)^n}{n!} z^n$

11. $\displaystyle\sum_{n=3}^{\infty} \binom{n}{3} (3z + i)^n$

12. $\displaystyle\sum_{n=1}^{\infty} 6^n (\tanh n^2) z^{2n}$

13. CAS PROJECT. Graphs of Partial Sums. (a) Figure 354. Produce this exciting figure using your software and adding further curves, say, those of s_{256}, s_{1024}, etc.

 (b) Power series. Study the nonuniformity of convergence experimentally by plotting partial sums near the endpoints of the convergence interval for real $z = x$.

14. TEAM PROJECT. Uniform Convergence. (a) Weierstrass M-test. Give a proof.

 (b) Termwise differentiation. Derive Theorem 4 from Theorem 3.

 (c) Subregions. Prove that uniform convergence of a series in a region G implies uniform convergence in any portion of G. Is the converse true?

 (d) Example 2. Find the precise region of convergence of the series in Example 2 with x replaced by a complex variable z.

 (e) Figure 355. Show that $x^2 \sum_{m=1}^{\infty} (1 + x^2)^{-m} = 1$ if $x \neq 0$ and 0 if $x = 0$. Verify by computation that the partial sums s_1, s_2, s_3 look as shown in Fig. 355.

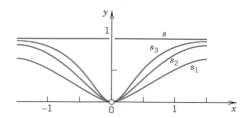

Fig. 355. Sum s and partial sums in Team Project 14(e)

Heat Equation. Show that (10), Sec. 11.5, with coefficients (11) is a solution of the heat equation for $t > 0$, assuming that $f(x)$ is continuous on the interval $0 \leq x \leq L$ and has one-sided derivatives at all interior points of that interval. Proceed as follows.

15. Show that $|B_n|$ is bounded, say, $|B_n| < K$ for all n. Conclude that

$$|u_n| < K e^{-\lambda_n^2 t_0} \qquad \text{if} \qquad t \geq t_0 > 0$$

and, by the Weierstrass test, the series (10) converges uniformly with respect to x and t for $t \geq t_0$, $0 \leq x \leq L$. Using Theorem 2, show that $u(x, t)$ is continuous for $t \geq t_0$ and thus satisfies the boundary conditions (2) for $t \geq t_0$.

16. Show that $|\partial u_n/\partial t| < \lambda_n^2 K e^{-\lambda_n^2 t_0}$ if $t \geq t_0$ and the series of the expressions on the right converges, by the ratio test. Conclude from this, the Weierstrass test, and Theorem 4 that the series (10) can be differentiated term by term with respect to t and the resulting series has the sum $\partial u/\partial t$. Show that (10) can be differentiated twice with respect to x and the resulting series has the sum $\partial^2 u/\partial x^2$. Conclude from this and the result of Prob. 15 that (10) is a solution of the heat equation for all $t \geq t_0$. (The proof that (10) satisfies the given initial condition can be found in Ref. [C7] listed in Appendix 1.)

CHAPTER 14 REVIEW

1. What is a convergence test for a series? State two of them (from memory). Give examples.
2. What is a power series? Why are these series important in complex analysis?
3. What do you know about the convergence behavior of power series? About the radius of convergence?
4. What is absolute convergence? Conditional convergence? Uniform convergence?
5. What do we mean by saying that two power series may be added term by term? What about multiplication of two power series?
6. What have power series to do with analytic functions?
7. Can we integrate a power series termwise? Differentiate termwise? Explain, giving details and examples.
8. What is a Taylor series? A Maclaurin series? Give examples.
9. How did we obtain Taylor's formula from Cauchy's formula?
10. List complex analogs of Maclaurin series in calculus.
11. Does Ln z have a Maclaurin series? Explain.
12. Explain some ways of obtaining a Taylor or Maclaurin series in practice.
13. Write a short essay on the ideas discussed in connection with uniform convergence.
14. Can properties of a function be discovered from its Maclaurin series? Give examples.
15. Does every function have Taylor series developments?

Find the radius of convergence of the following power series. Can you identify the sum as a familiar function in some of the problems? (Show the details of your work.)

16. $\displaystyle\sum_{n=1}^{\infty} n z^n$ **17.** $\displaystyle\sum_{n=1}^{\infty} \frac{z^n}{n}$ **18.** $\displaystyle\sum_{n=0}^{\infty} \frac{n^5}{n!} (z + 1 - i)^n$ **19.** $\displaystyle\sum_{n=0}^{\infty} \frac{(-1)^n (\pi z)^{2n+1}}{(2n + 1)!}$

20. $\displaystyle\sum_{n=1}^{\infty} \frac{n(n + 3)}{4^n(n + 1)} (z - i)^n$ **21.** $\displaystyle\sum_{n=1}^{\infty} \frac{(-2)^{n+1}}{2n} z^n$ **22.** $\displaystyle\sum_{n=0}^{\infty} \frac{(-1)^n}{n!} z^{2n}$

23. $\displaystyle\sum_{n=0}^{\infty} (2\pi)^n (z - i)^{2n}$ **24.** $\displaystyle\sum_{n=0}^{\infty} \frac{(z - 2i)^n}{(4 + 3i)^n}$ **25.** $\displaystyle\sum_{n=0}^{\infty} \frac{z^n}{(2n)!}$

Find the Taylor series of the following functions with the given point as center and determine the radius of convergence. (Show details.)

26. e^{2z}, $\frac{1}{2}\pi i$ **27.** $(e^{z^4} - 1)/z^3$, 0 **28.** $(z + 3 - 4i)^{-2}$, 0

29. $\sin^2 z$, 0 **30.** $\text{Ln } z$, 3 **31.** $1/z$, $3i$

32. z^4, i **33.** $1/(1 - z)^3$, 0 **34.** $\cos z$, $\frac{1}{2}\pi$

35. $\sinh 2z$, πi

36. Does there exist a power series in powers of z that converges at $z = 30 + 10i$ and diverges at $z = 31 - 6i$? (Give a reason.)

Where do the following power series converge uniformly?

37. $\displaystyle\sum_{n=0}^{\infty} \frac{(z - i)^{2n}}{2^n}$ **38.** $\displaystyle\sum_{n=1}^{\infty} \frac{(-1)^n}{7^n n} z^n$ **39.** $\displaystyle\sum_{n=3}^{\infty} \binom{n}{3} (4z + i)^n$ **40.** $\displaystyle\sum_{n=1}^{\infty} \frac{n!}{n^2} (z + 1)^n$

SUMMARY OF CHAPTER 14
POWER SERIES, TAYLOR SERIES

Sequences, series, and convergence tests are discussed in Sec. 14.1. A **power series** is of the form (Sec. 14.2)

$$(1) \qquad \sum_{n=0}^{\infty} a_n(z - z_0)^n = a_0 + a_1(z - z_0) + a_2(z - z_0)^2 + \cdots;$$

z_0 is its *center*. The series (1) converges for $|z - z_0| < R$ and diverges for $|z - z_0| > R$, where R is the **radius of convergence.** Some power series converge for all z (then we write $R = \infty$). Also, $R = \lim |a_n/a_{n+1}|$ if this limit exists. The series (1) converges absolutely (Sec. 14.2) and **uniformly** (Sec. 14.5) in every closed disk $|z - z_0| \leqq r < R$ ($R > 0$). It represents an analytic function $f(z)$ for $|z - z_0| < R$. The derivatives $f'(z)$, $f''(z), \cdots$ are obtained by termwise differentiation of (1), and these series have the same radius of convergence R as (1). See Sec. 14.3.

Conversely, *every* analytic function $f(z)$ can be represented by power series. These **Taylor series** of $f(z)$ are of the form (Sec. 14.4)

$$(2) \qquad f(z) = \sum_{n=0}^{\infty} \frac{1}{n!} f^{(n)}(z_0)(z - z_0)^n \qquad (|z - z_0| < R),$$

as in calculus. They converge for all z in the open disk with center z_0 and radius generally equal to the distance from z_0 to the nearest **singularity** of $f(z)$ (point at which $f(z)$ ceases to be analytic as defined in Sec. 14.4). If $f(z)$ is **entire** (analytic for all z; see Sec. 12.6), then (2) converges for all z. The functions e^z, $\cos z$, $\sin z$, etc. have Maclaurin and Taylor series similar to those in calculus (Sec. 14.4).

Laurent Series
Residue Integration

A **Laurent series** is a series of positive **and negative** integer powers of $z - z_0$ by which we can represent a given function $f(z)$ in an annulus (a circular ring with center z_0) in which $f(z)$ is analytic. $f(z)$ may have singularities outside the ring as well as in its "hole." The series (or finite sum) of the negative powers is called the **principal part** of the Laurent series.

An important special case is a Laurent series converging for $0 < |z - z_0| < r$, in "a ring whose inner circle is degenerated to a point." The principal part of this series serves to classify the singularity of $f(z)$ at z_0 (Sec. 15.2). The coefficient of the power $1/(z - z_0)$ of this Laurent series is called the **residue** of $f(z)$ at z_0. Residues are used in an elegant integration method for complex contour integrals (Sec. 15.3) as well as for certain complicated real integrals (Sec. 15.4).

Prerequisites for this chapter: Chaps. 12, 13, Sec. 14.2.
Sections that may be omitted in a shorter course: Secs. 15.2, 15.4.
References: Appendix 1, Part D.
Answers to problems: Appendix 2.

15.1 Laurent Series

In applications it will often be necessary to expand a function $f(z)$ around a point where $f(z)$ has a *singularity* (*is singular,* is no longer analytic; see the next section). Then we can no longer use Taylor series; instead, we need a new kind of series, called **Laurent series.**[1] This is a series of nonnegative integer powers of $z - z_0$—like a Taylor series—*and negative integer powers* of $z - z_0$—this is the new feature. It is used for two main purposes: for the classification of singularities (Sec. 15.2) and as the key to a powerful, elegant integration method (*residue integration;* Sec. 15.3). A Laurent series converges in an annulus (in the "hole" of which $f(z)$ may have singularities), as follows.

---

[1]PIERRE ALPHONSE LAURENT (1813—1854), French ingenieur and mathematician, published the theorem in 1843.

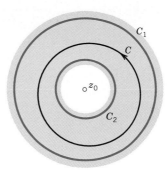

Fig. 356.   Laurent's theorem

THEOREM 1   **Laurent's theorem**

If $f(z)$ is analytic on two concentric circles[2] C_1 and C_2 with center z_0 and in the annulus between them, then $f(z)$ can be represented by the **Laurent series**

(1)

$$f(z) = \sum_{n=0}^{\infty} a_n(z - z_0)^n + \sum_{n=1}^{\infty} \frac{b_n}{(z - z_0)^n}$$

$$= a_0 + a_1(z - z_0) + a_2(z - z_0)^2 + \cdots$$

$$\cdots + \frac{b_1}{z - z_0} + \frac{b_2}{(z - z_0)^2} + \cdots$$

consisting of nonnegative powers and the **principal part** *(the negative powers). The coefficients of this Laurent series are given by the integrals[3]*

(2) $$a_n = \frac{1}{2\pi i} \oint_C \frac{f(z^*)}{(z^* - z_0)^{n+1}} \, dz^*, \quad b_n = \frac{1}{2\pi i} \oint_C (z^* - z_0)^{n-1} f(z^*) \, dz^*,$$

taken counterclockwise around any simple closed path C that lies in the annulus and encircles the inner circle (Fig. 356).

   This series converges and represents $f(z)$ in the open annulus obtained from the given annulus by continuously increasing the outer circle C_1 and decreasing C_2 until each of the two circles reaches a point where $f(z)$ is singular.

   In the important special case that z_0 is the only singular point of $f(z)$ inside C_2, this circle can be shrunk to the point z_0, giving convergence in a disk except at the center.

Comment. Obviously, instead of (1), (2) we may write (denoting b_n by a_{-n})

(1′)

$$f(z) = \sum_{n=-\infty}^{\infty} a_n(z - z_0)^n$$

where all the coefficients are given by a single integral formula, namely,

---

[2]Recall that by the definition of analyticity, this means that $f(z)$ is analytic in some domain containing the annulus as well as its boundary circles.

[3]We denote the variable of integration by z^* because z is used in $f(z)$.

$$(2') \qquad \boxed{a_n = \frac{1}{2\pi i} \oint_C \frac{f(z^*)}{(z^* - z_0)^{n+1}} \, dz^*} \qquad (n = 0, \pm 1, \pm 2, \cdots).$$

PROOF OF LAURENT'S THEOREM. (a) The nonnegative powers. We start from Cauchy's integral formula (3), Sec. 13.3, denoting the functions represented by the two integrals (times $1/2\pi i$) by $g(z)$ and $h(z)$,

$$(3) \qquad f(z) = g(z) + h(z) = \frac{1}{2\pi i} \oint_{C_1} \frac{f(z^*)}{z^* - z} \, dz^* - \frac{1}{2\pi i} \oint_{C_2} \frac{f(z^*)}{z^* - z} \, dz^*$$

where z is any point in the given annulus and we integrate counterclockwise over C_1 and also over C_2 (hence the minus sign!). We transform each of these two integrals as in Sec. 14.4. The first integral is *precisely* as in Sec. 14.4. Hence we get precisely the same result, namely, the Taylor series of $g(z)$,

$$(4) \qquad g(z) = \frac{1}{2\pi i} \oint_{C_1} \frac{f(z^*)}{z^* - z} \, dz^* = \sum_{n=0}^{\infty} a_n (z - z_0)^n$$

with coefficients [see (2), Sec. 14.4, counterclockwise integration]

$$(5) \qquad a_n = \frac{1}{2\pi i} \oint_{C_1} \frac{f(z^*)}{(z^* - z_0)^{n+1}} \, dz^*.$$

Here we can replace C_1 by C (see Fig. 356), by the principle of deformation of path, since z_0, the point where the integrand in (5) is not analytic, is not a point of the annulus. This proves the formula for the a_n in (2).

    (b) The negative powers (the principal part) in (1) and the formula for b_n in (2) are obtained if we consider $h(z)$ (the second integral times $-1/2\pi i$) in (3). Since z lies in the annulus, it lies *outside* the path C_2. Hence the situation differs from that for the first integral. The essential point is that instead of [see (7*) in Sec. 14.4]

$$(6) \qquad \text{(a)} \quad \left| \frac{z - z_0}{z^* - z_0} \right| < 1 \qquad \text{we now have} \qquad \text{(b)} \quad \left| \frac{z^* - z_0}{z - z_0} \right| < 1.$$

Consequently, we must develop the expression $1/(z^* - z)$ in the integrand of the second integral in (3) in powers of $(z^* - z_0)/(z - z_0)$ [instead of the reciprocal of this] to get a *convergent* series. We find

$$\frac{1}{z^* - z} = \frac{1}{z^* - z_0 - (z - z_0)} = \frac{-1}{(z - z_0)\left(1 - \dfrac{z^* - z_0}{z - z_0}\right)}.$$

Compare this for a moment with (7) in Sec. 14.4, to really understand the difference. Then go on and apply formula (8), Sec. 14.4, for a finite geometric sum, obtaining

$$\frac{1}{z^* - z} = -\frac{1}{z - z_0}\left\{ 1 + \frac{z^* - z_0}{z - z_0} + \left(\frac{z^* - z_0}{z - z_0}\right)^2 + \cdots + \left(\frac{z^* - z_0}{z - z_0}\right)^n \right\}$$

$$- \frac{1}{z - z^*}\left(\frac{z^* - z_0}{z - z_0}\right)^{n+1}.$$

Multiplication by $-f(z^*)/2\pi i$ and integration over C_2 on both sides now yield

$$h(z) = -\frac{1}{2\pi i} \oint_{C_2} \frac{f(z^*)}{z^* - z}\, dz^*$$

$$= \frac{1}{2\pi i} \left\{ \frac{1}{z - z_0} \oint_{C_2} f(z^*)\, dz^* + \frac{1}{(z - z_0)^2} \oint_{C_2} (z^* - z_0)f(z^*)\, dz^* \right.$$

$$\left. + \cdots + \frac{1}{(z - z_0)^{n+1}} \oint_{C_2} (z^* - z_0)^n f(z^*)\, dz^* \right\} + R_n^*(z)$$

with the last term on the right given by

$$(7) \qquad\qquad R_n^*(z) = \frac{1}{2\pi i(z - z_0)^{n+1}} \oint_{C_2} \frac{(z^* - z_0)^{n+1}}{z - z^*} f(z^*)\, dz^*.$$

As before, we can integrate over C instead of C_2 in the integrals on the right. We see that on the right, the power $1/(z - z_0)^n$ is multiplied by b_n as given in (2). This establishes Laurent's theorem, provided

$$(8) \qquad\qquad \lim_{n \to \infty} R_n^*(z) = 0.$$

(c) Convergence proof for the principal part [Proof of (8)]. The principal part will often consist of finitely many terms only. Then there is nothing to be proved. Otherwise, we begin by noting that $f(z^*)/(z - z^*)$ in (7) is bounded in absolute value, say,

$$\left| \frac{f(z^*)}{z - z^*} \right| < \tilde{M} \qquad\qquad \text{for all } z^* \text{ on } C_2$$

because $f(z^*)$ is analytic in the annulus and on C_2, and z^* lies on C_2 and z outside, so that $z - z^* \neq 0$. From this and the ML-inequality (Sec. 13.1) applied to (7) we get the inequality (L = length of C_2)

$$|R_n^*(z)| \leqq \frac{1}{2\pi|z - z_0|^{n+1}} |z^* - z_0|^{n+1} \tilde{M}L = \frac{\tilde{M}L}{2\pi} \left| \frac{z^* - z_0}{z - z_0} \right|^{n+1}.$$

From (6) we see that the expression on the right approaches zero as n approaches infinity. This proves (8). The representation (1) with coefficients (2) is now established in the given annulus.

(d) Convergence of (1) in the larger annulus. The first series in (1) is a Taylor series [representing $g(z)$]; hence it converges in the disk D with center z_0 whose radius equals the distance of that singularity of $g(z)$ which is closest to z_0. Also, $g(z)$ must be singular at all points outside C_1 where $f(z)$ is singular. (For an explanation of the notion of a singularity of an analytic function, see Sec. 14.4.)

The second series in (1) [the principal part of (1)], representing $h(z)$, is a power series in $Z = 1/(z - z_0)$. Let the given annulus be $r_2 < |z - z_0| < r_1$, where r_1 and r_2 are the radii of C_1 and C_2, respectively (Fig. 356). This corresponds to $1/r_2 > |Z| > 1/r_1$. Hence this power series in Z must converge at least in the disk $|Z| < 1/r_2$. This corresponds to the exterior $|z - z_0| > r_2$ of C_2, so that $h(z)$ is analytic for all z outside C_2. Also, $h(z)$ must be singular inside C_2 where $f(z)$ is singular, and the principal part converges for all z in the exterior E of the circle with center z_0 and radius equal to the maximum distance from z_0 to the singularities of $f(z)$ inside C_2. The domain common to D and E is the open annulus characterized near the end of Laurent's theorem, whose proof is now complete. ◄

Uniqueness. *The Laurent series of a given analytic function $f(z)$ in its annulus of convergence is unique* (see Team Project 22). *However, $f(z)$ may have different Laurent series in two annuli with the same center;* see the examples below. The uniqueness is essential. As for a Taylor series, to obtain the coefficients of Laurent series, we do not generally use the integral formulas (2); instead, we use various other methods, some of which we shall illustrate in our examples. If a Laurent series has been found by any such process, the uniqueness guarantees that it must be *the* Laurent series of the given function in the given annulus.

EXAMPLE 1   **Use of Maclaurin series**

Find the Laurent series of $z^{-5} \sin z$ with center 0.

Solution.   By (14), Sec. 14.4, we obtain

$$z^{-5} \sin z = \sum_{n=0}^{\infty} \frac{(-1)^n}{(2n+1)!} z^{2n-4} = \frac{1}{z^4} - \frac{1}{6z^2} + \frac{1}{120} - \frac{1}{5040} z^2 + - \cdots \qquad (|z| > 0).$$

Here the "annulus" of convergence is the whole complex plane without the origin. ◄

EXAMPLE 2   **Substitution**

Find the Laurent series of $z^2 e^{1/z}$ with center 0.

Solution.   From (12) in Sec. 14.4 with z replaced by $1/z$ we obtain

$$z^2 e^{1/z} = z^2 \left(1 + \frac{1}{1!z} + \frac{1}{2!z^2} + \cdots \right) = z^2 + z + \frac{1}{2} + \frac{1}{3!z} + \frac{1}{4!z^2} + \cdots \qquad (|z| > 0). \quad ◄$$

EXAMPLE 3   **Develop $1/(1 - z)$ (a) in nonnegative powers of z, (b) in negative powers of z.**

Solution.

(a)
$$\frac{1}{1 - z} = \sum_{n=0}^{\infty} z^n \qquad \text{(valid if } |z| < 1\text{).}$$

(b)
$$\frac{1}{1 - z} = \frac{-1}{z(1 - z^{-1})} = -\sum_{n=0}^{\infty} \frac{1}{z^{n+1}} = -\frac{1}{z} - \frac{1}{z^2} - \cdots \qquad \text{(valid if } |z| > 1\text{).} \quad ◄$$

EXAMPLE 4   **Laurent expansions in different concentric annuli**

Find all Laurent series of $1/(z^3 - z^4)$ with center 0.

Solution.   Multiplying by $1/z^3$, we get from Example 3

(I)
$$\frac{1}{z^3 - z^4} = \sum_{n=0}^{\infty} z^{n-3} = \frac{1}{z^3} + \frac{1}{z^2} + \frac{1}{z} + 1 + z + \cdots \qquad (0 < |z| < 1),$$

(II)
$$\frac{1}{z^3 - z^4} = -\sum_{n=0}^{\infty} \frac{1}{z^{n+4}} = -\frac{1}{z^4} - \frac{1}{z^5} - \cdots \qquad (|z| > 1). \quad ◄$$

EXAMPLE 5   **Use of partial fractions**

Find all Taylor and Laurent series of $f(z) = \dfrac{-2z + 3}{z^2 - 3z + 2}$ with center 0.

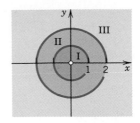

Fig. 357. Regions of convergence in Example 5

Solution. In terms of partial fractions,

$$f(z) = -\frac{1}{z-1} - \frac{1}{z-2}.$$

(a) and (b) in Example 3 take care of the first fraction. For the second fraction,

(c) $$-\frac{1}{z-2} = \frac{1}{2\left(1 - \frac{1}{2}z\right)} = \sum_{n=0}^{\infty} \frac{1}{2^{n+1}} z^n \qquad (|z| < 2),$$

(d) $$-\frac{1}{z-2} = -\frac{1}{z\left(1 - \frac{2}{z}\right)} = -\sum_{n=0}^{\infty} \frac{2^n}{z^{n+1}} \qquad (|z| > 2).$$

(I) From (a) and (c), valid for $|z| < 1$ (see Fig. 357),

$$f(z) = \sum_{n=0}^{\infty} \left(1 + \frac{1}{2^{n+1}}\right) z^n = \frac{3}{2} + \frac{5}{4}z + \frac{9}{8}z^2 + \cdots.$$

(II) From (c) and (b), valid for $1 < |z| < 2$,

$$f(z) = \sum_{n=0}^{\infty} \frac{1}{2^{n+1}} z^n - \sum_{n=0}^{\infty} \frac{1}{z^{n+1}} = \frac{1}{2} + \frac{1}{4}z + \frac{1}{8}z^2 + \cdots - \frac{1}{z} - \frac{1}{z^2} - \cdots.$$

(III) From (d) and (b), valid for $|z| > 2$,

$$f(z) = -\sum_{n=0}^{\infty} (2^n + 1) \frac{1}{z^{n+1}} = -\frac{2}{z} - \frac{3}{z^2} - \frac{5}{z^3} - \frac{9}{z^4} - \cdots. \qquad ◀$$

If $f(z)$ in Laurent's theorem is analytic inside C_2, the coefficients b_n in (2) are zero by Cauchy's integral theorem, so that the Laurent series reduces to a Taylor series. Examples 3(a) and 5(I) illustrate this.

PROBLEM SET 15.1

Laurent Series Near a Singularity at 0
In Probs. 1–8 find the Laurent series that converges for $0 < |z| < R$ and determine the precise region of convergence. (Show the details of your work.)

1. $\dfrac{\cos z}{z^4}$ **2.** $\dfrac{\sin \pi z}{z^2}$ **3.** $\dfrac{e^{z^2}}{z^3}$ **4.** $\dfrac{1}{z(z-1)}$

5. $z^3 \cosh \dfrac{1}{z}$ **6.** $\dfrac{e^{-5z}}{z^5}$ **7.** $\dfrac{e^{-1/z^2}}{z^2}$ **8.** $\dfrac{e^z}{z - z^2}$

Laurent Series Near a Singularity at z_0

Find the Laurent series that converges for $0 < |z - z_0| < R$ and determine the precise region of convergence. (Show details.)

9. $\dfrac{1}{z^2 + 1}$, $z_0 = i$ **10.** $\dfrac{e^z}{z(1 - z)}$, $z_0 = 1$ **11.** $\dfrac{\cos z}{(z - \pi)^2}$, $z_0 = \pi$

12. $\dfrac{z^2 - 4}{(z - 1)^2}$, $z_0 = 1$ **13.** $\dfrac{z^4}{(z + 2i)^4}$, $z_0 = -2i$ **14.** $\dfrac{\sin z}{(z - \frac{1}{4}\pi)^2}$, $z_0 = \frac{1}{4}\pi$

15. $\dfrac{\cosh z}{(z + \pi i)^2}$, $z_0 = -\pi i$ **16.** $\dfrac{e^{az}}{z - b}$, $z_0 = b$

Taylor and Laurent Series. Showing details, develop $f(z) = \dfrac{2z - 3i}{z^2 - 3iz - 2}$ in a series valid for

17. $0 < |z| < 1$ **18.** $1 < |z| < 2$ **19.** $|z| > 2$ **20.** $0 < |z + i| < 2$

21. CAS PROJECT. Partial Fractions. Write a program for obtaining Laurent series by the use of partial fractions. Using the program, verify the calculations in Example 5 of the text. Apply the program to two other functions of your choice.

22. TEAM PROJECT. Laurent Series. (a) **Uniqueness.** Prove that the Laurent expansion of a given analytic function in a given annulus is unique.

    (b) **Accumulation of singularities.** Does $\tan(1/z)$ have a Laurent series that converges in a region $0 < |z| < R$? (Give a reason.)

    (c) **Integrals.** Expand the following functions in a Laurent series that converges for $|z| > 0$:

$$\frac{1}{z^2} \int_0^z \frac{e^t - 1}{t} \, dt, \qquad \frac{1}{z^3} \int_0^z \frac{\sin t}{t} \, dt.$$

15.2 Singularities and Zeros. Infinity

Roughly, a *singular point* of an analytic function $f(z)$ is a z at which $f(z)$ ceases to be analytic, and a *zero* is a z at which $f(z) = 0$. Precise definitions follow below. Singularities may be discussed and classified by means of Laurent series. This is what we do first. Then we show that zeros may be discussed by means of Taylor series.

We say that a function[4] $f(z)$ **is singular** or **has a singularity** at a point $z = z_0$ if $f(z)$ is not analytic (perhaps not even defined) at $z = z_0$, but every neighborhood of $z = z_0$ contains points at which $f(z)$ is analytic. We also say that $z = z_0$ is a **singular point** of $f(z)$.

We call $z = z_0$ an **isolated singularity** of $f(z)$ if $z = z_0$ has a neighborhood without further singularities of $f(z)$. *Example:* $\tan z$ has isolated singularities at $\pm\pi/2$, $\pm 3\pi/2$, etc.; $\tan(1/z)$ has a nonisolated singularity at 0. (Explain!) Isolated singularities of $f(z)$ at $z = z_0$ can be classified by the Laurent series

(1) $$f(z) = \sum_{n=0}^{\infty} a_n(z - z_0)^n + \sum_{n=1}^{\infty} \frac{b_n}{(z - z_0)^n} \qquad \text{(Sec. 15.1)}$$

---

[4] We recall that, by definition, a function is a *single-valued* relation. (See Sec. 12.3.)

valid *in the immediate neighborhood* of the singular point $z = z_0$, except at z_0 itself, that is, in a region of the form

$$0 < |z - z_0| < R.$$

The sum of the first series is analytic at $z = z_0$, as we know from the last section. The second series, containing the negative powers, is called the **principal part** of (1). If it has only finitely many terms, it is of the form

$$\text{(2)} \qquad \frac{b_1}{z - z_0} + \cdots + \frac{b_m}{(z - z_0)^m} \qquad (b_m \neq 0).$$

Then the singularity of $f(z)$ at $z = z_0$ is called a **pole,** and m is called its **order.** Poles of the first order are also known as **simple poles.**

If the principal part of (1) has infinitely many terms, we say that $f(z)$ has at $z = z_0$ an **isolated essential singularity.**

We leave aside nonisolated singularities (see the example above).

EXAMPLE 1   **Poles. Essential singularities**

The function

$$f(z) = \frac{1}{z(z - 2)^5} + \frac{3}{(z - 2)^2}$$

has a simple pole at $z = 0$ and a pole of fifth order at $z = 2$. Examples of functions having an isolated essential singularity at $z = 0$ are

$$e^{1/z} = \sum_{n=0}^{\infty} \frac{1}{n! z^n} = 1 + \frac{1}{z} + \frac{1}{2! z^2} + \cdots$$

and

$$\sin \frac{1}{z} = \sum_{n=0}^{\infty} \frac{(-1)^n}{(2n + 1)! z^{2n+1}} = \frac{1}{z} - \frac{1}{3! z^3} + \frac{1}{5! z^5} - + \cdots.$$

Section 15.1 provides further examples. For instance, Example 1 shows that $z^{-5} \sin z$ has a fourth-order pole at 0. Example 4 shows that $1/(z^3 - z^4)$ has a third-order pole at 0 and a Laurent series with infinitely many negative powers. This is no contradiction, since this series is valid for $|z| > 1$; it merely tells us that it is quite important to consider the Laurent series valid *in the immediate neighborhood* of a singular point.   ◀

The classification of singularities into poles and essential singularities is not merely a formal matter, because the behavior of an analytic function in a neighborhood of an essential singularity is entirely different from that in the neighborhood of a pole.

EXAMPLE 2   **Behavior near a pole**

The function $f(z) = 1/z^2$ has a pole at $z = 0$, and $|f(z)| \to \infty$ as $z \to 0$ in any manner. This illustrates the following theorem.   ◀

THEOREM 1   **(Poles)**

If $f(z)$ is analytic and has a pole at $z = z_0$, then $|f(z)| \to \infty$ as $z \to z_0$ in any manner. (See Prob. 20.)

EXAMPLE 3   **Behavior near an essential singularity**

The function $f(z) = e^{1/z}$ has an essential singularity at $z = 0$. It has no limit for approach along the imaginary axis; it becomes infinite if $z \to 0$ through positive real values, but it approaches zero if $z \to 0$ through negative real values. It takes on any given value $c = c_0 e^{i\alpha} \neq 0$ in an arbitrarily small neighborhood of $z = 0$. In fact, setting $z = re^{i\theta}$, we must solve the equation

$$e^{1/z} = e^{(\cos \theta - i \sin \theta)/r} = c_0 e^{i\alpha}$$

for r and θ. Equating the absolute values and the arguments, we have $e^{(\cos \theta)/r} = c_0$, that is,

$$\cos \theta = r \ln c_0 \qquad \text{and} \qquad -\sin \theta = \alpha r.$$

From these two equations and $\cos^2 \theta + \sin^2 \theta = r^2(\ln c_0)^2 + \alpha^2 r^2 = 1$ we obtain the formulas

$$r^2 = \frac{1}{(\ln c_0)^2 + \alpha^2} \qquad \text{and} \qquad \tan \theta = -\frac{\alpha}{\ln c_0}.$$

Hence r can be made arbitrarily small by adding multiples of 2π to α, leaving c unaltered. This illustrates the very famous *Picard's theorem* (with $z = 0$ as the exceptional value). For the rather complicated proof, see Ref. [D8].   ◀

THEOREM 2   **(Picard's[5] theorem)**

If $f(z)$ is analytic and has an isolated essential singularity at a point z_0, it takes on every value, with at most one exceptional value, in an arbitrarily small neighborhood of z_0.

Removable singularities. We say that a function $f(z)$ has a *removable singularity* at $z = z_0$ if $f(z)$ is not analytic at $z = z_0$, but can be made analytic there by assigning a suitable value $f(z_0)$. Such singularities are of no interest since they can be removed as just indicated. *Example:* $f(z) = (\sin z)/z$ becomes analytic at $z = 0$ if we define $f(0) = 1$.

Zeros of Analytic Functions

A **zero** of an analytic function $f(z)$ in a domain D is a $z = z_0$ in D such that $f(z_0) = 0$. A zero has **order** n if not only f but also the derivatives f', f'', $\cdots$, $f^{(n-1)}$ are all 0 at $z = z_0$ but $f^{(n)}(z_0) \neq 0$. A first-order zero is also called a **simple zero.** For a second-order zero, $f(z_0) = f'(z_0) = 0$ but $f''(z_0) \neq 0$. And so on.

EXAMPLE 4   **Zeros**

The function $1 + z^2$ has simple zeros at $\pm i$. The function $(1 - z^4)^2$ has second-order zeros at ± 1 and $\pm i$. The function $(z - a)^3$ has a third-order zero at $z = a$. The function e^z has no zeros (see Sec. 12.6). The function $\sin z$ has simple zeros at 0, $\pm \pi$, $\pm 2\pi$, $\cdots$, and $\sin^2 z$ has second-order zeros at these points. The function $1 - \cos z$ has second-order zeros at 0, $\pm 2\pi$, $\pm 4\pi$, $\cdots$, and the function $(1 - \cos z)^2$ has fourth-order zeros at these points.   ◀

Taylor Series at a Zero. At an nth-order zero $z = z_0$ of $f(z)$, the derivatives $f'(z_0)$, $\cdots$, $f^{(n-1)}(z_0)$ are zero, by definition. Hence the first few coefficients a_0, $\cdots$, a_{n-1} of the Taylor series (1), Sec. 14.4, are zero, too, whereas $a_n \neq 0$, so that this series takes the form

(3)
$$
\begin{aligned}
f(z) &= a_n(z - z_0)^n + a_{n+1}(z - z_0)^{n+1} + \cdots \\
&= (z - z_0)^n [a_n + a_{n+1}(z - z_0) + a_{n+2}(z - z_0)^2 + \cdots]
\end{aligned}
\qquad (a_n \neq 0).
$$

This is characteristic of such a zero, because if $f(z)$ has such a Taylor series, it has an nth-order zero at $z = z_0$, as follows by differentiation.

Whereas nonisolated singularities may occur, for zeros we have

---

[5]See the footnote in Sec. 1.9.

THEOREM 3 **(Zeros)**

The zeros of an analytic function $f(z)$ ($\not\equiv 0$) are isolated; that is, each of them has a neighborhood that contains no further zeros of $f(z)$.

PROOF. In (3), the factor $(z - z_0)^n$ is zero only at $z = z_0$. The power series in the brackets $[\cdots]$ represents an analytic function (by Theorem 5 in Sec. 14.3), call it $g(z)$. Now $g(z_0) = a_n \neq 0$, since an analytic function is continuous, and because of this continuity, also $g(z) \neq 0$ in some neighborhood of $z = z_0$. Hence the same holds for $f(z)$. ◀

This theorem is illustrated by the functions in Example 4.

Poles are often caused by zeros in the denominator. (*Example:* $\tan z$ has poles where $\cos z$ is zero.) This is a major reason for the importance of zeros. The key to the connection is the following theorem, whose proof follows from (3) (see Team Project 8).

THEOREM 4 **(Poles and zeros)**

Let $f(z)$ be analytic at $z = z_0$ and have a zero of nth order at $z = z_0$. Then $1/f(z)$ has a pole of nth order at $z = z_0$.
The same holds for $h(z)/f(z)$ if $h(z)$ is analytic at $z = z_0$ and $h(z_0) \neq 0$.

Analytic or Singular at Infinity

Infinity (∞) was attached to the complex plane in Sec. 12.9, resulting in the **extended complex plane.** (This was suggested by properties of linear fractional transformations, which are conformal mappings.) The extended complex plane can be mapped onto the sphere of diameter 1 in Fig. 358 touching the plane at $z = 0$. The image P^* of a complex number P: z is the intersection of the sphere with the segment from P to the "North Pole" N. The point ∞ has the image N, by definition. The sphere representing the extended complex plane in this way is called the **Riemann number sphere.** The mapping of the sphere onto the plane is called a **stereographic projection** *with center N.* (What is the image of the Northern Hemisphere? Of the Western Hemisphere? Of a straight line through the origin?)

If we want to investigate a function $f(z)$ for large $|z|$, we may now set $z = 1/w$ and investigate $f(z) = f(1/w) \equiv g(w)$ in a neighborhood of $w = 0$. We define $f(z)$ to be **analytic** or **singular at infinity** if $g(w)$ is analytic or singular, respectively, at $w = 0$.

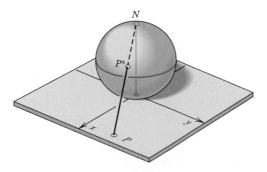

Fig. 358.   Riemann number sphere

We also define

(4) $$g(0) = \lim_{w \to 0} g(w)$$

if this limit exists.

Furthermore, we say that $f(z)$ has an *nth-order zero at infinity* if $f(1/w)$ has such a zero at $w = 0$. Similarly for poles and essential singularities.

EXAMPLE 5   **Functions analytic or singular at infinity. Entire and meromorphic functions**

The function $f(z) = 1/z^2$ is analytic at ∞ since $g(w) = f(1/w) = w^2$ is analytic at $w = 0$, and $f(z)$ has a second-order zero at ∞. The function $f(z) = z^3$ is singular at ∞ and has a third-order pole there since the function $g(w) = f(1/w) = 1/w^3$ has such a pole at $w = 0$. The function e^z has an essential singularity at ∞ since $e^{1/w}$ has such a singularity at $w = 0$. Similarly, $\cos z$ and $\sin z$ have an essential singularity at ∞.

Recall that an **entire function** is one that is analytic everywhere in the (finite) complex plane. Liouville's theorem (Sec. 13.4) tells us that the only *bounded* entire functions are the constants, hence any nonconstant entire function must be unbounded. Hence it has a singularity at ∞, a pole if it is a polynomial or an essential singularity if it is not. The functions just considered are typical in this respect.

An analytic function whose only singularities in the finite plane are poles is called a **meromorphic function.** Examples are rational functions with nonconstant denominator, $\tan z$, $\cot z$, $\sec z$, and $\csc z$.   ◀

In this section we used Laurent series for investigating singularities. In the next section we shall use these series for an elegant integration method.

PROBLEM SET 15.2

Zeros
Determine the location and order of the zeros of the following functions.

1. $\tan \pi z$ **2.** $(z^4 - 1)^4$ **3.** $\cos^2 \frac{1}{2} z$ **4.** $\cosh^2 2z$
5. $(z^2 + 1)(e^z - 1)$ **6.** $(\sin z - 1)^5$ **7.** $(z^4 - z^2 - 6)^3$

8. TEAM PROJECT. Zeros. (a) Derivative. Show that if $f(z)$ has a zero of order $n > 1$ at $z = z_0$, then $f'(z)$ has a zero of order $n - 1$ at z_0.
     (b) Poles and zeros. Prove Theorem 4.
     (c) Isolated k-points. Show that the points at which a nonconstant analytic function $f(z)$ has a given value k are isolated.
     (d) Identical functions. If $f_1(z)$ and $f_2(z)$ are analytic in a domain D and equal at a sequence of points z_n in D that converges in D, show that $f_1(z) \equiv f_2(z)$ in D.

Singularities
Determine the location and type of singularities of the following functions, including those at infinity. In the case of poles also state the order. (Give reasons.)

9. $z^2 - 1/z^2$ **10.** $\tan \frac{1}{2} \pi z$ **11.** $\cot 2z$ **12.** $\sin 3z - \cos 3z$
13. $2z^{-3} - z^{-1}$ **14.** $\cosh[(z^2 + 4)^{-1}]$ **15.** $(z^2 + a^2)^{-2}$ **16.** $(z + i)e^{1/(z+i)}$
17. $z^{-2} \sin^2 z$ **18.** $(\cos z - \sin z)^{-1}$ **19.** $(z - \pi i)^{-2} \sinh z$

20. Verify Theorem 1 for $f(z) = z^{-3} - z^{-1}$. Prove Theorem 1.
21. (Riemann number sphere) Assuming that we let the image of the x-axis be the meridians $0°$ and $180°$, describe and plot the images of the following regions on the Riemann number sphere: **(a)** $|z| > 100$, **(b)** the lower half-plane, **(c)** $\frac{1}{2} \leq |z| \leq 2$.

15.3 Residue Integration Method

The purpose of Cauchy's residue integration method is the evaluation of integrals

$$\oint_C f(z)\,dz$$

taken around a simple closed path C. The idea is as follows.

If $f(z)$ is analytic everywhere on C and inside C, such an integral is zero by Cauchy's integral theorem (Sec. 13.2), and we are done.

If $f(z)$ has a singularity at a point $z = z_0$ inside C, but is otherwise analytic on C and inside C, then $f(z)$ has a Laurent series

$$f(z) = \sum_{n=0}^{\infty} a_n(z - z_0)^n + \frac{b_1}{z - z_0} + \frac{b_2}{(z - z_0)^2} + \cdots$$

that converges for all points near $z = z_0$ (except at $z = z_0$ itself), in some domain of the form[6] $0 < |z - z_0| < R$. Now comes the key idea. The coefficient b_1 of the first negative power $1/(z - z_0)$ of this Laurent series is given by the integral formula (2), Sec. 15.1, with $n = 1$, that is,

$$b_1 = \frac{1}{2\pi i} \oint_C f(z)\,dz,$$

but since we can obtain Laurent series by various methods, without using the integral formulas for the coefficients (see the examples in Sec. 15.1), we can find b_1 by one of those methods and then use the formula for b_1 for evaluating the integral:

(1)
$$\oint_C f(z)\,dz = 2\pi i b_1.$$

Here we integrate counterclockwise around the simple closed path C that contains $z = z_0$ in its interior (but no other singular points of $f(z)$ on or inside C!).

The coefficient b_1 is called the **residue** of $f(z)$ at $z = z_0$ and we denote it by

(2)
$$b_1 = \operatorname*{Res}_{z=z_0} f(z).$$

EXAMPLE 1   **Evaluation of an integral by means of a residue**

Integrate the function $f(z) = z^{-4} \sin z$ counterclockwise around the unit circle C.

Solution.   From (14) in Sec. 14.4 we obtain the Laurent series

$$f(z) = \frac{\sin z}{z^4} = \frac{1}{z^3} - \frac{1}{3!\,z} + \frac{z}{5!} - \frac{z^3}{7!} + - \cdots$$

which converges for $|z| > 0$ (that is, for all $z \neq 0$). This series shows that $f(z)$ has a pole of third order at

---

[6]Sometimes called a **deleted neighborhood,** an old-fashioned term that we shall not use.

$z = 0$ and the residue

$$b_1 = -1/3!.$$

From (1) we thus obtain the answer

$$\oint_C \frac{\sin z}{z^4} \, dz = 2\pi i b_1 = -\frac{\pi i}{3} \,. \qquad \blacktriangleleft$$

EXAMPLE 2   **Be careful to use the *right* Laurent series!**

Integrate $f(z) = 1/(z^3 - z^4)$ clockwise around the circle C: $|z| = 1/2$.

Solution. $z^3 - z^4 = z^3(1 - z)$ shows that $f(z)$ is singular at $z = 0$ and $z = 1$. Now $z = 1$ lies outside C. Hence it is of no interest here. So we need the residue of $f(z)$ at 0. We find it from the Laurent series that converges for $0 < |z| < 1$. This is series (I) in Example 4, Sec. 15.1,

$$\frac{1}{z^3 - z^4} = \frac{1}{z^3} + \frac{1}{z^2} + \frac{1}{z} + 1 + z + \cdots \qquad (0 < |z| < 1).$$

We see from it that this residue is 1. Clockwise integration thus yields

$$\oint_C \frac{dz}{z^3 - z^4} = -2\pi i \operatorname*{Res}_{z=0} f(z) = -2\pi i.$$

Caution! Had we used the wrong series (II) in Example 4, Sec. 15.1,

$$\frac{1}{z^3 - z^4} = -\frac{1}{z^4} - \frac{1}{z^5} - \frac{1}{z^6} - \cdots \qquad (|z| > 1),$$

we would have obtained the wrong answer, 0, because this series has no power $1/z$. $\blacktriangleleft$

Two Formulas for Residues at Simple Poles

To calculate a residue at a pole we need not produce a whole Laurent series, but, more economically, we can derive formulas for residues once and for all. For a *simple pole* at $z = z_0$ the Laurent series (1), Sec. 15.1, is

$$f(z) = \frac{b_1}{z - z_0} + a_0 + a_1(z - z_0) + a_2(z - z_0)^2 + \cdots \qquad (0 < |z - z_0| < R).$$

Here $b_1 \neq 0$. (Why?) Multiplying both sides by $z - z_0$ we have

$$(z - z_0)f(z) = b_1 + (z - z_0)[a_0 + a_1(z - z_0) + \cdots].$$

We now let $z \to z_0$. Then the right side approaches b_1 because of continuity (Theorem 1, Sec. 14.3). This gives a first formula

(3)
$$\boxed{\operatorname*{Res}_{z=z_0} f(z) = b_1 = \lim_{z \to z_0} (z - z_0)f(z).}$$

EXAMPLE 3   **Residue at a simple pole**

$$\operatorname*{Res}_{z=i} \frac{9z + i}{z(z^2 + 1)} = \lim_{z \to i} (z - i) \frac{9z + i}{z(z + i)(z - i)} = \left[\frac{9z + i}{z(z + i)} \right]_{z=i} = \frac{10i}{-2} = -5i. \qquad \blacktriangleleft$$

Another, sometimes simpler formula for the residue at a simple pole is obtained if we start from

$$f(z) = \frac{p(z)}{q(z)} \qquad (p, q \text{ analytic})$$

where $p(z_0) \neq 0$ and $q(z)$ has a simple zero at z_0, so that $f(z)$ has a simple pole at z_0, by Theorem 4 in the last section. By the definition of a simple zero, the Taylor series of $q(z)$ with center z_0 is of the form

$$q(z) = (z - z_0) q'(z_0) + \frac{(z - z_0)^2}{2!} q''(z_0) + \cdots .$$

Substituting this into $f = p/q$ and then f into (3), we obtain

$$\operatorname*{Res}_{z=z_0} f(z) = \lim_{z \to z_0} (z - z_0) \frac{p(z)}{q(z)} = \lim_{z \to z_0} \frac{(z - z_0)p(z)}{(z - z_0)[q'(z_0) + (z - z_0)q''(z_0)/2 + \cdots]} .$$

$z - z_0$ cancels. The denominator has the limit $q'(z_0)$, again by continuity. Hence our second formula for the residue at a simple pole is

(4)
$$\operatorname*{Res}_{z=z_0} f(z) = \operatorname*{Res}_{z=z_0} \frac{p(z)}{q(z)} = \frac{p(z_0)}{q'(z_0)} .$$

EXAMPLE 4   **Residue at a simple pole calculated by formula (4)**

$$\operatorname*{Res}_{z=i} \frac{9z + i}{z(z^2 + 1)} = \left[\frac{9z + i}{3z^2 + 1} \right]_{z=i} = \frac{10i}{-2} = -5i \qquad \text{(see Example 3).} \quad \blacktriangleleft$$

Formula for the Residue at a Pole of Any Order

If $f(z)$ has a pole of order $m > 1$ at $z = z_0$, its Laurent series converging near z_0 (except at z_0 itself) is (see Sec. 15.2)

$$f(z) = \frac{b_m}{(z - z_0)^m} + \frac{b_{m-1}}{(z - z_0)^{m-1}} + \cdots + \frac{b_2}{(z - z_0)^2} + \frac{b_1}{z - z_0} + a_0 + a_1(z - z_0) + \cdots$$

where $b_m \neq 0$. The residue of $f(z)$ at $z = z_0$ is b_1. Multiplying both sides by $(z - z_0)^m$, we have

$$(z - z_0)^m f(z) = b_m + b_{m-1}(z - z_0) + \cdots$$
$$+ b_2(z - z_0)^{m-2} + b_1(z - z_0)^{m-1} + a_0(z - z_0)^m + \cdots .$$

We see that the residue b_1 of $f(z)$ at $z = z_0$ is now the coefficient of the power $(z - z_0)^{m-1}$ in the Taylor series of the function

$$g(z) = (z - z_0)^m f(z)$$

on the left, with center $z = z_0$. Thus, by Taylor's theorem (Sec. 14.4),

$$b_1 = \frac{1}{(m - 1)!} g^{(m-1)}(z_0).$$

Hence if $f(z)$ has a pole of mth order at $z = z_0$, the residue is given by

(5)
$$\operatorname*{Res}_{z=z_0} f(z) = \frac{1}{(m-1)!} \lim_{z \to z_0} \left\{ \frac{d^{m-1}}{dz^{m-1}} \left[(z - z_0)^m f(z) \right] \right\}.$$

In particular, for a second-order pole ($m = 2$),

(5*)
$$\operatorname*{Res}_{z=z_0} f(z) = \lim_{z \to z_0} \left\{ \left[(z - z_0)^2 f(z) \right]' \right\}.$$

EXAMPLE 5   **Residue at a pole of higher order**

The function

$$f(z) = \frac{50z}{(z+4)(z-1)^2}$$

has a pole of second order at $z = 1$, and from (5*) we obtain the corresponding residue

$$\operatorname*{Res}_{z=1} f(z) = \lim_{z \to 1} \frac{d}{dz} \left[(z-1)^2 f(z) \right] = \lim_{z \to 1} \frac{d}{dz} \left(\frac{50z}{z+4} \right) = 8. \qquad \blacktriangleleft$$

Several Singularities Inside the Contour. Residue Theorem

Residue integration can be extended from the case of a single singularity to the case of several singularities within the contour C. This is the purpose of the residue theorem. The extension is surprisingly simple.

THEOREM 1   **Residue theorem**

Let $f(z)$ be analytic inside a simple closed path C and on C, except for finitely many singular points $z_1, z_2, \cdots, z_k$ inside C. Then the integral of $f(z)$ taken counterclockwise around C equals $2\pi i$ times the sum of the residues of $f(z)$ at $z_1, \cdots, z_k$:

(6)
$$\oint_C f(z)\, dz = 2\pi i \sum_{j=1}^{k} \operatorname*{Res}_{z=z_j} f(z).$$

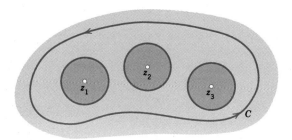

Fig. 359.   Residue theorem

PROOF.   We enclose each of the singular points z_j in a circle C_j with radius small enough that those k circles and C are all separated (Fig. 359). Then $f(z)$ is analytic in the multiply connected domain D bounded by C and $C_1, \cdots, C_k$ and on the entire boundary of D. From Cauchy's integral theorem we thus have

$$(7) \qquad \oint_C f(z)\,dz + \oint_{C_1} f(z)\,dz + \oint_{C_2} f(z)\,dz + \cdots + \oint_{C_k} f(z)\,dz = 0,$$

the integral along C being taken *counterclockwise* and the other integrals *clockwise* (as in Figs. 337 and 338, Sec. 13.2). We take the integrals over $C_1, \cdots, C_k$ to the right and compensate the resulting minus sign by reversing the sense of integration. Thus,

$$(8) \qquad \oint_C f(z)\,dz = \oint_{C_1} f(z)\,dz + \oint_{C_2} f(z)\,dz + \cdots + \oint_{C_k} f(z)\,dz$$

where all the integrals are now taken counterclockwise. By (1),

$$\oint_{C_j} f(z)\,dz = 2\pi i \operatorname*{Res}_{z=z_j} f(z), \qquad\qquad j = 1, \cdots, k,$$

so that (8) gives (6) and the residue theorem is proved.   ◀

   This important theorem has various applications in connection with complex and real integrals. Let us first consider some complex integrals. (Real integrals follow in the next section.)

EXAMPLE 6   **Integration by the residue theorem**

Evaluate the following integral counterclockwise around any simple closed path such that (a) 0 and 1 are inside C, (b) 0 is inside, 1 outside, (c) 1 is inside, 0 outside, (d) 0 and 1 are outside.

$$\oint_C \frac{4 - 3z}{z^2 - z}\,dz$$

Solution.   The integrand has simple poles at 0 and 1, with residues [by (3)]

$$\operatorname*{Res}_{z=0} \frac{4 - 3z}{z(z - 1)} = \left[\frac{4 - 3z}{z - 1} \right]_{z=0} = -4, \qquad \operatorname*{Res}_{z=1} \frac{4 - 3z}{z(z - 1)} = \left[\frac{4 - 3z}{z} \right]_{z=1} = 1.$$

[Confirm this by (4).] *Ans.* (a) $2\pi i(-4 + 1) = -6\pi i$, (b) $-8\pi i$, (c) $2\pi i$, (d) 0.   ◀

EXAMPLE 7   **Simplification of an earlier integration**

For counterclockwise integration around the circle $C: |z| = 3/2$, the residue theorem, together with the residue formula (4), gives

$$\oint_C \frac{\tan z}{z^2 - 1}\,dz = 2\pi i \left(\operatorname*{Res}_{z=1} \frac{\tan z}{z^2 - 1} + \operatorname*{Res}_{z=-1} \frac{\tan z}{z^2 - 1} \right)$$

$$= 2\pi i \left(\frac{\tan z}{2z} \bigg|_{z=1} + \frac{\tan z}{2z} \bigg|_{z=-1} \right) = 2\pi i \tan 1,$$

in agreement with Example 4 in Sec. 13.3. Note that we have avoided partial fraction reduction.   ◀

EXAMPLE 8   **Poles and essential singularities**

Evaluate the following integral, where C is the ellipse $9x^2 + y^2 = 9$ (counterclockwise).

$$\oint_C \left(\frac{ze^{\pi z}}{z^4 - 16} + ze^{\pi/z} \right) dz$$

Solution. Since $z^4 - 16 = 0$ at $\pm 2i$ and ± 2, the first term of the integrand has simple poles at $\pm 2i$ inside C, with residues [by (4); note that $e^{2\pi i} = 1$]

$$\operatorname*{Res}_{z=2i} \frac{ze^{\pi z}}{z^4 - 16} = \left[\frac{ze^{\pi z}}{4z^3} \right]_{z=2i} = -\frac{1}{16}, \qquad \operatorname*{Res}_{z=-2i} \frac{ze^{\pi z}}{z^4 - 16} = \left[\frac{ze^{\pi z}}{4z^3} \right]_{z=-2i} = -\frac{1}{16}$$

and simple poles at ± 2, which lie outside C, so that they are of no interest here. The second term of the integrand has an essential singularity at 0, with residue $\pi^2/2$ as obtained from

$$ze^{\pi/z} = z\left(1 + \frac{\pi}{z} + \frac{\pi^2}{2!z^2} + \frac{\pi^3}{3!z^3} + \cdots \right) = z + \pi + \frac{\pi^2}{2} \cdot \frac{1}{z} + \cdots.$$

Ans. $2\pi i(-\frac{1}{16} - \frac{1}{16} + \pi^2/2) = \pi(\pi^2 - \frac{1}{4})i = 30.221i$ by the residue theorem. ◀

PROBLEM SET 15.3

Residues

Find the residues at the singular points from the Laurent series or from one of the formulas (3)−(5). (Show details of your work.)

1. $\dfrac{4}{1 + z^2}$ 2. $\dfrac{\cos z}{z^4}$ 3. $\dfrac{\sin 2z}{z^6}$ 4. $\tan z$ 5. $\dfrac{1}{1 - e^z}$

6. $\sec z$ 7. $\dfrac{1}{(z^2 - 1)^2}$ 8. $\dfrac{z^4}{z^2 - iz + 2}$ 9. $\cot \pi z$ 10. $\dfrac{e^z}{(z - \pi i)^3}$

 11. CAS PROJECT. Residue at a Pole. Write a program for calculating the residue at a pole of any order. Use it for solving Probs. 1−10.

Residue Integration

Evaluate the following integrals (counterclockwise). (Show details.)

12. $\displaystyle\oint_C \frac{z - 23}{z^2 - 4z - 5} \, dz, \quad C: |z - 2| = 4$ 13. $\displaystyle\oint_C \tan \pi z \, dz, \quad C: |z| = 1$

14. $\displaystyle\oint_C \frac{z^2 \sin z}{4z^2 - 1} \, dz, \quad C: |z| = 2$ 15. $\displaystyle\oint_C \frac{e^z}{\cos z} \, dz, \quad C: |z| = 3$

16. $\displaystyle\oint_C \frac{e^z + z}{z^3 - z} \, dz, \quad C: |z| = \tfrac{1}{2}\pi$ 17. $\displaystyle\oint_C \frac{z + 1}{z^4 - 2z^3} \, dz, \quad C: |z| = \tfrac{1}{2}$

18. $\displaystyle\oint_C \frac{\sinh z}{2z - i} \, dz, \quad C: |z - i| = 1$ 19. $\displaystyle\oint_C \frac{z \cosh \pi z}{z^4 + 13z^2 + 36} \, dz, \quad C: |z| = \pi$

20. $\displaystyle\oint_C \frac{e^{-z^2}}{\sin 4z} \, dz, \quad C: |z| = 1$

15.4 Evaluation of Real Integrals

We shall now show the very surprising fact that the residue theorem also yields a very elegant and simple method for evaluating certain classes of complicated *real* integrals.

Integrals of Rational Functions of $\cos \theta$ and $\sin \theta$

We first consider integrals of the type

$$(1) \qquad I = \int_0^{2\pi} F(\cos \theta, \sin \theta) \, d\theta$$

where $F(\cos \theta, \sin \theta)$ is a real rational function of $\cos \theta$ and $\sin \theta$ [for example, $(\sin^2 \theta)/(5 - 4 \cos \theta)$] and is finite (does not become infinite) on the interval of integration. Setting $e^{i\theta} = z$, we obtain

$$(2) \qquad \boxed{\begin{aligned} \cos \theta &= \frac{1}{2}(e^{i\theta} + e^{-i\theta}) = \frac{1}{2}\left(z + \frac{1}{z}\right) \\[2mm] \sin \theta &= \frac{1}{2i}(e^{i\theta} - e^{-i\theta}) = \frac{1}{2i}\left(z - \frac{1}{z}\right). \end{aligned}}$$

Since F is rational in $\cos \theta$ and $\sin \theta$, Eq. (2) shows that it is now a rational function in z, say, $f(z)$. Since $dz/d\theta = ie^{i\theta}$, we have $d\theta = dz/iz$ and the given integral takes the form

$$(3) \qquad I = \oint_C f(z) \frac{dz}{iz}$$

and, as θ ranges from 0 to 2π in (1), the variable $z = e^{i\theta}$ ranges counterclockwise once around the unit circle $|z| = 1$. (Review Sec. 12.6 if needed.)

EXAMPLE 1 **An integral of the type (1)**

Show by the present method that $\displaystyle\int_0^{2\pi} \frac{d\theta}{\sqrt{2} - \cos \theta} = 2\pi$.

Solution. We use $\cos \theta = \frac{1}{2}(z + 1/z)$ and $d\theta = dz/iz$. Then the integral becomes

$$\oint_C \frac{dz/iz}{\sqrt{2} - \frac{1}{2}\left(z + \frac{1}{z}\right)} = \oint_C \frac{dz}{-\frac{i}{2}(z^2 - 2\sqrt{2}z + 1)} = -\frac{2}{i} \oint_C \frac{dz}{(z - \sqrt{2} - 1)(z - \sqrt{2} + 1)}.$$

We see that the integrand has a simple pole at $z_1 = \sqrt{2} + 1$ outside the unit circle C, so that it is of no interest here, and another simple pole at $z_2 = \sqrt{2} - 1$ (where $z - \sqrt{2} + 1 = 0$) inside C with residue [by (3), Sec. 15.3]

$$\operatorname*{Res}_{z=z_2} \frac{1}{(z - \sqrt{2} - 1)(z - \sqrt{2} + 1)} = \left[\frac{1}{z - \sqrt{2} - 1}\right]_{z=\sqrt{2}-1} = -\frac{1}{2}.$$

Answer: $2\pi i (-2/i)(-1/2) = 2\pi$. ◀

Improper Integrals of Rational Functions

We now consider real integrals of the type

(4)
$$\int_{-\infty}^{\infty} f(x)\, dx.$$

Such an integral, for which the interval of integration is not finite, is called an **improper integral,** and it has the meaning

(5′)
$$\int_{-\infty}^{\infty} f(x)\, dx = \lim_{a \to -\infty} \int_{a}^{0} f(x)\, dx + \lim_{b \to \infty} \int_{0}^{b} f(x)\, dx.$$

If both limits exist, we may couple the two independent passages to $-\infty$ and ∞, and write[7]

(5)
$$\int_{-\infty}^{\infty} f(x)\, dx = \lim_{R \to \infty} \int_{-R}^{R} f(x)\, dx.$$

We assume that the function $f(x)$ in (4) is a real rational function whose denominator is different from zero for all real x and is of degree at least two units higher than the degree of the numerator. Then the limits in (5′) exist, and we may start from (5). We consider the corresponding contour integral

(5*)
$$\oint_{C} f(z)\, dz$$

around a path C in Fig. 360. Since $f(x)$ is rational, $f(z)$ has finitely many poles in the upper half-plane, and if we choose R large enough, then C encloses all these poles. By the residue theorem we then obtain

$$\oint_{C} f(z)\, dz = \int_{S} f(z)\, dz + \int_{-R}^{R} f(x)\, dx = 2\pi i \sum \operatorname{Res} f(z)$$

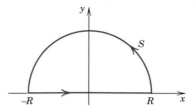

Fig. 360.   Path C of the contour integral in (5*)

---

[7]The expression on the right side of (5) is called the **Cauchy principal value** of the integral; it may exist even if the limits in (5′) do not exist. For instance,

$$\lim_{R \to \infty} \int_{-R}^{R} x\, dx = \lim_{R \to \infty} \left(\frac{R^2}{2} - \frac{R^2}{2} \right) = 0, \quad \text{but} \quad \lim_{b \to \infty} \int_{0}^{b} x\, dx = \infty.$$

where the sum consists of all the residues of $f(z)$ at the points in the upper half-plane at which $f(z)$ has a pole. From this we have

(6)
$$\int_{-R}^{R} f(x) \, dx = 2\pi i \sum \text{Res } f(z) - \int_{S} f(z) \, dz.$$

We prove that, if $R \to \infty$, the value of the integral over the semicircle S approaches zero. If we set $z = Re^{i\theta}$, then S is represented by $R = const$, and as z ranges along S, the variable θ ranges from 0 to π. Since, by assumption, the degree of the denominator of $f(z)$ is at least two units higher than the degree of the numerator, we have

$$|f(z)| < \frac{k}{|z|^2} \qquad (|z| = R > R_0)$$

for sufficiently large constants k and R_0. By the *ML*-inequality in Sec. 13.1,

$$\left| \int_{S} f(z) \, dz \right| < \frac{k}{R^2} \pi R = \frac{k\pi}{R} \qquad (R > R_0).$$

Hence, as R approaches infinity, the value of the integral over S approaches zero, and (5) and (6) yield the result

(7)
$$\boxed{\int_{-\infty}^{\infty} f(x) \, dx = 2\pi i \sum \text{Res } f(z),}$$

where we sum over all the residues of $f(z)$ corresponding to the poles of $f(z)$ in the upper half-plane.

EXAMPLE 2   **An improper integral from 0 to ∞**

Using (7), show that

$$\int_{0}^{\infty} \frac{dx}{1 + x^4} = \frac{\pi}{2\sqrt{2}}.$$

Solution.   Indeed, $f(z) = 1/(1 + z^4)$ has four simple poles at the points

$$z_1 = e^{\pi i/4}, \qquad z_2 = e^{3\pi i/4}, \qquad z_3 = e^{-3\pi i/4}, \qquad z_4 = e^{-\pi i/4}.$$

The first two of these poles lie in the upper half-plane (Fig. 361). From (4) in the last section we find the residues

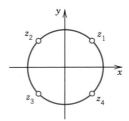

Fig. 361.   Example 2

$$\operatorname*{Res}_{z=z_1} f(z) = \left[\frac{1}{(1 + z^4)'}\right]_{z=z_1} = \left[\frac{1}{4z^3}\right]_{z=z_1} = \frac{1}{4}e^{-3\pi i/4} = -\frac{1}{4}e^{\pi i/4},$$

$$\operatorname*{Res}_{z=z_2} f(z) = \left[\frac{1}{(1 + z^4)'}\right]_{z=z_2} = \left[\frac{1}{4z^3}\right]_{z=z_2} = \frac{1}{4}e^{-9\pi i/4} = \frac{1}{4}e^{-\pi i/4}.$$

(Here we used $e^{\pi i} = -1$ and $e^{-2\pi i} = 1$.) By (1) in Sec. 12.7 and (7) in this section,

$$\int_{-\infty}^{\infty} \frac{dx}{1 + x^4} = -\frac{2\pi i}{4}(e^{\pi i/4} - e^{-\pi i/4}) = -\frac{2\pi i}{4} \cdot 2i \cdot \sin\frac{\pi}{4} \pi \sin\frac{\pi}{4} = \frac{\pi}{\sqrt{2}}.$$

Since $1/(1 + x^4)$ is an even function, we thus obtain, as asserted,

$$\int_0^{\infty} \frac{dx}{1 + x^4} = \frac{1}{2}\int_{-\infty}^{\infty} \frac{dx}{1 + x^4} = \frac{\pi}{2\sqrt{2}}. \qquad \blacktriangleleft$$

Fourier Integrals

The method of evaluating (4) by creating a closed contour (Fig. 360) and "blowing it up" extends to integrals

$$(8) \qquad\qquad \int_{-\infty}^{\infty} f(x) \cos sx\, dx \qquad \text{and} \qquad \int_{-\infty}^{\infty} f(x) \sin sx\, dx \qquad\qquad (s \text{ real})$$

as they occur in connection with the Fourier integral (Sec. 10.8).

If $f(x)$ is a rational function satisfying the assumption on the degree as for (4), we may consider the corresponding integral

$$\oint_C f(z)e^{isz}\, dz \qquad\qquad (s \text{ real and positive})$$

over the contour C in Fig. 360. Instead of (7) we now get

$$(9) \qquad\qquad \int_{-\infty}^{\infty} f(x)e^{isx}\, dx = 2\pi i \sum \operatorname{Res}\left[f(z)e^{isz}\right] \qquad\qquad (s > 0)$$

where we sum the residues of $f(z)e^{isz}$ at its poles in the upper half-plane. Equating the real and the imaginary parts on both sides of (9), we have

$$(10) \qquad \begin{aligned} \int_{-\infty}^{\infty} f(x) \cos sx\, dx &= -2\pi \sum \operatorname{Im} \operatorname{Res}\left[f(z)e^{isz}\right], \\[2mm] \int_{-\infty}^{\infty} f(x) \sin sx\, dx &= 2\pi \sum \operatorname{Re} \operatorname{Res}\left[f(z)e^{isz}\right]. \end{aligned} \qquad\qquad (s > 0)$$

To establish (9), we must show [as for (4)] that the value of the integral over the semicircle S in Fig. 360 approaches 0 as $R \to \infty$. Now $s > 0$ and S lies in the upper half-plane $y \geqq 0$. Hence

$$\left|e^{is(x+iy)}\right| = \left|e^{isx}\right|\left|e^{-sy}\right| = e^{-sy} \leq 1 \qquad (s > 0, \quad y \geq 0).$$

From this we obtain the inequality

$$\left|f(z)e^{isz}\right| = \left|f(z)\right|\left|e^{isz}\right| \leq \left|f(z)\right| \qquad (s > 0, \quad y \geq 0).$$

This reduces our present problem to that for (4). Continuing as before gives (9), which implies (10). ◀

EXAMPLE 3   **An application of (10)**

Show that

$$\int_{-\infty}^{\infty} \frac{\cos sx}{k^2 + x^2}\, dx = \frac{\pi}{k}\, e^{-ks}, \qquad \int_{-\infty}^{\infty} \frac{\sin sx}{k^2 + x^2}\, dx = 0 \qquad (s > 0, \quad k > 0).$$

Solution. In fact, $e^{isz}/(k^2 + z^2)$ has only one pole in the upper half-plane, namely, a simple pole at $z = ik$, and from (4) in Sec. 15.3 we obtain

$$\operatorname*{Res}_{z=ik} \frac{e^{isz}}{k^2 + z^2} = \left[\frac{e^{isz}}{2z}\right]_{z=ik} = \frac{e^{-ks}}{2ik}. \qquad \text{Thus} \qquad \int_{-\infty}^{\infty} \frac{e^{isx}}{k^2 + x^2}\, dx = 2\pi i\, \frac{e^{-ks}}{2ik} = \frac{\pi}{k}\, e^{-ks}.$$

Since $e^{isx} = \cos sx + i \sin sx$, this yields the above results [see also (15) in Sec. 10.8.] ◀

Another Kind of Improper Integral

We consider an improper integral

$$(11) \qquad \int_A^B f(x)\, dx$$

whose integrand becomes infinite at a point a in the interval of integration,

$$\lim_{x \to a} |f(x)| = \infty.$$

By definition, this integral (11) means

$$(12) \qquad \int_A^B f(x)\, dx = \lim_{\epsilon \to 0} \int_A^{a-\epsilon} f(x)\, dx + \lim_{\eta \to 0} \int_{a+\eta}^B f(x)\, dx$$

where both ϵ and η approach zero independently and through positive values. It may happen that neither limit exists if ϵ, $\eta \to 0$ independently, but

$$(13) \qquad \lim_{\epsilon \to 0}\left[\int_A^{a-\epsilon} f(x)\, dx + \int_{a+\epsilon}^B f(x)\, dx\right]$$

exists. This is called the **Cauchy principal value** of the integral. It is written

$$\text{pr. v.} \int_A^B f(x)\, dx.$$

For example,

$$\text{pr. v.} \int_{-1}^{1} \frac{dx}{x^3} = \lim_{\epsilon \to 0} \left[\int_{-1}^{-\epsilon} \frac{dx}{x^3} + \int_{\epsilon}^{1} \frac{dx}{x^3} \right] = 0;$$

the principal value exists, although the integral itself has no meaning.

In the case of simple poles on the real axis we shall obtain a formula for the principal value of an integral from $-\infty$ to ∞. This formula will result from the following theorem.

THEOREM 1 **(Simple poles on the real axis)**

If $f(z)$ has a simple pole at $z = a$ on the real axis, then (Fig. 362)

$$\lim_{r \to 0} \int_{C_2} f(z)\, dz = \pi i \operatorname*{Res}_{z=a} f(z).$$

PROOF. By the definition of a simple pole (Sec. 15.2) the integrand $f(z)$ has at $z = a$ the Laurent series

$$f(z) = \frac{b_1}{z - a} + g(z), \qquad b_1 = \operatorname*{Res}_{z=a} f(z)$$

where $g(z)$ is analytic on the semicircle of integration (Fig. 362)

$$C_2: \quad z = a + re^{i\theta}, \qquad 0 \leq \theta \leq \pi$$

and for all z between C_2 and the x-axis. By integration,

$$\int_{C_2} f(z)\, dz = \int_{0}^{\pi} \frac{b_1}{re^{i\theta}}\, ire^{i\theta}\, d\theta + \int_{C_2} g(z)\, dz.$$

The first integral on the right equals $b_1 \pi i$. The second cannot exceed $M\pi r$ in absolute value, by the *ML*-inequality (Sec. 13.1), and $M\pi r \to 0$ as $r \to 0$. ◀

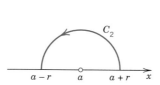

Fig. 362.   Theorem 1

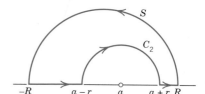

Fig. 363.   Application of Theorem 1

Figure 363 shows the idea of applying Theorem 1 to obtain the principal value of the integral of a rational function $f(x)$ from $-\infty$ to ∞. For sufficiently large R the integral over the contour in Fig. 363 has the value J given by $2\pi i$ times the sum of the residues of $f(z)$ at the singularities in the upper half-plane. We assume that $f(x)$ satisfies the degree condition imposed in connection with (4). Then the value of the integral over the large semicircle S approaches 0 as $R \to \infty$. For $r \to 0$ the integral over C_2 (clockwise!) approaches the value $K = -\pi i \operatorname{Res}_{z=a} f(z)$, by Theorem 1. Together this shows that the principal value P of the integral from $-\infty$ to ∞ plus K equals J; hence

$P = J - K = J + \pi i \operatorname{Res}_{z=a} f(z)$. If $f(z)$ has several simple poles on the real axis, then K will be $-\pi i$ times the sum of the corresponding residues. Hence the desired formula is

(14)
$$\text{pr. v.} \int_{-\infty}^{\infty} f(x)\, dx = 2\pi i \sum \operatorname{Res} f(z) + \pi i \sum \operatorname{Res} f(z)$$

where the first sum extends over all poles in the upper half-plane and the second over all poles on the real axis, the latter being simple by assumption.

EXAMPLE 4   **Poles on the real axis**

Find the principal value

$$\text{pr. v.} \int_{-\infty}^{\infty} \frac{dx}{(x^2 - 3x + 2)(x^2 + 1)}.$$

Solution.   Since $x^2 - 3x + 2 = (x - 1)(x - 2)$, the integrand $f(x)$, considered for complex z, has simple poles at

$$z = 1, \qquad \operatorname{Res}_{z=1} f(z) = \left[\frac{1}{(z - 2)(z^2 + 1)} \right]_{z=1} = -\frac{1}{2},$$

$$z = 2, \qquad \operatorname{Res}_{z=2} f(z) = \left[\frac{1}{(z - 1)(z^2 + 1)} \right]_{z=2} = \frac{1}{5},$$

$$z = i, \qquad \operatorname{Res}_{z=i} f(z) = \left[\frac{1}{(z^2 - 3z + 2)(z + i)} \right]_{z=i} = \frac{1}{6 + 2i} = \frac{3 - i}{20},$$

and at $z = -i$ in the lower half-plane, which is of no interest here. From (14) we get the answer

$$\text{pr. v.} \int_{-\infty}^{\infty} \frac{dx}{(x^2 - 3x + 2)(x^2 + 1)} = 2\pi i \left(\frac{3 - i}{20} \right) + \pi i \left(-\frac{1}{2} + \frac{1}{5} \right) = \frac{\pi}{10}. \qquad \blacktriangleleft$$

For further integrals of the kind considered in this section, see the problem set.

PROBLEM SET 15.4

Integrals Involving Cosine and Sine

Evaluate the following integrals. (Show the details of your work.)

1. $\displaystyle\int_0^{\pi} \frac{d\theta}{k + \cos \theta} \quad (k > 1)$

2. $\displaystyle\int_0^{2\pi} \frac{d\theta}{25 - 24 \cos \theta}$

3. $\displaystyle\int_0^{2\pi} \frac{1 + \sin \theta}{3 + \cos \theta}\, d\theta$

4. $\displaystyle\int_0^{2\pi} \frac{\cos \theta}{3 + \sin \theta}\, d\theta$

5. $\displaystyle\int_0^{\pi} \frac{\cos \theta}{17 - 8 \cos \theta}\, d\theta$

6. $\displaystyle\int_0^{2\pi} \frac{d\theta}{5 - 3 \sin \theta}\, d\theta$

7. $\displaystyle\int_0^{2\pi} \frac{\cos \theta}{13 - 12 \cos 2\theta}\, d\theta$

8. $\displaystyle\int_0^{2\pi} \frac{1 + 4 \cos \theta}{17 - 8 \cos \theta}\, d\theta$

Improper Integrals: Infinite Interval of Integration
Evaluate (showing details)

9. $\displaystyle\int_{-\infty}^{\infty} \frac{dx}{(1 + x^2)^2}$

10. $\displaystyle\int_{-\infty}^{\infty} \frac{dx}{x^4 + 16}$

11. $\displaystyle\int_{-\infty}^{\infty} \frac{x^3}{1 + x^8}\, dx$

12. $\displaystyle\int_{-\infty}^{\infty} \frac{dx}{(1 + x^2)^3}$

13. $\displaystyle\int_{-\infty}^{\infty} \frac{x^2}{(x^2 + 1)(x^2 + 4)}\, dx$

14. $\displaystyle\int_{-\infty}^{\infty} \frac{dx}{(x^2 - 2x + 5)^2}$

15. $\displaystyle\int_{-\infty}^{\infty} \frac{dx}{(x^2 + 1)(x^2 + 4)^2}$

16. $\displaystyle\int_{-\infty}^{\infty} \frac{dx}{(x^2 + 1)(x^2 + 9)}$

17. $\displaystyle\int_{0}^{\infty} \frac{\cos sx}{x^2 + 1}\, dx$

18. $\displaystyle\int_{-\infty}^{\infty} \frac{\sin 2x}{x^2 + x + 1}\, dx$

19. $\displaystyle\int_{-\infty}^{\infty} \frac{\sin 3x}{1 + x^4}\, dx$

20. $\displaystyle\int_{-\infty}^{\infty} \frac{\cos 2x}{(x^2 + 1)^2}\, dx$

Improper Integrals: Poles on the Real Axis
Find the **Cauchy principal value** of the following integrals.

21. $\displaystyle\int_{-\infty}^{\infty} \frac{dx}{x^2 - ix}$

22. $\displaystyle\int_{-\infty}^{\infty} \frac{dx}{x^2 - 2ix}$

23. $\displaystyle\int_{-\infty}^{\infty} \frac{x}{8 - x^3}\, dx$

24. $\displaystyle\int_{-\infty}^{\infty} \frac{dx}{x^4 - 1}$

 25. **CAS PROJECT. Check your CAS.** Find out to what extent your CAS can integrate integrals of the form (1), (4), and (8) correctly. Do this by comparing the results of direct integration (which may come out false) with that by using residues.

26. **TEAM PROJECT. Comments on Real Integrals. (a) Formula (10)** follows from (9). Give the details.

    (b) Use of auxiliary results. Integrating e^{-z^2} around the boundary of the rectangle with vertices $-a, a, a + ib, -a + ib$, letting $a \to \infty$, and using

$$\int_{0}^{\infty} e^{-x^2}\, dx = \frac{\sqrt{\pi}}{2},$$

show that

$$\int_{0}^{\infty} e^{-x^2} \cos 2bx\, dx = \frac{\sqrt{\pi}}{2}\, e^{-b^2}.$$

(This integral is needed in heat conduction in Sec. 11.6.)

    (c) Inspection. Solve Probs. 11 and 19 without calculation.

CHAPTER 15 REVIEW

1. What is a Laurent series? By what idea did we obtain it?

2. What do you know about the uniqueness of Laurent series? Of Taylor series?

3. List some methods of obtaining Laurent series. Give examples.

4. In what kind of domains do Laurent series converge?

5. What is the principal part of a Laurent series? What is it used for?

6. What is a singularity of an analytic function? What kinds of singularities did we discuss?

7. What is the residue of a function at some point? What do we mean by residue integration? Give examples.

8. What is a zero of an analytic function? What do you know about zeros?

9. What is the extended complex plane? What do we mean by saying that a function is analytic at infinity? Give typical examples.

10. Sketch the Riemann number sphere from memory. Explain it.

11. What is a meromorphic function? An entire function?

12. Write down two formulas for the residue at a simple pole. A formula for the residue at a pole of any order m.

13. State the residue theorem and explain the idea of proof from memory.

14. What are improper integrals? How did we evaluate some of them by residue integration?

15. Can the residue at a singular point be zero? Give a reason.

16. Can the residue at a simple pole be zero?

17. Can we apply residue integration in the case of a function that is not analytic but merely continuous?

18. Can we use residue integration for evaluating the integral of $\tan(1/z)$ around a closed contour containing $z = 0$ in its interior? Answer the same question for $e^{1/z}$.

19. In residue integration we need *closed* paths. How were we nevertheless able to evaluate (real) integrals over intervals?

20. What is the Cauchy principal value of an integral, and why did it occur in this chapter?

Integrate the given function counterclockwise over the given path C, using residue integration or one of the methods discussed in Chap. 13, and indicate whether residue integration can be used. (Show the details of your work.)

21. $\dfrac{z}{z^2 + 4}$, $C: |z| = 3$

22. $\dfrac{15z + 9}{z^3 - 9z}$, $C: |z - 3| = 2$

23. $\dfrac{25z}{(z + 4)(z - 1)^2}$, $C: |z - 1| = 2$

24. $\dfrac{15z + 9}{z^3 - 9z}$, $C: |z| = 4$

25. Re z, C the triangle with vertices $0, 1, 1 + i$

26. $\dfrac{z^2 \sin z}{4z^2 - 1}$, $C: |z| = 1$

27. $\dfrac{\cosh 5z}{(z^2 + 4)^2}$, $C: |z| = \tfrac{1}{2}\pi$

28. $z^3 e^{z^4}$, C any path from $1 + i$ to 1

29. $\dfrac{\cos z}{z^n}$, $n = 0, 1, \cdots$, $C: |z| = 1$

30. $\dfrac{\sin z}{z^n}$, $n = 0, 1, \cdots$, $|z| = 1$

31. $z/|z|^2$, $C: |z| = 1$

32. $\coth z$, $C: |z| = 1$

33. $\dfrac{\tan \pi z}{z}$, $C: |z| = 1$

Evaluate by the methods in this chapter, showing the details of your work:

34. $\displaystyle\int_0^{2\pi} \dfrac{\sin \theta}{3 + \cos \theta}\, d\theta$

35. $\displaystyle\int_0^{2\pi} \dfrac{d\theta}{13 - 5 \sin \theta}$

36. $\displaystyle\int_0^{2\pi} \dfrac{\sin \theta}{34 - 16 \sin \theta}\, d\theta$

37. $\displaystyle\int_{-\infty}^{\infty} \dfrac{x}{(1 + x^2)^2}\, dx$

38. $\displaystyle\int_{-\infty}^{\infty} \dfrac{dx}{1 + 4x^2}$

39. $\displaystyle\int_{-\infty}^{\infty} \dfrac{dx}{1 + 4x^4}$

40. $\displaystyle\int_{-\infty}^{\infty} \dfrac{\sin x}{x^2 + x + 1}\, dx$

Summary of Chapter 15
Laurent Series. Residue Integration

A **Laurent series** is a series of the form

$$(1) \qquad f(z) = \sum_{n=0}^{\infty} a_n(z - z_0)^n + \sum_{n=1}^{\infty} \frac{b_n}{(z - z_0)^n} \qquad \text{(Sec. 15.1)}$$

or, more briefly written [but this means the same as (1)!]

$$(1^*) \qquad f(z) = \sum_{n=-\infty}^{\infty} a_n(z - z_0)^n, \quad a_n = \frac{1}{2\pi i} \oint_C \frac{f(z^*)}{(z^* - z_0)^{n+1}} \, dz^*$$

where $n = 0, \pm 1, \pm 2, \cdots$. This series converges in an open annulus (ring) A with center z_0. In A the function $f(z)$ is analytic. At points not in A it may have singularities. The first series in (1) is a power series. The second series is called the **principal part** of the Laurent series. In a given annulus, a Laurent series of $f(z)$ is unique, but $f(z)$ may have different Laurent series in different annuli with the same center.

If $f(z)$ has an isolated singularity at $z = z_0$, the Laurent series of $f(z)$ that converges for

$$0 < |z - z_0| < R \qquad \qquad (R \text{ suitable})$$

can be used for classifying this singularity. The latter is called a **pole** if the principal part of this Laurent series is a finite sum, otherwise an (isolated) **essential singularity.** See Sec. 15.2.

A pole is said to be of **order** n if $1/(z - z_0)^n$ is the highest negative power of the principal part in (1). A first-order pole is also called a *simple pole.*

A similar classification holds for zeros (Sec. 15.2). An analytic function $f(z)$ in a domain D has a *zero of nth order* at a point $z = z_0$ in D if f and its derivatives $f', f'', \cdots, f^{(n-1)}$ are zero at z_0, whereas $f^{(n)}(z_0) \neq 0$. A first-order zero is also called a *simple zero.* If $g(z)$ has an nth-order zero at z_0, then $f(z) = 1/g(z)$ has an nth-order pole at z_0.

Section 15.2 also includes a discussion of the **extended complex plane,** obtained from the complex plane by attaching an improper point ∞ (*"infinity"*).

The **residue** of an analytic function $f(z)$ at a point $z = z_0$ is the coefficient b_1 of the power $1/(z - z_0)$ in that Laurent series

$$f(z) = a_0 + a_1(z - z_0) + \cdots + \frac{b_1}{z - z_0} + \frac{b_2}{(z - z_0)^2} + \cdots$$

of $f(z)$ which converges near z_0 (except at z_0 itself). This residue is given by the integral

$$(2) \qquad\qquad b_1 = \frac{1}{2\pi i} \oint_C f(z)\, dz \qquad\qquad \text{(Sec. 15.3)}$$

but can be obtained in various other ways, so that one can use (2) for evaluating integrals over closed curves. More generally, the **residue theorem** (Sec. 15.3) states that if $f(z)$ is analytic in a domain D except at finitely many points z_j and C is a simple closed path in D such that no z_j lies on C and the full interior of C belongs to D, then

$$(3) \qquad\qquad \oint_C f(z)\, dz = \frac{1}{2\pi i} \sum_j \operatorname*{Res}_{z=z_j} f(z)$$

(summation only over those z_j that lie *inside* C).

This integration method is elegant and powerful. Formulas for the residue at **poles** are ($m =$ order of the pole (Sec. 15.3))

$$(4) \qquad\qquad \operatorname*{Res}_{z=z_0} f(z) = \frac{1}{(m-1)!} \lim_{z \to z_0} \left(\frac{d^{m-1}}{dz^{m-1}} \left[(z - z_0)^m f(z) \right] \right),$$

where $m = 1, 2, \cdots$. Hence for a simple pole ($m = 1$),

$$\operatorname*{Res}_{z=z_0} f(z) = \lim_{z \to z_0} (z - z_0) f(z);$$

also,

$$\operatorname*{Res}_{z=z_0} \frac{p(z)}{q(z)} = \frac{p(z_0)}{q'(z_0)}.$$

Residue integration can also be used for **_real integrals_** (Sec. 15.4)

$$(5a) \qquad\qquad \int_0^{2\pi} F(\cos\theta,\ \sin\theta)\, d\theta$$

or

$$(5b) \qquad\qquad \int_{-\infty}^{\infty} f(x)\, dx.$$

Here $F(\cos\theta,\ \sin\theta)$ is a rational function of $\cos\theta$ and $\sin\theta$, and a closed contour (as needed in residue integration!) is obtained by setting $z = e^{i\theta}$. The integrand $f = p/q$ in (5b) is rational with degree $q \geqq$ degree $p + 2$, and a closed contour is obtained by attaching a semicircle to a segment of the real line and "blowing this contour up."

CHAPTER 16

Complex Analysis Applied to Potential Theory

Laplace's equation $\nabla^2\Phi = 0$ is one of the most important partial differential equations in engineering mathematics, because it occurs in connection with gravitational fields (Sec. 8.9), electrostatic fields (Sec. 11.11), steady-state heat conduction (Secs. 9.8, 11.5), incompressible fluid flow (Sec. 16.4), and other areas. (These references are just for orientation, not as a prerequisite for this chapter.) The theory of the solutions of this equation is called **potential theory,** and solutions whose second partial derivatives are continuous are called **harmonic functions.**

If in a problem the potential Φ in a region of space depends only on two of the three Cartesian coordinates, say, on x and y, we call it a **two-dimensional potential problem.** Then Laplace's equation becomes

$$(1) \qquad\qquad \nabla^2\Phi = \Phi_{xx} + \Phi_{yy} = 0.$$

Any such problem can be solved by complex analysis because solutions of (1) are closely related to complex analytic functions (as we know from Sec. 12.4).[1] This transition from real to complex also has the advantage that by the **"complex potential"** $F = \Phi + i\Psi$ we can simultaneously handle equipotential lines $\Phi = const$ and their orthogonal trajectories (the lines of force or flow $\Psi = const$).

Furthermore, in solving the **Dirichlet problem** of finding a potential with given boundary values we may often use **conformal mapping** (Sec. 16.2). This concerns electrostatics (Secs. 16.1, 16.2), heat conduction (Sec. 16.3), and hydrodynamics (Sec. 16.4).

Poisson's integral formula for potentials in disks and some general properties of potentials will be discussed in Secs. 16.5 and 16.6.

Prerequisites for this chapter: Chaps. 12, 13.
References: Appendix 1, Part D.
Answers to problems: Appendix 2.

---

[1]No such close relation exists in the three-dimensional case.
  On notation. We write Φ and later $\Phi + i\Psi$ since u and $u + iv$ will be needed in conformal mapping from Sec. 16.2 on.

16.1 Electrostatic Fields

The electrical force of attraction or repulsion between charged particles is governed by Coulomb's law. This force is the gradient of a function Φ, called the **electrostatic potential.** At any points free of charges, Φ is a solution of Laplace's equation

$$\nabla^2 \Phi = 0.$$

The surfaces $\Phi = const$ are called **equipotential surfaces.** At each point P the gradient of Φ is perpendicular to the surface $\Phi = const$ through $P;$ that is, the electrical force has the direction perpendicular to the equipotential surface. (See also Secs. 8.9 and 11.11.)

The problems we shall discuss in this entire chapter are **two-dimensional** (for the reason just given in the chapter opening), that is, they concern physical systems that lie in three-dimensional space (of course!), but are such that the potential Φ is independent of one of the space coordinates, so that Φ depends only on two coordinates, which we call x and y. Then **Laplace's equation** becomes

(1)
$$\nabla^2 \Phi = \frac{\partial^2 \Phi}{\partial x^2} + \frac{\partial^2 \Phi}{\partial y^2} = 0.$$

Equipotential surfaces now appear as **equipotential lines** (curves) in the xy-plane.

EXAMPLE 1 **Potential between parallel plates**

Find the potential Φ of the field between two parallel conducting plates extending to infinity (Fig. 364), which are kept at potentials Φ_1 and Φ_2, respectively.

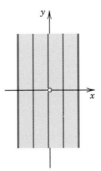

Fig. 364. Potential in Example 1

Solution. From the shape of the plates it follows that Φ depends only on x, and Laplace's equation becomes $\Phi'' = 0$. By integrating twice we obtain $\Phi = ax + b$, where the constants a and b are determined by the given boundary values of Φ on the plates. For example, if the plates correspond to $x = -1$ and $x = 1$, the solution is

$$\Phi(x) = \tfrac{1}{2}(\Phi_2 - \Phi_1)x + \tfrac{1}{2}(\Phi_2 + \Phi_1).$$

The equipotential surfaces are parallel planes. ◀

EXAMPLE 2   **Potential between coaxial cylinders**

Find the potential Φ between two coaxial conducting cylinders extending to infinity on both ends (Fig. 365) and kept at potentials Φ_1 and Φ_2, respectively.

Solution. Here Φ depends only on $r = \sqrt{x^2 + y^2}$, for reasons of symmetry, and Laplace's equation $r^2 u_{rr} + r u_r + u_{\theta\theta} = 0$ [(4), Sec. 11.9] with $u_{\theta\theta} = 0$ and $u = \Phi$ becomes (after division by r)

$$r\Phi'' + \Phi' = 0.$$

By separating variables and integrating we obtain

$$\frac{\Phi''}{\Phi'} = -\frac{1}{r}, \qquad \ln \Phi' = -\ln r + \tilde{a}, \qquad \Phi' = \frac{a}{r}, \qquad \Phi = a \ln r + b$$

and a and b are determined by the given values of Φ on the cylinders. Although no infinitely extended conductors exist, the field in our idealized conductor will approximate the field in a long finite conductor in that part which is far away from the two ends of the cylinders.   ◀

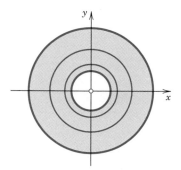

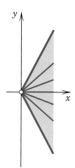

Fig. 365.   Potential
in Example 2

Fig. 366.   Potential
in Example 3

EXAMPLE 3   **Potential in an angular region**

Find the potential Φ between the conducting plates in Fig. 366, which are kept at potentials Φ_1 (the lower plate) and Φ_2, and make an angle α, where $0 < \alpha \leq \pi$. (In the figure we have $\alpha = 120° = 2\pi/3$.)

Solution. $\theta = \operatorname{Arg} z \ (z = x + iy \neq 0)$ is constant on rays $\theta = const$. It is harmonic since it is the imaginary part of an analytic function, $\operatorname{Ln} z$ (Sec. 12.8). Hence the solution is

$$\Phi(x, y) = a + b \operatorname{Arg} z$$

with a and b determined from the two boundary conditions (given values on the plates)

$$a + b(-\tfrac{1}{2}\alpha) = \Phi_1, \qquad a + b(\tfrac{1}{2}\alpha) = \Phi_2.$$

Thus $a = (\Phi_2 + \Phi_1)/2$, $b = (\Phi_2 - \Phi_1)/\alpha$. The answer is

$$\Phi(x, y) = \frac{1}{2}(\Phi_2 + \Phi_1) + \frac{1}{\alpha}(\Phi_2 - \Phi_1)\theta, \qquad \theta = \arctan \frac{y}{x}. \quad ◀$$

Complex Potential

Let $\Phi(x, y)$ be harmonic in some domain D and $\Psi(x, y)$ a conjugate harmonic of Φ in D (Sec. 12.4). Then[2]

$$\boxed{F(z) = \Phi(x, y) + i\Psi(x, y)}$$

---

[2]We write $F = \Phi + i\Psi$, reserving $f = u + iv$ for conformal mapping, as needed from the next section on.

is an analytic function of $z = x + iy$. This function F is called the **complex potential** corresponding to the real potential Φ. Recall from Sec. 12.4 that for given Φ, a conjugate Ψ is uniquely determined except for an additive real constant. Hence we may say ***the*** complex potential, without causing misunderstandings.

The use of F has two advantages, a technical one and a physical one. Technically, F is easier to handle than real or imaginary parts, in connection with methods of complex analysis. Physically, Ψ has a meaning. By conformality, the curves $\Psi = const$ intersect the equipotential lines $\Phi = const$ in the xy-plane at right angles [except where $F'(z) = 0$]. Hence they have the direction of the electrical force and, therefore, are called **lines of force.** They are the paths of moving charged particles (electrons in an electron microscope, etc.).

EXAMPLE 4   **Complex potential**

In Example 1, a conjugate is $\Psi = ay$. It follows that the complex potential is

$$F(z) = az + b = ax + b + iay,$$

and the lines of force are horizontal straight lines $y = const$ parallel to the x-axis. ◀

EXAMPLE 5   **Complex potential**

In Example 2 we have $\Phi = a \ln r + b = a \ln |z| + b$. A conjugate is $\Psi = a \operatorname{Arg} z$. Hence the complex potential is

$$F(z) = a \operatorname{Ln} z + b$$

and the lines of force are straight lines through the origin. $F(z)$ may also be interpreted as the complex potential of a source line whose trace in the xy-plane is the origin. ◀

EXAMPLE 6   **Complex potential**

In Example 3 we get $F(z)$ by noting that $i \operatorname{Ln} z = i \ln |z| - \operatorname{Arg} z$, multiplying this by $-b$, and adding a:

$$F(z) = a - ib \operatorname{Ln} z = a + b \operatorname{Arg} z - ib \ln |z|.$$

We see from this that the lines of force are concentric circles $|z| = const$. Can you sketch them? What do they tell you physically? ◀

Superposition

More complicated potentials can often be obtained by superposition.

EXAMPLE 7   **Potential of a pair of source lines**

Determine the potential of a pair of oppositely charged source lines of the same strength at the points $z = c$ and $z = -c$ on the real axis.

Solution. From Examples 2 and 5 it follows that the potential of each of the source lines is

$$\Phi_1 = K \ln |z - c| \qquad \text{and} \qquad \Phi_2 = -K \ln |z + c|,$$

respectively. Here the real constant K measures the strength (amount of charge). These are the real parts of the complex potentials

$$F_1(z) = K \operatorname{Ln} (z - c) \qquad \text{and} \qquad F_2(z) = -K \operatorname{Ln} (z + c).$$

Hence the complex potential of the combination of the two source lines is

$$(2) \qquad\qquad F(z) = F_1(z) + F_2(z) = K[\operatorname{Ln} (z - c) - \operatorname{Ln} (z + c)].$$

The equipotential lines are the curves

$$\Phi = \operatorname{Re} F(z) = K \ln \left| \frac{z - c}{z + c} \right| = const, \qquad \text{thus} \qquad \left| \frac{z - c}{z + c} \right| = const.$$

These are circles, as you may show by direct calculation. The lines of force are

$$\Psi = \operatorname{Im} F(z) = K[\operatorname{Arg}(z - c) - \operatorname{Arg}(z + c)] = const.$$

We write this briefly (Fig. 367)

$$\Psi = K(\theta_1 - \theta_2) = const.$$

Now $\theta_1 - \theta_2$ is the angle between the line segments from z to c and $-c$ (Fig. 367). Hence the lines of force are the curves along each of which the line segment S: $-c \leqq x \leqq c$ appears under a constant angle. These curves are the totality of circular arcs over S, as is known from elementary geometry. Hence the lines of force are circles. Figure 368 shows some of them together with some equipotential lines.

In addition to the interpretation as the potential of two source lines, this potential could also be thought of as the potential between two circular cylinders whose axes are parallel but do not coincide, or as the potential between two equal cylinders that lie outside each other, or as the potential between a cylinder and a plane wall. Explain this, using Fig. 368. ◀

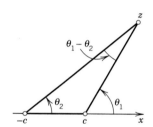

Fig. 367.   Arguments in Example 7

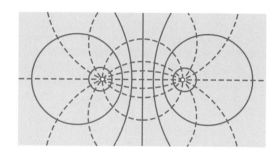

Fig. 368.   Equipotential lines and lines of force (dashed) in Example 7

The idea of the complex potential as just explained is a key to a close relation of potential theory to complex analysis and will recur in heat flow and fluid flow.

PROBLEM SET 16.1

Parallel Plates

1. Find and sketch the equipotential surfaces between two parallel plates at $x = -4$ and $x = 10$ having potentials 220 and 500 volts, respectively.

2. Find and sketch the complex potential in Prob. 1.

3. The plate $y = x$ is grounded (potential 0) and the plate $y = x + 1$ is kept at 110 volts. Find the corresponding real and complex potentials.

4. What is the potential between the plates $y = 2 - \frac{1}{2}x$ and $y = 5 - \frac{1}{2}x$ kept at potentials 200 and 500 volts, respectively? What is the complex potential?

Coaxial Cylinders

Find the potential Φ between two infinite coaxial cylinders of radii r_1 and r_2 $(> r_1)$ having potentials U_1 and U_2, respectively, where

5. $r_1 = 1$ cm, $r_2 = 5$ cm, $U_1 = 0$ volts, $U_2 = 100$ volts

6. $r_1 = 5$ mm, $r_2 = 2$ cm, $U_1 = -110$ volts, $U_2 = 110$ volts

7. $r_1 = 10$ cm, $r_2 = 1$ m, $U_1 = 10$ kV, $U_2 = 0$

8. If the cylinders $r_1 = 2$ cm and $r_2 = 4$ cm are kept at $U_1 = 10$ volts and $U_2 = 30$ volts, respectively, is the potential at $r = 3$ cm equal to 20 volts? Less? More? Answer without calculating. Then check by calculation. Explain the result.

Source Lines

9. Find the potential of two source lines at $z = a$ and $z = -a$ having the same charge.

10. Near each of the two source lines in Fig. 368 the circles $\Phi = const$ are almost concentric. Is this physically understandable?

11. Verify by calculation that the equipotential lines in Example 7 are circles.

Other Configurations

12. Find the potential in the first quadrant of the xy-plane between the axes (having potential 110 volts) and the hyperbola $xy = 1$ (having potential 60 volts).

13. Show that $F(z) = \cos^{-1} z$ may be interpreted as the complex potential of the configurations in Figs. 369 and 370.

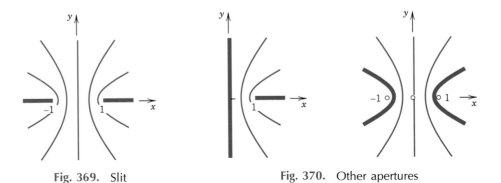

Fig. 369. Slit   **Fig. 370.** Other apertures

14. Show that $F(z) = \cosh^{-1} z$ may be interpreted as the complex potential between two confocal elliptic cylinders.

15. Find the real and complex potentials in the sector $-\pi/6 \leqq \theta \leqq \pi/6$ between the boundary $\theta = \pm \pi/6$ (kept at 0) and the curve $x^3 - 3xy^2 = 1$, kept at 220 volts.

16. CAS PROJECT. Complex Potential. Plot the equipotential lines and lines of force (on the same axes) for

   (a) $F(z) = z^2$,   **(b)** $F(z) = iz^2$,   **(c)** $F(z) = 1/z$ **(d)** $F(z) = i/z$.

   (e) Explore further complex potentials of your own choice (remembering the functions in Secs. 12.6–12.9) with the purpose of discovering configurations that might be of practical interest.

16.2 Use of Conformal Mapping

Complex potentials relate potential theory closely to complex analysis, as we have just seen. Another close relation results from the use of conformal mapping in solving **boundary value problems** for the Laplace equation, that is, in finding a solution of the equation in some domain assuming given values on the boundary (**"Dirichlet problem"**; see also Sec. 11.5). Then conformal mapping is used to map a given complicated domain onto a simpler one where the solution is known or can be found more easily. This solution is then mapped back to the given domain. This is the idea. That it works is due to the fact that harmonic functions remain harmonic under conformal mapping:

THEOREM 1   **(Harmonic functions under conformal mapping)**

Let Φ^* be harmonic in a domain D^* in the w-plane. Suppose that $w = u + iv = f(z)$ is analytic in a domain D in the z-plane and maps D conformally onto D^*. Then the function

$$(1) \qquad\qquad \Phi(x, y) = \Phi^*(u(x, y), v(x, y))$$

is harmonic in D.

PROOF.   The composite of analytic functions is analytic, as follows from the chain rule. Hence, taking a harmonic conjugate[3] $\Psi^*(u, v)$ of Φ^* and forming the analytic function $F^*(w) = \Phi^*(u, v) + i\Psi^*(u, v)$, we conclude that $F(z) = F^*(f(z))$ is analytic in D. Hence its real part, $\Phi(x, y) = \mathrm{Re}\, F(z)$ is harmonic in D.   ◄

EXAMPLE 1   **Potential between noncoaxial cylinders**

Find the potential between the cylinders C_1: $|z| = 1$ (grounded, that is, having potential $U_1 = 0$) and C_2: $|z - 2/5| = 2/5$ (having potential $U_2 = 110$ volts).

Solution.   We map the unit disk $|z| = 1$ onto the unit disk $|w| = 1$ in such a way that C_2 is mapped onto some cylinder C_2^*: $|w| = r_0$. By (7), Sec. 12.9, a linear fractional transformation mapping the unit disk onto the unit disk is

$$(2) \qquad\qquad w = \frac{z - b}{bz - 1}$$

where we have chosen $b = z_0$ real without restriction. z_0 is of no immediate help here because centers of circles do not map onto centers of the images, in general. However, we now have two free constants b and r_0 and shall succeed by imposing two reasonable conditions, namely, that 0 and 4/5 (Fig. 371) should be mapped onto r_0 and $-r_0$, respectively. This gives by (2)

$$r_0 = \frac{0 - b}{0 - 1} = b, \qquad \text{and with this,} \qquad -r_0 = \frac{4/5 - b}{4b/5 - 1} = \frac{4/5 - r_0}{4r_0/5 - 1},$$

a quadratic equation in r_0 with solutions $r_0 = 2$ (no good because $r_0 < 1$) and $r_0 = 1/2$. Hence our mapping function (2) with $b = 1/2$ becomes that in Example 5 of Sec. 12.9,

$$(3) \qquad\qquad w = f(z) = \frac{2z - 1}{z - 2}.$$

---

[3]See Sec. 12.4. We mention without proof that if D^* is simply connected (Sec. 13.2), then a harmonic conjugate of Φ^* exists. Another proof without the use of a harmonic conjugate is given in Appendix 4.

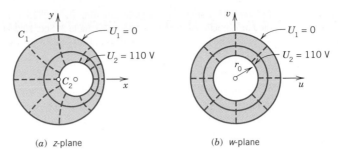

(a) z-plane   (b) w-plane

Fig. 371. Example 1

From Example 5 in Sec. 16.1, writing w for z we have as the complex potential in the w-plane the function $F^*(w) = a \operatorname{Ln} w + k$ and from this the real potential

$$\Phi^*(u, v) = \operatorname{Re} F^*(w) = a \ln |w| + k.$$

We determine a and k from the boundary conditions. If $|w| = 1$, then $\Phi^* = a \ln 1 + k = 0$, hence $k = 0$. If $|w| = r_0 = 1/2$, then $\Phi^* = a \ln (1/2) = 110$, hence $a = 110/\ln (1/2) = -158.7$. Substitution of (3) now gives the desired solutions in the given domain in the z-plane

$$F(z) = F^*(f(z)) = a \operatorname{Ln} \frac{2z - 1}{z - 2}.$$

The real potential is

$$\Phi(x, y) = \operatorname{Re} F(z) = a \ln \left| \frac{2z - 1}{z - 2} \right|, \qquad a = -158.7.$$

Can we "see" this result? Well, $\Phi(x, y) = const$ if and only if $|(2z - 1)/(z - 2)| = const$, that is, $|w| = const$ by (2) with $b = 1/2$. These circles are images of circles in the z-plane because the inverse of a linear fractional transformation is linear fractional (see (4), Sec. 12.9), and any such mapping maps circles onto circles (or straight lines), as mentioned in Example 1, Sec. 12.9. Similarly for the rays arg $w = const$. Hence the equipotential lines $\Phi(x, y) = const$ are circles, and the lines of force are circular arcs (dashed in Fig. 371). These two families of curves intersect orthogonally, that is, at right angles, as shown in Fig. 371. ◀

EXAMPLE 2   **Potential between two semicircular plates**

Find the potential between two semicircular plates P_1 and P_2 in Fig. 372a having potentials -3000 and 3000 volts, respectively. Use Example 3 in Sec. 16.1 and conformal mapping.

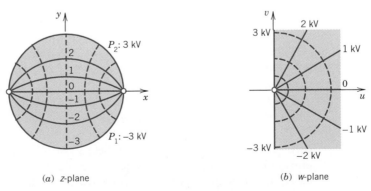

(a) z-plane   (b) w-plane

Fig. 372. Example 2

Solution. First step. We map the unit disk in Fig. 372a onto the right half of the w-plane (Fig. 372b) by using the linear fractional transformation in Example 3, Sec. 12.9:

$$w = f(z) = \frac{1+z}{1-z}.$$

The boundary $|z| = 1$ is mapped onto the boundary $u = 0$ (the v-axis), with $z = -1$, i, 1 going onto $w = 0$, i, ∞, respectively, and $z = -i$ onto $w = -i$. Hence the upper semicircle of $|z| = 1$ is mapped onto the upper half, and the lower semicircle onto the lower half of the v-axis, so that the boundary conditions in the w-plane are as indicated in Fig. 372b.

Second step. We determine the potential $\Phi^*(u, v)$ in the right half-plane of the w-plane. Example 3 in Sec. 16.1 with $\alpha = \pi$, $U_1 = -3000$, and $U_2 = 3000$ [with $\Phi^*(u, v)$ instead of $\Phi(x, y)$] yields

$$\Phi^*(u, v) = \frac{6000}{\pi} \varphi, \qquad\qquad \varphi = \text{arc tan } \frac{v}{u}.$$

On the positive half of the imaginary axis ($\varphi = \pi/2$), this equals 3000 and on the negative half -3000, as it should be. Φ^* is the real part of the complex potential

$$F^*(w) = -\frac{6000\, i}{\pi} \text{Ln } w.$$

Third step. We substitute the mapping function into F^* to get the complex potential $F(z)$ in Fig. 372a in the form

$$F(z) = F^*(f(z)) = -\frac{6000\, i}{\pi} \text{Ln } \frac{1+z}{1-z}.$$

The real part of this is the potential we wanted to determine:

$$\Phi(x, y) = \text{Re } F(z) = \frac{6000}{\pi} \text{Im Ln } \frac{1+z}{1-z} = \frac{6000}{\pi} \text{Arg } \frac{1+z}{1-z}.$$

As in Example 1 we conclude that the equipotential lines $\Phi(x, y) = const$ are circular arcs because they correspond to Arg $[(1 + z)/(1 - z)] = const$, hence to Arg $w = const$. Also, Arg $w = const$ are rays from 0 to ∞, the images of $z = -1$ and $z = 1$, respectively. Hence the equipotential lines all have -1 and 1 (the points where the boundary potential jumps) as their endpoints (Fig. 372a). The lines of force are circular arcs, too, and since they must be orthogonal to the equipotential lines, their centers can be obtained as intersections of tangents to the unit circle with the x-axis. (Explain!) ◀

Further examples can easily be constructed. Just take any mapping $w = f(z)$ in Secs. 12.6–12.9, a domain D in the z-plane, its image D^* in the w-plane, and a potential Φ^* in D^*. Then (1) gives a potential in D. Make up some examples of your own, involving, for instance, linear fractional transformations.

Basic Comment

We formulated the examples in this section in terms of the electrostatic potential. It is quite important to realize that this is accidental. We could equally well have phrased everything in terms of (time-independent) heat flow; then instead of voltages we would have had temperatures, the equipotential lines would have become isotherms (= lines of constant temperature), and the lines of the electrical force would have become lines along which heat flows from higher to lower temperatures (more on this in the next section). Or we could have talked about fluid flow; then the electrostatic lines of force would have become streamlines (more on this in Sec. 16.4). What we again see here is the **unifying power of mathematics:** different phenomena and systems from different areas in physics having the same types of model can be treated by the same mathematical methods. What differs from area to area is just the kinds of problems that are of practical interest.

PROBLEM SET 16.2

On Theorem 1 and Example 1

1. **(Second proof)** Carry out all steps of the proof (given in Appendix 4) in detail.

2. **(Verification)** Verify Theorem 1 for D: $x \leqq 0$, $0 \leqq y \leqq \pi$, $w = e^z$, $\Phi^* = 2uv$. Sketch D and D^*. What are the boundary values for D?

3. **(Rectangle)** Let D: $0 \leqq x \leqq \frac{1}{2}\pi$, $0 \leqq y \leqq 1$, D^* the image of D under $w = \sin z$, and $\Phi^* = u^2 - v^2$. Find the corresponding potential Φ in D and its boundary values. Sketch D and D^*.

4. **(Conjugate potential)** What happens in Prob. 3 if you replace the potential by a conjugate, $\Phi^* = 2uv$? Sketch or plot some curves $\Phi = const$.

5. Find the potential Φ in the region R in the first quadrant of the z-plane bounded by the axes (having potential 0) and the hyperbola $y = 1/x$ (having potential U_2) in two ways: (i) directly, (ii) by mapping R onto a suitable infinite strip.

6. **(Derivation)** Verify the steps of deriving (3) from (2).

7. **CAS PROJECT. Potential Fields.** (a) Plot equipotential lines in Example 1.
   (b) Plot equipotential lines if the complex potential is $F(z) = iz^2$, $F(z) = z^3$, $F(z) = e^z$.
   (c) Plot the equipotential surfaces corresponding to $F(z) = \ln z$ as cylinders in space.

8. **TEAM PROJECT. Extension of Example 1.** Find the potential between the noncoaxial cylinders C_1: $|z| = 1$ (potential $U_1 = 0$) and C_2: $|z - c| = c$ ($U_2 = 110$ volts), where $0 < c < \frac{1}{2}$. Sketch or plot the equipotential curves and their orthogonal trajectories for $c = 0.1, 0.2, 0.3, 0.4$. Try to think of the further extension C_1: $|z| = 1$, C_2: $|z - c| = \rho \neq c$.

Extensions of Example 2, Comments

9. **(Angular region)** Applying a suitable conformal mapping, obtain from Fig. 372b the potential Φ in the angular region $-\pi/4 < \text{Arg } z < \pi/4$, $\Phi = -3$ kV when $\text{Arg } z = -\pi/4$ and $\Phi = 3$ kV when $\text{Arg } z = \pi/4$.

10. **(Half-plane)** Find the complex and real potentials in the upper half-plane with boundary values 0 if $x < 1$ and 100 kV if $x > 1$ on the x-axis.

11. At $z = \pm 1$ in Fig. 372a the tangents to the equipotential lines shown make equal angles ($\pi/6$). Why?

12. Show that in Example 2, the y-axis is mapped onto the unit circle in the w-plane.

13. **(Disk)** Find the linear fractional transformation $z = g(Z)$ that maps $|Z| \leqq 1$ onto $|z| \leqq 1$ with $Z = i/2$ being mapped onto $z = 0$. Show that $Z_1 = (3 + 4i)/5$ is mapped onto $z = -1$, and $Z_2 = (-3 + 4i)/5$ onto $z = 1$, so that the equipotential lines of Example 2 look in $|Z| \leqq 1$ as shown in Fig. 373.

14. **(Quarter-disk)** In Example 2 set $z = Z^2$ ($Z = X + iY$) and show that the resulting potential $\Phi(X, Y) = \text{Re } F(Z^2)$ obtained from that in Fig. 372(a) is the potential in the portion of the unit disk $|Z| \leqq 1$ in the first quadrant having the boundary values 0 on the axes and 3 on $|Z| = 1$.

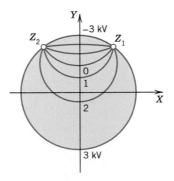

Fig. 373. Problem 13

16.3 Heat Problems

Laplace's equation also governs heat flow problems that are **steady**, that is, time-independent. Indeed, heat conduction in a body of homogeneous material is modeled by the **heat equation**

$$T_t = c^2 \nabla^2 T$$

where the function T is temperature, $T_t = \partial T / \partial t$, t is time, and c^2 is a positive constant (depending on the material of the body). Hence if a problem is **steady,** so that $T_t = 0$, and two-dimensional, then the heat equation reduces to the two-dimensional Laplace equation

(1) $$\nabla^2 T = T_{xx} + T_{yy} = 0,$$

so that the problem can be treated by our present methods.

$T(x, y)$ is called the **heat potential.** It is the real part of the **complex heat potential**

$$F(z) = T(x, y) + i\Psi(x, y).$$

The curves $T(x, y) = const$ are called **isotherms** (= lines of constant temperature) and the curves $\Psi(x, y) = const$ **heat flow lines,** because along them, heat flows from higher to lower temperatures.

It follows that all the examples considered so far (Secs. 16.1, 16.2) can now be reinterpreted as problems on heat flow. The electrostatic equipotential lines $\Phi(x, y) = const$ now become isotherms $T(x, y) = const$, and the lines of electrical force become lines of heat flow. Mathematically, the calculations remain the same. New problems may arise, involving boundary conditions that would make no sense physically in electrostatics or would be of no practical interest there. Examples 3 and 4 (below) illustrate this.

Physically, to have a steady problem, the boundary of the domain of heat flow must be kept at constant temperature, by heating or cooling.

EXAMPLE 1　**Temperature between parallel plates**

Find the temperature between two parallel plates $x = 0$ and $x = d$ in Fig. 374 having temperatures 0 and 100°C, respectively.

Solution.　As in Sec. 16.1 we conclude that $T(x, y) = ax + b$. From the boundary conditions, $b = 0$ and $a = 100/d$. The answer is

$$T(x, y) = \frac{100}{d} x \; [°C].$$

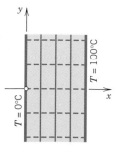

Fig. 374.　Example 1

The corresponding complex potential is $F(z) = (100/d)z$. Heat flows horizontally, in the negative x-direction, along the lines $y = const$.   ◀

EXAMPLE 2   **Temperature distribution between a wire and a cylinder**

Find the temperature field around a long thin wire of radius $r_1 = 1$ mm that is electrically heated to $T_1 = 500°$F and is surrounded by a circular cylinder of radius $r_2 = 100$ mm, which is kept at temperature $T_2 = 60°$F by cooling it with air. See Fig. 375.

Solution. T depends only on r, for reasons of symmetry. Hence, as in Sec. 16.1 (Example 2),

$$T(x, y) = a \ln r + b.$$

The boundary conditions are

$$T_1 = 500 = a \ln 1 + b, \qquad T_2 = 60 = a \ln 100 + b.$$

Hence $b = 500$ (since $\ln 1 = 0$) and $a = (60 - b)/\ln 100 = -95.54$. The answer is

$$T(x, y) = 500 - 95.54 \ln r \; [°\text{F}].$$

The isotherms are concentric circles. Heat flows from the wire radially outward to the cylinder. Sketch T as a function of r. Does it look physically reasonable?   ◀

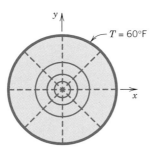

Fig. 375.   Example 2

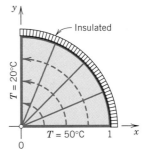

Fig. 376.   Example 3

EXAMPLE 3   **A mixed boundary value problem**

Find the temperature distribution in the region in Fig. 376 (cross section of a solid cylinder), whose vertical portion of the boundary is at 20°C, the horizontal portion at 50°C, and the circular portion is insulated.

Solution.   The insulated portion of the boundary must be a heat flow line, since by the insulation, heat is prevented from crossing such a curve, hence it must flow along it. Thus the isotherms must meet such a curve at right angles. Since T is constant along an isotherm, this means that

(2) $$\frac{\partial T}{\partial n} = 0 \qquad \text{along an insulated portion of the boundary.}$$

Here $\partial T/\partial n$ is the **normal derivative** of T, that is, the directional derivative (Sec. 8.9) in the direction normal (perpendicular) to the insulated boundary. Such a problem in which T is prescribed on one portion of the boundary and $\partial T/\partial n$ on the other portion is called a **mixed boundary value problem.**

In our case, the normal direction to the insulated circular boundary curve is the radial direction toward the origin. Hence (2) becomes $\partial T/\partial r = 0$, meaning that along this curve the solution must not depend on r. Now Arg $z = \theta$ satisfies (1), as well as this condition, and is constant (0 and $\pi/2$) on the straight portions of the boundary. Hence the solution is of the form

$$T(x, y) = a\theta + b.$$

The boundary conditions yield $a \cdot \pi/2 + b = 20$ and $a \cdot 0 + b = 50$. This gives

$$T(x, y) = 50 - \frac{60}{\pi} \theta, \qquad\qquad \theta = \arctan \frac{y}{x}.$$

The isotherms are portions of rays $\theta = const$. Heat flows from the x-axis along circles $r = const$ (dashed in Fig. 376) to the y-axis. ◀

EXAMPLE 4   **Another mixed boundary value problem in heat conduction**

Find the temperature field in the upper half-plane when the x-axis is at $T = 0°C$ for $x < -1$, insulated for $-1 < x < 1$, and at $T = 20°C$ for $x > 1$ (Fig. 377a).

Solution. We map the half-plane in Fig. 377a onto the vertical strip in Fig. 377b, find the temperature $T^*(u, v)$ there, and map it back to get the temperature $T(x, y)$ in the half-plane.

The idea of using that strip is suggested by Fig. 316 in Sec. 12.7 with the roles of $z = x + iy$ and $w = u + iv$ interchanged. The figure shows that $z = \sin w$ maps our present strip onto our half-plane in Fig. 377a. Hence the inverse function

$$w = f(z) = \sin^{-1} z$$

maps that half-plane onto the strip in the w-plane. This is the mapping function that we need according to Theorem 1 in Sec. 16.2.

The insulated segment $-1 < x < 1$ on the x-axis maps onto the segment $-\pi/2 < u < \pi/2$ on the u-axis. The rest of the x-axis maps onto the two vertical boundary portions $u = -\pi/2$ and $\pi/2$, $v > 0$, of the strip. This gives the transformed boundary conditions in Fig. 377b for $T^*(u, v)$, where on the insulated horizontal boundary, $\partial T^*/\partial n = \partial T^*/\partial v = 0$ because v is a coordinate normal to that segment.

Similarly to Example 1 we obtain

$$T^*(u, v) = 10 + \frac{20}{\pi} u$$

which satisfies all the boundary conditions. This is the real part of the complex potential $F^*(w) = 10 + (20/\pi)w$. Hence the complex potential in the z-plane is

$$F(z) = F^*(f(z)) = 10 + \frac{20}{\pi} \sin^{-1} z,$$

and $T(x, y) = \text{Re } F(z)$ is the solution. The isotherms are $u = const$ in the strip and the hyperbolas in the z-plane, perpendicular to which heat flows along the dashed ellipses from the 20°-portion to the cooler 0°-portion of the boundary, a physically very reasonable result. ◀

This section and the last one show the usefulness of conformal mappings and complex potentials. The latter will also play a role in the next section on fluid flow.

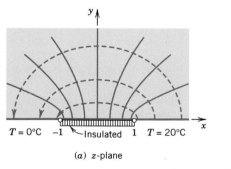

(a) z-plane

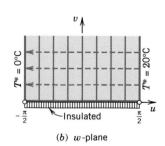

(b) w-plane

Fig. 377.   Example 4

PROBLEM SET 16.3

1. **(Parallel plates)** Find the temperature between two parallel plates $y = 0$ and $y = d$ kept at 0°C and 20°C, respectively. (i) Proceed directly. (ii) Use Example 1 and a suitable mapping.

2. **(Temperature in a plate)** Find the temperature and the complex potential in an infinite plate with edges $y = x - 4$ and $y = x + 4$ kept at -20°C and 40°C, respectively.

3. **(Mixed problem)** Find the temperature and the complex potential in Fig. 376 if $T = -40$°C on the y-axis, 100°C on the x-axis, and the circular portion of the boundary is insulated as before.

4. **(Sector)** Find the temperature T in the sector $0 \leqq \text{Arg } z \leqq 2\pi/3$, $|z| \leqq 1$ if $T = 10$°C on the x-axis, 80°C on $y = -\sqrt{3}x$, and the curved portion is insulated.

5. **(Quadrant)** Find the temperature and the complex potential in the first quadrant of the z-plane when the y-axis is kept at 100°C, the segment $0 < x < 1$ of the x-axis is insulated, and the portion $x > 1$ of the x-axis is kept at 300°C. *Hint.* Use Example 4.

6. **(Interpretation)** Interpret Prob. 13 in Problem Set 16.2 as a heat flow problem (with boundary temperatures, say, 20°C and 100°C). Along what curves does the heat flow?

7. **CAS PROJECT. Isotherms.** Plot isotherms and lines of heat flow in Examples 2–4. Can you see from these graphs where the temperature change is very rapid?

8. **TEAM PROJECT. Piecewise Constant Boundary Temperatures. (a) A basic building block** is shown in Fig. 378. Find the corresponding temperature and complex potential in the upper half-plane.

   (b) Superposition. Find the temperature T^* and the complex potential F^* in the upper half-plane satisfying the boundary conditions in Fig. 379.

   (c) Conformal mapping. What temperature in the first quadrant of the z-plane is obtained from (a) by the mapping $w = a + z^2$ and what are the transformed boundary conditions?

9. **(Semi-infinite strip)** Applying $w = \cosh z$ to Team Project 8(b), obtain the solution of the boundary value problem in Fig. 380.

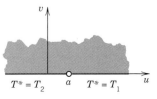

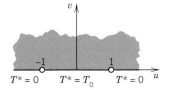

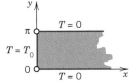

Fig. 378. Team Project 8(a)　　**Fig. 379.** Team Project 8(b)　　**Fig. 380.** Problem 9

Temperature Distributions in Plates

In Probs. 10–14 find the temperature $T(x, y)$ in the given thin metal plate whose faces are insulated and whose edges are kept at the temperatures shown in the figure.

10.

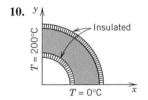

11.

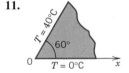

12.

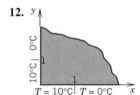

13.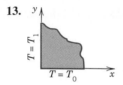

14.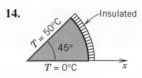

16.4 Fluid Flow

Laplace's equation also plays a basic role in hydrodynamics, in steady nonviscous fluid flow under physical conditions discussed later in this section. To keep in touch with complex analysis, our problems will be ***two-dimensional,*** so that the **velocity vector** V by which the motion of the fluid can be given depends only on two space variables x and y and the motion is the same in all planes parallel to the xy-plane.

Then we can use for V a complex function

$$(1) \qquad\qquad V = V_1 + iV_2$$

giving the magnitude $|V|$ and direction Arg V of the velocity at each point $z = x + iy$. Here V_1 and V_2 are the components of the velocity in the x and y directions. V is tangential to the path of the moving particles, called a **streamline** of the motion (Fig. 381).

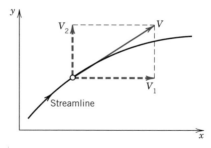

Fig. 381. Velocity

We show that under suitable assumptions (explained in detail following the examples), for a given flow there exists an analytic function

$$(2) \qquad\qquad \boxed{F(z) = \Phi(x,\, y) + i\Psi(x,\, y),}$$

called the **complex potential** of the flow, such that the streamlines are given by $\Psi(x,\, y) = const$, and the velocity vector or, briefly, the **velocity** is given by

$$(3) \qquad\qquad \boxed{V = V_1 + iV_2 = \overline{F'(z)}}$$

where the bar denotes the complex conjugate. Ψ is called the **stream function.** The function Φ is called the **velocity potential.**[4] The curves $\Phi(x,\, y) = const$ are called

---

[4]Some authors use $-\Phi$ (instead of Φ) as the velocity potential.

equipotential lines. V is the **gradient** of Φ; by definition, this means that

$$
(4) \qquad V_1 = \frac{\partial \Phi}{\partial x}, \qquad V_2 = \frac{\partial \Phi}{\partial y}.
$$

Indeed, for $F = \Phi + i\Psi$, Eq. (4) in Sec. 12.4 is $F' = \Phi_x + i\Psi_x$ with $\Psi_x = -\Phi_y$ by the second Cauchy–Riemann equation. Together,

$$
\overline{F'(z)} = \Phi_x - i\Psi_x = \Phi_x + i\Phi_y = V_1 + iV_2 = V.
$$

Furthermore, since $F(z)$ is analytic, Φ and Ψ satisfy Laplace's equation

$$
(5) \qquad \nabla^2 \Phi = \frac{\partial^2 \Phi}{\partial x^2} + \frac{\partial^2 \Phi}{\partial y^2} = 0, \qquad \nabla^2 \Psi = \frac{\partial^2 \Psi}{\partial x^2} + \frac{\partial^2 \Psi}{\partial y^2} = 0.
$$

Whereas in electrostatics the boundaries (conducting plates) are equipotential lines, in fluid flow a boundary across which fluid cannot flow must be a streamline. Hence in fluid flow the stream function is of particular importance.

Before discussing the conditions for the validity of the statements involving (2)–(5), let us consider two flows of practical interest, so that we first see what is going on from a practical point of view. Further flows follow in the problem set.

EXAMPLE 1   **Flow around a corner**

The complex potential $F(z) = z^2 = x^2 - y^2 + 2ixy$ describes a flow with

Equipotential lines	$\Phi = x^2 - y^2 = const$	(Hyperbolas)
Streamlines	$\Psi = 2xy = const$	(Hyperbolas).

From (3) we obtain the velocity vector

$$
V = 2\bar{z} = 2(x - iy), \qquad \text{that is,} \qquad V_1 = 2x, \qquad V_2 = -2y.
$$

The speed (magnitude of the velocity) is

$$
|V| = \sqrt{V_1{}^2 + V_2{}^2} = 2\sqrt{x^2 + y^2}.
$$

The flow may be interpreted as the flow in a channel bounded by the positive coordinates axes and a hyperbola, say, $xy = 1$ (Fig. 382). We note that the speed along a streamline S has a minimum at the point P where the cross section of the channel is large. ◀

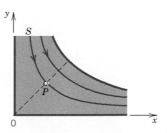

Fig. 382.   Flow around
a corner (Example 1)

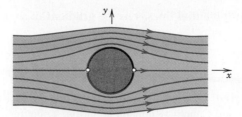

Fig. 383.   Flow around a cylinder
(Example 2)

EXAMPLE 2   **Flow around a cylinder**

Consider the complex potential

$$F(z) = \Phi(x, y) + i\Psi(x, y) = z + \frac{1}{z}.$$

Using the polar form $z = re^{i\theta}$, we obtain

$$F(z) = re^{i\theta} + \frac{1}{r} e^{-i\theta} = \left(r + \frac{1}{r}\right) \cos \theta + i \left(r - \frac{1}{r}\right) \sin \theta.$$

Hence the streamlines are

$$\Psi(x, y) = \left(r - \frac{1}{r}\right) \sin \theta = const.$$

In particular, $\Psi(x, y) = 0$ gives $r - 1/r = 0$ or $\sin \theta = 0$. Hence this streamline consists of the unit circle ($r = 1/r$ gives $r = 1$) and the x-axis ($\theta = 0$ and $\theta = \pi$). For large $|z|$ the term $1/z$ in $F(z)$ is small in absolute value, so that for these z the flow is nearly uniform and parallel to the x-axis. Hence we can interpret this as a flow around a long circular cylinder of unit radius. The flow has two **stagnation points** (that is, points at which the velocity V equals zero), at $z = \pm 1$. This follows from

$$F'(z) = 1 - \frac{1}{z^2}, \qquad \text{hence} \qquad z^2 - 1 = 0$$

and (3). See Fig. 383.   ◀

Assumptions and Theory Underlying (2)–(5)

If the domain of flow is simply connected and the flow is irrotational and incompressible, then the statements involving (2)–(5) hold. In particular, then the flow has a complex potential $F(z)$, which is an analytic function. (Explanation of terms below.)

We prove this, along with a discussion of basic concepts related to fluid flow. Consider any smooth curve C in the z-plane, given by $z(s) = x(s) + iy(s)$, where s is the arc length of C. Let the real variable V_t be the component of the velocity V tangent to C (Fig. 384).

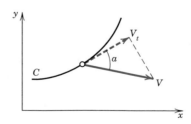

Fig. 384.   Tangential component of the velocity
with respect to a curve C

Then the value of the real line integral

(6)
$$\int_C V_t \, ds$$

taken along C in the sense of increasing values of s is called the **circulation** of the fluid along C. Dividing the circulation by the length of C, we obtain the mean velocity[5] of the flow along the curve C. Now

$$V_t = |V| \cos \alpha \qquad \text{(Fig. 384)}.$$

Hence V_t is the dot product (Sec. 8.2) of V and the tangent vector dz/ds of C (Sec. 12.5); thus in (6),

$$V_t \, ds = \left(V_1 \frac{dx}{ds} + V_2 \frac{dy}{ds} \right) ds = V_1 \, dx + V_2 \, dy.$$

The circulation (6) along C now becomes

(7)
$$\int_C V_t \, ds = \int_C (V_1 \, dx + V_2 \, dy).$$

As the next idea, let C be a *closed* curve, namely, the boundary of a simply connected domain D, and suppose that V has continuous partial derivatives in a domain containing D and C. Then we can use Green's theorem (Sec. 9.4) to represent the circulation around C by a double integral,

(8)
$$\oint_C (V_1 \, dx + V_2 \, dy) = \iint_D \left(\frac{\partial V_2}{\partial x} - \frac{\partial V_1}{\partial y} \right) dx \, dy.$$

The integrand of this double integral is called the **vorticity** of the flow. The vorticity divided by 2 is called the **rotation**

(9)
$$\omega(x, y) = \frac{1}{2} \left(\frac{\partial V_2}{\partial x} - \frac{\partial V_1}{\partial y} \right).$$

We assume the flow to be **irrotational,** that is, $\omega(x, y) \equiv 0$ throughout the flow; thus,

(10)
$$\frac{\partial V_2}{\partial x} - \frac{\partial V_1}{\partial y} = 0.$$

To understand the physical meaning of vorticity and rotation, take for C in (8) a circle.

---

[5]*Definitions:* $\dfrac{1}{b-a} \displaystyle\int_a^b f(x) \, dx$ = mean value of f on the interval $a \leq x \leq b$,

$\dfrac{1}{L} \displaystyle\int_C f(s) \, ds$ = mean value of f on C (L = length of C),

$\dfrac{1}{A} \displaystyle\iint_D f(x, y) \, dx \, dy$ = mean value of f on D (A = area of D).

Let r be the radius of C. Then the circulation divided by the length $2\pi r$ of C is the mean velocity of the fluid along C. Hence by dividing this by r we obtain the mean *angular* velocity ω_0 of the fluid about the axis of the circle:

$$\omega_0 = \frac{1}{2\pi r^2} \iint_D \left(\frac{\partial V_2}{\partial x} - \frac{\partial V_1}{\partial y} \right) dx\, dy = \frac{1}{\pi r^2} \iint_D \omega(x,\, y)\, dx\, dy.$$

If we now let $r \to 0$, the limit of ω_0 is the value of ω at the center of C. Hence, $\omega(x,\, y)$ is the limiting angular velocity of a circular element of the fluid as the circle shrinks to the point $(x,\, y)$. Roughly speaking, if a spherical element of the fluid were suddenly solidified and the surrounding fluid simultaneously annihilated, the element would rotate with the angular velocity ω.

Our second assumption is that the fluid is **incompressible.** (Examples are water and oil, whereas air is compressible.) Then

(11)
$$\frac{\partial V_1}{\partial x} + \frac{\partial V_2}{\partial y} = 0 \qquad \text{[see (7), Sec. 8.10]}$$

in every region that is free of **sources** or **sinks,** that is, points at which fluid is produced or disappears. [The expression in (11) is called the **divergence** of V and is denoted by div V.]

If the domain D of the flow is **simply connected** (Sec. 13.2) and the flow is irrotational, then (10) implies that the line integral (7) is independent of path in D (by Theorem 3 in Sec. 9.2, where $F_1 = V_1$, $F_2 = V_2$, $F_3 = 0$, and z is the third coordinate in space and has nothing to do with our present z). Hence if we integrate from a fixed point $(a,\, b)$ in D to a variable point $(x,\, y)$ in D, the integral becomes a function of the point $(x,\, y)$, say, $\Phi(x,\, y)$:

(12)
$$\Phi(x,\, y) = \int_{(a,\, b)}^{(x,\, y)} (V_1\, dx + V_2\, dy).$$

We claim that the flow has a velocity potential Φ, which is given by (12). To prove this, all we have to do is to show that (4) holds. Now since the integral (7) is independent of path, $V_1\, dx + V_2\, dy$ is exact (Sec. 9.2), namely, the differential of Φ, that is,

$$V_1\, dx + V_2\, dy = \frac{\partial \Phi}{\partial x}\, dx + \frac{\partial \Phi}{\partial y}\, dy.$$

From this we see that $V_1 = \partial \Phi / \partial x$ and $V_2 = \partial \Phi / \partial y$, which gives (4).

That Φ is harmonic follows at once by substituting (4) into (11), which gives the first Laplace's equation in (5).

We finally take a harmonic conjugate Ψ of Φ. Then the other equation in (5) holds. Also, assuming that the second partial derivatives of Φ and Ψ are continuous, we have that the complex function

$$F(z) = \Phi(x,\, y) + i\Psi(x,\, y)$$

is analytic in D. Since the curves $\Psi(x,\, y) = const$ are perpendicular to the equipotential curves $\Phi(x,\, y) = const$ (except where $F'(z) = 0$), we conclude that they are the streamlines. Hence Ψ is the stream function and $F(z)$ is the complex potential of the flow. ◀

PROBLEM SET 16.4

Flow Patterns: Streamlines, Complex Potential

These problems should encourage you to experiment with various functions $F(z)$, many of which model interesting flow patterns.

1. **(Parallel flow)** Show that $F(z) = Kz$ (K positive real) describes a uniform flow to the right, which can be interpreted as a uniform flow between two parallel lines (between two parallel planes in three-dimensional space). See Fig. 385. Find the velocity vector, the streamlines, and the equipotential lines.

2. **(Parallel flow)** What is the complex potential of a parallel flow in the direction of $y = x$ (upward)?

3. **(Extension of Example 1)** Sketch or plot the flow in Example 1 on the whole upper half-plane. Show that you can interpret it as a flow against a horizontal wall (the x-axis).

4. **(Corner)** Show that $F(z) = iz^2$ describes a flow around a corner. Find and sketch or plot the streamlines and equipotential lines. Find V.

5. **(Corner)** What $F(z)$ would be suitable in Example 1 if the angle of the corner were $\pi/3$ instead of $\pi/2$?

6. **(Potential $F = iz^3$)** Sketch or plot the streamlines and equipotential lines of $F(z) = iz^3$. Find V. Find all points at which V is parallel to the x-axis.

7. **(Conformal mapping)** Obtain the flow in Example 1 from that in Prob. 1 by a suitable conformal mapping.

8. **(Cylinder)** What happens to the flow in Fig. 383 if you replace z by $ze^{-i\alpha}$, where $\alpha = const$, for instance, $\alpha = \pi/4$?

9. **(Cylinder of radius r_0)** Change $F(z)$ in Example 2 slightly to obtain a flow around a cylinder of radius r_0 that gives the flow in Example 2 if $r_0 \to 1$.

10. **(Cylinder)** What happens in Example 2 if you replace z by z^2? Sketch and interpret the resulting flow in the first quadrant. Can you get help from Example 1?

11. **(Potential $1/z$)** Show that the streamlines obtained from $F(z) = 1/z$ are circles through the origin.

12. **(Aperture)** Show that $F(z) = \cosh^{-1} z$ gives confocal hyperbolas as streamlines, with foci at $z = \pm 1$, and the flow may be interpreted as a flow through an aperture (Fig. 386).

13. **(Elliptical cylinder)** Show that $F(z) = \cos^{-1} z$ gives confocal ellipses as streamlines, with foci at $z = \pm 1$, and the flow circulates around an elliptic cylinder or a plate (the segment from -1 to 1 in Fig. 387).

14. **TEAM PROJECT. Role of the Natural Logarithm in Flows. (a) Basic flows: Source and sink.** Show that $F(z) = (c/2\pi) \ln z$ with constant positive real c gives a flow directed radially outward (Fig. 388), so that F models a **point source** at $z = 0$ (that is, a **source line** $x = 0$, $y = 0$ in space) at which fluid is produced. c is called the **strength** or **discharge** of the source. If c is negative real, show that the flow is directed radially inward, so that F models

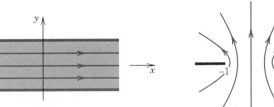

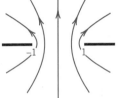

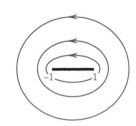

Fig. 385. Parallel flow in Problem 1

Fig. 386. Flow through an aperture in Problem 12

Fig. 387. Flow around a plate in Problem 13

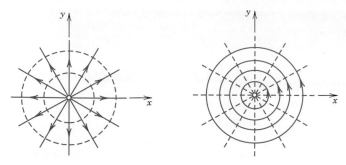

Fig. 388. Point source   **Fig. 389.** Vortex flow

a **sink** at $z = 0$, a point at which fluid disappears. Note that $z = 0$ is the singular point of $F(z)$.

(b) Basic flows: Vortex. Show that $F(z) = -(Ki/2\pi) \ln z$ with positive real K gives a flow circulating counterclockwise around $z = 0$ (Fig. 389). $z = 0$ is called a **vortex.** Note that each time we travel around the vortex, the potential increases by K.

(c) Addition of flows. Show that addition of the velocity vectors of two flows gives a flow whose complex potential is obtained by adding the complex potentials of those flows.

(d) Source and sink combined. Find the complex potentials of a flow with a source of strength 1 at $z = -a$ and of a flow with a sink of strength 1 at $z = a$. Add both and sketch or plot the streamlines. Show that for small $|a|$ these lines look similar to those in Prob. 11.

(e) Flow with circulation around a cylinder. Add the potential in (b) to that in Example 2. Show that this gives a flow for which the cylinder wall $|z| = 1$ is a streamline. Find the speed and show that the stagnation points are

$$z = \frac{iK}{4\pi} \pm \sqrt{\frac{-K^2}{16\pi^2} + 1};$$

if $K = 0$ they are at ± 1, as K increases they move up on the unit circle until they unite at $z = i$ ($K = 4\pi$, see Fig. 390), and if $K > 4\pi$ they lie on the imaginary axis (one lies in the field of flow and the other one lies inside the cylinder and has no physical meaning).

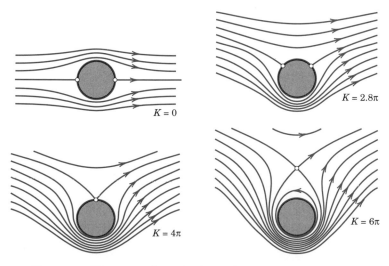

Fig. 390. Flow around a cylinder without circulation ($K = 0$) and with circulation

16.5 Poisson's Integral Formula

So far in this chapter we have seen that complex analysis offers powerful methods for discussing and solving two-dimensional potential problems based on conformal mappings and complex potentials. A further method results from complex integration. As a most important result it yields Poisson's integral formula (5) (below) for potentials in a standard domain (a circular disk) and from (5) a useful series (7) for these potentials.

Poisson's formula will follow from Cauchy's integral formula (Sec. 13.3)

$$(1) \qquad F(z) = \frac{1}{2\pi i} \oint_C \frac{F(z^*)}{z^* - z} \, dz^*.$$

Here C is the circle $z^* = Re^{i\alpha}$ (counterclockwise, $0 \leqq \alpha \leqq 2\pi$), and we assume that $F(z^*)$ is analytic in a domain containing C and its full interior. Since $dz^* = iRe^{i\alpha} \, d\alpha = iz^* \, d\alpha$, we obtain from (1)

$$(2) \qquad F(z) = \frac{1}{2\pi} \int_0^{2\pi} F(z^*) \frac{z^*}{z^* - z} \, d\alpha \qquad (z^* = Re^{i\alpha}, z = re^{i\theta}).$$

Now comes a little trick. If instead of z inside C we take a Z outside C, the integrals (1) and (2) are zero by Cauchy's integral theorem (Sec. 13.2). We choose $Z = z^*\bar{z}^*/\bar{z} = R^2/\bar{z}$, which is outside C because $|Z| = R^2/|z| = R^2/r > R$. From (2) we thus have

$$0 = \frac{1}{2\pi} \int_0^{2\pi} F(z^*) \frac{z^*}{z^* - Z} \, d\alpha = \frac{1}{2\pi} \int_0^{2\pi} F(z^*) \frac{z^*}{z^* - \dfrac{z^*\bar{z}^*}{\bar{z}}} \, d\alpha$$

$$= \frac{1}{2\pi} \int_0^{2\pi} F(z^*) \frac{\bar{z}}{\bar{z} - \bar{z}^*} \, d\alpha.$$

We subtract the last expression (which is zero!) from (2) and use the following formula that you can verify by direct calculation:

$$(3) \qquad \frac{z^*}{z^* - z} - \frac{\bar{z}}{\bar{z} - \bar{z}^*} = \frac{z^*\bar{z}^* - z\,\bar{z}}{(z^* - z)(\bar{z}^* - \bar{z})}.$$

We then have

$$(4) \qquad F(z) = \frac{1}{2\pi} \int_0^{2\pi} F(z^*) \frac{z^*\bar{z}^* - z\,\bar{z}}{(z^* - z)(\bar{z}^* - \bar{z})} \, d\alpha.$$

From the polar representations of z and z^* we see that the quotient in the integrand is real and equal to

$$\frac{R^2 - r^2}{(Re^{i\alpha} - re^{i\theta})(Re^{-i\alpha} - re^{-i\theta})} = \frac{R^2 - r^2}{R^2 - 2Rr \cos(\theta - \alpha) + r^2}.$$

We now write $F(z) = \Phi(r, \theta) + i\Psi(r, \theta)$ and take the real part on both sides of (4).

Then we obtain **Poisson's integral formula**[6]

$$(5) \qquad \Phi(r, \theta) = \frac{1}{2\pi} \int_0^{2\pi} \Phi(R, \alpha) \frac{R^2 - r^2}{R^2 - 2Rr \cos(\theta - \alpha) + r^2} \, d\alpha.$$

This formula represents the harmonic function Φ in the disk $|z| \leqq R$ in terms of its values $\Phi(R, \alpha)$ on the boundary (the circle) $|z| = R$.

Formula (5) is still valid if the boundary function $\Phi(R, \alpha)$ is merely piecewise continuous (as is practically often the case; see Fig. 372 in Sec. 16.2 for an example). Then (5) gives a function harmonic in the open disk, and on the circle $|z| = R$ equal to the given boundary function, except at points where the latter is discontinuous. A proof can be found in Ref. [D1] in Appendix 1.

Series for Potentials in Disks

From (5) we may obtain an important series development of Φ in terms of simple harmonic functions. We remember that the quotient in the integrand of (5) was derived from (3). We claim that the right side of (3) is the real part of

$$\frac{z^* + z}{z^* - z} = \frac{(z^* + z)(\bar{z}^* - \bar{z})}{(z^* - z)(\bar{z}^* - \bar{z})} = \frac{z^* \bar{z}^* - z\bar{z} - z^* \bar{z} + z\bar{z}^*}{|z^* - z|^2}.$$

Indeed, the denominator is real and so is $z^* \bar{z}^* - z\bar{z}$ in the numerator, whereas $-z^* \bar{z} + z\bar{z}^* = 2i \operatorname{Im}(z\bar{z}^*)$ in the numerator is pure imaginary. This verifies our claim. Now by the use of the geometric series we obtain

$$(6) \qquad \frac{z^* + z}{z^* - z} = \frac{1 + (z/z^*)}{1 - (z/z^*)} = \left(1 + \frac{z}{z^*}\right) \sum_{n=0}^{\infty} \left(\frac{z}{z^*}\right)^n = 1 + 2 \sum_{n=1}^{\infty} \left(\frac{z}{z^*}\right)^n.$$

Since $z = re^{i\theta}$ and $z^* = Re^{i\alpha}$, we have

$$\operatorname{Re}\left(\frac{z}{z^*}\right)^n = \operatorname{Re}\left[\frac{r^n}{R^n} e^{in\theta} e^{-in\alpha}\right] = \left(\frac{r}{R}\right)^n \cos(n\theta - n\alpha).$$

On the right, $\cos(n\theta - n\alpha) = \cos n\theta \cos n\alpha + \sin n\theta \sin n\alpha$. Hence from (6) we obtain

$$\operatorname{Re} \frac{z^* + z}{z^* - z} = 1 + 2 \sum_{n=1}^{\infty} \left(\frac{r}{R}\right)^n (\cos n\theta \cos n\alpha + \sin n\theta \sin n\alpha).$$

This expression is equal to the quotient in (5), as we have mentioned before, and by inserting the series into (5) and integrating term by term we find

$$(7) \qquad \Phi(r, \theta) = a_0 + \sum_{n=1}^{\infty} \left(\frac{r}{R}\right)^n (a_n \cos n\theta + b_n \sin n\theta)$$

---

[6]SIMÉON DENIS POISSON (1781—1840), French mathematician and physicist, professor in Paris from 1809. His work includes potential theory, partial differential equations (Poisson equation, Sec. 11.1), and probability (Sec. 22.7).

where the coefficients are

$$a_0 = \frac{1}{2\pi} \int_0^{2\pi} \Phi(R, \alpha) \, d\alpha, \qquad a_n = \frac{1}{\pi} \int_0^{2\pi} \Phi(R, \alpha) \cos n\alpha \, d\alpha,$$

(8)
$$n = 1, 2, \cdots,$$

$$b_n = \frac{1}{\pi} \int_0^{2\pi} \Phi(R, \alpha) \sin n\alpha \, d\alpha,$$

the Fourier coefficients of $\Phi(R, \alpha)$; see Sec. 10.2. Now for $r = R$ the series (7) becomes the Fourier series of $\Phi(R, \alpha)$. Hence the representation (7) will be valid whenever $\Phi(R, \alpha)$ can be represented by a Fourier series.

EXAMPLE 1   **Dirichlet problem for the unit disk**

Find the electrostatic potential $\Phi(r, \theta)$ in the unit disk $r < 1$ having the boundary values

$$\Phi(1, \alpha) = \begin{cases} -\alpha/\pi & \text{if} \quad -\pi < \alpha < 0 \\ \alpha/\pi & \text{if} \quad 0 < \alpha < \pi \end{cases} \qquad \text{(Fig. 391).}$$

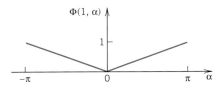

Fig. 391. Boundary values
in Example 1

Solution. Since $\Phi(1, \alpha)$ is even, $b_n = 0$, and from (8) we obtain $a_0 = \frac{1}{2}$ and

$$a_n = \frac{1}{\pi} \left[-\int_{-\pi}^0 \frac{\alpha}{\pi} \cos n\alpha \, d\alpha + \int_0^\pi \frac{\alpha}{\pi} \cos n\alpha \, d\alpha \right] = \frac{2}{n^2 \pi^2} (\cos n\pi - 1).$$

Hence, $a_n = -4/n^2\pi^2$ if n is odd, $a_n = 0$ if $n = 2, 4, \cdots$, and the potential is

$$\Phi(r, \theta) = \frac{1}{2} - \frac{4}{\pi^2} \left[r\cos\theta + \frac{r^3}{3^2} \cos 3\theta + \frac{r^5}{5^2} \cos 5\theta + \cdots \right].$$

Figure 392 shows the unit disk and some of the equipotential lines (curves $\Phi = const$). ◀

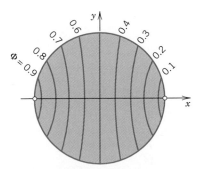

Fig. 392. Potential in Example 1

PROBLEM SET 16.5

Harmonic Functions in a Disk

Using (7), find the potential $\Phi(r, \theta)$ in the unit disk $r < 1$ having the given boundary values $\Phi(1, \theta)$. Using the first few terms of the series, compute some values of Φ and sketch a figure of the equipotential lines.

1. $\Phi(1, \theta) = \sin \theta$

2. $\Phi(1, \theta) = 2 - \cos \theta$

3. $\Phi(1, \theta) = \cos^2 \theta$

4. $\Phi(1, \theta) = \cos 4\theta - \cos 2\theta$

5. $\Phi(1, \theta) = \sin 5\theta$

6. $\Phi(1, \theta) = \cos^4 \theta$

7. $\Phi(1, \theta) = 4 \sin^3 \theta$

8. $\Phi(1, \theta) = \theta$ if $0 < \theta < 2\pi$

9. $\Phi(1, \theta) = \theta$ if $-\pi < \theta < \pi$

10. $\Phi(1, \theta) = 1$ if $0 < \theta < \pi$ and 0 otherwise

11. $\Phi(1, \theta) = \theta$ if $-\pi/2 < \theta < \pi/2$, $\Phi(1, \theta) = \pi - \theta$ if $\pi/2 < \theta < 3\pi/2$

12. Verify (3).

13. Show that each term in (7) is a harmonic function in the disk $r < R$.

14. TEAM PROJECT. Potential in a Disk. (a) Mean Value Property. Show that the value of a harmonic function Φ at the center of a circle C equals the mean of the value of Φ on C (see Sec. 16.4, footnote 5, for definitions of mean values).

    (b) Separation of variables. Show that the terms of (7) appear as solutions in separating the Laplace equation in polar coordinates.

    (c) Harmonic conjugate. Find a series for a harmonic conjugate Ψ of Φ from (7).

    (d) Power series. Find a series for $F(z) = \Phi + i\Psi$.

15. CAS PROJECT. Series (7). Write a program for series developments (7). Apply it to Example 1, namely, experiment on accuracy by computing values from partial sums and comparing them with values that you obtain from your CAS plot of Fig. 392. Do the same for the function in Prob. 10 (which is discontinuous on the circle!).

16.6 General Properties of Harmonic Functions

Analytic functions have various important general properties. We discuss two of them that follow from Cauchy's integral theorem and imply basic properties of **harmonic functions** (solutions of Laplace's equation whose second partial derivatives are continuous).

THEOREM 1   **(Mean value property of analytic functions)**

Let $F(z)$ be analytic in a simply connected domain D. Then the value of $F(z)$ at a point z_0 in D is equal to the mean value of $F(z)$ on any circle in D with center at z_0.

PROOF.   In Cauchy's integral formula (Sec. 13.3)

(1)
$$F(z_0) = \frac{1}{2\pi i} \oint_C \frac{F(z)}{z - z_0} \, dz$$

we choose for C the circle $z = z_0 + re^{i\alpha}$ in D. Then $z - z_0 = re^{i\alpha}$, $dz = ire^{i\alpha} \, d\alpha$, and (1) becomes

(2)
$$F(z_0) = \frac{1}{2\pi} \int_0^{2\pi} F(z_0 + re^{i\alpha}) \, d\alpha.$$

The right side is the mean value of F on the circle (= value of the integral divided by the length 2π of the interval of integration). This proves the theorem.   ◀

For harmonic functions, Theorem 1 implies

THEOREM 2   **(Two mean value properties of harmonic functions)**

The value of a harmonic function $\Phi(x, y)$ at a point (x_0, y_0) in D is equal to the mean value of $\Phi(x, y)$ on any circle with center at (x_0, y_0). It is also equal to the mean value of $\Phi(x, y)$ on any circular disk in D with center (x_0, y_0). [See footnote 5 in Sec. 16.4.]

PROOF.   The first part of the theorem follows from (2) by taking the real parts on both sides,

$$\Phi(x_0, y_0) = \mathrm{Re}\, F(x_0 + iy_0) = \frac{1}{2\pi} \int_0^{2\pi} \Phi(x_0 + r\cos\alpha, y_0 + r\sin\alpha)\, d\alpha.$$

The second part of the theorem follows by integrating this over r from 0 to r_0 (the radius of the disk) and division by $r_0^2/2$,

$$\Phi(x_0, y_0) = \frac{1}{\pi r_0^2} \int_0^{r_0} \int_0^{2\pi} \Phi(x_0 + r\cos\alpha, y_0 + r\sin\alpha) r\, d\alpha\, dr.$$

On the right this is the mean value (integral divided by the area of the region of integration).   ◀

Returning to analytic functions, we state and prove another famous consequence of Cauchy's integral formula. The corresponding indirect proof exhibits quite a nice idea of applying the *ML*-inequality.

THEOREM 3   **(Maximum modulus theorem for analytic functions)**

Let $F(z)$ be analytic and nonconstant in a domain containing a bounded[7] region D and its boundary. Then the absolute value $|F(z)|$ cannot have a maximum at an interior point of D. Consequently, the maximum of $|F(z)|$ is taken on the boundary of D. If $F(z) \neq 0$ in D, the same is true with respect to the minimum of $|F(z)|$.

PROOF.   We assume that $|F(z)|$ has a maximum at an interior point z_0 of D and show that this leads to a contradiction. Let $|F(z_0)| = M$ be this maximum. Since $F(z)$ is not constant, $|F(z)|$ is not constant, as follows from Example 3 in Sec. 12.4. Consequently, we can find a circle C of radius r with center at z_0 such that the interior of C is in D and $|F(z)|$ is smaller than M at some point P of C. Since $|F(z)|$ is continuous, it will be smaller than M on an arc C_1 of C that contains P (see Fig. 393 on the next page), say,

$$|F(z)| \leqq M - k \quad (k > 0) \qquad \text{for all } z \text{ on } C_1.$$

Let C_1 have the length L_1. Then the complementary arc C_2 of C has the length $2\pi r - L_1$. We now apply the *ML*-inequality (Sec. 13.1) to (1) and note that $|z - z_0| = r$.

---

[7]See Sec. 13.2, footnote 5.

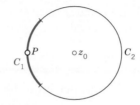

Fig. 393.   Proof of Theorem 3

We then obtain (using straightforward calculation in the second line of the formula)

$$M = |F(z_0)| \leqq \frac{1}{2\pi} \left| \int_{C_1} \frac{F(z)}{z - z_0} \, dz \right| + \frac{1}{2\pi} \left| \int_{C_2} \frac{F(z)}{z - z_0} \, dz \right|$$

$$\leqq \frac{1}{2\pi} \left(\frac{M - k}{r} \right) L_1 + \frac{1}{2\pi} \left(\frac{M}{r} \right) (2\pi r - L_1) = M - \frac{kL_1}{2\pi r} < M,$$

that is, $M < M$, which is impossible. Hence our assumption is false and the first statement is proved.

Next we prove the second statement. If $F(z) \neq 0$ in D, then $1/F(z)$ is analytic in D. From the statement already proved it follows that the maximum of $1/|F(z)|$ lies on the boundary of D. But this maximum corresponds to a minimum of $|F(z)|$. This completes the proof. ◀

This theorem has several fundamental consequences for harmonic functions, as follows.

THEOREM 4   (Harmonic functions)

Let $\Phi(x, y)$ be harmonic in a domain containing a simply connected bounded domain D and its boundary curve C. Then:

*(I) (**Maximum principle**) If $\Phi(x, y)$ is not constant, it has neither a maximum nor a minimum in D. Consequently, the maximum and the minimum are taken on the boundary of D.*

(II) If $\Phi(x, y)$ is constant on C, then $\Phi(x, y)$ is a constant.

(III) If $h(x, y)$ is harmonic in D and on C and if $h(x, y) = \Phi(x, y)$ on C, then $h(x, y) = \Phi(x, y)$ everywhere in D.

PROOF.   (I) Let $\Psi(x, y)$ be a conjugate harmonic function of $\Phi(x, y)$ in D. Then the complex function $F(z) = \Phi(x, y) + i\Psi(x, y)$ is analytic in D, and so is

$$G(z) = e^{F(z)}.$$

The absolute value is

$$|G(z)| = e^{\text{Re } F(z)} = e^{\Phi(x, y)}.$$

From Theorem 3 it follows that $|G(z)|$ cannot have a maximum at an interior point of D. Since e^Φ is a monotone increasing function of the real variable Φ, the statement about the maximum of Φ follows. From this, the statement about the minimum follows by replacing Φ by $-\Phi$.

(II) By (I) the function $\Phi(x, y)$ takes its maximum and its minimum on C. Thus, if $\Phi(x, y)$ is constant on C, its minimum must equal its maximum, so that $\Phi(x, y)$ must be a constant.

(III) If h and Φ are harmonic in D and on C, then $h - \Phi$ is also harmonic in D and on C, and by assumption, $h - \Phi = 0$ everywhere on C. By (II) we thus have $h - \Phi = 0$ everywhere in D, and (III) is proved. ◀

The last statement of Theorem 4 is very important. It means that a *harmonic function is uniquely determined in D by its values on the boundary of D.* Usually, $\Phi(x, y)$ is required to be harmonic in D and continuous on the boundary[8] of D. Under these circumstances the maximum principle (I) is still applicable. The problem of determining $\Phi(x, y)$ when the boundary values are given is called the Dirichlet problem for the Laplace equation in two variables, as we know. From (III) we thus have, as a highlight of our discussion,

THEOREM 5   **(Uniqueness theorem for the Dirichlet problem)**

If for a given region and given boundary values the Dirichlet problem for the Laplace equation in two variables has a solution, the solution is unique.

PROBLEM SET 16.6

Problems 1–9 Related to Theorems 1 and 2

Verify Theorem 1 for the given $F(z)$, z_0, and circle of radius 1.

1. $(z - 1)^2$, $\quad z_0 = \frac{1}{2}$ **2.** $5z^4$, $\quad z_0 = 0$ **3.** $(z + 2)^2$, $\quad z_0 = 1$

4. Integrate $|z|$ around the unit circle. Does the result contradict Theorem 1?

5. Obtain Theorem 2 from Poisson's integral formula.

Verify Theorem 2 for the given $\Phi(x, y)$, (x_0, y_0), and a circle of radius 1.

6. $x^2 - y^2$, $\quad (1, 0)$ **7.** $(x - 1)(y - 1)$, $\quad (3, -3)$ **8.** $3x^2y - y^3$, $\quad (1, 1)$

 9. CAS PROJECT. Plotting Potentials. Plot the potentials in Probs. 7 and 8 as well as three others of your choice as surfaces over a disk or a square in the xy-plane. Find the locations of maxima and minima by inspecting these plots.

Problems 10–15 Related to Theorems 3 and 4

10. TEAM PROJECT. Maximum Modulus of Analytic Functions. (a) Verify Theorem 3 for (i) $F(z) = z^2$ and the rectangle $1 \le x \le 4$, $3 \le y \le 7$,   (ii) $F(z) = e^z$ and any bounded domain.

(b) $F(z) = 1 + 5|z|^2$ is not zero in the disk D: $|z| \le 2$ and has a minimum at an interior point of D. Does this contradict Theorem 3?

(c) $F(x) = \sin x$ (x real) has a maximum at $x = \pi/2$. How does it follow that this cannot be a maximum of $|F(z)| = |\sin z|$ in a domain containing $z = \pi/2$?

(d) If $F(z)$ is analytic (not constant) in the closed disk D: $|z| \le 1$ and $|F(z)| = c = const$ on C: $|z| = 1$, show that $F(z)$ must have a zero in D. Can you extend this to an arbitrary simple closed curve C?

11. (Maximum principle) Verify the maximum principle for $\Phi(x, y) = xy$ and the disk $x^2 + y^2 \le 8$. Find the locations of the maxima and minima.

---

[8]That is, $\lim\limits_{\substack{x \to x_0 \\ y \to y_0}} \Phi(x, y) = \Phi(x_0, y_0)$, where (x_0, y_0) is on the boundary and (x, y) is in D.

12. **(Maximum principle)** Verify the principle for $\Phi(x, y) = e^x \sin y$ and the rectangle $a \leq x \leq b, 0 \leq y \leq 2\pi$.

13. **(Maximum)** Find the location and value of the maximum of $|\cos z|$ in the unit disk $|z| \leq 1$.

14. **(Conformal mapping)** Find the location (u_1, v_1) of the maximum of $\Phi^* = e^u \cos v$ in D^*: $|w| \leq 1$, $v \geq 0$, where $w = u + iv$. Find the region D that is mapped onto D^* by $w = f(z) = z^2$. Find the potential in D resulting from Φ^* and the location (x_1, y_1) of the maximum. Is (u_1, v_1) the image of (x_1, y_1)? If so, is this just by chance?

15. **(Conjugate)** Do Φ and a conjugate harmonic Ψ of Φ in a region D have their maximum at the same point of D? Prove or disprove.

CHAPTER 16 REVIEW

1. Why can two-dimensional potential problems be investigated and solved by methods of complex analysis?

2. What methods of complex analysis did we apply to potential problems?

3. Could three-dimensional potential problems be treated by complex analysis?

4. What fields of physics did we consider? Can you think of others?

5. What is a harmonic function? A harmonic conjugate?

6. What is the maximum modulus theorem? The maximum principle?

7. In some applications, equipotential lines were more important than their orthogonal trajectories; in other applications the situation was reversed. Explain.

8. Make a list of important potential functions, with applications (from memory).

9. What is the complex potential? Why is it important?

10. Write a short essay on potential theory in fluid flow (from memory).

11. What is the Dirichlet problem? Why did it occur in this chapter?

12. Why is steady-state heat flow related to potential theory?

13. What use of conformal mapping did we make in potential theory?

14. What is a mixed boundary value problem? Where did it occur?

15. List some remarkable properties common to all harmonic functions.

16. Find the potential and complex potential between the plates $y = x$ and $y = x + 10$ kept at 20 V and 220 V, respectively.

17. Find the potential between two coaxial cylinders of radii 1 and 10 kept at 100 V and 1000 V, respectively.

18. Find and sketch the equipotential lines of $F(z) = (1 + i)/z$.

19. Find the equipotential lines of $F(z) = i \operatorname{Ln} z$.

20. State the theorem on the behavior of harmonic functions under conformal mapping. Verify it for $\Phi^* = e^u \sin v$ and $w = z^2$.

21. Find the potential in the first quadrant of the xy-plane if the x-axis has potential 100 V and the y-axis is grounded.

22. Interpret Prob. 21 as a problem in heat conduction.

23. Find the temperature T in the upper half-plane if on the x-axis, $T = 20°C$ for $x > 1$ and $-20°C$ for $x < 1$.

24. If the region between two concentric cylinders of radii 2 cm and 10 cm contains water and the outer cylinder is kept at 20°C, to what temperature must we heat the inner cylinder in order to have 30°C at distance 5 cm from the axis?

25. Find the temperature and the complex potential in an infinite plate with edges $y = x - 2$ and $y = x + 2$ kept at $-10°C$ and 20°C, respectively.

26. Find the streamlines and the velocity for the complex potential $F(z) = (1 + i)z$. Describe the flow.

27. Describe the streamlines for $F(z) = \frac{1}{2}z^2 + z$.

28. Find V in Prob. 27 and verify that it gives tangent vectors to the streamlines.

29. What are the streamlines if $F(z) = i/z$?

30. What is the complex potential of a flow around a cylinder of radius 2 without circulation?

31. What is the complex potential of a source at $z = 2$? What are the streamlines?

32. Find the temperature in the unit disk $|z| \leq 1$ in the form of an infinite series if the left semicircle of $|z| = 1$ has the temperature of 100°C and the right semicircle the temperature zero.

33. Same task as in Prob. 32 if the upper semicircle is at 20°C and the lower at zero.

34. Find the stagnation points of the flow with $F(z) = z^2 + 1/z^2$.

35. Find a series for the potential in the unit disk with boundary values -1 if $-\pi < \theta < 0$ and 1 if $0 < \theta < \pi$.

SUMMARY OF CHAPTER 16
COMPLEX ANALYSIS APPLIED TO POTENTIAL THEORY

Potential theory is the theory of solutions of **Laplace's equation**

$$(1) \qquad\qquad\qquad \nabla^2 \Phi = 0.$$

Solutions whose second partial derivatives are *continuous* are called **harmonic functions.** Equation (1) is the most important partial differential equation in physics, where it is of interest in two and three dimensions. It appears in electrostatics (Sec. 16.1), steady-state heat problems (Sec. 16.3), fluid flow (Sec. 16.4), gravity, etc. Whereas the three-dimensional case requires other methods (see Chap. 11), two-dimensional potential theory can be handled by complex analysis, since the real and imaginary parts of an analytic function are harmonic (Sec. 12.4). They remain harmonic under conformal mapping (Sec. 16.2), so that **conformal mapping** becomes a powerful tool in solving boundary value problems for (1), as is illustrated in this chapter. With a real potential Φ in (1) we can associate a **complex potential**

$$(2) \qquad\qquad\qquad F(z) = \Phi + i\Psi \qquad\qquad (\text{Sec. 16.1}).$$

Then both families of curves $\Phi = const$ and $\Psi = const$ have a physical meaning. In electrostatics, they are equipotential lines and lines of electrical force (Sec. 16.1). In heat problems, they are isotherms (curves of constant temperature) and lines of heat flow (Sec. 16.3). In fluid flow, they are equipotential lines of the velocity potential and streamlines (Sec. 16.4).

For the disk, the solution of the Dirichlet problem is given by the **Poisson formula** (Sec. 16.5) or by a series that on the boundary circle becomes the Fourier series of the given boundary values (Sec. 16.5).

Harmonic functions, like analytic functions, have a number of general properties; particularly important are the **mean value property** and the **maximum modulus property** (Sec. 16.6), which implies the uniqueness of the solution of the Dirichlet problem (Theorem 5 in Sec. 16.6).

P A R T   E

Numerical Methods

Chapter 17 **Numerical Methods in General**

Chapter 18 **Numerical Methods in Linear Algebra**

Chapter 19 **Numerical Methods for Differential Equations**

No other field of mathematics has shown a recent increase in importance to the engineer comparable to that of numerical methods, nor has any other field developed as rapidly. Of course, the main reason for this evolution is the development of various computers, from the personal computer to the super computer, and we can see no end to it. Each major change in computer architecture and advancement in software engineering invites new research in numerical analysis. Small improvements in algorithms may have a great impact on time, storage demand, accuracy, and stability, which in turn lead to further development of well-structured numerical software.

Chapters $17-19$ concern the study and application of **numerical methods,** which provide the transition from the mathematical model of a problem (the equations or functions obtained in calculus or algebra, etc.) to an **algorithm** that we can program (or use directly on a calculator) to obtain the solution of the problem in the form of numbers or graphs. This includes the investigation of the range of applicability of numerical methods and their error analysis, stability, and properties in general.

We begin with numerical methods of a general nature in Chap. 17. In Chap. 18 we discuss numerical methods for problems in linear algebra, in particular, the solution of linear systems of equations and matrix eigenvalue problems. Chapter 19 is devoted to the numerical solution of ordinary and partial differential equations.

We give the algorithms in a form that seems best for showing how a method works and how to program it, even with little experience. The student is encouraged to program the given algorithms and try them out on the computer.[1] We also recommend strongly that the student make use of programs from public-domain or commercial software.

---

[1] No actual programs, FORTRAN, C, C^{++} or other, are given, because, in our experience, this could encourage some students to generate results without fully understanding the underlying numerical method.

Software

See also http://www.wiley.com/college/mat/kreyszig154962/

The following list will help you if you wish to obtain software. You may also obtain information on known and new software from magazines, such as *Byte Magazine* or *PC Magazine,* from articles published by the *American Mathematical Society* (see also their website at http://www.ams.org) or the *Society for Industrial and Applied Mathematics* (*SIAM*), and guidance to further literature from your library, Computer Science Department, or Mathematics Department.

DERIVE. Soft Warehouse, Inc., Honolulu, HI. Phone (808) 734-5801, website at http://www.derive.com.

EISPACK. See LAPACK.

GAMS (Guide to Available Mathematical Software). Website at http://gams.nist.gov. On-line cross-index of software development by NIST, with links to IMSL, NAG, and NETLIB.

IMSL (International Mathematical and Statistical Library). Visual Numerics, Inc., Houston, TX. Phone (713) 784-3131, website at http://www.vni.com. Mathematical and statistical FORTRAN subroutines with graphics.

LAPACK. FORTRAN 77 subroutines for linear algebra. This software package supersedes LINPACK and EISPACK. You can download the routines yourself from netlib@research.att.com or order them directly from NAG. The LAPACK User's Guide is available at http://www.netlib.com.

LINPACK see LAPACK.

MACSYMA. Macsyma, Inc., Arlington, MA. Phone 1-800-622-7962 or (617) 646-4550, website at http://www.macsyma.com.

MAPLE. Waterloo Maple, Inc., Waterloo, ON, Canada. Phone 1-800-267-6583 or (519) 747-2373. (Note that there is a MAPLE COMPUTER GUIDE to accompany this book. To order, call Wiley at 1-800-225-5945.)

MATHCAD. MathSoft, Inc., Cambridge, MA. Phone (617) 577-1017. Website at http://www.mathcad.com.

MATHEMATICA. Wolfram Research, Inc., Champaign, IL. Phone 1-800-965-3726 or (217) 398-0700. (Note that there is a MATHEMATICA COMPUTER GUIDE to accompany this book. To order, call Wiley at 1-800-225-5945.)

MATLAB. The Math Works, Inc., Natick, MA. Phone (508) 647-7000. Website at http://www.mathworks.com.

NAG. Numerical Algorithms Group, Inc., Downders Grove, IL. Phone (630) 971-2337, website at http://www.nag.com. Numerical routines in FORTRAN 77 and C.

NETLIB. Extensive library of public-domain software. Information on the Internet at netlib@research.att.com.

NIST. National Institute of Standards and Technology, Gaithersburg, MD. Phone (301) 975-2000, website at http://www.nist.gov. For Mathematical and Computational Science Division phone (301) 975-3800; see also http://math.nist.gov.

NUMERICAL RECIPES. Cambridge University Press, New York, NY. Phone (212) 924-3900. Books (also source codes on diskettes) containing numerical routines in C and FORTRAN. To order, call office at Port Chester, NY at 1-800-872-7423 or (212) 937-9600.

Numerical Methods in General

Numerical methods are methods for solving problems **numerically** (that is, in terms of numbers) on a **computer** or **calculator** (or in older times by hand). The computer has become very important in engineering work. It provides access to problems so large that they were out of reach in precomputer times. Much computing today is *"real-time";* it is done almost simultaneously with the process of generating data, for instance, in controlling ongoing chemical processes or guiding airplanes. Issues of speed, storage demand, and timing of portions of long programs then become very crucial.

Computers have changed, almost revolutionized, numerical methods—the field as a whole as well as many individual methods—and that development is continuing. Much research work is going on in creating new methods, adapting existing methods to new generations of computers, improving methods—in large-scale work even small improvements bring large savings in time or storage space—and investigating stability and accuracy of methods.

The purpose of this chapter is twofold. First, for the most important practical tasks, including solution of equations, interpolation, integration, and differentiation, the student should become familiar with the most basic (but not too complicated) numerical solution methods.[2] Such methods are needed because for many problems there is no solution formula (think of a complicated integral or of the roots of a polynomial of high degree) or in other cases a solution formula may be practically useless.

Second, the student should learn to understand some basic ideas and concepts that are important throughout the field, such as the idea of an algorithm, rounding errors, error estimation in general, ill-conditioning, order of convergence, and stability.

In the first section we explain some concepts that are basic in numerical work; this includes remarks on computing. Each of the other sections of the chapter is devoted to methods for one of the specific tasks already mentioned. These tasks are important throughout applied mathematics, regardless of the particular field of application.

Prerequisite for this chapter: Elementary calculus.
References: Appendix 1, Part E.
Answers to problems: Appendix 2.

---

[2]This will be continued with those for numerical linear algebra and differential equations in Chaps. 18 and 19.

17.1 Introduction

Numerical methods are used to solve problems on computers or calculators by numerical calculations, giving a table of numbers and/or graphical representations (figures). The steps from a given situation (in engineering, economics, etc.) to the final answer are usually as follows.

1. Modeling. We set up a mathematical model of the problem, such as an integral, a system of equations, or a differential equation.

2. Choice of mathematical methods, perhaps together with a preliminary error estimation, a choice of step sizes, etc.

3. Programming. From an algorithm we write a program, say, in FORTRAN, C, or C^{++}, and/or select suitable routines from a software system. Or we may decide to use a computing environment, such as MATHEMATICA, MAPLE, MATLAB, or MATHCAD.

4. Doing the computation.

5. Interpretation of results in physical or other terms, including decisions to rerun if further results are needed.

Steps 1 and 2 are related. A slight change of the model may often admit a more efficient method. To choose methods, we must first get to know them. Chapters $17-19$ contain efficient algorithms for the most important classes of problems occurring frequently in practice.

In Step 3 the program consists of the given data and a sequence of instructions to be executed by the computer in a certain order and eventually resulting in the answer in numerical or graphical form.

To create good understanding of the nature of numerical work, we continue in this section with some simple remarks on computing in general.

Floating-Point Form of Numbers

In the decimal notation, every real number is represented by a finite or infinite sequence of decimal digits. For machine computation the number must be replaced by a number of finitely many digits. Most digital computers have two ways of representing numbers, called *fixed point* and *floating point*. In a **fixed-point** system all numbers are given with a *fixed number of decimal places,* for example, 62.358, 0.013, 1.000. Fixed-point representations are impractical in most scientific computations because of their limited range (explain!) and will not concern us.

In a **floating-point** system used in computations, the number of significant digits is kept fixed (whereas the decimal point is "floating," as seen from the exponent); for instance,

$$0.6238 \times 10^3 \qquad 0.1714 \times 10^{-13} \qquad -0.2000 \times 10^1$$

also written[3]

$$0.6238E03 \qquad 0.1714E-13 \qquad -0.2000E01.$$

Significant digit of a number c is any given digit of c, except possibly for zeros to the left of the first nonzero digit that serve only to fix the position of the decimal point.

---

[3]One also uses a representation of the form 6.238×10^2, 1.1714×10^{-14}, etc.

(Thus, any other zero is a significant digit of c.) For instance, each of the numbers 1360, 1.360, 0.001 360 has 4 significant digits.[4]

Theoretically we can represent any nonzero number a as

$$(1) \qquad\qquad a = \pm m \cdot 10^e, \qquad 0.1 \leqq m < 1, \qquad e \text{ integer.}$$

On the computer, m is limited to t digits (e.g., $t = 8$) and e is limited, giving representations (for finitely many numbers only!)

$$(2) \qquad \bar{a} = \pm \bar{m} \cdot 10^e, \qquad \bar{m} = 0.d_1 d_2 \cdots d_t, \qquad d_1 > 0, \qquad |e| < M.$$

The fractional part m (or $\bar{m}$) is called the **mantissa**[5] and e is called the **exponent.**

Underflow and Overflow. The range of exponents that a typical computer can handle is very large. The IEEE (Institute of Electrical and Electronical Engineering) floating point standard for **single precision** (the usual number of digits in calculations) is about $-38 < e < 38$ (about $-125 < e^* < 125$ for the exponent in **binary representations,** i.e., representations in base 2). [For so-called **double precision** it is about $-308 < e < 308$ (about $-1020 < e^* < 1020$ for binary).] If in a computation a number outside that range occurs, this is called **underflow** when the number is smaller and **overflow** when it is larger. In the case of underflow the result is usually set to zero and computation continues. Overflow causes the computer to halt. Standard codes (by IMSL, NAG, etc.) are written to avoid overflow. Error messages on overflow may then indicate programming errors (incorrect input data, etc.).

Round-Off

An error is caused by **chopping** (= discarding all decimals from some decimal on) or **rounding.** This error is called **rounding error,** regardless of whether we chop or round. The rule for rounding off a number to k decimals is as follows. (The rule for rounding off to k significant figures is the same, with "decimal" replaced by "significant digit.")

Round-off rule. Discard the $(k + 1)$th and all subsequent decimals. (a) If the number thus discarded is less than half a unit in the kth place, leave the kth decimal unchanged (*"rounding down"*). (b) If it is greater than half a unit in the kth place, add one to the kth decimal (*"rounding up"*). (c) If it is exactly half a unit, round off to the nearest *even* decimal. (Example: Rounding off 3.45 and 3.55 to 1 decimal gives 3.4 and 3.6, respectively.)

The last part of the rule is supposed to ensure that in discarding exactly half a decimal, rounding up and rounding down happens about equally often, on the average.

If we round off 1.2535 to 3, 2, 1 decimals, we get 1.254, 1.25, 1.3, but if 1.25 is rounded off to one decimal, without further information, we get 1.2.

---

[4]In tables of functions showing k significant digits, it is conventionally assumed that any given value $\tilde{a}$ deviates from the corresponding exact value a by at most ± 0.5 unit of the last given digit, unless otherwise stated; for example, if $a = 1.1996$, then a table with 4 significant digits should show $\tilde{a} = 1.200$. Correspondingly, if 12 000 is correct to three digits only, we should write 120×10^2, etc. "Decimal" is abbreviated by D and "significant digit" by S. For example, 5D means 5 decimals, and 8S means 8 significant digits.

[5]This has nothing to do with "mantissa" as used in connection with logarithms.

Chopping is not recommended because the corresponding error can be larger than that in rounding, and is systematic. (Nevertheless, some computers use it because it is simpler and faster. On the other hand, some computers and calculators improve accuracy of results by doing intermediate calculations using one or more extra digits, called *guarding digits*.)

Error in rounding. Let $\bar{a} = fl(a)$ in (2) be the floating point computer approximation of a in (1) obtained by round-off, where fl suggests floating. Then the round-off rule gives (by dropping exponents) $|m - \bar{m}| \leq \frac{1}{2} \cdot 10^{-t}$. Since $|m| \geq 0.1$, this implies (when $a \neq 0$)

$$(3) \qquad \left| \frac{a - \bar{a}}{a} \right| \approx \left| \frac{m - \bar{m}}{m} \right| \leq \frac{1}{2} \cdot 10^{1-t}.$$

The right side $u = \frac{1}{2} \cdot 10^{1-t}$ is called the **rounding unit.** If we write $\bar{a} = a(1 + \delta)$, we have by algebra $(\bar{a} - a)/a = \delta$, hence $|\delta| \leq u$ by (3). *This shows that the rounding unit u is an error bound in rounding.*

Rounding errors may ruin a computation completely, even a small computation. In general, these errors become the more dangerous the more arithmetic operations (perhaps several millions!) we have to perform. It is therefore important to analyze computational programs for expected rounding errors and to find an arrangement of the computations such that the effect of rounding errors is as small as possible.

Algorithm. Stability

An **algorithm** is a finite sequence of rules for performing computations on a computer such that at each instant the rules determine exactly what the computer has to do next. These rules include a *"stopping rule"* that makes the computer stop, so it cannot run on indefinitely. Important algorithms follow in the next sections.

Stability. To be useful, an algorithm should be **stable;** that is, small changes in the initial data should give only correspondingly small changes in the final results. However, if small changes in the initial data produce large changes in the final results, we call the algorithm **unstable.**

This *"numerical instability,"* which can be avoided by choosing a better algorithm, must be distinguished from *"mathematical instability"* of a problem, which is called *"ill-conditioning,"* a concept we discuss in the next section.

Some algorithms are stable only for certain initial data, so that one must be careful in such a case.

Programming Errors

A common name for all kinds of errors in a computer program is **bugs,** and **debugging** means locating and removing bugs. Programming requires experience. Experience cannot be *taught* but must be *gained*. Nevertheless, a few general hints may be in order.

Prepare your program with the greatest care—it is easier to avoid errors this way than it is to discover them later. There are no general rules guaranteeing the detection of all bugs in all programs. Compilers provide *diagnostics* indicating all errors in a source program, except for errors in logic. Perhaps the best way to determine whether a program has bugs is to run it with data for which the answers are known or can be easily obtained in some other way.

If you are convinced that bugs exist because of nonsensical results but tests fail to actually find the bugs, apply selective (or even full) **tracing,** that is, printing out intermediate results and checking them over step by step.

Errors of Numerical Results

Final results of computations of unknown quantities generally are **approximations;** that is, they are not exact but involve errors. Such an error may result from a combination of the following effects. **Round-off errors** result from rounding, as discussed above. **Experimental errors** are errors of given data (probably arising from measurements). **Truncating errors** result from truncating (prematurely breaking off), for instance, if we replace a Taylor series by the sum of its first few terms. These errors depend on the computational method used and must be dealt with individually for each method. ["Truncating" is sometimes used as a term for chopping off (see before), a terminology that is not recommended.]

Formulas for errors. If $\tilde{a}$ is an approximate value of a quantity whose exact value is a, we call the difference

(4)
$$\epsilon = a - \tilde{a}$$

the **error** of $\tilde{a}$. Hence[6]

(4*)
$$a = \tilde{a} + \epsilon, \qquad \text{True value} = \text{Approximation} + \text{Error}.$$

For instance, if $\tilde{a} = 10.5$ is an approximation of $a = 10.2$, its error is $\epsilon = -0.3$. The error of an approximation $\tilde{a} = 1.60$ of $a = 1.82$ is $\epsilon = 0.22$.

The **relative error** ϵ_r of $\tilde{a}$ is defined by

(5)
$$\epsilon_r = \frac{\epsilon}{a} = \frac{a - \tilde{a}}{a} = \frac{\text{Error}}{\text{True value}} \qquad (a \neq 0).$$

This looks useless because a is unknown. But if $|\epsilon|$ is much less than $|\tilde{a}|$, then we can use $\tilde{a}$ instead of a and get

(5')
$$\epsilon_r \approx \frac{\epsilon}{\tilde{a}}.$$

This still looks problematic because ϵ is unknown—if it were known, we could get $a = \tilde{a} + \epsilon$ from (4) and we would be done. But what one often can obtain in practice is an **error bound** for $\tilde{a}$, that is, a number β such that

$$|\epsilon| \leq \beta, \qquad \text{hence} \qquad |a - \tilde{a}| \leq \beta.$$

This tells us how far away from our computed $\tilde{a}$ the unknown a can at most lie. Similarly, for the relative error, an error bound is a number β_r such that

$$|\epsilon_r| \leq \beta_r, \qquad \text{hence} \qquad \left| \frac{a - \tilde{a}}{a} \right| \leq \beta_r.$$

---

[6]This resembles standard situations in analysis, such as Series = Partial sum + Remainder, Integral = Riemann sum + Error, etc. However, in the literature, $\tilde{a} - a$ is sometimes also used as a definition of error.

Error Propagation

This is an important matter. It refers to how errors at the beginning and in later steps (round-off, for example) propagate into the computation and affect accuracy, sometimes very drastically. We state here what happens to error bounds. Namely, bounds for the *error* add under addition and subtraction, whereas bounds for the *relative error* add under multiplication and division. You do well to keep this in mind.

THEOREM 1 **(Error propagation)**

(**a**) *In addition and subtraction, an error bound for the results is given by the sum of the error bounds for the terms.*

(**b**) *In multiplication and division, a bound for the relative error of the results is given (approximately) by the sum of the bounds for the relative errors of the given numbers.*

PROOF. (**a**) We use the notations $x = \tilde{x} + \epsilon_1$, $y = \tilde{y} + \epsilon_2$, $|\epsilon_1| \leqq \beta_1$, $|\epsilon_2| \leqq \beta_2$. Then for the error ϵ of the *difference* we get

$$|\epsilon| = |x - y - (\tilde{x} - \tilde{y})|$$

$$= |x - \tilde{x} - (y - \tilde{y})|$$

$$= |\epsilon_1 - \epsilon_2| \leqq |\epsilon_1| + |\epsilon_2| \leqq \beta_1 + \beta_2.$$

The proof for the *sum* is similar and is left to the student.

(**b**) For the relative error ϵ_r of $\tilde{x}\tilde{y}$ we get from the relative errors ϵ_{r1} and ϵ_{r2} of $\tilde{x}$, $\tilde{y}$ and bounds β_{r1}, β_{r2}

$$|\epsilon_r| = \left| \frac{xy - \tilde{x}\tilde{y}}{xy} \right| = \left| \frac{xy - (x - \epsilon_1)(y - \epsilon_2)}{xy} \right| = \left| \frac{\epsilon_1 y + \epsilon_2 x - \epsilon_1 \epsilon_2}{xy} \right|$$

$$\approx \left| \frac{\epsilon_1 y + \epsilon_2 x}{xy} \right| \leqq \left| \frac{\epsilon_1}{x} \right| + \left| \frac{\epsilon_2}{y} \right| = |\epsilon_{r1}| + |\epsilon_{r2}| \leqq \beta_{r1} + \beta_{r2}.$$

This proof shows what "approximately" means: we neglected $\epsilon_1 \epsilon_2$ as small in absolute value compared to $|\epsilon_1|$ and $|\epsilon_2|$. The proof for the quotient is similar but slightly more tricky (see Prob. 19). ◀

Basic Error Principle

Every numerical method should be accompanied by an error estimate. If such a formula is lacking, is extremely complicated, or is impractical because it involves information (for instance, on derivatives) that is not available, the following may help.

Error Estimation by Comparison. *Do a calculation twice with different accuracy. Regard the difference $\tilde{a}_2 - \tilde{a}_1$ of the results $\tilde{a}_1$, $\tilde{a}_2$ as a (perhaps crude) estimate of the error ϵ_1 of the inferior result $\tilde{a}_1$. Indeed, $\tilde{a}_1 + \epsilon_1 = \tilde{a}_2 + \epsilon_2$ by (4*), hence $\tilde{a}_2 - \tilde{a}_1 = \epsilon_1 - \epsilon_2 \approx \epsilon_1$ because $\tilde{a}_2$ is generally more accurate than $\tilde{a}_1$, so that $|\epsilon_2|$ is small, compared to $|\epsilon_1|$.*

Loss of Significant Digits

This means that a result of a calculation has fewer correct digits than the numbers from which it was obtained. This happens if we subtract two numbers of about the same size, for example, $0.1439 - 0.1426$ ("subtractive cancellation"). It may occur in simple problems, but it can be avoided in most cases by simple changes of the algorithm—if one is aware of it! Let us illustrate this with the following basic problem.

EXAMPLE 1   **Quadratic equation. Loss of significant digits**

Find the roots of the equation

$$x^2 - 40x + 2 = 0,$$

using 4 significant digits in the computation.

Solution.   A formula for the roots x_1, x_2 of a quadratic equation $ax^2 + bx + c = 0$ is

$$(6) \qquad x_1 = \frac{1}{2a}\left(-b + \sqrt{b^2 - 4ac}\right), \qquad x_2 = \frac{1}{2a}\left(-b - \sqrt{b^2 - 4ac}\right).$$

Furthermore, since $x_1 x_2 = c/a$, another formula for those roots is

$$(7) \qquad x_1 \text{ as before}, \qquad x_2 = \frac{c}{ax_1}.$$

From (6) we obtain $x = 20 \pm \sqrt{398} = 20.00 \pm 19.95$. This gives $x_1 = 20.00 + 19.95 = 39.95$, involving no difficulty, whereas $x_2 = 20.00 - 19.95 = 0.05$ is poor because it involves loss of significant digits.

In contrast, (7) gives $x_1 = 39.95$, $x_2 = 2.000/39.95 = 0.05006$, in error by less than one unit of the last digit, as a computation with more digits shows.

Comment.   To avoid misunderstandings: 4S was used for convenience; (7) is better than (6) regardless of the number of digits used. For instance, the 8S-computation by (6) is $x_1 = 39.949\ 937$, $x_2 = 0.050\ 063$, which is poor, and by (7) it is x_1 as before, $x_2 = 2/x_1 = 0.050\ 062\ 657$.

In a quadratic equation with real roots, if x_2 is absolutely largest (because $b > 0$), use (6) for x_2 and then $x_1 = c/ax_2$.   ◀

PROBLEM SET 17.1

1. **(Floating point)** Write 23.49, -302.867, 0.000527532, -25700 in floating-point form, rounded to 4S (4 significant digits).

2. Write -89.216618, 500000, -0.002213675 in floating-point form, rounded to 5S.

3. **Small differences of large numbers** may be particularly strongly affected by rounding errors. Illustrate this by computing $0.81534/(35.724 - 35.596)$ as given with 5S, then rounding stepwise to 4S, 3S, and 2S, where 'stepwise' means: round the rounded numbers, not the given ones.

4. Do the work in Prob. 3 with numbers of your choice that give even more drastically differing results.

5. Write the quotient $a/(b - c)$ in Prob. 3 as $a(b + c)/(b^2 - c^2)$. Compute it first with 5S, then round the numerator 58.150 and the denominator 9.1290 stepwise as in Prob. 3. Compare and comment.

6. **(Quadratic equation)** Solve $x^2 - 30x + 1 = 0$ by (6) and by (7), using 6S in the computation. Compare and comment.

7. Do the computation in Prob. 6 with 4S and with 2S.

8. Solve $x^2 + 100x + 2 = 0$ by (6) and by (7) with 5S.

9. **(Change of formula)** How can we get good values of $\sqrt{9 + x^2} - 3$ if $|x|$ is small?

10. What is a good way to compute $\cos a - \cos b$ if a and b are nearly equal?

11. How can $\log a - \log b = \log (a/b)$ and $e^{a-b} = e^a/e^b$ be used to avoid loss of significant digits?

12. **Approximations to** $\pi = 3.141\ 592\ 653\ 589\ 79 \cdots$ are 22/7 and 355/113. Determine the corresponding errors and relative errors to 3S.

13. Compute π by Machin's approximation 16 arc tan (1/5) − 4 arc tan (1/239) to 10S. [Note that these digits are correct! For the first 100 000 digits of π, see D. Shanks and J. W. Wrench, *Mathematics of Computation* **16** (1962), pp. 76−99.]

14. **(Error of sum)** Let 65.43 and 17.0591 be correctly rounded to the number of digits shown. What is the smallest interval in which the exact sum of the numbers must lie?

15. **Addition** with a fixed number of significant digits generally depends on the order in which we add the numbers. Illustrate this with an example. Find an empirical rule for what is best.

16. What is the smallest interval in which the area of the surface of a rectangular box of edge lengths 10, 20, 30 (rounded to integer cm) must lie?

17. **(Rounding and adding)** Let $a_1, \cdots, a_n$ be numbers with a_j correctly rounded to D_j decimals. In calculating the sum $a_1 + \cdots + a_n$, retaining $D = \min D_j$ decimals, is it essential that we first add and then round the result or that we first round each number to D decimals and then add?

18. **(Theorems on errors)** Prove Theorem 1(a) for addition.

19. **(Quotient)** Prove Theorem 1(b) for division.

20. Show that in Example 1 the absolute value of the error of $x_2 = 2.000/39.95 = 0.05006$ is less than 0.00001.

21. Compute a two-decimal table[7] of $f(x) = x/16$, $x = 0(1)20$ and find out how the rounding error is distributed.

22. **(Nested form)** Evaluate $f(x) = x^3 - 7.5x^2 + 11.2x + 2.8 = ((x - 7.5)x + 11.2)x + 2.8$ at $x = 3.94$ using 3S arithmetic and rounding, in both of the given forms. The latter, called the *nested form*, is usually preferable since it minimizes the number of operations and thus the effect of rounding.

23. **Overflow and underflow** can sometimes be avoided by simple changes in a formula. Explain this in terms of $\sqrt{x^2 + y^2} = x\sqrt{1 + (y/x)^2}$ with $x^2 \geqq y^2$ and x so large that x^2 would cause overflow. Invent examples of your own.

24. **WRITING PROJECT. Numerical Analysis.** In your own words write about the overall role of numerical analysis in applied mathematics, why it is important, where and when it must be used or can be used, and how it is being influenced by the use of the computer in engineering and other work.

25. **CAS PROJECT. Chopping and Rounding.** (a) Let $x = 4/7$ and $y = 1/3$. Find the errors ϵ_{chop}, ϵ_{round} and the relative errors $\epsilon_{r,ch}$, $\epsilon_{r,rd}$ of $x + y$, $x - y$, xy, x/y in chopping and rounding to 5S.

   (b) Plot ϵ_{chop} and ϵ_{round} (for 5S) of $x = k/21$ as a function of $k = 1, 2, \cdots, 21$ on common axes. What average value can you read from the plot for ϵ_{chop}? For ϵ_{round}? What other integers (instead of 21) would give similar plots? Different types of plot? Can you characterize the different types in terms of prime factors?

   (c) How does the situation in (b) change if you take 4S instead of 5S?

   (d) Write programs for the work in (a)−(c).

---

[7]The notation $x = a(h)b$ means $x = a, a + h, a + 2h, \cdots, b$.

17.2 Solution of Equations by Iteration

From here on, each section will be devoted to some basic kind of problem and corresponding solution methods. We begin with methods of finding solutions of a single equation

(1)
$$f(x) = 0.$$

For this task there are practically no formulas (except in a few simple cases), so that one depends almost entirely on numerical algorithms. f in (1) is a given function. A **solution** of (1) is a number $x = s$ such that $f(s) = 0$. Here, s suggests "solution," but we shall also use other letters.

Examples are $x^2 - 3x + 2 = 0$, $x^3 + x = 1$, $\sin x = 0.5x$, $\tan x = x$, $\cosh x = \sec x$, $\cosh x \cos x = -1$, which can all be written in the form (1). The first two are **algebraic equations** because the corresponding f is a polynomial, and in this case the solutions are also called **roots** of the equations. The other equations are **transcendental equations** because they involve transcendental functions. Solving equations (1) is a task of prime importance because engineering applications abound: some occur in Chaps. 2, 3, 7 (characteristic equations), 5 (partial fractions), 11 (eigenvalues, zeros of Bessel functions), and 15 (integration), but there are many, many others.

To solve (1) when there is no formula for the exact solution, we can use an approximation method, in particular an **iteration method,** that is, a method in which we start from an initial guess x_0 (which may be poor) and compute step by step (in general better and better) approximations $x_1, x_2, \cdots$ of an unknown solution of (1). We discuss three such methods that are of particular practical importance and mention two others in the problem set.

In general, iteration methods are easy to program because the computational operations are the same in each step—just the data change from step to step—and, more important, if in a concrete case a method converges, it is stable (see Sec. 17.1) in general. (As a difficulty, it may sometimes not be too easy to decide when to terminate an iteration, as we shall see.)

Fixed-Point Iteration[8] for Solving Equations $f(x) = 0$

We transform (1) *algebraically* into the form

(2)
$$x = g(x).$$

Then we choose an x_0 and compute $x_1 = g(x_0)$, $x_2 = g(x_1)$, and in general

(3)
$$x_{n+1} = g(x_n)$$
$(n = 0, 1, \cdots).$

A solution of (2) is called a **fixed point** of g, motivating the name of the method. This is a solution of (1), since from $x = g(x)$ we can return to the original form $f(x) = 0$. From (1) we may get several different forms of (2). The behavior of corresponding iterative

---

[8]Our present use of the word "fixed point" has absolutely nothing to do with that in the last section.

sequences $x_0, x_1, \cdots$ may differ, in particular, with respect to their speed of convergence. Indeed, some of them may not converge at all. Let us illustrate these facts with a simple example.

EXAMPLE 1   **An iteration process (Fixed-point iteration)**

Set up an iteration process for the equation $f(x) = x^2 - 3x + 1 = 0$. Since we know the solutions

$$x = 1.5 \pm \sqrt{1.25}, \quad \text{thus} \quad 2.618\,034 \quad \text{and} \quad 0.381\,966,$$

we can watch the behavior of the error as the iteration proceeds.

Solution.   The equation may be written

(4a) $x = g_1(x) = \tfrac{1}{3}(x^2 + 1),$ thus $x_{n+1} = \tfrac{1}{3}(x_n^2 + 1).$

If we choose $x_0 = 1$, we obtain the sequence (Fig. 394a; computed with 6S and then rounded)

$$x_0 = 1.000, \quad x_1 = 0.667, \quad x_2 = 0.481, \quad x_3 = 0.411, \quad x_4 = 0.390, \cdots$$

which seems to approach the smaller solution. If we choose $x_0 = 2$, the situation is similar. If we choose $x_0 = 3$, we obtain the sequence (Fig. 394a, upper part)

$$x_0 = 3.000, \quad x_1 = 3.333, \quad x_2 = 4.037, \quad x_3 = 5.766, \quad x_4 = 11.415, \cdots$$

which seems to diverge.

Our equation may also be written (divide by x)

(4b) $x = g_2(x) = 3 - \dfrac{1}{x},$ thus $x_{n+1} = 3 - \dfrac{1}{x_n},$

and if we choose $x_0 = 1$, we obtain the sequence (Fig. 394b)

$$x_0 = 1.000, \quad x_1 = 2.000, \quad x_2 = 2.500, \quad x_3 = 2.600, \quad x_4 = 2.615, \cdots$$

which seems to approach the larger solution. Similarly, if we choose $x_0 = 3$, we obtain the sequence (Fig. 394b)

$$x_0 = 3.000, \quad x_1 = 2.667, \quad x_2 = 2.625, \quad x_3 = 2.619, \quad x_4 = 2.618, \cdots.$$

Our figures show the following. In the lower part of Fig. 394a the slope of $g_1(x)$ is less than the slope of $y = x$, which is 1, thus $|g_1'(x)| < 1$, and we seem to have convergence. In the upper part, $g_1(x)$ is steeper

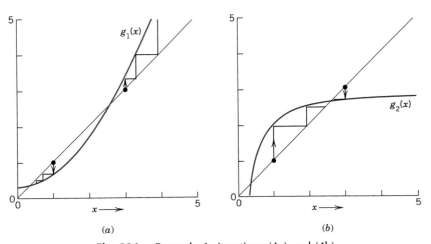

(a)   (b)

Fig. 394.   Example 1, iterations (4a) and (4b)

$(g_1'(x) > 1)$ and we have divergence. In Fig. 394b the slope of $g_2(x)$ is less near the intersection point ($x = 2.618$, fixed point of g_2, solution of $f(x) = 0$), and both sequences seem to converge. From all this we conclude that convergence seems to depend on the fact that in a neighborhood of a solution the curve of $g(x)$ is less steep than the straight line $y = x$, and we shall now see that this condition $|g'(x)| < 1$ ($= $ slope of $y = x$) is sufficient for convergence. ◄

An iteration process defined by (3) is called **convergent** for an x_0 if the corresponding sequence $x_0, x_1, \cdots$ is convergent.

A sufficient condition for convergence is given in the following theorem, which has various practical applications.

THEOREM 1   **(Convergence of fixed-point iteration)**

Let $x = s$ be a solution of $x = g(x)$ and suppose that g has a continuous derivative in some interval J containing s. Then if $|g'(x)| \leq K < 1$ in J, the iteration process defined by (3) converges for any x_0 in J.

PROOF.   By the mean value theorem of differential calculus there is a t between x and s such that

$$g(x) - g(s) = g'(t)(x - s) \qquad\qquad (x \text{ in } J).$$

Since $g(s) = s$ and $x_1 = g(x_0)$, $x_2 = g(x_1), \cdots$, we obtain from this and the condition on $|g'(x)|$ in the theorem

$$
\begin{aligned}
|x_n - s| &= |g(x_{n-1}) - g(s)| \\
&= |g'(t)|\,|x_{n-1} - s| \\
&\leq K|x_{n-1} - s| \\
&= K|g(x_{n-2}) - g(s)| \\
&= K|g'(\tilde{t})|\,|x_{n-2} - s| \\
&\leq K^2|x_{n-2} - s| \\
\cdots &\leq K^n|x_0 - s|.
\end{aligned}
$$

Since $K < 1$, we have $K^n \to 0$; hence $|x_n - s| \to 0$ as $n \to \infty$. ◄

We mention that a function g satisfying the condition in Theorem 1 is called a **contraction** because $|g(x) - g(v)| \leq K|x - v|$, where $K < 1$. Furthermore, K gives information on the speed of convergence. For instance, if $K = 0.5$, then the accuracy increases by at least 2 digits in only 7 steps because $0.5^7 < 0.01$.

EXAMPLE 2   **An iteration process. Illustration of Theorem 1**

Find a solution of $f(x) = x^3 + x - 1 = 0$ by iteration.

Solution.   A rough sketch shows that a real solution lies near $x = 1$. We may write the equation in the form

$$x = g_1(x) = \frac{1}{1 + x^2}, \qquad \text{so that} \qquad x_{n+1} = \frac{1}{1 + x_n^2}. \qquad \text{Also} \qquad |g_1'(x)| = \frac{2|x|}{(1 + x^2)^2} < 1$$

for any x because $4x^2/(1 + x^2)^4 = 4x^2/(1 + 4x^2 + \cdots) < 1$, so that we have convergence for any x_0. Choosing

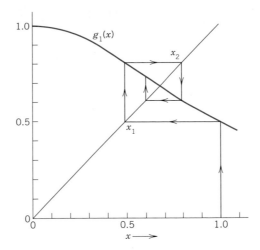

Fig. 395.   Iteration in Example 2

$x_0 = 1$, we obtain (Fig. 395)

$$x_1 = 0.500, \quad x_2 = 0.800, \quad x_3 = 0.610, \quad x_4 = 0.729, \quad x_5 = 0.653, \quad x_6 = 0.701, \cdots.$$

The solution exact to 6D is $s = 0.682\ 328$. The equation may also be written

$$x = g_2(x) = 1 - x^3. \qquad \text{Then} \qquad |g_2'(x)| = 3x^2$$

and this is greater than 1 near the solution, so that we cannot apply Theorem 1 and assert convergence. Try $x_0 = 1$, $x_0 = 0.5$, $x_0 = 2$ and see what happens.   ◀

Newton's Method for Solving Equations $f(x) = 0$

The **Newton method** or **Newton–Raphson**[9] **method** is another iteration method for solving equations $f(x) = 0$, where f is assumed to have a continuous derivative f'. The method is commonly used because of its simplicity and great speed. The underlying idea is that we approximate the graph of f by suitable tangents. Using an approximate value x_0 obtained from the graph of f, we let x_1 be the point of intersection of the x-axis and the tangent to the curve of f at x_0 (see Fig. 396). Then

$$\tan \beta = f'(x_0) = \frac{f(x_0)}{x_0 - x_1}, \qquad \text{hence} \qquad x_1 = x_0 - \frac{f(x_0)}{f'(x_0)}.$$

In the second step we compute $x_2 = x_1 - f(x_1)/f'(x_1)$, in the third step x_3 from x_2 again by the same formula, and so on. We thus have the algorithm shown in Table 17.1 on the next page. Formula (5) in this algorithm can also be obtained if we algebraically solve Taylor's formula

$$(5^*) \qquad f(x_{n+1}) \approx f(x_n) + (x_{n+1} - x_n)f'(x_n) = 0.$$

If it happens that $f'(x_n) = 0$ for some n (see line 2 of the algorithm), then try another starting value x_0. Line 3 is the heart of Newton's method.

---

[9]JOSEPH RAPHSON (1648—1715), English mathematician, who published a method similar to Newton's method. For historical details, see Ref. [2], p. 203, listed in Appendix 1.

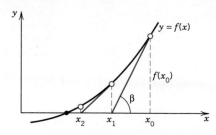

Fig. 396. Newton's method

The inequality in line 4 is a **termination criterion.** If the sequence of the x_n converges and the criterion holds, we have reached the desired accuracy and stop. In this line the factor $|x_n|$ is needed in the case of zeros of very small or very large absolute value.

Warning! The criterion by itself does not imply convergence. *Example.* The harmonic series diverges, although its partial sums $x_n = \sum_{k=1}^{n} 1/k$ satisfy the criterion because $\lim (x_{n+1} - x_n) = \lim (1/(n + 1)) = 0$.

Table 17.1
Newton's Method for Solving Equations $f(x) = 0$

ALGORITHM NEWTON (f, f', x_0, ϵ, N)

This algorithm computes a solution of $f(x) = 0$ given an initial approximation x_0 (starting value of the iteration). Here the function $f(x)$ is continuous and has a continuous derivative $f'(x)$.

INPUT: f, f', initial approximation x_0, tolerance $\epsilon > 0$, maximum number of iterations N.

OUTPUT: Approximate solution x_n ($n \leqq N$) or message of failure.

For $n = 0, 1, 2, \cdots, N - 1$ do:

1   Compute $f'(x_n)$.

2   If $f'(x_n) = 0$ then OUTPUT "Failure". Stop.

           [*Procedure completed unsuccessfully*]

3   Else compute

(5)
$$x_{n+1} = x_n - \frac{f(x_n)}{f'(x_n)}.$$

4   If $|x_{n+1} - x_n| \leqq \epsilon |x_n|$ then OUTPUT x_{n+1}. Stop.

           [*Procedure completed successfully*]

End

5   OUTPUT "Failure". Stop.

           [*Procedure completed unsuccessfully after N iterations*]

End NEWTON

Line 5 gives another termination criterion and is needed because Newton's method may diverge or, due to a poor choice of x_0, may not reach the desired accuracy by a reasonable number of iterations. Then we may try another x_0. If $f(x) = 0$ has more than one solution, different choices of x_0 may produce different solutions. Also, an iterative sequence may sometimes converge to a solution different from the expected one.

EXAMPLE 3   **Square root**

Set up a Newton iteration for computing the square root x of a given positive number c and apply it to $c = 2$.

Solution.   We have $x = \sqrt{c}$, hence $f(x) = x^2 - c = 0$, $f'(x) = 2x$, and (5) takes the form

$$x_{n+1} = x_n - \frac{x_n^2 - c}{2x_n} = \frac{1}{2}\left(x_n + \frac{c}{x_n}\right).$$

For $c = 2$, choosing $x_0 = 1$, we obtain

$$x_1 = 1.500\,000, \qquad x_2 = 1.416\,667, \qquad x_3 = 1.414\,216, \qquad x_4 = 1.414\,214, \cdots,$$

x_4 is exact to 6D.   ◀

EXAMPLE 4   **Iteration for a transcendental equation**

Find the positive solution of $2 \sin x = x$.

Solution.   Setting $f(x) = x - 2 \sin x$, we have $f'(x) = 1 - 2 \cos x$, and (5) gives

$$x_{n+1} = x_n - \frac{x_n - 2 \sin x_n}{1 - 2 \cos x_n} = \frac{2(\sin x_n - x_n \cos x_n)}{1 - 2 \cos x_n} = \frac{N_n}{D_n}.$$

From the graph of f we conclude that the solution is near $x_0 = 2$. We compute:

n	x_n	N_n	D_n	x_{n+1}
0	2.00000	3.48318	1.83229	1.90100
1	1.90100	3.12470	1.64847	1.89552
2	1.89552	3.10500	1.63809	1.89550
3	1.89550	3.10493	1.63806	1.89549

$x_4 = 1.89549$ is exact to 5D since the solution to 6D is 1.895 494.   ◀

EXAMPLE 5   **Newton's method applied to an algebraic equation**

Apply Newton's method to the equation $f(x) = x^3 + x - 1 = 0$.

Solution.   From (5) we have

$$x_{n+1} = x_n - \frac{x_n^3 + x_n - 1}{3x_n^2 + 1} = \frac{2x_n^3 + 1}{3x_n^2 + 1}.$$

Starting from $x_0 = 1$, we obtain

$$x_1 = 0.750\,000, \qquad x_2 = 0.686\,047, \qquad x_3 = 0.682\,340, \qquad x_4 = 0.682\,329, \cdots$$

where x_4 has the error $-1 \cdot 10^{-6}$. A comparison with Example 2 shows that the present convergence is much more rapid. This may motivate the concept of the *order of an iteration process,* to be discussed next.   ◀

Order of an Iteration Method. Speed of Convergence

We shall now see how we can characterize the quality of an iteration method by judging the speed of convergence, as follows.

Let $x_{n+1} = g(x_n)$ define an iteration method, and let x_n approximate a solution s of $x = g(x)$. Then $x_n = s - \epsilon_n$, where ϵ_n is the error of x_n. Suppose that g is differentiable a number of times, so that the Taylor formula gives

$$
\begin{aligned}
x_{n+1} = g(x_n) &= g(s) + g'(s)(x_n - s) + \tfrac{1}{2}g''(s)(x_n - s)^2 + \cdots \\
&= g(s) - g'(s)\epsilon_n + \tfrac{1}{2}g''(s)\epsilon_n^2 + \cdots .
\end{aligned}
$$

(6)

The exponent of ϵ_n in the first nonvanishing term after $g(s)$ is called the **order** of the iteration process defined by g. The order measures the speed of convergence.

To see this, subtract $g(s) = s$ on both sides of (6). Then on the left you get $x_{n+1} - s = -\epsilon_{n+1}$, where ϵ_{n+1} is the error of x_{n+1}. And on the right the remaining expression approximately equals its first nonzero term because $|\epsilon_n|$ is small in the case of convergence. Thus

(7)

$$
\begin{aligned}
&\text{(a)} \quad \epsilon_{n+1} \approx +g'(s)\epsilon_n && \text{in the case of first order,} \\
&\text{(b)} \quad \epsilon_{n+1} \approx -\tfrac{1}{2}g''(s)\epsilon_n^2 && \text{in the case of second order,} \quad \text{etc.}
\end{aligned}
$$

Thus if $\epsilon_n = 10^{-k}$ in some step, then for second order, $\epsilon_{n+1} = const \cdot 10^{-2k}$, so that the number of significant digits is about doubled in each step.

Convergence of Newton's Method

In Newton's method, $g(x) = x - f(x)/f'(x)$. By differentiation,

(8)
$$
g'(x) = 1 - \frac{f'(x)^2 - f(x)f''(x)}{f'(x)^2} = \frac{f(x)f''(x)}{f'(x)^2} .
$$

Since $f(s) = 0$, this shows that also $g'(s) = 0$. Hence Newton's method is at least of second order. If we differentiate again and set $x = s$, we find that

(8*)
$$
g''(s) = \frac{f''(s)}{f'(s)}
$$

which will not be zero in general. This proves

THEOREM 2   **(Second-order convergence of Newton's method)**

If $f(x)$ is three times differentiable and f' and f'' are not zero at a solution s of $f(x) = 0$, then for x_0 sufficiently close to s, Newton's method is of second order.

Comments. For Newton's method, (7b) becomes, by (8*),

(9)
$$
\epsilon_{n+1} \approx -\frac{f''(s)}{2f'(s)} \epsilon_n^2.
$$

For the rapid convergence of the method indicated in Theorem 2 it is important that s be a *simple* zero of $f(x)$ (thus $f'(s) \neq 0$) and that x_0 be close to s, because in Taylor's formula we took only the linear term [see (5*)], assuming the quadratic term to be negligibly small. (With a bad x_0 the method may even diverge!)

EXAMPLE 6 **Prior error estimate of the number of Newton iterations**

Use $x_0 = 2$ and $x_1 = 1.901$ in Example 4 for estimating how many iterations we need to produce the solution to 5D accuracy. This is an **a priori estimate** or **prior estimate** because we can compute it after only one iteration, prior to further iterations.

Solution.   We have $f(x) = x - 2 \sin x = 0$. Differentiation gives

$$\frac{f''(s)}{2f'(s)} \approx \frac{f''(x_1)}{2f'(x_1)} = \frac{2 \sin x_1}{2(1 - 2 \cos x_1)} \approx 0.57.$$

Hence (9) gives

$$|\epsilon_{n+1}| \approx 0.57 \epsilon_n^2 \approx 0.57^3 \epsilon_{n-1}^4 \approx \cdots \approx 0.57^M \epsilon_0^{M+1} \leqq 5 \cdot 10^{-6}$$

where $M = 2^n + 2^{n-1} + \cdots + 2 + 1 = 2^{n+1} - 1$. We show below that $\epsilon_0 \approx -0.11$. Consequently, our condition becomes

$$0.57^M 0.11^{M+1} \leqq 5 \cdot 10^{-6}.$$

Hence $n = 2$ is the smallest possible n, according to this crude estimate, in good agreement with Example 4.
 $\epsilon_0 \approx -0.11$ is obtained from $\epsilon_1 - \epsilon_0 = (\epsilon_1 - s) - (\epsilon_0 - s) = -x_1 + x_0 \approx 0.10$, hence $\epsilon_1 = \epsilon_0 + 0.10 \approx -0.57 \epsilon_0^2$ or $0.57 \epsilon_0^2 + \epsilon_0 + 0.10 \approx 0$, which gives $\epsilon_0 \approx -0.11$. ◀

Difficulties in Newton's Method. Difficulties may arise if $|f'(x)|$ is very small near a solution s of $f(x) = 0$, for instance, if s is a zero of $f(x)$ of second (or higher) order (so that Newton's method converges only linearly, as an application of l'Hôpital's rule to (8) shows). Geometrically, small $|f'(x)|$ means that the tangent of $f(x)$ near s almost coincides with the x-axis (so that double precision may be needed to get $f(x)$ and $f'(x)$ accurately enough). Then for values $x = \tilde{s}$ far away from s we can still have small function values

$$R(\tilde{s}) = f(\tilde{s}).$$

In this case we call the equation $f(x) = 0$ **ill-conditioned.** $R(\tilde{s})$ is called the **residual** of $f(x) = 0$ at $\tilde{s}$. Thus a small residual guarantees a small error of $\tilde{s}$ only if the equation is ***not*** ill-conditioned.

EXAMPLE 7 **An ill-conditioned equation.**

$f(x) = x^5 + 10^{-4} x = 0$ is ill-conditioned. $x = 0$ is a solution. $f'(0) = 10^{-4}$ is small. At $\tilde{s} = 0.1$ the residual $f(0.1) = 2 \cdot 10^{-5}$ is small, but the error -0.1 is larger in absolute value by a factor 5000. Invent a more drastic example of your own. ◀

Secant Method for Solving Equations $f(x) = 0$

Newton's method is very powerful, but the evaluation of the derivative involved may sometimes be difficult or expensive. This suggests the idea of replacing the derivative $f'(x_n)$ by the difference quotient

$$f'(x_n) \approx \frac{f(x_n) - f(x_{n-1})}{x_n - x_{n-1}}.$$

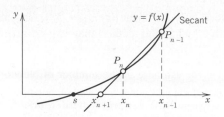

Fig. 397. Secant method

Then instead of (5) we have the popular **secant method**

(10)
$$x_{n+1} = x_n - f(x_n) \frac{x_n - x_{n-1}}{f(x_n) - f(x_{n-1})}.$$

Geometrically, we intersect the x-axis at x_{n+1} with the secant of $f(x)$ passing through P_{n-1} and P_n in Fig. 397. We need two starting values x_0 and x_1. Evaluation of derivatives is now avoided. It can be shown that convergence is **superlinear** (that is, more rapid than linear, $|\epsilon_{n+1}| \approx const \cdot |\epsilon_n|^{1.62}$), almost quadratic like Newton's method. The algorithm is similar to that of Newton's method, as the student may show.

It is **not** good to write (10) as

$$x_{n+1} = \frac{x_{n-1}f(x_n) - x_n f(x_{n-1})}{f(x_n) - f(x_{n-1})},$$

because this may lead to loss of significant digits if x_n and x_{n-1} are about equal. (Can you see this from the formula?)

EXAMPLE 8   **Secant method**

Find the positive solution of $f(x) = x - 2 \sin x = 0$ by the secant method, starting from $x_0 = 2$, $x_1 = 1.9$.

Solution.   Here, (10) is

$$x_{n+1} = x_n - \frac{(x_n - 2 \sin x_n)(x_n - x_{n-1})}{x_n - x_{n-1} + 2(\sin x_{n-1} - \sin x_n)} = x_n - \frac{N_n}{D_n}.$$

Numerical values are:

n	x_{n-1}	x_n	N_n	D_n	$x_{n+1} - x_n$
1	2.000 000	1.900 000	−0.000 740	−0.174 005	−0.004 253
2	1.900 000	1.895 747	−0.000 002	−0.006 986	−0.000 252
3	1.895 747	1.895 494	0		0

$x_3 = 1.895\,494$ is exact to 6D. See Example 4.   ◀

Summary of Methods. The methods for computing solutions s of $f(x) = 0$ with given continuous (or differentiable) $f(x)$ start with an initial approximation x_0 of s and generate a sequence $x_1, x_2, \cdots$ by **iteration. Fixed point methods** solve $f(x) = 0$ written as $x = g(x)$, so that s is a *fixed point* of g, that is, $s = g(s)$. For $g(x) = x - f(x)/f'(x)$ this is **Newton's method,** which for good x_0 and simple zeros converges quadratically (and for multiple zeros linearly). From Newton's method the **secant method** follows by replacing $f'(x)$ by a difference quotient. The **bisection method** and the **method of false position** in Problem Set 17.2 always converge, but often slowly.

PROBLEM SET 17.2

Fixed-Point Iteration

1. Why do we obtain a monotone sequence in Example 1, but not in Example 2?
2. Perform the iterations indicated at the end of Example 2. Sketch a figure similar to Fig. 395.
3. Sketch $f(x) = x^3 - 5.00x^2 + 1.01x + 1.88$. Conclude that roots are ± 1, 5, approximately. Write $x = g(x) = (5.00x^2 - 1.01x - 1.88)/x^2$. Find a root by starting from $x = 5, 4, 1, -1$.
4. Find a form $x = g(x)$ of $f(x) = 0$ in Prob. 3 that gives convergence to the root near 1.
5. **(Bessel functions, drumhead)** A partial sum of the Maclaurin series of $J_0(x)$ (Sec. 4.5) is $f(x) = 1 - \frac{1}{4}x^2 + \frac{1}{64}x^4 - \frac{1}{2304}x^6$. Conclude from a sketch that $f(x) = 0$ near $x = 2$. Write $f(x) = 0$ as $x = g(x)$ (by dividing $f(x)$ by $\frac{1}{4}x$ and taking the resulting x-term to the other side). Find the zero. (See Sec. 11.10 for the importance of these zeros.)
6. **(Elasticity)** Solve $x \cosh x = 1$. (Such equations appear in connection with vibrations of elastic beams.)
7. Find the smallest positive solution of $x = \tan x$, writing $x = \pi + \arctan x$ (why?).
8. Find the smallest positive solution of $\sin x = e^{-x}$.
9. Solve $x^4 - x - 0.12 = 0$ by starting from $x_0 = 1$.

 10. **CAS PROJECT. Fixed-Point Iteration.** **(a) Existence.** Prove that if g is continuous in a closed interval I and its range lies in I, then the equation $x = g(x)$ has at least one solution in I. Illustrate that it may have more than one solution in I.
 (b) Convergence. Let $f(x) = x^3 + 2x^2 - 3x - 4 = 0$. Write this as $x = g(x)$, for g choosing (1) $(x^3 - f)^{1/3}$, (2) $(x^2 - \frac{1}{2}f)^{1/2}$, (3) $x + \frac{1}{3}f$, (4) $x(1 + \frac{1}{4}f)$, (5) $(x^3 - f)/x^2$, (6) $(2x^2 - f)/2x$, (7) $x - f/f'$ and in each case $x_0 = 1.5$ and $x_0 = 2$. Find out about convergence and divergence and the number of steps to reach exact 6S-values of the roots.

Newton's Method

11. Design a Newton iteration for cube roots and compute $\sqrt[3]{7}$ (6D, $x_0 = 2$).
12. Design a Newton iteration for $\sqrt[k]{c}$ ($c > 0$). Use it to compute $\sqrt{2}, \sqrt[3]{2}, \sqrt[4]{2}, \sqrt[5]{2}$ (6D, $x_0 = 1$).
13. Obtain the formula for Newton's method by truncating the Taylor series.
14. **(Legendre polynomials)** Find the largest root of the Legendre polynomial $P_5(x)$ given by $P_5(x) = \frac{1}{8}(63x^5 - 70x^3 + 15x)$ (Sec. 4.3) to 6S-accuracy (to be needed in Gauss integration in Sec. 17.5).
15. **(Associated Legendre functions)** Find the smallest positive zero (to 6S-accuracy) of $P_4{}^2 = (1 - x^2)P_4'' = \frac{15}{2}(-7x^4 + 8x^2 - 1)$ (Sec. 4.3) (a) by Newton's method, (b) exactly, by solving a quadratic equation in x^2.
16. **(Heating, cooling)** At what time x (4S-accuracy) will the processes governed by $f_1(x) = 100(1 - e^{-0.2x})$ and $f_2(x) = 40e^{-0.01x}$ reach the same temperature? Also find the latter.
17. **(Vibrating beam)** Find the solution of $\cos x \cosh x = 1$ near $x = \frac{3}{2}\pi$ to 6S-accuracy. (This determines a frequency of a vibrating beam; see Problem Set 11.4.)
18. **Bessel functions** $J_{2n}(x) = 0$ approximately where $\cos x = -\sin x$ provided x is large (see Problem Set 4.5). Find a zero of J_{2n} near $x = 20$.
19. Solve Prob. 5 by Newton's method.

Secant Method

Solve the given problem by the secant method, using x_0 and x_1 as indicated.
20. Prob. 16, $x_0 = 2, x_1 = 3$
21. Prob. 17, $x_0 = 4, x_1 = 5$
22. $e^{-x} = \tan x$, $x_0 = 1, x_1 = 0.7$
23. Prob. 5, $x_0 = 2.0, x_1 = 2.5$

24. **TEAM PROJECT. Method of False Position (Regula falsi).** Figure 398 shows the idea. We assume that f is continuous. Compute the x-intercept c_0 of the line through $(a_0, f(a_0))$, $(b_0, f(b_0))$. If $f(c_0) = 0$, we are done. If $f(a_0)f(c_0) < 0$ (as in Fig. 398), set $a_1 = a_0, b_1 = c_0$

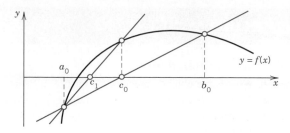

Fig. 398.   Method of false position

and repeat to get c_1, etc. If $f(a_0)f(c_0) > 0$, then $f(c_0)f(b_0) < 0$ and we set $a_1 = c_0$, $b_1 = b_0$, etc.

(a) Algorithm. Show that

$$c_0 = \frac{a_0 f(b_0) - b_0 f(a_0)}{f(b_0) - f(a_0)}$$

and write an algorithm for the method.

(b) Comparison. Solve $x^3 = 5x + 6$ by Newton's method, the secant method, and the method of false position. Compare.

(c) Solve $x^4 = 2$, $\cos x = \sqrt{x}$, and $x + \ln x = 2$ by the method of false position.

25. TEAM PROJECT. Bisection Method. This simple but slowly convergent method for finding a solution of $f(x) = 0$ with continuous f is based on the **intermediate value theorem,** which states that if a continuous function f has opposite signs at some $x = a$ and $x = b$ $(> a)$, that is, either $f(a) < 0$, $f(b) > 0$ or $f(a) > 0$, $f(b) < 0$, then f must be 0 somewhere on $[a, b]$. The solution is found by repeated bisection of the interval and in each iteration picking that half which also satisfies that sign condition.

(a) Algorithm. Write an algorithm for the method.

(b) Comparison. Solve $x = \cos x$ by Newton's method and by bisection. Compare.

(c) Solve $e^{-x} = \ln x$ and $e^x + x^4 + x = 2$ by bisection.

26. WRITING PROJECT. Solution of Equations. Compare the methods in this section and problem set, discussing advantages and disadvantages using simple examples of your own.

17.3 Interpolation

Interpolation means to find (approximate) values of a function $f(x)$ for an x between *different* x-values $x_0, x_1, \cdots, x_n$ at which the values of $f(x)$ are given. You know this for a table of logarithms, or you may think of recorded values (temperatures, results of censuses as in CAS Project 50 of Problem Set 1.6, etc.). So the given values

$$f_0 = f(x_0), \qquad f_1 = f(x_1), \qquad \cdots, \qquad f_n = f(x_n)$$

may come from a *"mathematical function"* given by a formula (which we may wish to approximate by a simpler function) or from an *"empirical function"* resulting from observations or experiments.

A standard idea in interpolation now is to find a polynomial $p_n(x)$ of degree n (or less) that assumes the given values; thus

(1) $$p_n(x_0) = f_0, \qquad p_n(x_1) = f_1, \qquad \cdots, \qquad p_n(x_n) = f_n.$$

We call this p_n an **interpolation polynomial** and $x_0, \cdots, x_n$ the **nodes.** And if $f(x)$ is a mathematical function, we call p_n an **approximation** of f (or a **polynomial approximation,** because there are other kinds of approximations, as we shall see later). We use p_n to get (approximate) values of f for x's between x_0 and x_n (**"interpolation"**) or sometimes outside that interval (**"extrapolation"**).

Motivation. Polynomials are convenient to work with because we can readily differentiate and integrate them, again obtaining polynomials. Moreover, they approximate *continuous* functions with any desired accuracy. That is, for any continuous $f(x)$ on an interval $J: a \leqq x \leqq b$ and error bound $\beta > 0$, there is a polynomial $p_n(x)$ (of sufficiently high degree n) such that $|f(x) - p_n(x)| < \beta$ for all x on J. This is the famous **Weierstrass approximation theorem** (for a proof see Ref. [9], p. 280; see Appendix 1).

Existence and uniqueness. p_n satisfying (1) for given data exists—we give formulas for it below. p_n is unique. Indeed, if a polynomial q_n also satisfies $q_n(x_0) = f_0, \cdots,$ $q_n(x_n) = f_n$, then $p_n(x) - q_n(x) = 0$ at $x_0, \cdots, x_n$, but a polynomial $p_n - q_n$ of degree n (or less) with $n + 1$ roots must be identically zero, as we know from algebra; thus $p_n(x) \equiv q_n(x)$ for all x, which means uniqueness. ◀

How to find p_n? This is the important practical question. We answer it by explaining several standard methods. For given data, these methods give the same polynomial, by the uniqueness just proved (which is thus of practical interest!) but in several forms suitable for different purposes.

Lagrange Interpolation

Given $(x_0, f_0), \cdots, (x_n, f_n)$ with arbitrarily spaced x_j, Lagrange had the idea of multiplying each f_j by a polynomial that is 1 at x_j and 0 at the other n nodes, and then to take the sum of these $n + 1$ polynomials to get the unique interpolation polynomial of degree n or less. Beginning with the simplest case, let us see how this works.

Linear interpolation is interpolation by the straight line through (x_0, f_0), (x_1, f_1); see Fig. 399. Thus by that idea, the linear Lagrange polynomial p_1 is a sum $p_1 = L_0 f_0 + L_1 f_1$ with L_0 the linear polynomial that is 1 at x_0 and 0 at x_1; similarly, L_1 is 0 at x_0 and 1 at x_1. Obviously,

$$L_0(x) = \frac{x - x_1}{x_0 - x_1}, \qquad L_1(x) = \frac{x - x_0}{x_1 - x_0}.$$

This gives the linear Lagrange polynomial

(2) $$p_1(x) = L_0(x)f_0 + L_1(x)f_1 = \frac{x - x_1}{x_0 - x_1} \cdot f_0 + \frac{x - x_0}{x_1 - x_0} \cdot f_1.$$

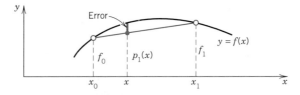

Fig. 399.   Linear interpolation

EXAMPLE 1　**Linear Lagrange interpolation**

Compute ln 9.2 from ln 9.0 = 2.1972, ln 9.5 = 2.2513 by linear Lagrange interpolation and determine the error from ln 9.2 = 2.2192 (4D).

Solution.　$x_0 = 9.0$, $x_1 = 9.5$, $f_0 = \ln 9.0$, $f_1 = \ln 9.5$.　In (2) we need

$$L_0(9.2) = \frac{9.2 - 9.5}{9.0 - 9.5} = 0.6, \qquad L_1(9.2) = \frac{9.2 - 9.0}{9.5 - 9.0} = 0.4$$

and we get the answer

$$\ln 9.2 \approx p_1(9.2) = L_0(9.2)f_0 + L_1(9.2)f_1 = 0.6 \cdot 2.1972 + 0.4 \cdot 2.2513 = 2.2188.$$

The error is $\epsilon = a - \tilde{a} = 2.2192 - 2.2188 = 0.0004$. Hence linear interpolation is not sufficient here to get 4D-accuracy; it would suffice for 3D-accuracy.　◀

Quadratic interpolation is interpolation of given (x_0, f_0), (x_1, f_1), (x_2, f_2) by a second-degree polynomial $p_2(x)$, which by Lagrange's idea is

(3a)　　　　　　　　$$p_2(x) = L_0(x)f_0 + L_1(x)f_1 + L_2(x)f_2$$

with $L_0(x_0) = 1$, $L_1(x_1) = 1$, $L_2(x_2) = 1$, and $L_0(x_1) = L_0(x_2) = 0$, etc. We claim that

$$L_0(x) = \frac{l_0(x)}{l_0(x_0)} = \frac{(x - x_1)(x - x_2)}{(x_0 - x_1)(x_0 - x_2)}$$

(3b)　　　　　　$$L_1(x) = \frac{l_1(x)}{l_1(x_1)} = \frac{(x - x_0)(x - x_2)}{(x_1 - x_0)(x_1 - x_2)}$$

$$L_2(x) = \frac{l_2(x)}{l_2(x_2)} = \frac{(x - x_0)(x - x_1)}{(x_2 - x_0)(x_2 - x_1)}.$$

How did we get this? Well, the numerator makes $L_k(x_j) = 0$ if $j \neq k$. And the denominator makes $L_k(x_k) = 1$ because it equals the numerator at $x = x_k$.

EXAMPLE 2　**Quadratic Lagrange interpolation**

Compute ln 9.2 by (3) from the data in Example 1 and ln 11.0 = 2.3979.

Solution.　In (3),

$$L_0(x) = \frac{(x - 9.5)(x - 11.0)}{(9.0 - 9.5)(9.0 - 11.0)} = x^2 - 20.5x + 104.5, \qquad L_0(9.2) = 0.5400,$$

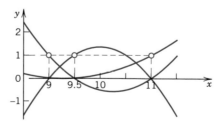

Fig. 400.　L_0 (highest at the left end),
L_1 (lowest at the left end), and L_2
in Example 2

$$L_1(x) = \frac{(x - 9.0)(x - 11.0)}{(9.5 - 9.0)(9.5 - 11.0)} = -\frac{1}{0.75}(x^2 - 20x + 99), \qquad L_1(9.2) = 0.4800,$$

$$L_2(x) = \frac{(x - 9.0)(x - 9.5)}{(11.0 - 9.0)(11.0 - 9.5)} = \frac{1}{3}(x^2 - 18.5x + 85.5), \qquad L_2(9.2) = -0.0200,$$

(see Fig. 400), so that (3a) gives, exact to 4D,

$$\ln 9.2 \approx p_2(9.2) = 0.5400 \cdot 2.1972 + 0.4800 \cdot 2.2513 - 0.0200 \cdot 2.3979 = 2.2192. \qquad \blacktriangleleft$$

General Lagrange interpolation polynomial. For general n we obtain

(4a)
$$f(x) \approx p_n(x) = \sum_{k=0}^{n} L_k(x)f_k = \sum_{k=0}^{n} \frac{l_k(x)}{l_k(x_k)} f_k$$

where $L_k(x_k) = 1$ and L_k is 0 at the other nodes, and the L_k are independent of the function f to be interpolated. We get (4a) if we take

$$l_0(x) = (x - x_1)(x - x_2) \cdots (x - x_n),$$

(4b)
$$l_k(x) = (x - x_0) \cdots (x - x_{k-1})(x - x_{k+1}) \cdots (x - x_n), \qquad 0 < k < n,$$

$$l_n(x) = (x - x_0)(x - x_1) \cdots (x - x_{n-1}).$$

We can easily see that $p_n(x_k) = f_k$. Indeed, inspection of (4b) shows that $l_k(x_j) = 0$ if $j \neq k$, so that for $x = x_k$, the sum in (4a) reduces to the single term $(l_k(x_k)/l_k(x_k))f_k = f_k$.

Error estimate. If f is itself a polynomial of degree n (or less), it must coincide with p_n because the $n + 1$ data $(x_0, f_0), \cdots, (x_n, f_n)$ determine a polynomial uniquely, so the error is zero. Now the special f has its $(n + 1)$st derivative identically zero. This makes it plausible that for a *general* f its $(n + 1)$st derivative $f^{(n+1)}$ should measure the error $\epsilon_n(x) = f(x) - p_n(x)$. It can be shown that this is true if $f^{(n+1)}$ exists and is continuous. Then, with a suitable t between x_0 and x_n (or between x_0, x_n, and x if we extrapolate),

(5)
$$\epsilon_n(x) = f(x) - p_n(x) = (x - x_0)(x - x_1) \cdots (x - x_n) \frac{f^{(n+1)}(t)}{(n + 1)!}.$$

Thus $|\epsilon_n(x)|$ is 0 at the nodes and small near them, because of continuity. The product $(x - x_0) \cdots (x - x_n)$ is large for x away from the nodes. This makes extrapolation risky and interpolation at an x best if we choose nodes on both sides of that x. Also, we get error bounds by taking the smallest and the largest value of $f^{(n+1)}(t)$ on the interval $x_0 \leqq t \leqq x_n$ (or on the interval also containing x if we *extra*polate). Most important: since p_n is unique, as we have shown, we have

THEOREM 1 **(Error of interpolation)**

Formula (5) *gives the error for **any** polynomial interpolation method if* $f(x)$ *has a continuous* $(n + 1)$*st derivative.*

Practical error estimate. If the derivative in (5) is difficult or impossible to obtain, apply the Error Principle (Sec. 17.1), that is, take another node and the Lagrange polynomial $p_{n+1}(x)$ and regard $p_{n+1}(x) - p_n(x)$ as a (crude) error estimate for $p_n(x)$.

EXAMPLE 3   **Error estimate (5) of linear interpolation. Damage by round-off. Error Principle**

Estimate the error in Example 1 by (5) and then by the Error Principle.

Solution.   **(A)** *Estimation by* **(5).** We have $n = 1$, $f(t) = \ln t$, $f'(t) = 1/t$, $f''(t) = -1/t^2$. Hence

$$\epsilon_1(x) = (x - 9.0)(x - 9.5)\frac{(-1)}{2t^2}, \qquad \text{thus} \qquad \epsilon_1(9.2) = \frac{0.03}{t^2}.$$

$t = 9.0$ gives the maximum $0.03/9^2 = 0.00037$ and $t = 9.5$ gives the minimum $0.03/9.5^2 = 0.00033$, so that we get $0.00033 \leqq \epsilon_1(9.2) \leqq 0.00037$, or better, 0.00038 because $0.3/81 = 0.003\,703 \cdots$.

But the error 0.0004 in Example 1 disagrees, and we can learn something! Repetition of the computation there with 5D instead of 4D gives

$$\ln 9.2 \approx p_1(9.2) = 0.6 \cdot 2.19722 + 0.4 \cdot 2.25129 = 2.21885$$

with an actual error $\epsilon = 2.21920 - 2.21885 = 0.00035$, which lies nicely near the middle between our two error bounds.

This shows that the discrepancy (0.0004 vs. 0.00035) was caused by round-off, which is not taken into account in (5).

(B) *Estimation by the Error Principle.* We calculate $p_1(9.2) = 2.21885$ as before and then $p_2(9.2)$ as in Example 2 but with 5D, obtaining

$$p_2(9.2) = 0.54 \cdot 2.19722 + 0.48 \cdot 2.25129 - 0.02 \cdot 2.39790 = 2.21916.$$

The difference $p_2(9.2) - p_1(9.2) = 0.00031$ is the approximate error of $p_1(9.2)$ that we wanted to obtain; this is an approximation of the actual error 0.00035 given above.   ◀

Newton's Divided Difference Interpolation

For given data $(x_0, f_0), \cdots, (x_n, f_n)$ the interpolation polynomial $p_n(x)$ satisfying (1) is unique, as we have shown. But for different purposes we may give $p_n(x)$ different forms. **Lagrange's form** just discussed is useful for interpolating tables and for deriving formulas for numerical differentiation (approximation formulas for derivatives) and integration. Equally important are Newton's forms of $p_n(x)$. They are more appropriate for computation. We shall also use them for solving differential equations (in Sec. 19.2). Moreover, they have the following advantage over Lagrange's form.

In practice we often do not know the degree of the interpolation polynomial that will give the required accuracy, so we should be prepared to increase the degree if necessary. Whereas in Lagrange interpolation this would need an entirely new polynomial, in Newton's interpolation we can use our previous work and simply add another term. This also simplifies the application of the Error Principle (used in Example 3 for Lagrange). The details of these ideas are as follows.

Let $p_{n-1}(x)$ be the $(n - 1)$st Newton polynomial (whose form we shall determine); thus $p_{n-1}(x_0) = f_0, \cdots, p_{n-1}(x_{n-1}) = f_{n-1}$. Furthermore, let us write the nth Newton polynomial as

(6) $$p_n(x) = p_{n-1}(x) + g_n(x),$$

with

(6′) $$g_n(x) = p_n(x) - p_{n-1}(x)$$

to be determined so that $p_n(x_0) = f_0, \cdots, p_n(x_n) = f_n$.

Since p_n and p_{n-1} agree at $x_0, \cdots, x_{n-1}$, we see that g_n is zero there. Also, g_n will generally be a polynomial of nth degree because so is p_n, whereas p_{n-1} can be of degree

$n - 1$ at most. Hence g_n must be of the form

$$(6'') \qquad\qquad g_n(x) = a_n(x - x_0)(x - x_1) \cdots (x - x_{n-1}).$$

We determine the constant a_n. For this we set $x = x_n$ and solve (6'') algebraically for a_n. Replacing $g_n(x_n)$ according to (6') and using $p_n(x_n) = f_n$, we see that this gives

$$(7) \qquad\qquad a_n = \frac{f_n - p_{n-1}(x_n)}{(x_n - x_0)(x_n - x_1) \cdots (x_n - x_{n-1})}.$$

We show that a_k equals the **kth divided difference**, recursively denoted and defined as follows:

$$a_1 = f[x_0, x_1] = \frac{f_1 - f_0}{x_1 - x_0},$$

$$a_2 = f[x_0, x_1, x_2] = \frac{f[x_1, x_2] - f[x_0, x_1]}{x_2 - x_0},$$

and in general

$$(8) \qquad a_k = f[x_0, \cdots, x_k] = \frac{f[x_1, \cdots, x_k] - f[x_0, \cdots, x_{k-1}]}{x_k - x_0}.$$

PROOF. If $n = 1$, then $p_{n-1}(x_n) = p_0(x_1) = f_0$, because $p_0(x)$ is constant and equal to f_0, the value of $f(x)$ at x_0. Hence (7) gives

$$a_1 = \frac{f_1 - p_0(x_1)}{x_1 - x_0} = \frac{f_1 - f_0}{x_1 - x_0} = f[x_0, x_1],$$

and (6) and (6'') give the Newton interpolation polynomial of the first degree

$$p_1(x) = f_0 + (x - x_0)f[x_0, x_1].$$

If $n = 2$, then this p_1 and (7) give

$$a_2 = \frac{f_2 - p_1(x_2)}{(x_2 - x_0)(x_2 - x_1)} = \frac{f_2 - f_0 - (x_2 - x_0)f[x_0, x_1]}{(x_2 - x_0)(x_2 - x_1)} = f[x_0, x_1, x_2]$$

where the last equality follows by straightforward calculation and comparison with the definition of the right side. (Verify it; be patient.) From (6) and (6'') we thus obtain the second Newton polynomial

$$p_2(x) = f_0 + (x - x_0)f[x_0, x_1] + (x - x_0)(x - x_1)f[x_0, x_1, x_2].$$

For $n = k$, formula (6) gives

$$(9) \qquad p_k(x) = p_{k-1}(x) + (x - x_0)(x - x_1) \cdots (x - x_{k-1})f[x_0, \cdots, x_k].$$

With $p_0(x) = f_0$ by repeated application with $k = 1, \cdots, n$ this finally gives **Newton's divided difference interpolation formula**

(10)

$$f(x) \approx f_0 + (x - x_0)f[x_0, x_1] + (x - x_0)(x - x_1)f[x_0, x_1, x_2]$$
$$+ \cdots + (x - x_0)(x - x_1) \cdots (x - x_{n-1})f[x_0, \cdots, x_n].$$

An algorithm is shown in Table 17.2. The first do-loop computes the divided differences and the second the desired value $p_n(\hat{x})$.

Example 4 shows how to arrange differences near the values from which they are obtained; the latter always stand a half-line above and a half-line below in the preceding column. Such an arrangement is called a (divided) **difference table.**   ◄

Table 17.2
Newton's Divided Difference Interpolation

ALGORITHM INTERPOL ($x_0, \cdots, x_n; f_0, \cdots, f_n; \hat{x}$)

This algorithm computes an approximation $p_n(\hat{x})$ of $f(\hat{x})$ at $\hat{x}$.

    INPUT: Data $(x_0, f_0), (x_1, f_1), \cdots, (x_n, f_n); \hat{x}$

    OUTPUT: Approximation $p_n(\hat{x})$ of $f(\hat{x})$

    Set $f[x_j] = f_j$ $(j = 0, \cdots, n)$.

    For $m = 1, \cdots, n - 1$ do:

            For $j = 0, \cdots, n - m$ do:

                $f[x_j, \cdots, x_{j+m}]$

$$= \frac{f[x_{j+1}, \cdots, x_{j+m}] - f[x_j, \cdots, x_{j+m-1}]}{x_{j+m} - x_j}$$

            End

    End

    Set $p_0(x) = f_0$.

    For $k = 1, \cdots, n$ do:

            $p_k(\hat{x}) = p_{k-1}(\hat{x}) + (\hat{x} - x_0) \cdots (\hat{x} - x_{k-1})f[x_0, \cdots, x_k]$

    End

    OUTPUT $p_n(\hat{x})$

End INTERPOL

EXAMPLE 4   **Newton's Divided Difference Interpolation Formula**

Compute $f(9.2)$ from the given values.

x_j	$f_j = f(x_j)$	$f[x_j, x_{j+1}]$	$f[x_j, x_{j+1}, x_{j+2}]$	$f[x_j, \cdots, x_{j+3}]$
8.0	2.079 442			
		0.117 783		
9.0	2.197 225		−0.006 433	
		0.108 134		0.000 411
9.5	2.251 292		−0.005 200	
		0.097 735		
11.0	2.397 895			

Solution.   We compute the divided differences as shown. Sample computation:

$$(0.097\ 735 - 0.108\ 134)/(11 - 9) = -0.005\ 200.$$

The values we need in (10) are circled. We have

$$f(x) \approx p_3(x) = 2.079\ 442 + 0.117\ 783(x - 8.0) - 0.006\ 433(x - 8.0)(x - 9.0)$$

$$+ 0.000\ 411(x - 8.0)(x - 9.0)(x - 9.5).$$

At $x = 9.2$,

$$f(9.2) \approx 2.079\ 442 + 0.141\ 340 - 0.001\ 544 - 0.000\ 030 = 2.219\ 208.$$

The value exact to 6D is $f(9.2) = \ln 9.2 = 2.219\ 203$. Note that we can nicely see how the accuracy increases from term to term:

$$p_1(9.2) = 2.220\ 782, \qquad p_2(9.2) = 2.219\ 238, \qquad p_3(9.2) = 2.219\ 208. \qquad \blacktriangleleft$$

Equal Spacing: Newton's Forward Difference Formula

Newton's formula (10) is valid for *arbitrarily spaced* nodes as they may occur in practice in experiments or observations. However, in many applications the x_j's are *regularly spaced*—for instance, in function tables or in measurements at regular intervals of time. Then we can write

(11) $$x_0, \quad x_1 = x_0 + h, \quad x_2 = x_0 + 2h, \quad \cdots, \quad x_n = x_0 + nh.$$

We show how (8) and (10) simplify in this case.

To get started, let us define the *first forward difference* of f at x_j by

$$\Delta f_j = f_{j+1} - f_j,$$

the *second forward difference* of f at x_j by

$$\Delta^2 f_j = \Delta f_{j+1} - \Delta f_j,$$

and, continuing in this way, the **kth forward difference** of f at x_j by

(12) $$\Delta^k f_j = \Delta^{k-1} f_{j+1} - \Delta^{k-1} f_j \qquad (k = 1, 2, \cdots).$$

Examples and an explanation of the name "forward" follow below. What is the point of this? We show that if we have regular spacing (11), then

(13)
$$f[x_0, \cdots, x_k] = \frac{1}{k!h^k} \Delta^k f_0.$$

We prove (13) by induction. It is true for $k = 1$ because $x_1 = x_0 + h$, so that

$$f[x_0, x_1] = \frac{f_1 - f_0}{x_1 - x_0} = \frac{1}{h}(f_1 - f_0) = \frac{1}{1!h}\Delta f_0.$$

Assuming (13) to be true for all forward differences of order k, we show that it holds for $k + 1$. We use (8) with $k + 1$ instead of k, then $x_{k+1} = x_0 + (k + 1)h$, resulting from (11), and finally (12) with $j = 0$. This gives

$$f[x_0, \cdots, x_{k+1}] = \frac{f[x_1, \cdots, x_{k+1}] - f[x_0, \cdots, x_k]}{(k + 1)h}$$

$$= \frac{1}{(k + 1)h}\left[\frac{1}{k!h^k}\Delta^k f_1 - \frac{1}{k!h^k}\Delta^k f_0\right]$$

$$= \frac{1}{(k + 1)!\,h^{k+1}}\Delta^{k+1} f_0$$

which is (13) with $k + 1$ instead of k. Formula (13) is proved. ◀

In (10) we finally set $x = x_0 + rh$. Then $x - x_0 = rh$, $x - x_1 = (r - 1)h$ since $x_1 - x_0 = h$, and so on. With this and (13), formula (10) becomes **Newton's** (or *Gregory*[10]*–Newton's*) **forward difference interpolation formula**

(14)
$$f(x) \approx p_n(x) = \sum_{s=0}^{n} \binom{r}{s} \Delta^s f_0 \qquad\qquad (x = x_0 + rh, \quad r = (x - x_0)/h)$$

$$= f_0 + r\Delta f_0 + \frac{r(r - 1)}{2!}\Delta^2 f_0 + \cdots + \frac{r(r - 1)\cdots(r - n + 1)}{n!}\Delta^n f_0$$

where the **binomial coefficients** in the first line are defined by

(15) $\displaystyle \binom{r}{0} = 1, \quad \binom{r}{s} = \frac{r(r - 1)(r - 2)\cdots(r - s + 1)}{s!}$ $(s > 0, \text{ integer})$

and $s! = 1 \cdot 2 \cdots s$.

Error. From (5) we get, with $x - x_0 = rh$, $x - x_1 = (r - 1)h$, etc.,

(16)
$$\epsilon_n(x) = f(x) - p_n(x) = \frac{h^{n+1}}{(n + 1)!} r(r - 1) \cdots (r - n)f^{(n+1)}(t)$$

with t as in (5). The next example explains the application.

---

[10]JAMES GREGORY (1638–1675), Scots mathematician, professor at St. Andrews and Edinburgh.
    Δ in (14) and ∇^2 (on p. 858) have nothing to do with the Laplacian.

Comments on accuracy. **(A)** The error $\epsilon_n(x)$ is about of the order of magnitude of the next difference not used in $p_n(x)$.

(B) One should choose $x_0, \cdots, x_n$ such that the x at which one interpolates is as well centered between $x_0, \cdots, x_n$ as possible.

The reason for (A) is that in (16),

$$f^{n+1}(t) \approx \frac{\Delta^{n+1}f(t)}{h^{n+1}}, \qquad \frac{|r(r-1)\cdots(r-n)|}{1 \cdot 2 \cdots (n+1)} \leqq 1 \quad \text{if} \quad |r| \leqq 1$$

(and actually for any r as long as we do not *extrapolate*). The reason for (B) is that $|r(r-1)\cdots(r-n)|$ becomes smallest for that choice.

EXAMPLE 5   **Newton's forward difference formula. Error estimation**

Compute cosh 0.56 from (14) and the four values in the following table and estimate the error.

j	x_j	$f_j = \cosh x_j$	Δf_j	$\Delta^2 f_j$	$\Delta^3 f_j$
0	0.5	1.127 626			
			0.057 839		
1	0.6	1.185 465		0.011 865	
			0.069 704		0.000 697
2	0.7	1.255 169		0.012 562	
			0.082 266		
3	0.8	1.337 435			

Solution. We compute the forward differences as shown in the table. The values we need are circled. In (14) we have $r = (0.56 - 0.50)/0.1 = 0.6$, so that (14) gives

$$\cosh 0.56 \approx 1.127\ 626 + 0.6 \cdot 0.057\ 839 + \frac{0.6(-0.4)}{2} \cdot 0.011\ 865 + \frac{0.6(-0.4)(-1.4)}{6} \cdot 0.000\ 697$$

$$= 1.127\ 626 + 0.034\ 703 - 0.001\ 424 + 0.000\ 039 = 1.160\ 944.$$

Error estimate. From (16), since the fourth derivative is $\cosh^{(4)} t - \cosh t$,

$$\epsilon_3(0.56) = \frac{0.1^4}{4!} \cdot 0.6\,(-0.4)(-1.4)(-2.4)\cosh t = A \cosh t,$$

where $A = -0.000\ 003\ 36$ and $0.5 \leqq t \leqq 0.8$. We do not know t, but we get an inequality by taking the largest and smallest cosh t in that interval:

$$A \cosh 0.8 \leqq \epsilon_3(0.62) \leqq A \cosh 0.5.$$

Since $f(x) = p_3(x) + \epsilon_3(x)$, this gives

$$p_3(0.56) + A \cosh 0.8 \leqq \cosh 0.56 \leqq p_3(0.56) + A \cosh 0.5.$$

Numerical values are

$$1.160\ 939 \leqq \cosh 0.56 \leqq 1.160\ 941.$$

The exact 6D-value is cosh $0.56 = 1.160\ 941$. It lies within these bounds. Such bounds are not always so tight. Also, we did not consider round-off errors, which will depend on the number of operations.   ◀

This example also explains the name "*forward* difference formula": we see that the differences in the formula slope forward in the difference table.

Equal Spacing: Newton's Backward Difference Formula

Instead of forward-sloping differences we may also employ backward-sloping differences. The difference table remains the same as before (same numbers, in the same positions), except for a very harmless change of the running subscript j (which we explain in Example 6, below). Nevertheless, purely for reasons of convenience it is standard to introduce a second name and notation for differences as follows. We define the *first backward difference* of f at x_j by

$$\nabla f_j = f_j - f_{j-1},$$

the *second backward difference* of f at x_j by

$$\nabla^2 f_j = \nabla f_j - \nabla f_{j-1},$$

and, continuing in this way, the **kth backward difference** of f at x_j by

(17)
$$\boxed{\nabla^k f_j = \nabla^{k-1} f_j - \nabla^{k-1} f_{j-1}} \qquad (k = 1, 2, \cdots).$$

A formula similar to (14) but involving backward differences is **Newton's** (or *Gregory–Newton's*) **backward difference interpolation formula**

(18)
$$f(x) \approx p_n(x) = \sum_{s=0}^{n} \binom{r + s - 1}{s} \nabla^s f_0 \qquad (x = x_0 + rh,\ r = (x - x_0)/h)$$

$$= f_0 + r\nabla f_0 + \frac{r(r + 1)}{2!} \nabla^2 f_0 + \cdots + \frac{r(r + 1) \cdots (r + n - 1)}{n!} \nabla^n f_0.$$

EXAMPLE 6   **Newton's forward and backward interpolations**

Compute a 7D-value of the Bessel function $J_0(x)$ for $x = 1.72$ from the four values in the following table, using (a) Newton's forward formula (14), (b) Newton's backward formula (18).

j_{for}	j_{back}	x_j	$J_0(x_j)$	1st Diff.	2nd Diff.	3rd Diff.
0	-3	1.7	0.397 9849			
				$-0.057\ 9985$		
1	-2	1.8	0.339 9864		$-0.000\ 1693$	
				$-0.058\ 1678$		$0.000\ 4093$
2	-1	1.9	0.281 8186		$0.000\ 2400$	
				$-0.057\ 9278$		
3	0	2.0	0.223 8908			

Solution.   The computation of the differences is the same in both cases. Only their notation differs.

(a) Forward. In (14) we have $r = (1.72 - 1.70)/0.1 = 0.2$, and j goes from 0 to 3 (see first column). In each column we need the first given number, and (14) thus gives

$$J_0(1.72) \approx 0.397\ 9849 + 0.2(-0.057\ 9985) + \frac{0.2(-0.8)}{2}(-0.000\ 1693)$$

$$+ \frac{0.2(-0.8)(-1.8)}{6} \cdot 0.000\ 4093$$

$$= 0.397\ 9849 - 0.011\ 5997 + 0.000\ 0135 + 0.000\ 0196$$

$$= 0.386\ 4183,$$

which is exact to 6D, the exact 7D-value being 0.386 4185.

(b) Backward. For (18) we use j shown in the second column, and in each column the last number. Since $r = (1.72 - 2.00)/0.1 = -2.8$, we thus get from (18)

$$J_0(1.72) \approx 0.223\ 8908 - 2.8(-0.057\ 9278) + \frac{-2.8(-1.8)}{2} \cdot 0.000\ 2400$$

$$+ \frac{-2.8(-1.8)(-0.8)}{6} \cdot 0.000\ 4093$$

$$= 0.223\ 8908 + 0.162\ 1978 + 0.000\ 6048 - 0.000\ 2750$$

$$= 0.386\ 4184. \qquad \blacktriangleleft$$

Central Difference Notation

This is a third notation for differences. The first central difference of $f(x)$ at x_j is defined by

$$\delta f_j = f_{j+1/2} - f_{j-1/2}$$

and the **kth central difference** of $f(x)$ at x_j by

(19)
$$\boxed{\delta^k f_j = \delta^{k-1} f_{j+1/2} - \delta^{k-1} f_{j-1/2}} \qquad (j = 2, 3, \cdots).$$

Thus in this notation a difference table, for example, for f_{-1}, f_0, f_1, f_2, looks as follows:

x_{-1}	f_{-1}			
		$\delta f_{-1/2}$		
x_0	f_0		$\delta^2 f_0$	
		$\delta f_{1/2}$		$\delta^3 f_{1/2}$
x_1	f_1		$\delta^2 f_1$	
		$\delta f_{3/2}$		
x_2	f_2			

Central differences are used in numerical differentiation (Sec. 17.5), differential equations (Chap. 19), and centered interpolation formulas (e.g., Everett's formula in Team Project 20). These are formulas that use function values "symmetrically" located on both sides of the interpolation point x. Such values are available near the middle of a given table, where centered interpolation formulas tend to give better results than those of Newton's formulas, which do not have that "symmetry" property.

Inverse Interpolation

The problem of finding x for given $f(x)$ is known as **inverse interpolation.** If f is differentiable and df/dx is not zero near the point at which the inverse interpolation is to be effected, the inverse $x = F(y)$ of $y = f(x)$ exists locally near the given value of f and it may happen that F can be approximated in that neighborhood by a polynomial of moderately low degree. Then we may effect the inverse interpolation by tabulating F as a function of y and applying methods of direct interpolation to F. If $df/dx = 0$ near or at the desired point, it may be useful to solve $p(x) = \tilde{f}$ by iteration; here $p(x)$ is a polynomial that approximates $f(x)$ and $\tilde{f}$ is the given value.

EXAMPLE 7 **Inverse interpolation**

Solve $e^x = 3.14$ by inverse interpolation using $x = F(y)$ with $F(3.0) = 1.0986$, $F(3.2) = 1.1632$, $F(3.4) = 1.2238$.

Solution. In Newton's forward formula (14) we have $r = (3.14 - 3.00)/0.2 = 0.7$. From this and the given data we obtain $\Delta x_0 = F(3.2) - F(3.0) = 0.0646$, $\Delta x_1 = F(3.4) - F(3.2) = 0.0606$, $\Delta^2 x_0 = \Delta x_1 - \Delta x_0 = -0.0040$, so that (14) gives, exact to 5S,

$$x = 1.0986 + 0.7 \cdot 0.0646 + \frac{0.7(-0.3)}{2} \cdot (-0.0040) = 1.1442.$$ ◀

PROBLEM SET 17.3

1. **(Linear interpolation)** Compute $p_1(x)$ in Example 1 and from it $\ln 9.3 \approx p_1(9.3)$.

2. **(Error bounds)** Estimate the error in Prob. 1 by (5).

3. **(Linear and quadratic interpolation)** Find $e^{-0.25}$ and $e^{-0.75}$ by linear interpolation with $x_0 = 0$, $x_1 = 0.5$ and $x_0 = 0.5$, $x_1 = 1$, respectively. Then find $p_2(x)$ interpolating e^{-x} with $x_0 = 0$, $x_1 = 0.5$, $x_2 = 1$ and from it $e^{-0.25}$ and $e^{-0.75}$. Compare the errors of these linear and quadratic interpolations.

4. **(Error bounds)** Derive error bounds for $p_2(9.2)$ in Example 2 from (5).

5. **(Quadratic interpolation)** Calculate the Lagrange polynomial $p_2(x)$ for the 4D-values of the Gamma function [(24), Appendix A3.1] $\Gamma(1.00) = 1.0000, \Gamma(1.02) = 0.9888, \Gamma(1.04) = 0.9784$ and from it approximations of $\Gamma(1.01)$ and $\Gamma(1.03)$.

6. **(Interpolation and extrapolation)** Calculate $p_2(x)$ in Example 2. Compute from it approximations of $\ln 9.4$, $\ln 10$, $\ln 10.5$, $\ln 11.5$, $\ln 12$, compute the errors by using exact 4D-values, and comment.

7. **(Extrapolation)** Does a sketch or plot of the product of the $(x - x_j)$ in (5) for the data in Prob. 6 indicate that extrapolation is likely to involve larger errors than interpolation does?

8. **(Cubic Lagrange interpolation)** Calculate and sketch or plot L_0, L_1, L_2, L_3 for $x = 0, 1, 2, 3$ on common axes. Find $p_3(x)$ for the data $(0, 1)$, $(1, 0.765198)$, $(2, 0.223891)$, $(3, -0.260052)$ [values of the Bessel function $J_0(x)$]. Find p_3 for $x = 0.5, 1.5, 2.5$ by interpolation.

9. **(Error function)** Calculate the Lagrange polynomial $p_2(x)$ for the 5D-values of the error function $f(x) = \text{erf } x = (2/\sqrt{\pi}) \int_0^x e^{-w^2} \, dw$, namely, $f(0.25) = 0.27633$, $f(0.5) = 0.52050$, $f(1) = 0.84270$, and from p_2 an approximation of $f(0.75)$ $(= 0.71116, 5D)$.

10. Derive an error bound in Prob. 9 from (5).

11. **(Newton's forward difference formula)** Set up Newton's forward difference formula (14) for the data in Prob. 5 and compute from it $\Gamma(1.01)$, $\Gamma(1.03)$, $\Gamma(1.05)$.

12. Using (14), find $f(1.25)$ by linear, quadratic, and cubic interpolation of the data $f(1.0) = 0.94608$, $f(1.5) = 1.32468$, $f(2.0) = 1.60541$, $f(2.5) = 1.77825$ and compare the errors. (These are values of the **sine integral** (40) in Appendix A3.1; exact (6S) $f(1.25) = 1.14645$.) For the linear interpolation use $f(1.0)$ and $f(1.5)$, for the quadratic $f(1.0)$, $f(1.5)$, and $f(2.0)$, etc.

13. **(Lower degree)** What is the degree of the interpolation polynomial for the data $(1, 5)$, $(2, 18)$, $(3, 37)$, $(4, 62)$, $(5, 93)$? Find the polynomial.

14. **(Newton's divided difference formula)** Set up Newton's divided difference formula for the data in Prob. 6 and derive from it $p_2(x)$ in Prob. 6.

15. Do the same task as in Prob. 14, for the data in Prob. 9.

16. **(Backward difference formula)** In Example 5, write down the backward difference formula with general x and Table A4 in Appendix 5, then use it to verify $\cosh 0.56$ in Example 5.

17. Using $p_2(x)$ in (18) and the values of erf x, $x = 0.2, 0.4, 0.6$ in Table A4 in Appendix 5, compute erf 0.3 and the error. [Exact 0.3286 (4S).]

18. **(Subtabulation)** Compute the Bessel function $J_1(x)$ for[11] $x = 0.1(0.2)0.9$ from $J_1(x)$ for

---

[11]The standard notation $x = a(h)b$ means $x = a, a + h, a + 2h, \cdots, b$.

$x = 0(0.2)1.0$ given by 0, 0.09950, 0.19603, 0.28670, 0.36884, 0.44005, respectively. Use (14) with $n = 5$.

19. **WRITING PROJECT. Interpolation: Comparison of Methods.** Make a list of $5-6$ ideas that you feel are most basic in this section. Arrange them in the best logical order. Discuss them in a $2-3$ page essay.

20. **TEAM PROJECT. Interpolation and Extrapolation. (a) Lagrange practical error estimate** (after Theorem 1). Apply this to $p_1(9.2)$ and $p_2(9.2)$ for the data $x_0 = 9.0$, $x_1 = 9.5$, $x_2 = 11.0$, $f_0 = \ln x_0$, $f_1 = \ln x_1$, $f_2 = \ln x_2$ (6S-values).

   (b) Extrapolation. Given $(x_j, \ f(x_j)) = (0.2, \ 0.9980)$, $(0.4, \ 0.9686)$, $(0.6, \ 0.8443)$, $(0.8, 0.5358)$, $(1.0, 0)$. Find $f(0.7)$ from the quadratic interpolation polynomials based on (α) 0.6, 0.8, 1.0, (β) 0.4, 0.6, 0.8, (γ) 0.2, 0.4, 0.6. Compare the errors and comment. [Exact $f(0.7) = \cos(\tfrac{1}{2}\pi \cdot 0.7^2) = 0.7181$ (4S).]

   (c) Plot the product of factors $(x - x_j)$ in the error formula (5) for $n = 2, \cdots, 10$ separately. What do these plots show regarding accuracy of interpolation and extrapolation?

   (d) Central differences. Show that $\delta^2 f_m = f_{m+1} - 2f_m + f_{m-1}$ and, furthermore, $\delta^3 f_{m+1/2} = f_{m+2} - 3f_{m+1} + 3f_m - f_{m-1}$, $\delta^n f_m = \Delta^n f_{m-n/2} = \nabla^n f_{m+n/2}$.

   (e) Everett's interpolation formula

   $$(20) \quad f(x) \approx (1 - r)f_0 + rf_1 + \frac{(2 - r)(1 - r)(-r)}{3!} \, \delta^2 f_0 + \frac{(r + 1)r(r - 1)}{3!} \, \delta^2 f_1$$

   is an example of a formula involving only even-order differences. Use it to compute the Bessel function $J_0(x)$ for $x = 1.72$ from $J_0(1.60) = 0.455\ 4022$ and $J_0(1.7)$, $J_0(1.8)$, $J_0(1.9)$ in Example 6.

17.4 Splines

One might expect the quality of interpolation to increase with increasing degree n of the polynomials used. Unfortunately, this is not generally true. Indeed, for various functions f the corresponding interpolation polynomials may tend to oscillate more and more between nodes as n increases. Hence we must be prepared for possible **numerical instability.** Figure 401 shows a famous example for which C. Runge[12] has proved that for equidistant nodes the maximum error even approaches infinity as $n \to \infty$. For a similar example, Fig. 402 shows how the error becomes large with increasing n.

Such oscillations are avoided by the method of splines, which was initiated by I. J. Schoenberg in 1946 (*Quarterly of Applied Mathematics* **4**, pp. $45-99$, $112-141$) and is now widely used in practice. The name is borrowed from a *draftman's spline,* which is an elastic rod, bent to conform to the points (and held in place by weights). The mathematical idea is this. Instead of a single high-degree polynomial over an interval J

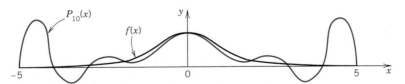

Fig. 401. Runge's example $f(x) = 1/(1 + x^2)$ and interpolating polynomial $P_{10}(x)$

---

[12]See the footnote in Sec. 19.1.

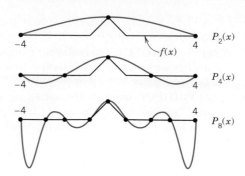

Fig. 402. Piecewise linear function $f(x)$ and interpolation polynomials
of increasing degrees

given by $a \leqq x \leqq b$ (which can oscillate considerably), we subdivide J and use several low-degree polynomials (which cannot oscillate much!), one over each subinterval, and get an interpolating function, called a **spline,** by fitting them together into a single curve. Thus **spline interpolation** *is piecewise polynomial interpolation.* In detail:

Given $f(x)$ on J, we **partition** J. That is, we subdivide J into subintervals with common endpoints, called **nodes,**

$$(1) \qquad\qquad a = x_0 < x_1 < \cdots < x_n = b.$$

We require the function $g(x)$, which will interpolate $f(x)$ on J, to be equal to $f(x)$ at the nodes. Thus $g(x_0) = f(x_0), \cdots, g(x_n) = f(x_n)$. On each of the n subintervals we require $g(x)$ to be given by a polynomial, such that at the nodes, $g(x)$ is several times differentiable. The function $g(x)$ thus obtained is called a spline. This method will in general be numerically stable since each such $g(x)$ will hardly oscillate, for the reason given.

The simplest continuous piecewise polynomial approximation would be by piecewise *linear* functions. But the graph of such a function has corners and would be of little practical interest—think of designing the body of a ship or a car. Hence it is preferable to use functions that have a certain number of derivatives *everywhere* on the interval $a \leqq x \leqq b$.

We shall consider cubic splines, which are perhaps the most important ones from a practical point of view. By definition, a **cubic spline** $g(x)$ on $a \leqq x \leqq b$ corresponding to the partition (1) is a continuous function $g(x)$ that has continuous first and second derivatives everywhere in that interval and, in each subinterval of that partition (1), is represented by a polynomial of degree not exceeding three. Hence $g(x)$ consists of n such polynomials, one in each subinterval.

If $f(x)$ is given on $a \leqq x \leqq b$ and a partition (1) has been chosen, we obtain a cubic spline $g(x)$ that approximates $f(x)$ by requiring that

$$(2) \qquad g(x_0) = f(x_0) = f_0, \quad g(x_1) = f(x_1) = f_1, \quad \cdots, \quad g(x_n) = f(x_n) = f_n,$$

as in the classical interpolation problem of the previous section. We claim that there is a cubic spline $g(x)$ satisfying these conditions (2). And if we also require that

$$(3) \qquad\qquad g'(x_0) = k_0, \qquad g'(x_n) = k_n$$

(k_0 and k_n given numbers, prescribing the tangent directions of $g(x)$ at the endpoints), then we have a uniquely determined cubic spline. This is the content of the following existence and uniqueness theorem, whose proof will also pave the way for the practical determination of splines. [Condition (3) will be discussed after the proof.]

THEOREM 1 **(Cubic splines)**

Let $f(x)$ be defined on the interval $a \leqq x \leqq b$, let a partition (1) be given, and let k_0 and k_n be any two given numbers. Then there exists one and only one cubic spline $g(x)$ corresponding to (1) and satisfying (2) and (3).

PROOF. By definition, on every subinterval I_j given by $x_j \leqq x \leqq x_{j+1}$, the spline $g(x)$ must agree with a polynomial $p_j(x)$ of degree not exceeding three, such that

(4) $$p_j(x_j) = f(x_j), \qquad p_j(x_{j+1}) = f(x_{j+1}) \quad (j = 0, 1, \cdots n - 1).$$

We write

$$\frac{1}{x_{j+1} - x_j} = c_j$$

and

(5) $$p_j'(x_j) = k_j, \qquad p_j'(x_{j+1}) = k_{j+1} \qquad (j = 0, 1, \cdots n - 1)$$

where k_0 and k_n are given, and $k_1, \cdots, k_{n-1}$ will be determined later. Equations (4) and (5) are four conditions for $p_j(x)$. By direct calculation we can verify that the unique cubic polynomial $p_j(x)$ satisfying (4) and (5) is

(6)
$$\begin{aligned}
p_j(x) = {} & f(x_j)c_j^2(x - x_{j+1})^2[1 + 2c_j(x - x_j)] \\
& + f(x_{j+1})c_j^2(x - x_j)^2[1 - 2c_j(x - x_{j+1})] \\
& + k_j c_j^2(x - x_j)(x - x_{j+1})^2 \\
& + k_{j+1}c_j^2(x - x_j)^2(x - x_{j+1}).
\end{aligned}$$

Differentiating twice, we obtain

(7) $$p_j''(x_j) = -6c_j^2 f(x_j) + 6c_j^2 f(x_{j+1}) - 4c_j k_j - 2c_j k_{j+1}$$

(8) $$p_j''(x_{j+1}) = 6c_j^2 f(x_j) - 6c_j^2 f(x_{j+1}) + 2c_j k_j + 4c_j k_{j+1}.$$

By definition, $g(x)$ has continuous second derivatives. This gives the conditions

$$p_{j-1}''(x_j) = p_j''(x_j) \qquad (j = 1, \cdots, n - 1).$$

If we use (8) with j replaced by $j - 1$, and (7), these $n - 1$ equations become

(9) $$\boxed{\, c_{j-1}k_{j-1} + 2(c_{j-1} + c_j)k_j + c_j k_{j+1} = 3\big[c_{j-1}^2 \nabla f_j + c_j^2 \nabla f_{j+1}\big] \,}$$

where $\nabla f_j = f(x_j) - f(x_{j-1})$ and $\nabla f_{j+1} = f(x_{j+1}) - f(x_j)$ and $j = 1, \cdots, n - 1$, as before. This linear system of $n - 1$ equations has a unique solution $k_1, \cdots, k_{n-1}$ since the coefficient matrix is strictly diagonally dominant (that is, in each row the (positive) diagonal entry is greater than the sum of the other (positive) entries). Hence by Theorem

3 in Sec. 18.7 the determinant of the matrix cannot be zero, so that we may determine unique values $k_1, \cdots, k_{n-1}$ of the first derivative of $g(x)$ at the nodes. This proves the theorem. ◀

Storage and time demands in solving (9) are modest since the matrix of (9) is **sparse** (has few nonzero entries) and **tridiagonal** (has nonzero entries only on the diagonal and on the two adjacent "parallels" above and below it). Pivoting (Sec. 6.3) is not necessary because of that dominance. This makes splines efficient in solving large problems with thousands of nodes.

Condition (3) includes the **clamped conditions**

(10) $$g'(x_0) = f'(x_0), \qquad g'(x_n) = f'(x_n),$$

in which the tangent directions $f'(x_0)$ and $f'(x_n)$ at the ends are given. Other conditions of practical interest are the **free** or **natural conditions**

(11) $$g''(x_0) = 0, \qquad g''(x_n) = 0$$

(geometrically: zero curvature at the ends, as for the draftman's spline), giving a **natural spline.** These names are motivated by Fig. 265 in Problem Set 11.4.

Determination of splines from (9) for equidistant nodes $x_0, x_1 = x_0 + h, \cdots,$ $x_n = x_0 + nh$. For these the idea is the same as for arbitrary nodes, but the formulas are simpler. Namely, in (9) we then have $c_j = 1/(x_{j+1} - x_j) = 1/h$. Hence (9), multiplied by h and with our previous notation $f(x_j) = f_j$ becomes simply

(12) $$k_{j-1} + 4k_j + k_{j+1} = \frac{3}{h}(f_{j+1} - f_{j-1}), \qquad j = 1, \cdots, n - 1,$$

k_0 and k_n are given, for instance, $k_0 = f'(a), k_n = f'(b)$ or otherwise, and $k_1, \cdots, k_{n-1}$ are obtained by solving the linear system (12) consisting of $n - 1$ equations. This was the first step.

Determination of splines. Second Step. We determine the coefficients of the spline $g(x)$. On the interval $x_j \leqq x \leqq x_{j+1} = x_j + h$ the spline $g(x)$ is given by a cubic polynomial, which we write in the form

(13) $$p_j(x) = a_{j0} + a_{j1}(x - x_j) + a_{j2}(x - x_j)^2 + a_{j3}(x - x_j)^3, \qquad (j = 0, \cdots, n - 1)$$

Using Taylor's formula, we obtain

(14)
$$a_{j0} = p_j(x_j) = f_j \qquad \text{by (2),}$$

$$a_{j1} = p_j'(x_j) = k_j \qquad \text{by (5),}$$

$$a_{j2} = \frac{1}{2}p_j''(x_j) = \frac{3}{h^2}(f_{j+1} - f_j) - \frac{1}{h}(k_{j+1} + 2k_j) \qquad \text{by (7),}$$

$$a_{j3} = \frac{1}{6}p_j'''(x_j) = \frac{2}{h^3}(f_j - f_{j+1}) + \frac{1}{h^2}(k_{j+1} + k_j)$$

with a_{j3} obtained by calculating $p_j''(x_{j+1})$ from (13) and equating the result to (8), that is,

$$p_j''(x_{j+1}) = 2a_{j2} + 6a_{j3}h = \frac{6}{h^2}(f_j - f_{j+1}) + \frac{2}{h}(k_j + 2k_{j+1}),$$

and now subtracting from this $2a_{j2}$ as given in (14) and simplifying.

EXAMPLE 1   **Spline interpolation (Fig. 403 on the next page)**

Interpolate $f(x) = x^4$ on the interval $-1 \leq x \leq 1$ by the cubic spline $g(x)$ corresponding to the partition $x_0 = -1$, $x_1 = 0$, $x_2 = 1$ and satisfying the clamped conditions $g'(-1) = f'(-1)$, $g'(1) = f'(1)$.

Solution.   We first write the given data in our standard notation, $f_0 = f(-1) = 1$, $f_1 = f(0) = 0$, $f_2 = f(1) = 1$. The given interval is partitioned into $n = 2$ parts of length $h = 1$. Hence our spline g consists of $n = 2$ polynomials (13),

$$p_0(x) = a_{00} + a_{01}(x + 1) + a_{02}(x + 1)^2 + a_{03}(x + 1)^3 \qquad (-1 \leq x \leq 0),$$

$$p_1(x) = a_{10} + a_{11}x + a_{12}x^2 + a_{13}x^3 \qquad (0 \leq x \leq 1).$$

Step 1.   Our goal is to determine the coefficients of p_0 and p_1, but (14) shows that they involve the k_j's, which we must thus first determine from (12). Since $n = 2$, we have $j = 1$ in (12) and get a single equation,

$$k_0 + 4k_1 + k_2 = \frac{3}{1}(f_2 - f_0) = 3(1 - 1) = 0.$$

Now (5) shows that $p_0'(x_0) = k_0$ and $p_1'(x_2) = k_2$. But $g = p_0$ at $x_0 = -1$ and $g = p_1$ at $x_2 = 1$. Hence, as given ($g' = f'$ at ± 1),

$$f'(-1) = -4 = g'(-1) = p_0'(-1) = k_0, \qquad f'(1) = 4 = g'(1) = p_1'(1) = k_2.$$

Substitution of $k_0 = -4$ and $k_2 = 4$ into (12) gives $k_1 = 0$.

Step 2.   From (14) we can now obtain the coefficients of p_0,

$$a_{00} = f_0 = 1, \qquad\qquad a_{01} = k_0 = -4$$

$$a_{02} = \frac{3}{1^2}(f_1 - f_0) - \frac{1}{1}(k_1 + 2k_0) = 3(0 - 1) - (0 - 8) = 5$$

$$a_{03} = \frac{2}{1^3}(f_0 - f_1) + \frac{1}{1^2}(k_1 + k_0) = 2(1 - 0) + (0 - 4) = -2.$$

Similarly, for the coefficients of p_1 we obtain from (14)

$$a_{10} = f_1 = 0, \qquad\qquad a_{11} = k_1 = 0$$

$$a_{12} = 3(f_2 - f_1) - (k_2 + 2k_1) = 3(1 - 0) - (4 + 0) = -1$$

$$a_{13} = 2(f_1 - f_2) + (k_2 + k_1) = 2(0 - 1) + (4 + 0) = 2.$$

This gives the polynomials

$$p_0(x) = 1 - 4(x + 1) + 5(x + 1)^2 - 2(x + 1)^3 = -x^2 - 2x^3,$$

$$p_1(x) = -x^2 + 2x^3$$

of which the spline $g(x)$ consists. Figure 403 shows $f(x)$ and this spline given by the formula

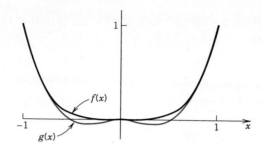

Fig. 403. Function $f(x) = x^4$ and cubic spline $g(x)$
in Example 1

$$g(x) = \begin{cases} -x^2 - 2x^3 & \text{if} \quad -1 \leqq x \leqq 0 \\ -x^2 + 2x^3 & \text{if} \quad 0 \leqq x \leqq 1. \end{cases}$$

Do you see that we could have saved over half of our work by using symmetry?   ◀

EXAMPLE 2   **Spline interpolation**

Interpolate $f_0 = f(0) = 1$, $f_1 = f(2) = 9$, $f_2 = f(4) = 41$, $f_3 = f(6) = 41$ by the cubic spline satisfying $k_0 = 0$, $k_3 = -12$.

Solution. $n = 3$, $h = 2$, so that (12) is

$$k_0 + 4k_1 + k_2 = \tfrac{3}{2}(f_2 - f_0) = 60$$
$$k_1 + 4k_2 + k_3 = \tfrac{3}{2}(f_3 - f_1) = 48.$$

Since $k_0 = 0$ and $k_3 = -12$, the solution is $k_1 = 12$, $k_2 = 12$.
   In (14) with $j = 0$ we have $a_{00} = f_0 = 1$, $a_{01} = k_0 = 0$,

$$a_{02} = \tfrac{3}{4}(9 - 1) - \tfrac{1}{2}(12 + 0) = 0$$
$$a_{03} = \tfrac{2}{8}(1 - 9) + \tfrac{1}{4}(12 + 0) = 1.$$

From this and, similarly, from (14) with $j = 1$ and $j = 2$ we get the spline $g(x)$ consisting of the three polynomials (see Fig. 404)

$$p_0(x) = 1 + x^3 \qquad\qquad\qquad\qquad\qquad\qquad\qquad (0 \leqq x \leqq 2)$$

$$p_1(x) = 9 + 12(x - 2) + 6(x - 2)^2 - 2(x - 2)^3 = 25 - 36x + 18x^2 - 2x^3 \quad (2 \leqq x \leqq 4)$$

$$p_2(x) = 41 + 12(x - 4) - 6(x - 4)^2 = -103 + 60x - 6x^2 \qquad\qquad (4 \leqq x \leqq 6). \quad ◀$$

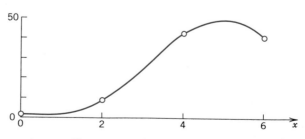

Fig. 404.   Spline in Example 2

PROBLEM SET 17.4

1. **WRITING PROJECT. Splines.** Write a short essay on the idea of spline interpolation, its motivation, its comparison with polynomial interpolation, and its applications. Try to say everything in your own words and with as few formulas as possible.

Derivations in the Text

2. Verify that (6) satisfies (4) and (5).
3. Obtain (7) and (8) from (6) as indicated in the text.
4. Verify the derivation of (9) indicated in the text.
5. Derive (12) from (9).
6. Give the details of the derivation of a_{j2} and a_{j3} in (14).
7. Verify the calculations in Example 1.
8. Compare the spline $g(x)$ in Example 1 with the quadratic interpolation polynomial $p(x)$ over the whole interval. What are the maximum deviations of $g(x)$ and $p(x)$ from $f(x)$? Comment.
9. Derive $p_1(x)$ and $p_2(x)$ in Example 2 and verify that $g(x)$ in Example 2 has continuous first and second derivatives.
10. If a cubic spline is three times continuously differentiable (that is, it has continuous first, second, and third derivatives), show that it must be a polynomial.

Determination of Splines

In Probs. 11–19 find the cubic spline $g(x)$ to the given data, with k_0 and k_n as indicated.

11. $f_0 = f(-1) = 0, f_1 = f(0) = 4, f_2 = f(1) = 0, \quad k_0 = 0, k_2 = 0$. Is $g(x)$ even? Why?
12. $f_0 = f(-2) = 1, f_1 = f(0) = 5, f_2 = f(2) = 17, \quad k_0 = -2, k_2 = -14$
13. $f_0 = f(0) = 3, f_1 = f(2) = 5, f_2 = f(4) = 31, \quad k_0 = 1, k_2 = 21$
14. $f_0 = f(0) = 0, f_1 = f(1) = 1, f_2 = f(2) = 6, f_3 = f(3) = 10, \quad k_0 = 0, k_3 = 0$
15. $f_0 = f(0) = 1, f_1 = f(1) = 0, f_2 = f(2) = -1, f_3 = f(3) = 0, \quad k_0 = 0, k_3 = -6$
16. $f(-2) = f(-1) = f(1) = f(2) = 0, f(0) = 1, \quad k_0 = k_4 = 0$. Is $g(x)$ even? Why? Also find the interpolation polynomial corresponding to the five f-values. Compare and comment. (See Fig. 405.)

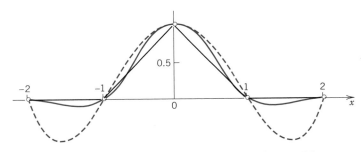

Fig. 405. Spline and interpolation polynomial in Problem 16

17. Could it happen that a spline is given by the same polynomial in adjacent intervals? First guess, then try $g(x)$ for $f(x) = \sin x$, partition $x_0 = -\frac{1}{2}\pi, x_1 = 0, x_2 = \frac{1}{2}\pi$, and $g' = f'$ at $\pm\frac{1}{2}\pi$.
18. $f(0) = 5, f(x) = 0$ if $x = -3, -2, -1, 1, 2, 3; k_0 = 0, k_6 = 0$. Also find the interpolation polynomial $p(x)$ for these data. Sketch or plot $g(x)$ and $p(x)$. Comment.
19. $f_0 = f(0) = 4, f_1 = f(2) = 0, f_2 = f(4) = 4, f_3 = f(6) = 80, \quad k_0 = 0, k_3 = 0$

20. TEAM PROJECT. A Short Look at Hermite Interpolation and Bezier Curves. In **Hermite interpolation** we are looking for a polynomial $p(x)$ (of degree $2n + 1$ or less) such that $p(x)$ and its derivative $p'(x)$ have given values at $n + 1$ nodes. (More generally, $p(x)$, $p'(x)$, $p''(x)$, $\cdots$ may be required to have given values at the nodes.)

(a) **Curves with given endpoints and tangents.** Let C be a curve in the xy-plane parametrically represented by $\mathbf{r}(t) = [x(t), y(t)]$, $0 \leqq t \leqq 1$ (see Sec. 8.5). Show that C has a given initial point A: $\mathbf{r}_0 = [x(0), y(0)] = [x_0, y_0]$ and tangent vector $\mathbf{v}_0 = [x'(0), y'(0)] = [x_0', y_0']$ and given terminal point B: $\mathbf{r}_1 = [x(1), y(1)] = [x_1, y_1]$ and tangent vector $\mathbf{v}_1 = [x'(1), y'(1)] = [x_1', y_1']$ if C is represented by

$$(15) \qquad \mathbf{r}(t) = \mathbf{r}_0 + \mathbf{v}_0 t + (3(\mathbf{r}_1 - \mathbf{r}_0) - (2\mathbf{v}_0 + \mathbf{v}_1))t^2 + (2(\mathbf{r}_0 - \mathbf{r}_1) + \mathbf{v}_0 + \mathbf{v}_1)t^3;$$

in components,

$$x(t) = x_0 + x_0't + (3(x_1 - x_0) - (2x_0' + x_1'))t^2 + (2(x_0 - x_1) + x_0' + x_1')t^3$$

$$y(t) = y_0 + y_0't + (3(y_1 - y_0) - (2y_0' + y_1'))t^2 + (2(y_0 - y_1) + y_0' + y_1')t^3.$$

We see that this is a cubic Hermite interpolation polynomial, and $n = 1$ because we have two nodes (the endpoints of C). (This has nothing to do with the Hermite polynomials in Sec. 4.8.) The two points G_A: $\mathbf{g}_0 = \mathbf{r}_0 + \mathbf{v}_0 = [x_0 + x_0', y_0 + y_0']$ and G_B: $\mathbf{g}_1 = \mathbf{r}_1 - \mathbf{v}_1 = [x_1 - x_1', y_1 - y_1']$ are called **guidepoints** because the segments AG_A and BG_B specify the tangents graphically. A, B, G_A, G_B determine C, and C can be changed quickly by moving the points. A curve consisting of such Hermite interpolation polynomials is called a **Bezier curve,** after the French engineer P. Bezier of the Renault Automobile Company, who introduced them in the early 1960s in designing car bodies. Bezier curves (and surfaces) are used in computer-aided design (CAD) and computer-aided manufacturing (CAM). (For more details, see Ref. [E12] in Appendix 1.)

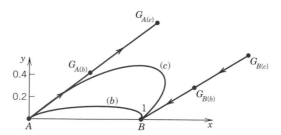

Fig. 406.   Team Project 20(b) and (c): Bezier curves

(b) Find and plot the Bezier curve and its guidepoints if A: $[0, 0]$, B: $[1, 0]$, $\mathbf{v}_0 = [\frac{1}{2}, \frac{1}{2}]$, $\mathbf{v}_1 = [-\frac{1}{2}, -\frac{1}{4}\sqrt{3}]$.

(c) **Changing guidepoints** changes C. Moving guidepoints farther away makes C "staying near the tangents for a longer time." Confirm this by changing $\mathbf{v}_0$ and $\mathbf{v}_1$ in (b) to $2\mathbf{v}_0$ and $2\mathbf{v}_1$ (see Fig. 406).

(d) Make experiments of your own. What happens if you change $\mathbf{v}_1$ in (b) to $-\mathbf{v}_1$. If you rotate the tangents? If you multiply $\mathbf{v}_0$ and $\mathbf{v}_1$ by positive factors less than 1?

17.5 Numerical Integration and Differentiation

Numerical integration means the numerical evaluation of integrals

$$J = \int_a^b f(x)\, dx$$

where a and b are given and f is a function given analytically by a formula or empirically by a table of values. Geometrically, J is the area under the curve of f between a and b (Fig. 407).

We know that if f is such that we can find a differentiable function F whose derivative is f, then we can evaluate J by applying the familiar formula

$$J = \int_a^b f(x)\, dx = F(b) - F(a) \qquad\qquad \left[F'(x) = f(x) \right].$$

Tables of integrals (e.g., Ref. [3] in Appendix 1) or a CAS (MATHEMATICA, MAPLE, etc.) may be helpful for this purpose.

However, applications often lead to integrals whose analytical evaluation would be very complicated or even impossible, or whose integrand is an empirical function given by recorded numerical values. Then we may obtain approximate numerical values of the integral by a numerical integration method.

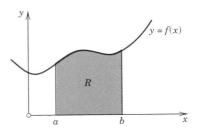

Fig. 407. Geometrical interpretation of a definite integral

Rectangular Rule. Trapezoidal Rule

Numerical integration methods are obtained by approximating the integrand f by functions that can easily be integrated. The simplest formula, the **rectangular rule,** is obtained if we subdivide the interval of integration $a \leqq x \leqq b$ into n subintervals of equal length $h = (b - a)/n$ and in each subinterval approximate f by the constant $f(x_j{}^*)$, the value of f at the midpoint $x_j{}^*$ of the jth subinterval (Fig. 408). Then f is approximated by a **step function** (piecewise constant function), the n rectangles in Fig. 408 have the areas $f(x_1{}^*)h, \cdots, f(x_n{}^*)h$, and the **rectangular rule** is

(1)
$$J = \int_a^b f(x)\, dx \approx h\left[f(x_1{}^*) + f(x_2{}^*) + \cdots + f(x_n{}^*) \right]$$

where $h = (b - a)/n$.

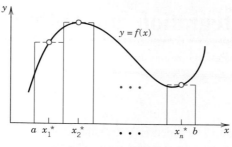
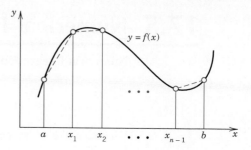

Fig. 408.   Rectangular rule   **Fig. 409.**   Trapezoidal rule

The **trapezoidal rule** is generally more accurate. We obtain it if we take the same subdivision as before and approximate f by a broken line of segments (chords) with endpoints $[a, f(a)], [x_1, f(x_1)], \cdots, [b, f(b)]$ on the curve of f (Fig. 409). Then the area under the curve of f between a and b is approximated by n trapezoids of areas

$$\tfrac{1}{2}[f(a) + f(x_1)]h, \qquad \tfrac{1}{2}[f(x_1) + f(x_2)]h, \qquad \cdots, \qquad \tfrac{1}{2}[f(x_{n-1}) + f(b)]h.$$

By taking their sum we obtain the **trapezoidal rule**

$$(2) \qquad J = \int_a^b f(x)\, dx \approx h\left[\tfrac{1}{2}f(a) + f(x_1) + f(x_2) + \cdots + f(x_{n-1}) + \tfrac{1}{2}f(b)\right]$$

where $h = (b - a)/n$, as in (1). The x_j's and a and b are called **nodes.**

EXAMPLE 1   **Trapezoidal rule**

Evaluate $J = \displaystyle\int_0^1 e^{-x^2}\, dx$ by means of (2) with $n = 10$.

Solution. $J \approx 0.1(0.5 \cdot 1.367\,879 + 6.778\,167) = 0.746\,211$ from Table 17.3.   ◀

Table 17.3
Computations in Example 1

j	x_j	x_j^2	$e^{-x_j^2}$	
0	0	0	1.000 000	
1	0.1	0.01		0.990 050
2	0.2	0.04		0.960 789
3	0.3	0.09		0.913 931
4	0.4	0.16		0.852 144
5	0.5	0.25		0.778 801
6	0.6	0.36		0.697 676
7	0.7	0.49		0.612 626
8	0.8	0.64		0.527 292
9	0.9	0.81		0.444 858
10	1.0	1.00	0.367 879	
Sums			1.367 879	6.778 167

Error Bounds and Estimate for the Trapezoidal Rule

This error estimate can be derived from (5), Sec. 17.3, with $n = 1$, as follows. Beginning with a single subinterval, we have

$$f(x) - p_1(x) = (x - x_0)(x - x_1) \frac{f''(t)}{2}$$

with a suitable t depending on x, between x_0 and x_1. Integration over x from $a = x_0$ to $x_1 = x_0 + h$ gives

$$\int_{x_0}^{x_0+h} f(x)\, dx - \frac{h}{2}[f(x_0) + f(x_1)] = \int_{x_0}^{x_0+h} (x - x_0)(x - x_0 - h) \frac{f''(t(x))}{2}\, dx.$$

Setting $x - x_0 = v$ and applying the mean value theorem of integral calculus, which we can use because $(x - x_0)(x - x_0 - h)$ does not change sign, we find that the right side equals

$$(3^*) \qquad \int_0^h v(v - h)\, dv\, \frac{f''(\tilde{t})}{2} = \left(\frac{h^3}{3} - \frac{h^3}{2}\right) \frac{f''(\tilde{t})}{2} = -\frac{h^3}{12} f''(\tilde{t})$$

where $\tilde{t}$ is a (suitable, unknown) value between x_0 and x_1. This is the error for the trapezoidal rule with $n = 1$, often called the *local error*.

Hence the **error** ϵ of (2) with any n is the sum of such contributions from the n subintervals; since $h = (b - a)/n$, $nh^3 = n(b - a)^3/n^3$, and $(b - a)^2 = n^2h^2$, we obtain

$$(3) \qquad \boxed{\epsilon = -\frac{(b - a)^3}{12n^2} f''(\hat{t}) = -\frac{(b - a)}{12} h^2 f''(\hat{t})}$$

with (suitable, unknown) $\hat{t}$ between a and b.

Because of (3) the trapezoidal rule (2) is also written

$$(2^*) \quad J = \int_a^b f(x)\, dx = h\left[\tfrac{1}{2}f(a) + f(x_1) + \cdots + f(x_{n-1}) + \tfrac{1}{2}f(b)\right] - \frac{b - a}{12} h^2 f''(\hat{t}).$$

Error bounds are now obtained by taking the largest value for f'', say, M_2, and the smallest value, $M_2{}^*$, in the interval of integration

$$(4) \qquad \boxed{KM_2 \leqq \epsilon \leqq KM_2{}^* \qquad \text{where} \qquad K = -\frac{(b - a)^3}{12n^2} = -\frac{b - a}{12} h^2.}$$

Error estimation by halving h (the Error Principle, Sec. 17.1) is advisable if h'' is very complicated or unknown (empirical data). We calculate by (2) with h, obtaining, say, $J = J_h + \epsilon_h$, and then with $\tfrac{1}{2}h$, obtaining $J = J_{h/2} + \epsilon_{h/2}$. Now if we replace h^2 in (3) by $(\tfrac{1}{2}h)^2$, the error is multiplied by $1/4$. Hence $\epsilon_{h/2} \approx \tfrac{1}{4}\epsilon_h$ (not exactly because $\hat{t}$ may differ). Together, $J_{h/2} + \epsilon_{h/2} = J_h + \epsilon_h \approx J_h + 4\epsilon_{h/2}$. Thus $J_{h/2} - J_h \approx (4 - 1)\epsilon_{h/2}$. This gives the error formula for $J_{h/2}$

$$(5) \qquad \boxed{\epsilon_{h/2} \approx \frac{1}{3}(J_{h/2} - J_h).}$$

EXAMPLE 2　　**Error estimations for the trapezoidal rule by (4) and (5)**

Estimate the error of the approximate value in Example 1 by (4) and (5).

Solution.　**(A)** *Error bounds by* **(4).** By differentiation, $f''(x) = 2(2x^2 - 1)e^{-x^2}$. Also, $f'''(x) > 0$ if $0 < x < 1$, so that the minimum and maximum occur at the ends of the interval. We compute $M_2 = f''(1) = 0.735\,759$ and $M_2^* = f''(0) = -2$. Furthermore, $K = -1/1200$, and (4) gives

$$-0.000\,614 \leqq \epsilon \leqq 0.001\,667.$$

Hence the exact value of J must lie between

$$0.746\,211 - 0.000\,614 = 0.745\,597 \qquad \text{and} \qquad 0.746\,211 + 0.001\,667 = 0.747\,878.$$

Actually, $J = 0.746\,824$, exact to 6D.

　(B) *Error estimate by* **(5).** $J_h = 0.746211$ in Example 1. Also,

$$J_{h/2} = 0.05 \left[\sum_{j=1}^{19} \exp\left[-(j/20)^2\right] + \tfrac{1}{2}(1 + 0.367879)\right] = 0.746671.$$

Hence $\epsilon_{h/2} = \tfrac{1}{3}(J_{h/2} - J_h) = 0.000153$ and $J_{h/2} + \epsilon_{h/2} = 0.746824$, exact to 6D.　　　◀

Simpson's Rule of Integration

Piecewise constant approximation of f led to the rectangular rule (1), piecewise linear approximation to the trapezoidal rule (2), and piecewise quadratic approximation will give Simpson's rule, which is of great practical importance because it is sufficiently accurate for most problems, but still sufficiently simple.

　　To derive Simpson's rule, we divide the interval of integration $a \leqq x \leqq b$ into an ***even number*** of equal subintervals, say, into $n = 2m$ subintervals of length $h = (b - a)/2m$, with endpoints $x_0 \, (= a), x_1, \cdots, x_{2m-1}, x_{2m} \, (= b)$; see Fig. 410. We now take the first two subintervals and approximate $f(x)$ in the interval $x_0 \leqq x \leqq x_2 = x_0 + 2h$ by the Lagrange polynomial $p_2(x)$ through $(x_0, f_0), (x_1, f_1), (x_2, f_2)$, where $f_j = f(x_j)$. From (4a) in Sec. 17.3 we obtain

$$(6) \quad p_2(x) = \frac{(x - x_1)(x - x_2)}{(x_0 - x_1)(x_0 - x_2)} f_0 + \frac{(x - x_0)(x - x_2)}{(x_1 - x_0)(x_1 - x_2)} f_1 + \frac{(x - x_0)(x - x_1)}{(x_2 - x_0)(x_2 - x_1)} f_2.$$

The denominators are $2h^2, -h^2$, and $2h^2$, respectively. Setting $s = (x - x_1)/h$, we have $x - x_0 = (s + 1)h, x - x_1 = sh, x - x_2 = (s - 1)h$, and we obtain

$$p_2(x) = \tfrac{1}{2}s(s - 1)f_0 - (s + 1)(s - 1)f_1 + \tfrac{1}{2}(s + 1)sf_2.$$

We now integrate with respect to x from x_0 to x_2. This corresponds to integrating with

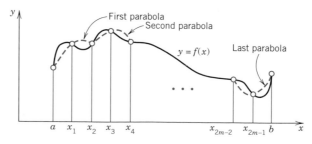

Fig. 410.　Simpson's rule

Table 17.4
Simpson's Rule of Integration

ALGORITHM SIMPSON (x_j, f_j, $j = 0, 1, \cdots, 2m$)

This algorithm computes the integral $J = \int_a^b f(x)\, dx$ from given values $f_j = f(x_j)$ at equidistant $x_0 = a$, $x_1 = x_0 + h, \cdots, x_{2m} = x_0 + 2mh = b$ by Simpson's rule (7), where $h = (b - a)/2m$.

$$\text{INPUT:} \quad a, b, m, f_0, \cdots, f_{2m}$$

$$\text{OUTPUT:} \quad \text{Approximate value } \tilde{J} \text{ of } J$$

$$\text{Compute} \quad s_0 = f_0 + f_{2m}$$

$$s_1 = f_1 + f_3 + \cdots + f_{2m-1}$$

$$s_2 = f_2 + f_4 + \cdots + f_{2m-2}$$

$$h = (b - a)/2m$$

$$\tilde{J} = \frac{h}{3}(s_0 + 4s_1 + 2s_2)$$

OUTPUT $\tilde{J}$. Stop.

End SIMPSON

---

respect to s from -1 to 1. Since $dx = h\, ds$, the result is

$$(7^*) \qquad \int_{x_0}^{x_2} f(x)\, dx \approx \int_{x_0}^{x_2} p_2(x)\, dx = h\left(\frac{1}{3} f_0 + \frac{4}{3} f_1 + \frac{1}{3} f_2\right).$$

A similar formula holds for the next two subintervals from x_2 to x_4, and so on. By summing all these m formulas we obtain **Simpson's rule**[13]

$$(7) \qquad \int_a^b f(x)\, dx \approx \frac{h}{3}(f_0 + 4f_1 + 2f_2 + 4f_3 + \cdots + 2f_{2m-2} + 4f_{2m-1} + f_{2m}),$$

where $h = (b - a)/2m$ and $f_j = f(x_j)$. Table 17.4 shows an algorithm for Simpson's rule.

Error of Simpson's rule (7). If the fourth derivative $f^{(4)}$ exists and is continuous on $a \leqq x \leqq b$, the **error** of (7) is

$$(8) \qquad \epsilon = -\frac{(b - a)^5}{180(2m)^4} f^{(4)}(\hat{t}) = -\frac{(b - a)}{180} h^4 f^{(4)}(\hat{t})$$

with (suitable, unknown) $\hat{t}$ between a and b. This is obtained similarly to (3). With this we may also write Simpson's rule (7) as

$$(7^{**}) \qquad \int_a^b f(x)\, dx = \frac{h}{3}(f_0 + 4f_1 + \cdots + f_{2m}) - \frac{(b - a)}{180} h^4 f^{(4)}(\hat{t}).$$

---

[13]THOMAS SIMPSON (1710—1761), self-taught English mathematician, author of several popular textbooks. Simpson's rule was used much earlier by Torricelli, Gregory (in 1668), and Newton (in 1676).

Error bounds. By taking for $f^{(4)}$ in (8) the maximum M_4 and minimum $M_4{}^*$ on the interval of integration we obtain from (8) the error bounds

$$(9) \qquad CM_4 \leqq \epsilon_S \leqq CM_4{}^* \qquad \text{where} \qquad C = -\frac{(b-a)^5}{180(2m)^4} = -\frac{(b-a)}{180}h^4.$$

Degree of precision (DP) *of an integration formula.* This is the maximum degree of arbitrary polynomials for which the formula gives exact values of integrals over any intervals. For the trapezoidal rule, DP = 1. For Simpson's rule we would expect DP = 2 (why?). Actually, DP = 3 by (9) because $f^{(4)}$ is identically zero for a cubic polynomial. This makes Simpson's rule sufficiently accurate for most practical problems and accounts for its popularity.

Numerical stability *with respect to round-off* is another important property of Simpson's rule. Indeed, for the sum of the rounding errors ϵ_j of the $2m + 1$ values f_j in (7) we obtain, since $h = (b - a)/2m$,

$$\frac{h}{3}|\epsilon_0 + 4\epsilon_1 + \cdots + \epsilon_{2m}| \leqq \frac{(b-a)}{3 \cdot 2m} 6mu = (b-a)u$$

where u is the round-off unit ($u = \frac{1}{2} \cdot 10^{-6}$ if we round to 6D; see Sec. 17.1). Also, 6 is the sum of the coefficients for a pair of intervals in (7). The bound $(b - a)u$ is independent of m, so that it cannot increase with increasing m (i.e., decreasing h). ◀

Newton–Cotes formulas. We mention that the trapezoidal and Simpson rules are special *closed Newton–Cotes formulas,* that is, integration formulas in which $f(x)$ is interpolated at equally spaced nodes by a polynomial of degree n ($n = 1$ for trapezoidal, $n = 2$ for Simpson), and **closed** means that a and b are nodes ($a = x_0$, $b = x_n$). $n = 3$ (the three-eighths rule; Prob. 21) and higher n are used occasionally. From $n = 8$ on, some of the coefficients become negative, so that a positive f_j could make a negative contribution to an integral, which is absurd. For more on this topic see Ref. [E13] in Appendix 1.

EXAMPLE 3 **Simpson's rule. Error estimate**

Evaluate $J = \int_0^1 e^{-x^2}\,dx$ by Simpson's rule with $2m = 10$ and estimate the error.

Solution. Since $h = 0.1$, Table 17.5 on the next page gives

$$J \approx \frac{0.1}{3}(1.367\ 879 + 4 \cdot 3.740\ 266 + 2 \cdot 3.037\ 901) = 0.746\ 825.$$

Estimate of error. Differentiation gives $f^{(4)}(x) = 4(4x^4 - 12x^2 + 3)e^{-x^2}$. By considering the derivative $f^{(5)}$ of $f^{(4)}$ we find that the largest value of $f^{(4)}$ in the interval of integration occurs at 0 and the smallest value at $x^* = (2.5 - 0.5\sqrt{10})^{1/2}$. Computation gives the values $M_4 = f^{(4)}(0) = 12$ and $M_4{}^* = f^{(4)}(x^*) = -7.419$. Since $2m = 10$ and $b - a = 1$, we obtain $C = -1/1\ 800\ 000 = -0.000\ 000\ 56$. Therefore, from (9),

$$-0.000\ 007 \leqq \epsilon_S \leqq 0.000\ 005.$$

Hence J must lie between $0.746\ 825 - 0.000\ 007 = 0.746\ 818$ and $0.746\ 825 + 0.000\ 005 = 0.746\ 830$, so that at least four digits of our approximate value are exact. Actually, $0.746\ 825$ is exact to 5D because $J = 0.746\ 824$ (exact to 6D).

Thus our result is much better than that in Example 1 obtained by the trapezoidal rule, whereas the number of operations is nearly the same in both cases. ◀

Instead of picking an $n = 2m$ and then estimating the error by (9), as in Example 3, it is better to require an accuracy (e.g., 6D) and then determine $n = 2m$ from (9).

Table 17.5
Computations in Example 3

j	x_j	x_j^2		$e^{-x_j^2}$	
0	0	0	1.000 000		
1	0.1	0.01		0.990 050	
2	0.2	0.04			0.960 789
3	0.3	0.09		0.913 931	
4	0.4	0.16			0.852 144
5	0.5	0.25		0.778 801	
6	0.6	0.36			0.697 676
7	0.7	0.49		0.612 626	
8	0.8	0.64			0.527 292
9	0.9	0.81		0.444 858	
10	1.0	1.00	0.367 879		
Sums			1.367 879	3.740 266	3.037 901

EXAMPLE 4   **Determination of $n = 2m$ in Simpson's rule from the required accuracy**

What n should we choose in Example 3 to get 6D-accuracy?

Solution. Using $M_4 = 12$ (which is bigger in absolute value than $M_4{}^*$), we get from (9), with $b - a = 1$ and the required accuracy,

$$|CM_4| = \frac{12}{180(2m)^4} = \frac{1}{2} \cdot 10^{-6}, \qquad \text{thus} \qquad m = \left[\frac{2 \cdot 10^6 \cdot 12}{180 \cdot 2^4} \right]^{1/4} = 9.55.$$

Hence we should choose $n = 2m = 20$. Do the computation, which parallels that in Example 3.

Note that the error bounds in (4) or (9) may sometimes be loose, so that in such a case a smaller $n = 2m$ may already suffice. ◀

Error estimation for Simpson's rule by halving h. The idea is the same as in (5) and gives

(10)
$$\epsilon_{h/2} \approx \frac{1}{15} (J_{h/2} - J_h).$$

J_h is obtained by using h and $J_{h/2}$ by using $\frac{1}{2}h$, and $\epsilon_{h/2}$ is the error of $J_{h/2}$.

Derivation. In (5) we had $\frac{1}{3}$ as the reciprocal of $3 = 4 - 1$ and $\frac{1}{4} = \left(\frac{1}{2}\right)^2$ resulted from h^2 in (3) by replacing h with $\frac{1}{2}h$. In (10) we have $\frac{1}{15}$ as the reciprocal of $15 = 16 - 1$ and $\frac{1}{16} = \left(\frac{1}{2}\right)^4$ results from h^4 in (8) by replacing h with $\frac{1}{2}h$.

EXAMPLE 5   **Error estimation for Simpson's rule by halving**

Integrate $f(x) = \frac{1}{4}\pi x^4 \cos \frac{1}{4}\pi x$ from 0 to 2 with $h = 1$ and apply (10).

Solution. $J_h = \frac{1}{3}[f(0) + 4f(1) + f(2)] = \frac{1}{3}(0 + 4 \cdot 0.555360 + 0) = 0.740480,$

$$J_{h/2} = \frac{1}{6} \left[f(0) + 4f\left(\frac{1}{2}\right) + 2f(1) + 4f\left(\frac{3}{2}\right) + f(2) \right]$$

$$= \frac{1}{6} [0 + 4 \cdot 0.045351 + 2 \cdot 0.555361 + 4 \cdot 1.521579 + 0] = 1.22974.$$

Hence (10) gives $\epsilon_{h/2} = \frac{1}{15}(1.22974 - 0.74048) = 0.032617$ and thus $J \approx J_{h/2} + \epsilon_{h/2} = 1.26236$, with an error -0.00283, which is less in absolute value than $\frac{1}{10}$ of the error 0.02979 of $J_{h/2}$. Hence the use of (10) was well worthwhile. Exact 5D-value: $J = 1.25953$. ◀

Adaptive Integration

The idea is to adapt step h to the variability of $f(x)$: small h where the variability is large, and large h where it is small. Changing h is done systematically, usually by halving h, and automatically (not "by hand") depending on the size of the (estimated) error over a subinterval. The subinterval is halved if the corresponding error is still too large, that is, larger than a given **tolerance** TOL (maximum admissible absolute error), or is not halved if the error is less than or equal to TOL.

Adaptive integration can be applied together with any integration formula. We explain it here in connection with Simpson's rule.

EXAMPLE 6   **Adaptive integration (with Simpson's rule)**

Integrate $f(x) = \frac{1}{4}\pi x^4 \cos \frac{1}{4}\pi x$ from $x = 0$ to 2 by adaptive integration and with Simpson's rule and TOL[0, 2] = 0.0002.

Solution.   Table 17.6 shows the calculations. Figure 411 shows the integrand $f(x)$ and the adapted intervals used. The first two intervals ([0, 0.5], [0.5, 1.0]) have length 0.5, hence $h = 0.25$ [because we use $2m = 2$ subintervals in Simpson's rule (7**)]. The next two intervals ([1.00, 1.25], [1.25, 1.50]) have length 0.25 (hence $h = 0.125$) and the last four intervals have length 0.125. *Sample computations.* For 0.740480 see Example 5. Formula (10) gives $(0.123716 - 0.122794)/15 = 0.000061$. Note that 0.123716 refers to [0, 0.5] and [0.5, 1],

Table 17.6
Computations in Example 6

Interval	Integral	Error (10)	TOL	Comment
[0, 2]	0.740480		0.0002	
[0, 1]	0.122794			
[1, 2]	1.10695			
	Sum = 1.22974	0.032617	0.0002	Divide further
[0.0, 0.5]	0.004782			
[0.5, 1.0]	0.118934			
	Sum = 0.123716*	0.000061	0.0001	TOL reached
[1.0, 1.5]	0.528176			
[1.5, 2.0]	0.605821			
	Sum = 1.13300	0.001803	0.0001	Divide further
[1.00, 1.25]	0.200544			
[1.25, 1.50]	0.328351			
	Sum = 0.528895*	0.000048	0.00005	TOL reached
[1.50, 1.75]	0.388235			
[1.75, 2.00]	0.218457			
	Sum = 0.606692	0.000058	0.00005	Divide further
[1.500, 1.625]	0.196244			
[1.625, 1.750]	0.192019			
	Sum = 0.388263*	0.000002	0.000025	TOL reached
[1.750, 1.875]	0.153405			
[1.875, 2.000]	0.065078			
	Sum = 0.218483*	0.000002	0.000025	TOL reached

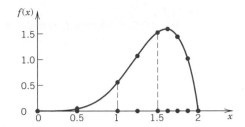

Fig. 411. Adaptive integration in Example 6

so that we must subtract the value corresponding to [0, 1] in the line before. Etc. TOL[0, 2] = 0.0002 gives 0.0001 for subintervals of length 1, 0.00005 for length 0.5, etc. The value of the integral obtained is the sum of the values marked by an asterisk (for which the error estimate has become less than TOL). This gives

$$J \approx 0.123716 + 0.528895 + 0.388263 + 0.218483 = 1.25936.$$

The exact 5D-value is $J = 1.25953$. Hence the error is 0.00017. This is about 1/200 of the absolute value of that in Example 5. Our more extensive computation has produced a much better result. ◀

Gauss Integration Formulas. Maximum Degree of Precision

Our integration formulas discussed so far use function values at *predetermined* (equidistant) x-values (nodes) and give exact results for polynomials not exceeding a certain degree [called the *degree of precision;* see after (9)]. But we can get much more accurate integration formulas as follows. We set

(11)
$$\int_{-1}^{1} f(t)\, dt \approx \sum_{j=1}^{n} A_j f_j \qquad\qquad [f_j = f(t_j)]$$

with fixed n (and $t = \pm 1$ obtained from $x = a, b$ by setting $x = \frac{1}{2}[a(t - 1) + b(t + 1)]$). Then we determine the n coefficients $A_1, \cdots, A_n$ and n nodes $t_1, \cdots, t_n$ so that (11) gives exact results for polynomials of degree k as high as possible. Since $n + n = 2n$ is the number of coefficients of a polynomial of degree $2n - 1$, it follows that $k \leqq 2n - 1$. Gauss has shown that exactness for polynomials of degree not exceeding $2n - 1$ (instead of $n - 1$ for predetermined nodes) can be attained, and he has given the location of the t_j (= the jth zero of the Legendre polynomial P_n in Sec. 4.3) and the coefficients A_j [which depend on n but not on $f(t)$]. With these t_j and A_j formula (11) is called a **Gauss integration formula** or *Gauss quadrature formula*. Table 17.7 on the next page gives the values for $n = 2, \cdots, 5$. (For larger n, see pp. 916–919 of Ref. [1] in Appendix 1.)

EXAMPLE 7   **Gauss integration formula with $n = 3$**

Evaluate the integral in Example 3 by the Gauss integration formula (11) with $n = 3$.

Solution.   We have to convert our integral from 0 to 1 into an integral from -1 to 1. We set $x = \frac{1}{2}(t + 1)$. Then $dx = \frac{1}{2}dt$, and (11) with $n = 3$ and the above values of the zeros and the coefficients yields

$$\int_{0}^{1} \exp(-x^2)\, dx = \frac{1}{2} \int_{-1}^{1} \exp\left(-\frac{1}{4}(t + 1)^2\right) dt$$

$$\approx \frac{1}{2}\left[\frac{5}{9} \exp\left(-\frac{1}{4}\left(1 - \sqrt{\frac{3}{5}}\right)^2\right) + \frac{8}{9} \exp\left(-\frac{1}{4}\right) + \frac{5}{9} \exp\left(-\frac{1}{4}\left(1 + \sqrt{\frac{3}{5}}\right)^2\right) \right] = 0.746\,815$$

(exact to 6D: 0.746 824), which is almost as accurate as the Simpson result obtained in Example 3 with a much

Table 17.7
Gauss Integration: Nodes t_j and Coefficients A_j

n	Nodes t_j	Coefficients A_j	Degree of Precision
2	$-0.57735\ 02692$	1	3
	$0.57735\ 02692$	1	
3	$-0.77459\ 66692$	$0.55555\ 55556$	5
	0	$0.88888\ 88889$	
	$0.77459\ 66692$	$0.55555\ 55556$	
4	$-0.86113\ 63116$	$0.34785\ 48451$	7
	$-0.33998\ 10436$	$0.65214\ 51549$	
	$0.33998\ 10436$	$0.65214\ 51549$	
	$0.86113\ 63116$	$0.34785\ 48451$	
5	$-0.90617\ 98459$	$0.23692\ 68851$	9
	$-0.53846\ 93101$	$0.47862\ 86705$	
	0	$0.56888\ 88889$	
	$0.53846\ 93101$	$0.47862\ 86705$	
	$0.90617\ 98459$	$0.23692\ 68851$	

larger number of arithmetical operations. With 3 function values (as in this example) and Simpson's rule we would get $\frac{1}{6}(1 + 4e^{-0.25} + e^{-1}) = 0.747\ 180$, with an error over 30 times that of the Gauss integration. ◄

EXAMPLE 8   **Gauss integration formula with $n = 4$ and 5**

Integrate $f(x) = \frac{1}{4}\pi x^4 \cos \frac{1}{4}\pi x$ from $x = 0$ to 2.

Solution. $x = t + 1$ gives $f(t) = \frac{1}{4}\pi(t + 1)^4 \cos (\frac{1}{4}\pi(t + 1))$, as needed in (11). For $n = 4$ we calculate (6S)

$$J \approx A_1 f_1 + \cdots + A_4 f_4 = A_1(f_1 + f_4) + A_2(f_2 + f_3)$$

$$= 0.347855\,(0.000290309 + 1.02570) + 0.652145\,(0.129464 + 1.25459) = 1.25950.$$

The error is 0.00003 because $J = 1.25953$ (6S). Calculating with 10S and $n = 4$ gives the same result; so the error is due to the formula, not rounding. For $n = 5$ and 10S we get $J \approx 1.25952\ 6185$, too large by 0.000000250 because $J = 1.25952\ 5935$ (10S). The accuracy is impressive, particularly if we compare the amount of work with that in Example 6. ◄

Gauss integration is popular in software. Its great accuracy outweighs the disadvantage of the complicated t_j and A_j (which have to be stored) whenever the integrand is given by a formula (not just by a table of numbers) or when experimental recordings can be set at times t_j (or whatever t represents) once and for all. Furthermore, Gauss coefficients A_j are positive for all n, in contrast to some of the Newton–Cotes coefficients for larger n. Of course, there are frequent applications with equally spaced nodes, so that Gauss integration does not apply (or has no great advantage if one first has to get the t_j in (11) by interpolation).

Since the endpoints -1 and 1 of the interval of integration in (11) are not zeros of P_n, they do not occur among $x_0, \cdots, x_n$, and the Gauss formula (11) is called, therefore, an **open formula,** in contrast to a **closed formula,** in which the endpoints of the interval of integration are x_0 and x_n. [For example, (2) and (7) are closed formulas.]

Numerical Differentiation

Numerical differentiation is the computation of values of the derivative of a function f from given values of f. Numerical differentiation should be avoided whenever possible, because, whereas *integration* is a *smoothing* process and is not affected much by small inaccuracies in function values, *differentiation* tends to make matters *rough* and generally give values of f' much less accurate than those of f—remember that the derivative is the limit of the difference quotient, and in the latter you usually have a small difference of large quantities that you then divide by a small quantity. However, the formulas to be obtained will be basic in the numerical solution of differential equations.

We use the notations $f'_j = f'(x_j)$, $f''_j = f''(x_j)$, etc., and may obtain rough approximation formulas for derivatives by remembering that

$$f'(x) = \lim_{h \to 0} \frac{f(x + h) - f(x)}{h} .$$

This suggests

(12)
$$f'_{1/2} \approx \frac{\delta f_{1/2}}{h} = \frac{f_1 - f_0}{h} .$$

Similarly, for the second derivative we obtain

(13)
$$f''_1 \approx \frac{\delta^2 f_1}{h^2} = \frac{f_2 - 2f_1 + f_0}{h^2} ,$$

and so on.

More accurate approximations are obtained by differentiating suitable Lagrange polynomials. Differentiating (6) and remembering that the denominators in (6) are $2h^2$, $-h^2$, $2h^2$, we have

$$f'(x) \approx p'_2(x) = \frac{2x - x_1 - x_2}{2h^2} f_0 - \frac{2x - x_0 - x_2}{h^2} f_1 + \frac{2x - x_0 - x_1}{2h^2} f_2.$$

Evaluating this at x_0, x_1, x_2, we obtain the "three-point formulas"

(14)

    (a) $f'_0 \approx \dfrac{1}{2h} (-3f_0 + 4f_1 - f_2),$

    (b) $f'_1 \approx \dfrac{1}{2h} (-f_0 + f_2),$

    (c) $f'_2 \approx \dfrac{1}{2h} (f_0 - 4f_1 + 3f_2).$

Applying the same idea to the Lagrange polynomial $p_4(x)$, we obtain similar formulas, in particular,

(15) $$f_2' \approx \frac{1}{12h}(f_0 - 8f_1 + 8f_3 - f_4).$$

Further details and formulas are included in Ref. [E3] listed in Appendix 1.

PROBLEM SET 17.5

1. **(Rectangular rule)** Show that (1) with $n = 5$ gives $J = 0.748\ 053$ for the integral in Example 1 (0.746 824 exact to 6S).
2. **(Error bounds for the rectangular rule)** Derive a formula for such lower and upper bounds and apply it to Prob. 1.
3. **(Trapezoidal rule)** To get a feeling for the increase in accuracy, integrate x^2 from 0 to 1 by (2) with $h = 1$, $h = 0.5$, $h = 0.25$.
4. **(Error estimate by halving)** Integrate $f(x) = x^4$ from 0 to 1 by (2) with $h = 1$, $h = 0.5$, $h = 0.25$ and estimate the error for $h = 0.5$ and for $h = 0.25$ by (5).
5. **(Error estimate)** Do the tasks in Prob. 4 for $f(x) = \sin \frac{1}{2}\pi x$.
6. **(Stability)** Prove that the trapezoidal rule is stable with respect to round-off.

Simpson's Rule
Evaluate the following integrals numerically as indicated in Probs. 7–12, and compare the result with the exact value obtained by a formula known from calculus.

$$A = \int_1^2 \frac{dx}{x}, \qquad B = \int_0^{0.4} x\,e^{-x^2}\,dx, \qquad J = \int_0^1 \frac{dx}{1+x^2}.$$

7. A, $2m = 4$ 8. A, $2m = 10$ 9. B, $2m = 4$
10. B, $2m = 10$ 11. J, $2m = 4$ 12. J, $2m = 10$
13. **(Error estimate for Simpson's rule)** Calculate the integral J (above) by Simpson's rule with $2m = 8$ and use the value and that in Prob. 11 to estimate the error by (10).
14. **(Error bounds and estimate)** Integrate e^{-x} from 0 to 2 by (7) with $h = 1$ and with $h = 0.5$. Give error bounds for the $h = 0.5$ value by (9) and an error estimate by (10).

Nonelementary Integrals
In Probs. 15–20 evaluate the following integrals numerically as indicated.

$$\text{Si}(x) = \int_0^x \frac{\sin \tilde{x}}{\tilde{x}}\,d\tilde{x}, \qquad C(x) = \int_0^x \cos(\tilde{x}^2)\,d\tilde{x}, \qquad J(x) = \int_0^x J_0(\tilde{x})\,d\tilde{x}.$$

$\text{Si}(x)$ is the **sine integral**, $C(x)$ a **Fresnel integral** (see Appendix A3.1), $J_0(x)$ the **Bessel function** of order 0 (use Table A1 in Appendix 5).
15. $\text{Si}(1)$ by (2), $n = 5$, $n = 10$ 16. $\text{Si}(1)$ by (7), $2m = 2$ and $2m = 10$

17. C(2) by (7), $2m = 10$ **18.** C(2) by (2), $n = 10$

19. J(1) by (2), $n = 10$ **20.** J(1) by (7), $2m = 10$

21. Three-eights rule:

$$\int_a^b f(x)\,dx = \frac{3}{8}\,h\,(f_0 + 3f_1 + 3f_2 + f_3) - \frac{(b-a)}{80}\,h^4 f^{(4)}(\hat{t}).$$

Apply this formula to the integral of $\cos x$ from 0 to $\frac{1}{2}\pi$ and give error bounds.

22. (Given TOL) Find the smallest n in computing A (see Probs. 7, 8) for which 5D-accuracy is guaranteed (a) by (4) in the use of (2), (b) by (9) in the use of (7). Compare and comment.

23. (Gauss integration) Solve Prob. 21 by Gauss integration with $n = 5$.

24. TEAM PROJECT. Romberg integration[14] uses the trapezoidal rule and gains precision stepwise by halving h and adding an error estimate. Do this for the integral of $f(x) = e^{-x}$ from $x = 0$ to $x = 2$ with TOL $= 10^{-3}$, as follows.

1st Step. Apply the trapezoidal rule (2) with $h = 2$ (hence $n = 1$) to get an approximation J_{11}. Halve h and use (2) to get J_{21} and an error estimate

$$\epsilon_{21} = \frac{1}{2^2 - 1}\,(J_{21} - J_{11}).$$

If $|\epsilon_{21}| \leqq$ TOL, stop. The result is $J_{22} = J_{21} + \epsilon_{21}$.

2nd Step. Show that $\epsilon_{21} = -0.066596$, hence $|\epsilon_{21}| >$ TOL and go on. Use (2) with $h/4$ (another halving!) to get J_{31} and add to it the error estimate $\epsilon_{31} = \frac{1}{3}(J_{31} - J_{21})$ to get the better $J_{32} = J_{31} + \epsilon_{31}$. Calculate

$$\epsilon_{32} = \frac{1}{2^4 - 1}\,(J_{32} - J_{22}) = \frac{1}{15}\,(J_{32} - J_{22}).$$

If $|\epsilon_{32}| \leqq$ TOL, stop. The result is $J_{33} = J_{32} + \epsilon_{32}$. (Why does $2^4 = 16$ come in?) Show that $\epsilon_{32} = -0.000266$, so that we can stop. Arrange your J- and ϵ-values in a kind of "difference table."

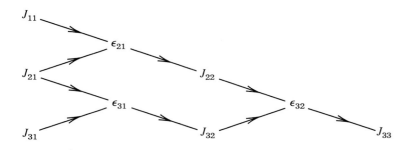

If $|\epsilon_{32}|$ were greater than TOL, you would have to go on and calculate in the next step J_{41} from (2) with $h = \frac{1}{4}$; then

[14]W. Romberg, *Kgl. Norske Videnskab. Selskab, Trondheim, Førh.* **28,** No. 7 (1955).

$$J_{42} = J_{41} + \epsilon_{41} \quad \text{with} \quad \epsilon_{41} = \frac{1}{3}(J_{41} - J_{31})$$

$$J_{43} = J_{42} + \epsilon_{42} \quad \text{with} \quad \epsilon_{42} = \frac{1}{15}(J_{42} - J_{32})$$

$$J_{44} = J_{43} + \epsilon_{43} \quad \text{with} \quad \epsilon_{43} = \frac{1}{63}(J_{43} - J_{33})$$

where $63 = 2^6 - 1$. (How does this come in?)

Apply the Romberg method to the integral of $f(x) = \frac{1}{4}\pi x^4 \cos \frac{1}{4}\pi x$ from $x = 0$ to 2 with TOL $= 10^{-4}$.

Differentiation

25. Consider $f(x) = x^4$ for $x_0 = 0$, $x_1 = 0.2$, $x_2 = 0.4$, $x_3 = 0.6$, $x_4 = 0.8$. Calculate f_2' from (14a), (14b), (14c), (15). Determine the errors. Compare and comment.

26. A **"four-point formula"** for the derivative is

$$f_2' \approx \frac{1}{6h}(-2f_1 - 3f_2 + 6f_3 - f_4).$$

Apply it to $f(x) = x^4$ with $x_1, \cdots, x_4$ as in Prob. 25, determine the error, and compare it with that in the case of (15).

27. The derivative $f'(x)$ can also be approximated in terms of first-order and higher order differences (see Sec. 17.3):

$$f'(x_0) \approx \frac{1}{h}\left(\Delta f_0 - \frac{1}{2}\Delta^2 f_0 + \frac{1}{3}\Delta^3 f_0 - \frac{1}{4}\Delta^4 f_0 + - \cdots\right).$$

Compute $f'(0.4)$ in Prob. 25 from this formula, using differences up to and including first order, second order, third order, fourth order.

28. Derive the formula in Prob. 27 from (14) in Sec. 17.3.

CHAPTER 17 REVIEW

1. What is a numerical method and why do we need such methods?
2. What is an error? A relative error? An error bound?
3. What are round-off errors? Why are they important? State the rules of rounding.
4. What does floating point mean? Overflow, underflow?
5. What do you know about stability?
6. What effect do computers have on the development of numerical methods?
7. Make a list of the main tasks that we have considered in this chapter.
8. What do you know about methods for solving equations?
9. What does polynomial interpolation mean? Spline interpolation?
10. What is an algorithm? Give basic examples.
11. What properties of an algorithm are important from the viewpoint of the computer?

12. List and compare numerical integration methods. When would you apply one of them?

13. What does adaptive integration mean? Give examples.

14. In what sense are Gauss integration formulas best possible?

15. What do we mean by the order of convergence of an iteration?

16. What do you remember about different error estimations for polynomial interpolation?

17. How did we use a Newton interpolation polynomial in the derivation of Simpson's rule?

18. What are closed and open integration formulas? Give examples.

19. What are Bezier curves?

20. How did we obtain formulas for numerical differentiation?

21. Write -4.268106, 8002.27, -0.00518789, 0.0178164, $1/3$, $-3/700$ in floating-point form with 4S (4 significant digits), with round-off.

22. Write 14.91009, -0.09184162, 3030.301, $-9/110$, $1/1024$ in floating-point form with 5S, with round-off.

23. Compute $(5.346 - 3.644)/(3.454 - 3.055)$ as given and then rounded stepwise to 3S, 2S, 1S. Comment. ("Stepwise" means rounding the four rounded numbers, not the given ones.)

24. Compute $0.29731/(4.1232 - 4.0872)$ as given and then rounded stepwise to 4S, 3S, 2S. Comment.

25. Let 9.1 and 15.84 be correctly rounded. Find the shortest interval in which the sum s of the true (unrounded) numbers must lie.

26. Do the same task as in Prob. 25 for the difference $43.2 - 16.29$.

27. What is the relative error of $n\tilde{a}$ in terms of that of $\tilde{a}$? (Here, $\tilde{a}$ is an approximation of a.)

28. Show that the relative error of $\tilde{a}^4$ is about 4 times that of $\tilde{a}$.

29. Solve $x^2 - 200x + 4 = 0$ by (6) and by (7), Sec. 17.1, using 5S in the computation. Compare and comment.

30. Find a good way to compute $\sqrt{x^2 + 16} - 4$ for small $|x|$.

31. Compute the solution of $x^4 = x + 0.12$ near $x = 0$ by transforming the equation algebraically into the form $x = g(x)$ and starting from $x_0 = 0$ (8S = 8 significant digits).

32. Solve $\cos x = x$ by iteration (6S, $x_0 = 1$), writing it as $x = (0.74x + \cos x)/1.74$, obtaining $x_4 = 0.739\,085$ (exact to 6S!). Why does this converge so rapidly?

33. Find all (real) solutions of $\sin x = \frac{1}{2}x$ by Newton's method.

34. Do the same task as in Prob. 33 for $\cos x = \sqrt{x}$.

35. Solve $x^4 - x^3 - 2x - 34 = 0$ by Newton's method.

36. Solve $\cos x = 2x$ by the method of false position.

37. Solve $\cosh x = 2 - x^2$ by the bisection method.

38. Let $f(0) = 1$, $f(1) = -1$, $f(2) = 0$. Find $f(0.3)$ by linear and by quadratic interpolation.

39. Compute $\sinh 0.3$ from $\sinh(-0.5) = -0.521$, $\sinh 0 = 0$, $\sinh 1 = 1.175$ by quadratic interpolation.

40. Find $f(1.3)$ by cubic interpolation from $f(0.5) = -2.325$, $f(1.0) = 0.300$, $f(1.5) = 5.425$, $f(2.0) = 14.700$.

41. Find the cubic spline for the data $f(0) = 0$, $f(1) = 0$, $f(2) = 4$, $k_0 = -1$, $k_2 = 5$.

42. Find the cubic spline for the data $f(-1) = 3$, $f(1) = 1$, $f(3) = 23$, $f(5) = 45$, $k_0 = k_3 = 3$.

43. Integrate x^3 from 0 to 1 by the trapezoidal rule with $n = 5$ and give error bounds.

44. Integrate $\cos(x^2)$ from 0 to 1 by Simpson's rule with $2m = 2$ and $2m = 4$ and give an error estimate by (10), Sec. 17.5. (The integral is a *Fresnel integral* $C(x)$ with $x = 1$; see (38), Appendix A3.1.)

45. Evaluate the integral in Prob. 44 by Gauss integration with $n = 3$ and $n = 5$.

SUMMARY OF CHAPTER 17
NUMERICAL METHODS IN GENERAL

In this chapter we discussed concepts that are relevant throughout numerical work as a whole and methods of a general nature, as opposed to methods for problems in linear algebra (Chap. 18) or in differential equations (Chap. 19).

In scientific computations we use the **floating-point** representation of numbers (Sec. 17.1); fixed-point representation is less suitable in most cases.

Numerical methods give approximate values $\tilde{a}$ of quantities. The **error** ϵ of $\tilde{a}$ is

$$(1) \qquad\qquad \epsilon = a - \tilde{a} \qquad\qquad \text{(Sec. 17.1)}$$

where a is the exact value. The *relative error* of $\tilde{a}$ is ϵ/a. Errors arise from rounding off, inaccuracy of measured values, truncation (that is, replacement of integrals by sums, derivatives by difference quotients, series by partial sums), and so on.

An algorithm is called **numerically stable** if small changes in the initial data give only correspondingly small changes in the final results. Unstable algorithms are generally useless because errors may become so large that results will be very inaccurate. Numerical instability of algorithms must not be confused with mathematical instability of problems (*"ill-conditioned problems,"* Sec. 17.2).

Fixed-point iteration is a method for solving equations $f(x) = 0$ in which the equation is first transformed algebraically to $x = g(x)$, an initial guess x_0 for the solution is made, and then approximations $x_1, x_2, \cdots$, are successively computed by iteration from (see Sec. 17.2)

$$(2) \qquad\qquad x_{n+1} = g(x_n) \qquad\qquad (n = 0, 1, \cdots).$$

Newton's method for solving equations $f(x) = 0$ is an iteration

$$(3) \qquad\qquad x_{n+1} = x_n - \frac{f(x_n)}{f'(x_n)} \qquad\qquad \text{(Sec. 17.2)}.$$

Here x_{n+1} is the x-intercept of the tangent of the curve $y = f(x)$ at the point x_n. This method is of second order (Theorem 2, Sec. 17.2). If we replace f' in (3) by a difference quotient (geometrically: we replace the tangent by a secant), we obtain the **secant method;** see (10) in Sec. 17.2. The *bisection method* (which converges slowly) and the *method of false position,* see Problem Set 17.2.

Polynomial interpolation means the determination of a polynomial $p_n(x)$ such that $p_n(x_j) = f_j$, where $j = 0, \cdots, n$ and $(x_0, f_0), \cdots, (x_n, f_n)$ are measured or observed values, values of a function, etc. $p_n(x)$ is called an *interpolation polynomial.* For given data, $p_n(x)$ of degree n (or less) is unique. However, it can be written in different forms, notably in **Lagrange's form** (4), Sec. 17.3, or in **Newton's divided difference form** (10), Sec. 17.3, which requires fewer operations. For regularly spaced $x_0, x_1 = x_0 + h, \cdots, x_n = x_0 + nh$ the latter becomes **Newton's forward difference formula** (Sec. 17.3)

$$f(x) \approx p_n(x) = f_0 + r\Delta f_0 + \frac{r(r-1)}{2!}\Delta^2 f_0 + \cdots$$

(4)

$$\cdots + \frac{r(r-1)\cdots(r-n+1)}{n!}\Delta^n f_0$$

where $r = (x - x_0)/h$ and the forward differences are $\Delta f_j = f_{j+1} - f_j$ and

$$\Delta^k f_j = \Delta^{k-1} f_{j+1} - \Delta^{k-1} f_j \qquad (k = 2, 3, \cdots).$$

A similar formula is *Newton's backward difference interpolation formula* (Sec. 17.3).

Interpolation polynomials may become numerically unstable as n increases, and instead of interpolating and approximating by a single high-degree polynomial it is preferable to use a cubic **spline** $g(x)$, that is, a twice continuously differentiable interpolation function [thus, $g(x_j) = f_j$], which in each subinterval $x_j \leqq x \leqq x_{j+1}$ consists of a cubic polynomial $p_j(x)$; see Sec. 17.4.

Simpson's rule of numerical integration is [see (7), Sec. 17.5]

$$(5) \quad \int_a^b f(x)\,dx \approx \frac{h}{3}(f_0 + 4f_1 + 2f_2 + 4f_3 + \cdots + 2f_{2m-2} + 4f_{2m-1} + f_{2m})$$

with equally spaced nodes $x_j = x_0 + jh$, $j = 1, \cdots, 2m$, $h = (b - a)/2m$, and $f_j = f(x_j)$. It is simple but accurate enough for many applications (degree of precision DP = 3 because the error (8), Sec. 17.5, involves h^4). A more practical error estimate is (10), Sec. 17.5,

$$\epsilon_{h/2} = \frac{1}{15}(J_{h/2} - J_h)$$

obtained by first computing with step h, then with step $h/2$, and then taking 1/15 of the difference of the results.

Simpson's rule is the most important of the **Newton–Cotes formulas,** which are obtained by integrating Lagrange interpolation polynomials, linear ones for the **trapezoidal rule** (2), Sec. 17.5, quadratic for Simpson's rule, cubic for the *three-eights rule,* etc.

Adaptive integration (Sec. 17.5, Example 6) is integration that adjusts (*"adapts"*) the step (automatically) to the variability of $f(x)$.

Romberg integration (Team Project 24, Problem Set 17.5) starts from the trapezoidal rule (2), Sec. 17.5, with h, $h/2$, $h/4$, etc. and improves results by systematically adding error estimates (a process called Richardson *extrapolation* because the improved values are no longer related to *interpolating* polynomials).

Gauss integration (11), Sec. 17.5, is important because of its great accuracy (DP = $2n - 1$, compared to Newton–Cotes's DP = $n - 1$ or n). This is achieved by an optimal choice of the nodes, which are not equally spaced; see Table 17.7, Sec. 17.5.

Numerical differentiation is discussed at the end of Sec. 17.5. (Its main application (to differential equations) follows in Chap. 19.)

Numerical Methods in Linear Algebra

In this chapter we consider some of the most important numerical methods for solving linear systems of equations (Secs. 18.1–18.4), for fitting straight lines or parabolas (Sec. 18.5), and for matrix eigenvalue problems (Secs. 18.6–18.9). These and similar methods are of great practical importance. Indeed, many engineering or other problems, for instance, in statistics, lead to mathematical models whose solution requires methods of numerical linear algebra.

This chapter is independent of Chap. 17 and can be studied immediately after Chap. 6 or 7.

Prerequisite for this chapter: Secs. 6.1, 6.2, 7.1.
Sections that may be omitted in a shorter course: Secs. 18.4–18.6, 18.9.
References: Appendix 1, Part E.
Answers to problems: Appendix 2.

18.1 Linear Systems: Gauss Elimination

A **linear system of n equations** in n *unknowns* $x_1, \cdots, x_n$ is a set of equations $E_1, \cdots, E_n$ of the form

$$
\begin{aligned}
E_1: &\quad a_{11}x_1 + \cdots + a_{1n}x_n = b_1 \\
E_2: &\quad a_{21}x_1 + \cdots + a_{2n}x_n = b_2 \\
&\quad \cdots\cdots\cdots\cdots\cdots\cdots \\
E_n: &\quad a_{n1}x_1 + \cdots + a_{nn}x_n = b_n
\end{aligned}
$$

(1)

where the **coefficients** a_{jk} and the b_j are given numbers. The system is called **homogeneous** if all the b_j are zero; otherwise it is **nonhomogeneous.** Using matrix multiplication (Sec. 6.2), we can write (1) as a single vector equation

(2)
$$
\mathbf{Ax} = \mathbf{b}
$$

where the **coefficient matrix** $\mathbf{A} = [a_{jk}]$ is the $n \times n$ matrix

$$\mathbf{A} = \begin{bmatrix} a_{11} & a_{12} & \cdots & a_{1n} \\ a_{21} & a_{22} & \cdots & a_{2n} \\ . & . & \cdots & . \\ a_{n1} & a_{n2} & \cdots & a_{nn} \end{bmatrix}, \quad \text{and} \quad \mathbf{x} = \begin{bmatrix} x_1 \\ \vdots \\ x_n \end{bmatrix} \quad \text{and} \quad \mathbf{b} = \begin{bmatrix} b_1 \\ \vdots \\ b_n \end{bmatrix}$$

are column vectors. The **augmented matrix** $\widetilde{\mathbf{A}}$ of the system (1) is

$$\widetilde{\mathbf{A}} = [\mathbf{A} \quad \mathbf{b}] = \begin{bmatrix} a_{11} & \cdots & a_{1n} & b_1 \\ a_{21} & \cdots & a_{2n} & b_2 \\ . & \cdots & . & . \\ a_{n1} & \cdots & a_{nn} & b_n \end{bmatrix}.$$

A **solution** of (1) is a set of numbers $x_1, \cdots, x_n$ that satisfy all the n equations, and a **solution vector** of (1) is a vector $\mathbf{x}$ whose components constitute a solution of (1).

The method of solving such a system by determinants (Cramer's rule in Sec. 6.6) is not practical, even with efficient methods for evaluating the determinants.

A practical method for the solution of a linear system is the so-called *Gauss elimination*, which we shall now discuss (proceeding independently of Sec. 6.3).

Gauss Elimination

This standard method for solving linear systems (1) is a systematic process of elimination that reduces (1) to **"triangular form"** because the system can then be easily solved by **"back substitution."** For instance, a triangular system is

$$3x_1 + 5x_2 + 2x_3 = 8$$
$$8x_2 + 2x_3 = -7$$
$$6x_3 = 3$$

and back substitution gives $x_3 = 3/6 = 1/2$ from the third equation, then

$$x_2 = \tfrac{1}{8}(-7 - 2x_3) = -1$$

from the second equation, and finally from the first equation

$$x_1 = \tfrac{1}{3}(8 - 5x_2 - 2x_3) = 4.$$

How do we reduce a given system (1) to triangular form? In the first step we eliminate x_1 from equations E_2 to E_n in (1). We do this by subtracting[1] suitable multiples of E_1

---

[1] This is better than using two names ("addition" and "subtraction") when referring to a single operation. ["Subtraction" seems more suggestive than "addition" (used in Chap. 6) in the numerical process of creating zeros.]

from equations $E_2, \cdots, E_n$ and taking the resulting equations, call them $E_2*, \cdots, E_n*$ as the new equations. The first equation, E_1, is called the **pivot equation** in this step, and a_{11} is called the **pivot.** This equation is left unaltered. In the second step we take the new second equation E_2* (which no longer contains x_1) as the pivot equation and use it to eliminate x_2 from E_3* to E_n*. And so on. After $n - 1$ steps this gives a triangular system that can be solved by back substitution as just shown. In this way we obtain precisely all solutions of the *given* system (as proved in Sec. 6.3).

The pivot a_{kk} (in step k) **_must be_** different from zero and **_should be_** large in absolute value, to avoid round-off magnification by the multiplication in the elimination. For this we choose as our pivot equation one that has the absolutely largest a_{jk} in column k on or below the main diagonal (actually, the uppermost if there are several such equations). This popular method is called **partial pivoting.**[2] It is used in CASs (e.g., in MATHEMATICA). Let us illustrate this method with a simple example.

EXAMPLE 1   **Gauss elimination. Partial pivoting**

Solve the system

$$E_1: \qquad\qquad 8x_2 + 2x_3 = -7$$
$$E_2: \quad 3x_1 + 5x_2 + 2x_3 = 8$$
$$E_3: \quad 6x_1 + 2x_2 + 8x_3 = 26.$$

Solution.   We must pivot since E_1 has no x_1-term. In column 1, equation E_3 has the largest coefficient. Hence we interchange E_1 and E_3,

$$6x_1 + 2x_2 + 8x_3 = 26$$
$$3x_1 + 5x_2 + 2x_3 = 8$$
$$8x_2 + 2x_3 = -7.$$

First Step. Elimination of x_1

It would suffice to show the augmented matrix and operate on it. We show both the equations and the augmented matrix. In the first step, the first equation is the pivot equation. Thus

$$
\begin{array}{ll}
\text{Pivot 6} \longrightarrow & \fbox{$6x_1$} + 2x_2 + 8x_3 = 26 \\
\text{Eliminate} \longrightarrow & \fbox{$3x_1$} + 5x_2 + 2x_3 = 8 \\
& \qquad\quad 8x_2 + 2x_3 = -7
\end{array}
\qquad
\begin{bmatrix}
6 & 2 & 8 & | & 26 \\
3 & 5 & 2 & | & 8 \\
0 & 8 & 2 & | & -7
\end{bmatrix}
$$

To eliminate x_1 from the other equations (here, from the second equation), do:

Subtract $3/6 = 1/2$ times the pivot equation from the second equation.

The result is

$$
\begin{array}{l}
6x_1 + 2x_2 + 8x_3 = 26 \\
\qquad\quad 4x_2 - 2x_3 = -5 \\
\qquad\quad 8x_2 + 2x_3 = -7
\end{array}
\qquad
\begin{bmatrix}
6 & 2 & 8 & | & 26 \\
0 & 4 & -2 & | & -5 \\
0 & 8 & 2 & | & -7
\end{bmatrix}
$$

---

[2]"Partial" pivoting as opposed to **total pivoting,** which involves both row *and* column interchanges, but is hardly used in practice.

Second Step. Elimination of x_2

The largest coefficient in column 2 is 8. Hence we take the *new* third equation as the pivot equation, interchanging equations 2 and 3,

$$
\begin{array}{rl}
6x_1 + 2x_2 + 8x_3 = 26 \\
\text{Pivot 8} \longrightarrow \quad \boxed{8x_2} + 2x_3 = -7 \\
\text{Eliminate} \longrightarrow \quad \boxed{4x_2} - 2x_3 = -5
\end{array}
\qquad
\begin{bmatrix}
6 & 2 & 8 & \vdots & 26 \\
0 & 8 & 2 & \vdots & -7 \\
0 & 4 & -2 & \vdots & -5
\end{bmatrix}
$$

To eliminate x_2 from the third equation, do:

Subtract 1/2 times the pivot equation from the third equation.

The resulting triangular system is shown below. This is the end of the forward elimination. Now comes the back substitution.

Back substitution. Determination of x_3, x_2, x_1

The triangular system obtained in Step 2 is

$$
\begin{array}{rl}
6x_1 + 2x_2 + 8x_3 = 26 \\
8x_2 + 2x_3 = -7 \\
- 3x_3 = -\tfrac{3}{2}
\end{array}
\qquad
\begin{bmatrix}
6 & 2 & 8 & \vdots & 26 \\
0 & 8 & 2 & \vdots & -7 \\
0 & 0 & -3 & \vdots & -\tfrac{3}{2}
\end{bmatrix}
$$

From this system, taking the last equation, then the second equation, and finally the first equation, we compute the solution

$$x_3 = \tfrac{1}{2}$$

$$x_2 = \tfrac{1}{8}(-7 - 2x_3) = -1$$

$$x_1 = \tfrac{1}{6}(26 - 2x_2 - 8x_3) = 4$$

This agrees with the values given above.   ◀

The general algorithm is shown in Table 18.1. To help explain the algorithm, we have numbered some of its lines. b_j is denoted by $a_{j,n+1}$, for uniformity. In lines 1 and 2 we look for a possible pivot. [For $k = 1$ we can always find one; otherwise x_1 would not occur in (1).] In line 2 we do pivoting if necessary, picking an a_{jk} of greatest absolute value (the one with the smallest j if there are several) and interchange the corresponding rows. If $|a_{kk}|$ is greatest, we do no pivoting. m_{jk} in line 3 suggests *multiplier,* since these are the factors by which we have to multiply the pivot equation $E_k{}^*$ in Step k before subtracting it from an equation $E_j{}^*$ below $E_k{}^*$ from which we want to eliminate x_k. Here we have written $E_k{}^*$ and $E_j{}^*$ to indicate that after Step 1 these are no longer the given equations in (1), but these underwent a change in each step, as indicated in line 4. Accordingly, a_{jk} etc. in lines 1–4 refer to the most recent equations, and $j \geqq k$ in line 1 indicates that we leave untouched all the equations that have served as pivot equations in previous steps. For $p = k$ in line 4 we get 0 on the right, as it should be in the elimination,

$$a_{jk} - m_j a_{kk} = a_{jk} - \frac{a_{jk}}{a_{kk}} a_{kk} = 0.$$

In line 5, if the last equation in the *triangular* system is $0 = b_n{}^* \neq 0$, we have no solution. If it is $0 = b_n{}^* = 0$, we have no unique solution because we then have fewer equations than unknowns.

Table 18.1
Gauss Elimination

Algorithm Gauss ($\widetilde{\mathbf{A}} = [a_{jk}] = [\mathbf{A} \quad \mathbf{b}]$)

This algorithm computes a unique solution $\mathbf{x} = [x_j]$ of the system (1) or indicates that (1) has no unique solution.

INPUT: Augmented $n \times (n + 1)$ matrix $\widetilde{\mathbf{A}} = [a_{jk}]$, where $a_{j,n+1} = b_j$

OUTPUT: Solution $\mathbf{x} = [x_j]$ of (1) or message that the system (1) has no unique solution

For $k = 1, \cdots, n - 1$, do:

1   If $a_{jk} = 0$ for all $j \geqq k$ then OUTPUT "No unique solution exists." Stop

   [*Procedure completed unsuccessfully; $\mathbf{A}$ is singular*]

2   Else exchange the contents of rows $\widetilde{j}$ and k of $\widetilde{\mathbf{A}}$ with $\widetilde{j}$ the smallest $j \geqq k$ such that $|a_{jk}|$ is maximum in column k.

3   For $j = k + 1, \cdots, n$, do:
$$m_{jk} := \frac{a_{jk}}{a_{kk}}$$

4   For $p = k + 1, \cdots, n + 1$, do:
$$a_{jp} := a_{jp} - m_{jk} a_{kp}$$
   End
   End
   End

5   If $a_{nn} = 0$ then OUTPUT "No unique solution exists."
   Stop
   Else

6    $$x_n = \frac{a_{n,n+1}}{a_{nn}} \qquad [\textit{Start back substitution}]$$

For $i = n - 1, \cdots, 1$, do:

7    $$x_i = \frac{1}{a_{ii}} \left(a_{i,n+1} - \sum_{j=i+1}^{n} a_{ij} x_j \right)$$
   End
   OUTPUT $\mathbf{x} = [x_j]$. Stop

End GAUSS

EXAMPLE 2   **Gauss elimination in Table 18.1, sample computation**

In Example 1 we had $a_{11} = 0$, so that pivoting was necessary. The greatest coefficient in column 1 was a_{31}. Thus $\widetilde{j} = 3$ in line 2, and we interchanged E_1 and E_3. Then in lines 3 and 4 we computed $m_{21} = 3/6 = \frac{1}{2}$ and

$$a_{22} = 5 - \tfrac{1}{2} \cdot 2 = 4, \qquad a_{23} = 2 - \tfrac{1}{2} \cdot 8 = -2, \qquad a_{24} = 8 - \tfrac{1}{2} \cdot 26 = -5,$$

and then $m_{31} = 0/6 = 0$, so that the third equation $8x_2 + 2x_3 = -7$ did not change in Step 1. In Step 2

($k = 2$) we had 8 as the greatest coefficient in column 2, hence $\tilde{j} = 3$. We interchanged equations 2 and 3, computed $m_{32} = 4/8 = \frac{1}{2}$ in line 4, and then $a_{33} = -2 - \frac{1}{2} \cdot 2 = -3$, $a_{34} = -5 - \frac{1}{2}(-7) = -\frac{3}{2}$. This produced the triangular form used in the back substitution. ◀

If $a_{kk} = 0$ in Step k, *we must pivot.* If $|a_{kk}|$ is small, *we should pivot* because of round-off error magnification that may seriously affect accuracy or even produce nonsensical results.

EXAMPLE 3   **Difficulty with small pivots**

The solution of the system

$$0.0004x_1 + 1.402x_2 = 1.406$$

$$0.4003x_1 - 1.502x_2 = 2.501$$

is $x_1 = 10$, $x_2 = 1$. We solve this system by the Gauss elimination, using four-digit floating-point arithmetic. (4D is for simplicity. Make an 8D-arithmetic example that shows the same.)

(a) Picking the first of the given equations as the pivot equation, we have to multiply this equation by $m = 0.4003/0.0004 = 1001$ and subtract the result from the second equation, obtaining

$$-1405x_2 = -1404.$$

Hence $x_2 = -1404/(-1405) = 0.9993$, and from the first equation, instead of $x_1 = 10$, we get

$$x_1 = \frac{1}{0.0004}(1.406 - 1.402 \cdot 0.9993) = \frac{0.005}{0.0004} = 12.5.$$

This failure occurs because $|a_{11}|$ is small compared to $|a_{12}|$, so that a small round-off error in x_2 led to a large error in x_1.

(b) Picking the second of the given equations as the pivot equation, we have to multiply this equation by $0.0004/0.4003 = 0.000\,9993$ and subtract the result from the first equation, obtaining

$$1.404x_2 = 1.404.$$

Hence $x_2 = 1$, and from the pivot equation $x_1 = 10$. This success occurs because $|a_{21}|$ is not very small compared to $|a_{22}|$, so that a small round-off error in x_2 would not lead to a large error in x_1. Indeed, for instance, if we had the value $x_2 = 1.002$, we would still have from the pivot equation the good value $x_1 = (2.501 + 1.505)/0.4003 = 10.01$. ◀

Error estimates for the Gauss elimination are discussed in Ref. [E3] listed in Appendix 1.

Row scaling means the multiplication of each row j by a suitable scaling factor s_j. It is done in connection with partial pivoting to get more accurate solutions. Despite much research (see Refs. [E6], [E8], [E17], [E19] in Appendix 1 and the proposition of several principles, scaling is still not well understood. As a possibility, one can scale for pivot choice only (not in the calculation, to avoid additional round-off) and take as first pivot the entry a_{j1} for which $|a_{j1}|/|A_j|$ is largest; here A_j is an entry of largest absolute value in row j. Similarly in the further steps of the Gauss elimination.

For instance, for the system

$$4.0000x_1 + 14020x_2 = 14060$$

$$0.4003x_1 - 1.502x_2 = 2.501$$

we might pick 4 as pivot, but dividing the first equation by 10^4 gives the system in Example 3, for which the second equation is a better pivot equation.

Operation Count

Quite generally, the quality of a numerical method is judged in terms of:

> Amount of storage
>
> Amount of time ($\equiv$ number of operations)
>
> Effect of round-off error.

For the Gauss elimination, the operation count for a full matrix is as follows. In Step k we eliminate x_k from $n - k$ equations. This needs $n - k$ divisions in computing the m_{jk} (line 3) and $(n - k)(n - k + 1)$ multiplications and as many subtractions (both in line 4). Since we do $n - 1$ steps, k goes from 1 to $n - 1$ and thus the total number of operations in this forward elimination is

$$f(n) = \sum_{k=1}^{n-1} (n - k) + 2 \sum_{k=1}^{n-1} (n - k)(n - k + 1) \qquad \text{(write } n - k = s\text{)}$$

$$= \sum_{s=1}^{n-1} s + 2 \sum_{s=1}^{n-1} s(s + 1) = \tfrac{1}{2}(n - 1)n + \tfrac{2}{3}(n^2 - 1)n \approx \tfrac{2}{3}n^3$$

where $2n^3/3$ is obtained by dropping lower powers of n. We see that $f(n)$ grows about proportional to n^3. We say that $f(n)$ is of *order* n^3 and write

$$f(n) = O(n^3)$$

where O suggests **order.** The general definition of O is as follows. We write

$$f(n) = O(h(n))$$

if the quotient $|f(n)/h(n)|$ remains bounded (does not trail off to infinity) as $n \to \infty$. In our present case, $h(n) = n^3$ and, indeed, $f(n)/n^3 \to 2/3$ because the omitted terms divided by n^3 go to zero as $n \to \infty$.

In the back substitution of x_i we make $n - i$ multiplications and as many subtractions, as well as 1 division. Hence the number of operations in the back substitution is

$$b(n) = 2\sum_{i=1}^{n} (n - i) + n = 2\sum_{s=1}^{n} s + n = n(n + 1) + n \approx n^2 = O(n^2).$$

We see that it grows more slowly than the number of operations in the forward elimination of the Gauss algorithm, so that it is negligible for large systems because it is smaller by a factor n, approximately. For instance, if an operation takes 10^{-9} sec, then the times needed are:

Algorithm	$n = 1000$	$n = 10\,000$
Elimination	0.7 sec	11 min
Back substitution	0.001 sec	0.1 sec

PROBLEM SET 18.1

For applications of linear systems see Secs. 6.3 and 7.2.
Systems of Two Equations. Solve graphically. Explain the result geometrically.

1. $3x_1 + 5x_2 = 5.9$
$x_1 - 4x_2 = 20.1$

2. $-2.50x_1 + 4.20x_2 = 0$
$20.50x_1 - 34.44x_2 = 0$

3. $25.38x_1 - 15.48x_2 = 30.60$
$-7.05x_1 + 4.30x_2 = -8.50$

Gauss Elimination

Solve the following linear systems by Gauss elimination, with partial pivoting if necessary (but without scaling). Check the result by substitution. If no solution or more than one solution exists, give a reason.

4. $4.12x_1 - 3.89x_2 = 33.000$
$6.04x_1 + 2.55x_2 = 5.464$

5. $1.5x_1 + 2.3x_2 = 16$
$-4.5x_1 - 6.9x_2 = 48$

6. $0.4x_1 + 20.0x_2 = 54$
$2.0x_1 - 65.5x_2 = 220.35$

7. $6x_2 + 13x_3 = 61$
$6x_1 - 8x_3 = -38$
$13x_1 - 8x_2 = 79$

8. $0.4x_1 + 1.6x_2 - 0.8x_3 = 3.80$
$3.2x_1 - 0.2x_2 + 1.2x_3 = 14.44$
$1.4x_1 - 4.2x_3 \quad 16.24$

9. $-3x_1 + 6x_2 - 9x_3 = -46.725$
$x_1 - 4x_2 + 3x_3 = 19.571$
$2x_1 + 5x_2 - 7x_3 = -20.073$

10. $4x_1 + 4x_2 + 2x_3 = 0$
$3x_1 - x_2 + 2x_3 = 0$
$3x_1 + 7x_2 + x_3 = 0$

11. $2x_1 + 5x_2 + 7x_3 = 25$
$-5x_1 + 7x_2 + 2x_3 = -4$
$x_1 + 22x_2 + 23x_3 = 71$

12. $5x_1 + 3x_2 + x_3 = 2$
$-4x_2 + 8x_3 = -3$
$10x_1 - 6x_2 + 26x_3 = 0$

13. $3.4x_1 - 6.12x_2 - 2.72x_3 = 0$
$-x_1 + 1.80x_2 + 0.80x_3 = 0$
$2.7x_1 - 4.86x_2 - 2.16x_3 = 0$

14. $x_1 + \frac{1}{2}x_2 = \frac{1}{2}$
$\frac{1}{2}x_1 + \frac{1}{2}x_2 + \frac{1}{2}x_3 = \frac{1}{3}$
$\frac{1}{3}x_2 + \frac{1}{3}x_3 = \frac{1}{4}$

15. $3x_2 + 5x_3 = 1.20736$
$3x_1 - 4x_2 = -2.34066$
$5x_1 + 6x_3 = -0.329193$

16. $-47x_1 + 4x_2 - 7x_3 = -118$
$19x_1 - 3x_2 + 2x_3 = 43$
$-15x_1 + 5x_2 = -25$

17. $3.2x_1 + 1.6x_2 = -0.8$
$1.6x_1 - 0.8x_2 + 2.4x_3 = 16.0$
$2.4x_2 - 4.8x_3 + 3.6x_4 = -39.0$
$3.6x_3 + 2.4x_4 = 10.2$

18. $2.2x_2 + 1.5x_3 - 3.3x_4 = -9.30$
$0.2x_1 + 1.8x_2 + 4.2x_4 = 9.24$
$-x_1 - 3.1x_2 + 2.5x_3 = -8.70$
$0.5x_1 - 3.8x_3 + 1.5x_4 = 11.94$

19. CAS PROJECT. Gauss Elimination. Write a program for the Gauss elimination with pivoting. Apply it to Probs. 9–12. Experiment with systems whose coefficient determinant is small in absolute value. Also investigate the performance of your program for larger systems of your choice, including sparse systems.

20. TEAM PROJECT. Linear Systems and Gauss Elimination. (a) Existence and uniqueness. Find a and b such that $ax_1 + x_2 = b$, $x_1 + x_2 = 3$ has (i) a unique solution, (ii) infinitely many solutions, (iii) no solutions.

(b) Gauss elimination and nonexistence. Apply the Gauss elimination to the following two systems and compare the calculations step by step. Explain why the elimination fails if no solution exists.

$$\begin{aligned}
x_1 + x_2 + x_3 &= 3 \\
4x_1 + 2x_2 - x_3 &= 5 \\
9x_1 + 5x_2 - x_3 &= 13
\end{aligned} \qquad\qquad
\begin{aligned}
x_1 + x_2 + x_3 &= 3 \\
4x_1 + 2x_2 - x_3 &= 5 \\
9x_1 + 5x_2 - x_3 &= 12.
\end{aligned}$$

(c) Zero determinant. Why may a computer program give you the result that a homogeneous linear system has only the trivial solution although you know its coefficient determinant to be zero?

(d) Pivoting. Solve

$$\begin{aligned}
\epsilon x_1 + x_2 &= 1 \\
x_1 + x_2 &= 2
\end{aligned}$$

by the Gauss elimination first without pivoting. Show that for any fixed machine word length and sufficiently small $\epsilon > 0$ the computer gives $x_2 = 1$ and then $x_1 = 0$. What is the exact solution? Its limit as $\epsilon \to 0$? Then solve the system by the Gauss elimination with pivoting. Compare and comment.

(e) Pivoting. Solve the following system by the Gauss elimination and three-digit rounding arithmetic, choosing (i) the first equation, (ii) the second equation as pivot equation. (Remember to round to 3S after each operation before doing the next, just as would be done on a computer!) Then use four-digit rounding arithmetic in those two calculations. Compare and comment.

$$\begin{aligned}
4.03x_1 + 2.16x_2 &= -4.61 \\
6.21x_1 + 3.35x_2 &= -7.19
\end{aligned}$$

18.2 Linear Systems: LU-Factorization, Matrix Inversion

We continue our discussion of numerical methods for solving linear systems of n equations in n unknowns $x_1, \cdots, x_n$,

(1) $$\mathbf{Ax} = \mathbf{b},$$

where $\mathbf{A} = [a_{jk}]$ is the $n \times n$ coefficient matrix and $\mathbf{x}^{\mathsf{T}} = [x_1 \cdots x_n]$ and $\mathbf{b}^{\mathsf{T}} = [b_1 \cdots b_n]$. We present three related methods that are modifications of the Gauss elimination. They are named after Doolittle, Crout, and Cholesky and use the idea of the LU-factorization of **A,** which we explain first.

An **LU-factorization** of a given square matrix $\mathbf{A}$ is of the form

(2)
$$\boxed{\mathbf{A} = \mathbf{LU}}$$

where $\mathbf{L}$ is lower triangular and $\mathbf{U}$ is upper triangular. For example,

$$\mathbf{A} = \begin{bmatrix} 2 & 3 \\ 8 & 5 \end{bmatrix} = \mathbf{LU} = \begin{bmatrix} 1 & 0 \\ 4 & 1 \end{bmatrix} \begin{bmatrix} 2 & 3 \\ 0 & -7 \end{bmatrix}.$$

It can be proved that for any nonsingular matrix (see Sec. 6.7) the rows can be reordered so that the resulting matrix $\mathbf{A}$ has an LU-factorization (2) in which $\mathbf{L}$ turns out to be the matrix of the **multipliers** m_{jk} of the Gauss elimination, with main diagonal $1, \cdots, 1$, and $\mathbf{U}$ is the matrix of the triangular system at the end of the Gauss elimination. (See Ref. [E3], pp. 155−156, listed in Appendix 1.)

The **crucial idea** now is that $\mathbf{L}$ and $\mathbf{U}$ in (2) can be computed directly, without solving simultaneous equations (thus, without using the Gauss elimination). As a count shows, this needs about $n^3/3$ operations, about half as many as the Gauss elimination, which needs about $2n^3/3$ (see Sec. 18.1). And once we have (2), we can use it for solving $\mathbf{Ax} = \mathbf{b}$ in two steps, involving only about n^2 operations, simply by noting that $\mathbf{Ax} = \mathbf{LUx} = \mathbf{b}$ may be written

(3)
$$\text{(a)} \quad \mathbf{Ly} = \mathbf{b} \qquad \text{where} \qquad \text{(b)} \quad \mathbf{Ux} = \mathbf{y}$$

and solving first (3a) for $\mathbf{y}$ and then (3b) for $\mathbf{x}$. This is called **Doolittle's method.** Both systems (3a) and (3b) are triangular, so their solution is the same as back substitution in the Gauss elimination.

A similar method, **Crout's method,** is obtained from (2) if $\mathbf{U}$ (instead of $\mathbf{L}$) is required to have main diagonal $1, \cdots, 1$. In either case the factorization (2) is unique.

EXAMPLE 1   **Doolittle's method**

Solve the system in Example 1 of Sec. 18.1 by Doolittle's method.

Solution.   The decomposition (2) is obtained from

$$\mathbf{A} = [a_{jk}] = \begin{bmatrix} 3 & 5 & 2 \\ 0 & 8 & 2 \\ 6 & 2 & 8 \end{bmatrix} = \begin{bmatrix} 1 & 0 & 0 \\ m_{21} & 1 & 0 \\ m_{31} & m_{32} & 1 \end{bmatrix} \begin{bmatrix} u_{11} & u_{12} & u_{13} \\ 0 & u_{22} & u_{23} \\ 0 & 0 & u_{33} \end{bmatrix}$$

by determining the m_{jk} and u_{jk}, using matrix multiplication. By going through $\mathbf{A}$ row by row we get successively

$a_{11} = 3 = u_{11}$	$a_{12} = 5 = u_{12}$	$a_{13} = 2 = u_{13}$
$a_{21} = 0 = m_{21}u_{11}$	$a_{22} = 8 = m_{21}u_{12} + u_{22}$	$a_{23} = 2 = m_{21}u_{13} + u_{23}$
$m_{21} = 0$	$u_{22} = 8$	$u_{23} = 2$
$a_{31} = 6 = m_{31}u_{11}$	$a_{32} = 2 = m_{31}u_{12} + m_{32}u_{22}$	$a_{33} = 8 = m_{31}u_{13} + m_{32}u_{23} + u_{33}$
$= m_{31} \cdot 3$	$= 2 \cdot 5 + m_{32} \cdot 8$	$= 2 \cdot 2 - 1 \cdot 2 + u_{33}$
$m_{31} = 2$	$m_{32} = -1$	$u_{33} = 6$

Thus the factorization (2) is

$$
\begin{bmatrix} 3 & 5 & 2 \\ 0 & 8 & 2 \\ 6 & 2 & 8 \end{bmatrix} = \begin{bmatrix} 1 & 0 & 0 \\ 0 & 1 & 0 \\ 2 & -1 & 1 \end{bmatrix} \begin{bmatrix} 3 & 5 & 2 \\ 0 & 8 & 2 \\ 0 & 0 & 6 \end{bmatrix}.
$$

We first solve $\mathbf{Ly} = \mathbf{b},$ determining $y_1,$ then $y_2,$ then $y_3,$ that is,

$$
\begin{bmatrix} 1 & 0 & 0 \\ 0 & 1 & 0 \\ 2 & -1 & 1 \end{bmatrix} \begin{bmatrix} y_1 \\ y_2 \\ y_3 \end{bmatrix} = \begin{bmatrix} 8 \\ -7 \\ 26 \end{bmatrix}. \qquad \text{Solution} \qquad \mathbf{y} = \begin{bmatrix} 8 \\ -7 \\ 3 \end{bmatrix}.
$$

Then we solve $\mathbf{Ux} = \mathbf{y},$ determining $x_3,$ then $x_2,$ then $x_1,$ that is,

$$
\begin{bmatrix} 3 & 5 & 2 \\ 0 & 8 & 2 \\ 0 & 0 & 6 \end{bmatrix} \begin{bmatrix} x_1 \\ x_2 \\ x_3 \end{bmatrix} = \begin{bmatrix} 8 \\ -7 \\ 3 \end{bmatrix}. \qquad \text{Solution} \qquad \mathbf{x} = \begin{bmatrix} 4 \\ -1 \\ 1/2 \end{bmatrix}.
$$

This agrees with the solution in Example 1 of Sec. 18.1.   ◄

Our formulas in Example 1 suggest that for general n the elements of the matrices $\mathbf{L} = [m_{jk}]$ (with main diagonal $1, \cdots, 1$ and m_{jk} suggesting "multiplier") and $\mathbf{U} = [u_{jk}]$ in the *Doolittle method* are computed from

$$
\begin{aligned}
u_{1k} &= a_{1k} && k = 1, \cdots, n \\[2mm]
m_{j1} &= \frac{a_{j1}}{u_{11}} && j = 2, \cdots, n \\[2mm]
u_{jk} &= a_{jk} - \sum_{s=1}^{j-1} m_{js} u_{sk} && k = j, \cdots, n; \quad j \geqq 2 \\[2mm]
m_{jk} &= \frac{1}{u_{kk}} \left(a_{jk} - \sum_{s=1}^{k-1} m_{js} u_{sk} \right) && j = k+1, \cdots, n; \quad k \geqq 2.
\end{aligned}
$$

(4)

Row Interchanges. Matrices, such as

$$
\begin{bmatrix} 0 & 1 \\ 1 & 1 \end{bmatrix} \qquad \text{or} \qquad \begin{bmatrix} 0 & 1 \\ 1 & 0 \end{bmatrix}
$$

have no LU-factorization (try!). This indicates that for obtaining an LU-factorization, row interchanges of $\mathbf{A}$ (and corresponding interchanges in $\mathbf{b}$) may be necessary.

Cholesky's Method

For a *symmetric, positive definite* matrix $\mathbf{A}$ (thus $\mathbf{A} = \mathbf{A}^{\mathsf{T}},$ $\mathbf{x}^{\mathsf{T}} \mathbf{A} \mathbf{x} > 0$ for all $\mathbf{x} \neq \mathbf{0}$) we can in (2) even choose $\mathbf{U} = \mathbf{L}^{\mathsf{T}},$ thus $u_{jk} = m_{kj}$ (but impose no conditions on the main diagonal entries). For example,

(5)
$$
\begin{bmatrix} 4 & 2 & 14 \\ 2 & 17 & -5 \\ 14 & -5 & 83 \end{bmatrix} = \begin{bmatrix} 2 & 0 & 0 \\ 1 & 4 & 0 \\ 7 & -3 & 5 \end{bmatrix} \begin{bmatrix} 2 & 1 & 7 \\ 0 & 4 & -3 \\ 0 & 0 & 5 \end{bmatrix}.
$$

The popular method of solving $\mathbf{Ax} = \mathbf{b}$ based on this factorization $\mathbf{A} = \mathbf{LL}^\mathsf{T}$ is called **Cholesky's method.** In terms of the entries of $\mathbf{L} = [l_{jk}]$ the formulas for the factorization are

(6)
$$
l_{11} = \sqrt{a_{11}}
$$

$$
l_{j1} = \frac{a_{j1}}{l_{11}} \qquad\qquad j = 2, \cdots, n
$$

$$
l_{jj} = \sqrt{a_{jj} - \sum_{s=1}^{j-1} l_{js}{}^2} \qquad\qquad j = 2, \cdots, n
$$

$$
l_{pj} = \frac{1}{l_{jj}} \left(a_{pj} - \sum_{s=1}^{j-1} l_{js} l_{ps} \right) \qquad\qquad p = j + 1, \cdots, n; \quad j \geqq 2.
$$

If $\mathbf{A}$ is symmetric but not positive definite, this method could still be applied, but then leads to a *complex* matrix $\mathbf{L}$, so that it becomes impractical.

EXAMPLE 2   Cholesky's method

Solve by Cholesky's method:

$$
\begin{aligned}
4x_1 + 2x_2 + 14x_3 &= 14 \\
2x_1 + 17x_2 - 5x_3 &= -101 \\
14x_1 - 5x_2 + 83x_3 &= 155.
\end{aligned}
$$

Solution.   From (6) or from the form of the factorization

$$
\begin{bmatrix} 4 & 2 & 14 \\ 2 & 17 & -5 \\ 14 & -5 & 83 \end{bmatrix} = \begin{bmatrix} l_{11} & 0 & 0 \\ l_{21} & l_{22} & 0 \\ l_{31} & l_{32} & l_{33} \end{bmatrix} \begin{bmatrix} l_{11} & l_{21} & l_{31} \\ 0 & l_{22} & l_{32} \\ 0 & 0 & l_{33} \end{bmatrix}
$$

we compute, in the given order,

$$
l_{11} = \sqrt{a_{11}} = 2 \qquad l_{21} = \frac{a_{21}}{l_{11}} = \frac{2}{2} = 1 \qquad l_{31} = \frac{a_{31}}{l_{11}} = \frac{14}{2} = 7
$$

$$
l_{22} = \sqrt{a_{22} - l_{21}{}^2} = \sqrt{17 - 1} = 4
$$

$$
l_{32} = \frac{1}{l_{22}} (a_{32} - l_{31} l_{21}) = \frac{1}{4}(-5 - 7 \cdot 1) = -3
$$

$$
l_{33} = \sqrt{a_{33} - l_{31}{}^2 - l_{32}{}^2} = \sqrt{83 - 7^2 - (-3)^2} = 5.
$$

This agrees with (5). We now have to solve $\mathbf{Ly} = \mathbf{b}$, that is,

$$\begin{bmatrix} 2 & 0 & 0 \\ 1 & 4 & 0 \\ 7 & -3 & 5 \end{bmatrix} \begin{bmatrix} y_1 \\ y_2 \\ y_3 \end{bmatrix} = \begin{bmatrix} 14 \\ -101 \\ 155 \end{bmatrix}. \qquad \text{Solution} \qquad \mathbf{y} = \begin{bmatrix} 7 \\ -27 \\ 5 \end{bmatrix}.$$

As the second step, we have to solve $\mathbf{Ux} = \mathbf{L}^{\mathsf{T}}\mathbf{x} = \mathbf{y}$, that is,

$$\begin{bmatrix} 2 & 1 & 7 \\ 0 & 4 & -3 \\ 0 & 0 & 5 \end{bmatrix} \begin{bmatrix} x_1 \\ x_2 \\ x_3 \end{bmatrix} = \begin{bmatrix} 7 \\ -27 \\ 5 \end{bmatrix}. \qquad \text{Solution} \qquad \mathbf{x} = \begin{bmatrix} 3 \\ -6 \\ 1 \end{bmatrix}. \qquad \blacktriangleleft$$

THEOREM 1 **(Stability of the Cholesky factorization)**

The Cholesky $\mathbf{LL}^{\mathsf{T}}$-factorization is numerically stable.

PROOF. We have $a_{jj} = l_{j1}{}^2 + l_{j2}{}^2 + \cdots + l_{jj}{}^2$ by squaring the third formula in (6) and solving it for a_{jj}. Hence for all l_{jk} (note that $l_{jk} = 0$ for $k > j$) we obtain (the inequality being trivial)

$$l_{jk}{}^2 \leqq l_{j1}{}^2 + l_{j2}{}^2 + \cdots + l_{jj}{}^2 = a_{jj}.$$

That is, $l_{jk}{}^2$ is bounded by an entry of $\mathbf{A}$, which means stability against round-off. $\blacktriangleleft$

The two methods that we have discussed were particularly popular in desk-computer times because of their relatively small number of intermediate results. They are still attractive since they can be implemented with accumulated inner products (see [E17]).

Gauss–Jordan Elimination. Matrix Inversion

Another variant of the Gauss elimination is the **Gauss–Jordan elimination,** introduced by W. Jordan in 1920, in which back substitution is avoided by additional computations that reduce the matrix to diagonal form, instead of the triangular form in the Gauss elimination. But this reduction from the Gauss triangular to diagonal form requires more operations than back substitution does, so that the method is *disadvantageous* for solving systems $\mathbf{Ax} = \mathbf{b}$. But it may be used for matrix inversion, where the situation is as follows.

The **inverse** of a nonsingular square matrix $\mathbf{A}$ may be determined in principle by solving the n systems

$$(7) \qquad\qquad\qquad \mathbf{Ax} = \mathbf{b}_j \qquad\qquad\qquad (j = 1, \cdots, n)$$

where $\mathbf{b}_j$ is the jth column of the $n \times n$ unit matrix.

However, it is preferable to produce $\mathbf{A}^{-1}$ by operating on the unit matrix $\mathbf{I}$ in the same way as the Gauss–Jordan algorithm, reducing $\mathbf{A}$ to $\mathbf{I}$. A typical illustrative example of this method is given in Sec. 6.7.

PROBLEM SET 18.2

Doolittle's Method

Solve the following linear systems by Doolittle's method, showing the details, in particular the LU-factorization.

1. $\quad 4x_1 + 5x_2 = 7$

$\quad 12x_1 + 14x_2 = 18$

2. $-3x_1 + 6x_2 = -34.77$

$\quad\quad 6x_1 - 8x_2 = 54.86$

3. $1.80x_1 + 2.60x_2 = 13.20$

$\quad 0.36x_1 + 3.72x_2 = 12.24$

4. $\quad 2x_1 + 2x_2 + 4x_3 = -2$

$\quad\quad 4x_1 + 5x_2 + 13x_3 = -7$

$\quad 10x_1 + 14x_2 + 43x_3 = -25$

5. $\quad 5x_1 + 4x_2 + x_3 = 3.4$

$\quad 10x_1 + 9x_2 + 4x_3 = 8.8$

$\quad 10x_1 + 13x_2 + 15x_3 = 19.2$

6. $5x_1 + 9x_2 + 2x_3 = 24$

$\quad 9x_1 + 4x_2 + x_3 = 25$

$\quad 2x_1 + x_2 + x_3 = 11$

Cholesky's Method

Solve by Cholesky's method. (Show details, in particular the LU-factorization.)

7. $\quad 9x_1 + 6x_2 + 12x_3 = 17.4$

$\quad\quad 6x_1 + 13x_2 + 11x_3 = 23.6$

$\quad 12x_1 + 11x_2 + 26x_3 = 30.8$

8. $0.01x_1 + 0.03x_3 = 0.14$

$\quad\quad\quad 0.16x_2 + 0.08x_3 = 0.16$

$\quad 0.03x_1 + 0.08x_2 + 0.14x_3 = 0.54$

9. $4x_1 + 6x_2 + 8x_3 = 0$

$\quad 6x_1 + 34x_2 + 52x_3 = -160$

$\quad 8x_1 + 52x_2 + 129x_3 = -452$

10. $\quad x_1 - x_2 + 3x_3 + 2x_4 = 15$

$\quad -x_1 + 5x_2 - 5x_3 - 2x_4 = -35$

$\quad 3x_1 - 5x_2 + 19x_3 + 3x_4 = 94$

$\quad 2x_1 - 2x_2 + 3x_3 + 21x_4 = 1$

11. $4x_1 + 2x_2 + 4x_3 = 10$

$\quad 2x_1 + 2x_2 + 3x_3 + 2x_4 = 18$

$\quad 4x_1 + 2x_2 + 6x_3 + 3x_4 = 30$

$\quad\quad\quad 2x_2 + 3x_3 + 9x_4 = 61$

12. **(Definiteness)** Let **A**, **B** be $n \times n$ and positive definite. Are $-\mathbf{A}$, $\mathbf{A}^\mathsf{T}$, $\mathbf{A} + \mathbf{B}$, $\mathbf{A} - \mathbf{B}$ positive definite?

13. **CAS PROJECT. Cholesky's Method.** (a) Write a program for solving linear systems by Cholesky's method and apply it to Example 2 in the text, Probs. 7 and 10, and to systems of your choice.

(b) **Splines.** Apply the factorization part of the program to the following matrices (as they occur in (9), Sec. 17.4 (with $c_j = 1$), in connection with splines).

$$\begin{bmatrix} 2 & 1 & 0 \\ 1 & 4 & 1 \\ 0 & 1 & 2 \end{bmatrix}, \quad \begin{bmatrix} 2 & 1 & 0 & 0 \\ 1 & 4 & 1 & 0 \\ 0 & 1 & 4 & 1 \\ 0 & 0 & 1 & 2 \end{bmatrix}.$$

14. TEAM PROJECT. Crout's method factorizes $\mathbf{A} = \mathbf{LU}$, where $\mathbf{L}$ is lower triangular and $\mathbf{U}$ is upper triangular with diagonal entries $u_{jj} = 1$, $j = 1, \cdots, n$. **(a) Formulas.** Obtain formulas for Crout's method similar to (4).
(b) Examples. Solve each system by Crout's method (showing details of your work).

$$3x_1 + 2x_2 = 18$$
$$18x_1 + 17x_2 = 123$$

$$x_1 - 4x_2 + 2x_3 = 81$$
$$-4x_1 + 25x_2 + 4x_3 = -153$$
$$2x_1 + 4x_2 + 24x_3 = 324$$

(c) Factorize the 3×3 coefficient matrix in (b) by the Doolittle, Crout, and Cholesky methods.
(d) Tridiagonal matrices can be factorized by Crout's method in the form

$$\mathbf{A} = \mathbf{LU} = \begin{bmatrix} a_1 & c_1 & 0 & 0 \\ b_2 & a_2 & c_2 & 0 \\ 0 & b_3 & a_3 & c_3 \\ 0 & 0 & b_4 & a_4 \end{bmatrix} = \begin{bmatrix} \alpha_1 & 0 & 0 & 0 \\ b_2 & \alpha_2 & 0 & 0 \\ 0 & b_3 & \alpha_3 & 0 \\ 0 & 0 & b_4 & \alpha_4 \end{bmatrix} \begin{bmatrix} 1 & \gamma_1 & 0 & 0 \\ 0 & 1 & \gamma_2 & 0 \\ 0 & 0 & 1 & \gamma_3 \\ 0 & 0 & 0 & 1 \end{bmatrix}$$

and similarly for general n. Show that

$$\alpha_1 = a_1, \qquad \alpha_j = a_j - b_j\gamma_{j-1}, \qquad j = 2, \cdots, n$$
$$\gamma_1 = c_1/\alpha_1, \quad \gamma_j = c_j/\alpha_j, \qquad j = 2, \cdots, n-1$$

Inverse of a Matrix

Find the inverse of $\mathbf{A}$ (if it exists) by the Gauss–Jordan method, where $\mathbf{A}$ is as follows. (Show the details of your work.)

15. The coefficient matrix in Prob. 1.
16. The coefficient matrix in Prob. 4.

17. $\begin{bmatrix} -2 & 4 & -1 \\ -2 & 3 & 0 \\ 7 & -12 & 2 \end{bmatrix}$ **18.** $\begin{bmatrix} 2 & 1 & 2 \\ -2 & 2 & 1 \\ 1 & 2 & -2 \end{bmatrix}$ **19.** $\begin{bmatrix} 1/3 & 1/4 & 2 \\ -1/9 & 1 & 1/7 \\ 4/63 & -3/28 & 13/49 \end{bmatrix}$

20. (Round-off) Find det $\mathbf{A}$ in Prob. 19 with $\mathbf{A}$ as given. What happens if you round the given entries to (a) 5S, (b) 4S, (c) 3S, (d) 2S, (e) 1S? What is the practical implication of your results?

18.3 Linear Systems: Solution by Iteration

The Gauss elimination and its variants in the last two sections belong to the **direct methods** for solving linear systems of equations; these are methods that give solutions after an amount of computation that can be specified in advance. In contrast, in an **indirect** or **iterative method** we start from an approximation to the true solution and, if successful, obtain better and better approximations from a computational cycle repeated as often as may be necessary for achieving a required accuracy, so that the amount of arithmetic depends upon the accuracy required and varies from case to case.

We apply iterative methods if the convergence is rapid (if matrices have large main diagonal entries, as we shall see), so that we save operations compared to a direct method. We also use iterative methods if a large system is **sparse,** that is, has very many zero coefficients, so that one would waste space in storing zeros, for instance, 9995 zeros per equation in a potential problem of 10^4 equations in 10^4 unknowns with typically only 5 nonzero terms per equation (more on this in Sec. 19.4).

Gauss–Seidel Iteration Method

This is an iterative method of great practical importance, which we can simply explain in terms of an example.

EXAMPLE 1　**Gauss–Seidel iteration**

We consider the linear system

$$
(1) \quad
\begin{aligned}
x_1 - 0.25x_2 - 0.25x_3 \qquad\quad &= 50 \\
-0.25x_1 + x_2 \qquad\quad - 0.25x_4 &= 50 \\
-0.25x_1 \qquad\quad + x_3 - 0.25x_4 &= 25 \\
-0.25x_2 - 0.25x_3 + x_4 &= 25.
\end{aligned}
$$

(Equations of this form arise in the numerical solution of partial differential equations and in spline interpolation.) We write the system in the form

$$
(2) \quad
\begin{aligned}
x_1 &= \quad\;\; 0.25x_2 + 0.25x_3 \qquad\quad + 50 \\
x_2 &= 0.25x_1 \qquad\qquad\quad + 0.25x_4 + 50 \\
x_3 &= 0.25x_1 \qquad\qquad\quad + 0.25x_4 + 25 \\
x_4 &= \quad\;\; 0.25x_2 + 0.25x_3 \qquad\quad + 25.
\end{aligned}
$$

We use these equations for iteration, that is, we start from a (possibly poor) approximation to the solution, say, $x_1^{(0)} = 100$, $x_2^{(0)} = 100$, $x_3^{(0)} = 100$, $x_4^{(0)} = 100$, and compute from (2) a presumably better approximation

Use "old" values

("New" values here not yet available)

$$
(3) \quad
\begin{aligned}
x_1^{(1)} &= \boxed{}\; 0.25x_2^{(0)} + 0.25x_3^{(0)} \qquad\quad\; + 50.00 = 100.00 \\
x_2^{(1)} &= 0.25x_1^{(1)} \;\boxed{} \qquad\quad + 0.25x_4^{(0)} \;+ 50.00 = 100.00 \\
x_3^{(1)} &= 0.25x_1^{(1)} \qquad\quad \boxed{}\; + 0.25x_4^{(0)} \;+ 25.00 = \;\;75.00 \\
x_4^{(1)} &= \quad\; 0.25x_2^{(1)} + 0.25x_3^{(1)} \;\boxed{} \qquad\quad + 25.00 = \;\;68.75.
\end{aligned}
$$

Use "new" values

We see that these equations are obtained from (2) by substituting on the right the ***most recent*** approximations. In fact, corresponding elements replace previous ones as soon as they have been computed, so that in the second and third equations we use $x_1^{(1)}$ (not $x_1^{(0)}$), and in the last equation of (3) we use $x_2^{(1)}$ and $x_3^{(1)}$ (not $x_2^{(0)}$ and $x_3^{(0)}$). The next step yields

$$
\begin{aligned}
x_1^{(2)} &= \quad\;\; 0.25x_2^{(1)} + 0.25x_3^{(1)} \qquad\quad\;\;\; + 50.00 = 93.75 \\
x_2^{(2)} &= 0.25x_1^{(2)} \qquad\qquad\quad + 0.25x_4^{(1)} \;+ 50.00 = 90.62 \\
x_3^{(2)} &= 0.25x_1^{(2)} \qquad\qquad\quad + 0.25x_4^{(1)} \;+ 25.00 = 65.62 \\
x_4^{(2)} &= \quad\;\; 0.25x_2^{(2)} + 0.25x_3^{(2)} \qquad\quad\;\;\; + 25.00 = 64.06.
\end{aligned}
$$

In practice, one would do further steps and obtain a more accurate approximate solution. The reader may show that the exact solution is $x_1 = x_2 = 87.5$, $x_3 = x_4 = 62.5$. ◀

To obtain an algorithm for the Gauss–Seidel iteration, let us derive the general formulas for this iteration.

We assume that $a_{jj} = 1$ for $j = 1, \cdots, n$. (Note that this can be achieved if we can rearrange the equations so that no diagonal coefficient is zero; then we may divide each equation by the corresponding diagonal coefficient.) We now write

(4) $$\mathbf{A} = \mathbf{I} + \mathbf{L} + \mathbf{U} \qquad (a_{jj} = 1)$$

where $\mathbf{I}$ is the $n \times n$ unit matrix and $\mathbf{L}$ and $\mathbf{U}$ are respectively lower and upper triangular matrices with zero main diagonals. If we substitute (4) into $\mathbf{Ax} = \mathbf{b}$, we have

$$\mathbf{Ax} = (\mathbf{I} + \mathbf{L} + \mathbf{U})\mathbf{x} = \mathbf{b}.$$

Taking $\mathbf{Lx}$ and $\mathbf{Ux}$ to the right, we obtain, since $\mathbf{Ix} = \mathbf{x}$,

(5) $$\mathbf{x} = \mathbf{b} - \mathbf{Lx} - \mathbf{Ux}.$$

Remembering from our computation in Example 1 that below the main diagonal we took "new" approximations and above the main diagonal "old" approximations, we obtain from (5) the desired iteration formulas

(6) $$\boxed{\mathbf{x}^{(m+1)} = \mathbf{b} - \mathbf{Lx}^{(m+1)} - \mathbf{Ux}^{(m)}} \qquad (a_{jj} = 1)$$

where $\mathbf{x}^{(m)} = \left[x_j^{(m)} \right]$ is the mth approximation and $\mathbf{x}^{(m+1)} = \left[x_j^{(m+1)} \right]$ is the $(m + 1)$st approximation. In components this gives the formula in line 1 in Table 18.2. The matrix

Table 18.2
Gauss–Seidel Iteration

Algorithm Gauss–Seidel ($\mathbf{A}, \mathbf{b}, \mathbf{x}^{(0)}, \epsilon, N$)

This algorithm computes a solution $\mathbf{x}$ of the system $\mathbf{Ax} = \mathbf{b}$ given an initial approximation $\mathbf{x}^{(0)}$, where $\mathbf{A} = [a_{jk}]$ is an $n \times n$ matrix with $a_{jj} \neq 0$, $j = 1, \cdots, n$.

INPUT: $\mathbf{A}, \mathbf{b}$, initial approximation $\mathbf{x}^{(0)}$, tolerance $\epsilon > 0$, maximum number of iterations N

OUTPUT: Approximate solution $\mathbf{x}^{(m)} = \left[x_j^{(m)} \right]$ or failure message that $\mathbf{x}^{(N)}$ does not satisfy the tolerance condition

For $m = 0, \cdots, N - 1$, do:
For $j = 1, \cdots, n$, do:

1 $\qquad x_j^{(m+1)} = \dfrac{1}{a_{jj}} \left(b_j - \displaystyle\sum_{k=1}^{j-1} a_{jk} x_k^{(m+1)} - \sum_{k=j+1}^{n} a_{jk} x_k^{(m)} \right)$

End

2 If $\max_j \left| x_j^{(m+1)} - x_j^{(m)} \right| < \epsilon$ then OUTPUT $\mathbf{x}^{(m+1)}$. Stop

[*Procedure completed successfully*]

End

OUTPUT: "No solution satisfying the tolerance condition obtained after N iteration steps." Stop

[*Procedure completed unsuccessfully*]

End GAUSS–SEIDEL

A must satisfy $a_{jj} \neq 0$ for all j. In Table 18.2 our assumption $a_{jj} = 1$ is no longer required, but is automatically taken care of by the factor $1/a_{jj}$ in line 1.

Convergence. Matrix Norms

An iteration method for solving $\mathbf{Ax} = \mathbf{b}$ is said to **converge** for an initial $\mathbf{x}^{(0)}$ if the corresponding iterative sequence $\mathbf{x}^{(0)}, \mathbf{x}^{(1)}, \mathbf{x}^{(2)}, \cdots$ converges to a solution of the given system. Convergence depends on the relation between $\mathbf{x}^{(m)}$ and $\mathbf{x}^{(m+1)}$. To get this relation for the Gauss–Seidel method, we use (6). We first have

$$(\mathbf{I} + \mathbf{L})\mathbf{x}^{(m+1)} = \mathbf{b} - \mathbf{U}\mathbf{x}^{(m)}$$

and by multiplying by $(\mathbf{I} + \mathbf{L})^{-1}$ from the left,

(7) $\qquad \boxed{\mathbf{x}^{(m+1)} = \mathbf{Cx}^{(m)} + (\mathbf{I} + \mathbf{L})^{-1}\mathbf{b}} \qquad$ where $\qquad \boxed{\mathbf{C} = -(\mathbf{I} + \mathbf{L})^{-1}\mathbf{U}.}$

The Gauss–Seidel iteration converges for every $\mathbf{x}^{(0)}$ if and only if all the eigenvalues (Sec. 7.1) of the "iteration matrix" $\mathbf{C} = [c_{jk}]$ have absolute value less than 1. (The proof is in Ref. [E3], p. 191, listed in Appendix 1.) *Caution!* If you want to get $\mathbf{C}$, first divide the rows of $\mathbf{A}$ by a_{jj} to have main diagonal $1, \cdots, 1$. If the **spectral radius** of $\mathbf{C}$ (= maximum of those absolute values) is small, then the convergence is rapid.

Sufficient convergence condition. A sufficient condition for convergence is

(8) $\qquad \boxed{\|\mathbf{C}\| < 1.}$

Here $\|\mathbf{C}\|$ is some **matrix norm,** such as

(9) $\qquad \|\mathbf{C}\| = \sqrt{\sum_{j=1}^{n}\sum_{k=1}^{n} c_{jk}^{2}} \qquad$ **(Frobenius norm)**

or

(10) $\qquad \|\mathbf{C}\| = \max_{k} \sum_{j=1}^{n} |c_{jk}| \qquad$ **(Column "sum" norm)**

or

(11) $\qquad \|\mathbf{C}\| = \max_{j} \sum_{k=1}^{n} |c_{jk}| \qquad$ **(Row "sum" norm).**

These are the most frequently used matrix norms in numerical work. In (9) we take the sum of the squares of all the entries and then the root of it. In (10) we take the sum of the $|c_{jk}|$ in column k, where $k = 1, \cdots, n$, and then the largest of these n sums. In (11) we sum the $|c_{jk}|$ in each row and then take the largest of these n sums.

EXAMPLE 2   **Test of convergence of the Gauss–Seidel iteration**

Test whether the Gauss–Seidel iteration converges for the system

$$\begin{aligned} 2x + y + z &= 4 \\ x + 2y + z &= 4 \qquad \text{written} \\ x + y + 2z &= 4 \end{aligned} \qquad \begin{aligned} x &= 2 - \tfrac{1}{2}y - \tfrac{1}{2}z \\ y &= 2 - \tfrac{1}{2}x - \tfrac{1}{2}z \\ z &= 2 - \tfrac{1}{2}x - \tfrac{1}{2}y. \end{aligned}$$

Solution.   The decomposition

$$
\begin{bmatrix} 1 & 1/2 & 1/2 \\ 1/2 & 1 & 1/2 \\ 1/2 & 1/2 & 1 \end{bmatrix} = \mathbf{I} + \mathbf{L} + \mathbf{U} = \mathbf{I} + \begin{bmatrix} 0 & 0 & 0 \\ 1/2 & 0 & 0 \\ 1/2 & 1/2 & 0 \end{bmatrix} + \begin{bmatrix} 0 & 1/2 & 1/2 \\ 0 & 0 & 1/2 \\ 0 & 0 & 0 \end{bmatrix}
$$

shows that

$$
\mathbf{C} = -(\mathbf{I} + \mathbf{L})^{-1}\mathbf{U} = - \begin{bmatrix} 1 & 0 & 0 \\ -1/2 & 1 & 0 \\ -1/4 & -1/2 & 1 \end{bmatrix} \begin{bmatrix} 0 & 1/2 & 1/2 \\ 0 & 0 & 1/2 \\ 0 & 0 & 0 \end{bmatrix} = \begin{bmatrix} 0 & -1/2 & -1/2 \\ 0 & 1/4 & -1/4 \\ 0 & 1/8 & 3/8 \end{bmatrix}.
$$

We compute the Frobenius norm of $\mathbf{C}$

$$
\|\mathbf{C}\| = \left(\frac{1}{4} + \frac{1}{4} + \frac{1}{16} + \frac{1}{16} + \frac{1}{64} + \frac{9}{64} \right)^{1/2} = \left(\frac{50}{64} \right)^{1/2} = 0.884 < 1
$$

and conclude from (8) that this Gauss–Seidel iteration converges. It is interesting that the other two norms would permit no conclusion, as you should verify. Of course, this points to the fact that (8) is sufficient for convergence rather than necessary.   ◀

Residual. Given a system $\mathbf{Ax} = \mathbf{b}$, the **residual r** of **x** is defined by

(12)
$$
\boxed{\mathbf{r} = \mathbf{b} - \mathbf{Ax}.}
$$

Clearly, $\mathbf{r} = \mathbf{0}$ if and only if **x** is a solution. Hence $\mathbf{r} \neq \mathbf{0}$ for an approximate solution. In the Gauss–Seidel iteration, at each stage we modify or *relax* a component of an approximate solution in order to reduce a component of **r** to zero. Hence the Gauss–Seidel iteration belongs to a class of methods often called **relaxation methods.** More about the residual follows in the next section.

Jacobi Iteration

The Gauss–Seidel iteration is a method of **successive corrections** because we replace approximations by corresponding new ones as soon as the latter have been computed. A method is called a method of **simultaneous corrections** if no component of an approximation $\mathbf{x}^{(m)}$ is used until *all* the components of $\mathbf{x}^{(m)}$ have been computed. A method of this type is the **Jacobi iteration,** which is similar to the Gauss–Seidel iteration but involves *not* using improved values until a step has been completed and then replacing $\mathbf{x}^{(m)}$ by $\mathbf{x}^{(m+1)}$ at once, directly before the beginning of the next cycle. Hence, if we write $\mathbf{Ax} = \mathbf{b}$ (*with $a_{jj} = 1$ as before!*) in the form $\mathbf{x} = \mathbf{b} + (\mathbf{I} - \mathbf{A})\mathbf{x}$, the Jacobi iteration in matrix notation is

(13)
$$
\boxed{\mathbf{x}^{(m+1)} = \mathbf{b} + (\mathbf{I} - \mathbf{A})\mathbf{x}^{(m)}}
$$
$(a_{jj} = 1)$.

This method converges for every choice of $\mathbf{x}^{(0)}$ if and only if the spectral radius of $\mathbf{I} - \mathbf{A}$ is less than 1. It has recently gained greater practical interest since on parallel processors all n equations can be solved simultaneously at each iteration step.

PROBLEM SET 18.3

Gauss–Seidel Iteration

Apply the Gauss–Seidel iteration to the following systems. Do 10 steps, starting from 1, 1, 1 and using 6 significant digits in the computation. *Hint.* Make sure that at the beginning you solve each equation for the variable that has the largest coefficient. (Why?) Show the details of your work.

1. $5x_1 + x_2 + 2x_3 = 19$

$x_1 + 4x_2 - 2x_3 = -2$

$2x_1 + 3x_2 + 8x_3 = 39$

2. $4x_1 - x_2 \qquad = 21$

$-x_1 + 4x_2 - x_3 = -45$

$\qquad - x_2 + 4x_3 = 33$

3. $10x_1 + x_2 + x_3 = 6$

$x_1 + 10x_2 + x_3 = 6$

$x_1 + x_2 + 10x_3 = 6$

4. $4x_1 \qquad + 5x_3 = 12.5$

$x_1 + 6x_2 + 2x_3 = 18.5$

$8x_1 + 2x_2 + x_3 = -11.5$

5. $x_1 + 9x_2 - 2x_3 = 36$

$2x_1 - x_2 + 8x_3 = 121$

$6x_1 + x_2 + x_3 = 107$

6. $3x_1 + 2x_2 + x_3 = 7$

$x_1 + 3x_2 + 2x_3 = 4$

$2x_1 + x_2 + 3x_3 = 7$

7. Apply the Gauss–Seidel iteration (3 steps) to the system in Prob. 3, starting from (a) 0, 0, 0, (b) 10, 10, 10. Compare and comment.

8. In Prob. 3, compute C (a) if you solve the first equation for x_1, the second for x_2, the third for x_3, proving convergence; (b) if you nonsensically solve the third equation for x_1, the first for x_2, the second for x_3, proving divergence.

 9. CAS PROJECT. Gauss–Seidel Iteration. (a) Write a program for Gauss–Seidel iteration.
(b) Apply the program to $A(t)x = b$, starting from $[0 \quad 0 \quad 0]^T$, where

$$A(t) = \begin{bmatrix} 1 & t & t \\ t & 1 & t \\ t & t & 1 \end{bmatrix}, \qquad b = \begin{bmatrix} 2 \\ 2 \\ 2 \end{bmatrix}.$$

For $t = 0.2, 0.5, 0.8, 0.9$ determine the number of steps to obtain the exact solution to 6S and the corresponding spectral radius of C. Plot the number of steps and the spectral radius as functions of t and comment.

(c) **Successive overrelaxation (SOR).** Show that by adding and subtracting $x^{(m)}$ on the right, formula (6) can be written

$$x^{(m+1)} = x^{(m)} + b - Lx^{(m+1)} - (U + I)x^{(m)} \qquad (a_{jj} = 1).$$

Anticipation of further corrections motivates the introduction of an **overrelaxation factor** $\omega > 1$ to get the SOR formula for Gauss–Seidel

(14) $$x^{(m+1)} = x^{(m)} + \omega(b - Lx^{(m+1)} - (U + I)x^{(m)}) \qquad (a_{jj} = 1)$$

intended to give more rapid convergence. A recommended value is $\omega = 2/(1 + \sqrt{1 - \rho})$, where ρ is the spectral radius of C in (7). Apply SOR to the matrix in (b) for $t = 0.5$ and 0.8 and notice the improvement of convergence. (Spectacular gains are made with larger systems. Convergence is obtained for $0 < \omega < 2$.)

Jacobi Iteration

Apply the Jacobi iteration as indicated. Do 5 steps, starting from $[1 \quad 1 \quad 1]^T$. Compare with the Gauss–Seidel iteration. Which of the two seems to converge faster? (Show the details of your work.)

10. The system in Prob. 2   **11.** The system in Prob. 1   **12.** The system in Prob. 4

13. (**Convergence**) Show convergence in Prob. 12 by verifying that $\mathbf{I} - \mathbf{A}$, where $\mathbf{A}$ is the matrix in Prob. 12 with the rows divided by the corresponding main diagonal elements, has the eigenvalues -0.519589 and $0.259795 \pm 0.246603i$.

14. (**Divergence**) Using eigenvalues, show that for the system in Example 2 in the text the Jacobi iteration diverges.

Norms

Compute the norms (9), (10), (11) for the following (square) matrices. Comment on the reasons for greater or smaller differences between the three numbers.

15. The matrix in Prob. 1   **16.** The matrix in Prob. 2   **17.** The matrix in CAS Project 9

18. $\begin{bmatrix} a & -a & a \\ -a & a & -a \\ a & -a & a \end{bmatrix}$ **19.** $\begin{bmatrix} 480 & -2 & -6 \\ 2 & 0 & 1 \\ 6 & 1 & 3 \end{bmatrix}$ **20.** $\begin{bmatrix} 0 & -10 & 0 \\ 10 & 0 & 0 \\ 0 & 0 & -10 \end{bmatrix}$

18.4 Linear Systems: Ill-Conditioning, Norms

One does not need much experience to observe that some systems $\mathbf{Ax} = \mathbf{b}$ are good, giving accurate solutions even under round-off or coefficient inaccuracies, whereas others are bad, so that these inaccuracies affect the solution strongly. We want to see what is going on and whether or not we can "trust" a linear system. Let us first formulate the two relevant concepts (ill- and well-conditioned) for general numerical work and then turn to linear systems and matrices.

A computational problem is called **ill-conditioned** (or *ill-posed*) if "small" changes in the data (the input) cause "large" changes in the solution (the output). The problem is called **well-conditioned** (or *well-posed*) if "small" changes in the data cause only "small" changes in the solution.

These concepts are qualitative. We would certainly regard a magnification of inaccuracies by a factor 100 as "large," but could debate where to draw the line between "large" and "small," depending on the kind of problem and on our viewpoint. Double precision may sometimes help, but if data are measured inaccurately, one should attempt *changing the mathematical setting* of the problem to a well-conditioned one.

Let us now turn to linear systems. Figure 412 explains that ill-conditioning occurs if and only if the two equations give two nearly parallel lines, so that their intersection point (the solution of the system) moves substantially if we raise or lower a line just a little. For larger systems the situation is similar in principle, although geometry no longer helps. We shall see that we may regard ill-conditioning as an approach to singularity of the matrix.

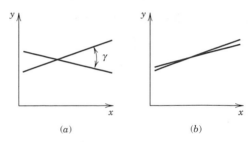

Fig. 412.   (*a*) Well-conditioned and (*b*) ill-conditioned linear system
of two equations in two unknowns

EXAMPLE 1   **An ill-conditioned system**

The reader may verify that the system

$$0.9999x - 1.0001y = 1$$
$$x - \quad y = 1$$

has the solution $x = 0.5$, $y = -0.5$, whereas the system

$$0.9999x - 1.0001y = 1$$
$$x - \quad y = 1 + \epsilon$$

has the solution $x = 0.5 + 5000.5\epsilon$, $y = -0.5 + 4999.5\epsilon$. This shows that the system is ill-conditioned because a change on the right of magnitude ϵ produces a change in the solution of magnitude 5000ϵ, approximately. We see that the lines given by the equations have nearly the same slope. ◄

Well-conditioning can be asserted if the main diagonal entries of **A** have large absolute values compared to those of the other entries. Similarly if $\mathbf{A}^{-1}$ and **A** have maximum entries of about the same absolute value.

Ill-conditioning is indicated if $\mathbf{A}^{-1}$ has entries of large absolute value compared to those of the solution [about 5000 in Examples 1 and 2 (below)] and if poor approximate solutions may still produce small residuals.

Residual. The *residual* **r** of an approximate solution $\tilde{\mathbf{x}}$ of $\mathbf{Ax} = \mathbf{b}$ is defined as

$$(1) \qquad\qquad \mathbf{r} = \mathbf{b} - \mathbf{A\tilde{x}}.$$

Now $\mathbf{b} = \mathbf{Ax}$, so that

$$(2) \qquad\qquad \mathbf{r} = \mathbf{A}(\mathbf{x} - \tilde{\mathbf{x}}).$$

Hence **r** is small if $\tilde{\mathbf{x}}$ has high accuracy, but the converse may be false:

EXAMPLE 2   **Poor approximate solution with a small residual**

The system

$$1.0001x_1 + \quad x_2 = 2.0001$$
$$x_1 + 1.0001x_2 = 2.0001$$

has the exact solution $x_1 = 1$, $x_2 = 1$. Can you see this by inspection? The very poor approximation

$\tilde{x}_1 = 2.0000$, $\tilde{x}_2 = 0.0001$ has the very small residual (to 4 decimals)

$$\mathbf{r} = \begin{bmatrix} 2.0001 \\ 2.0001 \end{bmatrix} - \begin{bmatrix} 1.0001 & 1.0000 \\ 1.0000 & 1.0001 \end{bmatrix} \begin{bmatrix} 2.0000 \\ 0.0001 \end{bmatrix} = \begin{bmatrix} 2.0001 \\ 2.0001 \end{bmatrix} - \begin{bmatrix} 2.0003 \\ 2.0001 \end{bmatrix} = \begin{bmatrix} -0.0002 \\ 0.0000 \end{bmatrix}.$$

From this, a naive person might draw the false conclusion that the approximation should be accurate to 3 or 4 decimals.

Our result is probably unexpected, but we shall see that it has to do with the fact that the system is ill-conditioned. ◀

Our goal is to show that ill-conditioning of a linear system and of its coefficient matrix **A** can be measured by a number, the *"condition number"* $\kappa(\mathbf{A})$. Its definition involves *norm,* a concept that is of great general interest throughout numerical analysis (and modern mathematics in general!). Other measures for ill-conditioning have also been proposed, but $\kappa(\mathbf{A})$ is probably most widely used.

We shall proceed in three steps, discussing

1. **Vector norms,**
2. **Matrix norms,**
3. **Condition number** κ of a square matrix.

Vector Norms

A **vector norm** for column vectors $\mathbf{x} = [x_j]$ with n components (n fixed) is a generalized length, is denoted by $\|\mathbf{x}\|$, and is defined by four properties of the usual length of vectors in three-dimensional space, namely,

(3)

    (a) $\|\mathbf{x}\|$ is a nonnegative real number.

    (b) $\|\mathbf{x}\| = 0$ if and only if $\mathbf{x} = \mathbf{0}.$

    (c) $\|k\mathbf{x}\| = |k| \, \|\mathbf{x}\|$ for all k.

    (d) $\|\mathbf{x} + \mathbf{y}\| \leqq \|\mathbf{x}\| + \|\mathbf{y}\|$ (Triangle inequality).

If we use several norms, we label them by a subscript. Most important in connection with computations is the ***p-norm*** defined by

(4)
$$\|\mathbf{x}\|_p = (|x_1|^p + |x_2|^p + \cdots + |x_n|^p)^{1/p}$$

where p is a fixed number and $p \geqq 1$. In practice, one usually takes $p = 1$ or 2 and, as a third norm, $\|\mathbf{x}\|_\infty$ (the latter as defined below), that is,

(5) $\|\mathbf{x}\|_1 = |x_1| + \cdots + |x_n|$ ("l_1-norm")

(6) $\|\mathbf{x}\|_2 = \sqrt{x_1^2 + \cdots + x_n^2}$ ("Euclidean" or "l_2-norm")

(7) $\|\mathbf{x}\|_\infty = \max_j |x_j|$ ("l_∞-norm")

For $n = 3$ the l_2-norm is the usual length of a vector in three-dimensional space. The l_1-norm and l_∞-norm are generally more convenient in computation.

EXAMPLE 3   **Vector norms**

If $\mathbf{x}^\mathsf{T} = [2 \quad -3 \quad 0 \quad 1 \quad -4]$, then

$$\|\mathbf{x}\|_1 = 10, \qquad \|\mathbf{x}\|_2 = \sqrt{30}, \qquad \|\mathbf{x}\|_\infty = 4. \qquad \blacktriangleleft$$

In three-dimensional space, two points with position vectors $\mathbf{x}$ and $\tilde{\mathbf{x}}$ have distance $|\mathbf{x} - \tilde{\mathbf{x}}|$ from each other. For a linear system $\mathbf{A}\mathbf{x} = \mathbf{b}$, this suggests that we take $\|\mathbf{x} - \tilde{\mathbf{x}}\|$ as a measure of accuracy and call it the **distance** between an exact and an approximate solution, or the **error** of $\tilde{\mathbf{x}}$.

Matrix Norms

If $\mathbf{A}$ is an $n \times n$ matrix and $\mathbf{x}$ any vector with n components, then $\mathbf{A}\mathbf{x}$ is a vector with n components. We now take a vector norm and consider $\|\mathbf{x}\|$ and $\|\mathbf{A}\mathbf{x}\|$. One can prove (see Ref. [9], p. 95, listed in Appendix 1) that there is a number c (depending on $\mathbf{A}$) such that

$$(8) \qquad \qquad \|\mathbf{A}\mathbf{x}\| \leqq c\,\|\mathbf{x}\| \qquad \qquad \text{for all } \mathbf{x}.$$

Let $\mathbf{x} \neq \mathbf{0}$. Then $\|\mathbf{x}\| > 0$ by (3b) and division gives

$$\frac{\|\mathbf{A}\mathbf{x}\|}{\|\mathbf{x}\|} \leqq c.$$

We obtain the smallest possible c valid for *all* $\mathbf{x}$ ($\neq \mathbf{0}$) by taking the maximum on the left. This smallest c is called the **matrix norm of** $\mathbf{A}$ *corresponding to the vector norm we picked* and is denoted by $\|\mathbf{A}\|$. Thus

$$(9) \qquad \qquad \boxed{\|\mathbf{A}\| = \max \frac{\|\mathbf{A}\mathbf{x}\|}{\|\mathbf{x}\|}} \qquad \qquad (\mathbf{x} \neq \mathbf{0}),$$

the maximum being taken over all $\mathbf{x} \neq \mathbf{0}$. Alternatively [see (c) in Team Project 20],

$$(10) \qquad \qquad \|\mathbf{A}\| = \max_{\|\mathbf{x}\|=1} \|\mathbf{A}\mathbf{x}\|$$

Note well that $\|\mathbf{A}\|$ depends on the vector norm that we picked.[3] In particular, one can show that we get

> for the l_1-norm (5) the column "sum" norm (10), Sec. 18.3,
>
> for the l_∞-norm (7) the row "sum" norm (11), Sec. 18.3.

By taking our best possible (our smallest) $c = \|\mathbf{A}\|$ we have from (8)

$$(11) \qquad \qquad \boxed{\|\mathbf{A}\mathbf{x}\| \leqq \|\mathbf{A}\|\,\|\mathbf{x}\|.}$$

---

[3]The maximum in (10) [hence in (9)] exists by Theorem 2.5-3 in Ref. [9], p. 77, since the norm is continuous (see Ref. [9], p. 60). The name "matrix *norm*" is justified since $\|\mathbf{A}\|$ satisfies (3) with $\mathbf{x}$ and $\mathbf{y}$ replaced by $\mathbf{A}$ and $\mathbf{B}$ (see Ref. [9], pp. 92–93).

The Frobenius norm (9), Sec. 18.3, cannot correspond to a vector norm.

This is the formula we need. Formula (9) also implies for two $n \times n$ matrices (see Ref. [9], p. 98)

(12) $\|\mathbf{AB}\| \leqq \|\mathbf{A}\| \, \|\mathbf{B}\|,$ thus $\|\mathbf{A}^n\| \leqq \|\mathbf{A}\|^n.$

See Refs. [9] and [E8] for other useful formulas on norms.

Before we go on, let us do a simple illustrative computation.

EXAMPLE 4   **Matrix norms**

Compute the matrix norms of the coefficient matrix $\mathbf{A}$ in Example 1 and of its inverse $\mathbf{A}^{-1}$ assuming that we use (a) the l_1-vector norm, (b) the l_∞-vector norm.

Solution.   We use (4*), Sec. 6.7, and then (10) and (11) in Sec. 18.3. Thus

$$\mathbf{A} = \begin{bmatrix} 0.9999 & -1.0001 \\ 1.0000 & -1.0000 \end{bmatrix}, \qquad \mathbf{A}^{-1} = \begin{bmatrix} -5000.0 & 5000.5 \\ -5000.0 & 4999.5 \end{bmatrix}.$$

(a) The l_1-vector norm gives the column "sum" norm (10), Sec. 18.3; from column 2 we thus obtain $\|\mathbf{A}\| = |-1.0001| + |-1.0000| = 2.0001$. Similarly, $\|\mathbf{A}^{-1}\| = 10000$.

(b) The l_∞-vector norm gives the row "sum" norm (11), Sec. 18.3; thus $\|\mathbf{A}\| = 2$, $\|\mathbf{A}^{-1}\| = 10000.5$ from row 1. We notice that $\|\mathbf{A}^{-1}\|$ is surprisingly large, which makes $\|\mathbf{A}\| \, \|\mathbf{A}^{-1}\|$ large (20001). We shall see below that this is typical of an ill-conditioned system.   ◄

Condition Number of a Matrix

We are now ready to introduce the key concept in our discussion of ill-conditioning, the **condition number** $\kappa(\mathbf{A})$ of a (nonsingular) square matrix $\mathbf{A}$, defined by

(13) $$\kappa(\mathbf{A}) = \|\mathbf{A}\| \, \|\mathbf{A}^{-1}\|.$$

The role of the condition number is seen from the following theorem.

THEOREM 1   **(Condition number)**

A linear system of equations $\mathbf{Ax} = \mathbf{b}$ whose condition number (13) is small is well-conditioned. A large condition number indicates ill-conditioning.

PROOF. $\mathbf{b} = \mathbf{Ax}$ and (11) give $\|\mathbf{b}\| \leqq \|\mathbf{A}\| \, \|\mathbf{x}\|$. Let $\mathbf{b} \neq \mathbf{0}$ and $\mathbf{x} \neq \mathbf{0}$. Then division by $\|\mathbf{b}\| \, \|\mathbf{x}\|$ gives

(14) $$\frac{1}{\|\mathbf{x}\|} \leqq \frac{\|\mathbf{A}\|}{\|\mathbf{b}\|}.$$

Multiplying (2) by $\mathbf{A}^{-1}$ from the left and interchanging sides, we have $\mathbf{x} - \tilde{\mathbf{x}} = \mathbf{A}^{-1}\mathbf{r}.$ Now (11) with $\mathbf{A}^{-1}$ and $\mathbf{r}$ instead of $\mathbf{A}$ and $\mathbf{x}$ yields

$$\|\mathbf{x} - \tilde{\mathbf{x}}\| = \|\mathbf{A}^{-1}\mathbf{r}\| \leqq \|\mathbf{A}^{-1}\| \, \|\mathbf{r}\|.$$

Division by $\|\mathbf{x}\|$ and use of (14) finally gives

(15) $$\frac{\|\mathbf{x} - \tilde{\mathbf{x}}\|}{\|\mathbf{x}\|} \leqq \frac{1}{\|\mathbf{x}\|} \|\mathbf{A}^{-1}\| \, \|\mathbf{r}\| \leqq \frac{\|\mathbf{A}\|}{\|\mathbf{b}\|} \|\mathbf{A}^{-1}\| \, \|\mathbf{r}\| = \kappa(\mathbf{A}) \frac{\|\mathbf{r}\|}{\|\mathbf{b}\|}.$$

Hence if $\kappa(\mathbf{A})$ is small, a small $\|\mathbf{r}\|/\|\mathbf{b}\|$ implies a small relative error $\|\mathbf{x} - \tilde{\mathbf{x}}\|/\|\mathbf{x}\|$, so that the system is well-conditioned. However, this does not hold if $\kappa(\mathbf{A})$ is large. ◄

EXAMPLE 5   **Condition numbers**

$$\mathbf{A} = \begin{bmatrix} 1 & 16 & 0 \\ 2 & 20 & 2 \\ 0 & 4 & 1 \end{bmatrix} \quad \text{has the inverse} \quad \mathbf{A}^{-1} = \begin{bmatrix} -0.6 & 0.8 & -1.6 \\ 0.1 & -0.05 & 0.1 \\ -0.4 & 0.2 & 0.6 \end{bmatrix}$$

so that (10) and (11) in Sec. 18.3 give the condition numbers $\kappa(\mathbf{A}) = 40 \cdot 2.3 = 92$ and $24 \cdot 3 = 72$, respectively, which are large. ◄

EXAMPLE 6   **Ill-conditioned linear system**

Example 4 gives by (10) or (11), Sec. 18.3, for the matrix in Example 1 the very large condition number $\kappa(\mathbf{A}) = 2.0001 \cdot 10\,000 = 2 \cdot 10\,000.5 = 20\,001$. This confirms that the system is very ill-conditioned. Similarly in Example 2, where by (4*), Sec. 6.7,

$$\mathbf{A}^{-1} = \frac{1}{0.0002} \begin{bmatrix} 1.0001 & -1.0000 \\ -1.0000 & 1.0001 \end{bmatrix} = \begin{bmatrix} 5000.5 & -5000.0 \\ -5000.0 & 5000.5 \end{bmatrix}$$

and (10), Sec. 18.3, gives

$$\kappa(\mathbf{A}) = (1.0001 + 1.0000)(5000.5 + 5000.0) \approx 20\,002.$$

This explains the surprising result in Example 2. ◄

In practice, $\mathbf{A}^{-1}$ will not be known, so that in computing the condition number $\kappa(\mathbf{A})$, one must estimate $\|\mathbf{A}^{-1}\|$. A method for this (proposed in 1979) is explained in Ref. [E8] listed in Appendix 1.

Inaccurate Matrix Entries. $\kappa(\mathbf{A})$ can be used for estimating the effect $\delta\mathbf{x}$ of an inaccuracy $\delta\mathbf{A}$ of $\mathbf{A}$ (errors of measurements of the u_{jk}, for instance). Instead of $\mathbf{Ax} = \mathbf{b}$ we then have

$$(\mathbf{A} + \delta\mathbf{A})(\mathbf{x} + \delta\mathbf{x}) = \mathbf{b}.$$

Multiplying out and subtracting $\mathbf{Ax} = \mathbf{b}$ on both sides, we obtain

$$\mathbf{A}\delta\mathbf{x} + \delta\mathbf{A}(\mathbf{x} + \delta\mathbf{x}) = \mathbf{0}.$$

Multiplication by $\mathbf{A}^{-1}$ from the left gives

$$\delta\mathbf{x} = -\mathbf{A}^{-1}\delta\mathbf{A}(\mathbf{x} + \delta\mathbf{x}).$$

Applying (11) with $\mathbf{A}^{-1}$ and vector $\delta\mathbf{A}(\mathbf{x} + \delta\mathbf{x})$ instead of $\mathbf{A}$ and $\mathbf{x}$, we get

$$\|\delta\mathbf{x}\| = \|\mathbf{A}^{-1}\delta\mathbf{A}(\mathbf{x} + \delta\mathbf{x})\| \leq \|\mathbf{A}^{-1}\| \ \|\delta\mathbf{A}(\mathbf{x} + \delta\mathbf{x})\|.$$

Again applying (11), with $\delta\mathbf{A}$ and $\mathbf{x} - \delta\mathbf{x}$ instead of $\mathbf{A}$ and $\mathbf{x}$, we obtain

$$\|\delta\mathbf{x}\| \leq \|\mathbf{A}^{-1}\| \ \|\delta\mathbf{A}\| \ \|\mathbf{x} + \delta\mathbf{x}\|.$$

Now $\|\mathbf{A}^{-1}\| = \kappa(\mathbf{A})/\|\mathbf{A}\|$, so that division by $\|\mathbf{x} + \delta\mathbf{x}\|$ shows that the relative inaccuracy of $\mathbf{x}$ is related to that of $\mathbf{A}$ via the condition number by the inequality

$$(16) \qquad \frac{\|\delta\mathbf{x}\|}{\|\mathbf{x}\|} \approx \frac{\|\delta\mathbf{x}\|}{\|\mathbf{x} + \delta\mathbf{x}\|} \leq \|\mathbf{A}^{-1}\| \ \|\delta\mathbf{A}\| = \kappa(\mathbf{A}) \ \frac{\|\delta\mathbf{A}\|}{\|\mathbf{A}\|} \ .$$

Conclusion. If the system is well-conditioned, small inaccuracies $\|\delta\mathbf{A}\| / \|\mathbf{A}\|$ can have only a small effect on the solution. However, in the case of ill-conditioning, if $\|\delta\mathbf{A}\| / \|\mathbf{A}\|$ is small, $\|\delta\mathbf{x}\| / \|\mathbf{x}\|$ *may* be large.

Inaccurate right side. You may show that, similarly, when $\mathbf{A}$ is accurate, an inaccuracy $\delta\mathbf{b}$ of $\mathbf{b}$ causes an inaccuracy $\delta\mathbf{x}$ satisfying

$$(17) \qquad \frac{\|\delta\mathbf{x}\|}{\|\mathbf{x}\|} \leq \kappa(\mathbf{A}) \ \frac{\|\delta\mathbf{b}\|}{\|\mathbf{b}\|} \ .$$

Hence $\|\delta\mathbf{x}\| / \|\mathbf{x}\|$ must remain relatively small whenever $\kappa(\mathbf{A})$ is small.

Further Comments on Condition Numbers. The following additional explanations may be helpful.

1. There is no sharp dividing line between "well-conditioned" and "ill-conditioned," but generally the situation will get worse as we go from systems with small $\kappa(\mathbf{A})$ to systems with larger $\kappa(\mathbf{A})$. Now always $\kappa(\mathbf{A}) \geq 1$, so that values of 10 or 20 or so give no reason for concern, whereas $\kappa(\mathbf{A}) = 100$, say, calls for caution, and systems such as those in Examples 1 and 2 are extremely ill-conditioned.

2. If $\kappa(\mathbf{A})$ is large (or small) in one norm, it will be large (or small, respectively) in any other norm. See Example 5.

3. The literature on ill-conditioning is extensive. For an introduction to it, see [E8].

This is the end of our discussion of numerical methods for solving linear systems. An important application follows in the next section.

PROBLEM SET 18.4

Vector Norms. Compute (5), (6), (7) for the following vectors. Find the corresponding **unit vectors** (vectors of norm 1) with respect to the l_∞-norm.

1. [7 −12 5 0]	**2.** [3 4 −5]	**3.** [0.8 −7.1 1.4]
4. [1 1 1 1 1]	**5.** [−6 6 −6 6]	**6.** [0 0 0 1 0]

Matrix Norms. For the following matrices compute the matrix norms and the condition numbers corresponding to the l_1- and l_∞-vector norms.

7. $\begin{bmatrix} 4 & 1 \\ 0 & 2 \end{bmatrix}$
    8. $\begin{bmatrix} 0.75 & 1.25 \\ 1.25 & 0.75 \end{bmatrix}$
    9. $\begin{bmatrix} -7 & 13 \\ 5 & -9 \end{bmatrix}$

10. $\begin{bmatrix} 1 & 3 & 1.5 \\ 0.75 & 1.5 & 1 \\ 0.6 & 1 & 0.75 \end{bmatrix}$ **11.** $\begin{bmatrix} -200 & 0 & 0 \\ 0 & 0.04 & 0 \\ 0 & 0 & 200 \end{bmatrix}$ **12.** $\begin{bmatrix} -2 & 4 & -1 \\ -2 & 3 & 0 \\ 7 & -12 & 2 \end{bmatrix}$

13. Verify (11) for $\mathbf{x}^T = \begin{bmatrix} 3 & 15 & -4 \end{bmatrix}$ taken with the l_∞-norm and the matrix in Prob. 12.

Ill-Conditioned Systems. Solve $\mathbf{Ax} = \mathbf{b}_1$, $\mathbf{Ax} = \mathbf{b}_2$, compare the solutions, and comment. Compute the condition number of $\mathbf{A}$.

14. $\mathbf{A} = \begin{bmatrix} 3 & 1.7 \\ 1.7 & 1 \end{bmatrix}$, $\mathbf{b}_1 = \begin{bmatrix} 4.7 \\ 2.7 \end{bmatrix}$, $\mathbf{b}_2 = \begin{bmatrix} 4.7 \\ 2.71 \end{bmatrix}$

15. $\mathbf{A} = \begin{bmatrix} 4.50 & 3.55 \\ 3.55 & 2.80 \end{bmatrix}$, $\mathbf{b}_1 = \begin{bmatrix} 5.2 \\ 4.1 \end{bmatrix}$, $\mathbf{b}_2 = \begin{bmatrix} 5.2 \\ 4.0 \end{bmatrix}$

16. (Residual) For $\mathbf{Ax} = \mathbf{b}_1$ in Prob. 15 guess what the residual of $\tilde{\mathbf{x}} = \begin{bmatrix} -10.0 & 14.1 \end{bmatrix}^T$, approximating $\begin{bmatrix} -2 & 4 \end{bmatrix}^T$, might be. Then calculate.

17. Try to construct an example of a small residual of a very poor approximate solution of Prob. 15 with $\mathbf{b}_1$.

18. (Minimum condition number) Show that $\kappa(\mathbf{A}) \geq 1$ for the matrix norms (10), (11), Sec. 18.3, and $\kappa(\mathbf{A}) \geq \sqrt{n}$ for the Frobenius norm (9), Sec. 18.3. *Hint.* Use $\mathbf{AA}^{-1} = \mathbf{I}$ and (12).

19. CAS PROJECT. Experimenting with Hilbert Matrices. The 3×3 Hilbert matrix is

$$\mathbf{H}_3 = \begin{bmatrix} 1 & \frac{1}{2} & \frac{1}{3} \\ \frac{1}{2} & \frac{1}{3} & \frac{1}{4} \\ \frac{1}{3} & \frac{1}{4} & \frac{1}{5} \end{bmatrix}.$$

The $n \times n$ Hilbert matrix is $\mathbf{H}_n = [h_{jk}]$, where $h_{jk} = 1/(j + k - 1)$. (Similar matrices occur in curve fitting by least squares.) Compute the condition number $\kappa(\mathbf{H}_n)$ for the matrix norm corresponding to the l_∞- (or l_1-) vector norm, for $n = 2, 3, \cdots, 6$ (or further if you wish). Try to find a formula that gives reasonable approximate values of these rapidly growing numbers. Solve a few linear systems involving an $\mathbf{H}_n$ of your choice.

20. TEAM PROJECT. Norms. **(a) Vector norms** in our text are **equivalent,** that is, they are related by double inequalities; for instance,

(18)   (a) $\|\mathbf{x}\|_\infty \leq \|\mathbf{x}\|_1 \leq n \|\mathbf{x}\|_\infty$ (b) $\dfrac{1}{n} \|\mathbf{x}\|_1 \leq \|\mathbf{x}\|_\infty \leq \|\mathbf{x}\|_1.$

Hence if for some **x,** one norm is large (or small), the other norm must also be large (or small). Thus in many investigations the particular choice of a norm is not essential. Prove (18).

(b) The **Cauchy–Schwarz inequality** is $|\mathbf{x}^T\mathbf{y}| \leq \|\mathbf{x}\|_2 \|\mathbf{y}\|_2$. It is very important. (Proof in Ref. [9] listed in Appendix 1.) Use it to prove

(19)   (a) $\|\mathbf{x}\|_2 \leq \|\mathbf{x}\|_1 \leq \sqrt{n} \|\mathbf{x}\|_2$ (b) $\dfrac{1}{\sqrt{n}} \|\mathbf{x}\|_1 \leq \|\mathbf{x}\|_2 \leq \|\mathbf{x}\|_1.$

(c) Formula (10) is often more practical than (9). Derive (10) from (9).

(d) Matrix norms. Illustrate (11) with examples. Give examples of (12) with equality as well as with strict inequality. Prove that the matrix norms (10), (11) in Sec. 18.3 satisfy the *axioms of a norm* $\|\mathbf{A}\| \geq 0$, $\|\mathbf{A}\| = 0$ if and only if $\mathbf{A} = \mathbf{0}$, $\|k\mathbf{A}\| = |k| \|\mathbf{A}\|$, $\|\mathbf{A} + \mathbf{B}\| \leq \|\mathbf{A}\| + \|\mathbf{B}\|$.

18.5 Method of Least Squares

In **curve fitting** we are given n points (pairs of numbers)

$$(x_1, y_1), \cdots, (x_n, y_n)$$

and we want to determine a function $f(x)$ such that $f(x_j) \approx y_j$, $j = 1, \cdots, n$. The type of function (for example, polynomials, exponential functions, sine and cosine functions) may be suggested by the nature of the problem (the underlying physical law, for instance), and in many cases a polynomial of a certain degree will be appropriate.

If we require strict equality $f(x_1) = y_1, \cdots, f(x_n) = y_n$ and use polynomials of sufficiently high degree, we may apply one of the methods discussed in Sec. 17.3 in connection with interpolation. However, in certain situations this would not be the appropriate solution of the actual problem. For instance, to the four points

(1) $(-1.3, 0.103), \qquad (-0.1, 1.099), \qquad (0.2, 0.808), \qquad (1.3, 1.897)$

there corresponds the Lagrange polynomial $f(x) = x^3 - x + 1$ (Fig. 413), but if we graph the points, we see that they lie nearly on a straight line. Hence if these values are obtained in an experiment and thus involve an experimental error, and if the nature of the experiment suggests a linear relation, we better fit a straight line through the points (Fig. 413). Such a line may be useful for predicting values to be expected for other values of x. A widely used principle for fitting straight lines is the **method of least squares** by Gauss. In the present situation it may be formulated as follows.

Method of least squares. *The straight line*

(2) $y = a + bx$

should be fitted through the given points $(x_1, y_1), \cdots, (x_n, y_n)$ so that the sum of the squares of the distances of those points from the straight line is minimum, where the distance is measured in the vertical direction (the y-direction).

The point on the line with abscissa x_j has the ordinate $a + bx_j$. Hence its distance from (x_j, y_j) is $\left| y_j - a - bx_j \right|$ (Fig. 414) and that sum of squares is

$$q = \sum_{j=1}^{n} (y_j - a - bx_j)^2.$$

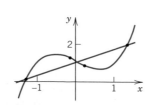

Fig. 413. Approximate fitting of a
straight line

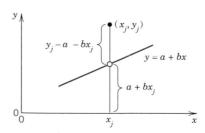

Fig. 414. Vertical distance of a point (x_j, y_j)
from a straight line $y = a + bx$

q depends on a and b. A necessary condition for q to be minimum is

(3)
$$\frac{\partial q}{\partial a} = -2 \sum (y_j - a - bx_j) = 0$$

$$\frac{\partial q}{\partial b} = -2 \sum x_j (y_j - a - bx_j) = 0$$

(where we sum over j from 1 to n). Writing each sum as three sums and taking one of them to the right, we obtain the result

(**4**)
$$\boxed{\begin{aligned} an \quad\quad + b \sum x_j &= \sum y_j \\ a \sum x_j + b \sum x_j^2 &= \sum x_j y_j. \end{aligned}}$$

These equations are called the **normal equations** of our problem.

EXAMPLE 1   **Straight line**

Using the method of least squares, fit a straight line to the four points given in formula (1).

Solution.   We obtain

$$n = 4, \quad \sum x_j = 0.1, \quad \sum x_j^2 = 3.43, \quad \sum y_j = 3.907, \quad \sum x_j y_j = 2.3839.$$

Hence the normal equations are

$$4a + 0.10b = 3.9070$$

$$0.1a + 3.43b = 2.3839.$$

The solution is $a = 0.9601$, $b = 0.6670$, and we obtain the straight line (Fig. 413)

$$y = 0.9601 + 0.6670x. \qquad\qquad \blacktriangleleft$$

Curve Fitting by Polynomials of Degree m

Our method of curve fitting can be generalized from a polynomial $y = a + bx$ to a polynomial degree m

(5)
$$p(x) = b_0 + b_1 x + \cdots + b_m x^m$$

where $m \leqq n - 1$. Then q takes the form

$$q = \sum_{j=1}^{n} (y_j - p(x_j))^2$$

and depends on $m + 1$ parameters $b_0, \cdots, b_m$. Instead of (3) we then have $m + 1$ conditions

(6)
$$\frac{\partial q}{\partial b_0} = 0, \quad \cdots, \quad \frac{\partial q}{\partial b_m} = 0$$

which give a system of $m + 1$ normal equations.

In the case of a quadratic polynomial

(7) $$p(x) = b_0 + b_1 x + b_2 x^2$$

the normal equations are (summation from 1 to n)

(8)
$$b_0 n \qquad + b_1 \sum x_j + b_2 \sum x_j^2 = \sum y_j$$
$$b_0 \sum x_j + b_1 \sum x_j^2 + b_2 \sum x_j^3 = \sum x_j y_j$$
$$b_0 \sum x_j^2 + b_1 \sum x_j^3 + b_2 \sum x_j^4 = \sum x_j^2 y_j.$$

The derivation of (8) is left to the reader.

For a general polynomial (5) the normal equations form a linear system of equations in the unknowns $b_0, \cdots, b_m$. When its matrix $\mathbf{M}$ is nonsingular, we can solve the system by Cholesky's method (Sec. 18.2) because then $\mathbf{M}$ is positive definite (and symmetric). When the equations are nearly linearly dependent, the normal equations may become ill-conditioned and should be replaced by other methods; see [E3], Sec. 5.7, listed in Appendix 1.

PROBLEM SET 18.5

Fitting by a Straight Line

Fit a straight line to the given points (x, y) by the method of least squares. Show the details of your work. Check your result by sketching the points and the line.

1. $(0, 3), (2, 1), (3, -1), (5, -2)$

2. How does the line in Prob. 1 change if you add a point far above it, say, $(1, 4)$? First guess and sketch. Then compute.

3. $(0, 1.8), (1, 1.6), (2, 1.1), (3, 1.5), (4, 2.3)$

4.

Time t	12:00	12:10	12:30	13:00	14:00
Temperature T [°C]	100	90	75	55	30

5. Do you get a better fit in Prob. 4 if you take $1/\ln T$ instead of T? Why? (Measure t in minutes, $t = 0$ corresponding to 12:00.)

6.

Revolutions per minute x	400	500	600	700	750
Power of a Diesel engine y [hp]	580	1030	1420	1880	2100

7. (**Ohm's law:** $U = Ri$) Estimate R from the least squares line that fits the following data. $(i, U) = (3.0, 162), (5.0, 255), (7.0, 360), (10.0, 495)$.

8. (**Hooke's law:** $F = ks$) Estimate the spring modulus k from the force F [lb] and elongation s [cm], where $(F, s) = (1, 0.3), (2, 0.7), (4, 1.3), (6, 1.9), (10, 3.2), (20, 6.3)$.

9. (**Speed**) Estimate the average speed $v = b_1$ [mph] of a motion in which the distance from the origin is given by $y(t) = b_0 + b_1 t$ [mi] (t = time [min]) and $y(t)$ fits the following data. $(t, y) = (0, 0), (20, 16), (40, 35), (60, 50), (80, 69)$.

Fitting by a Quadratic Parabola

Fit a parabola (7) to the given points (x, y) by the method of least squares. Show details. Check by sketching.

10. $(-1, 3), (1, 1), (2, 2), (3, 6)$

11. $(2, 0), (3, 3), (5, 4), (6, 3), (7, 1)$

12.

Speed of a snowplow x [mph]	5	7.5	10	12.5	15
Power of the plow y [lb]	4200	4600	5200	4800	4300

13. The data in Prob. 3. Plot the points, the line, and the parabola jointly, compare, and comment.

14. Fit (2) and (7) by least squares to $(-1, 7)$, $(1, 5)$, $(2, 6)$, $(3, 4)$, $(5, 7)$. Sketch jointly, compare, and comment.

15. Solve the normal equations in Prob. 13 by Cholesky's method.

Fitting by a Cubic Parabola

16. (**Normal equations**) Find the formula for the normal equations of a cubic least squares polynomial.

17. Fit a cubic parabola by least squares to $(-2, -8)$, $(-1, 0)$, $(0, 1)$, $(1, 2)$, $(2, 12)$, $(4, 80)$.

18. Fit (2), (7), and a cubic parabola by least squares to $(-2, -2)$, $(-1, 1)$, $(-\frac{1}{2}, \frac{1}{2})$, $(\frac{1}{4}, -\frac{1}{2})$, $(1, -1)$, $(2, 1)$

19. **CAS PROJECT.** Write programs for calculating and solving the normal equations (4) and (8). Apply the programs to Probs. 3, 6, 10, 11, and some data of your choice. If your CAS has a command for fitting (MATHEMATICA and MAPLE do), compare your results with those by your CAS command.

20. **TEAM PROJECT.** The **least squares approximation of a function** $f(x)$ on an interval $a \leqq x \leqq b$ by a function

$$F_m(x) = a_0 y_0(x) + a_1 y_1(x) + \cdots + a_m y_m(x) \qquad (y_0, \cdots, y_m \text{ given})$$

requires the determination of the coefficients $a_0, \cdots, a_m$ such that

$$(9) \qquad \int_a^b [f(x) - F_m(x)]^2 \, dx$$

becomes minimum. This integral is denoted by $\|f - F_m\|^2$, and $\|f - F_m\|$ is called the **L_2-norm** (L suggesting Lebesgue[4]). A necessary condition for that minimum is given by $\partial \|f - F_m\|^2 / \partial a_j = 0$, $j = 0, \cdots, m$ [the analog of (6)]. (**a**) Show that this leads to the $m + 1$ normal equations ($j = 0, \cdots, m$)

$$(10) \qquad \sum_{k=0}^m h_{jk} a_k = b_j \qquad \text{where} \qquad h_{jk} = \int_a^b y_j(x) y_k(x) \, dx, \quad b_j = \int_a^b f(x) y_j(x) \, dx.$$

(**b**) **Polynomial.** What form does (10) take if $F_m(x) = a_0 + a_1 x + \cdots + a_m x^m$? What is the coefficient matrix of (10) in this case when the interval is $0 \leqq x \leqq 1$?

(**c**) **Orthogonal functions.** What are the solutions of (10) if $y_0(x), \cdots, y_m(x)$ are orthogonal on the interval $a \leqq x \leqq b$? (For the definition, see Sec. 4.7. See also Sec. 4.8.)

18.6 Matrix Eigenvalue Problems: Introduction

In the remaining sections of this chapter we discuss some of the most important ideas and numerical methods for matrix eigenvalue problems. This very extensive part of numerical linear algebra is of great practical importance, with much research going on, particularly since 1945, and hundreds, if not thousands of papers published in various mathematical journals (see the references in [E7], [E8], [E10], [E16], [E17], [E19], [E20]). We begin with the concepts and general results we shall need in explaining and applying numerical methods for eigenvalue problems. (For typical applications, see Chap. 7.)

---

[4]HENRI LEBESGUE (1875—1941), great French mathematician, creator of a modern theory of measure and integration in his famous doctoral thesis of 1902.

An **eigenvalue** or **characteristic value** (or *latent root*) of a given $n \times n$ matrix $\mathbf{A} = [a_{jk}]$ is a real or complex number λ such that the vector equation

(1)
$$\mathbf{Ax} = \lambda\mathbf{x}$$

has a nontrivial solution, that is, a solution $\mathbf{x} \neq \mathbf{0}$, which is then called an **eigenvector** or **characteristic vector** of $\mathbf{A}$ corresponding to that eigenvalue λ. The set of all eigenvalues of $\mathbf{A}$ is called the **spectrum** of $\mathbf{A}$. Equation (1) can be written

(2)
$$(\mathbf{A} - \lambda\mathbf{I})\mathbf{x} = \mathbf{0}$$

where $\mathbf{I}$ is the $n \times n$ unit matrix. This homogeneous system has a nontrivial solution if and only if the **characteristic determinant** $\det(\mathbf{A} - \lambda\mathbf{I})$ is 0 (see Theorem 2 in Sec. 6.5). This gives (see Sec. 7.1)

THEOREM 1   **(Eigenvalues)**

The eigenvalues of $\mathbf{A}$ *are the solutions of the* **characteristic equation**

(3)
$$\det(\mathbf{A} - \lambda\mathbf{I}) = \begin{vmatrix} a_{11} - \lambda & a_{12} & \cdots & a_{1n} \\ a_{21} & a_{22} - \lambda & \cdots & a_{2n} \\ . & . & \cdots & . \\ a_{n1} & a_{n2} & \cdots & a_{nn} - \lambda \end{vmatrix} = 0.$$

Developing the characteristic determinant, we obtain the **characteristic polynomial** of $\mathbf{A}$, which is of degree n in λ. Hence $\mathbf{A}$ has at least one and at most n numerically different eigenvalues. If $\mathbf{A}$ is real, so are the coefficients of the characteristic polynomial. By familiar algebra it follows that then the roots (the eigenvalues of $\mathbf{A}$) are real or complex conjugates in pairs.

We shall usually denote the eigenvalues of $\mathbf{A}$ by

$$\lambda_1, \lambda_2, \cdots, \lambda_n$$

with the understanding that some (or all) of them may be numerically equal.

The sum of these n eigenvalues equals the sum of the entries on the main diagonal of $\mathbf{A}$, called the **trace** of $\mathbf{A}$; thus

(4)
$$\text{trace } \mathbf{A} = \sum_{j=1}^{n} a_{jj} = \sum_{k=1}^{n} \lambda_k.$$

Also, the product of the eigenvalues equals the determinant of $\mathbf{A}$,

(5)
$$\det \mathbf{A} = \lambda_1 \lambda_2 \cdots \lambda_n.$$

Both formulas follow from the product representation of the characteristic polynomial, which we denote by $f(\lambda)$,

$$f(\lambda) = (-1)^n (\lambda - \lambda_1)(\lambda - \lambda_2) \cdots (\lambda - \lambda_n).$$

If we take equal factors together and denote the *numerically distinct* eigenvalues of $\mathbf{A}$ by $\lambda_1, \cdots, \lambda_r$ ($r \leqq n$), then the product becomes

$$(6) \qquad f(\lambda) = (-1)^n (\lambda - \lambda_1)^{m_1} (\lambda - \lambda_2)^{m_2} \cdots (\lambda - \lambda_r)^{m_r}.$$

The exponent m_j is called the **algebraic multiplicity** of λ_j. The maximum number of linearly independent eigenvectors corresponding to λ_j is called the **geometric multiplicity** of λ_j. It is equal to or smaller than m_j.

A subspace S of R^n or C^n (if $\mathbf{A}$ is complex) is called an **invariant subspace** of $\mathbf{A}$ if for every $\mathbf{v}$ in S the vector $\mathbf{A}\mathbf{v}$ is also in S. **Eigenspaces** of $\mathbf{A}$ (spaces of eigenvectors; Sec. 7.1) are important invariant subspaces of $\mathbf{A}$.

Similarity. Spectral shift. Special matrices

An $n \times n$ matrix $\mathbf{B}$ is called **similar** to $\mathbf{A}$ if there is a nonsingular $n \times n$ matrix $\mathbf{T}$ such that

$$(7) \qquad \boxed{\mathbf{B} = \mathbf{T}^{-1}\mathbf{A}\mathbf{T}.}$$

Similarity is important for the following reason.

THEOREM 2   **(Similar matrices)**

Similar matrices have the same eigenvalues. If $\mathbf{x}$ is an eigenvector of $\mathbf{A}$, then $\mathbf{y} = \mathbf{T}^{-1}\mathbf{x}$ is an eigenvector of $\mathbf{B}$ in (7) *corresponding to the same eigenvalue.* (Proof in Sec. 7.5.)

Another theorem that has various applications in numerical work is as follows.

THEOREM 3   **(Spectral shift)**

If $\mathbf{A}$ has the eigenvalues $\lambda_1, \cdots, \lambda_n$, then $\mathbf{A} - k\mathbf{I}$ with arbitrary k has the eigenvalues $\lambda_1 - k, \cdots, \lambda_n - k$.

This theorem is a special case of the following **spectral mapping theorem.**

THEOREM 4   **(Polynomial matrices)**

If λ is an eigenvalue of $\mathbf{A}$, then

$$q(\lambda) = \alpha_s \lambda^s + \alpha_{s-1} \lambda^{s-1} + \cdots + \alpha_1 \lambda + \alpha_0$$

is an eigenvalue of the **polynomial matrix**

$$q(\mathbf{A}) = \alpha_s \mathbf{A}^s + \alpha_{s-1} \mathbf{A}^{s-1} + \cdots + \alpha_1 \mathbf{A} + \alpha_0 \mathbf{I}.$$

PROOF. $\mathbf{A}\mathbf{x} = \lambda \mathbf{x}$ implies $\mathbf{A}^2 \mathbf{x} = \mathbf{A}\lambda\mathbf{x} = \lambda \mathbf{A}\mathbf{x} = \lambda^2 \mathbf{x}$, $\mathbf{A}^3 \mathbf{x} = \lambda^3 \mathbf{x}$, etc. Thus

$$q(\mathbf{A})\mathbf{x} = (\alpha_s \mathbf{A}^s + \alpha_{s-1} \mathbf{A}^{s-1} + \cdots)\mathbf{x}$$
$$= \alpha_s \mathbf{A}^s \mathbf{x} + \alpha_{s-1} \mathbf{A}^{s-1} \mathbf{x} + \cdots$$
$$= \alpha_s \lambda^s \mathbf{x} + \alpha_{s-1} \lambda^{s-1} \mathbf{x} + \cdots = q(\lambda)\mathbf{x}. \qquad \blacktriangleleft$$

The eigenvalues of important special matrices can be characterized as follows.

THEOREM 5   **(Special matrices)**

The eigenvalues of Hermitian matrices (i.e., $\overline{\mathbf{A}}^{\mathsf{T}} = \mathbf{A}$), hence of real symmetric matrices (i.e., $\mathbf{A}^{\mathsf{T}} = \mathbf{A}$), are real. The eigenvalues of skew-Hermitian matrices (i.e., $\overline{\mathbf{A}}^{\mathsf{T}} = -\mathbf{A}$), hence of real skew-symmetric matrices (i.e., $\mathbf{A}^{\mathsf{T}} = -\mathbf{A}$) are pure imaginary or 0. The eigenvalues of unitary matrices (i.e., $\overline{\mathbf{A}}^{\mathsf{T}} = \mathbf{A}^{-1}$), hence of orthogonal matrices (i.e., $\mathbf{A}^{\mathsf{T}} = \mathbf{A}^{-1}$), have absolute value 1. (Proof in Sec. 7.4.)

The **choice of a numerical method** for matrix eigenvalue problems depends essentially on two circumstances, on the kind of matrix (real symmetric, real general, complex, sparse, or full) and on the kind of information to be obtained, that is, whether one wants to know all eigenvalues or merely specific ones, for instance, the largest eigenvalue, whether eigenvalues *and* eigenvectors are wanted, and so on. It is clear that we cannot enter into a systematic discussion of all these and further possibilities that arise in practice—look quickly into Ref. [E20] to get an idea—but we shall concentrate on some basic aspects and methods that will give us a general understanding of this fascinating field.

18.7 Inclusion of Matrix Eigenvalues

By **"inclusion"** we mean the determination of approximate values of eigenvalues and corresponding error bounds. As a first important "inclusion theorem," the theorem by Gerschgorin gives a region consisting of closed circular disks in the complex plane and including all the eigenvalues of a given matrix. Indeed, for each $j = 1, \cdots, n$ the inequality (1) in the theorem determines a closed circular disk in the complex λ-plane with center a_{jj} and radius given by the right side of (1); and Theorem 1 states that each of the eigenvalues of $\mathbf{A}$ lies in one of these n disks.

THEOREM 1   **(Gerschgorin's theorem)**

Let λ be an eigenvalue of an arbitrary $n \times n$ matrix $\mathbf{A} = [a_{jk}]$. Then for some integer j $(1 \leqq j \leqq n)$ we have

(1) $$\left| a_{jj} - \lambda \right| \leqq \left| a_{j1} \right| + \left| a_{j2} \right| + \cdots + \left| a_{j,j-1} \right| + \left| a_{j,j+1} \right| + \cdots + \left| a_{jn} \right|.$$

PROOF.   Let $\mathbf{x}$ be an eigenvector corresponding to an eigenvalue λ of $\mathbf{A}$. Then

(2) $$\mathbf{Ax} = \lambda\mathbf{x} \quad \text{or} \quad (\mathbf{A} - \lambda\mathbf{I})\mathbf{x} = \mathbf{0}.$$

Let x_j be a component of $\mathbf{x}$ that is largest in absolute value. Then we have $\left| x_m/x_j \right| \leqq 1$ for $m = 1, \cdots, n$. The vector equation (2) is equivalent to a system of n equations for the n components of the vectors on both sides. The jth of these n equations is

$$a_{j1}x_1 + \cdots + a_{j,j-1}x_{j-1} + (a_{jj} - \lambda)x_j + a_{j,j+1}x_{j+1} + \cdots + a_{jn}x_n = 0.$$

Division by x_j (which cannot be zero; why?) and reshuffling terms gives

$$a_{jj} - \lambda = -a_{j1}\frac{x_1}{x_j} - \cdots - a_{j,j-1}\frac{x_{j-1}}{x_j} - a_{j,j+1}\frac{x_{j+1}}{x_j} - \cdots - a_{jn}\frac{x_n}{x_j}.$$

By taking absolute values on both sides of this equation, applying the triangle inequality $|a + b| \leq |a| + |b|$ (where a and b are any complex numbers), and observing that because of the choice of j (which is crucial!),

$$\left| \frac{x_1}{x_j} \right| \leq 1, \quad \cdots, \quad \left| \frac{x_n}{x_j} \right| \leq 1,$$

we obtain (1), and the theorem is proved. ◀

EXAMPLE 1   **Gerschgorin's theorem**

For the eigenvalues of the matrix

$$\mathbf{A} = \begin{bmatrix} 0 & 1/2 & 1/2 \\ 1/2 & 5 & 1 \\ 1/2 & 1 & 1 \end{bmatrix}$$

we get the Gerschgorin disks (Fig. 415)

$$D_1: \quad \text{Center } 0, \quad \text{radius } 1, \quad D_2: \quad \text{Center } 5, \quad \text{radius } 1.5, \quad D_3: \quad \text{Center } 1, \quad \text{radius } 1.5.$$

Since $\mathbf{A}$ is symmetric, it follows from this and Theorem 5, Sec. 18.6, that the spectrum of $\mathbf{A}$ must actually lie in the intervals $[-1, 2.5]$ and $[3.5, 6.5]$ on the real axis. You may confirm this by showing that, to 3 decimals, $\lambda_1 = -0.209$, $\lambda_2 = 5.305$, $\lambda_3 = 0.904$.

It is interesting that here the Gerschgorin disks form two disjoint sets, namely, $D_1 \cup D_3$, which contains two eigenvalues, and D_2, which contains one eigenvalue. This is typical, as the following theorem shows. ◀

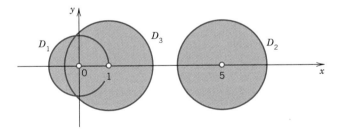

Fig. 415.   Gerschgorin disks in Example 1

THEOREM 2   **(Extension of Gerschgorin's theorem)**

If p Gerschgorin disks form a set S that is disjoint from the $n - p$ other disks of a given matrix $\mathbf{A}$, then S contains precisely p eigenvalues of $\mathbf{A}$ (each counted with its algebraic multiplicity, as defined in Sec. 18.6).

IDEA OF PROOF.   Set $\mathbf{A} = \mathbf{B} + \mathbf{C}$, where $\mathbf{B}$ is the diagonal matrix with entries a_{jj}, and apply Theorem 1 to $\mathbf{A}_t = \mathbf{B} + t\mathbf{C}$ with t growing from 0 to 1. ◀

EXAMPLE 2   **Another application of Gerschgorin's theorem. Similarity**

Suppose that we have diagonalized a matrix by some numerical method that left us with some off-diagonal entries of size 10^{-5}, say,

$$\mathbf{A} = \begin{bmatrix} 2 & 10^{-5} & 10^{-5} \\ 10^{-5} & 2 & 10^{-5} \\ 10^{-5} & 10^{-5} & 4 \end{bmatrix}.$$

What can we conclude about deviations of the eigenvalues from the main diagonal entries?

Solution. By Theorem 2, one eigenvalue must lie in the disk of radius $2 \cdot 10^{-5}$ centered at 4 and two eigenvalues (or an eigenvalue of algebraic multiplicity 2) in the disk of radius $2 \cdot 10^{-5}$ centered at 2. Actually, since the matrix is symmetric, these eigenvalues must lie in the intersections of these disks and the real axis, by Theorem 5 in Sec. 18.6.

We show how an isolated disk can always be reduced in size by a similarity transformation. The matrix

$$\mathbf{B} = \mathbf{T}^{-1}\mathbf{A}\mathbf{T} = \begin{bmatrix} 1 & 0 & 0 \\ 0 & 1 & 0 \\ 0 & 0 & 10^{-5} \end{bmatrix} \begin{bmatrix} 2 & 10^{-5} & 10^{-5} \\ 10^{-5} & 2 & 10^{-5} \\ 10^{-5} & 10^{-5} & 4 \end{bmatrix} \begin{bmatrix} 1 & 0 & 0 \\ 0 & 1 & 0 \\ 0 & 0 & 10^{5} \end{bmatrix}$$

$$= \begin{bmatrix} 2 & 10^{-5} & 1 \\ 10^{-5} & 2 & 1 \\ 10^{-10} & 10^{-10} & 4 \end{bmatrix}$$

is similar to $\mathbf{A}$. Hence by Theorem 2, Sec. 18.6, it has the same eigenvalues as $\mathbf{A}$. From row 3 we get the smaller disk of radius $2 \cdot 10^{-10}$. Note that the other disks got bigger, approximately by a factor of 10^5. And in choosing $\mathbf{T}$ we have to watch that the new disks do not overlap with the disk whose size we want to decrease. ◀

A **diagonally dominant matrix** $\mathbf{A} = [a_{jk}]$ is an $n \times n$ matrix such that

$$(3) \qquad\qquad |a_{jj}| \geq \sum_{k \neq j} |a_{jk}| \qquad\qquad j = 1, \cdots, n$$

where we sum over all off-diagonal entries in row j. The matrix is **strictly diagonally dominant** if $>$ in (3) for all j. Use Theorem 1 to prove the following basic property.

THEOREM 3   **(Strict diagonal dominance)**

Strictly diagonally dominant matrices are nonsingular.

Further Inclusion Theorems

An **inclusion theorem** is a theorem that specifies a set which contains at least one eigenvalue of a given matrix. Thus, Theorems 1 and 2 are inclusion theorems; they even include the whole spectrum. We now discuss some famous theorems that yield further inclusions of eigenvalues. We state the first two of them without proofs (which would exceed the level of this book).

THEOREM 4   **(Schur's theorem[5])**

Let $\mathbf{A} = [a_{jk}]$ be an $n \times n$ matrix. Then for each of its eigenvalues $\lambda_1, \cdots, \lambda_n$,

$$(4) \qquad\qquad |\lambda_m|^2 \leq \sum_{i=1}^{n} |\lambda_i|^2 \leq \sum_{j=1}^{n} \sum_{k=1}^{n} |a_{jk}|^2 \qquad \textbf{(Schur's inequality).}$$

In (4) the second equality sign holds if and only if $\mathbf{A}$ is such that

$$(5) \qquad\qquad \overline{\mathbf{A}}^{\mathsf{T}}\mathbf{A} = \mathbf{A}\overline{\mathbf{A}}^{\mathsf{T}}.$$

Matrices that satisfy (5) are called **normal matrices.** It is not difficult to see that

---

[5]ISSAI SCHUR (1875—1941), German mathematician, professor in Berlin, also known by his important work in group theory.

Hermitian, skew-Hermitian, and unitary matrices are normal, and so are real symmetric, skew-symmetric, and orthogonal matrices.

EXAMPLE 3   **Bounds for eigenvalues obtained from Schur's inequality**

For the matrix

$$A = \begin{bmatrix} 26 & -2 & 2 \\ 2 & 21 & 4 \\ 4 & 2 & 28 \end{bmatrix}$$

we obtain from Schur's inequality

$$|\lambda| \leqq \sqrt{1949} \approx 44.1475.$$

(The eigenvalues of **A** are 30, 25, and 20; thus $30^2 + 25^2 + 20^2 = 1925 < 1949$; in fact, **A** is not normal.) ◀

The preceding theorems are valid for every real or complex square matrix. There are other theorems that hold for special classes of matrices only. The following famous theorem is of this type.

THEOREM 5   **(Perron's theorem[6])**

*Let **A** be a real $n \times n$ matrix whose entries are all positive. Then **A** has at least one real positive eigenvalue λ, and the corresponding eigenvector can be chosen real and such that all its components are positive.*

This rather "theoretical" theorem has various applications, for instance, in economics. It is interesting that one can obtain from it a theorem that gives a numerical algorithm:

THEOREM 6   **(Collatz's inclusion theorem[7])**

*Let **A** $= [a_{jk}]$ be a real $n \times n$ matrix whose entries are all positive. Let **x** be any real vector whose components $x_1, \cdots, x_n$ are positive, and let $y_1, \cdots, y_n$ be the components of the vector **y** = **Ax**. Then the closed interval on the real axis bounded by the smallest and the largest of the n quotients $q_j = y_j / x_j$ contains at least one eigenvalue of **A**.*

PROOF.   We have **Ax** = **y** or

(6) $$\mathbf{y} - \mathbf{Ax} = \mathbf{0}.$$

The transpose $\mathbf{A}^\mathsf{T}$ satisfies the conditions of Theorem 5. Hence $\mathbf{A}^\mathsf{T}$ has a positive eigenvalue λ and, corresponding to this eigenvalue, an eigenvector **u** whose components u_j are all

---

[6]OSKAR PERRON (1880—1975), German mathematician, also known for his method in potential theory. Proof of Theorem 5 in Ref. [B2], vol. II, pp. 53–62). Extended to irreducible $n \times n$ matrices **A** with nonnegative entries by Frobenius (**"Perron–Frobenius theorem"**). **Irreducible** means that **A** cannot be brought into the form

$$\begin{bmatrix} \mathbf{B} & \mathbf{C} \\ \mathbf{0} & \mathbf{F} \end{bmatrix}$$

by interchanging rows or columns (or both); here **0** is a zero matrix and **B** and **F** are any $r \times r$ and $(n - r) \times (n - r)$ matrices.

[7]LOTHAR COLLATZ (1910—1990), German mathematician, known for his work in numerical analysis.

positive. Thus $\mathbf{A}^T\mathbf{u} = \lambda\mathbf{u}$, and by taking the transpose we obtain $\mathbf{u}^T\mathbf{A} = \lambda\mathbf{u}^T$. From this and (6) we have

$$\mathbf{u}^T(\mathbf{y} - \mathbf{A}\mathbf{x}) = \mathbf{u}^T\mathbf{y} - \mathbf{u}^T\mathbf{A}\mathbf{x} = \mathbf{u}^T\mathbf{y} - \lambda\mathbf{u}^T\mathbf{x} = \mathbf{u}^T(\mathbf{y} - \lambda\mathbf{x}) = 0$$

or written out

$$\sum_{j=1}^{n} u_j(y_j - \lambda x_j) = 0.$$

Since all the components u_j are positive, it follows that

(7)
$$y_j - \lambda x_j \geqq 0, \qquad \text{that is,} \qquad q_j \geqq \lambda \qquad \text{for at least one } j,$$
$$y_j - \lambda x_j \leqq 0, \qquad \text{that is,} \qquad q_j \leqq \lambda \qquad \text{for at least one } j.$$
and

Since $\mathbf{A}$ and $\mathbf{A}^T$ have the same eigenvalues, λ is an eigenvalue of $\mathbf{A}$, and from (7) the statement of the theorem follows. ◀

EXAMPLE 4   Bounds for eigenvalues from Collatz's theorem. Iteration

For a given matrix $\mathbf{A}$ with positive entries we choose an $\mathbf{x} = \mathbf{x}_0$ and **iterate,** that is, we compute $\mathbf{x}_1 = \mathbf{A}\mathbf{x}_0$, $\mathbf{x}_2 = \mathbf{A}\mathbf{x}_1, \cdots, \mathbf{x}_{20} = \mathbf{A}\mathbf{x}_{19}$. In each step, taking $\mathbf{x} = \mathbf{x}_j$ and $\mathbf{y} = \mathbf{A}\mathbf{x}_j = \mathbf{x}_{j+1}$ we compute an inclusion interval by Collatz's theorem. This gives (6S)

$$\mathbf{A} = \begin{bmatrix} 0.49 & 0.02 & 0.22 \\ 0.02 & 0.28 & 0.20 \\ 0.22 & 0.20 & 0.40 \end{bmatrix}, \quad \mathbf{x}_0 = \begin{bmatrix} 1 \\ 1 \\ 1 \end{bmatrix}, \quad \mathbf{x}_1 = \begin{bmatrix} 0.73 \\ 0.50 \\ 0.82 \end{bmatrix}, \quad \mathbf{x}_2 = \begin{bmatrix} 0.5481 \\ 0.3186 \\ 0.5886 \end{bmatrix},$$

$$\cdots, \quad \mathbf{x}_{19} = \begin{bmatrix} 0.00216309 \\ 0.00108155 \\ 0.00216309 \end{bmatrix}, \quad \mathbf{x}_{20} = \begin{bmatrix} 0.00155743 \\ 0.000778713 \\ 0.00155743 \end{bmatrix}$$

and the intervals $0.5 \leqq \lambda \leqq 0.82$, $0.3186/0.50 = 0.6372 \leqq \lambda \leqq 0.5481/0.73 = 0.750822$, etc. These intervals have length

j	1	2	3	10	15	20
Length	0.32	0.113622	0.0539835	0.0004217	0.0000132	0.0000004

You may show that the eigenvalues of $\mathbf{A}$ are 0.72, 0.36, 0.09, so that those intervals include the largest eigenvalue, 0.72. Their lengths decreased with j, so that the iteration was worthwhile. The reason will appear in the next section, where we discuss an iteration method for eigenvalues. ◀

PROBLEM SET 18.7

Gerschgorin Disks

Determine and sketch disks or intervals that contain the eigenvalues of the following matrices. If you have a CAS, find the spectrum and compare.

1. $\begin{bmatrix} 5.1 & 0.4 & -0.1 \\ -0.4 & 4.9 & 0.3 \\ 0.1 & -0.3 & -6.8 \end{bmatrix}$ 2. $\begin{bmatrix} 10 & 0.1 & -0.2 \\ 0.1 & 6 & 0 \\ -0.2 & 0 & 3 \end{bmatrix}$ 3. $\begin{bmatrix} 5 & 10^{-2} & 10^{-2} \\ 10^{-2} & 8 & 10^{-2} \\ 10^{-2} & 10^{-2} & 9 \end{bmatrix}$

$$
\textbf{4.} \begin{bmatrix} i & 0.5i & 1+i \\ 1-i & 0 & 2+i \\ 0 & 2i & 3+4i \end{bmatrix} \qquad
\textbf{5.} \begin{bmatrix} 5 & -2 & 2 \\ 2 & 0 & 4 \\ 4 & 2 & 7 \end{bmatrix} \qquad
\textbf{6.} \begin{bmatrix} 0 & 0.5 & -0.8 \\ -1.0 & 0.2 & 1.0 \\ 0.1 & 0 & 1.2 \end{bmatrix}
$$

7. (Similarity transformation) Find $\mathbf{T}^{-1}\mathbf{AT}$ such that the radius of the Gerschgorin circle with center 5 in Prob. 3 is reduced to 1/100 of its original value.

8. (Similarity) By what integer factor can you at most reduce the Gerschgorin circle with center 3 in Prob. 2?

9. If a symmetric $n \times n$ matrix $\mathbf{A} = [a_{jk}]$ has been diagonalized except for small off-diagonal entries of size 10^{-3}, what can you say about the eigenvalues?

10. (Eigenvalues on the circle) Illustrate with a 2×2 matrix that an eigenvalue may very well lie on a Gerschgorin circle (so that Gerschgorin disks can generally not be replaced by smaller disks without losing the inclusion property).

11. (Spectral radius $\rho(\mathbf{A})$) Show that $\rho(\mathbf{A})$ is not greater than the row sum norm of $\mathbf{A}$.

12. (Extended Gerschgorin theorem) Prove Theorem 2.

13. (Strict diagonal dominance) Prove Theorem 3.

14. (Normal matrices) Show that Hermitian, skew-Hermitian, and unitary matrices (hence real symmetric, skew-symmetric, and orthogonal matrices) are normal. Why is this of practical interest?

15. (Nonnormal matrix) Show that the matrix in Prob. 5 is not normal and has the eigenvalues $-1, 4, 9$.

16. (Normality) Let $\mathbf{A}, \mathbf{B}, \mathbf{C}$ be real $n \times n$ matrices and $\mathbf{A}$ and $\mathbf{B}$ normal. Are $\mathbf{A}^2$, $\mathbf{AB}$, $\mathbf{CC}^\mathsf{T} - \mathbf{C}^\mathsf{T}\mathbf{C}$ normal?

17. (Schur) Can (4) give a smaller upper bound for the spectral radius than Theorem 1 does? Can Theorem 1 give a lower bound for the spectral radius?

Collatz's Theorem. Apply Theorem 6 to the following matrices, choosing the given vectors as vectors $\mathbf{x}$.

$$
\textbf{18.} \begin{bmatrix} 20 & 8 & 1 \\ 8 & 21 & 8 \\ 1 & 8 & 20 \end{bmatrix}, \begin{bmatrix} 1 \\ 1 \\ 1 \end{bmatrix}, \begin{bmatrix} 1 \\ 2 \\ 1 \end{bmatrix}, \begin{bmatrix} 2 \\ 3 \\ 2 \end{bmatrix} \qquad
\textbf{19.} \begin{bmatrix} 5 & 1 & 1 \\ 1 & 5 & 1 \\ 1 & 1 & 5 \end{bmatrix}, \begin{bmatrix} 1 \\ 2 \\ 1 \end{bmatrix}, \begin{bmatrix} 1 \\ 1 \\ 1 \end{bmatrix}
$$

 20. CAS PROJECT. Collatz Iteration. (a) Write a program for the iteration in Example 4 (with any $\mathbf{A}$ and $\mathbf{x}_0$) that at each step prints the midpoint (why?), the endpoints, and the length of the inclusion interval.

(b) Apply the program to symmetric matrices of your choice. Explore how convergence depends on the choice of initial vectors. Can you construct cases in which the length of the inclusion intervals does not decrease monotone? Can you explain the reason? Can you experiment on the effect of round-off?

18.8 Eigenvalues by Iteration (Power Method)

A simple standard procedure for computing approximate values of the eigenvalues of an $n \times n$ matrix $\mathbf{A} = [a_{jk}]$ is the **power method.** In this method we start from any vector $\mathbf{x}_0 \; (\neq \mathbf{0})$ with n components and compute successively

$$
\mathbf{x}_1 = \mathbf{Ax}_0, \qquad \mathbf{x}_2 = \mathbf{Ax}_1, \qquad \cdots, \qquad \mathbf{x}_s = \mathbf{Ax}_{s-1}.
$$

For simplifying notation, we denote $\mathbf{x}_{s-1}$ by $\mathbf{x}$ and $\mathbf{x}_s$ by $\mathbf{y}$, so that $\mathbf{y} = \mathbf{Ax}$.

The method applies to any $n \times n$ matrix $\mathbf{A}$ that has a **dominant eigenvalue** (a λ such that $|\lambda|$ is greater than the absolute values of the other eigenvalues). If $\mathbf{A}$ is *symmetric*, it also gives the error bound (2), in addition to the approximation (1).

THEOREM 1   *Let $\mathbf{A}$ be an $n \times n$ real symmetric matrix. Let $\mathbf{x}$ ($\neq \mathbf{0}$) be any real vector with n components. Furthermore, let*

$$\mathbf{y} = \mathbf{Ax}, \qquad m_0 = \mathbf{x}^\mathsf{T}\mathbf{x}, \qquad m_1 = \mathbf{x}^\mathsf{T}\mathbf{y}, \qquad m_2 = \mathbf{y}^\mathsf{T}\mathbf{y}.$$

Then the quotient

(1)
$$q = \frac{m_1}{m_0}$$
(Rayleigh quotient[8])

is an approximation for an eigenvalue[9] λ of $\mathbf{A}$, and if we set $q = \lambda - \epsilon$, so that ϵ is the error of q, then

(2)
$$|\epsilon| \leqq \delta = \sqrt{\frac{m_2}{m_0} - q^2}.$$

PROOF. δ^2 denotes the radicand in (2). Since $m_1 = qm_0$ by (1), we have

$$(3) \qquad (\mathbf{y} - q\mathbf{x})^\mathsf{T}(\mathbf{y} - q\mathbf{x}) = m_2 - 2qm_1 + q^2m_0 = m_2 - q^2m_0 = \delta^2m_0.$$

Since $\mathbf{A}$ is real symmetric, it has an orthogonal set of n real unit eigenvectors $\mathbf{z}_1, \cdots, \mathbf{z}_n$ corresponding to the eigenvalues $\lambda_1, \cdots, \lambda_n$, respectively (some of which may be equal). (Proof in Ref. [B2], vol. 1, pp. 270−272, listed in Appendix 1.) Then $\mathbf{x}$ has a representation of the form

$$\mathbf{x} = a_1\mathbf{z}_1 + \cdots + a_n\mathbf{z}_n.$$

Now $\mathbf{Az}_1 = \lambda_1\mathbf{z}_1$, etc., and we obtain

$$\mathbf{y} = \mathbf{Ax} = a_1\lambda_1\mathbf{z}_1 + \cdots + a_n\lambda_n\mathbf{z}_n$$

and, since the $\mathbf{z}_j$ are orthogonal unit vectors,

$$(4) \qquad m_0 = \mathbf{x}^\mathsf{T}\mathbf{x} = a_1{}^2 + \cdots + a_n{}^2.$$

It follows that in (3),

$$\mathbf{y} - q\mathbf{x} = a_1(\lambda_1 - q)\mathbf{z}_1 + \cdots + a_n(\lambda_n - q)\mathbf{z}_n.$$

---

[8]LORD RAYLEIGH (JOHN WILLIAM STRUTT) (1842—1919), English physicist and mathematician, professor at Cambridge and London, known by his important contributions to various branches of applied mathematics and theoretical physics, in particular, the theory of waves, elasticity, and hydrodynamics.

[9]Ordinarily that λ which is greatest in absolute value, but no general statements are possible.

Since the $\mathbf{z}_j$ are orthogonal unit vectors, we thus obtain from (3)

(5) $\qquad \delta^2 m_0 = (\mathbf{y} - q\mathbf{x})^{\mathsf{T}} (\mathbf{y} - q\mathbf{x}) = a_1^2(\lambda_1 - q)^2 + \cdots + a_n^2(\lambda_n - q)^2.$

Now let λ_c be an eigenvalue of $\mathbf{A}$ to which q is closest. Then $(\lambda_c - q)^2 \leqq (\lambda_j - q)^2$ for $j = 1, \cdots, n$. From this and (5) we obtain the inequality

$$\delta^2 m_0 \geqq (\lambda_c - q)^2(a_1^2 + \cdots + a_n^2) = (\lambda_c - q)^2 m_0$$

Dividing by m_0, taking square roots, and recalling the meaning of δ^2 gives

$$\delta = \sqrt{\frac{m_2}{m_0} - q^2} \geqq |\lambda_c - q|.$$

This shows that δ is a bound for the error ϵ of the approximation q of an eigenvalue of $\mathbf{A}$ and completes the proof. ◀

The main advantage of the method is its simplicity. And it can handle *sparse matrices* too large to store as a full square array. Its disadvantage is its possibly slow convergence. From the proof of Theorem 1 we see that the speed of convergence depends on the ratio of the dominant eigenvalue to the next in absolute value ($2:1$ in Example 1, below). If the dominant eigenvalue is complex, we have no convergence since then its complex conjugate is an eigenvalue, too (because $\mathbf{A}$ is real), of the same absolute value. (More in Ref. [E17] listed in Appendix 1.)

If we want a convergent sequence of **eigenvectors,** then at the beginning of each step we **scale** the vector, say, by dividing its components by an absolutely largest one, as in Example 1.

EXAMPLE 1　**Application of Theorem 1. Scaling**

For the symmetric matrix $\mathbf{A}$ in Example 4, Sec. 18.7, and $\mathbf{x}_0 = [1 \quad 1 \quad 1]^{\mathsf{T}}$ we obtain from (1) and (2) and the indicated scaling

$$\mathbf{A} = \begin{bmatrix} 0.49 & 0.02 & 0.22 \\ 0.02 & 0.28 & 0.20 \\ 0.22 & 0.20 & 0.40 \end{bmatrix}, \quad \mathbf{x}_0 = \begin{bmatrix} 1 \\ 1 \\ 1 \end{bmatrix}, \quad \mathbf{x}_1 = \begin{bmatrix} 0.890244 \\ 0.609756 \\ 1 \end{bmatrix}, \quad \mathbf{x}_2 = \begin{bmatrix} 0.931193 \\ 0.541284 \\ 1 \end{bmatrix}$$

$$\mathbf{x}_5 = \begin{bmatrix} 0.990663 \\ 0.504682 \\ 1 \end{bmatrix}, \quad \mathbf{x}_{10} = \begin{bmatrix} 0.999707 \\ 0.500146 \\ 1 \end{bmatrix}, \quad \mathbf{x}_{15} = \begin{bmatrix} 0.999991 \\ 0.500005 \\ 1 \end{bmatrix}.$$

Here $\mathbf{A}\mathbf{x}_0 = [0.73 \quad 0.5 \quad 0.82]^{\mathsf{T}}$, scaled to $\mathbf{x}_1 = [0.73/0.82 \quad 0.5/0.82 \quad 1]^{\mathsf{T}}$, etc. The dominant eigenvalue is 0.72, an eigenvector $[1 \quad 0.5 \quad 1]^{\mathsf{T}}$. The corresponding q and δ are computed each time before the next scaling. Thus in the first step,

$$q = \frac{m_1}{m_0} = \frac{\mathbf{x}_0^{\mathsf{T}} \mathbf{A} \mathbf{x}_0}{\mathbf{x}_0^{\mathsf{T}} \mathbf{x}_0} = \frac{2.05}{3} = 0.683333$$

$$\delta = \left(\frac{m_2}{m_0} - q^2 \right)^{1/2} = \left(\frac{(\mathbf{A}\mathbf{x}_0)^{\mathsf{T}} \mathbf{A} \mathbf{x}_0}{\mathbf{x}_0^{\mathsf{T}} \mathbf{x}_0} - q^2 \right)^{1/2} = \left(\frac{1.4553}{3} - q^2 \right)^{1/2} = 0.134743.$$

This gives the following values of q, δ, and the error ϵ of q:

j	1	2	5	10	15
q	0.683333	0.716048	0.719944	0.720000	0.720000
δ	0.134743	0.038887	0.004499	0.000141	$4 \cdot 10^{-6}$
ϵ	0.036667	0.003952	0.000056	$5 \cdot 10^{-8}$	$5 \cdot 10^{-11}$

This is somewhat better than Collatz's method in Example 4, Sec. 18.7, at the expense of more operations. The error bounds are much larger than the actual errors. This is typical, although the bounds cannot be improved, that is, for special symmetric matrices they agree with the errors. ◀

Spectral shift, the transition from $\mathbf{A}$ to $\mathbf{A} - k\mathbf{I}$, shifts every eigenvalue by $-k$. Although finding a good k can hardly be made automatic, it may be helped by some other method or small preliminary computational experiments. In Example 1, Gerschgorin's theorem gives $-0.02 \leqq \lambda \leqq 0.82$ for the whole spectrum (verify!). Shifting by -0.4 might be too much (then $-0.42 \leqq \lambda \leqq 0.42$), so let us try -0.2.

EXAMPLE 2   **Power method with spectral shift**

For $\mathbf{A} - 2\mathbf{I}$ with $\mathbf{A}$ as in Example 1 we obtain the following improvements (where the index 1 refers to Example 1 and the index 2 to Example 2).

j	1	2	5	10
δ_1	0.134743	0.038887	0.004499	0.000141
δ_2	0.134743	0.034474	0.000693	$1.8 \cdot 10^{-6}$
ϵ_1	0.036667	0.003952	0.000056	$5 \cdot 10^{-8}$
ϵ_2	0.036667	0.002477	$1.3 \cdot 10^{-6}$	$9 \cdot 10^{-12}$

◀

PROBLEM SET 18.8

Power Method Without Scaling

Choosing $\mathbf{x}_0 = [1 \quad 1]^T$ or $[1 \quad 1 \quad 1]^T$ or $[1 \quad 1 \quad 1 \quad 1]^T$, respectively, apply the power method (3 steps) to the following symmetric matrices, computing Rayleigh quotients and error bounds. (Show the details of your work.)

1. $\begin{bmatrix} 9 & 4 \\ 4 & 3 \end{bmatrix}$ 2. $\begin{bmatrix} 6 & -3 \\ -3 & -2 \end{bmatrix}$ 3. $\begin{bmatrix} 3.6 & -1.8 & 1.8 \\ -1.8 & 2.8 & -2.6 \\ 1.8 & -2.6 & 2.8 \end{bmatrix}$ 4. $\begin{bmatrix} 3 & 2 & 3 \\ 2 & 6 & 6 \\ 3 & 6 & 3 \end{bmatrix}$

5. $\begin{bmatrix} 1 & 1 & 0 & 0 \\ 1 & -1 & 1 & 0 \\ 0 & 1 & -1 & 1 \\ 0 & 0 & 1 & 1 \end{bmatrix}$ 6. $\begin{bmatrix} 2 & 4 & 0 & 1 \\ 4 & 1 & 2 & 8 \\ 0 & 2 & 5 & 2 \\ 1 & 8 & 2 & 0 \end{bmatrix}$

7. **(Eigenvector)** Show that if $\mathbf{x}$ is an eigenvector, then $\delta = 0$ in (2).

8. **Collatz's theorem** (Sec. 18.7) applies to the matrix in Prob. 4. Use it with $\mathbf{x}_0 = [1 \quad 1 \quad 1]^T$, do 2 steps, and compare the results with those obtained from our Theorem 1.

9. **(Spectral shift, smallest eigenvalue)** $\lambda_1 \approx 7$ for $\mathbf{A}$ in Prob. 3. Apply the power method with spectral shift to $\mathbf{A} - 7\mathbf{I}$ (3 steps, $x_0 = [1, \quad 1, \quad 1]$). Why will this give inclusion intervals for the smallest eigenvalue ($\lambda_3 = 0.2$, shifted to -6.8)?

10. **(Error bound)** To understand the importance of the error bound (2), consider the matrix

$$
\mathbf{A} = \begin{bmatrix} 3 & 4 \\ 4 & -3 \end{bmatrix}, \qquad \text{choose} \qquad \mathbf{x}_0 = \begin{bmatrix} 3 \\ -1 \end{bmatrix},
$$

and show that $q = 0$ for all s. Find the eigenvalues and explain what happened. Start again, choosing another $\mathbf{x}_0$.

11. **(Rayleigh quotient)** Why is the Rayleigh quotient $q = m_1/m_0$ in general an approximation of that eigenvalue, λ_1, which is largest in absolute value? *Hint.* Let $\mathbf{z}_1, \cdots, \mathbf{z}_n$ be as in the proof of Theorem 1. Show that if

$$
\mathbf{x}_0 = \sum c_j \mathbf{z}_j, \quad \text{then} \quad \mathbf{x} = \mathbf{x}_{s-1} = \sum c_j \lambda_j^{s-1} \mathbf{z}_j, \quad \mathbf{y} = \mathbf{x}_s = \sum c_j \lambda_j^s \mathbf{z}_j,
$$

$$
q = m_1/m_0 = (c_1^2 \lambda_1^{2s-1} + \cdots)/(c_1^2 \lambda_1^{2s-2} + \cdots) \approx \lambda_1.
$$

Under what conditions will this be a good approximation?

Power Method with Scaling

Find approximations to an eigenvector in the following problems (3 steps, x_0 as before), with the component(s) of largest absolute value scaled to 1.

12. Prob. 2   13. Prob. 3   14. Prob. 6

15. **CAS PROJECT. Power Method with Scaling. (a)** Write a program for arbitrary $n \times n$ matrices that prints every step. Apply it to the (nonsymmetric!) matrix (20 steps)

$$
\mathbf{A} = \begin{bmatrix} -2.7 & 2.8 & 0.3 \\ -8.8 & 7.1 & 0.2 \\ -14.4 & 10.8 & 0.6 \end{bmatrix}.
$$

(b) Experiment in (a) with shifting. Which shift do you find optimal?

(c) Write a program as in (a) but for symmetric matrices that prints the vectors, scaled vectors, q, and δ. Apply it to the matrix in Prob. 14 (for which your CAS may have difficulties producing an eigenvector directly).

(d) Find a (nonsymmetric) matrix for which δ in (2) is no longer an error bound.

(e) Experiment systematically with convergence by choosing matrices with the second largest eigenvalue (i) almost equal to the largest, (ii) somewhat different, (iii) much different.

18.9 Tridiagonalization and QR-Factorization

We consider the problem of computing *all* the eigenvalues of a ***real symmetric*** matrix $\mathbf{A} = [a_{jk}]$. We discuss a method widely used in practice. In the first stage we apply Householder's method,[10] which reduces the given matrix to a **tridiagonal matrix,** that

---

[10]*Journal of the Association for Computing Machinery* **5** (1958), 335–342. See also Ref. [E19] in Appendix 1.

is, a matrix having all its nonzero entries on the main diagonal and in the positions immediately adjacent to the main diagonal (such as $\mathbf{A}_3$ in Fig. 416). In the second stage, the tridiagonal matrix is factorized in the form **QR,** where **Q** is orthogonal and **R** upper triangular, and the eigenvalues are actually determined (approximately); we discuss this afterward. (For extensions to general matrices, see Ref. [E20] listed in Appendix 1.)

Householder's Tridiagonalization Method

This method reduces a given real ***symmetric*** $n \times n$ matrix $\mathbf{A} = [a_{jk}]$ by $n - 2$ successive similarity transformations (see Sec. 18.6) to tridiagonal form. The matrices $\mathbf{P}_1, \mathbf{P}_2, \cdots, \mathbf{P}_{n-2}$ are orthogonal and symmetric matrices. Hence $\mathbf{P}_1^{-1} = \mathbf{P}_1^{\mathsf{T}} = \mathbf{P}_1$ and similarly for the others. The $n - 2$ similarity transformations that successively produce from the given $\mathbf{A}_0 = \mathbf{A} = [a_{jk}]$ the matrices $\mathbf{A}_1 = [a_{jk}^{(1)}]$, $\mathbf{A}_2 = [a_{jk}^{(2)}]$, etc. look as follows.

$$
\begin{aligned}
\mathbf{A}_1 &= \mathbf{P}_1\mathbf{A}_0\mathbf{P}_1 \\
\mathbf{A}_2 &= \mathbf{P}_2\mathbf{A}_1\mathbf{P}_2 \\
&\cdots\cdots\cdots\cdots \\
\mathbf{B} = \mathbf{A}_{n-2} &= \mathbf{P}_{n-2}\mathbf{A}_{n-3}\mathbf{P}_{n-2}.
\end{aligned}
\tag{1}
$$

These transformations create the necessary zeros, in the first step in row 1 and column 1, in the second step in row 2 and column 2, etc., as Fig. 416 illustrates for a 5×5 matrix. **B** is tridiagonal.

How do we determine $\mathbf{P}_1, \mathbf{P}_2, \cdots$? All these $\mathbf{P}_r$ are of the form

$$
\mathbf{P}_r = \mathbf{I} - 2\mathbf{v}_r\mathbf{v}_r^{\mathsf{T}} \qquad (r = 1, \cdots, n-2) \tag{2}
$$

where $\mathbf{v}_r = [v_{jr}]$ is a unit vector with its first r components 0; thus

$$
\mathbf{v}_1 = \begin{bmatrix} 0 \\ * \\ * \\ \vdots \\ * \end{bmatrix}, \quad
\mathbf{v}_2 = \begin{bmatrix} 0 \\ 0 \\ * \\ \vdots \\ * \end{bmatrix}, \quad \cdots, \quad
\mathbf{v}_{n-2} = \begin{bmatrix} 0 \\ 0 \\ \vdots \\ * \\ * \end{bmatrix} \tag{3}
$$

where the asterisks denote the other components (which will be nonzero in general).

$$
\begin{bmatrix} * & * & & & \\ * & * & * & * & * \\ & * & * & * & * \\ & * & * & * & * \\ & * & * & * & * \end{bmatrix}
\qquad
\begin{bmatrix} * & * & & & \\ * & * & * & & \\ & * & * & * & * \\ & & * & * & * \\ & & * & * & * \end{bmatrix}
\qquad
\begin{bmatrix} * & * & & & \\ * & * & * & & \\ & * & * & * & \\ & & * & * & * \\ & & & * & * \end{bmatrix}
$$

First Step	Second Step	Third Step
$\mathbf{A}_1 = \mathbf{P}_1\mathbf{A}\mathbf{P}_1$	$\mathbf{A}_2 = \mathbf{P}_2\mathbf{A}_1\mathbf{P}_2$	$\mathbf{A}_3 = \mathbf{P}_3\mathbf{A}_2\mathbf{P}_3$

Fig. 416. Householder's method for a 5×5 matrix. Positions left blank are zeros created by the method.

First Step. $\mathbf{v}_1$ has the components

(a)

$$v_{11} = 0$$

$$v_{21} = \sqrt{\frac{1}{2}\left(1 + \frac{|a_{21}|}{S_1}\right)}$$

(4) (b)

$$v_{j1} = \frac{a_{j1}\,\mathrm{sgn}\,a_{21}}{2v_{21}S_1} \qquad\qquad j = 3, 4, \cdots, n$$

where

(c)

$$S_1 = \sqrt{a_{21}^{\,2} + a_{31}^{\,2} + \cdots + a_{n1}^{\,2}}$$

where $S_1 > 0$, and $\mathrm{sgn}\,a_{21} = +1$ if $a_{21} \geqq 0$ and $\mathrm{sgn}\,a_{21} = -1$ if $a_{21} < 0$. With this we compute $\mathbf{P}_1$ by (2) and then $\mathbf{A}_1$ by (1). This was the first step.

Second Step. We compute $\mathbf{v}_2$ by (4) with all subscripts increased by 1 and the a_{jk} replaced by $a_{jk}^{(1)}$, the entries of $\mathbf{A}_1$ just computed. Thus [see also (3)]

$$v_{12} = v_{22} = 0$$

(4*)

$$v_{32} = \sqrt{\frac{1}{2}\left(1 + \frac{|a_{32}^{(1)}|}{S_2}\right)}$$

$$v_{j2} = \frac{a_{j2}^{(1)}\,\mathrm{sgn}\,a_{32}^{(1)}}{2v_{32}S_2} \qquad\qquad j = 4, 5, \cdots, n$$

where

$$S_2 = \sqrt{a_{32}^{(1)^2} + a_{42}^{(1)^2} + \cdots + a_{n2}^{(1)^2}}.$$

With this we compute $\mathbf{P}_2$ by (2) and then $\mathbf{A}_2$ by (1).

Third Step. We compute $\mathbf{v}_3$ by (4*) with all subscripts increased by 1 and the $a_{jk}^{(1)}$ replaced by the entries $a_{jk}^{(2)}$ of $\mathbf{A}_2$, and so on.

EXAMPLE 1   **Householder's method**

Tridiagonalize the real symmetric matrix

$$\mathbf{A} = \mathbf{A}_0 = \begin{bmatrix} 6 & 4 & 1 & 1 \\ 4 & 6 & 1 & 1 \\ 1 & 1 & 5 & 2 \\ 1 & 1 & 2 & 5 \end{bmatrix}.$$

Solution. ***First Step.*** We compute $S_1^{\,2} = 4^2 + 1^2 + 1^2 = 18$ from (4c). Since $a_{21} = 4 > 0$, we have $\mathrm{sgn}\,a_{21} = +1$ in (4b) and get from (4) by straightforward computation

$$\mathbf{v}_1 = \begin{bmatrix} 0 \\ v_{21} \\ v_{31} \\ v_{41} \end{bmatrix} = \begin{bmatrix} 0 \\ 0.985\,598\,56 \\ 0.119\,573\,16 \\ 0.119\,573\,16 \end{bmatrix}.$$

From this and (2),

$$
\mathbf{P}_1 = \begin{bmatrix} 1 & 0 & 0 & 0 \\ 0 & -0.942\,809\,04 & -0.235\,702\,27 & -0.235\,702\,27 \\ 0 & -0.235\,702\,27 & 0.971\,404\,52 & -0.028\,595\,48 \\ 0 & -0.235\,702\,27 & -0.028\,595\,48 & 0.971\,404\,52 \end{bmatrix}.
$$

From the first line in (1) we now get

$$
\mathbf{A}_1 = \mathbf{P}_1 \mathbf{A}_0 \mathbf{P}_1 = \begin{bmatrix} 6 & -\sqrt{18} & 0 & 0 \\ -\sqrt{18} & 7 & -1 & -1 \\ 0 & -1 & 9/2 & 3/2 \\ 0 & -1 & 3/2 & 9/2 \end{bmatrix}.
$$

Second Step. From (4*) we compute $S_2{}^2 = 2$ and

$$
\mathbf{v}_2 = \begin{bmatrix} 0 \\ 0 \\ v_{32} \\ v_{42} \end{bmatrix} = \begin{bmatrix} 0 \\ 0 \\ 0.923\,879\,53 \\ 0.382\,683\,43 \end{bmatrix}.
$$

From this and (2),

$$
\mathbf{P}_2 = \begin{bmatrix} 1 & 0 & 0 & 0 \\ 0 & 1 & 0 & 0 \\ 0 & 0 & -1/\sqrt{2} & -1/\sqrt{2} \\ 0 & 0 & -1/\sqrt{2} & 1/\sqrt{2} \end{bmatrix}.
$$

The second line in (1) now gives

$$
\mathbf{B} = \mathbf{A}_2 = \mathbf{P}_2 \mathbf{A}_1 \mathbf{P}_2 = \begin{bmatrix} 6 & -\sqrt{18} & 0 & 0 \\ -\sqrt{18} & 7 & \sqrt{2} & 0 \\ 0 & \sqrt{2} & 6 & 0 \\ 0 & 0 & 0 & 3 \end{bmatrix}.
$$

This matrix **B** is tridiagonal. Since our given matrix has order $n = 4$, we needed $n - 2 = 2$ steps to accomplish this reduction, as claimed. (Do you see that we got more zeros than we can expect in general?) **B** is similar to **A,** as we now show in general. This is essential because **B** thus has the same spectrum as **A,** by Theorem 2 in Sec. 18.6. ◀

We show that **B** in (1) is similar to $\mathbf{A} = \mathbf{A}_0$.
The matrix $\mathbf{P}_r$ is symmetric,

$$
\mathbf{P}_r{}^\mathsf{T} = (\mathbf{I} - 2\mathbf{v}_r\mathbf{v}_r{}^\mathsf{T})^\mathsf{T} = \mathbf{I}^\mathsf{T} - 2(\mathbf{v}_r\mathbf{v}_r{}^\mathsf{T})^\mathsf{T}
$$

$$
= \mathbf{I} - 2\mathbf{v}_r\mathbf{v}_r{}^\mathsf{T} = \mathbf{P}_r.
$$

Also, $\mathbf{P}_r$ is orthogonal because

$$
\mathbf{P}_r\mathbf{P}_r{}^\mathsf{T} = \mathbf{P}_r{}^2 = (\mathbf{I} - 2\mathbf{v}_r\mathbf{v}_r{}^\mathsf{T})^2 = \mathbf{I} - 4\mathbf{v}_r\mathbf{v}_r{}^\mathsf{T} + 4\mathbf{v}_r\mathbf{v}_r{}^\mathsf{T}\mathbf{v}_r\mathbf{v}_r{}^\mathsf{T}
$$

and $\mathbf{v}_r{}^T\mathbf{v}_r = 1$ in the last term because $\mathbf{v}_r$ is a unit vector (see above), so that the right side reduces to $\mathbf{I}$. This gives $\mathbf{P}_r{}^{-1} = \mathbf{P}_r{}^T = \mathbf{P}_r$. Hence from (1) we now obtain

$$\mathbf{B} = \mathbf{P}_{n-2}\mathbf{A}_{n-3}\mathbf{P}_{n-2} = \cdots$$

$$\cdots = \mathbf{P}_{n-2}\mathbf{P}_{n-3} \cdots \mathbf{P}_1\mathbf{A}\mathbf{P}_1 \cdots \mathbf{P}_{n-3}\mathbf{P}_{n-2}$$

$$= \mathbf{P}_{n-2}^{-1}\mathbf{P}_{n-3}^{-1} \cdots \mathbf{P}_1^{-1}\mathbf{A}\mathbf{P}_1 \cdots \mathbf{P}_{n-3}\mathbf{P}_{n-2}$$

$$= \mathbf{P}^{-1}\mathbf{A}\mathbf{P}$$

where $\mathbf{P} = \mathbf{P}_1\mathbf{P}_2 \cdots \mathbf{P}_{n-2}$. This proves our assertion.   ◀

QR-Factorization Method

In 1958 H. Rutishauser proposed the idea of using the LU-factorization (see Sec. 18.2; he called it LR-factorization) in solving eigenvalue problems. An improved version of Rutishauser's method (avoiding breakdown if certain submatrices become singular, etc.; see Ref. [E19]) is the **QR-method**,[11] which is based on the factorization **QR,** where **R** is upper triangular as before but **Q** is orthogonal (instead of lower triangular). We discuss the QR-method for a real symmetric matrix. (For extensions to general real or complex matrices, see Refs. [E19] and [E20] in Appendix 1.) In this method we start from a real *symmetric* tridiagonal matrix $\mathbf{B}_0 = \mathbf{B}$ (as obtained from a real symmetric matrix $\mathbf{A}$ by Householder's method). We compute stepwise $\mathbf{B}_1, \mathbf{B}_2, \cdots$ according to this rule:

First Step. Factor

$$\mathbf{B}_0 = \mathbf{Q}_0\mathbf{R}_0$$

where $\mathbf{Q}_0$ is orthogonal and $\mathbf{R}_0$ is upper triangular. Then compute

$$\mathbf{B}_1 = \mathbf{R}_0\mathbf{Q}_0.$$

Second Step. Factor $\mathbf{B}_1 = \mathbf{Q}_1\mathbf{R}_1$. Then compute $\mathbf{B}_2 = \mathbf{R}_1\mathbf{Q}_1$.
General Step. Factor

$$(5) \qquad \boxed{\mathbf{B}_s = \mathbf{Q}_s\mathbf{R}_s}$$

where $\mathbf{Q}_s$ is orthogonal and $\mathbf{R}_s$ is upper triangular. Then compute

$$(6) \qquad \boxed{\mathbf{B}_{s+1} = \mathbf{R}_s\mathbf{Q}_s.}$$

The method of obtaining the factorization (5) will be explained below.

$\mathbf{B}_{s+1}$ similar to B. Convergence to a diagonal matrix. From (5) we have $\mathbf{R}_s = \mathbf{Q}_s{}^{-1}\mathbf{B}_s$. Substitution into (6) gives

---

[11]Proposed independently by J. G. F. Francis, *Computer Journal* **4** (1961—1962), 265−271, 332−345, and V. N. Kublanovskaya, *Zhurnal Vych. Mat. i Mat. Fiz.* **1** (1961), 555−570.

(7) $$\mathbf{B}_{s+1} = \mathbf{R}_s \mathbf{Q}_s = \mathbf{Q}_s^{-1} \mathbf{B}_s \mathbf{Q}_s.$$

Thus $\mathbf{B}_{s+1}$ is similar to $\mathbf{B}_s$. Hence $\mathbf{B}_{s+1}$ is similar to $\mathbf{B}_0 = \mathbf{B}$ for all s. By Theorem 2, Sec. 18.6, this implies that $\mathbf{B}_{s+1}$ has the same eigenvalues as $\mathbf{B}$.

Also, $\mathbf{B}_{s+1}$ is symmetric. This follows by induction. Indeed, $\mathbf{B}_0 = \mathbf{B}$ is symmetric. Assuming $\mathbf{B}_s$ to be symmetric and using $\mathbf{Q}_s^{-1} = \mathbf{Q}_s^{\mathrm{T}}$ (since $\mathbf{Q}_s$ is orthogonal), we get from (7)

$$\mathbf{B}_{s+1}^{\mathrm{T}} = (\mathbf{Q}_s^{\mathrm{T}} \mathbf{B}_s \mathbf{Q}_s)^{\mathrm{T}} = \mathbf{Q}_s^{\mathrm{T}} \mathbf{B}_s^{\mathrm{T}} \mathbf{Q}_s = \mathbf{Q}_s^{\mathrm{T}} \mathbf{B}_s \mathbf{Q}_s = \mathbf{B}_{s+1}$$

as claimed.

If the eigenvalues of $\mathbf{B}$ are different in absolute value, say, $|\lambda_1| > |\lambda_2| > \cdots > |\lambda_n|$, then

$$\lim_{s \to \infty} \mathbf{B}_s = \mathbf{D}$$

where $\mathbf{D}$ is diagonal, with main diagonal entries $\lambda_1, \lambda_2, \cdots, \lambda_n$. (Proof in Ref. [E19] listed in Appendix 1.)

How do we get the QR-factorization, say, $\mathbf{B} = \mathbf{B}_0 = [b_{jk}] = \mathbf{Q}_0 \mathbf{R}_0$? The tridiagonal matrix $\mathbf{B}$ has $n - 1$ generally nonzero entries below the main diagonal. These are b_{21}, $b_{32}, \cdots, b_{n,n-1}$. We multiply $\mathbf{B}$ from the left by a matrix $\mathbf{C}_2$ such that $\mathbf{C}_2 \mathbf{B} = \left[b_{jk}^{(2)} \right]$ has $b_{21}^{(2)} = 0$. We multiply this by a matrix $\mathbf{C}_3$ such that $\mathbf{C}_3 \mathbf{C}_2 \mathbf{B} = \left[b_{jk}^{(3)} \right]$ has $b_{32}^{(3)} = 0$, etc. After $n - 1$ such multiplications we are left with an upper triangular matrix $\mathbf{R}_0$, namely,

(8) $$\mathbf{C}_n \mathbf{C}_{n-1} \cdots \mathbf{C}_3 \mathbf{C}_2 \mathbf{B}_0 = \mathbf{R}_0.$$

These $\mathbf{C}_j$ are very simple. $\mathbf{C}_j$ has the 2×2 submatrix

$$\begin{bmatrix} \cos \theta_j & \sin \theta_j \\ -\sin \theta_j & \cos \theta_j \end{bmatrix} \qquad (\theta_j \text{ suitable})$$

in rows $j - 1$ and j and columns $j - 1$ and j, entries 1 everywhere else on the main diagonal and all other entries 0. (This submatrix is the matrix of a plane rotation through the angle θ_j; see Team Project 18, Sec. 6.2.) For instance, if $n = 4$, writing $c_j = \cos \theta_j$, $s_j = \sin \theta_j$, we have

$$\mathbf{C}_2 = \begin{bmatrix} c_2 & s_2 & 0 & 0 \\ -s_2 & c_2 & 0 & 0 \\ 0 & 0 & 1 & 0 \\ 0 & 0 & 0 & 1 \end{bmatrix},$$

$$\mathbf{C}_3 = \begin{bmatrix} 1 & 0 & 0 & 0 \\ 0 & c_3 & s_3 & 0 \\ 0 & -s_3 & c_3 & 0 \\ 0 & 0 & 0 & 1 \end{bmatrix},$$

$$\mathbf{C}_4 = \begin{bmatrix} 1 & 0 & 0 & 0 \\ 0 & 1 & 0 & 0 \\ 0 & 0 & c_4 & s_4 \\ 0 & 0 & -s_4 & c_4 \end{bmatrix}.$$

These $\mathbf{C}_j$ are orthogonal. Hence their product in (8) is orthogonal, and so is the inverse of this product. We call this inverse $\mathbf{Q}_0$. Then from (8),

(9a) $$\mathbf{B}_0 = \mathbf{Q}_0 \mathbf{R}_0$$

where, with $\mathbf{C}_j^{-1} = \mathbf{C}_j^{\mathsf{T}}$,

(9b) $$\mathbf{Q}_0 = (\mathbf{C}_n \mathbf{C}_{n-1} \cdots \mathbf{C}_3 \mathbf{C}_2)^{-1} = \mathbf{C}_2^{\mathsf{T}} \mathbf{C}_3^{\mathsf{T}} \cdots \mathbf{C}_{n-1}^{\mathsf{T}} \mathbf{C}_n^{\mathsf{T}}.$$

This is our QR-factorization of $\mathbf{B}_0$. From it we have by (6) with $s = 0$

(10) $$\mathbf{B}_1 = \mathbf{R}_0 \mathbf{Q}_0 = \mathbf{R}_0 \mathbf{C}_2^{\mathsf{T}} \mathbf{C}_3^{\mathsf{T}} \cdots \mathbf{C}_{n-1}^{\mathsf{T}} \mathbf{C}_n^{\mathsf{T}}.$$

We do not need $\mathbf{Q}_0$ explicitly, but to get $\mathbf{B}_1$ from (10), we first compute $\mathbf{R}_0 \mathbf{C}_2^{\mathsf{T}}$, then $(\mathbf{R}_0 \mathbf{C}_2^{\mathsf{T}})\mathbf{C}_3^{\mathsf{T}}$, etc. Similarly in the further steps that produce $\mathbf{B}_2, \mathbf{B}_3, \cdots$.

Determination of $\cos \theta_j$ and $\sin \theta_j$. We finally show how to find the angles of rotation. $\cos \theta_2$ and $\sin \theta_2$ in $\mathbf{C}_2$ must be such that $b_{21}^{(2)} = 0$ in the product

$$\mathbf{C}_2 \mathbf{B} = \begin{bmatrix} c_2 & s_2 & 0 & \cdots \\ -s_2 & c_2 & 0 & \cdots \\ \cdot & \cdot & \cdot & \cdots \\ \cdot & \cdot & \cdot & \cdots \end{bmatrix} \begin{bmatrix} b_{11} & b_{12} & \cdots \\ b_{21} & b_{22} & \cdots \\ \cdot & \cdot & \cdots \\ \cdot & \cdot & \cdots \end{bmatrix}.$$

Now $b_{21}^{(2)}$ is obtained by multiplying the second row of $\mathbf{C}_2$ by the first column of $\mathbf{B}$,

$$b_{21}^{(2)} = -s_2 b_{11} + c_2 b_{21} = 0.$$

Hence $\tan \theta_2 = s_2/c_2 = b_{21}/b_{11}$, and

(11) $$\cos \theta_2 = \frac{1}{\sqrt{1 + (b_{21}/b_{11})^2}}, \qquad \sin \theta_2 = \frac{b_{21}/b_{11}}{\sqrt{1 + (b_{21}/b_{11})^2}}.$$

Similarly for $\theta_3, \theta_4, \cdots$.

The next example illustrates all this.

EXAMPLE 2   **QR-factorization method**

Compute all the eigenvalues of the matrix

$$\mathbf{A} = \begin{bmatrix} 6 & 4 & 1 & 1 \\ 4 & 6 & 1 & 1 \\ 1 & 1 & 5 & 2 \\ 1 & 1 & 2 & 5 \end{bmatrix}.$$

Solution. We first reduce $\mathbf{A}$ to tridiagonal form. Applying Householder's method, we obtain (see Example 1)

$$\mathbf{A_2} = \begin{bmatrix} 6 & -\sqrt{18} & 0 & 0 \\ -\sqrt{18} & 7 & \sqrt{2} & 0 \\ 0 & \sqrt{2} & 6 & 0 \\ 0 & 0 & 0 & 3 \end{bmatrix}.$$

From the characteristic determinant we see that $\mathbf{A_2}$, hence $\mathbf{A}$, has the eigenvalue 3. (Can you see this directly from $\mathbf{A_2}$?) Hence it suffices to apply the QR-method to the 3×3 matrix

$$\mathbf{B_0} = \mathbf{B} = \begin{bmatrix} 6 & -\sqrt{18} & 0 \\ -\sqrt{18} & 7 & \sqrt{2} \\ 0 & \sqrt{2} & 6 \end{bmatrix}.$$

First Step. We multiply $\mathbf{B}$ by

$$\mathbf{C_2} = \begin{bmatrix} \cos\theta_2 & \sin\theta_2 & 0 \\ -\sin\theta_2 & \cos\theta_2 & 0 \\ 0 & 0 & 1 \end{bmatrix} \quad \text{and then } \mathbf{C_2B} \text{ by} \quad \mathbf{C_3} = \begin{bmatrix} 1 & 0 & 0 \\ 0 & \cos\theta_3 & \sin\theta_3 \\ 0 & -\sin\theta_3 & \cos\theta_3 \end{bmatrix}.$$

Here $(-\sin\theta_2) \cdot 6 + (\cos\theta_2)(-\sqrt{18}) = 0$ gives [see (11)]

$$\cos\theta_2 = 0.816\,496\,58,$$

$$\sin\theta_2 = -0.577\,350\,27.$$

With these values, we compute

$$\mathbf{C_2B} = \begin{bmatrix} 7.348\,469\,23 & -7.505\,553\,50 & -0.816\,496\,58 \\ 0 & 3.265\,986\,32 & 1.154\,700\,54 \\ 0 & 1.414\,213\,56 & 6.000\,000\,00 \end{bmatrix}.$$

In $\mathbf{C_3}$ we get from $(-\sin\theta_3) \cdot 3.265\,986\,32 + (\cos\theta_3) \cdot 1.414\,213\,56 = 0$ the values

$$\cos\theta_3 = 0.917\,662\,94,$$

$$\sin\theta_3 = 0.397\,359\,71.$$

This gives

$$\mathbf{R_0} = \mathbf{C_3C_2B} = \begin{bmatrix} 7.348\,469\,23 & -7.505\,553\,50 & -0.816\,496\,58 \\ 0 & 3.559\,026\,08 & 3.443\,784\,13 \\ 0 & 0 & 5.047\,146\,15 \end{bmatrix}.$$

From this we compute

$$\mathbf{B_1} = \mathbf{R_0C_2}^\mathsf{T}\mathbf{C_3}^\mathsf{T} = \begin{bmatrix} 10.333\,333\,33 & -2.054\,804\,67 & 0 \\ -2.054\,804\,67 & 4.035\,087\,72 & 2.005\,532\,51 \\ 0 & 2.005\,532\,51 & 4.631\,578\,95 \end{bmatrix}$$

which is symmetric and tridiagonal. The off-diagonal entries in $\mathbf{B_1}$ are large in absolute value. Hence we have to go on.

Second Step. We do the same computations as in the first step, with $\mathbf{B}_0 = \mathbf{B}$ replaced by $\mathbf{B}_1$ and C_2 and C_3 changed accordingly, the new angles being $\theta_2 = -0.196\ 291\ 533$ and $\theta_3 = 0.513\ 415\ 589$. We obtain

$$\mathbf{R}_1 = \begin{bmatrix} 10.535\ 653\ 75 & -2.802\ 322\ 41 & -0.391\ 145\ 88 \\ 0 & 4.083\ 295\ 84 & 3.988\ 240\ 28 \\ 0 & 0 & 3.068\ 326\ 68 \end{bmatrix}$$

and from this

$$\mathbf{B}_2 = \begin{bmatrix} 10.879\ 879\ 88 & -0.796\ 379\ 18 & 0 \\ -0.796\ 379\ 18 & 5.447\ 386\ 64 & 1.507\ 025\ 00 \\ 0 & 1.507\ 025\ 00 & 2.672\ 733\ 48 \end{bmatrix}.$$

We see that the off-diagonal entries are somewhat smaller in absolute value than those of $\mathbf{B}_1$, but much too large for the diagonal entries to be good approximations of the eigenvalues of $\mathbf{B}$.

Further Steps. We list the main diagonal entries and the absolutely largest off-diagonal entry, which is $\left|b_{12}^{(j)}\right| = \left|b_{21}^{(j)}\right|$ in all steps. You may show that the given matrix $\mathbf{A}$ has the spectrum 11, 6, 3, 2.

Step j	$b_{11}^{(j)}$	$b_{22}^{(j)}$	$b_{33}^{(j)}$	$\max\limits_{j \neq k} \left\|b_{jk}^{(j)}\right\|$
3	10.966 892 9	5.945 898 56	2.087 208 51	0.585 235 82
5	10.997 087 2	6.001 815 41	2.001 097 38	0.120 653 34
7	10.999 742 1	6.000 244 39	2.000 013 55	0.035 911 07
9	10.999 977 2	6.000 022 67	2.000 000 17	0.010 684 77

◀

Looking back at our discussion, we recognize that the purpose of applying Householder's method before the QR-factorization method is a substantial reduction of cost in each QR-factorization, in particular if $\mathbf{A}$ is large.

Convergence acceleration can be achieved by a **spectral shift,** that is, if we take $\mathbf{B}_s - k_s\mathbf{I}$ instead of $\mathbf{B}_s$ with a suitable k_s. Possible choices of k_s are discussed in Ref. [E19], p. 510.

PROBLEM SET 18.9

Householder Tridiagonalization
Tridiagonalize the matrices in Probs. 1–6.

1. $\begin{bmatrix} 0.49 & 0.02 & 0.22 \\ 0.02 & 0.28 & 0.20 \\ 0.22 & 0.20 & 0.40 \end{bmatrix}$ **2.** $\begin{bmatrix} 6.0 & -3.6 & 3.6 \\ -3.6 & 4.4 & -5.2 \\ 3.6 & -5.2 & 4.4 \end{bmatrix}$ **3.** $\begin{bmatrix} 7 & 2 & 3 \\ 2 & 10 & 6 \\ 3 & 6 & 7 \end{bmatrix}$

4. $\begin{bmatrix} 6 & 4 & 0 & 1 \\ 4 & 5 & 2 & 8 \\ 0 & 2 & 9 & 2 \\ 1 & 8 & 2 & 4 \end{bmatrix}$
5. $\begin{bmatrix} 0.3 & 5.2 & 1.0 & 4.2 \\ 5.2 & 5.9 & 4.4 & 8.0 \\ 1.0 & 4.4 & 3.9 & 4.2 \\ 4.2 & 8.0 & 4.2 & 3.5 \end{bmatrix}$
6. $\begin{bmatrix} 5 & 4 & 1 & 1 \\ 4 & 5 & 1 & 1 \\ 1 & 1 & 4 & 2 \\ 1 & 1 & 2 & 4 \end{bmatrix}$

QR-Factorization Method

Do 3 QR-steps to find approximations to the eigenvalues of

7. $\begin{bmatrix} 9 & 1 & 0 \\ 1 & 4 & 1 \\ 0 & 1 & 1 \end{bmatrix}$
8. $\begin{bmatrix} 14.2 & -0.1 & 0 \\ -0.1 & -6.3 & 0.2 \\ 0 & 0.2 & 2.1 \end{bmatrix}$
9. $\begin{bmatrix} 7.0 & 0.5 & 0 \\ 0.5 & 3.5 & 0.1 \\ 0 & 0.1 & -1.5 \end{bmatrix}$

10. The matrix in the answer to Prob. 1   **11.** The matrix in the answer to Prob. 3

 12. CAS PROJECT. QR-Method. (a) Write a program for 3×3 matrices that first tridiagonalizes and then does QR-steps. Apply it to the matrices in Probs. 1 and 3.
(b) Do the same for 4×4 matrices and apply the program to the matrices in Probs. 4 and 5.
(c) Try to find out experimentally: On what properties of a matrix does the speed of decrease of off-diagonal entries in the QR-method depend?

CHAPTER 18 REVIEW

1. What are the main problem areas in numerical linear algebra?
2. What is pivoting? Why and how is it done?
3. When would you apply the Gauss–Seidel iteration and when the Gauss elimination?
4. What is the connection between Doolittle's method and Gauss elimination?
5. What do you know about the convergence of the Gauss–Seidel method?
6. What is ill-conditioning? What is the condition number and its significance?
7. What is least squares approximation? What are the normal equations?
8. What is an eigenvalue of a matrix? Why are eigenvalue problems important?
9. What numerical methods for eigenvalues and eigenvectors do you know?
10. What is a similarity transformation of a matrix and why is it important in designing numerical methods?
11. What is the power method for eigenvalues? State its advantages and disadvantages.
12. Why did we do scaling in the power method?
13. State Gerschgorin's theorem. Give an example. Try to prove the theorem.
14. What is tridiagonalization? When would you apply it?
15. What is the idea of the QR-method? Its purpose?

Solve the systems in Probs. 16–21 by the Gauss elimination. (Show the details of your work.)

16.
$$3x_2 - 6x_3 = 0$$
$$4x_1 - x_2 + 2x_3 = 16$$
$$-5x_1 + 2x_2 - 4x_3 = -20$$

17.
$$2.1x_1 + 3.7x_2 + 1.8x_3 = 4.8$$
$$-4.6x_1 - 1.2x_2 - 0.2x_3 = 8.2$$
$$0.3x_1 + 2.5x_2 + 0.5x_3 = 1.9$$

18. $5x_1 + x_2 - 3x_3 = 17$

$\qquad -5x_2 + 15x_3 = -10$

$\quad 2x_1 - 3x_2 + 9x_3 = 0$

19. $2x_1 + x_2 + 2x_3 = 5.6$

$\quad 8x_1 + 5x_2 + 13x_3 = 20.9$

$\quad 6x_1 + 3x_2 + 12x_3 = 11.4$

20. $x_1 + x_2 + x_3 = 5$

$\quad x_1 + 2x_2 + 2x_3 = 6$

$\quad x_1 + 2x_2 + 3x_3 = 8$

21. $\qquad 8x_2 - 6x_3 = 23.6$

$\quad 10x_1 + 6x_2 + 2x_3 = 68.4$

$\quad 12x_1 - 14x_2 + 4x_3 = -6.2$

22. Solve Prob. 20 by Doolittle's method and by Cholesky's method.

23. Solve Prob. 19 by Doolittle's method.

Find the inverse. (Show the details of your work.)

24. $\begin{bmatrix} 5 & 1 & 1 \\ 1 & 6 & 0 \\ 1 & 0 & 8 \end{bmatrix}$

25. $\begin{bmatrix} 1.5 & 2.0 & 1.0 \\ 2.0 & 3.5 & 1.5 \\ 1.0 & 1.5 & 9.0 \end{bmatrix}$

26. $\begin{bmatrix} 2.0 & 0.1 & 3.3 \\ 1.6 & 4.4 & 0.5 \\ 0.3 & -4.3 & 2.8 \end{bmatrix}$

Do 3 Gauss–Seidel steps, starting from $[1 \quad 1 \quad 1]$:

27. $10x_1 + x_2 - x_3 = 17$

$\quad 2x_1 + 20x_2 + x_3 = 28$

$\quad 3x_1 - x_2 + 25x_3 = 105$

28. $0.2x_1 + 4.0x_2 - 0.4x_3 = 32.0$

$\quad 0.5x_1 - 0.2x_2 + 2.5x_3 = -5.1$

$\quad 7.5x_1 + 0.1x_2 - 1.5x_3 = -12.7$

29. $4x_1 - x_2 \qquad = 11.0$

$\qquad 4x_2 - x_3 = 6.7$

$\quad -x_1 \qquad + 4x_3 = -1.2$

Compute the l_1-, l_2-, and l_∞-norms of the vectors:

30. $[-1 \quad 0 \quad 1 \quad 0]$ **31.** $[0 \quad -1 \quad -3]$ **32.** $[4 \quad -4 \quad 2 \quad 8 \quad 6]$

33. $[0 \quad 1 \quad 0 \quad 0]$ **34.** $[0.2 \quad -1.2 \quad 1.0]$ **35.** $[4, \quad 7, \quad 11]$

Compute the matrix norm corresponding to the l_∞-vector norm for the coefficient matrix:

36. In Prob. 16   **37.** In Prob. 27   **38.** In Prob. 28

Compute the condition number (corresponding to the l_1-vector norm) of the matrix:

39. In Prob. 25   **40.** In Prob. 26   **41.** In Prob. 24

42. Fit a least-squares straight line to (2, 3.8), (4, 5.1), (6, 5.8), (8, 7.0), (10, 7.9) and plot it. Show the details of your work.

43. What happens in Prob. 42 if you change (10, 7.9) to (10, 4.9)? First guess. Then compute and plot.

44. Fit a least-squares quadratic parabola to (0, 1.9), (1, 0.9), (3, 4.8), (4, 10.2).

45. Apply Gerschgorin's theorem to the matrix in Prob. 25. Then tridiagonalize the matrix and apply 3 QR-steps. (Spectrum to 6S: 9.65971, 4.07684, 0.263451.)

SUMMARY OF CHAPTER 18
NUMERICAL METHODS IN LINEAR ALGEBRA

This chapter deals with three large classes of numerical problems arising from linear algebra, namely, the numerical solution of linear systems (Secs. 18.1–18.4), the fitting of straight lines or parabolas through given data (represented as points in the plane; Sec. 18.5), and the numerical solution of eigenvalue problems (Secs. 18.6–18.9).

Thus, in Secs. 18.1–18.4 we solve systems $\mathbf{Ax} = \mathbf{b}$, where $\mathbf{A} = [a_{jk}]$ is an $n \times n$ matrix; written out,

(1)
$$
\begin{aligned}
\mathrm{E}_1: \quad & a_{11}x_1 + \cdots + a_{1n}x_n = b_1 \\
\mathrm{E}_2: \quad & a_{21}x_1 + \cdots + a_{2n}x_n = b_2 \\
& \cdots\cdots\cdots\cdots\cdots\cdots\cdots\cdots \\
\mathrm{E}_n: \quad & a_{n1}x_1 + \cdots + a_{nn}x_n = b_n.
\end{aligned}
$$

There are two types of numerical methods for this task, **direct methods,** such as the Gauss elimination, in which the amount of computation to get a solution can be specified in advance, and **indirect** or **iterative methods,** such as the Gauss–Seidel method, in which we start from a (possibly crude) approximation and improve it stepwise by repeatedly performing the same cycle of computation, with changing data (see below).

The **Gauss elimination** (Sec. 18.1) is a systematic elimination process that reduces (1) stepwise to triangular form. In Step 1 we eliminate x_1 from equations E_2 to E_n by subtracting $(a_{21}/a_{11})\,\mathrm{E}_1$ from E_2, then $(a_{31}/a_{11})\,\mathrm{E}_1$ from E_3, etc. Equation E_1 is called the **pivot equation** in this step and a_{11} the **pivot.** In Step 2 we take the new second equation as pivot equation and eliminate x_2, etc. If the triangular form is reached, we get x_n from the last equation, then x_{n-1} from the second last, etc. **Partial pivoting** (= interchange of equations) is *necessary* if candidates for pivots are zero, and *advisable* if they are small.

Doolittle's, Crout's, and **Cholesky's methods** in Sec. 18.2 are variants of the Gauss elimination. They use the idea of factoring

(2)
$$ \mathbf{A} = \mathbf{LU} $$

where $\mathbf{L}$ is a lower triangular and $\mathbf{U}$ an upper triangular matrix. One solves (2) by setting $\mathbf{Ux} = \mathbf{y}$ and then solving $\mathbf{Ax} = \mathbf{LUx} = \mathbf{Ly} = \mathbf{b}$ by first solving the triangular system $\mathbf{Ly} = \mathbf{b}$ for $\mathbf{y}$ and then the triangular system $\mathbf{Ux} = \mathbf{y}$ for $\mathbf{x}$.

Iterative methods are obtained by writing $\mathbf{Ax} = \mathbf{b}$ as

(3)
$$ \mathbf{x} = \mathbf{b} - (\mathbf{A} - \mathbf{I})\mathbf{x} $$

and then substituting approximations on the right to get new approximations on the left. In particular, in the **Gauss–Seidel method** we first divide each equation to make $a_{11} = a_{22} = \cdots = a_{nn} = 1$, then write $\mathbf{A} = \mathbf{I} + \mathbf{L} + \mathbf{U}$, so that (3) becomes

$x = b - Lx - Ux,$ and always take the most recent approximate x_j's on the right. This gives the iteration formula (Sec. 18.3)

(4) $$x^{(m+1)} = b - Lx^{(m+1)} - Ux^{(m)}.$$

If $\|C\| < 1,$ where $C = -(I + L)^{-1}U,$ then this process converges. Here, $\|C\|$ denotes any matrix norm (Sec. 18.3).

If the **condition number** $\kappa(A) = \|A\|\,\|A^{-1}\|$ of A is large, then the system $Ax = b$ is **ill-conditioned** (Sec. 18.4), and a small **residual** $r = b - A\tilde{x}$ does *not* imply that $\tilde{x}$ is close to the exact solution.

The fitting of a polynomial $p(x) = b_0 + b_1 x + \cdots + b_m x^m$ through given data (points in the xy-plane) $(x_1, y_1), \cdots, (x_n, y_n)$ by the method of **least squares** is discussed in Sec. 18.5.

An **eigenvalue** of an $n \times n$ matrix $A = [a_{jk}]$ is a number λ such that

(5) $$Ax = \lambda x, \qquad \text{thus} \qquad (A - \lambda I)x = 0$$

has a solution $x \neq 0,$ called an **eigenvector** of A corresponding to that $\lambda.$ In Sec. 18.6 we list basic facts about the eigenvalue problem needed in numerical methods.

Section 18.7 concerns theorems that give **inclusion sets** (sets in the complex λ-plane that contain one or several eigenvalues of A), notably the famous **Gerschgorin theorem,** which states that the **spectrum** (= set of all eigenvalues) of A lies in the n disks

(6) $$D_j: \qquad \text{Center } a_{jj}, \qquad \text{radius } \sum |a_{jk}| \qquad (j = 1, \cdots, n)$$

(sum over k from 1 to $n,$ $k \neq j$).

The **power method** (Sec. 18.8) gives approximations

(7) $$q = \frac{(Ax)^{\mathsf{T}}x}{x^{\mathsf{T}}x} \qquad \textit{(Rayleigh quotient),}$$

usually to the eigenvalue that is largest in absolute value, and, if A is symmetric, error bounds

(8) $$|\epsilon| \leqq \sqrt{\frac{(Ax)^{\mathsf{T}}Ax}{x^{\mathsf{T}}x} - q^2}.$$

Practically, one chooses any vector $x_0 \neq 0,$ computes the vectors

$$x_1 = Ax_0, \quad x_2 = Ax_1, \quad \cdots, \quad x_s = Ax_{s-1},$$

and takes $x = x_{s-1}$ and $Ax = x_s$ in (7) and (8).

For determining all the eigenvalues of a symmetric matrix $A,$ it is best to first **tridiagonalize** A and then to apply the **QR-method.** This method is based on a factorization $A = QR,$ where Q is orthogonal and R is upper triangular, and similarity transformations.

Numerical Methods for Differential Equations

Numerical methods for differential equations are of great importance to the engineer and physicist because practical problems often lead to differential equations that cannot be solved exactly by one of the methods in Chaps. 1–5, and 11 or by similar methods. Also, there are differential equations for which the solutions in terms of formulas are so complicated that one often prefers to apply a numerical method to such an equation.

This chapter includes basic methods for the numerical solution of ordinary differential equations (Secs. 19.1–19.3) and partial differential equations (Secs. 19.4–19.7).

Sections 19.1 and 19.2 may also be studied immediately after Chap. 1 and Sec. 19.3 after Chap. 2, since these sections are independent of Chaps. 17 and 18.

Sections 19.4–19.7 may also be studied immediately after Chap. 11, provided the reader has some knowledge of linear systems of algebraic equations.

Prerequisite for Secs. 19.1–19.3: Secs. 1.1–1.6, 2.1–2.3.
Prerequisite for Secs. 19.4–19.7: Secs. 11.1–11.3, 11.5, 11.11.
References: Appendix 1, Part E (see also Parts A and C).
Answers to problems: Appendix 2.

19.1 Methods for First-Order Differential Equations

From Chap. 1 we know that a *differential equation of the first order* is of the form $F(x, y, y') = 0$, and often it will be possible to write the equation in the *explicit form* $y' = f(x, y)$. An **initial value problem** consists of a differential equation and a condition the solution must satisfy (or several conditions referring to the same value of x if the equation is of higher order). In this section we shall consider initial value problems of the form

(1)
$$y' = f(x, y), \qquad y(x_0) = y_0$$

assuming f to be such that the problem has a unique solution on some interval containing x_0.

We shall discuss methods for computing numerical values of the solutions, which are needed if a formula for the solution of an equation is not available or is too complicated to be of practical use.

These methods are **step-by-step methods,** that is, we start from the given $y_0 = y(x_0)$ and proceed stepwise, computing approximate values of the solution $y(x)$ at the *"mesh points"*

$$x_1 = x_0 + h, \qquad x_2 = x_0 + 2h, \qquad x_3 = x_0 + 3h, \qquad \cdots,$$

where the **step size** h is a fixed number, for instance 0.2 or 0.1 or 0.01, whose choice we discuss later in this section.

The computation in each step is done by the same formula. Such formulas are suggested by the Taylor series

$$(2) \qquad\qquad y(x + h) = y(x) + h y'(x) + \frac{h^2}{2} y''(x) + \cdots.$$

Now for a small value of h, the higher powers $h^2, h^3, \cdots$ are very small. This suggests the crude approximation

$$y(x + h) \approx y(x) + h y'(x) = y(x) + h f(x, y)$$

(with the right side obtained from the given differential equation) and the following iteration process. In the first step we compute

$$y_1 = y_0 + h f(x_0, y_0)$$

which approximates $y(x_1) = y(x_0 + h)$. In the second step we compute

$$y_2 = y_1 + h f(x_1, y_1)$$

which approximates $y(x_2) = y(x_0 + 2h)$, etc., and in general

$$(3) \qquad\qquad \boxed{y_{n+1} = y_n + h f(x_n, y_n)} \qquad\qquad (n = 0, 1, \cdots).$$

This is called the **Euler method** or **Euler–Cauchy method.** Geometrically it is an approximation of the curve of $y(x)$ by a polygon whose first side is tangent to the curve at x_0 (see Fig. 417).

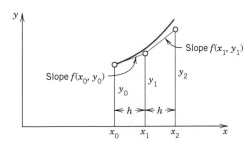

Fig. 417.   Euler method

This crude method is hardly ever used in practice, but since it is simple, it nicely explains the principle of methods based on the Taylor series.

Taylor's formula with remainder has the form $y(x + h) = y(x) + hy'(x) + \frac{1}{2}h^2y''(\xi)$ (where $x \leqq \xi \leqq x + h$). It shows that in Euler's method the *truncation error in each step* or **local truncation error** is proportional to h^2, written $O(h^2)$, where O suggests *order* (see also Sec. 18.1). Now over a fixed x-interval in which we want to solve an equation the number of steps is proportional to $1/h$. Hence the *total error* or **global error** is proportional to $h^2(1/h) = h^1$. For this reason, Euler's method is called a **first-order method.** In addition, there are **round-off errors** in this and other methods, which may affect the accuracy of the values $y_1, y_2, \cdots$ more and more as n increases, as we shall see.

EXAMPLE 1　**Euler method**

Apply Euler's method to the following initial value problem, choosing $h = 0.2$ and computing $y_1, \cdots, y_5$:

(4)
$$y' = x + y, \qquad y(0) = 0.$$

Solution.　Here $f(x, y) = x + y$, and we see that (3) becomes

$$y_{n+1} = y_n + 0.2(x_n + y_n).$$

Table 19.1 shows the computations, the values of the exact solution

$$y(x) = e^x - x - 1$$

obtained from (4) in Sec. 1.6, and the error. In practice the exact solution is unknown, but an indication of the accuracy of the values can be obtained by applying Euler's method once more with step $2h = 0.4$ and comparing corresponding approximations. This computation is:

x_n	y_n	$0.4(x_n + y_n)$	y_n in Table 19.1	Difference
0.0	0.000	0.000	0.000	0.000
0.4	0.000	0.160	0.040	0.040
0.8	0.160		0.274	0.114

Since the error is of order h^2, in a switch from h to $2h$ it is multiplied by $2^2 = 4$, but since we then need only half as many steps as before, it will only be multiplied by $4/2 = 2$. Hence the difference $2\epsilon_2 - \epsilon_2 = 0.040$ indicates the error ϵ_2 of y_2 in Table 19.1 (which actually is 0.052), and 0.114 that of y_4 (actual: 0.152). ◀

Table 19.1
Euler Method Applied to (4) in Example 1 and Error

n	x_n	y_n	$0.2(x_n + y_n)$	Exact Values	Error
0	0.0	0.000	0.000	0.000	0.000
1	0.2	0.000	0.040	0.021	0.021
2	0.4	0.040	0.088	0.092	0.052
3	0.6	0.128	0.146	0.222	0.094
4	0.8	0.274	0.215	0.426	0.152
5	1.0	0.489		0.718	0.229

Automatic Variable Step Size Selection in Modern Codes. These codes select variable step sizes h_n so that the error of the solution will not exceed a given maximum size TOL (suggesting *tolerance*). Now for Euler's method, when the step size is h_n, the local error at x_n is about $\frac{1}{2}h_n^2|y''(x_n)|$. We require that this be equal to a given tolerance TOL,

(5)   (a) $\frac{1}{2}h_n^2|y''(x_n)| = \text{TOL},$ thus   (b) $h_n = \sqrt{\dfrac{2\,\text{TOL}}{|y''(x_n)|}}.$

$y''(x)$ must not be zero on the interval $J\!:\, x_0 \leq x \leq x_N$ on which the solution is wanted. Let K be the minimum of $|y''(x)|$ on J and assume that $K > 0$. Minimum $|y''(x)|$ corresponds to maximum $h = H = \sqrt{2\,\text{TOL}/K}$ by (5). Thus, $\sqrt{2\,\text{TOL}} = H\sqrt{K}$. We can insert this into (5b), obtaining by straightforward algebra

(6) $h_n = \varphi(x_n)H$ where $\varphi(x_n) = \sqrt{\dfrac{K}{|y''(x_n)|}}.$

For other methods, automatic step size selection is the same in principle.

Improved Euler Method (Heun's Method)

By taking more terms in (2) into account we obtain numerical methods of higher order and precision. But there is a practical problem. If we substitute $y' = f(x, y(x))$ into (2), we have

(2*) $y(x + h) = y(x) + hf + \frac{1}{2}h^2 f' + \frac{1}{6}h^3 f'' + \cdots$

where, since y in f depends on x,

$$f' = f_x + f_y y' = f_x + f_y f$$

and the further derivatives f'', f''' become even much more cumbersome. The **general strategy** now is to avoid their computation and replace it by computing f for one or several suitably chosen auxiliary values of (x, y). "Suitably" means that they are chosen to make the order of the method as high as possible (to have high accuracy). Let us discuss two such methods that are of practical importance.

The first method is the so-called **improved Euler method** or **improved Euler–Cauchy method** (sometimes also called **Heun's method**). In each step of this method we compute first the auxiliary value

(7a) $$y_{n+1}^* = y_n + hf(x_n, y_n)$$

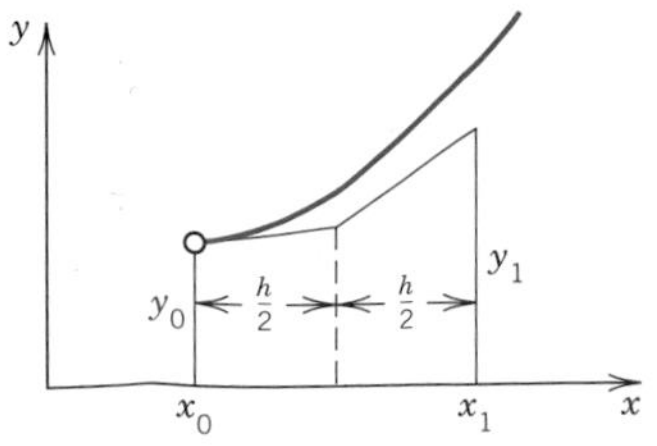

Fig. 418.   Improved Euler method

and then the new value

(7b)
$$y_{n+1} = y_n + \tfrac{1}{2}h\left[f(x_n, y_n) + f(x_{n+1}, y_{n+1}^*)\right].$$

This method has a simple geometric interpretation. In fact, we may say that in the interval from x_n to $x_n + \tfrac{1}{2}h$ we approximate the solution y by the straight line through (x_n, y_n) with slope $f(x_n, y_n)$, and then we continue along the straight line with slope $f(x_{n+1}, y_{n+1}^*)$ until x reaches x_{n+1} (see Fig. 418, where $n = 0$).

The improved Euler–Cauchy method is a **predictor–corrector method,** because in each step we first *predict* a value by (7a) and then *correct* it by (7b).

In algorithmic form, using the notations $k_1 = hf(x_n, y_n)$ in (7a) and $k_2 = hf(x_{n+1}, y_{n+1}^*)$ in (7b) we can write this method as shown in Table 19.2.

EXAMPLE 2   **Improved Euler method**

Apply the improved Euler method to the initial value problem (4), choosing $h = 0.2$, as before.

Solution.   For the present problem,

$$k_1 = 0.2\,(x_n + y_n)$$

$$k_2 = 0.2\,(x_n + 0.2 + y_n + 0.2\,(x_n + y_n))$$

$$y_{n+1} = y_n + \frac{0.2}{2}\,(2.2x_n + 2.2y_n + 0.2) = y_n + 0.22\,(x_n + y_n) + 0.02.$$

Table 19.3 shows that our present results are more accurate than those in Example 1; see also Table 19.6. ◀

Table 19.2
Improved Euler Method (Heun's Method)

ALGORITHM EULER (f, x_0, y_0, h, N)

This algorithm computes the solution of the initial value problem $y' = f(x, y)$, $y(x_0) = y_0$ at equidistant points $x_1 = x_0 + h$, $x_2 = x_0 + 2h$, $\cdots$, $x_N = x_0 + Nh$; here f is such that this problem has a unique solution on the interval $[x_0, x_N]$ (see Sec. 1.9).

    INPUT: Initial values x_0, y_0, step size h, number of steps N
    OUTPUT: Approximation y_{n+1} to the solution $y(x_{n+1})$ at $x_{n+1} = x_0 + (n + 1)h$, where $n = 0, \cdots, N - 1$

        For $n = 0, 1, \cdots, N - 1$ do:

             $x_{n+1} = x_n + h$

             $k_1 = hf(x_n, y_n)$

             $k_2 = hf(x_{n+1}, y_n + k_1)$

             $y_{n+1} = y_n + \tfrac{1}{2}(k_1 + k_2)$

             OUTPUT x_{n+1}, y_{n+1}

      End
      Stop
End EULER

Table 19.3
Improved Euler Method Applied to (4) and Error

n	x_n	y_n	$\begin{array}{c}0.22(x_n + y_n)\\ + 0.02\end{array}$	Exact Values	Error
0	0.0	0.0000	0.0200	0.0000	0.0000
1	0.2	0.0200	0.0684	0.0214	0.0014
2	0.4	0.0884	0.1274	0.0918	0.0034
3	0.6	0.2158	0.1995	0.2221	0.0063
4	0.8	0.4153	0.2874	0.4255	0.0102
5	1.0	0.7027		0.7183	0.0156

Local Error. *The local error of the improved Euler method is of order h^3.*

PROOF. Setting $\tilde{f}_n = f(x_n, y(x_n))$ and using (2*), we have

$$(8a) \qquad y(x_n + h) - y(x_n) = h\tilde{f}_n + \tfrac{1}{2}h^2\tilde{f}_n' + \tfrac{1}{6}h^3\tilde{f}_n'' + \cdots.$$

Approximating the expression in the brackets in (7b) by $\tilde{f}_n + \tilde{f}_{n+1}$ and again using the Taylor expansion, we obtain from (7b)

$$
\begin{aligned}
(8b) \qquad y_{n+1} - y_n &\approx \tfrac{1}{2}h\big[\tilde{f}_n + \tilde{f}_{n+1}\big] \\
&= \tfrac{1}{2}h\big[\tilde{f}_n + (\tilde{f}_n + h\tilde{f}_n' + \tfrac{1}{2}h^2\tilde{f}_n'' + \cdots)\big] \\
&= h\tilde{f}_n + \tfrac{1}{2}h^2\tilde{f}_n' + \tfrac{1}{4}h^3\tilde{f}_n'' + \cdots
\end{aligned}
$$

(where $' = d/dx_n$, etc.). Subtraction of (8b) from (8a) gives the error

$$\frac{h^3}{6}\,\tilde{f}_n'' - \frac{h^3}{4}\,\tilde{f}_n'' + \cdots = -\frac{h^3}{12}\,\tilde{f}_n'' + \cdots. \qquad \blacktriangleleft$$

Since the number of steps over a fixed x-interval is proportional to $1/h$, the global error is of order $h^3/h = h^2$. Hence the improved Euler method is a **second-order method.**

Runge–Kutta Methods

A still more accurate method of great practical importance is the *classical Runge–Kutta method of fourth order,* which we call briefly the **Runge–Kutta method.**[1] It is shown in Table 19.4. We see that in each step we first compute four auxiliary quantities k_1, k_2, k_3, k_4 and then the new value y_{n+1}. These formulas look complicated at first sight, but they are in fact very easy to program. Indeed, whereas in hand calculations the frequent calculation of $f(x, y)$ is somewhat laborious, on the computer it is not a problem. The method is well suited for computers because it needs no special starting procedure, makes

---

[1]Named after the German mathematicians CARL RUNGE (1856—1927), professor of applied mathematics at Göttingen, and WILHELM KUTTA (1867—1944). Runge [*Math. Annalen* **46** (1895), 167−178], KARL HEUN [*Zeitschr. Math. Phys.* **45** (1900), 23−38], and Kutta [*Zeitschr. Math. Phys.* **46** (1901), 435−453] developed various such methods. Theoretically, there are infinitely many fourth-order methods using four function values per step. The method in Table 19.4 is most popular from a practical viewpoint because of its "symmetrical" form and its simple coefficients. It was given by Kutta.

Table 19.4
Classical Runge–Kutta Method of Fourth Order

ALGORITHM RUNGE–KUTTA (f, x_0, y_0, h, N).

This algorithm computes the solution of the initial value problem $y' = f(x, y)$, $y(x_0) = y_0$ at equidistant points

$$x_1 = x_0 + h, \; x_2 = x_0 + 2h, \cdots, x_N = x_0 + Nh;$$

here f is such that this problem has a unique solution on the interval $[x_0, x_N]$ (see Sec. 1.9).

> INPUT: Initial values x_0, y_0, step size h, number of steps N
>
> OUTPUT: Approximation y_{n+1} to the solution $y(x_{n+1})$ at $x_{n+1} = x_0 + (n + 1)h$, where $n = 0, 1, \cdots, N - 1$

For $n = 0, 1, \cdots, N - 1$ do:

$$k_1 = hf(x_n, y_n)$$

$$k_2 = hf(x_n + \tfrac{1}{2}h, \; y_n + \tfrac{1}{2}k_1)$$

$$k_3 = hf(x_n + \tfrac{1}{2}h, \; y_n + \tfrac{1}{2}k_2)$$

$$k_4 = hf(x_n + h, \; y_n + k_3)$$

$$x_{n+1} = x_n + h$$

$$y_{n+1} = y_n + \tfrac{1}{6}(k_1 + 2k_2 + 2k_3 + k_4)$$

OUTPUT x_{n+1}, y_{n+1}

End

Stop

End RUNGE–KUTTA

---

light demand on storage, and repeatedly uses the same straightforward computational procedure. It is numerically stable.

Note that if f depends only on x, this method reduces to Simpson's rule of integration (Sec. 17.5).

EXAMPLE 3 **Classical Runge–Kutta method**

Apply the Runge–Kutta method to the initial value problem (4) in Example 1, choosing $h = 0.2$, as before, and computing five steps.

Solution. For the present problem we have $f(x, y) = x + y$. Hence

$$k_1 = 0.2(x_n + y_n), \qquad\qquad k_2 = 0.2(x_n + 0.1 + y_n + 0.5k_1),$$

$$k_3 = 0.2(x_n + 0.1 + y_n + 0.5k_2), \qquad k_4 = 0.2(x_n + 0.2 + y_n + k_3).$$

Since these expressions are so simple, we find it convenient to insert k_1 into k_2, obtaining the auxiliary quantity $k_2 = 0.22(x_n + y_n) + 0.02$, insert this into k_3, finding $k_3 = 0.222(x_n + y_n) + 0.022$, and finally insert this into k_4, finding $k_4 = 0.2444(x_n + y_n) + 0.0444$. If we use these expressions, the formula for y_{n+1} in Table 19.4 becomes

$$(9) \qquad\qquad y_{n+1} = y_n + 0.2214(x_n + y_n) + 0.0214.$$

Of course, our present inserting process is ***not typical*** of the Runge–Kutta method and should not be tried in general. Table 19.5 shows the computations. From Table 19.6 we see that the values are much more accurate than those in Examples 1 and 2.   ◄

Table 19.5
Runge–Kutta Method Applied to (4); Computations by the Use of (9)

n	x_n	y_n	$0.2214(x_n + y_n)$ $+ 0.0214$	Exact Values $y = e^x - x - 1$	$10^6 \times$ Error of y_n
0	0.0	0	0.021 400	0.000 000	0
1	0.2	0.021 400	0.070 418	0.021 403	3
2	0.4	0.091 818	0.130 289	0.091 825	7
3	0.6	0.222 106	0.203 414	0.222 119	11
4	0.8	0.425 521	0.292 730	0.425 541	20
5	1.0	0.718 251		0.718 282	31

Table 19.6
Comparison of the Accuracy of the Three Methods Under Consideration in the Case of the Initial Value Problem (4), with $h = 0.2$

x	$y = e^x - x - 1$	Error		
		Euler Method (Table 19.1)	Improved Euler (Table 19.3)	Runge–Kutta (Table 19.5)
0.2	0.021 403	0.021	0.0014	0.000 003
0.4	0.091 825	0.052	0.0034	0.000 007
0.6	0.222 119	0.094	0.0063	0.000 011
0.8	0.425 541	0.152	0.0102	0.000 020
1.0	0.718 282	0.229	0.0156	0.000 031

Error and Step Size Control. RKF (Rung–Kutta–Fehlberg)

The idea of adaptive integration (Sec. 17.5) has analogs for Runge–Kutta (and other) methods. In Table 19.4 for RK (Runge–Kutta), if we compute in each step with step sizes h and $2h$, the latter has error per step equal to $2^5 = 32$ times that of the former; however, since we have only half as many steps for $2h$, the actual factor is $2^5/2 = 16$. Hence the error ϵ of an approximation with step size h equals about $1/15$ times the difference $\delta = \tilde{y} - \tilde{\tilde{y}}$ of corresponding values for step sizes h and $2h$, respectively,

$$(10) \qquad\qquad \epsilon \approx \tfrac{1}{15}(\tilde{y} - \tilde{\tilde{y}}).$$

Table 19.7 illustrates (10) for the initial value problem

$$(11) \qquad\qquad y' = (y - x - 1)^2 + 2, \qquad y(0) = 1;$$

Table 19.7
Runge–Kutta Method Applied to the Initial Value Problem (11) and Error Estimate
(10). Exact Solution $y = \tan x + x + 1$

x	$\tilde{y}$ (Step size h)	$\tilde{\tilde{y}}$ (Step size $2h$)	Error Estimate (10)	Actual Error	Exact Solution (9D)
0.0	1.000 000 000	1.000 000 000	0.000 000 000	0.000 000 000	1.000 000 000
0.1	1.200 334 589			0.000 000 083	1.200 334 672
0.2	1.402 709 878	1.402 707 408	0.000 000 165	0.000 000 157	1.402 710 036
0.3	1.609 336 039			0.000 000 210	1.609 336 250
0.4	1.822 792 993	1.822 788 993	0.000 000 267	0.000 000 226	1.822 793 219

the step size is $h = 0.1$, and $0 \leq x \leq 0.4$. We see that the estimate is close to the actual error. This method of error estimation is simple but may be unstable.

RKF. E. Fehlberg [*Computing* **6** (1970), 61−71] proposed and developed error control by using two RK methods of different orders to go from (x_n, y_n) to (x_{n+1}, y_{n+1}). The difference of the computed y-values at x_{n+1} gives an error estimate to be used for step size control. Fehlberg discovered two RK formulas that together need only 6 function evaluations per step. We present these formulas here because RKF has become quite popular. For instance, MAPLE uses it (also for systems of differential equations).

Fehlberg's fifth-order RK method is

(12a)
$$y_{n+1} = y_n + \gamma_1 k_1 + \cdots + \gamma_6 k_6$$

with coefficient vector $\gamma = [\gamma_1 \cdots \gamma_6]$,

(12b)
$$\gamma = \begin{bmatrix} \frac{16}{135} & 0 & \frac{6656}{12825} & \frac{28561}{56430} & -\frac{9}{50} & \frac{2}{55} \end{bmatrix}.$$

His **fourth-order RK method** is

(13a)
$$y^*_{n+1} = y_n + \gamma^*_1 k_1 + \cdots + \gamma^*_5 k_5$$

with coefficient vector

(13b)
$$\gamma^* = \begin{bmatrix} \frac{25}{216} & 0 & \frac{1408}{2565} & \frac{2197}{4104} & -\frac{1}{5} \end{bmatrix}.$$

In both formulas we use only 6 different function evaluations altogether, namely,

(14)
$$
\begin{aligned}
k_1 &= hf(x_n, y_n) \\
k_2 &= hf(x_n + \tfrac{1}{4}h, \quad y_n + \tfrac{1}{4}k_1) \\
k_3 &= hf(x_n + \tfrac{3}{8}h, \quad y_n + \tfrac{3}{32}k_1 + \tfrac{9}{32}k_2) \\
k_4 &= hf(x_n + \tfrac{12}{13}h, \quad y_n + \tfrac{1932}{2197}k_1 - \tfrac{7200}{2197}k_2 + \tfrac{7296}{2197}k_3) \\
k_5 &= hf(x_n + h, \quad y_n + \tfrac{439}{216}k_1 - 8k_2 + \tfrac{3680}{513}k_3 - \tfrac{845}{4104}k_4) \\
k_6 &= hf(x_n + \tfrac{1}{2}h, \quad y_n - \tfrac{8}{27}k_1 + 2k_2 - \tfrac{3544}{2565}k_3 + \tfrac{1859}{4104}k_4 - \tfrac{11}{40}k_5)
\end{aligned}
$$

The difference of (12) and (13) gives the **error estimate**

(15)
$$\epsilon_{n+1} \approx y_{n+1} - y^*_{n+1} = \tfrac{1}{360}k_1 - \tfrac{128}{4275}k_3 - \tfrac{2197}{75240}k_4 + \tfrac{1}{50}k_5 + \tfrac{2}{55}k_6.$$

EXAMPLE 4   **Runge–Kutta–Fehlberg**

For the initial value problem (11) we obtain from (12)−(14) with $h = 0.1$ in the first step the 12S-values

$$k_1 = 0.200000\ 000000 \qquad k_2 = 0.200062\ 500000 \qquad k_3 = 0.200140\ 756867$$

$$k_4 = 0.200856\ 926154 \qquad k_5 = 0.201006\ 676700 \qquad k_6 = 0.200250\ 418651$$

$$y_1 = 1.200334\ 67253 \qquad y_1^* = 1.200334\ 66949 \qquad \epsilon_1 = 0.000000\ 00304.$$

The exact 12S-value is $y(0.1) = 1.200334\ 67209$. Hence the actual error of y_1 is $-4.4 \cdot 10^{-10}$, smaller than that in Table 19.7 by a factor 200.   ◀

Table 19.8 summarizes essential features of the methods in this section. It can be shown that these methods are *numerically stable* (definition in Sec. 17.1). They are **one-step methods** because in each step we use the data of just *one* preceding step, in contrast to **multistep methods** where in each step we use data from *several* preceding steps, as we shall see in the next section.

Table 19.8
Methods Considered and Their Order (= Their Global Error)

Method	Function Evaluation per Step	Global Error	Local Error
Euler	1	$O(h)$	$O(h^2)$
Improved Euler	2	$O(h^2)$	$O(h^3)$
RK (fourth order)	4	$O(h^4)$	$O(h^5)$
RKF	6	$O(h^5)$	$O(h^6)$

PROBLEM SET 19.1

Euler Method

Apply Euler's method to the following initial value problems. Do 10 steps. Solve the problem exactly. Compute the errors to see that the method is too inaccurate for practical purposes. (Show the details of your work.)

1. $y' + 0.1y = 0$, $y(0) = 2$, $h = 0.1$

2. $y' = \frac{1}{2}\pi \sqrt{1 - y^2}$, $y(0) = 0$, $h = 0.1$

3. $y' + 5x^4 y^2 = 0$, $y(0) = 1$, $h = 0.2$

4. $y' = (y + x)^2$, $y(0) = 0$, $h = 0.1$

Improved Euler Method

Solve the given initial value problems by the improved Euler method (10 steps). Find the exact solution and the error.

5. $y' = y$, $y(0) = 1$, $h = 0.1$

6. **(Logistic population)** $y' = y - y^2$, $y(0) = 0.5$, $h = 0.1$

7. $y' + 2xy^2 = 0$, $y(0) = 1$, $h = 0.2$

8. $y' = 2(1 + y^2)$, $y(0) = 0$, $h = 0.05$

9. Do Prob. 7 using Euler's method ($h = 0.2$) and compare the accuracy.

10. Do Prob. 7 using Euler's method (20 steps, $h = 0.1$) and compare.

Comparison of Euler and Runge–Kutta Methods

11. Solve $y' = 2x^{-1}\sqrt{y - \ln x} + x^{-1}$, $y(1) = 0$ for $1 \le x \le 1.8$ by Euler's method with $h = 0.1$. Verify that the exact solution is $y = (\ln x)^2 + \ln x$ and compute the error.

12. Solve Prob. 11 by the improved Euler method with $h = 0.2$, determine the error, compare with Prob. 11, and comment. Note that this is a fair comparison because here we evaluate $f(x, y)$ eight times (4 steps with 2 evaluations each), just as in Prob. 11.

13. Solve Prob. 11 by the classical Runge–Kutta method with $h = 0.4$, determine the error, and compare with Prob. 11. (Note that these 2 Runge–Kutta steps require 8 evaluations of $f(x, y)$, just as many as in Prob. 11.)

14. Solve Prob. 11 by the classical Runge–Kutta method with $h = 0.1$ and compare the error with that in Prob. 11.

15. Apply Euler's method and the improved Euler method with $h = 0.1$ and 10 steps to the initial value problem $y' = 2 - 2y$, $y(0) = 0$, and determine and compare the errors.

16. Apply the classical Runge–Kutta method with $h = 0.1$ and 10 steps to Prob. 15 and determine and compare the errors with those in Prob. 15.

17. Solve $y' = -0.2xy$, $y(0) = 1$ by the classical Runge–Kutta method (10 steps, $h = 0.2$). Determine the errors.

18. Solve Prob. 17 by the Euler and improved Euler methods (10 steps, $h = 0.2$) and compare the errors with those in Prob. 17.

19. Verify the error estimate (10) in Table 19.7.

20. **CAS PROJECT. RKF.** (**a**) Write a program for RKF that gives x_n, y_n, ϵ_n and, if the solution is known, the actual error.

(**b**) Apply the program to Example 4 in the text (10 steps, $h = 0.1$).

(**c**) ϵ_n in (b) gives a relatively good idea of the size of the actual error. Is this typical or accidental? Find out by experimentation with other problems on what properties of the equation or solution this might depend.

19.2 Multistep Methods

A **one-step method** is a method that in each step uses only values obtained in a single step, namely, in the preceding step. All the methods in Sec. 19.1 are one-step methods. In contrast, a **multistep method** is a method that in each step uses values from more than one of the preceding steps. The reason for using the multistep approach is that the additional information might increase accuracy. Such methods are obtained as follows.

Adams–Bashforth Methods

We consider an initial value problem

(**1**)
$$\boxed{y' = f(x, y), \qquad y(x_0) = y_0}$$

as before, with f such that the problem has a unique solution on some interval containing x_0. We integrate $y' = f(x, y)$ from x_n to $x_{n+1} = x_n + h$. This gives

$$\int_{x_n}^{x_{n+1}} y'(x)\, dx = y(x_{n+1}) - y(x_n) = \int_{x_n}^{x_{n+1}} f(x, y(x))\, dx.$$

Now comes the main idea. We replace $f(x, y(x))$ by an interpolation polynomial $p(x)$ (see Sec. 17.3), so that we can later integrate. This gives approximations y_{n+1} of $y(x_{n+1})$ and y_n of $y(x_n)$,

(2)
$$y_{n+1} = y_n + \int_{x_n}^{x_{n+1}} p(x)\, dx.$$

Different choices of $p(x)$ will now produce different methods. We explain the principle by taking a cubic polynomial, namely, the polynomial $p_3(x)$ that at (equidistant) x_n, x_{n-1}, x_{n-2}, x_{n-3} has the respective values

(3)
$$f_n = f(x_n, y_n), \quad f_{n-1} = f(x_{n-1}, y_{n-1}),$$
$$f_{n-2} = f(x_{n-2}, y_{n-2}), \quad f_{n-3} = f(x_{n-3}, y_{n-3}).$$

This will lead to a practically useful formula. We can obtain $p_3(x)$ from the Newton backward difference formula (18), Sec. 17.3:

$$p_3(x) = f_n + r\nabla f_n + \tfrac{1}{2}r(r + 1)\nabla^2 f_n + \tfrac{1}{6}r(r + 1)(r + 2)\nabla^3 f_n$$

where $r = (x - x_n)/h$. We integrate $p_3(x)$ over x from x_n to $x_{n+1} = x_n + h$, thus over r from 0 to 1. Since $x = x_n + hr$, we have $dx = h\,dr$. The integral of $\tfrac{1}{2}r(r + 1)$ is 5/12 and that of $\tfrac{1}{6}r(r + 1)(r + 2)$ is 3/8. We thus obtain

(4)
$$\int_{x_n}^{x_{n+1}} p_3\,dx = h\int_0^1 p_3\,dr = h\left(f_n + \frac{1}{2}\nabla f_n + \frac{5}{12}\nabla^2 f_n + \frac{3}{8}\nabla^3 f_n\right).$$

It is practical to replace these differences by their expressions in terms of f:

$$\nabla f_n = f_n - f_{n-1}$$
$$\nabla^2 f_n = f_n - 2f_{n-1} + f_{n-2}$$
$$\nabla^3 f_n = f_n - 3f_{n-1} + 3f_{n-2} - f_{n-3}.$$

We substitute this into (4) and collect terms. We then obtain the multistep formula of the **Adams–Bashforth method** *of fourth order*

(5)
$$y_{n+1} = y_n + \frac{h}{24}(55f_n - 59f_{n-1} + 37f_{n-2} - 9f_{n-3}).$$

It expresses the new value y_{n+1} [approximation of the solution y of (1) at x_{n+1}] in terms of 4 values of f computed from the y-values obtained in the preceding 4 steps. The local truncation error is of order h^5, as can be shown, so that the global error is of order h^4; hence (5) does define a fourth-order method.

Adams–Moulton Methods

These methods are obtained if for $p(x)$ in (2) we choose a polynomial that interpolates $f(x, y(x))$ at x_{n+1}, x_n, x_{n-1}, $\cdots$ (as opposed to x_n, x_{n-1}, $\cdots$ used before; this is the main point). We explain the principle for the cubic polynomial $\tilde{p}_3(x)$ that interpolates at x_{n+1}, x_n, x_{n-1}, x_{n-2}. (Before we had x_n, x_{n-1}, x_{n-2}, x_{n-3}.) Again using (18) in Sec. 17.3 and setting $r = (x - x_{n+1})/h$, we have

$$\tilde{p}_3(x) = f_{n+1} + r\nabla f_{n+1} + \tfrac{1}{2}r(r + 1)\nabla^2 f_{n+1} + \tfrac{1}{6}r(r + 1)(r + 2)\nabla^3 f_{n+1}.$$

We integrate over x from x_n to x_{n+1} as before. This now corresponds to integrating over r from -1 to 0. We obtain

$$\int_{x_n}^{x_{n+1}} \tilde{p}_3(x) \, dx = h \left(f_{n+1} - \frac{1}{2} \nabla f_{n+1} - \frac{1}{12} \nabla^2 f_{n+1} - \frac{1}{24} \nabla^3 f_{n+1} \right).$$

Replacing the differences as before gives

(6) $\quad y_{n+1} = y_n + \int_{x_n}^{x_{n+1}} \tilde{p}_3(x) \, dx = y_n + \frac{h}{24} (9f_{n+1} + 19f_n - 5f_{n-1} + f_{n-2}).$

This is usually called an **Adams–Moulton formula.** It is an **implicit formula** because $f_{n+1} = f(x_{n+1}, y_{n+1})$ appears on the right, so that it defines y_{n+1} only *implicitly,* in contrast to (5), which is an **explicit formula,** not involving y_{n+1} on the right. To use (6) we must *predict* a value y_{n+1}^*, for instance, by using (5), that is,

(7a) $$y_{n+1}^* = y_n + \frac{h}{24} (55f_n - 59f_{n-1} + 37f_{n-2} - 9f_{n-3}).$$

The corrected new value y_{n+1} is then obtained from (6) with f_{n+1} replaced by $f_{n+1}^* = f(x_{n+1}, y_{n+1}^*)$ and the other f's as in (6); thus,

(7b) $$y_{n+1} = y_n + \frac{h}{24} (9f_{n+1}^* + 19f_n - 5f_{n-1} + f_{n-2}).$$

This predictor–corrector method (7a), (7b) is often called the **Adams–Moulton method** *of fourth order.* Sometimes this name is reserved for the corresponding method with several corrections per step by (7b) until a specified accuracy is reached. Popular codes exist for these two versions of the method.

Getting started. Whereas one-step methods are *"self-starting"* (need no starting data beyond the given initial condition), multistep methods are not. This is an important point— and a disadvantage! In (5) we need f_0, f_1, f_2, f_3; thus, according to (3) we must first compute y_1, y_2, y_3 by some other method of great accuracy, for instance, by a Runge–Kutta or a Runge–Kutta–Fehlberg method. For other possibilities see Ref. [E14] listed in Appendix 1.

EXAMPLE 1   **Adams–Bashforth prediction (7a), Adams–Moulton correction (7b)**

Solve the initial value problem

(8) $$y' = x + y, \qquad y(0) = 0$$

by (7a), (7b) on the interval $0 \leqq x \leqq 2$, choosing $h = 0.2$.

Solution.   The problem is the same as in Examples 1–3, Sec. 19.1, so that we can compare the results. We compute starting values y_1, y_2, y_3 by the classical Runge–Kutta method. Then in each step we predict by (7a) and make one correction by (7b) before we execute the next step. The results are shown and compared with the exact values in Table 19.9. We see that the corrections improve the accuracy considerably. This is typical. ◀

Table 19.9
Adams–Moulton Method Applied to the Initial Value Problem (8);
Predicted Values Computed by (7a) and Corrected Values by (7b)

n	x_n	Starting y_n	Predicted y_n^*	Corrected y_n	Exact Values	$10^6 \times$ Error of y_n
0	0.0	0.000 000			0.000 000	0
1	0.2	0.021 400			0.021 403	3
2	0.4	0.091 818			0.091 825	7
3	0.6	0.222 107			0.222 119	12
4	0.8		0.425 361	0.425 529	0.425 541	12
5	1.0		0.718 066	0.718 270	0.718 282	12
6	1.2		1.119 855	1.120 106	1.120 117	11
7	1.4		1.654 885	1.655 191	1.655 200	9
8	1.6		2.352 653	2.353 026	2.353 032	6
9	1.8		3.249 190	3.249 646	3.249 647	1
10	2.0		4.388 505	4.389 062	4.389 056	−6

Comments on Comparison of Methods. An Adams–Moulton formula is generally much more accurate than an Adams–Bashforth formula of the same order. This justifies the greater complication and expense in using the former. The method (7a), (7b) is **numerically stable,** whereas the exclusive use of (7a) might cause instability. Step size control is relatively simple. If $|\text{Corrector} - \text{Predictor}| > \text{TOL}$, use interpolation to generate "old" results at half the current step size and then try $h/2$ as the new step.

Whereas the method (7a), (7b) needs only 2 evaluations per step, Runge–Kutta needs 4; however, with Runge–Kutta one may be able to take a step size more than twice as large, so that a comparison of this kind (widespread in the literature) is meaningless.

For more details, see Refs. [E13], [E14] listed in Appendix 1.

PROBLEM SET 19.2

Adams–Moulton Method (7a), (7b)

Solve the following initial value problems by the Adams–Moulton method (1 correction per step). Compute the errors by using the exact solution. (The given starting values should help you to spend your time entirely on the new method. Use the Runge–Kutta method where no such values are given.) Show the details of your work.

1. $y' = x + y$, $y(0) = 0$, 10 steps, $h = 0.1$ (0.00517083, 0.0214026, 0.0498585)
2. $y' = y$, $y(0) = 1$, 10 steps, $h = 0.1$ (1.105171, 1.221403, 1.349859)
3. $y' = (y - x - 1)^2 + 2$, $y(0) = 1$, 7 steps, $h = 0.1$ (1.20033, 1.40271, 1.60934)
4. $y' = 2x^{-1}\sqrt{y - \ln x} + x^{-1}$, $y(1) = 0$, $x = 1, \cdots, 1.8$, $h = 0.1$ (0.104 394, 0.215 563, 0.331 199)
5. $y' = -0.2xy$, $y(0) = 1$, 10 steps, $h = 0.2$.
6. $y' = x/y$, $y(1) = 3$, 10 steps, $h = 0.2$
7. Carry out and show the details leading to (4)−(7) in the text.
8. Solve the equation in Prob. 3 exactly and find the reason for the restriction to 7 steps.
9. **CAS PROJECT. Adams–Moulton.** (a) **Accurate starting** is important in (7a), (7b). Illustrate this in Example 1 of the text by using starting values from the improved Euler–Cauchy method and compare the results with those in Table 19.9.

(b) How much does the error in Prob. 1 decrease if you use exact starting values (instead of RK-values)?

(c) Experiment to find out for what differential equations poor starting is very damaging and for what equations it is not.

(d) The classical **RK method** often gives the same accuracy with step $2h$ as Adams–Moulton with step h, so that the total number of function evaluations is the same in both cases. Illustrate this with Prob. 6. (Hence corresponding comparisons in the literature in favor of Adams–Moulton are not valid.)

10. Show that by applying the method in the text to a polynomial of second degree we obtain the predictor and corrector formulas

$$y_{n+1}^* = y_n + \frac{h}{12}(23f_n - 16f_{n-1} + 5f_{n-2}),$$

$$y_{n+1} = y_n + \frac{h}{12}(5f_{n+1} + 8f_n - f_{n-1}).$$

11. Use Prob. 10 to solve $y' = 2xy$, $y(0) = 1$ (10 steps, $h = 0.1$, RK starting values). Compare with the exact solution and comment.

12. How much can you reduce the error in Prob. 11 by halving h (20 steps, $h = 0.05$)? First guess, then compute.

13. Use Prob. 10 to solve $y' = x + y$, $y(0) = 0$ (5 steps, $h = 0.2$). Compare with the exact solution.

14. Apply Prob. 10 to $y' = (x + y - 4)^2$, $y(0) = 4$ (7 steps, $h = 0.2$). Solve exactly and determine the error.

15. WRITING PROJECT. One-Step and Multistep Methods. Compare the two kinds of methods in a short essay, using your own formulations on relevant material from Secs. 19.1 and 19.2 and the problems.

19.3 Methods for Systems and Higher Order Equations

We consider initial value problems for first-order systems[2]

(1)
$$\boxed{\mathbf{y}' = \mathbf{f}(x, \mathbf{y}), \qquad \mathbf{y}(x_0) = \mathbf{y}_0,}$$

thus

$$y_1' = f_1(x, y_1, \cdots, y_m)$$

$$y_2' = f_2(x, y_1, \cdots, y_m)$$

$$\cdots\cdots\cdots\cdots\cdots\cdots\cdots$$

$$y_m' = f_m(x, y_1, \cdots, y_m).$$

We assume $\mathbf{f}$ to be such that the problem has a unique solution in some x-interval containing x_0. Equation (1) includes initial value problems for single mth order differential equations

---

[2]This section is practically independent of Chap. 3 on systems.

(2)
$$y^{(m)} = f(x, y, y', y'', \cdots, y^{(m-1)})$$

and initial conditions $y(x_0) = K_1$, $y'(x_0) = K_2, \cdots, y^{(m-1)}(x_0) = K_m$ as special cases. Indeed, we may set

(3)
$$y_1 = y, \qquad y_2 = y', \qquad y_3 = y'', \quad \cdots, \quad y_m = y^{(m-1)}.$$

Then we obtain

(4)
$$y_1' = y_2$$
$$y_2' = y_3$$
$$\vdots$$
$$y_{m-1}' = y_m$$
$$y_m' = f(x, y_1, \cdots, y_m)$$

and the initial conditions $y_1(x_0) = K_1$, $y_2(x_0) = K_2, \cdots, y_m(x_0) = K_m$.

Euler's Method for Systems

Methods for single first-order equations can be extended to systems (1) by writing vector functions $\mathbf{y}$ and $\mathbf{f}$ instead of scalar functions y and f, whereas x remains a scalar variable.

EXAMPLE 1 **Euler's method for a second-order equation. Mass–spring system**

Solve the initial value problem for a damped mass–spring system

$$y'' + 2y' + 0.75y = 0, \qquad y(0) = 3, \qquad y'(0) = -2.5$$

by Euler's method for systems with step $h = 0.2$ for x from 0 to 1.

Solution.   **Euler's method** (3), Sec. 19.1, generalizes to systems

(5)
$$\mathbf{y}_{n+1} = \mathbf{y}_n + h\,\mathbf{f}(x_n, \mathbf{y}_n),$$

in components

$$y_{1,n+1} = y_{1,n} + h f_1(x_n, y_{1,n}, y_{2,n})$$
$$y_{2,n+1} = y_{2,n} + h f_2(x_n, y_{1,n}, y_{2,n})$$

and similarly for systems of more than two equations. By (4) the given equation converts to the system

$$y_1' = f_1(x, y_1, y_2) = y_2$$
$$y_2' = f_2(x, y_1, y_2) = -2y_2 - 0.75y_1.$$

Hence (5) becomes

$$y_{1,n+1} = y_{1,n} + 0.2y_{2,n}$$
$$y_{2,n+1} = y_{2,n} + 0.2\,(-2y_{2,n} - 0.75y_{1,n}).$$

Table 19.10
Euler's Method for Systems in Example 1 (Mass–Spring System)

n	x_n	$y_{1,n}$	y_1 Exact (5D)	Error $\epsilon_1 = y_1 - y_{1,n}$	$y_{2,n}$	y_2 Exact (5D)	Error $\epsilon_2 = y_2 - y_{2,n}$
0	0.0	3.00000	3.00000	0.00000	-2.50000	-2.50000	0.00000
1	0.2	2.50000	2.55049	0.05049	-1.95000	-2.01606	-0.06606
2	0.4	2.11000	2.18627	0.76270	-1.54500	-1.64195	-0.09695
3	0.6	1.80100	1.88821	0.08721	-1.24350	-1.35067	-0.10717
4	0.8	1.55230	1.64183	0.08953	-1.01625	-1.12211	-0.10586
5	1.0	1.34905	1.43619	0.08714	-0.84260	-0.94123	-0.09863

The initial conditions are $y(0) = y_1(0) = 3$, $y'(0) = y_2(0) = -2.5$. The calculations are shown in Table 19.10. As for single equations, the results would not be accurate enough for practical purposes. Of course, the example merely serves to illustrate the method because the problem can be readily solved exactly,

$$y = y_1 = 2\,e^{-0.5x} + e^{-1.5x}, \qquad \text{thus} \qquad y' = y_2 = -e^{-0.5x} - 1.5\,e^{-1.5x}. \qquad \blacktriangleleft$$

Runge–Kutta Methods for Systems

These are obtained for initial value problems (1) simply by writing vector formulas for vectors with m components that for $m = 1$ reduce to the previous scalar formulas. Hence for the *classical* **Runge–Kutta method** *of fourth order* in Table 19.4 we obtain

(6a)
$$\mathbf{y}(x_0) = \mathbf{y}_0 \qquad \text{(Initial values)}$$

and for $n = 0, 1, \cdots, N - 1$ (N the number of steps) we obtain the auxiliary quantities

(6b)
$$\begin{aligned}
\mathbf{k}_1 &= h\,\mathbf{f}(x_n, \quad \mathbf{y}_n) \\
\mathbf{k}_2 &= h\,\mathbf{f}(x_n + \tfrac{1}{2}h, \quad \mathbf{y}_n + \tfrac{1}{2}\mathbf{k}_1) \\
\mathbf{k}_3 &= h\,\mathbf{f}(x_n + \tfrac{1}{2}h, \quad \mathbf{y}_n + \tfrac{1}{2}\mathbf{k}_2) \\
\mathbf{k}_4 &= h\,\mathbf{f}(x_n + h, \quad \mathbf{y}_n + \mathbf{k}_3)
\end{aligned}$$

and the new value [approximation of the solution $\mathbf{y}(x)$ at $x_{n+1} = x_0 + (n + 1)h$]

(6c)
$$\mathbf{y}_{n+1} = \mathbf{y}_n + \tfrac{1}{6}(\mathbf{k}_1 + 2\mathbf{k}_2 + 2\mathbf{k}_3 + \mathbf{k}_4).$$

EXAMPLE 2   **Runge–Kutta method for systems. Airy's equation. Airy function Ai(x)**

Solve the initial value problem

$$y'' = xy, \qquad y(0) = 1/(3^{2/3} \cdot \Gamma(2/3)) = 0.35502\,805, \qquad y'(0) = -1/(3^{1/3} \cdot \Gamma(1/3)) = -0.25881\,940$$

by the Runge–Kutta method for systems with $h = 0.2$; do 5 steps. This is **Airy's equation**,[3] which arose in optics (see Ref. [A7], p. 188, listed in Appendix 1). Γ is the gamma function (see Appendix A3.1). The initial

---

[3]Named after Sir GEORGE BIDELL AIRY (1801—1892), English mathematician, who is known for his work in elasticity and in partial differential equations.

conditions are such that we obtain a standard solution, the **Airy function** $\mathrm{Ai}(x)$, which has been tabulated and investigated (see Ref. [1], pp. 446, 475).

Solution. For $y'' = xy$, setting $y_1 = y$, $y_2 = y_1' = y'$ we obtain the system (4)

$$y_1' = y_2$$
$$y_2' = xy_1.$$

Hence $\mathbf{f} = [f_1 \quad f_2]^{\mathsf{T}}$ in (1) has the components

$$f_1(x, y) = y_2, \qquad f_2(x, y) = xy_1.$$

We now write (6) in components. The initial conditions (6a) are $y_{1,0} = 0.35502\,805$, $y_{2,0} = -0.25881\,940$. In (6b) we have fewer subscripts by simply writing

$$\mathbf{k}_1 = \mathbf{a}, \qquad \mathbf{k}_2 = \mathbf{b}, \qquad \mathbf{k}_3 = \mathbf{c}, \qquad \mathbf{k}_4 = \mathbf{d}$$

so that $\mathbf{a} = [a_1 \quad a_2]^{\mathsf{T}}$, etc. Then (6b) takes the form

(6b*)

$$\mathbf{a} = h \begin{bmatrix} y_{2,n} \\ x_n y_{1,n} \end{bmatrix}$$

$$\mathbf{b} = h \begin{bmatrix} y_{2,n} + \tfrac{1}{2}a_2 \\ (x_n + \tfrac{1}{2}h)(y_{1,n} + \tfrac{1}{2}a_1) \end{bmatrix}$$

$$\mathbf{c} = h \begin{bmatrix} y_{2,n} + \tfrac{1}{2}b_2 \\ (x_n + \tfrac{1}{2}h)(y_{1,n} + \tfrac{1}{2}b_1) \end{bmatrix}$$

$$\mathbf{d} = h \begin{bmatrix} y_{2,n} + c_2 \\ (x_n + h)(y_{1,n} + c_1) \end{bmatrix}.$$

For example, the second component of $\mathbf{b}$ is obtained as follows. $\mathbf{f}(x, \mathbf{y})$ has the second component $f_2(x, \mathbf{y}) = xy_1$. Now in $\mathbf{b}\ (= \mathbf{k}_2)$ the first argument is $x = x_n + \tfrac{1}{2}h$. The second argument in $\mathbf{b}$ is $\mathbf{y} = \mathbf{y}_n + \tfrac{1}{2}\mathbf{a}$, and the first component of this is $y_1 = y_{1,n} + \tfrac{1}{2}a_1$. Together, $xy_1 = (x_n + \tfrac{1}{2}h)(y_{1,n} + \tfrac{1}{2}a_1)$. Similarly for the other components in (6b*). Finally,

(6c*) $$\mathbf{y}_{n+1} = \mathbf{y}_n + \tfrac{1}{6}(\mathbf{a} + 2\mathbf{b} + 2\mathbf{c} + \mathbf{d}).$$

Table 19.11 shows the values $y(x) = y_1(x)$ of the Airy function $\mathrm{Ai}(x)$ and of its derivative $y'(x) = y_2(x)$ as well as of the (rather small!) error of $y(x)$. ◄

Table 19.11
RK Method for Systems: Values $y_{1,n}(x_n)$ of the Airy Function $\mathrm{Ai}(x)$ in Example 2

n	x_n	$y_{1,n}(x_n)$	$y_1(x_n)$ Exact (8D)	$10^8 \times$ Error of y_1	$y_{2,n}(x_n)$
0	0.0	0.35502 805	0.35502 805	0	−0.25881 940
1	0.2	0.30370 303	0.30370 315	12	−0.25240 464
2	0.4	0.25474 211	0.25474 235	24	−0.23583 073
3	0.6	0.20979 973	0.20980 006	33	−0.21279 185
4	0.8	0.16984 596	0.16984 632	36	−0.18641 171
5	1.0	0.13529 207	0.13529 242	35	−0.15914 687

Runge–Kutta–Nyström Methods (RKN Methods)

RKN methods are direct extensions of RK methods (Runge–Kutta methods) to second-order differential equations $y'' = f(x, y, y')$, as given by the Finnish mathematician E. J. Nyström [*Acta Soc. Sci. fenn.*, 1925, L, No. 13]. The best known of these uses the following formulas, where $n = 0, 1, \cdots, N - 1$ (N the number of steps):

(7a)
$$k_1 = \tfrac{1}{2}hf(x_n, y_n, y_n')$$
$$k_2 = \tfrac{1}{2}hf(x_n + \tfrac{1}{2}h, y_n + K, y_n' + k_1) \qquad \text{where } K = \tfrac{1}{2}h(y_n' + \tfrac{1}{2}k_1)$$
$$k_3 = \tfrac{1}{2}hf(x_n + \tfrac{1}{2}h, y_n + K, y_n' + k_2)$$
$$k_4 = \tfrac{1}{2}hf(x_n + h, y_n + L, y_n' + 2k_3) \qquad \text{where } L = h(y_n' + k_3).$$

From this we compute the approximation y_{n+1} of $y(x_{n+1})$ at $x_{n+1} = x_0 + (n + 1)h$,

(7b)
$$y_{n+1} = y_n + h(y_n' + \tfrac{1}{3}(k_1 + k_2 + k_3)),$$

and the approximation y_{n+1}' of the derivative $y'(x_{n+1})$ needed in the next step,

(7c)
$$y_{n+1}' = y_n' + \tfrac{1}{3}(k_1 + 2k_2 + 2k_3 + k_4).$$

The method is particularly advantageous for $y'' = f(x, y)$ with f not involving y' because then $k_2 = k_3$ in (7), so that the number of function evaluations is reduced. Indeed, then

(7*)
$$k_1 = \tfrac{1}{2}hf(x_n, y_n)$$
$$k_2 = \tfrac{1}{2}hf(x_n + \tfrac{1}{2}h, y_n + \tfrac{1}{2}h(y_n' + \tfrac{1}{2}k_1)) = k_3$$
$$k_4 = \tfrac{1}{2}hf(x_n + h, y_n + h(y_n' + k_2))$$
$$y_{n+1} = y_n + h(y_n' + \tfrac{1}{3}(k_1 + 2k_2))$$
$$y_{n+1}' = y_n' + \tfrac{1}{3}(k_1 + 4k_2 + k_4).$$

EXAMPLE 3 **Runge–Kutta–Nyström method. Airy's equation. Airy function Ai(x)**

For the problem in Example 2 and $h = 0.2$ as before we obtain from (7*) simply $k_1 = 0.1 x_n y_n$ and

$$k_2 = k_3 = 0.1(x_n + 0.1)(y_n + 0.1 y_n' + 0.05 k_1), \qquad k_4 = 0.1(x_n + 0.2)(y_n + 0.2 y_n' + 0.2 k_2).$$

Table 19.12 shows the results. The accuracy is the same as in Example 2, but the work was much less. ◀

Table 19.12
Runge–Kutta–Nyström Method Applied to Airy's Equation,
Computation of the Airy Function $y = $ Ai(x)

x_n	y_n	y_n'	$y(x)$ Exact (8D)	$10^8 \times$ Error of y_n
0.0	0.355 028 05	−0.258 819 40	0.355 028 05	0
0.2	0.303 703 04	−0.252 404 64	0.303 703 15	11
0.4	0.254 742 11	−0.235 830 70	0.254 742 35	24
0.6	0.209 799 74	−0.212 791 72	0.209 800 06	32
0.8	0.169 845 99	−0.186 411 34	0.169 846 32	33
1.0	0.135 292 18	−0.159 146 09	0.135 292 42	24

Our work in Examples 2 and 3 also illustrates that methods for differential equations are often useful in the tabulation of **"higher transcendental functions."**

PROBLEM SET 19.3

In each of these problems show the details of your work.

1. **(Euler for systems)** Solve $y_1' = 2y_1 - 4y_2$, $y_2' = y_1 - 3y_2$, $y_1(0) = 3$, $y_2(0) = 0$ by (5) (10 steps, $h = 0.1$). Plot the solution in the y_1y_2-plane.

2. **(Spiral)** Solve $y_1' = -y_1 + y_2$, $y_2' = -y_1 - y_2$, $y_1(0) = 0$, $y_2(0) = 4$, by Euler's method (5 steps, $h = 0.2$). Plot the solution in the y_1y_2-plane.

3. **(Second-order equation)** Apply Euler's method to $y'' - y = x$, $y(0) = 1$, $y'(0) = -2$ (4 steps, $h = 0.1$). Compare with the exact solution.

4. Apply Euler's method to $y'' = xy' - 3y$, $y(0) = 0$, $y'(0) = -3$ (5 steps, $h = 0.05$). Verify the solution $y = x^3 - 3x$. Compute the error.

5. **(RK for systems)** Solve Prob. 1 by (6) (2 steps, $h = 0.5$) and compare.

6. **(Second-order equation)** Solve Prob. 3 by RK in (6) (3 steps, $h = 0.1$), find the error, and compare with Prob. 3.

7. Apply RK to $x^2y'' - 2.5xy' - 2y = 0$, $y(1) = 0.3$, $y'(1) = 1.2$ (3 steps, $h = 0.2$). Find the error.

8. **(Bessel function J_0)** Apply RK in (6) to the (special) Bessel equation $xy'' + y' + xy = 0$, $y(1) = 0.765198$, $y'(1) = -0.440051$ (5 steps, $h = 0.5$). (These values give the standard solution $J_0(x)$ in Fig. 103, Sec. 4.5.)

9. **(Undamped pendulum)** Apply RK in (6) to $y'' + \sin y = 0$, $y(\pi) = 0$, $y'(\pi) = 1$, (5 steps, $h = 0.2$). How does your result fit into Fig. 89, Sec. 3.5?

10. **(Roughest method for second-order equations)** Show that for solving $y'' = f(x, y, y')$, $y(x_0) = y_0$, $y'(x_0) = y_0'$ the truncation of the Taylor series suggests

$$y_{n+1} = y_n + hy_n' + \tfrac{1}{2}h^2y_n'', \qquad y_{n+1}' = y_n' + hy_n'', \qquad y_{n+1}'' = f(x_{n+1}, y_{n+1}').$$

Apply this to $y'' = \tfrac{1}{2}(x + y + y' + 2)$, $y(0) = 0$, $y'(0) = 0$ (5 steps, $h = 0.2$) and calculate the error.

11. **(RKN)** Verify the formulas and computations for the Airy equation in Example 3.

12. Verify, to 3S-accuracy, the values y_0 and y_0' in Example 2, using (25) in Appendix A3.1 and Table A2 in Appendix 5.

13. Apply RKN to $y'' = xy' - 4y$, $y(0) = 3$, $y'(0) = 0$ (5 steps, $h = 0.2$). Exact: $y = x^4 - 6x^2 + 3$.

14. **(Bessel function)** Do Prob. 8 by RKN. Compare the accuracy.

15. **CAS PROJECT. Comparison of Methods.** **(a)** Write programs for RKN and RK for systems. **(b)** Try them out with second-order equations of your choice to find out empirically which is better in specific cases. **(c)** In using RKN, would it pay to first eliminate y' (see Prob. 29 in Problem Set 4.5)? Find out experimentally.

19.4 Methods for Elliptic Partial Differential Equations

The remaining sections of this chapter are devoted to numerical methods for partial differential equations, particularly for the Laplace, Poisson, heat, and wave equations, which are basic in applications and, at the same time, are model cases of elliptic, parabolic, and hyperbolic equations. The definitions are as follows.

A partial differential equation is called **quasilinear** if it is linear in the highest derivatives. Hence a second-order quasilinear equation in two independent variables x, y can be written

$$(1) \qquad au_{xx} + 2bu_{xy} + cu_{yy} = F(x, y, u, u_x, u_y).$$

u is the unknown function. This equation is said to be of

elliptic type	if $ac - b^2 > 0$	(example: *Laplace equation*)
parabolic type	if $ac - b^2 = 0$	(example: *heat equation*)
hyperbolic type	if $ac - b^2 < 0$	(example: *wave equation*)

(where in the heat and wave equations, y is time t). Here, the *coefficients a, b, c* may be functions of x, y, so that the type of (1) may be different in different regions of the xy-plane. This classification is not merely a formal matter but is of great practical importance because the general behavior of solutions differs from type to type and so do the additional conditions (boundary and initial conditions) that must be taken into account.

Applications involving *elliptic equations* usually lead to boundary value problems in a region R, called a *first boundary value problem* or **Dirichlet problem** if u is prescribed on the boundary curve C of R, a *second boundary value problem* or **Neumann problem** if $u_n = \partial u/\partial n$ (normal derivative of u) is prescribed on C, and a *third* or **mixed problem** if u is prescribed on a part of C and u_n on the remaining part. C usually is a closed curve (or sometimes consists of two or more such curves).

Difference Equations for the Laplace and Poisson Equations

In this section we consider the **Laplace equation**

$$(2) \qquad \boxed{\nabla^2 u = u_{xx} + u_{yy} = 0}$$

and the **Poisson equation**

$$(3) \qquad \boxed{\nabla^2 u = u_{xx} + u_{yy} = f(x, y).}$$

These are the most important elliptic equations in applications. To obtain methods of numerical solution, we replace the partial derivatives by corresponding difference quotients, as follows. By the Taylor formula,

$$(4) \quad \begin{aligned} &\text{(a)} \ \ u(x + h, y) = u(x, y) + hu_x(x, y) + \tfrac{1}{2}h^2 u_{xx}(x, y) + \tfrac{1}{6}h^3 u_{xxx}(x, y) + \cdots \\ &\text{(b)} \ \ u(x - h, y) = u(x, y) - hu_x(x, y) + \tfrac{1}{2}h^2 u_{xx}(x, y) - \tfrac{1}{6}h^3 u_{xxx}(x, y) + \cdots. \end{aligned}$$

We subtract (4b) from (4a), neglect terms in h^3, h^4, $\cdots$, and solve for u_x. Then

(5a)
$$u_x(x, y) \approx \frac{1}{2h}[u(x + h, y) - u(x - h, y)].$$

Similarly,

$$u(x, y + k) = u(x, y) + ku_y(x, y) + \tfrac{1}{2}k^2 u_{yy}(x, y) + \cdots$$

$$u(x, y - k) = u(x, y) - ku_y(x, y) + \tfrac{1}{2}k^2 u_{yy}(x, y) + \cdots.$$

By subtracting, neglecting terms in k^3, k^4, $\cdots$, and solving for u_y we obtain

(5b)
$$u_y(x, y) \approx \frac{1}{2k}[u(x, y + k) - u(x, y - k)].$$

We turn to second derivatives. Adding (4a) and (4b) and neglecting terms in h^4, h^5, $\cdots$, we obtain

$$u(x + h, y) + u(x - h, y) \approx 2u(x, y) + h^2 u_{xx}(x, y).$$

Solving for u_{xx} we have

(6a)
$$u_{xx}(x, y) \approx \frac{1}{h^2}[u(x + h, y) - 2u(x, y) + u(x - h, y)].$$

Similarly,

(6b)
$$u_{yy}(x, y) \approx \frac{1}{k^2}[u(x, y + k) - 2u(x, y) + u(x, y - k)].$$

We shall not need (see Prob. 1)

(6c) $u_{xy}(x, y) \approx \dfrac{1}{4hk}[u(x + h, y + k) - u(x - h, y + k)$

$$- u(x + h, y - k) + u(x - h, y - k)].$$

Figure 419 shows the points $(x + h, y)$, $(x - h, y)$, $\cdots$ in (5) and (6).

We now substitute (6a) and (6b) into the *Poisson equation* (3), choosing $k = h$ to obtain a simple formula:

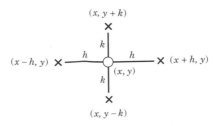

Fig. 419. Points in (5) and (6)

(7) $\boxed{u(x + h, y) + u(x, y + h) + u(x - h, y) + u(x, y - h) - 4u(x, y) = h^2 f(x, y).}$

This is a **difference equation** corresponding to (3). Hence for the **Laplace equation** (2) the corresponding difference equation is

(8) $\boxed{u(x + h, y) + u(x, y + h) + u(x - h, y) + u(x, y - h) - 4u(x, y) = 0.}$

h is called the **mesh size.** Equation (8) relates u at (x, y) to u at the four neighboring points shown in Fig. 420. For convenience, these neighboring points are often called E (East), N (North), W (West), S (South). Then Fig. 420 takes the form of Fig. 421 and (7) becomes

(7*) $\boxed{u(E) + u(N) + u(W) + u(S) - 4u(x, y) = h^2 f(x, y).}$

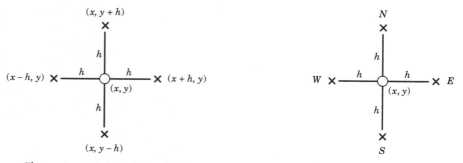

Fig. 420.   Points in (7) and (8)   **Fig. 421.**   Notation in (7*)

Our approximation of $h^2 \nabla^2 u$ in (7) and (8) is a 5-point approximation with the coefficient scheme or **stencil** (also called *pattern, molecule,* or *star*)

(9)
$$\left\{ \begin{matrix} & 1 & \\ 1 & -4 & 1 \\ & 1 & \end{matrix} \right\}.$$

And to grasp (7) at first sight, we may conveniently write it in terms of this stencil as

$$\left\{ \begin{matrix} & 1 & \\ 1 & -4 & 1 \\ & 1 & \end{matrix} \right\} u = h^2 f(x, y).$$

Note that (8) has a remarkable interpretation: u at (x, y) equals the mean of the values of u at the four neighboring points. This is an analog of the mean value property of harmonic functions (Sec. 16.6).

Dirichlet Problem

In the numerical solution of the Dirichlet problem (definition at the beginning of this section) in a region R we first choose h and introduce in R a grid consisting of equidistant horizontal and vertical straight lines of distance h. Their intersections are called **mesh points** (or *nodes* or *lattice points*). See Fig. 422. Then we use a difference equation approximating the given partial differential equation—formula (8) in the case of the Laplace equation—by which we relate the unknown values of u at the mesh points in R to each other and to the given boundary values, as will be discussed below. This yields a linear system of *algebraic* equations. By solving it we obtain approximations to the unknown values of u at the mesh points in R. We shall see that the number of equations equals the number of unknowns, that is, the number of mesh points in R. Since at each mesh point, u is only related to the values at the neighboring mesh points, the coefficients of the system form a **sparse matrix,** that is, a matrix with relatively few nonzero entries. In practice, this matrix will be large, since for obtaining high accuracy one needs many mesh points, and a 500×500 or larger matrix may cause a storage problem.[4] Hence an indirect method (see Sec. 18.3) is preferable to a direct one if the number of equations is large, say, greater than 50. In particular, we may use the **Gauss–Seidel method,** which in the present context is also called **Liebmann's method** in the older literature. The method is also convenient for storage because when a solution component has been computed, its old value is no longer needed and is overwritten.

We illustrate this approach with an example, keeping the number of equations small, for simplicity. As convenient *notations for mesh points and corresponding values of the solution* (and of approximate solutions) we use (see also Fig. 422)

$$(10) \qquad P_{ij} = (ih,\ jh), \qquad u_{ij} = u(ih,\ jh).$$

With this notation we can write (8) for any mesh point P_{ij} in the form

$$(11) \qquad \boxed{u_{i+1,j} + u_{i,j+1} + u_{i-1,j} + u_{i,j-1} - 4u_{ij} = 0.}$$

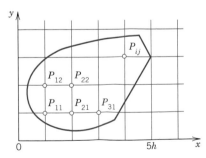

Fig. 422. Region in the xy-plane covered by a grid of mesh h, also showing mesh points $P_{11} = (h,\ h),\ \cdots,\ P_{ij} = (ih,\ jh),\ \cdots$

---

[4] The present matrix is *not* tridiagonal! (Definition below.) If it were, we could apply the Gauss elimination without causing that storage problem.

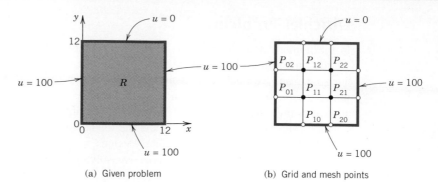

(a) Given problem　　　　(b) Grid and mesh points

Fig. 423.　Example 1

EXAMPLE 1　**Laplace equation. Liebmann's method**

The four sides of a square plate of side 12 cm made of homogeneous material are kept at constant temperature 0°C and 100°C as shown in Fig. 423a. Using a (very wide) grid of mesh 4 cm and applying Liebmann's method (that is, Gauss–Seidel iteration), find the (steady-state) temperature at the mesh points.

Solution.　In the case of independence of time, the heat equation (see Sec. 9.8)

$$u_t = c^2(u_{xx} + u_{yy})$$

reduces to the Laplace equation. Hence our problem is a Dirichlet problem for this equation. We choose the grid shown in Fig. 423b and consider the mesh points in the order $P_{11}, P_{21}, P_{12}, P_{22}$. We use (11) and, in each equation, take to the right all the terms resulting from the given boundary values. Then we obtain the system

(12)
$$
\begin{aligned}
-4u_{11} + u_{21} + u_{12} \qquad\quad &= -200 \\
u_{11} - 4u_{21} \qquad + u_{22} &= -200 \\
u_{11} \qquad\quad - 4u_{12} + u_{22} &= -100 \\
u_{21} + u_{12} - 4u_{22} &= -100.
\end{aligned}
$$

In practice, one would solve such a small system by the Gauss elimination, finding $u_{11} = u_{21} = 87.5$, $u_{12} = u_{22} = 62.5$.

More exact values (exact to 1D) of the solution of our problem are 88.1 and 61.9, respectively. (These were obtained by using Fourier series.) Hence the error is about 1%, which is surprisingly accurate for a grid of such a large mesh size h. If the system of equations were large, one would solve it by an indirect method, such as Liebmann's method. For (12) this is as follows. We write (12) in the form

$$
\begin{aligned}
u_{11} &= \qquad 0.25u_{21} + 0.25u_{12} \qquad\quad + 50 \\
u_{21} &= 0.25u_{11} \qquad\qquad\quad + 0.25u_{22} + 50 \\
u_{12} &= 0.25u_{11} \qquad\qquad\quad + 0.25u_{22} + 25 \\
u_{22} &= \qquad 0.25u_{21} + 0.25u_{12} \qquad\quad + 25.
\end{aligned}
$$

These equations are now used for the Gauss–Seidel iteration. They are identical with (2) in Sec. 18.3, where $u_{11} = x_1$, $u_{21} = x_2$, $u_{12} = x_3$, $u_{22} = x_4$, and the iteration is explained there, where 100, 100, 100, 100 are chosen as starting values. Using physical intuition on what the values at the mesh points might approximately be, one can save some work by choosing better starting values. You may verify that the exact solution of the system is $u_{11} = u_{21} = 87.5$, $u_{12} = u_{22} = 62.5$.

Remark.　It is interesting to note that if we choose mesh $h = L/n$ ($L =$ side of R) and consider the $(n-1)^2$ inner mesh points (i.e., mesh points not on the boundary) row by row in the order

$$P_{11}, P_{21}, \cdots, P_{n-1,1}, P_{12}, P_{22}, \cdots, P_{n-1,2}, \cdots,$$

then the system of equations has the $(n-1)^2 \times (n-1)^2$ coefficient matrix

$$(13) \quad \mathbf{A} = \begin{bmatrix} \mathbf{B} & \mathbf{I} & & & & \\ \mathbf{I} & \mathbf{B} & \mathbf{I} & & & \\ & & \cdot & & & \\ & & & \cdot & & \\ & & & & \cdot & \\ & & & \mathbf{I} & \mathbf{B} & \mathbf{I} \\ & & & & \mathbf{I} & \mathbf{B} \end{bmatrix} \quad \text{where} \quad \mathbf{B} = \begin{bmatrix} -4 & 1 & & & \\ 1 & -4 & 1 & & \\ & & \cdot & & \\ & & & \cdot & \\ & & & \cdot & \\ & & 1 & -4 & 1 \\ & & & 1 & -4 \end{bmatrix}$$

is an $(n-1) \times (n-1)$ matrix. (In (12) we have $n = 3$, $(n-1)^2 = 4$ inner mesh points, two submatrices **B**, and two submatrices **I**.) The matrix **A** is **irreducibly diagonally dominant,** that is, by definition, it is irreducible and diagonally dominant (see Sec. 18.7, text and footnote 6) with strict inequality $|a_{jj}| > \Sigma|a_{jk}|$ for at least one j (summation over all the off-diagonal entries in row j. It can be shown that this suffices for **A** to be nonsingular). (In Fig. 423 we have $n = 3$, hence $n - 1 = 2$, so that (13) involves two 2×2 matrices **B** and two unit matrices.) ◀

A matrix is called a **band matrix** if it has all its nonzero entries on the main diagonal and on sloping lines parallel to it (separated by sloping lines of zeros or not). For example, **A** in (13) is a band matrix. Although the Gauss elimination does not preserve zeros between bands, it does not introduce nonzero entries outside the limits defined by the original bands. Hence a band structure is advantageous. In (13) it has been achieved by carefully ordering the mesh points.

ADI Method

A matrix is called a **tridiagonal matrix** if it has all its nonzero entries on the main diagonal and on the sloping parallels immediately above or below the diagonal. In this case the Gauss elimination is particularly simple.

This raises the question of whether in the solution of the Dirichlet problem for the Laplace or Poisson equations one could obtain a system of equations whose coefficient matrix is tridiagonal. The answer is yes, and a popular method of that kind, called the **ADI method** (*alternating direction implicit method*) was developed by Peaceman and Rachford. The idea is as follows. The pattern in (9) shows that we could obtain a tridiagonal matrix if there were only the three points in a row (or only the three points in a column). This suggests that we write (11) in the form

$$(14a) \qquad u_{i-1,j} - 4u_{ij} + u_{i+1,j} = -u_{i,j-1} - u_{i,j+1}$$

so that the left side belongs to y-row j and the right side to x-column i. Of course, we can also write (11) in the form

$$(14b) \qquad u_{i,j-1} - 4u_{ij} + u_{i,j+1} = -u_{i-1,j} - u_{i+1,j}$$

so that the left side belongs to column i and the right side to row j. In the ADI method we proceed by iteration. At every mesh point we choose an arbitrary starting value $u_{ij}^{(0)}$. In each step we compute new values at all mesh points. In one step we use an iteration formula resulting from (14a) and in the next step an iteration formula resulting from (14b), and so on in alternating order.

In detail: suppose approximations $u_{ij}^{(m)}$ have been computed. Then, to obtain the next approximations $u_{ij}^{(m+1)}$, we substitute the $u_{ij}^{(m)}$ **on the right** side of (14a) and solve for the $u_{ij}^{(m+1)}$ on the left side; that is, we use

(15a)
$$u_{i-1,j}^{(m+1)} - 4u_{ij}^{(m+1)} + u_{i+1,j}^{(m+1)} = -u_{i,j-1}^{(m)} - u_{i,j+1}^{(m)}.$$

We use this for a fixed j, that is, **for a fixed row j,** and for all internal mesh points in this row. This gives a linear system of N algebraic equations (N = number of internal mesh points per row) in N unknowns, the new approximations of u at these mesh points. Note that (15a) involves not only approximations computed in the previous step but also given boundary values. We solve the system (15a) (j fixed!) by the Gauss elimination. Then we go to the next row, obtain another system of N equations and solve it by Gauss, and so on, until all rows are done. In the next step we **alternate direction,** that is, we compute the next approximations $u_{ij}^{(m+2)}$ column by column from the $u_{ij}^{(m+1)}$ and the given boundary values, using a formula obtained from (14b) by substituting the $u_{ij}^{(m+1)}$ **on the right:**

(15b)
$$u_{i,j-1}^{(m+2)} - 4u_{ij}^{(m+2)} + u_{i,j+1}^{(m+2)} = -u_{i-1,j}^{(m+1)} - u_{i+1,j}^{(m+1)}.$$

For each fixed i, that is, **for each column,** this is a system of M equations (M = number of internal mesh points per column) in M unknowns, which we solve by the Gauss elimination. Then we go to the next column, and so on, until all columns are done.

Let us consider an example that merely serves to explain the entire method. (In practice one would solve this problem directly by the Gauss elimination.)

EXAMPLE 2 **Dirichlet problem. ADI method**

Explain the procedure and formulas of the ADI method in terms of the problem in Example 1, using the same grid and starting values 100, 100, 100, 100.

Solution.   While working, we keep an eye on Fig. 423b, p. 966 and the given boundary values. We obtain first approximations $u_{11}^{(1)}$, $u_{21}^{(1)}$, $u_{12}^{(1)}$, $u_{22}^{(1)}$ from (15a) with $m = 0$. We write boundary values contained in (15a) without an upper index, for better identification and to indicate that these given values remain the same during the iteration. From (15a) with $m = 0$ we have for $j = 1$ (first row) the system

$$(i = 1) \qquad u_{01} - 4u_{11}^{(1)} + u_{21}^{(1)} \qquad\quad = -u_{10} - u_{12}^{(0)}$$
$$(i = 2) \qquad\qquad u_{11}^{(1)} - 4u_{21}^{(1)} + u_{31} = -u_{20} - u_{22}^{(0)}.$$

The solution is $u_{11}^{(1)} = u_{21}^{(1)} = 100$. For $j = 2$ (second row) we obtain from (15a) the system

$$(i = 1) \qquad u_{02} - 4u_{12}^{(1)} + u_{22}^{(1)} \qquad\quad = -u_{11}^{(0)} - u_{13}$$
$$(i = 2) \qquad\qquad u_{12}^{(1)} - 4u_{22}^{(1)} + u_{32} = -u_{21}^{(0)} - u_{23}.$$

The solution is $u_{12}^{(1)} = u_{22}^{(1)} = 66.667$.

Second approximations $u_{11}^{(2)}$, $u_{21}^{(2)}$, $u_{12}^{(2)}$, $u_{22}^{(2)}$ are now obtained from (15b) with $m = 1$ by using the first approximations just computed and the boundary values. For $i = 1$ (first column) we obtain from (15b) the system

$$(j = 1) \qquad u_{10} - 4u_{11}^{(2)} + u_{12}^{(2)} \qquad\quad = -u_{01} - u_{21}^{(1)}$$
$$(j = 2) \qquad\qquad u_{11}^{(2)} - 4u_{12}^{(2)} + u_{13} = -u_{02} - u_{22}^{(1)}.$$

The solution is $u_{11}^{(2)} = 91.11$, $u_{12}^{(2)} = 64.44$. For $i = 2$ (second column) we obtain from (15b) the system

$$(j = 1) \qquad u_{20} - 4u_{21}^{(2)} + u_{22}^{(2)} \qquad\quad = -u_{11}^{(1)} - u_{31}$$
$$(j = 2) \qquad\qquad u_{21}^{(2)} - 4u_{22}^{(2)} + u_{23} = -u_{12}^{(1)} - u_{32}.$$

The solution is $u_{21}^{(2)} = 91.11$, $u_{22}^{(2)} = 64.44$.

In this example, which merely serves to explain the practical procedure in the ADI method, the accuracy of the second approximations is about the same as of those of two Gauss–Seidel steps in Sec. 18.3 (where $u_{11} = x_1$, $u_{21} = x_2$, $u_{12} = x_3$, $u_{22} = x_4$), as the following table shows.

	u_{11}	u_{21}	u_{12}	u_{22}
ADI, 2nd approximations	91.11	91.11	64.44	64.44
Gauss–Seidel, 2nd approximations	93.75	90.62	65.62	64.06
Exact solution of (12)	87.50	87.50	62.50	62.50

◀

Improving convergence. Additional improvement of the convergence of the ADI method results from the following interesting idea. Introducing a parameter p, we can also write (11) in the form

$$(16a) \qquad u_{i-1,j} - (2 + p)u_{ij} + u_{i+1,j} = -u_{i,j-1} + (2 - p)u_{ij} - u_{i,j+1}$$

and

$$(16b) \qquad u_{i,j-1} - (2 + p)u_{ij} + u_{i,j+1} = -u_{i-1,j} + (2 - p)u_{ij} - u_{i+1,j}.$$

This gives the more general ADI iteration formulas

$$(17a) \qquad u_{i-1,j}^{(m+1)} - (2 + p)u_{ij}^{(m+1)} + u_{i+1,j}^{(m+1)} = -u_{i,j-1}^{(m)} + (2 - p)u_{ij}^{(m)} - u_{i,j+1}^{(m)}$$

and

$$(17b) \qquad u_{i,j-1}^{(m+2)} - (2 + p)u_{ij}^{(m+2)} + u_{i,j+1}^{(m+2)} = -u_{i-1,j}^{(m+1)} + (2 - p)u_{ij}^{(m+1)} - u_{i+1,j}^{(m+1)}.$$

For $p = 2$, this is (15). The parameter p may be used for improving convergence. Indeed, one can show that the ADI method converges for positive p, and that the optimum value for maximum rate of convergence is

$$(18) \qquad\qquad\qquad\qquad p_0 = 2 \sin \frac{\pi}{K}$$

where K is the larger of $M + 1$ and $N + 1$ (see above). Even better results can be achieved by letting p vary from step to step. More details of the ADI method and variants are discussed in Ref. [E13] listed in Appendix 1.

PROBLEM SET 19.4

1. Derive (5b), (6b), and (6c).

Gauss Elimination and Gauss–Seidel Method

2. Verify the calculations in Example 1. Find out experimentally how many steps are needed to obtain the solution of the linear system with an accuracy of 3S.

   For the grid in Fig. 424 on the next page compute the potential at the four interior points by Gauss and by 5 Gauss–Seidel steps, starting from 100, 100, 100, 100 and showing the details of your work, if the boundary values are

3. $u = 220$ on the upper and lower edges, $u = 110$ on the left edge, $u = -10$ on the right edge

4. $u = 0$ on the left edge, x^3 on the lower edge, $27 - 9y^2$ on the right edge, $x^3 - 27x$ on the upper edge

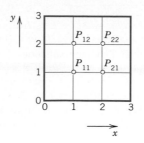

Fig. 424.   Problems 3−6

5. $u = \sin \frac{1}{3}\pi x$ on the upper edge, 0 on the other edges

6. **(Starting)** Do Prob. 5 by Gauss–Seidel, starting from **0.** Compare and comment on the substantial improvement.

7. **(Use of symmetry)** Conclude from the boundary values in Example 1 in the text that $u_{21} = u_{11}$ and $u_{22} = u_{12}$. Show that this leads to a system of two equations and solve it.

8. **(3 × 3 grid)** Solve Example 1, choosing $h = 3$ and starting values $100, 100, \cdots$.

9. For the square $0 \leqq x \leqq 4$, $0 \leqq y \leqq 4$ let the boundary temperatures be 0°C on the horizontal and 50°C on the vertical edges. Find the temperatures at the interior points of a square grid with $h = 1$.

10. Using the answer to Prob. 9, try to sketch the isotherms.

11. Find the isotherms for the square and grid in Prob. 9 if $u = \sin \frac{1}{4}\pi x$ on the horizontal and $-\sin \frac{1}{4}\pi y$ on the vertical edges. Try to sketch some isotherms.

12. Find the potential in Fig. 425 using (a) the coarse grid, (b) the fine grid, and Gauss elimination. *Hint.* In (b), use symmetry; take $u = 0$ as the boundary value at the two points at which the potential has a jump.

13. How many Gauss–Seidel steps would be needed to obtain the answer to Prob. 12, coarse grid, to 5S (5 significant figures) if one started from 0, 0? In the case of the fine grid in Prob. 12, the Gauss–Seidel method converges more slowly. Can you see the reason by inspecting the system of equations?

14. **(ADI)** Apply the ADI method to the Dirichlet problem in Prob. 5, using the grid in Fig. 424, as before, and starting values zero.

15. What p_0 in (18) should we choose for Prob. 14? Apply the ADI formulas (17) with $p_0 = 1.7$ to Prob. 14, performing 1 step. Illustrate the improved convergence by comparing with the corresponding values 0.077, 0.308 after the first step in Prob. 14. (Use the starting values zero.)

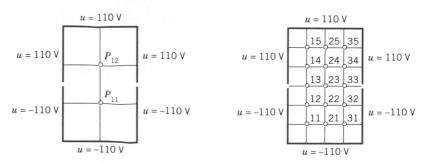

Fig. 425.   Region and grids in Problem 12

16. CAS PROJECT. Laplace Equation. (a) Write a program for Gauss–Seidel with 16 equations in 16 unknowns, composing the matrix (13) from the indicated 4×4 submatrices and including a transformation of the vector of the boundary values into the vector **b** of $\mathbf{Ax} = \mathbf{b}$.

(b) Apply the program to the square grid in $0 \leqq x \leqq 5$, $0 \leqq y \leqq 5$ with $h = 1$ and $u = 220$ on the upper and lower edges, $u = 110$ on the left edge and $u = -10$ on the right edge. Solve the linear system by Gauss elimination. What accuracy is reached in the 20th Gauss–Seidel step? Compare also with Prob. 3.

19.5 Neumann and Mixed Problems. Irregular Boundary

We continue our discussion of the numerical solution of boundary value problems for elliptic equations in a region R in the xy-plane. The Dirichlet problem was studied in the last section. In **Neumann** and **mixed problems** (defined in the last section) we are confronted with a new situation, because there are boundary points at which the (outer) **normal derivative** $u_n = \partial u / \partial n$ of the solution is given, but u itself is unknown since it is not given. To handle such points we need a new idea. This idea is the same for Neumann and mixed problems. Hence we may explain it in connection with one of these two types of problem. We shall do so and consider a typical example as follows.

EXAMPLE 1   **Mixed boundary value problem for a Poisson equation**

Solve the mixed boundary value problem for the Poisson equation

$$\nabla^2 u = u_{xx} + u_{yy} = f(x, y) = 12xy$$

shown in Fig. 426a.

Solution.   We use the grid shown in Fig. 426b, where $h = 0.5$. We recall that (7) in Sec. 19.4 has the right side $h^2 f(x, y) = 3xy$. From the formulas $u = 3y^3$ and $u_n = 6x$ given on the boundary we compute the boundary data

(1) $u_{31} = 0.375$, $u_{32} = 3$, $\dfrac{\partial u_{12}}{\partial n} = \dfrac{\partial u_{12}}{\partial y} = 6 \cdot 0.5 = 3$, $\dfrac{\partial u_{22}}{\partial n} = \dfrac{\partial u_{22}}{\partial y} = 6 \cdot 1 = 6$.

P_{11} and P_{21} are interior mesh points and can be handled as in the last section. Indeed, from (7), Sec. 19.4, with $h^2 = 0.25$ and $h^2 f(x, y) = 3xy$, and from the given boundary values, we obtain two equations corresponding to P_{11} and P_{21}, as follows.

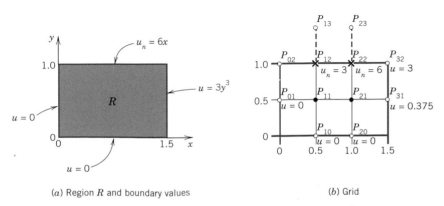

(a) Region R and boundary values   (b) Grid

Fig. 426. Mixed boundary value problem in Example 1

(2a)
$$-4u_{11} + u_{21} + u_{12} \qquad = 12(0.5 \cdot 0.5) \cdot \tfrac{1}{4} - 0 = 0.75$$
$$u_{11} - 4u_{21} \qquad + u_{22} = 12(1 \cdot 0.5) \cdot \tfrac{1}{4} - 0.375 = 1.125$$

The only difficulty with these equations seems to be that they involve the unknown values u_{12} and u_{22} of u at P_{12} and P_{22} on the boundary, where the normal derivative $u_n = \partial u/\partial n = \partial u/\partial y$ is given, instead of u; but we shall overcome this difficulty as we go on.

We consider P_{12} and P_{22}. The idea that will help us here is as follows. We imagine the region R to be extended above to the first row of external mesh points (corresponding to $y = 1.5$), and we assume that the differential equation also holds in the extended region. Then we can write down two more equations as before (Fig. 426b)

(2b)
$$u_{11} \qquad - 4u_{12} + u_{22} + u_{13} \qquad = 1.5 - 0 = 1.5$$
$$u_{21} + u_{12} - 4u_{22} \qquad + u_{23} = 3 - 3 = 0.$$

(On the right, 1.5 is $12xyh^2$ at (0.5, 1) and 3 is $12xyh^2$ at (1, 1).) We remember that we have not yet used the boundary condition on the upper part of the boundary of R, and we also notice that in (2b) we have introduced two more unknowns u_{13}, u_{23}. But we can now use that condition and get rid of u_{13}, u_{23} by applying the central difference formula for u_y. From (1) we then obtain (see Fig. 426b)

$$3 = \frac{\partial u_{12}}{\partial y} \approx \frac{u_{13} - u_{11}}{2h} = u_{13} - u_{11}, \qquad \text{hence} \qquad u_{13} = u_{11} + 3$$

$$6 = \frac{\partial u_{22}}{\partial y} \approx \frac{u_{23} - u_{21}}{2h} = u_{23} - u_{21}, \qquad \text{hence} \qquad u_{23} = u_{21} + 6.$$

Substituting these results into (2b) and simplifying, we have

$$2u_{11} \qquad - 4u_{12} + u_{22} = 1.5 - 3 = -1.5$$
$$2u_{21} + u_{12} - 4u_{22} = 3 - 3 - 6 = -6.$$

Together with (2a) this yields, written in matrix form,

(3)
$$\begin{bmatrix} -4 & 1 & 1 & 0 \\ 1 & -4 & 0 & 1 \\ 2 & 0 & -4 & 1 \\ 0 & 2 & 1 & -4 \end{bmatrix} \begin{bmatrix} u_{11} \\ u_{21} \\ u_{12} \\ u_{22} \end{bmatrix} = \begin{bmatrix} 0.75 \\ 1.125 \\ 1.5 - 3 \\ 0 - 6 \end{bmatrix} = \begin{bmatrix} 0.75 \\ 1.125 \\ -1.5 \\ -6 \end{bmatrix}.$$

(The entries 2 come from u_{13} and u_{23}, and so do -3 and -6 on the right.) The solution is as follows; the exact values of the problem are given in parentheses.

$$u_{12} = 0.866 \quad \text{(exact 1)} \qquad u_{22} = 1.812 \quad \text{(exact 2)}$$
$$u_{11} = 0.077 \quad \text{(exact 0.125)} \qquad u_{21} = 0.191 \quad \text{(exact 0.25)}.$$

Irregular Boundary

We continue our discussion of the numerical solution of boundary value problems for elliptic equations in a region R in the xy-plane. If R has a simple geometric shape, we can usually arrange for certain mesh points to lie on the boundary C of R, and then we can approximate partial derivatives as explained in the last section. However, if C intersects the grid at points that are not mesh points, then at points close to the boundary we must proceed differently, as follows.

The mesh point O in Fig. 427 is of that kind. For O and its neighbors A and P we obtain from Taylor's theorem

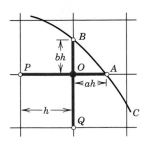

Fig. 427. Curved boundary C
of a region R, a mesh point O
near C, and neighbors A, B, P, Q

(4)
$\qquad$ (a) $\qquad u_A = u_O + ah \dfrac{\partial u_O}{\partial x} + \dfrac{1}{2}(ah)^2 \dfrac{\partial^2 u_O}{\partial x^2} + \cdots$

$\qquad$ (b) $\qquad u_P = u_O - h \dfrac{\partial u_O}{\partial x} + \dfrac{1}{2} h^2 \dfrac{\partial^2 u_O}{\partial x^2} + \cdots .$

We disregard the terms marked by dots and eliminate $\partial u_O / \partial x$. Equation (4b) times a plus equation (4a) gives

$$u_A + au_P \approx (1 + a)u_O + \frac{1}{2} a(a + 1)h^2 \frac{\partial^2 u_O}{\partial x^2} .$$

We solve this algebraically for the derivative, obtaining

$$\frac{\partial^2 u_O}{\partial x^2} \approx \frac{2}{h^2} \left[\frac{1}{a(1 + a)} u_A + \frac{1}{1 + a} u_P - \frac{1}{a} u_O \right].$$

Similarly, by considering the points O, B, and Q,

$$\frac{\partial^2 u_O}{\partial y^2} \approx \frac{2}{h^2} \left[\frac{1}{b(1 + b)} u_B + \frac{1}{1 + b} u_Q - \frac{1}{b} u_O \right].$$

By addition,

(5) $\qquad \nabla^2 u_O \approx \dfrac{2}{h^2} \left[\dfrac{u_A}{a(1 + a)} + \dfrac{u_B}{b(1 + b)} + \dfrac{u_P}{1 + a} + \dfrac{u_Q}{1 + b} - \dfrac{(a + b)u_O}{ab} \right].$

For example, if $a = \frac{1}{2}$, $b = \frac{1}{2}$, instead of the stencil (see Sec. 19.4)

$$\left\{ \begin{array}{ccc} & 1 & \\ 1 & -4 & 1 \\ & 1 & \end{array} \right\}, \qquad \text{we now have} \qquad \left\{ \begin{array}{ccc} & \frac{4}{3} & \\ \frac{2}{3} & -4 & \frac{4}{3} \\ & \frac{2}{3} & \end{array} \right\}.$$

The sum of all five terms still equals zero (which is useful for checking).

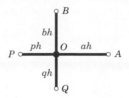

Fig. 428. Neighboring points
A, B, P, Q of a mesh point O
and notations in formula (6)

Using the same ideas, you may show that in the case of Fig. 428,

$$
\textbf{(6)} \quad \nabla^2 u_O \approx \frac{2}{h^2} \left[\frac{u_A}{a(a+p)} + \frac{u_B}{b(b+q)} + \frac{u_P}{p(p+a)} + \frac{u_Q}{q(q+b)} - \frac{ap+bq}{abpq} u_O \right],
$$

a formula that takes care of all conceivable cases.

EXAMPLE 2   **Dirichlet problem for the Laplace equation. Curved boundary**

Find the potential u in the region in Fig. 429 that has the boundary values given in that figure; here the curved portion of the boundary is an arc of the circle of radius 10 about $(0, 0)$. Use the grid in the figure.

Solution. u is a solution of the Laplace equation. From the given formulas for the boundary values $u = x^3$, $u = 512 - 24y^2, \cdots$ we compute the values at the points where we need them; the result is shown in the figure. For P_{11} and P_{12} we have the usual regular stencil, and for P_{21} and P_{22} we use (6), obtaining

$$
\textbf{(7)} \quad P_{11}, P_{12} \colon \begin{Bmatrix} & 1 & \\ 1 & -4 & 1 \\ & 1 & \end{Bmatrix}, \quad P_{21} \colon \begin{Bmatrix} & 0.5 & \\ 0.6 & -2.5 & 0.9 \\ & 0.5 & \end{Bmatrix}, \quad P_{22} \colon \begin{Bmatrix} & 0.9 & \\ 0.6 & -3 & 0.9 \\ & 0.6 & \end{Bmatrix}.
$$

We use this and the boundary values and take the mesh points in the order $P_{11}, P_{21}, P_{12}, P_{22}$. Then we obtain the system

$$
\begin{aligned}
-4u_{11} + u_{21} + u_{12} &= 0 - 27 &&= -27 \\
0.6u_{11} - 2.5u_{21} + 0.5u_{22} &= -0.9 \cdot 296 - 0.5 \cdot 216 &&= -374.4 \\
u_{11} - 4u_{12} + u_{22} &= 702 + 0 &&= 702 \\
0.6u_{21} + 0.6u_{12} - 3u_{22} &= 0.9 \cdot 352 + 0.9 \cdot 936 &&= 1159.2.
\end{aligned}
$$

In matrix form,

$$
\textbf{(8)} \qquad \begin{bmatrix} -4 & 1 & 1 & 0 \\ 0.6 & -2.5 & 0 & 0.5 \\ 1 & 0 & -4 & 1 \\ 0 & 0.6 & 0.6 & -3 \end{bmatrix} \begin{bmatrix} u_{11} \\ u_{21} \\ u_{12} \\ u_{22} \end{bmatrix} = \begin{bmatrix} -27 \\ -374.4 \\ 702 \\ 1159.2 \end{bmatrix}.
$$

The Gauss elimination yields the (rounded) values

$$
u_{11} = -55.6, \qquad u_{21} = 49.2, \qquad u_{12} = -298.5, \qquad u_{22} = -436.3.
$$

Clearly, from a grid with so few mesh points we cannot expect great accuracy. The exact solution having the given boundary values is $u = x^3 - 3xy^2$ and yields the values

$$
u_{11} = -54, \qquad u_{21} = 54, \qquad u_{12} = -297, \qquad u_{22} = -432.
$$

In practice one would use a much finer grid and solve the resulting large system by an indirect method.   ◀

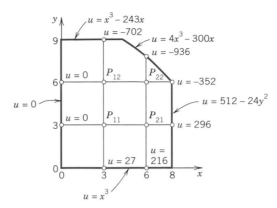

Fig. 429. Region, boundary values of the potential, and grid in Example 2

PROBLEM SET 19.5

Mixed Problems

1. Check the values given at the end of Example 1 by solving the system (3) by the Gauss elimination.

2. Solve the mixed boundary value problem for the Laplace equation $\nabla^2 u = 0$ in the rectangle in Fig. 426a (using the grid in Fig. 426b) and the boundary conditions $u_x = 0$ on the left edge, $u_x = 3$ on the right edge, $u = x^2$ on the lower edge, and $u = x^2 - 1$ on the upper edge.

3. Solve Prob. 2 when $u_n = 1$ on the upper edge and $u = 1$ on the other edges.

4. Solve the mixed boundary value problem for the Poisson equation $\nabla^2 u = 2(x^2 + y^2)$ in the region and for the boundary conditions shown in Fig. 430, using the indicated grid.

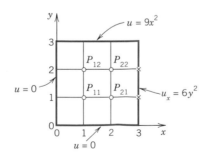

Fig. 430. Problem 4

5. Do Example 1 in the text for $\nabla^2 u = 0$ (instead of the Poisson equation) with boundary data and grid as before.

6. Solve $\nabla^2 u = -\pi^2 y \sin \frac{1}{3}\pi x$ for the grid in Fig. 430 and $u_y(1, 3) = u_y(2, 3) = \frac{1}{2}\sqrt{243}$, $u = 0$ on the other three sides of the square.

7. **CAS PROJECT. Mixed Problem.** Do Example 1 in the text with finer and finer grids of your choice and study the quality of the approximate values by comparing with the exact solution $u = 2xy^3$. (Verify the latter.)

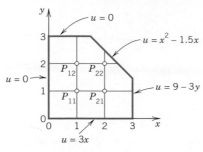

Fig. 431. Problem 13

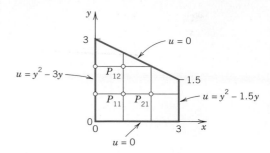

Fig. 432. Problem 16

Problems on Irregular Boundaries

8. We note that (4) can also be used to obtain an approximation formula for first partial derivatives if needed. Eliminating $\partial^2 u_O / \partial x^2$ from (4), derive

$$\frac{\partial u_O}{\partial x} \approx \frac{1}{h}\left[\frac{1}{a(1+a)}u_A - \frac{1-a}{a}u_O - \frac{a}{1+a}u_P\right].$$

9. Give the details of the derivation of (5). Verify the sample calculation after (5).
10. Give a detailed derivation of the basic general formula (6).
11. Verify the special stencil (7).
12. Solve (8) by the Gauss elimination. Show that the boundary values satisfy $u = x^3 - 3xy^2$. Show that $\nabla^2 u = 0$. Verify the given exact values.
13. Solve the Laplace equation in the region and for the boundary values shown in Fig. 431, using the indicated grid. (The sloping portion of the boundary is $y = 4.5 - x$.)
14. If in Prob. 13 the axes are grounded ($u = 0$), what constant potential must the other portion of the boundary have in order to produce 100 volts at P_{11}?
15. What potential do we have in Prob. 13 if $u = 190$ volts on the axes and $u = 0$ on the other portion of the boundary?
16. Solve the Poisson equation $\nabla^2 u = 2$ in the region and for the boundary values shown in Fig. 432, using the grid also shown in the figure.

19.6 Methods for Parabolic Equations

The last two sections concerned elliptic equations, and we now turn to parabolic equations. The definitions of elliptic, parabolic, and hyperbolic equations were given in Sec. 19.4. There it was also mentioned that the general behavior of solutions differs from type to type, and so do the problems of practical interest. This reflects on numerical methods as follows. For all three types, one replaces the equation by a corresponding difference equation, but for *parabolic* and *hyperbolic* equations this does not automatically guarantee the **convergence** of the approximate solution to the exact solution as the mesh $h \to 0$; in fact, it does not even guarantee convergence at all. For these two types of equation one needs additional conditions (inequalities) to assure convergence and **stability,** the latter meaning that small perturbations in the initial data (or small errors at any time) remain small at later times.

In this section we explain the numerical solution of the prototype of parabolic equations, the one-dimensional heat equation

$$u_t = c^2 u_{xx} \qquad\qquad (c \text{ constant}).$$

This equation is usually considered for x in some fixed interval, say, $0 \leqq x \leqq L$, and time $t \geqq 0$, and one prescribes the initial temperature $u(x, 0) = f(x)$ (f given) and boundary conditions at $x = 0$ and $x = L$ for all $t \geqq 0$, for instance $u(0, t) = 0$, $u(L, t) = 0$. We may assume $c = 1$ and $L = 1$; this can always be accomplished by a linear transformation of x and t (Prob. 1). Then the **heat equation** and those conditions are

(1)

$$u_t = u_{xx}$$

$0 \leqq x \leqq 1, t \geqq 0$

(2)

$$u(x, 0) = f(x)$$

(initial condition)

(3)

$$u(0, t) = u(1, t) = 0$$

(boundary conditions).

A simple finite difference approximation of (1) is [see (6a) in Sec. 19.4]

(4)

$$\frac{1}{k}(u_{i,j+1} - u_{ij}) = \frac{1}{h^2}(u_{i+1,j} - 2u_{ij} + u_{i-1,j}).$$

Figure 433 shows a corresponding grid and mesh points. The mesh size is h in the x-direction and k in the t-direction. Formula (4) involves the four points shown in Fig. 434. On the left we have used a *forward* difference quotient since we have no information for negative t at the start. From (4) we calculate $u_{i,j+1}$, which corresponds to time row $j + 1$, in terms of the three other u that correspond to time row j; solving (4) for $u_{i,j+1}$, we have

(5)

$$u_{i,j+1} = (1 - 2r)u_{ij} + r(u_{i+1,j} + u_{i-1,j}),$$

$r = \dfrac{k}{h^2}.$

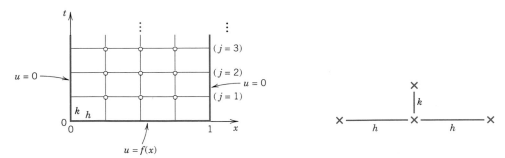

Fig. 433.   Grid and mesh points corresponding to (4), (5)

Fig. 434.   The four points in (4) and (5)

Computations by this **explicit method** based on (5) are simple. However, it can be shown that crucial to the convergence of this method is the condition

(6)

$$r = \frac{k}{h^2} \leqq \frac{1}{2},$$

that is, that u_{ij} have a positive coefficient in (5) or (for $r = \frac{1}{2}$) be absent from (5). Intuitively, (6) means that we should not move too fast in the t-direction. An example is given below.

Crank–Nicolson Method

Condition (6) is a handicap in practice. Indeed, to attain sufficient accuracy, we have to choose h small, which makes k very small by (6). For example, if $h = 0.1$, then $k \leqq 0.005$. And a switch to $\frac{1}{2}h$ quadruples the number of time steps needed to reach a certain t-value. Accordingly, we should look for a method based on a more satisfactory discretization of the heat equation.

Such a method that imposes no restriction on $r = k/h^2$ is the **Crank–Nicolson method,** which uses values of u at the six points in Fig. 435. The idea of the method is the replacement of the difference quotient on the right side of (4) by $\frac{1}{2}$ times the sum of two such difference quotients at two time rows (see Fig. 435). Instead of (4) we then have

(7)
$$\frac{1}{k}(u_{i,j+1} - u_{ij}) = \frac{1}{2h^2}(u_{i+1,j} - 2u_{ij} + u_{i-1,j})$$
$$+ \frac{1}{2h^2}(u_{i+1,j+1} - 2u_{i,j+1} + u_{i-1,j+1}).$$

Multiplying by $2k$ and writing $r = k/h^2$ as before, we collect the terms corresponding to time row $j + 1$ on the left and the terms corresponding to time row j on the right:

(8)　$(2 + 2r)u_{i,j+1} - r(u_{i+1,j+1} + u_{i-1,j+1}) = (2 - 2r)u_{ij} + r(u_{i+1,j} + u_{i-1,j}).$

How do we use (8)? In general, the three values on the left are unknown, whereas the three values on the right are known. If we divide the x-interval $0 \leqq x \leqq 1$ in (1) into n equal intervals, we have $n - 1$ internal mesh points per time row (see Fig. 433, where $n = 4$). Then for $j = 0$ and $i = 1, \cdots, n - 1$, formula (8) gives a linear system of $n - 1$ equations for the $n - 1$ unknown values $u_{11}, u_{21}, \cdots, u_{n-1,1}$ in the first time row in terms of the initial values $u_{00}, u_{10}, \cdots, u_{n0}$ and the boundary values $u_{01}, u_{n1}\ (= 0)$. Similarly for $j = 1, j = 2$, and so on; that is, for each time row we have to solve such a system of $n - 1$ linear equations resulting from (8).

Although $r = k/h^2$ is no longer restricted, smaller r will still give better results. In practice, one chooses a k by which one can save a considerable amount of work, without making r too large. For instance, often a good choice is $r = 1$ (which would be impossible in the previous method). Then (8) becomes simply

(9)　　　　　$4u_{i,j+1} - u_{i+1,j+1} - u_{i-1,j+1} = u_{i+1,j} + u_{i-1,j}.$

Fig. 435.　The six points in the Crank–Nicolson formulas (7) and (8)

EXAMPLE 1　**Temperature in a bar. Crank–Nicolson method, explicit method**

Consider a laterally insulated metal bar of length 1 and such that $c^2 = 1$ in the heat equation. Suppose that the ends of the bar are kept at temperature $u = 0°C$ and the temperature in the bar at some instant—call it $t = 0$—is $f(x) = \sin \pi x$. Applying the Crank–Nicolson method with $h = 0.2$ and $r = 1$, find the temperature $u(x, t)$ in the bar for $0 \leqq t \leqq 0.2$. Compare the results with the exact solution. Also apply (5) with an r satisfying (6), say, $r = 0.25$, and with values not satisfying (6), say, $r = 1$ and $r = 2.5$.

Crank–Nicolson Method

Condition (6) is a handicap in practice. Indeed, to attain sufficient accuracy, we have to choose h small, which makes k very small by (6). For example, if $h = 0.1$, then $k \leq 0.005$. And a switch to $\frac{1}{2}h$ quadruples the number of time steps needed to reach a certain t-value. Accordingly, we should look for a method based on a more satisfactory discretization of the heat equation.

Such a method that imposes no restriction on $r = k/h^2$ is the **Crank–Nicolson method,** which uses values of u at the six points in Fig. 435. The idea of the method is the replacement of the difference quotient on the right side of (4) by $\frac{1}{2}$ times the sum of two such difference quotients at two time rows (see Fig. 435). Instead of (4) we then have

$$
\begin{aligned}
\text{(7)} \quad \frac{1}{k}(u_{i,j+1} - u_{ij}) = {} & \frac{1}{2h^2}(u_{i+1,j} - 2u_{ij} + u_{i-1,j}) \\
& + \frac{1}{2h^2}(u_{i+1,j+1} - 2u_{i,j+1} + u_{i-1,j+1}).
\end{aligned}
$$

Multiplying by $2k$ and writing $r = k/h^2$ as before, we collect the terms corresponding to time row $j + 1$ on the left and the terms corresponding to time row j on the right:

$$
\text{(8)} \quad (2 + 2r)u_{i,j+1} - r(u_{i+1,j+1} + u_{i-1,j+1}) = (2 - 2r)u_{ij} + r(u_{i+1,j} + u_{i-1,j}).
$$

How do we use (8)? In general, the three values on the left are unknown, whereas the three values on the right are known. If we divide the x-interval $0 \leq x \leq 1$ in (1) into n equal intervals, we have $n - 1$ internal mesh points per time row (see Fig. 433, where $n = 4$). Then for $j = 0$ and $i = 1, \cdots, n - 1$, formula (8) gives a linear system of $n - 1$ equations for the $n - 1$ unknown values $u_{11}, u_{21}, \cdots, u_{n-1,1}$ in the first time row in terms of the initial values $u_{00}, u_{10}, \cdots, u_{n0}$ and the boundary values $u_{01}, u_{n1} \;(= 0)$. Similarly for $j = 1, j = 2$, and so on; that is, for each time row we have to solve such a system of $n - 1$ linear equations resulting from (8).

Although $r = k/h^2$ is no longer restricted, smaller r will still give better results. In practice, one chooses a k by which one can save a considerable amount of work, without making r too large. For instance, often a good choice is $r = 1$ (which would be impossible in the previous method). Then (8) becomes simply

$$
\text{(9)} \quad 4u_{i,j+1} - u_{i+1,j+1} - u_{i-1,j+1} = u_{i+1,j} + u_{i-1,j}.
$$

Fig. 435.   The six points in the Crank–Nicolson formulas (7) and (8)

EXAMPLE 1   **Temperature in a bar. Crank–Nicolson method, explicit method**

Consider a laterally insulated metal bar of length 1 and such that $c^2 = 1$ in the heat equation. Suppose that the ends of the bar are kept at temperature $u = 0°C$ and the temperature in the bar at some instant—call it $t = 0$— is $f(x) = \sin \pi x$. Applying the Crank–Nicolson method with $h = 0.2$ and $r = 1$, find the temperature $u(x, t)$ in the bar for $0 \leq t \leq 0.2$. Compare the results with the exact solution. Also apply (5) with an r satisfying (6), say, $r = 0.25$, and with values not satisfying (6), say, $r = 1$ and $r = 2.5$.

This equation is usually considered for x in some fixed interval, say, $0 \leq x \leq L$, and time $t \geq 0$, and one prescribes the initial temperature $u(x, 0) = f(x)$ (f given) and boundary conditions at $x = 0$ and $x = L$ for all $t \geq 0$, for instance $u(0, t) = 0$, $u(L, t) = 0$. We may assume $c = 1$ and $L = 1$; this can always be accomplished by a linear transformation of x and t (Prob. 1). Then the **heat equation** and those conditions are

(1)
$$u_t = u_{xx} \qquad 0 \leq x \leq 1, t \geq 0$$

(2)
$$u(x, 0) = f(x) \qquad \text{(initial condition)}$$

(3)
$$u(0, t) = u(1, t) = 0 \qquad \text{(boundary conditions)}.$$

A simple finite difference approximation of (1) is [see (6a) in Sec. 19.4]

(4)
$$\frac{1}{k}(u_{i,j+1} - u_{ij}) = \frac{1}{h^2}(u_{i+1,j} - 2u_{ij} + u_{i-1,j}).$$

Figure 433 shows a corresponding grid and mesh points. The mesh size is h in the x-direction and k in the t-direction. Formula (4) involves the four points shown in Fig. 434. On the left we have used a *forward* difference quotient since we have no information for negative t at the start. From (4) we calculate $u_{i,j+1}$, which corresponds to time row $j + 1$, in terms of the three other u that correspond to time row j; solving (4) for $u_{i,j+1}$, we have

(5)
$$u_{i,j+1} = (1 - 2r)u_{ij} + r(u_{i+1,j} + u_{i-1,j}), \qquad r = \frac{k}{h^2}.$$

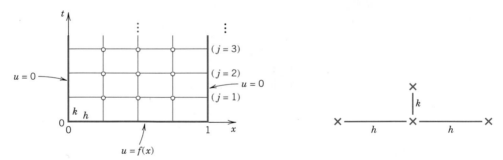

Fig. 433.   Grid and mesh points corresponding to (4), (5)

Fig. 434.   The four points in (4) and (5)

Computations by this **explicit method** based on (5) are simple. However, it can be shown that crucial to the convergence of this method is the condition

(6)
$$r = \frac{k}{h^2} \leq \frac{1}{2},$$

that is, that u_{ij} have a positive coefficient in (5) or (for $r = \frac{1}{2}$) be absent from (5). Intuitively, (6) means that we should not move too fast in the t-direction. An example is given below.

PROBLEM SET 19.6

1. **(Nondimensional form)** Show that the heat equation $\widetilde{u}_{\widetilde{t}} = c^2 \widetilde{u}_{\widetilde{x}\widetilde{x}}$, $0 \leq \widetilde{x} \leq L$, can be transformed to the "nondimensional" standard form $u_t = u_{xx}$, $0 \leq x \leq 1$, by setting $x = \widetilde{x}/L$, $t = c^2 \widetilde{t}/L^2$, $u = \widetilde{u}/u_0$, where u_0 is any constant temperature.

2. Derive the difference approximation (4) of the heat equation.

Explicit Method

3. Find the temperature at $t = 2$ in a laterally insulated bar of length 10 with ends kept at zero temperature and initial temperature $f(x) = x - 0.1x^2$. Use (5) with $h = 1$ and $k = 0.5$.

4. At what time will the maximum temperature in Prob. 3 have decreased to about 10% of its original value? *Hint.* Use Sec. 11.5.

5. In a laterally insulated bar of length 1 let the initial temperature be $f(x) = x$ if $0 \leq x \leq 0.2$, $f(x) = 0.25(1 - x)$ if $0.2 \leq x \leq 1$. Let $u(0, t) = 0$, $u(1, t) = 0$ for all t. Apply the explicit method with $h = 0.2$, $k = 0.01$. Do 5 steps.

6. Solve Prob. 5 for $f(x) = x$ if $0 \leq x \leq 0.5$, $f(x) = 1 - x$ if $0.5 \leq x \leq 1$, all the other data being as before. Can you expect the solution to satisfy $u(x, t) = u(1 - x, t)$ for all t?

7. If the left end of a laterally insulated bar extending from $x = 0$ to $x = 1$ is insulated, the boundary condition at $x = 0$ is $u_n(0, t) = u_x(0, t) = 0$. Show that in the application of the explicit method given by (5), we can compute $u_{0,j+1}$ by the formula

$$u_{0,j+1} = (1 - 2r)u_{0j} + 2ru_{1j}.$$

8. Applying the explicit method with $h = 0.2$ and $r = 0.25$, determine the temperature $u(x, t)$, $0 \leq t \leq 0.12$ in a laterally insulated bar extending from $x = 0$ to $x = 1$ if $u(x, 0) = 0$, the left end is insulated, and the right end is kept at temperature $g(t) = \sin \frac{50}{3}\pi t$. *Hint.* See Prob. 7.

Crank–Nicolson Method

9. Solve the heat equation (1) for the initial condition $f(x) = x$ if $0 \leq x \leq \frac{1}{2}$, $f(x) = 1 - x$ if $\frac{1}{2} < x \leq 1$, and boundary condition (3) by the Crank–Nicolson formula (9) with $h = 0.2$ for $0 \leq t \leq 0.20$. Compare with the exact values for $t = 0.20$ obtained from the series (2 terms) in Sec. 11.5.

10. Solve Prob. 3 by (9) with $h = 1$. Find exact values for $t = 2$ from the series in Sec. 11.5 (with suitable coefficients).

11. Solve Prob. 5 by (9) with $h = 0.2$, 2 steps. Compare with exact values (obtained similarly as in Prob. 9).

12. **CAS PROJECT. Comparison of Methods.** (a) Write programs for the explicit and the Crank–Nicolson methods.

    (b) Apply the programs to the heat problem of a laterally insulated bar of length 1 with $u(x, 0) = \sin \pi x$ and $u(0, t) = u(1, t) = 0$ for all t, using $h = 0.2$, $k = 0.01$ for the explicit method (20 steps), $h = 0.2$ and (9) for the Crank–Nicolson method (5 steps). Obtain exact 6D-values from a suitable series and compare.

    (c) Plot temperature curves in (b) in two figures similar to Fig. 271 in Sec. 11.6.

    (d) Experiment with smaller h (0.1, 0.05, etc.) for both methods to find out to what extent accuracy increases under systematic changes of h and k.

19.7 Methods for Hyperbolic Equations

In this section we consider the numerical solution of problems involving hyperbolic equations. We explain a standard method in terms of a typical setting for the prototype of a hyperbolic equation, the **wave equation:**

(1) $\qquad u_{tt} = u_{xx} \qquad\qquad 0 \leq x \leq 1,\, t \geq 0$

(2) $\qquad u(x,\, 0) = f(x) \qquad\qquad$ (Given initial displacement)

(3) $\qquad u_t(x,\, 0) = g(x) \qquad\qquad$ (Given initial velocity)

(4) $\qquad u(0,\, t) = u(1,\, t) = 0 \qquad\qquad$ (Boundary conditions).

Note that an equation $u_{tt} = c^2 u_{xx}$ and another x-interval can be reduced to the form (1) by a linear transformation of x and t. (This is similar to Sec. 19.6, Prob. 1.)

For instance, (1)–(4) is the model of a vibrating elastic string with fixed ends at $x = 0$ and $x = 1$ (see Sec. 11.2). Although an analytic solution of the problem is given in (13), Sec. 11.4, we use the problem for explaining basic ideas of the numerical approach that are also relevant for more complicated hyperbolic equations.

Replacing the derivatives by difference quotients as before, we obtain from (1) [see (6) in Sec. 19.4 with $y = t$]

(5) $\qquad \dfrac{1}{k^2}\, (u_{i,j+1} - 2u_{ij} + u_{i,j-1}) = \dfrac{1}{h^2}\, (u_{i+1,j} - 2u_{ij} + u_{i-1,j})$

where h is the mesh size in x, and k is the mesh size in t. This difference equation relates 5 points as shown in Fig. 438a. It suggests a rectangular grid similar to those for parabolic equations in the preceding section. We choose $r^* = k^2/h^2 = 1$. Then u_{ij} drops out and we have

(6) $\qquad \boxed{u_{i,j+1} = u_{i-1,j} + u_{i+1,j} - u_{i,j-1}} \qquad\qquad$ (Fig. 438b).

It can be shown that for $0 < r^* \leq 1$ the present **explicit method** is stable, so that from (6) we may expect reasonable results for initial data that have no discontinuities. (For a hyperbolic equation, the latter would propagate into the solution domain—a phenomenon that would be difficult to deal with on our present grid. For unconditionally stable **implicit methods** see [E1] in Appendix 1.)

Equation (6) still involves 3 time steps $j - 1, j, j + 1$, whereas the formulas in the parabolic case involved only 2 time steps. Furthermore, we now have 2 initial conditions. So we ask how we get started and how we can use the initial condition (3). This can be done as follows. From $u_t(x,\, 0) = g(x)$ we derive the difference formula

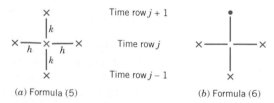

(a) Formula (5)   (b) Formula (6)

Fig. 438.   Mesh points used in (5) and (6)

(7) $\qquad \dfrac{1}{2k}(u_{i1} - u_{i,-1}) = g_i,$ hence $\qquad u_{i,-1} = u_{i1} - 2kg_i,$

where $g_i = g(ih)$. For $t = 0$, that is, $j = 0$, equation (6) is

$$u_{i1} = u_{i-1,0} + u_{i+1,0} - u_{i,-1}.$$

Into this we substitute $u_{i,-1}$ as given in (7). We obtain $u_{i1} = u_{i-1,0} + u_{i+1,0} - u_{i1} + 2kg_i$ and by simplification

(8) $\qquad \boxed{u_{i1} = \tfrac{1}{2}(u_{i-1,0} + u_{i+1,0}) + kg_i.}$

This expresses u_{i1} in terms of the initial data. It is for the beginning only. Now use (6).

EXAMPLE 1 **Vibrating string**

Apply the present method of numerical solution with $h = k = 0.2$ to the problem (1)–(4), where

$$f(x) = \sin \pi x, \qquad g(x) = 0.$$

Solution. The grid is the same as in Fig. 436, Sec. 19.6, except for the values of t, which now are 0.2, 0.4, · · · (instead of 0.04, 0.08, · · ·). The initial values u_{00}, u_{10}, · · · are the same as in Example 1, Sec. 19.6. From (8) and $g(x) = 0$ we have

$$u_{i1} = \tfrac{1}{2}(u_{i-1,0} + u_{i+1,0}).$$

From this we compute, using $u_{10} = u_{40} = \sin 0.2\pi = 0.587\,785$, $u_{20} = u_{30} = 0.951\,057$,

$$(i = 1) \qquad u_{11} = \tfrac{1}{2}(u_{00} + u_{20}) = \tfrac{1}{2} \cdot 0.951\,057 = 0.475\,528$$

$$(i = 2) \qquad u_{21} = \tfrac{1}{2}(u_{10} + u_{30}) = \tfrac{1}{2} \cdot 1.538\,842 = 0.769\,421,$$

and $u_{31} = u_{21}$, $u_{41} = u_{11}$ by symmetry as in Sec. 19.6, Example 1. From (6) with $j = 1$, using $u_{01} = u_{02} = \cdots = 0$, we now compute

$$(i = 1) \qquad u_{12} = u_{01} + u_{21} - u_{10} = 0.769\,421 - 0.587\,785 = 0.181\,636$$

$$(i = 2) \qquad u_{22} = u_{11} + u_{31} - u_{20} = 0.475\,528 + 0.769\,421 - 0.951\,057 = 0.293\,892,$$

and $u_{32} = u_{22}$, $u_{42} = u_{12}$ by symmetry; and so on. We thus obtain the following values of the displacement $u(x, t)$ of the string over the first half-cycle:

t	$x = 0$	$x = 0.2$	$x = 0.4$	$x = 0.6$	$x = 0.8$	$x = 1$
0.0	0	0.588	0.951	0.951	0.588	0
0.2	0	0.476	0.769	0.769	0.476	0
0.4	0	0.182	0.294	0.294	0.182	0
0.6	0	−0.182	−0.294	−0.294	−0.182	0
0.8	0	−0.476	−0.769	−0.769	−0.476	0
1.0	0	−0.588	−0.951	−0.951	−0.588	0

These values are exact to 3D (3 decimals), the exact solution of the problem being (see Sec. 11.3)

$$u(x, t) = \sin \pi x \cos \pi t.$$

The reason for the exactness follows from d'Alembert's solution (4), Sec. 11.4. (See Prob. 4, below.) ◀

This is the end of Chap. 19 on numerical methods for ordinary and partial differential equations, a rapidly developing field of interesting research.

PROBLEM SET 19.7

Solve the vibrating string problem (1)–(4) by the present numerical method with $h = k = 0.2$ on the given t-interval for initial velocity 0 and given initial deflection $f(x)$.

1. $0 \leq t \leq 2$, $f(x) = 0.1x(1 - x)$
2. $0 \leq t \leq 1$, $f(x) = x^2(1 - x)$
3. $0 \leq t \leq 1$, $f(x) = x$ if $0 \leq x \leq 0.2$, $f(x) = 0.25(1 - x)$ if $0.2 < x \leq 1$
4. Show that from d'Alembert's solution (14) in Sec. 11.4 with $c = 1$ it follows that (6) in the present section gives the exact value $u_{i,j+1} = u(ih, (j + 1)h)$.
5. If the string governed by the wave equation (1) starts from its equilibrium position with initial velocity $g(x) = \sin \pi x$, what is its displacement at time $t = 0.4$ and $x = 0.2, 0.4, 0.6, 0.8$? (Use the present method with $h = 0.2$, $k = 0.2$. Use (8). Compare with the exact values obtained from (13) in Sec. 11.4.)
6. Compute approximate values in Prob. 5, using a finer grid ($h = 0.1$, $k = 0.1$), and notice the increase in accuracy.
7. Illustrate the starting procedure for the present method in the case where both f and g are not identically zero, say,

$$f(x) = 1 - \cos 2\pi x, \qquad g(x) = x - x^2;$$

choose $h = k = 0.1$ and compute 2 time steps.
8. Show that (13) in Sec. 11.4 gives as another starting formula

$$u_{i1} = \frac{1}{2} (u_{i+1,0} + u_{i-1,0}) + \frac{1}{2} \int_{x_i - k}^{x_i + k} g(s) \, ds$$

(where one can evaluate the integral numerically if necessary). In what case is this identical with (8)?
9. Compute u in Prob. 7 for $t = 0.1$ and $x = 0.1, 0.2, \cdots, 0.9$, using the formula in Prob. 8, and compare the values.
10. Solve (1) numerically, subject to the conditions

$$u(x, 0) = x^2, \qquad u_t(x, 0) = 2x, \qquad u_x(0, t) = 2t, \qquad u(1, t) = (1 + t)^2,$$

choosing $h = k = 0.2$ (5 time steps).

CHAPTER 19 REVIEW

1. Explain Euler's method in geometrical terms.
2. What are the local and global orders of a method? Give examples.
3. What do you know about error estimates? Why are they important?
4. Is Euler's method accurate enough for practical purposes? Can it be improved? How?
5. How did we use the Taylor series for obtaining numerical methods?
6. Why did we compute auxiliary values in the classical Runge–Kutta method?
7. What is the Runge–Kutta–Fehlberg method?
8. What do you know about methods for systems of differential equations?
9. What is automatic step size control? When is it used?
10. What do we mean by single-step and multistep methods? Give examples.
11. What is a predictor–corrector method? Give examples.
12. What do we mean by saying that a method is self-starting? Not self-starting? Give examples.

13. What is the Runge–Kutta–Nyström method and for what kind of equations is it of practical interest?

14. Why and how did we use finite differences in this chapter?

15. By what difference equation did we approximate the Laplace equation in two variables? The Poisson equation?

16. Why does one need different methods for the different types of partial differential equations?

17. Can we expect a difference equation to give exact solutions of the corresponding differential equation?

18. In what method for partial differential equations did we have convergence problems?

19. Make a list of types of partial differential equations and the corresponding kinds of problems and methods of numerical solution.

20. How many initial conditions did we prescribe for the wave equation? For the heat equation?

21. Solve $y' = y$, $y(0) = 1$ by Euler's method (10 steps, $h = 0.1$).

22. Do the task in Prob. 21 with $h = 0.01$, 10 steps. Compute the errors. Compare with Prob. 21.

23. Solve $y' + 2y = e^x$, $y(0) = 0.5$ by the improved Euler method (10 steps, $h = 0.1$). Compute the errors, solving exactly.

24. Solve $y' + y = (x + 1)^2$, $y(0) = 3$ by the improved Euler method and determine the errors (10 steps, $h = 0.1$).

25. Compute $y = e^x$ for $x = 0, 0.1, 0.2, \cdots, 1.0$ by applying the classical Runge–Kutta method to $y' = y$, $y(0) = 1$ ($h = 0.1$). Show that the first five decimals of the result are correct.

26. Solve $y' = xy$, $y(0) = 1$ by the classical Runge–Kutta method (5 steps, $h = 0.2$). Compute the errors. Show details.

27. Solve $y' = (1 + x^{-1})y$, $y(1) = e$ by the classical Runge–Kutta method (5 steps, $h = 0.2$). Compute the errors. Show details.

28. $y_{n+1} = y_n + \frac{1}{6}(k_1 + 4k_2 + k_3^*)$ with k_1, k_2 as in Table 19.4 and $k_3^* = hf(x_{n+1}, y_n - k_1 + 2k_2)$ defines Kutta's third-order method. Apply it to (4) in Sec. 19.1 (5 steps, $h = 0.2$). Compare with Table 19.6.

29. $y_{n+1} = y_n + hf(x_n + \frac{1}{2}h, y_{n+1}^*)$, $y_{n+1}^* = y_n + \frac{1}{2}hf(x_n, y_n)$ gives another Euler–Cauchy method. Motivate it geometrically. Apply it to Prob. 21 (10 steps, $h = 0.1$). Compare the results.

30. Apply the Adams–Moulton method to $y' = (x + y - 4)^2$, $y(0) = 4$ ($h = 0.2$, $x = 0, \cdots, 1$; starting 4.00271, 4.02279, 4.08413).

31. Apply the Adams–Moulton method to $y' = \sqrt{1 - y^2}$, $y(0) = 0$ ($h = 0.2$, $x = 0, \cdots, 1$; starting 0.198668, 0.389416, 0.564637).

32. Show that by applying the method in Sec. 19.2 to a polynomial of first degree we obtain the multistep predictor and corrector formulas

$$y_{n+1}^* = y_n + \frac{h}{2}(3f_n - f_{n-1}), \qquad y_{n+1} = y_n + \frac{h}{2}(f_{n+1} + f_n).$$

33. Apply the multistep method in Prob. 32 to the initial value problem $y' = x + y$, $y(0) = 0$, choosing $h = 0.2$ and doing 5 steps. Compare with the exact values.

34. Apply Euler's method for systems to $y_1' = y_2$, $y_2' = -4y_1$, $y_1(0) = 2$, $y_2(0) = 0$ (10 steps, $h = 0.2$). Sketch the solution.

35. Apply Euler's method for systems to $y'' = x^2 y$, $y(0) = 1$, $y'(0) = 0$ (5 steps, $h = 0.1$).

36. Apply the Runge–Kutta method for systems to $y_1' = 6y_1 + 9y_2$, $y_2' = y_1 + 6y_2$, $y_1(0) = -3$, $y_2(0) = -3$ (3 steps, $h = 0.05$).

37. Apply Runge–Kutta for systems to $y'' + y = 2e^x$, $y(0) = 0$, $y'(0) = 1$ (5 steps, $h = 0.2$) and determine the errors.

38. Apply the method in Prob. 10 of Problem Set 19.3 to
$y'' = xy' - 3y$, $y(0) = 0$, $y'(0) = -3$ (5 steps, $h = 0.1$). Exact solution: $y = x^3 - 3x$ (verify!).

39. Apply the method in Prob. 10 of Problem Set 19.3 to
$(1 - x^2)y'' - 2xy' + 6y = 0$, $y(0) = -\frac{1}{2}$, $y'(0) = 0$ (5 steps, $h = 0.1$). Exact:
$y = \frac{1}{2}(3x^2 - 1)$.

40. Find the potential in the square in Fig. 424, Sec. 19.4, if $u(1, 0) = 60$, $u(2, 0) = 300$, $u = 100$ on the three other edges.

41. Verify that u_{11} and u_{21} in the answer to Prob. 5 of Problem Set 19.5 are the mean values of their four neighbors.

42. Find rough approximate values of the electrostatic potential at P_{11}, P_{12}, P_{13} in Fig. 439 that lie in a field between conducting plates (in Fig. 439 appearing as sides of a rectangle) kept at potentials 0 and 110 volts as shown. (Use the indicated grid.)

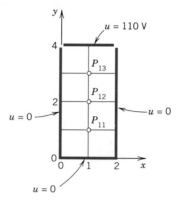

Fig. 439. Problem 42

43. Find the potential in Fig. 440, using the given grid and the boundary values $u(P_{10}) = u(P_{30}) = 960$, $u(P_{20}) = -480$, $u = 0$ elsewhere on the boundary.

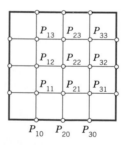

Fig. 440. Problem 43

44. A laterally insulated homogeneous bar with ends at $x = 0$ and $x = 1$ has initial temperature 0. Its left end is kept at 0, whereas the temperature at the right end varies sinusoidally according to

$$u(t, 1) = g(t) = \sin \tfrac{25}{3}\pi t.$$

Find the temperature $u(x, t)$ in the bar [solution of (1) in Sec. 19.6] by the explicit method with $h = 0.2$ and $r = 0.5$ (one period, that is, $0 \le t \le 0.24$).

45. Find the solution of the vibrating string problem $u_{tt} = u_{xx}$, $u(x, 0) = x(1 - x)$, $u_t = 0$, $u(0, t) = u(1, t) = 0$ by the method in Sec. 19.7 with $h = 0.1$ and $k = 0.1$ for $t = 0.3$.

Summary of Chapter 19
Numerical Methods
for Differential Equations

In this chapter we discussed numerical methods for ordinary differential equations (Secs. 19.1–19.3) and partial differential equations (Secs. 19.4–19.7). For first-order equations we considered initial value problems of the form (Sec. 19.1)

$$(1) \qquad y' = f(x, y), \qquad y(x_0) = y_0.$$

Numerical methods for solving such a problem can be obtained by truncating the Taylor series

$$y(x + h) = y(x) + hy'(x) + \frac{h^2}{2} y''(x) + \cdots$$

where, by (1), $y' = f$, $y'' = f' = \partial f/\partial x + (\partial f/\partial y)y'$, etc. Truncating after the term hy', we get *Euler's method*, in which we compute step by step

$$(2) \qquad y_{n+1} = y_n + hf(x_n, y_n) \qquad (n = 0, 1, \cdots).$$

Taking one more term into account, we obtain the *improved Euler* or *Heun's method*. Both methods show the basic idea in a simple form but are too inaccurate for most practical purposes.

Truncating after the term in h^4, we get the important classical **Runge–Kutta method** of fourth order. The crucial idea in this and similar methods is to replace the cumbersome evaluation of derivatives by the evaluation of $f(x, y)$ at suitable points (x, y); thus in each step we first compute four auxiliary quantities (Sec. 19.1)

$$(3a) \qquad \begin{aligned} k_1 &= hf(x_n, y_n) \\ k_2 &= hf(x_n + \tfrac{1}{2}h, \, y_n + \tfrac{1}{2}k_1) \\ k_3 &= hf(x_n + \tfrac{1}{2}h, \, y_n + \tfrac{1}{2}k_2) \\ k_4 &= hf(x_n + h, \, y_n + k_3) \end{aligned}$$

and then from them the new value

$$(3b) \qquad y_{n+1} = y_n + \tfrac{1}{6}(k_1 + 2k_2 + 2k_3 + k_4).$$

Error and step size control are possible by step halving or by **RKF** (Runge–Kutta–Fehlberg).

All the methods in Sec. 19.1 are **one-step methods** since y_{n+1} is computed from data of a single step. In contrast, a **multistep method** uses data from several preceding steps, thereby avoiding computations such as (3a). By integrating cubic interpolation polynomials we obtained an important multistep method consisting of an **Adams–Bashforth predictor** (Sec. 19.2)

$$(4a) \qquad y^*_{n+1} = y_n + \frac{1}{24} h(55f_n - 59f_{n-1} + 37f_{n-2} - 9f_{n-3}),$$

where $f_j = f(x_j, y_j)$, and an **Adams–Moulton corrector** (the actual new value)

$$(4b) \qquad y_{n+1} = y_n + \frac{1}{24} h(9f^*_{n+1} + 19f_n - 5f_{n-1} + f_{n-2}),$$

where $f^*_{n+1} = f(x_{n+1}, y^*_{n+1})$. Here, to get started, y_1, y_2, y_3 must be computed by the Runge–Kutta method or some other accurate method.

Section 19.3 concerned the extension of Euler's and RK methods to systems

$$\mathbf{y}' = \mathbf{f}(x, \mathbf{y}), \qquad \text{thus} \qquad y'_j = f_j(x, y_1, \cdots, y_m), \qquad j = 1, \cdots, m$$

This includes mth order differential equations, which are reduced to systems. Second-order equations can also be solved by **RKN** (Runge–Kutta–Nyström) **methods.** These are particularly advantageous for $y'' = f(x, y)$ with f not containing y'.

Numerical methods for **partial differential equations** are obtained by replacing partial derivatives by difference quotients. This leads to approximating difference equations, for the **Laplace equation** to

$$(5) \qquad u_{i+1,j} + u_{i,j+1} + u_{i-1,j} + u_{i,j-1} - 4u_{ij} = 0, \qquad \text{(Sec. 19.4)}$$

for the **heat equation** to

$$(6) \qquad \frac{1}{k}(u_{i,j+1} - u_{ij}) = \frac{1}{h^2}(u_{i+1,j} - 2u_{ij} + u_{i-1,j}), \qquad \text{(Sec. 19.6)}$$

and for the **wave equation** to

$$(7) \qquad \frac{1}{k^2}(u_{i,j+1} - 2u_{ij} + u_{i,j-1}) = \frac{1}{h^2}(u_{i+1,j} - 2u_{ij} + u_{i-1,j}) \qquad \text{(Sec. 19.7)};$$

here h and k are the mesh sizes of a grid in the x- and y-directions, respectively, where in (6) and (7) the variable y is time t.

These equations are *elliptic, parabolic,* and *hyperbolic,* respectively. Corresponding numerical methods differ, for the following reason. For elliptic equations we have boundary value problems, and we discussed for them the **Gauss–Seidel method** and the **ADI method** (Secs. 19.4, 19.5). For parabolic equations we are given one initial condition and boundary conditions, and we discussed an *explicit method* and the **Crank–Nicolson method** (Sec. 19.6). For hyperbolic equations, the problems are similar but we are given a second initial condition; in Sec. 19.7 we explained how to handle such problems numerically.

PART F

Optimization, Graphs

Chapter 20 **Unconstrained Optimization,**
Linear Programming

Chapter 21 **Graphs and Combinatorial Optimization**

The ideas of optimization and the application of graphs and digraphs (directed graphs) play an increasing role in engineering, computer science, systems theory, economics, and other areas. In this part we explain some basic concepts, methods, and results in unconstrained optimization (Sec. 20.1), linear programming (Secs. 20.1–20.4), graphs and digraphs, and optimization techniques for graphs and digraphs (Chap. 21). These so-called combinatorial optimization methods are relatively new and constitute an interesting area of ongoing applied and theoretical research.

CHAPTER 20

Unconstrained Optimization, Linear Programming

This chapter provides an introduction to the more important concepts, methods, and results of optimization. Optimization principles are of increasing importance in modern engineering design and systems operation in various areas. The recent development has been affected by computers capable of solving large-scale problems, and by the corresponding creation of new optimization techniques, so that the entire field is in the process of becoming a large area of its own.

Prerequisite: A modest knowledge of linear systems of equations.
References: Appendix 1, Part F.
Answers to problems: Appendix 2.

20.1 Basic Concepts Unconstrained Optimization

In an **optimization problem,** the objective is to *optimize* (*maximize* or *minimize*) some function f. This function f is called the **objective function.**

For example, an objective function f to be *maximized* may be the revenue in a production of TV sets, the yield per minute in a chemical process, the mileage per gallon of a certain type of car, the hourly number of customers served in a bank, the hardness of steel, or the tensile strength of a rope.

Similarly, we may want to *minimize* f if f is the cost per unit of producing certain cameras, the operating cost of some power plant, the daily loss of heat in a heating system, the idling time of some lathe, or the time needed to produce a fender.

In most optimization problems the objective function f depends on several variables

$$x_1, \cdots, x_n.$$

These are called **control variables** because we can "control" them, that is, choose their values.

For example, the yield of a chemical process may depend on pressure x_1 and temperature x_2. The efficiency of a certain air-conditioning system may depend on temperature x_1, air pressure x_2, moisture content x_3, cross-sectional area of outlet x_4, and so on.

Optimization theory develops methods for optimal choices of $x_1, \cdots, x_n$, which maximize (or minimize) the objective function f, that is, methods for finding optimal values of $x_1, \cdots, x_n$.

In many problems the choice of values of $x_1, \cdots, x_n$ is not entirely free but is subject to some **constraints,** that is, additional restrictions arising from the nature of the problem and the variables.

For example, if x_1 is production cost, then $x_1 \geqq 0$, and there are many other variables (time, weight, distance traveled by a salesman, etc.) that can take nonnegative values only. Constraints can also have the form of equations (instead of inequalities).

Let us first consider **unconstrained optimization** in the case of a real-valued function $f(x_1, \cdots, x_n)$. We also write $\mathbf{x} = (x_1, \cdots, x_n)$ and $f(\mathbf{x})$, for convenience.

By definition, f has a **minimum** at a point $\mathbf{x} = \mathbf{X}_0$ in a region R (where f is defined) if $f(\mathbf{x}) \geqq f(\mathbf{X}_0)$ for all $\mathbf{x}$ in R. Similarly, f has a **maximum** at $\mathbf{X}_0$ if $f(\mathbf{x}) \leqq f(\mathbf{X}_0)$ for all $\mathbf{x}$ in R. Minima and maxima together are called **extrema.**

Furthermore, f is said to have a **local minimum** at $\mathbf{X}_0$ if $f(\mathbf{x}) \geqq f(\mathbf{X}_0)$ for all $\mathbf{x}$ in a neighborhood of $\mathbf{X}_0$, say, for all $\mathbf{x}$ satisfying

$$|\mathbf{x} - \mathbf{X}_0| = \left[(x_1 - X_1)^2 + \cdots + (x_n - X_n)^2\right]^{1/2} < r,$$

where $\mathbf{X}_0 = (X_1, \cdots, X_n)$ and $r > 0$ is sufficiently small. A **local maximum** is defined similarly.

If f is differentiable and has an extremum at a point $\mathbf{X}_0$ in the *interior of R* (that is, not on the boundary), then the partial derivatives $\partial f / \partial x_1, \cdots, \partial f / \partial x_n$ must be zero at $\mathbf{X}_0$. These are the components of a vector that is called the **gradient** of f and denoted by grad f or ∇f. (For $n = 3$ this agrees with Sec. 8.9.) Thus

(1) $$\nabla f(\mathbf{X}_0) = \mathbf{0}.$$

A point $\mathbf{X}_0$ at which (1) holds is called a **stationary point** of f.

Condition (1) is necessary for an extremum of f at $\mathbf{X}_0$ in the interior of R, but is not sufficient. Indeed, if $n = 1$, then for $y = f(x)$, condition (1) is $y' = f'(X_0) = 0$; and, for instance, $y = x^3$ satisfies $y' = 3x^2 = 0$ at $x = X_0 = 0$ where f has no extremum but a point of inflection. Similarly, for $f(\mathbf{x}) = x_1 x_2$ we have $\nabla f(\mathbf{0}) = \mathbf{0},$ and f does not have an extremum but has a saddle point at $\mathbf{0}$. Hence after solving (1), one must still find out whether one has obtained an extremum. In the case $n = 1$ the conditions $y'(X_0) = 0$, $y''(X_0) > 0$ guarantee a local minimum at X_0 and the conditions $y'(X_0) = 0$, $y''(X_0) < 0$ a local maximum, as is known from calculus. For $n > 1$ there exist similar criteria. However, in practice even solving (1) will often be difficult. For this reason, one generally prefers solution by iteration, that is, by a search process that starts at some point and moves stepwise to points at which f is smaller (if a minimum of f is wanted) or larger (in the case of a maximum).

Cauchy's **method of steepest descent** or **gradient method** is of this type. It was introduced in 1847 and has enjoyed popularity ever since.[1] The idea is to find a minimum

---

[1]Convergence can sometimes be slow. For more refined methods and recent developments, see Ref. [F5] in Appendix 1.

of $f(\mathbf{x})$ by repeatedly computing minima of a function $g(t)$ of a single variable t, as follows. Suppose that f has a minimum at $\mathbf{X}_0$ and we start at a point $\mathbf{x}$. Then we look for a minimum of f closest to $\mathbf{x}$ along the straight line in the direction of

$$-\nabla f(\mathbf{x}),$$

the direction of steepest descent (= direction of maximum decrease) of f at $\mathbf{x}$; that is, we determine the value of t and the corresponding point

$$(2) \qquad\qquad\qquad \mathbf{z}(t) = \mathbf{x} - t\nabla f(\mathbf{x})$$

at which the function

$$(3) \qquad\qquad\qquad g(t) = f(\mathbf{z}(t))$$

has a minimum. We take this $\mathbf{z}(t)$ as our next approximation to $\mathbf{X}_0$.

EXAMPLE 1 **Method of steepest descent**

Determine a minimum of

$$(4) \qquad\qquad\qquad f(\mathbf{x}) = x_1{}^2 + 3x_2{}^2,$$

starting from $\mathbf{x}_0 = (6, 3) = 6\mathbf{i} + 3\mathbf{j}$ and applying the method of steepest descent.

Solution.   Clearly, inspection shows that $f(\mathbf{x})$ has a minimum at $\mathbf{0}$. Knowing the solution gives us a better feeling of how the method works.

We obtain $\nabla f(\mathbf{x}) = 2x_1\mathbf{i} + 6x_2\mathbf{j}$ and from this

$$\mathbf{z}(t) = \mathbf{x} - t\nabla f(\mathbf{x}) = (1 - 2t)x_1\mathbf{i} + (1 - 6t)x_2\mathbf{j}$$

$$g(t) = f(\mathbf{z}(t)) = (1 - 2t)^2 x_1{}^2 + 3(1 - 6t)^2 x_2{}^2.$$

We now calculate

$$g'(t) = 2(1 - 2t)x_1{}^2(-2) + 6(1 - 6t)x_2{}^2(-6),$$

set $g'(t) = 0$, and solve for t, finding

$$t = \frac{x_1{}^2 + 9x_2{}^2}{2x_1{}^2 + 54x_2{}^2}.$$

Starting from $\mathbf{x}_0 = 6\mathbf{i} + 3\mathbf{j}$, we compute the values in Table 20.1, which are shown in Fig. 441.   ◀

Table 20.1
Method of Steepest Descent, Computations in Example 1

n	$\mathbf{x}$		t	$1 - 2t$	$1 - 6t$
0	6.000	3.000	0.210	0.581	−0.258
1	3.484	−0.774	0.310	0.381	−0.857
2	1.327	0.664	0.210	0.581	−0.258
3	0.771	−0.171	0.310	0.381	−0.857
4	0.294	0.147	0.210	0.581	−0.258
5	0.170	−0.038	0.310	0.381	−0.857
6	0.065	0.032			

Figure 441 suggests that in the case of slimmer ellipses ("a long narrow valley"), convergence would be poor. You may confirm this by replacing the coefficient 3 in (4) with a large coefficient. For more sophisticated descent and other methods, some of them also applicable to vector functions of vector variables, we refer to the references listed in Part F of Appendix 1; see also [E6], [E13].

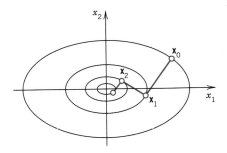

Fig. 441.   Method of steepest descent in Example 1

PROBLEM SET 20.1

1. Verify that in Example 1, successive gradients are orthogonal (perpendicular). Is this just by chance?

Apply 3 steepest descent steps to:

2. $f(\mathbf{x}) = x_1^2 + 4x_2^2 - 6x_1 - 8x_2;$ $\mathbf{x}_0 = [4 \quad 0]^T$

3. $f(\mathbf{x}) = 2(x_1^2 + x_2^2) + x_1x_2 - 5(x_1 + x_2);$ $\mathbf{x}_0 = [1 \quad -2]^T$

4. $f(\mathbf{x}) = 0.8x_1^2 + 0.35x_2^2 + 2.24x_1 + 0.21x_2;$ $\mathbf{x}_0 = [5 \quad 2]^T$

5. $f(\mathbf{x}) = 8x_1^2 - 20x_1x_2 + 17x_2^2 - 32x_1 + 40x_2;$ $\mathbf{x}_0 = [0 \quad 0]^T$

6. **(Parabolas)** Apply the present method to $f(\mathbf{x}) = x_1^2 - x_2$ (which has no minimum), $\mathbf{x}_0 = [1 \quad 1]^T$, 3 steps. Predict the outcome of further steps. Make a sketch.

7. **(Convergence)** Apply the present method to $f(\mathbf{x}) = x_1^2 + cx_2^2$. Starting from $[c \quad 1]^T$, show that subsequent approximations are $\mathbf{x}_m = a_m[c \quad (-1)^m]^T$, where $a_m = (c - 1)^m/(c + 1)^m$. Conclude that for large c (slim ellipses $f(\mathbf{x}) = const$) the convergence becomes poor.

8. **(Hyperbolas)** Apply the present method to $f(\mathbf{x}) = x_1^2 - x_2^2$, $\mathbf{x}_0 = [2 \quad 1]^T$, 5 steps. Make a sketch and guess the outcome. Then calculate.

9. What happens if you apply the method of steepest descent to $f(\mathbf{x}) = ax_1 + bx_2$? First guess, then calculate.

10. **CAS PROJECT. Steepest Descent. (a)** Write a program for the method.

   (b) Apply your program to $f(\mathbf{x}) = x_1^2 + 4x_2^2$, experimenting with respect to speed of convergence depending on the choice of $\mathbf{x}_0$.

   (c) Apply your program to $f(\mathbf{x}) = x_1^2 + x_2^4$ and to $f(\mathbf{x}) = x_1^4 + x_2^4$, $\mathbf{x}_0 = [2 \quad 1]^T$. Plot level curves and your path of descent. (Try to include plotting directly in your program.)

20.2 Linear Programming

Mathematical programming consists of methods for solving optimization problems **with constraints,** that is, for finding a maximum (or a minimum) of the objective function $z = f(x_1, \cdots, x_n)$ satisfying the constraints.

Linear programming (or **linear optimization**) means mathematical programming in which the objective function is a *linear function*

$$z = f(x_1, \cdots, x_n) = a_1x_1 + a_2x_2 + \cdots + a_nx_n,$$

and the constraints are **linear inequalities,** such as $3x_1 + 4x_2 \leq 36$, or $x_1 \geq 0$, etc. (examples below). Problems of this kind arise frequently, almost daily, for instance, in production, inventory management, bond trading, operation of power plants, routing delivery vehicles, airplane scheduling, and so on. Progress in computer technology has made it possible to solve programming problems involving hundreds or thousands of variables. Let us explain the setting of a linear programming problem and the idea of a "geometrical" solution, so that we shall see what is going on.

EXAMPLE 1   Production plan

Silvex Products produces gasoline tanks of types J and K. Two time constraints result from the use of two machines M_1 and M_2. On M_1 one needs 2 min for a J tank and 8 min for a K tank. On M_2 one needs 5 min for a J tank and 2 min for a K tank. A J tank sells for \$40 and a K tank for \$88 (because of higher material cost). Determine production figures x_1 for J and x_2 for K that maximize the hourly revenue

$$z = f(x_1, x_2) = 40x_1 + 88x_2.$$

Solution.   Production figures x_1 and x_2 must be nonnegative. Hence the objective function (to be maximized) and the four constraints are

$$(0) \qquad\qquad\qquad z = 40x_1 + 88x_2$$

$$(1) \qquad\qquad\qquad 2x_1 + 8x_2 \leq 60 \text{ min time on machine } M_1$$

$$(2) \qquad\qquad\qquad 5x_1 + 2x_2 \leq 60 \text{ min time on machine } M_2$$

$$(3) \qquad\qquad\qquad x_1 \geq 0$$

$$(4) \qquad\qquad\qquad x_2 \geq 0.$$

Figure 442 shows (0)–(4) as follows. Constancy lines $z = const$ are marked (0). These are **lines of constant revenue.** Their slope is $-40/88 = -5/11$. To increase z we must move the line upward (parallel to itself), as the arrow shows. Equation (1) with the equality sign is marked (1). It intersects the coordinate axes at $x_1 = 60/2 = 30$ (set $x_2 = 0$) and $x_2 = 60/8 = 7.5$ (set $x_1 = 0$). The arrow marks the side on which the points (x_1, x_2) lie that satisfy the inequality in (1). Similarly for Eqs. (2)–(4). The blue quadrangle thus obtained is called the **feasibility region.** It is the set of all **feasible solutions,** meaning solutions that satisfy all four constraints. The figure also lists the revenue at O, A, B, C. The optimal solution is obtained by moving the line of constant revenue up as much as possible without leaving the feasibility region. Obviously, this happens when that line passes through B, the intersection (10, 5) of (1) and (2). We see that the optimal revenue $z_{\max} = 40 \cdot 10 + 88 \cdot 5 = 840$ is obtained by producing twice as many J tanks as K tanks.   ◀

Note well that the problem in Example 1 or similar optimization problems *cannot* be solved by setting certain partial derivatives equal to zero, because crucial to such problems is the region in which the control variables are allowed to vary.

Furthermore, our "geometrical" or graphical method illustrated in Example 1 is confined to two variables x_1, x_2. However, most practical problems involve much more than two variables, so that we need other methods of solution.

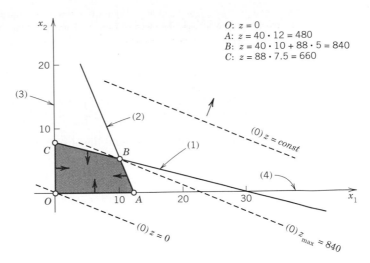

Fig. 442.   Linear programming in Example 1

Normal Form of a Linear Programming Problem

To prepare for those methods, we show that constraints can be written more uniformly. Let us explain the idea in terms of (1),

$$2x_1 + 8x_2 \leqq 60.$$

This inequality implies $60 - 2x_1 - 8x_2 \geqq 0$ (and conversely), that is, the quantity

$$x_3 = 60 - 2x_1 - 8x_2$$

is nonnegative. Accordingly, our original inequality can now be written

$$2x_1 + 8x_2 + x_3 = 60,$$

$$x_3 \geqq 0.$$

x_3 is a nonnegative auxiliary variable introduced for the purpose of converting inequalities to equations. Such a variable is called a **slack variable,** because it "takes up the slack" in the inequality.

EXAMPLE 2 **Conversion of inequalities by using slack variables**

With the help of two slack variables x_3, x_4 we can write the linear programming problem in Example 1 in the following form. *Maximize*

$$f = 40x_1 + 88x_2$$

subject to the constraints

$$2x_1 + 8x_2 + x_3 \qquad = 60$$

$$5x_1 + 2x_2 \qquad + x_4 = 60$$

$$x_i \geqq 0 \qquad (i = 1, \cdots, 4).$$

We now have $n = 4$ variables and $m = 2$ (linearly independent) equations, so that two of the four variables, for example, x_1, x_2, determine the others. Also note that each of the four sides of the quadrangle in Fig. 442 now has an equation of the form $x_i = 0$:

$$OA: x_2 = 0, \qquad AB: x_4 = 0, \qquad BC: x_3 = 0, \qquad CO: x_1 = 0.$$

A vertex of the quadrangle is the intersection of two sides. Hence at a vertex, $n - m = 4 - 2 = 2$ of the variables are zero and the others are nonnegative. Thus at A we have $x_2 = 0$, $x_4 = 0$, and so on. ◀

Our example suggests that a general linear optimization problem can be brought to the following **normal form.** *Maximize*

(5)
$$f = c_1 x_1 + c_2 x_2 + \cdots + c_n x_n$$

subject to the constraints

(6)
$$
\begin{aligned}
a_{11}x_1 + \cdots + a_{1n}\, x_n &= b_1 \\
a_{21}x_1 + \cdots + a_{2n}\, x_n &= b_2 \\
&\cdots\cdots\cdots\cdots\cdots \\
a_{m1}x_1 + \cdots + a_{mn}x_n &= b_m \\
x_i \geqq 0 \qquad (i = 1, &\cdots, n)
\end{aligned}
$$

with all b_j nonnegative. (If a $b_j < 0$, multiply the equation by -1.) Here $x_1, \cdots, x_n$ include the slack variables (for which the c's in f are zero). We assume that the equations in (6) are linearly independent. Then, if we choose values for $n - m$ of the variables, the system uniquely determines the others. Of course, since we must have $x_1 \geqq 0, \cdots, x_n \geqq 0$, this choice is not entirely free.

Our problem also includes the **minimization** of an objective function f since this corresponds to maximizing $-f$ and thus needs no separate consideration.

An n-tuple $(x_1, \cdots, x_n)$ that satisfies all the constraints in (6) is called a *feasible point* or **feasible solution.** A feasible solution is called an **optimal solution** if for it the objective function f becomes maximum, compared with the values of f at all feasible solutions.

Finally, by a **basic feasible solution** we mean a feasible solution for which at least $n - m$ of the variables $x_1, \cdots, x_n$ are zero. For instance, in Example 2 we have $n = 4$, $m = 2$, and the basic feasible solutions are the four vertices O, A, B, C in Fig. 442. Here B is an optimal solution (the only one in this example).

The following theorem is fundamental.

THEOREM 1   **(Optimal solution)**

Some optimal solution of a linear programming problem (5), (6) *is also a basic feasible solution of* (5), (6).

For a proof, see Ref. [F7], Chap. 3 (listed in Appendix 1). A problem can have many optimal solutions and not all of them may be *basic* feasible solutions; but the theorem guarantees that we can find an optimal solution by searching through the basic feasible solutions. This is a great simplification; but since there are $\binom{n}{n-m}$ different ways of equating $n - m$ of the n variables to zero, considering all these possibilities, dropping those which are not feasible and then searching through the rest would still involve very much work, even when n and m are relatively small. Hence a systematic search is needed. We shall explain an important method of this type in the next section.

PROBLEM SET 20.2

1. Could one find a profit $f(x_1, x_2) = a_1x_1 + a_2x_2$ whose maximum is at an interior point of the quadrangle in Fig. 442? (Give a reason for your answer.)
2. In Example 1, the solution is unique. Can we always expect uniqueness? (Give a reason.)
3. What is the meaning of the slack variables x_3, x_4 in Example 2 in terms of the problem in Example 1?

Regions, Constraints

Describe and graph the region in the first quadrant of the x_1x_2-plane determined by the given inequalities.

4. $x_1 - 3x_2 \geqq -6$

   $x_1 + x_2 \leqq 6$

5. $-0.5x_1 + x_2 \leqq 2$

   $x_1 + x_2 \geqq 2$

   $-x_1 + 5x_2 \geqq 5$

6. $2x_1 - x_2 \geqq 6$

   $8x_1 + 10x_2 \leqq 80$

   $x_1 - 2x_2 \geqq -3$

7. $x_1 - 2x_2 \leqq -2$

   $x_1 \leqq 4$

   $4x_1 + 5x_2 \geqq 20$

   $5x_1 - 4x_2 \geqq -6.4$

8. $x_1 + x_2 \geqq 2$

   $3x_1 + 5x_2 \geqq 15$

   $2x_1 - x_2 \geqq -2$

   $-x_1 + 2x_2 \leqq 10$

9. $-x_1 + x_2 \leqq -5$

   $2x_1 + x_2 \geqq 10$

   $x_2 \geqq 4$

   $10x_1 + 15x_2 \leqq 150$

Maximization, Minimization

10. Maximize $f = 30x_1 + 10x_2$ in the region in Prob. 9.
11. Minimize $f = 45.0x_1 + 22.5x_2$ in the region in Prob. 9.
12. Maximize $f = 5x_1 + 25x_2$ in the region in Prob. 5.
13. Minimize $f = 5x_1 + 25x_2$ in the region in Prob. 5.
14. Maximize $f = 20x_1 + 30x_2$ subject to $4x_1 + 3x_2 \geqq 12$, $x_1 - x_2 \geqq -3$, $x_2 \leqq 6$, $2x_1 - 3x_2 \leqq 0$.
15. Maximize $f = -10x_1 + 2x_2$ subject to $x_1 \geqq 0$, $x_2 \geqq 0$, $-x_1 + x_2 \geqq -1$, $x_1 + x_2 \leqq 6$, $x_2 \leqq 5$.
16. **(Maximum output)** Qualistep wants to maximize its daily total output of large step ladders by producing x_1 of them by a process P_1 and x_2 by a process P_2, where P_1 requires 2 hours of labor and 5 machine hours per ladder, and P_2 requires 4 hours of labor and 2 machine hours. For this kind of work, 800 hours of labor and 600 hours on the machines are at most available per day. Find the optimal x_1 and x_2.
17. **(Maximum profit)** Universal Electric manufactures and sells two models of lamps, L_1 and L_2, the profit being \$15 and \$10, respectively. The process involves two workers W_1 and W_2 who are available for this kind of work 100 and 80 hours per month, respectively. W_1 assembles L_1 in 20 minutes and L_2 in 30 minutes. W_2 paints L_1 in 20 minutes and L_2 in 10 minutes. Assuming that all lamps made can be sold without difficulty, determine production figures that maximize the profit.

18. **(Minimum cost)** Hardbrick Company has two kilns. Kiln I can produce 3000 grey bricks, 2000 red bricks, and 300 glazed bricks daily. For kiln II the corresponding figures are 2000, 5000, and 1500. Daily operating costs of kilns I and II are $400 and $600, respectively. Find the number of days of operation of each kiln so that the operation cost in filling an order of 9000 grey, 17000 red, and 4500 glazed bricks is minimized.

19. **(Maximum profit)** United Metal produces alloys B_1 (special brass) and B_2 (yellow tombac). B_1 contains 50% copper and 50% zinc. (Ordinary brass contains about 65% copper and 35% zinc.) B_2 contains 75% copper and 25% zinc. Net profits are $120 per ton of B_1 and $100 per ton of B_2. The daily copper supply is 15 tons. The daily zinc supply is 10 tons. Maximize the net profit of the daily production.

20. **(Nutrition)** Foods A and B have 700 and 500 calories, contain 10 g and 35 g of protein, and cost $1.50 and $2.00 per unit, respectively. Find the minimum cost diet of at least 3100 calories containing at least 100 g of protein.

20.3 Simplex Method

From the last section we recall the following. A linear optimization problem (linear programming problem) can be written in normal form; that is:

(1)

> **Maximize**
>
> $$z = f(\mathbf{x}) = c_1 x_1 + \cdots + c_n x_n$$
>
> **subject to the constraints**
>
> $$a_{11}x_1 + \cdots + a_{1n} x_n = b_1$$
>
> $$a_{21}x_1 + \cdots + a_{2n} x_n = b_2$$
>
> $$\cdots\cdots\cdots\cdots\cdots\cdots$$
>
> $$a_{m1}x_1 + \cdots + a_{mn}x_n = b_m$$
>
> $$x_i \geqq 0 \qquad (i = 1, \cdots, n).$$

(2)

For finding an optimal solution of this problem, we need to consider only the **basic feasible solutions** (defined in Sec. 20.2), but there are still so many that we have to follow a systematic search procedure. In 1948 G. B. Dantzig published an iterative method, called the **simplex method,** for that purpose. In this method, one proceeds stepwise from one basic feasible solution to another in such a way that the objective function f always increases its value. Let us explain this method in terms of the example in the last section.

In its original form the problem concerned the maximization of the objective function (giving the revenue)

$$z = 40x_1 + 88x_2$$

subject to

$$2x_1 + 8x_2 \leqq 60$$

$$5x_1 + 2x_2 \leqq 60$$

$$x_1 \qquad \geqq 0$$

$$x_2 \geqq 0.$$

Converting the first two inequalities to equations by introducing two slack variables x_3, x_4, we obtained the **normal form** of the problem in Example 2. Together with the objective function (written as an equation $z - 40x_1 - 88x_2 = 0$) this normal form is

$$
\begin{aligned}
z - 40x_1 - 88x_2 \qquad\qquad\qquad &= 0 \\
2x_1 + 8x_2 + x_3 \qquad\quad &= 60 \\
5x_1 + 2x_2 \qquad\quad + x_4 &= 60
\end{aligned}
$$

(3)

where $x_1 \geqq 0, \cdots, x_4 \geqq 0$. This is a linear system of equations. To find an optimal solution of it, we may consider its **augmented matrix** (see Sec. 6.3)

(4)
$$
\mathbf{T_0} = \begin{bmatrix}
z & x_1 & x_2 & x_3 & x_4 & b \\
1 & -40 & -88 & 0 & 0 & 0 \\
0 & 2 & 8 & 1 & 0 & 60 \\
0 & 5 & 2 & 0 & 1 & 60
\end{bmatrix}
$$

This matrix is called a **simplex tableau** or **simplex table** (the *initial simplex table*). These are standard names. The dashed lines and the letters z, $x_1, \cdots, b$ are for ease in further manipulation.

Every simplex table contains two kinds of variables x_j. By **basic variables** we mean those whose columns have only one nonzero entry. Thus x_3, x_4 in (4) are basic variables and x_1, x_2 are **nonbasic variables.**

Every simplex table gives a basic feasible solution. It is obtained by setting the nonbasic variables to zero. Thus (4) gives the basic feasible solution

$$
x_1 = 0, \qquad x_2 = 0, \qquad x_3 = 60/1 = 60, \qquad x_4 = 60/1 = 60, \qquad z = 0
$$

with x_3 obtained from the second row and x_4 from the third.

The optimal solution (its location and value) is now obtained stepwise by pivoting, designed to take us to basic feasible solutions with higher and higher values of z until the maximum of z is reached. Here, the choice of the **pivot equation** and **pivot** are quite different from what is done elsewhere in the book. The reason is that x_1, x_2, x_3, x_4 are restricted to nonnegative values.

1st Step. Operation O_1: Selection of the Column of the Pivot.
Select as the column of the pivot the first column with a negative entry in Row 1. In (4) this is Column 2 (because of the -40).

O_2: Selection of the Row of the Pivot.
Divide the right sides [60 and 60 in (4)] by the corresponding entries of the column just selected ($60/2 = 30$, $60/5 = 12$). Take as the pivot equation the equation that gives the *smallest* positive quotient. Thus the pivot is 5 because $60/5$ is smallest.

Operation O_3: Elimination by Row Operations.

This gives zeros above and below the pivot (as in Gauss–Jordan, Sec. 6.7).

With the notation for row operations as introduced in Sec. 6.3, the calculations in Step 1 give from the simplex table $\mathbf{T}_0$ in (4) the simplex table (augmented matrix)

$$
(5) \qquad \mathbf{T}_1 = \left[\begin{array}{c|cccc|c}
z & x_1 & x_2 & x_3 & x_4 & b \\
\hline
1 & 0 & -72 & 0 & 8 & 480 \\
\hline
0 & 0 & 7.2 & 1 & -0.4 & 36 \\
0 & 5 & 2 & 0 & 1 & 60
\end{array}\right]
\begin{array}{l}
\text{Row } 1 + 8 \text{ Row } 3 \\[1.5em]
\text{Row } 2 - 0.4 \text{ Row } 3 \\[1em]
\end{array}
$$

We see that basic variables are now x_1, x_3 and nonbasic variables are x_2, x_4. Setting the latter to zero, we obtain the basic feasible solution given by $\mathbf{T}_1$,

$$x_1 = 60/5 = 12, \qquad x_2 = 0, \qquad x_3 = 36/1 = 36, \qquad x_4 = 0, \qquad z = 480.$$

This is A in Fig. 442 (Sec. 20.2). We thus have moved from O: $(0, 0)$ with $z = 0$ to A: $(12, 0)$ with the greater $z = 480$. The reason for this increase is our elimination of a term $(-40x_1)$ with a negative coefficient. Hence **elimination is applied only to negative entries** in Row 1 but to no others (which are absent in our example). This motivates the selection of the *column* of the pivot.

We now motivate the selection of the *row* of the pivot. Had we taken the second row of $\mathbf{T}_0$ instead, we would have obtained $z = 1200$ (verify!), but this line of constant revenue $z = 1200$ lies entirely outside the feasibility region in Fig. 442. This motivates our cautious choice of the entry that gave the *smallest* quotient as our pivot, namely, 5.

2nd Step. The basic feasible solution given by (5) is not yet optimal because of the negative entry -72 in Row 1. Accordingly, we perform the operations O_1 to O_3 again, choosing a pivot in the column of -72.

Operation O_1. Select Column 3 of $\mathbf{T}_1$ in (5) as the column of the pivot (because $-72 < 0$).

Operation O_2. We have $36/7.2 = 5$ and $60/2 = 30$. Select 7.2 as the pivot (because $5 < 30$).

Operation O_3. Elimination by row operations gives

$$
(6) \qquad \mathbf{T}_2 = \left[\begin{array}{c|cccc|c}
z & x_1 & x_2 & x_3 & x_4 & b \\
\hline
1 & 0 & 0 & 10 & 4 & 840 \\
\hline
0 & 0 & 7.2 & 1 & -0.4 & 36 \\
0 & 5 & 0 & -\dfrac{1}{3.6} & \dfrac{1}{0.9} & 50
\end{array}\right]
\begin{array}{l}
\text{Row } 1 + 10 \text{ Row } 2 \\[1.5em]
\\[0.5em]
\text{Row } 3 - \dfrac{2}{7.2} \text{ Row } 2
\end{array}
$$

We see that now x_1, x_2 are basic and x_3, x_4 nonbasic. Setting the latter to zero, we obtain from $\mathbf{T}_2$ the basic feasible solution

$$x_1 = 50/5 = 10, \qquad x_2 = 36/7.2 = 5, \qquad x_3 = 0, \qquad x_4 = 0, \qquad z = 840.$$

This is B in Fig. 442 (Sec. 20.2). In this step, z has increased from 480 to 840, due to the elimination of -72 in $\mathbf{T}_1$. Since $\mathbf{T}_2$ contains no more negative entries in Row 1, we conclude that $z = f(10, 5) = 40 \cdot 10 + 88 \cdot 5 = 840$ is the maximum possible revenue. It is obtained if we produce twice as many J tanks as K tanks. This is the solution of our problem by the simplex method of linear programming.

Minimization. If we want to *minimize* $z = f(\mathbf{x})$ (instead of maximize), we take as the columns of the pivots those whose entry in Row 1 is *positive* (instead of negative). In such a Column k we consider only positive entries t_{jk} and take as pivot a t_{jk} for which b_j/t_{jk} is smallest (as before). For examples, see the problem set.

PROBLEM SET 20.3

In each case write the problem in normal form and solve it by the simplex method, assuming all the x_j to be nonnegative.

1. Maximize $z = 30x_1 + 20x_2$ subject to $-x_1 + x_2 \leqq 5$, $2x_1 + x_2 \leqq 10$.
2. Prob. 16 in Problem Set 20.2.
3. The problem in the text with the order of the constraints interchanged.
4. Maximize the daily output in producing x_1 chairs by process P_1 and x_2 chairs by process P_2 subject to $3x_1 + 4x_2 \leqq 550$ (machine hours), $5x_1 + 4x_2 \leqq 650$ (labor).
5. Maximize $z = 2x_1 + x_2 + 3x_3$ subject to $4x_1 + 3x_2 + 6x_3 \leqq 12$.
6. Maximize the daily profit in producing x_1 metal frames F_1 (profit \$90 per frame) and x_2 frames F_2 (profit \$50 per frame) subject to $x_1 + 3x_2 \leqq 18$ (material), $x_1 + x_2 \leqq 10$ (machine hours), $3x_1 + x_2 \leqq 24$ (labor).

7. Minimize $z = 5x_1 - 20x_2$ subject to $-2x_1 + 10x_2 \leqq 5$, $2x_1 + 5x_2 \leqq 10$.
8. Minimize $z = 4x_1 - 10x_2 - 20x_3$ subject to $3x_1 + 4x_2 + 5x_3 \leqq 60$, $2x_1 + x_2 \leqq 20$, $2x_1 + 3x_3 \leqq 30$.
9. Suppose that we produce x_1 batteries B_1 by process P_1 and x_2 by process P_2 and that we produce x_3 batteries B_2 by process P_3 and x_4 by process P_4. Let the profit per battery be \$10 for B_1 and \$20 for B_2. Maximize the total profit subject to the constraints

$$12x_1 + 8x_2 + 6x_3 + 4x_4 \leqq 120 \qquad \text{(machine hours)}$$

$$3x_1 + 6x_2 + 12x_3 + 24x_4 \leqq 180 \qquad \text{(labor hours)}.$$

 10. **CAS PROJECT. Simplex Method.** (a) Write a program for plotting a region R in the first quadrant of the x_1x_2-plane determined by linear constraints.
  (b) Write a program for maximizing $z = a_1x_1 + a_2x_2$ in R.
  (c) Write a program for maximizing $z = a_1x_1 + \cdots + a_nx_n$ subject to linear constraints.
  (d) Apply your programs to problems in this problem set and the previous one.

20.4 Simplex Method: Degeneracy, Difficulties in Starting

We recall from the last section that in the simplex method we proceed stepwise from one basic feasible solution to another, thereby increasing the value of the objective function f until we reach an optimal solution. Occasionally (but rather infrequently in practice), two kinds of difficulties may occur. The first of these is degeneracy, as follows.

A **degenerate feasible solution** is a feasible solution at which more than the usual number $n - m$ of variables are zero. Here n is the number of variables (slack and others) and m the number of constraints (not counting the $x_j \geqq 0$ conditions). In the last section, $n = 4$ and $m = 2$, and the occurring basic feasible solutions were nondegenerate; $n - m = 2$ variables were zero in each such solution.

In the case of a degenerate feasible solution we do an extra elimination step in which a basic variable that is zero for that solution becomes nonbasic (and a nonbasic variable becomes basic instead). We explain this in a typical case. For more complicated cases and techniques (rarely needed in practice) see Ref. [F7] in Appendix 1.

EXAMPLE 1   **Simplex method, degenerate feasible solution**

AB Steel, Inc., produces two kinds of iron I_1, I_2 by using three kinds of raw material R_1, R_2, R_3 (scrap iron and two kinds of ore) as shown. Maximize the daily profit.

Raw Material	Raw Material Needed per Ton		Raw Material Available per Day (tons)
	Iron I_1	Iron I_2	
R_1	2	1	16
R_2	1	1	8
R_3	0	1	3.5
Net profit per ton	$150	$300	

Solution.   Let x_1 and x_2 denote the amount (in tons) of iron I_1 and I_2, respectively, produced per day. Then our problem is as follows. Maximize

(1)
$$z = f(\mathbf{x}) = 150x_1 + 300x_2$$

subject to the constraints $x_1 \geqq 0$, $x_2 \geqq 0$ and

$$2x_1 + x_2 \leqq 16 \qquad \text{(raw material } R_1\text{)}$$
$$x_1 + x_2 \leqq 8 \qquad \text{(raw material } R_2\text{)}$$
$$x_2 \leqq 3.5. \qquad \text{(raw material } R_3\text{)}$$

By introducing slack variables x_3, x_4, x_5 we obtain the normal form of the constraints

(2)
$$2x_1 + x_2 + x_3 \qquad\qquad = 16$$
$$x_1 + x_2 \qquad + x_4 \qquad = 8$$
$$x_2 \qquad\qquad + x_5 = 3.5$$
$$x_i \geqq 0 \qquad (i = 1, \cdots, 5).$$

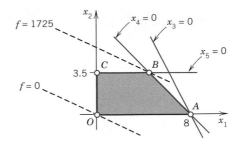

Fig. 443.   Example 1, where A is degenerate

As in the last section we obtain from (1) and (2) the initial simplex table

$$(3) \qquad \mathbf{T}_0 = \begin{bmatrix} \begin{array}{c|ccccc|c} z & x_1 & x_2 & x_3 & x_4 & x_5 & b \\ \hline 1 & -150 & -300 & 0 & 0 & 0 & 0 \\ \hline 0 & 2 & 1 & 1 & 0 & 0 & 16 \\ 0 & 1 & 1 & 0 & 1 & 0 & 8 \\ 0 & 0 & 1 & 0 & 0 & 1 & 3.5 \end{array} \end{bmatrix}.$$

We see that x_1, x_2 are nonbasic variables and x_3, x_4, x_5 are basic. With $x_1 = x_2 = 0$ we have from (3) the basic feasible solution

$$x_1 = 0, \qquad x_2 = 0, \qquad x_3 = 16/1 = 16, \qquad x_4 = 8/1 = 8, \qquad x_5 = 3.5/1 = 3.5, \qquad z = 0.$$

This is O: $(0, 0)$ in Fig. 443. We have $n = 5$ variables x_j, $m = 3$ constraints, and $n - m = 2$ variables equal to zero in our solution, which thus is nondegenerate.

1st Step of Pivoting

Operation O_1: Column Selection of Pivot. Column 2 (since $-150 < 0$).

Operation O_2: Row Selection of Pivot. $16/2 = 8$, $8/1 = 8$; $3.5/0$ is not possible. Hence we could choose Row 2 or Row 3. We choose Row 2. The pivot is 2.

Operation O_3: Elimination by Row Operations. This gives the simplex table

$$(4) \qquad \mathbf{T}_1 = \begin{bmatrix} \begin{array}{c|ccccc|c} z & x_1 & x_2 & x_3 & x_4 & x_5 & b \\ \hline 1 & 0 & -225 & 75 & 0 & 0 & 1200 \\ \hline 0 & 2 & 1 & 1 & 0 & 0 & 16 \\ 0 & 0 & \frac{1}{2} & -\frac{1}{2} & 1 & 0 & 0 \\ 0 & 0 & 1 & 0 & 0 & 1 & 3.5 \end{array} \end{bmatrix} \begin{array}{l} \text{Row } 1 + 75 \text{ Row } 2 \\ \\ \text{Row } 3 - \frac{1}{2} \text{ Row } 2 \\ \text{Row } 4 \end{array}$$

We see that the basic variables are x_1, x_4, x_5 and the nonbasic are x_2, x_3. Setting the nonbasic variables to zero, we obtain from $\mathbf{T}_1$ the basic feasible solution

$$x_1 = 16/2 = 8, \qquad x_2 = 0, \qquad x_3 = 0, \qquad x_4 = 0/1 = 0, \qquad x_5 = 3.5/1 = 3.5, \qquad z = 1200$$

This is A: $(8, 0)$ in Fig. 443. This solution is degenerate because $x_4 = 0$ (in addition to $x_2 = 0$, $x_3 = 0$); geometrically: the straight line $x_4 = 0$ also passes through A. This requires the next step, in which x_4 will become nonbasic.

2nd Step of Pivoting

Operation O_1: Column Selection of Pivot. Column 3 (since $-225 < 0$).

Operation O_2: Row Selection of Pivot. $16/1 = 16$, $0/\frac{1}{2} = 0$. Hence $\frac{1}{2}$ must serve as the pivot.

Operation O_3: Elimination by Row Operations. This gives the following simplex table.

$$(5) \quad \mathbf{T}_2 = \begin{bmatrix} 1 & 0 & 0 & -150 & 450 & 0 & 1200 \\ 0 & 2 & 0 & 2 & -2 & 0 & 16 \\ 0 & 0 & \frac{1}{2} & -\frac{1}{2} & 1 & 0 & 0 \\ 0 & 0 & 0 & 1 & -2 & 1 & 3.5 \end{bmatrix} \begin{matrix} \text{Row 1} + 450 \text{ Row 3} \\ \text{Row 2} - 2 \text{ Row 3} \\ \\ \text{Row 4} - 2 \text{ Row 3} \end{matrix}$$

Top header row: $z \quad x_1 \quad x_2 \quad x_3 \quad x_4 \quad x_5 \quad b$

We see that the basic variables are x_1, x_2, x_5 and the nonbasic are x_3, x_4. Hence x_4 has become nonbasic, as intended. By equating the nonbasic variables to zero we obtain from $\mathbf{T}_2$ the basic feasible solution

$$x_1 = 16/2 = 8, \quad x_2 = 0/\tfrac{1}{2} = 0, \quad x_3 = 0, \quad x_4 = 0, \quad x_5 = 3.5/1 = 3.5, \quad z = 1200.$$

This is still A: (8, 0) in Fig. 443 and z has not increased. But this opens the way to the maximum, which we reach in the next step.

3rd Step of Pivoting

Operation O_1: Column Selection of Pivot. Column 4 (since $-150 < 0$).

Operation O_2: Row Selection of Pivot. $16/2 = 8$, $0/(-\tfrac{1}{2}) = 0$, $3.5/1 = 3.5$. We can take 1 as the pivot. (With $-\tfrac{1}{2}$ as the pivot we would not leave A. Try it.)

Operation O_3: Elimination by Row Operations. This gives the simplex table

$$(6) \quad \mathbf{T}_3 = \begin{bmatrix} 1 & 0 & 0 & 0 & 150 & 150 & 1725 \\ 0 & 2 & 0 & 0 & 2 & -2 & 9 \\ 0 & 0 & \frac{1}{2} & 0 & 0 & \frac{1}{2} & 1.75 \\ 0 & 0 & 0 & 1 & -2 & 1 & 3.5 \end{bmatrix} \begin{matrix} \text{Row 1} + 150 \text{ Row 4} \\ \text{Row 2} - 2 \text{ Row 4} \\ \text{Row 3} + \tfrac{1}{2} \text{ Row 4} \\ \\ \end{matrix}$$

Top header row: $z \quad x_1 \quad x_2 \quad x_3 \quad x_4 \quad x_5 \quad b$

We see that basic variables are x_1, x_2, x_3 and nonbasic x_4, x_5. Equating the latter to zero we obtain from $\mathbf{T}_3$ the basic feasible solution

$$x_1 = 9/2 = 4.5, \quad x_2 = 1.75/\tfrac{1}{2} = 3.5, \quad x_3 = 3.5/1 = 3.5, \quad x_4 = 0, \quad x_5 = 0, \quad z = 1725.$$

This is B: (4.5, 3.5) in Fig. 443. Since Row 1 of $\mathbf{T}_3$ has no negative entries, we have reached the maximum daily profit $z_{\max} = f(4.5, 3.5) = 150 \cdot 4.5 + 300 \cdot 3.5 = 1725$. This is obtained by using 4.5 tons of iron I_1 and 3.5 tons of iron I_2. ◀

Difficulties in Starting

It may sometimes be difficult to find a basic feasible solution to start from. In such a case the idea of an **artificial variable** (or several such variables) is helpful. We explain this method in terms of a typical example.

EXAMPLE 2   **Simplex method: difficult start, artificial variable**

Maximize

$$(7) \qquad\qquad z = f(\mathbf{x}) = 2x_1 + x_2$$

subject to the constraints $x_1 \geqq 0$, $x_2 \geqq 0$ and (Fig. 444)

$$x_1 - \tfrac{1}{2}x_2 \geqq 1$$

$$x_1 - x_2 \leqq 2$$

$$x_1 + x_2 \leqq 4.$$

Solution.   By means of slack variables we achieve the normal form of the constraints

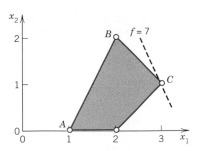

Fig. 444.   Feasibility region
in Example 2

$$
\begin{aligned}
x_1 - \tfrac{1}{2}x_2 - x_3 \qquad\qquad &= 1 \\
x_1 - \; x_2 \qquad + x_4 \qquad &= 2 \\
x_1 + \; x_2 \qquad\qquad + x_5 &= 4 \\
x_i \geqq 0 \quad (i = 1, \cdots, 5).
\end{aligned}
$$
(8)

Note that the first slack variable is negative (or zero), which makes x_3 nonnegative within the feasibility region (and negative outside). From (7) and (8) we obtain the simplex table

z	x_1	x_2	x_3	x_4	x_5	b
1	-2	-1	0	0	0	0
0	1	$-\tfrac{1}{2}$	-1	0	0	1
0	1	-1	0	1	0	2
0	1	1	0	0	1	4

x_1, x_2 are nonbasic, and we would like to take x_3, x_4, x_5 as basic variables. By our usual process of equating the nonbasic variables to zero we obtain from this table

$$
x_1 = 0, \qquad x_2 = 0, \qquad x_3 = 1/(-1) = -1, \qquad x_4 = 2/1 = 2, \qquad x_5 = 4/1 = 4, \qquad z = 0.
$$

$x_3 < 0$ indicates that $(0, 0)$ lies outside the feasibility region. Since $x_3 < 0$, we cannot proceed immediately. Now, instead of searching for other basic variables, we use the following idea. Solving the first equation in (8) for x_3, we have

$$
x_3 = -1 + x_1 - \tfrac{1}{2}x_2.
$$

We now add a variable x_6 on the right,

$$
x_3 = -1 + x_1 - \tfrac{1}{2}x_2 + x_6.
$$
(9)

x_6 is called an **artificial variable** and is subject to the constraint $x_6 \geqq 0$.

   We must take care that x_6 (which is not part of the given problem!) will disappear eventually. We shall see that we can accomplish this by adding a term $-Mx_6$ with very large M to the objective function. Because of (7) and (9) (solved for x_6) this gives the modified objective function for this **"extended problem"**

$$
\hat{z} = z - Mx_6 = 2x_1 + x_2 - Mx_6 = (2 + M)x_1 + (1 - \tfrac{1}{2}M)x_2 - Mx_3 - M.
$$
(10)

We see that the simplex table corresponding to (10) and (8) is

$$\mathbf{T}_0 = \begin{array}{c|ccc|cccc|c} \hat{z} & x_1 & x_2 & x_3 & x_4 & x_5 & x_6 & b \\ \hline 1 & -2-M & -1+\tfrac{1}{2}M & M & 0 & 0 & 0 & -M \\ \hline 0 & 1 & -\tfrac{1}{2} & -1 & 0 & 0 & 0 & 1 \\ 0 & 1 & -1 & 0 & 1 & 0 & 0 & 2 \\ 0 & 1 & 1 & 0 & 0 & 1 & 0 & 4 \\ 0 & 1 & -\tfrac{1}{2} & -1 & 0 & 0 & 1 & 1 \end{array}.$$

The last row of this table results from (9) written as $x_1 - \tfrac{1}{2}x_2 - x_3 + x_6 = 1$. We see that we can now start, taking x_4, x_5, x_6 as the basic variables and x_1, x_2, x_3 as the nonbasic variables. Column 2 has a negative first entry. We can take the second entry (1 in Row 2) as the pivot. This gives

$$\mathbf{T}_1 = \begin{array}{c|ccc|cccc|c} \hat{z} & x_1 & x_2 & x_3 & x_4 & x_5 & x_6 & b \\ \hline 1 & 0 & -2 & -2 & 0 & 0 & 0 & 2 \\ \hline 0 & 1 & -\tfrac{1}{2} & -1 & 0 & 0 & 0 & 1 \\ 0 & 0 & -\tfrac{1}{2} & 1 & 1 & 0 & 0 & 1 \\ 0 & 0 & \tfrac{3}{2} & 1 & 0 & 1 & 0 & 3 \\ 0 & 0 & 0 & 0 & 0 & 0 & 1 & 0 \end{array}.$$

This corresponds to $x_1 = 1$, $x_2 = 0$ (point A in Fig. 444), $x_3 = 0$, $x_4 = 1$, $x_5 = 3$, $x_6 = 0$. We can now drop Row 5 and Column 7. In this way we get rid of x_6, as wanted, and obtain

$$\mathbf{T}_2 = \begin{array}{c|ccc|ccc|c} z & x_1 & x_2 & x_3 & x_4 & x_5 & b \\ \hline 1 & 0 & -2 & -2 & 0 & 0 & 2 \\ \hline 0 & 1 & -\tfrac{1}{2} & -1 & 0 & 0 & 1 \\ 0 & 0 & -\tfrac{1}{2} & 1 & 1 & 0 & 1 \\ 0 & 0 & \tfrac{3}{2} & 1 & 0 & 1 & 3 \end{array}.$$

In Column 3 we choose 3/2 as the next pivot. We obtain

$$\mathbf{T}_3 = \begin{array}{c|ccc|ccc|c} z & x_1 & x_2 & x_3 & x_4 & x_5 & b \\ \hline 1 & 0 & 0 & -\tfrac{2}{3} & 0 & \tfrac{4}{3} & 6 \\ \hline 0 & 1 & 0 & -\tfrac{2}{3} & 0 & \tfrac{1}{3} & 2 \\ 0 & 0 & 0 & \tfrac{4}{3} & 1 & \tfrac{1}{3} & 2 \\ 0 & 0 & \tfrac{3}{2} & 1 & 0 & 1 & 3 \end{array}.$$

This corresponds to $x_1 = 2$, $x_2 = 2$ (this is B in Fig. 444), $x_3 = 0$, $x_4 = 2$, $x_5 = 0$. In Column 4 we choose 4/3 as the pivot, by the usual principle. This gives

$$\mathbf{T}_4 = \begin{array}{c|ccc|ccc|c} z & x_1 & x_2 & x_3 & x_4 & x_5 & b \\ \hline 1 & 0 & 0 & 0 & \tfrac{1}{2} & \tfrac{3}{2} & 7 \\ \hline 0 & 1 & 0 & 0 & \tfrac{1}{2} & \tfrac{1}{2} & 3 \\ 0 & 0 & 0 & \tfrac{4}{3} & 1 & \tfrac{1}{3} & 2 \\ 0 & 0 & \tfrac{3}{2} & 0 & -\tfrac{3}{4} & \tfrac{3}{4} & \tfrac{3}{2} \end{array}.$$

This corresponds to $x_1 = 3$, $x_2 = 1$ (point C in Fig. 444), $x_3 = \tfrac{3}{2}$, $x_4 = 0$, $x_5 = 0$. This is the maximum $f_{\mathrm{max}} = f(3, 1) = 7$.   ◀

PROBLEM SET 20.4

If in a step you have a choice between pivots, take the one that comes first in the column considered.

1. Maximize $z = f(\mathbf{x}) = 6x_1 + 12x_2$ subject to $0 \leq x_1 \leq 4$, $0 \leq x_2 \leq 4$, $6x_1 + 12x_2 \leq 72$.

2. Do Prob. 1 with the last two constraints interchanged.

3. Maximize the daily output in producing x_1 glass plates by a process P_1 and x_2 glass plates by a process P_2 subject to the constraints (labor hours, machine hours, raw material supply)

$$2x_1 + 3x_2 \leq 130, \qquad 3x_1 + 8x_2 \leq 300, \qquad 4x_1 + 2x_2 \leq 140.$$

4. Maximize $z = 300x_1 + 500x_2$ subject to $2x_1 + 8x_2 \leq 60$, $2x_1 + x_2 \leq 30$, $4x_1 + 4x_2 \leq 60$.

5. Do Prob. 4 with the last two constraints interchanged. Comment on the resulting simplification.

6. Maximize the total output $f = x_1 + x_2 + x_3$ (production figures of three different production processes) subject to input constraints (limitation of machine time)

$$4x_1 + 5x_2 + 8x_3 \leq 12, \qquad 8x_1 + 5x_2 + 4x_3 \leq 12.$$

7. Maximize $f = 6x_1 + 6x_2 + 9x_3$ subject to $x_j \geq 0$ ($j = 1, \cdots, 5$), and $x_1 + x_3 + x_4 = 1$, $x_2 + x_3 + x_5 = 1$.

8. Using an artificial variable, minimize $f = 2x_1 - x_2$ subject to $x_1 \geq 0$, $x_2 \geq 0$, $x_1 + x_2 \geq 5$, $-x_1 + x_2 \leq 1$, $5x_1 + 4x_2 \leq 40$.

9. Maximize $f = 4x_1 + x_2 + 2x_3$ subject to $x_1 \geq 0$, $x_2 \geq 0$, $x_3 \geq 0$, $x_1 + x_2 + x_3 \leq 1$, $x_1 + x_2 - x_3 \leq 0$.

10. If one uses the method of artificial variables in a problem without solution, this nonexistence will become apparent by the fact that one cannot get rid of the artificial variable. Illustrate this by trying to maximize $f = 2x_1 + x_2$ subject to $x_1 \geq 0$, $x_2 \geq 0$, $2x_1 + x_2 \leq 2$, $x_1 + 2x_2 \geq 6$, $x_1 + x_2 \leq 4$.

CHAPTER 20 REVIEW

1. What do we mean by constrained optimization? By unconstrained optimization?

2. Explain the idea of the method of steepest descent. Write down the corresponding formulas from memory.

3. On what does the speed of convergence of the method of steepest descent depend?

4. Write an algorithm for the method of steepest descent.

5. What is an objective function in linear programming?

6. What is the basic idea of linear programming?

7. What are slack variables? Artificial variables? Why did we introduce them?

8. Why can we not use a method from calculus in a linear programming problem?

9. What is a degenerate feasible solution? Give an example.

10. What do you know about difficulties that may arise in linear programming?

11. What happens in Example 1 of Sec. 20.1 if you replace the function $f(\mathbf{x}) = x_1^2 + 3x_2^2$ by $f(\mathbf{x}) = x_1^2 + 5x_2^2$? Do 5 steps, starting from $\mathbf{x}_0 = [6 \quad 3]^T$. Is the convergence faster or slower?

12. Apply the method of steepest descent to $f(\mathbf{x}) = 9x_1^2 + x_2^2 + 18x_1 - 4x_2$, 5 steps, starting from $\mathbf{x}_0 = [2 \quad 4]^T$.

13. In Prob. 12, could you start from $[0 \quad 0]^T$ and do 5 steps?

14. Show that the gradients in Prob. 13 are orthogonal. Give a reason.

15. What does the method of steepest descent amount to in the case of a function of a single variable?

16. Design a "method of steepest ascent" for determining maxima.

Sketch the region in the first quadrant of the x_1x_2-plane determined by the following inequalities.

17. $x_1 + 3x_2 \leqq 6$

$2x_1 + x_2 \leqq 4$

18. $x_1 - 2x_2 \leqq -2$

$0.8x_1 + x_2 \leqq 6$

19. $-x_1 + x_2 \geqq 0$

$x_1 + x_2 \leqq 4$

$-x_1 + x_2 \leqq 3.2$

20. $x_1 + x_2 \geqq 3$

$x_1 + x_2 \leqq 9$

$-x_1 + x_2 \geqq -3$

$-x_1 + x_2 \leqq 3$

21. $3x_1 - 7x_2 \geqq -28$

$x_1 + x_2 \geqq 6$

$9x_1 - 2x_2 \leqq 36$

$5x_1 + 6x_2 \leqq 48$

22. $-2x_1 + 3x_2 \leqq 9$

$5x_1 + x_2 \leqq 25$

$2x_1 + 3x_2 \geqq 3$

$x_1 - 4x_2 \leqq 4$

23. Maximize $z = 5x_1 + 10x_2$ subject to $0 \leqq x_1 \leqq 5$, $x_1 + x_2 \leqq 6$, $0 \leqq x_2 \leqq 4$.

24. Maximize $z = 20x_1 + 20x_2$ subject to $x_1 \geqq 0$, $x_2 \geqq 0$, $-x_1 + x_2 \leqq 1$, $x_1 + 3x_2 \leqq 15$, $3x_1 + x_2 \leqq 21$.

25. Minimize $z = 2x_1 - 10x_2$ subject to $x_1 \geqq 0$, $x_2 \geqq 0$, $x_1 - x_2 \leqq 4$, $2x_1 + x_2 \leqq 14$, $x_1 + x_2 \leqq 9$, $-x_1 + 3x_2 \leqq 15$.

SUMMARY OF CHAPTER 20
UNCONSTRAINED OPTIMIZATION,
LINEAR PROGRAMMING

In optimization problems we maximize or minimize an **objective function** $z = f(\mathbf{x})$ depending on control variables $x_1, \cdots, x_m$ whose domain is either unrestricted (**"unconstrained optimization,"** Sec. 20.1) or restricted by constraints in the form of inequalities or equations or both (**"constrained optimization,"** Sec. 20.2).

If the objective function is *linear* and the constraints are *linear inequalities* in $x_1, \cdots, x_m$, then by introducing **slack variables** $x_{m+1}, \cdots, x_n$ we can write the optimization problem in **normal form** with the objective function given by

(1) $$f_1 = c_1x_1 + \cdots + c_nx_n$$

(where $c_{m+1} = \cdots = c_n = 0$) and the constraints given by

(2)
$$a_{11}x_1 + a_{12}x_2 + \cdots + a_{1n}x_n = b_1$$
$$a_{21}x_1 + a_{22}x_2 + \cdots + a_{2n}x_n = b_2$$
$$\cdots\cdots\cdots\cdots\cdots\cdots\cdots\cdots\cdots\cdots$$
$$a_{m1}x_1 + a_{m2}x_2 + \cdots + a_{mn}x_n = b_m$$
$$x_1 \geqq 0, \cdots, x_n \geqq 0.$$

In this case we can then apply the widely used **simplex method** (Sec. 20.3), a systematic stepwise search through a very much reduced subset of all feasible solutions. Section 20.4 shows how to overcome difficulties that may arise in connection with this method.

Graphs and Combinatorial Optimization

Graphs and **digraphs** (= directed graphs) are presently developing into more and more powerful tools in electrical and civil engineering, communication networks, industrial management, operations research, computer science, economics, management science, marketing sociology, and in other areas. An accelerating factor of this growth is the impact of computers and their use in large-scale optimization problems that can be modeled in terms of graphs and solved by algorithms provided by graph theory. This approach yields models of general applicability and economic importance. It lies at the center of **"combinatorial optimization,"** a term introduced about thirty years ago for denoting optimization problems that have pronounced discrete or combinatorial structures.

This chapter gives an introduction to this wide area, which is full of new ideas as well as unsolved problems—in connection, for instance, with efficient computer algorithms and computational complexity. The classes of problems we shall consider include problems on transportation of minimum cost or time, best assignment of workers to jobs, most efficient use of telephone networks, and many others. These classes often form the core of larger and more involved practical problems.

Prerequisites: None.
References: Appendix 1, Part F.
Answers to problems: Appendix 2.

21.1 Graphs and Digraphs

Roughly, a *graph* consists of points, called *vertices,* and lines connecting them, called *edges.* For example, these may be four cities and five highways connecting them, as in Fig. 445. Or the points may represent some people, and we connect by an edge those who do business with each other. Or the vertices may represent computers and the edges connections between them. Let us now give a formal definition.

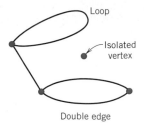

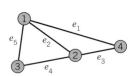

Fig. 445.   Graph consisting of   Fig. 446.   Isolated vertex, loop, double
4 vertices and 5 edges   edge. (Excluded by definition.)

Definition of a graph

A **graph** G consists of two finite sets (sets having finitely many elements), a set V of points, called **vertices,** and a set E of connecting lines, called **edges,** such that each edge connects two vertices, called the *endpoints* of the edge. We write

$$G = (V, E).$$

For simplicity we exclude[1] *isolated vertices* (vertices that are not endpoints of any edge), *loops* (edges whose endpoints coincide), and *multiple edges* (edges that have both endpoints in common). See Fig. 446. This is practical for our purpose.   ◀

We denote vertices by letters, $u, v, \cdots$ or $v_1, v_2, \cdots$ or simply by numbers $1, 2, \cdots$ (as in Fig. 445). We denote edges by $e_1, e_2, \cdots$ or by their two endpoints; for instance, $e_1 = (1, 4)$, $e_2 = (1, 2)$ in Fig. 445.

We say that an edge (v_i, v_j) is **incident** with the vertex v_i (and conversely); similarly, (v_i, v_j) is *incident* with v_j. The number of edges incident with a vertex v is called the **degree** of v. We call two vertices **adjacent** in G if they are connected by an edge in G (that is, if they are the two endpoints of some edge in G).

We meet graphs in different fields under different names: as "networks" in electrical engineering, "structures" in civil engineering, "molecular structures" in chemistry, "organizational structures" in economics, "sociograms," "road maps," "telecommunication networks," and so on.

Digraphs

Nets of one-way streets, pipeline networks, sequences of jobs in construction work, flows of computation in a computer, producer–consumer relations, and many other applications suggest the idea of a "digraph" (= directed graph) in which each edge has a direction (indicated by an arrow, as in Fig. 447 on the next page).

---

[1]As many authors do, but there is no uniformity, and one must be careful. Some authors permit multiple edges and call graphs without them **"simple graphs."** Others permit loops or isolated vertices, depending on the purpose.

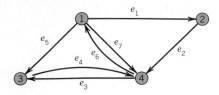

Fig. 447.   Digraph

Definition of a digraph (directed graph)

A **digraph** $G = (V, E)$ is a graph in which each edge $e = (i, j)$ has a direction from its *"initial point"* i to its *"terminal point"* j. ◀

Two edges connecting the same two points i, j are now permitted, provided they have opposite directions, that is, they are (i, j) and (j, i). *Example.* (1, 4) and (4, 1) in Fig. 447.

A **subgraph** or subdigraph of a given graph or digraph $G = (V, E)$, respectively, is a graph or digraph obtained by deleting some of the edges and vertices of G, retaining the other edges of G (together with their pairs of endpoints). For instance, e_1, e_3 (together with the vertices 1, 2, 4) form a subgraph in Fig. 445, and e_3, e_4, e_5 (together with the vertices 1, 3, 4) form a subdigraph in Fig. 447.

Computer Representation of Graphs and Digraphs

Drawings of graphs are useful to people in explaining or illustrating specific situations. Here one should be aware that a graph may be sketched in various ways (see Fig. 448). For handling graphs and digraphs in computers, one uses matrices or lists, as follows.

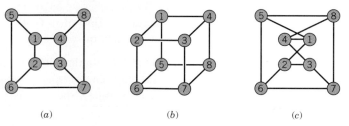

      (a)   (b)   (c)

Fig. 448.   Different sketches of the same graph

Adjacency matrix $\mathbf{A} = [a_{ij}]$ of a graph G

$$a_{ij} = \begin{cases} 1 & \text{if } G \text{ has an edge } (i, j), \\ 0 & \text{else.} \end{cases}$$

Thus $a_{ij} = 1$ if and only if two vertices i and j are adjacent in G. Here, by definition, no vertex is considered to be adjacent to itself; thus, $a_{ii} = 0$. $\mathbf{A}$ is symmetric, $a_{ij} = a_{ji}$. (Why?)

The adjacency matrix of a graph is generally much smaller than the so-called *incidence matrix* (see Prob. 18) and is preferred over the latter if one decides to store a graph in a computer in matrix form.

EXAMPLE 1   **Adjacency matrix of a graph**

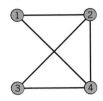

Vertex	1	2	3	4
Vertex 1	0	1	0	1
2	1	0	1	1
3	0	1	0	1
4	1	1	1	0

◀

Adjacency matrix $\mathbf{A} = [a_{ij}]$ of a digraph G

$$a_{ij} = \begin{cases} 1 & \text{if } G \text{ has a directed edge } (i, j), \\ 0 & \text{else.} \end{cases}$$

This matrix $\mathbf{A}$ is not symmetric. (Why?)

EXAMPLE 2   **Adjacency matrix of a digraph**

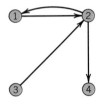

To vertex	1	2	3	4
From vertex 1	0	1	0	0
2	1	0	0	1
3	0	1	0	0
4	0	0	0	0

◀

Lists. The **vertex incidence list** of a graph shows for each vertex the incident edges. The **edge incidence list** shows for each edge its two endpoints. Similarly for a *digraph;* in the vertex list, outgoing edges then get a minus sign, and in the edge list we now have *ordered* pairs of vertices.

EXAMPLE 3   **Vertex incidence list and edge incidence list of a graph**

This graph is the same as in Example 1, except for notation.

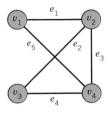

Vertex	Incident Edges
v_1	e_1, e_5
v_2	e_1, e_2, e_3
v_3	e_2, e_4
v_4	e_3, e_4, e_5

Edge	Endpoints
e_1	v_1, v_2
e_2	v_2, v_3
e_3	v_2, v_4
e_4	v_3, v_4
e_5	v_1, v_4

◀

"Sparse graphs" are graphs with few edges (far fewer than the maximum possible number $n(n - 1)/2$, where n is the number of vertices). For these graphs, matrices are not efficient. *Lists* then have the advantage of requiring much less storage and being easier to handle; they can be ordered, sorted, or manipulated in various other ways directly within the computer. For instance, in tracing a "walk" (a connected sequence of edges with pairwise common endpoints), one can easily go back and forth between the two lists just discussed, instead of scanning a large column of a matrix for a single 1.

Computer science has developed more refined lists, which, in addition to the actual content, contain "pointers" indicating the preceding item or the next item to be scanned or both items (in the case of a "walk": the preceding edge or the subsequent one). For details, see Ref. [E11] in Appendix 1.

This section was devoted to basic concepts and notations needed throughout this chapter, in which we shall discuss some of the most important classes of combinatorial optimization problems. This will at the same time help us to become more and more familiar with graphs and digraphs.

PROBLEM SET 21.1

1. Explain how the following can be regarded as graphs or digraphs: a family tree, air connections between given cities, relations between chapters in a book, trade relations between countries, and a hockey tournament.
2. Sketch the graph consisting of the vertices and edges of a triangle. Of a pentagon. Of a tetrahedron.
3. How would you represent a net of one-way and two-way streets by a digraph?
4. Worker W_1 can do jobs J_1, J_2, J_4, worker W_2 job J_3, worker W_3 jobs J_2, J_3. Represent this by a graph.

Adjacency Matrices

Find the adjacency matrix of the given graph or digraph.

5.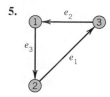

6.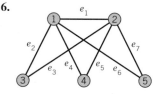

7.

8.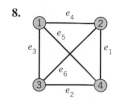

9.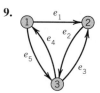

10.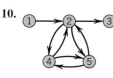

Sketch the graph for each adjacency matrix.

11. $\begin{bmatrix} 0 & 1 & 1 \\ 1 & 0 & 1 \\ 1 & 1 & 0 \end{bmatrix}$

12. $\begin{bmatrix} 0 & 1 & 0 & 1 \\ 1 & 0 & 1 & 0 \\ 0 & 1 & 0 & 0 \\ 1 & 0 & 0 & 0 \end{bmatrix}$

13. $\begin{bmatrix} 0 & 1 & 0 & 0 \\ 1 & 0 & 0 & 0 \\ 0 & 0 & 0 & 1 \\ 0 & 0 & 1 & 0 \end{bmatrix}$

14. Show that the adjacency matrix of a graph is symmetric.
15. When will the adjacency matrix of a digraph be symmetric?
16. **(Complete graph)** Show that a graph G with n vertices can have at most $n(n-1)/2$ edges, and G has exactly $n(n-1)/2$ edges if G is *complete*, that is, if every pair of vertices of G is joined by an edge. (Recall that loops and multiple edges are excluded.)
17. In what case are all the off-diagonal elements of the adjacency matrix of a graph G equal to 1?

18. Incidence matrix B of a graph. The definition is $\mathbf{B} = [b_{jk}]$, where

$$b_{jk} = \begin{cases} 1 & \text{if vertex } j \text{ is an endpoint of edge } e_k, \\ 0 & \text{otherwise.} \end{cases}$$

Find the incidence matrix of the graph in Prob. 6.

19. Incidence matrix $\widetilde{\mathbf{B}}$ of a digraph. The definition is $\widetilde{\mathbf{B}} = [\widetilde{b}_{jk}]$, where

$$\widetilde{b}_{jk} = \begin{cases} -1 & \text{if edge } e_k \text{ leaves vertex } j, \\ 1 & \text{if edge } e_k \text{ enters vertex } j, \\ 0 & \text{otherwise.} \end{cases}$$

Find the incidence matrix of the digraph in Prob. 7.

20. Make the vertex incidence list of the digraph in Prob. 7.

21.2 Shortest Path Problems Complexity

Beginning in this section, we shall discuss some of the most important classes of optimization problems that concern graphs and digraphs as they arise in applications. Basic ideas and algorithms will be explained and illustrated by small graphs, but you should keep in mind that real-life problems may often involve many ***thousands or even millions of vertices and edges*** (think of telephone networks, worldwide air travel, companies that have offices and stores in all larger cities). Then reliable and efficient systematic methods are an absolute necessity—solution by inspection or by trial and error would no longer work, even if "nearly optimal" solutions are acceptable.

We begin with **shortest path problems,** as they arise, for instance, in designing shortest (or least expensive, or fastest) routes for a traveling salesman, for a cargo ship, etc. Let us first explain what we mean by a path.

In a graph $G = (V, E)$ we can walk from a vertex v_1 along some edges to some other vertex v_k. Here we can

(A) make no restrictions, or

(B) require that each *edge* of G be traversed at most once, or

(C) require that each *vertex* be visited at most once.

In case (A) we call this a **walk.** Thus a walk from v_1 to v_k is of the form

(1) $(v_1, v_2), (v_2, v_3), \cdots, (v_{k-1}, v_k),$

where some of these edges or vertices may be the same. In case (B), where each *edge* may occur at most once, we call the walk a **trail.** Finally, in case (C), where each *vertex* may occur at most once (and thus each edge automatically occurs at most once), we call the trail a **path.**

We admit that a walk, trail, or path may end at the vertex it started from, in which case we call it **closed;** then $v_k = v_1$ in (1).

A closed path is called a **cycle.** *A cycle has at least three edges* (because we do not have double edges; see Sec. 21.1). Figure 449 on the next page illustrates all these concepts.

Fig. 449.   Walk, trail, path, cycle

1 − 2 − 3 − 2 is a walk (not a trail).
4 − 1 − 2 − 3 − 4 − 5 is a trail (not a path).
1 − 2 − 3 − 4 − 5 is a path (not a cycle).
1 − 2 − 3 − 4 − 1 is a cycle.

Shortest Path

Having said what a path is, we should say next what we mean by a shortest path in a graph $G = (V, E)$. For this, each edge (v_i, v_j) in G must have a given "length" $l_{ij} > 0$. Then a **shortest path** $v_1 \rightarrow v_k$ (with fixed v_1 and v_k) is a path (1) such that the sum of the lengths of its edges

$$l_{12} + l_{23} + l_{34} + \cdots + l_{k-1,k}$$

(l_{12} = length of (v_1, v_2), etc.) is minimum (as small as possible among all paths from v_1 to v_k). Similarly, a **longest path** $v_1 \rightarrow v_k$ is one for which that sum is maximum.

Shortest (and longest) path problems are among the most important optimization problems. Here, "length" l_{ij} (often also called "cost" or "weight") can be an actual length measured in miles or travel time or gasoline expenses, but it may also be something entirely different.

For instance, the *"traveling salesman problem"* requires the determination of a shortest **Hamiltonian[2] cycle** in a given graph, that is, a cycle that contains all the vertices of the graph.

As another example, by choosing the "most profitable" route $v_1 \rightarrow v_k$, a salesman may want to maximize Σl_{ij}, where l_{ij} is his expected commission minus his travel expenses for going from town i to town j.

In an investment problem, i may be the day an investment is made, j the day it matures, and l_{ij} the resulting profit, and one gets a graph by considering the various possibilities of investing and reinvesting over a given period of time.

Shortest Path If All Edges Have Length 1

Obviously, if all edges have length 1, then a shortest path $v_1 \rightarrow v_k$ is one that has the smallest number of edges among all paths $v_1 \rightarrow v_k$ in a given graph G. For this problem we discuss a BFS algorithm. BFS stands for **Breadth First Search.** This means that in each step the algorithm visits *all neighboring* (all adjacent) vertices of a vertex reached, as opposed to a DFS algorithm (**Depth First Search** algorithm), which makes a long trail (as in a maze). This widely used algorithm is shown in Table 21.1.

[2]WILLIAM ROWAN HAMILTON (1805—1865), Irish mathematician, known for his work in dynamics.

Table 21.1
Moore's BFS for Shortest Path (All Lengths One)[3]

ALGORITHM MOORE $[G = (V, E), s, t]$

This algorithm determines a shortest path in a connected graph $G = (V, E)$ from a vertex s to a vertex t.

INPUT: Connected graph $G = (V, E)$, in which one vertex is denoted by s and one by t, and each edge (i, j) has length $l_{ij} = 1$. Initially all vertices are unlabeled.

OUTPUT: A shortest path $s \rightarrow t$ in $G = (V, E)$

1. Label s with 0.
2. Set $i = 0$.
3. Find all *unlabeled* vertices adjacent to a vertex labeled i.
4. Label the vertices just found with $i + 1$.
5. If vertex t is labeled, then "backtracking" gives the shortest path

$$k \; (= \text{label of } t), k - 1, k - 2, \cdots, 0$$

   OUTPUT $k, k - 1, k - 2, \cdots, 0$. Stop
Else increase i by 1. Go to Step 3.
End MOORE

We want to find a shortest path in G from a vertex s (**start**) to a vertex t (**terminal**). To guarantee that there is a path from s to t, we make sure that G does not consist of separate portions. Thus we assume that G is **connected,** that is, for any two vertices v and w there is a path $v \rightarrow w$ in G. (Recall that a vertex v is called **adjacent** to a vertex u if there is an edge (u, v) in G.)

EXAMPLE 1   **Application of Moore's BFS algorithm**

Find a shortest path $s \rightarrow t$ in the graph G shown in Fig. 450.

Solution.   Figure 450 shows the labels. The blue edges form a shortest path (length 4). There is another shortest path $s \rightarrow t$. (Can you find it?) Hence in the program we must introduce a rule that makes backtracking unique because otherwise the computer would not know what to do next if at some step there is a choice (for instance, in Fig. 450 when it got back to the vertex labeled 2). The following rule seems to be natural.

Backtracking rule.   Using the numbering of the vertices from 1 to n (not the labeling!), at each step, if a vertex labeled i is reached, take as the next vertex that with the smallest number (not label!) among all the vertices labeled $i - 1$.   ◀

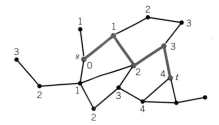

Fig. 450.   Example 1, given graph and result of labeling

[3]*Proceedings of the International Symposium for Switching Theory,* Part II, pp. 285–292. Cambridge: Harvard University Press, 1959.

Complexity of an Algorithm

Complexity *of Moore's algorithm.* To find the vertices to be labeled 1, we have to scan all edges incident with s. Next, when $i = 1$, we have to scan all edges incident with vertices labeled 1, etc. Hence each edge is scanned twice. These are $2m$ operations (m = number of edges of G). This is a function $c(m)$. Whether it is $2m$ or $5m + 3$ or $12m$ is not so essential; it *is* essential that $c(m)$ is proportional to m (not m^2, for example); it is of the "order" m. We write for any function $am + b$ simply $O(m)$, for any function $am^2 + bm + d$ simply $O(m^2)$, and so on; here, O suggests "order." The underlying idea and practical aspect are as follows.

In judging an algorithm, we are mostly interested in its behavior for very large problems (large m in the present case), since these are going to determine the limits of the applicability of the algorithm. Thus, the essential item is the fastest growing term (am^2 in $am^2 + bm + d$, etc.) since it will overwhelm the others when m is large enough. Also, a constant factor in this term is not very essential; for instance, the difference between two algorithms of orders, say, $5m^2$ and $8m^2$ is generally not very essential and can be made irrelevant by a modest increase in the speed of computers. However, it does make a great practical difference whether an algorithm is of order m or m^2 or of a still higher power m^p. And the biggest difference occurs between these "polynomial orders" and "exponential orders," such as 2^m.

For instance, on a computer that does 10^9 operations per second, a problem of size $m = 50$ will take 0.3 second with an algorithm that requires m^5 operations, but 13 days with an algorithm that requires 2^m operations. But this is not our only reason for regarding polynomial orders as good and exponential orders as bad. Another reason is the gain in using a faster computer. For example let two algorithms be $O(m)$ and $O(m^2)$. Then, since $1000 = 31.6^2$, an increase in speed by a factor 1000 has the effect that per hour we can do problems 1000 and 31.6 times as big, respectively. But since $1000 = 2^{9.97}$, with an algorithm that is $O(2^m)$, all we gain is a relatively modest increase of 10 in problem size because $2^{9.97} \cdot 2^m = 2^{m+9.97}$.

The **symbol** O is quite practical and commonly used whenever the order of growth is essential, but not the specific form of a function. Thus if a function $g(m)$ is of the form

$$g(m) = kh(m) + \text{more slowly growing terms} \qquad (k \neq 0, \text{constant}),$$

we say that $g(m)$ is of the *order* $h(m)$ and write

$$g(m) = O(h(m)).$$

For instance,

$$am + b = O(m), \qquad am^2 + bm + d = O(m^2), \qquad 5 \cdot 2^m + 3m^2 = O(2^m),$$

and so on.

We want an algorithm $\mathcal{A}$ to be "efficient," that is, "good" with respect to

        (i) *Time* (number $c_{\mathcal{A}}(m)$ of computer operations)

or

        (ii) *Space* (storage needed in the internal memory)

or both. Here $c_{\mathcal{A}}$ suggests **"complexity"** of $\mathcal{A}$. Two popular choices for $c_{\mathcal{A}}$ are

(*Worst case*) $c_{\mathcal{A}}(m)$ = longest time $\mathcal{A}$ takes for a problem of size m,

(*Average case*) $c_{\mathcal{A}}(m)$ = average time $\mathcal{A}$ takes for a problem of size m.

In problems on graphs, the "size" will often be m (number of edges) or n (number of vertices). For our present simple algorithm, $c_{\mathcal{A}}(m) = 2m$ in both cases.

For a "good" algorithm $\mathcal{A}$, we want that $c_{\mathcal{A}}(m)$ does not grow too fast. Accordingly, we call $\mathcal{A}$ **efficient** if $c_{\mathcal{A}}(m) = O(m^k)$ for some integer $k \geqq 0$; that is, $c_{\mathcal{A}}$ may contain only powers of m (or functions that grow even more slowly, such as $\ln m$), but no exponential functions. Furthermore, we call $\mathcal{A}$ **polynomially bounded** if $\mathcal{A}$ is efficient when we choose the "worst case" $c_{\mathcal{A}}(m)$. These conventional concepts have intuitive appeal, as our discussion shows.

Complexity should be investigated for every algorithm, so that one can also compare different algorithms for the same task. This may often exceed the level in this chapter; accordingly, we shall confine ourselves to a few occasional comments in this direction.

PROBLEM SET 21.2

Shortest Paths

Find a shortest path $P: s \rightarrow t$ and its length by Moore's BFS algorithm; sketch the graph with the labels and indicate P by heavier lines (as in Fig. 450).

1.   **2.**   **3.**

4. (**Maximum length**) How many edges can a shortest path between any two vertices in a graph with n vertices at most have? Give a reason. In a complete graph with all edges of length 1?
5. (**Moore's algorithm**) Show that if vertex v has label $\lambda(v) = k$, then there is a path $s \rightarrow v$ of length k.
6. (**Moore's algorithm**) Call the length of a shortest path $s \rightarrow v$ briefly the *distance* of v from s. Show that if v has distance l, it has label $\lambda(v) = l$.
7. (**Nonuniqueness**) Find another shortest path from s to t in Example 1 of the text.

Hamiltonian Cycle

8. Does the graph in Prob. 2 have a Hamiltonian cycle? (Give a reason.)
9. Find and sketch a Hamiltonian cycle in Prob. 1.
10. Find and sketch a Hamiltonian cycle in the graph of a dodecahedron, which has 12 pentagonal faces and 20 vertices (Fig. 451). This is a problem Hamilton considered.

Fig. 451.   Problem 10

11. Find and sketch a Hamiltonian cycle in Fig. 448, Sec. 21.1.

Euler Graph. Postman Problem

12. **(Euler graph)** An *Euler graph* G is a graph that has a closed Euler trail. An **Euler trail** is a trail that contains every edge of G exactly once. Which subgraph with four edges of the graph in Example 1, Sec. 21.1, is an Euler graph?

13. Is the graph in Fig. 452 an Euler graph? (Give a reason.)

14. Find two different closed Euler trails in Fig. 453.

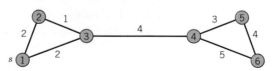

Fig. 452.   Problems 13, 17

Fig. 453.   Problem 14

15. **(Postman problem)** The *postman problem* (or *Chinese postman problem*[4]) is the problem of finding a closed walk $W: s \to s$ (s is the post office) in a graph G with edges (i, j) of length $l_{ij} > 0$ such that every edge of G is traversed at least once and the length of W is minimum. Explain some other applications in which this problem is essential.

16. Show that the length of a shortest postman trail is the same for every starting vertex.

17. Find a solution of the postman problem in Fig. 452 by inspection.

18. **(Order)** Show that $O(m^3) + O(m^3) = O(m^3)$ and $kO(m^p) = O(m^p)$, $0.02e^m + 100m^2 = O(e^m)$.

19. **(Order)** If we switch from one computer to another that is 100 times as fast, what is our gain in problem size per hour in the use of an algorithm that is $O(m)$, $O(m^2)$, $O(m^5)$, $O(e^m)$?

 20. **CAS PROBLEM. Moore's Algorithm.** Write a computer program for the algorithm in Table 21.1. Apply it to the graphs in Example 1, Probs. 1–3, and to some graphs of your own choice.

21.3 Bellman's Optimality Principle Dijkstra's Algorithm

We continue our discussion of the shortest path problem in a graph G. The last section concerned the special case that all edges had length 1. But in most applications the edges (i, j) will have any lengths $l_{ij} > 0$, and we now turn to this general case, which is of greater practical importance. We write $l_{ij} = \infty$ for any edge (i, j) that does not exist in G (setting $\infty + a = \infty$ for any number a, as usual).

We consider the problem of finding shortest paths from a given vertex, denoted by 1 and called the **origin,** to *all* other vertices 2, 3, $\cdots$, n of G. We let L_j denote the length of a shortest path $1 \to j$ in G.

---

[4]Since it was first considered in the journal *Chinese Mathematics* **1** (1962), 273–77.

Bellman's minimality principle (or optimality principle)
If P: $1 \to j$ is a shortest path from 1 to j in G and (i, j) is the last edge of P (Fig. 454), then P_i: $1 \to i$ [obtained by dropping (i, j) from P] is a shortest path $1 \to i$.

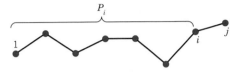

Fig. 454. Paths P and P_i in Bellman's minimality principle

PROOF.   Suppose that the conclusion is false. Then there is a path P_i^*: $1 \to i$ that is shorter than P_i. Hence if we now add (i, j) to P_i^*, we get a path $1 \to j$ that is shorter than P. This contradicts our assumption that P is shortest.   ◀

From Bellman's principle we can derive basic equations as follows. For fixed j we may obtain various paths $1 \to j$ by taking shortest paths P_i for various i for which there is in G an edge (i, j), and add (i, j) to the corresponding P_i. These paths obviously have lengths $L_i + l_{ij}$ (L_i = length of P_i). We can now take the minimum over i, that is, pick an i for which $L_i + l_{ij}$ is smallest. By the Bellman principle, this gives a shortest path $1 \to j$. It has the length

(1)
$$
\begin{aligned}
L_1 &= 0 \\
L_j &= \min_{i \neq j} (L_i + l_{ij}),
\end{aligned}
\qquad j = 2, \cdots, n.
$$

These are the **Bellman equations.** Since $l_{ii} = 0$ by definition, instead of $\min_{i \neq j}$ we can simply write $\min_i$. These equations suggest the idea of one of the best-known algorithms for the shortest path problem, as follows.

Dijkstra's Algorithm for Shortest Paths

Dijkstra's algorithm is shown in Table 21.2 on the next page. It is a labeling procedure. At each stage of the computation, each vertex v gets a label, either

(PL)   a *permanent label* = length L_v of a shortest path $1 \to v$

or

(TL)   a *temporary label* = upper bound $\widetilde{L}_v$ for the length of a shortest path $1 \to v$.

We denote by $\mathcal{PL}$ and $\mathcal{TL}$ the sets of vertices with a permanent label and with a temporary label, respectively. The algorithm has an initial step in which vertex 1 gets the permanent label $L_1 = 0$ and the other vertices get temporary labels, and then the algorithm alternates between Steps 2 and 3. In Step 2 the idea is to pick k "minimally." In Step 3 the idea is that the upper bounds will in general improve (decrease) and must be updated accordingly.

Table 21.2
Dijkstra's Algorithm for Shortest Path[5]

ALGORITHM DIJKSTRA $\left[G = (V, E), V = \{1, \cdots, n\}, l_{ij} \text{ for all } (i, j) \text{ in } E \right]$

Given a connected graph $G = (V, E)$ with vertices $1, \cdots, n$ and edges (i, j) having lengths $l_{ij} > 0$, this algorithm determines the lengths of shortest paths from vertex 1 to the vertices $2, \cdots, n$.

     INPUT: Number of vertices n, edges (i, j), and lengths l_{ij}
     OUTPUT: Lengths L_j of shortest paths $1 \rightarrow j, j = 2, \cdots, n$

   1. *Initial step*

      Vertex 1 gets PL: $L_1 = 0$.
      Vertex j ($= 2, \cdots, n$) gets TL: $\widetilde{L}_j = l_{1j}$ ($= \infty$ if there is no edge $(1, j)$ in G).
      Set $\mathscr{PL} = \{1\}$, $\mathscr{TL} = \{2, 3, \cdots, n\}$.

   2. *Fixing a permanent label*

      Find a k in $\mathscr{TL}$ for which $\widetilde{L}_k$ is minimum, set $L_k = \widetilde{L}_k$. Take the smallest k if there are several. Delete k from $\mathscr{TL}$ and include it in $\mathscr{PL}$.
      If $\mathscr{TL} = \varnothing$ (that is, $\mathscr{TL}$ is empty) then
          OUTPUT $L_2, \cdots, L_n$. Stop
      Else continue (that is, go to Step 3).

   3. *Updating temporary labels*

      For all j in $\mathscr{TL}$, set[6] $\widetilde{L}_j = \min_k \{\widetilde{L}_j, L_k + l_{kj}\}$.
      Go to Step 2.

End DIJKSTRA

EXAMPLE 1   **Application of Dijkstra's algorithm**

Applying Dijkstra's algorithm to the graph in Fig. 455*a*, find shortest paths from vertex 1 to vertices 2, 3, 4.

Solution.   We list the steps and computations.

   1. $L_1 = 0, \widetilde{L}_2 = 8, \widetilde{L}_3 = 5, \widetilde{L}_4 = 7,$ $\mathscr{PL} = \{1\},$ $\mathscr{TL} = \{2, 3, 4\}$

   2. $L_3 = \min \{\widetilde{L}_2, \widetilde{L}_3, \widetilde{L}_4\} = 5, k = 3,$ $\mathscr{PL} = \{1, 3\},$ $\mathscr{TL} = \{2, 4\}$

   3. $\widetilde{L}_2 = \min \{8, L_3 + l_{32}\} = \min \{8, 5 + 1\} = 6$

      $\widetilde{L}_4 = \min \{7, L_3 + l_{34}\} = \min \{7, \infty\} = 7$

   2. $L_2 = \min \{\widetilde{L}_2, \widetilde{L}_4\} = 6, k = 2,$ $\mathscr{PL} = \{1, 2, 3\},$ $\mathscr{TL} = \{4\}$

   3. $\widetilde{L}_4 = \min \{7, L_2 + l_{24}\} = \min \{7, 6 + 2\} = 7$

   2. $L_4 = 7, k = 4$ $\mathscr{PL} = \{1, 2, 3, 4\},$ $\mathscr{TL} = \varnothing.$

Figure 455*b* shows the resulting shortest paths, of lengths $L_2 = 6, L_3 = 5, L_4 = 7$.   ◀

---

[5]*Numerische Mathematik* **1** (1959), 269−271.

[6]That is, take the number $\min_k \{L_j, L_k + l_{kj}\}$ as your new $\widetilde{L}_j$.

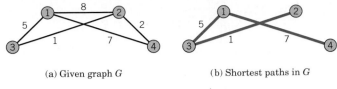

(a) Given graph G (b) Shortest paths in G

Fig. 455.   Example 1

Complexity. *Dijkstra's algorithm is $O(n^2)$.*

PROOF.   Step 2 requires comparison of elements, first $n - 2$, the next time $n - 3$, etc., a total of $(n - 2)(n - 1)/2$. Step 3 requires the same number of comparisons, a total of $(n - 2)(n - 1)/2$, as well as additions, first $n - 2$, the next time $n - 3$, etc., again a total of $(n - 2)(n - 1)/2$. Hence the total number of operations is $3(n - 2)(n - 1)/2 = O(n^2)$. ◄

PROBLEM SET 21.3

1. The net of roads in Fig. 456 connecting four villages is to be reduced to minimum length, but so that one can still reach every village from every other village. Which of the roads should be retained? Find the solution (a) by inspection, (b) by Dijkstra's algorithm.

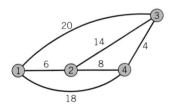

Fig. 456.   Problem 1

2. Show that in Dijkstra's algorithm, for L_k there is a path $P: 1 \to k$ of length L_k.

3. Show that in Dijkstra's algorithm, at each instant the demand on storage is light (data for less than n edges).

Dijkstra's Algorithm. In each of the graphs in Probs. 4–9 find the shortest paths.

4.
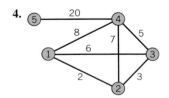

5.
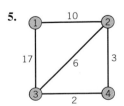

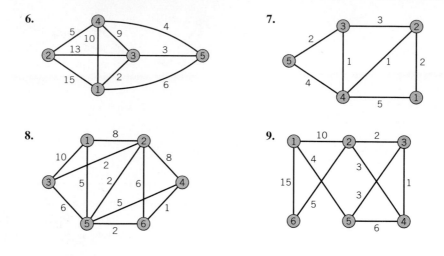

10. CAS PROBLEM. Write a program for Dijkstra's algorithm and apply it to Probs. 4–6.

21.4 Shortest Spanning Trees Kruskal's Greedy Algorithm

So far we have discussed shortest path problems. We now turn to a particularly important kind of graph, called a tree, along with related optimization problems that arise quite often in practice.

By definition, a **tree** T is a graph that is connected and has no cycles. **"Connected"** means that there is a path from any vertex in T to any other vertex in T. A **cycle**[7] is a path $s \to t$ of at least three edges that is closed ($t = s$); see also Sec. 21.2. Figure 457a shows an example.

A **spanning tree** T in a given connected graph $G = (V, E)$ is a tree containing *all* the n vertices of G. See Fig. 457b. Such a tree has $n - 1$ edges. (Proof?)

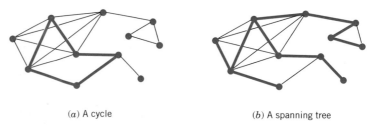

(a) A cycle   (b) A spanning tree

Fig. 457.   Example of (a) a cycle, (b) a spanning tree in a graph

A **shortest spanning tree** T in a connected graph G (whose edges (i, j) have lengths $l_{ij} > 0$) is a spanning tree for which Σl_{ij} (sum over all edges of T) is minimum compared to Σl_{ij} for any other spanning tree in G.

---

[7]Or **circuit.** Caution! The terminology varies considerably.

Trees are among the most important types of graphs, and they occur in various applications. Familiar examples are family trees and organization charts. Trees can be used to exhibit, organize, or analyze electrical networks, producer–consumer and other business relations, information in database systems, syntactic structure of computer programs, etc. We mention a few specific applications that need no lengthy additional explanations.

The set of shortest paths from vertex 1 to the vertices $2, \cdots, n$ in the last section forms a spanning tree.

Railway lines connecting a number of cities (the vertices) can be set up in the form of a spanning tree, the "length" of a line (edge) being the construction cost, and one wants to minimize the total construction cost. Similarly for bus lines, where "length" may be the average annual operating cost. Or for steamship lines (freight lines), where "length" may be profit and the goal is the maximization of total profit. Or in a network of telephone lines between some cities, a shortest spanning tree may simply represent a selection of lines that connect all the cities at minimal cost. In addition to these examples we could mention others from distribution networks, and so on.

We shall now discuss a simple algorithm for the shortest spanning tree problem, which is particularly suitable for sparse graphs (graphs with very few edges; see Sec. 21.1).

Table 21.3
Kruskal's Greedy Algorithm for Shortest Spanning Trees[8]

ALGORITHM KRUSKAL $[G = (V, E), l_{ij}$ for all (i, j) in $E]$

Given a connected graph $G = (V, E)$ with edges (i, j) having length $l_{ij} > 0$, the algorithm determines a shortest spanning tree T in G.

INPUT: Edges (i, j) of G and their lengths l_{ij}
OUTPUT: Shortest spanning tree T in G

1. Order the edges of G in ascending order of length.
2. Choose them in this order as edges of **T**, rejecting an edge only
     if it forms a cycle with edges already chosen.

  If $n - 1$ edges have been chosen, then
          OUTPUT T (= the set of edges chosen). Stop
End KRUSKAL

EXAMPLE 1   **Application of Kruskal's algorithm**

Using Kruskal's algorithm, we shall determine a shortest spanning tree in the graph in Fig. 458.

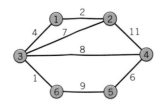

Fig. 458.   Graph in Example 1

[8]*Proceedings of the American Mathematical Society* **7** (1956), 48−50.

Solution. See Table 21.4. In some of the intermediate stages the edges chosen form a *disconnected* graph (see Fig. 459); this is typical. We stop after $n - 1 = 5$ choices since a spanning tree has $n - 1$ edges. In our problem the edges chosen are in the upper part of the list. This is typical of problems of any size; in general, edges farther down in the list have a smaller chance of being chosen. ◄

Table 21.4
Solution in Example 1

Edge	Length	Choice
(3, 6)	1	1st
(1, 2)	2	2nd
(1, 3)	4	3rd
(4, 5)	6	4th
(2, 3)	7	Reject
(3, 4)	8	5th
(5, 6)	9	
(2, 4)	11	

The efficiency of Kruskal's method is greatly increased by

Double labeling of vertices

Each vertex i carries a double label (r_i, p_i), where

     $r_i =$ *Root of the subtree to which i belongs,*
     $p_i =$ *Predecessor of i in its subtree,*
     $p_i = 0$ *for roots.*

This simplifies

Rejecting. *If (i, j) is next in the list to be considered, reject (i, j) if $r_i = r_j$ (that is, i and j are in the same subtree, so that they are already joined by edges and (i, j) would thus create a cycle). If $r_i \neq r_j$, include (i, j) in T.*

If there are several choices for r_i, choose the smallest. If subtrees merge (become a single tree), retain the smallest root as the root of the new subtree.

For Example 1, the double-label list is shown in Table 21.5. In storing it, at each instant one may retain only the latest double label. We show all double labels in order to exhibit the process in all its stages. Labels that remain unchanged are not listed again. Underscored are the two 1's that are the common root of vertices 2 and 3, the reason for rejecting the edge (2, 3). By reading for each vertex the latest label we can read from this list that 1 is

Table 21.5
List of Double Labels in Example 1

Vertex	Choice 1 (3, 6)	Choice 2 (1, 2)	Choice 3 (1, 3)	Choice 4 (4, 5)	Choice 5 (3, 4)
1		(1, 0)			
2		(1̲, 1)			
3	(3, 0)		(1̲, 1)		
4				(4, 0)	(1, 3)
5				(4, 4)	(1, 4)
6	(3, 3)		(1, 3)		

the vertex we have chosen as a root and the tree is as shown in the last part of Fig. 459. This is made possible by the predecessor label that each vertex carries. Also, for accepting or rejecting an edge we have to make only one comparison (the roots of the two endpoints of the edge).

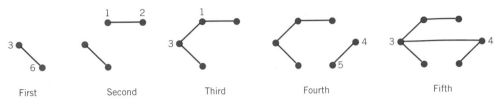

Fig. 459.   Choice process in Example 1

Ordering is the more expensive part of the algorithm. It is a standard process in data processing for which various methods have been suggested (see **Sorting** in Chap. 8 of Ref. [E13] listed in Appendix 1). For a complete list of m edges, an algorithm would be $O(m \log_2 m)$, but since the $n - 1$ edges of the tree are most likely to be found earlier, by inspecting the q ($< m$) topmost edges, for such a list of q edges one would have $O(q \log_2 m)$.

PROBLEM SET 21.4

Shortest Spanning Trees
Find a shortest spanning tree by Kruskal's algorithm. Sketch it.

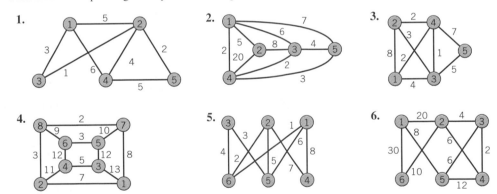

7. Find a shortest spanning tree in the complete graph of all possible 30 air connections between the six cities given (distances in miles, rounded). Can you think of a practical application of the result?

	Dallas	Denver	Los Angeles	New York	Washington, DC
Chicago	800	900	1800	700	650
Dallas		650	1300	1350	1200
Denver			850	1650	1500
Los Angeles				2500	2350
New York					200

8. To get a minimum spanning tree, instead of adding shortest edges, one could think of deleting longest edges. For what graphs would this be feasible? Describe an algorithm for this.

9. Apply the method suggested in Prob. 8 to the graph in Example 1. Do you get the same tree?

10. Design an algorithm for obtaining longest spanning trees.

11. Apply the algorithm in Prob. 10 to the graph in Example 1. Compare with the result in Example 1.

12. **CAS PROBLEM. Kruskal's Algorithm.** Write a corresponding program. (Sorting is discussed in Chap. 8 of Ref. [E13] listed in Appendix 1.)

General Properties of Trees

13. If in a graph G, any two vertices are connected by a unique path, show that G is a tree.

14. **(Uniqueness)** Show that the path connecting any two vertices u and v in a tree T is unique.

15. Show that a tree T with exactly two vertices of degree 1 (see Sec. 21.1) must be a path.

16. Show that a tree with n vertices has $n - 1$ edges.

17. **(Forest)** A (not necessarily connected) graph without cycles is called a *forest*. Give typical examples of applications in which graphs occur that are forests or trees.

18. Show that if a graph G has no cycles, then G must have at least 2 vertices of degree 1.

19. Show that if one joins two vertices in a tree T by a new edge, then a cycle is formed.

20. Show that a graph G with n vertices is a tree if and only if G has $n - 1$ edges and has no cycles.

21.5 Prim's Algorithm for Shortest Spanning Trees

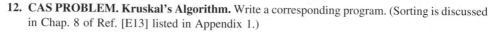

Prim's algorithm shown in Table 21.6 is another popular algorithm for the shortest spanning tree problem (see Sec. 21.4). This algorithm gives a tree T at each stage, a property that Kruskal's algorithm in the last section did not have (look back at Fig. 459 if you did not notice it).

In Prim's algorithm, starting from any single vertex, which we call 1, we "grow" the tree T by adding edges to it, one at a time, according to some rule (below) until T finally becomes a *spanning* tree, which is shortest.

We denote by U the set of vertices of the growing tree T and by S the set of its edges. Thus, initially $U = \{1\}$ and $S = \varnothing$; at the end, $U = V$, the vertex set of the given graph $G = (V, E)$, whose edges (i, j) have length $l_{ij} > 0$, as before.

Thus at the beginning (Step 1) the labels $\lambda_2, \cdots, \lambda_n$ of the vertices $2, \cdots, n$ are the lengths of the edges connecting them to vertex 1 (or ∞ if there is no such edge in G). And we pick (Step 2) the shortest of these as the first edge of the growing tree T and include its other end j in U (choosing the smallest j if there are several, to make the process unique). Updating labels in Step 3 (at this stage and at any later stage) concerns each vertex k not yet in U. Vertex k has label $\lambda_k = l_{i(k),k}$ from before. If $l_{jk} < \lambda_k$, this means that k is closer to the new member j just included in U than k is to its old "closest neighbor" $i(k)$ in U. Then we update the label of k, replacing $\lambda_k = l_{i(k),k}$ by $\lambda_k = l_{jk}$ and setting $i(k) = j$. If, however, $l_{jk} \geqq \lambda_k$ (the *old* label of k), we don't touch the old label. Thus the label λ_k always identifies the closest neighbor of k in U, and this is updated in Step 3 as U and the tree T grow. From the final labels we can backtrack the final tree, and from their numerical values we compute the total length (sum of the lengths of the edges) of this tree.

Table 21.6
Prim's Algorithm for Shortest Spanning Tree[9]

ALGORITHM PRIM $[G = (V, E), V = \{1, \cdots, n\}, l_{ij} \text{ for all } (i, j) \text{ in } E]$

Given a connected graph $G = (V, E)$ with vertices $1, 2, \cdots, n$ and edges (i, j) having length $l_{ij} > 0$, this algorithm determines a shortest spanning tree T in G and its length $L(T)$.

INPUT: n, edges (i, j) of G and their lengths l_{ij}
OUTPUT: Edge set S of a shortest spanning tree T in G; $L(T)$
[*Initially, all vertices are unlabeled.*]

1. *Initial step*
     Set $i(k) = 1$, $U = \{1\}$, $S = \varnothing$.
     Label vertex $k (= 2, \cdots, n)$ with $\lambda_k = l_{ik}$ [$= \infty$ if G has no edge $(1, k)$].

2. *Addition of an edge to the tree T*
     Let λ_j be the smallest λ_k for k not in U. Include vertex j in U and edge $(i(j), j)$ in S.
     If $U = V$ then compute
          $L(T) = \Sigma l_{ij}$ (sum over all edges in S)
          OUTPUT S, $L(T)$. Stop
          [S *is the edge set of a shortest spanning tree T in G.*]
     Else continue (that is, go to Step 3).

3. *Label updating*
     For every k not in U, if $l_{jk} < \lambda_k$, then set $\lambda_k = l_{jk}$ and $i(k) = j$.
     Go to Step 2.
End PRIM

EXAMPLE 1   **Application of Prim's algorithm**

Find a shortest spanning tree in the graph in Fig. 460 (which is the same as in Example 1, Sec. 21.4, so that we can compare).

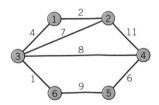

Fig. 460.   Graph in Example 1

The solution is shown on the next page.

---

[9]*Bell System Technical Journal* **36** (1957), 1389–1401. For an improved version of the algorithm, see Cheriton and Tarjan, *SIAM Journal on Computation* **5** (1976), 724–742.

Solution.   The steps are as follows.

1. $i(k) = 1$, $U = \{1\}$, $S = \varnothing$, initial labels see Table 21.7.

2. $\lambda_2 = l_{12} = 2$ is smallest, $U = \{1, 2\}$, $S = \{(1, 2)\}$

3. Update labels as shown in Table 21.7, column (I).

2. $\lambda_3 = l_{13} = 4$ is smallest, $U = \{1, 2, 3\}$, $S = \{(1, 2), (1, 3)\}$

3. Update labels as shown in Table 21.7, column (II).

2. $\lambda_6 = l_{36} = 1$ is smallest, $U = \{1, 2, 3, 6\}$, $S = \{(1, 2), (1, 3), (3, 6)\}$

3. Update labels as shown in Table 21.7, column (III).

2. $\lambda_4 = l_{34} = 8$ is smallest, $U = \{1, 2, 3, 4, 6\}$, $S = \{(1, 2), (1, 3), (3, 4), (3, 6)\}$

3. Update labels as shown in Table 21.7, column (IV).

2. $\lambda_5 = l_{45} = 6$ is smallest, $U = V$, $S = (1, 2), (1, 3), (3, 4), (3, 6), (4, 5)$. Stop.

The tree is the same as in Example 1, Sec. 21.4. Its length is 21. You will find it interesting to compare the growth process of the present tree with that in Sec. 21.4.   ◀

Table 21.7
Labeling of Vertices in Example 1

Vertex	Initial Label	Relabeling			
		(I)	(II)	(III)	(IV)
2	$l_{12} = 2$	—	—	—	—
3	$l_{13} = 4$	$l_{13} = 4$	—	—	—
4	∞	$l_{24} = 11$	$l_{34} = 8$	$l_{34} = 8$	—
5	∞	∞	∞	$l_{65} = 9$	$l_{45} = 6$
6	∞	∞	$l_{36} = 1$	—	—

PROBLEM SET 21.5

Prim's Algorithm

1. In what case will at the end $S = E$ in Prim's algorithm?

2. **(Complexity)** Show that Prim's algorithm has complexity $O(n^2)$.

3. How does Prim's algorithm prevent the generation of cycles as one grows T?

4. For a complete graph (or one that is almost complete), if our data is an $n \times n$ distance table (as in Prob. 7, Sec. 21.4), show that the present algorithm [which is $O(n^2)$] cannot easily be replaced by an algorithm of order less than $O(n^2)$.

5. What would the result be if one applied Prim's algorithm to a graph that is not connected?

In Probs. 6–13 find a shortest spanning tree T using Prim's algorithm. Sketch T.

6.
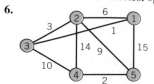

7.
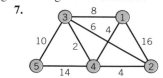

8.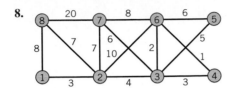

9.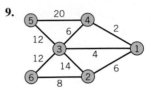

10. For the graph in Prob. 6, Sec. 21.4

12. For the graph in Prob. 4, Sec. 21.4

11. For the graph in Prob. 2, Sec. 21.4

13. For the graph in Prob. 1, Sec. 21.4

14. TEAM PROJECT. Center of a Graph and Related Concepts. (a) Distance, Eccentricity. Call the length of a shortest path $u \rightarrow v$ in a graph $G = (V, E)$ the *distance* $d(u, v)$ from u to v. For fixed u, call the greatest $d(u, v)$ as v ranges over V the *eccentricity* $\epsilon(u)$ of u. Find the eccentricity of vertices 1, 2, 3 in the graph in Prob. 9.

(b) Diameter, Radius, Center. The *diameter* $d(G)$ of a graph $G = (V, E)$ is the maximum of $d(u, v)$ as u and v vary over V, and the *radius* $r(G)$ is the smallest eccentricity $\epsilon(v)$ of the vertices v. A vertex v with $\epsilon(v) = r(G)$ is called a *central vertex*. The set of all central vertices is called the *center* of G. Find $d(G)$, $r(G)$ and the center of the graph in Prob. 9.

(c) What are the diameter, radius, and center of the spanning tree in Example 1?

(d) Explain how the idea of a center can be used in setting up an emergency service facility on a transportation network. In setting up a fire station, a shopping center. How would you generalize the concepts in the case of two or more such facilities?

(e) Show that a tree T whose edges all have length 1 has a center consisting of either one vertex or two adjacent vertices.

(f) Set up an algorithm of complexity $O(n)$ for finding the center of a tree T.

15. CAS PROBLEM. Prim's Algorithm. Write a program and apply it to Probs. 6–9.

21.6 Networks.
Flow Augmenting Paths

After shortest path problems and problems for trees, as a third large area in combinatorial optimization we discuss **flow problems in networks** (electrical, water, communication, traffic, business connections, etc.), turning from graphs to digraphs (directed graphs; see Sec. 21.1).

By definition, a **network** is a digraph $G = (V, E)$ in which each edge (i, j) has assigned to it a **capacity** $c_{ij} > 0$ [= maximum possible flow along (i, j)], and at one vertex, s, called the **source,** a flow is produced that flows along the edges to another vertex, t, called the **target** or **sink,** where the flow disappears.

In applications, this may be the flow of electricity in wires, of water in pipes, of cars on roads, of people in a public transportation system, of goods from a producer to consumers, of letters from senders to recipients, and so on.

We denote the flow along a (directed!) edge (i, j) by f_{ij} and impose two conditions:

1. For each edge (i, j) in G the flow does not exceed the capacity c_{ij},

(1) $$0 \leqq f_{ij} \leqq c_{ij} \quad \text{("Edge condition")}.$$

2. For each vertex i, not s or t,

  Inflow = Outflow *("Vertex condition," "Kirchhoff's law")*;

in a formula,

(2) $$\underbrace{\sum_{k} f_{ki}}_{\text{Inflow}} - \underbrace{\sum_{j} f_{ij}}_{\text{Outflow}} = \begin{cases} 0 \text{ if vertex } i \neq s, \ i \neq t, \\ -f \text{ at the source } s, \\ f \text{ at the target (sink) } t, \end{cases}$$

where f is the total flow (and at s the inflow is zero, whereas at t the outflow is zero). Figure 461 illustrates the notation (for some hypothetical figures).

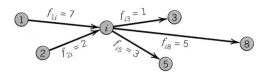

Fig. 461. Notation in (2): inflow and outflow for a vertex i (not s or t)

Paths

By a **path** $v_1 \rightarrow v_k$ from a vertex v_1 to a vertex v_k in a digraph G we mean a sequence of edges

$$(v_1, v_2), (v_2, v_3), \cdots, (v_{k-1}, v_k),$$

regardless of their directions in G, that forms a path as in a graph (see Sec. 21.2). Hence when we travel along this path from v_1 to v_k we may traverse some edge *in* its given direction—then we call it a **forward edge** of our path—or *opposite to* its given direction—then we call it a **backward edge** of our path. In other words, our path consists of one-way streets, and forward edges (backward edges) are those that we travel *in the right direction (in the wrong direction).* Figure 462 shows a forward edge (u, v) and a backward edge (w, v) of a path $v_1 \rightarrow v_k$.

  Caution! Each edge in a network has a given direction, *which we cannot change.* Accordingly, if (u, v) is a forward edge in a path $v_1 \rightarrow v_k$, then (u, v) can become a backward edge only in another path $x_1 \rightarrow x_j$ in which it is an edge and is traversed in the opposite direction as one goes from x_1 to x_j; see Fig. 463. Keep this in mind, to avoid misunderstandings.

Fig. 462. Forward edge (u, v) and backward edge (w, v) of a path $v_1 \rightarrow v_k$

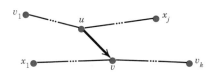

Fig. 463. Edge (u, v) as forward edge in the path $v_1 \rightarrow v_k$ and as backward edge in the path $x_1 \rightarrow x_j$

Flow Augmenting Paths

Our goal will be to **maximize the flow** from the source s to the target t of a given network. We shall do this by developing methods for increasing an existing flow (including the special case in which the latter is zero). The idea then is to find a path $P: s \to t$ all of whose edges are not fully used, so that we can push additional flow through P. This suggests the following concept.

Definition

A **flow augmenting path** in a network with a given flow f_{ij} on each edge (i, j) is a path $P: s \to t$ such that

(i) no forward edge is used to capacity; thus $f_{ij} < c_{ij}$ for these;

(ii) no backward edge has flow 0; thus $f_{ij} > 0$ for these.

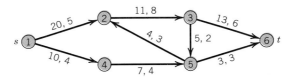

Fig. 464. Network in Example 1
First number = Capacity, Second number = Given flow

EXAMPLE 1 **Flow augmenting paths**

Find flow augmenting paths in the network in Fig. 464, where the first number is the capacity and the second number a given flow.

Solution. In practical problems, networks are large and one needs a *systematic method for augmenting flows, which we discuss in the next section.* In our small network, which should help to illustrate and clarify the concepts and ideas, we can find flow augmenting paths by inspection and augment the existing flow $f = 9$ in Fig. 464. (The outflow from s is $5 + 4 = 9$, which equals the inflow $6 + 3$ into t.)

We use the notation

$$\Delta_{ij} = c_{ij} - f_{ij} \quad \text{for forward edges}$$

$$\Delta_{ij} = f_{ij} \qquad\quad \text{for backward edges}$$

$$\Delta = \min \Delta_{ij} \quad \text{taken over all edges of a path.}$$

From Fig. 464 we see that a flow augmenting path $P_1: s \to t$ is $P_1:\ 1 - 2 - 3 - 6$ (Fig. 465), with $\Delta_{12} = 20 - 5 = 15$, etc., and $\Delta = 3$. Hence we can use P_1 to increase the given flow 9 to $f = 9 + 3 = 12$. All three edges of P_1 are forward edges. We augment the flow by 3. Then the flow in each of the edges of P_1 is increased by 3, so that we now have $f_{12} = 8$ (instead of 5), $f_{23} = 11$ (instead of 8), and $f_{36} = 9$ (instead of 6). Edge (2, 3) is now used to capacity. The flow in the other edges remains as before.

We shall now try to increase the flow in this network (Fig. 464) beyond $f = 12$.

There is another flow augmenting path $P_2: s \to t$, namely, $P_2:\ 1 - 4 - 5 - 3 - 6$ (Fig. 465). It shows how a backward edge comes in and how it is handled. Edge (3, 5) is a backward edge. It has flow 2, so that $\Delta_{35} = 2$. We compute $\Delta_{14} = 10 - 4 = 6$, etc. (Fig. 465) and $\Delta = 2$. Hence we can use P_2 for another augmentation to get $f = 12 + 2 = 14$. The new flow is shown in Fig. 466. No further augmentation is possible. We shall confirm later that $f = 14$ is maximum. (The "cut" in Fig. 466 will be explained below.) ◀

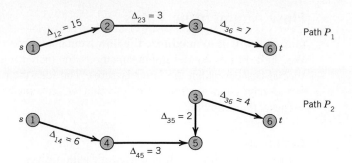

Fig. 465. Flow augmenting paths in Example 1

Cut Sets

A "cut set" is a set of edges in a network. The underlying idea is simple and natural. If we want to find out what is flowing from s to t in a network, we may cut the network somewhere between s and t (Fig. 466 shows an example) and see what is flowing in the edges hit by the cut, because any flow from s to t must sometimes pass through some of these edges. These form what is called a **cut set.** [In Fig. 466, the cut set consists of the edges (2, 3), (5, 2), (4, 5).] We denote this cut set by (S, T). Here S is the set of vertices on that side of the cut on which s lies ($S = \{s, 2, 4\}$ for the cut in Fig. 466) and T is the set of the other vertices ($T = \{3, 5, t\}$ in Fig. 466). We say that a cut *"partitions"* the vertex set V into two parts S and T. Obviously, the corresponding cut set (S, T) consists of all the edges in the network with one end in S and the other end in T.

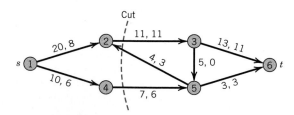

Fig. 466. Maximum flow in Example 1

By definition, the **capacity** cap (S, T) of a cut set (S, T) is the sum of the capacities of all **forward edges** in (S, T) (forward edges only!), that is, the edges that are directed *from S to T,*

$$(3) \qquad \text{cap } (S, T) = \Sigma c_{ij} \qquad \text{[sum over the forward edges of } (S, T)].$$

Thus, cap $(S, T) = 11 + 7 = 18$ in Fig. 466.

The other edges (directed *from T to S*) are called **backward edges** of the cut set (S, T), and by the **net flow** through a cut set we mean the sum of the flows in the forward edges minus the sum of the flows in the backward edges of the cut set.

Caution! Distinguish well between forward and backward edges in a cut set and in a path: (5, 2) in Fig. 466 is a backward edge for the cut shown but a forward edge in the path $1 - 4 - 5 - 2 - 3 - 6$.

For the cut in Fig. 466 the net flow is $11 + 6 - 3 = 14$. For the same cut in Fig. 464 (not indicated there), the net flow is $8 + 4 - 3 = 9$. In both cases it equals the flow f. We claim that this is not just by chance, but cuts do serve the purpose for which we have introduced them:

THEOREM 1   **(Net flow in cut sets)**

Any given flow in a network G is the net flow through any cut set (S, T) of G.

PROOF.   By Kirchhoff's law (2), multiplied by -1, at a vertex i,

$$(4) \qquad \underbrace{\sum_j f_{ij}}_{\text{Outflow}} - \underbrace{\sum_l f_{li}}_{\text{Inflow}} = \begin{cases} 0 & \text{if } i \neq s, t, \\ f & \text{if } i = s. \end{cases}$$

Here we can sum over j and l from 1 to n ($=$ number of vertices) by putting $f_{ij} = 0$ for $j = i$ and also for edges without flow or nonexisting edges; hence we can write the two sums as one,

$$\sum_j (f_{ij} - f_{ji}) = \begin{cases} 0 & \text{if } i \neq s, t, \\ f & \text{if } i = s. \end{cases}$$

We now sum over all i in S. Since $s \in S$, this sum equals f:

$$(5) \qquad \sum_{i \in S} \sum_{j \in V} (f_{ij} - f_{ji}) = f.$$

We claim that in this sum, only the edges belonging to the cut set contribute. Indeed, edges with both ends in T cannot contribute, since we sum only over i in S; but edges (i, j) with both ends in S contribute $+f_{ij}$ at one end and $-f_{ij}$ at the other, a total contribution of 0. Hence the left side of (5) equals the net flow through the cut set. By (5), this is equal to the flow f and proves the theorem.   ◀

This theorem has the following consequence, which we shall need below.

THEOREM 2   **(Upper bound for flows)**

A flow f in a network G cannot exceed the capacity of any cut set (S, T) in G.

PROOF.   By Theorem 1, the flow f equals the net flow through the cut set, $f = f_1 - f_2$, where f_1 is the sum of the flows through the forward edges and f_2 ($\geqq 0$) is the sum of the flows through the backward edges of the cut set. Thus $f \leqq f_1$. Now f_1 cannot exceed the sum of the capacities of the forward edges; but this sum equals the capacity of the cut set, by definition. Together, $f \leqq \text{cap}\,(S, T)$, as asserted.   ◀

Cut sets will now bring out the full importance of augmenting paths:

THEOREM 3   **Main Theorem (Augmenting path theorem for flows)**

A flow from s to t in a network G is maximum if and only if there does not exist a flow augmenting path s → t in G.

PROOF.   **(a)** If there is a flow augmenting path $P: s \to t$, we can use it to push through it an additional flow. Hence the given flow cannot be maximum.

(b) On the other hand, suppose that there is no flow augmenting path $s \to t$ in G. Let S_0 be the set of all vertices i (including s) such that there is a flow augmenting path $s \to i$, and let T_0 be the set of the other vertices in G. Consider any edge (i, j) with i in S_0 and j in T_0. Then we have a flow augmenting path $s \to i$ since i is in S_0, but $s \to i \to j$ is not flow augmenting because j is not in S_0. Hence we must have

(6) $$f_{ij} = \begin{cases} c_{ij} \\ 0 \end{cases} \text{ if } (i, j) \text{ is a } \begin{cases} \text{forward} \\ \text{backward} \end{cases} \text{edge of the path } s \to i \to j.$$

Otherwise we could use (i, j) to get a flow augmenting path $s \to i \to j$. Now (S_0, T_0) defines a cut set (since t is in T_0; why?). Since by (6), forward edges are used to capacity and backward edges carry no flow, the net flow through the cut set (S_0, T_0) equals the sum of the capacities of the forward edges, which is cap (S_0, T_0) by definition. This net flow equals the given flow f by Theorem 1. Thus $f = $ cap (S_0, T_0). Also, $f \leq $ cap (S_0, T_0) by Theorem 2. Hence f must be maximum since we have reached equality.   ◄

The end of this proof yields another basic result (by Ford and Fulkerson, *Canadian Journal of Mathematics* **8** (1956), 399−404), namely, the so-called

THEOREM 4   **Max-flow min-cut theorem**

The maximum flow[10] *in any network G equals the capacity of a "**minimum cut set**" (= a cut set of minimum capacity) in G.*

PROOF.   We have just seen that $f = $ cap (S_0, T_0) for a maximum flow f and a suitable cut set (S_0, T_0). Now by Theorem 2 we also have $f \leq $ cap (S, T) for this f and any cut set (S, T) in G. Together, cap $(S_0, T_0) \leq $ cap (S, T). Hence (S_0, T_0) is a minimum cut set, and the theorem is proved.   ◄

The two basic tools in connection with networks are flow augmenting paths and cut sets. In the next section we show how flow augmenting paths can be used as the basic tool in an algorithm for maximum flows.

---

[10]The existence of a maximum flow follows in the case of rational capacities from the algorithm in the next section and in the case of arbitrary capacities from a modification of this algorithm mentioned in footnote 11 in the next section.

PROBLEM SET 21.6

Cut Sets, Capacity
Find T and cap (S, T) for:

1. Fig. 466, $S = \{1, 2, 3\}$ **2.** Fig. 466, $S = \{1, 3, 5\}$

3. Fig. 466, $S = \{1, 2, 4, 5\}$ **4.** Fig. 467, $S = \{1, 2\}$

5. Fig. 467, $S = \{1, 2, 4, 5\}$ **6.** Fig. 467, $S = \{1, 2, 3\}$

7. **(Minimum cut set)** Find a minimum cut set in Fig. 464. Verify that its capacity equals the maximum flow f.

8. Why are backward edges not considered in the definition of the capacity of a cut set?

9. Find a minimum cut set in Fig. 467 and its capacity.

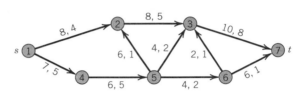

Fig. 467.   Problems 4–6, 9

Flow Augmenting Paths, Maximum Flow
Try to find flow augmenting paths:

10. In Fig. 464 (another path)   **11.** In Fig. 466

12.   **13.**

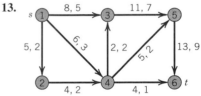

14.   **15.**

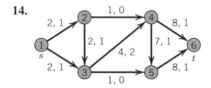   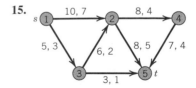

Find the maximum flow by inspection:

16. In Prob. 12   **17.** In Prob. 13

18. In Prob. 14   **19.** In Prob. 15

20. In which case can an edge (i, j) be used both as a forward edge or a backward edge in augmenting a given flow?

21.7 Ford–Fulkerson Algorithm for Maximum Flow

Flow augmenting paths, as discussed in the last section, are used as the basic tool in the Ford–Fulkerson algorithm in Table 21.8 in which a given flow (for instance, zero flow in all edges) is increased until it is maximum. The algorithm accomplishes the increase by a stepwise construction of flow augmenting paths, one at a time, until no further such paths can be constructed, which happens precisely when the flow is maximum.

Table 21.8
Ford–Fulkerson Algorithm for Maximum Flow

ALGORITHM FORD–FULKERSON

$[G = (V, E)$, vertices $1 \; (= s), \cdots, n \; (= t)$, edges (i, j), $c_{ij}]$

This algorithm computes the maximum flow in a network G with source s, sink t, and capacities $c_{ij} > 0$ of the edges (i, j).

    INPUT: n, $s = 1$, $t = n$, edges (i, j) of G, c_{ij}

    OUTPUT: Maximum flow f in G

1. Assign an initial flow f_{ij} (for instance, $f_{ij} = 0$ for all edges), compute f.

2. Label s by $\varnothing$. Mark the other vertices *"unlabeled."*

3. Find a labeled vertex i that has not yet been scanned. Scan i as follows.
   For every unlabeled adjacent vertex j, if $c_{ij} > f_{ij}$, compute

$$\Delta_{ij} = c_{ij} - f_{ij} \quad \text{and} \quad \Delta_j = \begin{cases} \Delta_{1j} & \text{if } i = 1 \\ \min (\Delta_i, \Delta_{ij}) & \text{if } i > 1 \end{cases}$$

   and label j with a *"forward label"* (i^+, Δ_j); or if $f_{ji} > 0$, compute

$$\Delta_j = \min (\Delta_i, f_{ji})$$

   and label j by a "backward label" (i^-, Δ_j).
   If no such j exists then OUTPUT f. Stop
       [f *is the maximum flow.*]
   Else continue (that is, go to Step 4).

4. Repeat Step 3 until t is reached.
       [*This gives a flow augmenting path P: $s \rightarrow t$.*]
   If it is impossible to reach t then OUTPUT f. Stop
       [f *is the maximum flow.*]
   Else continue (that is, go to Step 5).

5. Backtrack the path P, using the labels.

6. Using P, augment the existing flow by Δ_t. Set $f = f + \Delta_t$.

7. Remove all labels from vertices $2, \cdots, n$. Go to Step 3.

End FORD–FULKERSON

In Step 1, an initial flow may be given. In Step 3, a vertex j can be labeled if there is an edge (i, j) with i labeled and

$$c_{ij} > f_{ij} \qquad \textbf{\textit{("forward edge")}}$$

or if there is an edge (j, i) with i labeled and

$$f_{ji} > 0 \qquad \textbf{\textit{("backward edge")}}.$$

To **scan** a labeled vertex i means to label every unlabeled vertex j adjacent to i that can be labeled. Before scanning a labeled vertex i, scan all the vertices that got labeled before i. This BFS (Breadth First Search) strategy was suggested by Edmonds and Karp[11] in 1972. It has the effect that one gets shortest possible augmenting paths. The computational advantage of this is illustrated in Prob. 15.

EXAMPLE 1   **Ford–Fulkerson algorithm**

Applying the Ford–Fulkerson algorithm, determine the maximum flow for the network in Fig. 468 (which is the same as that in Example 1, Sec. 21.6, so that we can compare).

Solution.   The algorithm proceeds as follows.

1. An initial flow $f = 9$ is given.

2. Label $s \ (= 1)$ by $\varnothing$. Mark 2, 3, 4, 5, 6 "unlabeled."

3. Scan 1.
   Compute $\Delta_{12} = 20 - 5 = 15 = \Delta_2$. Label 2 by $(1^+, 15)$.
   Compute $\Delta_{14} = 10 - 4 = 6 = \Delta_4$. Label 4 by $(1^+, 6)$.

4. Scan 2.
   Compute $\Delta_{23} = 11 - 8 = 3$, $\Delta_3 = \min(\Delta_2, 3) = 3$. Label 3 by $(2^+, 3)$.
   Compute $\Delta_5 = \min(\Delta_2, 3) = 3$. Label 5 by $(2^-, 3)$.
   Scan 3.
   Compute $\Delta_{36} = 13 - 6 = 7$, $\Delta_6 = \Delta_t = \min(\Delta_3, 7) = 3$. Label 6 by $(3^+, 3)$.

5. P: $1 - 2 - 3 - 6 \ (= t)$ is a flow augmenting path.

6. $\Delta_t = 3$. Augmentation gives $f_{12} = 8$, $f_{23} = 11$, $f_{36} = 9$, other f_{ij} unchanged. Augmented flow $f = 9 + 3 = 12$.

7. Remove labels on vertices 2, $\cdots$, 6. Go to Step 3.

3. Scan 1.
   Compute $\Delta_{12} = 20 - 8 = 12 = \Delta_2$. Label 2 by $(1^+, 12)$.
   Compute $\Delta_{14} = 10 - 4 = 6 = \Delta_4$. Label 4 by $(1^+, 6)$.

4. Scan 2.
   Compute $\Delta_5 = \min(\Delta_2, 3) = 3$. Label 5 by $(2^-, 3)$.
   Scan 4. [*No vertex left for labeling.*]   (*Continued*)

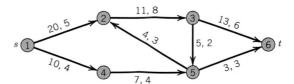

Fig. 468.   Network in Example 1 with capacities (first numbers) and given flow

---

[11]*Journal of the Association for Computing Machinery* **19** (1972), 248−64. The Ford–Fulkerson algorithm was first published in *Canadian Journal of Mathematics* **9** (1957), 210−218.

Scan 5.

Compute $\Delta_3 = \min(\Delta_5, 2) = 2$. Label 3 by $(5^-, 2)$.

Scan 3.

Compute $\Delta_{36} = 13 - 9 = 4$, $\Delta_6 = \min(\Delta_3, 4) = 2$. Label 6 by $(3^+, 2)$.

5. P: $1 - 2 - 5 - 3 - 6 (= t)$ is a flow augmenting path.

6. $\Delta_t = 2$. Augmentation gives $f_{12} = 10$, $f_{52} = 1$, $f_{35} = 0$, $f_{36} = 11$, other f_{ij} unchanged. Augmented flow $f = 12 + 2 = 14$.

7. Remove labels on vertices $2, \cdots, 6$. Go to Step 3.

One can now scan 1 and then scan 2, as before, but in scanning 4 and then 5 one finds that no vertex is left for labeling. Thus one can no longer reach t. Hence the flow obtained (Fig. 469) is maximum, in agreement with our result in the last section. ◀

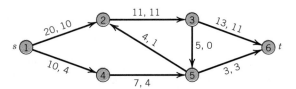

Fig. 469. Maximum flow in Example 1

PROBLEM SET 21.7

Ford–Fulkerson Algorithm

Find the maximum flow:

1.
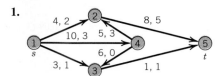

2.
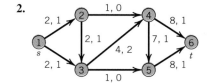

3.
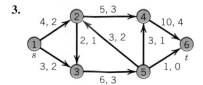

4.
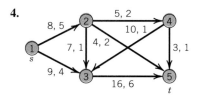

5. Find a minimum cut set in Fig. 468 and its capacity.

6. In Prob. 5, the cut set contains precisely all forward edges used to capacity by the maximum flow (Fig. 469). Is this just by chance?

7. Which are the "bottleneck" edges by which the flow in Example 1 in the text is actually limited? Hence which capacities could be decreased without affecting the maximum flow?

8. Apply the Ford–Fulkerson algorithm to Example 1 in the text with initial flow 0. Comment on the amount of work compared to that in Example 1.

9. How can one see from the algorithm that Ford and Fulkerson follow a BFS technique?

10. If the Ford–Fulkerson algorithm stops without reaching t, show that the edges with one end labeled and the other end unlabeled form a cut set (S, T) whose capacity equals the maximum flow.

11. CAS PROBLEM. Ford–Fulkerson. Write a program and apply it to Probs. 1–4.

General Properties of Flows

12. Show that in a network G with all $c_{ij} = 1$, the maximum flow equals the number of edge-disjoint paths $s \to t$.

13. In the augmentation of a flow by the use of a flow augmenting path, Kirchhoff's law is preserved. What is the (simple) reason for this?

14. Show that in a network G with capacities all equal to 1, the capacity of a minimum cut set (S, T) equals the minimum number q of edges whose deletion destroys all directed paths $s \to t$. (A **directed path** $v \to w$ is a path in which each edge has the direction in which it is traversed in going from v to w.)

15. How many augmentations would you need to obtain the maximum flow in the network in Fig. 470 if you started from zero flow and alternatingly used the paths P_1: $s - 2 - 3 - t$ and P_2: $s - 3 - 2 - t$? How does the Ford–Fulkerson algorithm prevent this poor choice?

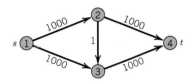

Fig. 470. Problem 15

21.8 Assignment Problems
Bipartite Matching

From digraphs we return to graphs and discuss another important class of combinatorial optimization problems that arises in **assignment problems** of workers to jobs, jobs to machines, goods to storage, ships to piers, classes to classrooms, exams to time periods, and so on. To explain the problem, we need the following concepts.

A **bipartite graph** $G = (V, E)$ is a graph in which the vertex set V is partitioned into two sets S and T (without common elements, by the definition of a partition) such that every edge of G has one end in S and the other in T. Hence there are no edges in G that have both ends in S or both ends in T. Such a graph $G = (V, E)$ is also written $G = (S, T; E)$.

Figure 471 shows an illustration. V consists of seven elements, three workers a, b, c, making up the set S, and four jobs 1, 2, 3, 4, making up the set T. The edges indicate that worker a can do the jobs 1 and 2, worker b the jobs 1, 2, 3, and worker c the job 4. The problem is to assign one job to each worker so that every worker gets one job to do. This suggests the next concept, as follows.

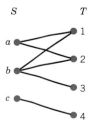

Fig. 471. Bipartite graph in the assignment
of a set $S = \{a, b, c\}$ of workers
to a set $T = \{1, 2, 3, 4\}$ of jobs

Definition

A **matching** in $G = (S, T; E)$ is a set M of edges of G such that no two of them have a vertex in common. If M consists of the greatest possible number of edges, we call it a **maximum cardinality matching**[12] in G.   ◄

For instance, a matching in Fig. 471 is $M_1 = \{(a, 2), (b, 1)\}$. Another is $M_2 = \{(a, 1), (b, 3), (c, 4)\}$; obviously, this is of maximum cardinality.

A vertex v is **exposed** (or *not covered*) by a matching M if v is not an endpoint of an edge of M. This concept, which always refers to some matching, will be of interest when we begin to augment given matchings (below). If a matching leaves no vertex exposed, we call it a **complete matching.** Obviously, a complete matching can exist only if S and T consist of the same number of vertices.

We now want to show how one can stepwise increase the cardinality of a matching M until it becomes maximum. Central in this task is the concept of an augmenting path:

An **alternating path** is a path that consists alternately of edges in M and not in M (Fig. 472A). An **augmenting path** is an alternating path both of whose endpoints (a and b in Fig. 472B) are exposed. By dropping from the matching M the edges that are on an augmenting path P (two edges in Fig. 472B) and adding to M the other edges of P (three in the figure), we get a new matching, with one more edge than M. This is how we use an augmenting path in augmenting a given matching by one edge. We assert that this will always lead, after a number of steps, to a maximum cardinality matching. Indeed, the basic role of augmenting paths is expressed in the following theorem.

(A) Alternating path

(B) Augmenting path P

Fig. 472.   Alternating and augmenting paths.
Heavy edges are those belonging to a matching M.

THEOREM 1   **Augmenting path theorem for bipartite matching**

A matching M in a bipartite graph $G = (S, T; E)$ is of maximum cardinality if and only if there does not exist an augmenting path P with respect to M.

PROOF.   **(a)** We show that if such a path P exists, then M is not of maximum cardinality. Let P have q edges belonging to M. Then P has $q + 1$ edges not belonging to M. (In Fig. 472B we have $q = 2$.) The endpoints a and b of P are exposed, and all the other vertices on P are endpoints of edges in M, by the definition of an alternating path. Hence if an edge of

---

[12]Or simply a **maximum matching,** but this term is sometimes also used in a different sense, which does not interest us here.

M is not an edge of P, it cannot have an endpoint on P since then M would not be a matching. Consequently, the edges of M not on P, together with the $q + 1$ edges of P not belonging to M form a matching of cardinality one more than the cardinality of M because we omitted q edges from M and added $q + 1$ instead. Hence M cannot be of maximum cardinality.

(b) We now show that if there is no augmenting path for M, then M is of maximum cardinality. Let M^* be a maximum cardinality matching and consider the graph H consisting of all edges that belong to M or to M^*, but not to both. Then it is possible that two edges of H have a vertex in common, but three edges cannot have a vertex in common since then two of the three would have to belong to M (or to M^*), violating that M and M^* are matchings. So every v in V can be in common with two edges of H or with one or none. Hence we can characterize each "component" (= maximal *connected* subset) of H; a component can be:

(A) A closed path with an *even* number of edges (in the case of an *odd* number, two edges from M or two from M^* would meet, violating the matching property). See (A) in Fig. 473.

(B) An open path P with the same number of edges from M and edges from M^*, for the following reason. P must be alternating, that is, an edge of M is followed by an edge of M^*, etc. (since M and M^* are matchings). Now if P had an edge more from M^*, then P would be augmenting for M [see (B2) in Fig. 473], contradicting our assumption that there is no augmenting path for M. If P had an edge more from M, it would be augmenting for M^* [see (B3) in Fig. 473], violating the maximum cardinality of M^*, by part (a) of this proof. Hence in each component of H, the two matchings have the same number of edges. Adding to this the number of edges that belong to both M and M^* (which we left aside when we made up H), we conclude that M and M^* must have the same number of edges. Since M^* is of maximum cardinality, this shows that the same holds for M, as we wanted to prove. ◀

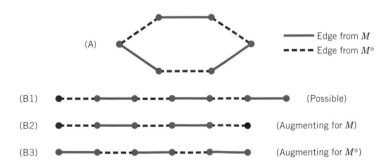

Fig. 473. Proof of the augmenting path theorem for bipartite matching

This theorem suggests an algorithm for obtaining augmenting paths in which vertices are labeled for the purpose of backtracking paths. Such a label is *in addition* to the number of the vertex, which is also retained. Clearly, to get an augmenting path, one must start from an *exposed* vertex, and then trace an alternating path until one arrives at another *exposed* vertex. Table 21.9 shows such an algorithm. After Step 3 all vertices in S are labeled. In Step 4, the set T contains at least one exposed vertex, since otherwise we would have stopped at Step 1.

Table 21.9
Bipartite Maximum Cardinality Matching

ALGORITHM MATCHING $[G = (S, T; E), M, n]$

This algorithm determines a maximum cardinality matching M in a bipartite graph G by augmenting a given matching in G.

INPUT: Bipartite graph $G = (S, T; E)$ with vertices $1, \cdots, n$, matching M in G (for instance, $M = \varnothing$)

OUTPUT: Maximum cardinality matching M in G

1. If there is no exposed vertex in S then
   OUTPUT M. Stop
   [M is of maximum cardinality in G.]
   Else label all *exposed* vertices *in S* with $\varnothing$.

2. For each i in S and edge (i, j) *not* in M, label j with i, unless already labeled.

3. For each *nonexposed* j in T, label i with j, where i is the other end of the unique edge (i, j) in M.

4. Backtrack the alternating paths P ending on an exposed vertex in T by using the labels on the vertices.

5. If no P in Step 4 is augmenting then
   OUTPUT M. Stop
   [M is of maximum cardinality in G.]
   Else augment M by using an augmenting path P.
   Remove all labels.
   Go to Step 1.

End MATCHING

EXAMPLE 1   **Maximum cardinality matching**

Is the matching M_1 in Fig. 474a of maximum cardinality? If not, augment it until maximum cardinality is reached.

Solution. We apply the algorithm.

1. Label 1 and 4 with $\varnothing$.

2. Label 7 with 1. Label 5, 6, 8 with 3.

3. Label 2 with 6, and 3 with 7.
   [*All vertices are now labeled as shown in Fig. 474a.*]

4. P_1: $1 - 7 \blacksquare 3 - 5$. [*By backtracking, P_1 is augmenting.*]
   P_2: $1 - 7 \blacksquare 3 - 8$. [*P_2 is augmenting.*]

5. Augment M_1 by using P_1, dropping $(3, 7)$ from M_1 and including $(1, 7)$ and $(3, 5)$. Remove all labels. Go to Step 1.
   Figure 474b shows the resulting matching $M_2 = \{(1, 7), (2, 6), (3, 5)\}$.

1. Label 4 with $\varnothing$.

2. Label 7 with 2. Label 6 and 8 with 3.

3. Label 1 with 7, and 2 with 6, and 3 with 5.

4. P_3: $5 \blacksquare 3 - 8$. [*P_3 is alternating but not augmenting.*]

5. Stop. M_2 is of maximum cardinality (namely, 3). ◀

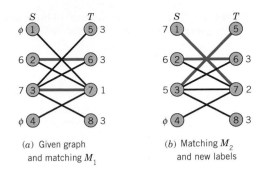

(a) Given graph
and matching M_1

(b) Matching M_2
and new labels

Fig. 474. Example 1

PROBLEM SET 21.8

Bipartite or Not?
If your answer is yes, find S and T.

1.

2.

3.

4.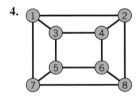

5.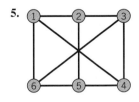

6.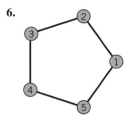

Augmenting Paths. Matching
Find an augmenting path:

7.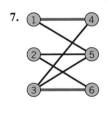

8.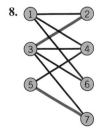

9.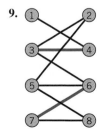

By augmenting the given matching, find a maximum cardinality matching of the graph:
10. In Prob. 8   **11.** In Prob. 7   **12.** In Prob. 9

Applications. Further Topics
13. (**Computer storage**) In some computations, temporary storage of frequently used variables $v_1, \cdots, v_6$ during overlapping time intervals (0, 3), (2, 4), (3, 6), (1, 4), (5, 7), (3, 6), respectively, is required. How many index registers (storage locations) are needed? *Hint.* Join v_i and v_j by an edge if their intervals overlap. Then color vertices.

14. What would be the answer to Prob. 13 if not only overlapping but also common endpoints are excluded?

15. **(Complete bipartite graphs)** A bipartite graph $G = (S, T; E)$ is called *complete* if every vertex in S is joined to every vertex in T by an edge, and is denoted by K_{n_1, n_2}, where n_1 and n_2 are the numbers of vertices in S and T, respectively. How many edges does this graph have?

16. **(Timetabling and matching)** Teacher T_1 teaches classes c_1, c_2, c_4. Teacher T_2 teaches classes c_1, c_2, c_3, c_4. Teacher T_3 teaches classes c_2, c_3, c_4. Represent this by a bipartite graph G. Show that a teaching schedule for one period corresponds to a matching in G. Set up a teaching schedule with the smallest possible number of periods.

17. **(Vertex coloring and exam scheduling)** What is the smallest number of exam periods for six subjects a, b, c, d, e, f if some of the students take a, b, f, some c, d, e, some a, c, e, and some c, e? Solve this as follows. Sketch a graph with six vertices $a, \cdots, f$ and join vertices if they represent subjects simultaneously taken by some students. Color the vertices so that adjacent vertices receive different colors. (Use numbers $1, 2, \cdots$ instead of actual colors if you want.) What is the minimum number of colors you need? For any graph G, this minimum number is called the (vertex) **chromatic number** $\chi_v(G)$. Why is this the answer to the problem? Write down a possible schedule.

18. Show that all trees can be vertex colored with two colors.

19. **(Planar graph)** A *planar graph* is a graph that can be drawn on a sheet of paper so that no two edges cross. Show that the complete graph K_4 with four vertices is planar. The complete graph K_5 with five vertices is not planar. Make this plausible by attempting to draw K_5 so that no edges cross. Interpret the result in terms of a net of roads between five cities.

20. **(Bipartite graph $K_{3,3}$ not planar)** Three factories 1, 2, 3 are each supplied underground by water, gas, and electricity, from points A, B, C, respectively. Show that this can be represented by $K_{3,3}$ (the complete bipartite graph $G = (S, T; E)$ with S and T consisting of three vertices each) and that eight of the nine supply lines (edges) can be laid out without crossing. Make it plausible that $K_{3,3}$ is not planar by attempting to draw the ninth line without crossing the others.

CHAPTER 21 REVIEW

1. What is a graph? A digraph? A tree? A cycle? A path?

2. State several possibilities of handling graphs on computers. (From memory!)

3. Give basic examples of situations and problems that can be modeled in terms of graphs or digraphs.

4. How did we handle flows in terms of graphs or digraphs?

5. What is a bipartite graph and for what kind of applications is it appropriate?

6. Why did we use matrices in this chapter?

7. What kind of optimization problems did we consider in this chapter? Give typical examples for each class of problems.

8. What is a list and how did we use lists in this chapter?

9. What is BFS? DFS? In what connection did these concepts occur?

10. What is the traveling salesman problem?

11. In what applications do spanning trees play a role?

12. What is the idea of Kruskal's greedy algorithm?

13. What is the intuitive idea of Bellman's optimality principle? How is the principle used in Dijkstra's algorithm?

14. Where did cut sets occur? Can you remember a famous theorem on them?

15. What is a flow augmenting path? A forward edge? A backward edge?

Find the adjacency matrix of the given graph or digraph.

16.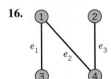

17.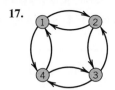

18.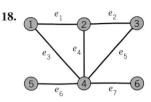
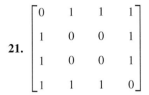

Sketch the graph whose adjacency matrix is

19.
$$\begin{bmatrix} 0 & 1 & 1 & 1 \\ 1 & 0 & 1 & 1 \\ 1 & 1 & 0 & 1 \\ 1 & 1 & 1 & 0 \end{bmatrix}$$

20.
$$\begin{bmatrix} 0 & 1 & 0 & 1 \\ 1 & 0 & 0 & 1 \\ 0 & 0 & 0 & 1 \\ 1 & 1 & 1 & 0 \end{bmatrix}$$

21.
$$\begin{bmatrix} 0 & 1 & 1 & 1 \\ 1 & 0 & 0 & 1 \\ 1 & 0 & 0 & 1 \\ 1 & 1 & 1 & 0 \end{bmatrix}$$

22. Make a vertex incidence list of the graph in Prob. 18.

23. Find a shortest path of the graph below and its length by Moore's BFS algorithm, assuming that all the edges have length 1.

24. Apply Dijkstra's algorithm to the graph below.

25. Find a shortest spanning tree in the graph below.

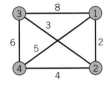

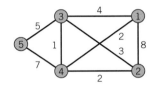

Problem 23   **Problem 24**   **Problem 25**

26. What is the difference between the traveling salesman problem and the postman problem?

27. **Cayley's theorem** states that the number of spanning trees in a complete graph with n vertices is n^{n-2}. Verify this for $n = 2, 3, 4$.

28. Find flow augmenting paths and the maximum flow in the graph below.

29. How does the Ford–Fulkerson algorithm prevent the formation of cycles?

30. Are the consecutive flow augmenting paths produced by the Ford–Fulkerson algorithm unique?

31. **(Integer flow theorem)** Show that if the capacities in a network G are integers, there is a maximum flow that is an integer.

32. Company A has offices in Chicago, Los Angeles, and New York; Company B in Boston and New York; Company C in Chicago, Dallas, and Los Angeles. Represent this by a bipartite graph.

33. Find an augmenting path in the graph below.

34. Extend the given matching in the graph below to a maximum cardinality matching.

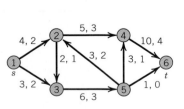

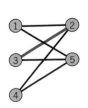

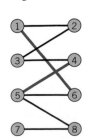

Problem 28   **Problem 33**   **Problem 34**

SUMMARY OF CHAPTER 21
GRAPHS AND COMBINATORIAL OPTIMIZATION

Combinatorial optimization concerns optimization problems of a discrete or combinatorial structure. It uses graphs and digraphs (Sec. 21.1) as basic tools.

A **graph** $G = (V, E)$ consists of a set V of **vertices** $v_1, v_2, \cdots, v_n$, (often simply denoted by $1, 2, \cdots, n$) and a set E of **edges** $e_1, e_2, \cdots, e_m$, each of which connects two vertices. We also write (i, j) for an edge with vertices i and j as endpoints. A **digraph** (= directed graph) is a graph in which each edge has a direction (indicated by an arrow). For handling graphs and digraphs in computers, one can use *matrices* or *lists* (Sec. 21.1).

This chapter is devoted to important classes of optimization problems for graphs that all arise from practical applications, and corresponding algorithms, as follows.

In a **shortest path problem** (Sec. 21.2) we determine a path of minimum length (consisting of edges) from a vertex s to a vertex t in a graph whose edges (i, j) have a "length" $l_{ij} > 0$, which may be an actual length or a travel time or cost or an electrical resistance [if (i, j) is a wire in a net], and so on. **Dijkstra's algorithm** (Sec. 21.3) or, when all $l_{ij} = 1$, **Moore's algorithm** (Sec. 21.2) are suitable for these problems.

A **tree** is a graph that is connected and has no **cycles** (no closed paths). Trees are very important in practice. A *spanning tree* in a graph G is a tree containing *all* the vertices of G. If the edges of G have lengths, we can determine a **shortest spanning tree,** for which the sum of the lengths of all its edges is minimum. Corresponding algorithms are those by **Kruskal** (Sec. 21.4) and by **Prim** (Sec. 21.5).

A **network** (Sec. 21.6) is a digraph in which each edge (i, j) has a *capacity* $c_{ij} > 0$ [= maximum possible flow along (i, j)] and at one vertex, the *source s,* a flow is produced that flows along the edges to a vertex t, the *sink* or *target,* where the flow disappears. The problem is to maximize the flow, for instance, by applying the **Ford–Fulkerson algorithm** (Sec. 21.7), which uses **flow augmenting paths** (Sec. 21.6). Another related concept is that of a **cut set,** as defined in Sec. 21.6.

A **bipartite graph** $G = (V, E)$ (Sec. 21.8) is a graph whose vertex set V consists of two parts S and T such that every edge of G has one end in S and the other in T, so that there are no edges connecting vertices in S or vertices in T. A **matching** in G is a set of edges, no two of which have an endpoint in common. The problem then is to find a **maximum cardinality matching** in G, that is, a matching M that has a maximum number of edges. For an algorithm, see Sec. 21.8.

PART G

Probability and Statistics

Chapter 22 **Data Analysis.**
Probability Theory

Chapter 23 **Mathematical Statistics**

Modern **mathematical statistics** has various engineering applications, for instance in testing materials, performance tests of systems, robotics and automatization in general, control of production processes, and so on. To this we could add a long list of other applications, for instance, in agriculture, biology, computer science, demography, economics, geography, management of natural resources, medicine, meteorology, politics, psychology, sociology, traffic control, etc. Although these applications are heterogeneous, we shall see that most statistical methods are rather universal in the sense that they can be applied in various fields.

Probability theory (Chap. 22) will provide models of probability distributions (theoretical models of the observable reality) to be tested by statistical tests, and it will also furnish the mathematical foundation of those tests and other methods, as we explain in Chap. 23.

CHAPTER 22

Data Analysis. Probability Theory

We first show how to handle data numerically or graphically (in terms of figures), in order to see what properties they may have and what kind of information we can extract from them. If data are influenced by chance effect (e.g., weather data, properties of steel, stock prices, etc.), they may suggest and motivate concepts and rules of probability theory because this is the theoretical counterpart of the observable reality whenever "chance" is at work. This theory gives us mathematical models of such chance processes (briefly called "experiments"; Sec. 22.2). In any such experiment we observe a "random variable" X, a function whose values in the experiment occur "by chance" (Sec. 22.5), which is characterized by a probability distribution (Secs. 22.5–22.8). Or we observe more than one random variable, for example, height and weight of persons, hardness and tensile strength of copper. This is discussed in Sec. 22.9, which will also give the basis for the mathematical justification of statistical methods in Chap. 23.

Prerequisite for this chapter: Calculus.
References: Appendix 1, Part G.
Answers to problems: Appendix 2.

22.1 Data: Representation, Average, Spread

Data can be represented numerically or graphically in various ways. For instance, your daily newspaper may contain tables of stock prices and money exchange rates, curves or bar charts illustrating economical or political developments, or pie charts showing how your tax dollar is spent. And there are numerous other representations of data for special purposes.

In this section we discuss the use of standard representations of data in statistics. (For these, software packages, such as DATA DESK and MINITAB, are available, and MAPLE or MATHEMATICA may also be helpful.) We explain corresponding concepts and methods in terms of typical examples, beginning with

(1)   89 84 87 81 89 86 91 90 78 89 87 99 83 89.

These are $n = 14$ measurements of the tensile strength of sheet steel in kg/mm^2, recorded in the order obtained and rounded to integer values. To see what is going on, we **sort** these data, that is, we order them by size,

(2)   78   81   83   84   86   87   87   89   89   89   89   90   91   99.

Sorting is a standard process on the computer; see Ref. [E13], Chap. 8, listed in Appendix 1.

Graphical Representation of Data

We shall now discuss standard graphical representations used in statistics for obtaining information on properties of data.

Stem-and-Leaf Plot

This is one of the simplest but most useful representations of data. For (1) it is shown in Fig. 475. The numbers in (1) range from 78 to 99 [see (2)]. We divide these numbers into 5 groups, 75–79, 80–84, 85–89, 90–94, 95–99. The integers in the tens position of the groups are 7, 8, 8, 9, 9. These form the *stem* in Fig. 475. The first *leaf* is 8 (representing 78). The second leaf is 134 (representing 81, 83, 84), and so on.

Leaf unit = 1.0

1	7	8
4	8	134
11	8	6779999
13	9	01
14	9	9

Fig. 475.   Stem-and-leaf plot of the data in (1) and (2)

The number of times a value occurs is called its **absolute frequency.** Thus 78 has absolute frequency 1, the value 89 has absolute frequency 4, etc. The column to the extreme left in Fig. 475 shows the **cumulative absolute frequencies,** that is, the sum of the absolute frequencies of the values up to the line of the leaf. Thus, the number 4 in the second line on the left shows that (1) has 4 values up to and including 84. The number 11 in the next line shows that there are 11 values not exceeding 89, etc. Dividing the cumulative absolute frequencies by n (= 14 in Fig. 475) gives the **cumulative relative frequencies.**

Histogram

For large sets of data, histograms are better in displaying the distribution of data than stem-and-leaf plots. The principle is explained in Fig. 476. (An application to a larger

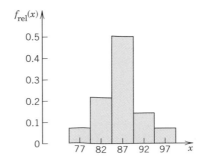

Fig. 476.   Histogram of the data in (1) and (2) (grouped as in Fig. 475)

data set is shown in Sec. 23.7). The bases of the rectangles in Fig. 476 are the x-intervals (known as **class intervals**) 74.5–79.5, 79.5–84.5, 84.5–89.5, 89.5–94.5, 94.5–99.5, whose midpoints (known as **class marks**) are $x = 77, 82, 87, 92, 97$, respectively. The height of a rectangle with class mark x is the **relative class frequency** $f_{rel}(x)$, defined as the number of data values in that class interval, divided by n (= 14 in our case). Hence the areas of the rectangles are proportional to these relative frequencies, so that histograms give a good impression of the distribution of data.

Center and Spread of Data: Median, Quartiles

As a center of the location of data values we can simply take the **median,** the data value that falls in the middle when the values are ordered. In (2) we have 14 values. The seventh of them is 87, the eighth is 89, and we split the difference, obtaining the median 88. (In general, we would get a fraction.)

The spread (variability) of the data values can be measured by the **range** $R = x_{max} - x_{min}$, the largest minus the smallest data values, $R = 99 - 78 = 21$ in (2).

Better information gives the **interquartile range** IQR $= q_U - q_L$. Here the **upper quartile** q_U is the middle value among the data values *above* the median. The **lower quartile** q_L is the middle value among the data values *below* the median. Thus in (2) we have $q_U = 89$ (the fourth value from the end), $q_L = 84$ (the fourth value from the beginning), and IQR $= 89 - 84 = 5$. The median is also called the **middle quartile** and is denoted by q_M. The rule of "splitting the difference" (just applied to the middle quartile) is equally well used for the other quartiles if necessary.

Boxplot

The **boxplot** of (1) in Fig. 477 is obtained from the five numbers $x_{min}, q_L, q_M, q_U, x_{max}$ just determined. The box extends from q_L to q_U. Hence it has the height IQR. The position of the median in the box shows that the data distribution is not symmetric. The two lines extend from the box to x_{min} below and to x_{max} above. Hence they mark the range R.

Boxplots are particularly suitable for making comparisons. For example, Fig. 477 shows boxplots of the data sets (1) and

(3)   91 89 93 91 87 94 92 85 91 90 96 93 89

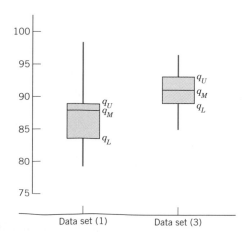

Fig. 477. Boxplots of data sets (1) and (3)

(consisting of $n = 13$ values). Ordering gives

(4)   85 87 89 89 90 91 91 91 92 93 93 94 96

(tensile strength, as before). From the plot we immediately see that the box of (3) is shorter than the box of (1) (indicating the higher quality of the sheets!) and that q_M is located in the middle of the box (showing the more symmetric form of the distribution). Finally, x_{max} is closer to q_U for (3) than for (1), a fact that we shall discuss later.

For plotting the box of (3) we took from (4) $x_{min} = 85$, $q_L = 89$, $q_M = 91$, $q_U = 93$, $x_{max} = 96$.

Outliers

An **outlier** is a value that appears to be uniquely different from the rest of the data set. It might indicate that something went wrong with the data collection process. In connection with quartiles an outlier is conventionally defined as a value more than a distance of $1.5\,\text{IQR}$ from either end of the box.

For the data in (1) we have $\text{IQR} = 5$, $q_L = 84$, $q_U = 89$. Hence outliers are smaller than $84 - 7.5$ or larger than $89 + 7.5$, so that 99 is an outlier [see (2)]. The data (3) have no outliers, as you can readily verify.

Mean. Standard Deviation. Variance

Medians and quartiles are easily obtained by ordering and counting, practically without calculation. But they do not give full information on data: you can move data values around to some extent without changing the median. Similarly for the quartiles.

The average size of the data values can be measured in a more refined way by the **mean**

(5)
$$\bar{x} = \frac{1}{n} \sum_{j=1}^{n} x_j = \frac{1}{n} (x_1 + x_2 + \cdots + x_n).$$

This is the arithmetic mean of the data values, obtained by taking their sum and dividing by the data *size* n. Thus in (1),

$$\bar{x} = \tfrac{1}{14} (89 + 84 + \cdots + 89) = \tfrac{611}{7} \approx 87.3.$$

Every data value contributes, and changing one of them will change the mean.

Similarly, the spread (variability) of the data values can be measured in a more refined way by the **standard deviation** s or by its square, the **variance**

(6)
$$s^2 = \frac{1}{n-1} \sum_{j=1}^{n} (x_j - \bar{x})^2 = \frac{1}{n-1} \left[(x_1 - \bar{x})^2 + \cdots + (x_n - \bar{x})^2 \right].$$

Thus, to obtain the variance of the data, take the difference $x_j - \bar{x}$ of each data value from the mean, square it, take the sum of these n squares, and divide it by $n - 1$ (not n, as we motivate in Sec. 23.2). To get the standard deviation s, take the square root of s^2.

For example, using $\bar{x} = 611/7$, we get for the data (1) the variance

$$s^2 = \tfrac{1}{13} \left[\left(89 - \tfrac{611}{7}\right)^2 + \left(84 - \tfrac{611}{7}\right)^2 + \cdots + \left(89 - \tfrac{611}{7}\right)^2 \right] = \tfrac{176}{7} \approx 25.14.$$

Hence the standard deviation is $s = \sqrt{176/7} \approx 5.014$. Note that the standard deviation has the same dimension (kg/mm^2) as the data values. On the other hand, the variance is more advantageous than the standard deviation in developing statistical methods.

Caution! Your CAS (MAPLE, for instance) may use $1/n$ instead of $1/(n-1)$ in (6), but the latter is better when n is small (see Sec. 23.2).

PROBLEM SET 22.1

Representation of Data

Represent the following data by a stem-and-leaf plot, a histogram, and a boxplot.

1. 12   11   9   5   12   6   7   9   11   11
2. 17   18   17   16   17   16   18   16
3. 46   48   44   23   31   20   34   27   41   36   46   28   28   39   29
4. −0.51   0.12   −0.47   0.95   0.25   −0.18   −0.54
5. 50.6   50.9   49.1   51.3   50.5   49.7   51.5   49.8   51.1   48.9   50.3   49.2   51.2   50.4   52.8
6. 13.1   11.0   13.4   11.5   10.2   18.2   12.4   12.8   15.7   10.9
7. Release time [sec] of a relay

$$1.3 \quad 1.2 \quad 1.4 \quad 1.5 \quad 1.3 \quad 1.3 \quad 1.4 \quad 1.1 \quad 1.5 \quad 1.4$$

$$1.6 \quad 1.3 \quad 1.5 \quad 1.1 \quad 1.4 \quad 1.2 \quad 1.3 \quad 1.5 \quad 1.4 \quad 1.4$$

8. Carbon content [%] of coal

$$86 \quad 87 \quad 86 \quad 81 \quad 77 \quad 85 \quad 87 \quad 86 \quad 85 \quad 87$$

$$82 \quad 84 \quad 83 \quad 79 \quad 82 \quad 73 \quad 86 \quad 84 \quad 83 \quad 83$$

9. Miles per gallon of gasoline required by six cars of the same make

$$15.0 \quad 15.5 \quad 14.5 \quad 15.0 \quad 15.5 \quad 15.0$$

10. Weight of filled bags [grams] in an automatic filling process

$$203 \quad 199 \quad 198 \quad 201 \quad 200 \quad 201 \quad 201$$

Average and Spread

Find the mean and compare it with the median. Find the standard deviation and compare it with the interquartile range.

11. The data in Prob. 1
12. The data in Prob. 2
13. The data in Prob. 5
14. 5   22   7   23   6. Why is $|\bar{x} - q_M|$ so large?
15. Construct the simplest possible data with $\bar{x} = 100$ but $q_M = 0$.
16. **(Mean)** Prove that $\bar{x}$ must always lie between the smallest and the largest data values.
17. **(Outlier, reduced data)** Calculate s for the data 4, 1, 3, 10, 2. Then reduce the data by deleting the outlier and calculate s. Comment.
18. **WRITING PROJECT. Average and Spread.** Compare Q_M, IQR and $\bar{x}$, s, illustrating the advantages and disadvantages with examples and plots of your own.

22.2   Experiments, Outcomes, Events

We now turn to **probability theory.** This theory has the purpose of providing mathematical models of situations affected or even governed by "chance effects," for instance, in weather forecasting, life insurance, quality of technical products (computers, batteries, steel sheets, etc.), traffic problems, and, of course, games of chance with cards or dice. And the accuracy of these models can be tested by suitable observations or experiments—this is the purpose of **statistics** in Chap. 23.

We begin by defining some standard terms. An **experiment** is a process of measurement or observation, in a laboratory, in a factory, on the street, in nature, or wherever; so "experiment" is used in a rather general sense. Our interest is in experiments that involve **randomness,** chance effects, so that we cannot predict a result exactly. A **trial** is a single performance of an experiment. Its result is called an **outcome** or a **sample point.** n trials then give a **sample** of **size** n consisting of n sample points. The **sample space** S of an experiment is the set of all possible outcomes. Examples are

EXAMPLES 1–6

(1) Inspecting a lightbulb. $S = \{$Defective, Nondefective$\}$.

(2) Rolling a die. $S = \{1, 2, 3, 4, 5, 6\}$.

(3) Measuring tensile strength of wire. S the numbers in some interval.

(4) Measuring copper content of brass. S: 50% to 90%, say.

(5) Counting daily traffic accidents in New York. S the integers in some interval.

(6) Asking for opinion about a new car model. $S = \{$Like, Dislike, Undecided$\}$.

The subsets of S are called **events** and the outcomes **simple events.**

EXAMPLE 7   **Events**

In (2), events are $A = \{1, 3, 5\}$ (*"Odd number"*), $B = \{2, 4, 6\}$ (*"Even number"*), $C = \{5, 6\}$, etc. Simple events are $\{1\}, \{2\}, \cdots, \{6\}$.   ◀

If in a trial an outcome a happens and $a \in A$ (*a is an element of A*), we say that A happens. For instance, if a die turns up a 3, the event A: *Odd number* happens. Similarly, if C happens (meaning 5 or 6 turns up), then $D = \{4, 5, 6\}$ happens. Also note that S happens in each trial, meaning that *some* event of S always happens. All this is quite natural.

Unions, Intersections, Complements of Events

In connection with basic probability laws we shall need the following concepts and facts about events (subsets) $A, B, C, \cdots$ of a given sample space S.

The **union** $A \cup B$ of A and B consists of all points in A or B or both.

The **intersection** $A \cap B$ of A and B consists of all points that are in both A and B.

If A and B have no points in common, we write

$$A \cap B = \varnothing$$

where $\varnothing$ is the *empty set* (set with no elements) and we call A and B **mutually exclusive** (or **disjoint**) because the occurrence of A *excludes* that of B (and conversely)—if your die turns up an odd number, it cannot turn up an even number in the same trial. Similarly, a coin cannot turn up *Head* and *Tail* at the same time.

The **complement**[1] A^c of A consists of all the points of S *not* in A. Thus,

$$A \cap A^c = \varnothing, \qquad A \cup A^c = S.$$

In Example 7 we have $A^c = B$, hence $A \cup A^c = \{1, 2, 3, 4, 5, 6\} = S$.

Unions and intersections of more events are defined similarly. The **union**

$$\bigcup_{j=1}^{m} A_j = A_1 \cup A_2 \cup \cdots \cup A_m$$

of events $A_1, \cdots, A_m$ consists of all points that are in at least one A_j. Similarly for the union $A_1 \cup A_2 \cup \cdots$ of infinitely many subsets $A_1, A_2, \cdots$ of an *infinite* sample space S (that is, S consists of infinitely many points). The **intersection**

$$\bigcap_{j=1}^{m} A_j = A_1 \cap A_2 \cap \cdots \cap A_m$$

of $A_1, \cdots, A_m$ consists of the points of S that are in each of these events. Similarly for the intersection $A_1 \cap A_2 \cap \cdots$ of infinitely many subsets of S.

Working with events can be illustrated and facilitated by **Venn diagrams**[2] for showing unions, intersections, and complements, as in Figs. 478 and 479, which are typical examples that give the idea.

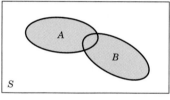

Union $A \cup B$

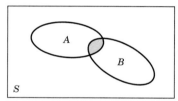

Intersection $A \cap B$

Fig. 478. Venn diagrams showing two events A and B in a sample space S and their union $A \cup B$ (colored) and intersection $A \cap B$ (colored)

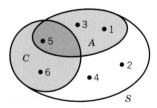

Fig. 479. Venn diagram for the experiment of rolling a die, showing S, $A = \{1, 3, 5\}$, $C = \{5, 6\}$, $A \cup C = \{1, 3, 5, 6\}$, $A \cap C = \{5\}$

---

[1]Or $\overline{A}$, but we shall not use this because in set theory it is used to denote the closure of A.
[2]JOHN VENN (1834—1923), English mathematician.

EXAMPLE 8   **Unions and intersections of 3 events**

In rolling a die, consider the events

   A: *Number greater than* 3,   *B:*   *Number less than* 6,   *C:*   *Even number.*

Then $A \cap B = \{4, 5\}$, $B \cap C = \{2, 4\}$, $C \cap A = \{4, 6\}$, $A \cap B \cap C = \{4\}$. Can you sketch a Venn diagram of this? Furthermore, $A \cup B = S$, hence $A \cup B \cup C = S$ (why?), etc. ◀

PROBLEM SET 22.2

Sample Spaces, Events

Graph a sample space for the following experiments.

1. Drawing three screws from a lot of right-handed and left-handed screws
2. Rolling two dice
3. Tossing two coins
4. Rolling a die until the first 6 appears
5. Drawing bolts from a lot of 10, containing 1 defective *D*, until *D* is drawn, assuming **sampling without replacement,** that is, bolts drawn are not returned to the lot.
6. In Prob. 1 let *A, B, C, D* mean 1 right-handed, 1 left-handed, 2 right-handed, 2 left-handed, respectively, among the 3 screws drawn. Are *A* and *B* mutually exclusive? *C* and *D?*
7. In rolling two dice, are *A: Sum divisible by* 3, *B: Sum divisible by* 5 mutually exclusive? Answer the same question for rolling three dice.
8. In Prob. 2 circle and mark the events *A: Faces are equal, B: Sum of faces less than* 5, $A \cup B$, $A \cap B$, A^c, B^c.
9. List all eight subsets of $S = \{a, b, c\}$.
10. In Prob. 4 list the outcomes that make up the event *E: First "Six" in rolling at most* 5 *times.* Describe E^c.

Venn Diagrams

11. In connection with a trip to Europe by some students, consider the events *P* that they see Paris, *G* that they have a good time, and *M* that they run out of money, and describe in words the events $1, \cdots, 7$ in the diagram.

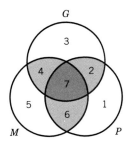

Problem 11

12. In a lot of 20 gaskets, 7 have no defect, 3 have a *T*-defect (too thin), 6 have an *L*-defect (too large), and 4 have both defects. Show this in a Venn diagram, also giving the number in each set.
13. (**De Morgan's laws**) Using Venn diagrams, graph and check *De Morgan's laws*

$$(A \cup B)^c = A^c \cap B^c$$

$$(A \cap B)^c = A^c \cup B^c.$$

14. Using Venn diagrams, graph and check the rules

$$A \cup (B \cap C) = (A \cup B) \cap (A \cup C)$$
$$A \cap (B \cup C) = (A \cap B) \cup (A \cap C).$$

15. Show that, by the definition of complement, for any subset A of a sample space S,

$$(A^c)^c = A, \qquad S^c = \varnothing, \qquad \varnothing^c = S, \qquad A \cup A^c = S, \qquad A \cap A^c = \varnothing.$$

22.3 Probability

The "probability" of an event A in an experiment is supposed to measure how frequently A is about to occur if we make many trials. If we flip a coin, then heads H and tails T will appear *about* equally often—we say that H and T are **"equally likely."** Table 22.1 confirms this. Similarly, for a regularly shaped die of homogeneous material (**"fair die"**) each of the six outcomes $1, \cdots, 6$ will be equally likely. These are examples of experiments in which the sample space S consists of finitely many outcomes (points) that for reasons of some symmetry can be regarded as equally likely. This suggests the following definition.

Definition 1. Probability

If the sample space S of an experiment consists of finitely many outcomes (points) that are equally likely, then the probability $P(A)$ of an event A is

(1)
$$P(A) = \frac{\text{Number of points in } A}{\text{Number of points in } S}.$$

Thus, in particular,

(2)
$$P(S) = 1$$

as follows directly from (1).

EXAMPLE 1   **Fair die**

In rolling a fair die, what is the probability $P(A)$ of A of obtaining at least a 5? The probability of B: *"Even number"?*

Solution. The six outcomes are equally likely, so that each has probability 1/6. Thus $P(A) = 2/6 = 1/3$ because $A = \{5, 6\}$ has 2 points, and $P(B) = 3/6 = 1/2$. ◀

Table 22.1
Coin Tossing

Experiments by	Number of Throws	Number of Heads	Relative Frequency of Heads
BUFFON	4,040	2,048	0.5069
K. PEARSON	12,000	6,019	0.5016
K. PEARSON	24,000	12,012	0.5005

Definition 1 takes care of many games as well as some practical applications, as we shall see, but certainly not of all experiments, simply because in many problems we do not have finitely many equally likely outcomes. To arrive at a more general definition of probability, we regard ***probability as the counterpart of relative frequency.*** Recall from Sec. 22.1 that the **absolute frequency** $f(A)$ of an event A in n trials is the number of times A occurs, and the **relative frequency** of A in these trials is $f(A)/n$; thus

(3)
$$f_{\text{rel}}(A) = \frac{f(A)}{n} = \frac{\text{Number of times } A \text{ occurs}}{\text{Number of trials}}$$

Now if A did not occur, then $f(A) = 0$. If A always occurred, then $f(A) = n$. These are the extreme cases. Division by n gives

(4*)
$$0 \leqq f_{\text{rel}}(A) \leqq 1.$$

In particular, for $A = S$ we have $f(S) = n$ because S always occurs (meaning that some event always occurs; if necessary, see Sec. 22.2, after Example 7). Division by n gives

(5*)
$$f_{\text{rel}}(S) = 1.$$

Finally, if A and B are mutually exclusive, they cannot occur together. Hence the absolute frequency of their union $A \cup B$ must equal the sum of the absolute frequencies of A and B. Division by n gives the same relation for the relative frequencies,

(6*)
$$f_{\text{rel}}(A \cup B) = f_{\text{rel}}(A) + f_{\text{rel}}(B) \qquad (A \cap B = \varnothing).$$

We are now ready to extend the definition of probability to experiments in which equally likely outcomes are not available. Of course, the extended definition should include Definition 1. Since probabilities are supposed to be the theoretical counterpart of relative frequencies, we choose the properties in (4*), (5*), (6*) as axioms. (Historically, such a choice is the result of a long process of gaining experience on what might be best and most practical.)

Definition 2. Probability

Given a sample space S, with each event A of S (subset of S) there is associated a number $P(A)$, called the **probability** of A, such that the following **axioms of probability** are satisfied.

1. For every A in S,

(4)
$$0 \leqq P(A) \leqq 1.$$

2. The entire sample space S has the probability

(5)
$$P(S) = 1.$$

3. For mutually exclusive events A and B ($A \cap B = \varnothing$; see Sec. 22.2),

(6)
$$P(A \cup B) = P(A) + P(B) \qquad (A \cap B = \varnothing).$$

If S is infinite (has infinitely many points), Axiom 3 has to be replaced by[3]

3′. For mutually exclusive events $A_1, A_2, \cdots$,

(6′) $$P(A_1 \cup A_2 \cup \cdots) = P(A_1) + P(A_2) + \cdots.$$

Basic Theorems for Probability

We shall see that the axioms of probability will enable us to build up probability theory and its application to statistics. We begin with three basic theorems. The first of them is useful if we can get the probability of the complement A^c more easily than $P(A)$ itself.

THEOREM 1 **(Complementation rule)**

For an event A and its complement A^c in a sample space S,

(7) $$P(A^c) = 1 - P(A).$$

PROOF. By the definition of complement (Sec. 22.2) we have $S = A \cup A^c$ and $A \cap A^c = \varnothing$. Hence by Axioms 2 and 3,

$$1 = P(S) = P(A) + P(A^c), \qquad \text{thus} \qquad P(A^c) = 1 - P(A). \qquad \blacktriangleleft$$

EXAMPLE 2 **Coin tossing**

Five coins are tossed simultaneously. Find the probability of the event A: *At least one head turns up.* Assume that the coins are fair.

Solution. Since each coin can turn up heads or tails, the sample space consists of $2^5 = 32$ outcomes. Since the coins are fair, we may assign the same probability $(1/32)$ to each outcome. Then the event A^c (*No heads turn up*) consists of only 1 outcome. Hence $P(A^c) = 1/32$, and the answer is $P(A) = 1 - P(A^c) = 31/32$. $\blacktriangleleft$

The next theorem is a simple extension of Axiom 3, which you can readily prove by induction:

THEOREM 2 **(Addition rule for mutually exclusive events)**

For mutually exclusive events $A_1, \cdots, A_m$ in a sample space S,

(8) $$P(A_1 \cup A_2 \cup \cdots A_m) = P(A_1) + P(A_2) + \cdots + P(A_m).$$

EXAMPLE 3 **Mutually exclusive events**

If the probability that on any workday a garage will get 10–20, 21–30, 31–40, over 40 cars to service is 0.20, 0.35, 0.25, 0.12, respectively, what is the probability that on a given workday the garage gets at least 21 cars to service?

Solution. Since these are mutually exclusive events, Theorem 2 gives the answer $0.35 + 0.25 + 0.12 = 0.72$. Check this by the complementation rule. $\blacktriangleleft$

---

[3]In the infinite case, for a *theoretical* restriction of the subsets of S, of no *practical* consequence to us, see "σ-algebra," for example, in Ref. [8] listed in Appendix 1.

In many cases, events will not be mutually exclusive. Then we have

THEOREM 3   **(Addition rule for arbitrary events)**

For events A and B in a sample space,

(9) $$P(A \cup B) = P(A) + P(B) - P(A \cap B).$$

PROOF.   *C, D, E* in Fig. 480 make up $A \cup B$ and are mutually exclusive (disjoint). Hence by Theorem 2,

$$P(A \cup B) = P(C) + P(D) + P(E).$$

This gives (9) because on the right $P(C) + P(D) = P(A)$ by Axiom 3 and disjointness; and $P(E) = P(B) - P(D) = P(B) - P(A \cap B)$, also by Axiom 3 and disjointness.   ◀

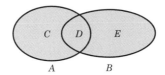

Fig. 480.   Proof of Theorem 3

Note that for mutually exclusive events *A* and *B* we have $A \cap B = \varnothing$ by definition and, by comparing (9) and (6),

(10) $$\boxed{P(\varnothing) = 0.}$$

(Can you also prove this by (5) and (7)?)

EXAMPLE 4   **Union of arbitrary events**

In tossing a fair die, what is the probability of getting an odd number or a number less than 4?

Solution.   Let *A* be the event *"Odd number"* and *B* the event *"Number less than* 4." Then Theorem 3 gives the answer

$$P(A \cup B) = \tfrac{3}{6} + \tfrac{3}{6} - \tfrac{2}{6} = \tfrac{2}{3}$$

because $A \cap B = $ *"Odd number less than* 4" $= \{1, 3\}$.   ◀

Conditional Probability. Independent Events

Often it is required to find the probability of an event *B* under the condition that an event *A* occurs. This probability is called the **conditional probability** *of B given A* and is denoted by $P(B|A)$. In this case *A* serves as a new (reduced) sample space, and that probability is the fraction of $P(A)$ which corresponds to $A \cap B$. Thus

(11) $$\boxed{P(B|A) = \frac{P(A \cap B)}{P(A)}}$$ $[P(A) \neq 0].$

Similarly, the *conditional probability of A given B* is

(12) $$P(A|B) = \frac{P(A \cap B)}{P(B)}$$ $[P(B) \neq 0]$.

Solving (11) and (12) for $P(A \cap B)$, we obtain

THEOREM 4   **(Multiplication rule)**

If A and B are events in a sample space S and $P(A) \neq 0$, $P(B) \neq 0$, then

(13) $$P(A \cap B) = P(A)P(B|A) = P(B)P(A|B).$$

EXAMPLE 5   **Multiplication rule**

In producing screws, let A mean "screw too slim" and B "screw too short." Let $P(A) = 0.1$ and let the conditional probability that a slim screw is also too short be $P(B|A) = 0.2$. What is the probability that a screw that we pick randomly from the lot produced will be both too slim and too short?

Solution. $P(A \cap B) = P(A)P(B|A) = 0.1 \cdot 0.2 = 0.02 = 2\%$, by Theorem 4.   ◀

Independent events. If events A and B are such that

(14) $$P(A \cap B) = P(A)P(B),$$

they are called **independent events.** Assuming $P(A) \neq 0$, $P(B) \neq 0$, we see from (11)–(13) that in this case

$$P(A|B) = P(A), \qquad P(B|A) = P(B).$$

This means that the probability of A does not depend on the occurrence or nonoccurrence of B, and conversely. This justifies the term "independent."

Independence of *m* events. Similarly, m events $A_1, \cdots, A_m$ are called **independent** if

(15a) $$P(A_1 \cap \cdots \cap A_m) = P(A_1) \cdots P(A_m)$$

as well as for every k different events $A_{j_1}, A_{j_2}, \cdots, A_{j_k}$

(15b) $$P(A_{j_1} \cap A_{j_2} \cap \cdots \cap A_{j_k}) = P(A_{j_1})P(A_{j_2}) \cdots P(A_{j_k})$$

where $k = 2, 3, \cdots, m - 1$.

Accordingly, three events A, B, C are independent if

(16)
$$P(A \cap B) = P(A)P(B),$$
$$P(B \cap C) = P(B)P(C),$$
$$P(C \cap A) = P(C)P(A),$$
$$P(A \cap B \cap C) = P(A)P(B)P(C).$$

Sampling. Our next example has to do with randomly drawing objects, *one at a time,* from a given set of objects. This is called **sampling from a population,** and there are two ways of sampling, as follows.

   1. In **sampling with replacement,** the object that was drawn at random is placed back to the given set and the set is mixed thoroughly. Then we draw the next object at random.

   2. In **sampling without replacement** the object that was drawn is put aside.

EXAMPLE 6   **Sampling with and without replacement**

A box contains 10 screws, three of which are defective. Two screws are drawn at random. Find the probability that none of the two screws is defective.

Solution.   We consider the events

$$A: \textit{First drawn screw nondefective.}$$

$$B: \textit{Second drawn screw nondefective.}$$

Clearly, $P(A) = \frac{7}{10}$ because 7 of the 10 screws are nondefective and we sample at random, so that each screw has the same probability ($\frac{1}{10}$) of being picked. If we sample with replacement, the situation before the second drawing is the same as at the beginning, and $P(B) = \frac{7}{10}$. The events are independent, and the answer is

$$P(A \cap B) = P(A)P(B) = 0.7 \cdot 0.7 = 0.49 = 49\%.$$

If we sample without replacement, then $P(A) = \frac{7}{10}$, as before. If A has occurred, then there are 9 screws left in the box, 3 of which are defective. Thus $P(B|A) = \frac{6}{9} = \frac{2}{3}$, and Theorem 4 yields the answer

$$P(A \cap B) = \frac{7}{10} \cdot \frac{2}{3} \approx 47\%.$$

Is it intuitively clear that this value must be smaller than the preceding one?   ◀

PROBLEM SET 22.3

   1. In rolling two fair dice, what is the probability of obtaining a sum greater than 3 but not exceeding 6?
   2. In Prob. 1, what is the probability of obtaining a sum not exceeding 10?
   3. If a box contains 10 left-handed and 20 right-handed screws, what is the probability of obtaining at least one right-handed screw in drawing 2 screws with replacement?
   4. Will the probability in Prob. 3 increase or decrease if we draw without replacement. First guess, then calculate.
   5. Three screws are drawn at random from a lot of 100 screws, 10 of which are defective. Find the probability of the event that all 3 screws drawn are nondefective, assuming that we draw (a) with replacement, (b) without replacement.
   6. What is the probability of obtaining at least one Six in rolling three fair dice?
   7. Under what conditions will it make *practically* no difference whether we sample with or without replacement?
   8. Two boxes contain ten chips each, numbered from 1 to 10, and one chip is drawn from each box. Find the probability of the event E that the sum of the numbers on the drawn chips is greater than 4.
   9. If a certain kind of tire has a life exceeding 30 000 miles with probability 0.90, what is the probability that a set of these tires on a car will last longer than 30 000 miles?
   10. A batch of 200 iron rods consists of 50 oversized rods, 50 undersized rods, and 100 rods of the desired length. If two rods are drawn at random without replacement, what is the probability

of obtaining (a) two rods of the desired length, (b) exactly one of the desired length, (c) none of the desired length, (d) two undersized rods?

11. If we inspect envelopes by drawing 4 of them without replacement from every lot of 100, what is the probability of getting 4 clean envelopes although 3% of them contain spots?

12. If a circuit contains four automatic switches and we want that, with a probability of 99%, during a given time interval the switches to be all working; what probability of failure per time interval can we admit for a single switch?

13. A pressure control apparatus contains 3 electronic tubes. The apparatus will not work unless all tubes are operative. If the probability of failure of each tube during some interval of time is 0.04, what is the corresponding probability of failure of the apparatus?

14. Suppose that in a production of spark plugs the fraction of defective plugs has been constant at 2% over a long time and that this process is controlled every half hour by drawing and inspecting two just produced. Find the probabilities of getting (a) no defectives, (b) 1 defective, (c) 2 defectives. What is the sum of these probabilities?

15. What gives the greater probability of hitting at least once: (a) hitting with probability 1/2 and firing 1 shot, or (b) hitting with probability 1/4 and firing 2 shots? First guess. Then calculate.

16. Suppose that we draw cards repeatedly and with replacement from a file of 200 cards, 100 of which refer to male and 100 to female persons. What is the probability of obtaining the second "female" card before the third "male" card?

17. What is the complementary event of the event considered in Prob. 16? Calculate its probability and use it to check your result in Prob. 16.

18. Show that if B is a subset of A, then $P(B) \leqq P(A)$.

19. Extending Theorem 4, show that $P(A \cap B \cap C) = P(A)P(B|A)P(C|A \cap B)$.

20. You may wonder whether in (16) the last relation follows from the others, but the answer is no. To see this, imagine that a chip is drawn from a box containing 4 chips numbered $000, 011, 101, 110$, and let A, B, C be the events that the first, second, and third digit, respectively, on the drawn chip is 1. Show that then the first three formulas in (16) hold but the last one does not hold.

22.4 Permutations and Combinations

Permutations and combinations help in finding probabilities $P(A) = a/k$ by **systematically counting** the number a of points of which an event A consists; here, k is the number of points of the sample space S. The practical difficulty is that a may often be surprisingly large, so that actual counting becomes hopeless. For example, if in assembling some instrument you need 10 different screws in a certain order and you want to draw them randomly from a box (which contains nothing else) the probability of obtaining them in the required order is only 1/3 628 800 because there are

$$10! = 1 \cdot 2 \cdot 3 \cdot 4 \cdot 5 \cdot 6 \cdot 7 \cdot 8 \cdot 9 \cdot 10 = 3\ 628\ 800$$

orders in which they can be drawn. Similarly, in many other situations the numbers of orders, arrangements, etc. are often incredibly large. (If you are unimpressed, take 20 screws—how much bigger will the number be?)

Permutations

A **permutation** of given things (*elements* or *objects*) is an arrangement of these things in a row in some order. For example, for three letters a, b, c there are $3! = 1 \cdot 2 \cdot 3 = 6$ permutations: *abc, acb, bac, bca, cab, cba*. This illustrates (a) in the following theorem.

THEOREM 1 **(Permutations)**

(a) *Different things.* *The number of permutations of n different things taken all at a time is*

(1)
$$n! = 1 \cdot 2 \cdot 3 \cdots n$$
(read "*n factorial*").

(b) *Classes of equal things.* *If n given things can be divided into c classes of alike things differing from class to class, then the number of permutations of these things taken all at a time is*

(2)
$$\frac{n!}{n_1! n_2! \cdots n_c!}$$
$(n_1 + n_2 + \cdots + n_c = n)$

where n_j is the number of things in the jth class.

PROOF. (a) There are n choices for filling the first place in the row. Then $n - 1$ things are still available for filling the second place, etc.

(b) n_1 alike things in class 1 make $n_1!$ permutations collapse into a single permutation (those in which class 1 things occupy the same n_1 positions), etc., so that (2) follows from (1). ◀

EXAMPLE 1 **Illustration of Theorem 1(b)**

If a box contains 6 red and 4 blue balls, the probability of drawing first the red and then the blue balls is

$$P = 6!4!/10! = 1/210 \approx 0.5\%.$$ ◀

A **permutation of *n* things taken *k* at a time** is a permutation containing only k of the n given things. Two such permutations consisting of the same k elements, in a different order, are different, by definition. For example, there are 6 different permutations of the three letters *a, b, c,* taken two letters at a time, *ab, ac, bc, ba, ca, cb.*

A **permutation of *n* things taken *k* at a time with repetitions** is an arrangement obtained by putting any given thing in the first position, any given thing, including a repetition of the one just used, in the second, and continuing until k positions are filled. For example, there are $3^2 = 9$ different such permutations of *a, b, c* taken 2 letters at a time, namely, the preceding 6 permutations and *aa, bb, cc.* You may prove (see Team Project 16)

THEOREM 2 **(Permutations)**

The number of different permutations of n different things taken k at a time without repetitions is

(3a)
$$n(n - 1)(n - 2) \cdots (n - k + 1) = \frac{n!}{(n - k)!}$$

and with repetitions is

(3b)
$$n^k.$$

EXAMPLE 2   **Illustration of Theorem 2**

In a coded telegram the letters are arranged in groups of five letters, called *words*. From (3b) we see that the number of different such words is

$$26^5 = 11\ 881\ 376.$$

From (3a) it follows that the number of different such words containing each letter no more than once is

$$26!/(26 - 5)! = 26 \cdot 25 \cdot 24 \cdot 23 \cdot 22 = 7\ 893\ 600.\qquad \blacktriangleleft$$

Combinations

In a permutation, the order of the selected things is essential. In contrast, a **combination** of given things means any selection of one or more things *without regard to order*. There are two kinds of combinations, as follows.

The number of **combinations of n different things, taken k at a time, without repetitions** is the number of sets that can be made up from the n given things, each set containing k different things and no two sets containing exactly the same k things.

The number of **combinations of n different things, taken k at a time, with repetitions** is the number of sets that can be made up of k things chosen from the given n things, each being used as often as desired.

For example, there are three combinations of the three letters a, b, c, taken two letters at a time, without repetitions, namely, ab, ac, bc, and six such combinations with repetitions, namely, ab, ac, bc, aa, bb, cc.

THEOREM 3   **(Combinations)**

The number of different combinations of n different things, k at a time, without repetitions, is

(4a)
$$\binom{n}{k} = \frac{n!}{k!(n-k)!} = \frac{n(n-1)\cdots(n-k+1)}{1 \cdot 2 \cdots k},$$

and the number of those combinations with repetitions is

(4b)
$$\binom{n+k-1}{k}.$$

The statement involving (4a) follows from the first part of Theorem 2 by noting that there are $k!$ permutations of k things from the given n things that differ by the order of the elements (see Theorem 1), but there is only a single combination of those k things of the type characterized in the first statement of Theorem 3. The last statement of Theorem 3 can be proved by induction (see Team Project 16).

EXAMPLE 3   **Illustration of Theorem 3**

The number of samples of five lightbulbs that can be selected from a lot of 500 bulbs is [see (4a)]

$$\binom{500}{5} = \frac{500!}{5!495!} = \frac{500 \cdot 499 \cdot 498 \cdot 497 \cdot 496}{1 \cdot 2 \cdot 3 \cdot 4 \cdot 5} = 255\ 244\ 687\ 600.\qquad \blacktriangleleft$$

Factorial Function

In (1)–(4) the **factorial function** is basic. By definition,

$$(5) \qquad\qquad 0! = 1.$$

Values may be computed recursively from given values by

$$(6) \qquad\qquad (n + 1)! = (n + 1)n!.$$

For large n the function is very large (see Table A3 in Appendix 5). A convenient approximation for large n is the **Stirling formula**[4]

$$(7) \qquad\qquad \boxed{n! \sim \sqrt{2\pi n}\left(\frac{n}{e}\right)^n} \qquad\qquad (e = 2.718\cdots)$$

where $\sim$ is read *"asymptotically equal"* and means that the ratio of the two sides of (7) approaches 1 as n approaches infinity.

EXAMPLE 4　**Stirling formula**

To get a feeling for the Stirling formula, verify the following approximations:

$n!$	By (7)	Exact Value	Relative Error
4!	23.5	24	2.1%
10!	3 598 696	3 628 800	0.8%
20!	$2.422\ 79 \cdot 10^{18}$	2 432 902 008 176 640 000	0.4%

◀

Binomial Coefficients

The binomial coefficients are defined by the formula

$$(8) \qquad\qquad \binom{a}{k} = \frac{a(a - 1)(a - 2)\cdots(a - k + 1)}{k!} \qquad (k \geqq 0, \text{ integer}).$$

The numerator has k factors. Furthermore, we define

$$(9) \qquad\qquad \binom{a}{0} = 1, \qquad \text{in particular,} \qquad \binom{0}{0} = 1.$$

For integer $a = n$ we obtain from (8)

$$(10) \qquad\qquad \binom{n}{k} = \binom{n}{n - k} \qquad\qquad (n \geqq 0, 0 \leqq k \leqq n).$$

---

[4]JAMES STIRLING (1692—1770), Scots mathematician.

Binomial coefficients may be computed recursively, because

(11)
$$\binom{a}{k} + \binom{a}{k+1} = \binom{a+1}{k+1} \qquad (k \geqq 0, \text{ integer}).$$

Formula (8) also yields

(12)
$$\binom{-m}{k} = (-1)^k \binom{m+k-1}{k} \qquad \begin{array}{l} (k \geqq 0, \text{ integer}) \\ (m > 0). \end{array}$$

There are numerous further relations; we mention

(13)
$$\sum_{s=0}^{n-1} \binom{k+s}{k} = \binom{n+k}{k+1} \qquad \begin{array}{l} (k \geqq 0, n \geqq 1, \\ \text{both integer}) \end{array}$$

and

(14)
$$\sum_{k=0}^{r} \binom{p}{k} \binom{q}{r-k} = \binom{p+q}{r}.$$

PROBLEM SET 22.4

1. List all permutations of four digits 1, 2, 3, 4, taken all at a time.
2. List (a) all permutations, (b) all combinations without repetitions, (c) all combinations with repetitions, of 5 letters *a, e, i, o, u* taken 2 at a time.
3. How many different samples of 4 objects can be drawn from a lot of 50 objects?
4. In how many ways can we assign 7 workers to 7 jobs (one worker to each job and conversely)?
5. In how many ways can we choose a committee of 3 from 8 persons?
6. In how many different ways can we select a committee consisting of 3 engineers, 2 chemists, and 2 mathematicians from 8 engineers, 6 chemists, and 5 mathematicians? (First guess, then compute.)
7. An urn contains 2 green, 3 yellow, and 5 red balls. We draw 1 ball at random and put it aside. Then we draw the next ball, and so on. Find the probability of drawing at first the 2 green balls, then the 3 yellow ones, and finally the red ones.
8. If a cage contains 100 mice, 3 of which are male, what is the probability that the 3 male mice will be included if 10 mice are randomly selected?
9. How many different license plates showing 6 symbols, namely, 3 letters followed by 3 digits, could be made?
10. In how many different ways can 6 people be seated at a round table?
11. How many automobile registrations may the police have to check in a hit-and-run accident if a witness reports KDP7 and cannot remember the last two digits on the license plate but is certain that all three were different?
12. If 3 suspects who committed a burglary and 6 innocent persons are lined up, what is the probability that a witness who is not sure and has to pick three persons will pick the three suspects by chance? That the witness picks 3 innocent persons by chance? First guess. Then calculate.
13. In a lot of 8 items, 2 are defective. (a) Find the number of different samples of 3. Find the number of samples containing (b) no defectives, (c) 1 defective, (d) 2 defectives.

14. (Birthday problem) What is the probability that in a group of 20 people (that includes no twins) at least two have the same birthday, if we assume that the probability of having birthday on a given day is 1/365 for every day. First guess, then calculate.

15. CAS PROJECT. Stirling formula. (a) Using (7), compute approximate values of $n!$ for $n = 1, \cdots, 20$.

(b) Determine the relative error in (a). Find an empirical formula for that relative error.

(c) An upper bound for that relative error is $e^{1/12n} - 1$. Try to relate your empirical formula to this.

(d) Search through the literature for further information on Stirling's formula. Write a short essay about your findings, arranged in logical order and illustrated with numerical examples.

16. TEAM PROJECT. Permutations, Combinations. (a) Prove Theorem 2.

(b) Prove the last statement of Theorem 3.

(c) Derive (11) from (8).

(d) By the **binomial theorem,**

$$(a + b)^n = \sum_{k=0}^{n} \binom{n}{k} a^k b^{n-k},$$

so that $a^k b^{n-k}$ has the coefficient $\binom{n}{k}$. Can you conclude this from Theorem 3 or is this a mere coincidence?

(e) Prove (14) by using the binomial theorem.

(f) Collect further formulas for binomial coefficients from the literature and illustrate them numerically.

22.5 Random Variables, Probability Distributions

In Sec. 22.1 we considered frequency distributions of data. These distributions show the absolute or relative frequency of the data values. Similarly, a **probability distribution** or, briefly, a **distribution,** shows the probabilities of events in an experiment. The quantity that we observe in an experiment will be denoted by X and called a **random variable** (or **stochastic variable**) because the value it will assume in the next trial depends on chance, on **randomness**—if you roll a dice, you get one of the numbers from 1 to 6, but you don't know which one will show up next. Thus $X = $ *Number a die turns up* is a random variable. So is $X = $ *Elasticity of rubber* (elongation at break). ("Stochastic" means related to chance.)

If we **count** (cars on a road, deaths by cancer, tosses until a die shows the first Six), we have a **discrete random variable and distribution.** If we **measure** (electric voltage, rainfall, hardness of steel), we have a **continuous random variable and distribution.** Precise definitions follow. In both cases the distribution of X is determined by the **distribution function**[5]

(1)
$$F(x) = P(X \leq x);$$

this is the probability that X will assume any value not exceeding x. For (1) to make sense in both cases we impose conditions given in the following definition.

---

[5]***Caution!*** The terminology is not uniform! $F(x)$ is sometimes also called the **cumulative distribution function.**

Definition (Random variable)

A **random variable** X is a function defined on the sample space S of an experiment. Its values are real numbers. For every number a the probability $P(X = a)$ with which X assumes a is defined. Similarly, for any interval I the probability $P(X \in I)$ with which X assumes any value in I is defined.

Although this definition is very general, practically only a very small number of distributions will occur over and over again in applications.

From (1) we obtain the fundamental formula for the probability corresponding to an interval $a < x \leqq b$,

$$(2) \qquad \boxed{P(a < X \leqq b) = F(b) - F(a).}$$

This follows because $X \leqq a$ (*"X assumes any value not exceeding a"*) and $a < X \leqq b$ (*"X assumes any value in the interval $a < x \leqq b$"*) are mutually exclusive events, so that by (1) and Axiom 3,

$$F(b) = P(X \leqq b) = P(X \leqq a) + P(a < X \leqq b) = F(a) + P(a < X \leqq b)$$

and subtraction of $F(a)$ on both sides gives (2). ◀

Discrete Random Variables and Distributions

By definition, a random variable X and its distribution are **discrete** if X assumes only finitely many or at most countably many values $x_1, x_2, x_3, \cdots$, called the **possible values** of X, with positive probabilities $p_1 = P(X = x_1)$, $p_2 = P(X = x_2)$, $p_3 = P(X = x_3)$, $\cdots$, whereas the probability $P(X \in I)$ is zero for any interval I containing no possible value.

Obviously, the discrete distribution is also determined by the **probability function** $f(x)$ of X, defined by

$$(3) \qquad \boxed{f(x) = \begin{cases} p_j & \text{if } x = x_j \\ 0 & \text{otherwise} \end{cases}} \qquad (j = 1, 2, \cdots),$$

From this we get the values of the **distribution function** $F(x)$ by taking sums,

$$(4) \qquad \boxed{F(x) = \sum_{x_j \leqq x} f(x_j) = \sum_{x_j \leqq x} p_j}$$

where for any given x we sum all the probabilities p_j for which x_j is smaller than or equal to that x. This is a **step function** with upward jumps of size p_j at the possible values x_j of X and constant in between.

EXAMPLE 1 **Probability function and distribution function**

Figure 481 shows the probability function $f(x)$ and the distribution function $F(x)$ of the discrete random variable

$$X = \text{Number a fair die turns up.}$$

X has the possible values $x = 1, 2, 3, 4, 5, 6$ with probability 1/6 each. At these x the distribution function has upward jumps of magnitude 1/6. Hence from the graph of $f(x)$ we can construct the graph of $F(x)$, and conversely.

In Figure 481 (and the next one) at each jump the *fat dot* indicates the *function value at the jump!* ◀

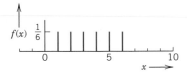

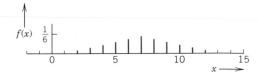

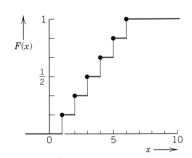

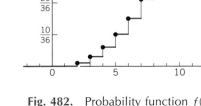

Fig. 481. Probability function $f(x)$ and distribution function $F(x)$ of the random variable $X = $ *Number obtained in tossing a fair die once*

Fig. 482. Probability function $f(x)$ and distribution function $F(x)$ of the random variable $X = $ *Sum of the two numbers obtained in tossing two fair dice once*

EXAMPLE 2   **Probability function and distribution function**

The random variable $X = $ *Sum of the two numbers two fair dice turn up* is discrete and has the possible values $2 (= 1 + 1), 3, 4, \cdots, 12 (= 6 + 6)$. There are $6 \cdot 6 = 36$ equally likely outcomes $(1, 1), (1, 2), \cdots, (6, 6)$, where the first number is that shown on the first die and the second number that on the other die. Each such outcome has probability 1/36. Now $X = 2$ occurs in the case of the outcome $(1, 1)$; $X = 3$ in the case of the two outcomes $(1, 2)$ and $(2, 1)$; $X = 4$ in the case of the three outcomes $(1, 3), (2, 2), (3, 1)$; and so on. Hence $f(x) = P(X = x)$ and $F(x) = P(X \leq x)$ have the values

x	2	3	4	5	6	7	8	9	10	11	12
$f(x)$	1/36	2/36	3/36	4/36	5/36	6/36	5/36	4/36	3/36	2/36	1/36
$F(x)$	1/36	3/36	6/36	10/36	15/36	21/36	26/36	30/36	33/36	35/36	36/36

Figure 482 shows a bar chart of this function and the graph of the distribution function, which is again a step function, with jumps (of different height!) at the possible values of X. ◀

Two useful formulas for discrete distributions are readily obtained as follows. For the probability corresponding to intervals we have from (2) and (3)

$$(5) \qquad \boxed{P(a < X \leq b) = \sum_{a < x_j \leq b} p_j} \qquad (X \text{ discrete}).$$

This is the sum of all probabilities p_j for which x_j satisfies $a < x_j \leq b$. (Be careful about $<$ and $\leq$!) From this and $P(S) = 1$ (Sec. 22.3) we obtain the following formula.

(6)
$$\sum_j p_j = 1$$
(sum of all probabilities).

EXAMPLE 3   **Illustration of formula (5)**

In Example 2, compute the probability of a sum of at least 4 and at most 8.

Solution. $P(3 < X \leqq 8) = F(8) - F(3) = \frac{26}{36} - \frac{3}{36} = \frac{23}{36}$.   ◄

EXAMPLE 4   **Waiting problem. Countably infinite sample space**

In tossing a fair coin, let X = *Number of trials until the first head appears.* Then, by independence of events (Sec. 22.3),

$$P(X = 1) = P(H) \quad\;\; = \tfrac{1}{2} \qquad\qquad\qquad (H = \text{Head})$$

$$P(X = 2) = P(TH) \;\;\; = \tfrac{1}{2} \cdot \tfrac{1}{2} \quad\; = \tfrac{1}{4} \qquad\qquad (T = \text{Tail})$$

$$P(X = 3) = P(TTH) = \tfrac{1}{2} \cdot \tfrac{1}{2} \cdot \tfrac{1}{2} = \tfrac{1}{8}, \quad \text{etc.}$$

and in general $P(X = n) = \left(\tfrac{1}{2}\right)^n$, $n = 1, 2, \cdots$. Also, (6) can be confirmed by the sum formula for the geometric series,

$$\tfrac{1}{2} + \tfrac{1}{4} + \tfrac{1}{8} + \cdots = -1 + \frac{1}{1 - \tfrac{1}{2}} = -1 + 2 = 1. \qquad\qquad ◄$$

Continuous Random Variables and Distributions

Discrete random variables appear in experiments in which we **count** (defectives in a production, days of sunshine in Chicago, customers standing in a line, etc.). Continuous random variables appear in experiments in which we **measure** (lengths of screws, voltage in a power line, Brinell hardness of steel, etc.). By definition, a random variable X and its distribution are *of continuous type* or, briefly, **continuous,** if its distribution function $F(x)$ [defined in (1)] can be given by an integral

(7)
$$F(x) = \int_{-\infty}^{x} f(v)\, dv$$

(we write v because x is needed as the upper limit of the integral) whose integrand $f(x)$, called the **density** of the distribution, is nonnegative, and is continuous, perhaps except for finitely many x-values. Differentiation gives the relation of f to F as

(8)
$$f(x) = F'(x)$$

for every x at which $f(x)$ is continuous.

From (2) and (7) we obtain the very important formula for the probability corresponding to an interval:

(9)
$$P(a < X \leqq b) = F(b) - F(a) = \int_{a}^{b} f(v)\, dv.$$

This is the analog of (5) for the discrete case.

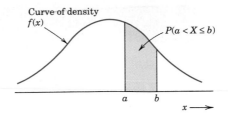

Fig. 483. Example illustrating formula (9)

From (7) and $P(S) = 1$ (Sec. 22.3) we also have the analog of (6):

$$(10) \qquad\qquad \int_{-\infty}^{\infty} f(v)\, dv = 1.$$

Continuous random variables are **simpler than discrete ones** with respect to intervals. Indeed, in the continuous case the four probabilities corresponding to $a < X \leq b$, $a < X < b$, $a \leq X < b$, and $a \leq X \leq b$ with any fixed a and b $(> a)$ are all the same. Can you see why? (*Answer*. This probability is the area under the density curve, as in Fig. 483, and does not change by adding or subtracting a single point in the interval of integration.) This is different from the discrete case! (Explain.)

The next example illustrates notations and typical applications of our present formulas.

EXAMPLE 5   **A continuous distribution**

Let X have the density function $f(x) = 0.75(1 - x^2)$ if $-1 \leq x \leq 1$ and zero otherwise. Find the distribution function. Find the probabilities $P(-\frac{1}{2} \leq X \leq \frac{1}{2})$ and $P(\frac{1}{4} \leq X \leq 2)$. Find x such that $P(X \leq x) = 0.95$.

Solution.   From (7) we obtain $F(x) = 0$ if $x \leq -1$,

$$F(x) = 0.75 \int_{-1}^{x} (1 - v^2)\, dv = 0.5 + 0.75x - 0.25x^3 \qquad \text{if } -1 < x \leq 1,$$

and $F(x) = 1$ if $x > 1$. From this and (9) we get

$$P(-\tfrac{1}{2} \leq X \leq \tfrac{1}{2}) = F(\tfrac{1}{2}) - F(-\tfrac{1}{2}) = 0.75 \int_{-1/2}^{1/2} (1 - v^2)\, dv = 68.75\%$$

(because $P(-\frac{1}{2} \leq X \leq \frac{1}{2}) = P(-\frac{1}{2} < X \leq \frac{1}{2})$ for a continuous distribution) and

$$P(\tfrac{1}{4} \leq X \leq 2) = F(2) - F(\tfrac{1}{4}) = 0.75 \int_{1/4}^{1} (1 - v^2)\, dv = 31.64\%.$$

(Note that the upper limit of integration is 1, not 2. Why?) Finally,

$$P(X \leq x) = F(x) = 0.5 + 0.75x - 0.25x^3 = 0.95.$$

Algebraic simplification gives $3x - x^3 = 1.8$. A solution is $x = 0.73$, approximately.

Sketch $f(x)$ and mark $x = -\frac{1}{2}, \frac{1}{2}, \frac{1}{4}$, and 0.73, so that you can see the results (the probabilities) as areas under the curve. Sketch also $F(x)$.   ◀

Further examples of continuous distributions are included in the next problem set and in later sections.

PROBLEM SET 22.5

1. Sketch the probability function $f(x) = x^2/14$ ($x = 1, 2, 3$) and the distribution function.
2. Find k in the probability function $f(x) = k\binom{3}{x}$, $x = 0, 1, 2, 3$, and sketch f and the distribution function F.
3. Sketch f and F when $f(0) = f(3) = 1/6$, $f(1) = f(2) = 1/3$. Can f have further positive values?
4. Let X be the number of years before a certain kind of pump needs replacement. Let X have the probability function $f(x) = kx^3$, $x = 0, 1, 2, 3, 4$. Find k. Sketch f and F.
5. If X has the probability function $f(x) = k/x!$ ($x = 0, 1, 2, \cdots$), what are k and $P(X \geqq 3)$?

6. Sketch the density $f(x) = \frac{1}{4}$ ($2 < x < 6$) and the distribution function. Find $P(X \geqq 4)$, $P(X \leqq 3)$.
7. In Prob. 6 find c such that (a) $P(X \leqq c) = 90\%$, (b) $P(X \geqq c) = \frac{1}{2}$, (c) $P(X \leqq c) = 5\%$.
8. Let $F(x) = 0$ if $x < 0$, $F(x) = 1 - e^{-0.1x}$ if $x > 0$. Sketch F and the density f. Find c such that $P(X \leqq c) = 95\%$.
9. Let X [millimeters] be the thickness of washers a machine turns out. Assume that X has the density $f(x) = kx$ if $0.9 < x < 1.1$ and 0 otherwise. Find k. What is the probability that a washer will have thickness between 0.95 mm and 1.05 mm?

10. Two screws are randomly drawn without replacement from a box containing 7 right-handed and 3 left-handed screws. Let X be the number of left-handed screws drawn. Find $P(X = 0)$, $P(X = 1)$, $P(X = 2)$, $P(1 < X < 2)$, $P(0.5 < X < 5)$.
11. Find the probability that none of three bulbs in a traffic signal will have to be replaced during the first 1500 hours of operation if the lifetime X of a bulb is a random variable with the density $f(x) = 6[0.25 - (x - 1.5)^2]$ when $1 \leqq x \leqq 2$ and $f(x) = 0$ otherwise, where x is measured in multiples of 1000 hours.
12. If the diameter X of axles has the density $f(x) = k$ if $119.9 \leqq x \leqq 120.1$ and 0 otherwise, how many defectives will a lot of 500 axles approximately contain if defectives are axles slimmer than 119.91 or thicker than 120.09?
13. If the life of ball bearings has the density $f(x) = ke^{-0.2x}$ if $0 \leqq x \leqq 10$ and 0 otherwise, what is k? What is the probability $P(X \geqq 5)$?
14. Find the probability function of $X = $ *Number of times a fair die is rolled until the first Six appears* and show that it satisfies (6).
15. Suppose that certain bolts have length $L = 400 + X$ mm, where X is a random variable with density $f(x) = \frac{3}{4}(1 - x^2)$ if $-1 \leqq x \leqq 1$ and 0 otherwise. Determine c so that with a probability of 95% a bolt will have any length between $400 - c$ and $400 + c$.
16. Suppose that in an automatic process of filling oil into cans, the content of a can (in gallons) is $Y = 100 + X$, where X is a random variable with density $f(x) = 1 - |x|$ when $|x| \leqq 1$ and 0 when $|x| > 1$. Sketch $f(x)$ and $F(x)$. In a lot of 1000 cans, about how many will contain 100 gallons or more? What is the probability that a can will contain less than 99.5 gallons? Less than 99 gallons?

17. Let $f(x) = kx^2$ if $0 \leqq x \leqq 2$ and 0 otherwise. Find k. Find c_1 and c_2 such that $P(X \leqq c_1) = 0.1$ and $P(X \leqq c_2) = 0.9$.
18. Let X be the ratio of sales to profits of some firm. Assume that X has the distribution function $F(x) = 0$ if $x < 2$, $F(x) = (x^2 - 4)/5$ if $2 \leqq x < 3$, $F(x) = 1$ if $x \geqq 3$. Find and sketch the density. What is the probability that X is between 2.5 (40% profit) and 5 (20% profit)?
19. Let X be a random variable that can assume every real value. What are the complements of the events $X \leqq b$, $X < b$, $X \geqq c$, $X > c$, $b \leqq X \leqq c$, $b < X \leqq c$?
20. Show that $b < c$ implies $P(X \leqq b) \leqq P(X \leqq c)$.

22.6   Mean and Variance of a Distribution

The mean μ and variance σ^2 of a random variable X and of its distribution are the theoretical counterparts of the mean $\bar{x}$ and variance s^2 of a frequency distribution in Sec. 22.1 and serve a similar purpose. Indeed, the mean characterizes the central location and the variance the spread (the variability) of the distribution. The **mean** μ (mu) is defined by

(1)

$$\textbf{(a)} \quad \mu = \sum_j x_j f(x_j) \qquad \text{(Discrete distribution)}$$

$$\textbf{(b)} \quad \mu = \int_{-\infty}^{\infty} x f(x)\, dx \qquad \text{(Continuous distribution)}.$$

and the **variance** σ^2 (sigma square) by

(2)

$$\textbf{(a)} \quad \sigma^2 = \sum_j (x_j - \mu)^2 f(x_j) \qquad \text{(Discrete distribution)}$$

$$\textbf{(b)} \quad \sigma^2 = \int_{-\infty}^{\infty} (x - \mu)^2 f(x)\, dx \qquad \text{(Continuous distribution)}.$$

σ (the positive square root of σ^2) is called the **standard deviation** of X and its distribution. f is the probability function or the density, respectively, in (a) and (b).

The mean μ is also denoted by $E(X)$ and is called the **expectation** *of* X because it gives the average value of X to be expected in many trials. Quantities such as μ and σ^2 that measure certain properties of a distribution are called **parameters.** μ and σ^2 are the two most important ones. From (2) we see that

(3) $\sigma^2 > 0$

(except for a discrete "distribution" with only one possible value, so that $\sigma^2 = 0$). We assume that μ and σ^2 exist (are finite), as is the case for practically all distributions that are useful in applications.

EXAMPLE 1   **Mean and variance**

The random variable $X = $ *Number of heads in a single toss of a fair coin*   has the possible values $X = 0$ and $X = 1$ with probabilities $P(X = 0) = \frac{1}{2}$ and $P(X = 1) = \frac{1}{2}$. From (1a) we thus obtain the mean $\mu = 0 \cdot \frac{1}{2} + 1 \cdot \frac{1}{2} = \frac{1}{2}$, and (3a) yields the variance

$$\sigma^2 = (0 - \tfrac{1}{2})^2 \cdot \tfrac{1}{2} + (1 - \tfrac{1}{2})^2 \cdot \tfrac{1}{2} = \tfrac{1}{4}. \qquad \blacktriangleleft$$

EXAMPLE 2   **Uniform distribution. Variance measures spread**

The distribution with the density

$$f(x) = \frac{1}{b - a} \quad \text{if} \quad a < x < b$$

and $f = 0$ otherwise is called the **uniform distribution** on the interval $a < x < b$. From (1b) (or from Theorem 1, below) we find that $\mu = (a + b)/2$, and (2b) yields the variance

$$\sigma^2 = \int_a^b \left(x - \frac{a + b}{2} \right)^2 \frac{1}{b - a}\, dx = \frac{(b - a)^2}{12}.$$

Figure 484 on p. 1076 illustrates that the spread is large if and only if σ^2 is large. $\blacktriangleleft$

Note that in Fig. 484 the increase of the length of the interval by a factor 3 leads to a large increase of the variance (by what factor?).

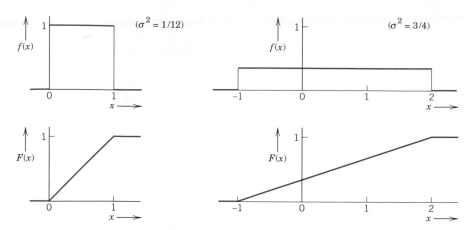

Fig. 484.   Uniform distributions having the same mean (0.5)
but different variances σ^2

Symmetry. We can obtain the mean μ without calculation if a distribution is symmetric. Indeed, you may prove

THEOREM 1   **(Mean of a symmetric distribution)**

If a distribution is **symmetric** *with respect to $x = c$, that is, $f(c - x) = f(c + x)$, then $\mu = c$. (Examples 1 and 2 illustrate this.)*

Transformation of Mean and Variance

Given a random variable X with mean μ and variance σ^2, we want to calculate the unknown mean and variance of $X^* = a_1 + a_2X$, where a_1 and a_2 are given constants. This problem is important in statistics, where it appears often. Notably, we shall need it like our daily bread in connection with the "normal distribution," beginning in Sec. 22.8.

THEOREM 2   **(Transformation of mean and variance)**

(a) *If a random variable X has mean μ and variance σ^2, then the random variable*

$$(4) \qquad\qquad\qquad X^* = a_1 + a_2X \qquad\qquad\qquad (a_2 > 0)$$

has the mean μ^* and variance σ^{*2}, where

$$(5) \qquad\qquad \mu^* = a_1 + a_2\mu \qquad \text{and} \qquad \sigma^{*2} = a_2{}^2\sigma^2.$$

(b) *In particular, the* **standardized random variable** *Z corresponding to X, given by*

$$(6) \qquad\qquad\qquad Z = \frac{X - \mu}{\sigma}$$

has the mean 0 and the variance 1.

PROOF.　We prove (5) for a continuous distribution. To a small interval I of length Δx on the x-axis there corresponds the probability $f(x)\Delta x$ [approximately; the area of a rectangle of base Δx and height $f(x)$]. Then the probability $f(x)\Delta x$ must equal that for the corresponding interval on the x^*-axis, that is, $f^*(x^*)\Delta x^*$, where f^* is the density of X^* and Δx^* is the length of the interval on the x^*-axis corresponding to I. Hence for differentials we have $f^*(x^*)\,dx^* = f(x)\,dx$. Also, $x^* = a_1 + a_2 x$ by (4), so that (1b) applied to X^* gives

$$\mu^* = \int_{-\infty}^{\infty} x^* f^*(x^*)\,dx^* = \int_{-\infty}^{\infty} (a_1 + a_2 x) f(x)\,dx = a_1 \int_{-\infty}^{\infty} f(x)\,dx + a_2 \int_{-\infty}^{\infty} x f(x)\,dx.$$

On the right the first integral equals 1, by (10) in Sec. 22.5. The second integral is μ. This proves (5) for μ^*. It implies

$$x^* - \mu^* = (a_1 + a_2 x) - (a_1 + a_2 \mu) = a_2(x - \mu).$$

From this and (2) applied to X^*, again using $f^*(x^*)\,dx^* = f(x)\,dx$, we obtain the second formula in (5),

$$\sigma^{*2} = \int_{-\infty}^{\infty} (x^* - \mu^*)^2 f^*(x^*)\,dx^* = a_2{}^2 \int_{-\infty}^{\infty} (x - \mu)^2 f(x)\,dx = a_2{}^2 \sigma^2.$$

For a discrete distribution the proof of (5) is similar.

Choosing $a_1 = -\mu/\sigma$ and $a_2 = 1/\sigma$ we obtain (6) from (4), writing $X^* = Z$. For these a_1, a_2 formula (5) gives $\mu^* = 0$ and $\sigma^{*2} = 1$, as claimed in (b).　◀

Expectation, Moments

Recall that (1) defines the expectation (the mean) of X, the value of X to be expected on the average, written $\mu = E(X)$. More generally, if $g(x)$ is nonconstant and continuous for all x, then $g(X)$ is a random variable. Hence its *mathematical expectation* or, briefly, its **expectation** $E(g(X))$ is the value of $g(X)$ to be expected on the average, defined [similarly to (1)] by

(7)　　　　$E(g(X)) = \sum_j g(x_j) f(x_j)$　　or　　$E(g(X)) = \int_{-\infty}^{\infty} g(x) f(x)\,dx.$

In the first formula, f is the probability function of the discrete random variable X. In the second formula, f is the density of the continuous random variable X. Important special cases are the **kth moment** of X $(k = 1, 2, \cdots)$

(8)　　　　$E(X^k) = \sum_j x_j{}^k f(x_j)$　　or　　$\int_{-\infty}^{\infty} x^k f(x)\,dx$

and the **kth central moment** of X $(k = 1, 2, \cdots)$

(9)　　　　$E([X - \mu]^k) = \sum_j (x_j - \mu)^k f(x_j)$　　or　　$\int_{-\infty}^{\infty} (x - \mu)^k f(x)\,dx.$

This includes the first moment, the **mean** of X

(10)　　　　　　　　　　$\boxed{\mu = E(X)}$　　　　　　　　[(8) with $k = 1$].

It also includes the second central moment, the **variance** of X

(11)
$$\sigma^2 = E([X - \mu]^2)$$
[(9) with $k = 2$].

For later use you may prove

(12)
$$E(1) = 1.$$

PROBLEM SET 22.6

Mean and Variance

Find the mean and the variance of the random variable X, where $f(x)$ is the probability function or density.

1. $f(x) = k\binom{3}{x}$, $x = 0, 1, 2, 3$

2. $X = $ *Number a fair die turns up*

3. $f(x) = 2x$ $(0 \leq x \leq 1)$

4. $f(x) = e^{-x}$ $(x > 0)$

5. Uniform distribution on $[0, 10]$

6. $Y = 4X - 2$ with X as in Prob. 4

7. If the diameter X [cm] of certain bolts has the density $f(x) = k(x - 0.9)(1.1 - x)$ for $0.9 < x < 1.1$ and 0 for other x, what are k, μ, and σ^2? Sketch $f(x)$.

8. If in Prob. 7 a defective bolt is one that deviates from 1.00 cm by more than 0.06 cm, what percentage of defectives should we expect?

9. For what choice of the maximum possible deviation from 1.00 cm shall we obtain 10% defectives in Probs. 7 and 8?

10. What total sum can you expect in rolling a fair die 20 times? Do the experiment. Repeat it a number of times and record how the sum varies.

11. A small filling station is supplied with gasoline every Saturday afternoon. Assume that its volume X of sales in ten thousands of gallons has the probability density $f(x) = 6x(1 - x)$ if $0 \leq x \leq 1$ and 0 otherwise. Determine the mean, the variance, and the standardized variable.

12. What capacity must the tank in Prob. 11 have in order that the probability that the tank will be emptied in a given week be 5%?

13. If the life of certain tires (in thousands of miles) has the density $f(x) = \theta e^{-\theta x}$ $(x > 0)$, what mileage can you expect to get on one of these tires? Let $\theta = 0.05$ and find the probability that a tire will last at least 30 000 miles.

14. If in rolling a fair die, Jack wins as many dimes as the die shows, how much per game should Jack pay to make the game fair?

15. What is the expected daily profit if a small grocery store sells X turkeys per day with probabilities $f(5) = 0.1$, $f(6) = 0.3$, $f(7) = 0.4$, $f(8) = 0.2$ and the profit per turkey is $3.50?

16. **TEAM PROJECT. Means, Variances, Expectations. (a)** Show that $E(X - \mu) = 0$,
$\sigma^2 = E(X^2) - \mu^2$.

(b) Prove (10)–(12).

(c) Find all the moments of the uniform distribution on an interval $a \leq x \leq b$.

(d) The **skewness** γ of a random variable X is defined by

$$\gamma = \frac{1}{\sigma^3} E([X - \mu]^3).$$

Show that for a symmetric distribution (whose third central moment exists) the skewness is zero.

   (e) Find the skewness of the distribution with density $f(x) = xe^{-x}$ when $x > 0$ and $f(x) = 0$ otherwise. Sketch $f(x)$.

   (f) Calculate the skewness of a few simple discrete distributions of your own choice.

   (g) Find a nonsymmetric discrete distribution with 3 possible values, mean 0, and skewness 0.

22.7 Binomial, Poisson, and Hypergeometric Distributions

These are the three most important *discrete* distributions, with numerous applications.

Binomial Distribution

The **binomial distribution** occurs in games of chance (rolling a die, see below, etc.), quality inspection (e.g., count of the number of defectives), opinion polls (number of employees favoring certain schedule changes, etc.), medicine (e.g., number of patients recovered by a new medication), and so on. The conditions of its occurrence are as follows.

We are interested in the number of times an event A occurs in n independent trials. In each trial the event A has the same probability $P(A) = p$. Then in a trial, A will not occur with probability $q = 1 - p$. In n trials the random variable that interests us is

$$X = \textit{Number of times A occurs in n trials.}$$

X can assume the values $0, 1, \cdots, n$, and we want to determine the corresponding probabilities. Now $X = x$ means that A occurs in x trials and in $n - x$ it does not occur. This may look as follows.

$$\text{(1)} \qquad \underbrace{AA \cdots A}_{x \text{ times}} \quad \underbrace{BB \cdots B.}_{n - x \text{ times}}$$

Here $B = A^c$ is the complement of A, meaning that A does not occur. We now use the assumption that the trials are independent, that is, they do not influence each other. Hence (1) has the probability (see Sec. 22.3 on independent events)

$$\text{(1*)} \qquad \underbrace{pp \cdots p}_{x \text{ times}} \quad \underbrace{qq \cdots q}_{n - x \text{ times}} = p^x q^{n-x}.$$

Now (1) is just one order of arranging x A's and $n - x$ B's. We now use Theorem 1(b) in Sec. 22.4, which gives the number of permutations of n things (the n outcomes of the n trials) consisting of 2 classes, class 1 containing the $n_1 = x$ A's and class 2 containing the $n - n_1 = n - x$ B's. This number is

$$\frac{n!}{x!(n-x)!} = \binom{n}{x}.$$

Accordingly, (1*) multiplied by this number gives the probability $P(X = x)$ of $X = x$, that is, of obtaining precisely x A's in n trials. Hence X has the probability function

(2)
$$f(x) = \binom{n}{x} p^x q^{n-x}$$
$(x = 0, 1, \cdots, n)$

and $f(x) = 0$ otherwise. The distribution of X with probability function (2) is called the **binomial distribution** or *Bernoulli distribution*. The occurrence of A is called *success* (regardless of what it actually is; it may mean that you miss your plane or lose your watch) and the nonoccurrence of A is called *failure*. Figure 485 shows typical examples. Numerical values are given in Table A5 in Appendix 5.

The mean of the binomial distribution is (see Team Project 14)

(3)
$$\mu = np$$

and the variance is (see Team Project 14)

(4)
$$\sigma^2 = npq.$$

For the symmetric case of equal chance of success and failure ($p = q = 1/2$) this gives the mean $n/2$, the variance $n/4$, and the probability function

(2*)
$$f(x) = \binom{n}{x} \left(\frac{1}{2}\right)^n$$
$(x = 0, 1, \cdots, n).$

EXAMPLE 1   **Binomial distribution**

Compute the probability of obtaining at least two *"Six"* in rolling a fair die 4 times.

Solution. $p = P(A) = P(\text{"Six"}) = 1/6, q = 5/6, n = 4.$ *Answer:*

$$P = f(2) + f(3) + f(4) = \binom{4}{2}\left(\frac{1}{6}\right)^2\left(\frac{5}{6}\right)^2 + \binom{4}{3}\left(\frac{1}{6}\right)^3\left(\frac{5}{6}\right) + \binom{4}{4}\left(\frac{1}{6}\right)^4$$

$$= \frac{1}{6^4}(6 \cdot 25 + 4 \cdot 5 + 1) = \frac{171}{1296} = 13.2\%.$$ ◀

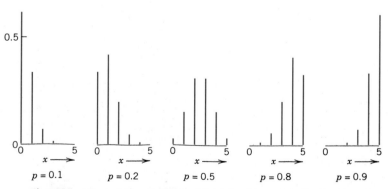

$p = 0.1$ $p = 0.2$ $p = 0.5$ $p = 0.8$ $p = 0.9$

Fig. 485.   Probability function (2) of the binomial distribution
for $n = 5$ and various values of p

Poisson Distribution

The discrete distribution with infinitely many possible values and probability function

$$(5) \qquad f(x) = \frac{\mu^x}{x!}\, e^{-\mu} \qquad\qquad (x = 0, 1, \cdots)$$

is called the **Poisson distribution,** named after S. D. Poisson (Sec. 16.5). Figure 486 shows (5) for some values of μ. It can be proved that this distribution is obtained as a limiting case of the binomial distribution, if we let $p \to 0$ and $n \to \infty$ so that the mean $\mu = np$ approaches a finite value. (For instance, $\mu = np$ may be kept constant.) The Poisson distribution has the mean μ and the variance (see Team Project 14)

$$(6) \qquad\qquad \sigma^2 = \mu.$$

Figure 486 gives the impression that with increasing mean the spread of the distribution increases, thereby illustrating formula (6), and that the distribution becomes more and more (approximately) symmetric.

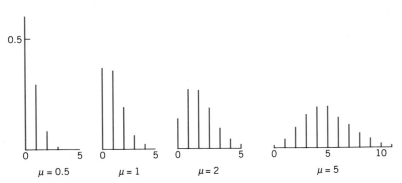

Fig. 486.   Probability function (5) of the Poisson distribution for various values of μ

EXAMPLE 2   **Poisson distribution**

If the probability of producing a defective screw is $p = 0.01$, what is the probability that a lot of 100 screws will contain more than 2 defectives?

Solution.   The complementary event is A^C: *Not more than 2 defectives.* For its probability we get from the binomial distribution with mean $\mu = np = 1$ the value [see (2)]

$$P(A^C) = \binom{100}{0} 0.99^{100} + \binom{100}{1} 0.01 \cdot 0.99^{99} + \binom{100}{2} 0.01^2 \cdot 0.99^{98}.$$

Since p is very small, we can approximate this by the much more convenient Poisson distribution with mean $\mu = np = 100 \cdot 0.01 = 1$, obtaining [see (5)]

$$P(A^C) \approx e^{-1}\left(1 + 1 + \frac{1}{2}\right) = 91.97\%.$$

Thus $P(A) = 8.03\%$. Show that the binomial distribution gives $P(A) = 7.94\%$, so that the Poisson approximation is quite good.   ◀

EXAMPLE 3   **Parking problems. Poisson distribution**

If on the average, 2 cars enter a certain parking lot per minute, what is the probability that during any given minute 4 or more cars will enter the lot?

Solution.   To understand that the Poisson distribution is a model of the situation, we imagine the minute to be divided into very many short time intervals, let p be the (constant) probability that a car will enter the lot during any such short interval, and assume independence of the events that happen during those intervals. Then we are dealing with a binomial distribution with very large n and very small p, which we can approximate by the Poisson distribution with $\mu = np = 2$. Thus the complementary event "3 *cars or fewer enter the lot*" has the probability

$$f(0) + f(1) + f(2) + f(3) = e^{-2}\left(\frac{2^0}{0!} + \frac{2^1}{1!} + \frac{2^2}{2!} + \frac{2^3}{3!}\right) = 0.857. \qquad \textit{Answer: } 14.3\%. \quad \blacktriangleleft$$

Sampling with Replacement

This means that we draw things from a given set one by one, and after each trial we replace the thing drawn (put it back to the given set and mix) before we draw the next thing. This guarantees independence of trials and leads to the **binomial distribution.** Indeed, if a box contains N things, for example, screws, M of which are defective, the probability of drawing a defective screw in a trial is $p = M/N$. Hence the probability of drawing a nondefective screw is $q = 1 - p = 1 - M/N$, and (2) gives the probability of drawing x defectives in n trials in the form

(7)
$$f(x) = \binom{n}{x}\left(\frac{M}{N}\right)^x\left(1 - \frac{M}{N}\right)^{n-x} \qquad (x = 0, 1, \cdots, n).$$

Sampling Without Replacement. Hypergeometric Distribution

Sampling without replacement means that we return no screw to the box. Then we no longer have independence of trials (why?), and instead of (7) the probability of drawing x defectives in n trials is

(8)
$$f(x) = \frac{\binom{M}{x}\binom{N-M}{n-x}}{\binom{N}{n}} \qquad (x = 0, 1, \cdots, n).$$

The distribution with this probability function is called the **hypergeometric distribution.**[6]

Derivation of (8). By (4a) in Sec. 22.4 there are

(a) $\binom{N}{n}$ different ways of picking n things from N,

(b) $\binom{M}{x}$ different ways of picking x defectives from M,

---

[6]Because the moment generating function (see Team Project 14) of this distribution can be expressed in terms of the hypergeometric function.

(c) $\binom{N-M}{n-x}$ different ways of picking $n-x$ nondefectives from $N-M$,

and each way in (b) combined with each way in (c) gives the total number of mutually exclusive ways of obtaining x defectives in n drawings without replacement. Since (a) is the total number of outcomes and we draw at random, each such way has the probability $1/\binom{N}{n}$. From this, (8) follows.   ◀

The hypergeometric distribution has the mean (Team Project 14)

(9)
$$\mu = n\frac{M}{N}$$

and the variance

(10)
$$\sigma^2 = \frac{nM(N-M)(N-n)}{N^2(N-1)}.$$

EXAMPLE 4   **Sampling with and without replacement**

We want to draw random samples of two gaskets from a box containing 10 gaskets, three of which are defective. Find the probability function of the random variable $X = $ *Number of defectives in the sample*.

Solution.   We have $N = 10$, $M = 3$, $N - M = 7$, $n = 2$. For sampling with replacement, (7) yields

$$f(x) = \binom{2}{x}\left(\frac{3}{10}\right)^x\left(\frac{7}{10}\right)^{2-x}, \qquad f(0) = 0.49, \quad f(1) = 0.42, \quad f(2) = 0.09.$$

For sampling without replacement we have to use (8), finding

$$f(x) = \binom{3}{x}\binom{7}{2-x}\Big/\binom{10}{2}, \qquad f(0) = f(1) = \frac{21}{45} \approx 0.47, \quad f(2) = \frac{3}{45} \approx 0.07.$$ ◀

If N, M, and N − M are large compared with n, then it does not matter too much whether we sample with or without replacement, and in this case the hypergeometric distribution may be approximated by the binomial distribution (with $p = M/N$), *which is somewhat simpler.*

*Hence in sampling from an indefinitely large population (***"infinite population"***) we may use the binomial distribution, regardless of whether we sample with or without replacement.*

PROBLEM SET 22.7

1. Five fair coins are tossed simultaneously. Find the probability function of the random variable $X = $ *Number of heads* and compute the probabilities of obtaining no heads, precisely 1 head, at least 1 head, not more than 4 heads.
2. If the probability of hitting a target is 25% and 4 shots are fired independently, what is the probability that the target will be hit at least once?
3. In Prob. 2, if the probability of hitting would be 5% and we fired 20 shots, would the probability of hitting at least once be less than, equal to, or greater than that in Prob. 2? Guess first.

4. Suppose that 4% of steel rods made by a machine are defective, the defectives occurring at random during production. If the rods are packaged 100 per box, what is the Poisson approximation of the probability that a given box will contain $x = 0, 1, \cdots, 5$ defectives?

5. Classical experiments by E. Rutherford and H. Geiger in 1910 showed that the number of alpha particles emitted per second in a radioactive process is a random variable X having a Poisson distribution. If X has mean 0.5, what is the probability of observing two or more particles during any given second?

6. Let $p = 2\%$ be the probability that a certain type of lightbulb will fail in a 24-hour test. Find the probability that a sign consisting of 15 such bulbs will burn 24 hours with no bulb failures.

7. Guess how much less the probability in Prob. 6 would be if the sign consisted of 100 bulbs. Then calculate.

8. If a ticket office can serve at most 4 customers per minute and the average number of customers is 120 per hour, what is the probability that during a given minute customers will have to wait? (Use the Poisson distribution, Table 6 in Appendix 5.)

9. Suppose that in the production of 60-ohm radio resistors, nondefective items are those that have a resistance between 58 and 62 ohms and the probability of a resistor's being defective is 0.1%. The resistors are sold in lots of 200, with the guarantee that all resistors are nondefective. What is the probability that a given lot will violate this guarantee? (Use the Poisson distribution.)

10. A carton contains 20 fuses, 5 of which are defective. Find the probability that, if a sample of 3 fuses is chosen from the carton by random drawing without replacement, x fuses in the sample will be defective.

11. Suppose that a test for extrasensory perception consists of naming (in any order) 3 cards randomly drawn from a deck of 13 cards. Find the probability that by chance alone, the person will correctly name (a) no cards, (b) 1 card, (c) 2 cards, (d) 3 cards.

12. A distributor sells rubber bands in packages of 100 and guarantees that at most 10% are defective. A consumer controls each package by drawing 10 bands without replacement. If the sample contains no defective rubber bands, he accepts the package. Otherwise he rejects it. Find the probability that in this process the consumer rejects a package that contains 10 defective bands (so that it still satisfies the guarantee).

13. If X is the number of cars per minute passing a certain point of a country road between 11 P.M. and 12 P.M. and X has a Poisson distribution with mean 5, what is the probability of observing fewer than 5 cars during any given minute?

14. TEAM PROJECT. Moment Generating Function $G(t)$. This function is defined by

$$G(t) = E(e^{tX}) = \sum_j e^{tx_j} f(x_j)$$

or

$$G(t) = E(e^{tX}) = \int_{-\infty}^{\infty} e^{tx} f(x)\, dx$$

where X is a discrete or continuous random variable, respectively.

(a) Assuming that termwise differentiation and differentiation under the integral sign is permissible, show that $E(X^k) = G^{(k)}(0)$, where $G^{(k)} = d^k G/dt^k$, in particular, $\mu = G'(0)$.

(b) Show that the binomial distribution has the moment generating function

$$G(t) = \sum_{x=0}^{n} e^{tx} \binom{n}{x} p^x q^{n-x}$$

$$= \sum_{x=0}^{n} \binom{n}{x} (pe^t)^x q^{n-x} = (pe^t + q)^n.$$

(c) Using (b), prove (3).

(d) Prove (4).

(e) Show that the Poisson distribution has the moment generating function $G(t)$ and prove (6), where

$$G(t) = e^{-\mu}e^{\mu e^t}.$$

(f)   Prove $x\begin{pmatrix} M \\ x \end{pmatrix} = M\begin{pmatrix} M-1 \\ x-1 \end{pmatrix}$.   Using this, prove (9).

15. By definition, the **multinomial distribution** has the probability function

$$f(x_1, \cdots, x_k) = \frac{n!}{x_1! \cdots x_k!} p_1{}^{x_1} \cdots p_k{}^{x_k}$$

where $0 \leqq x_j \leqq n,\ j = 1, \cdots, k,$ and $x_1 + \cdots + x_k = n;$ also, $p_1 + p_2 + \cdots + p_n = 1.$ Show that this is the probability of obtaining in n independent trials precisely x_j A_j's, $j = 1, \cdots, k,$ where p_j is the probability of A_j in a single trial.

22.8 Normal Distribution

Turning from discrete to continuous distributions, in this section we discuss the normal distribution. This is the most important continuous distribution because in applications many random variables are **normal random variables** (that is, they have a normal distribution) or they are approximately normal or can be transformed into normal random variables in a relatively simple fashion. Furthermore, the normal distribution is a useful approximation of more complicated distributions, and it also occurs in the proofs of various statistical tests.

The **normal distribution** or *Gauss distribution* is defined as the distribution with the density

(1)
$$f(x) = \frac{1}{\sigma\sqrt{2\pi}} \exp\left[-\frac{1}{2}\left(\frac{x-\mu}{\sigma}\right)^2\right] \qquad (\sigma > 0)$$

where exp is the exponential function with base $e = 2.718 \cdots$. This is simpler than it may at first look. $f(x)$ has these features (see also Fig. 487 on the next page).

1. μ is the mean and σ the standard deviation.

2. $1/(\sigma\sqrt{2\pi})$ is a constant factor that makes the area under the curve equal to 1, as it must be by (10), Sec. 22.5.

3. The curve of $f(x)$ is symmetric with respect to $x = \mu$ because the exponent is quadratic. Hence for $\mu = 0$ it is symmetric with respect to the y-axis $x = 0$ (Fig. 487, *"bell-shaped curves"*).

4. The exponential function in (1) goes to zero very fast—the faster the smaller the standard deviation σ is, as it should be (Fig. 487).

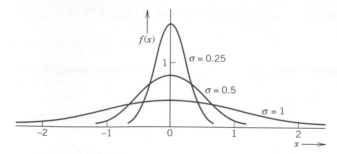

Fig. 487.   Density (1) of the normal distribution with $\mu = 0$
for various values of σ

Distribution Function $F(x)$

From (7) in Sec. 22.5 and (1) we see that the normal distribution has the **distribution
function**

$$
(2) \qquad F(x) = \frac{1}{\sigma\sqrt{2\pi}} \int_{-\infty}^{x} \exp\left[-\frac{1}{2}\left(\frac{v-\mu}{\sigma}\right)^2\right] dv.
$$

Here we needed x as the upper limit of integration and wrote v (instead of x) in the
integrand.

For the corresponding **standardized normal distribution** with mean 0 and standard
deviation 1 we denote $F(x)$ by $\Phi(z)$. Then we simply have from (2)

$$
(3) \qquad \Phi(z) = \frac{1}{\sqrt{2\pi}} \int_{-\infty}^{z} e^{-u^2/2}\, du.
$$

This integral cannot be integrated by one of the methods of calculus. But this is no serious
handicap because the integral has been tabulated (Table A7 in Appendix 5) since one
needs its values in working with the normal distribution. The curve of $\Phi(z)$ is S-shaped.
It increases monotone (why?) from 0 to 1 and intersects the vertical axis at 1/2 (why?),
as shown in Fig. 488.

It is now of greatest practical importance that the general $F(x)$ in (2) with any μ and
σ can be expressed in terms of the tabulated standard $\Phi(z)$:

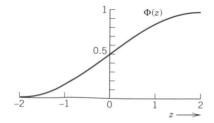

Fig. 488.   Distribution function $\Phi(z)$ of the normal distribution
with mean 0 and variance 1

THEOREM 1   **(Use of the normal table (Table A7) in Appendix 5)**

The distribution function $F(x)$ of the normal distribution with any μ and σ [see (2)] is related to the standardized distribution function $\Phi(z)$ in (3) by the formula

(4)
$$F(x) = \Phi\left(\frac{x - \mu}{\sigma}\right).$$

PROOF.   Comparing (2) and (3), we see that we should set

$$u = \frac{v - \mu}{\sigma}. \qquad \text{Then } v = x \text{ gives} \qquad u = \frac{x - \mu}{\sigma}$$

as the new upper limit of integration. Also $v - \mu = \sigma u$, thus $dv = \sigma\, du$. Together, since σ drops out,

$$F(x) = \frac{1}{\sigma\sqrt{2\pi}} \int_{-\infty}^{(x-\mu)/\sigma} e^{-u^2/2}\, \sigma\, du = \Phi\left(\frac{x - \mu}{\sigma}\right). \qquad \blacktriangleleft$$

Probabilities corresponding to intervals will be needed quite frequently in statistics in Chap. 23. These are obtained as follows.

THEOREM 2   **(Normal probabilities for intervals)**

The probability that a normal random variable X with mean μ and standard deviation σ assume any value in an interval $a < x \leq b$ is

(5)
$$P(a < X \leq b) = F(b) - F(a) = \Phi\left(\frac{b - \mu}{\sigma}\right) - \Phi\left(\frac{a - \mu}{\sigma}\right).$$

PROOF.   Formula (2) in Sec. 22.5 gives the first equality in (5), and (4) in this section gives the second equality. $\blacktriangleleft$

Numerical Values

In practical work with the normal distribution it is good to remember that about 2/3 of all values of X to be observed will lie between $\mu \pm \sigma$, about 95% between $\mu \pm 2\sigma$, and practically all between the **three-sigma limits** $\mu \pm 3\sigma$. More precisely, by Table A7 in Appendix 5

(6)

(a) $\quad P(\mu - \sigma < X \leq \mu + \sigma) \approx 68\%$

(b) $\quad P(\mu - 2\sigma < X \leq \mu + 2\sigma) \approx 95.5\%$

(c) $\quad P(\mu - 3\sigma < X \leq \mu + 3\sigma) \approx 99.7\%.$

These formulas are illustrated in Fig. 489 on the next page.

The formulas in (6) show that a value deviating from μ by more than σ, 2σ, or 3σ will occur in one of about 3, 20, and 300 trials, respectively.

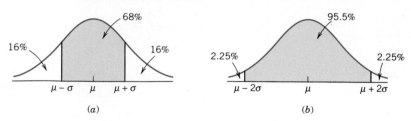

Fig. 489.   Illustration of formula (6)

In tests (Chap. 23) we shall ask conversely for the intervals that correspond to certain given probabilities; practically most important are probabilities of 95%, 99%, and 99.9%. For these, Table A8 in Appendix 5 gives the answers $\mu + 2\sigma$, $\mu \pm 2.5\sigma$, and $\mu \pm 3.3\sigma$, respectively. More precisely,

(7)

	(a)	$P(\mu - 1.96\sigma < X \le \mu + 1.96\sigma) = 95\%$
	(b)	$P(\mu - 2.58\sigma < X \le \mu + 2.58\sigma) = 99\%,$
	(c)	$P(\mu - 3.29\sigma < X \le \mu + 3.29\sigma) = 99.9\%.$

Working with the Normal Tables in Appendix 5

There are two normal tables in Appendix 5, Tables A7 and A8. If you want probabilities, use Table A7. If probabilities are given and corresponding intervals or x-values are wanted, use Table A8. The following examples are typical. Do them with care, verifying all values, and don't just regard them as dull exercises for your software. Make sketches of the density to see whether the results look reasonable.

EXAMPLE 1   **Reading entries from Table A7**

If X is standardized normal (so that $\mu = 0$, $\sigma = 1$), then

$$P(X \le 2.44) = 0.9927 \approx 99\tfrac{1}{4}\%$$

$$P(X \le -1.16) = 0.1230 \approx 12\tfrac{1}{2}\%$$

$$P(X \ge 1) = 1 - P(X \le 1) = 1 - 0.8413 = 0.1587 \text{ by (7), Sec. 22.3}$$

$$P(1.0 \le X \le 1.8) = \Phi(1.8) - \Phi(1.0) = 0.9641 - 0.8413 = 0.1228.$$ ◄

EXAMPLE 2   **Probabilities for given intervals, Table A7**

Let X be normal with mean 0.8 and variance 4 (so that $\sigma = 2$). Then by (4) and (5)

$$P(X \le 2.44) = F(2.44) = \Phi\left(\frac{2.44 - 0.80}{2}\right) = \Phi(0.82) = 0.7939 \approx 80\%$$

or if you like it better (similarly in the other cases)

$$P(X \leq 2.44) = P\left(\frac{X - 0.80}{2} \leq \frac{2.44 - 0.80}{2}\right) = P(Z \leq 0.82) = 0.7939$$

$$P(X \geq 1) = 1 - P(X \leq 1) = 1 - \Phi\left(\frac{1 - 0.8}{2}\right) = 1 - 0.5398 = 0.4602$$

$$P(1.0 \leq X \leq 1.8) = \Phi(0.5) - \Phi(0.1) = 0.6915 - 0.5398 = 0.1517.$$ ◄

EXAMPLE 3 **Unknown values c for given probabilities, Table A8**

Let X be normal with mean 5 and variance 0.04 (hence standard deviation 0.2). Find c or k corresponding to the given probability

$$P(X \leq c) = 95\%, \qquad \Phi\left(\frac{c - 5}{0.2}\right) = 95\%, \qquad \frac{c - 5}{0.2} = 1.645, \qquad c = 5.329$$

$$P(\ 5 - k \leq X \leq 5 + k) = 90\%, \qquad 5 + k = 5.329 \qquad \text{(as before; why?)}$$

$$P(X \geq c) = 1\%, \qquad \text{thus } P(X \leq c) = 99\%, \qquad \frac{c - 5}{0.2} = 2.326, \qquad c = 5.4652.$$ ◄

EXAMPLE 4 **Defectives**

In a production of iron rods let the diameter X be normally distributed with mean 2 in. and standard deviation 0.008 in.

(a) What percentage of defectives can we expect if we set the tolerance limits at 2 ± 0.02 in.?

(b) How should we set the tolerance limits to allow for 4% defectives?

Solution. **(a)** $1\frac{1}{4}\%$ because from (5) and Table A7 we obtain for the complementary event the probability

$$P(1.98 \leq X \leq 2.02) = \Phi\left(\frac{2.02 - 2.00}{0.008}\right) - \Phi\left(\frac{1.98 - 2.00}{0.008}\right)$$

$$= \Phi(2.5) - \Phi(-2.5) = 0.9938 - (1 - 0.9938) = 0.9876 \approx 98\tfrac{3}{4}\%.$$

(b) 2 ± 0.0164 because for the complementary event we have

$$0.96 = P(2 - c \leq X \leq 2 + c) \qquad \text{or} \qquad 0.98 = P(X \leq 2 + c)$$

so that Table A8 gives

$$0.98 = \Phi\left(\frac{2 + c - 2}{0.008}\right), \qquad \frac{2 + c - 2}{0.008} = 2.054, \qquad c = 0.0164.$$ ◄

Binomial Distribution Approximated by Normal Distribution

The probability function of the binomial distribution is (Sec. 22.7)

$$(8) \qquad\qquad f(x) = \binom{n}{x} p^x q^{n-x} \qquad\qquad (x = 0, 1, \cdots, n).$$

If n is large, the binomial coefficients and powers become very inconvenient. It is of great practical (and theoretical) importance that in this case the normal distribution provides a good approximation of the binomial distribution, according to the following theorem, one of the most important theorems in all probability theory.

THEOREM 3   **(Limit theorem of De Moivre and Laplace)**

For large n,

$$(9) \qquad\qquad\qquad f(x) \sim f^*(x) \qquad\qquad\qquad (x = 0, 1, \cdots, n).$$

Here f is given by (8). The function

$$(10) \qquad\qquad f^*(x) = \frac{1}{\sqrt{2\pi}\sqrt{npq}}\, e^{-z^2/2}, \qquad z = \frac{x - np}{\sqrt{npq}}$$

*is the density of the normal distribution with mean $\mu = np$ and variance $\sigma^2 = npq$ (the mean and variance of the binomial distribution). The symbol $\sim$ (read **asymptotically equal**) means that the ratio of both sides approaches 1 as n approaches ∞. Furthermore, for any nonnegative integers a and b ($> a$),*

$$(11) \qquad\qquad P(a \leqq X \leqq b) = \sum_{x=a}^{b} \binom{n}{x} p^x q^{n-x} \sim \Phi(\beta) - \Phi(\alpha),$$

$$\alpha = \frac{a - np - 0.5}{\sqrt{npq}}, \qquad \beta = \frac{b - np + 0.5}{\sqrt{npq}}.$$

     A proof of this theorem can be found in [G3] listed in Appendix 1. The proof shows that the term 0.5 in α and β is a correction caused by the change from a discrete to a continuous distribution.

PROBLEM SET 22.8

1. Let X be normal with mean 10 and variance 4. Find $P(X > 12)$, $P(X < 10)$, $P(X < 11)$, $P(9 < X < 13)$.

2. Let X be normal with mean 105 and variance 25. Find $P(X \leqq 112.5)$, $P(X > 100)$, $P(110.5 < X < 111.25)$.

3. Let X be normal with mean 50 and variance 9. Determine c such that $P(X < c) = 5\%$, $P(X > c) = 1\%$, $P(50 - c < X < 50 + c) = 50\%$.

4. Let X be normal with mean 3.6 and variance 0.01. Find c such that $P(X \leqq c) = 50\%$, $P(X > c) = 10\%$, $P(-c < X - 3.6 \leqq c) = 99.9\%$.

5. If the lifetime X of a certain kind of automobile battery is normally distributed with a mean of 5 years and a standard deviation of 1 year, and the manufacturer wishes to guarantee the battery for 4 years, what percentage of the batteries will he have to replace under the guarantee?

6. If the standard deviation in Prob. 5 were smaller, would that percentage be larger or smaller?

7. If the resistance X of certain wires in electrical networks is normal with mean 0.01 ohm and standard deviation 0.001 ohm, how many of 1000 wires will meet the specification that they have resistance between 0.009 and 0.011 ohm?

8. What is the probability of obtaining at least 2048 heads if a coin is tossed 4040 times and heads and tails are equally likely? (See Table 22.1 in Sec. 22.3.)

9. If the mathematics scores of the SAT college entrance exams are normal with mean 480 and standard deviation 100 (these are about the actual values over the past years) and if some college sets 500 as the minimum score for new students, what percent of students would not reach that score?

10. A producer sells electric bulbs in cartons of 1000 bulbs. Using (11), find the probability that any given carton contains not more than 1% defective bulbs, assuming the production process to be a Bernoulli experiment with $p = 1\%$ (= probability that any given bulb will be defective). First guess. Then calculate.

11. If the monthly machine repair and maintenance cost X in a certain factory is known to be normal with mean \$12000 and standard deviation \$2000, what is the probability that the repair cost for the next month will exceed the budgeted amount of \$15000?

12. The breaking strength X [kg] of a certain type of plastic block is normally distributed with a mean of 1500 kg and a standard deviation of 50 kg. What is the maximum load such that we can expect no more than 5% of the blocks to break?

13. If sick-leave time X used by employees of a company in one month is (very roughly) normal with mean 1000 hours and standard deviation 100 hours, how much time t should be budgeted for sick leave during the next month if t is to be exceeded with probability of only 20%?

14. **TEAM PROJECT. Normal Distribution.** (a) Derive the formulas in (6) and (7) from the appropriate normal table.

   (b) Show that $\Phi(-z) = 1 - \Phi(z)$. Give an example.

   (c) Find the points of inflection of the curve of (1).

   (d) Considering $\Phi^2(\infty)$ and introducing polar coordinates in the double integral (a standard trick worth remembering), prove

   $$\text{(12)} \qquad \Phi(\infty) = \frac{1}{\sqrt{2\pi}} \int_{-\infty}^{\infty} e^{-u^2/2}\, du = 1.$$

   (e) Show that σ in (1) is indeed the standard deviation of the normal distribution. [Use (12).]

   (f) **Bernoulli's law of large numbers.** In an experiment let an event A have probability p $(0 < p < 1)$, and let X be the number of times A happens in n independent trials. Show that for any given $\epsilon > 0$,

   $$P\left(\left| \frac{X}{n} - p \right| \leq \epsilon \right) \to 1 \qquad\qquad \text{as } n \to \infty.$$

   (g) **Transformation.** If X is normal with mean μ and variance σ^2, show that $X^* = c_1 X + c_2$ $(c_1 > 0)$ is normal with mean $\mu^* = c_1\mu + c_2$ and variance $\sigma^{*2} = c_1^2 \sigma^2$.

15. **WRITING PROJECT. Use of Tables.** Give a systematic discussion of the use of Tables A7 and A8 for obtaining $P(X < b)$, $P(X > a)$, $P(a < X < b)$, $P(X < c) = k$, $P(X > c) = k$, $P(\mu - c < X < \mu + c) = k$; include simple examples.

22.9 Distributions of Several Random Variables

Distributions of two or more random variables are of interest for two reasons:

1. They occur in experiments in which we observe several random variables, for example, carbon content X and hardness Y of steel, amount of fertilizer X and yield of corn Y, height X_1, weight X_2, and blood pressure X_3 of persons, and so on.

2. They are needed in the mathematical justification of the methods of statistics in Chap. 23.

In this section we consider two random variables X and Y or, as we also say, a **two-dimensional random variable** (X, Y). For (X, Y) the outcome of a trial is a pair of numbers $X = x$, $Y = y$, briefly $(X, Y) = (x, y)$, which we can plot as a point in the XY-plane.

The **two-dimensional probability distribution** of the random variable (X, Y) is given by the **distribution function**

(1)
$$\boxed{F(x, y) = P(X \leqq x, Y \leqq y).}$$

This is the probability that in a trial, X will assume any value not greater than x and in the same trial, Y will assume any value not greater than y. This corresponds to the blue region in Fig. 490, which extends to $-\infty$ to the left and below. $F(x, y)$ determines the probability distribution uniquely, because in analogy to the formula $P(a < X \leqq b) = F(b) - F(a)$ we now have for a rectangle (see Prob. 14)

(2) $\quad P(a_1 < X \leqq b_1, \quad a_2 < Y \leqq b_2) = F(b_1, b_2) - F(a_1, b_2) - F(b_1, a_2) + F(a_1, a_2).$

As before, in the two-dimensional case we shall also have discrete and continuous random variables and distributions.

Discrete Two-Dimensional Distributions

In analogy to the case of a single random variable we call (X, Y) and its distribution **discrete** if (X, Y) can assume only finitely many or at most countably infinitely many pairs of values (x_1, y_1), (x_2, y_2), $\cdots$ with positive probabilities, whereas the probability for any domain containing none of those values of (X, Y) is zero.

Let (x_i, y_j) be any of those pairs and let $P(X = x_i, Y = y_j) = p_{ij}$ (where we admit that p_{ij} may be 0 for certain pairs of subscripts i, j). Then we define the **probability function** $f(x, y)$ of (X, Y) by

(3) $\qquad f(x, y) = p_{ij} \quad \text{if} \quad x = x_i, y = y_j \quad \text{and} \quad f(x, y) = 0 \text{ otherwise};$

here, $i = 1, 2, \cdots$ and $j = 1, 2, \cdots$ independently. In analogy to (4), Sec. 22.5, we now have for the distribution function the formula

(4)
$$F(x, y) = \sum_{x_i \leqq x} \sum_{y_j \leqq y} f(x_i, y_j).$$

Instead of (6) in Sec. 22.5 we now have the condition

(5)
$$\sum_i \sum_j f(x_i, y_j) = 1.$$

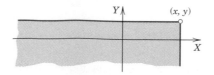

Fig. 490. Formula (1)

EXAMPLE 1   **Two-dimensional discrete distribution**

If we simultaneously toss a dime and a nickel and consider

$$X = \textit{Number of heads the dime turns up,}$$

$$Y = \textit{Number of heads the nickel turns up,}$$

then X and Y can have the values 0 or 1, and the probability function is

$$f(0, 0) = f(1, 0) = f(0, 1) = f(1, 1) = \tfrac{1}{4}, \quad f(x, y) = 0 \text{ otherwise.}$$ ◀

Continuous Two-Dimensional Distributions

In analogy to the case of a single random variable we call (X, Y) and its distribution **continuous** if the corresponding distribution function $F(x, y)$ can be given by a double integral

$$(6) \qquad\qquad F(x, y) = \int_{-\infty}^{y} \int_{-\infty}^{x} f(x^*, y^*) \, dx^* \, dy^*$$

whose integrand f, called the **density,** is nonnegative everywhere, and is continuous, possibly except on finitely many curves.

From (6) we obtain the probability that (X, Y) assume any value in a rectangle (Fig. 491), given by the formula

$$(7) \qquad\qquad P(a_1 < X \leqq b_1, \quad a_2 < Y \leqq b_2) = \int_{a_2}^{b_2} \int_{a_1}^{b_1} f(x, y) \, dx \, dy.$$

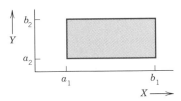

Fig. 491.   Notion of a two-dimensional distribution

Figures 492 and 493 give an illustration that will be discussed on the next page.

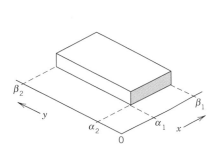

Fig. 492.   Probability density function (8)
of the uniform distribution

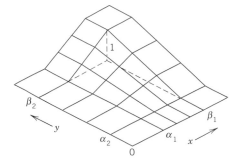

Fig. 493.   Distribution function of the
uniform distribution defined by (8)

EXAMPLE 2   **Two-dimensional uniform distribution in a rectangle**

Let R be the rectangle $\alpha_1 < x \leqq \beta_1$, $\alpha_2 < y \leqq \beta_2$. The density (see Fig. 492)

(8) $f(x, y) = 1/k$ if (x, y) is in R, $f(x, y) = 0$ otherwise

defines the so-called **uniform distribution** *in the rectangle R;* here $k = (\beta_1 - \alpha_1)(\beta_2 - \alpha_2)$ is the area of R.
The distribution function is shown in Fig. 493.   ◀

Marginal Distributions of a Discrete Distribution

This is a rather natural idea, without counterpart for a single random variable. It amounts
to being interested only in one of the two variables in (X, Y), say, X, and asking for its
distribution, called the **marginal distribution** of X in (X, Y). So we ask for the probability
$P(X = x, Y$ arbitrary$)$. Since (X, Y) is discrete, so is X. We get its probability function,
call it $f_1(x)$, from the probability function $f(x, y)$ of (X, Y) by summing over y:

(9)
$$f_1(x) = P(X = x, Y \text{ arbitrary}) = \sum_y f(x, y)$$

where we sum all the values of $f(x, y)$ that are not 0 for that x.

From (9) we see that the distribution function of the marginal distribution of X is

(10) $F_1(x) = P(X \leqq x, Y \text{ arbitrary}) = \sum_{x^* \leqq x} f_1(x^*).$

Similarly, the probability function

(11)
$$f_2(y) = P(X \text{ arbitrary}, Y = y) = \sum_x f(x, y)$$

determines the **marginal distribution** of Y in (X, Y). Here we sum all the values of
$f(x, y)$ that are not zero for the corresponding y. The distribution function of this marginal
distribution is

(12) $F_2(y) = P(X \text{ arbitrary}, Y \leqq y) = \sum_{y^* \leqq y} f_2(y^*).$

The next example illustrates all this. It also motivates the name "marginal distributions"
because they appear on the margins of Table 22.2.

EXAMPLE 3   **Marginal distributions of a discrete two-dimensional random variable**

In drawing 3 cards with replacement from a bridge deck let us consider

$(X, Y),$ $X = $ *Number of queens,* $Y = $ *Number of kings or aces.*

The deck has 52 cards. These include 4 queens, 4 kings, and 4 aces. Hence in a single trial a queen has probability
$4/52 = 1/13$ and a king or ace $8/52 = 2/13$. This gives the probability function of (X, Y),

$$f(x, y) = \frac{3!}{x!\, y!\, (3 - x - y)!}\left(\frac{1}{13}\right)^x \left(\frac{2}{13}\right)^y \left(\frac{10}{13}\right)^{3-x-y} \qquad (x + y \leqq 3)$$

and $f(x, y) = 0$ otherwise. Table 22.2 shows the values of $f(x, y)$ in the center and, on the right and lower margins, the values of the probability functions $f_1(x)$ and $f_2(y)$ of the marginal distributions of X and Y, respectively.　◀

Table 22.2
Values of the Probability Functions $f(x, y)$, $f_1(x)$, $f_2(y)$ in Drawing Three Cards with Replacement from a Bridge Deck, where X is the Number of Queens Drawn and Y is the Number of Kings or Aces Drawn

y x	0	1	2	3	$f_1(x)$
0	$\frac{1000}{2197}$	$\frac{600}{2197}$	$\frac{120}{2197}$	$\frac{8}{2197}$	$\frac{1728}{2197}$
1	$\frac{300}{2197}$	$\frac{120}{2197}$	$\frac{12}{2197}$	0	$\frac{432}{2197}$
2	$\frac{30}{2197}$	$\frac{6}{2197}$	0	0	$\frac{36}{2197}$
3	$\frac{1}{2197}$	0	0	0	$\frac{1}{2197}$
$f_2(y)$	$\frac{1331}{2197}$	$\frac{726}{2197}$	$\frac{132}{2197}$	$\frac{8}{2197}$	

Marginal Distributions of a Continuous Distribution

This is conceptually the same as for discrete distributions, with probability functions and sums replaced by densities and integrals. For a continuous random variable (X, Y) with density $f(x, y)$ we now have the **marginal distribution** of X in (X, Y), defined by the distribution function

(13)
$$F_1(x) = P(X \leq x, \ -\infty < Y < \infty) = \int_{-\infty}^{x} f_1(x^*) \, dx^*$$

with the density f_1 of X obtained from $f(x, y)$ by integration over y,

(14)
$$f_1(x) = \int_{-\infty}^{\infty} f(x, y) \, dy.$$

Interchanging the roles of X and Y, we obtain the **marginal distribution** of Y in (X, Y) with the distribution function

(15)
$$F_2(y) = P(-\infty < X < \infty, \ Y \leq y) = \int_{-\infty}^{y} f_2(y^*) \, dy^*$$

and density

(16)
$$f_2(y) = \int_{-\infty}^{\infty} f(x, y) \, dx.$$

Independence of Random Variables

X and Y in a (discrete or continuous) random variable (X, Y) are said to be **independent** if

(17)
$$F(x, y) = F_1(x)F_2(y)$$

holds for all (x, y). Otherwise these random variables are said to be **dependent.** These definitions are suggested by the corresponding definitions for events in Sec. 22.3.

Necessary and sufficient for independence is

(18)
$$f(x, y) = f_1(x)f_2(y)$$

for all x and y. Here the f's are the above probability functions if (X, Y) is discrete or those densities if (X, Y) is continuous. (See Prob. 16.)

EXAMPLE 4 **Independence and dependence**

In tossing a dime and a nickel, X = *Number of heads on the dime*, Y = *Number of heads on the nickel* may assume the values 0 or 1 and are independent. The random variables in Table 22.2 are dependent. ◀

Extension of independence to ***n*-dimensional random variables** will be needed throughout Chap. 23. The distribution of such a random variable $\mathbf{X} = (X_1, \cdots, X_n)$ is determined by a **distribution function** of the form

$$F(x_1, \cdots, x_n) = P(X_1 \leqq x_1, \cdots, X_n \leqq x_n).$$

The random variables $X_1, \cdots, X_n$ are said to be **independent** if

(19)
$$F(x_1, \cdots, x_n) = F_1(x_1)F_2(x_2) \cdots F_n(x_n)$$

for all $(x_1, \cdots, x_n)$. Here $F_j(x_j)$ is the distribution function of the marginal distribution of X_j in **X,** that is,

$$F_j(x_j) = P(X_j \leqq x_j, X_k \text{ arbitrary}, k \neq j).$$

Otherwise these random variables are said to be **dependent.**

Functions of Random Variables

Let (X, Y) be a two-dimensional random variable. Let $g(x, y)$ be a continuous function (not a constant) defined for all (x, y). Then $Z = g(X, Y)$ is a random variable, too. For example, if we roll two dice and X is the number that the first die turns up whereas Y is the number that the second die turns up, then $Z = X + Y$ is the sum of those two numbers (see Fig. 482 in Sec. 22.5).

If $(X_1, \cdots, X_n)$ is an n-dimensional random variable and $g(x_1, \cdots, x_n)$ is a continuous function that is defined for all $(x_1, \cdots, x_n)$ and is not constant, then $Z = g(X_1, \cdots, X_n)$ is a random variable, too.

In the case of a *discrete* random variable (X, Y) we may obtain the probability function $f(z)$ of $Z = g(X, Y)$ by summing all $f(x, y)$ for which $g(x, y)$ equals the value of z considered; thus

$$(20) \qquad f(z) = P(Z = z) = \sum \sum_{g(x,y)=z} f(x, y).$$

Hence the distribution function of Z is

$$(21) \qquad F(z) = P(Z \leqq z) = \sum \sum_{g(x,y)\leqq z} f(x, y)$$

where we sum all values of $f(x, y)$ for which $g(x, y) \leqq z$.

In the case of a *continuous* random variable (X, Y) we similarly have

$$(22) \qquad F(z) = P(Z \leqq z) = \iint\limits_{g(x,y)\leqq z} f(x, y)\, dx\, dy$$

where for each z we integrate over the region $g(x, y) \leqq z$ in the xy-plane.

Addition of Means

The number

$$(23) \qquad E(g(X, Y)) = \begin{cases} \displaystyle\sum_x \sum_y g(x, y)f(x, y) & [(X, Y) \text{ discrete}] \\[2em] \displaystyle\int_{-\infty}^{\infty} \int_{-\infty}^{\infty} g(x, y)f(x, y)\, dx\, dy & [(X, Y) \text{ continuous}] \end{cases}$$

is called the **mathematical expectation** or, briefly, the **expectation of** $g(X, Y)$. Here it is assumed that the double series converges absolutely and the integral of $|g(x, y)|f(x, y)$ over the xy-plane exists (is finite). Since summation and integration are linear processes, we have from (23)

$$(24) \qquad E(ag(X, Y) + bh(X, Y)) = aE(g(X, Y)) + bE(h(X, Y)).$$

An important special case is

$$E(X + Y) = E(X) + E(Y),$$

and by induction we have the following result.

THEOREM 1   **(Addition of means)**

The mean (expectation) of a sum of random variables equals the sum of the means (expectations), that is,

$$(25) \qquad \boxed{E(X_1 + X_2 + \cdots + X_n) = E(X_1) + E(X_2) + \cdots + E(X_n).}$$

Furthermore we readily obtain

THEOREM 2   **(Multiplication of means)**

*The mean (expectation) of the product of **independent** random variables equals the product of the means (expectations), that is,*

(26) $$E(X_1 X_2 \cdots X_n) = E(X_1)E(X_2) \cdots E(X_n).$$

PROOF.   If X and Y are independent random variables (both discrete or both continuous), then $E(XY) = E(X)E(Y)$. In fact, in the discrete case we have

$$E(XY) = \sum_x \sum_y xy f(x, y) = \sum_x x f_1(x) \sum_y y f_2(y) = E(X)E(Y),$$

and in the continuous case the proof of the relation is similar. Extension to n independent random variables gives (26), and Theorem 2 is proved.   ◀

Addition of Variances

This is another matter of practical importance that we shall need. As before, let $Z = X + Y$ and denote the mean and variance of Z by μ and σ^2. Then we first have (see Team Project 16 in Problem Set 22.6)

$$\sigma^2 = E([Z - \mu]^2) = E(Z^2) - [E(Z)]^2.$$

From (24) we see that the first term on the right equals

$$E(Z^2) = E(X^2 + 2XY + Y^2) = E(X^2) + 2E(XY) + E(Y^2).$$

For the second term on the right we obtain from Theorem 1

$$[E(Z)]^2 = [E(X) + E(Y)]^2 = [E(X)]^2 + 2E(X)E(Y) + [E(Y)]^2.$$

By substituting these expressions into the formula for σ^2 we have

$$\sigma^2 = E(X^2) - [E(X)]^2 + E(Y^2) - [E(Y)]^2$$
$$+ 2[E(XY) - E(X)E(Y)].$$

From Team Project 16, Sec. 22.6, we see that the expression in the first line on the right is the sum of the variances of X and Y, which we denote by σ_1^2 and σ_2^2, respectively. The quantity in the second line (except for the factor 2), that is,

(27) $$\boxed{\sigma_{XY} = E(XY) - E(X)E(Y)}$$

is called the **covariance** of X and Y. Consequently, our result is

(28) $$\sigma^2 = \sigma_1^2 + \sigma_2^2 + 2\sigma_{XY}.$$

If X and Y are independent, then

$$E(XY) = E(X)E(Y);$$

hence $\sigma_{XY} = 0$, and

(29) $$\sigma^2 = \sigma_1{}^2 + \sigma_2{}^2.$$

Extension to more than two variables gives the basic

THEOREM 3   **(Addition of variances)**

*The variance of the sum of **independent** random variables equals the sum of the variances of these variables.*

Caution! In the numerous applications of Theorems 1 and 3 we must always remember that Theorem 3 holds only for **independent** variables.

    This is the end of Chap. 22 on probability theory. Most of the concepts, methods, and special distributions discussed in this chapter will play a fundamental role in the next chapter, which deals with methods of **statistical inference,** that is, conclusions from samples to populations, whose unknown properties we want to know and try to discover by looking at suitable properties of samples that we have obtained.

PROBLEM SET 22.9

1. Let $f(x, y) = k$ when $4 \leq x \leq 10$ and $0 \leq y \leq 5$ and zero elsewhere. Find k. Find $P(X \leq 8, 3 \leq Y \leq 4)$ and $P(9 \leq X \leq 13, Y \leq 1)$.
2. Find $P(X > 4, Y > 4)$ and $P(X \leq 1, Y \leq 1)$ if (X, Y) has the density $f(x, y) = 1/32$ if $x \geq 0$, $y \geq 0$, $x + y \leq 8$.
3. Let $f(x, y) = k$ if $x > 0$, $y > 0$, $x + y < 3$ and 0 otherwise. Find k. Sketch $f(x, y)$. Find $P(X + Y \leq 1)$, $P(Y > X)$.
4. Find the density of the marginal distribution of X in Prob. 2.
5. Find the density of the marginal distribution of Y in Fig. 492.
6. If certain sheets of paper have a mean weight of 2 grams each, with a standard deviation of 0.03 gram, what are the mean weight and the standard deviation of a pack of 10 000 sheets?
7. What are the mean thickness and the standard deviation of transformer cores each consisting of 50 layers of sheet metal and 49 insulating paper layers if the metal sheets have mean thickness 0.5 mm each with a standard deviation of 0.05 mm and the paper layers have mean 0.05 mm each with a standard deviation of 0.02 mm?
8. Let X[cm] and Y[cm] be the diameters of a pin and a hole, respectively. Suppose that (X, Y) has the density

$$f(x, y) = 625 \quad \text{if} \quad 0.98 < x < 1.02, \quad 1.00 < y < 1.04$$

and 0 otherwise. (a) Find the marginal distributions. (b) What is the probability that a pin chosen at random will fit a hole whose diameter is 1.00?

9. An electronic device consists of two components. Let X and Y [years] be the times to failure of the first and second components, respectively. Assume that (X, Y) has the density $f(x, y) = 4e^{-2(x+y)}$ if $x > 0$ and $y > 0$ and 0 otherwise. (a) Are X and Y dependent or independent? (b) Find the densities of the marginal distributions. (c) What is the probability that the first component will have a lifetime of 2 years or longer?

10. If the weight of certain (empty) containers has 5 lb and standard deviation 0.2 lb, and if the filling of the containers has mean weight 100 lb and standard deviation 0.5 lb, what are the mean weight and the standard deviation of filled containers?

11. Using Theorems 1 and 3, obtain the formulas for the mean and the variance of the binomial distribution.

12. Using Theorem 1, obtain the formula for the mean of the hypergeometric distribution. Can you use Theorem 3 to obtain the variance of that distribution?

13. Give an example of two different discrete distributions that have the same marginal distributions.

14. Prove (2).

15. Show that the two-dimensional random variables with the densities $f(x, y) = x + y$ and $g(x, y) = (x + \frac{1}{2})(y + \frac{1}{2})$ if $0 \leq x \leq 1$, $0 \leq y \leq 1$, have the same marginal distributions.

16. Prove the statement involving (18).

CHAPTER 22 REVIEW

1. What are stem-and-leaf plots, histograms, and boxplots? Compare their advantages.

2. Why did we begin the chapter with a section on data?

3. What do you know about outliers?

4. What possibilities do we have for measuring the average size of data values? Of their spread?

5. Why did we consider probability theory? What is its role in statistics?

6. What do we mean by an experiment? By a random variable related with it? By outcomes and events?

7. What is a Venn diagram and how can we use it?

8. State the definition of probability from memory.

9. What is relative frequency and to what extent does it motivate probability?

10. State the main theorems on probability and illustrate them by simple examples.

11. What are independent events? Independent random variables?

12. What are permutations and combinations? Why did we consider them?

13. What is a random variable? Its distribution function?

14. There are two kinds of random variables of practical importance. Explain their difference.

15. How are mean and variance defined and what properties of a probability distribution do they characterize?

16. What is a standardized random variable? Why is this concept of interest?

17. Under what conditions will an experiment involve a binomial distribution?

18. What do you know about the Poisson and hypergeometric distributions?

19. Explain how the tables of the normal distribution are used.

20. State the most important facts about distributions of two random variables.

21. If $P(A) = P(B)$ and $A \subseteq B$, can $A \neq B$? If $E \neq S$, can $P(E) = 1$?

22. How is the density of a continuous distribution related to the distribution function?

23. Can the probability function of a discrete random variable have infinitely many values? (Give a reason.)

24. What is $P(X = a)$ (a arbitrary) if X is a continuous random variable?

25. Under what condition does the addition formula for variances hold? The addition formula for means?

26. Make a stem-and-leaf plot, histogram, and boxplot of the data 110, 113, 109, 118, 110, 115, 104, 111, 116, 113.

27. Same task as in Prob. 26, for the data 13.5, 13.2, 12.1, 13.6, 13.3.

28. Find the mean, standard deviation, and variance in Prob. 26.

29. Find the mean, standard deviation, and variance in Prob. 27.

30. Show that the mean always lies between the smallest and the largest data value.

31. Plot a histogram of the data 8, 2, 4, 10 and guess $\bar{x}$ and s by inspecting the histogram. Then calculate $\bar{x}$, s^2, and s.

32. What are the outcomes in the sample space of the experiment of simultaneously tossing three coins?

33. What are the outcomes of the sample space of X: *Tossing a coin until the first Head appears?*

34. Using a Venn diagram, show that $A \subseteq B$ if and only if $A \cap B = A$.

35. Using a Venn diagram, show that $A \subseteq B$ if and only if $A \cup B = B$.

36. Of a lot of 12 items, 3 are defective. (a) Find the number of different samples of 3 items. Find the number of samples of 3 items containing (b) no defectives, (c) 1 defective, (d) 2 defectives, (e) 3 defectives.

37. Obtain $P(\varnothing) = 0$ from the complementation rule.

38. Find the probability function of X = *Number of times of tossing a fair coin until the first head appears.* Find the mean.

39. Determine the number of different bridge hands. (A bridge hand consists of 13 cards selected from a full deck of 52 cards.)

40. What is the probability that in simultaneously rolling 6 dice all the numbers $1, 2, \cdots, 6$ will show up?

41. Sketch the probability function $f(x) = x^2/30$ ($x = 1, 2, 3, 4$) and the distribution function.

42. Sketch $F(x) = 0$ if $x \leq 0$, $F(x) = 0.2x$ if $0 < x \leq 5$, $F(x) = 1$ if $x > 5$, and its density $f(x)$.

43. If the life of ball bearings has the density $f(x) = ke^{-x}$ if $0 \leq x \leq 2$ and 0 otherwise, what is k? What is the probability $P(X \geq 1)$?

44. Find the mean and variance of a discrete random variable X having the probability function $f(0) = \frac{1}{4}$, $f(1) = \frac{1}{2}$, $f(2) = \frac{1}{4}$.

45. Find the mean and the variance of the distribution having the density $f(x) = \frac{1}{2}e^{-|x|}$.

46. Find the skewness of the distribution with density $f(x) = 2(1 - x)$ if $0 < x < 1$, $f(x) = 0$ otherwise.

47. Suppose that a certain type of plastic tape contains, on the average, 2 defects per 100 meters. What is the probability that a roll of tape 300 meters long will contain (a) x defects, (b) no defects?

48. Let X be normal with mean 80 and variance 9. Find $P(X > 83)$, $P(X < 81)$, $P(X < 80)$, and $P(78 < X < 82)$.

49. Let X be normal with mean 14 and variance 4. Determine c such that $P(X \leq c) = 95\%$, $P(X \leq c) = 5\%$, $P(X \leq c) = 99.5\%$.

50. A five-gear assembly is put together with spacers between the gears. The mean thickness of the gears is 5.030 cm with a standard deviation of 0.008 cm. The mean thickness of the spacers is 0.140 cm with a standard deviation of 0.005 cm. Find the mean and the standard deviation of the thickness of the assembled units consisting of five randomly selected gears and four randomly selected spacers.

SUMMARY OF CHAPTER 22
DATA ANALYSIS. PROBABILITY THEORY

A *random experiment,* briefly called **experiment,** is a process in which the result (**"outcome"**) depends on "chance" (effects of factors unknown to us). Examples are games of chance with dice or cards, measuring the hardness of steel, observing weather conditions, or recording the number of accidents in a city. (Thus the word "experiment" is used here in a much wider sense than in common language.) The outcomes are regarded as points (elements) of a set S, called the **sample space,** whose subsets are called **events.** For events E we define a **probability** $P(E)$ by the axioms (Sec. 22.3)

$$0 \leq P(E) \leq 1$$
$$(1) \qquad P(S) = 1$$
$$P(E_1 \cup E_2 \cup \cdots) = P(E_1) + P(E_2) + \cdots \quad (E_j \cap E_k = \varnothing).$$

These axioms are motivated by properties of frequency distributions of data (Sec. 22.1).

The complement E^c of E has the probability

$$(2) \qquad P(E^c) = 1 - P(E).$$

The **conditional probability** of an event B under the condition that an event A happens is (Sec. 22.3)

$$(3) \qquad P(B|A) = \frac{P(A \cap B)}{P(A)} \qquad [P(A) > 0].$$

Two events A and B are called **independent** if the probability of their simultaneous appearance in a trial equals the product of their probabilities, that is, if

$$(4) \qquad P(A \cap B) = P(A)P(B).$$

With an experiment we associate a **random variable X.** This is a function defined on S whose values are real numbers; furthermore, X is such that the probability $P(X = a)$ with which X assumes any value a, and the probability $P(a < X \leq b)$ with which X assumes any value in an interval $a < X \leq b$ are defined (Sec. 22.5). The **probability distribution** of X is determined by the distribution function

$$(5) \qquad F(x) = P(X \leq x).$$

There are two practically important kinds of random variables: those of **discrete** type, which appear if we count (defective items, customers in a bank, etc.) and those of **continuous** type, which appear if we measure (length, speed, temperature, weight, etc.).

A discrete random variable has a **probability function**

$$(6) \qquad\qquad f(x) = P(X = x).$$

Its **mean** μ and **variance** σ^2 are (Sec. 22.6)

$$(7) \qquad \mu = \sum_j x_j f(x_j) \qquad \text{and} \qquad \sigma^2 = \sum_j (x_j - \mu)^2 f(x_j)$$

where the x_j are the values for which X has a positive probability. Important discrete random variables and distributions are the **binomial, Poisson,** and **hypergeometric distributions** discussed in Sec. 22.7.

A continuous random variable has a **density**

$$(8) \qquad\qquad f(x) = F'(x) \qquad\qquad \text{[see (5)].}$$

Its mean and variance are (Sec. 22.6)

$$(9) \qquad \mu = \int_{-\infty}^{\infty} x f(x)\, dx \qquad \text{and} \qquad \sigma^2 = \int_{-\infty}^{\infty} (x - \mu)^2 f(x)\, dx.$$

Very important is the **normal distribution** (Sec. 22.8), whose density is

$$(10) \qquad f(x) = \frac{1}{\sigma\sqrt{2\pi}} \exp\left[-\frac{1}{2} \left(\frac{x - \mu}{\sigma} \right)^2 \right]$$

and whose distribution function is (Sec. 22.8; Tables A7, A8 in Appendix 5)

$$(11) \qquad F(x) = \Phi\left(\frac{x - \mu}{\sigma} \right).$$

A **two-dimensional random variable** (X, Y) occurs if we simultaneously observe two quantities (for example, height X and weight Y of adults). Its distribution function is (Sec. 22.9)

$$(12) \qquad\qquad F(x, y) = P(X \leqq x, Y \leqq y).$$

X and Y have the distribution functions (Sec. 22.9)

$$(13) \quad F_1(x) = P(X \leqq x, Y \text{ arbitrary}) \quad \text{and} \quad F_2(y) = P(x \text{ arbitrary}, Y \leqq y)$$

respectively; their distributions are called **marginal distributions.** If both X and Y are discrete, then (X, Y) has a probability function

$$f(x, y) = P(X = x, Y = y).$$

If both X and Y are continuous, then (X, Y) has a density $f(x, y)$.

Mathematical Statistics

In **probability theory** we set up mathematical models of processes and systems that are affected by "chance." In mathematical statistics or, briefly, **statistics,** we check these models against the reality, to determine whether they are faithful and accurate enough for practical purposes. This is done mainly as a basis for predictions, decisions, and actions, for instance, in analyzing markets, planning productions, buying equipment, investing in business projects, and so on. The process of checking models is called **statistical inference.**

In this process we draw random samples, briefly called **samples.** These are sets of data values from a much larger set of data values that could be studied, called the **population.** Examples are 10 diameters of screws from a large lot of screws, 100 household incomes in your community, 5 values a die turns up in 5 trials (here the population is *hypothetical,* consisting of an infinite sequence of outcomes of trials). Such an inference from samples to a population holds true, not absolutely, but with some high probability, that we can choose (95%, for instance) or at least compute.

Methods of statistical inference are based on drawing samples ("sampling," Sec. 23.1). Most important are **estimation of parameters** (Secs. 23.2, 23.3) and **hypothesis testing** (Sec. 23.4, 23.7, 23.8) with application to *quality control* (Sec. 23.5) and *acceptance sampling* (Sec. 23.6). The last two sections (Secs. 23.9, 23.10) give an introduction to **regression** and **correlation analysis,** which concern experiments involving two variables.

Prerequisites for this chapter: Chap. 22.
Sections that may be omitted in a shorter course: 23.5, 23.6, 23.8, 23.10.
References: Appendix 1, Part G.
Answers to problems: Appendix 2.
Statistical tables: Appendix 5.

23.1 Introduction. Random Sampling

Mathematical statistics consists of methods for designing and evaluating random experiments to obtain information about practical problems, such as exploring the relation between iron content and density of iron ore, the quality of raw material or manufactured products, the efficiency of air-conditioning systems, the performance of certain cars, the effect of advertising, consumer reactions to a new product, etc.

Random variables occur more frequently in engineering (and elsewhere) than one would think. For example, properties of mass-produced articles (screws, lightbulbs, etc.) always show **random variation,** due to small (uncontrollable!) differences in raw material or manufacturing processes. Thus the diameter of screws is a random variable X and we have *nondefective screws,* with diameter between tolerance limits, and *defective screws,* with diameter outside those limits. We can ask for the distribution of X, for the percentage of defective screws to be expected, and for necessary improvements (by using an automatic screw machine with higher precision).

Samples are selected from populations—20 screws from a lot of 1000, 100 of 5000 voters, 8 beavers in a wildlife conservation project—because inspecting the entire population would be too expensive, time-consuming, impossible or even senseless (think of destructive testing of lightbulbs or dynamite). To obtain meaningful conclusions, samples must be **random selections.** Each of the 1000 screws must have the same chance of being sampled (of being drawn when we sample), at least approximately. Only then will the sample mean $\bar{x} = (x_1 + \cdots + x_{20})/20$ (Sec. 22.1) of a sample of size $n = 20$ (or any other n) be a good approximation of the population mean μ (Sec. 22.6); and the accuracy of the approximation will generally improve with increasing n, as we shall see. Similarly for other parameters (standard deviation, variance, etc.).

Independent sample values will be obtained in experiments with an infinite sample space S (Sec. 22.2), certainly for the normal distribution. This is also true in sampling with replacement. It is approximately true in drawing *small* samples from a large finite population (for instance, 5 or 10 of 1000 items). However, if we sample without replacement from a small population, the effect of dependence of sample values may be considerable.

Random numbers help in obtaining samples that are in fact random selections. This is sometimes not easy to accomplish because there are many subtle factors that can bias sampling (by personal interviews, by poorly working machines, by the choice of nontypical observation conditions, etc.). Random numbers can be obtained from a **random number generator** in MAPLE, MATHEMATICA, or other systems. (The numbers are not truly random, as they would be produced in flipping coins or rolling dice, but are calculated by a tricky formula that produces numbers that do have practically all the essential features of true randomness.)

EXAMPLE 1   **Random numbers from a generator and by dice**

To select a sample of size $n = 10$ from 80 given ball bearings, we number the latter from 1 to 80. We then let the generator randomly produce 10 of the integers from 1 to 80 and include the bearings with the numbers obtained in our sample, for example,

$$44 \quad 55 \quad 53 \quad 03 \quad 52 \quad 61 \quad 67 \quad 78 \quad 39 \quad 54$$

or whatever.

If you have no generator, take two fair dice. Roll them. Each time take the sum minus 3, ignoring trials resulting in -1. Take the remaining results in pairs in the order obtained. These pairs give you the ones and tens of your random numbers. Omit $81, \cdots, 99$ and fill up by replacements until you have 10 numbers.

Random numbers are also contained in (older) statistical tables.   ◀

Representing and processing data were considered in Sec. 22.1 in connection with frequency distributions. These are the empirical counterparts of probability distributions and helped motivating axioms and properties in probability theory. The new aspect in this chapter is *randomness:* the data are samples selected *randomly* from a population. Accordingly, we can immediately make the connection to Sec. 22.1, using stem-and-leaf plots, box plots, and histograms for representing samples graphically.

Also, we now call the mean $\bar{x}$ in (5), Sec. 22.1, the **sample mean**

$$
(1) \qquad \bar{x} = \frac{1}{n} \sum_{j=1}^{n} x_j = \frac{1}{n} (x_1 + x_2 + \cdots + x_n).
$$

We call n the **sample size,** the variance s^2 in (6), Sec. 22.1, the **sample variance**

$$
(2) \qquad s^2 = \frac{1}{n-1} \sum_{j=1}^{n} (x_j - \bar{x})^2 = \frac{1}{n-1} \left[(x_1 - \bar{x})^2 + \cdots + (x_n - \bar{x})^2 \right],
$$

and its positive square root s the **sample standard deviation.** These parameters will be needed throughout this chapter.

23.2 Estimation of Parameters

Beginning in this section, we shall discuss the most basic practical tasks in statistics and corresponding statistical methods. The first of them is point estimation of **parameters,** that is, quantities appearing in distributions, such as p in the binomial distribution and μ and σ in the normal distribution.

A **point estimate** of a parameter is a number (point on the real line), which is computed from a given sample and serves as an approximation of the unknown exact value of the parameter. An **interval estimate** is an interval (*"confidence interval"*) obtained from a sample; such estimates will be considered in the next section. Estimation of parameters is of great practical importance in many applications.

As an approximation of the mean μ of a population we may take the mean $\bar{x}$ of a corresponding sample. This gives the estimate $\hat{\mu} = \bar{x}$ for μ, that is,

$$
(1) \qquad \hat{\mu} = \bar{x} = \frac{1}{n} (x_1 + \cdots + x_n)
$$

where n is the sample size. Similarly, an estimate $\hat{\sigma}^2$ for the variance of a population is the variance s^2 of a corresponding sample, that is,

$$
(2) \qquad \hat{\sigma}^2 = s^2 = \frac{1}{n-1} \sum_{j=1}^{n} (x_j - \bar{x})^2.
$$

Clearly, (1) and (2) are estimates of parameters for distributions in which μ or σ^2 appear explicitly as parameters, such as the normal and Poisson distributions. For the binomial distribution, $p = \mu/n$ [see (3) in Sec. 22.7]. From (1) we thus obtain for p the estimate

$$
(3) \qquad \hat{p} = \frac{\bar{x}}{n}.
$$

We mention that (1) is a special case of the so-called **method of moments.** In this method the parameters to be estimated are expressed in terms of the moments of the distribution (see Sec. 22.6). In the resulting formulas those moments are replaced by the corresponding moments of the sample. This gives the estimates. Here the **kth moment of a sample** $x_1, \cdots, x_n$ is

$$m_k = \frac{1}{n} \sum_{j=1}^{n} x_j^k.$$

Maximum Likelihood Method

Another method for obtaining estimates is the so-called **maximum likelihood method** of R. A. Fisher [*Messenger Math.* **41** (1912), 155−160]. To explain it, we consider a discrete (or continuous) random variable X whose probability function (or density) $f(x)$ depends on a single parameter θ. We take a corresponding sample of n independent values $x_1, \cdots, x_n$. Then in the discrete case the probability that a sample of size n consists precisely of those n values is

(4) $$l = f(x_1)f(x_2) \cdots f(x_n).$$

In the continuous case the probability that the sample consists of values in the small intervals $x_j \leqq x \leqq x_j + \Delta x$ $(j = 1, 2, \cdots, n)$ is

(5) $$f(x_1)\Delta x \, f(x_2)\Delta x \cdots f(x_n)\Delta x = l(\Delta x)^n.$$

Since $f(x_j)$ depends on θ, the function l depends on $x_1, \cdots, x_n$ and θ. We imagine $x_1, \cdots, x_n$ to be given and fixed. Then l is a function of θ, which is called the **likelihood function.** The basic idea of the maximum likelihood method is very simple, as follows. We choose that approximation for the unknown value of θ for which l is as large as possible. If l is a differentiable function of θ, a necessary condition for l to have a maximum in an interval (not at the boundary) is

(6) $$\frac{\partial l}{\partial \theta} = 0.$$

(We write a *partial* derivative, because l depends also on $x_1, \cdots, x_n$.) A solution of (6) depending on $x_1, \cdots, x_n$ is called a **maximum likelihood estimate** for θ. We may replace (6) by

(7) $$\frac{\partial \ln l}{\partial \theta} = 0,$$

because $f(x_j) > 0$, a maximum of l is in general positive, and $\ln l$ is a monotone increasing function of l. This often simplifies calculations.

Several parameters. If the distribution of X involves r parameters $\theta_1, \cdots, \theta_r$, then instead of (6) we have the r conditions $\partial l/\partial \theta_1 = 0, \cdots, \partial l/\partial \theta_r = 0$, and instead of (7) we have

(8) $$\frac{\partial \ln l}{\partial \theta_1} = 0, \qquad \cdots, \qquad \frac{\partial \ln l}{\partial \theta_r} = 0.$$

EXAMPLE 1   **Normal distribution**

Find maximum likelihood estimates for μ and σ in the case of the normal distribution.

Solution.   From (1), Sec. 22.8, and (4) we obtain the likelihood function

$$l = \left(\frac{1}{\sqrt{2\pi}}\right)^n \left(\frac{1}{\sigma}\right)^n e^{-h} \qquad \text{where} \qquad h = \frac{1}{2\sigma^2} \sum_{j=1}^n (x_j - \mu)^2.$$

Taking logarithms, we have

$$\ln l = -n \ln \sqrt{2\pi} - n \ln \sigma - h.$$

The first equation in (8) is $\partial \ln l / \partial \mu = 0$, written out

$$\frac{\partial \ln l}{\partial \mu} = -\frac{\partial h}{\partial \mu} = \frac{1}{\sigma^2} \sum_{j=1}^n (x_j - \mu) = 0, \qquad \text{hence} \qquad \sum_{j=1}^n x_j - n\mu = 0.$$

The solution is the desired estimate $\hat{\mu}$ for μ; we find

$$\hat{\mu} = \frac{1}{n} \sum_{j=1}^n x_j = \bar{x}.$$

The second equation in (8) is $\partial \ln l / \partial \sigma = 0$, written out

$$\frac{\partial \ln l}{\partial \sigma} = -\frac{n}{\sigma} - \frac{\partial h}{\partial \sigma} = -\frac{n}{\sigma} + \frac{1}{\sigma^3} \sum_{j=1}^n (x_j - \mu)^2 = 0.$$

Replacing μ by $\hat{\mu}$ and solving for σ^2, we obtain the estimate

$$\tilde{\sigma}^2 = \frac{1}{n} \sum_{j=1}^n (x_j - \bar{x})^2$$

which we shall use in Sec. 23.7. Note that this differs from (2). We cannot discuss criteria for the goodness of estimates but want to mention that for small n, formula (2) is preferable.   ◀

PROBLEM SET 23.2

1. Apply the maximum likelihood method to the normal distribution with $\mu = 0$.
2. Find the maximum likelihood estimate for the parameter μ of the normal distribution with known variance $\sigma^2 = \sigma_0^2$.
3. **(Poisson distribution)** Apply the maximum likelihood method to the Poisson distribution.
4. **(Uniform distribution)** Show that in the case of the parameters a and b of the uniform distribution (see Sec. 22.6), the maximum likelihood estimate cannot be obtained by equating the first derivative to zero. How can we obtain maximum likelihood estimates in this case?
5. Find the maximum likelihood estimate of θ in the density $f(x) = \theta e^{-\theta x}$ if $x \geqq 0$ and $f(x) = 0$ if $x < 0$.
6. In Prob. 5, find the mean μ, substitute it in $f(x)$, find the maximum likelihood estimate of μ, and show that it is identical with the estimate for μ which can be obtained from that for θ in Prob. 5.
7. Compute $\hat{\theta}$ in Prob. 5 from the sample 1.9, 0.4, 0.7, 0.6, 1.4. Graph the sample distribution function $\hat{F}(x)$ and the distribution function $F(x)$ of the random variable, with $\theta = \hat{\theta}$, on the same axes. Do they agree reasonably well? (We consider goodness of fit systematically in Sec. 23.7.)
8. Do the same task as in Prob. 7 if the given sample is 0.4, 0.7, 0.2, 1.1, 0.1.

9. **(Binomial distribution)** Derive a maximum likelihood estimate for p.

10. Extend Prob. 9 as follows. Suppose that m times n trials were made and in the first n trials A happened k_1 times, in the second n trials A happened k_2 times, $\cdots$, in the mth n trials A happened k_m times. Find a maximum likelihood estimate of p based on this information.

11. Suppose that in Prob. 10 we made 3 times 4 trials and A happened 2, 3, 2 times, respectively. Estimate p.

12. Consider $X = $ *Number of independent trials until an event A occurs.* Show that X has the probability function $f(x) = pq^{x-1}$, $x = 1, 2, \cdots$, where p is the probability of A in a single trial and $q = 1 - p$. Find the maximum likelihood estimate of p corresponding to a single observed value x of X.

13. In Prob. 12, find the maximum likelihood estimate of p resulting from a sample $x_1, \cdots, x_n$.

14. In rolling a die, suppose that we get the first Six in the 7th trial and in doing it again we get it in the 6th trial. Estimate the probability p of getting a Six in rolling that die once.

23.3　Confidence Intervals

Confidence intervals[1] for an unknown parameter θ of some distribution (e.g., $\theta = \mu$) are intervals $\theta_1 \leqq \theta \leqq \theta_2$ that contain θ, not with certainty but with a high probability γ, which we can choose (95% and 99% are popular). Such an interval is calculated from a sample. $\gamma = 95\%$ means probability $1 - \gamma = 5\% = 1/20$ of being wrong—one of about 20 such intervals will not contain θ. Instead of writing $\theta_1 \leqq \theta \leqq \theta_2$, we denote this more distinctly by writing

$$\text{(1)} \qquad\qquad\qquad \text{CONF}_\gamma \{\theta_1 \leqq \theta \leqq \theta_2\},$$

where γ is called the **confidence level,** and θ_1 and θ_2 are the **lower** and **upper confidence limits.** They depend on γ. The larger we choose γ, the smaller the error probability $1 - \gamma$ is, but the longer the confidence interval will be; if $\gamma \to 1$, then its length goes to infinity. The choice of γ depends on the kind of application. In taking no umbrella, a 5% chance of getting wet is not tragic. In a medical matter of life and death, a 1% chance of being wrong ($\gamma = 99\%$) may be better.

Confidence intervals are more valuable than point estimates (Sec. 23.2). Indeed, we can take the midpoint of (1) as an approximation of θ and half the length of (1) as an "error bound" (not in the strict sense of numerical analysis, but except for an error whose probability we know).

θ_1 and θ_2 in (1) are calculated from a sample $x_1, \cdots, x_n$. These are n observations of a random variable X. Now comes a **standard trick.** *We regard $x_1, \cdots, x_n$ as **single** observations of n random variables $X_1, \cdots, X_n$ (with the same distribution, namely, that of X).* Then $\theta_1 = \theta_1(x_1, \cdots, x_n)$ and $\theta_2 = \theta_2(x_1, \cdots, x_n)$ in (1) are observed values of two random variables $\Theta_1 = \Theta_1(X_1, \cdots, X_n)$ and $\Theta_2 = \Theta_2(X_1, \cdots, X_n)$. The condition involving γ can now be written

$$\text{(2)} \qquad\qquad\qquad P(\Theta_1 \leqq \theta \leqq \Theta_2) = \gamma.$$

Let us see what all this means in concrete practical cases.

---

[1]The modern theory and terminology of confidence intervals were developed by J. Neyman [*Annals of Mathematical Statistics* **6** (1935), $111-116$]. See also footnote 3 in the next section.

In mathematics, $\theta_1 \leqq \theta \leqq \theta_2$ means that θ lies between θ_1 and θ_2, and, to avoid misunderstandings, it seems worthwhile to characterize a confidence interval by a special symbol, such as CONF.

Confidence Interval for μ of the Normal Distribution with Known σ^2

Table 23.1
Determination of a Confidence Interval for the Mean μ of a Normal Distribution with Known Variance σ^2

1st Step. Choose a confidence level γ (95%, 99%, or the like).

2nd Step. Determine the corresponding c:

γ	0.90	0.95	0.99	0.999
c	1.645	1.960	2.576	3.291

3rd Step. Compute the mean $\bar{x}$ of the sample $x_1, \cdots, x_n$.

4th Step. Compute $k = c\sigma/\sqrt{n}$. The confidence interval for μ is

$$(3) \qquad \mathrm{CONF}_\gamma\{\bar{x} - k \leqq \mu \leqq \bar{x} + k\}.$$

EXAMPLE 1   **Confidence interval for μ of the normal distribution with known σ^2**

Determine a 95% confidence interval for the mean of a normal distribution with variance $\sigma^2 = 9$, using a sample of $n = 100$ values with mean $\bar{x} = 5$.

Solution. 1st Step. $\gamma = 0.95$ is required.

2nd Step.   The corresponding c equals 1.960; see Table 23.1.

3rd Step. $\bar{x} = 5$ is given.

4th Step.   We need $k = 1.960 \cdot 3/\sqrt{100} = 0.588$. Hence $\bar{x} - k = 4.412$, $\bar{x} + k = 5.588$ and the confidence interval is

$$\mathrm{CONF}_{0.95}\ \{4.412 \leqq \mu \leqq 5.588\}.$$

This is sometimes written $\mu = 5 \pm 0.588$, but we shall not use this notation, which can be misleading.   ◀

Theory for Table 23.1

We show that the method in Table 23.1 follows from the basic

THEOREM 1   **(Sum of independent normal random variables)**

Let $X_1, \cdots, X_n$ be **independent** normal random variables each of which has mean μ and variance σ^2. Then the following holds.

(a) *The sum $X_1 + \cdots + X_n$ is normal with mean $n\mu$ and variance $n\sigma^2$.*

(b) *The following random variable $\overline{X}$ is normal with mean μ and variance σ^2/n.*

$$(4) \qquad \overline{X} = \frac{1}{n}(X_1 + \cdots + X_n)$$

(c) *The following random variable Z is normal with mean 0 and variance 1.*

$$(5) \qquad Z = \frac{\overline{X} - \mu}{\sigma/\sqrt{n}}$$

PROOF. The statements about the mean and variance in (a) follow from Theorems 1 and 3 in Sec. 22.9. From this and Theorem 2 in Sec. 22.6 we see that $\overline{X}$ has the mean $(1/n)n\mu = \mu$ and the variance $(1/n)^2 n\sigma^2 = \sigma^2/n$. This implies that Z has the mean 0 and variance 1, by Theorem 2(b) in Sec. 22.6. The normality of $X_1 + \cdots + X_n$ is proved in Ref. [G3] listed in Appendix 1. This implies the normality of (4) and (5). ◀

Derivation of (3) in Table 23.1. Sampling from a normal distribution gives independent sample values (see Sec. 23.1), so that Theorem 1 applies. Hence we can choose γ and then determine c such that

$$P(-c \leqq Z \leqq c) = P\left(-c \leqq \frac{\overline{X} - \mu}{\sigma/\sqrt{n}} \leqq c\right)$$

(6)

$$= \Phi(c) - \Phi(-c) = \gamma.$$

For the value $\gamma = 0.95$ we obtain $z(D) = 1.960$ from Table A8 in Appendix 5, as used in Example 1. For $\gamma = 0.9, 0.99, 0.999$ we get the other values of c listed in Table 23.1. Finally, all that we have to do is convert the inequality in (6) into one for μ and insert observed values obtained from the sample. We multiply $-c \leqq Z \leqq c$ by -1 and then by $\sigma/\sqrt{n}$, writing $c\sigma/\sqrt{n} = k$ (as in Table 23.1),

$$P(-c \leqq Z \leqq c) = P(c \geqq -Z \geqq -c)$$

$$= P\left(c \geqq \frac{\mu - \overline{X}}{\sigma/\sqrt{n}} \geqq -c\right)$$

$$= P(k \geqq \mu - \overline{X} \geqq -k) = \gamma.$$

Adding $\overline{X}$ gives

$$P(\overline{X} + k \geqq \mu \geqq \overline{X} - k) = \gamma$$

or

(7)
$$P(\overline{X} - k \leqq \mu \leqq \overline{X} + k) = \gamma.$$

Inserting the observed value $\bar{x}$ of $\overline{X}$ gives (3). Here we have regarded $x_1, \cdots, x_n$ as single observations of $X_1, \cdots, X_n$ (the standard trick!), so that $x_1 + \cdots + x_n$ is an observed value of $X_1 + \cdots + X_n$ and $\bar{x}$ is an observed value of $\overline{X}$. Note further that (7) is of the form (2) with $\Theta_1 = \overline{X} - k$ and $\Theta_2 = \overline{X} + k$. ◀

EXAMPLE 2 **Sample size needed for a confidence interval of prescribed length**

How large must n be in Example 1 if we want to obtain a 95% confidence interval of length $L = 0.4$?

Solution. The interval (3) has the length $L = 2k = 2c\sigma/\sqrt{n}$. Solving for n, we obtain

$$n = (2c\sigma/L)^2.$$

In the present case the answer is $n = (2 \cdot 1.960 \cdot 3/0.4)^2 \approx 870$.

Figure 494 on the next page shows how L decreases as n increases and that for $\gamma = 99\%$ the confidence interval is substantially longer than for $\gamma = 95\%$ (and the same sample size n). ◀

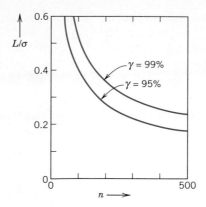

Fig. 494. Length of the confidence interval (3) (measured in multiples of σ) as a function of the sample size n for $\gamma = 95\%$ and $\gamma = 99\%$

Confidence Interval for μ of the Normal Distribution with Unknown σ^2

So far the variance σ^2 was assumed to be known. If this is no longer the case, as in most applications, the entire theory changes, although the practical technicalities remain quite similar. Now using less information (σ^2 is not known), we should expect somewhat longer intervals (which is true). We may hope that the sample variance will help us out (which is also true). Table 23.2 shows the steps. k differs from that in Table 23.1, and c now depends on n and must be determined from Table A9 in Appendix 5. That table contains values z corresponding to given values of the distribution function

$$(8) \qquad F(z) = K_m \int_{-\infty}^{z} \left(1 + \frac{u^2}{m}\right)^{-(m+1)/2} du$$

Table 23.2
**Determination of a Confidence Interval for the Mean μ
of a Normal Distribution with Unknown Variance σ^2**

> **1st Step.** Choose a confidence level γ (95%, 99%, or the like).
>
> **2nd Step.** Determine the solution c of the equation
>
> $$(9) \qquad F(c) = \tfrac{1}{2}(1 + \gamma)$$
>
> from the table of the t-distribution with $n - 1$ degrees of freedom (Table A9 in Appendix 5; n = sample size).
>
> **3rd Step.** Compute the mean $\bar{x}$ and the variance s^2 of the sample $x_1, \cdots, x_n$.
>
> **4th Step.** Compute $k = sc/\sqrt{n}$. The confidence interval is
>
> $$(10) \qquad \text{CONF}_\gamma \{\bar{x} - k \leq \mu \leq \bar{x} + k\}.$$

of the so-called **t-distribution** of Student.[2] Here, m ($= 1, 2, \cdots$) is a parameter, called the **number of degrees of freedom** of the distribution. The constant K_m is such that $F(\infty) = 1$. By integration it turns out that $K_m = \Gamma(\frac{1}{2}m + \frac{1}{2}) / \left[\sqrt{m\pi}\,\Gamma(\frac{1}{2}m)\right]$, where Γ is the gamma function (see (24) in Appendix A3.1).

EXAMPLE 3   **Confidence interval for μ of the normal distribution with unknown σ^2**

Five independent measurements of the flash point (°F) of Diesel oil (D-2) gave the values

$$144 \quad 147 \quad 146 \quad 142 \quad 144.$$

Assuming normality, determine a 99%-confidence interval for the mean.

Solution. 1st Step. $\gamma = 0.99$ is required.

2nd Step. $F(c) = \frac{1}{2}(1 + \gamma) = 0.995$, and Table A9 in Appendix 5 with $n - 1 = 4$ degrees of freedom gives $c = 4.60$.

3rd Step. $\bar{x} = 144.6$, $s^2 = 3.8$.

4th Step. $k = \sqrt{3.8} \cdot 4.60/\sqrt{5} = 4.01$. The confidence interval is

$$\mathrm{CONF}_{0.99}\,\{140.5 \leqq \mu \leqq 148.7\}.$$

If the variance σ^2 were known and equal to the sample variance s^2, thus $\sigma^2 = 3.8$, then Table 23.1 would give

$$k = c\sigma/\sqrt{n} = 2.576\sqrt{3.8}/\sqrt{5} = 2.25$$

and

$$\mathrm{CONF}_{0.99}\,\{142.35 \leqq \mu \leqq 146.85\}.$$

The length is 4.5, whereas the previous interval is almost twice as long, 8.1. Hence for small samples the difference is considerable! See also Fig. 495.   ◀

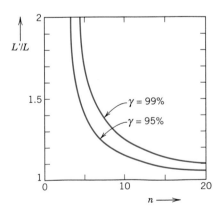

Fig. 495.   Ratio of the lengths L' and L
of the confidence intervals (10) and (3)
with $\gamma = 95\%$ and $\gamma = 99\%$ as a function
of the sample size n for equal s and σ

---

[2]Pseudonym for WILLIAM SEALY GOSSET (1876—1937), English statistician, who discovered the t-distribution in 1907–1908.

Theory for Table 23.2

For deriving (10) in Table 23.2 we need

THEOREM 2   *Let $X_1, \cdots, X_n$ be independent normal random variables with the same mean μ and the same variance σ^2. Then the random variable*

(11)
$$T = \frac{\overline{X} - \mu}{S/\sqrt{n}}$$

has a t-distribution [see (8)] with $n - 1$ degrees of freedom; here $\overline{X}$ is given by (4) and

(12)
$$S^2 = \frac{1}{n-1} \sum_{j=1}^{n} (X_j - \overline{X})^2.$$

Proof in Ref. [G3], listed in Appendix 1.

Derivation of (10). This is similar to the derivation of (3). We choose a number γ between 0 and 1 and determine a number c from Table A9, Appendix 5, with $n - 1$ degrees of freedom such that

(13)
$$P(-c \leq T \leq c) = F(c) - F(-c) = \gamma.$$

Since the *t*-distribution is symmetric, we have $F(-c) = 1 - F(c)$, and (13) assumes the form (9). Substituting (11) into (13) and transforming the result as before, we obtain

(14)
$$P(\overline{X} - K \leq \mu \leq \overline{X} + K) = \gamma \qquad \text{where} \qquad K = cS/\sqrt{n}.$$

By inserting the observed values $\bar{x}$ of $\overline{X}$ and s^2 of S^2 into (14) we finally obtain (10). ◀

Confidence Interval for the Variance σ^2 of the Normal Distribution

EXAMPLE 4   **Confidence interval for the variance of the normal distribution according to Table 23.3**

Determine a 95% confidence interval (16) for the variance, using the sample (tensile strength of sheet steel in kg/mm^2, rounded to integer values)

$$89 \quad 84 \quad 87 \quad 81 \quad 89 \quad 86 \quad 91 \quad 90 \quad 78 \quad 89 \quad 87 \quad 99 \quad 83 \quad 89.$$

Solution. 1st Step. $\gamma = 0.95$ is required.

2nd Step.   For $n - 1 = 13$ we find $c_1 = 5.01$ and $c_2 = 24.74$.

3rd Step. $13s^2 = 326.9$.

4th Step. $13s^2/c_1 = 65.25$, $13s^2/c_2 = 13.21$. The confidence interval is

$$\text{CONF}_{0.95} \{13.21 \leq \sigma^2 \leq 65.25\}. \qquad ◀$$

Table 23.3
Determination of a Confidence Interval for the Variance σ^2
of a Normal Distribution, Whose Mean Need Not Be Known

> **1st Step.** Choose a confidence level γ (95%, 99%, or the like).
>
> **2nd Step.** Determine solutions c_1 and c_2 of the equations
>
> (15) $F(c_1) = \frac{1}{2}(1 - \gamma), \qquad F(c_2) = \frac{1}{2}(1 + \gamma)$
>
> from the table of the chi-square distribution with $n - 1$ degrees of freedom (Table A10 in Appendix 5; n = sample size).
>
> **3rd Step.** Compute $(n - 1)s^2$, where s^2 is the variance of the sample $x_1, \cdots, x_n$.
>
> **4th Step.** Compute $k_1 = (n - 1)s^2/c_1$ and $k_2 = (n - 1)s^2/c_2$. The confidence interval is
>
> (16) $\text{CONF}_\gamma\{k_2 \leqq \sigma^2 \leqq k_1\}.$

Theory for Table 23.3

The steps in Table 23.3 are quite similar to those in Tables 23.1 and 23.2. But we now have to determine two numbers c_1 and c_2. Both are obtained from Table A10 in Appendix 5, which contains values z corresponding to given values of the distribution function $F(z) = 0$ if $z < 0$ and

$$F(z) = \begin{cases} 0 & \text{if} \quad z < 0 \\ C_m \displaystyle\int_0^z e^{-u/2}u^{(m-2)/2}\, du & \text{if} \quad z \geqq 0. \end{cases}$$

This is the distribution function of the so-called χ^2**-distribution** (*chi-square distribution*); here, $m\ (= 1, 2, \cdots)$ is a parameter, called the **number of degrees of freedom** of the distribution, and $C_m = 1/\left[2^{m/2}\Gamma(\frac{1}{2}m)\right]$.

For deriving (16) in Table 23.3 we need

THEOREM 3   *Under the assumptions in Theorem 2 the random variable*

(17) $Y = (n - 1)\dfrac{S^2}{\sigma^2}$

with S^2 given by (12) has a chi-square distribution with $n - 1$ degrees of freedom.

Proof in Ref. [G3], listed in Appendix 1.

Derivation of (16). This is similar to the derivation of (3) and (10). We choose a number γ between 0 and 1 and determine c_1 and c_2 from Table A10, Appendix 5, such that [see (15)]

$$P(Y \leqq c_1) = F(c_1) = \tfrac{1}{2}(1 - \gamma), \qquad P(Y \leqq c_2) = F(c_2) = \tfrac{1}{2}(1 + \gamma).$$

Subtraction yields

$$P(c_1 \leqq Y \leqq c_2) = P(Y \leqq c_2) - P(Y \leqq c_1) = \gamma.$$

Transforming $c_1 \leqq Y \leqq c_2$ with Y given by (17) into an inequality for σ^2, we obtain

$$\frac{n-1}{c_2} S^2 \leqq \sigma^2 \leqq \frac{n-1}{c_1} S^2.$$

By inserting the observed value s^2 of S^2 we obtain (16).   ◀

Confidence Intervals for Other Distributions

The methods in Tables 23.1–23.3 for confidence intervals for μ and σ^2 are designed for the normal distribution. We now show that they can also be applied to other distributions if we use large samples.

We know that if $X_1, \cdots, X_n$ are independent random variables with the same mean μ and the same variance σ^2, then their sum $Y_n = X_1 + \cdots + X_n$ has the following properties.

    (A) Y_n has the mean $n\mu$ and the variance $n\sigma^2$ (by Theorems 1 and 3 in Sec. 22.9).

    (B) If those variables are normal, then Y_n is normal (by Theorem 1).

If those random variables are not normal, then **(B)** fails to hold. However, for large n the random variable Y_n is still *approximately* normal. This follows from the central limit theorem, which is one of the most fundamental results in probability theory.

THEOREM 4   **(Central limit theorem)**

Let $X_1, \cdots, X_n, \cdots$ be independent random variables that have the same distribution function and therefore the same mean μ and the same variance σ^2. Let $Y_n = X_1 + \cdots + X_n$. Then the random variable

(18)
$$Z_n = \frac{Y_n - n\mu}{\sigma\sqrt{n}}$$

*is **asymptotically normal** with mean 0 and variance 1; that is, the distribution function $F_n(x)$ of Z_n satisfies*

$$\lim_{n \to \infty} F_n(x) = \Phi(x) = \frac{1}{\sqrt{2\pi}} \int_{-\infty}^{x} e^{-u^2/2} \, du.$$

A proof can be found in Ref. [G3] listed in Appendix 1.

Hence when applying Tables 23.1–23.3 to a nonnormal distribution we must use *sufficiently large samples.* As a rule of thumb, if the sample indicates that the skewness of the distribution (the asymmetry; see Team Project 16(d), Problem Set 22.6) is small, use at least $n = 20$ for the mean and at least $n = 50$ for the variance.

PROBLEM SET 23.3

1. Why are interval estimates in most cases more useful than point estimates?
2. What happens in Example 1 in the text if you use 99% instead of 95%? If the mean is 8 instead of 5?
3. Determine a 99% confidence interval for the mean of a normal population with standard deviation 2.5, using the sample 30.8, 30.0, 29.9, 30.1, 31.7, 34.0.
4. Find a 95% confidence interval for the mean μ of a normal population with standard deviation 1.2, using the sample 10, 10, 8, 12, 10, 11, 10, 11.
5. Determine a 95% confidence interval for the mean μ of a normal population with variance $\sigma^2 = 16$, using a sample of size 200 with mean 74.81.
6. What will happen to the length of the interval in Prob. 5 if we reduce the sample size to 50?
7. What sample size would be needed to produce a 95% confidence interval (3) of length (a) 2σ, (b) σ?
8. Obtain a 99% confidence interval for the mean of a normal population with variance $\sigma^2 = 0.36$ from Fig. 494, using a sample of size 290 with mean 16.30. (This problem should merely help you to understand the meaning of the figure.)

Assuming that the populations from which the following samples are taken are normal, determine a 99% confidence interval for the mean μ of the population.

9. A sample of lengths of 20 bolts with mean 15.50 cm and variance 0.09 cm^2
10. Melting point of aluminum (°C) 658, 665, 652, 661, 660
11. Copper content [%] of brass 65, 65, 64, 63, 65, 66, 63, 64, 62, 63

12. Find a 99% confidence interval for the parameter p of the binomial distribution, using Pearson's result in the last row of Table 22.1 in Sec. 22.3.
13. Find a 95% confidence interval for the percentage of cars on a certain highway that have poorly adjusted brakes, using a random sample of 500 cars stopped at a roadblock on that highway, 87 of which had poorly adjusted brakes.

Assuming that the populations from which the following samples are taken are normal, determine a 95% confidence interval for the variance σ^2 of the population.

14. Ultimate tensile strength (kpsi) of alloy steel (Maraging H) at room temperature:

$$251, 255, 258, 253, 253, 252, 250, 252, 255, 256$$

15. Mean energy (keV) of delayed neutron group (Group 3, half-life 6.2 sec) for uranium U^{235} fission: 435, 451, 430, 444, 438
16. Carbon monoxide emission (grams per mile) of a certain type of passenger car (cruising at 55 mph):

$$17.3, 17.8, 18.0, 17.7, 18.2, 17.4, 17.6, 18.1$$

17. If X is normal with mean 40 and variance 4, what distributions do $3X$ and $5X - 2$ have? (Use a part of the team project in Sec. 22.8.)
18. If X_1 and X_2 are independent normal random variables with mean 16 and 12 and variance 8 and 2, respectively, what distribution does $4X_1 - X_2$ have?
19. A machine fills boxes weighing Y lb with X lb of salt, where X and Y are normal with mean 200 lb and 10 lb and standard deviation 2 lb and 0.5 lb, respectively. What percent of filled boxes weighing between 208 lb and 212 lb are to be expected?
20. If the weight X of bags of cement is normally distributed with a mean of 40 kg and a standard deviation of 2 kg, how many bags can a delivery truck carry so that the probability of the total load exceeding 2000 kg will be 5%?

23.4 Testing of Hypotheses, Decisions

The ideas of confidence intervals and of tests[3] are perhaps the two most important ideas in modern statistics. In a statistical **test** we make inference from sample to population through testing a **hypothesis,** resulting from experience or observations, from a theory or a quality requirement, and so on. In many cases the result of a test is then used as a basis for a **decision,** for instance, to buy or not to buy a certain model of car, depending on a test of the gasoline mileage; to apply some medication, depending on a test of its effect; to proceed with a marketing strategy, depending on a test of consumer reactions, etc. Let us explain such a test in terms of a typical example and introduce the corresponding standard notions of statistical testing.

EXAMPLE 1   **Test of a hypothesis. Alternative. Significance level α**

We want to buy 100 coils of a certain kind of wire provided we can verify the manufacturer's claim that the wire has a breaking limit $\mu = \mu_0 = 200$ lb (or more). This is a test of the **hypothesis** (also called *null hypothesis*) $\mu = \mu_0 = 200$. We shall not buy the wire if the (statistical) test shows that actually $\mu = \mu_1 < \mu_0$, the wire is weaker, the claim does not hold. μ_1 is called the **alternative** (or *alternative hypothesis*) of the test. We shall **accept** the hypothesis if the test suggests that it is true, except for a small error probability α, called the **significance level** of the test. Otherwise we **reject** the hypothesis. Hence α is the probability of rejecting a hypothesis although it is true. The choice of α is up to us. 5% and 1% are popular values.

For the test we need a sample. We randomly select 25 coils of the wire, cut a piece from each coil, and determine the breaking limit experimentally. Suppose that this sample of $n = 25$ values of the breaking limit has the mean $\bar{x} = 197$ lb (somewhat less than the claim!) and the standard deviation $s = 6$ lb.

At this point we could only speculate whether this difference $197 - 200 = -3$ is due to randomness, is a chance effect, or whether it is **significant,** due to the actually inferior quality of the wire. To continue beyond speculation requires probability theory, as follows.

We assume that the breaking limit is normally distributed. (This assumption could be tested by the method in Sec. 23.7. Or we could remember the central limit theorem (Sec. 23.3) and take a still larger sample.) Then

$$T = \frac{\bar{X} - \mu_0}{S/\sqrt{n}}$$

in (11), Sec. 23.3, with $\mu = \mu_0$ has a *t*-**distribution** with $n - 1$ degrees of freedom ($n - 1 = 24$ for our sample). Also $\bar{x} = 197$ and $s = 6$ are observed values of $\bar{X}$ and S to be used later. We can now choose a significance level, say, $\alpha = 5\%$. From Table A10 in Appendix 5 we then obtain a critical value c such that $P(T \le c) = \alpha = 5\%$. For $P(T \le \tilde{c}) = 1 - \alpha = 95\%$ the table gives $\tilde{c} = 1.71$, so that $c = -\tilde{c} = -1.71$ because of the symmetry of the distribution (Fig. 496).

---

[3]A systematic theory of tests was developed by the American statistician JERZY NEYMAN (1894—1981) and the English statistician EGON SHARPE PEARSON (1895—1980), the son of Karl Pearson (see footnote 6 near the end of this section), beginning around 1930.

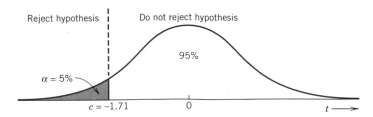

Fig. 496.　t-distribution in Example 1

We now reason as follows—this is the ***crucial idea*** of the test. If the hypothesis is true, we have a chance of only α (= 5%) that we observe a value t of T (calculated from a sample) that will fall between $-\infty$ and -1.71. Hence if we nevertheless do observe such a t, we assert that the hypothesis cannot be true and we reject it. Then we accept the alternative. If, however, $t \geqq c$, we accept the hypothesis.

A simple calculation finally gives $t = (197 - 200)/(6/\sqrt{25}) = -2.5$ as an observed value of T. Since $-2.5 < -1.71$, we reject the hypothesis (the manufacturer's claim) and accept the alternative $\mu = \mu_1 < 200$.　　　　　　　　　　　　　　　◀

This example illustrates the **steps of a test:**

1. Formulate the hypothesis $\theta = \theta_0$ to be tested. ($\theta_0 = \mu_0$ in the example.)

2. Formulate an alternative $\theta = \theta_1$. ($\theta_1 = \mu_1$ in the example.)

3. Choose a significance level α (5%, 1%, 0.1%).

4. Use a random variable $\hat{\Theta} = g(X_1, \cdots, X_n)$ whose distribution depends on the hypothesis and on the alternative and is known in both cases. Determine a critical value c from the distribution of $\hat{\Theta}$, assuming the hypothesis to be true. ($\hat{\Theta} = T$, c from $P(T \leqq c) = \alpha$ in the example.)

5. Use a sample $x_1, \cdots, x_n$ to determine an observed value $\hat{\theta} = g(x_1, \cdots, x_n)$ of $\hat{\Theta}$. (t in the example.)

6. Accept or reject the hypothesis, depending on the size of $\hat{\theta}$ relative to c. ($t < c$ in the example, rejection of the hypothesis.)

Two important facts require further discussion and careful attention. The first is the choice of an alternative. In the example, $\mu_1 < \mu_0$, but other applications may require $\mu_1 > \mu_0$ or $\mu_1 \neq \mu_0$. The second fact has to do with errors. We know that α (the significance level of the test) is the probability of rejecting a true hypothesis. And we shall discuss the probability β of accepting a false hypothesis.

Kinds of Alternatives (Fig. 497 on Page 1120)

Let θ be an unknown parameter in a distribution, and suppose that we want to test the hypothesis $\theta = \theta_0$. Then there are three main kinds of alternatives, namely,

(1)　　　　　　　　　　　　　　　$\theta > \theta_0$

(2)　　　　　　　　　　　　　　　$\theta < \theta_0$

(3)　　　　　　　　　　　　　　　$\theta \neq \theta_0$.

(1) and (2) are **one-sided alternatives,** and (3) is a **two-sided alternative.**

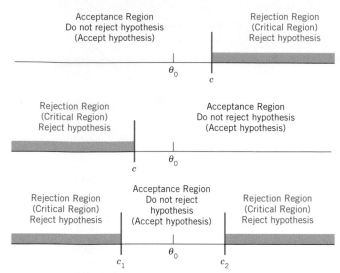

Fig. 497. Test in the case of alternative (1) (upper part of the figure), alternative (2) (middle part), and alternative (3)

In (1) the critical c lies to the right of θ_0 because the alternative lies to the right of θ_0; see Fig. 497, upper part. Hence the rejection region extends to the right. This is called a **right-sided test.** In (2) the critical c lies to the left of θ_0 (as in Example 1), the rejection region extends to the left, and we have a **left-sided test** (Fig. 497, middle part). These are **one-sided tests.** In (3) we have two rejection regions. This is called a **two-sided test** (Fig. 497, lower part).

All three kinds of alternatives occur in practical problems. For example, (1) may arise if θ_0 is the maximum tolerable inaccuracy of a voltmeter or some other instrument. Alternative (2) may occur in testing strength of material, as in Example 1. Finally, θ_0 in (3) may be the diameter of axle-shafts, and shafts that are too thin or too thick are equally undesirable, so that we have to watch for deviations in both directions.

Types of Errors in Tests

Tests always involve **risks of making false decisions:**

(I) Rejecting a true hypothesis (**Type I error**). α = probability of making a Type I error.

(II) Accepting a false hypothesis (**Type II error**). β = probability of making a Type II error.

Clearly, we cannot avoid these errors because no absolutely certain conclusions about populations can be drawn from samples. But we show that there are ways and means of choosing suitable levels of risks, that is, of values α and β. The choice of α depends on the nature of the problem (e.g., a small risk $\alpha = 1\%$ is used if it is a matter of life or death).

Let us discuss this systematically for a test of a hypothesis $\theta = \theta_0$ against an alternative that is a single number[4] θ_1, for simplicity. We let $\theta_1 > \theta_0$, so that we have a right-sided

---

[4]This standard notation has absolutely nothing to do with the use of the notation θ_1 in connection with confidence intervals in Sec. 23.3.

test. For a left-sided or a two-sided test the discussion is quite similar.

We choose a critical $c > \theta_0$ (as in the upper part of Fig. 497, by methods discussed below). From a given sample $x_1, \cdots, x_n$ we then compute a value

$$\hat{\theta} = g(x_1, \cdots, x_n)$$

with a suitable g (whose choice will be a main point of our further discussion; for instance, take $g = (x_1 + \cdots + x_n)/n$ in the case in which θ is the mean). If $\hat{\theta} > c$, we reject the hypothesis. If $\hat{\theta} \leq c$, we accept it. Here, the value $\hat{\theta}$ can be regarded as an observed value of the random variable

(4) $$\hat{\Theta} = g(X_1, \cdots, X_n)$$

because x_j may be regarded as an observed value of X_j, $j = 1, \cdots, n$. In this test there are two possibilities of making an error, as follows.

Type I error (see Table 23.4). The hypothesis is true but is rejected (hence the alternative is accepted) because Θ assumes a value $\hat{\theta} > c$. Obviously, the probability of making such an error equals

(5) $$\boxed{P(\hat{\Theta} > c)_{\theta = \theta_0} = \alpha.}$$

α is called the **significance level** of the test, as mentioned before.

Type II error (see Table 23.4). The hypothesis is false but is accepted because $\hat{\Theta}$ assumes a value $\hat{\theta} \leq c$. The probability of making such an error is denoted by β; thus

(6) $$\boxed{P(\hat{\Theta} \leq c)_{\theta = \theta_1} = \beta.}$$

$\eta = 1 - \beta$ is called the **power** of the test. Obviously, this is the probability of avoiding a Type II error.

Formulas (5) and (6) show that both α and β depend on c, and we would like to choose c so that these probabilities of making errors are as small as possible. But Fig. 498 shows that these are conflicting requirements because to let α decrease we must shift c to the right, but then β increases. In practice we first choose α (5%, sometimes 1%), then determine c, and finally compute β. If β is large so that the power $\eta = 1 - \beta$ is small, we should repeat the test, choosing a larger sample, for reasons that will appear shortly.

Table 23.4
Type I and Type II Errors in Testing a Hypothesis $\theta = \theta_0$
Against an Alternative $\theta = \theta_1$

		Unknown Truth	
		$\theta = \theta_0$	$\theta = \theta_1$
Accepted	$\theta = \theta_0$	True decision $P = 1 - \alpha$	Type II error $P = \beta$
	$\theta = \theta_1$	Type I error $P = \alpha$	True decision $P = 1 - \beta$

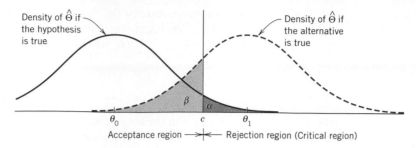

Density of $\hat{\Theta}$ if the hypothesis is true

Density of $\hat{\Theta}$ if the alternative is true

Acceptance region $\longrightarrow$|$\longleftarrow$ Rejection region (Critical region)

Fig. 498.   Illustration of Type I and II errors in testing a hypothesis $\theta = \theta_0$ against an alternative $\theta = \theta_1$ ($> \theta_0$, right-sided test)

If the alternative is not a single number but is of the form (1)–(3), then β becomes a function of θ. This function $\beta(\theta)$ is called the **operating characteristic** (OC) of the test and its curve the **OC curve.** Clearly, in this case $\eta = 1 - \beta$ also depends on θ. This function $\eta(\theta)$ is called the **power function** of the test. (Examples will follow.)

Of course, from a test that leads to the acceptance of a certain hypothesis θ_0, it does **not** follow that this is the only possible hypothesis or the best possible hypothesis. Hence the terms **"not reject"** or **"fail to reject"** are perhaps better than the term **"accept."**

Test for μ of the Normal Distribution with Known σ^2

The following examples will explain tests of practically important hypotheses.

EXAMPLE 2   **Test for the mean of the normal distribution with known variance**

Let X be a normal random variable with variance $\sigma^2 = 9$. Using a sample of size $n = 10$ with mean $\bar{x}$, test the hypothesis $\mu = \mu_0 = 24$ against the three kinds of alternatives, namely,

$$(a) \quad \mu > \mu_0 \qquad (b) \quad \mu < \mu_0 \qquad (c) \quad \mu \neq \mu_0.$$

Solution.   We choose the significance level $\alpha = 0.05$. An estimate of the mean will be obtained from

$$\bar{X} = \frac{1}{n}(X_1 + \cdots + X_n).$$

If the hypothesis is true, $\bar{X}$ is normal with mean $\mu = 24$ and variance $\sigma^2/n = 0.9$, see Theorem 1, Sec. 23.3. Hence we may obtain the critical value c from Table A8 in Appendix 5.

Case (a).   We determine c from $P(\bar{X} > c)_{\mu=24} = \alpha = 0.05$, that is,

$$P(\bar{X} \leq c)_{\mu=24} = \Phi\left(\frac{c - 24}{\sqrt{0.9}}\right) = 1 - \alpha = 0.95.$$

Table A8 in Appendix 5 gives $(c - 24)/\sqrt{0.9} = 1.645$, and $c = 25.56$, which is greater than μ_0, as in the upper part of Fig. 497. If $\bar{x} \leq 25.56$, the hypothesis is accepted. If $\bar{x} > 25.56$, it is rejected. The power of the test is (Fig. 499)

$$\eta(\mu) = P(\bar{X} > 25.56)_\mu = 1 - P(\bar{X} \leq 25.56)_\mu$$

(7)

$$= 1 - \Phi\left(\frac{25.56 - \mu}{\sqrt{0.9}}\right) = 1 - \Phi(26.94 - 1.05\mu).$$

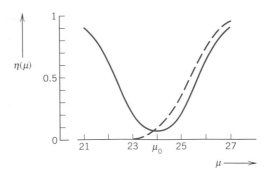

Fig. 499. Power $\eta(\mu)$ in Example 2, case (a) (dashed)
and case (c)

Case (b). The critical value c is obtained from the equation

$$P(\overline{X} \leq c)_{\mu=24} = \Phi\left(\frac{c-24}{\sqrt{0.9}}\right) = \alpha = 0.05.$$

Table A8 in Appendix 5 yields $c = 24 - 1.56 = 22.44$. If $\bar{x} \geq 22.44$, we accept the hypothesis. If $\bar{x} < 22.44$, we reject it. The power of the test is

$$(8) \qquad\qquad \eta(\mu) = P(\overline{X} \leq 22.44)_{\mu} = \Phi\left(\frac{22.44-\mu}{\sqrt{0.9}}\right) = \Phi(23.65 - 1.05\mu).$$

Case (c). Since the normal distribution is symmetric, we choose c_1 and c_2 equidistant from $\mu = 24$, say, $c_1 = 24 - k$ and $c_2 = 24 + k$, and determine k from

$$P(24 - k \leq \overline{X} \leq 24 + k)_{\mu=24} = \Phi\left(\frac{k}{\sqrt{0.9}}\right) - \Phi\left(-\frac{k}{\sqrt{0.9}}\right) = 1 - \alpha = 0.95.$$

Table A8 in Appendix 5 gives $k/\sqrt{0.9} = 1.960$, $k = 1.86$. Hence $c_1 = 24 - 1.86 = 22.14$ and $c_2 = 24 + 1.86 = 25.86$. If $\bar{x}$ is not smaller than c_1 and not greater than c_2, we accept the hypothesis. Otherwise we reject it. The power of the test is (Fig. 499)

$$\eta(\mu) = P(\overline{X} < 22.14)_{\mu} + P(\overline{X} > 25.86)_{\mu} = P(\overline{X} < 22.14)_{\mu} + 1 - P(\overline{X} \leq 25.86)_{\mu}$$

$$(9) \qquad\qquad = 1 + \Phi\left(\frac{22.14-\mu}{\sqrt{0.9}}\right) - \Phi\left(\frac{25.86-\mu}{\sqrt{0.9}}\right)$$

$$= 1 + \Phi(23.34 - 1.05\mu) - \Phi(27.26 - 1.05\mu).$$

Consequently, the operating characteristic $\beta(\mu) = 1 - \eta(\mu)$ (see before) is (Fig. 500 on the next page)

$$\beta(\mu) = \Phi(27.26 - 1.05\mu) - \Phi(23.34 - 1.05\mu).$$

If we take a larger sample, say, of size $n = 100$ (instead of 10), then $\sigma^2/n = 0.09$ (instead of 0.9) and the critical values are $c_1 = 23.41$ and $c_2 = 24.59$, as can be readily verified. Then the operating characteristic of the test is

$$\beta(\mu) = \Phi\left(\frac{24.59-\mu}{\sqrt{0.09}}\right) - \Phi\left(\frac{23.41-\mu}{\sqrt{0.09}}\right)$$

$$= \Phi(81.97 - 3.33\mu) - \Phi(78.03 - 3.33\mu).$$

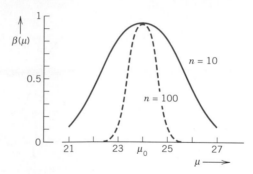

Fig. 500.   Curves of the operating characteristic (OC curves)
in Example 2, case (c), for two different sample sizes n

Figure 500 shows that the corresponding OC curve is steeper than that for $n = 10$. This means that the increase of n has led to an improvement of the test. In any practical case, n is chosen as small as possible but so large that the test brings out deviations between μ and μ_0 that are of practical interest. For instance, if deviations of ± 2 units are of interest, we see from Fig. 500 that $n = 10$ is much too small because when $\mu = 24 - 2 = 22$ or $\mu = 24 + 2 = 26$, β is almost 50%. On the other hand, we see that $n = 100$ is sufficient for that purpose. ◀

Tests for μ when σ^2 is Unknown, and for σ^2

EXAMPLE 3   **Test for the mean of the normal distribution with unknown variance**

The tensile strength of a sample of $n = 16$ manila ropes (diameter 3 in.) was measured. The sample mean was $\bar{x} = 4482$ kg, and the sample standard deviation was $s = 115$ kg (N. C. Wiley, 41st Annual Meeting of the American Society for Testing Materials). Assuming that the tensile strength is a normal random variable, test the hypothesis $\mu_0 = 4500$ kg against the alternative $\mu_1 = 4400$ kg. Here μ_0 may be a value given by the manufacturer, while μ_1 may result from previous experience.

Solution.   We choose the significance level $\alpha = 5\%$. If the hypothesis is true, it follows from Theorem 2 in Sec. 23.3, that the random variable

$$T = \frac{\bar{X} - \mu_0}{S/\sqrt{n}} = \frac{\bar{X} - 4500}{S/4}$$

has a t-distribution with $n - 1 = 15$ degrees of freedom. The test is left-sided. The critical value c is obtained from $P(T < c)_{\mu_0} = \alpha = 0.05$. Table A9 in Appendix 5 gives $c = -1.75$. As an observed value of T we obtain from the sample $t = (4482 - 4500)/(115/4) = -0.626$. We see that $t > c$ and accept the hypothesis. For obtaining numerical values of the power of the test, we would need tables called noncentral Student t-tables; we shall not discuss this question here. ◀

EXAMPLE 4   **Test for the variance of the normal distribution**

Using a sample of size $n = 15$ and sample variance $s^2 = 13$ from a normal population, test the hypothesis $\sigma^2 = \sigma_0^2 = 10$ against the alternative $\sigma^2 = \sigma_1^2 = 20$.

Solution.   We choose the significance level $\alpha = 5\%$. If the hypothesis is true, then

$$Y = (n - 1) \frac{S^2}{\sigma_0^2} = 14 \frac{S^2}{10} = 1.4 S^2$$

has a chi-square distribution with $n - 1 = 14$ degrees of freedom, by Theorem 3, Sec. 23.3. From

$$P(Y > c) = \alpha = 0.05; \qquad \text{that is,} \qquad P(Y \leq c) = 0.95,$$

and Table A10 in Appendix 5 with 14 degrees of freedom we obtain $c = 23.68$. This is the critical value of Y. Hence to $S^2 = \sigma_0^2 Y/(n - 1) = 0.714 Y$ there corresponds the critical value $c^* = 0.714 \cdot 23.68 = 16.91$. Since $s^2 < c^*$, we accept the hypothesis.

If the alternative is true, the variable

$$Y_1 = 14 \frac{S^2}{\sigma_1^2} = 0.7 S^2$$

has a chi-square distribution with 14 degrees of freedom. Hence our test has the power

$$\eta = P(S^2 > c^*)_{\sigma^2=20} = P(Y_1 > 0.7c^*)_{\sigma^2=20} = 1 - P(Y_1 \leqq 11.84)_{\sigma^2=20}.$$

From a more extensive table of the chi-square distribution (e. g. in Ref. [G3] or [G8]) we see that $\eta \approx 62\%$. Hence the Type II risk is very large, namely, 38%. To make this risk smaller, we would have to increase the sample size.　◀

Comparison of Means and of Variances

EXAMPLE 5　　**Comparison of the means of two normal distributions**

Using a sample $x_1, \cdots, x_{n_1}$ from a normal distribution with unknown mean μ_1 and a sample $y_1, \cdots, y_{n_2}$ from another normal distribution with unknown mean μ_2, we want to test the hypothesis that the means are equal, $\mu_1 = \mu_2$, against an alternative, say, $\mu_1 > \mu_2$. The variances need not be known but are assumed to be equal.[5] Two cases are of practical importance:

Case A.　*The samples have the **same size**. Furthermore, each value of the first sample corresponds to precisely one value of the other,* because corresponding values result from the same person or thing (**paired comparison**)— for example, two measurements of the same thing by two different methods or two measurements from the two eyes of the same animal. More generally, they may result from pairs of *similar* individuals or things, for example, identical twins, pairs of used front tires from the same car, etc. Then we should form the differences of corresponding values and test the hypothesis that the population corresponding to the differences has mean 0, using the method in Example 3. If we have a choice, this method is better than the following.

Case B.　*The two samples are independent and not necessarily of the same size.* Then we may proceed as follows. Suppose that the alternative is $\mu_1 > \mu_2$. We choose a significance level α. Then we compute the sample means $\bar{x}$ and $\bar{y}$ as well as $(n_1 - 1)s_x^2$ and $(n_2 - 1)s_y^2$, where s_x^2 and s_y^2 are the sample variances. Using Table A9 in Appendix 5 with $n_1 + n_2 - 2$ degrees of freedom, we now determine c from

$$(10) \qquad\qquad P(T \leqq c) = 1 - \alpha.$$

We finally compute

$$(11) \qquad\qquad t_0 = \sqrt{\frac{n_1 n_2 (n_1 + n_2 - 2)}{n_1 + n_2}} \; \frac{\bar{x} - \bar{y}}{\sqrt{(n_1 - 1)s_x^2 + (n_2 - 1)s_y^2}} \, .$$

It can be shown that this is an observed value of a random variable that has a t-distribution with $n_1 + n_2 - 2$ degrees of freedom, provided the hypothesis is true. If $t_0 \leqq c$, the hypothesis is accepted. If $t_0 > c$, it is rejected.
If the alternative is $\mu_1 \neq \mu_2$, then (10) must be replaced by

$$(10^*) \qquad\qquad P(T \leqq c_1) = 0.5\alpha, \qquad\qquad P(T \leqq c_2) = 1 - 0.5\alpha.$$

Note that for samples of equal size $n_1 = n_2 = n$, formula (11) reduces to

$$(12) \qquad\qquad t_0 = \sqrt{n} \, \frac{\bar{x} - \bar{y}}{\sqrt{s_x^2 + s_y^2}} \, .$$

---

[5]If the test in the next example shows that the variances differ significantly, then choose two not too small samples of the same size $n_1 = n_2 = n$ (> 30, say), use the fact that (12) is an observed value of an approximately normal random variable with mean 0 and variance 1, and proceed as in Example 2.

To illustrate the computations, let us consider the two samples

| 105 | 108 | 86 | 103 | 103 | 107 | 124 | 105 |

and

| 89 | 92 | 84 | 97 | 103 | 107 | 111 | 97 |

showing the relative output of tin plate workers under two different working conditions [J. J. B. Worth, *Journal of Industrial Engineering* **9**, 249–253). Assuming that the corresponding populations are normal and have the same variance, let us test the hypothesis $\mu_1 = \mu_2$ against the alternative $\mu_1 \neq \mu_2$. (Equality of variances will be tested in the next example.)

Solution. We find

$$\bar{x} = 105.125, \qquad \bar{y} = 97.500, \qquad s_x^2 = 106.125, \qquad s_y^2 = 84.000.$$

We choose the significance level $\alpha = 5\%$. From (10*) with $0.5\alpha = 2.5\%$, $1 - 0.5\alpha = 97.5\%$ and Table A9 in Appendix 5 with 14 degrees of freedom we obtain $c_1 = -2.14$ and $c_2 = 2.14$. Formula (12) with $n = 8$ gives the value

$$t_0 = \sqrt{8} \cdot 7.625/\sqrt{190.125} = 1.56.$$

Since $c_1 \leqq t_0 \leqq c_2$, we accept the hypothesis $\mu_1 = \mu_2$ that under both conditions the mean output is the same.

Case A applies to the example because the two first sample values correspond to a certain type of work, the next two were obtained in another kind of work, etc. So we may use the differences

| 16 | 16 | 2 | 6 | 0 | 0 | 13 | 8 |

of corresponding sample values and the method in Example 3 to test the hypothesis $\mu = 0$, where μ is the mean of the population corresponding to the differences. As a logical alternative we take $\mu \neq 0$. The sample mean is $\bar{d} = 7.625$, and the sample variance is $s^2 = 45.696$. Hence

$$t = \sqrt{8}\,(7.625 - 0)/\sqrt{45.696} = 3.19.$$

From $P(T \leqq c_1) = 2.5\%$, $P(T \leqq c_2) = 97.5\%$ and Table A9 in Appendix 5 with $n - 1 = 7$ degrees of freedom we obtain $c_1 = -2.36$, $c_2 = 2.36$ and reject the hypothesis because $t = 3.19$ does not lie between c_1 and c_2. Hence our present test, in which we used more information (but the same samples), shows that the difference in output is significant. ◄

EXAMPLE 6 **Comparison of the variances of two normal distributions**

Using the two samples in the last example, test the hypothesis $\sigma_1^2 = \sigma_2^2$; assume that the corresponding populations are normal and the nature of the experiment suggests the alternative $\sigma_1^2 > \sigma_2^2$.

Solution. We find $s_x^2 = 106.125$, $s_y^2 = 84.000$. We choose the significance level $\alpha = 5\%$. Using $P(V \leqq c) = 1 - \alpha = 95\%$ and Table A11 in Appendix 5, with $(n_1 - 1, n_2 - 1) = (7, 7)$ degrees of freedom, we determine $c = 3.79$. We finally compute $v_0 = s_x^2 / s_y^2 = 1.26$. Since $v_0 \leqq c$, we accept the hypothesis. If $v_0 > c$, we would reject it.

This test is justified by the fact that v_0 is an observed value of a random variable that has a so-called **F-distribution** with $(n_1 - 1, n_2 - 1)$ degrees of freedom, provided the hypothesis is true. (Proof in Ref. [G3] listed in Appendix 1.) The F-distribution with (m, n) degrees of freedom was introduced by R. A. Fisher[6] and has the distribution function $F(z) = 0$ if $z < 0$ and

$$\textbf{(13)} \qquad F(z) = K_{mn} \int_0^z t^{(m-2)/2} (mt + n)^{-(m+n)/2}\, dt \qquad (z \geqq 0),$$

where $K_{mn} = m^{m/2} n^{n/2} \Gamma(\frac{1}{2}m + \frac{1}{2}n) / \Gamma(\frac{1}{2}m)\Gamma(\frac{1}{2}n)$. (For Γ see Appendix A3.1.) ◄

---

[6]After the pioneering work of the English statistician and biologist, KARL PEARSON (1857—1936), the founder of the English school of statistics, and W. S. GOSSET (see the footnote in Sec. 23.3), the English statistician Sir RONALD AYLMER FISHER (1890—1962), professor of eugenics in London (1933—1943) and professor of genetics in Cambridge, England (1943—1957), had great influence on the further development of modern statistics.

This long section contained the basic ideas and concepts of testing, along with typical applications, and you may perhaps want to review it quickly before going on, because the next sections concern an adaption of these ideas to tasks of great practical importance and resulting tests in connection with quality control, acceptance (or rejection) of goods produced, and so on.

PROBLEM SET 23.4

1. Test $\mu = 0$ against $\mu > 0$, assuming normality and using the sample 1, -1, 1, 3, -8, 6, 0 (deviations of the azimuth [multiples of 0.01 radian] in some revolution of a satellite). Choose $\alpha = 5\%$.

2. Using the data of Buffon in Table 22.1 (Sec. 22.3), test the hypothesis that the coin is fair against the alternative that heads are more likely than tails. (Choose $\alpha = 5\%$.)

3. Do the same test as in Prob. 2, using the data of Pearson in Table 22.1.

4. Assuming normality and known variance $\sigma^2 = 9$, test the hypothesis $\mu = 60.0$ against the alternative $\mu = 57.0$ using a sample of size 20 with mean $\bar{x} = 58.05$ and choosing $\alpha = 5\%$.

5. How does the result in Prob. 4 change if we use a smaller sample, say, of size 5, the other data ($\bar{x} = 58.05$, $\alpha = 5\%$, etc.) remaining as before?

6. Determine the power of the test in Prob. 4.

7. What is the rejection region in Prob. 4 in the case of a two-sided test with $\alpha = 5\%$?

8. Sketch a figure similar to Fig. 498 for a left-sided test and a two-sided test.

9. Verify the calculations for Fig. 500. If you have a CAS, plot Fig. 500 and further OC curves for values of n smaller than 10 and larger than 100.

10. If a sample of 25 tires of a certain kind has a mean life of 37 000 miles and a standard deviation of 5000 miles, can the manufacturer claim that the true mean life of such tires is greater than 35 000 miles? Set up and test a corresponding hypothesis at the 5% level, assuming normality.

11. A firm sells oil in cans containing 5000 g oil per can and is interested to know whether the mean weight differs significantly from 5000 g at the 5% level, in which case the filling machine has to be adjusted. Set up a hypothesis and an alternative and perform the test, assuming normality and using a sample of 50 fillings with mean 4990 g and standard deviation 20 g.

12. If a standard medication cures about 75% of patients with a certain disease and a new medication cured 310 of the first 400 patients on whom it was tried, can we conclude that the new medication is better? Choose $\alpha = 5\%$. First guess. Then calculate.

13. If simultaneous measurements of electric voltage by two different types of voltmeter yield the differences (in volts) 0.4, -0.6, 0.2, 0.0, 1.0, 1.4, 0.4, 1.6, can we assert at the 5% level that there is no significant difference in the calibration of the two types of instruments? (Assume normality.)

14. Suppose that in operating battery-powered electrical equipment, it is less expensive to replace all batteries at fixed intervals than to replace each battery individually when it breaks down, provided the standard deviation of the lifetime is less than a certain limit, say, less than 5 hours. Set up and apply a suitable test, using a sample of 28 values of lifetimes with standard deviation $s = 3.5$ hours and assuming normality; choose $\alpha = 5\%$.

15. Suppose that in the past the standard deviation of weights of certain 100.0-oz packages filled by a machine was 0.8 oz. Test the hypothesis H_0: $\sigma = 0.8$ against the alternative H_1: $\sigma > 0.8$ (an undesirable increase), using a sample of 20 packages with standard deviation 1.0 oz and assuming normality. (Choose $\alpha = 5\%$.)

16. The two samples 70, 80, 30, 70, 60, 80 and 140, 120, 130, 120, 120, 130, 120 are values of the differences of temperatures (°C) of iron at two stages of casting, taken from two different crucibles. Is the variance of the first population larger than that of the second? (Assume normality. Choose $\alpha = 5\%$.)

17. Brand *A* gasoline was used in 16 similar automobiles under identical conditions. The corresponding sample of 16 values (miles per gallon) had mean 19.6 and standard deviation 0.4. Under the same conditions, high-power brand *B* gasoline gave a sample of 16 values with mean 20.2 and standard deviation 0.6. Is the mileage of *B* significantly better than that of *A?* (Test at the 5% level; assume normality. First guess. Then calculate.)

18. Test for equality of population means against the alternative that the means are different, assuming normality, choosing $\alpha = 5\%$ and using two samples of sizes 12 and 18, with means 10 and 14, respectively, and equal standard deviation 3.

19. Show that for a normal distribution the two types of errors in a test of a hypothesis H_0: $\mu = \mu_0$ against an alternative H_1: $\mu = \mu_1$ can be made as small as one pleases (not zero) by taking the sample sufficiently large.

20. Graph the OC curves in Example 2, cases (*a*) and (*b*).

23.5 Quality Control

The ideas on testing can be adapted and extended in various ways to serve basic practical needs in engineering and other fields. We show this in the remaining sections for some of the most important tasks solvable by statistical methods. As a first such area of problems, we discuss industrial quality control.

No production process is so perfect that all the products are completely alike. There is always a small variation that is caused by a great number of small, uncontrollable factors and must therefore be regarded as a chance variation. It is important to make sure that the products have required values (for example, length, strength, or whatever property may be essential in a particular case). For this purpose one makes a test of the hypothesis that the products have the required property, say, $\mu = \mu_0$, where μ_0 is a required value. If this is done after an entire lot has been produced (for example, a lot of 100,000 screws), the test will tell us how good or how bad the products are, but it is obviously too late to alter undesirable results. It is much better to test during the production run. This is done at regular intervals of time (for example, every hour or half-hour) and is called **quality control.** Each time a sample of the same size is taken, in practice 3 to 10 items. If the hypothesis is rejected, we stop the production and look for the cause of the trouble.

If we stop the production process even though it is progressing properly, we make a Type I error. If we do not stop the process even though something is not in order, we make a Type II error (see Sec. 23.4). The result of each test is marked in graphical form on what is called a **control chart.** This was proposed by W. A. Shewhart in 1924 and makes quality control particularly effective.

Control Chart for the Mean

An illustration and example of a control chart is given in the upper part of Fig. 501. This control chart for the mean shows the **lower control limit** LCL, the **center control line** CL, and the **upper control limit** UCL. The two **control limits** correspond to the critical values c_1 and c_2 in case (*c*) of Example 2 in Sec. 23.4. As soon as a sample mean falls outside the range between the control limits, we reject the hypothesis and assert that the production process is "out of control"; that is, we assert that there has been a shift in process level. Action is called for whenever a point exceeds the limits.

If we choose control limits that are too loose, we shall not detect process shifts. On the other hand, if we choose control limits that are too tight, we shall be unable to run the process because of frequent searches for nonexistent trouble. The usual significance level is $\alpha = 1\%$. From Theorem 1 in Sec. 23.3 and Table A8 in Appendix 5 we see that in the case of the normal distribution the corresponding control limits for the mean are

$$
\text{(1)} \qquad \text{LCL} = \mu_0 - 2.58 \, \frac{\sigma}{\sqrt{n}} \, , \qquad\qquad \text{UCL} = \mu_0 + 2.58 \, \frac{\sigma}{\sqrt{n}} \, .
$$

Here σ is assumed to be known. If σ is unknown, we may compute the standard deviations of the first 20 or 30 samples and take their arithmetic mean as an approximation of σ. The broken line connecting the means in Fig. 501 is merely to display the results.

Additional, more subtle controls are often used in industry. For instance, one observes the motions of the sample means above and below the centerline, which should happen frequently. Accordingly, long runs (conventionally of length 7 or more) of means all above (or all below) the centerline could indicate trouble.

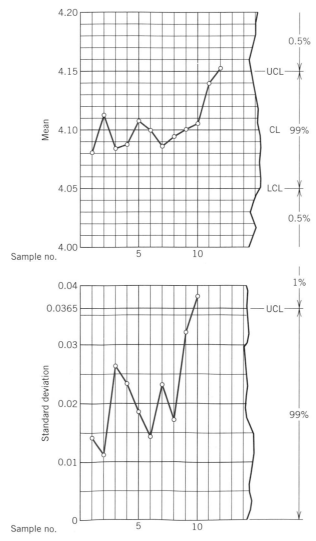

Fig. 501. Control charts for the mean (upper part of figure) and the standard deviation in the case of the sample in Table 23.5, p. 1130

Table 23.5
Twelve Samples of Five Values Each
(Diameter of Small Cylinders, Measured in Millimeters)

Sample Number	Sample Values					$\bar{x}$	s	R
1	4.06	4.08	4.08	4.08	4.10	4.080	0.014	0.04
2	4.10	4.10	4.12	4.12	4.12	4.112	0.011	0.02
3	4.06	4.06	4.08	4.10	4.12	4.084	0.026	0.06
4	4.06	4.08	4.08	4.10	4.12	4.088	0.023	0.06
5	4.08	4.10	4.12	4.12	4.12	4.108	0.018	0.04
6	4.08	4.10	4.10	4.10	4.12	4.100	0.014	0.04
7	4.06	4.08	4.08	4.10	4.12	4.088	0.023	0.06
8	4.08	4.08	4.10	4.10	4.12	4.096	0.017	0.04
9	4.06	4.08	4.10	4.12	4.14	4.100	0.032	0.08
10	4.06	4.08	4.10	4.12	4.16	4.104	0.038	0.10
11	4.12	4.14	4.14	4.14	4.16	4.140	0.014	0.04
12	4.14	4.14	4.16	4.16	4.16	4.152	0.011	0.02

Control Chart for the Variance

In addition to the mean, one often controls the variance, the standard deviation, or the range. To set up a control chart for the variance in the case of a normal distribution, we may employ the method in Example 4 of Sec. 23.4 for determining control limits. It is customary to use only one control limit, namely, an upper control limit. Now from Example 4 of Sec. 23.4 we have $S^2 = \sigma_0^2 Y/(n - 1)$, where because of our normality assumption the random variable Y has a chi-square distribution with $n - 1$ degrees of freedom. Hence the desired control limit is

$$(2) \qquad \boxed{\text{UCL} = \frac{\sigma^2 c}{n - 1}}$$

where c is obtained from the equation

$$P(Y > c) = \alpha, \qquad \text{that is,} \qquad P(Y \leq c) = 1 - \alpha$$

and the table of the chi-square distribution (Table A10 in Appendix 5) with $n - 1$ degrees of freedom; here α (5% or 1%, say) is the probability that in a properly running process an observed value s^2 of S^2 is greater than the upper control limit.

If we wanted a control chart for the variance with both an upper control limit UCL and a lower control limit LCL, these limits would be

$$(3) \qquad \text{LCL} = \frac{\sigma^2 c_1}{n - 1} \qquad \text{and} \qquad \text{UCL} = \frac{\sigma^2 c_2}{n - 1}$$

where c_1 and c_2 are obtained from the equations

(4) $$P(Y \leqq c_1) = \frac{\alpha}{2} \quad \text{and} \quad P(Y \leqq c_2) = 1 - \frac{\alpha}{2}$$

and Table A10 in Appendix 5 with $n - 1$ degrees of freedom.

Control Chart for the Standard Deviation

Similarly, to set up a control chart for the standard deviation, we need an upper control limit

(5) $$\text{UCL} = \frac{\sigma \sqrt{c}}{\sqrt{n-1}}$$

obtained from (2). For example, in Table 23.5 we have $n = 5$. Assuming that the corresponding population is normal with standard deviation $\sigma = 0.02$ and choosing $\alpha = 1\%$, we obtain from the equation

$$P(Y \leqq c) = 1 - \alpha = 99\%$$

and Table A10 in Appendix 5 with 4 degrees of freedom the critical value $c = 13.28$ and from (5) the corresponding value

$$\text{UCL} = \frac{0.02\sqrt{13.28}}{\sqrt{4}} = 0.0365,$$

which is shown in the lower part of Fig. 501.

A control chart for the standard deviation with both an upper and a lower control limit is obtained from (3).

Control Chart for the Range

Instead of the variance or standard deviation, one often controls the **range** R (= largest sample value minus smallest sample value). It can be shown that in the case of the normal distribution, the standard deviation σ is proportional to the expectation of the random variable R^* for which R is an observed value, say, $\sigma = \lambda_n E(R^*)$, where the factor of proportionality λ_n depends on the sample size n and has the values

n	2	3	4	5	6	7	8	9	10
$\lambda_n = \sigma/E(R^*)$	0.89	0.59	0.49	0.43	0.40	0.37	0.35	0.34	0.32

n	12	14	16	18	20	30	40	50
$\lambda_n = \sigma/E(R^*)$	0.31	0.29	0.28	0.28	0.27	0.25	0.23	0.22

Since R depends on two sample values only, it gives less information about a sample than s does. Clearly, the larger the sample size n is, the more information we lose in using R instead of s. A practical rule is to use s when n is larger than 10.

PROBLEM SET 23.5

1. Suppose a machine for filling cans with lubricating oil is set so that it will generate fillings which form a normal population with mean 1 gal and standard deviation 0.02 gal. Set up a control chart of the type shown in Fig. 501 for controlling the mean (that is, find LCL and UCL), assuming that the sample size is 4.

2. **(Three-sigma control chart)** Show that in Prob. 1, the requirement of the significance level $\alpha = 0.3\%$ leads to LCL $= \mu - 3\sigma/\sqrt{n}$ and UCL $= \mu + 3\sigma/\sqrt{n}$, and find the corresponding numerical values.

3. What sample size should we choose in Prob. 1 if we want LCL and UCL somewhat closer together, say, UCL $-$ LCL $= 0.02$, without changing the significance level?

4. What effect on UCL $-$ LCL does it have if we double the sample size? If we switch from $\alpha = 1\%$ to $\alpha = 5\%$?

5. Eight samples of size 2 were taken from a lot of screws. The values (length in inches) are

Sample No.	1	2	3	4	5	6	7	8
Length	3.50	3.51	3.49	3.52	3.53	3.49	3.48	3.52
	3.51	3.48	3.50	3.50	3.49	3.50	3.47	3.49

   Assuming that the population is normal with mean 3.500 and variance 0.0004 and using (1), set up a control chart for the mean and graph the sample means on the chart.

6. Graph the means of the following 10 samples (thickness of gaskets, coded values) on a control chart, assuming that the population is normal with mean 5 and standard deviation 1.16.

Time	10:00	11:00	12:00	13:00	14:00	15:00	16:00	17:00	18:00	19:00
	5	7	7	4	5	6	5	5	3	3
Sample	2	5	3	4	6	4	5	2	4	6
values	5	4	6	3	4	6	6	5	8	6
	6	4	5	6	6	4	4	3	4	8

7. Graph the ranges of the samples in Prob. 6 on a control chart for ranges.

8. Graph $\lambda_n = \sigma/E(R^*)$ as a function of n. Why is λ_n a monotone decreasing function of n?

9. **(Number of defectives)** Find formulas for the UCL, CL and LCL (corresponding to 3σ-limits) in the case of a control chart for the number of defectives, assuming that in a state of statistical control the fraction of defectives is p.

10. **(Attribute control charts).** Fifteen samples of size 100 were taken from a production of containers. The numbers of defectives (leaking containers) in those samples (in the order observed) were

$$1 \quad 4 \quad 5 \quad 4 \quad 9 \quad 7 \quad 0 \quad 5 \quad 6 \quad 13 \quad 0 \quad 2 \quad 1 \quad 12 \quad 8$$

   From previous experience it was known that the average fraction defective is $p = 4\%$ provided that the process of production is running properly. Using the binomial distribution, set up a *fraction defective chart* (also called a **p-chart**), that is, choose the LCL $= 0$ and determine the UCL for the fraction defective (in percent) by the use of 3-sigma limits, where σ^2 is the variance of the random variable $\overline{X}$ = *Fraction defective in a sample of size* 100. Is the process in control?

11. **(Number of defects per unit)** A so-called *c-chart* or *defects-per-unit chart* is used for the control of the number X of defects per unit (for instance, the number of defects per 100 meters of paper, the number of rivets missing from an airplane wing, etc.). (a) Set up formulas for CL and LCL, UCL corresponding to $\mu \pm 3\sigma$, assuming that X has a Poisson distribution. (b) Compute CL, LCL, and UCL in a control process of the number of imperfections in sheet glass; assume that this number is 3.6 per sheet on the average when the process is in control.

12. How should we change the sample size in controlling the mean of a normal population if we want UCL $-$ LCL to decrease to half its original value?

13. Since the presence of a point outside control limits for the mean indicates trouble, how often would we be making the mistake of looking for nonexistent trouble if we used (a) 1-sigma limits, (b) 2-sigma limits? (Assume normality.)

14. What LCL and UCL should we use instead of (1) if instead of $\bar{x}$ we use the sum $x_1 + \cdots + x_n$ of the sample values? Determine these limits in the case of Fig. 501.

15. How would progressive tool wear in an automatic lathe operation be indicated by a control chart for the mean? Answer the same question for a sudden change in the position of the tool in that operation.

23.6 Acceptance Sampling

Acceptance sampling is usually done when products leave the factory (or in some cases even within the factory). The standard situation in acceptance sampling is that a **producer** supplies to a **consumer** a lot of N items (a carton of screws, for instance). The decision to **accept** or **reject** the lot is made by determining the number x of **defectives** ($=$ defective items) in a sample of size n from the lot. The lot is accepted if $x \leq c$, where c is called the **acceptance number,** giving the allowable number of defectives. If $x > c$, the consumer rejects the lot. Clearly, producer and consumer must agree on a certain **sampling plan** giving n and c.

From the hypergeometric distribution (Sec. 22.7) we see that the event A: *"Accept the lot"* has probability

$$(1) \qquad P(A) = P(X \leq c) = \sum_{x=0}^{c} \binom{M}{x} \binom{N-M}{n-x} \Big/ \binom{N}{n}$$

where M is the number of defectives in a lot of N items. In terms of the **fraction defective** $\theta = M/N$ we can write (1) as

$$(2) \qquad P(A; \theta) = \sum_{x=0}^{c} \binom{N\theta}{x} \binom{N-N\theta}{n-x} \Big/ \binom{N}{n}.$$

$P(A; \theta)$ can assume $n + 1$ values corresponding to $\theta = 0, 1/N, 2/N, \cdots, N/N$; here, n and c are fixed. A smooth curve through these points is called the **operating characteristic curve (OC curve)** of the sampling plan considered.

EXAMPLE 1   Sampling plan

Suppose that certain tool bits are packaged 20 to a box, and the following sampling plan is used. A sample of two tool bits is drawn, and the corresponding box is accepted if and only if both bits in the sample are good. In this case, $N = 20$, $n = 2$, $c = 0$, and (2) takes the form (a factor 2 drops out)

$$P(A; \theta) = \binom{20\theta}{0} \binom{20 - 20\theta}{2} \Big/ \binom{20}{2} = \frac{(20 - 20\theta)(19 - 20\theta)}{380}.$$

The values of $P(A, \theta)$ for $\theta = 0, 1/20, 2/20, \cdots, 20/20$ and the resulting OC curve are shown in Fig. 502 on the next page. (Verify!) ◀

In most practical cases θ will be small (less than 10%). Then if we take small samples compared to N, we can approximate (2) by the Poisson distribution (Sec. 22.7); thus

$$(3) \qquad P(A; \theta) \sim e^{-\mu} \sum_{x=0}^{c} \frac{\mu^x}{x!} \qquad (\mu = n\theta).$$

EXAMPLE 2　**Sampling plan. Poisson distribution**

Suppose that for large lots the following sampling plan is used. A sample of size $n = 20$ is taken. If it contains not more than one defective, the lot is accepted. If the sample contains two or more defectives, the lot is rejected. In this plan, we obtain from (3)

$$P(A; \theta) \sim e^{-20\theta}(1 + 20\theta).$$

The corresponding OC curve is shown in Fig. 503. ◀

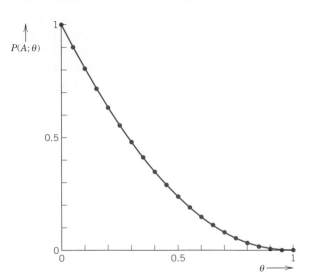

Fig. 502. OC curve of the sampling plan with $n = 2$ and $c = 0$ for lots of size $N = 20$

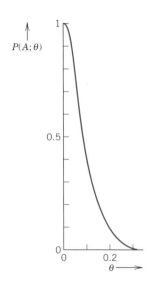

Fig. 503. OC curve in Example 2

Errors in Acceptance Sampling

We show how acceptance sampling fits into general test theory (Sec. 23.4) and what this means from a practical point of view. The producer wants the probability α of rejecting an **acceptable lot** (a lot for which θ does not exceed a certain number θ_0 on which the two parties agree) to be small. θ_0 is called the **acceptable quality level** (AQL). Similarly, the consumer (the buyer) wants the probability β of accepting an **unacceptable lot** (a lot for which θ is greater than or equal to some θ_1) to be small. θ_1 is called the **lot tolerance percent defective** (LTPD) or the **rejectable quality level** (RQL). α is called **producer's risk.** It corresponds to a Type I error in Sec. 23.4. β is called **consumer's risk** and corresponds to a Type II error. Figure 504 shows an example. We see that the points $(\theta_0, 1 - \alpha)$ and (θ_1, β) lie on the OC curve. It can be shown that for large lots we can choose θ_0, θ_1 ($> \theta_0$), α, β and then determine n and c such that the OC curve runs very close to those prescribed points. Table 23.6 shows the analogy between acceptance sampling and hypothesis testing in Sec. 23.4.

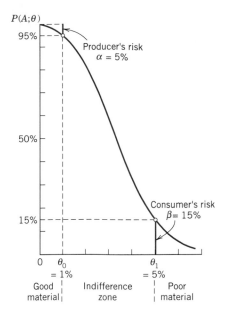

Fig. 504. OC curve, producer's and consumer's risks

Table 23.6
Acceptance Sampling and Hypothesis Testing

Acceptance Sampling	Hypothesis Testing
Acceptable quality level (AQL) $\theta = \theta_0$	Hypothesis $\theta = \theta_0$
Lot tolerance percent defectives (LTPD) $\theta = \theta_1$	Alternative $\theta = \theta_1$
Allowable number of defectives c	Critical value c
Producer's risk α of rejecting a lot with $\theta \leqq \theta_0$	Probability α of making a Type I error (significance level)
Consumer's risk β of accepting a lot with $\theta \geqq \theta_1$	Probability β of making a Type II error

Rectification

Rectification of a *rejected* lot means that the lot is inspected item by item and all defectives are removed and replaced by nondefective items. (This may be too expensive if the lot is cheap; in that case it may be sold at a cut-rate price or scrapped.) If a production turns out $100\theta\%$ defectives, then in K lots of size N each, $KN\theta$ of the KN items are defectives. Now $KP(A; \theta)$ of these lots are accepted. These contain $KPN\theta$ defectives, whereas the rejected and rectified lots contain no defectives, because of the rectification. Hence after the rectification the fraction defective in all K lots equals $KPN\theta/KN = \theta P(A; \theta)$. This is called the **average outgoing quality** (AOQ); thus

$$\text{(4)} \qquad \boxed{\text{AOQ}(\theta) = \theta P(A; \theta).}$$

Figure 505 shows an example. Since $\text{AOQ}(0) = 0$ and $P(A; 1) = 0$, the AOQ curve has a maximum at some $\theta = \theta^*$, giving the **average outgoing quality limit** (AOQL). This is the worst average quality that may be expected to be accepted under rectification.

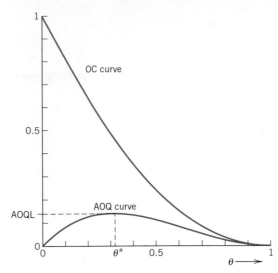

Fig. 505.　OC curve and AOQ curve
for the sampling plan in Fig. 502

PROBLEM SET 23.6

1. Lots of knives are inspected by a sampling plan that uses a sample of size 20 and the acceptance number $c = 1$. What are the probabilities of accepting a lot with 1%, 2%, 10% defectives (dull knives)? Use Table A6 in Appendix 5. Graph the OC curve.

2. What happens in Prob. 1 if the sample size is increased to 50? First guess. Then calculate. Graph the OC curve and compare.

3. How will the probabilities in Prob. 1 with $n = 20$ change (up or down) if we decrease c to zero? First guess.

4. What are the producer's and consumer's risks in Prob. 1 if the AQL is 1.5% and the RQL is 7.5%?

5. Lots of copper pipes are inspected according to a sample plan that uses sample size 30 and acceptance number 1. Graph the OC curve of the plan, using the Poisson approximation.

6. Graph the AOQ curve in Prob. 5. Determine the AOQL, assuming that rectification is applied.

7. In Example 1 in the text, what are the producer's and consumer's risks if the AQL is 0.1 and the RQL is 0.6?

8. Find the binomial approximation of the hypergeometric distribution in Example 1 in the text and compare the approximate and the accurate values.

9. Graph and compare sampling plans with $c = 1$ and increasing values of n, say, $n = 2, 3, 4$. (Use the binomial distribution.)

10. What happens in Example 1 in the text if we increase the sample size to $n = 3$, leaving the other data as before? Compute $P(A; 0.1)$ and $P(A; 0.2)$ and compare.

11. Samples of 3 fuses are drawn from lots and a lot is accepted if in the corresponding sample we find no more than 1 defective fuse. Criticize this sampling plan. In particular, find the probability of accepting a lot that is 50% defective. (Use the binomial distribution (7), Sec. 22.7.)

12. A lot of batteries for watches is accepted if and only if a sample of 20 contains at most 1 defective. Graph the OC and AOQ curves. Find AOQL. [Use (3).]

13. Graph the OC curve and the AOQ curve for the sampling plan for large lots with $n = 5$ and $c = 0$, and find the AOQL.

14. If in a sampling plan for large lots of spark plugs, the sample size is 100 and we want the AQL to be 5% and the producer's risk 2%, what acceptance number c should we choose? (Use the normal approximation.)

15. What is the consumer's risk in Prob. 14 if we want the RQL to be 12%?

23.7 Goodness of Fit. χ^2-Test

So far we have discussed tests for unknown parameters (μ, σ, etc.) in distributions whose kind we knew or did not care about because we had a large sample, so that by the central limit theorem (Sec. 23.3) we could also apply methods designed for the normal distribution to other distributions (with proper caution!). However, what can we do if we want to test whether a distribution is of a certain kind, for instance, normal? This means we wish to test that a certain function $F(x)$ is the distribution function of a distribution from which we have a sample $x_1, \cdots, x_n$. This important task is called testing for **goodness of fit:** if the **sample distribution function** $\widetilde{F}(x)$ defined by

$$\widetilde{F}(x) = \textit{Sum of the relative frequencies of all sample values } x_j \textit{ not exceeding } x$$

fits $F(x)$ "sufficiently well", we shall accept the hypothesis that $F(x)$ is the distribution function of the population; if not, we shall reject it.

To decide in this fashion, we have to know how much $\widetilde{F}(x)$ can differ from $F(x)$ if the hypothesis is true. Hence we must first introduce a quantity that measures the deviation of $\widetilde{F}(x)$ from $F(x)$, and we must know the probability distribution of this quantity under the assumption that the hypothesis is true. Then we proceed as follows. We determine a number c such that if the hypothesis is true, a deviation greater than c has a small preassigned probability. If, nevertheless, a deviation greater than c occurs, we have reason to doubt that the hypothesis is true and we reject it. On the other hand, if the deviation does not exceed c, so that $\widetilde{F}(x)$ approximates $F(x)$ sufficiently well, we accept the hypothesis. Of course, if we accept the hypothesis, this means that we have insufficient evidence to reject it and does not exclude the possibility that there are other functions that would not be rejected in the test. In this respect the situation is quite similar to that in Sec. 23.4.

Table 23.7 on the next page shows a test of that type, which was introduced by R. A. Fisher. This test is justified by the fact that if the hypothesis is true, then $\chi_0{}^2$ is an observed value of a random variable whose distribution function approaches that of the chi-square distribution with $K - 1$ degrees of freedom (or $K - r - 1$ degrees of freedom if r parameters are estimated) as n approaches infinity. The requirement that at least five sample values lie in each interval in Table 23.7 results from the fact that for finite n that random variable has only *approximately* a chi-square distribution. A proof can be found in Ref. [G3] listed in Appendix 1. If the sample is so small that the requirement cannot be satisfied, one may continue with the test, but the result should then be used with caution.

Table 23.7

Chi-square Test for the Hypothesis That $F(x)$ is the Distribution Function of a Population from Which a Sample $x_1, \cdots, x_n$ is Taken

1st Step. Subdivide the x-axis into K intervals $I_1, I_2, \cdots, I_K$ such that each interval contains at least 5 values of the given sample $x_1, \cdots, x_n$. Determine the number b_j of sample values in the interval I_j, where $j = 1, \cdots, K$. If a sample value lies at a common boundary point of two intervals, add 0.5 to each of the two corresponding b_j.

2nd Step. Using $F(x)$, compute the probability p_j that the random variable X under consideration assumes any value in the interval I_j, where $j = 1, \cdots, K$. Compute

$$e_j = np_j.$$

(This is the number of sample values theoretically expected in I_j if the hypothesis is true.)

3rd Step. Compute the deviation

$$(1) \qquad \chi_0{}^2 = \sum_{j=1}^{K} \frac{(b_j - e_j)^2}{e_j}.$$

4th Step. Choose a significance level (5%, 1%, or the like).

5th Step. Determine the solution c of the equation

$$P(\chi^2 \leqq c) = 1 - \alpha$$

from the table of the chi-square distribution with $K - 1$ degrees of freedom (Table A10 in Appendix 5). If r parameters of $F(x)$ are unknown and their maximum likelihood estimates (Sec. 23.2) are used, then use $K - r - 1$ degrees of freedom (instead of $K - 1$). If $\chi_0{}^2 \leqq c$, accept the hypothesis. If $\chi_0{}^2 > c$, reject the hypothesis.

Table 23.8

Sample of 100 Values of the Splitting Tensile Strength (lb/in.^{2}) of Concrete Cylinders

320	380	340	410	380	340	360	350	320	370
350	340	350	360	370	350	380	370	300	420
370	390	390	440	330	390	330	360	400	370
320	350	360	340	340	350	350	390	380	340
400	360	350	390	400	350	360	340	370	420
420	400	350	370	330	320	390	380	400	370
390	330	360	380	350	330	360	300	360	360
360	390	350	370	370	350	390	370	370	340
370	400	360	350	380	380	360	340	330	370
340	360	390	400	370	410	360	400	340	360

D. L. IVEY, Splitting tensile tests on structural lightweight aggregate concrete. Texas Transportation Institute, College Station, Texas.

EXAMPLE 1   **Test of normality**

Test whether the population from which the sample in Table 23.8 was taken is normal.

Solution.   Table 23.8 shows the values (column by column) in the order obtained in the experiment. Table 23.9 gives the frequency distribution and Fig. 506 the histogram. It is hard to guess the outcome of the test—does the histogram resemble a normal density curve sufficiently well or not?

The maximum likelihood estimates for μ and σ^2 are $\hat{\mu} = \bar{x} = 364.7$ and $\tilde{\sigma}^2 = 712.9$. The computation in Table 23.10 yields $\chi_0^2 = 2.942$. It is very interesting that the interval $375 \cdots 385$ contributes over 50% of χ_0^2. From the histogram we see that the corresponding frequency looks much too small. The second largest

Table 23.9
Frequency Table of the Sample in Table 23.8

1 Tensile Strength x [lb/in.^{2}]	2 Absolute Frequency	3 Relative Frequency $\tilde{f}(x)$	4 Cumulative Absolute Frequency	5 Cumulative Relative Frequency $\tilde{F}(x)$
300	2	0.02	2	0.02
310	0	0.00	2	0.02
320	4	0.04	6	0.06
330	6	0.06	12	0.12
340	11	0.11	23	0.23
350	14	0.14	37	0.37
360	16	0.16	53	0.53
370	15	0.15	68	0.68
380	8	0.08	76	0.76
390	10	0.10	86	0.86
400	8	0.08	94	0.94
410	2	0.02	96	0.96
420	3	0.03	99	0.99
430	0	0.00	99	0.99
440	1	0.01	100	1.00

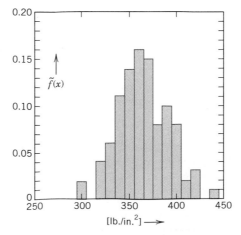

Fig. 506.   Frequency histogram
of the sample in Table 23.8

contribution comes from $395 \cdots 405$, and the histogram shows that the frequency seems somewhat too large, which is perhaps not obvious from inspection.

We choose $\alpha = 5\%$. Since $K = 10$ and we estimated $r = 2$ parameters we have to use Table A10 in Appendix 5 with $K - r - 1 = 7$ degrees of freedom. We find $c = 14.07$ as the solution of $P(\chi^2 \leqq c) = 95\%$. Since $\chi_0^2 < c$, we accept the hypothesis that the population is normal. ◀

Table 23.10
Computations in Example 1

x_j	$\dfrac{x_j - 364.7}{26.7}$	$\Phi\left(\dfrac{x_j - 364.7}{26.7}\right)$	e_j	b_j	Terms in (1)
$-\infty \cdots 325$	$-\infty \quad \cdots -1.49$	$0.0000 \cdots 0.0681$	6.81	6	0.096
$325 \cdots 335$	$-1.49 \cdots -1.11$	$0.0681 \cdots 0.1335$	6.54	6	0.045
$335 \cdots 345$	$-1.11 \cdots -0.74$	$0.1335 \cdots 0.2296$	9.61	11	0.201
$345 \cdots 355$	$-0.74 \cdots -0.36$	$0.2296 \cdots 0.3594$	12.98	14	0.080
$355 \cdots 365$	$-0.36 \cdots \quad 0.01$	$0.3594 \cdots 0.4960$	13.66	16	0.401
$365 \cdots 375$	$0.01 \cdots \quad 0.39$	$0.4960 \cdots 0.6517$	15.57	15	0.021
$375 \cdots 385$	$0.39 \cdots \quad 0.76$	$0.6517 \cdots 0.7764$	12.47	8	1.602
$385 \cdots 395$	$0.76 \cdots \quad 1.13$	$0.7764 \cdots 0.8708$	9.44	10	0.033
$395 \cdots 405$	$1.13 \cdots \quad 1.51$	$0.8708 \cdots 0.9345$	6.37	8	0.417
$405 \cdots \infty$	$1.51 \cdots \quad \infty$	$0.9345 \cdots 1.0000$	6.55	6	0.046

$$\chi_0^2 = 2.942$$

PROBLEM SET 23.7

1. If 100 flips of a coin result in 40 heads and 60 tails, can we assert on the 5% level that the coin is fair?

2. If in 10 flips of a coin we get the same ratio as in Prob. 1 (4 heads and 6 tails), is the conclusion the same as in Prob. 1? First conjecture, then compute.

3. Can you claim on a 5% level that a die is fair if 60 trials give $1, \cdots, 6$ with absolute frequencies 10, 13, 9, 11, 9, 8?

4. In a classical experiment, R. Wolf rolled a die 20 000 times and obtained $1, \cdots, 6$ with the absolute frequencies 3407, 3631, 3176, 2916, 3448, 3422. Test whether the die was fair, using $\alpha = 5\%$.

5. If a service station has served 60, 49, 56, 46, 68, 39 cars from Monday through Friday between 1 P.M. and 2 P.M., can one claim on a 5% level that the differences are due to randomness? First guess. Then calculate.

6. Three samples of 200 rivets each were taken from a large production of each of three machines. The numbers of defective rivets in the samples were 13, 3, and 8. Is this difference significant? (Use $\alpha = 5\%$.)

7. Using the given sample, test that the corresponding population has a Poisson distribution. x is the number of alpha particles per 7.5-second intervals observed by E. Rutherford and H. Geiger in one of their classical experiments in 1910, and $a(x)$ is the absolute frequency (= number of time periods during which exactly x particles were observed).

x	0	1	2	3	4	5	6	7	8	9	10	11	12	$\geqq 13$
a	57	203	383	525	532	408	273	139	45	27	10	4	2	0

8. Test for normality at the 1% level using a sample of $n = 79$ (rounded) values x (tensile strength $[\text{kg/mm}^2]$ of steel sheets of 0.3 mm thickness). $a = a(x)$ = absolute frequency. (Take the first two values together, also the last three, to get $K = 5$.)

x	57	58	59	60	61	62	63	64
a	4	10	17	27	8	9	3	1

9. In a sample of 100 patients having a certain disease 45 are men and 55 women. Does this support the claim that the disease is equally common among men and women? Choose $\alpha = 5\%$.

10. In Prob. 9 find the smallest number (>50) of women that leads to the rejection of the hypothesis on the levels 5%, 1%, 0.5%.

11. If it is known that 25% of certain steel rods produced by a standard process will break when subjected to a load of 5000 lb, can we claim that a new process yields the same breakage rate if we find that in a sample of 160 rods produced by the new process, 50 rods broke when subjected to that load? Choose $\alpha = 5\%$.

12. A manufacturer claims that in a process of producing razor blades, only 2.5% of the blades are dull. Test the claim against the alternative that more than 2.5% of the blades are dull, using a sample of 400 blades containing 17 dull blades. (Use $\alpha = 5\%$.)

13. Can we assert that the traffic on the three lanes of an expressway (in one direction) is about the same on each lane if a count gives 920, 870, 750 cars on the right, middle, and left lanes, respectively, during the same interval of time?

14. TEAM PROJECT. Difficulty with Random Selection. Seventy-seven students were asked to select three of the numbers 11, 12, 13, $\cdots$, 30 in a completely arbitrary fashion. The amazing result was as follows.

Number	11	12	13	14	15	16	17	18	19	20
Frequency	11	10	20	8	13	9	21	9	16	8

Number	21	22	23	24	25	26	27	28	29	30
Frequency	12	8	15	10	10	9	12	8	13	9

If the selected numbers were completely random, each of the following hypotheses should be true.

(*a*) The 20 numbers are equally likely.

(*b*) The 10 even numbers together are as likely as the 10 odd numbers together.

(*c*) The 6 prime numbers together have probability 0.3 and the other numbers together have probability 0.7.

Test these hypotheses.

    Design further experiments that illustrate the difficulties of random selection.

15. CAS PROJECT. Random Number Generator. Check your generator experimentally by imitating results of n trials of rolling a fair die, with a convenient n (e.g., 60 or 300 or the like). Do this many times and see whether you can notice any "nonrandomness" features, for example, too few Sixes, too many even numbers, etc., or whether your generator seems to work properly. Design and perform other kinds of checks.

23.8 Nonparametric Tests

Nonparametric tests, also called **distribution-free tests,** are valid for any distribution. Hence they are practical in cases when the kind of distribution is unknown, or is known but such that no tests specifically designed for it are available. In this section we shall explain the basic idea of these tests, which are based on **"order statistics"** and are very simple. If there is a choice, then tests designed for a specific distribution generally give better results than do nonparametric tests. For instance, this applies to the tests in Sec. 23.4 for the normal distribution.

We shall discuss two tests in terms of typical examples. In deriving the distributions used in the test, it is essential that the distributions from which we sample are continuous. (Nonparametric tests can also be derived for discrete distributions, but this is slightly more complicated.)

EXAMPLE 1 **Sign test for the median**

A **median** of the population is a solution $x = \tilde{\mu}$ of the equation $F(x) = 0.5$, where F is the distribution function.

Suppose that eight radio operators were tested, first in rooms without air-conditioning and then in air-conditioned rooms over the same period of time, and the difference of errors (unconditioned minus conditioned) were

$$9 \quad 4 \quad 0 \quad 6 \quad 4 \quad 0 \quad 7 \quad 11.$$

Test the hypothesis $\tilde{\mu} = 0$ (that is, air-conditioning has no effect) against the alternative $\tilde{\mu} > 0$ (that is, inferior performance in unconditioned rooms).

Solution. We choose the significance level $\alpha = 5\%$. If the hypothesis is true, the probability p of a positive difference is the same as that of a negative difference. Hence in this case, $p = 0.5$, and the random variable

$$X = Number\ of\ positive\ values\ among\ n\ values$$

has a binomial distribution with $p = 0.5$. Our sample has eight values. We omit the values 0, which do not contribute to the decision. Then six values are left, all of which are positive. Since

$$P(X = 6) = \binom{6}{6} (0.5)^6 (0.5)^0 = 0.0156 = 1.56\% < \alpha,$$

we reject the hypothesis and assert that the number of errors made in unconditioned rooms is significantly higher, so that installing air-conditioning should be considered. ◀

EXAMPLE 2 **Test for arbitrary trend**

A certain machine is used for cutting lengths of wire. Five successive pieces had the lengths

$$29 \quad 31 \quad 28 \quad 30 \quad 32.$$

Using this sample, test the hypothesis that there is **no trend,** that is, the machine does not have the tendency to produce longer and longer pieces or shorter and shorter pieces. Assume that the type of machine suggests the alternative that there is *positive trend,* that is, there is the tendency of successive pieces to get longer.

Solution. We count the number of **transpositions** in the sample, that is, the number of times a larger value precedes a smaller value:

$$29 \text{ precedes } 28 \qquad (1 \text{ transposition}),$$

$$31 \text{ precedes } 28 \text{ and } 30 \qquad (2 \text{ transpositions}).$$

The remaining three sample values follow in ascending order. Hence in the sample there are $1 + 2 = 3$ transpositions. We now consider the random variable

$$T = Number\ of\ transpositions.$$

If the hypothesis is true (no trend), then each of the 5! = 120 permutations of five elements 1 2 3 4 5 has the same probability (1/120). We arrange these permutations according to their number of transpositions:

$T = 0$	$T = 1$	$T = 2$	$T = 3$
1 2 3 4 5	1 2 3 5 4	1 2 4 5 3	1 2 5 4 3
	1 2 4 3 5	1 2 5 3 4	1 3 4 5 2
	1 3 2 4 5	1 3 2 5 4	1 3 5 2 4
	2 1 3 4 5	1 3 4 2 5	1 4 2 5 3
		1 4 2 3 5	1 4 3 2 5
		2 1 3 5 4	1 5 2 3 4
		2 1 4 3 5	2 1 4 5 3
		2 3 1 4 5	2 1 5 3 4
		3 1 2 4 5	2 3 1 5 4
			2 3 4 1 5
			2 4 1 3 5
			3 1 2 5 4
			3 1 4 2 5
			3 2 1 4 5
			4 1 2 3 5

etc.

From this we obtain

$$P(T \leqq 3) = \tfrac{1}{120} + \tfrac{4}{120} + \tfrac{9}{120} + \tfrac{15}{120} = \tfrac{29}{120} = 24\%.$$

Hence we accept the hypothesis.

Values of the distribution function of T in the case of no trend are shown in Table A12, Appendix 5. For instance, if $n = 3$, then $F(0) = 0.167$, $F(1) = 0.500$, $F(2) = 1 - 0.167$. If $n = 4$, then $F(0) = 0.042$, $F(1) = 0.167$, $F(2) = 0.375$, $F(3) = 1 - 0.375$, $F(4) = 1 - 0.167$, and so on.

Our method and those values refer to *continuous* distributions. Theoretically, we may then expect that all the values of a sample are different. Practically, some sample values may still be equal, because of rounding off. If m values are equal, add $m(m - 1)/4$ (= mean value of the transpositions in the case of the permutations of m elements), that is, $\tfrac{1}{2}$ for each pair of equal values, $\tfrac{3}{2}$ for each triple, etc. ◀

PROBLEM SET 23.8

1. What would change in Example 1, had we observed only 5 positive values? Only 4?

2. Test $\widetilde{\mu} = 0$ against $\widetilde{\mu} > 0$, using 1, −1, 1, 3, −8, 6, 0 (deviations of the azimuth [multiples of 0.01 radian] in some revolution of a satellite).

3. Are oil filters of type A better than type B filters if in 11 trials, A gave cleaner oil than B in 7 cases, B gave cleaner oil than A in 1 case, whereas in 3 of the trials the results for A and B were practically the same?

4. Each of 10 patients were given two different sedatives A and B. The following table shows the effect (increase of sleeping time, measured in hours). Using the sign test, find out whether the difference is significant.

A	1.9	0.8	1.1	0.1	−0.1	4.4	5.5	1.6	4.6	3.4
B	0.7	−1.6	−0.2	−1.2	−0.1	3.4	3.7	0.8	0.0	2.0
Difference	1.2	2.4	1.3	1.3	0.0	1.0	1.8	0.8	4.6	1.4

5. Assuming that the populations corresponding to the samples in Prob. 4 are normal, apply the test explained in Example 3, Sec. 23.4. Use $\alpha = 5\%$.

6. Thirty new employees were grouped into 15 pairs on the basis of intelligence and previous experience and were then instructed in data processing. Two methods of instruction were applied, an old method (*A*) to one (randomly selected) person of each pair and a new, presumably better, method (*B*) to the other. Test the hypothesis that the methods are equally effective against the alternative that (*B*) is better, using the following scores obtained at the end of the training period.

A	60	70	80	85	75	40	70	45	95	80	90	60	80	75	65
B	65	85	85	80	95	65	100	60	90	85	100	75	90	60	80

7. Assuming normality, solve Prob. 6 by a suitable test from Sec. 23.4.

8. Test whether a thermostatic switch is properly set to 20°C against the alternative that its setting is too low. Use a sample of 8 values, 7 of which are less than 20°C and 1 is greater than 20°C.

9. Complete the table in Example 2 and make a similar table for $n = 4$.

10. Do the readings of a certain type of voltmeter tend to increase with temperature?

Temperature T[°C]	10	20	30	40	50
Reading V [volts]	110.4	111.1	110.9	111.0	111.3

11. Test the following gains in weight [kg] of 10 animals in a swine feeding experiment (ordered according to increasing daily amounts of food) for a positive trend:

$$22 \quad 19 \quad 21 \quad 20 \quad 25 \quad 18 \quad 27 \quad 30 \quad 26 \quad 24.$$

12. Does the amount of fertilizer increase the yield of wheat *X* [kg/plot]? Use a sample of values ordered according to increasing amounts of fertilizer:

$$33.4 \quad 35.3 \quad 31.6 \quad 35.0 \quad 36.1 \quad 37.6 \quad 36.5 \quad 38.7.$$

13. Does an increase in temperature cause an increase of the yield of a chemical reaction from which the following sample was taken?

Temperature [°C]	10	20	30	40	60	80
Yield [kg/min]	1.6	2.1	1.9	2.6	2.2	3.0

14. Apply the test explained in Example 2 to the following data (*x* = diastolic blood pressure [mm Hg], *y* = weight of heart [in grams] of 10 patients who died of cerebral hemorrhage).

x	121	120	95	123	140	112	92	100	102	91
y	521	465	352	455	490	388	301	395	375	418

15. Apply the test in Example 2 to the following data (*x* = disulfide content of a certain type of wool, measured in percent of the content in unreduced fibers; *y* = saturation water content of the wool, measured in percent).

x	10	15	30	40	50	55	80	100
y	51	47	44	43	37	40	38	34

23.9　Regression Analysis Fitting Straight Lines

So far we have been concerned with random experiments in which we observed a single quantity (random variable) and got samples whose values were single numbers. In this section we discuss experiments in which we observe or measure two quantities simultaneously, so that we get samples of *pairs* of values $(x_1, y_1), (x_2, y_2), \cdots, (x_n, y_n)$. In practice we may distinguish between two kinds of experiments, as follows.

1. In **correlation analysis** (Sec. 23.10) both quantities are random variables and we are interested in relations between them. Examples are the relation (one says "correlation") between wear X and wear Y of the front tires of cars, between grades X and Y of students in mathematics and in physics, respectively, between the hardness X of steel plates in the center and the hardness Y near the edges of the plates, etc.

2. In **regression analysis** (this section) one of the two variables, call it x, can be regarded as an ordinary variable because we can measure it without substantial error or we can even give it values we want. x is called the **independent variable,** or sometimes the **controlled variable** because we can control it (set it at values we choose). The other variable, Y, is a random variable, and we are interested in the dependence of Y on x. Typical examples are the dependence of the blood pressure Y on the age x of a person or, as we shall now say, the regression of Y on x, the regression of the gain of weight Y of certain animals on the daily ration of food x, the regression of the heat conductivity Y of cork on the specific weight x of the cork, etc.

In general, in an experiment we usually choose $x_1, \cdots, x_n$ and then observe corresponding values $y_1, \cdots, y_n$ of Y, so that we get a sample $(x_1, y_1), \cdots, (x_n, y_n)$. Now in regression analysis the dependence of Y on x is a dependence of the mean μ of Y on x, so that $\mu = \mu(x)$ is a function in the ordinary sense. The curve of $\mu(x)$ is called the **regression curve** of Y on x.

In this section we discuss the simplest case, namely, that of a straight line

(1)
$$\mu(x) = \kappa_0 + \kappa_1 x.$$

Then we may want to plot the sample values as n points in the xY-plane, fit a straight line through them, and use it for estimating $\mu(x)$ at values of x that interest us, so that we know what values of Y we can expect for those x. Fitting that line by eye would not be good because it would be subjective; that is, different persons' results would come out differently, particularly if the points are scattered. So we need a mathematical method that gives a unique result depending only on the n points. A widely used procedure is the following.

Method of Least Squares

For the present task we may formulate Gauss's least squares method as follows.

Least squares principle. *The straight line should be fitted through the given points so that the sum of the squares of the distances of those points from the straight line is minimum, where the distance is measured in the vertical direction (the y-direction).* (Formulas below.)

To get uniqueness, we need some extra condition. To see this, take the sample $(0, 1)$, $(0, -1)$. Then all the lines $y = k_1 x$ with any k_1 satisfy the principle. (Can you see it?) The following assumption will imply uniqueness, as we shall find out.

General assumption (A1)

The x-values $x_1, \cdots, x_n$ in our sample $(x_1, y_1), \cdots, (x_n, y_n)$ are not all equal.

From a given sample $(x_1, y_1), \cdots, (x_n, y_n)$ we shall now determine a straight line by least squares. We write the line as

(2)
$$y = k_0 + k_1 x$$

and call it the **sample regression line** because it will be the counterpart of the population regression line (1).

Now a sample point (x_j, y_j) has the vertical distance (distance measured in the y-direction) from (2) given by

$$\left| y_j - (k_0 + k_1 x_j) \right| \qquad \text{(see Fig. 507)}.$$

Hence the sum of the squares of these distances is

(3)
$$q = \sum_{j=1}^{n} (y_j - k_0 - k_1 x_j)^2.$$

In the method of least squares we now have to determine k_0 and k_1 such that q is minimum. From calculus we know that a necessary condition for this is

(4)
$$\frac{\partial q}{\partial k_0} = 0 \qquad \text{and} \qquad \frac{\partial q}{\partial k_1} = 0.$$

We shall see that from this condition we obtain for the **sample regression line** the formula

(5)
$$y - \bar{y} = k_1(x - \bar{x})$$

where $\bar{x}$ and $\bar{y}$ are the means of the x- and the y-values in our sample,

(6) $$\bar{x} = \frac{1}{n}(x_1 + \cdots + x_n) \qquad \text{and} \qquad \bar{y} = \frac{1}{n}(y_1 + \cdots + y_n)$$

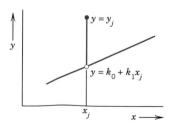

Fig. 507. Vertical distance of a point (x_j, y_j)
from a straight line $y = k_0 + k_1 x$

and the slope k_1, called the **regression coefficient,** is given by

(7)
$$k_1 = \frac{s_{xy}}{s_x^2}$$

with the **"sample covariance"** s_{xy} given by

(8) $\quad s_{xy} = \dfrac{1}{n-1} \sum_{j=1}^{n} (x_j - \bar{x})(y_j - \bar{y}) = \dfrac{1}{n-1} \left[\sum_{j=1}^{n} x_j y_j - \dfrac{1}{n} \left(\sum_{i=1}^{n} x_i \right) \left(\sum_{j=1}^{n} y_j \right) \right]$

and

(9) $\quad\quad s_x^2 = \dfrac{1}{n-1} \sum_{j=1}^{n} (x_j - \bar{x})^2 = \dfrac{1}{n-1} \left[\sum_{j=1}^{n} x_j^2 - \dfrac{1}{n} \left(\sum_{j=1}^{n} x_j \right)^2 \right].$

From (5) we see that the sample regression line passes through the point $(\bar{x}, \bar{y})$, by which it is determined, together with the regression coefficient (7). We may call s_x^2 the *variance* of the x-values in our sample [analogous to (2), Sec. 23.1], but we should keep in mind that x is an ordinary variable, not a random variable.

Derivation of (5) and (7). Differentiating (3) and using (4), we first obtain

$$\frac{\partial q}{\partial k_0} = -2 \sum (y_j - k_0 - k_1 x_j) = 0, \qquad \frac{\partial q}{\partial k_1} = -2 \sum x_j (y_j - k_0 - k_1 x_j) = 0$$

where we sum over j from 1 to n. We now divide by 2, write each of the two sums as three sums, and take those containing y_j and $x_j y_j$ over to the right. Then we get the **"normal equations"**

(10)
$$k_0 n + k_1 \sum x_j = \sum y_j$$
$$k_0 \sum x_j + k_1 \sum x_j^2 = \sum x_j y_j.$$

This is a linear system of two equations in the two unknowns k_0 and k_1. Its coefficient determinant is [see (9)]

$$\begin{vmatrix} n & \sum x_j \\ \sum x_j & \sum x_j^2 \end{vmatrix} = n \sum x_j^2 - \left(\sum x_j \right)^2 = n(n-1)s_x^2 = n \sum (x_j - \bar{x})^2$$

and is not zero because of Assumption (A1). Hence the system has a unique solution. Dividing the first equation of (10) by n and using (6), we get $k_0 = \bar{y} - k_1 \bar{x}$. Together with $y = k_0 + k_1 x$ in (2) this gives (5). To get (7), we solve the system (10) by Cramer's rule or elimination, finding

(11)
$$k_1 = \frac{n \sum x_j y_j - \sum x_i \sum y_j}{n(n-1)s_x^2}.$$

This gives (7)−(9) and completes the derivation. [The equality of the two expressions in (8) and in (9) may be shown by the student; see Prob. 10.] ◀

EXAMPLE 1   **Regression line**

The decrease of volume y [%] of leather for certain fixed values of high pressure x [atmospheres] was measured. The results are shown in the first two columns of Table 23.11. Find the regression line of y on x.

Solution.   We see that $n = 4$ and obtain the values $\bar{x} = 28\,000/4 = 7000$, $\bar{y} = 19.0/4 = 4.75$, and from (9) and (8)

$$s_x^2 = \frac{1}{3}\left(216\,000\,000 - \frac{28\,000^2}{4}\right) = \frac{20\,000\,000}{3}$$

$$s_{xy} = \frac{1}{3}\left(148\,400 - \frac{28\,000 \cdot 19}{4}\right) = \frac{15\,400}{3}.$$

Hence $k_1 = 15\,400/20\,000\,000 = 0.000\,77$ from (7), and the regression line is

$$y - 4.75 = 0.000\,77(x - 7000) \qquad \text{or} \qquad y = 0.000\,77x - 0.64.$$

Note that $y(0) = -0.64$, which is physically meaningless, but typically indicates that a linear relation is merely an approximation valid on some restricted interval. ◀

Table 23.11
Regression of the Decrease of Volume y [%] of
Leather on the Pressure x [Atmospheres]

Given Values		Auxiliary Values	
x_j	y_j	x_j^2	$x_j y_j$
4,000	2.3	16,000,000	9,200
6,000	4.1	36,000,000	24,600
8,000	5.7	64,000,000	45,600
10,000	6.9	100,000,000	69,000
28,000	19.0	216,000,000	148,400

Confidence Intervals in Regression Analysis

If we want to get confidence intervals, we have to make assumptions about the distribution of Y (which we have not made so far; least squares is a "geometric principle," nowhere involving probabilities!). We assume normality and independence in sampling:

Assumption (A2)

For each fixed x the random variable Y is normal with mean

(12) $$\mu(x) = \kappa_0 + \kappa_1 x$$

and variance σ^2 independent of x.

Assumption (A3)

The n performances of the experiment by which we obtain a sample

$$(x_1, y_1), \qquad (x_2, y_2), \qquad \cdots, \qquad (x_n, y_n)$$

are independent. (This is similar to the one-dimensional case in Sec. 23.1.)

κ_1 in (12) is called the *population* **regression coefficient** because it can be shown that under Assumptions (A1)−(A3) the maximum likelihood estimate of κ_1 is the sample regression coefficient k_1 given by (11).

Under Assumptions (A1)−(A3) we may now obtain a confidence interval for κ_1, as shown in Table 23.12.

Table 23.12
Determination of a Confidence Interval for κ_1 in (1)
under Assumptions (A1)−(A3)

1st Step. Choose a confidence level γ (95%, 99%, or the like).

2nd Step. Determine the solution c of the equation

$$(13) \qquad\qquad F(c) = \tfrac{1}{2}(1 + \gamma)$$

from the table of the t-distribution with $n - 2$ degrees of freedom (Table A9 in Appendix 5; $n =$ sample size).

3rd Step. Using a sample $(x_1, y_1), \cdots, (x_n, y_n)$, compute $(n - 1)s_x^2$ from (9), $(n - 1)s_{xy}$ from (8), k_1 from (7),

$$(14) \qquad\qquad (n - 1)s_y^2 = \sum_{j=1}^{n} y_j^2 - \frac{1}{n}\left(\sum_{j=1}^{n} y_j\right)^2,$$

and

$$(15) \qquad\qquad q_0 = (n - 1)(s_y^2 - k_1^2 s_x^2).$$

4th Step. Compute

$$K = c\sqrt{\frac{q_0}{(n - 2)(n - 1)s_x^2}}.$$

The confidence interval is

$$(16) \qquad\qquad \text{CONF}_\gamma\{k_1 - K \leqq \kappa_1 \leqq k_1 + K\}.$$

EXAMPLE 2   **Confidence interval for the regression coefficient**

Using the sample in Table 23.11, determine a confidence interval for κ_1 by the method in Table 23.12.

Solution. 1st Step.   We choose $\gamma = 0.95$.

2nd Step.   Equation (13) takes the form $F(c) = 0.975$, and Table A9 in Appendix 5 with $n - 2 = 2$ degrees of freedom gives $c = 4.30$.

3rd Step.   From Example 1 we have $3s_x^2 = 20\,000\,000$ and $k_1 = 0.00077$. From Table 23.11 we compute

$$3s_y^2 = 102.2 - \frac{19^2}{4} = 11.95, \qquad q_0 = 11.95 - 20\,000\,000 \cdot 0.00077^2 = 0.092.$$

4th Step.   We thus obtain $K = 4.30\sqrt{0.092/(2 \cdot 20\,000\,000)} = 0.000\,206$ and

$$\text{CONF}_{0.95}\{0.00056 \leqq \kappa_1 \leqq 0.00098\}. \qquad\qquad \blacktriangleleft$$

PROBLEM SET 23.9

Sample Regression Line

Find and sketch or plot the sample regression line of y on x and plot the given data on the same axes.

1. (2, 12), (5, 24), (9, 33), (14, 50)
2. $(-2, 3.5)$, (0, 1.5), (2, 1.0), $(4, -0.5)$, $(6, -1.0)$
3. **(Ohm's law; Sec. 1.7)** Also find the resistance.

Voltage x [volts]	30	30	60	60	90	90
Current y [amperes]	3.1	3.2	6.3	6.5	10.0	10.1

4. **(Hooke's law; Sec. 2.5)** Also find the spring modulus.

Force x [lb]	1	2	3	4
Extension y [in.]	3.1	5.9	8.8	12.1

5. **(Thermal conductivity of water)** Also find y at room temperature 66°F.

Temperature x [°F]	32	50	100	150	212
Conductivity y [Btu/hr·ft·°F]	0.337	0.345	0.365	0.380	0.395

6. **(Stopping distance of a passenger car)** Also find y at 35 mph.

Speed x [mph]	30	40	50	60
Stopping distance y [ft]	160	240	330	435

Confidence Interval

Using the given sample and assuming that assumptions (A2) and (A3) hold, find a 95% confidence interval for the regression coefficient κ_1.

7. The sample in Prob. 1   8. The sample in Prob. 4   9. The sample in Prob. 6

10. Derive the second expression for s_x^2 in (9) from the first one.

23.10 Correlation Analysis

This section serves as an introduction to correlation analysis; for proofs see Ref. [G8] in Appendix 1. **Correlation analysis** is concerned with the relation between X and Y in a two-dimensional random variable (X, Y) (Sec. 22.9). A sample consists of n ordered pairs of values $(x_1, y_1), \cdots, (x_n, y_n)$. From Sec. 23.9 we shall use the sample means

$$\bar{x} = \frac{1}{n} \sum_{j=1}^{n} x_j, \qquad \bar{y} = \frac{1}{n} \sum_{j=1}^{n} y_j,$$

the sample variances

$$s_x{}^2 = \frac{1}{n-1} \sum_{j=1}^{n} (x_j - \bar{x})^2 \qquad s_y{}^2 = \frac{1}{n-1} \sum_{j=1}^{n} (y_j - \bar{y})^2$$

and the sample **covariance**

$$s_{xy} = \frac{1}{n-1} \sum_{j=1}^{n} (x_j - \bar{x})(y_j - \bar{y}).$$

The sample **correlation coefficient** is

(1)
$$r = \frac{s_{xy}}{s_x s_y}.$$

Both s_{xy} and r measure the interrelation between the x and y values. But r has the advantage that it does not change under a multiplication of these values by a factor (in going from feet to inches, etc.):

THEOREM 1   **(Sample correlation coefficient)**

r satisfies $-1 \leqq r \leqq 1$, and $r = \pm 1$ if and only if the sample values lie on a straight line. (See Fig. 508.)

The theoretical counterpart of r is the **correlation coefficient** ρ of X and Y,

(2)
$$\rho = \frac{\sigma_{XY}}{\sigma_X \sigma_Y}$$

where $\mu_X = E(X)$, $\mu_Y = E(Y)$, $\sigma_X{}^2 = E([X - \mu_X]^2)$, $\sigma_Y{}^2 = E([Y - \mu_Y]^2)$ (the means and variances of the marginal distributions of X and Y; see Sec. 22.9), and σ_{XY} is the **covariance** of X and Y given by (see Sec. 22.9)

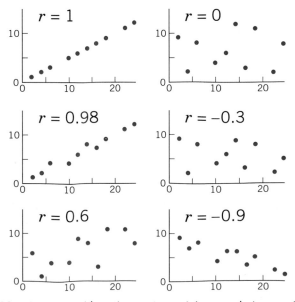

Fig. 508. Samples with various values of the correlation coefficient r

(3) $$\sigma_{XY} = E([X - \mu_X][Y - \mu_Y]) = E(XY) - E(X)E(Y).$$

The analog of Theorem 1 is

THEOREM 2   **(Correlation coefficient)**

ρ satisfies $-1 \leq \rho \leq 1$, and X and Y are linearly related $(Y = \gamma X + \delta, X = \gamma^* Y + \delta^*)$ if and only if $\rho = \pm 1$.

X and Y are called **uncorrelated** if $\rho = 0$.

THEOREM 3   **(Independence. Normal distribution)**

(a) Independent X and Y (see Sec. 22.9) are uncorrelated.

(b) If (X, Y) is normal (see below), then uncorrelated X and Y are independent.

Here the two-dimensional normal distribution can be introduced by taking two independent standardized normal random variables X^*, Y^*, whose joint distribution thus has the density

(4) $$f^*(x^*, y^*) = \frac{1}{2\pi} e^{-(x^{*2} + y^{*2})/2}$$

(representing a surface of revolution over the $x^* y^*$-plane with a bell-shaped curve as cross section) and setting

$$X = \mu_X + \sigma_X X^*$$

$$Y = \mu_Y + \rho \sigma_Y X^* + \sqrt{1 - \rho^2}\, \sigma_Y Y^*.$$

This gives the general **two-dimensional normal distribution** with the density

(5a) $$f(x, y) = \frac{1}{2\pi \sigma_X \sigma_Y \sqrt{1 - \rho^2}} e^{-h(x,y)/2}$$

where

(5b) $$h(x, y) = \frac{1}{1 - \rho^2} \left[\left(\frac{x - \mu_X}{\sigma_X} \right)^2 - 2\rho \left(\frac{x - \mu_X}{\sigma_X} \right) \left(\frac{y - \mu_Y}{\sigma_Y} \right) + \left(\frac{y - \mu_Y}{\sigma_Y} \right)^2 \right].$$

As we see from the following example, normality is important in Theorem 3(b).

EXAMPLE 1   **Uncorrelated but dependent random variables**

If X assumes $-1, 0, 1$ with probability 1/3 and $Y = X^2$, then $E(X) = 0$ and in (3)

$$\sigma_{XY} = E(XY) = E(X^3) = (-1)^3 \cdot \frac{1}{3} + 0^3 \cdot \frac{1}{3} + 1^3 \cdot \frac{1}{3} = 0,$$

so that $\rho = 0$ and X and Y are uncorrelated. But they are certainly not independent since they are even functionally related.   ◀

Test for the Correlation Coefficient ρ

Table 23.13 shows a test for ρ in the case of the two-dimensional normal distribution. t is an observed value of a random variable that has a t-distribution with $n - 2$ degrees of freedom. This was shown by R. A. Fisher in 1915 (*Biometrika* **10**, 507−521).

Table 23.13
Test of the Hypothesis $\rho = 0$ Against the Alternative $\rho > 0$ in the Case of the Two-Dimensional Normal Distribution

1st Step. Choose a significance level α (5%, 1%, or the like).

2nd Step. Determine the solution c of the equation

$$P(T \le c) = 1 - \alpha$$

from the t-distribution (Table A9 in Appendix 5) with $n - 2$ degrees of freedom.

3rd Step. Compute r from (1), using a sample $(x_1, y_1), \cdots, (x_n, y_n)$.

4th Step. Compute

$$t = r\left(\sqrt{\frac{n-2}{1-r^2}}\right).$$

If $t \le c$, accept the hypothesis. If $t > c$, reject the hypothesis.

EXAMPLE 2   **Test for the correlation coefficient ρ**

Test the hypothesis $\rho = 0$ (independence of X and Y, because of Theorem 3) against the alternative $\rho > 0$, using the data in the lower left corner of Fig. 508, where $r = 0.6$ (manual soldering errors on 10 two-sided circuit boards done by 10 workers; $x =$ front, $y =$ back of the boards).

Solution.   We choose $\alpha = 5\%$; thus $1 - \alpha = 95\%$. Since $n = 10$, $n - 2 = 8$, the table gives $c = 1.86$. Also, $t = 0.6\sqrt{8/0.64} = 2.12 > c$. We reject the hypothesis and assert that there is a **positive correlation.** A worker making few (many) errors on the front side also tends to make few (many) errors on the reverse side of the board. ◄

CHAPTER 23 REVIEW

1. What is a sample? A population? Why do we sample in statistics?
2. If we have several samples from the same population, do they have the same sample distribution function? The same mean and variance?
3. What role does probability theory play in statistics?
4. What is a parameter? Why are parameters important?
5. Which results from Chap. 22 did we use in Chap. 23? (Answer from memory.)
6. Can we apply a statistical method without having a sample?
7. We have discussed two kinds of estimating parameters. What are they? How do they compare?

8. What is the idea of the maximum likelihood method? Why do we say "likelihood" rather than "probability"?

9. Couldn't we make the error of interval estimation zero simply by choosing the confidence level 1?

10. What is testing? Why do we test? What are the errors involved?

11. Make a list of concepts related to testing (from memory) and explain them.

12. What will generally happen to errors in testing if we take larger samples?

13. What will generally happen to confidence intervals if we take larger samples?

14. When did we use the t-distribution? The F-distribution?

15. What is the chi-square (χ^2) test? Give a simple example from memory.

16. What is a nonparametric test? What are its advantages?

17. What are one-sided and two-sided tests? Give typical examples.

18. How do we test in quality control?

19. Acceptance sampling uses principles of testing. Explain.

20. What is the power of a test and what do you do when it is low?

21. Can you also apply tests for the normal distribution to other distributions? Explain.

22. What is Gauss's least squares principle (which he found at age 18)?

23. What is regression? Give two typical examples.

24. What is correlation? Explain how it differs from regression.

25. Are independent random variables uncorrelated? What about the converse? Explain.

Find the mean, variance and standard deviation of the samples.

26. 21.0   21.6   19.9   19.6   15.6   20.6   22.1   22.2

27. 0.28   −1.5   −1.2   0.15   2.0   0.69   1.2   0.50   0.47   0.087

Assuming normality, find the maximum likelihood estimates of mean and variance from the sample:

28. In Prob. 26.   29. In Prob. 27.

30. Determine a 95% confidence interval for the mean μ of a normal population with variance $\sigma^2 = 25$, using a sample of size 500 with mean 22.

31. What will happen to the length of the interval in Prob. 30 if we reduce the sample size to 125?

32. Determine a 99% confidence interval for the mean of a normal population with standard deviation 3.2, using the sample 32, 33, 32, 34, 35, 29, 29, 27.

33. What confidence interval do we obtain in Prob. 32 if we assume the variance to be unknown?

34. Find a 95% confidence interval for the mean from a sample of 25 values with $\bar{x} = 116$ and $s = 7$, assuming normality.

35. Assuming normality, find a 95% confidence interval for the variance in Prob. 34.

36. Assuming normality, find a 95% confidence interval for the variance from the sample 145.3, 145.1, 145.4, 146.2.

37. Using a sample of 10 values with mean 14.5 from a normal population with variance $\sigma^2 = 0.25$, test the hypothesis $\mu_0 = 15.0$ against the alternative $\mu_1 = 14.5$ on the 5% level.

38. In Prob. 37, change the alternative to $\mu \neq 15.0$ and test as before.

39. Find the power in Prob. 37.

40. Using a sample of 20 values with mean 29.8 and variance 1.2, test the hypothesis $\mu_0 = 28.0$ against the alternative $\mu_1 = 30.0$, assuming normality and taking $\alpha = 1\%$.

41. If a sample of 100 tires of a certain kind has a mean life of 26 000 km and a standard deviation of 2000 km, can the manufacturer claim that the true mean life of such tires is greater than 25 000 km? Set up and test a corresponding hypothesis, assuming normality and choosing $\alpha = 1\%$.

42. Three specimens of high-quality concrete had compressive strength 357, 359, 413 $[\text{kg/cm}^2]$, and for three specimens of ordinary concrete the values were 346, 358, 302. Test for equality of the population means, $\mu_1 = \mu_2$, against the alternative $\mu_1 > \mu_2$. (Assume normality and equality of variances. Choose $\alpha = 5\%$.)

43. Assume the thickness X of washers to be normal with mean 2.75 mm and variance 0.00024 mm^2. Set up a control chart for μ and graph the means of the five samples (2.74, 2.76), (2.74, 2.74), (2.79, 2.81), (2.78, 2.76), (2.71, 2.75) on the chart.

44. The OC curve in acceptance sampling cannot have a strictly vertical portion. Why?

45. Samples of 5 screws are drawn from a lot with fraction defective θ. The lot is accepted if the sample contains (a) no defective screws, (b) at most 1 defective screw. Using the binomial distribution, find, graph, and compare the OC curves.

46. Find the risks in the sampling plan with $n = 6$ and $c = 0$, assuming that the AQL is $\theta_0 = 1\%$ and the RQL is $\theta_1 = 15\%$. How do the risks change if we increase n?

47. The number of books borrowed from a public library during a particular week was 500 on Monday, 450 on Tuesday, 480 on Wednesday, 460 on Thursday, and 510 on Friday. Using this sample, test the hypothesis that the number of books borrowed does not depend on the day of the week. Use $\alpha = 5\%$.

48. Does a process of producing plastic rods of length $\tilde{\mu} = 2$ meters need adjustment if in a sample, 2 rods have the exact length and 15 are shorter and 3 longer than 2 meters? (Use the sign test.)

49. Find the regression line of y (expansion of gelatin [%]) on x (humidity of air [%]), using $(x, y) = (10, 0.8), (20, 1.6), (30, 2.3), (40, 2.8)$.

50. Find the regression line of y on x for the data $(x, y) = (0, 4), (2, 0), (4, -5), (6, -9), (8, -10)$.

SUMMARY OF CHAPTER 23
MATHEMATICAL STATISTICS

We recall from Chap. 22 that with an experiment in which we observe some quantity (number of defectives, height of persons, etc.) there is associated a random variable X whose probability distribution is given by a distribution function

$$(1) \qquad\qquad F(x) = P(X \leqq x) \qquad\qquad (\text{Sec. 22.5})$$

which for each x gives the probability that X assumes any value not exceeding x. In statistics we take random samples $x_1, \cdots, x_n$ of size n by performing that experiment n times (Sec. 23.1) and draw conclusions from properties of samples about properties of the distribution of the corresponding X. We do this by calculating *point estimates* or *confidence intervals* or by performing a *test* for **parameters** (μ and σ^2 in the normal distribution, p in the binomial distribution, etc.) or by a test for distribution functions.

A **point estimate** (Sec. 23.2) is an approximate value for a parameter in the distribution of X obtained from a sample. For instance, the **sample mean** (Sec. 23.1)

$$(2) \qquad \bar{x} = \frac{1}{n} \sum_{j=1}^{n} x_j = \frac{1}{n} (x_1 + \cdots + x_n)$$

is an estimate of the mean μ of X, and the **sample variance** (Sec. 23.1)

$$(3) \qquad s^2 = \frac{1}{n-1} \sum_{j=1}^{n} (x_j - \bar{x})^2 = \frac{1}{n-1} \left[(x_1 - \bar{x})^2 + \cdots + (x_n - \bar{x})^2 \right]$$

is an estimate of the variance σ^2 of X. Point estimation can be done by the basic **maximum likelihood method** (Sec. 23.2).

Confidence intervals (Sec. 23.3) are intervals $\theta_1 \leq \theta \leq \theta_2$ with endpoints calculated from a sample such that with a high probability γ we obtain an interval that contains the unknown true value of the parameter θ in the distribution of X. Here, γ is chosen at the beginning, usually 95% or 99%. We denote such an interval by $\text{CONF}_\gamma \{ \theta_1 \leq \theta \leq \theta_2 \}$.

In a **test** for a parameter we test a *hypothesis* $\theta = \theta_0$ against an *alternative* $\theta = \theta_1$ [this notation has nothing to do with the θ_1 above] and then, on the basis of a sample, accept the hypothesis, or we reject it in favor of the alternative (Sec. 23.4). Like any conclusion about X from samples, this may involve errors leading to a false decision. There is a small probability α (which we can choose, 5% or 1%, for instance) that we reject a true hypothesis, and there is a probability β (which we can compute and decrease by taking larger samples) that we accept a false hypothesis. α is called the **significance level** and $1 - \beta$ the **power** of the test. Among many other engineering applications, testing is used in **quality control** (Sec. 23.5) and **acceptance sampling** (Sec. 23.6).

If not merely a parameter but the kind of distribution of X is unknown, we can use the **chi-square test** (Sec. 24.7) for testing the hypothesis that some function $F(x)$ is the unknown distribution function of X. This is done by determining the discrepancy between $F(x)$ and the distribution function $\tilde{F}(x)$ of a given sample.

"Distribution-free" or **nonparametric tests** are tests that apply to any distribution, since they are based on combinatorial ideas. These tests are usually very simple. Two of them are discussed in Sec. 23.8.

The last two sections deal with samples of **pairs of values,** which arise in an experiment when we simultaneously observe two quantities. In **regression analysis** (Sec. 23.9), one of the quantities, x, is an ordinary variable and the other, Y, is a random variable whose mean μ depends on x, say, $\mu(x) = \kappa_0 + \kappa_1 x$. In **correlation analysis** (Sec. 23.10) the relation between X and Y in a two-dimensional random variable (X, Y) is investigated, notably in terms of the **correlation coefficient** ρ.

APPENDIX 1

References

General References

Software *see* **p. 829.**

[1] Abramowitz, M. and I. A. Stegun (eds.), *Handbook of Mathematical Functions.* 10th printing, with corrections. Washington, DC: National Bureau of Standards, 1972. (Also New York: Dover, 1965.)

[2] Cajori, F., *A History of Mathematics.* 3rd ed. New York: Chelsea, 1980.

[3] *CRC Handbook of Mathematical Sciences.* 6th ed. Boca Raton, FL: CRC Press, 1987.

[4] Courant, R. and D. Hilbert, *Methods of Mathematical Physics.* 2 vols. New York: Wiley-Interscience, 1989.

[5] Courant, R., *Differential and Integral Calculus.* 2 vols. New York: Wiley, 1988.

[6] Erdélyi, A., W. Magnus, F. Oberhettinger and F. G. Tricomi, *Higher Transcendental Functions.* 3 vols. New York: McGraw-Hill, 1953, 1955.

[7] Graham, R. L., D. E. Knuth and O. Patashnik, *Concrete Mathematics.* 2nd ed. Reading, MA: Addison-Wesley, 1994.

[8] Itô, K. (ed.), *Encyclopedic Dictionary of Mathematics.* 4 vols. 2nd ed. Cambridge, MA: MIT Press, 1987.

[9] Kreyszig, E., *Introductory Functional Analysis with Applications.* New York: Wiley, 1989.

[10] Kreyszig, E., *Differential Geometry.* Mineola, NY: Dover, 1991.

[11] Magnus, W., F. Oberhettinger and R. P. Soni, *Formulas and Theorems for the Special Functions of Mathematical Physics.* 3rd ed. New York: Springer, 1966.

[12] Szegö, G., *Orthogonal Polynomials.* 4th ed. New York: American Mathematical Society, 1975. (9th printing 1995.)

[13] Thomas, G. B. and R. L. Finney, *Calculus and Analytic Geometry.* 9th ed. Reading, MA: Addison-Wesley, 1996.

Part A.   Ordinary Differential Equations (Chaps. 1−5)

See also **Part E: Numerical Methods.**

[A1] Birkhoff, G. and G.-C. Rota, *Ordinary Differential Equations.* 4th ed. New York: Wiley, 1989.

[A2] Churchill, R. V., *Operational Mathematics.* 3rd ed. New York: McGraw-Hill, 1972.

[A3] Coddington, E. A. and N. Levinson, *Theory of Ordinary Differential Equations.* New York: McGraw-Hill, 1955.

[A4] Erdélyi, A., W. Magnus, F. Oberhettinger and F. Tricomi, *Tables of Integral Transforms.* 2 vols. New York: McGraw-Hill, 1954.

[A5] Ince, E. L., *Ordinary Differential Equations.* New York: Dover, 1956.

[A6] Oberhettinger, F. and L. Badii, *Tables of Laplace Transforms.* New York: Springer, 1973.

[A7] Watson, G. N., *A Treatise on the Theory of Bessel Functions.* 2nd ed. Cambridge: University Press, 1944. (Reprinted 1966.)

[A8] Widder, D. V., *The Laplace Transform.* Princeton, NJ: Princeton University Press, 1941.

[A9] Zwillinger, D., *Handbook of Differential Equations.* 3rd ed. New York: Academic Press, 1998.

Part B.   Linear Algebra, Vector Calculus (Chaps. 6—9)

For books on *numerical* linear algebra *see also* Part E: Numerical Methods.

[B1] Chatelin, F., *Eigenvalues of Matrices.* New York: Wiley-Interscience, 1993.

[B2] Gantmacher, F. R., *The Theory of Matrices.* 2 vols., 2nd ed. New York: Chelsea, 1990, 1988.

[B3] Gohberg, I., P. Lancaster and L. Rodman, *Invariant Subspaces of Matrices with Applications.* New York: Wiley, 1986.

[B4] Herstein, I. N., *Topics in Algebra.* 2nd ed. Lexington, MA: Xerox College Publishing, 1975.

[B5] Lang, S., *Linear Algebra.* 3rd ed. New York: Springer, 1987.

[B6] MacDuffee, C. C., *The Theory of Matrices.* New York: Chelsea, 1946.

[B7] Nef, W., *Linear Algebra.* 2nd ed. New York: Dover, 1988.

[B8] Wilkinson, J. H., *The Algebraic Eigenvalue Problem.* Oxford: Clarendon, 1965.

Part C.   Fourier Analysis and Partial Differential Equations (Chaps. 10, 11)

For books on *numerical* methods for partial differential equations, *see also* Part E: Numerical Methods.

[C1] Carslaw, H. S. and J. C. Jaeger, *Conduction of Heat in Solids.* 2nd ed. Oxford: Clarendon, 1959. (Reprinted 1986.)

[C2] Churchill, R. V. and J. W. Brown, *Fourier Series and Boundary Value Problems.* 4th ed. New York: McGraw-Hill, 1987.

[C3] Erdélyi, A., W. Magnus, F. Oberhettinger and F. Tricomi, *Tables of Integral Transforms.* 2 vols. New York: McGraw-Hill, 1954.

[C4] Hanna, J. R. and J. H. Rowland, *Fourier Series, Transforms and Boundary Value Problems.* 2nd ed. New York: Wiley, 1990.

[C5] John, F., *Partial Differential Equations.* 4th ed. New York: Springer, 1982.

[C6] Tolstov, G. P., *Fourier Series.* New York: Dover, 1976.

[C7] Widder, D. V., *The Heat Equation.* New York: Academic Press, 1975.

[C8] Zauderer, E., *Partial Differential Equations of Applied Mathematics.* 2nd ed. New York: Wiley, 1989.

[C9] Zygmund, A., *Trigonometric Series.* 2nd ed., reprinted with corrections. Cambridge: University Press, 1988.

Part D.   Complex Analysis (Chaps. 12—16)

[D1] Ahlfors, L. V., *Complex Analysis.* 3rd ed. New York: McGraw-Hill, 1979.

[D2] Ahlfors, L. V. and L. Sario, *Riemann Surfaces.* Princeton, NJ: Princeton University Press, 1971.

[D3] Bieberbach, L., *Conformal Mapping.* New York: Chelsea, 1964.

[D4] Henrici, P., *Applied and Computational Complex Analysis.* 3 vols. New York: Wiley, 1993.

[D5] Hille, E., *Analytic Function Theory.* 2 vols. 2nd ed. New York: Chelsea, 1990, 1987.

[D6] Knopp, K., *Elements of the Theory of Functions.* New York: Dover, 1952.

[D7] Knopp, K., *Theory of Functions.* 2 parts. New York: Dover, 1945, 1947.

[D8] Titchmarsh, E. C., *The Theory of Functions.* 2nd ed. London: Oxford University Press, 1939. (Reprinted 1975.)

Part E.   Numerical Methods (Chaps. 17—19)

[E1] Ames, W. F., *Numerical Methods for Partial Differential Equations.* 3rd ed. New York: Academic Press, 1992.

[E2] Anderson, E. et al., *LAPACK User's Guide.* 2nd ed. Philadelphia: SIAM, 1995.

[E3] Dahlquist, G. and Å. Björck, *Numerical Methods.* Englewood Cliffs, NJ: Prentice-Hall, 1974.

[E4] DeBoor, C., *A Practical Guide to Splines*. New York: Springer, 1978. (Reprinted 1991.)

[E5] Dongarra, J. J. et al., *LINPACK Users Guide*. Philadelphia: SIAM Publications, 1978. (*See also p.* 829.)

[E6] Forsythe, G. E., M. A. Malcolm and C. B. Moler, *Computer Methods for Mathematical Computations*. Englewood-Cliffs, NJ: Prentice-Hall, 1977.

[E7] Garbow, B. S. et al., *Matrix Eigensystem Routines: EISPACK Guide Extension*. New York: Springer, 1972. (Reprinted 1990.) (*See also p.* 829.)

[E8] Golub, G. H. and C. F. Van Loan, *Matrix Computations*. 3rd ed. Baltimore: Johns Hopkins University Press, 1996.

[E9] Higham, N. J., *Accuracy and Stability of Numerical Algorithms*. Philadelphia: SIAM, 1996.

[E10] IMSL (International Mathematical and Statistical Libraries), *FORTRAN Subroutines for Mathematics and Statistics. User's Manuals, Version 2.0*. 6 vols. Houston, TX: Visual Numerics, 1991. (*See also p.* 829.)

[E11] Knuth, D. E., *The Art of Computer Programming*. 3 vols. 2nd ed. Reading, MA: Addison-Wesley, 1981.

[E12] Mortensen, M. E., *Geometric Modeling*. New York: Wiley, 1985.

[E13] Press, W. H. et al., *Numerical Recipes in C: The Art of Scientific Computing*. 2nd ed. Cambridge: University Press, 1992.

[E14] Shampine, L. F., *Numerical Solution of Ordinary Differential Equations*. New York: Chapman and Hall, 1994.

[E15] Shampine, L. F. et al., *Fundamentals of Numerical Computing*. New York: Wiley, 1996.

[E16] Smith, B. T. et al., *Matrix Eigensystem Routines–EISPACK Guide*. 2nd ed. New York: Springer, 1976. (Reprinted 1990.) (*See also p.* 829.)

[E17] Stewart, G. W., *Introduction to Matrix Computations*. New York: Academic Press, 1973.

[E18] Strikwerda, J. C., *Finite Difference Schemes and Partial Differential Equations*. Pacific Grove, CA: Brooks/Cole, 1989.

[E19] Wilkinson, J. H., *The Algebraic Eigenvalue Problem*. Oxford: Clarendon, 1988.

[E20] Wilkinson, J. H. and C. Reinsch, *Linear Algebra*. New York: Springer, 1971.

Part F.   Optimization, Graphs (Chaps. 20, 21)

[F1] Berge, C., *Graphs*. 3rd ed. New York: North-Holland, 1985.

[F2] Bondy, J. A. and U. S. R. Murty, *Graph Theory with Applications*. New York: North-Holland, 1976.

[F3] Chvátal, V., *Linear Programming*. San Francisco: Freeman, 1983.

[F4] Cook, W. J. et al., *Combinatorial Optimization*. New York: Wiley, 1993.

[F5] Dixon, L. C. W., *Nonlinear Optimization*. London: English Universities Press, 1972.

[F6] Ford, L. R., Jr. and D. R. Fulkerson, *Flows in Networks*. Princeton, NJ: Princeton University Press, 1962. (Reprinted 1992.)

[F7] Gass, S. I., *Linear Programming, Methods and Applications*. 3rd ed. New York: McGraw-Hill, 1969.

[F8] Gondran, M. and M. Minoux, *Graphs and Algorithms*. New York: Wiley, 1984.

[F9] Harari, F., *Graph Theory*. Reading, MA: Addison-Wesley, 1969.

[F10] Papadimitriou, C. H. and K. Steiglitz, *Combinatorial Optimization: Algorithms and Complexity*. Englewood Cliffs, NJ: Prentice-Hall, 1982.

Part G.   Probability and Statistics (Chaps. 22, 23)

[G1] American Society for Testing and Materials, *Manual on Presentation of Data and Control Chart Analysis*. 6th ed. Philadelphia: ASTM, 1992.

[G2] Anderson, T. W., *An Introduction to Multivariate Statistical Analysis.* 2nd ed. New York: Wiley, 1984.

[G3] Cramér, H., *Mathematical Methods of Statistics.* Princeton: Princeton University Press, 1946. (Reprinted 1991.)

[G4] Gibbons, J. D., *Nonparametric Statistical Inference.* 2nd ed. New York: Dekker, 1985.

[G5] Grant, E. L. and R. S. Leavenworth, *Statistical Quality Control.* 5th ed. New York: McGraw-Hill, 1980.

[G6] IMSL, *FORTRAN Subroutines for Mathematics and Statistics. User's Manuals, Version 2.0.* 6 vols. Houston, TX: Visual Numerics, 1991. (*See also p.* 829.)

[G7] Kendall, M. and A. Stuart, *The Advanced Theory of Statistics.* 3 vols. 4th ed. New York: Macmillan, 1977, 1979, 1983. (Vol. 3 coauthored by K. Ord.) (*See also* [G9].)

[G8] Kreyszig, E., *Introductory Mathematical Statistics. Principles and Methods.* New York: Wiley, 1970.

[G9] Stuart, A. and J. K. Orth, *Kendall's Advanced Theory of Statistics.* Vols. 1, 2. 5th ed. Kent, U.K.: Arnold, 1987, 1991.

APPENDIX 2

Answers to Odd-Numbered Problems

PROBLEM SET 1.1, page 8

1. $x^3/3 + c$

3. $\frac{1}{6}x^2 + ax + b$

5. 1

7. 3

9. 1

11. Nothing

13. $1.4 - 0.4e^{-2x}$

15. $y^2 - 2x^2 = 1$

17. $x^2/2^2 + y^2 = 2$

19. $\exp(-1.4 \cdot 10^{-11}t) = 1/2, t = 10^{11}(\ln 2)/1.4$ [sec] $= 1570$ years

21. $e^{-3.636k} = 1/2, 0.8264, 6.040\,77 \cdot 10^{-31}$

23. 4.5 sec, 6.4 sec

25. $y_0 = 5.3, k = \ln(13/5.3)/30 = 0.03, y = 5.3e^{0.03t}$ gives 32 for 1860 ($t = 60$), but the other values are much too large. For a better model, see Sec. 1.6.

PROBLEM SET 1.2, page 12

1. $y = ce^{x^2/2}, c = \pm 1, \pm 2$

3. $y = ce^{-x}$

5. $y = \dfrac{1}{\pi}\sin \pi x + c$

17. $v(0) < 3.13$

PROBLEM SET 1.3, page 18

1. If you add a constant later, you may not get a solution. Example: $y' = y, \ln |y| = x + c$, $y = e^{x+c} = \tilde{c}e^x$ but not $e^x + c$ (with $c \neq 0$).

3. $10 \tan (0.1x + c)$

5. $ce^{x^2/4}$

7. $x/(c - x)$

9. $x\sqrt{2 \ln |x| + c}$

11. $\arccos (x + c)$

13. $xy = -4$

15. $1/(2 + (4 + 2x)e^{-x})$

17. $\cot y = -\tanh x$

19. $I = I_0 e^{-(R/L)t}$

21. $x \arctan (x^3 - 1)$

23. $(2x^4 - 4x^2)^{1/2}$

25. $2y - 2x + (y + 2x)^2 = c$

27. CAS PROJECT. See (36) in Appendix A3.1 for the Maclaurin series.

PROBLEM SET 1.4, page 23

1. $2^3 y_0, 2^7 y_0 = 128 y_0$

3. 45.4 sec, 281 km/hr. Don't forget to add the initial speed!

5. 69.6% of y_0

7. $1/2^6 = 0.016 > 0.01 > 1/2^7 = 0.0078$

9. $pV = c = const$, the law empirically found by Boyle (1662) and Mariotte (1676).

11. $y(t) = 0.01e^{-0.275t}$

13. $T = 22 - 17e^{-0.5306t} = 21.9°C$ when $t = 9.68$ min.

15. Ellipses $y^2 + 4x^2 = c$

17. $y' = y/x, y = cx$

19. PROJECT. $y_0 e^{kt_1} = y_1$, $y_0 e^{kt_2} = y_2$, $e^{k(t_1 - t_2)} = \dfrac{y_1}{y_2}$, $t_H = \dfrac{1}{k} \ln \dfrac{1}{2} = \dfrac{t_1 - t_2}{\ln(y_1/y_2)} \ln \dfrac{1}{2}$

$\quad = \dfrac{1 - 4}{\ln(0.2/0.05)} \ln \dfrac{1}{2} = 1.5$

PROBLEM SET **1.5**, page 31

1. $2x\,dx + 8y\,dy = 0$ **3.** $e^{x^2/y}(2xy^{-1}\,dx - x^2 y^{-2}\,dy) = 0$

5. $(2y\,dy - 3x^2\,dx)/\cos^2(y^2 - x^3) = 0$, semicubical parabolas

7. $x^2 y = c$ **9.** $\cosh x \cos y = c$

11. $r^2 e^{-2\theta} = c$ **13.** No, $1/(3 \ln |x| + 2)$

15. No, $y = \sqrt{x^2 + 3x}$. Use $u = y/x$.   **17.** Yes, $(e^x - e^y)x = e - 1$

19. Yes, $\sinh y \cos 2x = \sinh 1$ **21.** $b = k$, $ax^2 + 2kxy + ly^2 = c$

23. $xy^2 = c$ **25.** $e^x \sin y = c$

27. $y^{b+1} x^{a+1} = c$ **29.** $x^2 y^2 e^x = c$

31. $F = \sinh x$, $\sinh^2 x \cos y = c$ **33.** $F = x$, $x^2 \cos y + x^4 = c$

35. $F = e^{x^2}$, $e^{x^2} \tan y = c$ **37.** $F = x$, $x \cosh y = c$

PROBLEM SET **1.6**, page 38

3. $y = ce^x - 4$ **5.** $y = ce^{-3x^2/2}$ **7.** $y = (c + x)e^{-kx}$

9. $y = x^2(c + e^x)$ **11.** $y = 2 + c \sin x$ **13.** $y = (c + x)e^{\cos x}$

15. $y = -3e^{-4x} + 5$ **17.** $y = 1 + 3 \cosh 2x$ **19.** $y = -0.1 \cos x + 0.3 \sin x$

21. $y = \tan x$ **31.** $y = 2/(1 + 2ce^{2x})$ **33.** $y^{-3} = ce^x - 2x - 1$

35. $x = ce^{-2y} + 2e^y$ **37.** $y^2 = 1 + ce^{-x^2}$

39. $y(t) = 1000 + e^{-0.1t}(-2.494 \cos t + 49.88 \sin t) - 797.51e^{-0.05t}$

41. $T = 240e^{kt} + 60$, $T(10) = 200$, $k = -0.0539$, $t = 102$ min

45. Substitute $y = w + v$, $y' = w' + v'$. Since v is a solution, there remains
$\quad w' = -pw + g(w^2 + 2wv)$, a Bernoulli equation.

47. $y' = y' + xy'' + (dg/ds)y''$, $y''(x + dg/ds) = 0$, etc.

PROBLEM SET **1.7**, page 47

3. $t = L/R$ gives $1 - e^{-\alpha t} = 1 - e^{-1} = 0.63$.

5. $e^{-t/\tau_L} = \frac{1}{2}$ gives $t = \tau_L \ln 2 = 0.7\tau_L$ [sec].

7. $e^{-Rt/L} = 0.75$, $L = Rt/\ln(1/0.75) = 0.348$ because $Rt = 1/10$.

9. (a) $I(t) = ce^{-(R/L)t} + e^{-t}/(R - L)$, (b) $I(t) = (c + t/L)e^{-t}$

13. $RQ' + Q/C = E_0$, $Q(t) = E_0 C(1 - e^{-t/RC})$, $V(t) = Q(t)/C = 24(1 - e^{-0.025t})$

15. $Q = Q_0 e^{-t/RC}$

17. $Q = 1.5(e^{-t} - e^{-3t})$, $Q' = 0$ gives $t_m = 0.549$, $Q_m = 0.577$ coulomb.

19. $Q = \sin 2t + \sin 4t$

PROBLEM SET **1.8**, page 51

1. $x^2/(c^2 + 4) + y^2/c^2 - 1 = 0$

3. $y - \cosh(x - c) + c = 0$

5. $y' = 2y$ **7.** $xy' = y$ **9.** $y' = xy/(x^2 - 1)$

11. $y = \sqrt{2x + c^*}$ **13.** $x^2 + y^2/2 = c^*$ **15.** $2x^2 + 3y^2 = c^*$

17. $x^2 - 2y^2 = c^*$ **19.** $y = c^* x$ **21.** $x^2 - y^2 = c^*$

PROBLEM SET 1.9, page 58

1. General solution $y = cx^4$. No; $f(x, y) = 4y/x$ is not defined when $x = 0$.

3. $\alpha = a$ because we can take b as large as we please.

5. Two solutions would then satisfy the same initial condition $y(x_1) = y_1$, where (x_1, y_1) is the common point.

7. $f(x, y) = r(x) - p(x)y,\ \partial f/\partial y = -p(x)$

9. $K = 1 + b^2$ is smallest. The derivative of $b/(1 + b^2)$ is zero for $b = 1$.
Ans. $\alpha = 1/(1 + 1) = 1/2$.

13. $y_0 = -1,\ y_n = -1 - x + x^{n+1}/(n + 1)!,\ y = -1 - x$

15. $y_n = x^2 - x^{2n+2}/2^n(n + 1)!,\ y = x^2$

17. $\dfrac{1}{2}, \dfrac{1}{2} + \dfrac{x}{4}, \dfrac{1}{2} + \dfrac{x}{4} - \dfrac{x^3}{48}, \dfrac{1}{2} + \dfrac{x}{4} - \dfrac{x^3}{48} + \dfrac{x^5}{480} - \dfrac{x^7}{16128},$ etc., $y = 1/(1 + e^{-x})$

CHAPTER 1 REVIEW, page 59

15. $y = ce^{-4x} - \cos x + 4 \sin x$

17. $9x^2 - 25y^2 = c$

19. $\cos x \sin 2y = c$

21. $y^2 + x^2/3 = c\sqrt{x}$

23. $F = x,\ x^3 e^y + x^2 y = c$

25. $y = \pi \cosh x$

27. $y = (3x - x^2)^{-1}$

29. $9 \sin x + \sin y = 0$

31. $F = y,\ x^3 y^2 = 9$

33. $y = +(1 + 3e^{-x^2})^{1/2}$

35. $y = ce^{-3x}$

37. $y = ce^x - 0.1 \cos 10x - 0.01 \sin 10x$

39. $y = -\frac{1}{2} \ln |x| + c^*$

41. $y = \sqrt{2} \ln |x| + c^*$

43. $1, 1 + 2x, 1 + 2x + (2x)^2/2$, etc.

45. 1.7 days, 2.7 days

47. $e^k = 0.9$, 6.6 days, 43.7 days from $e^{kt} = 0.5,\ e^{kt} = 0.01$

49. $I = I_0 e^{-(t - t_2)/RC}$ with $I_0 = -0.99 E_0/R$ from (7), Sec. 1.7.

51. $a = 99 \cdot 10^8$ meters/sec^2, $s = \frac{1}{2}at^2 + v_0 t = 50.5$ [meters]

53. $y^2 - x^2 = c^*$

55. $r^2 = a^2 \cos 2\theta$

PROBLEM SET 2.1, page 71

1. $F(x, z, z') = 0$

3. $c_1 e^x + c_2$

5. $(c_1 x + c_2)^{-1}$

7. $x = e^y + c_1 y + c_2$

9. $y_2 = x^3 \ln |x|$

11. $y_2 = x^{-1/2} \sin x$

13. $y = [(2t + 4)^{3/2} - 2]/3,\ y(6) = 62/3,\ y'(6) = 4$

15. $y = (e^{2x} - 1)/2$

17. $4 \cos 3x - 2 \sin 3x$

19. $x^{-1/2} + 2x^{3/2}$

PROBLEM SET 2.2, page 75

1. $c_1 e^{x/2} + c_2 e^{-3x/2}$

3. $c_1 + c_2 e^{9x/2}$

5. $c_1 e^{-4x} + c_2 e^{-5x}$

7. $(c_1 + c_2 x)e^{5x/3}$

9. $(c_1 + c_2 x)e^{-kx}$

11. $(1 + 3x)e^{-2x}$

13. $0.3e^{-x/4} - 0.5e^{x/2}$

15. $2e^{-1.3x}$

17. $e^{-0.5x}$

19. Linearly dependent

21. Linearly dependent

23. Linearly independent

25. Linearly independent

PROBLEM SET 2.3, page 80

1. $e^x(A \cos x + B \sin x)$

3. $e^{-x/2}\left(A \cos \dfrac{3x}{2} + B \sin \dfrac{3x}{2}\right)$

5. II, $(c_1 + c_2 x)e^{-0.8x}$

7. III, $e^{x/4}\left(A \cos \dfrac{x}{2} + B \sin \dfrac{x}{2}\right)$

9. I, $c_1 e^{3\pi x} + c_2 e^{-3\pi x}$ **11.** II, $(c_1 + c_2 x)e^{x\sqrt{2}}$ **13.** $(4 - 3x)e^{-x/3}$

15. $2e^{5x} - 2e^{-5x}$ **17.** $3e^{-x} - 7e^{2x}$ **19.** $3\cos 2x + c_2 \sin 2x$

21. $e^{-x}\cos x$

23. If $y \not\equiv 0$, $y(P_1) = y(P_2) = 0$ and y_1 satisfies (1), (11), so does $y_2 = y_1 + y$, and $y_2 \not\equiv y_1$. Conversely, if y_1 and y_2 are different solutions of (1), (11), then $y = y_1 - y_2 \not\equiv 0$ is a solution of (1) and $y(P_1) = y(P_2) = 0$.

PROBLEM SET 2.4, page 83

1. $9e^{3x}$, $10e^{2x} - 2e^{-x}$, 0 **3.** $0, 3e^{2x}, 0, -3e^{-x}$ **5.** $c_1 e^{-x} + c_2 e^{2x}$

7. $c_1 e^{4x} + c_2$ **9.** $(c_1 + c_2 x)e^{-kx}$ **11.** $(c_1 + c_2 x)e^{-x/8}$

13. $(c_1 + c_2 x)e^{-0.6x}$

PROBLEM SET 2.5, page 90

1. $y = (y_0^2 + v_0^2 \omega_0^{-2})^{1/2} \cos\left[\omega_0 t - \arctan(v_0/(y_0\omega_0))\right]$ **3.** Lower by $\sqrt{2}$, higher

7. $0.3183, 0.4775, k = k_1 + k_2, 0.5738$

9. $my'' = -a\gamma y$, where $m = 1$ kg, $ay = \pi \cdot 0.01^2 \cdot 2y$ meter^3 is the volume of the water that causes the restoring force $a\gamma y$ with $\gamma = 9800$ nt ($=$ weight/meter^3). $y'' + \omega_0^2 y = 0$, $\omega_0^2 = a\gamma/m = a\gamma = 0.000628\gamma$. Frequency $\omega_0/2\pi = 0.4$ [sec^{-1}].

13. $y = [(v_0 + \alpha y_0)t + y_0]e^{-\alpha t}$

15. The positive solutions of $\tan t = 1$, that is, $\pi/4$ (max), $5\pi/4$ (min), etc.

17. $0.0231 = (\ln 2)/30$ [kg/sec] from $\exp(-10 \cdot 3c/2m) = \frac{1}{2}$.

19. See (9). Oscillations eventually disappear as c increases.

PROBLEM SET 2.6, page 96

3. $c_1 x^{-4} + c_2 x^5$ **5.** $(c_1 + c_2 \ln x)x^{-1.8}$ **7.** $A\cos(\ln x) + B\sin(\ln x)$

9. $c_1 x^{-1/2} + c_2 x^{-3/2}$ **11.** $\sqrt{x}\,[A\cos(\ln x) + B\sin(\ln x)]$

13. $(c_1 + c_2 \ln x)x^{-3}$ **15.** $(2 - \ln x)x^{-5/2}$

17. $3x^2 \ln x$ **19.** $570 - 2700/r$

PROBLEM SET 2.7, page 100

1. $(\lambda_2 - \lambda_1)\exp(\lambda_1 x + \lambda_2 x)$ **3.** $3e^{-ax}$

5. x^7 **7.** $2x^{2\mu - 1}$

9. $y'' - 6y' + 9y = 0$, $W = e^{6x}$ **11.** $x^2 y'' - 3xy' + 4y = 0$, $W = x^3$

13. $x^2 y'' - 1.5xy' + y = 0$, $W = -\frac{3}{2}x^{3/2}$ **15.** $y'' + 4\pi^2 y = 0$, $W = 2\pi$

17. $4x^2 y'' + 4xy' - 9y = 0$, $W = -3/x$

PROBLEM SET 2.8, page 103

1. $c_1 e^{-x} + c_2 e^x + e^{-3x}$ **3.** $c_1 e^{-2x} + c_2 e^{-x} + 2x^2 - 6x + 7$

5. $c_1 e^{-4x} + c_2 e^x - \cos 2x$ **7.** $A\cos x + B\sin x + \ln \pi x$

9. $-\cos x + 6\sin x + 2x$ **11.** $-e^x + xe^x$

13. $\ln x$ **15.** $(x - \ln x)e^{-2x}$

PROBLEM SET **2.9, page 107**

1. $A\cos 2x + B\sin 2x - \frac{1}{5}\sin 3x$ **3.** $c_1 e^{-3x} + c_2 + \frac{1}{2}e^{4x} + \frac{7}{2}e^{-4x}$

5. $e^{-x}(A\cos 3x + B\sin 3x) + \frac{5}{2}x^2 - x$ **7.** $c_1 e^{-3x} + c_2 e^{2x} + x^3$

9. $c_1 e^{-7x} + c_2 e^{5x} + xe^{5x} - 0.1\cos 5x - 0.6\sin 5x$

11. $(c_1 + c_2 x)e^{-5x} + \frac{1}{2}x^2 e^{-5x}$

13. $(c_1 + c_2 x)e^{-4x} + \frac{1}{2}e^{4x} + 16x^2 e^{-4x}$ **15.** $4e^{-2x} + x^4$

17. $-\frac{1}{8}\sinh 2x + \frac{1}{2}x - \frac{1}{4}xe^{-2x}$ **19.** $(x + 2x^2)e^{-0.6x}$

21. $3e^{-x} + 5 + 2x + \frac{1}{3}x^3$

PROBLEM SET **2.10, page 111**

1. $(c_1 + c_2 x + x\ln|x| - x)e^{2x}$ **3.** $(c_1 + c_2 x - \cos x)e^{-x}$

5. $(c_1 + c_2 x)e^x + \frac{1}{2}e^x/x$ **7.** $(c_1 + c_2 x + \frac{12}{35}x^{7/2})e^x$

9. $(c_1 + c_2 x - 2\ln|x|)e^{-2x}$ **11.** $c_1 x^2 + c_2 x^3 + \frac{1}{2}x^{-4}$

13. $c_1 x^{1/2} + c_2 x^{-3/2} + (x^2 - x^3)/3$ **15.** $c_1 x + c_2 x^2 - x\cos x$

17. $c_1 x^3 + c_2 x^{-3} + 3x^5$

PROBLEM SET **2.11, page 117**

1. $-\cos 2t + 3\sin 2t$ **3.** $-0.02525\cos 0.2t + 0.25\sin 0.2t$

5. $e^t/3$ **7.** $c_1 e^{-t} + c_2 e^{-2t} - 6\cos 4t - 7\sin 4t$

9. $A\cos t + B\sin t + (1 - \omega^2)^{-1}\cos \omega t$ **11.** $c_1 e^{-t} + c_2 e^{-3t} + 3.2\sin 2t - 0.4\cos 2t$

13. $\cos 5t + \sin t$ **15.** $-13\sin 2t - 16\cos 2t$

17. $\cos 2t + \frac{1}{3}\sin t - \frac{1}{15}\sin 3t - \frac{1}{105}\sin 5t$

PROBLEM SET **2.12, page 122**

1. $R > R_{\text{crit}} = 2\sqrt{L/C}$ is Case 1, etc.

3. I_0 in (5) is maximum when $S = 0$, thus $C = 1/(\omega^2 L)$.

5. $10(\cos 5t + \sin 5t)$

7. $c_1 e^{-50t} + c_2 e^{-30t} + \frac{1}{109}(32\cos 100t + 34\sin 100t)$

9. $e^{-50t}(A\cos 30t + B\sin 30t) + e^{-t}\cos t$

11. $5e^{-2t}\sin t$ **13.** $1 - \cos t$ **15.** $\frac{110}{21}(\cos 4t - \cos 10t)$

17. $R = 10$ ohms, $C = 1/29$ farad, $E = 11\sin 5t$ volts (so that $E' = 55\cos 5t$)

PROBLEM SET **2.13, page 131**

1. $W = 12$, $y = 1 - \frac{1}{2}x^2 + 5x^3$ **3.** $W = 2e^{-9x}$, $y = (4 - x + 2x^2)e^{-3x}$

5. $W = 32$, $y = 2\cosh x \sin x$ **7.** $W = 2/x^3$, $y = 2 - 3x + 5/x$

9. $W = -100$, $y = 4\cosh 2x - 4\sin x$ **11.** Linearly dependent

13. Linearly independent   **15.** Linearly dependent

17. Linearly independent

PROBLEM SET **2.14, page 137**

1. $c_1 e^{2x} + c_2 e^{-2x} + A\cos 2x + B\sin 2x$ **3.** $(c_1 + c_2 x)e^{-x} + (c_3 + c_4 x)e^x$

5. $c_1 e^{-x} + c_2 e^x + c_3 e^{2x}$ **7.** $c_1 e^{-x} + (c_2 + c_3 x)e^x$

9. $c_1e^{x/2} + c_2e^{-x/2} + c_3e^{3x/2} + c_4e^{-3x/2}$

11. $4x^3 - 2x^2 + 16x + 1$

13. $(2 - x)e^x$

15. $25\sinh x + 5e^{-5x}$

17. $(1 + x)e^{-x}\cos x$

19. $c_1/x + c_2x + c_3x^2$

PROBLEM SET 2.15, page 141

1. $(c_1 + c_2x + c_3x^2)e^{-x} + e^x + x$

3. $c_1x^{-1} + c_2x + c_3x^2 - \frac{1}{12}x^{-2}$

5. $c_1e^x + c_2e^{2x} + c_3e^{-2x} + 2e^{-x}$

7. $c_1x^{-1} + c_2 + c_3x + x^{-1}e^x$

9. $(1 + x)^2e^{-x} + e^{-x}\cos x$

11. $x - x^3 + x^5$

13. $2 - 2\sin x + \cos x$

CHAPTER 2 REVIEW, page 142

17. $c_1e^{2x} + c_2e^{-x/2} + x^2 - 3x$

19. $(c_1 + c_2\ln|x|)x^2 + 3$

21. $c_1x^2 + c_2x^4 + c_3x^6$

23. $e^{-x}(A\cos x + B\sin x - \cos 2x)$

25. $c_1e^x + c_2e^{-x} + c_3e^{2x} + c_4e^{-2x} + \cos 2x$

27. $(c_1 + c_2x + c_3x^2)e^{-x} - 2\cos x - 2\sin x$

29. $(c_1 + c_2\ln|x|)x^{0.7} + \frac{1}{7}$

31. $5\cos 4x - \frac{3}{4}\sin 4x + e^x$

33. $10x^2 - 5x^3 - x^2\sin\pi x$

35. $\cos 2x + 2\sin 2x + e^{-2x} + x^2$

37. $x - 4x^{-1} + 2x^3$

39. $-1/2x - 3x + 10x^2 - 3x^3/2$

41. $I(t) = -0.01093\cos 415t + 0.05273\sin 415t$

43. $I(t) = \frac{1}{73}(50\sin 4t - 110\cos 4t)$

45. RLC-circuit with $R = 20$ ohms, $L = 4$ henrys, $C = 0.1$ farad, $E = -25\cos 4t$ volts

47. $\frac{1}{2}\sin 2t$

49. $e^{-t}\cos\sqrt{5}t + \frac{1}{2}\sin 2t$

PROBLEM SET 3.1, page 158

1. $y_1 = 75 - 75e^{-0.08t}$, $y_2 = 75 + 75e^{-0.08t}$. Faster approach to the limit.

3. Same effect as in Prob. 1. How can this best be seen?

5. $I_1 = 16e^{-2t} - 10e^{-0.8t} + 3$, $I_2 = 8e^{-2t} - 8e^{-0.8t}$

7. $\mathbf{y} = c_1[1 \quad 1]^Te^t + c_2[1 \quad -1]^Te^{-t}$

9. $\mathbf{y} = c_1[1 \quad 3]^Te^{3t} + c_2[1 \quad -3]^Te^{-3t}$

11. $\mathbf{y} = c_1[1 \quad 0]^T + c_2[1 \quad 4]^Te^{4t}$

13. (a) For example, $C = 1000$ gives -2.39993, $-0.000\,167$. (b) -2.4, 0.
(c) Use $1/2.4 = 1/4 + 1/6$ (see Fig. 76). (d) $a_{22} = -4 + 2\sqrt{6.4} = 1.05964$ gives the critical case. C about 0.18506.

PROBLEM SET 3.3, page 169

1. $y_1 = c_1e^{-t} + c_2e^t$, $y_2 = -c_1e^{-t} + c_2e^t$, $y_1^2 - y_2^2 = const$

3. $y_1 = c_1e^{-2t} + c_2e^{2t}$, $y_2 = -3c_1e^{-2t} + c_2e^{2t}$

5. $y_1 = e^t(A\cos t + B\sin t)$, $y_2 = e^t(A\sin t - B\cos t)$

7. $y_1 = c_1e^{-18t} + 2c_2e^{9t} + 2c_3e^{18t}$, $y_2 = 2c_1e^{-18t} + c_2e^{9t} - 2c_3e^{18t}$,
$y_3 = 2c_1e^{-18t} - 2c_2e^{9t} + c_3e^{18t}$

9. $y_1 = -2c_1 + (c_2 - c_3)e^{3t}$, $y_2 = c_1 + c_3e^{3t}$, $y_3 = c_1 + 2c_2e^{3t}$

11. $y_1 = \cosh t$, $y_2 = \sinh t$

13. $y_1 = 10e^{-t/2}$, $y_2 = -5e^{-t/2}$

15. $y_1 = -4e^{-9t} + 3e^{-4t}$, $y_2 = -2e^{-9t} + 3e^{-4t}$

17. $I_1 = c_1e^{-t} + 3c_2e^{-3t}$, $I_2 = -3c_1e^{-t} - c_2e^{-3t}$

PROBLEM SET 3.4, page 174

1. Unstable improper node, $y_1 = c_1e^t$, $y_2 = c_2e^{2t}$

3. Saddle point, always unstable, $y_1 = c_1e^{-t} + c_2e^{3t}$, $y_2 = -c_1e^{-t} + c_2e^{3t}$

5. Stable spiral, $y_1 = e^{-2t}(A \cos 2t + B \sin 2t)$, $y_2 = e^{-2t}(-A \sin 2t + B \cos 2t)$

7. Center, always stable, $y_1 = A \cos 3t + B \sin 3t$, $y_2 = -3A \sin 3t + 3B \cos 3t$

9. Saddle point, always unstable, $y_1 = 3c_1e^{-10t} + c_2e^{4t}$, $y_2 = 4c_1e^{-10t} - c_2e^{4t}$

11. Saddle point, always unstable, $y_1 = c_1e^{(2-\sqrt{6})t} + c_2e^{(2+\sqrt{6})t}$,

     $y_2 = -c_1\sqrt{6}\, e^{(2-\sqrt{6})t} + c_2\sqrt{6}\, e^{(2+\sqrt{6})t}$

13. $y = c_1 + c_2e^{-at}$, $y_2 + ay_1 = const$; parallel straight lines

17. $p = 0.2 \neq 0$ (was 0), $\Delta < 0$; spiral point, unstable

19. For instance, $b = 1$ (spiral point), $b = -1$ (degenerate node), $b = -2$ (saddle point).

PROBLEM SET 3.5, page 183

5. At $(0, 0)$, $y_1' = y_2$, $y_2' = y_1$; saddle point. At $(1, 0)$, set $y_1 = 1 + \tilde{y}_1$.
Then $y_1 - y_1^2 = -\tilde{y}_1 - \tilde{y}_1^2 \approx -\tilde{y}_1$; thus $\tilde{y}_1' = y_2$, $y_2' = -\tilde{y}_1$; center.

7. Center at $(0, 0)$. At $(2, 0)$ set $y_1 = \tilde{y}_1 + 2$. Then $\tilde{y}_2' = \tilde{y}_1$. Saddle point at $(2, 0)$.

9. $(0, 0)$ saddle point, $(-3, 0)$ and $(3, 0)$ centers

11. $(\frac{1}{2}\pi \pm 2n\pi, 0)$ saddle points, $(-\frac{1}{2}\pi \pm 2n\pi, 0)$ centers

13. Hyperbolas $y_1y_2 = const$

PROBLEM SET 3.6, page 189

1. $\mathbf{y}^{(h)}$ exists on J because of the continuity assumption. Let $\tilde{\mathbf{y}}$ be any solution of (1) on J containing no arbitrary constants. Then $\mathbf{Y} = \tilde{\mathbf{y}} - \mathbf{y}^{(p)}$ satisfies the homogeneous system,

$$\mathbf{Y}' = \tilde{\mathbf{y}}' - \mathbf{y}^{(p)'} = \mathbf{A}\tilde{\mathbf{y}} - \mathbf{g} - (\mathbf{A}\mathbf{y}^{(p)} - \mathbf{g}) = \mathbf{A}(\tilde{\mathbf{y}} - \mathbf{y}^{(p)}) = \mathbf{A}\mathbf{Y}.$$

Hence $\mathbf{Y}$ is obtained if we assign suitable values to the arbitrary constants in $\mathbf{y}^{(h)}$, and for $\tilde{\mathbf{y}}$ this gives the representation $\tilde{\mathbf{y}} = \mathbf{Y} + \mathbf{y}^{(p)}$, as claimed.

3. $y_1 = c_1e^{-t} + c_2e^t$, $y_2 = -c_1e^{-t} + c_2e^t - e^{3t}$

5. $y_1 = c_1e^{-4t} + c_2e^{4t} + \sin 3t$, $y_2 = -7c_1e^{-4t} + c_2e^{4t} + 3 \cos 3t$

7. $y_1 = e^{-t}(c_1 - c_2 + c_2t) + \frac{1}{4}e^t$, $y_2 = e^{-t}(c_1 + c_2t) + \frac{3}{4}e^t$

9. $y_1 = 8 \cosh t + 3e^{2t}$, $y_2 = 8 \sinh t$

11. $y_1 = \cos 5t + 2 \sin 5t + 3t$, $y_2 = -\sin 5t + 2 \cos 5t - 4$

13. $y_1 = \cos 2t + \sin 2t + 4 \cos t$, $y_2 = -2 \sin 2t + 2 \cos 2t + \sin t$

15. $y_1 = 4e^{6t} + e^t + t^2 - 5$, $y_2 = e^{6t} - e^t - t$

17. $I_1 = 2c_1e^{\lambda_1 t} + 2c_2e^{\lambda_2 t} + 100$,

     $I_2 = (1.1 + \sqrt{0.41})c_1e^{\lambda_1 t} + (1.1 - \sqrt{0.41})c_2e^{\lambda_2 t}$,

     $\lambda_1 = -0.9 + \sqrt{0.41}$, $\lambda_2 = -0.9 - \sqrt{0.41}$

19. $c_1 = 17.948$, $c_2 = -67.948$

CHAPTER 3 REVIEW, page 190

17. $y_1 = c_1 e^{-5t} + c_2 e^t$, $y_2 = -c_1 e^{-5t} + 2c_2 e^t$; saddle point

19. $y_1 = e^{-4t}(A \cos t + B \sin t)$, $y_2 = \frac{1}{5} e^{-4t}[(B - 2A) \cos t - (A + 2B) \sin t]$;
asymptotically stable spiral point

21. $y_1 = 4c_1 e^{5t} + c_2 e^{-2t}$, $y_2 = 3c_1 e^{5t} - c_2 e^{-2t}$; saddle point

23. $y_1 = e^{-t}(2A \cos t + 2B \sin t)$,
$y_2 = e^{-t}[A(5 \cos t + \sin t) + B(5 \sin t - \cos t)]$; asymptotically stable spiral point

25. $y_1 = c_1 e^{-t} + 3c_2 e^t + te^{-t} + e^{-t}$, $y_2 = -c_1 e^{-t} - c_2 e^t - te^{-t}$

27. $y_1 = c_1 e^{-4t} + c_2 e^{4t} - 1 - 8t^2$, $y_2 = -c_1 e^{-4t} + c_2 e^{4t} - 4t$

29. Saddle point

31. $I_1 = (19 + 32.5t)e^{-5t} - 19 \cos t + 62.5 \sin t$,
$I_2 = (-6 - 32.5t)e^{-5t} + 6 \cos t + 2.5 \sin t$

33. $y_1 = 30e^{-0.03t} - 30e^{-0.15t}$, $y_2 = 45e^{-0.03t} + 45e^{-0.15t}$

35. $(0, 0)$ saddle point; $(-1, 0)$, $(1, 0)$ centers

37. $(n\pi, 0)$ center when n is even and saddle point when n is odd

PROBLEM SET 4.2, page 204

1. $y = a_0(1 - x^2 + x^4/2! - x^6/3! + - \cdots) = a_0 e^{-x^2}$

3. $y = a_3 x^3 - k/3$

5. $y = a_0 + a_1 x + (\frac{3}{2}a_1 - a_0)x^2 + (\frac{7}{6}a_1 - a_0)x^3 + \cdots$. Setting $a_0 = A + B$ and $a_1 = A + 2B$, we obtain $y = Ae^x + Be^{2x}$. This illustrates the fact that even if the solution of an equation is a known function, the power series method may not yield it immediately in the usual form. Practically, this does not matter because the main interest concerns equations for which the power series solutions define *new* functions.

7. $y = a_0(1 - 2x^2 + \frac{2}{3}x^4 - + \cdots) + a_1(x - \frac{2}{3}x^3 + \frac{2}{15}x^5 - + \cdots) = a_0 \cos 2x + \frac{1}{2}a_1 \sin 2x$

9. $y = a_0 \left[1 + kt + \frac{(kt)^2}{2!} + \cdots \right] = a_0 \left[1 + k(x - 1) + \frac{k^2}{2!}(x - 1)^2 + \cdots \right]$

11. $(t + 1)\dot{y} - y = t + 1$, $y = a_0(1 + t) + (1 + t)(t - \frac{1}{2}t^2 + - \cdots)$
$= a_0 x + x \ln x$ $(\dot{y} = dy/dt)$

13. ∞ **15.** $\sqrt{3}$ **17.** $\sqrt{|k|}$

19. $\sqrt{3/2}$ **21.** $\sqrt{2}$

23. $\displaystyle\sum_{m=1}^{\infty} \frac{(m + 1)(m + 2)}{(m + 1)^2 + 1} x^m$;   1   **25.** $\displaystyle\sum_{m=1}^{\infty} \frac{m^3}{(m + 1)!} x^m$, ∞

PROBLEM SET 4.3, page 209

1. Use $\ln(1 + x) = x - \frac{1}{2}x^2 + \frac{1}{3}x^3 - + \cdots$, etc.

7. Set $x = az$; $y = P_3(x/a)$.

9. Use (4).

PROBLEM SET 4.4, page 216

3. $y_1 = 1/(1 - x)$, $y_2 = 1/x$

5. $y_1 = x^{-1} \cos x$, $y_2 = x^{-1} \sin x$

7. $y_1 = e^x$, $y_2 = e^x \ln x$

9. $y_1 = x + 2$, $y_2 = 1/(x + 2)$

11. $y_1 = \sqrt{x}$, $y_2 = 1 + x$

13. $y_1 = x^2$, $y_2 = x^{-3}$

15. $y_1 = x^{-2} \sin(x^2)$, $y_2 = x^{-2} \cos(x^2)$

17. $A(1 - x)^{-1/4} + B\sqrt{x}\, F(1, \frac{3}{4}, \frac{3}{2}; x)$

19. $A(1 - 8x + \frac{32}{5}x^2) + Bx^{3/4}F(\frac{7}{4}, -\frac{5}{4}, \frac{7}{4}; x)$

PROBLEM SET 4.5, page 226

1. $AJ_{1/3}(x) + BJ_{-1/3}(x)$

3. $AJ_{1/4}(x^2) + BJ_{-1/4}(x^2)$

5. $AJ_{2/3}(x^2) + BJ_{-2/3}(x^2)$

7. $x^{-2}J_2(x)$; see Theorem 2.

9. $x^{1/3}(AJ_{1/3}(x^{1/3}) + BJ_{-1/3}(x^{1/3}))$

13. Use (24) with $\nu = 1$.

15. $J_2(x) = 2x^{-1}J_1(x) - J_0(x)$; see (26) with $\nu = 1$.

17. Use J_3 from Example 2. 0.1289, 0.1623, 0.1981, 0.2353, 0.2727.

19. Start from (25), with ν replaced by $\nu - 1$, and replace $J_{\nu-1}$ by using (24).

21. Let $J_n(x_1) = J_n(x_2) = 0$. Then $x_1^{-n}J_n(x_1) = x_2^{-n}J_n(x_2) = 0$, and $[x^{-n}J_n(x)]' = 0$ somewhere between x_1 and x_2 by Rolle's theorem. Now use (25). Then use (24) with $\nu = n + 1$.

25. $-2J_2(x) - J_0(x) + c$ by (27)

27. $-2J_4 - 2J_2 - J_0 + c$ by (27)

29. $u'' + u = 0$, $y = ux^{-1/2} = x^{-1/2}(A \cos x + B \sin x)$

PROBLEM SET 4.6, page 232

3. $AJ_{3/2}(5x) + BY_{3/2}(5x)$ is elementary.

5. $\sqrt{x}\left[AJ_{1/4}(\frac{1}{2}x^2) + BY_{1/4}(\frac{1}{2}x^2)\right]$

7. $\sqrt{x}\left[AJ_{1/4}(\frac{1}{2}kx^2) + BY_{1/4}(\frac{1}{2}kx^2)\right]$

9. $x^{1/4}(AJ_{1/4}(x^{1/4}) + BJ_{-1/4}(x^{1/4}))$

11. 1.1

13. Set $H_\nu^{(1)} = kH_\nu^{(2)}$, use (10), obtain a contradiction.

15. For $x \neq 0$ all the terms of the series (13) are real and positive.

PROBLEM SET 4.7, page 238

3. $\lambda = [(2n + 1)\pi/2]^2$, $n = 0, 1, \cdots$; $y_n(x) = \sin(\frac{1}{2}(2n + 1)\pi x)$

5. $\lambda = [(2n + 1)\pi/2L]^2$, $n = 0, 1, \cdots$; $y_n(x) = \sin((2n + 1)\pi x/2L)$

7. $\lambda = ((2n + 1)\pi/2)^2$, $n = 0, 1, \cdots$; $y_n(x) = \sin((n + \frac{1}{2})\pi \ln|x|)$

9. $\lambda = n^2\pi^2$, $n = 1, 2, \cdots$; $y_n(x) = x \sin(n\pi \ln|x|)$; Euler–Cauchy equation

11. $y'' + \lambda y = 0$, $y(0) = y(2L)$, $y'(0) = y'(2L)$

13. Set $x = ct + k$.

15. $x = \cos\theta$, $dx = -\sin\theta\, d\theta$, etc.

PROBLEM SET 4.8, page 246

1. $16(P_0 - P_2 + P_4)$

3. $\frac{4}{5}P_0 - \frac{4}{7}P_2 - \frac{8}{35}P_4$

5. $0.6366P_0 - 0.6871P_2 + 0.0518P_4 - 0.0013P_6 + \cdots$

7. $\frac{1}{4}P_0 + \frac{1}{2}P_1 + \frac{5}{16}P_2 - \frac{3}{32}P_4 + \frac{13}{256}P_6 + \cdots$

9. (c) $a_m = 2/(\alpha_{m0}J_1(\alpha_{m0}))$

CHAPTER 4 REVIEW, page 247

17. $\cos\frac{1}{2}x$, $\sin\frac{1}{2}x$

19. $(x + 1)^{3/4}$, $(x + 1)^{1/4}$

21. $J_{\sqrt{3}}(x)$, $J_{-\sqrt{3}}(x)$

23. $x^{-1}\cos 2x$, $x^{-1}\sin 2x$

25. e^{-x}, x^2

27. $\sqrt{\omega/\pi}\sin n\omega x$, $n = 1, 2, \cdots$

29. $\sqrt{2/\pi}$, $(2/\sqrt{\pi})\cos 4nx$, $(2/\sqrt{\pi})\sin 4nx$

31. $\lambda = n^2\pi^2/L^2$, $n = 0, 1, \cdots$, $y_n = 1$, $\cos(n\pi x/L)$, $\sin(n\pi x/L)$

33. $\lambda = n^2$, $y_n = \cos nx$ $(n = 1, 3, 5, \cdots)$, $\sin nx$ $(n = 2, 4, 6, \cdots)$

35. $8(P_0 - P_2 + P_4)$

37. $\sqrt{x}\left[AJ_{1/3}(\frac{4}{3}x^{3/2}) + BY_{1/3}(\frac{4}{3}x^{3/2})\right]$

39. $x^2\left[AJ_2(x^2) + BY_2(x^2)\right]$

PROBLEM SET 5.1, page 257

1. $2/s^2 + 6/s$

3. $\pi/(s^2 + \pi^2)$

5. $e^a/(s + b)$

7. $(\omega \cos \delta + s \sin \delta)/(s^2 + \omega^2)$

9. $1/s + (e^{-s} - 1)/s^2$

11. $k(1 - e^{-cs})/s$

13. $(1 - e^{-ks})/s^2 - ke^{-ks}/s$

15. $\frac{1}{2}(e^{-s} - 1)/s^2 - e^{-s}/2s + 1/s$

17. $0.1 \cos 1.8t + 0.5 \sin 1.8t$

19. $3e^{-t} - 4e^{2t}$

21. $0.4t^3 - 1.9t^5$

23. $[\cos (n\pi t/L)]/L^2$

25. $\displaystyle\sum_{k=1}^{5} a_k e^{-k^2 t}$

27. $(e^{\sqrt{3}t} - e^{-\sqrt{2}t})/(\sqrt{3} + \sqrt{2})$

29. $\dfrac{2}{(s + 3)^3}$

31. $\dfrac{10}{(s - 2)^2 - 4}$

33. $\dfrac{s^2 - 2}{s^4 + 4}$

35. te^{-t}

37. $e^{-3t} \sin 3t$

39. $e^{-t/2}(\cos t - \frac{1}{2} \sin t)$

PROBLEM SET 5.2, page 264

1. $y = e^{-3t} - \cos t + 3 \sin t$

3. $y = 0.05t - 0.25$

5. $y = 2e^{-2at} + 4e^{at}$

7. $y = 2t + e^t - e^{3t}$

9. $y = -2e^{-2t} + \frac{11}{2}e^{-3t} - \frac{3}{2}e^t$

11. $(s^2 + 2)/(s^3 + 4s)$

13. $\frac{1}{4}(1 - e^{-4t})$

15. $(1 - \cos \omega t)/\omega^2$

17. $\cosh t - 1$

19. $1 + t - \cos 3t - \frac{1}{3} \sin 3t$

PROBLEM SET 5.3, page 273

3. e^{-s}/s^2

5. $e^{-s}(2s^{-3} + 2s^{-2} + s^{-1})$

7. $-4e^{-\pi s}s/(s^2 + 1)$

9. $\omega(1 + e^{-\pi s/\omega})/(s^2 + \omega^2)$

11. $(1 - e^{1-s})/(s - 1)$

13. $-10s(e^{-s} + e^{-2s})/(s^2 + \pi^2)$

15. $\frac{1}{2}(t - 3)^2 u(t - 3)$

17. $[1 + u(t - \pi)] \sin 3t$

19. $(\cos \pi t)u(t - 2)$

21. $\frac{1}{3}(e^t - 1)^3 e^{-5t}$

23. $\sin 3t + \sin t$ $(0 < t < \pi)$; $\frac{4}{3} \sin 3t$ $(t > \pi)$

25. $e^t - \sin t$ $(0 < t < 2\pi)$; $e^t - \frac{1}{2} \sin 2t$ $(t > 2\pi)$

27. $\sin t$ $(0 < t < \pi)$; 0 $(\pi < t < 2\pi)$; $-\sin t$ $(t > 2\pi)$

29. $3e^{-3t} - 2e^{-t} + 2e^t$ $(0 < t < \frac{1}{2})$; $3e^{-3t} - 2e^{-t} + 2e^t + \frac{1}{4}(-e^{-3t+3/2} + e^{t-1/2})$ $(t > \frac{1}{2})$

31. $L(e^{-Rt/L} - 1)/R^2 + t/R$ $(0 < t < 4\pi)$; $Le^{-Rt/L}/R^2 + 4\pi/R - (L/R^2)e^{-(t-4\pi)R/L}$ $(t > 4\pi)$

33. $sI + I/s = (1 - e^{-s})/s^2$. Ans. $1 - \cos t$ $(0 < t < 1)$; $\cos (t - 1) - \cos t$ $(t > 1)$

35. $\frac{1}{2}(e^{-t} - \cos t + \sin t)$ $(0 < t < \pi)$; $\frac{1}{2}[-(1 + e^{-\pi}) \cos t + (3 - e^{-\pi}) \sin t]$ $(t > \pi)$

37. 0 $(t < 1)$; $e^{-0.1(t-1)}$ $(1 < t < 2)$; $e^{-0.1(t-1)} - e^{-0.1(t-2)}$ $(t > 2)$

39. 0 $(t < 2)$; $(10e^{-(t-2)} - e^{-0.1(t-2)})/900e^2$ $(t > 2)$

PROBLEM SET 5.4, page 278

1. $\dfrac{1}{(s-1)^2}$

3. $\dfrac{2s^3 + 6\pi^2 s}{(s^2 - \pi^2)^3}$

5. $\dfrac{s^2 - \omega^2}{(s^2 + \omega^2)^2}$

7. $\dfrac{2s+2}{(s^2 + 2s + 2)^2}$

9. $\frac{1}{2}t^2 e^{3t}$

11. $t \cos \pi t$

13. $2t^{-1}(e^t - \cos t)$

15. $\frac{1}{4}t \sin 2t$

PROBLEM SET 5.5, page 283

1. t

3. $\sinh t$

5. $\frac{1}{2}t \sin \omega t$

7. $e^t - t - 1$

9. $\frac{1}{2}(1 - e^{-2(t-3)})u(t-3)$

11. $e^t - t - 1$

13. $\frac{1}{4} - \frac{1}{4}\cos 2t$

15. $(t \sin \pi t)/2\pi$

17. $(\omega t - \sin \omega t)/\omega^2$

19. $y = \cos t - \cos 2t$

21. $y = \frac{1}{4}(1 + 3\cos 2t) + \frac{1}{4}[\cos(2t - 2) - 1]u(t - 1)$

23. $y = e^{-t}(\cos t + \sin t) + [-2\cos t + \sin t + e^{-(t-2\pi)}(2\cos t + \sin t)]u(t - 2\pi)$

25. $y = [t - \cos(t - 1) - \sin(t - 1)]u(t - 1) + [-t + 2\cos(t - 2) + \sin(t - 2)]u(t - 2)$

27. e^t

29. $\cos t$

31. $\sinh t$

33. $\cosh t$

PROBLEM SET 5.6, page 289

1. $e^{4t} - e^{-2t}$

3. $1 + 3\sinh 3t$

5. $\cos 3t + \cosh 3t$

7. $\frac{1}{2}(t^2 e^{-t} + t^3)$

9. $e^{-2t}(1 + t^2 - 2t^3)$

PROBLEM SET 5.7, page 294

1. $y_1 = e^{-t}\cos t,\ y_2 = -e^{-t}\sin t$

3. $y_1 = 4e^{2t} - e^{-5t},\ y_2 = 3e^{2t} + e^{-5t}$

5. $y_1 = \sin t,\ y_2 = \cos t$

7. $y_1 = e^t + e^{2t},\ y_2 = e^{2t}$

9. $y_1 = e^t,\ y_2 = e^{-t},\ y_3 = e^t - e^{-t}$

11. $y_1 = -1 + \cos t + \sin t + u(t - 1)[1 - \cos(t - 1) - \sin(t - 1)],$
$y_2 = 1 - \cos t + \sin t + u(t - 1)[-1 + \cos(t - 1) - \sin(t - 1)]$

13. $y_1 = -e^{-2t} + 4e^t + \frac{1}{3}u(t - 1)(-e^{3-2t} + e^t),\ y_2 = -e^{-2t} + e^t + \frac{1}{3}u(t - 1)(-e^{3-2t} + e^{+t})$

17. $y_1 = 50 - 49.955e^{-0.12t} + 0.015e^{-0.04t} - 0.060\cos 2t - 1.497\sin 2t,$
$y_2 = 50 + 99.910e^{-0.12t} + 0.030e^{-0.04t} + 0.060\cos 2t - 0.005\sin 2t$

19. $i_1 = 2e^{-8t} + 13e^{-2t} - 15\cos t + 42\sin t,\ i_2 = -e^{-8t} + 13e^{-2t} - 12\cos t + 18\sin t$

CHAPTER 5 REVIEW, page 299

17. $\dfrac{s^2 + 2}{s(s^2 + 4)}$

19. $\dfrac{e^{-2s+2}}{s - 1}$

21. $\dfrac{1}{s^2(s + 3)}$

23. $\dfrac{s}{s^2 - 0.01}$

25. $\cos 3t + \sin 3t$

27. $tu(t - 1)$

29. $e^t - 2e^{2t} + 2$

31. $\cosh 2t - \sinh t$

33. $e^{-2t}(3\cos t - 2\sin t)$

35. $\cos 2t + \frac{1}{2}u(t - 3)\sin^2(t - 3)$

37. $(1 - u(t - \pi))\sin^3 t$

39. $y_1 = \cos t,\ y_2 = \sin t$

41. $y_1 = 4e^t - e^{-2t},\ y_2 = e^t - e^{-2t}$

45. $1 - e^{-t}$ if $0 < t < 4$, $(e^4 - 1)e^{-t}$ if $t > 4$

47. $i(t) = e^{-4t}(\frac{3}{26} \cos 3t - \frac{10}{39} \sin 3t) - \frac{3}{26} \cos 10t + \frac{8}{65} \sin 10t$

49. $i_1 = e^{-4t} + t, i_2 = -4e^{-4t} + \frac{1}{5}$

PROBLEM SET 6.1, page 309

1. $\begin{bmatrix} 0 & 6 \\ 1 & 15 \end{bmatrix}, \begin{bmatrix} 0 & 6 \\ 1 & 15 \end{bmatrix}$, undefined   **3.** Undefined, **C**, undefined

5. $\begin{bmatrix} 2 & 0 & -29 \\ -18 & 20 & 60 \end{bmatrix}, \begin{bmatrix} 2 & -18 \\ 0 & 20 \\ -29 & 60 \end{bmatrix}$ **7.** $\begin{bmatrix} -4 & 4 \\ -1 & 1 \end{bmatrix}, \begin{bmatrix} -4 & 4 \\ -1 & 1 \end{bmatrix}, \mathbf{0}$

9. $[26 \quad -40 \quad 18], [26 \quad -40 \quad 18]^T$ **11.** $[3 \quad 39 \quad -51], [3 \quad 39 \quad -51]$

13. $[25 \quad 45 \quad -25], [-25 \quad -45 \quad 25]$ **15.** $\mathbf{a}, \mathbf{0}$, undefined

19. (b) $\mathbf{C} = \mathbf{S} + \mathbf{T} = \frac{1}{2}(\mathbf{C} + \mathbf{C}^T) + \frac{1}{2}(\mathbf{C} - \mathbf{C}^T)$

PROBLEM SET 6.2, page 319

1. Undefined, $[2 \quad 8]$, undefined   **3.** $\begin{bmatrix} 56 & 30 & 24 \\ 30 & 45 & 9 \\ 24 & 9 & 14 \end{bmatrix}, \mathbf{C}^2, \mathbf{C}^2$

5. Undefined, 16, 16, $\begin{bmatrix} 4 & 3 & 0 \\ 16 & 12 & 0 \\ 12 & 9 & 0 \end{bmatrix}$

7. $\begin{bmatrix} 13 & -6 & -3 \\ -6 & 4 & 2 \\ -3 & 2 & 1 \end{bmatrix}, \begin{bmatrix} 4 & -6 \\ -6 & 14 \end{bmatrix}, \begin{bmatrix} 26 & -54 \\ -12 & 28 \\ -6 & 14 \end{bmatrix}$

9. $\begin{bmatrix} 24 & 9 & 10 \\ 9 & 22.5 & 1.5 \\ 10 & 1.5 & 8 \end{bmatrix}$, same, $\mathbf{0}$ **11.** $\mathbf{p} = [85 \quad 62 \quad 30]^T$

13. $[0.7 \quad 0.7], [0.49 \quad 0.91], [0.427 \quad 0.973], [0.4081 \quad 0.9919]$

PROBLEM SET 6.3, page 329

1. $x = -3, y = 5$ **3.** $x = 0.4, y = 1.2$ **5.** $x = 1, y = 3, z = -5$

7. $x = 6, y = -7$ **9.** No solution   **11.** $x = 0, z = 3 - 2y$

13. $w = 1, y = 2z - x$ **15.** $w = 4, x = 0, y = 2, z = 6$

17. $I_1 = 2, I_2 = 6, I_3 = 8$

19. $I_1 = (R_1 + R_2)E_0/R_1R_2, I_2 = E_0/R_1, I_3 = E_0/R_2$ (amperes)

21. $x_2 = 1600 - x_1, x_3 = 600 + x_1, x_4 = 1000 - x_1$. No.

PROBLEM SET 6.4, page 336

1. Linearly independent   **3.** Linearly dependent   **5.** Linearly dependent

7. Linearly independent   **9.** 1   **11.** 2

13. 2   **15.** 2   **17.** 4

19. 2, [3   0   −2], [0   1   0]   **21.** No

23. 1, 1   **25.** No

29. For instance, [3   1   4], [0   5   8] and [1   0   −1 $\frac{1}{3}$]^{T}, [1   2   1   1]^{T}

PROBLEM SET **6.6, page 349**

5. 36.18   **7.** 0   **9.** $m^3 + n^3 + p^3 - 3mnp$

11. −477   **13.** 72   **15.** 3

17. $x = 5, y = -4$ **19.** $x = -7, y = 0, z = 1$

PROBLEM SET **6.7, page 357**

1. $\begin{bmatrix} 3 & -1 & 1 \\ -15 & 6 & -5 \\ 5 & -2 & 2 \end{bmatrix}$ **3.** Singular   **5.** $\begin{bmatrix} 29 & -11 & 10 \\ -160 & 61 & -55 \\ 55 & -21 & 19 \end{bmatrix}$

7. $\frac{1}{9}\mathbf{A}$ **9.** $\begin{bmatrix} 1 & -8 & 31 \\ 0 & 1 & -3 \\ 0 & 0 & 1 \end{bmatrix}$

11. Note that the matrix in Prob. 5 is the square of that in Prob. 1.

15. $\begin{bmatrix} 3 & -1 & 1 \\ -15 & 6 & -5 \\ 5 & -2 & 2 \end{bmatrix}$ **17.** $\begin{bmatrix} 3 & 0 & -1 \\ 0 & 5 & 1 \\ 1 & 0 & -1 \end{bmatrix}$ **19.** $\begin{bmatrix} 1 & 1 & 2 \\ 1 & 2 & 3 \\ 2 & 3 & 3 \end{bmatrix}$

PROBLEM SET **6.8, page 364**

1. 2; [3   1   0], [2   0   −1]   **3.** 4; 1, x, x^2, x^3

5. 1; $\begin{bmatrix} 0 & 1 \\ -1 & 0 \end{bmatrix}$ **7.** 3; $\begin{bmatrix} 0 & 1 \\ 0 & 0 \end{bmatrix}, \begin{bmatrix} 0 & 0 \\ 1 & 0 \end{bmatrix}, \begin{bmatrix} 1 & 0 \\ 0 & -1 \end{bmatrix}$

9. No

11. For instance, [1,   0], [0,   1] and [1,   1], [1,   −1] and [1,   0], [0,   −1]

13. 2.58650   **15.** $\sqrt{13/18}$

17. $\sqrt{33}$ **21.** $[v_1, -\frac{1}{2}v_1, v_3]$; v_1, v_3 arbitrary. Yes.

23. $\| [2.4 \quad 4.3 \quad -7.2]^{\mathsf{T}} \| = \sqrt{76.09} = 8.72 \leqq \sqrt{6.69} + \sqrt{38} = 8.75$

25. $x_1 = -0.2y_1 + 0.4y_2$
$x_2 = \quad 0.8y_1 - 0.6y_2$

27. $x_1 = \quad 2y_1 - \quad 3y_2$
$x_2 = -10y_1 + 16y_2 + y_3$
$x_3 = \quad -7y_1 + 11y_2 + y_3$

29. $x_1 = \quad 4y_1 - 2y_2 + 2y_3$
$x_2 = -2y_1 - 4y_2 + 4y_3$
$x_3 = -4y_1 + 2y_2 + 8y_3$

CHAPTER 6 REVIEW, page 365

11. $\begin{bmatrix} -1 & 6 & 1 \\ -18 & 8 & -7 \\ -13 & -2 & -7 \end{bmatrix}$ **13.** $\begin{bmatrix} 0 & 5 & 4 \\ -5 & 0 & 7 \\ -4 & -7 & 0 \end{bmatrix}$ **15.** $\begin{bmatrix} -17 & 2 & -8 \\ 2 & -20 & -4 \\ -8 & -4 & -5 \end{bmatrix}$

17. $\begin{bmatrix} 17 & -2 & 8 \\ -2 & 20 & 4 \\ 8 & 4 & 5 \end{bmatrix}$ **19.** $\begin{bmatrix} 21 \\ -8 \\ -31 \end{bmatrix}$ **21.** -1

23. $[2.1 \quad -0.8 \quad -3.1]$ **25.** 0

27. $x = 3, y = 2z + 1$ **29.** $x = 0.4, y = -1.3, z = 1.7$

31. $x = 0, y = 0, z = 0$ **33.** $x = 9, y = -9, z = 1$

35. 2, 2   **37.** 3, 3

39. 3, 3

41. $\begin{bmatrix} -16 & 11 & -14 \\ 11 & -6 & 9 \\ -14 & 9 & -11 \end{bmatrix}$ **43.** No inverse   **45.** $\begin{bmatrix} 4 & 7 & -2 \\ 7 & 10 & 1 \\ -2 & 1 & 1 \end{bmatrix}$

47. $I_1 = 16.5, I_2 = 11, I_3 = 5.5$ [amps]

49. The current $i_1 = i_2$ causes a voltage drop $Zi_1 = Zi_2$ across the resistor (Ohm's law); hence $u_1 = u_2 + Zi_2, i_1 = i_2$, which gives the indicated matrix.

PROBLEM SET 7.1, page 375

1. $4, \begin{bmatrix} 1 \\ 0 \end{bmatrix}, -6, \begin{bmatrix} 0 \\ 1 \end{bmatrix}$ **3.** $-8, \begin{bmatrix} 2 \\ 9 \end{bmatrix}, 6, \begin{bmatrix} 1 \\ 1 \end{bmatrix}$

5. $-5, \begin{bmatrix} 1 \\ -2 \end{bmatrix}, 5, \begin{bmatrix} 2 \\ 1 \end{bmatrix}$ **7.** $1, \begin{bmatrix} 1 \\ 0 \end{bmatrix}, 3, \begin{bmatrix} 1 \\ 1 \end{bmatrix}$

9. $3, \begin{bmatrix} 1 \\ 0 \\ 0 \end{bmatrix}, -8, \begin{bmatrix} 0 \\ 1 \\ 0 \end{bmatrix}, 4, \begin{bmatrix} 0 \\ 0 \\ 1 \end{bmatrix}$ **11.** $1, \begin{bmatrix} 7 \\ -4 \\ 2 \end{bmatrix}, 3, \begin{bmatrix} 1 \\ 0 \\ 0 \end{bmatrix}, 4, \begin{bmatrix} 5 \\ 1 \\ 0 \end{bmatrix}$

13. $\frac{1}{2}, \begin{bmatrix} 0 \\ 1 \\ 0 \end{bmatrix}, 3, \begin{bmatrix} 1 \\ 0 \\ -1 \end{bmatrix}$

15. $\begin{bmatrix} 1 & 0 \\ 0 & -1 \end{bmatrix}$, (a) $1, \begin{bmatrix} 1 \\ 0 \end{bmatrix}$, (b) $-1, \begin{bmatrix} 0 \\ 1 \end{bmatrix}$. (a) Any point on the x-axis is mapped onto itself.

(b) Any point $(0, y)$ on the y-axis is mapped onto $(0, -y)$, so that $[0 \quad y]^T$ is an eigenvector corresponding to $\lambda = -1$.

17. $\begin{bmatrix} 4 & 0 \\ 0 & 4 \end{bmatrix}$. $\lambda = 4$ is the uniform stretch in every direction, which is preserved, so that it corresponds to an eigenvector.

19. $\begin{bmatrix} 0 & 0 \\ 0 & 1 \end{bmatrix}, 1, \begin{bmatrix} 0 \\ 1 \end{bmatrix}, 0, \begin{bmatrix} 1 \\ 0 \end{bmatrix}$. A point on the y-axis goes onto itself, a point on the x-axis onto the origin.

PROBLEM SET 7.2, page 379

1. $-35.26°$, 2; $54.74°$, 8
3. $-45°$, 1.5; $45°$, 4.5
5. $-54.74°$, 0.5; $35.26°$, 2
7. $[5 \quad 8]^{\mathsf{T}}$
9. $[2 \quad 3 \quad 6]^{\mathsf{T}}$
11. 1.5
13. $c[10 \quad 18 \quad 25]^{\mathsf{T}}$, $c > 0$
15. $\mathbf{x} = (\mathbf{I} - \mathbf{A})^{-1}\mathbf{y} = [0.55 \quad 0.64375 \quad 0.6875]^{\mathsf{T}}$

PROBLEM SET 7.3, page 384

1. Orthogonal, $0.96 \pm 0.28i$
3. Not skew-symmetric!, $1 \pm 4i$
5. Skew-symmetric, $\pm 25i$, 0
7. $a_{jj} = -a_{jj}$ implies $a_{jj} = 0$.
9. Rotation about the x-axis
11. $\mathbf{A}^{-1} = (-\mathbf{A}^{\mathsf{T}})^{-1} = -(\mathbf{A}^{-1})^{\mathsf{T}}$
13. No, since $\det \mathbf{A} = 0$ if $\mathbf{A}^{\mathsf{T}} = -\mathbf{A}$ and $n = 3$

PROBLEM SET 7.4, page 390

3. $[2 + i \quad i]^{\mathsf{T}}$, $[2 + i \quad -5i]^{\mathsf{T}}$
5. Hermitian; -3, $[-3 - 4i \quad 5]^{\mathsf{T}}$; 7, $[3 + 4i \quad 5]^{\mathsf{T}}$
7. Unitary; $\frac{1}{2} - i\sqrt{\frac{3}{4}}$, $[1 \quad -1]^{\mathsf{T}}$; $\frac{1}{2} + i\sqrt{\frac{3}{4}}$, $[1 \quad 1]^{\mathsf{T}}$
9. Skew-Hermitian, unitary; $-i$, $[0 \quad -1 \quad 1]^{\mathsf{T}}$; i, $[0 \quad 1 \quad 1]^{\mathsf{T}}$, $[1 \quad 0 \quad 0]^{\mathsf{T}}$
11. $\mathbf{A} = \mathbf{H} + \mathbf{S}$, $\mathbf{H} = \frac{1}{2}(\mathbf{A} + \overline{\mathbf{A}}^{\mathsf{T}})$, $\mathbf{S} = \frac{1}{2}(\mathbf{A} - \overline{\mathbf{A}}^{\mathsf{T}})$

13. $\begin{bmatrix} 4 & -4 \\ -4 & 5 \end{bmatrix}$
15. $\begin{bmatrix} 0.5 & 0.4 & 0 \\ 0.4 & 0 & -0.7 \\ 0 & -0.7 & 0 \end{bmatrix}$
17. $\begin{bmatrix} 1 & -1 & 0 \\ -1 & 1 & 0 \\ 0 & 0 & -4 \end{bmatrix}$

21. Hermitian, 4
23. Skew-Hermitian, $17i$
25. Skew-Hermitian, $16i$

PROBLEM SET 7.5, page 397

1. $\begin{bmatrix} 0 & 0 \\ \frac{7}{3} & 5 \end{bmatrix}$, 0, $\begin{bmatrix} 15 \\ -7 \end{bmatrix}$, $\mathbf{x} = \begin{bmatrix} -6 \\ 3 \end{bmatrix}$; 5, $\begin{bmatrix} 0 \\ 1 \end{bmatrix}$, $\mathbf{x} = \begin{bmatrix} 3 \\ 6 \end{bmatrix}$

3. $\begin{bmatrix} -25 & 12 \\ -50 & 25 \end{bmatrix}$, -5, $\begin{bmatrix} 3 \\ 5 \end{bmatrix}$, $\mathbf{x} = \begin{bmatrix} -2 \\ 4 \end{bmatrix}$; 5, $\begin{bmatrix} 2 \\ 5 \end{bmatrix}$, $\mathbf{x} = \begin{bmatrix} 2 \\ 1 \end{bmatrix}$

5. $\begin{bmatrix} 355 & -42 & 560 \\ 0 & 1 & 0 \\ -225 & 27 & -355 \end{bmatrix}$, -5, $\begin{bmatrix} -14 \\ 0 \\ 9 \end{bmatrix}$, $\mathbf{x} = \begin{bmatrix} -1 \\ 0 \\ 3 \end{bmatrix}$; 1, $\begin{bmatrix} -14 \\ 2 \\ 9 \end{bmatrix}$, $\mathbf{x} = \begin{bmatrix} -1 \\ 2 \\ 3 \end{bmatrix}$;

$\qquad$ 5, $\begin{bmatrix} -8 \\ 0 \\ 5 \end{bmatrix}$, $\mathbf{x} = \begin{bmatrix} -1 \\ 0 \\ 1 \end{bmatrix}$

7. $[1/\sqrt{2} \quad -i/\sqrt{2}]^{\mathsf{T}}$, $[1/\sqrt{2} \quad i/\sqrt{2}]^{\mathsf{T}}$
9. $[1/\sqrt{2} \quad i/\sqrt{2}]^{\mathsf{T}}$, $[1/\sqrt{2} \quad -i/\sqrt{2}]^{\mathsf{T}}$

11. $\begin{bmatrix} 1 \\ 1 \end{bmatrix}$, $\begin{bmatrix} 1 \\ -2 \end{bmatrix}$, $\begin{bmatrix} 3 & 0 \\ 0 & 0 \end{bmatrix}$
13. $\begin{bmatrix} 1 \\ 2 \\ 3 \end{bmatrix}$, $\begin{bmatrix} 0 \\ 1 \\ 2 \end{bmatrix}$, $\begin{bmatrix} 0 \\ 0 \\ 1 \end{bmatrix}$, $\begin{bmatrix} 16 & 0 & 0 \\ 0 & -8 & 0 \\ 0 & 0 & 4 \end{bmatrix}$

15. $\begin{bmatrix} 1 \\ 1 \\ 0 \end{bmatrix}, \begin{bmatrix} 2 \\ 0 \\ 1 \end{bmatrix}, \begin{bmatrix} 0 \\ 1 \\ 1 \end{bmatrix}, \begin{bmatrix} 15 & 0 & 0 \\ 0 & 0 & 0 \\ 0 & 0 & -15 \end{bmatrix}$

17. Ellipse $4y_1^2 + 10y_2^2 = 200$, $x_1 = (y_1 + y_2)/\sqrt{2}$, $x_2 = (-y_1 + y_2)/\sqrt{2}$

19. Straight lines $y_2 = \pm 2$, $x_1 = (y_1 + 3y_2)/\sqrt{10}$, $x_2 = (3y_1 - y_2)/\sqrt{10}$

21. Hyperbola $y_1^2 - 4y_2^2 = 4$, $x_1 = (2y_1 + 3y_2)/\sqrt{13}$, $x_2 = (3y_1 - 2y_2)/\sqrt{13}$

CHAPTER 7 REVIEW, page 398

11. $3, \begin{bmatrix} 1 \\ 1 \end{bmatrix}; 2, \begin{bmatrix} 1 \\ -1 \end{bmatrix}$ **13.** $\frac{1}{2}, \begin{bmatrix} 2 \\ 1 \end{bmatrix}; -\frac{1}{4}, \begin{bmatrix} 1 \\ -2 \end{bmatrix}$

15. $2, \begin{bmatrix} 3 \\ 0 \\ 1 \end{bmatrix}; -2, \begin{bmatrix} 1 \\ 1 \\ 0 \end{bmatrix}; 1, \begin{bmatrix} 2 \\ 0 \\ 1 \end{bmatrix}$ **17.** $\begin{bmatrix} 1 \\ 1 \end{bmatrix}, \begin{bmatrix} 1 \\ 5 \end{bmatrix}, \begin{bmatrix} 7 & 0 \\ 0 & 3 \end{bmatrix}$

19. $\begin{bmatrix} 1 \\ 1 \\ 2 \end{bmatrix}, \begin{bmatrix} -1 \\ -1 \\ 1 \end{bmatrix}, \begin{bmatrix} 2 \\ -1 \\ 1 \end{bmatrix}, \begin{bmatrix} 9 & 0 & 0 \\ 0 & 0 & 0 \\ 0 & 0 & -12 \end{bmatrix}$

21. Ellipse $8y_1^2 + 18y_2^2 = 72$, $x_1 = (3y_1 + y_2)/\sqrt{10}$, $x_2 = (y_1 - 3y_2)/\sqrt{10}$

23. Ellipse $y_1^2 + 9y_2^2 = 9$, $x_1 = (y_1 + 2y_2)/\sqrt{5}$, $x_2 = (-2y_1 + y_2)/\sqrt{5}$

PROBLEM SET 8.1, page 407

1. 3, 4, 0; 5   **3.** 1, 2, 3; $\sqrt{14}$ **5.** $-4, 4, 0$; $\sqrt{32}$

7. 0, -18, 0; 18   **9.** (3, 3, 0); $\sqrt{13}$ **11.** (0, 0, 0); $\sqrt{74}$

13. $(\frac{5}{4}, 1, -\frac{9}{2}); \frac{3}{4}$ **15.** [7, -1, 0]   **17.** [7, 2, 0]

19. [44, 0, -4]   **21.** [18, -48, 6]   **25. 0**

27. $[-6, -10, -14]$; $\sqrt{332}$ **29.** $[-1, -6, 0]$ **31.** $p_3 = -17$

33. $|\mathbf{w}|/2 \sin \alpha$

PROBLEM SET 8.2, page 413

1. -8 **3.** $4\sqrt{14}$, $\sqrt{59}$ **5.** 30

7. $-22, 22$ **9.** $-8, -8\mathbf{c}$ **13.** 28

15. 0   **19.** 19.1°   **21.** 38.0°

23. 18.4°   **25.** 22.2°   **27.** 55.15°, 62.425°, 62.425°

29. 54.6°, 125.4°   **31.** $1/\sqrt{3}$ **33.** $-14/\sqrt{20}$

35. 4.6

PROBLEM SET 8.3, page 421

1. $8\mathbf{k}, -8\mathbf{k}, 1, 1$ **3.** $[8, -4, -1]$, 9, 9, 8   **5.** 120$\mathbf{k}$

7. $[16, 8, -14]$ **9.** $[-24, 16, 0]$, $[-26, 13, -4]$

11. c, 32

13. 1, -1

15. $-256, -256$

17. $-16, 16$

21. $[72, 0, -27]$, $\sqrt{5913}$

23. $[-20, 20, 0]$

25. $[42, -9, 21]$

27. $\sqrt{4014}$

29. $\frac{1}{2}\sqrt{3081}$

31. $\frac{27}{2}x - 9y + 18z = 0$

33. 50

35. 79

37. Linearly dependent

PROBLEM SET 8.4, page 427

1. 100, 44.5, 153.667

3. Between the ellipses $x^2/4 + y^2/9 = 1$ and $x^2/16 + y^2/36 = 1$

5. $y/x = c$, straight lines

7. Hyperbolas

9. Circles

11. Planes

13. Ellipsoids

15. Cones

25. $[0, 0, 2x]$, $[2y, 0, 0]$, $[0, 2z, 0]$

27. $[y, 0, z]$, $[x, z, 0]$, $[0, y, x]$

29. $[-\sin x \cosh y, -\cos x \sinh y]$, $[\cos x \sinh y, -\sin x \cosh y]$

PROBLEM SET 8.5, page 433

1. $(4 + t)\mathbf{i} + (2 + t)\mathbf{j}$

3. $[3 + 4t, 1 + 7t, 5 - t]$

5. $[2 + 3t, 3 - 4t, 0]$

7. $[1 + 2t, 2, 3 - 3t]$

9. $y = x^3 + 2$, $z = 0$

11. Circle in the yz-plane

13. Hyperbola $x^2 - y^2 = 1$, $z = 0$

15. Counterclockwise. $t = -\tilde{t}$

17. $[0, 3 \cos t, 3 + 3 \sin t]$

19. Ellipse $[\cos t, \sin t, \sin t]$

21. (a) $\mathbf{r}' = [1, 3t^2, 0]$, $\mathbf{u} = (9t^4 + 1)^{-1/2}\mathbf{r}'$,   (b) $[1, 3, 0]$, $[1/\sqrt{10}, 3/\sqrt{10}, 0]$,   (c) $[1 + w, 1 + 3w, 0]$

23. (a) $\mathbf{r}' = [-\sin t, 2 \cos t, 0]$, $(\sin^2 t + 4 \cos^2 t)^{-1/2}\mathbf{r}'$,   (b) $[-\sqrt{3}/2, 1, 0]$, $[-\sqrt{3/7}, \sqrt{4/7}, 0]$,   (c) $[\frac{1}{2} - w\sqrt{3}/2, \sqrt{3} + w, 0]$

25. $\mathbf{r}' = \sinh t\, \mathbf{i} + \cosh t\, \mathbf{j}$, $\mathbf{u} = (\sinh^2 t + \cosh^2 t)^{-1/2}\mathbf{r}'$,   (b) $[\frac{4}{3}, \frac{5}{3}, 0]$, $[4/\sqrt{41}, 5/\sqrt{41}, 0]$,   (c) $(1/3)[5 + 4w, 4 + 5w, 0]$

27. $\sinh 1 = 1.175$

29. $8(\sqrt{1000} - 1)/27 = 9.073$

31. Start from $\mathbf{r}(t) = t\mathbf{i} + f(t)\mathbf{j}$.

PROBLEM SET 8.6, page 439

1. $\mathbf{a} = \mathbf{0}$

3. $\mathbf{a} = 10\mathbf{k} = \mathbf{a}_{\text{tang}}$, $\mathbf{a}_{\text{norm}} = \mathbf{0}$

5. $\mathbf{a} = -b \cos t\, \mathbf{i} - b \sin t\, \mathbf{j} = \mathbf{a}_{\text{norm}}$, $\mathbf{a}_{\text{tang}} = \mathbf{0}$. Circle

7. $\mathbf{a} = e^t\mathbf{i} + e^{-t}\mathbf{j}$, $\mathbf{a}_{\text{tang}} = \tanh 2t\,(e^t\mathbf{i} - e^{-t}\mathbf{j})$. Hyperbola $xy = 1$

9. $\mathbf{v}(0) = 2\omega R\mathbf{i}$, $\mathbf{a}(0) = -\omega^2 R\mathbf{j}$, $\mathbf{v}(\pi/\omega) = \mathbf{0}$, $\mathbf{a}(\pi/\omega) = \omega^2 R\mathbf{j}$

11. $\mathbf{w} = \omega\mathbf{k} = \text{const}$, $\mathbf{a} = \mathbf{v}' = (\mathbf{w} \times \mathbf{r})' = \mathbf{w} \times \mathbf{r}' = \mathbf{w} \times \mathbf{v} = \omega\mathbf{k} \times (\omega\mathbf{k} \times \mathbf{r})$
    $= \omega^2\mathbf{k} \times [\mathbf{k} \times (x\mathbf{i} + y\mathbf{j})] = \omega^2\mathbf{k} \times (x\mathbf{j} - y\mathbf{i}) = \omega^2(-x\mathbf{i} - y\mathbf{j}) = -\omega^2\mathbf{r}$

13. 1 year $= 365 \cdot 86400$ sec, $R = 30 \cdot 365 \cdot 86400/2\pi = 151.10^6$[km],
    $|\mathbf{a}| = \omega^2 R = |\mathbf{v}|^2/R = 30^2/R = 5.98 \cdot 10^{-6}$ [km/sec^2]

15. $R = 3960 + 80$ mi $= 2.133 \cdot 10^7$ ft, $g = |\mathbf{a}| = \omega^2 R = |\mathbf{v}|^2/R$,
    $|\mathbf{v}| = \sqrt{gR} = \sqrt{6.61 \cdot 10^8} = 25700$ [ft/sec] $= 17500$ [mph]

PROBLEM SET 8.7, page 443

3. $\mathbf{r}(t) = [t, \quad y(t), \quad 0], \mathbf{r}' = [1, \quad y', \quad 0], \mathbf{r}' \cdot \mathbf{r}' = 1 + y'^2, \mathbf{r}'' = [0, \quad y'', \quad 0]$, etc.

5. $2/(1 + 4x^2)^{3/2}$ **7.** $a/(a^2 + c^2)$

9. $6/(\sqrt{t}\,(4 + 9t)^{3/2})$ **15.** $3/(1 + 9t^2 + 9t^4)$

PROBLEM SET 8.8, page 446

1. $4\sqrt{2}\,(\sinh 8t)/(\cosh 8t)^{1/2}$ **3.** $(\cos t)^{(\sin t)-1}[\cos^2 t \ln (\cos t) - \sin^2 t]$

5. $-t(1 + t^2)^{-3/2}$ **7.** $e^{2u} \sin 2v, e^{2u} \cos 2v$

9. $-(u^2 + v^2)^{-3}u, -(u^2 + v^2)^{-3}v$

PROBLEM SET 8.9, page 452

1. $[2x, -2y], [-2, -6]$ **3.** $(x^2 + y^2)^{-1}[2x, 2y], [1, 0]$

5. $-(x^2 + y^2 + z^2)^{-3/2}[x, y, z], -14^{-3/2}[2, 1, 3]$

7. $(x^2 + y^2)^{-2}[2xz, 2yz, -x^2 - y^2], [0, 4, -1]$

9. $[\sin x \cosh y, -\cos x \sinh y], [\cosh 1, 0]$

11. $\left[-\dfrac{1}{y}, \dfrac{x}{y^2}\right], [1, 8]$ **15.** $[0.8, -0.6]$ **17.** $[0.6, 0.8]$

19. $(1/\sqrt{50})[3, 4, -5]$ **21.** $x^2 + 2y^2 + 4z^2$ **23.** No potential

25. xy/z **29.** $-2/\sqrt{5}$ **31.** $-7/(125\sqrt{3})$

33. $7/3$ **35.** $-2e^2/\sqrt{13}$

PROBLEM SET 8.10, page 456

1. 3   **3.** $2e^x \cos y$ **5.** 0

7. $2e^x$

9. (a) Parallel flow, 0; (b) outflow on the left and right, no flow across the other sides, hence div $\mathbf{v} > 0$; (c) outflow left and right, inflow from above and below, balance perhaps zero; by calculation, div $\mathbf{v} = 0$. Etc.

13. (b) $(fv_1)_x + (fv_2)_y + (fv_3)_z = f[(v_1)_x + (v_2)_y + (v_3)_z] + f_x v_1 + f_y v_2 + f_z v_3$, etc.
(c) Use (b) with $\mathbf{v} = \nabla g$.

15. 28   **17.** $2xy/z^3$ **19.** 0

PROBLEM SET 8.11, page 459

3. $[y - z, z - x, x - y]$

5. **0**

7. $x(z^2 - y^2)\mathbf{i} + y(x^2 - z^2)\mathbf{j} + z(y^2 - x^2)\mathbf{k}$

9. div $\mathbf{v} = 0$, curl $\mathbf{v} = [0, -3x^2, 0]$, $\mathbf{v} = [x', y', z'] = [0, 0, x^3]$, $x' = 0, x = c_1, y' = 0$, $y = c_2, z' = x^3 = c_1^3, z = c_1^3 t + c_3$

11. curl $\mathbf{v} = -2\mathbf{k}$, incompressible, $x' = y, y' = -x, x^2 + y^2 = c_1, z = c_2$

13. curl $\mathbf{v} = \mathbf{0}$, div $\mathbf{v} = 1$, compressible, $\mathbf{r} = [c_1 e^t, c_2 e^t, c_3 e^{-t}]$

15. $[x^2 z - 2xyz, y^2 x - 2xyz, z^2 y - 2xyz]$

17. **0**, $[xy - zx, yz - xy, zx - yz]$

19. $-xy - yz - zx, 0$

CHAPTER 8 REVIEW, page 459

15. $36\mathbf{k}, -36\mathbf{k}$　　　　　　**17.** $-24, 24$　　　　　　**19.** $[266, 182, -184]$

21. $-1696, -1696$　　　　**23.** $[-5040, -3360, 3480]$

25. $\sqrt{270}, \sqrt{278} - \sqrt{52}$　　**27.** $178.2°$　　　　　　**29.** -12

31. If $|\mathbf{a}| = |\mathbf{b}|$ or if $\mathbf{a}$ and $\mathbf{b}$ are orthogonal　　　　**35.** $7x - 8y + 6z = 5$

37. $[0, -5, 0]$　　　　　**39.** $1, -2y$　　　　　　**41.** 0

43. $[0, -2, 0]$　　　　　**45.** $-7/11$　　　　　　**47.** $\mathbf{0}$

49. 0. Why?

PROBLEM SET 9.1, page 470

1. 4　　　　　　　　　**3.** $8/5$　　　　　　　　**5.** $891/64$

7. 9π　　　　　　　　**9.** $-1 - e^{-1} + 2e \approx 4.07$　　**13.** $80\sqrt{10}/3$

15. $\sqrt{5}(4\pi + 256\pi^3/3) \approx 5944$　　　　　　**17.** $2\sqrt{2}\pi$

19. $3\pi/8$

PROBLEM SET 9.2, page 477

3. 183　　　　　　　　**5.** 25　　　　　　　　**7.** $-2 \sinh 2$

9. $\cosh 1 - 2$　　　　**11.** Independent, $a^2b^2 + c$　　**13.** Dependent

15. Dependent　　　　**17.** Independent, $\sin(a + bc)$　**19.** Independent, $ae^c + b^2$

PROBLEM SET 9.3, page 484

1. $160/3$　　　　　　　**3.** 54　　　　　　　　**5.** $1 - 1/\sqrt{2}$

7. $\frac{1}{2}\sinh 4 - \sinh 2$　　**9.** $e^{30} - 9e^5 - e^2 + e$　　**11.** 144

13. $a^3/3$　　　　　　　**15.** $4a/3\pi$　　　　　　**17.** $I_x = bh^3/12, I_y = b^3h/4$

19. $I_x = (a + b)h^3/24, I_y = h(a^4 - b^4)/(48(a - b))$

PROBLEM SET 9.4, page 490

1. $9(e^2 - 1) - \frac{8}{3}(e^3 - 1)$　　**3.** $-\pi/2$　　　　　　**5.** 0

7. $2 \cdot 5^7/7$　　　　　**9.** $16/5$　　　　　　　**11.** $a^2(3\pi/8 - 1)$

13. $2 + 3\pi/4$　　　　　**15.** $\frac{1}{2}(\cosh 4 - 1)$　　　**17.** 10π

19. Set $F_1 = -ww_y, F_2 = ww_x$ in Green's theorem. Then $F_1\, dx + F_2\, dy = w(\text{grad } w)\bullet\mathbf{n}\, ds$, etc.

PROBLEM SET 9.5, page 495

1. Straight lines, $\mathbf{k}$

3. $z = c\sqrt{x^2 + y^2}$, circles, straight lines, $[-cu \cos v, \quad -cu \sin v, \quad u]$

5. $z = x^2/a^2 + y^2/b^2$, ellipses, parabolas, $[-2bu^2 \cos v, \quad -2au^2 \sin v, \quad abu]$

7. $z = x^2/a^2 - y^2/b^2$, hyperbolas, parabolas, $[-2bu^2 \cosh v, \quad 2au^2 \sinh v, \quad abu]$

9. $x^2/a^2 + y^2/b^2 = 1$, ellipses, straight lines, $[-b \cos v, \quad -a \sin v, \quad 0]$

13. $[8u, \quad 6v, \quad 4(1 - u - v)], [3, 4, 6]$

15. $[\cos v \cos u, \quad \cos v \sin u, \quad 2 \sin v], [2 \cos u \cos^2 v, \quad 2 \sin u \cos^2 v, \quad \sin v \cos v]$

17. $[\cosh u, \quad \sinh u, \quad v], [\cosh u, \quad -\sinh u, \quad 0]$

19. $[u \cos v, \quad u \sin v, \quad 9u^2], [-18u^2 \cos v, \quad -18u^2 \sin v, \quad u]$
21. $\tilde{\mathbf{r}} = [u, \quad v, \quad 9(u^2 + v^2)], \tilde{\mathbf{N}} = [-18u, -18v, 1]$
25. $(1/9)[4, -4, 7]$ **27.** $(1/a)[0, y, z]$ **29.** $(1/6)[x, y, z]$

PROBLEM SET 9.6, page 503

1. -36 **3.** $2 \sinh 6$ **5.** 0
7. $\mathbf{F(r)} \cdot \mathbf{N} = \cos^3 v \cos u \sin u$ from (3), Sec. 9.5. *Ans.* 1/3
9. $\mathbf{F(r)} \cdot \mathbf{N} = -u^3$. *Ans.* -128π
13. $dA = \sqrt{26}\, dx\, dy$. Integrate over $1 \leqq x < \infty, 1 \leqq y < \infty$. *Ans.* $\sqrt{26}/e$
15. $G(\mathbf{r}) = (1 + 9u^4)^{3/2}, |\mathbf{N}| = (1 + 9u^4)^{1/2}$. *Ans.* 54.4   **17.** 27π
21. $2\pi h$ **23.** $\pi h^4/\sqrt{2}$ **25.** $\pi h + 2\pi h^3/3$

PROBLEM SET 9.7, page 509

1. 224   **3.** $4 + 8/\pi^2$ **5.** 1/10
7. 16/3   **9.** $2a^5/3$ **11.** $\pi a^4 h/2$
13. 0   **15.** 0   **17.** 2/3
19. $2916\pi/5$

PROBLEM SET 9.8, page 514

1. Integrals $8(-2)$ $(x = 1)$, $4(-4)$ $(y = 2)$, $2 \cdot 16$ $(z = 4)$. Sum 0. $x = 0, y = 0, z = 0$ no contribution.
3. $16\pi/3$
5. 480π
7. $3/2 - \pi^2/4$
9. $\mathbf{F} = [x, y, z]$, div $\mathbf{F} = 3$. In (2), Sec. 9.7, $\mathbf{F} \cdot \mathbf{n} = |\mathbf{F}||\mathbf{n}| \cos \phi = \sqrt{x^2 + y^2 + z^2} \cos \phi = r \cos \phi$.
11. $\mathbf{F} = [x, 0, 0]$, div $\mathbf{F} = 1$, use (3), Sec. 9.7. Etc.

PROBLEM SET 9.9, page 520

1. 5. Line integral $= 1 + 5 - 1 + 0 = 5$
3. $\pm 4(1 - e^4) \approx \mp 214$. ± 4 from $(0, 0, 0)$ to $(4, 0, 0)$, $\mp 4e^4$ from $(4, 2, 4)$ to $(0, 2, 4)$. The integrals over the parabolas cancel each other.
5. $\pm 4/3$. $\pm 4/3$ over the semicircle, 0 over the parabola
7. curl $\mathbf{F} = 9\mathbf{k}$. *Ans.* 36π **9.** curl $\mathbf{F} = 2\mathbf{j} - 2\mathbf{k}$. *Ans.* -4π
11. (curl $\mathbf{F}) \cdot \mathbf{N} = -xy + y(1 - x - y)$. *Ans.* 0   **13.** $-\sqrt{3}/10$

CHAPTER 9 REVIEW, page 521

17. Exact. e **19.** Not exact, ± 2 **21.** 4, by Stokes
23. Exact, $1 + \sqrt{2}$ **25.** Exact, $2 + e^2 - e$ **27.** 0 since curl $\mathbf{F} = \mathbf{0}$
29. 0, by Green   **31.** 4/5, 8/15
33. $M = 2 \ln 2 - 1, \bar{x} = (2 \ln 2 - \frac{3}{4})/M, \bar{y} = (\ln 2 - 1)^2/M$
35. 8/7, 118/49   **37.** -8 **39.** 63
41. $-4/3$ **43.** 4π **45.** 72π

PROBLEM SET 10.1, page 528

1. 2π, 2π, π, π, 2, 2, 1, 1

PROBLEM SET 10.2, page 536

1. $\dfrac{1}{2} + \dfrac{2}{\pi}\left(\cos x - \dfrac{1}{3}\cos 3x + \dfrac{1}{5}\cos 5x - + \cdots\right)$

3. $\dfrac{k}{2} + \dfrac{2k}{\pi}\left(\sin x + \dfrac{1}{3}\sin 3x + \dfrac{1}{5}\sin 5x + \cdots\right)$

5. $2\left(\sin x - \dfrac{1}{2}\sin 2x + \dfrac{1}{3}\sin 3x - \dfrac{1}{4}\sin 4x + - \cdots\right)$

7. $\dfrac{\pi^2}{3} - 4\left(\cos x - \dfrac{1}{4}\cos 2x + \dfrac{1}{9}\cos 3x - \dfrac{1}{16}\cos 4x + - \cdots\right)$

9. $2\left[\left(\dfrac{\pi^2}{1} - \dfrac{6}{1^3}\right)\sin x - \left(\dfrac{\pi^2}{2} - \dfrac{6}{2^3}\right)\sin 2x + \left(\dfrac{\pi^2}{3} - \dfrac{6}{3^3}\right)\sin 3x - + \cdots\right]$

11. $-\dfrac{4}{\pi}\left(\sin x + \dfrac{1}{3}\sin 3x + \dfrac{1}{5}\sin 5x + \cdots\right)$

13. $\dfrac{4}{\pi}\left(\cos x - \dfrac{1}{3}\cos 3x + \dfrac{1}{5}\cos 5x - + \cdots\right)$

15. $\dfrac{2}{\pi}\sin x + \dfrac{1}{2}\sin 2x - \dfrac{2}{9\pi}\sin 3x - \dfrac{1}{4}\sin 4x + \dfrac{2}{25\pi}\sin 5x + \cdots$

PROBLEM SET 10.3, page 540

1. $\dfrac{4}{\pi}\left(\sin \pi x + \dfrac{1}{3}\sin 3\pi x + \dfrac{1}{5}\sin 5\pi x + \cdots\right)$

3. $1 + \dfrac{4}{\pi}\left(\sin \dfrac{\pi x}{2} + \dfrac{1}{3}\sin \dfrac{3\pi x}{2} + \dfrac{1}{5}\sin \dfrac{5\pi x}{2} + \cdots\right)$

5. $\dfrac{4}{\pi}\left(\sin \pi x - \dfrac{1}{2}\sin 2\pi x + \dfrac{1}{3}\sin 3\pi x - + \cdots\right)$

7. $1 - \dfrac{12}{\pi^2}\left(\cos \pi x - \dfrac{1}{4}\cos 2\pi x + \dfrac{1}{9}\cos 3\pi x - \dfrac{1}{16}\cos 4\pi x + - \cdots\right)$

9. $\dfrac{1}{4} - \dfrac{2}{\pi^2}\left(\cos \pi x + \dfrac{1}{9}\cos 3\pi x + \cdots\right) + \dfrac{1}{\pi}\left(\sin \pi x - \dfrac{1}{2}\sin 2\pi x + - \cdots\right)$

11. $4\left(\dfrac{1}{2} - \dfrac{1}{1\cdot 3}\cos 2\pi x - \dfrac{1}{3\cdot 5}\cos 4\pi x - \dfrac{1}{5\cdot 7}\cos 6\pi x - \cdots\right)$

PROBLEM SET 10.4, page 546

1. Even: $|x^3|$, $x^2\cos nx$, $\cosh x$. Odd: $x\cos nx$, $\sinh x$, $x|x|$.

3. Neither   **5.** Even   **7.** Odd

9. Neither

11. $\dfrac{k}{2} + \dfrac{2k}{\pi}\left(\cos x - \dfrac{1}{3}\cos 3x + \dfrac{1}{5}\cos 5x - + \cdots\right)$

13. $\dfrac{4}{\pi} \left(\sin x - \dfrac{1}{9} \sin 3x + \dfrac{1}{25} \sin 5x - + \cdots \right)$

15. $\dfrac{\pi^2}{6} - 2 \left(\cos x - \dfrac{1}{4} \cos 2x + \dfrac{1}{9} \cos 3x - \dfrac{1}{16} \cos 4x + - \cdots \right)$

21. $\dfrac{L}{2} - \dfrac{4L}{\pi^2} \left(\cos \dfrac{\pi x}{L} + \dfrac{1}{9} \cos \dfrac{3\pi x}{L} + \dfrac{1}{25} \cos \dfrac{5\pi x}{L} + \cdots \right);$

$\dfrac{2L}{\pi} \left(\sin \dfrac{\pi x}{L} - \dfrac{1}{2} \sin \dfrac{2\pi x}{L} + \dfrac{1}{3} \sin \dfrac{3\pi x}{L} - + \cdots \right)$

23. $\dfrac{\pi}{2} + \dfrac{4}{\pi} \left(\cos x + \dfrac{1}{3^2} \cos 3x + \dfrac{1}{5^2} \cos 5x + \cdots \right);$

$2 \left(\sin x + \dfrac{1}{2} \sin 2x + \dfrac{1}{3} \sin 3x + \cdots \right)$

25. $a_0 = \dfrac{1}{L} (e^L - 1), \quad a_n = \dfrac{2L}{L^2 + n^2\pi^2} \left[(-1)^n e^L - 1 \right];$

$b_n = \dfrac{2n\pi}{L^2 + n^2\pi^2} (1 - (-1)^n e^L)$

PROBLEM SET 10.5, page 549

3. $i \displaystyle\sum_{\substack{n=-\infty \\ n\neq 0}}^{\infty} \dfrac{(-1)^n}{n} e^{inx}$ **5.** $\pi + i \displaystyle\sum_{\substack{n=-\infty \\ n\neq 0}}^{\infty} \dfrac{1}{n} e^{inx}$ **7.** Use (7).

PROBLEM SET 10.6, page 552

3. $y = C_1 \cos \omega t + C_2 \sin \omega t + A(\omega) \sin t, \quad A(\omega) = 1/(\omega^2 - 1), \quad A(0.5) = -1.33,$
$A(0.7) = -1.96, \quad A(0.9) = -5.3, \quad A(1.1) = 4.8, \quad A(1.5) = 0.8, \quad A(2) = 0.33,$
$A(10) = 0.01$

5. $y = C_1 \cos \omega t + C_2 \sin \omega t + \displaystyle\sum_{n=1}^{N} \dfrac{b_n}{\omega^2 - n^2} \sin nt$

7. $y = C_1 \cos \omega t + C_2 \sin \omega t + \dfrac{\pi}{2\omega^2} + \dfrac{4}{\pi} \left(\dfrac{1}{\omega^2 - 1} \cos t + \dfrac{1}{9(\omega^2 - 9)} \cos 3t + \cdots \right)$

11. $y = \displaystyle\sum_{n=1}^{N} (A_n \cos nt + B_n \sin nt),$

$A_n = \left[(1 - n^2)a_n - ncb_n \right]/D_n, \quad B_n = \left[(1 - n^2)b_n + nca_n \right]/D_n, \quad D_n = (1 - n^2)^2 + n^2c^2$

13. $y = \displaystyle\sum_{n=1}^{\infty} \left[\dfrac{(-1)^n c}{n^2 D_n} \cos nt - \dfrac{(-1)^n(1 - n^2)}{n^3 D_n} \sin nt \right], \quad D_n = (1 - n^2)^2 + n^2c^2$

15. $I = \displaystyle\sum_{n=1}^{\infty} (A_n \cos nt + B_n \sin nt), \quad A_n = (-1)^{n+1} \dfrac{240(10 - n^2)}{n^2 D_n},$

$B_n = \dfrac{(-1)^{n+1} 2400}{n D_n}, \quad D_n = (10 - n^2)^2 + 100n^2$

PROBLEM SET 10.7, page 556

1. $F = \dfrac{4}{\pi} \left[\sin x + \dfrac{1}{3} \sin 3x + \cdots + \dfrac{1}{N} \sin Nx \right]$ for odd N;

$E^* = 1.19, 1.19, 0.62, 0.62, 0.42$

3. $F = 2 \left(\sin x - \dfrac{1}{2} \sin 2x + \cdots + \dfrac{(-1)^{N+1}}{N} \sin Nx \right)$; $E^* \approx 8.1, 5.0, 3.6, 2.8, 2.3$

5. $F = 2 \left[(\pi^2 - 6) \sin x - \dfrac{1}{8} (4\pi^2 - 6) \sin 2x + \dfrac{1}{27} (9\pi^2 - 6) \sin 3x - + \cdots \right]$;

863, 675, 455, 326, 266, 219

7. $F = \dfrac{2}{\pi} \sin x + \dfrac{1}{2} \sin 2x - \dfrac{2}{9\pi} \sin 3x - \dfrac{1}{4} \sin 4x + \dfrac{2}{25\pi} \sin 5x + \cdots$,

$E^* = \dfrac{\pi^3}{12} - \pi \left[\dfrac{4}{\pi^2} + \dfrac{1}{4} + \dfrac{4}{81\pi^2} - \dfrac{1}{16} + \dfrac{4}{625\pi^2} + \cdots \right]$;

1.311, 0.525, 0.509, 0.313, 0.311

9. $N = 13$

15. Use the Fourier series $\cos^3 x = \tfrac{3}{4} \cos x + \tfrac{1}{4} \cos 3x$.

PROBLEM SET 10.8, page 563

7. $\dfrac{2}{\pi} \displaystyle\int_0^\infty \dfrac{\sin w \cos xw}{w}\, dw$

9. $\dfrac{2}{\pi} \displaystyle\int_0^\infty \left[\dfrac{a \sin aw}{w} + \dfrac{\cos aw - 1}{w^2} \right] \cos xw\, dw$

11. $A = \dfrac{2}{\pi} \displaystyle\int_0^\infty \dfrac{\cos wv}{1 + v^2}\, dv = e^{-w} \; (w > 0), \quad f(x) = \displaystyle\int_0^\infty e^{-w} \cos wx\, dw$

13. $\dfrac{2}{\pi} \displaystyle\int_0^\infty \dfrac{1 - \cos aw}{w} \sin xw\, dw$

15. $\dfrac{2}{\pi} \displaystyle\int_0^\infty \dfrac{\sin \pi w}{1 - w^2} \sin xw\, dw$

17. $\dfrac{2}{\pi} \displaystyle\int_0^\infty \dfrac{w - e(w \cos w - \sin w)}{1 + w^2} \sin xw\, dw$

19. For $n = 1, 2, 11, 12, 31, 32, 49, 50$ the value of $\mathrm{Si}(n\pi) - \pi/2$ equals $0.28, -0.15, 0.029, -0.026, 0.0103, -0.0099, 0.0065, -0.0064$ (rounded).

PROBLEM SET 10.9, page 568

1. $\sqrt{2/\pi}\, (2 \sin w - \sin 2w)/w$

3. $\sqrt{2/\pi}\, (\cos aw + aw \sin aw - 1)/w^2$

5. $e^{-w}\sqrt{\pi/2}$

7. $\sqrt{2/\pi}\, (2w \cos w + (w^2 - 2) \sin w)/w^3$

11. $\sqrt{2/\pi}\, w/(a^2 + w^2)$

13. $\sqrt{2/\pi}\, ((2 - w^2) \cos w + 2w \sin w - 2)/w^3$

15. $\sqrt{2/\pi}\, w/(1 + w^2)$

17. $\sqrt{\pi/2}\, e^{-w} \cos w$

19. No

PROBLEM SET 10.10, page 575

1. $i(e^{-ibw} - e^{-iaw})/(w\sqrt{2\pi})$

3. $(e^{a-iaw} - e^{-a+iaw})/((1 - iw)\sqrt{2\pi})$

5. $(e^{-iaw}(1 + iaw) - 1)/(w^2\sqrt{2\pi})$

7. $1/((1 + iw)^2\sqrt{2\pi})$

9. $i\sqrt{2/\pi}\, (\cos w - 1)/w$

11. $e^{-w^2/2}$

13. Use $e^{-ic(a-w)} - e^{ic(a-w)} = -2i \sin c(a - w)$.

15. $f(x) = g(x) = e^{-x} \; (x > 0)$. Show that $(f*g)(x) = xe^{-x}$, etc.

CHAPTER 10 REVIEW, page 579

17. $\dfrac{\pi^2}{3} - 4 \left(\cos x - \dfrac{1}{4} \cos 2x + \dfrac{1}{9} \cos 3x - + \cdots \right)$

19. $1 - \dfrac{8}{\pi^2}\left(\cos\dfrac{\pi x}{2} + \dfrac{1}{9}\cos\dfrac{3\pi x}{2} + \dfrac{1}{25}\cos\dfrac{5\pi x}{2} + \cdots\right)$

21. $\dfrac{1}{2} - \dfrac{2}{\pi}\left(\sin\pi x + \dfrac{1}{3}\sin 3\pi x + \dfrac{1}{5}\sin 5\pi x + \cdots\right)$

23. $\dfrac{2}{\pi} + \dfrac{4}{\pi}\left(\dfrac{1}{1\cdot 3}\cos 2x - \dfrac{1}{3\cdot 5}\cos 4x + \dfrac{1}{5\cdot 7}\cos 6x - + \cdots\right)$

25. $\dfrac{\pi}{2} - \dfrac{4}{\pi}\left(\cos x + \dfrac{1}{9}\cos 3x + \dfrac{1}{25}\cos 5x + \cdots\right)$

27. $\dfrac{8}{\pi}\left(\cos x + \dfrac{1}{9}\cos 3x + \dfrac{1}{25}\cos 5x + \cdots\right)$

29. $\dfrac{4}{3} + \dfrac{4}{\pi^2}\left(\cos\pi x + \dfrac{1}{4}\cos 2\pi x + \dfrac{1}{9}\cos 3\pi x + \cdots\right)$

$\qquad\qquad\qquad\qquad - \dfrac{4}{\pi}\left(\sin\pi x + \dfrac{1}{2}\sin 2\pi x + \dfrac{1}{3}\sin 3\pi x + \cdots\right)$

31. Prob. 19   **33.** 5.16771, same, 0.074755, same, 0.011879, same

35. $y = C_1\cos\omega t + C_2\sin\omega t + \dfrac{\pi^2}{12\omega^2} - \dfrac{1}{\omega^2 - 1}\cos t + \dfrac{1}{4(\omega^2 - 4)}\cos 2t - + \cdots$

37. $\sqrt{2/\pi}\,(\cos aw - \cos w + aw\sin aw - w\sin w)/w^2$

39. $1/((2 + iw)\sqrt{2\pi})$

PROBLEM SET 11.1, page 584

15. $u = c(x)e^y$ **17.** $u = h(x)y + k(x)$ **19.** $u = c(x)e^y + h(y)$

21. $u = c(x)\exp(xy^2)$ **23.** $3\ln(x^2 + y^2)/\ln 4$ **25.** $axy + bx + cy + k$

PROBLEM SET 11.3, page 594

3. $k(\cos t\sin x - \tfrac{1}{2}\cos 2t\sin 2x)$

5. $1.2\left(\cos t\sin x - \dfrac{1}{2^3}\cos 2t\sin 2x + \dfrac{1}{3^3}\cos 3t\sin 3x - + \cdots\right)$

7. $\dfrac{4}{\pi^2}\left(\cos t\sin x - \dfrac{1}{9}\cos 3t\sin 3x + \dfrac{1}{25}\cos 5t\sin 5x - + \cdots\right)$

9. $\dfrac{1.6}{\pi^2}\left((2 - \sqrt{2})\cos t\sin x - \dfrac{1}{9}(2 + \sqrt{2})\cos 3t\sin 3x\right.$

$\qquad\qquad\left. + \dfrac{1}{25}(2 + \sqrt{2})\cos 5t\sin 5x - + \cdots\right)$

13. $u = ce^{k(x+y)}$ **15.** $u = c\exp\left[\tfrac{1}{2}(x^2 + y^2) + k(x - y)\right]$

17. $u = c\exp(kx + y/k)$ **19.** $u = cx^k e^{-y^2/k}$

PROBLEM SET 11.4, page 597

3. $(1/2\pi)(n\pi/2)70 = 17.5n$ [cycles/sec]

13. Parabolic, $v = x$, $z = 2x - y$, $u_{vv} = 0$, $u = xf_1(2x - y) + f_2(2x - y)$

15. Hyperbolic, $v = 3x + y$, $z = x + y$, $u_{vz} = 0$, $u = f_1(3x + y) + f_2(x + y)$

17. Hyperbolic, $v = x + 2y$, $z = x - 2y$, $u_{vz} = 0$, $u = f_1(x + 2y) + f_2(x - 2y)$

PROBLEM SET 11.5, page 608

3. $u = \sin 0.1\pi x\, e^{-1.752\pi^2 t/100}$

5. $u = \dfrac{800}{\pi^3}\left(\sin 0.1\pi x\, e^{-0.01752\pi^2 t} + \dfrac{1}{3^3}\sin 0.3\pi x\, e^{-0.01752(3\pi)^2 t} + \cdots\right)$

7. $u_I = U_1' + (U_2 - U_1)x/L$, the solution of (1) with $\partial u/\partial t = 0$ satisfying the boundary conditions.

11. $u = k$

13. $u = \dfrac{1}{2} + \dfrac{4}{\pi^2}\left(\cos x\, e^{-t} + \dfrac{1}{9}\cos 3x\, e^{-9t} + \dfrac{1}{25}\cos 5x\, e^{-25t} + \cdots\right)$

17. $u = \dfrac{80}{\pi}\displaystyle\sum_{n=1}^{\infty}\dfrac{1}{2n-1}\sin\dfrac{(2n-1)\pi x}{24}\dfrac{\sinh[(2n-1)\pi y/24]}{\sinh(2n-1)\pi}$

19. $u(x, y) = \dfrac{A_0}{24}x + \displaystyle\sum_{n=1}^{\infty}A_n\dfrac{\sinh(n\pi x/24)}{\sinh n\pi}\cos\dfrac{n\pi y}{24}$,

$A_0 = \dfrac{1}{24}\displaystyle\int_0^{24}f(y)\,dy,\qquad A_n = \dfrac{1}{12}\displaystyle\int_0^{24}f(y)\cos\dfrac{n\pi y}{24}\,dy$

PROBLEM SET 11.6, page 615

3. $A(p) = \dfrac{2\sin p}{\pi p},\ B(p) = 0,\quad u = \dfrac{2}{\pi}\displaystyle\int_0^{\infty}\dfrac{\sin p}{p}\cos px\, e^{-c^2 p^2 t}\,dp$

5. $u(x, t) = \displaystyle\int_0^1\cos px\, e^{-c^2 p^2 t}\,dp$

9. Set $w = s/\sqrt{2}$ in (21).

PROBLEM SET 11.8, page 626

1. c increases and so does the frequency.

5. $c\pi\sqrt{85}$ (F_{29}, F_{67}, F_{76}, F_{92}), $c\pi\sqrt{221}$, $c\pi\sqrt{260}$, etc.

7. $B_{mn} = (-1)^{n+1}8/(mn\pi)$ (m odd), 0 (m even)

9. $B_{mn} = 64/(m^3 n^3 \pi^2)$ (m, n both odd), 0 otherwise

11. $u = 0.1\cos 5\pi t\sin 3\pi x\sin 4\pi y$

13. $u = \dfrac{64k}{\pi^6}\displaystyle\sum_{\substack{m=1 \\ m,n\ \text{odd}}}^{\infty}\sum_{n=1}^{\infty}\dfrac{1}{m^3 n^3}\cos(\pi t\sqrt{m^2 + n^2})\sin m\pi x\sin n\pi y$

17. $A = ab$, $b = A/a$, $(m^2 a^{-2} + n^2 a^2 A^{-2})' = 0$ gives $a/b = m/n$.

19. $\cos\left(\pi\sqrt{\left(\dfrac{4}{a^2} + \dfrac{9}{b^2}\right)}\,t\right)\sin\dfrac{2\pi x}{a}\sin\dfrac{3\pi y}{b}$

PROBLEM SET 11.9, page 628

7. $u = \dfrac{3}{4}r\sin\theta - \dfrac{1}{4}r^3\sin 3\theta$

9. $u = \dfrac{2}{\pi}r\sin\theta + \dfrac{1}{2}r^2\sin 2\theta - \dfrac{2}{9\pi}r^3\sin 3\theta - \dfrac{1}{4}r^4\sin 4\theta + \cdots$

11. $u = \dfrac{\pi}{2} \pm \dfrac{4}{\pi}\left(r + \dfrac{1}{9}r^3 + \dfrac{1}{25}r^5 + \cdots\right)$

13. $u = \dfrac{4u_0}{\pi}\left(r \sin\theta + \dfrac{1}{3}r^3 \sin 3\theta + \dfrac{1}{5}r^5 \sin 5\theta + \cdots\right)$

15. $\nabla^2 u = u_{x^*x^*} + u_{y^*y^*}$

PROBLEM SET 11.10, page 634

3. $T = 6.828\rho R^2 f_1{}^2$, f_1 the fundamental frequency

5. c increases, $\lambda_m = ck_m$ increases, and so does the frequency.

7. Differentiation brings in a factor $1/\lambda_m = R/c\alpha_m$ in (15).

9. No

PROBLEM SET 11.11, page 641

7. $u = 320/r + 60$

9. $v = F(r)G(t)$, $F'' + k^2 F = 0$, $\dot{G} + c^2 k^2 G = 0$, $F_n = \sin(n\pi r/R)$,

$$G_n = B_n \exp(-c^2 n^2 \pi^2 t/R^2), \qquad B_n = \dfrac{2}{R}\int_0^R rf(r)\sin\dfrac{n\pi r}{R}\,dr$$

11. $u = 1$

13. $\cos 2\phi = 2\cos^2\phi - 1$, $2w^2 - 1 = \frac{4}{3}P_2(w) - \frac{1}{3}$, $u = \frac{4}{3}r^2 P_2(\cos\phi) - \frac{1}{3}$

15. $u = B_0/r$ in (19) since $B_1 = B_2 = \cdots = 0$ in (20) because of orthogonality.

19. This is the analog of Example 1 with 55 replaced by 10.

25. TEAM PROJECT (b) $a(x^3 - 3xy^2) + k(-3x^2y + y^3)$ with arbitrary constants a and k.

PROBLEM SET 11.12, page 646

5. $U(x, s) = \dfrac{c(s)}{x^s} + \dfrac{x}{s^2(s+1)}$, $U(0, s) = 0$, $c(s) = 0$, $u(x, t) = x(t - 1 + e^{-t})$

9. Set $x^2/(4c^2\tau) = z^2$. Use z as a new variable of integration. Use $\mathrm{erf}(\infty) = 1$.

CHAPTER 11 REVIEW, page 647

21. $A(y)\cos 3x + B(y)\sin 3x$

23. $c_1(x)e^{-4y} + c_2(x)e^y - 3$

25. $cy^k e^{kx}$

27. $f_1(y + 2x) + f_2(y - 2x)$

29. $f_1(x) + f_2(y + x)$

31. $\cos 10t \sin 5x$

33. $\sin 0.01\pi x\, e^{-0.001143t}$

35. $\frac{3}{4}\sin 0.01\pi x\, e^{-0.001143t} - \frac{1}{4}\sin 0.03\pi x\, e^{-0.01029t}$

37. $250\cos 2x\, e^{-4t}$

39. $\sin\dfrac{\pi x}{12}\left(\sinh\dfrac{\pi y}{12}\right)\Big/\sinh\pi$

41. $\sin\dfrac{\pi x}{4}\left(\sinh\dfrac{\pi y}{4}\right)\Big/\sinh 3\pi$

49. $u = (u_1 - u_0)(\ln r)/\ln(r_1/r_0) + (u_0 \ln r_1 - u_1 \ln r_0)/\ln(r_1/r_0)$

PROBLEM SET 12.1, page 656

3. $23 - 14i$

5. $0.16 - 0.12i$

7. $-44, 64$

9. $-7 + 26i, -7 - 26i$

11. $(-7 - 26i)/29$

13. $-y/(x^2 + y^2)$

15. 256

17. $(x^3 - 3xy^2)/(x^2 + y^2)$

PROBLEM SET 12.2, page 662

1. $\sqrt{2}(\cos \frac{1}{4}\pi + i \sin \frac{1}{4}\pi)$

3. $5(\cos 2.2143 - i \sin 2.2143)$

5. $3(\cos \frac{1}{2}\pi \pm i \sin \frac{1}{2}\pi)$

7. $4(\cos \pi + i \sin \pi)$

9. $\sqrt{5/34}(\cos 1.00407 + i \sin 1.00407)$

11. $-\pi/4$

13. ± 0.9273

15. $-3\pi/4$

17. $2 + 2i$

19. $-3 + 3i$

21. $\sqrt[6]{2}\left(\cos \dfrac{1}{12}k\pi + i \sin \dfrac{1}{12}k\pi\right), k = 1, 9, 17$

23. $6, -3 \pm 3\sqrt{3}\,i$

25. $\pm(3 + 4i)$

27. $\pm 1, \pm i, \pm(1 \pm i)/\sqrt{2}$

29. $3 + 4i, 4 - 3i$

31. $i, -1 - i$

33. Equation (5) holds when $z_1 + z_2 = 0$. Let $z_1 + z_2 \neq 0$ and $c = a + ib = z_1/(z_1 + z_2)$. By (19) in Prob. 36, $|a| \leq |c|, |a - 1| \leq |c - 1|$. Thus $|a| + |a - 1| \leq |c| + |c - 1|$. Clearly $|a| + |a - 1| \geq 1$. Together we have the inequality below; multiply by $|z_1 + z_2|$ to get (5).

$$1 \leq |c| + |c - 1| = \left|\frac{z_1}{z_1 + z_2}\right| + \left|\frac{z_2}{z_1 + z_2}\right|$$

PROBLEM SET 12.3, page 668

1. Closed disk, center $-2 - 5i$, radius $\frac{1}{2}$

3. Unit disk without its center

5. Horizontal infinite strip of width 2π

7. Region between the branches of the hyperbola $x^2 - y^2 = 1$

11. $-3/20, 1/20$

13. No

15. Yes

17. $-i/2$

19. $(-111 + 423i)/625$

21. $15(1 - i)/2$

23. 0

PROBLEM SET 12.4, page 673

1. Yes

3. Yes

5. No

7. Yes

9. No

11. No

13. $r_x = x/r = \cos \theta, r_y = \sin \theta, \theta_x = -(\sin \theta)/r, \theta_y = (\cos \theta)/r$,
   (a) $0 = u_x - v_y = u_r \cos \theta + u_\theta(-\sin \theta)/r - v_r \sin \theta - v_\theta(\cos \theta)/r$.
   (b) $0 = u_y + v_x = u_r \sin \theta + u_\theta(\cos \theta)/r + v_r \cos \theta + v_\theta(-\sin \theta)/r$.
   Multiply (a) by $\cos \theta$, (b) by $\sin \theta$, and add. Etc.

17. $1/z + ic$

19. $\ln |z| + i \operatorname{Arg} z + ic$

21. $e^{-x}(\cos y - i \sin y) + ic$

23. No

25. $a = b = 0, v = c$

27. $a = \pm 5, v = \pm e^{\pm 5x} \sin 5y + c$

PROBLEM SET 12.5, page 678

1. See Example 1.

3. $|w| = 3, 2, \cdots, \operatorname{Arg} w = 0, \mp\pi/4, \cdots$

7. $|w| > 4$

9. $\pi < \operatorname{Arg} w < 3\pi/2$

11. a

13. ± 1, $\pm i$

15. $-b/2$

17. $3 + 2 \cos t + i(-1 + 2 \sin t)$

19. $4 \cosh t + 2i \sinh t$

PROBLEM SET 12.6, page 682

1. $-e^2$, e^2

3. $e^{2\pi}$, $e^{2\pi}$

5. $-i$, 1

7. $5 \exp (i \arctan \frac{3}{4})$

9. $4e^{\pi i}$

11. $z = \pm 2n\pi i$

13. $z = \ln 3 + (1 \pm 2n)\pi i$

17. First quadrant

19. Annulus $3 < |w| < 5$

PROBLEM SET 12.7, page 686

1. Use (11), then (5) for e^{iy}, and simplify.

3. Use (11) and simplify.

5. $11.5487i$

7. $9.66667 - 2.79915i$

9. $\frac{1}{2}(2n + 1)\pi - (-1)^n 1.818i$

11. $\pm(\pi/3)i \pm 2n\pi i$, $n = 0, 1, \cdots$

13. $\pi/2 \pm 2n\pi \pm 3i$

15. Interior of the ellipse $u^2/(\cosh^2 2) + v^2/(\sinh^2 2) = 1$ in the first quadrant

17. Lower half of the interior of the ellipse $u^2/\cosh^2 1 + v^2/\sinh^2 1 = 1$

19. Lower half-plane $v < 0$

PROBLEM SET 12.8, page 691

5. $\ln 5 + \pi i$

7. $\frac{1}{2} \ln 2 \pm \frac{1}{4}\pi i$

9. $1.3701 \pm (1.571 \mp 1.096)i$

11. 0, $\pm 2\pi i$, $\pm 4\pi i$, $\cdots$

13. $(\pi - 1 \pm 2n\pi)i$, $n = 0, 1, \cdots$

15. $-i$

17. $-54.05 - 7.70i$

19. $\ln 2 \leqq u \leqq \ln 3$, $\pi/4 \leqq v \leqq \pi/2$

21. $81(\cos (\ln 3) - i \sin (\ln 3)) = 36.84 - 72.14i$

23. $2^{-1/2}e^{-\pi/4}(\cos (\frac{1}{4}\pi - \ln \sqrt{2}) - i \sin (\frac{1}{4}\pi - \ln \sqrt{2})) = 0.2919 - 0.1370i$

25. $(1 + i)/\sqrt{2}$

27. $\sqrt[3]{5}(\cos (\frac{1}{3} \arctan \frac{4}{3}) + i \sin (\frac{1}{3} \arctan \frac{4}{3})) = 1.629 + 0.520i$

29. $e^{(3-i) \ln (-3)} = e^{(3-i)(\ln 3 + \pi i)} = 27e^{\pi} (\cos (3\pi - \ln 3) + i \sin (3\pi - \ln 3))$
$= -284.179 + 556.431i$

PROBLEM SET 12.9, page 698

7. $\dfrac{3z + 1}{z - 1}$

9. $\dfrac{1}{z}$

11. $-\dfrac{i}{2}z$

13. $\dfrac{z - i}{z + i}$

15. $\dfrac{2z - i}{-iz - 2}$

19. Equation (5) gives no fixed points if and only if $c = 0$, $a - d = 0$, $b \neq 0$; thus $w = z + b/a$, a translation.

PROBLEM SET 12.10, page 700

1. w moves once around the unit circle $|w| = 1$.

5. $|z| = 1$; $\ln z = \ln |z| + i\theta = i\theta$ moves up the v-axis by 2π each time.

7. ± 1, 2 sheets

9. $-\frac{1}{2}i$, 3 sheets

CHAPTER 12 REVIEW, page 701

17. 16

19. $i, -i$

21. $\sqrt{32}\, e^{3\pi i/4}$

23. $25e^{-\pi i/2}$

25. $6.9044e^{0.2903i}$

27. $\pm(2-i)$

29. $\pm(1.4142 - 4.2426i)$

31. z^3

33. $z^2 - 2z$

35. 1

37. $1.6837 - 0.3805i$

39. $0.2718 + 1.0839i$

41. $u > 0$

43. $1/e < |w| < e,\ -1 < \text{Arg}\, w < 1$

45. $\ln 2 < u < \ln 4,\ 0 < v < \tfrac{1}{2}\pi$

PROBLEM SET 13.1, page 711

1. $(4 - 7i)t\ (0 \leqq t \leqq 1)$

3. $4 - 2i + 3e^{it}\ (0 \leqq t \leqq \pi)$

5. $t + i/t\ (1 \leqq t \leqq 4)$

7. $-3 + i + 5e^{it}\ (0 \leqq t \leqq 2\pi)$

9. Straight segment from 0 to $6 + 8i$

11. Ellipse (semi-axes 4, 1, center $2i$)

13. Hyperbola $x^2 - y^2 = 1$ from 1 to $27.31 + 27.29i$

15. $4 + 2i$

17. $(\pi - \tfrac{1}{2}\sinh 2\pi)i$

19. 0

21. $-\sinh 1$ by (9)

23. $2i\sinh \pi$ by (9)

25. $1 - i\tanh\tfrac{1}{4}\pi$ by (9)

PROBLEM SET 13.2, page 720

3. Yes, by the deformation principle.

5. $-i\sinh \pi$

7. 0, yes

9. 0, no

11. πi, no

13. 0, yes

15. $-\pi$, no

17. $2\pi i$

19. 0

21. $2\pi i$

23. $-\pi, \pi$

25. $-4\pi i$

PROBLEM SET 13.3, page 724

1. $-\pi i/2$

3. 0

5. $\pi/8$

7. πi

11. $\pi/2$

13. $2\pi i\, \text{Ln}\, 5 \approx 10.11i$

15. $2\pi i\cosh(-\pi^2 - \pi i) = -2\pi i\cosh \pi^2 \approx -60739i$

17. πe^{2i}

19. Use partial fractions.

PROBLEM SET 13.4, page 729

1. $8\pi i/3$

3. $\pi i/4$

5. $4\pi i$

7. 0

9. $-9\pi(1 + i)/(64\sqrt{2})$

11. $2\pi^2 i$

13. πi

15. $\pi(-5 + 8i)$

17. $\tfrac{1}{2}\pi i(\sin\tfrac{1}{2} + \tfrac{3}{2}\cos\tfrac{1}{2})$

19. 0 (why?)

CHAPTER 13 REVIEW, page 730

17. $56i/15$

19. $-\tfrac{1}{2}\sinh(\pi^2)$

21. 0

23. 0

25. $2\pi(11e^4 + 5e^{-4})$

27. $-8\pi i$

29. π

PROBLEM SET 14.1, page 740

1. Let l_1 and l_2 be two limits. Let $|l_1 - l_2| = d$ and $\epsilon = d/3$. Then by the definition of a limit there is an $N(\epsilon)$ such that $|z_n - l_1| < \epsilon$, $|z_n - l_2| < \epsilon$ for all $n > N(\epsilon)$. But this is impossible because the disks $|z - l_1| < d/3$ and $|z - l_2| < d/3$ are disjoint.

3. Yes, yes, 0   **5.** Yes, yes, πi **7.** Yes, no, $\pm 1 + 100i$

9. Yes, yes, $-\pi i/3$ **13.** Convergent   **15.** Divergent

17. Divergent

PROBLEM SET 14.2, page 745

1. $-i\sqrt{2}$, 1   **3.** i, ∞ **5.** 0, 0

7. $-2 + i$, $\sqrt{2}$ **9.** 0, $\sqrt{1.3}$ **11.** 0, 2/27

13. $-\pi$, e **15.** $2i$, ∞ **17.** 0, 1/3

PROBLEM SET 14.3, page 750

3. 5   **5.** 1/6   **7.** $1/\sqrt{2}$

9. ∞ **11.** 1   **13.** 1

PROBLEM SET 14.4, page 757

1. $1 - 2z^4 + \frac{2}{3}z^8 - \frac{4}{45}z^{12} + - \cdots$; $R = \infty$

3. $\frac{1}{2} - \frac{1}{2}\cos 2z = z^2 - \frac{1}{3}z^4 + \frac{2}{45}z^6 - \frac{1}{315}z^8 + - \cdots$; $R = \infty$

5. $2 + z + 2z^2 + z^3 + 2z^4 + \cdots$; $R = 1$

7. $-\dfrac{i}{3} + \dfrac{1}{9}z + \dfrac{i}{27}z^2 - \dfrac{1}{81}z^3 - \dfrac{i}{243}z^4 + \cdots$; $R = 3$

9. $(2/\sqrt{\pi})(z - z^3/3 + z^5/2!5 - z^7/3!7 + \cdots)$; $R = \infty$

11. $z^3/1!3 - z^7/3!7 + z^{11}/5!11 - + \cdots$; $R = \infty$

15. $1 + \frac{1}{3}z^3 + \frac{2}{15}z^5 + \frac{17}{315}z^7 + \cdots$; $R = \frac{1}{2}\pi$

17. $\frac{1}{2} - \frac{1}{4}(z - 2) + \frac{1}{8}(z - 2)^2 - \frac{1}{16}(z - 2)^3 + \frac{1}{32}(z - 2)^4 - + \cdots$; $R = 2$

19. $e^a\left(1 + (z - a) + \dfrac{1}{2!}(z - a)^2 + \dfrac{1}{3!}(z - a)^3 + \cdots\right)$; $R = \infty$

21. $(z - 1) - \frac{1}{2}(z - 1)^2 + \frac{1}{3}(z - 1)^3 - \frac{1}{4}(z - 1)^4 + - \cdots$; $R = 1$

23. $1 - \dfrac{1}{2!}(z - \frac{1}{2}\pi)^2 + \dfrac{1}{4!}(z - \frac{1}{2}\pi)^4 - \dfrac{1}{6!}(z - \frac{1}{2}\pi)^6 + - \cdots$; $R = \infty$

25. $2(z - \frac{1}{2}i) + \dfrac{2^3}{3!}(z - \frac{1}{2}i)^3 + \dfrac{2^5}{5!}(z - \frac{1}{2}i)^5 + \dfrac{2^7}{7!}(z - \frac{1}{2}i)^7 + \cdots$; $R = \infty$

27. $\dfrac{1}{e}\left[1 + (z - 1)^2 + \dfrac{1}{2!}(z - 1)^4 + \dfrac{1}{3!}(z - 1)^6 + \dfrac{1}{4!}(z - 1)^8 + \cdots\right]$; $R = \infty$

PROBLEM SET 14.5, page 766

1. Use Theorem 1.   **3.** $|\cos^n |z|| \le 1$; $\Sigma\, 1/n^2$ converges.

5. $R = 4$; use Theorem 1.   **7.** $|z + i| \le \sqrt{5} - \delta$, $\delta > 0$

9. $|z - 1| \le 2 - \delta$, $\delta > 0$ **11.** $|z + \frac{1}{3}i| \le \frac{1}{3} - \delta$, $\delta > 0$

CHAPTER 14 REVIEW, page 767

17. $\text{Ln}\,[1/(1-z)]$, $R = 1$

19. $\sin \pi z$, $R = \infty$

21. $\text{Ln}\,(1 + 2z)$, $R = \frac{1}{2}$

23. $1/[1 - 2\pi(z - i)^2]$, $R = 1/\sqrt{2\pi}$

25. $\cosh \sqrt{z}$, $R = \infty$

27. $z + \frac{1}{2}z^5 + \frac{1}{6}z^9 + \cdots$; $R = \infty$

29. $\frac{1}{2} - \frac{1}{2}\cos 2z = z^2 - (2^3/4!)z^4 + (2^5/6!)z^6 - + \cdots$; $R = \infty$

31. $-\frac{1}{3}i + \frac{1}{9}(z - 3i) + \frac{1}{27}i(z - 3i)^2 - \frac{1}{81}(z - 3i)^3 - \cdots$; $R = 3$

33. $1 + 3z + 6z^2 + 10z^3 + 15z^4 + 21z^5 + \cdots$; $R = 1$

35. $2(z - \pi i) + \frac{8}{6}(z - \pi i)^3 + \frac{32}{120}(z - \pi i)^5 + \cdots$; $R = \infty$

37. $|z - i| \leq \sqrt{2} - \delta$, $\delta > 0$

39. $|z + \frac{1}{4}i| \leq \frac{1}{4} - \delta$, $\delta > 0$

PROBLEM SET 15.1, page 775

1. $z^{-4} - \frac{1}{2}z^{-2} + \frac{1}{24} - \frac{1}{720}z^2 + - \cdots$; $\ 0 < |z| < \infty$

3. $z^{-3} + z^{-1} + \frac{1}{2}z + \frac{1}{6}z^3 + \cdots$; $\ 0 < |z| < \infty$

5. $z^3 + \frac{1}{2}z + \frac{1}{24}z^{-1} + \frac{1}{720}z^{-3} + \cdots$; $\ 0 < |z| < \infty$

7. $z^{-2} - z^{-4} + \frac{1}{2}z^{-6} - \frac{1}{6}z^{-8} + \frac{1}{24}z^{-10} - + \cdots$; $\ 0 < |z| < \infty$

9. $-\frac{1}{2}i(z - i)^{-1} + \frac{1}{4} + \frac{1}{8}i(z - i) - \frac{1}{16}(z - i)^2 - \frac{1}{32}i(z - i)^3 + \cdots$; $\ 0 < |z - i| < 2$

11. $-(z - \pi)^{-2} + \frac{1}{2} - \frac{1}{24}(z - \pi)^2 + \frac{1}{720}(z - \pi)^4 - + \cdots$; $\ |z - \pi| > 0$

13. $1 - 8i(z + 2i)^{-1} - 24(z + 2i)^{-2} + 32i(z + 2i)^{-3} + 16(z + 2i)^{-4}$

15. $\cosh z = -\cosh(z + \pi i)$, $-(z + \pi i)^{-2} - \frac{1}{2} - \frac{1}{24}(z + \pi i)^2 - \cdots$; $\ |z + \pi i| > 0$

17. $\frac{3}{2}i + \frac{5}{4}z - \frac{9}{8}iz^2 - \frac{17}{16}z^3 + \frac{33}{32}iz^4 + \frac{65}{64}z^5 + \cdots$

19. $\dfrac{2}{z} + \dfrac{3i}{z^2} - \dfrac{5}{z^3} - \dfrac{9i}{z^4} + \dfrac{17}{z^5} + \dfrac{33i}{z^6} - \dfrac{65}{z^7} + \cdots$

PROBLEM SET 15.2, page 780

1. $0, \pm 1, \pm 2, \cdots$; simple

3. $\pm \pi, \pm 3\pi, \pm 5\pi, \cdots$; second order

5. $\pm i, 0, \pm 2\pi i, \pm 4\pi i, \cdots$, simple

7. $\pm \sqrt{3}, \pm i\sqrt{2}$; third order

9. $0, \infty$ (second-order poles)

11. $\pm \frac{1}{2}n\pi$, $n = 0, 1, \cdots$ (simple poles); ∞ (essential singularity)

13. 0 (third-order pole)

15. $\pm ia$ (second-order poles)

17. ∞ (essential singularity)

19. πi (simple pole); ∞ (essential singularity)

PROBLEM SET 15.3, page 786

1. $\pm 2i$ (at $\mp i$)

3. $4/15$ (at 0)

5. -1 (at $\pm 2n\pi i$)

7. $\pm \frac{1}{4}$ (at ∓ 1)

9. $1/\pi$ (at $0, \pm 1, \cdots$)

13. $-4i$

15. $-4\pi i \sinh \frac{1}{2}\pi$

17. $-\frac{3}{4}\pi i$

19. $4\pi i/5$

PROBLEM SET 15.4, page 793

1. $\dfrac{\pi}{\sqrt{k^2 - 1}}$

3. $\dfrac{\pi}{\sqrt{2}}$

5. $\dfrac{\pi}{60}$

7. 0

9. $\dfrac{\pi}{2}$

11. 0

13. $\dfrac{\pi}{3}$ **15.** $\frac{5}{144}\pi$ **17.** $\frac{1}{2}\pi e^{-s}$

19. 0 **21.** π **23.** $-\sqrt{3}\,\pi/6$

CHAPTER 15 REVIEW, page 794

21. $2\pi i$ **23.** $8\pi i$ **25.** $i/2$, no

27. 0 **29.** 0 for even n, $(-1)^{(n-1)/2}2\pi i/(n-1)!$ for odd n

31. 0, no   **33.** 0 **35.** $\pi/6$

37. 0 **39.** $\pi/2$

PROBLEM SET 16.1, page 802

1. $20x + 300$ **3.** $\Phi = 110(y - x)$, $F = -110(1 + i)z$

5. $100 \ln r/\ln 5$ **7.** $(10/\ln 10)(\ln 100 - \ln r) = 20.00 - 4.34 \ln r$

9. $\Phi = c \,\text{Re}\, [\text{Ln}\,(z - a) + \text{Ln}\,(z + a)] = c \ln |z^2 - a^2|$

13. Use Fig. 316, Sec. 12.7, with the z- and w-planes interchanged, and $\cos z = \sin (z + \frac{1}{2}\pi)$.

15. $\Phi = 220(x^3 - 3xy^2)$, $F = 220z^3$

PROBLEM SET 16.2, page 807

1. Straightforward calculation involving the chain rule and the Cauchy–Riemann equations.

3. See Fig. 317, Sec. 12.7. $\Phi = \sin^2 x \cosh^2 y - \cos^2 x \sinh^2 y$.

5. $\Phi(x, y) = U_2 xy$. $w = u + iv = iz^2$ maps R onto $-2 \leqq u \leqq 0$.

9. Apply $w = z^2$.

11. Corresponding rays in the w-plane make equal angles, and the mapping is conformal.

13. $z = (2Z - i)/(-iZ - 2)$ by (7) in Sec. 12.9.

PROBLEM SET 16.3, page 811

1. $(20/d)y$. Rotate through $\pi/2$.   **3.** $100 - 280\theta/\pi$; $100 + (280/\pi)i \,\text{Ln}\, z$

5. $\text{Re}\, F(z) = 100 + (400/\pi)\,\text{Re}\,(\sin^{-1} z)$

9. $T = \dfrac{T_0}{\pi} \text{Arg}\, \dfrac{\cosh z - 1}{\cosh z + 1}$. The boundary in Fig. 380 is mapped onto the u-axis;

$\cosh x$, $0 \leqq x < \infty$, gives $u \geqq 1$; $\cosh (x + i\pi) = -\cosh x$, $0 \leqq x < \infty$, gives $u < -1$; the vertical segment $(x = 0)$ is mapped onto $-1 < u < 1$.

11. $(120/\pi)\,\text{Arg}\, z$ **13.** $T_0 + 2\pi^{-1}(T_1 - T_0)\,\text{Arg}\, z$

PROBLEM SET 16.4, page 817

1. $V = V_1 = K$, $Ky = const$, $Kx = const$ **5.** $F(z) = z^3$

7. $f(z) = z^2/K$, $F^*(w) = Kw$, $F(z) = F^*(f(z))$

9. $F(z) = z/r_0 + r_0/z$ **11.** $y/(x^2 + y^2) = c$ or $x^2 + (y - k)^2 = k^2$.

13. Use that $w = \cos^{-1} z$ is the mapping $w = \cos z$ with the roles of the z- and w-planes interchanged.

PROBLEM SET 16.5, page 822

1. $\Phi = r \sin \theta$　　　　　　**3.** $\Phi = \frac{1}{2} + \frac{1}{2}r^2 \cos 2\theta$　　　**5.** $\Phi = r^5 \sin 5\theta$
7. $\Phi = 3r \sin \theta - r^3 \sin 3\theta$
9. $\Phi = 2(r \sin \theta - \frac{1}{2}r^2 \sin 2\theta + \frac{1}{3}r^3 \sin 3\theta - + \cdots)$
11. $\Phi = (4/\pi)(r \sin \theta - \frac{1}{9}r^3 \sin 3\theta + \frac{1}{25}r^5 \sin 5\theta - + \cdots)$

PROBLEM SET 16.6, page 825

1. Use (2). $F(\frac{1}{2}) = \frac{1}{4}$.　　　　　　　　　**3.** Use (2). $F(1) = 9$.
5. Set $r = 0$ in that formula.　　　　　　　**7.** $\Phi(3, -3) = -8$
11. $\pm(2, 2)$ max, $\pm(2, -2)$ min by inspection since $|xy|$ is the area of a rectangle.
13. $\cos(\pm i) = (1 + \sinh^2 1)^{1/2} = 1.543081$　　　**15.** No; make up counterexample.

CHAPTER 16 REVIEW, page 826

17. $(900 \ln r) \ln 10 + 100$　　　　　　　**19.** Arg $z = const$
21. $100(1 - (2/\pi) \text{Arg } z)$　　　　　　　**23.** $20[1 - (2/\pi) \text{Arg }(z - 1)]$
25. $T(x, y) = 5 + 7.5(y - x)$　　　　　　**27.** Hyperbolas $(x + 1)y = c$
29. Circles $(x - c)^2 + y^2 = c^2$

31. $F(z) = \dfrac{c}{2\pi} \ln(z - 2)$, Arg $(z - 2) = c$

33. $10 + (40/\pi)(r \sin \theta + \frac{1}{3}r^3 \sin 3\theta + \frac{1}{5}r^5 \sin 5\theta + \cdots)$

35. $\Phi(r, \theta) = \dfrac{4}{\pi} \left(r \sin \theta + \dfrac{r^3}{3} \sin 3\theta + \dfrac{r^5}{5} \sin 5\theta + \cdots \right)$

PROBLEM SET 17.1, page 836

1. 0.2349×10^2, -0.3029×10^3, 0.5275×10^{-3}, 0.2570×10^5
3. 6.3698, 6.794, 8.15, impossible　　　　**5.** 6.3698, 6.370, 6.37, 6.4, 6
7. 29.96, 0.04;　29.96, 0.03338.　　30, 0.0;　30, 0.033
9. Multiply and divide by $\sqrt{9 + x^2} + 3$.
15. Small terms first. $(0.0004 + 0.0004) + 1.000 = 1.001$ but $(1.000 + 0.0004) + 0.0004 = 1$ (4S).
17. Add first, then round.
19. $\dfrac{a_1}{a_2} = \dfrac{\tilde{a}_1 + \epsilon_1}{\tilde{a}_2 + \epsilon_2} = \dfrac{\tilde{a}_1 + \epsilon_1}{\tilde{a}_2} \left(1 - \dfrac{\epsilon_2}{\tilde{a}_2} + \dfrac{\epsilon_2^{\,2}}{\tilde{a}_2^{\,2}} - + \cdots \right) \approx \dfrac{\tilde{a}_1}{\tilde{a}_2} + \dfrac{\epsilon_1}{\tilde{a}_2} - \dfrac{\epsilon_2}{\tilde{a}_2} \cdot \dfrac{\tilde{a}_1}{\tilde{a}_2},$

　　　hence $\left| \left(\dfrac{a_1}{a_2} - \dfrac{\tilde{a}_1}{\tilde{a}_2} \right) \middle/ \left| \dfrac{a_1}{a_2} \right| \right| \approx \left| \dfrac{\epsilon_1}{a_1} - \dfrac{\epsilon_2}{a_2} \right| \leqq |\epsilon_{r1}| + |\epsilon_{r2}| \leqq \beta_{r1} + \beta_{r2}$

25. (a) $19/21 = 0.90476\ 1905$, $\epsilon_{chop} = \epsilon_{round} = 0.1905 \times 10^{-5}$,
　　　$\epsilon_{r,chop} = \epsilon_{r,round} = 0.2106 \times 10^{-5}$, etc.

PROBLEM SET 17.2, page 847

3. Convergence to 4.7 for all starting values.
5. 2, 2.44444, 2.39774, 2.39221, 2.39170, 2.39165 (exact to 6S). From the series, 2.40483.

7. To get $|g'(x)| < 1$. Starting from $\frac{3}{2}\pi$ (why?) gives 4.49341 (6S exact) with 4 steps.

9. $x = \sqrt[4]{x + 0.12}$; 1, 1.02874, $\cdots$, 1.03717 (6S exact, 7 steps)

11. $x_{n+1} = (2x_n + 7/x_n^2)/3$, 1.912 931 $(= x_4)$

13. $f(x_{n+1}) \approx f(x_n) + (x_{n+1} - x_n)f'(x_n) = 0$; solving this algebraically gives (5).

15. (a) 0.5, 0.375, 0.377968, 0.377 964; (b) $1/\sqrt{7}$

17. 4.5, 4.80388, 4.73492, 4.73006, 4.73004 (6S exact)

19. 2, 2.38095, 2.39163, 2.39165 (6S exact)

21. 5, 4.48457, 4.66728, 4.74888, 4.72885, 4.73002, 4.73004

23. 2.5, 2.39635, 2.39157, 2.39165

25. (a) **ALGORITHM BISECT** (f, a_0, b_0, N) **Bisection Method**
   This algorithm computes an interval $[a_n, b_n]$ containing a solution of $f(x) = 0$
   (f continuous), or it computes a solution c_n, given an initial interval $[a_0, b_0]$ such
   that $f(a_0)f(b_0) < 0$. Here N is determined by $(b - a)/2^N \leqq \beta$, β the required accuracy.
   INPUT: Initial interval $[a_0, b_0]$, maximum number of iterations N.
   OUTPUT: Interval $[a_N, b_N]$ containing a solution, or a solution c_n.
   For $n = 0, 1, \cdots, N - 1$ do:

   > Compute $c_n = \frac{1}{2}(a_n + b_n)$.
   > If $f(c_n) = 0$ then OUTPUT c_n. Stop. [*Procedure completed*]
   > Else continue.
   > If $f(a_n)f(c_n) < 0$ then $a_{n+1} = a_n$ and $b_{n+1} = c_n$.
   > Else set $a_{n+1} = c_n$ and $b_{n+1} = b_n$.

   End
   OUTPUT $[a_N, b_N]$. Stop.
   [*Procedure completed*]
   End BISECT

   (b) 0.739085; (c) 1.30980, 0.429494

PROBLEM SET 17.3, page 860

1. $L_0(x) = -2x + 19$, $L_1(x) = 2x - 18$, $p_1(x) = 0.1082x + 1.2234$, $p_1(9.3) = 2.2297$

3. 0.8033 $(\epsilon = -0.0245)$, 0.4872 $(\epsilon = -0.0148)$; $p_2(x) = 0.3096x^2 - 0.9418x + 1$,
   0.7839 $(\epsilon = -0.0051)$, 0.4679 $(\epsilon = 0.0045)$

5. $p_2(x) = x^2 - 2.580x + 2.580$, $\Gamma(1.01) = 0.9943$, $\Gamma(1.03) = 0.9835$ (exact to 4D)

9. $p_2(x) = -0.44304x^2 + 1.30896x - 0.02322$, $p_2(0.75) = 0.70929$

11. $p_2(x) = 1.0000 - 0.0112r + 0.0008r(r - 1)/2 = x^2 - 2.580x + 2.580$, $r = (x - 1)/0.02$;
   0.9943, 0.9835, 0.9735

13. $3x^2 + 4x - 2$

15. $p_2(x) = 0.27633 + 0.97668(x - 0.25) - 0.44304(x - 0.25)(x - 0.5)$
   $= -0.44304x^2 + 1.30896x - 0.02322$

17. $r = -1.5$, $p_2(0.3) = 0.6039 + (-1.5) \cdot 0.1755 + \dfrac{(-1.5)(-0.5)}{2} \cdot (-0.0302) = 0.3293$.
   Error -0.0007.

PROBLEM SET 17.4, page 867

11. $4 - 12x^2 - 8x^3$, $4 - 12x^2 + 8x^3$. Yes

13. $3 + x - 2x^2 + x^3$, $5 + 5(x - 2) + 4(x - 2)^2$

15. $1 - x^2$, $-2(x - 1) - (x - 1)^2 + 2(x - 1)^3$, $-1 + 2(x - 2) + 5(x - 2)^2 - 6(x - 2)^3$

17. $g(x) = -4x^3/\pi^3 + 3x/\pi$

19. $4 + x^2 - x^3$, $-8(x - 2) - 5(x - 2)^2 + 5(x - 2)^3$,
   $4 + 32(x - 4) + 25(x - 4)^2 - 11(x - 4)^3$

PROBLEM SET 17.5, page 880

3. 0.5, 0.375, 0.34375

5. $\epsilon_{0.5} \approx 0.03452$ ($\epsilon_{0.5} = 0.03307$), $\epsilon_{0.25} \approx 0.00829$ ($\epsilon_{0.25} = 0.00820$)

7. 0.693254 (exact 0.693147)   **9.** 0.073930 (exact 0.073928)

11. 0.785 392 (exact 0.785 398)

13. $\frac{1}{15}(0.785398 - 0.785392) \approx 4 \cdot 10^{-7}$

15. 0.94508, 0.94583 (exact 0.94608)   **17.** 0.4612 (exact 0.4615)

19. 0.91936 (exact 0.91973)   **21.** 1.001005, $-0.001476 \leqq \epsilon \leqq 0$

23. $\displaystyle\int_0^{\pi/2} \cos x \, dx = \frac{1}{2} \int_{-\pi/2}^{\pi/2} \cos x \, dx = \frac{\pi}{4} \int_{-1}^{1} \cos \frac{1}{2}\pi t \, dt$

$$\approx \frac{\pi}{4}\left[2A_1 \cos \tfrac{1}{2}\pi t_1 + 2A_2 \cos \tfrac{1}{2}\pi t_2 + A_3 \cdot 1\right] = 1.00000\ 00552$$

25. 0.08, 0.32, 0.176, 0.256 (exact)

27. $5(0.1040 - \frac{1}{2} \cdot 0.1760 + \frac{1}{3} \cdot 0.1344 - \frac{1}{4} \cdot 0.0384) = 0.256$

CHAPTER 17 REVIEW, page 882

21. -0.4268×10^1, 0.8002×10^4, -0.5188×10^{-2}, 0.1782×10^{-1}, 0.3333×10^0, -0.4286×10^{-2}

23. 4.266, 4.38, 6.0, impossible   **25.** $24.885 \leqq s \leqq 24.995$

27. Same   **29.** 199.98, 0.02; 199.98, 0.020002

31. -0.12, $-0.11979\ 264$, $-0.11979\ 407$ **33.** ±1.89549

35. 3, 2.82278, 2.80167, 2.80139; -2, -2.13043, -2.12066, -2.12060

37. 0.8090   **39.** 0.334 [exact 0.305 (3D)]

41. $-x + x^3$, $2(x-1) + 3(x-1)^2 - (x-1)^3$ **43.** 0.26, $-0.02 \leqq \epsilon \leqq 0$

45. 0.90296, 0.90453 (exact 0.904524)

PROBLEM SET 18.1, page 893

1. $x_1 = 7.3$, $x_2 = -3.2$ **3.** $x_1 = (30.6 + 15.48x_2)/25.38$, x_2 arbitrary

5. No solution   **7.** $x_1 = 3$, $x_2 = -5$, $x_3 = 7$

9. $x_1 = 3.908$, $x_2 = -1.998$, $x_3 = 2.557$ **11.** $x_1 = 2 + x_2$, x_2 arbitrary, $x_3 = 3 - x_2$

13. x_1 arbitrary, x_2 arbitrary, $x_3 = 1.25x_1 - 2.25x_2$

15. $x_1 = 0.142857$, $x_2 = 0.692308$, $x_3 = -0.173913$

17. $x_1 = 1.5$, $x_2 = -3.5$, $x_3 = 4.5$, $x_4 = -2.5$

PROBLEM SET 18.2, page 899

1. $\begin{bmatrix} 1 & 0 \\ 3 & 1 \end{bmatrix} \begin{bmatrix} 4 & 5 \\ 0 & -1 \end{bmatrix}$, $\begin{aligned} x_1 &= -2 \\ x_2 &= 3 \end{aligned}$ **3.** $\begin{bmatrix} 1 & 0 \\ 0.2 & 1 \end{bmatrix} \begin{bmatrix} 1.8 & 2.6 \\ 0 & 3.2 \end{bmatrix}$, $\begin{aligned} x_1 &= 3 \\ x_2 &= 3 \end{aligned}$

5. $\begin{bmatrix} 1 & 0 & 0 \\ 2 & 1 & 0 \\ 2 & 5 & 1 \end{bmatrix} \begin{bmatrix} 5 & 4 & 1 \\ 0 & 1 & 2 \\ 0 & 0 & 3 \end{bmatrix}$, $\begin{aligned} x_1 &= 0.2 \\ x_2 &= 0.4 \\ x_3 &= 0.8 \end{aligned}$

7. $\begin{bmatrix} 3 & 0 & 0 \\ 2 & 3 & 0 \\ 4 & 1 & 3 \end{bmatrix} \begin{bmatrix} 3 & 2 & 4 \\ 0 & 3 & 1 \\ 0 & 0 & 3 \end{bmatrix}$, $\begin{matrix} x_1 = 0.6 \\ x_2 = 1.2 \\ x_3 = 0.4 \end{matrix}$

9. $\begin{bmatrix} 2 & 0 & 0 \\ 3 & 5 & 0 \\ 4 & 8 & 7 \end{bmatrix} \begin{bmatrix} 2 & 3 & 4 \\ 0 & 5 & 8 \\ 0 & 0 & 7 \end{bmatrix}$, $\begin{matrix} x_1 = 8 \\ x_2 = 0 \\ x_3 = -4 \end{matrix}$

11. $\begin{bmatrix} 2 & 0 & 0 & 0 \\ 1 & 1 & 0 & 0 \\ 2 & 1 & 1 & 0 \\ 0 & 2 & 1 & 2 \end{bmatrix} \begin{bmatrix} 2 & 1 & 2 & 0 \\ 0 & 1 & 1 & 2 \\ 0 & 0 & 1 & 1 \\ 0 & 0 & 0 & 2 \end{bmatrix}$, $\begin{matrix} x_1 = 3 \\ x_2 = -1 \\ x_3 = 0 \\ x_4 = 7 \end{matrix}$

15. $\begin{bmatrix} -3.5 & 1.25 \\ 3 & -1 \end{bmatrix}$ **17.** $\begin{bmatrix} 6 & 4 & 3 \\ 4 & 3 & 2 \\ 3 & 4 & 2 \end{bmatrix}$ **19.** No inverse

PROBLEM SET 18.3, page 905

1. Exact 2, 1, 4   **3.** Exact 0.5, 0.5, 0.5   **5.** Exact 15, 5, 12

7. (a) $\mathbf{x}^{(3)T} = [0.49982 \quad 0.50001 \quad 0.50002]$, (b) $\mathbf{x}^{(3)T} = [0.50333 \quad 0.49985 \quad 0.49968]$

9. 6, 15, 46, 96 steps; spectral radius 0.09, 0.35, 0.72, 0.85

11. $[1.99934 \quad 1.00043 \quad 3.99684]^T$ (Jacobi, step 5); $[2.00004 \quad 0.998059 \quad 4.00072]^T$ (Gauss–Seidel)

15. $\sqrt{128} = 11.3$, 12, 13   **17.** $\sqrt{3 + 6t^2}$, $1 + 2t$, $1 + 2t$

19. $\sqrt{230\,491} = 480.09$, 488, 488

PROBLEM SET 18.4, page 912

1. 24, $\sqrt{218} \approx 14.8$, 12, $[7/12 \quad -1 \quad 5/12 \quad 0]$

3. 9.3, $\sqrt{53.01} \approx 7.3$, 7.1, $[0.113 \quad -1 \quad 0.197]$

5. 24, 12, 6, $[-1 \quad 1 \quad -1 \quad 1]$

7. 4, 5; $4 \cdot 0.625 = 2.5$, $5 \cdot 0.5 = 2.5$

9. 22, 20; $22 \cdot 10 = 220$, $20 \cdot 11 = 220$

11. 200, 200; $200 \cdot 25 = 5000$, $200 \cdot 25 = 5000$

15. -2, 4; -144, 184; 25921

17. Fix a, set $\tilde{\mathbf{x}} = [a \quad y]^T$, find y that makes $\mathbf{r}$ small.

19. 27, 748, 28375, 943656, 29070279

PROBLEM SET 18.5, page 916

1. $2.846 - 1.038x$ **3.** $1.48 + 0.09x$

5. $0.215 + 0.000642t$ **7.** $U = 18.7 + 47.9i$; 47.9 ohms

9. $0.86 \cdot 60 = 51.6$ mph   **11.** $-8.36 + 5.45x - 0.589x^2$

13. $1.89 - 0.739x + 0.207x^2$ **17.** $0.92 + 0.23x + 0.24x^2 + 1.16x^3$

PROBLEM SET 18.7, page 924

1. 5.1, 4.9, −6.8; radii 0.5, 0.7, 0.4   **3.** 5 ± 0.02, 8 ± 0.02, 9 ± 0.02

5. 5, 0, 7; radii 4, 6, 6   **7.** $t_{11} = 100$, $t_{22} = t_{33} = 1$

9. They lie in intervals with endpoints $a_{jj} \pm (n - 1)10^{-3}$. (Why?)

11. $\| A \|_{\infty} = \max_{j} \sum_{k} |a_{jk}| = \max_{j} (|a_{jj}| + \text{Gerschgorin radius})$

13. 0 lies in no Gerschgorin disk, by (3) with >; hence det $\mathbf{A} = \lambda_1 \cdots \lambda_n \neq 0$.

19. $6 \leqq \lambda \leqq 8$, $\lambda = 7$

PROBLEM SET 18.8, page 928

1. $q = 10, 10.9908, 10.9999$; $|\epsilon| \leqq 3, 0.3028, 0.0275$

3. $q = 1.33333, 6.57377, 7.15811$; $|\epsilon| \leqq 2.1746, 1.74223, 0.47378$

5. $q = 1.5, 1.6, 1.61538$; $|\epsilon| \leqq 0.5, 0.2, 0.07692$

7. $\mathbf{y} = \mathbf{Ax} = \lambda\mathbf{x}$, $\mathbf{y}^{\mathsf{T}}\mathbf{x} = \lambda\mathbf{x}^{\mathsf{T}}\mathbf{x}$, $\mathbf{y}^{\mathsf{T}}\mathbf{y} = \lambda^2\mathbf{x}^{\mathsf{T}}\mathbf{x}$, $\epsilon^2 \leqq \mathbf{y}^{\mathsf{T}}\mathbf{y}/\mathbf{x}^{\mathsf{T}}\mathbf{x} - (\mathbf{y}^{\mathsf{T}}\mathbf{x}/\mathbf{x}^{\mathsf{T}}\mathbf{x})^2 = \lambda^2 - \lambda^2 = 0$

9. $q = -5.66667, -6.53818, -6.63628$ approximates $-6.8 = 0.2 - 7$; $|\epsilon| \leqq 2.1746, 0.59575, 0.48492$

13. $[1 \quad -0.444444 \quad 0.555556]^{\mathsf{T}}$, $[1 \quad -0.831276 \quad 0.835391]^{\mathsf{T}}$, $[1 \quad -0.954483 \quad 0.954608]^{\mathsf{T}}$

PROBLEM SET 18.9, page 937

1.
$$\begin{bmatrix} 0.49 & -0.220907 & 0 \\ -0.220907 & 0.435082 & 0.185902 \\ 0 & 0.185902 & 0.244918 \end{bmatrix}$$

3.
$$\begin{bmatrix} 7 & -3.60555 & 0 \\ -3.60555 & 13.4615 & 3.69231 \\ 0 & 3.69231 & 3.53846 \end{bmatrix}$$

5.
$$\begin{bmatrix} 0.3 & -6.7587 & 0 & 0 \\ -6.7587 & 14.3532 & 4.53518 & 0 \\ 0 & 4.53518 & 2.3342 & 0.312618 \\ 0 & 0 & 0.312618 & -3.38744 \end{bmatrix}$$

7.
$$\begin{bmatrix} 9.1585 & 0.4409 & 0 \\ 0.4409 & 4.1452 & 0.1801 \\ 0 & 0.1801 & 0.6963 \end{bmatrix}, \begin{bmatrix} 9.1893 & 0.1983 & 0 \\ 0.1983 & 4.1236 & 0.0300 \\ 0 & 0.0300 & 0.6871 \end{bmatrix}, \begin{bmatrix} 9.1955 & 0.0888 & 0 \\ 0.0888 & 4.1177 & 0.0050 \\ 0 & 0.0050 & 0.6868 \end{bmatrix}$$

9.
$$\begin{bmatrix} 7.0533 & 0.2463 & 0 \\ 0.2463 & 3.4483 & -0.0435 \\ 0 & -0.0435 & -1.5016 \end{bmatrix}, \begin{bmatrix} 7.0661 & 0.1200 & 0 \\ 0.1200 & 3.4358 & 0.0190 \\ 0 & 0.0190 & -1.5019 \end{bmatrix}, \begin{bmatrix} 7.0691 & 0.0583 & 0 \\ 0.0583 & 3.4329 & -0.0083 \\ 0 & -0.0083 & -1.5020 \end{bmatrix}$$

11.
$$\begin{bmatrix} 11.2903 & -5.0173 & 0 \\ -5.0173 & 10.6144 & 0.7499 \\ 0 & 0.7499 & 2.0952 \end{bmatrix}, \begin{bmatrix} 14.9028 & -3.1265 & 0 \\ -3.1265 & 7.0883 & 0.1966 \\ 0 & 0.1966 & 2.0089 \end{bmatrix}, \begin{bmatrix} 15.8298 & -1.2932 & 0 \\ -1.2932 & 6.1692 & 0.0625 \\ 0 & 0.0625 & 2.0010 \end{bmatrix}$$

CHAPTER 18 REVIEW, page 938

17. $[-2 \quad 0 \quad 5]^{\mathsf{T}}$ **19.** $[2.2 \quad 3.0 \quad -0.9]^{\mathsf{T}}$ **21.** $[3.9 \quad 4.3 \quad 1.8]^{\mathsf{T}}$

23. $u_{11} = 2$, $u_{12} = 1$, $u_{13} = 2$, $u_{22} = 1$, $u_{23} = 5$, $u_{33} = 6$, $m_{21} = 4$, $m_{31} = 3$, $m_{32} = 0$

25. $\begin{bmatrix} 2.8193 & -1.5904 & -0.0482 \\ -1.5904 & 1.2048 & -0.0241 \\ -0.0482 & -0.0241 & 0.1205 \end{bmatrix}$

27. $\begin{bmatrix} 1.700 \\ 1.1800 \\ 4.043 \end{bmatrix}, \begin{bmatrix} 1.986 \\ 0.999 \\ 4.002 \end{bmatrix}, \begin{bmatrix} 2.000 \\ 1.000 \\ 4.000 \end{bmatrix}$

29. $\begin{bmatrix} 3.000 \\ 1.925 \\ 0.450 \end{bmatrix}, \begin{bmatrix} 3.231 \\ 1.788 \\ 0.508 \end{bmatrix}, \begin{bmatrix} 3.197 \\ 1.801 \\ 0.499 \end{bmatrix}$

31. $4, \sqrt{10}, 3$

33. 1, 1, 1   **35.** $22, \sqrt{186}, 11$ **37.** 29

39. 51.27   **41.** 2.469   **43.** $4.09 + 0.205x$

45. $\begin{bmatrix} 1.5 & -2.23607 & 0 \\ -2.23607 & 5.8 & -3.1 \\ 0 & -3.1 & 6.7 \end{bmatrix}$, Step 3: $\begin{bmatrix} 9.44973 & 1.06216 & 0 \\ 1.06216 & 4.28682 & -0.00308 \\ 0 & -0.00308 & 0.26345 \end{bmatrix}$

PROBLEM SET 19.1, page 951

1. $y = 2e^{-0.1x}$; 0.00048, 0.00091 (errors of y_5, y_{10})

3. $y = 1/(x^5 + 1)$; -0.0541, 0.0076 (errors of y_5, y_{10})

5. $y = e^x$; 0.001 275, 0.004 201 (errors of y_5, y_{10})

7. $y = 1/(1 + x^2)$; -0.0033, -0.0030 (errors of y_5, y_{10})

9. -0.007, 0.014 (errors of y_5, y_{10})

11. $y = 0, 0.1000, 0.2034, 0.3109, 0.4217, 0.5348, 0.6494, 0.7649, 0.8806$;
error 0, 0.0044, 0.0122, $\cdots$, 0.0527

13. $y = 0, 0.4352, 0.9074$; error 0.0145, 0.0258 (about 50% of that in Prob. 11)

15. For instance, $x = 0.5$, $y = 0.672$ (error -0.040), 0.6293 (error 0.0029);
$x = 1$, $y = 0.893$ (error -0.028), 0.8626 (error 0.0021)

17. $y = e^{-0.1x^2}$; errors of the order 10^{-9} to 10^{-7}

PROBLEM SET 19.2, page 955

1. Errors between $-2 \cdot 10^{-6}$ and $+3 \cdot 10^{-7}$

3. Positive and negative errors of order 10^{-6} to 10^{-5}

5. $y = e^{-0.1x^2}$; errors of the same order as the round-off errors of the starting values

9. (a) 0, 0.02, 0.0884, 0.215 848, $y_4 = 0.417\,818$, $y_5 = 0.708\,887$ (poor). (b) By 30–50%.

11. Errors $\times 10^5$ from $x = 0.3$ to 1: $-5, -11, -19, -31, -47, -69, 102, 150$

13. $y_1 = 0.021400$, $y_2 = 0.091818$ (starting values), $y_3 = 0.222146$, $y_4 = 0.425619$, $y_5 = 0.718431$

PROBLEM SET 19.3, page 961

1. $y_1 = 3, 3.6, 4.2, \cdots, 10.27$; $y_2 = 0, 0.3, \cdots, 2.49$

3. $y = y_1 = 1, 0.8, 0.61, 0.429, 0.2561$

5. $y_1 = 4e^x - e^{-2x}$, $y_2 = e^x - e^{-2x}$ (verify!). 6.21875, 1.27344 ($x = 0.5$, error 0.00826, 0.00740; 6.114, 1.283 in Prob. 1). 10.7288, 2.57672 ($x = 1$, error 0.00903, 0.00623; 10.26, 2.486 in Prob. 1)

7. $y = 0.3x^4$, $y = 0.621\,895, 1.15197, 1.96505$. Errors $\times 10^4$: 2, 5, 10.

9. $y = 0.198669, 0.389494, 0.565220, 0.719632, 0.847790$;
$y' = 0.980132, 0.922062, 0.830020, 0.709991, 0.568572$

13. You get the exact solution, except for a round-off error [e.g., $y_1 = 2.761\ 608$, $y(0.2) = 2.7616$ (exact), etc.]. Why?

PROBLEM SET 19.4, page 969

3. $u_{11} = 150$, $u_{21} = 120$, $u_{12} = 150$, $u_{22} = 120$; 149.927, 119.963, 149.963, 119.982 (Step 5)

5. $u_{11} = u_{21} = 0.108253$, $u_{12} = u_{22} = 0.324759$; 0.400, 0.254, 0.471, 0.398 (Step 5)
is very poor.

7. $-3u_{11} + u_{12} = -200$, $u_{11} - 3u_{12} = -100$

9. $u_{12} = u_{32} = 31.25$, $u_{21} = u_{23} = 18.75$, $u_{jk} = 25$ at the others

11. $u_{21} = u_{23} = 0.25$, $u_{12} = u_{32} = -0.25$, $u_{jk} = 0$ else

13. Only 5 steps

15. $\sqrt{3}$, $u_{11} = u_{21} = 0.0849$, $u_{12} = u_{22} = 0.3170$. (0.1083, 0.3248 are 4S-values of the solution
of the linear equations of the problem.)

PROBLEM SET 19.5, page 975

3. **A** as in Example 1, right sides -2, -2, -2, -2.
Solution $u_{11} = u_{21} = 1.14286$, $u_{12} = u_{22} = 1.42857$

5. $u_{11} = 0.766$, $u_{21} = 1.109$, $u_{12} = 1.957$, $u_{22} = 3.293$

13. $-4u_{11} + u_{21} + u_{12} = -3$, $u_{11} - 4u_{21} + u_{22} = -12$, $u_{11} - 4u_{12} + u_{22} = 0$,
$2u_{21} + 2u_{12} - 12u_{22} = -14$, $u_{11} = u_{22} = 2$, $u_{21} = 4$, $u_{12} = 1$

15. $\mathbf{b} = [-380 \quad -190, \quad -190, \quad 0]^\mathsf{T}$; $u_{11} = 140$, $u_{21} = u_{12} = 90$, $u_{22} = 30$

PROBLEM SET 19.6, page 981

3. 0, 0.6625, 1.25, 1.7125, 2, 2.1, 2, etc.

5. Step 5 gives 0, 0.06279, 0.09336, 0.08364, 0.04707, 0.

7. Use (5) and $0 = \partial u_{0j}/\partial x = (u_{1j} - u_{-1,j})/2h$.

9. 0.1636, 0.2545 ($t = 0.04$, $x = 0.2$, 0.4), 0.1074, 0.1752 ($t = 0.08$), 0.0735, 0.1187
($t = 0.12$), 0.0498, 0.0807 ($t = 0.16$), 0.0339, 0.0548 ($t = 0.2$; exact 0.0331, 0.0535)

11. Step 2: 0 (exact 0), 0.0453 (0.0422), 0.0672 (0.0658), 0.0671 (0.0628), 0.0394 (0.0373),
0 (0)

PROBLEM SET 19.7, page 984

1. For $x = 0.2$, 0.4 we obtain 0.012, 0.02 ($t = 0.2$), 0.004, 0.008 ($t = 0.4$), -0.004,
-0.008 ($t = 0.6$), etc.

3. $u(x, 1) = 0$, -0.05, -0.10, -0.15, -0.20, 0

5. 0.190, 0.308, 0.308, 0.190 (0.178, 0.288, 0.288, 0.178 exact to 3D)

7. 0, 0.354, 0.766, 1.271, 1.679, 1.834, $\cdots$ ($t = 0.1$); 0, 0.575, 0.935, 1.135, 1.296,
1.357, $\cdots$ ($t = 0.2$)

CHAPTER 19 REVIEW, page 984

21. $y = e^x$; 0.038, 0.125 (errors of y_5, y_{10})

23. $y = (e^{-2x} + 2e^x)/6$; -0.00138, -0.00218 (errors of y_5 and y_{10})

27. $y = xe^x$; errors 0, 0.000190, 0.000461, 0.000846, 0.001385, 0.002131

29. 0.0013, 0.0042 (errors of y_5, y_{10})

31. $y = \sin x$; $y_4 = 0.717366$, $y_5 = 0.841496$ (errors -1×10^{-5}, -2.5×10^{-5})

33. $y = e^x - x - 1$; 0.021400 (starting), 0.092322, 0.223342, 0.427788, 0.721945. Too inaccurate.

35. $y = 1, 1, 1, 1.0001, 1.0006, 1.002$

37. $y_1' = y_2$, $y_2' = 2e^x - y_1$, $y_1 = 0, 0.241, 0.571, \cdots$; errors between 10^{-6} and 10^{-5}.

39. You get values of the exact solution. Why?

43. $u(P_{11}) = u(P_{31}) = 270$, $u(P_{21}) = u(P_{13}) = u(P_{23}) = u(P_{33}) = 30$, $u(P_{12}) = u(P_{32}) = 90$, $u(P_{22}) = 60$

45. 0, 0.04, 0.08, 0.12, 0.15, 0.16, 0.15, 0.12, 0.08, 0.04, 0

PROBLEM SET 20.1, page 993

3. Step 3: $[1.03311 \quad 0.984418]^T$. Fast convergence (almost a circle)

5. $[0.688 \quad -0.860]^T$, $[1.764 \quad 0]^T$, $[1.845 \quad -0.102]^T$ (Exact $[2 \quad 0]$)

PROBLEM SET 20.2, page 997

1. No   **3.** x_3, x_4 unused time on M_1, M_2, respectively

11. Not unique; $f = 225$ on the segment from (3, 4) to (0, 10)

13. $f(5/6, 7/6) = 100/3$ **15.** $f(0, 5) = 10$

17. $f_{max} = f(210, 60) = 3750$

19. $0.5x_1 + 0.75x_2 \leq 15$ (copper), $0.5x_1 + 0.25x_2 \leq 10$, $f = 120x_1 + 100x_2$, $f_{max} = f(15, 10) = 2800$

PROBLEM SET 20.3, page 1001

1. $f(\frac{5}{3}, \frac{20}{3}) = 183\frac{1}{3}$

3. Matrices with rows 2 and 3 and columns 4 and 5 interchanged

5. $f_{max} = 6$ on the segment from (3, 0, 0) to (0, 0, 2)

7. $f_{min} = -10$ at $x_1 = 0$, $x_2 = 5/10 = \frac{1}{2}$

9. $f_{max} = 2200/7 \approx 314$ at $x_1 = 60/21 \approx 3$, $x_2 = 0$, $x_3 = 1500/105 \approx 14$, $x_4 = 0$

PROBLEM SET 20.4, page 1007

1. $f(4, 4) = 72$ **3.** $f(20, 30) = 50$ **5.** $f(10, 5) = 5500$

7. $f(1, 1, 0) = 12$ **9.** $f(\frac{1}{2}, 0, \frac{1}{2}) = 3$

CHAPTER 20 REVIEW, page 1007

11. Step 5: $[0.353 \quad -0.028]^T$. Slower   **13.** Of course! Step 5: $[-1.003 \quad 1.897]^T$

23. $f(2, 4) = 50$ **25.** $f(3, 6) = -54$

PROBLEM SET 21.1, page 1014

5. $\begin{bmatrix} 0 & 1 & 0 \\ 0 & 0 & 1 \\ 1 & 0 & 0 \end{bmatrix}$ **7.** $\begin{bmatrix} 0 & 1 & 1 & 1 \\ 0 & 0 & 0 & 0 \\ 1 & 0 & 0 & 0 \\ 0 & 0 & 0 & 0 \end{bmatrix}$ **9.** $\begin{bmatrix} 0 & 1 & 1 \\ 0 & 0 & 1 \\ 1 & 1 & 0 \end{bmatrix}$

11.

13.

17. If G is complete.

19.

Edge	e_1	e_2	e_3	e_4
Vertex 1	-1	-1	1	-1
2	1	0	0	0
3	0	1	-1	0
4	0	0	0	1

PROBLEM SET 21.2, page 1019

1. 5   **3.** 4

5. The idea is to go backward. There is a v_{k-1} adjacent to v_k and labeled $k - 1$, etc. Now the only vertex labeled 0 is s. Hence $\lambda(v_0) = 0$ implies $v_0 = s$, so that $v_0 - v_1 - \cdots - v_{k-1} - v_k$ is a path $s \to v_k$ that has length k.

13. No; there is no way of traveling along (3, 4) only once.

15. Police patrol, track repair crew, farmer's best route for seeding his fields.

17. $1 - 2 - 3 - 4 - 5 - 6 - 4 - 3 - 1, L = 25$

19. From m to $100m$, $10m$, $2.5m$, $m + 4.6$

PROBLEM SET 21.3, page 1023

1. (1, 2), (2, 4), (4, 3); $L_2 = 6$, $L_3 = 18$, $L_4 = 14$

5. (1, 2), (2, 4), (3, 4); $L_2 = 10$, $L_3 = 15$, $L_4 = 13$

7. (1, 2), (2, 4), (3, 4), (3, 5); $L_2 = 2$, $L_3 = 4$, $L_4 = 3$, $L_5 = 6$

9. (1, 5), (2, 3), (2, 6), (3, 4), (3, 5); $L_2 = 9$, $L_3 = 7$, $L_4 = 8$, $L_5 = 4$, $L_6 = 14$

PROBLEM SET 21.4, page 1027

1. $1 - 3 - 2 \begin{smallmatrix} 4 \\ \\ 5 \end{smallmatrix}$ **3.** $\begin{smallmatrix} 2 \\ \\ 1 \end{smallmatrix} 4 - 3 - 5$ **5.** $5 - 3 - 6 \begin{smallmatrix} 1 \\ \\ 2 - 4 \end{smallmatrix}$

7. New York $-$ Washington $-$ Chicago $-$ Dallas $-$ Denver $-$ Los Angeles

9. Yes   **11.** $1 - 3 - 4 \begin{smallmatrix} 2 \\ \\ 5 - 6, \end{smallmatrix}$ $L = 38$

13. G is connected. If G were not a tree, it would have a cycle, but this cycle would provide two paths between any pair of its vertices, contradicting the uniqueness.

19. If we add an edge (u, v) to T, then since T is connected, there is a path $u \to v$ in T which, together with (u, v), forms a cycle.

PROBLEM SET 21.5, page 1030

1. If G is a tree

5. A shortest spanning tree of the largest connected graph that contains vertex 1

7. (1, 4), (4, 3), (4, 2), (3, 5); $L = 20$

9. (1, 4), (1, 3), (1, 2), (2, 6), (3, 5); $L = 32$

11. $(1, 4), (4, 3), (4, 5), (1, 2); L = 12$
13. $(1, 3), (3, 2), (2, 5), (2, 4); L = 10$

PROBLEM SET **21.6, page 1037**

1. $\{4, 5, 6\}, 10 + 5 + 13 = 28$ **3.** $\{3, 6\}, 11 + 3 = 14$
5. $\{3, 6, 7\}, 8 + 4 + 4 = 16$ **7.** $T = \{3, 6\}, f = 14$
9. $S = \{1, 4\}, \text{cap}(S, T) = 6 + 8 = 14$ **11.** None
13. $\Delta_{13} = 3, \Delta_{35} = 4, \Delta_{56} = 4$, etc., max $= 17$ **15.** $\Delta_{13} = 2, \Delta_{35} = 2$, etc.
17. 17   **19.** 15

PROBLEM SET **21.7, page 1040**

1. $1 - 2 - 5, \Delta_t = 2; 1 - 4 - 2 - 5, \Delta_t = 1; f = 6 + 2 + 1 = 9$
3. $1 - 2 - 4 - 6, \Delta_t = 2; 1 - 3 - 5 - 6, \Delta_t = 1; f = 4 + 2 + 1 = 7$
5. $S = \{1, 2, 4, 5\}, T = \{3, 6\}, \text{cap}(S, T) = 14$
7. $(2, 3)$ and $(5, 6)$ **15.** 2000

PROBLEM SET **21.8, page 1045**

1. $S = \{1, 4\}, T = \{2, 3\}$ **3.** No
5. $S = \{1, 3, 5\}, T = \{2, 4, 6\}$ **7.** $5 - 1 - 4 - 3 - 6 - 2$
9. $1 - 4 - 3 - 6 - 7 - 8$ **11.** $(1, 5), (2, 6), (3, 4)$
13. 4   **15.** $n_1 n_2$
17. 3

CHAPTER **21 REVIEW, page 1046**

17. $\begin{bmatrix} 0 & 1 & 0 & 1 \\ 1 & 0 & 1 & 0 \\ 0 & 1 & 0 & 1 \\ 1 & 0 & 1 & 0 \end{bmatrix}$ **19.** 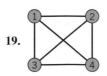   **21.**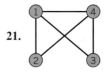

23. 4   **25.** $1 - 4 \left\langle \begin{array}{l} 2 \\ 3 - 5 \end{array} \right.$

29. By considering only edges with one labeled end and one unlabeled
33. $1 - 2 - 3 - 5$

PROBLEM SET **22.1, page 1054**

1. $q_L = 7, q_M = 10, q_U = 11$ **3.** $q_L = 28, q_M = 34, q_U = 44$
5. $q_L = 49.7, q_M = 50.5, q_U = 51.2$ **7.** $q_L = 1.3, q_M = 1.4, q_U = 1.45$
9. $q_L = q_M = 15, q_U = 15.5$ **11.** $\bar{x} = 9.3, s = 2.54, IQR = 4$
13. $\bar{x} = 50.5, s = 1.05, IQR = 1.5$ **15.** $0 \quad 0 \quad 300$
17. 3.54, 1.29

PROBLEM SET 22.2, page 1057

1. 2^3 outcomes: *RRR, RRL, RLR, LRR, RLL, LRL, LLR, LLL*
3. 4 outcomes: *HH, HT, TH, TT* (*H* = head, *T* = tail)
5. 10 outcomes: *D, ND, NND*, etc.   **7.** Yes, no
9. Include *S* and $\varnothing$

PROBLEM SET 22.3, page 1063

1. 1/3   **3.** 8/9
5. (a) $0.9^3 = 72.9\%$, (b) $\frac{90}{100} \cdot \frac{89}{99} \cdot \frac{88}{98} = 72.65\%$ **9.** $0.9^4 = 65.6\%$
11. 88.36%   **13.** $1 - 0.96^3 = 11.5\%$
15. $1 - 0.75^2 = 0.4375 < 0.5$
17. $P(MMM) + P(MMFM) + P(MFMM) + P(FMMM) = \frac{1}{8} + 3 \cdot \frac{1}{16} = \frac{5}{16}$

PROBLEM SET 22.4, page 1068

3. 230300   **5.** 56
7. $2!3!5!/10! \approx 0.04\%$ (by Theorem 1)   **9.** $26^3 \cdot 10^3$
11. 72   **13.** 56, 20, 30, 6
15. (b) $1/12n$ or $8/n$ [%]

PROBLEM SET 22.5, page 1074

3. No. Why?
5. $k = 1/e$, $1 - P(X < 3) = f(0) + f(1) + f(2) = 8.03\%$
7. 5.6, 4, 2.2   **9.** $k = 5$; 50%
11. $0.5^3 = 12.5\%$ **13.** 0.2313, 0.2689
15. $\frac{3}{4}(2c - \frac{2}{3}c^3) = 0.95$, $c = 0.8114$ **17.** $k = 3/8$; $x = 0.9283, 1.9310$
19. $X > b$, $X \geqq b$, $X < c$, $X \leqq c$, etc.

PROBLEM SET 22.6, page 1078

1. $k = 1/8$, $\mu = 1.5$, $\sigma^2 = 0.75$ **3.** 2/3, 1/18
5. 5, 100/12   **7.** 750, 1, 0.002
9. 0.073   **11.** $\frac{1}{2}$, $\frac{1}{20}$, $(X - \frac{1}{2})\sqrt{20}$
13. $\mu = 1/\theta$; 22.3%
15. $23.45

PROBLEM SET 22.7, page 1083

1. $\binom{5}{x} 0.5^5$, 0.03125, 0.15625, $1 - f(0) = 0.96875$, 0.96875
3. $1 - 0.95^{20} = 64.15\%$ **5.** $0.5^x e^{-0.5}/x!$, $1 - 0.91 = 9\%$
7. $13\frac{1}{4}\%$ **9.** $1 - e^{-0.2} = 18\%$
11. 42%, 47.2%, 10.5%, 0.3%   **13.** 44%

PROBLEM SET 22.8, page 1090

1. 0.1587, 0.5, 0.6915, 0.6247   **3.** 45.065, 56.978, 2.022
5. 15.9%   **7.** About 680 (Fig. 489*a*)

9. About 58%   **11.** 0.067

13. $t = 1084$

PROBLEM SET 22.9, page 1099

1. $k = 1/30$; 2/15, 1/30   **3.** 2/9, 1/9, 1/2

5. $f_2(y) = 1/(\beta_2 - \alpha_2)$ if $\alpha_2 < y < \beta_2$ **7.** 27.45 mm, 0.38 mm

9. Independent, $f_1(x) = 2e^{-2x}$ if $x > 0$, $f_2(y) = 2e^{-2y}$ if $y > 0$, 1.83%

13. Example 1 and $f(0, 0) = f(1, 1) = 1/8$, $f(0, 1) = f(1, 0) = 3/8$

CHAPTER 22 REVIEW, page 1100

27. $Q_L = 13.2$, $Q_M = 13.3$, $Q_U = 13.5$ **29.** $\bar{x} = 13.14$, $s = 0.602$, $s^2 = 0.363$

31. $\bar{x} = 6$, $s = 3.65$ **33.** H, TH, TTH, etc.

35. Always $B \subseteq A \cup B$. If also $A \subseteq B$, then $B = A \cup B$. Etc.

39. $\binom{52}{13} = 635\ 013\ 559\ 600$ **43.** $k = 1.1565$; 26.9%

45. 0, 2   **47.** $6^x e^{-6}/x!$, e^{-6}

49. 17.29, 10.71, 19.152

PROBLEM SET 23.2, page 1108

3. $\hat{\mu} = \bar{x}$ **5.** $\hat{\theta} = n/\Sigma\, x_j = 1/\bar{x}$ **7.** $\hat{\theta} = 1$

9. $l = p^k(1 - p)^{n-k}$, $\hat{p} = k/n$, k = number of successes in n trials

11. 7/12   **13.** $\hat{p} = 1/\bar{x}$

PROBLEM SET 23.3, page 1117

3. $\text{CONF}_{0.99}\{28.45 \leqq \mu \leqq 33.71\}$ **5.** $\text{CONF}_{0.95}\{74.25 \leqq \mu \leqq 75.37\}$

7. 4, 16   **9.** $\text{CONF}_{0.99}\{15.308 \leqq \mu \leqq 15.692\}$

11. $\text{CONF}_{0.99}\{62.71 \leqq \mu \leqq 65.29\}$

13. $c = 1.96$, $\bar{x} = 87$, $s^2 = 71.86$, $k \approx cs/\sqrt{n} = 0.742$, $\text{CONF}_{0.95}\{86 \leqq \mu \leqq 88\}$, $\text{CONF}_{0.95}\{0.17 \leqq p \leqq 0.18\}$

15. $\text{CONF}_{0.95}\{23 \leqq \sigma^2 \leqq 553\}$

17. Normal, means 120, 198, variances 36, 100

19. $Z = X + Y$ is normal with mean 210 and variance 4.25. *Ans.* $P(208 \leqq Z \leqq 212) = 67\%$.

PROBLEM SET 23.4, page 1127

1. $t = (0.286 - 0)/(4.31/\sqrt{7}) = 0.18 < c = 1.94$; accept the hypothesis.

3. $c = 6090 > 6019$, $c = 12\ 127 > 12\ 012$; accept the hypothesis.

5. $\sigma^2/n = 1.8$, $c = 57.8$, accept the hypothesis.

7. $\mu < 58.69$ or $\mu > 61.31$

11. Alternative $\mu \neq 5000$, $t = (4990 - 5000)/(20/\sqrt{50}) = -3.54 < c = -2.01$ (Table A9, Appendix 5). Reject the hypothesis $\mu = 5000$ g.

13. Two-sided, $t = (0.55 - 0)/\sqrt{0.546/8} = 2.11 < c = 2.37$ (Table A9, Appendix 5), no difference

15. $19 \cdot 1.0^2/0.8^2 = 29.69 < c = 30.14$ (Table A10, Appendix 5); accept the hypothesis.

17. By (12), $t_0 = \sqrt{16}\,(20.2 - 19.6)/\sqrt{0.16 + 0.36} > c = 1.70$. Assert that B is better.

PROBLEM SET 23.5, page 1132

1. LCL $= 1 - 2.58 \cdot 0.02/2 = 0.974$, UCL $= 1.026$

3. 27

5. $2.58\sqrt{0.0004}/\sqrt{2} = 0.036$, LCL $= 3.464$, UCL $= 3.536$

9. LCL $= np - 3\sqrt{np(1-p)}$, CL $= np$, UCL $= np + 3\sqrt{np(1-p)}$

11. LCL $= \mu - 3\sqrt{\mu}$ is negative in (b) and we set LCL $= 0$, CL $= \mu = 3.6$,
UCL $= \mu + 3\sqrt{\mu} = 9.3$.

13. In about 30% (5%) of the cases

15. Continuous increase of means. Abrupt change

PROBLEM SET 23.6, page 1136

1. 0.9825, 0.9384, 0.4060   **3.** 0.8187, 0.6703, 0.1353   **5.** $e^{-30\theta}(1 + 30\theta)$

7. 19%, 15%   **9.** $(1 - \theta)^n + n\theta(1 - \theta)^{n-1}$

11. $(1 - \frac{1}{2})^3 + 3 \cdot \frac{1}{2}(1 - \frac{1}{2})^2 = \frac{1}{2}$

13. $(1 - \theta)^5$, $[\theta(1 - \theta)^5]' = 0$, $\theta = 1/6$, AOQL $= 6.7\%$

15. $\displaystyle\sum_{x=0}^{9} \binom{100}{x} 0.12^x 0.88^{100-x} = 22\%$ (by the normal approximation)

PROBLEM SET 23.7, page 1140

1. $\chi_0^2 = (40 - 50)^2/50 + (60 - 50)^2/50 = 4 > c = 3.84$; no

3. $\chi_0^2 = 16/10 < 11.07$; yes   **5.** $\chi_0^2 = 10.264 < 11.07$; yes

7. Combining the last three nonzero values, we have $K - r - 1 = 9$ ($r = 1$ since we estimated the mean, $\frac{10094}{2608} \approx 3.87$). $\chi_0^2 = 12.8 < c = 16.92$. Accept the hypothesis.

9. $\chi_0^2 = 1 < 3.84$; yes

11. $\chi_0^2 = 10^2/40 + 10^2/120 < c = 3.84$; yes

13. $(73^2 + 23^2 + 97^2)/847 = 18.02 > c = 5.99$; reject

PROBLEM SET 23.8, page 1143

3. $(\frac{1}{2})^8 + 8 \cdot (\frac{1}{2})^8 = 3.5\%$ is the probability that 7 or 8 in 8 trials favor A under the hypothesis that A and B are equally good. Reject.

5. Hypothesis $\mu = 0$. Alternative $\mu > 0$, $\bar{x} = 1.58$,
$t = 1.58/(1.23/\sqrt{10}) = 4.06 > c = 1.83$ ($\alpha = 5\%$). Hypothesis rejected.

7. $\bar{x} = 9.67$, $s = 11.87$, $t_0 = 9.67/(11.87/\sqrt{15}) = 3.15 > c = 1.76$ ($\alpha = 5\%$)

11. $P(T \leq 15) = 10.8\%$ from Table 12 in Appendix 5 with $n = 10$. Accept that there is no trend.

13. $P(T \leq 2) = 2.8\%$. Assert positive trend.   **15.** $P(T \leq 2) = 0.1\%$; assert negative trend.

PROBLEM SET 23.9, page 1150

1. $y = 6.74074 + 3.0679x$ **3.** $y = -0.367 + 0.115x$, $R = 8.70$

5. $y = 0.32923 + 0.00032x$, $y(66) = 0.35035$

7. $q_0 = 6.377$, $c = 4.30$, $K = 0.853$, $CONF_{0.95}\{2.215 \leqq \kappa_1 \leqq 3.921\}$

9. $q_0 = 157.65$, $c = 4.30$, $K = 1.707$, $CONF_{0.95}\{7.44 \leqq \kappa_1 \leqq 10.86\}$

CHAPTER 23 REVIEW, page 1153

27. 0.268, 1.052, 1.026   **29.** 0.268, 0.947

31. It doubles.

33. $CONF_{0.99}\{27.94 \leqq \mu \leqq 34.81\}$ is 15% longer.

35. $CONF_{0.95}\{29.85 \leqq \mu \leqq 94.84\}$ **37.** $c = 14.74 > 14.5$; reject μ_0.

39. $\Phi\left(\dfrac{14.74 - 14.50}{\sqrt{0.025}}\right) = 0.9355$

41. $t = (26\,000 - 25\,000)/(2000/\sqrt{100}) = 5 > c = 2.37$ (Table A9, Appendix 5). Reject the hypothesis $\mu_0 = 25\,000$ and assert that the manufacturer's claim is justified.

43. $2.58 \cdot \sqrt{0.00024}/\sqrt{2} = 0.028$, LCL = 2.722, UCL = 2.778

45. $(1 - \theta)^5$, $(1 - \theta)^5 + 5\theta(1 - \theta)^4$

47. $\chi_0^2 = (20^2 + 30^2 + 0^2 + 20^2 + 30^2)1/480 < c = 9.49$. Accept.

49. $y = 0.2 + 0.067x$

Appendix 3

Auxiliary Material

A3.1 Formulas for Special Functions

For tables of numerical values, see Appendix 5.

Exponential function e^x (Fig. 509)

$$e = 2.71828\ 18284\ 59045\ 23536\ 02874\ 71353$$

$$(1) \qquad e^x e^y = e^{x+y}, \qquad e^x/e^y = e^{x-y}, \qquad (e^x)^y = e^{xy}$$

Natural logarithm (Fig. 510)

$$(2) \qquad \ln(xy) = \ln x + \ln y, \qquad \ln(x/y) = \ln x - \ln y, \qquad \ln(x^a) = a \ln x$$

$\ln x$ is the inverse of e^x, and $e^{\ln x} = x$, $e^{-\ln x} = e^{\ln(1/x)} = 1/x$.

Logarithm of base ten $\log_{10} x$ or simply $\log x$

$$(3) \qquad \log x = M \ln x, \quad M = \log e = 0.43429\ 44819\ 03251\ 82765\ 11289\ 18917$$

$$(4) \qquad \ln x = \frac{1}{M} \log x, \quad \frac{1}{M} = 2.30258\ 50929\ 94045\ 68401\ 79914\ 54684$$

$\log x$ is the inverse of 10^x, and $10^{\log x} = x$, $10^{-\log x} = 1/x$.

Sine and cosine functions (Figs. 511, 512). In calculus, angles are measured in radians, so that $\sin x$ and $\cos x$ have period 2π.

$\sin x$ is odd, $\sin(-x) = -\sin x$, and $\cos x$ is even, $\cos(-x) = \cos x$.

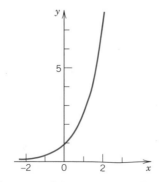

Fig. 509. Exponential function e^x

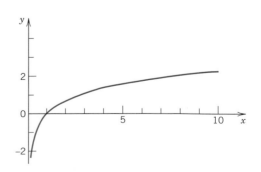

Fig. 510. Natural logarithm $\ln x$

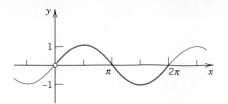

Fig. 511. sin x

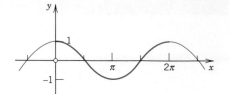

Fig. 512. cos x

$$1° = 0.01745\ 32925\ 19943\ \text{radian}$$

$$1\ \text{radian} = 57°\ 17'\ 44.80625''$$

$$= 57.29577\ 95131°$$

$$(5) \qquad \sin^2 x + \cos^2 x = 1$$

$$(6) \quad \begin{cases} \sin(x + y) = \sin x \cos y + \cos x \sin y \\ \sin(x - y) = \sin x \cos y - \cos x \sin y \\ \cos(x + y) = \cos x \cos y - \sin x \sin y \\ \cos(x - y) = \cos x \cos y + \sin x \sin y \end{cases}$$

$$(7) \qquad \sin 2x = 2 \sin x \cos x, \qquad \cos 2x = \cos^2 x - \sin^2 x$$

$$(8) \quad \begin{cases} \sin x = \cos\left(x - \dfrac{\pi}{2}\right) = \cos\left(\dfrac{\pi}{2} - x\right) \\ \cos x = \sin\left(x + \dfrac{\pi}{2}\right) = \sin\left(\dfrac{\pi}{2} - x\right) \end{cases}$$

$$(9) \qquad \sin(\pi - x) = \sin x, \qquad \cos(\pi - x) = -\cos x$$

$$(10) \qquad \cos^2 x = \tfrac{1}{2}(1 + \cos 2x), \qquad \sin^2 x = \tfrac{1}{2}(1 - \cos 2x)$$

$$(11) \quad \begin{cases} \sin x \sin y = \tfrac{1}{2}[-\cos(x + y) + \cos(x - y)] \\ \cos x \cos y = \tfrac{1}{2}[\cos(x + y) + \cos(x - y)] \\ \sin x \cos y = \tfrac{1}{2}[\sin(x + y) + \sin(x - y)] \end{cases}$$

$$(12) \quad \begin{cases} \sin u + \sin v = 2 \sin \dfrac{u + v}{2} \cos \dfrac{u - v}{2} \\[2mm] \cos u + \cos v = 2 \cos \dfrac{u + v}{2} \cos \dfrac{u - v}{2} \\[2mm] \cos v - \cos u = 2 \sin \dfrac{u + v}{2} \sin \dfrac{u - v}{2} \end{cases}$$

$$(13) \qquad A \cos x + B \sin x = \sqrt{A^2 + B^2} \cos(x \pm \delta), \quad \tan \delta = \dfrac{\sin \delta}{\cos \delta} = \mp \dfrac{B}{A}$$

$$(14) \qquad A \cos x + B \sin x = \sqrt{A^2 + B^2} \sin(x \pm \delta), \quad \tan \delta = \dfrac{\sin \delta}{\cos \delta} = \pm \dfrac{A}{B}$$

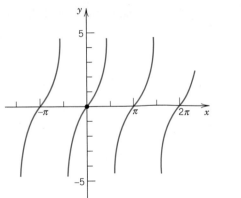

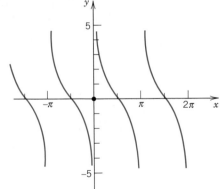

Fig. 513. tan x **Fig. 514.** cot x

Tangent, cotangent, secant, cosecant (Figs. 513, 514)

(15) $\tan x = \dfrac{\sin x}{\cos x}$, $\cot x = \dfrac{\cos x}{\sin x}$, $\sec x = \dfrac{1}{\cos x}$, $\csc x = \dfrac{1}{\sin x}$

(16) $\tan (x + y) = \dfrac{\tan x + \tan y}{1 - \tan x \tan y}$, $\tan (x - y) = \dfrac{\tan x - \tan y}{1 + \tan x \tan y}$

Hyperbolic functions (hyperbolic sine sinh x, etc.; Figs. 515, 516)

(17) $\sinh x = \tfrac{1}{2}(e^x - e^{-x}),\qquad \cosh x = \tfrac{1}{2}(e^x + e^{-x})$

(18) $\tanh x = \dfrac{\sinh x}{\cosh x}$, $\coth x = \dfrac{\cosh x}{\sinh x}$

(19) $\cosh x + \sinh x = e^x,\qquad \cosh x - \sinh x = e^{-x}$

(20) $\cosh^2 x - \sinh^2 x = 1$

(21) $\sinh^2 x = \tfrac{1}{2}(\cosh 2x - 1),\qquad \cosh^2 x = \tfrac{1}{2}(\cosh 2x + 1)$

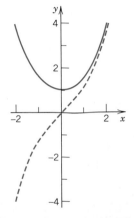

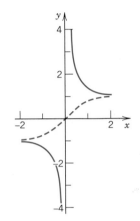

Fig. 515. sinh x (dashed) and cosh x **Fig. 516.** tanh x (dashed) and coth x

$$(22) \qquad \begin{cases} \sinh (x \pm y) = \sinh x \cosh y \pm \cosh x \sinh y \\[2mm] \cosh (x \pm y) = \cosh x \cosh y \pm \sinh x \sinh y \end{cases}$$

$$(23) \qquad \tanh (x \pm y) = \frac{\tanh x \pm \tanh y}{1 \pm \tanh x \tanh y}$$

Gamma function (Fig. 517 and Table A2 in Appendix 5). The gamma function $\Gamma(\alpha)$ is defined by the integral

$$(24) \qquad \Gamma(\alpha) = \int_{0}^{\infty} e^{-t} t^{\alpha - 1} \, dt \qquad\qquad (\alpha > 0)$$

which is meaningful only if $\alpha > 0$ (or, if we consider complex α, for those α whose real part is positive). Integration by parts gives the important *functional relation of the gamma function,*

$$(25) \qquad \Gamma(\alpha + 1) = \alpha \Gamma(\alpha).$$

From (24) we readily have $\Gamma(1) = 1$; hence if α is a positive integer, say k, then by repeated application of (25) we obtain

$$(26) \qquad \Gamma(k + 1) = k! \qquad\qquad (k = 0, 1, \cdots).$$

This shows that *the gamma function can be regarded as a generalization of the elementary factorial function.* [Sometimes the notation $(\alpha - 1)!$ is used for $\Gamma(\alpha)$, even for noninteger values of α, and the gamma function is also known as the **factorial function.**]

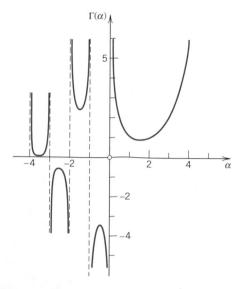

Fig. 517. Gamma function

By repeated application of (25) we obtain

$$\Gamma(\alpha) = \frac{\Gamma(\alpha + 1)}{\alpha} = \frac{\Gamma(\alpha + 2)}{\alpha(\alpha + 1)} = \cdots = \frac{\Gamma(\alpha + k + 1)}{\alpha(\alpha + 1)(\alpha + 2) \cdots (\alpha + k)}$$

and we may use this relation

(27) $\qquad \Gamma(\alpha) = \dfrac{\Gamma(\alpha + k + 1)}{\alpha(\alpha + 1) \cdots (\alpha + k)} \qquad\qquad (\alpha \neq 0, -1, -2, \cdots)$

for defining the gamma function for negative α ($\neq -1, -2, \cdots$), choosing for k the smallest integer such that $\alpha + k + 1 > 0$. *Together with* (24), *this then gives a definition of* $\Gamma(\alpha)$ *for all* α *not equal to zero or a negative integer* (Fig. 517).

It can be shown that the gamma function may also be represented as the limit of a product, namely, by the formula

(28) $\qquad \Gamma(\alpha) = \lim_{n \to \infty} \dfrac{n! \, n^\alpha}{\alpha(\alpha + 1)(\alpha + 2) \cdots (\alpha + n)} \qquad\qquad (\alpha \neq 0, -1, \cdots).$

From (27) or (28) we see that, for complex α, the gamma function $\Gamma(\alpha)$ is a meromorphic function with simple poles at $\alpha = 0, -1, -2, \cdots$.

An approximation of the gamma function for large positive α is given by the **Stirling formula**

(29) $$\Gamma(\alpha + 1) \approx \sqrt{2\pi\alpha} \left(\frac{\alpha}{e}\right)^\alpha$$

where e is the base of the natural logarithm. We finally mention the special value

(30) $$\Gamma(\tfrac{1}{2}) = \sqrt{\pi}.$$

Incomplete gamma functions

(31) $\qquad P(\alpha, x) = \displaystyle\int_0^x e^{-t} t^{\alpha-1} \, dt, \qquad Q(\alpha, x) = \displaystyle\int_x^\infty e^{-t} t^{\alpha-1} \, dt \qquad (\alpha > 0)$

(32) $$\Gamma(\alpha) = P(\alpha, x) + Q(\alpha, x)$$

Beta function

(33) $\qquad B(x, y) = \displaystyle\int_0^1 t^{x-1} (1 - t)^{y-1} \, dt \qquad\qquad (x > 0, y > 0)$

Representation in terms of gamma functions:

(34) $$B(x, y) = \frac{\Gamma(x)\Gamma(y)}{\Gamma(x + y)}$$

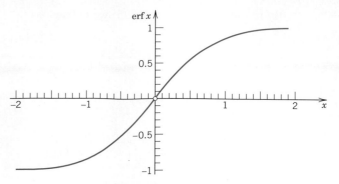

Fig. 518. Error function

Error function (Fig. 518 and Table A4 in Appendix 5)

$$\text{(35)} \qquad \text{erf } x = \frac{2}{\sqrt{\pi}} \int_0^x e^{-t^2} \, dt$$

$$\text{(36)} \qquad \text{erf } x = \frac{2}{\sqrt{\pi}} \left(x - \frac{x^3}{1!3} + \frac{x^5}{2!5} - \frac{x^7}{3!7} + - \cdots \right)$$

erf $(\infty) = 1$, *complementary error function*

$$\text{(37)} \qquad \text{erfc } x = 1 - \text{erf } x = \frac{2}{\sqrt{\pi}} \int_x^{\infty} e^{-t^2} \, dt$$

Fresnel integrals[1] (Fig. 519)

$$\text{(38)} \qquad C(x) = \int_0^x \cos (t^2) \, dt, \qquad S(x) = \int_0^x \sin (t^2) \, dt$$

$C(\infty) = \sqrt{\pi/8}$, $S(\infty) = \sqrt{\pi/8}$, *complementary functions*

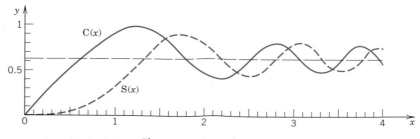

Fig. 519. Fresnel integrals

---

[1]AUGUSTIN FRESNEL (1788—1827), French physicist and mathematician. For tables see Ref. [1].

(39)
$$c(x) = \sqrt{\frac{\pi}{8}} - C(x) = \int_x^\infty \cos(t^2)\, dt$$
$$s(x) = \sqrt{\frac{\pi}{8}} - S(x) = \int_x^\infty \sin(t^2)\, dt$$

Sine integral (Fig. 520 and Table A4 in Appendix 5)

(40)
$$\text{Si}(x) = \int_0^x \frac{\sin t}{t}\, dt$$

$\text{Si}(\infty) = \pi/2$, *complementary function*

(41)
$$\text{si}(x) = \frac{\pi}{2} - \text{Si}(x) = \int_x^\infty \frac{\sin t}{t}\, dt$$

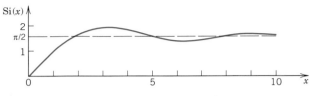

Fig. 520. Sine integral

Cosine integral (Table A4 in Appendix 5)

(42)
$$\text{ci}(x) = \int_x^\infty \frac{\cos t}{t}\, dt \qquad\qquad (x > 0)$$

Exponential integral

(43)
$$\text{Ei}(x) = \int_x^\infty \frac{e^{-t}}{t}\, dt \qquad\qquad (x > 0)$$

Logarithmic integral

(44)
$$\text{li}(x) = \int_0^\infty \frac{dt}{\ln t}$$

A3.2 Partial Derivatives

For differentiation formulas, see inside of front cover

Let $z = f(x, y)$ be a real function of two independent real variables, x and y. If we keep y constant, say, $y = y_1$, and think of x as a variable, then $f(x, y_1)$ depends on x alone. If the derivative of $f(x, y_1)$ with respect to x for a value $x = x_1$ exists, then the value of this derivative is called the **partial derivative** of $f(x, y)$ *with respect to x at the point* (x_1, y_1) and is denoted by

$$\frac{\partial f}{\partial x}\bigg|_{(x_1, y_1)} \qquad \text{or by} \qquad \frac{\partial z}{\partial x}\bigg|_{(x_1, y_1)}.$$

Other notations are

$$f_x(x_1, y_1) \qquad \text{and} \qquad z_x(x_1, y_1);$$

these may be used when subscripts are not used for another purpose and there is no danger of confusion.

We thus have, by the definition of the derivative,

$$(1) \qquad \frac{\partial f}{\partial x}\bigg|_{(x_1, y_1)} = \lim_{\Delta x \to 0} \frac{f(x_1 + \Delta x, y_1) - f(x_1, y_1)}{\Delta x}.$$

The partial derivative of $z = f(x, y)$ with respect to y is defined similarly; we now keep x constant, say, equal to x_1, and differentiate $f(x_1, y)$ with respect to y. Thus

$$(2) \qquad \frac{\partial f}{\partial y}\bigg|_{(x_1, y_1)} = \frac{\partial z}{\partial y}\bigg|_{(x_1, y_1)} = \lim_{\Delta y \to 0} \frac{f(x_1, y_1 + \Delta y) - f(x_1, y_1)}{\Delta y}.$$

Other notations are $f_y(x_1, y_1)$ and $z_y(x_1, y_1)$.

It is clear that the values of those two partial derivatives will in general depend on the point (x_1, y_1). Hence the partial derivatives $\partial z/\partial x$ and $\partial z/\partial y$ at a variable point (x, y) are functions of x and y. The function $\partial z/\partial x$ is obtained as in ordinary calculus by differentiating $z = f(x, y)$ with respect to x, *treating y as a constant*, and $\partial z/\partial y$ is obtained by differentiating z with respect to y, *treating x as a constant*.

EXAMPLE 1 Let $z = f(x, y) = x^2 y + x \sin y$. Then

$$\frac{\partial f}{\partial x} = 2xy + \sin y, \qquad \frac{\partial f}{\partial y} = x^2 + x \cos y. \qquad \blacktriangleleft$$

The partial derivatives $\partial z/\partial x$ and $\partial z/\partial y$ of a function $z = f(x, y)$ have a very simple **geometric interpretation.** The function $z = f(x, y)$ can be represented by a surface in space. The equation $y = y_1$ then represents a vertical plane intersecting the surface in a curve, and the partial derivative $\partial z/\partial x$ at a point (x_1, y_1) is the slope of the tangent (that is, $\tan \alpha$ where α is the angle shown in Fig. 521) to the curve. Similarly, the partial derivative $\partial z/\partial y$ at (x_1, y_1) is the slope of the tangent to the curve $x = x_1$ on the surface $z = f(x, y)$ at (x_1, y_1).

The partial derivatives $\partial z/\partial x$ and $\partial z/\partial y$ are called *first partial derivatives* or *partial derivatives of first order*. By differentiating these derivatives once more, we obtain the four *second partial derivatives* (or *partial derivatives of second order*)[2]

---

[2]Caution! In the subscript notation the subscripts are written in the order in which we differentiate, whereas in the "∂" notation the order is opposite.

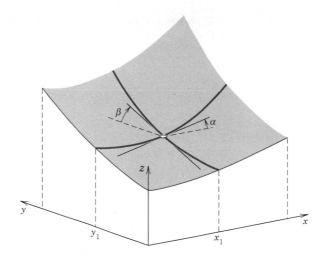

Fig. 521.   Geometrical interpretation of first partial derivatives

(3)
$$\frac{\partial^2 f}{\partial x^2} = \frac{\partial}{\partial x}\left(\frac{\partial f}{\partial x}\right) = f_{xx}$$

$$\frac{\partial^2 f}{\partial x\,\partial y} = \frac{\partial}{\partial x}\left(\frac{\partial f}{\partial y}\right) = f_{yx}$$

$$\frac{\partial^2 f}{\partial y\,\partial x} = \frac{\partial}{\partial y}\left(\frac{\partial f}{\partial x}\right) = f_{xy}$$

$$\frac{\partial^2 f}{\partial y^2} = \frac{\partial}{\partial y}\left(\frac{\partial f}{\partial y}\right) = f_{yy}.$$

It can be shown that if all the derivatives concerned are continuous, then the two mixed partial derivatives are equal, so that the order of differentiation does not matter (see Ref. [5] in Appendix 1), that is,

(4)
$$\frac{\partial^2 z}{\partial x\,\partial y} = \frac{\partial^2 z}{\partial y\,\partial x}.$$

EXAMPLE 2   For the function in Example 1,

$$f_{xx} = 2y, \qquad f_{xy} = 2x + \cos y = f_{yx}, \qquad f_{yy} = -x\sin y. \qquad \blacktriangleleft$$

By differentiating the second partial derivatives again with respect to x and y, respectively, we obtain the *third partial derivatives* or *partial derivatives of the third order* of f, etc.

If we consider a function $f(x, y, z)$ of **three independent variables,** then we have the three first partial derivatives $f_x(x, y, z)$, $f_y(x, y, z)$, and $f_z(x, y, z)$. Here f_x is obtained by differentiating f with respect to x, *treating both y and z as constants.* Thus, analogous to (1), we now have

$$\left.\frac{\partial f}{\partial x}\right|_{(x_1, y_1, z_1)} = \lim_{\Delta x \to 0} \frac{f(x_1 + \Delta x,\, y_1,\, z_1) - f(x_1,\, y_1,\, z_1)}{\Delta x},$$

etc. By differentiating f_x, f_y, f_z again in this fashion we obtain the second partial derivatives of f, etc.

EXAMPLE 3   Let $f(x, y, z) = x^2 + y^2 + z^2 + xy\, e^z$. Then

$$f_x = 2x + y\, e^z, \qquad f_y = 2y + x\, e^z, \qquad f_z = 2z + xy\, e^z,$$

$$f_{xx} = 2, \qquad f_{xy} = f_{yx} = e^z, \qquad f_{xz} = f_{zx} = y\, e^z,$$

$$f_{yy} = 2, \qquad f_{yz} = f_{zy} = x\, e^z, \qquad f_{zz} = 2 + xy\, e^z.$$ ◀

A3.3 Sequences and Series

See also Chap. 14.

Monotone Real Sequences

We call a real sequence $x_1, x_2, \cdots, x_n, \cdots$ a **monotone sequence** if it is either **monotone increasing,** that is,

$$x_1 \leqq x_2 \leqq x_3 \leqq \cdots$$

or **monotone decreasing,** that is,

$$x_1 \geqq x_2 \geqq x_3 \geqq \cdots.$$

We call $x_1, x_2, \cdots$ a **bounded sequence** if there is a positive constant K such that $|x_n| < K$ for all n.

THEOREM 1   *If a real sequence is bounded and monotone, it converges.*

PROOF.   Let $x_1, x_2, \cdots$ be a bounded monotone increasing sequence. Then its terms are smaller than some number B and, since $x_1 \leqq x_n$ for all n, they lie in the interval $x_1 \leqq x_n \leqq B$, which will be denoted by I_0. We bisect I_0; that is, we subdivide it into two parts of equal length. If the right half (together with its endpoints) contains terms of the sequence, we denote it by I_1. If it does not contain terms of the sequence, then the left half of I_0 (together with its endpoints) is called I_1. This is the first step.

In the second step we bisect I_1, select one half by the same rule, and call it I_2, and so on (see Fig. 522).

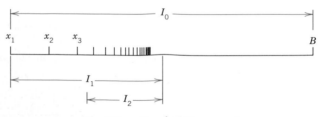

Fig. 522.   Proof of Theorem 1

In this way we obtain shorter and shorter intervals $I_0, I_1, I_2, \cdots$ with the following properties. Each I_m contains all I_n for $n > m$. No term of the sequence lies to the right of I_m, and, since the sequence is monotone increasing, all x_n with n greater than some number N lie in I_m; of course, N will depend on m, in general. The lengths of the I_m approach zero as m approaches infinity. Hence there is precisely one number, call it L, that lies in all those intervals,[3] and we may now easily prove that the sequence is convergent with the limit L.

In fact, given an $\epsilon > 0$, we choose an m such that the length of I_m is less than ϵ. Then L and all the x_n with $n > N(m)$ lie in I_m, and, therefore, $|x_n - L| < \epsilon$ for all those n. This completes the proof for an increasing sequence. For a decreasing sequence the proof is the same, except for a suitable interchange of "left" and "right" in the construction of those intervals. ◀

Real Series

THEOREM 2   **Leibniz test for real series**

Let $x_1, x_2, \cdots$ be real and monotone decreasing to zero, that is,

$$(1) \qquad (a) \quad x_1 \geqq x_2 \geqq x_3 \geqq \cdots, \qquad (b) \quad \lim_{m \to \infty} x_m = 0.$$

Then the series with terms of alternating signs

$$x_1 - x_2 + x_3 - x_4 + - \cdots$$

converges, and for the remainder R_n after the nth term we have the estimate

$$(2) \qquad\qquad |R_n| \leqq x_{n+1}.$$

PROOF.   Let s_n be the nth partial sum of the series. Then, because of (1a),

$$s_1 = x_1, \qquad\qquad s_2 = x_1 - x_2 \leqq s_1,$$

$$s_3 = s_2 + x_3 \geqq s_2, \qquad s_3 = s_1 - (x_2 - x_3) \leqq s_1,$$

so that $s_2 \leqq s_3 \leqq s_1$. Proceeding in this fashion, we conclude that (Fig. 523)

$$(3) \qquad\qquad s_1 \geqq s_3 \geqq s_5 \geqq \cdots \geqq s_6 \geqq s_4 \geqq s_2$$

which shows that the odd partial sums form a bounded monotone sequence, and so do the even partial sums. Hence, by Theorem 1, both sequences converge, say,

---

[3]This statement seems to be obvious, but actually it is not; it may be regarded as an axiom of the real number system in the following form. Let $J_1, J_2, \cdots$ be closed intervals such that each J_m contains all J_n with $n > m$, and the lengths of the J_m approach zero as m approaches infinity. Then there is precisely one real number that is contained in all those intervals. This is the so-called **Cantor–Dedekind axiom,** named after the German mathematicians GEORG CANTOR (1845—1918), the creator of set theory, and RICHARD DEDEKIND (1831—1916), known for his fundamental work in number theory. For further details see Ref. [2] in Appendix 1. (An interval I is said to be **closed** if its two endpoints are regarded as points belonging to I. It is said to be **open** if the endpoints are not regarded as points of I.)

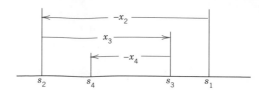

Fig. 523. Proof of the Leibniz test

$$\lim_{n\to\infty} s_{2n+1} = s, \qquad\qquad \lim_{n\to\infty} s_{2n} = s^*.$$

Now, since $s_{2n+1} - s_{2n} = x_{2n+1}$, we readily see that (1b) implies

$$s - s^* = \lim_{n\to\infty} s_{2n+1} - \lim_{n\to\infty} s_{2n} = \lim_{n\to\infty} (s_{2n+1} - s_{2n}) = \lim_{n\to\infty} x_{2n+1} = 0.$$

Hence $s^* = s$, and the series converges with the sum s.

We prove the estimate (2) for the remainder. Since $s_n \to s$, it follows from (3) that

$$s_{2n+1} \geqq s \geqq s_{2n} \qquad \text{and also} \qquad s_{2n-1} \geqq s \geqq s_{2n}.$$

By subtracting s_{2n} and s_{2n-1}, respectively, we obtain

$$s_{2n+1} - s_{2n} \geqq s - s_{2n} \geqq 0, \qquad 0 \geqq s - s_{2n-1} \geqq s_{2n} - s_{2n-1}.$$

In these inequalities, the first expression is equal to x_{2n+1}, the last is equal to $-x_{2n}$, and the expressions between the inequality signs are the remainders R_{2n} and R_{2n-1}. Thus the inequalities may be written

$$x_{2n+1} \geqq R_{2n} \geqq 0, \qquad 0 \geqq R_{2n-1} \geqq -x_{2n}$$

and we see that they imply (2). This completes the proof.   ◀

A3.4 Grad, Div, Curl, ∇^2 in Curvilinear Coordinates

To simplify formulas we write Cartesian coordinates $x = x_1$, $y = x_2$, $z = x_3$. We denote curvilinear coordinates by q_1, q_2, q_3. Through each point P there pass three coordinate surfaces $q_1 = const$, $q_2 = const$, $q_3 = const$. They intersect along coordinate curves. We assume the three coordinate curves through P to be **orthogonal** (perpendicular to each other). We write coordinate transformations as

$$(1) \qquad x_1 = x_1(q_1, q_2, q_3), \qquad x_2 = x_2(q_1, q_2, q_3), \qquad x_3 = x_3(q_1, q_2, q_3).$$

Corresponding transformations of grad, div, curl, and ∇^2 can all be written by using

$$(2) \qquad h_j{}^2 = \sum_{k=1}^{3} \left(\frac{\partial x_k}{\partial q_j} \right)^2.$$

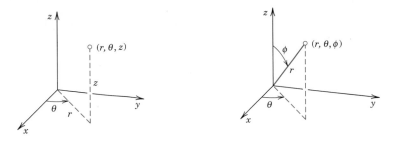

(a) Cylindrical coordinates   (b) Spherical coordinates

Fig. 524.   Special curvilinear coordinates

Next to Cartesian coordinates, most important are **cylindrical coordinates** $q_1 = r$, $q_2 = \theta$, $q_3 = z$ (Fig. 524a) defined by

$$(3) \qquad x_1 = q_1 \cos q_2 = r \cos \theta, \qquad x_2 = q_1 \sin q_2 = r \sin \theta, \qquad x_3 = q_3 = z$$

and **spherical coordinates** $q_1 = r$, $q_2 = \theta$, $q_3 = \phi$ (Fig. 524b) defined by[4]

$$(4) \qquad \begin{aligned} x_1 &= q_1 \cos q_2 \sin q_3 = r \cos \theta \sin \phi, \qquad x_2 = q_1 \sin q_2 \sin q_3 = r \sin \theta \sin \phi, \\ x_3 &= q_1 \cos q_3 = r \cos \phi. \end{aligned}$$

In addition to the general formulas for any orthogonal coordinates q_1, q_2, q_3, we shall give additional formulas for these important special cases.

Linear Element ds. In Cartesian coordinates,

$$ds^2 = dx_1{}^2 + dx_2{}^2 + dx_3{}^2 \qquad \text{(Sec. 8.5).}$$

For the q-coordinates,

$$(5) \qquad\qquad ds^2 = h_1{}^2 \, dq_1{}^2 + h_2{}^2 \, dq_2{}^2 + h_3{}^2 \, dq_3{}^2.$$

$$(5') \qquad\qquad ds^2 = dr^2 + r^2 \, d\theta^2 + dz^2 \qquad\qquad \text{(Cylindrical coordinates).}$$

For polar coordinates set $dz^2 = 0$.

$$(5'') \qquad\qquad ds^2 = dr^2 + r^2 \sin^2 \phi \, d\theta^2 + r^2 \, d\phi^2 \qquad\qquad \text{(Spherical coordinates).}$$

Gradient. grad $f = \nabla f = \begin{bmatrix} f_{x_1}, & f_{x_2}, & f_{x_3} \end{bmatrix}$ (partial derivatives; Sec. 8.9). In the q-system, with **u, v, w** denoting unit vectors in the positive directions of the q_1, q_2, q_3 coordinate curves, respectively,

$$(6) \qquad\qquad \text{grad } f = \nabla f = \frac{1}{h_1} \frac{\partial f}{\partial q_1} \mathbf{u} + \frac{1}{h_2} \frac{\partial f}{\partial q_2} \mathbf{v} + \frac{1}{h_3} \frac{\partial f}{\partial q_3} \mathbf{w}$$

---

[4]This is the notation used in calculus and in many other books. It is logical since in it, θ plays the same role as in polar coordinates. *Caution!* Some books interchange the roles of θ and ϕ.

(6') $$\text{grad } f = \nabla f = \frac{\partial f}{\partial r}\mathbf{u} + \frac{1}{r}\frac{\partial f}{\partial \theta}\mathbf{v} + \frac{\partial f}{\partial z}\mathbf{w}$$ (Cylindrical coordinates)

(6'') $$\text{grad } f = \nabla f = \frac{\partial f}{\partial r}\mathbf{u} + \frac{1}{r\sin\phi}\frac{\partial f}{\partial \theta}\mathbf{v} + \frac{1}{r}\frac{\partial f}{\partial \phi}\mathbf{w}$$ (Spherical coordinates).

Divergence $\text{div } \mathbf{F} = \nabla\cdot\mathbf{F} = (F_1)_{x_1} + (F_2)_{x_2} + (F_3)_{x_3}$ ($\mathbf{F} = [F_1, F_2, F_3]$, Sec. 8.10):

(7) $$\text{div } \mathbf{F} = \nabla\cdot\mathbf{F} = \frac{1}{h_1 h_2 h_3}\left[\frac{\partial}{\partial q_1}(h_2 h_3 F_1) + \frac{\partial}{\partial q_2}(h_3 h_1 F_2) + \frac{\partial}{\partial q_3}(h_1 h_2 F_3)\right]$$

(7') $$\text{div } \mathbf{F} = \nabla\cdot\mathbf{F} = \frac{1}{r}\frac{\partial}{\partial r}(rF_1) + \frac{1}{r}\frac{\partial F_2}{\partial \theta} + \frac{\partial F_3}{\partial z}$$ (Cylindrical coordinates)

(7'') $$\text{div } \mathbf{F} = \nabla\cdot\mathbf{F} = \frac{1}{r^2}\frac{\partial}{\partial r}(r^2 F_1) + \frac{1}{r\sin\phi}\frac{\partial F_2}{\partial \theta} + \frac{1}{r\sin\phi}\frac{\partial}{\partial \phi}(\sin\phi\, F_3)$$ (Spherical coordinates).

Laplacian $\nabla^2 f = \nabla\cdot\nabla f = \text{div (grad } f) = f_{x_1 x_1} + f_{x_2 x_2} + f_{x_3 x_3}$ (Sec. 8.10):

(8) $$\nabla^2 f = \frac{1}{h_1 h_2 h_3}\left[\frac{\partial}{\partial q_1}\left(\frac{h_2 h_3}{h_1}\frac{\partial f}{\partial q_1}\right) + \frac{\partial}{\partial q_2}\left(\frac{h_3 h_1}{h_2}\frac{\partial f}{\partial q_2}\right) + \frac{\partial}{\partial q_3}\left(\frac{h_1 h_2}{h_3}\frac{\partial f}{\partial q_3}\right)\right]$$

(8') $$\nabla^2 f = \frac{\partial^2 f}{\partial r^2} + \frac{1}{r}\frac{\partial f}{\partial r} + \frac{1}{r^2}\frac{\partial^2 f}{\partial \theta^2} + \frac{\partial^2 f}{\partial z^2}$$ (Cylindrical coordinates)

(8'') $$\nabla^2 f = \frac{\partial^2 f}{\partial r^2} + \frac{2}{r}\frac{\partial f}{\partial r} + \frac{1}{r^2 \sin^2\phi}\frac{\partial^2 f}{\partial \theta^2} + \frac{1}{r^2}\frac{\partial^2 f}{\partial \phi^2} + \frac{\cot\phi}{r^2}\frac{\partial f}{\partial \phi}$$ (Spherical coordinates).

Curl (Sec. 8.11):

(9) $$\text{curl } \mathbf{F} = \nabla\times\mathbf{F} = \frac{1}{h_1 h_2 h_3}\begin{vmatrix} h_1\mathbf{u} & h_2\mathbf{v} & h_3\mathbf{w} \\ \dfrac{\partial}{\partial q_1} & \dfrac{\partial}{\partial q_2} & \dfrac{\partial}{\partial q_3} \\ h_1 F_1 & h_2 F_2 & h_3 F_3 \end{vmatrix}.$$

For cylindrical or spherical coordinates, use (9) with

$$h_1 = h_r = 1, \qquad h_2 = h_\theta = q_1 = r, \qquad h_3 = h_z = 1$$ (Cylindrical coordinates)

or

$$h_1 = h_r = 1, \qquad h_2 = h_\theta = q_1 \sin q_3 = r\sin\phi, \qquad h_3 = h_\phi = q_1 = r$$ (Spherical coordinates).

Additional Proofs

SECTION 2.7, PAGE 97

PROOF OF THEOREM 1 (Uniqueness)[1]

Assuming that the problem consisting of the differential equation

$$y'' + p(x)y' + q(x)y = 0 \tag{1}$$

and the two initial conditions

$$y(x_0) = K_0, \qquad y'(x_0) = K_1 \tag{3}$$

has two solutions $y_1(x)$ and $y_2(x)$ on the interval I in the theorem, we show that their difference

$$y(x) = y_1(x) - y_2(x)$$

is identically zero on I; then $y_1 \equiv y_2$ on I, which implies uniqueness.

Since (1) is homogeneous and linear, y is a solution of that equation on I, and since y_1 and y_2 satisfy the same initial conditions, y satisfies the conditions

$$y(x_0) = 0, \qquad y'(x_0) = 0. \tag{10}$$

We consider the function

$$z(x) = y(x)^2 + y'(x)^2$$

and its derivative

$$z' = 2yy' + 2y'y''.$$

From the differential equation we have

$$y'' = -py' - qy.$$

By substituting this in the expression for z' we obtain

$$z' = 2yy' - 2py'^2 - 2qyy'. \tag{11}$$

---

[1]This proof was suggested by my colleague, Prof. A. D. Ziebur. In this proof we use formula numbers that have not yet been used in Sec. 2.7.

Now, since y and y' are real,

$$(y \pm y')^2 = y^2 \pm 2yy' + y'^2 \geqq 0.$$

From this we immediately obtain the two inequalities

(12)   (a) $2yy' \leqq y^2 + y'^2 = z,$ (b) $-2yy' \leqq y^2 + y'^2 = z.$

From (12b) we have $2yy' \geqq -z$. Together, $|2yy'| \leqq z$. For the last term in (11) we now obtain

$$-2qyy' \leqq |-2qyy'| = |q||2yy'| \leqq |q|z.$$

Using this result as well as $-p \leqq |p|$ and applying (12a) to the term $2yy'$ in (11), we find

$$z' \leqq z + 2|p|y'^2 + |q|z.$$

Since $y'^2 \leqq y^2 + y'^2 = z$, from this we obtain

$$z' \leqq (1 + 2|p| + |q|)z$$

or, denoting the function in parentheses by h,

(13a) $z' \leqq hz$ for all x on I.

Similarly, from (11) and (12) it follows that

$$-z' = -2yy' + 2py'^2 + 2qyy'$$

(13b)

$$\leqq z + 2|p|z + |q|z = hz.$$

The inequalities (13a) and (13b) are equivalent to the inequalities

(14) $z' - hz \leqq 0,$ $z' + hz \geqq 0.$

Integrating factors for the two expressions on the left are

$$F_1 = e^{-\int h(x)\, dx} \qquad \text{and} \qquad F_2 = e^{\int h(x)\, dx}.$$

The integrals in the exponents exist because h is continuous. Since F_1 and F_2 are positive, we thus have from (14)

$$F_1(z' - hz) = (F_1 z)' \leqq 0 \qquad \text{and} \qquad F_2(z' + hz) = (F_2 z)' \geqq 0,$$

which means that $F_1 z$ is nonincreasing and $F_2 z$ is nondecreasing on I. Since $z(x_0) = 0$ by (10), when $x \leqq x_0$ we thus obtain

$$F_1 z \geqq (F_1 z)_{x_0} = 0, \qquad F_2 z \leqq (F_2 z)_{x_0} = 0$$

and similarly, when $x \geqq x_0,$

$$F_1 z \leqq 0, \qquad F_2 z \geqq 0.$$

Dividing by F_1 and F_2 and noting that these functions are positive, we have altogether

$$z \leqq 0, \qquad z \geqq 0 \qquad\qquad \text{for all } x \text{ on } I.$$

This implies that $z = y^2 + y'^2 \equiv 0$ on I. Hence $y \equiv 0$ or $y_1 \equiv y_2$ on I.   ◀

SECTION 4.4, PAGE 213

PROOF OF THEOREM 2 (Frobenius method. Basis of solutions. Three cases)

The formula numbers in this proof are the same as in the text of Sec. 4.4. An additional formula not appearing in Sec. 4.4 will be called (A) (see below).

The differential equation in Theorem 2 is

(1)
$$y'' + \frac{b(x)}{x} y' + \frac{c(x)}{x^2} y = 0,$$

where $b(x)$ and $c(x)$ are analytic functions. We can write it

(1′)
$$x^2 y'' + x b(x) y' + c(x) y = 0.$$

The indicial equation of (1) is

(4)
$$r(r - 1) + b_0 r + c_0 = 0.$$

The roots r_1, r_2 of this quadratic equation determine the general form of a basis of solutions of (1), and there are three possible cases as follows.

Case 1 (Distinct roots not differing by an integer). A first solution of (1) is of the form

(5)
$$y_1(x) = x^{r_1}(a_0 + a_1 x + a_2 x^2 + \cdots)$$

and can be determined as in the power series method. For a proof that in this case, equation (1) has a second independent solution of the form

(6)
$$y_2(x) = x^{r_2}(A_0 + A_1 x + A_2 x^2 + \cdots),$$

see Ref. [A5] listed in Appendix 1.

Case 2 (Double root). The indicial equation (4) has a double root r if and only if $(b_0 - 1)^2 - 4c_0 = 0$, and then $r = \frac{1}{2}(1 - b_0)$. A first solution

(7)
$$y_1(x) = x^r(a_0 + a_1 x + a_2 x^2 + \cdots), \qquad\qquad r = \tfrac{1}{2}(1 - b_0),$$

can be determined as in Case 1. We show that a second independent solution is of the form

(8) $$y_2(x) = y_1(x) \ln x + x^r(A_1 x + A_2 x^2 + \cdots) \qquad (x > 0).$$

We use the method of reduction of order (see Sec. 2.1), that is, we determine $u(x)$ such that $y_2(x) = u(x)y_1(x)$ is a solution of (1). By inserting this and the derivatives

$$y_2' = u'y_1 + uy_1', \qquad y_2'' = u''y_1 + 2u'y_1' + uy_1''$$

into the differential equation (1') we obtain

$$x^2(u''y_1 + 2u'y_1' + uy_1'') + xb(u'y_1 + uy_1') + cuy_1 = 0.$$

Since y_1 is a solution of (1'), the sum of the terms involving u is zero, and this equation reduces to

$$x^2 y_1 u'' + 2x^2 y_1' u' + xby_1 u' = 0.$$

By dividing by $x^2 y_1$ and inserting the power series for b we obtain

$$u'' + \left(2\frac{y_1'}{y_1} + \frac{b_0}{x} + \cdots\right)u' = 0.$$

Here and in the following the dots designate terms that are constant or involve positive powers of x. Now from (7) it follows that

$$\frac{y_1'}{y_1} = \frac{x^{r-1}[ra_0 + (r+1)a_1 x + \cdots]}{x^r[a_0 + a_1 x + \cdots]}$$

$$= \frac{1}{x}\left(\frac{ra_0 + (r+1)a_1 x + \cdots}{a_0 + a_1 x + \cdots}\right) = \frac{r}{x} + \cdots.$$

Hence the previous equation can be written

(A) $$u'' + \left(\frac{2r + b_0}{x} + \cdots\right)u' = 0.$$

Since $r = (1 - b_0)/2$, the term $(2r + b_0)/x$ equals $1/x$, and by dividing by u' we thus have

$$\frac{u''}{u'} = -\frac{1}{x} + \cdots.$$

By integration we obtain $\ln u' = -\ln x + \cdots$, hence $u' = (1/x)e^{(\cdots)}$. Expanding the exponential function in powers of x and integrating once more, we see that u is of the form

$$u = \ln x + k_1 x + k_2 x^2 + \cdots.$$

Inserting this into $y_2 = uy_1$, we obtain for y_2 a representation of the form (8).

Case 3 (Roots differing by an integer). We write $r_1 = r$ and $r_2 = r - p$ where p is a *positive* integer. A first solution

(9)
$$y_1(x) = x^{r_1}(a_0 + a_1 x + a_2 x^2 + \cdots)$$

can be determined as in Cases 1 and 2. We show that a second independent solution is of the form

(10)
$$y_2(x) = k y_1(x) \ln x + x^{r_2}(A_0 + A_1 x + A_2 x^2 + \cdots)$$

where we may have $k \neq 0$ or $k = 0$. As in Case 2 we set $y_2 = u y_1$. The first steps are literally as in Case 2 and give equation (A),

$$u'' + \left(\frac{2r + b_0}{x} + \cdots \right) u' = 0.$$

Now by elementary algebra, the coefficient $b_0 - 1$ of r in (4) equals minus the sum of the roots,

$$b_0 - 1 = -(r_1 + r_2) = -(r + r - p) = -2r + p.$$

Hence $2r + b_0 = p + 1$, and division by u' gives

$$\frac{u''}{u'} = -\left(\frac{p + 1}{x} + \cdots \right).$$

The further steps are as in Case 2. Integrating, we find

$$\ln u' = -(p + 1) \ln x + \cdots, \qquad \text{thus} \qquad u' = x^{-(p+1)} e^{(\cdots)}$$

where dots stand for some series of nonnegative integer powers of x. By expanding the exponential function as before we obtain a series of the form

$$u' = \frac{1}{x^{p+1}} + \frac{k_1}{x^p} + \cdots + \frac{k_{p-1}}{x^2} + \frac{k_p}{x} + k_{p+1} + k_{p+2} x + \cdots.$$

We integrate once more. Writing the resulting logarithmic term first, we get

$$u = k_p \ln x + \left(-\frac{1}{px^p} - \cdots - \frac{k_{p-1}}{x} + k_{p+1} x + \cdots \right).$$

Hence, by (9) we get for $y_2 = u y_1$ the formula

$$y_2 = k_p y_1 \ln x + x^{r_1 - p} \left(-\frac{1}{p} - \cdots - k_{p-1} x^{p-1} + \cdots \right) (a_0 + a_1 x + \cdots).$$

But this is of the form (10) with $k = k_p$ since $r_1 - p = r_2$ and the product of the two series involves nonnegative integer powers of x only.　◀

SECTION 4.7, PAGE 234

THEOREM   **(Reality of Eigenvalues)**

If p, q, r, and r' in the Sturm–Liouville equation (1) *of Sec. 4.7 are real-valued and continuous on the interval $a \leqq x \leqq b$ and $p(x) > 0$ throughout that interval (or $p(x) < 0$ throughout that interval), then all the eigenvalues of the Sturm–Liouville problem* (1), (2), *Sec. 4.7, are real.*

PROOF   Let $\lambda = \alpha + i\beta$ be an eigenvalue of the problem and let

$$y(x) = u(x) + iv(x)$$

be a corresponding eigenfunction; here α, β, u, and v are real. Substituting this into (1), Sec. 4.7, we have

$$(ru' + irv')' + (q + \alpha p + i\beta p)(u + iv) = 0.$$

This complex equation is equivalent to the following pair of equations for the real and the imaginary parts:

$$(ru')' + (q + \alpha p)u - \beta p v = 0$$

$$(rv')' + (q + \alpha p)v + \beta p u = 0.$$

Multiplying the first equation by v, the second by $-u$ and adding, we get

$$-\beta(u^2 + v^2)p = u(rv')' - v(ru')'$$

$$= [(rv')u - (ru')v]'.$$

The expression in brackets is continuous on $a \leqq x \leqq b$, for reasons similar to those in the proof of Theorem 1, Sec. 4.7. Integrating over x from a to b, we thus obtain

$$-\beta \int_a^b (u^2 + v^2)p \, dx = \left[r(uv' - u'v) \right]_a^b.$$

Because of the boundary conditions the right side is zero; this is as in that proof. Since y is an eigenfunction, $u^2 + v^2 \not\equiv 0$. Since y and p are continuous and $p > 0$ (or $p < 0$) on the interval $a \leqq x \leqq b$, the integral on the left is not zero. Hence, $\beta = 0$, which means that $\lambda = \alpha$ is real. This completes the proof.   ◀

SECTION 6.6, PAGE 344

PROOF THAT THE DEFINITION OF A DETERMINANT IN SEC. 6.6 IS UNAMBIGUOUS

We show that the definition of a determinant

$$(7) \qquad D = \det \mathbf{A} = \begin{vmatrix} a_{11} & a_{12} & \cdots & a_{1n} \\ a_{21} & a_{22} & \cdots & a_{2n} \\ \cdot & \cdot & \cdots & \cdot \\ \cdot & \cdot & \cdots & \cdot \\ a_{n1} & a_{n2} & \cdots & a_{nn} \end{vmatrix}$$

as given in Sec. 6.6 is unambiguous, that is, it yields the same value of D no matter which rows or columns we choose. (Here we shall use formula numbers not yet used in Sec. 6.6.)

We shall prove first that the *the same value is obtained no matter which row is chosen.*

The proof is by induction. The statement is true for a second-order determinant, for which the developments by the first row $a_{11}a_{22} + a_{12}(-a_{21})$ and by the second row $a_{21}(-a_{12}) + a_{22}a_{11}$ give the same value $a_{11}a_{22} - a_{12}a_{21}$. Assuming the statement to be true for an $(n - 1)$th-order determinant, we prove that it is true for an nth-order determinant.

For this purpose we expand D in terms of each of two arbitrary rows, say, the ith and the jth, and compare the results. Without loss of generality let us assume $i < j$.

First expansion. We expand D by the ith row. A typical term in this expansion is

$$(19) \qquad a_{ik}C_{ik} = a_{ik} \cdot (-1)^{i+k}M_{ik}.$$

The minor M_{ik} of a_{ik} in D is an $(n - 1)$th-order determinant. By the induction hypothesis we may expand it by any row. We expand it by the row corresponding to the jth row of D. This row contains the entries a_{jl} $(l \neq k)$. It is the $(j - 1)$th row of M_{ik}, because M_{ik} does not contain entries of the ith row of D, and $i < j$. We have to distinguish between two cases as follows.

Case I. If $l < k$, then the entry a_{jl} belongs to the lth column of M_{ik} (see Fig. 525). Hence the term involving a_{jl} in this expansion is

$$(20) \qquad a_{jl} \cdot (\text{cofactor of } a_{jl} \text{ in } M_{ik}) = a_{jl} \cdot (-1)^{(j-1)+l}M_{ikjl}$$

where M_{ikjl} is the minor of a_{jl} in M_{ik}. Since this minor is obtained from M_{ik} by deleting the row and column of a_{jl}, it is obtained from D by deleting the ith and jth rows and the kth and lth columns of D. We insert the expansions of the M_{ik} into that of D. Then it follows from (19) and (20) that the terms of the resulting representation of D are of the form

$$(21a) \qquad\qquad a_{ik}a_{jl} \cdot (-1)^{b}M_{ikjl} \qquad\qquad (l < k)$$

where

$$b = i + k + j + l - 1.$$

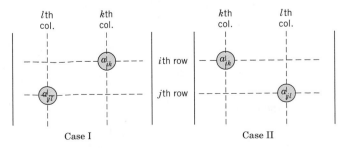

Case I   Case II

Fig. 525. Cases I and II of the two expansions of D

Case II. If $l > k$, the only difference is that then a_{jl} belongs to the $(l-1)$th column of M_{ik}, because M_{ik} does not contain entries of the kth column of D, and $k < l$. This causes an additional minus sign in (20), and, instead of (21a), we therefore obtain

(21b) $$-a_{ik}a_{jl} \cdot (-1)^b M_{ikjl} \qquad\qquad (l > k)$$

where b is the same as before.

Second expansion. We now expand D at first by the jth row. A typical term in this expansion is

(22) $$a_{jl}C_{jl} = a_{jl} \cdot (-1)^{j+l} M_{jl}.$$

By the induction hypothesis we may expand the minor M_{jl} of a_{jl} in D by its ith row, which corresponds to the ith row of D, since $j > i$.

Case I. If $k > l$, the entry a_{ik} in that row belongs to the $(k-1)$th column of M_{jl}, because M_{jl} does not contain entries of the lth column of D, and $l < k$ (see Fig. 525). Hence the term involving a_{ik} in this expansion is

(23) $$a_{ik} \cdot (\text{cofactor of } a_{ik} \text{ in } M_{jl}) = a_{ik} \cdot (-1)^{i+(k-1)} M_{ikjl},$$

where the minor M_{ikjl} of a_{ik} in M_{jl} is obtained by deleting the ith and jth rows and the kth and lth columns of D [and is, therefore, identical with M_{ikjl} in (20), so that our notation is consistent]. We insert the expansions of the M_{jl} into that of D. It follows from (22) and (23) that this yields a representation whose terms are identical with those given by (21a) when $l < k$.

Case II. If $k < l$, then a_{ik} belongs to the kth column of M_{jl}, we obtain an additional minus sign, and the result agrees with that characterized by (21b).

We have shown that the two expansions of D consist of the same terms, and this proves our statement concerning rows.

The proof of the statement concerning *columns* is quite similar; if we expand D in terms of two arbitrary columns, say, the kth and the lth, we find that the general term involving $a_{jl}a_{ik}$ is exactly the same as before. This proves that not only all column expansions of D yield the same value, but also that their common value is equal to the common value of the row expansions of D.

This completes the proof and shows that *our definition of an nth-order determinant is unambiguous.* ◀

SECTION 8.3, PAGE 414

PROOF OF FORMULA (2)

We prove that in right-handed Cartesian coordinates, the vector product

$$\mathbf{v} = \mathbf{a} \times \mathbf{b} = [a_1 \quad a_2, \quad a_3] \times [b_1, \quad b_2, \quad b_3]$$

has the components

(2) $$\boxed{v_1 = a_2b_3 - a_3b_2, \qquad v_2 = a_3b_1 - a_1b_3, \qquad v_3 = a_1b_2 - a_2b_1.}$$

We need only consider the case $\mathbf{v} \neq \mathbf{0}$. Since $\mathbf{v}$ is perpendicular to both $\mathbf{a}$ and $\mathbf{b}$, Theorem 1 in Sec. 8.2 gives $\mathbf{a} \cdot \mathbf{v} = 0$ and $\mathbf{b} \cdot \mathbf{v} = 0$; in components [see (2), Sec. 8.2],

(3)
$$a_1 v_1 + a_2 v_2 + a_3 v_3 = 0,$$
$$b_1 v_1 + b_2 v_2 + b_3 v_3 = 0.$$

Multiplying the first equation by b_3, the last by a_3, and subtracting, we obtain

$$(a_3 b_1 - a_1 b_3)v_1 = (a_2 b_3 - a_3 b_2)v_2.$$

Multiplying the first equation by b_1, the last by a_1, and subtracting, we obtain

$$(a_1 b_2 - a_2 b_1)v_2 = (a_3 b_1 - a_1 b_3)v_3.$$

We can easily verify that these two equations are satisfied by

(4) $\qquad v_1 = c(a_2 b_3 - a_3 b_2), \qquad v_2 = c(a_3 b_1 - a_1 b_3), \qquad v_3 = c(a_1 b_2 - a_2 b_1)$

where c is a constant. The reader may verify by inserting that (4) also satisfies (3). Now each of the equations in (3) represents a plane through the origin in $v_1 v_2 v_3$-space. The vectors $\mathbf{a}$ and $\mathbf{b}$ are normal vectors of these planes (see Example 6 in Sec. 8.2). Since $\mathbf{v} \neq \mathbf{0}$, these vectors are not parallel and the two planes do not coincide. Hence their intersection is a straight line L through the origin. Since (4) is a solution of (3) and, for varying c, represents a straight line, we conclude that (4) represents L, and every solution of (3) must be of the form (4). In particular, the components of $\mathbf{v}$ must be of this form, where c is to be determined. From (4) we obtain

$$|\mathbf{v}|^2 = v_1{}^2 + v_2{}^2 + v_3{}^2 = c^2 \big[(a_2 b_3 - a_3 b_2)^2 + (a_3 b_1 - a_1 b_3)^2 + (a_1 b_2 - a_2 b_1)^2 \big].$$

This can be written

$$|\mathbf{v}|^2 = c^2 \big[(a_1{}^2 + a_2{}^2 + a_3{}^2)(b_1{}^2 + b_2{}^2 + b_3{}^2) - (a_1 b_1 + a_2 b_2 + a_3 b_3)^2 \big],$$

as can be verified by performing the indicated multiplications in both formulas and comparing. Using (2) in Sec. 8.2, we thus have

$$|\mathbf{v}|^2 = c^2 \big[(\mathbf{a} \cdot \mathbf{a})(\mathbf{b} \cdot \mathbf{b}) - (\mathbf{a} \cdot \mathbf{b})^2 \big].$$

By comparing this with the formula in Team Project 38(a) of Problem Set 8.3 we conclude that $c = \pm 1$.

We show that $c = +1$. This can be done as follows.

If we change the lengths and directions of $\mathbf{a}$ and $\mathbf{b}$ continuously and so that at the end $\mathbf{a} = \mathbf{i}$ and $\mathbf{b} = \mathbf{j}$ (Fig. 170a in Sec. 8.3), then $\mathbf{v}$ will change its length and direction continuously, and at the end, $\mathbf{v} = \mathbf{i} \times \mathbf{j} = \mathbf{k}$. Obviously we may effect the change so that both $\mathbf{a}$ and $\mathbf{b}$ remain different from the zero vector and are not parallel at any instant. Then $\mathbf{v}$ is never equal to the zero vector, and since the change is continuous and c can only assume the values $+1$ or -1, it follows that at the end c must have the same value as before. Now at the end $\mathbf{a} = \mathbf{i}$, $\mathbf{b} = \mathbf{j}$, $\mathbf{v} = \mathbf{k}$ and, therefore, $a_1 = 1$, $b_2 = 1$, $v_3 = 1$, and the other components in (4) are zero. Hence from (4) we see that $v_3 = c = +1$. This proves Theorem 1.

For a left-handed coordinate system, $\mathbf{i} \times \mathbf{j} = -\mathbf{k}$ (see Fig. 170*b* in Sec. 8.3), resulting in $c = -1$. This proves the statement right after formula (2). ◄

SECTION 8.11, page 459

PROOF OF THEOREM 1 (Invariance of the curl)

This proof will follow from two theorems (A and B), which we prove first.

THEOREM A   **(Transformation law for vector components)**

For any vector $\mathbf{v}$ *the components* v_1, v_2, v_3 *and* $v_1{}^*$, $v_2{}^*$, $v_3{}^*$ *in any two systems of Cartesian coordinates* x_1, x_3, x_3 *and* $x_1{}^*$, $x_2{}^*$, $x_3{}^*$, *respectively, are related by*

$$
\begin{aligned}
v_1{}^* &= c_{11}v_1 + c_{12}v_2 + c_{13}v_3 \\
v_2{}^* &= c_{21}v_1 + c_{22}v_2 + c_{23}v_3 \\
v_3{}^* &= c_{31}v_1 + c_{32}v_2 + c_{33}v_3,
\end{aligned}
\tag{1}
$$

and conversely

$$
\begin{aligned}
v_1 &= c_{11}v_1{}^* + c_{21}v_2{}^* + c_{31}v_3{}^* \\
v_2 &= c_{12}v_1{}^* + c_{22}v_2{}^* + c_{32}v_3{}^* \\
v_3 &= c_{13}v_1{}^* + c_{23}v_2{}^* + c_{33}v_3{}^*
\end{aligned}
\tag{2}
$$

with coefficients

$$
\begin{array}{lll}
c_{11} = \mathbf{i}^* \cdot \mathbf{i} & c_{12} = \mathbf{i}^* \cdot \mathbf{j} & c_{13} = \mathbf{i}^* \cdot \mathbf{k} \\
c_{21} = \mathbf{j}^* \cdot \mathbf{i} & c_{22} = \mathbf{j}^* \cdot \mathbf{j} & c_{23} = \mathbf{j}^* \cdot \mathbf{k} \\
c_{31} = \mathbf{k}^* \cdot \mathbf{i} & c_{32} = \mathbf{k}^* \cdot \mathbf{j} & c_{33} = \mathbf{k}^* \cdot \mathbf{k}
\end{array}
\tag{3}
$$

satisfying

$$
\sum_{j=1}^{3} c_{kj} c_{mj} = \delta_{km}
\tag{4}
\qquad (k, m = 1, 2, 3),
$$

where the **Kronecker delta**[2] *is given by*

$$
\delta_{km} = \begin{cases} 0 & (k \neq m) \\ 1 & (k = m), \end{cases}
$$

---

[2]LEOPOLD KRONECKER (1823—1891), German mathematician at Berlin, who made important contributions to algebra, group theory, and number theory.

We shall keep our discussion completely independent of Chap. 6, but readers familiar with matrices should recognize that we are dealing with **orthogonal transformations and matrices** and that our present theorem follows from Theorem 2 in Sec. 7.3.

and **i, j, k** *and* **i*, j*, k*** *denote the unit vectors in the positive* x_1-, x_2-, x_3- *and* x_1^*-, x_2^*-, x_3^*-*directions, respectively.*

PROOF.　The representations of **v** in the two systems are

(5)　　　　(a)　$\mathbf{v} = v_1\mathbf{i} + v_2\mathbf{j} + v_3\mathbf{k}$　　　　(b)　$\mathbf{v} = v_1^*\mathbf{i}^* + v_2^*\mathbf{j}^* + v_3^*\mathbf{k}^*.$

Since $\mathbf{i}^* \cdot \mathbf{i}^* = 1$, $\mathbf{i}^* \cdot \mathbf{j}^* = 0$, $\mathbf{i}^* \cdot \mathbf{k}^* = 0$, we get from (5b) simply $\mathbf{i}^* \cdot \mathbf{v} = v_1^*$ and from this and (5a)

$$v_1^* = \mathbf{i}^* \cdot \mathbf{v} = \mathbf{i}^* \cdot v_1\mathbf{i} + \mathbf{i}^* \cdot v_2\mathbf{j} + \mathbf{i}^* \cdot v_3\mathbf{k} = v_1\mathbf{i}^* \cdot \mathbf{i} + v_2\mathbf{i}^* \cdot \mathbf{j} + v_3\mathbf{i}^* \cdot \mathbf{k}.$$

Because of (3), this is the first formula in (1), and the other two formulas are obtained similarly, by considering $\mathbf{j}^* \cdot \mathbf{v}$ and then $\mathbf{k}^* \cdot \mathbf{v}$. Formula (2) follows by the same idea, taking $\mathbf{i} \cdot \mathbf{v} = v_1$ from (5a) and then from (5b) and (3)

$$v_1 = \mathbf{i} \cdot \mathbf{v} = v_1^*\mathbf{i} \cdot \mathbf{i}^* + v_2^*\mathbf{i} \cdot \mathbf{j}^* + v_3^*\mathbf{i} \cdot \mathbf{k}^* = c_{11}v_1^* + c_{21}v_2^* + c_{31}v_3^*,$$

and similarly for the other two components.

We prove (4). We can write (1) and (2) briefly as

(6)　　　　(a)　$v_j = \sum_{m=1}^{3} c_{mj}v_m^*,$　　　　(b)　$v_k^* = \sum_{j=1}^{3} c_{kj}v_j.$

Substituting v_j into v_k^*, we get

$$v_k^* = \sum_{j=1}^{3} c_{kj} \sum_{m=1}^{3} c_{mj}v_m^* = \sum_{m=1}^{3} v_m^* \left(\sum_{j=1}^{3} c_{kj}c_{mj} \right),$$

where $k = 1, 2, 3$. Taking $k = 1$, we have

$$v_1^* = v_1^*\left(\sum_{j=1}^{3} c_{1j}c_{1j} \right) + v_2^*\left(\sum_{j=1}^{3} c_{1j}c_{2j} \right) + v_3^*\left(\sum_{j=1}^{3} c_{1j}c_{3j} \right).$$

For this to hold for *every* vector **v,** the first sum must be 1 and the other two sums 0. This proves (4) with $k = 1$ for $m = 1, 2, 3$. Taking $k = 2$ and then $k = 3$, we obtain (4) with $k = 2$ and 3, for $m = 1, 2, 3$.　　　◀

The most general transformation of a Cartesian coordinate system into another such system may be decomposed into a transformation of the type just considered and a translation. Under a translation, corresponding coordinates differ merely by a constant. We thus obtain

THEOREM B　**(Transformation law for Cartesian coordinates)**

The transformation of any Cartesian $x_1x_2x_3$-coordinate system into any other Cartesian $x_1^*x_2^*x_3^*$-coordinate system is of the form

(7)　　　　　　　　　　$x_m^* = \sum_{j=1}^{3} c_{mj}x_j + b_m,$　　　　　　　　$m = 1, 2, 3,$

with coefficients (3) *and constants* b_1, b_2, b_3; *conversely,*

$$(8) \qquad\qquad x_k = \sum_{n=1}^{3} c_{nk} x_n{}^* + \tilde{b}_k, \qquad\qquad k = 1, 2, 3.$$

PROOF OF THEOREM 1 (Invariance of the curl)

We write again x_1, x_2, x_3 instead of x, y, z, and similarly $x_1{}^*$, $x_2{}^*$, $x_3{}^*$ for other Cartesian coordinates, assuming that both systems are right-handed. Let a_1, a_2, a_3 denote the components of curl $\mathbf{v}$ in the $x_1 x_2 x_3$-coordinates, as given by (1), Sec. 8.11, with

$$x = x_1, \qquad y = x_2, \qquad z = x_3.$$

Similarly, let $a_1{}^*$, $a_2{}^*$, $a_3{}^*$ denote the components of curl $\mathbf{v}$ in the $x_1{}^* x_2{}^* x_3{}^*$-coordinate system. We prove that the length and direction of curl $\mathbf{v}$ are independent of the particular choice of Cartesian coordinates, as asserted in the theorem. We do this by showing that the components of curl $\mathbf{v}$ satisfy the transformation law (2), which is characteristic of vector components. We consider a_1. We use (6a), and then the chain rule for functions of several variables (Sec. 8.8). This gives

$$a_1 = \frac{\partial v_3}{\partial x_2} - \frac{\partial v_2}{\partial x_3} = \sum_{m=1}^{3} \left(c_{m3} \frac{\partial v_m{}^*}{\partial x_2} - c_{m2} \frac{\partial v_m{}^*}{\partial x_3} \right)$$

$$= \sum_{m=1}^{3} \sum_{j=1}^{3} \left(c_{m3} \frac{\partial v_m{}^*}{\partial x_j{}^*} \frac{\partial x_j{}^*}{\partial x_2} - c_{m2} \frac{\partial v_m{}^*}{\partial x_j{}^*} \frac{\partial x_j{}^*}{\partial x_3} \right).$$

From this and (7) we obtain

$$a_1 = \sum_{m=1}^{3} \sum_{j=1}^{3} (c_{m3} c_{j2} - c_{m2} c_{j3}) \frac{\partial v_m{}^*}{\partial x_j{}^*}$$

$$= (c_{33} c_{22} - c_{32} c_{23}) \left(\frac{\partial v_3{}^*}{\partial x_2{}^*} - \frac{\partial v_2{}^*}{\partial x_3{}^*} \right) + \cdots$$

$$= (c_{33} c_{22} - c_{32} c_{23}) a_1{}^* + (c_{13} c_{32} - c_{12} c_{33}) a_2{}^* + (c_{23} c_{12} - c_{22} c_{13}) a_3{}^*.$$

Note what we did. The double sum had $3 \times 3 = 9$ terms, 3 of which were zero (when $m = j$), and the remaining 6 terms we combined in pairs as we needed them in getting $a_1{}^*$, $a_2{}^*$, $a_3{}^*$.

We now use (3), Lagrange's identity (see Team Project 38(d) in Problem Set 8.3) and $\mathbf{k}^* \times \mathbf{j}^* = -\mathbf{i}^*$ and $\mathbf{k} \times \mathbf{j} = -\mathbf{i}$. Then

$$c_{33} c_{22} - c_{32} c_{23} = (\mathbf{k}^* \cdot \mathbf{k})(\mathbf{j}^* \cdot \mathbf{j}) - (\mathbf{k}^* \cdot \mathbf{j})(\mathbf{j}^* \cdot \mathbf{k})$$

$$= (\mathbf{k}^* \times \mathbf{j}^*) \cdot (\mathbf{k} \times \mathbf{j}) = \mathbf{i}^* \cdot \mathbf{i} = c_{11}, \qquad \text{etc.}$$

Hence $a_1 = c_{11} a_1{}^* + c_{21} a_2{}^* + c_{31} a_3{}^*$. This is of the form of the first formula in (2), p. A74, and the other two formulas of the form (2) are obtained similarly. This proves the theorem for right-handed systems. If the $x_1 x_2 x_3$-coordinates are left-handed, then $\mathbf{k} \times \mathbf{j} = +\mathbf{i}$, but then there is a minus sign in front of the determinant in (1), Sec. 8.11.   ◀

SECTION 9.2, PAGE 472

PROOF OF THEOREM 1, PART (b). We prove that if

$$(1) \qquad \int_C \mathbf{F(r)} \cdot d\mathbf{r} = \int_C (F_1\, dx + F_2\, dy + F_3\, dz)$$

with continuous F_1, F_2, F_3 in a domain D is independent of path in D, then F = grad f in D for some f; in components

$$(2') \qquad F_1 = \frac{\partial f}{\partial x}, \qquad F_2 = \frac{\partial f}{\partial y}, \qquad F_3 = \frac{\partial f}{\partial z}\,.$$

We choose any fixed A: (x_0, y_0, z_0) in D and any B: (x, y, z) in D and define f by

$$(3) \qquad f(x, y, z) = f_0 + \int_A^B (F_1\, dx^* + F_2\, dy^* + F_3\, dz^*),$$

with any constant f_0 and any path from A to B in D. Since A is fixed and we have independence of path, the integral depends only on the coordinates x, y, z, so that (3) defines a function $f(x, y, z)$ in D. We show that $\mathbf{F} = \text{grad } f$ with this f, beginning with the first of the three relations $(2')$. Because of independence of path, we may integrate from A to B_1: (x_1, y, z) and then parallel to the x-axis along the segment B_1B in Fig. 526 with B_1 chosen so that the whole segment lies in D. Then

$$f(x, y, z) = f_0 + \int_A^{B_1} (F_1\, dx^* + F_2\, dy^* + F_3\, dz^*) + \int_{B_1}^B (F_1\, dx^* + F_2\, dy^* + F_3\, dz^*).$$

We now take the partial derivative with respect to x on both sides. On the left we get $\partial f/\partial x$. We show that on the right we get F_1. The derivative of the first integral is zero because A: (x_0, y_0, z_0) and B_1: (x_1, y, z) do not depend on x. We consider the second integral. Since on the segment B_1B, both y and z are constant, the terms $F_2\, dy^*$ and $F_3\, dz^*$ do not contribute to the derivative of the integral. The remaining part can be written as a definite integral,

$$\int_{B_1}^B F_1\, dx^* = \int_{x_1}^x F_1(x^*, y, z)\, dx^*.$$

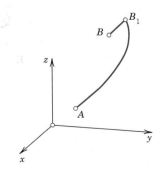

Fig. 526. Proof of Theorem 1

Hence its partial derivative with respect to x is $F_1(x, y, z)$, and the first of the relations $(2')$ is proved. The other two formulas in $(2')$ follow by the same argument. ◀

SECTION 12.4, PAGE 671

PROOF OF THEOREM 2 (Cauchy–Riemann equations)

We prove that the Cauchy–Riemann equations

$$(1) \qquad\qquad u_x = v_y, \qquad\qquad u_y = -v_x$$

are sufficient for a complex function $f(z) = u(x, y) + iv(x, y)$ to be analytic; precisely, *if the real part u and the imaginary part v of f(z) satisfy* (1) *in a domain D in the complex plane and if the partial derivatives in* (1) *are* **continuous** *in D, then f(z) is analytic in D.*

In this proof we write $\Delta z = \Delta x + i\Delta y$ and $\Delta f = f(z + \Delta z) - f(z)$. The idea of proof is as follows.

(a) We express Δf in terms of first partial derivatives of u and v, by applying the mean value theorem of Sec. 8.8.

(b) We get rid of partial derivatives with respect to y by applying the Cauchy–Riemann equations.

(c) We let Δz approach zero and show that then $\Delta f/\Delta z$ as obtained approaches a limit, which is equal to $u_x + iv_x$, the right side of (4) in Sec. 12.4, regardless of the way of approach to zero.

(a) Let $P\colon (x, y)$ be any fixed point in D. Since D is a domain, it contains a neighborhood of P. We can choose a point $Q\colon (x + \Delta x, y + \Delta y)$ in this neighborhood such that the straight-line segment PQ is in D. Because of our continuity assumptions we may apply the mean value theorem in Sec. 8.8. This yields

$$u(x + \Delta x, y + \Delta y) - u(x, y) = (\Delta x)u_x(M_1) + (\Delta y)u_y(M_1)$$

$$v(x + \Delta x, y + \Delta y) - v(x, y) = (\Delta x)v_x(M_2) + (\Delta y)v_y(M_2).$$

where M_1 and M_2 ($\neq M_1$ in general!) are suitable points on that segment. The first line is Re Δf and the second is Im Δf, so that

$$\Delta f = (\Delta x)u_x(M_1) + (\Delta y)u_y(M_1) + i[(\Delta x)v_x(M_2) + (\Delta y)v_y(M_2)].$$

(b) $u_y = -v_x$ and $v_y = u_x$ by the Cauchy–Riemann equations, so that

$$\Delta f = (\Delta x)u_x(M_1) - (\Delta y)v_x(M_1) + i[(\Delta x)v_x(M_2) + (\Delta y)u_x(M_2)].$$

Also $\Delta z = \Delta x + i\Delta y$, so that we can write $\Delta x = \Delta z - i\Delta y$ in the first term and $\Delta y = (\Delta z - \Delta x)/i = -i(\Delta z - \Delta x)$ in the second term. This gives

$$\Delta f = (\Delta z - i\Delta y)u_x(M_1) + i(\Delta z - \Delta x)v_x(M_1) + i[(\Delta x)v_x(M_2) + (\Delta y)u_x(M_2)].$$

By performing the multiplications and reordering we obtain

$$\Delta f = (\Delta z)u_x(M_1) - i\Delta y\{u_x(M_1) - u_x(M_2)\}$$

$$+ i[(\Delta z)v_x(M_1) - \Delta x\{v_x(M_1) - v_x(M_2)\}].$$

Division by Δz now yields

(A) $\quad \dfrac{\Delta f}{\Delta z} = u_x(M_1) + iv_x(M_1) - \dfrac{i\Delta y}{\Delta z}\{u_x(M_1) - u_x(M_2)\} - \dfrac{i\Delta x}{\Delta z}\{v_x(M_1) - v_x(M_2)\}.$

(c) We finally let Δz approach zero and note that $|\Delta y/\Delta z| \leqq 1$ and $|\Delta x/\Delta z| \leqq 1$ in (A). Then $Q\colon (x + \Delta x, y + \Delta y)$ approaches $P\colon (x, y)$, so that M_1 and M_2 must approach P. Also, since the partial derivatives in (A) are assumed to be continuous, they approach their value at P. In particular, the differences in the braces $\{\cdots\}$ in (A) approach zero. Hence the limit of the right side of (A) exists and is independent of the path along which $\Delta z \to 0$. We see that this limit equals the right side of (4) in Sec. 12.4. This means that $f(z)$ is analytic at every point z in D, and the proof is complete.   ◀

SECTION 13.2, PAGES 714 AND 715

GOURSAT'S PROOF OF CAUCHY'S INTEGRAL THEOREM. Goursat proved Cauchy's integral theorem without assuming that $f'(z)$ is continuous, as follows.

We start with the case when C is the boundary of a triangle. We orient C counterclockwise. By joining the midpoints of the sides we subdivide the triangle into four congruent triangles (Fig. 527). Let C_I, C_II, C_III, C_IV denote their boundaries. We claim that (see Fig. 527).

(1) $$\oint_C f\,dz = \oint_{C_\mathrm{I}} f\,dz + \oint_{C_\mathrm{II}} f\,dz + \oint_{C_\mathrm{III}} f\,dz + \oint_{C_\mathrm{IV}} f\,dz.$$

Indeed, on the right we integrate along each of the three segments of subdivision in both possible directions (Fig. 527), so that the corresponding integrals cancel out in pairs, and the sum of the integrals on the right equals the integral on the left. We now pick an integral on the right that is biggest in absolute value and call its path C_1. Then, by the triangle inequality (Sec. 12.2),

$$\left|\oint_C f\,dz\right| \leqq \left|\oint_{C_\mathrm{I}} f\,dz\right| + \left|\oint_{C_\mathrm{II}} f\,dz\right| + \left|\oint_{C_\mathrm{III}} f\,dz\right| + \left|\oint_{C_\mathrm{IV}} f\,dz\right| \leqq 4\left|\oint_{C_1} f\,dz\right|.$$

We now subdivide the triangle bounded by C_1 as before and select a triangle of subdivision with boundary C_2 for which

$$\left|\oint_{C_1} f\,dz\right| \leqq 4\left|\oint_{C_2} f\,dz\right|. \qquad \text{Then} \qquad \left|\oint_C f\,dz\right| \leqq 4^2\left|\oint_{C_2} f\,dz\right|.$$

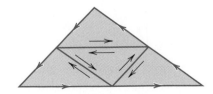

Fig. 527.   Proof of Cauchy's integral theorem

Continuing in this fashion, we obtain a sequence of triangles $T_1, T_2, \cdots$ with boundaries $C_1, C_2, \cdots$ that are similar and such that T_n lies in T_m when $n > m$, and

$$\text{(2)} \qquad \left| \oint_C f \, dz \right| \leq 4^n \left| \oint_{C_n} f \, dz \right|, \qquad n = 1, 2, \cdots.$$

Let z_0 be the point that belongs to all these triangles. Since f is differentiable at $z = z_0$, the derivative $f'(z_0)$ exists. Let

$$\text{(3)} \qquad h(z) = \frac{f(z) - f(z_0)}{z - z_0} - f'(z_0).$$

Solving this algebraically for $f(z)$ we have

$$f(z) = f(z_0) + (z - z_0)f'(z_0) + h(z)(z - z_0).$$

Integrating this over the boundary C_n of the triangle T_n gives

$$\oint_{C_n} f(z) \, dz = \oint_{C_n} f(z_0) \, dz + \oint_{C_n} (z - z_0)f'(z_0) \, dz + \oint_{C_n} h(z)(z - z_0) \, dz.$$

Since $f(z_0)$ and $f'(z_0)$ are constants and C_n is a closed path, the first two integrals on the right are zero, as follows from Cauchy's proof, which is applicable because the integrands do have continuous derivatives (0 and *const,* respectively). We thus have

$$\oint_{C_n} f(z) \, dz = \oint_{C_n} h(z)(z - z_0) \, dz.$$

Since $f'(z_0)$ is the limit of the difference quotient in (3), for given $\epsilon > 0$ we can find a $\delta > 0$ such that

$$\text{(4)} \qquad |h(z)| < \epsilon \qquad \text{when} \qquad |z - z_0| < \delta.$$

We may now take n so large that the triangle T_n lies in the disk $|z - z_0| < \delta$. Let L_n be the length of C_n. Then $|z - z_0| < L_n$ for all z on C_n and z_0 in T_n. From this and (4) we have $|h(z)(z - z_0)| < \epsilon L_n$. The *ML*-inequality in Sec. 13.1 now gives

$$\text{(5)} \qquad \left| \oint_{C_n} f(z) \, dz \right| = \left| \oint_{C_n} h(z)(z - z_0) \, dz \right| \leq \epsilon L_n \cdot L_n = \epsilon L_n{}^2.$$

Now let L be the length of C. Then the path C_1 has the length $L_1 = L/2$, the path C_2 has the length $L_2 = L_1/2 = L/4$, etc., and C_n has the length $L_n = L/2^n$. Hence $L_n{}^2 = L^2/4^n$. From (2) and (5) we thus obtain

$$\left| \oint_C f \, dz \right| \leq 4^n \left| \oint_{C_n} f \, dz \right| \leq 4^n \epsilon L_n{}^2 = 4^n \epsilon \frac{L^2}{4^n} = \epsilon L^2.$$

By choosing $\epsilon \, (> 0)$ sufficiently small we can make the expression on the right as small as we please, while the expression on the left is the definite value of an integral.

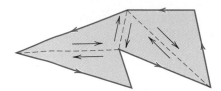

Fig. 528.   Proof of Cauchy's integral theorem for a polygon

Consequently, this value must be zero, and the proof is complete.

The proof for *the case in which C is the boundary of a polygon* follows from the previous proof by subdividing the polygon into triangles (Fig. 528). The integral corresponding to each such triangle is zero. The sum of these integrals is equal to the integral over C, because we integrate along each segment of subdivision in both directions, the corresponding integrals cancel out in pairs, and we are left with the integral over C.

The case of a general simple closed path C can be reduced to the preceding one by inscribing in C a closed polygon P of chords, which approximates C "sufficiently accurately," and it can be shown that there is a polygon P such that the integral over P differs from that over C by less than any preassigned positive real number $\tilde{\epsilon}$, no matter how small. The details of this proof are somewhat involved and can be found in Ref. [D7] listed in Appendix 1.   ◀

SECTION 14.1, PAGE 735

PROOF OF THEOREM 4 (Cauchy's convergence principle for series)

(a) In this proof we need two concepts and a theorem, which we list first.

1. A **bounded sequence** $s_1, s_2, \cdots$ is a sequence whose terms all lie in a disk of (sufficiently large, finite) radius K with center at the origin; thus $|s_n| < K$ for all n.

2. A **limit point** a of a sequence $s_1, s_2, \cdots$ is a point such that, given an $\epsilon > 0$, there are infinitely many terms satisfying $|s_n - a| < \epsilon$. (Note that this does *not* imply convergence, since there may still be infinitely many terms that do not lie within that circle of radius ϵ and center a.)

Example: $\frac{1}{4}, \frac{3}{4}, \frac{1}{8}, \frac{7}{8}, \frac{1}{16}, \frac{15}{16}, \cdots$ has the limit points 0 and 1 and diverges.

3. A bounded sequence in the complex plane has at least one limit point. (Bolzano–Weierstrass theorem; proof below. Recall that "sequence" always means *infinite* sequence.)

(b) We now turn to the actual proof that $z_1 + z_2 + \cdots$ converges if and only if for every $\epsilon > 0$ we can find an N such that

$$(1) \qquad |z_{n+1} + \cdots + z_{n+p}| < \epsilon \qquad \text{for every } n > N \text{ and } p = 1, 2, \cdots.$$

Here, by the definition of partial sums,

$$s_{n+p} - s_n = z_{n+1} + \cdots + z_{n+p}.$$

Writing $n + p = r$, we see from this that (1) is equivalent to

$$(1^*) \qquad |s_r - s_n| < \epsilon \qquad \text{for all } r > N \text{ and } n > N.$$

Suppose that $s_1, s_2, \cdots$ converges. Denote its limit by s. Then for a given $\epsilon > 0$ we can find an N such that

$$|s_n - s| < \frac{\epsilon}{2} \qquad \text{for every } n > N.$$

Hence, if $r > N$ and $n > N$, then by the triangle inequality (Sec. 12.2),

$$|s_r - s_n| = |(s_r - s) - (s_n - s)| \leqq |s_r - s| + |s_n - s| < \frac{\epsilon}{2} + \frac{\epsilon}{2} = \epsilon,$$

that is, (1*) holds.

(c) Conversely, assume that $s_1, s_2, \cdots$ satisfies (1*). We first prove that then the sequence must be bounded. Indeed, choose a fixed ϵ and a fixed $n = n_0 > N$ in (1*). Then (1*) implies that all s_r with $r > N$ lie in the disk of radius ϵ and center s_{n_0} and only *finitely many* terms $s_1, \cdots, s_N$ may not lie in this disk. Clearly, we can now find a circle so large that this disk and these finitely many terms all lie within this new circle. Hence the sequence is bounded. By the Bolzano–Weierstrass theorem, it has at least one limit point, call it s.

We now show that the sequence is convergent with the limit s. Let $\epsilon > 0$ be given. Then there is an N^* such that $|s_r - s_n| < \epsilon/2$ for all $r > N^*$ and $n > N^*$, by (1*). Also, by the definition of a limit point, $|s_n - s| < \epsilon/2$ for *infinitely many* n, so that we can find and fix an $n > N^*$ such that $|s_n - s| < \epsilon/2$. Together, for *every* $r > N^*$,

$$|s_r - s| = |(s_r - s_n) + (s_n - s)| \leqq |s_r - s_n| + |s_n - s| < \frac{\epsilon}{2} + \frac{\epsilon}{2} = \epsilon;$$

that is, the sequence $s_1, s_2, \cdots$ is convergent with the limit s. ◀

BOLZANO–WEIERSTRASS THEOREM[3]

A bounded infinite sequence $z_1, z_2, z_3, \cdots$ in the complex plane has at least one limit point.

PROOF.   It is obvious that we need both conditions: a finite sequence cannot have a limit point, and the sequence $1, 2, 3, \cdots$, which is infinite but not bounded, has no limit point. To prove the theorem, consider a bounded infinite sequence $z_1, z_2, \cdots$ and let K be such that $|z_n| < K$ for all n. If only finitely many values of the z_n are different, then, since the sequence is infinite, some number z must occur infinitely many times in the sequence, and, by definition, this number is a limit point of the sequence.

We may now turn to the case when the sequence contains infinitely many *different* terms. We draw a large square Q_0 that contains all z_n. We subdivide Q_0 into four congruent squares, which we number 1, 2, 3, 4. Clearly, at least one of these squares (each taken with its complete boundary) must contain infinitely many terms of the sequence. The square of this type with the lowest number (1, 2, 3, or 4) will be denoted by Q_1. This is the first step. In the next step we subdivide Q_1 into four congruent squares and select a square Q_2 by the same rule, and so on. This yields an infinite sequence of squares

---

[3]BERNARD BOLZANO (1781—1848), Austrian mathematician and professor of religious studies, was a pioneer in the study of point sets, the foundations of analysis, and mathematical logic.
For Weierstrass, see footnote 8 in Sec. 14.5.

$Q_0, Q_1, Q_2, \cdots, Q_n, \cdots$ with the property that the side of Q_n approaches zero as n approaches infinity, and Q_m contains all Q_n with $n > m$. It is not difficult to see that the number which belongs to all these squares,[4] call it $z = a$, is a limit point of the sequence. In fact, given an $\epsilon > 0$, we can choose an N so large that the side of the square Q_N is less than ϵ and, since Q_N contains infinitely many z_n, we have $|z_n - a| < \epsilon$ for infinitely many n. This completes the proof.   ◀

SECTION 14.3, page 749

PART (b) OF THE PROOF OF THEOREM 5

We have to show that

$$\sum_{n=2}^{\infty} a_n \left[\frac{(z + \Delta z)^n - z^n}{\Delta z} - nz^{n-1} \right]$$

$$= \sum_{n=2}^{\infty} a_n \, \Delta z [(z + \Delta z)^{n-2} + 2z(z + \Delta z)^{n-3} + \cdots + (n - 1)z^{n-2}],$$

thus,

$$\frac{(z + \Delta z)^n - z^n}{\Delta z} - nz^{n-1}$$

$$= \Delta z [(z + \Delta z)^{n-2} + 2z(z + \Delta z)^{n-3} + \cdots + (n - 1)z^{n-2}].$$

If we set $z + \Delta z = b$ and $z = a$, thus $\Delta z = b - a$, this becomes simply

(7a) $$\frac{b^n - a^n}{b - a} - na^{n-1} = (b - a)A_n \qquad (n = 2, 3, \cdots),$$

where A_n is the expression in the brackets on the right,

(7b) $$A_n = b^{n-2} + 2ab^{n-3} + 3a^2b^{n-4} + \cdots + (n - 1)a^{n-2};$$

thus, $A_2 = 1$, $A_3 = b + 2a$, etc. We prove (7) by induction. When $n = 2$, then (7) holds, since then

$$\frac{b^2 - a^2}{b - a} - 2a = \frac{(b + a)(b - a)}{b - a} - 2a = b - a = (b - a)A_2.$$

Assuming that (7) holds for $n = k$, we show that it holds for $n = k + 1$. By adding and subtracting a term in the numerator and then dividing we first obtain

$$\frac{b^{k+1} - a^{k+1}}{b - a} = \frac{b^{k+1} - ba^k + ba^k - a^{k+1}}{b - a} = b \frac{b^k - a^k}{b - a} + a^k.$$

---

[4]The fact that such a unique number $z = a$ exists seems to be obvious, but it actually follows from an axiom of the real number system, the so-called *Cantor–Dedekind axiom;* see footnote 3 in Appendix A3.3.

By the induction hypothesis, the right side equals $b[(b - a)A_k + ka^{k-1}] + a^k$. Direct calculation shows that this is equal to

$$(b - a)\{bA_k + ka^{k-1}\} + aka^{k-1} + a^k.$$

From (7b) with $n = k$ we see that the expression in the braces $\{\cdots\}$ equals

$$b^{k-1} + 2ab^{k-2} + \cdots + (k - 1)ba^{k-2} + ka^{k-1} = A_{k+1}.$$

Hence our result is

$$\frac{b^{k+1} - a^{k+1}}{b - a} = (b - a)A_{k+1} + (k + 1)a^k.$$

Taking the last term to the left, we obtain (7) with $n = k + 1$. This proves (7) for any integer $n \geqq 2$ and completes the proof. ◄

SECTION 16.2, PAGE 804

ANOTHER PROOF OF THEOREM 1, *without the use of a harmonic conjugate*

We show that if $w = u + iv = f(z)$ is analytic and maps a domain D conformally onto a domain D^* and $\Phi^*(u, v)$ is harmonic in D^*, then

(1) $$\Phi(x, y) = \Phi^*(u(x, y), v(x, y))$$

is harmonic in D, that is, $\nabla^2 \Phi = 0$ in D. We make no use of a harmonic conjugate of Φ^*, but use straightforward differentiation. By the chain rule,

$$\Phi_x = \Phi_u^* u_x + \Phi_v^* v_x.$$

We apply the chain rule again, underscoring the terms that will drop out when we form $\nabla^2 \Phi$:

$$\Phi_{xx} = \underline{\Phi_u^* u_{xx}} + (\Phi_{uu}^* u_x + \underline{\Phi_{uv}^* v_x}) u_x$$

$$+ \underline{\Phi_v^* v_{xx}} + (\underline{\Phi_{vu}^* u_x} + \Phi_{vv}^* v_x) v_x.$$

Φ_{yy} is the same with each x replaced by y. We form the sum $\nabla^2 \Phi$. In it, $\Phi_{vu}^* = \Phi_{uv}^*$ is multiplied by

$$u_x v_x + u_y v_y,$$

which is 0 by the Cauchy–Riemann equations. Also $\nabla^2 u = 0$ and $\nabla^2 v = 0$. There remains

$$\nabla^2 \Phi = \Phi_{uu}^* (u_x^2 + u_y^2) + \Phi_{vv}^* (v_x^2 + v_y^2).$$

By the Cauchy–Riemann equation this becomes

$$\nabla^2 \Phi = (\Phi_{uu}^* + \Phi_{vv}^*)(u_x^2 + v_x^2)$$

and is 0 since Φ^* is harmonic. ◄

Tables

Tables of Laplace transforms in Secs. 5.8 and 5.9
Tables of Fourier transforms in Sec. 10.11

Table A1
Bessel Functions

For more extensive tables see Ref. [1] in Appendix 1.

x	$J_0(x)$	$J_1(x)$	x	$J_0(x)$	$J_1(x)$	x	$J_0(x)$	$J_1(x)$
0.0	1.0000	0.0000	3.0	−0.2601	0.3391	6.0	0.1506	−0.2767
0.1	0.9975	0.0499	3.1	−0.2921	0.3009	6.1	0.1773	−0.2559
0.2	0.9900	0.0995	3.2	−0.3202	0.2613	6.2	0.2017	−0.2329
0.3	0.9776	0.1483	3.3	−0.3443	0.2207	6.3	0.2238	−0.2081
0.4	0.9604	0.1960	3.4	−0.3643	0.1792	6.4	0.2433	−0.1816
0.5	0.9385	0.2423	3.5	−0.3801	0.1374	6.5	0.2601	−0.1538
0.6	0.9120	0.2867	3.6	−0.3918	0.0955	6.6	0.2740	−0.1250
0.7	0.8812	0.3290	3.7	−0.3992	0.0538	6.7	0.2851	−0.0953
0.8	0.8463	0.3688	3.8	−0.4026	0.0128	6.8	0.2931	−0.0652
0.9	0.8075	0.4059	3.9	−0.4018	−0.0272	6.9	0.2981	−0.0349
1.0	0.7652	0.4401	4.0	−0.3971	−0.0660	7.0	0.3001	−0.0047
1.1	0.7196	0.4709	4.1	−0.3887	−0.1033	7.1	0.2991	0.0252
1.2	0.6711	0.4983	4.2	−0.3766	−0.1386	7.2	0.2951	0.0543
1.3	0.6201	0.5220	4.3	−0.3610	−0.1719	7.3	0.2882	0.0826
1.4	0.5669	0.5419	4.4	−0.3423	−0.2028	7.4	0.2786	0.1096
1.5	0.5118	0.5579	4.5	−0.3205	−0.2311	7.5	0.2663	0.1352
1.6	0.4554	0.5699	4.6	−0.2961	−0.2566	7.6	0.2516	0.1592
1.7	0.3980	0.5778	4.7	−0.2693	−0.2791	7.7	0.2346	0.1813
1.8	0.3400	0.5815	4.8	−0.2404	−0.2985	7.8	0.2154	0.2014
1.9	0.2818	0.5812	4.9	−0.2097	−0.3147	7.9	0.1944	0.2192
2.0	0.2239	0.5767	5.0	−0.1776	−0.3276	8.0	0.1717	0.2346
2.1	0.1666	0.5683	5.1	−0.1443	−0.3371	8.1	0.1475	0.2476
2.2	0.1104	0.5560	5.2	−0.1103	−0.3432	8.2	0.1222	0.2580
2.3	0.0555	0.5399	5.3	−0.0758	−0.3460	8.3	0.0960	0.2657
2.4	0.0025	0.5202	5.4	−0.0412	−0.3453	8.4	0.0692	0.2708
2.5	−0.0484	0.4971	5.5	−0.0068	−0.3414	8.5	0.0419	0.2731
2.6	−0.0968	0.4708	5.6	0.0270	−0.3343	8.6	0.0146	0.2728
2.7	−0.1424	0.4416	5.7	0.0599	−0.3241	8.7	−0.0125	0.2697
2.8	−0.1850	0.4097	5.8	0.0917	−0.3110	8.8	−0.0392	0.2641
2.9	−0.2243	0.3754	5.9	0.1220	−0.2951	8.9	−0.0653	0.2559

$J_0(x) = 0$ for $x = 2.405, 5.520, 8.654, 11.792, 14.931, \cdots$
$J_1(x) = 0$ for $x = 0, 3.832, 7.016, 10.173, 13.324, \cdots$

Table A1 *(continued)*

x	$Y_0(x)$	$Y_1(x)$	x	$Y_0(x)$	$Y_1(x)$	x	$Y_0(x)$	$Y_1(x)$
0.0	$(-\infty)$	$(-\infty)$	2.5	0.498	0.146	5.0	−0.309	0.148
0.5	−0.445	−1.471	3.0	0.377	0.325	5.5	−0.339	−0.024
1.0	0.088	−0.781	3.5	0.189	0.410	6.0	−0.288	−0.175
1.5	0.382	−0.412	4.0	−0.017	0.398	6.5	−0.173	−0.274
2.0	0.510	−0.107	4.5	−0.195	0.301	7.0	−0.026	−0.303

Table A2
Gamma Function [see (24) in Appendix A3.1]

α	$\Gamma(\alpha)$	α	$\Gamma(\alpha)$	α	$\Gamma(\alpha)$	α	$\Gamma(\alpha)$	α	$\Gamma(\alpha)$
1.00	1.000 000	1.20	0.918 169	1.40	0.887 264	1.60	0.893 515	1.80	0.931 384
1.02	0.988 844	1.22	0.913 106	1.42	0.886 356	1.62	0.895 924	1.82	0.936 845
1.04	0.978 438	1.24	0.908 521	1.44	0.885 805	1.64	0.898 642	1.84	0.942 612
1.06	0.968 744	1.26	0.904 397	1.46	0.885 604	1.66	0.901 668	1.86	0.948 687
1.08	0.959 725	1.28	0.900 718	1.48	0.885 747	1.68	0.905 001	1.88	0.955 071
1.10	0.951 351	1.30	0.897 471	1.50	0.886 227	1.70	0.908 639	1.90	0.961 766
1.12	0.943 590	1.32	0.894 640	1.52	0.887 039	1.72	0.912 581	1.92	0.968 774
1.14	0.936 416	1.34	0.892 216	1.54	0.888 178	1.74	0.916 826	1.94	0.976 099
1.16	0.929 803	1.36	0.890 185	1.56	0.889 639	1.76	0.921 375	1.96	0.983 743
1.18	0.923 728	1.38	0.888 537	1.58	0.891 420	1.78	0.926 227	1.98	0.991 708
1.20	0.918 169	1.40	0.887 264	1.60	0.893 515	1.80	0.931 384	2.00	1.000 000

Table A3
Factorial Function

n	n!	log (n!)	n	n!	log (n!)	n	n!	log (n!)
1	1	0.000 000	6	720	2.857 332	11	39 916 800	7.601 156
2	2	0.301 030	7	5 040	3.702 431	12	479 001 600	8.680 337
3	6	0.778 151	8	40 320	4.605 521	13	6 227 020 800	9.794 280
4	24	1.380 211	9	362 880	5.559 763	14	87 178 291 200	10.940 408
5	120	2.079 181	10	3 628 800	6.559 763	15	1 307 674 368 000	12.116 500

Table A4
Error Function, Sine and Cosine Integrals [see (35), (40), (42) in Appendix A3.1]

x	erf x	Si(x)	ci(x)	x	erf x	Si(x)	ci(x)
0.0	0.0000	0.0000	∞	2.0	0.9953	1.6054	−0.4230
0.2	0.2227	0.1996	1.0422	2.2	0.9981	1.6876	−0.3751
0.4	0.4284	0.3965	0.3788	2.4	0.9993	1.7525	−0.3173
0.6	0.6039	0.5881	0.0223	2.6	0.9998	1.8004	−0.2533
0.8	0.7421	0.7721	−0.1983	2.8	0.9999	1.8321	−0.1865
1.0	0.8427	0.9461	−0.3374	3.0	1.0000	1.8487	−0.1196
1.2	0.9103	1.1080	−0.4205	3.2	1.0000	1.8514	−0.0553
1.4	0.9523	1.2562	−0.4620	3.4	1.0000	1.8419	0.0045
1.6	0.9763	1.3892	−0.4717	3.6	1.0000	1.8219	0.0580
1.8	0.9891	1.5058	−0.4568	3.8	1.0000	1.7934	0.1038
2.0	0.9953	1.6054	−0.4230	4.0	1.0000	1.7582	0.1410

Table A5
Binomial Distribution

Probability function $f(x)$ [see (2), Sec. 22.7] and distribution function $F(x)$

n	x	$p = 0.1$ $f(x)$	$F(x)$	$p = 0.2$ $f(x)$	$F(x)$	$p = 0.3$ $f(x)$	$F(x)$	$p = 0.4$ $f(x)$	$F(x)$	$p = 0.5$ $f(x)$	$F(x)$
		0.		0.		0.		0.		0.	
1	0	9000	0.9000	8000	0.8000	7000	0.7000	6000	0.6000	5000	0.5000
	1	1000	1.0000	2000	1.0000	3000	1.0000	4000	1.0000	5000	1.0000
2	0	8100	0.8100	6400	0.6400	4900	0.4900	3600	0.3600	2500	0.2500
	1	1800	0.9900	3200	0.9600	4200	0.9100	4800	0.8400	5000	0.7500
	2	0100	1.0000	0400	1.0000	0900	1.0000	1600	1.0000	2500	1.0000
3	0	7290	0.7290	5120	0.5120	3430	0.3430	2160	0.2160	1250	0.1250
	1	2430	0.9720	3840	0.8960	4410	0.7840	4320	0.6480	3750	0.5000
	2	0270	0.9990	0960	0.9920	1890	0.9730	2880	0.9360	3750	0.8750
	3	0010	1.0000	0080	1.0000	0270	1.0000	0640	1.0000	1250	1.0000
4	0	6561	0.6561	4096	0.4096	2401	0.2401	1296	0.1296	0625	0.0625
	1	2916	0.9477	4096	0.8192	4116	0.6517	3456	0.4752	2500	0.3125
	2	0486	0.9963	1536	0.9728	2646	0.9163	3456	0.8208	3750	0.6875
	3	0036	0.9999	0256	0.9984	0756	0.9919	1536	0.9744	2500	0.9375
	4	0001	1.0000	0016	1.0000	0081	1.0000	0256	1.0000	0625	1.0000
5	0	5905	0.5905	3277	0.3277	1681	0.1681	0778	0.0778	0313	0.0313
	1	3281	0.9185	4096	0.7373	3602	0.5282	2592	0.3370	1563	0.1875
	2	0729	0.9914	2048	0.9421	3087	0.8369	3456	0.6826	3125	0.5000
	3	0081	0.9995	0512	0.9933	1323	0.9692	2304	0.9130	3125	0.8125
	4	0005	1.0000	0064	0.9997	0284	0.9976	0768	0.9898	1563	0.9688
	5	0000	1.0000	0003	1.0000	0024	1.0000	0102	1.0000	0313	1.0000
6	0	5314	0.5314	2621	0.2621	1176	0.1176	0467	0.0467	0156	0.0156
	1	3543	0.8857	3932	0.6554	3025	0.4202	1866	0.2333	0938	0.1094
	2	0984	0.9841	2458	0.9011	3241	0.7443	3110	0.5443	2344	0.3438
	3	0146	0.9987	0819	0.9830	1852	0.9295	2765	0.8208	3125	0.6563
	4	0012	0.9999	0154	0.9984	0595	0.9891	1382	0.9590	2344	0.8906
	5	0001	1.0000	0015	0.9999	0102	0.9993	0369	0.9959	0938	0.9844
	6	0000	1.0000	0001	1.0000	0007	1.0000	0041	1.0000	0156	1.0000
7	0	4783	0.4783	2097	0.2097	0824	0.0824	0280	0.0280	0078	0.0078
	1	3720	0.8503	3670	0.5767	2471	0.3294	1306	0.1586	0547	0.0625
	2	1240	0.9743	2753	0.8520	3177	0.6471	2613	0.4199	1641	0.2266
	3	0230	0.9973	1147	0.9667	2269	0.8740	2903	0.7102	2734	0.5000
	4	0026	0.9998	0287	0.9953	0972	0.9712	1935	0.9037	2734	0.7734
	5	0002	1.0000	0043	0.9996	0250	0.9962	0774	0.9812	1641	0.9375
	6	0000	1.0000	0004	1.0000	0036	0.9998	0172	0.9984	0547	0.9922
	7	0000	1.0000	0000	1.0000	0002	1.0000	0016	1.0000	0078	1.0000
8	0	4305	0.4305	1678	0.1678	0576	0.0576	0168	0.0168	0039	0.0039
	1	3826	0.8131	3355	0.5033	1977	0.2553	0896	0.1064	0313	0.0352
	2	1488	0.9619	2936	0.7969	2965	0.5518	2090	0.3154	1094	0.1445
	3	0331	0.9950	1468	0.9437	2541	0.8059	2787	0.5941	2188	0.3633
	4	0046	0.9996	0459	0.9896	1361	0.9420	2322	0.8263	2734	0.6367
	5	0004	1.0000	0092	0.9988	0467	0.9887	1239	0.9502	2188	0.8555
	6	0000	1.0000	0011	0.9999	0100	0.9987	0413	0.9915	1094	0.9648
	7	0000	1.0000	0001	1.0000	0012	0.9999	0079	0.9993	0313	0.9961
	8	0000	1.0000	0000	1.0000	0001	1.0000	0007	1.0000	0039	1.0000

Table A6
Poisson Distribution

Probability function $f(x)$ [see (5), Sec. 22.7] and distribution function $F(x)$

x	$\mu = 0.1$ $f(x)$	$F(x)$	$\mu = 0.2$ $f(x)$	$F(x)$	$\mu = 0.3$ $f(x)$	$F(x)$	$\mu = 0.4$ $f(x)$	$F(x)$	$\mu = 0.5$ $f(x)$	$F(x)$
0	**0.** 9048	0.9048	**0.** 8187	0.8187	**0.** 7408	0.7408	**0.** 6703	0.6703	**0.** 6065	0.6065
1	0905	0.9953	1637	0.9825	2222	0.9631	2681	0.9384	3033	0.9098
2	0045	0.9998	0164	0.9989	0333	0.9964	0536	0.9921	0758	0.9856
3	0002	1.0000	0011	0.9999	0033	0.9997	0072	0.9992	0126	0.9982
4	0000	1.0000	0001	1.0000	0003	1.0000	0007	0.9999	0016	0.9998
5							0001	1.0000	0002	1.0000

x	$\mu = 0.6$ $f(x)$	$F(x)$	$\mu = 0.7$ $f(x)$	$F(x)$	$\mu = 0.8$ $f(x)$	$F(x)$	$\mu = 0.9$ $f(x)$	$F(x)$	$\mu = 1$ $f(x)$	$F(x)$
0	**0.** 5488	0.5488	**0.** 4966	0.4966	**0.** 4493	0.4493	**0.** 4066	0.4066	**0.** 3679	0.3679
1	3293	0.8781	3476	0.8442	3595	0.8088	3659	0.7725	3679	0.7358
2	0988	0.9769	1217	0.9659	1438	0.9526	1647	0.9371	1839	0.9197
3	0198	0.9966	0284	0.9942	0383	0.9909	0494	0.9865	0613	0.9810
4	0030	0.9996	0050	0.9992	0077	0.9986	0111	0.9977	0153	0.9963
5	0004	1.0000	0007	0.9999	0012	0.9998	0020	0.9997	0031	0.9994
6			0001	1.0000	0002	1.0000	0003	1.0000	0005	0.9999
7									0001	1.0000

x	$\mu = 1.5$ $f(x)$	$F(x)$	$\mu = 2$ $f(x)$	$F(x)$	$\mu = 3$ $f(x)$	$F(x)$	$\mu = 4$ $f(x)$	$F(x)$	$\mu = 5$ $f(x)$	$F(x)$
0	**0.** 2231	0.2231	**0.** 1353	0.1353	**0.** 0498	0.0498	**0.** 0183	0.0183	**0.** 0067	0.0067
1	3347	0.5578	2707	0.4060	1494	0.1991	0733	0.0916	0337	0.0404
2	2510	0.8088	2707	0.6767	2240	0.4232	1465	0.2381	0842	0.1247
3	1255	0.9344	1804	0.8571	2240	0.6472	1954	0.4335	1404	0.2650
4	0471	0.9814	0902	0.9473	1680	0.8153	1954	0.6288	1755	0.4405
5	0141	0.9955	0361	0.9834	1008	0.9161	1563	0.7851	1755	0.6160
6	0035	0.9991	0120	0.9955	0504	0.9665	1042	0.8893	1462	0.7622
7	0008	0.9998	0034	0.9989	0216	0.9881	0595	0.9489	1044	0.8666
8	0001	1.0000	0009	0.9998	0081	0.9962	0298	0.9786	0653	0.9319
9			0002	1.0000	0027	0.9989	0132	0.9919	0363	0.9682
10					0008	0.9997	0053	0.9972	0181	0.9863
11					0002	0.9999	0019	0.9991	0082	0.9945
12					0001	1.0000	0006	0.9997	0034	0.9980
13							0002	0.9999	0013	0.9993
14							0001	1.0000	0005	0.9998
15									0002	0.9999
16									0000	1.0000

Table A7
Normal Distribution

Values of the distribution function $\Phi(z)$ [see (3), Sec. 22.8]. $\Phi(-z) = 1 - \Phi(z)$

z	$\Phi(z)$	z	$\Phi(z)$	z	$\Phi(z)$	z	$\Phi(z)$	z	$\Phi(z)$	z	$\Phi(z)$
	0.		0.		0.		0.		0.		0.
0.01	5040	0.51	6950	1.01	8438	1.51	9345	2.01	9778	2.51	9940
0.02	5080	0.52	6985	1.02	8461	1.52	9357	2.02	9783	2.52	9941
0.03	5120	0.53	7019	1.03	8485	1.53	9370	2.03	9788	2.53	9943
0.04	5160	0.54	7054	1.04	8508	1.54	9382	2.04	9793	2.54	9945
0.05	5199	0.55	7088	1.05	8531	1.55	9394	2.05	9798	2.55	9946
0.06	5239	0.56	7123	1.06	8554	1.56	9406	2.06	9803	2.56	9948
0.07	5279	0.57	7157	1.07	8577	1.57	9418	2.07	9808	2.57	9949
0.08	5319	0.58	7190	1.08	8599	1.58	9429	2.08	9812	2.58	9951
0.09	5359	0.59	7224	1.09	8621	1.59	9441	2.09	9817	2.59	9952
0.10	5398	0.60	7257	1.10	8643	1.60	9452	2.10	9821	2.60	9953
0.11	5438	0.61	7291	1.11	8665	1.61	9463	2.11	9826	2.61	9955
0.12	5478	0.62	7324	1.12	8686	1.62	9474	2.12	9830	2.62	9956
0.13	5517	0.63	7357	1.13	8708	1.63	9484	2.13	9834	2.63	9957
0.14	5557	0.64	7389	1.14	8729	1.64	9495	2.14	9838	2.64	9959
0.15	5596	0.65	7422	1.15	8749	1.65	9505	2.15	9842	2.65	9960
0.16	5636	0.66	7454	1.16	8770	1.66	9515	2.16	9846	2.66	9961
0.17	5675	0.67	7486	1.17	8790	1.67	9525	2.17	9850	2.67	9962
0.18	5714	0.68	7517	1.18	8810	1.68	9535	2.18	9854	2.68	9963
0.19	5753	0.69	7549	1.19	8830	1.69	9545	2.19	9857	2.69	9964
0.20	5793	0.70	7580	1.20	8849	1.70	9554	2.20	9861	2.70	9965
0.21	5832	0.71	7611	1.21	8869	1.71	9564	2.21	9864	2.71	9966
0.22	5871	0.72	7642	1.22	8888	1.72	9573	2.22	9868	2.72	9967
0.23	5910	0.73	7673	1.23	8907	1.73	9582	2.23	9871	2.73	9968
0.24	5948	0.74	7704	1.24	8925	1.74	9591	2.24	9875	2.74	9969
0.25	5987	0.75	7734	1.25	8944	1.75	9599	2.25	9878	2.75	9970
0.26	6026	0.76	7764	1.26	8962	1.76	9608	2.26	9881	2.76	9971
0.27	6064	0.77	7794	1.27	8980	1.77	9616	2.27	9884	2.77	9972
0.28	6103	0.78	7823	1.28	8997	1.78	9625	2.28	9887	2.78	9973
0.29	6141	0.79	7852	1.29	9015	1.79	9633	2.29	9890	2.79	9974
0.30	6179	0.80	7881	1.30	9032	1.80	9641	2.30	9893	2.80	9974
0.31	6217	0.81	7910	1.31	9049	1.81	9649	2.31	9896	2.81	9975
0.32	6255	0.82	7939	1.32	9066	1.82	9656	2.32	9898	2.82	9976
0.33	6293	0.83	7967	1.33	9082	1.83	9664	2.33	9901	2.83	9977
0.34	6331	0.84	7995	1.34	9099	1.84	9671	2.34	9904	2.84	9977
0.35	6368	0.85	8023	1.35	9115	1.85	9678	2.35	9906	2.85	9978
0.36	6406	0.86	8051	1.36	9131	1.86	9686	2.36	9909	2.86	9979
0.37	6443	0.87	8078	1.37	9147	1.87	9693	2.37	9911	2.87	9979
0.38	6480	0.88	8106	1.38	9162	1.88	9699	2.38	9913	2.88	9980
0.39	6517	0.89	8133	1.39	9177	1.89	9706	2.39	9916	2.89	9981
0.40	6554	0.90	8159	1.40	9192	1.90	9713	2.40	9918	2.90	9981
0.41	6591	0.91	8186	1.41	9207	1.91	9719	2.41	9920	2.91	9982
0.42	6628	0.92	8212	1.42	9222	1.92	9726	2.42	9922	2.92	9982
0.43	6664	0.93	8238	1.43	9236	1.93	9732	2.43	9925	2.93	9983
0.44	6700	0.94	8264	1.44	9251	1.94	9738	2.44	9927	2.94	9984
0.45	6736	0.95	8289	1.45	9265	1.95	9744	2.45	9929	2.95	9984
0.46	6772	0.96	8315	1.46	9279	1.96	9750	2.46	9931	2.96	9985
0.47	6808	0.97	8340	1.47	9292	1.97	9756	2.47	9932	2.97	9985
0.48	6844	0.98	8365	1.48	9306	1.98	9761	2.48	9934	2.98	9986
0.49	6879	0.99	8389	1.49	9319	1.99	9767	2.49	9936	2.99	9986
0.50	6915	1.00	8413	1.50	9332	2.00	9772	2.50	9938	3.00	9987

Table A8
Normal Distribution

Values of z for given values of $\Phi(z)$ [see (3), Sec. 22.8] and $D(z) = \Phi(z) - \Phi(-z)$
Example: $z = 0.279$ if $\Phi(z) = 61\%$; $z = 0.860$ if $D(z) = 61\%$.

%	$z(\Phi)$	$z(D)$	%	$z(\Phi)$	$z(D)$	%	$z(\Phi)$	$z(D)$
1	-2.326	0.013	41	-0.228	0.539	81	0.878	1.311
2	-2.054	0.025	42	-0.202	0.553	82	0.915	1.341
3	-1.881	0.038	43	-0.176	0.568	83	0.954	1.372
4	-1.751	0.050	44	-0.151	0.583	84	0.994	1.405
5	-1.645	0.063	45	-0.126	0.598	85	1.036	1.440
6	-1.555	0.075	46	-0.100	0.613	86	1.080	1.476
7	-1.476	0.088	47	-0.075	0.628	87	1.126	1.514
8	-1.405	0.100	48	-0.050	0.643	88	1.175	1.555
9	-1.341	0.113	49	-0.025	0.659	89	1.227	1.598
10	-1.282	0.126	50	0.000	0.674	90	1.282	1.645
11	-1.227	0.138	51	0.025	0.690	91	1.341	1.695
12	-1.175	0.151	52	0.050	0.706	92	1.405	1.751
13	-1.126	0.164	53	0.075	0.722	93	1.476	1.812
14	-1.080	0.176	54	0.100	0.739	94	1.555	1.881
15	-1.036	0.189	55	0.126	0.755	95	1.645	1.960
16	-0.994	0.202	56	0.151	0.772	96	1.751	2.054
17	-0.954	0.215	57	0.176	0.789	97	1.881	2.170
18	-0.915	0.228	58	0.202	0.806	97.5	1.960	2.241
19	-0.878	0.240	59	0.228	0.824	98	2.054	2.326
20	-0.842	0.253	60	0.253	0.842	99	2.326	2.576
21	-0.806	0.266	61	0.279	0.860	99.1	2.366	2.612
22	-0.772	0.279	62	0.305	0.878	99.2	2.409	2.652
23	-0.739	0.292	63	0.332	0.896	99.3	2.457	2.697
24	-0.706	0.305	64	0.358	0.915	99.4	2.512	2.748
25	-0.674	0.319	65	0.385	0.935	99.5	2.576	2.807
26	-0.643	0.332	66	0.412	0.954	99.6	2.652	2.878
27	-0.613	0.345	67	0.440	0.974	99.7	2.748	2.968
28	-0.583	0.358	68	0.468	0.994	99.8	2.878	3.090
29	-0.553	0.372	69	0.496	1.015	99.9	3.090	3.291
30	-0.524	0.385	70	0.524	1.036			
31	-0.496	0.399	71	0.553	1.058	99.91	3.121	3.320
32	-0.468	0.412	72	0.583	1.080	99.92	3.156	3.353
33	-0.440	0.426	73	0.613	1.103	99.93	3.195	3.390
34	-0.412	0.440	74	0.643	1.126	99.94	3.239	3.432
35	-0.385	0.454	75	0.674	1.150	99.95	3.291	3.481
36	-0.358	0.468	76	0.706	1.175	99.96	3.353	3.540
37	-0.332	0.482	77	0.739	1.200	99.97	3.432	3.615
38	-0.305	0.496	78	0.772	1.227	99.98	3.540	3.719
39	-0.279	0.510	79	0.806	1.254	99.99	3.719	3.891
40	-0.253	0.524	80	0.842	1.282			

Table A9
t-Distribution

Values of z for given values of the distribution function $F(z)$ (see p. 1112)
Example: For 9 degrees of freedom, $z = 1.83$ when $F(z) = 0.95$.

$F(z)$	Number of Degrees of Freedom									
	1	2	3	4	5	6	7	8	9	10
0.5	0.00	0.00	0.00	0.00	0.00	0.00	0.00	0.00	0.00	0.00
0.6	0.32	0.29	0.28	0.27	0.27	0.26	0.26	0.26	0.26	0.26
0.7	0.73	0.62	0.58	0.57	0.56	0.55	0.55	0.55	0.54	0.54
0.8	1.38	1.06	0.98	0.94	0.92	0.91	0.90	0.89	0.88	0.88
0.9	3.08	1.89	1.64	1.53	1.48	1.44	1.41	1.40	1.38	1.37
0.95	6.31	2.92	2.35	2.13	2.02	1.94	1.89	1.86	1.83	1.81
0.975	12.7	4.30	3.18	2.78	2.57	2.45	2.36	2.31	2.26	2.23
0.99	31.8	6.96	4.54	3.75	3.36	3.14	3.00	2.90	2.82	2.76
0.995	63.7	9.92	5.84	4.60	4.03	3.71	3.50	3.36	3.25	3.17
0.999	318.3	22.3	10.2	7.17	5.89	5.21	4.79	4.50	4.30	4.14

$F(z)$	Number of Degrees of Freedom									
	11	12	13	14	15	16	17	18	19	20
0.5	0.00	0.00	0.00	0.00	0.00	0.00	0.00	0.00	0.00	0.00
0.6	0.26	0.26	0.26	0.26	0.26	0.26	0.26	0.26	0.26	0.26
0.7	0.54	0.54	0.54	0.54	0.54	0.54	0.53	0.53	0.53	0.53
0.8	0.88	0.87	0.87	0.87	0.87	0.86	0.86	0.86	0.86	0.86
0.9	1.36	1.36	1.35	1.35	1.34	1.34	1.33	1.33	1.33	1.33
0.95	1.80	1.78	1.77	1.76	1.75	1.75	1.74	1.73	1.73	1.72
0.975	2.20	2.18	2.16	2.14	2.13	2.12	2.11	2.10	2.09	2.09
0.99	2.72	2.68	2.65	2.62	2.60	2.58	2.57	2.55	2.54	2.53
0.995	3.11	3.05	3.01	2.98	2.95	2.92	2.90	2.88	2.86	2.85
0.999	4.03	3.93	3.85	3.79	3.73	3.69	3.65	3.61	3.58	3.55

$F(z)$	Number of Degrees of Freedom									
	22	24	26	28	30	40	50	100	200	∞
0.5	0.00	0.00	0.00	0.00	0.00	0.00	0.00	0.00	0.00	0.00
0.6	0.26	0.26	0.26	0.26	0.26	0.26	0.25	0.25	0.25	0.25
0.7	0.53	0.53	0.53	0.53	0.53	0.53	0.53	0.53	0.53	0.52
0.8	0.86	0.86	0.86	0.85	0.85	0.85	0.85	0.85	0.84	0.84
0.9	1.32	1.32	1.31	1.31	1.31	1.30	1.30	1.29	1.29	1.28
0.95	1.72	1.71	1.71	1.70	1.70	1.68	1.68	1.66	1.65	1.65
0.975	2.07	2.06	2.06	2.05	2.04	2.02	2.01	1.98	1.97	1.96
0.99	2.51	2.49	2.48	2.47	2.46	2.42	2.40	2.36	2.35	2.33
0.995	2.82	2.80	2.78	2.76	2.75	2.70	2.68	2.63	2.60	2.58
0.999	3.50	3.47	3.43	3.41	3.39	3.31	3.26	3.17	3.13	3.09

Table A10
Chi-square Distribution

Values of x for given values of the distribution function $F(z)$ (see p. 1115)
Example: For 3 degrees of freedom, $z = 11.34$ when $F(z) = 0.99$.

$F(z)$	Number of Degrees of Freedom									
	1	2	3	4	5	6	7	8	9	10
0.005	0.00	0.01	0.07	0.21	0.41	0.68	0.99	1.34	1.73	2.16
0.01	0.00	0.02	0.11	0.30	0.55	0.87	1.24	1.65	2.09	2.56
0.025	0.00	0.05	0.22	0.48	0.83	1.24	1.69	2.18	2.70	3.25
0.05	0.00	0.10	0.35	0.71	1.15	1.64	2.17	2.73	3.33	3.94
0.95	3.84	5.99	7.81	9.49	11.07	12.59	14.07	15.51	16.92	18.31
0.975	5.02	7.38	9.35	11.14	12.83	14.45	16.01	17.53	19.02	20.48
0.99	6.63	9.21	11.34	13.28	15.09	16.81	18.48	20.09	21.67	23.21
0.995	7.88	10.60	12.84	14.86	16.75	18.55	20.28	21.96	23.59	25.19

$F(z)$	Number of Degrees of Freedom									
	11	12	13	14	15	16	17	18	19	20
0.005	2.60	3.07	3.57	4.07	4.60	5.14	5.70	6.26	6.84	7.43
0.01	3.05	3.57	4.11	4.66	5.23	5.81	6.41	7.01	7.63	8.26
0.025	3.82	4.40	5.01	5.63	6.26	6.91	7.56	8.23	8.91	9.59
0.05	4.57	5.23	5.89	6.57	7.26	7.96	8.67	9.39	10.12	10.85
0.95	19.68	21.03	22.36	23.68	25.00	26.30	27.59	28.87	30.14	31.41
0.975	21.92	23.34	24.74	26.12	27.49	28.85	30.19	31.53	32.85	34.17
0.99	24.73	26.22	27.69	29.14	30.58	32.00	33.41	34.81	36.19	37.57
0.995	26.76	28.30	29.82	31.32	32.80	34.27	35.72	37.16	38.58	40.00

$F(z)$	Number of Degrees of Freedom									
	21	22	23	24	25	26	27	28	29	30
0.005	8.0	8.6	9.3	9.9	10.5	11.2	11.8	12.5	13.1	13.8
0.01	8.9	9.5	10.2	10.9	11.5	12.2	12.9	13.6	14.3	15.0
0.025	10.3	11.0	11.7	12.4	13.1	13.8	14.6	15.3	16.0	16.8
0.05	11.6	12.3	13.1	13.8	14.6	15.4	16.2	16.9	17.7	18.5
0.95	32.7	33.9	35.2	36.4	37.7	38.9	40.1	41.3	42.6	43.8
0.975	35.5	36.8	38.1	39.4	40.6	41.9	43.2	44.5	45.7	47.0
0.99	38.9	40.3	41.6	43.0	44.3	45.6	47.0	48.3	49.6	50.9
0.995	41.4	42.8	44.2	45.6	46.9	48.3	49.6	51.0	52.3	53.7

$F(z)$	Number of Degrees of Freedom							
	40	50	60	70	80	90	100	>100 (Approximation)
0.005	20.7	28.0	35.5	43.3	51.2	59.2	67.3	$\frac{1}{2}(h - 2.58)^2$
0.01	22.2	29.7	37.5	45.4	53.5	61.8	70.1	$\frac{1}{2}(h - 2.33)^2$
0.025	24.4	32.4	40.5	48.8	57.2	65.6	74.2	$\frac{1}{2}(h - 1.96)^2$
0.05	26.5	34.8	43.2	51.7	60.4	69.1	77.9	$\frac{1}{2}(h - 1.64)^2$
0.95	55.8	67.5	79.1	90.5	101.9	113.1	124.3	$\frac{1}{2}(h + 1.64)^2$
0.975	59.3	71.4	83.3	95.0	106.6	118.1	129.6	$\frac{1}{2}(h + 1.96)^2$
0.99	63.7	76.2	88.4	100.4	112.3	124.1	135.8	$\frac{1}{2}(h + 2.33)^2$
0.995	66.8	79.5	92.0	104.2	116.3	128.3	140.2	$\frac{1}{2}(h + 2.58)^2$

In the last column, $h = \sqrt{2m - 1}$, where m is the number of degrees of freedom.

Table A11
F-Distribution with (m, n) Degrees of Freedom

Values of z for which the distribution function $F(z)$ [see (13), Sec. 23.4] has the value **0.95**
Example: For (7, 4) degrees of freedom, $z = 6.09$ if $F(z) = 0.95$.

n	$m=1$	$m=2$	$m=3$	$m=4$	$m=5$	$m=6$	$m=7$	$m=8$	$m=9$
1	161	200	216	225	230	234	237	239	241
2	18.5	19.0	19.2	19.2	19.3	19.3	19.4	19.4	19.4
3	10.1	9.55	9.28	9.12	9.01	8.94	8.89	8.85	8.81
4	7.71	6.94	6.59	6.39	6.26	6.16	6.09	6.04	6.00
5	6.61	5.79	5.41	5.19	5.05	4.95	4.88	4.82	4.77
6	5.99	5.14	4.76	4.53	4.39	4.28	4.21	4.15	4.10
7	5.59	4.74	4.35	4.12	3.97	3.87	3.79	3.73	3.68
8	5.32	4.46	4.07	3.84	3.69	3.58	3.50	3.44	3.39
9	5.12	4.26	3.86	3.63	3.48	3.37	3.29	3.23	3.18
10	4.96	4.10	3.71	3.48	3.33	3.22	3.14	3.07	3.02
11	4.84	3.98	3.59	3.36	3.20	3.09	3.01	2.95	2.90
12	4.75	3.89	3.49	3.26	3.11	3.00	2.91	2.85	2.80
13	4.67	3.81	3.41	3.18	3.03	2.92	2.83	2.77	2.71
14	4.60	3.74	3.34	3.11	2.96	2.85	2.76	2.70	2.65
15	4.54	3.68	3.29	3.06	2.90	2.79	2.71	2.64	2.59
16	4.49	3.63	3.24	3.01	2.85	2.74	2.66	2.59	2.54
17	4.45	3.59	3.20	2.96	2.81	2.70	2.61	2.55	2.49
18	4.41	3.55	3.16	2.93	2.77	2.66	2.58	2.51	2.46
19	4.38	3.52	3.13	2.90	2.74	2.63	2.54	2.48	2.42
20	4.35	3.49	3.10	2.87	2.71	2.60	2.51	2.45	2.39
22	4.30	3.44	3.05	2.82	2.66	2.55	2.46	2.40	2.34
24	4.26	3.40	3.01	2.78	2.62	2.51	2.42	2.36	2.30
26	4.23	3.37	2.98	2.74	2.59	2.47	2.39	2.32	2.27
28	4.20	3.34	2.95	2.71	2.56	2.45	2.36	2.29	2.24
30	4.17	3.32	2.92	2.69	2.53	2.42	2.33	2.27	2.21
32	4.15	3.29	2.90	2.67	2.51	2.40	2.31	2.24	2.19
34	4.13	3.28	2.88	2.65	2.49	2.38	2.29	2.23	2.17
36	4.11	3.26	2.87	2.63	2.48	2.36	2.28	2.21	2.15
38	4.10	3.24	2.85	2.62	2.46	2.35	2.26	2.19	2.14
40	4.08	3.23	2.84	2.61	2.45	2.34	2.25	2.18	2.12
50	4.03	3.18	2.79	2.56	2.40	2.29	2.20	2.13	2.07
60	4.00	3.15	2.76	2.53	2.37	2.25	2.17	2.10	2.04
70	3.98	3.13	2.74	2.50	2.35	2.23	2.14	2.07	2.02
80	3.96	3.11	2.72	2.49	2.33	2.21	2.13	2.06	2.00
90	3.95	3.10	2.71	2.47	2.32	2.20	2.11	2.04	1.99
100	3.94	3.09	2.70	2.46	2.31	2.19	2.10	2.03	1.97
150	3.90	3.06	2.66	2.43	2.27	2.16	2.07	2.00	1.94
200	3.89	3.04	2.65	2.42	2.26	2.14	2.06	1.98	1.93
1000	3.85	3.00	2.61	2.38	2.22	2.11	2.02	1.95	1.89
∞	3.84	3.00	2.60	2.37	2.21	2.10	2.01	1.94	1.88

Table A11
F-Distribution with (*m, n*) Degrees of Freedom *(continued)*

Values of z for which the distribution function $F(z)$ [see (13), Sec. 23.4] has the value **0.95**

n	$m = 10$	$m = 15$	$m = 20$	$m = 30$	$m = 40$	$m = 50$	$m = 100$	∞
1	242	246	248	250	251	252	253	254
2	19.4	19.4	19.4	19.5	19.5	19.5	19.5	19.5
3	8.79	8.70	8.66	8.62	8.59	8.58	8.55	8.53
4	5.96	5.86	5.80	5.75	5.72	5.70	5.66	5.63
5	4.74	4.62	4.56	4.50	4.46	4.44	4.41	4.37
6	4.06	3.94	3.87	3.81	3.77	3.75	3.71	3.67
7	3.64	3.51	3.44	3.38	3.34	3.32	3.27	3.23
8	3.35	3.22	3.15	3.08	3.04	3.02	2.97	2.93
9	3.14	3.01	2.94	2.86	2.83	2.80	2.76	2.71
10	2.98	2.85	2.77	2.70	2.66	2.64	2.59	2.54
11	2.85	2.72	2.65	2.57	2.53	2.51	2.46	2.40
12	2.75	2.62	2.54	2.47	2.43	2.40	2.35	2.30
13	2.67	2.53	2.46	2.38	2.34	2.31	2.26	2.21
14	2.60	2.46	2.39	2.31	2.27	2.24	2.19	2.13
15	2.54	2.40	2.33	2.25	2.20	2.18	2.12	2.07
16	2.49	2.35	2.28	2.19	2.15	2.12	2.07	2.01
17	2.45	2.31	2.23	2.15	2.10	2.08	2.02	1.96
18	2.41	2.27	2.19	2.11	2.06	2.04	1.98	1.92
19	2.38	2.23	2.16	2.07	2.03	2.00	1.94	1.88
20	2.35	2.20	2.12	2.04	1.99	1.97	1.91	1.84
22	2.30	2.15	2.07	1.98	1.94	1.91	1.85	1.78
24	2.25	2.11	2.03	1.94	1.89	1.86	1.80	1.73
26	2.22	2.07	1.99	1.90	1.85	1.82	1.76	1.69
28	2.19	2.04	1.96	1.87	1.82	1.79	1.73	1.65
30	2.16	2.01	1.93	1.84	1.79	1.76	1.70	1.62
32	2.14	1.99	1.91	1.82	1.77	1.74	1.67	1.59
34	2.12	1.97	1.89	1.80	1.75	1.71	1.65	1.57
36	2.11	1.95	1.87	1.78	1.73	1.69	1.62	1.55
38	2.09	1.94	1.85	1.76	1.71	1.68	1.61	1.53
40	2.08	1.92	1.84	1.74	1.69	1.66	1.59	1.51
50	2.03	1.87	1.78	1.69	1.63	1.60	1.52	1.44
60	1.99	1.84	1.75	1.65	1.59	1.56	1.48	1.39
70	1.97	1.81	1.72	1.62	1.57	1.53	1.45	1.35
80	1.95	1.79	1.70	1.60	1.54	1.51	1.43	1.32
90	1.94	1.78	1.69	1.59	1.53	1.49	1.41	1.30
100	1.93	1.77	1.68	1.57	1.52	1.48	1.39	1.28
150	1.89	1.73	1.64	1.54	1.48	1.44	1.34	1.22
200	1.88	1.72	1.62	1.52	1.46	1.41	1.32	1.19
1000	1.84	1.68	1.58	1.47	1.41	1.36	1.26	1.08
∞	1.83	1.67	1.57	1.46	1.39	1.35	1.24	1.00

Table A11
F-Distribution with (*m*, *n*) Degrees of Freedom *(continued)*

Values of z for which the distribution function $F(z)$ [see (13), Sec. 23.4] has the value **0.99**

n	$m = 1$	$m = 2$	$m = 3$	$m = 4$	$m = 5$	$m = 6$	$m = 7$	$m = 8$	$m = 9$
1	4052.	4999	5403	5625	5764	5859	5928	5981	6022
2	98.5	99.0	99.2	99.2	99.3	99.3	99.4	99.4	99.4
3	34.1	30.8	29.5	28.7	28.2	27.9	27.7	27.5	27.3
4	21.2	18.0	16.7	16.0	15.5	15.2	15.0	14.8	14.7
5	16.3	13.3	12.1	11.4	11.0	10.7	10.5	10.3	10.2
6	13.7	10.9	9.78	9.15	8.75	8.47	8.26	8.10	7.98
7	12.2	9.55	8.45	7.85	7.46	7.19	6.99	6.84	6.72
8	11.3	8.65	7.59	7.01	6.63	6.37	6.18	6.03	5.91
9	10.6	8.02	6.99	6.42	6.06	5.80	5.61	5.47	5.35
10	10.0	7.56	6.55	5.99	5.64	5.39	5.20	5.06	4.94
11	9.65	7.21	6.22	5.67	5.32	5.07	4.89	4.74	4.63
12	9.33	6.93	5.95	5.41	5.06	4.82	4.64	4.50	4.39
13	9.07	6.70	5.74	5.21	4.86	4.62	4.44	4.30	4.19
14	8.86	6.51	5.56	5.04	4.69	4.46	4.28	4.14	4.03
15	8.68	6.36	5.42	4.89	4.56	4.32	4.14	4.00	3.89
16	8.53	6.23	5.29	4.77	4.44	4.20	4.03	3.89	3.78
17	8.40	6.11	5.18	4.67	4.34	4.10	3.93	3.79	3.68
18	8.29	6.01	5.09	4.58	4.25	4.01	3.84	3.71	3.60
19	8.18	5.93	5.01	4.50	4.17	3.94	3.77	3.63	3.52
20	8.10	5.85	4.94	4.43	4.10	3.87	3.70	3.56	3.46
22	7.95	5.72	4.82	4.31	3.99	3.76	3.59	3.45	3.35
24	7.82	5.61	4.72	4.22	3.90	3.67	3.50	3.36	3.26
26	7.72	5.53	4.64	4.14	3.82	3.59	3.42	3.29	3.18
28	7.64	5.45	4.57	4.07	3.75	3.53	3.36	3.23	3.12
30	7.56	5.39	4.51	4.02	3.70	3.47	3.30	3.17	3.07
32	7.50	5.34	4.46	3.97	3.65	3.43	3.26	3.13	3.02
34	7.44	5.29	4.42	3.93	3.61	3.39	3.22	3.09	2.98
36	7.40	5.25	4.38	3.89	3.57	3.35	3.18	3.05	2.95
38	7.35	5.21	4.34	3.86	3.54	3.32	3.15	3.02	2.92
40	7.31	5.18	4.31	3.83	3.51	3.29	3.12	2.99	2.89
50	7.17	5.06	4.20	3.72	3.41	3.19	3.02	2.89	2.78
60	7.08	4.98	4.13	3.65	3.34	3.12	2.95	2.82	2.72
70	7.01	4.92	4.07	3.60	3.29	3.07	2.91	2.78	2.67
80	6.96	4.88	4.04	3.56	3.26	3.04	2.87	2.74	2.64
90	6.93	4.85	4.01	3.54	3.23	3.01	2.84	2.72	2.61
100	6.90	4.82	3.98	3.51	3.21	2.99	2.82	2.69	2.59
150	6.81	4.75	3.91	3.45	3.14	2.92	2.76	2.63	2.53
200	6.76	4.71	3.88	3.41	3.11	2.89	2.73	2.60	2.50
1000	6.66	4.63	3.80	3.34	3.04	2.82	2.66	2.53	2.43
∞	6.63	4.61	3.78	3.32	3.02	2.80	2.64	2.51	2.41

Table A11
F-Distribution with (*m, n*) Degrees of Freedom *(continued)*

Values of z for which the distribution function $F(z)$ [see (13), Sec. 23.4] has the value **0.99**

n	$m = 10$	$m = 15$	$m = 20$	$m = 30$	$m = 40$	$m = 50$	$m = 100$	∞
1	6056	6157	6209	6261	6287	6303	6330	6366
2	99.4	99.4	99.4	99.5	99.5	99.5	99.5	99.5
3	27.2	26.9	26.7	26.5	26.4	26.4	26.2	26.1
4	14.5	14.2	14.0	13.8	13.7	13.7	13.6	13.5
5	10.1	9.72	9.55	9.38	9.29	9.24	9.13	9.02
6	7.87	7.56	7.40	7.23	7.14	7.09	6.99	6.88
7	6.62	6.31	6.16	5.99	5.91	5.86	5.75	5.65
8	5.81	5.52	5.36	5.20	5.12	5.07	4.96	4.86
9	5.26	4.96	4.81	4.65	4.57	4.52	4.42	4.31
10	4.85	4.56	4.41	4.25	4.17	4.12	4.01	3.91
11	4.54	4.25	4.10	3.94	3.86	3.81	3.71	3.60
12	4.30	4.01	3.86	3.70	3.62	3.57	3.47	3.36
13	4.10	3.82	3.66	3.51	3.43	3.38	3.27	3.17
14	3.94	3.66	3.51	3.35	3.27	3.22	3.11	3.00
15	3.80	3.52	3.37	3.21	3.13	3.08	2.98	2.87
16	3.69	3.41	3.26	3.10	3.02	2.97	2.86	2.75
17	3.59	3.31	3.16	3.00	2.92	2.87	2.76	2.65
18	3.51	3.23	3.08	2.92	2.84	2.78	2.68	2.57
19	3.43	3.15	3.00	2.84	2.76	2.71	2.60	2.49
20	3.37	3.09	2.94	2.78	2.69	2.64	2.54	2.42
22	3.26	2.98	2.83	2.67	2.58	2.53	2.42	2.31
24	3.17	2.89	2.74	2.58	2.49	2.44	2.33	2.21
26	3.09	2.82	2.66	2.50	2.42	2.36	2.25	2.13
28	3.03	2.75	2.60	2.44	2.35	2.30	2.19	2.06
30	2.98	2.70	2.55	2.39	2.30	2.25	2.13	2.01
32	2.93	2.66	2.50	2.34	2.25	2.20	2.08	1.96
34	2.89	2.62	2.46	2.30	2.21	2.16	2.04	1.91
36	2.86	2.58	2.43	2.26	2.18	2.12	2.00	1.87
38	2.83	2.55	2.40	2.23	2.14	2.09	1.97	1.84
40	2.80	2.52	2.37	2.20	2.11	2.06	1.94	1.80
50	2.70	2.42	2.27	2.10	2.01	1.95	1.82	1.68
60	2.63	2.35	2.20	2.03	1.94	1.88	1.75	1.60
70	2.59	2.31	2.15	1.98	1.89	1.83	1.70	1.54
80	2.55	2.27	2.12	1.94	1.85	1.79	1.66	1.49
90	2.52	2.24	2.09	1.92	1.82	1.76	1.62	1.46
100	2.50	2.22	2.07	1.89	1.80	1.73	1.60	1.43
150	2.44	2.16	2.00	1.83	1.73	1.66	1.52	1.33
200	2.41	2.13	1.97	1.79	1.69	1.63	1.48	1.28
1000	2.34	2.06	1.90	1.72	1.61	1.54	1.38	1.11
∞	2.32	2.04	1.88	1.70	1.59	1.52	1.36	1.00

Table A12
Distribution Function $F(x) = P(T \le x)$ of the Random Variable T in Section 23.8

Values given are $0.\text{xxx}$

$n=3$

x	
0	167
1	500

$n=4$

x	
0	042
1	167
2	375

$n=5$

x	
0	008
1	042
2	117
3	242
4	408

$n=6$

x	
0	001
1	008
2	028
3	068
4	136
5	235
6	360
7	500

$n=7$

x	
1	001
2	005
3	015
4	035
5	068
6	119
7	191
8	281
9	386
10	500

$n=8$

x	
2	001
3	003
4	007
5	016
6	031
7	054
8	089
9	138
10	199
11	274
12	360
13	452

$n=9$

x	
4	001
5	003
6	006
7	012
8	022
9	038
10	060
11	090
12	130
13	179
14	238
15	306
16	381
17	460

$n=10$

x	
6	001
7	002
8	005
9	008
10	014
11	023
12	036
13	054
14	078
15	108
16	146
17	190
18	242
19	300
20	364
21	431
22	500

$n=11$

x	
8	001
9	002
10	003
11	005
12	008
13	013
14	020
15	030
16	043
17	060
18	082
19	109
20	141
21	179
22	223
23	271
24	324
25	381
26	440
27	500

$n=20$

x	
50	001
51	002
52	002
53	003
54	004
55	005
56	006
57	007
58	008
59	010
60	012
61	014
62	017
63	020
64	023
65	027
66	032
67	037
68	043
69	049
70	056
71	064
72	073
73	082
74	093
75	104
76	117
77	130
78	144
79	159
80	176
81	193
82	211
83	230
84	250
85	271
86	293
87	315
88	339
89	362
90	387
91	411
92	436
93	462
94	487

$n=19$

x	
43	001
44	002
45	002
46	003
47	003
48	004
49	005
50	006
51	008
52	010
53	012
54	014
55	017
56	021
57	025
58	029
59	034
60	040
61	047
62	054
63	062
64	072
65	082
66	093
67	105
68	119
69	133
70	149
71	166
72	184
73	203
74	223
75	245
76	267
77	290
78	314
79	339
80	365
81	391
82	418
83	445
84	473
85	500

$n=18$

x	
38	001
39	002
40	003
41	003
42	004
43	005
44	007
45	009
46	011
47	013
48	016
49	020
50	024
51	029
52	034
53	041
54	048
55	056
56	066
57	076
58	088
59	100
60	115
61	130
62	147
63	165
64	184
65	205
66	227
67	250
68	275
69	300
70	327
71	354
72	383
73	411
74	441
75	470
76	500

$n=17$

x	
32	001
33	002
34	002
35	003
36	004
37	005
38	007
39	009
40	011
41	014
42	017
43	021
44	026
45	032
46	038
47	046
48	054
49	064
50	076
51	088
52	102
53	118
54	135
55	154
56	174
57	196
58	220
59	245
60	271
61	299
62	328
63	358
64	388
65	420
66	452
67	484

$n=16$

x	
27	001
28	002
29	002
30	003
31	004
32	006
33	008
34	010
35	013
36	016
37	021
38	026
39	032
40	039
41	048
42	058
43	070
44	083
45	097
46	114
47	133
48	153
49	175
50	199
51	225
52	253
53	282
54	313
55	345
56	378
57	412
58	447
59	482

$n=15$

x	
23	001
24	002
25	003
26	004
27	006
28	008
29	010
30	014
31	018
32	023
33	029
34	037
35	046
36	057
37	070
38	084
39	101
40	120
41	141
42	164
43	190
44	218
45	248
46	279
47	313
48	349
49	385
50	423
51	461
52	500

$n=14$

x	
18	001
19	002
20	002
21	003
22	005
23	007
24	010
25	013
26	018
27	024
28	031
29	040
30	051
31	063
32	079
33	096
34	117
35	140
36	165
37	194
38	225
39	259
40	295
41	334
42	374
43	415
44	457
45	500

$n=13$

x	
14	001
15	001
16	002
17	003
18	005
19	007
20	011
21	015
22	021
23	029
24	038
25	050
26	064
27	082
28	102
29	126
30	153
31	184
32	218
33	255
34	295
35	338
36	383
37	429
38	476

$n=12$

x	
11	001
12	002
13	003
14	004
15	007
16	010
17	016
18	022
19	031
20	043
21	058
22	076
23	098
24	125
25	155
26	190
27	230
28	273
29	319
30	369
31	420
32	473

INDEX

Page numbers A1, A2, A3, · · · refer to Appendix 1 to Appendix 5 at the end of the book.

A

Absolute
  convergence 735
  frequency 1051, 1059
  value 657
Acceleration 436
Acceptable quality level 1134
Acceptance sampling 1133
Adams–Bashforth methods 952
Adams–Moulton methods 954
Adaptive integration 876
Addition of
  complex numbers 653
  matrices 308
  means 1097
  normal random variables 1110
  power series 747
  variances 1099
  vectors 308, 359, 403
Addition rule 1060, 1061
Adiabatic 608
ADI method 967
Adjacency matrix 1012
Adjacent vertices 1011
Airfoil 677
Airy's equation 598, 958, 960
Algebraic multiplicity 374, 919
Algorithm 833
  Dijkstra 1021
  efficient 1019
  Ford–Fulkerson 1038
  Greedy 1025
  Kruskal 1025
  Moore 1017
  polynomially bounded 1019
  Prim 1029
  stable, unstable 833

Allowable number of defectives 1133
Alternating
  direction implicit method 967
  path 1042
Alternative hypothesis 1119
Ampère 42
Amplification 116
Amplitude spectrum 557
Analytic function 203, 667
Analytic at infinity 779
Angle
  between curves 48
  between vectors 409
Angular speed 418, 436
Annulus 663
Anticommutative 417
AOQ, AOQL 1135
Approximate solution of
  differential equations 10, 56, 942–988
  eigenvalue problems 920–938
  equations 838–848
  systems of equations 886–913
Approximation
  least squares 914, 1145
  trigonometric 553
A priori estimate 845
AQL 1134
Arc of a curve 430
Archimedes's principle 91
Arc length 432
Arc tan 692
Area 481, 488, 503
Argand diagram 655
Argument 657
Artificial variable 1004
Assignment problem 1041
Associated Legendre functions 210
Astroid 40, 435

Asymptotically
    equal 1090
    normal 1116
    stable 172
Attractive 172
Augmented matrix 322
Augmenting path 1033, 1042
    theorem 1036
Autonomous 175, 180
Average (*see* Mean value)
Average outgoing quality 1135
Axioms of probability 1059

B

Backward
    differences 858
    edge 1032, 1034
Band matrix 967
Bashforth method 953
Basic
    feasible solution 996, 998
    variables 999
Basis 68, 97, 125, 161, 334, 340, 393
Beam 130, 598
Beats 115
Bellman
    equations 1021
    optimality principle 1021
Bell-shaped curve 15, 16, 1085
Bernoulli 36, 93
    distribution 1080
    equation 36
    law of large numbers 1091
    numbers 758
Bessel 218
    equation 218, 631
    functions 219, 231, 238, 631, A85
    functions, tables A85
    inequality 245, 555
Beta function A55
Bezier curves 868
BFS 1016
Bijective mapping 674
Binary 832
Binomial
    coefficients 856, 1067
    distribution 1080, 1089, A87
    series 757
    theorem 1069
Binormal 441
Bipartite
    graph 1041
    matching 1042

Birthday problem 1069
Bisection method 848
Bolzano–Weierstrass theorem A82
Bonnet 210
Boundary
    conditions 80, 233, 587
    point 664
    value problem 80, 96, 234
Bounded
    domain 713
    function 53
    region 478
    sequence A60, A81
Boxplot 1052
Boyle–Mariotte's law 23
Branch point 699
Breadth first search 1016
Buffon 1058
Bugs 833

C

Cable 71, 227, 642
Cancellation law 356
Cantor–Dedekind axiom A61
Capacitance 42
Capacitive time constant 46
Capacitor 42
Capacity
    of a cut set 1034
    of an edge 1031
Cardano 652
Cardioid 435, 489
Cartesian coordinates 402, 655
CAS x, 829
Catenary 71, 434
Cauchy 93
    convergence principle 735
    determinant 134
    –Hadamard formula 744
    inequality 728
    integral formula 722
    integral theorem 714
    method of steepest descent 991
    principal value 788, 791
    product 748
    –Riemann equations 52, 669, 672
Cayley transformation 695
Center 166
    of a graph 1031
    of gravity 481, 504
Central
    differences 859
    limit theorem 1116

Central (*Cont.*)
  moments 1077
Centrifugal force 436
Centripetal acceleration 436
Cgs system: Front cover
Chain rule 444
Characteristic
  determinant 373, 918
  equation 72, 132, 373, 918
  functions 590, 622, 632
  polynomial 373
  value 371, 590, 622, 918
  vector 371, 918
Chebyshev polynomials 239
Chinese postman problem 1020
Chi-square
  distribution 1115, A92
  test 1138
Cholesky's method 897
Chopping 832
Chromatic number 1046
Circle 429, 663
  of convergence 743
Circuit 41, 118, 274, 301
Circular
  disk 663
  helix 429, 433, 441
  membrane 629
Circulation 519, 815, 818
Cissoid 435
Clairaut equation 40
Class intervals 1052
Closed
  disk 663
  integration formula 878
  interval A61
  point set 664
  region 478
Coefficient matrix 322, 887
Coefficients of a
  differential equation 65
  power series 199
  system of equations 322, 886
Cofactor 344
Coin tossing 1058
Collatz's theorem 923
Column 306, 344
  space 335
  sum norm 903
  vector 306
Combination 1066
Combinatorial optimization 1010
Comparison test 736
Complement 664, 1056

Complementary
  error function A56
  Fresnel integrals A56–A57
  sine integral A57
Complementation rule 1060
Complete
  graph 1014
  matching 1042
  orthonormal set 244
Complex
  conjugate 655
  exponential function 77, 679
  Fourier integral 570
  Fourier series 548
  function 664
  hyperbolic functions 684
  impedance 123, 124
  indefinite integral 717
  line integral 704
  logarithm 687, 756
  number 652, 657, 680
  number sphere 779
  plane 655
  plane, extended 693, 779
  potential 798, 802, 808, 812
  sequence 732
  series 734
  trigonometric functions 682, 755
  variable 665
  vector space 358, 359, 389
Complexity 1018, 1030
Component 306, 402, 411
Composite transformation 317
Compound interest 10
Compressible fluid 454
Computer graphics 321
Conchoid 435
Condition number 910
Conditionally convergent 735
Conditional probability 1061
Conduction of heat 512, 600, 808
CONF 1109
Confidence
  intervals 1109–1116, 1148
  level 1109
  limits 1109
Conformal mapping 676, 804
Conic sections 396
Conjugate
  complex numbers 655
  harmonic function 672
Connected
  graph 1017
  set 664
Conservative 450, 458, 474

Consistent equations 326
Constraints 991
Consumer's risk 1134
Continuity
        of a complex function 666
        equation 455
        of a vector function 425
Continuous
        distribution 1072, 1093
        random variable 1072
Contour integral 714
Contraction 840
Control
        chart 1128
        limit 1128
        variables 990
Convergence
        absolute 735
        circle of 743
        conditional 735
        interval 200, 743
        of an iterative process 840, 903
        mean 244
        mean-square 244
        in norm 244
        principle 735
        radius 200, 743
        of a sequence 425, 733
        of a series 199, 734
        tests 735–739
        uniform 760
Conversion to a system 156
Convolution 279, 574, 614
Cooling 21, 39
Coordinate transformations A62, A75
Coordinates
        Cartesian 402, 655
        curvilinear A62
        cylindrical 636, A63
        polar 483, 488, 626, 657, 680
        spherical 637, A63
Coriolis acceleration 437
Corrector 946, 954
Correlation analysis 1150
Cosecant 683, A53
Cosine
        of a complex variable 682, 755
        hyperbolic 684
        integral A57
        of a real variable A52
Cotangent 683
Coulomb 42
        law 451
Covariance 1098, 1151

Cramer's rule 342, 343, 347
Crank–Nicolson method 978
Critical
        damping 88
        point 164, 171, 676
        region 1120
Cross product 414
Crout's method 895
Cubic spline 862
Cumulative
        distribution function 1069
        frequency 1051
Curl 457, 475, 486, 515, A64
Curvature 440, 443
Curve 428
        arc length of 432
        fitting 915
        orientation of 428
        piecewise smooth 465, 706
        rectifiable 432
        simple 430
        smooth 465, 705
Curvilinear coordinates A62
Cut set 1034
Cycle 1015
Cycloid 439, 490
Cylinder, flow around 814, 818
Cylindrical coordinates 636, A63

D

D'Alembert's solution 596
Damping 86, 115
Dantzig 998
Debugging 833
Decay 6, 9
Decreasing sequence A60
Decrement 92
Dedekind A61
Defect 374
Defective item 1133
Definite complex integral 704
Definiteness 391
Deformation of path 717
Degenerate feasible solution 1002
Degree of
        precision 675
        a vertex 1011
Degrees of freedom 1113, 1115, 1126
Deleted neighborhood 781
Delta
        Dirac 271
        Kronecker 240

De Moivre 660
    formula 660
    limit theorem 1090
De Morgan's laws 1057
Density 1072, 1093
Dependent
    linearly 68, 97, 125, 332, 360, 406
    random variables 1096
Depth first search 1016
Derivative
    of a complex function 666, 670, 725, 749
    directional 447
    left-hand 535
    right-hand 535
    of a vector function 425
DERIVE 829
Descartes 402
Determinant 341, 343
    Cauchy 134
    characteristic 373, 918
    of a matrix 343
    of a matrix product 356
    Vandermonde 134
DFS 1016
Diagonalization 187, 394
Diagonal matrix 314
Diagonally dominant matrix 922
Diameter of a graph 1031
Differences 853, 855, 858, 859
Difference table 854
Differentiable complex function 666
Differential 26, 474
Differential equation
    Airy 598, 958, 960
    Bernoulli 36
    Bessel 218, 631
    Cauchy–Riemann 52, 669, 672
    with constant coefficients 72, 132
    elliptic 597, 962
    Euler–Cauchy 93, 111, 214
    exact 26
    homogeneous 16, 33, 65, 124, 160, 583
    hyperbolic 597, 962, 982
    hypergeometric 216
    Laguerre 278
    Laplace 512, 605, 672, 799, 962
    Legendre 205, 638
    linear 33, 64, 124, 583
    nonhomogeneous 33, 65, 101, 124, 138, 160, 583
    nonlinear 64, 124, 175
    numerical methods for 942–988
    ordinary 2

Differential equation (*Cont.*)
    parabolic 597, 962, 976
    partial 583, 962
    Poisson 583, 962, 971
    separable 14
    Sturm–Liouville 233
    of vibrating beam 598
    of vibrating mass 85, 87, 111, 158, 173, 293, 378, 550
    of vibrating membrane 618, 629
    of vibrating string 585
Differential
    form 26, 474
    operator 82
Differentiation
    analytic functions 749
    complex functions 666
    Laplace transforms 275
    numerical 879
    power series 201, 748
    series 764
    vector functions 425
Diffusivity 512, 600
Digraph 1012
Dijkstra's algorithm 1021
Dimension of vector space 334, 360, 406
Diocles 435
Dirac's delta 271
Directed
    graph 1012
    line segment 401
    path 1041
Directional derivative 447
Direction field 10
Direct method 900
Dirichlet 514
    discontinuous factor 560
    problem 514, 606, 636, 825, 962, 968
Discharge of a source 817
Discrete
    Fourier transform 575
    random variable 1070, 1092
    spectrum 572
Disjoint events 1055
Disk 663
Dissipative 474
Distribution 1069
    Bernoulli 1080
    binomial 1080, 1089, A87
    chi-square 1115, A92
    continuous 1072, 1093
    discrete 1070, 1092
    Fisher's F- 1126, A93

Distribution (*Cont.*)
-free test 1142
function 1069, 1096
Gauss 1085
hypergeometric 1082
marginal 1094
multinomial 1085
normal 616, 1085, 1108–1116,
1122–1127, 1139, 1152, A89
Poisson 1081, 1134, A88
Student's *t*- 1113, A91
two-dimensional 1092
uniform 1075, 1093
Divergence
theorem of Gauss 506
of vector fields 453, 510, 816, A64
Divergent
sequence 733
series 199, 734
Divided differences 853
Division of complex numbers 654, 659
Domain 444, 664, 713
Doolittle's method 895
Dot product 315, 408
Double
Fourier series 624
integral 479
labeling 1026
precision 832
Driving force 112
Drumhead 629
Duffing equation 183
Duhamel's formula 646

E

Eccentricity of a vertex 1031
Echelon form 328
Edge 1011
incidence list 1013
Efficient algorithm 1019
Eigenfunction 234, 590, 602, 622, 632
expansion 240
Eigenspace 371, 919
Eigenvalue 151, 234, 371, 590, 622, 632, 918
problem 162, 371, 917
Eigenvector 371, 918
EISPACK 829
Elastic membrane 619, 629
Electrical network (*see* Networks)
Electric circuit (*see* Circuit)
Electromechanical analogies 119
Electromotive force 41

Electrostatic
field 799
potential 799
Element of matrix 305
Elementary
matrix 331
operations 326
Elimination of first derivative 227
Ellipse 429, 488
Ellipsoid 495
Elliptic
cylinder 495
differential equation 597, 962
paraboloid 495
Empty set 1055
Engineering system: Front cover
Entire function 679, 728, 780
Entry 305, 344
Equally likely 1058
Equality of
complex numbers 652
matrices 307
vectors 307, 401
Equipotential
lines 673, 799
surfaces 799
Equivalence relation 331
Equivalent linear systems 326
Erf 616, 758, A56, A86
Error 834
bound 834
function 616, 758, A56, A86
propagation 835
Type I, Type II 1120
Essential singularity 777
Estimation of parameters 1106–1117
Euclidean
norm 362
space 362
Euler 93
beta function A55
–Cauchy equation 93, 111, 214
–Cauchy method 943, 957
constant 230
formula 680, 683
formulas for Fourier coefficients 531,
537, 624
graph 1020
numbers 758
method for systems 957
trail 1020
Evaporation 23
Even function 541

Event 1055
Everett interpolation formula 861
Exact
    differential 26
    differential equation 26
    differential form 26, 474
Existence theorem
    differential equations 53, 97, 126, 159
    Fourier integral 559
    Fourier series 535
    Laplace transforms 256
    Linear equations 338
Expectation 1077, 1097
Experiment 1055
Explicit solution 4
Exponential
    decay 6, 9, 23
    function, complex 77, 679, 755
    function, real A51
    growth 9, 23
    integral A57
Exposed vertex 1042
Extended complex plane 693, 779
Extension, periodic 544
Extrapolation 849
Extremum 991

F

Factorial function 1067, A54, A86
Failure 1080
Fair die 1058
Falling body 3, 9
False position 847
Family of curves 5, 18, 49
Faraday 42
Fast Fourier transform 575
F-distribution 1126, A93
Feasible solution 994, 996
Fibonacci 751
Field
    conservative 450, 458, 474
    of force 424
    gravitational 424, 450, 636
    irrotational 458, 815
    scalar 423
    vector 423
    velocity 424
Finite complex plane 693
First
    fundamental form 504
    Green's formula 513
    shifting theorem 253

Fisher 1107, 1126, 1137
    F-distribution 1126, A93
Fixed
    decimal point 831
    point 693, 831, 838
Flat spring 91
Floating point 831
Flow augmenting path 1033
Flows in networks 1031
Fluid flow 454, 812
Flux 454, 497
    integral 496
Folium of Descartes 435
Forced oscillations 90, 111, 286, 550, 594
Ford–Fulkerson algorithm 1038
Forest 1028
Form
    Hermitian 389
    quadratic 388
    skew-Hermitian 389
Forward
    differences 855
    edge 1032, 1034
Fourier 526
    –Bessel series 243, 633
    coefficients 532, 537
    coefficients, complex 548
    constants 240
    cosine integral 562
    cosine series 542
    cosine transform 565, 576
    double series 624
    half-range expansions 544
    integral 559, 610, 790
    integral, complex 569
    Legendre series 242, 639
    series 241, 532, 537
    series, complex 548
    series, generalized 240
    sine integral 562
    sine series 542
    sine transform 565, 577, 615
    transform 570, 578, 612
Four-terminal network 367
Fractional linear transformation 692
Fraction defective 1133
Free
    fall 3, 9
    oscillations 83, 90
Frenet formulas 442
Frequency 86
    of values in samples 1051, 1059
Fresnel integrals 758, A56

Friction 24
Frobenius 211
    method 211
    norm 903
    theorem 923
Fulkerson 1038
Full-wave rectifier 290
Function
    analytic 203, 667
    Bessel 219, 231, 238, 631, A85
    beta A55
    bounded 53
    characteristic 590, 622, 632
    complex 664
    conjugate harmonic 672
    entire 679, 728, 780
    error 616, 758, A56, A86
    even 541
    exponential 77, 679, 755, A51
    factorial 1067, A54, A86
    gamma 221, A54, A86
    Hankel 232
    harmonic 512, 636, 672, 804, 824
    holomorphic 667
    hyperbolic 684, 755, A53
    inverse hyperbolic 692
    inverse trigonometric 692
    Legendre 205, 210
    logarithmic 687, A51
    meromorphic 780
    Neumann 230
    odd 541
    orthogonal 235, 534
    orthonormal 235
    periodic 527
    probability 1070, 1092
    rational 668
    scalar 423
    staircase 290
    step 265
    trigonometric 682, 755, A52
    unit step 265
    vector 423
Fundamental
    form 504
    matrix 161
    mode 590
    period 527
    system 68, 125, 161
    theorem of algebra 729

G

Galilei 24

Gamma function 221, A54, A86
GAMS 829
Gauss 216
    distribution 1085
    divergence theorem 506
    elimination method 324, 887
    hypergeometric equation 216
    integration formula 877
    –Jordan elimination 352, 898
    least squares 914, 1145
    quadrature 877
    –Seidel iteration 901, 965
General
    powers 687
    solution 5, 68, 101, 125, 161
Generalized
    Fourier series 240
    function 271
    solution 592
    triangle inequality 659
Generating function 209, 246
Geometric
    multiplicity 374
    series 195, 736, 755, 760
Gerschgorin's theorem 920
Gibbs phenomenon 540, 561
Global error 944
Golden Rule 29
Goodness of fit 1137
Gosset 1113
Goursat 715, A79
Gradient 446, 813, A63
    method 991
Graph 1011
    bipartite 1041
    complete 1014
    Euler 1020
    planar 1046
Gravitation 424, 450, 636
Greedy algorithm 1025
Greek alphabet: Back cover
Green 485
    formulas 513
    theorem 485, 518
Gregory–Newton formulas 856, 858
Guldin's theorem 505

H

Hadamard's formula 744
Half-life time 9, 24
Half-plane 663
Half-range Fourier series 544
Half-wave rectifier 290, 539

Hamiltonian cycle 1016
Hanging cable 227
Hankel functions 232
Hard spring 183
Harmonic
    conjugate 672
    function 512, 636, 672, 804, 824
    oscillation 85
    series 738
Heart pacemaker 60
Heat
    equation 512, 600, 610, 767, 808, 962, 977
    potential 808
Heaviside 251
    formulas 289
    function 266
Helicoid 495
Helix 429, 433, 441
Helmholtz equation 620
Henry 42
Hermite
    interpolation 868
    polynomials 246
Hermitian 385, 389
Hertz 86
Hesse's normal form 413
Heun's method 945
High-frequency line equations 643
Hilbert 361
    matrix 358, 913
    space 361
Histogram 1051
Holomorphic 667
Homogeneous
    differential equation 16, 33, 65, 124, 160, 583
    system of equations 322, 340
Hooke's law 84
Householder's tridiagonalization 930
Hyperbolic
    differential equation 597, 962, 982
    functions, complex 684, 755
    functions, real A53
    paraboloid 495
    spiral 435
Hyperboloid 495
Hypergeometric
    differential equation 216
    distribution 1082
    functions 216
    series 217
Hypocycloid 435
Hypothesis 1118

I

Idempotent matrix 320
Identity
    matrix (see Unit matrix)
    of Lagrange 422
    theorem for power series 747
    transformation 693
    trick 392
Ill-conditioned 906
Image 362, 674
Imaginary
    axis 655
    part 652
    unit 653
Impedance 120, 124
Implicit solution 4
Improper integral 788, 791
IMSL 829
Impulse 270
Incidence
    list 1013
    matrix 1015
Incomplete gamma function A55
Incompressible 456, 816
Inconsistent equations 326
Increasing sequence A60
Indefinite
    integral 704, 717
    integration 707
Independence of path 471, 520, 715
Independent
    events 1062
    random variables 1096
Indicial equation 213
Indirect method 900
Inductance 42
Inductive time constant 44
Inductor 42
Inequality
    Bessel 245, 555
    Cauchy 728
    ML- 711
    Schur 922
    triangle 361, 409, 658
Infinite
    dimensional 360
    population 1105
    sequence 732
    series (see Series)
Infinity 693, 779
Initial
    condition 7, 67, 126, 159, 587
    value problem 7, 52, 67, 126, 159, 942, 956

Injective mapping 674
Inner product 315, 361, 389, 408
    space 361
Input 38, 112, 260
Integral
    contour 714
    definite 704
    double 479
    equation 282
    Fourier 559, 610, 790
    improper 788, 791
    indefinite 704, 717
    line 465, 704
    surface 496, 501
    theorems, complex 714, 722
    theorems, real 485, 506, 516
    transform 564
    triple 506
Integrating factor 29
Integration
    complex functions 704–731, 781–794
    Laplace transforms 276
    numerical 869–878
    power series 748
    series 764
Interest 10
Interlacing of zeros 226
Intermediate value theorem 848
Interpolation 848–868
    Hermite 868
    inverse 859
    Lagrange 849
    Newton 854, 856, 858
    spline 862
Interquartile range 1052
Intersection of events 1055
Interval
    closed A61
    of convergence 200
    estimate 1106
    open 4, A61
Invariant subspace 919
Inverse
    hyperbolic functions 692
    interpolation 859
    of a matrix 350, 352, 898
    trigonometric functions 692
Inversion 692
Irreducible 923
Irregular boundary 972
Irrotational 458, 815
Isocline 11
Isolated singularity 776
Isotherms 808

Iteration
    for eigenvalues 925
    for equations 838–848
    Gauss–Seidel 901
    Jacobi 904
    Picard 56

J

Jacobian 482
Jacobi iteration 904
Jordan 352
Joukowski airfoil 677

K

Kirchhoff's laws 43, 119, 324
Kronecker delta 240
Kruskal's algorithm 1025
Kutta 947, 960

L

l_1, l_2, l_∞ 908
L_2 917
Labeling 1026
Lagrange 69, 108
    identity of 422
    interpolation 849
Laguerre polynomials 239, 278
Lambert's law 23
Lamé 435
LAPACK 829
Laplace 251
    equation 512, 605, 672, 799, 962
    integrals 562
    limit theorem 1090
    operator 451
    transform 251, 643
Laplacian 451, 489, 626, 628
Latent root 371
Laurent series 771, 776, 781
Law of
    absorption 23
    cooling 21
    gravitation 23
    large numbers 1091
    mass action 61
    the mean (*see* Mean value theorem)
LC-circuit 122
LCL 1128
Least squares 914, 1145
Lebesgue 917

Left-hand
    derivative 535
    limit 535
Left-handed 415
Legendre 205
    differential equation 205, 638
    functions 205, 210
    polynomials 207, 237, 639, 877
Leibniz 21
    convergence test A61
Lemniscate 61
Length
    of a curve 432
    of a vector 401
Leonardo of Pisa 751
Leontief 380
Leslie model 378
Level curve 673
Liapunov 172
Libby 19
Liebmann's method 965
Likelihood function 1107
Limaçon 435, 490
Limit
    of a complex function 665
    cycle 181
    left-hand 535
    point A81
    right-hand 535
    of a sequence 733
    vector 425
    of a vector function 425
Lineal element 10
Linear
    algebra 303–399
    combination 125, 332, 360
    dependence 68, 97, 125, 332, 360, 406
    differential equation 33, 64, 124, 583
    element 432, A63
    fractional transformation 692
    independence 68, 97, 125, 332, 360, 406
    interpolation 849
    optimization 994
    programming 994
    space (*see* Vector space)
    system of equations 321, 886
    transformation 316, 362
Linearization 176
Line integral 465, 704
Lines of force 801
LINPACK 829
Liouville 233
    theorem 728
Lipschitz condition 55

List 1013
Local
    error 944
    minimum 991
Logarithm 687, 756, A51
Logarithmic
    decrement 92
    integral A57
    spiral 435
Logistic population law 13, 37
Longest path 1016
Loss of significant digits 836
Lotka-Volterra 178
Lot tolerance per cent defective 1134
Lower
    control limit 1128
    triangular matrix 314
LTPD 1134
LU-factorization 894

M

Maclaurin 752
    series 752
    trisectrix 435
MACSYMA 829
Magnitude of a vector (*see* Length)
Main diagonal 306, 344
Malthus's law 9, 37
MAPLE 829
Mapping 362, 674, 692, 804
Marginal distributions 1094
Mariotte 23
Markov process 318, 377
Mass–spring systems 84, 111, 158, 173, 293, 378, 550
Matching 1042
MATHCAD 829
MATHEMATICA 829
Mathematical expectation 1077, 1097
MATLAB 829
Matrix 305, 306
    addition 308
    augmented 322
    band 967
    diagonal 314
    eigenvalue problem 162, 371, 917
    Hermitian 385, 389
    identity (*see* Unit matrix)
    inverse 350, 353, 898
    inversion 352, 898
    multiplication 311
    nonsingular 350
    norm 903, 909

Matrix (*Cont.*)
    normal 391, 922
    null (*see* Zero matrix)
    orthogonal 381
    scalar 314
    singular 350
    skew-Hermitian 385
    skew-symmetric 307
    sparse 965
    square 306
    stochastic 318
    symmetric 307
    transpose 307, 315
    triangular 314
    tridiagonal 929, 967
    unit 314
    unitary 385
    zero 309
Max-flow min-cut theorem 1036
Maximum 991
    likelihood method 1107
    matching 1042
    modulus theorem 823
    principle 824
Mean convergence 244
Mean-square convergence 244
Mean value of a (an)
    analytic function 822
    distribution 1075
    function 815
    harmonic function 823
    sample 1053
Mean value theorem 445, 479, 510
Median 1052, 1142
Membrane 616–626, 629–635
Meromorphic function 780
Mesh incidence matrix 311
Method
    of false position 847
    of least squares 914, 1145
    of moments 1106
    of steepest descent 991
    of undetermined coefficients 104, 138, 184
    of variation of parameters 108, 140, 186
Middle quartile 1052
Minimum 991
Minor 344
Mixed
    boundary value problem 606, 636, 809, 962, 971
    triple product 419
Mixing problem 20, 23 35, 152, 169, 291
Mks system: Front cover
ML-inequality 711

Möbius 499
    strip 499
    transformation 692
Mode 590, 632
Modeling 1, 6, 19, 46, 83, 118, 585, 616
Modified Bessel functions 232
Modulus 657
Molecule 964
Moivre's formula 660
Moment
    of a distribution 1077
    of a force 417
    generating function 1084
    of inertia 481, 502, 510
    of a sample 1107
    vector 418
Monotone sequence A60
Moore's shortest path algorithm 1017
Morera's theorem 728
Moulton 954
Moving trihedron (*see* Trihedron)
M-test 765
Multinomial distribution 1085
Multiple point 430
Multiplication of
    complex numbers 653, 659
    matrices 312
    means 1098
    power series 202, 747
    vectors 312, 315, 408, 414
Multiplication rule for events 1062
Multiplicity 374, 919
Multiply connected 713
Multistep method 951, 952
"Multivalued function" 665
Mutually exclusive events 1055

N

Nabla 447
NAG 829
Natural
    equations 442
    frequency 113
    logarithm 687, 756, A51
Neighborhood 425, 663
Nested form 837
NETLIB 829
Networks 154, 170, 190, 292, 295, 310, 324, 330
    in graph theory 1031
Neumann
    functions 230
    problem 606, 636, 962, 971

Newton 21
  –Cotes formulas 874
  interpolation formulas 854, 856, 858
  law of cooling 21, 39
  law of gravitation 22, 424
  method 841
  –Raphson method 841
  second law 13, 84
Neyman 1109, 1118
Nicolson 978
Nicomedes 435
Nilpotent matrix 320
NIST 829
Nodal
  incidence matrix 310
  line 622
Node 164, 590, 849, 870
Nonbasic variables 999
Nonconservative 474
Nonhomogeneous
  differential equation 33, 65, 101, 124,
    138, 160, 583
  system of equations 322
Nonlinear differential equations 64, 124, 175
Nonorientable surface 499
Nonparametric test 1142
Nonsingular matrix 350
Norm 235, 361, 382, 401, 903, 908
Normal
  acceleration 436
  asymptotically 1116
  to a curve 441
  derivative 490
  distribution 616, 1085, 1108–1116,
    1122–1127, 1139, A89
    two-dimensional 1152
  equations 915, 1147
  matrix 391, 922
  mode 590, 632
  form (differential equation) 597
  plane 440 (Fig. 194)
  to a plane 412
  random variable 1085
  to a surface 449, 494
  vector 412, 494
Null
  hypothesis 1118
  matrix (see Zero matrix)
  space 340
  vector (see Zero vector)
Nullity 340
Number sphere 779
Numerical methods 828–988
  differentiation 879

Numerical methods (*Cont.*)
  eigenvalues 920–938
  equations 838–848
  integration 869–878
  interpolation 848–868
  linear equations 886–913
  matrix inversion 352, 898
  optimization 990–1048
  ordinary differential equations 942–961
  partial differential equations 962–984
Nyström method 960

O

O 1018
Objective function 990
OC curve 1133
Odd function 541
Ohm's law 42
One-dimensional
  heat equation 600
  wave equation 586, 587
One-sided test 1120
One-step method 951
One-to-one mapping 674
Open
  disk 663
  integration formula 878
  interval 4, A61
  point set 444, 664
Operating characteristic 1133
Operational calculus 81, 250
Operation count 892
Operator 81, 362
Optimal solution 996
Optimality principle, Bellman's 1021
Optimization 990–1048
Orbit 163
Order 944, 1018
  of a determinant 343
  of a differential equation 4, 583
  of an iteration process 844
Ordering 1027
Ordinary differential equations 1–302, 942–961
  (*see also* Differential equation)
Orientable surface 499
Orientation of a
  curve 428
  surface 498
Orthogonal
  coordinates A62
  curves 48
  eigenvectors 393
  expansion 240

Orthogonal (*Cont.*)
functions 235, 534
matrix 381
trajectories 48
transformation 382
vectors 361, 408
Orthonormal 235, 383, 412
Oscillations
in circuits 45, 120
damped 86, 115
forced 111, 286, 550, 594
free 83
harmonic 85
of a beam 598
of a cable 227
of a mass on a spring 83, 111, 158, 173,
293, 378, 550
of a membrane 616, 629
self-sustained 181
of a string 585, 983
undamped 85, 113
Osculating plane 440 (Fig. 194)
Outcome 1055
Outlier 1053
Output 38, 112, 260
Overdamping 87
Overrelaxation 905
Overdetermined system 326
Overflow 832
Overtone 590

P

Paired comparison 1125
Pappus's theorem 505
Parabolic differential equation 597, 962,
976
Paraboloid 495
Parachutist 13
Parallelepiped 420
Parallel flow 817
Parallelogram
equality 361, 410
law 404
Parameter of a distribution 1075
Parametric representation 428, 492
Parseval's equality 245, 556
Partial
derivative 426, A57
differential equation 583, 962
fractions 284
pivoting 325, 888
sum 199, 734
Particular solution 5, 68, 101, 125

Pascal 490
Path 163
in a digraph 1032
in a graph 1015
of integration 465, 705
Peaceman-Rachford method 967
Pearson, E. S. 1118
Pearson, K. 1058
Pendulum 91, 176, 180
Period 527
Periodic
extension 544
function 527
Permutation 1064
Perron–Frobenius theorem 381, 923
Phase
angle 45, 116
of complex number (*see* Argument)
lag 116
plane 163, 170
portrait 163, 171
Picard 56
iteration method 56
theorem 778
Piecewise
continuous 255
smooth 465, 494, 706
Pivot 325, 888, 999
Planar graph 1046
Plane 350, 412
Plane curve 429
Point
estimate 1106
at infinity 693
set 663
source 817
spectrum 572
Poisson 820
distribution 1081, 1134, A88
equation 583, 962, 971
integral formula 820
Polar
coordinates 483, 488, 626, 657, 680
form of complex numbers 657, 680
moment on inertia 481
Pole 777
Polynomially bounded 1019
Polynomial matrix 919
Polynomials 668
Chebyshev 239
Hermite 246
Laguerre 239, 278
Legendre 207, 237, 639, 877
trigonometric 554

Population 1104
Population models 9, 13, 37, 178, 378
Position vector 403
Positive definite 361, 391, 409, 896
Possible values 1070
Postman problem 1020
Postmultiplication 313
Potential 450, 473, 640, 799
    complex 798, 802, 808, 812
    theory 512, 672, 798–827
Power 690
    method 925
    of a test 1121
    series 195, 198, 741
    series method 194
Predator–prey 178
Precision 832
Predictor–corrector 946, 954
Pre-Hilbert space 361
Premultiplication 313
Prim's algorithm 1029
Principal
    axes theorem 396
    diagonal 306, 344
    directions 376
    normal 441
    part 771, 777
    value 658, 661, 687, 690, 788, 791
Prior estimate 845
Probability 1058, 1059
    conditional 1061
    density 1072, 1093
    distribution 1069, 1092
    function 1070, 1092
Producer's risk 1134
Product (*see* Multiplication)
Programming errors 833
Projection of a vector 411
Pure imaginary number 653

Q

QR-factorization method 933
Quadratic
    equation 836
    form 388
    interpolation 850
Qualitative methods 175
Quality control 1128
Quartile 1052
Quasilinear 962
Quotient of complex numbers 654, 659

R

Rachford method 967
Radiation 6
Radiocarbon dating 19
Radius
    of convergence 200, 743
    of a graph 1031
Random
    experiment 1055
    numbers 1105
    variable 1070, 1092
Range of a
    function 665
    sample 1052, 1131
Rank of a matrix 333, 347
Raphson 841
Rational function 668
Ratio test 737
Rayleigh 183
    equation 183
    quotient 926
RC-circuit 45, 269
Reactance 120
Real
    axis 655
    part 652
    sequence 732
    vector space 359
Rectangular
    membrane 619
    pulse 269, 272
    rule 869
    wave 532, 538
Rectifiable curve 432
Rectification of a lot 1135
Rectifier 290, 539
Rectifying plane 440 (Fig. 194)
Reduction of order 69, 137
Region 664
Regression 1145
    coefficient 1149
    line 1146
Regula falsi 847
Regular
    point 212
    Sturm-Liouville problem 236
Relative
    class frequency 1052
    error 834
    frequency 1059
Relaxation 904
Remainder 199, 734, 751
Removable singularity 778
Representation 363, 535

Residual 904, 907
Residue 781
Residue theorem 784
Resistance 42
Resonance 114, 288
Response 38, 112
Restoring force 85
Resultant of forces 404
Riccati equation 40
Riemann 669
    number sphere 779
    surface 699
Right-hand
    derivative 535
    limit 535
Right-handed 415
Risk 1134
RC-circuit 45
RL-circuit 43
RLC-circuit 118, 121, 553
Rodrigues's formula 209
Romberg integration 881
Root 660
Root test 739
Rotation 320, 382, 418, 458, 692, 815
Rounding 832
Row 306, 344
    -equivalent 326
    operations 326
    scaling 891
    space 335
    sum norm 903
    vector 306
Runge–Kutta methods 947, 958
Runge–Kutta–Fehlberg 949
Runge–Kutta–Nyström 960

S

Saddle point 165
Sample 1055
    covariance 1147
    distribution function 1137
    mean 1106
    moments 1107
    point 1055
    range 1052
    size 1055, 1106
    space 1055
    standard deviation 1053
    variance 1053, 1106
Sampling 1063, 1082
    plan, 1133
Sawtooth wave 290, 543, 556

Scalar 308, 359, 401
    field 423
    function 423
    matrix 314
    multiplication 308, 359, 405
    triple product 419
Scaling 891, 927
Schrödinger 271
Schur's inequality 922
Schwartz 271
Schwarz inequality 361, 409
Secant 683, A53
    method 846
Second
    Green's formula 513
    shifting theorem 267
Sectionally continuous
    (*see* Piecewise continuous)
Seidel 765, 901, 965
Self-starting 954
Self-sustained oscillations 181
Semicubical parabola 434
Separable differential equation 14
Separation of variables 14, 587
Sequence 732, A60
Series 734
    addition of 747
    of Bessel functions 243, 633
    binomial 757
    convergence of 199, 734
    differentiation of 748, 764
    double Fourier 624
    of eigenfunctions 240
    Fourier 241, 532, 537, 548
    geometric 195, 736, 755, 760
    harmonic 738
    hypergeometric 217
    infinite 734, A61
    integration of 748, 764
    Laurent 771, 776, 781
    Maclaurin 752
    multiplication of 202, 747
    of orthogonal functions 240
    partial sums of 199, 734
    power 198, 741
    real A61
    remainder of 199, 734, 751
    sum of 199, 734
    Taylor 751
    trigonometric 528
    value of 199, 734
Serret–Frenet formulas
    (*see* Frenet formulas)
Set of points 663

Shifted data problem 263
Shifting theorems 253, 267
Shortest
        path 1016
        spanning tree 1024
Significance level 1118, 1121
Significant
        digit 831
        in statistics 1118
Sign test 1142
Similar matrices 392, 919
Simple
        curve 430
        graph 1011
        pole 777
        zero 778
Simplex
        method 998
        table 999
Simply connected 475, 707, 713
Simpson's rule 873
Simultaneous
        corrections 904
        differential equations 146
        linear equations (*see* Linear systems)
Sine
        of a complex variable 682, 755
        hyperbolic 684, A53
        integral 560, 758, A57
        of a real variable A52
Single precision 832
Single-valued relation 665
Singular
        at infinity 779
        matrix 350
        point 212, 754, 776
        solution 5, 70, 99, 129
        Sturm–Liouville problem 236
Singularity 754, 776
Sink 454, 511, 817, 1031
SI system: Front cover
Size of a sample 1055, 1106
Skew-Hermitian 385, 389
Skewness 1078
Skew-symmetric matrix 307, 381
Skydiver 13
Slack variable 995
Slope field 10
Smooth
        curve 465, 705
        piecewise 465, 494, 706
        surface 494
Sobolev 271
Soft spring 183

Software 829
Solution
        of a differential equation 4, 65, 124, 583
        general 5, 68, 101, 125, 161
        particular 5, 68, 101, 125
        singular 5, 70, 99, 129
        space 340
        steady-state 45, 115
        of a system of differential equations 159
        of a system of equations 322
        vector 322
SOR 905
Source 454, 511, 817, 1031
Span 335
Spanning tree 1024
Sparse
        graph 1013
        matrix 965
        system of equations 900
Spectral
        density 571
        mapping theorem 381, 919
        radius 371
        representation 571
        shift 381, 919, 937
Spectrum 371, 571, 590, 918
Speed 418, 435
Sphere 493, 501
Spherical coordinates 637, A63
Spline 862
Spiral 435
        point 166
Spring 84, 91
Square
        error 554
        matrix 306
        membrane 622
        root 662
        wave 241, 538, 557
Stability 172, 833, 874, 976
        chart 173
Stagnation point 814
Staircase function 290
Standard
        basis 363, 406
        deviation 1053, 1075
        form 65, 124
Standardized random variable 1076
Stationary point 991
Statistical
        inference 1104
        tables A87–A97
Steady 455, 605
        state 45, 115

Steepest descent 991
Steiner 504
Stem-and-leaf plot 1051
Stencil 964
Step-by-step method 943
Step function 265
Step size control 949
Stereographic projection 779
Stirling formula 1067, A55
Stochastic
    matrix 318
    variable 1069
Stokes's theorem 516
Straight line 412, 429
Stream function 812
Streamline 812
Strength of a source 817
Strictly diagonally dominant 922
String 585, 643
Student's t-distribution 1113, A91
Sturm–Liouville problem 234
Subgraph 1012
Submarine cable equations 643
Submatrix 338
Subsidiary equation 260
Subspace 335
    invariant 919
Success 1080
Successive
    corrections 904
    overrelaxation 905
Sum (see Addition)
Sum of a series 199, 734
Superlinear 846
Superposition principle 66, 125, 160
Surface 491
    area 481, 488, 503
    integral 496, 501
    normal 449, 494
Surjective mapping 674
Symmetric matrix 307, 381
System of
    differential equations 159, 291, 956
    linear equations (see Linear system)
    units: Front cover

T

Tables
    of Fourier transforms 576–578
    of functions A85–A97
    of Laplace transforms 297–299
    statistical A87–A97

Tangent 683, 697, A53
    to a curve 431, 441
    hyperbolic 684, A53
    plane 449, 494
    vector 431
Tangential acceleration 436
Target 1031
Taylor 751
    formula 752
    series 751
Tchebichef (see Chebyshev)
t-distribution 1113
Telegraph equations 643
Termination criterion 842
Termwise
    differentiation 764
    integration 764
    multiplication 748
Test
    chi-square 1138
    for convergence 735–739
    of hypothesis 1118–1126
    nonparametric 1142
Tetrahedron 420
Thermal
    conductivity 512, 600
    diffusivity 512, 600
Three
    -eights rule 881
    -sigma limits 1087
Timetabling 1046
Torricelli's law 24
Torsion of a curve 441
Torsional vibrations 91
Torus 501
Total
    differential 25
    pivoting 325, 888
    square error 554
Trace of a matrix 380, 918
Tracing 833
Trail 1015
Trajectories 48, 163, 170
Transfer function 260
Transformation 362
    of Cartesian coordinates A75
    by a complex function 674
    of integrals 482, 485, 506, 516
    linear 316, 362
    orthogonal 382
    similarity 392
    unitary 389
    of vector components A74

Transient state 45, 115
Translation 401, 692
Transmission
    line equations 642
    matrix 367
Transpose of a matrix 307, 315
Transpositions 1142, A97
Trapezoidal rule 870
Traveling salesman problem 1016
Tree 1024
Trend 1142
Trial 1055
Triangle inequality 361, 409, 658
Triangular matrix 314
Tricomi equation 598
Tridiagonalization 930
Tridiagonal matrix 929, 967
Trigonometric
    approximation 553
    form of complex numbers 657, 680
    functions, complex 682, 755
    functions, real A52
    polynomial 554
    series 528, 538
    system 528, 534
Trihedron 440
Triple
    integral 506
    product 419
Trisectrix 435
Trivial solution 33, 39, 322, 340
Truncation error 834, 944
Tuning 122, 590
Twisted curve 429
Two-dimensional
    distribution 1092
    normal distribution 1152
    random variable 1092
    wave equation 618
Two-sided test 1120
Type of a differential equation 597
Type I and II errors 1121

U

UCL 1128
Unconstrained optimization 991
Uncorrelated 1152
Undamped system 85, 113
Underdamping 89
Underdetermined system 326
Underflow 832
Undetermined coefficients 104, 138, 184

Uniform
    convergence 760
    distribution 1075, 1093
Union of events 1055
Uniqueness
    differential equations 54, 97, 126, 159
    Dirichlet problem 514
    Laurent series 774
    linear equations 338, 347
    power series 746
Unit
    binormal vector 441
    circle 661, 663
    impulse 271
    matrix 314
    normal vector 494
    principal normal vector 441
    step function 265
    tangent vector 431, 441
    vector 361, 401
Unitary
    matrix 385
    system of vectors 389
    transformation 389
Unstable 172, 833
Upper control limit 1128

V

Value of a series 199, 734
Vandermonde 134
Van der Pol equation 181
Variable
    complex 665
    random 1070, 1092
    standardized random 1076
    stochastic 1069
Variance of a
    distribution 1075, 1098
    sample 1053
Variation of parameters 108, 140, 186
Vector 306, 359, 401
    addition 308, 359, 403
    field 423
    function 423
    moment 418
    norm 908
    product 414
    space 334, 359, 389
    subspace 335
Velocity
    field 424
    potential 812

Velocity (*Cont.*)
    vector 435
Venn diagram 1056
Verhulst 13, 37, 41
Vertex 1011
    exposed 1042
    incidence list 1013
Vibrations (*see* Oscillations)
Violin string 585
Volta 42
Voltage drop 41
Volterra 178
Volume 481
Vortex 818
Vorticity 815

W

Walk 1015
Wave equation 583, 587, 618, 630, 962, 982

Weber 247
    equation 247
    functions 230
Website *see* Preface
Weierstrass 765, A82
    approximation theorem 849
    M-test 765
Weight function 235
Well-conditioned 906
Wessel 655
Wheatstone bridge 330
Work 410, 468, 521,
    integral 466, 468
Wronskian 98, 110, 127, 161

Z

Zero
    of analytic function 778
    matrix 309
    vector 403

Some Constants

$e = 2.71828\ 18284\ 59045\ 23536$
$\sqrt{e} = 1.64872\ 12707\ 00128\ 14685$
$e^2 = 7.38905\ 60989\ 30650\ 22723$

$\pi = 3.14159\ 26535\ 89793\ 23846$
$\pi^2 = 9.86960\ 44010\ 89358\ 61883$
$\sqrt{\pi} = 1.77245\ 38509\ 05516\ 02730$

$\log_{10} \pi = 0.49714\ 98726\ 94133\ 85435$
$\ln \pi = 1.14472\ 98858\ 49400\ 17414$
$\log_{10} e = 0.43429\ 44819\ 03251\ 82765$
$\ln 10 = 2.30258\ 50929\ 94045\ 68402$

$\sqrt{2} = 1.41421\ 35623\ 73095\ 04880$
$\sqrt[3]{2} = 1.25992\ 10498\ 94873\ 16477$
$\sqrt{3} = 1.73205\ 08075\ 68877\ 29353$
$\sqrt[3]{3} = 1.44224\ 95703\ 07408\ 38232$
$\ln 2 = 0.69314\ 71805\ 59945\ 30942$
$\ln 3 = 1.09861\ 22886\ 68109\ 69140$

$\gamma = 0.57721\ 56649\ 01532\ 86061$
$\ln \gamma = -0.54953\ 93129\ 81644\ 82234$
(see Sec. 4.6)
$1° = 0.01745\ 32925\ 19943\ 29577$ rad
1 rad $= 57.29577\ 95130\ 82320\ 87680°$
$= 57°17'44.806''$

Polar Coordinates

$x = r \cos \theta \qquad r = \sqrt{x^2 + y^2}$

$y = r \sin \theta \qquad \theta = \arc\tan \dfrac{y}{x}$

$dx\, dy = r\, dr\, d\theta$

Series

$$\frac{1}{1-x} = \sum_{m=0}^{\infty} x^m \quad (|x| < 1)$$

$$e^x = \sum_{m=0}^{\infty} \frac{x^m}{m!}$$

$$\sin x = \sum_{m=0}^{\infty} \frac{(-1)^m x^{2m+1}}{(2m+1)!}$$

$$\cos x = \sum_{m=0}^{\infty} \frac{(-1)^m x^{2m}}{(2m)!}$$

$$\ln(1-x) = -\sum_{m=0}^{\infty} \frac{x^m}{m} \quad (|x| < 1)$$

$$\arc\tan x = \sum_{m=0}^{\infty} \frac{(-1)^m x^{2m+1}}{2m+1} \quad (|x| < 1)$$

Greek Alphabet

α	Alpha		ν	Nu
β	Beta		ξ	Xi
γ, Γ	Gamma		o	Omicron
δ, Δ	Delta		π	Pi
ϵ, ε	Epsilon		ρ	Rho
ζ	Zeta		σ, Σ	Sigma
η	Eta		τ	Tau
$\theta, \vartheta, \Theta$	Theta		υ, Υ	Upsilon
ι	Iota		ϕ, φ, Φ	Phi
κ	Kappa		χ	Chi
λ, Λ	Lambda		ψ, Ψ	Psi
μ	Mu		ω, Ω	Omega

Vectors

$$\mathbf{a} \cdot \mathbf{b} = a_1 b_1 + a_2 b_2 + a_3 b_3$$

$$\mathbf{a} \times \mathbf{b} = \begin{vmatrix} \mathbf{i} & \mathbf{j} & \mathbf{k} \\ a_1 & a_2 & a_3 \\ b_1 & b_2 & b_3 \end{vmatrix}$$

$$\operatorname{grad} f = \nabla f = \frac{\partial f}{\partial x} \mathbf{i} + \frac{\partial f}{\partial y} \mathbf{j} + \frac{\partial f}{\partial z} \mathbf{k}$$

$$\operatorname{div} \mathbf{v} = \nabla \cdot \mathbf{v} = \frac{\partial v_1}{\partial x} + \frac{\partial v_2}{\partial y} + \frac{\partial v_3}{\partial z}$$

$$\operatorname{curl} \mathbf{v} = \nabla \times \mathbf{v} = \begin{vmatrix} \mathbf{i} & \mathbf{j} & \mathbf{k} \\ \dfrac{\partial}{\partial x} & \dfrac{\partial}{\partial y} & \dfrac{\partial}{\partial z} \\ v_1 & v_2 & v_3 \end{vmatrix}$$